KU-523-984

ENCYCLOPEDIA
OF
SPECTROSCOPY
AND
SPECTROMETRY

ENCYCLOPEDIA
OF
SPECTROSCOPY
AND
SPECTROMETRY

Editor-in-Chief

JOHN C. LINDON

Editors

GEORGE E. TRANTER

JOHN L. HOLMES

ACADEMIC PRESS

A Harcourt Science and Technology Company

San Diego San Francisco New York Boston
London Sydney Tokyo

This book is printed on acid-free paper

The following articles are US Government works in the public domain and not subject to copyright:

Food and Dairy Products, Applications of Atomic Spectroscopy
Proton Affinities

ACADEMIC PRESS
A Harcourt Science and Technology Company
24–28 Oval Road
London NW1 7DX, UK
http.hbuk.co.uk/ap/

ACADEMIC PRESS
525 B Street, Suite 1900,
San Diego, CA 92101–4495, USA
http://www.apnet.com

ISBN 0-12-226680-3

A catalogue record for this Encyclopedia is available from the British Library

Library of Congress Catalog Card Number: 98–87952

Access for a limited period to an on-line version of the Encyclopedia of Spectroscopy and Spectrometry is included in the purchase price of the print edition.
This on-line version has been uniquely and persistently identified by the Digital Object Identifier (DOI)

10.1006/rwsp.2000

By following the link

http://dx.doi.org/10.1006/rwsp.2000

from any Web Browser, buyers of the Encyclopedia of Spectroscopy and Spectrometry will find instructions on how to register for access.

Typeset by Macmillan India Limited, Bangalore, India
Printed and bound in Great Britian by The University Printing House, Cambridge, UK.
00 01 02 04 05 CU 9 8 7 6 5 4 3 2 1

Editors

EDITOR-IN-CHIEF

John C. Lindon
Biological Chemistry
Division of Biomedical Sciences
Imperial College of Science, Technology and Medicine
Sir Alexander Fleming Building
South Kensington
London SW7 2AZ, UK

EDITORS

George E. Tranter
Glaxo Wellcome Medicines Research
Physical Sciences Research Unit
Gunnells Wood Road
Stevenage
Hertfordshire SG1 2NY, UK

John L. Holmes
University of Ottawa
Department of Chemistry
PO Box 450
Stn 4, Ottawa, Canada KIN 6N5

Editorial Advisory Board

Preface

This encyclopedia provides, we believe, a comprehensive and up-to-date explanation of the most important spectroscopic and related techniques together with their applications.

The *Encyclopedia of Spectroscopy and Spectrometry* is a cumbersome title but is necessary to avoid misleading readers who would comment that a simplified title such as the "Encyclopedia of Spectroscopy" was a misnomer because it included articles on subjects other than spectroscopy. Early in the planning stage, the editors realized that the boundaries of spectroscopy are blurred. Even the expanded title is not strictly accurate because we have also deliberately included other articles which broaden the content by being concerned with techniques which provide localized information and images. Consequently, we have tried to take a wider ranging view on what to include by thinking about the topics that a professional spectroscopist would conveniently expect to find in such a work as this. For example, many professionals use spectroscopic techniques, such as nuclear magnetic resonance, in conjunction with chromatographic separations and also make use of mass spectrometry as a key method for molecular structure determination. Thus, to have an encyclopedia of spectroscopy without mass spectrometry would leave a large gap. Therefore, mass spectrometry has been included. Likewise, the thought of excluding magnetic resonance imaging (MRI) seemed decidedly odd. The technique has much overlap with magnetic resonance spectroscopy, it uses very similar equipment and the experimental techniques and theory have much in common. Indeed, today, there are a number of experiments which produce multidimensional data sets of which one dimension might be spectroscopic and the others are image planes. Again the subject has been included.

This led to the general principle that we should include a number of so-called spatially-resolved methods. Some of these, like MRI, are very closely allied to spectroscopy but others such as diffraction experiments or scanning probe microscopy are less so, but have features in common and are frequently used in close conjunction with spectroscopy. The more peripheral subjects have, by design, not been treated in the same level of detail as the core topics. We have tried to provide an overview of as many as possible techniques and applications which are allied to spectroscopy and spectrometry or are used in association with them. We have endeavoured to ensure that the core subjects have been treated in substantial depth. No doubt there are omissions and if the reader feels we got it wrong, the editors take the blame.

The encyclopedia is organized conventionally in alphabetic order of the articles but we recognize that many readers would like to see articles grouped by spectroscopic area. We have achieved this by providing separate contents lists, one listing the articles in an intuitive alphabetical form, and the other grouping the articles within specialities such as mass spectrometry, atomic spectroscopy, magnetic resonance, etc. In addition each article is flagged as either a "Theory", "Methods and Instrumentation" or "Applications" article. However, inevitably, there will be some overlap of all of these categories in some articles. In order to emphasize the substantial overlap which exists among the spectroscopic and spectrometric approaches, a list has been included at the end of each article suggesting other articles in this encyclopedia which are related and which may provide relevant information for the reader. Each article also comes with a "Further Reading" section which provides a source of books and major reviews on the topic of the article and in some cases also provides details of seminal research papers. There are a number of colour plates in each volume as we consider that the use of colour can add greatly to the information content in many cases, for example for imaging studies. We have also included extensive Appendices of tables of useful reference data and a contact list of manufacturers of relevant equipment.

We have attracted a wide range of authors for these articles and many are world recognized authorities in their fields. Some of the subjects covered are relatively static, and their articles provide a distillation of the established knowledge, whilst others are very fast moving areas and for these we have aimed at presenting up-to-date summaries. In addition, we have included a number of entries which are retrospective in nature, being historical reviews of particular types of spectroscopy. As with any work of this magnitude some of the articles which we desired and commissioned to include did not make it for various reasons. A selection of these will appear in a separate section in the on-line version of the encyclopedia, which will be available to all purchasers of the print version and will have extensive hypertext links and advanced search tools. In this print version there are 281 articles contributed by more than 500 authors from 24 countries. We have persuaded authors from Australia, Belgium, Canada, Denmark, Finland, France, Germany, Hungary, India,

Israel, Italy, Japan, Mexico, New Zealand, Norway, Peru, Russia, South Africa, Spain, Sweden, Switzerland, The Netherlands, the UK and the USA to contribute.

The encyclopedia is aimed at a professional scientific readership, for both spectroscopists and non-spectroscopists. We intend that the articles provide authoritative information for experts within a field, enable spectroscopists working in one particular field to understand the scope and limitations of other spectroscopic areas and allow scientists who may not primarily be spectroscopists to grasp what the various techniques comprise in considering whether they would be applicable in their own research. In other words we tried to provide something for everone, but hope that in doing so, we have not made it too simple for the expert or too obscure for the non-specialist. We leave the reader to judge.

John Lindon
John Holmes
George Tranter

Acknowledgements

Without a whole host of dedicated people, this encyclopedia would never have come to completion. In these few words I, on behalf of my co-editors, can hope to mention the contributions of only some of those hard working individuals.

Without the active co-operation of the hundreds of scientists who acted as authors for the articles, this encyclopedia would not have been born. We are very grateful to them for endeavouring to write material suitable for an encyclopedia rather than a research paper, which has produced such high-quality entries. We know that all of the people who contributed articles are very busy scientists, many being leaders in their fields, and we thank them. We, as editors, have been ably supported by the members of the Editorial Advisory Board. They made many valuable suggestions for content and authorship in the early planning stages and provided a strong first line of scientific review after the completed articles were received. This encyclopedia covers such a wide range of scientific topics and types of technology that the very varied expertise of the Editorial Advisory Board was particularly necessary.

Next, this work would not have been possible without the vision of Carey Chapman at Academic Press who approached me about 4 years ago with the excellent idea for such an encyclopedia. Four years later, am I still so sure of the usefulness of the encyclopedia? Of course I am, despite the hard work and I am further bolstered by the thought that I might not ever have to see another e-mail from Academic Press. For their work during the commissioning stage and for handling the receipt of manuscripts and dealing with all the authorship problems, we are truly indebted to Lorraine Parry, Colin McNeil and Laura O'Neill who never failed to be considerate, courteous and helpful even under the strongest pressure. I suspect that they are now probably quite expert in spectroscopy. In addition we need to thank Sutapas Bhattacharya who oversaw the project through the production stages and we acknowledge the hard work put in by the copy-editors, the picture researcher and all the other production staff coping with very tight deadlines.

Finally, on a personal note, I should like to acknowledge the close co-operation I have received from my co-editors George Tranter and John Holmes. I think that we made a good team, even if I say it myself.

John Lindon
Imperial College of Science, Technology and Medicine
London
22 April 1999

Guide to Use of the Encyclopedia

Structure of the Encyclopedia

The material in the Encyclopedia is arranged as a series of entries in alphabetical order.

There are 4 categories of entry:

- Historical Overview
- Theory
- Methods and Instrumentation
- Applications

To help you realize the full potential of the material in the Encyclopedia we have provided the following features to help you find the topic of your choice.

1. Contents lists

Your first point of reference will probably be the main alphabetical contents list. The complete contents list appearing in each volume will provide you with both the volume number and the page number of the entry.

Alternatively you may choose to browse through a volume using the alphabetical order of the entries as your guide. To assist you in identifying your location within the Encyclopedia a running headline indicates the current entry. Furthermore, a "reference box" is provided on the opening page of each entry so that it is immediately clear whether it is a theory, methods & instrumentation, applications, or historical entry, and which of the following areas of spectroscopy it covers.

- Atomic Spectroscopy
- Electronic Spectroscopy
- Fundamentals in Spectroscopy
- High Energy Spectroscopy
- Magnetic Resonance
- Mass Spectrometry
- Spatially Resolved Spectroscopic Analysis
- Vibrational, Rotational, & Raman Spectroscopies

Example:

NUCLEAR OVERHAUSER EFFECT 1643

Nuclear Overhauser Effect

Anil Kumar and **R Christy Rani Grace**,
Indian Institute of Science, Bangalore, India

Copyright © 1999 Academic Press

MAGNETIC RESONANCE
Theory

You will find "dummy entries" in the following instances:

1. where obvious synonyms exist for entries. For example, a dummy entry appears for ESR Imaging which directs you to EPR Imaging where the material is located.
2. where we have grouped together related topics. For example, a dummy entry appears for Arsenic, NMR Applications which leads you to **Heteronuclear NMR Applications (As, Sb, Bi)** where the material is located.

3. where there is debate over whether an entry title begins with the application of a technique, or with the technique itself. For example, a dummy entry appears for Raman Spectroscopy in Biochemistry which directs you to Biochemical Applications of Raman Spectroscopy where the material is located.

Dummy entries appear in both the contents list and the body of the text.

Example:

If you were attempting to locate material on the application of spectroscopic techniques in astronomy via the contents list the following information would be provided.

Astronomy, Applications of Spectroscopy *See* Interstellar Molecules, Spectroscopy of; Stars, Spectroscopy of.

The page numbers for these entries are given at the appropriate location in the contents list.

If you were trying to locate the material by browsing through the text and you looked up Astronomy then the following would be provided.

Astronomy, Applications of Spectroscopy

See **Interstellar Molecules, Spectroscopy of; Stars, Spectroscopy of.**

Alternatively if you looked up Stars the following information would be provided.

STARS, SPECTROSCOPY OF 2199

Stars, Spectroscopy of

AGGM Tielens, Rijks Universiteit, Groningen, The Netherlands

Copyright © 1999 Academic Press

ELECTRONIC SPECTROSCOPY
Applications

Further to aid the reader to locate material the main alphabetical Contents list is followed by a list of the entries grouped within their relevant subject area. (Subject areas follow each other alphabetically). Within each subject area the entries are further broken down into those covering historical aspects, theory, methods & instrumentation, or applications. The entries are listed alphabetically within these categories, and their relevant page numbers are given.

2. Cross References

To direct the reader to other entries on related topics a "see also" section is provided at the end of each entry

Example:

The entry Nuclear Overhauser Effect includes the following cross-references:

See also: **Chemical Exchange Effects in NMR; Macromolecule–Ligand Interactions Studied By NMR; Magnetic Resonance, Historical Perspective; NMR Pulse Sequences; NMR Relaxation Rates; Nucleic Acids Studied Using NMR; Proteins Studied Using NMR Spectroscopy; Structural Chemistry Using NMR Spectroscopy, Organic Molecules; Structural Chemistry Using NMR Spectroscopy, Peptides; Structural Chemistry Using NMR Spectroscopy, Pharmaceuticals; Two-Dimensional NMR Methods.**

3. Index

The index appears in each volume. Any topic not found through the Contents list can be located by referring to the index. On the opening page of the index detailed notes on its use are provided.

4. Colour plates

The colour figures for each volume have been grouped together in a plate section. The location of this section is cited at the end of the contents list.

5. Appendices

The appendices appear in volume 3.

6. Contributors

A full list of contributors appears at the beginning of each volume.

Contributors

Adams, Fred
Department of Chemistry
University of Instelling Antwerp
University Pleim 1, B-2610, Antwerp, Belgium

Aime, S
University of Torino
Department of Chemistry
via Giuria 7, 10125, Torino, Italy

Andersson, L A
Vassar College
Poughkeepsie, Box 589, NY 12604-0589, USA

Ando, Isao
Department of Polymer Chemistry
Tokyo Institute of Technology
Meguro Ku, Tokyo, 152, Japan

Andrenyak, David M
University of Utah
Center for Human Toxicology
Salt Lake City, Utah 84112, USA

Andrews, David L
School of Chemical Sciences
University of East Anglia
Norwich, NR4 7TJ, UK

Andrews, Lester
Chemistry Department
University of Virginia
McCormick Road,
Charlottesville, VA 22901, USA

Appleton, T G
Department of Chemistry
The University of Queensland
Brisbane, Queensland 4072, Australia

Arroyo, C M
USA Medical Research Institute for Chemical
Defense, Drug Assessment Division
Advanced Assessment Branch
3100 Ricketts Point Road, Aberdeen Proving
Ground, Maryland, MD 21010, USA

Artioli, Gilberto
Dipartimento di Scienze della Terra
Universita degli Studi di Milano
via Botticelli 23, I-20133, Milan, Italy

Ashfold, Michael N R
University of Bristol
School of Chemistry
Bristol, BS8 1TS, UK

Aubery, M
Laboratorie de Glycobiologie et Reconnaissance
Cellulaire
Université Paris V - UFR Biomédicale
45, rue des Saints-Péres, 75006, Paris, France

Baer, Tom
Department of Chemistry
University of North Carolina
Chapel Hill, NC 27599-3290, USA

Bain, A D
Department of Chemistry
McMaster University
1280 Main Street W., Hamilton,
Ontario L8S 4M1, Canada

Baker, S A
Beltsville Human Nutrition Research Center
U.S. Department of Agriculture
Food Composition Lab
Beltsville, MD 21054, USA

Baldwin, Mike
Mass Spectrometry Facility
University of California
San Francisco, CA 94143-0446, USA

Bateman, R
Micromass LTD
Manchester, M23 9LZ, UK

Batsanov, Andrei
Department of Chemistry
University of Durham
South Road, Durham, DH1 3LE, UK

Beauchemin, Diane
Department of Chemistry
Queen's University
Kingston, ONT K7L 3N6, Canada

Bell, Jimmy D
The Robert Steiner MR Unit, MRC Clinical
Sciences Centre
Imperial College School of Medicine
Hammersmith Hospital
Du Cane Road, London, W12 0HS, UK

Belozerski, G N
Post Box 544, B-155, 199155, St. Petersburg,
Russia

Bernasek, S L
Princeton University
Department of Chemistry
Princeton, NJ 05844, USA

Berova, Nina
Columbia University
Department of Chemistry
New York, NY 10027, USA

Berthezene, Y
Hospital L. PradelUMR CNRS 5515
Dept de Imagerie Diagnostique et Therapeutique
BP Lyon Montchat, F-69394,
Lyon 03, France

Boesl, Ulrich
Institut für Physikalische und Theoretische Chemie
Technische Universität München
Lichtenbergstrasse 4,
D-85748, München, Germany

Bogaerts, Annemie
Department of Chemistry
University of Antwerp
Universiteitsplein 1, B-2610, Wilrijk, Belgium

Bohme, D
Department of Chemistry & Centre for Research in Earth & Space Science
York University
North York, Ontario M3J 1P3, Canada

Bonchin, Sandra L
Los Alamos National Laboratory
Nuclear Materials Technology-Analytical Chemistry
NMT-1, MS G740, Los Alamos,
NM 87545, USA

Bowie, John H
The University of Adelaide
Department of Organic Chemistry
South Australia 5005, Australia

Brand, Willi A
Max-Planck-Institute for Biochemistry
P.O. Box 100164, 07701, Jena, Germany

Braslavsky, Silvia E
Max Planck-Institut für Strahlenchemie
Postfach 101365,
D-45470, Mülheim an der Ruhr, Germany

Braut-Boucher, F
Laboratorie de Glycobiologie et Reconnaissance Cellulaire
Université Paris V - UFR Biomédicale
45, rue des Saints-Peres,
75006, Paris, France

Brittain, H G
Center for Pharmaceutical Physics
10 Charles Road, Milford,
NJ 08848, USA

Brumley, W C
National Exposure Research Laboratory
US EPA, Division of Environmental Science
PO Box 93478, Las Vegas,
Nevada 89193, USA

Bryce, David L
Dalhousie University
Department of Chemistry
Halifax, Nova ScotiaCanada

Bunker, Grant
Department of Physics
Illinois Institute of Technology
3101 S. Dearborn, Chicago, IL 60616, USA

Burgess, C
Rose Rae
Startforth
Barnard Castle, Durham, DL12 9AB, UK

Buss, Volker
University of Duisberg
Department of Theoretical Chemistry
D-47048, Duisberg, Germany

Callaghan, P T
Department of Physics
Massey University
Palmerston North, New Zealand

Calucci, Lucia
Dipartimento di Chimica e Chimica Industriale
via Risorgimento 35, 56126, Pisa, Italy

Cammack, Richard
Division of Life Sciences, Kings College
University of London
Campden Hill Road, London, W8 7AH, UK

Canè, E
Universita di Bologna
Dipartimento di Chimica Fisica e Inorganica
Viale Risorgimento 4,
40136, Bologna, Italy

Canet, D
Laboratorie Methode RMN
Universite de Nancy 1
FU CNRS E008, INCM,
F-S4506, Vandoeuvre, Nancy, France

Carter, E A
University of Bradford
Chemical and Forensic Sciences
Bradford, BD7 1DP, UK

Caruso, Joseph A
University of Cincinnati
Department of Chemistry
Cincinnati, OH 45221-0037, USA

Cerdan, Sebastian
Instituto de Investigaciones Biomedicas,
C.S.I.C
c/ Arturo Duperier 4, 28029,
Madrid, Spain

Chakrabarti, C L
Chemistry Department
Carlton University
Ottawa, Ontario K1S 5B6, Canada

Chen, Peter C
Department of Chemistry
Spelman College
Spelman Lane, Atlanta, Georgia 30314-4399,
USA

Cheng, H N
University of Delaware
Department of Chemistry
Newark, DE 19176, USA

Chichinin, A I
Institute of Chemical Kinetics and Combustion
Institutskaya 3, 630090, Novosibirsk, Russia

Claereboudt, Jan
University of Antwerp
Department of Pharmaceutical Sciences
Universiteitsplein 1, B-2610, Antwerp, Belgium

Claeys, Magda M
University of Antwerp
Department of Pharmaceutical Sciences
Universiteitsplein 1, B-2610, Antwerp, Belgium

Colarusso, Pina
National Institues of Digestive and Diabetes and
Kidney Diseases, National Institutes of Health
Laboratory of Chemical Physics
Bethesda, MD 20892, USA

Conrad, Horst
Fritz Haber Institute of the
Max Planck Gessellschaft
Faradayweg 4-6, D14195, Berlin, Germany

Cory, D G
Department of Nuclear Engineering
MIT
Cambridge, MA 02139, USA

Crouch, Dennis J
University of Utah
Center for Human Toxicology
Salt Lake City, Utah 84112, USA

Cruz, Fatima
Instituto de Investigaciones Biomedicas, C.S.I.C
c/ Arturo Duperier 4, 28029,
Madrid, Spain

Curbelo, Raul
Digilab Division
Bio-Rad Laboratories
237 Putnam Avenue, Cambridge, MA 02139, USA

Dåbakk, Eigil
Foss Sverige
Turebergs Torg 1, Box 974, SE 191 92,
Sollentuna, Sweden

Davies, M C
The University of Nottingham
Laboratory of Biophysics and Surface Analysis,
School of Pharmaceutical Siences
University Park, Nottingham,
NG7 2RD, UK

Dawson, P H
Iridian Spectral Technologies Ltd
Industry Partnership Facility [M5O]
1200, Montreal Road,
Ottawa, K1A 0R6, Ontario Canada

Demtroder, W
Fachbereich Physik
Universtität Kaiserslautern
D-6750, Kaiserslautern, Germany

Di, Qiao Qing
University of Florida
College of Pharmacy
P.O. Box 100485, Gainesville,
FL 32610, USA

Dirl, Rainer
Centre for Computational Material Science
Institut für Theoretische Physik
Tu Wien, Wiedner Hauptstrabe 8-10,
A-1040, Vienna, Austria

Dixon, Ruth M
MRC Biochemical and Clinical Magnetic
Resonance Unit
Department of Biochemistry
South Parks Road, Oxford, OX1 3QU, UK

Docherty, John
Institute of Biodiagnostics
436 Ellice Avenue, Winnipeg,
Manitoba R3B 1Y6, Canada

Dong, R Y
Department of Physics and Astronomy
Brandon University
Brandon, Manitoba R7A 6A9, Canada

Douglas, D
University of British Columbia
Department of Chemistry
2036 Main, Mall,
Vancouver, BC, V6T 1Z1, Canada

Dua, Suresh
The University of Adelaide
Department of Chemistry
South Australia 5005, Australia

Dugal, Robert
The Canadian Pharmaceutical Manufacturers Association
Doping Control Laboratory
Ottawa, Canada

Durig, J R
University of Missouri-Kansas City
5100 Rockhill Road, Kansas City,
Missouri 64110-2499, USA

Dworzanski, Jacek
Center for Micro Analysis and Reaction Chemistry
University of Utah
110 South Central Campus Drive, Room 214,
Salt Lake City, Utah 84112, USA

Dybowski, Cecil R
University of Delaware
Department of Chemistry
Newark, DE 19716, USA

Eastwood, DeLyle
Air Force Institution of Technology
MS AFIT/ENP
Wright-Patterson AFB,
OH 45433-7765, USA

Edwards, H G M
Chemistry and Forensic Sciences
University of Bradford
Bradford,
West Yorkshire BD7 1DP, UK

Eggers, L
Gerhard-Mercator-Universität
Institut für Physikalische und Theoretische Chemie
D-47048, Duisburg, Germany

Emsley, James W
Department of Chemistry
University of Southampton
Highfield, Southampton, UK

Endo, I
Department of Physics
Hiroshima University
1-3-1 Kagamiyama,
Higashi Hiroshima, 739, Japan

Ens, W
Department of Physics
University of Manitoba
Winnipeg,
Manitoba R3T2N2, Canada

Farley, J W
University of Nevada
Department of Physics
Las Vegas, NV 89154, USA

Farrant, R D
Physical Sciences
GlaxoWellcome R & D
Gunnels Wood Road, Stevenage, SG1 2NY, UK

Feeney, J
National Institute for Medical Research
Medical Research Council
The Ridgeway, Mill Hill, London, NW7 1AA, UK

Fennell, Timothy R
Chemical Industry Institute of Toxicology
6 Davis Drive, PO Box 12137, Research Triangle Park, North Carolina NC 27709-2137, USA

Ferrer, N
Serv. Cientif. Tecn. University of Barcelona
Lluis Sole Sabaris 1, E-08028, Barcelona, Spain

Fisher, A J
Department of Physics and Astronomy
University College London
Gower Street, London, WC1E 6BT, UK

Flack, H D
University of Geneva
Laboratory of Crystallography
24 Quai Ernest Ansernet,
CH 1211, Geneva, Switzerland

Flytzanis, Chr.
Laboratorie d'Optique Quantique
CNRS - Ecole Polytechnique
F-91128, Palaiseau, Cedex France

Foltz, Rodger L
Center for Human Toxicology
University of Utah
20 S 2030 ERM 490,
Salt Lake City, UT 84112-9457, USA

Ford, Mark
University of York
Department of Chemistry
Heslington, York Y010 5DD, UK

Friedrich, J
Lehrstuhl für Physik, Weihenstephan
Technische Universität München
D-85350, Freisling, Germany

Fringeli, Urs
Insitute of Physical Chemistry
University of Vienna
Althanstrasse 14/UZA II,
A-1090, Vienna, Austria

Frost, T
GlaxoWellcome R & D
Temple Hill, Dartford, Kent DA1 5AH, UK

Fuller, Watson
Keele University
Department of Physics
Keele, Staffs ST5 5BG, UK

Futrell, J H
Department of Chemistry & Biochemistry
University of Delaware
Newark, Delaware 19716, USA

Geladi, Paul
Department of Chemistry
Umeå University
SE 901 87, Umeå, Sweden

Gensch, Thomas
Katholieke Universiteit of Leuven
Department of Organic Chemistry
Molecular Dynamics and Spectroscopy
Celestijnenlaan 200F, B-3001,
Heverlee, Belgium

Gerothanassis, I P
Department of Chemistry
University of Iannina
GR-45110, Iannina, Greece

Gilbert, A S
19 West Oak, Beckenham,
Kent BR3 5EZ, UK

Gilchrist, Alison
University of Leeds
Department of Colour Chemistry
Leeds, LS2 9JT, UK

Gilmutdinov, AKh
Kazan Lenin State University
Department of Physics
Kazan, 420008, Russia

Gorenstein, D G
Sealy Centre for Structural Biology
Medical Branch
University of Texas
Galveston, Texas 77555-1157, USA

Grace, R Christy Rani
Department of Physics
Indian Institute of Science
Bangalore, India

Green-Church, Kari B
Department of Chemistry
Louisiana State University
Baton Rouge, LA 70803, USA

Greenfield, Norma J
Department of Neuroscience and Cell Biology,
Robert Wood Johnson Medical School
University of Medicine and Dentistry of
New Jersey
675 Hoes Lane, Piscataway,
NJ 08854, USA

Grime, G
Department of Materials
University of Oxford
Parks Road, Oxford, UK

Grutzmacher, Hans
Universität Bielefeld
Fakultat für Chemie
Postfach 100131,
D-33501, Bielefeld, Germany

Guillot, G
Unite de Recherche en Resonance,
Magnetique Medicale, CNRS URS 2212
Bat.220 Universite Paris-Sud
91405, ORSAY, Cedex France

Hallett, F R
Department of Physics
University of Guelph
Guelph, Ontario N1G 2W1, Canada

Hannon, A C
ISIS Facility, Rutherford Appleton Laboratory
Didcot, Oxon OX11 0QX, UK

Harada, Noboyuki
Tohoku University
Institute of Chemical Reaction Science
Sendai, 980 77, Japan

Hare, John F
SmithKline Beecham Pharmaceuticals
The Frythe, Welwyn,
Herts, AL6 9AR, UK

Harmony, Marlin D
Department of Chemistry, Marlott Hall
University of Kansas
Lawrence, Kansas 66045, USA

Harrison, A G
Chemistry Department
University of Toronto
80 St George Street, Toronto,
Ontario M5S 3H6, Canada

Hawkes, G E
Department of Chemistry,
Queen Mary and Westfield College
University of London
Mile End Road, London, E1 4NS, UK

Hayes, Cathy
Department of Botany
Trinity College
Dublin 2, Eire

Heck, Albert J R
Bijvoet Center for Biomolecular Research,
Utrecht University
Department of Chemistry and Pharmacy
Sorbonnelaan 16, 3584 CA, Utrecht,
The Netherlands

Herzig, Peter
Institut für Physikalische Chemie
Universität Wien
Währingerstraße 42,
A-1090, Wien, Austria

Hess, Peter
Physikalisch-Chemisches Institut
Universität Heidelberg
Im Neuenheimer Feld 253,
D-69120, Heidelberg, Germany

Hicks, J M
Department of Chemistry
Georgetown University
Washington DC, 20057, USA

Hildebrandt, Peter
Max-Planck-Institut für Strahlenchemie
Postfach 101365,
D-45413, Mülheim/Ruhr, Germany

Hill, Steve J
Department of Environmental Science
University of Plymouth
Drake Circus, Plymouth PL4 8AA, UK

Hills, Brian P
Institute of Food Research
Norwich Laboratory
Norwich Research Park, Colney,
Norwich NR4 7UA, UK

Hockings, P D
SmithKline Beecham Pharmaceuticals
Analytical Sciences Department
The Frythe, Welwyn,
Herts, AL6 9AR, UK

Hofer, Tatiana
Universität Kaiserslautern
Fachbereich Chemie der
D-67663, Kaiserslautern, Germany

Hoffmann, G G
Hoffmann Datentechnik
Postfach 10 06 31,
D-46006, Oberhausen, Germany

Holcombe, James A
Department of Chemistry
University of Texas
Austin, Texas7871-1167, USA

Holliday, Keith
University of San Francisco
Department of Physics
2130 Fulton Street, San Francisco,
CA 94117, USA

Holmes, John L
Department of Chemistry
University of Ottawa
PO Box 450, Stn 4, Ottawa, K1N 6N5, Canada

Homer, J
Chemical Engineering and Applied Chemistry,
School of Engineering and Applied Science
Aston University
Aston Triangle, Birmingham, B4 7ET, UK

Hore, P J
Physical and Theoretical Chemistry Laboratory
University of Oxford
South Parks Road, Oxford, OX1 3QZ, UK

Huenges, Martin
Technische Universität Munchen
Institut für Organische Chemie and Biochemie -
Leharul II
Lichtenbergatrahe 4,
D-85747, Garching, Germany

Hug, W
Institut de Chimie Physique
Universite de Fribourg
CH-1700, Fribourg, Switzerland

Hunter, Edward P L
Physical and Chemistry Properties Division (838)
Physics Building (221), Room A 113
NIST, PHY A 111, Gaithersburg,
Maryland 20899, USA

Hurd, Ralph
GE Medical Systems
47697 Westinghouse Drive, Fremont,
California 94539, USA

Imhof, Robert E
Department of Physics and Applied Physics
Strathclyde University
Glasgow, G4 0NG, UK

Jackson, Michael
National Research Council Canada
Institute for Biodiagnostics
435 Ellice Avenue, Winnipeg,
Manitoba R3B 1Y6, Canada

Jalsovszky, G
Chemistry Research Centre, Institute of Chemistry
Hungarian Academy of Sciences
PO Box 17, H-1525, Budapest, Hungary

Jellison, G E
Oak Ridge National Laboratory
Solid State Division
POB 2008, Oak Ridge, Tennessee, TN 37831, USA

Jokisaari, J
Department of Physical Sciences
University of Oulu
P O Box 3000, Oulu, FIN-90Y01, Finland

Jonas, J
School of Chemical Sciences
University of Illinois
Urbana, Illinois, 61801, USA

Jones, J R
Department of Chemistry
University of Surrey
Guildford, Surrey GU2 5XH, UK

Juchum, John
University of Florida
College of Pharmacy
P.O.Box 100485, Gainsville, FL 32610, USA

Katoh, Etsuko
National Institute of Agrobiological Resources
2-1-2, Kannondai, Tsukuba, Ibaraki 305-0856, Japan

Kauppinen, J
University of Turku
Department of Applied Physics
FIN-20014, Turku 50, Finland

Kessler, Horst
Institut für Organische Chemie und Biochemie
Technische Universität München
Lichtenbergstrabe 4,
D-85747, Garching, Germany

Kettle, S F
School of Chemical Sciences
University of East Anglia
Norwich, NR4 7TJ, UK

Kidder, Linda H
National Institues of Digestive and Diabetes and Kidney Diseases, National Institutes of Health
Laboratory of Chemical Physics
Bethesda, MD 20892, USA

Kiefer, Wolfgang
Institut für Physikalische Chemie
Der Universität Wurzburg
Am Hubland,
D-97074, Wurzburg, Germany

Kiesewalter, Stefan
Universität Kaiserslautern
Fachbereich Chemie der
D-67663, Kaiserslautern, Germany

Kimmich, Rainer
Universität Ulm
Sektion Kernresonanzspektroskopie
D-89069, Ulm, Germany

Klinowski, J
Department of Chemistry
University of Cambridge
Lensfield Road,
Cambridge, CB2 1EW, UK

Koenig, J L
Case Western Reserve University
Department of Macromolecular Science
10900 Euclid Avenue, Cleveland,
Ohio 44106-7202, USA

Kolemainen, E
Department of Chemistry
University of Jyvaskyla
Jyvaskyla, FIN-40351, Finland

Kooyman, R P H
University of Twente
Department of Applied Physics
Enschede, NL 7500 AE, The Netherlands

Kordesch, Martin E
Department of Physics and Astronomy
Ohio University
Athens, Ohio 45701, USA

Kotlarchyk, M
Department of Physics
3242 Gosnell, Rochester Institute of Technology
85 Lomb Memorial Drive, Rochester,
NY 14623-5603, USA

Kramar, U
Institute of Petrography and Geochemistry
University of Karlsruhe
Kaiserstrasse 12,
D-76128, Karlsruhe, Germany

Kregsamer, P
Atominstitut der Osterreichischen Universitaten
Stadionallee 2, 1020, Wien, Austria

Kruppa, Alexander I
Institute of Chemical Kinetics and Combustion
Novosibirsk-90, 630090, Russia

Kuball, Hans-Georg
Universität Kaiserslautern
Fachbereich Chemie der
D-67653, Kaiserslautern, Germany

Kumar, A
Department of Physics and Sophisticated Instruments Facility
Indian Institute of Science
Bangalore, 560012,
Karnataka, India

Kushmerick, J G
The Pennsylvania State University
Department of Chemistry
University Park, PA 16802-6300, USA

Kvick, Ake
European Synchrotron Radiation Facility
BP 220, Avenue des Martyrs, F-38043,
Grenoble, France

Laeter, J Rde
Curtin University of Technology
Bentley, Western Australia 6102, Australia

Latosińska, Jolanta N
Insitute of Physics
Adam Mickiewicz University
Umultowska 85, 61-614, Poznań, Poland

Leach, M O
Clinical Magnetic Resonance Research Group
Institute of Cancer Research
Royal Marsden Hospital
Sutton, Surrey SM2 5PT, UK

Lecomte, S
CNRS-Université Paris VI
Thiais, France

Leshina, T V
Russian Academy of Sciences
Institute of Chemical Kinetics and Combustion
Novosibirsk-90, Russia

Levin, Ira W
National Institues of Digestive and Diabetes and Kidney Diseases, National Institutes of Health
Laboratory of Chemical Physics
Bethesda, MD 20892, USA

Lewen, Nancy S
Bristol Myers Squibb, 1 Squibb Dr., Bldg. 101 Rm B18, New Brunswick, NJ 08903, USA

Lewiński, J
Department of Chemistry
Warsaw University of Technology
Noakowskiego 3, PL-00664, Warsaw, Poland

Lewis, Neil
National Institutes of Digestivo and Diabetes and Kidney Diseases, National Institute of Health
Laboratory of Chemical Physics
Bethesda, MD 20892, USA

Leyh, Bernard
F.N.R.S. and University of Leige
Department of Chemistry (B6)
B.4000, Sart Tilman, Belgium

Lias, S
Physical and Chemical Properties Division (838)
Physics Building (221), Room A 113
NIST, PHY A 111, Gaithersburg,
Maryland 20899, USA

Lifshitz, Chava
Department of Physical Chemistry
The Farkas Centre for Light-induced Processes
The Hebrew University of Jerusalem
Jerusalem, 91904, Israel

Limbach, Patrick A
Louisiana State University
Department of Chemistry
Baton Rouge, LA 70803, USA

Lindon, John C
Biological Chemistry
Division of Biomedical Sciences
Imperial College School of Science, Technology and Medicine
Sir Alexander Fleming Building
South Kensington, London SW7 2AZ, UK

Linuma, Masataka
Hiroshima University
Department of Physics
1-3-1 Kagamiyama, Higashi,
Hiroshima, 739, Japan

Liu, Maili
The Chinese Academy of Sciences
Wuhan Institute of Physics and Mathematics
Laboratory of Magnetic Resonance and Atomic and Molecular Physics
Wuhan, 430071,
Peoples' Republic of China

Lorquet, J C
Departement of Chemie
Universite de Liege
Sart-Tilman B6 (Batiment. B6), B-4000, Liege 1, Belgium

Louer, D
Groupe Crystallochimique, Chemical Solids and Inorganic Molecules Laboratory
University of Rennes
CNRS UMR 6511 Ave. Gen. Leclerc,
F-35042, Rennes, France

Luxon, Bruce A
Sealy Center for Structural Biology
University of Texas Medical Branch
Galveston, Texas 77555, USA

Maccoll, Allan
10, The Avenue, Claygate, Surrey KT10 0RY, UK

Macfarlane, R D
Chemistry Department
Texas A & M University
College Station, TX 77843-3255, USA

Maerk, T D
Institut fuer Ionenphysik
Leopold Franzens Universitaet
Technikerstr. 25, A-6020, Innsbruck, Austria

Magnusson, Robert
Department of Electronic Engineering
University of Texas
Arlington, Texas TX 76019, USA

Mahendrasingam, A
Keele University
Physics Department
Staffordshire, ST5 5BG, UK

Maier, J P
Institut für Physikalische Chemie
Universitat Basel
Klingelbergstrasse, CH 4056, Basel, Switzerland

Makriyannis, A
School of Pharmacy
University of Connecticut
Storrs, CT 06269, USA

Malet-Martino, Myriam
Universite Paul Sabatier
Groupe de RMN Biomedicale, Laboratorie des IMRCP (UMR CNRS 5623)
31062, Toulouse, Cedex, France

Mamer, O
Mass Spectrometry Unit
McGill University
1130 Pine Avenue West, Montreal,
Quebec H3A 1A3, Canada

Mandelbaum, Asher
Technion-Israel Institute of Technology
Department of Chemistry
Technion, Haifa 32000, Israel

Mantsch, H H
National Research Council of Canada
Institute for Biodiagnostics
435 Ellice Avenue, Winnipeg, R3B 1Y6, Canada

Mao, Xi-an
The Chinese Academy of Sciences
Wuhan Institute of Physics and Mathematics,
Laboratory of Magnetic Resonance and Atomic and Molecular Physics
Wuhan, 430071, Peoples' Republic of China

March, Raymond E
Department of Chemistry
Trent University
Peterborough, Ontario K9J 7B8, Canada

Marchetti, Fabio
Universita di Camerino
Dipartmento di Scienze Chemiche
via S. Agostino 1, 62032, Camerino MC, Italy

Mark, Tilmann D
Loepold Franzens Universitat
Institut für Ionenphysik
Technikerstrasse 25, A-6020, Innsbruck, Austria

Marsmann, H C
University Gesemthsch. Paderborn
Fachbereich Chem.
Warburger Str. 100,
D-33095, Paderborn, Germany

Martino, Robert
Universite Paul Sabatier
Groupe de RMN Biomedicale
Laboratore des IMRCP (UMR CNRS 5623)
31062, Toulouse Cedex, France

Maupin, Christine L
Michigan Technological University
Department of Chemistry
Houghton, MI 4993, USA

McClure, C K
Department of Chemistry
Montana State University
Bozeman, MT 59171, USA

McLaughlin, D
Kodak Research Laboratories
Eastman Kodak Co.
Rochester, New York NY 14650, USA

McNab, Iain
Department of Physics
University of Newcastle
Newcastle-upon-Tyne, NE1 7RU, UK

McNesby, K L
6735 Indian River Drive, Citrus Heights,
CA 95621, USA

Meuzelaar, H L C
Center for Micro Analysis & Reaction Chemistry
University of Utah
110 South Central Campus Drive, Room 214,
Salt Lake City, Utah 84112, USA

Michl, Josef
Department of Chemistry & Biochemistry
University of Colorado
Boulder, CO 80309-0215, USA

Miklos, Andras
Institute of Physical Chemistry
University of Heidelberg
Im Neuenheimer Feld 253,
D-69120, Heidelberg, Germany

Miller, S A
Princeton University
Department of Chemistry
Princeton, NJ 08544, USA

Miller-Ihli, Nancy J
U.S. Department of Agriculture
Food Composition Laboratory
Building 161, Rm. 1, BARC-East, Beltsville,
MD 20705, USA

Morris, G A
Department of Chemistry
University of Manchester
Oxford Road, Manchester, M13 9PL, UK

Mortimer, R J
Department of Chemistry
Loughborough University
Loughborough, Leics LE11 3TU, UK

Morton, Thomas H
Department of Chemistry
University of California
Riverside, CA 92521-0403, USA

Muller-Dethlefs, K
University of York
Department of Chemistry
Heslington, York YO1 5DD, UK

Mullins, Paul G M
SmithKline Beecham Pharmaceuticals
Analytical Sciences Department
The Frythe, Welwyn, Herts, AL6 9AR, UK

Murphy, Damien M
Cardiff University
National ENDOR Centre, Department of Chemistry
Cardiff, CF1 3TB, UK

Nafie, L A
Department of Chemistry
Syracuse University
Syracuse, New York 13244-4100, USA

Naik, Prasad A
Room #201, R & D Block "D"
Centre for Advanced Technology
Indore, 452013, Madhya Pradesh, India

Nakanishi, Koji
Department of Chemistry
Columbia University
New York NY 10027, USA

Nicholson, J K
Biological Chemistry, Division of Biomedical Sciences
Imperial College of Science, Technology & Medicine
Sir Alexander Fleming Building,
South Kensington, London,
SW7 2AZ, UK

Nibbering, N M M
Institute of Mass Spectrometry
University of Amsterdam
Nieuwe Achtergracht 129, 1018 WS,
Amsterdam, The Netherlands

Niessen, W M A
hyphen MassSpec Consultancy
De Wetstraat 8, 2332 XT, Leiden,
The Netherlands

Nobbs, Jim
University of Leeds
Department of Colour Chemistry
Leeds, LS2 9JT, UK

Norden, B
Department of Physical Chemistry
Chalmers University of Technology
S-41296, Gothenburg, Sweden

Norwood, T
Department of Chemistry
University of Leicester
Leicester, LE1 7RH, UK

Olivieri, A C
Facultad de Ciencias Biochimicas y Farmaceuticas, Departamento Quimica Analitica
Universita Nacional Rosario
Suipacha 531,
RA-2000, Rosario, Santa Fe, Argentina

Omary, Mohammed A
University of Maine
Department of Chemistry
Orono, Maine ME 04469, USA

Parker, S F
Rutherford Appleton Laboratory, ISIS Facility
Oxon, Didcot OX11 0QX, UK

Partanen, Jari O
University of Turku
Department of Applied Phyics
FIN 20014,
Turku, Finland

Patterson, Howard H
Department of Chemistry
University of Maine
Orono, Maine ME 04469, USA

Pavlopoulos, Spiro
University of Connecticut
Institute of Materials Science and School of Pharmacy
Storrs, Connecticut 06269, USA

Pettinari, C
Università degli Studi di Camerino
Scienze Chimiche
Via S. Agostino 1, 62032, Camerino, MC, Italy

Poleshchuk, O K
Department of Inorganic Chemistry
Tomsk Pedagogical University
Komsomolskii 75, 634041, Tomsk,
Russian Federation

Rafaiani, Giovanni
Università di Camerino
Dipartimento di Scienze Chemie
Via S Agostino 1, 62032, Camerino MC, Italy

Ramsey, Michael H
Imperial College of Science Technology and Medicine
TH Huxley School of Environmental, Earth Science and Engineering
London, SW7 2BP, UK

Randall, Edward W
Department of Chemistry, Queen Mary & Westfield College
University of London
Mile End Road,
London, E1 4NS, UK

Rehder, D
Institute of Inorganic Chemistry
University of Hamburg
Martin-Luther-King Platz 6,
D-20146, Hamburg, Germany

Reid, David G
Smithkline Beecham Pharmaceuticals
Analytical Sciences Department
The Frythe, Welwyn, Herts, AL6 9AR, UK

Reid, Ivan D
Paul Scherrer Institute
CH-5232, Villigen PSI, Switzerland

Reynolds, William F
Lash Miller Chemical Laboratories
University of Toronto
80 George Street, Toronto,
Ontario M5S 1A1, Canada

Richards-Kortum, Rebecca
Dept. of Elec. and Computer Engin.
University of Texas at Austin
Austin, TX 78712, Inter-Office C0803, USA

Riddell, Frank G
Department of Chemistry
University of St Andrews
The Purdie Building, St Andrews, Fife KY16 9ST, Scotland

Riehl, J P
Michigan Technological University
Department of Chemistry
Houghton, MI 49931, USA

Rinaldi, Peter L
Department of Chemistry
University of Akron
Akron, Ohio OH 44325-3061, USA

Roberts, C J
The University of Nottingham
Department of Pharmaceutical Sciences
University Park, Nottingham, NG7 2RD, UK

Rodger, Alison
University of Warwick
Department of Chemistry
Coventry, CV4 7AL, UK

Rodger, C
University of Strathclyde
Department of Pure & Applied Chemistry
Glasgow, G1 1XL, UK

Roduner, Emil
Universität Stuttgart
Physikalisch-Chemisches Institut
Pfaffenwaldring 55, D-70550, Stuttgart, Germany

Rost, F W D
45 Charlotte Street, Ashfield, NSW 2131, Australia

Rowlands, C C
Cardiff University
National ENDOR Centre, Department of Chemistry
Cardiff, CF1 3TB, UK

Rudakov, Taras N
7 Reen Street, St James, WA 6102, Australia

Salamon, Z
University of Arizona
Department of Biochemistry
Tuscon, AZ 85721, USA

Salman, S R
Chemistry Department
University of Qatar
PO Box 120174, Doha, Qatar

Sanders, Karen
University of Warwick
Department of Chemistry
Coventry, CV4 7AL, UK

Sanderson, P N
Protein Science Unit
GlaxoWellcome Medicines Research Centre
Gunnels Wood Road, Stevenage,
Herts, SG1 2NY, UK

Santini, Carlo
Università di Camerino
Dipartimento di Scienze Chemiche
Via s. Agostino 1, 62032, Camerino MC, Italy

Santos Gómez, J
Instituto de Estructura de la Materia, CSIC
28006, Madrid, Spain

Schafer, Stefan
Institute of Physical Chemistry
University of Heidelberg
Im Neuenheimer Feld 253,
D-69120, Heidelberg, Germany

Schenkenberger, Martha M
Bristol-Myers Squibb
Pharmaceutical Research Institute
1 Squibb Dr., New Brunswick,
NJ 08903, USA

Schrader, Bernhard
Institut für Physikalische und Theoretisch Chemie
Universität Essen
Fachbereich 8, D-45117, Essen, Germany

Schulman, Stephen G
Department of Medicinal Chemistry
College of Pharmacy
University of Florida
Gainesville, FL 32610-0485, USA

Schwarzenbach, D
University of Lausanne
Institute of Crystallography
BSP Dorigny,
CH-1015, Lausanne, Switzerland

Seitter, Ralf-Oliver
Universität Ulm
Sektion Kernresonanzspektroskopie
89069, Ulm Germany

Seliger, J
Department of Physics
Faculty of Mathematics and Physics
University of Ljubljana
Jadranska, 19, 1000,
Ljubljana Slovenia

Shaw, R Anthony
National Research Council Canada
Institute for Biodiagnostics
435 Ellice Avenue, Winnipeg,
Manitoba R3B 1Y6, Canada

Shear, Jason B
Department of Chemistry
University of Texas
Austin, TX 78712, USA

Sheppard, Norman
School of Chemical Sciences
University of East Anglia
Norwich, NR4 7TJ, UK

Shluger, A L
Department of Physics and Astronomy
University College London
Gower Street, London, WC1E 6BT, UK

Shockcor, J P
Bioanalysis and Drug Metabolism
Stine-Haskell Research Center
Dupont-Merck, P O Box 30, Elkton Road, Newark,
Delaware DE 19714, USA

Shukla, Anil K
University of Delaware
Department of Chemistry and Biochemistry
Newark,
DE 19176, USA

Shulman, R G
MR Center, Department of Molecular Biophysics
Yale University
New Haven, Connecticut 06520, USA

Sidorov, Lev N
Physical Chemistry Division
Chemistry Department
Moscow State University
119899, Moscow, Russia

Sigrist, Markus
Swiss Federal Insitute of Technology (ETH)
Insitute of Quantum Electronics, Laboratory for
Laser Spectroscopy and Environmental Sensing
Hoenggerberg,
CH-8093, Zurich, Switzerland

Simmons, Tracey A
Department of Chemistry
Louisiana State Universty
Baton Rouge, LA 70803, USA

Smith, W E
University of Strathclyde
Department of Pure & Applied Chemistry
Glasgow, G1 1XL, UK

Smith, David
The University of Keele
Department of Biomedical Engineering and
Medical Physics, Hospital Centre
Thornburrow Drive, Hartshill,
Stoke-on-Trent ST4 7QB, UK

Snively, C M
Department of Macromolecular Science
Case Western Reserve University
Cleveland, OH 44106, USA

Somorjai, J
Institute for Biodiagnostics
435 Ellice Avenue, Winnipeg,
R3B 1Y6, Canada

Španěl, Patrick
Keele University
Department of Biomedical Engineering and
Medical Physics
Thornburrow Drive, Hartshill,
Stoke-on-Trent ST4 7QB, UK

Spanget-Larsen, Jens
Department of Life Sciences and Chemistry
Roskilde University
POB 260, DK-4000, Roskilde, Denmark

Spiess, H W
Max Plank Institute of Polymer Research
Postfach 3148, D-55021, Mainz, Germany

Spinks, Terence
PET Methodology Group, MRC Clinical Sciences
Centre, Royal Postgraduate Medical School
Hammersmith Hospital
Du Cane Road,
London, W12 0NN, UK

Spragg, R A
Perkin-Elmer Analytical Instruments
Post Office Lane, Beaconsfield,
Bucks HP9 1QA, UK

Standing, K G
Department of Physics
University of Manitoba
Winnipeg, Manitoba R3T 2N2, Canada

Steele, Derek
The Centre for Chemical Sciences
Royal Holloway
University of London
Egham, Surrey TW40 0EX, UK

Stephens, Philip J
University of Southern California
Department of Chemistry
Los Angeles, CA 90089-0482, USA

Stilbs, Peter
Royal Institute of Technology Physical Chemistry
S-10044, Stockholm, Sweden

Streli, C
Atominstitut of the Austrian Universities
Stadionallee 2, 1020, Wien, Austria

Styles, Peter
John Radcliffe Hospital
MRL Biochemistry & Chemical Magnetic
Resonance Unit
Headington, Oxford,
OX3 9DU, UK

Sumner, Susan C J
Chemical Industry Institute of Toxicology
6 Davis Drive, PO Box 12137, Research Triangle
Park, North Carolina
NC 27709-2137, USA

Sutcliffe, L H
Institute of Food Research
Norwich NR4 7UA, UK

Szepes, Laszlo
Eotvos Lorand University
Department of General and Inorganic Chemistry
Pazmany Peter Satany 2, 1117,
Budapest Hungary

Taraban, Marc B
Institute of Chemical Kinetics and Combustion
Novosibirsk-90, 630090, Russia

Tarczay, Gyorgy
Eotvos University
Department of General and Inorganic Chemistry
Pazmany Peter S. 2,
Budapest H-1117, Hungary

Taylor, A
Robens Institute of Health and Safety
Trace Elements Laboratory
University of Surrey
Guildford, Surrey GU2 5XH, UK

Tendler, S J B
University of Nottingham
Department of Pharmaceutical Sciences
University Park, Nottingham, NG7 2RD, UK

Terlouw, J
McMaster University
Department of Chemistry
ABB-455, 1280 Main Street West,
Hamilton, ON L8S 4M1, Canada

Thompson, Michael
University of London
Birkbeck College, Department of Chemistry
Gordon House, 29 Gordon Square,
London, WC1H 0PP, UK

Thulstrup, Erik W
Department of Chemistry & Life Sciences
Roskilde University (RUC)
Building 17.2, PO Box 260,
DK-4000, Roskilde, Denmark

Tielens, A G G M
Kapteyn Astronomical Institute
PO Box 800, 9700 AV, Groningen, The Netherlands

Tollin, Gordon
University of Arizona
Department of Biochemistry
Tucson, AZ 85721, USA

Traeger, John C
Department of Chemistry
La Trobe University
Bundoora, Victoria 3083, Australia

Tranter, George E
Glaxo Wellcome Medicines Research Centre
Gunnels Wood Road, Stevenage,
Herts, SG1 2NY, UK

Trombetti, A
Università di Bologna
Dipartimento Chimica Fisica e Inorganica
Viale Risorgimento 4, I-40136, Bologna, Italy

True, N S
Department of Chemistry
University of California Davis
Davis, CA 95616, USA

Ulrich, Anne S
Institute of Molecular Biology
University of Jena
Winzerlaer Strasse 10, D-07745, Jena, Germany

Utzinger, Urs
The University of Texas at Austin
Texas USA

Van Vaeck, Luc
Department of Chemistry
University Instelling Antwerp
University Pleim 1, B-2610, Antwerp, Belgium

Vandell, Victor E
Louisiana State University
Department of Chemistry
Baton Rouge, LA 708023, USA

Varmuza, Klaus
Department of Chemometrics
Technical University of Vienna
A-1060, Vienna Austria

Veracini, C A
Dipartimento Chemica
University of Pisa
I-56126, Pisa, Italy

Viappiani, Christiano
Dipartimento di Scienze Ambientali
Universita degli Studi di Parma
viale delle Scienze, I-43100, Parma, Italy

Vickery, Kymberley
Chalmers University of Technology
Department of Physical Chemistry
S-412 96, Gothenburg, Sweden

Wagnière, Georges H
Physikalisch-Chemisches Institut der Universitat Zurich
Winterhurerstrasse 190,
CH-8057, Zurich, Switzerland

Waluk, Jacek
Institute of Physical Chemistry
Polish Academy of Sciences
01-224 Warszawa, Kasprzaka, 44/52, Poland

Wasylishen, Roderick E
Department of Chemistry
Dalhousie University
Halifax, Nova Scotia B3H 4J3, Canada

Watts, Anthony
Department of Biochemistry
University of Oxford
Oxford, OX1 3QU, UK

Webb, G A
University of Surrey
Department of Chemistry
Guildford, GU2 5XH, UK

Weiss, P S
Penn State University
Department of Chemistry
University Park,
Pennsylvania PA 16802-6300, USA

Weller, C T
School of Biomedical Sciences
University of St Andrews
St Andrews, KY16 9ST, UK

Wenzel, Thomas J
Department of Chemistry
Bates College
Lewiston, Maine ME 04240, USA

Wesdemiotis, Chrys
Chemistry Department
University of Akron
Akron, OH 44325-3601, USA

Western, Colin M
University of Bristol
School of Chemistry
Bristol, BS8 1TS, UK

White, R L
Department of Chemistry
University of Oklahoma
Norman, Oklahoma OK 73019-0390, USA

Wieser, Michael E
University of Calgary
2500 University Drive NW, Calgary,
Alberta T2N 1N4, Canada

Wilkins, John
Unilever Research
Colworth, Sharnbrook Beds. MK44 1LQ, UK

Williams, P M
The University of Nottingham
Laboratory of Biophysics and Surface Analysis,
School of Pharmaceutical Sciences
University Park, Nottingham, NG7 2RD, UK

Williams, Antony J
Wobrauschek P Atominstitut der Osterreichischen Universitaten
Stadionallee 2, 1020, Wien, Austria

Wlodarczak, G
Laboratorie de Spectroscopie Hertzienne
URA CNRS 249, Univ. de Lille 1,
F59655, Villeneured'Aacg Cedese, France

Woźniak, Stanisław
Mickiewicz University
Warsaw, Poland

Young, Ian
Robert Steiner MR Unit
Royal Postgraduate Medical School
Hammersmith Hospital
Ducane Road, London, W12 0HS, UK

Zagorevskii, Dimitri
Department of Chemistry
University of Missouri-Columbia
Columbia MO 65201, USA

Zoorob, Grace K
Biosouth Research Laboratories Inc.
5701 Crawford Street, Harahan,
LA 70123, USA

Zwanziger, J W
Department of Chemistry
Indiana University
Bloomington, Indiana 47405, USA

Contents

Volume 1

A

B

C

D

E

F

G

H

Volume 2

I

L

M

N

Volume 3

O

P

Q

R

S

X

Y

Z

Entry Listing by Subject Area

Subject areas follow each other alphabetically in this list. Within each subject area entries are categorised into those covering historical aspects, theory, methods and instrumentation, or applications. The entries are listed alphabetically within these categories.

Atomic Spectroscopy

Historical Overview

Theory

Methods & Instrumentation

Applications

Electronic Spectroscopy

Theory

Fundamentals of Spectroscopy

Theory

Methods & Instrumentation

High Energy Spectroscopy

Theory

Methods & Instrumentation

Applications

Magnetic Resonance

Historical Overview

Theory

Methods & Instrumentation

Applications

Mass Spectrometry

Historical Overview

Theory

Methods & Instrumentation

Applications

Spatially Resolved Spectroscopic Analysis

Theory

Methods & Instrumentation

Applications

Vibrational, Rotational & Raman Spectroscopies

Historical Overview

Theory

Methods & Instrumentation

Applications

I

In Vivo NMR, Applications, ^{31}P

Ruth M Dixon and **Peter Styles**, University of Oxford, UK

MAGNETIC RESONANCE
Applications

Introduction

The phosphorus nucleus has occupied centre stage in the development of *in vivo* NMR spectroscopy for two important reasons. First, the NMR properties of the nucleus are well suited to *in vivo* NMR (100% natural abundance, reasonable sensitivity and adequate but not excessive chemical shift). Second, key metabolites, particularly those involved in the production of energy, contain phosphorus and are present in substantial concentrations in living tissue. It is for these reasons that the first NMR spectra of living cells (a suspension of erythrocytes), and later of excised frog muscle, were obtained using ^{31}P NMR. In both instances, resonances were observed from adenosine triphosphate and inorganic phosphate, the position of the latter being an indicator of intracellular pH. Additional peaks were seen from 2,3-diphosphoglycerate (2,3-DPG) in red blood cells and creatine phosphate in skeletal muscle. Despite the comparative simplicity of ^{31}P NMR tissue spectra, the technique offers unique insight into the regulation of critical energetic pathways. The early studies of *ex vivo* samples were soon extended to *in vivo* examination of intact animals and later to humans, experiments which were made possible by the development of large superconducting magnets, the use of local surface coil probes and the invention of a variety of techniques for localizing the source of the NMR signal.

Techniques for ^{31}P spectroscopy *in vivo*

Signal acquisition and spatial localization

Perhaps the most critical step in the development of *in vivo* spectroscopy was the realization that it was not necessary to surround the 'sample' with a conventional NMR probe, but that a small local coil (a surface coil) could be placed against an animal to preferentially receive signals from the tissues adjacent to this coil. The surface coil has two advantages, namely excellent sensitivity close to the coil and inherent localizing properties owing to the rapid fall-off of the magnetic flux away from the coil. However, when deep tissues are to be investigated, additional localization is required. In choosing an appropriate technique for obtaining ^{31}P NMR spectra from a live animal or man, careful consideration must be given to the spatial resolution and signal-to-noise ratio which can reasonably be expected. As the phosphorus nucleus has only 6% of the sensitivity of the proton, it follows that these expectations will be somewhat more modest than for ^{1}H spectroscopy. The usual approach is to adopt the least demanding methodology appropriate for any particular application. As a consequence, in many instances simple surface coil localization is used to give adequate localization and provide the best signal-to-noise ratio in short acquisition times. When this simple approach is inadequate, a range of single and multiple voxel methods can be considered, but care must be taken to choose methods which work well at short echo times to avoid excessive T_2 relaxation and minimize problems due to the J modulation of peaks such as adenosine triphosphate (ATP).

Dealing first with the single voxel approach, image selected *in vivo* spectroscopy (ISIS) is frequently the method of choice because the spins are kept predominantly along the z axis, thus minimizing T_2 and J coupling effects. Other localizing schemes (e.g. stimulated echo acquisition mode spectroscopy (STEAM) and point resolved spectroscopy (PRESS)) which are commonly used for proton spectroscopy can give reasonable results, but are more sensitive to the aforementioned problems.

Where multivoxel localization is advantageous, chemical shift imaging techniques can be applied in one, two or three dimensions, but relatively poor spatial resolution is obtainable and long acquisition times are usually required.

Finally, it will often be advantageous to implement hybrid protocols in which, for example, two dimensions might be defined by an ISIS sequence, and phase encoded chemical shift imaging (CSI) implemented along the third axis, thus producing a set of spectra that represent slices through a well-defined column of tissue.

Proton decoupling

The quality of *in vivo* ^{31}P spectra can be improved by the use of proton decoupling. Two benefits are available. Some of the phosphorus spectral lines are significantly broadened by scalar coupling to hydrogen. Irradiation of the protons during acquisition removes this interaction and sharpens the spectra. The effect is particularly advantageous in the phosphomonoester (PME) and phosphodiester (PDE) regions of the spectrum where conventional acquisition will fail to resolve, for example, glycerophosphocholine and glycerophosphoethanolamine. Together with the increased resolution, proton decoupling also increases the signal-to-noise ratio due to peak narrowing. A further increase in signal is available by low power decoupling during the interpulse relaxation period. Here, the nuclear Overhauser effect (NOE) can result in improved sensitivity of up to about 50% owing to interactions between the phosphorus nuclei and neighbouring protons.

Magnetisation transfer

Under certain conditions, rates of chemical exchange can be measured using saturation- or inversion-transfer methods. These work by magnetically labelling one chemical species and observing the transfer of this magnetic signature to a second compound which is coupled by chemical exchange. For the technique to be viable, there needs to be significant exchange during a time scale which is of the order of the longitudinal relaxation time T_1. Such reactions include the synthesis and hydrolysis of ATP (by measuring exchange between ATP and inorganic phosphate) and the fast exchange catalysed by the creatine kinase enzyme.

The *in vivo* ^{31}P NMR spectrum

Figure 1 shows spectra from human skeletal muscle, heart, liver and brain. Each contains the three peaks from ATP and one from inorganic phosphate. ATP shows three signals due to the different chemical environments of its three phosphorus atoms. These signals are further split by ^{31}P–^{31}P *J*-coupling, but this splitting is not always detectable *in vivo* owing to the relatively broad lines. These peaks are normally designated 'ATP' but also contain minor contributions from other nucleotides, such as guanosine triphosphate, NADH, NADP, UDP-sugars and ADP. The signal which contains the least contamination from other compounds is the one from the central (beta) phosphate, which appears at the upfield end of the spectrum, and so this peak is generally used for the measurement of ATP levels. In addition, spectra from muscle, heart and brain contain a prominent peak from creatine phosphate. This compound is absent from liver. Liver and brain spectra also contain substantial peaks designated as phosphomonoester (PME) and phosphodiester (PDE). These regions of the spectra can contain metabolites involved in membrane metabolism and also phosphorylated sugars involved in glycolysis. An important feature of the ^{31}P spectrum is that the position of the inorganic phosphate (P_i) peak is pH dependent, thus providing a sensitive marker of intracellular acidity/alkalinity.

Biological relevance of the peaks

Adenosine triphosphate (ATP) ATP is the universal energy carrier in all cells. When energy is required for cell functions such as preserving ionic gradients, muscular contraction, etc., the terminal phosphate is cleaved by ATPase enzymes to form adenosine diphosphate (ADP) and inorganic phosphate, an

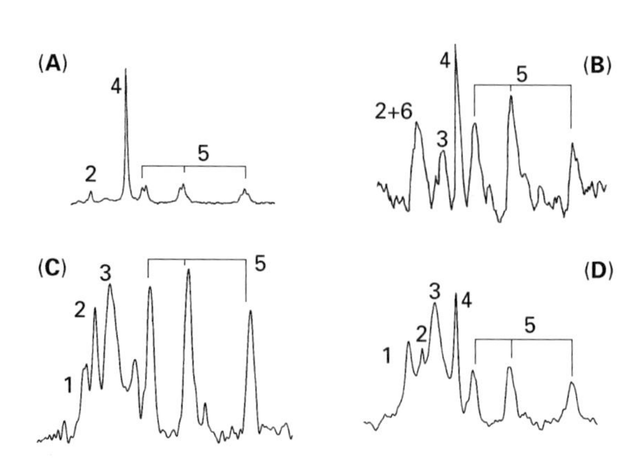

Figure 1 ^{31}P spectra of human organs (A) calf muscle, (B) heart, (C) liver, (D) brain. Peak assignments: 1. PME, 2. Pi, 3. PDE, 4. PCr, 5. ATP, 6. 2,3-DPG. Reproduced with permission of Academic Press from Fu, Schroeder, Jones, Santini and Gorenstein (1998) *Journal of Magnetic Resonance* **77**: 577–582.

energy producing reaction:

$$ATP + H_2O \rightarrow ADP + P_i$$

ATP is resynthesized by one of several pathways, most usually by ATP synthesis using energy obtained from oxidative or anaerobic metabolism.

Inorganic phosphate (P_i) This compound may be considered as the waste product of ATP hydrolysis or conversely as a substrate for ATP synthesis. It may also play a role in regulating the supply of energy via oxidative metabolism in the cells' mitochondria. At physiological pH, P_i exists in two ionic forms in fast equilibrium via the reaction:

$$H_2PO_4^- \rightleftharpoons HPO_4^{2-} + H^+$$

This reaction has a pK_a of 6.75, and so the relative concentrations of P_i^- and P_i^{2-} vary with the pH. Owing to the fast exchange between these two species, the position of the P_i peak in the spectrum reflects the weighted average of the two chemical shifts, and is thus a direct measure of pH.

Creatine phosphate (PCr) This molecule is present in abundance in muscle, heart and brain cells. Through the fast equilibrium reaction catalysed by creatine kinase, PCr buffers the ATP concentration by phosphorylating ADP, the reaction being:

$$PCr + ADP + H^+ \rightleftharpoons ATP + Cr$$

PCr is depleted during sudden energetic demands such as vigorous muscle contraction.

Adenosine diphosphate (ADP) ADP is a substrate for ATP production, and a product of its hydrolysis. ADP stimulates respiration (oxidative metabolism) – breakdown of ATP increases ADP which in turn activates the processes which provide energy for ATP synthesis. In tissues such as muscle, much of the ADP is bound to macromolecules and therefore invisible to NMR. However, the free (and therefore metabolically active) ADP usually present in only micromolar concentrations, may be deduced using the creatine kinase equilibrium (see above).

Phosphomonoesters (PME) Certain intermediaries in the glycolytic pathway such as glucose and fructose phosphates resonate in this region of the spectrum and are sometimes detected. More usually, the PME peak comprises the lipid precursors phosphocholine and phosphoethanolamine.

Phosphodiester (PDE) This region also contains compounds involved in membrane metabolism, specifically glycerophosphocholine (GPC) and glycerophosphoethanolamine (GPE).

Broad resonances In addition to the sharp peaks that have just been mentioned, there will often be underlying broad 'humps' in the phosphorus spectrum, arising from phosphorus moieties that are immobile or tumbling very slowly. Phospholipid head groups in cell membranes produce a broad and asymmetric signal centred under the PDE part of the spectrum. Bone also produces a broad resonance, several kilohertz wide, which is easily visible in surface coil spectra of the head, but is not usually present in localized spectra due to its very short T_2.

Applications in skeletal muscle

Skeletal muscle is a tissue that is particularly suited to study by ^{31}P NMR. From a technical standpoint, muscle is an ideal target because it exists close to the body surface and is not obscured by any other pool of phosphorus-containing metabolites. From a biochemical standpoint, muscle is especially interesting because its energetic demands vary by at least two orders of magnitude (from rest to exercise) and so it provides an example of exquisite metabolic control which can be studied non-invasively. Finally, muscle can be stressed readily by exercise within the NMR instrument, thus facilitating the real-time investigation of the dynamics of energetic regulation.

Animal studies of skeletal muscle often employ direct stimulation of the sciatic nerve to induce contraction and associated metabolic response. Basic metabolic control can be investigated using such a model, as can a wide range of disease-related processes. Genetically abnormal animals provide models of diseases such as muscular dystrophy and ^{31}P NMR spectroscopy provides a tool for investigating metabolism, pH regulation and possible therapeutic interventions.

In a typical human examination, a subject will perform an exercise regime within the magnet, perhaps squeezing a handgrip or pressing down a loaded pedal. The NMR signals are usually collected using a surface coil placed over the exercising muscle and a series of spectra are collected to follow the metabolic state before, during and after the exercise. A typical time course is shown in **Figure 2**. The several metabolite concentrations and pH are then used to quantitatively assess the metabolic response to the stress, and, by using models of metabolic control, unique insight can be gained into the complementary processes of mitochondrial respiration and anaerobic metabolism.

Muscle metabolism

Figure 3 is a block diagram of the main energy-producing pathways in skeletal muscle (and other tissues, except that the creatine kinase pathways may not be present). In resting muscle, ATP is synthesized only by oxidative metabolism, but during exercise glycolysis (anaerobic metabolism) and PCr breakdown via the creatine kinase reaction provide additional ATP production. Although only three compounds are directly detected by ^{31}P NMR, others may be deduced by making assumptions based on data derived from biopsy analysis. In particular, an estimate of free [ADP] is obtained using assumed values of ATP and total creatine concentrations, and applying the known value for the creatine kinase equilibrium constant.

In using these data to investigate metabolic control *in vivo*, two periods within an exercise regime are of particular importance. At the start of exercise, energy is produced primarily by glycolysis and a breakdown of creatine phosphate. The latter is measured directly while the former produces lactate which is reflected in the pH. At the start of recovery, glycolysis stops, the energy demands fall dramatically and oxidative metabolism restores the PCr. Using models for proton efflux and metabolic control by ADP, it is possible to assess the relative contributions of the various energy pathways in both normal and diseased muscle.

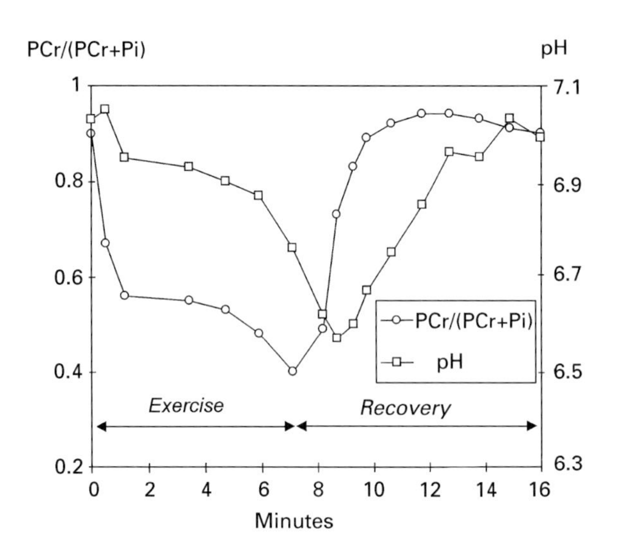

Figure 2 A typical time course for changes in PCr and pH during exercise and recovery in the human forearm muscle.

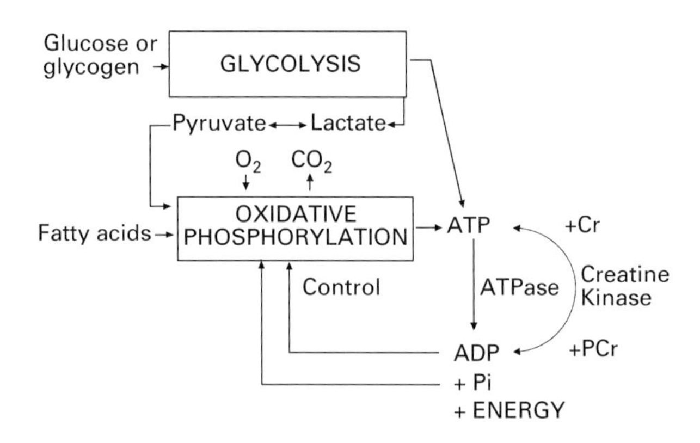

Figure 3 A schematic diagram of energy metabolism in skeletal muscle, heart and brain. The creatine kinase pathway is absent in liver.

Diseases of skeletal muscle

In man, the understanding and quantification of muscle energetics has facilitated the study and, sometimes, diagnosis of a wide range of muscle diseases. Often these stem from inherited genetic abnormalities, including the dystrophies, mitochondrial myopathies and so-called energy storage diseases. Muscle can also be affected by arterial disease and systemic abnormalities secondary to, for example, renal and cardiac failure. Many studies have looked at these phenomena and monitored the response to training or therapeutic intervention.

Applications in the heart

Basic NMR research of the heart has concentrated on studies of the organ *ex vivo* (the excised heart, perfused with physiological buffer, can be maintained in a viable and beating condition for many hours). *In vivo* studies in both animals and man are more difficult – the heart is a moving target, is filled with blood and overlaid with skeletal muscle which gives a similar ^{31}P NMR signature. The approaches that have been applied in animal systems often involve exposing the organ and suturing a surface coil to the heart wall, thus facilitating the measurement of localized metabolic events. In man, single voxel ISIS is sometimes used, or chemical shift imaging, usually as a hybrid experiment (see above) where spectra are obtained from parallel slabs of tissue. The biochemistry of the heart is in many ways similar to skeletal muscle. However, the dynamic range of energetic demand is less extreme than in muscle and in normal controls it is difficult to induce significant changes in the ^{31}P spectrum from the heart. In the diseased heart, the main findings have been that cardiac energetics in the resting subject are substantially stable until the onset of heart failure, when PCr falls significantly. In patients with coronary artery disease, metabolic levels can be modulated by exercise. As an alternative to exercise, the positive inotrope atropine-dobutamine has been administered to normal subjects, and at high cardiac output a reduction in the PCr:ATP ratio has been observed.

Intracellular pH is difficult to measure because the P_i peaks can be obscured by 2,3-DPG in the blood. However, it is possible to overcome this problem using proton decoupling to sharpen the resonances, or saturation transfer to identify the intracellular P_i unambiguously.

Applications in brain

Using animal models, many disorders of the brain have been studied, including the effects of ischaemia (e.g. models of stroke and birth asphyxia), systemic disturbances (e.g. hepatic encephalopathy) and genetic abnormalities (e.g. models of Duchenne dystrophy). Spatial localization is usually kept to a minimum – often the scalp is removed to avoid contamination by signals from the cranial musculature, and a single surface coil is used to obtain the ^{31}P NMR signal either from the whole brain or, by appropriate placing of the coil, predominantly from a single hemisphere. Almost invariably it is the energy metabolism of the brain which is of interest.

There have been many studies using ^{31}P NMR to assess disease processes in man. Perhaps the most interesting have been the investigation of birth asphyxia in neonates where energy metabolite levels and intracellular pH give direct measures of brain damage and provide a good prognostic index. Other disorders which have been investigated include stroke, multiple sclerosis, epilepsy, hydrocephalus, tumours, dementias and systemic encephalopathies. In all cases there have been reports of altered energy metabolism (reduced PCr or increased P_i) and, in multiple sclerosis, changes have been observed in the broad phospholipid peak. The pH data are somewhat more variable – acidic values have been reported in acute stroke and hydrocephalus, but alkaline values have been observed in chronic stroke and brain tumours. One of the main limitations in performing ^{31}P NMR spectroscopy in the human brain is that the spatial resolution is inadequate for interrogating focal lesions of moderate size. Nevertheless the global pictures that have emerged give insight into the way that brain energetics respond in disease.

Applications in liver

The liver is relatively simple to study *in vivo* by ^{31}P spectroscopy. On a macroscopic level it is homogeneous, and localization is generally straightforward. Since hepatocytes do not express creatine kinase, the liver contains no detectable phosphocreatine and so the degree of contamination of the liver spectra by signals from overlying muscle can be readily assessed (and in some cases corrected for).

Localization strategies have ranged from simple surface coil detection (with a hard 180° inversion pulse at the surface to suppress the muscle signal) through Fourier series window approaches and ISIS to one-, two- or three-dimensional chemical shift imaging. Proton decoupling and NOE irradiation have occasionally been employed, but the technical requirements and power deposition considerations have limited these to a few cases. In animal studies, localization methods are often avoided by exposing the liver and using a surface coil placed directly on the organ.

Liver metabolism

The liver's role is to maintain chemical homeostasis and it can, in certain circumstances, be stressed by means of a metabolic load. One example has been to administer an oral fructose load which causes minimal spectral changes in normal subjects but, at very low dosage, results in an increase in the fructose-1-phosphate peak in patients with fructose intolerance. Other metabolic stresses have been administered to investigate galactose intolerance, glycogen storage diseases and cirrhosis. The rat liver (in situ or perfused) has been studied to elucidate details of carbohydrate and fatty acid metabolism, sometimes by a combination of ^{13}C and ^{31}P NMR. The control of pH in the liver has also been studied.

Liver diseases

Some common diseases that involve the liver have also been studied by ^{31}P NMR. These include alcoholic liver disease, hepatitis, cirrhosis of various etiologies and liver tumours. The findings have often allowed the diseased liver to be distinguished from the healthy one, but the changes have not usually been specific to a particular disease and the interpretation of the changes is often speculative. Overall, these studies have shown that the major long-term changes in liver spectra involve the PME and the PDE signals, which probably signify changes in phospholipid metabolism (see below) while the energetics of the liver (assessed from the P_i:ATP ratio and the pH) remains normal until a very late stage of disease. Another area that shows promise is in assessing the transplanted liver, either before or after transplantation.

Spectral interpretation

The PME and the PDE signals in the liver spectrum are each composed of several components, which cannot generally be resolved, except with the use of proton decoupling. Aqueous extracts show that the PME signal contains primarily phosphoethanolamine and phosphocholine, with contributions from 2,3-DPG from blood and other minor components such as AMP, glucose-6-phosphate, *sn*-glycerol-1-phosphate and 3-phosphoglycerate. The PDE signal (at the low magnetic field strengths used for whole-body studies) is mostly composed of signals from phospholipid headgroups, broadened by chemical shift anisotropy and proton–phosphorus dipolar coupling, and also contains signals from GPE and GPC. These can be resolved by proton decoupling as can the PME signal, but this has rarely been undertaken.

Applications in cancer

From the first ^{31}P NMR studies of tumours *in vivo*, the differences between tumour spectra and those from other tissues were obvious. The tumour spectra were dominated by the PME signal, and the pH was found to be higher than in most other tissues. While these findings had potential use in distinguishing cancerous from normal tissue, it was hoped that subtle differences in the spectra would also help in determining the type or grade of the tumour, but these hopes have to a large extent not been realized. Indeed some non-cancerous tissue, such as the regenerating liver or the neonatal brain, also show these characteristics.

The PME signal in most tumour types is predominantly composed of phosphoethanolamine, with variable contributions from phosphocholine. These molecules are precursors (and breakdown products) of membrane phospholipids, and it has been suggested that high levels may indicate a rapid rate of membrane turnover, although this has not been shown experimentally. Where proton decoupling has been applied, the relative amounts of phosphoethanolamine, phosphocholine, GPE and GPC can be assessed separately, and this may lead to a better biochemical explanation of the changes.

pH of tumour cells

Because tumour cells tend to be glycolytic, their pH was expected to be low. It was therefore a surprise to find that the NMR-determined pH was high (about 7.4, similar to that in blood plasma). These findings have been explained by the intracellular pH being kept high by the extrusion of protons into the extracellular space, which has a low pH, as shown by microelectrode studies. This 'reverse pH gradient' (opposite to that found in most cells) has implications for the uptake of drugs, and drug design.

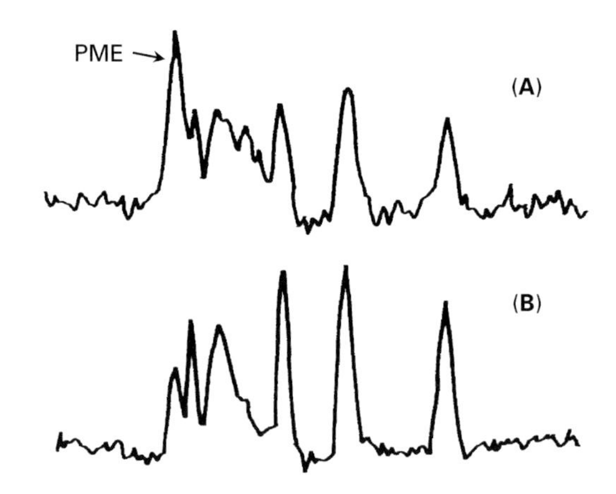

Figure 4 An example of response to chemotherapy in the lymphomatous liver. Note the dramatic change in the PME peak: (A) before chemotherapy, (B) following chemotherapy.

Response to therapy

Various groups have studied the response of tumours to therapy, both in patients and in experimental models. The results of such studies are often affected by partial volume effects, i.e. apparent metabolic changes in the tumour may simply be the result of the tumour shrinking, and the spectra containing more signal from the surrounding tissue. The most consistent findings are that tumours that respond to chemotherapy show a greater decrease in the PME:ATP ratio than do non-responders (**Figure 4**). In some cases, these changes take place before any measurable change in tumour size. Changes in pH and Pi:ATP ratios have also been reported. Experimental manipulation of tumour pH and blood flow, together with radiotherapy or chemotherapy, in implanted tumours in animals have led to a greater understanding of the way in which tumours respond to therapy.

List of symbols

J = coupling constant; T_1 = longitudinal relaxation time; T_2 = transverse relaxation time.

See also: **Biofluids Studied By NMR; Cells Studied By NMR; *In Vivo* NMR, Methods; *In Vivo* NMR,**

Applications – Other Nuclei; Isotopic Labelling in Biochemistry, NMR Studies; Medical Applications of Mass Spectrometry; Nuclear Overhauser Effect; ^{31}P NMR; Perfused Organs Studied Using NMR Spectroscopy.

Further reading

Bottomley PA (1994) MR spectroscopy of the human heart: the status and the challenges. *Radiology* **191**: 593–612.

Gadian DG (1995) *NMR and its Applications to Living Systems*. Oxford: Oxford University Press.

Gillies RJ (ed) (1994) *NMR in Physiology and Biomedicine*. San Diego: Academic Press.

Kemp GJ and Radda GK (1994) Quantitative interpretation of bioenergetic data from ^{31}P and ^{1}H magnetic resonance spectroscopy studies of skeletal muscle: an analytic review. *Magnetic Resonance Quarterly* **10**: 43–63.

Negendank W (1992) Studies of human tumours by MRS: a review. *NMR in Biomedicine* **5**: 303–324.

In Vivo NMR, Applications, Other Nuclei

Jimmy D Bell, **E Louise Thomas** and **K Kumar Changani**, Hammersmith Hospital, London, UK

MAGNETIC RESONANCE
Applications

Introduction

In vivo nuclear magnetic resonance spectroscopy (or *in vivo* MRS as it has become known) is increasingly being used to provide direct, localized biochemical information on animal and human metabolism in health and disease. Dedicated systems for animal studies use magnetic fields ranging from 2.0 to 7.5 T. Most human MRS systems range from 1.5 to 2.0 T, although 4 T systems are available. Clearly these systems offer spectra with a quality and sensitivity no better than that obtained with high-resolution NMR systems in the mid-1960s. However, even within these constraints, *in vivo* MRS has contributed greatly to our understanding of the metabolic processes associated with a wide range of clinical and metabolic disorders by providing researchers with information that could not be obtained previously even with tissue biopsy. The application of *in vivo* MRS has expanded rapidly in recent years and its value as a research tool depends on the organ under investigation as well as the nuclei being utilized. This article reviews the work carried out to date in *in vivo* MRS using ^{13}C, ^{19}F, ^{15}N, ^{23}Na and other less common nuclei.

A list of nuclei currently being used for *in vivo* biomedical research is shown in **Table 1**. Although many of these nuclei are routinely used in high-resolution *in vitro* NMR spectrometers, their implementation in *in vivo* MRS systems required substantial technical development and many are available only to a handful of research groups around the world. This is partly due to the considerable cost associated with developing multinuclear capabilities into commercial systems. Accordingly, most commercial MR systems have very restricted spectroscopic capabilities, with ^{1}H and ^{31}P being the nuclei generally available. This has, of course, restricted the application of less common nuclei in biomedical research. A summary of the main areas of research and application of some of these nuclei is shown in **Table 2**. Results of many of these studies have made considerable contributions to important areas of clinical and biochemical research, while other studies have concentrated on demonstrating technical feasibility, although they suggest the prospect of major areas of future research.

In vivo ^{13}C MRS

Whole-body ^{13}C MRS is potentially the most powerful tool available for the study of human metabolism *in vivo*. The nuclide's low natural abundance (1.1%) allows the study of important biological molecules such as glycogen and lipids, while at the same time allowing the use of ^{13}C-enriched compounds in dynamic measurements of intermediate metabolism. However, several technical issues need to be addressed before its implementation *in vivo*, including a second transmit channel, optimum localization and proton decoupling. The latter has been one of the major stumbling blocks for the application of *in vivo*

Table 1 NMR characteristics of some nuclei of biomedical interest

Nucleus	*Spin*	*Relative sensitivity*	*Natural abundance (%)*	*Observation frequency for ¹H observation at 100 MHz*
^{2}H	1	9.65×10^{-3}	1.5×10^{-2}	15.35
^{7}Li	$\frac{3}{2}$	2.93×10^{-1}	92.58	38.86
^{11}B	$\frac{3}{2}$	1.65×10^{-1}	80.42	32.08
^{13}C	$\frac{1}{2}$	1.59×10^{-2}	1.11	25.14
^{15}N	$\frac{1}{2}$	1.03×10^{-3}	0.37	10.13
^{17}O	$\frac{5}{2}$	2.91×10^{-2}	0.04	13.56
^{19}F	$\frac{1}{2}$	8.3×10^{-1}	100	94.08
^{23}Na	$\frac{3}{2}$	9.25×10^{-1}	100	26.45
^{27}Al	$\frac{5}{2}$	2.06×10^{-1}	100	26.06
^{29}Si	$\frac{1}{2}$	7.84×10^{-3}	4.70	19.86
^{87}Rb	$\frac{3}{2}$	1.7×10^{-1}	27.85	32.72
^{129}Xe	$\frac{1}{2}$	2.12×10^{-2}	26.44	27.66

^{13}C MRS, because of the potential problem of power deposition during proton decoupling. This problem has now been addressed by the use of the newer pulse train decoupling sequences, which allow full decoupling of signal with low power deposition. The need for a second transmit channel with frequency synthesizer, phase modulator and power amplifier, together with the need for specifically designed ^{1}H/^{13}C coil systems, means that the actual cost of a standard *in vivo* scanner is substantially increased, hence limiting its availability. Thus, although many of the technical problems associated with the use of ^{13}C MRS *in vivo* have been overcome, it is still not routinely available to the great majority of research groups.

Table 2 *In vivo* applications of NMR spectroscopy

Nucleus	*Application*	*Organ/tissue*
^{13}C	Glycogen	Liver, muscle
	Amino acids, glucose	Brain
	Lipids	Adipose tissue, liver
^{19}F	Drug metabolism	Liver, brain, muscle
	Anaesthetics	Brain
	Phospholipids	Tumour
	Intracellular Ca^{2+}	Brain, kidney, spleen
	Glucose metabolism	Brain
^{15}N	Glutamine metabolism	Brain
^{23}Na	Electrolyte balance	Muscle, brain, liver, heart
^{2}H (D)	D_2O, methionine, glucose	Brain, adipose tissue, liver
	Choline	Kidney
^{27}Al	Gastric emptying	Gut
^{129}Xe	Lung volume	Lungs
^{17}O	Blood flow	Brain, liver
^{87}Ru	Na^{+}/K^{+}-ATPase	Muscle
^{7}Li	Psychiatric disorders	Brain, muscle
^{11}B	Therapy agents	Liver, brain, tumour
^{29}Si	Breast implants	Liver

Application of *in vivo* ¹³C MRS to carbohydrate metabolism

Carbohydrate metabolism is one area of research that has significantly benefited from *in vivo* ^{13}C MRS since signals from sugar carbons occur in a region of the spectrum that is relatively unpopulated by other resonances. Using specific carbon labelling on different substrates, various metabolic pathways have been studied. Further information has been gained by labelling two adjacent carbon atoms within the same molecule; this produces a characteristic ^{13}C–^{13}C spin–spin coupling, helping to elucidate important information about particular enzyme pathways.

Liver The first *in vivo* experiments, performed in 1981, explored the potential application of ^{13}C MRS to glucose/glycogen metabolism. Following infusion of ^{13}C-enriched glucose in rats, well-resolved resonances from both glucose and glycogen could be observed, enabling the time course of the relative changes in concentration to be established simultaneously. One area of debate at the time was the degree of glycogen visibility in liver. It was initially thought that glycogen, being such a large macromolecule, would be only partly NMR 'visible'. However, the most recent consensus is that glycogen is in fact 100% visible and that ^{13}C MRS can therefore provide reliable noninvasive measurements of hepatic glycogen concentrations over physiological ranges.

Glycogen repletion studies have been carried out in animals and humans. Infusions of [1-^{13}C]glucose and [3-^{13}C]alanine to fasted rats suggested that 30% of the total glycogen formed originated from the direct conversion of glucose in the short term. However, over the longer term, active glycogen synthesis from [3-^{13}C]alanine occurred for a longer period of time, with a rate constant 3-fold higher than from [1-^{13}C]glucose, implying greater efficiency of glycogen synthesis from the indirect route. These studies have also highlighted potential problems associated with substrate cycling, since following entry into the Krebs cycle the label may be assimilated into other pathways and then re-introduced into the gluconeogenic pathway. Definitive mechanisms of glycogen repletion therefore become difficult to assess *in vivo*, unless measurements are performed immediately following administration.

Human studies have concentrated on direct observation of glycogen repletion using natural abundance ^{13}C or after administration of [1-^{13}C]glucose. Early work compared pre- and post-meal hepatic glycogen repletion rates in individuals starved for 30 hours, showing repletion rates 3.8 times greater in the post-meal states compared with the fasted basal rate. Hepatic gluconeogenesis in healthy subjects and those with non-insulin-dependent diabetes mellitus (NIDDM) was also investigated by ^{13}C MRS and conventional methods. Concentration of glycogen was lower in NIDDM patients compared to controls (**Figure 1**). Furthermore, following infusion of [6-^{3}H]glucose, gluconeogenesis in the NIDDM subjects was found to be 60% greater than in control subjects and accounted for 88% of the total glucose production compared with 70% in the normal subjects. This suggests that increased gluconeogenesis accounts for the increase in whole-body glucose production in overnight-fasted NIDDM subjects.

^{13}C MRS has also been applied to determine the metabolic alterations caused by glycogen storage diseases. Patients with glycogen type IIIA storage disease were shown to have increased levels of hepatic glycogen compared with normal volunteers. This disease manifests owing to the inactivation of amylo-1,6-glucosidase, which is required for glycogen hydrolysis. It was concluded that *in vivo* ^{13}C MR spectroscopy could play an important role in the assessment of other glycogen storage diseases where current clinical practice requires needle biopsies to be performed.

^{13}C MRS has also been used to assess enzyme capacities and kinetics in animals. Following infusion of [1-^{13}C]butyrate (90% enriched), rates of ketogenesis could be monitored by measuring the formation of acetoacetate and β-hydroxybutyrate within the carbonyl region of the ^{13}C MR spectrum. Further studies of ^{13}C-labelled fatty acids, such as [1,3-^{13}C]butyrate, assessed the metabolic pathways of ketone body production in fasted and diabetic rats. Increased β-hydroxybutyrate production was seen in diabetic rats; it was thought that a large fraction of butyrate evaded β-oxidation to acetyl-CoA and was converted directly to acetoacetyl-CoA. Different physiological states can therefore be monitored by *in vivo* ^{13}C MRS since labelling patterns following ^{13}C-enriched metabolite administration would be different.

In vivo ^{13}C MR spectroscopy has also been applied to monitor the pharmokinetics of ^{13}C-labelled xenobiotic products in hepatic metabolism. One such study used a phenacyl imidaolium compound that effects glycogen synthesis. It was shown that glycogen synthesis increases 2-fold in drug-treated rats and it was suggested that augmentation of the indirect pathway is the probable mechanism. Further studies have shown that long-term monitoring of drug concentrations is possible using ^{13}C-labelled compounds, which may be important for monitoring of hepatic tumour treatments.

Muscle Most human ^{13}C MRS studies in muscle have concentrated on the study of glycogen synthesis and its role in NIDDM, glycogen storage diseases and muscle myopathies. Studies of insulin action on muscle glycogen synthesis have shown that there appears to be an impairment of muscle's ability to either take up glucose or assimilate it as glycogen in NIDDM.

Recent studies have used *in vivo* ^{13}C MRS to study muscles following intense exercise. This, together with interleaved ^{31}P MRS, has provided important information on the phosphorylation states of

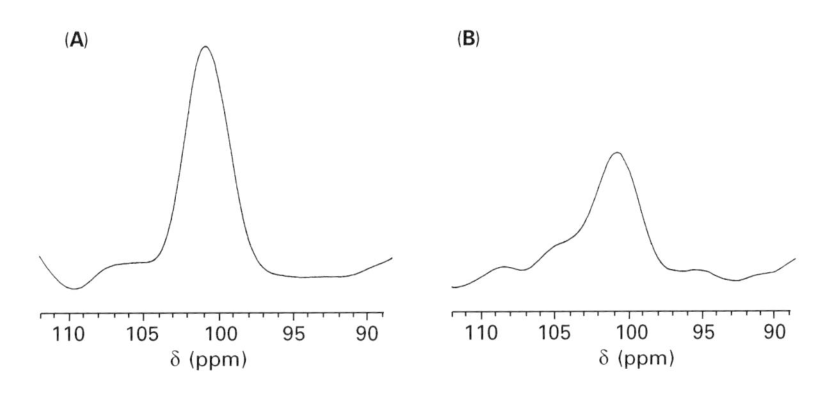

Figure 1 Typical *in vivo* ^{1}H-decoupled ^{13}C MRS spectra at 22.5 MHz of the C-1 position of liver glycogen from a control subject (A) and from a type II diabetic patient (B) 4 hours after a liquid meal. Reproduced with permission from Magnusson I, Rothman DL, Katz LD, Shulman RG and Shulman GI (1992) Increased rate of gluconeogenesis in type II diabetes mellitus. *Journal of Clinical Investigation* **90**: 1323–1327.

glycogen synthase that could be correlated with post-exercise glycogen synthesis rates. Studies of this kind could provide important information on enzyme pathways that may be defective in particular muscle myopathies. Acetate metabolism has also been assessed in rabbit muscles to determine rates of anaplerotic flux (i.e. determination of label of glutamate and glutamine) compared with basal flux of the Krebs cycle. Different muscle types (muscle fibre/oxidative or nonoxidative) have dissimilar anaplerotic flux rates, which may again be relevant for particular disease types.

Heart *In vivo* ^{13}C MR spectroscopy of the heart is difficult owing to continuous contraction of the heart. For this reason, together with the relative insensitivity of the ^{13}C nucleus, which requires a large number of averages, *in vivo* MRS studies are few. Consideration of the metabolites within blood must also be made so that the spectral acquisitions contain information purely reflecting heart metabolism. The bulk of studies on the heart have therefore been conducted using isolated, perfused organs. However, there have been a few *in vivo* studies that have looked at glycogen synthesis rates following infusions of acetate and glucose. These studies can provide information about physiological conditions that may effect carbohydrate utilization in favour of fatty acid metabolism, with implications for coronary heart disease.

Brain *In vivo* ^{13}C MR spectroscopy of the brain has been used to determine *in vivo* levels of glutamate, glutamine, γ-aminobutyrate (GABA), and lactate following intravenous infusions of enriched precursors. Brain tumours in rats have been studied to investigate carbohydrate metabolism following administration of [1-^{13}C]glucose. Rat gliomas produced lactate and glutamate/glutamine as well as glycogen. Further studies used combinations of labelled substrates such as glucose and acetate to determine the relative flux through the tricarboxylic acid cycle together with their relative conversion rates into the cerebral amino acids, important for neurotransmission. Both these studies suggested that *in vivo* ^{13}C MRS is ideal for studying the individual carbons of different metabolites in the brain. Furthermore, cerebral compartmentation of glial and neuronal cells may also be studied because of their different metabolism. Two separate glutamate pools and two separate tricarboxylic acid cycles, one preferentially metabolizing acetate and the other mainly using unlabelled acetyl-CoA and a predominant production of GABA in the glutamate pool lacking glutamine synthase, have been identified, accounted for by either the glial or neuronal cells.

A further application of *in vivo* ^{13}C MRS has been to study the metabolite changes associated with mild or severe hypoxia as well as electroshock. Changes associated with lactate and the cerebral amino acids were correlated with both electroshock and degree of hypoxia. This work may be useful in characterizing the metabolic state following epilepsy, stroke, trauma, tumours and other pathological conditions in humans. In humans, total brain glutamate concentration has been assessed following [1-^{13}C]glucose infusions, while in stroke it has helped to determine metabolic turnover. Some of these studies will benefit from the recent introduction of proton-observed carbon-edited (POCE) techniques, which can provide better relative sensitivity and resolution. This technique has been very useful, for example, in determining the effects on brain metabolism of drugs such as inhibitors of succinate dehydrogenase and the GABA-transaminase inhibitor vigabatrin.

Application of *in vivo* ^{13}C MRS to lipids

The main application of *in vivo* ^{13}C MRS to the field of lipids has been the study of adipose tissue composition. Although the natural abundance of the ^{13}C isotope is low, the high content of lipids in adipose tissue has enabled acquisition of ^{13}C MR spectra *in vivo* in a matter of minutes, with or without proton decoupling. This has allowed the use of this technique on subjects (e.g. neonates and in pregnancy) where power deposition from decoupling, though small, it is still unacceptable.

Adipose tissue Adipose tissue is principally composed of triglycerides and it has been shown to play an important role in major diseases, including NIDDM and cardiovascular disease. Preliminary studies in the early 1980s showed that *in vivo* ^{13}C MR spectroscopy could be used to detect differences in the lipid composition of adipose tissue and liver in rats fed a diet high in polyunsaturated fatty acids. The first human studies showed that linoleic acid (18:2*n*–6), usually the principal polyunsaturated fatty acid present in human adipose tissue, dominated the polyunsaturated signal observed *in vivo*, allowing estimation of this stored essential fatty acid. Further work showed that fatty acid composition of adipose tissue closely correlated with dietary fat content, with subjects on long-term vegan diet showing significantly higher levels of polyunsaturated fatty acids and reduced saturated fatty acids compared to omnivores (general public) and vegetarians. Indeed, there was no difference in the adipose tissue composition of vegetarians compared to omnivores, suggesting that the former have substituted the animal fat (mainly saturated) with saturated fat in dairy products.

In vivo ^{13}C MR spectroscopy has also been used to study adipose tissue composition in babies. The work showed that the influence of maternal diet on infant adipose tissue composition and changes due to gestational age and during postnatal development could all be detected noninvasively by *in vivo* ^{13}C MR spectroscopy. Preterm infants had significantly lower levels of unsaturated adipose tissue fatty acids than the infants born at full term. Furthermore, the level of unsaturated fatty acids continued to increase as the infants matured (from birth to 6 weeks, to 6 months). Interestingly, the authors also found that infants breast-fed by vegan mothers had 70% more polyunsaturated fatty acids than infants breast-fed by omnivore mothers, the long-term consequences of which (beneficial or harmful) are unknown. The same authors showed that *in vivo* ^{13}C MRS could also measure the effect of intensive exercise on adipose tissue composition of adult human volunteers. A decrease in polyunsaturated fatty acids was observed, suggesting that exercise is an independent factor for adipose tissue composition.

In addition, *in vivo* ^{13}C MR spectroscopy has been applied to the study of adipose tissue composition in disease. Children with cystic fibrosis were shown to have lower levels of polyunsaturated adipose tissue fatty acids than healthy children, possibly owing to a disorder in essential fatty acid metabolism that may be partly responsible for the development of the disease. Further studies with *in vivo* ^{13}C MRS in disease have shown a significant increase in saturated adipose tissue fatty acids following transplantation and subsequent weight gain in malnourished patients with liver cirrhosis. It was suggested that this increase in saturated fatty acids may be secondary to a general repletion of membrane polyunsaturated fatty acids or the use of essential fatty acids for biosynthesis of eicosanoids in the postoperative period.

Liver ^{13}C MRS has also been used for noninvasive quantification of hepatic triglyceride content. Diagnosis of hepatic steatosis (fatty liver) is important because this is a reversible condition and early detection can help to prevent irreversible liver damage. The condition is normally diagnosed by a liver biopsy, but the invasive nature of this procedure and the problems associated with biopsies make it difficult to monitor patients at risk of developing hepatic steatosis. In a study of 15 patients with varying degrees of fatty infiltration it was shown that by quantifying the intensity of the CH_2 resonances from the *in vivo* ^{13}C MR spectrum of the liver it was possible to determine hepatic lipid content. Excellent correlation with conventional liver biopsy measurements was observed. As the authors suggested, this technique will in future become a valuable tool in the diagnosis and follow-up in patients with hepatic steatosis.

Brain The application of *in vivo* ^{13}C MR spectroscopy to the study of brain lipids has been rather limited, partly because most of the lipids in the brain are present in the form of membranes. This gives rise to broad and uninformative signals. An interesting possibility for improving the discriminatory power of MRS, as applied to lipids, is the use of ^{13}C-enriched fatty acids. However, at present such studies would be prohibitively expensive as the subjects would be required to consume gram quantities of ^{13}C-enriched fatty acid.

In vivo ^{19}F MRS

Fluorine-19 is a highly NMR-sensitive and naturally abundant nucleus. An important advantage of ^{19}F MRS is the fact that it allows direct observation of fluorinated compounds and their metabolites in the human body without background signal from tissue. *In vivo* ^{19}F MR spectroscopy has been used to detect noninvasively anticancer and psychoactive drugs and other fluorinated compounds and to study their metabolism. It has also been used to monitor the effects of anaesthetics and to measure intracellular pH and calcium levels.

One of the major applications of *in vivo* ^{19}F MR spectroscopy has been in the field on oncology, detecting and monitoring the metabolism of fluorine-containing anticancer drugs. An important chemotherapeutic compound is the fluorinated drug 5-fluorouracil (5-FU). Much of the work with ^{19}F MRS has concentrated on determining optimal drug administration schedules and studying the effects of this drug on both tumour and healthy tissue. ^{19}F MRS has allowed detection and serial measurement of 5-FU and its metabolites in patients undergoing chemotherapy, increasing our understanding of the variation in efficacy of chemotherapy in different subjects (**Figure 2**). For example, levels of 5-FU 'trapping' by tumours following intravenous administration of the drug have been measured. The results reveal that patients whose tumours trapped the drug had a significantly better response to chemotherapy than those patients whose tumours did not trap 5-FU. This differential in response to therapy could not be ascertained from the standard plasma concentration measurements. From this study it was concluded that ^{19}F MRS can be used to identify patients who are likely to respond to chemotherapy with 5-fluorouracil and can therefore aid selection of the optimal

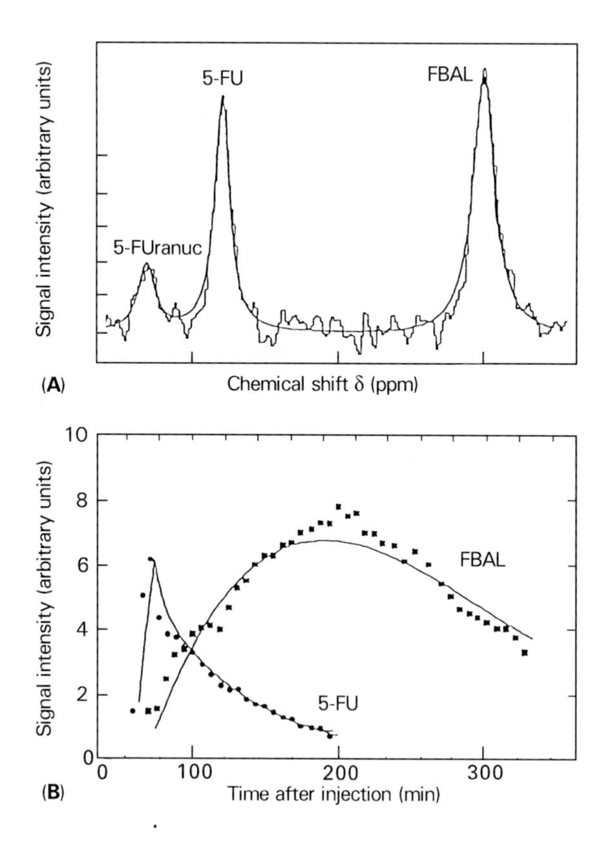

Figure 2 (A) Liver *in vivo* ^{19}F MR spectrum at 60 MHz following administration of 5-fluorouracil (5-FU). (B) Time course of 5-FU and its major catabolite α-fluoro-β-alanine (FBAL). (FUranuc = 5-fluorouracil nucleoside + nucleotides). Reproduced with permission from Semmler W, Bacher-Baumann P and Guckel F (1990) Real time follow-up of 5-fluorouracil metabolism in the liver of tumor patients by means of F-19 MR spectroscopy. *Radiology* **174**: 141–145.

treatment for individual patients. Studies of 5-FU metabolism in patients with protein calorie malnutrition (a common condition in patients with cancer) have shown increased toxicity from this drug. Therefore, identifying alterations in 5-FU metabolism may be an important tool in reducing toxicity during chemotherapy. Interestingly, like ^{31}P MRS, *in vivo* ^{19}F MRS may also be used to measure pH noninvasively. This may be useful for ^{19}F MRS studies of tumours, as the uptake and retention of 5-fluorouracil is dependent on tumour pH.

An important area of research using *in vivo* ^{19}F MRS has been the study of the pharmacokinetics of fluvoxamine and fluoxetine, drugs used to treat obsessive-compulsive disorder and depression, respectively. Several groups have shown that it is possible to detect these compounds noninvasively in the brain. Brain fluvoxamine levels were shown to reach a steady state 30 days after consistent daily dosing, substantially more rapidly than reported for fluoxetine, which was shown to plateau after 6–8 months of treatment. Brain concentrations of both drugs were shown to correlate with plasma levels, though the absolute concentrations were substantially higher in the brain. Comparison of the *in vivo* measurement of fluoxetine with *in vitro* samples taken at autopsy suggest that fluorinated drugs may not be 100% visible *in vivo*. These are based on the *in vitro* findings from just one brain that had been fixed in formaldehyde, so further studies would be required to confirm these findings.

Some of the other applications of *in vivo* ^{19}F MRS to the study of fluorinated drugs have included the study of fluorinated antibiotics and antifungal agents in animals and humans. Orally administered fleroxacin (a fluoroquinolone antibiotic agent) could be detected in the liver and calf muscle by *in vivo* ^{19}F MRS. Furthermore, its washout and metabolism were studied over a 24-hour period. There are important implications of this work, and it will in future be possible to ensure that drugs reach their target site in appropriate concentrations.

In vivo ^{19}F MRS has also been used to study the pharmacokinetics and metabolism of anaesthetic agents such as halothane, isoflurane and desflurane in experimental animals. It has also been shown that anaesthetic agents can be observed noninvasively in the brains of patients immediately following surgery. The principal aim of these studies was to elucidate the mechanism of action of these compounds and eventually help in the development of more effective and selective anaesthetics. Unfortunately, factors such as sensitivity and localization of the region of interest (which are particularly relevant to the study of anaesthetics) have greatly limited the impact of this research.

Animal studies have shown that *in vivo* ^{19}F MRS can be used in metabolic studies. It has been shown that 3-fluoro-3-deoxy-D-glucose in the brain is metabolized primarily in the aldose reductase sorbitol pathway, suggesting that ^{19}F MRS may be a useful tool for the further elucidation of this pathway. Further studies have shown that it is also possible to study the metabolism of fluorinated galactose, tryptophan and protein *in vivo* and determine intracellular calcium levels. The concentration of this cation was shown to be the same in the kidney, spleen and brain of rats, at approximately 200 nM. These measurements were obtained by infusing the calcium indicator 5F-BAPTA (=5,5-difluoro-1,2-bis(*o*-aminophenoxy)ethane-*N*,*N*,*N*^{1},*N*^{1}-tetraacetic acid) intravenously or intraventricularly into the animal, and so the potential application to human subjects appears limited.

It has also been possible to follow the time course of recovery from nerve injury in skeletal muscle using *in vivo* ^{19}F MRS. Muscle injury leads to significant decrease in the energy state and local circulation dynamics, which appear to reverse during recovery. While ^{31}P MRS was used to determine the energy status of the cells, ^{19}F MRS was utilized to determine local circulation dynamics by measuring blood volume. The results indicate that the recovery processes of circulation precede those of energy state.

Many of the above studies have shown that in the future ^{19}F MRS is likely to become a valuable tool for monitoring drug metabolism at the site of action, though at the moment this is limited by the availability of *in vivo* MR scanners with ^{19}F capabilities.

Less common nuclei

In the last decade there has been a significant increase in the number of 'less common' nuclei that are being utilized for *in vivo* MRS (see **Table 1**). The use of some of these nuclei was initially limited by technical constraints and/or inherent NMR characteristics. However, recent studies have shown that these nuclei can be used to increase our understanding of some important physiological processes.

A number of studies have utilized *in vivo* MRS to assess the extent of delivery of pharmacological substances to target tissue. Nitrogen-15 MRS in combination with intravenous $^{15}NH_4^+$ infusion into animal models of hepatic encephalopathy (**Figure 3**) enabled the determination of glutamine synthetase activity *in vivo*. The results show that *in situ* the activity of this enzyme is kinetically limited by the levels of some substrates and cofactors, and that the activity of phosphate-activated glutaminase was maintained at a low level by a suboptimal concentration of P_i and the strong inhibitory effect of glutamine. The introduction of the heteronuclear multiple quantum coherence (HMQC) transfer technique is likely to further these studies by probing the kinetics of glutamine synthesis at physiological concentrations.

Similarly, lithium-7 MRS has been used to examine the pharmacokinetics of lithium, a drug that is widely used in the treatment of a number of psychiatric disorders. Initial studies showed that there was a slow accumulation of lithium in the brain, which may be responsible for the delay in therapeutic response that is often seen following the initiation of therapy. Subsequent studies showed that lithium levels in the brain and muscle were lower than the average serum concentration. From the ratio of these *in vivo* measurements, the authors hypothesized that the minimal effective concentration of brain lithium level for maintenance treatment

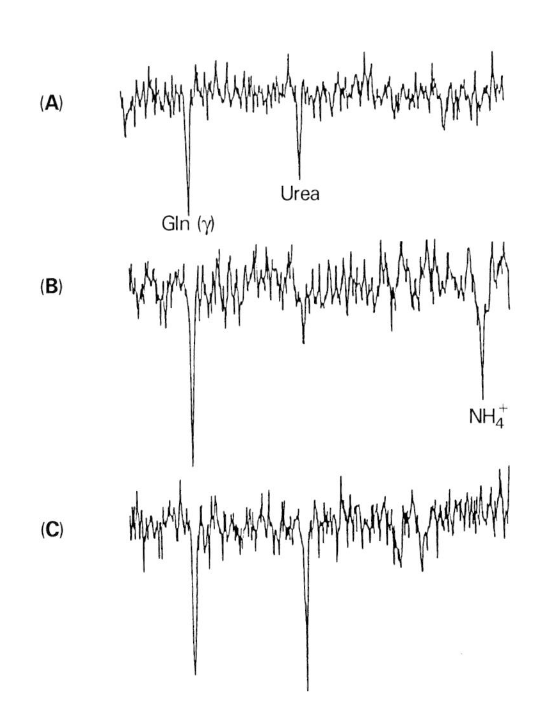

Figure 3 *In vivo* ^{15}N spectra at 24.3 MHz from cerebral region of (A) a control rat given low-rate $^{15}NH_4^+$ infusion, showing cerebral [γ-^{15}N]glutamine at a near physiological concentration; (B) a portacaval-shunt rat showing cerebral and blood $^{15}NH_4^+$, in addition to [γ-^{15}N]glutamine; (C) a control rat showing cerebral and blood [^{15}N]urea and lower [γ-^{15}N]glutamine. Peaks are downgoing following NMR convention for ^{15}N spectroscopy. Reproduced with permission from Kanamori K, Parivar F and Ross BD (1993) A ^{15}N NMR study of *in vivo* cerebral glutamine synthesis in hyperammonemic rats. *NMR in Biomedicine* **6**: 21–26.

of bipolar disorder could be determined. However, in a similar study, researchers found that in a subgroup of patients there was only a weak correlation between serum and brain lithium levels, which may account for the failure of lithium therapy in some patients with serum lithium levels within the therapeutic range. This suggests that measurement of *in vivo* brain lithium concentration by MRS may be more clinically relevant than simply assessing serum concentration.

Boron-11 MRS has been used to assess the delivery of boron neutron capture therapy (BNCT) agents to a target lesion. Uptake and retention of these compounds was determined *in vivo* in humans and animals and the results show that there is differential retention between normal and abnormal tissue. Furthermore, ^{1}H-observed ^{10}B-edited MRS has been shown to provide much higher sensitivity and could be utilized to monitor the distribution and excretion of boron agents noninvasively in patients about to undergo BNCT.

Sodium-23 magnetic resonance spectroscopy has been used to yield information on electrolyte balance, which appears to reflect physiological changes associated with cell function. In myotonic dystrophy, an inherited multisystem disease that leads to muscle malfunction, ^{23}Na MRS was shown to be a sensitive method for quantification of disease progression. Quantification of intracellular and extracellular sodium by MRS has also generated important information on organ function following transplantation or injury. In heart transplantation experiments in rats, ^{23}Na MRS showed that there was a steady increase in intracellular sodium 3 days prior to rejection, followed by a sharp increase after rejection. In the liver, intracellular sodium levels were shown to be a sensitive indicator of hepatic dysfunction after major systemic injury. In the kidney, sodium compartmentation (intracellular, vascular and intraluminal) was determined using the shift reagent TmDOTP5, while in the brain this shift reagent was shown to be a useful tool in determining compartmental sodium concentration and measuring blood flow kinetics.

The activity of ion transport systems, such as sodium- and potassium-activated adenosine triphosphatase, has been studied *in vivo* by the use of rubidium-87 MRS. This has been possible because rubidium has been shown to substitute for potassium in a number of transmembrane transport systems, accumulating in the intracellular space. Standard *in vitro* methods of determining Na^+/K^+-ATPase activity, which also use rubidium, give highly variable measurements and in some cases contradictory results. Many of these problems appear to have been overcome by the use of *in vivo* ^{87}Rb MRS, especially when sequential measurements are required. In a longitudinal study of spontaneously hypertensive rats, ^{87}Rb MRS showed that skeletal muscle rubidium rose at a faster rate in hypertensive rats than in control animals, which is consistent with a marked increase in Na^+/K^+-ATPase activity. This type of experiment emphasizes the value of *in vivo* MRS since rubidium kinetics can be determined sequentially on the same animal, minimizing inter/intra-subject variability as well as the number of animals required in the study.

In vivo deuterium (^{2}H) MRS has been used to characterize amino acid metabolism, body iron content, brain and kidney metabolism and body fat utilization rates in rodents. These studies rely on the use of deuterium labelling or the existence of natural-abundance deuterium in water or lipids. For example, deuterium-labelled methionine was used to confirm the dominant contribution of the glycine/sarcosine shuttle to the metabolism of excess methionine, while deuterium-labelled glucose was used to show that systemic glucose level influences brain metabolic recovery following partial ischaemia. The renal distribution and metabolism of methyl groups was assessed by the use of deuterium-labelled choline. The results show that the concentration of trimethylamines, metabolic products of choline, was higher in the cortex and inner medulla than in the outer medulla, but that the inner medullary fraction was more liable to diuresis.

By combining D_2O dilution techniques with ^{2}H MRS it has been possible to study fat turnover *in vivo*. The rate of loss of deuterium from body fat, after a short period of D_2O administration, has been shown to reflect utilization of body fat. This technique was used successfully to demonstrate that induction of diabetes in mice did not affect the utilization of fat as a metabolic fuel. Given that D_2O administration is regularly used with human subjects, this opens up the possibility, for the first time, of determining fat turnover noninvasively in human subjects.

A number of studies have used nuclei such as oxygen-17, aluminium-27 and xenon-129 in conjunction with *in vivo* magnetic resonance spectroscopy for structural or dynamic measurements in different organs. ^{129}Xe signals were detected in the lung and head from human volunteers following inhalation of hyperpolarized ^{129}Xe gas, allowing the determination of the temporal evolution of the gas-phase and dissolved-phase components. Most of this work, however, was used to demonstrate the feasibility of obtaining xenon images of the lung. ^{129}Xe imaging now allows the possibility of imaging the lung at resolution equivalent to that of conventional imaging of normal tissue. ^{27}Al MRS has been applied to the visualization of gastric emptying and measurement of gastric phospholipids, while ^{17}O MRS has been used to explore cerebral blood flow. Although these studies have again concentrated mainly on their technical feasibility, they do open up important areas of clinical and biochemical research.

Conclusions

The application of *in vivo* multinuclear MRS to biological sciences is an ever-expanding area of research. Present-day clinical NMR systems routinely allow the use of ^{1}H and ^{31}P MR spectroscopy in clinical and scientific research; however, the use of other nuclei is still restricted to a relatively small number of research groups around the world. This is principally for commercial reasons rather than scientific or technical considerations. Demands for an increase in the multinuclear capability of commercial systems is of course increasing as the usefulness of nuclei such as ^{13}C, ^{19}F and ^{15}N continues to be demonstrated both in animal models and in human subjects. It is conceivable that in the near future these technical capabilities will be routinely available in most clinical systems as they are today available in most *in vivo* animal systems. This will no doubt greatly help to further elucidate metabolic mechanisms in health and disease and in the development and monitoring of the efficacy of novel clinical treatments, including gene therapy.

See also: **^{13}C NMR, Methods; Drug Metabolism Studied Using NMR Spectroscopy; ^{19}F NMR, Applications, Solution State; Heteronuclear NMR Applications (B, Al, Ga, In, Tl); Heteronuclear NMR Applications (Ge, Sn, Pb); *In vivo* NMR, Applications – ^{31}P; *In vivo* NMR, Methods; Labelling Studies in Biochemistry Using NMR; Nitrogen NMR; NMR Spectrometers; Nuclear Overhauser Effect; Perfused Organs Studied Using NMR Spectroscopy; Xenon NMR Spectroscopy.**

Further reading

Arai T, Mori K, Nakao S *et al* (1991) *In vivo* oxygen-17 nuclear magnetic resonance for the estimation of cerebral blood flow and oxygen consumption. *Biochemical and Biophysical Research Communications* **179**: 954–961.

Bartels M and Albert K (1995) Detection of psychoactive drugs using ^{19}F MR spectroscopy. *Journal of Neural Transmission, General Section* **99**: 1–6.

Cox IJ (1996) Development and applications of *in vivo* clinical magnetic resonance spectroscopy. *Prog Biophys Mol Biol* **65**: 45–81.

Eng J, Berkowitz BA and Balaban RS (1990) Renal distribution and metabolism of [$^{2}H_9$] choline. A ^{2}H NMR and MRI study. *NMR in Biomedicine* **3**: 173–177 (and references therein).

Gadian DG (1995) *NMR and its Applications to Living Systems*. Oxford: Oxford Science Publications, Oxford University Press.

Kabalka GW, Davis M and Bendel P (1988) Boron-11 MRI and MRS of intact animals infused with a boron neutron capture agent. *Magnetic Resonance in Medicine* **8**: 231–237.

Kanamori K, Parivar F and Ross BD (1993) A ^{15}N NMR study of *in vivo* cerebral glutamine synthesis in hyperammonemic rats. *NMR in Biomedicine* **6**: 21–26.

Kaspar A, Bilecen D, Scheffler K and Seelig J (1996) Aluminium-27 nuclear magnetic resonance spectroscopy and imaging of the human gastric lumen. *Magnetic Resonance in Medicine* **36**: 177–192.

Kushnir T, Knubovets T, Itzchak Y *et al* (1997) *In vivo* ^{23}Na NMR studies of myotonic dystrophy. *Magnetic Resonance in Medicine* **37**: 192–196 (and references therein).

Leach MO (1994) Magnetic resonance spectroscopy applied to clinical oncology. *Technology and Health Care* **2**: 235–246.

Magnusson I, Rothman DL, Katz LD, Shulman RG and Shulman GI (1992) Increased rate of gluconeogenesis in type II diabetes mellitus. *Journal of Clinical Investigation* **90**: 1323–1327.

Mugler JP, Driehuys B, Brookman JR *et al* (1997) MR imaging and spectroscopy using hyperpolarized ^{129}Xe gas: preliminary human results. *Magnetic Resonance in Medicine* **37**: 809–905.

Semmler W, Bacher-Baumann P and Guckel F (1990) Real time follow-up of 5-fluorouracil metabolism in the liver of tumor patients by means of F-19 MR spectroscopy. *Radiology* **174**: 141–145.

Soares JC, Krishnan KR and Keshavan MS (1996) Nuclear magnetic resonance spectroscopy: new insights into the pathophysiology of mood disorders. *Depression* **4**: 14–30 (and references therein).

Syme PD, Dixon RM, Allis JL, Aronson JK, Grahame-Smith DG and Radda GK (1990) A non-invasive method of measuring concentrations of rubidium in rat skeletal muscle *in vivo* by ^{87}Rb nuclear magnetic resonance spectroscopy: implications for the measurement of cation transport activity *in vivo*. *Clinical Science* **78**: 303–309 (and references therein).

Yoshida SH, Swan S, Teuber SS and Gershwin ME (1995) Silicone breast implants: immunotoxic and epidemiologic issues. *Life Science* **56**: 1299–1310.

Young IR and Charles HC (1996) *MR Spectroscopy: Clinical Applications and Techniques*. London: Martin Dunitz.

In Vivo NMR, Methods

John C Lindon, Imperial College of Science, Technology and Medicine, London, UK

MAGNETIC RESONANCE
Methods & Instrumentation

Introduction

NMR spectroscopy is a technique that has found widespread applicability in many areas of science. Although originally a physics method for measuring nuclear magnetic moments, it soon became invaluable in studies of identification and structuring of organic and inorganic chemicals with the discovery of the chemical shift and the resolution of spin–spin coupling splittings. Later the application of NMR spectroscopy was amplified by the realization that pieces of tissue could yield reasonably resolved NMR spectra (initially of the ^{31}P nucleus) and this sparked a major research effort in the use of NMR spectroscopy to study biochemical systems such as intact cell suspensions and perfused tissues.

A logical next step was the investigation of whole living animals and humans and nowadays there is much research activity in these fields. Many of the *in vivo* NMR studies have been in the area of clinical medicine and by analogy with magnetic resonance imaging (MRI), where the 'nuclear' part has been dropped, *in vivo* NMR spectroscopy is often referred to as simply magnetic resonance spectroscopy (MRS), really a misnomer as there are other types of magnetic resonance spectroscopy such as electron paramagnetic resonance (EPR).

Methods

In order to obtain high resolution NMR spectra from a subject *in vivo*, it is necessary to localize the NMR excitation and detection to a particular region of interest. There are two principal ways of achieving this. Of course, the NMR magnet has to be large enough to accommodate the animal or human subject, or at least that part which is being studied (e.g. leg or arm muscles) and therefore *in vivo* NMR experiments are usually conducted using horizontal bore magnets designed for MRI investigations. Some small animals and species such as invertebrates can be studied using conventional high resolution NMR spectrometers with specially developed MRI probes.

The simplest approach to NMR detection uses a RF transmitter and receiver coil placed on the surface of the subject inside the NMR magnet. The size of the coil and its siting determines which tissue will be examined, and the depth of penetration of the detection region can be effected by the strength of the RF pulse used. This method has been used extensively for ^{31}P NMR spectroscopy. Surface coil methods when combined with field gradients can also give greater spatial discrimination. One example is the DRESS technique where a field gradient is applied perpendicular to the plane of the surface coil and a frequency selective RF pulse is used to excite only the region of the tissue where the Larmor frequency is achieved.

The second approach makes use of pulsed field gradients and techniques adapted from MRI. In MRI, a three-dimensional region of tissue (usually a slice of square cross-section) is excited selectively by a combination of RF pulses and field gradients and, according to the method used, an image is produced based upon parameters such as the proton density, relaxation times or diffusion coefficients of the water in that region. For *in vivo* spectroscopy, a different pulse sequence is used which gives the high resolution NMR spectrum from that active region. One of the first methods was termed ISIS (image selected *in vivo* spectroscopy). This could be the ^{31}P or ^{1}H NMR spectrum. If a ^{1}H NMR spectrum is required then it is also necessary to suppress the strong resonance from water. In addition, if ^{1}H NMR spectra are obtained by this means then the pulse sequence is usually based on some form of the spin–echo experiment in order to suppress the peaks from fast relaxing protons which occur in the less mobile components of the tissue such as proteins or membrane lipids. One of the popular methods is called PRESS (point resolved spectroscopy). More complicated methods are also used in which information on spectroscopy and imaging is obtained simultaneously, techniques such as chemical shift imaging (CSI) where up to 4-dimensional data sets are obtained in which there are up to three spatial axes and one spectroscopic axis. From such data it is possible to obtain spectra at any voxel in the image or alternatively to determine the spatial distribution of any given metabolite detected spectroscopically.

Applications

^{31}P NMR spectroscopy

The first *in vivo* experiments were carried out using ^{31}P NMR spectroscopy. This nuclide is particularly favourable because it has 100% natural abundance, high intrinsic sensitivity, the number of organic phosphorus compounds is relatively few and such molecules are important in cellular processes particularly for energy balance.

Most studies have concentrated on muscle with much emphasis on the heart. However, a typical ^{31}P NMR spectrum obtained from a human forearm is shown in **Figure 1** using a surface coil. This shows the main features which comprise the three signals of nucleotide phosphates, in this case mainly ATP, and phosphocreatine. Saturation transfer experiments have been carried out on these resonances to determine the rate of phosphate exchange between the phosphocreatine and ADP which yields creatine and ATP. This reaction is catalysed by the important enzyme creatine kinase. In addition, a major area of research is the study of the effects of brain ischaemia, such as occurs after a stroke, on the ^{31}P NMR spectrum. In the brain, unlike muscle, peaks are also seen from a wider variety of phosphorus-containing species such as inorganic phosphate, phosphomonoesters such as AMP or glucose phosphate and phosphodiesters such as ADP or glycerophosphorylethanolamine.

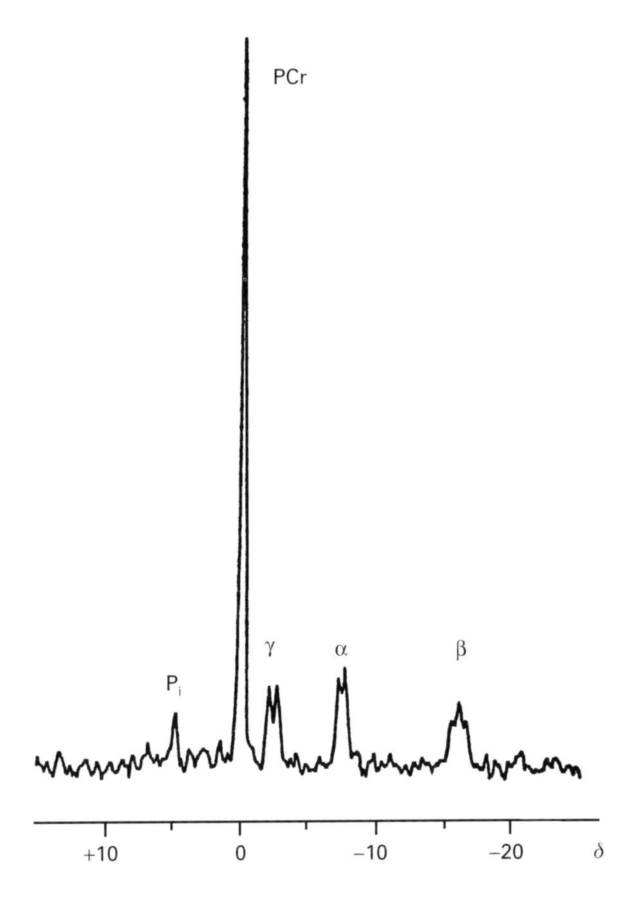

Figure 1 40.5 MHz ^{31}P NMR spectrum from a human forearm measured using a 4 cm diameter surface coil. P_i – inorganic phosphate, PCr – creatine phosphate, α, β, and γ – phosphate groups at the α, β and γ positions in adenosine triphosphate. ^{31}P–^{31}P spin–spin coupling causes the splittings on the α, β and γ peaks. Chemical shifts are references to PCr at 0 ppm. Reproduced with permission from Friebolin H (1998) *Basic One- and Two-Dimensional NMR Spectroscopy*, 3rd edn. Weinheim: Wiley-VCH.

The chemical shift of inorganic phosphate is pH dependent and is thus a marker of intracellular pH. Typically, this is measured using an equation of the form

$$\mathrm{pH} = \mathrm{p}K_\mathrm{a} + \log\left(\frac{\delta - \delta_\mathrm{a}}{\delta_\mathrm{b} - \delta}\right)$$

where $\mathrm{p}K_\mathrm{a}$ is $-\log K_\mathrm{a}$, the proton dissociation constant for the monoanionic species of phosphate, δ is the chemical shift difference between inorganic phosphate and phosphocreatine, and δ_a and δ_b are the chemical shifts of the diprotonated and monoprotonated forms of phosphate respectively. Usually, $\mathrm{p}K_\mathrm{a}$ is taken as 6.77 with δ_a and δ_b being 3.27 ppm and 5.78 ppm respectively, although different research groups use slightly different values.

^{1}H NMR spectroscopy

Nowadays, the bulk of publications on *in vivo* NMR spectroscopy are concerned with ^{1}H NMR spectroscopy of human brain. Many small molecule metabolites have been detected and much effort has gone into attempts to quantify these components. Many studies are concerned with altered levels of metabolites in various brain diseases. A typical ^{1}H NMR spectrum of human brain acquired *in vivo* is shown in **Figure 2**.

^{19}F NMR spectroscopy

In vivo ^{19}F NMR spectroscopy can be a very useful technique for monitoring the distribution of fluorinated drugs and their metabolites. These include the PET agent 2-fluoro-2-deoxyglucose, the study of its 3-fluoro-isomer as a probe of aldose reductase activity in brain, the elimination from the brain of fluorinated anaesthetics, metabolism of halothane in the liver, the distribution and catabolism of the anticancer drug 5-fluorouracil and the uptake of trifluoromethylthymidine into mouse tumours.

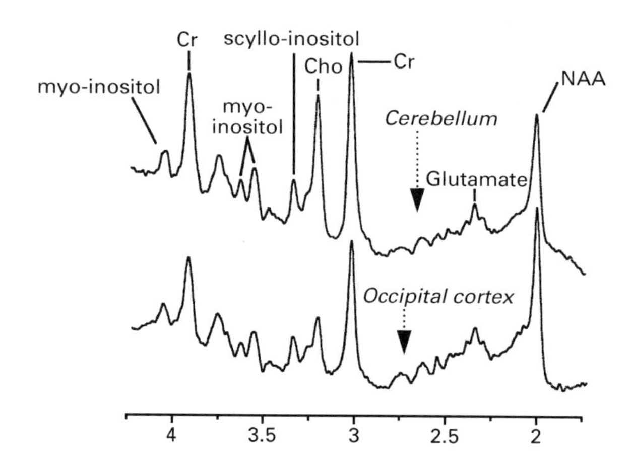

Figure 2 ^{1}H NMR spectra from the brain of a healthy 43 year old male human volunteer. The upper spectrum is from the cerebellum and the lower spectrum is from the occipital cortex. The spectrum was measured at 170 MHz using a 4T magnet with a 15 cm diameter semi-volume coil. Localization was achieved using a pulse sequence based on the STEAM method with a nominal volume of 27 mL. Water peak suppression used selective Gaussian pulses. Cho – choline groups, Cr – creatine, NAA – *N*-acetylaspartate. Reproduced with permission from Seaquist ER and Gruetter R (1998) Identification of a high concentration of scyllo-inositol in the brain of a healthy human subject using ^{1}H and ^{13}C NMR. *Magnetic Resonance in Medicine* **39**: 313–316.

A number of fluorinated molecules which have pH-sensitive chemical shifts can be taken up by cells and used to monitor intracellular pH.

^{13}C NMR spectroscopy

A number of reports of the use of *in vivo* ^{13}C NMR spectroscopy are in the literature. These include studies of lipid deposits, glycogen and its breakdown in the liver and the monitoring of the metabolism of ^{13}C-labelled substances such as ^{13}C-glucose in various tissues such as the brain.

See also: **Biofluids Studied By NMR; ^{13}C NMR, Methods; Cells Studied By NMR; ^{19}F NMR, Applications, Solution State; MRI Applications, Biological; MRI Applications, Clinical; MRI Applications, Clinical Flow Studies; MRI Instrumentation; ^{31}P NMR; Solvent Suppression Methods in NMR Spectroscopy.**

Further reading

Bailey IA, Williams SR, Radda GK and Gadian DG (1981) A ^{31}P NMR study of the effects of reflow on the ischaemic rat heart. *Biochemical Journal* **196**: 171–178.

Berliner LJ and Reuben J (eds) (1992) *In Vivo Spectroscopy, Biological Magnetic Resonance*, vol 11. New York: Plenum.

Bottomley PA, Foster TB and Darrow RD (1984) Depth-resolved surface coil spectroscopy (DRESS) for *in vivo* ^{1}H, ^{31}P and ^{13}C NMR. *Journal of Magnetic Resonance* **59**: 338–342.

Duyn JH, Frank JA and Moonen CTW (1995) Incorporation of lactate measurement in multi-spin-echo proton spectroscopic imaging. *Magnetic Resonance in Medicine* **33**: 101–107.

Hoult DI, Busby SJW, Gadian DG, Radda GK, Richards RE and Seeley PJ (1974) Observation of tissue metabolites using ^{31}P nuclear magnetic resonance. *Nature (London)* **252**: 285–287.

Ordidge RJ, Connelly A and Lohman JAB (1986) Image-selected *in vivo* spectroscopy (ISIS) – a new technique for spatially selective NMR spectroscopy. *Journal of Magnetic Resonance* **66**: 283–294.

Pettegrew JW (ed) (1990) *NMR: Principles and Applications to Biomedical Research*. New York: Springer-Verlag.

Ross BD, Radda GK, Gadian DG, Rocker G, Esiri M and Falconer-Smith J (1981) Examination of a case of suspected McArdle's syndrome by ^{31}P nuclear magnetic resonance. *New England Journal of Medicine* **304**: 1338–1342.

Sillerud LO and Shulman RG (1983) Structure and metabolism of mammalian liver glycogen monitored by ^{13}C nuclear magnetic resonance. *Biochemistry* **22**: 1087–1094.

van Zijl PCM, Chesnick AS, Despres D, Moonen CTW, Ruiz-Cabello J and van Gelderen P (1993) *In vivo* proton spectroscopy and spectroscopic imaging of (1-^{13}C)-glucose and its metabolic products. *Magnetic Resonance in Medicine* **30**: 544–551.

Indium NMR, Applications

See **Heteronuclear NMR Applications (B, Al, Ga, In, Tl)**

Induced Circular Dichroism

Kymberley Vickery and **Bengt Nordén**, Chalmers University of Technology, Gothenburg, Sweden

ELECTRONIC SPECTROSCOPY
Applications

Induced circular dichroism (ICD) is the CD observed in an optically inactive (achiral) chromophore due to its interaction with an optically active (chiral) moiety. Early applications exploited ICD just as evidence for such an interaction. For example, the ICD in the absorption band of an achiral drug may be used to prove its binding to a (chiral) protein or nucleic acid. More recent use of ICD includes the structural information that theoretical analysis of the inducing mechanism can provide. ICD has been an important complement to linear dichroism (LD) for the assessment of binding geometries in drug–nucleic acid systems.

The CD induced in an achiral chromophore upon interaction with a chiral substrate can be ascribed to one or several mechanisms, each corresponding to a more or less idealized situation: the chromophore molecule may be perturbed, by mainly steric interactions, to adopt a chiral (equilibrium) conformation or it may gain optical activity through electric and/or magnetic interactions with surrounding chromophores.

Different types of ICD

Strictly, ICD encompasses all kinds of circular dichroism effects except for those of inherently chiral chromophores. This is so because most systems (molecules) may be more or less arbitrarily divided into one achiral chromophore in which the ICD is observed and a remaining component that includes the chirality. Thus, the CD exhibited by a chiral molecule may be considered ICD if the chromophoric entity itself is achiral. With this broader definition of ICD, molecules such as carbonyl compounds (in which the carbonyl chromophore is achiral) display an intramolecular ICD in the first allowed $n \rightarrow \pi^*$ transition around 300 nm due to perturbation from a surrounding chiral (e.g. steroid) skeleton. Distinction between this type of molecule and one which is an inherently chiral chromophore could be difficult to make as the boundaries of the chromophore are often diffuse. For example, when a diene is in immediate proximity of a carbonyl, a composite chromophore is obtained, which, if skewed, is inherently chiral. Therefore, it is practical to restrict the definition of ICD to cases in which the achiral part is distinctly separable from the chiral part, generally by being a separate molecular species.

ICD could arise in an achiral chromophore mainly through two mechanisms: (1) ICD of an electric dipole-allowed (eda) transition due to coupling with other eda transitions of the surrounding (chirally arranged) chromophore(s) or (2) ICD of a magnetic dipole-allowed (mda) transition due to mixing with eda transitions in the same chromophore under the perturbation of the electronic states of the chiral surrounding. The latter mechanism generally gives more easily detectable ICD as a combined effect of typically strong CD and weak absorption of mda transitions (CD/absorbance sets the limit for the signal-to-noise ratio of CD detection). In addition to carbonyl groups in organic compounds, d–d transitions of transition metals provide examples where strong ICD from interaction with chiral surroundings are frequently observed. For example, the first d–d transition of $[Co(NH_3)_6]^{3+}$ displays ICDs of varying sign and magnitude due to hydrogen-bonded outer-sphere coordination with tartrate, amino acids and other chiral oxoanions. These ICD effects have been exploited to assess binding constants of the outer-sphere complexes formed, the strength of the ICD and of the binding constants suggesting that they are due to relatively well ordered structures. However, attempts towards more detailed structural assignments have been unsuccessful owing to the complicated nature of this mechanism of induced CD. For this reason, the focus here will be on the first (electric dipole) ICD mechanism, which, despite its weaker ICD signals and consequently more demanding experiments, can be more simply related to structural models.

Interactions between electric dipole-allowed transitions

This mechanism of ICD is related to the strong exciton CD that is observed for a chirally arranged set of identical intense absorbers such as in a chiral homodimer (or polymer) species. A characteristic feature is the strong bisignate CD doublet observed at the frequency of the transition. The dominant CD bands centred around 200 nm in proteins and at

260 nm in nucleic acids are both of exciton-type $\pi \rightarrow \pi^*$ transitions. Exciton CD will not be considered an ICD as the transition is not localized on a single achiral chromophore but is spread over the whole dimer (polymer) system and thus constitutes an inherently chiral chromophore. Instead, the ICD arises from the interaction between two different (nondegenerate) chromophores. It is generally of a much weaker magnitude than the exciton CD, but becomes gradually more intense as the energy between the two transitions decreases. This is seen from the energy difference in the denominator of the expression (Eqn [1]) for the ICD rotational strength $\boldsymbol{R}(A)$ for an eda transition, with transition moment $\mu(A)$ occurring at frequency $\nu(A)$ on chromophore A, due to interaction with another transition $\mu(B)$ at frequency $\nu(B)$ on chromophore B of the chiral environment. A requirement for non-vanishing $\boldsymbol{R}(A)$ is that the transition moments $\mu(A)$ and $\mu(B)$ are chirally arranged relative to each other, that is, they must not be coplanar (**Figure 1**).

$$\boldsymbol{R}(\mathrm{A}) = \frac{-2\pi\nu_\mathrm{A}\nu_\mathrm{B}}{hc(\nu_\mathrm{B}^2 - \nu_\mathrm{A}^2)} V_\mathrm{BA} O_\mathrm{BA} \qquad [1]$$

where V_{BA} is the corresponding dipole–dipole interaction energy:

$$V_\mathrm{BA} = \frac{1}{4\pi\varepsilon_0}\frac{1}{|\boldsymbol{R}_\mathrm{BA}|^3}\left[(\boldsymbol{\mu}_\mathrm{A}\cdot\boldsymbol{\mu}_\mathrm{B}) - \frac{3}{|\boldsymbol{R}_\mathrm{BA}|^2} \times (\boldsymbol{\mu}_\mathrm{A}\cdot\boldsymbol{R}_\mathrm{BA})(\boldsymbol{\mu}_\mathrm{B}\cdot\boldsymbol{R}_\mathrm{BA})\right] \qquad [2]$$

where μ_A and μ_B are the two (unperturbed) electronic transition moments, ε_0 is the permittivity of vacuum, and $\boldsymbol{R}_{BA}$ is the vector from the centre of A to the centre of B. O_{BA} in Equation [1] is an 'optical factor' which depends on the geometrical arrangements of the two electronic transition moments relative to each other:

$$O_\mathrm{BA} = \boldsymbol{R}_\mathrm{BA}\cdot(\boldsymbol{\mu}_\mathrm{A}\times\boldsymbol{\mu}_\mathrm{B}) \qquad [3]$$

A decreased difference between $\nu(A)$ and $\nu(B)$ in the denominator of Equation (1) would make $\boldsymbol{R}(A)$ increase; consequently, ICD of bands close to the inducing band would be strong. The ICD further depends on the geometry of the A–B pair, both through V_{BA} (Eqn [2]) and O_{BA} (Eqn [3]), which could be described by four coordinates: the length of the separation vector $\boldsymbol{R}_{BA}$ and three angles α, β, and γ defining how $\mu(A)$ and $\mu(B)$ are oriented relative to $\boldsymbol{R}_{BA}$. As seen from Equations [2] and [3], the interaction factor V_{BA} and the optical factor O_{BA} of Equation [1] are both dependent on all four coordinates. The ICD drops off as $\boldsymbol{R}_{AC}^{-2}$ as a combined effect of the $\boldsymbol{R}_{BA}$ dependences in Equations [2] and [3]. Furthermore, the ICD is proportional to the dipole strengths μ_A^2 and μ_B^2, i.e. to absorption of both A and B chromophores:

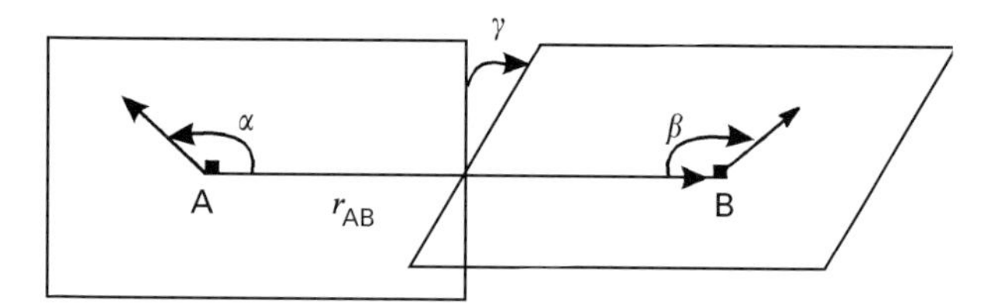

Figure 1 Geometry and coordinates of the A–B system. Transition moments μ_A and μ_B of chromophores A and B and the separation vector $\boldsymbol{R}_{BA}$.

$$\boldsymbol{R}(\mathrm{A}) = -\frac{2\pi\nu_\mathrm{A}\nu_\mathrm{B}(\mu_\mathrm{A}\mu_\mathrm{B})^2}{hc(\nu_\mathrm{B}^2 - \nu_\mathrm{A}^2)\boldsymbol{R}^2{}_\mathrm{BA}}(\sin\alpha\sin\beta\cos\gamma + 2\cos\alpha\cos\beta)\sin\alpha\sin\beta\sin\gamma \qquad [4]$$

Generally, the achiral chromophore A interacts with several chromophores (B_i) of the chiral substrate. If $\nu_A << \nu_B$, the ICD may be regarded as a sum of pairwise contributions, each corresponding to one term in Equation (1) with a given set of B_i, V_{B_iA} and O_{B_iA}. In many molecular systems, such as DNA and cyclodextrins, the symmetric distribution of B chromophores makes it possible to deduce simple expressions for the net ICD.

The following examples will focus on studies of DNA-induced circular dichroism; however the principles are general, based on the application of Equations [1]–[4].

Detection and quantitation of DNA–drug complex

Many DNA-binding molecules (usually referred to as ligands or drugs although they may have little or no therapeutic value) are themselves achiral. However, upon binding to DNA, they acquire an ICD that is characteristic of their interaction. When considering CD applications of proteins and DNA, the ligand ICD can be used at a number of levels. The simplest use is to note that it exists, and therefore that the molecule *does* bind to the chiral substrate; however, more information can usually be extracted.

Empirical use of ICD for binding site analysis

It is often useful to measure a series of spectra where some variable such as temperature, ionic strength, or mixing ratio (drug: DNA ratio) is changed. If the CD intensity changes but the shape of the spectrum remains constant during such an experiment, this indicates that the ligand binding mode is unchanged although the amount of bound ligand may have changed. If, however, there is a change in the shape of the spectrum as an experimental parameter is changed, then a change in the DNA–drug interaction as a function of that parameter is implied. The change is frequently due to occupancy of more than one binding site as the drug load of the DNA increases. Alternatively, it may also be due to changes in DNA conformation or to drug–drug interactions. In other words, if the experiment is conducted with a constant ligand concentration and varying DNA concentration, then unless the ligand binding mode is changing, there should be no change in the observed ligand ICD. A particularly dramatic change, due to exciton CD, is observed if, as the DNA concentration is decreased, the ligands begin to stack or self-associate with each other.

Extended treatment for structural analysis

The important case of coupled oscillator (exciton) CD of interacting, identical chromophores (i.e. degenerate transitions) has already been mentioned. Exciton CD was early observed for dye molecules upon association to biopolymers, attributed to interactions between adjacently bound dimolecules with chiral orientation relative to each other. The less exploited case of coupling of electric dipole transition of the ligand molecule A due to interaction with one or several *nondegenerate* transitions of the chiral molecular system B has been considered in terms of Equations [1]–[4]. Because of the nondegeneracy, this CD is generally very small and should be measured at low drug:biopolymer ratios to avoid dominance by the much more intense exciton CD from drug–drug interactions, DNA-induced monomer CD is exhibited by, for example, intercalating dyes such as methylene blue, acridine orange and proflavin. The DNA-induced CD of several of these intercalators depends sensitively on both salt conditions and on base sequence. These effects have been interpreted in terms of varying binding geometries in the intercalation pocket or to varying inducing transitions of the nucleobases. For eda drug transitions which are energetically isolated from the inducing transitions of the DNA bases, a simple relation has been derived between the DNA-induced monomer CD (rotational strength $\boldsymbol{R}$) and the 'azimuthal' angle γ, defining the orientation of the intercalator transition in the plane of the intercalation pocket (angle to dyad axis):

$$\boldsymbol{R} = f|\boldsymbol{\mu}|^2(1 - 2\cos^2\gamma) \qquad [5]$$

Where f is a function comprising contributions of the various transitions of the surrounding nucleobases. The DNA-induced CD of an intercalator has also been calculated using the more rigorous 'matrix method'. The orientation dependence obtained in this manner (**Figure 2**) was found in good comparison with the $(1-2\cos^2\gamma)$ dependence predicted by Equation [2]. The CD amplitudes were also in reasonable agreement with measured ICD spectra of various intercalators which can be seen in **Table 1**.

The situation is more complicated for a nonintercalator than for an intercalator in that there will be more degrees of geometrical freedom (**Figure 2**) and, as a result of the external position, interactions with a larger number of DNA bases must to be considered.

Groove-bound and intercalator ligand–DNA complex geometries

The relative geometries of DNA and the transition moment of an externally bound chromophore are defined by the parameters r, α, β, and γ, in **Figure 2B**. r and γ define the *position* of the external chromophore's transition moment relative to the DNA helical axis and α and β describe the *orientation* of the transition moment; α is the angle between the z-axis (DNA helical axis) and the transition moment and β is the angle between the projection of the transition moment onto the xy plane and the vector r the distance from the helix axis to the position of the transition moment.

In **Figure 3**, an example is shown of the ICD calculated for a nonintercalator (r = 7 Å) sitting outside a GC step in the minor groove, as a function of the α and β angles. $\alpha = 45°$ and $\beta = 90°$ correspond to a minor groove binding geometry (indicated by an ×), with a transition moment pointing along the groove. This geometry should give a strong positive ICD. However, $\alpha = 90°$ and $\beta = 0°$, which correspond to a transition moment along the duplex dyad axis of an intercalator which has been withdrawn from the centre, should produce a negative ICD. A strong positive ICD appears to be a common feature for minor groove binders with transition moments parallel with the groove. However, depending on the DNA

Table 1 Calculated and observed ICD rotational strengths for DNA intercalators and externally bound drugs

Drug system[a]	*Energy of transition* (10^{0} *cm* $^{-1}$)	*Transition moment angles (Fig. 2)/°*			*Rotational strength (DBM units)*	
		α	β	γ	*Expt.*	*Calc./predicted*
Intercalators						
DAPI–[Poly(dC–dG)]$_2$	30	80	0	0	+0.01	+0.01
Acridine orange–DNA	20	90	0	90	–	–
Pseudoisocyanine–DNA	18	90	0	90	–	–
APF–[poly(dC–dG)]$_2$	28	80	0	0	+	+
Minor groove binders						
DAPI–[poly(dA–dT)]$_2$	30	45	0	90	+1.8	+0.5
Netropsin–DNA	32	45	0	90	+1.3	+0.4
Hoechst 33258–DNA	26	45	0	90	+2.8	+0.4
APF–[poly(dA–dT)]$_2$	28	45	0	90	+	+
Major groove binders						
Methyl green–DNA	23	90	180	0	−0.6	−0.1
	15	45	180	270	+0.8	+0.2

[a] DAPI = 4,6-diamidino-2-phenylindole and APF = 2,5-bis(4-amidinophenyl)furan.

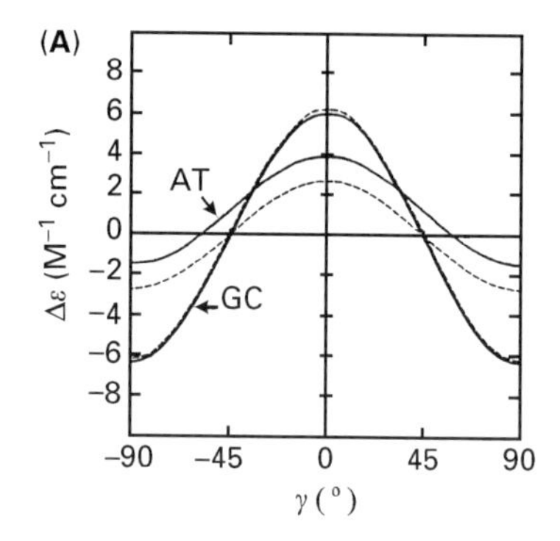

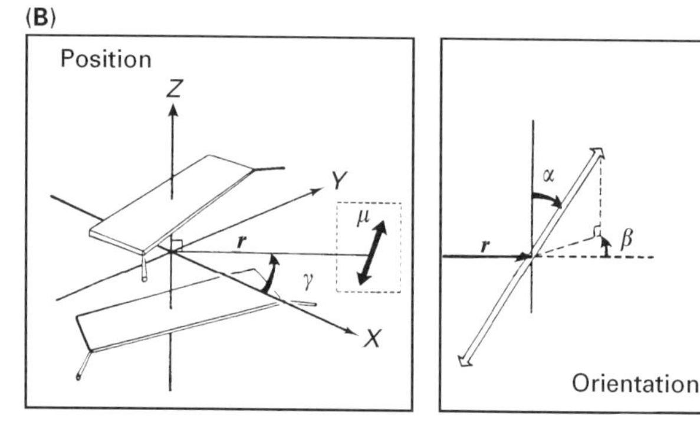

Figure 2 (A) ICD of a DNA intercalator calculated as a function of the transition moment in the plane of the intercalation pocket and centred on the helix axis. γ is the angle between the transition moment and the DNA dyad axis. Reproduced with permission from Lyng, Hard and Nordén (1987) *Biopolymer* **26**: 1327. (B) definition of geometrical parameters of a DNA–ligand system: with $r = 0$, $\alpha = 90°$ and $\beta = 0°$, then γ corresponds to the orientation of an intercalator transition moment (see top figure). Typical minor groove binding geometry parameters are $r = 7$ Å, $\alpha = 45°$, $\beta = 0°$ and $\gamma = 0$.

sequence, both the sign and amplitude are predicted to vary for other nonintercalated and semi-intercalated geometries.

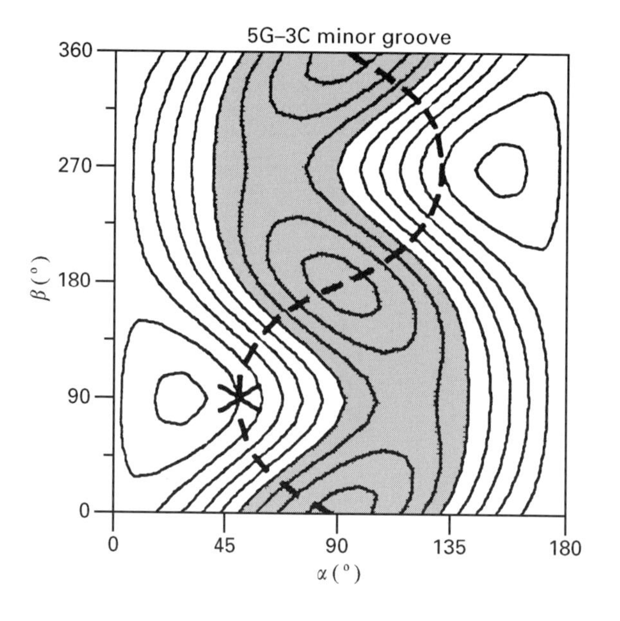

Figure 3 Example of ICD calculated for a minor groove binder (at 5′G–3′C step) as a function of angles α and β (see **Figure 2**). Reproduced with permission of Wiley from Lyng, Hard and Nordén (1991) *Biopolymers* **31**: 1709.

As can seen in **Table 1**, DAPI and APF, which are both known to bind preferentially, in the minor groove at AT-rich regions of duplex DNA, can also bind by intercalation in pure GC tracts (this difference is most clearly seen in the flow linear dichroism spectrum, which probes the α angle). Thus, **Figure 4** shows three types of DNA ICD. Spectrum (a) shows the APF [2,5 bis(4-amidinophenyl)furan] ligand bound in the minor groove of [poly(dAdT)$_2$]. Spectrum (f), obtained at a higher drug: DNA binding ratio, also contains a contribution from exciton CD

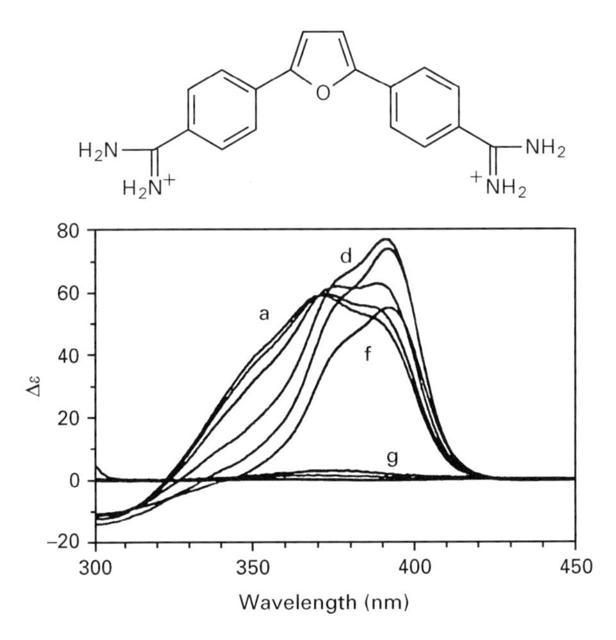

Figure 4 Measured ICD of APF (see formula) bound to $[poly(dAdT)]_2$ [spectra (a)–(f)] and to $[poly(dGdC)]_2$ [spectrum(g)]. The strongly positive ICD in spectrum (a) (drug: DNA phosphate ratio $r = 0.01$) is consistent with minor groove binding (see **Figure 3**), while increasing binding ratios ($r = 0.02$, 0.04, 0.08, 0.16 and 0.31) give a gradually increasing contribution of CD due to exciton coupling between adjacently bound APF molecules. The weak positive ICD in $[poly(dGdC)_2]$ is consistent with intercalation (transition moment along the DNA dyad axis, see **Figure 2**). Reproduced with permission of the American Chemical Society from Jansen, Lincoln and Nordén (1993) *Biochemistry* **32**: 6605.

due to drug–drug interactions between adjacent sites in the minor groove. Spectrum (g), which refers to APF in $[poly(dGdC)_2]$, is the weak positive CD corresponding to the intercalated drug with the transition moment preferentially directed along the short (dyad) axis of the intercalation pocket.

Conformational ICD in DNA systems

When chromophore A could adopt a chiral conformation and B could change its conformation upon binding, some additional sources of ICD induced by chromophore B should be considered. The ICD is generally assumed to be the difference

$$\mathrm{ICD} = \Delta\mathrm{CD} = \mathrm{CD(B+A)} - \mathrm{CD(B)} - \mathrm{CD(A)} = 0 \quad [6]$$

where A = achiral molecule (drug) and B = chiral molecule (DNA). However, the net ΔCD may also have a contribution due to a change of conformation in B:

$$\Delta\mathrm{CD} = \mathrm{ICD(A)} + \delta\mathrm{CD(conf\ B)} \quad [7]$$

where δCD(conf B) = CD change of B due to conformational change of B upon interaction with A.

Since ICD(A) is the true ICD, providing information about the interaction, it is recommended that one corrects, if possible, for any contributions due to conformational change of B[ΔCD − δCD(conf B)]. Three contributions to the ICD of A may be distinguished:

$$\mathrm{ICD(A)} = \mathrm{ICD(A)_{el}} + \delta\mathrm{CD(conf\ A)} + \delta\mathrm{CD(exciton\ A)} \quad [8]$$

where $\mathrm{ICD(A)_{el}}$ = induced CD in A's transitions due to electronic coupling with B's transition's, δCD(conf A) = CD of A due to adopted chiral conformation, and δCD(exciton A) = exciton CD of A interaction with neighbour A molecules.

Special situation (a). When A is a chiral molecule added in enantiomerically pure form, then ICD(A) = induced CD in A's transitions due to electronic coupling with B's transition + δCD(chir) of A due to any changed (chiral) conformation of A + δCD(Abs) of A due to any changed absorption property (hypo-/hyperchromism) of A. The third contribution scales directly as the changed absorption, δAbs:

$$\delta\mathrm{CD(Abs)} = \frac{\delta\mathrm{Abs} \times \mathrm{CD(A)}}{\mathrm{Abs}}$$

Hence if CD(A) is large, δCD(Abs) will be large, even for small δAbs.

Special situation (b). A is a chiral molecule added as a racemate: then, in addition to the contributions in situation (*a*), there is additionally the effect of any enantiopreferentiality of binding A to B, which adds the following extra term to ICD(A): + δCD of A due to that the enantiomer bound may have changed its absorption properties (hypo/hyperchromism) or conformation.

Special situation (c). A is a chiral molecule subject to inversion. The enantiopreferential binding to B can then affect the equilibrium between the enantio-mers to make the total of A non-racemic (an excess of one enantiomer). Generally, the preferred enantiomer is thermodynamically stabilized. This is the so-called 'Pfeiffer effect'. An example is given by the addition of labile $[Fe(phen)_3]^{2+}$ to DNA where the observed

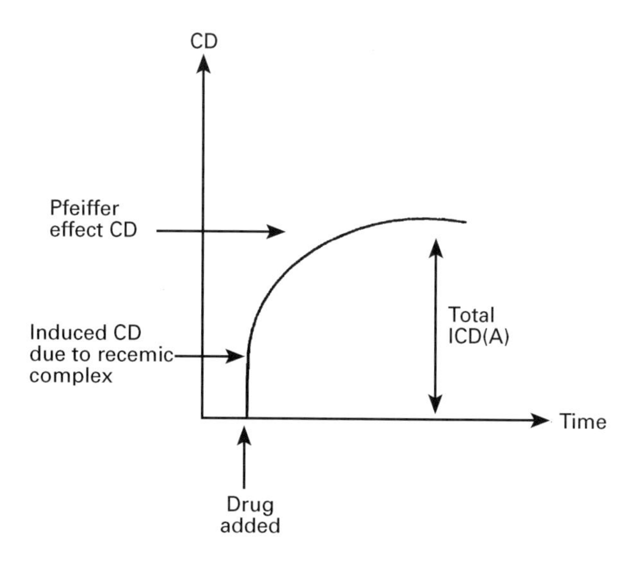

Figure 5 The 'Pfeiffer' ICD of inversion labile $[Fe(phen)_3]^{2+}$ upon binding to DNA. Following DNA addition, a small ICD is immediately seen due to the preferential association of one enantiomer and electronic perturbation of its CD [*special situations (a) and (b)*]. Therefore, a strong CD is more slowly evolving as a result of a shifted diastereomeric equilibrium giving net excess to one enantiomer.

ICD is comprised of two components. There is an immediate effect according to situation (b) and a slower developing (stronger) Pfeiffer ICD due to an inversion yielding a net excess of the enantiomer in the solution, as shown in **Figure 5**.

ICD in cyclodextrins

The CD of large, formally rather complicated systems is sometimes as simple as that of much smaller ones, particularly if the system has high symmetry. One such system is a β-cyclodextrin host–guest complex. Cyclodextrins are α-1,4-linked D-glucose oligomers, where α-cyclodextrins have six glucose units and β-systems have seven and so on. Since cyclodextrins have been found to catalyse reactions involving their guests, it is of interest to know the orientation of the guest inside the sugar ring (**Figure 6**). The CD induced into the transitions of known polarization of the guest may show that immediately. Most experiments have been conducted using the β-cyclodextrin framework, which has a cavity of the correct size to accommodate a single aromatic molecule such as benzene or naphthalene. Another application of cyclodextrin-induced CD is for the assignment of transitions in C_{2v} or D_{2h} symmetric molecules: ICD of opposite signs correspond to transitions with mutually perpendicular polarizations.

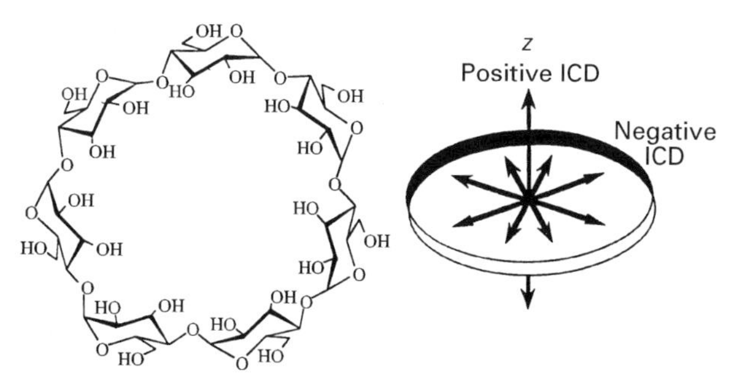

Figure 6 β-Cyclodextrin (left) and the signs of the CD observed in the transitions of a guest molecule depending on their polarization directions: axially polarized transitions appear with positive ICD, whereas transitions polarized in the plane of the cyclodextrin ring appear with negative ICD. Reproduced with permission from Roger A and Nordén B (1997) *Circular Dichroism and Linear Dichroism*, Oxford: Oxford University Press.

List of symbols

O = optical factor; r = distance from helix axis to position of transition moment; r = drug:DNA phosphate ratio; $\boldsymbol{R}$ = rotational strength; V = dipole–dipole interaction energy; α = angle; β = angle; γ = angle; ε_0 = permittivity of vacuum; μ = transition moment; ν = frequency.

See also: **Biomacromolecular Applications of Circular Dichroism and ORD; Chiroptical Spectroscopy, Emission Theory; Chiroptical Spectroscopy, Oriented Molecules and Anisotropic Systems; Chiroptical Spectroscopy, General Theory; Macromolecule-Ligand Interactions Studied By NMR; Nucleic Acids and Nucleotides Studied Using Mass Spectrometry; Nucleic Acids Studied Using NMR; Pharmaceutical Applications of Atomic Spectroscopy.**

Further reading

Harada N and Nakanishi K (1983) *Circular Dichroism Spectroscopy: Exciton Coupling in Organic Stereochemistry.* Mill Valley, CA: California University Science Books.

Mason SF (1982) *Molecular Optical Activity and the Chiral Discrimination.* Cambridge: Cambridge University, Press.

Nakanishi K, Berova N and Woody RW (eds) (1994) *Circular Dichroism Principles and Applications*, New York: VCH.

Nordén B, Kubista M and Kurucsev T (1992) *Quarterly Reviews of Biophysics* **25**: 51–170.

Rodger A, and Nordén B (1997) *Circular Dichroism and Linear Dichroism.* Oxford: Oxford University Press.

Inductively Coupled Plasma Mass Spectrometry, Methods

Diane Beauchemin, Queen's University, Kingston, Ontario, Canada

MASS SPECTROMETRY
Methods & Instrumentation

Inductively coupled plasma mass spectrometry (ICP-MS) combines an ICP with a mass spectrometer. It features a multielemental capability, good precision, a long linear dynamic range, simple spectra, low detection limits and the ability to do rapid isotopic analysis. With such advantages, ICP-MS has found wide application to the analysis of a variety of samples, including water and food, as well as geochemical, environmental and biological samples. However, it has a number of limitations, such as matrix effects, which can be important, and the need to keep the concentration of dissolved solids low to avoid clogging problems. These complicate the analysis by requiring sample pre-treatment and/or more involved calibration strategies. Nonetheless, with the appropriate method development and/or optional accessories, these limitations can be largely circumvented, thereby allowing ICP-MS to be applied to the analysis of virtually any type of sample.

Description of a basic ICP-MS instrument

Figure 1 shows a block diagram of an ICP-MS instrument. A 1–2.5 kW Ar ICP is the typical ion source for MS. The ICP is a high-temperature (5000–10 000 K) electrodeless discharge that is sustained in argon flowing within a torch. In general, the ICP uses samples in solution, which presents the advantages of good control over homogeneity and ease of calibration. A sample solution is delivered to a nebulizer using a peristaltic pump to minimize physical interferences (from, for instance, changes in viscosity). The nebulizer converts this solution into an aerosol that is then carried by the nebulizer gas through a spray chamber where large droplets condense and drain out. The resulting fine aerosol is injected into the heart of the plasma, and undergoes several sequential processes as it moves deeper into the plasma: desolvation, vapourization, atomization and ionization.

In an Ar ICP, the degree of ionization of 52 elements is expected to be $\geq 90\%$. Only three elements (He, F and Ne) which possess a first ionization potential greater than that of Ar would not be ionized and could therefore not be determined by ICP-MS with an Ar ICP. Similarly, the highest degree of double ionization is 10% and occurs only for a few elements. The ICP is therefore an efficient elemental ion source for MS since the majority of elements in the periodic table are singly ionized.

A portion of the plasma (which is typically three sampler orifice diameters wide and two sampler orifice diameters deep) is sampled from the central channel of the plasma, and extracted through a differentially pumped interface (through two water-cooled metal cones with orifice diameter ~1 mm, the sampler and the skimmer), which is maintained at approximately 1 torr using a mechanical roughing pump. This portion is then transmitted into the mass spectrometer where ions are separated according to their mass-to-charge ratios (m/z) and counted.

Two different categories of ICP-MS instruments are currently commercially available: low-resolution instruments, using either a quadrupole or a time-of-flight (TOF) tube to achieve ion separation, and double-focusing high-resolution instruments that use an electrostatic and a magnetic analyser in series. Both the quadrupole-based and the double-focusing instruments allow a sequential multielement measurement, whereas TOF-ICP-MS allows simultaneous multielement detection.

Selected specifications for these two categories are summarized in **Table 1**. Aside from resolution, one distinctive feature of high-resolution instruments is the negligible background which is observed across the mass range. This results from the curved path that the ion beam must follow, which efficiently prevents photons from reaching the detector, as well as from the operating pressure that is considerably lower than that of low-resolution instruments, and thereby minimizes collisional processes in the ion beam.

Capabilities of ICP-MS

Table 2 shows that ICP-MS has numerous advantageous features: large linear dynamic range,

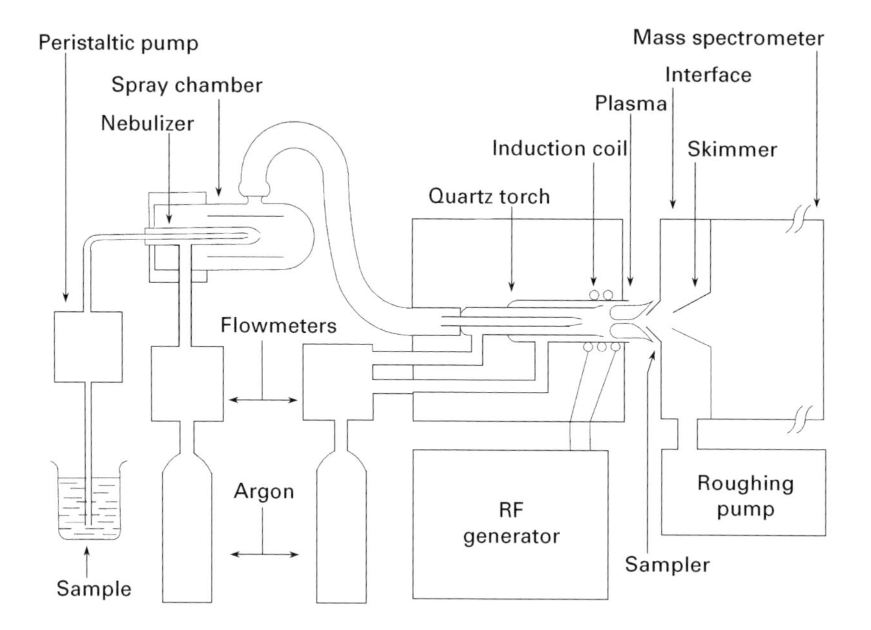

Figure 1 Schematic representation of a conventional ICP-MS instrument.

Table 1 Typical specifications of ICP-MS instruments

	Type of instrument	
Parameter	*Low resolution*	*High resolution*
Resolution	1 amu[a]	300–10 000 amu[b]
Mass range	1–300 amu	1–500 amu
Background level across the mass range	< 2–50 cps	<< 0.1 cps

[a] Peak width at 10% from the peak maximum.
[b] 10% valley between two adjacent peaks.

Table 2 Attributes of ICP-MS instruments

	Type of instrument	
Feature	*Low resolution*	*High resolution*
Orders of linear dynamic range	5–9	9
Short term precision for 10 repeats (%)	1–3	2
Sensitivity (Mcps/ppm)	1–30	20 at $R = 300$ 1 at $R = 3\,000$
Detection limit[a] (ng/L)	<1–10	<0.01–1 ($R = 300$–500)

[a] Based on three times the standard deviation of the blank.

advantage of working with samples in solution, good precision, low detection limits. The linear dynamic range for many elements is such that major, minor and trace levels may all be determined at once. Furthermore, the multielement capability of ICP-MS allows the determination of, for instance, 70 elements in less than 2 min, using less than 2 mL of solution at an uptake rate of 1 mL min^{-1}. Detection limits are typically ≤ 1 ng L^{-1} for most elements with all low-resolution instruments, and even lower with high-resolution instruments. This difference arises from the lower background typically obtained on high-resolution instruments, as well as the higher reliability in peak hopping which results from the flat-top shape of the peaks observed in high resolution. Sensitivity is similar on all instruments in low resolution, but is degraded by an increase in resolution because this is typically achieved by decreasing the slit width of the entrance and exit slits on double-focusing instruments, which in turn leads to a reduction in ion transmission.

In any case, since specific isotopes are detected, isotope ratios can be measured. Also, calibration by the very accurate isotope dilution analysis method can be done (where internal standardization is essentially carried out using an enriched isotope of the analyte). The flat-top peaks, which are typical of high-resolution instruments, should allow the measurement of ratios with better precision than that achievable with quadrupole-based instruments. However, because these mass spectrometers cannot be scanned as fast as quadrupoles, the resulting precision can be degraded compared to that obtained with quadrupole MS. The best precision is expected with TOF instruments since they allow the simultaneous detection of all masses.

All these features explain why even the basic ICP-MS instrument described so far is being applied to analyses in a variety of areas. For instance, **Table 3** shows that ICP-MS is widely used for environmental,

Table 3 Examples of applications of ICP-MS

General category	*Samples*	*Sample pretreatment*
Water	Sea water, industrial waste water, pore water, river water, fresh water, rain, ground water, lake water, spring water, snow, tap water, high-purity water	Filtration and any of the following: acidification and dilution; ion exchange separation to remove interferents and/or preconcentrate analytes; preconcentration by evaporation
Environmental	Soil, dust, leaves, sediments, sewage, industrial effluents, paint, atmospheric aerosols, suspended particulate matter, tobacco smoke condensate, domestic sludge, glass, spent nuclear fuel, leachates of high-activity waste, rocks, metals, ores, alloys, ceramic materials	Crushing or grinding, and either digestion with high-purity acids or fusion and dissolution, followed by dilution
Food	Prawn parts, wheat flour, spinach, cabbage, wine, mussels, beef kidney, milk	Digestion with high-purity acids followed by dilution
Biological	Faecal material, urine, human serum, plasma, red blood cells, whole blood, cerebrospinal fluids	Dilution or drying, ashing and/or digestion with high-purity acids followed by dilution or extraction
Organic	Crude oil, gasoline	Extraction into aqueous acidic solution

clinical and geochemical studies, with appropriate sample preparation procedures. These are required not only to put solid samples in solution, but also to avoid or circumvent limitations of ICP-MS by, for instance, removing sources of interference.

Limitations of ICP-MS

Table 2 indicates that, whatever the resolution, detection limits vary with the element being determined. There are, indeed, several factors affecting detection limits in ICP-MS. The degree of ionization of the analyte, the natural abundance of the isotope that is used for the determination of a particular element, spectroscopic interferences from the background and/or the sample matrix, and mass discrimination are the most notorious sources of degradation. Obviously, if an element is only 50% ionized in the plasma, then its detection limit will be reduced by a factor of 2 compared to an element which is completely ionized. Similarly, if only one isotope of a multi-isotope element is detected, the resulting detection limit will be less than that of a monoisotopic element.

Furthermore, the sample introduction system is the Achilles' heel of all ICP-based instruments. Indeed, only about 2% of the sample solution actually reaches the plasma with a conventional system (composed of a pneumatic nebulizer and spray chamber, as shown in **Figure 1**), most of it going down the drain. This means, therefore, that only a small fraction of the total possible signal is obtained.

As mentioned earlier, because the mass spectrum is simple, the likelihood of spectroscopic interferences is reduced, particularly with high-resolution instruments. However, when they occur, the number of alternative lines is small (there are indeed a number of monoisotopic elements). One example of an isobaric interference that cannot be resolved even at high resolution is that of $^{40}Ar^{+}$ on $^{40}Ca^{+}$, the major isotope of Ca. Other severe interferences arise from Ar-containing polyatomic ions as well as oxides of matrix elements. Several of these can be resolved at higher resolution but with a concurrent sacrifice in sensitivity and detection limits. On low-resolution instruments, a 'cold plasma', where the ICP is operated at lower power and higher nebulizer gas flow rate, can be used to reduce important interferences from the background but, again, a lower sensitivity concurrently results across the mass range.

Mass discrimination is most important for quadrupole-based instruments where the transmission of ions is not uniform over the mass range. In particular, the transmission of light ions is reduced compared to heavier ions. Another parameter which may also affect the detection capability of ICP-MS is the plasma sampling position, which may not be optimal for all the analytes being determined simultaneously. This position is indeed important to maximize the density of analyte ions extracted from the ICP. Even the chemical form (i.e. speciation) of an element in solution may influence where it undergoes the various processes (desolvation, vapourization, etc) in the plasma, ultimately determining where the maximum ion density will be.

The sampling process itself is the source of another limitation: because part of the plasma passes through a small orifice, this orifice can become clogged by dissolved solids. The maximum concentration of dissolved solids which can be tolerated on a continuous basis is 0.2%. Finally, even if this level is kept reasonably low, effects of concomitant elements (also called matrix effects) can occur, whereby the analyte signal is either suppressed or enhanced by the presence of other elements. Some trends have been observed and partly explained. For example, a heavy matrix element will in general induce a more important suppression than a light matrix element on quadrupole-based instruments because of space–charge effects. However, these matrix effects are still largely unpredictable, which complicates the calibration process for samples that do not possess a relatively simple matrix. Whenever possible, a simple dilution of the sample is recommended to minimize the effects of concomitant elements, since they depend on the absolute amount of concomitant element.

The various calibration strategies available with ICP-MS are summarized in **Table 4**, along with implications of the above-mentioned limitations. (More elaborate chemometric calibration procedures have also been developed by a few groups.) In general, unless the sample matrix is simple, internal standardization (or frequent calibration) will be required for quantitative analysis using external calibration with a series of standard solutions. The choice of the internal standard(s) depends on the instrument, the sample, etc. For the best results, an element with properties similar to that of the analyte (so that it will behave similarly in ICP-MS) is needed, so several internal standards may be required for a multielemental analysis spanning the entire mass range.

Table 4 Calibration strategies in ICP-MS

Method	*Requirements*	*Quality of results*
External calibration	One isotope free of spectroscopic interference per analyte Matrix matching and/or correction for drift (using internal standardization or frequent calibration)	Good accuracy and precision
Standard additions	One isotope free of spectroscopic interference per analyte May require a correction for drift (using internal standardization)	Good accuracy and precision
Isotope dilution	Two isotopes free of spectroscopic interference per analyte Good equilibrium between isotopic spike and sample	High accuracy and precision

The method of standard additions efficiently compensates for matrix effects but not for drift. If the latter is also present (for instance because of gradual clogging), internal standardization may also be required. The best method to compensate for both matrix effects and drift is isotope dilution analysis. However, it is only applicable to elements with at least two isotopes (stable and/or long-lived radioactive) free of spectroscopic interferences. Because of mass discrimination, this normally standardless technique requires a prior determination of mass bias using standards of known isotopic composition in order to correct the measured isotope ratios for mass discrimination.

Accessories available to expand the capabilities of ICP-MS

Major modifications can be made to the ICP-MS instrument, such as adding one or two quadrupoles, and operating one quadrupole in an RF-only mode so that it can serve as a cell where gas-phase collisions are performed to break polyatomic ions. However, several more or less expensive accessories that can be readily interfaced to ICP-MS are available to circumvent one or more of the above-mentioned limitations. The most frequently used are summarized in **Table 5**.

The sample introduction system can simply be replaced by a high-efficiency one, such as an ultrasonic nebulizer coupled to a desolvation system. Analyte transport efficiencies of 10–20% are typically achieved using such a system. Other high-performance nebulization systems include direct injection nebulization, which introduces 100% of the sample directly into the plasma (i.e. no spray chamber is needed), thermospray, hydraulic high-pressure nebulization and monodisperse dried microparticulate injection.

The characteristics of the plasma can also be modified by adding another gas to argon. For instance, an addition of O_2 to the plasma is beneficial for the analysis of organic solutions as it prevents the deposition of soot on the interface. The addition of gases with a higher thermal conductivity than argon (such as H_2) can be used to enhance sensitivity (by improving the ionization efficiency), reduce and/or eliminate spectroscopic and non-spectroscopic interferences, and reduce mass discrimination.

Table 5 demonstrates that various approaches can be used with a basic ICP-MS instrument to overcome spectroscopic interferences. Many of these approaches consist in separating the problematic element(s) from the analyte during sample preparation

Table 5 Options widely used to enhance the capabilities of ICP-MS

Method	*Advantages*	*Disadvantages*	*Examples of applications*
Ultrasonic nebulization with desolvation	High sample introduction efficiency	Prone to memory effects	Analysis of river water, lake water, rain and other samples with simple matrix
	Reduction of oxides through removal of solvent	Cone blockage and matrix effects from concentrated matrix	
Mixed-gas plasmas	Reduced spectroscopic interferences	Higher incidence of torch melting	Analysis of fuel oil, vegetable oil, sea water, urine, slurries
	Reduced matrix effects		
Flow injection analysis	Reduced memory effects	System must be modified and optimized for each type of application	Analysis of samples with high level of dissolved solids (sea water, brines, urine, etc), concentrated acids, volume-limited samples
	Increased sample throughput		
	Reduction of matrix effects		
	Enhancement of sensitivity		
	On-line pretreatment chemistry		
Chromatographic techniques	Speciation of aqueous and gaseous samples	Different interfaces required for different techniques	Determination of toxic and non-toxic species in biological materials, food, environmental samples
	Separation of analyte from matrix	Chromatographic resolution dependent on interface to ICP	
Chemical vapour generation	Selective separation of analyte from matrix	Different chemistries and/or conditions required for different analytes	Determination of hydride-forming elements and/or species in a variety of samples
	Up to 100% sample introduction efficiency		
Electrothermal vapourization	Small sample volume	Only one or a few elements can be determined at once	Analysis of waters, biological materials, food, volume-limited samples
	Direct solid analysis	Matrix modifiers often required to prevent loss of analyte	
	In-situ pre-treatment of sample		
	Up to 100% sample introduction efficiency		
Slurry nebulization	Eliminates digestion and fusion steps	No available sample blanks	Analysis of rocks, soils, hard-to-dissolve samples (such as coal)
	Calibration with aqueous standards	Efficiency of calibration dependent on particle size and sample type	
Laser ablation	Direct solid analysis	Limited number of reference materials available for calibration	Analysis of geological samples, ceramic materials, glass
	In-situ analysis of interstitial fluid	No available sample blanks	
	Microscopic profiling of solid samples		

(by extraction, chelation, selective precipitation, etc). Although they can be quite effective at reducing both spectroscopic and non-spectroscopic interferences (if the source of the latter is also removed during sample processing), they are nonetheless time-consuming and increase the likelihood of contamination.

Most of these separations can be carried out on-line with ICP-MS using a flow injection manifold. A big advantage of this approach is that all the chemistry is done in a closed system, often on a reduced scale, thereby reducing sources of contamination, as well as sample and reagent consumption. Chemical vapourization approaches can also selectively isolate the analyte and transform it into a volatile species, in either continuous or flow injection fashion. Alternatively, selected species of the analyte can be determined by coupling gas, liquid or supercritical chromatography, or capillary electrophoresis, to ICP-MS.

As mentioned earlier, solid samples must first be put in solution for analysis by conventional ICP-MS, using digestion, fusion, dry-chlorination, etc, which all introduce potential sources of contamination. Slurry nebulization, which simply consists in grinding the sample and suspending it in a suitable dispersing agent, can significantly reduce these contaminations. Alternatively, solids can be analysed directly using various techniques coupled to ICP-MS: electrothermal vapourization, laser ablation (as well as arc/spark ablation), direct sample insertion, fluidized bed introduction, etc. These techniques are even more interesting for samples which are difficult to dissolve. Some of them (electrothermal vapourization and direct sample insertion) can also be used for the analysis of solutions.

All these approaches enhance the capabilities of ICP-MS but complicate the optimization process. The full potential of some of them (such as ablation techniques, flow injection, etc), which generate short, transient signals, will be better realized if an instrument allowing simultaneous multielement detection (such as TOF) is used. In any case, no single method can be used for the direct analysis of every sample. Some method development is therefore required for each new sample type, whether ICP-MS is used with or without one or more of the available options.

See also: **Calibration and Reference Systems (Regulatory Authorities); Chromatography-MS, Methods; Food Science, Applications of Mass Spectrometry; Forensic Science, Applications of Mass Spectrometry; Glow Discharge Mass Spectrometry, Methods; Isotope Ratio Studies Using Mass Spectrometry; Quadrupoles, Use of in Mass Spectrometry; Sector Mass Spectrometers; Time of Flight Mass Spectrometers.**

Further reading

Brenner IB and Taylor HE (1992) A critical review of inductively coupled plasma mass spectrometry for geoanalysis, geochemistry, and hydrology. Part I. Analytical performance. *CRC Critical Reviews in Analytical Chemistry* **23**: 355–367.

Date AR and Gray AL (eds) (1989) *Applications of Inductively Coupled Plasma Mass Spectrometry.* Glasgow: Blackie.

Gray AL (1988) Inductively coupled plasma source mass spectrometry. In: Adams F, Gijbels R and Van Grieken R (eds) *Inorganic Mass Spectrometry, Chemical Analysis Series*, Vol 95, pp 257–300. London: Wiley.

Hall GEM (1989) Inductively coupled plasma mass spectrometry. In: Thompson M and Walsh JN (eds) *Handbook of Inductively Coupled Plasma Spectrometry*, 2nd edn. Glasgow: Blackie.

Houk RS (1986) Mass spectrometry of inductively coupled plasmas. *Analytical Chemistry* **58**: 97A–105A.

Houk RS (1994) Elemental and isotopic analysis by inductively coupled plasma mass spectrometry. *Accounts of Chemical Research* **27**: 333–339.

Jarvis KE, Gray AL and Houk RS (1992) *Handbook of Inductively Coupled Plasma Mass Spectrometry.* Glasgow: Blackie.

Montaser A (ed) (1998) *Inductively Coupled Plasma Mass Spectrometry.* New York: Wiley.

Montaser A and Golightly DW (eds) (1992) *Inductively Coupled Plasmas in Analytical Atomic Spectrometry*, 2nd edn. New York: VCH.

Olesik JW (1991) Elemental analysis using ICP-OES and ICP/MS. An evaluation and assessment of remaining problems. *Analytical Chemistry* **63**: 12A–21A.

Industrial Applications of IR and Raman Spectroscopy

AS Gilbert, Beckenham, Kent, UK
RW Lancaster, Glaxo-Wellcome plc,
Medicines Research Centre, Stevenage, Herts, UK

VIBRATIONAL, ROTATIONAL & RAMAN SPECTROSCOPIES
Applications

Extensive use is made of IR and Raman spectroscopy in industry, because these techniques have the capacity to provide information at the chemical molecular level on virtually any material of interest be it solid, liquid or gas. They are well understood, can be easily applied and results can be obtained quickly and cheaply. Instruments utilized range from simple single-channel IR filtometers to sophisticated computerized interferometers and spectrographs supported by flexible data handling software. In addition to process functions they are employed in quality assurance and troubleshooting in the quality control (QC) environment and in analysis (both qualitative and quantitative) in research and development (R&D). Areas of application are pre-eminently in the chemical, oil and gas, plastics and pharmaceutical industries, but there is also considerable use in materials industries, mining, agriculture and food processing.

Applications of near-infrared (NIR) spectroscopy will not be described in this article but can be found elsewhere in the Encyclopedia. The examples of IR and Raman spectroscopy given below represent, of course, only a tiny fraction of what is done in industry and are necessarily a rather arbitrary selection.

Brief history

With the introduction of commercially available, automatically recording, spectrometers during World War 2, IR spectroscopy began to be utilized in industrial laboratories and by the 1960s dedicated instruments had reached the factory floor. The advent of FT-IR machines designed specifically for the analytical market and a much wider range of accessories then saw a considerable expansion in use from the early 1980s onwards, particularly in QC and development laboratories.

Expense and technical difficulty meant that it was not until the 1970s that anything more than a handful of Raman spectrometers were employed in industry. Take-up, even by industrial research laboratories, continued to be slow until the arrival of relatively cheap computerized spectrographs, FT-Raman and advanced methods of sampling during the 1980s.

The special conditions of the industrial environment

Employment of instrumentation is much the same in industrial research laboratories as it is in academic institutions, although of course the aims of the; former are necessarily more focused to particular objectives. But elsewhere in industry both time and money, within the constraints of safety, are of prime importance and the usefulness of information provided by any technique must be judged primarily by these criteria. The nearer the production line, the greater is the degree of rigour with which these criteria are applied. Rapid measurement, high degree of automation and reliability, all at minimum cost, are therefore demanded of manufacturers of spectroscopic equipment.

A major cost element of any activity is personnel. This predicates dedicated operation with as much 'black box' control of instrumentation as possible and incorporating robust protocols.

The factors described above tend inevitably to induce conservative attitudes in industrial practitioners so that it is important that any technique and its particular area of application be well understood, widely accepted and validated by longtime usage. The activities of organizations such as ASTM (American Society for Testing Materials) which have overseen the establishment of standard procedures for many common but specific spectroscopic measurements have been highly important in cementing acceptance of such methodology. Scientific forums that bring together both industrial and academic workers are extremely helpful in circulating ideas. The long-standing British-based IRDG (Infrared and Raman Discussion Group) is a good example of a body that organizes regular lecture meetings, often at industrial locations.

Instrumentation

As vibrational spectroscopy was originally a research technique, many spectrometers used in industry in the past have been derivatives of research-oriented machines. However, most equipment is now produced for the 'analytical' market so this relationship (even for Raman spectrometers) has been turned around and now many so-called 'research' machines are just the standard 'analytical' instrument with add-on features. Outside R&D most instrumentation is designed for dedication to specific tasks and a great many IR machines sold are single-channel filtometers.

Such instruments are non-dispersive and utilize a narrow bandpass filter to monitor one compound of interest, commonly carbon dioxide or carbon monoxide, in a process environment. The bandwidth of the filter is closely matched to the bandwidth (for carbon dioxide the unresolved envelope of a number of rotational sidebands) of the analyte and a reference signal is obtained from another filter which passes radiation from a transparent area of the spectrum.

Full-spectrum machines are now, in the case of IR, mostly interferometers. The benefits of full computerization, high sensitivity, ease of operation and 'quantitative analysis' software packages have rendered the classical dispersive spectrometers obsolete. Portable operating software has meant that industry standard PCs and workstations can be used as 'front ends' on most if not all commercial instruments now sold.

Until relatively recently, Raman spectroscopy was not generally suited to industrial use with the exception of specific problems in R&D. Although the classical double grating dispersive scanning spectrometers could be computerized, they were slow and often subject to problems such as sample fluorescence. This has been changed by the advent first of simple and rugged spectrographs with multichannel (usually charge-coupled device (CCD)) detectors and second of FT-Raman modules that can be 'bolted on' to standard FT-IR machines. Spectrographs have addressed the speed problem while FT-Raman, which uses near-IR excitation, is not troubled by fluorescence.

Sampling methods

Many of the sampling accessories now available for vibrational spectroscopy were developed for specific industrial use or with industry in mind. Some of the most important recent developments have been in the production of fibre-optic attachments that can allow data to be obtained *in situ* with remote spectrometers. Optical fibres are available for the Raman region but though suitable materials (e.g. chalcogenides) that are transparent in parts of the mid-IR region can now be manufactured, fibre-optic probes in this region are somewhat limited in usefulness. **Figure 1** demonstrates the quality of a Raman spectrum that can be obtained from inside a container, in this example an amber vial holding acetone. The flexible fibre-optic probe was merely placed on the outside of the glass so that sampling was completely non-invasive.

Raman microscopy imaging, the spectroscopic mapping of a sample placed on a computer-controlled motorized stage, has applications in many areas. A TV attachment gives visual guidance to the

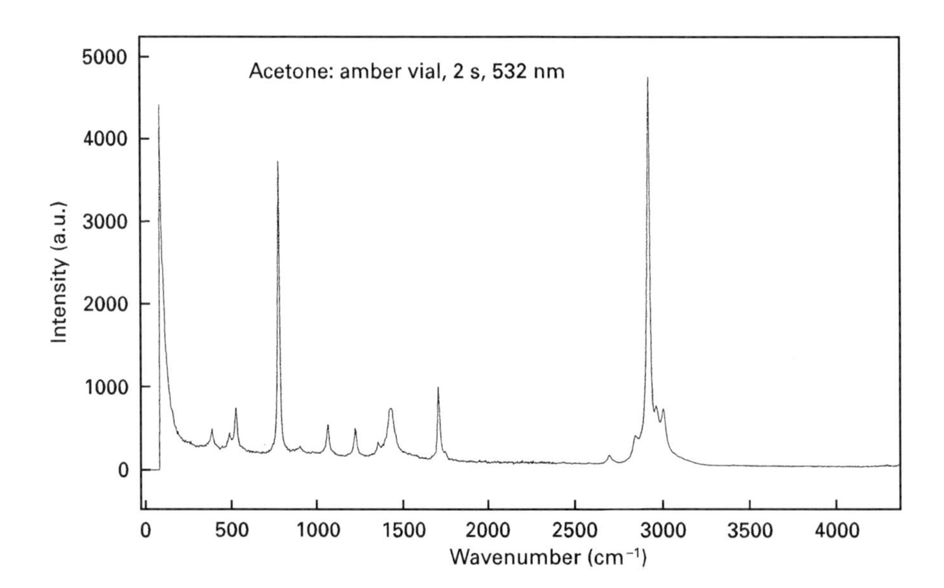

Figure 1 The Raman spectrum of acetone contained within an amber vial. A HoloProbe (Kaiser Optical Systems, USA) fibre-optic probe transmitting a visible excitation line at 532 nm was utilized. The length of the fibre optic can be 100 m or more while, by changing attachments, the head can collect the Raman signal from distances of between about 0.03 and 40 cm.

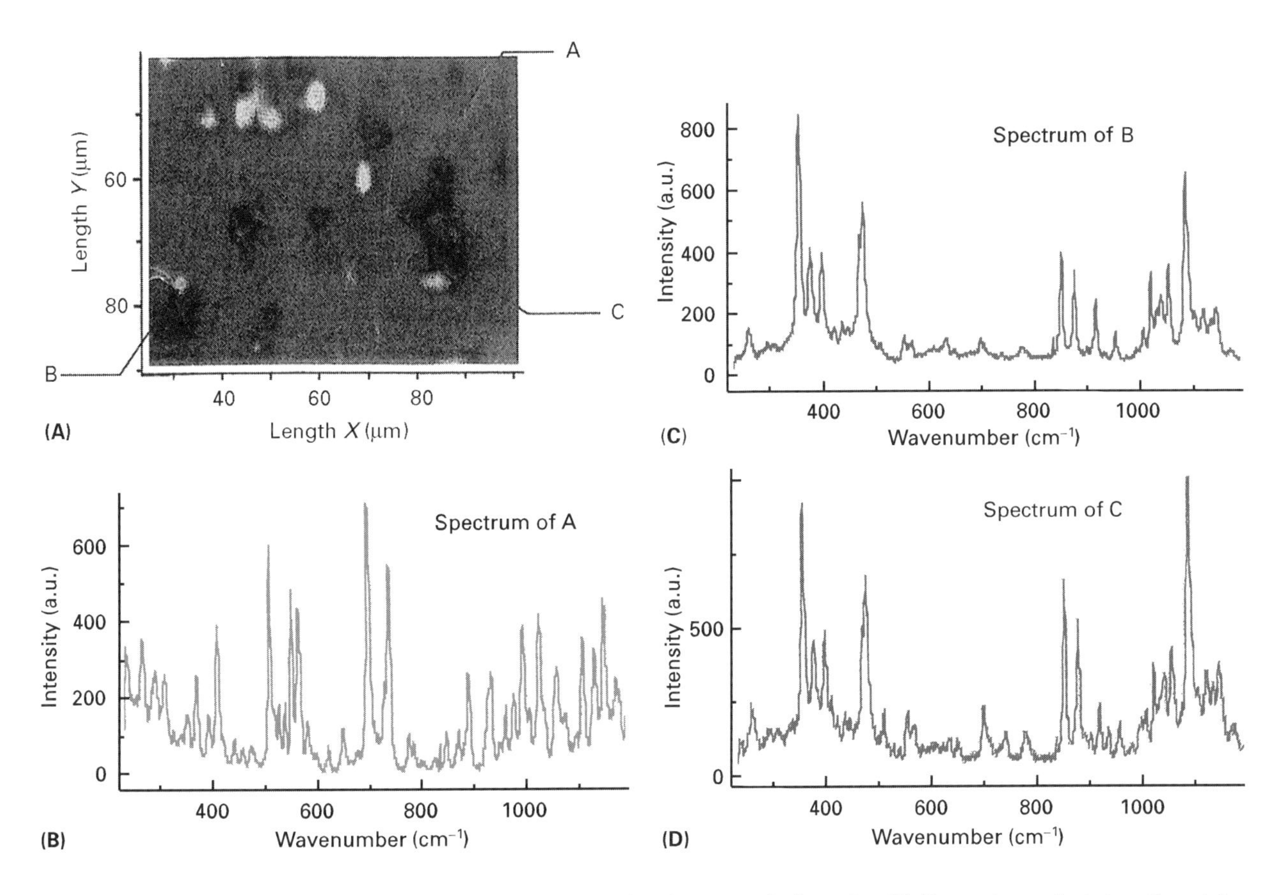

Figure 2 Raman microscope image and associated spectra of a pharmaceutical powder. (A) Raman image that gives the relative importance of each compound (A, B, C), and spectra of (B) compound A, (C) compound B and (D) compound C. (See Colour Plate 23)

operator for location of an area of interest and real-time spectra are produced. A particular application, in this example from the pharmaceutical industry, is the examination of powders, particles down to 1 μm (or less) being distinguished. This is an order of magnitude better than IR microscopy. **Figure 2A** illustrates a colour-coded Raman map that offers chemical information on the disposition of different components of a mixture, the three distinct spectra shown in **Figures 2B**, **C** and **D** being obtained by 'specifying' three particular points in the image. The Raman image analysis software can also be used to assess particle size distribution.

Attenuated total reflection (ATR) probes can be used to follow chemical reactions *in situ*. An example is the react-IR (Applied Systems) instrument. Although ZnSe is commonly used as the optical element in ATR, it is not robust enough to deal with certain chemical reactions, e.g. bromination, so a diamond tip is utilized instead. **Figure 3** shows the spectrum of a fused-ring β-lactam which yields a carbonyl band at 1785 cm^{-1}. This can be seen reacting via an intermediate to give a less strained β-lactam with a carbonyl vibration near 1760 cm^{-1}.

The production of potassium bromide discs for IR spectroscopy has never been automated, although at least one scheme has indicated the possibility. When dispersive spectrometers were used, this did not matter so much as spectrum acquisition was of the same order of duration as sample preparation, but FT-IR machines are much more rapid. Thus diffuse reflectance has recently become a very popular technique for monitoring the IR spectra of organic compounds that are soluble in volatile solvents as it only requires the placing of a drop of solution on a small layer of ground alkali metal halide powder.

For dealing with large numbers of compounds from chromatographic or other types of separation, the so-called 'hyphenated' methods are used. The most popular are the combinations with gas chromatography (GC-IR), thermogravimetry (TG-IR) and gel-permeation chromatography (GPC-IR). **Figure 4** is the IR chromatogram of part of a GC separation of fractions in a typical gasoline (petrol). One of the aims was to detect aromatic components the levels of which are strictly controlled by various 'clean fuel' legislations (in the USA and elsewhere). Only IR absorption from the 900–600 cm^{-1} region has been

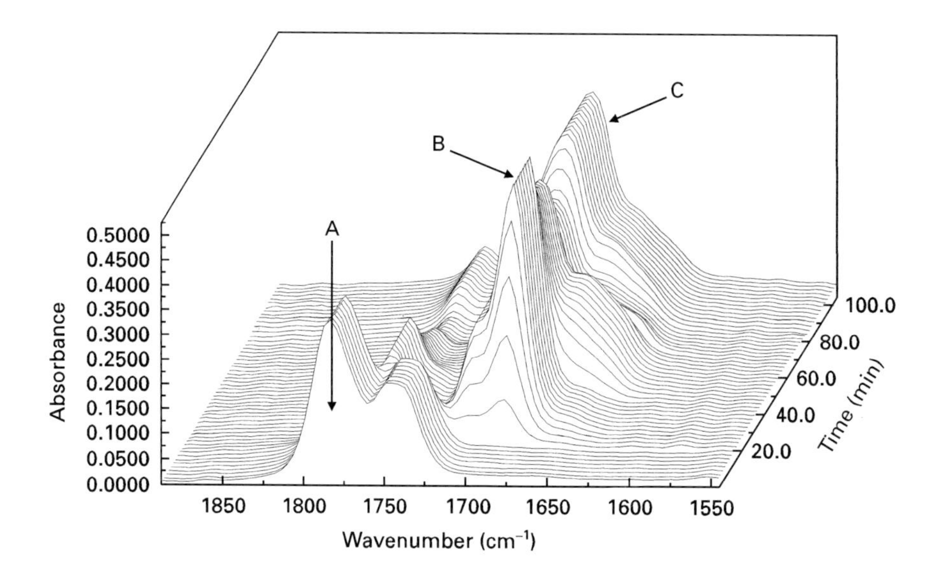

Figure 3 IR stack plot of the progress of a reaction of a β-lactam reactant A giving intermediate B and product C.

utilized in this example, this being the region of the characteristic C–H out-of-plane bending modes of aromatic compounds. The level of benzene in this sample, which was of the order of 0.02% by mass, could be determined by the GC-IR method to an accuracy of within 1% of the measured value.

Data handling

Data transfer to computer networks such as LIMS (laboratory information management systems) is now straightforward using the JCAMP world standard data format and modern operational software for vibrational spectrometers is usually set up for this. Sophisticated quantitative analysis packages are generally available for use in analytical laboratories and designed for largely 'black box' operation, which is obviously attractive to industry. Some of the chemometrics methodology (e.g. partial least squares (PLS)) behind the software has only been developed relatively recently, however, and still draws criticism from some quarters, which might cause problems when regulatory authorities are involved.

Research and development

Many substances of interest to industry are highly complex but vibrational spectra can still provide useful insights into their nature. **Figures 5A, B and C** show the IR spectra of a series of coal tar fractions which were obtained by carbonization of a wet charge of coal in an industrial coke oven at a mean flue temperature of 1220°C. The fractions are principally composed of complex multi-ring aromatics. While the spectra clearly yield many bands characteristic of aromatic systems, for instance the substitution sensitive C–H out-of-plane bending vibrations below 1000 cm^{-1}, other functionalities are obviously present, such as O–H and N–H. Interestingly, spectrum (B) shows a band at about 2220 cm^{-1}

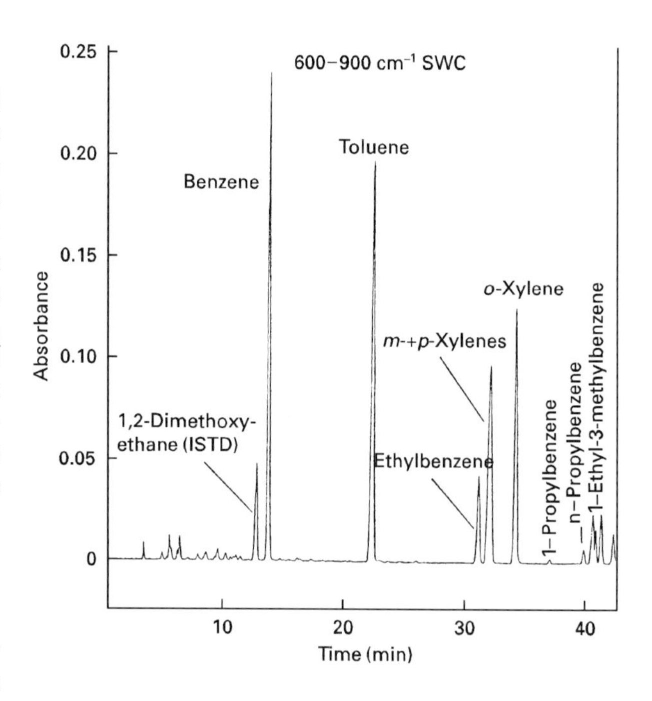

Figure 4 IR chromatogram of the first 40 min of a GC-IR separation of aromatic components in a sample of gasoline. The peaks represent the total integrated absorbance intensity in the region 900–600 cm^{-1}. Reprinted with permission from Diehl JW, Finkbeiner JW and DiSanzo FP (1995) Determination of aromatic hydrocarbons in gasolines by gas chromatography/Fourier transform infrared spectroscopy. *Analytical Chemistry* **67**: 2015–2019. Copyright 1995 American Chemical Society.

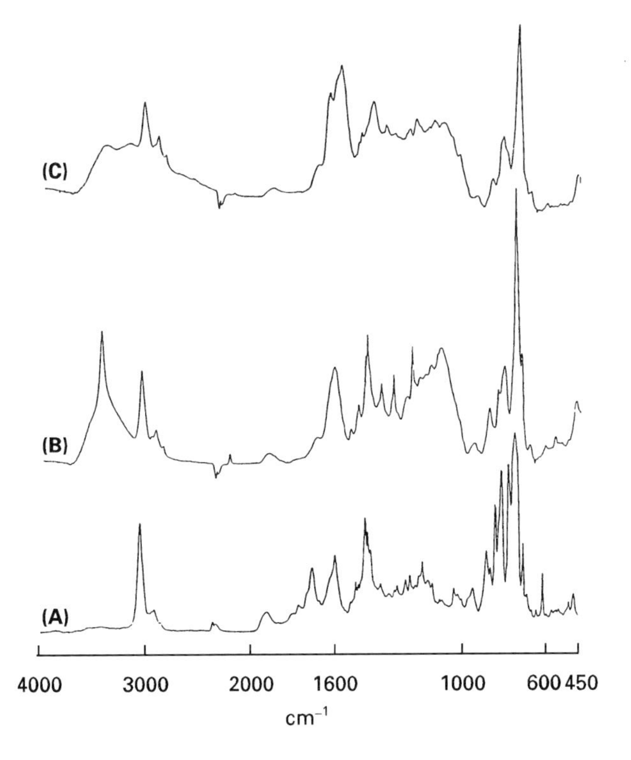

Figure 5 FT-IR spectra of coal tar fractions examined as either (A) a film or (B, C) KBr discs. Reprinted from Diez MA, Alvarez R, Gonzalez AI, Menendez R, Moinelo SR and Bermejo J (1994) *Fuel* **73**: 139–142, Copyright 1994, with permission from Elsevier Science.

(just below the negative, instrument-uncompensated, CO_2 peak at about 2300 cm^{-1}) indicating nitrile, possibly from a naphthocarbonitrile.

Vibrational bands are very sensitive to intermolecular effects, thus different crystal modifications of a single compound usually yield discrete spectra. The occurrence of differing crystal forms is known as polymorphism and is of interest as the physical properties are dissimilar. This can have considerable ramifications in the pharmaceutical industry as the dissolution and absorption rates of polymorphic forms of a drug may differ significantly, with serious consequences for therapeutic action and patient tolerance.

Figures 6A and **B** display the IR spectra of forms I and II, respectively, of lamivudine, a 1,3-oxathiolane nucleoside which has anti-viral activity:

The two forms are not strict polymorphs because I is a hydrate (one H_2O per five lamivudines), but have different crystal structures and spectra. The O–H stretching mode from the hydrate water in I can clearly be seen at 3545 cm^{-1}. The strong bands just below 3000 cm^{-1} are from the hydrocarbon mulling agent used to prepare the material for spectroscopic measurement.

Combinatorial chemistry

Schemes for combinatorial syntheses of compounds, particularly candidate pharmaceuticals, now attract great interest because of the potential for greatly speeding up the discovery of new drug molecules. They are based on carrying out chemical reactions on solid artifacts, the starting material, intermediates and final product being attached to the solid phase; on completion the product is cleaved away into solution. Typically the solid substrate is a mass of small (of the order of tens or 100 μm or so) porous beads made of polymer or silica resin. They are suspended, by agitation, in reactant solution, which is removed on completion by filtration. The beads are washed with solvent between stages of synthesis.

A considerable challenge is thereby presented to analytical technology if the progress of the reaction is to be monitored without going through the time-consuming business of analyte removal. Mass spectrometry cannot be used and NMR only with difficulty because of sensitivity problems. However, IR spectroscopy is fairly easy to utilize, either by diffuse reflectance or, as in the cases below of two drug intermediates, ATR sampling methods. **Figure 7** is the spectrum of

attached to polystyrene beads, the carbonyl stretching band being clearly seen at 1690 cm^{-1}. There is little interference with this highly characteristic band from the polymer. The silica resin (Tentagel) presents a greater problem, although as shown in **Figure 8**, a number of bands from attached

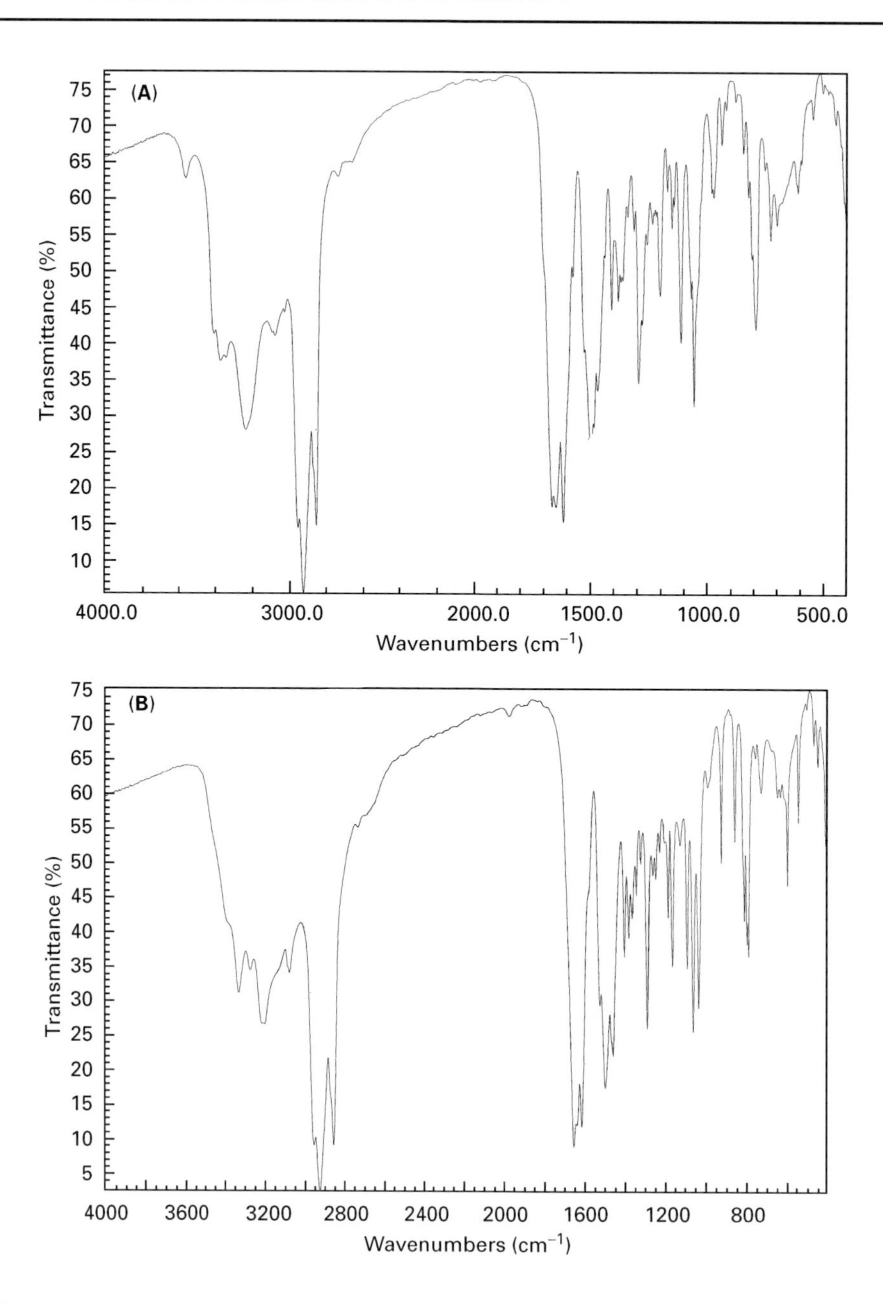

Figure 6 FT-IR spectra of forms (A) I and (B) II of lamivudine as Nujol mulls. Resolution, 2 cm^{-1}. Reproduced with permission from Harris RK, Yeung RR, Lamont RB, Lancaster RW, Lynn SM and Staniforth SE (1997) *Journal of the Chemical Society, Perkin Transactions* **2**: 2653–2659.

can still be observed, e.g. the carbonyl vibration at 1741 cm^{-1}. The very strong band near 1100 cm^{-1} is, of course, from Si–O stretching. These spectra were obtained from a small cluster of beads, but it is possible using microscopy techniques to examine a single bead at a time.

Quality assurance

Pharmaceuticals

As visible radiation is unaffected by passage through clear, colourless glass (apart from inducing minor fluorescence), Raman spectroscopy can be used to monitor gases contained within the headspace of sealed vials. **Figure 9** is a photograph of equipment that performs automated headspace analysis of vials (1 mm thick borosilicate glass) containing the pharmaceutical product Flolan (sodium prostacyclin) that is freeze-dried and sealed (with rubber stoppers secured by aluminium caps) under nitrogen. The product, although buffered, is highly sensitive to degradation by ingress of carbon dioxide if admission of air occurs, leading to a severe reduction in shelf-life. Any leakage, which might very occasionally occur before final capping, is detectable, however, by

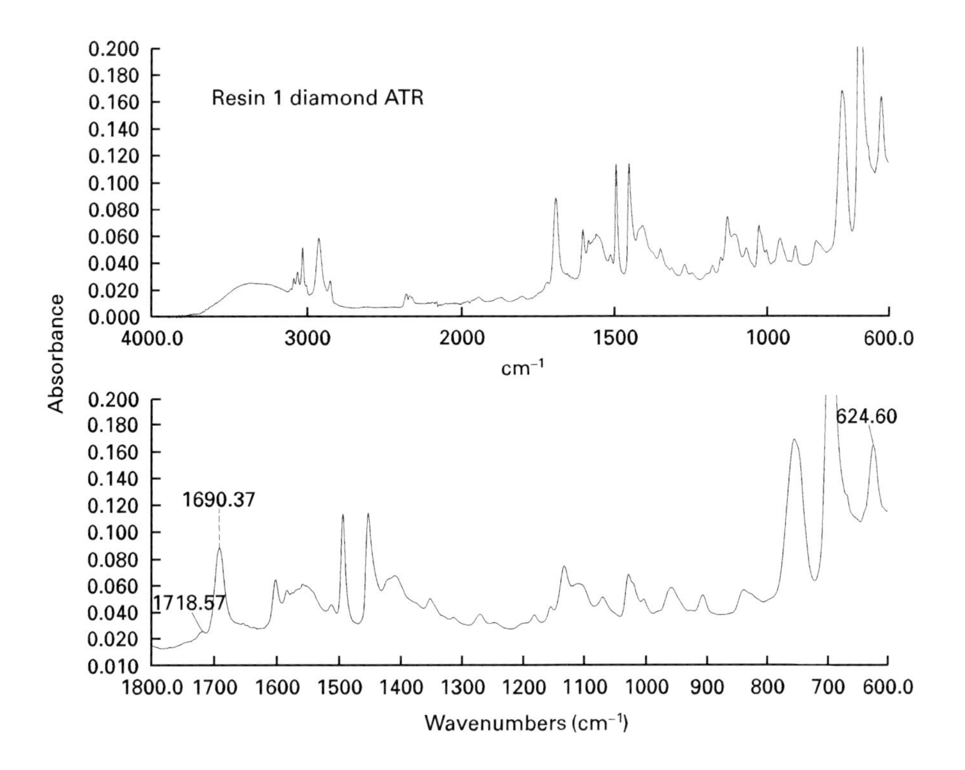

Figure 7 Part of the FT-IR/ATR spectrum of polystyrene beads with attached compound (see text). Courtesy of Perkin-Elmer Ltd.

assaying for oxygen with the nitrogen as reference, using the Raman bands from the stretching modes of these molecules at about 1550 and 2330 cm^{-1}, respectively (**Figure 10**).

The analyser is built around a spectrograph with a CCD multichannel detector and the spectra are generated by laser radiation at 488 nm from an argon ion laser. Oxygen percentages can be measured to

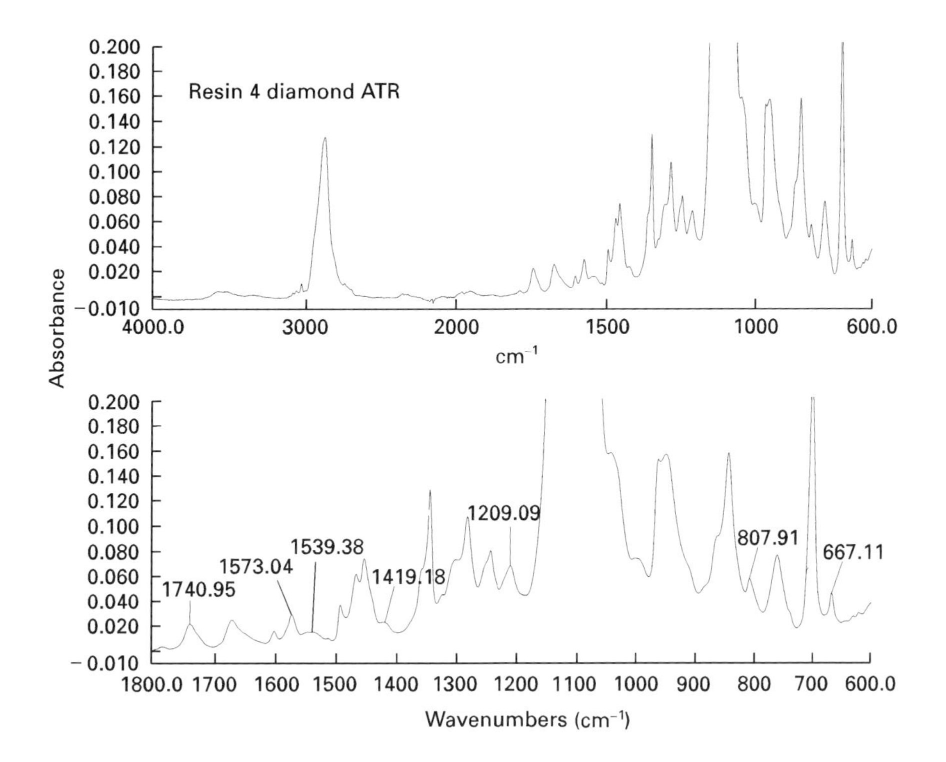

Figure 8 Part of the FT-IR/ATR spectrum of Tentagel beads with attached compound (see text). Courtesy of Perkin-Elmer Ltd.

Figure 9 Automated headspace analyser. A Model HR640 MSL (Instruments SA) Raman spectrograph with an Astromed CCD multichannel detector is utilized. Laser excitation at 488 nm. Courtesy of Hobbs KW, Glaxo-Wellcome plc.

±0.3% O_2 (± one standard deviation) for a measurement time of 1 s. Mechanical restrictions mean that sample throughput is slower than this, but still sufficiently fast that total batch inspection can be performed.

Semiconductors

An important requirement in the semiconductor industry is to be able to measure various undesirable impurities in silicon wafers which can interfere with their electrical properties. Two significant impurities are oxygen which can incorporate interstitially and carbon which can substitute into lattice sites. Although pure silicon itself yields a very weak IR spectrum, fairly thick samples can be examined in transmission and therefore trace levels of the impurities can be detected and routinely quantified. Standard methods for IR spectroscopic determination of the impurities have been published by ASTM.

Oxygen forms Si–O–Si bonds which yield a characteristic band at 515 cm^{-1} from Si–O–Si deformation (a stronger band from Si–O stretching occurs at 1107 cm^{-1}). Carbon can be detected by the Si–C stretching band at 607 cm^{-1}. **Figure 11** shows some room temperature difference spectra of CZ silicon using FZ silicon as reference (CZ and FZ refer to two different methods of growing silicon crystals, Czochralski process and float zone, respectively). It is interesting to note that the high-carbon CZ sample contains so much oxygen that weak bands from carbon–oxygen complexes are observed. Quantitative determination of carbon and oxygen is possible down to low ppm levels.

Dedicated FT-IR spectrometers are extensively employed to measure the thickness of epitaxial layers

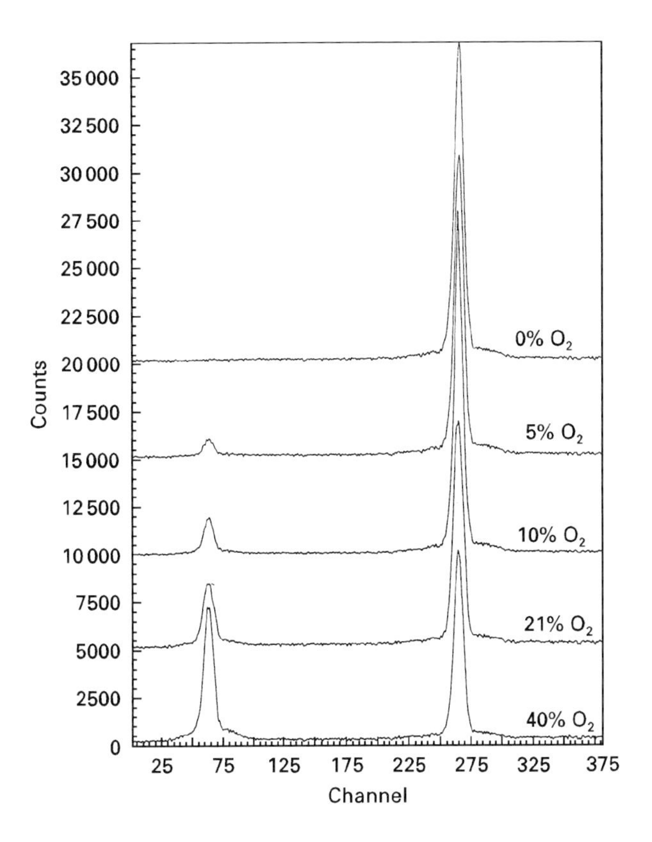

Figure 10 Raman spectra of various mixtures of nitrogen and oxygen obtained with the spectrograph in **Figure 9**. For the sake of clarity the spectra have been vertically displaced from one another. The N_2 peak is on the right. Band intensity is essentially from the pure vibrational mode. The rotational–vibrational side bands form weak overlapped 'wings' on either side of the main peaks. Reproduced with permission of the Society of Photo-Optical Instrumentation Engineers (SPIE) from Gilbert AS, Hobbs KW, Reeves AH and Jobson PP (1994) Automated headspace analysis for quality assurance of pharmaceutical vials by laser Raman spectroscopy. *Proceedings of the SPIE – Society of Photo-Optical Instrumentation Engineers* **2248**: 391–398.

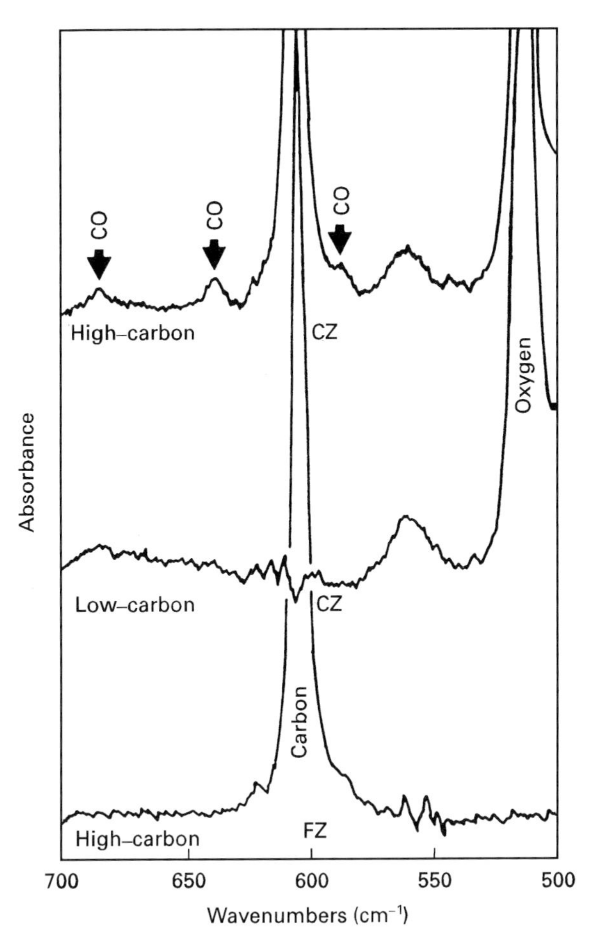

Figure 11 IR difference spectra, CZ–FZ, using FZ (bottom) as reference. Samples were 2 mm thick. Carbon–oxygen complex bands are indicated by arrows. See text for definition of CZ and FZ. Reproduced by permission of Academic Press from Krishnan K, Stout PJ and Watanabe M (1990) Characterisation of semiconductor silicon using Fourier transform infrared spectrometry. In: Ferraro JR and Krishnan K (eds) *Practical Fourier Transform Infrared Spectroscopy,* pp 286–351. New York: Academic Press.

on silicon wafers. The measurements are made in reflectance mode and what is observed is not the spectrum but the interference pattern from multiple back and forth reflections within the layer. The regular distance between the maxima (or minima) gives the thickness and the fringes can be directly perceived from the interferogram as two 'side-bursts' on either side of the centre.

Process control

Analysis of edible oils and fats is an important activity in the food industry. Quantification of *cis* and *trans* unsaturation levels is one part of this activity. Many edible oils are found in the *cis* form and are liquids. They are often partially hydrogenated to improve stability and hardness (e.g. for 'healthy' alternatives to butter), but in the process some *cis* bonds are converted to *trans*. Unfortunately, *trans* unsaturated fatty acids have been implicated in heart disease, but they can easily be detected by the presence of the characteristic out-of-plane olefinic C–H bending vibration at 980–965 cm^{-1}. *Cis* bonds do not interfere as they yield the corresponding vibrational mode at 730–650 cm^{-1}.

The standard IR spectroscopic procedure (American Oil Chemists Society) involves dissolution in carbon disulfide with chemical modification if *trans* levels are $< 15\%$. Peak height only is used for calculation and the overall procedure is slow. By taking advantage of modern multivariate chemometric methods, a more rapid and convenient means of analysis has recently been developed utilizing a

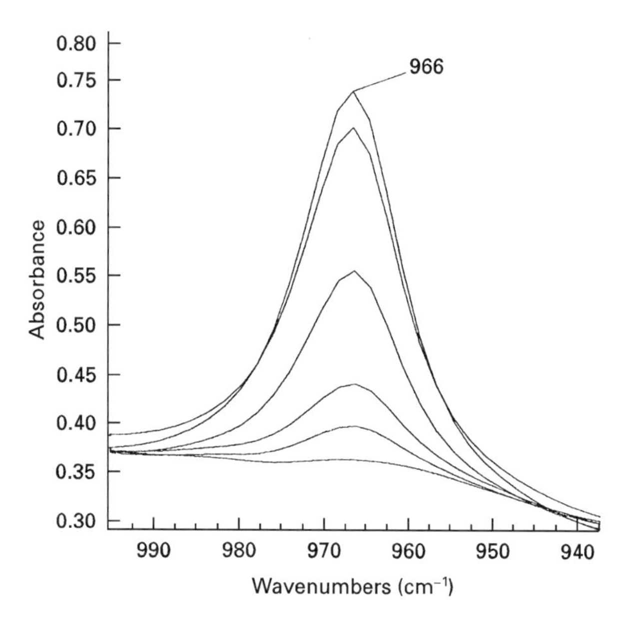

Figure 12 FT-IR calibration standard spectra of the '*trans*' region. The standards are a '*trans*' free base oil with various known amounts of trielaidin, a triglyceride of carbon chain length 18, containing a *trans* double bond at the 1-position. Reproduced with permission of the AOCS from Sedman J, van de Voort FR, Ismail AA and Maes P (1998) *Journal of the American Oil Chemists Society* **75**: 33–39.

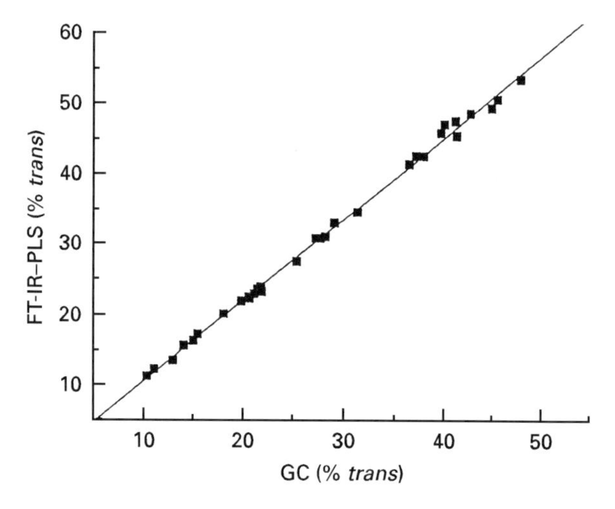

Figure 13 Comparison of results from FT-IR and GC methods for determination of the '*trans*' contents of edible oils. Reproduced with permission of the AOCS from Sedman J, van de Voort FR, Ismail AA and Maes P (1998) *Journal of the American Oil Chemists Society* **75**: 33–39.

heated flow cell (thermostated to 80±0.2°C) to analyse the neat sample in its liquid state. **Figure 12** is the IR spectrum of the *trans* region for a series of calibration standards and **Figure 13** is a plot comparing measured values by the new IR method with gas chromatography, demonstrating high correlation between the two techniques. Calculation for the new IR procedure is carried out by the PLS method using multiple spectroscopic points and is thus very 'black box'. The sample measurement time is claimed to be 2 min, sufficiently rapid, therefore, for process monitoring.

A mundane but very common requirement is the on-line measurement of the water content of industrial solvents and feedstocks in process streams. Chemical methods (e.g. Karl Fischer) are not suited to continuous monitoring. Relatively high levels (> 0.1%) can be dealt with using NIR methodology, but for very low levels the much stronger (fundamental) O–H stretching bands in the mid-IR region need to be observed. **Figure 14** shows the IR spectrum of traces of water in 1,2-dichloroethane (ethylene dichloride, EDC). Because of the low concentration there is no significant hydrogen bonding between water molecules so that the 'free' asymmetric and symmetric stretching modes appear at 3685 and 3595 cm^{-1}, respectively.

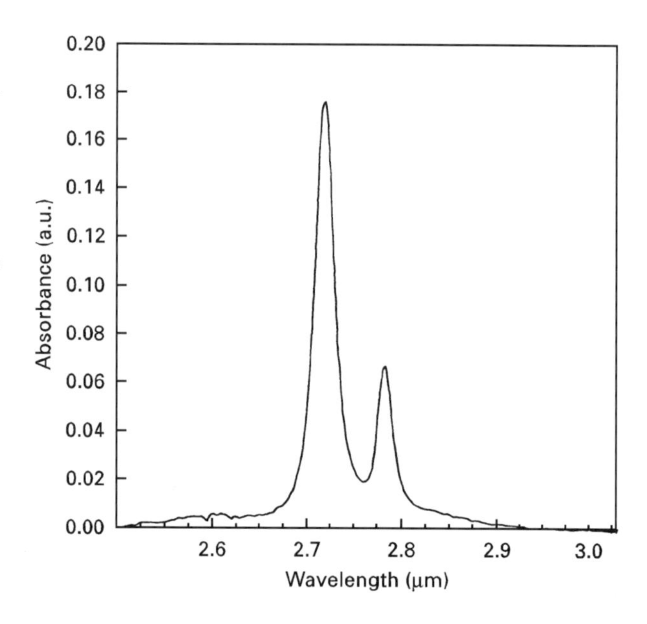

Figure 14 IR transmission spectrum of traces of water in 1,2-dichloroethane. Note that the scale is in microns, which can be converted to wavenumbers by taking the reciprocal and multiplying by 10 000; e.g. 2.7 μm is equivalent to 3703.7 cm^{-1}. Reproduced with permission of the Royal Society of Chemistry from Bruce SH and Dhaliwal HK (1991) On-line moisture analysis by IR, in Davies AMC and Creaser CS (eds) *Analytical Applications of Spectroscopy II*, pp 46–52. Cambridge: Royal Society of Chemistry.

EDC is the feedstock for the manufacture of PVC, being first cracked in a furnace to produce vinyl chloride followed by polymerization. The moisture measurement is usually done in the transmission

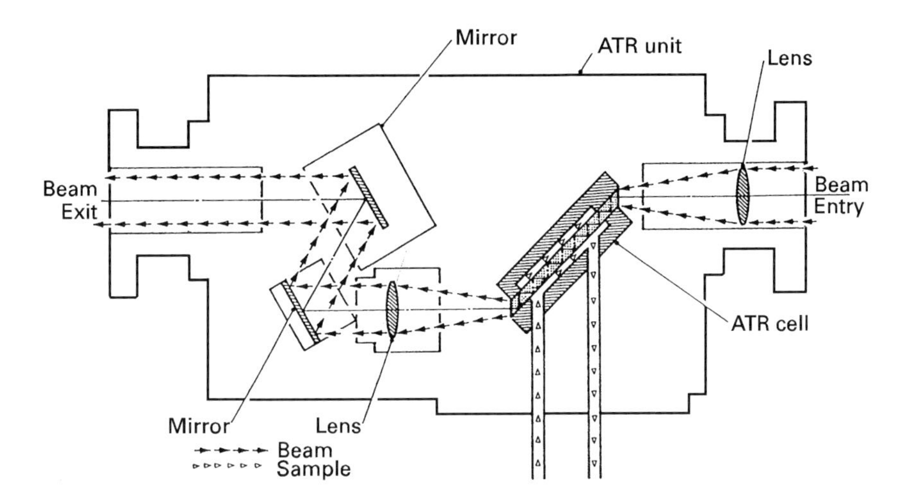

Figure 15 Diagram of process ATR cell for on-line moisture analysis. Reproduced with permission of the Royal Society of Chemistry from Bruce SH and Dhaliwal HK (1991) On-line moisture analysis by IR, in Davies AMC and Creaser CS (eds) *Analytical Applications of Spectroscopy II*, pp 46–52. Cambridge: Royal Society of Chemistry.

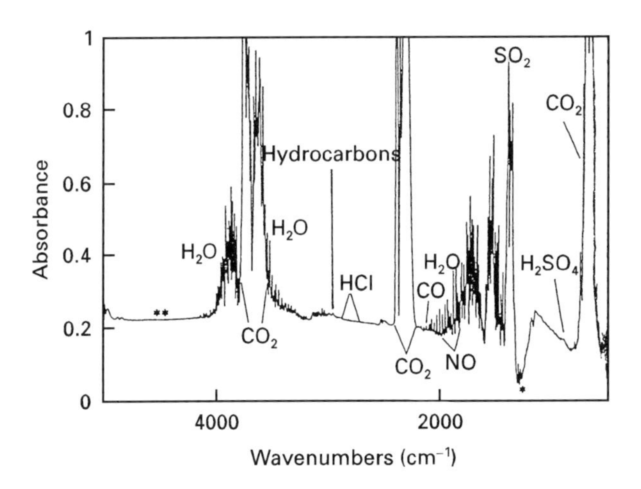

Figure 16 FT-IR spectrum of a flue gas stream, resolution 4 cm^{-1}. Reproduced with permission of the Society of Photo-Optical Instrumentation Engineers from Solomon PR, Morrison PW Jr, Serio MA *et al* (1992) Fourier transform infrared spectroscopy for process monitoring and control. *Proceedings of the SPIE – Society of Photo-Optical Instrumentation Engineers* **1681**: 264–273.

mode, with a pathlength of around 2 mm, and as there are no other bands in the O–H stretching region a simple IR filtometer can be used. In other cases where the pathlength is too restrictive, for instance if the solvent/feedstock is too viscous, an ATR method can be employed instead. **Figure 15** shows an on-line ruggedized (150 psig, 150°C) ATR cell for this purpose (Servomex, UK) which can be fitted with a variety of IR crystals (for chemical compatibility) and adjusted to give the requisite effective pathlength.

The high sensitivity of FT-IR to many small gas molecules can be exploited to monitor flue gases. **Figure 16** shows the make-up of exhaust gas from a coal-fired boiler. To see the minor components it is necessary to expand the spectral scale as in **Figure 17**. Of importance are obviously the levels of the acid gases, NO and (unburned) hydrocarbons, which can all be measured at ppm levels.

Most industrial polymers contain additives which are there as stabilizers (e.g. cross-linkers, antioxidants) or property modifiers (e.g. slip agents). The diagram of a system (Automatik, Germany) for molten polymer process control using FT-IR to monitor additives on-line is given in **Figure 18**. The flow cell can withstand high pressures (300 bar) and temperatures (400°C) and has an adjustable pathlength. The spectrum shown in **Figure 19** is of various amounts of the slip agent oleamide in low-density polyethylene the monitored band being the amide carbonyl stretch near 1715 cm^{-1}.

Conclusion

The wide application of vibrational spectroscopy to industrial problems is long established and its use is growing, as evidenced by steadily increasing sales of instrumentation. In many instances other techniques can do the same job but are either less sensitive or slower or more expensive, or are ruled out through other circumstances. In certain cases vibrational spectroscopy may be the only method practicable, such as in the semiconductor applications described above.

The fact that vibrational spectroscopy is amenable to almost any chemical substance is responsible for its general utility, but this 'non-specificity' can be a

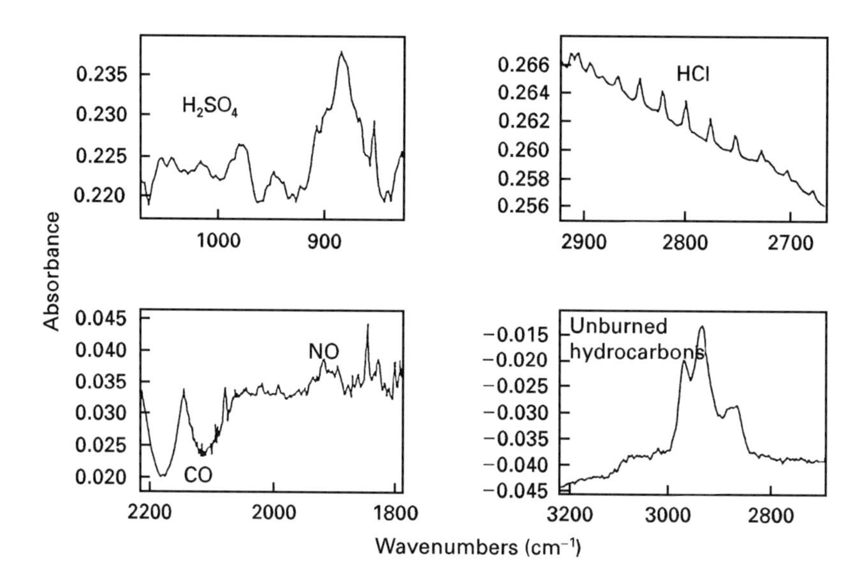

Figure 17 Expanded regions of spectrum in **Figure 16** to show trace gases. Reproduced with permission of the Society of Photo-Optical Instrumentation Engineers from Solomon PR, Morrison PW Jr, Serio MA *et al* (1992) Fourier transform infrared spectroscopy for process monitoring and control. *Proceedings of the SPIE – Society of Photo-Optical Instrumentation Engineers* **1681**: 264–273.

weakness where it is necessary to 'pick out' one component from a background. There are one or two special techniques in Raman spectroscopy where this can be done (e.g. resonance Raman and surface-enhanced Raman spectroscopy), but they are not of general applicability.

A particular strength compared with other main-line analytical techniques is in the monitoring of gases from the points of view of both the sensitivity of IR spectroscopy and the non-invasive capability of Raman spectroscopy, coupled with the capacity of both techniques for remote viewing.

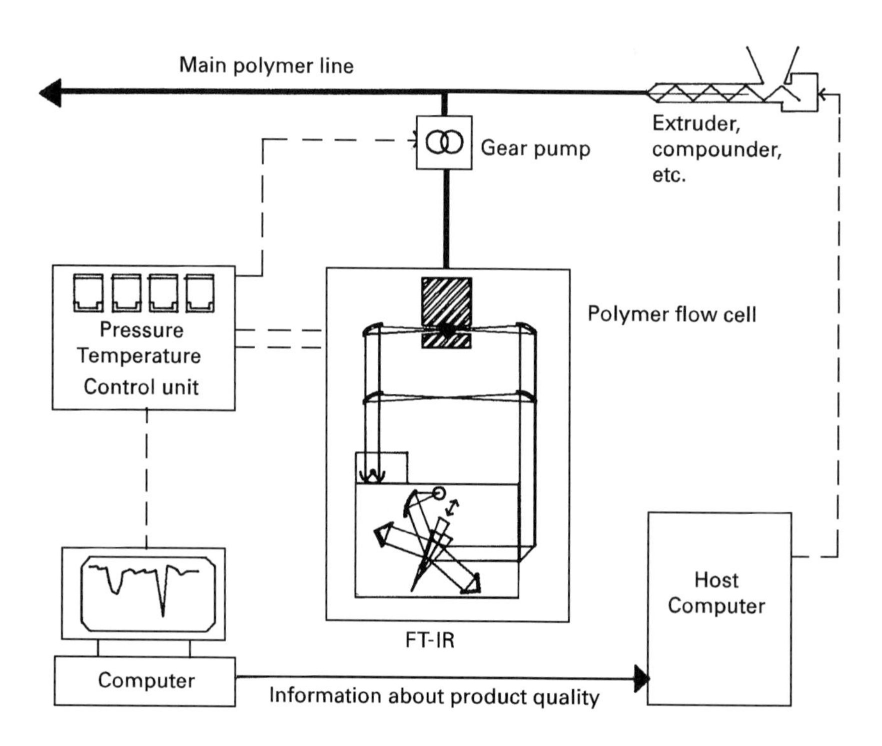

Figure 18 Diagram of an on-line IR process control system for polymer production. Reproduced with permission of the Society of Photo-Optical Instrumentation Engineers from Stengler RK and Weis G (1992) Infrared process control on molten polymers using a high pressure, high temperature flow cell. *Proceedings of the SPIE – Society of Photo-Optical Instrumentation Engineers* **1681**: 33–38.

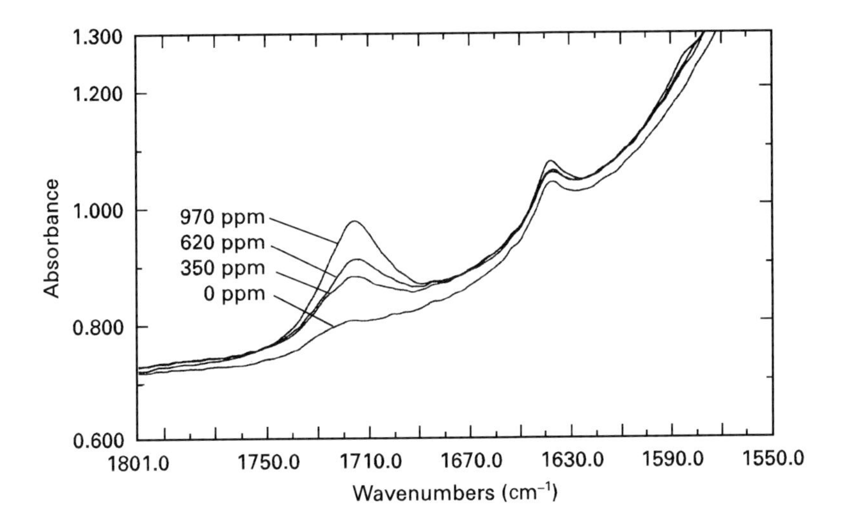

Figure 19 FT-IR spectra of various concentrations of oleamide in molten low-density polyethylene. Reproduced with permission of the Society of Photo-Optical Instrumentation Engineers from Stengler RK and Weis G (1992) Infrared process control on molten polymers using a high pressure, high temperature flow cell. *Proceedings of the SPIE – Society of Photo-Optical Instrumentation Engineers* **1681**: 33–38.

See also: **ATR and Reflectance IR Spectroscopy, Applications; Biochemical Applications of Raman Spectroscopy; Food Science, Applications of Mass Spectrometry; Food Science, Applications of NMR Spectroscopy; Fourier Transformation and Sampling Theory; FT-Raman Spectroscopy, Applications; IR Spectrometers, IR Spectroscopy Sample Preparation Methods; IR Spectroscopy, Theory; IR Spectral Group Frequencies of Organic Compounds; Nonlinear Optical Properties; Raman Optical Activity, Spectrometers; Raman Spectrometers.**

Further reading

Bower DI and Maddams WF (1989) *The Vibrational Spectroscopy of Polymers*. Cambridge: Cambridge University Press.

Chalmers JM and Dent G (1997) *Industrial Analysis with Vibrational Spectroscopy*. Cambridge: Royal Society of Chemistry.

Durig JR (1985) *Chemical, Biological and Industrial Applications of Infrared Spectroscopy*. New York: Wiley.

Ferraro JR and Krishnan K (1990) *Practical Fourier Transform Infrared Spectroscopy*. New York: Academic Press.

White R (1990) *Chromatography/Fourier Transform Infrared Spectroscopy and its Applications*. New York: Marcel Dekker.

Inelastic Neutron Scattering, Applications

Stewart F Parker, Rutherford Appleton Laboratory, Didcot, UK

HIGH ENERGY SPECTROSCOPY
Applications

Inelastic neutron scattering (INS) spectroscopy was pioneered in the 1950s by the American physicist Bertram Brockhouse. So it is not surprising that most of the early applications were physics based; particularly the determination of phonon dispersion curves of metals which are critical to the understanding of their thermodynamic properties. However, like many other spectroscopic methods, of which nuclear magnetic resonance is probably the best example, the use of the technique has greatly broadened, and it is now used by researchers in virtually any field that involves the condensed state. The wide energy (or, equivalently, timescale) range available to INS means that different kinds of motion can be probed in the same system, although not usually simultaneously. The applications span from biology through to materials science and to engineering, in addition to physics and chemistry. Particularly active areas include: catalysis, polymers, magnetism, hydrogen-in-metals, hydrogen bonding, glasses, geology and quantum fluids.

Coherent and incoherent scattering

Experimentally, the quantity that is usually measured is the scattering function, $S(Q, \omega)$ [Q is the momentum transfer (Å^{-1}) and ω the energy transfer (meV)]. The scattering can be coherent, $S^{coh}(Q, \omega)$, or incoherent, $S^{inc}(Q, \omega)$. The definition of $S^{coh}(Q, \omega)$ involves the correlation between the positions of different nuclei at different times and thus gives interference effects. This gives information on collective motions in the sample. $S^{inc}(Q, \omega)$ only involves the correlation between the position of the same nucleus, so there are no interference effects, and the motions of a single particle are probed. This is a powerful method for studying the local environment of the scattering atom.

Whether the scattering is coherent or incoherent depends on the relative size of the coherent and incoherent scattering cross sections of the scattering nuclei; see **Table 1** for a selection of elements. In practice, the signal is a mixture of coherent and incoherent scattering, unless the experiment is designed to select one type of scattering. The cross section is both element and isotope dependent. The most important feature is the enormous incoherent cross section of hydrogen; in contrast, deuterium and most other elements have a modest incoherent cross section and a significant coherent cross section. This means that the scattering from a protonic sample will be mostly incoherent, while that from its deuterated isotopomer will be largely coherent. Isotopic substitution therefore provides a very powerful method to change the scattering characteristics of a sample.

Table 1 The coherent and incoherent scattering cross sections of selected elements

		Scattering cross section (barns = 10^{-28} m^{-2})	
Element	*Mass[a] (amu)*	*Coherent*	*Incoherent*
Hydrogen	1.0078	1.758	79.7
Deuterium	2.0141	5.597	2.04
Carbon	12.011	5.6	0.001
Nitrogen	14.007	11.0	0.49
Oxygen	15.999	4.23	0.000
Aluminium	26.982	1.492	0.0085
Silicon	28.066	2.163	0.015
Chlorine	35.453	11.531	5.2
Iron	55.847	11.4	0.22
Copper	63.546	7.486	0.52
Platinum	195.08	11.65	0.13

[a] The atomic mass given refers to the natural isotopic composition of the element except for hydrogen and deuterium where the mass is that of the isotopes ^{1}H and ^{2}H respectively.

The energy range ±10 meV (±80 cm^{-1}): quasielastic scattering and tunnelling spectroscopy

Quasielastic scattering

Quasielastic scattering is concerned with scattered neutrons that have undergone small energy transfers, typically ≤ 2 meV (≤ 16 cm^{-1}). This originates from neutrons scattered by atoms that are undergoing diffusion or reorientation on a timescale of 10^{-12}–10^{-7} s. Since the motions are not quantized, they form a continuous distribution, and the effect is manifested as a broadening of the peak due to the elastically scattered neutrons. This is entirely analogous to the Doppler broadening of an audio signal by reflection from a moving object. The neutron has a magnetic moment so it can be scattered magnetically and the random fluctuations of unpaired electron spins of stationary atoms can also cause quasielastic broadening.

In general, a molecule in a condensed phase will be simultaneously undergoing translation, rotation and vibration and the scattering law is a convolution of the scattering laws corresponding to the different types of motion. In practice, the vibrational term only affects the intensity as a Debye–Waller factor, $\exp(-Q^2\langle u^2\rangle)$, where $\langle u^2\rangle$ is the mean square displacement in the vibrational modes. For hydrogenous materials the scattering will be incoherent and only the motion of the hydrogen atoms need be considered. In this case the scattering law can be written as the sum of an elastic and an inelastic component:

$$S(Q,\omega) = \text{Elastic} + \text{Inelastic} = A_0(Q)\delta(\omega) + S^{\text{inc}}(Q,\omega) \quad [1]$$

Provided that the experimental data can be resolved into elastic and inelastic contributions, the $A_0(Q)$ term can be determined from:

$$A_0(Q) = \frac{\text{Elastic intensity}}{\text{Total intensity}} \quad [2]$$

and is called the elastic incoherent structure factor (EISF). The elastic intensity can be considered to arise from motions that are localized in space and the inelastic arise from motions that are not localized. Thus for purely translational motion, the EISF is zero since there is no elastic peak. For rotational motion the EISF is unity at $Q = 0$ and falls to a minimum at a Q value that is related to the radius of gyration. At higher Q the EISF is oscillatory and the shape gives information on the geometry of the rotational process. This is most characteristic beyond the first minimum and shows the importance of obtaining data at as wide a range of Q as possible.

Rotational motions in many different systems have been studied. In particular, methyl group rotations in polymers have been extensively studied. The rotational motion is represented by instantaneous jumps between three equivalent positions on a circle of radius R. When all orientations are equally likely, Equation [1] may be written as:

$$S(Q,\omega) = A_0(Q)\delta(\omega) + \frac{1}{\pi}[1 - A_0(Q)]L(\omega) \quad [3]$$

with the EISF given by:

$$A_0(Q) = \frac{1}{3}\left[1 + 2J_0(QR\sqrt{3})\right] \quad [4]$$

where $J_0(x)$ is the zeroth-order spherical Bessel function. $L(\omega)$ is a Lorentzian line shape:

$$L(\omega) = \frac{2\tau/3}{1 + \omega^2(2\tau/3)^2} = \frac{\Gamma}{\Gamma^2 + \omega^2} \quad [5]$$

whose full width at half maximum is given by 2Γ or $3/\tau$ (τ being the average time between jumps). **Figure 1A** shows the decomposition of the experimental data into an elastic and a quasielastic part for selectively deuterated poly(methyl methacrylate) at 25°C. The deuterated part of the molecule gives considerable coherent scattering and after correction for this, the experimental data, **Figure 1B**, shows excellent agreement with the EISF calculated from Equation [4].

Recent work has shown that the data can be better modelled by a distribution of jump rates, since for amorphous polymers the methyl groups are in a variety of environments.

For translational diffusion, the simplest model is to consider that the diffusing particle starts at the origin at time zero. Since this is the dynamics of a single particle, the scattering is incoherent. In this case the scattering law takes the form:

$$S(Q,\omega) = \frac{1}{\pi}\frac{D_t Q^2}{(D_t Q^2)^2 + \omega^2} \quad [6]$$

where D_t is the tracer diffusion coefficient. This is valid at small Q (≤ 0.5 Å^{-1}) where the diffusion is being studied over long (tens of Å) distances, since in this case the details of the individual jumps (≤ 2 Å) are lost by averaging. The function is a Lorentzian with full width at half maximum of $D_t Q^2$, allowing D_t to be extracted from the width of the quasielastic peak at low Q.

For coherent scattering, there will only be intensity at low Q if the scattering atoms are in a compressible fluid. The intensity arises because at $t = 0$ there is an atom at the origin and hence excess concentration. This diffuses away in the same manner as for D_t, except that it is a concentration distribution that is studied. This gives rise to the chemical or Fick's law diffusion coefficient, D_{chem}, which is generally larger than D_t.

The diffusion of methane in the zeolite ZSM-5 can be treated using a more sophisticated version of the same idea that assumes that the jumps can be described by a Gaussian distribution of jump lengths. Physically, this corresponds to the molecule making jumps along the channel segments as well as across the channel intersections. Using this model,

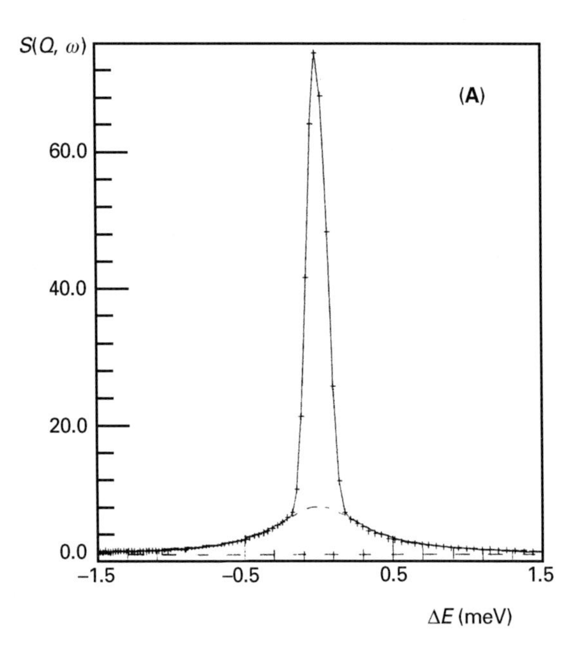

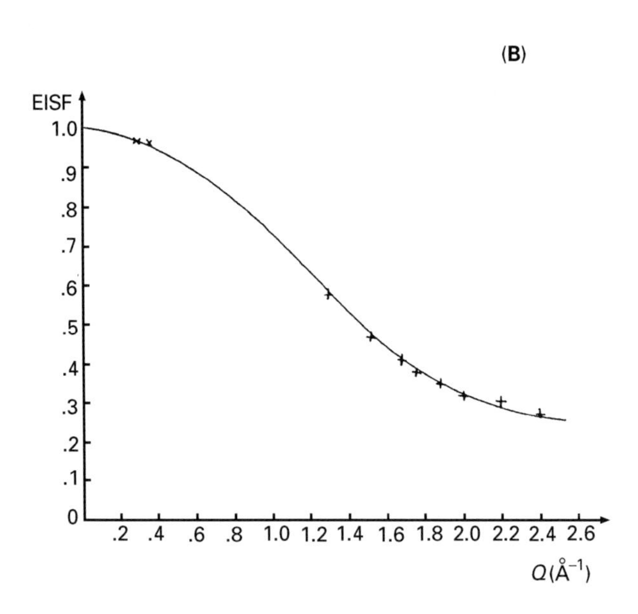

Figure 1 (A) Decomposition of the quasielastic scattering at $Q = 2.06$ Å^{-1} from selectively deuterated syndiotactic poly(methyl methacrylate) at 25°C into a delta function (sharp component) superimposed on a Lorentzian (broad component), due to the ester methyl rotation. Both components are resolution broadened. (B) EISF for the same sample. The data points have been corrected for coherent scattering and the solid line is a fit using Equation [4]. (Data recorded with IN5 and IN6 at the ILL). Reproduced with permission from Gabrys B, Higgins JS, Ma KT and Roots JE (1984) Rotational motion of the ester methyl group in stereoregular poly(methyl methacrylate): A neutron scattering study. *Macromolecules* **17**: 560–566.

there is a more complex Q dependence, as shown in **Figure 2**. At small values of Q Equation [6] is obtained and at large values of Q the width tends asymptotically towards the mean jump rate. From the initial slope of the data in **Figure 2**, $D_t = 5 \times 10^5$ cm^2 s^{-1} and the mean time between jumps is 3.6×10^{-11} s.

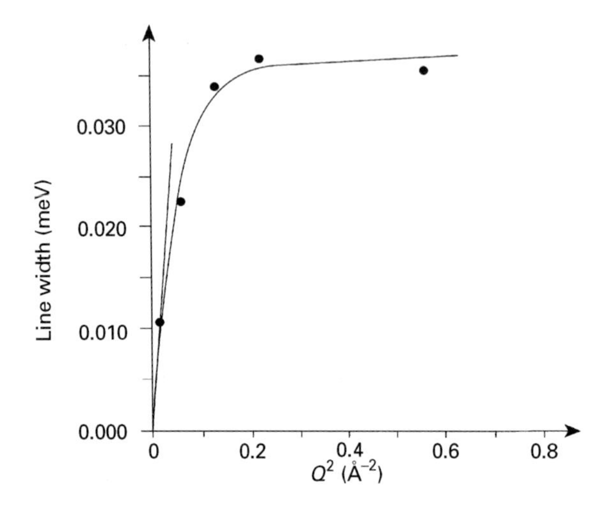

Figure 2 The broadening of the elastic peak as a function of Q^2 for methane at a loading of 1.5 molecules per unit cell in ZSM-5 at 250 K. (Data recorded with IN5 at the ILL). Reproduced with permission from Jobic H, Bée M and Kearley GJ (1989) Translational and rotational dynamics of methane in ZSM-5 zeolite: A quasielastic neutron scattering study. *Zeolites* **9**: 312–317.

A traditional strength, and still a major application, of INS is the study of hydrogen-in-metal systems. In these systems hydrogen occupies a well-defined lattice site, but is often free to move between sites. This proton mobility has led to numerous applications in battery technology, particularly for $LaNi_5H_x$ and the AB_2 Laves phase compounds. The simplest model is to consider that: the jumps are all of the same length (l), that jumps in any direction are equally probable, and that the jump time is negligibly small compared to the residence time (τ). With these assumptions Equation [6] becomes:

$$S(Q,\omega) = \frac{1}{\pi}\frac{F(Q)}{[F(Q)]^2 + \omega^2} \qquad [7]$$

with

$$F(Q) = \frac{1}{\tau}\left(1 - \frac{\sin Ql}{Ql}\right) = \frac{6D_t}{l^2}\left(1 - \frac{\sin Ql}{Ql}\right) = \frac{6D_t}{l^2}(1 - \operatorname{sinc} Ql) \qquad [8]$$

since $D_t = l_2/6\tau$. As shown in **Figure 3A** for $ZrTi_2H_{3.6}$, the scattering function is still Lorentzian

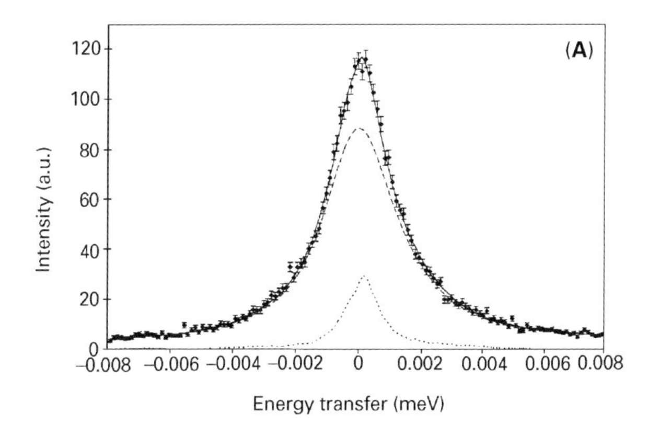

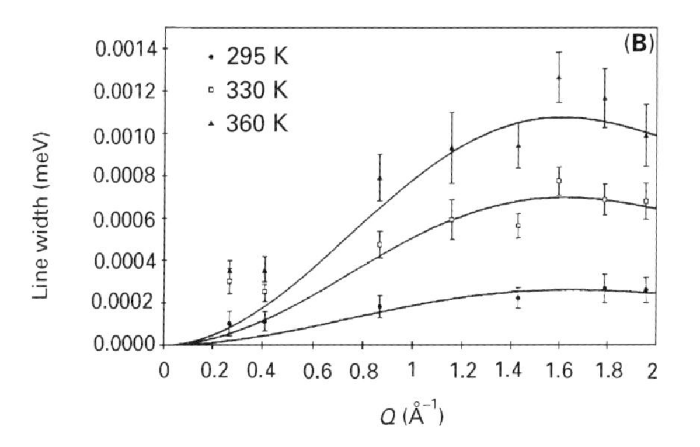

Figure 3 (A) Scattering at $Q = 0.87$ Å^{-1} and 360K from $ZrTi_2H_{3.6}$. The fit (solid line) to the data (•) is the sum of the quasielastic scattering with a Lorentzian line shape, (dashed line) and elastic scattering from the metal atoms (dotted line). (B) Line width of the Lorentzian component as a function of Q for $T = 295$, 330 and 360 K. The solid lines are fits from the Chudley–Elliott model (Eqn [8]) with $l = 2.8$ Å. (Data recorded with IN10 at the ILL). Reproduced with permission from Fernandez JF, Kemali M, Johnson MR and Ross DK (1997) Quasielastic neutron scattering measurements on the $ZrTi_2H_{3.6}$ C-15 Laves phase compound. *Physica B* **234–236**: 903–905.

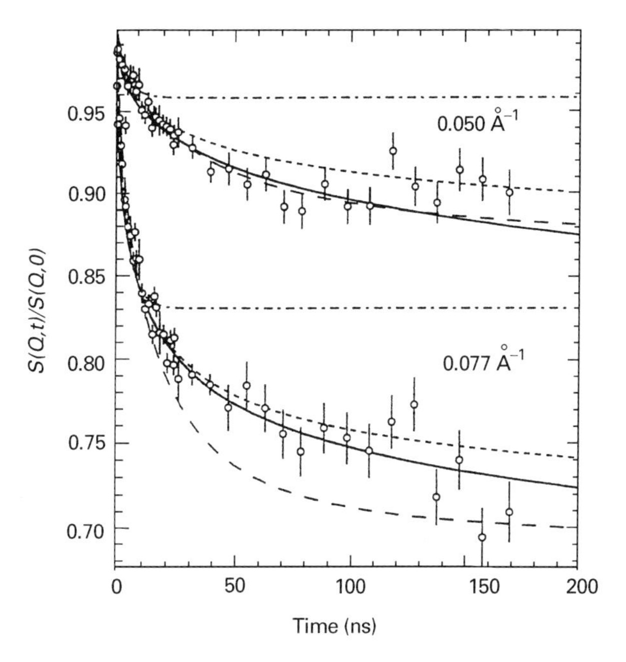

Figure 4 Plot of the normalized intermediate scattering function $[S(Q,t)/S(Q,0)]$ versus time for hydrogenous polyethylene (12% w/w) in deuterated polyethylene at 509 K for $Q = 0.050$ and 0.077 Å^{-1}. Also shown are the predictions from the reptation model (solid line) and competing (dashed and dot-dashed lines). (Data recorded with IN15 at the ILL). Reproduced with permission from Schleger P, Farago B, Lartigue C, Kollmar A and Richter D (1998) Clear evidence for reptation in polyethylene from neutron spin echo spectroscopy. *Physical Review Letters* **81**: 124–127.

but the line width has a more complex Q dependence (**Figure 3B**). This model, known as the Chudley–Elliott model, was originally developed for the case of diffusion in liquids. It has since been extended to take account of sites that are different both crystallographically and energetically.

Neutron spin echo has been largely used to look at the main chain dynamics of polymers in concentrated solution or in the melt. In these systems the entanglements between chains determine key physical properties, particularly the viscosity, which is an area of central concern to the processing of polymers. The most successful theory to explain the properties of such systems is that of reptation. This considers the polymer to be confined to a 'tube' formed by its neighbours and that the only motion is a snake-like creep along the tube, lateral motion being prevented by the walls of the tube. However, there are competing theories and there was no direct experimental evidence that the reptation model is correct. For the case of hydrogenous polyethylene (12% w/w) in deuterated polyethylene, recent improvements in instrumentation have allowed the models to be tested. **Figure 4** compares the experimental data and the predictions of the various models. Clearly only the reptation model (solid line) accurately describes the data over the entire time range, thus providing compelling evidence that reptation is the best model.

Tunnelling spectroscopy

Tunnelling is the movement of a particle from one site to another at temperatures where, classically, it does not have sufficient energy for the transition. Thus tunnelling is intrinsically a quantum phenomenon and is essentially restricted to motion of hydrogen (and its isotopes). It can be rotational or translational in origin. **Figure 5** illustrates a model for the hindered rotation in a system with three equivalent hydrogens such as $-CH_3$ or NH_3. The crucial factors are the barrier height, V, and the rotational constant, B, of the molecule. As the ratio V/B decreases, the overlap of the ground-state wavefunctions increases. Since the

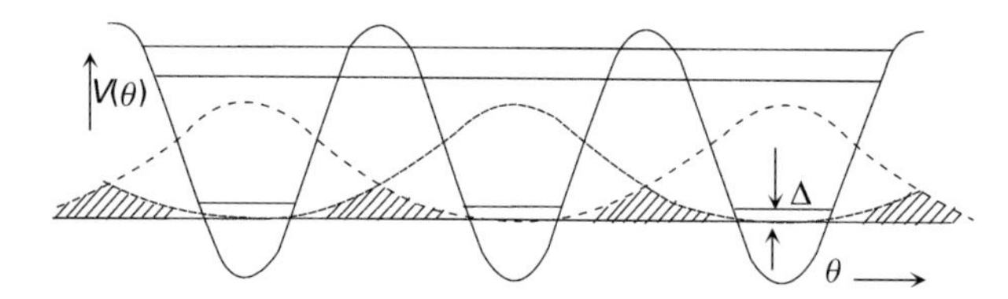

Figure 5 Model of the tunnel splitting (Δ) of the rotational ground state and of the first excited torsional state. The splitting is due to the overlap of the wavefunctions (hatched region) in neighbouring wells.

energy levels of the protons in each well are the same, they are degenerate and they split to remove the degeneracy. The degree of splitting is the tunnelling splitting. Since the wavefunctions are approximately Gaussian, the splitting depends exponentially on the barrier height, making the tunnel splitting a very sensitive probe of the immediate environment of the molecule. INS is sensitive to tunnel splittings in the range 0.1 μeV (10^{-4} cm^{-1}) to 5 meV (40 cm^{-1}).

Intercalation of graphite with small molecules is used to modify its physical and chemical (including catalytic) properties. **Figure 6A** shows the rotational tunnelling of ammonia in a caesium–graphite intercalate. $C_{28}Cs(NH_3)_x$. As x varies from 0.0 to 1.90 the tunnelling peaks can be seen to grow and decay as the layers of caesium atoms in the layers become coordinatively saturated by ammonia ligands. The peaks occur in pairs at ± energies because even at the low temperature of the experiment, the first excited state is populated, thus both neutron energy gain and neutron energy loss peaks are observed. The staging behaviour is reproduced in the diffraction pattern of the intercalate which was recorded simultaneously (**Figure 6B**). The ability to obtain structural and dynamic information simultaneously is a unique aspect of neutron studies. Tunnelling is very strongly temperature dependent and is only usually observed at very low temperatures. As the temperature is raised the intensity and transition energy of the lines decrease, while simultaneously a broad quasielastic feature appears.

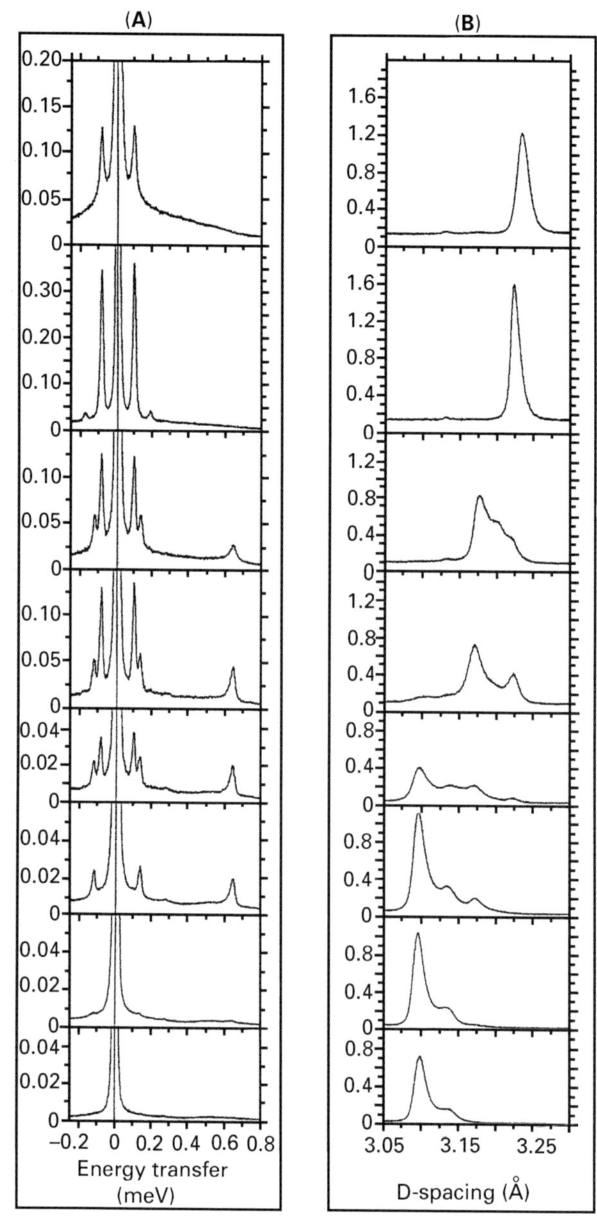

Figure 6 Neutron studies of $C_{28}Cs(NH_3)_x$ (from top to bottom x = 1.90, 1.54, 1.25, 0.94, 0.74, 0.42, 0.20, 0.0). (A) Rotational tunnelling spectra and (B) neutron diffraction patterns recorded simultaneously at 4.2 K as a function of x. (Data recorded with IRIS at ISIS). Reproduced with permission from Carlile CJ, Jamie IM, Lockhart G and White JW (1992) Two-dimensional caesium–ammonia solid solutions in $C_{28}Cs(NH_3)_x$. *Molecular Physics* **76**: 173–200.

One of the major discoveries in chemistry in recent years is the characterization of a series of complexes where dihydrogen, H_2, acts as a ligand. For the compounds $M(CO)_3(\eta^2\text{-}H_2)(PR_3)$ (M = Mo, R = cyclohexyl; M = W, R = cyclohexyl, isopropyl) the tunnel spectra are shown in **Figure 7**. The observation of the tunnel splitting is proof that the H–H bond is intact. It can be seen that changing the metal has a large (300%) effect on the tunnel splitting (compare **Figure 7A** and **Figures 7B** and **7C**), whereas changing the phosphine has a minor (20%) effect (compare **Figure 7B** and **Figure 7C**), indicating that the electronic component of the M–H_2 bonding is more significant than the steric influences of the phosphines. (Two peaks in the gain and loss spectra are observed because of crystallographic disorder.)

Translational tunnelling is conceptually different since the hydrogen may be in a different crystallographic site after the tunnelling event. Hydrogen-in-metal systems may show this type of

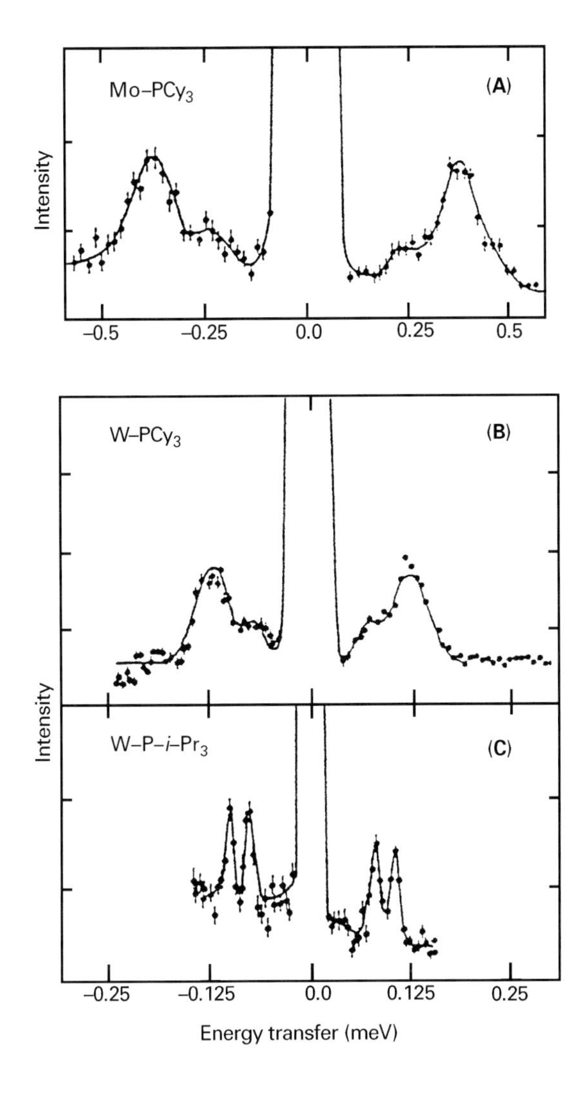

Figure 7 Rotational tunnel spectra of dihydrogen complexes at 4 K: (A) $Mo(CO)_3(\eta^2\text{-}H_2)[P(C_6D_{11})_3]$, (Mo–$PCy_3$), (B) $W(CO)_3(\eta^2\text{-}H_2)[P(C_6D_{11})_3]$, (W–$PCy_3$) and (C) $W(CO)_3(\eta^2\text{-}H_2)$ $[P(\textit{iso}\text{-}C_3D_7)_3]$, (W–P-*i*-$Pr_3$). Note that the energy scale in (B) and (C) is different from that in (A). (Data recorded with IN5 at the ILL). Reproduced with permission from Eckert J, Kubas GJ, Hall JH, Hay J and Boyle CM (1990) Molecular hydrogen complexes 6. The barrier to rotation of η^2-H_2 in $M(CO)_3(PR_3)(\eta^2\text{-}H_2)$ (M = W, Mo; R = Cy, *i*-Pr) inelastic neutron scattering, theoretical and molecular mechanics studies. *Journal of the American Chemical Society* **112**: 2324–2332.

tunnelling. **Figures** 8 shows the tunnelling transition of hydrogen in α-Mn. The tunnelling in this instance is between two sites of the same symmetry separated by 0.68 Å. This is a very unusual system in that the tunnelling energy is 6.3 meV (51 cm^{-1}), which is ~15 times higher than commonly found for hydrogen-in-metal systems, and the integrated intensity is very large when compared to the optic modes at 80–120 meV (645 – 968 cm^{-1}).

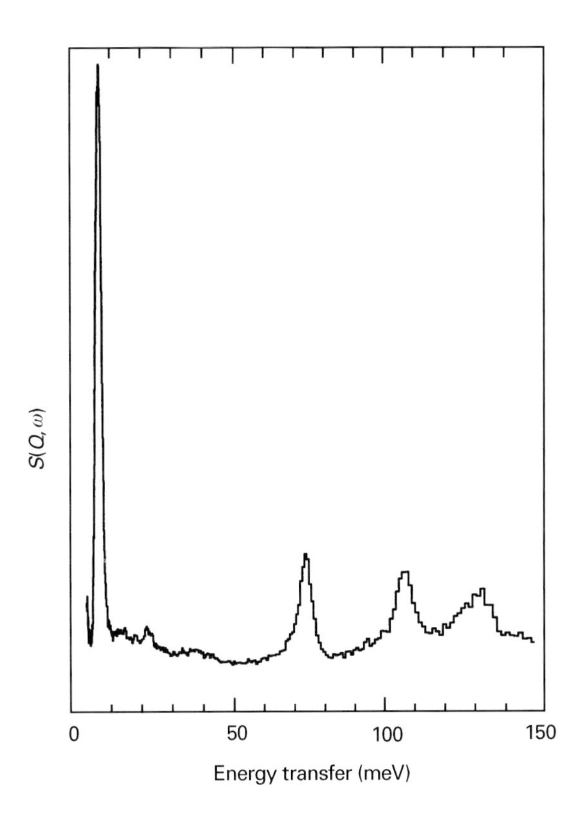

Figure 8 Hydrogen tunnelling in α-Mn. The intensity of the tunnelling line at 6.3 meV (51 cm^{-1}) is unusually large when compared to the optic modes at 80–120 meV (645–968 cm^{-1}). (Data recorded with TFXA at ISIS).

The energy range 0–1 eV (0–8000 cm^{-1}): vibrational spectroscopy and magnetic excitations

Coherent INS spectroscopy

Coherent INS spectroscopy gives information on collective motions in crystals. These are important in understanding phenomena as diverse as magnetism, phase transitions and thermodynamic properties. Neutron diffraction is coherent elastic scattering from a notionally static lattice; coherent one-phonon neutron scattering can be considered to be 'diffraction' from a lattice where the atoms are undergoing sinusoidal motion of frequency given by the phonon frequency. This means that the scattering occurs in specific directions at specific energies. It is to be emphasized that coherent INS is the most general method of measuring the excitations (which can be vibrational or magnetic in origin) across the entire Brillouin zone.

Palladium hydride has been extensively studied because of its ready take-up and large capacity for hydrogen. This has lead to interest in its use as a

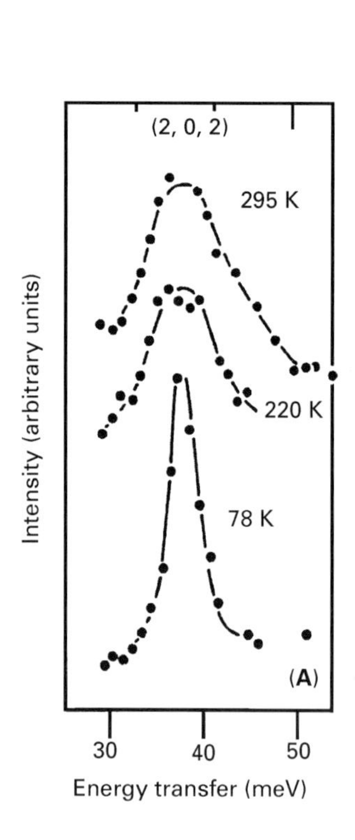

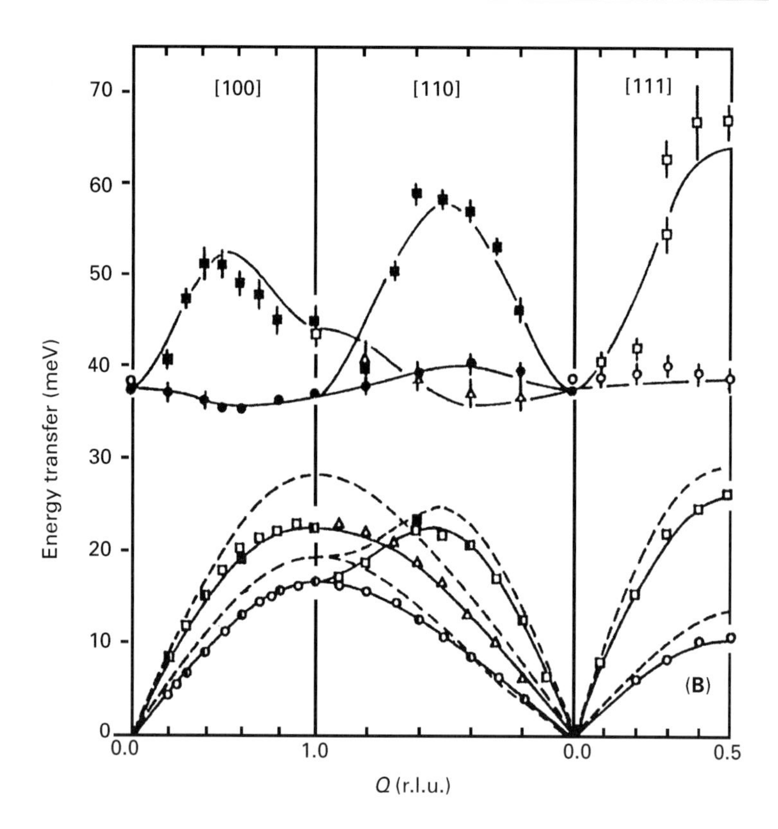

Figure 9 (A) Example of a phonon group in $PdD_{0.63}$ and (B) the dispersion curves for $PdD_{0.63}$. The dashed lines are the dispersion curves for pure Pd (r.l.u. = reciprocal lattice unit) (data recorded with HB3 at Oak Ridge). Reproduced with permission from: Rowe JM, Rush JJ, Smith HG, Mostoller M and Flotow HE (1974) Lattice dynamics of a single crystal of $PdD_{0.63}$. *Physical Review Letters* **21**: 1297–1300.

hydrogen storage material. **Figure 9A** shows one of the measured transitions for $PdD_{0.63}$ and **Figure 9B** the dispersion curves derived from the data. The lower set of curves are the acoustic modes, where all the atoms in the unit cell move in the same direction (as is the case for sound waves) and largely reflect the dynamics of the metal atoms. The higher energy curves are the optic modes and different types of atom move in different directions. This generates a changing dipole, and these modes would give rise to infrared or Raman active modes. Infrared and Raman spectroscopies only observe modes at approximately zero wave vector and from **Figure 9B** only the optic modes have nonzero frequencies at $Q = 0$. It is one of the great strengths of INS that modes at all wave vectors can be studied.

The comparatively low energy of the optic modes indicates that the Pd–D force constant is small. For $PdD_{0.63}$ the variation of energy transfer with wave vector (dispersion) of the optic modes shows that there are strong interactions between the H atoms. To fit the data fourth-nearest neighbours have to be included. Comparision of the acoustic modes for $PdD_{0.63}$ and Pd (dashed lines) shows that the former are lower in energy so the bonding is weaker. This is consistent with the lattice expansion observed on hydrogen (deuterium) take-up.

The unique properties of water are essential to life but are still not understood in detail. They derive from the hydrogen bonding present in water which is manifested in the complex phase diagram and anomalously high melting point of ice. Coherent INS studies of ice I_h provide a stringent test of models of the dynamics. **Figure 10A** shows one of the measured spectra for a single crystal of D_2O ice I_h and **Figure 10B** shows the derived dispersion curves. It can be seen that while the longitudinal acoustic (LA) and transverse acoustic (TA) modes, which mainly involve oxygen motion are sharp, the transverse optic (TO) mode is broad. The TO mode is largely due to deuteron motion and the width probably reflects the highly disordered arrangement of deuterons (protons) in ice I_h.

Virtually all of our understanding of magnetism at the atomic level is based on neutron scattering; neutron diffraction allows the orientation of the spins in the lattice to be determined and inelastic scattering allows their dynamics to be studied.

Magnetoresistance is a phenomenon observed in some materials where the electrical resistivity

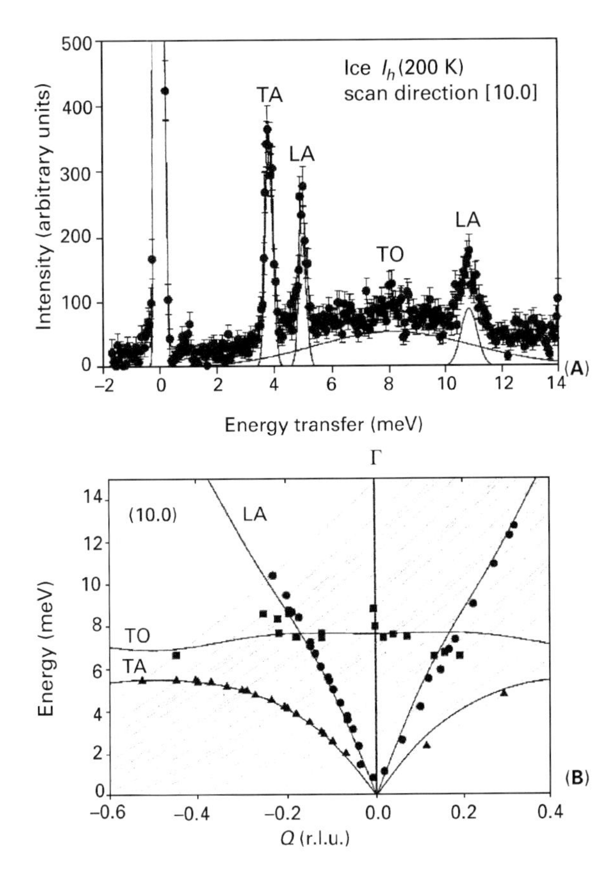

Figure 10 (A) One of the measured spectra of the phonon dispersion along the (10.0) direction about the (11.0) reciprocal lattice point of D_2O ice I_h. (B) The derived dispersion curves; the solid lines are fits to the TA and LA modes. The dashed lines show the Q,ω trajectories available in this experiment (r.l.u. = reciprocal lattice unit) (data recorded with PRISMA at ISIS).

changes in an applied magnetic field. In recent years there has been a huge revival in the study of manganese perovskites which, depending on the addition of other elements, can change their resistance by several orders of magnitude in fields of a few tesla; this is known as colossal magnetoresistance (CMR). The interest in these materials is not just because of possible technological applications, particularly read-out heads for hard disks that will allow much higher storage densities and faster access, but also because of fundamental questions relating to the transition between metallic and insulating phases.

The CMR manganites are metallic and derived from the insulating ionic compound $LaMnO_3$. Partial substitution of La^{3+} by a divalent metal ion e.g. Ba^{2+}, Sr^{2+}, Pb^{2+}, introduces mobile charge carriers and these are responsible for the unusual behaviour. The materials are ferromagnets in which the magnetic moments associated with the manganese ions are aligned in the same direction below the Curie temperature. In 1930 Bloch showed that the lowest energy excitations in a ferromagnet are spin waves; coherent precessions of the spins around their average orientation. **Figure 11** shows an image of the inelastic magnetic scattering by a single crystal of $La_{0.7}Pb_{0.3}MnO_3$. The spin wave is clearly seen as a ring of scattering around the (1,0,0) reciprocal lattice point. From the spectra the dispersion relation for the spin wave can be obtained by plotting the position of maximum intensity along the main directions of crystal symmetry. Fitting the data shows that the dynamics can be described by only considering nearest neighbours, a surprising result for such a complex material.

Incoherent INS spectroscopy

Incoherent INS spectroscopy probes the local environment of the scattering atom and is thus complementary to infrared and Raman spectroscopies. However there are important differences as illustrated in **Figure 12A–C** which shows the infrared, Raman and INS spectra respectively, of *N*-phenylmaleimide, a model compound for bismaleimide composites that are used in aerospace applications. It can be seen that all three spectra are very different. There are two main reasons for this. Firstly, whether a mode is infrared and/or Raman active (or is inactive in both) is determined by the symmetry of the molecule. Note that even when a mode is allowed it may still have little or no intensity. In contrast, there are no selection rules for INS spectroscopy and all modes are allowed.

The strongest bands in the infrared and Raman spectra are the out-of-phase carbonyl stretch at 212 meV (1707 cm^{-1}) and the in-phase carbonyl stretch at 219 meV (1770 cm^{-1}) respectively. In contrast, the carbonyl bands are completely absent in the INS spectrum, but there are several strong bands that do not have any obvious counterparts in the infrared and Raman spectra.

The differences arise because the intensity of the ith INS band at a frequency ω_i is proportional to:

$$S(Q,\omega_i) \propto \sum_d \frac{|QU_d^i|^2}{m_d\,\omega_i}\sigma_d^{\text{inc}}\exp(-Q^2U_{\text{total}}^2) \qquad [9]$$

The exponential term is the Debye–Waller factor, U_{total}^2 is the mean square displacement of the molecule and its magnitude is in part determined by the thermal motion of the molecule. This can be reduced by cooling the sample and so spectra are typically recorded below 30 K.

U_d^i is the amplitude of vibration of the dth atom in the ith mode, σ_d^{inc} and m_d are its incoherent cross section and mass respectively. From **Table 1** it can be seen that if hydrogen is present, the combination

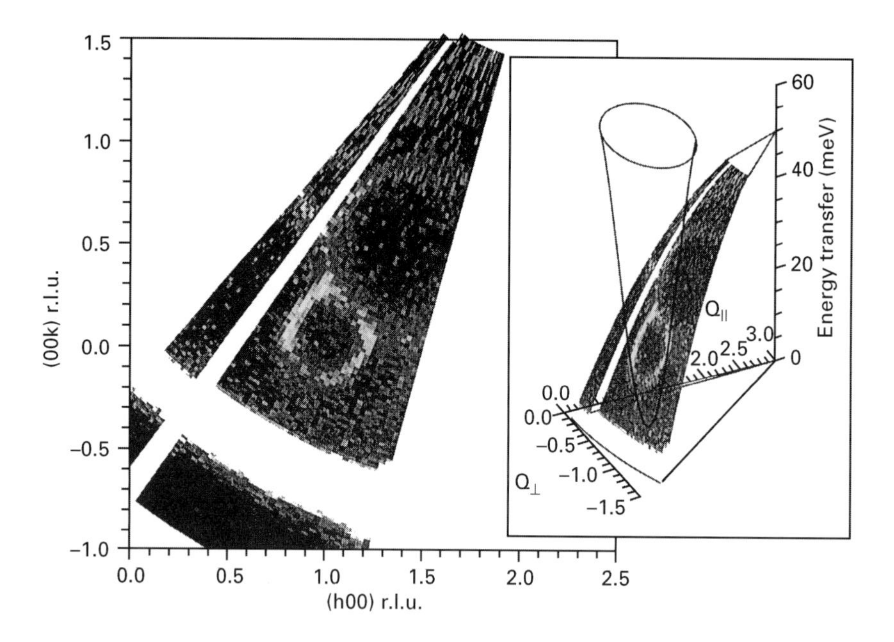

Figure 11 (Left) Image of the magnetic excitations in $La_{0.7}Pb_{0.3}MnO_3$. (Right) Data plotted in a three-dimensional space defined by two coordinates of momentum in the horizontal plane and the energy vertically. The shaded surface indicates the possible energies and momenta of the scattered neutrons that are accessible in this experiment; maximum intensity occurs where this surface intersects the dispersion surface of the spin waves (r.l.u. = reciprocal lattice unit) (data recorded with HET at ISIS).

of large incoherent cross section and small mass means that the scattering from hydrogen will dominate the spectrum. For hydrogenous materials the scattering from all other atoms can be neglected. This dependence on the cross section is why the INS spectrum is so different from the optical spectroscopies. There, the intensity derives from changes in the electronic properties of the molecule that occur as the vibration is executed (the dipole moment and the polarizability for infrared and Raman spectroscopy respectively).

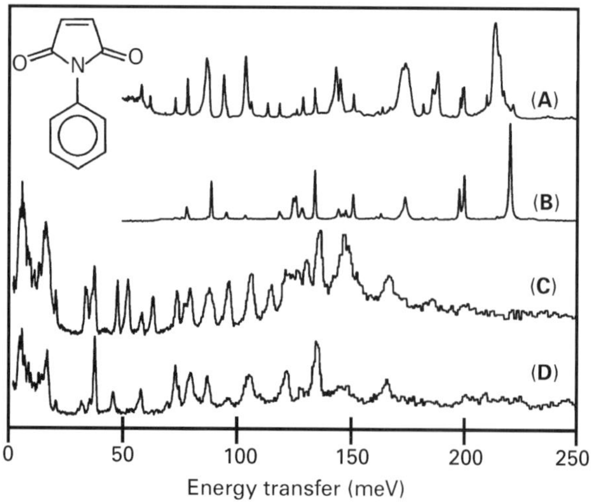

Figure 12 Vibrational spectra of *N*-phenylmaleimide: (A) infrared, (B) Raman, (C) INS; (D) INS spectrum of *N*-(perdeuterophenyl)maleimide (INS data recorded with TFXA at ISIS).

The cross-section dependence can be exploited to 'eliminate' parts of the molecule from the spectrum. **Figure 12D** is again *N*-phenylmaleimide but with the phenyl ring deuterated. This results in a considerable simplification of the spectrum and all of the modes can be assigned to vibrations of the maleimide ring with the phenyl group treated as a point mass.

Vibrational spectroscopy is a commonly used tool for the study of catalysts. The neutron is a particle with zero charge which means that it is highly penetrating and is not a surface-sensitive probe. This can be circumvented by using high surface area materials and hydrogenous adsorbates that offer contrast to the substrate. The advantages are that common catalyst-supports such as silica, alumina and carbon are weak scatterers, see **Table 1**, and the entire 0–500 meV (0–4000 cm^{-1}) 'mid-infrared' range is accessible, and in particular the low-energy range below 125 meV (1000 cm^{-1}). This range is very difficult to study with optical methods; the supports tend to be highly absorbing and the important metal–adsorbate vibrations are often very weak.

Formate (HCOO) species adsorbed on a copper-based catalyst are a key intermediate in the industrial synthesis of methanol from CO_2 and H_2. INS spectra were recorded for a number of different copper surfaces; spectra were taken before and after formic

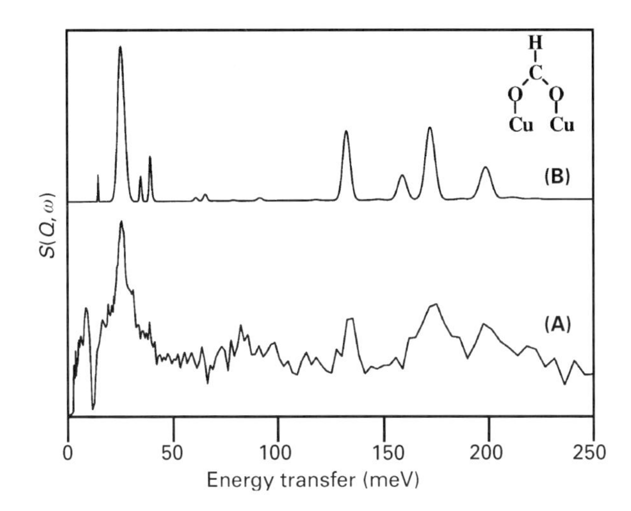

Figure 13 Experimental (A) and calculated (B) spectra of bridging formate on CuO. The experimental spectrum is the difference spectrum [CuO + formate] – [CuO]. (Data recorded with TFXA at ISIS).

acid adsorption. In all the spectra adsorbed, formate could be clearly distinguished even though it represented less than 1% of the total mass in the beam. The difference spectrum ([CuO + formate] – [CuO]) is shown in **Figure 13A**.

Equation [9] is noteworthy in that the intensity of a mode is determined only by the cross section, the momentum transfer and the amplitude of vibration, hence it is purely dynamic. From a conventional Wilson–Decius–Cross normal coordinate analysis it is possible to calculate both the energies (from the eigenvalues) and intensities (from the eigenvectors) for INS spectra of molecular species. The requirement to fit the INS intensities is a stringent test of any force field. The spectrum calculated for a bidentate configuration of formate bound to two Cu atoms (**Figure 13B**) gives good agreement with the experimental data, supporting this as the mode of binding on this surface. It also demonstrated that the mode at 26.8 meV (216 cm^{-1}) is the Cu–O torsion, which was the first time this mode was detected on a surface.

The rapid growth in computer power means that increasingly complex systems can be treated by *ab initio* quantum chemistry techniques. The vibrational spectrum of the object of interest is often calculated, since this offers a check of whether the structure is a true minimum or a saddle point on the potential surface, as well as being a rigorous test of the results. The amplitudes of vibration are calculated as part of this process and it is relatively straightforward to derive the INS spectrum from these via Equation [9]. In contrast, the infrared and Raman intensities are poorly reproduced for anything but the simplest systems. The use of *ab initio* methods is undoubtedly the way that INS data will be routinely analysed in the near future. The power of the method is illustrated in **Figure 14** where it allows an unambiguous differentiation between two possible modes of water binding to the Brönsted acid sites of a zeolite. Clearly, the hydrogen-bonded water model fits the data better than the hydroxonium ion, H_3O^+, model.

The energy range >1 eV (> 8000 cm^{-1}): neutron Compton scattering

Neutron Compton scattering (NCS) is the measurement of atomic momentum distributions, $n(p)$, in condensed matter systems by high-energy inelastic neutron scattering. NCS measurements are particularly important in the study of quantum fluids.

For protons, $n(p)$ is related to the Fourier transform of the proton wavefunction and an NCS measurement of $n(p)$ can be used to determine the wavefunction in a way analogous to the determination of a real space structure from a diffraction pattern. If $n(p)$ is known then, in principle, both the proton wavefunction and the exact form of the potential energy well can be reconstructed.

For accurate NCS measurements energy transfers much greater than the maximum vibrational frequency are required for the impulse approximation to be valid. In the impulse approximation, momentum and energy are conserved in the scattering event and from a measurement of the momentum and energy change of the neutron, the momentum of the proton before the collision can be determined. This requires neutrons with energies >1 eV and only spallation sources provide a sufficient flux of high-energy neutrons for the measurement to be feasible. At lower energies the impulse approximation is no longer valid and $n(p)$ is not related in a simple way to the observed intensities.

An important feature is that the neutron energy loss depends on the atomic mass of the scattering atom and hence different masses occur at different times-of-flight. In the harmonic approximation, the peak width depends on the Gaussian average of the vibrational frequencies. The peak intensity depends only on the number of scatterers and their cross section. Thus NCS functions as a 'neutron mass spectrometer', albeit with poor mass resolution. Hydrogen is usually well separated from other elements and allows its quantification, even in the presence of deuterium.

This aspect of NCS has been exploited in the study of the battery material $LaNi_5$. For a well-annealed polycrystalline sample of α-$LaNi_5$ after hydrogen

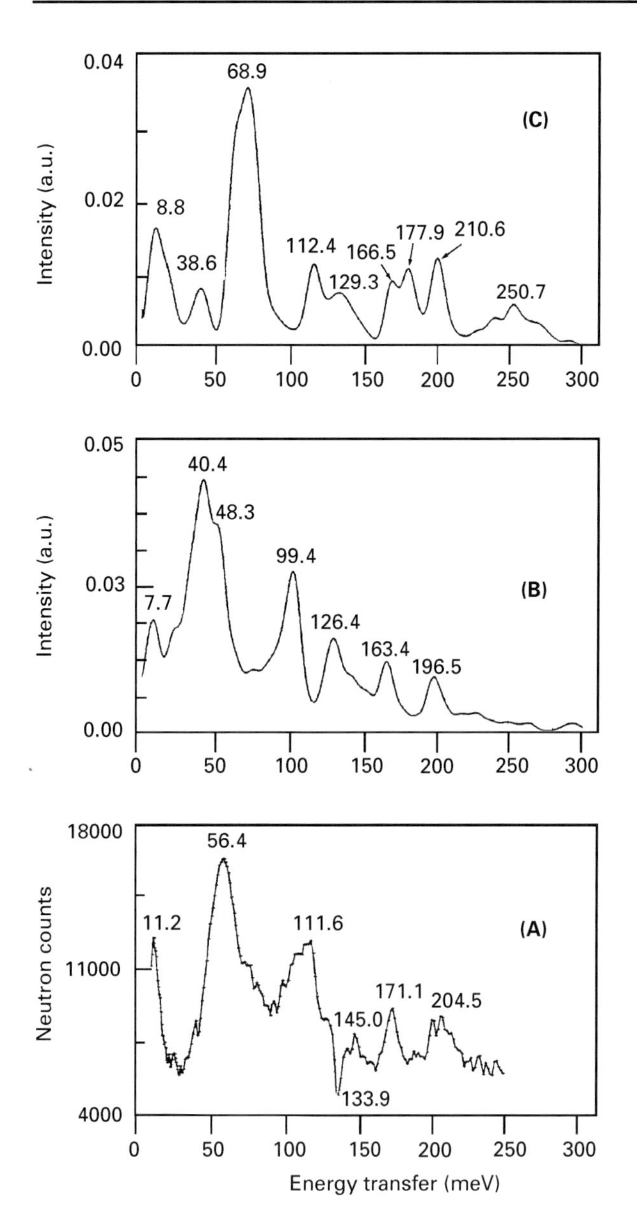

Figure 14 INS spectrum of water adsorbed at low loading on H-ZSM5. (A) Experimental spectrum. *Ab initio* simulation for: (B) water hydrogen-bonded to a bridging hydroxyl and (C) a hydroxonium ion, H_3O^+. (Data recorded with IN1BeF at ILL). Reproduced with permission from Jobic H, Tuel A, Krossner M and Sauer J (1996) Water in interaction with acid sites in H-ZSM-5 zeolite does not form hydroxonium ions. A comparison between neutron scattering results and *ab initio* calculations. *Journal of Physical Chemistry* **100**: 19545–19550.

desorption, there should be no residual hydrogen. **Figure 15A** shows an intense peak at 375 μs due to the metal atoms and a weak feature at 275 μs assigned to hydrogen. This is confirmed by the absorption of hydrogen, resulting in an increase of the scattering at 275 μs (**Figure 15B**). Analysis shows that the nominally hydrogen-free material has a

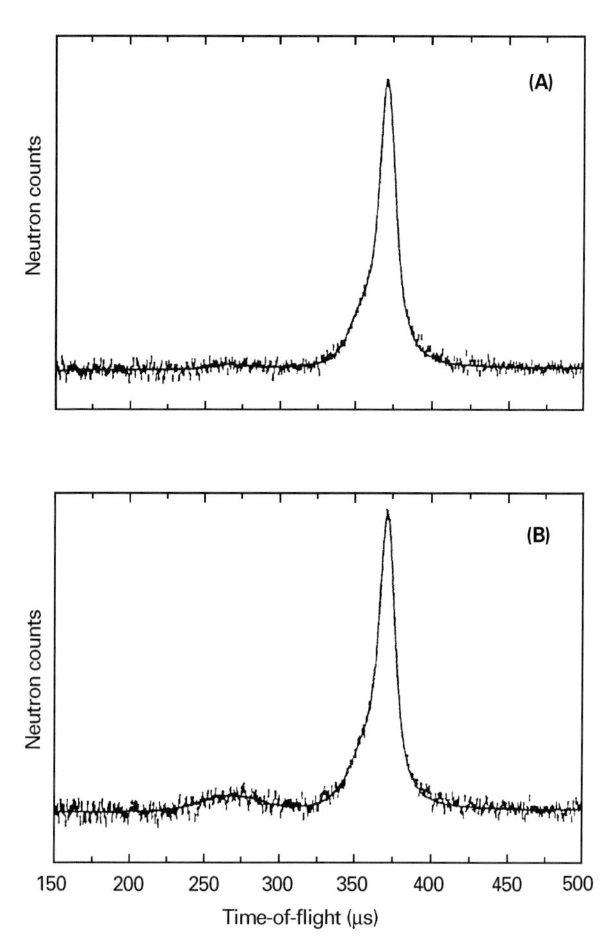

Figure 15 Time-of-flight spectra for α-$LaNi_5$: (A) nominally hydrogen-free (however, note the small peak at 275 μs due to hydrogen), (B) after absorption of hydrogen. (Data recorded with eVS at ISIS). Reproduced with permission from Gray EMA, Kemali M, Mayers J and Norland J (1997) Interstitial site occupation in α-$LaNi_5$ studied by deep inelastic neutron scattering. *Journal of Alloys and Compounds* **253–254**: 291–294.

hydrogen-to-metal ratio of 0.017 ± 0.002, indicating that there is a significant number of trap sites, which are probably detrimental to battery performance.

List of symbols

$A_0(Q)$ = elastic incoherent structure factor (EISF); B = rotational constant; D_{chem} = chemical diffusion coefficient; D_t = tracer diffusion coefficient; $J_0(x)$ = zeroth-order spherical Bessel function; $L(\omega)$ = Lorentzian line shape; m_d = mass of dth atom; $n(p)$ = atomic momentum distribution; Q = momentum transfer; $S^{\mathrm{coh}}(Q, \omega)$ = coherent scattering function; $S^{\mathrm{inc}}(Q, \omega)$ = incoherent scattering function; $\langle u^2 \rangle$ = mean square displacement in the vibrational modes; U^2_{total} = mean square displacement; U^i_d =

amplitude of vibration of the dth atom in the ith mode; V = barrier height; σ_d^{inc} = incoherent cross section of dth atom; τ = average time between jumps; ω = energy transfer.

See also: **Inelastic Neutron Scattering, Instrumentations; IR Spectral Group Frequencies of Organic Compounds; IR and Raman Spectroscopy of Inorganic, Coordination and Organometallic Compounds; Scattering Theory; Vibrational, Rotational and Raman Spectroscopy, Historical Perspective.**

Further reading

Bée M (1988) *Quasielastic Neutron Scattering*. Bristol, UK: Adam Hilger.

Eckert J and Kearley GJ (eds) Spectroscopic applications of inelastic neutron scattering: Theory and practice (1992) *Spectrochimica Acta, Part A* 48: 269–478.

Furrer A (ed) (1994) *Neutron Scattering from Hydrogen in Materials*. Singapore: World Scientific.

Higgins JS and Benoît HC (1994) *Polymers and Neutron Scattering*. Oxford: Oxford University Press.

Newport RJ, Rainford BD and Cywinski R (eds) (1988) *Neutron Scattering at a Pulsed Source*. Bristol, UK: Adam Hilger.

Prager M and Heidemann A (1997) Rotational tunnelling and neutron spectroscopy. *Chemical Reviews* 97: 2933–2966.

Ross DK (1997) Neutron scattering studies on metal hydrogen systems In: Wipf H (ed) *Hydrogen in Metals III: Properties and Applications (Topics in Applied Physics. Vol. 73)*, pp. 153–214. Berlin: Springer-Verlag.

Squires GL (1996) *Introduction to the Theory of Thermal Neutron Scattering*. New York: Dover Publications.

Inelastic Neutron Scattering, Instrumentation

Stewart F Parker, Rutherford Appleton Laboratory, Didcot, UK

HIGH ENERGY SPECTROSCOPY

Methods & Instrumentation

Introduction

The prime requirement for inelastic neutron scattering (INS) spectroscopy is a source of neutrons. The only methods that produce a sufficient flux of neutrons (> 10^{14} neutrons cm^{-2} s^{-1}) for spectroscopy are: ^{235}U fission in a nuclear reactor and spallation. The latter process involves accelerating a beam of protons to near-light-speed and impacting them onto a heavy-metal target, usually made of either depleted uranium or tantalum. The target atoms absorb the protons, generating highly excited nuclei that decay, in part, by emission of neutrons. In the case of a uranium target, fission also occurs, increasing the neutron yield. Typically, a reactor yields one neutron per fission event that is available for spectroscopy, while spallation yields up to 30 neutrons per proton absorbed. At present, the most powerful reactor source is the Institut Laue Langevin (ILL) (Grenoble, France) and the most powerful spallation source is the ISIS Facility at the Rutherford Appleton Laboratory (Chilton, UK). Usually, fission sources operate continuously and spallation sources are pulsed, although this is not always the case. The reactor at the Franck Laboratory of Nuclear Physics (Dubna, Russia) consists of two pieces of sub-critical uranium, one of which is mounted on a rotating disc; as the two pieces come into proximity, the mass goes critical, producing neutrons, and as it goes sub-critical again, neutron production ceases. The net result is a pulsed reactor. The SINQ facility at the Paul Scherrer Institute (Villigen, Switzerland) is a continuous spallation source.

By whatever process they are produced, the 'newborn' neutrons are very energetic, $> 2 \times 10^9$ meV (1 meV = 8.067 cm^{-1} = 0.241 THz) and must be brought to useful energies by multiple inelastic collisions in a moderator; this is usually a hydrogenous material. By suitable design it is possible to establish a quasi-thermal equilibrium between the temperature of the moderator and the energy of the neutrons, which provides a means of tailoring the neutrons' energy to the requirements of the experiment.

Neutrons are quantum entities, so they exhibit both particle-like and wave-like properties. For INS spectroscopy, the particle-like properties are relevant since the neutron–sample interaction occurs on a femtosecond timescale and the interaction can be considered to be analogous to billiard-ball scattering.

Table 1 Relationship between neutron properties and typical values

Neutron energy				Wavelength	Wave vector
meV	cm^{-1}	THz	K	(Å)	($Å^{-1}$)
E			$=k_BT$	$=\dfrac{h^2}{2m\lambda^2}$	$=\dfrac{\hbar^2k^2}{2m}$
E	$=8.066\,7E$	$=\dfrac{E}{4.135\,4}$	$=0.086\,165T$	$=\dfrac{81.787}{\lambda^2}$	$=2.071\,7k^2$
0.01	0.081	0.002 42	0.116	90.4	0.069 5
0.1	0.867	0.024 2	1.16	28.6	0.220
1.0	8.067	0.241 8	11.16	9.04	0.695
10.0	80.7	2.418	116.1	2.86	2.197
100.0	807	24.18	1 161	0.904	6.948
1 000.0	8 067	241.8	11 606	0.286	21.97
10 000.0	80 667	2 418	116 056	0.090	69.48

meV = millielectronvolts, cm^{-1} = wavenumbers, THz = terahertz (10^{12} Hz), K = kelvins, Å = angstroms (10^{-10} m), h = Planck's constant, $\hbar = h/2\pi$, m = neutron mass, k_B = Boltzmann's constant.

However, the selection of the incident energy or the analysis of the scattered neutrons' energy often uses Bragg diffraction from a single crystal, making use of the wave-like properties. **Table 1** shows the relationships between the units commonly used.

INS spectroscopy spans an enormous energy range, which is equivalent to a wide range of timescales. **Figure 1** illustrates the ranges available and some of the applications. It is convenient to divide the whole energy range into three sectors, while recognizing that there is significant overlap between them. In the lowest energy range, ±10 meV, the applications are principally quasielastic scattering and tunnelling spectroscopy. The second range, 0–1000 meV, covers the regions of vibrational spectroscopy and magnetic excitations. The highest range above 1000 meV is the province of neutron Compton scattering.

The major difference between INS spectroscopy and the photon-based spectroscopies that cover the same energy range is that the neutron has a significant mass (1.009 amu). Thus an inelastic collision

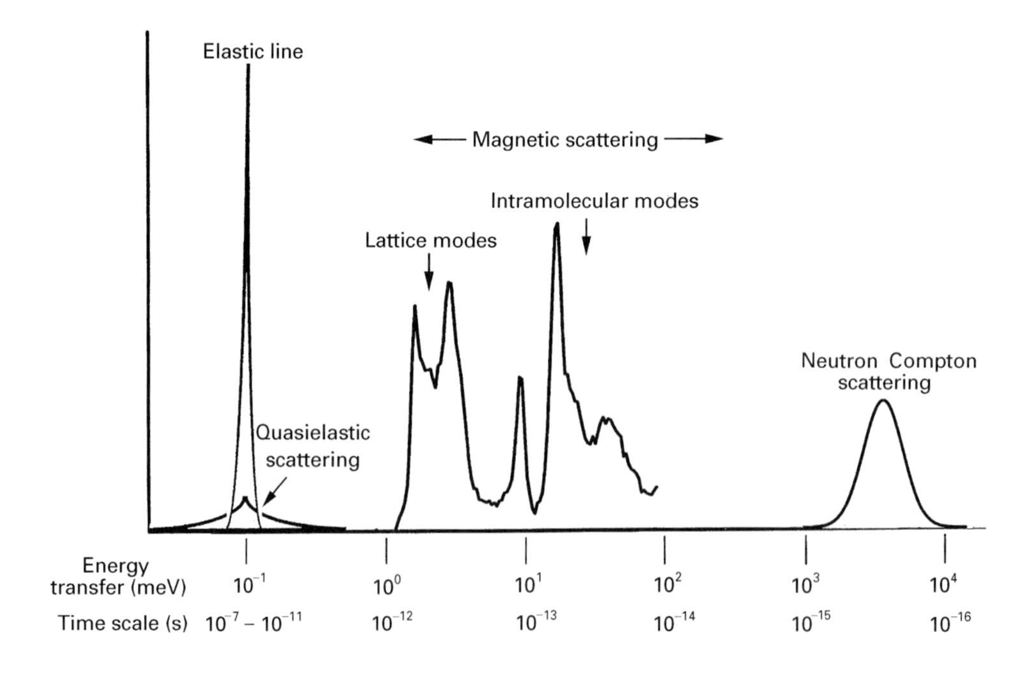

Figure 1 Illustration of the energy range and the equivalent timescales and length scales available with inelastic neutron scattering. The type of information available is also indicated.

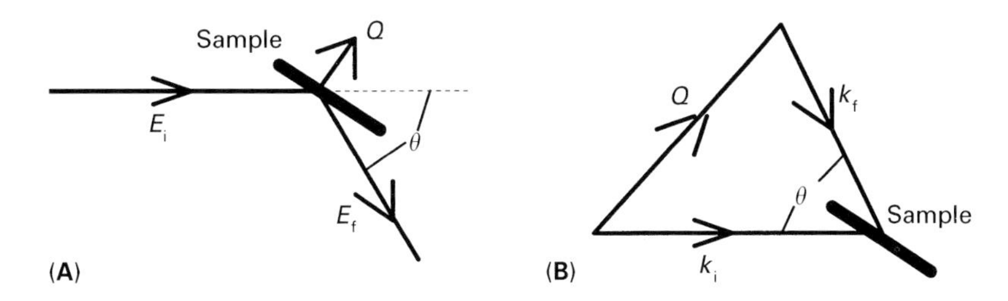

Figure 2 Neutron scattering in (A) real space and (B) reciprocal space.

results in transfer of both momentum (Q, Å^{-1}) and energy (E, meV). In contrast, optical processes result in essentially zero momentum transfer. The aim of an INS experiment is, usually, to measure scattered intensity at each energy transfer as a function of the momentum transfer. In **Figure 2** the scattering process is shown in both real space and reciprocal space.

The energy transfer, $\hbar\omega$, is given by the difference in the incident (i) and final (f) neutron energies:

$$\hbar\omega = E_\mathrm{i} - E_\mathrm{f} \qquad [1]$$

where E is the kinetic energy of the neutron and

$$E = mv^2/2 \qquad [2]$$

where m is the neutron mass and v its velocity. The momentum transfer, Q, is given by:

$$Q = \sqrt{(k_\mathrm{i}^2 + k_\mathrm{f}^2 - 2k_\mathrm{i}k_\mathrm{f}\cos\theta)} \qquad [3]$$

where θ is the scattering angle and k (Å^{-1}) is the wavevector:

$$k = 2\pi/\lambda \qquad [4]$$

associated with the wavelength, λ (Å). Thus INS offers the possibility of exploring both momentum and energy transfer and can be considered to be a 'two-dimensional' form of spectroscopy. This is illustrated in **Figure 3** which shows the Q, E coverages of some of the instruments at ISIS and the ILL.

The energy range ±10 meV: quasielastic scattering and tunnelling spectroscopy

Quasielastic scattering is a very low energy inelastic process. The term is usually taken to mean a broadening of the elastic line and is most commonly the result of diffusional (translational or rotational) motion of atoms. (It can also be caused by the randon fluctuations of unpaired electronic spins of

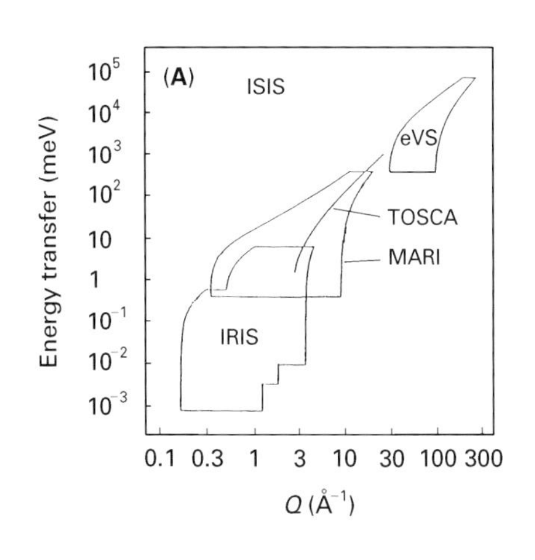

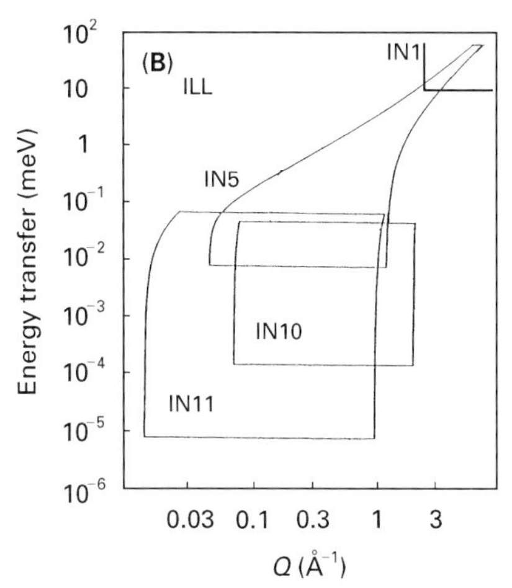

Figure 3 The Q, E coverages of some of the instruments at (A) the ILL and (B) ISIS.

stationary atoms). Translational and rotational diffusion occur simultaneously but on different timescales. The energy resolution of a spectrometer, ΔE, and the timescale τ of the motion, are related by the Heisenberg uncertainty principle:

$$\Delta E \times \tau \sim h/2\pi \quad [5]$$

Thus rapid motions require relaxed resolution, while slower motions require high resolution. Note that these are relative terms; typically an energy resolution of better than 1% $\Delta E/E$ and 100 µeV is required. The timescale to be probed can be separated into three regimes, each of which uses a different technique: for $\tau \sim 10^{-11}$ s, ΔE is 10–100 µeV and direct-geometry time-of-flight is used, for $\tau \sim 10^{-9}$ s, ΔE is 0.3–20 µeV and a backscattering crystal analyser is used, for $\tau \sim 10^{-7}$ s, ΔE is 0.005–1 µeV and neutron spin echo is used. Examples of each type of spectrometer will be considered.

Direct-geometry time-of-flight

For time-of-flight INS spectroscopy it is necessary to determine either the incident energy or the final energy of the neutron. Spectrometers that define the neutron's final energy are indirect-geometry instruments and those that determine the neutron's incident energy are direct-geometry instruments. In a direct-geometry quasielastic spectrometer, a monochromatic (single energy), pulsed beam of neutrons is produced by a rotating mechanical chopper system. This pulse of neutrons is incident upon the sample and, after interacting with the atoms of the sample, the scattered neutrons are timed over a known distance to the detectors which are set at a range of scattering angles. The time-of-flight, T_1, from the sample to the detectors is given by $T_1 = L_1/\nu_1$, where L_1 is the sample-to-detector distance. Since the incident neutron velocity, ν_0, is known it follows from Equations [1] and [2] that the energy transfer, $\hbar\omega$ is:

$$\hbar\omega = \frac{m}{2}(\nu_1^2 - \nu^2) = \frac{mL_1^2}{2}\left(\frac{1}{T_1^2} - \frac{1}{T^2}\right) \quad [6]$$

where T_0 is the time-of-flight of the elastically scattered neutrons. Using Equation [3] and the known angular position of the detectors the momentum transfer Q of the detected neutrons can also be determined.

Figure 4 illustrates the principle for the multichopper time-of-flight spectrometer IN5 at the ILL. The incoming neutron distribution spreads from ~36 meV (1.5 Å) to ~0.25 meV (18 Å) with a maximum flux at 3.3 meV (5 Å). As chopper 1 opens, a pulse of neutrons is admitted; the only neutrons that pass through the last chopper (number 4) are those with the correct velocity to arrive at chopper 4 as it opens. Thus the energy of the neutrons incident on the sample is determined by the phase angle between the first and the last choppers. Neutrons with velocities that are one-half of the desired value, would also pass through chopper 4

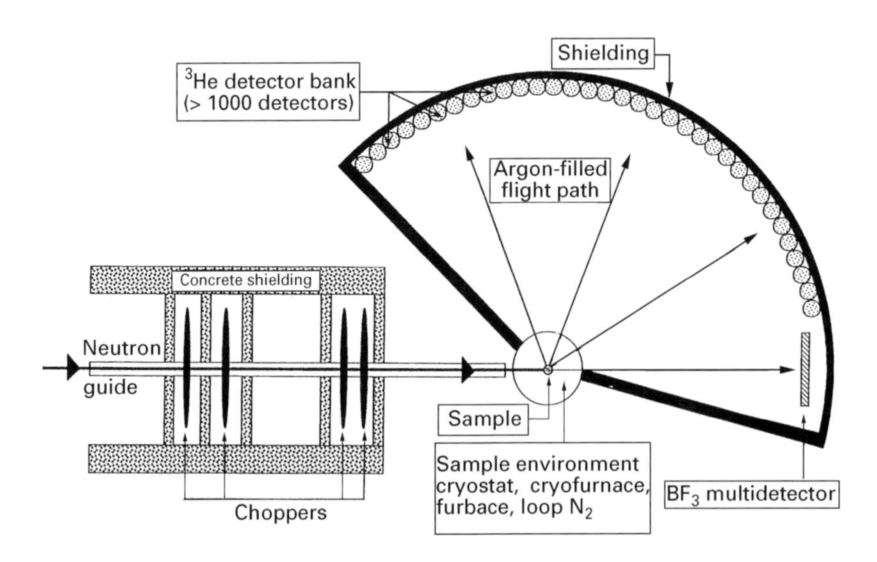

Figure 4 Layout of the cold-neutron multi-chopper spectrometer IN5 at the ILL. Reproduced with permission of Adenine Press from Ferrand M (1997) Neutron instrumentation in studying dynamics of biomolecules. In: Cusack S, Büttner H, Ferrand M, Langan P and Timmins P (eds). *Biological Macromolecular Dynamics*. Schenectady: Adenine Press.

after a further revolution; these are eliminated by chopper 2. Chopper 3 spins at an integral fraction of the speed of the other choppers and eliminates a certain number of pulses, e.g. 1 in 2, or 3 in 4. This is to prevent frame-overlap, where fast neutrons from a following pulse overtake slow neutrons from the previous pulse. This is a problem since the time-of-flight method depends on knowing when a given neutron was created; from Equation [6] any uncertainty in the time-of-flight translates into a corresponding uncertainty in the energy transfer.

Backscattering crystal analyser

The backscattering crystal analyser is an example of an indirect-geometry spectrometer and is used when higher resolution is required. Neutrons with a band of energies whose width (in energy) is greater than the maximum energy transfer to be measured, are incident on the sample. From the scattered neutrons, one neutron energy is selected by Bragg diffraction from a single crystal. Bragg's law states:

$$\lambda = 2d \sin\theta \tag{7}$$

where λ is the incident neutron wavelength, d is the interplanar spacing in the analyser crystal and θ is the angle of incidence. The resolution is determined by the differential:

$$\frac{\Delta E}{E} = 2\frac{\Delta\lambda}{\lambda} = 2\left([\Delta\theta \cot\theta]^2 + \left[\frac{\Delta d}{d}\right]^2\right)^{0.5} \tag{8}$$

The dominant term in most cases is cot θ since for most single crystals the uncertainty in the lattice parameter is ~10^{-4} and can be ignored. In the limit as the Bragg angle approaches 90°, cot θ tends to zero and the energy resolution of the analyser becomes extremely good, being 0.5 μeV for silicon and 10 μeV for graphite. Hence, the term 'backscattering spectrometer'.

The implementation of a backscattering spectrometer is somewhat different at a reactor than at a spallation source. **Figure 5** shows a schematic view of the spectrometer IN10 at the ILL and **Figure 6** shows the equivalent spectrometer IRIS at ISIS.

IN10 sits at the end of a cold-neutron guide. For low-energy (long-wavelength) neutrons, total external reflection occurs with a corresponding increase in transmission as compared to a tube of the same diameter. This provides a means of transporting neutrons long distances, several tens of metres, without the intensity decrease that would result from solid-angle considerations. High-energy neutrons do

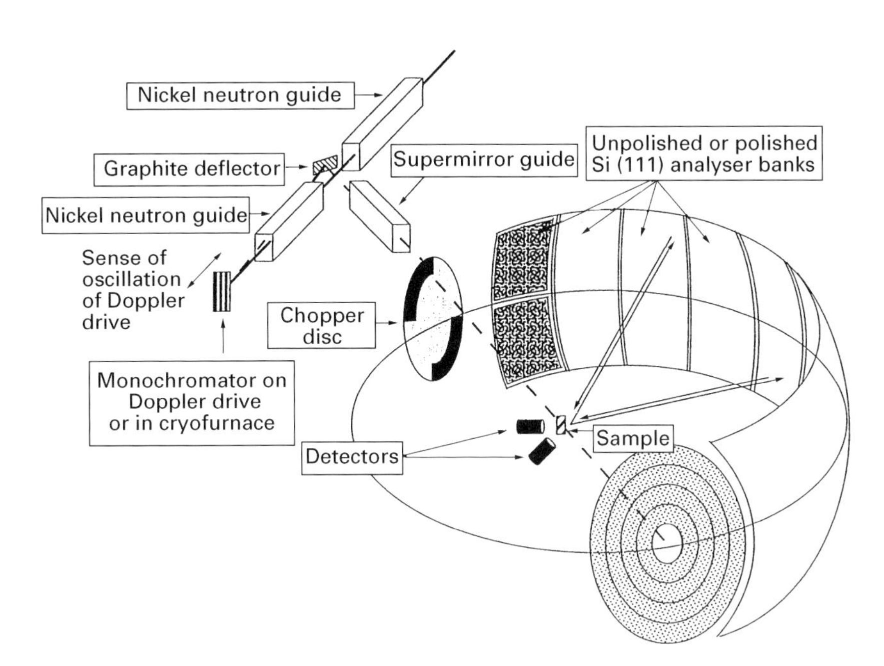

Figure 5 Schematic view of the cold-neutron spectrometer IN10 at the ILL. The incident energy is changed either by Doppler-shifting the neutron wavelengths (IN10A) or by thermal expansion of the monochromator (IN10B). Reproduced with permission of Adenine Press from Ferrand M (1997) Neutron instrumentation in studying dynamics of biomolecules. In: Cusack S, Büttner H, Ferrand M, Langan P and Timmins P (eds). *Biological Macromolecular Dynamics*. Schenectady: Adenine Press.

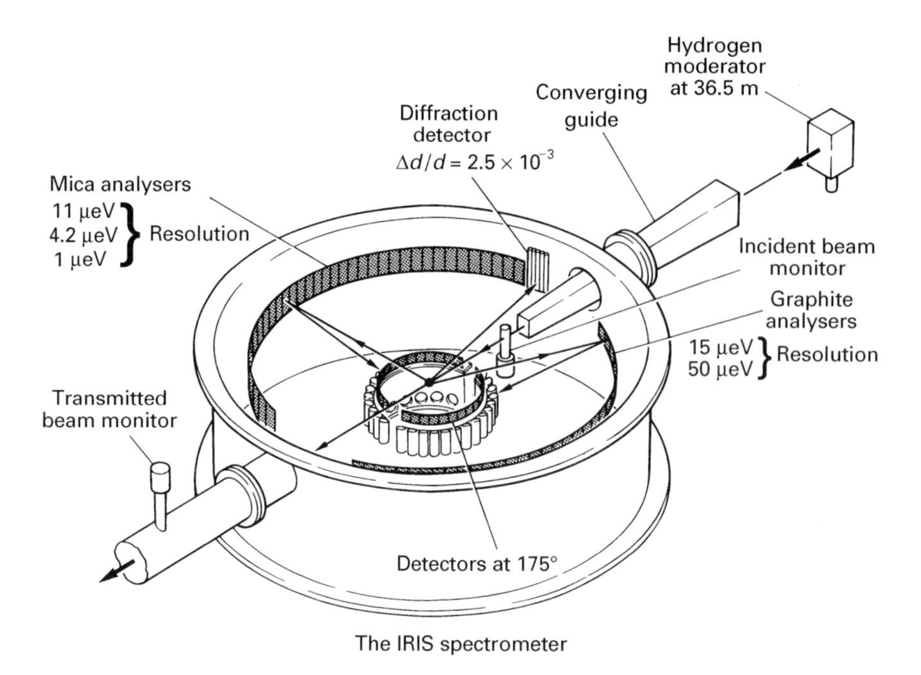

Figure 6 Schematic of the cold-neutron backscattering spectrometer IRIS at ISIS. Reprinted from *Physica*, 182B, Carlile CJ and Adams MA, 431–440, 1992, with kind permission of Elsevier Science – NL, Sara Burgerhartstraat 25, 1055 KV Amsterdam, The Netherlands.

not undergo total reflection and pass through the walls of the guide. By making the guide gently curved, with a radius of curvature of the order of a kilometre, a guide can also provide a means of selecting the neutrons of interest.

On IN10 the energy scan is performed by varying the energy of the incident neutrons. One may do this scan, which typically covers ±15 μeV for 2 meV incident neutrons, by oscillating the monochromator to produce a Doppler shift in the velocity of the neutrons impinging on the sample. An alternative scanning procedure is to slowly vary the temperature and hence the lattice constant of the monochromator.

It can be seen in **Figure 6** that the analyser system on IRIS is similar to that of IN10 but that the incident white beam is generated differently. From the neutron pulse emitted by the liquid-hydrogen moderator, a broad band of neutron energies is selected by a rotating disc chopper at 6.4 m. The required energy resolution in the incident beam is achieved by allowing this band of neutrons to disperse in time over a long distance as it drifts along a curved neutron guide to the sample situated at 36 m from the moderator. Neutrons whose energies change to precisely match the analyser energy after scattering from the sample are selected by Bragg scattering from the pyrolytic graphite crystal array in near-backscattering geometry. A continuous range of scattering angles from 15° to 165° is covered simultaneously by the analyser–detector array. Using the 002 reflection from the graphite analyser an elastic resolution of 15 μeV is provided over an energy transfer range from +4 meV to −1 meV.

An advantage of all indirect-geometry instruments at spallation sources is that a relatively broad range of neutron wavelengths is incident on the sample. This means that it is possible to add a diffraction capability simply by the addition of detectors that have a clear view of the sample; see **Figure 6**. This allows information on both structure and dynamics to be collected simultaneously. Inelastic scattering is much weaker than elastic scattering, so over the timescale of the experiment a few diffraction detectors will be sufficient to record a diffraction pattern of good quality.

Neutron spin echo

The highest energy resolution is obtained by using the neutron spin echo technique. In this type of instrument, the velocities of neutrons incident on, and scattered from, a sample are coded as the number of Larmor precessions that the neutron spins undergo in a well-defined applied magnetic field. The method requires a beam of polarized neutrons with a distribution of velocities. As indicated in **Figure 7**, these neutrons enter the precession-field region with their spins perpendicular to the field. As a neutron

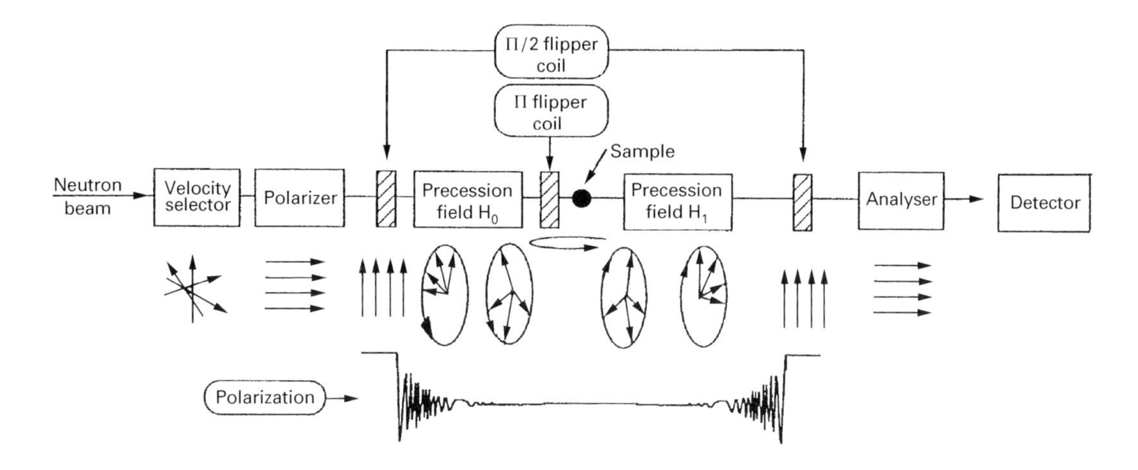

Figure 7 Schematic diagram of a neutron spin echo spectrometer. The direction of the neutron spin is shown as it traverses the spectrometer. Reproduced with permission of Adenine Press from Ferrand M (1997) Neutron instrumentation in studying dynamics of biomolecules. In: Cusack S, Büttner H, Ferrand M, Langan P and Timmins P (eds) *Biological Macromolecular Dynamics*. Schenectady: Adenine Press.

traverses this field, it precesses and the number of precessions is directly proportional to the time it spends in the magnetic field and thus inversely proportional to its velocity. After scattering by the sample, the spins 'unwind' in a reverse-precession-field region. The method is unique in using the magnetic moment of the neutron to detect the velocity change rather than the wave (as in the backscattering spectrometer) or particle properties (as in the direct-geometry spectrometer).

If the scattering at the sample is elastic and the precession fields before and after the sample are identical in magnitude and spatial extent, all neutrons will be in phase at the point in space where they leave the second precession region. This alignment of spins is the 'echo'. However, if the scattering is not purely elastic, the spins will not be completely in phase, weakening the echo. The distribution of phases is then a measure of the distribution of the atomic velocities in the sample.

The high resolution of this technique stems from the large number of precessions the neutron undergoes, $\sim 10^5$ precessions m^{-1}. As a result, the velocity can be measured to an accuracy of better than 10^{-5} and hence the energy exchange can be measured to similar accuracy.

Another feature that is unique to neutron spin echo is that the signal, which is the neutron polarization at the echo point, is directly proportional to the intermediate scattering function, $I(Q, t)$, which is the Fourier transform of the scattering function $S(Q, \omega)$. This is in contrast to the other types of spectrometer that measure $S(Q, \omega)$. It is often advantageous to calculate $I(Q, t)$ rather than $S(Q, \omega)$.

The energy range 0–1000 meV: vibrational spectroscopy and magnetic excitations

This energy range corresponds to the mid- and near-infrared regions of the electromagnetic spectrum. This is the province of vibrational spectroscopy where the forces between atoms in condensed matter can be investigated directly. It is also the region where magnetic excitations are observed. These are electronic transitions between the energy levels of a metal atom (or ion) and usually involve d or f orbitals. They are often optically forbidden and the transitions can be observed by INS because of their interaction with the magnetic moment of the neutron.

Neutrons in this energy range are known as thermal neutrons, because typical energies correspond to room temperature; see **Table 1**. Thermal-neutron INS spectroscopy was the first type to be studied and the instrument invented by Bertram Brockhouse in 1952, the triple-axis spectrometer, is still a mainstay of inelastic instrumentation at reactor sources. In 1994, Brockhouse, together with the pioneer of elastic neutron scattering (diffraction) Clifford Shull, received the Nobel prize for physics for his invention and subsequent work with it.

The triple-axis spectrometer is designed to study coherent excitations, which gives information about the relative motions of the nuclei. This approach mandates the use of single crystals and these need to be very large, of the order of several cubic centimetres. The scattering from powders and liquids is largely incoherent, which gives information about the motions of the individual nuclei. For these types of materials different spectrometers are used. Spectrometers for coherent and incoherent INS spectroscopy will be considered separately, although this division is one of convenience only.

Spectrometers for coherent INS spectroscopy

The principle of the triple-axis spectrometer is displayed in **Figure 8**. The prerequisite is a neutron source with a constant flux, which can be a steady reactor or a continuous spallation source. A monochromatic neutron beam is selected from the neutron source by using Bragg reflection from a single crystal, the monochromator. This beam is then incident on the sample. In a selected direction another single crystal, the analyser, is positioned in such a way that only a given wavelength can be reflected; a neutron detector is placed after this analyser. If neutrons are counted in the detector, this means that there has been some process in the sample which has induced the specific changes of trajectory and wavelength of the incident neutrons.

The important point is that a given configuration of the spectrometer corresponds to a single point in (Q, ω) space. Changing step-by-step either k_i or k_f, the crystal is rotated and the scattering angle varied so as to scan the (Q, ω) space and detect all possible neutron-scattering processes, which can be vibrational (phonons) or magnetic (magnons) in origin. Furthermore, a given (Q, ω) point may be obtained in different ways (varying, for instance, either k_i or k_f). This gives the experimentalist the possibility to adjust conditions to the specific requirements of a given problem.

The great strength of triple-axis spectroscopy is simultaneously its Achilles heel. The instrumentation is enormously flexible and, in principle, any point in (Q, ω) space is accessible. However, it is a point-by-point method and this means that it is slow and to map the phonon dispersion in one sample can require weeks of measurement time.

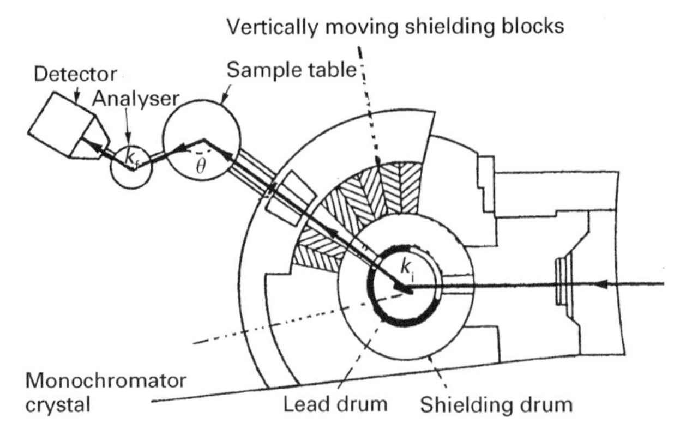

Figure 8 General layout of the hot-neutron triple-axis spectrometer IN1 at the ILL.

The pulsed-source equivalent of a triple-axis spectrometer is demonstrated by PRISMA at ISIS. The difference is that a white beam of neutrons is incident on the sample, rather than a monochromated beam. As in the triple-axis spectrometer, Bragg reflection from analyser crystals is used to select the neutrons scattered by the sample that will ultimately be detected. However, PRISMA has an array of 16 independent analyser and detector arms instead of just one detector. Each arm of PRISMA measures along a parabolic path in (Q, ω) space, so generating a two-dimensional scan through (Q, ω) space for each ISIS pulse, in a single setting of the instrument and sample.

Spectrometers for incoherent INS spectroscopy

The simplest form of instrument is a filter spectrometer which consists of a monochromator to define the incident neutron energy on the sample and a filter after the sample that only allows neutrons of a given energy to reach the detector. This type of instrument is used at a reactor and is exemplified by IN1BeF at the ILL. The monochromator is the same as for the triple-axis instrument IN1, see **Figure 8**, but the analyser is replaced by a cooled beryllium filter. Beryllium only transmits neutrons with energy of less than 5 meV; higher-energy neutrons are Bragg-scattered out of the beam. The transmission and the sharpness of the cut-off are improved by cooling the filter to below 100 K and it is routinely operated at liquid-nitrogen temperature. Since both the incident and final energies are known, the energy transfer is obtained from Equation [1]. The resolution of the instrument is determined by the bandpass of the filter and the detector response at low energies and by the monochromator at higher energies. Typically the resolution increases from ~3 to ~8% of the energy transfer between 20 and 300 meV. The resolution can be improved by using a graphite filter which has a narrower bandpass of ~1.5 meV but at the cost of a large decrease in detected flux.

At a spallation source, both indirect- and direct-geometry instruments are used. A schematic of the indirect-geometry spectrometer TOSCA at ISIS is shown in **Figure 9**. A small fraction of the neutrons from the incident white beam (which extends from 2 meV to beyond 2500 meV) are inelastically scattered by the sample; those that are backscattered through an angle of 135° impinge on a graphite crystal. Only one wavelength (and its harmonics) are Bragg-scattered by the crystal; the remainder pass through the

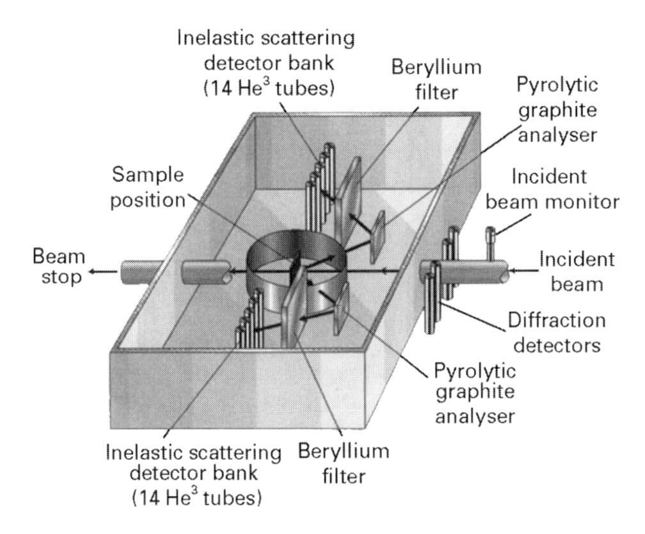

Figure 9 Principle of operation of the indirect-geometry spectrometer TOSCA at ISIS. Only two of the 10 analyser/detector modules are shown.

graphite crystal to be absorbed by the shielding. The neutrons at multiples of the fundamental wavelength are removed by the beryllium filter, so only neutrons of ~4 meV reach the detectors. The net effect of the combination of the graphite crystal and beryllium filter is to act as a narrow bandpass filter.

The use of a small final energy has two important consequences. Since for most energies $k_i >> k_f$ it follows that both the magnitude and direction of Q are largely determined by k_i. Thus Q is almost parallel to the incoming neutron beam and perpendicular to the sample plane across virtually the entire energy transfer range. The significance of this is that in order to observe an INS transition, the vibration must have a component of motion parallel to Q. For a randomly oriented (i.e. polycrystalline) sample this condition will be satisfied for all the vibrations and all will be observed. This is not the case for an oriented sample and experiments directly analogous to optical polarization measurements are possible. Further, because Q is independent of k_f there is a unique value of Q for each energy transfer and $Q^2(\text{Å}^{-2}) \approx E(\text{meV})/2$. Thus TOSCA (and IN1BeF since it also uses the same final energy) trace out a parabola in (Q, ω) space as shown in **Figure 3B**.

The resolution function varies with energy transfer but is typically 2% of the energy transfer between 2 and 500 meV. The advantages of this system are that it is mechanically simple (there are no moving parts) and it offers excellent resolution across a wide energy range with very low background. Further, the reflection geometry means that the spectrometer is relatively insensitive to multiple scattering (where the neutron undergoes more than one inelastic collision in the sample) so quite large samples (several grams) can be used to improve the count rate.

The pre-eminent direct-geometry spectrometer for this energy range is MARI at ISIS. **Figure 10** shows a schematic of the instrument. Two background-suppression choppers are used, the first is a Nimonic chopper which is used to suppress the prompt pulse of very high energy neutrons and gamma rays that are produced when the proton pulse hits the target. The second is a disc chopper made of borated resin with a single hole. This is designed to suppress the flux of delayed neutrons that are a significant fraction of the background when using a depleted-uranium target.

The core of the instrument is the Fermi chopper. This is a metal drum with a series of slots cut through it. It is magnetically suspended in a vacuum perpendicular to the beam and able to rotate at speeds up to 600 Hz. Thus for most of a rotation, the incoming neutrons are blocked, but at one particular time the slots are parallel to the incoming neutrons and they pass through to the sample. The incident neutron energy is selected by phasing the opening time of the slots with respect to the neutron pulse from the target station. Incident energies in the range 10 to 2000 meV can be selected. The detector bank continuously covers the angular range from 3° to 135° and so is able to map large regions of (Q, ω) space in a single measurement. The energy resolution

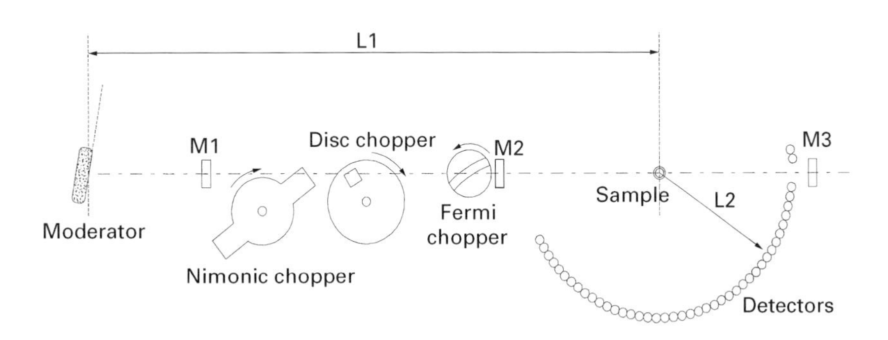

Figure 10 Schematic diagram of the direct-geometry spectrometer MARI at ISIS. M1, M2 and M3 are incident beam monitors.

is between 1–2% of the incident energy, and with all the detectors at the same secondary flight path, this resolution is constant for all the detector banks. The large range in Q is essential because phonons and magnons have different Q-dependencies and this is the best way to distinguish between the two sorts of scattering processes.

Three low-efficiency (because of the high neutron flux) detectors are placed in the main beam. The first is placed before the background choppers to monitor the incident flux for the purposes of normalization. The second and third are placed just after the Fermi chopper and behind the sample respectively. These are used to accurately determine the incident energy of the neutrons.

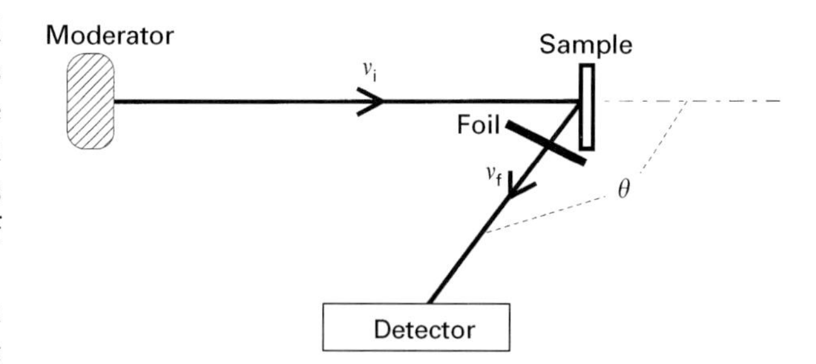

Figure 11 Schematic diagram of the eVS spectrometer at ISIS. The incident neutrons have velocity ν_i, that of the scattered neutrons is ν_f and the scattering angle is θ. The moderator-to-sample distance is L_i and the sample-to-detector distance is L_f. The final neutron energy E_f is defined by the analyser foil, which can be either gold or uranium.

The energy range >1000 meV: neutron Compton scattering

In this energy range only spallation sources can provide a sufficient flux of high-energy neutrons. The electron volt spectrometer (eVS) at ISIS is a neutron spectrometer, which uses the high intensity of neutrons in the electronvolt energy range to measure atomic-momentum distributions in a variety of condensed matter systems. When the momentum transfer in the scattering process is sufficiently large (at least 10 times that of the mean atomic momentum), the scattering can be interpreted within the impulse approximation. In this approximation, the scattering is essentially single-atom billiard-ball scattering, which is determined entirely by conservation of momentum and kinetic energy of the atom and the neutron. By measuring the energy and momentum change of the neutron, one can determine the momentum component of the target atom along the direction of the scattering vector. This process is neutron Compton scattering.

The instrument is shown schematically in **Figure 11**. eVS is an indirect-geometry inelastic spectrometer. Energy transfers in the 1000–30000 meV region and momentum transfers between 30 and 200 Å^{-1} are achieved using a filter–difference technique. The filter is a thin foil which has a strong nuclear resonance that absorbs neutrons over a narrow band of energy centered at E_1. Two time-of-flight spectra are taken, one with the foil between the sample and detectors and the second with the foil removed. The analyser foils are moved in and out of the scattered beam by computer-controlled pneumatic pistons. One complete cycle is performed every 10 minutes, i.e. 5 min with foil in and 5 min with foil out. This is to avoid effects due to long-term changes in the relative efficiency of the detectors, since typical measurement times are 2–12 hr.

The difference between these spectra is due to those neutrons absorbed in the foil and is effectively the spectrum for neutrons scattered with energy E_1. If E_1 is known, the momentum and energy transfer can be determined from a measurement of the total time-of-flight t of the neutron from moderator to detector. The velocity ν_1 of the scattered neutrons can be calculated from the kinetic energy. The velocity ν_i of the incident neutron is determined from a measurement of the neutron time-of-flight t:

$$t = \frac{L_i}{\nu_i} + \frac{L_f}{\nu_f} + t_0 \qquad [9]$$

where L_i is the distance from source to sample and L_f that from sample to detector. t_0 is an electronic time delay constant. The energy transfer is then given by Equation [6] and the momentum transfer is:

$$Q = m\left(\nu_i^2 + \nu_f^2 + 2\nu_i\nu_f\cos\theta\right)^{0.5} \qquad [10]$$

where θ is the scattering angle.

Future developments

In terms of instrumentation, developments in guide technology will lead to modest gains (factors of two to five) in the flux at the sample. The prevent trend is towards increasing the area covered by the detectors with corresponding gains in sensitivity and data rate.

INS is flux-limited technique. Thus major instrumental advances can only be expected when more-intense neutron sources become available. These will be spallation sources, since reactor sources, such as the ILL, are already close to the limit of the heat load generated by the fission process in the reactor core that can be handled. In addition, concerns about nuclear-weapon proliferation (enriched uranium is required for the reactor core) mean that these are politically and environmentally contentious. In the near future, it is planned to increase the ISIS proton beam current by 50% with a corresponding flux increase. For the year 2005, the USA is proposing to build a spallation source with several times the ISIS flux. In the longer term, the most powerful source will be the European Spallation Source, which will have a time-averaged flux equal to that of the ILL and a peak intensity 30 times that of ISIS.

List of symbols

d = analyser crystal interplanar spacing; E = neutron kinetic energy; h = Planck's constant; INS = inelastic neutron scattering; I = intermediate scattering function; k = wave vector; k_B = Boltzmann's constant; L = sample–source/detector distance; m = neutron mass; Q = neutron momentum transfer; S = scattering function; T, t = time of flight; $\hbar = h/2\pi$; λ = wavelength; v = neutron velocity; θ = scattering angle; τ = timescale; ω = frequency.

See also: **Inelastic Neutron Scattering, Applications; Neutron Diffraction, Theory; Structure Refinement (Solid State Diffraction).**

Further reading

The most up-to-date and useful source of information on neutron scattering centres and their instrumentation is the World Wide Web. The site: http://www.neutron.anl.gov/neutronf.htm provides an extensive list of websites. The ISIS website is at: http://www.isis.rl.ac.uk and the ILL website is at: http://www.ill.fr/

Bée M (1988) *Quasielastic Neutron Scattering*. Bristol: Adam Hilger.

Eckert J and Kearley GJ eds (1992) Spectroscopic applications of inelastic neutron scattering: theory and practice. *Spectrochimica Acta* **48A**: 269–478.

Newport RJ, Rainford BD and Cywinski R eds (1988) *Neutron Scattering at a Pulsed Source*. Bristol: Adam Hilger.

Nicholson LK (1981) The neutron spin-echo spectrometer: a new high resolution technique in neutron scattering. *Contemporary Physics* **22**: 451–475.

Squires GL (1978) *Introduction to the Theory of Thermal Neutron Scattering*. Mineola: Dover Publications.

Windsor CG (1981) *Pulsed Neutron Scattering*. London: Taylor and Francis.

Inorganic Chemistry, Applications of Mass Spectrometry

Lev N Sidorov, Moscow State University, Russia

MASS SPECTROMETRY
Applications

As a new research field, high-temperature mass spectrometry was established in the early 1950s. Of special value was the combined use of the classic Knudsen effusion method and mass spectral study of the evaporation products of inorganic compounds (the KCMS technique). The first review appeared as early as 1959 and contained data for approximately 120 systems and compounds which had been studied by that time.

Currently, high temperature mass spectrometry is one of the most powerful methods in high temperature chemistry. A particular feature of this method is a high temperature molecular beam source, namely a Knudsen (or effusion) cell. A Knudsen cell is an isothermal enclosure with a small orifice of precisely known dimensions. Vapour molecules escape through the orifice at a rate assumed to equal the equilibrium rate of collisions with the wall of the cell. Total vapour pressure inside the Knudsen cell has to be less than 10 Pa to ensure molecular flow conditions. The effusion cell with the substance under investigation is placed inside the mass spectrometer near an ionization chamber. By increasing the temperature of the Knudsen cell, the equilibrium pressure of the sample also increases, and usually a measurable signal can be detected from a pressure of 10^{-7} Pa inside the cell. Thus, the working range of the sample pressures in high temperature mass spectrometry is 10–10^{-7} Pa. After passing collimating slits, the molecular beam goes through the ionization chamber where electron impact ionization takes place. As usual in mass spectrometry, the electron energy is approximately 40–60 eV and can be changed to measure appearance energies.

Measurements of the ion current intensities provides the partial pressures of individual species inside the Knudsen cell and equilibrium constants of gas-phase and heterogeneous chemical reactions. Thermal ionization of samples inside the Knudsen cell also takes place at high temperatures. These thermal ions are in equilibrium with the neutral species and can be drawn out from the cell by a weak electrostatic field and their partial pressures measured. This provides the equilibrium constants of gas-phase ion molecule reactions. As a result, the following thermodynamic data can be obtained: molecular and ion composition of the vapour; enthalpies and Gibbs free energy of gas-phase and heterogeneous chemical reactions; enthalpies of formation, Gibbs free energies, electron affinities and ionization energies for vapour species (molecules, positive and negative ions); activities of the components in mixtures, partial and integral Gibbs free energies of mixing and Gibbs free energies of formation of solid compounds in the mixture.

Knudsen cell–mass spectrometer equipment

Mass analysers

In principle, any type of mass analyser can be coupled with the Knudsen cell system. As a rule, a mass range up to 1000 a.m.u. and resolution near 800 are sufficient for refractory material investigations. A single focusing magnetic sector field is used in most cases for mass analysis of effusion beams. Quadrupole and time-of-flight mass spectrometers are also applied. More expensive double focusing instruments provide high resolution and help to avoid interfering background ion intensities arising from ionization of organic impurities.

Knudsen cells, heating and temperature measurement systems

Knudsen cells are mostly heated by resistance heaters or by electron bombardment. In the first case, the limiting working temperature of the cell does not exceed 1900 K and can be measured by thermocouples. Platinum–platinum/rhodium is applied up to 1300 K and tungsten–rhenium for higher temperatures. When electron bombardment is used the upper temperature is limited by melting and evaporation of the cell material. Optical pyrometers of various types are the instruments of choice for temperature measurements in the range 2200–3000 K. Various types of Knudsen cell are used depending on the particular aim of the investigation. Stainless steel, quartz, nickel, graphite, tungsten or molybdenum may be applied as the cell materials. Protecting layers on the inner walls of a cell are also used to avoid chemical interactions between the sample and the cell material.

Some of the cells are shown in **Figure 1**. Two-compartment Knudsen cells (**Figure 1D**) are used if the vapour species formed from two different compounds with very different vapour pressures have to be investigated. They can also be used to get unsaturated vapours. The second compartment is replaced by a gas inlet system if gases are involved in the chemical reactions to be studied. 'Russian doll' type and a twin cell with a connecting channel (**Figures 1B** and **1C**) allow generation of saturated and unsaturated vapours of the sample.

Multiple Knudsen cells (**Figures 1E** and **1F**) are used for the determination of chemical activities. One cell contains a reference compound, whereas the others contain mixtures having different compositions. A comparison of the vapour pressures in the different cells gives the activities of the components and differences between the sublimation enthalpies of the samples.

Local heating of the sample to higher temperatures can be provided by a laser. Containerless heating and free evaporation take place under such conditions. Local heating is widely used in cluster formation studies.

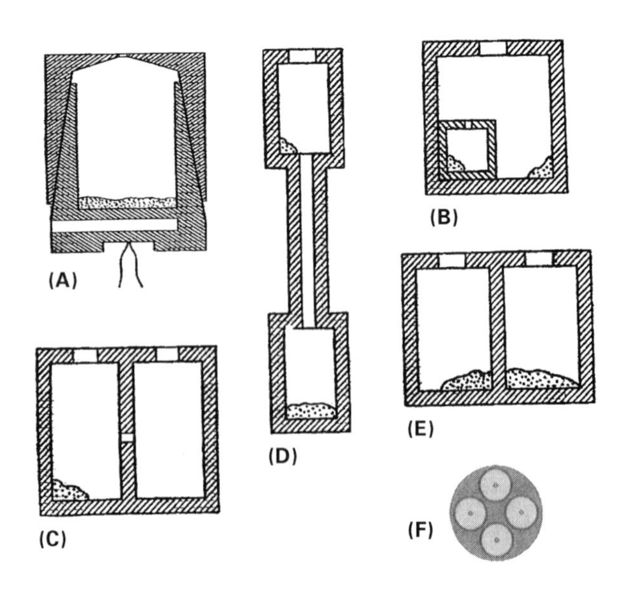

Figure 1 (A) Knudsen cell; (B) Russian doll; (C) twin cell with connecting channel; (D) two compartment cells; (E) twin cell; (F) four Knudsen cell units (multiple cells).

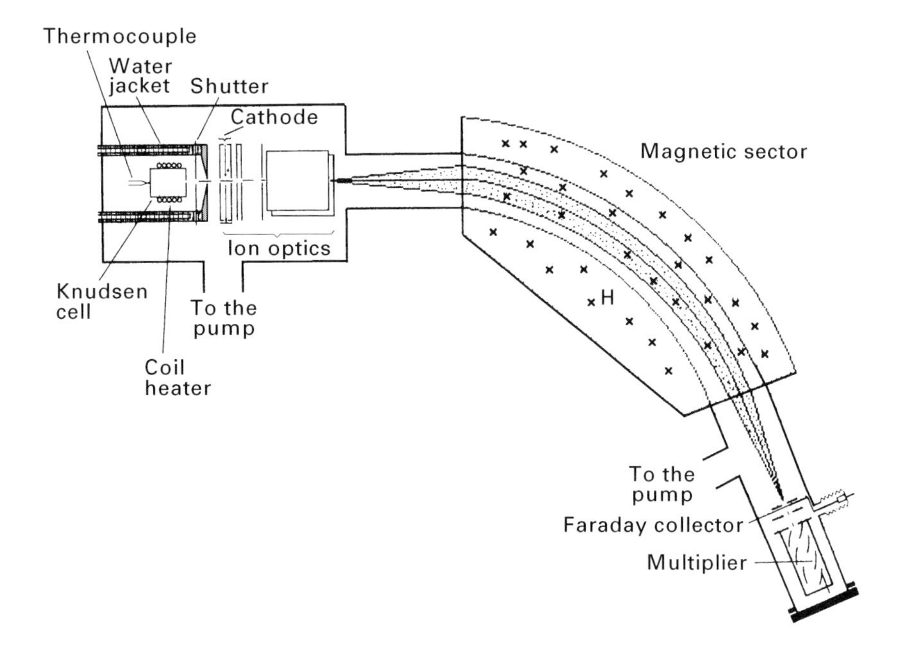

Figure 2 Knudsen cell-sector magnetic mass spectrometer with coaxial ion and molecular beams.

Ion sources

Conventional ion sources with electron impact ionization are utilized. Photo-ionization and different types of electron energy selectors are applied to improve the accuracy of ion appearance energy measurements. The electron beam is always perpendicular to the effusion beam. The ion beam can be coaxial with the effusion direction or perpendicular to both the electron and the effusion directions. Both types of KSMC instruments are shown in **Figures 2** and **3**.

Thermal ionization of samples inside the Knudsen cell takes place at high temperatures and the Knudsen cell can itself be a source of thermal ions. The charge and quantity of escaping ions depend on the draw-out potential in front of the effusion hole. A combined ion source can operate in two modes, to register either neutral or charged particles escaping from the cell.

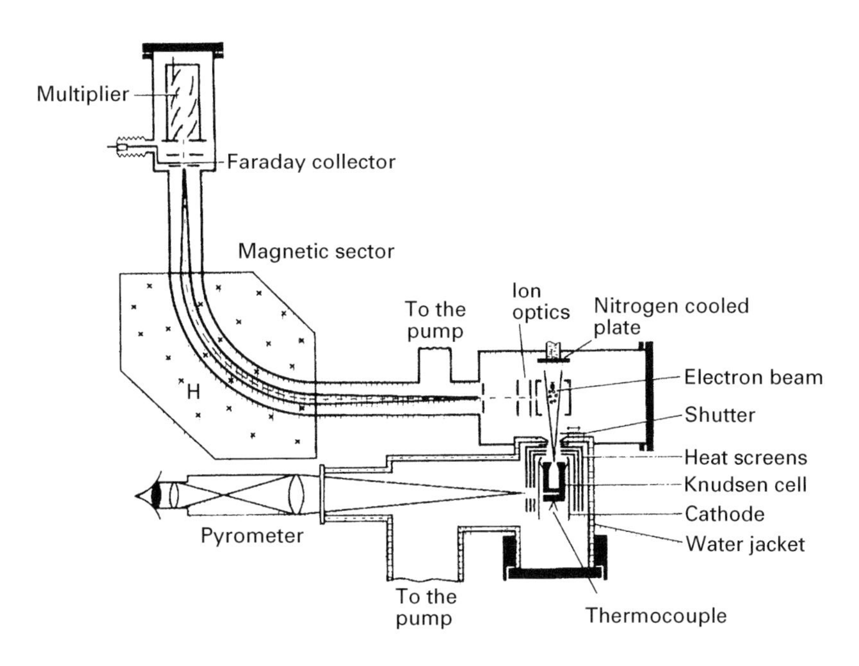

Figure 3 Knudsen cell-sector magnetic mass spectrometer with perpendicular ion and molecular beams.

At an ion pressure of less than 10^{-8} Pa, the Debye shielding radius becomes larger than the linear dimension of the cell and the concentrations of positive and negative ions in the cell are not equal. Changing the electron work function of the inner surface of the cell can increase the concentration of the ions of interest.

Deflection plates are placed behind the exit slit of the ion source. The vector of the electrostatic field is coaxial with the sector magnetic field vector. The plates permit measurement of the ion kinetic energy distribution in the direction coinciding with the vector of the magnetic analyser field strength.

Background and legitimate signal

A problem which is equally important for both neutral and thermal ion registration is separation of the legitimate signal from the background. In neutral species studies, the legitimate signal is separated from the background by use of a movable shutter impeding the passage of the molecular beam to the ionization cell. In thermal ion studies, the signal is separated and located by means of deflecting plates. Because the electrostatic draw-out field has an axial symmetry and the axis of symmetry passes through the centres of the effusion hole and of the draw-out lens, the ions generated away from the axis (in the space between the cell and the lens) have an additional velocity in the direction normal to the axis of symmetry. This additional velocity component can be readily fixed by using the deflection plates. The use of these plates allows one to discriminate between the ions drawn out from the effusion hole (the legitimate signal) and those appearing on the outer surface of the effusion cell, protective screens, etc. (the background). Moreover, the method makes it possible to monitor the surface and to locate the place of sample evaporation (**Figures** 4 and 5).

Due to the high temperature in the surroundings of the Knudsen cell some additional units are installed. There is a water jacket around the heating assembly and a target cooled by liquid nitrogen to condense the effusion beam escaping from the Knudsen cell.

A valve between the Knudsen cell compartment and the ion source is useful and makes possible a fast change of sample in the cell without changing the parameters of the ion source. The Knudsen cell and heating assembly are fixed on a moving plate and can be shifted relative to the ionization chamber entrance slit. This allows one to find the appropriate positions of the electron and effusion beams which give the highest ionization efficiency.

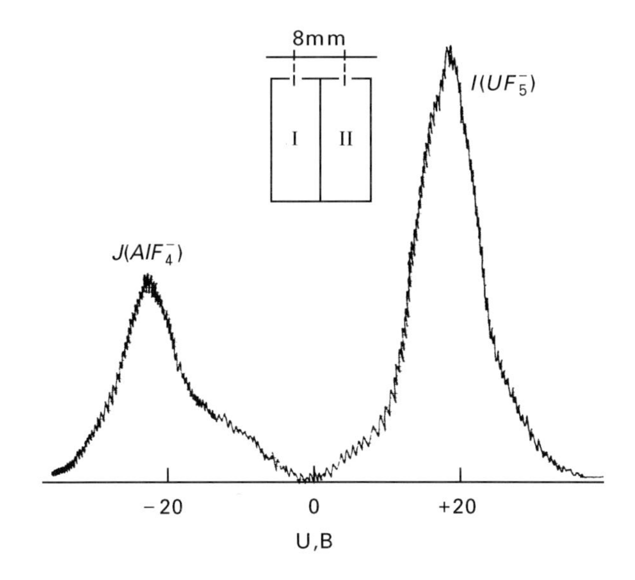

Figure 4 Ion current intensity as a function of the deflecting plates potential. Twin cell with NaF–AlF_3 in one section and KF–UF_4 in the other. The distance between effusion holes is 8 mm. Distance resolution is 5.2 V mm^{-1}.

Ionization efficiency curves and cross-sections

Ionization efficiency curves are plotted in the electron energy range 5–50 eV and ion appearance energies are determined in most cases. The threshold law is written as $I = \text{const}(E - E_{\text{th}})^{n-1}$ where n is the number of electrons leaving the collision area. In the case of the formation of singly charged positive ions,

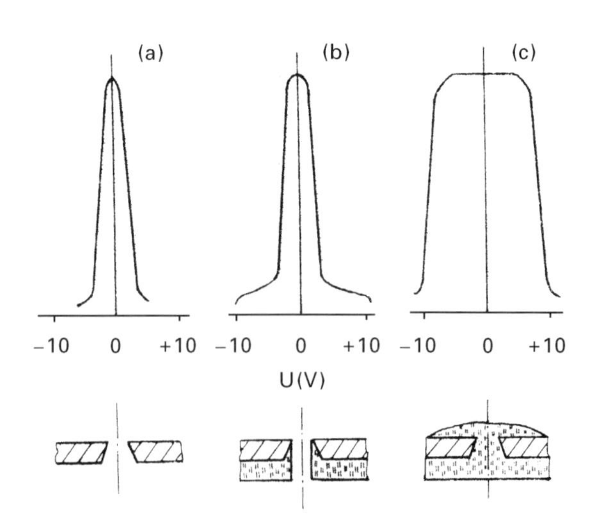

Figure 5 Ion current intensity as a function of the deflecting plates potential. (A) clean effusion hole; (B) condensation on the inner surface around the effusion hole; (C) condensation on the outer surface around effusion hole (creeping).

the linear extrapolation of the linear part of the ionization efficiency curves is widely used to find the ion appearance energy. The ionization cross-sections of atoms can be measured for different ionization energies between 0 and 220 V in steps of 1 eV and their maximal values are located in the range 40–60 eV. These data are used in Knudsen effusion mass spectrometry; the calculated maximum values are considered to be in agreement within 20% with the experimental data if the ionization cross-sections are compared at the same energy.

Ionization cross-sections of molecules can be obtained from those of atoms by applying the additivity rule postulated by Otvos and Stevenson in 1956. The ionization cross-section calculated in such a way is a total for the electron energy in the range 40–60 eV. Fragmentation always takes place at such electron energies, and the mass spectrum of a molecule gives the ratios of partial cross-sections with respect to the base peak. The total ionization cross-section is the sum of the partial ones. It is generally accepted that the above additivity rule gives a value which differs from the experimental one by not more than a factor of 2. This rough estimate is only correct for the total ionization cross-section. There are two limiting cases in Knudsen cell mass spectrometry: the first is when the molecular ion peak is the base peak and its partial ionization cross-section with respect to the total is more than 80%. In the second case, the molecular ion peak is small and its partial ionization cross-section is less than 10% of the total. Treatment of the experimental data and the experimental approaches are essentially different in these cases. In the first case, measurements of the molecular ion intensities are sufficient for the reliable interpretation of the experimental data. In the second case, one has to know the fragmentation pattern (mass spectrum) of each vapour species and calculate the total ion current intensities for each vapour species. This task is rather complicated because a particular feature of high temperature vapour is the existence of vapour species which cannot be isolated as individual compounds and whose mass spectra cannot be measured in the previous runs. The superposition of the same fragments formed from different species takes place and ion current intensities have to be apportioned between each species.

The total ionization cross-section is a function of the electronic states of a molecule. It does not significantly depend on temperature in the one- to two-hundreds degree temperature range. It is not a case of partial cross-sections which depend on vibrational states. The temperature dependence of the mass spectrum of the individual molecules is well known and has to be taken into account before the experimental data are interpreted.

Relationships between ion intensities and partial pressures of vapour species

The relation between the measured ion current intensities and partial pressures of the neutral species inside the effusion cell can be written as follows:

$$p_j = \frac{k}{\sigma_{ij} I_{ij} T} \qquad [1]$$

$$p_j = \frac{k_i}{\sigma_j \sum I_{ij} T} = \frac{k}{\sigma_j I_j T} \qquad [2]$$

where p_j is the partial pressure of molecule j, k is the sensitivity constant, σ_j is the total ionization cross-section of molecule j, σ_{ij} is the partial ionization cross-section of molecule j, I_{ij} is the measured ion current of ion i originating from molecule j (I is calculated allowing for the isotopic composition of the elements and the secondary multiplier gain, $I_j = \sum_i I_{ij}$ is the total ion current from molecule j, T is the temperature of the effusion cell. While the thermal ions escaping the effusion cell are recorded the corresponding equation relating the measured ion currents and ion pressures inside the effusion cell is given as:

$$p_i = C\, I_i\, M_i^{1/2}\, T^{1/2} \qquad [3]$$

where P_i is the partial pressure of ion i, C is the sensitivity constant and M_i is the mass of ion i.

Mass spectra and gaseous species

The chemical formula of the ions in the mass spectrum is obtained from the mass of the ions and their isotopic distribution. The next step is the assignment of the observed ions to their neutral precursors in the equilibrium vapour and the calculation of the vapour species, partial pressures. Equations [1] to [3] reveal a set of tasks which have to be completed if calculated partial pressure values are to be reliable:

determination of sensitivity, k and C in Equations [1], [2] and [3] (calibration of an instrument);
determination of molecular weights of vapour species, j;

apportioning the measured ion currents $I_i = \sum_i I_{ij}$, i.e. determination of values I_{ij};
calculation of total σ_j or partial σ_{ij} ionization cross-sections.

The last three points relate to neutral species only.

Molecular weight of neutral species

Molecular ions and ion fragments have to be identified. The main approach is based on efficiency curve measurements and appearance potential estimates. For example, consider bismuth oxide. The ionization energy (IE) of Bi is 7.3 eV, IE of Bi_2O_3 is 9.5 and IE of Bi_4O_6 is 9.0 eV. The expected IE of molecular ions with atomic composition shifted to bismuth have to be in the range 7.3–9.0. There are 6 ions in the mass spectrum of saturated bismuth oxide vapour of Bi_4O_5, Bi_4O_4, Bi_4O_2, Bi_3O_4, Bi_2O_2 and Bi_2O. The first three ions have IE 9.1, 10.0 and 10.3 eV and are treated as fragments, the next have IE 7.5, 7.9 and 7.5 eV and are considered to be molecular ions.

Apportion of ion currents

When molecular ions give the base peaks in mass spectra, one can decrease the electron energy and obtain a mass spectrum which mainly consists of molecular ions. The estimated total ionization cross-section can be utilized and the partial pressures of vapour species calculated. When molecular ions give only a minor peak in the mass spectrum, decreasing the electron energy does not give satisfactory results and therefore some thermodynamic approaches have been developed. These are based on the variation of ion current intensities with the isothermal change of vapour composition.

The isothermal change of vapour composition is governed by the thermodynamic laws, and their analytical expressions can be included in the equation that relates measured and apportioned ion currents. A variation of partial pressures at constant temperature can be achieved by changing the composition of the solid (liquid) phase equilibrated with the vapour. When an individual compound is under study, Knudsen cells of **Figures 1B**, C and D allow one to compare the saturated and unsaturated vapours at the same temperature and apportion the measured ion current.

Instrument calibration

For thermal ions, the ratio of ion partial pressures is measured without calculation of absolute values. In the case of neutral species, the ratio and absolute values of partial pressures have to be determined in most cases. Two methods of calibration can be used: (1) the sensitivity k is determined by comparison with a reference compound having a known vapour pressure. The ratio of the ionization cross-sections of a sample and a reference compound must be calculated according to the additivity rule and theoretical values of atomic cross-sections; (2) The sensitivity k_j is determined directly by complete evaporation of a weighed amount of a sample, and if complete evaporation of two samples is carried out the ratio of their ionization cross-sections can be determined. In the first case

$$p_j = \frac{k}{\sigma_j} I_j T \quad [4]$$

$$p_{\text{ref}} = \frac{k}{\sigma_{\text{ref}}} I_{\text{ref}} T \quad [5]$$

and

$$p_j = p_{\text{ref}} \frac{\sigma_{\text{ref}}}{\sigma_j} \frac{I_j}{I_{\text{ref}}} \quad [6]$$

where p_{ref} and p_j are the pressures of the reference and investigated samples respectively.

In the second case, the Herz–Knudsen equation is utilized:

$$q[\text{mol}] = s_{\text{eff}}(2\pi M_j R)^{-1/2} \int_0^t p_j T^{-1/2}\, \text{d}t = B_j \int_0^t p_j T^{-1/2} \text{d}t \quad [8]$$

where s_{eff} is the orifice area, q = weight of sample; M_j the mass of vapour specimen and R is the gas constant. Taking into account Equation [2] it can be written as follows:

$$q[\text{mol}] = B_j k_j \int_0^t I_j T^{1/2}\, \text{d}t \quad [9]$$

Measurements of temperature, ion current intensities as a function of time, weight of sample and the orifice area make it possible to calculate k_j and P_j at any time during evaporation. If there are several species in the vapour, Equation [9] has to be written as

$$q = \sum_j B_j k_j \int_0^t I_j T^{1/2}\, \text{d}t \quad [10]$$

Dynamic range

The upper limit of vapour pressure in the cell is determined by the molecular flow through the effusion orifice. When the orifice diameter is 0.3–1.0 mm, the upper limit lies in the pressure range 1–10 Pa. A vapour pressure equal to 10^{-7} Pa is accepted as the lower limit for neutral species. In the case of thermal ions, the sensitivity of the instrument is 5–6 orders of magnitude higher, and the data fall in the range 10^{-6}–10^{-12} Pa. On the other hand, the concentration of ions in the vapour is 6–7 orders of magnitude lower than that of the neutral species, and as a result the ion currents from neutral species and thermal ions are of the same order.

Second and third law calculations

When partial pressures are measured, the equilibrium constants of the gas phase and heterogeneous reactions can be calculated and their temperature dependence plotted.

The equation $d/dT(\ln K_p) = \Delta H^o/RT^2$ allows the determination of reaction enthalpies from the slope of the curve $\ln K_p$ versus $1/T$ (second law calculation). As $K_p = \Pi_i P_j^{v_j} = \text{const}\ \Pi_i I_j^{v_j} T^{\Delta v_i} = \text{const}\ K_p(I)$ one can determine the enthalpy change from the slope of the curve $\ln K_p(I)$ versus $1/T$.

The equation $\Delta H^o_o = -RT \ln K_p + \Delta\Phi^o T$ allows one to calculate reaction enthalpies from the absolute value of K_p (third law calculation), where $\Phi^o_T = (G^o_T - H^o_o)/T$.

Determination of activities

According to the definition of activity, a_j is the ratio p_j/p_j^0, and putting ion current intensities I_j instead of partial pressures one has $a_j = I_j/I_j^0$. Even in the case of a binary mixture, not less than 10 samples of different composition have to be tested to obtain reliable data for the activities of both components in the complete range of composition. Two approaches have been developed to increase the efficiency of this type of measurement.

One of them is the multiple Knudsen cell which allows one to obtain activities for several mixture compositions during a run. Another is the isothermal evaporation method (IEM) which gives activities as a continuous function of the melt composition. The isothermal evaporation process should be carried out rather slowly (1–3% per hour) and can be considered a sequence of equilibrium states under such conditions. Simultaneously, ion current intensities are recorded as a function of time. Experimental data taken in such a way give ion current intensities as a function of time, integrals $\int_0^t I_{ij} T^{1/2} dt$ and derivatives of two types $dI_i(t)/dt$ and $dI_i(t)/dI_u(t)$. These data can be transposed into partial pressure and activity depending on the melt composition. The mathematical part of IEM which makes this transformation possible is based on the Herz–Knudsen and Gibbs–Duhem equations.

There are many binary systems whose vapour contains mixed associates and a large number of vapour species. The virtue of the KSMC method is the possibility of directly determining the partial pressures of vapour species, rather than the total pressure in the system. These advantages of mass spectrometry are widely used in thermodynamic research into two-component systems. If there are mixed species A_kB_l and A_rB_s in the saturated vapour of system A–B, the Gibbs–Duhem equation $\Sigma_j N_j d \ln a_j = 0$ can be written as

$$\frac{d\ln P(A_rB_s)}{d\ln P(A_kB_l)} = \frac{rN_B - sN_A}{kN_B - lN_A}$$

where N_A (N_B) is the mole fraction of A(B) in the solution. Partial pressures can be replaced by ion currents and any pair of vapour species can be treated in accordance with the Gibbs–Duhem equation. In the simplest cases ($r = 1$, $s = 0$, $k = 0$, $l = 1$) it will be

$$\frac{d\ln p_A}{d\ln p_B} = -\frac{N_B}{N_A} = \frac{d\ln I_A}{d\ln I_B}$$

and for $r = 1$, $s = -1$, $k = 0$, $l = 1$

$$N_A d\ln\frac{p_A}{p_B} + d\ln p_B = 0 = N_A d\ln\frac{I_A}{I_B} + d\ln I_B$$

Different forms of Gibbs–Duhem equation are used and the choice of independent variables is determined by the vapour species being easily registered. The replacement of low intensities by the ratio of measured ones is a general approach in KCMS. Any type of equilibrium involving molecules j can be used for this purpose. For example, consider the reactions; A + B $\leftrightharpoons$ AB

$$K_p = \frac{p_{AB}}{p_A\, p_B}$$

and p_A may be replaced by the ratio p_{AB}/p_B.

Applications

Vapour species and their thermodynamic properties

During the last four decades all major classes of inorganic compounds have been studied by KSMC. The discovery of new gas phase molecules and ions is considered a main achievement of KCMS. Thermodynamic properties of the newly discovered vapour species were also obtained and most of them can be found in reference books. Metal vapours consist of atoms and small amounts of dimers. The main vapour species of carbon is a trimer and a series of C_n ($n = 1$–8) molecules were found in the saturated vapour. The main components of P, As and Sb vapours are tetramers. In the vapour of S and Se, a series of polymers with n = 1–12 were observed with the most abundant species being S_8 and Se_6. Heteroatomic molecules are always present in the vapour of mixtures of elements. The saturated vapour of inorganic compounds that evaporate in the temperature range 700–3000 K, generally contains species formed by the dissociation of the base compound and a number of associates. Many interesting and unexpected examples can be found in the extensive database on the species in inorganic vapours. The existence of the dimers and trimers of oxides V_2O_5, WO_3 and MoO_3 has been reported. Salts of oxygen-containing acids can be evaporated without decomposition to give small amounts of dimers. Mixtures of halides of alkali metals and halides of polyvalent metals give associates in the vapour and the electrical properties of such associates have been investigated.

Heterogeneous reactions and gas phase reactions

Gas phase reactions of any type can be successfully studied. The magnitude of the equilibrium constant is important. *Isomolecular* reactions $A + B \leftrightharpoons C + D$ do not require an absolute instrument calibration which makes them the most convenient and frequently observed type of reaction. This also concerns ion–molecule and electron exchange reactions. A bond dissociation energy is calculated from the equilibrium constants of the reaction $MX_n \leftrightharpoons MX_{n-1} + X$. The sublimation enthalpy and dissociation enthalpy are determined from the equilibrium constants of the heterogeneous reactions $A_{solid} \leftrightharpoons A_{gas}$ and $AB_{solid} \leftrightharpoons B_{solid} + A_{gas}$.

Ion/molecule equilibrium in the saturated vapour

A study of the complex halogenides and salts of oxygen-containing acids by KCMS modified for ion/molecule equilibrium study (KCMS-IME) showed a very high concentration of ions in the saturated vapour (only 5 orders of magnitude less than the vapour pressure). These salts of alkali metals have large anions and as a result the enthalpies of heterolytic dissociation of the complex salts are much lower than those of alkali metal fluorides. For example, for the system KF–33 mol% UF_4, $T = 1000$ K, $p(KUF_5) = 0.35$ Pa; $p(K^+) = p(UF_5) = 10^{-6}$ Pa. At such ion pressures the Debye shielding radius R_d (0.2 mm) is significantly smaller than the linear dimension of the Knudsen cell (10 mm). An electronless plasma is formed and the main vapour species are:

Molecules KF, K_2F_2, KUF_5, UF_4
Positive ions K^+, K_2F^+, $K_2UF_5{}^+$
Negative ions KF_2^-, UF_5^-, UF_6^-, $U_2F_9^-$

Electron affinity determination

Addition of complex alkali metal salts to a high temperature system leads to negative ion formation in the vapour. KCMS-IME can be applied to electron affinities and the estimation of negative ion enthalpies of formation. The EA of fluorides and oxides of transition metals in their higher oxidation states were determined. Almost all of these compounds have EA in the range 3.5–7.0 eV. Application of this method to fullerene vapours has yielded the electron affinities for the series of higher fullerenes C_n ($n = 70, 72, \ldots, 106$).

Activities in mixtures and the Gibbs free energy of formation of solid compounds

The determination of activities of components allows calculation of the partial and integral Gibbs free energy for the mixing of liquid and solid solutions. If solid compounds A_nB_m are formed in the mixture A–B, the Gibbs free energy of formation of these compounds from A and B may be calculated.

P–T–X diagram study

Measurements of the dependence of partial pressures on temperature along the liquidus (two-phase equilibrium between liquid and solid phases) and solidus (solid–solid) lines and partial pressure dependences on system composition at constant temperature give the total pressure and vapour brutto composition as function of temperature and system composition. After these data are obtained, a T–X diagram can be transposed to a P–T–X diagram. Such diagrams are used in applied science e.g. the growth of crystals and films by the vapour deposition method.

Homogeneous range, Partial pressures of oxygen

KCMS is often used for the study of the homogeneous range of solid compounds. The partial pressure change can be easily selected. Isothermal evaporation is sometimes used to identify solid phase composition changes. Diffusion in the solid phase and enrichment of a surface layer by a hard volatile component demands special attention. When oxides, glasses or ceramics are studied the oxygen partial pressure is of particular interest. One advance in the last decade is that very low partial pressures of oxygen can be calculated and controlled by measurement of the ratio of two negative ions (KCMS-IME). Oxides of variable valency metals are the source of these two ions and the equilibrium constant of the reactions $MeO_n^- + \frac{1}{2}O_2 \leftrightharpoons MeO_{(n+1)}^-$ have to be previously obtained. The same approach is used for activity determinations and is based on similar reactions $A + B^- \leftrightharpoons AB^-$ and the measurements of negative ion ratios $I(AB^-)$ to $I(B^-)$.

List of symbols

A_kB_l and A_rB_s are vapour species in the binary system A–B; $B_j = \rho_{eff}(2\pi M_j R)^{-1/2}$; E_{th} = threshold energy; I_{ij} = measured ion current of ion i originating from molecule j; $I_i = \sum_j I_{ij}$ = measured current of ion i; $I_j = \sum_i I_{ij}$ = total ion current from molecule j; M_i = mass of ion i; M_j = mass of vapour specimen j; $N_A(N_B)$ = mole fraction of A(B); p_i = partial pressure of ion j; P_i = partial pressure of ion i; p_j = partial pressure of molecule j; p_{ref}, p_j = pressure of reference and investigated samples; q = weight of sample; s_{eff} = orifice area; σ_j = total ionization cross-section of molecule j; σ_{ij} = partial ionization cross-section of molecule j; $\Phi_T^o = -(G_T^o - H_0^o)/T$; ν = stoichiometric coefficient of chemical reaction.

See also: **Cluster Ions Measured Using Mass Spectrometry; Flame and Temperature Measurement Using Vibrational Spectroscopy; Gas Phase Applications of NMR Spectroscopy; Ionization Theory; Ion Molecule Reactions in Mass Spectrometry; Negative Ion Mass Spectrometry, Methods; Photoionization and Photodissociation Methods in Mass Spectrometry; Quadrupoles, Use of in Mass Spectrometry; Time of Flight Mass Spectrometers.**

Further reading

Boltalina OB, Ponomarev DB and Sidorov LN (1997) Thermochemistry of fullerene anions in the gas phase. *Mass Spectrometry Reviews* **16**(N6): 333–352.

Droward J (1964) In: Mass spectrometric studies of the vaporation of inorganic substances at high temperature. Rutner E, Goldfinger F and Hirtz JP (eds) *Condensation and Evaporation of Solids*, pp 255–310. New York: Gordon and Breach.

Droward J and Goldfinger P (1967) Investigation of inorganic systems at high temperature by mass spectrometry. *Angewandte Chemie* **79**: 589–604.

Droward J (1985) Mass spectrometric studies at high temperature. In: Todd JFJ (ed) *Advances in Mass Spectrometry, Part A*, pp 195–214. New York: Wiley.

Hastie JW (1984) New techniques and opportunities in high temperature mass spectrometry. *Pure and Applied Chemistry* **56**: 1583–1600.

Hilpert K (1990) Chemistry of Inorganic Vapors. *Structure and Bonding* **73**: 97–198.

Inghram MG and Droward J (1960) Mass spectrometry applied to high temperature chemistry. In *Proceedings of an International Symposium on High Temperature Technology*, pp 219–240. New York: McGraw-Hill.

Sidorov LN and Sholtz VB (1972) Mass spectrometric investigation of two component systems of complex vapour composition by the isothermal evaporation method. *International Journal of Mass Spectrometry and Ion Physics* **8**: 437–458.

Sidorov LN, Zhuravleva LV and Sorokin ID (1986) High temperature mass spectrometry and studies of ion–ion, ion–molecule and molecule–molecule equilibrium. *Mass Spectrometry Reviews* **5**: 73–97.

Stolyarova VL and Semenov GA (1994) *Mass Spectrometry Study of the Vaporization of Oxides Systems.* New York: Wiley.

Inorganic Chemistry, Applications of NMR Spectroscopy

See **Structural Chemistry Using NMR Spectroscopy, Inorganic Molecules.**

Inorganic Compounds and Minerals Studied Using X-Ray Diffraction

Gilberto Artioli, Università degli Studi di Milano, Italy

HIGH ENERGY SPECTROSCOPY
Applications

Introduction

X-ray diffraction from the solid state is a fundamental technique for the characterization of synthetic and natural materials. Since the discovery of Bragg-type X-ray diffraction from periodic crystal lattices, the technique has proven an essential tool for (1) the identification of crystalline compounds, (2) the quantitative analysis of polyphasic mixtures, (3) the study of the long-range atomic structure of crystals, including detailed charge density studies, and (4) the physical analysis of crystalline aggregates, including orientation texture, crystallite size distribution, and lattice microstrain effects. Such routine applications have developed during the first half of the 20th century into standard analysis procedures, and during the last decades they became so widely utilized that data are now available for most known inorganic compounds and minerals, to the extent that electronic databases are now accessible for automatic identification and rapid retrieval of crystallographic and structural information of crystalline substances.

The focus of the modern crystallographic research on inorganic solids is the interpretation of the physical and chemical properties of materials in terms of their ideal or defective atomic structure, and the transfer of the acquired crystal chemical knowledge to the engineering of solid-state compounds with novel technological properties. The years since the mid 1980s have also witnessed the development of dedicated second and third generation synchrotron radiation sources in the region of hard X-rays. The availability of brilliant, polarized and collimated synchrotron radiation beams with a wavelength that is tunable over a wide spectral range has made a number of interesting advances possible in materials characterization. State-of-the-art investigations of inorganic compounds and minerals by synchrotron X-ray diffraction include: *in situ* dynamic diffraction of kinetic processes and phase transformations, structural characterization of compounds at ultra-high pressures, and use of resonant scattering effects for element-selective structural analysis.

X-ray diffraction analysis in the laboratory

X-ray diffraction (XRD) data collection and analysis is routinely performed in any laboratory dealing with inorganic or mineral compounds. Samples are commonly in the form of small single crystals, polycrystalline aggregates (powders), or oriented specimens (fibres, thin films, polished surfaces). The experimental techniques and the data analysis methods to be applied may therefore vary depending on the physical state and the chemical composition of the sample, besides of course the nature of the requested information. Typical laboratory applications in inorganic chemistry, solid-state physics and mineralogy may involve: (1) identification of unknown crystalline phases, (2) quantification of phase abundance in polycrystalline mixtures, (3) crystal structure determination and refinement, (4) physical analysis of the sample in terms of crystallite size, lattice microstrain, or orientation texture of the crystal domains. On one hand the mineralogist usually employs X-ray diffraction to identify and classify mineral specimens, to quantify mineral phases in raw materials for industrial use, or to characterize the temperature–pressure–time history of a rock through the detailed analysis of the crystal structure of the phases present in the mineral assemblage. On the other hand the inorganic chemist frequently characterizes by X-ray diffraction the products of synthesis reactions in an attempt to refine the synthesis processes or produce new materials with novel structural and physical properties.

The experimental apparatus commonly involve a sealed-tube or rotating-anode X-ray source, a 2-circles (for powders) or a 4-circles (for single crystals) automated diffractometer for sample orientation with respect to the incident beam, and one or more detector banks for detection of the diffracted signal. The angular position, the integrated intensity, and the peak profile shape of the diffracted signal are important for the subsequent analysis.

Phase identification

Powder X-ray diffraction patterns are the fingerprint of all crystalline substances. The technique is a rapid and powerful tool for identification of all crystalline compounds, and it is especially valuable in the presence of polymorphs or quasi-isochemical compounds, where chemical techniques cannot be applied. The data collection time of a medium-resolution powder diffraction pattern for identification purposes is typically of the order of 20–30 min. Identification of the unknowns in a compound composed of a few phases is a straight-forward automatic procedure in which the Bragg diffraction peaks in the observed pattern are matched against the reported powder diffraction patterns of all known crystalline substances in the powder diffraction file. Advanced computer software is available for such applications.

If the investigated material is composed of a large number of phases, then the identification process can be more complicated due to substantial overlap of the Bragg peaks related to the different phases. Chemical information on the compound of interest may in this case help in defining subsets of phases from the database containing particular chemical elements. Subsets search may be more successful than the search performed on the whole database, although it requires prior knowledge of information on the sample.

Quantitative phase analysis

Quantification of the relative abundance of crystalline phases in a multiphase mixture is an everyday problem in a wide range of applications. Common examples are: evaluation of the yield in inorganic synthesis and catalytic processes, characterization of raw mineral materials for industrial processes, quality check of fired ceramic products, and many more. While in most cases the required accuracy level of the analysis is a few percent at best, in particular cases such as in the quantification of phase contaminants in technologically important materials, or of hazardous and toxic phases in environmentally dispersed aerosols, the required level of accuracy must be substantially lower than 1 wt% relative abundance. Accuracy levels of 2–3 wt% are commonly reached if standard procedures of quantitative phase analysis by diffraction data are properly performed. Generally employed analytical methods include: the internal or external standard method, the matrix flushing method, and the reference intensity ratio method. Very recently, the availability of analysis techniques of powder diffraction data based on full-profile (Rietveld method), originally developed for structural analysis, had a great impact in the area of quantitative analysis: if the crystallographic structure of the phases present in the polycrystalline mixture are known, then the complete simulation of the diffraction pattern provides an easy standardless method for accurate quantification of all phases. The major advantage is that only one powder diffraction pattern is needed, with no need for standard addition or laborious external calibration of the peak intensities. **Figure 1** shows the observed, calculated and difference powder diffraction patterns relative to the quantitative analysis of a polycrystalline specimen composed of three standard phases, performed using the standardless Rietveld technique. Estimated standard deviations of relative proportions of all phases are in the range 0.1–0.2 wt%.

If the diffraction data are of above-average quality, the full-profile method can easily allow quantification of crystalline phases well below the 1 wt% accuracy level. Advanced nonroutine application of the full-profile technique involve modelling of anisotropic Bragg peak broadening for the quantification of defective or disordered crystal phases. The

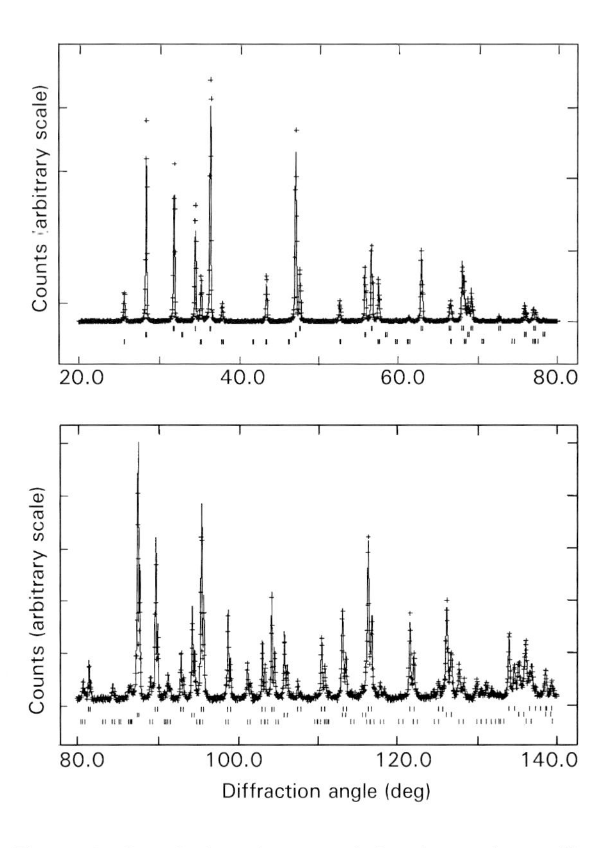

Figure 1 Quantitative phase analysis of a polycrystalline mixture containing three phases performed by the Rietveld method. Selected portions of the observed (crosses) and calculated (solid line) X-ray powder diffraction patterns are shown. The vertical marks indicate the position of the single Bragg peaks.

combination of the full-profile Rietveld analysis with the addition of a known quantity of an internal standard in the mixture provides a simple and reliable method for estimation of the amorphous phase in samples containing noncrystalline components. The accuracy and reliability levels reached by modern methods of quantitative phase analysis based on powder diffractometry commonly exceed the specifications required by phase quantification in traditional and advanced industrial materials.

Crystal structure analysis

The analysis of the atomic structure of crystals by diffraction is one of the more developed applications in the whole field of diffraction. Although often used as a simple, rapid technique for molecular structure determination, the full analysis of the crystal structure by X-ray diffraction indeed represents a powerful tool for the detailed interpretation of the relationship between the long-range atomic arrangement and the physicochemical properties in any solid-state material. Recent advances in the preparation, engineering and characterization of high-technology materials such as high-T_c superconducting oxides and fullerenes, could hardly have been achieved without the accurate study of their crystal structure; and the long list of presently known synthetic microporous compounds widely used in catalysis would be much shorter were it not for the detailed structural data obtained on aluminosilicate zeolite minerals.

The standard crystallographic description of the atomic arrangement in a crystal includes the definition of a translation unit cell with its symmetry properties (Bravais lattice type and space group symmetry), the fractional coordinates of each crystallographic position together with the occupancy factor of each chemical species located on it, and its atomic displacement parameters (related to the probability density function of the electron distribution). Standard crystallographic descriptions as such are now available for most known solid-state compounds: the primary reference sources for inorganic crystals and minerals are the Structure Reports (in printed form) and the Inorganic Crystal Structure Database (in CD-ROM form). A number of comprehensive monographs review the essential concepts of chemical bonding in crystals and the classification of basic structural types. The vast majority of structural information was derived from single-crystal diffraction studies performed with standard laboratory apparatus at room temperature and pressure, although in recent years a growing number of inorganic crystal structure data has resulted from diffraction studies performed on polycrystalline samples using the full-profile Rietveld techniques of analysis. Equilibrium studies at nonambient temperature or pressure, and studies of slow kinetic processes are also accessible in modern X-ray laboratories, both using single-crystal and powder specimens.

The standard crystal structure description of a compound is commonly aimed at the geometrical analysis of the long-range atomic arrangement in terms of interatomic bond distances and angles, coordination polyhedra, and topological bond connectivity. This is the minimal information required for the correct interpretation of the macroscopic physicochemical properties of the compound at the atomic level. Common applications include the understanding of the solid-state reactivity of inorganic compounds, and the reconstruction of the crystallization and transformation processes, i.e. the geological record, in minerals.

Sometimes subtle or elusive physical effects in crystals require a deeper understanding of the chemical bonding and the crystallochemical role of the atoms. Examples are the oxygen site vacancies and cation substitutions in high-T_c superconducting materials, the atomic or molecular polarization effects in nonlinear optical materials, the cation partitioning among sublattice sites in microporous framework structures, or the distinction between atomic static (i.e. site disorder) and dynamic (i.e. anisotropic motion) effects in minerals exhibiting nonideal solid solution behaviour. To face such complex problems, nonstandard single crystal structure analysis techniques are available. For example, detailed mapping of the electron density in a crystal can be performed by means of charge density studies, commonly performed by low-temperature (down to a few K) X-ray diffraction in order to minimize atomic thermal vibrations smearing out the density effects due to localized electrons. **Figure 2** shows the detailed mapping of the electron density of the H–O–H plane in the water molecule enclosed in the channel of a porous hydrous sodium aluminosilicate, the natural zeolite natrolite, whose complete structure is shown in **Figure 3**. The charge density distribution is derived from the experimental structure factor amplitudes measured by single-crystal X-ray diffraction. Also shown is the map of the Laplacian of the electron charge density in the same plane. On the basis of a Bader-type analysis the map allows detailed interpretation of the character of the chemical bonds in the crystal structure.

Accurate crystal structure studies can be performed on the same crystal in a wide temperature range and allow extrapolation of the true atomic

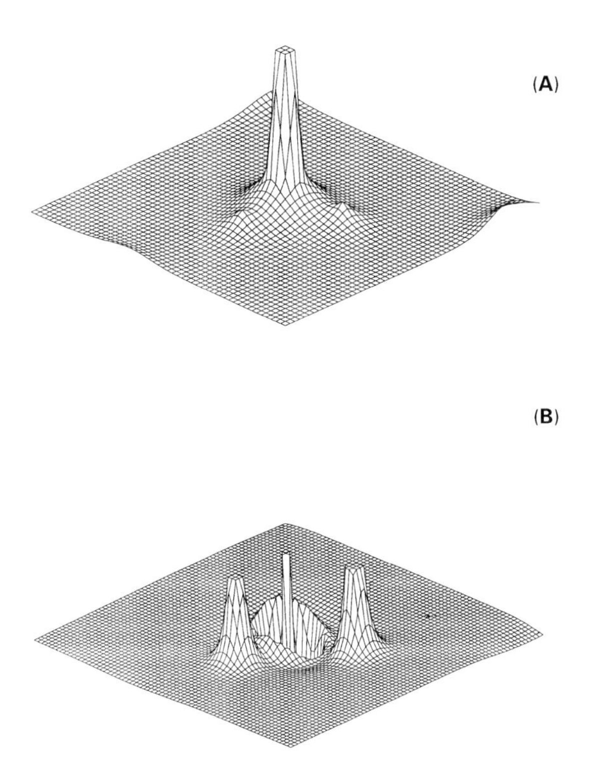

Figure 2 Experimental distribution of the electron charge density (A) in the H–O–H plane of the water molecule in the zeolitic channel of natrolite obtained from single-crystal X-ray diffraction data. Detailed interpretation of the chemical bond features, including the position and character of the critical points, is possible through the analysis of the maps of the Laplacian of the charge density (B).

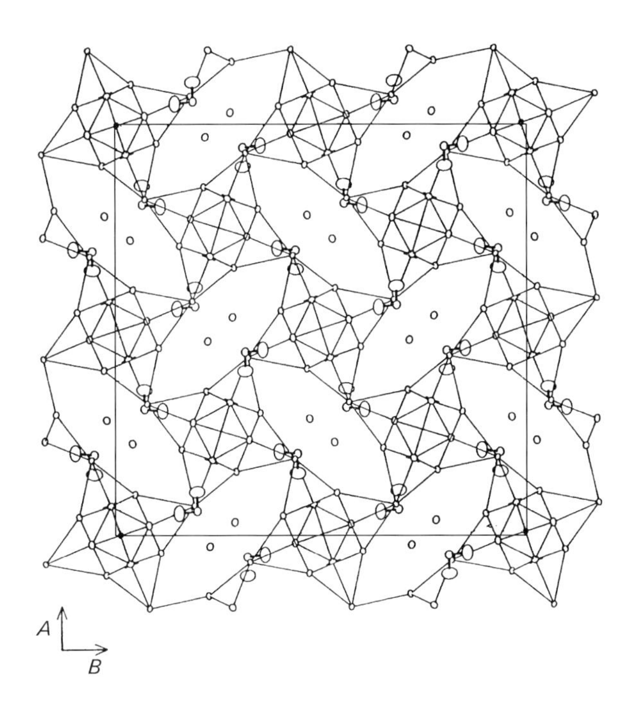

Figure 3 Projection of the structure of the zeolite natrolite along the orthorhombic *c* axis, showing the zeolitic channels containing the Na atoms (small circles) and the water molecules forming hydrogen bonds to framework oxygen atoms.

displacement parameters down to the 0 K temperature region, evaluation of the zero-point energy of the atom, and control of the presence of static subsite disorder effects. **Figure 4** shows selected experimentally determined values of the atomic displacement parameters for the Mg, Si, Al and O atoms in a synthetic pyrope garnet of composition $Mg_3Al_2Si_3O_{12}$. The deviation of the values at 30 K from the linear fit performed on the atomic displacement values measured above 298 K is due to the contribution of the zero-point energy to the atomic motion. Accurate diffraction data of this kind allow separation of the static and dynamic contributions to the measured atomic displacement, and many also indicate the presence of anharmonic vibrational components, although these effects are more appropriately evaluated using temperature-dependent neutron diffraction data.

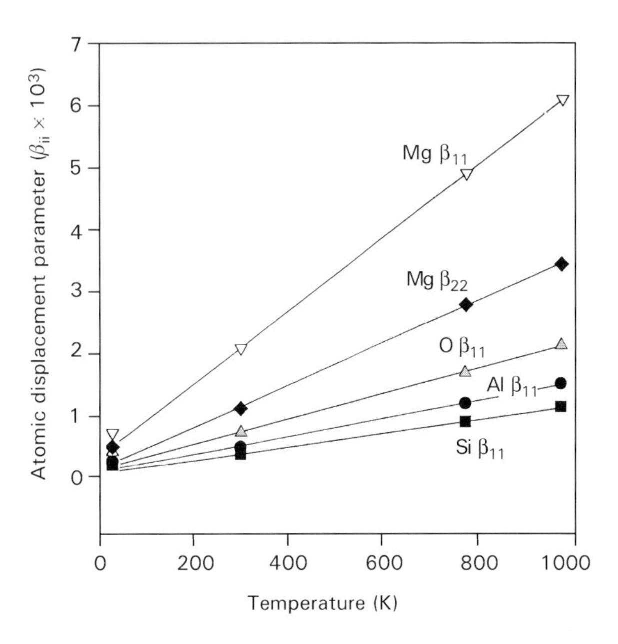

Figure 4 Temperature dependance of selected atomic displacement parameters for the Mg, Si, Al and O atoms in pyrope garnet. Analysis of the displacement parameters obtained from accurate single-crystal diffraction data allow evaluation of the zero-point energy contribution, separation of static and dynamic effects, and detection of anharmonic vibrational contributions to the atomic motion.

Physical analysis of crystals and crystal aggregates

The basic information concerning the nature of the phases present in a synthetic or natural material, and the ideal models of their crystallographic structures are frequently known. The information of interest for the characterization of the material and the understanding of its crystallochemical behaviour then reside in the physical status of the crystal or crystal aggregate. It is known that many material properties of interest in technological applications are directly related to what we might call the deviation from the 'ideal' crystal, i.e. the presence in the material of a variety of structural defects at the atomic level, such as doping elements, chemical modulations and stacking faults. These perturbations of the periodic crystal lattice produce effects of various kinds observable in the diffraction signal, i.e. diffuse scattering, superlattice reflections, commensurate or noncommensurate Bragg peak modulations, peak shifts, peak profile distortions, etc. A substantial part of X-ray crystallography deals with the measurement of the diffraction effects caused by lattice disorder, and their interpretation in terms of structural defects. The whole branch of 'quasicrystallography' is devoted to alloys and solid materials exhibiting diffraction effects incompatible with the standard three-dimensional space group symmetry.

A rapidly developing area of study is the measurement of lattice microstrain in polycrystalline aggregates. It is based on the interpretation of the angular dependence of the broadening of Bragg peak profiles in a powder diffraction pattern. The surface or volume mapping of the distribution of lattice strain gradients in the material yields a variety of information on the presence of residual stresses, developing tensions due to microcracking, local imperfections due to coprecipitated phases, and so on. Furthermore the analysis of the peak profile broadening as a function of crystallographic direction (anisotropy of the broadening effects) allows determination of the mean particle size and shape of the crystallites in the compound.

The analysis of the statistical or preferred orientation of the crystallites in solid polycrystalline materials is commonly referred to as texture analysis. Again, the diffraction technique allows the definition of the relationship between a microscopic property, i.e. the orientation of the crystallites defined as coherent diffraction domains, and the macroscopic physical properties of the crystal aggregate. Texture studies are of course crucial in the characterization of oriented synthetic materials such as cold-rolled metals or oxide thin films, but they are also of great relevance in the study of the formation processes of mineral assemblages. As an example, the texture features of olivine or pyroxene minerals in meteoritic chondrules yield information on the early condensation sequence and cooling history of the meteorite body they belong to. **Figure 5** shows the heavily textured diffraction pattern of chondrule from the Mafra L-4 chondritic meteorite collected using synchrotron radiation and a charge-coupled device (CCD) area detector. The pattern displays at the same time sharp diffraction spots produced by the olivine and pyroxene microcrystals present within the chondrule, and the inhomogeneous intensity distribution on the diffraction rings produced by the polycrystalline matrix. The example indicates how hard X-rays from synchrotron sources can effectively be used to investigate precious materials in a nondestructive way.

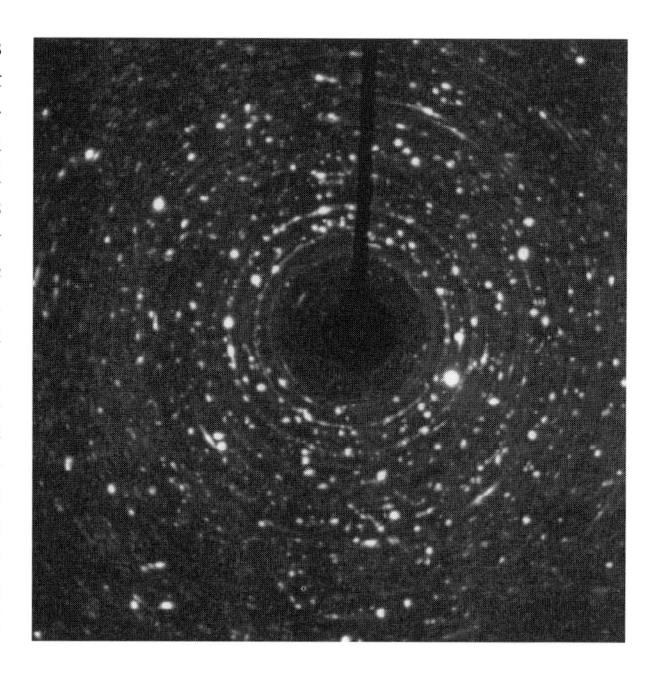

Figure 5 Textured diffraction pattern from a meteoritic chondrule collected using short wavelength synchrotron radiation (λ = 0.5 Å) and a CCD area detector. The image displays sharp diffraction spots from olivine and pyroxene single crystals, and powder diffraction rings from the polycrystalline matrix of the chondrule.

Synchrotron X-ray diffraction

The availability in recent years of the intense, collimated and energy-tunable X-ray beams emitted by synchrotrons and electron storage rings has had a spectacular impact on most fields related to materials characterization. As a consequence, a number of theoretical and instrumental aspects of spectroscopic techniques are rapidly developing in order to take full advantage of the peculiar properties of synchrotron radiation. All kinds of laboratory X-ray diffraction experiments described above can be performed at synchrotron sources as well, and the results are of significantly higher quality in terms of angle- or energy-resolution and signal-to-noise ratio. Furthermore, since the incident radiation beam is several orders of magnitude more brilliant than the laboratory X-ray beams, the experiments can usually be performed using a much smaller volume of sample (down to few μm^3) or, alternatively, by collecting diffraction data with extremely fast acquisition times (down to the ms time range). Finally, several completely new types of experiment, such as those outlined in the following sections, are only possible at synchrotron sources.

Dynamic diffraction

The study of the evolution of the diffraction signal in a sample during a phase transformation or a chemical reaction involving crystalline phases is defined as dynamic diffraction. The term implies that the sample is studied *in situ* under operating conditions of temperature, pressure, vacuum or controlled atmosphere; and that the diffraction pattern is sampled using acquisition times much shorter than the time the process takes to reach completion. Diffraction experiments under nonambient conditions can be performed on laboratory instrumentation only at thermodynamic equilibrium. Direct determination of the metastable and transient phases present during the reaction path, and measurement of fast kinetics are only possible if dynamic diffraction using synchrotron radiation is employed. Commonly the resolution in reciprocal space (defined as $\Delta d/d$, or the ability to distinguish two closely spaced Bragg diffraction peaks) is not a critical experimental parameter when studying the qualitative aspects of the phase transformations, or for the quantification of phase proportions during the process (for example in kinetics studies). On other hand, resolution is a crucial factor in the investigation of structural phase changes, where high-quality powder spectra collected in time-resolved mode are required.

A typical dynamic diffraction experiment involves (a) a high X-ray flux incident on the sample, (b) a conditioning chamber around the specimen controlling the environmental variables, and (c) a fast data acquisition system (i.e. position sensitive or area detectors). Depending on the require $\Delta d/d$ resolution, at synchrotron sources these experimental conditions are easily met either using angle-dispersive or energy-dispersive geometries, i.e. using monochromatic or polychromatic incident radiation, respectively. The flexible design of a synchrotron experiment also allows simultaneous *in situ* experiments to be carried out, such as XRD and small-angle X-ray scattering, XRD and differential scanning calorimetry, or XRD and X-ray absorption spectroscopy in fast scanning mode. Combined experiments of this kind provide a wealth of information, as they furnish the means of simultaneously investigating the process at different space- or momentum-transfer scales. For example, simultaneous time-resolved X-ray powder diffraction and X-ray absorption spectroscopy performed during a solid-state reaction catalysed by a specific chemical element can provide identification and quantification of all the crystalline phases involved, and at the same time also provide short-range information on the coordination environment of the catalyser element at each stage of the reaction.

Recent applications of the dynamic diffraction experimental techniques were mostly in the areas of temperature-induced dehydration processes in natural zeolites, hydrothermal crystallization of aluminosilicate and aluminophosphate microporous materials, thermal decomposition of layer silicate minerals, high-temperature synthesis of advanced ceramics, and hydrothermal ion-exchange and conversion processes in synthetic molecular sieves. The time-resolved powder diffraction patterns relative to the isothermal nucleation and growth process of zeolite-A by the hydrothermal treatment of activated metakaolinite is shown in **Figure 6**.

Although most of the cited studies only imply the qualitative analysis of the transformation, or the simple quantification of the crystalline phases involved in the process in order to properly extract the kinetic parameters, a number of full structural studies have been performed, which allow the complete determination of the structure response to the external gradient applied to the sample.

The case study of the temperature-induced phase transformation in the Ca-rich zeolite mineral garronite is illustrated here to show the large amount of detailed information to be gained by dynamic diffraction studies. The time evolution of the experimental powder diffraction pattern is shown in **Figure 7**. The image was obtained by synchrotron radiation using a gas flow heater, a polycrystalline sample in a glass capillary, and an image plate detector installed on a translating camera. Constant-temperature integration of the image pixels yield a number of powder diffraction histograms for the structure analysis. **Figure 8** shows the Rietveld analysed powder spectra of phase A at room temperature, and phase B at 170°C. Full structural analysis of all powder data sets produce a detailed picture of the structure response to dehydration, involving the positional shift and the change in coordination of the Ca cation in zeolitic channels, the progressive distortion of the aluminosilicate tetrahedral framework

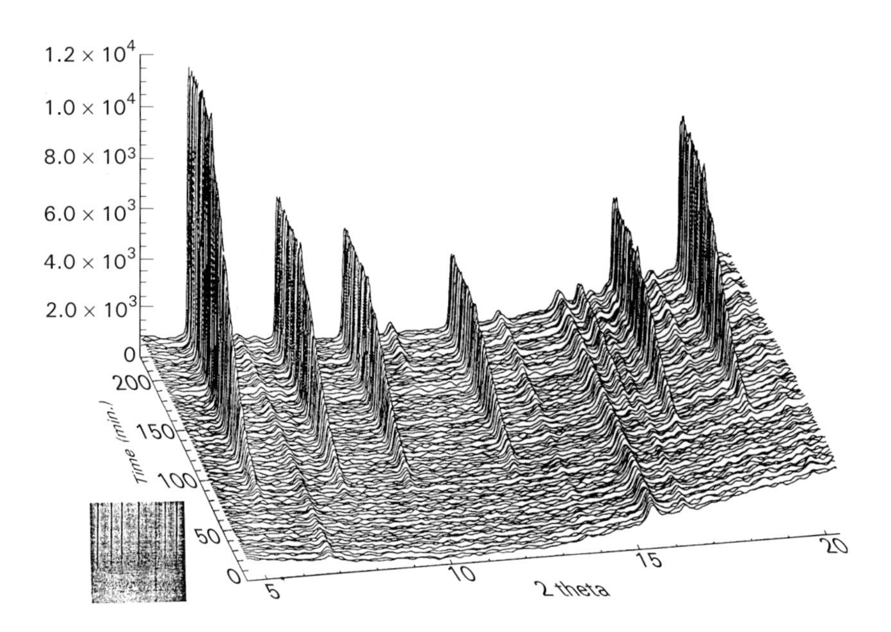

Figure 6 Time evolution of the powder diffraction pattern of zeolite-A crystallizing from activated metakaolinite under hydrothermal conditions. The original image plate picture is shown as a small figure at the bottom left.

inducing a space-group change from monoclinic to tetragonal, and ultimately the collapse and amorphization of the structure. **Figure 9** describes the progression of the process by the measured unit cell parameters and volume, and **Figure 10** shows the distortion of the zeolite channels caused by the contraction of the tetrahedral framework along the *c* tetragonal direction in phase B, which corresponds to the *b* monoclinic axis of phase A.

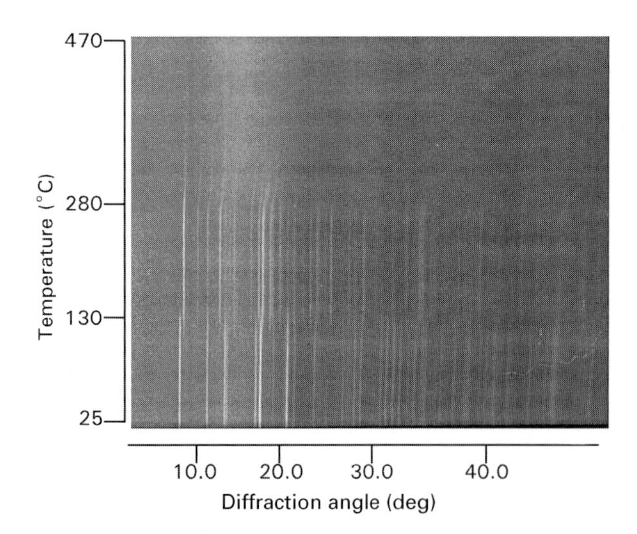

Figure 7 Evolution of the X-ray powder diffraction pattern of the natural zeolite garronite during the temperature-induced dehydration process. The date were collected using synchrotron radiation, a gas flow heater, a capillary sample, and an image plate detector installed on a translating camera.

High pressure studies

Small volume samples are required to reach ultrahigh pressures (i.e. above 10^2 GPa). Therefore, high-brilliance synchrotron radiation is needed to get a useful diffraction signal out of microsamples. This is the reason why the new techniques of high-pressure diffraction were mostly developed with synchrotron radiation sources. Physics, materials science, and especially mineral physics greatly benefited from such developments, because of the possibility of studying high-pressure-stabilized crystal structures using *in situ* diffraction techniques. Since the earth's and planetary interiors are not directly reachable for sampling, we need laboratory techniques to stabilize and characterize the materials satisfying the observed large-scale geophysical (density distribution within the planetary interior, seismic waves velocity) and geochemical (abundance of chemical elements in the solar system, element distribution within the planetary bodies) parameters governing the global planetary processes, such as the thermal flow, the mantle convection regime and the crust plate tectonics.

High-pressure mineral physics therefore is concerned with the characterization of the physical properties and stability fields of planetary materials under extreme pressure and temperature conditions. For example, conditions present at the centre of the earth are thought to be in excess of 300 GPa and 5000 K. Such conditions can at present be obtained by laser-heated diamond anvil cells or multianvil presses. Emphasis of current *in situ* diffraction

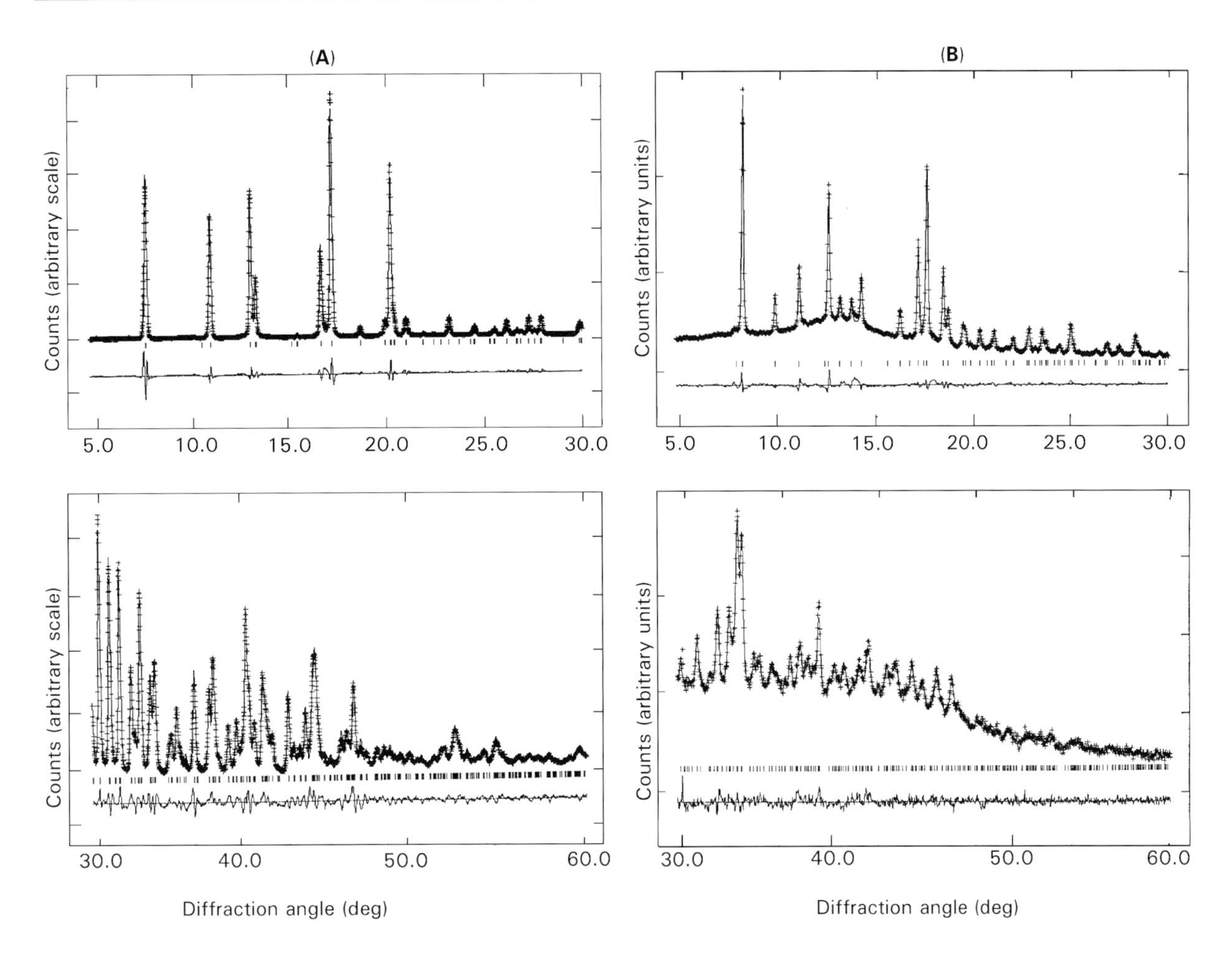

Figure 8 Selected portions of the Rietveld-analysed powder diffraction pattern of the zeolite garronite at 25 (A) and 170 (B)°C. The patterns were integrated from the image plate data shown in **Figure 2**. A full structure analysis of the zeolite as a function of temperature was performed.

studies using synchrotron radiation under extreme conditions is on the definition of the thermodynamic phase stability fields of the crystalline phases possibly present in the earth's mantle and core, on the experimental measurement of the equations of state in P–V–T space, and in the crystal structure determination of the high-pressure phases.

Resonant scattering experiments

A noticeable advantage of synchrotron radiation is the possibility of fine tuning the incident energy over a wide range, typically 5–50 keV for bending magnet radiation from high-energy third-generation synchrotrons (up to 100 keV for wiggler radiation from the same machines) and 2–20 keV for bending magnet radiation from low-energy third-generation synchrotrons. The available energy range covers absorption K-edges of chemical elements from P to the rare earth elements, and L-edges from Rb to the actinides. The wavelength tunability is not only of obvious advantage for X-ray absorption spectroscopy, but it also offers diffraction an increased flexibility through proper exploitation of the resonant scattering effect. The excitation of resonant scattering takes place when diffraction is performed using an incident energy close to that of an absorption edge of a chemical element present in the sample. Under these conditions the atomic scattering factor of the element is modified by a factor (called the anomalous dispersion correction) amounting to several electrons. It is obvious that by performing diffraction studies at different wavelengths in the vicinity of an absorption edge, it is possible to change to scattering contrast between the chemical elements present in the sample, and therefore increase the ability of diffraction to (1) determine the location of the particular element in the structure, or (2) quantify the statistical amount of that element in specific crystallographic sites.

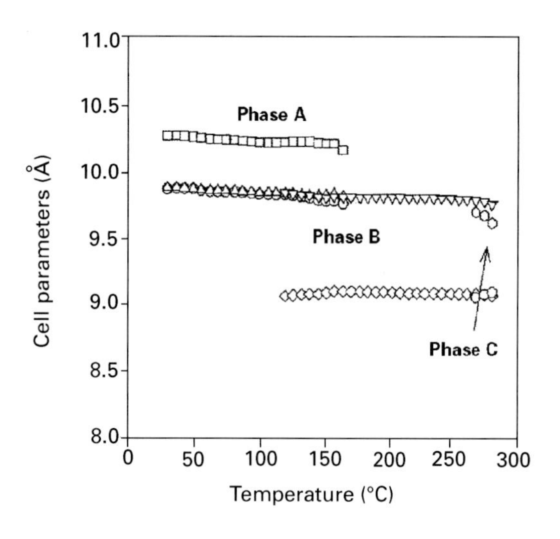

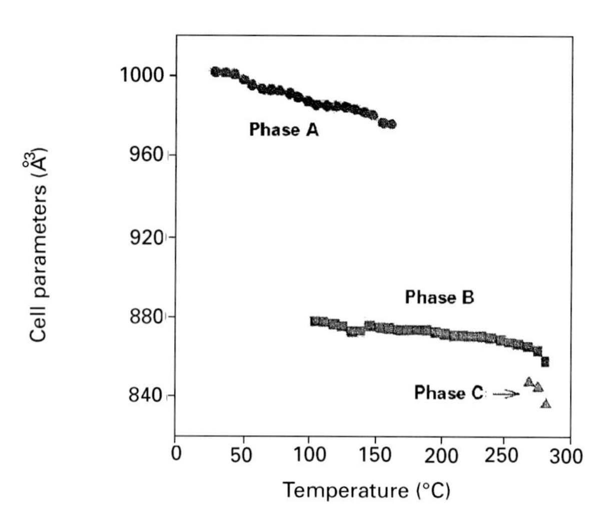

Figure 9 Structure evolution of the zeolite garronite during the dehydration process. The unit cell parameters and volume resulted from the Rietveld refinements of all powder diffraction data shown in **Figure 2**. Phase A is monoclinic *I*2/*a* (circles = cell *a*; squares = cell *b*; triangles = cell *c*), phase B is tetragonal $P4_12_12$ (triangles = cell *a* = *b*; diamonds = cell *c*) and phase C is also tetragonal $P4_12_12$ (hexagons = cell *a* = *b*; circles = cell *c*). It may be noted that the cell volume contraction during the monoclinic–tetragonal phase transition is related to the shortening of the monoclinic *b* axis alone (corresponding to the *c* axis of the tetragonal cell). The structure analysis indicates that the framework distortion is related to the change in bond coordination of the Ca cation in the zeolitic cages.

Since an energy shift of the absorption edge is observed depending on the oxidation state of the element, the accurate determination of the energy dependence of the anomalous dispersion term yields information on the valence state of the element present in a crystallographic site, much the same way this information is obtained from the chemical shift of the edge in X-ray absorption spectroscopy. However, the information obtained from absorption spectroscopy is averaged over the sample volume, and cannot discriminate between different valence states of the same element simultaneously present on different structural sites, whereas diffraction under resonant conditions intrinsically yields site-resolved information. The resonance effect is also exploited for the resolution of crystal structures, especially organic macromolecules, using the multiwavelength anomalous dispersion technique.

Diffraction exploiting the resonant scattering effect is generally used in inorganic chemistry and mineralogy to determine the cation partitioning of specific elements among different crystallographic sites, or to evaluate the oxidation state of chemical elements present in specific structural sites. A promising

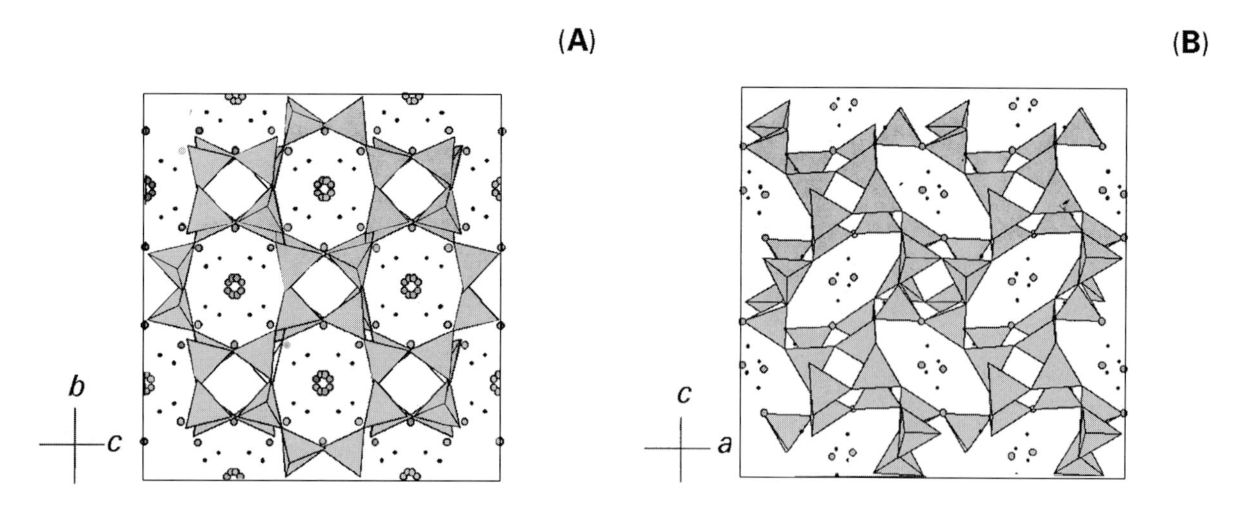

Figure 10 Projections of the crystal structures of garronite: (A) phase A at 25°C along the [100] monoclinic direction; (B) phase B at 170°C along the [010] tetragonal direction. The distortion of the zeolitic channels caused by the change in coordination of the Ca atoms is evident.

extension of the technique is called diffraction anomalous fine structure spectroscopy, and it involves measurement of the energy dependence of the integrated intensity of particular Bragg diffraction peaks across an absorption edge.

See also: **Fibres and Films Studied Using X-Ray Diffraction; Materials Science Applications of X-Ray Diffraction; Neutron Diffraction, Instrumentation; Neutron Diffraction, Theory; Powder X-Ray Diffraction, Applications; Quantitative Analysis; Small Molecule Applications of X-Ray Diffraction; Structure Refinement (Solid State Diffraction).**

Further reading

Bader RFW (1994) *Atoms in Molecules, a Quantum Theory*. Oxford: Clarendon Press.

Coppens P (1992) *Synchrotron Radiation Crystallography*. London: Academic Press.

Guinier A (1963) *X-ray Diffraction in Crystals, Imperfect Crystals, and Amorphous Bodies*. San Francisco: W.H. Freeman.

Hyde BG and Andersson S (1989) *Inorganic Crystal Structures*. New York: John Wiley.

ICSD, *Inorganic Crystal Structure Database. Karlsruhe*: FIZ Fachinformationszentrum.

Materlik G, Spark CJ and Fischer K (eds) (1994) *Anomalous X-ray Scattering. Theory and Applications*. Amsterdam: North Holland.

PDF, Powder Diffraction File. Swarthmore, PA: ICDD International Center for Diffraction Data.

Structure Reports (1940–1990) *Metals and Inorganic Sections*. IUCr International Union of Crystallography Geneva: IUCr; Dordrecht: Kluwer.

Wells AF (1984) *Structural Inorganic Chemistry*. Oxford: Clarendon Press.

Willis BTM and Pryor AW (1975) *Thermal Vibrations in Crystallography*. Cambridge: University Press.

Young RA (ed) (1993) *The Rietveld Method. IUCr Monographs on Crystallography*. Oxford: Oxford University Press.

Zevin LS and Kimmel G (1995) *Quantitative X-ray Diffractometry*. New York: Springer-Verlag.

Inorganic Condensed Matter, Applications of Luminescence Spectroscopy

Keith Holliday, University of Strathclyde, Glasgow, UK

ELECTRONIC SPECTROSCOPY
Applications

The Interaction of impurities in condensed matter with light

The optical spectroscopy of impurities in solids and liquids is similar to that of free particles except that interpretation requires an extra interaction to be taken into account, i.e. the influence of the host on the energy levels of the impurity. In other words, the presence of the impurity particle in a particular environment adds an extra term to the Hamiltonian: the crystal field.

There are two components of the crystal field; static, referring to the average position of nearest neighbour particles, and dynamic, referring to their motion. The static crystal field shifts energy levels and lifts degeneracies to cause splittings whilst the dynamic crystal field allows coupling of electronic energy levels to vibrations. In terms of luminescence spectroscopy, the static crystal field is principally responsible for shifting emission wavelengths whilst the dynamic component broadens emission lines into bands of widely varying widths and shapes. Both static and dynamic effects can influence the oscillator strength of the transition and thus the luminescence lifetime. Excited state lifetimes can also be influenced by factors such as nonradiative decay and concentration quenching.

Insulators provide a useful trap for optically active ions due to their transparency in the visible region. Nuclear (vibrational) resonances with electromagnetic radiation take place in the infrared part of the spectrum whilst resonances with bound electrons (from the valence band to the conduction band) are usually in the ultraviolet. Visible radiation therefore interacts with the impurities alone. The role of the host is simply to influence the optical behaviour of the impurity.

Spectroscopic characteristics of impurities in condensed matter

Lanthanide ion doping

The optically active electrons in lanthanide ions are 4f electrons and they are shielded from the influence of the crystal field by the complete 5s and 5p subshells that surround them. Therefore, the static crystal field has only a small effect on the energy levels of lanthanides and is treated as a perturbation to the free ion states. This perturbation is significant, however. Electric dipole transitions between pure 4f states are parity forbidden (though some of the much weaker magnetic dipole transitions are allowed). The addition of the crystal field term causes weak mixing of 4f states, most importantly with 5d states, so that some electric dipole transitions become weakly allowed. Reducing the symmetry of the crystal field tends to cause greater mixing thus making the optical transitions more intense. At the same time, the *J*-degeneracy of the state is lifted so that the luminescence spectrum of a particular transition appears as a number of closely separated lines such as those shown in **Figure 1A**. The dynamic crystal field is also weakly interacting with lanthanide ion impurities so that phonon-assisted transitions are rarely significant features.

Luminescence studies are particularly useful for lanthanide-activated matter as transitions are often weak thus rendering absorption difficult to measure. Each lanthanide ion has a unique energy level structure so that excitation of such a material will result in a luminescence spectrum characteristic of that impurity ion. Only a small proportion of energy levels luminesce, however. Most decay nonradiatively through multiphonon emission to states at slightly lower energies. The excited electron rapidly tumbles between states until it reaches one that is separated by more than about 1000–3000 cm^{-1} (1 eV ≈ 8000 cm^{-1}), depending on the system, from the next lowest level. The probability of multiphonon decay is an exponential function of separation so that states with wide gaps to next lowest levels are unlikely to decay nonradiatively. Luminescence is thus emitted from such states. As an example, the energy structure of Nd^{3+} is shown in **Figure 1B**.

Transition metal ion doping

The active electrons in transition metal ions are not shielded by outer subshells and this leads to a much

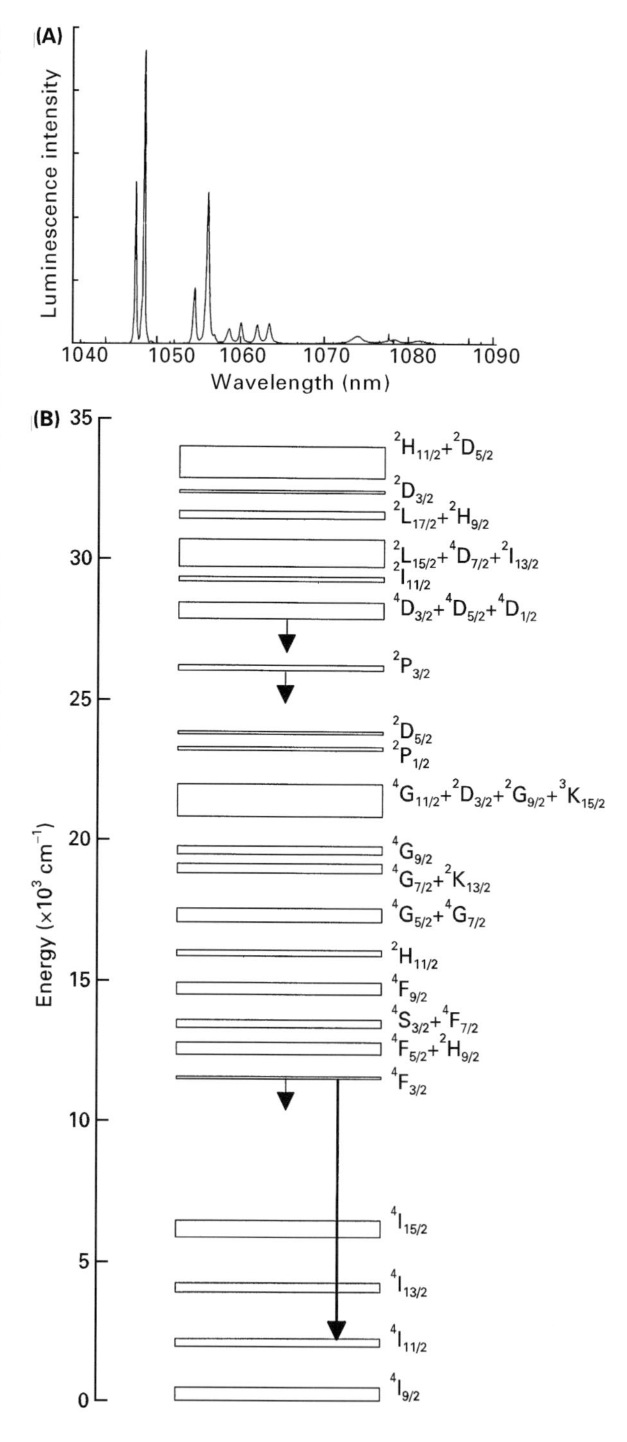

Figure 1 Luminescence from a typical lanthanide-doped crystal, Nd^{3+}-doped $KLiYF_5$. The $^4F_{3/2} \rightarrow {}^4I_{11/2}$ luminescence multiplet, most often used as the gain transition in Nd-based lasers, is shown in (A) and the energy level diagram is shown in (B). The total Nd^{3+} multiplet splittings for this host are indicated by the thickness of the boxes, levels that radiate significantly are denoted by small arrows and the $^4F_{3/2} \rightarrow {}^4I_{11/2}$ transition is indicated by a long arrow. This luminescence multiplet would normally be expected to be split into no more than six components but the 12 observed indicates that Nd^{3+} substitutes into two different sites in this crystal. The similarity of the spectra due to different site substitution is typical for lanthanides and the energy level diagram in (B) varies very little from host to host.

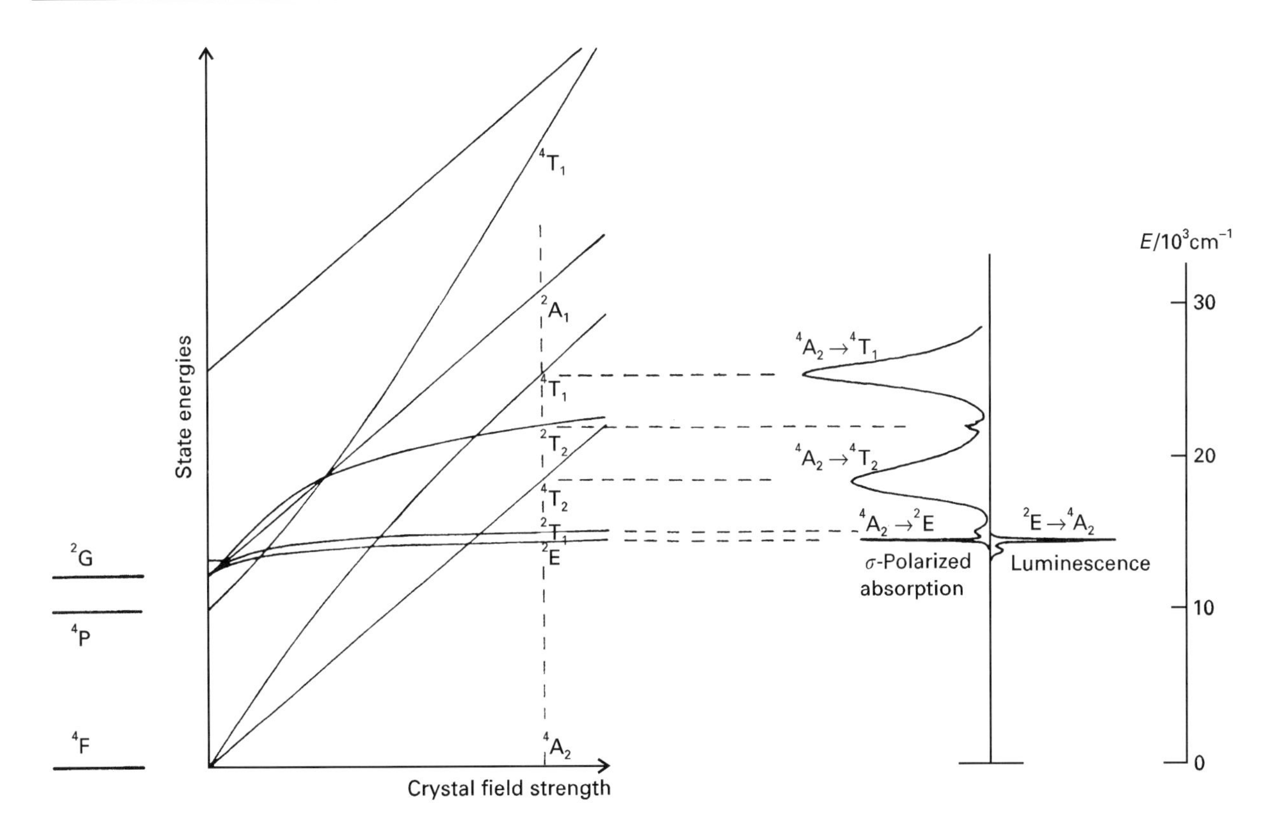

Figure 2 A Tanabe–Sugano diagram for a $3d^3$ impurity ion such as Cr^{3+}. The octahedral crystal field increases along the *x*-axis and the energy levels split and shift as indicated. For zero crystal field the free ion LS states are indicated on the left. A dashed line indicates the crystal field strength relevant to ruby. The energy levels indicated appear in the absorption spectrum shown on the right. Luminescence is only from the lowest lying excited state, ^2E. For materials that provide weaker field environments the $^4T_2$ state is lower and much broader bandwidth emission is obtained. Reproduced with permission of Oxford University Press from Henderson B and Imbusch GF (1989) *Optical Spectroscopy of Inorganic Solids.* Oxford: Clarendon Press.

stronger interaction with the crystal field. The Hamiltonian contains the same components as for lanthanides but now the crystal field cannot be treated as a perturbation to the free ion states as this term is similar in strength to the electron–electron interactions and stronger than spin–orbit coupling terms. The resulting wavefunctions are denoted by a nomenclature that derives from group theory.

Plots, known as Tanabe–Sugano diagrams, show how the state energies of transition metal ions vary as a function of crystal field strength. From these plots absorption and luminescence characteristics can be predicted or explained. **Figure 2** shows the Tanabe-Sugano diagram for Cr^{3+}. The dashed line indicates the crystal field strength for aluminium oxide ($Al_2O_3:Cr^{3+}$ is ruby) and it indicates where the excited state energy levels lie relative to the ground state ($^4A_2$). Absorption bands appear resonant with the energy levels, broad in the case of states that are strongly vibronically coupled and narrow for weak electron–phonon coupling. The emission spectrum consists of a single line with some weak vibronically assisted structure regardless of which absorption transition is excited. This is because, as for lanthanides, rapid nonradiative decay prevents luminescence from states that have other levels at energies close below. In this case, narrow band luminescence is seen (the ruby laser transition) but for materials that provide a weaker crystal field environment for Cr^{3+} ions, the $^4T_2$ state may be at a lower energy than the ^2E state making it the emitting level. Much stronger electron–phonon coupling occurs for this transition (as for the reverse $^4A_2 \rightarrow {}^4T_2$ (absorption) transition indicated in **Figure 2**) and broadband emission results.

Colour centres

Colour centres (or F centres) result from the presence of defects in the structure of a solid, often intentionally induced, for instance by high-energy-photon irradiation. Such treatment causes disruption to the lattice and leads to the production of particular defects. For instance, a single electron can become trapped in an anion vacancy. The Hamiltonian for such a system is considerably different from that of an impurity ion but the resulting eigenenergies give rise to transitions in the optical region of the spectrum. Such a centre is influenced very strongly by the crystal field and so transitions are always strongly coupled to vibrations resulting in very broad absorption and luminescence spectra (~1000–4000 cm^{-1}). There is a wide range of different optically active defects and determining the structure of colour centres is difficult using only optical techniques. However, by applying external fields and stresses and making measurements of properties such as polarization and magnetic circular dichroism much can be understood. Of great assistance are magnetic resonance measurements, in particular electron paramagnetic resonance. Such techniques reveal information on the number of electrons present and the symmetry of the colour centre from which structure can be deduced.

Crystals, glasses and liquids as hosts

It might be expected that the various types of condensed matter host will give rise to quite different optical properties for the impurities that they contain. In fact, the impurity has a much stronger influence than the host. This is especially true for lanthanide ions. These are not greatly influenced by their surroundings so that a change of environment does not make very much difference to optical behaviour. Splittings between *J*-levels will vary between hosts and transition strengths will also vary whilst remaining relatively weak.

Transition metal ions are more strongly influenced by environment, for instance Cr^{3+} ions decay via entirely different transitions depending on crystal field strength. Generally, transition metal ions will have an energetically preferred orientation. The energy of solvation of the Cr^{3+} ion is lowest when octahedrally coordinated, in solids and liquids. In aqueous solution six water molecules are arranged about the chromium ion. In glasses lack of long-range order means that the solvation energy of the Cr^{3+} ion can be minimized without influencing the overall structure of the host so that, again, Cr^{3+} ions become octahedrally coordinated, this time surrounded by the dominant anions (often fluoride or oxide ions). In crystals the situation is somewhat different. The lattice energy of the host is minimized by long-range order so that considerations of the solvation energy of the Cr^{3+} ion are of much less significance. If there is a suitable octahedral site (or somewhat distorted octahedral site) for the Cr^{3+} ion then substitution can occur. If not, then the chromium ion may enter as an alternative valence (e.g. Cr^{4+} in a tetragonal site) or may not enter the lattice at all. In fluids and glasses it is more likely that a transition metal ion can find its energetically preferred orientation due to the impurity ion's greater influence on its own neighbourhood. In crystals, transition metal ions often enter more distorted sites according to availability and this sometimes leads to more unusual optical properties such as large energy level splittings or transition strengths.

The main difference between spectra obtained in different classes of hosts is in the broadening of features. In a perfect crystal every impurity ion would experience the same potential due to neighbour particles so that each ion would behave in a spectroscopically identical way. Spectra would not be broadened, other than by homogeneous and vibrational effects, and would consist of very narrow zero phonon lines and phonon side bands. Perfect crystals do not exist and inhomogeneous broadening due to variation in crystal field contributions to the Hamiltonian is always present (see **Figure 3**). This is not always apparent at room temperature when homogeneous broadening dominates in most systems but below about 100 K, in solids, inhomogeneous broadening has a significant influence. In glasses and liquids inhomogeneous broadening is usually more significant. The local environment of the impurity ion is influenced less strictly by the host and the lack of order leads to a much greater range of environments and thus greater broadening of spectral features.

Lanthanide doped crystals typically give rise to luminescence multiplets with spreads in energy of around 100–1000 cm^{-1}. Line widths of individual features at low temperatures tend to be around 100 times narrower than this. In glasses and liquids, however, the effect of variation in environment causes luminescence to be emitted in bands that encompass the whole of the multiplet and individual lines cannot usually be resolved. Such bandwidths are not greatly broadened beyond the multiplet widths of crystals, however.

The zero phonon lines of transition metal ions behave similarly to those of lanthanides but phonon side bands are more evident. Such is the extent of broadening due to electron–phonon coupling that the influence of environment is proportionally much smaller. It is unusual to see disorder-induced broadening of phonon-assisted bands by more than a factor of two.

The greatest broadening effects are actually observed in substitutionally disordered crystals. Substitutional disorder in crystals occurs when two or more constituent ions have interchangeable lattice positions. Whilst the stoichiometry and periodicity of the crystal is maintained, the local environment varies throughout the structure. Optically active impurity ions experience different crystal fields as a

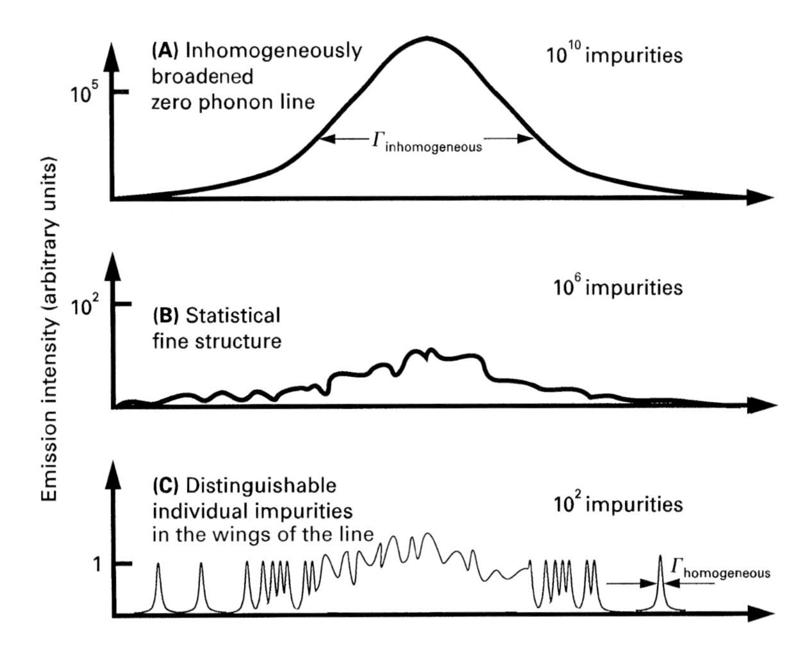

Figure 3 A typical zero phonon luminescence line from a bulk sample would appear as in (A) with a typical inhomogeneous width (Γ) of 1–10 cm^{-1}. As the number of emitting impurities decreases, the smooth line shape typical of a large distribution is lost (B) and eventually the contributions of individual homogeneously broadened contributions appear in the wings (C). The homogenous width is exaggerated by a factor of about 10^4 for temperatures of less than about 4 K.

result of this local variation. Inhomogeneous broadening through substitutional disorder is strongest when it is the nearest neighbour ions that are randomly occupied. For instance, the $^7F_0 \rightarrow {}^5D_0$ transition of Sm^{2+} in BaFCl is broadened by a factor of 30 at low temperatures through the addition of bromine to the melt to produce $BaFCl_{0.5}Br_{0.5}$. Such is the magnitude of the effect that inhomogeneous broadening dominates homogenous broadening up to temperatures in excess of 300 K. This discovery prompted a successful search for room temperature persistent spectral hole-burning materials (see below).

Techniques used to study optically activated condensed matter

Very simple methods can be used to obtain useful information on the optical behaviour of impurity doped condensed matter. The first step in the characterization of such a material is usually to measure its absorption spectrum in the region of about 0.3–1 μm. Such measurements are often sufficient to identify the active impurities and to provide some information on their environment. For crystalline hosts, the polarization of absorption spectra can be analysed to draw conclusions on site symmetries and crystal field parameters (see **Figure 4**).

As well as ground state absorption measurements, it is often also useful to know the absorption characteristics of excited metastable states, particularly for potential laser gain media. Such measurements are considerably more difficult than ground state measurements because it is necessary to maintain a significant excited state population. The experimental arrangement required to obtain such spectra requires two excitation sources, a powerful beam to excite the material and a weaker probe to measure the absorption immediately after the excitation pulse.

There are some materials for which even ground state absorption is too weak to measure. A method that provides equivalent information is zero-order excitation spectroscopy. Emission at all wavelengths is directed to a detector without dispersion. A time delay is required to prevent excitation light from swamping the signal. Ideally, this is achieved using a pulsed laser and gated detector but other systems also work, such as asynchronous choppers in a continuous beam. By varying the wavelength of the exciting beam and recording the strength of emission at all wavelengths a spectrum equivalent to that of the absorption of all luminescent centres is recorded.

The zero-order excitation spectrum reveals the wavelengths at which the material can be excited to obtain luminescence. The luminescence spectrum can be recorded by exciting at one of these wavelengths and dispersing the emission using a monochromator. When emission is weak a chopper can be placed in the excitation beam to produce a modulated intensity that a lock-in amplifier can detect.

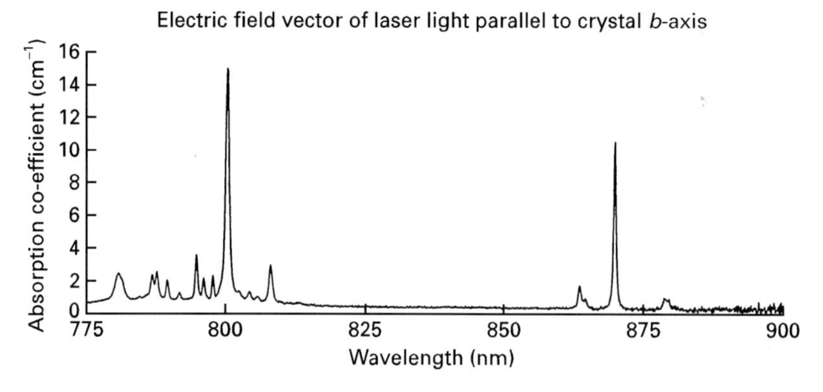

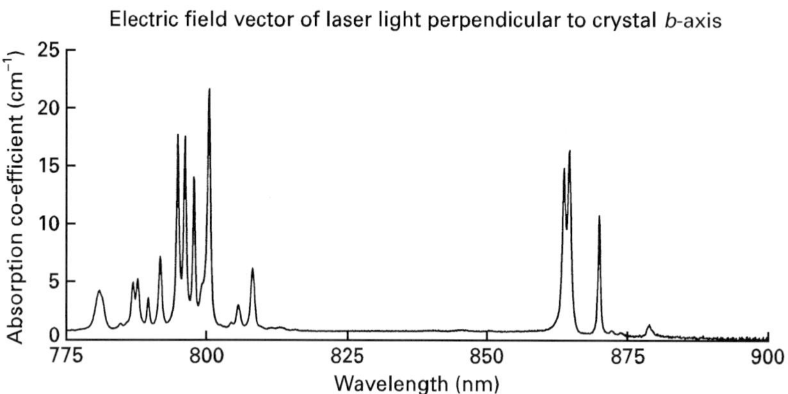

Figure 4 The absorption spectrum of the $^4I_{9/2} \rightarrow {}^4F_{5/2} + {}^2H_{9/2}$ and $^4I_{9/2} \rightarrow {}^4F_{3/2}$ transitions (see **Figure 1B**) in Nd^{3+}-doped $KLiYF_5$ measured in two different polarizations at 77 K. Similar variations in intensities within multiplets are observed in luminescence spectra.

Generally, excited ions step between energetically nearby states nonradiatively until a large gap is encountered from which radiative transitions occur. In a crystal in which all impurity ions occupy equivalent sites similar spectra are obtained regardless of which absorption transition is excited, unless there are two or more widely separated emitting states. In the latter case a different luminescence spectrum is obtained for each emitting state when excitation is resonant with levels between it and the next highest emitting state (see **Figure 1**).

Occasionally, the same impurity ion may substitute into two or more crystallographically different impurity sites within a single host. Each will normally have its own luminescence characteristics so that excitation at particular wavelengths may excite one set of ions and not the other when absorption bands do not overlap. This occurs most often for materials with narrow band absorption spectra, in particular lanthanides, and different luminescence spectra will result depending on the selected absorption feature (see **Figure 5**). Energy transfer between sites may cause both sites to emit under some circumstances.

In hosts in which a distribution of sites exist, shifts in emission bands result from excitation at different energies within an absorption band. The site distribution may be broadened by variation in crystal field strength or in electron–phonon coupling strength and measurements of the shift of the emission band as a function of excitation energy allow the contributions of each distribution to be quantified. **Figure 6** shows the luminescence spectrum of Cr^{3+}-doped strontium gallogermanate, a substitutionally disordered crystal that gives rise to luminescence via both the $^2E \rightarrow {}^4A_2$ and $^4T_2 \rightarrow {}^4A_2$ transitions due to a large variation in crystal field strength that causes the emitting state to vary between sites (see **Figure 2**).

Selective excitation techniques may be used to make high-resolution measurements. At low temperatures where inhomogeneous broadening dominates, it is possible to use a highly monochromatic excitation source such as a single frequency dye laser to select a subset of impurities, each of which contributes to the same portion of a spectral line. The resolution of spectral measurements is then, in principle, limited by the homogeneous line width. Below 10 K this is typically of the order of 10^6 times smaller than the inhomogeneous line width.

In the absence of energy transfer, exciting a subset of impurities will result in luminescence from only this subset. Single mode dye lasers can routinely

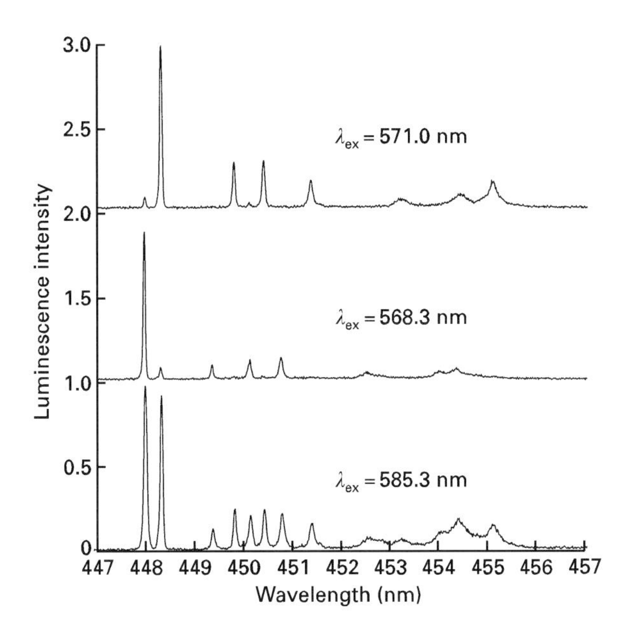

Figure 5 Selective excitation spectra for the $^2P_{3/2} \rightarrow {}^4I_{13/2}$ transition in Nd^{3+}-doped $KLiYF_5$. The top two spectra are produced by exciting the material at wavelengths that result in luminescence from only one of the two Nd^{3+} substitutional sites whereas the bottom spectrum is excited at a wavelength that excites both. This is an example of upconversion as the exciting photon energy is lower than the output photon energy. Site selection in this case occurs during excited state absorption from the $^4F_{3/2}$ state. Reproduced with permission of Elsevier Science from Russell DL, Henderson B, Chai BH, Nicholls JFH and Holliday K (1997) *Optics Communications* **134**: 398–406.

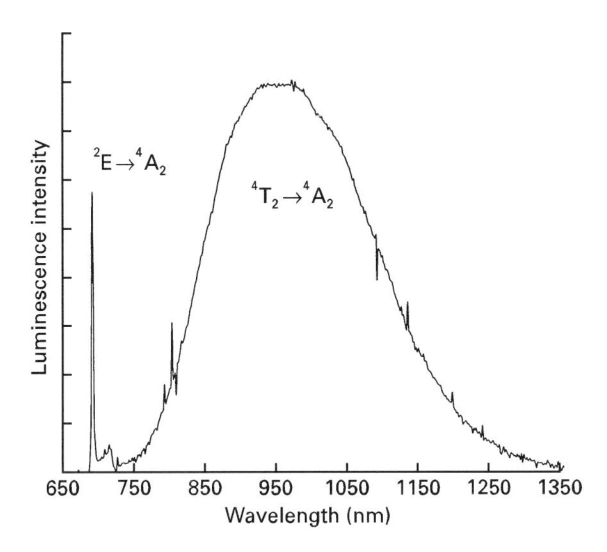

Figure 6 The luminescence spectrum of Cr^{3+}-doped $Sr_3Ga_2Ge_4O_{14}$ which decays via two different transitions due to the large variation in crystal fields experienced by the Cr^{3+} ions (see **Figure 2**). The structure in the $^4T_2 \rightarrow {}^4A_2$ band is instrumental noise.

provide excitation line widths of 1 MHz and solid state tunable lasers can do considerably better. A typical monochromator or scanning Fabry–Perot etalon cannot reach such resolutions and so the detected emission is limited by the resolution of the dispersion apparatus. Nevertheless, fluorescence line narrowing (see **Figure 7E**) can reveal information on, for instance, ground state splittings, that are normally hidden beneath the inhomogeneous broadening of the transition. Fluorescence line narrowing is usually employed on purely electronic transitions but is quite generally applicable to these.

For certain materials the excitation process results in a subsequent decrease in absorption at the same wavelength. This selective bleaching may persist indefinitely and can occur through a variety of mechanisms, all of which rely on the removal of a proportion of absorbing impurities from resonance. The spectral hole line shape may then be recorded by scanning the excitation wavelength and the spectral resolution is determined by the laser line width alone, making possible measurements of homogenous line widths. Persistent spectral hole-burning (see **Figure 7C**) is not a universal phenomenon but does occur for a wide range of materials. Amongst the most common mechanisms for impurities to be removed from resonance are oxidation (e.g. $Sm^{2+} \rightarrow Sm^{3+}$), motion of the impurity ion within the lattice (or alternatively a redistribution of the local environment about the impurity ion) and redistribution of ground state nuclear spins. The persistence of the spectral hole allows perturbation experiments (e.g. Stark effect, Zeeman effect, stress effects) to be undertaken with much greater sensitivity than is possible when observing the effects of very high fields on entire inhomogeneously broadened spectral lines.

The ultimate way to unambiguously examine the behaviour of impurities and defects in host matrices is to detect single chromophores (see **Figure 3**). The principle underpinning this luminescence excitation technique relies, like spectral hole-burning, on the reduction of the homogenous line width of the zero phonon electronic transition with decreasing temperature. Inhomogeneous broadening is insensitive to temperature and so the overlap of spectral lines from individual chromophores decreases with decreasing temperature. If the sample is then diluted until the concentration is sufficiently low, the superposition of absorption contributions is no longer smooth. This is known as statistical fine structure. As the sample is diluted further, single chromophores will have no effective absorptive spectral overlap with other chromophores in the wings of the absorption line and any emission detected will correspond to this single atom, ion, molecule or defect centre. With continuing dilution, single chromophores may be detected closer to the inhomogeneous line centre. To date, this technique has only been

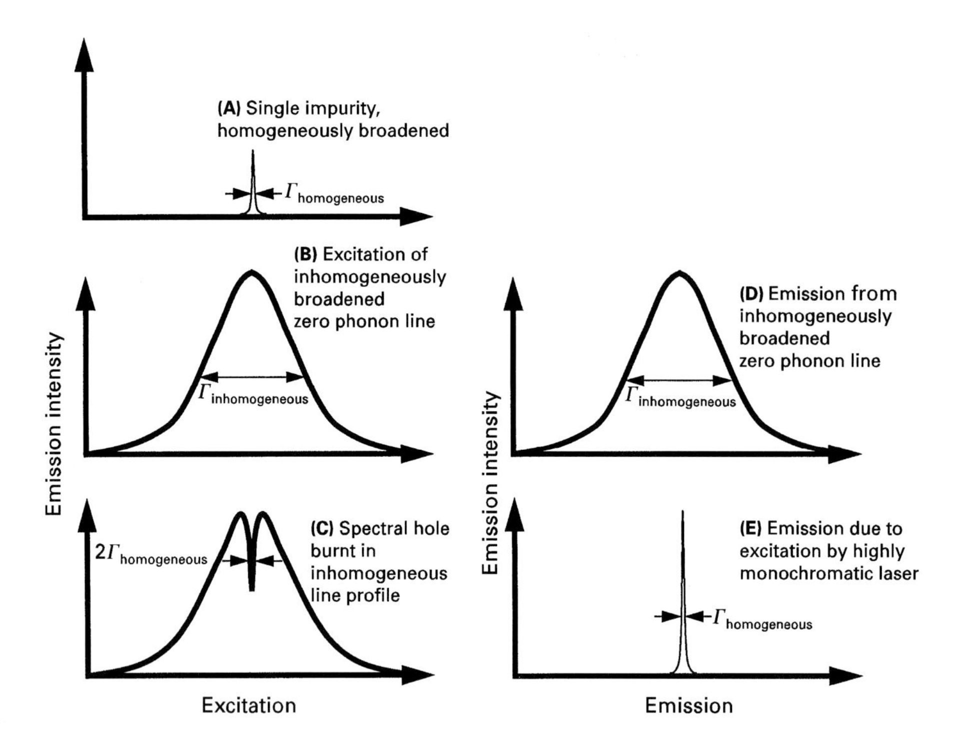

Figure 7 Schematic illustration of two high-resolution laser techniques. On the left is a technique that relies on excitation (or absorption) spectroscopy, spectral hole-burning, which is therefore instrumentally limited only by laser line width. On the right, fluorescence line narrowing is indicated, a technique that involves recording an emission spectrum which is therefore limited both by laser line width and dispersion resolution.

applied to dye molecules in organic hosts but could in principle be applied to any impurity doped solid that has a zero phonon electronic transition.

Luminescence phenomena studied in optically activated condensed matter

The interaction between impurity (or defect) and host determines the energy levels of the optically active species and so the most obvious application of luminescence spectroscopy to condensed matter is determination of site structure. The magnitude of the splittings of electronic levels is determined by the strength of the crystal field and the extent to which degeneracies are lifted is determined by the site symmetry of the crystal field. Simple absorption, excitation or luminescence spectra can thus reveal such information through application of group theoretical analytical techniques and this can be enhanced with data from perturbation experiments. Clearly these methods are most useful for materials in which the electron–phonon coupling is weak so that zero phonon lines dominate and hence lanthanide (and some transition metal) doped crystals are most appropriate for such a study. Doping impurities into hosts is not a sensible method to study the pure host as the presence of the impurity can distort the local lattice structure, but more general conclusions on the host (such as the extent of crystallinity) can be reached by studying parameters such as the broadening of transitions.

When electron–phonon coupling is strong, so that zero phonon lines are weak, other techniques can be used to determine information about site structure. Polarized measurements of features in absorption, excitation and luminescence spectra provide direct evidence of the symmetries of the wavefunctions involved in these transitions and this leads to information on site structure. Usually this information is best used in conjunction with other spectroscopic techniques such as electron paramagnetic resonance (detected either directly or optically). The form of a phonon-assisted sideband can itself reveal information on site structure, especially the extent of the local deformation that occurs when the impurity is excited to a higher electronic state. **Figure 8** compares the luminescence spectra of Cr^{3+} impurity ions in $LiCaAlF_6$ and $LiSrAlF_6$, crystals that are isomorphous but with different until cell sizes. The decreased bandwidth and increased structure associated with the $LiCaAlF_6$ spectrum indicates weaker electron–phonon coupling and a smaller excited state deformation. Again, group theoretical analysis can be applied to such results to obtain detailed information. For

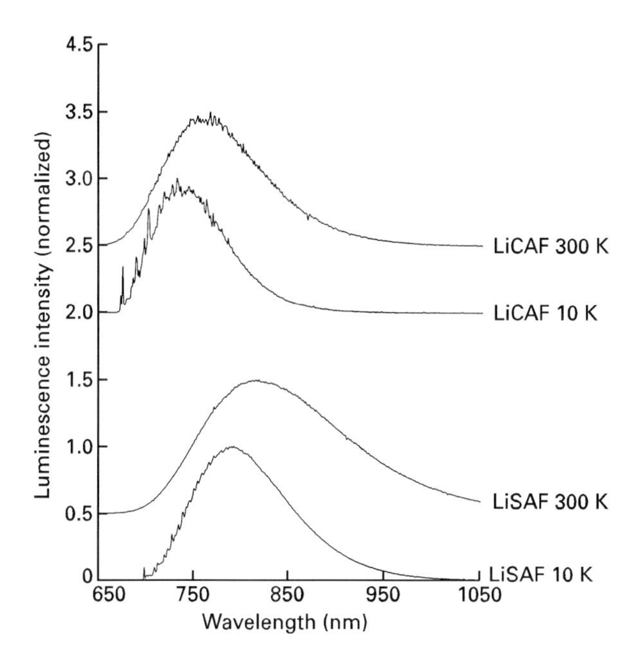

Figure 8 Luminescence spectra of the $^4T_2 \rightarrow {}^4A_2$ transition at indicated temperatures in Cr^{3+}-doped $LiCaAlF_6$ (LiCAF) and $LiSrAlF_6$ (LiSAF), host crystals that are isostructural but with slightly different unit cell sizes.

$LiCaAlF_6$ specific phonon transitions can be identified at low temperatures. Application of polarized spectroscopy to these features and comparison with Raman spectra can provide further information on the details of electron–phonon coupling and site structure.

Upconversion refers to a process in which a material emits photons of higher energy than those with which it is excited (see **Figure** 5). It is a phenomenon most commonly associated with lanthanide-doped materials and can take place through a variety of mechanisms including cross relaxation, energy transfer, excited state absorption and two-photon absorption. Studying the dynamics of such systems using time resolved techniques is an excellent method for exploring these processes within solids. Variation of doping levels and pump intensities help to determine exactly which processes dominate and the way in which they operate.

Fluorescence line narrowing is also a useful technique for studying energy transfer, in particular when measurements are time resolved. When a subsection of the impurities in a material is excited via energy selection it is the same impurities that luminesce, except when energy transfer takes place. If time-resolved spectra are taken the rate at which energy transfer processes progress can be determined as the line broadens. The line narrowing that results from energy selection can also be used to resolve small splittings in either the ground or excited state of the material. Cr^{3+}-doped strontium gallogermanate provides a nice example of this. **Figure 9** shows spectra obtained by exciting the $^4A_2 \rightarrow {}^2E$ electronic transition and recording emission from the reverse transition. By using relatively low resolution dispersion (a monochromator) the relatively large excited state splitting (~ 20 cm^{-1}) can be measured, but when detecting using a Fabry–Perot etalon the ground state splitting (~ 2 cm^{-1}) is revealed.

Spectral hole-burning has been used to investigate a wide variety of phenomena that includes impurity and defect structure, electron–nuclear interactions, spin-spin cross relaxation, tunnelling processes, optical dephasing, spectral diffusion in glasses, electron–phonon coupling and the behaviour of special materials such as biological compounds (involved in photosynthesis for example) and thin films. Much fuller details are given elsewhere in this encyclopedia.

A range of phenomena can be studied in solution, though generally only the simpler techniques discussed above are available due to the dynamics of the liquid phase. Measurements of luminescent lifetime can be interpreted in terms of the coordination number of the optically active solvated ion. Quenching and sensitizing of luminescence by the manipulation of the solution (e.g. addition of other species, change of acidity, etc.) is another common area of investigation. Generally, such effects are chemical in nature. The luminescent ion forms a bond, either loosely with surrounding solvent molecules or more strongly with ligand species. Quenching or sensitization takes place depending on the interaction between the optically active ion and its immediate surroundings. Much of this area of study is related to organic chemistry in that large organometallic ions are formed in solution and it is their behaviour that is being probed in optical experiments. This is discussed further elsewhere in the encyclopedia.

Technological applications of optically activated condensed matter

To investigate materials such as laser gain media only simple experimental techniques are usually required. To evaluate the potential of a new material in a technological application it is often sufficient to measure absorption and emission spectra along with luminescence lifetimes, though it is desirable to also have an indication of relevant loss mechanisms. These include nonradiative decay, excited state absorption and energy transfer. The result of significant nonradiative decay is indicated by variations in luminescence intensity and lifetime as a function of temperature

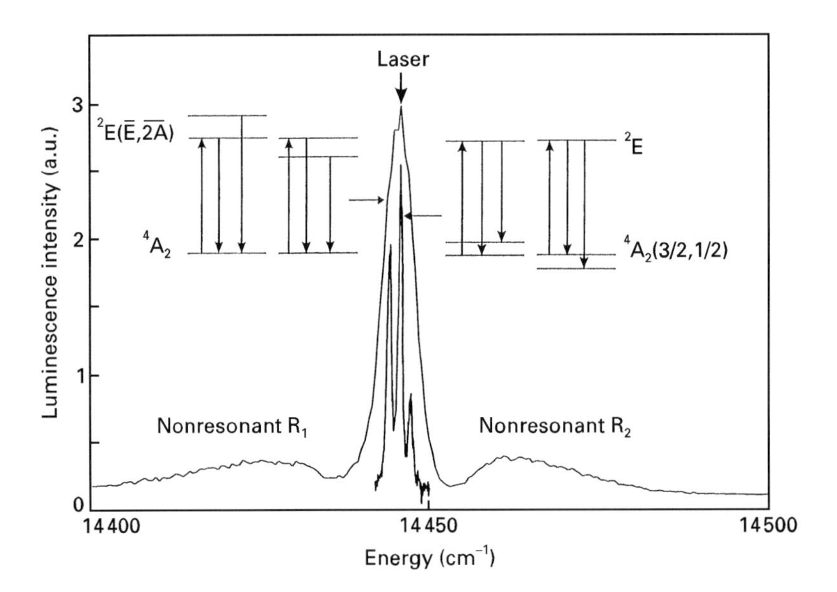

Figure 9 Fluorescence line narrowed spectra of the $^2E \rightarrow {}^4A_2$ transition in Cr^{3+}-doped $Sr_3Ga_2Ge_4O_{14}$ (see **Figure 6**). When dispersing emission using a monochromator (long spectrum) the excited state (^2E) splitting of ~20 cm^{-1} is observed. When dispersing emission using a Fabry–Perot etalon (internal spectrum) the ground state ($^4A_2$) splitting of ~2 cm^{-1} is observed. Reproduced with permission of the American Physical Society from Grinberg M, Macfarlane PI, Henderson B and Holliday K (1995) *Physical Review* B **52**: 3917–3929.

whereas energy transfer can be investigated using techniques such as time-resolved fluorescence line narrowing and upconversion. The results of studies of laser gain media and related materials has led to the concept of 'crystal field engineering' where the ideal combination of impurity and crystal field for any particular application can be calculated and then created through careful choice of host. In reality there is always feedback between spectroscopic interpretation and technological implementation.

The best known materials that have been developed over the last four decades are for implementation in lasers, either as gain media or as saturable absorbers. The first laser gain medium was ruby (see **Figure 2**), providing single wavelength output in the visible part of the spectrum. Neodymium based lasers have now largely replaced ruby as the industrial laser of choice. The $^4F_{3/2} \rightarrow {}^4I_{11/2}$ electronic transition (see **Figure 1**) provides laser emission just beyond 1 μm in a number of hosts, most notably in yttrium aluminium garnet ($Y_3Al_5O_{12}$). More recently, tunable laser gain media such as Ti^{3+}-doped sapphire (Al_2O_3) and Cr^{3+}-doped $LiSrAlF_6$ (**Figure 8**) have been developed and these provide variable wavelength outputs across the near-infrared. Both systems are based on vibronically assisted luminescence bands that arise due to strong electron–phonon coupling. Many other laser systems have been developed, including those that use lanthanide ion doped optical fibres as the gain medium and others that are able to sustain oscillation on upconversion transitions.

The much greater inhomogeneous broadening that results from disorder in glasses has also been exploited, for instance in erbium-doped fibre amplifiers. In these devices a small part of an optical fibre communication system is doped with erbium. A diode laser is coupled into the fibre and pumps the erbium ions to an excited metastable state. As the signal passes through this part of the fibre it is amplified through stimulated emission. Using a glass host for this purpose broadens the transition and allows broad bandwidth operation.

There are more speculative applications associated with some of the more advanced techniques. For instance, spectral hole-burning has been suggested as a high density optical storage medium whereby binary digits are indicated by the presence or absence of a hole at any given wavelength thus adding wavelength to spatial position in a compact disc-like memory. Extensions to this idea include the storage of images as spectral holes through the production of wavelength multiplexed holographic gratings and the combination of these images through Stark effect splittings of spectral holes to produce an optical parallel processor. These ideas have been demonstrated in principle but, to date, are only feasible when the sample is maintained at very low temperatures.

Amongst the more important applications of luminescence spectroscopy in the liquid phase is the detection of small amounts of impurities in naturally occurring solutions such as rivers and groundwater. A full knowledge of quenching and sensitization effects are required to be able to solve these problems. Of particular importance in this area is the detection of pollution and especially radioactive pollution that can be dangerous in very small quantities. Actinides are generally strongly luminescent and so spectroscopic techniques are very effective.

See also: **EPR, Methods; Fluorescence Microscopy, Applications; Fluorescent Molecular Probes; Hole Burning Spectroscopy, Methods; Laser Magnetic Resonance; Laser Applications in Electronic Spectroscopy; Laser Spectroscopy Theory; Light Sources and Optics; Luminescence Theory; Near-IR Spectrometers; Raman Optical Activity, Applications; Symmetry in Spectroscopy, Effects of; UV-Visible Absorption and Fluorescence Spectrometers; Zeeman and Stark Methods in Spectroscopy, Applications.**

Further reading

Bartram RH and Henderson B (1999) *Crystal Field Engineering of Solid State Laser Materials.* Cambridge: Cambridge University Press.

Choppin GR and Peterman DR (1998) Applications of lanthanide luminescence spectroscopy to solution studies of coordination chemistry. *Coordination Chemistry Reviews* **174**: 283–299.

Gan F (1995) *Laser Materials.* Philadelphia: World Scientific.

Henderson B and Imbusch GF (1989) *Optical Spectroscopy of Inorganic Solids.* Oxford: Clarendon Press.

Holliday K and Wild UP (1993) Spectral hole-burning. In: Schulman S (ed) *Molecular Luminescence Spectroscopy: Methods and Applications Part III*, Chapter 5, pp 149–228.

Jorgensen CK and Reisfeld R (1982) Uranyl photophysics. *Structure and Bonding* **50**: 121–171.

Kaminskii AA (1990) *Laser Crystals* Berlin: Springer.

Kaplyanskii AA and Macfarlane RM (eds) (1987) *Spectroscopy of Solids Containing Rare Earth Ions.* Amsterdam: North-Holland.

Sugano S, Tanabe Y and Kamimura H (1970) *Multiplets of Transition Metal Ions in Crystals.* New York: Academic Press.

Wybourne BG (1965) *Spectroscopic Properties of Rare Earths.* London: Interscience.

Interstellar Molecules, Spectroscopy of

AGGM Tielens, Kapteyn Astronomical Institute, Groningen, The Netherlands

ELECTRONIC SPECTROSCOPY
Applications

Introduction

The origin and evolution of life has long fascinated mankind. It is directly tied to the abiotic formation and chemical history of the biogenic elements, in particular H, C, N, O, P and S. The evolution of most elements starts with nucleosynthesis inside the fiery cauldrons of stellar interiors. Eventually, these elements are injected into the interstellar medium, either through violent supernova explosions or through more gentle stellar winds which terminate the life of most stars. Thus, stellar formation, evolution and death are intimately connected, as the ashes of one generation of stars become the building blocks of the next generation. Much of the expelled material is in the form of molecules, ranging from the simple diatomics such as molecular hydrogen (H_2) and carbon monoxide (CO) to more complex species such as polycyclic aromatic hydrocarbons (PAHs) and fullerenes (C_{60}), or in the form of small ($\approx$ 100 nm) dust grains – carbon-based (i.e. soot and silicon carbide) and silicates (aluminium oxide, olivine, enstatite). In their evolution from their birth sites, through the interstellar medium, until their incorporation into new planetary systems, these elements undergo a complex chemical history. Various processes cycle the material from the gas phase to the solid phase and from one chemical compound to

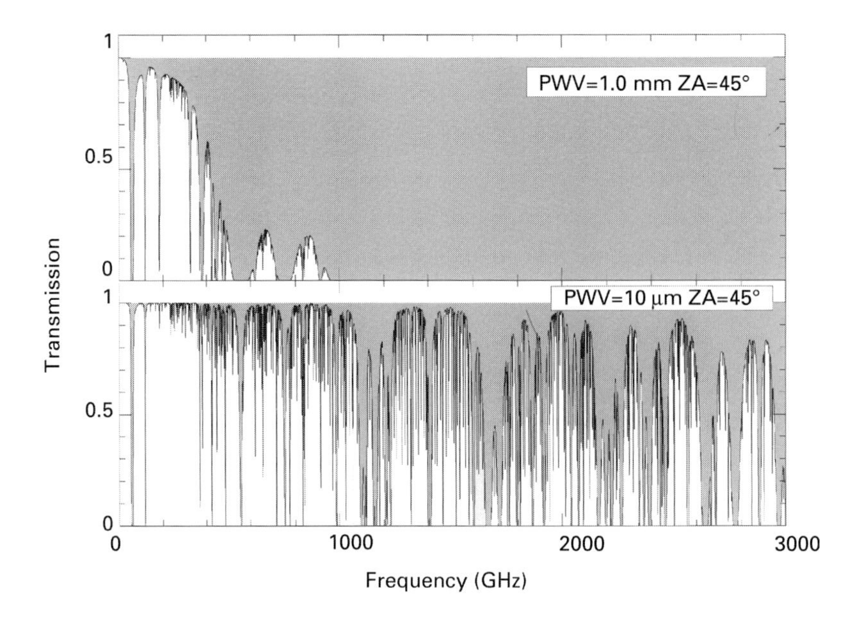

Figure 1 Atmospheric transmission in the submillimetre and far-infrared for typical observing conditions on a dry mountain top with 1 mm of precipitable water vapour and under a zenith angle of 45 degrees (top panel) and on an airborne platform with 10 μm of precipitable water vapour flying at an altitude of about 40 000 feet also under a zenith angle of 45 degrees (bottom panel). Courtesy of TG Philips.

another. Understanding this history and its relation to the origin of life is one of the key questions of modern astrophysics.

The most important tool for astronomers to understand the origin and evolution of the molecular universe is spectroscopy. This includes studies of the rotational emission spectrum of the cold interstellar gases in the microwave wavelength region. Near bright stars, vibrational modes can be excited through fluorescence. Electronic spectra are predominantly measured through absorption spectroscopy against a bright background source. Each of these wavelength regions has its own techniques and provides unique information on interstellar molecules. Together they point to a chemically diverse and active molecular universe. Each of these techniques and their results will be briefly summarized.

Rotational spectroscopy

Hundreds of pure rotational lines of a variety of molecules have been detected in the interstellar medium, most of them in emission but some also in absorption. These lines occur throughout the millimetre, centimetre and decametre wavelength region of the spectrum and are detected using radio antennas and heterodyne detection techniques at spectral resolutions of typically 300 000. At submillimetre and far-infrared wavelengths, the atmosphere is not very transparent (**Figure 1**) and telescopes are located on high, dry sites, on airborne platforms or in space. **Figure 2** shows the results of a line survey of a star-forming region in the Orion constellation, illustrating the richness of the spectrum and the diversity of the molecules contributing. Many of these rotational lines are bright enough to map the spatial distribution of the molecular emission. Such maps can be intercompared to study the interrelationship of various molecules. Also, they can be compared to maps of (newborn) stars to determine the interaction of stars with the surrounding (natal) molecular gas.

Similar spectral and spatial studies have been performed for the ejecta of stars in the later stages of their life to probe their molecular composition and evolution. Because the central star heats the gas to warm temperatures (up to 1000 K), rotational lines in the far-infrared can be excited as well. **Figure 3** shows the far-infrared spectrum of such a star, IRC 10216, obtained by the Infrared Space Observatory. For these objects, molecules can also be detected through their rovibrational transitions against the bright far-red, and near- and mid-infrared continuum of the star.

Rotational transitions are very characteristic for molecules and, at the high resolution allowed by heterodyne techniques, precise molecular identifications can be made using laboratory measured rotational constants. Over a hundred different molecules have been detected in interstellar space (**Table 1**) and new ones are found at a rate of a few per year. Some of these species are very simple well-known molecules such as water and ammonia; however, most are

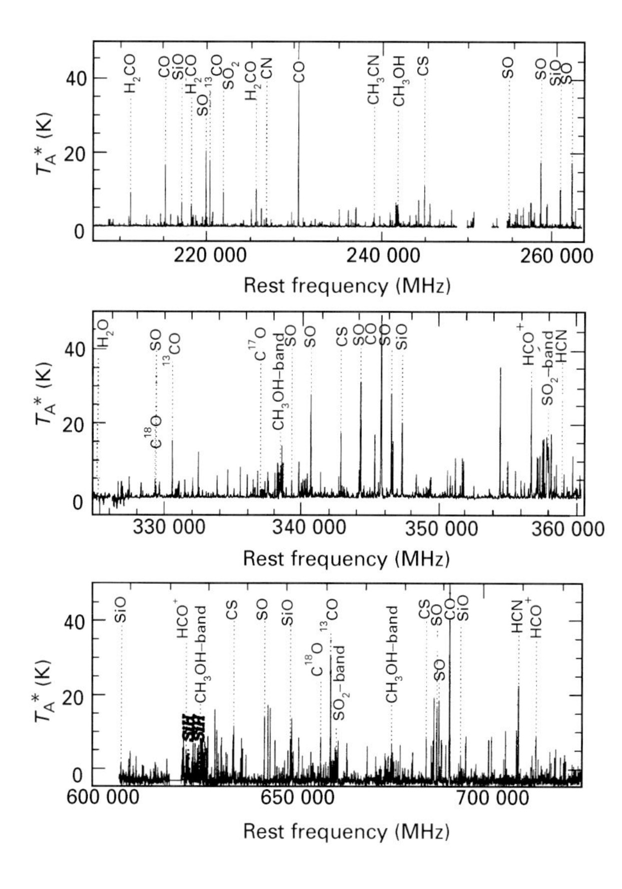

Figure 2 Ground-based line surveys of a star-forming region in the Orion constellation in three atmospheric windows centred at 240, 330 and 700 GHz. Note the richness of the spectrum. Some of the more prominent lines are labelled by the molecule responsible. Courtesy of TG Philips.

Table 1 Identified interstellar and circumstellar molecules

Simple hydrides, oxides, sulfides, halogens and related molecules				
H_2 (IR,UV)	CO	NH_3	CS	NaCl*[a]
HCl	SiO	SiH_4* (IR)	SiS	AlCl*
H_2O	SO_2	C_2 (IR)	H_2S	KCl*
N_2O	OCS	CH_4 (IR)	PN	AlF*
HF				
Nitriles and acetylene derivatives				
C_3 (IR,UV)	HCN	CH_3CN	HNC	C_2H_4* (IR)
C_5* (IR)	HC_3N	CH_3C_3N	HNCO	C_2H_2 (IR)
C_3O	HC_5N	CH_3C_5N ?	HNCS	
C_3S	HC_7N	CH_3C_2H	HNCCC	
C_4Si*	HC_9N	CH_3C_4H	CH_3NC	
	$HC_{11}N$	CH_3CH_2CN	HCCNC	
	HC_2CHO	CH_2CHCN		
Aldehydes, alcohols, ethers, ketones, amides and related molecules				
H_2CO	CH_3OH	HCOOH	CH_2NH	CH_2CC
H_2CS	CH_3CH_2OH	$HCOOCH_3$	CH_3NH_2	CH_2CCC
CH_3CHO	CH_3SH	$(CH_3)_2O$	NH_2CN	
NH_2CHO	$(CH_3)_2CO$	H_2CCO	CH_3COOH	
Cyclic molecules				
C_3H_2	SiC_2	c-C_3H	CH_2OCH_2	
Molecular ions				
CH^+(VIS)	HCO^+	$HCNH^+$	H_3O^+	HN_2^+
HCS^+	$HOCO^+$	HC_3NH^+	HOC^+	H_3^+(IR)
CO^+	H_2COH^+	SO^+		
Radicals				
OH	C_2H	CN	C_2O	C_2S
CH	C_3H	C_3N	NO	NS
CH_2	C_4H	HCCN*	SO	SiC*
NH (U V)	C_5H	CH_2CN	HCO	SiN*
NH_2	C_6H	CH_2N	MgNC	CP*
HNO	C_7H	NaCN	MgCN	
C_6H_2	C_8H	C_5N*		

[a] Species denoted with * have only been detected in the circumstellar envelope of carbon-rich stars, notably IRC 10216. Most molecules have been detected at radio and millimetre wavelengths, unless otherwise indicated (IR, VIS or UV). Isotopomers of many of these species – including D, ^{13}C, ^{15}N, and ^{18}O – have been detected as well.

rather exotic, unsaturated carbon chain molecules, radicals and ions. To a large extent this reflects observational selection effects. The detection of these species is heavily favoured because of their large dipole moments and their relatively small partition function, which allows detection of even trace amounts. Also, while in terrestrial laboratories many of these species are very transient, they can persist much longer under the extreme vacuum conditions of the interstellar medium when properly shielded from stellar ultraviolet photons by the ubiquitous dust grains. Many of the ions are protonated versions of stable molecules such as water and carbon monoxide. Finally, isotopomers of many of these species have been discovered as well.

The intensity of rotational line emission depends on the density and temperature of the gas as well as the abundance of the molecular species. Because many lines of the rotational ladder of a species can be measured, these spectra provide excellent probes of the physical conditions of the emitting gas. Most of the molecules are present in so-called molecular clouds, which are dense for interstellar standards (1000 molecules cm^{-3}), cold (10 K), large scale (30 light-years) structures. About 40% of the interstellar gas is in such molecular clouds. The large column densities of dust associated with these molecular clouds shields the molecules from the dissociating radiation from surrounding bright stars and allows them to survive. At the same time, these molecules allow the clouds to cool down to low temperatures and be compressed. Gravity can then overcome the thermal and magnetic support of these clouds and new stars can form. Molecular clouds are indeed the exclusive sites for the formation of new stars. The molecular inventory of these clouds is therefore directly fed into newly formed stars and their planetary systems.

The most abundant species in molecular clouds is H_2, simply because hydrogen is the most abundant element. Carbon monoxide is second at an abundance relative to hydrogen of 10^{-4}. Most of the other species in **Table 1** are trace species at abundances of 10^{-8} or less. Many processes contribute to the formation and evolution of these molecules. Molecular hydrogen is thought to be formed on the surfaces of dust grains (mainly because gas-phase routes towards this molecule are measured to be extremely slow). Once molecular hydrogen is present, other molecules can be formed through gas-phase ion–molecule reactions. These reactions are fast and, unlike neutral–neutral reactions, often do not have reaction barriers because of the Coulomb interaction. A trace amount (10^{-7}) of ionization is provided by penetrating high energy (MeVs) cosmic ray

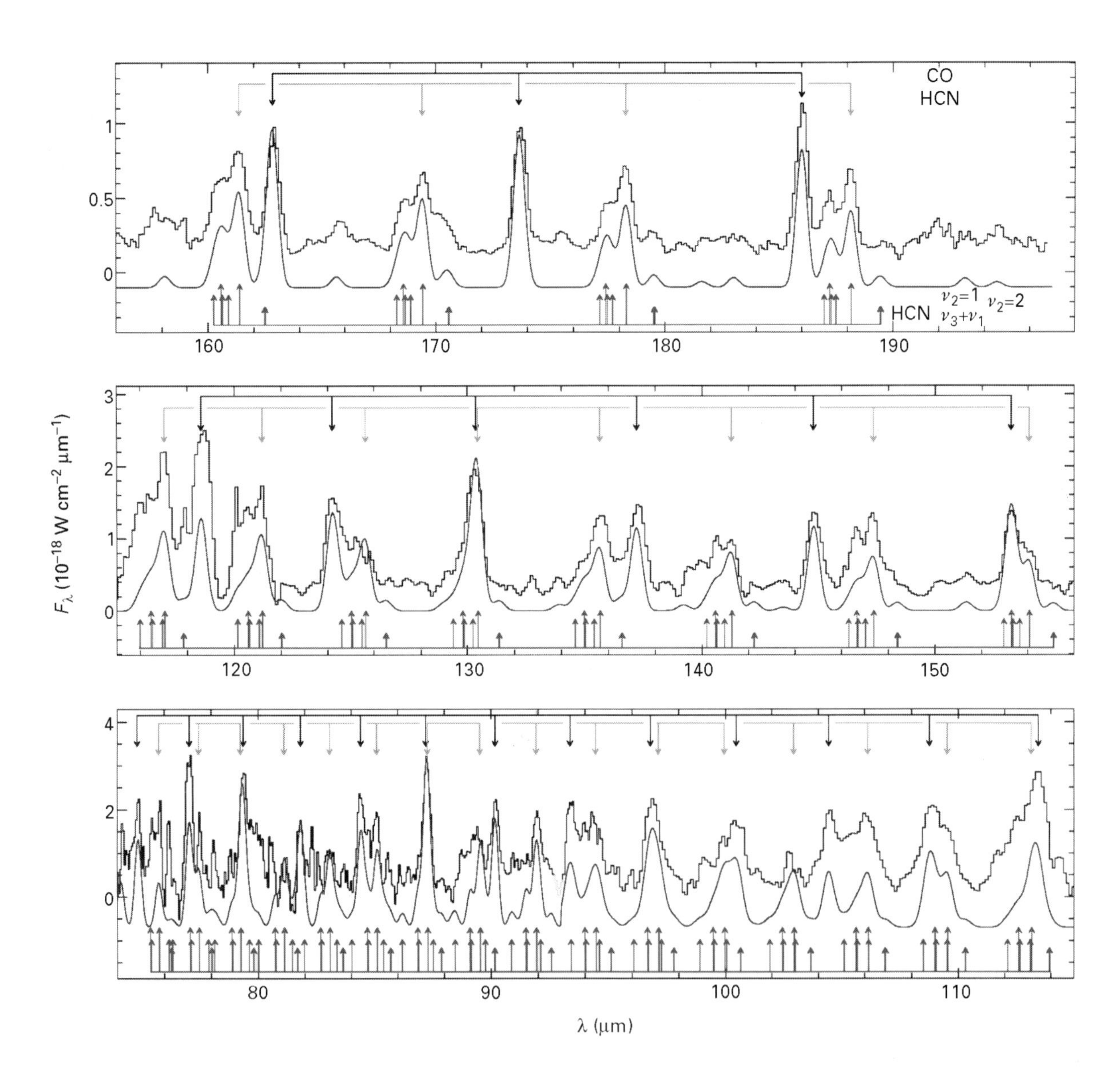

Figure 3 Far-infrared spectrum of the ejecta from the star IRC 10216 obtained with the Long-Wavelength Spectrometer on board the Infrared Space Observatory at a resolution of about 100. Most of the lines in this spectrum can be identified with pure rotational transitions in CO, HCN and their isotopomers (top arrows). Rotational transitions associated with the lowest vibrational states of HCN are also present (bottom arrows). Reproduced with permission from Cernicharo J, Barlow MJ, Gonzaléz-Alfonso *et al* (1996) *Astronomy and Astrophysics* **315**: L201.

protons. The cosmic rays ionize H_2 producing H_2^+ which reacts with another H_2 to form protonated molecular hydrogen, H_3^+. This species can transfer its proton to other species, in particular atomic oxygen, to start the chain that will lead to the formation of OH, H_2O and their protonated forms. Reaction of these species with atomic C and C^+ leads then eventually to the formation of CO. This very stable molecule locks up most of the available carbon, but trace amounts of atomic C and C^+ present can drive the build up of more complex carbon molecules including the carbon chains. Finally, the low temperatures (10 K) of molecular clouds will enhance isotopic fractionation. In particular, because of the zero-point energy difference between hydrogenated and deuterated molecules (typically 100–500 K), deuterium fractionation is important and the abundance of deuterated molecules can reach levels comparable to their hydrogenated counterparts despite the very low elemental abundance of deuterium (10^{-5} relative to H).

Vibrational spectroscopy

Ices in star-forming regions

Gas and dust in molecular clouds are very cold and hence vibrational spectroscopy is limited to absorption spectroscopy, where a newly formed star inside the cloud or a chance superposition of the cloud against a background star provides the continuum against which the spectra can be measured. Such spectra show a variety of broad absorption features due to simple molecular species: H_2O, CH_3OH, CO, CO_2, OCS and CH_4. **Figure 4** shows as an example the 2–20 μm spectrum of the luminous protostar W33A. These species are very volatile and are therefore never seen in emission in the infrared. At this resolution, it is easy to separate absorption due to gaseous molecules from that of solid ices (**Figure 5**). **Table 2** summarizes the observed composition of interstellar ice towards this source. H_2O dominates the column density of ice by a large factor. The carbon-bearing species, CO, CO_2 and CH_3OH, are less abundant by a factor 10 to 20, while CH_4 is at the 1% level compared to H_2O. This

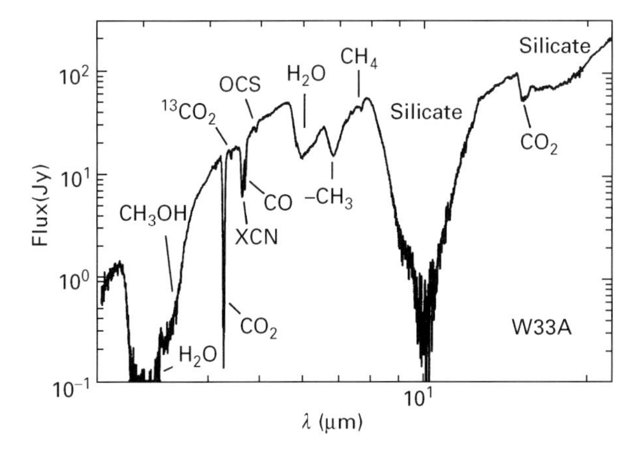

Figure 4 The 2–20 μm spectrum of the protostar W33A obtained at a spectral resolution of 250 using the Short-Wavelength Spectrometer on board the Infrared Space Observatory. Except for the 10 μm silicate band, all features are due to simple molecules in ice mantles.

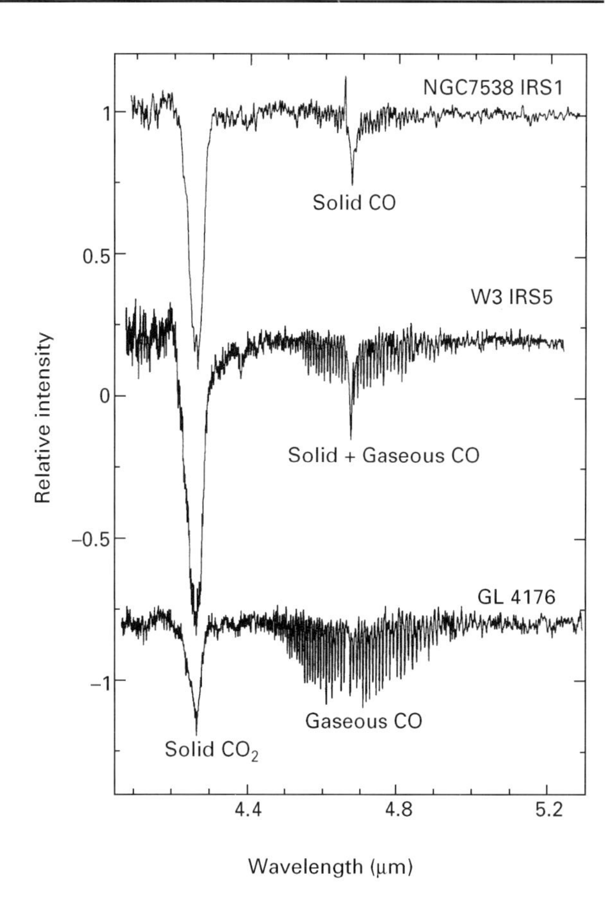

Figure 5 Absorption spectra of a number of protostars in the stretching region of CO and CO_2 obtained by the Short-Wavelength Spectrometer on board the Infrared Space Observatory. The broad band at about 4.25 μm is due to the stretching mode of solid CO_2. The narrow features in the 4.5–4.9 μm range are the rovibrational transitions of gaseous CO. The broader feature in the Q-band gap of this linear molecule is the stretching mode of solid CO. The narrow emission feature at about 4.66 μm in the spectrum of NGC 7538 IRS 1 is a hydrogen recombination line. These different sources are characterized by different temperatures of the absorbing gas and ices. The relative amounts of the rather-volatile molecule CO in the gas phase and the ice reflect this rather strikingly. Reproduced with permission of the Royal Society of Chemistry from van Dishoeck EF (1998) *Faraday Discussions* **109**: 31.

Table 2 Observed abundances in interstellar ices

Species	*W33A*	*NGC 7538-IRS 9*	*Comets*
H_2O	100	100	100
NH_3	–	12	1
CH_4	0.4	2	0.2–1.2
CH_3OH	4	5	0.3–5
CO	2	12	6
CO_2	3	15	3
OCS	0.04	–	0.1
HCOOH	3	3	0.2

Observed abundances normalized to that of water ice (=100). W33A and NGC 7538-IRS 9 are two luminous protostars which span the observed range in interstellar ice composition. The abundances for the comets are an average of those observed for comets Hyakutake and Hale–Bopp.

dominance of H_2O is directly evident from the complete IR spectrum of this source (**Figure 4**), since the H_2O absorption features are more conspicuous than any of the other ice features and yet the intrinsic strength of absorption features are very similar for most molecules.

The most striking aspect of the IR spectrum of interstellar ices is the apparent simplicity of the composition. Nevertheless, this global ice inventory is somewhat misleading. The peak position and profile of absorption bands are sensitive to the molecular environment of the absorbing species in the ice. Thus, H_2O is mainly located in an H_2O-rich ice, as evidenced by the width and peak position of the 3 and 6 μm bands representing OH stretching and bending modes. While some of the solid CO is embedded in the H_2O ice, most of the absorption is due to an apolar ice component dominated by O_2, N_2 or CO itself. Solid CO_2 studies also reveal the presence of multiple ice mantle components along many lines of sight. **Figure 6** shows the bending mode of CO_2 in the spectra of a number of protostars embedded in their natal molecular clouds. The spectra show weak structure in the depth of the band at 15.1 and 15.25 μm as well as a shoulder at about 15.4 μm. These spectra are well fitted by laboratory spectra of mixtures of H_2O, CH_3OH and CO_2 heated to about 100 K. At about that temperature, these mixtures segregate into separate H_2O, CH_3OH and CO_2 ices and this is reflected in the absorption spectra. Thus, the features at 15.1 and 15.25 μm are due to the bending mode in pure solid CO_2. This mode is doubly degenerate and splits when the axial symmetry of the molecule is broken in the CO_2 matrix. The observed peak positions are slightly shifted from those of a pure CO_2 ice film because, in these warmed-up mixtures, the CO_2 ice clusters are still embedded in the H_2O and CH_3OH ices. The strength of the 15.1 and 15.25 μm features in the interstellar spectra reflect the timescale of this thermal processing; i.e. the lifetime of the newly formed star. The CO_2 stretching mode is not sensitive to these effects and neither the laboratory nor the interstellar spectra show much variation upon warm up. However, the stretching mode of the ^{13}C isotope of solid CO_2 does show these effects and the interstellar spectra are in good agreement with the conclusions derived from the bending mode of $^{12}CO_2$.

Interstellar ices are formed through the accretion of simple gas-phase species on surfaces of small dust grains injected into the interstellar medium by stars. Typical accretion rates are about one species a day.

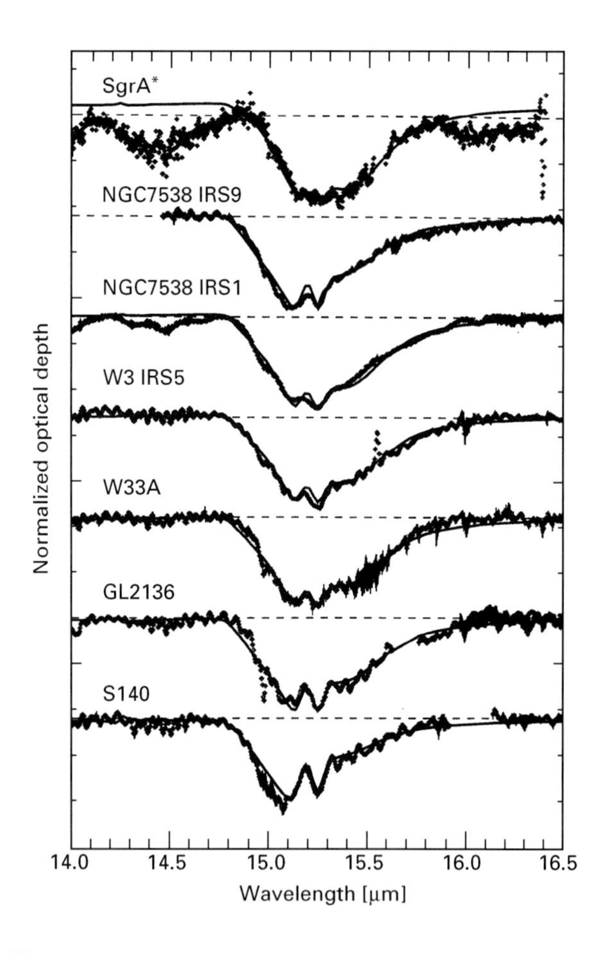

Figure 6 Absorption spectra of the bending mode of CO_2 ice towards a number of protostars embedded in their natal molecular clouds obtained by the Short-Wavelength Spectrometer on board the Infrared Space Observatory at a spectral resolution of about 2000. The observed profiles show weak features at about 15.1 and 15.25 μm as well as a shoulder at about 15.4 μm. The solid lines are laboratory spectra of the H_2O:CH_3OH:CO_2 = 1:1:1 mixture deposited at 10 K and then warmed to temperatures in the range of 114 to 118 K. The sharp structure near 14.97 μm in some sources is due to the rovibrational transitions in the Q branch of gaseous CO_2 which at this resolution pile up in one spectral resolution element. The individual P and R branch lines are too weak to be observable at this resolution.

Interstellar grain surface chemistry is dominated by hydrogenation and oxidation reactions. Of particular interest are those reactions involving the dominant gas-phase molecule, CO, leading to H_2CO, CH_3OH and CO_2. Atomic oxygen, carbon and nitrogen are readily hydrogenated to H_2O, NH_3 and CH_4. In the general interstellar medium, the omnipresent stellar radiation photodesorbs these species keeping the grain surfaces clean. However, the shielding provided by molecular clouds allows thin (10 s of nm) ice mantles to form. During the gravitational collapse associated with the formation of a new star, the accretion rate of gas-phase species on a grain becomes so high that essentially all species can freeze out in these ice mantles. Such ice grains entering the cold outskirts of protoplanetary disks can survive and agglomerate into large bodies, leading to the formation of comets. The great similarity in molecular inventory and abundances of interstellar ices and comets (cf. **Table 2**) supports such a scenario. Besides grain surface chemistry, the composition of interstellar ices is also affected by thermal processing when the ices are too close to a newborn star, much like the outgassing of comets when they approach the inner solar system. The difference in volatility between apolar molecules, such as CO and CH_4, and hydrogen-bonding species such as H_2O and CH_3OH can then lead to thermal fractionation.

Interstellar and circumstellar PAHs

Ground-based, airborne and space-borne observations have shown that the IR spectra of bright sources with associated dust and gas are dominated by relatively broad emission features at 3.3, 6.2, 7.7, 8.6 and 11.3 μm, which always appear together. Because the carriers of these bands remained unidentified for almost a decade, these bands have become collectively known as the unidentified infrared (UIR) bands. These bands are now unequivocally identified with an aromatic carrier and this name is somewhat of a misnomer. Nevertheless, the abbreviation has stuck and is still widely used. As an example, **Figure 7** shows three mid-infrared spectra of carbon-rich outflows from stars in the latest stages of their evolution. It is now known that these emission features are not limited to stellar outflows but are also present in the spectra of disks of newly formed

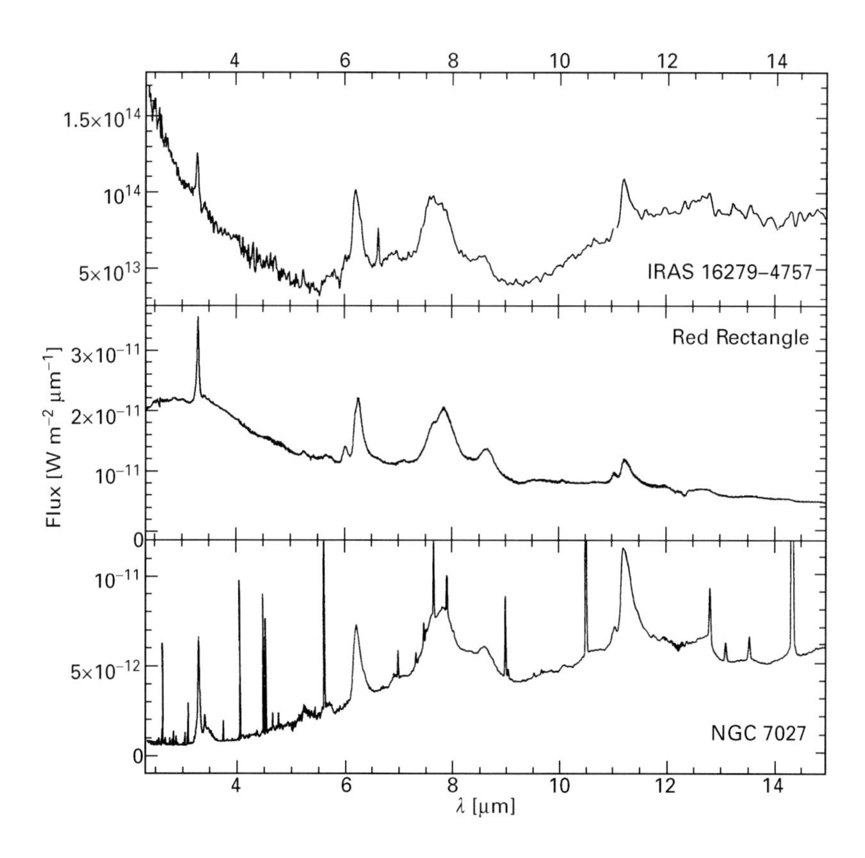

Figure 7 The 3–15 μm emission spectra of three stars in the latest stages of their evolution illustrating the ubiquitous nature and the richness of the UIR spectrum. These spectra were obtained with the Short-Wavelength Spectrometer on board the Infrared Space Observatory at a spectral resolution ranging from 250 to 2000.

stars, the spectra of the surfaces of interstellar gas clouds, the spectra of reflection nebulae, the spectra of ionized nebulae around massive stars, and the spectra of the general interstellar medium of the Milky Way and other nearby galaxies. The carrier is thus ubiquitous and able to withstand the harsh environment of interstellar space.

These bands are very characteristic for aromatic hydrocarbon species in which carbon atoms are arranged in hexagons decorated by hydrogens at the edges. These features are observed far from illuminating sources where interstellar gases are cold (10–100 K) and exciting collisions infrequent (once a day). The emission is therefore attributed to infrared fluorescence from a collection of vibrationally excited PAH species electronically pumped through the absorption of one stellar UV photon (about 10 eV). During the vibrational cascade (about 1 s), the molecule emits in its C–H and C–C vibrational modes (the UIR bands) and, afterwards, the molecule remains cold (about 10 K) until the next UV photon is absorbed (about a day). Analysis of this process leads then to a determination of the size of the emitting species of about 50 carbon atoms. Thus the carriers of the UIR bands are PAH molecules such as hexabenzocoronene ($C_{48}H_{24}$) and circumcoronene ($C_{54}H_{18}$).

Typical abundances of PAHs are of the order of 10^{-7} relative to hydrogen, which makes these species the most abundant polyatomic molecules present in space. Vibrational spectroscopy is eminently suited for detecting the presence of classes of molecules, but it is much more difficult to identify specific molecules within a collection of species. However, while the gross characteristics of these stellar and interstellar spectra are very similar, when examined in detail, they vary from source to source. Analysis of these variations may well provide us with a tool to identify specific molecules within the circumstellar and interstellar PAH family and such efforts, supported by extensive laboratory studies, are now underway.

The 11–15 μm region of the spectrum is rich in details showing bands at 11.0, 11.3, 11.9, 12.8, 13.6, 14.2 and 14.8 μm plus a number of weaker shoulders (**Figure 8**). Emission features in this spectral region are due to a C–H out-of-plane deformation mode of aromatic hydrocarbons and the observed pattern is very characteristic for the edge structure of the emitting aromatic species. Due to coupling between the vibrating H atoms, the exact peak position of these modes depends on the number of directly adjacent H atoms (i.e. bonded to neighbouring C atoms on a ring). The 11.3 μm band is attributed to this mode for isolated Hs in PAHs while the 11.9 and 12.8 μm bands are due to 2s and 3s. The weak bands at 13.6 and 14.2 μm are likely due to phenyl (5s) groups; i.e. one ring connected by a single bond to the rest of the PAH. In that case, the 14.2 μm band is actually a ring-bending mode.

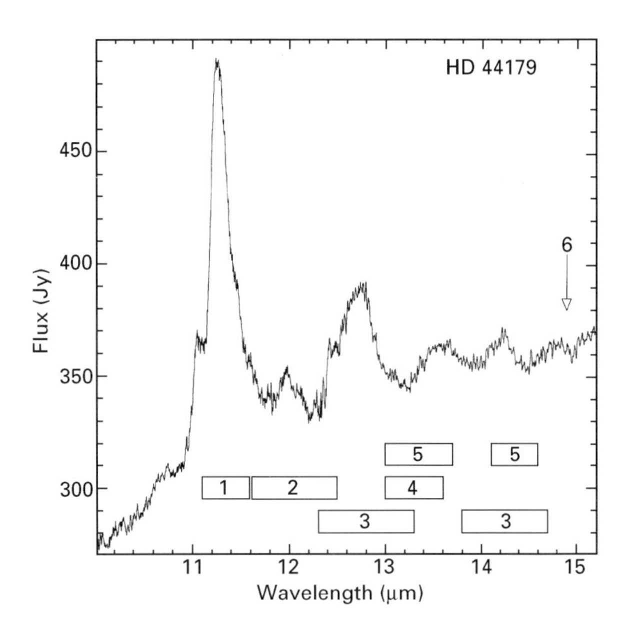

Figure 8 The 10–15 μm spectrum of the outflow from a star known as the Red Rectangle because of the rectangular appearance of its associated reflection nebula on red plates. Bands in this spectral region are largely due to the C–H out-of-plane bending mode whose exact peak position is sensitive to the number of adjacent H atoms involved. The boxes underneath the spectra give the ranges where PAHs with the indicated number of adjacent H atoms emit. The arrow labelled 6 indicates the position for benzene. These spectra were obtained with the Short-Wavelength Spectrometer on board the Infrared Space Observatory at a spectral resolution ranging from 250 to 2000.

These PAHS are thought to form in carbon-rich stellar winds through chemical processes akin to those occurring in sooting flames and other terrestrial combustion environments. Acetylene is the precursor molecule from which larger species, and eventually soot, are formed. This process starts with the formation of the first aromatic ring (e.g. benzene). Rapid chemical growth of this ring through the addition of acetylene then forms larger PAHs. Further chemical growth as well as clustering of these PAHs leads to the formation of small (≈100 nm) soot particles. These soot particles and the gas are driven out by the radiation pressure of the luminous star, thereby enriching the interstellar medium.

Electronic spectroscopy

Elemental abundances in the galaxy

With a few notable exceptions, most atomic species possess strong absorption lines at far-ultraviolet

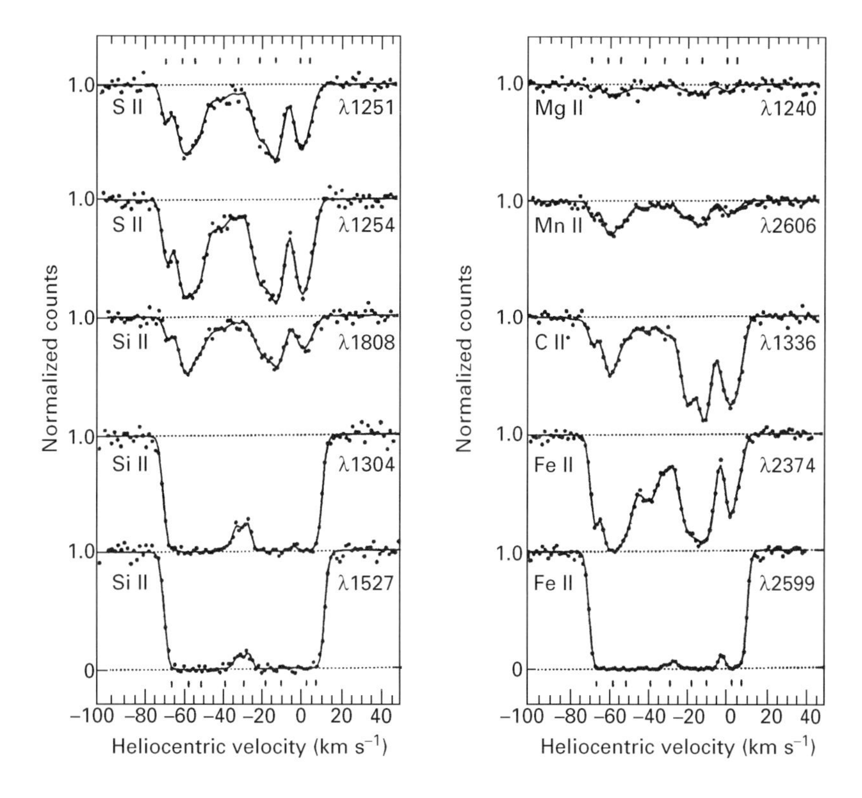

Figure 9 Normalized intensity versus heliocentric velocity for a number of lines of different interstellar atoms towards the star HD 93521. These were obtained with the Goddard High-Resolution Spectrograph on the Hubble Space Telescope at a resolution of about 100 000. Tick marks at the top and bottom of the panels indicate individual absorption components present along the line of sight to this star.

wavelengths and their detection requires space-borne instrumentation. Since the 1970s, a number of missions have measured electronic absorptions of many elements as well as a few diatomic molecules. These lines are due to gas in the interstellar medium measured in absorption against a bright background star. A sample of such is shown in **Figure 9** obtained by the Goddard High-Resolution Spectrograph on the Hubble Space Telescope. At the high spectral resolution of this instrument, these absorption lines show components due to many individual clouds along the line of sight which have different velocities with respect to the Sun because of galactic rotation. The lines shown cover a range of line strengths, due to differences in intrinsic oscillator strength and column density of the absorbing species involved. These observations allow the determination of the abundance of elements in the interstellar medium with respect to atomic hydrogen. Typically, the measured abundances fall short of the elemental abundances measured in the Sun, in solar system objects such as meteorites, and in stellar photospheres (**Figure 10**). Because the Sun and other stars are formed from interstellar gas and, hence, should have very similar abundances, these large differences are attributed to the depletion of these elements from the gas phase in the form of small solid dust grains. This is supported by the loose correlation of the amount of material depleted from the gas phase with condensation temperature of the likely condensate of these elements (**Figure 10**). These dust grains are formed in the high-density, high-temperature environments of stars and injected into the interstellar medium in a supernova explosion or a stellar wind.

The diffuse interstellar bands

The diffuse interstellar bands (DIBs) are weak absorption features which typically appear in the visual range of stellar spectra. These bands were discovered over 60 years ago and their origin in interstellar material along the line of sight towards these stars was quickly established, because they do not share the stellar velocity variations in spectroscopic binaries and their strength increases with distance and the column of absorbing interstellar material along the line of sight. **Figure 11** shows a compilation of these absorption features. The richness of the DIB spectrum is apparent: over 200 DIBs have been discovered. Most DIBs have a width in the range 0.5–3 Å,

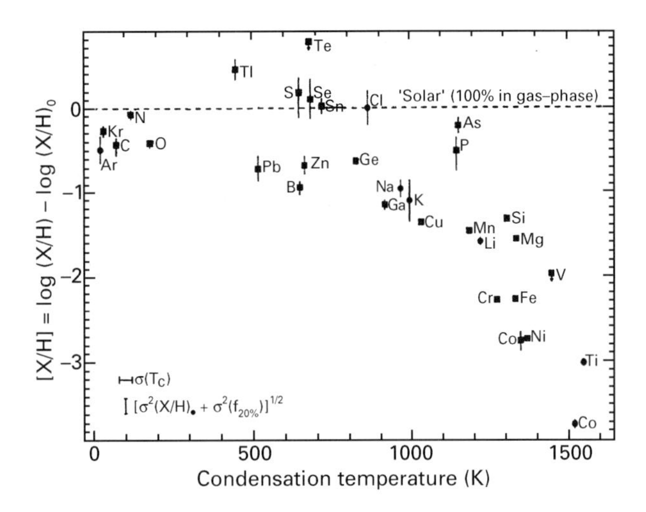

Figure 10 Gas-phase abundances measured towards the star z Oph of various elements relative to atomic hydrogen (X/H) and normalized with respect to solar abundances $(X/H)_0$ plotted versus condensation temperature of these elements. The condensation temperature is the temperature at which 50% of that element is depleted from the gas phase into the solid phase. The error bars on the squares indicate measurement errors only. Representative uncertainties in the solar abundances, oscillator strength and condensation temperature are indicated in the lower left hand corner.

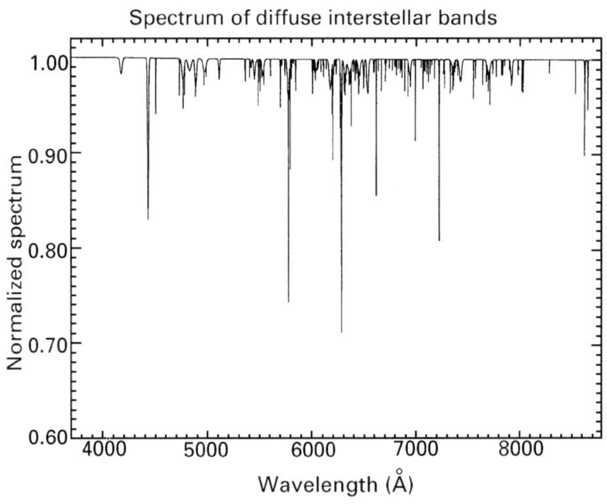

Figure 11 A compilation of the diffuse interstellar bands spectrum. Individual DIBs have been smoothed to a spectral resolution of 2 Å and their strength normalized to a column corresponding to line-of-sight dust reddening equal to unity. Reproduced with permission from Jenniskens P and Désert F-X (1994). *Astronomy and Astrophysics Supplement Series* **106**: 39.

but bands as broad as 20 Å exist as well. The strengths of individual DIBs correlate reasonably well with each other but not perfectly, indicating that more than one carrier is involved. Some DIBs show substructure at high spectral resolution (**Figure 12**). In recent years, a few DIBs have been observed in the ultraviolet and far-ultraviolet part of the spectrum. But such studies are severely hampered by atmospheric absorption and by interstellar absorption due to solid dust grains in the interstellar medium, which increases rapidly with increasing frequency.

At present, none of the DIBs has been identified with certainty – not for lack of trying, though. The DIBs are too broad to be due to electronic transitions in atoms. For decades, the DIBs were thought to be associated with small (100 nm) dust grains which are responsible for the reddening and extinction of starlight. In that view, the DIBs were just thought to represent a small substructure in the extinction properties of the grains likely due to electronic transitions in impurities in an otherwise dielectric material such as silicates or oxides. However, the constant peak position and width of the DIBs in all directions argues against that since, theoretically, the profile is expected to be very sensitive to the grain size as well as the detailed composition of the matrix material. Also, these interstellar dust grains are observed to cause a small amount of polarization of starlight but the DIBs are unpolarized.

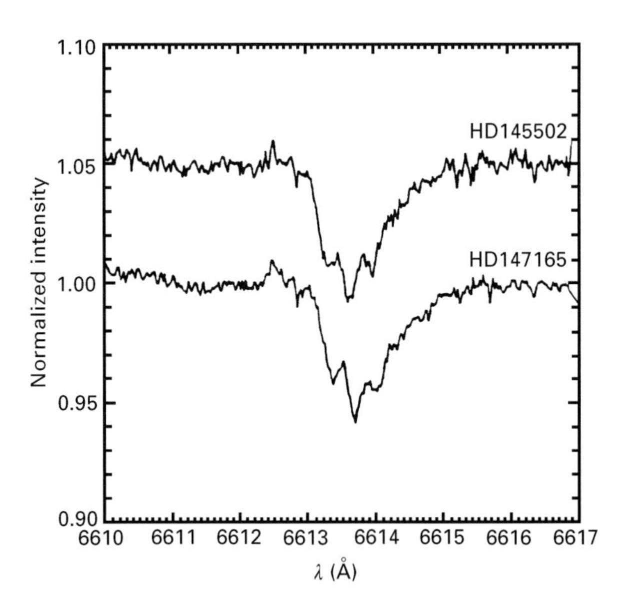

Figure 12 Line profile for the 16 613 Å DIB at a resolution of about 100 000 revealing a three-peaked structure characteristic for rotational contours of electronic transitions in molecular species. Reproduced with permission from Ehrenfreund P and Foing B (1996) *Astronomy and Astrophysics* **307**: L25

Instead, it is now generally accepted that the DIBs are due to absorption by interstellar gaseous molecules. Absorption by molecules (CH, CN) in the visible part of the spectrum was detected only shortly after the discovery of the DIBs, but these species are too small and have too low an abundance to be responsible for the DIBs. This and the perceived difficulty to form an abundant concentration of molecules in the harsh environment of the interstellar medium has long driven the field away from a molecular carrier. However, presently, there is abundant evidence from other wavelength regions for the presence of large and complex molecules such as PAHs in the interstellar medium and this argument has lost much of its persuasion. There is also direct observational support for a molecular origin of the DIBs. One of the key pieces of evidence for this was found in the visible spectrum of the red rectangle. The spectra of the nebulosity associated with this object (known to be bright in PAH emission features in the infrared) show emission features near the wavelengths of prominent DIBs, including those at 5797 and 6613 Å, and, with increasing distance from the central illuminating star, their peak position shifts closer and closer to that of the interstellar DIB absorption features. These visual emission bands also become progressively narrower with distance from the star. This is the behaviour expected for fluorescence of electronic transitions in molecules excited by the radiation of the central star. The gas flowing away from the star cools and, consequently, fluorescence is dominated by lower and lower rotational lines. In the cold interstellar medium, absorption only occurs from the lowest rotational levels. A molecular origin is also supported by the substructure observed at high resolution in some DIBs, which is characteristic for rotational contours of electronic transitions in molecules (cf. **Figure 12**). Analysis of these line profiles shows that, depending on the class of species responsible (i.e. the rotational constants of the electronic states involved), the carrier contains between 10 and 60 C atoms. Finally, two new DIBs have been detected in the far-red (9577 and 9632 Å), which are close to laboratory measured absorption features of the fullerene cation, C_{60}^{+}, in a neon matrix. Unfortunately, no gas-phase spectra of this cation are presently available to confirm this identification unambiguously.

The carrier of the DIBs has still not been identified unequivocally. Part of this is, likely, that there is not one molecular carrier, or even one class of molecular carriers, responsible for all the DIBs, but rather a variety of species contribute. Among the likely candidates are PAHs, fullerenes and carbon chains (such as C_7). Vigorous studies on the visible absorption properties of such species are underway at various laboratories.

See also: **Cosmochemical Applications Using Mass Spectrometry; Microwave and Radiowave Spectroscopy, Applications; Rotational Spectroscopy, Theory; Solid State NMR, Rotational Resonance; Vibrational, Rotational and Raman Spectrocopy, Historical Perspective.**

Further reading

Allamandola LJ, Tielens AGGM and Barker JR (1989) Interstellar polycyclic aromatic hydrocarbons: the infrared emission bands, the excitation/emission mechanism, and the astrophysical implications. *The Astrophysical Journal Supplement Series* **71**: 733–775.

Bakes ELO (1997) *The Astrochemical Evolution of the Interstellar Medium.* Vledder: Twin Press.

Kaler JB (1997) *Cosmic Clouds: Birth, Death, and Recycling in the Galaxy.* New York: Scientific American Library.

Latter WB, Radford SJE, Jewell PR, Mangum JG and Bally J (eds) (1997) *CO: Twenty-Five Years of Millimeter-Wave Spectroscopy.* Dordrecht: Kluwer.

Millar TJ and Williams DA (1993) *Dust and Chemistry in Astronomy.* Bristol: Institute of Physics.

Savage BD and Sembach KR (1996) Interstellar abundances from absorption line observations with the Hubble space telescope. *Annual Reviews of Astronomy and Astrophysics* **34**: 279–329.

Tielens AGGM and Snow TP (eds) (1995) *The Diffuse Interstellar Bands.* Dordrecht: Kluwer.

van Dishoeck EF (ed) (1997) *Molecules in Astrophysics.* Dordrecht: Kluwer.

Ion Beam Analysis, Methods

See **High Energy Ion Beam Analysis.**

Ion Collision Theory

Anil K Shukla and **Jean H Futrell**, University of Delaware, Newark, DE, USA

MASS SPECTROMETRY
Theory

Introduction

Collisions between an ion and neutral species result in a number of possible outcomes depending upon the chemical and physical properties of the two reactants, their relative velocities, and the impact parameter of their trajectories. These include elastic and inelastic scattering of the colliding particles, charge transfer (including dissociative charge transfer), atom abstraction, complex formation and dissociation of the colliding ion. Each of these reactions may be characterized in terms of their energy-dependent rate coefficients, cross sections and reaction kinematics. This article outlines a theoretical framework for discussing these processes that emphasizes simple models and classical mechanics. We divide the discussion of collision processes into the two categories of low-energy and high-energy collisions. Experiments under thermal or quasi-thermal conditions – swarms, drift tubes, chemical ionization and ion cyclotron resonance – are strongly influenced by long-range forces and often involve 'capture collisions' in which atom exchange and extensive energy exchange are common characteristics. High-energy collisions are typically impulsive, involve short-range intermolecular forces and are direct, fast process.

Low-energy collisions

For historical reasons, and because the majority of investigations of ion–molecule reactions have involved quasi-thermal collisions of relatively low-mass ions and neutrals, we begin with the simplest case of a point charge interacting with a polarizable neutral. It is recognized that this Langevin model applies only at low collision energies where long-range forces dominate collision dynamics. Despite these restrictions, this simplistic picture provides a near-quantitative rationalization for the rates of thousands of ion–molecule reactions.

Ion–molecule collisions at low energies are dominated by the attractive long-range polarization force described in the simplest case by

$$V(r) = \frac{\alpha q^2}{2r^4} \quad [1]$$

where the ion is tacitly assumed to be a point charge, q, and the neutral is a point polarizable atom or molecule having a polarizability α at a distance r from the ion. At relative velocities where repulsive forces can be neglected (that is, they are sensed at much closer distances than the impact parameter for orbiting collisions, as discussed below), Equation [1] is a plausible approximation. Conservation of angular momentum in a collision is imposed by expressing the effective potential of the ion–molecule system as the sum of the central potential energy $V(r)$ and the centrifugal potential energy:

$$V_{\text{eff}}(r) = -\frac{q^2\alpha}{2r^4} + \frac{L^2}{2\mu r^2} \quad [2]$$

where L is the classical orbital angular momentum of the two particles and μ is the reduced mass of the system. Here L is given by $\mu v b$ where b is the impact parameter and v is relative velocity. The total relative energy of the system is given by

$$E_{\text{r}} = E_{\text{trans}}(r) + V_{\text{eff}}(r) \quad [3]$$

where $E_{\text{trans}}(r)$ is the translational energy along the line of centres of the collision.

A plot of $V_{\text{eff}}(r)$ versus r at constant E_{r} for several values of the impact parameter is shown in **Figure 1** for the collisions of N_2^+ with N_2. For the special case in which the centrifugal barrier height equals E_{r} (at $b = b_{\text{c}}$ in **Figure 1**), $E_{\text{trans}}(r) = 0$, and the particles will orbit the scattering centre with a constant separation distance r_{c}. For $b > b_{\text{c}}$ the distance of closest approach $\geq b_{\text{c}}/\sqrt{2}$ and for $b < b_{\text{c}}$ the particles spiral into small radii before they separate. The cross section for orbiting collisions is

$$\sigma(v) = \pi b_{\text{c}}^2(v) \quad [4]$$

The rate coefficient corresponding to the orbiting or capture cross section is given by

$$k_{\text{c}} = v\sigma_{\text{c}} = v\pi q(2\alpha/E_{\text{r}})^{1/2} = 2\pi q(\alpha/\mu)^{1/2} \quad [5]$$

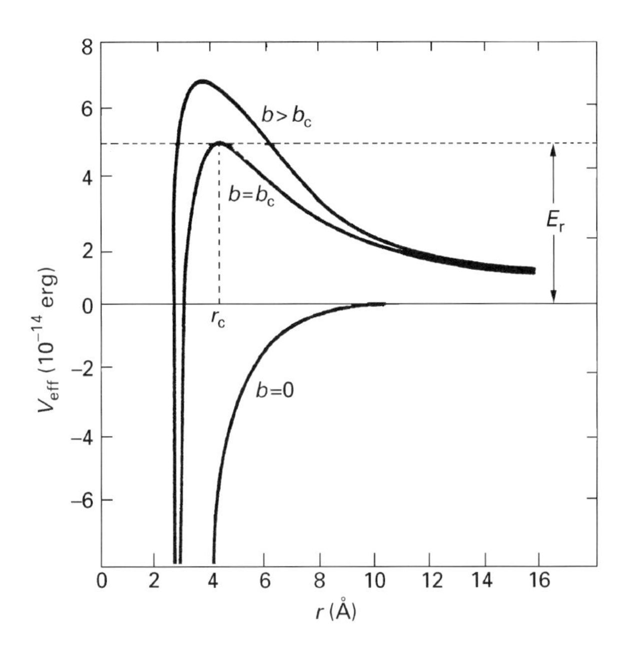

Figure 1 Plot of effective potential, V_{eff}, versus r for N_2^+ colliding with N_2 at a fixed energy of E_r and several values of the impact parameter. Reproduced with permission of Academic Press from Su T and Bowers MT (1979) In Bowers MT (ed) *Gas Phase Ion Chemistry*, Vol. 1.

which is the commonly cited energy (velocity)-independent Langevin rate constant for ion–molecule reactions. Use of these expressions implicitly assumes there is a reactive cross section σ_{rx} for real systems which is smaller than the capture cross section given by Equation [4].

The reaction cross section also sets a conceptual limit of relative velocity beyond which treatments of only long-range critical potentials cannot apply. For a hypothetical example of $\sigma_{rx} = 50$ Å^2, polarizability of 10 Å^3 and reduced mass of 50 amu, the capture cross section equals the σ_{rx} at a centre-of-mass (CM) collision energy of approximately 0.7 eV. We may infer that collision energies of the order of 1 eV roughly define the boundary between low-energy collisions – e.g., reactions in high-pressure ion sources, drift tubes, ion cyclotron resonance and plasma swarm experiments – and higher-energy collisions, considered later. Considering these limitations, it is remarkable that this simplistic theory accurately correlates collision cross section and rate coefficients for hundreds of reactions of ions with nonpolar molecules.

Ion–dipole and ion–quadrupole interactions

Over time, increasingly sophisticated modifications to the Langevin capture cross-section model have been developed to rationalize cross sections larger than Langevin and the temperature/energy dependences of cross sections. Considering first ion–dipole interactions, the classical expression for the effective potential is

$$V(r,\theta) = \frac{L^2}{2\mu r^2} - \frac{\alpha q^2}{2r^4} - \frac{\mu_D q}{r^2}\cos\theta \qquad [6]$$

where μ_D is the dipole moment of the molecule and θ is the angle between the dipole and the line of centres of the collision. The first such treatment by Hamill and co-workers assumed that the dipole 'locks in' on the ion, i.e. $\theta = 0$. This upper limit, locked-dipole approximation is obviously unrealistic for reactions at and above room temperature, for which rotation of the polar molecule tends to average out the dipolar potential.

Addressing this limitation, Su and Bowers developed the average dipole orientation (ADO) theory in which the orientation angle of the dipole with respect to the line of centres of collision is averaged as a function of ion–molecule separation distance. This concept leads to a 'locking constant', c, which is multiplied by the $\cos\theta$ term; this constant has been parametrized as a function of $\mu_D/\alpha^{1/2}$ for the range of temperatures that are of general interest in low-energy ion–molecule reactions. Further elaboration of this approach by Su and Bowers modified the treatment of angular momentum to formalize angle-averaged dipole orientation (AADO) theory. These workers also considered the interaction with a point charge of the quadrupole moment of molecules with $D_{\infty h}$ symmetry and formulated the average quadrupole orientation (AQO) theory for these types of long-range interaction.

Barker and Ridge developed a statistical model for ion–polar molecule collisions based on similar concepts. They considered the average interaction energy for a statistical ensemble of ion–neutral pairs as a function of their separation. The effective potential for such an ensemble is

$$V(r) = -\frac{\alpha q^2}{2r^4} - \frac{\mu q}{r^2}\mathscr{L}\left(\frac{\mu q}{r^2 k_B T_R}\right) + \frac{Eb^2}{r^2} \qquad [7]$$

where k_B is the Boltzmann constant, T_R is the rotational temperature of the neutral, E is the relative translational energy, b is the impact parameter and $\mathscr{L}(X)$ is the Langevin function defined as $\mathscr{L}(X) = \text{cotanh}(X) - 1/X$. The momentum transfer collision frequencies calculated from this model or from the family of ADO theories described above are in reasonable agreement with experimental measurements of rate coefficients. They rationalize both the larger cross sections found for ion–molecule reactions involving highly polar neutrals and the increase in reaction rate coefficients for polar molecules at low temperatures.

Quantum-mechanical approaches

Quantum-mechanical calculations of ion–dipole capture rate constants have been modelled successfully for low-temperature experiments and correlate well with the classical collision theory and capture cross sections just described. Clary has calculated rate constants of several ion–molecule reactions using the adiabatic capture centrifugal sudden approximation (ACCSA). Since the rotational motion of the neutral is strongly hindered by the ion at low velocity, he chose collinear reaction geometry, leading to a basis set that localizes the rotational motion of the neutral about this collinear configuration. The partial wave Hamiltonian for the entrance reaction channel is given by

$$H = -\frac{\hbar^2}{2\mu r}\frac{d^2}{dr^2}r + Bj^2 + \frac{|J-j|^2}{2\mu r^2} + V(r,\theta) \qquad [8]$$

where $\hbar$ is Planck's constant divided by 2π, μ is the reduced mass, B is the rotational constant, j is the rotational angular momentum operator, J is the total angular momentum operator, V is the ion–molecule potential energy surface, θ is the orientation angle and r is the position vector of the particle measured from the centre-of-mass. As an approximation, $|J-j|^2$ is replaced by the diagonal value $[J(J+1)\hbar^2+j^2-2\hbar^2\Omega^2]$, where Ω is the projection of both rotational and total angular momentum along the Z axis (centre-of-mass vector). Thus for a fixed value of r, the Hamiltonian becomes

$$H = Bj^2 + \frac{1}{2\mu r^2}(j^2 - 2\Omega^2\hbar^2) + V(r,\theta) \qquad [9]$$

The reaction cross section is obtained from partial wave expansion and the capture approximation discussed earlier; that is, the reaction takes place if there is enough energy to surmount the centrifugal barrier. It follows that

$$\sigma(j) = \frac{\pi}{k_j^2(2j+1)} \sum_{J=0}^{J_{\max}(j,\Omega)} \sum_{\Omega=-\min(J,j)}^{\min(J,j)} (2J+1)P_j^{J\Omega} \qquad [10]$$

where $P_j^{J\Omega}$ is the reaction probability. When $J_{\max}(j, \Omega)$ is larger than j, the sum over J in this equation gives the relationship

$$\sigma(j) = \frac{\pi}{k_j^2(2j+1)}\left[\left(\sum_{\Omega=-j}^{j}(J_{\max}(j,\Omega)+1)^2\right) - (2j+1)(j+1)^{(j/3)}\right] \qquad [11]$$

with the condition that $\sigma(j) = 0$ if the energy is below the centrifugal barrier. Rate constants calculated by this formalism are in excellent agreement with experimental room-temperature constants for such systems as positive and negative ion–molecule reactions of HCN with H^-, D^- and H_3^+ ions.

Troe has used a statistical adiabatic channel model to calculate thermal ion–molecule capture rates using a pure long-range potential. The adiabatic channel potentials $V(r)$ are calculated from perturbation theory, and analytical expressions for channel threshold energies, activated complex partition functions and capture rate constants are obtained. His results are in good agreement with trajectory calculations at temperatures above 10 K for such systems as $H_3^+ + HCN$. Additionally, Sakimoto has developed a time-independent quantum-mechanical approach for collinear triatomic systems A+BC in the energy range from thermal to several eV. Using hyperspherical coordinates, the time-independent Schrödinger equation is solved numerically using the discrete variable representation algorithm for collision energies of 1–6 eV. Partial cross sections for reactive scattering and ion dissociation for $He + H_2^+$ (plus isotopes) were calculated that generally agree with limited experimental data for this system.

High-energy ion–neutral collisions: Beam scattering

Low-energy ion–neutral collisions considered thus far often result in random angular scattering with extensive mixing of orbital and angular momentum. Beam experiments at the low-energy limit demonstrate that persistent complexes are formed but provide no further information on reaction mechanisms. In contrast, conservation of angular momentum in high-energy collisions leads to specific angular scattering, which can provide detailed information on reaction mechanisms. Deducing this information is a complex exercise and we begin with a description of elastic scattering. After these scattering characteristics have been described – both classically and quantum mechanically – we shall outline briefly how reactive scattering is treated in this framework. A collision is better described by CM coordinates, which permit us to describe kinetic energy T as the motion of the centre-of-mass of the system in the laboratory frame and the relative motion of the particles in the CM frame

$$T = \frac{1}{2}MV_{\mathrm{CM}}^2 + \frac{1}{2}\mu v^2 \qquad [12]$$

where M is the sum of the masses, V_{CM} is the velocity of the CM and the reduced mass and relative velocity have been defined previously. Since the motion of the CM is conserved in any collision, we can focus our attention on the relative motion in which the two particles are described as a single particle having mass μ, angular momentum $\mu(\boldsymbol{r} \times \boldsymbol{v})$ and kinetic energy $\frac{1}{2}\mu v^2$.

For central potentials considered previously, the angular momentum is constant and the total CM energy at separation r is

$$E = \frac{1}{2}\mu\left(\frac{\mathrm{d}r}{\mathrm{d}t}\right)^2 + V_{\mathrm{eff}}(r) \qquad [13]$$

where the effective potential

$$V_{\mathrm{eff}}(r) = V(r) + \frac{L^2}{2\mu r^2} \qquad [14]$$

includes the centrifugal potential $L^2/2\mu r^2$ described previously. The centrifugal potential is included in the equation for radial motion to account for angular motion of the system.

The observable results of a collision are the velocity of the particle after collision (deduced from its kinetic energy) and its direction of motion. For ordinary collisions between spherical particles, total kinetic energy must be conserved, so the initial and final velocities must be equal at distances large enough that $V(r)$ is negligible. **Figure 2** depicts how the deflection angle – the quantity of interest in these elastic and nonreactive collisions – relates to the interaction potential. Here a particle of reduced mass μ approaches parallel to the x axis with kinetic energy $\frac{1}{2}\mu v^2$ and angular momentum μvb. When the effective potential energy V_{eff} equals the initial kinetic energy, the particle reaches its distance of closest approach, r_c, the turning point of the trajectory. The particle then recedes, and at a large distance from the scattering centre no longer senses the interaction potential. The change in direction of motion is the deflection angle θ, and is measured from the negative x axis in **Figure 2**. We note parenthetically that deflection through $+\theta$ and $-\theta$ are experimentally indistinguishable and that deflection by $\pm 2\pi N\theta$ (where N is an integer) is indistinguishable from the deflection θ.

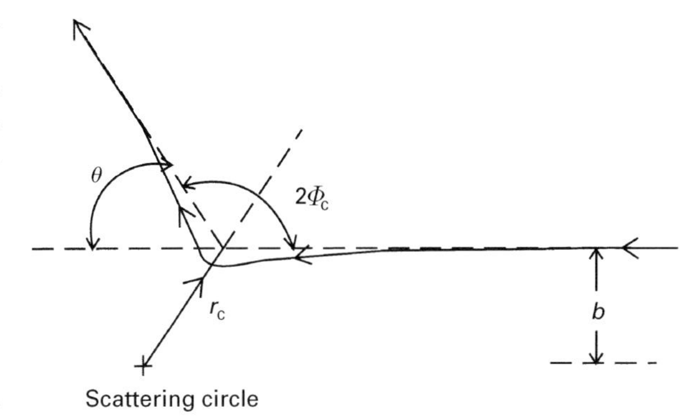

Figure 2 Deflection of particles interacting through a potential $V(r)$. A single particle of reduced mass μ and velocity v moves toward the stationary scattering centre with impact parameter b and is scattered by a central potential $V(r)$ through an angle θ. Reproduce with permission of John Wiley from Shirts RB (1986) In Futrell JH (eds) *Gaseous Ion Chemistry and Mass Spectrometry.*

Figure 2 illustrates that the deflection is symmetric about the turning point; that is, the deflection of the particle in the outgoing trajectory equals its deflection in the approach trajectory for elastic scattering. Consequently, the polar scattering angle $2\phi_c$ and the final scattering angle θ is $\pi-2\phi_c$ (see **Figure 2**). This leads to the expression for the deflection angle as a function of initial energy and impact parameter:

$$\phi(b, E) = \pi - 2b\int_{r_c}^{\infty}\frac{\mathrm{d}r}{r^2}\left(1 - \frac{V(r)}{E} - \frac{b^2}{r^2}\right)^{-1/2} \qquad [15]$$

where r_c is the outermost solution of $E = L^2/2\mu r_c^2 + V(r_c)$. Note that if $b = 0$ (head-on collision), then $\theta = \pi$ (unless $V = 0$). Further, as b increases from zero, θ decreases from π, and θ decreases to zero as b increases without limit.

A reasonable central force potential for ion–neutral elastic scattering is the Morse potential given by

$$V(r) = D(\mathrm{e}^{-2x} - 2\mathrm{e}^{-x}) \qquad [16]$$

where $x = \alpha(r - r_e)$, with α specifying the curvature of the potential at its minimum. This potential has an attractive well of depth D at the equilibrium separation, $r = r_e$. For $r < r_e$, the repulsive part of the potential rises steeply to a high value, for $r > r_e$ the attractive part of the potential is dominant. Because the Morse potential goes to zero exponentially, it does not correctly describe the potential at large r. If necessary, it can be modified by adding a term for long-range polarization forces. It nicely describes the short-range behaviour and the balance between attractive and repulsive forces.

Differential scattering

To illustrate the relationship between potential and scattering, **Figures** 3 to 5 show respectively the effective potential, deflection function and scattering intensity for the Morse potential, (Eqn [16]), parameterized to describe scattering of low-energy protons by argon atoms. The potential for zero impact parameter (hence zero angular momentum, $L = 0$) in **Figure** 3 has the parameters $r_e = 1.28$ Å, $D = 4.17$ eV and $\alpha = 1.85$ Å^3; the Morse potential with these parameters matches the experimentally determined scattering of protons by argon to better than 1% accuracy over the interval from 0 to 5 Å.

The curves in **Figure** 3 are labelled with the angular momentum, L, in units of $\hbar$. At high L, the centrifugal potential overpowers the attractive well and the scattering is always repulsive. At $L = 75.95\hbar$ there is an inflection point at $V = 1.153$ eV, and for lower angular momenta collisions the effective potential has two regions of repulsion separated by an attractive well. The outermost root is the classical turning point for the potential and a plot of the turning point r as a function of impact parameter (or, equivalently, of L) leads to a singularity where $\sqrt{c} = b$ for all collision energies. This corresponds to the point where the potential $V(r) = 0$ [~1 Å for the potential shown in **Figure** 3] and is a measure of the 'size' of the interacting particles, analogous to but different from a hard-sphere radius for the collision.

Figure 4 shows the deflection function calculated from Equation [15] for the effective potential given in **Figure** 3 for a number of collision energies. For zero impact parameter, the deflection angle is π, just as it would be for hard-sphere collisions, with positive scattering angle corresponding to repulsive collisions. The deflection angle goes through zero and becomes negative as the collision senses the attractive potential at longer distances. As b increases, the deflection angle goes smoothly to zero. For moderate energy collisions, the centrifugal potential dominates scattering at all impact parameters. For very low energies, < 1.153 eV for the potential illustrated, there are three roots of the deflection function. (The reader is reminded that positive and negative deflections are experimentally equivalent.) The 'piling up'

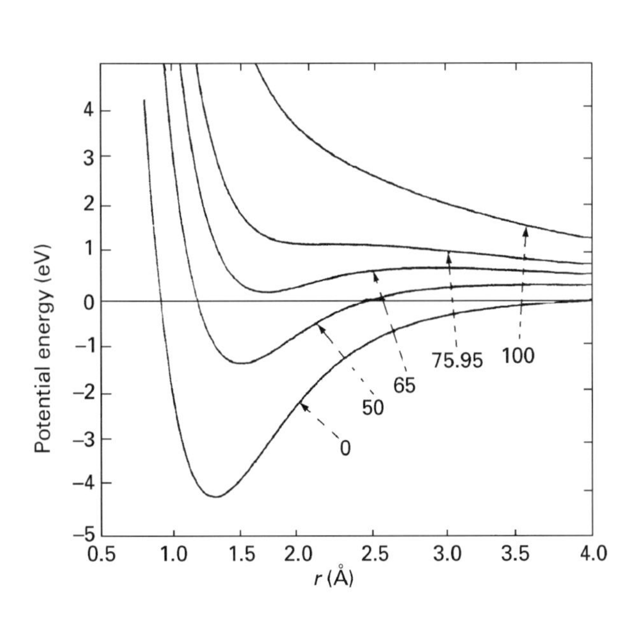

Figure 3 Plot of effective potentials derived using the Morse potential for several values of angular momentum, L, as marked on each curve. See text for the parameters used for the Morse potential. Reproduced with permission of John Wiley from Shirts RB (1986) In Futrell JH (eds) *Gaseous Ion Chemistry and Mass Spectrometry*.

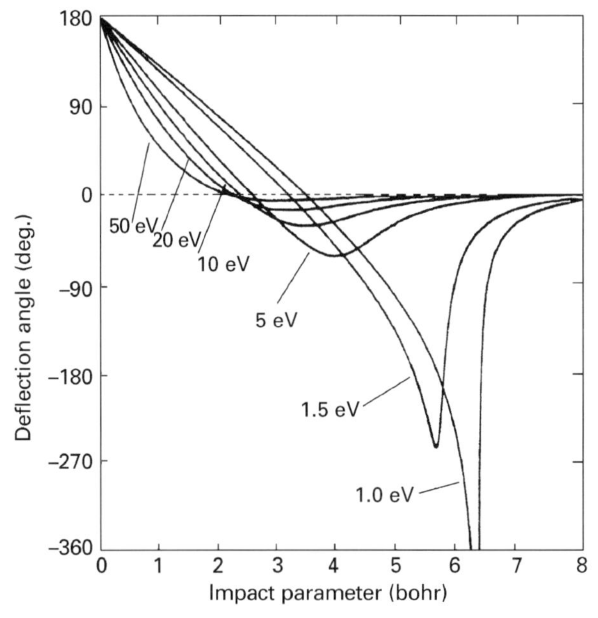

Figure 4 The relationship between deflection function and impact parameter for interactions governed by the Morse potential of **Figure 3** at the indicated collision energies. 1 bohr = 0.529 Å. Reproduced with permission of John Wiley from Shirts RB (1986) In Futrell JH (eds) *Gaseous Ion Chemistry and Mass Spectrometry*.

of intensity at a particular scattering angle is called the rainbow angle. Measurement of rainbow scattering is an important experimental parameter for deducing the scattering potential. For specific L values below this critical energy, two of the roots coincide at the maximum of the effective potential. For these trajectories, orbiting occurs and the deflection angle becomes infinite. This phenomenon is readily detected in careful beam scattering experiments; it corresponds to the capture cross section in our description of low-energy collisions.

Figure 5 illustrates the connection of crossed-beam measurements of ion scattering with the interaction potential of the ion–neutral collision, again using the Morse potential. The quantity measured experimentally is the differential scattering cross section for a given CM collision energy. The differential cross section, $I(\theta)$, at a specified energy is given by

$$I(\theta)\mathrm{d}\Omega = \frac{I(b)}{I_0} = I(\theta)[2\pi \sin\theta\, \mathrm{d}\theta] = 2\pi b\, \mathrm{d}b \quad [17]$$

$$I(\theta) = \frac{b}{|\sin\theta\, (\mathrm{d}\theta/\mathrm{d}b)|} \quad [18]$$

where $\mathrm{d}\Omega$ is the differential of solid angle. The cross section calculated for the Morse potential shown in

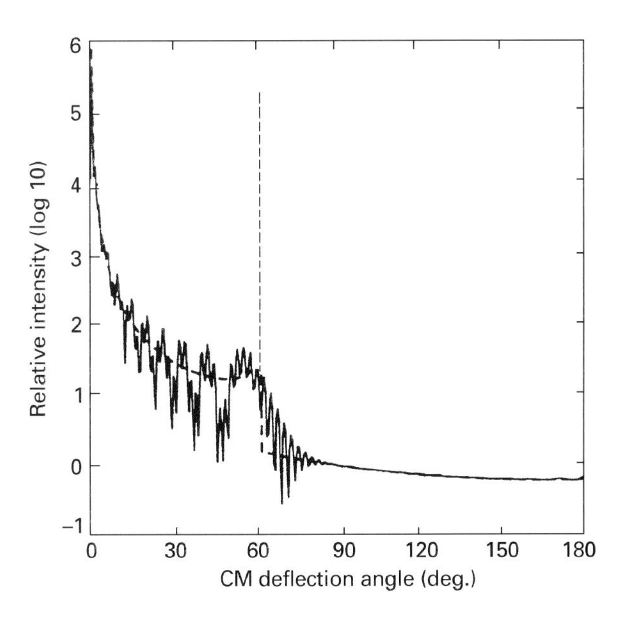

Figure 5 Plot of scattering intensity (differential cross section) versus centre-of-mass scattering angle for interactions with Morse potential of **Figure 3** using classical treatment. Reproduced with permission of John Wiley from Shirts RB (1986) In Futrell JH (eds) *Gaseous Ion Chemistry and Mass Spectrometry.*

Figure 5 illustrates for several energies the singularities determined by the two terms in the denominator of Equation [18]. The sin θ term gives infinity at CM angles of zero and π. The rainbow angle gives rise to a singularity when $\mathrm{d}\theta/\mathrm{d}b$ goes to zero; this occurs when several impact parameters give the same deflection angle. At higher collision energies this occurs at small angles; with decreasing collision energy the rainbow angle increases smoothly with decreasing energy and, for 1.5 eV is greater than 180°. Numerically it is 254° for this potential, leading to an observed scattering angle of 360° –254° = 106°. Two details distinguish this class of trajectories experiment from high energy collisions – namely, the singularity at 180° and the reversal of the 'bright side' of the rainbow. The backscattering of particles corresponding to the sin θ singularity at 180° is the experimental demonstration that particles are scattered by a potential sufficiently attractive at that collision energy to hold them together for at least half a revolution.

Reactive encounters

We have explained in some detail the relationship of measured variables in elastic scattering to the interaction potential between the ion and neutral. This framework is the starting point for describing inelastic and reactive scattering. Inelastic scattering may be introduced as an instantaneous conversion of kinetic energy of motion into internal energy of either or both of the collision partners, while reactive collisions involve both energy exchange and reaction. It is plausible that these physical and chemical conversions occur at or near the distance of closest approach in the respective trajectories. This breaks the symmetry of the elastic scattering formalism and the retreat trajectory of products involves both a different energy inventory in the collision partners and a change in the potential governing the trajectory as they separate from the collision centre. This approach has been reasonably successful in rationalizing the differential scattering cross sections of a variety of proton transfer reactions. The time-dependent quantum-mechanical treatment of collisions described in the next section automatically takes into account energy transfer and reaction – at least in principle.

Newton diagrams

As noted previously, describing collision processes in the CM frame is more informative than in the laboratory (LAB) frame. Explicitly the motion of the CM is conserved in collisions and is unavailable to the reactants. The proper framework for displaying collisions is the CM relative velocity diagram, frequently called the Newton diagram. The collision process (for simplicity assumed to occur at right angles) of

$$M_1^+ + M_2 \rightarrow M_3^+ + M_4 \qquad [19]$$

with M_1^+ and M_2, having velocities of V_1 and V_2, is depicted in **Figure 6**. The collision centre is the origin of the laboratory reference frame. In the CM frame, M_1^+ and M_2 move collinearly towards each other with velocities U_1 and U_2 and collide at the CM located on their relative velocity vector at the point defined by conservation of momentum such that $M_1U_1 = M_2U_2$. After collision, the two particles recoil from the CM in opposite direction with velocities $U_1'$and U_2'. The velocity of the CM is given by

$$V_{CM} = \frac{M_1 V_1}{M_1 + M_2} + \frac{M_2 V_2}{M_1 + M_2} \qquad [20]$$

and is constant. The relative kinetic energy of the collision partners, E_{CM}, is available for the reaction and is given by

$$E_{CM} = \frac{1}{2} M_1 U_1^2 \left(1 + \frac{M_1}{M_2}\right) \qquad [21]$$

Similarly, the postcollision relative kinetic energy is given by

$$E'_{CM} = \frac{1}{2} M_1 U_1'^2 \left(1 + \frac{M_1}{M_2}\right) \qquad [22]$$

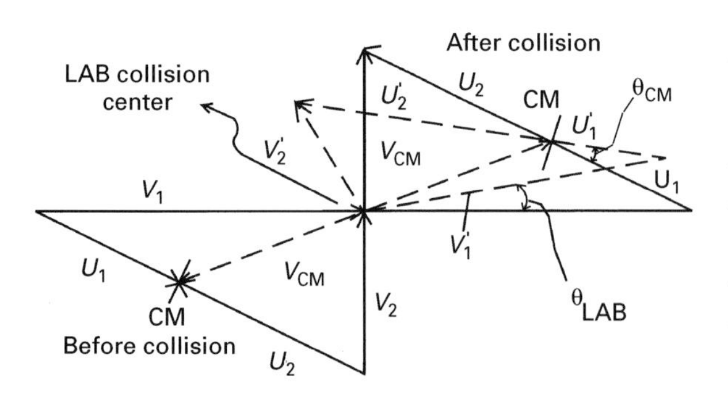

Figure 6 Newton diagram showing pre- and postcollision velocity vectors in the laboratory and centre-of-mass reference frames for the collision of M_1^+ and M_2 having velocities of V_1 and V_2.

The difference $\Delta T = E'_{CM} - E_{CM}$, the translational exoergicity of the collision process, is a direct measure of the energy used in the process. Consequently, experimental measurements and transformation of mass, intensity, energy and angular distributions of product ions as Newton diagrams provides quantitative information on the energetics and dynamics of the collision process.

Two limiting cases of ion–molecule reaction dynamics will serve to illustrate the utility of Newton diagrams. If a collision between an ion and a molecule proceeds via an intermediate whose lifetime is larger than several rotational periods ($\sim 10^{-12}$ s), internal energy redistribution is usually complete and the complex loses its memory of the reactant velocity vectors. The products will separate with equal probability on both sides of a plane passing through the CM and normal to the collision axis, giving rise to the forward–backward symmetry. Such orbiting complexes are usually formed at relative kinetic energies that are of the order of or less than the well depth of the interaction potential governing the collision.

As relative kinetic energy increases, the lifetime of the intermediate complex decreases and there is a transition to a direct reaction. The interaction time is too short for the intermediate to rotate and the Newton diagram contour plot becomes asymmetric with respect to the CM. The dependence of reaction probability on impact parameter and the effective potential for the approach and retreat trajectories determine the scattering angle and the form of the Newton diagram.

Semiclassical theory of collisions, the WKB phase shift

In three dimensions and for central potentials, the Schrödinger equation separates in spherical coordinates analogously to corresponding classical equations of motion. The incoming particle is described by a wavefunction with a local wavenumber $k(r)$. Repulsive potentials decrease the velocity of the wave packet and the resulting scattered wavefunction has fewer nodes, resulting in a negative phase shift. Conversely, attractive potentials result in an increase in velocity; the wavefunction acquires more nodes and this results in a positive phase shift. The overall phase shift for scattered particles is

$$\delta(E,b) = \underbrace{\int k(r)\,\mathrm{d}r}_{\text{actual path}} - \underbrace{\int k_0(r)\,\mathrm{d}r}_{\text{unscattered path}} \qquad [23]$$

Integrating this expression from the distance of closest approach to infinity (this gives half the total difference in phase for the entire trajectory; this is the convention for phase shift) gives

$$\delta(E,b) = \frac{\mu v}{h}\lim_{R\to\infty}\left[\int_{r_c}^{R}\left(1-\frac{V(r)}{E}-\frac{b^2}{r^2}\right)^{1/2} \mathrm{d}r - \int_{b}^{R}\left(1-\frac{b^2}{r^2}\right)^{1/2} \mathrm{d}r\right] \quad [24]$$

For both integrals, the lower limit of integration is the (outermost) zero of the associated integrand. Since both integrals are divergent, δ depends on cancellation between the two. This difficulty is repaired by subtracting $\int_{r_c}^{R}\mathrm{d}r$ from both terms:

$$\delta(E,b) = \frac{\mu v}{\hbar}\left[\int_{r_c}^{\infty}\left(1-\frac{v(r)}{E}-\frac{b^2}{r^2}\right)^{1/2} - 1\right]\mathrm{d}r - r_c + \frac{\pi b}{2} \quad [25]$$

This may be recast into a form incorporating the definition of phase shift from Equation [23]:

$$\delta(E,b) = \int_{r_c}^{\infty}(k(r)-k)\,\mathrm{d}r - k\left(r_c - \frac{\pi b}{2}\right) \quad [26]$$

where the integral term gives the phase difference resulting from the interaction potential and a correction for the difference in turning points of the unscattered and scattered particle includes the effect of the centrifugal potential.

Finally, it can be shown that the phase shift method leads to expressions for the collision time (different between $V(n) \neq 0$ and $V(n) = 0$) and for scattering angle.

$$\tau(E,b) = 2\hbar\left(\frac{\partial\delta(E,b)}{\partial E}\right)_b \quad [27]$$

$$\theta(E,b) = \frac{2\hbar}{\mu\nu}\left(\frac{\partial\delta(E,b)}{\partial E}\right)_E \quad [28]$$

This makes the precise connection we seek between WKB scattering theory and the classical model developed previously.

Quantum scattering

Moving from a semiclassical towards a full quantum treatment we note that, at large distances, the wavefunction for a scattered particle is a combination of an incoming plane wave and a scattered wave:

$$\psi(r,\theta) = \mathrm{e}^{\mathrm{i}kz} + \frac{f(\theta)}{r}\mathrm{e}^{\mathrm{i}kr} \quad [29]$$

where the z direction is the direction of travel. The first term describes a plane wave and the second term is an outgoing spherical wave whose amplitude is $f(\theta)$. This simple form of the wavefunction is valid for r large enough that no interference between the incoming and outgoing waves is important. This applies experimentally to a collimated beam of finite width.

Since the wavefunction in Equation [29] obeys the Schrödinger equation, it can be expanded in any complete set of orthogonal functions, for example Legendre polynomials:

$$\psi(r\ \theta) = \sum_{l=0}^{\infty}\frac{\psi_2(r)}{kr}P_l(\cos\theta)(2l+1)i^l \quad [30]$$

where P_l is Legendre polynomial. Multiplying by $P_{l'}(\cos\theta)$, and integrating over θ we obtain for large values of r the asymptotic result

$$\left(\frac{\mathrm{d}^2}{\mathrm{d}r^2}+k^2\right)\psi_l(r) = 0 \quad [31]$$

which has solutions $\psi_1(r) = a_1\sin(kr - l\pi/2 + \delta_l)$. Here the phase shift, δ_l, for the lth partial wave contains all the information about the potential, and the term $l\pi/2$ comprises the centrifugal potential. The expression for scattering amplitude is obtained as

$$f(\theta) = \sum_{l=0}^{\infty}\frac{2l+1}{k}\mathrm{e}^{\mathrm{i}\delta_l}\sin\delta_l\,P_l(\cos\theta) \quad [32]$$

The differential cross skection $I(\theta)$ is the probability density for observing a particle at an angle $\theta, |f(\theta)|^2$ and the total cross section is obtained by integration

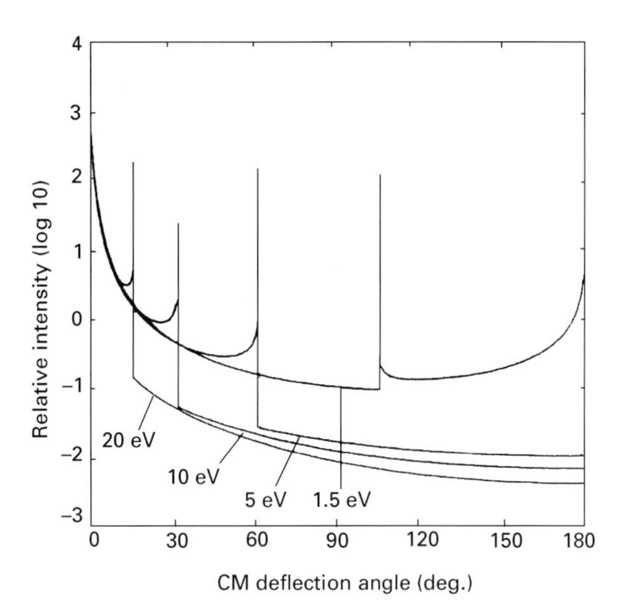

Figure 7 Plot of scattering intensity versus centre-of-mass scattering angle at 5 eV collision energy using quantum-mechanical treatment. The calculations were performed using 500 phase shifts and the Morse potential of **Figure 3**. Reproduced with permission of John Wiley from Shirts RB (1986) In Futrell JH (ed) *Gaseous Ion Chemistry and Mass Spectrometry.*

of $I(\theta)$ over all angles as given below.

$$\sigma = \frac{4\pi}{k^2}\sum_{l=0}^{\infty}(2l+1)\sin^2\delta_l \qquad [33]$$

The phase shifts for calculating scattering as a function of angle are obtained by starting at $r = 0$ and increasing r until $V(r)$ becomes negligible.

Figure 7 illustrates the differential cross section for the Morse potential (Eqn [17]) calculated using Equation [32] and 500 phase shifts that were calculated using Equation [26]. The relative energy is 5 eV. The classical differential cross section from **Figure 5** is also shown. The quantum-mechanical form of $I(\theta)$ removes the classical singularity. The additional peaks occurring at angles smaller than the rainbow angle are interferences from the three different values of the impact parameter that contribute to scattering into these angles. The quantum result tends to oscillate about the classical mechanical result, which cannot recover the interferences of the quantum result.

For noncentral potentials – e.g. molecular scattering – the Schrödinger equation does not separate into partial waves. The solution can still be expanded in partial waves, but then the individual partial-wave radial Schrödinger equations will have additional terms that couple them together. Sophisticated computer programs are now available to solve these complex equations and calculate the detailed dynamics of a collision process.

List of symbols

b = impact parameter; B = rotational constant; c = locking constant; D = depth of attractive potential well; E = energy; $f()$ = function; H = Hamiltonian; $I(\theta)$ = differential cross section; j = rotational angular momentum operator; J = total angular momentum operator; k = rate coefficient; local wavenumber; k_B = Boltzmann constant; $\mathscr{L}$ = Langevin function; L = classical orbital angular momentum; P = Legendre polynomial reaction probability; q = ionic charge; r = distance; magnitude of position vector; polar coordinate; r = position vector; r_e = equilibrium distance; T = kinetic energy; T_R = rotational temperature; v = relative velocity vector; v = magnitude of relative velocity vector; $V(r)$ = central potential energy; V, V', U, U' = velocity in centre-of-mass frame; V_{CM} = velocity of centre-of-mass; α = atomic/molecular polarizability; curvature of potential energy curve at minimum; δ = phase shift on scattering; θ = angle between dipole and line of centres; final scattering angle; polar coordinate; μ = reduced mass; μ_D = dipole moment; σ = collision cross section; τ = collision time; φ_c = polar scattering angle; ψ = wavefunction; Ω = projection of rotational and total angular momentum on Z axis; solid angle.

Subscripts

c = capture; c = closest approach; CM = centre of mass; e = equilibrium; eff = effective; r = relative; trans = translational

See also: **Ionization Theory; Quadrupoles, Use of in Mass Spectrometry.**

Further reading

Chesnavich WJ, Su T and Bowers MT (1979) Ion-dipole collisions: recent theoretical advances. In Ausloos P (ed) *Kinetics of Ion Molecule Reactions*, pp. 31–53. New York: Plenum.

Child MS (1973) *Molecular Collision Theory.* New York: Academic Press.

Fluendy MAD and Lawley KP (1973) *Chemical Applications of Molecular Beam Scattering.* London: Chapman and Hall.

Franklin JL (ed) (1972) *Ion Molecule Reactions.* New York: Plenum Press.

Futrell JH (ed) (1986) *Gaseous Ion Chemistry and Mass Spectrometry.* New York: Wiley.

Levine RD and Bernstein RB (1987) *Molecular Reaction Dynamics and Chemical Reactivity*. Oxford: Oxford University Press.
McDaniel EW (1964) *Collision Phenomena in Ionized Gases*. New York: Wiley.
McDaniel EW, Cermak V, Dalgarno A, Ferguson EE and Friedman L (1970) *Ion–Molecule Reactions*. New York: Wiley-Interscience.
Shirts RB (1986) Collision theory and reaction dynamics. In Futrell JH (ed) *Gaseous Ion Chemistry and Mass Spectrometry*, pp. 25–57.New York: Wiley.
Steinfeld JI, Francisco JS and Hase WL (1989) *Chemical Kinetics and Dynamics*. Englewood Cliffs, NJ: Prentice-Hall.
Su T and Bowers MT (1979) Classical ion-molecule collision theory. In Bowers MT (ed) *Gas Phase Ion Chemistry*, Vol. 1, pp. 83–118. New York: Academic Press.
Vestal ML, Wahrhaftig AL and Futrell JH (1976) Application of a modified elastic spectator model to proton transfer reactions in polyatomic system. *Journal of Physical Chemistry*, 80: 2892–2899.

Ion Dissociation Kinetics, Mass Spectrometry

Bernard Leyh, F.N.R.S. and Université de Liège, Belgium

MASS SPECTROMETRY
Theory

The ionization process taking place in the source of a mass spectrometer or the activation step of a tandem mass spectrometry experiment leads to ions with a range of internal energies extending usually well above the first dissociation threshold, i.e. the molecular ion has enough energy to dissociate. Dissociation is a dynamic process characterized by its associated lifetime. The whole story of kinetics in general and of ion dissociation kinetics in particular is to measure how fast a dissociation proceeds under given conditions (state selection, well-defined energy or temperature, for example) and to determine the reaction mechanisms. Both aspects will be addressed.

Unimolecular rate constant

Dissociation kinetics of state-selected reactant ions has been investigated only in a limited number of relatively small systems. However, a large amount of data are now available on ionic reactions investigated with internal energy selection.

The dissociation of an energy-selected molecular ion is a unimolecular process, the rate of which, $R(E,t)$ is defined by the following equation, where N_0 is the initial number of ions formed:

$$\mathrm{ABC^+\ (internal\ energy = E) \rightarrow AB^+ + C}$$

$$R(E,t) = \frac{\mathrm{d}N_{\mathrm{AB^+}}}{\mathrm{d}t} = k_{\mathrm{AB^+}}(E)N_{\mathrm{ABC^+}}(E,t) = k_{\mathrm{AB^+}}(E)N_0 \exp\left(-\,k_{\mathrm{AB^+}}(E)t\right)$$

This represents the most simple situation where a single dissociation process occurs for ions possessing a single total energy, *E*. $k(E)$ is called the unimolecular rate constant. However, in practice, the parent ion can be formed with a broad distribution of internal energies, $P(E)$, and competitive reactions can take place. If the different competitive channels are characterized by rate constants denoted $k_i(E)$, then the following equation holds for the rate of production of the *j*th fragment ion:

$$R_j(E,t) = \frac{\mathrm{d}N_j}{\mathrm{d}t} = N_0 \int_E \mathrm{d}E\, k_j(E)P(E) \exp\left(-\sum_i k_i(E)t\right)$$

If further dissociations occur, the rates must be calculated by solving the set of differential equations resulting from the appropriate kinetic scheme.

If the dissociating ion is characterized by a welldefined internal energy, i.e. if $P(E)$ reduces to a single value, then each ion of the ionized sample can be considered to be one of the replicas of a microcanonical ensemble. For this reason, $k(E)$ is often referred to as the 'microcanonical rate constant'. If, however, $P(E)$

is a thermal Boltzmann distribution characterized by a temperature T, then the corresponding rate constant, $k(T)$ is called the 'canonical rate constant'.

One of the most widespread approaches to model the experimental microcanonical rate constant is the RRKM-QET statistical theory, which postulates that a rapid internal energy randomization takes place before dissociation and that a transition state can be defined. This leads to the following well-known equation:

$$k(E) = \frac{\sigma N^{\ddagger}(E - E_0)}{h\rho(E)}$$

where σ is the degeneracy of the reaction path (e.g. $\sigma = 4$ for the loss of H from CH_4), $N^{\ddagger}(E - E_0)$ is the number of states of the transition state up to an energy $E - E_0$ above the critical energy E_0 and $\rho(E)$ is the density of states of the parent ion at the internal energy E.

If $k(E)$ is averaged over a thermal internal energy distribution at temperature T, the equation of transition state theory for the canonical rate constant is recovered. It can be expressed as a function of the activation entropy, $\Delta S^{\circ\ddagger}$, and of the activation enthalpy, $\Delta H^{\circ\ddagger}$.

$$k(T) = \frac{k_B T}{h} \exp\left(\frac{\Delta S^{\circ\ddagger}}{R}\right) \exp\left(-\frac{\Delta H^{\circ\ddagger}}{RT}\right)$$

k_B and h are respectively Boltzmann's and Planck's constants. The value of the activation entropy gives insight into the looseness or the tightness of the transition state. Loose transition states are usually associated with simple bond cleavages, whereas rearrangement processes are expected to involve tighter transition states. Competitive processes, multistep fragmentations, nonadiabatic processes and tunnelling effects affect the variation of the rate constant with internal energy, which then forms a diagnostic tool to investigate such effects. Furthermore, as in neutral physical organic chemistry, kinetic isotope effects (mostly upon deuteration) are one of the most powerful tools to investigate the mechanisms of a chemical reaction.

Experimental determination of the microcanonical rate constant

Various experimental techniques are now available to measure the unimolecular rate constant over quite an extended range, typically from 10^2 to 10^{10} s^{-1}. Broadly speaking, most experiments can be classified in one or the other of the three following groups, depending on the rate constant range accessible: microsecond time window ($10^5 < k < 10^7$ s^{-1}), nanosecond time window ($10^7 < k < 10^{10}$ s^{-1}) and millisecond time range ($10^2 < k < 10^5$ s^{-1}). Trapping techniques are instrumental in this last range. The techniques pertaining to the first two groups are based on the fact that the kinetic energy acquired by a fragment ion produced within an accelerating electric field depends on the position (or, equivalently, on the time) at which dissociation took place. The rate constant can then be extracted from the shape of an ion kinetic energy spectrum or of a time-of-flight distribution. The rate constant range accessible in a given experiment depends on the magnitude of the electric field.

Microsecond time window

Ions dissociating in the microsecond time window, i.e. with a rate constant in the 10^5–10^7 s^{-1} range, have received considerable attention. This range is very often referred to as the metastable time frame and processes occurring within this time window are called metastable Photoelectron–photoion coincidence (PEPICO) spectroscopy is the technique of choice in this range. The most useful version of this technique, called threshold PEPICO (TPEPICO), initiated by Stockbauer and by Baer, uses a variable-wavelength photon source, nowadays in most cases a synchrotron or a tunable laser, and detects ions in delayed coincidence with zero-kinetic-energy electrons (threshold electrons). Under such conditions, the ion internal energy is equal to the photon energy less the adiabatic ionization energy of the molecule. By combining angular and time-of-flight discriminations of energetic electrons, a resolution of about 10 meV can be obtained. Even better resolutions are reached by using recent zero-kinetic-energy techniques. The data have to be corrected for the analyser apparatus function, for the thermal distribution of the parent ions and for the possible release of translational energy to the dissociation fragments. Extracting rate constant information in the presence of an appreciable translational energy release is somewhat hazardous.

The upper limit in the rate constant range accessible by threshold PEPICO is dictated by the extraction field which cannot exceed a few V cm^{-1} to maintain a reasonable energy resolution.

Another way of selecting ions with a well-defined internal energy involves multiphoton ionization (MPI) where the vibrationless ground state of the molecular ion is first prepred and then photodissociated by absorption of an additional photon. In practice, the upper rate constant limit is about the same as in PEPICO. The major advantage of MPI is it makes rotational state selection possible, so that the influence of angular momentum on ionic fragmentations can be investigated.

Nanosecond time window

In field ionization, much higher electric fields can be achieved (of the order of 10^5 V cm^{-1}): the ions are then much more strongly accelerated, and dissociation decays in the nanosecond to picosecond range can be monitored. No internal energy selection is performed in such experiments. Nontrivial data handling procedures assuming some reasonable form for the internal energy distribution have been developed to obtain $k(E)$ data.

Photodissociation mass-analysed ion kinetic energy spectrometry (PD-MIKE) has been designed by Kim and co-workers to monitor nanosecond range decays. Laser photodissociation of mass-selected ions is induced in the presence of a strong electric field within the region located between the magnet and the electrostatic analyser of a reverse-geometry sector mass spectrometer (a region which would be otherwise field-free). The internal energy range of the parent ion prior to photodissociation is relatively broad but can be narrowed by collisional cooling inside the ion source.

Millisecond time window

Slow dissociation rates (10^2–10^4 s^{-1}) have been measured in Dunbar's laboratory by time-resolved photodissociation, which consists of trapping ions in an ICR cell during a variable delay time after a photodissociating photon pulse. The technique called 'time-resolved photoionization mass spectrometry', developed by Lifshitz, consists of trapping photoions in a cylindrical trap at very low pressure to avoid bimolecular collisions, and then ejecting them into a mass filter after a variable delay covering the microsecond to millisecond range. When the dissociation rate constant becomes lower than *ca.* 10^3 s^{-1}, competition with infrared fluorescence takes place and limits the lifetime of the decomposition process. This has to be taken into account to extract the dissociation rate constant from the experimental data.

The rate constant for the $C_6H_5Br^{+\bullet} \rightarrow C_6Hc + Br^{\bullet}$ reaction has been measured by all these experimental techniques, so that a broad range could be sampled (**Figure 1**). Note that very good agreement is obtained between independent measurements in the metastable domain.

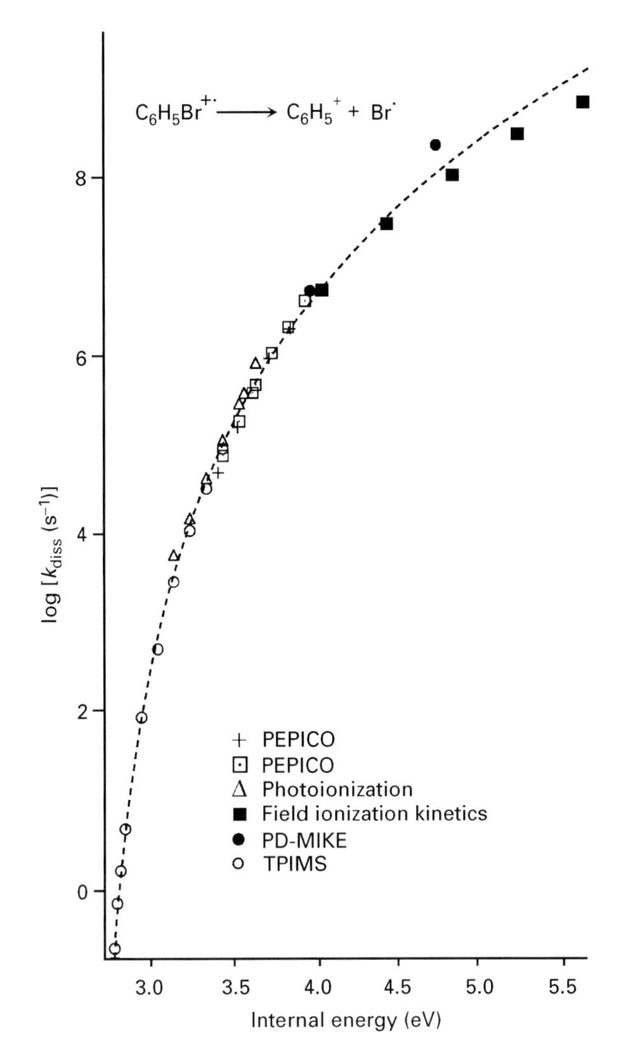

Figure 1 Microcanonical rate constant for the $C_6H_5Br^{+\bullet} \rightarrow C_6H_5^+$ + Br• decomposition obtained using various experimental techniques. (+) Baer *et al* (1976) (+) Baer *et al* (1982) (⊡) Rosenstock *et al* (1980) (Δ) Pratt and Chupka (1981) (■) Brand *et al* (1984) (●) Lim *et al* (1998) (○) Malinovich *et al* (1985).

Canonical rate constant

Flow tube techniques (flowing afterglow, selected ion flow tube) have been used extensively to study ion–molecule reactions and to measure their rate coefficients. There are far fewer studies on the thermal decomposition of gaseous ions. Recently, cluster ions and noncovalent complexes have, however, received considerable attention.

Dissociation rate constants of ions at atmospheric pressure have been obtained by ion mobility spectrometry. The time frame of such experiments is in the millisecond range. Dissociation rates of proton-bound amine dimer ions, for example, have been measured recently.

In the technique called 'blackbody infrared radiation dissociation' (BIRD), developed in E.R. Williams' laboratory, ions trapped in an ICR cell at very low pressure absorb infrared photons emitted by the cell walls. Dissociation decays in the second to minute time frame can be followed. In both techniques, the analysis of the temperature dependence of the rate constant leads to activation parameters. BIRD has been applied to biologically important non-covalent complexes like proton-bound dimers of amino acids or double-stranded DNA complexes.

Kinetic isotope effects

Isotope effects, mostly involving hydrogen–deuterium, are one of the most powerful tools for chemical kinetics investigations. The kinetic isotope effect is defined as the ratio of the rate constants corresponding to the unlabelled and labelled species at a given internal energy. A distinction is made between intermolecular and intramolecular kinetic isotope effects. An intramolecular isotope effect refers to the competition, from a unique partially labelled molecular ion, between two channels differing only by the location of the heavier isotope, e.g.:

$$\mathrm{CHDOH^+} \rightarrow \begin{cases} \mathrm{DCO^+ + H_2} \\ \mathrm{HCO^+ + HD} \end{cases}$$

An intermolecular isotope effect refers to two equivalent dissociation pathways from two differently labelled molecules. As an example:

$$\mathrm{C_2H_5Cl^{+\bullet} \rightarrow C_2H_4^{+\bullet} + HCl}$$
$$\mathrm{C_2D_5Cl^{+\bullet} \rightarrow C_2D_4^{+\bullet} + DCl}$$

For intramolecular isotope effects, as one starts from a unique precursor, the two processes correspond to the same internal energy distribution, so that useful information can be inferred even from ionabundances measured without internal energy selection (like metastable dissociations or field ionization kinetics). This is not the case for intermolecular effects, where there is no warranty that the two internal energy distributions are the same. Within the RRKM framework, the intramolecular kinetic isotope effect depends only on the transition state properties (critical energy, rotational constants and vibrational frequencies) and not on the reactant properties.

In primary isotope effects, the isotopically substituted atom is directly involved in bond breaking or bond formation. In secondary isotope effects, the substituted atom is remote from the incriminating bond. Primary isotope effects occur largely because of changes in the critical energy E_0 upon labelling. Secondary isotope effects result from changes in the densities of states. Intermolecular secondary isotope effects can be larger than primary effects. The larger the critical energy, the larger the secondary isotope effect. A kinetic isotope effect, where k_H/k_D is larger than unity, is said to be normal. If $k_H/k_D < 1$, the effect is called inverse.

Isotope effects increase when internal energy decreases. Large effects are therefore expected in the metastable time range which corresponds to relatively low internal energies. Isotope effects are instrumental in (i) localizing the particular atoms involved in H-shift reactions, (ii) providing evidence of hydrogen scrambling, (iii) distinguishing between stepwise or concerted processes, (iv) postulating a reasonable structure for the transition state, (v) deciding whether tunnelling is involved or not. Examples will be given in the next sections.

Reaction mechanisms from kinetic data

We will consider, successively, reactions involving only one potential energy surface and those where atleast two electronic states of the same or of a different multiplicity interact. These latter reactions are called nonadiabatic. Among the adiabatic reactions, we will move from the simple to the more complex, from one-step dissociations to competitive processes and finally to multistep fragmentations. While convenient, this classification is oversimplified. A multistep fragmentation, for example, can involve both adiabatic and nonadiabatic intermediate steps.

One-step dissociations

Dissociations of substituted benzene ions are probably among the most extensively studied ionic reactions. Most of them were satisfactorily modelled by the RRKM equation. The looseness or tightness of the transition state is usually inferred from the value of the activation entropy (at a somewhat arbitrary temperature, usually 1000 K) resulting from the RRKM fit. A clear-cut positive value for $\Delta S^{\circ\ddagger}_{1000\ \mathrm{K}}$ (usually not exceeding 35 J K^{-1} mol^{-1}) indicates a loose transition state, whereas a slightly positive or a negative value (usually not smaller than −30 J K^{-1} mol^{-1}) is the signature of a tighter transition state. **Table 1** illustrates this for a few selected reactions,

Table 1 Entropies of activation for selected one-step dissociations

Reaction	$\Delta S^{\circ\ddagger}_{1000\,K}$ (J K^{-1} mol^{-1})
$C_6H_5Br^{+\bullet} \rightarrow C_6H_5^+ + Br^\bullet$	33.8[a]
$C_6H_5I^{+\bullet} \rightarrow C_6H_5^+ + I^\bullet$	31.1[b]
$C_6H_5OH^{+\bullet}$(phenol)$\rightarrow$ c-$C_5H_6^{+\bullet}$ + CO	9.2[c]
p-$C_6H_4Cl_2^{+\bullet}$ + $C_6H_4Cl^+$ + $Cl^\bullet$	7.5[d]
$C_6H_5CN^{+\bullet} \rightarrow C_6H_4^{+\bullet}$+ HCN	1.7[e]
$C_8H_8^{+\bullet}$(styrene) $\rightarrow C_6H_6^{+\bullet} + C_2H_2$	−26.8[f]

[a] Malinovich Y, Arakawa R, Haase G and Lifshitz C (1985) Time-dependent mass spectra and breakdown graphs. 6. Slow unimolecular dissociatioon of bromobenzene ions at near threshold energies. *Journal of Physical Chemistry* **89**: 2253–2260.
[b] Malinovich Y, and Lifshitz C (1986) Time-dependent mass spectra and breakdown graphs. 7. Time-resolved photoionization mass spectrometry of iodobenzene. The heat of formation of $C_6H_5^+$. *Journal of Physical Chemistry* **90**: 2200–2203.
[c] Malinovich Y and Llifshitz C (1986) Time-dependent mass spectra and breakdown graphs. 8. Dissociative photoionization of phenol. *Journal of Physical Chemistry* **90**: 4311–4317.
[d] Olesik S, Baer T and Morrow JC (1986) Dissociation rates of energy-selected dichloro- and dibromobenzene ions. *Journal of Physical Chemistry* **90**: 3563–3568.
[e] Rosenstock HM, Stockbauer R and Parr AC (1980) Photoelectron-photoion coincidence study of benzonltrile. *Journal de Chimie Physique* **77**: 745–750.
[f] Dunbar RC (1989) Time-resolved unimolecular dissociation of styrene ion. Rates and activation parameters. *Journal of the American Chemical Society* **111**: 5572–5576.

which are believed to involve a single dissociative step, i.e. to take place on a single-well potential energy surface.

At first sight, it might seem inconsistent to use a thermal quantity to characterize a microcanonical rate constant. However, the canonical rate constant is related, via a thermal average, to the microcanonical rate constant with the same transition state. The advantage of the canonical formulation is twofold. First, the essential information is compacted into two factors, the critical energy and the entropy of activation, rather than being scattered among many individual vibrational frequencies. Second, it makes a comparison possible with thermal data for the neutral counterpart of an ionic reaction. Alternatively, some authors calculate the microcanonical entropy of activation at infinite internal energy, so that the critical energy becomes negligible, according to:

$$\Delta S^{\ddagger}_{\mu}(E \rightarrow +\infty) = R \ln \frac{N^{\ddagger}(E - E_0 \rightarrow +\infty)}{h\nu_R \rho(E \rightarrow +\infty)}$$

ν_R is the frequency associated with the reaction coordinate. For the $C_6H_6^{\bullet +} \rightarrow C_6H_5^+ + H^\bullet$ reaction, e.g. $\Delta S^{\circ\ddagger}_{1000K}$ = 11.3 J K^{-1} mol^{-1} and $\Delta S^{\ddagger}_{\mu}(E \rightarrow +\infty)$ = 13.2 J K^{-1} mol^{-1}. For the $C_2H_4^{+\bullet} \rightarrow C_2H_3^+ + H^\bullet$ reaction, the canonical and microcanonical entropies of activation amount to 32 and 36 J K^{-1} mol^{-1} respectively.

The large activation entropies observed for the dissociations of the bromo- and iodobenzene cations are compatible with a maximum entropy analysis of the product energy distributions, which leads to a nearly complete (about 80%) phase sample sampling. As illustrated in **Figure 1**, an excess energy of about 0.7 eV, called the 'kinetic shift' is necessary to bring the rate constant above 10^5 s^{-1} so that dissociation can be detected in the metastable window.

The dissociation of the styrene ion involves a transition state very similar to the reactant. The large negative entropy of activation results from the fact that the internal energy content of the transition state is reduced, compared to the reactant ion, by an amount equal to the critical energy. The transition state structure for the CO loss from the phenol cation is also close to the precursor but some vibrational modes loosen up, resulting in a slightly positive activation entropy.

Activation entropies result from a fitting procedure including the critical energy E_0, too. In many cases, the limited range of internal energies investigated, as well as the lack of information on the structure and/or the thermochemistry of the fragments, and on a possible reverse activation barrier, preclude any reliable control over E_0. One should therefore be aware of energy–entropy trade-offs.

Competitive reactions

Ions generally have low isomerization barriers so that decomposition channels leading to different product ions with the same molecular formula can compete. Two examples will illustrate this point.

In contrast to the simple C–I bond cleavage observed with the iodobenzene cation, the corresponding reaction with the iodotoluene cation is much more complex, leading to a mixture of three $C_7H_7^+$ isomers with tolyl, benzyl and tropylium structures. The experimental data available have been analysed by Dunbar and co-workers using a two-channel model (**Figure 2**). The lowest energy channel leads to the benzyl and tropylium structures via a rearrangement characterized by a tight transition state ($\Delta S^{\circ\ddagger}_{1000\,K}$ = −29 J K^{-1} mol^{-1}), whereas the highest energy channel is a direct bond cleavage ($\Delta S^{\circ\ddagger}_{1000\,K}$ = 32 J K^{-1} mol^{-1}). As expected, the branching ratio between the two channels depends dramatically on the internal energy content.

Loss of HF from the 1,1-difluoroethylene cation at low internal energy leads to the fluoroacetylene fragment ion with a rate constant in the metastable range and a large reverse activation barrier of nearly 1 eV. At higher internal energy, collisional activation experiments and TPEPICO data show that a competitive channel becomes open, leading probably to the fluoroethenylidene cation. This is associated with a decrease in the kinetic energy released to the fragments.

Multistep fragmentations

A dissociation can very seldom be described as a one-step process starting from a unique potential energy minimum and leading to the fragments via a unique transition state. One or more reversible isomerization steps prior to dissociation have been frequently proposed to interpret experimental data in the framework of a classical kinetic scheme coupled with a RRKM evaluation of the appropriate rate constants. *Ab initio* calculations have revealed the importance of unconventional species like bridged structures and of distonic ions as stable ionic structures. Distonic ions are radical ions for which the ionic site and the radical site are formally separated, like in $^{\bullet}CH_2CH_2OH_2^+$. Furthermore, ion–neutral complexes have been suggested in many instances to play an important role in the disso-ciation mechanism. The production of HCO^+ from $CH_3OH^{+\bullet}$ at threshold has been shown to proceed via the following multistep sequence:

$$\begin{aligned}CH_3OH^{\bullet+} &\rightarrow \overset{\bullet}{C}H_2\,\overset{+}{O}H_2 \rightarrow [CH_2OH\ldots H]^{\bullet+}\\ &\rightarrow [CH_2O\ldots H_2]^{\bullet+} \rightarrow CH_2O^{\bullet+} + H_2\\ &\qquad\qquad\qquad\qquad \downarrow\\ &\qquad\qquad\qquad HCO^+ + H^{\bullet} + H_2\end{aligned}$$

where both a distonic ion and two ion–neutral complexes (bracketed species) are involved. The typical ion–neutral separation in such a complex is in the 3–5 Å range. Rearrangements or ion–neutral reactions can take place within this entity. They are favoured at low internal energy, whereas simple cleavages become more probable at higher internal energy. A spectacular ion–molecule reaction within an ion-neutral complex has been discovered by Botter and Longevialle: it results in the transfer of a proton from the D side to the A side of a steroid molecular ion.

The decomposition of the propanol cation is a nice example of a concerted investigation by coincidence methods, including selective deuteration, and *ab*

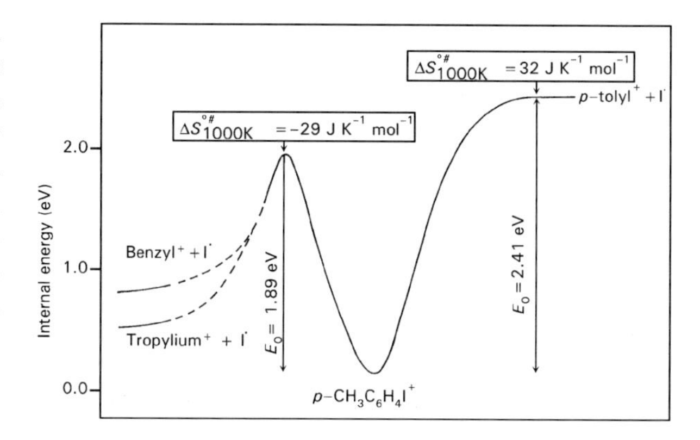

Figure 2 Reaction mechanism for the loss of iodine from the *p*-iodotoluene cation, after the work of Lin and Dunbar: Lin CY and Dunbar RC (1994) Time-resolved photodissociation rates and kinetic modelling for unimolecular dissociations of iodotoluene Ions. *The Journal of Physical Chemistry* **98**: 1369–1375.

initio calculations. **Figure 3**, adapted from work of Baer and Booze, describes the multistep process involved in the competition between the H loss and H_2O loss channels. Intermolecular isotope effects were studied with partially deuterated propanol cations, $CD_3CH_2CH_2OH^+$. A very small isotope effect is observed for the H loss, which, according to the model, is a secondary effect. Water loss, on the other hand, involves a 1,4-hydrogen shift as an intermediate step. Deuteration brings about a primary isotope effect, characterized by a 0.15 eV increase in the critical energy.

The intermediate steps which have just been mentioned correspond to successive minima of the potential energy surface. What about the transition state? It is not unique either. To find its position is a particularly critical problem if no saddle point appears along the dissociation pathway. As suggested by Bowers and co-workers, two transition states exist in this case, respectively at small and large values of the reaction coordinate. The internal transition state results from the competition between the conversion of kinetic into potential energy and the decrease of the vibration frequencies as the system moves along the reaction coordinate. The second transition state, which is called the orbiting transition state, is associated with the centrifugal barrier appearing at larger values of the reaction coordinate. The internal transition state is expected to play a significant role at high internal energy, whereas the external transition state governs the dissociation rate at energies closer to the threshold. The experimental observation of this transition state switching is still sought for single-well potential energy surfaces in ionic systems. Other kinds of transition states can also be invoked. A strong

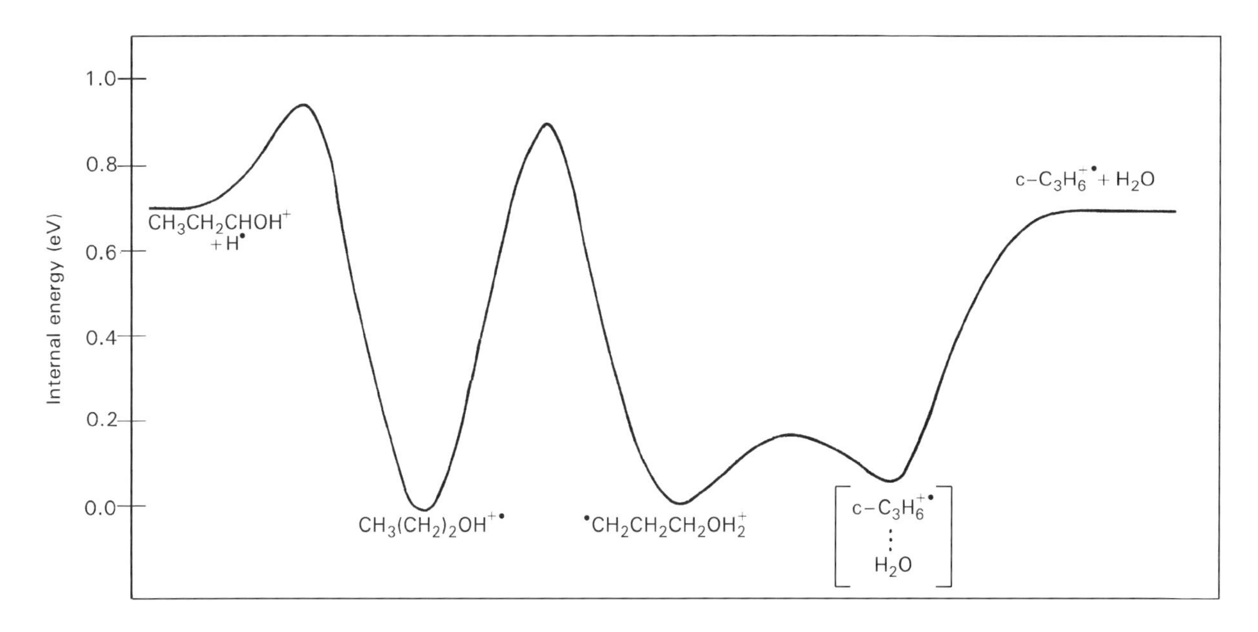

Figure 3 Multistep reaction pathway for the loss of H and of H_2O from ionized propanol. The figure is adapted from a study by Booze and Baer: Booze JA and Baer T (1992) Dissociation dynamics of energy selected propanol ions from a σ-type ion structure. *The Journal of Physical Chemistry* **96**: 5715–5719.

curvature of the reaction path brings about a bottleneck in the energetically allowed phase space and thus a decrease of the reaction rate. An avoided crossing between two potential energy surfaces limits also the reactive flux due to a branching between the two electronic states. This latter aspect will be discussed in the next subsection.

Nonadiabatic reactions

The excited electronic states of the molecular ion which are populated by the ionization process in general decay very rapidly (lifetime of the order of 0.1 ps) to the ground electronic state by internal conversion of electronic energy into vibrational energy. If the electronic ground state does not correlate adiabatically to a given pair of fragments, the production of these fragments requires a transition to a potential energy surface correlating to the appropriate dissociation asymptote, at some stage of the decomposition process. A reaction requiring such an interstate transition is referred to as nonadiabatic. Nonadiabatic behaviour is always expected when a competition involves two close-lying dissociation asymptotes, differing only by the location of the positive charge on one or the other fragments. In general, this will always be the case for charge-transfer reactions. The statistical theories can be adapted to accommodate nonadiabatic transitions occurring along the reaction path. Briefly, the non-adiabatic transition probability, i.e. the probability of switching from one potential energy surface to the other, is introduced as a kind of transmission coefficient in a RRKM-like equation.

The two electronic states involved in a non-adiabatic transition have either the same or a different multiplicity. In the former case, the coupling of the electronic and nuclear motions is involved. If multiplicity changes, the transition is due to the spin–orbit coupling. A recent example deals with the unimolec-ular loss of singlet H_2 from triplet CH_3O^+ to produce singlet HCO^+. Based on the measurement of intramolecular kinetic isotope effects and on *ab initio* calculations, Schwarz and co-workers rejected the usually accepted stepwise mechanism in favour of a mechanism in which loss of H_2 and intersystem crossing are concerted.

Tunnelling

If a reverse activation barrier exists, tunnelling is possible. The probability of tunnelling depends dramatically on the reduced mass. The smaller the 'reduced mass, the larger the probability. Furthermore, the largest H–D isotope effects are expected to occur near the threshold. This corresponds generally to the metastable time scale, which is the most easily accessible to mass spectrometric experiments. If very large H/D kinetic isotope effects are observed, sometimes larger than 1000, then tunnelling is often proposed as the most probable explanation. To take tunnelling into account in the RRKM equation, a transmission coefficient is introduced. It can be calculated analytically provided an appropriate model is used for the potential energy barrier. The asymmetric Eckart potential is often chosen.

Ab initio calculations performed by different groups on the loss of H_2 from the ethane cation agree that the appearance energy for this rearrange-ment lies below the top of the barrier. On the other hand, any RRKM calculation neglecting tunnelling leads to a rate constant at threshold larger than 10^8 s^{-1}, which is inconsistent with the metastability of this reaction. RRKM calculations by Weitzel including tunnelling are in good agreement with $k(E)$ measurements by the TPEPICO technique.

The intermolecular kinetic isotope effect k_H/k_D for the loss of HCl or DCl from $C_2H_5Cl^{+\bullet}$ and $C_2D_5Cl^{+\bullet}$ respectively lies between 10^2 and 10^5. It has been well accounted for by RRKM calculations including tunnelling (Booze and co-workers).

As shown by Lifshitz, Schwarz and co-workers, the methane elimination from the acetone cation proceeds via two successive intermediates, a hydrogen-bridged complex and a ketene–methane ion–neutral complex. The barrier separating these two entities lies 14.2 kJ mol^{-1} above the $CH_2CO^{+\bullet} + CH_4$ asymptote. Tunnelling through this barrier is required to account for the appearance of the $CH_2CO^{+\bullet}$ fragment at its thermochemical threshold.

Isolated state decay

In most ionic unimolecular reactions, internal conversion to the ground electronic state is rapid compared to dissociation. In a few cases, however, evidence has been found of a rapid and specific dissociation process occurring on the potential energy surface of an excited electronic state. When such an 'isolated state decay' occurs, a correlation exists between the branching ratio for the specific channel and the photoelectron spectrum. Ionized difluoroethylenes for example, dissociate statistically below the $\tilde{C}$ state but this latter state favours strongly the F loss channel.

Future prospects

A fairly important amount of kinetic data is nowadays available, which in the vast majority of cases, have been accounted for by a classical kinetic scheme coupled to a RRKM evaluation of the rate constants involved, possibly including tunelling. The story is, however, far from being complete.

1. The rate constant is often known only in a limited range so that being able to fit the data to a statistical equation like the RRKM-QET one is not a definitive proof of statisticity by itself. The translational energy distribution of the fragment ions is complementary piece of information which is more sensitive to any deviation from a purely ergodic situation. Rather few detailed analyses of experimental distributions are available for the time being.
2. Nonadiabatic reactions are probably much more widespread than usually thought. In particular the role of spin–orbit coupling in unimolecular chemistry deserves careful investigation. In addition to the experimental component, this raises nontrivial *ab initio* computational problems.
3. The interpretation of kinetic experiments is almost always performed on the basis of a classical kinetic scheme. A quantum approach studying the decay of quasibound states, called resonances, could be more appropriate under given circumstances. Large fluctuations of nearly two orders of magnitude have been found, for example, for the decay rate of state-selected formaldehyde.

List of symbols

E = total energy; h = Planck constant; k_B = Boltzmann constant; $k(E)$ = unimolecular microcanonical rate constant; $k(T)$ = canonical rate constant; N_0 = initial number of ions formed; $N^{\ddagger}(E - E_0)$ = number of states of the transition state up to $E - E_0$ above the critical energy E_0; $P(E)$ = distribution of internal energies; $R(E,t)$ = rate of dissociation; T = temperature; $\Delta H^{\circ\ddagger}$ = activation enthalpy; $\Delta S^{\circ\ddagger}$ = activation entropy; $\Delta S^{\ddagger}_{\mu}$ = microcanonical entropy of activation; ν_R = reaction coordinate frequency; $\rho(E)$ = density of states of the parent ion at internal energy E; σ = degeneracy of reaction path.

See also: **Fragmentation in Mass Spectrometry; Metastable Ions; Multiphoton Excitation in Mass Spectrometry; Photoelectron–Photoion Coincidence Methods in Mass Spectrometry (PEPICO); Statistical Theory of Mass Spectra.**

Further reading

Baer T and Hase WL (1996) *Unimolecular Reaction Dynamics. Theory and Experiments.* Oxford: Oxford University Press.

Baer T, Booze J and Weitzel KM (1991) Photoelectron–photoion coincidence studies of ion dissociation dynamics. In: Ng CY (ed), *Vacuum Ultraviolet Photoionization and Photodissociation of Molecules and Clusters*. Singapore: World Scientific.

Derrick PJ and Donchi KF (1983) Mass spectrometry. In: Bamford CH and Tipper CFH (eds) *Comprehensive Chemical Kinetics*. Vol. 24. *Modern Methods in Kinetics*. Amsterdam: Elsevier.

Forst W (1973) *Theory of Unimolecular Reactions*. New York: Academic Press.
Gilbert RG and Smith SC (1990) *Theory of Unimolecular and Recombination Reactions*. Oxford: Blackwell Scientific Publications.
Lifshitz C (1992) Recent developments in applications of RRKM-QET. *International Journal of Mass Spectrometry and Ion Processes* 118/119: 315–337.
Longevialle P (1992) Ion–neutral complexes in the unimolecular reactivity of organic cations in the gas phase. *Mass Spectrometry Reviews* **11**: 157–192.
Lorquet JC and Leyh B (1984) Potential energy surfaces and theory of unimolecular dissociation. *Ionic Processes in the Gas Phase. NATO ASI SERIES* **C118**: 1–6
Steinfeld JI, Francisco JS and Hase WL (1999) *Chemical Kinetics and Dynamics*, 2nd edn. Upper Saddle River: Prentice Hall.

Ion Energetics in Mass Spectrometry

John Holmes, University of Ottawa, Canada

MASS SPECTROMETRY
Theory

Neutral thermochemistry

For neutral organic compounds there are only some 3000 good quality data for molecular heats of formation, $\Delta_f H^0$[M]. These have been carefully selected by Pedley and co-workers over an extended period. The standard heat of formation of an entity, $\Delta_f H^0$, is defined as the heat absorbed or released when 1 mole of the species is formed from its constituent elements at 298 K and at 1 atm pressure. This is represented by an equation, e.g.

$$C(s) + 2H_2(g) \rightarrow CH_4(g), \Delta_f H^0 = -17.78\,\text{kcal mol}^{-1}$$

The terms in brackets represent the phase in which the reactants and products are stable in their standard state at 298 K and 1 atm pressure (1 cal = 4.184 J).

Although so few organic compounds have accurate experimental $\Delta_f H^0$ values, the knowledge that $\Delta_f H^0$ for homologous organic compounds can be represented as an additive property has resulted in the production of additivity coefficients, the terms which are most closely associated with the work of Benson. The coefficients now available allow one to estimate $\Delta_f H^0$ for a very wide range of organic compounds via simple arithmetical calculations. Sufficient data exist for the estimation of many geometric effects such as ring strain energies, geometric isomers, etc.

Ionization energies and ionic heats of formation, $\Delta_f H^0[M^{\bullet+}]$

The $\Delta_f H^0$ value for a molecular ion $M^{\bullet+}$ is obtained from the measurement of the ionization energy (IE) of neutral M. The reaction

$$M \rightarrow M^{\bullet+} + e^-$$

has

$$\Delta H^0_{\text{reaction}} = \text{IE[M]} = \Delta_f H^0[M^{\bullet+}] - \Delta_f H^0[M] \quad [1]$$

$\Delta_f H^0$[M] is obtained from reference data, from an additivity estimation or, more recently, from high level *ab initio* molecular orbital theory calculations. The IE is derived from an experimental measurement. In Equation [1] the electron is given an enthalpy of zero (i.e. it is assumed to be stationary). Ionization energies can be obtained from photoelectron spectra or from direct measurement of the ionization threshold using monochromated photons or energy-selected electrons. Whatever method is chosen, the most important factor in determining a molecular IE value is the relative geometry of the molecular ion and the neutral molecule, both in their ground states. The terms vertical IE and adiabatic IE are defined in **Figure 1**.

If the geometry difference between the molecule and its ion is particularly large then the adiabatic IE (that necessary for the determination of $\Delta_f H^0$ [$M^{\bullet+}$] in Eqn [1]) will be difficult to measure. Photoionization is only a vertical process; electron impact is less so because the approach of the ionizing electron and its effects on the molecule are not instantaneous. For moderate geometry differences, the adiabatic IE can be measured using energy-selected electron beams, as a reasonably well-defined asymptote in the ionization efficiency curve, see **Figure 1**.

Molecular adiabatic ionization energies are to be found in the NIST database (see Further reading section) and other reference books.

A number of attempts have been made to produce empirical equations for the estimation of molecular ion enthalpies, notably by Mouvier and co-workers and by Holmes, Fingas and Lossing. The latter is a simple combination of Benson-type additivity terms for the neutral's $\Delta_f H^0$ and a reciprocal term that incorporates the IE; This arises from the observation that for homologous molecules their IE values fall linearly with the reciprocal of the ion size, represented as $1/n$, where n is the number of atoms in the molecule. The equation

$$\Delta_f H^0[\mathrm{M}^{\bullet+}] = A - Bn + (C/n)$$

where A, B and C are derived from experimental data for a given homologous series works well for various organic compounds that contain C, H, O and halogen. The secondary empirical relationship, that IE for homologous molecules are proportional to $1/n$, has proved useful in many applications where an unmeasured IE needs to be estimated or where an experimental value appears to be anomalous.

The empirical equation of Mouvier is more complex:

$$\log\left[\frac{\mathrm{IE(R^1XR^2)} - \mathrm{IE}_\infty}{\mathrm{IE}_0 - \mathrm{IE}_\infty}\right] = 0.106[\mathrm{I(R^1)} + \mathrm{IR^2})]$$

where X is a functional group and R^1 and R^2 are attached alkyl groups, IE_∞ is a constant for each compound type and IE_0 is that for a reference compound. $I(R^n)$ are empirical constants for each R group. The IE values derived and the $\Delta_f H^0$[ion] values obtained using the appropriate neutral enthalpies provide constant ionic heats of formation.

Perhaps the most important use of all such schemes is their ability to reveal misfits in reference data or to highlight an ion having unusual stabilizing or destabilizing properties.

The $\Delta_f H^0$ values for even-electron ions (i.e. ionized free radicals) can be obtained from direct measurement provided that a clean source of the appropriate free radical can be devised. The most versatile method has been that pioneered by Lossing in which the flash pyrolysis of nitrite esters was used to generate the radicals, e.g.

$$(\mathrm{CH_3})_2\mathrm{CHCH_2ONO} \xrightarrow{\mathrm{Heat}} (\mathrm{CH_3})_2\mathrm{CH}^\bullet + \mathrm{CH_2O} + \mathrm{NO}^\bullet$$

Direct IE measurements by these means remove the uncertainties which arise in appearance energy (AE) measurements (described below).

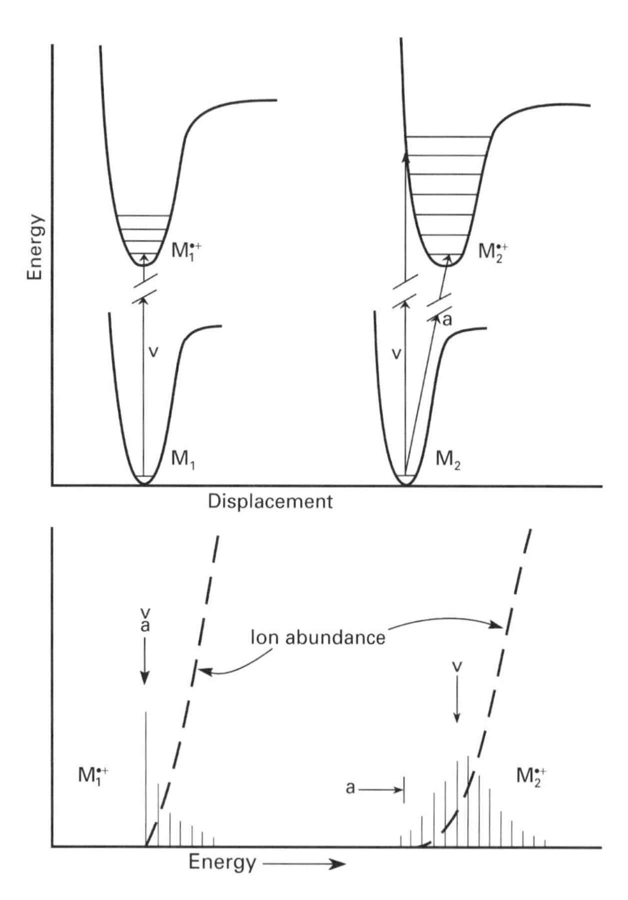

Figure 1 Vertical (v) and adiabatic (a) ground state to ground state ionization energies. Upper curves are the potential energy diagrams for two molecules, M_1 and M_2, and their molecular ions. For M_1 there is no significant geometry change on ionization and v and a have the same value. For molecule M_2, the geometry of the ion is not the same as that of the neutral and so v > a. The lower figure shows the ion onset curves (broken lines) and the vibrational modes accessed in the ion with the lines' length showing their relative importance in the ionization process. Note, the more poorly defined onset for $M_2^{\bullet+}$.

Appearance energies

The measurement of the $\Delta_f H^0$ for a fragment ion is an important aid to ion structure determination. The energy change associated with the general equation

$$M_1 \rightarrow M_2{}^+ + M_3{}^\bullet + e^-$$

is called the appearance energy (AE) (for the fragment ion M_2^+). The simple thermochemical relation, Equation [2], corresponding to this reaction:

$$\Delta H_2^0 = AE[M_2^+] = \Delta_f H^0[M_2^+] + \Delta_f H^0[M_3{}^\bullet] - \Delta_f H^0[M_1] \quad [2]$$

is insufficient to describe the full result of the AE measurement for several reasons:

(i) The reverse reaction may have an energy barrier, E_{rev}, and so the measured AE will exceed the thermochemical minimum for the reaction.

(ii) The measured AE may be too high because the dissociation of $M_1^{\bullet+}$ involves a significant kinetic shift. A kinetic shift arises as follows. The time-scale of the experiment for measuring the AE is usually very short, typically 10^{-4} to 10^{-6} s, and so for the unimolecular dissociation of $M_1^{\bullet+}$ ions to achieve rate constants of 10^4–10^6 s^{-1} they must contain excess energy above the minimum required for the reaction to proceed. For many dissociations, particularly simple bond cleavages in small ions, the kinetic shift effect is negligible. For dissociations of large ions or molecular elimination reactions, the effect can be appreciable, as much as 0.5 eV. In addition, for small molecular ions which lie in deep potential wells, the large density of states at the dissociation limit may give rise to a significant kinetic shift. A good example is provided by the phenyl halides, where on the μs time-scale the mean kinetic shift for the simple bond cleavage reaction

$$C_6H_5Br^{\bullet+} \longrightarrow C_6H_5{}^+ + Br^\bullet$$

is 0.7 eV. Note that if the AE values for fragment ions from *metastable ion* precursors are measured, the limiting observational rate constant reduces to ~10^4 s^{-1} and the above kinetic shift is halved. Delayed observations, on the ms time-scale, reduce the kinetic shift even further.

(iii) In general, the internal energy of the reactant molecule is taken to be that corresponding to the ambient temperature of the ion source and that of the products is essentially unknown. In photoionization AE measurements, Traeger has supposed that the products are at 0 K, perhaps an unlikely outcome for polyatomic species. Proposing that the products' energy differs little from those of the neutral reactant does not appear to introduce major errors.

(iv) The caveats regarding $\Delta_f H^0[M_1]$ apply also for $\Delta_f H^0[M_3{}^\bullet]$, which as a free radical may not have a well-assigned enthalpy of formation. Indeed, radical $\Delta_f H^0$ values can reliably be obtained from AE measurements when the ionic heat of formation $\Delta_f H^0[M_2{}^+]$ is well established and $\Delta_f H^0[M_3{}^\bullet]$ is the unknown. The results of this approach equal those derived from activation energies determined from the kinetics of neutral species' reactions.

(v) Hidden rearrangements: in many cases the structure of the ion and the accompanying neutral fragment are not in doubt. However, a significant number of molecular ions undergo unpredicted rearrangements before they fragment. If, therefore, there is any doubt as to the structure of the fragmentation products, then appropriate experiments should be performed. Two examples suffice to illustrate the problem: the $C_2H_6O^{\bullet+}$ ion generated by loss of CH_2O from ionized propane-1,3-diol is neither $CH_3CH_2OH^{\bullet+}$ nor $CH_3OCH_3^{\bullet+}$ (molecular ions of the only two stable C_2H_6O compounds) but has the structure $\dot{C}H_2CH_2\overset{+}{O}H_2$. This ion behaves as an electrostatically bound water molecule – ethene ion; its $\Delta_f H^0$ value is *lower* than that of either of the above two conventional ions and, indeed, is the lowest energy $C_2H_6O^{\bullet+}$ isomer.

$$HOCH_2{-}CH_2{-}CH_2OH^{\bullet+} \longrightarrow CH_2O + \dot{C}H_2CH_2\overset{+}{O}H_2$$

The dissociation of ionized methyl acetate appears to be trivial

$$CH_3CO_2CH_3{}^{\bullet+} \longrightarrow CH_3\overset{+}{C}O + [C, H_3, O]^\bullet$$

with the formation of the acetyl ion and a methoxy radical as the products. Astonishingly, the ion is as expected but the lowest energy dissociation of this molecular ion produces the isomeric hydroxymethyl radical, $^\bullet CH_2OH$, instead of the predicted $CH_3O^\bullet$ species. In this case, profound rearrangements have preceded dissociation.

In summary, AE measurements, although relatively easy to perform, may be difficult to evaluate because of the above uncertainties. However, a reverse energy barrier will often be manifest by the magnitude of the kinetic energy release observed in the metastable peak accompanying the fragmentation. A small kinetic energy release of say < 10 meV likely indicates neither a reverse energy barrier nor a significant kinetic shift.

Except for the existence of a reverse energy barrier, most of the above problems are avoided in photoelectron photoion coincidence experiments, where the dissociation of only internal energy selected species are observed.

Estimation of ion enthalpies

It was described above how empirical expressions can be used to estimate $\Delta_f H^0$ values for odd-electron (molecular) ions.

No such simple formulae have been derived for even-electron (fragment) ions, but some useful general semiquantitative relationships have been found. The simplest is that for successive substitution of a single group (e.g. CH_3, OH, etc.) at a formal charge-bearing site. $\Delta_f H^0$[Ion] decreases linearly with log(ion size), the latter usually being represented by the number of atoms in the ion. This is illustrated in **Figure 2** for successive substitution of H by X, beginning with the methyl cation. Note that the OH substitutions (steep lines) and CH_3 substitutions produce linear plots of similar slope.

Charge density in ions

It is worth emphasizing that thermochemical data give insights as to the distribution of charge in related ions. Thus, for example, the effect of methyl substitution at carbon in $[CH_2OH]^+$ is much greater than for the substitution at oxygen; $\Delta_f H^0$ $[CH_2OH]^+$ = 169 kcal mol^{-1}. $\Delta_f H^0[CH_3CHOH]^+$ = 139 kcal mol^{-1} [$\Delta(\Delta_f H^0) = -30$ kcal mol^{-1}] whereas $\Delta_f H^0[CH_2OCH_3]^+$ = 157 kcal mol^{-1} [$\Delta(\Delta_f H^0)$ = −12 kcal mol^{-1}]. Thus the hydroxymethyl cation (protonated formaldehyde) is better represented as $^+CH_2OH$ (a $^+$C hydroxy substituted carbocation) than as $CH_2 = O^+H$ (an oxonium ion). This is also borne out by the ion's reactivity and its behaviour in ion–molecule complexes.

Table 1 The effect of OH substitution at various positions in alkenes and their ions (energies in kcal mol^{-1})

Alkene	$\Delta_f H^0$[M]	$\Delta_f H^0[M^{\bullet+}]$	$\Delta(\Delta_f H^0)$ Neutral	$\Delta(\Delta_f H^0)$ Ion
CH_2=$CHCH_2CH_3$	0	222	–	–
HOCH=$CHCH_2CH_3$	−42	150	−42	−72
CH_2=C(OH) CH_2CH_3	−43	150	−43	−72
CH_2=CHCH(OH)CH_3	−38	181	−38	−41
CH_2=$CHCH_2CH_2OH$	−36	184	−36	−38
CH_2=$C(CH_3)_2$	−4	209	–	–
HOCH=$C(CH_3)_2$	−45	145	−41	−64
CH_2=$C(CH_3)CH_2OH$	−38	176	−34	−33

The effects of, for example, OH substitution at, near and away from a formal charge-bearing site are clearly and significantly different, a good example being provided by the isomeric ions shown in **Table 1**. Note that the heat of formation of the ion, when OH is *not* at the formal charge-bearing site (here the π-system), is governed by the difference in $\Delta_f H^0$ values for the neutral alkene and alkenol. When OH is at the charge site, the stabilization effect is much greater.

Proton affinities

It is predicted that there should be a linear relationship between the proton affinities of homologous molecules and their ionization energies. A simple thermochemical cycle gives

$$\mathrm{PA[M] = HA[M^+] + IE[H] - IE[M]}$$

If the hydrogen atom affinity (HA) of homologous ions is a constant, then

$$\mathrm{PA[M]} \approx \mathrm{constant - IE[M]}$$

This equation indeed works quite well for homologous alcohols, ethers, carbonyl compounds, amines, etc.

Electron affinities

The electron affinity (EA) of a molecule is defined as

$$\mathrm{EA[M]} = \Delta_f H^0\mathrm{[M]} - \Delta_f H^0\mathrm{[M^-]}$$

Again the problems of adiabatic and vertical values

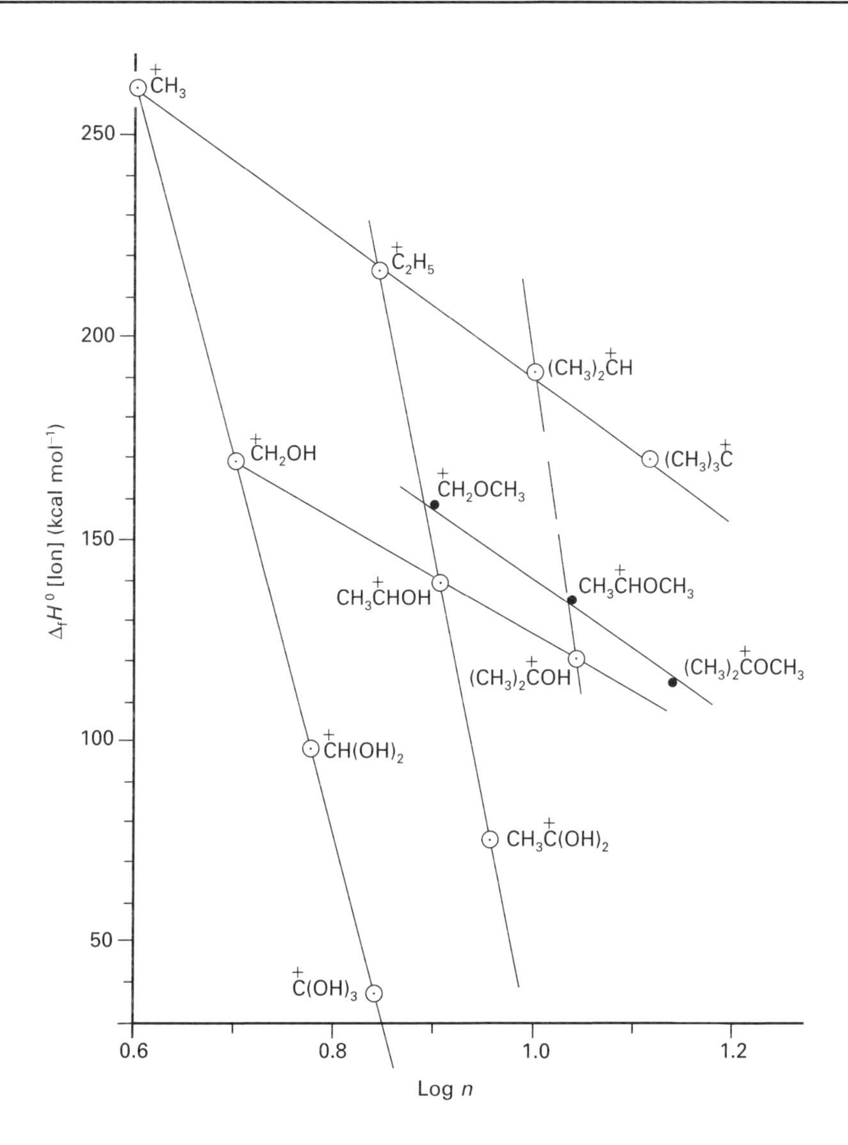

Figure 2 Plots of $\Delta_f H^0$ [ion]$^+$ versus log (ion size) the latter being represented by n, the number of atoms in the ion. The steep curves show the effect of successive OH substitution in alkyl ions and the shallow curves are for CH_3 substitution therein.

arise. The best values probably are determined by laser photoelectron spectroscopy and laser photodetachment. Both of these methods directly measure a threshold energy for the removal of an electron from the anion. EA values are listed in the NIST inventory of data. In general, electron-binding energies to molecules and radicals are less precisely known than cationic $\Delta_f H^0$ values. In thermochemical calculations, the term 'gas phase acidity' $\Delta_{acid} H^0[AH]$ is often employed

$$\Delta_{acid} H^0[AH] = BDE(A-H) - EA[A] + IE[H]$$

where BDE(A–H) is the homolytic bond strength (A–H). Alternatively

$$\Delta_{acid} H^0[AH] = \Delta_f H^0[A^-] + \Delta_f H^0[H^+] - \Delta_f H^0[AH]$$

List of symbols

E_{rev} = energy barrier of reverse reaction; n = number of atoms in a molecule; $\Delta_f H[M]$ = molecular heat of formation.

See also: **Ion Dissociation Kinetics, Mass Spectrometry; Ion Structures in Mass Spectrometry; Metastable Ions; Photoelectron–Photoion Coincidence Methods in Mass Spectrometry (PEPICO); Proton Affinities; Thermospray Ionization in Mass Spectrometry.**

Further reading

Bachiri M, Mouvier G, Carlier P and Dubois PE (1980) *Journal de Chimie Physique* 77: 899–914.
Bensen SW (1976) *Thermochemical Kinetics*, 2nd edn. New York: Wiley.
Holmes, JL, Fingas M and Lossing FP (1981) *Canadian Journal of Chemistry* **59**: 80–93.
Hunter PL and Lias SG (1988) Evaluated gas phase basicities and proton affinities of molecules. *Journal of Physical Chemistry Reference Data* **27**: 413–656.
Lias SG, Bartmess JE, Liebman JF, Holmes JL, Levin RD and Mallard WG (1988) Gas phase ion and neutral thermochemistry. *Journal of Physical Chemistry Reference Data* **17**(suppl 1).
NIST Chemistry WebBook: **http://webbook.nist.gov/chemistry/**
Pedley JB, Naylor RD and Kirby SP (1986) *Thermochemical Data of Organic Compounds*. London: Chapman & Hall.
Traegar JC, McLoughlin RG (1981) *Journal of the American Chemical Society* **103**: 3647–3652.

Ion Imaging Using Mass Spectrometry

Albert JR Heck, University of Utrecht,
The Netherlands, UK

MASS SPECTROMETRY
Applications

Ion imaging, as described here, is a multiplex detection technique used in gas-phase molecular reaction dynamics to visualize the breaking and formation of chemical bonds. The primary aim of the ion imaging technique is to reveal quantitatively the way in which energy associated with the processes that govern the breaking and formation of chemical bonds is released into the reaction products. The ion imaging technique is based on a combination of two more established techniques: optical spectroscopy (resonance-enhanced multiphoton ionization, REMPI) and two-dimensional time-of-flight mass spectrometry. Ion imaging measures the three-dimensional recoil (velocity, i.e. speed and angular distribution) of the ensuing fragments or reaction products and uni- and bimolecular reactions. Using resonance-enhanced multi-photon ionization, the recoil of individual quantum states (electronic, vibrational and/or rotational) of a selected fragment or reaction product can be measured. More extensive reviews covering various aspects of the ion imaging technique are available.

Principal aspects

One of the major goals of molecular reaction dynamics, and of ion imaging, is to elucidate how the energy deposited into products of a molecular dissociation or bimolecular reaction is distributed over all available degrees of freedom. Consider a unimolecular dissociation initiated by absorption of a photon. Following dissociation, the nascent fragments will recoil into space, away from the point where the reaction was initiated. The speed of the products will depend on the amount of energy deposited in the translational degrees of freedom, and on the masses of the products. The recoil of the fragments or reaction products does not have to be isotropic. In a laser-induced dissociation, the angular distribution of the photofragments depends on the polarization vector of the photolysis source, on the symmetry of the electronic states involved, and on the lifetime of the activated complex. In the case of a bimolecular reaction, the products will be scattered at various angles with respect to the centre-of-mass velocity vector. Measurement of such recoil will reveal directly whether reaction products are, for example, forward or backward scattered. Additionally, reaction products can be formed with excess internal energy distributed over the electronic, vibrational and rotational degrees of freedom. In ion imaging, the internal energy distributions are typically measured using laser-based techniques, i.e. resonance-enhanced multiphoton ionization, although laser-induced fluorescence imaging has been employed as well.

Simultaneous measurement of all the above quantities, as achieved by ion imaging, is important so that correlated quantities (e.g. the quantum state of one product correlated with that of the other, or the velocities of fragments correlated with the particular quantum state populated) can be determined. A further elegant aspect of the imaging technique is that the recoil of the fragments and/or reaction products is visualized by the ion images appearing on a

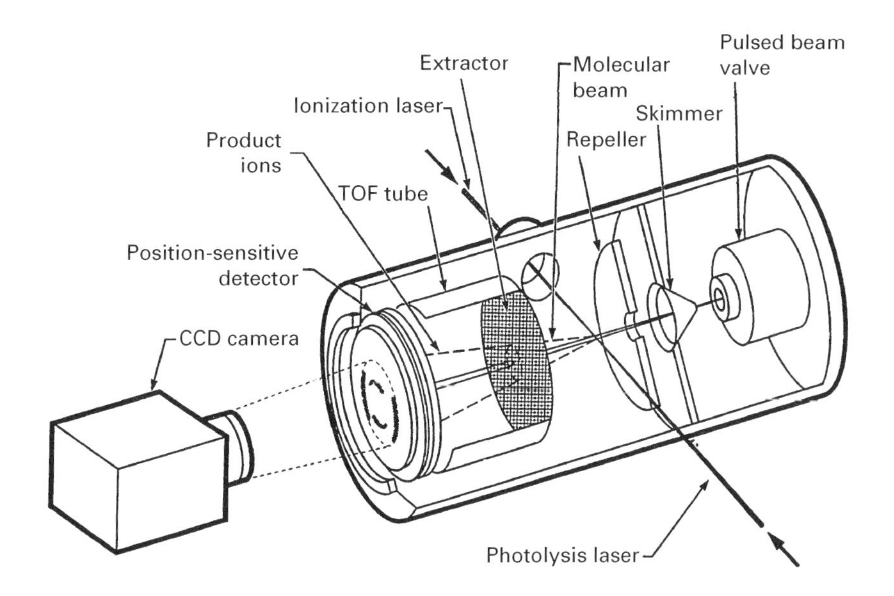

Figure 1 Schematic diagram of an ion imaging apparatus (TOF = time of flight). Reproduced with permission from Heck AJR and Chandler DW (1995) Imaging techniques for the study of chemical reaction dynamics. *Annual Review of Physical Chemistry* **46**: 335–378. © 1995 Annual Reviews Inc.

phosphor screen. In this way the ion imaging technique provides instructive and direct 'snapshots' of the breaking and making of chemical bonds.

Experimental aspects

Figure 1 shows a schematic of an ion imaging apparatus. The basic experimental components of an ion imaging instrument are: (1) molecular beams used to collisionally cool the reactants and to generate reactants with a well-defined velocity; (2) lasers used to photolyse or generate the primary reactants, and to ionize the reaction product quantum-state-selectively; (3) two-dimensional time-of-flight mass spectrometry used to measure the recoil of the reaction products; (4) a position-sensitive detector used to detect two-dimensional time-of-flight profiles. The position-sensitive detector typically consists of a pair of chevron-type microchannel plates (MCP) coupled to a phosphor screen and a charge-coupled-device (CCD) camera. The CCD camera records the images appearing on the phosphor screen.

A typical photofragment imaging experiment starts when a jet-cooled molecule absorbs a laser photon, eventually leading to dissociation. The usual short (~3 ns) and focused laser pulse provides a well-defined starting point in time as well as in space for the experiment. Following the bond cleavage, the nascent fragments start to fly away from that point on three-dimensional spheres. Before the fragments recoil far away, the products of interest are selectively ionized using REMPI. Ionization does not change the recoil of the fragments significantly. The ions are extracted from the reaction region and transferred to the position-sensitive detector through a field-free time-of-flight region. During their flight to the detector, the spheres expand significantly. The flight-times of the fragments, and thus also the physical size of the spheres upon reaching the detector, can be adjusted experimentally by varying the extraction field strengths in the time-of-flight region. On arrival at the detector, the three-dimensional distribution of the ionized fragments is crushed onto the plane of the two-dimensional detector. The position where the ion impinges on the detector is directly fibreoptically coupled to a small spot on the phosphor screen. The light emitted by the phosphor screen is detected by a CCD camera. The measured ion images are therefore two-dimensional projections of the three-dimensional velocity distributions.

Extraction of information from the images

Several methods can be used to analyse the ion images. The most straightforward approach, analysing the 'raw' images, already provides insight into the dynamical processes involved in the photofragmentation. The two-dimensional projections may reveal different channels involved in the reaction; however, branching ratios cannot be determined quantitatively when contributions of competing channels spatially overlap in the image. Two-dimensional projections show qualitatively the angular distributions of the recoiling fragments. Knowledge of the time of flight of the ionized fragment, the masses of the two concomitant fragments, the photolysis laser wavelength and the bond energy in the parent molecule provides information about the translational and internal energy of the complementary fragment.

More quantitative information can be obtained from the images when the full three-dimensional speed and angular distributions are reconstructed using mathematical transformations of the crushed two-dimensional images, or alternatively by using forward convolution simulation techniques. If the initial three-dimensional distribution has cylindrical symmetry, a *unique* transformation – the inverse Abel transform – can be used to reconstruct the initial three-dimensional velocity distribution. As the photolysis laser vector defines automatically an axis of cylindrical symmetry, the inverse Abel transformation can usually be used, as long as the plane of the position-sensitive detector is placed parallel to the laser polarization vector.

The angular distribution of photofragments depends particularly on the symmetry of the electronic transition and the direction of the photolysis laser polarization vector. The angular distribution of photofragments is usually described by the anisotropy parameter β. In particular, the recoil distribution, which is related to the velocity of the fragment v and the polarization vector of the electromagnetic field $\boldsymbol{E}$, is given by

$$I(\theta) = (4\pi)^{-1}[1 + \beta P_2(\cos\theta)] \qquad [1]$$

where θ is the angle between v and $\boldsymbol{E}$, P_2 is the second-order Legendre polynomial and β is the anisotropy parameter. This anisotropy parameter β may range from −1 for a perpendicular transition, i.e. μ perpendicular to $\boldsymbol{E}$, resulting in a $\sin^2(\theta)$ distribution of the photofragments and 2 for a parallel transition (which results in a $\cos^2(\theta)$ distribution). $\beta = 0$ corresponds to an isotropic distribution of the fragments. **Figure 2** shows schematic representations of three-dimensional distributions of photofragments resulting from prompt dissociations following a parallel transition (**Figure 2A**) and a perpendicular transition (**Figure 2B**). Following a parallel transition, the fragments will recoil primarily along the laser polarization vector, whereas following a perpendicular transition, the fragments recoil perpendicular to the laser polarization vector.

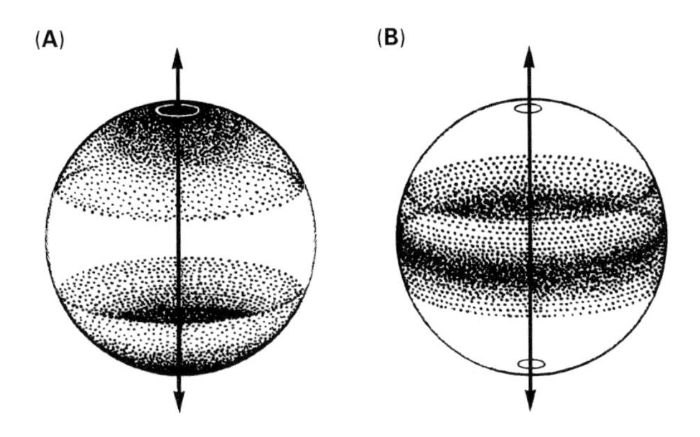

Figure 2 Schematic diagram of the three-dimensional distributions of photofragments formed following a (A) pure parallel transition in the precursor molecule and (B) a pure perpendicular transition in the precursor molecule. The arrows indicate the direction of the laser polarization vector.

Figure 3 shows a series of reconstructed ion images of nascent D (deuterium) atoms generated via the photolysis of DI, at three different photolysis wavelengths. These images can be used to illustrate many of the points made above. The upper panels show the inverse Abel transformed ion images, the lower panels the corresponding D atom speed distributions. What information is inherent in the ion images of **Figure 3**? If the nascent D fragments had not acquired any kinetic energy following the photodissociation the ionized D^+ ions would have shown up at the centre of the image. From the measurement of the size of the rings and arrival time of the ions at the detector, the speed of the D atoms may be determined. It is immediately clear that the speed of the nascent D atoms is much higher when DI is photolysed using $\lambda = 205$ nm photons instead of $\lambda = 308$ nm photons, since the ion image obtained at $\lambda = 308$ nm is smaller. From these measurements and knowledge of the photolysis energy (or laser wavelength), the bond dissociation energy of the DI molecule may be determined. The photodissociation of HI or DI is possibly one of the simplest, but also most-studied, dissociation processes. At ultraviolet wavelengths two product channels are energetically accessible:

$$\begin{aligned} \mathrm{DI} &\rightarrow \mathrm{D} + \mathrm{I}\,(^2\mathrm{P}_{3/2}) \\ \mathrm{DI} &\rightarrow \mathrm{D} + \mathrm{I}^*(^2\mathrm{P}_{1/2}) \end{aligned} \qquad [2]$$

where I ($^2P_{3/2}$), or I, is ground-state atomic iodine and I ($^2P_{1/2}$), or I*, is the spin–orbit excited state of iodine, which is 0.94 eV higher in energy than the ground state. Therefore, two distinct D atom translational energies are possible corresponding to the concomitant formation of the two different spin–orbit states of the iodine atom. From analysis of the inverse Abel-transformed ion images we can directly obtain the I/I* branching ratios for the dissociation, and moreover the angular distributions of the D atoms for the two channels individually. The three images presented in **Figure 3** reveal that the branching ratio between the I and I* channel changes quite dramatically with photolysis energy. At low energy, only the I channel is observed and a single ring is seen that corresponds with a pure $\sin^2(\theta)$ angular distribution. As the energy is increased, the I* channel becomes energetically accessible and an additional component, corresponding to a $\cos^2(\theta)$ distribution, is seen in the images. This example of imaging of D atoms from the photolysis of DI reveals the general capabilities of the technique. In a single measurement, many valuable parameters that characterize a molecular dissociation, such as product branching ratios, angular distributions of the nascent fragments and bond dissociation energies, are revealed. At the time of writing, photofragment ion imaging has already been applied to more than 30 molecular systems, in most cases revealing exciting new aspects of the dynamics governing molecular photodissociations.

New developments

Ion imaging is now, some ten years after its introduction, a well-established tool for the study of molecular reaction dynamics. The most recent years have seen some remarkable progress in the field, in particular in the areas of velocity resolution and photoelectron imaging, complex photodissociations of polyatomic molecules exhibiting many competing reaction pathways, and reaction product imaging of bimolecular reactions.

Energy resolution, 'velocity mapping'

Parker and Eppink have introduced a modification to the ion optics in ion imaging instruments employing a set of electrostatic lenses. The modified setup does not require grids, which were typically used to cover the large holes (of approximately 20 mm diameter) in the extractor plates, to ensure parallel field lines (see **Figure 1**). Such grids may cause 'grid

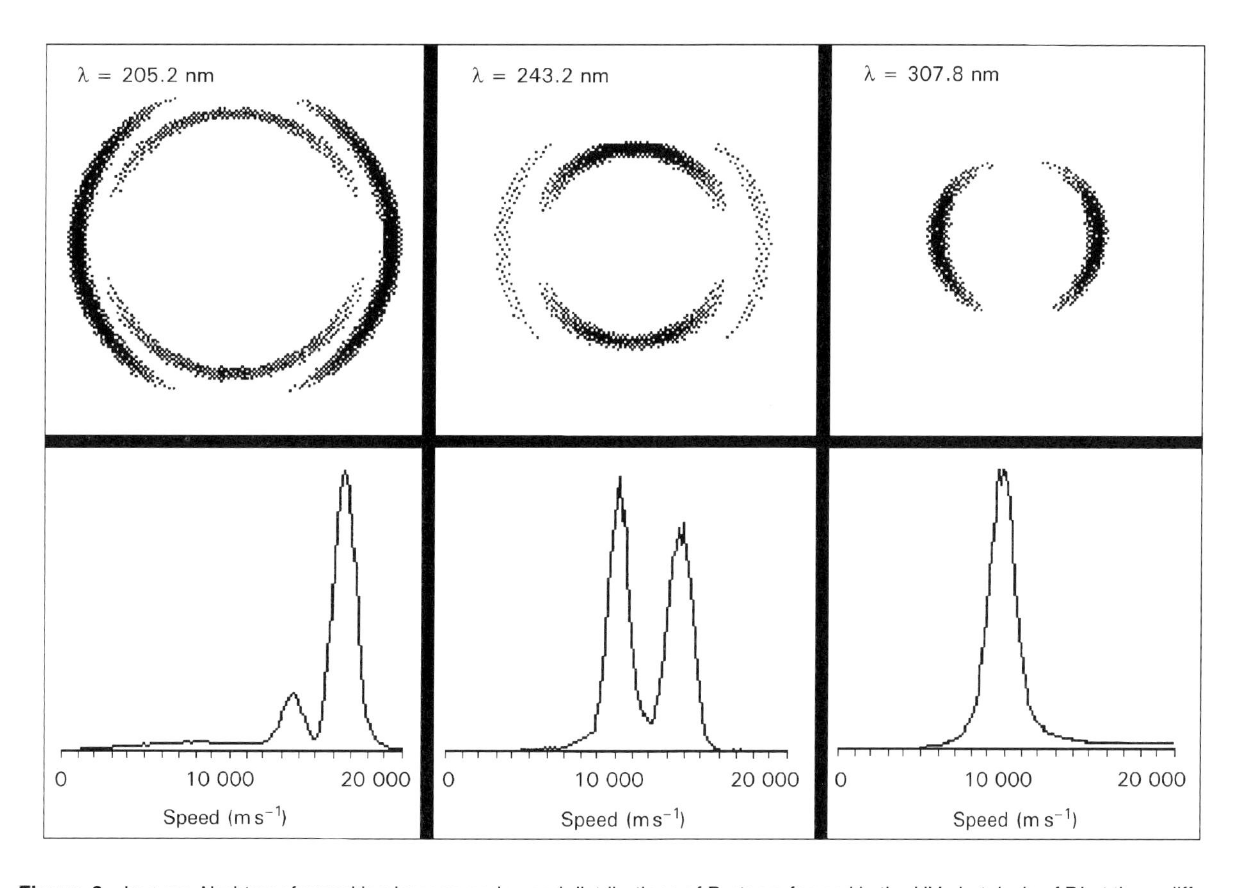

Figure 3 Inverse Abel-transformed ion images and speed distributions of D atoms formed in the UV photolysis of DI at three different photolysis wavelengths. The faster fragments will have recoiled farther from the centre of the image. For the images obtained at photolysis wavelengths λ = 205 nm and 243 nm, the outer rings correspond with the formation of D atoms concomitantly with ground-state I atoms, whereas the inner rings correspond with the formation of D atoms concomitantly with spin–orbit excited-state I* atoms. At a photolysis wavelength λ = 308 nm, only ground-state I is formed concomitantly with the D atoms. Intensities are shown in a linear grey scale (darker corresponds to higher intensities). Reproduced with permission from Heck AJR and Chandler DW (1995) Imaging techniques for the study of chemical reaction dynamics. *Annual Review of Physical Chemistry* **46**: 335–378. © 1995 Annual Reviews Inc.

distortions' in the measured ion images. Another problem encountered in the original photofragment ion imaging instruments stemmed from the interaction region between the molecular beam and the photolysis laser often being not, as desired, a point source but rather more a line source defined by the physical size of the interaction volume of the molecular beam and the laser beam. Line sources may cause 'blurring' of ion images. To reduce 'blurring' of the images, several computational 'deblurring' algorithms have been introduced. The experimental approach introduced by Parker and Eppink greatly diminishes blurring as it focuses all ions with the same initial velocity to the same spot (i.e. pixel) on the MCP detector and therefore overcomes the need for computational deblurring routines. Focusing conditions are defined by the time-of-flight path length and the ratio of the voltages applied to the repeller and extractor plates. Parker and colleagues demonstrated the improved capabilities of the modified ion imaging apparatus investigating complex competing pathways in photoionization and dissociation of molecular oxygen, excited to the $v=2$, $n = 2$ level of the 3dp $^3\Sigma^-_{1g}$ Rydberg state. By absorbing a subsequent photon, these excited intermediate states of molecular oxygen may (a) dissociate generating ground-state and excited-state oxygen atoms (the latter being ionized by a subsequent photon), (b) autoionize forming O_2^+ ions in ground or excited states, (c) autoionize and dissociate forming O^+ and ground-state and excited-state O atoms, or (d) autoionize and dissociate after absorption of one more photon. In total, more than ten different channels are expected, which normally would hamper any sort of analysis.

An interesting advantage of imaging instruments is that they may be converted, without major modifications, to photoelectron imaging instruments as described by Helm and co-workers, providing an alternative method of angle-resolved photoelectron spectroscopy. Owing to the enhanced resolution and by combining results of the O^+ images with the images of the corresponding photoelectrons, Parker and Eppink were able to unravel the relative importance of most of the anticipated channels, all exhibiting different kinetic energy releases and different angular recoil. **Figure 4** shows on the left the ion image of the O^+ ions and on the right the corresponding image of the photoelectrons. The positions of the rings corresponding to the formation of O* atoms are indicated, as well as the positions of the rings involving intermediate O_2^+ (ν) states. Channels separated by less than 100 meV in kinetic energy are still quite well resolved. This improvement, brought about by a relatively simple modification, increases the resolution in the images tremendously and is expected to enhance the capabilities of photofragment ion imaging ever more.

Ion imaging of complex polyatomic photodissociations

The first ion imaging study, by Chandler and Houston, focused on the molecular dissociation of CH_3I. Although, this is a relatively large polyatomic molecule, its dissociation behaviour is comparatively simple as far as possible reaction products are concerned, in that the H_3C–I bond cleavage is by far the most dominant process. One of the most notable features of ion imaging is the capability to distinguish different competing reaction pathways even when they produce the same products. The study of complex vacuum ultraviolet (VUV) photolysis of methane benefited greatly from the multiplex detection capabilities of ion imaging described by Zare. Considering only spin-conserving dissociation channels, reactions [3] are energetically accessible following VUV absorption:

$$CH_4(^1A_1) \rightarrow \begin{cases} CH_3\,(^2A_2'') + H \\ CH_2\,(a\,^1A_1) + H_2 \\ CH_2\,(^3B_1) + H + H \\ CH_2\,(a\,^1A_1) + H + H \\ CH\,(^2\Pi) + H_2 + H \end{cases} \quad [3]$$

Many of these five channels lead to one or more identical reaction products, and the contribution of all these pathways to the VUV photolysis could only be unravelled using the ion imaging technique. $H_2(v, J)$ photofragment images revealed that the two channels leading to molecular hydrogen in the photolysis of methane were indeed active. The rovibrational distributions of the $H_2(v, J)$ products were determined using REMPI, and by combining REMPI with the imaging technique for each of these two channels individually. Similarly, H atom photofragment images revealed the relative contributions of the pathways leading to atomic hydrogen products.

The group of Reisler has employed ion imaging to study the dynamics evolving on the S_0 and S_1 excited states of isocyanic acid (HNCO). The unimolecular decomposition of HNCO is equally complex as it involves coupling between different electronic surfaces and competition among different reaction pathways. Following ultraviolet absorption, the HNCO molecule dissociates via the pathways in Equation [4].

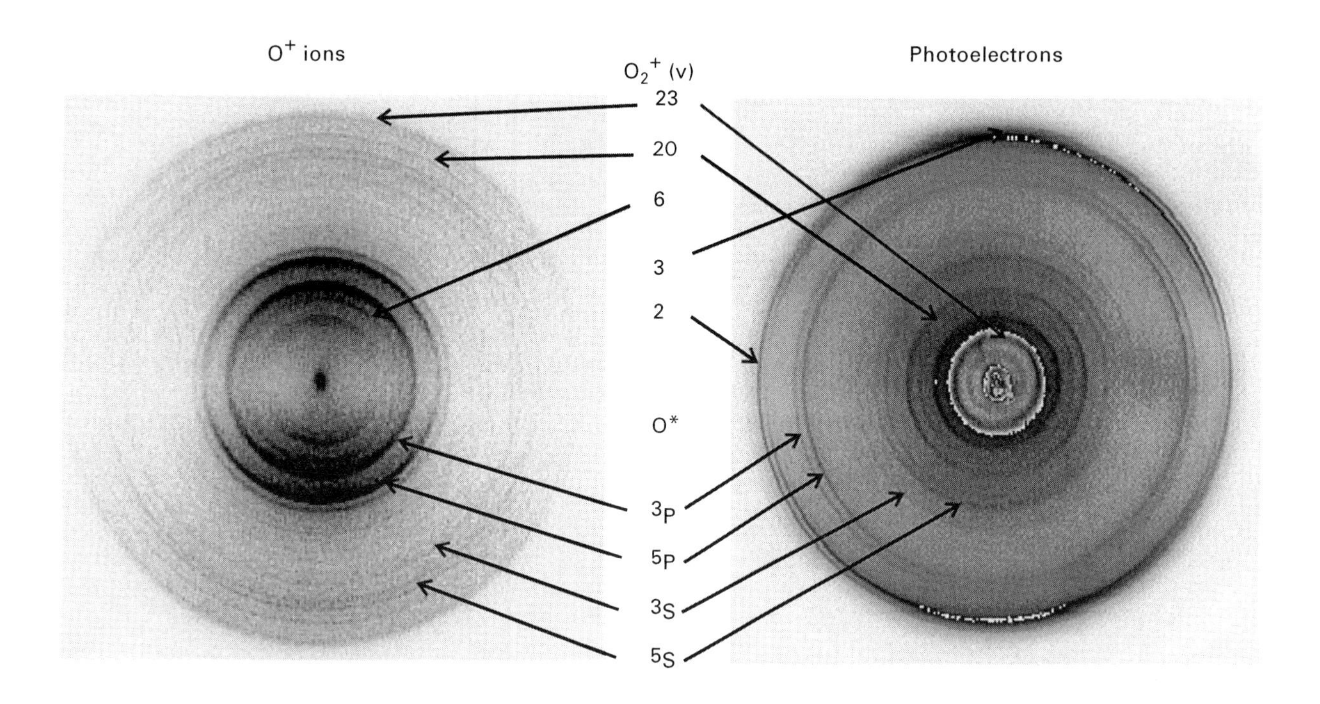

Figure 4 Photolysis and ionization of O_2 through the $v = 2$, $n = 2$ level of the $3dp\,^3\Sigma^-_{1g}$ Rydberg intermediate state. The figure on the left shows a raw O^+ ion image, whereas the figure on the right shows the corresponding photoelectron image. The positions of rings that correspond with O_2^+ vibrational levels and excited O* atoms are indicated in both images. The laser polarization vector was vertical for both these images. Intensities are shown in a linear grey scale (darker corresponds to higher intensities). Reproduced with permission from Parker DH and Eppink AJJB (1997) Photoelectron and photofragment velocity map imaging of state-selected molecular oxygen dissociation/ionization dynamics. *Journal of Chemical Physics* **107**: 2357–2362.

$$\text{HNCO} \rightarrow \begin{cases} \text{H}(^2\text{S}) & + \text{NCO}(\text{X}\,^2\Pi) \\ \text{NH}(\text{X}\,^3\Sigma^-) & + \text{CO}(\text{X}\,^1\Sigma^+) \\ \text{NH}(\text{a}\,^1\Delta) & + \text{CO}(\text{X}\,^1\Sigma^+) \end{cases} \qquad [4]$$

Although the formation of triplet NH is the least endothermic reaction path, this channel is formally not allowed. However, this spin-forbidden channel has been observed. Initially, the two latter channels contributing to the formation of ground-state CO molecules, concomitant with triplet and singlet NH respectively, were identified using laser-induced fluorescence of the singlet and triplet NH separately. However, such an approach does not allow the identification of the relative branching ratios of these two channels, unless the transition strengths of the probed transitions are known quantitatively. Therefore, Reisler and colleagues studied the formation of ground-state CO by photofragment ion imaging, using state-selective REMPI on CO ($v = 0, J$). Probing only the CO fragments, the two channels are not resolved spectroscopically, but they can be resolved in velocity space using ion imaging: the two different reaction pathways lead to different velocities of the CO molecules. **Figure 5** shows an ion image obtained for CO with zero quanta of vibration and 10 quanta

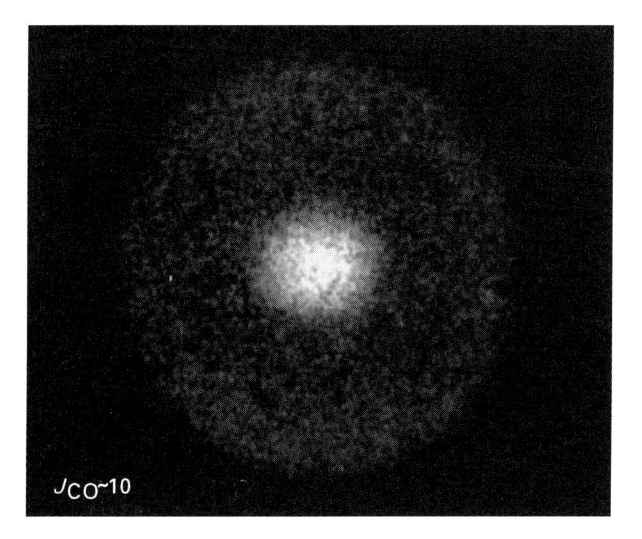

Figure 5 Ion image of CO ($v = 0$, $J = 10$) formed in the photolysis of HNCO at $\lambda = 230.1$ nm. The outer part of the image originates from CO molecules formed concomitantly with triplet NH, whereas most of the CO molecules in the inner, more dense area of the image originate from CO molecules formed concomitantly with singlet NH. Intensities are shown in an inverse linear grey scale (darker corresponds to lower intensities). Reprinted with permission of Elsevier Science-NL from Droz-Georget T, Zyrianov M, Reislev H and Chandler DW (1997) Correlated distributions in the photodissociation of HNCO to $NH(a^1\Delta + CO(X_1\Sigma^+)$ near the barrier on S_1. *Chemical Physics Letters* **276**: 316–324.

of rotation. CO ($v = 0, J = 10$) fragments are formed concomitantly with triplet as well as singlet NH. Singlet NH is approximately 12 500 cm^{-1} higher in energy than triplet NH. The singlet pathway leads to the smaller, somewhat more intense image, as less energy is available following this decomposition pathway, whereas the triplet channel leads to the bigger image, i.e. higher average speed of the CO molecules. As in the methane study, inverse Abel transformation of the images allowed the determination of rotational distributions of the CO fragments formed via these separate channels.

Reaction product ion imaging

Photofragment ion imaging is now an established method for the measurement of photofragment velocity (speed and angle) distributions from unimolecular dissociation processes. Recently, the domain of the ion imaging technique has been expanded to investigate bimolecular reactions as well, with the aim of determining differential cross sections in a single measurement. Houston and colleagues have studied the inelastic scattering reaction of NO molecules off argon atoms. In this study, a molecular beam of NO was crossed with a molecular beam of argon. The scattering of NO molecules was investigated for NO products in selected rotational and spin–orbit levels. The scattering of NO was visualized directly using the reaction product imaging technique, through selective ionization of the selected NO quantum state, and appeared to vary quite dramatically depending on the J level of the NO product.

Chandler and colleagues studied the bimolecular reaction $H+D_2 \rightarrow D+HD$. In their study, an atomic hydrogen beam (generated through the photolysis of HI) of well-defined velocity was crossed with a molecular beam of D_2. The velocity distribution of the deuterium atoms formed in the reaction was studied at two different centre-of-mass collision energies of 0.54 and 1.29 eV. **Figure 6** shows the D

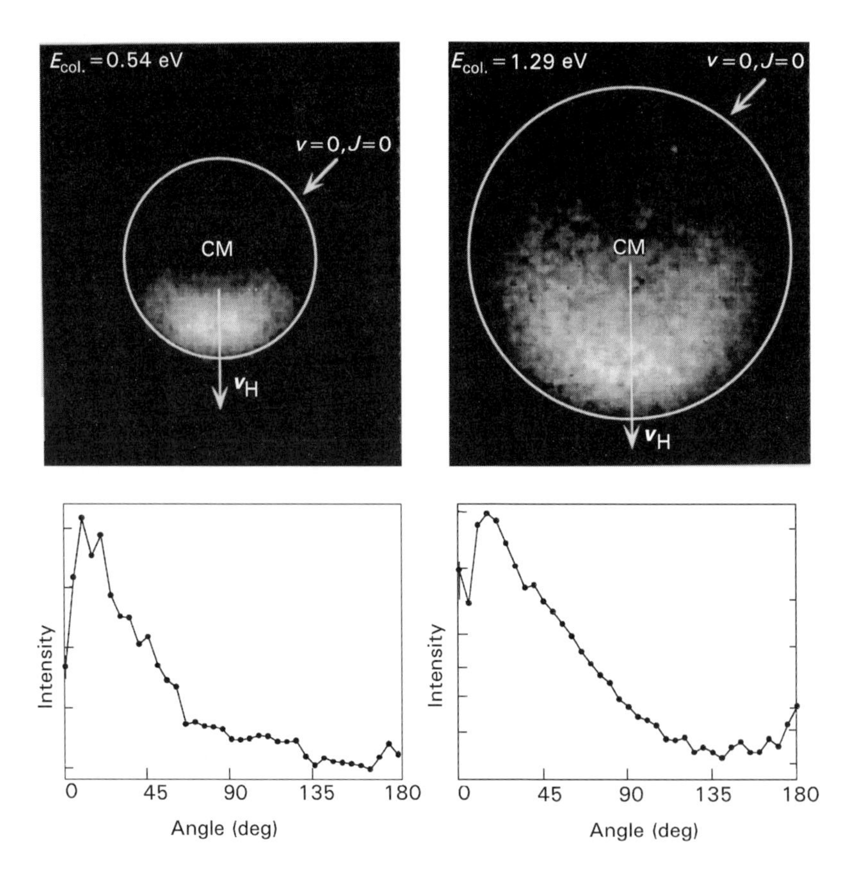

Figure 6 Raw images of D atoms formed in the reaction of $H+D_2$. Intensities are shown in an inverse linear grey scale (darker corresponds to lower intensities). The left and right columns show results obtained for this reaction at centre-of-mass collision energies of 0.54 eV and 1.29 eV, respectively. In the images, the centre-of-mass (CM) of the reaction and the direction of the H atoms are indicated. The circles indicate the calculated positions of the fastest D atoms corresponding to formation of HD in the $v = 0$, $J = 0$ state. At the bottom of the figure the differential cross-sections, obtained from the inverse Abel-transformed images are displayed. Reproduced with permission from Heck AJR and Chandler DW (1995) Imaging techniques for the study of chemical reactions dynamics. *Annual Review of Physical Chemistry* **46**: 335–378 © 1995 Annual Reviews Inc.

atom reaction product images at these two collision energies. The circles represent the maximum speed the D atoms could have, corresponding to the concomitant formation of HD in its rotational and vibrational ground states. It is directly obvious from these images that the nascent D atoms are predominately forward scattered with respect to the direction of the H atom beam velocity. The differential cross sections obtained from the inverse Abel-transformed images for the $H+D_2 \rightarrow D+HD$ reaction at these two different collision energies agreed fairly well with theoretically predicted differential cross sections. Reaction product imaging offers several advantages over more traditional methods used to study bimolecular reactive scattering in that the differential cross sections of the products, possibly in a single selected quantum state, are obtained in a single measurement.

List of symbols

$I(\theta)$ = photo-fragment angular distribution; J = rotational quantum number; P_n = nth order Legendre polynomial; β = anisotropy parameter; μ = mass; v = vibrational quantum number; θ = angle between v and E; λ = photon wavelength.

See also: **Fragmentation in Mass Spectrometry; Gas Phase Applications of NMR Spectroscopy; Ion Dissociation Kinetics, Mass Spectrometry; Ion Energetics in Mass Spectrometry; Photoionization and Photodissociation Methods in Mass Spectrometry; Time of Flight Mass Spectrometers.**

Further reading

Bontuyan LS, Suits AG, Houston PL and Whitaker BJ (1993) State-resolved differential cross-sections for crossed-beam Ar-NO inelastic scattering by direct imaging. *Journal of Physical Chemistry* **97**: 6342–6350.

Bordas C, Paulig FH, Helm H, Huestis DL (1996) Photoelectron imaging spectrometry: Principle and inversion method. *Review of Scientific Instruments* **67**: 2257–2268.

Chandler DW and Houston PL (1987) Two-dimensional imaging of state-selected photodissociation products detected by multiphoton ionization. *Journal of Chemical Physics* **87**: 1445–1447.

Droz-Georget T, Zyrianov M, Reisler H and Chandler DW (1997) Correlated distributions in the photodissociation of HNCO to $NH(a^1\Delta) + CO(X_1\Sigma^+)$ near the barrier on S_1. *Chemical Physics Letters* **276**: 316–324.

Eppink ATJB and Parker DH (1997) Velocity map imaging of ions and electrons using electrostatic lenses: Application in photoelectron and photofragment ion imaging of molecular oxygen. *Review of Scientific Instruments* **68**: 3477–3484.

Heck AJR (1997) Throwing light on molecules falling apart: photofragment ion imaging. *European Mass Spectrometry* **3**: 171–183.

Heck AJR and Chandler DW (1995) Imaging techniques for the study of chemical reaction dynamics. *Annual Reviews of Physical Chemistry* **46**: 335–372.

Heck AJR, Zare RN and Chandler DW (1996) Photofragment imaging of methane. *Journal of Chemical Physics* **104**: 4019–4030.

Houston PL (1995) Snapshots of chemistry: product imaging of molecular reactions. *Accounts of Chemical Research* **28**: 453–460.

Houston PL (1996) New laser-based and imaging methods for studying the dynamics of molecular collisions. *Journal of Physical Chemistry* **100**: 12757–12770.

Kitsopoulos TN, Buntine MA, Baldwin DP, Zare RN and Chandler DW (1993) Reaction product imaging: The $H + D_2$ reaction. *Science* **260**: 1605–1610.

Parker DH and Eppink ATJB (1997) Photoelectron and photofragment velocity map imaging of state-selected molecular oxygen dissociation/ionization dynamics. *Journal of Chemical Physics* **107**: 2357–2362.

Whitaker BJ (1993) Photo-ion imaging and future directions in reactive scattering. In Compton RG and Hancock G (eds) *Research in Chemical Kinetics*, Vol 1, pp 307–346. Oxford: Elsevier.

Ion Molecule Reactions in Mass Spectrometry

Diethard K Böhme, York University, Toronto, Ontario, Canada

MASS SPECTROMETRY
Methods & Instrumentation

Mass spectrometry has played a dominant role in measurements of ion–molecule reactions. These measurements have had broad impact, in both fundamental and applied areas of science. For example, they have provided extensive thermodynamic information about ions and molecules including standard enthalpies of ion formation and the electron affinities, ionization energies and proton affinities of neutral molecules. They have found important applications in analytical and biomedical mass spectrometry and in plasma chemistry. In physical-organic chemistry they have provided fundamental insight into the intrinsic efficiencies of ion–molecule reactions with analogues in solution (e.g. proton transfer (acid–base), nucleophilic substitution (S_N2) and metal–ion coordination reactions) and into the influence of solvation on these efficiences. Also, they have provided a deeper understanding of the chemistry of the earth's ionosphere and the ionospheres of other planets, of chemical synthesis in interstellar and circumstellar clouds, of flame chemistry, of radiolysis and of ion-induced polymerization.

Ion–molecule reaction measurements

Ion–molecule reactions have been studied using a large variety of mass-spectrometric techniques for over 3 decades over a large range in pressure and temperature.

Low pressure (from ~10^{-5} to 10^{-7} torr) measurements have been performed with Fourier transform (FT) and ion-cyclotron resonance (ICR) mass spectrometers and with ion traps. Moderate pressure (from 0.1 to 5 torr) techniques used to investigate ion–molecule reactions include high-pressure mass spectrometry, the flowing afterglow (FA) technique, the selected-ion flow tube (SIFT) technique and the variable-temperature selected-ion flow-drift tube (VT-SIFDT) technique. The latter technique operates over a temperature range from 85 to 550 K and at average kinetic energies up to at least 1 eV. Measurements at atmospheric pressure have been restricted largely to ion mobility spectrometry and flame-ion mass spectrometry. The latter provides data in the high-temperature range from 1800 to 2400 K. Measurements at temperatures below 160 K have been achieved in a cooled Penning trap, a drift tube and a flow reactor that utilizes cooling by adiabatic expansion. A free jet-flow reactor has been employed to study ion–molecule reaction rate coefficients at temperatures as low as 0.3 to 30 K.

Rates of ion–molecule reactions

Reactions between ions and molecules are among the fastest chemical reactions known. This is because the electrostatic interaction between the charge on the ion and the polarizable (or polar) molecule draws these reactants together at long range. Furthermore, the resulting interaction energy at short range is often sufficiently large to overcome intrinsic energy barriers to chemical change. As a consequence, exothermic ion–molecule reactions often proceed very rapidly at room temperature *at the capture rate*, without an activation energy. This is in sharp contrast to molecule–molecule reactions that normally involve an activation energy. Generalized profiles for the dependence of reactant energy on the reaction coordinate for ion–molecule and molecule–molecule reactions are shown in **Figure 1**. Ion–molecule reactions are often best described by a double-minimum, rather than a single-minimum, potential energy profile.

The reaction rate coefficient, k_r, for an exothermic ion–molecule reaction proceeding without an activation energy is equal to the capture or collision rate coefficient, k_c, which is typically ~1×10^{-9} cm^3 molecule^{-1} s^{-1}. This is some one to two orders of magnitude larger than that for an exothermic molecule–molecule reaction proceeding without an activation energy. Several theories are available that allow calculation of the capture or collision rate coefficient, k_c, for an ion–molecule collision. According to the most popular theory, the 'average dipole orientation' (ADO) theory, the magnitude of k_c is determined by the physical properties of the reactants: the polarizability, α, and permanent dipole, μ_D, of the reacting molecule, the charge on the ion, q, and the reduced mass, μ, of the colliding reactants. The collision-rate coefficient k_c can be calculated using ADO theory from the following equation:

$$k_c = \frac{2\pi q}{\mu^{1/2}}\left[\alpha^{1/2} + C\mu_D \frac{2^{1/2}}{(\pi k_B T)^{1/2}}\right]$$

where C is a correction factor that is a function of α and μ_D, and k_B is Boltzmann's constant. The ratio of the reaction rate coefficient to the collision rate coefficient, k_r/k_c, is a measure of reaction probability per collision or reaction efficiency. Although many exothermic ion–molecule reactions proceed with unit reaction probability at room temperature, others exhibit reaction probabilities much less than unity.

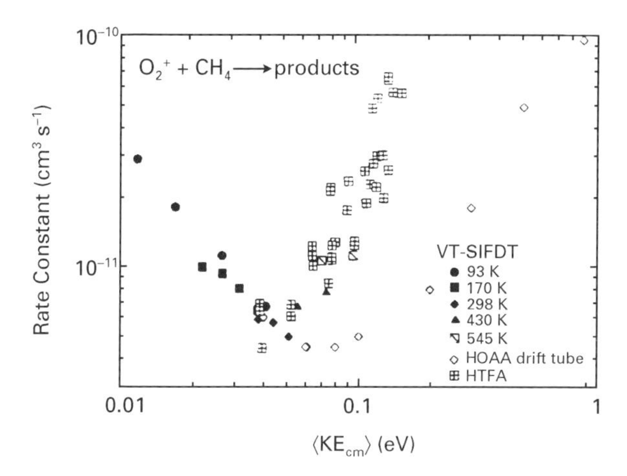

Figure 2 Rate coefficients for the reaction of O_2^+ with methane measured as a function of average kinetic energy at several temperatures in a variable-temperature selected-ion flow drift tube (VT-SIFDT). Also shown are data taken with a high-temperature flowing afterglow (HFTA) and a drift tube. Reproduced with permission from Viggiano AA and Morris MA (1996) *Journal of Physical Chemistry* **100**: 19227–19240.

Temperature/kinetic energy dependence

Ion–molecule reactions that are rapid and efficient at room temperature frequently remain rapid as the temperature of the reactants is varied and follow the temperature dependence of the collision rate. Slow reactions at room temperature almost always increase in rate with decreasing or increasing temperature or kinetic energy (see **Figure 2**). This can be understood in terms of the double-minimum potential shown in **Figure 1** and the following reaction mechanism:

$$A^+ + B \rightleftharpoons (AB)^{+*} \rightarrow C^+ + D$$

As the reactants approach each other they are attracted by electrostatic ion–induced dipole and ion–dipole forces and form the 'entrance channel' complex $(AB)^{+*}$. Further advance along the minimum potential energy path ultimately leads to an increase of potential energy due to bond redisposition. Then the complex exits along the 'product channel' and finally dissociates into products. The two potential energy minima (or 'wells') are separated by a barrier that controls the overall reaction efficiency, even though the barrier often lies below the initial energy of the reactants. A rate of dissociation back to reactants that decreases more rapidly with decreasing temperature than the rate of the dissociation of $(AB)^{+*}$ into products leads to an increase in overall rate with decreasing temperature. An increase in rate at higher temperatures often results from the emergence of new vibrational, electronic or chemical channels.

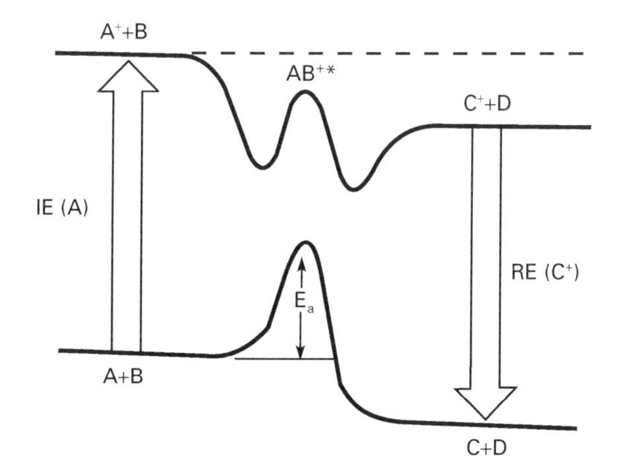

Figure 1 Generalized profiles for the dependence of reactant energy on the reaction coordinate for ion–molecule and molecule–molecule reactions. IE and RE represent ionization energy and recombination energy, respectively. Some ion–molecule reactions are better represented by a single- rather than a double-minimum potential.

Endothermic ion–molecule reactions and ion–molecule reactions with an activation energy have been examined over a wide range of the kinetic energy of the reactant ion (from ~0.1 to 600 eV in the laboratory frame). They exhibit a reaction threshold with increasing ion kinetic energy, rapidly increase in reaction rate and then slow down again at higher kinetic energies (see **Figure 3**). An analysis of the kinetic energy dependence of the reaction cross-section can provide thermodynamic information about the reacting ions and neutral molecules.

Rotational and translational energy have been found to be equally effective in driving endothermic reactions. For exothermic reactions large rotational effects appear mainly when one or both of the

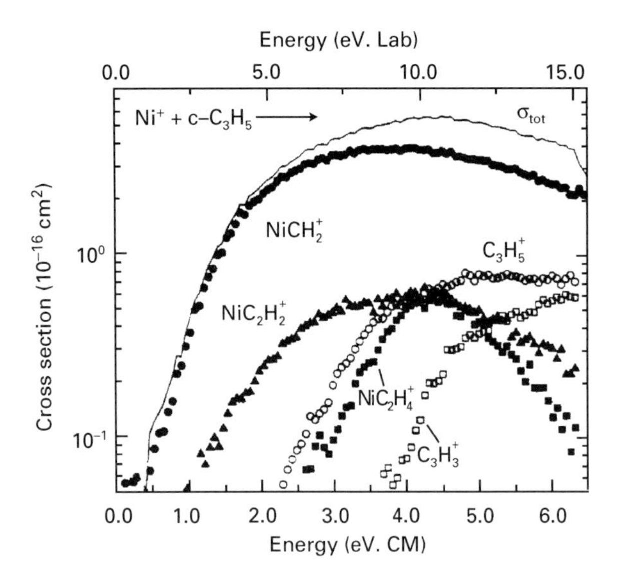

Figure 3 Observed variations of product cross-sections with translational energy in the centre-of-mass frame of reference (lower scale) and the laboratory frame (upper scale) for reactions of cyclopropane with Ni^+. Reproduced with permission from Fisher ER and Armentrout PB (1990) *Journal of Physical Chemistry* **94**: 1674–1683.

reactants have a large rotational constant. Vibrational effects are more varied: in some reactions vibrational excitation in the anticipated reaction coordinate can strongly affect reaction probability.

Types of ion–molecule reaction

Reactions between ions and molecules can lead to a variety of consequences ranging from the simple transfer of an electron between an ion and a molecule to the nominal transfer of a large molecular fragment accompanied by extensive chemical bond redisposition in both the molecule and the ion. **Table 1** summarizes various types of ion–molecule reaction that have been investigated involving singly charged ions, but recently many also have been observed with multiply charged ions.

Table 1 Types of ion–molecule reaction

Reaction type	
Electron (charge) transfer	$A^{\pm} + B \rightarrow B^{\pm} + A$
Proton transfer	$AH^+ + B \rightarrow BH^+ + A$
	$A^- + BH \rightarrow AH + B^-$
H-atom transfer	$A^{\pm} + BH \rightarrow AH^{\pm} + B$
R^+ transfer	$AR^+ + B \rightarrow BR^+ + A$
Nucleophilic displacement	$A^- + RX \rightarrow AR + X^-$
Bond redisposition	$AB^{\pm} + C \rightarrow AC^{\pm} + B$
Radiative association	$A^{\pm} + B \rightarrow AB^+ + h\nu$
Collisional association	$A^{\pm} + B + M \rightarrow AB^{\pm} + M$
Switching	$A^{\pm}.B + C \rightarrow A^{\pm}.C + B$
Associative detachment	$A^- + B \rightarrow AB + e$

Electron-transfer reactions

Electron transfer between ions and molecules at thermal energies is subject to energy resonance and Franck–Condon effects. When small ions are involved, electron transfer is generally inefficient unless there exists a near-resonance between an energy level of the product ion and an available recombination energy of the reactant ion. In addition there must be large Franck–Condon factors connecting the upper and lower states involved in the transition. Energy resonance is necessary, but not always sufficient. For electron-transfer reactions involving triatomic or polyatomic species, the density of states is much larger so that an energy match and Franck–Condon requirement can almost always be obtained. Vibrational-energy dependence is often observed in electron-transfer reactions presumably due to energy resonance and Franck–Condon effects. Low energy electron-transfer collisions often produce products with large amounts of internal energy that can lead to the dissociation of the product ion. The dissociation products provide a signature of the reactant molecule so that electron transfer is useful in the analysis of gases by chemical ionization mass spectrometry. Electron transfer with multiply charged ions leads to charge separation according to the following reaction:

$$A^{n+} + B \rightarrow A^{(n-1)+} + B^+$$

Such reactions involve a 'reverse activation energy' resulting from the Coulombic repulsion between the product ions and as a consequence exhibit a delayed onset.

Proton-transfer reactions

Proton transfer, either from a protonated molecule to a neutral molecule or from a neutral molecule to a negative ion, often proceeds at or near the collision rate when exothermic (see **Figure 4**). When the exothermicity is large, internal energy appearing in the protonated product may be sufficient to cause dissociation with the loss of one or more neutral fragments. Thus, as was the case with dissociative electron-transfer reactions, the product ions again provide a signature of the reactant molecule so that these reactions also are useful in the analysis of gases by chemical ionization mass spectrometry. Virtually any molecule can be chemically ionized with an appropriate proton donor and often this is possible

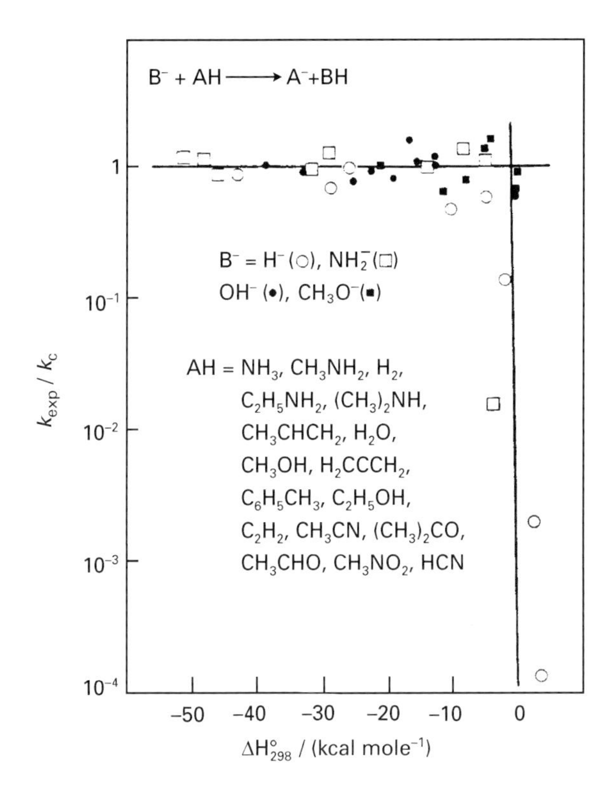

Figure 4 A correlation between the efficiency of proton transfer, k_{exp}/k_c, and its overall change in enthalpy, ΔH°, at room temperature. The measurements were taken using the flowing-afterglow technique. Reproduced with permission from Bohme DK (1981) *Transactions of the Royal Society of Canada* **19**: 265–284.

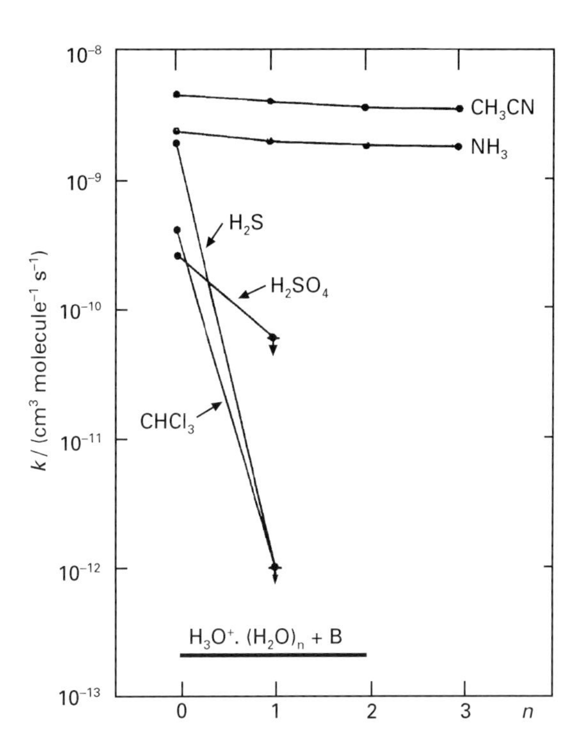

Figure 5 Observed variations with the number of water molecules in the rate coefficient for proton-transfer reactions between hydrated hydronium ions and various molecules B at room temperature. The measurements were taken using the flowing-afterglow technique. Adapted with permission from Bohme DK, Mackay GI and Tanner SD (1979) *Journal of the American Chemical Society* **101**: 3724–3730.

with high efficiency and selectivity. Solvation can influence the rate of proton transfer as is illustrated in **Figure 5**, but proton transfer generally remains fast until solvation renders the reaction endothermic. There is a growing interest in the ion chemistry of multiply protonated biological ions such as multiply protonated peptides and proteins. Proton transfer from multiply protonated cations may occur according to the following reaction:

$$AH_n^{n+} + B \rightarrow AH_{n-1}^{(n-1)+} + BH^+$$

Again, a 'reverse activation energy' is involved in such reactions as a consequence of the Coulombic repulsion between the product ions.

H-atom and R$^+$ transfer and bond redisposition reactions

Successive H-atom transfer reactions can lead to the hydrogenation of unsaturated positive or negative ions. They find importance, for example, in the conversion of atomic ions to hydrogenated polyatomic ions in interstellar clouds rich in molecular hydrogen. There appears to be no systematic or easy way to predict the efficiency of such ion–molecule reactions other than to say that hydrogen-atom transfer is rarely observed when electron-transfer or proton-transfer channels are exothermic. Similar comments apply to the transfer of larger R$^+$ species and more general bond redisposition reactions.

Nucleophilic displacement reactions

Nucleophilic displacement reactions, reactions that formally also involve the transfer of R$^+$, have been explored in great depth in the gas phase with the majority of available mass spectrometric techniques. In part this has been because of the importance of S_N2 reactions in solution chemistry and their extreme sensitivity to solvent, both in solution and in the gas phase. Unsolvated S_N2 reactions are characterized by double minima in the reaction energy profile as shown in **Figure 1** and the following reaction mechanism:

$$A^- + RX \rightleftharpoons A^- \ldots RX \rightleftharpoons AR \ldots X^- \rightarrow AR + X^-$$

The central barrier in the double-minimum potential energy profile, a Walden-inversion transition state, may lie above or below the initial energy of the reactants. The rate coefficients for S_N2 reactions show a slight negative temperature dependence.

Gas-phase measurements also have provided information on the dependence of reaction efficiency on the stepwise molecular solvation of the nucleophile A^- in reactions of the following type, where X can be a halogen atom, R=CH_3 and S a solvent molecule:

$$A^-S_n + RX \rightarrow AR + X^-S_m + S_{n-m}$$

Such reactions are also characterized by double minima in the reaction-energy profiles. Because of the lower stabilization of the charge-delocalized transition state relative to the charge-localized reactant and product ions, these profiles converge to a single-maximum profile typical for S_N2 reactions in solution as the number of solvent molecules on the reactant and product ions is increased. Gas-phase measurements of the influence of stepwise solvation on reaction efficiency have shown that solvation by even one or two molecules can reduce the specific rate for nucleophilic displacement precipitously (see **Figure 6**). Retention of all the solvent molecules in the product ion appears to be the exception rather than the rule. Limited measurements indicate that increasing solvation increases the magnitude of the negative temperature dependence.

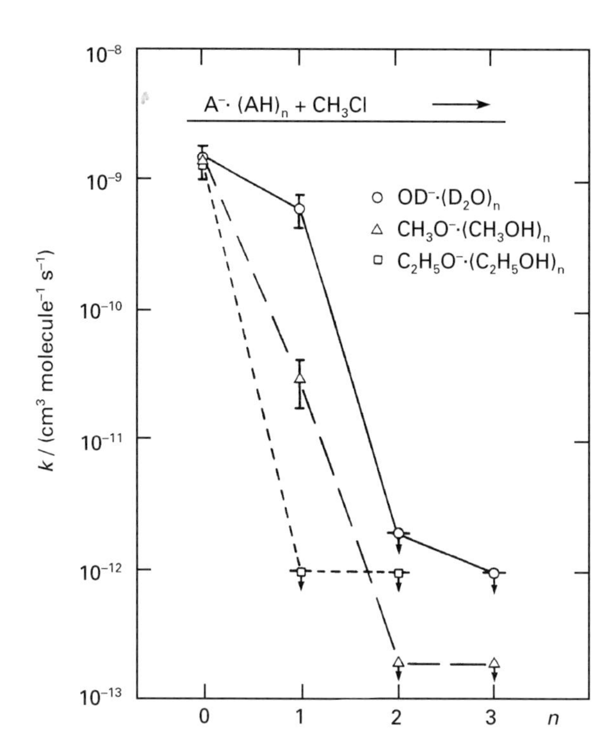

Figure 6 Observed variations with the number of solvent molecules in the rate coefficient for nucleophilic-displacement reactions between various solvated nucleophiles and methyl chloride at room temperature. The measurements were taken using the flowing-afterglow technique. Reproduced with permission from Bohme DK and Raksit AB (1984) *Journal of the American Chemical Society* **106**: 3447–3452.

Association reactions

Association reactions lead to the growth of an ion though the formation of new bonds ranging from weak electrostatic to strong covalent bonds. For example, association reactions may lead to the solvation of an ion by weak electrostatic or hydrogen bonding, to ion ligation involving bonds of intermediate strength, or even to strong chemical bond formation. Two steps are required for association as shown below: the formation of an intermediate excited adduct ion $(AB)^{\pm *}$ followed by its de-excitation.

$$A^{\pm} + B \rightleftharpoons (AB)^{\pm *}$$
$$(AB)^{\pm *} \rightarrow AB^{\pm} + h\nu$$
$$(AB)^{\pm *} + M \rightarrow AB^{\pm} + M$$

Radiative association occurs when the intermediate becomes stabilized by the emission of a photon ($h\nu$), usually in the infrared spectral region. Collisional association results when the intermediate is stabilized by collision with a third molecule M. Although radiative association is a bimolecular and therefore second-order process, termolecular collisional association is third-order at low pressures and becomes second-order at high pressures when the frequency of collisions with M exceeds the frequency of the unimolecular dissociation of the intermediate ion $(AB)^{\pm *}$. The rate coefficient for collisional association exhibits a *negative* temperature dependence. Also, FT-ICR measurements have shown that the rate coefficient for radiative association has a strong negative temperature dependence.

Radiative association finds importance in high-vacuum ion-trapping instruments and the low-pressure environments of interstellar space. In contrast, collisional association predominates in high-pressure, low-temperature environments. When they occur *sequentially*, collisional association reactions can play a vital role in gas-phase ion solvation (as occurs naturally, for example, in the earth's lower ionosphere), in ion-cluster formation generally, and in the coordination or multiple derivatization of the type important in organometallic chemistry or in ion-induced polymerization, respectively. The following are examples of ion solvation, ion coordination and ion-induced polymerization, respectively:

$$H_3O^+(H_2O)_n + H_2O + M \rightarrow H_3O^+(H_2O)_{n+1} + M$$

$$Fe^+(C_2H_2)_n + C_2H_2 + M \rightarrow Fe^+(C_2H_2)_{n+1} + M$$

$$C_{60}^{2+}(\text{allene})_n + \text{allene} + M \rightarrow C_{60}^{2+}(\text{allene})_{n+1} + M$$

In the latter reaction involving a multiply-charged reactant ion, intramolecular charge separation is expected to occur and, in this case, may even act to direct the polymerization away from the surface of C_{60}.

Switching reactions

Adduct ions formed by radiative or collisional association may be transformed by bimolecular switching reactions in which one solvent or ligand molecule is replaced by another. Such reactions may occur at any extent of solvation or ligation and are usually efficient at room temperature when they are exothermic.

Associative detachment reactions

Associative detachment reactions are unique to negative ions. The energy profiles for such reactions involve one or more curve-crossings between one or more of the attractive potential curves of the reactants and the potential curve for the product AB. AB is formed by the rapid autodetachment of an electron from $(AB)^{\pm *}$ at short interaction distances. Exothermic associative detachment for atomic reagents is inevitably fast. Inhibiting factors can arise for molecular systems.

Sources of information

For up-to-date accounts in the rapidly evolving field of ion chemistry, the reader is directed to volumes 1 to 3 and future volumes in the series *Advances in Gas Phase Ion Chemistry* edited by NG Adams and LM Babcock. For earlier reports on various aspects of ion–molecule reactions see the NATO ASI Series edited by P Ausloos, by P Ausloos and SG Lias and by MA Almoster Ferreira. Another source of earlier information is the *Gas Phase Ion Chemistry* series edited by MT Bowers.

There have been a number of tabulations of rate coefficients for gas-phase ion–molecule reactions that include leading references to individual ion–molecule reactions. For the most part they are restricted to limited areas of interest. They include tabulations of rate constants measured with flow reactors (DL Albritton), rate constants of interest in modelling the chemistry of planetary atmospheres, cometary comae and interstellar clouds (VG Anicich), rate constants for reactions involving organic ions containing only C and H (LW Sieck) and those containing more than C and H (LW Sieck and SG Lias). An early, more general, tabulation has been published by Talroze. The most comprehensive tabulation of rate constants for both bimolecular and termolecular reactions determined up to 1986 has been published by Ikezoe and co-workers.

See also: **Chemical Ionization in Mass Spectrometry; Cosmochemical Applications using Mass Spectrometry; Interstellar Molecules, Spectroscopy of; Ion Energetics in Mass Spectrometry; Ion Trap Mass Spectrometers; Mass Spectrometry, Historical Perspective; SIFT Applications in Mass Spectrometry.**

Further reading

Adams NG and Babcock LM (eds) (1992) *Advances in Gas Phase Ion Chemistry*, Vol 1. Greenwich, CT: JAI Press.

Adams NG and Babcock LM (eds) (1996)*Advances in Gas Phase Ion Chemistry*, Vol 2. Greenwich, CT: JAI Press.

Adams NG and Babcock LM (eds) (1998) *Advances in Gas Phase Ion Chemistry*, Vol 3. Greenwich, CT: JAI Press.

Albritton DL (1978) *Ion–Molecule Reaction-Rate Coefficients Measured in Flow Reactors Through 1977.* Atomic Data Nuclear Data Tables 22: 1–101. Academic Press.

Almoster Ferreira MA (ed) (1984) *Ionic Processes in the Gas Phase,* NATO ASI Series C, Vol 118. Dordrecht: Reidel.

Ausloos P (ed) (1975) *Interactions between Ions and Molecules*, NATO ASI Series B, Vol 6. New York: Plenum.

Ausloos P (ed) (1979) *Kinetics of Ion-Molecule Reactions,* NATO ASI Series B, Vol 40. New York: Plenum.

Anicich VC (1993) A survey of bimolecular ion-molecule reactions for use in modeling the chemistry of planetary atmospheres, cometary comae and interstellar clouds. *Astrophysical Journal. Supplement Series* **84**: 215–315.

Anicich VC (1993) Evaluated bimolecular ion-molecule gas phase kinetics of positive ions for use in modeling the chemistry of planetary atmospheres, cometary comae and interstellar clouds. *Journal of Physical Chemistry Reference Data* **22**: 1469–1569.

Ausloos P and Lias SG (eds) (1987) *Structure/Reactivity and Thermochemistry of Ions,* NATO ASI Series C, Vol 193. Dordrecht: Reidel.

Bowers MT (ed) (1979) *Gas Phase Ion Chemistry*. New York: Academic.

Sieck LW and Lias SG (1976) Rate coefficients for ion-molecule reactions: ions containing C and H. *J. Phys. Chem. Ref. Data* 5, 1123–1146.

Sieck LW (1979) *Rate Coefficients for Ion-Molecule Reactions: Ions Other Than Those Containing Only C and H.* National Standards Reference Data System – National Bureau of Standards 64, pp 1–20.

Ikezoe Y, Matsuoka S, Takebe M and Viggiano A (1987) *Gas Phase Ion–Molecule Reaction Rate Constants Through 1986*. Tokyo: Maruzen (Distributors).

Virin LI, Dzhagatspanyan RB, Karachevtsev GV and Talroze VL (1979) *Rate Coefficients for Ion-Molecule Reactions* (in Russian). Moscow: Nauka.

Ion Structures in Mass Spectrometry

Peter C Burgers, Hercules European Research Center B.V., Barneveld, The Netherlands
Johan K Terlouw, McMaster University, Hamilton, Ontario, Canada

MASS SPECTROMETRY
Theory

Motivation and methods

During the early days in the development and application of mass spectrometry to the analysis of organic molecules, electron ionization (EI) was the only ionization method available. Today, many other ionization techniques are available for the analysis of large (bio)molecules, but EI remains a popular method for the ionization of organic molecules that are not too involatile; it is, for example, the method of choice for identifying components in complex mixtures using the powerful technique of gas chromatography–mass spectrometry (GC-MS) in conjunction with fast searches in large EI spectral databases. Although EI produces both positively and negatively charged ions, usually only the singly charged positive ions, which make up conventional EI mass spectra, are used for interpretive purposes. It is not surprising, then, that most of the work published on ion structures concerns the properties of singly charged positive ions.

Almost all organic molecules contain an even number of electrons and so removal of an electron by EI leads to a radical cation, $M^{\bullet+}$, the molecular ion. The conventional mass spectrum reflects the competitive and consecutive fragmentations with rate constants $k > 10^6\ s^{-1}$ of the molecular ion induced by EI. The interpreter's task then is to deduce the structure of the original neutral molecule from the occurrence or absence of ionic fragmentations. Obviously, a good understanding of which ion structures can be formed and how they dissociate is critical to the interpreter's success.

All experimental methods for ion structure elucidation give information only about the ion's constitution or connectivity, i.e. which atoms are joined together and the probable (formal) locations of charge (and radical) sites. Details of the ion's configuration, such as bond lengths and angles, are not yet accessible to experiment for polyatomic ions. In the mass spectrometer, ions are gaseous isolated species: they are free from solvent molecules and this makes them especially amenable to molecular orbital (MO) calculations. Indeed, the growing synergism between experiment and theory in gas-phase ion chemistry has greatly facilitated our understanding in this area. State-of-the-art *ab initio* MO calculations can be as accurate as experiment and as often as not disagreements that emerge are solved by reinterpretation of the experimental data.

The most important methods for ion structure determination are:

- Ion dissociation characteristics obtained by tandem mass spectrometry: both unimolecular, as observed in the metastable ion (MI) mass spectrum and collision-induced, as observed in the collision-induced dissociation (CID) mass spectrum.
- Reactivity, as studied by ion–molecule reactions.
- Ion thermochemistry, that is heat of formation (ΔH_f) data obtained by ionization energy (IE) and appearance energy (AE) measurements, often in conjunction with theoretical calculations.

These may be complemented by isotope labelling experiments, a time-honoured method of unravelling fragmentation mechanisms. With these methods, ions of different internal energies and ion lifetimes are sampled. MI characteristics, as obtained with a sector mass spectrometer, reflect ions that decompose in a field-free region with a narrow range of internal energies just above the threshold for decomposition, ~10 μs after their formation. CID mass spectra, also measured with a sector mass spectrometer, and ion–molecule reactions as studied in an ion cyclotron resonance (ICR) cell involve nondecomposing ions with internal energies up to the lowest threshold for decomposition; these ions have lifetimes >10 μs and >1 ms, respectively. Finally, ΔH_f data and MO calculations characterize ions in their electronic and vibrational ground states having, in principle, infinite lifetimes. These different techniques sample ions of different internal energies and lifetimes and, depending on the heights of possible isomerization barriers, they may also sample ions of different structure.

Isomerization of solitary ions

Structure elucidation of an organic ion may be rendered difficult by isomerization reactions; this is especially true for the low-energy (metastable) ions. This is illustrated in **Figure 1** for the isomeric ions ABC^+ and ACB^+ whose dissociations of lowest energy requirement from the unrearranged species lead to AB^+ and AC^+, respectively.

In **Figure 1A** the isomerization barrier is higher than either dissociation threshold and the isomeric ions will yield different ion products in their mass spectra, so that differentiation is possible on the basis of MI spectra alone. In the case of **Figure 1B** the isomerization barrier is lower than either dissociation threshold and now, prior to dissociation, ions ACB^+ and ABC^+ interconvert. As a consequence, prior to dissociation, the metastable ions will consist of an equilibrium mixture $ACB^+ \rightleftharpoons ABC^+$ that will choose the exit channel of lowest energy, namely to AB^+. In this case the isomers will yield identical MI spectra. Ions with energies below the transition state for isomerization will have retained their original structure, but these do not dissociate. These ions can be induced to decompose by collisional activation with an inert gas, such as helium, to dissociation products of higher enthalpy, resulting in the CID mass spectrum. These dissociations take place rapidly (<0.1 μs) and so interconversion $ACB^+ \rightleftharpoons ABC^+$ will be slow by comparison and the isomers can be differentiated on the basis of their respective CID spectra. The situation pertinent to ion ACB^+ in **Figure 1B** is frequently observed: because they dissociate slowly, metastable ions can undergo more or less extensive rearrangement reactions to form the most 'economical', but usually non-structure-characteristic, set of products. In CID ('sledge-hammer' approach), the energized species fragment rapidly and so more of the structure-characteristic direct bond cleavages are seen. Finally, the isomerization barrier can be intermediate between the dissociation levels of $AB^+ + C^•$ and $AC^+ + B^•$. In this case ACB^+ isomerizes slowly to ABC^+ which then fragments rapidly. Now the metastable ions will yield the same products but with different internal energies and kinetics. In particular, the kinetic energy released (T value) upon dissociation will be larger for ACB^+, resulting in broadening of the metastable peak.

Usually CID experiments serve to increase the internal energy of the ions, which then fragment. Under certain conditions, however, such collisions may lead not only to dissociation but also to association. Thus collisions of $C_{60}^{•+}$ with He at high translational energies (~3×10^4 m s^{-1}) leads to the capture complex $C_{60}He^{•+}$. This complex survives a neutralization–reionization (NR) experiment, showing that the ionic and neutral complexes have an endohedral structure He@C_{60}. Complexes

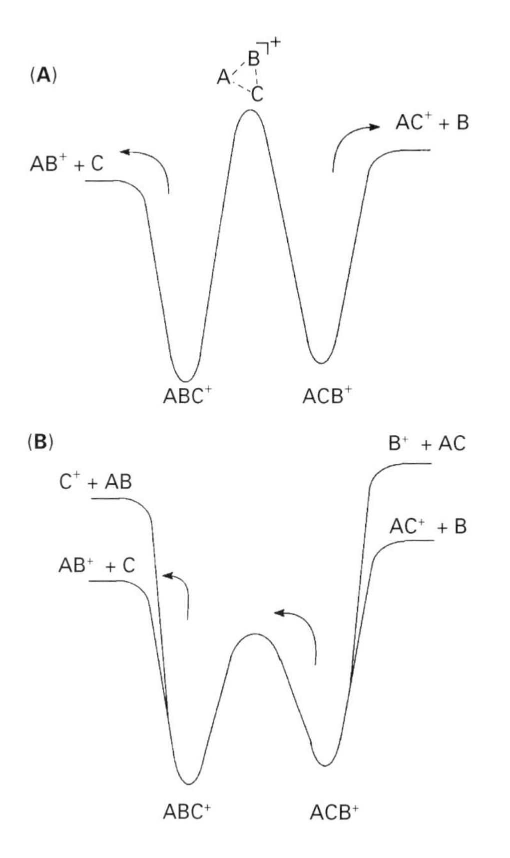

Figure 1 Schematic potential energy diagrams showing the relationship between the heights of isomerization barriers and observed products.

containing two noble-gas atoms have also been prepared by collision experiments.

Ion structures

The molecular radical cation, an odd-electron ion, can dissociate in two basic ways. It can undergo a direct homolytic cleavage to produce an even-electron ion and a radical, or it can produce another odd-electron ion by multiple cleavage (as in a cyclic compound) or via rearrangement (as in the McLafferty rearrangement).

Even electron ions

Ionized radicals

Even-electron ions are formed by loss of a radical from the molecular ions or by addition of a reagent ion to a molecule in an ion–molecule reaction. Most, but not all, even-electron ions formed by dissociative ionization can be viewed as ionized radicals. Such species may be stabilized by resonance or aromaticity. Thus $C_3H_5^+$ ions are abundant ions in many mass spectra, even in cases where the C_3H_5 substructure is not present in the original molecule, and this can be attributed to stabilization by resonance: $CH_2{=}CH{-}CH_2^+ \Leftrightarrow {}^+CH_2{-}CH{=}CH_2$. Similarly, the $C_7H_7^+$ ion, *m/z* 91, is a ubiquitous peak in the mass spectra of aromatic compounds. Elegant ^{13}C labelling experiments performed in the 1950s have shown that this species in the spectrum of toluene and substituted aromatics is not the expected benzyl ion, $C_6H_5CH_2^+$ (even though allylically stabilized) but rather the thermodynamically more stable (by ~40 kJ mol^{-1}) cyclic tropylium ion. The reverse situation holds for radicals; here the benzyl structure is the more stable one. The tropylium ion obeys Hückel's $4n+2$ rule ($n = 1$) on aromaticity, whereas the tropyl radical does not, and indeed it was Hückel who predicted 70 years ago that the cation should enjoy considerable stability. A similar, but more dramatic, situation occurs for the C_3H_3 structures. Here, for the neutrals, the cyclopropenyl radical is ~96 kJ mol^{-1} less stable than the propargyl radical, $^{\bullet}CH_2{-}C{\equiv}CH$ whereas, for the ions, the cyclopropenium ion ($n = 0$) is the more stable isomer (by 105 kJ mol^{-1}). For the cations, the large destabilization due to ring strain is more than offset by the energy gain due to aromaticity. These examples serve to illustrate an important phenomenon in mass spectrometry: upon ionization a stability reversal of isomeric structures may take effect. This becomes particularly important for odd-electron ions (see below)

The $C_2H_3O^+$ system is one of the earliest examples where experiment and theory have gone hand in hand to produce a detailed description of the potential energy surface. Four stable $C_2H_3O^+$ ions have been generated and identified by MI, CID and isotope labelling experiments and all are calculated by theory to be stable species. These are (with relative calculated ΔH_f values in kJ mol^{-1} in parentheses):

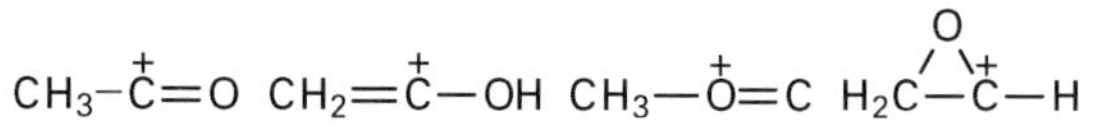

$$CH_3{-}\overset{+}{C}{=}O \quad CH_2{=}\overset{+}{C}{-}OH \quad CH_3{-}\overset{+}{O}{=}C \quad H_2C{-}\overset{+}{C}{-}H \text{ (O bridging)}$$

(0) (181) (216) (244)

In particular the isomer $CH_3{-}O^+{=}C$ presented an experimental challenge, but after theory predicted it should be stable, it was subsequently made. Not surprisingly, the most stable ion, the acetylium ion $CH_3C{=}O^+$, is the base peak in the mass spectra of methyl ketones. As shown by CID experiments, the ion is also abundantly formed from radical cations that do not contain the acetyl substructure, for example from ionized vinyl ethers, $CH_2{=}C(H)OR^{\bullet+}$, which do not produce the intrinsically unstable α-formylcarbenium ion $^+CH_2C(H){=}O$ via a simple bond-breaking reaction but rather $CH_3C{=}O^+$. Such 'pseudo-simple cleavage reactions' – which proceed via a 'hidden' hydrogen migration in the molecular ion – often occur, and if they are not recognized for what they are, various erroneous conclusions may be drawn from the experiments. This is exemplified by the behaviour of the dimethyl sulfoxide (DMSO) radical cation, $CH_3{-}S({=}O)CH_3^{\bullet+}$, which is a case par excellence of the situation sketched in **Figure 1B**, where ACB^+ represents the DMSO radical cation. Loss of $^{\bullet}CH_3$ had always, quite reasonably, been viewed as a direct bond-cleavage reaction to produce $CH_3S{=}O^+$, and indeed CID experiments on the product ions formed in the ion source (i.e. those ions formed at relatively high internal energies) showed that this is indeed the case. From AE measurements a $\Delta H_f = 736$ kJ mol^{-1} was derived for the product ions and, logically then, this value was proposed to correspond to the structure $CH_3S{=}O^+$. However, it followed from CID experiments on the metastably generated product ions (i.e. those ions formed at lower internal energies, corresponding more closely to the AE measurements) that these ions did not have the $CH_3S{=}O^+$ connectivity, but rather the carbenium structure $^+CH_2{-}S{-}OH$, formed via a low-lying ('hidden') 1,3-H shift in the molecular ions, yielding the intermediate $CH_2{=}S(OH)CH_3^{\bullet+}$ ions, ABC^+ in **Figure 1B**, and the measured ΔH_f was now assigned to the structure $^{+}CH_2$–S–OH. Interestingly, these CID experiments were inspired by the *ab initio* MO calculations which correctly predicted that $CH_2=S^{+}OH$ should have a lower ΔH_f than $CH_3S=O^{+}$ (by 84 kJ mol^{-1}) in sharp contrast to the carbon analogues $CH_2=C^{+}OH$ and $CH_3–C=O^{+}$ (see above). In contrast to ketene, $CH_2=C=O$, where protonation takes place at the carbon to produce the stable structure $CH_3C=O^{+}$, the protonation of sulfine $CH_2=S=O$, takes place at the oxygen to produce $^{+}CH_2$–S–OH.

Proton-bound molecule pairs

Even-electron proton-bound molecule pairs ($M_1\cdots H^{+}\cdots M_2$) can conveniently be made by ion–molecule reactions and their properties (for example, bond dissociation energies) can be obtained by ICR-based experiments. Such experiments on symmetric ($M\cdots H^{+}\cdots M$) proton-bound dimers and asymmetric ($M_1\cdots H^{+}\cdots M_2$) proton-bound molecular pairs indicate a remarkable constancy in hydrogen bond energies. It appears that symmetric proton-bound dimers have hydrogen bond energies of 130 ± 10 kJ mol^{-1} and so ΔH_f of a proton-bound dimer is given by $\Delta H_f = \Delta H_f[MH^{+}] + \Delta H_f[M] - 130$; for asymmetric proton-bound molecule pairs, the hydrogen bond energy depends on the difference in proton affinities (PA) of the bases involved, and for the case $PA[M_1] > PA[M_2]$ the following empirical equation (Eqn [1]) applies for their ΔH_f:

$$\Delta H_f = 0.54\{\Delta H_f[M_2H^{+}] + \Delta H_f[M_1]\} + 0.46\{\Delta H_f[M_1H^{+}] + \Delta H_f[M_2]\} - 130 \quad [1]$$

Thus, the constancy in hydrogen bond energy allows the assessment of ΔH_f of the proton-bound molecule pair, provided that the PA values of M_1 and M_2 are known. Alternatively, the relative PA values of M_1 and M_2 may be assessed from MI experiments of the proton-bound molecule pair using a sector-type instrument: the intensity ratio of the observed M_1H^{+} and M_2H^{+} in the MI spectrum is a measure of the relative PA values of the molecules M_1 and M_2.

Odd-electron Ions

Enol and related radical cations

The McLafferty rearrangement, discovered in 1952, is the only named reaction in mass spectrometry. For 2-pentanone, the reaction is represented by Equation [2].

$$CH_3C(=O^{\bullet+})CH_2CH_2CH_3 \longrightarrow [CH_3C(OH)=CH_2]^{\bullet+} + CH_2=CH_2 \quad [2]$$

$$[CH_3C(OH)=CH_2]^{\bullet+} \not\rightleftharpoons CH_3C(=O^{\bullet+})CH_3$$

At first it was thought that the driving force of this reaction was the expulsion of a stable neutral species (such as $CH_2=CH_2$) rather than formation of a stable neutral (keto) ionic product, but from subsequent AE measurements it was concluded – and these findings have been amply corroborated – that the product ions had an enol and not the tautomeric keto structure and that, in contrast to the corresponding neutral molecules, enol radical cations are more stable thermodynamically than the keto tautomers. Thus, as with the C_7H_7 and C_3H_3 structures discussed above, enol structures undergo a stability reversal upon ionization; this is illustrated in **Figure 2** for the acetone radical cation and its enol structures.

Because of their enhanced stability, enol radical cations are produced from many compounds (*m/z* 58 from methyl ketones; *m/z* 59 from amides; *m/z* 60 from acids) and they are thus highly structure-characteristic. They provide a rich source for the generation of neutral enols via NR (neutralization–reionization) mass spectrometry. It is found from such experiments that gaseous neutral enols do not isomerize to the more stable keto counterparts, in agreement with results from pyrolysis mass spectrometry. The class of enol radical cations can be regarded as the harbinger of all ions of unexpected stability (see below). Imidic acids RC(OH)=NR (iminols) are the 'enol' equivalents of amides, RC(=O)NHR, and the prototype imidic acid radical cation, $HN=C(H)OH^{\bullet+}$, ionized formimidic acid, has been generated by an appropriate dissociative ionization and identified by CID type experiments. Neutral imidic acids have been proposed as intermediates in the acid-catalysed proton exchange of amides, peptides and proteins, a process of fundamental importance in biochemistry (Eqn [3] as opposed to the classical mechanism, Eqn [4]). Simple imidic acids have not yet been seen in solution.

$$\underset{\text{amide}}{R^1{-}C(=O){-}NH{-}R^2} \overset{H^+}{\rightleftharpoons} R^1{-}C(OH){=}\overset{+}{N}H{-}R^2 \overset{-H^+}{\rightleftharpoons} \underset{\text{imidic acid}}{R^1{-}C(OH){=}N{-}R^2} \qquad [3]$$

$$R^1{-}C(=O){-}NH{-}R^2 \overset{H^+}{\rightleftharpoons} R^1{-}C(=O){-}\overset{+}{N}H{-}R^2 \qquad [4]$$

NR experiments on ionized formimidic acid showed that a significant fraction of the ions survive neutralization, but, because of interference from neutral products arising from dissociation of the neutralized species, it was not possible, from straightforward NR experiments alone, to assign a unique structure to the neutralized species. Such problems plague other systems too, but neutral identification may still be feasible by selectively analysing the ions recovered after the NR process. Such a NR-CID type of multiple collision experiment is demanding in terms of sensitivity, but nevertheless revealing. Such experiments showed that the formimidic acid structure was maintained throughout the NR experiment, and so neutral formimidic acid is a stable species on the microsecond timescale, which makes proton exchange via the imidic acid mechanism (Eqn [3]) in principle possible. NR-CID experiments are also a good check of the (isomeric) purity of the primary ion beam.

Another isomer of ionized formamide is ionized aminohydroxycarbene, $H_2NC^{\bullet +}OH$. This ion was also generated in the gas phase by an appropriate dissociative ionization and distinguished from its isomers by CID experiments on multiply labelled ions. This followed earlier results indicating that carbene radical cations are remarkably stable species.

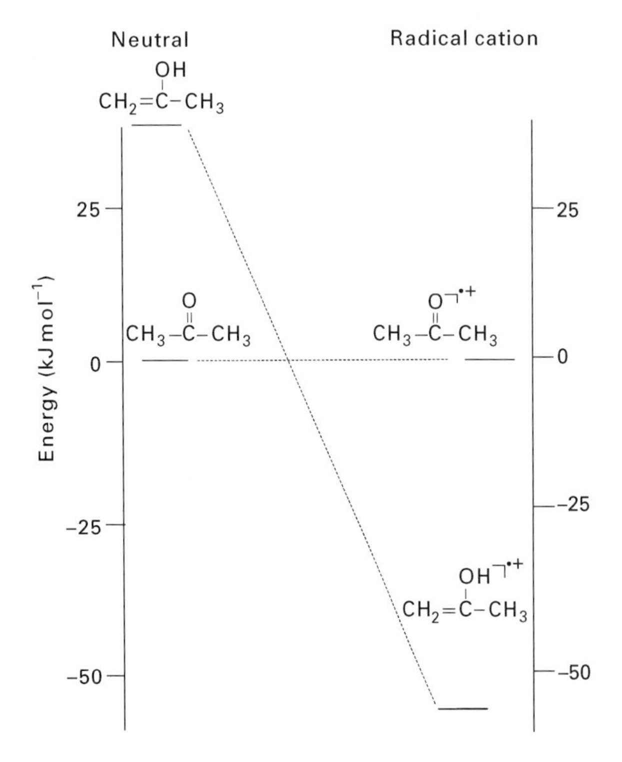

Figure 2 Heats of formation of neutral and ionized acetone and its enol showing stability reversal upon ionization. Arbitrary energy scales. $\Delta H_f\,[CH_3C(=O)CH_3] = -218\ kJ\ mol^{-1}$, $\Delta H_f\,[CH_3(=O)CH_3]^{\bullet +} = 720\ kJ\ mol^{-1}$.

Carbene radical cations

In the early days of mass spectrometry little attention was paid to carbene-type radical cations, $R_1CR_2^{\bullet +}$, probably because it was thought that such species, like their neutral counterparts, would be thermodynamically too unstable to be formed. For example, neutral hydroxycarbene, $HC^{\bullet\bullet}OH$ lies 218 kJ mol^{-1} above its isomer formaldehyde, $CH_2{=}O$. However, from experiments and theoretical calculations from the early 1980s, it appeared that the radical cations of simple carbenes could be as stable thermodynamically as their better-known isomers: for example, $HC^{\bullet +}OH$ is only 13 kJ mol^{-1} less stable than $CH_2{=}O^{\bullet +}$, compared to 218 kJ mol^{-1} for the neutral system. This behaviour parallels that of keto–enol structures (see above). The generation of $HC^{\bullet +}OH$ is simple: it is formed by a 1,1-H_2 loss from ionized methanol, but other ionized carbenes are not so easily made. Several oxacarbene radical cations, $R_1C^{\bullet +}OR_2$(R = H or alkyl) have been generated and they all possess a surprising property: direct cleavage of R_2 leads to stable acylium ions (see above), but without exception this reaction has a large barrier (~100 kJ mol^{-1}) for the reverse reaction. This is illustrated in **Figure 3** for the prototype oxacarbene $HC^{\bullet +}OH$.

Many ionized carbenes have been generated by appropriate dissociative ionizations and these ions provide a convenient source for the generation of neutral carbenes by NRMS, for example the prototype dioxacarbene, $HOC^{\bullet\bullet}OH$.

Keto–enol tautomerization

In the rarefied gas phase, keto–enol tautomerization of neutral species can only occur via intramolecular

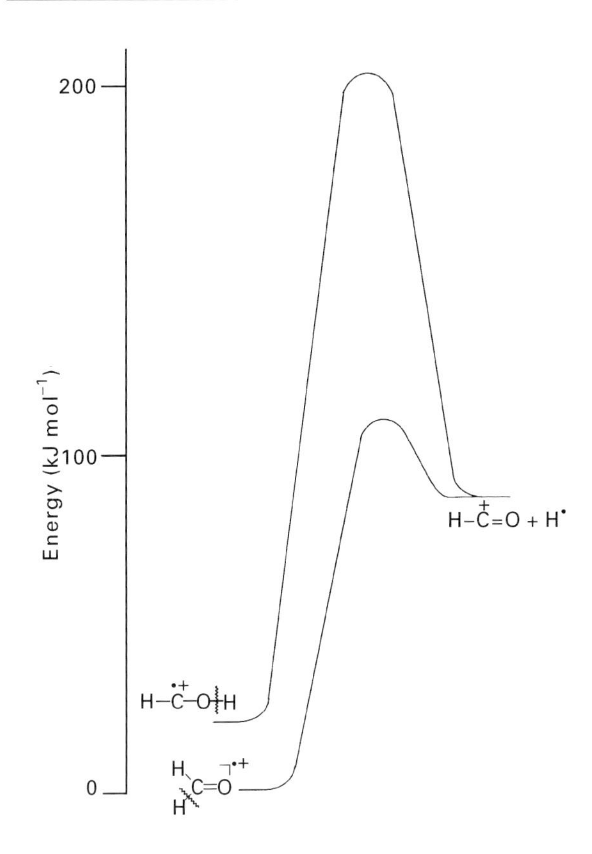

Figure 3 Energy level diagram for H• loss from the isomers CH_2=O•+ and HC•+–OH showing a larage reverse energy for the direct O–H cleavage.

hydrogen transfers. For simple systems, such as methyl acetate, $CH_3C(=O)OCH_3$, tautomerization is only possible via an energy-demanding 1,3-H shift ($CH_3C(=O)OCH_3 \longrightarrow CH_2=C(OH)OCH_3$) and thus, in the gas phase, enol structures, despite their unfavourable heats of formation, remain as they are (see above). The situation is much different for the corresponding radical cations. Here keto–enol tautomerization is mediated by a surprisingly stable third structure, for which no neutral counterpart exists. Instead of a one-step energy-consuming and thus slow 1,3-H shift, the tautomerization takes place via two less energy-demanding fast 1,4-H shifts. This is why deuterium-labelled methyl acetate molecular ions show H/D randomization reactions ('scrambling'). The stable intermediate structure is called a 'distonic ion'.

keto ⇌ distionic ⇌ enol [5]

Distonic ions

In *Tetrahedron Report* number 280 (1990) the question was addressed: 'One electron more, one electron less. What does it change?' The authors critically evaluated the perception that 'comparing the reactivity of molecules whose only difference was the electron number was like comparing apples and oranges'. However, if one thing has become clear over the past decade, it is that for gas phase species 'one electron less' can change the chemistry of the species beyond recognition. One of the great discoveries in the 1980s with respect to ion structures was the finding that certain ion structures that did not correspond to those of stable, neutral molecules could nevertheless be very stable, both thermodynamically and kinetically. For example, even a simple radical cation such as $CH_3OH^{\bullet+}$ has a stable isomer, $CH_2OH_2^{\bullet+}$, the methyleneoxonium radical cation. It has been said, not without exaggeration, that $CH_2OH_2^{\bullet+}$ represents a triumph for theory. This is because, theory (*ab initio* MO calculations) predicted not only that $CH_2OH_2^{\bullet+}$ is more stable than $CH_3OH^{\bullet+}$ but also that a large barrier separates these ions; this is shown in **Figure 4A**.

Soon after this prediction was made, the ion was indeed generated and identified. These new species all have in common that, in contrast to conventional isomers, the charge and radical site are formally at separate atoms and as such they are referred to as 'distonic' ions from the Greek 'diestos' and Latin 'distans' meaning 'separate'. If the radical and charge sites are on adjacent atoms such as in $CH_2OH_2^{\bullet+}$, the distonic ion can be viewed as an ionized ylid, and as such they are also referred to as 'ylidion'. Distonic

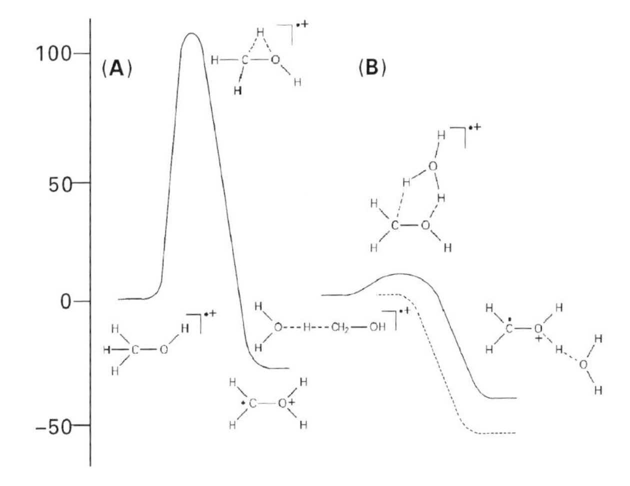

Figure 4 (A) Energy diagram for the (unassisted 1,2-H shift) isomerization $CH_3OH^{\bullet+}$ to $CH_2OH_2^{\bullet+}$. (B) Energy diagram for the rearrangement $CH_3OH^{\bullet+}$ to $CH_2OH_2^{\bullet+}$, catalysed by water. Broken lines indicate catalysis by formaldehyde. Energies in kJ mol^{-1}.

ions can also be viewed as charged or protonated radicals, but, because they form a well-defined group of radical cations having characteristic properties, the term distonic ion is a useful one. Distonic ions can be made by judiciously chosen dissociative ionizations and they can be differentiated from their conventional isomers by MI and CID experiments and by ion–molecule reactions. In **Figure 5** are shown the CID spectra of ionized methanol, $CH_3OH^{\bullet+}$ and of the $CH_2OH_2^{\bullet+}$ product ion generated by the loss of $CH_2{=}O$ from ionized 1,2-ethanediol (Eqn [6]):

$$[HOCH_2CH_2OH]^{\bullet+} \longrightarrow {}^{\bullet}CH_2{-}\overset{+}{O}H_2 + CH_2{=}O \quad [6]$$

The major dissociation products in the spectrum of **Figure 5B** are $CH_2^{\bullet+}$ (*m*/*z* 14) and $H_2O^{\bullet+}$ (*m*/*z* 18), and this is compatible only with the $CH_2OH_2^{\bullet+}$ structure. This spectrum also reveals that the doubly charged ion $CH_2OH_2^{++}$, *m*/*z* 16, which can be viewed as doubly protonated formaldehyde, also is a remarkably stable species, despite the large coulombic repulsion of the charges. Such doubly charged ions have been invoked in the field of super-acid chemistry pioneered by Olah, and in solution these species behave as 'super electrophiles'. Thus $^+CH_2OH_2^+$ is proposed to be a key intermediate in the formation of methane from formaldehyde under 'super-acid' conditions.

In the mass spectrometer, many initial molecular ions rearrange to distonic ions by way of a hydrogen shift; these intermediate distonic ions may rearrange further and their reactions involve both the radical and charge sites: often a charged moiety can easily migrate to the radical site, thereby generating an isomeric distonic ion that in turn can shift its ionized part to the new radical site. Such a sequence can rationalize many otherwise unintelligible reactions. Distonic ions can serve as precursors for the genesis of novel neutral species via the techniques of NRMS. This is illustrated by the generation of the Hammick intermediate [1], a previously unobserved species that has been held responsible for the accelerated decarboxylation of 2-picolinic acid [3] (Eqn [7]).

$[3]^{\bullet+} \longrightarrow [1]^{\bullet+} \xrightarrow{NR} [1] \quad [7]$

$[2]^{\bullet+}$

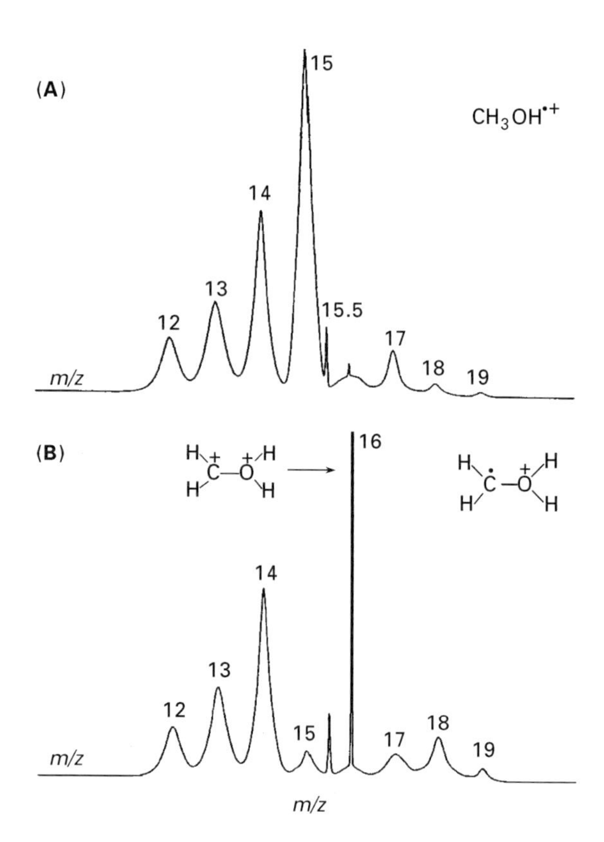

Figure 5 Collision induced dissociation mass spectra of (A) $CH_3OH^{\bullet+}$ and (B) $CH_2OH_2^{\bullet+}$. Note the intense signal for the doubly charged ion in (B).

In the mass spectrometer, the molecular ions of 2-picolinic acid eliminate CO_2 and CID experiments indicate that the resulting *m*/*z* 79 product ions are the ylidion isomer of pyridine $[1]^{\bullet+}$ and not ionized pyridine $[2]^{\bullet+}$ itself. (The possibility of differing descriptive formalisms, i.e. ylidion or ionized carbene, exists for $[1]^{\bullet+}$). NR experiments showed that a significant fraction of the ylidions survive neutralization experiments, but because of interference from neutral products arising from dissociation of the neutralized species, it was not possible to assign a structure to the neutralized species. The same problem plagues the formidic acid experiment (see above). However, the NR-CID mass spectrum clearly showed that the ylid structure was maintained throughout the NR experiments and so the Hammick intermediate is a stable species on the microsecond timescale.

The first homologue of the methyleneoxonium ion is the distonic ion $^{\bullet}CH_2CH_2OH_2^+$, an isomer of ionized ethanol. *Ab initio* calculations show that the distonic ion is indeed a low-energy species lying ~40 kJ mol^{-1} beneath ionized ethanol, but also that the C–O bond readily stretches to form the ion–dipole (ion–neutral) complex $CH_2{=}CH_2^{\bullet+}/H_2O$. This species can best be viewed as a positively charged ethene rod to which the water molecule is attached by its dipole, but which can be moved more or less freely along and about the positive rod.

Ion–neutral complexes

Traditionally, unimolecular rearrangements of gas-phase (radical) cations have been rationalized in terms of transformations within the covalently bound ion. Increasingly, ion–neutral complexes with a relatively strong electrostatic bond between the incipient fragments are being invoked as intermediates to account for reactions that would otherwise be considered implausible for the intact ion. In an ion–neutral complex the charged and neutral components are sufficiently separated that they show reactivities similar to those expected for the isolated species. Two distinct but related phenomena may be interpreted by means of ion–neutral interactions: the ionic component may react individually by isomerization to a more stable structure (Eqn [8a]) or the ionic and neutral species may react with each other, for example by proton transfer (Eqn [8b]).

$$I_1N^+ \rightarrow I_1^+ \cdots N \rightarrow I_2^+ \cdots N \rightarrow I_2^+ + N \qquad [8a]$$

$$\rightarrow [I_1\text{–}H] \cdots HN^+ \rightarrow [I_1\text{–}H] + HN^+ \qquad [8b]$$

Although the earliest suggestion of a complex-mediated dissociation dates back to 1956, it was not until the early 1980s that it became evident that their involvement in the unimolecular dissociations of odd and even electron ions is a widespread phenomenon.

The formation of $CH_3CH^+CH_3$ from the onium ion $CH_2{=}O^+CH_2CH_2CH_3$ is illustrative of isomerization (via a 1,2-hydride shift) of the cation within the complex followed by dissociation, that is: $CH_2{=}O\cdots{}^+CH_2CH_2CH_3 \rightarrow CH_2{=}O\cdots{}^+CH(CH_3)_2$.

The rearrangement of carbocations, such as the archetypal cation isomerization $^+CH_2CH_2CH_3 \rightarrow {}^+CH(CH_3)_2$, is a central theme of organic chemistry; the ion in an ion–neutral complex occupies a position between that of the 'bare' ion and that corresponding to the ion in solution. Thus the influence of a single solvent molecule on the rearrangement may be studied.

A classical case of proton transfer between the formal components of a complex concerns the dissociation of steroidal diamines, for example 3,20-diaminopregnane. Here, a long-distance intramolecular and interfunctional proton transfer was proposed to occur in an ion–neutral complex, where upon separation the incipient fragments rotate more or less freely, thus allowing proton transfer between functional groups that originally were distant in the intact molecular ion (Eqn [9]).

The maximum stabilization energy (SE) of an ion–dipole complex is achieved when the dipole moment vector points away from the charge; SE (kJ mol^{-1}) = $288\ \mu/r^2$ where μ is the dipole moment (in Debye) and r is the distance (in Å) between the point charge and the dipole. For typical values of μ (1.5D) and r (2 Å) one obtains SEs of ~100 kJ mol^{-1} with concomitantly smaller values for lower μ. Thus, a strong ion–dipole bond is energetically on a par with a weak covalent bond.

Criteria have been proposed for the intermediacy of ion–neutral complexes, but probably the only good criteria follow from the proposal that such complexes dissociate exceedingly close to threshold. These are (1) very small kinetic energy releases and (2) unprecedented isotope effects. Thus, extreme isotopic fractionation due to differences in vibrational frequencies in the products has been identified in ion–dipole dissociations.

One of the puzzling observations seen in MI spectra is that metastable ions invariably dissociate to the products of lowest enthalpy, despite the obvious mechanistic complexities involved from a conventional point of view; the particular virtue of intermediate ion–neutral complexes is that they facilitate the formation of such low-energy products.

Hydrogen-bridged radical cations

Hydrogen-bridged radical cations are the odd-electron counterparts of proton-bound molecule pairs, $M_1\cdots H^+M_2$. Formally they can be represented as proton-bound molecule–radical pairs ($M\cdots H^+\cdots R^\bullet$), but they are better viewed as H-bridged ion–neutral complexes of the type $MH^+\cdots R^\bullet$ or $M\cdots HR^{\bullet+}$, with the bridging H closer to the partner of higher PA, where most of the stabilization energy is provided by ion–dipole attractions. The H-bridge furnishes additional stabilization by about 20 kJ mol^{-1} but its main function is to direct the course of isomerization by allowing a facile proton transfer. Hydrogen-bridged radical cations are harder to make than

$M^{\bullet+}$ → … → $[M-43]^{\bullet+}$ [9]

proton-bound molecule pairs via ion-molecule reactions. They are usually generated by elimination of a stable molecule from a molecular ion, such as loss of CO from ionized β-hydroxypyruvic acid (Eqn [10]).

$$\xrightarrow{-CO} H_2\overset{+}{C}{-}O{-}H{\cdots}O{=}\dot{C}{-}OH \quad [10]$$

There remains an inherent problem with regard to the identification of a H-bridged radical cation as a stable product ion: the ion may have isomers with closely similar reactivity and dissociation characteristics. This is a well-recognized problem in general, but it becomes particularly acute for H-bridged radical cations because they often dissociate very much like distonic isomers. A case in point is the structure of the ion generated by the loss of C_2H_4 from ionized 1,4-butanediol, $HOCH_2CH_2CH_2CH_2OH^{\bullet+}$. Its dissociation characteristics rule out a simple extrusion reaction yielding $HOCH_2CH_2OH^{\bullet+}$ (ionized 1,2-ethanediol), but they are compatible with the formation of species comprising vinyl alcohol and water: i.e. with both the H-bridged ion $H_2O{\cdots}HOC(H){=}CH_2^{\bullet+}$ and the distonic ions $H_2O^+CH_2C^{\bullet}HOH$ and $^{\bullet}CH_2CH(OH)OH_2^+$. Analysis of *ab initio* computational results on the isomers, stabilities and conversion barriers in conjunction with energetic information from experiment (AE measurements) resolved the ambiguity and led to the identification of the H-bridged ion.

In some cases, H-bridged radical cations have been generated by ion–molecule reactions. Electron impact ionization of a mixture of H_2O and CO_2 leads to the H-bridged species $HOH{\cdots}O{=}C{=}O^{\bullet+}$. Although the ion could be differentiated from its covalently bounded isomer ionized carbonic acid, $(HO)_2C{=}O^{\bullet+}$, through its different CID spectrum, the NR mass spectrum provided definitive evidence of its structure. As expected, and in contrast to $(HO)_2C{=}O^{\bullet+}$, the H-bridged species does not survive neutralization and dissociates completely to CO_2 and H_2O.

Hydrogen-bridged radical cations, like ion–neutral complexes, consist of two separate entities and as such they may be considered the smallest type of clusters. Dimer radical cations $M_2^{\bullet+}$ constitute an interesting class of complexes within the mainstream of cluster chemistry. They can be made by ligand exchange reactions at low pressures in an FT-ICR cell. Noble gas dimer radical cations such as $Xe_2^{\bullet+}$ can exchange, in a three-body collision, a Xe atom for a molecule M to form a $XeM^{\bullet+}$ radical cation. The latter can exchange, in a similar process, the second Xe atom for another molecule to form the dimer radical cation $M_2^{\bullet+}$. In this way, the water dimer radical cation $(H_2O)_2^{\bullet+}$ has been generated. With this technique it is possible to measure thermochemical properties of such complexes. Thus, bracketing of the electron transfer processes $(H_2O)_2^{\bullet+} + M \longrightarrow M^{\bullet+} + H_2O + H_2O$ leads to an energy difference of 1030–1038 kJ mol^{-1} between $(H_2O)_2^{\bullet+}$ and $H_2O + H_2O$. *Ab initio* calculations show that the H-bridged radical cation $H_2O{-}H^+{\cdots}OH^{\bullet}$ is 38 kJ mol^{-1} more stable than the 'true' ion-dipole complex $H_2O^{\bullet+}{\cdots}OH_2$ and that it lies 1021 kJ mol^{-1} below $H_2O + H_2O$, indicating that the observed species is in fact H-bridged.

Hydrogen-bridged radical cations, in particular O···H···O bonded ions and their less stable C···H···O bonded counterparts, are important intermediates in the dissociation chemistry of many oxygen-containing radical cations. They can interconvert via a so-called proton-transport catalysis. It was mentioned above that distonic radical cations are often more stable than their conventional counterparts and that they may be separated from those isomers by large barriers; see **Figure 4A** for the 1,2-H shift separating the conventional ion $CH_3OH^{\bullet+}$ and its distonic isomer $CH_2OH_2^{\bullet+}$. However, experiments and *ab initio* calculations show that interaction with an appropriately basic molecule, such as water or formaldehyde, greatly lowers the barrier. Thus the isomerization $CH_3OH^{\bullet+} \rightarrow CH_2OH_2^{\bullet+}$, which does not occur for the bare ions, is greatly accelerated by the addition of water; see **Figure 4B**. Here the water molecule attracts one of the methanol C-protons and then via a 5-membered ring donates one of its original protons to the O atom, thus producing $CH_2OH_2^{\bullet+}$. The barrier, which for the unassisted reaction is 108 kJ mol^{-1} is reduced to a mere 10 kJ mol^{-1} for the water-catalysed rearrangement. Proton transport catalysis has been shown to occur in many ionic rearrangement/dissociation processes. For example, the rearrangement $CH_2OH_2^{\bullet+} \rightarrow CH_3OH^{\bullet+}$, but now catalysed by formaldehyde, plays a key role in the low-energy rearrangement of ionized 1,2-ethanediol. These are cases par excellence of effects induced by a single solvent molecule on the intrinsic chemistry of the ion.

Charge transfer complexes

Although charge (or electron) transfer complexes are not stable entities as such, *ab initio* calculations indicate that they do play a crucial role in the rearrangement of a variety of molecular ions. The situation often arises that a molecular ion $M–DH^{\bullet+}$ (D = dipole) rearranges to an ion–dipole complex of the type $M^{\bullet+} \rightarrow D–H$, where the arrow represents the dipole vector, and that a hydrogen needs to be transferred from the dipole molecule D–H to $M^{\bullet+}$ in order to obtain the observed products $MH^+ + D^\bullet$. This, however, is not possible because in order to transfer its hydrogen to $M^{\bullet+}$, D–H must rotate to such an extent that the resulting ion–dipole repulsion would lead to dissociation rather than to hydrogen transfer. In such cases charge (electron) transfer through orbital interaction ($M^{\bullet+} \rightarrow D–H \rightleftharpoons M \leftarrow D–H^{\bullet+}$) provides an alternative because now, and in contrast to the situation before charge transfer, the charged $D–H^{\bullet+}$ moiety is more or less free to rotate to produce $M\cdots H–D^{\bullet+}$ after which proton transfer takes place to produce $MH^+ + D^\bullet$. Thus, the charge transfer process kills two birds with one stone: it allows rotation of D–H and subsequent proton transfer rather than hydrogen atom transfer.

Dications

Historically, dications were regarded as a curiosity and were observed only incidentally in mass spectrometric studies. Small polyatomic dications are remarkable species indeed; usually they are thermodynamically unstable with respect to dissociation into two monocations, but significantly kinetic stability may result if large barriers impede dissociation. This is illustrated by the methyleneoxonium dication $CH_2OH_2^{++}$ discussed above. This species is less stable than its dissociation products $CH_2OH^+ + H^+$ (by 105 kJ mol^{-1}), but a barrier of 250 kJ mol^{-1} makes the ion experimentally accessible, see **Figure 5**. In many cases, deprotonation of a dication is best viewed as a two-stage process. Initially, the departing unit is a hydrogen atom rather than a proton, and later an electron transfer takes place to form the products: $AH^{++} \rightarrow A^{++} - H \rightarrow A^+\cdots H^+ \rightarrow A^+ + H^+$. The first step, a homolytic cleavage, is responsible for the large barrier. Hence, a dication is truly a tiger in a cage.

List of symbols

k = rate constants.

See also: **Fragmentation in Mass Spectrometry; Hydrogen Bonding and other Physicochemical Interactions Studied By IR and Raman Spectroscopy; Ion Collision Theory; Ion Energetics in Mass Spectrometry; Ion Molecule Reactions in Mass Spectrometry; Metastable Ions; Negative Ion Mass Spectrometry, Methods; Neutralization–Reionization in Mass Spectrometry; Proton Affinities; Sector Mass Spectrometers; Stereochemistry Studied Using Mass Spectrometry.**

Further reading

Bouchoux G (1988) Keto-enol tautomers and distonic ions: the chemistry of $[C_nH_{2n}O]$ radical cations. *Mass Spectrometry Reviews* 7: 1–39, 203–255.

Burgers PC and Terlouw JK (1989) Structures and reactions of gas-phase organic ions. In Rose ME (ed) *Specialist Periodical Reports Mass Spectrometry* 10, Chapter 2, pp 35–74. Cambridge, UK: The Royal Society of Chemistry.

Freiser B (1996) Gas-phase metal ion chemistry. *Journal of Mass Spectrometry* 31: 703–715.

Hammerum S (1988) Distonic radical cations in gaseous and condensed phase. *Mass Spectrometry Reviews* 7: 123–202.

Heinrich N and Schwarz H (1989) Ion/molecule complexes as central intermediates in unimolecular decompositions of metastable radical cations of some keto/enol tautomers: theory and experiment in concert. In Maier JP (ed) *Ion and Cluster Ion Spectroscopy and Structure*, pp 329–370. Amsterdam: Elsevier.

Holmes JL (1985) Assigning structures to ions in the gas-phase. *Organic Mass Spectrometry* **20**: 169–183.

Longevialle P (1992) Ion–neutral complexes in the unimolecular reactivity of organic cations in the gas phase. *Mass Spectrometry Reviews* **11**: 157–192.

McLafferty FW and Turecek F (1993) *Interpretation of Mass Spectra.* Mill Valley, CA: University Science Books.

Morton TH (1992) The reorientation criterion and positive ion–neutral complexes. *Organic Mass Spectrometry* **27**: 353–368.

Radom L (1991) Chemistry by computer: a theoretical approach to gas-phase ion chemistry. *Organic Mass Spectrometry* **26**: 359–373.

Splitter JS (1994) Mass-spectral intermediate-ion structures. In Splitter JS and Turecek F (ed) *Applications of Mass Spectrometry to Organic Stereochemistry*, Chapter 3, pp 39–81. Weinheim: Verlag Chemie.

Ion Trap Mass Spectrometers

Raymond E March, Trent University, Peterborough, Ontario, Canada

MASS SPECTROMETRY
Methods & Instrumentation

Introduction

An ion trap mass spectrometer functions both as a mass spectrometer of considerable mass range and variable mass resolution and as an ion store in which gaseous ions can be confined. Unlike other mass spectrometers that operate at pressures $< 10^{-6}$ Torr, the ion trap operates at 1 mTorr of helium. As a storage device, the ion trap acts as an 'electric-field test-tube' for the confinement of gaseous ions either positively charged or negatively charged. The confining capacity of the ion trap arises from a trapping potential well formed when appropriate potentials are applied to the ion trap electrodes. The ion trap functions as a mass spectrometer when the trapping field is changed, so that the trajectories of simultaneously trapped ions of consecutive specific mass/charge ratio become sequentially unstable, and ions leave the trapping field in order of mass/charge ratio. Upon ejection from the ion trap, ions strike a detector and provide an output signal.

With the advent of new methods by which gaseous ions can be formed from polar molecules and injected into an ion trap, a wider range of ion trap applications is possible. The coupling of liquid chromatography (LC) with electrospray (ES) ionization and with mass spectrometry (MS) in the early 1980s led to the development of new ion trap instruments for the analysis of nonvolatile, polar and thermally labile compounds. In 1995, new ion trap instruments (Finnigan's LCQ and GCQ, and Bruker-Franzen's ESQUIRE) were introduced, which employ external ion sources with injection of ions into the ion trap. The major focus for the application of these instruments, using LC-ES/MS, is the examination of high-molecular-mass biopolymers such as proteins, peptides and oligodeoxyribonucleotides.

The quadrupole ion trap mass spectrometer

The quadrupole ion trap consists of three electrodes which are shown in open array in **Figure 1**. Two of the electrodes are virtually identical and, while having hyperboloidal geometry, resemble small inverted saucers; these saucers are end-cap electrodes and are distinguishable by the number of holes in the centre of each electrode. One end-cap electrode has a single aperture through which electrons and/or ions can be gated, while the other has several apertures arranged centrally and through which ions pass to a detector. The third electrode, also of hyperboloidal geometry but of two sheets rather than one, is the ring electrode; it resembles a napkin holder and is of similar size, since the radius, r_0, in the central plane is

Figure 1 The three electrodes of the quadrupole ion trap shown in open array.

~1 cm. The ring electrode is positioned symmetrically between two end-cap electrodes as shown in **Figure 2**; **Figure 2A** is a photograph of an ion trap cut in half along the axis of cylindrical symmetry, while **Figure 2B** is a cross-section of an ideal ion trap showing the asymptotes and the dimensions r_0 and z_0, where $2z_0$ is the separation of the end-cap electrodes along the axis of the ion trap.

The electrodes in **Figure 2** are truncated for practical purposes but, in theory, they extend to infinity and meet the asymptotes shown. The electrode geometrics are defined so that, when an RF potential is applied to the ring electrode with the end-cap electrodes grounded, a near-ideal quadrupole field is produced, which creates a parabolic potential well for ion confinement, **Figure 3**. As shown in **Figure 3**, the potential well in the axial direction is of depth $\bar{D}_z$, while that in the radial direction $\bar{D}_r$; since $\bar{D}_z \approx 2\bar{D}_r$, the potential well resembles more a flower vase than a bowl.

Mass-selective ejection of ions from the potential well is accomplished by linearly ramping the amplitude of the RF potential; each ion species is ejected at a specific RF amplitude and, since the initial amplitude and ramping rate are known, the mass/charge ratio can be determined for each ion species. This method for measuring mass/charge ratios of confined ions was developed by Stafford and is known as the 'mass-selective axial instability mode'; this method made possible the commercialization of the ion trap in the early 1980s. A prerequisite is that ions be focused initially to the ion trap centre by momentum-dissipating collisions with helium atoms.

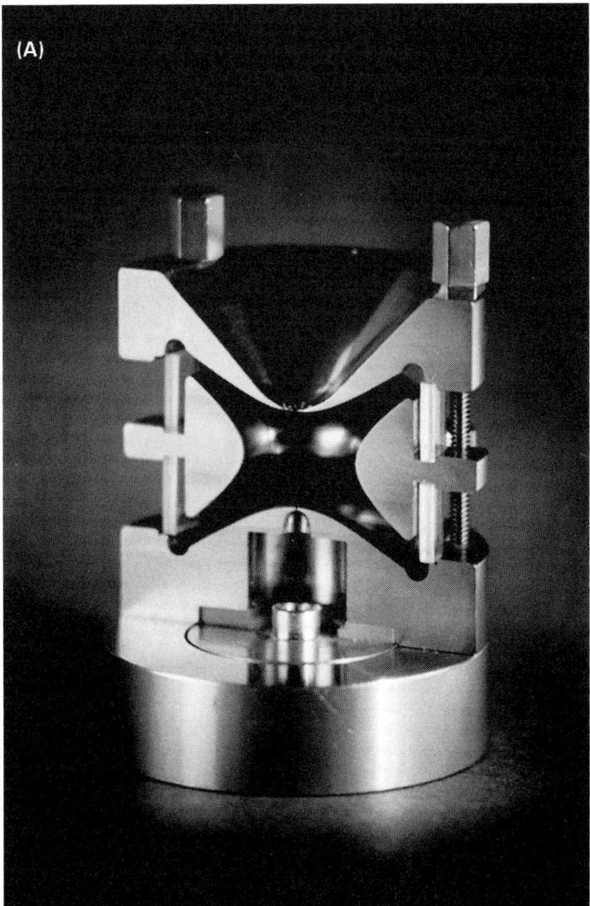

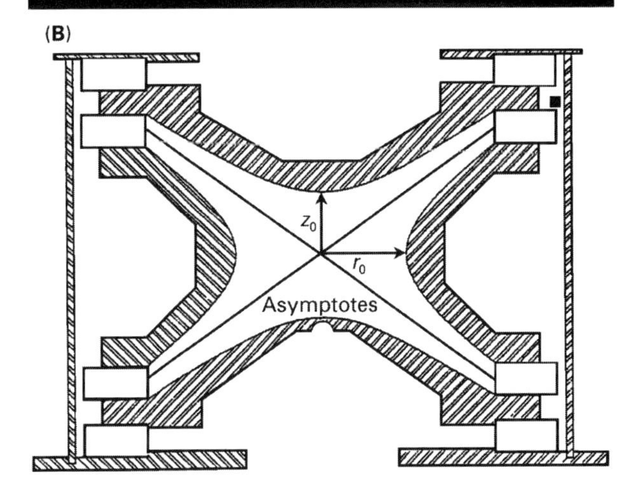

Figure 2 Quadrupole ion trap. (A) Photograph of an ion trap cut in half along the axis of cylindrical symmetry. (B) Schematic diagram of the three-dimensional ideal ion trap showing the asymptotes and the dimensions r_0 and z_0.

History and literature

The history of the quadrupole ion trap originates in the pioneering work of Paul and Steinwedel in the mid-1950s; their work was recognized by the award

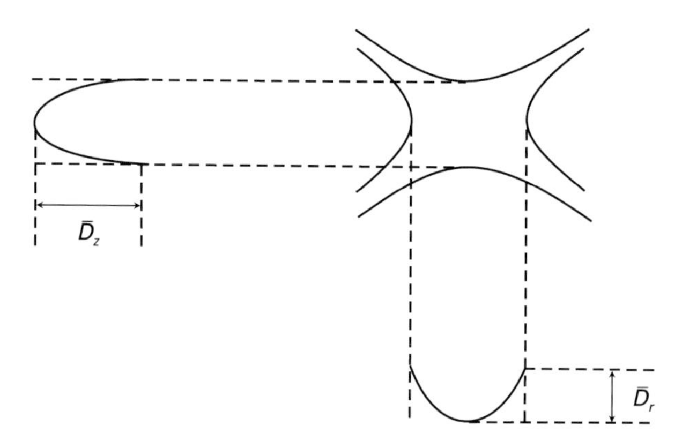

Figure 3 Representation of the parabolic trapping potential wells of depths $\bar{D}_z$ and $\bar{D}_r$.

of the 1989 Nobel Prize in Physics to Wolfgang Paul. Yet the basis of the theory of operation of quadrupole devices was laid down over 140 years ago by Mathieu, from his investigation of vibrating stretched skins.

Those papers that are landmarks in this field are listed in the bibliography together with the five texts in which are described a wide variety of applications of the ion trap. The history of this device has been expressed as several 'ages', beginning with mass-selective detection; this age, which covered the 1950s and began with disclosure of quadrupolar devices by Paul and Steinwedel, included the storage of microparticles by Wuerker, Shelton and Langmuir and the first use of the ion trap as a mass spectrometer by Fischer. The second age is described as mass-selective storage and covers the period 1962–82. Detailed accounts of the development of the quadrupole-type devices during the early part of this period have been published by Dawson and Whetten and by Dawson; this latter publication has been reprinted in the series of 'American Vacuum Society Classics' by the American Institute of Physics under ISBN 1563964554. The third age (1983–88) is described as mass-selective ejection; this age ushered in the commercialization of the ion trap as a mass-selective detector for gas chromatograph and was a period of intense activity. In 1989, the present age of recent advances and developments commenced with the recognition of the work of Wolfgang Paul and Hans G. Dehmelt by the award, in part, of the Nobel Prize in Physics. In the same year, an account of the ion trap together with a full treatment of ion trap theory appeared in the text *Quadrupole Storage Mass Spectrometry* by March, Hughes and Todd; the historical account in this text by Todd was expanded into a full-scale review.

Other reviews have been contributed by Cooks and co-workers and a special collection reporting upon recent developments has also appeared. Ion trap mass spectrometry was reviewed for the 12th International Mass Spectrometry Conference in 1991. In 1995, three volumes entitled *Practical Aspects of Ion Trap Mass Spectrometry* were published in the CRC Series, Modern Mass Spectrometry. Volume 1 of this series, subtitled 'Fundamentals of Ion Trap Mass Spectrometry', covers the history of the ion trap, nonlinear ion traps, ion activation, ion–molecule reactions and ion trajectory simulations; the reader is referred to chapter 1 for a discussion of the Ages of the Quadrupole Ion Trap and to chapter 2 for an exposition of the mathematical basis of ion trap operation. Volume 2, subtitled 'Ion Trap Instrumentation', deals with enhancement of ion trap performance, confinement of externally generated ions, ion structure differentiation, ion photodissociation, lasers and the ion trap, and ion traps in the study of Physics. Volume 3, subtitled 'Chemical, Environmental and Biomedical Applications', includes a revisitation of fundamentals and expositions on gas chromatography–ion trap tandem mass spectrometry (GC–MS/MS) and liquid chromatography–ion trap tandem mass spectrometry (LC–MS/MS). Recently, an introduction to the quadrupole ion trap written in a tutorial form has been published. An excellent source of information is the publication entitled *Proceedings of the Annual Conference of the American Society for Mass Spectrometry and Allied Topics*; almost 1000 extended abstracts on ion traps can be found in the Proceedings from the past 15 years.

The trapping potential well

The trapping potential well created within the electrode assembly of an ion trap is of parabolic cross section; ion species are confined in layers in the well rather like an exotic drink of several liqueurs arranged carefully in horizontal layers according to their density, as shown in **Figure 4A**. However, the ions of lowest mass/charge ratio reside at the ion trap centre (bottom of the well) surrounded, like the centre of an onion, by layers of ions of increasing mass/charge ratio. Upon tilting the bowl to the right, equivalent to ramping the RF trapping potential, the layer of least density, corresponding to ions of highest mass/charge ratio, will be poured from the well. To withdraw the layer of greatest density, that is, ions of lowest mass/charge ratio, a 'straw' is introduced and the bottom layer is sucked out as shown in **Figure 4B**; the straw represents axial modulation (see below) and the liqueur glass in **Figure 4B** corresponds to the detector.

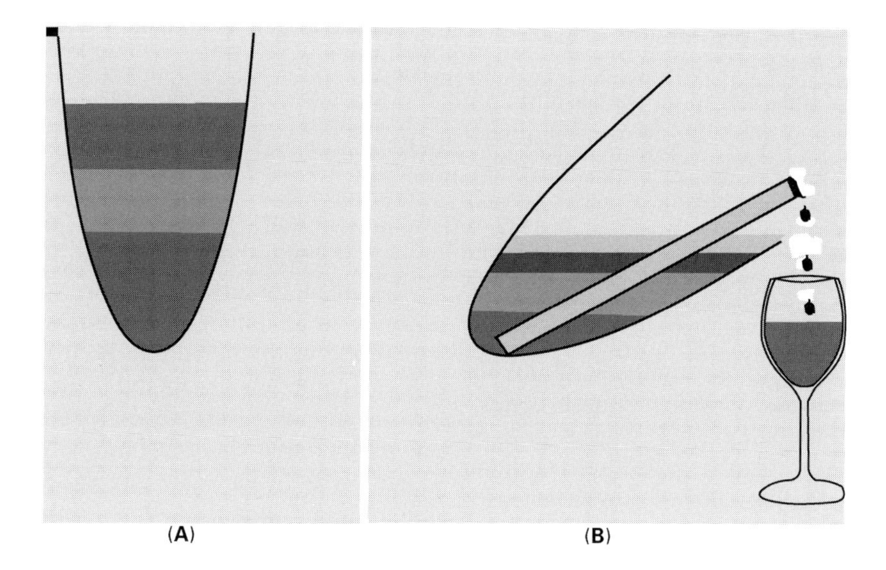

Figure 4 (A) Schematic presentation of a trapping parabolic potential well where the three liquids differing in density represent ions differing in mass/charge ratio. (B) The tilting of the well corresponds to ramping of the RF potential while the straw, with which ions are withdrawn in order of increasing mass/charge ratio, represents axial modulation.

For an ideal quadrupole field, the following identity is given,

$$r_0^2 = 2z_0^2 \quad [1]$$

so that once the magnitude of r_0 is given the sizes of all three electrodes and the electrode spacings are fixed; in the majority of commercial ion traps, r_0 lies in the range 0.7–1.0 cm.

The theory of ion trap operation

The motions of ions in quadrupole devices differ markedly from those in magnetic and electrostatic sectors. The quadrupole ion trap is described as a *dynamic* instrument since ion trajectories are influenced by time-dependent forces.

An ion in a quadrupole field

An ion in a quadrupole field experiences strong focusing in that the restoring force, which drives the ion back towards the centre of the device, increases as the ion deviates from the centre. The resulting ion trajectories resemble Lissajous figures (figures-of-eight) and the motion of ions is described mathematically by the solutions to the second-order linear differential equation described originally by Mathieu from his investigation of the mathematics of vibrating stretched skins. He described solutions in terms of regions of stability and instability; these solutions and the criteria for stability and instability also describe the trajectories of ions confined in quadrupole devices and define the limits to trajectory stability.

The Mathieu equation

The canonical form of the Mathieu equation is

$$\frac{d^2u}{d\xi^2} + (a_u - 2q_u \cos 2\xi)u = 0 \quad [2]$$

where u represents the coordinate axes r and z, ξ is a dimensionless parameter equal to $\Omega t/2$, Ω (for the ion trap) is the radial frequency of the RF potential applied to the ring and a_u and q_u are dimensionless trapping parameters. For the quadrupole ion trap, the trapping parameters are expressed as

$$a_r = \frac{4eU}{mr_0^2\Omega^2}; \quad q_r = \frac{-2eV}{mr_0^2\Omega^2} \quad [3]$$

$$a_z = \frac{-8eU}{mr_0^2\Omega^2}; \quad q_z = \frac{4eV}{mr_0^2\Omega^2} \quad [4]$$

where U is a DC potential and V is the amplitude of the RF potential of the form $V\cos \Omega t$. It is seen that $a_z = -2a_r$ and $q_z = -2q_r$.

Since $U = 0$, a_r and a_z are equal to zero and the common mode of ion trap operation corresponds to operation on the q_z axis of the stability diagram. The expression for q_z contains the mass/charge ratio for a given ion, the size of the ion trap, r_0, the amplitude V of the RF potential and the radial frequency Ω, that is, all of the parameters that are needed to understand the operations of the ion trap.

The 'stretched' ion trap

The ion trap electrodes are truncated in order to obtain a practical instrument, but this truncation introduces higher-order multipole components to the potential. To compensate for these multipole components, the electrodes of commercial devices prior to 1995 were assembled with a 'stretched' separation of the end-cap electrodes; the value of z_0 was increased by 10.6%. An account of the 'stretching' of the ion trap is given by Syka in chapter 4 of volume 1 in the CRC books, while in chapter 3 is presented an account by Franzen, Gabling, Schubert and Wang of nonlinear ion traps. The immediate consequences of stretching are that the asymptotes to the end-cap electrodes no longer coincide with those for the ring electrode, $r_0^2 \neq 2z_0^2$ and the values of the trapping parameters are changed. The trapping parameters are now expressed as

$$a_r = \frac{8eU}{m(r_0^2 + 2z_0^2)\Omega^2}; \quad q_r = \frac{-4eV}{m(r_0^2 + 2z_0^2)\Omega^2} \qquad [5]$$

and

$$a_z = \frac{-16eU}{m(r_0^2 + 2z_0^2)\Omega^2}; \quad q_z = \frac{8eV}{m(r_0^2 + 2z_0^2)\Omega^2} \qquad [6]$$

For the ion trap in the LCQ and GCQ instruments, $r_0 = 0.707$ cm and $z_0 = 0.785$ cm, such that the geometry has been stretched by ~57%.

Regions of ion trajectory stability

Quadrupole ion trap operation is concerned with the criteria that govern the stability of an ion trajectory in both r and z directions within the field, that is, the experimental conditions that determine whether or not an ion is stored.

The solutions to Mathieu's equation are of two types, (i) periodic but unstable, and (ii) periodic and stable. Solutions of type (i) form the boundaries of unstable regions on the stability diagram and correspond to those values of a trapping parameter, β_z that are integers, that is, 0, 1, 2, 3, ... ; β_z is a complex function of a_z and q_z that is approximated as

$$\beta_z = \sqrt{a_z + \frac{q_z^2}{2}} \qquad [7]$$

for $q_z < 0.4$. The boundaries represent, in practical terms, the point at which an ion trajectory becomes unbounded. Solutions of type (ii) determine ion motion in an ion trap. The stability regions corresponding to stable solutions in the z-direction are shaded and labelled z-stable in **Figure 5A**, while those corresponding to stable solutions in the r-direction are shaded and labelled r-stable in **Figure 5B**; the latter are doubled in magnitude along the ordinate and inverted.

Ions are confined in the ion trap provided their trajectories are stable in the r and z directions simultaneously; such trajectory stability is obtained in the region closest to the origin, that is, region A in **Figure 6**, which is plotted in a_u, q_u space, that is, where a_u is plotted against q_u. Regions A and B are referred to as stability regions; region A is shown in detail in **Figure 7**. The coordinates of the stability region in **Figure 7** are the parameters a_z and q_z. In **Figure 7**, the $\beta_z = 1$ stability boundary intersects with the q_z axis at $q_z = 0.908$; this intersection is the working point of the ion of lowest mass/charge ratio that can be stored.

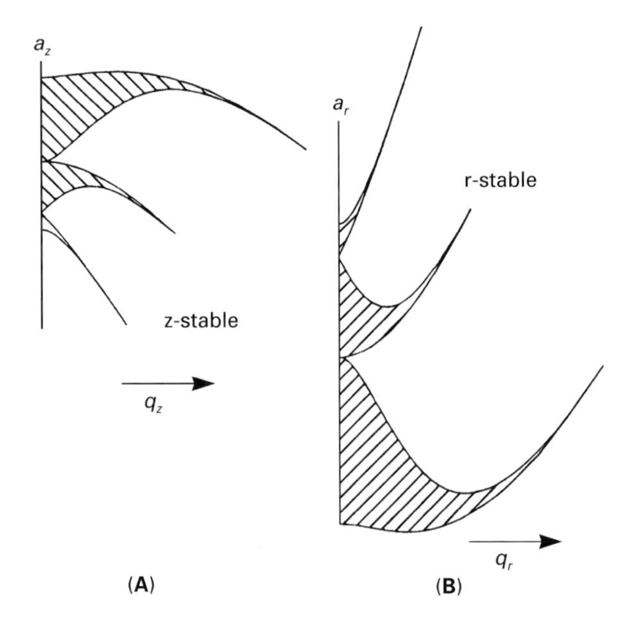

Figure 5 Several Mathieu stability regions for the three-dimensional quadrupole field. (A) Diagrams for the z direction of (a_z,q_z) space. (B) Diagrams for the r direction of (a_z,q_z) space.

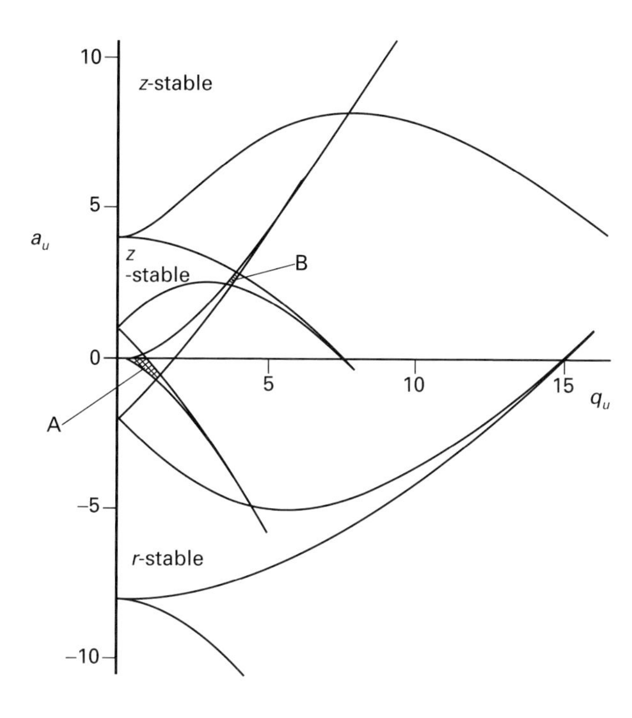

Figure 6 The Mathieu stability diagram in (a_z,q_z) space for the quadrupole ion trap in both the *r* and *z* directions. Regions of simultaneous overlap are labelled A and B.

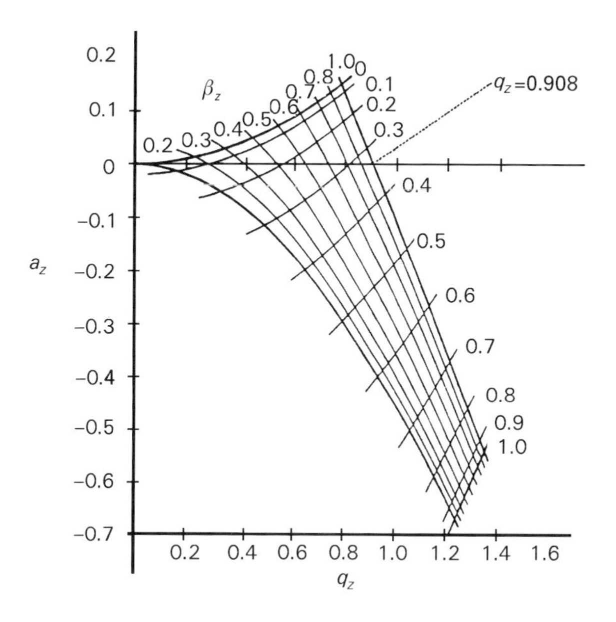

Figure 7 Stability diagram in (a_z,q_z) space for the region of simultaneous stability in both *r* and *z* directions near the origin for the three-dimensional quadrupole ion trap; the iso-β_r and iso-β_z lines are shown. The q_z axis intersects the $\beta_z = 1$ boundary at $q_z = 0.908$, which corresponds to q_{max} in the mass-selective instability mode.

Secular frequencies

A three-dimensional representation of an ion trajectory, shown in **Figure 8**, has the general appearance of a Lissajous curve composed of two fundamental frequency components, $\omega_{r,0}$ and $\omega_{z,0}$ of the secular motion. Higher-order (n) frequencies exist and the family of frequencies is described by $\omega_{r,n}$ and $\omega_{z,n}$ as given by

$$\omega_{u,n} = (n + \beta_u/2)\Omega; \quad 0 \le n < \infty \qquad [8]$$

and

$$\omega_{u,n} = -(n + \beta_u/2)\Omega; \quad -\infty < n < 0 \qquad [9]$$

The simulated ion trajectory shown in **Figure 8** resembles a roller-coaster ride and depicts the motion of an ion on the potential shown in **Figure 9**. The oscillatory motion of the ion results from the undulations of the potential surface, which can be envisaged as rotating. The simulation of the ion trajectory was carried out using ITSIM simulation program, while the potential surface was generated by calculating the potential for increments of 1 mm in both radial and axial directions.

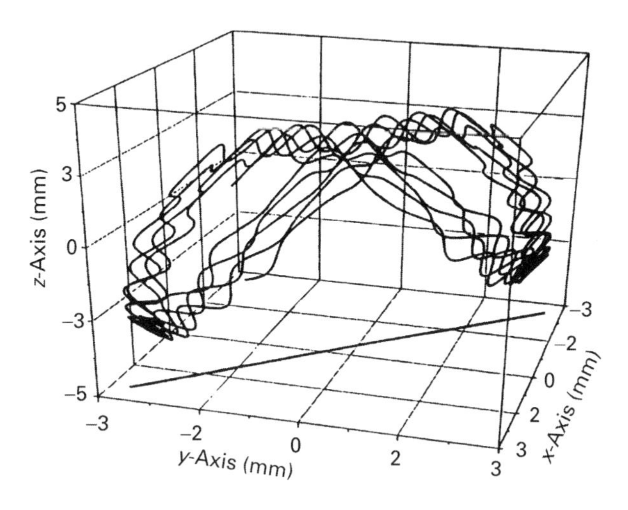

Figure 8 Trajectory of a trapped ion of *m*/*z* 105. The initial position was selected randomly from a population with an initial gaussian distribution (FWHM of 1 mm); $q_z = 0.3$; zero initial velocity. The projection onto the *xy* plane illustrates planar motion in three-dimensional space. The trajectory develops a shape that resembles a flattened boomerang. Reproduced from Nappi *et al* (1997) *International Journal of Mass Spectrometry and Ion Processes* **161**: 77–85.

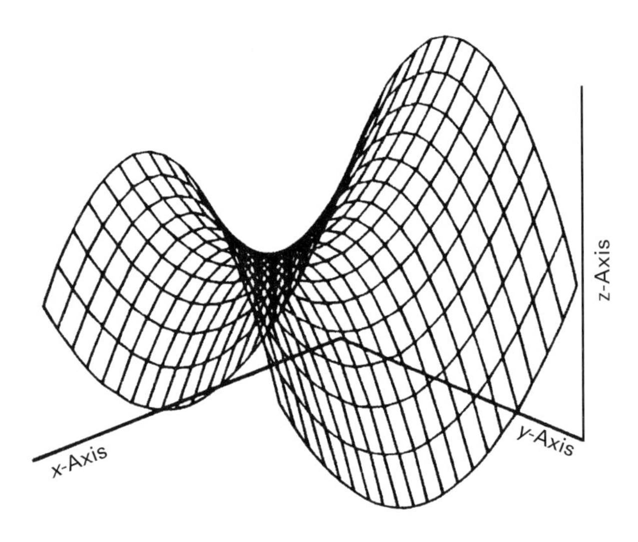

Figure 9 Pure quadrupole field, or potential surface, for a quadrupole ion trap. Note the four poles of the surface and the similarity of the field shape to the trajectory in **Figure 8**.

Resonant excitation

The motion of confined ions can be excited upon resonant irradiation at ω_z; resonant excitation is a powerful technique in ion trap mass spectrometry since predetermined waveforms composed of specified frequencies or frequency ranges can be used. Irradiation is effected by applying a supplementary potential of some hundreds of millivolts across the end-cap electrodes. Prior to resonant excitation, ions are focused collisionally to the ion trap centre by collisions with helium atoms. This process is described as 'ion cooling' in that ion kinetic energies are reduced to ~0.1 eV, corresponding to ~800 K. Resonant excitation, or 'tickling' of cooled ions causes ions to move away from the ion trap centre so that they experience the trapping field and are accelerated to kinetic energies of tens of electronvolts.

Resonant excitation is used to increase ion kinetic energy for the following purposes. (1) To eject unwanted ions during ionization and to isolate a narrow range of mass/charge ratios. (2) To promote endothermic ion/molecule reactions. (3) To increase ion internal energy through momentum-exchange collisions with helium atoms; in the limit, ions dissociate. (4) To move ions towards an end-cap electrode where an image current can be detected for their nondestructive measurement and re-measurement. (5) To eject ions either for ion isolation or for mass-selective ejection while the applied frequency is swept. (6) To eject ions while the amplitude V of the main RF potential is ramped up; this mode, known as axial modulation, uses a fixed frequency to eject ions just before their trajectories become unstable. In axial modulation, the resonant frequency is slightly less than half the main drive frequency Ω. Resonant excitation at lower frequencies has been used with great success to extend the normal mass range of the ion trap.

Operation of the ion trap as a mass spectrometer

In the ion trap, gaseous molecules are bombarded with 50–80 eV electrons emitted from a heated filament and gated into the trap, as shown in **Figure 10A**. Under automatic gain control (AGC); the number of ions formed during 200 μs is used to scale the ionization time and produce the required number of ions. During ionization, an RF voltage V_0 is applied to the ring electrode so as to confine ions in a given range of mass/charge ratio. Nascent ions are subjected to about 20 collisions per millisecond with helium at a pressure of 10^{-3} torr; those ions that are not ejected become focused near the trap centre. In **Figure 10B**, an RF amplitude is ramped over the period 30–85 ms, during which mass-selective ion ejection and mass analysis occur.

Each ion species confined within the ion trap is associated with a q_z value that lies on the q_z axis on the stability diagram; ions of high mass/charge ratio have q_z values near the origin, while ions of lower mass/charge ratio have q_z values that extend towards the $\beta_z = 1$ stability boundary, as shown diagrammatically using stick-people of various sizes in **Figure 11A**. At the intersection of the $\beta_z = 1$ stability boundary and the q_z axis, where $q_z = 0.908$ (see **Figure 7**), the trajectories of trapped ions become unstable axially and ions leave the ion trap. Once the

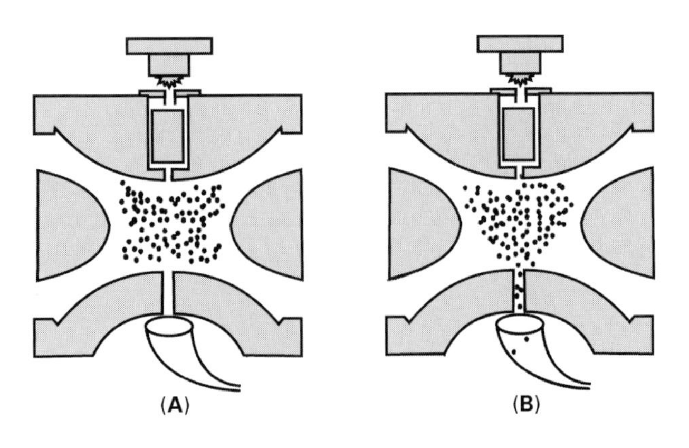

Figure 10 An overview of MS-in-time. (A) Step 1: a trapping RF amplitude is applied for 0–30 ms during which ions are formed from sample molecules and stored. (B) Step 2: an RF amplitude is ramped over the period 30–85 ms during which mass-selective ion ejection and mass analysis occurs. Reproduced from *An Overview of MS-in-Time*.

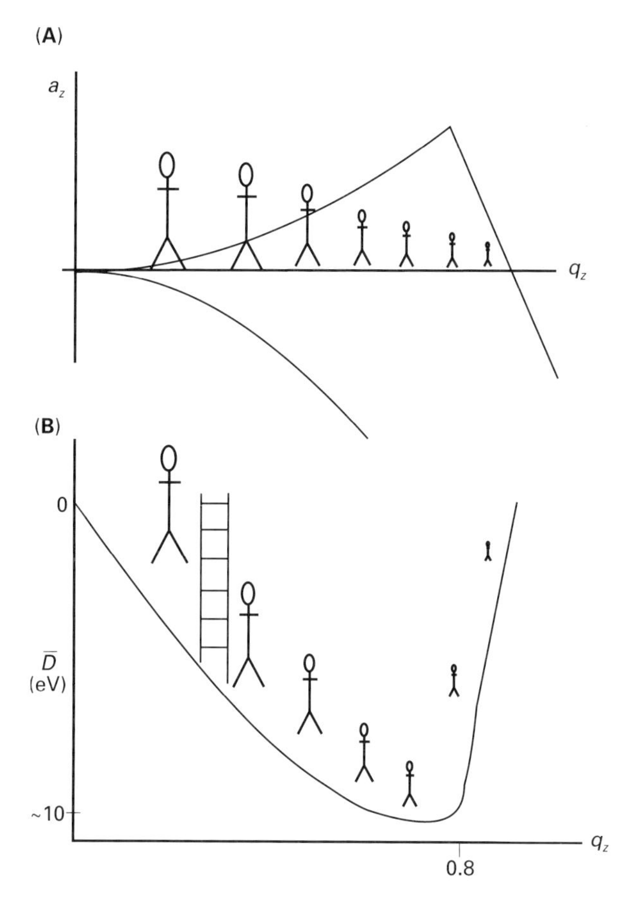

Figure 11 (A) Schematic representation of working points (that is, coordinates in a_z,q_z space) in the stability diagram for several species of ions stored concurrently. The arrangement of working points with respect to mass/charge ratio is depicted by figures that differ in size. (B) Ions are shown residing near the bottom of their respective axial potential wells of depth $\overline{D}_z$; the ladder represents the opportunity for resonant ejection of an ion species.

ion cloud has been focused collisionally to the ion trap centre, the amplitude of the RF potential is ramped; this operation, which is described as an analytical ramp or analytical scan, increases the q_z values of all ion species and, when $q_z = 0.908$ for each ion species, the ions are ejected axially through the end-cap electrodes. This mass-selective axial instability method of ion ejection has been supplanted by the use of axial modulation. Originally, axial modulation described resonant ejection of ions at a frequency of 485 kHz and at a q_z value slightly less than 0.908. When the RF amplitude is ramped, ions come into resonance at 485 kHz as their q_z values approach 0.908 and are ejected axially in order of increasing mass/charge ratio. Since ions are focused near the ion trap centre in the fashion of an onion, resonance ejection has the effect of removing the ions of low mass/charge ratio residing in the innermost layer of the onion from the influence of space-charge perturbations induced by other ion species, so that ions are ejected free of space-charge and with enhanced mass resolution. Resonant ejection, which can be carried out over a wide frequency range, is depicted pictorially in **Figure 11**B, which shows ions residing near the bottom of their respective axial potential wells of depth $\overline{D}_z$; the ladder represents resonant ejection of an ion species. For axial modulation, the ladder is positioned at a q_z value slightly less than 0.908. Resonantly ejected ions pass through holes in both end-cap electrodes so that ~50% impinge upon an electron multiplier located behind one of the end-cap electrodes; ion signals are created that produce a mass spectrum in order of increasing mass/charge ratio.

A mass spectrum is obtained by running the scan function (see below) a number of times specified by the number of microscans; the signals from each microscan are averaged and yield a mass spectrum.

Scan function

The above operation can be expressed succinctly in a scan function, which shows the temporal variation of the potentials applied to the electrodes such that a scan function is a visual representation of the sequence of program segments in the software that controls ion trap operation. The scan function for the above mass spectrometric operation is shown in **Figure 12**.

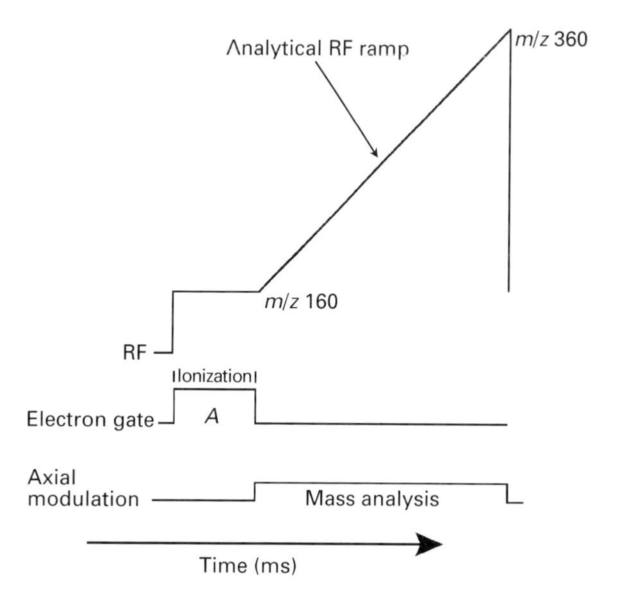

Figure 12 Scan function for obtaining an EI mass spectrum. The scan function shows the ionization period, *A*, followed immediately by the analytical ramp with concurrent axial modulation. Note that the pre-scan for the automatic gain control algorithm is not shown.

Collision-induced dissociation

Collision-induced dissociation (CID) of an isolated ion species in a quadrupole ion trap is a powerful technique for both the determination of ion structures and the analytical identification of compounds with high specificity. Under the influence of the tickle voltage, ions are moved from the centre to a region of higher potential, whereupon they are accelerated and their kinetic energies are increased. Subsequent collisions with helium atoms lead to enhancement of ion internal energy. In analytical applications, where the objective is to dissociate all isolated ions and to maximize the trapping of fragment ions produced, ion kinetic energy uptake must be balanced with incremental accumulation of internal energy so that ejection of isolated ions and fragment ions is avoided.

Tandem mass spectrometry

Tandem (Latin: "at length") mass spectrometry, MS/MS, is the practice of performing one mass-selective operation after another, much as the riders are seated on a tandem bicycle. The first mass-selective operation isolates an ion species designated as the parent ion, while the second determines the mass/charge ratios of the fragment, or product, ions formed by CID of the parent ions. MS/MS with a quadrupole ion trap, where successive mass-selective operations are carried out in time, offers a number of advantages. First, since the ion trap operates in a pulsed mode, mass-selected ions can be accumulated over time. Second, since CID is wrought by many collisions of mass-selected ions with helium atoms wherein the energy transferred per collision is small, dissociation channels of lowest activation energy are accessed almost exclusively. Third, all isolated ions can be dissociated and fragment ions arising from some 90% of them can be confined. Fourth, a sequence of several mass-selective operations can be performed as in MS*n*.

When gas chromatography is interfaced with an ion trap tandem mass spectrometer, individual compounds can be detected at the hundreds of femtograms level. The high specificity or informing power obtainable with GC–MS/MS is achieved by observation of specific fragment ion signals from an isolated molecular ion species $M^{\bullet +}$ formed from M that elutes within a specified retention-time window.

Chemical ionization and ion molecule reactions

In the quadrupole ion trap, several types of reactions can and do occur simultaneously and spontaneously once electron ionization of a compound has occurred. Ion–molecule reactions involving charge transfer, proton transfer and clustering occur sequentially, resulting in the formation of stable even-electron ions. Proton transfer chemical ionization (CI) involves the transfer to a neutral species of a proton (or other even-electron charged particles) from an ion that has been formed in an ion–molecule reaction. Common CI reagents are CH_5^+ and, $C_2H_5^+$, which are formed rapidly and can be isolated prior to reaction. CI reagent ions can be created externally and injected into the ion trap, isolated mass-selectively and allowed to react with sample molecules.

Conclusions

Ion trap mass spectrometry is a versatile technique of high sensitivity and high specificity. The relatively low cost of commercial instrumentation has permitted a substantial growth in the practice of mass spectrometry. The theory of ion trap operation differs from those of other mass spectrometers and presents an exciting challenge to the mass spectrometry community.

List of symbols

a_u = dimensionless trapping parameter ($u = r$ or z); $\bar{D}_r$ = depth of potential well in radial direction; $\bar{D}_z$ = depth of potential well in axial direction; m = ion mass [amu]; n = order of frequency component; q_u = dimensionless trapping parameter ($u = r$ or z); r_0 = radius of ring electrode; u = collective coordinate axes r and z; U = DC potential; V = 0-to-peak amplitude of the RF potential; z_0 = half the separation of the end-cap electrodes, along the axis of cylindrical symmetry; β_z = trapping parameter = $(a_z + q_z^2/2)^{1/2}$; ξ = a dimensionless parameter = $\Omega t/2$; ω = fundamental frequency component; Ω = radial frequency of the applied RF potential.

See also: **Chemical Ionization in Mass Spectrometry; Chromatography-MS, Methods; Proton Affinities.**

Further reading

Cooks RG and Kaiser RE Jr (1990) Quadrupole ion trap mass spectrometry. *Accounts of Chemical Research* **23**: 213–224.

Dawson PH (1976) *Quadrupole Mass Spectrometry and Its Applications*. Amsterdam: Elsevier.

Dawson PH and Whetten NR (1969) Radiofrequency quadrupole mass spectroscopy. *Electronics and Electron Physics* **27**: 58–158.

Fischer E (1959) Three-dimensional stabilization of charge carriers in a quadrupole field. *Zeitschrift für Physik* **156**: 1–26.

Glish GL and McLuckey SA (eds) (1991) *Quadrupole Ion Traps* (*International Journal of Mass Spectrometry and Ion Processes* **106**).

March RE (1992) Ion trap mass spectrometry. *International Journal of Mass Spectrometry and Ion Processes* **118/119**: 71.

March RE (1997) An introduction to quadrupole ion trap mass spectrometry. *Journal of Mass Spectrometry*, **32**: 351.

March RE, Hughes RJ and Todd JFJ (1989) *Quadrupole Storage Mass Spectrometry*. New York: Wiley Interscience.

March RE and Todd JFJ (eds) (1995) *Practical Aspects of Ion Trap Mass Spectrometry*, Modern Mass Spectrometry Series; Vol. 1, Fundamentals, ISBN 0-8493-4452-2; Vol. 2, Instrumentation, ISBN 0-8493-8253-X; Vol. 3, Chemical, Biomedical, and Environmental Applications, ISBN 0-8493-8251-3. Boca Raton, FL: CRC Press.

Mathieu E (1868) *Journal de Mattematiques Pures et Applies (J. Liouville)* **13**: 137. (See also McLachan NW (1947) *Theory and Applications of Mathieu Functions*. Oxford: Clarendon Press and Campbell R (1955) *Théorie Générale de l'Equation de Mathieu*. Paris: Masson.

Nappi M, Weil C, Cleven CD, Horn LA, Wollnik H and Cooks RG (1997) Visual representations of simulated three-dimensional ion trajectories in an ion trap mass spectrometer. *International Journal of Mass Spectrometry and Ion Processes* **161**: 77–85.

Nourse BD and Cooks RG (1990) Aspects of recent developments in ion trap mass spectrometry. *Analytica Chimica Acta* **228**: 1–11.

Paul W (1990) Electromagnetic traps for charged and neutral particles (Nobel Lecture). *Angewandte Chemie* **29**: 739–748.

Paul W and Steinwedel H (1960) Apparatus for separating charged particles of different specific charges. *German Patent* 944 900, 1056; *US Patent* 2 939 952.

Reiser HP, Kaiser RE Jr and Cooks RG (1992) A versatile method of simulation of the operating of ion trap mass spectrometers. *International Journal of Mass Spectrometry and Ion Processes* **121**: 49–63.

Stafford GC Jr, Kelley PE, Syka JEP, Reynolds WE and Todd JFJ (1984) Recent improvements in and analytical applications of advanced ion trap technology. *International Journal of Mass Spectrometry and Ion Processes* **60**: 85–98.

Todd JFC (1991) *Mass Spectrometry Reviews* **10**: 3.

Wuerker RF, Shelton H and Langmuir RV (1959) *Journal of Applied Physics* **30**: 324.

Ionization Methods in MS

See **Atmospheric Pressure Ionization in Mass Spectrometry; Chemical Ionization in Mass Spectrometry; Fast Atom Bombardment Ionization in Mass Spectrometry; Ionization Theory; Plasma Desorption Ionization in Mass Spectrometry; Thermospray Ionization in Mass Spectrometry.**

Ionization Theory

C Lifshitz, The Hebrew University of Jerusalem, Israel
TD Märk, Leopold Franzens Universität, Innsbruck, Austria

MASS SPECTROMETRY
Theory

Introduction: Electron ionization, photoionization

At the heart of mass spectrometry lies the formation of the ions to be analysed. A thorough understanding of the production of the ions is extremely important, since the method used to ionize a sample markedly affects its mass spectrum (fragmentation pattern). Besides electron impact ionization (EI) and photoionization (PI), a number of different methods are used in mass spectrometry including chemical ionization (CI), field ionization, fast atom bombardment, surface ionization, electrospray ionization, laser ablation and other plasma methods. EI is by far the most common method as it is simple to use, easy to set up and very effective. For instance the maximum ionization cross section (see definition below) for the hydrogen atom is about 10^{-20} m^2 (at 100 eV) for electrons as compared to about 5×10^{-22} m^2 (at 14 eV) for photons. On the other hand, PI has yielded basic and accurate data required for the understanding of the energetics and dynamics of ionization and fragmentation.

Because of its overwhelming importance in mass spectrometry, only EI (and, to a lesser extent, PI) will be treated here. We will consider the ionization mechanism, the various types of ions produced and the kinetics and dynamics of the ionization process and subsequent fragmentation reactions.

Electron ionization (EI) is the process by which an atom or a molecule M is ionized by electron impact to form a positive ion (Eqn [1]).

$$\mathrm{M} + \mathrm{e}^- \rightarrow \mathrm{M}^{\bullet+} + 2\mathrm{e}^- \qquad [1]$$

Photoionization (PI) is the process by which M is ionized through photon impact.

$$\mathrm{M} + h\nu \rightarrow \mathrm{M}^{\bullet+} + \mathrm{e}^- \qquad [2]$$

The *ionization energy* (IE) is the minimum electron energy or photon energy required to produce ionization from M and is related to the binding energy of the most loosely bound valence electron in the molecule. When the electron energy or photon energy is varied continuously there is a threshold energy below which the ion current is zero and above which it rises with increasing energy, according to a particular threshold law (see below). This threshold is an experimental measure of the IE. The IE often (but not always) can be determined spectroscopically as the limit of a series of Rydberg states in which the principal quantum number n of the electron reaches increasingly higher values. The IE can be determined mass spectrometrically even in those cases for which a Rydberg series has not been observed spectroscopically.

The majority of neutral, chemically stable, molecules possess a closed shell. As a result, ionization leads to a radical cation, i.e. to an odd-electron ion ($\mathrm{OE}^{\bullet+}$). In spectroscopic terms, a neutral singlet state leads to an ionic doublet state.

Ionization mechanism

EI or PI involves the collision of an electron or photon of sufficient energy with a neutral (or ionized) target particle and the subsequent production of an ion and the respective ejected electron(s). The term 'electron impact ionization' is somewhat misleading, because an electron is small compared to the size of a molecule and thus would have difficulty 'hitting' any part of an atomic target. It is better to think of the electron as passing close to or even through the atomic target, while in quantum-mechanical terms the wave of the electron interacts with and distorts the electric field of the atomic system. Electrons accelerated through a potential of several tens of volts have a de Broglie wavelength of ~0.1 nm. In this case the wavelength and the molecular dimensions are similar and mutual quantum effects (distortions) occur. The wavelength of a photon required for PI is about 100 nm which is much larger than the usual molecular dimension and almost no molecular distortion occurs during ionization. Because of this difference in interaction in the case of EI, the ionizing transition is not strictly vertical (see below) and more states (also taking into account spin conservation from the target to the product ion) can be reached relative to PI.

In addition, electron impact can also give rise to the birth of negative ions (anions). In contrast to the formation of positive ions (cations), direct attachment of the incident electron to a polyatomic target to give a stable anion is rather improbable. The reason is that the energy of the attaching electron and the binding energy (electron affinity) must be taken up (accommodated) in the emerging product (anion). Usually, the excess energy leads either to fragmentation of the anion or to shake-off of the electron (autodetachment); radiative stabilization of this excited anion takes place only with very low probability. One way to produce stable anions by direct electron attachment is to have sufficient gas pressure in the ion source for three-body reactions to occur that remove the excess energy (three-body stabilization).

Ionization channels

Ionization of atoms results only in the production of singly and multiply charged atomic ions. As the energy deposited by the ionizing agent is increased, the variety and abundance of the ions produced from a specific molecular target will increase, because the ionization process may proceed via different reaction channels, each of which gives rise to characteristic ionized and neutral products. For the simple case of a diatomic molecule AB the corresponding reaction channels are as shown in Equations [3]–[13] where e_s is a 'scattered' electron and e_e is an 'ejected' electron.

$$AB + e \rightarrow$$

$$AB^{\bullet+} + e_s + e_e \qquad \text{(single ionization)} \qquad [3]$$

$$AB^{2+} + e_s + 2e_e \qquad \text{(double ionization)} \qquad [4]$$

$$AB^{z+} + e_s + z\,e_e \qquad \text{(multiple ionization)} \qquad [5]$$

$$AB^{K+} + e_s + e_e \qquad \text{(K-shell (inner) ionization)} \qquad [6]$$

$$AB^{**} + e_s \rightarrow AB^{\bullet+} + e_s + e_e \qquad \text{(autoionization)} \qquad [7]$$

$$AB^{+*} + e_s + e_e \rightarrow \begin{cases} A^{\bullet+} + B^{\bullet} + e_s + e_e \\ \text{(metastable fragmentation)} \quad [8] \\ AB^{2+} + e_s + 2e_e \\ \text{(autoionization)} \quad [9] \\ AB^{\bullet+} + e_s + e_e + h\nu \\ \text{(radiative ionization)} \quad [10] \end{cases}$$

$$A^{+} + B + e_s + e_e \qquad \text{(prompt dissociative ionization)} \qquad [11]$$

$$AB^{-*} \rightarrow A^{\bullet-} + B^{\bullet} \qquad \text{(dissociative attachment)} \qquad [12]$$

$$AB^{**} + e_s \rightarrow A^{+} + B^{-} + e_s \qquad \text{(ion pair formation)} \qquad [13]$$

Similar reaction channels are possible for PI but there are no scattered electrons and dissociative attachment is only possible for EI. Different and more complex reactions may occur when more complex targets are involved, i.e. polyatomic molecules or clusters (even including multiple electron collisions and subsequent reactions within the molecular target). Clusters and fullerenes have also been demonstrated to undergo a process of delayed ionization which is akin to thermionic emission in bulk material.

Most of the ionization reactions shown above (e.g. Eqns [3]–[6], [11]) can be classified as being due to a direct ionization mechanism in which the ejected and the scattered electrons leave the ion within 10^{-16} s of each other. Conversely, there exist alternative ionization channels (competing with direct ionization) in which either the electrons are ejected one after the other (autoionization) or the molecular ion decays at a later stage (see **Figure 1** and below). The autoionization event (e.g. Eqns [7] and [9]) can be described as a two-step reaction: first, a neutral molecule (or atom) is raised to a superexcited state, which can exist for some finite time. Then, radiationless transition into the underlying ionization continuum occurs. Superexcited states exist in principle for every target system except for the hydrogen atom. For molecules, the upper autoionization rate (and hence the ionization cross section) is limited by the characteristic energy-storage mode frequency. In addition, if predissociation (into two neutrals) is faster than autoionization, the latter will not occur at an appreciable rate.

Autoionization by photons is a resonance process. Autoionization complicates the ionization cross section function (see definition below) at low as well as at high electron energies. PI has been extremely valuable in the quantitative study of autoionization processes both in atomic and in molecular systems. Classical examples involve the photoionization mass spectra of noble gas atoms such as xenon (**Figure 2**) and molecular hydrogen, H_2 (**Figure 3**). Xenon has filled 5s and 5p shells. Ionization is initially observed at $h\nu$ = 12.130 eV with the ejection of a 5p electron and the formation of the ground $^2P_{3/1}$ state of Xe^+. In the interval between $^2P_{3/1}$ and its spin–orbit partner $^2P_{1/2}$ (12.12–13.436 eV), autoionization structure is prominent (see **Figure 2**) owing to Rydberg states converging to $^2P_{1/2}$. One notes sharp $p \rightarrow ns$

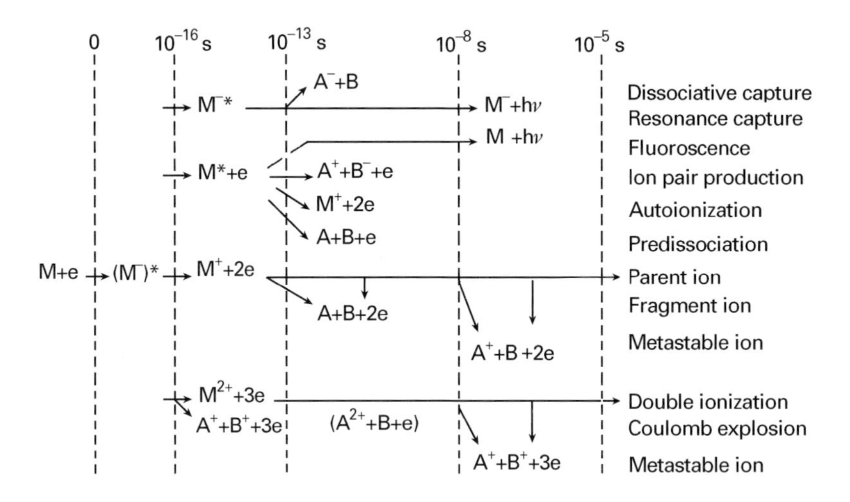

Figure 1 Schematic time evolution of EI of a molecule.

transitions and broad ones corresponding to p $\rightarrow$ *n*d transitions. Autoionization takes place by virtue of a 'spin–flip' mechanism in which J changes from $\frac{1}{2}$ to $\frac{3}{2}$ and Rydberg electron is ejected. Autoionization in the case of molecular hydrogen (**Figure 3**) involves conversion of vibrational energy into additional electronic excitation of the Rydberg electron. Predissociation competes with autoionization of H_2.

Franck–Condon principle

Inelastic collisions between electrons/photons and molecules involve transitions between defined (electronic, vibrational and rotational) molecular states. The energy deposited into the excitation of molecular vibration and rotation is usually rather small compared with the energy of the electronic transition, at least if one assumes that only one vibrational quantum is excited in a single collision. For example, the greatest separation between a ground and a first excited vibrational state of any molecular ion is that of H_2^+ (0.27 eV), which is small compared with the ionization energy of 15.426 eV. The changes in the vibrational levels from v'' to v' that result in ionization can be described in terms of the Franck–Condon principle.

Qualitatively, the Franck–Condon principle may be summarized as follows. During an electronic transition no (or only negligible) changes occur in the nuclear separation and the velocity of relative nuclear motion. Owing to the large ratio of nuclear to electronic mass and the short interaction time (~10^{-17} s, as compared with ~10^{-13} s for a vibration), the point on the upper potential-energy curve (corresponding to the configuration after the transition) lies directly above the starting point on the initial potential-energy curve (vertical transition). This leads to a number of possible electronic transitions, which depend on the relative shapes of the potential-energy curves available in a specific system. Several

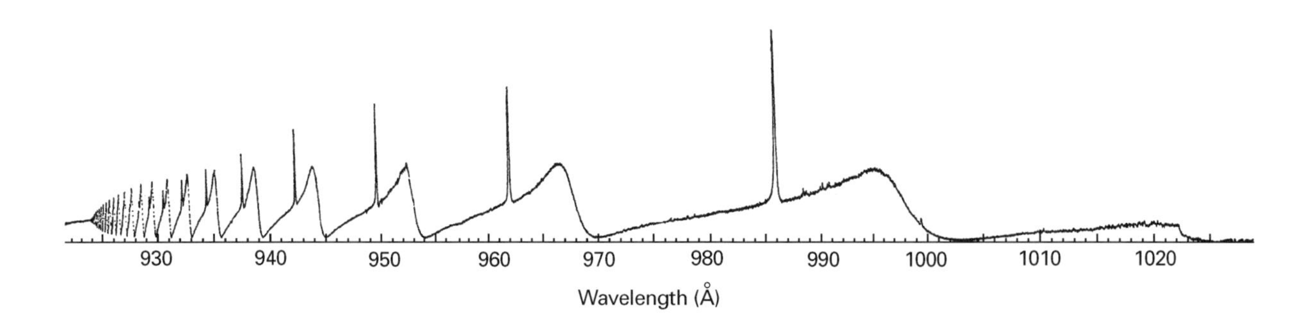

Figure 2 Photoionization mass spectrum of xenon between the $^2P_{3/2}$ (ground) and $^2P_{1/2}$ (excited) ionic states, displaying sharp 's' like and broad 'd'-like resonances. Reproduced with permission of Academic Press from Berkowitz J (1979) *Photoabsorption, Photoionization, and Photoelectron Spectroscopy*, Chapter VI.

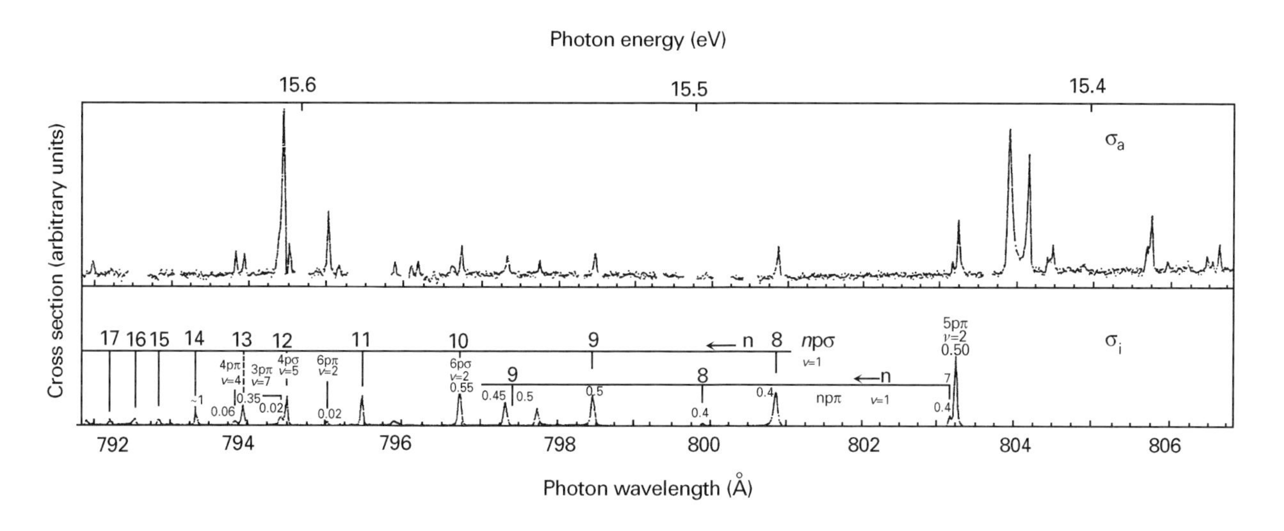

Figure 3 Relative photoionization and photoabsorption cross sections σ_i and σ_a respectively, for parahydrogen at 78 K. The numbers above peaks in the photoionization data give the value of η_i the quantum yield of ionization. Reproduced with permission of Academic Press from Berkowitz J (1979) *Photoabsorption, Photoionization and Photoelectron Spectroscopy*, Chapter VI.

cases can be distinguished. (1) The final accessible level lies within the region of discrete vibrational states of the upper potential-energy curve. The probability that the vibrational quantum number will change depends on the relative positions of the potential-energy curves. (2) The final accessible level not only lies within the region of discrete vibrational states, but includes some part of the continuum. Hence, some of the transitions will lead to dissociation. (3) The final accessible level lies within the continuum of a repulsive state and all transitions lead to dissociation.

The Franck–Condon principle can be used to treat quantitatively the fragmentation of diatomic or pseudodiatomic molecules at low energies. The cross section for ionization from the vibrational level v'' of the neutral molecule to a vibrational level v' in the ionized molecule is given by the Franck–Condon factor (FCF) which is equal to the square of the overlap integral between the respective vibrational wavefunctions, i.e. FCF = $|\langle\psi_{v''}|\psi_{v'}\rangle|^2$. In polyatomic molecules, the two-dimensional potential-energy curves have to be replaced by multidimensional potential-energy hypersurfaces. Although EI and PI proceed without nuclear displacement (Franck–Condon principle), the resulting (excited) polyatomic ion will usually subsequently undergo further internal transitions (e.g. radiationless transfer of energy, that is transitions from one surface to another surface), possibly leading to subsequent unimolecular decomposition (see below). Even with an accurate knowledge of the potential energy surfaces, a detailed description of the evolution of the ionized system is virtually impossible; it is necessary to use statistical theory methods.

Koopman's theorem; photoelectron spectroscopy

The photoionization process (Eqn [2]) may be studied by using a fixed photon wavelength $h\nu$ and recording the kinetic energy distribution of the emitted photoelectrons. This is called photoelectron spectroscopy (PES). The photoelectron carries away the difference between the photon energy and the binding energy of the ejected electron (Eqn [14]).

$$\varepsilon = h\nu - \mathrm{IE(M)} \qquad [14]$$

where ε is the kinetic energy of the electron and IE(M) is the ionization energy of the molecule, equal to the binding energy of the photoionized electron. Each ionization energy is approximately equal to the negative eigenvalue of the molecular orbital (MO) from which the electron is ejected. This approximation is known as Koopman's theorem, which is based on the self-consistent field (SCF) model.

The photoelectron spectrum of N_2 (**Figure 4**), obtained with $h\nu = 21.21$ eV, demonstrates what happens upon ionization from the three outermost MOs of the molecule, whose electronic structure is $(1\sigma_g)^2(1\sigma_u)^2(2\sigma_g)^2(2\sigma_u)^2(1\pi_u)^4(3\sigma_g)^2$. Ionization of an electron from a nonbonding orbital leads to an ionic state whose equilibrium internuclear distance is not shifted appreciably with respect to the neutral. The vertical transition, according to the Franck–Condon principle, leads to very little, if any vibrational excitation. The outermost σ_g orbital is very weakly bonding. As a result, the strongest transition observed is from $v'' = 0$ of the neutral to $v' = 0$ of the ion.

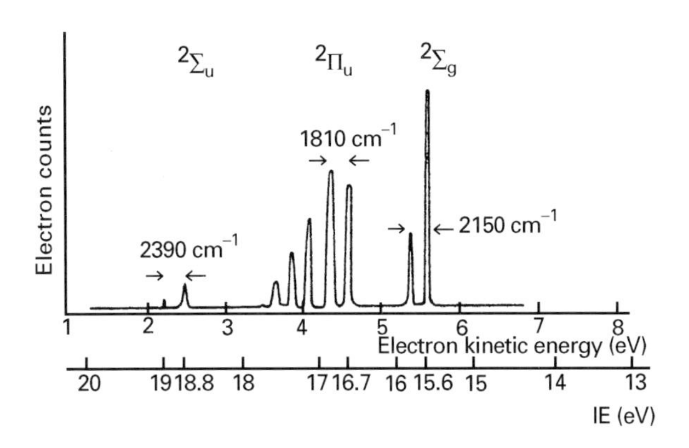

Figure 4 Photoelectron spectrum of N_2 obtained with 21.21 eV incident radiation; electron counts versus electron kinetic energy (upper) and versus IE of N_2 (lower). Reproduced with permission of Oxford University Press from Atkins PW (1986) *Physical Chemistry*, 3rd edn, p 483.

Furthermore, the vibrational spacing in the $^2\Sigma_g$ ionic state that is produced is 2150 cm^{-1}, only slightly less than in the neutral (2345 cm^{-1}). The second band envelope reached, which is due to the $^2\Pi_u$ state, has a much richer vibrational structure, because the π_u electron is ionized from a strongly bonding orbital. This causes the position of the minimum in the potential energy curve to be shifted to a larger internuclear distance in the ion with respect to the neutral, to excitation of a number of vibrational levels in the ion within the vertical Franck–Condon region, and to a rather small vibrational spacing of 1810 cm^{-1} in the ion. Finally, the third band, which is due to formation of the $^2\Sigma_u$ state of the ion, demonstrates the ionization of a weakly antibonding electron, with a strong propensity for excitation of the 0–0 transition and with a slightly larger vibrational spacing (2390 cm^{-1}) in the ion as compared to the neutral.

Types of ions produced

Parent ions

Parent ions are positively charged ions produced by reaction [3] through removal of one electron from the neutral precursor. The production of these parent ions relative to that of other ions originating from the same neutral precursor (molecule) depends on the electron or photon energy and on the properties of the neutral molecule. At and just above the IE, only singly charged (parent) ions are produced, but at higher electron energies, other ions (see below) are also observed.

In general, for small molecules the parent ion is the dominant ion at all impact energies, although there are exceptions, such as CCl_4 and CF_4 which have no stable parent ions. Conversely, for large molecules the relative parent-ion abundance usually decreases with increasing molecular mass and increasing incident energy. Again there are exceptions to this rule, the most notable one concerning the fullerenes, where the singly charged $C_{60}^{\bullet+}$ parent ion is the most abundant ion in the mass spectrum (see **Figure 5**).

The parent-ion abundance depends also on the temperature of the molecular gas target. At constant ion-source gas pressure a decrease in all ion intensities is noted with increasing temperature, while an increase in vibrational energies of the molecular ion (due to the increase in the vibrational energies of the

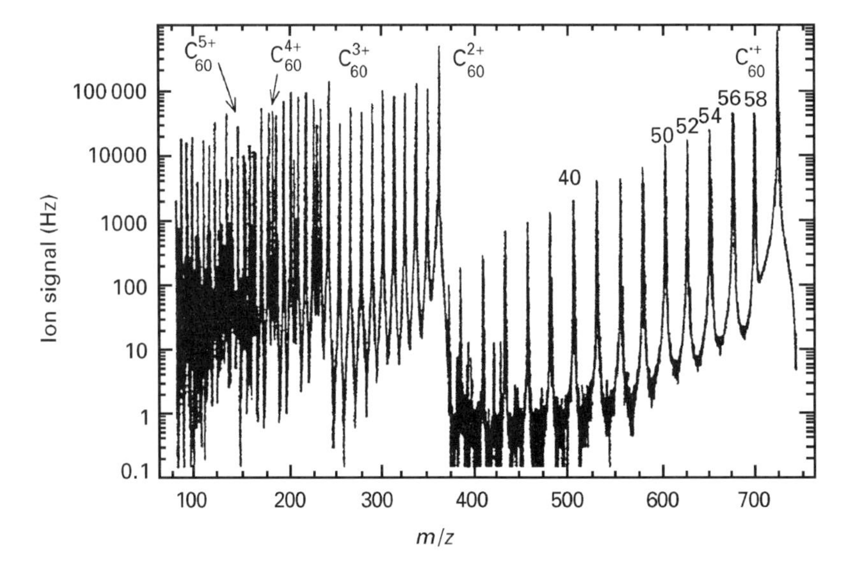

Figure 5 Mass spectrum of C_{60} ionized with 200 eV electrons.

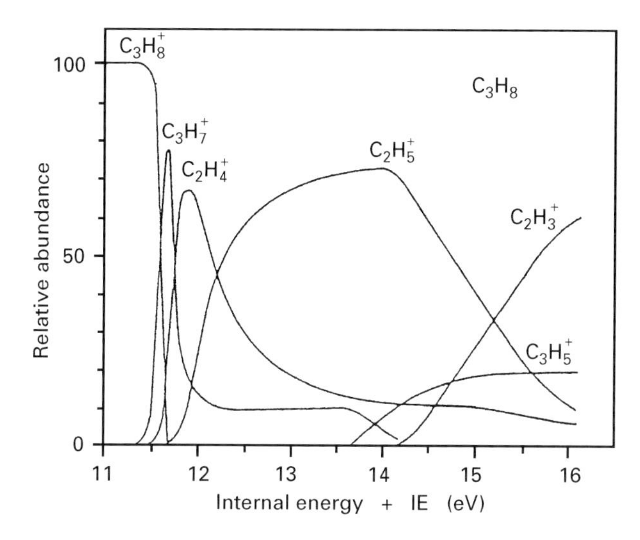

Figure 6 Breakdown graph of propane. The relative abundances of the ions indicated are plotted as functions of the sum of the IE of propane and the internal energy of the parent propane ion.

neutral precursor) may lead to an appreciable decrease in the relative parent-ion intensity. For polyatomic molecules, this latter effect can be explained in terms of statistical theories.

Fragment ions

If the incident energy is increased above the IE of the molecule, fragment ions appear owing to reaction paths discussed above in relation to the Franck–Condon principle. In the case of polyatomic molecules, a wide variety of fragmentation channels is available; that is, the molecular ion can immediately decay into a fragment ion with an even or odd number of electrons (primary fragment ion). These primary fragment ions may also be produced in (unstable) excited states and immediately decay into further fragments. The number of fragment ions and their relative cross-section functions are characteristic of the corresponding parent molecule. The relative importance of the various ions as a function of internal energy can be demonstrated with the help of breakdown curves (see **Figure 6** showing the breakdown graph of propane).

In the case of diatomic parent molecules, the fragmentation can be treated quantitatively in terms of the Franck–Condon principle. For the dissociation of small polyatomic molecules, spectroscopic and quantum-theoretical ideas (correlation rules) can be used to determine the dissociation path and predict the resulting electronic states. With the advent of *ab initio* CI (configuration interaction) calculations, detailed information on the potential energy surfaces of the different states and on their interactions is becoming available. However, to deal with large polyatomic molecules, statistical theories have to be used, i.e. the so-called quasiequilibrium (QET) theory or an almost identical approach, the so-called RRKM theory (named after Rice, Ramsperger, Kassel and Marcus).

In general, the relative abundance of any fragment ion is related to its rate of formation and its rate of dissociation by unimolecular decomposition. Hence, the measured fragmentation pattern of a molecule is a record in time of the 'quasiequilibrium' balance of these rates. In other words, because the fragmentation pattern is a slice of the three-dimensional plot of ion current as a function of incident energy and mass-to-charge ratio, the respective partial ionization cross-section functions will depend on the time after formation of the primary ion. Trapped-ion mass spectrometry has been used to investigate this phenomenon for EI and PI. Typical results for short and long delay times are shown for $C_6H_6^{\bullet+}$, $C_6H_5^+$ and $C_6H_4^{\bullet+}$ in **Figure 7**. The increased fragmentation of $C_6H_6^{\bullet+}$ at long delay times (due to the presence of metastable ions, see below) is obvious.

In this context it is interesting to point out that the reason for running mass spectra at relatively high electron energies (50–100 eV) is that fragmentation patterns do not vary very much with electron energy in this energy range and that the partial ionization cross-section functions (and thus the detection efficiency) have their respective maxima in this energy range (see below). On the other hand, much energy can be transferred to the molecular ion in this energy range, resulting in extensive fragmentation (see below), sometimes making it difficult to identify the parent ion. Because of this, EI is considered a 'non-soft' ionization technique compared with PI or CI. However, if similar energies are used for PI and EI, very similar mass spectra are often observed.

Finally, it should be pointed out that dissociative ionization yields fragment ions with small to moderate kinetic energies. To describe the dissociative ionization process completely, these kinematic properties must be known.

Metastable ions

Ions produced with lifetimes longer than those of excited ions that decompose in the ion source (prior to about 10^{-6} s) are called metastable ions. The existence of metastable ions can be explained by different mechanisms depending on the size and property of the precursor ion. Normally, dissociation of an excited ion occurs during the first vibration (prompt dissociation), or predissociation that involves a transition from one potential energy hypersurface to another, which also occurs rapidly;

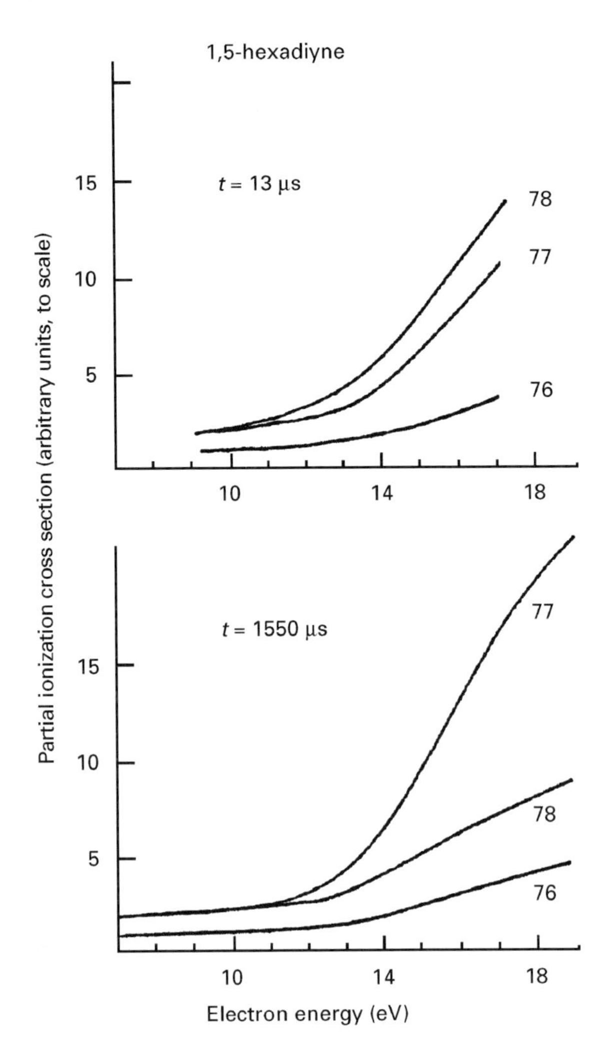

Figure 7 Partial ionization cross-section functions for ions of 1,5-hexadiyne (a linear isomer of benzene) at two different detection times, *t*.

therefore, the product ions are produced in the ion source (see fragment ions). However, if crossings of potential hypersurfaces are spin-forbidden, some ions will nevertheless undergo electronic predissociation but will do so at a reduced rate. Thus, electronic (forbidden) predissociation via repulsive states is one origin of metastable ions. A second possible mechanism is dissociation by tunnelling. In this case the initial excitation is to bound levels of an excited state of the molecular ion that are above the dissociation limit of this state, but below the top of either an intrinsic or centrifugal barrier. The existence of large metastable ions can be rationalized in the framework of QET and is termed vibrational predissociation. After ionization, radiationless transitions yield highly vibrationally excited ground-state ions with a distribution of internal energies and hence a distribution of decay rate constants

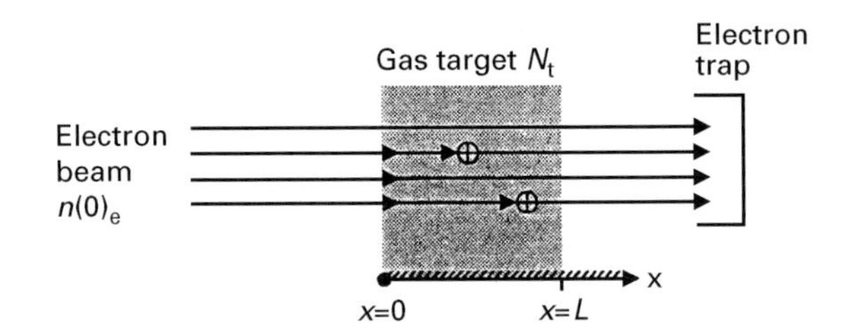

Figure 8 Schematic view of an electron impact ionization experiment.

including those that lead to unimolecular decomposition in the metastable time range.

If the metastable decomposition occurs in flight before the ion reaches the analyser of a mass spectrometer, a typical metastable peak may be observed in the mass spectrum depending on the type of mass analyser used. Moreover, using appropriate methods, it is also possible to deduce the kinetic energy distribution of the nascent fragment ions at these late times after the initial excitation.

Multiply charged ions

Multiply charged atomic and molecular ions were observed and identified as early as 1912. Subsequent observation of numerous doubly charged ions followed. Triply charged molecular ions have been detected in low abundances in the mass spectra of some species (e.g. CO_2, CS_2, C_2N_2 and aromatic compounds). Moreover, EI of free van der Waals clusters can lead to multiply charged ions with up to five and more elementary charges. Conversely, with certain molecules (e.g. H_2O and CH_4), it is not possible to detect any stable doubly charged molecular parent ions by EI.

The partial ionization cross sections for the production of multiply charged atoms and molecular ions rarely exceed 1–5% of that of the dominant singly charged ion. Certain atoms and compounds, however, have been found to possess an increased ability to sustain two or even more positive charges, an especially intriguing example being C_{60} (see **Figure** 5).

Ionization cross sections

Definitions

Consider, as shown in **Figure** 8, a parallel, homogeneous and monoenergetic beam of electrons crossing a semi-infinite medium containing N_t target particles per cubic centimetre at rest. If $n(0)_e$ represents the initial electron intensity (number of electrons per area and per second), the intensity of the electron beam at depth x is given by the exponential absorption law (Eqn [15]):

$$n(x)_{\mathrm{e}} = n(0)_{\mathrm{e}} \exp(-N_{\mathrm{t}}\sigma x) \quad [15]$$

If $N_t \sigma x << 1$ (single-collision condition), the number of ions generated per second along the collision interaction path $x = L$ (over which the ions are collected and analysed) is

$$n(L)_{\mathrm{i}} = n(0)_{\mathrm{e}} N_{\mathrm{t}} \sigma_{\mathrm{c}} L \quad [16]$$

where σ_c is the counting ionization cross section. The total positive-ion current i_t produced in this interaction volume is given by

$$i_{\mathrm{t}} = n(0)_{\mathrm{e}} e N_{\mathrm{t}} \sigma_{\mathrm{t}} L \quad [17]$$

where σ_t is the total ionization cross section. If the ions produced are analysed with respect to their mass m and charge ze, the respective individual ion currents measured with a Faraday cup are given by

$$i_{\mathrm{ms}} = n(0)_{\mathrm{e}} z e N_{\mathrm{t}} \sigma_{zi} L \quad [18]$$

where i_{ms} is the mass selected ion current and σ_{zi} is the partial ionization cross section for the production of a specific ion i with charge ze.

Total and counting ionization cross sections of a specific target system are the weighted and the simple sum of the various single and multiple partial cross sections, respectively:

$$\sigma_{\mathrm{t}} = \Sigma \sigma_{zi} z \qquad \text{and} \qquad \sigma_c = \Sigma \sigma_{zi} \quad [19]$$

Sometimes the macroscopic cross section $s = \sigma N_t$, which represents the total effective cross-sectional area for ionization of all target molecules in 1 cm^3 of the target medium, is used.

Partial and total ionization cross sections are of great importance for practical applications, but they give only a limited insight into the ionization process itself. Conversely, a great deal of information about the kinematic aspects can be obtained from differential electron ionization cross sections, including the single, double and triple differential cross sections. The single differential cross section $d\sigma(E,W)/dW$, where W is the energy of the electrons after the ionizing collision, measured at the collision energy E represents the electron energy distribution (integrated over all angles) of the two outgoing electrons. The energy distribution is in principle symmetrical with respect to $\frac{1}{2}(W_1 + W_2) = \frac{1}{2}(E - \mathrm{IE})$, since for each fast electron a corresponding slow electron of complementary energy is ejected. If the incident energy E is large enough, the low-energy part of the distribution is mainly due to 'ejected' electrons and the high-energy part is due to 'scattered' electrons. Although in principle the two outgoing electrons are indistinguishable, this interpretation is supported by the different angular behaviour of these two groups of electrons. Usually, the single differential cross section has a minimum at $W = \frac{1}{2}(W_1 + W_2)$ and maxima close to $W = 0$ and $W = (E - \mathrm{IE})$. At very small incident energies, the cross section becomes independent of W (at least for atoms).

The double differential cross section $d^2\sigma\,(E, W, \delta)/dW\,d\Omega$ is higher by one order with respect to the number of kinematic parameters, i.e. the energy W and the polar angle δ of the outgoing electrons have to be measured. In practice, one of the three experimental parameters E, W or δ is varied while the others are kept constant. This allows the determination of either angular distributions (E and W constant) of ejected electrons or energy loss spectra. In general, the higher the energy of the scattered electron, the more their intensity is peaked forward, whereas the angular distribution of very low-energy electrons is nearly isotropic. The triple differential cross section $d^3\sigma\,(E, W_1, \delta_1, \delta_2, \varphi_2)\,dW\,d\Omega_1\,d\Omega_2$ contains all available kinematic information. It has to be measured in coincidence ('e,2e experiments') and full graphical representation is impossible.

Ionization threshold law

Following the postulates of Wigner that the ionization probability for EI to a first approximation depends critically on the likelihood of the separation step in the exit channel (that is, separation of the collision complex AB^{-*} into a stable ion and two free electrons depends on the probability of disposing of the excess energy), Wannier, using a very simple phase space argument, showed that this probability depends on the number of degrees of freedom n for sharing the excess energy between the electrons, thus giving a threshold law

$$\sigma = k(E - \mathrm{IE})^n \quad [20]$$

For single ionization and thus emission of two electrons, $n = 1$ and the coefficients k is determined by the electronic transition probability, which is

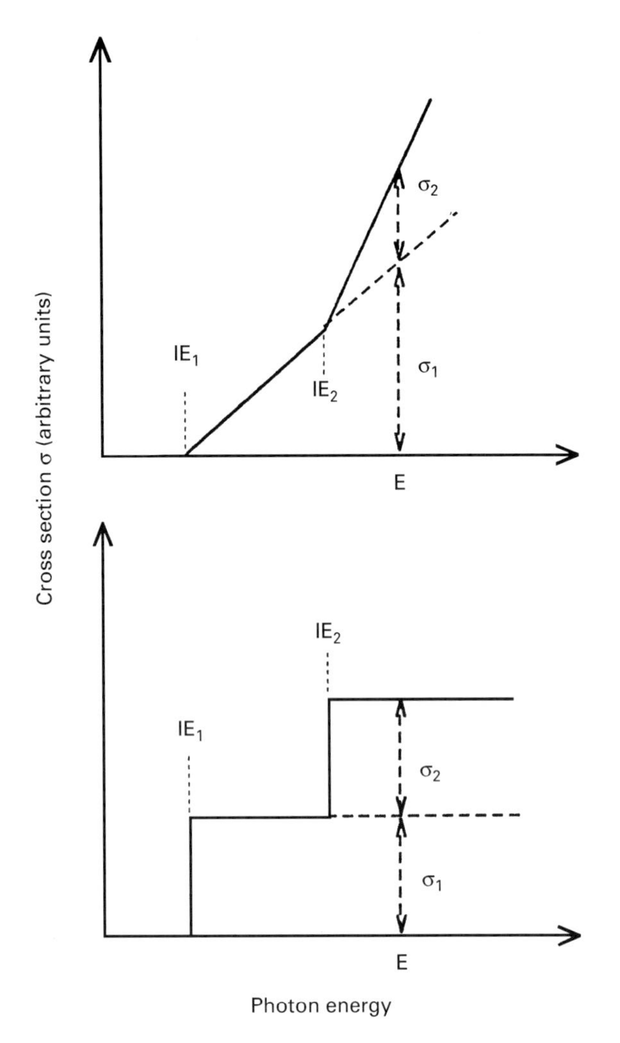

Figure 9 Idealized threshold laws for electron impact ionization and photoionization.

proportional to the overlap of the vibronic wavefunctions for the neutral ground state and the state of the ion produced. This argument can be extended to z-fold ionization, giving $n = z$. More detailed calculations predict that these power laws should all be increased by 0.127. Using the same arguments for PI, it is then possible to conclude that

$$\frac{d\sigma_{\text{electron}}}{dE} = \sigma_{\text{photon}} \qquad [21]$$

Therefore, for single ionization a linear increase of the cross section with electron energy is expected, whereas for direct PI a step-function cross-section law is predicted. This simple theory gives no prediction of the energy range of validity and in practice the threshold laws are slightly different for several reasons, including influence of competitive channels, tailing due to rotational levels, and autoionization

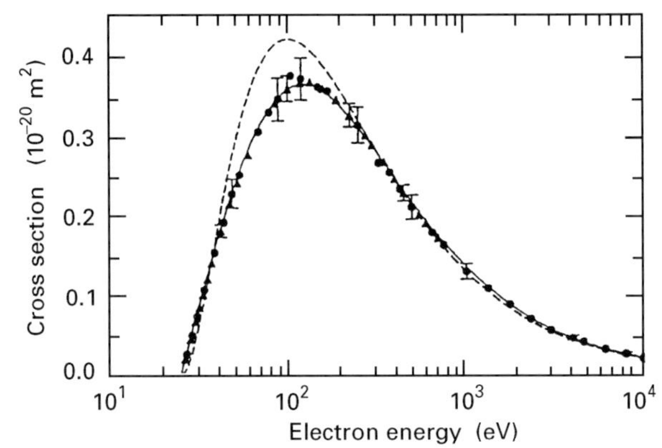

Figure 10 Experimental (solid symbols) and theoretical (dashed line, distorted-wave Born approximation; full line, Deutsch–Maerk formula and BEB (Binary–Encounter–Bethe) approximation) electron ionization cross section for helium.

which will obscure the true threshold behaviour. Nevertheless, it is customary to assume in accordance with many experimental examples, that for each state of the ion there exists a separate ionization probability described by Equation [20] and that to a first approximation these probabilities are additive (see **Figure 9**). Finally, it is interesting to note that the shape of the threshold law is especially important for the determination of the ionization and appearance energies, because owing to the low ion signal close to the onset it is necessary to invoke an extrapolation method and knowledge of the threshold behaviour facilitates this procedure.

Partial and total ionization cross sections

As discussed above, there exist different threshold laws for EI and PI. This difference pertains also for higher energies. Whereas in the case of EI the

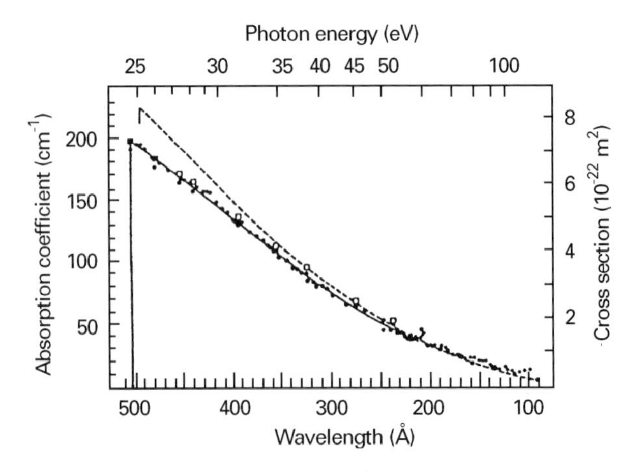

Figure 11 Experimental and theoretical (dashed line) photoionization cross sections of helium.

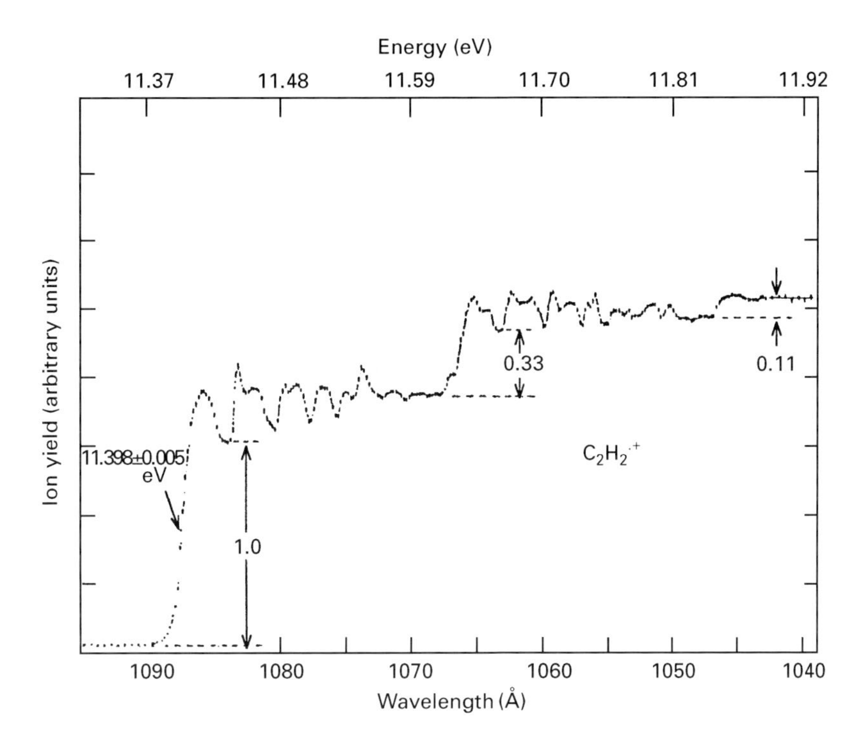

Figure 12 Mass-selected photoion yield curve of $C_2H_2^{\cdot+}$ in the vicinity of the ionization threshold. Three steps are observed owing to direct ionization from the vibrational ground state of the neutral to $v' = 0$, 1 and 2 of the C–C stretch of the ion. Autoionization structure is superimposed upon the steplike vibrational structure. Reproduced with permission of Academic Press from Berkowitz J (1979) *Photoabsorption, Photoionization, and Photoelectron Spectroscopy*, Chapter VI.

ionization cross-section function (**Figure 10**) rises more or less linearly over some tens of eV, reaches a maximum at an energy of about 10 eV and then decreases monotonically over some thousands of eV (a logarithmic decrease is predicted in the limit of the Born–Bethe approximation), in the case of PI the abrupt rise at threshold is followed by a monotonic, more or less exponential decline to higher energies (**Figure 11**).

In the limit of zero excess photon energy above threshold, the details of the absorption probability characteristic of the individual targets do not enter and the cross-section function is only dependent on the long-range interaction of the departing particles (electron and positive ion). The Coulomb force then leads to the abrupt onset to a finite crosssection value with a post-threshold behaviour independent of excess photon energy. But even for atomic targets this step function description is inappropriate (see as an example helium in **Figure 11**, which has been studied both experimentally and theoretically). For the heavier noble gases, autoionization fine structures due to spin–orbit interaction immediately set in beyond the threshold and thus prevent examination of the isolated continuum (see above and **Figure 2**). For more complicated targets, such as molecular species, additional effects thwart the observation and analysis of the unperturbed continuum. Even the abrupt rise at threshold is not always apparent and therefore step-function behaviour, while a convenient idealization, has little correspondence with reality. A nice example where the step-function behaviour does occur is in the case of the vibrational staircase behaviour of acetylene PI (**Figure 12**).

Inelastic collisions of the electrons with atomic targets are usually divided into two regimes, fast collisions and slow collisions, depending on the

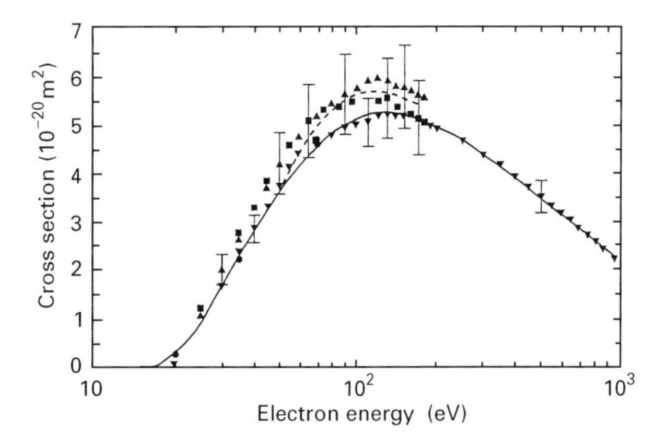

Figure 13 Experimental (solid symbols) and theoretical (dashed line, Deutsch-Maerk formula; full line, Binary–encounter–Bethe approximation) total electron ionization cross section for CF_4.

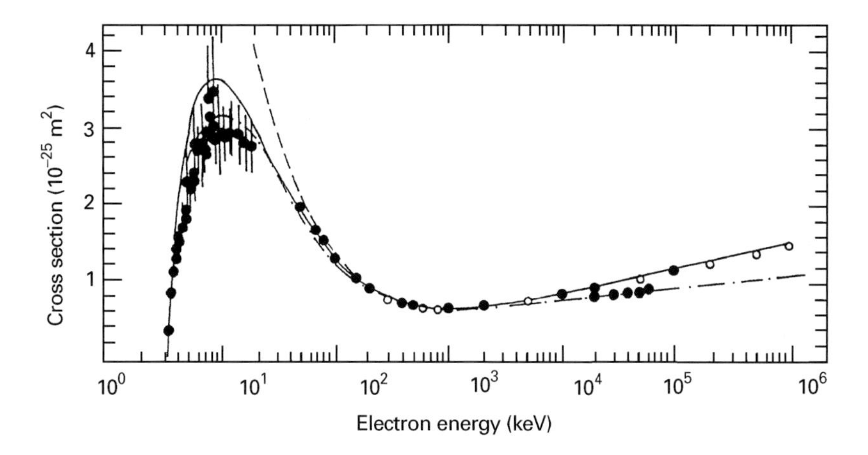

Figure 14 Experimental (solid circles and full line) and theoretical K-shell electron ionization cross section for argon: dashed line, Born–Bethe approximation; dash-dotted line, Deutsch–Maerk formula; open circles, Born–Bethe including relativistic corrections.

relation between the impacting electron velocity to the mean orbital velocity of the electron(s) in the (sub)shell responsible for the inelastic process under consideration. For fast collisions, the influence of the incoming electron upon the target can be viewed as a sudden and small external perturbation. This concept leads to a cross-section function derived in the Born–Bethe approximation, where the ionization cross section is given in good accordance with experimental determinations asymptotically by the simple expression

$$\sigma = (4\pi a_0{}^2/u)\, M_{\rm i}{}^2 \ln(4cu) \qquad [22]$$

with a_0 the Bohr radius, $u = E/R$ the reduced electron energy, R the ionization energy of H, $M_i^2 = \int(\mathrm{d}f/\mathrm{d}E(1/u)\,\mathrm{d}E$ with $\mathrm{d}f/\mathrm{d}E$ the differential oscillator strength and c a collision parameter. It is possible to extract M_i^2 from the slope of a plot of σu versus $\ln(u)$ (Fano-plot).

Conversely, for slow collisions the combined system of incoming electron and target molecule has to be considered, leading in the exit channel to a full three-body problem. Quantum-mechanical (approximate) calculations are difficult and have been carried out only for a few selected examples. Therefore, other methods have been developed with the goal of obtaining reasonably accurate cross sections using either classical or semiclassical theories and by devising semiempirical formulae. Some of these concepts are based on the Born–Bethe formula [22] and on the observation that the ejection of an atomic electron with quantum numbers (n,l) is approximately proportional to the mean-square radius of the electron shell (n,l). This leads also to proposed correlations of the ionization cross section with polarizability, diamagnetic susceptibility, number of electrons, and the additivity of atomic cross sections to give molecular cross sections. Moreover, there exist simple empirical relations such as between the energy position $u_{\rm max}$ of the maximum cross section and the ionization energy $\sigma_{\rm max}u_{\rm max} \approx (R/\mathrm{IE})^2$, which is the high-energy limit of the classical Thomson formula.

A particularly useful, simple and versatile formula is the semi-classical Deutsch–Maerk (DM) approach giving the ionization cross section for single ionization σ_i in terms of $\xi_{n,l}$ (the number of equivalent electrons in the (n,l) subshell), $r_{n,l}$ (the electron radius of the (n,l) shell), and some weighting factors $g_{n,l}$. It has been demonstrated recently that this formula may be applied not only to calculate ionization cross sections for ground-state atoms (**Figure 10**), but also for the calculation of cross sections for excited atomic species, for molecules (**Figure 13**), for radicals, for inner-shell ionization (**Figure 14**), for ionization of ions and even clusters, thereby spanning a range of cross-section values from about 10^{-34} m^2 for 14-fold ionization of Si all the way up to about 5×10^{-19} m^2 for C_{60} and in the case of inner-shell ionization energies from threshold up to 10^9 eV.

List of symbols

a_0 = Bohr radius; $\mathrm{d}f/\mathrm{d}E$ = differential oscillator strength; E = electron collision energy; i = ionization current (t, total; ms, mass selected); IE = ionization energy; $N_{\rm t}$ = number of target particles cm^{-3}; $n(0)_{\rm e}$ = initial electron intensity [cm^{-2} s^{-1}]; u (= E/R) = reduced electron energy [R = H ionization energy]; W = electron energy after ionization collision; σ = ionization cross section (t, total; c, counting; zi for ion i of charge ze); δ = polar angle of outgoing electron.

See also: **Chemical Ionization in Mass Spectrometry; Fragmentation in Mass Spectrometry; Metastable Ions; Photoelectron Spectroscopy; Photoionization and Photodissociation Methods in Mass Spectrometry.**

Further reading

Berkowitz J (1979) *Photoabsorption, Photoionization and Photoelectron Spectroscopy*. New York: Academic Press.
Beckey HD (1977) *Principles of Field Ionization and Field Desorption Mass Spectrometry*. Oxford: Pergamon Press.
Christophorou LG (1984) *Electron–Molecule Interactions and Their Applications*. Orlando: Academic Press.
Christophorou LG, Illenberger E and Schmidt WF (1994) *Linking the Gaseous and Condensed Phases of Matter: The Behaviour of Slow Electrons*. New York: Plenum.
Duncan ABF (1971) *Rydberg Series in Atoms and Molecules*. New York: Academic Press.
Field FH and Franklin JL (1970) *Electron Impact Phenomena and the Properties of Gaseous Ions*. New York: Academic Press.
Futrell JH (1986) *Gaseous Ion Chemistry and Mass Spectrometry*. New York: Wiley-Interscience.
Hasted JB (1972) *Physics of Atomic Collisions*. London: Butterworths.
Illenberger E and Momigny J (1992) *Gaseous Molecular Ions*. Darmstadt: Steinkopff.
Levsen K (1978) *Fundamental Aspects of Organic Mass Spectrometry*. Weinheim: Verlag Chemie.
Maerk TD and Dunn GH (1985) *Electron Impact Ionization*. Wien: Springer
Marr GV (1967) *Photoionization Processes in Gases*. New York: Academic Press.
Massey HSW, Burhop EHS and Gilbody HB (1969) *Electronic and Ionic Impact Phenomena*. Oxford: Clarendon Press.
Rose ME and Johnstone RAW (1982) *Mass Spectrometry for Chemists and Biochemists*. London: Cambridge University Press.

IR and Raman Spectroscopy of Inorganic, Coordination and Organometallic Compounds

Claudio Pettinari and **Carlo Santini**, Università di Camerino, Italy

VIBRATIONAL, ROTATIONAL & RAMAN SPECTROSCOPIES
Applications

Infrared and Raman spectroscopy techniques are being employed to an ever-increasing extent for the recognition and the quantitative analysis of structural units in unknown compounds. The infrared region of the electromagnetic spectrum extends from the red end of the visible spectrum to the microwave region. It includes radiation at wavelengths between 14 000 and 20 cm^{-1}, but the spectral range used most is the mid-infrared one, which covers frequencies from 4000 to 200 cm^{-1}. Infrared spectrometry involves examination of the twisting, bending, rotating and vibrational motions of atoms in a molecule. Upon interaction with infrared radiation, portions of the incident radiation are absorbed at specific wavelengths. The multiplicity of vibrations occurring simultaneously produces a highly complex absorption spectrum that is uniquely characteristic of the functional groups that make up the molecule and of the overall configuration of the molecule as well.

When monochromatic radiation is scattered by molecules, a small fraction of the scattered radiation is observed to have a different frequency from that of the incident radiation; this is known as the *Raman effect*. Since its discovery in 1928, the Raman effect has been an important method for the elucidation of molecular structure, for locating various functional groups or chemical bonds in molecules and for the quantitative analysis of complex mixtures. Although vibrational Raman spectra are related to infrared absorption spectra, a Raman spectrum arises in a different manner and thus often provides complementary information. Vibrations that are active in Raman scattering may be inactive in infrared, and vice versa. A unique feature of Raman scattering is that each line has a characteristic polarization, and polarization data provide additional information about molecular structure.

Raman spectra can be used to study materials in aqueous solutions, a medium that transmits infrared radiation very poorly. Another advantage is the ability to examine the entire vibrational spectrum with one instrument, unlike infrared spectroscopy in which the far-infrared is usually scanned separately from the mid-infrared. For quantitative determinations, the Raman scattering power is directly proportional to concentration, whereas, in infrared spectroscopy, it is the logarithm of the ratio of incident to transmitted power, the absorbance, that is proportional to concentration. Furthermore, the application of infrared spectroscopy to the identification of inorganic and coordination compounds has been somewhat unsuccessful because many simple inorganic compounds such as the borides, nitrides and oxides do not exhibit absorption in the region between 4000 and 600 cm^{-1}, which for many years was the range of the infrared region covered by most commercially available spectrometers.

In addition, the Raman technique has proved to be particularly valuable in the study of single crystals where the infrared technique has greater limitations on sample size and geometry. Polarization data obtained from Raman spectra allow unambiguous classification of fundamentals and lattice modes into the various symmetry classes. Although Raman spectroscopy will never challenge X-ray diffraction as a tool for quantitative structural analysis, it is the preferred technique when qualitative information is sufficient because it is faster and less expensive.

Experimental aspects

The spectrum obtained for a given sample of inorganic, coordination or organometallic compounds depends upon the physical state of the sample. Gaseous samples usually exhibit a rotational fine structure which is damped in solution because collisions of molecules in the condensed phase occur before a rotation is completed. In addition to the difference in resolved fine structure, the number of absorption bands and the frequencies of the vibrations vary in the different states. Often there are more bands in the liquid state than in the gaseous state of a substance, and frequently new bands below 300 cm^{-1} caused by *lattice vibrations*, i.e. translational and torsional motions of the molecules in the lattice (*phonon modes*) appear in the solid state spectrum. The stronger intermolecular forces that exist in the solid and liquid states compared with those in the gaseous state are often the cause of slight shifts in the frequencies.

An additional complication arises if the unit cell of the crystal contains more than one chemically equivalent molecule. When this is the case, the vibrations in the individual molecules can couple with each other. This intermolecular coupling can give rise to frequency shifts and band splitting.

Sample handling also presents a number of problems in the infrared region. For example, there is no rugged window material for cuvettes that is transparent and also inert over this region. The alkali halides are widely used, particularly sodium chloride, which is transparent at wavelengths as long as 625 cm^{-1}. For frequencies less than 600 cm^{-1}, polyethylene cells are frequently used.

Samples that are liquid at room temperature are usually scanned in their neat form or in solution. Unfortunately, not all substances are soluble to a reasonable concentration in a solvent not absorbing in the regions of interest. In some cases use is made of solvents susceptible to hydrogen-bonding effects which can alter the vibrational frequency characteristic of the compound examined; the stronger the hydrogen bonding, the more the fundamental frequency is lowered.

Powders or solids can be examined as a thin paste or mull (mineral oil, nujol, hexachlorobutadiene, perfluorokerosene) or through the pellet technique which allows quantitative analyses to be performed since accurate measurements can be made of the ratio of weight of sample to internal standard in each disk. Modern technique using diamond ATR is removing the need for mulls and disks.

The use of laser excitation allows Raman spectroscopy to be performed on specimens in almost any state: liquid, solution, transparent solid, translucent solid, powder, pellet or gas. Water is a weak scatterer and therefore an excellent solvent for Raman work. This has important consequences in studies of biochemical interest and in the pharamaceutical industry. Water is a bad solvent in FT Raman because the Raman light is self absorbed.

Group frequencies and band assignments

The infrared spectrum of a compound is essentially the superposition of absorption bands of specific functional groups: even small interactions with the surrounding atoms of the molecule impose the stamp of individuality on the spectrum of each compound. For qualitative analysis, one of the best features of an infrared spectrum is that the absorption or the *lack of absorption* in specific frequency regions can be correlated with specific stretching and bending motions and, in some cases, with the relationship of these groups to the rest of the molecule.

In the near-infrared region (NIR), which meets the visible region at about 12 500 cm^{-1} and extends to about 4000 cm^{-1}, generally there are many absorption bands that result from harmonic overtones of fundamental and combination bands often associated with hydrogen atoms.

Many useful correlations have been found in the mid-infrared region. This region is divided into the 'group frequency' region, 4000–1300 cm^{-1}, and the fingerprint region, 1300–650 cm^{-1}. In the group frequency region the principal absorption bands are assigned to vibration units consisting of only two atoms of a molecule: in the interval from 4000 to 2500 cm^{-1}, the absorption is characteristic of hydrogen stretching vibrations with elements of mass 19 or less. The intermediate frequency range, 2500–1540 cm^{-1} (*unsaturated* region) contains triple bond frequencies which appear from 2500 to 2000 cm^{-1} and double bond frequencies from 2000 to 1540 cm^{-1}. In the region between 1300 and 650 cm^{-1} there are single bond stretching frequencies and bending vibrations (skeletal frequencies) of polyatomic systems that involve motions of bonds linking a substituent group to the remainder of the molecule.

The region 667–10 cm^{-1} contains the bending vibrations of carbon, nitrogen, oxygen and fluorine with atoms heavier than mass 19, and additional bending motions in cyclic or unsaturated systems. The low-frequency molecular vibrations in the far-infrared are particularly sensitive to changes in the overall structure of the molecule; thus the far-infrared bands often differ in a predictable manner for different isomeric forms of the same basic compound.

The far-infrared frequencies of organometallic compounds are often sensitive to the metal ion or atom, and this can be used advantageously in the study of coordination bonds. Moreover, this region is particularly well suited to the study of organometallic or inorganic compounds whose atoms are heavy and whose bonds are inclined to be weak.

In Raman spectroscopy those vibrations that originate in relatively nonpolar bonds with symmetrical charge distributions and that are symmetrical in nature produce the greatest polarizability changes and are the most intense. Vibrations from –C=C–, –C≡C–, –C≡N, –C=S, –C–S–, –S–S–, –N=N– and –S–H bonds are readily observed. Absorption bands due to the skeletal vibrations of finite chains and rings of saturated and unsaturated hydrocarbons are generally sharper in Raman than in infrared spectra. Aromatic compounds exhibit particularly strong spectra with a strong ring deformation mode at 1600 ± 30 cm^{-1}. Skeletal motions are very characteristic and highly useful for cyclic and aromatic rings, steroids and long chains of methylenes.

In solid samples, sharpening and intensifying of certain bands appear to be a function of crystallinity. For all molecules that have a centre of symmetry, a band allowed in the infrared is forbidden in the Raman, and vice versa. In molecules with symmetry elements other than a centre of symmetry, certain bands may be active in the Raman, infrared, both or neither. For a complex molecule that has no symmetry, all of the normal vibrational modes are allowed in both the infrared and Raman spectra. One other strong difference is the tendency for peaks in a Raman spectrum to have a greater range of intensities, from very weak to very strong. These factors, plus the contrasting relative intensities for a given group, are the main basis for making more confident assignment of chemical structure through the combination of infrared and Raman data. For example, in infrared spectra the 3300 cm^{-1} region is badly obscured by intense OH absorption, whereas the same region in Raman is very helpful in the assignment of NH and CH stretchings because the OH absorption is weak.

Applications

Selection rules

Molecular symmetry is very important in determining the infrared activity and degeneracy of a molecular vibration. When a molecule is present in a crystal, the symmetry of the surroundings of the molecule in the unit cell, the so called *site symmetry*, determines the selection rules. Often, bands forbidden in the gaseous state or in solution appear in the solid, and degenerate vibrations in the gaseous state are split in the solid. To show this effect, let us consider the infrared spectra of carbonate derivatives. The infrared (**Figure 1A**) and Raman spectra of $CaCO_3$ in calcite, where the carbonate ion is in a site of D_3 symmetry, contain the following bands (in cm^{-1}): ν_1, 1087 (R); ν_2, 879 (IR); ν_3, 11 432 (IR, R); ν_4, 710 (IR, R). The infrared spectra of $CaCO_3$ (**Figure 1B**) in aragonite, where the site symmetry for the carbonate is C_8 is different: ν_1 is infrared active and both ν_3 and ν_4 split into two bands.

In **Figure 2** some of the possible structures for a compound with an empirical formula NSF_3 are shown whereas the calculated number and symmetry of bands for the illustrated structures are listed in **Table 1**. In the infrared spectrum of this compound, six intense bands are found, in agreement with a C_{3v} structure but certainly not in conflict with other possible structures. It was found that the four bands ν_1, ν_2, ν_3 and ν_5, have *P*, *Q* and *R* branches indicating that the molecule investigated is a symmetric top molecule and supports a C_{3v} structure. The spectral data are contained in **Table 2**.

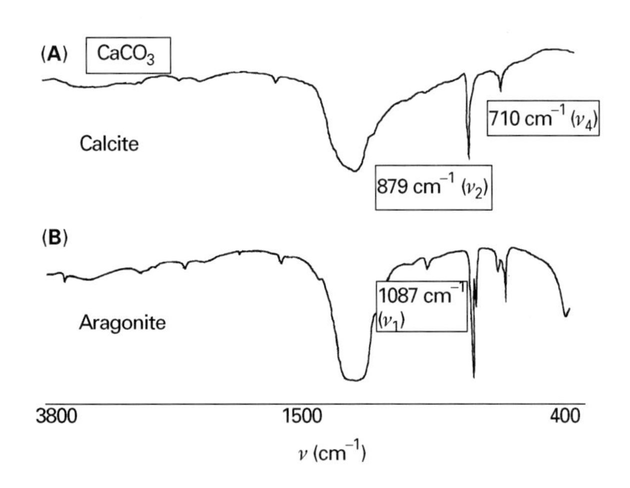

Figure 1 Infrared spectra of $CaCO_3$ in calcite and in aragonite.

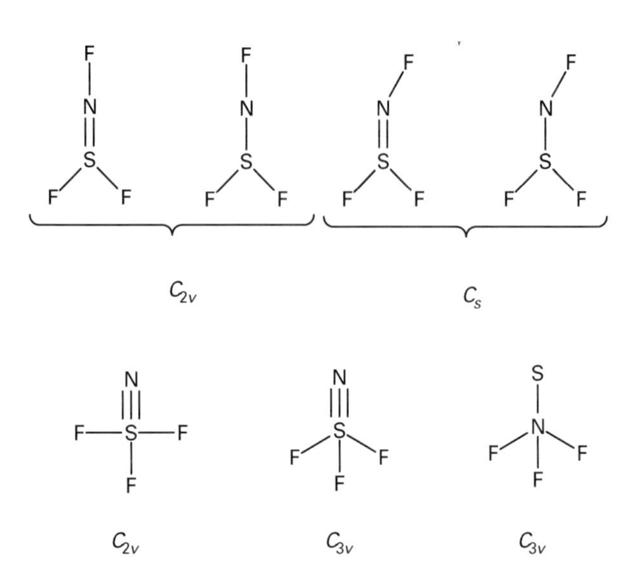

Figure 2 Possible structures for NSF_3.

Table 1 Calculated number of bands for some of the possible structures for NSF_3

C_{3v}	C_{2v}	C_s
$3A_1$ (IR active)	$4A_1$ (IR active)	$7A'$ (IR active)
	$3B_1$ (IR active)	$2A''$ (IR active)
$3E$ (IR active)	$2B_2$ (IR active)	

Table 2 Fundamental vibration frequencies (cm^{-1}) for NSF_3

ν_1	1515	ν_4	811
ν_2	775	ν_4	429
ν_3	521	ν_6	342

These data, combined with an NMR study, have been employed to support the C_{3v} structure $F_3S{-}N$. The F_3NS structure has not been eliminated by these data; however, the other observed bands support F_3SN: the S–N stretch is the only fundamental vibration that could be expected at a frequency of 1515 cm^{-1}. The NF stretching vibrations in NF_3 occur at 1031 cm^{-1} and would not be expected to be as high as 1515 cm^{-1} in $F_3N{-}S$. The F_3SN structure has been demonstrated on the basis of microwave and mass spectroscopic studies.

Change in the spectra of donor molecules upon coordination

In the infrared spectrum of *N,N*-dimethylacetamide in CCl_4, a strong absorption band at 1662 cm^{-1} has been found due to a highly coupled carbonyl absorption. The low frequency with respect to that found for acetone (1715 cm^{-1}) is attributed to a resonance interaction with the lone pair on the nitrogen (**Figure 3**).

A decrease in the frequency of this band has generally been found upon complexation of the molecule with Lewis acids. This has been attributed to coordination of the oxygen to the acid. Oxygen coordination produces a decrease in the carbonyl force constant by draining the π electron density out of the carbonyl group, which causes the observed decrease in the carbonyl frequency. The absence of any absorption in the carbonyl region on the high-frequency side of the uncomplexed carbonyl band further supports coordination of this molecule through the oxygen. On the other hand, in $Pd(NH_2CONH_2)_2Cl_2$, and $Pt(NH_2CONH_2)_2Cl_2$, coordination of the ligand NH_2CONH_2 through the nitrogen occurs, which results in a decrease of the C–N vibration frequency and in a higher frequency carbonyl absorption.

Analogously, a decrease in the S–O stretching frequency, indicative of oxygen coordination, is observed when dimethyl sulfoxide or tetramethylene sulfoxide is complexed to many metal ions, iodine and phenol, whereas the S–O stretching frequency increases in the palladium complex of dimethyl sulfoxide, compared to free sulfoxide, which suggests sulfur coordination.

Figure 3 Resonance interaction between the carbonyl and the lone pair on the nitrogen in *N,N′*-dimethylacetamide.

The infrared spectra of ethylenediaminetetraacetic acid metal complexes have been used to distinguish between tetradentate, pentadentate or hexadentate coordination of the ligand on the basis of the absorption bands in the carbonyl region corresponding to free and complexed carbonyl groups.

The change in C≡N stretching frequency of nitriles and metal cyanides resulting from their interaction with Lewis acids has also attracted considerable interest. When such molecules are coordinated to Lewis acids not generally involved in π back-bonding, the C≡N stretching frequency increases. A combined molecular orbital and normal coordinate analysis of acetonitrile and some of its adducts indicates that a slight increase is to be expected from this effect, but that the principal contribution to the observed shift arises from an increase in the C≡N force constant. This increase is mainly due to an increase in the C≡N σ bond strength from nitrogen rehybridization. In those systems where there is extensive π back-bonding from the acid into the π^* orbitals of the nitrile group, the decreased π bond energy accounts for the decreased frequency.

Change in symmetry upon coordination

Some of the most useful applications of infrared spectroscopy in the area of coordination and organometallic chemistry originate from the change in the symmetry of a ligand upon coordination. For example, when small molecules (e.g. N_2, O_2 and H_2) are linked to transition metal ions a symmetry change occurs which has a strong influence on the infrared spectra. The stretching vibration of free N_2 is infrared inactive but Raman active (2331 cm^{-1}). If the azide ion is added to $Ru(NH_3)_5H_2O^{3+}$, $Ru(NH_3)_5N_2^{2+}$ is obtained, which exhibits in infrared a sharp absorption due to ν(N–N) at 2130 cm^{-1}.

Infrared spectroscopy is a very effective tool for determining the nature of the bonding of the sulfate group in its derivatives. The sulfate ion (T_d symmetry) has two infrared bands in the 1200–600 cm^{-1} region, one assigned to ν_3 at 1104 cm^{-1} and one to ν_4 at 613 cm^{-1}. In the complex $[Co(NH_3)_5OSO_3]Br$ the coordinated sulfate group has lower symmetry, C_{3v}, with six new bands at 970 (ν_1), 438 (ν_2), 1032 to 1044, and 1117 to 1143 (from ν_3), and 645 and 604 cm^{-1} (from ν_4). In a bridged sulfate group, the symmetry is lowered to C_{2v} and even more bands appear: the ν_3 band is split into three peaks and the ν_4 band into three other peaks.

Five-coordinate addition compounds such as $(CH_3)_3SnCl \cdot (CH_3)_2SO$ had a structure in which the three methyl groups were in the equatorial positions: in fact, the symmetric Sn–C stretch present in

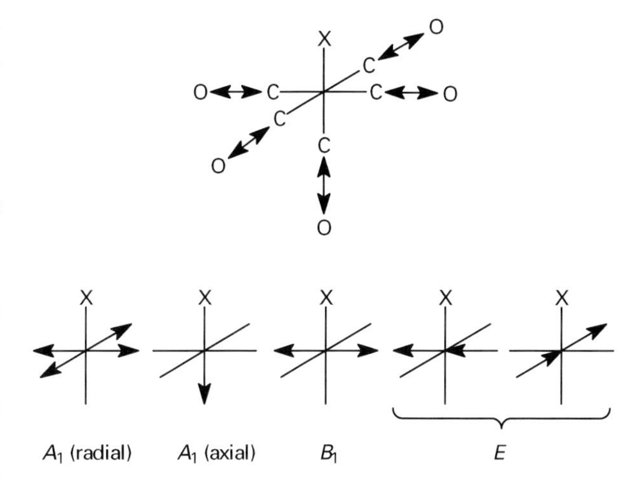

Figure 4 Carbonyl stretching vibrations of $M(^{12}CO)_5X$ derivatives.

$(CH_3)_3SnCl$ at 545 cm^{-1} disappeared in the addition compound where a single Sn–C vibration due to the asymmetric Sn–C stretch was detected at 551 cm^{-1}. This is in accordance with an isomer with three methyl groups in the equatorial position: in fact, a small dipole moment change is associated with the symmetric stretch in compounds having this structure. All the other possible structures possess at least two different Sn–C vibrations.

Also, the carbonyl stretching vibrations of $M(CO)_5$ X (and their ^{13}C-substituted) derivatives can be employed to study change in symmetry upon coordination. The operations or the C_{4v} point group for all-^{12}CO compounds such as those in **Figure 4** produce 2A_1 (one radial and one axial), B_1 (radial) and E (radial) vibrations, three of which are infrared active and four Raman active.

If the molecule has an axial ^{13}CO, the symmetry is still C_{4v} and the B_1 and E modes (which have no contribution from the axial group) will be unaffected by the mass change. The two A_1 modes having the same symmetry will mix, so isotopic substitution could affect both; but it will have a major effect on A_1 (axial). If a ^{13}CO is located in the radial position, the symmetry is lowered to C_s, leading to the possible vibrations shown in **Figure 5**.

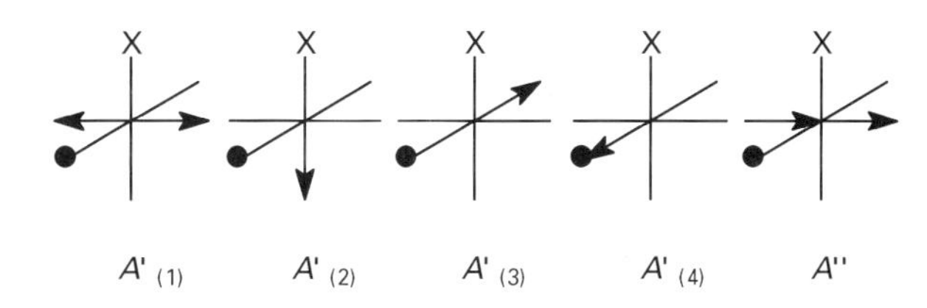

Figure 5 Possible vibrations for a $M(^{12}CO)_4(^{13}CO)X$ derivative, containing the ^{13}CO in the radial position.

Only the A″ will be unaffected by the mass change, and it resembles the E mode in the all-^{12}CO molecule. All of the other four modes can mix and could be shifted by distributing the mass effect over all four modes. This would shift the vibrations to lower frequencies than those in the all-^{12}CO molecule. Several normal coordinate analyses on such systems, employing a large number of isotopes, have been made which indicates that as the formal positive charge on the central atom increases in forming $M(CO)_5Br$ from $M(CO)_6$, the σ-bonding increases and the extent of π back-bonding decreases.

Metal-isotope spectroscopy

The metal–ligand vibrations appear in the low-frequency region (600–100 cm^{-1}) and provide direct information about the structure of the coordination sphere and the nature of the metal–ligand bond. It is difficult to make unequivocal assignments of metal–ligand vibrations since the interpretation of the low-frequency spectrum is complicated by the appearance of ligand vibrations, as well as lattice vibrations in the case of solid-state spectra. However, different methods can be employed: for example, the metal–ligand vibrations are absent in the spectrum of the free donor (it should be taken into account that in some cases new ligand vibrations, activated by complex formation, can appear).

Metal–ligand vibrations are also metal sensitive and are shifted by changing the metal or its oxidation state: this method is applicable only to isostructural metal complexes and it does not provide definitive assignments since some ligand vibrations (for example the chelate ring deformations) are also metal sensitive.

If the ligand is isotopically substituted, the metal–ligand vibrations present an isotope shift: the ν(Ni–N) in $[Ni(NH_3)_6]Cl_2$ at 334 cm^{-1} is shifted to 318 cm^{-1} upon deuteration of the ammonia groups. It is possible, with this method, to assign the metal–ligand vibrations of chelate compounds such as oxamido ($^{14}N/^{15}N$) and acetylacetonato ($^{16}O/^{17}O$) complexes. However isotopic substitution of the α-atom (an atom directly bonded to the metal) causes shifts not only of metal–ligand vibrations but also of ligand vibrations involving the motion of the α-atom. Finally, the frequency of a metal–ligand vibration may also be predicted if the metal–ligand stretching and other force constants are known as a priori.

The 'metal-isotope technique' is the best one for obtaining reliable metal–ligand assignments. Isotope pairs such as (H/D) and ($^{16}O/^{17}O$) had been used in the past routinely by many spectroscopists. Isotopic pairs of heavy metals such as ($^{58}Ni/^{62}Ni$) and ($^{104}Pd/^{110}Pd$) were not employed until 1969 when the first report on the assignments of the Ni–P vibrations of *trans*-$Ni(PEt_3)_2X_2$ (X = Cl and Br) was made. The delay in their use was probably due to the high cost of the pure metal isotopes.

The magnitudes of metal isotope shifts are generally of the order of 2–10 cm^{-1} for stretching modes and 0–2 cm^{-1} for bending modes, whereas the experimental error in measuring the frequency could be as small as ±0.2 cm^{-1}.

In highly symmetrical molecules (T_d, O_h, etc.), the central atom does not move during the totally symmetric vibration and no metal-isotope shifts are expected. When the central atom is coordinated by several donor atoms, multiple isotope labelling is necessary to distinguish different coordinate bond-stretching vibrations. For example, complete assignments of bis(β-diketonato)nickel(II) derivatives require $^{16}O/^{18}O$ as well as $^{58}Ni/^{62}Ni$ isotope shift data. The metal-isotope technique is often indispensable not only in assigning the metal–ligand vibrations but also in refining metal–ligand stretching force constants in normal coordinate analysis. The presence of vibrational coupling between metal–ligand and other vibrations can also be detected by combining metal-isotope data with normal coordinate calculations, since both experimental and theoretical isotope shift values will be smaller when such couplings occur.

The metal-isotope technique is very important in biological molecules such as haem proteins: structural and bonding information about the active site (iron porphyrin) can be obtained through definitive assignments of coordinate bond-stretching vibrations around the iron atom. Using resonance Raman techniques, it is possible to observe iron porphyrin and iron–axial ligand vibrations without interference from peptide chain vibrations. Thus, these vibrations can be assigned by comparing a resonance Raman spectrum of a natural haem protein with that of a ^{54}Fe-reconstituted haem protein.

Inorganic compounds

This and the following paragraphs are devoted to a general resumé of the more common and important spectroscopic and structural information on inorganic compounds: the assignment of the fundamental vibrations are based on point group symmetry.

Diatomic molecules have only one vibration along the chemical bond: in homopolar AA($D_{\infty h}$) molecules, this vibration is Raman active but not infrared active, whereas the stretching vibration is both Raman and infrared active in heteropolar AB($D_{\infty v}$) molecules.

Table 3 Observed frequencies of diatomic molecules (cm^{-1})

Molecule	ν	*Molecule*	ν
N_2	2360	HF	4138
CO	2138	HCl	2991
NO	1880	HBr	2650
O_2^+	1865	HI	2309
O_2	1580	AlH	1593[a]
O_2^-	1097	$[OH]^-$	3637[b]
O_2^{2-}	766	$[CN]^-$	2080[c]

[a] Matrix.
[b] Solid state.
[c] Solution.

Table 3 lists some relevant observed frequencies for diatomic molecules, ions and radicals: the effect of the charge on the frequency in the series of the dioxygen species is worthy of note and also the vibrational frequencies characteristic of hydrogen halides which generally polymerize in the condensed phases. The HX stretching bands are shifted markedly to lower frequencies upon polymerization.

IR and Raman spectra can be properly used to select between bent and linear structures in triatomic inorganic molecules (**Table 4**): the linear $X_3(D_{\infty h})$ and XAX ($D_{\infty h}$) type molecules have three normal modes (**Figure 6**), ν_1, Raman active but not infrared active, and ν_2 and ν_3, infrared active but not Raman active. In the case of bent X_3 (C_{2v}) and XAX (C_{2v}) type molecules all three vibrations are infrared and Raman active.

Pyramidal XY_3 (C_{3v}) molecules have four normal modes of vibration (**Figure 7**) (**Table 5**), all infrared and Raman active, which have been used, for example, to confirm the presence of the hydronium (H_3O^+) ion in hydrated acids. Substitution of one of the Y atoms of a pyramidal XY_3 molecule by a Z atom lowers the symmetry from C_{3v} to C_s. Then the degenerate vibrations split into two bands [$\nu_d(XY) \rightarrow \nu_s(XY) + \nu_a(XY)$; $\delta_d(YXY) \rightarrow \delta_s(YXY) + \delta_a(YXZ)$] ($\delta$ = bending) and all six vibrations become infrared and Raman active (**Table 5**).

Table 4 Vibrational frequencies of triatomic molecules (cm^{-1})

Compound	*Structure*	ν_1	ν_2	ν_3
$BeCl_2$	Linear	(680)	345	1555
$MgCl_2$	Linear	327	93	601
$SnCl_2$	Bent	354	(120)	334
$[CuCl_2]^-$	Linear	300	109	405
$BaCl_2$	Bent	225	–	260
O_3	Bent	1135	716	1089
$S_{3(g)}$	Bent	585	490/310	651
$ClOCl_{(s)}$	Bent	631	296	671
SCS	Linear	658	397	1533
$O^{13}CO$	Linear	–	(649)	2284
NNO	Linear	2224	589	1285

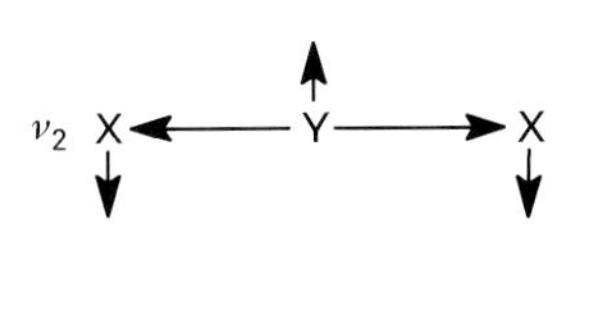

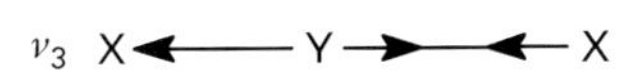

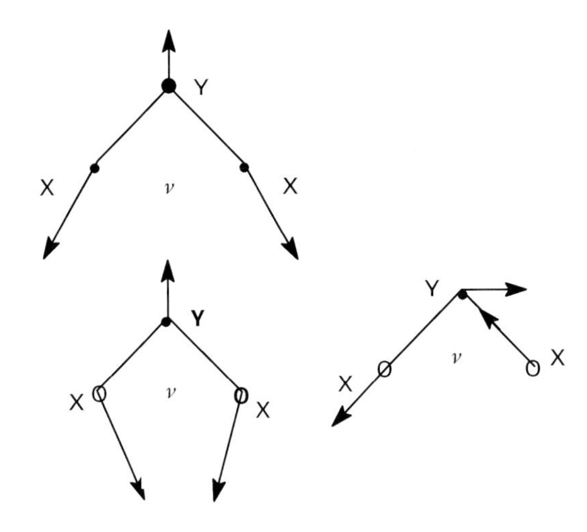

Figure 6 Normal modes of vibration for triatomic inorganic molecules.

The four normal modes of vibration of planar XY_3 (D_{3h}) molecules are shown in **Figure 8**, whereas significant vibrational frequencies of planar XY_3 molecules are listed in **Table 5**. Pyramidal and planar

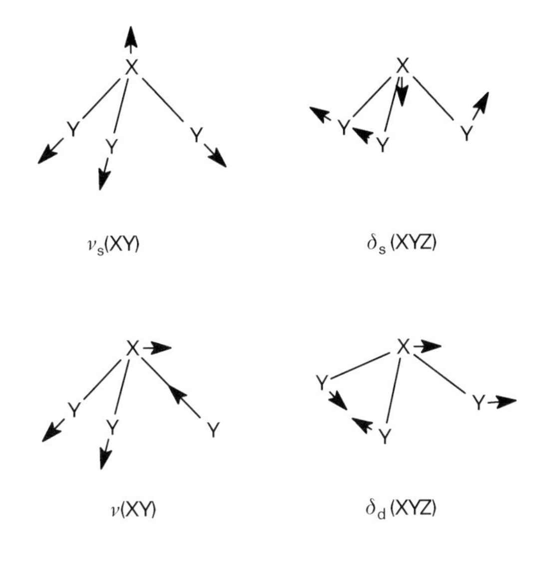

Figure 7 Normal modes of vibration for a pyramidal XY_3 molecule.

Table 5 Vibrational frequencies of tetratomic molecules (cm^{-1})

Compound	*Structure*	*State*	ν_1	ν_2	ν_3	ν_4
NH_3	Pyramidal	Solid	3223	1060	3378	1646
$[OH_3]^+$	Pyramidal	Solution	3560	1095	3510	1600
PH_3	Pyramidal	Gas	2327	990	2421	1121
NF_3	Planar	Gas	1035	649	910	500
$[SnCl_3]^-$	Planar	Solution	297	128	256	103
$BiCl_3$	Planar	Gas	342	123	322	107
$BiBr_3$	Planar	Gas	220	77	214	63
BiI_3	Planar	Solid	146	90	115	71

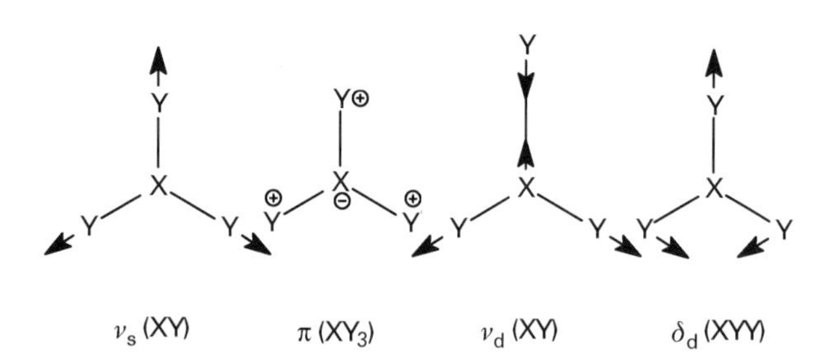

Figure 8 Normal modes of vibration for a planar XY_3 molecule.

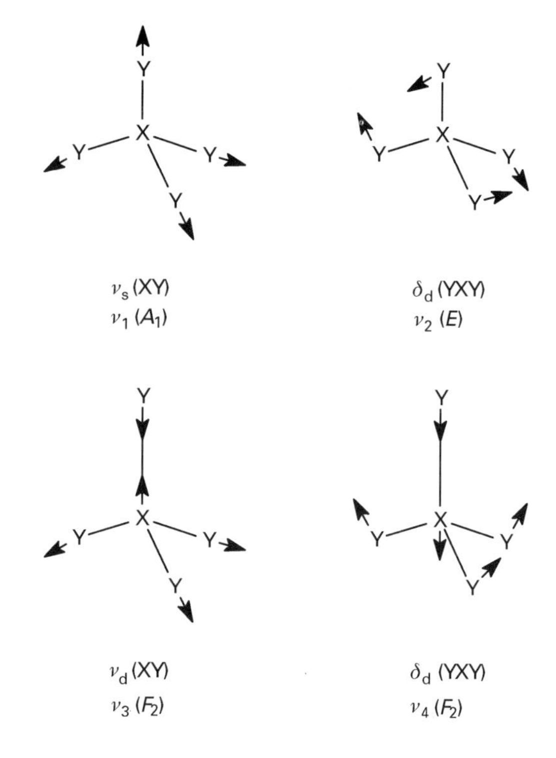

Figure 9 Normal modes of vibration for a tetrahedral XY_4 molecule.

structures can be distinguished easily on the basis of the difference in selection rules.

Tetrahedral XY_4 molecules have four normal modes of vibration (**Figure 9**) which are all Raman active, whereas only ν_3 and ν_4 are infrared active. In XH_4-type molecules (**Table 6**) the following trend is observed: $\nu_3 > \nu_1$ and $\nu_2 > \nu_4$. In MX_4 derivatives the MX stretching frequency generally decreases on going from F to I. The average values of ν(MBr)/ν(MCl) and ν(MI)/ν(MCl) calculated from selected compounds are 0.76 and 0.62, respectively, for ν_3, and 0.61 and 0.42, respectively, for ν_1. These values are very useful in the assignment of the MX stretching bands in halogeno complexes.

The effect of changing the oxidation state on the MX stretching frequency has been extensively studied with $[FeX_4]^-$ and $[FeX_4]^{2-}$ (X = Cl and Br) species: it has been found that the MX stretching frequency increases as the oxidation state of the metal increases. In tetrahedral MO_4^-, MS_4^- and MSe_4^- compounds, the ν_1/ν_3 ratio increases as the negative charge of the anion increases; for anions having the same negative charge and the central atom belonging to the same group of the periodic table, the ν_1/ν_3 ratio increases with increasing mass of the central atom, and finally the ν_1/ν_3 ratio increases with increasing negative charge of the anion in isoelectronic ions, with the mass of the central atom approximately constant.

Table 6 Vibrational frequencies of tetrahedral compounds (cm^{-1})

Compound	ν_1	ν_2	ν_3	ν_4	ν_5	ν_6
CF_4	908	434	1283	631		
$SiCl_4$	423	145	616	220		
$[MnCl_4]^{2-}$	256	–	278, 301	120		
$[BrO_4]^-$	801	331	878	410		
$FSiCl_3$	465	948	239	640	282	167
$FCCl_3$	538	1080	351	848	243	395
ONF_3	743	1691	528	883	558	400
$[IF_4]^{+a}$	728	614	345	263	–	–

[a] Distorted, ν_7: 388 ; ν_8: 719; ν_9: 319

Figure 10 shows the seven normal modes of vibration of square-planar XY_4 molecules. Vibra-tions ν_3, ν_6 and ν_7 are infrared active, whereas ν_1, ν_2 and ν_4 are Raman active. Some vibrational frequencies for selected square-planar XY_4 molecules are reported in **Table 7**.

XY_5 molecules may be trigonal bipyramidal (D_{3h}) or tetragonal pyramidal (C_{4v}). Trigonal bipyramidal have eight normal modes of vibration, six of which (A_1', E' and E'') are Raman active and five (A_2'' and E') are infrared active. Normal coordinate studies have demonstrated that in neutral trigonal bipyramidal MX_5 compounds, the equatorial bonds are stronger than axial ones.

In tetragonal XY_5 molecules, the axial stretching frequency (ν_1) is generally higher than equatorial

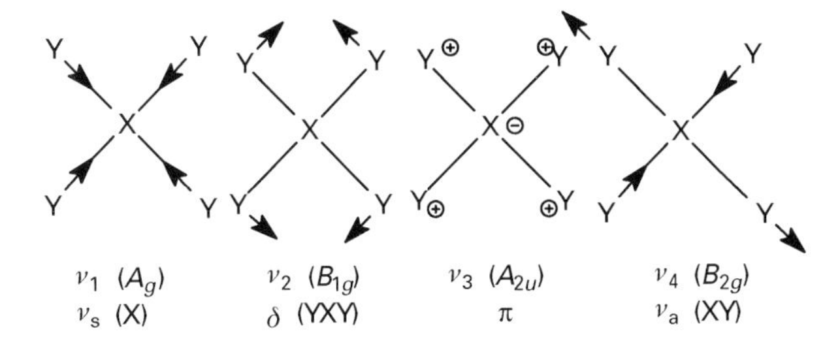

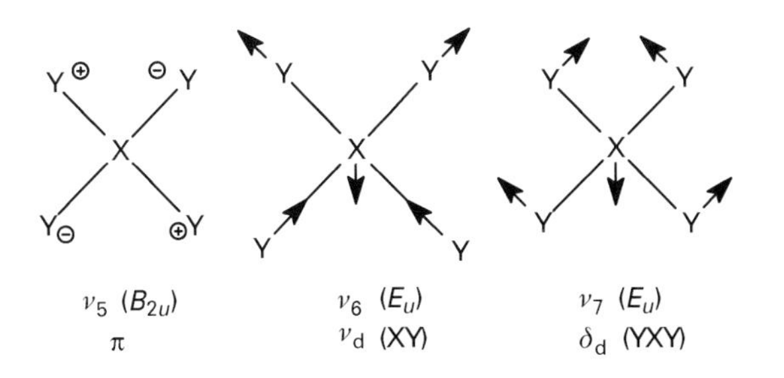

Figure 10 Normal modes of vibration for a square planar XY_4 molecule.

stretching frequencies (ν_2, ν_4 and ν_7). This is the opposite of what is found for trigonal bipyramidal XY_5 molecules.

Figure 11 shows the six normal modes of vibration of an octahedral XY_6 molecule. Vibrations ν_1, ν_2 and ν_5 are Raman active, whereas only ν_3 and ν_4 are infrared active; ν_6 is inactive in both, and its frequency is estimated from an analysis of combination

Table 7 Vibrational frequencies of square planar XY_4 molecules (cm^{-1})

Compound	ν_1	ν_2	ν_3	ν_4	ν_6	ν_7
XeF_4	554	218	291	524	586	(161)
$[AuCl_4]^-$	347	171	–	324	350	179
$[PtCl_4]^-$	330	171	147	312	313	165
ClF_4	505	288	425	417	680–500	–

and overtone bands. Vibrational frequencies of several octahedral XY_6 molecules are reported in **Table 8**. The order of the stretching frequencies can be $\nu_1 > \nu_3 >> \nu_2$ or $\nu_1 < \nu_3 >> \nu_2$, depending on the compound.

The order of the bending frequencies is $\nu_4 > \nu_5 > \nu_6$ in most cases. In the same group of the periodic table, the stretching frequencies decrease as the mass of the central atom increases.

Several XY_6-type ions show splitting of degenerate vibrations due to lowering of symmetry in the crystalline state, an effect which has been attributed to a static Jahn–Teller effect.

Characteristic infrared frequencies and their assignment for some other important anions are summarized in **Table 9**.

In some cases where there is high point group symmetry, vibrations normally infrared inactive often appear as a weak band and doubly and triply degenerate vibrations can split into two and three compo-

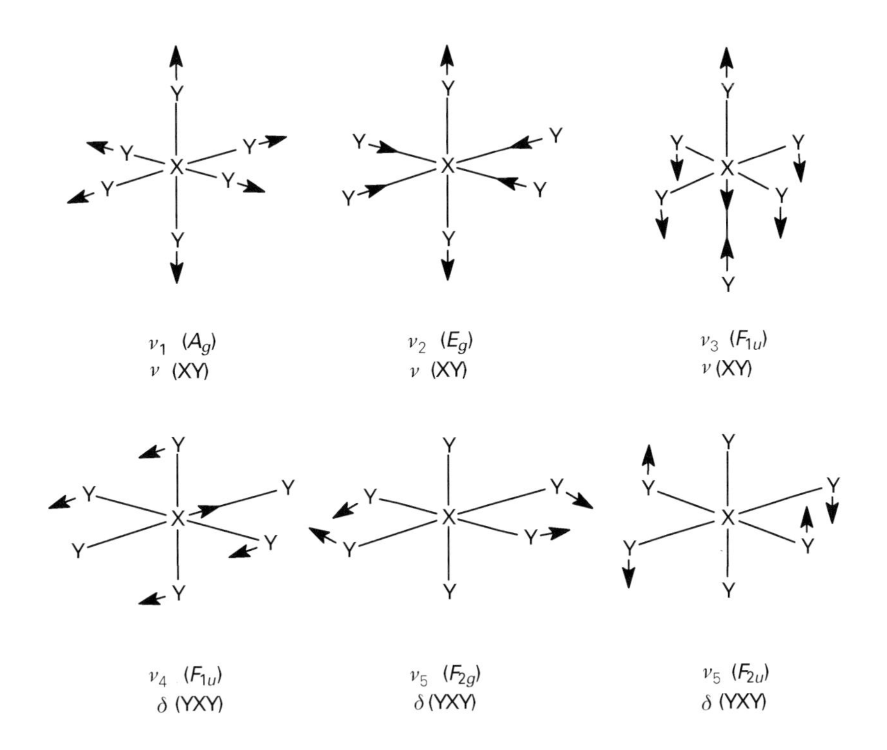

Figure 11 Normal modes of vibration for an octahedral XY_6 molecule.

Table 8 Vibrational frequencies of octahedral XY_6 molecules (cm^{-1})

Compound	ν_1	ν_2	ν_3	ν_4	ν_5	ν_6
SF_6	775	643	939	614	534	(347)
$[BiCl_6]^{3-}$	259	215	172	130	115	–
$[SnI_6]^{2-}$	311	229	303	166	158	–
$[TiCl_6]^{2-}$	320	271	316	183	173	–
$[BrF_6]^+$	658	660	–	–	405	–

Table 9 Fundamental vibrations of selected inorganic ions (cm^{-1})

Formula	*Ion*	*Point group*	*Vibrations*
AlH_4^-	Tetrahydroaluminate	T_d	1790 (ν_1), 799 (ν_2), 1740 (ν_3), 764 (ν_4)
CN^-	Cyanide		2080–2239 ($\nu_{C\equiv N}$)
SCN^-	Thiocyanate	$C_{\infty v}$	1293 (ν_1), 470 (ν_2), 2066 (ν_3)
CO_3^{2-}	Carbonate	D_{3h}	1087 (ν_1), 874 (ν_2), 1432 (ν_3), 706 (ν_4)
PO_4^{3-}	Orthophosphate	T_d	935 (ν_1), 420 (ν_2), 1080 (ν_3), 550 (ν_4)
MoO_4^{2-}	Molybdate(III)	T_d	940 (ν_1), 220 (ν_2), 895 (ν_3), 365 (ν_4)
$SnCl_6^{2-}$	Hexachlorostannate	O_d	311 (ν_1), 229 (ν_2), 158 (ν_3)
IO_3^-	Iodate	C_{4v}	779 (ν_1), 390 (ν_2), 826 (ν_3), 330 (ν_4)
ClO_4^-	Perchlorate	T_d	935 (ν_1), 460 (ν_2), 1050–1170 (ν_3), 630 (ν_4)
$PtCl_4^{2-}$	Tetrachloroplatinate	D_{4h}	335 (ν_1), 164 (ν_2), 304 (ν_4)
NO_2^-	Nitrite	$C_{\omega h}$	1320–1365 (ν_1), 807–818 (ν_2), 1221–1251 (ν_3)
NO_3^-	Nitrate	D_{3h}	1018–1050 (ν_1), 807–850 (ν_2), 1310–1405 (ν_3), 697–716 (ν_4)
MnO_4^-	Permanganate	T_d	840 (ν_1), 340–350 (ν_2), 900 (ν_3), 387 (ν_4)

nents, respectively. These effects derive either from lowering of the point group symmetry or from factor group splitting as a result of different crystalline environments.

Coordination compounds

Ammine complexes

The NH_3 stretching frequencies in hexammine complexes (**Figure 12**) are generally lower than those in the free NH_3 molecule due to the fact that upon coordination the N–H bond is weakened: the stronger the M–N bond, the weaker is the N–H bond and the lower are the NH_3 stretching frequencies. For this

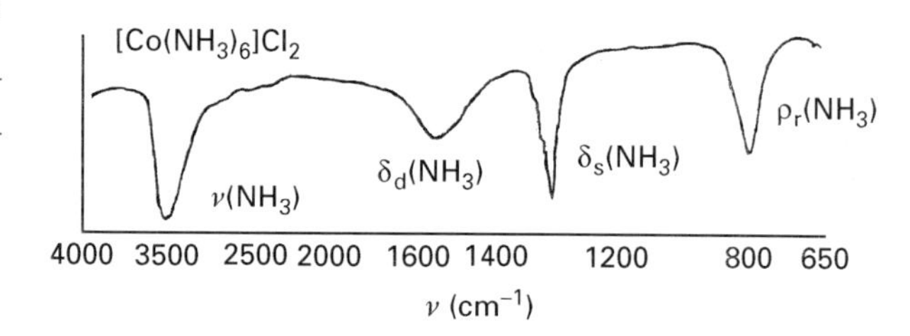

Figure 12 Infrared spectrum of the hexammine cobalt complex $[Co(NH_3)_6]Cl_2$.

reason, the NH_3 stretching frequencies may be used as a rough measure of the M–N bond strength. The effect of the counterion should be considered: the NH_3 stretching frequencies in chloride derivatives are much lower than those in the perchlorate ones and the weakening of the N–H bond is due to the formation of the N–H$\cdots$Cl-type hydrogen bond in the former. The main effect of coordination and hydrogen bonding is the shift to higher frequencies of the NH_3 deformation and rocking modes. The NH_3 rocking mode is most sensitive whereas the degenerate deformation is least sensitive to these effects.

Nitro, nitrito and nitrato complexes

The NO_2^- ion is able to coordinate metals in a variety of ways (**Figure 13**), and vibrational spectroscopy is very useful in distinguishing the possible structures.

If the NO_2 group is bonded to a metal through its N atom (nitro form) it exhibits $\nu_a(NO_2)$ and $\nu_s(NO_2)$ in the 1470–1370 and 1340–1320 cm^{-1} regions, respectively. The free NO_2^- ion exhibits the same modes at 1250 and 1335 cm^{-1}, respectively; thus $\nu_a(NO_2)$ shifts markedly to a higher frequency, whereas $\nu_s(NO_2)$ changes very little upon coordination.

On the other hand, if the NO_2 group is bonded to a metal through one of its O atoms (nitrito form) it shows two $\nu(NO_2)$ well separated at 1485–1400 cm^{-1}, ν(N=O), and 1110–1050 cm^{-1}, ν(NO), respectively. The nitrito complexes lack the wagging modes near 620 cm^{-1} which appear in all nitro derivatives and exhibit ν(MO) (M = Cr(III), Rh(III) and Ir(III)) in the 360–340 cm^{-1} region.

If the NO_2 group is O_2-chelating, both antisymmetric and symmetric NO_2 stretching frequencies are lower and the ONO bending frequency is higher than that for unidentate N-bonded nitro complexes. In this case, $\nu_a(NO_2)$ depends on the degree of asymmetry of the coordinated nitro group: it is lowest when two N–O bonds are equivalent and becomes higher as the degree of asymmetry increases.

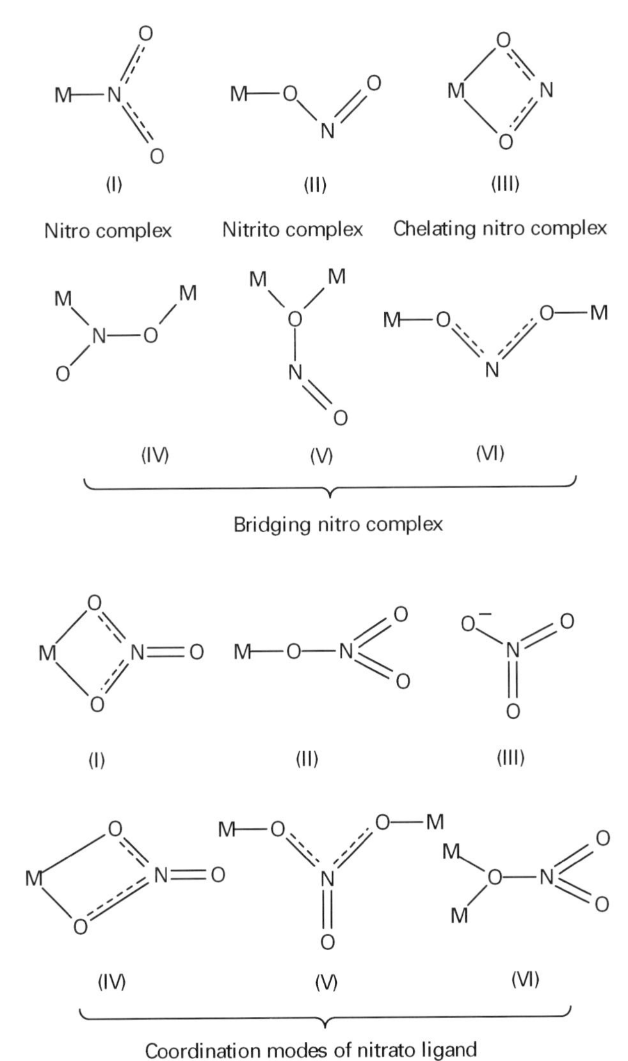

Figure 13 Coordination modes of the NO_2^- and NO_3^- ions.

The nitro group is also known to form a bridge between two metal atoms (**Figure 13**) with NO_2 stretching bands appearing at 1492 and 1180 cm^{-1}. Upon isotopic substitution of the bridging oxygen, the latter is shifted by −10 cm^{-1} while the former is almost unchanged. For this reason, these bands have been assigned to the ν(N=O) (outside the bridge) and ν(N–O) (bridge), respectively.

The NO_3^- ion is able to coordinate in unidentate, symmetric and asymmetric chelating bidentate, and bridging bidentate fashion (**Figure 13**). It is very difficult to distinguish the possible structures by vibrational spectroscopy since the symmetry of the nitrate ion differs very little among them (C_{2v} or C_s). The unidentate NO_3 group exhibits three NO stretching bands, as expected due to its C_{2v} symmetry. For example, in $[Ni(en)_2(NO_3)_2]$ containing unidentate nitrato groups the three bands at 1420, 1305 and 1008 cm^{-1} have been identified as $\nu_a(NO_2)$, $\nu_s(NO_2)$ and ν(NO), respectively; whereas $[Ni(en)_2NO_3]ClO_4$ which contains a chelating bidentate nitrate, exhibits the three bands at 1476 cm^{-1}, 1290 cm^{-1} and 1025 cm^{-1}, due to ν(N=O), $\nu_a(NO_2)$ and $\nu_s(NO_2)$, respectively.

The separation of the two highest frequency bands is larger for bidentate than for unidentate coordination, a rule which is not applicable if the complexes are markedly different. For structural diagnosis, it is possible to use the combination band, $\nu_1+\nu_4$, of free NO_3^- ion which appears in the 1800–1700 cm^{-1} region. Upon coordination the ν_4 (E', in-plane bending) near 700 cm^{-1} splits into two bands, and the magnitude of this splitting is expected to be larger for bidentate than for unidentate ligands. In some cases, this produces the separation of two ($\nu_1+\nu_4$) bands in the 1800–1700 cm^{-1} region, the NO_3^- ion being bidentate if the separation is 70–20 cm^{-1} and unidentate if it is 25–5 cm^{-1}.

Lattice water

It is possible to classify water in inorganic salts and coordination compounds such as lattice or coordinated water, but no definite borderline exists between the two forms. Vibrational spectra are often useful for providing information which allows the two forms to be distinguished.

Lattice water exhibits strong absorptions at 3550–3200 cm^{-1} (antisymmetric and symmetric OH stretchings) and at 1630–1600 cm^{-1} (HOH bending). In the region 600–200 cm^{-1}, lattice water exhibits 'libration modes' that are due to rotational oscillations of the water molecule.

Aquo and hydroxo complexes

The structure of aquo complexes is often assigned on the basis of their vibrational spectra: $TiCl_3 \cdot 6H_2O$ has been formulated as *trans*-$[Ti(H_2O)_4Cl_2]Cl \cdot 2H_2O$ since it exhibits one TiO stretching (500 cm^{-1}, E_u) and one TiCl stretching (336 cm^{-1}, A_{2u}) mode.

In the Raman spectra of aqueous solutions of nitrate and sulfate zinc(II) and mercury(II) salts, the polarized Raman bands in the 400–360 cm^{-1} region have been assigned to the metal–O stretching modes of hexacoordinated aquo complex ions.

The hydroxo group can be distinguished from the aquo group since the former lacks the HOH bending mode near 1600 cm^{-1}, and hydroxo complexes exhibit MOH bending modes below 1200 cm^{-1} (for example, 1150 cm^{-1} for the $[Sn(OH)_6]^{2-}$ ion). The OH group is able to bridge two metal atoms: in $[(bipy)Cu(OH)_2Cu(bipy)]SO_4 \cdot 5H_2O$ the bridging OH bending mode is at 955 cm^{-1} and is shifted to 710 cm^{-1} upon deuteration.

Perchlorato complexes

Infrared and Raman spectroscopy techniques have been used extensively to determine the mode of coordination of the ClO_4^- ligand. For example: the ionic $K[ClO_4]$ shows only two characteristic bands at 1170–1050 and 935 cm^{-1}, whereas $Cu(ClO_4)_2 \cdot 2H_2O$, where the perchlorate is coordinated in unidentate fashion, exhibits three strong absorptions at 1158, 1030 and 920 cm^{-1}. In the bidentate perchlorate complexes, the first absorption is split into three bands (1270–1245, 1130 and 948–920 cm^{-1}) whereas the second band appears at 1030 cm^{-1}.

β-Diketones

β-Diketones form metal chelates of type I, II, III and IV (**Figure 14**). The ν(MO) of these chelates, assigned by using the metal-isotope technique, provides direct information about the M–O bond strength. Vibrational spectroscopy can be employed to differentiate *cis* and *trans* octahedral complexes $M(acac)_2X_2$. For example, $Ti(acac)_2F_2$ is '*cis*' with two ν(TiF) at 633 and 618 cm^{-1} whereas *cis*- and *trans*-$Re(acac)_2Cl_2$ can be isolated. In the infrared spectrum the *trans* isomer exhibits ν(ReO) and ν(ReCl) at 464 and 309 cm^{-1}, respectively, while each of these bands splits into two in the *cis* isomer: 472 and 460 cm^{-1} for ν(ReO) and 346 and 333 cm^{-1} for ν(ReCl).

In some derivatives, the keto form of acac forms a chelate ring (II). This kind of coordination has been observed in $M(acacH)Cl_2$ (M=Co and Zn) which exhibits a strong ν(C=O) band near 1700 cm^{-1}.

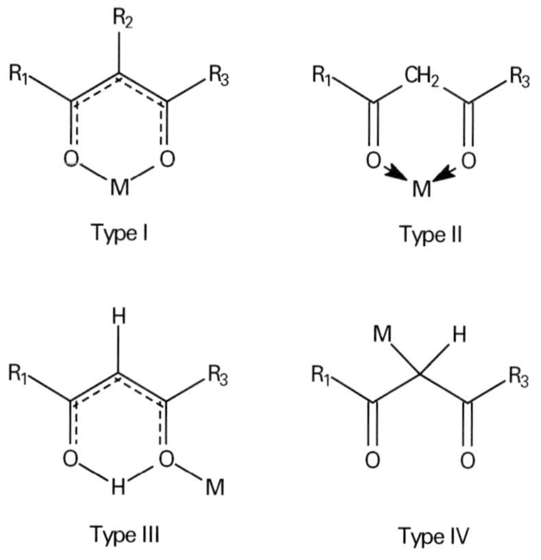

Figure 14 Coordination modes of β-diketone ligands.

In $Mn(acacH)_2Br_2$, the acacH coordinates in the unidentate enolic form through only one O atom (III). The CO and CC stretching bands of the enol ring were found at 1627 and 1564 cm^{-1}, respectively.

In $[Pt(acac)_2Cl_2]^{2-}$, the metal is bonded to the γ-carbon atom of the acetylacetonato ion (IV). The infrared spectra show two ν(C=O) at 1652 and 1626 cm^{-1}, two ν(C–C) at 1350 and 1193 cm^{-1}, and one ν(Pt–C) at 567 cm^{-1}.

Carbonyl complexes

Carbonyl complexes generally exhibit strong sharp ν(CO) bands between 2100 and 1800 cm^{-1}. Studies of ν(CO) provide valuable information about the structure and bonding of carbonyl complexes because ν(CO) is generally free from coupling with other modes and is not obscured by the presence of other vibrations. The ν(CO) is generally at lower frequencies in the complexes than in the free CO (2155 cm^{-1}). However, the opposite trend is observed when CO is bonded to metal halides with the metals in a relatively higher oxidation state. Frequencies for bridging CO groups are much lower (1900–1800 cm^{-1}) than those of the terminal CO group (2100–2000 cm^{-1}), and extremely low ν(CO) (~1300 cm^{-1}) are observed when the bridging CO group forms an adduct through its O atom.

Vibrational studies on $Fe(CO)_5$ and $M(CO)_6$ (M = Cr, Mo and W) derivatives, including their ^{13}C and ^{18}O species, have indicated that the M–C bond strength increases in the order Mo < Cr < W, an order also supported by a Raman study.

It has been found that in the $M(CO)_4$ complexes (T_d), ν(CO) decreases and ν(MC) increases on going from $Ni(CO)_4$ to $[Co(CO)_4]^-$ to $[Fe(CO)_4]^{2-}$. This result indicates that the M→CO π back-donation increases in the order Ni(0) < Co(-I) < Fe(-II). Highly reduced species such as $[W(CO)_4]^{4-}$ exhibit very low ν(CO) values (1530–1460 cm^{-1}).

Vibrational spectroscopy has been a subject useful for elucidating structures of polynuclear carbonyls. For $CO_2(CO)_8$, which has a structure containing six terminal and two bridging CO groups, five terminal and two bridging ν(CO) are expected to be infrared active: the former are in fact observed at 2075, 2064, 2047, 2035 and 2028 cm^{-1}, whereas the latter are at 1867 and 1859 cm^{-1}.

In $MX(CO)_X$ derivatives (X = Cl, Br or I), ν(CO) is highest for the chloro compound and lowest for the iodo compound (the bromo compound being between the two) in accordance with their electronegativities.

Dioxygen complexes

Dioxygen (O_2) adducts of metal complexes have been extensively investigated with vibrational spectroscopy because of their importance as models for oxygen carriers in biological systems and as catalytic intermediates in oxidation reactions of organic compounds.

The bond order of the O–O linkage decreases as the number of electrons in the antibonding 2pπ^* orbital increases: the bond order in $[O_2^+]AsF_6$, O_2, $K[O_2^-]$ and $Na_2[O_2^{2-}]$ is 2.5, 2.0, 1.5 and 1.0, respectively. This decrease causes an increase in the O–O distance and a decrease in the $\nu(O_2)$, which are at 1858, 1555, 1108 and 760 cm^{-1}, respectively.

An interesting example of dioxygen adducts of transition-metal complexes is the *trans*-planar $[IrCl(CO)(PPh_3)_2]$ (Vaska's salt), which binds dioxygen reversibly to form a trigonal bipyramidal complex. The O–O distance of this compound (1.30 Å) is close to that of the O_2^- ion (1.28 Å): its $\nu(O_2)$ (857 cm^{-1}) is typical of peroxo adducts.

Halogeno complexes

Halogens (X) are the most common ligands in coordination chemistry. Terminal MX stretching bands appear in the regions of 750–500 cm^{-1} for MF, 400–200 cm^{-1} for MCl, 300–200 cm^{-1} for MBr, and 200–100 cm^{-1} for MI. The ν(MX) generally is higher with a higher oxidation state of the metal. In some cases, such as the $[M(dias)_2Cl_2]^{n+}$, ν(MCl) changes rather drastically on going from Ni(III) (d^7) to Ni(IV) (d^6) (240 and 421 cm^{-1}, respectively) while very little change is observed between Fe(III) (d^5) and Fe(IV) (d^4) (384 and 390 cm^{-1}, respectively). This was attributed to the presence of one electron in the antibonding e_g^* orbital in the Ni(III) complex.

The SnX stretching force constants of halogenotin compounds are approximately proportional to the oxidation number of the metal divided by the coordination number of the complex. It is interesting to note that the ν(SnCl) of free $SnCl_3^-$ ion [289 (A_1) and 252 (E) cm^{-1}] are shifted to higher frequencies upon coordination to a metal: the ν(SnCl) of $[Rh_2Cl_2(SnCl_3)_4]^{2-}$ are at 339 and 323 cm^{-1}.

The number of ν(MX) vibrations observed is often very useful for determining the stereochemistry of the complex: in the vibrational and Raman spectra of planar $M(NH_3)_2X_2$ [M = Pt(II) and Pd(II)] the *trans* isomer (D_{2h}) exhibits one ν(MX) (B_{3u}), whereas the *cis* isomer (C_{2v}) exhibits two ν(MX) (A_1 and B_2) bands in the infrared.

Some derivatives, such as $Ni(PPh_2R)_2Br_2$ (R = alkyl) exist in two isomeric forms, tetrahedral and trans-planar. Distinction between these two can be made easily since the numbers and frequencies of infrared-active ν(NiBr) and ν(NiP) are different for each isomer: in fact, ν(NiBr) and ν(NiP) are at ~330 and 260 cm^{-1}, respectively, for the planar form, and at ~270–230 and 200–160 cm^{-1}, respectively, for the tetrahedral form.

IR data allow us to differentiate between *fac* and *mer* $[MX_3(L)_3]$ isomers: for example *fac*-$[RhCl_3(Py)_3]$ gives two bands at 341 and 325 cm^{-1} whereas *mer*-$[RhCl_3(Py)_3]$ shows three bands at 355, 322 and 295 cm^{-1}.

Bridging MX stretching frequencies [ν_b(MX)] are generally lower than terminal MX stretching frequencies [ν_t(MX)].

Complexes containing metal–metal bonds

Complexes containing metal–metal bonds show ν(MM) in the low-frequency region (250–100 cm^{-1}) because the M–M bonds are relatively weak and the masses of metals are relatively large. In the infrared, ν(MM) is forbidden if the complex is perfectly centrosymmetric with respect to the M–M. If the derivative is not centrosymmetric, weak ν(MM) appear.

In heteronuclear complexes the ν(MM′) vibration is more strong than ν(MM) because of the presence of a dipole moment along the M–M′ bond.

In Raman spectroscopy both ν(MM) and ν(MM) appear strong since large changes in polarizabilities are expected as a result of stretching long, covalent M–M bonds. Several ν(MM) of polynuclear carbonyls have been reported, and the MM stretching force constants obtained from normal coordinate analysis have been used to discuss the nature of the M–M bond. For example, $(CO)_5Mn–W(CO_5)$ exhibits the ν(M–M) in the Raman spectra at 153 cm^{-1}.

Compounds containing unusually short M–M bonds exhibit unusually high ν(MM). The Mo–Mo distance in $Mo_2(OAc)_4$ is 2.09 Å, and the ν(MM) appear at 406 cm^{-1} because the Mo–Mo bond consists of one σ-bond, two π-bonds and a δ-bond (bond order 4).

Organometallic compounds

Metal alkanes

The methyl group bonded to a metal (M–CH_3) exhibits the six normal vibrations expected for tetrahedral ZXY_3 molecules. In addition, CMC bending and CH_3 torsional modes are expected for $M(CH_3)_n$ ($n \geq 2$) compounds. In $M(CH_3)_4$ derivatives (M = Si or Sn) the CH_3 rocking, MC stretching and CMC bending frequencies are very sensitive to the change in metals, and the number of these skeletal infrared or Raman active modes provides direct information about the structure of the MC skeleton.

Some metal alkyls are polymerized in the solid state: $Li(CH_3)$ forms a tetramer containing $Li–CH_3–Li$ bridges in the solid state and its CH_3 frequencies are lower than those of nonbridging compounds. The infrared spectra of $Li[Al(CH_3)_4]$ and $Li_2[Zn(CH_3)_4]$ have been interpreted on the basis of linear polymeric chains in which the Al (or Zn) atom and the Li atom are bonded alternately through two CH_3 groups.

Metal alkenes and alkynes

A vinyl group σ-bonded to a metal ($M–CH=CH_2$) exhibits, in addition to the MC stretching and CMC bending modes, strong C=C stretching vibrations in the Raman spectra. Their intensity and position in infrared spectra depend on the metal: for example $Hg(CH=CH_2)_2$ exhibits ν(C=C) at 1603, and ν(MC) at 541 and 513 cm^{-1}.

Also, the C≡C stretching frequencies of acetylenic compounds are strong in the Raman, but vary from strong to weak in the infrared, depending on the metal involved. In contrast to σ-bonded complexes, the C=C and C≡C stretching bands of π-bonded complexes show marked shifts to lower frequencies relative to those of free ligands.

Alkenes form π-complexes with transition metals. Zeise's salt, $K[Pt(C_2H_4)Cl_3]\cdot H_2O$ has been extensively investigated, but assignments have been often controversial. Two types of bonding are involved in the $Pt–C_2H_4$ bond: a σ-bond formed by the overlap of the filled 2pπ bonding orbital of the alkene with the vacant dsp^2 bonding orbital of the metal and a π-bond formed by the overlap of the 2pπ* antibonding orbital of the alkene with a filled dp hybrid orbital of the metal. If the σ-bond is predominant, bonding scheme I (**Figure 15**), which predicts one Pt(II)-alkene stretching mode, should be preferred.

Some workers assigned the absorption at 407 cm^{-1} band of Zeise's salt to this mode. On the other hand, if bonding scheme II is operating, a five coordinate Pt atom is formed and the two bands at 491 and 403 cm^{-1} can be assigned to the symmetric and antisymmetric PtC stretching modes, respectively. The real bonding is somewhere between I and II, and the latter may become more significant as the oxidation state of the metal decreases. In Zeise's salt X-ray analysis and MO calculations, together with infrared studies suggests bonding scheme I.

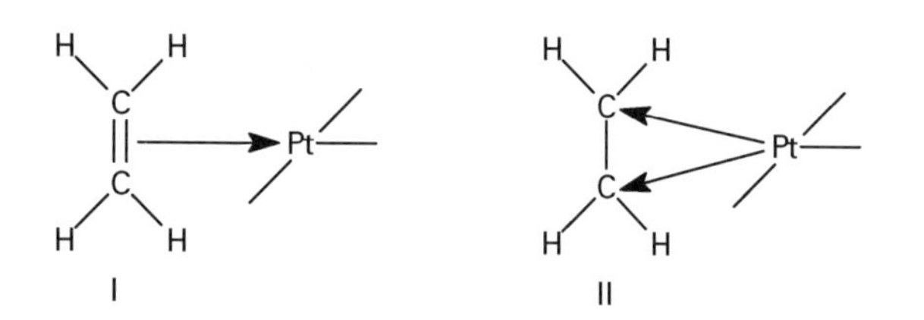

Figure 15 Possible bonding schemes for the $Pt–C_2H_4$ bond in Zeise's salt.

Free HC≡CH exhibits a CC stretching band at 1974 cm^{-1}. When coordinated as in $(HC≡CH)Co_2(CO)_6$, the CC stretching band was observed at 1402 cm^{-1}, which is ~570 cm^{-1} lower than the value for free acetylene. The spectrum of the coordinated acetylene in this complex is similar to that of free acetylene in its first excited state, at which the molecule takes on a *trans*-bent structure. Free $HC≡C(C_6H_5)$ exhibits the C≡C stretching band at 2111 cm^{-1}. In the case of σ-bonded complexes, this band shifts slightly to a lower frequency (2036–2017 cm^{-1}), while in $M[–C≡C(C_6H_5)]_2$ [M = Cu(I) and Ag(I)], it shifts to 1926 cm^{-1} due to the formation of both σ-and π-type bonding.

See also: **Biochemical Applications of Raman Spectroscopy; Far-IR Spectroscopy, Applications; IR Spectroscopy, Theory; IR Spectrometers; IR Spectroscopy Sample Preparation Methods; Raman Spectrometers; Rayleigh Scattering and Raman Spectroscopy, Theory.**

Further reading

Degen IA and Newman GA (1993) Raman spectra of inorganic ions. *Spectrochimica Acta Part A (Molecular Spectroscopy)* 49: 859–887.

Ferraro JR and Basile LJ (eds) (1978) *Fourier Transform Infrared Spectroscopy*. New York: Academic Press.

Greenwood NN (ed) (1968) *Spectroscopy Properties of Inorganic and Organometallic Compounds*. London: The Chemical Society.

Herzberg G (1945) *Infrared and Raman Spectra of Polyatomic Molecules*. Princeton, New York: Van Nostrand.

Maslowsky E Jr (1977) *Vibrational Spectra of Organometallic Compounds*. New York: Wiley.

Nakamoto K (1969) Characterization of organometallic compounds by infrared spectroscopy. In: Tsutsui (ed) *Characterization of Organometallic Compounds*, Part I. New York: Wiley Interscience.

Nakamoto K (ed) (1986) *Infrared and Raman Spectra of Inorganic and Coordination Compounds*, IVth edn. Toronto: Wiley.

Nyquist RA and Kagel RO (1971) *Infrared Spectra of Inorganic Compounds*. San Diego: Academic Press.

Nyquist RA, Kagel RO, Putzig CL and Lugers MA (eds) (1996) *The Handbook of Infrared Spectra of Inorganic Compounds and Organic Salts*. San Diego: Academic Press.

IR and Raman Spectroscopy Studies of Works of Art

See **Art Works Studied Using IR and Raman Spectroscopy.**

IR Microspectroscopy

See **Raman and IR Microspectroscopy.**

IR Spectral Group Frequencies of Organic Compounds

AS Gilbert, Beckenham, Kent, UK

VIBRATIONAL, ROTATIONAL & RAMAN SPECTROSCOPIES
Applications

The vibrational spectra of organic compounds are typically complex, possessing many individual bands of variable intensity and shape. It is observed that certain bands are associated with specific chemical groups, that is, certain vibrational motions are largely concentrated within a portion of the molecule. Such bands are identifiable primarily by their occurrence within a narrow range of frequencies and their presence or absence from a spectrum allows empirical judgments to be made concerning the chemical structure. No chemical group, however, is completely isolated in vibrational or electronic terms, and therefore the exact frequency and intensity is determined by and can provide information about the adjacent chemical environment. Examination of many thousands of organic compounds has enabled numerous correlations to be established between elements of IR spectra and chemical structure and extensive tables of group frequency data are available in many publications. Although it is rarely possible to assign most of the bands in an IR spectrum, sufficient information can often be gleaned that, when allied with data from Raman spectra and other techniques, allows complete structural determination to a high degree of confidence. The concept of group frequencies underpins much of vibrational spectroscopy, though it is not necessary for quantitative analysis.

Initial considerations and definitions

Frequency and wavenumber

IR absorption (or Raman shift) is caused by a quantum transition between two energy levels, and so band positions ought to be defined in terms of energy units. However, atoms can be treated as oscillating under the influence of spring-like bonds obeying Hooke's law, so the terms vibration and vibrational frequency are habitually used. The vibrational frequency is considered to be equal to the frequency of absorbed radiation (or frequency difference in the case of Raman) and is proportional to the energy difference, but it is customary to define its value not in hertz but in a frequency-dependent unit, the wavenumber, cm^{-1}, which is the reciprocal of the spectral wavelength as measured in centimetres.

Differences between IR and Raman spectroscopies

For a variety of reasons, IR spectroscopy is more fruitful than Raman for applying group frequencies to structural analysis. Some of these reasons are technical, for instance access to IR equipment is much easier and many compounds are not amenable to Raman spectroscopy with visible excitation (as used in the majority of instruments) owing to fluorescence or decomposition under laser illumination.

Chemical moieties that yield good group IR frequencies tend to be polar in nature and such entities tend not to give prominent bands in Raman spectra. Conversely, some elements of structure such as C=C bonds, particularly if situated in largely symmetrical and nonpolar surroundings, absorb very weakly in the IR, but in contrast are relatively strong in the Raman.

This complementary feature of Raman spectroscopy arises because it is change in *polarizability* during vibration that is important for determining intensity, rather than dipole change as in the IR. Confidence in assigning group frequencies is therefore often considerably strengthened by utilizing Raman spectroscopy if possible.

Overall molecular symmetry is rarely of importance in studying organic compounds as the vast majority of compounds do not possess any. However, local group symmetry usually has a bearing on the relative intensities of certain bands.

Physical state

In what follows, the listed frequencies are mostly for samples measured in liquid or solid phases. These can often be found to vary by a few wavenumbers owing to environmental conditions so that it is usually pointless to quote exact values. Care should be exercised with gas-phase frequencies as they can differ quite substantially, owing to the removal of intermolecular association effects. Hydrogen-bonded groups such as hydroxyl are particularly susceptible.

The basis of vibrational frequency

For simple harmonic oscillation of a diatomic molecule the value of the vibrational frequency ν of the fundamental mode, in cm^{-1}, is given by the relationship

$$\nu = 130.3\left(f\left[\frac{1}{m_1}+\frac{1}{m_2}\right]\right)^{1/2} \quad [1]$$

$$= 130.3(f/\mu)^{1/2} \quad [2]$$

where f is the force constant (stiffness) of the bond in newtons per metre, m_1 and m_2 are the individual atomic masses (in atomic mass units) and μ is the 'reduced' atomic mass, which is given by

$$\mu = \frac{m_1 m_2}{m_1 + m_2} \quad [3]$$

A lone X–H bond (where X represents C, N, O etc.) in a polyatomic compound vibrates almost as if it were a diatomic molecule. As μ in this case is always close to 1 and X–H bonds have broadly similar values of f, then the X–H moiety will yield a band in the same general region. For C–H, f is about 490 N m^{-1}, which yields a frequency of 3000 cm^{-1}; actual values are usually 3100–2800 cm^{-1}. Frequencies for O–H and N–H are usually slightly higher owing to greater values of f.

The reduced mass of X–X′ is rather larger than that of X–H, for instance for C–O μ is about 6.86 and since f here is roughly similar to that of C–H (both being single bonds) ν can be expected to be about √6.86 times less than 3000 cm^{-1}, i.e. about 1150 cm^{-1}. Double bonds are approximately twice as strong (and stiff) as single bonds, so C=O should 'vibrate' approximately √2 times more rapidly than C–O in the region around 1600 cm^{-1}. In fact, the 'average' carbonyl frequency is around 1700 cm^{-1} and for a variety of reasons, discussed below, can be found at least 100 cm^{-1} or so on either side of this value.

Likewise, triple bonds have values of f roughly three times greater than single bonds, thus C≡C and C≡N groups would be expected around 2000 cm^{-1}; they are usually seen between 2300 and 2100 cm^{-1} or so.

The simplest case of a deformation or bending mode (given the symbol δ as opposed to ν which refers to stretching modes) involves a three-atom group with the atoms undergoing a scissors-like motion. Geometric considerations mean that the expression for μ is more complicated than, but not too different from, that for the diatomic oscillator, but the force constants are rather lower. Thus δ(XH_2) vibrations are broadly positioned in the region around 1500 cm^{-1} while δ(XYH) occurs from 1500 to 1000 cm^{-1} and beyond. Deformations of XYZ (all C, N, O or heavier atoms) are below 1000 cm^{-1}.

The frequencies of stretching vibrations can usually be roughly predicted from knowledge of the bond order and the atomic masses. However, the values of the force constants involved in deformations are not obvious, so that their frequencies often appear to be rather arbitrary. The total number of vibrations for a nonlinear molecule is given by the relation, $3N$-6, where N is the number of atoms.

These simple considerations explain the general appearance of the vibrational spectra of organic chemicals. Essentially, most bands are found in two regions. First, from about 3600 to 2800 cm^{-1}, wherein lie features resulting from X–H stretching modes. Second, from about 1800 to below 400 cm^{-1}, a region densely occupied by bands from most other vibrations. The upper part, 1800 to about 1500 cm^{-1}, is often referred to as the double bond region, the rest as the fingerprint region. The region between 2800 and 1800 cm^{-1} is generally quite barren unless triple (or fused double) carbon–carbon or carbon–nitrogen bonds or one or two less common groups are present. Strongly hydrogen-bonded O–H and N–H can also produce features here. **Figure 1** provides an overview in chart form.

The IR spectra of inorganic compounds have a rather different aspect, consisting mostly of broad features at low frequencies. Although O–H and N–H are sometimes present, the spectrum down to about 1300 cm^{-1} is usually empty except for weak overtones and combination bands.

Finally, it will be obvious from the equations above that ν(X–D), δ(XD_2) and δ(XYD), where D is deuterium, will be considerably different in frequency from the corresponding vibrations with hydrogen.

Vibrational interactions

No bond vibrates independently of any other in a polyatomic molecule, and where adjacent oscillators vibrate with similar or identical frequency, strong interaction takes place. This can be illustrated by considering the simple case of a symmetric linear triatomic molecule or group Y–X–Y, joined together by two equivalent bonds. When analysed in terms of

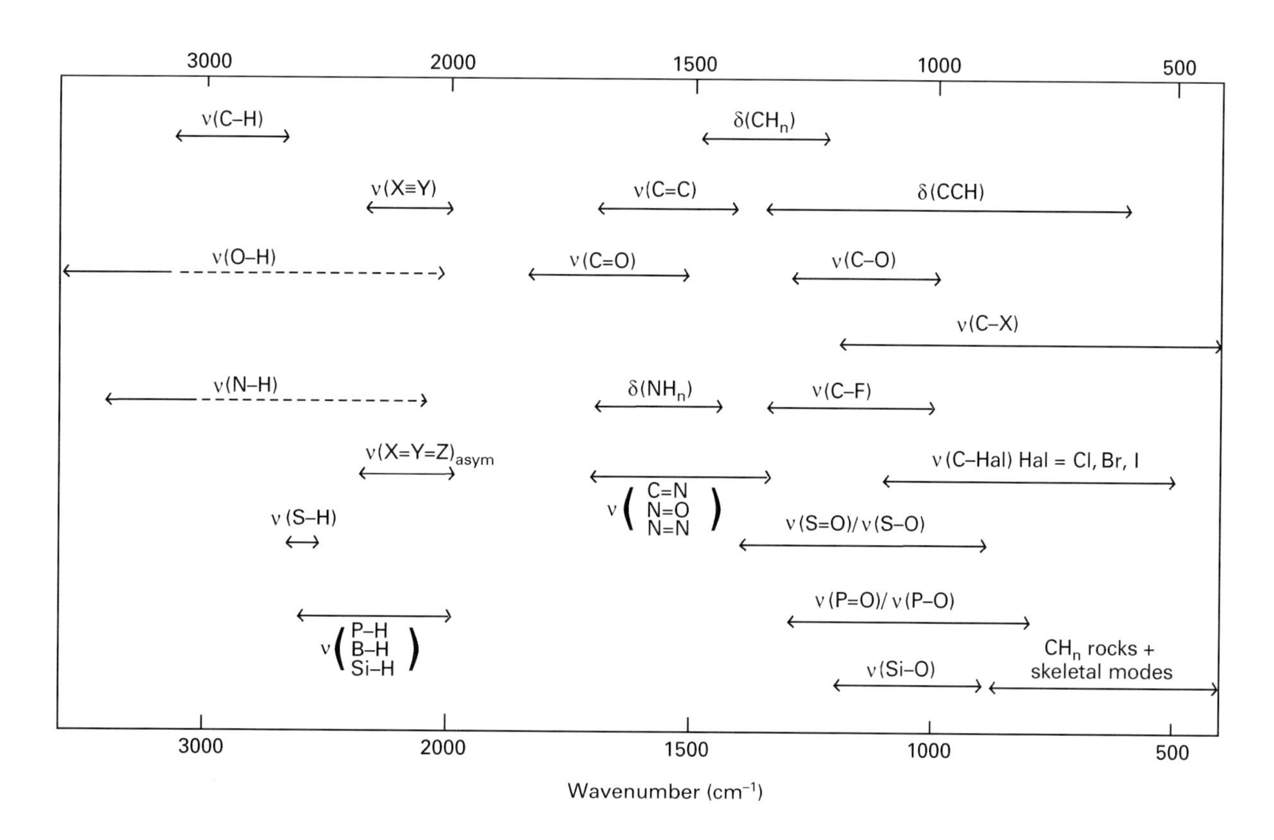

Figure 1 Correlation chart for vibrational spectra of organic compounds. The symbols ν and δ and X, Y and Z are defined in the text.

Table 1 Observed and calculated frequencies (cm^{-1}) for triatomic molecules Y=X=Y: carbon disulfide (CS_2) and allene (C_3H_4)

observed	*ν(asym) observed*	*ν(sym) observed*	ν (asym)[a] / ν (sym)	ν (asym)[b] / ν (sym)	ν *(X=Y)*[c]	ν *(X=Y) observe*
CS_2	1522	650	2.34	2.52	1158, 1254	1250–1020
C_3H_4	1980	1071	1.85	1.83	1596,1576	1680–1630

[a] Observed ratio.
[b] Theoretical ratio.
[c] Calculated from ν(asym) and ν(sym) using theoretical ratios.

classical mechanics the two individual oscillators couple mechanically together to yield two distinct modes of vibration with different frequencies. Representations of these modes are given in **Figure 2**.

The frequencies of each mode can be derived in a simplified fashion as follows. For the in-phase (or symmetric) mode, the central mass remains fixed in space so that each end atom is effectively joined to an infinite mass, thus:

$$\nu = 130.3\left(f\frac{1}{m_{\mathrm{Y}}}\right)^{1/2} \qquad [4]$$

For the out-of-phase (antisymmetric or asymmetric) mode, the central mass is shared so that each end atom effectively vibrates independently with a central atom of half mass, thus:

$$\nu = 130.3\left(f\left[\frac{1}{m_{\mathrm{Y}}}+\frac{2}{m_{\mathrm{X}}}\right]\right)^{1/2} \qquad [5]$$

The frequency of the isolated diatomic fragment Y–X is given by,

$$\nu = 130.3\left(f\left[\frac{1}{m_{\mathrm{Y}}}+\frac{1}{m_{\mathrm{X}}}\right]\right)^{1/2} \qquad [6]$$

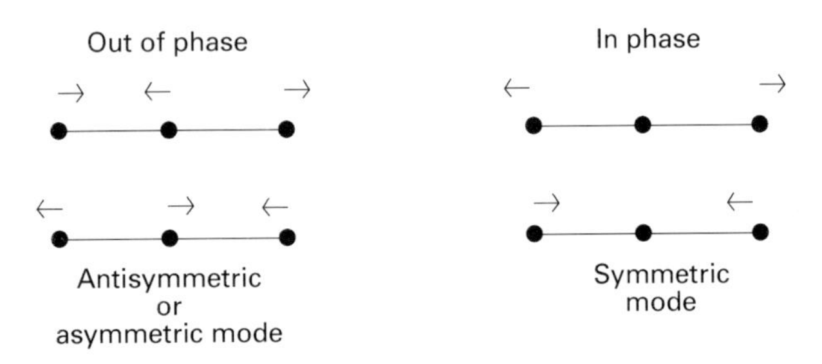

Figure 2 Representation of the two stretching modes of vibration of a linear triatomic molecule.

By substituting some real values into the equations, it may be seen that the individual oscillators couple to produce two new frequencies, one considerably above and one considerably below the isolated oscillator frequency. **Table 1** gives the relevant bands of carbon disulfide (S=C=S) and allene ($H_2C{=}C{=}CH_2$) and the theoretical ratios of their frequency values as calculated from Equations [4], [5] and [6]. For the case of allene, Y is treated as CH_2 with a mass of 14.

The strong coupling shown above takes place because the individual vibrators are identical. The less similar they are in terms of frequency, the lower will be the degree of coupling. This is revealed by consideration of the frequencies of simple triatomic linear cyanide molecules shown in **Table 2**. For HCN and the halogenides, the 'natural' (isolated) C–H or C–halogen frequencies are very different from ν(C≡N) and there is little interaction, as demonstrated by the roughly constant values of the latter. But this is not so for DCN and TCN (T = tritium), where by extrapolation from ν(C–H) the C–D and C–T oscillators would have natural frequencies of about 2430 and 2050 cm^{-1} respectively.

The reality of group frequencies

It should be evident from the above that where two or more oscillators are strongly coupled, they will be highly sensitive to replacement or substitution of any individual units. The associated frequencies will

Table 2 Stretching frequencies of linear triatomic nitriles

X	ν (Ξ−X) ($\chi\mu^{-1}$)	ν (X≡N) ($\chi\mu^{-1}$)
H	3312	2089
D	2629	1906
T	2460	1724
F	1077	2290
Cl	729	2201
Br	580	2187
I	470	2158

[a] Strongly coupled and therefore mixed modes.

therefore not be reliable indicators of particular items of chemical structure.

The observation that many simple groups are associated with bands that occur regularly in fairly narrow regions of the spectrum implies that they are largely isolated in vibrational terms. This means that the frequencies are insensitive to gross changes in molecular structure elsewhere. This is understandable for a carbonyl group or a C=C bond positioned in a singly bonded chain of carbon atoms, for instance. While individual vibrators within a group may couple strongly with each other, as long as they only couple weakly with external moieties, they retain their status as group frequencies.

Such characteristic and reliable frequencies provide a powerful means for identification of molecular subunits. By definition, they are largely insensitive to mechanical (vibrational) interactions and electronic effects from different adjacent atoms or groups. However, these perturbations can produce consistent shifts that are not so great as to destroy the characteristic nature of the group frequency, so that they provide information on the nature of the immediate chemical environment.

Mixed-character vibrational modes

The modes of the linear triatomic molecule, described in the previous section, are purely bond stretching in character. In nonlinear molecules, such as H_2O, bond stretching leads to small changes in bond angles through electron redistribution, and the converse occurs for deformation modes. In these simple cases the degree of mixing of bend and stretch is not significant, so that the modes are easily distinguishable from each other.

However, where atoms are joined together in a small ring, neither bond stretching nor bond angle bending can take place without each significantly affecting the other so that many of the modes are highly mixed in character. Such vibrations of the molecular framework are often termed 'skeletal' modes. Some idea of their complexity can be gained from **Figure 3** which shows how the individual atoms oscillate for the various modes of the benzene framework.

There are few good identifiable group frequencies below about 1000 cm^{-1}, spectra here are usually dominated by very mixed carbon skeletal and $C(H)_n$ rocking modes.

Some examples of group frequencies

A few of the more common submolecular units are now discussed below. Space permits only a cursory treatment and the frequency values given are necessarily a compromise between figures quoted in various publications, which tend to differ slightly.

Hydrocarbons

The methylene group This is shown schematically in [1] together with a representation of its associated

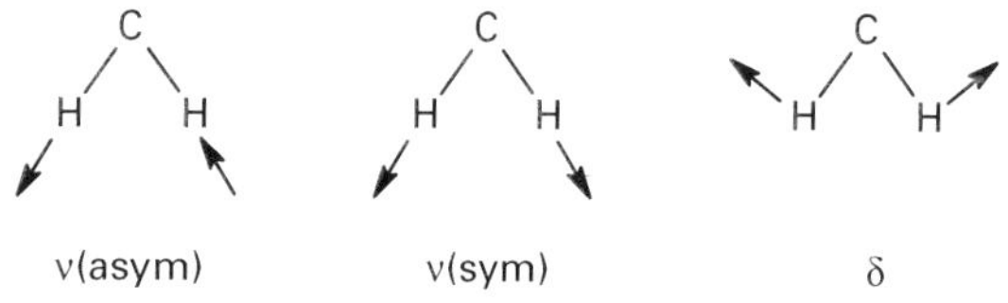

ν(asym) ν(sym) δ

[1]

local modes of vibration. The two C–H bonds are coupled to give two discrete stretching modes of vibration. Simple analysis shows that typically both overall dipole fluctuation and frequency are greater for ν(asym), so that in IR spectra the higher component of such a band doublet has the greater intensity.

The exact position of each band and its relative intensity is dependent on the nature of the atoms to which the methylene group is connected. In simple hydrocarbons the bands are situated in very narrow ranges, 2925 ± 10, 2855 ± 10 and 1465 ± 15 cm^{-1}, respectively. Shifts occur from the presence of other groups and Fermi resonance can often bring some confusion to the region.

When methylene is part of a cyclopropane ring, distortion of bond angles leads to increasing sp^2 character in the C–H bonds and thereby increase in force constant. This leads to a rise in both ν(asym) and ν(sym), to over 3000 cm^{-1}.

Methylene also engages in external vibrations where the group, as an effectively rigid unit, oscillates with respect to other atoms to which it is connected. One such mode, called a 'rock', is shown below, and where more than four adjacent CH_2 groups form an extended hydrocarbon chain, a characteristic band appears in the IR spectrum near to 725 cm^{-1}. The band is essentially due to a mode where all the CH_2 groups are rocking in phase, see [2].

The methyl group There are three stretching modes, two normally identical (degenerate) asymmetric modes, and one symmetric, with a similar relation of frequency and intensity as for methylene but displaced to higher wavenumber. Like methylene, again, they are situated in narrow ranges, 2960 ± 10 and 2870 ± 10 cm^{-1}, respectively, in simple hydrocarbons.

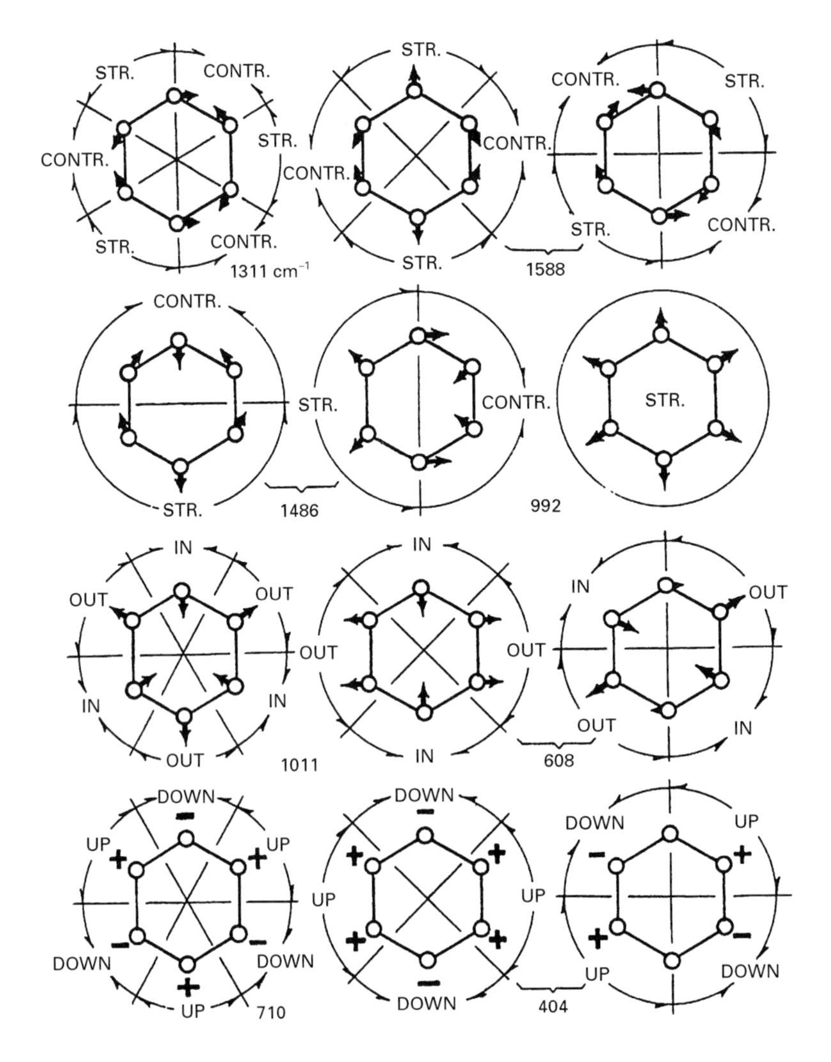

Figure 3 Representation of the skeletal modes of the benzene molecule. Reproduced with permission from Colthup NB, Daly LH and Wiberley SE (1975) *Introduction to Infrared and Raman Spectroscopy*. New York, San Francisco and London: Academic Press.

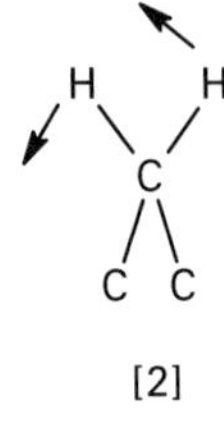

[2]

As with the stretching modes, there are two normally or nearly degenerate asymmetric deformation modes that usually overlap with δ(CH_2) when both groups are present, and a symmetric bending mode in the region 1380–1370 cm^{-1} for simple hydrocarbons. Substituents other than carbon can shift the latter frequency quite markedly (**Table 3**), while carbonyl instead raises the intensity.

Table 3 Symmetrical δ(XH_3) frequencies for various X–CH_3

X	*δ (CH_3)*(cm^{-1})
F	~1475
O	1455–1435
N	1430–1370
CF_3	~1410
CCl_3	~1385
Cl	~1355
B	1320–1290
Br	~1305
S	1325–1300
P	1310–1290
Se	~1280
Si	1280–1250

Table 4 Group frequencies of triatomics

Group[a]	*ν (asym)* (cm^{-1})	*ν (asym)* (cm^{-1})	*δ* (cm^{-1})	*External vibrations* (cm^{-1})
NH_2	3500–3350	3400–3300	1650–1590	<900
aph NO_2	1550–1530	1380–1350 }	630–610	<700
aro NO_2	1550–1470	1350–1310		
CO_2^-	1610–1550	1420–1380	?	
aph SO_2	1330–1300	1150–1110 }	610–545	<525
aro SO_2	1370–1350	1170–1150		
R-$SO_2(NH_2)$	1380–1310	1180–1140	?	

[a] aph = aliphatic; aro = aromatic.

Triatomic groups

The presence of these groups is signalled by a number of characteristic bands, as shown in **Table 4**. The respective bands are well separated and stably positioned in their respective regions, and therefore provide easily identifiable group frequencies. Typically, in IR spectra, the highest frequency band, from ν(asym), is the strongest or is roughly equal in intensity to ν(sym), while the deformation mode or modes are the weakest. In Raman spectra, however, both δ and ν(asym) are weak while ν(sym) is very prominent.

The carbonyl group

The carbonyl group provides one of the best indicators for the IR spectrum yielding a strong band typically in the region between 1800 and 1650 cm^{-1}, though occasionally up to 100 cm^{-1} more on either side. It is also an exemplar for the various electronic factors that influence group frequencies.

The C=O bond is polarized owing to the tendency of the oxygen to attract electrons from carbon (inductive effect) and this gives the group some single-bond character, ($^+$C–O$^-$) when the substituents are, for instance, methyls. If one of these is replaced by an electron-withdrawing atom such as chlorine, then the effect is largely cancelled out so that bond order is more nearly 2. The bond is in consequence stiffer and its vibrational frequency higher (see [3]).

Nitrogen, like chlorine, is also more electronegative than carbon but it affects the carbonyl frequency in the opposite sense because of the additional and

H_3C—C(=O)—CH_3 ν(C=O): ~1720 cm^{-1}

H_3C—C(=O)—Cl ~1790 cm^{-1}

[3]

dominant effect of mesomerism. Here the C=O loses some of its double-bond character, and is thereby weakened, owing to redistribution of electrons into the C–N bond, ($^-$O–C=N$^+$).

In the case of esters, the mesomeric effect is weak so that the inductive effect of electronegative oxygen dominates to raise the frequency, and for vinyl and aryl substituents on oxygen it is higher still.

Conjugation reduces frequency because of reduction of double-bond character. Single conjugation usually generates a lowering of about 20–30 cm^{-1}, double conjugation a little more; for example, diaryl ketones yield ν(C=O) near 1660 cm^{-1}.

Noncyclic amides yield ν(C=O) between 1690 and 1620 cm^{-1} and, if primary or secondary, have also one, slightly weaker, band from N–H bending at 1650–1620 cm^{-1} or 1550–1510 cm^{-1}, respectively.

Where carbonyl is part of a strained ring, its frequency is raised, probably as a consequence of several different effects, not just that of electron redistribution. For instance, simple lactams display in the regions illustrated in [4].

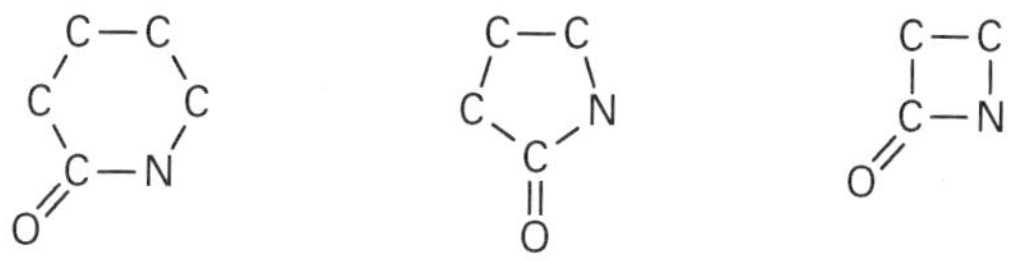

1670–1640 cm^{-1} 1750–1700 cm^{-1} 1760–1730 cm^{-1}

[4]

Table 5 lists the basic frequencies for the carbonyl group in some of its more commonly encountered chemical forms. It can be seen that the carboxylic acid frequency observed depends on the technique. This is because they are almost always (in liquid and solid states) dimerized (see [5]) so that the IR band observed is from the coupled asymmetric mode while that in the Raman is from the symmetric one.

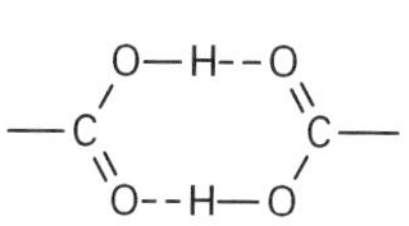

[5]

Where two carbonyls are coupled together as in anhydrides or imides the symmetric stretching mode is, unexpectedly, at higher frequency and the relative intensities of the IR bands are strongly dependent on the relative orientation of the groups. For noncyclic anhydrides the higher mode is the most intense, but for cyclic anhydrides and imides it is the weaker.

Table 5 The carbonyl bond as a component of various chemical groups

Group[a]	ν (C=O) (cm^{-1})
R–(C=O)–OR	1750–1730
Ar–(C=O)–OR	1730–1715
R–(C=O)–O–C=C	1775–1750
R–(C=O)–O–(C=O)–R	1825–1815, 1755–1745
5-membered ring cyclic anhydride	1870–1845, 1800–1775
5-membered ring cyclic imide	1790–1735, 1745–1680
Lactone (6 mem ring)	1750–1735
Lactone (5 mem ring)	1795–1765
Lactone (4 mem ring)	> 1800
R–(C=O)–NH_2	1670–1650
R–(C=O)–NH–	1690–1620
R–(C=O)–N	1670–1630
R–(C=O)–R	1725–1705
Ar–(C=O)–Ar	1670–1660
R–(C=O)–H	1740–1730
Ar–(C=O)–H	1715–1695
R–(C=O)–OH	1725–1700 (IR)
R–(C=O)–OH	1670–1640 (Raman)
Ar–(C=O)–OH	1715–1680 (IR)
Ar–(C=O)–OH	1690–1620 (Raman)
HN–(C=O)–OR	1740–1680
N–(C=O)–N	1670–1610

[a] R = alkyl; Ar = aromatic.

Table 6 Group frequencies of some unconjugted multiple bonds

Mode	*Frequency* (cm^{-1})
ν(C=C)	2140–2100
ν(C≡N)	2260–2240
ν(C=C)	1680–1630[a]
ν(C=N)	1690–1630
ν(S=O)	1060–1040[b]
ν(R–N≡C)	2145–2135[c]
ν(R–N=C=N–R) asym	2150–2100
ν(–N=N=N) asym	2170–2080
ν(N=N)	1580–1550
ν(P=O)	1300–1140

[a] Olefin. [b] Sulfone. [c] Isonitrile.

Alcohols, ethers and esters

The C–O stretching bands of these units, while coupled with and potentially obscured by other vibrations, can be identified by their relatively high intensity in IR spectra. For ethers, ν(C–O) occurs over the range 1140–1080 cm^{-1}; aliphatic alcohols can be anywhere between 1100 and 1000 cm^{-1}; while aromatic examples are higher in frequency, but usually weaker in intensity, at 1260–1180 cm^{-1}. For esters, ν(C–O) is between 1300 and 1150 cm^{-1}, the exact position being dependent on the nature of the substituents both on oxygen and on the carbon. A second band is also observed around about 1100 cm^{-1} or so and both can probably be regarded as having C–O stretching character coupled with adjacent C–C stretching vibrations. With regard to the higher frequency, simple acetates are generally 1250–1230 cm^{-1}, while longer-chain homologues are 1200–1190 cm^{-1}; conjugation gives greater values, so that unsaturated compounds such as phthalates and benzoates yield a strong band between 1300 and 1260 cm^{-1}.

Multiple bonds

Table 6 gives information for some multiple bond groups. Bands from the CC and CN groups, where isolated, show rather variable intensity in IR spectra and, particularly those from C=C, are often difficult to identify. By contrast, these moieties feature strongly in the Raman. The asymmetric stretching mode band of a fused double-bond group, however, is always strong in the IR and occurs in the triple-bond region; for example, CO_2 which absorbs at just above 2300 cm^{-1}.

Frequencies are rarely lowered by more than 50 cm^{-1} on conjugation, but an exception is the azo group, for which it can be reduced by almost 200 cm^{-1}.

Usually P=O is present as part of a larger group such as a phosphorous acid or a phosphate ester and this is often the case as well for S=O as in sulfonic acids. The double bond can be delocalized somewhat into adjacent P–O or S–O single bonds (particularly on ionization in salts) and coupling of their vibrations with neighbouring C–O or C–C bonds leads to several bands between 1300 and 800 cm^{-1}. These often provide good diagnostic information.

Hydrogens attached directly to doubly or triply bonded carbons give stretching bands between 3100 and 3000 cm^{-1}, owing to the sp^2 nature of the C–H bond, and are therefore a good indicator of unsaturation. In olefins, out-of-plane δ(C=C–H) modes are substitution dependent, *trans* substitution yields a strong IR band at 980–965 cm^{-1}, while for *cis* an IR band is seen at 730–650 cm^{-1}.

The benzene ring

Although the benzene ring is a relatively large entity, it nevertheless yields a number of very useful group frequencies. As in olefins, C–H stretching absorptions appear from 3100 to 3000 cm^{-1}. Between 1600 and 1450 cm^{-1} a number of bands appear; often as many as four can be distinguished. They are produced by what are essentially stretching modes of conjugated C=C bonds with some contributions from in-plane C–C–H bending motions.

The most prominent IR bands are substitution sensitive and occur below 1000 cm^{-1}. **Table 7** lists the relevant frequencies. Bands above 730 cm^{-1} are from out-of-plane C–C–H bending (C–H wag) modes, while those below are from ring bending.

Raman spectra of benzene rings show a strong very characteristic band in the 1000 cm^{-1} region. For benzene itself this comes from a ring breathing mode, but in substituted rings it is derived from other modes.

Heterocyclic aromatic compounds show some similarity to benzene and its homologues.

Halogens

Halogenated organics do not give good group frequencies. Bands from carbon–halogen stretching are spread across quite wide ranges; ν(C–F) is anywhere from 1350 to 1000 cm^{-1}, while for C–Cl, C–Br and C–I stretching bands are 1100–500 cm^{-1}. Where carbon is aromatic, the frequencies tend to be higher than for aliphatic in each case. With the exception of fluorine (in the IR), the bands are of medium intensity so that it is difficult to differentiate them from the multitude of others in these regions.

Hydroxyl and N–H stretching bands

Both O–H and N–H moieties generally form intra- and intermolecular hydrogen bonds and their bands therefore occur over fairly wide range of wavenumbers below the position of the unbonded frequencies. Usually this is no more than 200–300 cm^{-1}, but in extreme cases such as carboxylic acids or ionized N–H or where tautomerism occurs, the shift can be rather larger.

In this sense they are not good group frequencies, however, hydrogen bonding broadens the width and raises the intensity of the bands in the IR so they can normally be recognized though not definitively assigned to one or other. **Figure 4** shows part of theIR spectrum of *meta*-chlorobenzoic acid where ν(O–H) is represented by the diffuse underlying absorption between about 3500 and 2500 cm^{-1}.

Table 7 Substitution sensitive bands of benzene below 1000 cm^{-1}

Substitution pattern	*Band (s)* (cm^{-1})
Mono	770–730 and 710 – 690
Ortho	765–735
Meta	790–770 and 705–685
Para	860–800

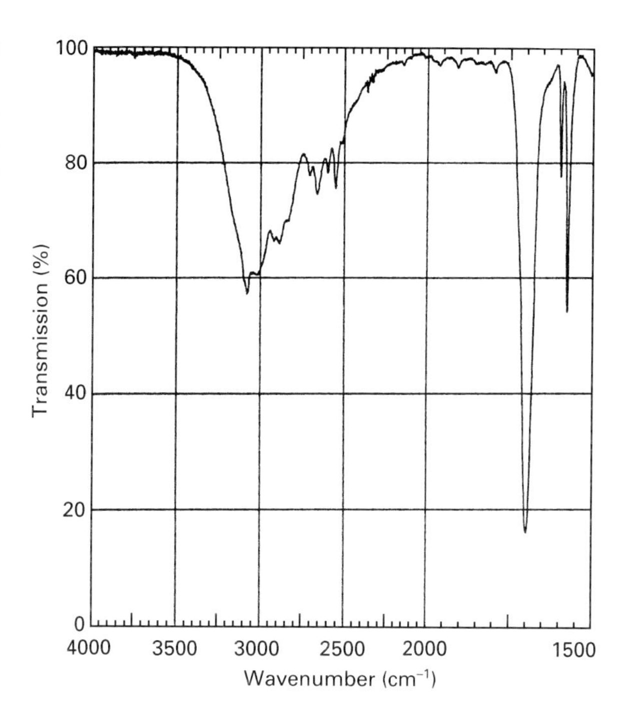

Figure 4 Part of the IR spectrum of *meta*-chlorobenzoic acid. The bandlets between 2700 and 2500 cm^{-1} are a typical part of such broad bands and are derived from interactions (Fermi-resonance) with combinations and overtones from lower-frequency vibrations.

Group frequencies in spectral interpretation and structure determination

Some IR spectra are presented below as examples. Band assignments are given in the figure captions.

Figure 5 shows the spectrum of isobutyl methyl ketone. The position of the carbonyl, while suiting a ketone, is also consistent with conjugated ester and carboxylic acid, but other characteristic bands from these two groups are not present.

Figure 6 is from benzyl alcohol (Ph–CH_2–OH). The hydroxyl bands are dominant despite only one O–H to seven C–H vibrators; this is due mainly to the effects of hydrogen bonding. The aromatic bands are all consistent with or indicative of single substitution, including four from combinations of lower-frequency modes (not labelled) between 2000 and 1700 cm^{-1}. These latter are usually too weak to be of use in any other than simple molecules.

Figure 7 is from cinnamonitrile (Ph–CH=CH–C≡N). The position of the nitrile band at 2220 cm^{-1} is indicative of conjugation.

Figure 8 shows *para*-nitroaniline. The polar substituents dominate, but some aromatic bands can still be seen.

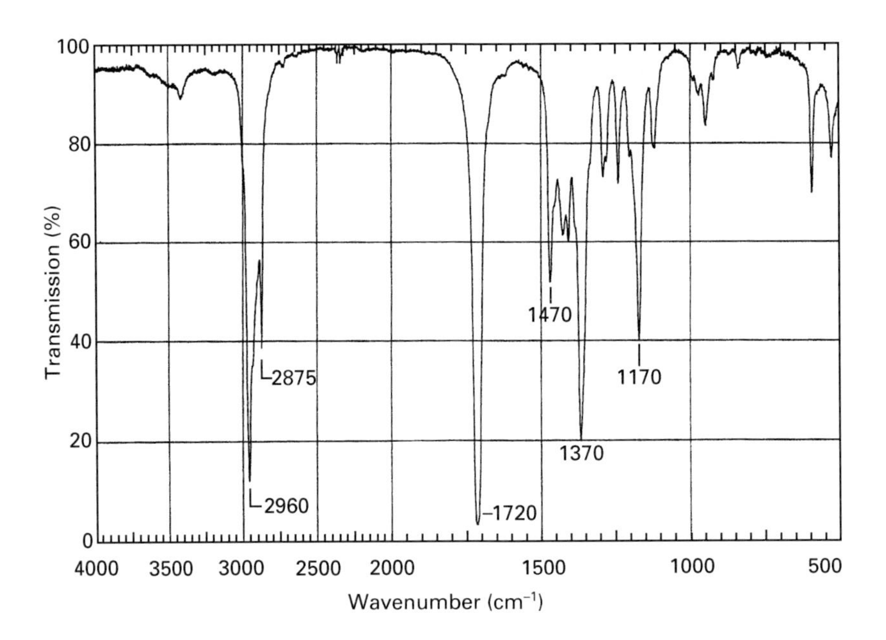

Figure 5 Band assignments for the IR spectrum of isobutyl methyl ketone: 2960, 2875, ν(CH_3) asym, sym; 1720, ν(C=O)$_g$; 1470, δ(CH_3) asym, δ(CH_2); 1370, δ(CH_3) sym; 1170, ν(CCC) asym of –CH_2–C(=O)–CH_3 unit.

Finally, **Figure 9** shows acetanilide (Ph–NH–C(=O)–CH_3). The 3300 cm^{-1} band could suit ν(O–H) but there is no obvious ν(C–O) band. The strength of the 1600 cm^{-1} band is typical of anilino N–H bending rather than a purely benzene ring mode.

In these cases, structure has been assumed and band assignment is therefore definitive. Where a compound is unknown, assignments must initially be tentative and an iterative procedure has to be followed until all attributions are consistent with one another and with any other relevant information. With luck, confirmation may be possible by matching to a spectrum in a database, bearing in mind minor but not significant spectral differences due to different measurement conditions.

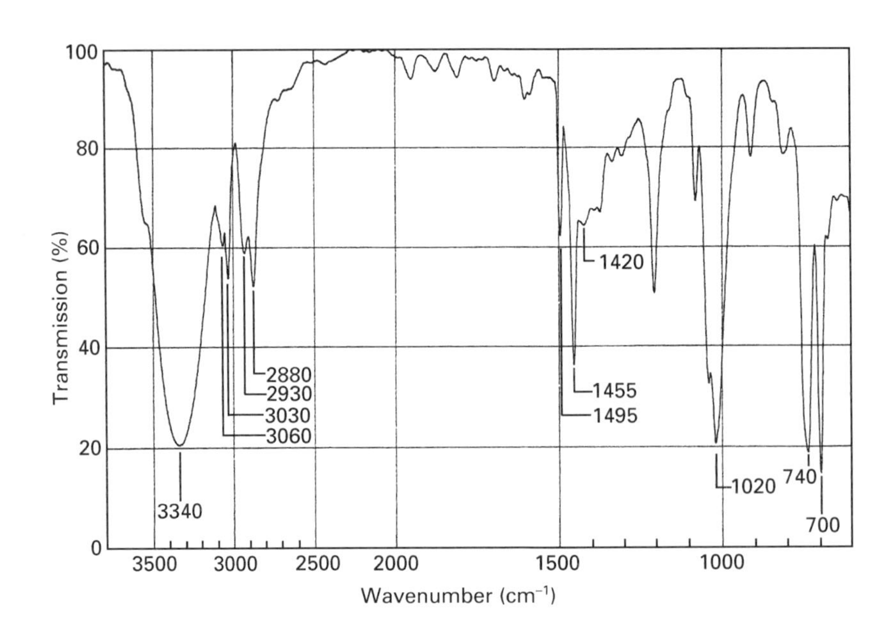

Figure 6 Band assignments for the IR spectrum of benzyl alcohol (Ph–CH_2–OH): 3340, ν(O–H); 3060, 3030, ν(C–H) benzene ring; 2930, 2880, ν(CH_2) asym, sym; 1495, ν(C=C) benzene ring; 1455, δ(CH_2); 1420, δ(COH); 1020, ν(C–O); 740, 700, δ(CCH) monosubstituted benzene ring.

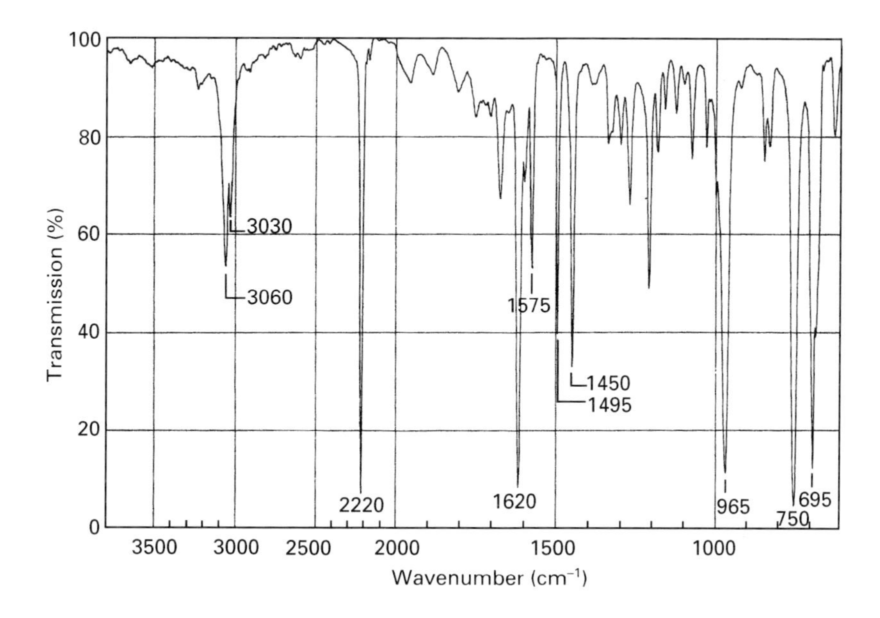

Figure 7 Band assignments for the IR spectrum of cinnamonitrile(Ph–CH=CH–C≡N): 3060, 3030, ν(C–H) benzene ring; 2220, ν(C≡N); 1620, ν(C=C) exocyclic; 1575, 1495, 1450, ν(C=C) benzene ring; 965, ν(C=C) *trans* substituted; 750, 695, δ(CCH) monosubstituted benzene ring.

There are number of commercial databases with facilities for computer searching. The Sadtler collection of organic compounds contains nearly 100 000 IR and several thousand Raman spectra.

Some other uses of group frequencies

Group frequencies are also useful in examining secondary effects and structure.

$$\text{Ph—CH}_2\text{—CH(CO}_2\text{H)(NH}_2\text{)}$$

[6]

A simple example of physicochemical influence is provided by phenylalanine, whose IR spectrum is shown in **Figure 10**. This compound is a β-aminocarboxylic acid [6], but while the general appearances of the 3300–2500 cm^{-1} region could be suggestive of the broad diffuse absorption yielded by $-CO_2H$ (**Figure 4**), there is no obvious carbonyl frequency in the right place nor are there any distinct ν(NH_2) bands in the 3500–3350 cm^{-1} region. Instead, carboxylate is indicated by bands at 1565 and 1410 cm^{-1}. The broad higher-frequency absorptions around 3000 cm^{-1} are thus consistent with, and evidently from, ionized N(H) groups, in this case NH_3^+. The compound therefore exists as an internal salt, or zwitterion.

A more complex case is that of secondary structure in large polypeptides and proteins. These molecules can assume a wide variety of conformations that often include large sections of regular structure where the backbone is arranged either in the form of the alpha-helix or beta-sheet. Vibrational spectra of proteins typically show that the amide carbonyl group is represented by several broad and overlapping bands that are spread over a range of frequencies, 1690–1610 cm^{-1}, rather larger than can be explained by the presence of different amino acids. Vibrational coupling/electronic effects from the different side chains would only be expected to shift ν(C=O) by a few wavenumbers.

This multiplicity of bands arises instead because of coupling along the polypeptide backbone, dipole–dipole interactions between carbonyl groups and hydrogen bonding between C=O and N–H groups, all of which are modified by differences in the secondary structure. Thus, amidic ν(C=O) from alpha-helix is found about 1660–1645 cm^{-1} and from beta-sheet between 1640 and 1625 cm^{-1}. Disordered structures produce bands above 1670 and below 1630 cm^{-1}.

Comparison of IR and Raman spectroscopy with other techniques

Vibrational spectroscopy provides particular insights into molecular structure but rarely a completely clear view. X-ray crystallography can provide absolute determination of structure but is expensive and

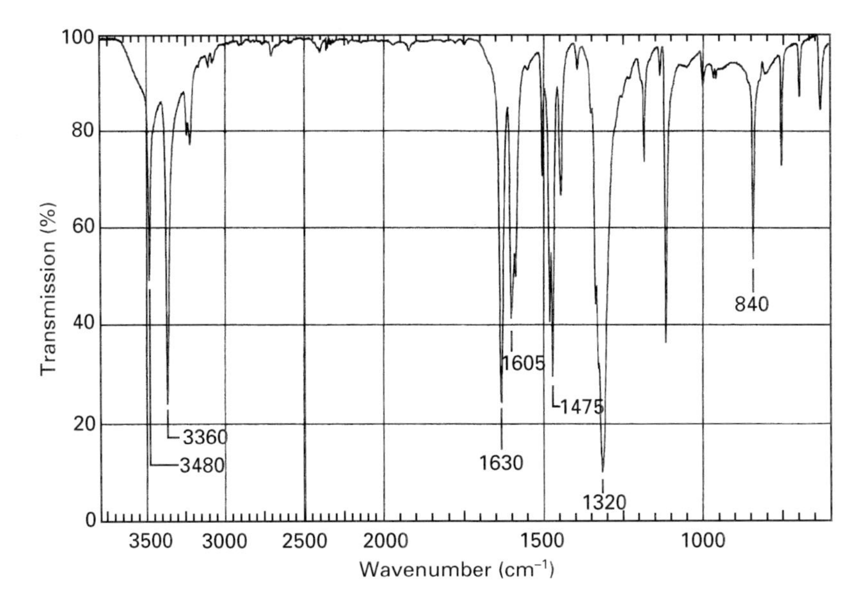

Figure 8 Band assignments for the IR spectrum of *p*-nitroaniline: 3480, 3360, ν(NH_2) asym, sym; 1630, δ(NH_2); 1605, ν(C=C) benzene ring; 1475, 1320, ν(NO_2) asym, sym; 840, δ(CCH) *para*-disubstituted benzene.

limited to small compounds that will crystallize. Mass spectrometry (MS) can give the empirical formula, but identification of groups is dependent on the somewhat arbitrary way the molecule fragments. Nuclear magnetic resonance spectroscopy (NMR), the most powerful and popular structural technique nowadays, while very good at determining many of the finer points of structure, is limited in practical terms by the types of nuclei that will give useful spectra.

Fortunately, these spectroscopic techniques complement each other and it is therefore sensible to utilize them all for complete structural determination of unknowns. A particular strength of IR spectroscopy is that it provides easily the best method for initially surveying a completely unknown material. Physical state or solvation properties are not a problem (as they can be for MS or NMR in obtaining a spectrum) and it is immediately apparent whether

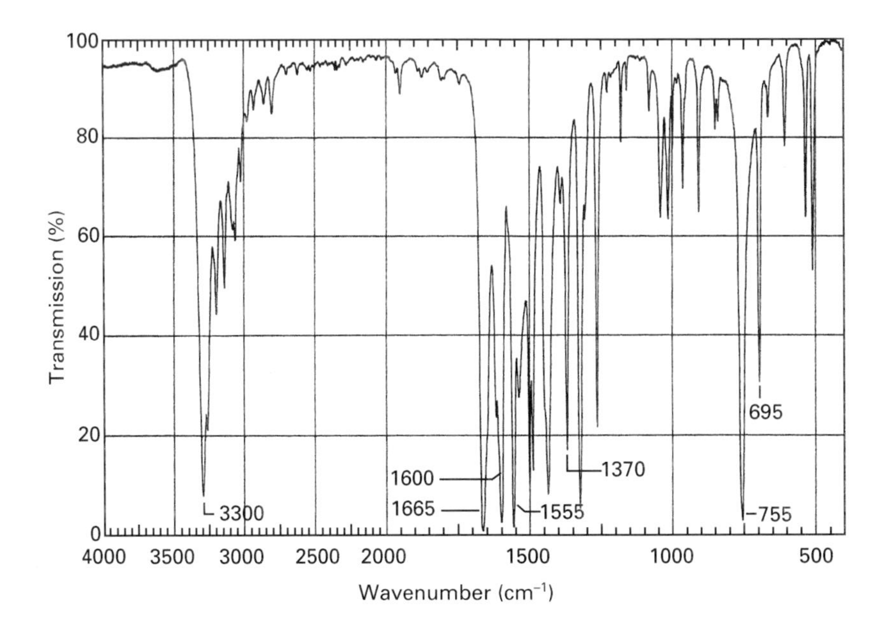

Figure 9 Band assignments for the IR spectrum of acetanilide (Ph–NH–C(=O) CH_3): 3300, ν(N–H); 1665, 1555, ν(C=O), δ(CNH) secondary amide; 1600, δ(CNH); 1370, δ(CH_3) sym; 755, 695, δ(CCH) monosubstituted benzene ring.

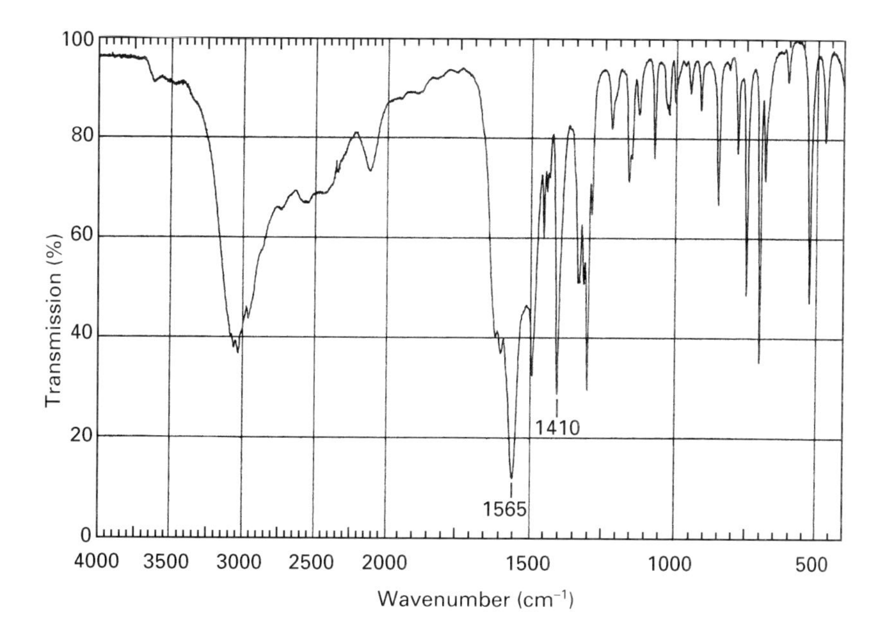

Figure 10 IR spectrum of phenylalanine. Band assignments are given in the text.

the sample is organic or inorganic (or a mixture) and what the major groups are likely to be.

List of symbols

f = bond force constant (N m^{-2}); δ = bending mode frequency (cm^{-1}); μ = reduced mass; ν = stretching mode frequency (cm^{-1}).

See also: **ATR and Reflectance IR Spectroscopy, Applications; Chromatography-IR Methods and Instrumentation; Far IR Spectroscopy, Applications; Forensic Science, Applications of IR Spectroscopy; High Resolution IR Spectroscopy (Gas Phase), Applications; High Resolution IR Spectroscopy (Gas Phase), Instrumentation; Hydrogen Bonding and other Physicochemical Interactions Studied By IR and Raman Spectroscopy; Industrial Applications of IR and Raman Spectroscopy; IR and Raman Spectroscopy of Inorganic, Coordination and Organometallic Compounds; IR Spectrometers; IR Spectroscopy Sample Preparation Methods; Medical Science Applications of IR; Near IR Spectrometers; Polymer Applications of IR and Raman Spectroscopy; Surface Studies By IR Spectroscopy.**

Further reading

Bellamy LJ (1975) *The Infrared Spectra of Complex Molecules*. London: Chapman and Hall.

Bellamy LJ (1975) *Advances in Infrared Group Frequencies*. London: Chapman and Hall.

Colthup NB, Daly LH and Wiberley SE (1975, 1990) *Introduction to Infrared and Raman Spectroscopy*, 2nd and 3rd editions. New York, San Francisco and London: Academic Press.

Dollish FR, Fatelely WG and Bentley FF (1974) *Characteristic Raman Frequencies of Organic Compounds*. New York: Wiley.

Lin-Vien D, Colthup NB, Fateley WG and Graselli JG (1991) *The Handbook of Infrared and Raman Characteristic Frequencies of Organic Molecules*. London: Academic Press.

Roeges NPG (1994) *A Guide to the Complete Interpretation of Infrared Spectra of Organic Structures*. New York: Wiley.

Silverstein RM and Webster FX (1998) *Spectrometric Identification of Organic Compounds*. New York: Wiley.

Socrates G (1994) *Infrared Characteristic Group Frequencies*. New York: Wiley.

IR Spectrometers

RA Spragg, Perkin-Elmer Analytical Instruments, Beaconsfield, Bucks, UK

VIBRATIONAL, ROTATIONAL & RAMAN SPECTROSCOPIES
Methods & Instrumentation

Introduction

From the 1980s onwards Fourier-transform infrared (FT-IR) spectrometers almost entirely replaced dispersive instruments. The most significant feature of FT spectrometers is that radiation from all wavelengths is measured simultaneously. This is much more efficient than the operation of the dispersive spectrometer, in which different wavelengths are measured successively. This difference, known as the Fellgett advantage, leads to much faster or more sensitive measurement. For example, a dispersive spectrometer typically takes at least 3 min to scan from 4000 to 400 cm^{-1}, while most FT instruments can acquire a spectrum in 1 s or less. The first commercial mid-IR Fourier transform spectrometer was introduced in 1969, but initially the cost of the necessary computing power confined the use of such instruments to research applications. More recently, FT-IR spectrometers have become less expensive in real terms than the dispersive instruments they have replaced. As this article will deal with instruments that are capable of recording complete spectra, it will not consider laser-based measurements.

The IR region

The IR region covers wavelengths from about 800 nm to 1 nm (**Figure 1**). However, spectroscopists generally divide this into three, near IR (NIR) from the visible to 2.5 µm, mid IR from 2.5 to 25 µm, and far IR (FIR) beyond 25 µm. The main significance of this division is that most fundamental molecular vibrations occur in the mid IR, making this region the richest in chemical information. The near-IR region contains overtones and combinations of the fundamental vibrations, especially those involving hydrogen atoms. Some electronic absorptions are also found in this region. The far-IR region is mostly of interest for vibrations involving heavy atoms, lattice modes of solids and some rotational absorptions of small molecules.

This article will concentrate on the mid-IR region, which covers the widest range of applications.

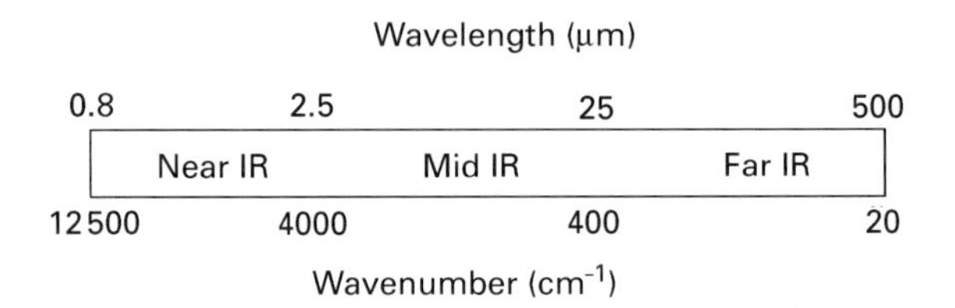

Figure 1 The divisions of the IR region.

However, it is worth noting that interest in the near-IR region has increased steadily in recent years.

Dispersive IR spectrometers

The principle of operation of a dispersive spectrometer is very simple. A beam from a broad-band source of radiation is dispersed by a prism or grating. The dispersed radiation falls on a slit through which a narrow range of wavelengths passes to reach a detector. Rotating the prism or grating brings successive wavelengths onto the slit, so scanning the spectrum. For the mid IR, the source is a ceramic rod or wire heated to around 1300 K. In practice, instruments use one or more gratings to cover the mid-IR region, with filters to separate different orders of diffraction. The usual detector is a thermocouple. Instruments operate in a double-beam mode (**Figure 2**) switching rapidly between two equivalent optical paths, one of which contains the sample. The transmission of the sample is measured from the ratio between the detector signals from the two beams directly (ratio recording) or by moving an attenuator into the reference beam until the detector signal matches that through the sample.

Fourier-transform spectrometers

Most commercial Fourier-transform (FT) spectrometers are based on the Michelson interferometer (**Figure 3**). A collimated beam of radiation falls on a beam splitter that divides it into two beams with approximately equal intensities. These two beams strike mirrors that return them to the beam splitter where they recombine, creating interference, and

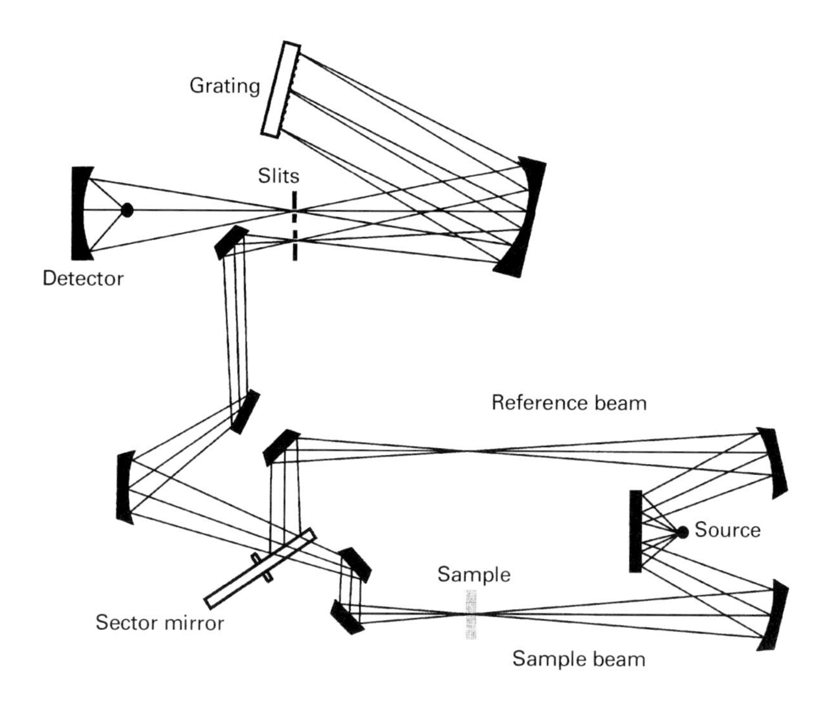

Figure 2 The optical arrangement of a double-beam dispersive IR spectrometer.

then proceed to a detector. Moving one of the mirrors creates a varying optical path difference (OPD) between the two beams. As the path difference changes, the two beams move in and out of phase with each other, so that there is alternating constructive and destructive interference.

Monochromatic radiation produces a sinusoidal signal, the frequency of which is directly proportional to the wavenumber. The signal from a broadband source consists of a combination of overlapping sinusoidal signals, called an interferogram. A typical interferogram has a strong signal, called the centre burst, at the point where the paths of the two beams are equal, and decays as the path difference increases (**Figure** 4). The interferogram can be converted into the required spectrum of intensities at different wavenumbers by a Fourier transformation (**Figure** 5).

Unlike dispersive spectrometers, FT instruments operate in a single-beam mode. The sample is normally placed between the interferometer and the detector, in contrast to the arrangement in dispersive spectrometers where the sample is placed between the source and the monochromator. Two spectra are recorded, first of the instrument alone and then with the sample in place. The ratio of these, calculated automatically, is the transmission spectrum of the sample (**Figures** 6 and 7).

Range and resolution in FT-IR spectrometers

The detector in a FT spectrometer sees a range of frequencies that is limited by the output of the source

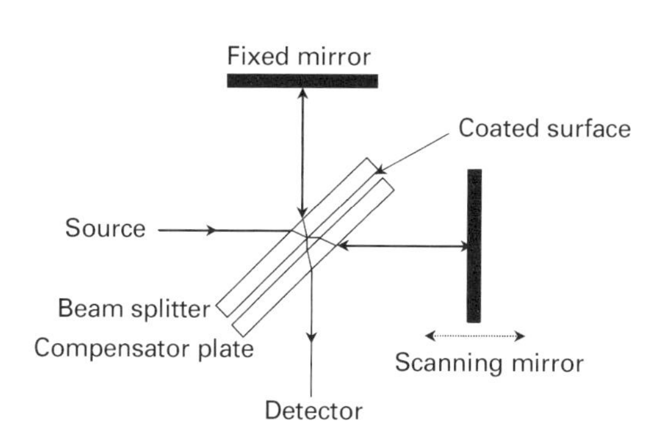

Figure 3 The basic optics of a Michelson interferometer.

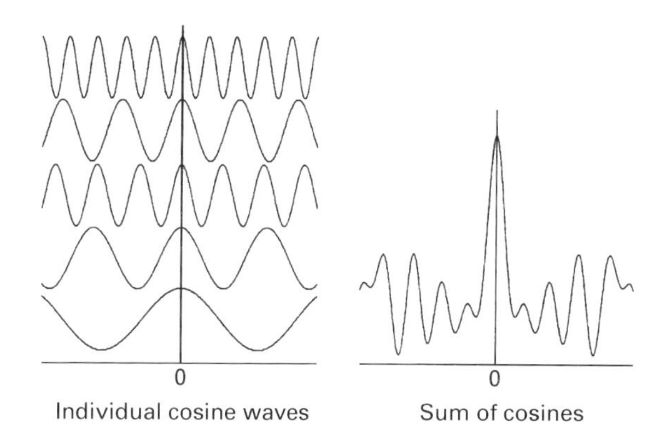

Figure 4 The interference of overlapping cosine waves.

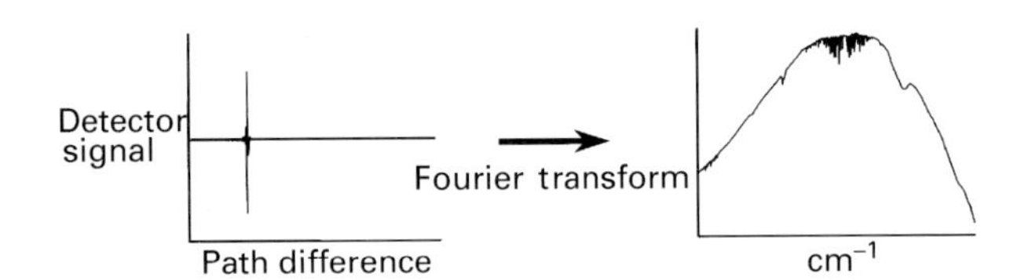

Figure 5 The interferogram from a broad-band source and the corresponding spectrum.

and the transmission of the optics. The interferogram is digitized prior to the Fourier transformation. The range of frequencies that can be identified in the Fourier analysis depends on the spacing of the data points in the digitized signal. There must be at least two data points for each cycle of the highest frequency in the analysis. This is known as the Nyquist criterion. If there are fewer than two data points per cycle, the signal is indistinguishable from that of another, lower frequency, a phenomenon known as aliasing. Optical or electrical filtering is therefore needed to remove any frequencies that are above the Nyquist limit (**Figure 8**).

The resolution in the spectrum depends on the spacing of the frequencies provided by the Fourier transform. This is determined by the length of the interferogram. The frequencies used in a Fourier transform are those such that an exact number of cycles fits the length of the interferogram from zero to the maximum OPD, Δ (in cm). Two adjacent frequencies in the analysis, corresponding to wavenumbers ν_1 and ν_2 (cm^{-1}), have n and $(n+1)$ cycles in this length. The wavenumbers are given by

$$\nu_1 = n/\Delta \qquad \text{and} \qquad \nu_2 = (n+1)/\Delta \qquad [1]$$

From this it follows that the difference between the wavenumbers is

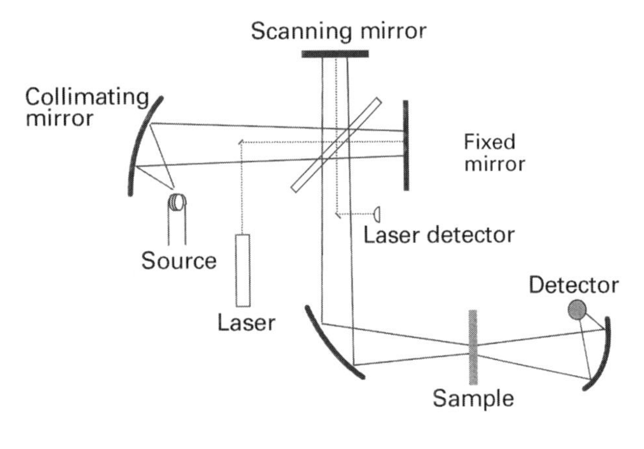

Figure 6 The optical arrangement of a FT-IR spectrometer.

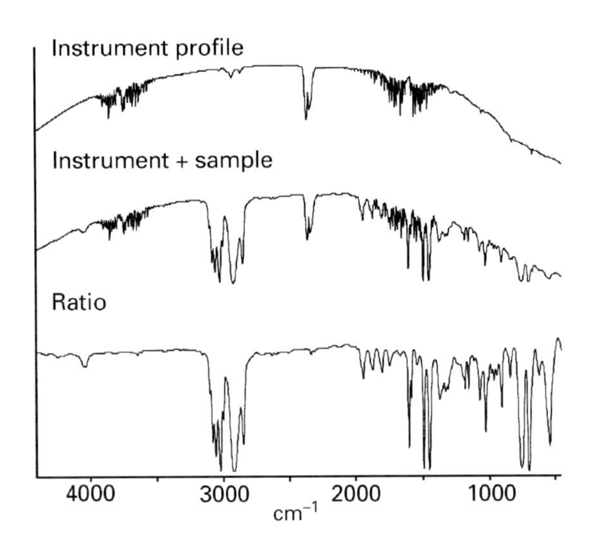

Figure 7 Single-beam spectra and the resulting sample spectrum.

$$(\nu_2 - \nu_1) = 1/\Delta \qquad [2]$$

Thus, the spectrum generated by the Fourier transform contains wavenumbers separated by the inverse of the maximum path difference in centimetres. Instruments for routine analytical work typically have highest resolution of 1 or 0.5 cm^{-1}, requiring a maximum path difference of 1 or 2 cm. This resolution is constant across the spectrum. Research instruments have been built with resolutions as high as 0.0009 cm^{-1}, which requires a path difference in excess of 10 m.

In a practical instrument, the beam passing through the interferometer cannot be perfectly parallel. There are rays at different angles and these experience different changes in path length for a given displacement of the moving mirror. This

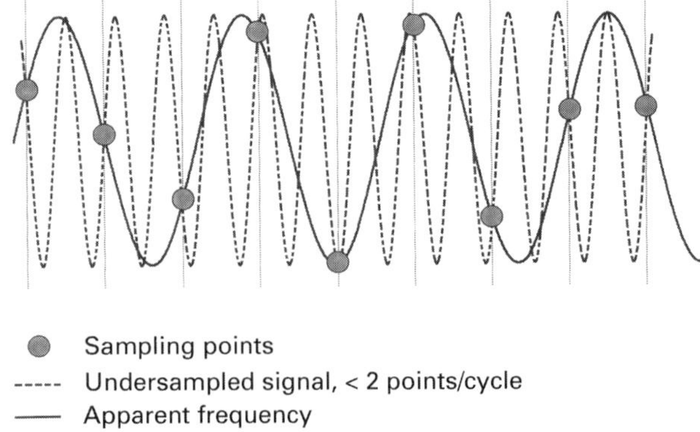

Figure 8 A signal sampled at less than two points per cycle and a lower frequency with identical values at the sampling points.

causes line broadening by an amount that is proportional to the wavenumber. The range of angles admitted to the interferometer has to be restricted so that the required resolution is achieved. Incorporating an aperture, called a Jacquinot stop, at an appropriate point in the optical system controls this. In some ways this aperture is analogous to the slit in a dispersive spectrometer. For a given resolution, the optical throughput of a FT spectrometer, limited by a circular aperture, can be higher than that of a dispersive spectrometer with a rectangular slit. This is referred to as the Jacquinot advantage and contributes to the higher sensitivity of FT instruments. The size of this aperture depends on the resolution and the wavenumber range. Lowering the resolution by a factor of 2 allows the energy reaching the detector to be doubled. However, in most spectrometers it is not possible to increase the energy above that permitted for 4 cm^{-1} resolution because of other limitations such as the range of the analogue-to-digital converter.

Measuring the interferogram

To establish the wavelength scale of the spectrometer, the interferogram has to be measured at precisely known intervals of the path difference. Generally this is achieved by using a helium–neon laser at 0.6328 μm (15 800 cm^{-1}) to provide a reference signal. The laser radiation passes through the interferometer to a separate detector. This produces a sinusoidal signal with a period equal to the wavelength of the laser (**Figure 9**). Typically the interferogram signal is measured at each point where the laser signal passes through zero. With this spacing of the data points in the interferogram, all wavenumbers below 15 800 cm^{-1} can be distinguished. The laser wavelength is known very accurately and is very stable. As a result, the wavelength scale of FT spectrometers is much more accurate than that of dispersive instruments. This superiority is known as the Connes advantage after the originator of the laser referencing technique. The absolute accuracy of the scale depends on the spectrometer resolution, while the scan-to-scan precision is of the order of 0.01 cm^{-1}.

The time taken for an interferometer to scan from zero to the maximum path difference is typically of the order of 1 s. For most routine measurements, the interferogram signal is averaged over several scans to reduce random noise that is largely generated by the detector. By averaging N scans the noise level in the final spectrum can be reduced by a factor of $N^{1/2}$ relative to that from a single scan. For the signal averaging to work properly, the interferogram must be digitized with sufficient precision to record the noise

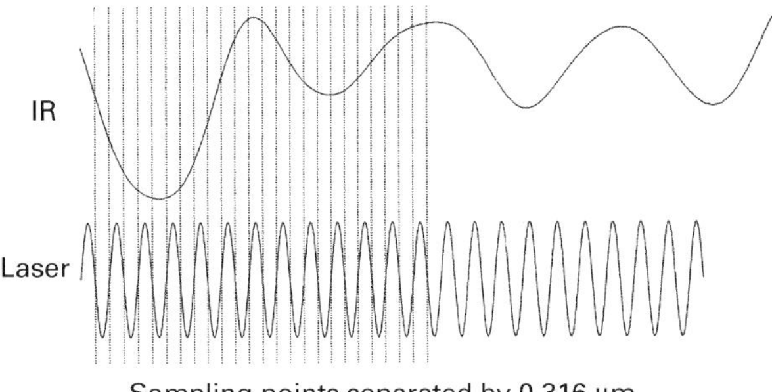

Figure 9 A short length of an interferogram with the sampling points corresponding to zero-crossings of the laser signal.

in a single scan. The analogue-to-digital converter must be able to handle the very large centre burst and the much smaller noise component. The precision needed for this is at least 16 bits.

In an ideal interferometer, the complete spectral information could be obtained by scanning the path difference from zero to the value needed to achieve the required resolution. However, data have to be measured for a limited distance on both sides of the nominal point of zero path difference in order to carry out phase correction. For signal averaging, the data points corresponding to the same path difference for successive scans have to be combined. The laser signal measures changes in the path difference but not its absolute value. In early FT instruments this was measured by including a broad-band white light source, the interferogram from which is a very sharp spike at zero path difference. The optics were arranged so that zero path difference for the white light was offset from that for the IR radiation. The white-light signal was used to trigger data collection (**Figure 10**). The interferometer scanned in one direction and collected a complete interferogram on

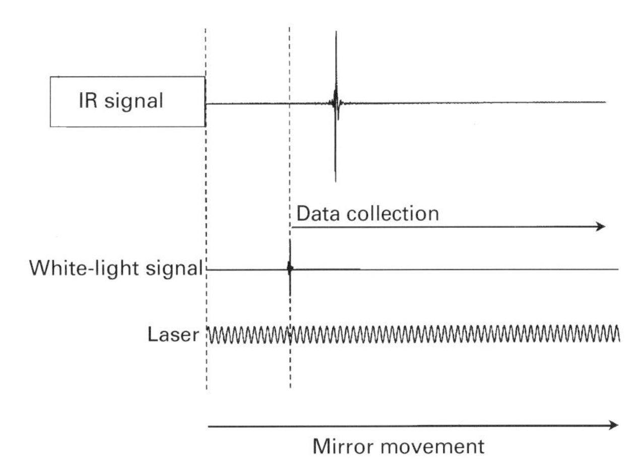

Figure 10 The use of a white-light interferogram to trigger data collection.

one side of zero path difference only. After each scan, the interferometer would move back past the start position to begin the next scan. Such a scheme is described as unidirectional single-sided scanning (**Figure 11**). It was necessary to trigger each scan separately because the path difference could not be tracked through the change of direction at the end of the scan. Most systems now use two detectors with a phase shift between them to monitor the laser signal, allowing forward and reverse motion to be distinguished. This quadrature detection arrangement can track the path difference continuously and so control the co-addition of successive scans without individual triggering. It also allows the interferometer to collect data from scans in both directions. This bidirectional scanning achieves more efficient data collection as there is no fly-back period between scans (**Figure 11**).

Processing the interferogram signal

The signal processing that converts the interferogram into a spectrum usually involves phase correction, zero-filling and apodization as well as the Fourier transform. An ideal interferogram would be perfectly symmetrical about zero path difference. However, in practice there is some asymmetry because of phase differences between the signals from different wavenumbers. A phase correction routine is applied before the Fourier transform. Phase correction can be avoided by scanning the full length of the interferogram on both sides of zero path difference. This allows a magnitude spectrum to be calculated without phase correction. The Fourier transform itself uses the fast Fourier transform (FFT) algorithm devised by Cooley and Tukey. This algorithm requires that the number of points to be transformed is equal to a power of 2. However, the number of data points collected by scanning to a path difference of 1 cm, for example, does not correspond to a power of 2. The interferogram is therefore extended by additional data points with the value of zero to give the required number of points. This process is called zero-filling. It is a common practice to apply further zero-filling to double or quadruple the number of points used. This improves the appearance of the final spectrum by increasing the density of the data points, although it does not improve the resolution. In some systems, interpolation in the spectrum is used to give results that are equivalent to the additional zero-filling.

Apodization is an operation that is necessary because of the finite length of the interferogram. Monochromatic radiation produces a sinusoidal interferogram, but the length of this is limited by the maximum path difference in the interferometer scan. The Fourier transform of such a signal is a line of finite width with oscillations on either side. This line has the form of a sinc function ($(\sin x/x)$) (**Figure 12**). The oscillations (or side-lobes) that

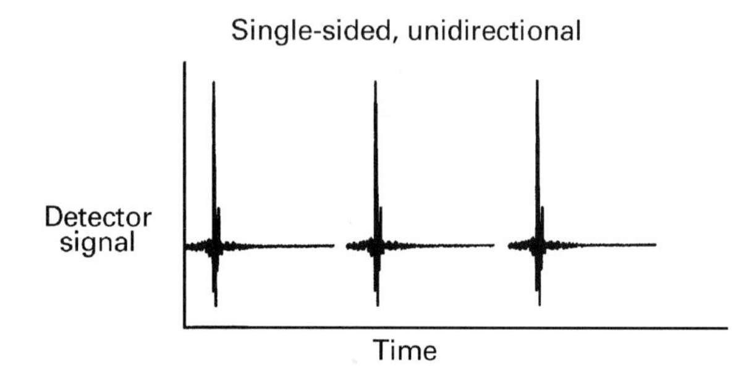

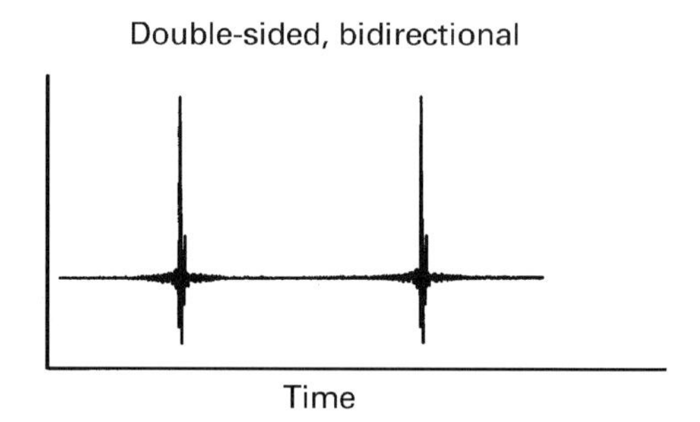

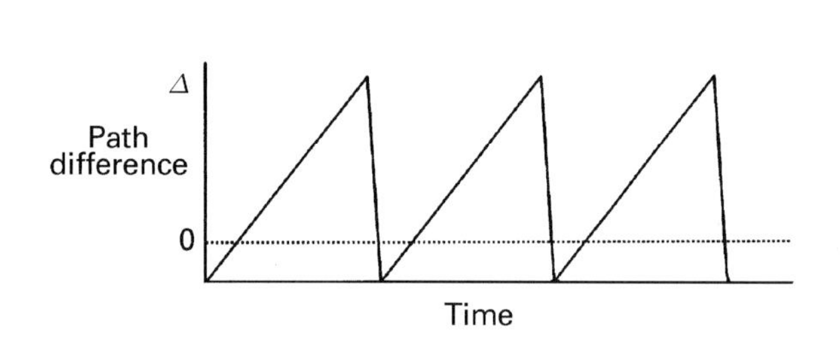

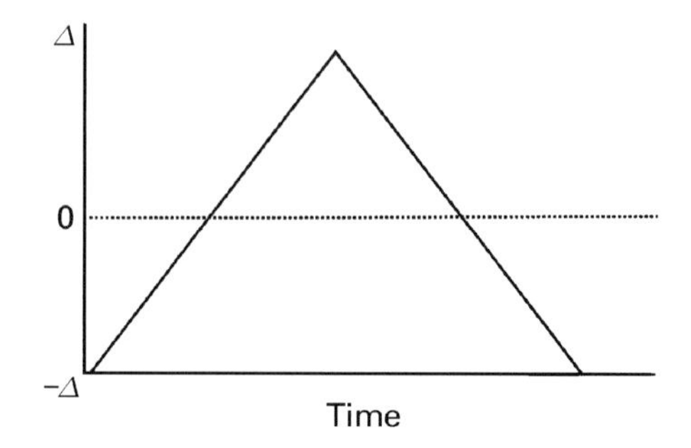

Figure 11 Alternative scanning and data collection schemes.

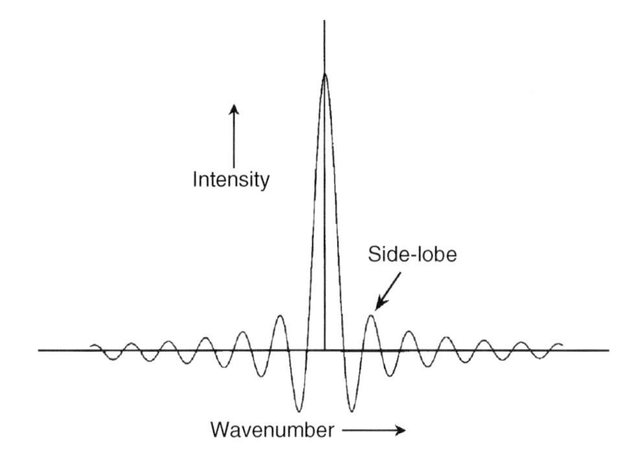

Figure 12 The sinc function generated by Fourier transformation of a cosine wave of finite length.

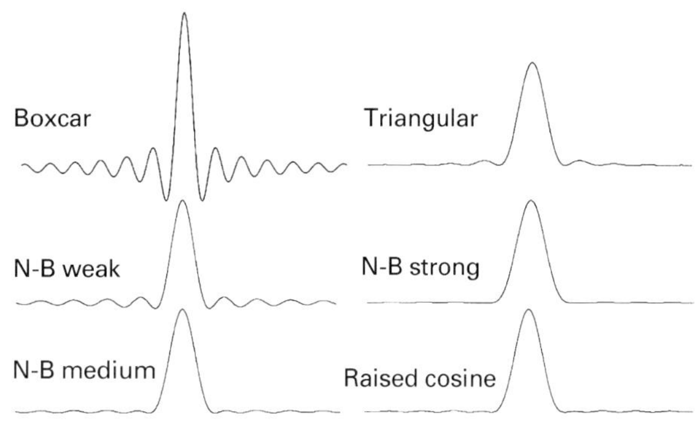

Figure 13 Some common apodization functions and the resulting line shapes.

accompany each line are obviously a problem because they could obscure smaller features in the spectrum. They can be reduced at the expense of some broadening of the lines by multiplying the interferogram with a function that decreases with increasing path difference. This process is called apodization. Various functions are used for this, representing different compromises between line broadening and the reduction of the side-lobes. Historically, a simple triangular function was often used, but other functions can reduce the side-lobes further with less broadening. By examining a very large number of alternative functions, Norton and Beer established empirical limits for the reduction in side-lobe intensity that could be achieved for a given increase in line width. Three functions identified by Norton and Beer are in common use, weak, medium and strong (**Figure 13**), corresponding to increasing degrees of side-lobe reduction and line broadening. The use of no apodization in order to achieve the narrowest possible lines is called boxcar truncation.

Scanning mechanisms

To create interference, the beams returned from the two mirrors in the interferometer must overlap exactly as they meet at the beam splitter. Any change in the alignment of the mirrors reduces the efficiency of the interference by an amount that increases with the square of the wavenumber. Small mechanical changes, for example associated with thermal expansion, can cause the alignment to change slowly. Changes that occur between measurement of the background and sample spectra result in a curved baseline (**Figure 14**).

The scanning mechanisms of interferometers have to be designed to minimize any change in alignment during the scan that would introduce noise or distortion into the spectrum. In systems where a plane mirror moves linearly, the problem is to avoid any tilting of the moving mirror. In some spectrometers the alignment is monitored and corrected continuously in a process called dynamic alignment. Multiple detectors measure the laser reference signal from different points in the beam. Any mirror tilt introduces a phase shift between the signals from the detectors because the path differences are not equal. This is automatically corrected in real time by using piezoelectric transducers to tilt the non-scanning mirror. The problem can also be addressed by using a cube-corner reflector instead of a plane mirror. Tilting the cube corner does not alter the direction of the reflected radiation but lateral movement would affect the alignment. There are also scanning arrangements that completely avoid alignment errors

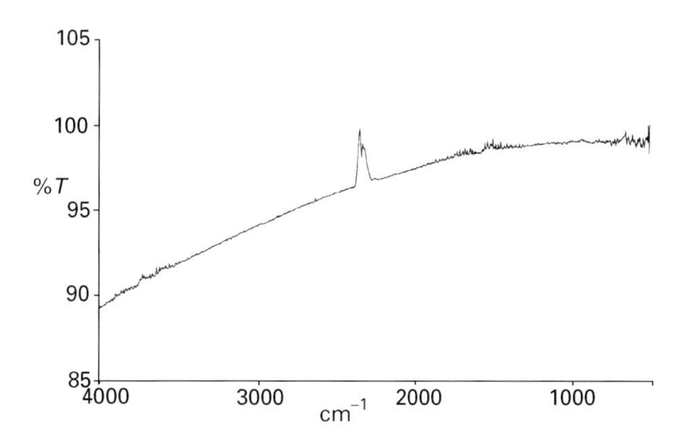

Figure 14 A parabolic baseline caused by misalignment of the interferometer.

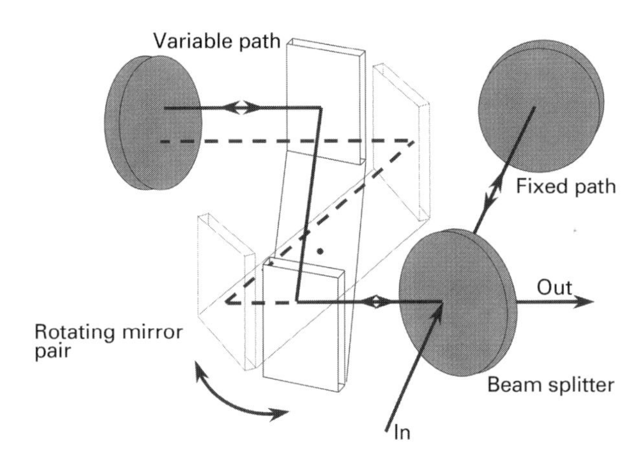

Figure 15 A scanning mechanism with a rotating mirror pair.

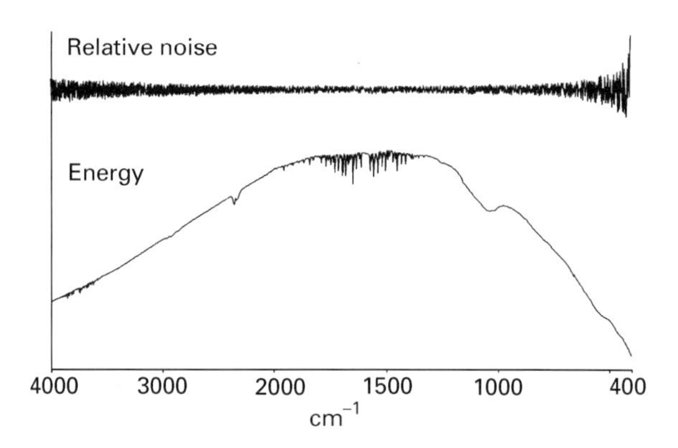

Figure 16 The instrument profile for a mid-IR spectrometer and the resulting variation of the noise level.

during scanning. One of the simplest generates the path difference by rotating a pair of parallel mirrors in one arm of the interferometer. The mirror pair does not alter the direction of radiation passing through it, whatever its orientation, so that it cannot affect the alignment as it rotates (**Figure 15**).

Some commercial instruments use a scanning mechanism that does not involve moving mirrors. Changing the refractive index instead of the physical length can alter the optical path. By moving a wedge of a material such as potassium bromide (KBr) into one arm of an interferometer and displacing air, the optical path is increased. Because the mechanical tolerances are less severe than for simple moving-mirror systems, these instruments are mostly used in non-laboratory applications. One disadvantage is that the refractive index varies with wavelength so that the wavenumber scale is not inherently linear as it is in conventional instruments.

The range covered by a spectrometer depends primarily on the properties of the source and the beam splitter, and is sometimes limited by the detector. The sources used in mid-IR spectrometers typically operate at temperatures around 1300 K. Their output has a fairly similar profile to a black body and so provides significant energy across the entire mid-IR region. An ideal beam splitter would provide uniform reflectivity of 50% and no absorption across the region of interest. In practice, a single beam splitter made from KBr with a multilayer coating can be used from the near IR to about 400 cm^{-1}, where the KBr absorbs strongly. Most interferometer designs incorporate a compensator plate so that both beams pass through the same thickness of KBr. A FT spectrometer can cover the entire mid-IR region in a single measurement. As can be seen from **Figure 16** the single-beam energy falls off towards the ends of the range. In consequence, the noise level in the final spectrum increases at each end.

Detectors

For routine measurements the most common detector material is deuterated triglycine sulfate (DTGS). This is a thermal detector, the signal being generated in response to the change in temperature caused by absorption of the IR radiation. It is generally operated at a temperature close to ambient. The response time of these detectors is less than 1 ms, allowing them to follow changes occurring at rates up to several kilohertz. In spectrometers with DTGS detectors, the OPD is typically scanned at about 0.3 cm s^{-1} because faster scanning would reduce the response of the detector. The time taken to scan the 0.25 cm needed for a spectrum at 4 cm^{-1} resolution is therefore approximately 1 s. These detectors have excellent linearity over a wide range of signal levels. Lithium tantalate is an alternative room-temperature detector material with somewhat lower performance but also lower cost.

When higher speed or sensitivity is needed, liquid nitrogen-cooled MCT (mercury cadmium telluride) detectors are used. These are photon detectors in which electrons are excited directly by the absorption of radiation. Cooling to liquid-nitrogen temperature is needed to avoid excitation of electrons by thermal motion. These detectors do not cover the complete IR range because there is a minimum energy needed to excite an electron, corresponding to a long-wavelength limit. Changing the composition can vary this limit, but extending the range reduces the sensitivity. The highest sensitivity is obtained with narrow-band detectors, which have a limit of about 750 cm^{-1}, while wide-band detectors

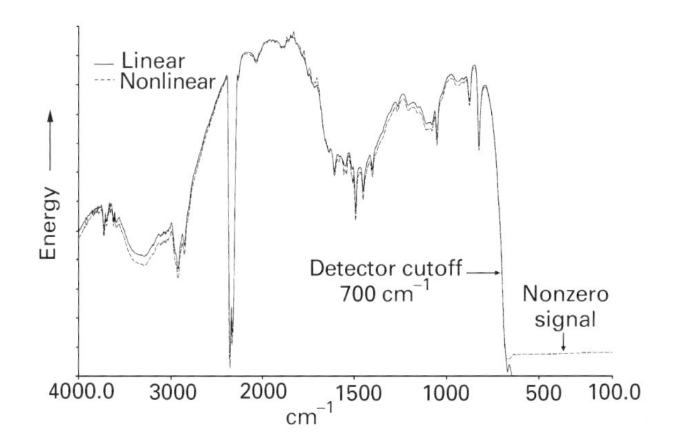

Figure 17 The instrument profile with a cooled MCT detector showing linear and nonlinear behaviour.

Table 1

Region	*Range* (cm^{-1})	*Source*	*Beam splitter*	*Detector*
Mid IR	7 000–400	Ceramic, wire,Sic	KBr, ZnSe	DTGS, LiTa, MCT, InSb
Near IR	15 000–4 000	Tungsten–halogen	CaF_2, quartz	DTGS, PbS, InGaAs, InSb
Far IR	400–20	Ceramic, wire, mercury arc	Mylar, wire grid, CsI (>200 cm^{-1})	DTGS, Si

reach 450 cm^{-1}, close to the limit for KBr optics. Scan rates are not limited by the detector response, and indeed the detector performance improves as the scan rate is increased. This allows data to be acquired at rates up to 100 scans per second in commercial instruments. The response of MCT detectors is inherently nonlinear (see **Figure 17**). This restricts their use to low signal levels, typically less than 10% of the energy available. However this is not a problem in applications where the signal level is restricted, such as microscopy. Because of the nonlinearity the centre burst signal is compressed relative to the rest of the interferogram. After Fourier transformation, the primary consequence is an offset in transmission, so that totally absorbing bands do not appear at zero. The presence of nonlinearity can be detected by examining the single-beam signal in the region beyond the long-wavelength limit of the detector. Any nonlinearity causes this signal to be nonzero. Various methods, both in electronics and in software, have been employed to minimize the effects of nonlinearity. None of these is entirely successful. For this reason, room-temperature detectors are preferred for quantitative measurements.

The combinations of components found in use for the three regions are summarized in **Table 1**.

Step-scan instruments

In most spectrometers the path difference is scanned continuously, but so-called step-scan instruments have become important for some research applications (see **Figure 18**). In these instruments the path difference is held constant at one point while the interferogram signal is measured before being moved to the next point. The IR signal has to be modulated to match the frequency response of the detector. This is achieved by applying a small-amplitude oscillation to the fixed mirror. In practice, a constant path difference can be maintained either by moving the two mirrors synchronously or by holding both stationary. Step-scanning instruments can address a much wider range of dynamic measurements than continuously scanning systems because the signal modulation frequencies are not restricted by the velocity of the moving mirror.

The principal applications have been to very rapid kinetics, photoacoustic spectroscopy and spectroscopic imaging with array detectors. The kinetic applications involve processes that are repeated for each step of the path difference. The signal can be measured at time intervals limited only by the detector response and electronics. At each path difference measurements are made at the same times after initiation of the process. The data are then reshuffled to create a series of interferograms in which each point was measured at the same time into the process. This approach has allowed FT-IR measurements to reach the nanosecond timescale. In photoacoustic spectroscopy there are advantages because the modulation frequency is constant across the spectrum, making it feasible to measure the spectra of subsurface layers. In the imaging application, the sample is focused onto an array detector. The signal from each pixel in

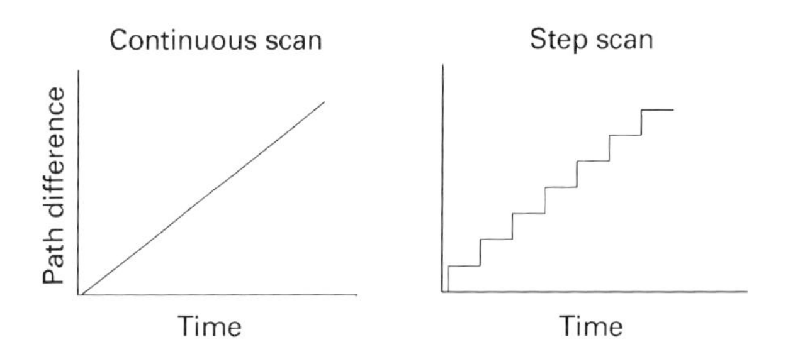

Figure 18 The difference between continuous and step scanning.

the array can be read and averaged at each path difference. This is not feasible with continuous scanning spectrometers because the array cannot be read out sufficiently fast. With a step-scan spectrometer a 64 × 64-pixel image containing 4096 spectra can be generated in a few minutes.

Commercial instruments

Commercial spectrometers can be grouped into those intended for routine measurements and research instruments. In general, the routine instruments are designed to simplify operation while research instruments aim to offer versatility. In virtually all cases the instrument is controlled by a personal computer that handles an increasing amount of the digital signal processing in the system.

A typical routine mid-IR spectrometer has KBr optics, best resolution of around 1 cm^{-1}, and a room-temperature DTGS detector. Noise levels below 0.1%T peak-to-peak can be achieved in a few seconds. To reduce the problem of fluctuating levels of atmospheric water vapour and carbon dioxide, much of the optical path is sealed and desiccated. The sample compartment will accommodate a variety of sampling accessories such as those for ATR (attenuated total reflection) and diffuse reflection. In most cases the output from the interferometer can be directed through an external port, principally for coupling to an IR microscope. Other external accessories provide coupling to a gas chromatograph or thermogravimetric analyser. These instruments often have versions for near-IR operation in which the source, beam splitter and detector are different.

Research spectrometers are generally larger, permitting higher optical throughput and hence better signal-to-noise performance, and higher resolution. There may be multiple external ports for the interferometer output so that several experiments can be installed simultaneously. The major components – source, beam splitter and detector – are generally exchangeable, so that a wide spectral range can be covered. A common arrangement is to have alternative sources and detectors permanently installed and selectable by switching mirrors. Different ranges can then be selected simply by changing the beam splitter. For far-IR operation, the beam splitters are generally Mylar films, which have a narrow range, or metal grids on a polymer substrate. Cooled silicon bolometers are the detectors of choice for high resolution and sensitivity. Some instruments have a vacuum version that removes the very strong water absorption in this region, allowing measurement out to 5 cm^{-1}. Although there have been relatively few reported applications, some spectrometers are also capable of measuring throughout the visible region.

The range of scan speeds in research instruments is greater than in routine instruments. The higher speeds provide better performance with cooled detectors and for kinetics, and the lower speeds are used with photoacoustic detectors. Digital signal processing is increasingly being used to facilitate experiments involving external modulation in which two channels of information have to be acquired simultaneously. Examples include polarization modulation in reflection–absorption of species on metal surfaces and in measuring the dynamic dichroism associated with polymer stretching. Research instruments generally have both continuous-scanning and step-scan modes as discussed earlier.

Since the introduction of the first commercial FT instrument in the late 1960s, there have been great improvements in speed and sensitivity. Routine measurements using the common sampling techniques now take only seconds, so that further improvements in performance will bring little benefit. Future advances are more likely to be seen in the development of smaller and more robust spectrometers. There are areas where improved performance is desirable, such as the sensitivity of chromatographic interfaces and the spatial resolution of IR microscopes. However, there has been little recent progress and there is little prospect of significant advances in these areas. The advent of step-scan spectrometers has led to great advances, but these have been in rather specialized applications, apart from imaging. Thus FT-IR instrumentation can be classified as mature. There is little prospect of any alternative methods of measuring IR spectra replacing FT instruments for most purposes.

List of symbols

n = number of cycles in Δ; N = number of scans that are averaged; %T = percent transmission; Δ = maximum optical path difference; ν = frequency.

See also: **ATR and Reflectance IR Spectroscopy, Applications; Chromatography-IR Applications; Chromatography-IR, Methods and Instrumentation; Far IR Spectroscopy, Applications; Fourier Transformation and Sampling Theory; High Resolution IR Spectroscopy (Gas Phase), Applications; High Resolution IR Spectroscopy (Gas Phase), Instrumentation; IR Spectral Group Frequencies of Organic Compounds; IR and Raman Spectroscopy of Inorganic, Coordination and Organometallic Compounds; IR**

Spectroscopy, Theory; Near-IR Spectrometers; Surface Studies by IR Spectroscopy.

Further reading

Analytical Chemistry (1995) **67**: 381A–385A, (1998) **70**: 273A–276A. (Reviews of currently available commercial laboratory spectrometers appear every few years in this journal. These are the most recent.)

Chalmers JM and Dent G (1997) *Industrial Analysis with Vibrational Spectroscopy*, RSC Analytical Spectroscopy Monographs. Cambridge: Royal Society of Chemistry. (More practically oriented than Griffiths and de Haseth, this describes the optical layouts of both dispersive and FT commercial spectrometers.)

Griffiths PR and de Haseth JA (2000) *Fourier Transform Infrared Spectroscopy*. 2nd edn. New York: Wiley. (This is the standard book on the subject with detailed discussion of the principles of operation and design of FT-IR spectrometers.)

Mackenzie MW (ed) (1989) *Advances in Applied Fourier Transform Spectroscopy*. Chichester: Wiley.

Persky MJ (1995) *Review of Scientific Instruments* **66**: 4763–4797. (A review of spaceborne Fourier transform spectrometers for remote sensing. This is an excellent review of non-laboratory instrumentation with a very clear summary of the difficulties. Numerous designs are described.)

IR Spectroscopic Studies of Hydrogen Bonding

See **Hydrogen Bonding and other Physicochemical Interactions Studied By IR and Raman Spectroscopy.**

IR Spectroscopy in Forensic Science

See **Forensic Science, Applications of IR Spectroscopy.**

IR Spectroscopy in Medicine

See **Medical Science Applications of IR.**

IR Spectroscopy of Surfaces

See **Surface Studies By IR Spectroscopy.**

IR Spectroscopy Sample Preparation Methods

RA Spragg, Perkin-Elmer Analytical Instruments,
Beaconsfield, Bucks, UK

Introduction

One of the strengths of IR spectroscopy as an analytical technique is the ability to obtain spectra from a very wide range of materials. However, in many cases some form of sample preparation is required in order to obtain a good quality spectrum. For direct transmission spectra, samples have to be no more than a few tens of micrometres thick. This means that reflection measurements are often needed if materials are to be measured in their original form. Fortunately, the choice of methods for obtaining spectra by reflection has expanded significantly in recent years. When neither transmission nor reflection spectra are satisfactory, spectra can generally be obtained by photoacoustic spectroscopy, although this is a relatively expensive option.

Modern spectrometers achieve very low noise levels and have a range of data processing options that can improve the appearance of spectra. However, these are not sufficient to generate a good spectrum from a sample that has not been prepared satisfactorily. In the context of this article, a good spectrum is considered as one that is suitable for comparison with reference data. This requires that the relative intensities, shapes and positions of both strong and weak bands should be observable. With modern spectrometers the intensities of bands at 1% transmission can be measured adequately for qualitative purposes, although for quantitative analysis it is wise to avoid bands at below 10% transmission. Although noise levels can be well below 0.1%, it should be remembered that the measurement of very weak bands is more likely to be hindered by interferences such as matrix impurities or instrument artefacts than by high-frequency noise.

Transmission measurements

Liquids

Liquids are measured as thin films between IR-transmitting windows. The choice of window material depends on solubility, chemical compatibility and the range needed. The properties of common windows are listed in **Table 1**. For organic materials KBr is the most common choice. Zinc selenide has advantages for aqueous solutions because of the wide range, but the high refractive index leads to interference fringes that can be a problem for quantitative work, and calcium fluoride and barium fluoride therefore continue to be preferred. KBr and other alkali metal halide windows deteriorate with continued exposure to moist atmospheres and are easily scratched, but can be repolished fairly easily. Gloves should be worn to handle them to avoid fogging. The windows used for aqueous samples are more robust and are not generally repolished.

There are three main ways to measure liquids, in cells with spacers that may be fixed or demountable, or as a film directly between windows. In so-called sealed cells the spacer is a metal amalgam which provides a fixed pathlength. These are generally preferred for quantitative work. Demountable cells use a spacer made from metal foil or PTFE. The windows can be separated for cleaning or to change the pathlength. However, the pathlength generally changes slightly when the cell is reassembled, creating problems for quantitative work. When assembling demountable cells it is important to avoid dust particles on the spacer since these affect the spacing and can cause leaks. The pathlengths used for mid-IR measurements range from about 15 µm for aqueous samples to 500 µm or more for measuring components at low concentrations in weakly absorbing solvents. The pathlength of an empty cell is easily determined by measuring interference fringes. The thickness is equal to $(2 \times \text{fringe spacing})^{-1}$.

Care must be taken in filling short-pathlength cells because it is possible to generate enough internal pressure to damage the windows. For this reason it is better practice to pull the liquid into the cell by attaching a reservoir to one port and a syringe to the other, than to push liquid into the cell from a syringe. The cell should be tilted for filling to avoid trapped bubbles that cause stray light. A demountable cell is shown in **Figure 1**.

Neat liquids can be measured by placing one drop on a window and putting a second window on top of this. The pathlength can be varied by applying pressure. This is very convenient for qualitative work, but cannot be used for very volatile materials.

Table 1 Properties of IR-transmitting materials

Material	*Range* (cm^{-1})[a]	*Refractive index*	*Properties*
NaCl	>600	1.52	Soluble in water and some alcohols, easily polished
KBr	>400	1.54	As NaCl, hygroscopic
CsI	>200	1.74	Soluble in water and some alcohols, very hygroscopic
CaF_2	>1100	1.40	Insoluble, attacked by NH_4^+ salts, robust
BaF_2	>750	1.45	As CaF_2, but sensitive to shock
ZnSe	>500	2.43	Insoluble but attacked by acids and strong bases
ZnS	>750	2.25	Insoluble, attacked by acids and strong oxidizing agents, good thermal shock resistance
Diamond	>2200, <1900	2.38	Highest chemical and mechanical resistance
KRS–5	>250	2.38	Slightly soluble, attacked by bases, very soft, toxic
Sapphire	>1600	1.62	Very hard, slightly soluble in acids and alkalis
Ge	>600	4.01	Hard and inert, brittle
Si	>650, <450	3.42	Very hard and inert
AMTIR	>725	2.5	Insoluble, attacked by bases, brittle
Polyethylene	<625	1.5	Insoluble, swells in some organic solvents
Fused silica	>2500	1.46	Insoluble
AgBr	>300	2.30	Insoluble, attacked by acids and NH_4^+ salts, soft

[a] This is for windows. The longer pathlength in ATR elements generally reduces the range, e.g. to 650 cm^{-1} for ZnSe.

For non-volatile liquids porous polymer films are a convenient alternative to conventional windows. These '3M cards' (**Figure 2**) are made of polyethylene or PTFE. The polyethylene films absorb in rather narrow regions and, except for the region near 3000 cm^{-1}, their absorptions can be removed by subtraction. The PTFE cards are useful for obtaining spectra in the CH stretching region around 3000 cm^{-1}.

Solids

The two most common forms of sample preparation for solids both involve grinding the material to a fine powder and dispersing it in a matrix. Scattering and reflection effects are reduced when the particle size is less than the wavelength of incident radiation, which in this case is a few micrometres. Surrounding the particles by a matrix with a similar refractive index further reduces the effects. Most samples can be ground either mechanically with a small vibrating mill or by hand with a pestle and mortar, the choice being a matter of personal preference. For manual grinding the amount of sample should be no more than around 20 mg. The sample should be ground until it is adhering to the sides of the mortar and has a glossy appearance. This process may take several minutes.

The ground material can be dispersed in a liquid to form a mull. The most commonly used liquid is

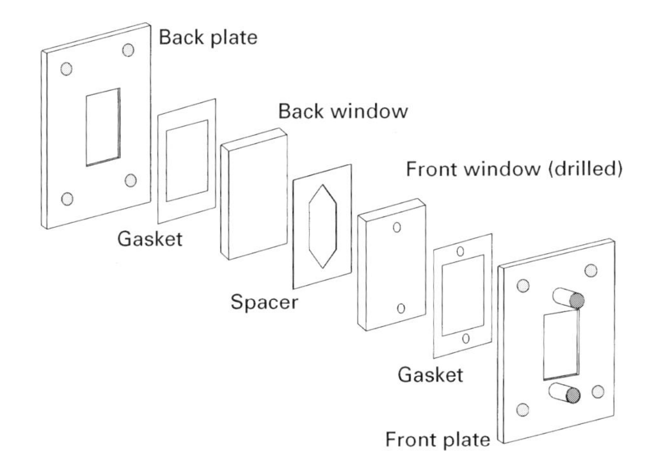

Figure 1 Construction of a demountable cell for liquids.

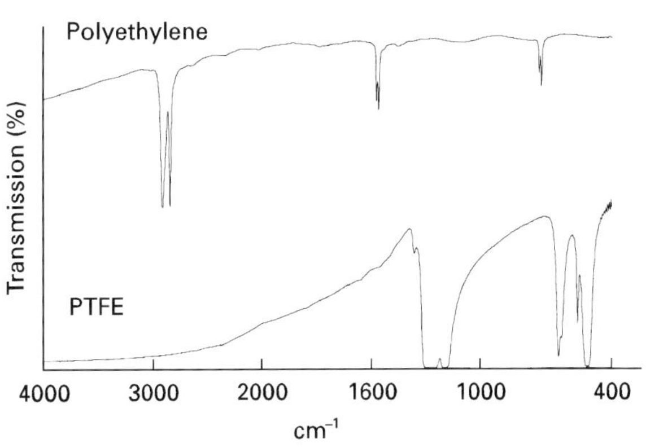

Figure 2 Spectra of 3M cards.

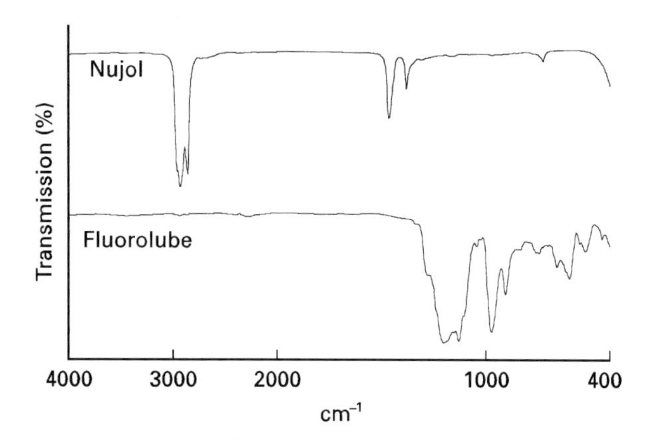

Figure 3 Spectra of mulling agents.

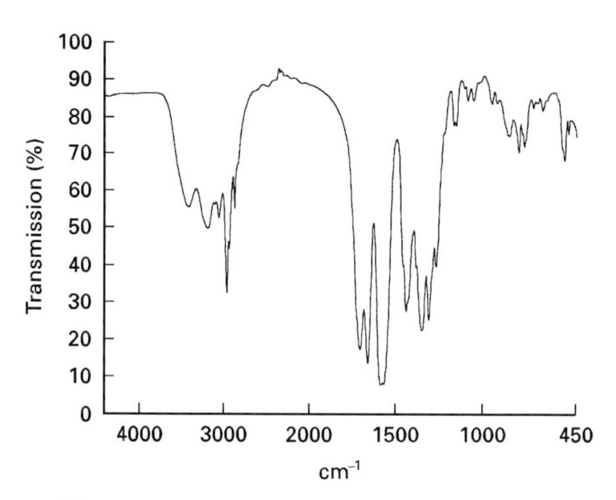

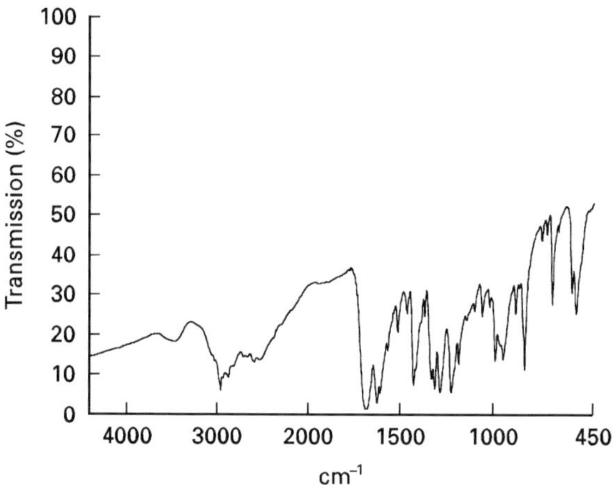

Figure 4 Spectra from poorly (bottom) and well (top) ground discs.

mineral oil (Nujol), while a fluorinated hydrocarbon (Fluorolube) can be used to observe the 3000 cm^{-1} region (**Figure 3**). No more than one or two drops are needed to produce a paste that can be spread between windows. The consistency of this paste is a compromise between the ease of spreading and the interference from the absorptions of the mulling agent.

Potassium bromide (KBr) is probably the most widely used matrix material, even though preparing a good KBr disc is time consuming and requires practice. Between 1 and 3 mg of ground material is mixed thoroughly with about 350 mg of ground KBr. The mixture is transferred to a die that has a barrel diameter of 13 mm. This is placed in a suitable press, evacuated and then pressed at around 12 000 psi for 1–2 min. Recrystallization of the KBr results in a clear glassy disc about 1 mm thick. Discs can be made with manual presses and without evacuation, but the best results are obtained with the more elaborate procedure.

Provided that the sample, the KBr and the disc are weighed, this technique can be used for quantitative work. It has the advantage that there are no matrix absorptions above 400 cm^{-1}, although adsorbed water in the KBr has to be avoided. Potential limitations are that the grinding and pressing can modify the crystalline form of the sample and that ion exchange can occur. To avoid ion exchange it is common to use potassium chloride for amine hydrochlorides. Caesium iodide is used for the region to 200 cm^{-1} while for far-IR measurements polyethylene is a suitable matrix. Grinding is the main issue in disc preparation (**Figure 4**). Inadequate grinding results in sloping baselines caused by light scattering and in asymmetric bands. The asymmetry, known as the Christiansen effect (**Figure 5**) results from refractive index differences.

Very few solid samples arrive in a suitable form for direct transmission measurement, the principal exception being polymer films. Unfortunately, interference fringes often distort the spectra. There are several techniques for eliminating interference fringes by reducing the surface reflections. The simplest are to abrade the surface or to sandwich the film between windows. Another possibility is to smear one surface with mineral oil, provided that this does not obscure

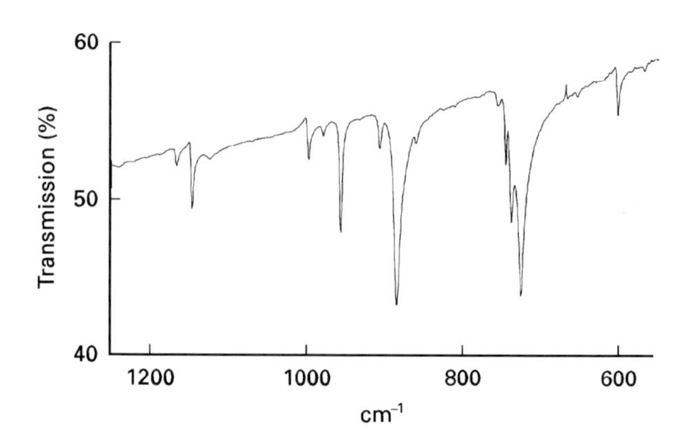

Figure 5 Christiansen effect.

important regions of the spectrum. Polymer samples that are too thick for direct measurements can often be pressed to a suitable thickness. Heated platens for making films of controlled thickness are available from accessory suppliers.

Simple compression cells with diamond windows have appeared recently. Although they generate lower pressures than traditional diamond anvil cells, they can provide excellent spectra from a range of samples that are too thick for direct measurements.

Soluble materials can be prepared as cast films by evaporating them from solution on a suitable substrate such as a KBr window. An infrared lamp is a convenient means of warming the film to remove residual solvent. This approach is especially useful for those polymers that can be cast on to a glass block from which they can be removed for measurement without a support. A wide range of organic compounds can be prepared on the porous polymer substrates known as 3M cards. The best spectra are obtained when the compound forms an amorphous deposit. The spectra of crystalline samples may be distorted by scattering and reflection effects.

Gases

Measurements on gases require much longer pathlengths than those for condensed phases; 10 cm is usual for compounds at relatively high concentrations. For trace concentrations multiple reflection cells provide pathlengths of several metres in compact designs. Detection limits are well below 1 ppm for many compounds.

Internal reflection methods

In recent years, internal reflection methods have developed into perhaps the most versatile approach to obtaining IR spectra. Total internal reflection can occur at the interface between media with different refractive indices (**Figure 6**). If the angle of incidence exceeds a critical value, radiation cannot pass into the medium with the lower refractive index; instead, it is totally reflected. Although there is total reflection, an electric field, sometimes referred to as an evanescent wave (**Figure 7**), extends a short distance into the second medium. This distance is of the same order of magnitude as the wavelength. In an internal reflection measurement the radiation enters a high refractive index material and emerges after undergoing internal reflection. The high refractive index element is usually called the crystal, even though the material may be a glass. An absorbing sample placed on the surface of the crystal interacts with the electric field to generate a spectrum. This is called an attenuated total reflection or ATR spectrum. The band positions and shapes are slightly different from those in transmission spectra, but not enough to affect qualitative identification.

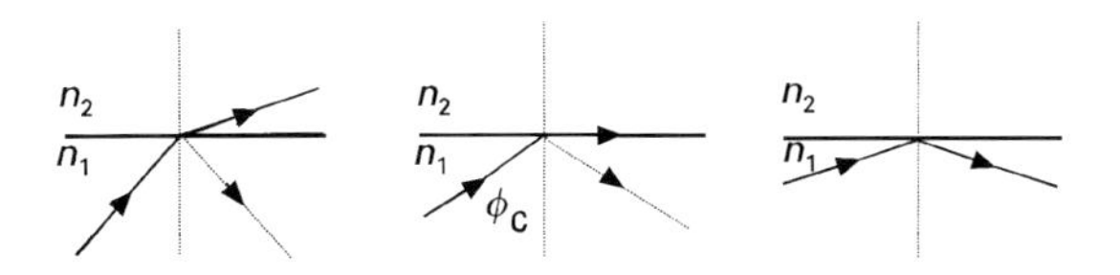

Figure 6 Total internal reflection at interface where $n_1 > n_2$.

The penetration depth, d_p, of the light into the lower refractive index material is defined as the depth at which the field amplitude has fallen to 1/e of its value at the surface. For two materials with refractive indices n_1 and n_2 it is given by

$$d_p = \frac{\lambda_1}{2\pi(\sin^2 \alpha - n_{21}^2)^{1/2}}$$

where λ_1 is the wavelength in the higher refractive index material and is related to the incident wavelength λ by the expression $\lambda_1 = (\lambda/n_1)$, α is the angle of incidence and $n_{21} = n_2/n_1$.

The intensity of an ATR spectrum depends on the distance that the electric field extends into the sample and on the number of reflections. The most commonly used materials are zinc selenide and diamond, both with a refractive index of 2.3, and germanium, with a refractive index of 4.0. The higher refractive index reduces the distance that the electric field penetrates into the sample, giving a weaker spectrum. The depth of penetration is proportional to the wavelength, so that when ATR spectra are compared with transmission the intensities at longer

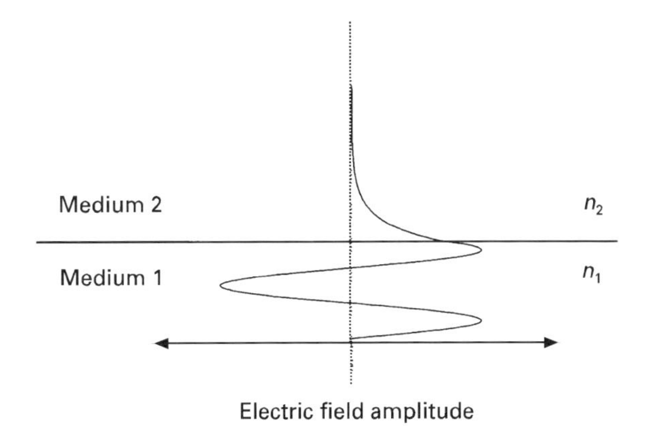

Figure 7 Evanescent wave at an interface where $n_1 > n_2$.

wavelengths appear enhanced. For an ATR crystal with a refractive index of 2.2 and a 45° angle of incidence, the depth of penetration in a typical organic sample is about 4μm at 1000 cm^{-1}.

Because ATR measurements are obtained from a thin surface layer, the spectra may not be representative of the bulk material. For example, spectra of polymer laminates such as packaging materials generally show only the surface layer. However, the spectra of emulsions and suspensions may be unrepresentative of the bulk material because of segregation at the surface.

There are two basic designs of ATR accessory in common use. These have replaced traditional designs in which a thin sample was clamped against the vertical face of the crystal. In horizontal ATR (HATR) units the crystal is a parallel-sided plate, typically about 5 cm by 1 cm, with the upper surface exposed (**Figure 8**). Radiation traverses the length of the crystal, undergoing internal reflection at the upper and lower surfaces several times. The number of reflections at each surface is usually between five and ten, depending on the length and thickness of the crystal and the angle of incidence. Liquids are simply poured on to the crystal, which is recessed within a metal plate to retain the sample. Pastes and other semisolid samples are readily measured by spreading them on the crystal. Horizontal ATR units are often used for quantitative work in preference to transmission cells because they are easier to clean and maintain.

When measuring solids by ATR, there is the problem of ensuring good optical contact between the sample and the crystal. The accessories have devices that clamp the sample to the crystal surface and apply pressure. This works well with elastomers and other deformable materials, and also with fine powders, but many solids give very weak spectra because the contact is confined to small areas. The effects of poor contact are greatest at shorter wavelengths where the depth of penetration is lowest.

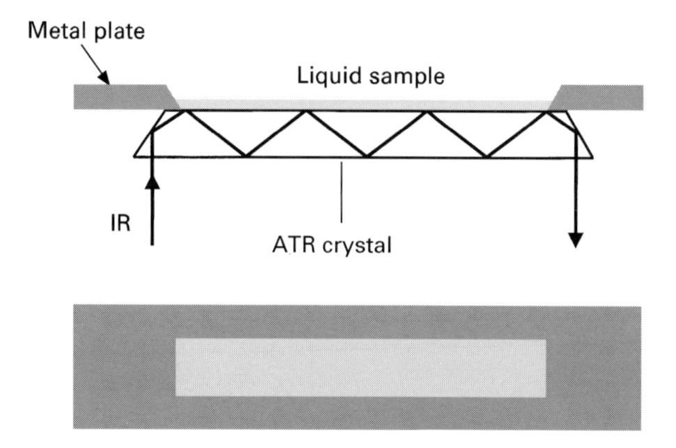

Figure 8 Schematic of an HATR top plate for liquids.

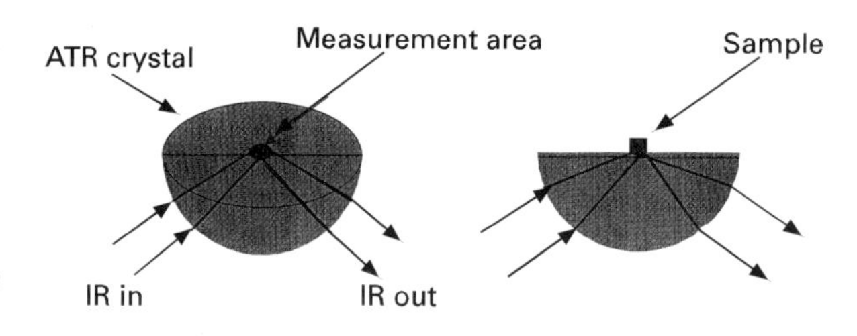

Figure 9 Schematic of a single-reflection ATR accessory.

The problem of contact has been overcome to a great extent by the introduction of accessories with very small crystals, typically about 2 mm across. The crystal is often diamond, which offers durability and chemical inertness. Germanium and silicon are lower cost alternatives. The accessories generally provide only a single reflection (**Figure 9**), but this is generally sufficient, given the very low noise levels of modern instruments. Because of the small area, much higher pressure can be generated with limited force. A much smaller area of contact is required than in conventional HATR units. As a result, spectra can be obtained from a wide variety of solid samples, even from minerals using systems designed for high pressures. Some accessories use a thin diamond element in combination with zinc selenide. This approach can provide multiple reflections, giving spectra equivalent to those from HATR accessories from very small sample volumes. Because of their versatility, accessories of this type are sometimes described as universal sampling devices. However, it must always be remembered that ATR only measures the surface of the sample.

The ATR technique is also employed with the unit inserted into bulk liquids, for example for process monitoring. The principle is exactly the same, the difference being that the form of the crystal is a prism so that radiation enters and leaves in the same direction. Radiation is coupled into the prism with light guides or optical fibres. There are also ATR probes in which the fibre itself forms the ATR element. The fibres are either chalcogenide glasses or mixed silver halides. One design of contact probe has a U-shaped section of exposed fibre that is pressed against the sample. This approach has been used for *in vivo* measurements of skin.

External reflection methods

Reflection measurements can be conveniently divided into surface or specular reflection and diffuse

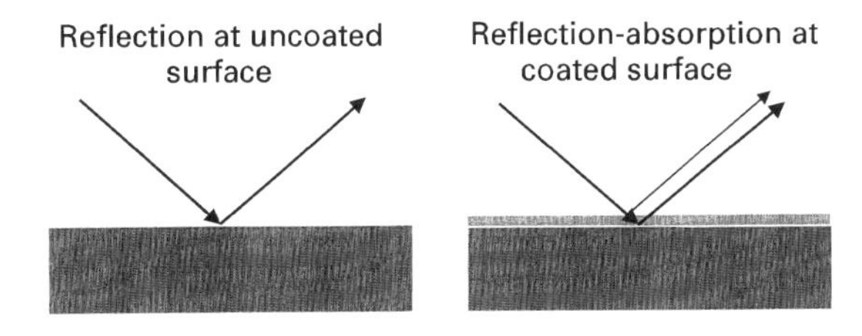

Figure 10 Reflection and reflection–absorption at surfaces.

reflection. The reflection spectrum from the surface of a continuous solid, such as a large crystal or a block of polymer, does not resemble an absorption spectrum, being determined by the refractive index according to the Fresnel equations. However, it can be converted into an absorbance spectrum by a mathematical process called the Kramers–Kronig transform which is provided in the software from instrument suppliers. This is a common approach for obtaining spectra of single crystals and from polished surfaces of minerals. It is also very effective for carbon-filled polymers. Very simple accessories using two flat mirrors can be used to obtain reflection spectra from flat surfaces. The same accessories can be used to measure the spectra of coatings on metal surfaces such as protective or decorative polymer films on food and beverage cans. In this case what is measured is a transmission spectrum of the coating, the radiation having passed twice through the coating and having been reflected at the metal surface. This is sometimes called a transflectance measurement. Examples of the above are shown in **Figures 10** and **11**.

Diffuse reflection is seen when radiation penetrates below the surface before being scattered or reflected back from within the sample. Spectra of this kind have the character of transmission spectra since there is absorption as the radiation travels inside the sample. Samples for diffuse reflection are generally powders (**Figure 12**). Accessories for this kind of measurement have to collect the light that emerges from the sample over a wide range of angles. Integrating spheres are little used because suitable large detectors are not available for the mid-IR region. Instead, commercial accessories focus the radiation on to a small region, typically 1–2 mm across, and collect over a wide but limited range of angles (**Figure 13**). The spectra obtained with these accessories represent a mixture of reflection from the surface and from within the sample. They can be used to measure pure surface reflection from samples that have no internal reflection or scattering.

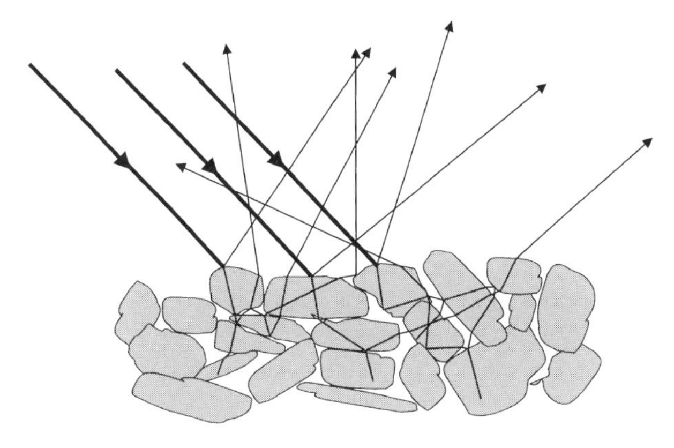

Figure 12 Representation of diffuse reflection by a powder.

In spectra obtained directly from powdered organic compounds, the stronger bands often appear flattened and are distorted by surface reflection. The diffuse reflection spectrum can be thought of as a transmission spectrum in which there is a range of different pathlengths. This results in enhancement of

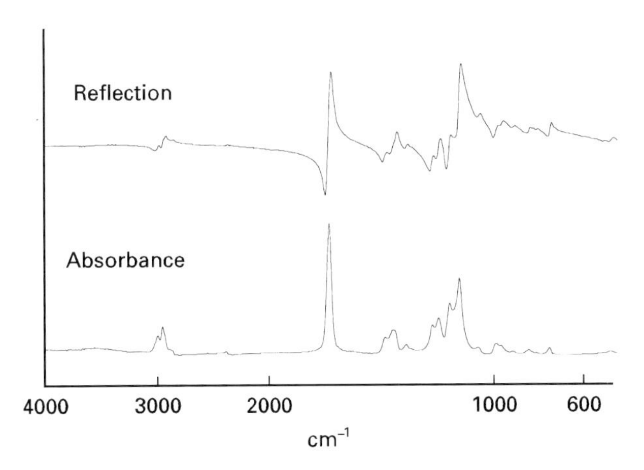

Figure 11 Top, reflection spectrum of poly(methyl methacryiate); bottom, absorbance calculated by Kramers–Kronig transformation.

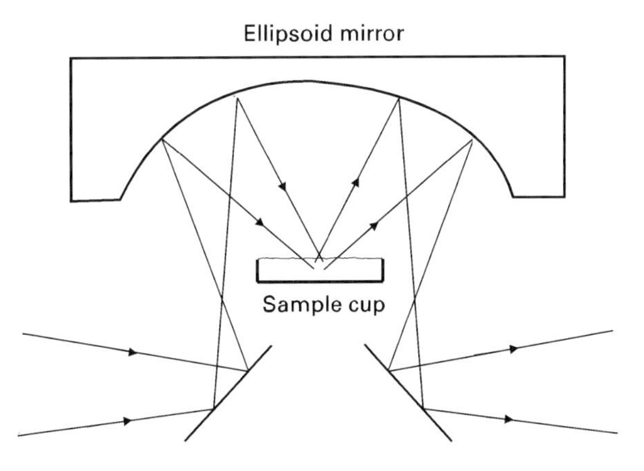

Figure 13 Optics of diffuse reflection accessory.

the relative intensities of the weaker bands. In general, the aim is to obtain a result that resembles a true transmission spectrum. The standard technique is to grind the sample to a fine powder, ideally to a particle size below 10 μm, and to mix this with a nonabsorbing matrix such as finely ground potassium bromide at a concentration of about 1%. The small particle size and dilution together ensure that absorption is weak for all pathlengths. This means that relative band intensities in transmission are similar for all contributing pathlengths and so are correct in the final spectrum. The presence of the non-absorbing matrix reduces reflection at the surfaces of the particles by better refractive index matching, and so minimizes the surface reflection component. Sample preparation is quicker than for a potassium bromide disc. It is also gentler, making it less likely to induce changes in the crystalline form. However, it has proved difficult to achieve sufficiently consistent sample preparation to allow good quantitative measurements. Representative spectra of finely ground aspirin are shown in **Figure 14**.

Diffuse reflection spectra are often transformed to Kubelka–Munk units as defined below, just as spectra measured in transmission are converted to absorbance:

$$f(R_\infty) \propto (1 - R_\infty)^2 / R_\infty$$

where R_∞ is the ratio of the sample reflectance to that of a reference such as potassium bromide. The reason is that under certain conditions the band intensities in these units are proportional to concentration. However, the necessary conditions are not met in typical mid-IR measurements, so there is little reason to prefer this presentation to the alternative log(1/*R*), which is a direct equivalent of absorbance (**Figure 15**). It is worth noting that near-IR diffuse reflection spectra are used extensively for quantitative work, almost invariably using log(1/*R*) rather than Kubelka–Munk units.

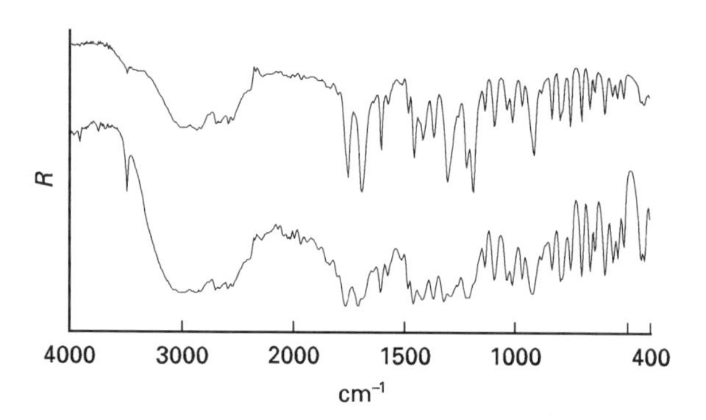

Figure 14 Diffuse reflection spectra of finely ground aspirin. Top, diluted in KBr; bottom, neat.

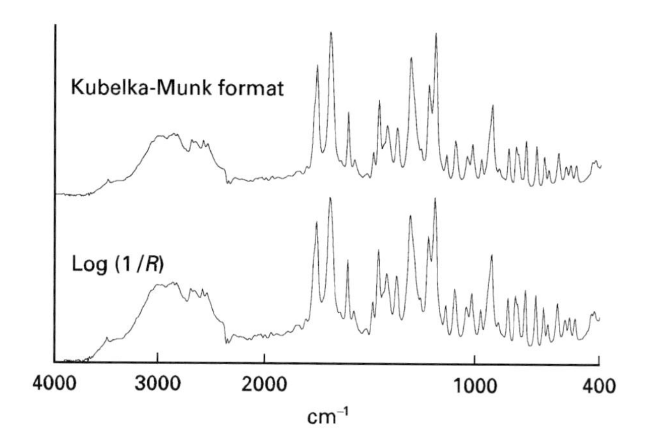

Figure 15 Diffuse reflection spectra of aspirin in Kubelka–Munk and log(1/*R*) units.

Another way of presenting samples for diffuse reflection is as a thin film of fine particles. This also achieves the objective of ensuring weak absorptions. Samples can be taken from the surface of large objects by abrading them with silicon-carbide-coated paper. The spectrum of the material removed can be measured directly on the abrasive substrate. This method has proved ideal for painted surfaces and moulded plastic objects. It can be extended to very hard materials by using abrasive devices such as metal rods tipped with diamond powder. The main limitation to this technique is that abrading soft and rubbery materials generally does not produce particles that are small enough to give good spectra.

Photoacoustic spectroscopy

Photoacoustic measurements are unique in that they depend directly on the energy absorbed by the sample, rather than on what is transmitted or reflected (**Figure 16**). The sample is placed in a sealed chamber. When radiation is absorbed it is converted into heat which can be detected as a change of pressure in the gas above the sample. This heat transfer process is sufficiently fast to follow the modulation of the radiation produced in an FT-IR spectrometer. The change in pressure is detected with a microphone. A spectrum is obtained by ratioing the detector signal to that generated with a totally absorbing material such as carbon black. As with ATR, the spectrum comes from a surface layer, but in this case the effective penetration can be more than 100 μm. This can result in total absorption for strong bands, so that relative intensities are distorted. The effective

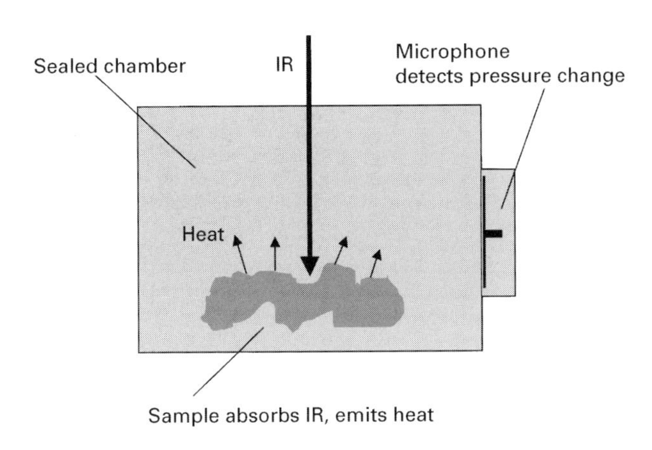

Figure 16 The principle of a photoacoustic measurement.

depth of penetration depends on the modulation frequency which varies with wavelength in rapid-scanning FT-IR spectrometers. In step-scan spectrometers the modulation frequency is constant for all wavelengths, so simplifying interpretation of the spectra. Some information about the depth distribution of different components can be obtained by varying the modulation frequency. However, depth profiling, measuring a concentration quantitatively as a function of depth, has not been achieved.

In practice, the main value of photoacoustic measurements has been in obtaining spectra from strongly absorbing samples such as carbon-filled polymers and for materials that cannot be ground to a fine powder or be prepared with a flat surface. Sample preparation is minimal, requiring only that the specimen is small enough to fit in the chamber. Commercial accessories can accommodate samples several millimeters across.

Microsampling

In recent years, the approach to measuring small samples has been transformed by the widespread use of IR microscopes. The advantages over traditional beam condensers lie in much greater convenience and versatility, as well as in the ability to measure much smaller samples. The region for measurement can be selected while a relatively large area of the sample is viewed. Spectra can be obtained from isolated samples that are only a few micrometres across. However, in extended samples diffraction limits spatial resolution to about 10 µm. Because of this, it is generally preferable to remove small inclusions from a matrix rather than to try to measure them *in situ*. One consequence of having very small samples is that it is often easy to compress them to a suitable thickness for transmission measurement. Compression cells are therefore more useful than they are with large samples. Sample preparation is greatly helped by having standard microscopy tools and a microtome available. Reflection measurements are more generally useful than they are with larger samples. Because of the small sample area and large collection angle, it is easier to find an area that gives a pure surface reflection spectrum. This can then be converted into an absorbance spectrum by the Kramers–Kronig transform.

Miscellaneous techniques

Emission spectra can be obtained without any sample preparation, but they are of very limited use. If the sample is opaque, the information provided is identical with that in the reflection spectrum since the sum of transmission, reflection and emission at any wavelength adds up to one. The reflection spectra are usually measured more easily than emission. Emission is very useful to obtain spectra from thin coatings on uneven metal surfaces when diffuse reflection measurements would require a small sample.

Pyrolysis can provide qualitative information from intractable samples without requiring expensive accessories. The simplest approach is to heat the sample in the bottom of a long test-tube so that decomposition products condense at the top of the tube and can be measured by the usual methods. Commercial devices also allow gaseous products to be collected. There are collections of the spectra of decomposition products from polymers. The most sophisticated version of this method is to couple a gas cell in the spectrometer to a thermogravimetric analyser and measure the products evolved at different temperatures.

Separation techniques are needed in order to identify individual components in mixtures. The most important of these is gas chromatography. To maximize sensitivity, the sample has to be confined to a small area. The usual approach is to pass the gas flow from the GC column through a light pipe ~1 mm in diameter and 10 cm long. This achieves detection limits of ~10 ng for strongly absorbing compounds. For higher sensitivity a more elaborate deposition method is available. The effluent from the column is directed on to a moving cold window where the peaks in the chromatogram can be condensed as a series of spots about 100 µm across. The detection limits are below 100 ng.

Coupling to liquid chromatography is much less successful because of the problem of eliminating the solvent and modifiers. In the most widely used system the eluate is deposited in a circular track on a rotating heated window which is subsequently transferred to the spectrometer. This is used especially in combination with size-exclusion chromatography.

List of symbols

d_p = penetration depth; n = refractive index; R_∞ = ratio of sample reflectance to that of a reference; α = angle of incidence; λ = wavelength.

See also: **ATR and Reflectance IR Spectroscopy, Applications; Chromatography-IR, Applications; Chromatography-IR, Methods and Instrumentation; High Resolution IR Spectroscopy (Gas Phase), Applications; High Resolution IR Spectroscopy (Gas Phase), Instrumentation; Photoacoustic Spectroscopy, Applications; Photoacoustic Spectroscopy, Theory; Surface Studies by IR Spectroscopy.**

Further reading

Chalmers JM and Dent G (1998) *Industrial Analysis with Vibrational Spectroscopy*. Cambridge: Royal Society of Chemistry.

Griffiths PR and de Haseth JA (1986) *Fourier Transform Infrared Spectroscopy*. New York: Wiley.

Harrick NJ (1977) *Internal Reflection Spectroscopy*. Ossining, NY: Harrick Scientific.

Messerschmidt RG and Harthcock MA (1988) *Infrared Microspectroscopy*. New York: Marcel Dekker.

Nishikida K, Nishio E and Hannah RW (1995) *Selected Applications of Modern FT-IR Techniques*, Tokyo: Kodanasha.

IR Spectroscopy, Theory

Derek Steele, Royal Holloway College, Egham, UK

VIBRATIONAL, ROTATIONAL & RAMAN SPECTROSCOPIES
Theory

Factors which control absorption

Absorption in the infrared region arises predominantly from excitation of molecular vibrations. For N atoms there are $3N–6$ vibrational degrees of freedom ($3N–5$ for a linear molecule), which is equivalent to stating that all vibrational distortions can be described as the sum of $3N–6$ '*fundamental*' vibrational modes. Where any symmetry exists in a molecule, the vibrations will reflect that symmetry in that the vibrational wavefunctions will be symmetric or antisymmetric with respect to any simple symmetry element. This statement has to be modified slightly for situations where degeneracy of rotations occurs. For a detailed discussion of groups and their relevance in spectroscopy the reader is referred to the Further reading section. In the event that a rigid point group is inappropriate, as for free rotors (e.g. CH_3BF_2), then the invariance of energy to certain distortions requires that permutation group theory be used. What we are seeking to explore with group theory is the invariance of energy and the consequences to certain transformations of the molecular configuration. In free rotors, by definition, the energy is unchanged by rotation about a bond. This then makes rigid group theory inappropriate.

For excitation of any single mode i, the energy, E_i, can be expressed as

$$\frac{E_i}{hc} = \omega_i(v+\tfrac{1}{2}) - \omega_i\, x_i\, (v+\tfrac{1}{2})^2 + \omega_i\, y_i\, (v+\tfrac{1}{2})^3 + \cdots \qquad [1]$$

where v is the vibrational quantum number and ω, the wavenumber, is ν/c. Restricting the discussion initially to a harmonic oscillator with no molecular rotation, we can then examine the conditions required for vibrations to be excited through interaction with electromagnetic radiation. For absorption the following conditions must be satisfied:

(i) the photon energy, $h\nu_i$, must be equal to the change in vibrational energy. We should qualify this by giving due allowance for the Doppler effect. In a low-density gas the spread of molecular velocities relative to the probing beam leads to a natural line width, $\Delta\nu$, at a temperature T of

$$\Delta\nu = \frac{\nu_i}{c}\left(\frac{2\,k_B\,T\ln 2}{M}\right)^{\frac{1}{2}}$$

where M is the relative molecular mass. Such widths are very small and relevant only when probing fine structure. Thus, for methane at room temperature and for a band at $\omega = \nu_i/c = 1500\ \mathrm{cm}^{-1}$ then $\Delta\omega = 0.003\ \mathrm{cm}^{-1}$. By using molecular beams moving at right-angles to the probing beam resolutions substantially less than this are attainable. Other techniques exist for exploring intrinsic line widths which are less than the Doppler width. These include two, or more, photon excitation and saturation spectroscopy.

(ii) Interaction of electromagnetic radiation and matter can only occur through electric and magnetic fields. For vibrational spectroscopy the electric field interaction is by far the greater. In classical terms there must be a change in dipole moment for the vibration to be excited. Thus for stretching of the bond of a homonuclear diatomic no absorption at the vibrational frequency occurs. In general, the absorption is proportional to the square of the dipole change. In terms of group theory only those vibrations which form representations of the same classes as the electric dipole vector (or simple translations) can be excited by EMR. Thus, for CF_3CF_3 (D_{3h} symmetry), only vibrations of the A_2'' and E' species can absorb electromagnetic energy of the appropriate frequency since a dipole vector forms such representations. We must note that it is unnecessary for an electric dipole to exist in the equilibrium configuration.

According to quantum mechanics, the probability of an induced transition per unit time per unit radiation density between states $|i\rangle$ and $|f\rangle$ is given by Fermi's 'Golden Rule':

$$B_{if} = B_{fi} = 2\pi^2/h^2\varepsilon_0|\langle f|\hat{e}_r\cdot\mu|i\rangle|^2\ \delta(\nu - \nu_f + \nu_i) \qquad [2]$$

where ε_0 is the permittivity of free space, $\hat{e}_r$ is a unit vector in the direction of the electric field vector of the radiation and $\delta(\ )$ is the Kronecker delta, which takes the value zero unless its argument is zero, in which case it has a value of unity. In the above equation the Kronecker delta requires that the radiation frequency is exactly equal to the difference in energies of the initial and final states divided by Planck's constant.

For an ensemble of randomly oriented molecules, the power absorption is given by

$$\mathrm{d}I_\nu = (N_iB_{if} - N_fB_{fi})\rho_\nu\,h\nu_{fi} \qquad [3]$$

where N_i is the number of molecules in state $|i\rangle$ and ρ_ν is the radiation density. Defining the intensity over the kth band as

$$\Gamma_k = \frac{1}{Cl}\oint \ln\frac{I_0^\nu}{I^\nu}\frac{\mathrm{d}\nu}{\nu} \qquad [4]$$

and assuming harmonic oscillator wavefunctions, then Equations [2], [3] and [4] yield

$$\Gamma_k = \frac{N\pi}{3\nu_k}\left(\frac{\partial\mu}{\partial Q_k}\right)^2$$

where C is the concentration of the absorbing species, l is the pathlength, I^ν and I_0^ν are the intensities of radiation at the frequency ν with and without the sample absorption and $\partial\mu/\partial Q_k$ is the dipole change with respect to the vibrational motion in normal, or fundamental, mode k. ν_k is the fundamental transition frequency. The definition of band intensity in Equation [4] differs from the usual integral (the integrated absorbance, A) in the presence of ν in the denominator, but is much more satisfactory in relation to theory and to isotopic intensity sum rules. Since the band widths are generally much smaller than the frequency band centre, the integrated absorbance, A, is related to Γ by $A \approx \Gamma\nu$.

(iii) The dipole moment can be written as a Taylor series expansion in the vibrational motion Q:

$$\mu = \mu_0 + \frac{\partial\mu}{\partial Q}Q + \frac{1}{2}\frac{\partial^2\mu}{\partial^2 Q}Q^2 + \text{higher terms}$$

For a harmonic oscillator, quantum theory requires that the transition integral $\langle f|Q|i\rangle$ disappears unless the vibrational quantum numbers v_i and v_f are related by $v_i - v_f = 0$ or ± 1. This requires that overtones are forbidden if only the linear term in Q exists. It is known, however, that the vibrational spectrum contains weak features due to overtones and combination bands in which two quantum numbers change simultaneously. These occur because the dipole expansion in general has finite terms quadratic in the deformation and because true wavefunctions are nonharmonic. Occasionally combination bands appear with considerable strength due to mixing with a nearby fundamental transition of the same symmetry (Fermi resonance). Interaction leads to mutual repulsion of the mixed energy states. In the event that the combination transition has zero or much less intensity than the fundamental, then the unperturbed fundamental transition can be estimated as the intensity weighted mean of the two bands. A classical example of Fermi resonance is seen in the infrared spectrum of CCl_4. The very strong doublet at 768/797 cm^{-1} is due to the triply degenerate CCl stretch interacting with the combination state $\nu_1 + \nu_4$ (305 + 458 cm^{-1}).

There is considerable interest in the factors that determine the dipole derivative. The stretching of a highly polar bond, such as C–F, might well be expected to lead to a large dipole change. In practice, the dipole change is much greater than would be expected for the charges on the atoms in their equilibrium positions. Charge flows as the nuclei move, in this case towards the situation where the fluorine atom carries a complete extra electron. In certain cases where a *trans* lone pair of electrons exist, such as for the hydrogen atom in the aldehyde unit, extension of the bond results in a flow of electrons from the lone pair into an antibonding orbital of the C–H bond, thereby greatly enhancing the dipole change. This is the same phenomenon that is responsible for bond weakening in such cases. In bending motions, nonbonding interactions, as well as rehybridization of the electrons at the nonterminal atom, cause substantial deviations from the dipole change expected from the equilibrium charge distribution. This explains the remarkably high intensity associated with movement of a hydrogen atom from the plane of benzene or ethene derivatives. As the electrons rehybridize around the carbon (moving from sp^2 towards sp^3), so electrons build up on the opposite side of the skeletal plane to the hydrogen, thereby enhancing its positive charge.

Many studies have been made of the absorption intensities associated with bands arising from group vibrations. It is recognized that the transition moments are far more sensitive to neighbouring group effects than are the frequencies themselves, and intensities are not particularly useful in evaluating the concentration of functional groups. However, some qualification of this statement is required in view of the widespread application of principal factor (or principal component) analysis (PFA) in industrial spectroscopy. This has been mainly used with near-infrared spectra, but this is probably more to do with ease of generating suitable spectra than to fundamental limitations. The application to the mid-infrared spectra of coals is an excellent example. The relevance to group intensities is that PFA is a statistical technique of fitting data to a progressively increasing number of unknown components until the fit reaches an acceptable level. As such it requires the data to consist of a finite number of linearly additive components. The application to very complex mixtures such as coals, wheat and petroleum, both refined and unrefined, strongly suggests additivity is much better than used to be believed. This has been rationalized by demonstrating that group spectra are characteristic provided that the neighbouring groups are the same.

Band shapes in liquids

In low-density vapours an absorption band consists of lines arising from a large number of transitions arising from rovibrational transitions, perhaps from excited vibrational states as well as the ground state. In high-density fluid states the interaction between different molecules will affect the spectrum by causing perturbations of the rovibrational energy states and by reducing the lifetimes of the excited states through energy transfer processes. The individual transitions initially broaden and then merge into a smooth band contour. It is sometimes stated that the rotational transitions are suppressed, but this cannot be so. The equipartition theorem states that energy must be equally distributed between each degree of freedom, and this must extend to the condensed state. It has been demonstrated that the contours still contain the rotational contribution. Thus for a diatomic molecule of moment of inertia I, the integrated second moment of the intensity (Γ as defined in Eqn [1]) about the vibrational band centre, divided by the integrated intensity itself, is, apart from a vibrational relaxation contribution, $k_B T/I_X$, where I_X is the moment of inertia. This is the same for vapour and for liquid. It is simply the relaxation of the excited states that has changed. There is a vibrational contribution to the width (see below). It is possible in certain situations to separate these contributions using Raman spectroscopic methods. If we designate the transition dipole moment at a time t of a given transition as $\mu(t)$ then its development as a function of the time lapse $t = t'' - t'$ is related to Fourier transform about the band centre of the intensity as

$$C(t'' - t') = \langle \mu(t'') \cdot \mu(t') \rangle = \langle \mu(0) \cdot \mu(t'' - t') \rangle$$
$$= \int_{-\infty}^{+\infty} \Gamma(\nu_0 + \Delta\nu) \exp(i2\pi \, \Delta\nu t) \, d\nu$$

The brackets $\langle \, \rangle$ designate an ensemble average and, since the transition dipole is a vector, this transform yields the ensemble time decay of the transition dipole moment from an arbitrary time of origin. The time scale of the decay will be of the order of the inverse of the band width. Thus for a band of half-band width 10 cm^{-1} the half-life of the decay will be ~1/($3 \times 10^{10} \times 10$) ≈$3 \times 10^{-12}$ s. If we discount vibrational relaxation and consider the transition moment as being fixed to the molecule, then the correlation function $\langle\mu(0)\cdot\mu(t' - t'')\rangle$ will describe the rotation of the molecule as defined by the transition dipole axis. In the condensed state after an initial period in which the motion is governed by the rotational energy, then decay becomes exponential due to collisional processes. In the vapour the dipole correlation shows a minimum, perhaps becoming negative, in a period after which most molecules are facing in the opposite direction to that in which they started (see **Figure 1**). As in magnetic resonance, the decay time will consist of isotropic and anisotropic contributions. In a fluid the transition dipole will also decay by transfer of energy between molecules and by multi-body collision deactivation of the vibration. These are responsible for the isotropic component. Owing to the very short time-scale of these processes, separation of the relaxation mechanisms is generally impossible. However, some progress has been made with laser optical nutation experiments and with Raman studies.

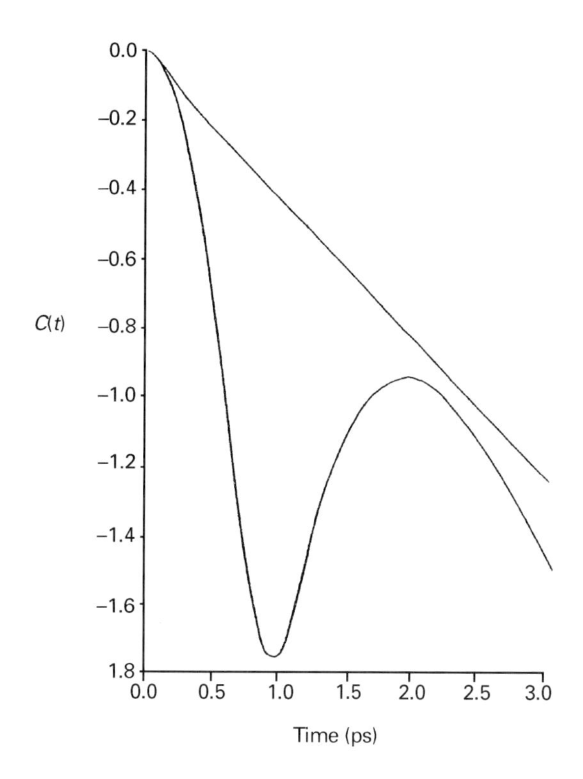

Figure 1 The dipole autocorrelation function of the ν_{11} band of benzene. The top line is the dipole correlation function for benzene dissolved in cyclohexane and the bottom line is that for benzene vapour. Hill IR and Steele D, unpublished work.

Calculation of vibrational spectra

Although energetic aspects of vibrational spectroscopy, such as zero point energy and crystal lattice energies, are strictly quantum phenomena, much can be achieved by classical-type calculations based on inter-atomic potential functions. Assuming that the potential energy for deformation about an equilibrium configuration is strictly quadratic in deformations then the vibrational frequencies are given as the eigenvalues of the matrix product ***GF*** where ***F*** is the matrix of quadratic force constants and ***G*** (so-called inverse kinetic energy matrix) is such that

$$G_{ij} = \sum_k \frac{\partial R_i}{\partial \xi_k}\frac{\partial R_j}{\partial \xi_k}\frac{1}{m_k} \quad \xi = x,\ y,\ z$$

where R represents an internal coordinate of the molecule, usually chosen from bond stretches, angle bends and bond torsions. The sum is over all atoms. The methodology of carrying out these classical calculations are well documented. One of the principle advantages of such calculations is that the eigen-vectors of ***GF*** yield the nature of the molecular vibration.

Up to this point we have assumed harmonic oscillations. In reality we cannot ignore the impact of cubic and higher terms in Equation [1]. For a change in quantum number of $\Delta\nu$, then from Equation [1] we have for a transition $|0\rangle \rightarrow |\nu\rangle$

$$\frac{\Delta E}{hc} = \omega_i \Delta\nu - \omega_i x_i (2\nu + 1)\Delta\nu + \text{higher terms}$$

A fundamental transition (ν = 0→1) then has a wavenumber of $\omega_i - 2\omega_i x_i$. The anharmonicity constant x is much less than 1. For the C–H vibrations it is relatively large because of the considerable amplitude of vibration, and it has a value of 0.021. This means that for a transition seen at 3000 cm^{-1} the fundamental frequency is at 3126 cm^{-1}. The first overtone would be expected at $2\omega_i - 6\omega_i x_i$ = 5874 cm^{-1}. In fact the C–H stretching overtone and combination frequencies deviate from these expectations. The cause of this lies in the large amplitudes of motion which result in a breakdown of local symmetry. A methylene group in an environment which conserves the equivalence of the two bonds will have vibrations which are symmetric and antisymmetric with respect to the bonds, and in the limit of small amplitude vibrations point group symmetry applies. As the amplitudes of all the vibrations increase, so the symmetry breaks down and the bonds tend towards

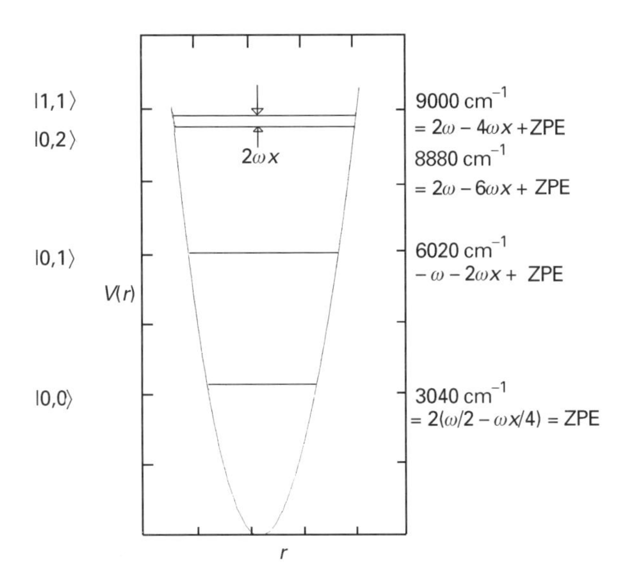

Figure 2 Relative energies in the local mode model of ground, fundamental, first harmonic and combination states for a system with two degenerate vibrators. A fundamental harmonic wavenumber of 3100 cm^{-1} and a vibrational anharmonicity of $\omega x = 60$ cm^{-1} have been assumed. ZPE represents the zero point energy.

independent oscillators. In this event the theoretical model needed becomes that of the local mode. Application of Equation [1] shows that the separation of states in which one of two equivalent bonds is doubly excited and that in which both bonds are singly excited is $2x_i\omega_i$. The higher transition will be at twice the observed fundamental frequency (see **Figure 2**).

Quantum theoretical calculations of spectra

Applied quantum theory has reached the stage where infrared and Raman spectra can be predicted to a very useful degree of accuracy for molecules of fairly substantial size. The ability to programme the calculations so that force constants and dipole gradients are calculated analytically from the wavefunctions is one of the features which has greatly enhanced these computations. The calculations as routinely performed are of the harmonic wavenumbers and the band intensities. At the Hartree–Fock level the wavenumbers are too high when compared with experiment owing to the harmonic approximation, but also due to lack of electron correlation and the restriction of electron pairing. The latter leads to dissociation to states in which electron pairing is still present, and these are generally of much higher energy. Very good results are obtained by simply multiplying the wavenumbers (frequencies) by a scaling factor. Improved results are generated by scaling the force constants, keeping one factor for C–H stretching, another for the C–H bending constants, etc. Improved results, with scaling constants nearer to unity, are obtained with perturbation calculations of the electron correlation correction (Möller–Plesset theory). Density functional theory (DFT) now permits very good frequencies to be obtained in times which are close to the Hartree–Fock times. Both Raman and infrared intensities are much more difficult to compute. This is not surprising when it is seen that dipole moments are extremely sensitive to the basis sets. Although there are cases where the results are as yet disappointing, DFT appears to be yielding much improved intensities and dipole moments over earlier Hartree–Fock and Möller–Plesset calculations. Although these calculations have led to much improved force fields and spectra, it must be said that some difficulties still exist.

Vibration rotation spectra

When a vibrational quantum state changes, other quantum states of lower energy will also generally change. Thus, for a heteronuclear diatomic molecule the $|0\rangle$ to $|1|$ transition is accompanied by a change in rotational quantum number of ±1 leading to the well known P and R branches. For polyatomic molecules the selection rules required so as to allow the nondisappearance of the transition integral are more complex, and for details the reader is referred to a specialist text. In the situation where individual rovibrational transitions cannot be resolved in an asymmetric top, then the overall contour can be related to the direction of the transition moment with respect to the inertial axes and to the moments themselves. Thus transitions in which the dipole change occurs along the direction of greatest moment of inertia are characterized by a strong Q branch (type C bands), corresponding to no change in the rotational quantum number change (see **Figure 3**).

The contours have been calculated as a function of the moments of inertia and presented in easily used figures. With laser and high-resolution Fourier instruments, an increasing amount of detail of rovibrational information is becoming available. Here we shall refer to a few general points.

The parameters of fitting include the moments of inertia. Even with small molecules and isotopically substituted molecules it is very difficult to generate accurate molecular geometries. The inertial constants refer to the ground and vibrationally excited states, but even in the ground state at the absolute zero of temperature each quantum state has a half quantum of energy (zero point energy, see **Figure 2**). The average of the root mean square bond length

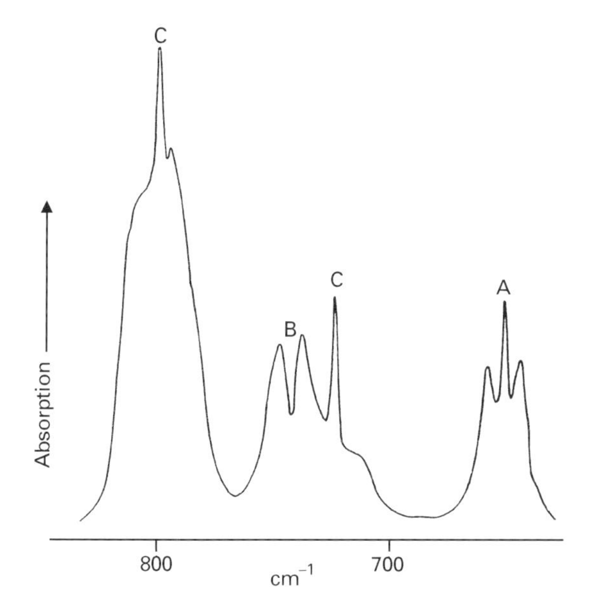

Figure 3 Part of the infrared absorption spectrum of 2,6-difluoropyridine in the vapour phase showing typical A, B and C contours (transition moments along axes of least, intermediate and greatest moments of inertia, respectively.) Reproduced with permission from Bailey RT and Steele D (1967) *Spectrochimica Acta, Part A* **23**: 2997.

during a vibration differs significantly from the equilibrium bond length. *Ab initio* calculations, such as referred to in the previous section, may be used to estimate the necessary corrections.

One feature unique to vibration rotation interaction is due to Coriolis interaction. There is a component of force on a moving particle due to rotation about a perpendicular axis equal to

$$\omega \otimes (\omega \otimes m_i r_i)$$

where $\otimes$ represents the vector product operator, ω is the molecular angular velocity and m_i and r_i are the mass and coordinate of the ith particle. Such coupling leads to interaction between two states, the product of whose symmetry representations belongs to the same species as a molecular rotation. For degenerate vibrations, this can be a first-order effect and has a dramatic impact on the overall vibrational rotational contour. The same phenomenon, of course, controls the contrary direction of rotation of air masses in the northern and southern hemispheres.

List of symbols

A = integrated absorbance; B_{if} = probability of an induced transition per unit time per unit radiation density; c = velocity of light; C = concentration; E_i = energy of ith mode; F = quadratic force constant matrix; G = inverse kinetic energy matrix; h = Planck's constant; k_B = Boltzmann's constant; l = pathlength; M = molecular mass; Q = normal mode; R = internal coordinate; v = vibrational quantum number; Γ = integrated absorption intensity; μ = dipole moment; ν = frequency; ρ = radiation density; ω = wavenumber.

See also: **ATR and Reflectance IR Spectroscopy, Applications; Far IR Spectroscopy, Applications; Fast Atom Bombardment Ionization in Mass Spectrometry; High Resolution IR Spectroscopy (Gas Phase), Applications; High Resolution IR Spectroscopy (Gas Phase), Instrumentation; Hydrogen Bonding and other Physicochemical Interactions Studied By IR and Raman Spectroscopy; IR Spectral Group Frequencies of Organic Compounds; IR and Raman Spectroscopy of Inorganic, Coordination and Organometallic Compounds; IR Spectrometers; IR Spectroscopy Sample Preparation Methods; Near-IR Spectrometers; Rotational Spectroscopy, Theory; Vibrational, Rotational and Raman Spectroscopy, Historical Perspective.**

Further reading

Allen HC and Cross PC (1963) *Molecular Vib-Rotors.* New York: Wiley.

Badger RM and Zumvalt LR (1938) *Journal of Chemical Physics* **6**: 711.

Berne BJ and Pecora R (1976) *Dynamic Light Scattering.* New York: Wiley.

Brewer RG and Shoemaker RL (1971) *Physical Review Letters* **27**: 631.

Bright Wilson E, Decius JC and Cross PC (1955) *Molecular Vibrations.* New York: McGraw-Hill.

Cagnac B (1982) *Philosophical Transactions of the Royal of the Society of London, Series A* **307**: 633.

Cotton, FA (1990) *Chemical Applications of Group Theory*, 3rd edn. New York: Wiley.

Crawford B Jr (1952) *Journal of Chemical Physics* **20**: 977.

Fredericks PM, Lee JB, Osborn PR and Swinkels DAJ (1985) *Applied Spectroscopy* **39**: 303, 311.

Gordon RG (1965) *Journal of Chemical Physics* **42**: 3658.

Gordon RG (1965) *Journal of Chemical Physics* **43**: 1307.

Jennings DE (1978) *Applied Physics Letters* **33**: 493.

Longuet Higgins HC (1963) *Molecular Physics* **6**: 445.

Martens H and Naes T (1989) *Multivariate Calibration.* New York: Wiley.

Mills I and Robiette AG (1985) *Molecular Physics* **56**: 743.

Parker ME, Steele D and Smith MJC (1997) *Journal of Physical Chemistry, Section A* **101**: 9618.

Steele D (1971) *Theory of Vibrational Spectroscopy.* Philadelphia: WB Saunders.

Wexler AS (1967) *Applied Spectroscopy* **1**: 29.

Iridium NMR, Applications

See **Heteronuclear NMR Applications (La–Hg).**

Iron NMR, Applications

See **Heteronuclear NMR Applications (Sc–Zn).**

Isotope Ratio Studies Using Mass Spectrometry

Michael E Wieser, University of Calgary, Alberta, Canada
Willi A Brand, Max-Planck-Institute for Biogeochemistry, Jena, Germany

MASS SPECTROMETRY
Applications

Introduction

Isotopes of an element contain the same number of protons but different numbers of neutrons. Whereas the former means that isotopically different compounds undergo the same reactions, the latter means that they differ in mass (e.g. $^{12}_{6}C$ and $^{13}_{6}C$ are 12 and 13 atomic mass units, respectively). As a result of their different masses, isotopes of an element participate in chemical, biological and physical processes at different rates. Hence, the isotope composition of the element in a given compound depends on the history and origins of the sample. Isotope abundance data can provide information concerning the source of a material or processes responsible for its synthesis and conversion. Variations in the isotope abundance ratios of many elements of biogeochemical importance are subtle, but significant. To resolve these differences, isotope ratio mass spectrometers must accurately measure variations of 20 to 50 parts per million. In the case of carbon, the average $^{13}C/^{12}C$ isotope abundance ratio of 0.011 200 ranges over ±0.000 450 with biogeochemically important variations of ±0.000 000 5. Such requirements have resulted in a branch of mass spectrometry that has developed its own specialized instrumentation and analytical methods.

Isotope ratio studies are employed in a variety of multidisciplinary research projects encompassing chemistry, physics, biology, medicine, geology, archaeology and environmental technology. This article will describe mechanisms responsible for isotope abundance variations, the essential components of the isotope ratio mass spectrometer (with particular reference to electron impact ionization) and isotope abundance variations of H, C, N, O and S. Finally, the development of recent isotope ratio monitoring techniques to extract isotope information from transient signals is presented.

Natural isotope ratio variations

To understand how mass differences give rise to isotope abundance variations, it is useful to consider chemical bonds as harmonic oscillators with fundamental vibrational frequencies inversely proportional to the square root of the molecular masses. The zero point energy of a molecule is defined as the finite energy (above $h\nu$ where h is Planck's constant and ν is the frequency of the oscillation) possessed by that molecule at 0 K. Isotope substitution changes the mass of the molecule and results in a shift in the zero point energy. A molecule containing a lighter isotope of an element will have a higher vibrational frequency and hence higher zero point energy compared to the molecule with the heavier isotope. The bonds in the molecule with the higher zero point energy are more easily broken. For all elements except hydrogen, the vibrational terms have the largest isotope effect. The magnitudes of these effects are proportional to the relative mass difference between the isotopic species.

Processes that result in isotope ratio variations can be divided into three general categories: (1) isotope exchange reactions such that isotopes of an element are redistributed among different molecules or different phases of the same molecule at thermodynamic equilibrium; (2) unidirectional processes where the reaction rates of chemical or physical reactions of isotopic species differ; and (3) radiogenic decay.

Equilibrium isotope effects

Equilibrium isotope exchanges involve the redistribution of an element's isotopes among different chemical compounds or different phases of a molecule. While chemical equilibrium is necessary in order to attain isotopic equilibrium, isotope exchange will occur without any net change in the distribution of the chemical species in the system. Generally, the heavier isotope is favoured in the molecule that has the lowest zero point energy or the more condensed phase. The equilibrium constant of the isotope exchange reaction is related to the partition function ratios of the products and reactants. Partition functions control the distribution of internal energies within molecules. It is possible to calculate reduced partition function ratios and hence equilibrium constants for simple gaseous molecules (and to some extent condensed phases) based on spectroscopic and thermodynamic data. The dependence of the equilibrium constant on temperature is such that at higher temperatures the extent of isotope fractionation decreases.

The relationship between the isotope abundance ratios of the different species in the reaction is expressed as the fractionation factor 'α'. For example, consider the oxygen isotope exchange of H_2O between condensed and vapour phases:

$$H_2{}^{18}O_{\text{vapour}} + H_2{}^{16}O_{\text{liquid}} \leftrightarrow H_2{}^{16}O_{\text{vapour}} + H_2{}^{18}O_{\text{liquid}}$$

The corresponding isotope fractionation factor α is 1.0092 at 25°C, i.e. the liquid phase is enriched in ^{18}O over the gaseous phase by 0.92 percent (Eqn [1]):

$$\alpha_{\text{liquid-vapour}} = \frac{(^{18}O/^{16}O)_{\text{liquid}}}{(^{18}O/^{16}O)_{\text{vapour}}} = 1.0092 \qquad [1]$$

Kinetic isotope effects

Kinetic isotope effects are observed in unidirectional processes (such as diffusion, evaporation, and bacterial conversions) where the reaction rate depends on the masses of the reacting molecules. Given the same amount of kinetic energy, molecules of lighter mass m_1 will react at a rate $\sqrt{(m_2/m_1)}$ faster than the heavier molecules of mass m_2.

Biochemical reaction pathways are often dependent on the mass of an element at a particular position in a molecule. The result mostly is to favour the lighter isotopic species in the formation of the product. Often, exact information about the intermediate activated complex between reactants and products is not available. This makes calculation of rate constants and hence isotope fractionations from first principles difficult.

Radiogenic isotopes

The transmutation of elements due to radioactive decay is another important cause of isotope abundance variations. Over time, the numbers of radioactive parent isotopes will decrease as they decay into the daughter products. The abundance determination of radiogenic daughter isotopes is applied extensively for age determinations of geological and organic materials. The ages of minerals and rocks (t) can be calculated from the measurement of the number of daughter nuclides formed in a mineral that decayed under closed conditions (D^*) and the parent atoms (N) where the half-life of the parent (probability of a decay, λ) is known (Eqn [2]).

$$D^* = N[\exp(\lambda t) - 1] \qquad [2]$$

Successful age determinations are possible providing (1) the rock or mineral has remained closed and changes in the numbers of parent and daughter atoms are due to nuclear decay only; (2) the decay constant is known accurately; (3) the number of daughter nuclides originally present (at time $t = 0$) is known; and (4) the numbers of daughter and parent atoms can be measured accurately.

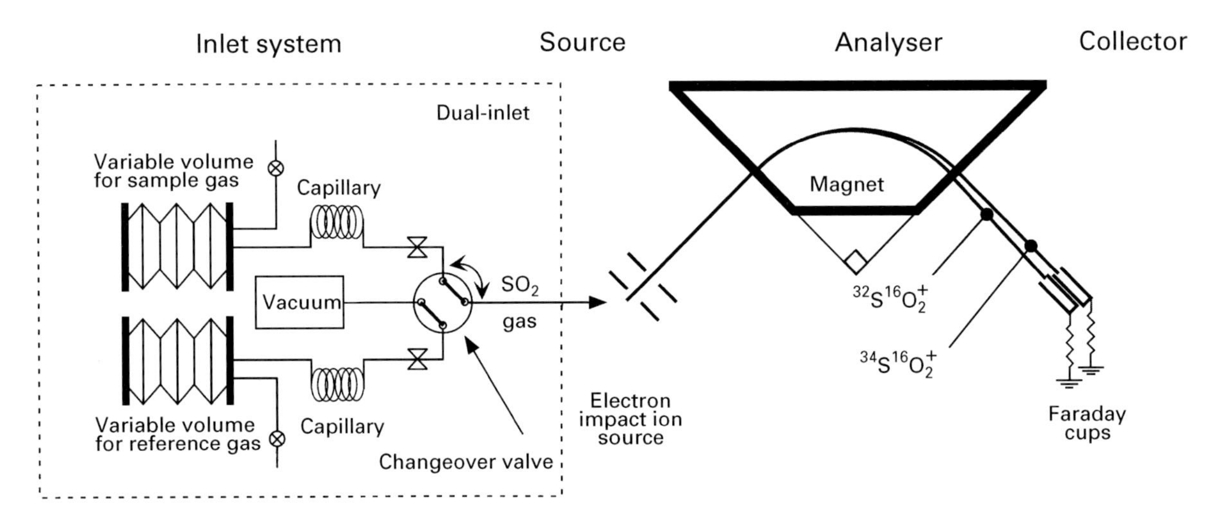

Figure 1 Typical mass spectrometer and inlet system for isotopic comparison of an unknown sample with a reference gas of precisely known isotopic composition. Clean gases are needed for such systems.

Isotope ratio mass spectrometer

The majority of isotope abundance determinations are performed using magnetic sector mass spectrometers based on a design from Alfred Nier dating back to the 1940s. The isotope ratio mass spectrometer consists of an ion source, magnetic sector and Faraday cup ion detector(s) (**Figure 1**). The source is responsible for the production of ions from either solid or gaseous material. The magnetic sector separates ions according to their mass to charge ratio and brings them into focus at the collector. The detector records and converts the charge of the ions into (computer readable) electrical signals. Neutral species are removed from the interior of the mass spectrometer (maintained at less than 10^{-8} mbar or 10^{-6} Pa) to minimize the number of collisions between ions and residual molecules in the ionization and analyser regions. These interactions would otherwise cause the ions to lose variable amounts of kinetic energy, resulting in a defocused ion beam.

Isotope ratio mass spectrometers tend to be of low resolution (~ 100, 10% valley) to resolve the isotopomers in the simple gases introduced (i.e. in the case of CO_2 gas; $^{12}C^{16}O_2^+$, $^{13}C^{16}O_2^+$ and $^{12}C^{16}O^{18}O^+$ have masses 44, 45 and 46). The mass spectrometer provides high precision isotope abundance data, typically of the order of 0.001%. This is achieved by optimizing the number of ions collected at the detector to improve the counting statistics.

Magnetic sector

The task of the ion optics of the mass spectrometer (of which the magnetic sector is the critical component) is to bring all of the ions produced at the ion source into focus at the collector. Ions formed in the source are collimated and accelerated through an electric potential difference (V) from 2000 to 10000 V, depending on instrument type. The accelerated ions enter the magnetic field and are separated along radial paths dependent on their charge to mass ratio and kinetic energy. The radius (r) of the path traced by an ion while in the magnetic field is related to its mass (M), charge (q), the potential of the source (V), and the magnetic field flux density (B) according to Equation [3]:

$$r = \sqrt{\frac{2MV}{qB^2}} \qquad [3]$$

It is possible to change either the accelerating potential of the source or the magnetic field strength to bring an ion of a given mass into focus at the collector. A unique feature of the magnetic sector isotope ratio mass spectrometer is the trapezoidal peak shape produced by the selection of ion source (entrance) and collector (exit) slit widths (**Figure 2**). As the ion beam is scanned across the exit slit of the mass spectrometer, the ion current increases until the entire width of the ion image is contained inside the collector. The resulting flat-top peak ensures that the measured ion current does not change if there are small fluctuations in the accelerating potential (V) and/or magnetic flux density (B).

Faraday cup detectors

The ion currents realized with most applications are of the order of 10^{-9} A for the major (most abundant) isotopic species and can be measured by a Faraday

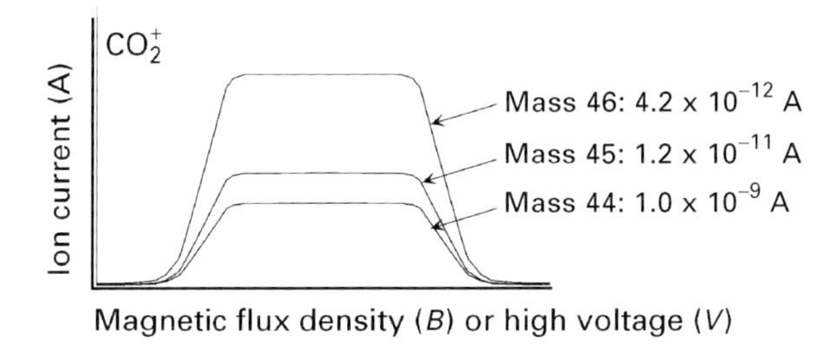

Figure 2 Flat-topped peaks ensure that small fluctuations in the magnet current or high voltage regulation do not hamper precision. For clarity, the peaks are recorded with different sensitivity factors (1:100 : 300 for 44, 45 and 46).

cup type detectors (**Figure** 3). The Faraday cup is connected to ground via a high ohmic resistor (10^8 to 10^{12} Ω). The incident ions produce a current (I) resulting in a voltage drop (V) across the resistor (R) proportional to the ion current incident at the collector by Ohm's law ($V = IR$). Either voltage to frequency converters or analog to digital converters digitize the voltage for computer-assisted processing of the data.

The accurate Faraday cup detector must neutralize the incoming singly charged particle with exactly one electron. However, due to the energy of the ion (2 to 10 keV), the surface of the detector is sputtered and secondary electrons and ions are ejected. These must be suppressed to prevent a false measurement of the ion current. Therefore, Faraday cup detectors are mostly thin, deep boxes with graphite surfaces and an electric field (~100 V) at their entrance to force secondary electrons back into the cup. The noise of the detector system can be no greater than 1.5×10^{-16} A or 1000 ions per second in order to measure the ratio of $^1H^2H$ to $^1H_2$ with a precision of 0.01% (where a sample of ocean water contains about 156 ppm ^2H and the $^1H_2^+$ signal is 5×10^{-9} A).

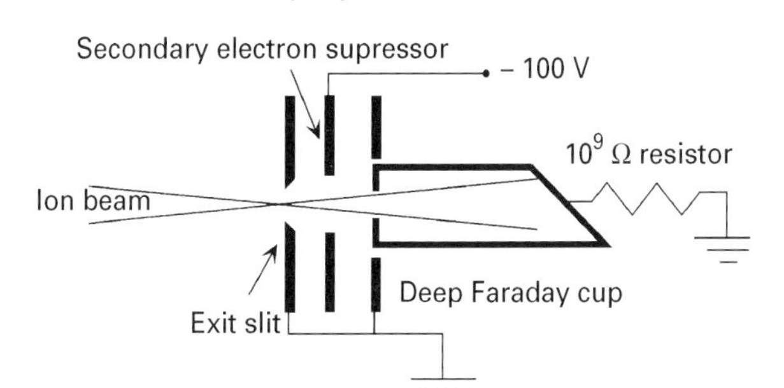

Figure 3 Schematic layout of a Faraday cup assembly. The cups are deep and have secondary electron suppression in order to precisely record one charge per incoming singly charged ion. Different resistor values are used depending on the ion current in question.

As suggested by the name isotope *ratio* mass spectrometry, more than one ion current is measured in order to determine isotope abundance ratios. Most modern isotope ratio mass spectrometers have from 3 to 8 collectors carefully positioned along the focal plane of the instrument to measure several isotopic ion currents simultaneously, typically a few mass units apart. This configuration is known as simultaneous collection and has several advantages over designs employing a single detector. Since multiple detectors are measuring the different isotopic ion currents at the same time, any changes in the total ion current affect the signal in all detectors. Fluctuations in the production of ions by the ion source effectively cancel and the stability of the ion current is not a limiting factor.

Ion source

Two ion production methods are employed for isotope abundance ratios. Thermal ionization is best suited for the analysis of nanogram quantities of metals. Electron impact is suitable for the measurement of gaseous samples and is the most widely used ionization process.

Thermal ionization Thermal ionization sources produce ions by heating sample-coated metallic filaments. Samples in solution are first deposited as microlitre drops on the filaments and dried at low temperatures forming thin salt or oxide layers (**Figure 4**). The prepared filament is transferred into the mass spectrometer source, the source is evacuated, and the filament is then heated to temperatures ranging from 800 to 2000°C. The probability of forming an ion in thermal equilibrium with the surface of the hot metal filament depends on the temperature (T), work function of the filament (W), and the ionization potential associated with the production of a given ion species (IP). This is described by the Langmuir–Saha equation as the relative number of ionized particles (N^+) to neutral species (N^0) evaporated from the filament (Eqn [4])

$$\frac{N^+}{N^0} \propto \exp\left[\frac{(W - \mathrm{IP})}{kT}\right] \qquad [4]$$

Typically, a filament material is chosen such that the work function of the metal is higher than the ionization potential of the sample. The filament is just heated to temperatures that allow efficient ionization without melting the metal.

Figure 4 Deposition of sample on a Re filament for thermal ionization analysis.

Using similar techniques, it is also possible to produce and analyse negative ions. This method has the advantage of allowing a number of elements with high ionization potentials to be more easily ionized as negative atomic or oxide species (i.e. B, Se and Re). Again, the number of ions relative to the number of neutral species produced can be described by the Langmuir–Saha equation. In this instance, the relative number of negative ions (N^{-}) produced depends on the electron affinity of the sample (EA) as well as the filament's temperature (T) and work function (W) (Eqn [5]:

$$\frac{N^{-}}{N^{0}} \propto \exp\left[\frac{(\mathrm{EA} - W)}{kT}\right] \qquad [5]$$

Mass-dependent fractionation processes that occur during the ionization process limit the ultimate accuracy and precision of isotope ratio measurements. As the sample is heated and ions are formed, the lighter isotopic species will evolve from the filament at a faster rate. The remaining (unionized) sample becomes relatively more enriched in the heavier isotopes and no longer has a representative isotope composition. Corrections to this fractionation are based on empirical calculations according to exponential or Rayleigh distillation models.

Principal applications of thermal ionization mass spectrometry (TIMS) are in the earth sciences to measure radiogenic isotopes for age determinations (i.e. U-Pb, Pb-Pb, Rb-Sr, Re-Os, Nd-Sm), absolute isotope abundance measurements, tracing of Pb and other metals in the environment, and isotope dilution mass spectrometry. The high efficiency of the ion source for many elements (i.e. 1 ion collected for every 200 rubidium atoms on the filament) allows very small samples (10^{-9} to 10^{-12} g) to be analysed.

Electron impact Electron impact sources ionize gases by collisions with transverse electron beams of up to 1 mA current and 100 eV energy (**Figure 5**). The mostly singly charged ions produced are extracted from the ionization region by an applied electric field that further serves to collimate and accelerate the ions. The resulting ion current depends on the amount of sample gas inside the ion source. The influx of the sample gas into the ion source is of the order of 10^{-12} mol s^{-1} under viscous flow conditions. The electron impact ion source will result in 1 ion being collected for every 1000 to 10 000 molecules introduced, depending upon the sensitivity of the ion source and the focusing properties of the mass spectrometer. Collisions in the ion source between ions and neutral species can lead to a nonlinear dependence of the ion current ratios with the source pressure. To prevent this, ions must be extracted immediately after formation. For isotope abundance measurements, simple gases (e.g. CO_2, N_2, H_2, SO_2, CO) are prepared from raw materials being careful to preserve the isotope composition of the sample.

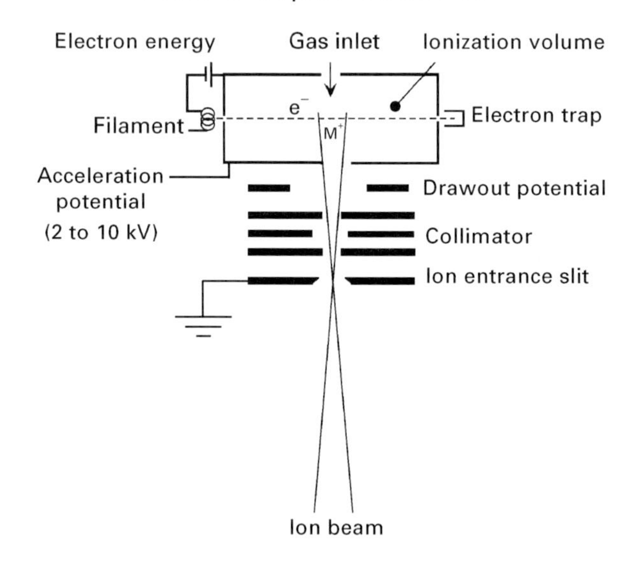

Figure 5 Schematic diagram of electron impact ion source for the ionization of gaseous samples. The whole ion source is closed, leaving only small holes for the electrons to enter and the ions to exit. This ensures the highest sensitivity by having the gas molecules pass through the ionization region several times before being pumped away.

Inlet systems: the changeover concept Mass spectrometers do not measure isotope abundance ratios accurately due to inherent biases caused by the ion optical design, ion current detectors and electronics, and method of sample preparation. In practice, to correct for machine bias, the measured ion current ratios of the sample are compared to those from a reference gas (of known isotopic composition) analysed under identical operating conditions. This can be achieved using a *dual inlet* system with two stainless steel variable volumes (1 to 100 mL) for reference and sample gases (**Figure 1**). The volume of these reservoirs is adjusted to change the pressure of the gas (10 to 1000 mbar) and obtain matched major ion current intensities of both reference and sample gases. Each measurement of the sample ion currents is immediately followed by an analysis of the standard's ion currents.

Stainless steel capillaries (1 m length and 0.1 to 0.2 mm i.d.) transport the gas from the variable volumes to the ion source. The *changeover* valve alternately connects the reference and sample bellows to either the ion source or vacuum and the isotope compositions of the two gases are measured in turn. This method ensures a constant flow of gas through the capillaries at all times. To maintain viscous flow conditions, a pressure of ~20 mbar in the inlet system is required. This constitutes a lower limit of gas suitable for analysis with the dual inlet system. The smallest practical volume of an inlet system is ~250 μL. This represents 200 nmol of gas at 20 mbar. In some cases, there is not enough sample material available to produce sufficient quantities of gas and alternative analytical methods must be employed, e.g. isotope ratio monitoring.

Delta (δ) values

In order to realize inter- and intra-laboratory consistency, measured isotope abundance ratios of the sample and reference gases are compared using a δ (delta) scale. In the case of carbon isotope ratio measurements, δ values are calculated as shown in Equation [6]. The δ values for other elements are calculated based on the isotope abundance ratios listed in **Table 1**.

$$\delta^{13}\mathrm{C} = \left[\frac{(^{13}\mathrm{C}/^{12}\mathrm{C})_{\mathrm{sample}}}{(^{13}\mathrm{C}/^{12}\mathrm{C})_{\mathrm{reference}}} - 1\right] \times 1000 \qquad [6]$$

This quantity is the deviation of the isotope abundance ratio of the sample from that of a standard in parts per thousand or per mill (‰). Positive δ values indicate that the sample is relatively enriched in the heavier isotope of the element with respect to the reference. Negative δ values represent samples that are relatively depleted in the heavier isotope compared to the reference material. The National Institute of Standards and Technology (NIST) in Gaithersburg, Maryland, USA and the International Atomic Energy Agency (IAEA) in Vienna, Austria both distribute and continuously calibrate isotope reference materials for a number of elements. The reference materials chosen are representative chemical compounds and reflect nominal terrestrial isotope abundances.

Table 1 The isotopes that are most commonly compared for the determination of δ values for the elements H, C, N, O and S

Element	*Isotope abundance ratio*	*δ value*
Hydrogen	D/H or $^{2}H/^{1}H$	$\delta\Delta$ or $\delta^{2}H$
Carbon	$^{13}C/^{12}C$	$\delta^{13}C$
Nitrogen	$^{15}N/^{14}N$	$\delta^{15}N$
Oxygen	$^{18}O/^{16}O$	$\delta^{18}O$
Sulfur	$^{34}S/^{32}S$	$\delta^{34}S$

Analysis of H, C, N, O and S isotope ratio variations

Hydrogen and oxygen

Hydrogen exhibits the largest natural isotope abundance variations due to the relatively large mass difference between its two stable isotopes; $^{1}_{1}H$ and $^{2}_{1}H$ (also written commonly as H and D, respectively). H is approximately 99.985% and D 0.015% naturally abundant. The three isotopes of oxygen $^{16}_{8}O$ $^{17}_{8}O$ and $^{18}_{8}O$ are ~99.763%, ~0.0375% and ~0.1995% abundant, respectively. Examples of isotope abundance variations measured for H and O are summarized in **Figures 6** and **7**.

Hydrogen and oxygen combine to form water and the isotopic species of H_2O vary in mass from 18 to 22 (i.e. $H_2{}^{16}O$, $HD^{16}O$, $D_2{}^{18}O$). The decreasing vapour pressure with increasing mass of the different isotopomers causes evaporated water vapour in equilibrium with the condensed phase to be relatively enriched in the lighter isotopes of both oxygen and hydrogen. Conversely, the condensation of water from the vapour phase results in a precipitate enriched in the heavier isotopes of O and H (higher $\delta^{18}O$ and δD values). The condensation of water vapour in equilibrium with its liquid phase – a situation that occurs in clouds – can be described by

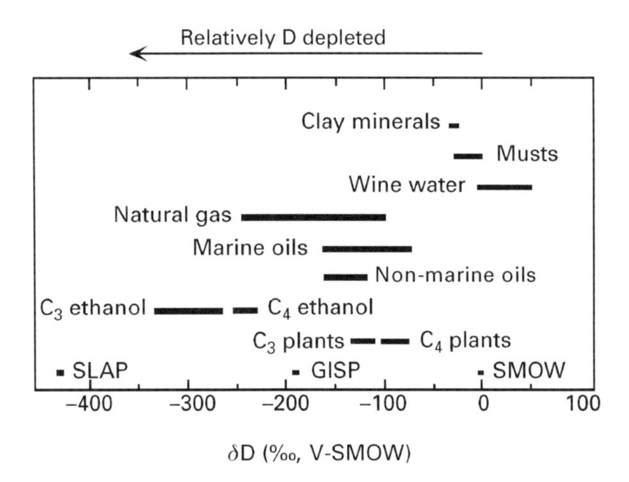

Figure 6 Natural variations in hydrogen isotope abundances. V-SMOW, SLAP, and GISP are international reference water samples available from the International Atomic Energy Agency (IAEA) in Vienna, Austria.
V-SMOW = Vienna Standard Mean Ocean Water
SLAP = Standard Light Antarctic Precipitation
GISP = Greenland Ice Standard Precipitation.

the Rayleigh fractionation law (Eqn [7]).

$$R/R_0 = f^{\alpha-1} \qquad [7]$$

Here, R_0 is the initial $^{18}O/^{16}O$ ratio of the gas phase, R is the actual ratio as a function of the remaining molar fraction in the gas phase (f). The term $\alpha_{\text{liquid-vapour}}$ is the fractionation factor between the liquid and vapour phases at a given temperature. With the increasing amount of ^{18}O-enriched liquid formed and removed from the vapour, the gas phase and liquid phase formed from it becomes isotopically lighter. The $\delta^{18}O$ values for liquid and vapour as a result of this process are plotted in **Figure 8**. As the precipitation event continues, the immediate removal of the condensed phase from the cloud without any isotope exchange or reevaporation continuously depletes the vapour in D and ^{18}O. As a result, precipitation formed at higher latitudes becomes progressively more depleted in the heavier isotopes of hydrogen and oxygen.

Because the extent of isotope fractionation between the condensed and vapour phases is temperature dependent, variations in the δD and $\delta^{18}O$ values recorded at a particular site reflect seasonal and possibly long-term fluctuations. Ices cores from Greenland, Antarctica, and elsewhere are excellent archives of past climate change, the information recorded mainly in the $^{18}O/^{16}O$ ratio of the ice. $\delta^{18}O$ measurements indicate that very rapid changes from cold to warm and vice versa within a decade have occurred over the past 160 000 years. Such data have challenged our understanding of the underlying processes and caught the attention of the media because of the social and political implications changing climatic events could have today.

Hydrogen and oxygen are also found in many minerals and are fractionated during geochemical processes. The hydrogen and oxygen isotope compositions of the mineral are controlled by the isotope

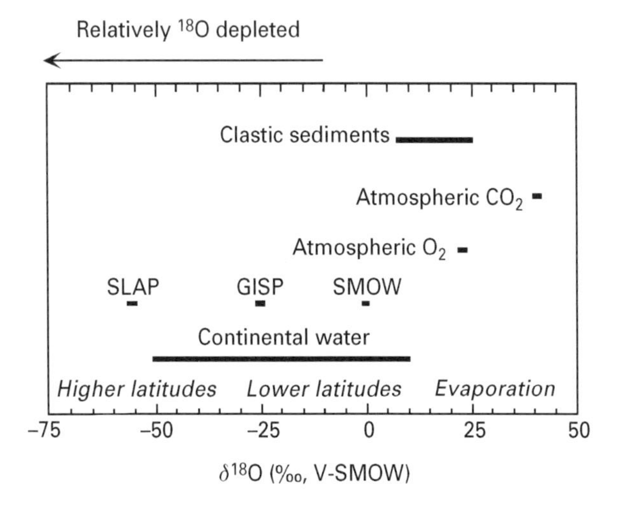

Figure 7 Natural variations in oxygen isotope abundances. The $\delta^{18}O$ values in ancient precipitation (Greenland Ice, Antarctic Ice, etc.) and in forameniferal carbonate from sediments provide an excellent record of the temperatures of the past.

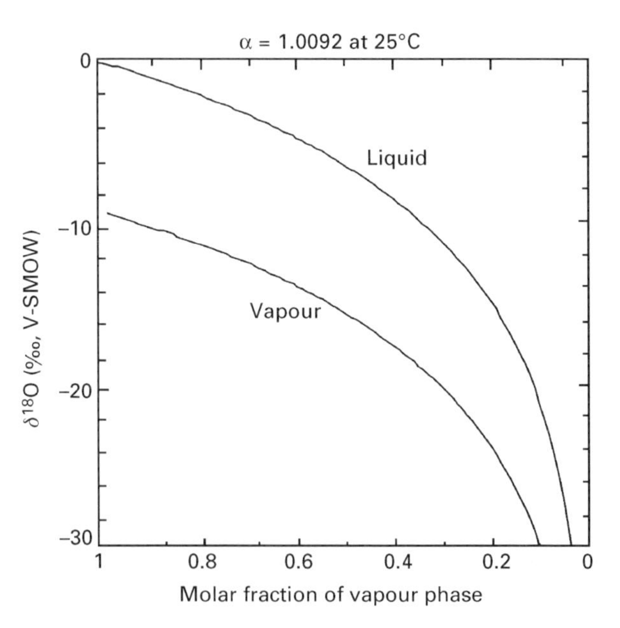

Figure 8 Rayleigh fractionation during water condensation. Condensation starts with the liquid enriched in $\delta^{18}O$ by 9.2‰ over the gas phase. As the condensation process proceeds, the remaining vapour becomes lighter and, consequently, also the precipitate formed from it.

composition of the fluid and the fractionation factor appropriate for the temperature during formation. For example, the $\delta^{18}O$ value of oxygen in $CaCO_3$ differs from that of the water as the carbonate precipitates under equilibrium conditions. The extent of fractionation is temperature dependent and at 25 °C the mineral is enriched in ^{18}O by about 28‰ compared to the water. The $\delta^{18}O$ value of carbonates and fluids are used to assess water–rock interactions. Whereas the δD value of hot spring waters may be similar to that of local precipitation, the $\delta^{18}O$ value may be enriched in ^{18}O indicating oxygen isotope exchange with nearby rocks.

Standards The standard for both hydrogen and oxygen is V-SMOW (Vienna Standard Mean Ocean Water) distributed by the IAEA in Vienna, Austria. A secondary standard often employed is SLAP (Standard Light Antarctic Precipitation) which has a δD values of −428‰ and $\delta^{18}O$ values of −55.5‰ with respect to V-SMOW.

Sample preparation Hydrogen isotopes in water are measured from H_2 gas formed by reducing water with a suitable reducing agent like U, Zn, Cr or C. It is also possible to exchange hydrogen isotopes between H_2 gas and H_2O in the presence of Pt catalysts.

Ratios of HD to H_2 (mass 3 to 2) of the sample hydrogen gas are measured and reported as δD values. Ion–molecule reactions in the mass spectrometer source will result in the production of H_3^+, which is isobaric with HD^+ at mass 3. The number of H_3^+ ions produced is proportional to the H_2 pressure and to the number of H_2^+ ions in the source, necessitating an empirical correction factor to the measured ratios.

Oxygen isotope compositions of water samples are determined from the measurement of $^{12}C^{16}O_2^+$ and $^{12}C^{16}O^{18}O^+$ at masses 44 and 46, respectively. CO_2 gas is equilibrated with 5 mL of water at 25°C for 8 h and isotope exchange between CO_2 gas and water makes the oxygen isotope composition of the CO_2 gas reflect that of the water.

Carbon

Carbon has two stable isotopes ($^{12}_{6}C$ and $^{13}_{6}C$) of ~98.89% and ~1.11% abundance, respectively. Carbon compounds are exchanged between the oceans, atmosphere, biosphere and lithosphere. Significant isotope ratio variations exist within these groups due to both kinetic and equilibrium isotope effects. Representative carbon isotope abundance variations are shown in **Figure 9**.

During photosynthesis, the metabolized products become depleted in ^{13}C compared to the CO_2. The initial assimilation and intracellular diffusion of

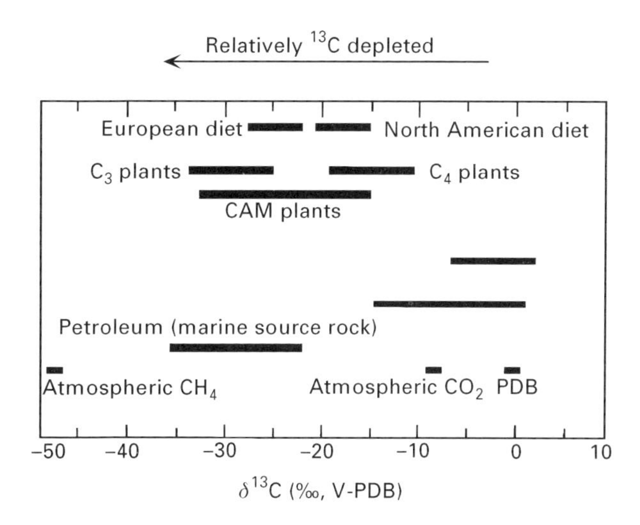

Figure 9 Natural variations in carbon isotope abundances. North American diet to a large extent is derived from maize, a C_4 plant. In countries like Japan, where fish is a major part of the average diet, $\delta^{13}C$ values are in between the European and North American values.

CO_2 in the plant produces a 4‰ depletion in its $\delta^{13}C$ value. Isotope fractionation during the decarboxylation of the absorbed CO_2 is dependent on the plant species and causes a much larger shift in the $\delta^{13}C$ value. Plants that form a 3-carbon phosphoglyceric acid as the first product of carbon fixation according to the Calvin cycle (commonly referred to as C_3 plants) have $\delta^{13}C$ values averaging −28‰ V-PDB. Examples of C_3 plants include cereal grains, peanuts, rice, tobacco, beans, sugar beets and all evergreen and deciduous trees. Plants that photosynthesize dicarboxylic acid by the Hatch–Slack mechanism (referred to as C_4 plants) are more enriched in ^{13}C and have $\delta^{13}C$ values closer to −13% V-PDB. C_4 plants include corn, millet, sorghum, sugar cane and many grasses such as crab grass and bermuda grass. Crassulacean acid metabolism (CAM) plants have intermediate $\delta^{13}C$ values. This arises because they metabolize CO_2 via C_3 pathways in light and C_4 in darkness. The large isotope fractionations that arise in plants are generally retained as the organic material is incorporated into sediments and eventually as hydrocarbon deposits.

The carbon isotope composition of food ingested by animals is generally reflected in the metabolized products. For example, it is possible to differentiate between animals that have primarily C_3 versus C_4 diets based on the measurement of the $\delta^{13}C$ value of hair, finger nails, bone collagen, etc. This information is useful in food web studies to determine primary nutrient sources and to discover who is the prey and who the hunter. Nitrogen isotopes are often combined with $\delta^{13}C$ data for a comprehensive trophic level analysis.

Oceans play a significant role in the exchange of CO_2 between the atmosphere and biosphere. Carbon isotope exchange between CO_2 in the air and HCO_3^- in the oceans results in ^{13}C enrichment of the latter. CO_2 in air is well mixed globally and currently has a $\delta^{13}C$ value of −8‰ V-PDB compared to marine limestones at ~0‰ V-PDB. Freshwater carbonates have much more variable and lower $\delta^{13}C$ values due to the oxidation of ^{13}C depleted organic matter and subsequent formation of HCO_3^-.

The CO_2 content of the atmosphere has increased dramatically due to the combustion of fossil fuels and oxidation of soil and organic matter. Fossil fuels are generally depleted in ^{13}C because they were derived from organic material. Consequently, the global $\delta^{13}C$ value of atmospheric CO_2 has seen a change from 6.5‰ in preindustrial times to about 8.0‰ today.

Standards The standard for carbon isotope abundance measurements is based on a Cretaceous belemnite sample from the Peedee formation in South Carolina, USA. The original material is no longer available. It has been replaced by the convention that NBS 19, a carbonate material, has a value of +1.95‰ versus PDB. This new scale is termed V-PDB (Vienna-PDB). The IAEA distributes a number of secondary standards including graphite (USGS24) with a $\delta^{13}C$ value of −15.99‰ V-PDB, oil (NBS-22) at −29.74‰ V-PDB, and calcium carbonate (NBS-18) with a value of −5.01‰ V-PDB.

The oxygen isotope composition of carbonates is also commonly referenced to the $^{18}O/^{16}O$-isotope ratio of V-PDB. It is not used as a reference for $\delta^{18}O$ analyses of igneous or metamorphic rocks. The difference between V-SMOW and V-PDB $\delta^{18}O$ values is approximately 30‰ ($\delta^{18}O_{V\text{-}SMOW} = 1.03091 \times \delta^{18}O_{V\text{-}PDB} + 30.91$).

Sample preparation CO_2 gas is produced from carbonate minerals by reaction with 100% H_3PO_4 or by thermal decomposition. The carbon *and* oxygen isotope compositions of the carbonate can be determined simultaneously from the CO_2 produced. $\delta^{13}C$ values from organic samples are analysed by combusting the material at high temperature with an oxygen source (commonly CuO) in sealed quartz tubes. Ion currents at mass 44, 45 and 46 produced from CO_2 gas are measured in order to determine a $\delta^{13}C$ value. The isobaric interference at mass 45 from ^{17}O (i.e. both $^{13}C^{16}O^{16}O^+$ and $^{12}C^{17}O^{16}O^+$ will be collected by the same Faraday detector) requires an empirical correction to enable the calculation of the $^{13}C/^{12}C$ ratio from the measured mass 45 (^{17}O correction).

Nitrogen

Nitrogen is present primarily as N_2 gas in the atmosphere and dissolved N_2 in oceans. In terrestrial systems, nitrogen is found in minor amounts bound to H, C and O in both reduced and oxidized states. Nitrogen has two stable isotopes $^{14}_{7}N$ and $^{15}_{7}N$ of ~99.64% and ~0.36% abundance, respectively. Isotope abundance variations measured for nitrogen are summarized in **Figure 10.**

Biochemical reactions mediated by bacteria are responsible for many nitrogen isotope fractionations. The fixation of N_2 gas to compounds such as NH_3 at the nodules of plant roots requires substantial energy to break the N_2 bonds and results in negligible isotope fractionation. However, the conversion of ammonia to nitrates (nitrification) initially to NO_2^- by *Nitrosomonas* ssp. and then to NO_2^- by *Nitrobacter* yields nitrates that can have $\delta^{15}N$ values 20‰ more negative than the ammonia. The bacterial conversion of nitrates to N_2 under anaerobic conditions (denitrification) by *Pseudomonas denitrificans* or *Thiobacillus denitrificans* enriches the ^{15}N content of the residual nitrate increasing the $\delta^{15}N$ value by as much as 30‰ as the reaction progresses. Generally, the organic matter of the biosphere is enriched in ^{15}N compared to the atmosphere and $\delta^{15}N$ values for marine environments are greater than those measured in continental regions.

Increasing populations and agricultural activities have a significant influence on the amount of nitrate in groundwater. Within a given region, nitrate concentrations and $\delta^{15}N$ values can be used to trace the source of anthropogenic nitrogen compounds into a system and monitor its occurrence over time.

The relationship between the nitrogen isotope composition of an animal and its position in the food

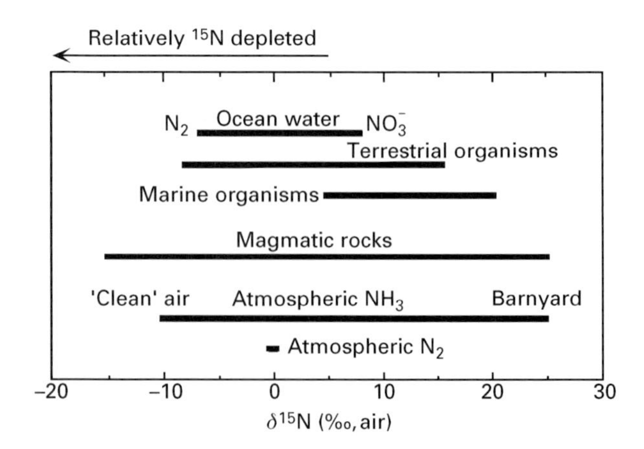

Figure 10 Natural variations in nitrogen isotope abundances. $\delta^{15}N$ values are often used to establish the position of a particular species within the food chain of an ecosystem (trophic level).

chain is such that one observes an increase in the $\delta^{15}N$ value of approximately 3‰ per trophic level. Compared to the food source, excreted nitrogen is relatively depleted in ^{15}N due to fractionation during urea formation. The analysis of contemporary and archaeological specimens of feathers, bone, hair or skin can provide information about the diets of people and animals and changes that may have occurred in response to past climatic or historical events.

Standards Nitrogen in the atmosphere is well mixed and $\delta^{15}N$ values are reported relative to the isotope composition of N_2 in air.

Sample preparation Nitrogen isotope abundances are determined by measuring ions at masses 28 ($^{14}N^{14}N^+$) and 29 ($^{14}N^{15}N^+$). Nitrogen compounds are converted to N_2 gas by high temperature combustion in an elemental analyser. N_2 gas introduced into the ion source must be free from CO or CO_2 that produce an isobaric interference at masses 28 and 29.

Sulfur

Sulfur has four stable isotopes; $^{32}_{16}S$, $^{33}_{16}S$, $^{34}_{16}S$ and $^{36}_{16}S$ which are ~95.02%, ~0.75%, ~4.21% and 0.02% abundant, respectively. Sulfur is ubiquitous in the global environment and major sulfur reservoirs include sulfate in the oceans, evaporites, sulfide ore deposits, and organic sulfur compounds. Sulfur isotope abundance variations are shown in **Figure 11**.

Large kinetic isotope effects occur during the reduction of sulfate by anaerobic bacteria such as *Desulfovibrio* ssp. resulting in the product sulfide having $\delta^{34}S$ values ranging from +3 to −46‰ with respect to the remaining sulfate. Another potential mechanism for sulfur isotope fractionation includes exchange reactions among sulfur compounds with S in different valence states. For example, sulfur isotope exchange between sulfate and sulfide (and resulting change in oxidation state from +6 to −2) is predicted to yield fractionations as large as 75‰ between the H_2S and SO_4^{2-} at 25 °C. However, sulfur isotope exchange between sulfates and sulfides proceeds very slowly and probably occurs at high temperature where the equilibrium constant is close to unity. Sulfur oxidation, mineral precipitation, and high temperature processes result in less extensive isotope fractionations.

Sulfur isotope compositions are employed to assess the effect of industrial activities in pristine environments, particularly in regions where the pollutant sulfur has a distinct $\delta^{34}S$ value compared to that of the environmental receptors. For example, in the province of Alberta, Canada, emissions from sour gas processing plants have $\delta^{35}S$ values 20‰ higher compared to sulfur isotope compositions of unaffected soils. The distinct isotope composition of the anthropogenic sulfur allows it to be followed as it enters the environment via biogeochemical reactions.

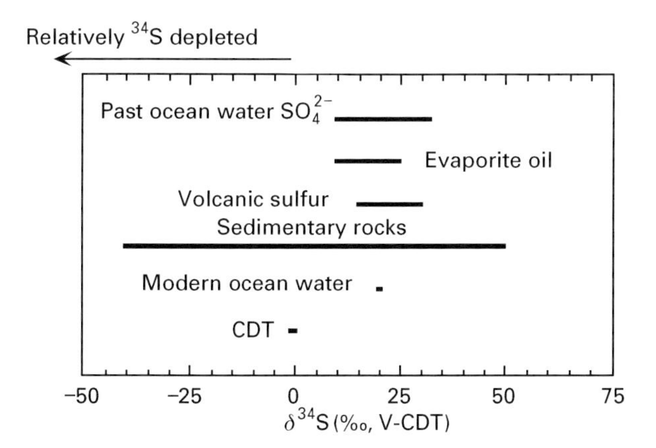

Figure 11 Natural variations in sulfur isotope abundances. Large sulfur isotope fractionations are possible during the reduction of sulfates to sulfides by anaerobic bacteria that produce ^{32}S enriched H_2S relative to the original sulfate.

Standards The international sulfur isotope scale is based on a troilite sample from the Canyon Diablo meteorite, termed CDT. The International Atomic Energy Agency (IAEA) in Vienna has promoted the use of a more accurate sulfur isotope scale, V-CDT, to replace CDT because of isotopic inhomogenities of ±0.4‰ in the original sample. A new Ag_2S standard, IAEA S-1, has a defined $\delta^{34}S$ value of −0.30‰ V-CDT.

Sample preparation Sulfur isotope abundances are measured commonly from isotopic ion currents of SO_2^+ at masses 64, 65 and 66. SO_2 gas is generated by reacting sulfur or sulfide minerals with V_2O_5 and SiO_2 in a ratio of 1 : 10 : 10 in sealed quartz tubes heated to 900 °C. Isobaric interference from oxygen isotopes necessitates the preparation of SO_2 from standards and samples with identical oxygen sources. SO_2 is easily adsorbed onto glass and metal surfaces and care is required in its handling to avoid fractionation and memory effects among different samples. An alternative (and lesser used) approach is to form the more stable SF_6 gas and measure SF_5^+ ions.

Isotope ratio monitoring

A recent analytical technique for high precision, high sensitivity isotope ratio measurements is isotope ratio

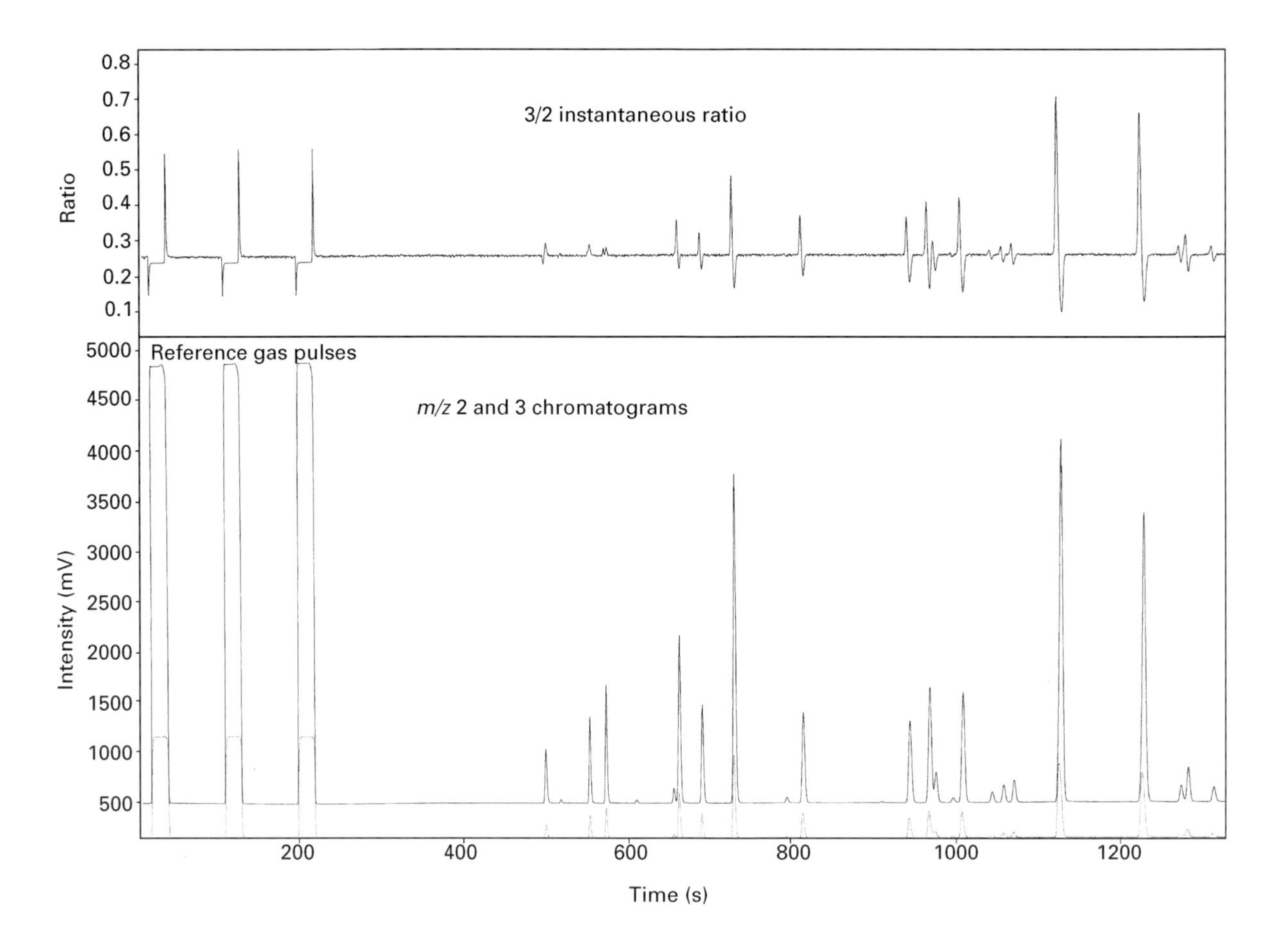

Figure 12 *irm*-GCMS chromatogram from a crude oil extract. The isotope ratio of deuterium versus hydrogen from a post column high temperature (1400°C) pyrolysis reaction is studied. The lower trace shows the mass 2 intensity whereas the upper trace exhibits the instantaneous intensity ratio of 3/2 as a function of time. The deuterated compounds elute earlier than the nondeuterated ones with a time shift between 1 and 5 s. This gives rise to the marked isotope swing in the upper 'ratio' trace. The first 3 peaks are reference gas pulses injected behind the GC, thus they do not exhibit this behaviour. Here, differences in the amplifier time constants give rise to the sharp edges displayed on the ratio trace.

monitoring (*irm*). The *irm* instrument is a further development of the classical isotope ratio mass spectrometer including improvements in source design, vacuum pump technology, high speed detection electronics and powerful personal computers for data processing. This method for isotope analysis does not employ a dual inlet (**Figure 1**) and hence there is no requirement to generate large amounts of sample gas to maintain viscous flow conditions in the capillaries. Instead, isotope abundance data are extracted from time-varying ('transient') signals as opposed to measurements of prepared gases stored in reservoirs that produce near constant signal intensities. The entire sample entrained in a helium carrier gas enters the ion source as a transient and is ionized for isotope abundance measurements. A typical *irm* chromatogram for D/H analysis from GC eluates is shown in **Figure 12**. All ion currents of interest are integrated as the He carrier sweeps the sample through the source. A pulse of reference gas is admitted to the source and its isotope ratio determined in a similar manner. From the two measurements the δ value of the respective sample peak is calculated. Advantages of the *irm* technique include the smaller amount of sample that must be prepared for analysis (1 nmol compared to 200 nmol). In addition, reduced sample handling and preparation is possible because raw materials are converted to simple gases for analysis by automated sample preparation devices (such as elemental analysers). It should be noted that there are a number of terms in current use in the literature to describe this method, including continuous flow isotope ratio mass spectrometry or CF-IRMS.

Configuration of *irm* inlet systems

The *irm* system consists of three components; the sample preparation device, isotope ratio mass spectrometer and the gas and electronic interface between the two. Depending on the nature of the sample and type of analysis required, there are two approaches to assembling an *irm* inlet system (**Figure 13**). For isotope abundance ratio determinations from bulk samples, the entire sample is either oxidized or pyrolysed, producing a mixture of reaction products (CO_2, N_2, SO_2 and H_2O for the oxidation, CO and H_2 for the pyrolysis) that must be separated prior to introduction into the mass spectrometer. In contrast, compound specific analysis is the separation of a particular component from some complex mixture followed by conversion to a gas suitable for isotope analysis (by oxidation or pyrolysis) and subsequent introduction into the mass spectrometer ion source. Here, reaction products are isolated by chemical means on-line when possible.

The amount of sample and carrier gas that enters the ion source is limited to a maximum of about

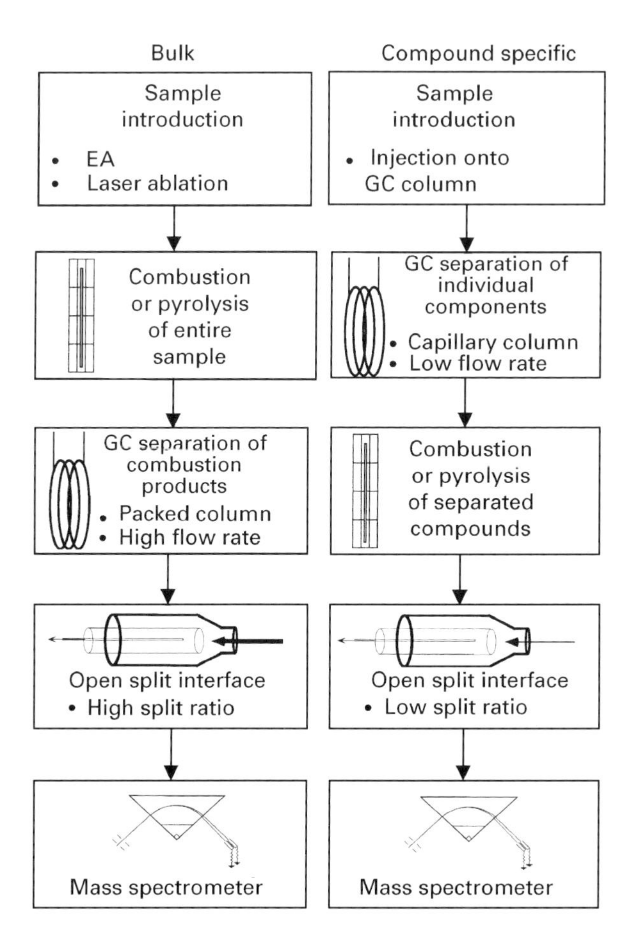

Figure 13 Comparison of isotope ratio monitoring (*irm*) sample preparation and introduction for bulk versus compound specific sample analysis. In bulk sample analysis, the GC follows the combustion step whereas in the compound specific *irm* system CO_2 and N_2 from a single peak both enter the mass spectrometer at the same time. In both cases the amount of gas flowing to the ion source is comparable. The difference in sensitivity comes from the difference in carrier gas flow rate (about 100 ml min^{-1} versus ~1.5 ml min^{-1}).

0.5 mL min^{-1}. Therefore, only a portion of the gas from the sample preparation device is admitted, requiring an open split. A high gas flow from the preparation system will require a higher split ratio and reduce the relative amount of sample available for ionization. To maximize sensitivity the split ratio should be as low as practically possible thereby maximizing the amount of sample ionized.

Bulk composition isotope ratio monitoring The determination of isotope compositions from bulk samples is one of the most widespread applications of *irm* techniques primarily because of its ease of use, possibility for automated analysis, and the variety of sample types that can be analysed. Raw samples are most often converted to simple gases by an elemental analyser (EA) where submilligram quantities of sample are packed into Sn or Ag capsules (2 × 5 mm) and loaded into the autosampler of the EA. Carrier gas flow rates are typically 80 to 120 mL min^{-1}, necessitating a high split ratio at the mass spectrometer. Individual samples are dropped into an oxygen-enriched He stream sweeping a reactor at 1050°C where they are combusted instantaneously. A second copper-filled furnace at 600°C in-line reduces nitrogen oxides produced in the combustion process to N_2 gas. It also scavenges excess oxygen not used in the combustion. Water is removed from the combustion products and the CO_2, N_2 and SO_2 (if any) are separated by a packed column (**Figure 14**).

Compound specific isotope ratio monitoring The isotope composition of individual components constituting a complex mixture can be determined by

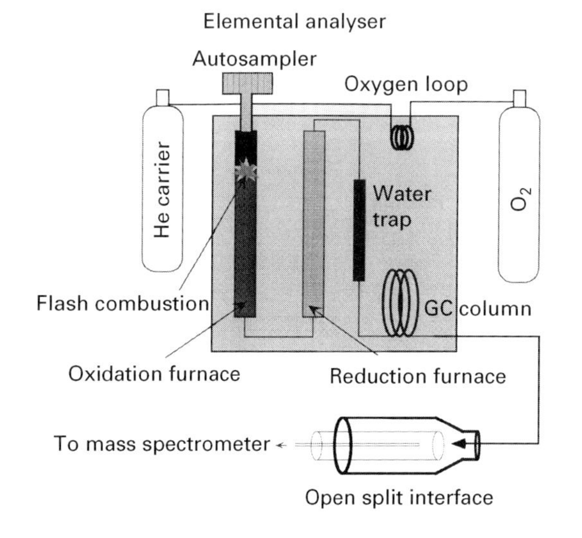

Figure 14 Bulk sample isotope ratio monitoring interface. The means for reference gas injection and for effluent dilution are not shown.

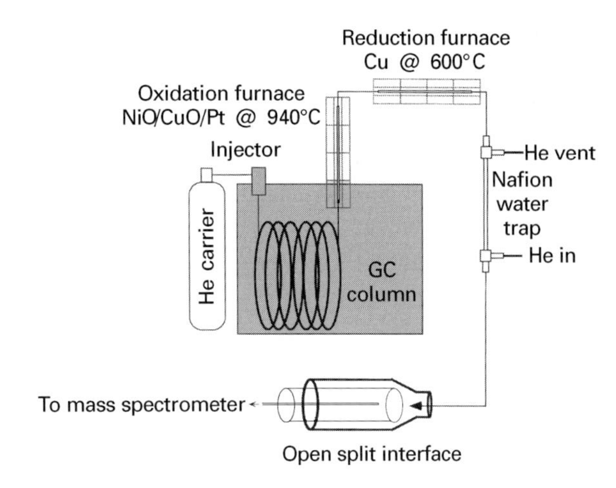

Figure 15 Compound specific isotope ratio monitoring interface. For clarity, the devices for backflushing the solvent, for effluent diversion from the ion source, and for reference gas injection are not shown.

separating out the individual fractions followed by conversion into simple gases suitable for analysis. The basic arrangement is a gas (or liquid) chromatograph followed by an oxidation (or pyrolysis) furnace and open split to the mass spectrometer (**Figure 15**). A capillary chromatographic column first separates the mixtures.

For oxidation and carbon or nitrogen isotopic analysis the components are then swept sequentially through a furnace converting them to N_2, CO_2 and H_2O. Water is removed and CO_2 is removed cryogenically if $\delta^{15}N$ measurements are to be made. For $\delta^{13}C$ analysis alone, nitrogen, if present, also enters the ion source. It does not interfere with the measurement.

For pyrolysis at high temperatures (δD or $\delta^{18}O$ analysis) the gaseous reaction products H_2 and CO are allowed to enter the mass spectrometer simultaneously. For D/H it is possible to remove CO cryogenically using, for instance, a piece of capillary molecular sieve column immersed in liquid nitrogen.

The low carrier gas flow (<3 mL min^{-1}) allows for a low split ratio or high sensitivity. This technique termed *irm*-GCMS is sometimes referred to as gas chromatography–combustion mass spectrometry (GCCMS).

Characteristics of the *irm* mass spectrometer

The *irm* mass spectrometer must be able to handle the resulting high pressures in the ion source region (10^{-5} mbar) while processing data from transient signals with precisions comparable to, or better than, dual-inlet performance. Differential pumping of the source and analyser regions enables the ion source of the *irm* instrument to operate under a high continuous He load.

To measure nanomoles of gas, the mass spectrometer ion source must maximize the number of ions detected per number of gas molecules introduced. Typical *irm* instruments are capable of producing 1 ion for every 1500 molecules admitted to the source. In practice, this is achieved by using a 'closed' source design and allowing a large angle of the ion beam to enter the analyser region of the mass spectrometer. The ion optical design of the instrument ensures that the vertical height of the ion beam is identical at the entrance and exit slits so that loss of ions due to collisions with the walls of the flight tube is minimized.

The transient signals encountered in *irm* mass spectrometry vary over several orders of magnitude during an analysis. In order to measure accurate isotope abundance ratios, the measured ion current ratios must not depend on the ion current intensities. This characteristic is defined as source linearity. Note that with dual-inlet systems, major ion current intensities of the sample and reference gases were matched by varying the pressure of the gas in the ion source to counter errors introduced by nonlinear source behaviour. Two major causes of nonlinear source behaviour are the buildup of space charge due to the large numbers of He atoms in the source and the reaction of neutrals with protonating agents leading to isobaric interference.

The Faraday cups receive a constant stream of time-varying ion currents that must be processed without loss of data. This capability was only possible recently with the development of high speed, low cost personal computers. Ion currents are converted to voltages by high gain, low noise operational amplifiers. The voltages are then digitized into signals that can be processed by a computer. Typically, the voltage is integrated over 0.25 s time slices and converted to a frequency pulse such that the number of pulses within the interval is directly proportional to the measured ion current. The time constants of the ion current amplifiers (i.e. how quickly the device responds to a changing signal) are optimized and matched for all detectors.

Examples of *irm* applications

Doping in athletics Testosterone is a potent performance-enhancing drug that is used in both human and equestrian sports. The current method to detect testosterone abuse is based on measurement of the testosterone to epi-testosterone ratio using GC or LC/MS techniques. However, this method is not reliable because some 'clean' individuals may naturally register a 'positive' result and the measurement must be made no more than 5 h after ingestion of the drug. Successful detection of drug abuse by *irm* is possible because commercially available testosterone has a lower $\delta^{13}C$ value than that metabolized by the body (approximately 5 ‰ lower). The isotope ratio determination of a subject's testosterone and metabolites by *irm*-GCMS can detect this shift in $\delta^{13}C$ values and detect testosterone abuse 24 h after the initial dose.

Food adulteration Quality assurance of food products is an important activity in the food industry where the substitution of lower cost alternatives constitutes fraud as well as a public health risk. Stable isotope measurements have been successfully applied to many instances where 'all-natural' fruit juices, honey and maple syrup have been artificially sweetened. This is possible when the added sugar is from a plant species which follows a different photosynthetic pathway from that of the bulk material (C_3 versus C_4 plants), leading to a $\delta^{13}C$ composition of the food product intermediate to the 12‰ difference between the two plant types.

The determination of the $\delta^{13}C$ value of individual fatty acids of adulterated and unadulterated maize oils by *irm*-GCMS allows as little as 5% of an adulterant C_3 oil to be detected. More recently, the measurement of $\delta^{18}O$ value of olive oils that were pyrolysed over nickelized carbon demonstrated a geographical dependence between ^{18}O and latitude. Lower $\delta^{18}O$ values were measured for olive oils coming from Italy as compared to Greece, allowing the origin of the product to be verified.

Detection of *Heliobacter pylorii* The *Heliobacter pylorii* bacterium is found in the mucous membrane of the stomach and is the major cause of gastritis, gastric ulcers and cancer of the stomach and duodenum. It produces urease by cleaving urea to form ammonia and CO_2. The detection of this dangerous bacterium is possible by administering a small amount (100 mg) of ^{13}C-labelled urea to the patient and measuring the $\delta^{13}C$ value of breath CO_2. If after 30 min an increase in the $\delta^{13}C$ of the breath-CO_2 is detected, the bacterium is present and actively forming ^{13}C-enriched CO_2 from the labelled urea. This noninvasive procedure is simple to administer and is widely applied as a ^{13}C urea breath test by a number of medical laboratories.

Trace gas analysis in air Greenhouse gases such as methane, CO_2 and N_2O are causing the temperature of the planet to increase, or they are interfering with ozone chemistry in the stratosphere. Isotope analysis of these gases may help to understand their cycling in the atmosphere or between global pools. For CH_4 and N_2O, dual inlet techniques require in excess of 70 L of air to achieve the precision (±0.2‰ for $\delta^{13}C$, $\delta^{18}O$ and $\delta^{15}N$) required to discriminate among different sources. Such large samples are difficult to process and are of low spatial and temporal resolution. To obtain meaningful, high resolution data from modern and archived air samples, *irm* techniques have been developed which are capable of analysing the isotope composition of CH_4 (1.7 ppm) and N_2O (0.3 ppm) in 25 mL and 100 mL air samples, respectively.

List of symbols

B = magnetic field flux density; h = Planck's constant; I = current; m = mass of molecule; N = number of parent nuclides; q = charge; R = path radius; t = time; T = temperature; W = work function; α = isotope fractionation factor; ν = frequency of oscillation; λ = probability of decay; ‰ = per mille; $\delta^{13}C$, $\delta^{18}O$, δD, $\delta^{15}N$ and $\delta^{34}S$ = Deviations of isotope ratios from a reference.

See also: **Biochemical Applications of Mass Spectrometry; Cosmochemical Applications Using Mass Spectrometry; Food Science, Applications of Mass Spectrometry; Food Science, Applications of NMR Spectroscopy; Forensic Science, Applications of IR Spectroscopy; Forensic Science, Applications of Mass Spectrometry; Inductively Coupled Plasma Mass Spectrometry, Methods; Interstellar Molecules, Spectroscopy of; IR Spectroscopy, Theory; Isotopic Labelling in Mass Spectrometry; Mass Spectrometry, Historical Perspective; Medical Applications of Mass Spectrometry; Scattering Theory; Sector Mass Spectrometers.**

Further reading

Bowen R (1988) *Isotopes in the Earth Sciences*. Essex, England: Elsevier.

Bowen R (1991) *Isotopes and Climate*. London: Elsevier.

Brand WA (1996) High precision isotope ratio monitoring techniques in mass spectrometry. *Journal of Mass Spectrometry* 31: 225–235.

Brand WA (1998) Isotope ratio mass spectrometry: precision from transient signals. In: *Advances in Mass Spectrometry*. Karajalainen EJ, Hesso AE, Jalonen JE and Karjalainen UP (eds), Vol 14, pp 655–680. Amsterdam: Elsevier.

Dickin AP (1995) *Radiogenic Isotope Geology*. Cambridge: Cambridge University Press.

Faure G (1986) *Principles of Isotope Geology*, 2nd edn. New York: Wiley.

Fritz P and Fontes J Ch (eds) (1980)*Handbook of Environmental Isotope Geochemistry*. Amsterdam: Elsevier.

Heumann KG (1988) Isotope dilution mass spectrometry. In: Adams F, Gijbels R and van Grieken R (eds) *Inorganic Mass Spectrometry (Chemical Analysis Series, vol. 95)*, pp 301–376. New York: Wiley.

Hoefs J (1997) *Stable Isotope Geochemistry (Minerals and Rocks; 9)*, 4th edn. Berlin: Springer-Verlag.

Krouse HR and Grinencko VA (eds) (1991) *Stable Isotopes: Natural and Anthropogenic Sulphur in the Environment (Scope/Scientific Committee on Problems of the Environment, No 43)*. Wiley.

Platzner IT (ed) (1997) *Modern Isotope Ratio Mass Spectrometry*, Vol 145 in Chemical Analysis. Chichester: Wiley.

Isotopic Labelling in Biochemistry, NMR Studies

See **Labelling Studies in Biochemistry Using NMR.**

Isotopic Labelling in Mass Spectrometry

Thomas Hellman Morton, University of California, Riverside, CA, USA

MASS SPECTROMETRY
Applications

As is well known, an atomic nucleus is characterized by its positive charge Z (an integer that designates the number of fundamental electric charges) and its mass. Nuclear masses are so close to being multiples of the proton rest mass that they are also characterized by integers, A. The integers A and Z characterize a nuclide. Most elements as found in nature contain mixtures of nuclides with different values of A, known as isotopes.

Pure isotopes have masses that differ measurably from the integers A (based on a scale on which ^{12}C has a mass defined as 12.00000 atomic mass units or amu). None of the stable nuclides (i.e. those with nuclear half-lives $>10^9$ years) deviates from integer mass by more than 0.1 amu. The nominal mass of a molecule is the sum of the A values of its constituent atoms.

Labelling nuclides

In 1914 F.W. Aston demonstrated mass spectrometric separation of the two most abundant isotopes of neon. Since that time, careful measurement has established that 20 chemical elements are monoisotopic (i.e. exhibit only one naturally occurring stable nuclide to any significant extent). From the remaining members of the periodic table, the relative abundances for hundreds of isotopic pairs (same Z, different A) have been determined by mass spectrometry. The 163 stable nuclides for elements lighter than xenon give rise to slightly more than 300 isotopic pairs.

Labelling embraces those experimental designs in which two nuclides have been mixed under conditions in which they might be expected to behave identically (or nearly so), followed by an analysis that distinguishes them. Usually the mass spectrometer acts as the analyser, but it can play the role of the mixer instead. A measurement of the ^{44}Ti half-life (a piece of data pivotal for investigating astronomical cataclysms) illustrates one such application in nuclear physics. Bare nuclei with the same A/Z ratio will coincide in a mass separator. At gigavolt kinetic energies, where ions undergo nuclear reactions with target atoms and emerge without any bound electrons, ions from different elements having $A = 2Z$ will not be separated but can be isolated from those for which $A \neq 2Z$. This technique has been used to implant mixtures of ^{22}Na and ^{44}Ti simultaneously into solid matrices, where the radioactive sodium provides a standard for measuring the decay rate of the radioactive titanium. In this example, one nuclide has been used to label another that has very different chemistry, so as to provide a well-calibrated mixture of the two.

More conventionally, mass spectrometry is used to separate isotopes of a given element or isotopically tagged compounds. Four stable isotopic pairs have found application in the vast majority of molecular labelling experiments examined by mass spectrometry – $^{1}H/^{2}H$, $^{12}C/^{13}C$, $^{14}N/^{15}N$, and $^{16}O/^{18}O$. In each of these cases, the heavier isotope is the rarer, with natural abundances of 0.15% (^{2}H), 1.1% (^{13}C), 0.37% (^{15}N), and 0.20% (^{18}O). The survey below will confine itself to molecular labelling experiments that utilize hydrogen, carbon, nitrogen and oxygen isotopes.

Isomers, isotopomers and isobars

The term isomer has two different meanings, only one of which will pertain here. The meaning that is not relevant is employed by physicists, to whom *nuclear* isomers designate states of atomic nuclei with the same A and Z (generally having different net spins), which have different masses. For example, two nuclear isomers of cadmium, ^{113}Cd and ^{113m}Cd, differ in mass by 0.0003 amu. Nuclear isomers will be omitted from further mention. Isomerism will refer below only to molecules that possess the same connectivities among their atoms but which are not superimposable.

The term isotopomer designates a variant of a molecule with a different net isotopic content. Except for the isotopic difference, the two molecules must otherwise have identical structures. Three isotopomers of phenol provide an example. The (^{13}C)phenol drawn as structure [1] represents an isotopomer of phenol containing only ^{12}C. But [1] is not an isotopomer of structure [2], which also contains one ^{13}C. Instead [1] and [2], properly speaking, are *isotopic isomers*, differing only by the placement of the labelled atom.

OH H^{13}C [1]   OH H^{13}C [2]   OD [3]

Ions with different elemental compositions may display the same nominal mass-to-charge ratio (m/z), such as the m/z 28 positive ions $N_2^{\bullet+}$, $CO^{\bullet+}$, $C_2H_4^{\bullet+}$, $HCNH^{+}$, Fe^{2+}, and Ce^{5+}. These are sometimes called isobars. Among physicists, *isobars* denote two nuclides with the same value of A but different values of Z. Chemists use the term less restrictively. The monodeuterated phenol [3] is an isotopomer of [1] and [2]. If all of its carbons are ^{12}C, then the mass of [3] will have the same integer value as the masses of [1] and [2], and all three molecular ions will have m/z 95. [3] is an *isobaric isotopomer* of [1] and [2]. The deuterated ion does not have the same exact mass as the ^{13}C-containing ions, and they can be distinguished by high-resolution mass spectrometry (HRMS).

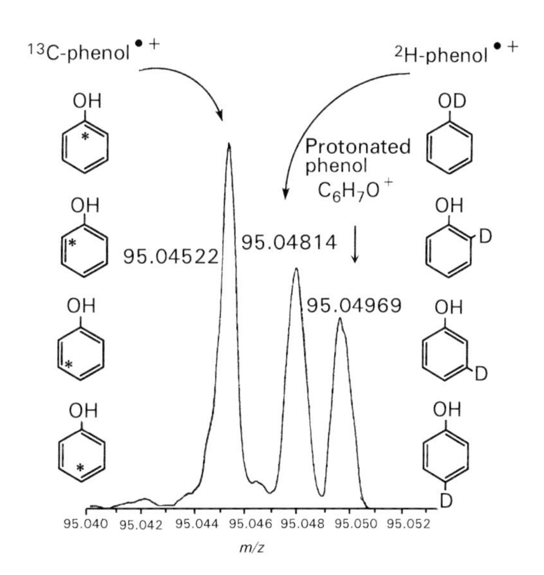

Figure 1 HRMS of m/z 95 ions from 70 eV electron ionization of $C_6H_5OCH_2CD_2OH$, illustrating the resolution of isobaric ions on a double focusing reverse Nier–Johnson (B-E) instrument with resolving power of 70 000. Each of the peaks that contains a rare isotope is a radical ion (odd number of electrons), which is a mixture of isomeric structures, all the possibilities for which are drawn. Asterisks designate ^{13}C-containing positions of the ring. The unlabelled ion is an even-electron species. Reproduced a from Nguyen V, Bennett JS and Morton TH (1997) *Journal of the American Chemical Society* **119**: 8342 with permission of the American Chemical Society with modifications.

Mass spectrometry cannot separate the molecular ions of isomers, although two isotopic isomers might well give rise to different fragmentation patterns (see below). Mass spectrometry can separate isotopomers, even if they happen to have the same nominal mass. **Figure 1** illustrates a published example, an experimental measurement of the abundances of m/z 95 fragment ions from a larger molecule. Phenol (C_6H_5OH) tagged with a single ^{13}C atom corresponds to any one of four isomers (including [1] and [2]) with identical masses, or a mixture thereof. Monodeuterated phenol has the same nominal mass as (^{13}C)phenol. The positive-ion mass spectrum reproduced in **Figure 1** displays the resolution of the two isotopomers (which differ by 0.003 amu) as well as protonated phenol ($C_6H_7O^+$), which has a different elemental composition.

At the time of writing, the techniques that give the highest resolution (such as Ion Cyclotron Resonance (ICR) MS) do not enjoy a reputation for accurate quantitation of isobars. Double focusing sector mass spectrometers (such as was used for the mass spectrum in **Figure 1**) appear to be the current method of choice for experiments that require high-resolution measurements of relative ion abundances, even though they have lower resolving power. It is important to bear in mind the distinction between high-resolution mass spectrometry (as exemplified by **Figure 1**) and exact mass determination (for which ICR gives excellent results based on accurate frequency measurements).

Variations of natural abundances

Precise measurement of relative natural abundances finds application in chemistry. The food industry routinely assays $^{12}C/^{13}C$ ratios of flavourings in order to ascertain their origin. For instance, that isotopic ratio is slightly greater for synthetic vanillin than for the identical molecule from the vanilla bean. As the commercial value of natural vanillin is approximately one million times greater than that of the synthetic, isotope ratio mass spectrometry (IRMS) plays a major role in quality control.

Isotopes do not fractionate identically among the products of most chemical reactions. If a reaction proceeds in 100% yield, the net balance of isotopes in the products ought to be identical to that of the reactants, though the distribution may differ between two products. When a reaction does not run to completion, the natural abundance in the products may vary appreciably from that of the reactants. Measurement of the distribution of natural-abundance isotopes represents an important tool in studying reaction pathways, but the partitioning does not always depend solely upon isotopic mass. In 1978 B. Kraeutler and N.J. Turro used mass spectrometry to determine the fractionation of ^{13}C in carbon monoxide from decomposition of the radical pair produced by photolysis of dibenzyl ketone (Eqn [1]).

$$(C_6H_5CH_2)_2CO \xrightarrow{h\nu} [C_6H_5CH_2\cdot \quad C_6H_5CH_2CO\cdot] \rightarrow (C_6H_5CH_2)_2 + CO \qquad [1]$$

The nuclear magnetism of the heavier carbon isotope perturbed the branching ratio under the reaction conditions, favouring return of the intermediate caged radical pair to the starting material. Hence, recovered CO was measurably depleted in ^{13}C. In part, the elegance of this experiment derives from the fact that it required no synthesis or separation of isotopically enriched starting material. In 1980 R.G. Lawler pointed out that conclusive demonstration of nuclear spin isotope effects demands bracketing of the magnetic isotope between two nonmagnetic ones (e.g. depletion of ^{13}C relative to ^{14}C as well as depletion relative to ^{12}C); however, few experiments have been reported that fulfil this rigorous requirement.

Isotopic dilution and tracer techniques

Isotopic dilution and tracer techniques represent different methodologies, but they overlap to a degree that obviates any clear-cut categorization. Both are techniques for quantitative analysis. In an isotopic dilution experiment one adds a known amount of an isotopomer (sometimes called the 'spike') of a known compound (usually containing a rare isotope) and measures the abundance of ions derived from it relative to those from the analyte (which is usually the more common isotopomer). If, for example, one wished to quantitate phenol (C_6H_5OH) in a mixture, one might add a known quantity of pentadeuterated phenol (C_6D_5OH) as a spike and measure the *m/z* 94:*m/z* 99 intensity ratio by mass spectrometry.

This conceptually simple approach, called a mass isotopomer abundance ratio (MIAR) measurement, requires a number of refinements. First of all, other components of the mixture might give rise to ions of the same mass – for instance, all alkyl phenyl ethers larger than anisole give very prominent *m/z* 94 ions. Therefore, MIAR experiments ordinarily require chromatographic separation prior to mass spectrometry (such as GC-MS), an approach sometimes referred to as isotope ratio monitoring (IRM) chromatography MS. Since different isotopic variants may have slightly different retention times, one must monitor intensity ratios over the entire elution profile of the component under analysis. Second, isotope effects may alter the relative abundances of fragment ions; hence, calibration with authentic samples is necessary. Finally, MIAR measurements give best results when the peaks to be compared have comparable intensities. When one ion has abundance more than an order of magnitude greater than the other, interferences and nonlinearities begin to introduce uncertainty and the measured MIAR may depend upon the amount of sample introduced into the mass spectrometer. C.K. Fagerquist and J.-M. Schwarz have surveyed the origins of systematic deviations and list ion–molecule reactions within the mass spectrometer and detector nonlinearities as possible sources of error.

The most reliable abundance measurements involving stable isotopes make use of carbon, nitrogen or oxygen labelling, with chromatographic separation followed by quantitative conversion to light gases prior to mass spectrometric measurement. GC–combustion–isotopic dilution MS (GC-C-IDMS) depends upon the high precision with which isotopic abundances for CO_2 and N_2 can be measured by means of isotope ratio mass spectrometry. Oxygen has three stable isotopes, and ^{13}C natural abundance interferes if one wishes to analyse ^{17}O content in CO_2. For such purposes, CO_2 can be converted to molecular oxygen with 5% molecular fluorine in helium. The basic equation for IDMS (regardless of whether the mass analysis is performed on the intact molecule or on its degradation products) is

$$N^{\text{sample}} = N^{\text{spike}} \frac{(R^{\text{spike}} - R^{\text{blend}})\Sigma R_i^{\text{sample}}}{(R^{\text{blend}} - R^{\text{sample}})\Sigma R_i^{\text{spike}}} \quad [2]$$

where N^{spike} stands for the (known) number of moles of spike; N^{sample} stands for the number of moles of analyte (to be quantitated); R^{spike} stands for the observed peak intensity ratio for the pure spike substance; R^{sample} stands for the observed peak intensity ratio for the pure analyte; R^{blend} stands for the observed peak intensity ratio for the mixture of spike and sample; and the R_i stand for the observed ratios for all isotopomers.

The following worked example illustrates how to apply Equation [2]. Consider an unknown number of moles of analyte (N^{sample}) for which the observed ^{13}C:^{12}C peak intensity ratio is $R^{\text{sample}} = 1/9$, to which one adds a known number of moles of spike (N^{spike}), for which the observed ^{13}C:^{12}C peak intensity ratio is $R^{\text{spike}} = 2/1$. Suppose the observed ^{13}C:^{12}C peak intensity ratio for the mixture is $R^{\text{blend}} = 23/37$. The values to be summed in the numerator are $\Sigma R_i^{\text{sample}} = 1/9 + 1$, the sum of the ^{13}C:^{12}C peak intensity ratio and the ^{13}C:^{13}C peak intensity ratio for the pure analyte. The values to be summed in the denominator are $\Sigma R_i^{\text{spike}} = 2/1 + 1$, the sum of the ^{13}C:^{12}C peak intensity ratio and the ^{13}C:^{13}C peak intensity ratio of the pure spike. Inserting these values into Equation [2] gives

$$\begin{aligned} N^{\text{sample}} &= N^{\text{spike}} \left[\frac{(2/1 - 23/37)(10/9)}{(23/37 - 1/9)(3)}\right] \\ &= N^{\text{spike}}[1] = N^{\text{spike}} \end{aligned} \quad [3]$$

In other words, the example corresponds to the outcome for equal amounts of sample and spike.

The above instance might obtain when the sample has nine carbon atoms (containing natural-abundance ^{13}C) and the spike has one partially labelled atom (though the spike might contain a mixture of isotopic isomers). Quantitation becomes more reliable if naturally occurring isotopomers do not overlap the peaks from the spike, so that $R^{\text{sample}} \rightarrow 0$. Multiple labelling can adequately address that issue. Also, precision increases as the isotopic purity of the spike is increased, so that $R^{\text{spike}} \rightarrow \infty$. Strictly speaking, it is not the values of R^{sample} and R^{spike} that introduce error, but rather the uncertainty in measuring those ratios when they do not have their limiting values of zero and infinity.

In tracer studies, the analyte is not present at the time the spike is added. Diagnosis of human infection by *Helicobacter pylori* (the bacteria implicated in digestive tract ulcers) exemplifies such a technique. This microorganism survives at low pH in the stomach by generating locally high concentrations of ammonia to buffer its immediate environment. When a patient swallows a sample of ^{13}C-labelled urea, a colony of these bacteria (if present) will hydrolyse the urea to NH_3 and CO_2, giving a characteristic rate of production of ^{13}C-carbon dioxide in the patient's breath.

The distinction between tracer and isotope dilution methodologies is often blurred. For instance, IDMS sometimes requires a series of chemical transformations. A recently published protocol for quantitating nitrate in environmental samples dilutes the sample with $^{15}NO_3^-$ followed by a series of steps that reduce it to nitrite and then convert it to an azo dye, Sudan I, as **Scheme 1** summarizes. Derivatization of label-containing Sudan I to its *t*-butyldimethylsilyl (TBDMS) ether then renders the labelled molecule sufficiently volatile to be introduced into a mass spectrometer. The ratio of M–57 fragment ion intensities quantitates the $^{15}N/^{14}N$ ratio, once correction has been made for ^{29}Si (natural abundance 5% that of ^{28}Si) and ^{13}C natural abundance of the 18 carbons in the ion.

A physiological labelling technique that applies tracer and isotopic dilution methodologies concurrently involves the use of doubly labelled water (containing ^{18}O in addition to a heavy isotope of hydrogen) for metabolic studies. After an animal is dosed with doubly labelled water, part of the labelled oxygen turns up in exhaled CO_2. From the amount of ^{18}O tracer in the CO_2 and the proportions of ^{18}O and isotopic hydrogen in excreted water (the isotopic dilution), one can infer the rate of energy expenditure of a living organism.

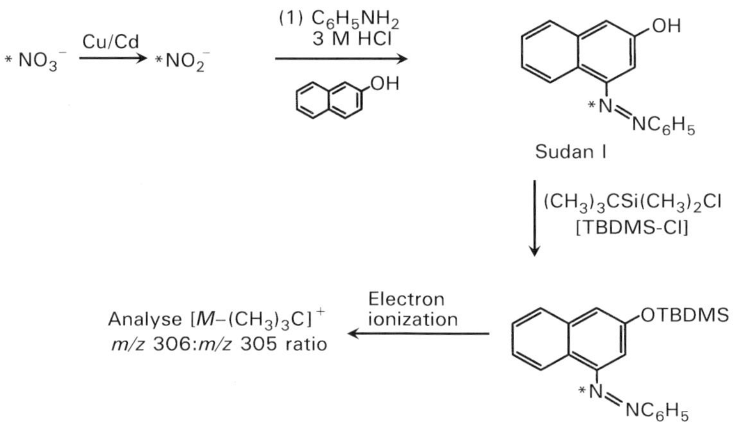

$*NO_3^-$ Cu/Cd → $*NO_2^-$ (1) $C_6H_5NH_2$ 3 M HCl, [2-naphthol] OH → Sudan I ($*N{=}NC_6H_5$, OH)

$(CH_3)_3CSi(CH_3)_2Cl$ [TBDMS-Cl] → OTBDMS, $*N{=}NC_6H_5$

Electron ionization → Analyse $[M-(CH_3)_3C]^+$ m/z 306:m/z 305 ratio

Scheme 1

Quantitative analysis of tracer experiments can be more complicated than for isotopic dilution. Consider a hypothetical outcome for pure $H_2{}^{18}O$ mixed thoroughly with a 10-fold excess of pure $H_2{}^{16}O$, where all of the oxygen atoms (and only those oxygen atoms) turn up in recovered CO_2. Statistically, the carbon dioxide should exhibit a distribution of zero, 1, or 2 labelled oxygens in the proportions m/z 44:m/z 46:m/z 48 = 100:20:1. It is clear that Equation [2] is not the right expression for analysing that result. Moreover, experiment has shown that observed ratios often deviate from statistical proportions because of fractionation factors (i.e. isotope effects that preferentially partition the heavier isotope into one chemical form versus another). The mathematical analysis for doubly labelled water studies, which includes the effects of fractionation factors, is well documented elsewhere.

Accelerator mass spectrometry

The sensitivity of tracer experiments increases as the naturally occurring isotopic background decreases. The natural abundance of ^{14}C in the biosphere (standardized to the year 1950) is approximately 1.2 parts per trillion (a level designated as Modern), a factor of 10^{-10} lower than the natural abundance of ^{13}C. Moreover, when an organism dies, it ceases to equilibrate with atmospheric CO_2 and its ^{14}C decays away. Many commercially available compounds are petroleum derived and routinely contain ^{14}C levels < 1% Modern. Therefore, mass spectrometric determination of ^{14}C has much greater sensitivity for tracer experiments than ^{13}C, with a usable detection limit of 0.001 fmole for ^{14}C versus 0.25 fmole for ^{13}C.

Accelerator mass spectrometry (AMS) is designed to analyse unstable isotopes, and a greatly simplified schematic is drawn in **Figure 2**. Historically, AMS measurements of isotopic ratios have been employed to measure half-lives of radionuclides by monitoring the disappearance of an unstable nuclide relative to a stable one (e.g. ^{44}Ti relative to ^{46}Ti) over a period of months. Contemporary applications address more chemical problems. The general technique involves two stages of ionization alternating with two stages of m/z selection. In the first ionization, atomic negative ions are produced from a solid sample, and the first mass spectrometer selects all ions with a given nominal mass. The negative ion is then accelerated to high velocity and transmitted through a foil stripper (a thin sheet of solid material), a passage that removes the outer electrons from the projectile ions. This second stage of ionization converts atomic ions with a single negative charge into positive ions with multiple charges, which are analysed by a second mass spectrometer.

When applied to ^{14}C, AMS makes use of a sample that has been converted to graphite. Since all chemical elements are present in the sample in concentrations greater than one part in 10^{12}, a major obstacle is to avoid isobaric ions. Bombarding the graphite with a fast ion beam produces C^- among the stable atomic ions. Nitrogen atoms do not form stable negative ions, so the only atomic negative ion produced having m/z 14 is $^{14}C^-$. Along with this are produced a number of isobaric m/z 14 molecular ions, such as $^{13}CH^-$ and $^{12}CH_2{}^-$. The m/z 14 negative ions are mass-selected by a low-energy mass spectrometer. The ion beam is then accelerated to megavolt kinetic energies and passed through a thin carbon foil, which strips off all the valence electrons. The molecular ions fragment as a consequence, while the atomic carbon ions simply acquire up to six positive charges. The mass filtering of AMS relies upon selection of m/z 3.5 positive ions ($^{14}C^{4+}$) formed from previously selected m/z 14 negative ions. One might

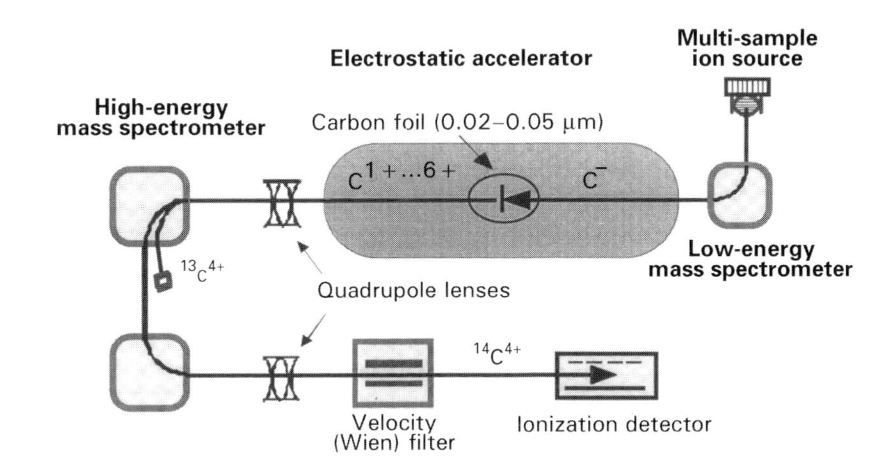

Figure 2 Schematic of the accelerator mass spectrometer at the Centre for AMS at Lawrence Livermore National Laboratory. Negative ions from a multi-sample ion source are separated by a low-energy mass spectrometer, accelerated to 6 MeV in an electrostatic accelerator, converted to positive ions by passage through a foil stripper, and accelerated again. Quadruply charged carbon atomic ions (30 MeV kinetic energy) are focused by quadrupole lenses and resolved by high-energy mass spectrometers, followed by velocity selection and identification of charge state in an ionization detector.

think that two stages of mass analysis would suffice to separate the ^{14}C atomic ions from all other isobars, but there is a substantial interference from $^{7}Li_2^-$ in the source that is converted to $^{7}Li^{2+}$ by passage through the foil stripper. Therefore, the final stage of identification of ^{14}C requires that the ion beam be velocity selected and then passed through an ionization chamber, where collisions with neutral gas molecules produce secondary ions with a rate that distinguishes between a doubly charged projectile ($^{7}Li^{2+}$) and a quadruply charged projectile ($^{14}C^{4+}$). The high kinetic energy of the ions in AMS permits sensitive single-ion counting, and samples containing $< 10^6$ atoms of ^{14}C in < 1 mg total carbon can routinely be analysed.

In practice AMS is used to measure $^{14}C/^{13}C$ isotopic ratios. For most applications the variation of ^{13}C abundance in the sample does not significantly affect the accuracy of an AMS measurement. AMS has been widely used for samples whose ^{14}C content is below Modern (as in carbon dating) and is beginning to find major applications for samples where ^{14}C content greatly exceeds Modern (as in biomedical research). The latter is exemplified by a recent study of covalent modification of proteins. When the enzyme aldolase is treated with acetoacetic ester, the protein irreversibly attaches a small but reproducible amount of the ester. Such a low level of tagging is sometimes referred to as background labelling, since (even with carrier-free (^{14}C)acetoacetic ester) there is not enough incorporated radioactivity to give a decay rate very much higher than the background for detecting nuclear disintegrations. The isotopic label stands out well above natural ^{14}C levels, but decay counting is an intrinsically inefficient method of detection. No matter what type of counter is used, only fourteen ^{14}C nuclei will decay per hour for every 10^9 that are present in the sample. In the case of tagged aldolase, the amount of incorporated radioactivity was too low for decay counting to be used for fragments of the tagged enzyme, and AMS presented the only method for analysing ^{14}C content after cleavage of the protein.

AMS takes advantage of the inherent efficiency of mass spectrometry and can readily measure the amount of label in protein fragments purified by analytical gel electrophoresis. Site-specific chemical cleavage of tagged aldolase gave four electrophoresis bands, each containing $\leq$0.05 mg of polypeptide. Individual gel slices (containing 5–10 mg of polyacrylamide each) were converted to CO_2 and then reduced to pure graphite. AMS gave analyses that showed $^{14}C/^{13}C$ ratios in the range from 5 to 50 times Modern reproducibly, depending on the identity and quantity of labelled polypeptide in the gel slice, which (when corrected for the great excess of stable carbon isotopes contributed by the polyacrylamide) corresponded to 0.03–3 fmol ^{14}C per polypeptide fraction. The extent of labelling was found to be nonuniform: it did not correlate with the molecular mass of the fragments but was proportional to the number of tyrosine residues. No correlation was observed for any other amino acid. Therefore, what had hitherto been dismissed as background labelling turns out to be an amino acid-specific covalent modification.

Tracer studies for characterizing reaction pathways

Mass spectrometry is typically used to look at products from reactions where a reactant has intentionally been enriched in a rare stable isotope. This is illustrated by a 'crossover experiment' reported by P.D. Bartlett and T.G. Traylor in 1963, who reported the labelling of molecular oxygen liberated by chain termination of free *t*-butyl hydroperoxy radicals. When a radical doubly labelled with ^{18}O reacted with an unlabelled radical to yield di-*t*-butyl peroxide (Eqn [4]), the predominant gaseous product gave *m/z* 34 ($^{16}O^{18}O$). That outcome indicates that each radical contributes one oxygen atom and implicates a tetroxide intermediate, which could not be directly observed.

$$\begin{aligned}(CH_3)_3C^{18}O^{18}O\cdot + (CH_3)_3C^{16}O^{16}O\cdot &\rightarrow \\ [(CH_3)_3C^{18}O^{18}O^{16}O^{16}OC(CH_3)_3] &\rightarrow \\ {}^{16}O^{18}O + (CH_3)_3COOC(CH_3)_3 & \qquad [4]\end{aligned}$$

An experiment that fails to show crossover can be equally meaningful. The 1953 report by F.A. Loewus, F.H. Westheimer and B. Vennesland on the enzyme alcohol dehydrogenase represents one of the earliest and most profound applications of isotopic labelling. These experiments demonstrated the facial selectivity of the enzyme-catalysed reaction, which is depicted schematically in Equation [5]. They performed mass spectrometric analyses of mixtures of HD and H_2 produced by reduction of water from combusted samples of ethanol and acetaldehyde, and they established that the (*R*) enantiomer of deuterated ethanol transfers only deuterium to the coenzyme NAD^+ (a partial structure of which is drawn in Eqn [5]). The reduced coenzyme transfers only deuterium back to acetaldehyde. Under conditions where the reaction goes back and forth many times, no deuterated acetaldehyde ($CH_3CD{=}O$) is detected. If, on the other hand, $CH_3CD{=}O$ reacts with the unlabelled, reduced coenzyme (known as NADH), the reduction gives (*S*)-1-deuterioethanol. No label transfers between deuterated acetaldehyde and the coenzyme. The remarkable conclusion (confirmed for many enzymes subsequently) is that biological catalysts distinguish the two faces of planar carbon atoms that bear three nonidentical substituents (e.g. the top and bottom faces of the planar ring of NAD^+ and of the carbonyl carbon of acetaldehyde).

Alcohol dehydrogenase

(NAD^+) + (*R*)-1-Deuterioethanol ⇌ H^+ + Acetaldehyde [5]

MS/MS – tandem versus ion storage

In the gas phase, a neutral molecule can be ionized by removal or addition of an electron via an electron beam (electron ionization or EI), by reaction with another gaseous ion (chemical ionization or CI), or by interacting with ionizing radiation. Most methods of depositing charge are so energetic that they lead to fragmentation. By close examination of fragment ions, one can sometimes draw conclusions regarding the placement of isotopic label and the mechanism of ionic decomposition. Quite often the source mass spectrum is confusing, since isotopic label frequently randomizes under typical ionization conditions. Sequential mass spectrometry separates a portion of the ion beam (usually a single nominal *m/z*) and reanalyses it after it has undergone subsequent ion decompositions. Two general types of MS/MS apparatus enjoy wide usage: those that have mass spectrometers in spatial sequence (tandem instruments) and those that examine ions using a single mass spectrometer, in which the two stages of mass selection occur in temporal sequence (ion storage instruments).

Ions that fragment between two stages of mass selection are said to undergo metastable ion decompositions. These often show patterns different from a source mass spectrum. The spectrum from which **Figure 1** is taken provides an example. Electron ionization of $C_6H_5OCH_2CD_2OH$ shows a $C_6H_7O^{\bullet+}$:$C_6H_5DO^{\bullet+}$:$C_6H_6DO^+$ ratio of about 0.75:1:0.6. Metastable ion decomposition of the corresponding molecular ion $C_6H_5OCH_2CD_2OH^{\bullet+}$ shows a ratio $< 0.05:1:2$. The high-resolution spectrum in **Figure 1** required a two-sector instrument. In metastable studies, the same two sectors were used in tandem. Since isobaric *m/z* 95 ions cannot be resolved by a single sector, the $C_6H_7O^{\bullet+}$: $C_6H_5DO^{\bullet+}$ abundance ratio from metastable ion decompositions had to be inferred from a comparison of several additional isotopic variants, $C_6H_5OCH_2CD_2OD$, $C_6H_5OCD_2CH_2OH$, $C_6H_5OCD_2CD_2OD$ and $C_6D_5OCH_2CH_2OH$.

Determination of isotopic stereochemistry

In favourable cases, configuration of an isotopically substituted stereogenic centre can be determined by means of mass spectrometric fragmentation. The labelled analyte must contain a ring or else must pass through a cyclic transition state in the course of decomposition. Acyclic secondary alcohols with a single deuterium adjacent to the oxygen-bearing carbon (general formula $R^1CHOHCHDR^2$) have two diastereomeric forms as a consequence of isotopic substitution. While the mass spectra of the alcohols present major difficulties of interpretation, several derivatives undergo mass spectrometric eliminations by which relative stereochemistry can be assigned. The ionized phenyl ethers, for instance, expel neutral alkene. The $C_6H_5OD^{\bullet+}$: $C_6H_5OH^{\bullet+}$ ratios differ from one stereoisomer to the other to such an extent that ion storage techniques can be used to assay the proportions of isotopic diastereomers. While chromatography cannot separate isotopic isomers, GC-MS/MS for quantitating mixtures of isotopic diastereomers has been reduced to practice.

In 1979 Knowles, McLafferty and their co-workers made dramatic use of metastable ion mass spectrometry in solving the stereochemistry of phosphate monoesters $ROP^{16}O^{17}O^{18}O^{2-}$ that are chiral by virtue of substitution with oxygen isotopes. Reaction with optically active 1,2-propanediol affords the mixture of three isotopomeric cyclic triesters shown to the left in **Scheme 2**. One $ROP^{16}O^{17}O^{18}O^{2-}$ enantiomer gives the mixture depicted; the other $ROP^{16}O^{17}O^{18}O^{2-}$ enantiomer gives products in which ^{17}O and ^{18}O are transposed. Ring opening with methanol gives a mixture of three isotopomeric triesters (middle of **Scheme 2**) that exhibits mass spectrometric fragment ions corresponding to trimethyl phosphate. The isotopomeric trimethyl phosphate radical ions display prominent metastable ion decompositions from loss of formaldehyde. For one $ROP^{16}O^{17}O^{18}O^{2-}$ enantiomer the *m/z* 142 fragment loses only $CH_2{}^{16}O$, as **Scheme 2** portrays, while *m/z* 141 loses either $CH_2{}^{16}O$ or $CH_2{}^{17}O$ and *m/z* 143 loses either $CH_2{}^{16}O$ or $CH_2{}^{18}O$. These metastable ion decompositions are summarized beneath their respective trimethylphosphate ions to the right in **Scheme 2**. The other $ROP^{16}O^{17}O^{18}O^{2-}$ enantiomer leads to ions (not shown in **Scheme 2**) for which *m/z* 141 loses only $CH_2{}^{16}O$; *m/z* 142 loses $CH_2{}^{16}O$ or $CH_2{}^{18}O$; and *m/z* 143 loses $CH_2{}^{16}O$ or $CH_2{}^{17}O$.

Elucidation of fragmentation pathways and isotope effects

Isotopic labelling provides an excellent method for probing how ions decompose in the mass spectrometer. Frequently the pathway followed by an ion resembles a cognate reaction from solution chemistry, but sometimes the outcome differs quite unexpectedly. Many elimination reactions – both in solution and in the gas phase – operate via loss of groups attached to adjacent atoms. But EI on $(CH_3)_2CDCH_2CH_2Br$ yields an M–DBr daughter ion that is much more abundant than the M–HBr fragment. The structure of the ion (inferred from subsequent surface neutralization and analysis of the C_5H_{10} neutral) is $(CH_3)_2C{=}CHCH_3$. Clearly some

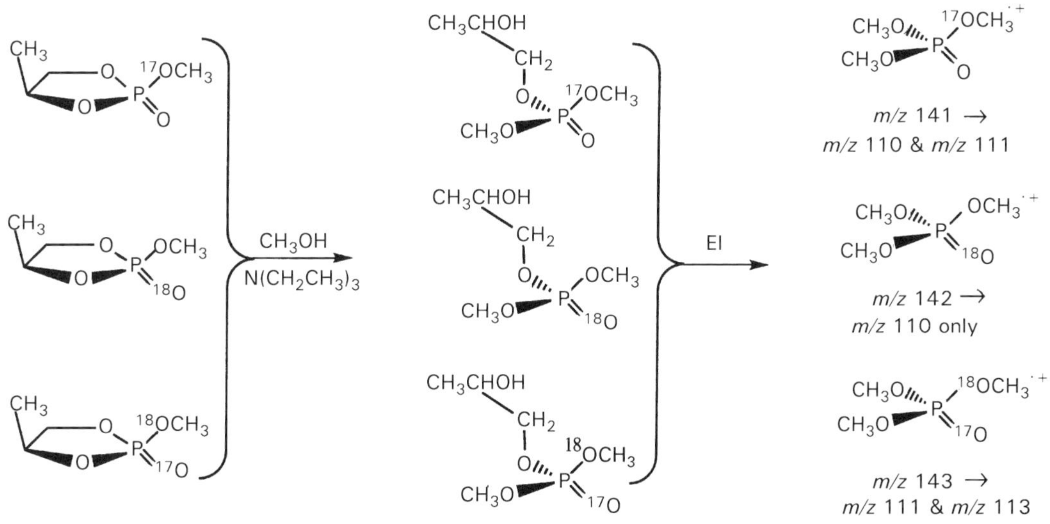

CH3OH
N(CH2CH3)3
EI
m/z 141 → m/z 110 & m/z 111
m/z 142 → m/z 110 only
m/z 143 → m/z 111 & m/z 113

Scheme 2

transposition of hydrogen has taken place. Investigating the rearrangements of gaseous ions represents a major area of research in mass spectrometry.

Deuterium substitution often affects the competition between two otherwise identical fragmentation pathways. EI on 1,4-dimethoxybutane shows no molecular ion and an intense peak from loss of methyl radical (a fragmentation not seen in the shorter or longer homologous α,ω-dimethoxy-*n*-alkanes). When ionizing energy is lowered, the contribution from methyl loss increases until it represents > 70% of the total ionization. Deuterating one methyl ($CD_3OCH_2CH_2CH_2CH_2OCH_3$) exerts a comparatively small isotope effect: loss of CD_3 is 0.7 as intense as loss of CH_3. Deuterating the nearest methylene as well ($CD_3OCD_2CH_2CH_2CH_2OCH_3$) lowers the loss of CD_3 to 0.5 the intensity of loss of CH_3. In other words, replacing two H atoms with D on the other side of the oxygen has as great an effect as replacing all three H atoms with D on the methyl itself. This has been taken to imply a hidden hydrogen rearrangement, in which a proton transfers to the remote oxygen simultaneously with a C–O bond cleavage.

If two reactions exhibit different isotope effects in a product-determining step, they cannot be proceeding via a common intermediate. For example, elimination of propene produces the most intense fragment ion in the EI and CI mass spectra of *n*-propyl phenyl ether, with extensive scrambling of all of the hydrogens in the side-chain. Could it be that the *n*-propyl ether ions isomerize to isopropyl ether ions prior to dissociating? If so, photoionization of $CD_3CH_2CH_2OPh$ and $CH_3(CD_3)CHOPh$ should manifest the same isotope effects on propene expulsion. But the isotope effect (measured as the $C_6H_5OH^{\bullet+}/C_6H_5OD^{\bullet+}$ intensity ratio) is about 1.2 for *n*-propyl, significantly less than the value (1.4) for isopropyl. Therefore, the two isomeric ions dissociate via distinct pathways.

Discrimination of isobars and isomers

Isotopic labelling often shifts the mass of one isobar relative to another for the purposes of identification. In ionized mixtures of methyl fluoride with a trace of ammonia, the weakly hydrogen-bonded cluster ion $CH_3F\text{–}HNH_3^+$ (*m/z* 52) has the same mass as a much more stable ammonia cluster ion, $(NH_3)_3H^+$. With $^{15}NH_3$ the mass-to-charge ratios become *m/z* 53 and *m/z* 55, respectively. Moreover, the isotopic substitution identifies the product of the *m/z* 52 → *m/z* 53 fragmentation as FCH_2^+.

Ion structures can be deduced from isotopic labelling. One of the first distonic ions ever demonstrated, portrayed in Equation [6],

$$\overset{D^+}{CH_3OCH_2\dot{C}H_2} \xrightarrow{B} BD^+ \qquad [6]$$

was characterized by its reaction with gaseous base to transfer only D^+. If the ion had had the ionized methyl ethyl ether structure, it would have transferred H^+ at least as often as D^+. A similar type of experiment demonstrated that the distonic ion in Equation [7] converts to (or interconverts with) ionized methoxycyclopropane, since the two isotopomers [4] and [5] transfer H^+ and D^+ to gaseous bases in the same proportions.

$$\dot{C}H_2CD_2CH{=}\overset{+}{O}CH_3 \rightleftharpoons \text{(2,2-dideuterio-cyclopropyl)}\text{–}\overset{+\bullet}{O}CH_3$$
[4]

$$\dot{C}D_2CH_2CD{=}\overset{+}{O}CH_3 \rightleftharpoons \text{(1,2,2-trideuterio-cyclopropyl)}\text{–}\overset{+\bullet}{O}CH_3$$
[5]

$$\xrightarrow{B} BH^+/BD^+ = 1.6$$

Nuclear magnetic resonance and mass spectrometry

Neutral products from decomposition of gaseous ions can be isolated and subsequently characterized by gas chromatography or magnetic resonance spectroscopy. Fluorine NMR has been especially useful in this regard, since the ^{19}F chemical shift scale is so wide that isotopic shifts of NMR peak positions permit assessment of the extents and positions of deuteration in isomers/isotopomer mixtures.

Figure 3 reproduces ^{19}F NMR spectra of fluoroethylenes recovered from EI of two different isomers of dideuterated β-fluorophenetole. The isomers give slightly different mass spectrometric fragmentation patterns, but the decomposition mechanism could not be fully understood without the analysis of the expelled neutrals. As the NMR shows, the distribution of deuterated fluoroethylenes is not the same for the two isomers. One isomer yields a product not formed by the other, and vice versa. $C_6H_5OCD_2CH_2F$ produces CHD=CHF, but $C_6H_5OCD_2CH_2F$ does not. $C_6H_5OCH_2CD_2F$ produces CHD=CDF, but $C_6H_5OCD_2CH_2F$ does not. Hence hydrogens cannot be randomizing within ionized β-fluorophenetole prior to elimination.

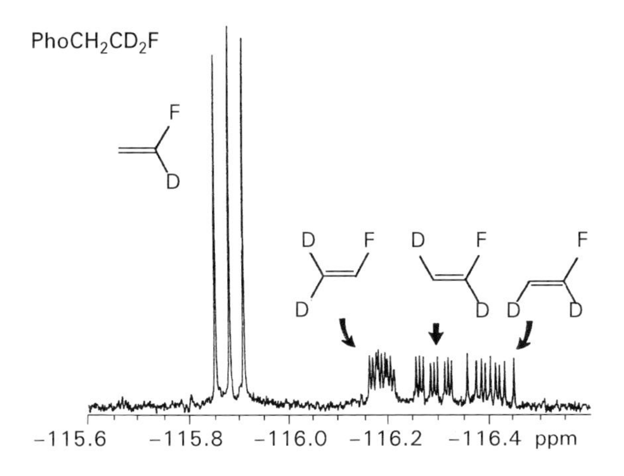

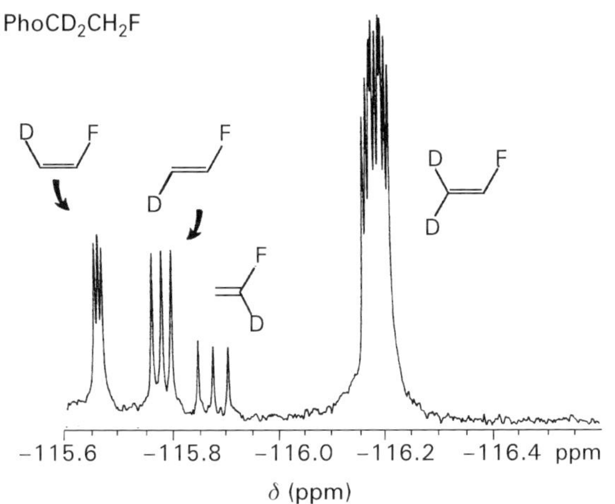

Figure 3 ^{19}F NMR spectra (proton noise-decoupled) of deuterated fluoroethylenes expelled from the indicated dideuterated β-fluorophenetoles after 70 eV electron ionization. Observed peak splittings come from deuterium–fluorine spin–spin coupling, by which isomers/isotopomers are identified. Reproduced with permission of the American Chemical Society from Nguyen V, Cheng X and Morton TH (1992) *Journal of the American Chemical Society* **114:** 7127.

It turns out that ionized β-fluorophenetole dissociates via two competing pathways, illustrated in Equation [8], which are indistinguishable but for the isotopic label. Both involve cleavage of the sp^3-carbon oxygen bond to form ion–neutral complexes, which live of the order of a nanosecond before a proton transfers from the fluorine-containing cation to the phenoxy radical. The upper pathway drawn in Equation [8] has the fluorine bridging between two carbons, yielding a symmetrical cation structure. The lower pathway involves a hydrogen shift to form the most stable of the possible $C_2H_5F^+$ structures. The two competing mechanisms operate in about a 1:2 ratio.

$$PhOCH_2CH_2F^{\cdot +} \longrightarrow \left[\text{(bridged } F^+\text{)} \quad PhO^{\cdot} \right] \text{ or } \left[CH_3CH{=}F^+ \quad PhO^{\cdot} \right] \quad [8]$$

Ion–neutral complexes

$$\longrightarrow CH_2{=}CHF + PhOH^{\cdot +}$$

Chemists have long dreamed of combining mass spectrometry with NMR, so as to probe the structures of gaseous ions directly. While that vision remains to be fulfilled, NMR analysis of neutral fragments expelled by gaseous ions represents a step towards that goal.

List of symbols

N^{sample} = number of moles of sample; N^{spike} = number of moles of spike; R^{sample} = peak intensity ratio for pure analyte; R^{spike} = peak intensity ratio for pure spike substance; R^{blend} = peak intensity ratio for mixture of sample and spike.

See also: **Biochemical Applications of Mass Spectrometry; Chromatography-MS, Methods; Cosmochemical Applications Using Mass Spectrometry; Food Science, Applications of Mass Spectrometry; Fragmentation in Mass Spectrometry; Isotope Ratio Studies Using Mass Spectrometry; Labelling Studies in Biochemistry Using NMR; Medical Science Applications of IR; MS–MS and MS^n; Metastable Ions; Sector Mass Spectrometers; Stereochemistry Studied Using Mass Spectrometry.**

Further reading

Bennett JS, Bell DW, Buchholz BA, Kwok ESC, Vogel JS and Morton TH (1996) Accelerator mass spectrometry for assaying irreversible covalent modification of aldolase by acetoacetic ester. *International Journal of Mass Spectrometry* **179/180:** 185–193.

Brenna JT (1997) High-precision continuous flow isotope ratio mass spectrometry. *Mass Spectrometry Reviews* **16:** 227–258.

DeBièvre P (1993) Isotope dilution mass spectrometry as a primary method of analysis. *Analytical Proceedings* **30:** 328–333.

Holmes JL (1975) The elucidation of mass spectral fragmentation mechanisms by isotopic labelling. In: Buncel E and Lee CC (eds) *Isotopes in Organic Chemistry*, Vol 1. Amsterdam: Elsevier, pp 61–133.

McAdoo DJ and Morton TH (1993) Gas phase analogues of cage effects. *Accounts of Chemical Research* **26**: 295–302.

Morton TH and Beauchamp JL (1975) Chainlength effects upon the interaction of remote functional groups. *Journal of the American Chemical Society* **97**: 2355–2362.

Shaler TA, Borchardt D and Morton TH (1999) Competing 1,3- and 1,2-hydrogen shifts in gaseous fluoropropyl ions. *Journal of the American Chemical Society* **121**: 7907–7913.

Speakman JR (1997) *Doubly Labelled Water*. London: Chapman and Hall.

Taphanel MH, Morizur JP, Leblanc D, Borchardt D and Morton TH (1997) Stereochemical analysis of monodeuterated isomers by GC/MS/MS. *Analytical Chemistry* **69**: 4191–4196.

L

Labelling Studies in Biochemistry Using NMR

Timothy R Fennell and **Susan CJ Sumner**,
Chemical Industry Institute of Toxicology,
Research Triangle Park, NC, USA

MAGNETIC RESONANCE
Applications

The elucidation of the pathways involved in the metabolism of exogenous and endogenous materials has been advanced considerably by the use of labelled chemicals. The transformation of these labelled intermediates can be followed by a variety of techniques. Much of the early work in the elucidation of pathways of intermediary metabolism involved the use of radioactive isotopes, and ^{14}C in particular, as tracers. More recently, the focus of research has been on the control of metabolism, and flux through pathways. Nuclear magnetic resonance (NMR) spectroscopy has been used in the study of metabolic processes to follow the conversion of compounds with natural abundance isotopes such as ^{1}H, ^{19}F, or ^{31}P or isotopically labelled compounds containing ^{2}H, ^{15}N, or ^{13}C. This discussion focuses on the application of isotope labels to track the fate of chemicals in biological systems. The majority of the NMR applications involving labels use ^{13}C-enriched chemicals and in general are designed to describe the pathways of metabolism or to evaluate flux through pathways. Although less frequently used, ^{2}H and ^{15}N can also provide useful information. An isotope effect on metabolism resulting from the increase in mass with ^{2}H compared with ^{1}H may be a significant limitation. ^{15}N labels are limited to the study of nitrogen-containing compounds and have generally been limited by the low sensitivity for detection of ^{15}N. Two types of applications can be identified: those in which the metabolism of endogenous materials is assessed by addition of labelled materials, with the fate of the label indicating the fate of the pool of both labelled and unlabelled material and those in which the metabolism of labelled exogenous materials is assessed.

Detection of ^{13}C-labelled metabolites

^{1}H NMR spectroscopy, which is widely used for structure elucidation, is very sensitive as a result of both the high natural abundance and NMR sensitivity of the proton. As a means of tracking the fate of chemicals in biological systems, ^{1}H NMR is widely used and works well with chemicals with functional groups whose NMR signals can be distinguished readily from endogenous materials (see also other sections). However, ^{1}H NMR is, in general, not very selective, and frequently lacks the ability to discriminate among compounds of interest in complex biological mixtures. In addition, water in biological systems causes dynamic range issues that must be circumvented by suppression techniques during the acquisition of ^{1}H NMR spectra. Carbon-13 occurs with a low natural abundance (1.1%), and has a spin quantum number of 1/2; ^{13}C has an NMR sensitivity 1.6% that of ^{1}H, but is highly selective. Signals in ^{13}C spectra are dispersed over a wide frequency range, depending on the substituents of the carbon atom from which the signal arises. The wide dispersion of ^{13}C NMR signals is one of the major reasons for using ^{13}C-enriched chemicals in the study of metabolism. Using ^{13}C-enriched materials permits detection over the low natural abundance of unlabelled materials. Compared with natural abundance of 1.1%, enrichment of a carbon atom to 99%

would cause an increase in the signal-to-noise ratio of 90-fold. The use of NMR for the detection of labelled materials has the advantage over other techniques that the enrichment at a particular carbon atom in a molecule can readily be determined without degradation of the molecule.

Proton-decoupled ^{13}C NMR signals for natural abundance materials occur as singlets since the probability of two adjacent ^{13}C nuclei is 0.0123%. An isolated enriched carbon atom in an otherwise natural abundance material gives rise to a single resonance. Adjacent enriched carbon atoms give multiplet patterns as a result of carbon–carbon coupling. Two adjacent ^{13}C atoms each give rise to a doublet as shown in **Figure 1**. The NMR spectrum of three adjacent ^{13}C atoms (C_a–C_b–C_c) will give rise to a doublet for each of the external carbons (C_a and C_c) and a doublet of doublets for the centre carbon atom (C_b), excluding the effects of long-range ^{13}C–^{13}C coupling. This type of splitting pattern permits the distinction of exogenous and endogenous compounds. In mixtures, a singly labelled compound may not readily be distinguished from endogenous signals, whereas the multiplet patterns associated with multiply labelled materials readily stand out.

The ^{1}H-decoupled 1D ^{13}C NMR spectrum is the general starting place for the detection of labelled metabolic intermediates or products. In the study of intermediary metabolism, assignments can frequently be made by comparison of the chemical shifts of signals with those from known standards. With multiply labelled metabolites, more detailed structural information can be obtained by the acquisition of additional spectral information. The measurements of carbon–carbon coupling constants from the 1D spectrum may be used to determine the connection between carbon resonances. In mixtures, however, coupling patterns for carbon atoms from different compounds can be similar. Carbon–carbon connectivity experiments (e.g. incredible natural abundance double quantum transfer experiment; INADEQUATE) provide an unambiguous determination of carbon–carbon connectivity. The two-dimensional spectra generated from these experiments contain a dimension for chemical shift and a dimension for double quantum frequency. The presence of two contours with the same double quantum frequency (i.e. aligned horizontally along the F_1 axis) but with different chemical shifts (F_2 axis) indicates coupling between the carbons giving rise to these signals (**Figure 2**). These experiments can be conducted to determine direct and long-range carbon–carbon coupling.

A number of techniques permit the detection of the number of protons attached to specific carbon atoms. The attached proton test (APT) experiment and related pulse sequences can be used to distinguish between CH, CH_2, and CH_3 carbon signals. However, they are not always able to distinguish between CH and CH_3 groups. The 2D carbon proton experiment (e.g. heteronuclear 2D *J*-resolved or HET2DJ) permits the identification of the number of hydrogen atoms attached to each carbon atom. The 2D spectrum generated from these pulse sequences contains a dimension for carbon chemical shift and a dimension indicating the multiplets produced by the attached protons ($2nI+1$ rule). Carbons with three, two, or one attached protons will have four, three, or two contour peaks, respectively, aligned at the chemical shift of the carbon signal in the 2D spectrum (**Figure 3**).

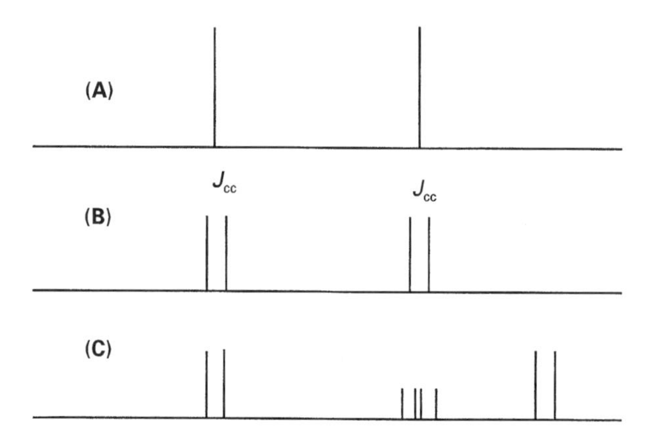

Figure 1 Diagrammatic representation of ^{13}C NMR spectra illustrating carbon–carbon coupling. (A) The spectrum of two connected natural abundance carbon atoms. (B) The spectrum of two directly connected ^{13}C atoms. The separation between each doublet is equal to the carbon–carbon coupling constant (J_{CC}) (C) The spectrum of three directly connected ^{13}C atoms.

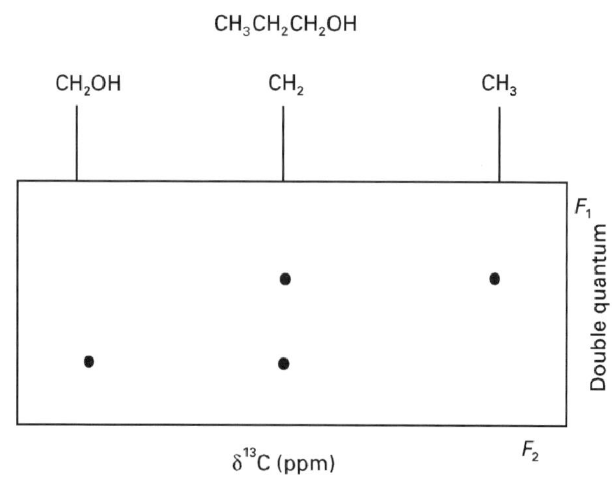

Figure 2 Diagrammatic representation of an INADEQUATE spectrum illustrating carbon–carbon coupling.

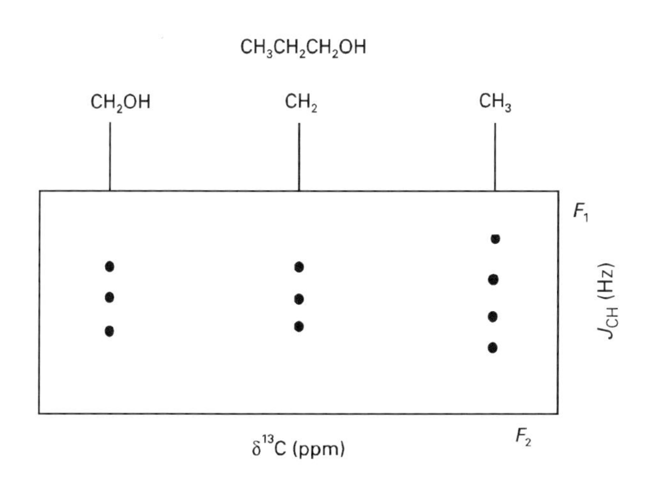

Figure 3 Diagrammatic representation of a HET2DJ spectrum illustrating carbon–proton coupling.

Two-dimensional ^{13}C–^1H heteronuclear shift correlation allows the detection of the proton signals associated with the labelled carbon atoms. Indirect detection methods are heteronuclear techniques with detection of the nucleus with higher gyromagnetic ratio. Detection of carbon signals through attached protons provides a considerable increase in sensitivity over that obtained in the 1D ^1H-decoupled ^{13}C spectrum. For biological samples, the increase in sensitivity can be outweighed by other factors, as recently described by Gemmecker and Kessler. In a complex mixture, for example, a large number of increments (second dimension of a 2D experiment) is required to obtain sufficient resolution of carbon cross peaks to allow the identification of the carbon chemical shift. Also, water suppression techniques must be employed with biological samples.

The chemical shifts of carbon atoms in various environments can be estimated through the use of substituent increments that have been tabulated over many years. Computational programs are also available, some of which use the same logic in the estimation of shift. In addition, database programs can be used to search existing spectral data and to calculate chemical shifts by interpolating from compounds in the database.

In the investigation of intermediary metabolism, the position of labelling in a newly synthesized molecule provides important information about the pathway of its formation. The potential labelling possibilities and the impact that these will have on the NMR spectra need to be considered. For a molecule with N carbon atoms, the number of carbon isotopomers (isomers with differing distributions of ^{13}C and ^{12}C) is 2^N. In biochemical studies, the potential number of isotopomers possible may not be realized. However, even a small number of possibilities can add considerably to the complexity of the signals detected. For many studies, critical information is extracted from the analysis of isotopomer distribution by NMR and requires the interpretation of complex signals for individual metabolites. Frequently computer simulation of the expected spectra is conducted to enable interpretation.

Detection of other labels

Deuterium-labelled materials are readily available and can be applied to studying metabolic processes. However, ^2H has a spin quantum number of 1 and is relatively insensitive compared with ^1H (9.7×10^{-3}). It has a low natural abundance of 0.016%.

Of the two isotopes of nitrogen, ^{15}N with a spin quantum number of 1/2 is the most widely used. It is very insensitive (1×10^{-3}), and has a low natural abundance of 0.37%. While some studies have used the direct detection of ^{15}N-labelled compounds, the indirect detection of ^{15}N through ^1H by techniques such as INEPT (insensitive nuclei enhanced by polarization transfer) and HMQC (heteronuclear multiple quantum coherence), which considerably enhance sensitivity, have increased the applicability of ^{15}N in biological systems.

Applications of labels

Exogenous metabolites

^{13}C-labelled tracers have been used in the elucidation of the pathways of metabolism of a number of exogenous chemicals, with labels incorporated in one or several carbon atoms of the materials of interest. The metabolism of [^{13}C]formaldehyde, a chemical with a single labelled carbon, provides an uncomplicated illustration of the use of a labelled material in the study of metabolism. The metabolism of formaldehyde to formate, methanol, and several additional metabolites could be determined from the 1D ^1H-decoupled ^{13}C spectrum of *Escherichia coli* incubated with [^{13}C]formaldehyde (**Figure 4**). From ^1H-coupled spectra, two of these metabolites with signals at 64 and 68 ppm appeared to contained labelled carbons present as CH_2 groups. ^{13}C–^1H correlation spectra indicated that these additional metabolites contained protons in the range 3.4–3.7 ppm, with nonequivalent protons on each labelled carbon atom. The metabolism of formaldehyde to produce 1,2-propanediol, 1,3-propanediol, and glycerol was established by culturing *E. coli* in medium containing 14.3% uniformly labelled [^{13}C]glucose, administering 90% labelled [^{13}C]formaldehyde, and analysing

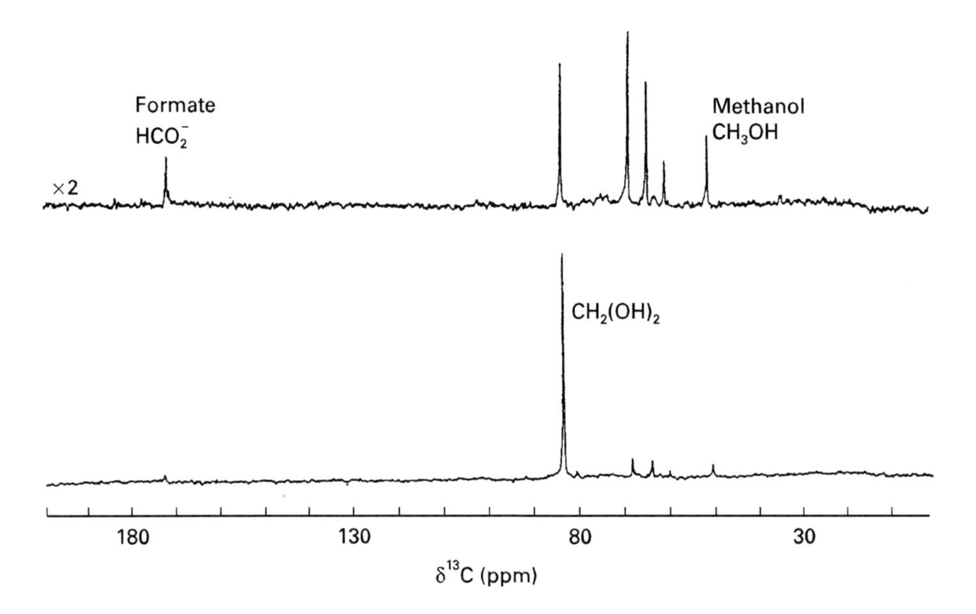

Figure 4 1D ^1H-decoupled ^{13}C spectrum of *E. coli* incubated with [^{13}C]formaldehyde. Reprinted with permission from Hunter BK, Nicholls KM and Sanders JKM (1984) Formaldehyde metabolism by *Escherichia coli. In vivo* carbon, deuterium, and two-dimensional NMR observations of multiple detoxifying pathways. *Biochemistry* **23**: 508–514. Copyright 1984 American Chemical Society.

the products formed by ^{13}C NMR spectroscopy. The ^{13}C–^{13}C couplings produced permitted the detection of the signals from the glucose-derived carbon atoms adjacent to the formaldehyde-derived carbons. An additional metabolite was determined to be *S*-[^{13}C]hydroxymethylglutathione, formed by addition of [^{13}C]formaldehyde to glutathione.

^{13}C-labelled materials have been used in the elucidation of pathways of metabolism of many chemicals. Information can be obtained from chemicals containing more than one carbon atom but only a single labelled carbon. Following administration of [3-^{13}C]-1,2-dibromo-3-chloropropane (81 mg kg^{-1}) to rats, 15 biliary metabolites and 12 urinary metabolites were observed. Metabolites were assigned based on chemical shift, proton multiplicity, ^{13}C–^1H correlation spectra, and comparison with synthetic standards. The metabolic profile proposed for this material is shown in **Figure 5**.

The incorporation of multiple labels into a compound with multiple carbon atoms provides an opportunity for using additional methods for characterizing metabolites. Following administration of [1,2,3-^{13}C]acrylonitrile to rats and mice, a number of metabolites in urine were detected and assigned structures based on chemical shift, carbon–carbon multiplicity, carbon–carbon connectivity, and proton multiplicity. Examples of the spectra obtained in urine from rats administered [1,2,3-^{13}C]acrylonitrile are shown in **Figure 6**. The multiplets assigned to the labelled carbons of the metabolites result from carbon–carbon coupling (**Figure 6C**). The HET2DJ spectrum shown in **Figure 7** illustrates the determination of the number of protons attached to each labelled carbon atom. The HET2DJ spectra obtained from multiply-labelled compounds provide an additional means of distinguishing signals derived from labelled compounds from natural abundance carbon signals in that the carbon–carbon coupling displaces the multiplets from the centre of the F_1 axis. For example, each peak of the doublet labelled 3a (**Figure 7**) has three contours aligned along the *x*-axis, indicating that the signals are from a labelled methylene carbon. The nature of the labelled carbon atoms in metabolites derived from acrylonitrile could be assigned. The identity of the unlabelled substituents of the metabolites could be inferred from the substituent effects on the chemical shifts of the labelled carbons. Most of the metabolites detected were derived from conjugation of acrylonitrile or its epoxide metabolite, cyanoethylene oxide, with glutathione and further metabolism of the glutathione-derived portion of the metabolites. The metabolism of acrylonitrile concluded from this study is shown in **Figure 8**. Quantitation of metabolites was conducted with dioxane added as an internal standard by measuring peak areas under conditions that ensured adequate relaxation of the carbon signals of interest. This combination of techniques has been used to study the metabolism of a number of compounds.

The metabolism of mixtures has received little attention, in part because of the complexity of the problem. NMR spectroscopy with labelled materials provides an ideal tool for studying the metabolism of

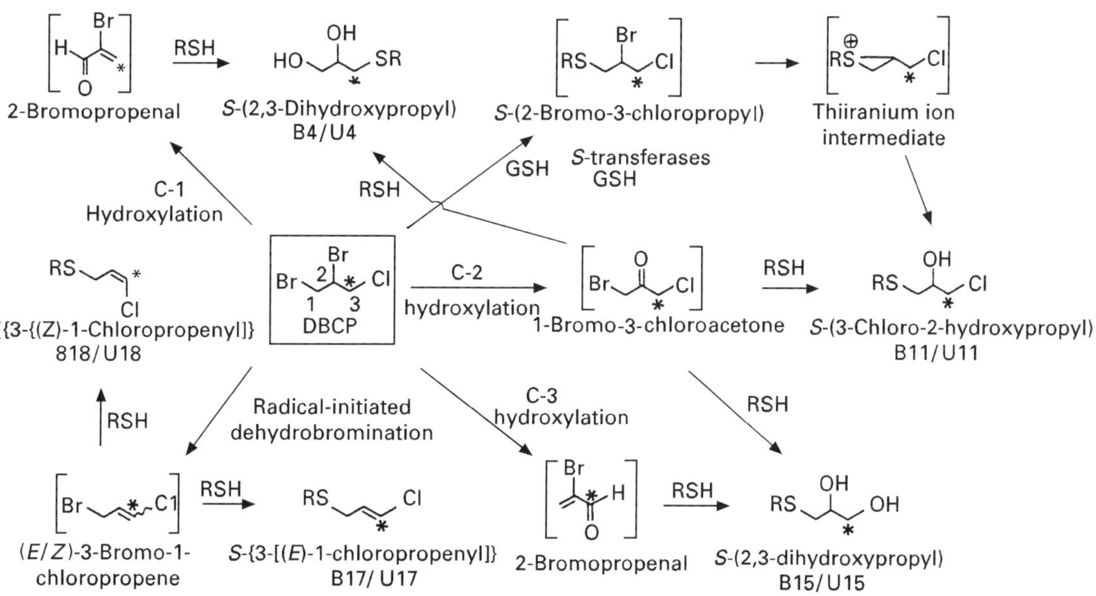

Figure 5 Metabolism of [3-^{13}C]-1,2-dibromo-3-chloropropane in rats. Reprinted with permission from Dohn DR, Graziano MJ and Casida JE (1988) Metabolites of [3-^{13}C]-1,2-dibromo-3-chloropropane in male rats studied by ^{13}C and ^{1}H–^{13}C correlated two-dimensional NMR spectroscopy. *Biochemical Pharmacology* **37**: 3485–3495. Copyright 1988, Elsevier Science.

several components of a mixture. On coadministration of [1,2,3-^{13}C]acrylonitrile and [1,2,3-^{13}C]acrylamide to rats and mice, the metabolites of both compounds could be resolved readily in the urine.

Endogenous metabolites

NMR spectroscopy has been widely used in the investigation of metabolism. Therefore, only a very general review of this area can be provided. For additional information, the reader is directed to the references contained in the Further reading section. Most of the recently reported studies involve investigation of the regulation of metabolism. Generally, one or more labelled starting materials are administered in a biological system: subcellular fractions, cells, isolated perfused organs, or whole animals. NMR spectroscopy has been applied to the investigation of metabolism in a wide range of experimental systems from plants to animals and people. Analysis can be conducted at the level of cell extracts, whole cells, body fluids, whole tissues, and *in vivo*. Mammalian tissues that are widely studied include brain, liver, muscle, and heart. The use of multiply labelled compounds aids in distinguishing high levels of naturally occurring unlabelled materials (which give singlet signals) from low levels of added labelled material (which yield multiplet signals) and can provide a means for distinguishing added from endogenous material in determining the size of pools by reverse isotope dilution.

Many pathways of metabolism have been studied using labelled chemicals, including the conversion of glucose to glycogen, glycolysis, the Krebs cycle, fatty acid metabolism, amino acid metabolism, and the pentose phosphate pathway. Among the considerations in conducting studies involving the use of labels in metabolism is the introduction of the label into the biological system, the concentration of label required, distinction of pathways, and isotopomer distribution. Some of these issues can be illustrated with consideration of specific examples.

Oxaloacetate, malate, fumarate, and α-ketoglutarate are intermediates in the Krebs cycle that are present in very low concentration and are generally not directly detectable by NMR spectroscopy. Measurement of surrogates that accumulate to measurable levels such as glutamate, alanine, aspartate, and lactate is used for the determination of isotope distribution. Oxaloacetate, malate, and fumarate exist virtually in equilibrium. Glutamate and α-ketoglutarate are in fast equilibrium via transamination. Determination of the labelling pattern in glutamate is thought to be reflective of that of α-ketoglutarate and, in turn, of oxaloacetate. Similarly, aspartate and oxaloacetate are in equilibrium by transamination, and the labelling pattern of aspartate is thought to reflect that of oxaloacetate.

Succinate and fumarate are symmetrical molecules, and label that passes through these intermediates can become scrambled. Introduction of label from α-ketoglutarate into succinate (**Figure 9**)

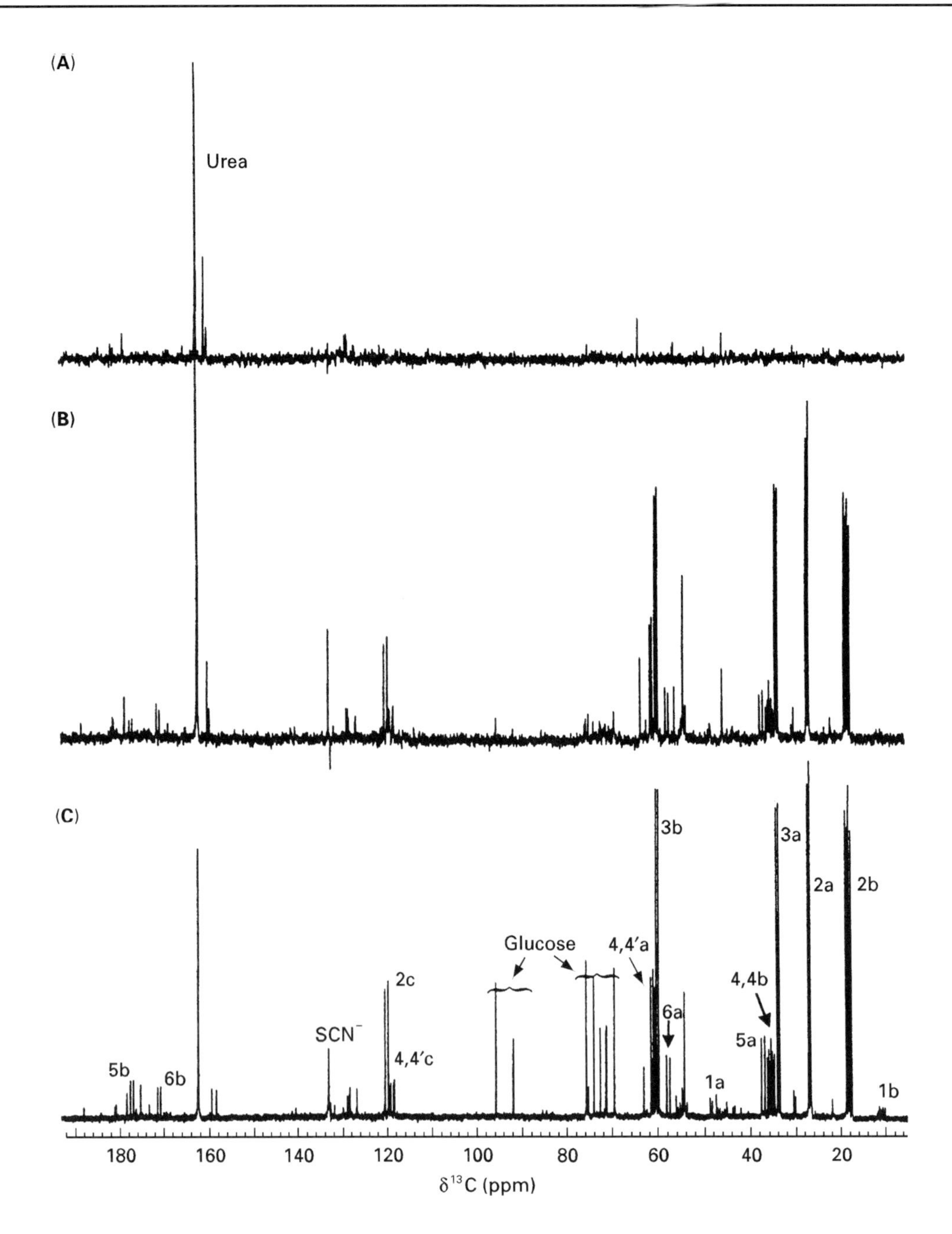

Figure 6 The ^{1}H-decoupled ^{13}C NMR spectrum of (A) control rat urine and urine collected for 24 h following administration of (B) 10 or (C) 30 mg kg^{-1} [1,2,3-^{13}C]acrylonitrile. Signals are labelled according to metabolite number (see Figure 8, and the letter of carbon-derived form acrylonitrile ($_aCH_2{=}_bCH_2{-}_cCN$). Reprinted with permission from Fennell TR, Kedderis GL and Sumner SC (1991) Urinary metabolites of [1,2,3-^{13}C]acrylonitrile in rats and mice detected by ^{13}C nuclear magnetic resonance spectroscopy. *Chemical Research in Toxicology* **4**: 678–687. Copyright 1991, American Chemical Society.

would result in the presence of label at more than one site in a molecule. Such mixing can be indicative that the label has passed through a symmetrical intermediate.

Quantitative information can be obtained from isotopomer analysis in one of several ways: modelling with differential equations, modelling with infinite convergent series, and modelling with input–output equations (see the review by Künnecke). Simplified models that focus on a limited number of isotopomers can be used to estimate the contribution of competing pathways.

The incorporation of [1-^{13}C]glucose into glycogen can be followed by NMR spectroscopy. Glucose and glycogen give prominent signals in the natural abundance ^{13}C NMR spectrum of liver. The C-1

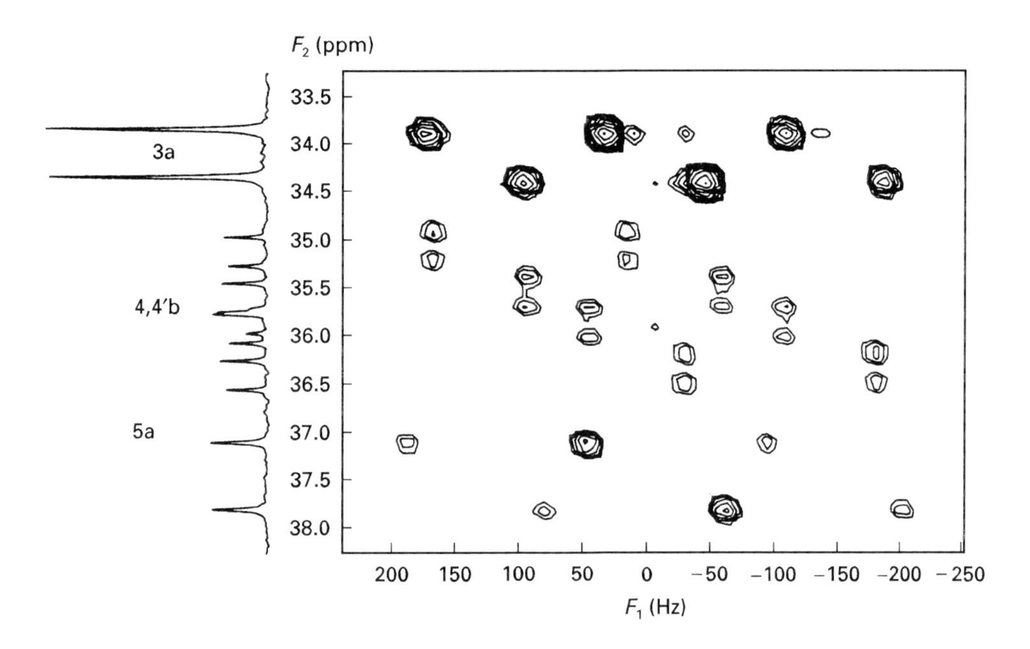

Figure 7 HET2DJ spectrum of urine from a rat administered [1,2,3-^{13}C]acrylonitrile. Reprinted with permission from Fennell TR, Kedderis GL and Sumner SC (1991) Urinary metabolites of [1,2,3-^{13}C]acrylonitrile in rats and mice detected by ^{13}C nuclear magnetic resonance spectroscopy, *Chemical Research In Toxicology* **4**: 678–687. Copyright 1991, American Chemical Society.

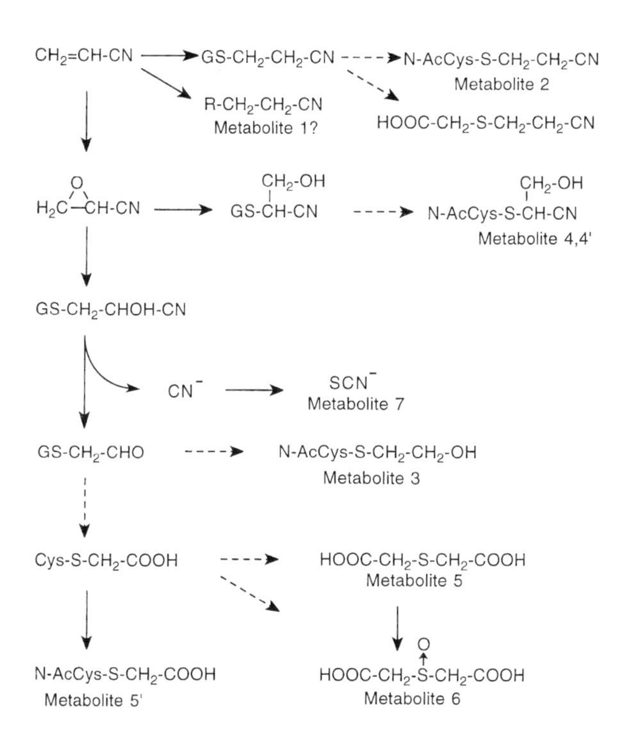

Figure 8 Proposed metabolism scheme for acrylonitrile in the rat and mouse. GS– represents a glutathionyl residue and Cys–S– and NAcCys–S– a cysteinyl and an *N*-acetylcysteinyl residue, respectively. The broken arrows represent processes that involve several transformations. Reprinted with permission from Fennell TR, Kedderis GL and Sumner SC (1991). Urinary metabolites of [1,2,3-^{13}C]acrylonitrile in rats and mice detected by ^{13}C nuclear magnetic resonance spectroscopy. *Chemical Research in Toxicology* **4**: 678–687. Copyright 1991, American Chemical Society.

carbon signal of glycogen at 100.4 ppm can be readily distinguished from the C-1 signals of α- and β-glucose at 92.6 and 96.7 ppm. Although glycogen is a high molecular mass (10^9) molecule with extensive branching, the resonances of glycogen can be readily detected. The internal motion of the glucopyranose

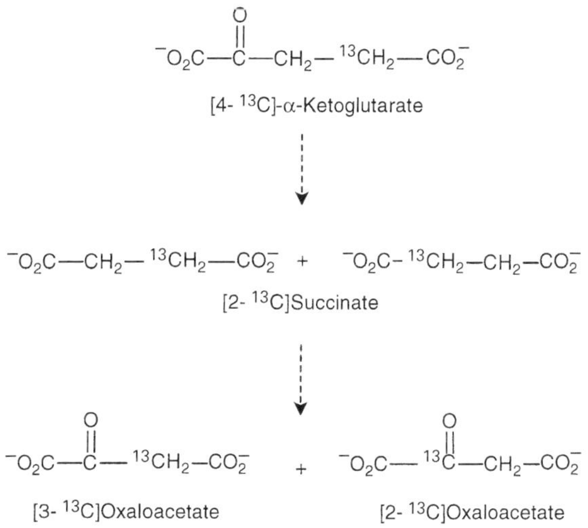

$^-O_2C—C(=O)—CH_2—{}^{13}CH_2—CO_2^-$

[4-^{13}C]-α-Ketoglutarate

$^-O_2C—CH_2—{}^{13}CH_2—CO_2^-$ + $^-O_2C—{}^{13}CH_2—CH_2—CO_2^-$

[2-^{13}C]Succinate

$^-O_2C—C(=O)—{}^{13}CH_2—CO_2^-$ + $^-O_2C—{}^{13}C(=O)—CH_2—CO_2^-$

[3-^{13}C]Oxaloacetate [2-^{13}C]Oxaloacetate

Figure 9 An illustration of the scrambling of label by passage through the symmetrical intermediate succinate. Incorporation of ^{13}C in the 4-position of α-ketoglutarate results in the formation of [2-^{13}C]succinate, and in turn [2-^{13}C]- and [3-^{13}C]oxaloacetate. Since no molecules labelled at both the 2- and 3-position would be produced, no coupling would be observed between the resonances.

residues of glycogen is thought to be virtually unrestricted resulting in the observation of narrow line widths for glycogen signals. The incorporation of glucose into glycogen can take place by several routes, and not just by the direct conversion of glucose → glucose 6-phosphate → glucose 1-phosphate → UDP glucose → glycogen. Additional pathways involving the conversion of glucose → C_3 units → glucose 6-phosphate → glycogen had been extensively studied by NMR spectroscopy. Injection of [1,2-^{13}C]glucose into fed rats resulted in the formation of liver glycogen labelled at both the C-1 and C-2 carbons, with spin–spin coupling (**Figure 10**).

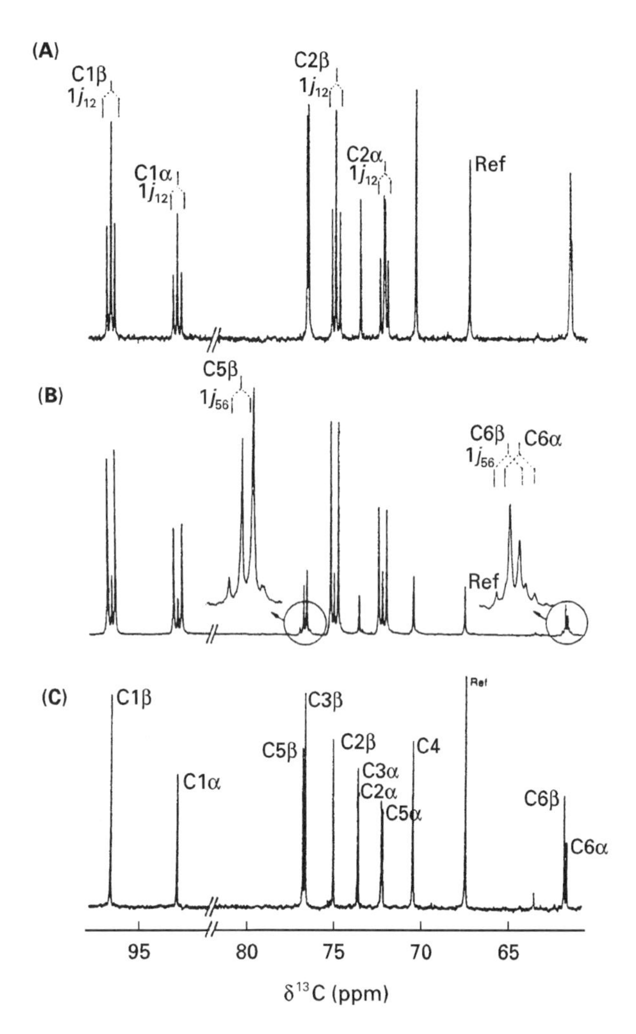

Figure 10 High-resolution ^{13}C NMR spectra (100.16 MHz) of enzymatically hydrolysed liver glycogen. Extracted and hydrolysed liver glycogen of (A) a fed and (B) a fasted rat after injection of [1,2-$^{13}C_2$]glucose. Reprinted with permission from Künnecke B and Seelig J (1991). Glycogen metabolism as detected by *in vivo* and *in vitro* ^{13}C-NMR spectroscopy using [1,2-$^{13}C_2$]glucose as substrate. *Biochimica et Biophysica Acta* **1095**: 103–113. Copyright 1991, Elsevier Science.

In fasted rats, the additional presence of label at C-5 and C-6 indicates the synthesis of glycogen from newly formed glucose. The absence of 1-^{13}C- and 1,3-^{13}C-isotopomers ruled out the involvement of the pentose phosphate pathway in the synthesis of glucose. In the indirect pathway for glycogen synthesis, glucose is converted in several steps to dihydroxyacetone phosphate, which is rapidly isomerized to glyceraldehyde 3-phosphate. Glyceraldehyde 3-phosphate, in turn, can be used for gluconeogenesis via fructose 6-phosphate or via pyruvate and oxaloacetate. In either eventuality, the ^{13}C labels can appear simultaneously at the 5- and 6-positions of glucose in glycogen, giving rise to [1,2-^{13}C]-, [5,6-^{13}C]-, and [1,2,5,6-^{13}C]glucose.

Areas of development

Although few studies have used indirect detection methods for the analysis of labels in biological systems, their use will become more widespread as the technology develops. Examples of the application of these techniques can be found in the literature. Metabolites of [1-^{13}C]glucose have been detected in maize root tips using heteronuclear multiple quantum coherence (HMQC) NMR. Labelling of the amide group of glutamine could be detected by ^{1}H–^{15}N HMQC in root tips labelled with $^{15}NH_4^+$. Following infusion of $^{15}NH_4^+$, ^{1}H–^{15}N HMQC NMR has been applied to the detection of [5-^{15}N]glutamine protons in rat brain.

See also: **Biofluids Studied By NMR; ^{13}C NMR, Methods; ^{13}C NMR, Parameter Survey; Carbohydrates Studied By NMR; Cells Studied By NMR; Drug Metabolism Studied Using NMR Spectroscopy; ^{19}F NMR Applications, Solution State; *In Vivo* NMR, Applications, Other Nuclei; Nucleic Acids Studied Using NMR; Perfused Organs Studied Using NMR Spectroscopy; ^{31}P NMR; Proteins Studied Using NMR Spectroscopy; Structural Chemistry Using NMR Spectroscopy, Pharmaceuticals; Tritium NMR, Applications; Two-Dimensional NMR, Methods.**

Further reading

Beckmann N (1995) *Carbon-13 NMR Spectroscopy of Biological Systems*. San Diego: Academic Press.

Brainard JR, Downey RS, Bier DM and London RE (1989) Use of multiple ^{13}C-labeling strategies and ^{13}C NMR to detect low levels of exogenous metabolites in the presence of large endogenous pools: measurement of glucose turnover in a human subject. *Analytical Biochemistry* **176**: 307–312.

Cohen JS, Lyon RC and Daly PF (1989) Monitoring intracellular metabolism by nuclear magnetic resonance. In: Oppenheimer NJ and James TL (eds) *Nuclear Magnetic Resonance. Part B. Structure and Mechanism*, pp. 435–452. San Diego: Academic Press.

Cohen SM (1989) Enzyme regulation of metabolic flux. In: Oppenheimer NJ and James TL (eds) *Nuclear Magnetic Resonance. Part B. Structure and Mechanism*, pp. 417–434. San Diego: Academic Press.

Gemmecker G and Kessler H (1995) Methodology and applications of heteronuclear and multidimensional ^{13}C NMR to the elucidation of molecular structure and dynamics in the liquid state. In: Beckmann N (ed.) *Carbon-13 NMR Spectroscopy of Biological Systems*, pp. 7–64. San Diego: Academic Press.

Jans AW and Kinne RK (1991) ^{13}C NMR spectroscopy as a tool to investigate renal metabolism. *Kidney International* **39**: 430–437.

Jeffrey FM, Rajagopal A, Malloy CR and Sherry AD (1991) ^{13}C-NMR: a simple yet comprehensive method for analysis of intermediary metabolism. *Trends in Biochemical Science* **16**: 5–10.

Kanamori K, Ross BD and Tropp J (1995) Selective, *in vivo* observation of [5-^{15}N]glutamine amide protons in rat brain by ^{1}H-^{15}N heteronuclear multiple-quantum-coherence transfer NMR. *Journal of Magnetic Resonance B* **107**: 107–115.

Künnecke B (1995) Application of ^{13}C NMR spectroscopy to metabolic studies on animals. In: Beckmann N (ed.) *Carbon-13 NMR Spectroscopy of Biological Systems*, pp. 159–257. San Diego: Academic Press.

London RE (1988) ^{13}C Labeling in studies on metabolic regulation. *Progress in Nuclear Magnetic Resonance Spectroscopy* **20**: 337–383.

Lundberg P, Harmsen E, Ho C and Vogel HJ (1990) Nuclear magnetic resonance studies of cellular metabolism. *Analytical Biochemistry* **191**: 193–222.

Nicholson JK and Wilson ID (1987) High resolution nuclear magnetic resonance spectroscopy of biological samples as an aid to drug development. *Progress in Drug Research* **31**: 427–479.

Laboratory Information Management Systems (LIMS)

David R McLaughlin, Eastman Kodak Company, Rochester, NY, USA
Antony J Williams, Advanced Chemistry Development Inc., Toronto, Ontario, Canada

FUNDAMENTALS OF SPECTROSCOPY
Methods & Instrumentation

The purpose of an analytical laboratory is to provide measurement support, generate information, and solve problems. This often involves processing many samples and requests and produces large numbers of test results and reports. In an attempt to increase efficiency, many analytical laboratories have computerized their logbooks, focusing on tracking jobs, samples, tests and results. With this information accessible via a computer, it is then possible to provide many additional functions such as methods management, test results, archiving, calculation, comparison to specifications, control charting and verification of instrument performance. The information is also readily available and valuable for reference, problem solving and modelling. Inclusion of spectroscopic data in such systems has been hampered by a lack of standards and cooperation between instrument vendors and software developers. Some companies have developed successful systems by building customized interfaces and data manipulation software. Advances in Web-based user interfaces suggest that spectroscopy laboratories may soon be able to take full advantage of integrated spectroscopy software, chemical structure handling software and information systems.

The product of an analytical laboratory

The main product of an analytical laboratory is information. Some of this information is in the form of new analytical techniques or methods that extend a laboratory's ability to analyse new materials. For most laboratories, however, the largest amount of information is in the form of measurements and reports. These measurements are key to understanding and controlling manufacturing processes, solving manufacturing and development problems and developing new high-performance materials. Preparing samples, making measurements, analysing data and generating reports is the common process analytical laboratories use to generate information. Properly managing this process and tracking the flow of work through the laboratory is critical to the efficient operation of a laboratory.

When computers became available, analytical laboratories began to use them to help automate many of their processes, including the information-handling and information-generating processes. Analytical chemists and information specialists working to improve their own operations were the first to develop laboratory information management systems, or LIMS. They typically supported only one technique or small unit of a laboratory and did not communicate with other information systems.

The first commercial LIMS arrived on the market in about 1980. Today, there are dozens of LIMS vendors providing packages with many capabilities. There is even a World Wide Web site devoted to providing information on LIMS resources, analyses of the latest technology, and the products and services available from vendors and consultants. In 1994 the American Society for Testing and Materials published a standard guide on LIMS. The guide is intended for anyone interested in LIMS, including users, developers, implementers and laboratory managers. It defines standard terminology, a concept model, primary functions, an implementation guide and a checklist. These are useful for discussing and understanding LIMS as well as developing specifications and cost justification. The guide is a good reflection of what it was thought a traditional LIMS should be and of what most vendors have marketed.

Types of laboratories

The information-handling requirements of analytical laboratories vary from the research to the manufacturing environments. A standard testing laboratory that supports manufacturing quality control and regulatory compliance uses analysis procedures that are well defined and strictly followed. These laboratories analyse relatively large numbers of samples and produce numerical or tabular results that are often plotted against time. Spectral analysis is usually used in this environment to determine concentration or verify identification relative to known reference spectra.

In contrast, a research analytical laboratory supporting the discovery of new materials uses analysis procedures that are defined in the mind of the analyst. While the analyst may use analytical techniques that are well defined, the process applied to a particular sample is not predetermined, but is developed in response to the questions that need to be answered. These laboratories analyse fewer samples and regularly produce results that combine graphics and images with textual reports. In this environment, spectroscopy is also used to elucidate chemical structure and determine three-dimensional spatial relationships.

Other types of analytical laboratory environments vary between these two idealized cases. Process control environments model a fully automated standard testing laboratory. High-throughput screening is similar, but often includes special processing for unusual results or potentially interesting materials. Like research, problem solving does not follow a fixed analytical process, but often results must be expressed in manufacturing terms. Development environments reflects the need to move research testing for understanding to standard testing for manufacturing. Problems arise when a LIMS vendor tries to sell one generalized solution, customized to fit all environments.

Selecting a LIMS

Many factors drive analytical laboratories to implement LIMS. These include the need to demonstrate value, comply with detailed regulatory requirements, manage large amounts of samples and tests from automated analysis and screening systems, automate laboratory processes, improve accuracy and deal with increased demands for efficiency and documentation. Implementing a LIMS can be expensive and have significant impact on laboratory workflow and productivity. For successful implementation of a LIMS, it is important to know which factors are most important and where productivity will be affected, both positively and negatively. It is also extremely important that the laboratory workflow is supported by the LIMS in a simple and appropriate manner. It may be appropriate to utilize a re-engineering process to simplify a laboratory's workflow before implementing a LIMS.

Even with useful guides and books and the availability of highly functional LIMS and capable consultants, successful installation of a LIMS is a difficult and frustrating task for some laboratories. Vendors have generally focused on developing generic systems that provide a great deal of functionality but have grown to be quite complex. These products have worked quite well in some laboratory environments and have completely failed in others. They work best in laboratories where the operation is well understood and standardized and where the processes do not have to be changed to match the LIMS. This is generally not the case in research and development environments, where microscopy and molecular spectroscopy laboratories usually reside. The need to manage images, various types of spectra and chemical structures adds to the problem. Since LIMS vendors generally do not consider these types of data, significant customization and software development is required. This task is easier with LIMS that utilize recent improvements in computer technology.

Managers may still have a hard time finding a system that will work for their operations and meet the expectations of users. Often, the laboratory will have to change its workflow to match that of a LIMS package or deal with additional expense when significant customization is required. The business case for the buy-and-customize route versus the build route can be unclear. While the general trend has been towards purchasing commercial systems, recent surveys examining potential year-2000 problems indicate that most LIMS are still internally developed and maintained.

Functions of a LIMS

Whether they support manufacturing, research, or problem solving, all analytical laboratories receive and prepare samples, make measurements, analyse data, and report results and information. A simplified illustration of the analytical support process is shown in **Figure 1**. Improving the overall efficiency of this process is the primary function of a LIMS. It can be used to support only a few aspects of this process or it can support all aspects. The most common and perhaps most important use of a LIMS is to maintain proper records of samples and the tests requested on them. This electronic logbook or sample management type function assigns sample numbers or bar codes and records administrative information, often including an indication of the work requested. Most LIMS also allow numerical and tabular results to be entered for tests performed on the samples. Once data have been collected and properly recorded, the LIMS may be used for data manipulation and automated report generation. Significant productivity enhancements are usually obtained from this last step. The following sections describe how a LIMS is used to support the steps in **Figure 1**.

Definition of problem or job

For the laboratory supporting research, problem solving or method development, the definition of a problem or job is the first step of the analysis process. Job requests could include determining what is happening to a material as it degrades, how the properties of a material might be improved, or whether a fabrication process is producing devices with the desired composition and geometry. When a new chemical is to be commercialized, the request may be to develop tests and fitness for use criteria needed for production monitoring. The job usually defines the context for investigative analytical work performed on many samples. This information is usually stored in laboratory notebooks with only a job title stored in LIMS. When the work is complete and a report is written, this information may be made available in an electronic library. There is an opportunity for the field of knowledge management to improve productivity in this area.

In a standard testing laboratory, the problems or jobs are predefined by the time materials are analysed. Given a type of material and its point-in-process, the suite of required tests, testing methods, specification limits and resulting actions and reports are clearly defined. This information is often stored in LIMS as master or static data and used to automate the workflow for these materials.

Sample log-in

A primary function of the LIMS is to track the samples of materials to be analysed by a laboratory. This can be as simple as tracking when a sample enters and leaves a laboratory or as detailed as tracking every action that is performed on the sample while it is in the laboratory. The process normally begins with the client or analyst recording general descriptive information about each new sample or batch of samples and the work request. This is ideally entered into or created by information systems used by the clients of the laboratory and interfaced to LIMS, but is normally entered manually on an input screen. Client systems include electronic laboratory notebooks used by researchers or manufacturing scheduling and control systems. Typical information includes the name of the requester of the work, an account number to be charged for the work, the date the request is made, the client's identifier for the sample, the date the sample is received, the job or project to which this sample belongs, the priority, and the hazard information. Test requests vary from a general description of the problem to be solved to a suite of specific point-in-process tests using very specific methods.

Sample receipt and review

Samples may be logged into a LIMS before they arrive in the laboratory. This is quite common in

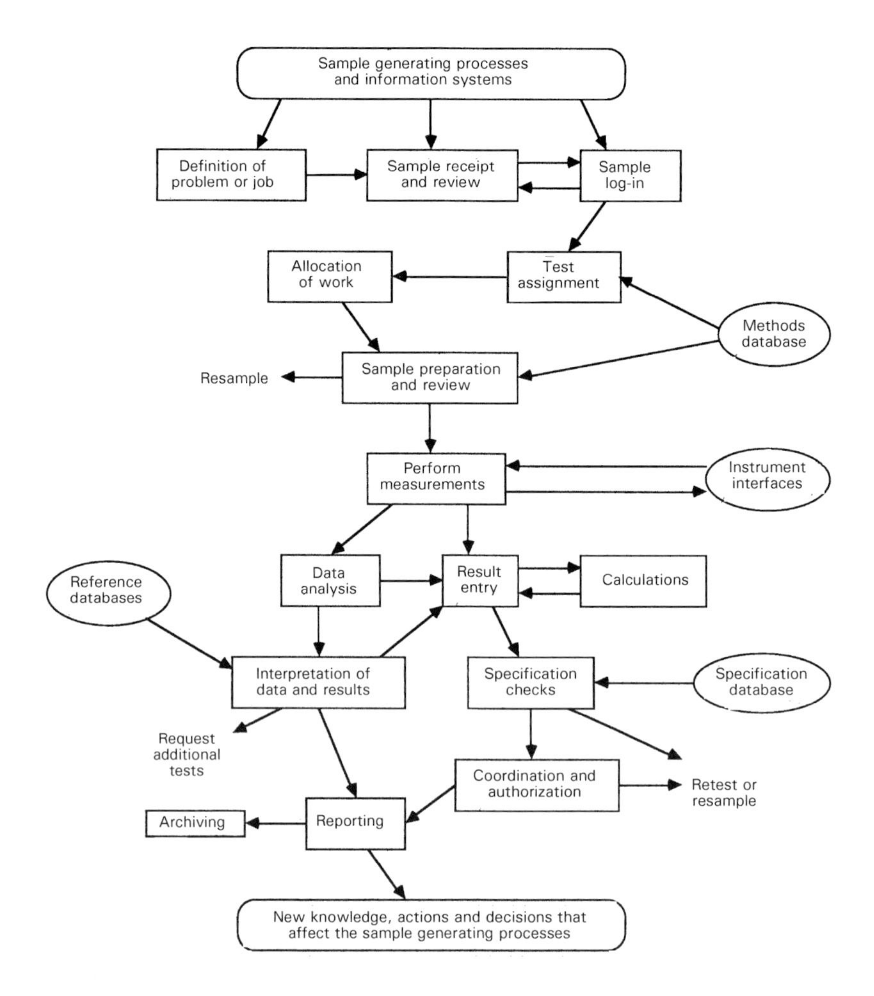

Figure 1 Generic analytical laboratory workflow supported by LIMS.

standard testing laboratories with a LIMS that is integrated with their client's manufacturing resource-planning system. Good-quality operating procedures dictate that when a sample arrives it should be reviewed to ensure it is suitable for analysis, that the requested work is appropriate and that it has not visibly degraded. The LIMS sample receipt function allows the time a sample arrives and any anomalies to be recorded. If anomalies are observed in the sample, then a decision to proceed, resample or request other tests is made, often after consultation with the client.

Test assignment

The client may request a specific collection of tests when the sample is logged in. In a standard testing environment, the suite of tests is often assigned automatically by the LIMS. If a general test is requested or the request is for some problem to be solved or investigated, then the appropriate specific tests and course of investigation are selected by the analyst.

Allocation of work

Work may be allocated among specific instruments, analysts or support groups. This is done to match the test requirements to appropriately skilled analysts and equipment and ensure the best utilization of resources. The differing priority levels of samples received in a laboratory can also be accommodated in this step. When work is allocated, it is recorded in LIMS so that it can generate work lists for the analysts and instruments. LIMS can often make default assignments automatically. In other cases, analysts simply check out the next pending sample from the queue in LIMS. In a research or problem-solving area, this may not be recorded in a LIMS until the results are entered.

Method management

Methods are the procedures and protocols by which materials are analysed. LIMS designed for standard testing laboratories often provide some support for managing methods. This can allow complete automation of the analytical process, as exemplified by on-line analysis.

Sample preparation

Most samples and analysis techniques require some preparation such as dissolving, diluting or mounting before measurements can be made. This step and the subsequent measurement step are a typical focus for laboratory automation using robotics. Such systems follow detailed procedures that may vary for every sample. Associating the correct procedure with each sample and supplying that information to the robotic systems is logically a job for LIMS. Most automated sample preparation and measurement systems were initially isolated, performing very focused tasks and requiring manual data transfer at the beginning and end of a run. This is changing as automation and LIMS continue to mature.

Perform measurements

Measurements are performed using instruments. In an automated laboratory, the LIMS can download data acquisition instructions to the instruments and receive data and analysis results for entry into its database. Most LIMS provide some capabilities for interfacing with instruments. This has been particularly successful in chromatography, where good standards have been developed and followed by both LIMS and instrument vendors.

Result entry

After the tests have been performed, the results are often entered into the LIMS. This is a manual process, unless a specific instrument interface has been built or purchased. Many instrument vendors now provide such interfaces, but this has been limited in spectroscopy. Results entered into LIMS are generally derived from processing raw analysis data on the instruments. There can be simple yes or no phrase results to tables of amounts of various materials. Usually these tables have to be predefined by the method that is used to perform the analysis. Interfacing efforts in spectroscopy have been limited partly by the lack of widely accepted standard spectral formats and partly by the lower demand for general spectroscopy applications in standard testing areas.

Calculations and data analysis

Most LIMS systems provide simple capabilities for performing calculation on the results that are entered. These calculations may simply be transformations into standard units or the generation of averages. More sophisticated analysis is generally performed using instrument vendor data systems or specialized data analysis software. Most LIMS provide customization functions to facilitate entry of these results.

Specification checking

Many standard testing laboratories run analyses to determine whether a material is within manufacturing specification limits. The manufacturing process and analysis method usually define these limits. Most LIMS provide a mechanism for these specifications to be stored or retrieved and compared to the results. The specifications are often stored in a database controlled by the clients of the analytical laboratory. Materials that are within specifications can be automatically released. Those that are not can be flagged or operators can be automatically notified regarding the test results.

Coordination and authorization

In some environments, it is important to validate results before they are released. This usually involves a person other than the analyst checking the results and verifying the procedures. The LIMS can enforce such rules to help ensure that the business processes are followed. In environments where several tests are performed on one material, this role may be to coordinate reporting to the client and to identify and resolve inconsistencies. This activity may result in requests for materials to be retested or resampled.

Interpretation of data and results

The result of research processes is new knowledge. This is generally in the form of written documents with graphics and images that may involve many samples. A LIMS that supports analytical research needs to support the storage of these kinds of results as well as numerical and tabular results.

Report generation

The output of an analytical laboratory is information that is reported to the client and often affects further actions and decisions. A major function of a LIMS is to improve the efficiency of report generation. In order to facilitate this process, report templates can be established that are automatically triggered when a step in the workflow is completed. Analysis results, billing, turnaround time, instrument calibration and control charts, justification and inventory are just a few of the common reports that can be triggered. Because of the variety of clients that analytical laboratories interact with, report generation is an area where significant customization is invested to make a LIMS effective.

Archiving

The results and reports from an analytical laboratory can be of great economic value. Reference information, historical understanding of processes, solutions to problems and assigned spectra from structure elucidation efforts represent a few of the valuable resources that need to be maintained, sometimes indefinitely. This information is usually best stored in a knowledge management system specifically designed for this purpose. Laboratories also produce information needed to support regulatory compliance. Organizations may be required to save this information for many years. Archiving of this data is generally supported by LIMS.

Summary

The analytical support process exists to aid and enhance the sample generation process. If it cannot be measured, then it is hard to improve it. Making the information produced by this process available electronically can significantly enhance its value to an organization. Having measurements available and logically organized is required for any material or process modelling. Quick and easy access avoids duplication of work and speeds research and problem solving. Computerization of the process requires consideration of interfaces to other information-storing or information-generating systems. These include methods management, manufacturing resource planning and control, electronic laboratory notebooks, instrument interfaces and reference databases.

The power of LIMS lies in its ability to automate mundane tasks and to work with other systems to enhance the access of authorized users to information and the proper maintenance of that information for future reference.

LIMS for spectroscopy

Ever since computers became of a practical size to fit in a spectroscopy laboratory, they have had an ever more intimate relationship with the spectrometers themselves. Spectrometer vendors have applied them to instrument control, data collection, library searching and information management. The information management functions of these systems could be considered a LIMS with a focus limited to a particular type of testing. With the exception of chromatography, these systems are limited in scope and do not interface well with laboratory-wide LIMS. Vendors of spectral analysis packages such as *Grams* (Galactic) and *Spectacle* (LabControl) similarly provide some limited LIMS functions or the ability to link with spreadsheets where such functions may be built.

To aid in sharing spectra obtained using equipment from different vendors, a number of efforts have been made to establish standard exchange formats. In 1987, the Joint Committee on Atomic and Molecular Physical Data (JCAMP) published the JCAMP-DX format as a standard for exchange of infrared spectra. This general format was subsequently extended to include mass and NMR spectra. Although JCAMP-DX files are created with small variations from vendor to vendor, it is supported as an export format by most infrared and mass spectroscopy instrument vendors.

The Analytical Instrument Association (AIA) created a netCDF-based Analytical Data Interchange (ANDI) format for chromatography that received widespread acceptance. After this success, the AIA adopted a standard for mass spectroscopy in 1993 and began definitions for infrared and NMR. This mass spectroscopy standard, although supported by a few vendors, has not received wide usage and the infrared and NMR definitions have not been implemented.

There is currently no generally used spectral format in NMR. However, the International Union of Pure and Applied Chemistry (IUPAC) has a working party that expects to produce a JCAMP-based multidimensional NMR data standard by March 2000.

Despite the existence of these standard formats, spectrometer vendors have been slow to fully support them as export options. This has been a hurdle for spectral analysis and database software vendors, who have been required to deliver format converters. Spectrometer manufacturers often consider these vendors as competition. Until spectrometer users demand better integration by the vendors, progress will be slow and the lack of widely accepted formats for spectral data will remain a barrier to the general integration of spectroscopy into LIMS.

Some corporate solutions

In the early 1970s, Wolfgang Bremser and others at BASF began to develop software to aid in spectral interpretation and chemical structure elucidation. Their efforts were focused on improving speed, cost and reliability by using databases of reference spectra to provide reasonable suggestions and dialogue for the interpreting spectroscopist. This work resulted in the very useful SpecInfo system, which is still being enhanced at BASF. This system and associated databases were the basis for formation of the company Chemical Concepts in 1990, which now sells a variety of advanced interpretation tools.

Eastman Kodak Company began a similar effort in the late 1970s. The initial goal was to improve the efficiency and quality of structure elucidation using NMR, IR, MS and UV-visible data with the ultimate aim of completely automating the analysis of routine samples. Such a system would automatically determine whether the structure a chemist proposed for a sample was consistent with the spectral data obtained. It was soon recognized that high-quality databases relevant to Kodak materials were needed and that interpretation software operated in isolation would be of limited utility. The proposed structure and resulting spectral data needed to be available from a LIMS. Integrated sample management software that incorporated chemical structures and spectral data was developed to solve this problem and to help minimize the task of building an in-house database. The complete system, called QUANTUM (**Figure 2**), combines the spectral and structural analysis software with a sample management system (SoftLog) and a spectral database (SDM).

SoftLog was probably the first LIMS designed specifically for spectroscopists. It incorporated several important features not typically available in LIMS. These included full and substructure searches, spectral display and search, easy incorporation of results into the reference database and a logical interface to fully automated analysis software. In addition, the user was automatically informed of previous sample or material analyses, of inconsistent results, of reference data and of related compounds such as impurities, models, by-products or precursors. The notification of previous analyses of the same material included data generated by spectroscopy laboratories dispersed across the entire corporation. This was particularly useful, because it made analysis information initially obtained in a research environment available to analytical laboratories supporting development.

The Web is the way

Since the mid-1990s the world has experienced explosive growth of the World Wide Web or Internet. This growth was primarily caused by the introduction of graphical web browsers that provide a simple and intuitive point-and-click environment for access to information. The Web is based on widely accepted and robust standards that are open and simple enough that everyone can participate. It supports most kinds of information (text, graphics, audio, video, binary, etc.) and provides a simple mechanism for adding specialized support. Inexpensive or free graphical browsers are available on virtually all computer platforms commonly used in

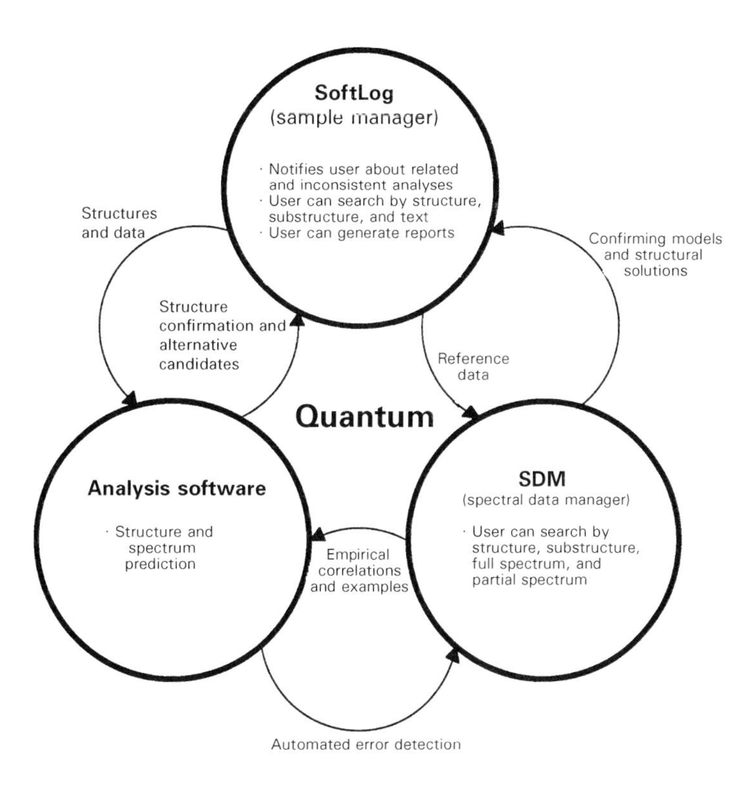

Figure 2 Components of an integrated spectroscopy information management environment.

analytical laboratories. This is valuable in spectroscopy or other laboratories that typically utilize multiple computer platforms to control spectrometers and other instruments. With this paradigm shift in user interface design has come a new level of comfort for chemists, researchers, and analysts in managing information electronically.

A few companies have reported on their efforts to automate their LIMS and spectroscopy environments. Scientists in the analytical research laboratory at Parke-Davis Pharmaceuticals provide NMR and mass spectral analysis support for synthetic chemistry by combining robotic automation with internally developed interfaces to their Beckman Lab Manager LIMS. The LIMS provides preparation conditions to the sample preparation robot and a work list for batch operation of the spectrometer. These workers have used Web technology to provide information on the status of sample analyses and other functions of the analytical laboratory. Web technology has also been used to develop sample tracking and data analysis in support of genome mapping. SmithKline Beecham Pharmaceuticals, Eli Lilly, NASA Lewis Research Center and some oil companies have reported on integration of LIMS with spectroscopy.

The analytical laboratories at Kodak have converted a major portion of its LIMS to a Web-based system. This work began with the development of their Web-based Information Management System, WIMS, in the molecular spectroscopy areas. This integrated Web-based system provides access to sample information including reports, structures and spectra. Sample information is integrated with a Web interface to QUANTUM for structure management needs. Spectra from various instruments are copied as they are processed to a server. Structures and spectra are retrieved and displayed as GIF images by simple C programs and are available for download to the desktop for further analysis by local applications. Reports of analysis can be attached as text, as HTML or as standard word-processor or spreadsheet documents that can be viewed using helper applications. Numerical and tabular results are stored in a relational database linked to the rest of the system with a Web interface developed with one of the several good tools available.

Using the many distinct advantages of Web-based technology as well as its fast growth curve, commercial LIMS vendors are presently developing Web interfaces for their LIMS. The one product with a focus on spectroscopy laboratories is the S^{3}LIMS product from Advanced Chemistry Development. This LIMS, inspired by WIMS, is a fully Web-based integration of sample, structure and spectral management utilizing Java-based structure drawing and spectral display applets. The Java implementation and compatibility with Netscape and Microsoft browsers ensures a platform independence for the system and marries the system with the proton NMR, carbon NMR and other structure-based prediction engines also available from Advanced Chemistry Development. Chemical Concepts, now owned by John Wiley & Sons, has also developed a fully web-based 'version' of their interpretation tools.

Conclusions

When selecting or developing a LIMS, a laboratory should be clear why the system is needed and what workflow and business functions it must support. Keeping the installation simple in scope, design and, most importantly, user interface will help ensure success. As the laboratory's business changes over time, so must its LIMS. Continued investment in information management must be expected.

LIMS is still an evolving concept. Most commercial LIMS have focused on standard testing, perhaps because the processes are more clearly defined or standardized and perhaps because the market is larger. Recent developments in knowledge management and electronic notebooks systems may help develop better systems for research environments. Web technology allows information system interfaces to be developed more simply and more quickly. There is increasing interest from pharmaceutical companies in using spectroscopy with high-throughput screening. With these technology developments and increased market demand, LIMS with better integration and simpler interfaces for spectroscopy may become ubiquitous.

See also: **Structural Chemistry Using NMR Spectroscopy, Inorganic Molecules; Structural Chemistry Using NMR Spectroscopy, Organic Molecules; Structural Chemistry Using NMR Spectroscopy, Pharmaceuticals.**

Further reading

ASTM (1994) *E1578-93e1 Standard Guide for Laboratory Information Management Systems (LIMS).* West Conshohocken, PA: American Society for Testing and Materials.

Hinton MD (1995) *Laboratory Information Management Systems: Development and Implementation for a Quality Assurance Laboratory.* New York: Marcel Dekker.

LIMSource (1998) *LIMSource: the best site for LIMS and lab data management system info.* http//:www.limsource.com.

Mahaffey RR (1990) *LIMS: Applied Information Technology for the Laboratory*. New York: Van Nostrand Reinhold.
McDowall RD (ed) (1987) *Laboratory Information Management Systems, Concepts, Integration, Implementation*. Wilmslow, UK: Sigma Press.
Nakagawa AS (1994) *LIMS: Implementation and Management*. Cambridge, UK: The Royal Society of Chemistry.
Stafford J (ed) (1995) *Advanced LIMS Technology: Case Studies and Business Opportunities*. Andover, UK: Chapman & Hall.

Lanthanum NMR, Applications

See **Heteronuclear NMR Applications (La–Hg).**

Laser Applications in Electronic Spectroscopy

Wolfgang Demtröder, University of Kaiserslautern, Germany

ELECTRONIC SPECTROSCOPY
Applications

Introduction

The applications of lasers to atomic and molecular spectroscopy has revolutionized this field. Although methods in classical spectroscopy of electronic transitions using large grating spectrographs or interferometers have brought a wealth of information on molecular structure and dynamics, the new techniques of laser spectroscopy have by far surpassed the capabilities of former methods with regard to sensitivity and spectral or time resolution. This article reviews these new spectroscopic techniques, their applications and some of the results obtained so far; it will be restricted to the spectroscopy of free atoms and molecules and does not include solid-state spectroscopy.

Various techniques of sensitive absorption spectroscopy, including nonlinear techniques, which allow a spectral resolution below the Doppler width are described first. These techniques are termed sub-Doppler-spectroscopy and include linear spectroscopy in collimated molecular beams, nonlinear saturation and polarization spectroscopy, and Doppler-free two-photon spectroscopy. Emission spectroscopy, which covers laser-induced fluorescence as well as stimulated emission methods, is described next. The assignment of complex molecular spectra is greatly facilitated by optical-double resonance methods, which are treated together with optical pumping. The development of ultrashort laser pulses in the picosecond to femtosecond time domain has opened a wide area for investigations of molecular dynamics, intramolecular and intermolecular energy transfer, fast collision-induced relaxations, and detailed real-time studies of chemical reactions by observation of the forming and breaking of chemical bonds. This fascinating and rapidly expanding field of femtosecond chemistry is covered briefly. The final section discusses some aspects of coherent spectroscopy with pulsed and CW lasers that rely on the coherence properties of the laser light and on the coherent preparation of atomic or molecular states. Techniques such as quantum beat spectroscopy, correlation spectroscopy and STIRAP are described and their relevance for the electronic spectroscopy of atoms and molecules is outlined.

Sensitive absorption spectroscopy

In classical absorption spectroscopy, the intensity I_T of a light beam transmitted over a path length L through a sample of absorbing molecules with absorption coefficient α (cm^{-1}) is compared with the incident intensity I_0. According to Beer's law of linear absorption,

$$I_r = I_0 \exp(-\alpha L) \approx I_0(1 - \alpha L) \qquad \text{for } \alpha L \ll 1 \quad [1]$$

the absorption can be written as

$$\alpha L \approx \frac{I_0 - I_T}{T_0} \quad [2]$$

The smallest still detectable absorption is limited by intensity fluctuations of the incident intensity I_0 and by detector noise. The absorption coefficient $\alpha = N_i\sigma_{ik}$ is the product of number density N_i of molecules in the absorbing level $|i\rangle$ and absorption cross section $\sigma_{ik}(\nu)$ for the transition $|k\rangle \leftarrow |i\rangle$ The minimum detectable number N_i of absorbing molecules

$$N_i = \frac{\Delta I_{\min}}{I_0\, L\, \sigma} \quad [3]$$

is then given by the minimum difference $\Delta I_{\min} = I_0 - I_T$ that still can be safely measured. For a given value of $\Delta I_{\min}$, the detection sensitivity can be increased by making the product $I_0 L\sigma$ as large as possible. The various sensitive techniques of laser absorption spectroscopy attempt to minimize $\Delta I_{\min}$ as well as to maximize $I_0 L\sigma$.

High-frequency modulation spectroscopy

For high-resolution laser absorption spectroscopy, a narrow-band tuneable CW laser can be used. Possible candidates are dye lasers, Ti:sapphire or other vibronic solid-state lasers, and the variety of tuneable semiconductor diode lasers, which cover a spectral range from the blue to the mid-infrared. Of particular importance are the optical parametric oscillators that have been brought to reliable operation in the pulsed as well as in the CW mode with single-mode performance. They can now span the spectral region from 500 to 5000 nm.

When the output of such a tuneable monochromatic laser is sent through an electrooptic modulator, the phase and thus the frequency of the transmitted laser wave is modulated and the frequency spectrum consists of the carrier and two side-band frequencies, where the side-bands have opposite phases (**Figure 1**). At a modulation frequency of $\Omega = 1$ GHz, the separation of the sidebands from the carrier is about equal to the Doppler width of the absorption lines. If this modulated laser beam is detected after transmission through the absorption cell, the

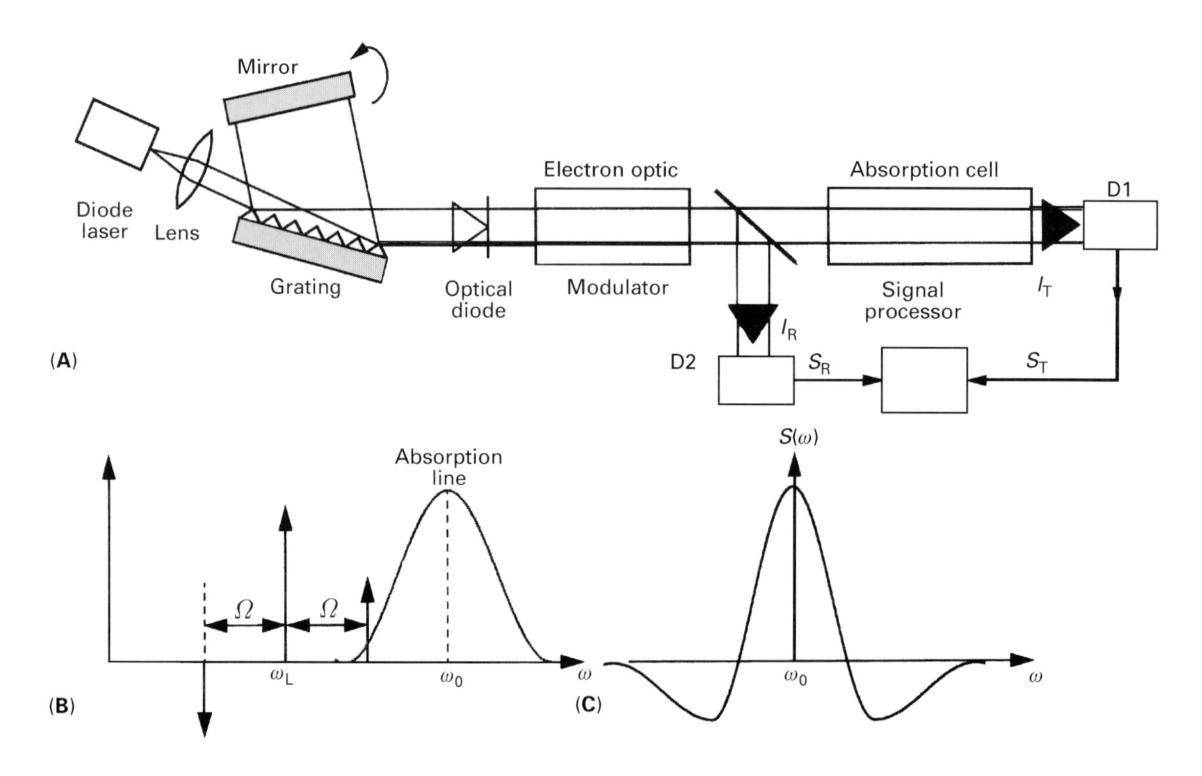

Figure 1 Principle of modulation spectroscopy. (A) apparatus; (B) sidebands in of phase modulation; (C) line profile of the signal $S(\omega)$.

output of a lock-in, tuned to the frequency Ω, will not show any signal as long as neither the carrier nor the sidebands overlap with an absorption line, since the beat signals between carrier and the two sidebands exactly cancel. Note also that any fluctuations of the laser intensity are automatically subtracted. If, however, one of the three frequencies coincides with a molecular absorption line, the exact balance is perturbed and a detector signal is generated. When the laser frequency ω is tuned over an absorption line, the measured signal profile $S(\omega)$ is similar to the second derivative of the absorption profile $\alpha(\omega)$.

Since lock-in detectors now available cannot measure frequencies above 100 MHz, the output of the detector is heterodyned with a stable local oscillator of frequency Ω_0 and the beat signal with frequency $(\Omega - \Omega_0)$ is sent to the lock-in.

The sensitivity of this modulation spectroscopy is two to three orders of magnitude higher than in conventional spectroscopy and reaches a minimum detectable absorption of $\alpha L < 10^{-7}$. Using multiple-path absorption cells (Herriot cells) and balanced detectors for detection of reference and transmitted beams, absorption coefficients as low as $\alpha \leq 10^{-10}$ cm^{-1} can still be measured.

Excitation spectroscopy

Instead of measuring the transmitted laser intensity, the photons absorbed by the sample molecules can be directly detected because they excite the absorbing molecules into a state $|k\rangle$ that emits fluorescence (**Figure 2**). The rate of detected fluorescence photons is

$$\dot{n}_{\mathrm{Fl}} = \dot{n}_{\mathrm{abs}} \cdot \eta_{\mathrm{m}} \cdot \varepsilon \cdot \eta_{\mathrm{k}} \quad [4]$$

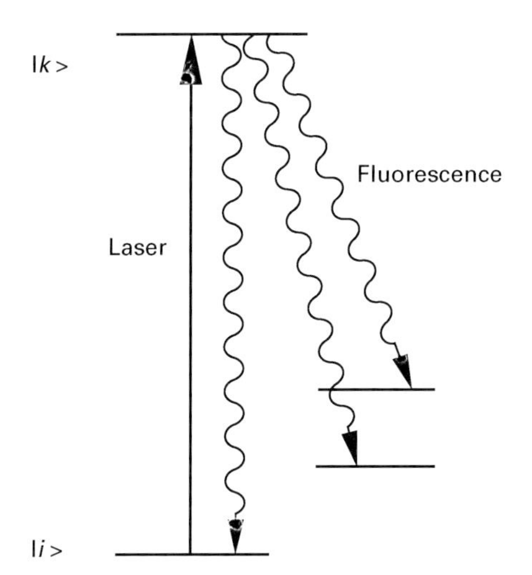

Figure 2 Level scheme of excitation spectroscopy.

where $\dot{n}_{\mathrm{abs}}$ is the rate of absorbed laser photons, $\eta_{\mathrm{m}} \leq 1$ is the quantum efficiency of the excited atom or molecule, ε is the collection probability of the emitted fluorescence and η_{k} is the quantum efficiency of the detector. With 1W incident laser power at $\lambda = 500$ nm, a collection efficiency of $\varepsilon = 0.1$, quantum efficiencies $\eta_{\mathrm{m}}=1$ and $\eta_{\mathrm{k}} = 0.2$ and an absorption probability of $\alpha L = 10^{-15}$, we obtain a photon counting rate of 60 photons s^{-1}, which gives a signal-to-noise ratio of 6 at a time constant of 1 s and a dark current of 10 photons s^{-1}.

For selected atomic transitions, the upper level may decay only into the initial lower level (true two-level system). In this case the same atom can be excited many times during its flight time T through the laser beam. If its upper-state lifetime is τ, the number of fluorescence photons emitted per second may be as high as T/τ (photon burst), if the laser is intense enough to saturate the transition.

Ionization spectroscopy

The absorption of a photon in the transition $|i\rangle \rightarrow |k\rangle$ can be also detected by ionizing the molecule in the excited state $|k\rangle$ by a second photon (**Figure 3**). If the intensity of the second laser is sufficiently high, the ionizing transition may be saturated, which implies that every excited molecule is ionized. Since the ions are extracted from the ionization volume by an electric field that focuses them onto the cathode of an ion multiplier, the detection efficiency can reach 100%.

This makes the technique very sensitive and in favourable cases single absorbed photons can be monitored. Since the first transition $|i\rangle \rightarrow |k\rangle$ can be readily saturated, this also implies that single molecules can be detected.

In case of pulsed lasers, the laser power is sufficiently high to allow saturation of the two transitions even without focusing or with only weak focusing. If the ionization energy is high, two or more photons might be required for the ionizing step (REMPI, resonant multiphoton ionization). With CW lasers, tight focusing is required to reach the necessary high intensity.

The further advantage of ionization spectroscopy is the mass-selective detection of the ions by means of a mass spectrometer. This allows the detection of specific isotopes in the presence of other species. With pulsed laser ionization, a time-of-flight mass spectrometer offers the best choice; for CW lasers, a quadrupole mass spectrometer is suitable.

Absorption spectroscopy in molecular beams

The spectral resolution for absorption spectroscopy in the gas phase is generally restricted by the Doppler

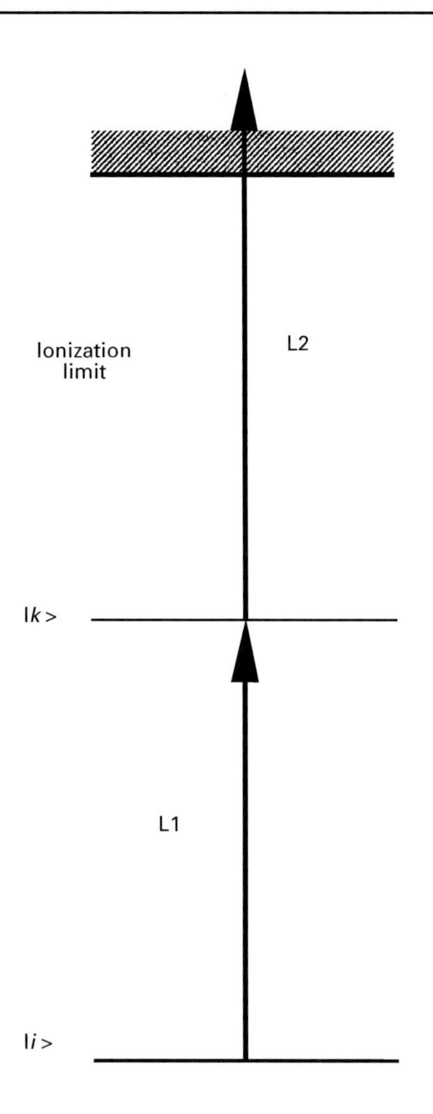

Figure 3 Level scheme of ionization spectroscopy.

width of the absorption lines. If the laser beam crosses perpendicularly a molecular beam that is formed by molecules passing from a reservoir through a narrow nozzle A into vacuum and is collimated by a slit of width b at a distance d from the nozzle (**Figure 4**), the velocity components of the beam molecules parallel to the direction of the laser beam are reduced, owing to the geometric collimation. This reduces the Doppler width by a factor $\varepsilon = b/d$, which equals the collimation ratio $b/d \approx 0.01$ of the molecular beam.

Besides the increase in spectral resolution, supersonic molecular beams have the additional advantage that during the adiabatic expansion the molecules transfer a large fraction of their internal energy E_{int} into directed flow energy. This means that the molecules are cooled and their population distribution $N(E_{\text{int}})$ is compressed into the lowest rotational–vibrational levels. This reduces the number of absorbing levels and therefore simplifies the absorption spectra considerably and facilitates the assignment of complex spectra of larger molecules.

Nonlinear sub-Doppler spectroscopy

The Doppler limit in spectral resolution can be also overcome by nonlinear spectroscopy, which is based on saturation of absorbing transitions resulting in a partial or complete depletion of the population in the absorbing level.

The scheme of saturation spectroscopy is depicted in **Figure 5**. The output beam of a single-mode tuneable laser is split into a strong pump beam and a weak probe beam, which pass in opposite $\pm z$ directions through the sample cell with length L. Only those molecules with velocity components v_z can absorb the pump photons on the transition $|i\rangle \rightarrow |k\rangle$ with centre frequency ω_0, which are Doppler shifted into resonance with the laser frequency ω. This gives the relation

$$|\omega_0 - \omega \pm kv_z| \leq \gamma$$

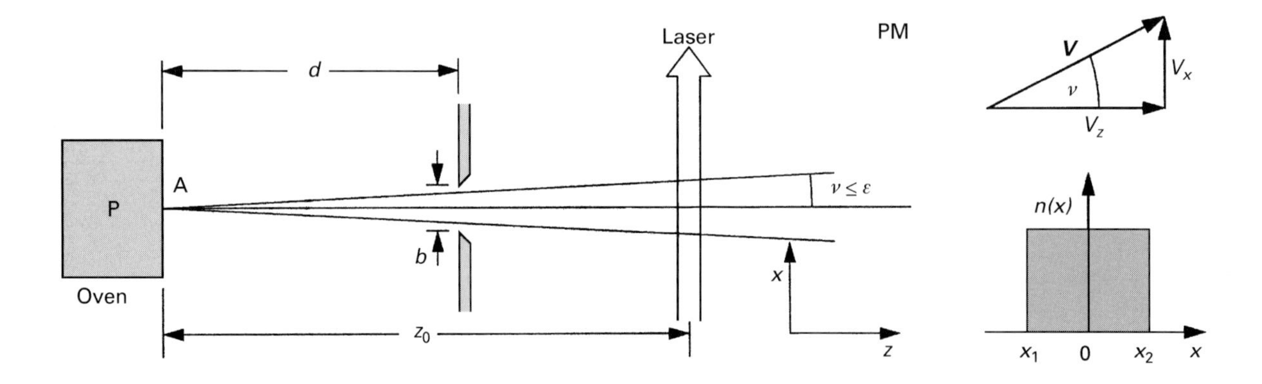

Figure 4 Sub-Doppler spectroscopy in a collimated molecular beam.

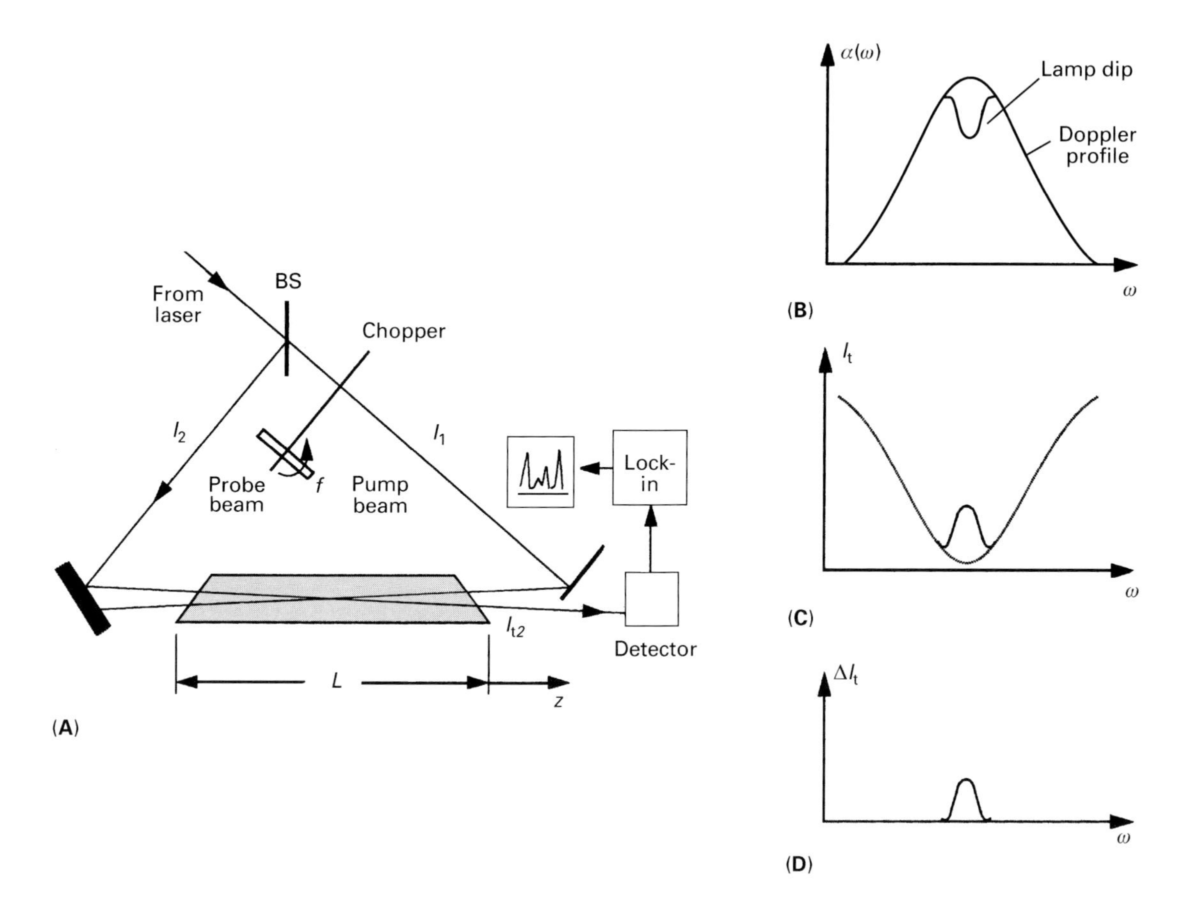

Figure 5 Saturation spectroscopy: (A) experimental scheme; (B) Lamb dip at the centre of the Doppler-broadened absorption profile; (C) Lamb peak; (D) difference of the transmitted probe intensity with and without pump.

where γ is the homogeneous line width of the transition. Since the wave vector $\boldsymbol{k}$ of the probe is opposite to that of the pump beam, the Doppler shifts for pump or probe absorption are opposite. A molecule can only interact with both beams if its Doppler shift is smaller than the homogeneous width γ, i.e. $|k{\cdot}v_z| < \gamma/2$. In this case the probe absorption is decreased due to the saturation by the pump, which depletes the number N_i of absorbing molecules. When the laser frequency ω is tuned over the Doppler broadened absorption profile, the absorption $\Delta I = \alpha L$ of the probe intensity shows a dip at the line centre (Lamb dip) with a spectral width γ that may be two orders of magnitude smaller than the Doppler width (**Figure 5B**). This results in a corresponding peak in the transmitted probe intensity $I_T \approx I_0(1 - \alpha L)$ (**Figure 5C**).

The disadvantage of this simple arrangement is its low sensitivity, since a tiny change of the transmitted intensity has to be detected. If the pump intensity is chopped and the probe transmission is recorded with a lock-in detector, tuned to the chopping frequency, the difference of the transmitted intensities with and without saturation is monitored. This eliminates the Doppler-broadened background and yields Doppler-free profiles (i.e. the Lamb peaks) with higher sensitivity (**Figure 5D**).

Still higher sensitivities can be reached with polarization spectroscopy, in which the sample is placed between two crossed polarizers (**Figure 6**). The circularly polarized pump beam with intensity I_2 induces transitions with $\Delta M = 1$ (M = quantum number of the projection of the total angular momentum $\boldsymbol{J}$ on the laser beam axis), causing a partial orientation of the otherwise randomly orientated molecules, owing to non-uniform saturation of the different M-sublevels. This makes the otherwise isotropic gaseous medium birefringent and causes a rotation of the polarization vector of a linearly polarized probe wave with intensity $I_1 << I_2$, passing in the opposite direction through the sample. However, this happens only if the probe wave interacts with the same molecules as the pump wave, i.e. if these molecules have velocity components $|v_z| \leq \gamma/2k$, just as in the case of saturation spectroscopy.

Since without saturation no signal is detected through the crossed polarizers, the method suppresses the background and therefore reduces

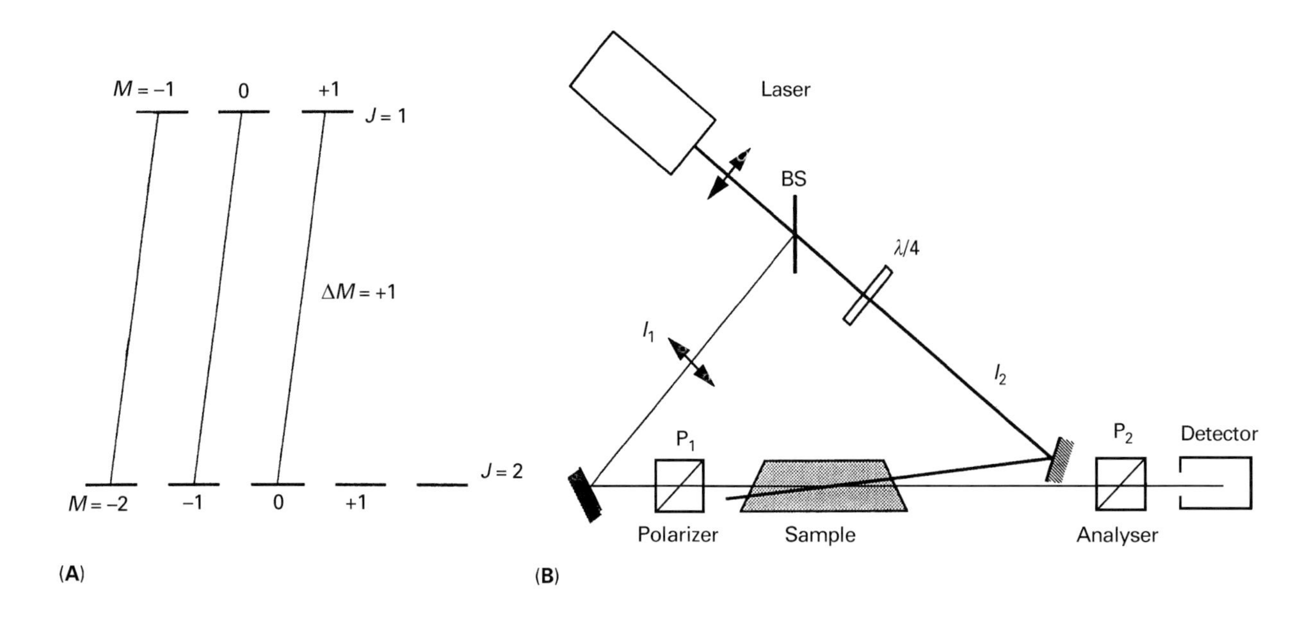

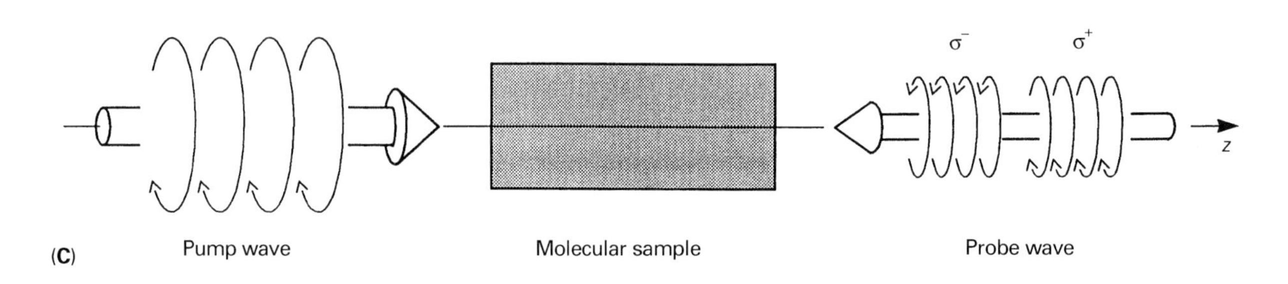

Figure 6 Polarization spectroscopy.

much of the noise, resulting in a higher signal-to-noise ratio.

Doppler-free two-photon spectroscopy

When two photons interact simultaneously with a molecule, a two-photon transition $|i\rangle \xrightarrow{h\nu_1+h\nu_2} |k\rangle$ can occur if the two levels $|i\rangle$ and $|k\rangle$ have the same parity and differ in their angular momenta J by $\Delta J = 0, \pm 2$. This allows the excitation of high-lying electronic states in atoms or molecules, such as Rydberg states. Since the transition probability for two-photon transitions is much smaller than for one-photon transitions, generally pulsed lasers with high peak powers are used. However, two-photon transitions have been observed even with CW lasers. If the two photons come from two opposite beams of the same tuneable lasers, the two-photon transition in a molecule moving with velocity component v_z requires the energy conservation

$$E_K - E_i = (\hbar\omega_1 - kv_z) + (\hbar\omega_2 + kv_z) \\ = 2\hbar\omega \quad \text{for} \quad \omega_1 = \omega_2 = \omega \qquad [5]$$

independent of the molecular velocity v_z! This means that all molecules in level $|i\rangle$, independent of their velocities, contribute to the two-photon signal at a laser frequency $\omega = (E_k - E_i)/2\hbar$. This is different from saturation or polarization spectroscopy where only a small subgroup of molecules with $|v_z| \leq \gamma/2$ can absorb both laser beams. Despite the smaller transition probability, it therefore yields comparable signal magnitudes as for the other nonlinear techniques. This is illustrated by **Figure 7**, which shows the Doppler-free two-photon transition $5S_{1/2} \leftarrow 3S_{1/2}$ of sodium-atoms, measured by Cagnac and co-workers.

Emission spectroscopy

Classical emission spectroscopy is based on excitation of atoms or molecules into higher electronic states by electron impact (in gas discharges), photon absorption or thermal excitation at high temperatures (in star atmospheres). Excitation by narrow-band lasers may result in the selective

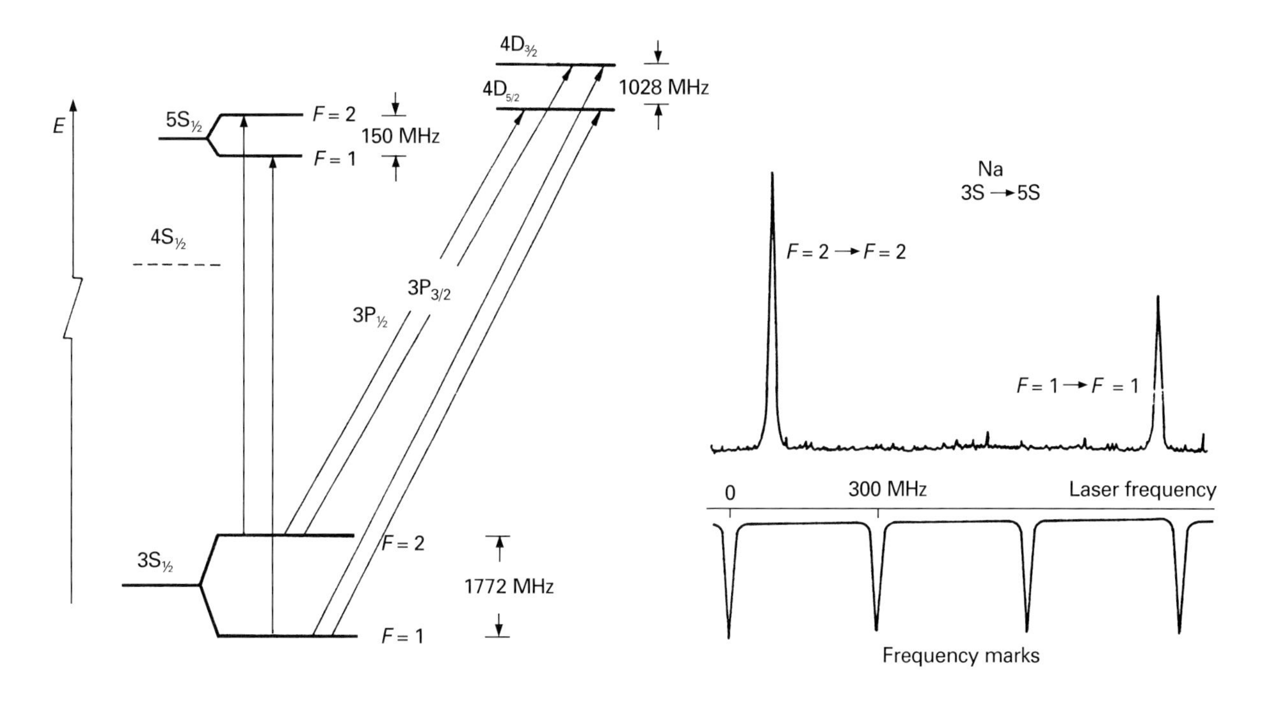

Figure 7 Doppler-free two-photon spectroscopy of the $5S_{1/2} \leftarrow 3S_{1/2}$ transition of the sodium atom. Reproduced from Cagnac and co-workers.

population of wanted levels, which emit their excitation energy as fluorescence photons. Such a laser-induced fluorescence (LIF) spectrum is much simpler than the emission spectrum of a gas discharge, where the superposition of fluorescence from many emitting levels is observed.

Measurements of spectrally resolved LIF allows the determination of the final levels into which the fluorescence is emitted (**Figure 8A**). This gives access, for example, to high vibrational–rotational levels in the electronic ground state that are not thermally populated and are therefore not accessible to absorption spectroscopy. Examples are levels closely below the dissociation energy, where the vibrating atoms reach very large internuclear distances at their outer turning point.

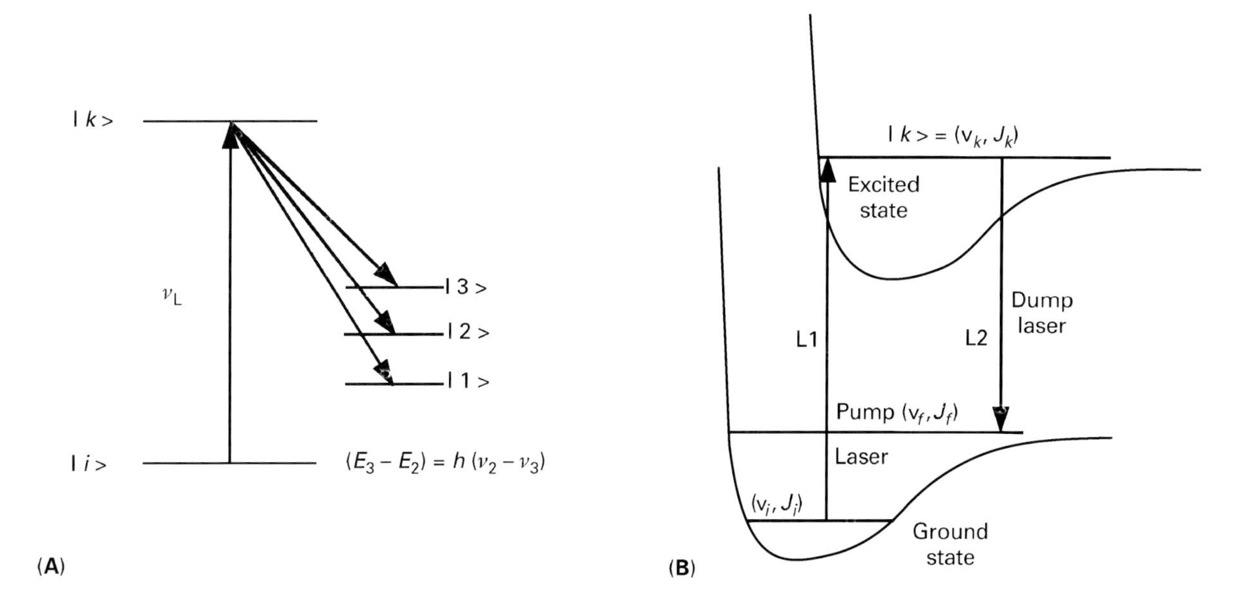

Figure 8 (A) Laser-induced fluorescence and (B) stimulated emission pumping.

The accuracy of wavelength measurements is limited by that of the spectrograph or interferometer used for dispersing the LIF spectrum. Higher spectral resolution can be achieved by stimulated emission pumping, where a second tuneable laser L2 is used to induce stimulated emission from the excited level $|f\rangle$ (**Figure 8B**). The transition can be detected by the corresponding decrease of the population $N(k)$, which may be monitored by LIF or by photoionization of level $|k\rangle$.

Double-resonance spectroscopy

Double-resonance spectroscopy is based on the simultaneous resonant interaction of two electromagnetic waves, which induce two transitions sharing a common level (**Figure 9**). In the V-type double resonance (**Figure 9A**) this common level of both transitions is the lower level $|i\rangle$ of the two transitions. The laser L1, called the pump laser, depletes the population of level $|i\rangle$ and increases that of $|k\rangle$. If L1 is chopped, the populations N_i and N_k are modulated with opposite phases. The probe laser L2 is tuned through the spectrum of allowed transitions. Each time its wavelength λ_2 coincides with a transition $|i\rangle \rightarrow |m\rangle$ or $|k\rangle \rightarrow |n\rangle$, its absorption will be modulated. This can be monitored either on the transmitted intensity I_T(L2), or with LIF or RTPI. Since the starting level, marked by the pump, is known, these optical double resonance signals can be more readily assigned than the much more abundant lines in a conventional absorption spectrum. The OODR technique therefore is a very convenient method for facilitating the assignment of complex and perturbed spectra.

The Λ-type OODR scheme (**Figure 9B**) can be regarded as stimulated resonance Raman scattering. It can be performed in cells or in molecular beams and yields Doppler-free signals if the two lasers involved are operated as single-mode lasers. The resonance signal appears when L2 is tuned to resonance $\nu_1 - \nu_2 = (E_i - E_f)/h$. The Λ-type scheme can be used, for instance, to investigate with sub-Doppler resolution the structure and dynamics of levels closely below the dissociation limit. One example is the mixing of *gerade* and *ungerade* singlet and triplet states at large internuclear distances, where the energy differences between the two electronic states becomes comparable to the hyperfine splittings of the atomic dissociation products. In these cases the indirect coupling of the nuclear spins by the electron spins can mix both states. This process may become important in cold atomic clouds kept in magnetooptical traps where collisions between atoms can lead to spin flips, which correspond in the molecular picture to transitions between triplet and singlet states.

The double-resonance scheme of **Figure 9C** corresponds to stepwise excitation, which may be regarded as the special resonance case of two-photon absorption. The first laser populates the excited level $|k\rangle$ and the second laser induces further excitation into high energy levels, such as Rydberg states. They can be monitored by field ionization or photoionization with high efficiency. These Rydberg levels may also be autoionizing levels above the ionization limit, which can be directly detected through the ions. Since all transitions induced by L2 start from the known level $|k\rangle$, their assignment is greatly facilitated. This enables detailed investigation of structure and dynamics of Rydberg levels and their autoionization probabilities.

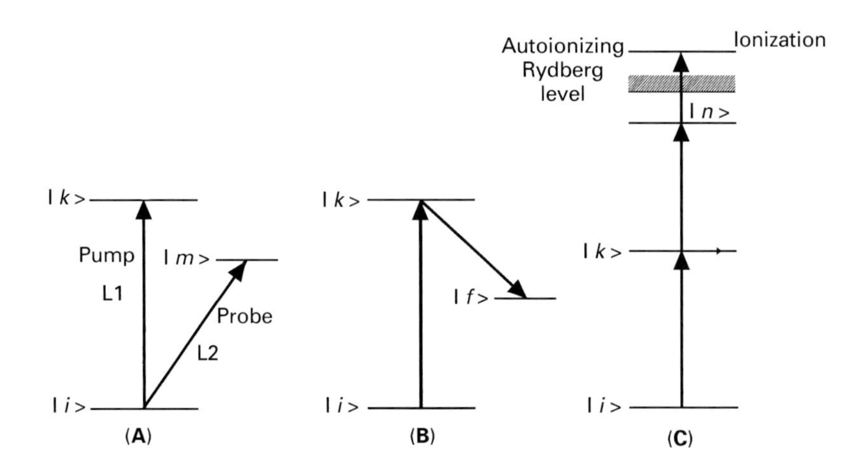

Figure 9 Double-resonance schemes: (A) V-type; (B) Λ-type; (C) stepwise excitation.

Time-resolved electronic spectroscopy

The development of ultrashort laser pulses down to pulse durations of 5×10^{-15} s has opened access to studies of extremely fast transient phenomena. Examples are the relaxation of electrons in semiconductors after their excitation by a short light pulse. The electrons excited with a definite energy $\hbar\omega$ into the conduction band thermalize within 10^{-13} s by electron–phonon collisions (**Figure 10**). With a much longer decay constant, they recombine with holes in the valence band before they can be excited again by the next pulse. Such time-resolved studies give important information on the limiting processes for the maximum speed of computers.

In molecular physics, elementary processes such as the vibrations of molecules, the formation or breaking of chemical bonds or electronic isomerization processes can now be visualized by femtosecond spectroscopy. This will be illustrated by two examples. The first is concerned with the wavepacket dynamics of vibrating diatomic molecules. The schematic principle is shown in **Figure 11**. The molecule is excited by a short pulse with duration T from its vibrational ground state into vibrational levels of an electronically excited state. If the spectral width $\Delta\nu \approx 1/T$ is large compared to the vibrational spacings, several vibrational levels are simultaneously and coherently excited. This coherent superposition of vibrational wavefunctions produces a vibrational wavepacket in the upper state that moves with the average velocity of the vibrating nuclei towards the outer turning point, where it is reflected. When a second laser pulse with a variable time delay Δt brings the excited molecule into a higher electronic state, the transition probability depends on the location of the wavepacket, because the overlap with the vibrational wavefunctions of this higher state depends on the internuclear distance. For the example of **Figure 11**, transitions from the inner turning point produce the molecular ion M^+ plus the ejected electron e^-, which carries away the excess energy. From the outer turning point, however, the repulsive ion potential can be reached and the molecular ion M^+ dissociates into $A^+ + B$ or $A + B^+$. When the time delay between pump and probe is continuously increased, the ratio M^+/A^+, or respectively M^+/B^+, will show a periodic modulation, where the period reflects the periodic movement of the wavepacket. The important point is that this pump-and-probe technique allows a time resolution on the femtosecond scale, although the detectors may be much slower. With such controlled time delay of femtosecond pulses, the resultant reaction products may be selected, and in such favourable cases channel-selective photochemistry can be realized.

A second example illustrates the dissociation of a molecule following optical excitation into an electronic state with a repulsive potential (**Figure 12**). The dissociating molecule is excited by a second laser pulse into a fluorescing bound state. The wavelength of this second transition depends on the internuclear separation R. Choosing the wavelength $\lambda(R_1)$, the correct time delay between pump and probe for exciting the fluorescence reflects the time that the dissociating molecule needs to reach R_1. This method is not restricted to diatomic molecules but has been demonstrated also for polyatomic molecules.

Measurements of lifetimes of excited states of atoms and molecules give direct information on the

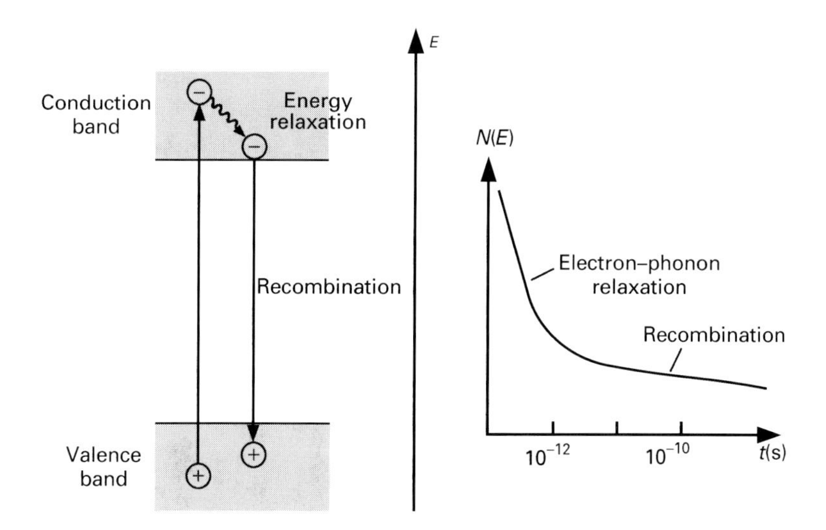

Figure 10 Thermalization and recombination of electrons after excitation into the energy E within the conduction band in a semiconductor with a femtosecond laser pulse.

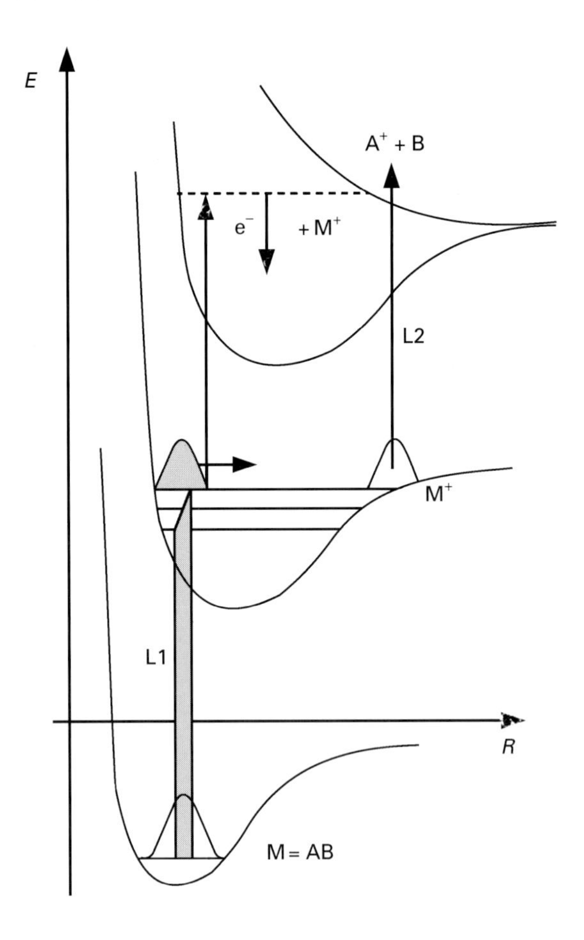

Figure 11 Wavepacket dynamics of a vibrating diatomic molecule.

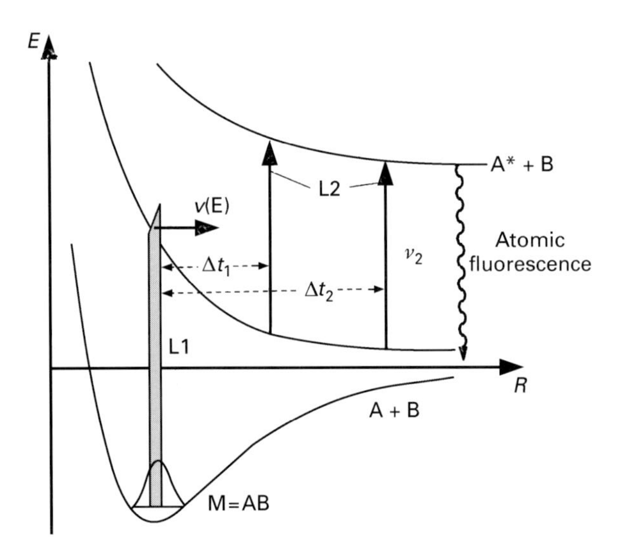

Figure 12 Level scheme of the pump-and-probe technique for a dissociating molecule.

absolute transition probabilities and are therefore important for many fields in atomic physics, chemistry and astrophysics. Accordingly, several experimental methods have been developed that cover different time domains. In the nanosecond to microsecond range, the delayed coincidence photon counting technique has proved to be very useful. The molecules are excited by a short laser pulse, which also triggers a linear voltage ramp. The first fluorescence photon emitted by the excited molecules produces an output voltage in the photomultiplier that stops the voltage ramp. The output voltage of this time-to-pulse height converter is proportional to the time delay between the exciting laser pulse and the emission of the fluorescence photon. Measuring many excitation fluorescence cycles yields a voltage pulse height distribution that is stored in a multichannel analyser (or a computer) and reflects the exponential decay of the excited molecules.

Since the probability per cycle of detecting a fluorescence photon has to be smaller than unity, the cycle frequency should be large and is only limited by the requirement that the time between successive excitation pulses should be larger than the lifetime of the excited molecules. Therefore, mode-locked synchronously pumped cavity-dumped CW dye lasers are ideal excitation sources. For the excitation of higher electronic states, the visible output pulses can be converted into ultraviolet pulses by optical frequency doubling in optically nonlinear crystals.

In the picosecond and femtosecond range, the pump-and-pulse technique is more suitable where the time delay between pump and probe can be accurately realized by an optical delay line (**Figure 13**) where one of the laser pulses travels a variable path length before it is again superimposed on the pump pulse direction.

Time-resolved investigations have been applied to measurements of fast relaxation processes, such as the primary steps in the visual process or in photosynthesis, or to collision-induced relaxation of excited states in liquids or gases under high pressures.

List of symbols

E_{int} = internal energy; E_i, E_k = energies of levels i, k; $\hbar$ = Planck constant/2π; I = light intensity; J = total angular momentum quantum number; $\boldsymbol{k}$ = wave vector, $k = |\boldsymbol{k}|$; L = path length; M = quantum number of the projection of the total angular momentum on the laser beam axis; $\dot{n}_{\text{abs}}$ = rate of absorption of laser photons; N_i = number density of molecules in level $|i\rangle$; R = internuclear separation; $S(\omega)$ = signal profile; T = flight time, pulse duration; $v_{i,k}$ = vibrational quantum number of level i, k; $v_z = z$ direction

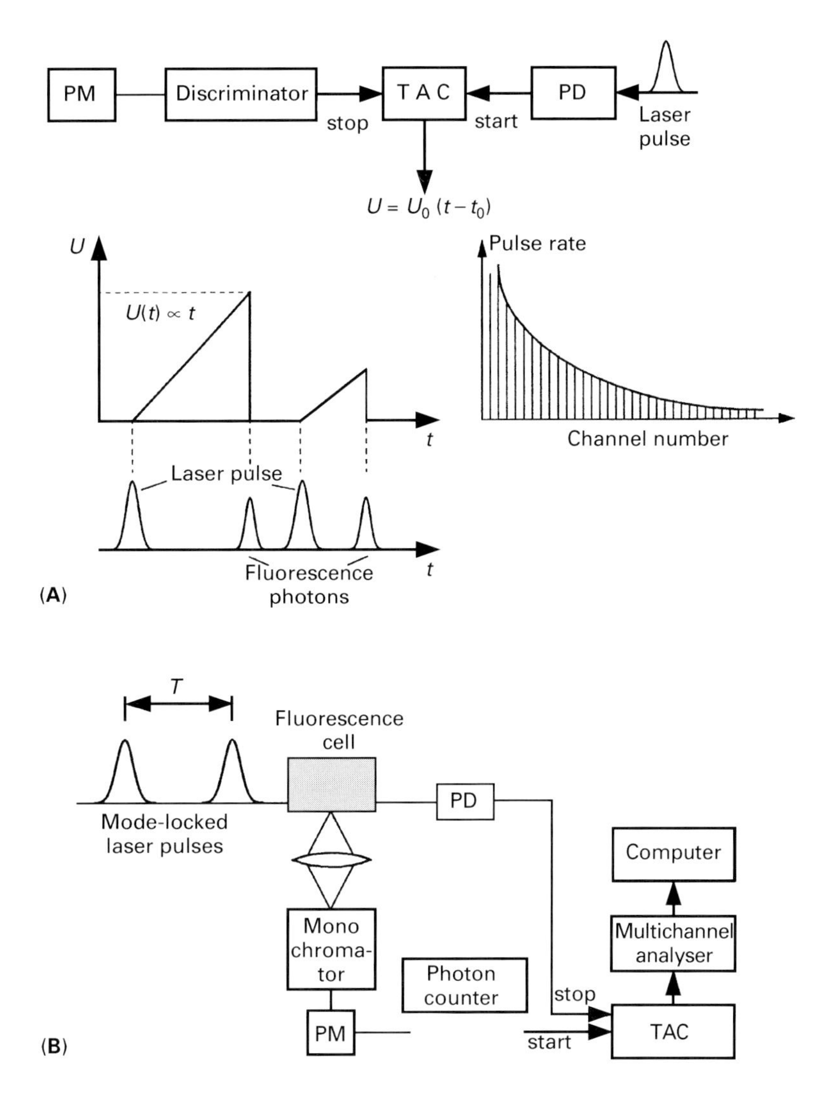

Figure 13 Delayed coincidence photon counting: (A) principal scheme; (B) experimental setup for lifetime measurements.

velocity component; α = absorption coefficient; $\alpha(\omega)$ = absorption profile; Δt = laser pulse time delay; γ = homogeneous line width; ε = collection probability of emitted fluorescence; η_k = quantum efficiency of detector; η_m = quantum efficiency of excited species; λ = wavelength; ν = transition frequency; $\sigma_{ik}(\nu)$ = cross section for transition $|k\rangle \leftarrow |i\rangle$; τ = upper-state lifetime; ω = laser (pump) frequency; Ω = modulation frequency; Ω_0 = local oscillator frequency.

See also: **Laser Spectroscopy Theory; Multiphoton Spectroscopy, Applications; Nonlinear Optical Properties; X-Ray Emission Spectroscopy, Applications; X-Ray Emission Spectroscopy, Methods.**

Further reading

Demtröder W (1998) *Laser Spectroscopy*, 2nd edn. Berlin: Springer-Verlag.

Hurst GS and Payne MG (1988) *Principles and Applications of Resonance Ionization Spectroscopy*. Bristol: Hilger.

Levenson MD (1986) *Introduction to Nonlinear Laser Spectroscopy*, 2nd edn. New York: Academic Press.

Scoles G (ed) (1991, 1992) *Atomic and Molecular Beam Methods*, Vols 1 + 2. Oxford: Oxford University Press.

Ye Jun, Ma-Long-Sheng and Hall JL (1988) Ultrasensitive detections in atomic and molecular physics. *Journal of the Optical Society of America* B15: 6.

Zewail AH (1994) *Femtochemistry*, Vols I, II. Singapore: World Scientific.

Laser Induced Optoacoustic Spectroscopy

Thomas Gensch, Katholieke Universiteit of Leuven, Heverlee, Belgium
Cristiano Viappiani, Università di Parma and Instituto Nazionale per la Fisica della Materia, Italy
Silvia E Braslavsky, Max-Planck-Institut für Strahlenchemie, Mülheim an der Ruhr, Germany

ELECTRONIC SPECTROSCOPY
Applications

Introduction

Time-resolved laser-induced optoacoustic spectroscopy (LIOAS) is a sensitive photobaric method for nondestructive probing of materials that monitors the pressure changes induced in a liquid sample after photoexcitation with a pulsed laser. LIOAS delivers information generally not available from optical studies and therefore complements other methods. After excitation, commonly with nanosecond pulsed lasers, the time evolution of the pressure pulse is monitored by fast piezoelectric transducers frequently located in a plane parallel to the direction of the laser beam (right-angle arrangement). Front-face geometry has also been used, especially with strongly absorbing samples. This article concentrates mainly on the former arrangement.

So far, time-resolved information can be used only in the case of pseudo-first-order reactions, either parallel or sequential, including thermal equilibria, since the LIOAS data analysis has been developed for schemes complying with a time-dependent pressure evolution represented by a sum of single-exponential decays.

Measurement of the heat evolved is used for determination of energy levels of intermediates (including the energy associated with chromophore–solvent interactions), provided that the quantum yields for the reactions are known. Conversely, if the energy level of the intermediates is known from other spectroscopic techniques, the determination of quantum yields is possible, for example for isomerization, charge transfer, fluorescence, intersystem crossing and energy transfer processes.

The structural volume changes are essentially density changes and reflect movements in the chromophore such as photoinduced changes in charge distribution and in the strength of the interactions with solvent molecules. Time-resolved LIOAS data also afford the decay rate constants of the transients, and therefore provide a check for the assignment of the observed events to a particular intermediate. A further identification of the transients involved is offered by the action spectrum of the pressure evolved.

LIOAS is an interesting tool for investigating photoinduced 'optically silent' intermediates, i.e. displaying no absorption or emission. A phenomenological approach to the equations is employed here and a description is given of the equipment and data handling as well as a few examples of applications of LIOAS to small molecules. Applications to biological complex systems can be found in the work by Schulenberg and Braslavsky listed in the Further reading section.

Phenomenological approach

Pressure pulses in the illuminated sample arise from the volume changes produced by radiationless relaxation (ΔV_{th}) and structural solute–solvent rearrangements (ΔV_{r}). Relaxation originates either from nonradiative decay of excited states or release of heat (enthalpy change) in photoinitiated reactions, including the heat involved in the rearrangement of the solvent. The structural volume changes reflect movements of the photoexcited molecules and/or the surrounding solvent in response to such events as dipole moment change, charge transfer, and photoisomerization. The overall pressure change is given by Equation [1].

$$\Delta p = -\frac{1}{\kappa_T}\left(\frac{\Delta V}{V_0}\right)_T = -\frac{1}{\kappa_T}\left(\frac{\Delta V_{\mathrm{th}} + \Delta V_{\mathrm{r}}}{V_0}\right)_T \quad [1]$$

where p is the pressure change, V_0 is the volume of the illuminated sample, κ_T is the isothermal compressibility (see below). The pressure change induces an electrical signal H from the transducer, proportional to the pressure change through an instrumental factor k (Eqn [2]):

$$H = k \cdot \Delta p \qquad [2]$$

The voltage is given by

$$H = \frac{1}{\kappa_T}\frac{k}{V_0}(\Delta V_{\rm th} + \Delta V_{\rm r}) = k'(\Delta V_{\rm th} + \Delta V_{\rm r}) \qquad [3]$$

Since the signal is proportional to the number of moles of absorbed photons, n,

$$n = \frac{N_{\rm ph}}{N_{\rm A}}\left(1 - 10^{-A}\right) \qquad [4]$$

where A is the absorbance of the solution at the excitation wavelength λ, N_A is the Avogadro number, and N_{ph} is the number of incident photons, the total signal is given by

$$H^{\rm S} = k'n\left(\alpha\frac{\beta}{c_{\rm p}\rho}E_\lambda + \Delta V_{\rm e}\right) \qquad [5]$$

where $E_\lambda = N_A\ (h.c/\lambda)$ is the energy of an Einstein of wavelength λ, $\alpha = q/E_\lambda$ is the fraction of absorbed energy E_λ released as heat q, $\Delta V_e = \Delta V_r/n$ is the structural volume change per absorbed Einstein, c_p is the specific heat capacity, ρ is the density, and β is the cubic thermal expansion coefficient of the solution [for dilute solutions (< 1 mM) these properties are those of the solvent]. To obtain α and ΔV_e, the instrumental constant, k', must be determined by comparing the signal of the sample with that of a calorimetric reference, i.e. a system that releases all the absorbed energy as heat in a time shorter than the instrumental response. The calorimetric reference and the sample solutions should be measured under identical experimental conditions such as solvent, temperature, excitation wavelength and geometry.

Since the isothermal compressibility κ_T is related to the adiabatic compressibility, κ_S by

$$\kappa_T = \kappa_S + T\frac{\beta^2}{c_{\rm p}\rho} = \frac{1}{\rho\nu_{\rm a}^2} + T\frac{\beta^2}{c_{\rm p}\rho} \qquad [6]$$

(with v_a the sound velocity in the medium) care should be taken to use the proper values of κ_T in Equation [3], especially when determining the thermoelastic parameters using a calorimetric reference in a different solvent. In neat water $\kappa_T = \kappa_S$ and the situation is simpler.

The signal for the calorimetric reference is given (α=1) by

$$H^{\rm R} = k'n\frac{\beta}{c_{\rm p}\rho}E_\lambda \qquad [7]$$

When permanent products or transient species living much longer (~5 times) than the pressure integration time are the only products of the photoreaction, the amplitudes of the LIOAS signals serve to determine the heat and structural volume changes through application of Equation [8] (obtained from the ratio of Eqns [7] and [5]) relating the fluence-normalized LIOAS signal amplitudes for sample and reference solutions, H_n^S and H_n^R, respectively:

$$\frac{H_{\rm n}^{\rm S}}{H_{\rm n}^{\rm R}} = \alpha + \frac{\Delta V_{\rm e}}{E_\lambda}\frac{c_{\rm p}\rho}{\beta} \qquad [8]$$

For transient species with lifetimes within the pressure integration time, kinetic information is obtained by considering that the signal is a time convolution between the transfer function $H^R(t)$ of the instrument, obtained with the calorimetric reference, and the time derivative $H(t)$, of the time-dependent total volume change:

$$H^{\rm S}(t) = \int_0^t H^{\rm R}(t)\,H(t - t')\,{\rm d}t' \qquad [9]$$

Numerical reconvolution has been used to retrieve the kinetic, enthalpic and volumetric parameters. Direct deconvolution methods have been developed as well. The reconvolution methods assume a sum of single-exponential decay functions for the time evolution of the pressure:

$$H(t) = \sum_i \frac{\varphi_i}{\tau_i}{\rm e}^{-t/\tau_i} \qquad [10]$$

where τ_i is the lifetime of transient i and φ_i is its fractional contribution to the measured volume change. This assumption bears no implications regarding the mechanism involved, since several mechanisms lead to a function such as Equation [10].

A quadratic approximation to the convolution integral has proved to be the best method for optimizing the results. A Levenberg–Marquardt algorithm performing a least-squares minimization is used.

The number of transients is successively increased until no further (10–20%) improvements is observed in the value of χ^2 and in the residuals trend. A single-exponential decay did not correctly describe the signal of benzophenone in acetonitrile, whereas the sum of two single-exponential decays yielded a perfect fit (**Figure 1**). Upon addition of a third exponential no significant improvement in χ^2 was observed (not shown).

Up to three lifetimes have been resolved in pulsed photoacoustic spectroscopy experiments, provided that they were separated by at least a factor of 4. In the case of closer values, a strong correlation between amplitudes and lifetimes appears, preventing a reliable recovery of the parameters. Lifetimes below a few nanoseconds are integrated in the prompt response, for which it is only possible to obtain the fractional amplitude.

Lifetimes between a few nanoseconds and some microseconds can readily be deconvoluted. By applying simulations based on a theoretical treatment of the signal, restoration limits have been derived for amplitudes and lifetimes in the case of a sum of two single-exponential decays. The influences of variable geometry, energy distribution and different deconvolution methods, as well as that of the signal-to-noise ratio on the restoration of the amplitudes and lifetimes, have been analysed. Since a general feature of the microsecond transients is a strong correlation between their lifetime and amplitude, their evaluation is more critical. Simulations show that the front-face geometry in principle covers a broader frequency range than does the perpendicular arrangement at the same signal-to-noise level. However, the polymeric detector used in this scheme has a smaller sensitivity compared to the piezoceramic detectors, which defeats the purpose of a higher signal-to-noise ratio. Experimental validation of these restoration limits has not yet been done. The high absorbance needed and other technical problems constitute a problem with the front-face geometry.

The fractional amplitudes φ_i in Equation [10] contain the heat release q_i and the structural volume change $\Delta V_{r,i}$ of the ith step. Both contributions can be considered additive provided the time evolution of both is the same (Eqn [11]). This assumption has been validated in several systems.

$$\varphi_i = \alpha_i + \frac{\Delta V_{e,i}}{E_\lambda}\frac{c_p \rho}{\beta}, \quad \alpha_i = \frac{q_i}{E_\lambda}, \quad \Delta V_{e,i} = \Phi_i \, \Delta V_{r,i} \qquad [11]$$

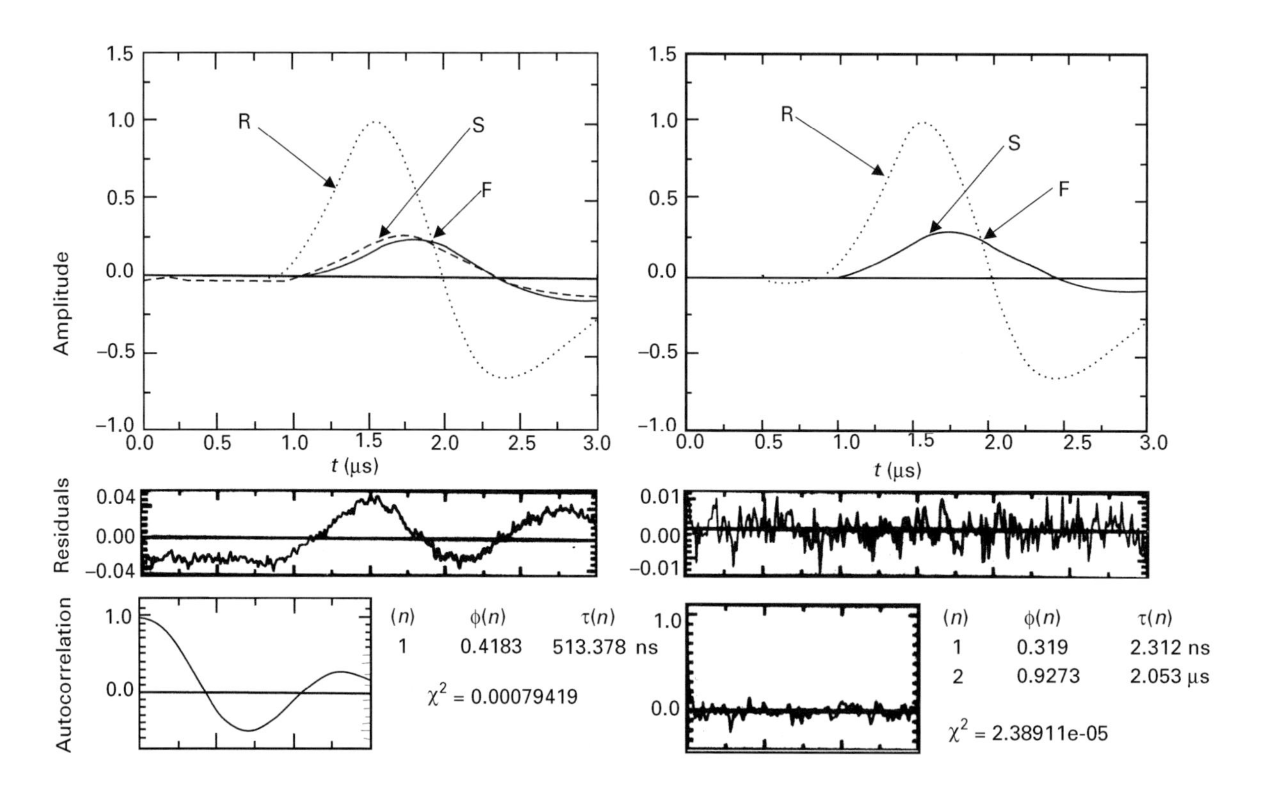

Figure 1 LIOAS signals of benzophenone (S, -----) and 2-hydroxybenzophenone (R, ·····) in N_2-bubbled acetonitrile. Fitted curves (F,——) obtained by reconvolution of the reference signal R with single-exponential (left) and two-exponential (right) decay model functions $H(t)$. Residuals and autocorrelation function are given for inspection of the goodness of the fits. Index n corresponds to subindex i in text.

where Φ_i is the quantum yield of step *i*. q_i is related to the enthalpy of the species produced, including the reorganization energy of the medium and to the value of Φ_i. $\Delta V_{r,i}$ represents the molar structural volume change of the *i*th step, which includes intrinsic changes in the chromophore as well as solute–solvent interactions. Equation [8] is thus a particular case of Equation [11] for $i = 1$.

A mechanism needs to be determined for the proper interpretation of φ_i. In the case of a sequential reaction scheme such as depicted in equation [12],

$$A \xrightarrow{h\nu} A^* \xrightarrow{k_1} B \xrightarrow{k_2} C \xrightarrow{k_3} D \quad [12]$$

the amplitudes φ_i^{app} derived from the reconvolution analysis are related to the parameters of the elementary steps (i.e., the weight of each step φ_i and the lifetimes τ_i) by Equations [13]–[15]:

$$\varphi_3 = \frac{(\tau_3 - \tau_1)(\tau_3 - \tau_2)}{\tau_3^2} \varphi_3^{\text{app}} \quad [13]$$

$$\varphi_2 = \frac{\tau_2 - \tau_1}{\tau_2} \varphi_2^{\text{app}} + \frac{\tau_2}{\tau_3 - \tau_2} \varphi_3 \quad [14]$$

$$\varphi_1 = \varphi_1^{\text{app}} + \frac{\tau_1}{\tau_2 - \tau_1} \varphi_2 + \frac{\tau_1^2}{(\tau_2 - \tau_1)(\tau_1 - \tau_3)} \varphi_3 \quad [15]$$

It is obvious that for a system with $\tau_3 >> \tau_2 >> \tau_1$, the equations simplify and $\varphi_i = \varphi_i^{\text{app}}$.

The equations for the reaction scheme [16], including a thermal equilibrium,

$$A \xrightarrow{h\nu} A^* \xrightarrow{k_1} B \underset{k_{32}}{\overset{k_{23}}{\rightleftarrows}} C \xrightarrow{k_{34}} D, \; k_1 < 0.1 \text{ ns} \quad [16]$$

are given by Equations [17]–[19]:

$$\varphi_3 = \frac{1}{k_{23}} \left[\left(k_{23} - \frac{1}{\tau_2} \right) \varphi_2^{\text{app}} + \left(k_{23} - \frac{1}{\tau_3} \right) \varphi_3^{\text{app}} \right] \quad [17]$$

$$\varphi_2 = \frac{1}{k_{23}} \left(\frac{1}{\tau_2} \varphi_2^{\text{app}} + \frac{1}{\tau_3} \varphi_3^{\text{app}} \right) \quad [18]$$

$$\varphi_1 \approx \varphi_1^{\text{app}} \quad [19]$$

Separation of heat release and structural volume changes

Separation of the heat dissipation and structural terms in Equation [11] is achieved by measuring the LIOAS signals as a function of the solvent ratio $c_p\rho/\beta$. In aqueous solutions the ratio $c_p\rho/\beta$ depends strongly on temperature (mainly owing to the changes in β). Thus, the separation of the contributions is obtained by performing measurements at various temperatures in a relatively small range (e.g. 8–20°C). A prerequisite is that the temperature does not significantly affect the kinetics of the various processes. The ratio $(c_p\rho/\beta)_T$ for neat water is obtained from tabulated values. In particular, $\beta_{\text{water}} = 0$ to 3.9°C ($T_{\beta=0}$), is positive above and negative below this temperature. At $T_{\beta=0}$, heat release gives rise to no signal. This peculiar feature allows the straightforward assessment of the existence of structural volume changes. For aqueous solutions containing salts or other additives at millimolar concentrations or higher, $(c_p\rho/\beta)_T$ and $T_{\beta=0}$ must be determined by comparison of the signal obtained for a calorimetric reference in the solvent of interest and in water.

The precision of the recovered slope of plots with Equation [11], $\Delta V_{e,i}/E_\lambda$, is generally good, whereas the intercept α_i is affected by larger errors. In case the main goal is the determination of the structural volume change, a two-temperature method may be applied. Essentially, the LIOAS wave is measured for the sample at $T_{\beta=0}$ and at a slightly higher temperature $T_{\beta\neq0}$ for which $\beta \neq 0$. The parameters $\Delta V_{e,i}/E_\lambda$ and α_i are then calculated with Equations [20] and [21]:

$$\Delta V_{e,i} = \varphi_i(T_{\beta=0}) \left(\frac{\beta}{c_p\rho} \right) E_\lambda \quad [20]$$

$$\alpha_i = \varphi_i(T_{\beta\neq0}) - \varphi_i(T_{\beta=0}) \quad [21]$$

These equations hold only for values of $T_{\beta=0}$ and $T_{\beta\neq0}$ that are near to each other, such that the rise time of the LIOAS signal and the compressibility of the solution are similar. The signal of the reference at $T_{\beta\neq0}$ is used together with that of the sample at $T_{\beta=0}$ and at $T_{\beta\neq0}$ to obtain the values of $\varphi_i(T_{\beta=0})$ and $\varphi_i(T_{\beta\neq0})$. A time shift of the sample waveform with respect to the reference waveform at $T_{\beta\neq0}$ is necessary to compensate for the change in the speed of sound, which leads to a slightly different arrival time at both temperatures. Thermal instability of the sample holder and trigger arrival fluctuations make the shifting an important option in the fitting routines.

The temperature dependence of the $c_p\rho/\beta$ ratio in organic solvents is poor, rendering the above procedures inapplicable in these solvents. An alternative approach has been used that relies on the determination of the LIOAS signals in a series of homologous solvents such as linear alkanes, cycloalkanes, or, alternatively, in solvent mixtures. This procedure is valid provided that no significant perturbations are introduced in the photophysics and photochemistry of the solute and that the solvation enthalpy and structural volume change do not vary with change of solvent. These are strong assumptions that need verification in each case.

Calorimetric calculations

The heat released in each step, q_i, is interpreted in terms of enthalpy changes by using the energy balance

$$E_\lambda = \Phi_f E_f + \sum_i q_i + E_{stored} \qquad [22]$$

where the input energy, E_λ, is equal to the sum of the energy released as emission (e.g. fluorescence; Φ_f and E_f are the quantum yield and energy constant of the emitting state) and the heat dissipated (changes in the states' energy content plus energy associated with solute–solvent interactions) in each of the steps. E_{stored} represents the total energy trapped in species living much longer than the pressure-integration time of the experiment. Analysis of the calorimetric information requires knowledge of the reaction scheme and the parameters associated with the process such as quantum yields and energy levels derived, for example, from optical measurements.

Instrumentation

Figure 2 shows a typical LIOAS setup. A calibrated beam splitter guides a fraction of each laser pulse to an energy meter for normalization purposes. The sample holder contains a normal quartz cuvette inside a thermostated metal block.

The unfocused beam is shaped by passing it through a slit (200 μm–2 mm) placed in front of the cuvette. With fast transducers, the beam size determines the rise time of signal and, therefore, the time resolution. The piezoelectric transducer is pressed against the cuvette (**Figure** 3).

The higher sensitivity of the slit-shaping geometry allows the use of the lower photon densities. The transducer is either a crystal such as lead zirconate titanate (PZT) or a film such as β-poly(vinylidene difluoride) (PVF). PZT is more sensitive, whereas the wide band PVF films provide a better temporal resolution. PZTs are generally resonant, damped transducers, the highest usable peak frequencies being about 100 MHz, with typical values around 1 MHz. Very high impedance is common to both types of transducers. After impedance matching, the electrical signal is fed into a high gain amplifier (typically × 100 and of bandwidth compatible with the transducer in use). The amplified signal is stored by a transient recorder.

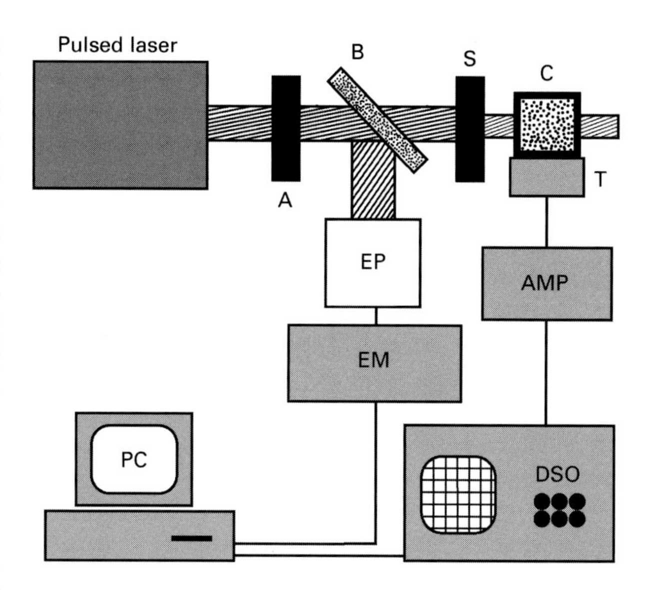

Figure 2 Block diagram of a typical LIOAS setup. A, variable attenuator; B, beam splitter; S, slit; C, quartz cuvette; T, transducer; AMP, amplifier; EP, energy probe; EM, energy meter; DSO, digital sampling oscilloscope; PC, personal computer.

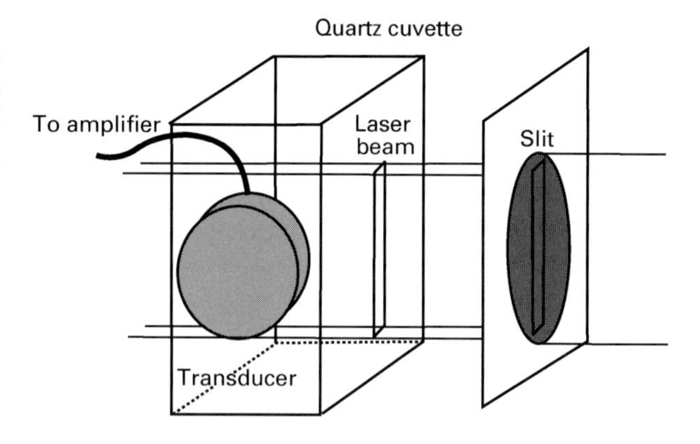

Figure 3 Close-up of right-angle detection geometry with slit-shaping of the laser beam.

Applications to selected systems

The triplet state of a porphyrin ($TPPS_4$)

The high triplet yield of porphyrins is important in a number of photobiological and photochemical problems, for example in the photodestruction of the photosynthetic apparatus and in photodynamic therapy (PDT). 5,10,15,20-Tetrakis(4-sulfonatophenyl)porphyrin ($TPPS_4$) is a water-soluble porphyrin derivative used as standard compound in PDT research. The LIOAS signal upon excitation of $TPPS_4$ in 10 mM Tris buffer (pH 7.8) at 30°C in the presence of O_2 (**Figure 4**) was fitted with a sum of three exponential decays (Eqn [10]), corresponding to the porphyrin triplet formation within a few nanoseconds, its decay (lifetime of about 2 μs) and the decay of singlet molecular oxygen ($O_2(^1\Delta_g)$, lifetime 3–3.5 μs) formed by energy transfer from triplet $TPPS_4$ to ground-state O_2 (see **Figure 6**). Measurements at various temperatures (5–30°C) and in D_2O showed that the triplet formation and decay exhibit large structural volume changes, whereas $O_2(^1\Delta_g)$ does not.

Owing to the closeness of the values of the triplet and the $O_2(^1\Delta_g)$ lifetimes, it was not possible to fit the signal with a three-exponential function in the whole temperature range. A sum of two single-exponential functions was used instead. The slow component (fixed at 2.5 μs) represented the triplet plus the $O_2(^1\Delta_g)$ decay. The plot of the parameters φ_i (Eqn [12]) vs $c_p\rho/\beta$ (**Figure 5**) shows that triplet formation (φ_1) is accompanied by a contraction $\Delta V_{r,1} = -10 \pm 3$ mL mol^{-1}, whereas triplet decay (φ_2) is concomitant with an expansion of similar magnitude, $\Delta V_{r,2} = 10 \pm 3$ mL mol^{-1}. This structural volume change indicates that triplet $TPPS_4$ has a stronger hydrogen-binding ability than the ground state, inducing a contraction of the surrounding hydration shell. Further LIOAS experiments with $TPPS_4$ and other mesosubstituted porphyrins enabled attribution of the structural volume changes to the photoinduced changes in the interactions with water (via hydrogen bonds) of the N atoms in the pyrrol units of the porphyrin ring.

The molar energy absorbed by the system, E_λ, is dissipated throughout the four possible processes undergone by the excited molecules, i.e. fluorescence, prompt heat by vibrational relaxation and internal conversion ($S_1 \rightarrow S_0$), intersystem crossing ($S_1 \rightarrow T_1$; $T_1 \rightarrow S_0$), loss during energy transfer to molecular oxygen, and stored heat (Eqn [23]). Phosphorescence from $TPPS_4$ triplet and from $O_2(^1\Delta_g)$ are known to

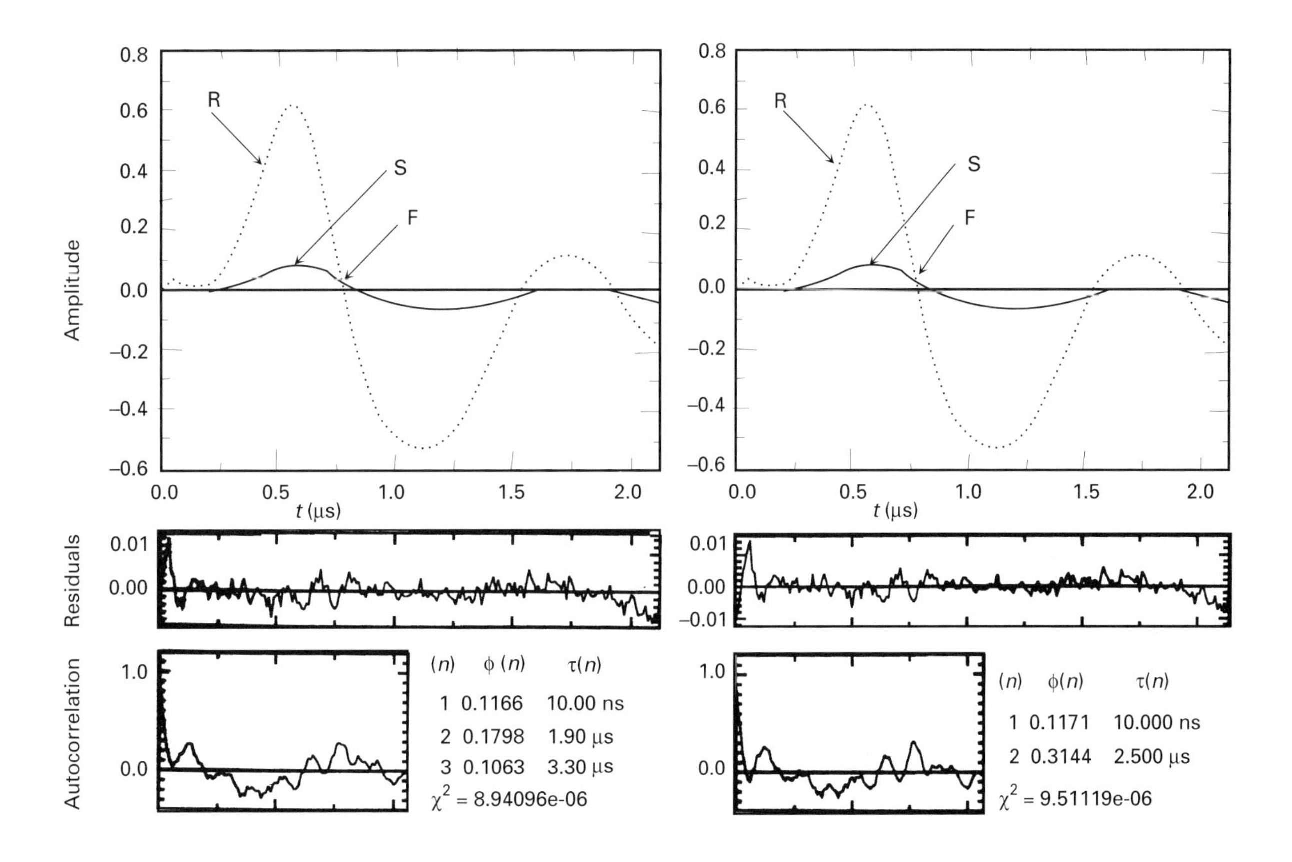

Figure 4 Fitted curves (F, ——) obtained from deconvolution of the LIOAS signal of $TPPS_4$ (S, -----) in air-satured 10 mM Tris buffer (pH 7.8) at 303 K with two-exponential (right) and three-exponential (left) decay model functions. The reference compound was bromocresol purple (R, ·····). Residuals and autocorrelation function are given for inspection of the goodness of the fits. Index *n* corresponds subindex *i* in text.

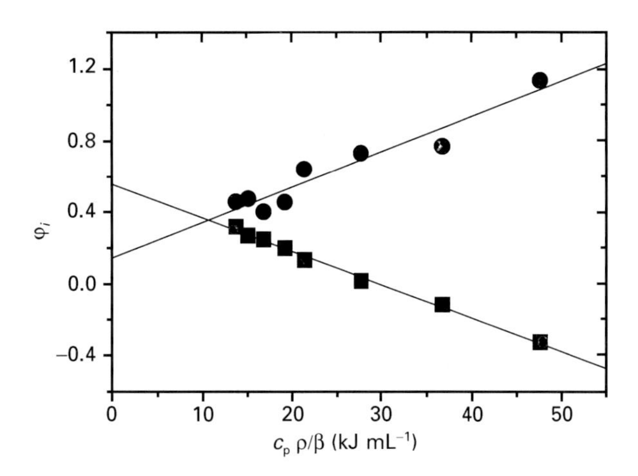

Figure 5 Plot of $\varphi_1(\tau_1 < 10\text{ns})$ and φ_2 (τ_2 fixed at 2.5 μs) as a function of $c_p\rho/\beta$ (equation [11]) for $TPPS_4$ in air-satured 10 mM Tris buffer (pH 7.8) with bromocresol purple as a calorimetric reference. Reproduced with permission from Gensch T and Braslavsky SE (1997) *Journal of Physical Chemistry B* **101**: 101–108. Copyright 1997, American Chemical Society.

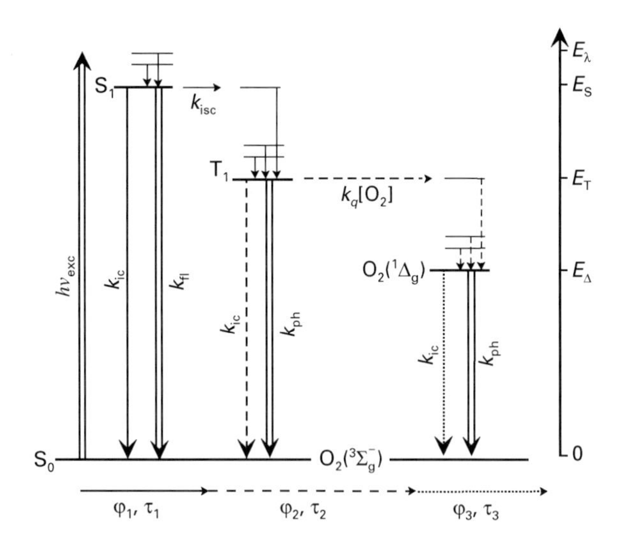

Figure 6 Term diagram for the photophysical processes occurring after excitation of $TPPS_4$ in aerated aqueous solutions. Double arrows denote light absorption and radiative decay process not detectable with LIOAS. Single full, dashed and broken arrows denote radiationless processes contributing to φ_1, φ_2, and φ_3. Subscript fl denotes fluorescence, ic internal conversion, isc intersystem crossing, ph phosphorescence, K_q [O_2] is the quenching constant of triplet T_1 by oxygen.

have yields below 10^{-3} and were neglected.

$$E_\lambda = \Phi_f \cdot E_f + q_1 + q_2 + q_3 \quad [23]$$

$$\begin{aligned} q_1 &= E_\lambda - E_S + \Phi_f(E_S - E_f) \\ &\quad + (1 - \Phi_f - \Phi_T)E_S + \Phi_T(E_S - E_T) \\ q_2 &= (\Phi_T - \Phi_\Delta)E_T + \Phi_\Delta(E_T - E_\Delta) \\ q_3 &= \Phi_\Delta E_\Delta \end{aligned}$$

The heat released during the formation of the triplet state as calculated from the intercept of the plot (**Figure 5**, Eqn [12]), $\alpha_1 = 0.58 \pm 0.05$, was used to determine the triplet quantum yield $\Phi_T = 0.61$, by using the literature values $E_f = 174$ kJ mol^{-1}, $E_S = E_{0-0}$ = 186 kJ mol^{-1} $\Phi_f = 0.06$, and $E_T = 139$ kJ mol^{-1} (at 77 K) and a simple energy balance equation (see **Figure 6** for the meaning of the symbols). Alternatively, the average $\Phi_T = 0.67$ (from literature) can be used to calculate $E_T = 126$ kJ mol^{-1}, which is the first report of a triplet energy value for this porphyrin at room temperature.

Porphyrin to quinone electron transfer

The interest in understanding the molecular basis of the biologically fundamental electron transfer process from the chlorophyll derivative pheophytin to a quinone in photosynthesis has led to the development of a large number of models, such as synthetic systems with porphyrin derivatives as donor and quinones as acceptor. Feitelson and Mauzerall demonstrated the power of LIOAS for the determination of rate constants, quantum yields and structural volume changes in solution of such systems. The triplet yields and electron transfer efficiencies were determined in various solvents. Electrostriction after formation of the ion pair was suggested as the source of structural volume change leading to a contraction in the charge-separated state. Electrostriction describes the contraction of the solvent around a charged molecule. The magnitude of the effect is inversely proportional to the radius of the molecule and to the dielectric constant of the solution. A problem in this connection is that the theoretically derived values using a solvent continuum model are far to small compared to the measured effects. This problem needs further attention.

The structural volume changes for triplet formation of metal-free uroporphyrin (−0.45 cm^3 mol^{-1}) and of its zinc complex (−1.1 cm^3 mol^{-1}), as well as the structural volume changes due to electrostriction after electron transfer from Zn-uroporphyrin to a quinone (−4.4 cm^3 mol^{-1}), were estimated in aqueous solutions. The entropy change due to ion formation was then estimated as −202 kJ mol^{-1}K for the latter system.

Ru(II) bipyridyl complexes

Ruthenium(II) bipyridyl (bpy) complexes constitute interesting compounds for LIOAS studies since their photophysical parameters are known and their transient lifetimes are within the integration time of the LIOAS experiment. In all these complexes, the metal-to-ligand charge transfer (MLCT) state in aqueous solution is formed with unity quantum yield and relaxes within 100–600 ns (depending on the complex) to the ground state.

In particular, the structural volume changes for the formation of the MLCT state of $Ru(bpy)(CN)_4^{2-}$, $Ru(bpy)(CN)_3(CNCH_3)^-$, $Ru(bpy)_2(CN)_2$, and $Ru(bpy)_3^{2+}$ of 15, 10, 5, and −3.5 cm³ mol⁻¹ correlate with the number of CN groups. Thus, the structural volume changes are interpreted as expansion of the strong second-sphere donor–acceptor interaction of the CN groups with water after weakening of the π backbone in the Lewis-base cyano substituents concomitant with the ruthenium photooxidation. The complex $Ru(bpy)_3^{2+}$ (with no CN groups) shows a small contraction of −3.5 cm³ mol⁻¹ due to intrinsic Ru–bpy bond shortening. The lifetimes derived by deconvolution were similar to those obtained by emission.

For the cyano-Ru(II) complexes in neat water, the heats derived from the intercepts of plots with Equation [11] afforded values of the energy content of the MLCT states similar to those derived from emission data. Both for intra- and intermolecular electron transfer reactions of cyano-Ru(II) complexes in aqueous solutions in the presence of salts, the structural volume change correlates with the entropy change of the reaction. An enthalpy–entropy compensation effect occurs upon perturbation of the water structure by the addition of the salts.

$Ru(bpy)(CN)_4$ inside the water pools of sodium bis(2-ethylhexyl)sulfo-succinate (AOT) reverse micelles with a large water pool inside the micelle showed a structural volume change similar to that in neat water, whereas in micelles with a small water pool the structural volume changes were smaller.

These findings support the model of a change in the CN–water interaction as the origin of the expansion upon MLCT formation. In a large water pool $Ru(bpy)(CN)_4$ interacts with free water (i.e. the $Ru(bpy)(CN)_4$ environment is similar to that in neat water), whereas in a small water pool the interactions involve 'bound water' (water molecules interacting with the charged detergent molecules), and the expansion is impaired and reduced.

Proton transfer and the reaction volume for water formation

The structural volume changes photoinduced in solution during the proton transfer between *o*-nitrobenzaldehyde and hydroxyls in water have been recently studied by means of LIOAS. Upon photoexcitation, *o*-nitrobenzaldehyde is converted to *o*-nitrosobenzoic acid and protons are detached with a quantum yield of about 0.4. At pH around and below neutrality, solvation of the charges is accompanied by a fast, subresolution contraction of the medium of approximately −5.2 ± 0.4 cm³ mol⁻¹.

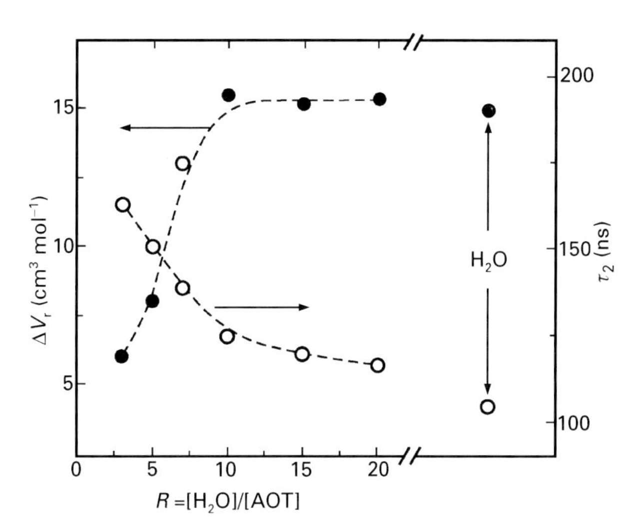

Figure 7 Structural volume change (ΔV_r) associated with the formation of the MLCT state of $Ru(bpy)(CN)_4^{2-}$ and its decay (lifetime $\tau_2 = \tau_{MLCT}$) in 0.1 M AOT–water–*n*-alkane reverse micelles at various values of R = [water]/[AOT]. Reproduced with permission from Borsarelli CD and Braslavsky SE (1997). Nature of the water structure inside the pools of reverse micelles sensed by laser-induced optoacoustic spectroscopy. *Journal of Physical Chemistry B* **101**: 6036–6042 Copyright 1997 American Chemical Society.

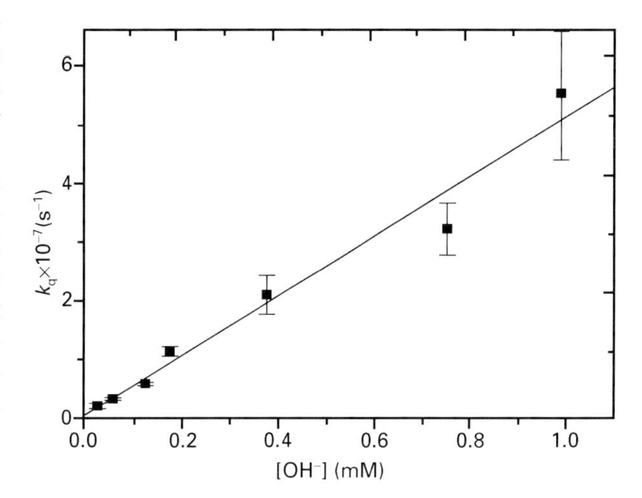

Figure 8 Rate constant for the reaction between photodetached protons and hydroxyls in water at various OH⁻ concentrations. Reproduced with permission of the Editors of *Chemical Physics Letters* from Bonetti G, Vecli A and Viappini C (1997) Reaction volume of water formation detected by time-resolved photoacoustics – photoinduced proton transfer between *o*-nitrobenzaldehyde and hydroxyls in water. *Chemical Physics Letters* **209**: 268–273.

This contraction is at least qualitatively accountable for on the basis of the electrostrictive effects described by the Drude–Nernst equation, considering the negative charge localized on the carboxylic oxygen atom of *o*-nitrosobenzoic acid and using the molar volume of the proton in water. At pH > 9, the reaction with hydroxyls becomes relevant and the formation of water molecules produces an expansion of 24.5 ± 0.4 cm^3 mol^{-1}. The plot of the quenching rate constant (k_q, **Figure 8**), derived from LIOAS data vs the concentration of hydroxyls is linear with a slope of $(4.9 \pm 0.4) \times 10^{10}$ M^{-1} s^{-1}, indicating that the process obeys pseudo-first-order kinetics.

List of symbols

A = absorbance; c_p = specific heat capacity; E_λ = energy of an Einstein of wavelength λ; H = transducer response; k = instrumental factor; n = number of moles of absorbed photons; N_A = Avogadro number; N_{ph} = number of incident photons; q = energy released as heat; T = temperature; v_a = sound velocity; V = volume; V_0 = volume of illuminated sample; $\alpha_i = q_i/E_\lambda$; β = cubic expansion coefficient; Δp = pressure change; ΔV_e = structural volume change per absorbed Einstein; ΔV_r = volume change due to solute–solvent rearrangements; ΔV_{th} = volume change due to radiationless relaxation; κ_S = adiabatic compressibility; κ_T = isothermal compressibility; λ = excitation wavelength; ρ = density; τ_i = lifetime of transient i; Φ_i = quantum yield of step i;φ_i = fractional contribution to volume change of transient i.

See also: **Laser Applications in Electronic Spectroscopy; Laser Microprobe Mass Spectrometers; Laser Spectroscopy Theory; Photoacoustic Spectroscopy, Applications; Photoacoustic Spectroscopy, Theory.**

Further reading

Bonetti G, Vecli A and Viappiani C (1997) Reaction volume of water formation detected by time-resolved photoacoustic – photoinduced proton transfer between *o*-nitrobenzaldehyde and hydroxyls in water. *Chemical Physics Letters* **269**: 268–273.

Borsarelli CD and Braslavsky SE (1997) Nature of the water structure inside the pools of reverse micelles sensed by laser-induced optoacoustic spectroscopy. *Journal of Physical Chemistry B* **101**: 6036–6042.

Borsarelli CD and Braslavsky SE (1999) Enthalpy, volume, and entropy changes associated with the electron transfer reaction between the ^{3}MLCT state of $Ru(bpy)_3^{2+}$ and methyl viologen in aqueous solutions. *Journal of Physical Chemistry A* **103**: 1719–1727.

Braslavsky SE and Heibel GE (1992) Time-resolved photothermal and photoacoustic methods applied to photoinduced processes in solution. *Chemical Reviews* **92**: 1381–1410.

Churio MS, Angermund KP and Braslavsky SE (1994) Combination of laser-induced optoacoustic spectroscopy (LIOAS) and semiempirical calculations for the determinations of molecular volume changes: the photoisomerization of carbocyanines. *Journal of Physical Chemistry* **98**: 1776–1782.

Crippa PR, Vecli A and Viappiani C (1994) Time resolved photoacoustic spectroscopy: new developments of an old idea. *Journal of Photochemistry and Photobiology B: Biology* **24**: 3–15.

Feitelson J and Mauzerall D (1996) Photoacoustic evaluation of volume and entropy changes in energy and electron transfer. Triplet state porphyrin with oxygen and naphthoquinone-2-sulfonate. *Journal of Physical Chemistry* **100**: 7698–7703.

Gensch T and Braslavsky SE (1997) Volume changes related to triplet formation of water soluble porphyrins. A laser-induced optoacoustic spectroscopy (LIOAS) study. *Journal of Physical Chemistry B* **101**: 101–108.

Gensch T, Strassburger JM, Gärtner W and Braslavsky SE (1998) Volume and enthalpy changes upon photoexcitation of bovine rhodopsin derived from optoacoustic studies by using an equilibrium between bathorhodopsin and blue-shifted intermediate. *Israel Journal of Chemistry* **38**: 231–236.

Mauzerall D, Feitelson J and Prince R (1995) Wide-band time-resolved photoacoustic study of electron-transfer reactions: difference between measured enthalpies and redox free energies. *Journal of Physical Chemistry* **99**: 1090–1093.

Puchenkov OV (1995) Photoacoustic diagnostics of fast photochemical and photobiological processes. *Biophysical Chemistry* **56**: 241–261.

Schmidt R and Schütz M. (1996) Determination of reaction volumes and reaction enthalpies by photoacoustic calorimetry. *Chemical Physics Letters* **263**: 795–802.

Schmidt R and Schütz M (1997) Methodical studies on the time resolution of photoacoustic calorimetry. *Journal of Photochemistry and Photobiology A: Chemistry* **103**: 39–44.

Schulenberg P and Braslavsky SE (1997) Time-resolved photothermal studies with biological supramolecular systems. In: Mandelis A and Hess P (eds) *Progress in Photothermal and Photoacoustic Science and Technology*, pp 58–81. Bellingham WA: SPIE Press.

Small JR, Libertini LJ and Small EW (1992) Analysis of photoacoustic waveforms using the non-linear least square method. *Biophysical Chemistry* **42**: 24–48.

Wegewijs B, Verhoeven JW and Braslavsky SE (1996) Volume changes associated with intramolecular exciplex formation in a semiflexible donor-bridge-acceptor compound. *Journal of Physical Chemistry* **100**: 8890–8894.

Laser Magnetic Resonance

Al Chichinin, Institute of Chemical Kinetics and Combustion, Novosibirsk, Russia

MAGNETIC RESONANCE
Methods & Instrumentation

Laser magnetic resonance (LMR) is a sensitive technique for studying rotational, vibrational-rotational, or electronic Zeeman spectra of paramagnetic atoms and molecules, using fixed-frequency infrared lasers. High resolution and sensitivity are important characteristics of LMR. In a pioneering experiment in 1968, Evenson and his co-workers succeeded in demonstrating the feasibility of this technique when they detected a rotational transition of O_2 using a HCN laser. Several years later LMR was extended to the mid-infrared by using CO and CO_2 lasers with similar success. Since then, throughout 15 years, much of the increase in understanding of the structure and properties of short-lived paramagnetic molecules has come from the application of LMR. The method has allowed the detection of more than 100 free radicals, including elusive species such as HO_2, CH_2, FO and $Cl(^2P_{1/2})$ which cannot easily be detected by other means. The high sensitivity and versatility of LMR has resulted in numerous applications of the method to chemical kinetics.

Theory of LMR

Atoms

As a simple illustration let us consider the LMR spectra of Cl atoms. The energy level diagram at the top of **Figure 1** shows the Zeeman effect in Cl atoms.

In a magnetic field, the $|J, I, F\rangle$ levels of the atom split into components labelled by M_J and M_F quantum numbers (here $\boldsymbol{F} = \boldsymbol{J} + \boldsymbol{I} = \boldsymbol{L} + \boldsymbol{S} + \boldsymbol{I}$). The magnetic field is swept, and at suitable values of the field the frequencies of the transitions $|J, M_J, M_F\rangle \rightarrow \langle J', M'_J, M'_F|$ come into coincidence with the frequency of a fixed-frequency laser. These resonances result in a decrease in detected laser power. In practice, a small modulation (up to 50 G) is added to the magnetic field and the change in transmitted laser power is recorded as a first derivative signal by processing the detector output in a phase-sensitive amplifier.

LMR spectra of Cl atoms are shown at the bottom of **Figure 1**. Note that $^2P_{3/2}$–$^2P_{1/2}$ magnetic dipole transitions with $\Delta M_F = \pm 1$ and $\Delta M_F = 0$ occur at $\boldsymbol{E} \perp \boldsymbol{B}$ and $\boldsymbol{E} \parallel \boldsymbol{B}$ polarizations, respectively; here $\boldsymbol{E}$ is the electric field of the laser radiation and $\boldsymbol{B}$ is the magnetic field of the electromagnet.

The Hamiltonian may be written as

$$H = A_{\mathrm{so}}\boldsymbol{LS} + H_{\mathrm{hfs}}(a, b, \boldsymbol{J}, \boldsymbol{I}) + \mu_0(g_J\boldsymbol{JB} + g_I\boldsymbol{IB})$$

where H_{hfs} is the hyperfine interaction operator, A_{so} is a spin-orbit interaction constant, a and b are hyperfine constants, $\mu_0 = e\hbar/2mc$ is the Bohr magneton, and g_J and g_I are electron and nuclear g-factors. All these atomic constants can be obtained from the analysis of LMR spectra; the precise values for A_{so} have been determined in this way for the first time.

Diatomic radicals

A more common application is the measurement of LMR spectra of linear radicals with both spin and orbital angular momentum (SnH, NiH, GeH, CH, SD, SeH, ... etc). The effective Hamiltonian is a summation of spin-orbit (H_{so}), vibrational (H_{v}), rotational (H_{rot}), spin-rotational (H_{sr}), spin-spin (H_{ss}), lambda-doubling (H_{ld}), hyperfine (H_{hfs}) and Zeeman (H_{z}) terms:

$$H = H_{\mathrm{so}} + H_{\mathrm{v}} + H_{\mathrm{rot}} + H_{\mathrm{sr}} + H_{\mathrm{ss}} + H_{\mathrm{ld}} + H_{\mathrm{hfs}} + H_{\mathrm{z}} \qquad [1]$$

The choice of significant terms and representation of the terms are determined by the particular type of radical. Some usual representations are presented below

$$\begin{aligned}
H_{\mathrm{so}} &= A_{\mathrm{so}}\,\boldsymbol{LS} \\
H_{\mathrm{v}} &= \hbar\omega_{\mathrm{e}}(v + 1/2) - \omega_{\mathrm{e}}x_{\mathrm{e}}\hbar(v + 1/2)^2 \\
H_{\mathrm{rot}} &= B_v\boldsymbol{N}^2 - D_v\boldsymbol{N}^4 \\
H_{\mathrm{sr}} &= \gamma\boldsymbol{NS} \\
H_{\mathrm{ss}} &= \frac{2}{3}\lambda(3S_z^2 - \boldsymbol{S}^2) \\
H_{\mathrm{z}} &= \mu_0(g_{\mathrm{s}}\boldsymbol{S} + g_L\boldsymbol{L} + g_{\mathrm{r}}\boldsymbol{N})\boldsymbol{B}
\end{aligned}$$

Here v denotes the vibrational quantum number and $\boldsymbol{N} = \boldsymbol{J} - \boldsymbol{L} - \boldsymbol{S}$ is the nuclear rotational angular

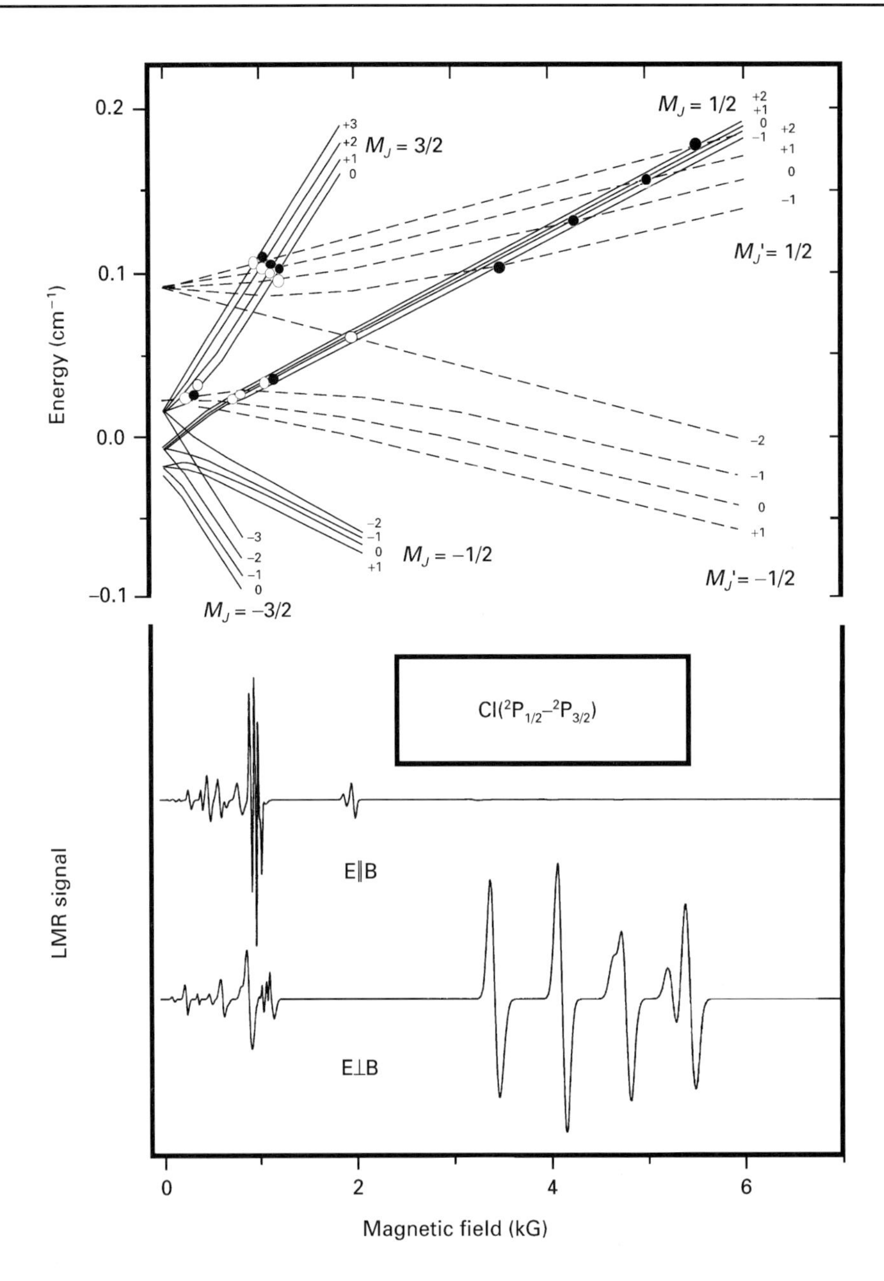

Figure 1 Top: Zeeman effect in ^{35}Cl atoms. All sublevels are labelled by M_F values. The lower $^2P_{3/2}$ state is represented by solid lines. Upper $^2P_{1/2}$ state (dashed lines) is downshifted by an amount of the laser photon energy. Intersections of solid and dashed lines with $\Delta M_F = \pm 1$ (open circles) or with $\Delta M_F = 0$ (solid circles) represent possible LMR transitions. Bottom: LMR spectra of Cl atoms (both ^{35}Cl and ^{37}Cl).

momentum. Centrifugal corrections to some parameters are often used as well as a tensor expression for H_z.

Polyatomic radicals

The most often encountered type of radical observed by LMR is the three-atomic asymmetric-top radical, such as NH_2, PH_2, HCO, HO_2, etc., Hund's case (b). In this case, the Hamiltonian [1] is used. Without spin-orbit and lambda-doubling terms, H_{rot}, H_{ss} and H_{sr} are usually expressed as

$$H_{rot} = A_v N_a^2 + B_v N_b^2 + C_v N_c^2 + H_{cd}(\boldsymbol{N}, \Delta_N, \Delta_{NK}, \Delta_K, \delta_N, \delta_K)$$

$$H_{ss} = \frac{1}{3} D_v (2S_a^2 - S_b^2 - S_c^2) + E_v (S_b^2 - S_c^2)$$

$$H_{sr} = \varepsilon_{aa} N_a S_a + \varepsilon_{bb} N_b S_b + \varepsilon_{cc} N_c S_c$$

where H_{cd} is the Watson's centrifugal distortion operator; the molecule-fixed components of N, S, and ε are labelled by a, b, and c.

Finally note that the representation of the Hamiltonian is very dependent on the particular radical. Some terms are often omitted, and some new ones are added. For example, Jahn–Teller distortion should be taken into account for symmetric top radicals, such as CH_3O and SiH_3.

The LMR detection of polyatomic radicals with heavy (not hydrogen) atoms is hindered for several reasons: (1) the population of the lower state is low, it decreases with rotational partition function; (2) Zeeman splitting is small, it decreases both with moments of inertia and rotational quantum numbers of the radical; hence LMR spectra are observable only at low magnetic fields because of a weak coupling of S with N. Moreover, the spectra are often unresolved because of a large number of components.

An important exception was found for the first time by Uehara and Hakuta who have observed high-field spectra in the ν_1 band of ClO_2 in spite of its small spin-orbit interaction. Later, similar high-field far-infrared LMR spectra of FO_2, ClSO, FSO and NF_2 were detected. These spectra were assigned to transitions induced by avoided crossings between Zeeman levels having the same value of M_J but differing by one in N. This type of transition offers a new means to study radicals which might otherwise be inaccessible to the LMR technique.

LMR spectrometer

Basic features

The essential features of an LMR spectrometer are illustrated schematically in **Figure 2**, which shows three of the most commonly encountered experimental configurations.

A conventional intracavity LMR spectrometer is shown in **Figure 2A**. Placing a free-radical absorption cell inside the laser cavity results in a gain in sensitivity due to multipassing of the laser radiation. A further sensibility gain may also be obtained when the modulation frequency is close to the frequency of the laser intensity relaxation oscillations. Another benefit of the intracavity configuration is that it favours the observation of saturation Lamb dips. This is especially important in the mid-infrared region, where Doppler widths are greater and the high resolution of saturation spectroscopy is needed more. The laser power is coupled out by the zeroth order of the grating and is detected by a photoresistor. Coupling through the hole in one of the mirrors or by a variable coupler inserted into the laser cavity are also used.

An intracavity LMR spectrometer based on an optically pumped laser is shown in **Figure 2B**. The only difference is the change of the CO (or CO_2) laser in **Figure 2A** by an optically pumped laser. The pump radiation is multiply reflected between two metallic-coated flats parallel to the far-infrared laser axis. Longitudinal pumping in which the pump radiation is introduced along the laser axis through a hole in one of the far-infrared laser mirrors is also employed. The cavity is divided by a beam splitter (e.g. thin polypropylene).

Figure 2C gives the block diagram for the Faraday LMR setup, in which there is an extracavity arrangement and detection of paramagnetic species is via their polarization effects. The laser light propagates along the z-axis which coincides with the magnetic field direction. The polarization, determined by the polarizer P_x points along the x-axis and the polarizer P_y points along the y-axis. The multipass absorption cell is placed between these crossed polarizers. The absorbing radicals produce a change in the polarization; this effect is used for sensitive detection of the radicals.

In essence, the Faraday arrangement is extracavity LMR with two crossed polarizers. These polarizers reduce the laser power by a large factor and the LMR signal by the square of that factor. Hence the signal-to-noise ratio for the Faraday arrangement is significantly better (more than two orders of magnitude) than that for conventional extracavity LMR.

Note that time resolution of the intracavity configurations is determined by the medium of the laser: it is usually ~2–4 μs; the time resolution of the Faraday arrangement is considerably better.

The Faraday configuration requires the magnetic field of the electromagnet (solenoid) to be parallel to the Poynting vector, $B \parallel S_p$; An alternative (Voigt) configuration is proposed in which $B \perp S_p$ and a polarization angle relative to B is 45°. Although the Voigt configuration is expected to have similar sensitivity to that of Faraday LMR, it is rarely encountered.

Usually, the configurations in **Figure 2A** and **2C** are used in mid-infrared spectroscopy. The configuration in **Figure 2B** is used in far-infrared spectroscopy only. In intracavity configurations, the polarization of the laser with respect to the magnetic field is determined by the intracavity beam splitter in **Figure 2B** or by windows in **Figure 2A** set at Brewster's angle. The splitter (or the windows) are rotatable about the laser axis so the polarization can be rotated. When the electric vector of the laser lies perpendicular to the magnetic field, $E \perp B$, then

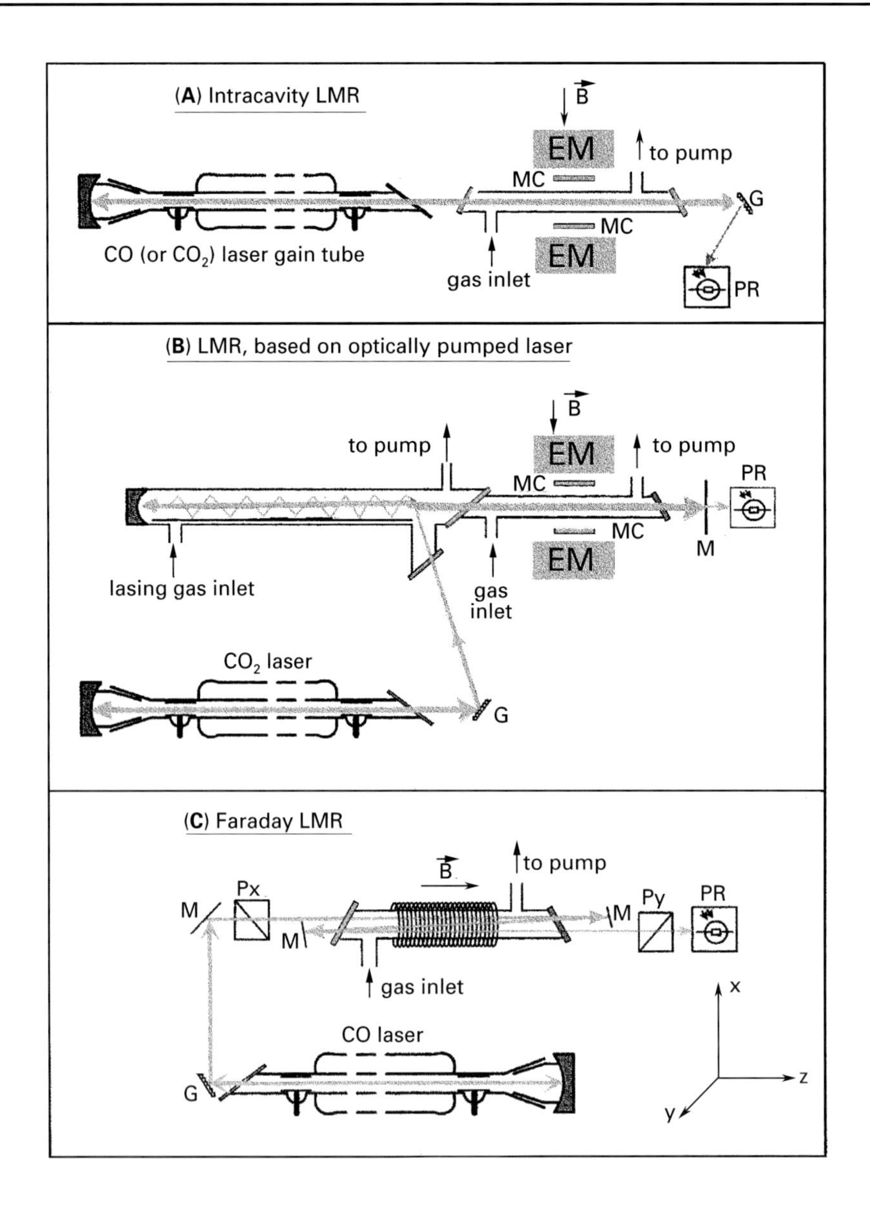

Figure 2 Schematic representation of a LMR spectrometer: (**A**) mid-infrared intracavity arrangement, (**B**) far-infrared intracavity arrangement with optically pumped laser, (**C**) mid-infrared Faraday extracavity arrangement. EM = electromagnet pole, MC = modulation coil, G = diffraction grating, PR = photoresistor, M = mirror, P_x and P_y = polarizers.

$\Delta M_J = \pm 1$ (σ) electric dipole transitions are induced; the $\Delta M_J = 0$ (π) transitions appear with parallel polarization, $E \parallel B$. Note that the latter case is impossible in Faraday LMR.

Infrared lasers

The first far-infrared LMR spectra were recorded by using HCN, H_2O and D_2O lasers. Since then, optically pumped lasers are usually used with a line-tunable continuous-wave CO_2 laser as a light source. By use of these lasers, over 1000 laser frequencies are available in the far-infrared region (30–1200 μm); some 500 have been used in LMR.

In medium-infrared, most of the radical spectra are observed with CO (1200–2000 cm^{-1}) and CO_2 (875–1110 cm^{-1}) lasers. In the former case, the spectral region is now strongly extended by using a CO-overtone laser which provides 200 ($\Delta v = 2$) transitions in the region of 2500–3500 cm^{-1}. In the latter case, a great increase in capacity of LMR (several hundreds of lines) was attained by using $^{13}C^{16}O_2$, $^{13}C^{18}O_2$, $^{13}C^{16}O^{18}O$ and $C^{18}O_2$ isotope modifications; an N_2O laser has also been used.

A colour centre laser (2–4 μm), and spin-flip Raman laser (which consisted of a InSb crystal pumped by a CO laser) have also been employed.

Infrared detectors

The first far-infrared LMR spectra were recorded by using Golay cells. These have the advantage of room temperature operation and are easy to use, but their response is slow and hence only low modulation frequencies (approximately 10–100 Hz) can be used. A further gain in sensitivity was realized by use of a helium-cooled Ge bolometer: the much faster response of the bolometer allows much higher modulation frequencies (~1 kHz), and this, together with the lower noise equivalent power, means that the sensitivity is limited mainly by inherent noise in the laser source. By far the most commonly encountered far-infrared detectors are helium-cooled photoresistors (Ge:B, Ge:In, Ge:Ga, Ge:As, nIn:Sb, etc.); modulation frequencies of up to hundreds of kHz can be used.

In mid-infrared, the most commonly used photoresistors are Ge: Au (77 K), Ge:Hg (53 K), Ge: (Zn, Sb) (64 K) and Hg:Cd:Te (77 K) for CO_2 laser output and Ge:Ga (77 K), In:Sb (77 K), Hg:Cd:Te (77 K) for CO laser output.

In general the main noise source is the laser rather than the photoresistor. However, in the mid-infrared region attenuation of laser light is often required since detectors are easily saturated by laser light. For example, for InSb (77 K) detector saturation starts near 1 mW, while typical CO laser output power reaches 1 W. Note that in Faraday and Voigt LMR arrangements, the saturation is absent.

Modulation

The modulation frequency f is determined as a compromise between two factors. First, at low frequencies the noise of the LMR spectrometer increases due to vibrations of the spectrometer. Second, the modulation amplitude decreases inversely with the square root of f at a fixed power of the generator for modulation coils. Finally, for extracavity LMR the best frequency is 15–20 kHz. For intracavity LMR spectrometers, a gain of sensibility can be obtained when the modulation frequency is close to the frequency of the laser intensity relaxation oscillations (50–150 kHz).

A simple method to improve the sensitivity of LMR is to increase the integration time constant of the lock-in amplifier. However, the long-term instability of the LMR spectrometer usually makes useless time constants larger than 3 s. This problem can be solved by using double modulation. The magnetic field is modulated at high (~100 kHz) and low (~2 Hz) frequencies. The high-frequency signal is demodulated by a phase-sensitive lock-in amplifier. The resulting low-frequency output of the lock-in is again demodulated by another phase-sensitive lock-in. As a result, the effective time constant of the system is ~200 s and the LMR sensitivity increases considerably. Note that the best peak-to-peak amplitude of the low-frequency modulation is equal to the spectral line width. The drawback of this double modulation system is the slow response.

Intracavity systems are very sensitive to acoustic cross talk between the modulation unit and the laser resonator, thus increasing the noise just at the modulation frequency f. Hence another method of improving the sensitivity of intracavity LMR is to use a lock-in detector at $2f$ (the second harmonic), where the noise is considerably smaller. In some cases, this change can increase the signal-to-noise ratio of LMR.

Preparation of radicals

The great majority of radicals studied so far by LMR have been generated by atom–molecule reactions in the gas phase, using discharge-flow techniques or by pumping the products of a microwave discharge rapidly into the sample region of the spectrometer. In addition, a time-resolved arrangement allows generation of radicals either directly by UV photolysis (or multiphoton dissociation), or by the reactions of appropriate molecules with the species prepared photolytically.

Detection of molecular ions

Molecular ions (DCl^+, DBr^+, etc.) are generated by discharge-flow techniques. Two difficulties arise in experiments with intracavity LMR detection of these ions. First is the rapid deflection of the ions to the reactor walls in the magnetic field of the electromagnet; the second is the very high noise of a DC discharge situated inside the laser cavity. These problems have been solved by using the Faraday extracavity LMR arrangement, which does not suffer to the same extent from modulation pickup via discharge plasma as the intracavity arrangement. In the Faraday arrangement the discharge is stable, since the magnetic field of the solenoid and the electric field of the discharge are collinear.

An additional advantage of this arrangement is the possibility of tracing hot-band transitions not only for open-shell ions but also for free radicals.

Sensitivity of LMR, comparison with EPR

The technique of LMR is very similar to other magnetic resonance methods such as EPR and NMR. While NMR uses radiofrequency radiation to produce transitions between nuclear spin levels, and EPR uses microwave radiation to produce transitions between electron spin levels, LMR uses radiation of a laser to produce transitions between rotational (far-infrared) or vibrational–rotational (mid-infrared) levels in paramagnetic molecules. Note that the sensitivity γ_{min} is nearly equal for EPR and LMR: $\gamma_{min} \simeq 10^{-10}$–$10^{-9}$ cm^{-1}. The concentration sensitivity is given by

$$N_{min} = \frac{\gamma_{min}}{\sigma d}$$

where d is the difference of relative populations of the upper and lower levels and σ is the cross section of the transition. The superiority of LMR over EPR in sensitivity is determined by the d factor (2–3 orders of magnitude) and by the cross section σ which is proportional to the transition frequency. The factor in favour of EPR is the high Q-factor of an EPR resonator. When EPR is compared with mid-infrared LMR, an additional factor in favour of EPR is the dipole transition matrix element, which is an order of magnitude higher for rotational transitions in radicals than that for vibrational–rotational transitions. In general, LMR is much more sensitive than EPR.

The sensitivities of far-infrared intracavity LMR and mid-infrared Faraday LMR are as high as that of laser-induced fluorescence (LIF); the sensitivity of mid-infrared intracavity LMR is usually one order of magnitude less. These estimates, however, are very dependent on the particular radical. The sensitivity achieved in practice also depends on the refinement of the apparatus. The typical values for up-to-date spectrometers are listed in **Table 1**.

For applications in chemical kinetics, combined EPR/LMR spectrometers for both far- and mid-infrared regions have been developed. Three advantages are gained from this arrangement: (i) the versatility of the combined spectrometer, (ii) the possibility of the absolute calibration of the LMR spectrometer by comparison of the LMR and EPR signals of the same radicals, (iii) usage of commercial EPR hardware for LMR detection. In the mid-infrared, EPR and LMR make use of a common detection zone. In the far-infrared, the detection zones of EPR and LMR are placed side by side; they cannot coincide, because of the large wavelength of the far-infrared laser.

Applications of LMR

Spectroscopy studies

LMR has allowed the spectroscopic study of many free radicals that could not be detected by more conventional infrared techniques.

Table 1 Typical sensitivities of LMR and EPR

Radical	*Laser*	*Wavelength* (μm)	*Sensitivity* (cm^{-3})
Cl	CO_2	11.3	2(9)[a]
O	CH_3OD	145.7	1(10)
Hg	CO	5.67	5(9)
NO	CO	5.33	1(7)
	CH_3OH	1224	1(11)
OH	H_2O	118.6	1(6)
	D_2O	84.3	2(7)
	EPR	30 000	1(10)
O_2	CH_3OH	699.5	5(10)
	EPR	30 000	3(13)
ClO	CD_3I	556.9	2(8)
	H_2O	118.6	4(8)
HO_2	CO_2	10	1(10)
	EPR	30 000	1(13)
NO_2	H_2O	118.6	1(11)
CH_2	$^{13}CH_3OH$	157.9	3(8)
NF_2	CO_2	10	1(10)
	EPR	30 000	1(14)

[a] $a(b) = a \times 10^b$.

In **Table 2** a list of the species detected by LMR to 1998 is presented. However, the pace of developments has slowed, since many of the better-known free radicals accessible for LMR have now been observed. As for lesser-known free radicals, LMR is a difficult method for obtaining initial knowledge (at least for polyatomics) because the Zeeman effect must be analysed as well as the zero-field spectrum. In the mid-infrared, the tunable semiconductor diode lasers are another factor in the development of infrared spectroscopy of high sensitivity and resolution, although LMR provides magnetic parameters inaccessible to the diode laser spectroscopy. In **Figure 3** the rise and the fall of LMR are illustrated.

A typical spectroscopic study of a radical includes: recording of numerous LMR spectra, assignment of the observed resonances to quantum numbers, and determination of the parameters of the Hamiltonian (for both ground and vibrationally excited states in the case of vibrational–rotational spectra). Although the accuracy of these parameters (up to MHz) is somewhat lower than that obtained by microwave spectroscopy, it is not restricted to rotational transitions but includes vibrational, fine structure and hyperfine effects.

The main requirements for a successful LMR study are: (i) the atom or molecule must be paramagnetic ($S > 0$ or $L > 0$); (ii) the closeness of laser and transition frequencies (<1 cm^{-1}); (iii) usually, no more than two heavy (not hydrogen) atoms [exceptions are non-hydride triatomics which are either linear (NCO, N_3, NCN), or which have

Table 2 Free radicals detected by LMR[a]

H_2O, D_2O, HCN, InSb[b] lasers	OH, OD, CH, O_2, NO, PH, HO_2, NH_2, NO_2, HCO, PH_2
Optically pumped lasers	C, O, Si, CH, CD, CF, O_2, $O^{17}O$, $O^{18}O$, OH, ^{17}OH, OD, NH, ^{15}NH, ND, ClO, SiH, PH, PD, PO, SH, SD, NS, S_2, FeH, CoH, GeH, SeH, SeD, NiH, NiD, CH_2, CD_2, CCH, CH_2F, CH_2Br, CD_2Br, NH_2, NCO, NHD, HO_2, DO_2, FO_2, PH_2, PO_2, ClSO, FSO, FeD_2, CH_3, NH_2O, CH_3O, CH_2OH, N*, Mg*, NF*, O_2*, CO*, CH*, AsH*, CH_2*, OH^+, OD^+, HCl^+, HBr^+
CO_2 laser	Cl, FO, BrO, SH, SD, SO, SiC, NiH, AsO, SeO, NSe, CH_2, $^{13}CH_2$, NH_2, ND_2, PH_2, HO_2, DO_2, HCO, HSO, FCO, FO_2, ClO_2, NO_2, NF_2, SiH_3, SiF_3, CH_3O, Kr*, Xe*, SO*
CO laser	NO, ClO, CD, CF, MgO, SiH, SiD, PH, PD, SD, FeH, CoH, CrH, NiH, NiD, GeH, GeD, AsH, SeH, SeD, SnH, SbH, TeD, C_2H, C_2D, NCN, NCO, NH_2, NO_2, $^{15}NO_2$, N_3, HCO, DCO, HO_2, DO_2, FO_2, FeH_2, Hg*, CO*, NCO*, DCl^+, DBr^+, SD^+

[a] Electronically excited species are marked by *.
[b] Spin-flip Raman laser: InSb crystal pumped by CO laser.

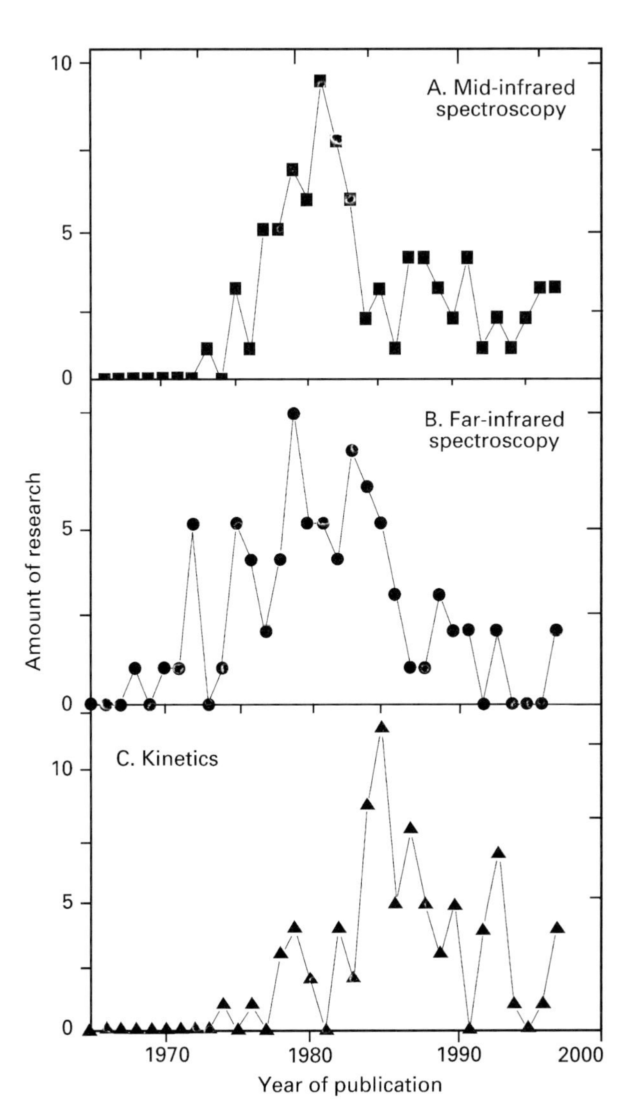

Figure 3 Amount of research (number of publications) against year of publication. C. Kinetics = chemical kinetics.

high-field LMR spectra due to level anticrossing mechanisms (FO_2, ClSO, FSO, ClO_2)]; (iv) a moderate transition dipole moment; (v) numerous resonances at numerous laser lines ought to be detected for successful extraction of the Hamiltonian parameters. If the spectra are complex and energy level predictions are not available, the data from other spectroscopic techniques are especially valid.

LMR in chemical kinetics

The combination of high sensitivity and versatility makes LMR attractive in chemical kinetics studies. Three experimental arrangements of LMR are usually employed.

(i) The first are discharge-flow techniques in which the products of a microwave discharge and consequent chemical reactions are pumped rapidly through the detection zone of a LMR spectrometer. The kinetics are obtained by varying the distance between the radical injector and the detection zone.
(ii) The second is time-resolved LMR in which radicals are produced by either UV photolysis of multiphoton dissociation of a suitable precursor; the LMR signal kinetics is monitored.

These experimental arrangements for kinetics studies are not specific to LMR; they are widely used with other spectroscopic methods, such as EPR, LIF, mass spectrometry, etc.

(iii) The third arrangement is specific to LMR; it is used only to study vibrational relaxation of radicals; an intracavity LMR spectrometer based on a CO_2 laser is usually employed. Radicals are prepared in a discharge-flow system; the kinetics of the LMR signal saturation by a CO_2 laser radiation field after a fast magnetic field jump are monitored. The magnetic field jump provides fast adjustment to the absorption spectral line of the radical. The saturation kinetics are exponential, the exponent decay time being

$$1/\tau = 2\sigma J_0 + \sum_i k_i[\mathrm{M}_i]$$

where σ is the vibrational–rotational transition cross section, J_0 is the photon flux density, $[M_i]$ is the concentration of the ith gas-relaxator and k_i are the vibrational relaxation rate constants. This relation is used to obtain k_i by measuring τ at different $[M_i]$.

This method is applicable only to radicals that show sufficient saturation (>10%) of a vibrational transition by the radiation field. The prerequisites for successful detection of the saturation are: radiation intensity in the cavity of the CO_2 laser of $>100\ W\ cm^{-2}$, rather a small rotational partition function ($\leq 10^3$, i.e. nonlinear radicals with low moments of inertia, or linear radicals), and moderate transition dipole moment (~0.1 D).

List of symbols

A_{so} = spin-orbit interaction constant; A_v B_v, C_v = rotational constants of radical; a, b = hyperfine constants; $\boldsymbol{B}$ = external magnetic field; d = difference of relative populations; D_v = centrifugal constant for diatomic molecule; D_v, E_v = parameters of Hamiltonian H_{ss}; E = electric field of laser radiation; f = modulation frequency; g_J, g_I = electron and nuclear g-factors; g_s, g_L, g_r = spin, orbital and rotational g-factors; $\hbar$ = Planck's constant; H_{cd} = Watson's centrifugal distortion operator; H_{hfs} = hyperfine interaction operator; H_{ld} = lambda-doubling term in Hamiltonian; H_{rot} = rotational term in Hamiltonian; H_{so} = spin-orbit term in Hamiltonian; H_{sr} = spin-rotational term in Hamiltonian; H_{ss} = spin-spin term in Hamiltonian; H_v = vibrational term in Hamiltonian; H_z = Zeeman term in Hamiltonian; $\boldsymbol{J}$, $\boldsymbol{I}$, $\boldsymbol{F}$, $\boldsymbol{L}$, $\boldsymbol{S}$, $\boldsymbol{N}$, M_F, M_J = momenta and their projections; J_0 = photon flux density; k_i = vibrational relaxation rate constant; $[M_i]$ = concentration of the ith gas-relaxator; N_a, N_b, N_c = components of $\boldsymbol{N}$; N_{min} = concentrational sensitivity; $^2P_{3/2}$, $^2P_{1/2}$ = Cl atomic states; S_a, S_b, S_c, S_z = projections of $\boldsymbol{S}$; S_p = Poynting vector; v = vibrational quantum number; w_e, x_e = parameters of vibrational Hamiltonian; γ = parameter of spin-rotational term H_{sr}; γ_{min} = sensitivity in cm^{-1} units; Δ_N, Δ_{NK}, Δ_K, δ_N, δ_K = centrifugal constants of asymmetric rotor; λ = parameter of spin-spin term H_{ss}; ε_{aa}, ε_{bb}, ε_{cc} = components of spin-rotational tensor; $\mu_0 = e\hbar/2mc$ = Bohr magneton; σ = cross section of the transition; τ = decay time for the saturation kinetics.

See also: **Atomic Absorption, Theory; Electromagnetic Radiation; EPR, Methods; EPR Spectroscopy, Applications in Chemistry; EPR Spectroscopy, Theory; Far-IR Spectroscopy, Applications; High Resolution IR Spectroscopy (Gas Phase) Instrumentation; IR Spectroscopy, Theory; Laser Spectroscopy Theory; Near-IR Spectrometers; Rotational Spectroscopy, Theory; Spectroscopy of Ions; Zeeman and Stark Methods in Spectroscopy, Applications; Zeeman and Stark Methods in Spectroscopy, Instrumentation.**

Further reading

Davies PB (1981) Laser magnetic resonance spectroscopy. *Journal of Physical Chemistry* 85: 2599–2607.

Evenson KM (1981) Far-infrared laser magnetic resonance. *Faraday Discussions of the Chemical Society* 71: 7–14.

Evenson KM, Saykally RJ, Jennings DA, Curl RF and Brown JM (1980) In: Moore CB (ed) *Chemical and Biochemical Applications of Lasers*, Vol 5, p 95. New York: Academic Press.

Hills GW (1984) *Magnetic Resonance Review* 9: 15–64.

Russell DK (1983) *Laser Magnetic Resonance Spectroscopy Electron Spin Resonance: A Specialist Periodical Report the Royal Society of Chemistry*, Vol 8, pp 1–30. London: Royal Society of Chemistry, Burlington House.

Russell DK (1991) *Specialist Periodical Report of the Royal Society of Chemistry* 12B: 64–98.

Laser Microprobe Mass Spectrometers

Luc Van Vaeck and **Freddy Adams**, University of Antwerpen (UIA), Wilrijk, Belgium

MASS SPECTROMETRY
Methods and instrumentation

Introduction

The defining attribute of laser microprobe mass spectrometry (LMMS) is the use of a focused laser to irradiate a 5–10 μm spot of a solid sample at a power density above 10^6 W cm^{-2}. The photon–solid interaction yields ions which are mass analysed by time-of-flight (TOF) or Fourier transform (FT) MS. The technique is sometimes referred to as laser probe microanalysis (LPA or LPMA), laser ionization mass analysis (LIMA) and laser microprobe mass analysis (LAMMA).

The use of lasers to ionize solids in mass spectrometry dates back to the early 1960s, when the first high-power pulsed lasers became available. These highly directional and intense monochromatic beams can be easily introduced in the confined space of ion sources without disturbing the electrical fields. Moreover, upon the photon interaction with nonconducting materials, no charging of the sample occurs. First, lasers became an interesting alternative for spark-source MS to quantify elements in dielectrical specimens. Later, the ultrafast heating of the solid by laser irradiation was exploited for the desorption and ionization (DI) of labile organic compounds without thermal decomposition. The growing interest in microanalysis triggered the application of a focused laser in the 1970s. The initial aim was again elemental analysis in nonconducting samples but the main strength of LMMS was found to be the molecule-specific information on the local constituents. Organic and inorganic compounds are characterized by a combination of fragments and adduct ions. The former arise from the structure-specific breakdown of the analyte. The latter simply consist of the intact analyte attached to one or more of these stable fragments.

The understanding of material properties on a local level implies knowledge of the chemistry and, hence, of the molecules present. Therefore the molecular specificity of LMMS is a major advantage in comparison with most micro- and surface analysis techniques, which characterize relative elemental abundances, bonds present or functional groups. Quantitation in LMMS is difficult because adequate reference materials are not always available. However, qualitative identification of local constituents can often be achieved by deductive reasoning only. As a result, LMMS became appreciated as a versatile tool in diverse problem-solving applications in science and technology. The main sample requirement is its stability in the vacuum.

Instrumentation

The simple construction and operation, excellent transmission, and the registration of full mass spectra made a TOF analyser the obvious choice for the early LMMS instruments. Later it was found that the time definition of the ion production from the laser pulse was less suited to TOF MS than initially assumed. Also, the complex mass spectra required a significantly higher mass resolution and mass accuracy than that available in TOF LMMS. Hence, FT LMMS was developed. Although the laser microbeam ionization was retained, the species detected and their relative intensities differ significantly from TOF LMMS. The reason is that each MS may capture a different fraction of the initial ion population, depending on its acceptance with respect to the initial kinetic energy (E_{kin}), their time and place of formation in the ion source. Therefore, a clear distinction between TOF and FT LMMS is mandatory.

TOF LMMS

Figure 1 depicts a commercial instrument. The sample is mounted inside the vacuum chamber of the MS. The DI is commonly performed by 266 nm UV pulses from a frequency-quadrupled Nd:YAG (neodymium–yttrium–aluminium–garnet) laser. This instrument allows repositioning of the sample and the optics for analysis in transmission or in reflection. In the former case, the laser hits the lower surface of the sample while the upper surface faces the MS. This suits thin films or particles of about 1 μm on a polymer film. Reflection means that the beam impinges on the sample side facing the MS. The surface of bulk samples can thus be characterized. Micropositioners allow one to move the spot of interest on the sample into the waist of the ionizing beam, which is visualized by a collinear He–Ne laser.

The UV-irradiated spot is typically 1–3 μm in TOF LMMS so that power densities between 10^6 and 10^{11} W cm^{-2} can be attained with mJ pulses of 10–15 ns. Refractive objectives allow spot sizes down to the diffraction limit of 0.5 μm. Reflective optics (as in **Figure 1**) allow larger working distances and are free from chromatic aberrations so that use of other wavelengths is facilitated. Ionization with a tunable dye laser allows resonant one-step DI or postionization of the laser ablated neutrals. Specificity and sensitivity can thus be improved substantially but wavelength selection becomes cumbersome for unknown compounds.

The principles of a TOF mass analyser are covered in another article. Commercial TOF LMMS permits a mass resolution of about 500, a much lower value than theory predicts, and this figure strongly depends on the analyte. The reason is quite fundamental. The application of TOF implies that ions with different *m*/*z* and therefore different velocities must arrive at the same time at the entrance to the drift tube. The pulsed laser ionization does not meet this requirement as ion formation may continue long after the laser pulse, especially for organic compounds but also for inorganic analytes. As a result, mass resolution and mass accuracy may become problematic. The asset of an unlimited *m*/*z* range is cancelled by the low mass resolution. A major advantage is the inherent panoramic registration.

FT LMMS

Figure 2 illustrates an instrument with an external ion source and shows the different microprobe related devices such as the optical interface, the micropositioners, the sample observation and the exchange system. This set-up features laser irradiation of the sample in reflection with a spot of 5 μm. Electrostatic fields transport ions through a differentially pumped transfer line from the source at 10^{-6} torr to the FT MS cell at 10^{-10} torr inside a 4.7 tesla magnetic field (*B*). Inherent advantages of FT MS are the routinely obtainable high mass resolution of over 100 000 at *m*/*z* 1000, and up to a few million below *m*/*z* 100, and the mass accuracy of better than 1 ppm up *m*/*z* 1000. Internal calibration of the *m*/*z* scale by adding a reference compound is not necessary as is the case in magnetic high–resolution mass spectrometers. A remarkable feature of FT MS is that better mass resolution means better sensitivity. This contrasts with other types of MS where resolution is increased by selecting a central fraction of the ion beam, thereby sacrificing sensitivity. The direct link of sensitivity and mass resolution in FT MS results from the influence of space charge effects, field imperfections and pressure in the cell. All these cause a faster decay of the coherence in the ion packets after excitation, which decreases mass resolution, but they also limit the quantitative trapping and/or the radius increase of the ion orbit during excitation, which diminish detection sensitivity. FT MS is also limited in the

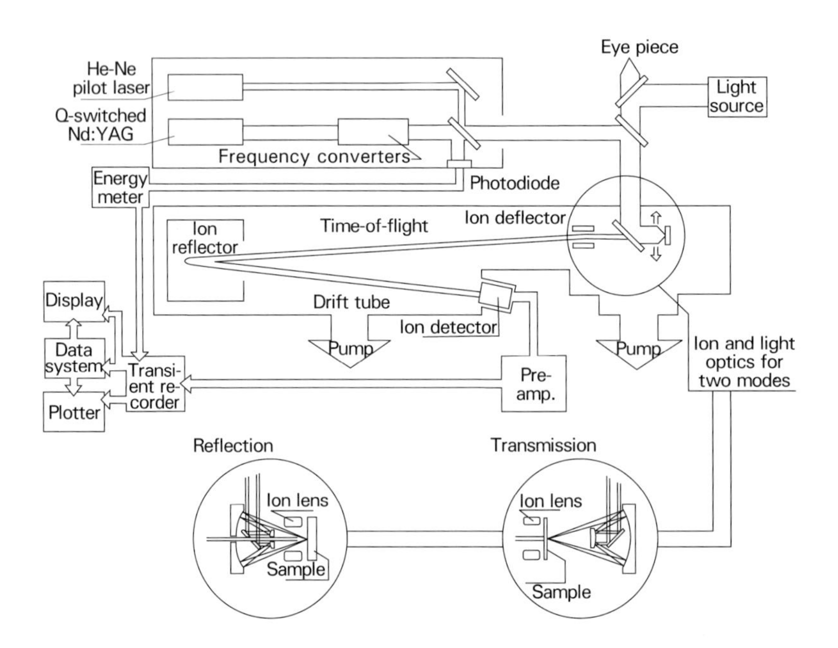

Figure 1 Schematic diagram of the LIMA 2A TOF LMMS. Reprinted from the LIMA technical documentation with permission of Kratos Analytical.

detection of low and high m/z by its detection bandwidth and B, respectively. A range between m/z 15 and 10 000 is common.

The use of FT MS in microanalysis implies that the technique is operated close to its detection limits. Therefore, the vacuum in the cell becomes a prime reason for using an external ion source be used for LMMS as opposed to a single- or dual-cell instrument, where the sample is placed in the FT MS cell. Additionally, the E_{kin} of laser ions can exceed the energy range imposed by the optimal trapping potentials of typically less than 2 V. Biasing the sample holder or placing additional electrodes around the sample inside the cell disturbs the trapping field. Cooling of the ions by ion–molecule interactions is not always efficient. The use of micropositioners, laser and observation optics inside the narrow bore of the strong magnetic field, is not obvious. However, injection of external ions requires temporary deactivation of the potential gradient in the hole of the first trapping plate. Upon reflection against the second trapping plate, ions come back and may escape unless the entrance is blocked. The use of an electrostatic transfer line causes a time dispersion of ions of different m/z as in TOF MS, i.e. the so-called TOF effect. To trap a high m/z ion together with a low m/z ion, the high m/z ion, which arrives later, must enter the cell before the low m/z ion (which arrived earlier and has been reflected by the second trapping plate) leaves the cell. In practice, the m/z ratio of high-to-low m/z ions that can be trapped simultaneously in the cell is about a factor of 3–4, depending on the time of ion formation and life time of the specific ions. Adapting the time between the laser pulse and the closure of the injection lenses, allows one to shift the m/z window along the m/z range in successive experiments.

Analytical characteristics

Table 1 surveys the common techniques for micro- and surface analysis of solids. Both static secondary ion mass spectrometry (SSIMS) and LMMS provide molecular information on local organic and inorganic compounds. However, the primary interaction of keV ions with the sample in SSIMS as opposed to eV-range photons in LMMS makes the direct structural linkage of the detected signals to the sample composition less obvious in SSIMS. Hence, comparison with reference spectra is usually required in SSIMS, while LMMS allows a deductive interpretation. In fact, both methods are complementary with respect to their typical applications. SSIMS can detect high m/z ions from polymers, while LMMS yields more detailed structural information on analytes of up to a few kDa. SSIMS generates primarily ions from the upper monolayer, making surface contamination a problem in real–life applications. In contrast, ions in LMMS originate from the upper 10–50 nm surface layer, although the crater depth goes up to 0.1–1 μm. SSIMS allows imaging, while LMMS performs spot analysis.

The reproducibility of LMMS strongly depends on the positioning of the sample surface in the waist of the UV beam. Hence, mapping by motorizing the

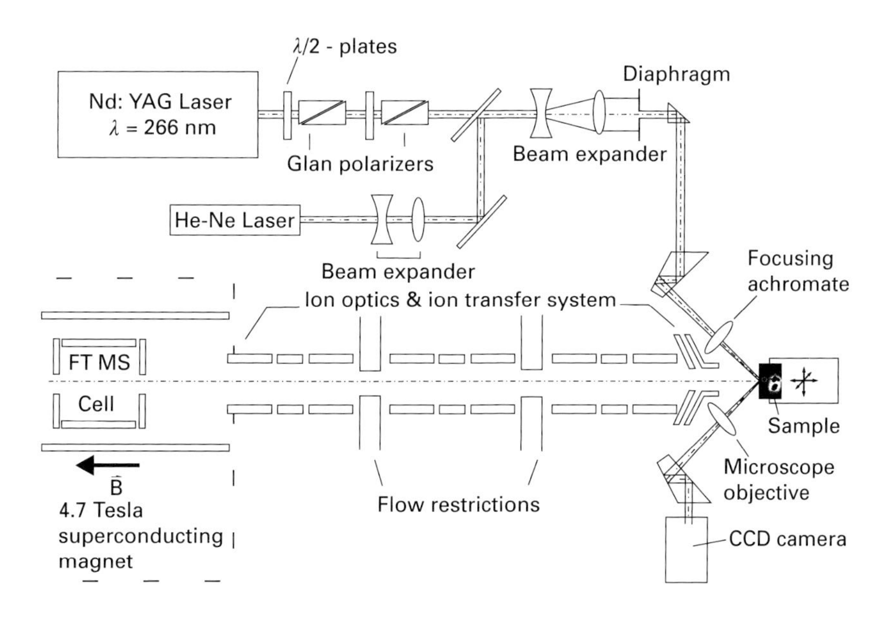

Figure 2 Schematic diagram of the FT LMMS with external ion source. Adapted from Struyf H, Van Roy W, Van Vaeck L, Van Grieken R, Gijbels R and Caravatti P (1993) *Analytica Chimica Acta*, **283**: 139–151 with permission of Elsevier Science.

Table 1 Overview of some important microanalytical techniques

						Dynamic SIMS		*Static SIMS*	*LMMS*	
	EPXMA	*Auger*	*ESCA*	*Raman*	*FT-IR*	*Stigmatic*	*Scanning*	*TOF*	*TOF*	*FT*
Probe input beam	20 keV electrons	Electrons ≤ 3 keV	Photons X-ray, UV	Photons visible	Photons IR	≤ 20 keV ions	40–60 keV ions	20 keV	UV photons (few eV)	
Detected beam	X-rays	Electrons	Electrons	Photons	Photons	+/– ions		+/– ions	+/– ions	
Parameter	WDS, EDS	Energy	Energy	Wavelength	Wavelength	*m/z*		*m/z*	*m/z*	
Resolution of detector	WDS 20 eV EDS 150 eV	1–15 eV	0.3–1 eV	0.7 cm^{-1} (sp) 8 cm^{-1} (im)	0.7 cm^{-1} (IR) 2 cm^{-1} (μIR)	400–10^4	250–500	10^3–10^4	≤850	10^4–10^6
Typically analysed area	±1 μm	0.2 μm	≥150 μm ≥5 μm (μESCA)	1 μm	5–10 μm	2–250 μm	≥ 20 nm	0.1 μm	1–3 μm	5 μm
Depth of information	≤1 μm	1–3 nm	1–10 nm	10 μm 5–50 nm (PAS)	10 μm	0.5 nm	1 nm	Monolayer	(0.1–1 μm)	n.a.
Image resolution	SEM/X ±1 μm STEM/X< μm	50–100 nm	No Yes (μESCA)	μm	n.a. 1 mm (μIR)	0.5 mm	20 nm	0.5 μm	1 μm	>1 μm
Detection limit	WDS 100 ppm EDS 1000 ppm	≥ 1 %	≥ 1 %	Major	ppm	≤ ppm	10–100 ppm	Monolayer	10^{-3}–10^{-15}	10^{-11}–10^{-12}g
Detection range	WDS $Z \geq 4$ EDS $Z \geq 11$[a]	All but H, He	All but H, He	n.a.	n.a.	H–U		H-unlimited	H-unlimited	15–15 000
Direct isotope information	No	No	No	No	No	Yes		Yes	Yes	
Compound speciation	No	No	Yes	Yes	Yes	No		Yes	Yes	
Organic characterization	No	No	No	Yes	Yes	Yes		Yes	Yes	
Destructive	(No)	No	No	No	No	Yes		Yes	Yes	
with sputtering	n.a.	Yes	Yes	n.a.	n.a.	n.a.		n.a.	n.a.	
In-depth profiling	No	No	No	No	No	Yes		No	Yes	
With sputtering	n.a.	Yes	Yes	n.a.	n.a.	n.a.		n.a.	n.a.	
Quantification	Yes	Yes	Yes	(Yes)	Yes	Difficult		Difficult	Very difficult	
Easy analysis of insulators	No	No	Yes	Yes	Yes	No		No	Yes	
Sample in vaccum requ red	Yes	Yes	Yes	No	No	Yes		Yes	Yes	

[a] EDS ranges to lower elements with windowless detector. PAS=photoacoustic single detection; WDS=wavelength dispersive spectrometry; EDS=energy dispersive spectrometry; SEM=scanning electron microscopy; STEM=scanning transmission microscopy; sp=spectrum; im=imaging mode; μIR=microscope FT-IR; μESCA=microESCA; n.a.=not applicable.

sample positioners is only feasible for extremely flat samples. As to the kind of materials analysed, LMMS requires stability in a vacuum of 10^{-6} torr but SSIMS needs a pressure of under 10^{-8} torr, which prevents the analysis of 'volatile' organic compounds. Charge build-up in dielectrical samples occurs in SSIMS, not in LMMS.

Quantification in MS requires adequate reference samples. For LMMS this means that not only the chemical composition but also the UV absorption, reflective and refractive properties of each microvolume must be comparable to ensure that the energy deposition and ion yield are similar. Hence, preparation of suitable reference materials often becomes the bottleneck. Biological sections or ambient aerosols are chemically and optically heterogeneous, so that quantification is not feasible. On the other hand, element diffusion in the wafers from semiconductor applications can be quantitatively studied.

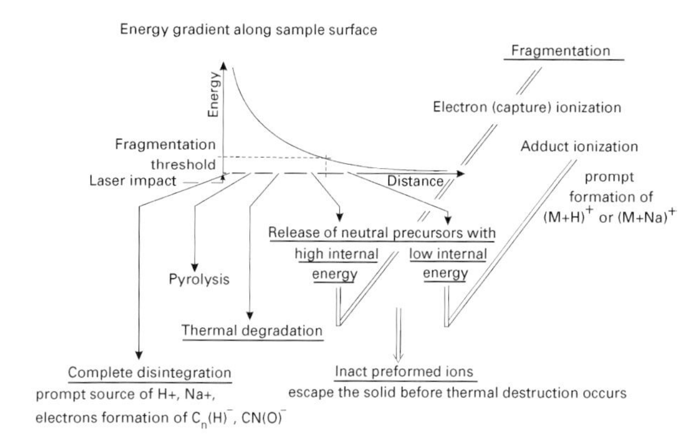

Figure 3 Relationship between the energy regimes at the surface and the ion formation process according to the tentative model for DI in LMMS. Reprinted from Van Vaeck L, Struyf H, Van Roy W and Adams F (1994) *Mass Spectrometry Reviews*, **13:** 189–208 with permission of John Wiley and Sons.

Diagnostic use of the mass spectra

Model of ion formation

Deduction of the analyte composition from the signals requires that one must be able to extrapolate the mass spectra from a limited database and ideally predict the ions from a given analyte. Therefore, practical concepts about ion formation are mandatory. **Figures 3–5** survey a tentative model for DI in LMMS, applicable to both organic and inorganic compounds. Basically, the ionization event is described in terms of the three prime parameters in MS, namely local energy, pressure and time.

First of all, laser impact is assumed to create different local energy regimes in physically distinct regions within and around the irradiated spot. **Figure 3** depicts schematically the occurrence of atomization and destructive pyrolysis in hot spots, direct ejection of fragments and molecular aggregates from surrounding regions. Ultrafast thermal processes allow preservation of the molecular structure of even thermolabile analytes. The initial species will undergo fragmentation or subsequent ion–molecule interactions. Thermionic emission of e.g. alkali ions, yields the necessary flux of ions to form adducts with the initially released neutrals and ion-pairs. This occurs primarily after the laser pulse in the selvedge, i.e. the gas phase just above the sample.

Secondly, the local pressure must be considered. As shown in **Figure 4**, the released neutrals and ions give rise to a gradually expanding microcloud. Its initial density depends on the 'volatility' of the analyte, i.e. the number of species released at a given power density. The importance of ion–molecule interactions depends on the local pressure in this cloud, which determines the collision probability. In low-density regions, ions follow unimolecular behaviour and fragmentation only depends on the internal energy.

Finally, the timescale of the ionization is an essential feature of the approach since it must match the time window of detectable ions by a given mass analyser. The time evolution of the DI process is assumed to include two distinct phases. The schematic picture in **Figure 5** shows the that the ion–generation process comprises two consecutive contributions. An initial intense ion current is produced during the first 10–25 ns, compatible with detection in TOF LMMS. However, selvedge ionization may already start in that initial period but will reach its maximum later and finally continues for microseconds after the end

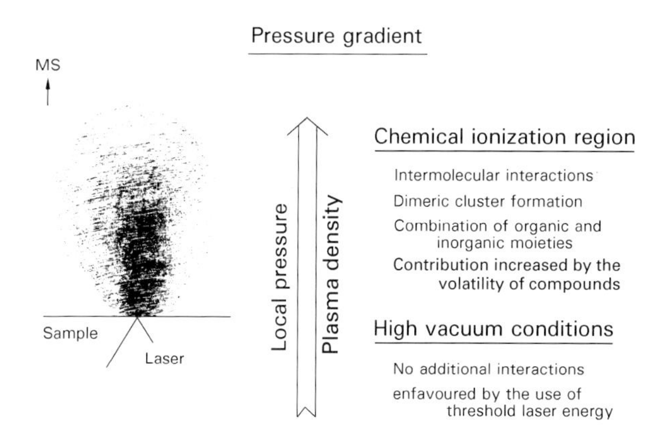

Figure 4 Effect of the local pressure on the type of ions formed according to the tentative model for DI in LMMS. Reprinted from Van Vaeck L, Struyf H, Van Roy W and Adams F (1994) *Mass Spectrometry Reviews* **13**: 189–208 with permission of John Wiley and Sons.

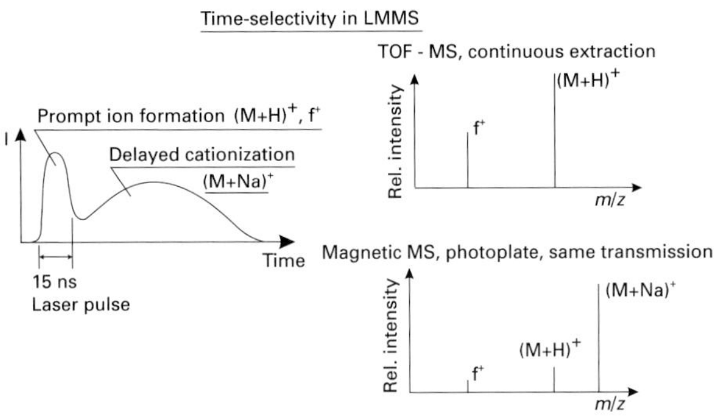

Figure 5 Influence of the time domain of ion formation and of the analyser on the mass spectra recorded according to the tentative model for DI in LMMS. Reprinted from Van Vaeck L, Struyf H, Van Roy W and Adams F (1994) *Mass Spectrometry, Reviews*, **13**: 189–208 with permission of John Wiley and Sons.

of the laser pulse. This process may be small, but integrated over time, this contribution often prevails over the prompt DI. This rather slow DI component is not compatible with detection in TOF LMMS but is inherently included in the signal when magnetic mass spectrometers or FT MS ion traps are used.

This tentative DI model does not give an accurate description of all the physical processes involved, but up to now, it allows a consistent explanation of the formation of the ions detected. Also, it unifies the processes of organic and inorganic ion formation and allows rationalization of the differences between TOF and FT LMMS spectra as to the species detected and their relative intensities. Moreover, it allows linkage of LMMS to the older literature on the soft ionization of organic compounds by defocused lasers and magnetic analysers, where fragmentation was virtually absent.

Inorganic speciation

Figure 6 illustrates how FT LMMS signals can be used for analyte identification. The low *m/z* fragments and their adducts to the original molecule readily allow tentative identification of an unknown compound by deductive reasoning alone. This represents a major advantage in e.g. industrial materials-science applications, where many not well characterized reagents are possible causes of anomalous behaviour. Recording reference spectra of all possible candidates for the given anomaly could be time consuming if one cannot select deductively the most likely one.

The most demanding speciation task is the distinction between analytes with the same elements but in different ratios, e.g. sodium sulfate, sulfite and thiosulfate. The positive fragments Na^+, Na_2O, $Na_2O{\cdot}H^+$ and $Na_2O{\cdot}Na^+$ as well as the signals from $Na_2SO_3{\cdot}Na^+$ and $Na_2SO_4{\cdot}Na^+$ are observed by FT LMMS for all three analogues. However, only thiosulfate produces additional peaks for cationized Na_2S, which characterize the presence of the

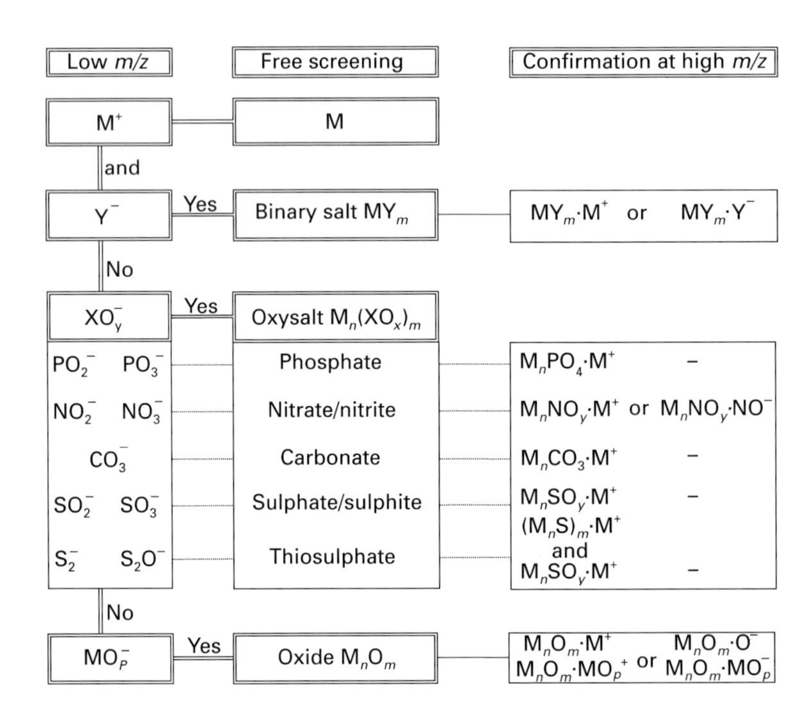

Figure 6 Scheme for the deductive identification of the analyte molecular composition from mass spectra taken by FT LMMS. Reprinted from Struyf H, Van Vaeck L, Poels K and Van Grieken R (1998) *Journal of the American Society for Mass Spectrometry* **9**: 482–497 with permission of Elsevier Science.

sodium–sulfur bond in this analogue. Adduct ions of $Na_2S_2O_3$ are also seen at low intensity. Distinction can be based on the ratio of the ions $Na_2SO_3{\cdot}Na^+$ to $Na_2SO_4{\cdot}Na^+$, which is 0.82 ± 0.13 for sulfite and 0.30 ± 0.06 for sulfate. The production of cationized sulfate from sulfite and vice versa is rationalized by the relatively high-energy regime in the hot spots, where the stability of the formed ion prevails. Nevertheless, the softer regime of the periphery is assumed to produce the cationized analyte and thereby restores the logical connection with the original composition. The major signals in the negative mode are found at the nominal *m/z* values of 64 and 80 and are common for all three analogues. However, these signals are uniquely due to SO_2^- and SO_3^- for sulfite and sulfate, while thiosulfate produces additional contributions from S_2^- and S_2O^- also at *m/z* 64 and 80 respectively. The FT LMMS high mass resolution is needed to exploit such distinctive features.

Identification of organic compounds

Deductive identification is even more important in organic microanalysis because of the numerous structures for each molecular weight. **Figure** 7 shows the positive and negative TOF LMMS data of a microscopic residue, obtained from the eluate of a single peak from analytical HPLC. The nanogram amount of available material was not consumed significantly after microprobe analysis although it was insufficient for conventional MS. The latter required the combination of fractions from numerous elutions to yield the mass spectrum of miconazole, the parent drug. In contrast, TOF LMMS detected the protonated molecules by the cluster of isotope peaks around *m/z* 503. Subsequent loss of HCOOH yields the fragments around *m/z* 457, while the accompanying lower *m/z* signals are readily rationalized by the known ring formation and rearrangement mechanisms in organic MS. Complementary information comes with the negative ions. Specifically the fragment at *m/z* 153 serves to specify further the presence of the imidazole methacrylate function in the molecule. This example highlights two features of the LMMS technique, namely the minute material consumption and the deductive identification of labile compounds without thermal destruction as a result of the ultrafast heating rate of the solid.

Comparison of mass spectra recorded by TOF LMMS and FT LMMS

Several characteristic differences are observed between the mass spectra, recorded by FT LMMS and TOF LMMS from the same analyte under the same experimental conditions with respect to laser wavelength, pulse duration and power density. While FT LMMS offers superior mass resolution and mass accuracy in comparison to TOF LMMS, the TOF effect in FT LMMS with an external source (see earlier sections on TOF LMMS and FT LMMS) means that only partial mass spectra in given *m/z* windows can be recorded (TOF LMMS always yields mass spectra that cover the entire *m/z* range). As well as these directly instrument-related data between FT LMMS and TOF LMMS, the mass spectral patterns still reflect additional distinctive features. Specifically, more intense signals from the adduct ions in comparison to the summed intensities of the fragments are seen in FT LMMS as compared to TOF LMMS. This is related to the increased sampling of the selvedge contribution to the initial ion population, created by the laser impact. Also the existence of the two contributions, i.e. direct ion emission from the solid and the selvedge recombination in the gas phase, could be observed by selectively tuning the ion source in FT LMMS so that only one of the contributions had initial energies within the 1 eV trapping energy of the cell. Note that this increased adduct ion detection in FT LMMS applies to both organic and inorganic compounds. For instance, cation and anion adducts of molecules and even dimers are generally detected in FT LMMS from virtually all salts such as Na_2SO_4, Na_3PO_4, oxides and binary salts, as opposed to TOF LMMS. Additionally, FT LMMS shows increased signal intensities specifically for fragment ions associated with the fragmentation of adduct ions with low internal energy. Such ions are observed to undergo the loss of small neutral molecules, while ions with high internal energy are subject to more drastic cleavages and rearrangement, which may reduce structural specificity. Finally, TOF LMMS is often handicapped in the analysis of high molecular mass and polar substances. These compounds are hard to desorb from the solid without the aid of matrix-assisted techniques, that require dissolution of the sample in a UV-absorbing matrix and, therefore, are incompatible with the direct local analysis of as-received solid samples. Especially for such strongly adsorbed compounds, the slow release of the analyte over a long time interval (up to several microseconds) occurs (continuing postlaser desorption). The adducts formed upon recombination with for instance Na^+ (also thermionically emitted for long periods after the laser pulse) no longer fulfil the TOF LMMS requirement of ion formation within 25 ns. As a result, these ions give rise to broad unresolved peaks or even to a continuous background. However, in FT LMMS, ions formed within a period of 50 μs to several hundreds of μs (depending on the *m/z*) after the laser pulse are still trapped in the FT LMMS cell

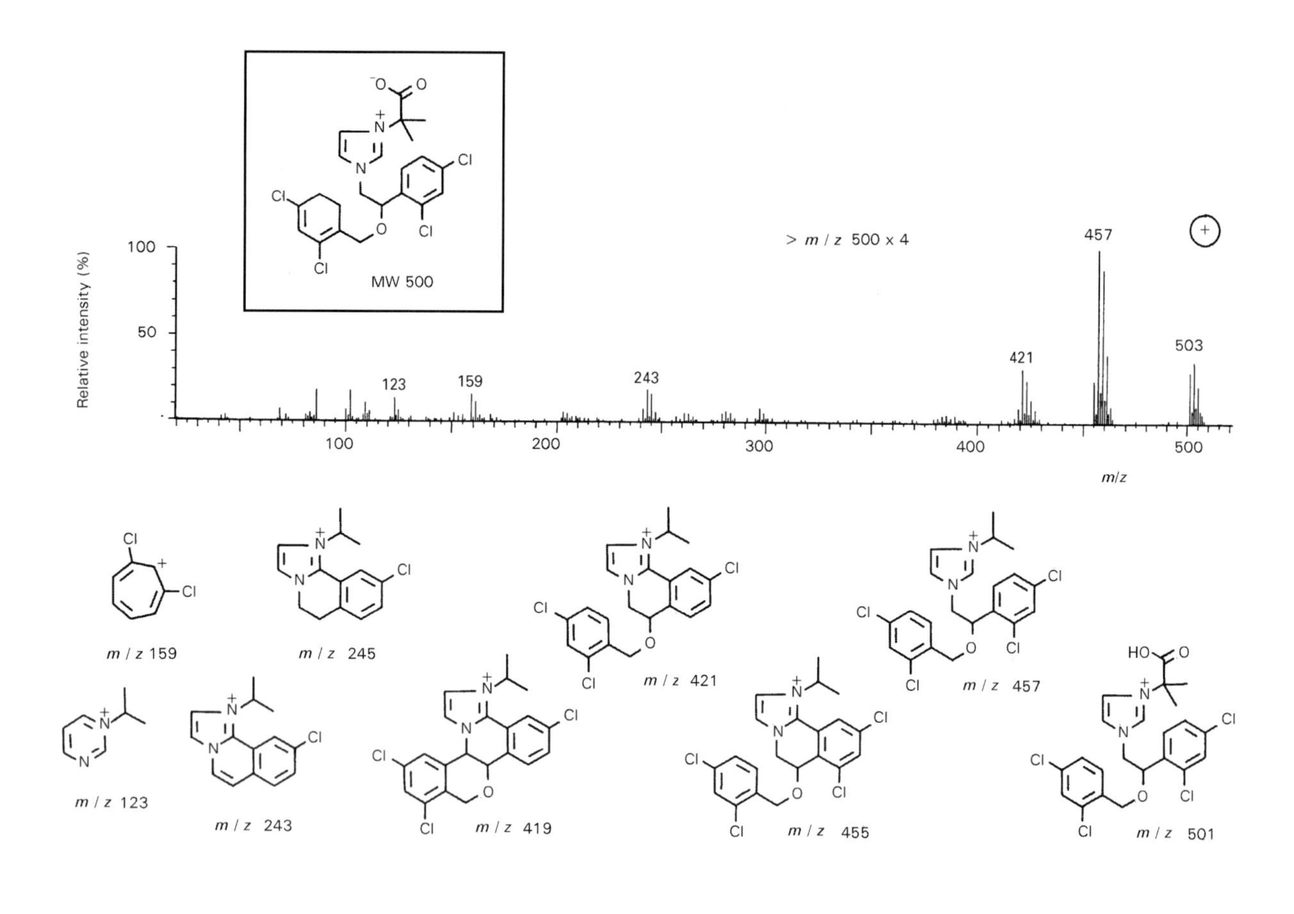

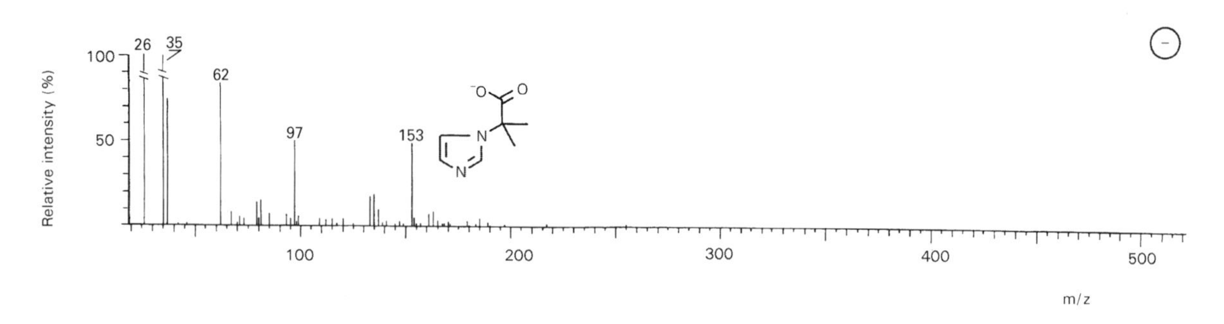

Figure 7 Positive and negative mass spectrum recorded by TOF LMMS from the residue of the eluate of a single peak in analytical HPLC. Reprinted from Van Vaeck L and Lauwers W (1989), *Advances in Mass Spectrometry*, **11a**: 348–349 with permission of Heyden and Son.

together with the ions formed during the laser pulse. As a result, the rang e of compounds to which FT LMMS can be applied, significantly extends towards higher molecular mass and polarity as compared to TOF LMMS.

Selected applications

Review of the literature reveals that LMMS is applied to a wide variety of problems in the field of bioscience, environmental chemistry and materials research. Complete coverage is not feasible in this contribution, which only aims to demonstrate the strengths of the method. Specifically, LMMS excels at yielding, within a relatively short period, qualitative information on the organic or inorganic local surface components from the most diverse samples, often with negligible sample preparation.

The transmission type TOF LMMS instruments especially suits biomedical section samples as commonly used for optical and electron microscopy.

The lateral resolution of 0.5 μm allows experiments at the subcellular level. Initial work has concentrated on the localization of elements in specific sites of the tissue. Typical examples involve the assessment of aluminium levels in the bones of chronic haemodialysis patients, the study of lead accumulation in kidneys and calcifications in intraperitoneal soft tissue as a result of chronic lead intoxication, and the detection of several heavy metals in the amalgam tattoos of the oral mucose membrane and human gingiva in direct contact with dental alloys. The relatively low lateral resolution, the lack of automated mapping and the almost impossible quantization strongly hinder the use of LMMS, especially in view of analytical electron microscopy (AEM) with X-ray analysis and the emerging possibilities of nuclear microscopy.

As a result, experiments have become gradually directed towards applications where the direct extraction of molecular information from the biological sector could be exploited. Interpretation of the mass spectra in TOF LMMS with low mass resolution may become problematic when the local microvolume constitutes of a complex multicomponent system. As a result, applications were especially successful when LMMS was used to characterize the micrometre-size microliths and foreign bodies, often found in histological sections. In this way, the spheroliths in the Bowman's membrane of patients suffering from primary atypical bandkeratopathy and the intrarenal microliths, formed after the administration of high doses of cyclosporin were proven to consist of hydroxyapatite. Implants often lead to long-term problems by dispersion of wear particles in tissues, leaching of specific compounds and sometimes subsequent chemical transformation. The pathogenesis of the aseptic loosening of joint prostheses was related to the presence of zirconium oxide in the granular foreign bodies of surrounding tissue. This compound was added to the bone cement to enhance the contrast in later X-ray radiographs.

Examples of characterization of organic molecules include the detection of deposits from an anti-leprosy drug in the spleen of treated mice. The study of biomedical sections is not the exclusive field of TOF LMMS, although the lower lateral resolution of FT LMMS limits the number of applications. However, the much better specificity from the superior mass resolution and mass accuracy pays off with increasing complexity of the local composition. FT LMMS is particularly appreciated because of the separation of isobaric ions in the tissue, such as CaO^+ and Fe^+ ions, the facile assignment of a signal as an 'organic' ion or an 'inorganic' cluster on the basis of accurate *m*/*z* values, and the increased contribution of ions from the selvedge, which decreases the influence of the local laser power density on the mass spectra. Practical applications widen the range, initiated by TOF LMMS. For instance, the foreign bodies in the inflammed tissue around implants were identified as wear particles from a titanium knee-implant. Other experiments involved the verification of the local molecular composition deduced from relative element abundances in AEM. The apoptotic cell death in the tissue around vein grafts was associated with the presence of hydroxyapatite because of the intense Ca and P signals in AEM. FT LMMS clearly showed the erroneous nature of this conclusion since both elements did not belong to the same molecule. This clearly illustrates the importance of molecular information in microanalysis.

Figure 8 illustrates the application of FT LMMS for the *in situ* identification of pigments in a lichen, *Haemmatomma ventosum*. The interest in such natural products relates to the understanding of their role in the interception of sunlight and their possible interaction in energy transfer. The optical micrograph shows the dark-red dish-like fruiting bodies or apothecia on the light coloured thallus. The combination of positive and negative ion mass spectra taken from the apothecia gives quite a detailed picture of the molecule. Specifically, the molecular mass is available from the potassium adduct and the numerous fragments serve to deduce structural features. The material is analysed as it is found in nature, without any prior sample preparation.

As to materials research, diverse problem-solving examples of LMMS are described. Typical examples involve the identification of local heterogeneities in poorly dispersed rubbers as one of the ingredients of the formulation, the tracing of the origin of occasional organic and inorganic contaminants at the surface of microelectronic devices and the study of segregation and formation of specific compounds in the joints of welded or heat-treated oxide-dispersion-strengthened alloys. Particularly interesting is the study of the dispersion of the magnetic elements inside the polyethylene terephthalate matrix at the surface of faulty floppy discs. Here the capability to detect both inorganic ions and organic structural fragments in the same spectrum is essential for trouble shooting.

A final example illustrates how LMMS can contribute to the fundamental understanding of processing of materials. Specific ally, aluminium strips and plates are produced by hot rolling the primary alloy ingots under high pressure and temperature. The lubricating emulsions contain numerous additives, whose composition is empirically optimized. Sometimes, given constituents adhere or interact with the

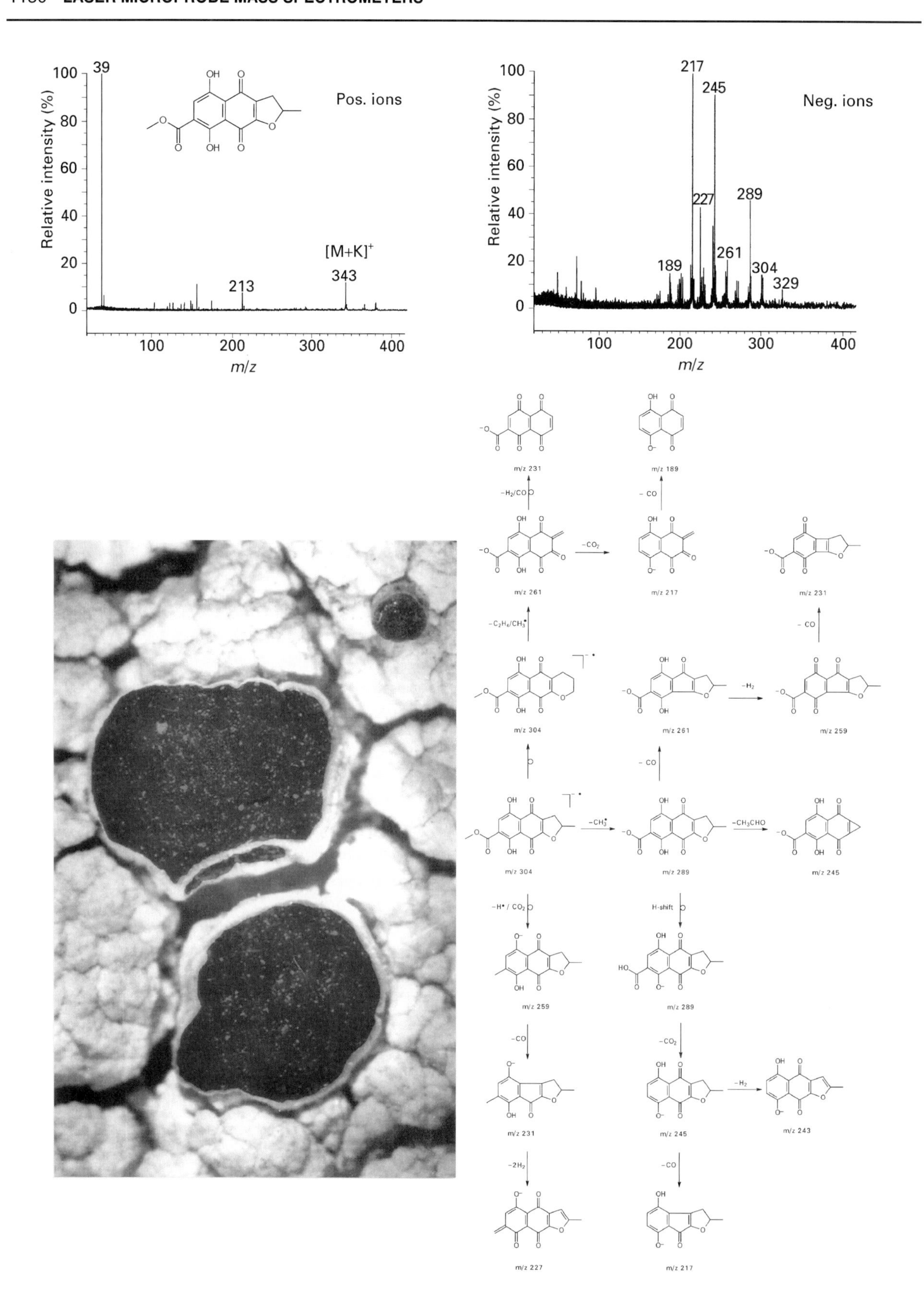

Figure 8 *In situ* FT LMMS analysis of a pigment in the apothecia of a microlichen. Reprinted from Van Roy W, Matthey A and Van Vaeck L (1996) *Rapid Communications in Mass Spectrometry* **10**: 562–572 with permission of John Wiley and Sons.

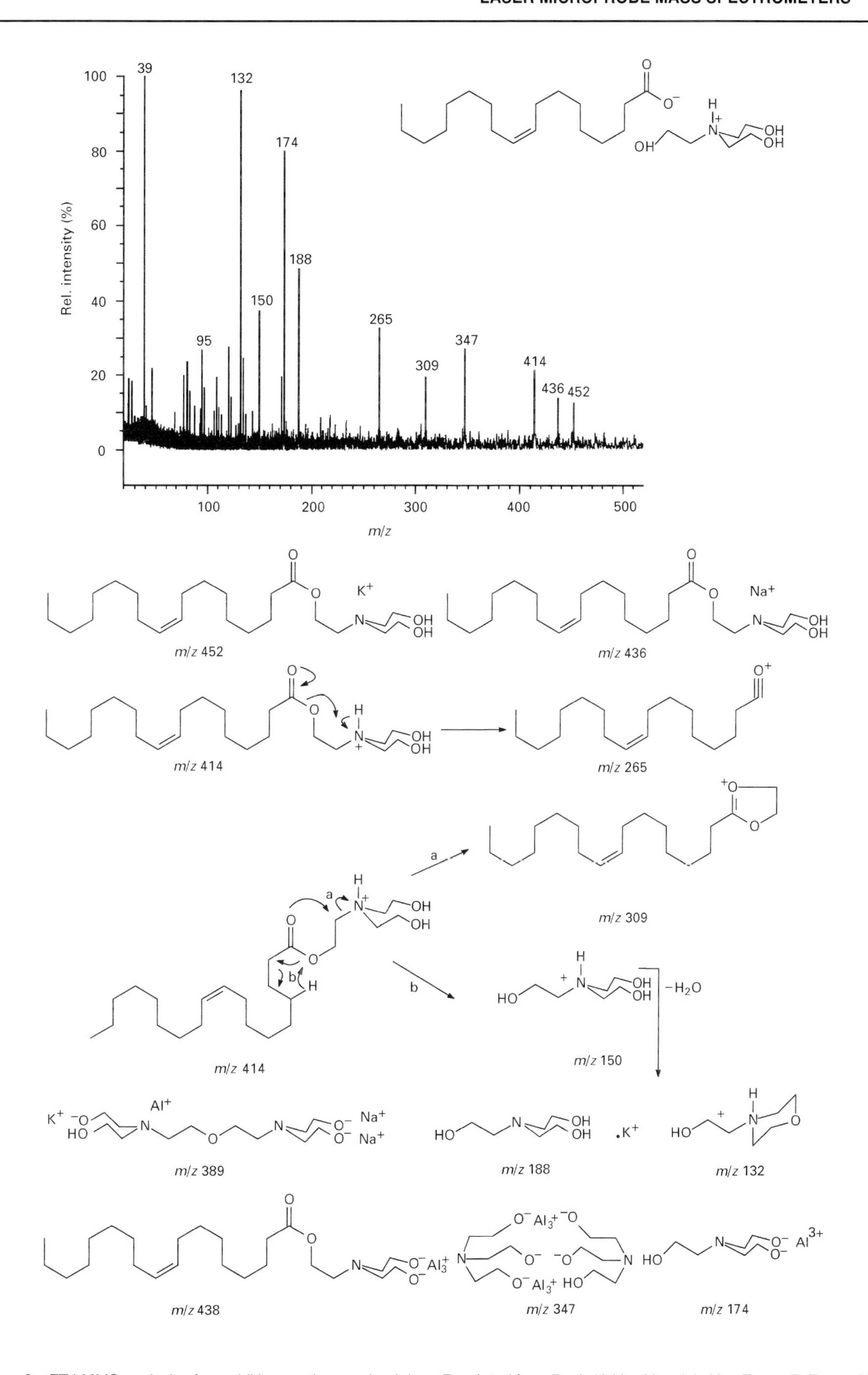

Figure 9 FT LMMS analysis of an additive coating on aluminium. Reprinted from Poels K, Van Vaeck L, Van Espen P, Terryn H and Adams F (1996) *Rapid Communications in Mass Spectrometry* **10**: 1351–1360 with permission of John Wiley and Sons.

surface and cause optical defects upon subsequent coating or surface treatment. FT LMMS was used to study the possible additive–metal interactions. The data in **Figure 9** for triethanolamine (TEA) oleate on aluminium show the normally expected simple adduct and fragments, as well as particularly interesting signals due to ions, which include multiply charged anions with Al^{3+}. Since only singly charged aluminium is formed in laser microbeam DI, these Al^{3+}-containing ions must exist as such in the solid and undergo direct ejection. This implies that TEA oleate is likely to bind effectively to the metal, while this additive is often considered as a chemically non-aggressive surfactant in the lubricating emulsions.

To conclude, it should be mentioned that one of the major breakthroughs in the field of organic MS emerged from the research on the initial TOF LMMS instruments. Indeed, the now booming field of matrix assisted laser desorption ionization (MALDI) for the characterization of high molecular mass compounds up to 230 kDa found its roots in the laboratory of Prof. Hillenkamp, which was one of the driving forces behind the development of the initial TOF LMMS instruments. It shows the ingenuity of chemists in exploiting the powerful combination of laser ionization and MS. As to local analysis, the second generation of LMMS, FT LMMS seems to represent a major step in the search for a versatile microprobe, enabling us to characterize the molecular composition of organic and inorganic compounds at the surface of almost any type of solid, electrically conducting or not, with minimal sample preparation.

List of symbols

E_{kin} = initial kinetic energy of ions.

See also: **FT-Raman Spectroscopy, Applications; Time of Flight Mass Spectrometers.**

Further reading

Eeckhaoudt S, Van Vaeck L, Gijbels R and Van Grieken R (1994) Laser microprobe mass spectrometry in biology and biomedicine. *Scanning Electron Microscopy supplement* 8: 335–358.

Poels K, Van Vaeck L, Van Espen P, Terryn H and Adams F (1996) Feasibility of Fourier transform laser microprobe mass spectrometry for the analysis of lubricating emulsions on rolled aluminium. *Rapid Communications in Mass Spectrometry* **10**: 1351–1360.

Struyf H, Van Roy W, Van Vaeck L, Van Grieken R, Gijbels R and Caravatti P (1993) A new laser microprobe Fourier transform mass spectrometer with external ion source for organic and inorganic microanalysis. *Analytica Chimica Acta* **283**: 139–151.

Struyf H, Van Vaeck L, Poels K and Van Grieken R (1998) Fourier transform laser microprobe mass spectromery for the molecular identification of inorganic compounds. *Journal of the American Society for Mass Spectrometry* **9**: 482–497.

Van Roy W, Matthey A and Van Vaeck L (1996) *In situ* analysis of lichen pigments by Fourier transform laser microprobe mass spectrometry with external ion source. *Rapid Communications in Mass Spectrometry* **10**: 562–572.

Van Vaeck L, Gijbels R and Lauwers W (1989) Laser microprobe mass spectrometry: an alternative for structural characterisation of polar and thermolabile organic compounds. *In:* Longiévalle P (ed) *Advances in Mass Spectrometry*, Vol. 11A, pp. 348–349. London: Heyden.

Van Vaeck L, Struyf H, Van Roy W and Adams F (1994) Organic and inorganic analysis with laser microprobe mass spectrometry. Part 1: Instrumentation and methodology. *Mass Spectrometry Reviews* **13**: 189–209.

Van Vaeck L, Struyf H, Van Roy W and Adams F (1994) Organic and inorganic analysis with laser microprobe mass spectrometry. Part 2: Applications. *Mass Spectrometry Reviews* **13**: 209–232.

Verbueken A, Bruynseels F, Van Grieken R and Adams F (1988) Laser Microprobe Mass Spectrometry. *In:* Adams F, Gijbels R and Van Grieken R (eds) *Inorganic Mass Spectrometry*, pp. 173–256. New York: Wiley.

Vertes A, Gijbels R and Adams F (eds) (1993) *Laser ionisation mass analysis*. Chemical Analysis Series, Vol. 124. New York: Wiley.

Laser Spectroscopy Theory

David L Andrews, University of East Anglia, Norwich, UK

ELECTRONIC SPECTROSCOPY
Theory

The theory underlying electronic spectroscopy with lasers is essentially the theory of visible or ultraviolet photon interactions. Any spectroscopic technique based on laser instrumentation might in principle be studied with some other kind of light source, but usually at the expense of data quality in terms of signal-to-noise or resolution. The distinctive features that arise with the deployment of laser light in electronic spectroscopy are principally those that relate to or exploit the qualities of the electric field produced by the laser beam. The customarily high level of monochromaticity affords the means to obtain high-resolution data, and high field strengths offer scope for the study of multiphoton processes. Another widely vaunted attribute of the laser, the coherence of its output, is only indirectly relevant to spectroscopic applications. In the so-called 'coherence spectroscopies' such as self-induced transparency and photon echo, laser coherence is significant only in enabling short and highly intense pulses to be created.

Electric field and intensity

Laser electronic spectroscopy is primarily based on coupling between the electron clouds of individual ions, atoms, chromophores or molecules of the sample with the electric field of the impinging laser radiation. The coupling is usually of dipolar character, and its detailed involvement in photonic interactions is to be discussed below. First, it is expedient to characterize the electric field, and to relate it to the laser intensity – as the latter is more directly amenable to experimental measurement. The electric field produced by a coherent, parallel, radially symmetric and plane polarized laser beam propagating in the z-direction, with a wavelength λ and a frequency ν, oscillates sinusoidally in time and space and has an amplitude A which depends on the radial displacement r from the beam centre:

$$E(z, r, t) = A(r)\cos(kz - \omega t + \phi) \qquad [1]$$

Here $k = 2\pi/\lambda$, the 'circular frequency' $\omega = 2\pi\nu$ and ϕ is the phase. The vector character of the electric field is in general determined by the state of polarization, as discussed later: in the case of plane polarization it is directed along a unit vector $\boldsymbol{i}$ perpendicular to the z-direction of propagation. In cases where the line width of the radiation is significant the amplitude additionally has a frequency dependence and, particularly in the case of pulsed radiation, it also varies with the time t.

The beam irradiance I (the power per element of cross sectional area), to which the electric and magnetic fields of the radiation contribute equally, is in general given by

$$I(r, \nu, t) = \tfrac{1}{2}c\varepsilon_0 A^2(r, \nu, t) \qquad [2]$$

The beam thus has radial, frequency and temporal intensity profiles. The radial profile is determined by the mode structure, reflecting the pattern of standing waves sustained within the laser cavity. In the simplest or ideal uniphase case, radiation tracks back and forth in an exactly axial fashion between the end-mirrors. With no hard edge, the beam is then described by an essentially Gaussian distribution, with a beam diameter $2w$ defined as the transverse distance between points at which the irradiance drops to $1/e^2$ (13.5%) of its central value, $I(0)$:

$$I(r) = I(0)\exp\left[-2(r/w)^2\right] \qquad [3]$$

Accordingly, the electric field amplitude falls away as $\exp[-(r/w)^2]$. For a beam of instantaneous power W, the beam centre irradiance $I(0)$ is given by

$$I(0) = 2W/\pi w^2 \qquad [4]$$

Fundamental considerations show that any laser beam can only be focused down to a limiting width whose diameter is of the same order as the wavelength – this is the diffraction limit. The focused beam diameter $2w_0$ is generally determined by the relation

$$w_0 = 2M^2\lambda/\pi\theta \qquad [5]$$

where λ is the wavelength and θ the angle of convergence of the focused beam. The quantity M^2 is a measure of beam quality and has a diffraction-limited value of unity for a fundamental Gaussian-mode beam as represented by Equation [3].

The frequency profile of the laser beam is usually characterized by its full-width at half-maximum (FWHM), $\Delta\nu$. The single parameter that most effectively characterizes the degree of monochromaticity is then the quality factor Q, defined as the ratio of the laser emission frequency to its line width. It is also expressible as the coherence length l_c (the distance over which photons remain effectively in phase) divided by the wavelength:

$$Q = \nu/\Delta\nu = l_c/\lambda \quad [6]$$

In absorption-based laser spectroscopy this parameter represents an upper limit on the achievable resolution, which in practice may be reduced by other features of the instrumentation.

The issue of temporal profile is one that primarily relates to pulsed lasers, whose high powers result from their delivery of energy over a short period of time rather than continuously. For such devices the detailed temporal profile of the intensity $I(t)$ is determined by the means of pulsed operation. Individual pulses from giant pulse lasers seldom have the smoothly symmetric shape needed to follow any simple analytic form – and although the more closely symmetric pulses from a mode-locked train are often characterized as having Gaussian shape, that more than anything reflects their characterization by averaging over a large number of pulses. The sech^2 intensity profile widely regarded as ideal is, however, achieved in pulses from many of the newer generation of ultrafast lasers.

Even continuous-wave laser light generally exhibits temporal fluctuations as a result of imperfect coherence, and the probability of finding a single photon in the volume of space V occupied by the species of spectroscopic interest generally approximates to a Poisson distribution. With a mean number of photons M given by

$$M = \frac{IV\lambda}{hc^2} \quad [7]$$

the probability P_N of finding N photons is

$$P_N = \frac{M^N e^{-M}}{N!} \quad [8]$$

Interaction of light and matter

In formulating the general theory it is helpful first to establish the dipolar response of each ion, atom, chromophore or molecule ξ to the oscillating electric field of the radiation. This is associated with an interaction operator:

$$V(\xi) = -\mu(\xi) \cdot \boldsymbol{E}(\xi) \quad [9]$$

where $\mu(\xi)$ is the appropriate electric dipole operator and $\boldsymbol{E}(\xi)$ the electric field vector at the position in space occupied by the species ξ. Electric dipole coupling accounts for the vast majority of the observations in electronic spectroscopy, though other kinds of multipolar interaction (such as electric quadrupole or magnetic dipole) can permit otherwise 'forbidden' transitions to occur, through a coupling which is typically weaker by a factor of 10^2 or 10^3.

Even with the high laser intensities employed for the study of multiphoton processes, the electric fields they produce are seldom comparable to intramolecular coulombic fields. As such, the interaction operator [9] can be treated as a perturbation on the molecular states. The result of these perturbations is to modify the form of the Schrödinger equation so that its usual eigenfunctions no longer represent stationary states, and hence transitions occur. The quantum amplitude for a transition between a given initial state i and a final state f is given by the following result from time-dependent perturbation theory, expressed in Dirac notation:

$$M_{fi}(t,t_0) = \langle f|U_I(t,t_0)|i\rangle = \left(\frac{1}{i\hbar}\right)^n \langle f| \int_{t_0}^{t} dt_1 \times \int_{t_0}^{t_1} dt_2 \cdots \int_{t_0}^{t_{n-1}} dt_n \tilde{V}(t_1)\tilde{V}(t_2)\cdots\tilde{V}(t_n)|i\rangle \quad [10]$$

where

$$\tilde{V}(t_i) = \exp\left(\frac{iH_0(t_i - t_0)}{\hbar}\right) V \exp\left(\frac{-iH_0(t_i - t_0)}{\hbar}\right) \quad [11]$$

and H_0 is the normal, unperturbed, Schrödinger operator for the molecule. Except for measurements with the most extreme time resolution, evaluation of the quantum amplitude given in Equation [10] leads to an essentially time-independent result and yields a corresponding transition rate Γ given by Fermi's 'Golden Rule':

$$\Gamma = \frac{2\pi}{\hbar} |M_{fi}|^2 \rho_f \quad [12]$$

Here ρ_f is the number of energy levels per unit energy interval, the so-called 'density of states'. Though often overlooked, the size of this latter factor contributes significantly to the ultrafast rates of transitions observed to occur in many large molecules, as a result of their tightly packed quasi-continua of vibronic levels. With increasing n, successive orders of the perturbation series [10] represent a progressively diminishing coupling, with a scale corresponding to the ratio of the radiative electric field in relation to intramolecular coulombic fields. Together, Equations [1], [2] and [9]–[12] correspondingly show that each additional photon interaction typically reduces the rate by a factor of the order (I/J), where J is a level of irradiance which would produce ionization or dissociation – its value is generally in the region $10^{18\pm4}$ W m^{-2}. This figure is substantially undershot in most laser excitation experiments, and as a result the higher orders of photon interaction demand increasingly sensitive detection.

Single-photon absorption

For a process of single-photon absorption, the quantum amplitude is simply given by

$$M_{fi} = -\mu^{fi} \cdot \boldsymbol{E} \quad [13]$$

using the convenient shorthand for the vector transition dipole moment, $\mu^{fi} = \langle f | \mu | i \rangle$. The rate Γ as given by Equation [12] depends quadratically on the electric field, and hence through Equations [1] and [2] exhibits a linear dependence on the laser intensity.

It follows that, since absorption produces a diminution of the laser intensity on passage through the absorbing sample, the rate of intensity loss is proportional to its instantaneous value, $-\mathrm{d}I(t)/\mathrm{d}t \propto I(t)$. In the case of pulsed irradiation, if the amount of absorption per pulse is small and decay processes from the excited state occur on a timescale that is slow compared with the pulse duration, the net absorption per pulse is proportional to the pulse energy. More generally, as the distance z the radiation travels through the sample is proportional to time t through the speed of light, we also have $-\mathrm{d}I(z)/\mathrm{d}z \propto I(z)$. Thus the amount of absorption depends on the path length through Beer's Law:

$$I(z) = I(0)\exp(-\alpha Cz) \quad [14]$$

where α is the absorption coefficient and C the concentration of (ground state) absorbing species. The former is calculable from secondary effects pursuant on the absorption, or more directly from the relative attenuation of the excitation beam. Weak absorption leads to an intensity loss which has an almost directly linear dependence on distance, as follows from Taylor series expansion of the right-hand side of Equation [14].

Laser excitation here offers few distinctive features at the molecular level, except that at high levels of intensity the increased flux can lead to saturation. Then, a significantly high proportion of the sample molecules undergoes transition to an excited state and, as C varies through depletion of the ground state population, departures from Beer's law arise. This is a phenomenon that is exploited for analytical purposes with pulsed radiation, in concentration-modulated absorption spectroscopy.

The selection rules associated with single-photon absorption stem from its dependence on the transition dipole moment, which must be non-zero for absorption to occur. To satisfy this fundamental requirement, the direct product symmetry species of the initial and final wavefunctions must contain the symmetry species of the dipole operator – the latter transforming like the translation vectors $\boldsymbol{x}$, $\boldsymbol{y}$ and $\boldsymbol{z}$. Probably the most familiar aspect is the Laporte selection rule for centrosymmetric molecules, which allows transitions only between states of opposite parity, gerade (g) $\longleftrightarrow$ ungerade (u).

For single-photon absorption in isotropic media, there is essentially no dependence on beam polarization. Such dependence as does exist arises only for optically active (chiral) compounds, and is associated with quantum interference between electric dipole and magnetic dipole (or electric quadrupole) interactions. These weak effects produce the characteristic polarization dependence that is manifest in the phenomena of circular dichroism and optical rotation.

Multiphoton absorption

For processes involving the absorption of n photons from a single laser beam, energy conservation dictates

$$nh\nu = E_f - E_i \quad [15]$$

Multiphoton absorption of this kind is essentially a concerted rather than a multistep process, as there is no physically identifiable intermediate stage between the absorption of successive photons. It is, therefore, necessary for the absorbing molecule to be intercepted by the necessary number of photons almost simultaneously, a condition which can seldom be satisfied other than by the use of pulsed laser radiation.

Determination of the rate of n-photon absorption from the Fermi rule [12] invokes the quantum amplitude for the process as given by Equations [9]–[11]. In the case of two-photon absorption, for example, we obtain a quadratic coupling with the impinging electric field:

$$M_{fi} = -S^{\mathrm{fi}}_{\alpha\beta}(\omega) : E_\alpha E_\beta \qquad [16]$$

a tensor counterpart to the scalar product of the single-photon result shown in Equation [13], in which α and β here represent the Cartesian indices x, y and z. Equation [16] is written in this concise form (with obvious extension to higher orders) through adoption of the Einstein convention for summation over repeated tensor indices, i.e. both α and β in the above expression are implicitly summed over x, y and z. The molecular response exhibited in Equation [16] is cast in terms of a two-photon tensor whose explicit structure is as follows

$$S^{\mathrm{fi}}_{\alpha\beta}(\omega) = \sum_r \left[\frac{\langle f|\mu_\alpha|r\rangle\langle r|\mu_\beta|i\rangle}{E_r - E_i - \hbar\omega - \mathrm{i}\Gamma_r} \right] \qquad [17]$$

in which there is summation over a set of contributions from each molecular state r, of energy E_r and FWHM Lorentzian line width $2\Gamma_r$. Again the extension to higher orders follows, each additional photon interaction introducing one further transition dipole in the numerator and also an additional energy denominator term. For all but the simplest cases the exact form of these is best determined through quantum electrodynamical methods using Feynman time-ordered diagrams.

In general the n-photon quantum amplitude exhibits an nth power dependence on the electric field and accordingly the rate given by Equation [12] depends on the nth power of the laser irradiance I. The lack of coherence which is manifest as intensity fluctuations in the beam generally means that the time-average value of I^n differs from $\bar{I}^n$, where $\bar{I}$ is the mean irradiance. Consequently the result is usually cast instead in terms of $g^{(n)}\bar{I}^n$, where $g^{(n)}$ is the degree of nth order coherence – having the value of unity for a perfect Poissonian source.

In the case of pulsed radiation the extent of absorption depends on the integral $\int I^n(t)\mathrm{d}t$ and so involves not only the energy of the pulse but also the temporal profile of the latter. Equally, the radial profile of the beam has a bearing on the extent of absorption. In particular, the nonlinear dependence on intensity means that if the laser beam is focused within a medium exhibiting multiphoton absorption, most of the absorption occurs at the focus. For example a Gaussian beam focused from a beam waist w to the diffraction-limited value w_0 as given by Equation [5] generates across its focus a net absorption which is $(w/w_0)^{2n}$ times larger than across the unfocused beam. This principle is exploited in imaging techniques based on the detection of fluorescence resulting from multiphoton absorption.

Through the nonlinear intensity dependence, Beer's law condition totally fails to be satisfied in processes involving the absorption of two or more photons, and results cannot be given in terms of conventional absorption coefficients. Again, for two-photon absorption the counterpart to Equation [14] takes the form

$$I(z) = I(0)/(1 + \beta C z) \qquad [18]$$

where β determines the strength of the two-photon transition. However, the weakness of the absorption generally means that the leading terms of a Taylor series for the right-hand side of Equation [18] provide a very good approximation for the beam attenuation $\Delta I = I(0) - I(z)$, which then displays the same essentially linear dependence on distance as weak single-photon absorption.

It is evident from the form of Equation [17] that the two-photon tensor and hence the extent of two-photon absorption, is significantly enhanced if the molecule possesses an electronic excited state of an energy close to $E_i + \hbar\omega$, for then, in the sum over r, the term that state contributes has an absolute minimum value for the denominator. This situation, typical of multiphoton processes, is known as resonance enhancement, and its physical basis can be understood in terms of the time–energy uncertainty principle:

$$\Delta E \Delta t \geq \tfrac{1}{4}\hbar \qquad [19]$$

In frequency regions well removed from resonance, absorption of the first laser photon leads the molecule into a state with a large ΔE, i.e. one that is energy non-conserving. Accordingly, the molecule can exist in such a state only if further absorption restores energy conservation within a correspondingly short time. However, if the first photon is near resonance with a real molecular state, the intermediate state has a small ΔE and can persist for much longer. This greatly increases the probability that the next photon will arrive within the necessary window. In the limit where exact resonance occurs and ΔE is zero there is no longer any temporal constraint save for the decay lifetime of the resonant state, as that state can be physically populated by single-photon absorption. Then the two-photon process can be completed in a second stage, without the need to fulfil the exacting conditions for both photons to arrive at the same instant. Under resonance conditions the power law for the intensity dependence is also in general modified to reflect the greater significance of whichever stage is the rate-determining step.

Each photon interaction in a multiphoton process governed by dipole coupling is subject to the Laporte selection rules. However, the product selection rules depart substantially from the normal results; for example, if an even number of photons is involved, as in Raman scattering, two-photon absorption and four-wave mixing, then the selection rules g $\longleftrightarrow$ g and u $\longleftrightarrow$ u apply. In general, for a given transition to occur it is necessary that the molecular response tensor possesses at least some non-zero elements. The criterion for this rule to be satisfied is that the product of the irreducible representations of the molecular initial and final states must be spanned by one or more components of the tensor. The representation of the (electric dipole) molecular response tensor $T^{(n)}$, for a process of n-photon absorption from a single beam, has the following general decomposition into irreducible parts or weights n, $(n-2)$ . . .;

$$\mathcal{D}(T^{(n)}) = \mathcal{D}^{(n\varphi)} \oplus \mathcal{D}^{((n-2)\varphi)} \oplus \cdots \qquad [20]$$

where for even n the parity $\varphi = +1$ and for odd $\varphi = -1$, and the series terminates with $\mathcal{D}^{(0\varphi)}$ or $\mathcal{D}^{(1\varphi)}$ respectively. For molecules of reasonably high symmetry, the operation of the multiphoton selection rule leads to far more states being accessible through multiphoton processes than through conventional single-photon absorption.

Polarization effects

Laser beam polarization plays an obvious role in the orientational effects displayed in photoabsorption by anisotropic media such as crystals, liquid crystals and surfaces. However, strong pumping with a plane polarized laser beam can induce anisotropy in an originally isotropic sample. This results from the fact that the probability of single-photon depends on $\cos^2\theta$, where θ is the angle between the relevant molecular transition moments and the photon electric field vector, as in Equation [13]. High intensity pulsed laser radiation can thus create a preferentially oriented population of excited molecules, a process termed photoselection. In many cases rotational relaxation nonetheless destroys this anisotropy within a matter of picoseconds after the inducing laser pulse.

The electric field of laser light need not oscillate in a single plane, and often optics are employed to produce other polarizations with a degree of circularity. Circular polarizations are important in forms of laser spectroscopy which exploit angular momentum selection rules, because the photons carry unit quanta of angular momentum. With chiral substances, a small degree of sensitivity to the handedness of the radiation is also manifest in the circular differential response. For two-photon and higher-order processes, however, even the spectra of reasonably symmetrical molecules display a marked dependence on polarization.

For generality the polarization state of the laser beam can be expressed in a form that can accommodate arbitrary plane, circular or elliptical states. Instead of the $\boldsymbol{i}$ which determines the direction of the electric field in Equation [1], we then have a unit polarization vector given by

$$\boldsymbol{e} = \mathrm{e}^{\mathrm{i}\phi} \cos \sigma \hat{\boldsymbol{X}} + \mathrm{e}^{\mathrm{i}\zeta} \sin \sigma \hat{\boldsymbol{Y}} \qquad [21]$$

where

$$(\sigma, \phi, \zeta) \mapsto \begin{cases} (\pi/4, 0, \pi/2) & \text{R-circular polarization} \\ (\pi/4, 0, -\pi/2) & \text{L-circular polarization} \\ (\pi/2, 0, 0) & \text{p-polarization} \\ (0, 0, 0) & \text{s-polarization} \end{cases} \qquad [22]$$

where the latter mutually orthogonal plane polarizations are given the labels that would usually be employed in connection with irradiation of a surface with normal vector Z.

List of symbols

A = amplitude of a plane polarized laser beam; c = speed of light; C = ground state concentration;

E = electric field; E_r = energy of level r; $g^{(n)}$ = degree of nth order coherence; I = beam irradiance; J = irradiance level producing ionization or dissociation; l_c = coherence length; M = mean number of photons; M_{fi} = quantum probability amplitude; Q = quality factor; r = radial displacement from the beam centre; t = time; $T^{(n)}$ = molecular response tensor; $V(\xi)$ = interaction operator for species ξ; V = volume; w = beam radius; w_0 = focused beam radius; W = instantaneous beam power; α = absorption coefficient; ε_0 = vacuum permittivity; Γ = transition rate; θ = angle of convergence of the focused beam; μ^{fi} = transition dipole moment; λ = wavelength; $\mu(\xi)$ = electric dipole operator of species ξ; $E(\xi)$ = electric field vector at the position occupied by species ξ; ρ_f = number of energy levels per unit energy interval; ν = frequency; ω = circular frequency; $\hbar$ = Planck's constant/2π.

See also: **Chemical Reactions Studied By Electronic Spectroscopy; Environmental Applications of Electronic Spectroscopy; Laser Applications in Electronic Spectroscopy; Multiphoton Excitation in Mass Spectrometry; Multiphoton Spectroscopy, Applications; Nonlinear Optical Properties.**

Further reading

Andrews DL and Demidov AA (1995) *An Introduction to Laser Spectroscopy*. New York: Plenum Press.

Craig DP and Thirunamachandran T (1984) *Molecular Quantum Electrodynamics: An Introduction to Radiation–Molecule Interactions*. London: Academic Press.

Demtröder W (1996) *Laser Spectroscopy: Basic Concepts and Instrumentation*, 2nd edn. Berlin: Springer-Verlag.

Kettle SFAK (1985) *Symmetry and Structure*. New York: Wiley.

Siegman AE (1986) *Lasers*. Oxford: Oxford University Press.

Lead NMR, Applications

See **Heteronuclear NMR Applications (Ge, Sn, Pb).**

Light Sources and Optics

R Magnusson, The University of Texas at Arlington, TX, USA

FUNDAMENTALS OF SPECTROSCOPY
Methods & Instrumentation

Optics

Light

Light emanates from sources with a wide variety of physical embodiments. Examples include the Sun, a sodium arc lamp, an incandescent light bulb, and a helium–neon laser. The physical processes producing light emission appear to be distinctly different for each type of source. On a fundamental level, however, light is generated as a result of acceleration of electric charges. Thus, a synchrotron generates light by accelerating free electrons in a circular storage ring. An incandescent light bulb radiates as a result of electric-current-induced electronic charge vibrations and subsequent emission of radiation. A laser operates under atomic transitions involving bound charges acting as miniature, light-radiating electric dipole antennas.

Electromagnetics

Light propagates as an electromagnetic wave. Maxwell's equations govern the propagation of light and other electromagnetic waves in free space, in material media, at interfaces, and through apertures. Visible light has wavelengths, λ, expressed in nm or μm (10^3 nm = 1 μm = 10^{-6} m), or frequency, ν, in Hz or s^{-1}, detectable by the human eye typically falling within the $380 < \lambda < 780$ nm range. The electromagnetic spectrum includes the ultraviolet (UV) region often defined to be $10 < \lambda < 380$ nm and the infrared (IR) region 780 nm $< \lambda <$ 1 mm, also of great importance in optics and spectroscopy.

Maxwell's equations yield a wave equation that predicts the propagation of electromagnetic radiation. The most elementary solution is the plane wave expressible as a function of space (x, y, z) and time (t) as

$$\boldsymbol{e}(z,t) = \hat{x}e_o \cos[\omega t - kz + \phi] \qquad [1]$$

This wave, propagating in the $+z$ direction, is monochromatic with angular frequency $\omega = 2\pi\nu$, wavenumber $k = 2\pi/\lambda$, and with $\lambda\nu = c$ where c is the speed of light in the medium in which the wave propagates. An arbitrary phase factor ϕ is included. This wave, expressed in terms of the electric field vector $\boldsymbol{e}(z,t)$, has amplitude e_o and is linearly polarized along the x direction, with $\hat{x}$ denoting a unit vector. An electromagnetic plane wave is a transverse wave without any field component along the direction of propagation. With $\boldsymbol{e}(z,t)$ representing the electric field (V m^{-1}), Maxwell's equations require that an orthogonal magnetic field (A m^{-1}) vector $\boldsymbol{h}(z,t)$ also be associated with the wave as shown in **Figure 1**. Since e_o = constant, the plane wave is transversely infinite and thus carries infinite energy; it is further infinite in time and exists everywhere in space. In spite of these idealizations, the plane-wave model is exceedingly successful in predicting a large range of optical phenomena such as reflection/refraction at metallic and dielectric interfaces, diffraction by spectroscopic gratings, wave propagation in nonlinear and anisotropic media, etc. A practical, transversely finite, laser beam may be 1 mm in diameter which is 2000λ for $\lambda = 500$ nm; it is well approximated as a plane wave since it is wide on the scale of λ.

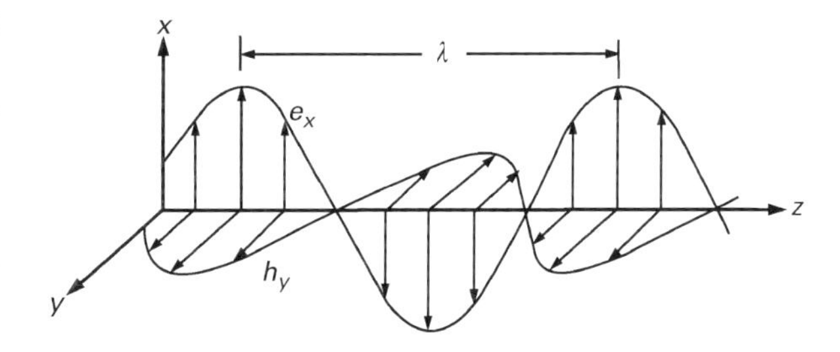

Figure 1 A snapshot of an electromagnetic plane wave with wavelength λ illustrating its spatial characteristics. The electric-field component e_x and the magnetic-field component h_y oscillate in phase but orthogonal to each other and to the direction of propagation. This wave is said to be linearly polarized along the x-direction.

Polarization

The vectorial nature of the electromagnetic fields representing light implies polarization, a fundamental property of light. The particular orientation of the e-vector of a plane wave incident on an interface has a profound effect on the detailed interaction that takes place and influences, for example, the amount of light that is reflected or transmitted. If the electric field vector is confined to oscillate in a plane (e.g. the x–z plane in **Figure 1**), the wave is said to be linearly polarized. If the tip of the electric field vector traverses an elliptical path around the direction of propagation, the state of polarization is elliptical. Circular polarization and linear polarization are special cases of the general state of elliptical polarization.

The influence of the state of polarization is illustrated in **Figure 2** where a polarized plane wave in air (refractive index, $n_1 = 1$) is incident on a piece of glass ($n_2 = 1.5$). The index of refraction is defined as $n = c_o/c$, where c_o is the speed of light in vacuum. For s-polarized light, with the e-field oscillating orthogonal to the plane of incidence as shown, the reflectance (R_s, defined as the ratio of optical power density (W cm^{-2}) reflected to that of the input wave) increases monotonically as the angle of incidence, θ_i, is increased. For a p-polarized incident wave, the reflectance, R_p, differs markedly, even exhibiting a reflection zero when $\theta_i = \theta_B$, which defines the Brewster angle given by $\theta_B = \tan^{-1}[n_2/n_1]$. The angle of incidence θ_i is related to the angle of refraction θ_t by Snell's law $n_1 \sin\theta_i = n_2 \sin\theta_t$ which holds for both polarization states. Additionally, the angle of reflection is $\theta_r = \theta_i$. If the wave traverses a rare-to-dense interface ($n_1 < n_2$) then $\theta_t < \theta_i$. Conversely, if $n_1 > n_2$, $\theta_t > \theta_i$. In particular, for θ_i such that $\theta_t = 90°$, total internal reflection occurs and $R_s = R_p = 1$ for all $\theta_i > \theta_c$, which is the critical angle given by $\theta_c = \sin^{-1}[n_2/n_1]$. Light propagation inside optical fibres in telecommunications systems occurs under conditions of total internal reflection.

Diffraction gratings

Diffraction gratings are widely applied on account of their dispersive properties. Thus a beam of polychromatic light incident on a grating spatially separates according to its spectral content. A diffraction grating with period Λ larger than the wavelength generally exhibits multiple diffracted waves excited by a single incident plane wave as illustrated in

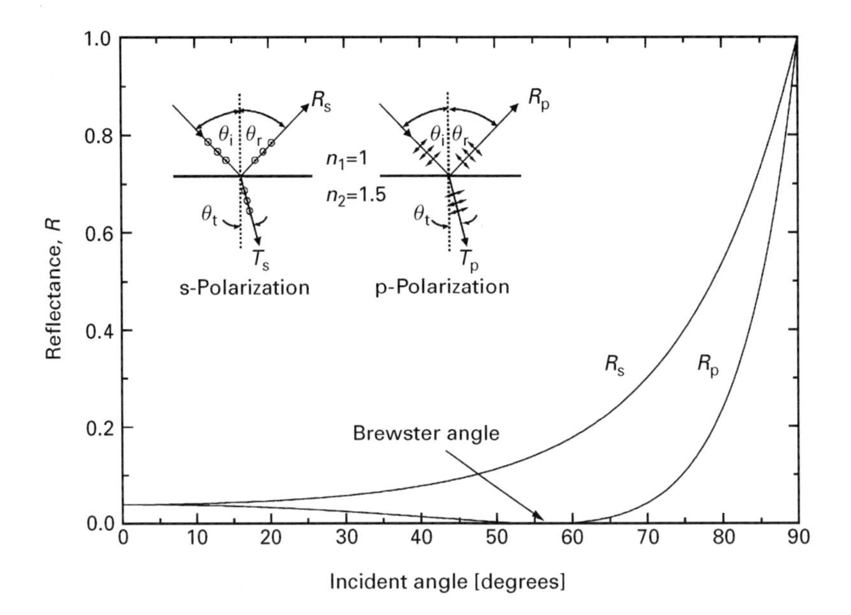

Figure 2 Light reflectance as a function of the angle of incidence of a planar air–glass interface. The influence of the state of polarization of the incident plane wave is illustrated.

Figure 3. Diffraction gratings operate in reflection or transmission. In spectroscopic devices, such as monochromators, reflection gratings play key roles.

With reference to **Figure 3**, the grating equation describes the spatial division of the incident wave of wavelength λ into distinct directions θ_ι for each order ι as

$$\Lambda(\sin\theta_i + \sin\theta_\iota) = \iota\lambda \qquad [2]$$

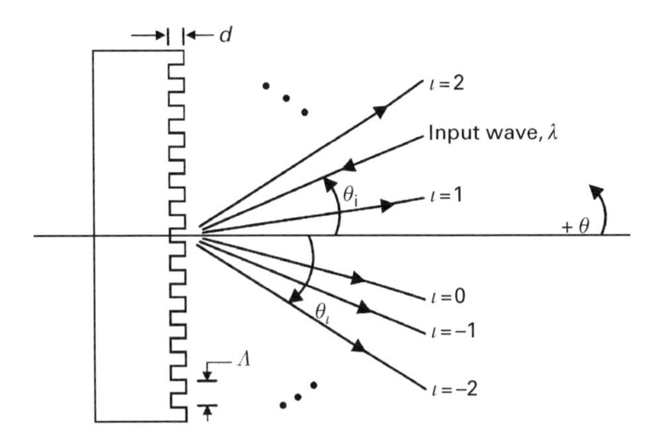

Figure 3 Illustration of the dispersion properties of a surface-relief reflection grating. The monochromatic incident wave is diffracted into several waves propagating in directions specified by the grating equation. A change in the wavelength, λ, of the incident wave will alter the directions θ_ι of the diffracted orders. Spectroscopic instruments often employ the first order ($\iota = 1$) to characterize the spectral content of the input light.

where $\iota = 0, \pm1, \pm2, \ldots$ labels the order of diffraction. Essentially the directions θ_ι correspond to conditions of constructive interference of waves emanating from the grating facets. For a polychromatic incident wave, each component wavelength propagates into directions specified by Equation [2]. Taking θ_i = constant and differentiating, this angular dispersion is

$$\frac{d\theta}{d\lambda} = \frac{\iota}{\Lambda\cos\theta_\iota} \qquad [3]$$

Thus, the angular spread per unit spectral interval is largest for higher-order (ι) diffracted waves, small grating periods (Λ), and large diffraction angles (θ_ι).

The diffraction efficiency, DE_ι, specifies the optical power carried by the diffracted waves normalized with the power of the incident wave. For the surface-relief reflection grating of **Figure 3** with a square-wave profile, the diffraction efficiency is given by

$$DE_\iota = (2/\iota\pi)^2\sin^2(2g) \quad \iota = 1, 3, 5, \ldots \qquad [4]$$

where $g = \pi n_1 d/\lambda \cos\theta_i$. This grating obtains a maximum DE_1 of 40.5%. A lossless grating with a saw-tooth (blazed) profile has a theoretical maximum DE_1 of 100%.

Monochromators

A typical monochromator is schematically illustrated in **Figure 4**. The input light contains a range of

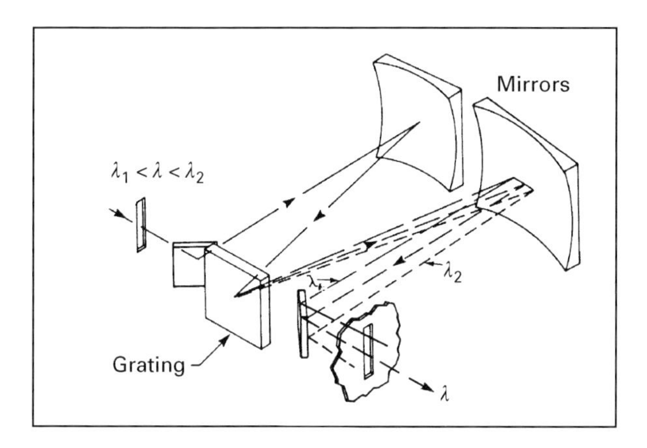

Figure 4 The basic elements of a monochromator. The input spectrum contains wavelengths in the range $\lambda_1 < \lambda < \lambda_2$. The grating spreads the light within the diffraction orders (say $\iota = 1$) in which the instrument is operated. No other diffracted orders are shown in the drawing for clarity. The width of the slit defines the spectral content of the output light. Reproduced with permission from Oriel Corporation (1989) *Oriel 1989 Catalog*, Volume II.

wavelengths that are spatially separated with the diffraction grating as shown. The exit slit controls the spectral content of the output light by blocking the undesired wavelengths surrounding the central one. The spectrograph is a similar instrument but without a slit. In this case, the output spectrum is picked up by a detector array such that a prescribed spectral band can be associated with a given detector element.

Thermal sources

Thermal sources supply light by heating the source material to a sufficiently high temperature. If an electric current is passed through a filament made of a metal with high melting point such as tungsten, the filament glows. Thermal radiation exhibits a characteristic, spectrally broad, continuous emission spectrum. The laws of blackbody radiation are useful for describing the emission spectra of thermal, or incandescent, sources. A blackbody consists of an ideal material that completely absorbs all incident radiation independent of wavelength. Conversely, the maximum radiation emitted by a hot solid material at a given temperature is that of a blackbody. The radiation properties of a blackbody are determined completely by the absolute temperature.

In practice, the characteristics of a blackbody can be obtained approximately by a cavity with a small hole in its side. Radiation entering the cavity through the hole is completely absorbed. As the cavity is heated, blackbody radiation emerges from the hole.

Planck's radiation law gives the spectral distribution of blackbody radiation. It may be expressed as

$$M_\lambda(T) = C_1\lambda^{-5}[\exp(C_2/\lambda T) - 1]^{-1} \qquad [5]$$

where M_λ is the spectral radiant emittance given in W cm^{-2} per unit wavelength. Thus M_λ is the radiated power density per wavelength, λ. Further, T is the absolute temperature (K), $C_1 = 3.7418 \times 10^{-12}$ W cm^2 and $C_2 = 1.4388$ cm K. These coefficients are expressed in terms of fundamental physical constants as $C_1 = 2\pi hc_o^2$ and $C_2 = hc_o/k_B$ where h is Planck's constant and k_B is Boltzmann's constant. **Figures 5** and **6** illustrate Planck's law on logarithmic and linear scales demonstrating the shift of the emission peak to shorter wavelengths with increasing temperature as well as increase of the radiated power density.

Taking $dM_\lambda/d\lambda = 0$ yields the wavelength, λ_{peak}, at which the radiation peak occurs as $\lambda_{peak}T = 0.28978$ cm K, which is Wien's displacement law traced by the dashed line in **Figure 5**. Finally, calculating the integral

$$M = \int_0^\infty M_\lambda \, d\lambda = \sigma T^4 \qquad [6]$$

gives the total power per unit area, M, emitted by the blackbody, where the Stefan–Boltzmann constant $\sigma = 5.6703 \times 10^{-12}$ W cm^{-2} K^{-4}.

As the temperature of the blackbody increases, its colour changes as evident in **Figures 5** and **6**. The colour of a blackbody is thus uniquely dependent on its temperature. Real sources such as tungsten lamps that have spectral distributions approximating those of blackbodies can be assigned a colour temperature at which the spectrum most nearly overlays that of the blackbody. Outside the Earth's atmosphere, the Sun can, for example, be approximated as a blackbody at 6200 K, which is, then, the Sun's colour temperature.

Figure 7A shows the spectral distribution of an incandescent lamp fitted with a tungsten filament inside a quartz bulb enclosing a rare gas and a halogen. The spectrum is smooth with a colour temperature exceeding 3000 K containing a significant amount of near-infrared light.

Electric-discharge sources

Electric discharge denotes the excitation of atomic states in a gaseous medium on passing an electric current through the medium. An ordinary household

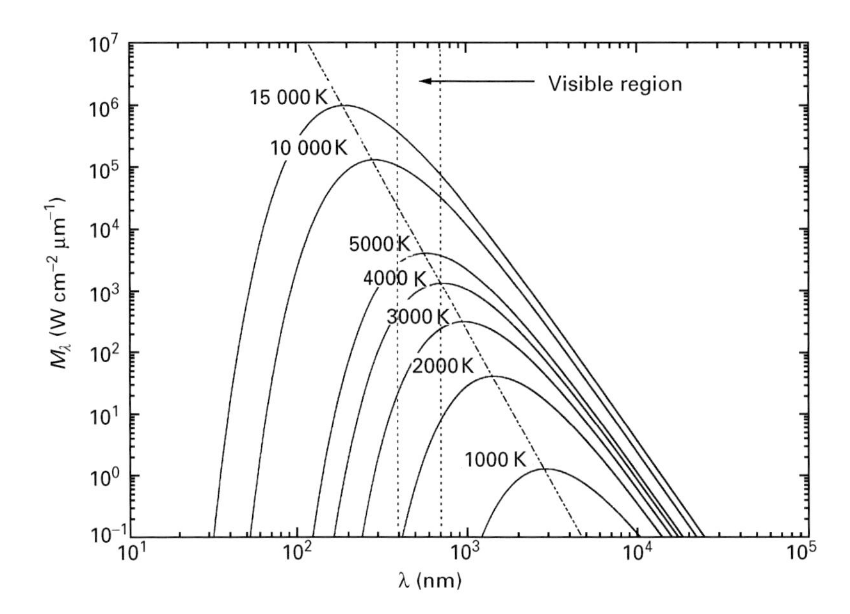

Figure 5 Spectral characteristics of blackbody radiation. The spectral radiant emittance is plotted as a function of the wavelength for several values of the absolute temperature. The slanted, dashed line indicates Wien's displacement law.

fluorescent lamp is an example of an electric-discharge source. The lamp tube, coated on the inside with fluorescent phosphors, contains a mixture of a rare gas and mercury vapour. On application of sufficiently high voltage across the lamp's terminals, a low-pressure mercury arc forms. Thus, excited Hg atoms emit UV radiation (a particularly strong line exists at 253.7 nm) which is absorbed by the phosphor and re-emitted as visible light. The phosphor radiation spectrum is concentrated in the visible spectral region; it is smooth with several superimposed discrete lines originating in the Hg atoms.

By increasing the lamp's internal operating pressures and temperatures, high-intensity light can be produced. High-intensity discharge lamps are classified as mercury-vapour, metal-halide, or high-pressure-sodium types. In these sources, the discharge is contained in arc-tubes placed inside the lamp envelope filled with inert gas. Additionally, short-arc lamps, such as xenon and mercury–xenon lamps, are capable of power output in the multi-kilowatt region. Typical spectral characteristics of such sources are shown in **Figures 7B** and **C**.

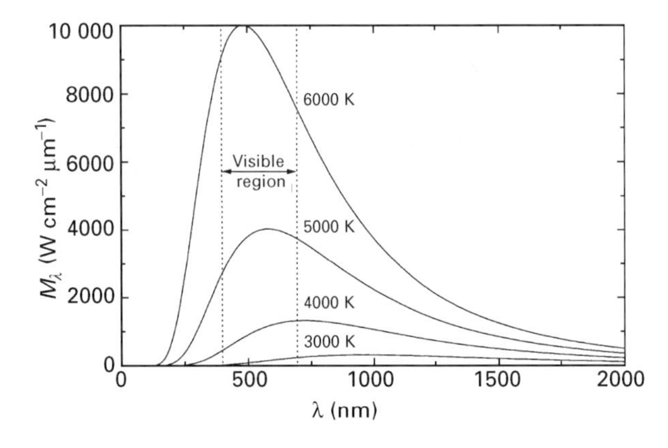

Figure 6 Similar to **Figure 5** on a linear scale.

Table 1 gives wavelength ranges and power levels of selected light sources.

Lasers

Key characteristics

In contrast to light emitted by traditional thermal or luminescent sources, the laser beam is directional with all of its power confined within a narrow angular range. As the beam expands slowly on propagation, its divergence, $\Delta\theta$, is small, often of the order of a milliradian (mrad). A direct result of this confinement of laser power within a small angular region is the high value of laser brightness, or radiance, measured in watts (W) per unit area (cm^2) per solid angle (steradian, sr). A typical argon laser beam with 1 mm diameter and 1 mrad divergence carrying a power of 1 W has brightness of ~10^8 W cm^{-2} sr^{-1} whereas a 100 W light bulb radiating uniformly into a solid angle of 4π obtains only ~1 W cm^{-2} sr^{-1}. In addition, laser light is highly monochromatic, meaning that the spectral wavelength spread, $\Delta\lambda$, is small. An associated property of monochromatic sources is temporal coherence; lasers are highly coherent

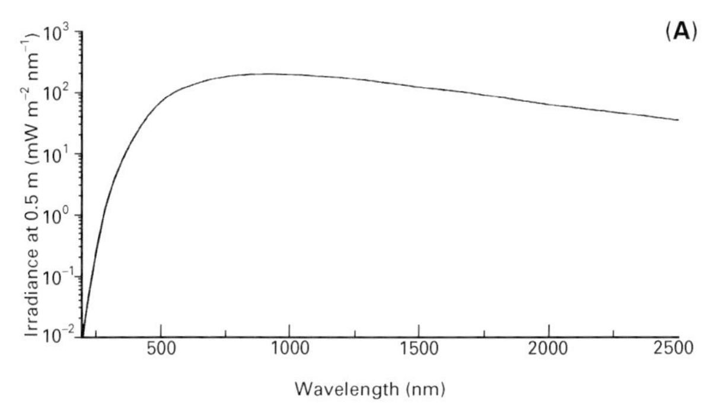

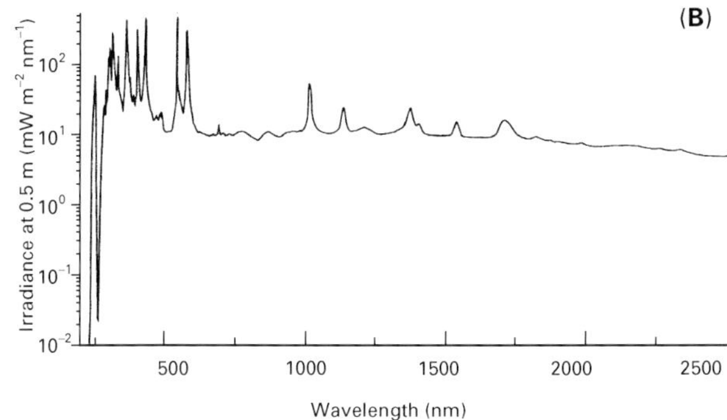

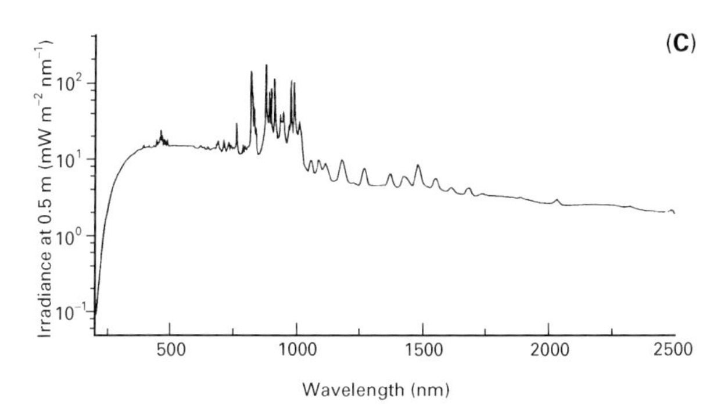

Figure 7 (A) Spectrum of a 1000 W tungsten–halogen lamp. Reproduced with permission from Oriel Corporation (1994) *Oriel 1994 Catalog*, Volume II. (B) Spectrum of a 200 W mercury arc lamp. Reproduced with permission from Oriel Corporation (1994) *Oriel 1994 Catalog*, Volume II. (C) Spectrum of a 150 W xenon arc lamp. Reproduced with permission from Oriel Corporation (1994) *Oriel 1994 Catalog*, Volume II.

sources. Further, lasers operate in continuous-wave (CW) or pulsed modes, and they may be polarized or unpolarized. While many lasers operate with fixed output wavelengths, some laser types are tunable over relatively large spectral regions. Finally, laser beams can be focused to tiny spots (~1 μm diameter) with extreme power densities obtainable even for CW lasers.

Applications

These unique characteristics of lasers make them exceedingly useful in numerous applications. Thus,

Table 1 Representative power levels and wavelength ranges for common light sources

Source material	*Wavelength range* (nm)	*Rated power* (W)
Mercury	200–2500	50–1500
Mercury–xenon	200–2500	30–5000
Xenon	200–2500	15–30 000
Deuterium	180–400	30
Sodium	400–800	250–400
Tungsten–halogen	240–2 700	50–1 000

lasers enable applications in diverse fields such as telecommunications, holography, spectroscopy, integrated optics, lithography, and medicine. Specific functions include alignment, distance measurements, holographic data storage, signal processing, image processing, velocity measurement, industrial process control, inspection, surgery, welding, fusion, and propagation of fibre-optic communication signals.

Atomic laser processes

The term 'laser' stands for 'light amplification by stimulated emission of radiation'. This stimulated emission occurs on interaction between a photon propagating in the laser's active medium and an atom in an excited state. The incident photon essentially stimulates the emission of an identical photon which joins the incident photon coherently at the same frequency and along the same direction. By viewing the photon as an electromagnetic wave packet, stimulated emission is a coherent wave process wherein the electromagnetic field of the incident photon drives the electrons of the excited atoms into coherent oscillation. The atom then undergoes energy level transition by radiating energy in the form of an optical wavefield that is spatially and temporally coherent with the incident, stimulating wavefield.

The fundamental laser process, thus, relies on quantum-mechanical energy levels in atomic systems. Media sustaining laser action possess discrete energy levels with discrete spectral components produced on transitions between particular levels. If the transition is from an upper level E_2 to a lower level E_1, the emitted photon will have frequency given by

$$\nu_{21} = (E_2 - E_1)/h \qquad [7]$$

where h is Planck's constant. The transition energy is $\Delta E = h\nu = hc/\lambda$.

Suitable laser energy levels have been discovered in thousands of atomic systems occurring in a variety of media. For example, discrete electron energy levels of isolated Ne atoms, originating in proton–electron Coulomb attraction, enable the ubiquitous HeNe laser. Vibrations of molecules with associated harmonic-oscillator type quantum levels are basic to the CO_2 laser. Organic dye molecules exhibit electronic, rotational, and vibrational energy levels on account of their size and complex structure. This provides a continuum of available transitions and yields broadband, tunable dye lasers. Semiconductor lasers operate on transitions based on energy-level bands that form on merging of the discrete levels associated with individual atoms in the solid.

In thermal equilibrium, an atomic system with N_1 atoms in level E_1 and N_2 atoms in $E_2 > E_1$ obeys the Boltzmann principle

$$N_2/N_1 = \exp[-(E_2 - E_1)/k_{\mathrm{B}}T] = \exp[-h\nu/k_{\mathrm{B}}T] \quad [8]$$

Therefore, in thermal equilibrium, $N_2 < N_1$ and, if ν is in the optical region, $N_2 << N_1$. Such a system is absorptive and not amplifying. To obtain the amplification effect associated with stimulated emission, an external pump mechanism is required to induce population inversion such that $N_2 > N_1$. Population inversion in the active laser medium is, thus, a non-equilibrium condition that enables lasing. The pumping approach employed depends on the laser medium. Optical flashlamps are used to invert some solid-state laser media (Nd:YAG, ruby), electric-current discharge is effective for many gas lasers (HeNe, argon), direct current injection pumps semiconductor lasers (GaAs, InP), and optical pumping with an auxiliary laser (e.g. argon laser) is used to operate dye lasers.

Figure 8 schematically illustrates a four-level laser system such as that for a Nd:YAG laser. Population inversion and lasing is established between levels E_2 and E_1. The optical pump populates level E_3 that may be a broad range of closely spaced levels in practice. Decay from E_3 down to the metastable upper laser level E_2 may occur via radiative or nonradiative relaxation processes. Random radiative emission, or spontaneous emission, occurs without a stimulating electric field, in contrast to stimulated emission. Nonradiative relaxation implies energy transfer via lattice vibration also called phonon modes.

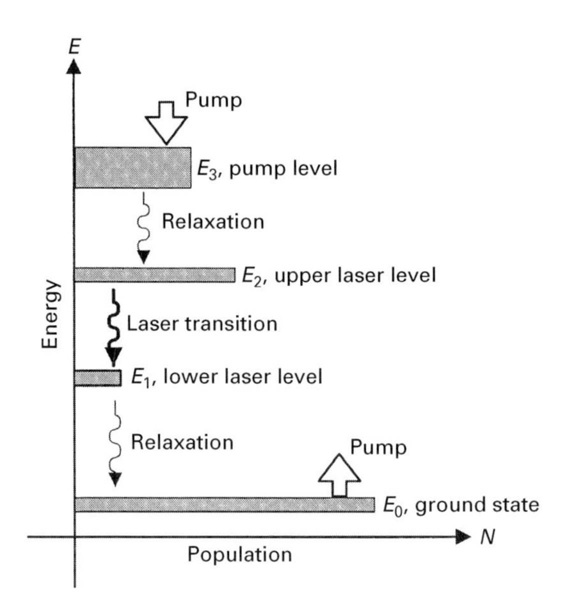

Figure 8 A four-level atomic energy system.

Spectral content

Spontaneous transitions in laser media possess a finite spectral bandwidth $\Delta\nu$. The concomitant frequency distribution is described by a line shape function, $g(\nu)$, taken to be normalized as

$$\int_{-\infty}^{\infty} g(\nu)\mathrm{d}\nu = 1 \quad [9]$$

The function $g(\nu)$ determines the spectral characteristics of the laser gain coefficient in inverted media. Light emission and absorption are both spectrally governed by $g(\nu)$.

The functional shape of $g(\nu)$ is defined by the particular line-broadening mechanisms of the atomic systems that constitute the active medium. If the atoms in the collection are indistinguishable, i.e. each atom acts the same, the line broadening is said to homogeneous. Lifetime broadening and collision broadening are examples of homogeneous broadening. Systems with distinguishable atoms suffer inhomogeneous broadening such as Doppler broadening and lattice-strain broadening.

Lifetime broadening is due to the finite lifetime of the excited atomic state. It can be modelled as an exponentially decaying oscillator of the form (see Eqn [1])

$$e(t) = e_0 \exp(-t/\tau)\cos(\omega_0 t), \quad t \geq 0 \quad [10]$$

emitting a wave of frequency ω_0 during the lifetime τ. The frequency content of the decaying oscillator is found by applying a Fourier transform

$$F(\omega) = \int_{-\infty}^{\infty} f(t)\mathrm{e}^{-\mathrm{i}\omega t}\,\mathrm{d}t \quad [11]$$

This leads to a normalized, Lorentzian, line shape function in the vicinity of $\omega \cong \omega_0 = 2\pi\nu_0$ expressed as

$$g(\nu) = \frac{\Delta\nu}{2\pi\left[(\nu - \nu_0)^2 - (\Delta\nu/2)^2\right]} \quad [12]$$

where $\Delta\nu = 1/\pi\tau$ is the line width (full width at half maximum, FWHM) of the lifetime-broadened system. This result applies to homogeneous broadening, where each atom in the collection is characterized by a bandspread $\Delta\nu$. Short pulses (τ small) yield broad spectra. Atomic collisions that interrupt the decay cause additional broadening by reducing the effective average value of τ.

Doppler broadening is a commonly cited example of inhomogeneous broadening. The atoms in a gas have a statistical distribution of velocities; the atoms are thus distinguishable by their differing velocities that correspond to different atomic resonance, or transition, frequencies. The resulting Gaussian line shape function is given by

$$g(\nu) = \frac{2\sqrt{\ln 2}}{\sqrt{\pi}\Delta\nu_{\mathrm{D}}}\exp\{-[4\ln 2][(\nu - \nu_o)/\Delta\nu_{\mathrm{D}}]^2\} \quad [13]$$

where the FWHM line width is $\Delta\nu_{\mathrm{D}} = 2\nu_0[2k_{\mathrm{B}}T \ln 2/mc^2]^{1/2}$ where m is the atomic mass. Thus, as expected, $\Delta\nu_{\mathrm{D}}$ increases with increasing T and decreasing mass. For example, the HeNe laser is Doppler broadened with $\Delta\nu_{\mathrm{D}}$ ~1.5 GHz whereas the Nd:YAG laser is homogeneously broadened with $\Delta\nu$ ~120 GHz.

The line shape function $g(\nu)$ spectrally governs the rate of spontaneous and stimulated emission. The stimulated transition rate per atom is given by $W_{\mathrm{i}} = \lambda^2 I_\nu g(\nu)/(8\pi n^2 h\nu\tau)$, where n is the refractive index and I_ν is the intensity (W cm^{-2}) of the monochromatic optical wave (at frequency ν) causing stimulated emission. The relation to laser gain is established by considering propagation along the z-direction in the laser medium, which can be described by

$$\frac{\mathrm{d}I_\nu}{\mathrm{d}z} = (N_2 - N_1)\left[\left(\lambda^2 g(\nu)\right)/(8\pi n^2\tau)\right]I_\nu \equiv \gamma(\nu)I_\nu \quad [14]$$

which defines the exponential gain coefficient. The solution is

$$I_\nu(z) = I_\nu(0)\exp[\gamma(\nu)z] \quad [15]$$

implying exponential growth of the laser radiation intensity for $\gamma(\nu) > 0$ or $N_2 > N_1$ (inverted medium) and a traditional decay if the system is absorbing, or noninverted, in thermal equilibrium.

Basic laser components

Figure 9 illustrates the building blocks of typical lasers. The active medium (gas, solid, or liquid) is placed between two mirrors constituting an optical resonator. The back mirror is highly reflective with reflectance, R, approaching 100%. The front mirror (output coupler) is also highly reflective but with a finite value of transmittance, T, typically a few percent, to allow the useful output laser beam to emerge from the resonator. The resonator-cavity mirrors may be flat or curved. A laser with two flat mirrors is difficult to align as small misalignment causes the oscillating light wave, making multiple round trips, to walk off and leave the cavity. It is thus a marginally stable configuration. Improved design is obtained with curved mirrors confining the laser light energy efficiently within the resonator. In particular, two mirrors with long radii of curvature constitute a practical, stable laser cavity. The laser medium is pumped with an external source, also shown in **Figure 9**, to maintain the inverted atomic population needed for laser power amplification to occur. Pumping methods include optical pumping of solid state lasers with incoherent (e.g. xenon) flash lamps, electric-discharge pumping of gas lasers, and direct-current-injection pumping of semiconductor lasers and light-emitting diodes. Collisions can also play a role

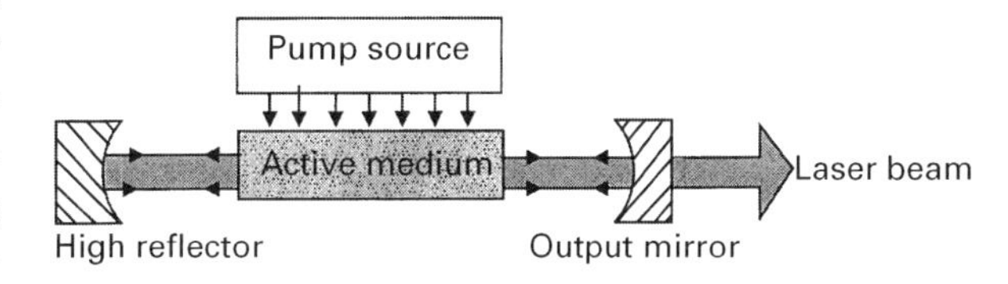

Figure 9 Chief components of a laser.

in achieving the population inversion. For example, in the HeNe laser, the electrical discharge (a few kV at a few tens of mA current) drives electrons that collide with He atoms, raising them to excited levels (2s level desired). The He atoms then interact with Ne atoms, raising them to $3s_2$ levels by collision energy transfer. The resulting metastable state with a few μs lifetime radiates the familiar red photons with 633 nm wavelength.

Laser modes

Laser oscillation can build up from noise or from a deliberately injected signal. Before lasing, spontaneous emission generates a flood of photons propagating in random directions. Those taking a path along the resonator axis are reflected back into the active medium (i.e. feedback) and initiate the stimulated emission and amplification process with power provided by the external pump. If the round-trip gain exceeds the round-trip loss, optical energy builds up in the resonator. Since the pump provides finite energy, this oscillating power buildup saturates with steady-state oscillation established when the round-trip gain equals the round-trip loss. Under these conditions, a steady beam may be emitted from the laser, since the useful output light power is taken as part of the loss.

Thus, laser oscillation occurs inside a resonator confining the laser light. The resonator defines the frequency distribution and the spatial distribution of the output light. A resonator with planar mirrors separated by distance L supports longitudinal modes with discrete frequencies

$$\nu_q = q(c/2L), \quad q = 1, 2, 3 \ldots \qquad [16]$$

Adjacent modes are thus separated by $c/2L$. A laser will support those longitudinal modes of frequency ν_q that fall under the gain curve of the laser medium.

Figure 10A shows an example atomic line and several longitudinal cavity modes with six modes having sufficient gain to oscillate, the rest being below threshold and not surviving. The spectral properties of the individual longitudinal modes can be understood by treating the resonator as a Fabry–Perot etalon. Their spectral width, $\delta\nu$, is given in terms of the resonator's free spectral range, FSR = $c/2L$, and its finesse, F, as $\delta\nu$ = FSR/F. If each of the cavity mirrors has a reflectance R, the finesse is $F = \pi R^{1/2}/(1-R)$.

Figure 11 depicts some of the main components in a large-frame laser such as a commercial argon-ion laser. The resonator supports multiple longitudinal modes as suggested by Equation [16]. The active medium, in the case of argon, supports multiple well-separated atomic lines any one of which, as shown in **Figure 10A**, can accommodate several longitudinal cavity modes. To select one of the atomic lines, a prism is inserted in the cavity as shown. By angularly tuning the prism, a principal line, or wavelength, is selected such that it emerges out of the prism at normal incidence on the high reflector. This line can then oscillate in the cavity with other lines cut off. To narrow the laser's spectrum further, an intracavity etalon is applied. The etalon is a planar Fabry–Perot cavity of thickness d_e with its FSR = $c/2d_e$ thus supporting multiple longitudinal modes separated by a much larger frequency spread than the laser cavity modes since $L \gg d_e$. The etalon is tuned by rotation and by varying its temperature. When an etalon mode coincides with one of the laser modes in **Figure 10A**, that mode oscillates by itself with the adjacent modes suppressed. The laser then oscillates in a single mode with a corresponding decrease in

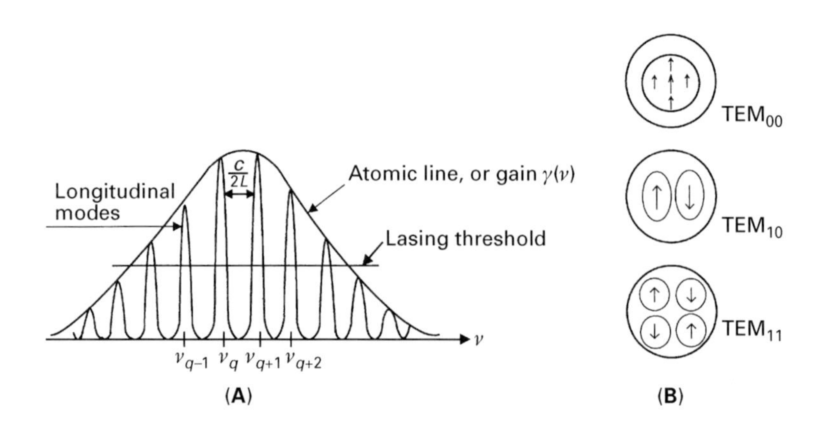

Figure 10 Spectral and spatial modes of a laser. (A) Atomic line enclosing several longitudinal resonator modes. (B) Transverse spatial modes of a laser beam. The fundamental TEM_{00} mode is the Gaussian laser beam profile. The arrows indicate the direction of the beam's electric field at a given instant of time.

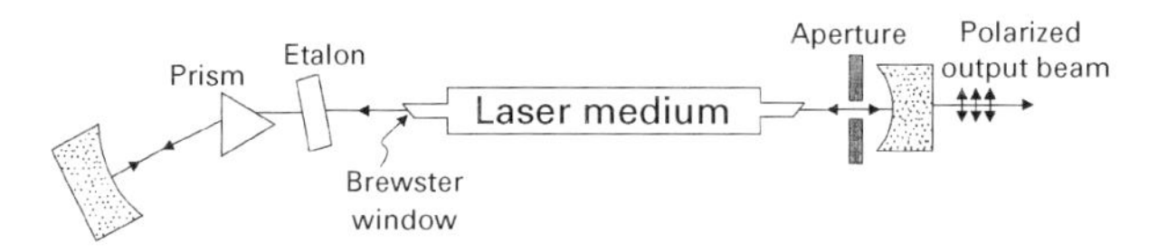

Figure 11 Principal components used to optimize the performance of a laser. The prism selects a particular atomic line and the wavelength of operation. The etalon selects a longitudinal mode and increases the laser's coherence. Reducing the aperture diameter favours the Gaussian spatial mode. The Brewster windows polarize the beam linearly.

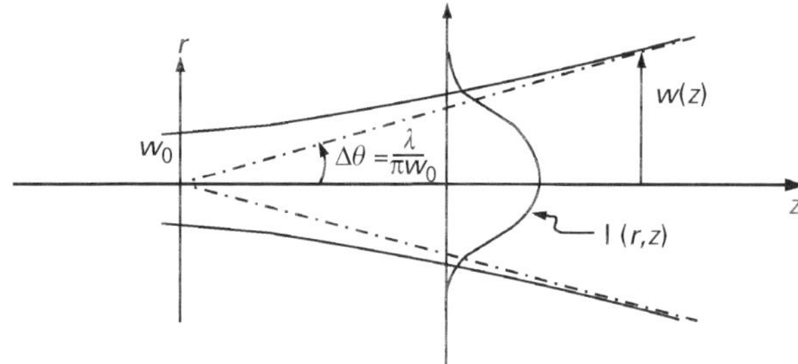

Figure 12 Parameters of a Gaussian laser beam.

the output spectral line width and associated increase in its level of coherence. The laser in **Figure 11** is equipped with Brewster windows (see **Figure 2**) that force the laser to favour a single polarization state as indicated.

In addition to the longitudinal modes discussed, the laser resonator admits multiple transverse modes that appear as spatial intensity variations across the output laser beam as shown in **Figure 10B**. Such transverse electromagnetic (TEM) modes are labelled with subscripts that count the number of zeros along the (x,y) transverse directions. The most useful mode has no zeros and thus the TEM_{00} mode is preferred. This mode is the ubiquitous Gaussian laser beam obtained when the laser aperture in **Figure 11** suppresses higher-order transverse modes. With reference to **Figure 12**, the intensity of a Gaussian laser beam carrying a total power P is expressible as

$$I(r,z) = \frac{2P}{\pi w^2(z)} \exp\left(-\frac{2r^2}{w^2(z)}\right) \qquad [17]$$

with $w(z) = w_0[1 + (z/z_0)^2]^{1/2}$ where $z_0 = \pi w_0^2/\lambda$, called the Rayleigh range, w_0 being the beamwaist radius occurring at $z = 0$, and $r^2 = x^2 + y^2$. If a thin lens of focal length f, with the lens placed at the waist, focuses this Gaussian beam, the radius of the focal point is

$$w_f = w_0/[1 + (z_0/f)^2]^{1/2} \qquad [18]$$

Often $z_0 >> f$ which gives $w_f \sim f\lambda/\pi w_0$. As an example, with $f = 10$ mm, $\lambda = 0.5$ μm, and $w_0 = 0.5$ mm, $w_f \sim 3$ μm. If $P = 1$ W, this yields a power density at focus of ~3 MW cm^{-2} or ~3 kW cm^{-2} if $P = 1$ mW (e.g. a small HeNe laser). In view of these high power densities, laser safety is an important concern. Eye protective goggles matched to the laser's wavelengths should be worn when working with lasers.

Pulsed lasers

Generation of intense short pulses of light is of interest in many fields, including nonlinear optics and spectroscopy. Pulse lengths in the femtosecond region with pulse power levels of terawatts have been achieved. Two common methods for short-pulse generation, Q-switching and mode locking, are summarized in this section.

In Q-switching, the laser is prevented from oscillating with the pump running such that a large inverted population builds up in the cavity. Oscillation occurs on removal of the blockage thereby switching the cavity Q back to its normally high value. A short, powerful optical pulse develops as the available gain rapidly depletes, typically on nanosecond timescales. Q-switching is accomplished by placing electro-optic or acousto-optic modulators inside the cavity or, in a passive manner, with a saturable absorber. The passive approach is particularly simple; a saturable absorber (organic dye solution) becomes transparent when sufficiently high gain has developed thus permitting a high-intensity pulse to escape. As an example, a Q-switched Nd:YAG laser may deliver ~1 J energy per pulse with pulse length of τ_p ~5 ns, power per pulse ~200 MW, and repetition rate of 10 Hz.

An ideal, inhomogeneously broadened laser oscillates on multiple longitudinal modes separated by $c/2L$ in frequency. These free-running modes oscillate independently with random, uncorrelated phases. If the phases (ϕ in Eqn [1]) can be locked together, a periodic laser-pulse train is obtained as output light. To accomplish this, an electro-optic or acousto-optic switching element can be placed in the cavity. It is modulated with a period equal to the resonator's roundtrip transit time $T = 2L/c$. Optimum conditions for light buildup correspond to minimum loss, that is an open switch. This switching action tends to lock the phases of the longitudinal modes such that their interference creates a pulse that passes through the open switch without loss. Wide atomic line widths, $\Delta\nu$, enclosing a large number of

Table 2 Representative figures characterizing common lasers

Laser type	*Wavelengths* (nm)	*Power*	*Beam dia* (mm)	*Divergence* (mrad)	*Output*
Helium–neon	633,1152	0.1–50 mW	0.5–3.0	0.5–6.0	CW
Argon–ion	351–529 several	5 mW–25 W	0.5–2.0	0.5–1.0	CW
Helium–cadmium	325, 442	1–50 mW	0.3–1.2	0.5–3.0	CW
Dye (Ar pump)	400–1000	to 5 W	0.5–1.0	1–2	CW, tunable
Nd:YAG	1064	0.1–500 W (ave)	1–10	1–10	Pulsed
Nd:YAG	1064	0.5–1000 W	1–10	1–10	CW
CO_2	10 600	1–100 W	3–7	1–5	CW
GaAlAs	750–900	1–100 mW	Divergent output	Large, ~30°	CW
InGaAsP	1100–1600	1–100 mW	Divergent output	Large, ~30°	CW
Ti:sapphire (Ar pump)	700–1100	0.1–1 W	0.5–1	1–2	CW, tunable

longitudinal modes produce the shortest mode-locked pulses since $\tau_p \sim \Delta\nu^{-1}$. For example, Ti:sapphire lasers, with their broad line widths, provide short pulses, with $\tau_p \sim 100$ fs, average power ~100 mW, and output pulse rate of ~100 MHz being representative numerical data.

Table 2 provides numerical parameters for some common lasers.

List of symbols

c = speed of light in medium; c_o = speed of light in vacuum; C_1, C_2 = coefficients in Planck's law; d = grating thickness; d_e = etalon thickness; DE_ι = diffraction efficiency of order ι; $\boldsymbol{e}$ = electric field vector; e_o = e-field amplitude; E_1, E_2 = energy levels; f = focal length; FWHM = full width at half maximum; FSR = free spectral range; $F(\omega)$ = Fourier transform of $f(t)$; g = normalized grating modulation strength; $g(\nu)$ = atomic line shape function; $\boldsymbol{h}$ = magnetic field vector; h = Planck's constant; $I(r,z)$ = intensity of Gaussian laser beam; I_ν = intensity of light of frequency ν; k = wavenumber; k_B = Boltzmann's constant; L = laser resonator length; m = atomic mass; M = emittance; M_λ = spectral radiant emittance; n = index of refraction; N_1, N_2 = atomic populations; P = power; q = integer mode index; $R_{s,p}$ = reflectance of s- or p-polarized light; $r^2 = x^2 + y^2$; t = time; T = absolute temperature; $T_{s,p}$ = transmittance of s- or p-polarized light; w_0 = radius of a Gaussian beam at the waist; w_f = focal point radius; $w(z)$ = beam radius at arbitrary z; x, y, z = space coordinates; z_o = Rayleigh range; $\gamma(\nu)$ = gain coefficient; $\delta\nu$ = FWHM of longitudinal cavity modes; $\Delta\theta$ = laser beam divergence; $\Delta\lambda$ = spectral line width, wavelength; $\Delta\nu$ = spectral line width, frequency; $\Delta\nu_D$ = Doppler line width; τ = lifetime; θ_i = angle of incidence; θ_r = angle of reflection; θ_B = Brewster angle; θ_c = critical angle; θ_ι = angle of diffraction; θ_t = angle of refraction; λ = wavelength of light; Λ = grating period; ν = frequency of light; ν_{12} = transition frequency; ν_q = frequency of longitudinal cavity mode; σ = Stefan–Boltzmann constant; ι = diffraction-order integer; τ_p = pulse length; ω = angular frequency of light; ϕ = phase.

See also: **Electromagnetic Radiation; Ellipsometry; Fibre Optic Probes in Optical Spectroscopy, Clinical Applications; Laser Applications in Electronic Spectroscopy; Laser Spectroscopy Theory; Optical Frequency Conversion; Raman Spectrometers.**

Further reading

Bass M ed (1995) *Handbook of Optics: Fundamentals, Techniques & Design*, 2nd edn, Vol I, Part 4: Optical sources. New York: McGraw-Hill.

Born M and Wolf E (1975) *Principles of Optics*, 5th edn. Oxford: Pergamon Press.

Hecht E (1987) *Optics*, 2nd edn. Reading, MA: Addison-Wesley.

Hecht J (1992) *The Laser Guidebook*, 2nd edn. New York: McGraw-Hill.

Proudfoot CN (ed) (1997) *Handbook of Photographic Science and Engineering*, 2nd edn, Section 1: Radiation sources. Springfield, VA: Society for Imaging Science and Technology.

Saleh BEA and Teich MC (1991) *Fundamentals of Photonics*. New York: Wiley.

Siegman AE (1986) *Lasers*. Mill Valley, CA: University Science Books.

Silfvast WT (1996) *Laser Fundamentals*. New York: Cambridge University Press.

Waynant RW and Ediger MN (eds) (1994) *Electro-Optics Handbook*. New York: McGraw-Hill.

Yariv A (1997) *Optical Electronics in Modern Communications*, 5th edn. New York: Oxford University Press.

Linear Dichroism, Applications

Erik W Thulstrup, Roskilde University, Denmark
Jacek Waluk, Polish Academy of Sciences, Warsaw, Poland
Jens Spanget-Larsen, Roskilde University, Denmark

ELECTRONIC SPECTROSCOPY
Applications

Light absorption is not only characterized by energy and intensity, it also has directional properties. These are often overlooked, even when they might provide essential information. Such information includes assignments of electronic and vibrational transitions and information on molecular structure, conformation and association. The directional properties of individual molecules cannot be observed in isotropic samples but require samples that are non-isotropic or aligned.

Although a natural (nonpolarized) light beam is nonisotropic in nature, since its electric and magnetic vectors are located exclusively in the plane perpendicular to the direction of the beam, the directional properties of aligned samples may best be studied using linearly polarized light. The term 'linear dichroism' refers to the change in absorption that may often be observed for nonisotropic samples when the direction of polarization of the light is changed by 90°.

In reality, most light in nature is partially polarized – scattering produces polarized light and most light around us is scattered sunlight. Historians believe that the Vikings used the polarization of light from the blue sky as a navigational aid when they travelled to North America about 1000 years ago. Similarly, except for common solutions, most naturally occurring samples are nonisotropic; this is especially true for samples with a biological origin. The first linear dichroism experiment was actually carried out on a dye-stained cell membrane in the 1880s.

The transition moment

Many observable properties in connection with electronic transitions between state $|0\rangle$ and $|f\rangle$ may be calculated from the two wavefunctions by means of an integral of the type

$$\langle 0|\hat{\boldsymbol{O}}|f\rangle = O_{0f} \quad [1]$$

where $\hat{O}$ is an operator describing the desired property. In many cases $\hat{O}$ is not a scalar but a vector or tensor. The operator that describes absorption probability, the electric dipole operator $\hat{M}$, is a vector. Thus the corresponding observable property, the transition moment $\boldsymbol{M}_{0f}$, will also be a vector.

$\boldsymbol{M}_{0f}$ may vanish for reasons of symmetry; then the transition between $|0\rangle$ and $|f\rangle$ is said to be symmetry forbidden. For example, if the molecule possesses a plane of symmetry, the product of $\langle 0|$, $\hat{M}$ and $|f\rangle$ will be either symmetric or antisymmetric with respect to the plane; if it is antisymmetric, the integral vanishes. The electric dipole operator $\hat{M}$ corresponds to the sum of charges on particles in the molecule times their coordinates and is therefore antisymmetric with respect to any symmetry plane, thus $\langle 0|$ and $|f\rangle$ must have different symmetry properties with respect to the plane in order for the integral to be nonzero.

In molecules with symmetry elements, the direction of $\boldsymbol{M}_{0f}$ is determined by the symmetry of the two states $|0\rangle$ and $|f\rangle$. In the important cases of C_{2v}, D_{2h} and D_2 symmetric molecules, only three different excited-state symmetries can be reached by an electric dipole transition from a totally symmetric ground state. For example, in the case of D_{2h}, each of the three 'allowed' excited state types are antisymmetric with respect to one, and only one, of the three symmetry planes of the molecule. The direction of $\boldsymbol{M}_{0f}$ for the three different allowed transitions is in each case perpendicular to the antisymmetry plane. For transitions to all other excited states, $\boldsymbol{M}_{0f}$ becomes zero; these transitions are said to be symmetry forbidden.

Production of linear dichroic samples

Although many samples occurring in nature are partially aligned from the start, most linear dichroism experiments are performed on samples that have been aligned in the laboratory. Single crystals may be grown of many molecules: if the crystal structure is known and is suitable, information may be obtained on transition moments for electronic and vibrational transitions. In practice, this is often experimentally difficult – very thin crystals are required for reasonably strong transitions and the interpretation of the spectra may be very complicated, primarily owing to strong intermolecular interactions in the crystal. Alignment in electric fields may be used, especially for large molecules (see the Kerr effect). Magnetic fields may be used to align molecules with permanent magnetic moments, but the effect is weak (see the Cotton–Mouton effect). Further details may be found in the article on linear dichroism instrumentation.

In the following we shall consider samples aligned either by photoselection or by dissolving the molecules in 'anisotropic solvents'. Ordinary photoselection is not based on a physical alignment of a set of molecules. Instead a partially aligned subset of an isotropic molecular assembly is selected by absorption of natural or polarized light. Since the relative transition probability for molecules with different spatial orientations is known, the orientation distribution of molecules in the photoselected sample will often be known. In order to preserve this alignment, the molecules must be kept in a high viscosity solvent, usually at low temperature. Anisotropic solvents may be single crystals of a suitable host molecule, membranes, a fast-flowing liquid, aligned liquid crystals or stretched polymers. With the possible exception of the first of these solvents, the orientation distribution of the sample molecules is not known, but, as we shall see, a complete interpretation of absorption spectra is possible even when only a few, simple properties of this distribution are known.

Description of molecular alignment: One-photon processes

The transition probability for linearly polarized light with its electric vector ε_U along sample (laboratory) axis U at the wavelength of the transition to state $|f\rangle$ in a single molecule is proportional to

$$(\varepsilon_U \cdot \boldsymbol{M}_{0f})^2 = (\boldsymbol{M}_{0f})^2 \cos^2(U, \boldsymbol{M}_{0f}) \qquad [2]$$

where $(U, \boldsymbol{M}_{0f})$ is the angle between U and $\boldsymbol{M}_{0f}$. The sample axis U is usually chosen according to the properties of the individual sample; most important samples are uniaxial (see below) and in these U is the unique axis. It is clear that the absorption depends on molecular alignment; if the molecule is aligned so that $\boldsymbol{M}_{0f}$ is perpendicular to U, the probability becomes zero, while the maximum absorption probability occurs when $\boldsymbol{M}_{0f}$ is parallel to U. In a typical partially aligned sample the angle $(U, \boldsymbol{M}_{0f})$ may take on all possible values between 0° and 90° for individual molecules in the sample. This means that an average value over all sample molecules must be used and we obtain

$$\langle(\varepsilon_U \cdot \boldsymbol{M}_{0f})^2\rangle = (\boldsymbol{M}_{0f})^2 \langle\cos^2(U, \boldsymbol{M}_{0f})\rangle \qquad [3]$$

where the brackets refer to average values for all molecules in the sample.

In the important case of a molecular symmetry (C_{2v}, D_{2h} and D_2) that only allows transition moments along three different perpendicular x, y and z directions within the molecule, we label the corresponding average cosine square values K_x, K_y and K_z:

$$\langle\cos^2(U, \boldsymbol{M}_{0f})\rangle = K_x, K_y \text{ or } K_z \qquad [4]$$

for x-, y- or z-polarized transitions, respectively. These three K values are not independent; their sum is equal to 1, which often provides a useful test. Note, however, that K values are useful for any direction in a molecule, for example a low-symmetry molecule; the transition to state $|f\rangle$ will, for a given sample alignment, be characterized by the K value K_f:

$$K_f = \langle\cos^2(U, \boldsymbol{M}_{0f})\rangle \qquad [5]$$

The molecular direction that corresponds to the largest possible K_f – in other words the direction that, on the average, is best aligned in the molecule with respect to U – is called the orientation axis. Since it is determined by averaging, the position of the orientation axis is not always obvious; as a result, the axis is sometimes called the effective orientation axis. In the symmetrical molecules mentioned above it will be one of the three directions determined by symmetry. It is often labelled z. The least well-aligned axis, with the lowest K value, is often called x and the intermediate axis is called y. The three axes are those that diagonalize a 'K-tensor' with elements $\langle\cos(u,U)\cos(v,U)\rangle$, where u and v are directions in the molecule.

The orientational dependence of the two linear dichroism spectra can easily be expressed in terms of the K values. If the electric vector is along the U axis, we obtain for the resulting absorbance $E_U(\nu)$:

$$E_U(\nu) = \sum K_f A_f(\nu) \qquad [6]$$

where the sum is over all relevant transitions f and $A_f(\nu)$ is equal to the contribution to the absorbance from transition f in a sample of perfectly aligned molecules (this is also equal to three times the contribution from transition f to the absorbance of an isotropic sample). Similarly, if the electric vector is along the V laboratory axis, perpendicular to U, we obtain for the resulting absorbance $E_V(\nu)$

$$E_V(\nu) = \sum K'_f A_f(\nu) \quad [7]$$

where $K'_f = \langle \cos^2 (V, \mathbf{M}_{0f}) \rangle$. The linear dichroism is equal to

$$\mathrm{LD}(\nu) = E_U(\nu) - E_V(\nu) \quad [8]$$

Uniaxial samples

The equations above are valid for any sample. Each transition moment direction is characterized by two quantities, K and K'. If the sample is uniaxial, i.e. all directions perpendicular to a given sample axis are equivalent, the equations simplify further. We chose U as the sample axis. Since V is now equivalent to the third laboratory axis, W, it can be shown that for all transitions f, $K'_f = (1 - K_f)/2$. We obtain

$$\begin{aligned} E_U(\nu) &= \sum K_f A_f(\nu) \\ E_V(\nu) &= \tfrac{1}{2}\sum(1 - K_f)A_f(\nu) \\ \mathrm{LD}(\nu) &= \tfrac{1}{2}\sum(3K_f - 1)A_f(\nu) \end{aligned} \quad [9]$$

When the two curves $E_U(\nu)$ and $E_V(\nu)$ differ very little, i.e. when the LD(ν) is small, a direct measurement of LD(ν) by means of polarization modulation provides a more accurate result. The second (linearly independent) curve is then either the isotropic spectrum

$$\begin{aligned} E_{\text{iso}}(\nu) = [E_U(\nu) + 2\,E_V(\nu)]/3, \\ E_V(\nu), \text{or} \\ [E_U(\nu) + E_V(\nu)]/2 \end{aligned}$$

Many important samples have been shown to be or are assumed to be uniaxial, for example solutes in stretched polyethylene or in nematic liquid crystals. It is now simple to determine K_f. If transition f dominates the absorption at ν_f (if it does not overlap with other transitions), the dichroic ratio at ν_f, $E_U(\nu_f)/E_V(\nu_f)$, can be written

$$\frac{E_U(\nu_f)}{E_V(\nu_f)} = \frac{2K_f}{1 - K_f} \quad [10]$$

Therefore, K_f can be directly determined:

$$K_f = \frac{E_Z(\nu_f)}{2E_V(\nu_f) + E_U(\nu_f)} \quad [11]$$

The situation for nonoverlapping transitions is common in IR spectroscopy, but less common in UV spectroscopy. If transition f is overlapped by other transitions, it is still often possible to determine K_f. This is done by a trial-and-error method (the TEM method) in which a series of linear combinations of the observed spectra are formed, using values of K near the expected value for K_f:

$$\tfrac{1}{2}(1 - K)E_U(\nu) + KE_V(\nu) \quad [12]$$

The spectral feature (peak, shoulder) due to transition f disappears from the linear combination when $K = K_f$. This may provide a very accurate determination of K_f, even in cases of strong overlap. **Figure 1** shows the observed spectra, the dichroic ratio ($E_U(\nu)/E_V(\nu)$), and linear combinations of the former that allow a determination of the two different K values present in the spectrum from the disappearance of the three major peaks in the spectrum.

Figure 2 shows a plot of K_z values against K_y values for a large number of symmetrical molecules aligned in stretched polyethylene. Obviously, shape is very important for the alignment. The triangle shows the theoretical limits for the K values, assuming that $K_z \geq K_y \geq K_x$. The lower left side corresponds to a rod-like alignment, $K_x = K_y$, and the lower right side to a disc-like alignment, $K_z = K_y$.

It is remarkable that the simple use of squares of directional cosines (the K values; sometimes equivalent quantities are used, the Saupe S matrices and the order parameters) provides a description of all directional properties of the sample, which can be observed in absorption spectroscopy, without any assumptions. Several alternative descriptions (the Fraser–Beer model, the Tanizaki model, the Popov model) involve assumptions that are rarely fulfilled. At the same time, some of these models tend to be mathematically complex. A simple extension of the

K values to averages of fourth powers of directional cosines (L values) makes it possible to describe two-photon processes (luminescence, two-photon absorption, Raman) in aligned samples.

Linear dichroism spectra: Interpretation

Photoselection

We shall in the following only consider the photoselection of a subset of molecules from an isotropic, 'stationary' sample. By stationary is meant that the sample molecules do not rotate on the timescale of the experiment. Descriptions of photoselection from an already aligned sample or, from a sample with molecular rotational movement, or of photoselection of each molecule with two photons, are more complicated but are available in the literature. Photoselection is usually performed with a collimated light beam with a well-defined wavelength; the light may be either linearly polarized or natural/circularly polarized, the two latter will produce similar results.

When an aligned subset of sample molecules is selected, another aligned subset, that consisting of the remaining molecules, is also created. When the sample depletion is negligible, the former set of molecules is small in number, but have a high degree of alignment, while the remaining molecules have negligible alignment. When the sample depletion is substantial, both the selected set and remaining set of molecules will be aligned. We shall in the following assume that only one transition is responsible for the photoselection of a small fraction of the sample. Other cases are also of interest and are described in the literature.

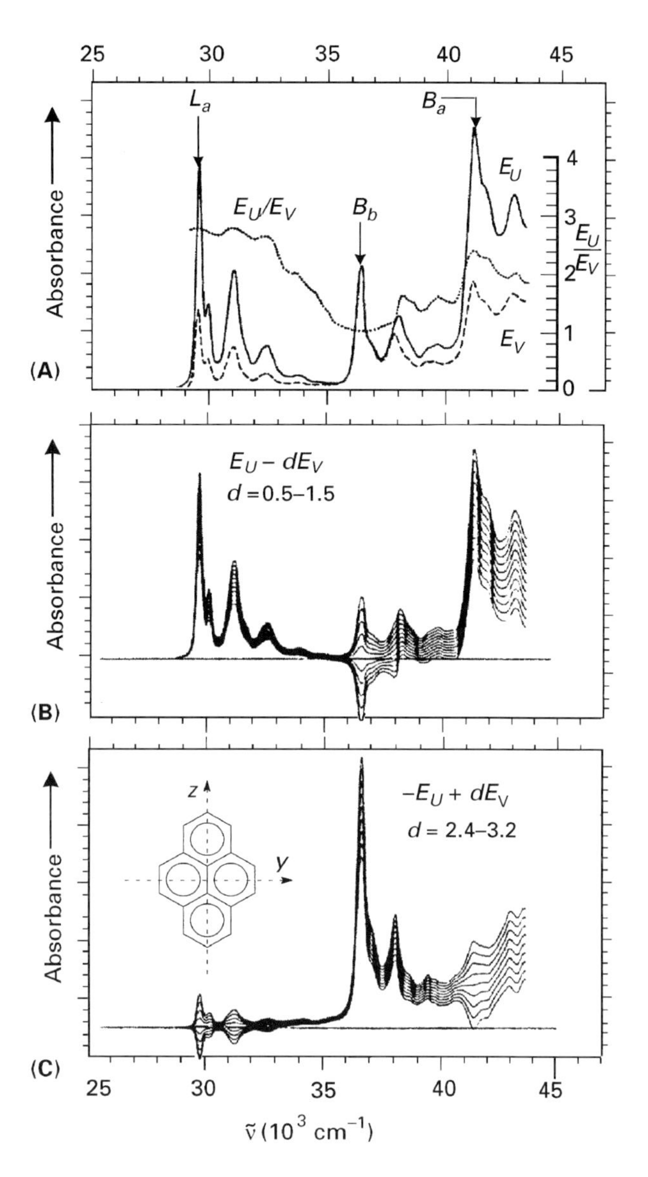

Figure 1 UV spectra of pyrene in stretched polyethylene at 77K. Absorbance in arbitrary units. (A) E_U and E_V (baseline-corrected) and the dichroic ratio E_U/E_V. (B) and (C) Linear combinations of E_U and E_V, in search of spectral curves corresponding to A_z and A_y, respectively (the TEM trial-and-error process). Adapted by permission from Langkilde FW, Gisin M, Thulstrup EW and Michl J (1983) *Journal of Physics and Chemistry* **87**: 2901–2911.

Photoselection: negligible depletion; linearly polarized light

The photoselected sample becomes uniaxial with respect to the electric vector of the light, which we label U. On the molecular level, the transition moment direction for the transition responsible for the photoselection becomes the orientation axis, which we label molecular axis z. All directions perpendicular to z will have the same K value. We obtain a rod-like alignment of the photoselected molecules: $K_z = 3/5$; $K_y = K_x = 1/5$.

Photoselection: negligible depletion; natural or circularly polarized light

The photoselected sample becomes uniaxial with respect to the direction of light propagation, which we label U. On the molecular level, the transition moment of the absorbing transitions now defines the axis that is least aligned with U, which we label x. We obtain a disc-like alignment of the photoselected molecules: $K_x = 1/5$; $K_y = K_z = 2/5$.

An interesting aspect of these orientation distributions obtained by photoselection is that not only the K values needed for interpretation of absorption spectra but the complete orientation distribution function is known, including the L values.

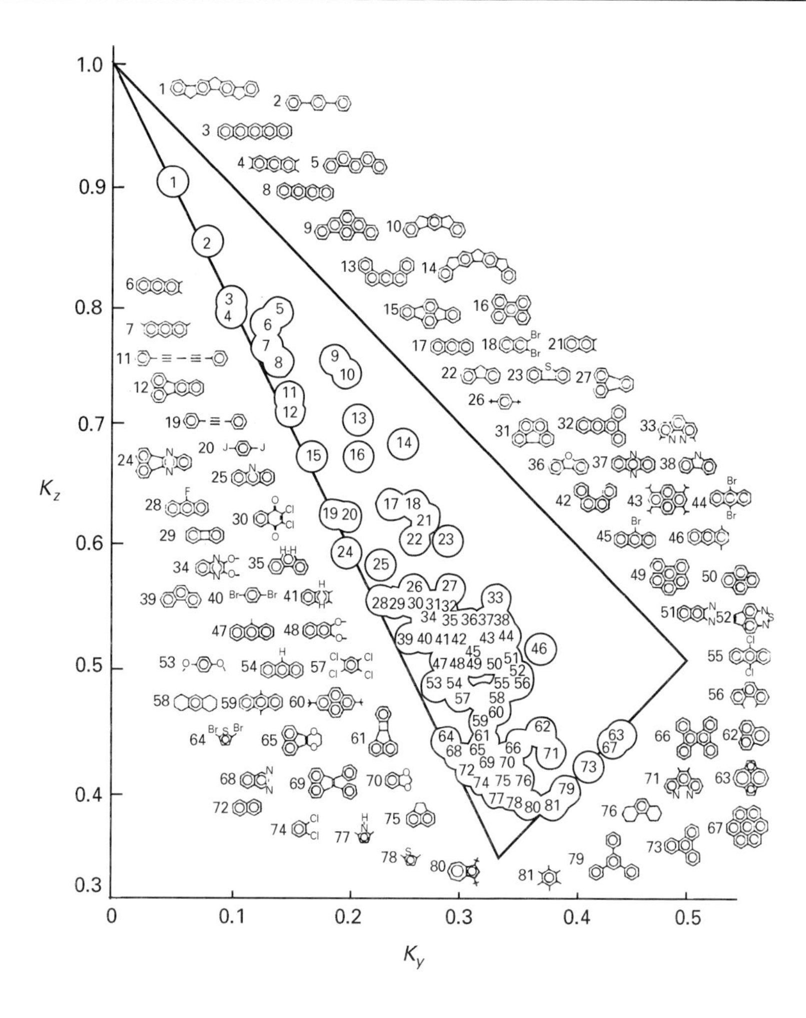

Figure 2 Values of (K_z, K_y) for aromatic molecules in stretched polyethylene at 293 K. The molecular z axes are horizontal in the structures, the y-axes are vertical. Adapted by permission from Thulstrup EW and Michl J (1982) *Journal of the American Chemical Society* **104**: 5594–5604.

Anisotropic (uniaxial) solvents

In the case of molecules aligned in stretched polymer sheets or nematic liquid crystals, the orientation distribution is unknown. Nevertheless, K values for many or all relevant transitions may be easily determined using the TEM method. For molecules with only two different K values, for example with a C_3 or higher axis, and in spectral regions where only two different K values exist, linear combinations of the two observed spectra may be formed that correspond to sums of contributions from transitions with the same K value.

One important example is that of π–π* transitions in a molecule of C_{2v} or D_{2h} symmetry. The transition moments for these are restricted to two different, perpendicular directions in the molecule; we shall label these directions y and z. The sum of contributions from z-polarized transitions is

$$A_z(\nu) = [(1 - K_y)E_U(\nu) - 2K_yE_V(\nu)]/(K_z - K_y)$$

while the sum of contributions from y-polarized transitions is

$$A_y(\nu) = [2K_zE_V(\nu) - (1 - K_z)E_U(\nu)]/(K_z - K_y)$$

Figure 3 illustrates the resulting curves for the π–π* transitions in pyrene. The separation of transitions into two groups may help identify excited-state symmetries, separate close-lying, differently polarized transitions and reveal hidden transitions. It thereby simplifies the assignment of both electronic and vibrational transitions considerably.

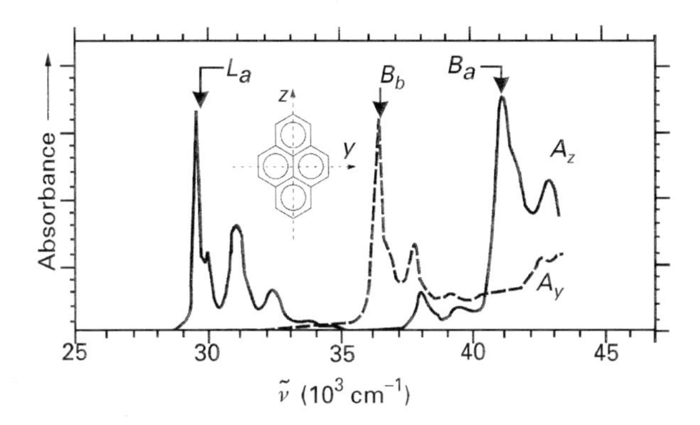

Figure 3 Curves corresponding to purely z-polarized (A_z) and purely y-polarized absorbance (A_y) obtained by the TEM method as illustrated in **Figure 1**. Adapted by permission from Michl J and Thulstrup EW (1995) *Spectroscopy with Polarized Light: Solute Alignment by Photoselection, in Liquid, Polymers, and Membranes.* New York: Wiley-VCH.

If the molecule is of lower symmetry, very useful, but less specific, results can still be obtained. Let us assume that the molecule has a symmetry plane as its only symmetry element and that the out-of-plane direction, x, corresponds to the lowest possible K value, K_x. Transitions must be polarized either along x or in the yz plane, perpendicular to x. We may therefore observe the following K values: $K = K_x$ or $K_y \leq K \leq K_z$.

For a transition f in the (y, z) plane, the angle between the molecular long axis, z, and the transition moment $\boldsymbol{M}_{0f}$ for transition f, $(z, \boldsymbol{M}_{0f})$, can now be determined from K_y, K_z and K_f:

$$\tan^2(z, \boldsymbol{M}_{0f}) = \frac{K_z - K_f}{K_f - K_y} \qquad [13]$$

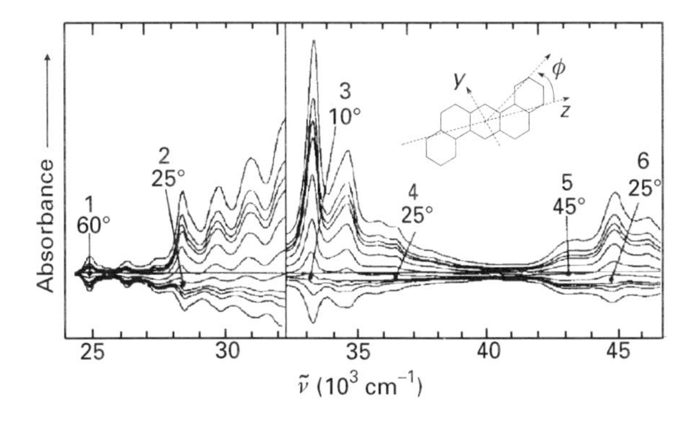

Figure 4 UV spectra of dibenz[*a,h*] anthracene. An example of a planar molecule for which π–π^* transition moments may be along any direction in the molecular plane. The curves are linear combinations of E_U and E_V and correspond to transitions that form angles of 0°, 15°, 30°, 45°, 60°, 75°, and 90° with the molecular z axis. Adapted by permission from Pedersen PB, Thulstrup EW and Michl J (1981) *Chemical Physics* **60**: 187–198.

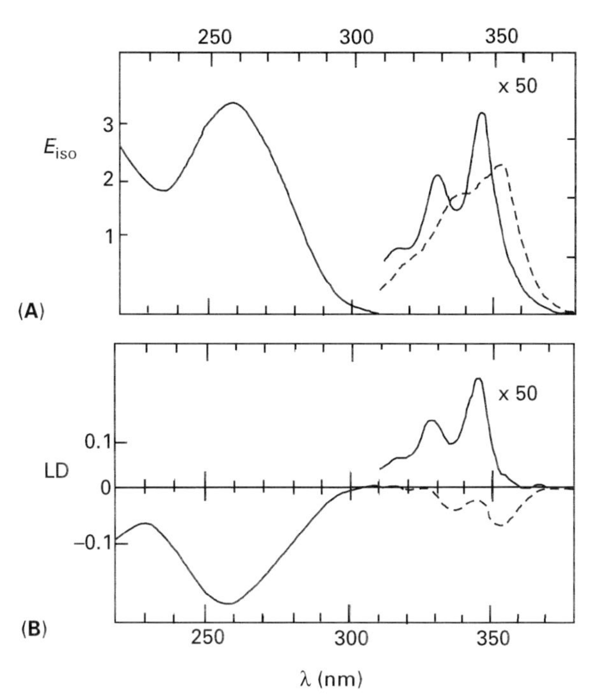

Figure 5 UV absorption (E_{iso}) (A) and LD spectra (B) of two derivatives of benz[a]pyrene, covalently bound to DNA aligned in a flow. Both derivatives are dihydroxyepoxytetrahydrobenzpyrenes, but their two hydroxy groups are positioned on different sides of the molecular plane (*syn*, broken line; anti, full line). The absorption below 300 nm is mainly due to the DNA bases, the planes of which are predominantly perpendicular to the DNA helix axis. This explains the negative linear dichroism below 300 nm. The absorption above 300 nm is mainly due to the two derivatives; it is seen that *anti* has a positive LD, corresponding to a small angle between its molecular plane and the helix axis, while *syn* has a negative LD, corresponding to a large angle between its molecular plane and the helix axis. The biological difference between the two molecules is considerable; *syn* is a weak carcinogen, while *anti* is a very strong one. Adapted by permission from Tjerneld F (1982) Thesis, Chalmers University of Technology, Gothenburg, Sweden.

An example of the determination of such angles in a molecule with only the molecular plane as symmetry element is shown in **Figure 4.**

Macromolecules, especially DNA, that are difficult to dissolve in stretched polymer sheets or nematic liquid crystals have often successfully been aligned in a flow. Flow LD is measured in a solution between a stationary and a fast-rotating cylinder using a Couette or Maxwell cell. It has proved very useful for the study of the interaction between macromolecules and smaller molecules, for example the attachment of carcinogenics to DNA. The long axis of the DNA molecules, the helix axis, becomes the orientation axis. So far such studies have primarily been performed in the UV-visible region. The two linearly independent spectra obtained are

$$\mathrm{LD}(\nu) = E_U(\nu) - E_V(\nu) \quad \text{and}$$
$$E_{\mathrm{iso}}(\nu) = [E_U(\nu) + 2E_V(\nu)]/3 \qquad [14]$$

The latter curve may be obtained when the flow is stopped. **Figure 5** shows $E_{\mathrm{iso}}(\nu)$ and LD(ν) for two metabolites of benzo[*a*]pyrene, of which one is a strong carcinogen, the other weak. The measurement was done in a Couette cell containing a mixture of a small amount of a metabolite with DNA in a weak solution of sodium cacodylate in water. The strong absorption of the DNA bases dominates the spectrum below 300 nm (with a negative LD because the DNA base absorption is polarized in the molecular planes of the bases, which are almost perpendicular to the helix axis).

The weak LD absorption of the metabolites correspond to an in-plane polarized transition; it can be studied above 300 nm. The stronger carcinogen of the metabolites has a positive LD, the other a negative. This means that the former is positioned with its molecular plane relatively parallel to the helix axis, while the molecular plane of the latter is more perpendicular to the helix axis.

List of symbols

$A_f(\nu)$ = contribution to the absorbance of transition f in a sample of perfectly aligned molecules; $E_U(\nu)$ = absorbance (at frequency ν) with electric vector in the specified direction (U, V, etc.); $|f\rangle$ = final state; K_f = average cosine square value = $\langle\cos^2(U, \mathbf{M}_{0f})$; $\hat{M}$ = electric dipole operator; $\mathbf{M}_{0f}$ = electric dipole transition moment; $\hat{O}$ = generic operator; u = a direction in the molecule; U = laboratory axis, unique axis; v = a direction in the molecule (= x, y, z); V = laboratory axis perpendicular to U; W = laboratory axis perpendicular to U and V; $|0\rangle$ = initial state; ε = electric vector; ν = light frequency.

See also: **Laser Spectroscopy Theory; Liquid Crystals and Liquid Crystal Solutions Studied By NMR; Nucleic Acids and Nucleotides Studied Using Mass Spectrometry; Nucleic Acids Studied Using NMR; Optical Spectroscopy, Linear Polarization Theory; Vibrational CD Spectrometers; Vibrational CD, Applications; Vibrational CD, Theory.**

Further reading

Kliger DS, Lewis JW, Randall CE (1990) *Polarized Light in Optics and Spectroscopy*. London: Academic Press.

Michl J and Thulstrup EW (1995) *Spectroscopy with Polarized Light: Solute Alignment by Photoselection, in Liquid Crystals, Polymers, and Membranes*. New York: Wiley-VCH.

Rodger A and Nordén B (1997) *Circular Dichroism and Linear Dichroism*, Vol. 106, Oxford Chemistry Masters. Oxford: Oxford University Press.

Samori' B and Thulstrup EW (eds) (1988) *Polarized Spectroscopy of Ordered Systems*, Vol. 242 of NATO ASI Series, Series C: Mathematical and Physical Sciences. Dordrecht: Kluwer Academic.

Thulstrup EW and Michl J (1989). *Elementary Polarization Spectroscopy*. New York: VCH.

Linear Dichroism, Instrumentation

Erik W Thulstrup and **Jens Spanget-Larsen**,
Roskilde University, Denmark
Jacek Waluk, Polish Academy of Sciences, Warsaw, Poland

ELECTRONIC SPECTROSCOPY
Methods & Instrumentation

All linear dichroism (LD) measurements are performed on anisotropic samples. Many samples occurring in nature are partially oriented from the start. Nevertheless, most LD experiments are performed on samples that have been aligned in the laboratory. Therefore, prior to learning the techniques of measurement, one should become familiar with different procedures used to align molecular ensembles.

Methods of alignment

The majority of LD studies are done on partially aligned samples, which exhibit uniaxial orientation. Such samples have a unique direction (optical axis, unique sample axis) with respect to which the solute molecules are oriented. All directions perpendicular to this axis are equivalent. The most widespread method for obtaining partial alignment is the use of anisotropic solvents: stretched polymers or liquid crystals.

Stretched polymers This is the simplest and the least expensive method. Solutes may be introduced into the polymers by (i) swelling the polymer with a solution containing the substrate; (ii) diffusion of the vapour of the substrate into the polymer; (iii) casting a film of the polymer from a solution containing the substrate; or (iv) dissolving the substrate in a molten polymer. The most commonly used polymers are low-density polyethylene, best for nonpolar solutes, and poly(vinyl alcohol), used for polar and nonpolar solutes. Both are transparent in the UV-visible regions, and in some regions of the infrared. Mechanical stretching produces uniaxial orientation, its degree increasing with the stretching ratio. The mechanism of orientation is still under debate, but it is evident that the molecular shape is an important factor: the more elongated molecules align on the average better than those that have more spherical shape.

Liquid crystals These are experimentally more demanding than polymers. Usually, thermotropic liquid crystals are used, either nematic or compensated nematic mixtures, obtained from cholesteric components. Many liquid crystals absorb in the UV, but some are transparent down to about 200 nm. The alignment of the liquid crystal sample can be controlled by external electric or magnetic fields, or by cell surface pretreatment such as rubbing or coating. These procedures make it possible to switch the direction of the unique axis from perpendicular to the surface (homeotropic alignment) to parallel (homogeneous alignment). The degree of orientation of the solute depends on its nature and concentration, as well as on the nature and temperature of the sample and the imperfections of the cell surface.

Lyotropic liquid crystals, Langmuir-Blodgett films and even monolayers adsorbed on surfaces can also be used as uniaxial orienting media.

Single crystals In principle, this classical technique makes it possible to obtain nearly perfect orientation and absolute polarization assignments. However, both the measurement and the interpretation are often difficult. Very thin crystals are required for electronic absorption studies. Intermolecular interactions (Davydov splitting) often complicate the spectra. The solution is to find a single crystal of a suitable host molecule, transparent in the region of interest. These two conditions are often very difficult to meet simultaneously.

Electric and magnetic fields Linear dichroism produced as a result of applying an electric field is known as electric dichroism (Kerr effect). The effect is usually quite small for normal-sized molecules, although polymers such as DNA can be oriented very well. The orientation imposed by a magnetic field is even smaller, and therefore its practical use is limited to large molecular ensembles (Cotton–Mouton effect).

Flow field Macromolecules, such as DNA, can become orientated by the hydrodynamic shear associated with liquid flow. This is realized experimentally in a Maxwell cell (sometimes referred to as Couette cell). The cell consists of a fixed cylinder and a rotating cylinder: a flow gradient is produced in the liquid solution contained in the annular gap between the two cylinders.

Photoselection and photoorientation This elegant method allows one to obtain a set of partially oriented molecules in a completely isotropic medium. It makes use of the fact that the photoexcited ensemble of molecules is always anisotropic, and thus exhibits LD (transient dichroism). Moreover, if the excitation is followed by chemical transformation, and the environment is rigid enough to prevent molecular rotation, a permanent alignment of both reactant and product is obtained. Such photooriented samples can be studied by conventional LD techniques. Particularly attractive media for use in photoorientation are low-temperature rare gas matrices, which are inert and transparent in both the UV-visible and IR regions.

Techniques of measurement

For partially oriented uniaxial samples, two linearly independent spectra can be obtained from an experiment, irrespective of the mode of measurement and the polarization state of the absorbed light. The obvious procedure is to record two absorption spectra, one with the electric vector of light parallel to U, the unique axis of the sample (E_U), the other with the electric vector perpendicular to the unique axis (E_V). The spectra must be baseline-corrected by subtracting the curves recorded on pure solvents. This yields the largest difference between the tow linearly independent spectra. Other alternatives are also possible. A combination of E_U or E_V with the spectrum of an isotropic sample, E_{iso} $(=[E_U + 2E_V]/3)$ is easily obtained in liquid crystal. In cases where the sample geometry precludes the measurement of both E_U and the isotropic spectrum (membranes), the two spectra are E_V and E_ω, the latter recorded with the incidence angle $\pi/2 - \omega$, and with the plane of polarization containing the U axis (the 'tilted plate method'). Another possibility is to combine either E_U or E_V with the unpolarized spectrum measured with a depolarizer in front of the sample and the light beam perpendicular to the U axis. Finally, one can use procedures that do not require the use of a polarizer at all, such as recording first the E_V spectrum for an aligned liquid solution (with light propagating along the U axis), and then E_{iso} for the same solution with temperature raised above the isotropic melting point.

All these measurements can be performed on a commercial double–beam spectrophotometers supplemented with one or two linear polarizers (**Figure 1**). Calcite Glan prisms are used for measurements down to about 220 nm. Sheet polarizers can tolerate larger apertures and are much cheaper. On the other hand, their transmission is lower and the wavelength range usually limited to the visible range, although the UV-transmitting polarizing sheets have become available.

Various potential sources of errors in LD measurements are: (i) scattering and depolarization by the sample, particularly significant in the UV region; (ii) sample inhomogeneity and irreproducibility, which often makes it difficult to reliably record the baseline; (iii) sample birefringence; and (iv) polarization bias of the instrument. For weakly dichroic samples, the separate recording of two spectra is not practical, and the LD has to be measured directly. This is achieved by modulation methods. The state of polarization of the light beam is modulated in time, using a combination of a linear polarizer and, most often, a photoelastic modulator. Since the transmission of the dichroic sample varies for different polarizations of the light, the detector produces an a.c. signal, superimposed on the d.c. signal, owing to the average intensity of the transmitted beam. The analysis of the two signals allows an evaluation of the LD and of the average optical density of the sample. For measurements of this type, one can adapt a spectropolarimeter normally used for circular dichroism measurements (**Figure 2**).

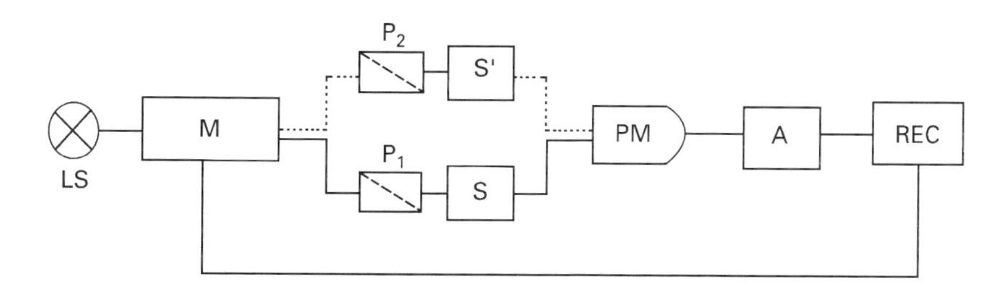

Figure 1 Schematic diagram of a double-beam spectrophotometer for polarized spectroscopy. LS, light source; M, monochromator; P_1, P_2, polarizers; S, sample; S′, reference sample (optional); PM, photomultiplier; A, amplifier; REC, recorder. Adapted by permission from Michl J and Thulstrup EW (1995) *Spectroscopy with Polarized Light: Solute Alignment by Photoselection, in Liquid, Polymers, and Membranes.* New York: Wiley-VCH.

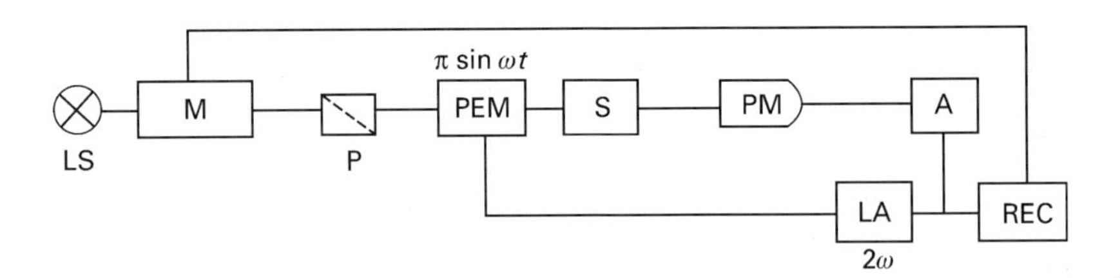

Figure 2 Schematic diagram of a spectropolarimeter adapted for linear dichroism measurements. LS, light source; M, monochromator; P, polarizer; PEM, photoelastic modulator producing time-dependent retardation $\pi \sin \omega t$; S, sample; PM, photomultiplier; A, d.c. amplifier; LA, lock-in amplifier operating at frequency 2ω; REC, recorder. Adapted by permission from Michl J and Thulstrup EW (1995) *Spectroscopy with Polarized Light: Solute Alignment by Photoselection, in Liquid, Polymers, and Membranes*. New York: Wiley-VCH.

Although most LD studies are performed by recording the intensity of transmitted light, other methods are also used. One of them detects the LD via fluorescence, the other by photoacoustic spectroscopy. Both methods are very sensitive and avoid some of the errors inherent in the transition mode, such as light scattering.

Most of the alignment techniques and methods of measurements used for electronic linear dichroism can be adapted for studies in the IR region. The merits of IR-LD make this technique a very attractive choice, complementing and in many respects surpassing the informational content provided by electronic linear dichroism.

List of symbols

E_{iso} = absorbance in an isotropic sample; E_U = absorbance with the electric field vector parallel to the unique axis, U; E_V = absorbance with the electric field vector perpendicular to the unique axis (parallel to the perpendicular axis, V); U = the unique optical axis; ω = angle complementary to the incidence angle in the 'tilted plate method'.

See also: **Linear Dichroism, Theory; Linear Dichroism, Applications; Nucleic Acids and Nucleotides Studied Using Mass Spectrometry; Nucleic Acids Studied Using NMR; Vibrational CD, Applications; Vibrational CD Spectrometers; Vibrational CD, Theory.**

Further reading

Kliger DS, Lewis JW and Randall CE (1990) *Polarized Light in Optics and Spectroscopy*. San Diego: Academic Press.

Michl J and Thulstrup EW (1995) *Spectroscopy with Polarized Light. Sample Alignment by Photoselection, in Liquid Crystal, Polymers, and Membranes*. New York: Wiley-VCH.

Rodger A and Nordén B (1997) *Circular Dichroism and Linear Dichroism*, Vol. 106. Oxford Chemistry Masters. Oxford: Oxford University Press.

Samori B and Thulstrup EW (1988) *Polarized Spectroscopy of Ordered Systems*, Vol. 242 of NATO ASI Series, Series C: Mathematical and Physical Sciences. Dordrecht: Kluwer Academic.

Thulstrup EW and Michl J (1989) *Elementary Polarization Spectroscopy*. New York: VCH.

Linear Dichroism Theory

See **Optical Spectroscopy, Linear Polarization Theory.**

Plate 22 (left) Radiocarbon dating. View of a linear accelerator used as part of an accelerator mass spectrometer (AMS). This device is capable of counting the relatively few carbon-14 atoms in a radioactive sample. The proportion of carbon-14 to carbon-12 atoms in the sample may be used to determine the radiocarbon age of an organic object. This is then adjusted by various corrections to give the true age. *See Isotope Ratio Studies Using Mass spectrometry*. Reproduced with permission from Science Photo Library.

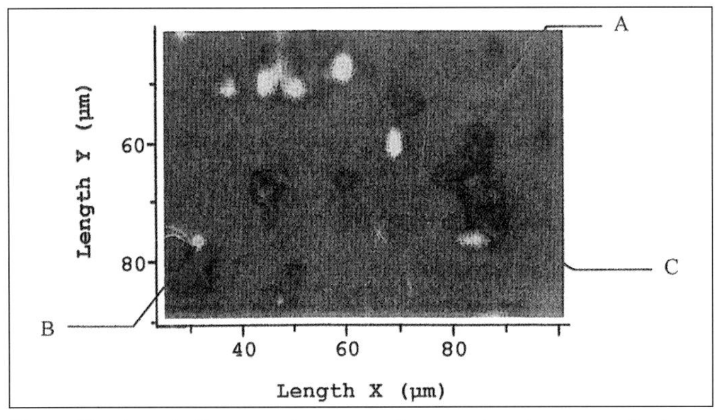

Plate 23 (above) Raman microscope image of a pharmaceutical powder showing the relative importance of three compound, A, B and C. *See Industrial Applications of IR and Raman Spectroscopy*. Courtesy of A. S. Gilbert and R. W. Lancaster.

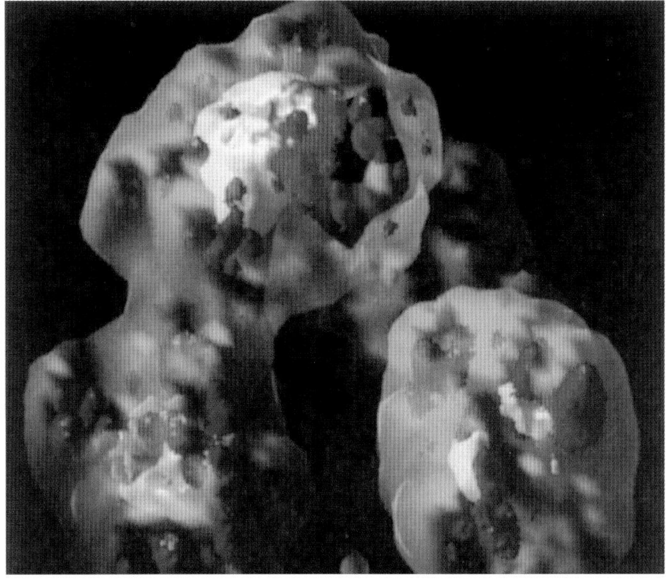

Plate 24 (above) 3-D reconstruction of a plant cell undergoing meiotic cell division, created from two-photon cross-sectioned images. *See Laser Spectroscopy Theory; Light sources and Optics*. Reproduced with permission from Spectra-Physics © W. Zipfel and C. Conley/Cornell University.

Plate 25 (below) Thorax and abdomen. Coloured Magnetic Resonance Imaging (MRI) scan of the thorax and abdomen of a woman aged 58 years, seen in posterior view. The arms are seen on either side, with the waist and hips at lower frame. The thorax contains the blue lung fields (upper frame) with bones of the thoracic spine at upper centre. In the abdomen are lobes of the liver (green, below the lungs) and a pair of kidneys (green, below the liver lobes). Bands of muscle can be seen around the lumbar spine (lower centre) and the shoulder (at top, magenta). MRI uses probes of radiowave energy in the presence of a magnetic field to create slices. *See MRI Applications, Clinical; MRI Instrumentation; MRI Theory; Contrast mechanisms in MRI*. Reproduced with permission from Science Photo Library.

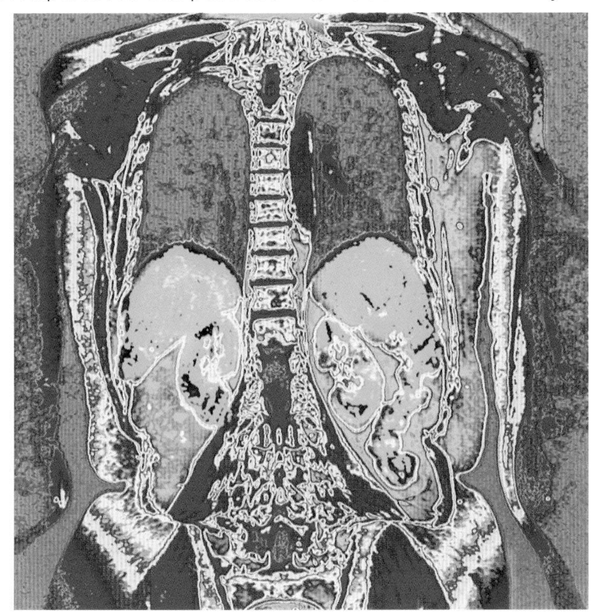

Plate 26 (right) False-colour magnetic resonance image (MRI) of a whole human body, a woman, taken in coronal (frontal) section. Various parts of the body are prominent: bone appears in black, with the hip and shoulder joints clearly defined. The lungs are also black. Vertebrae forming the spinal column are also obvious (in yellow and white). Part of the spinal cord is visible (in gold) beneath the base of the brain, the hemispherical structure of which is revealed in this section. This whole body image is the product of a number of MRI scans made along the length of the body. The complete image was rendered from data relating to the various sections stored on the scanner's computer. *See MRI Applications, Clinical; MRI Instrumentation; MRI Theory; Contrast mechanisms in MRI*. Reproduced with permission from Science Photo Library.

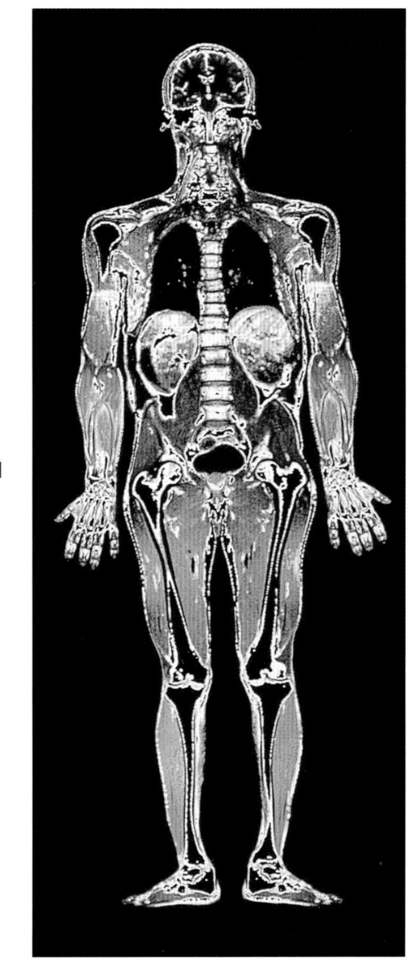

Plate 27 (left) Portrait of Francis Aston (1877–1945) British physicist and Nobel Laureate. After WW1, Aston helped Thomson in his studies of the deflection of ions in magnetic fields. He went on to improve Thomson's apparatus, designing it so as to make all atoms of a given mass fall on the same part of a photographic plate. Working with neon, he found that two lines were isotopes. He repeated this with chlorine with similar results. The device, called the mass spectrometer, showed that most stable elements had isotopes. His work earned him the 1922 Nobel Prize for Chemistry, and introduced a powerful new analytical tool to science. *See Mass Spectrometry, Historical Perspective*. Reproduced with permission from Science Photo Library.

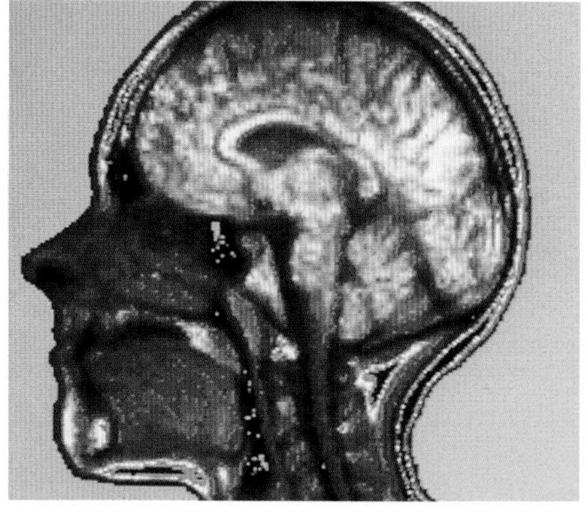

Plate 28 (right) Magnetic Resonance Image of human female brain. *See MRI Applications, Clinical*. Reproduced with permission from © Medipics/Dan McCoy/ Rainbow.

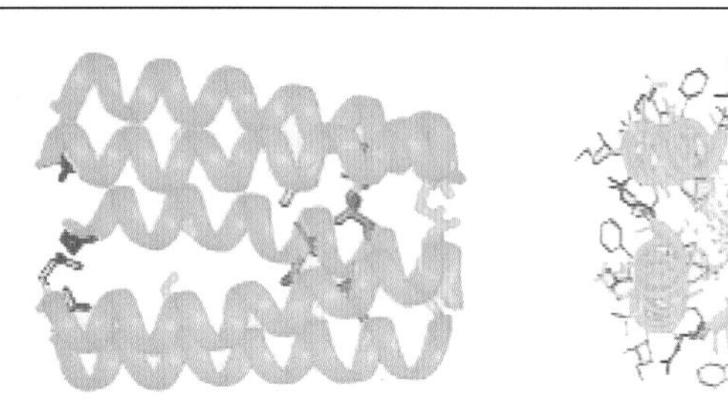

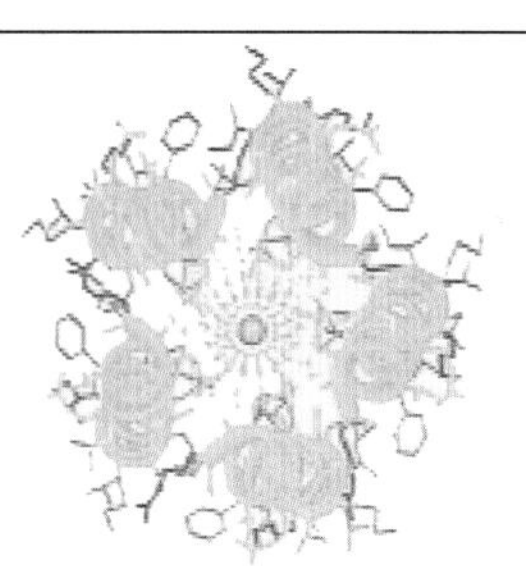

Plate 29 (above left) Structure of membrane proteins. Membrane peptides often self assemble into controlled oligomeric forms to make molecular selective channels whose structure can be very difficult to resolve at the molecular level by most methods, although solid state NMR methods can make a contribution to their functional and structural description. Here the pentameric funnel-like bundle of the M2-peptide from the nicotinic acetly choline receptor has been resolved from ^{15}N NMR studies of orientated M2 peptides in lipid bilayers. The funnel has a wide mouth at the N-terminal, intracellular side of the pore. The pore lining residues has also been modelled and distances between residues in the channel estimated. The a-carbon backbone is in cyan, acidic residues in red and basic residues in blue, polar residues in yellow and lipophilic residues in purple. (Figure adapted from Opella et al., (1999) Nature St. Biology, 6: 374–379). *See Membranes Studied by NMR Spectroscopy*.

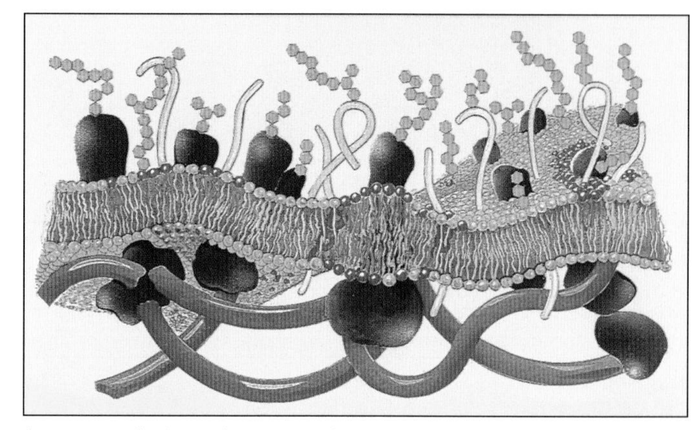

Plate 30 (above right) Biological membranes which enclose all living cells, and are also present within the cells of higher life forms, are very complex and heterogeneous in their chemical make-up. Between 20 and 40% of components encoded by the genome in any one living cell eventually end up in the membranes of the cell. Any one typical membrane may be composed of over 100 different bilayer lipids and sterols, several hundreds of different proteins, some of which have polysaccharides attached to them, and the whole assembly may be supported by a protein scaffold underlying the membrane on the cytoplasmic side. The whole molecular complex is in dynamic equilibrium, with lipids and sterols rotating fast (about 10^9 times per second) around their long axis, and diffusing laterally (covering about 10^8 cm^2 in one second), whilst at the same time maintaining a regular and relatively ordered structure, on average. Such dynamic information has come from a range of spectroscopic methods, including NMR, in which it is usual to study just one, or a limited number of such components in a simplified model system, and try and understand them and their interactions with a limited number of other components. (Figure by Ove Broo Sorensen of the Technical University of Copenhagen, Denmark). *See Membranes Studied by NMR Spectroscopy*.

Plate 31 (left) Richard Ernst, Nobel Laureate in Chemistry 1991. Notes for this contribution to NMR spectroscopy and MRI. *See Magnetic Resonance, Historical Perspective*. Reproduced with permission from The Nobel Foundation.

Plate 32 (right) Felix Bloch who was awarded the Nobel prize for physics in 1952 jointly with Edward Purcell. They led two independent research groups which, in 1945, first detected the nuclear magnetic resonance phenomenon in bulk matter. *See Magnetic Resonance, Historical Perspective*. Reproduced with permission from The Nobel Foundation.

Plate 33 The first fully functioning electrospray mass spectrometer built at Yale by Masamichi Yamashita in 1983. Reproduced with permission from John B. Fen.

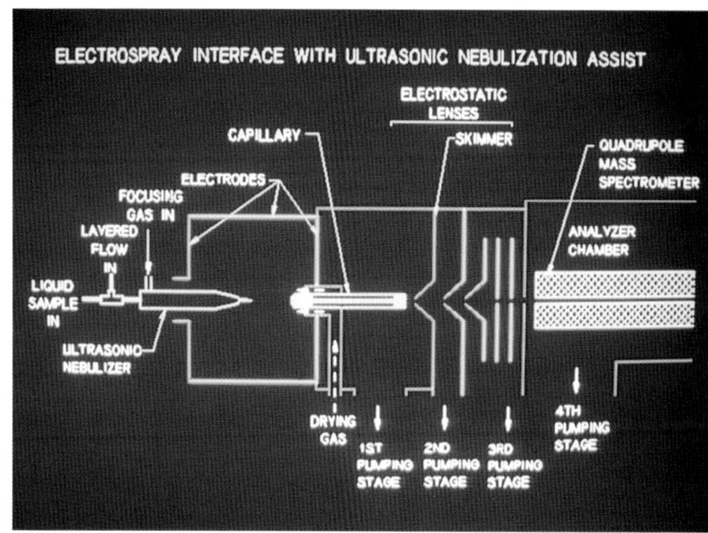

Plate 34 A schematic representation of a typical electrospray mass spectrometer in which the mass analyser is a quadrupole mass filter. Reproduced with permission from John B. Fen.

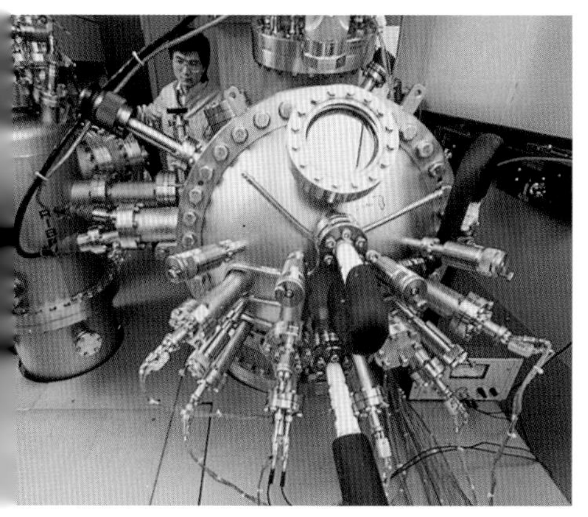

Plate 35 A Bell Lab ZAB high field mass spectrometer. Reproduced with permission from © Medipics/Dan McCoy/Rainbow.

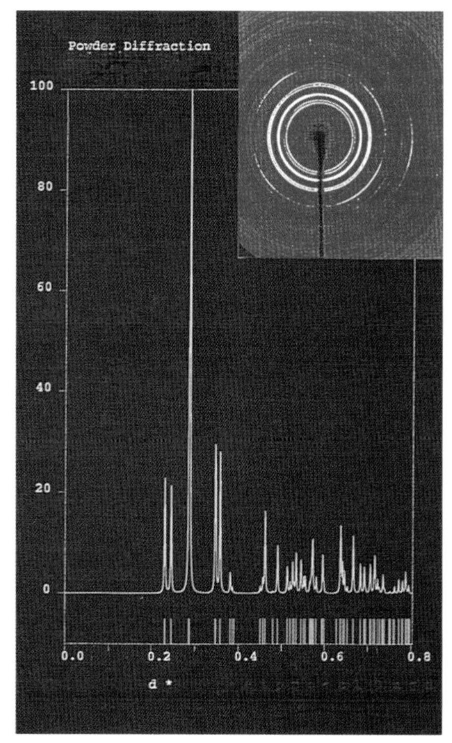

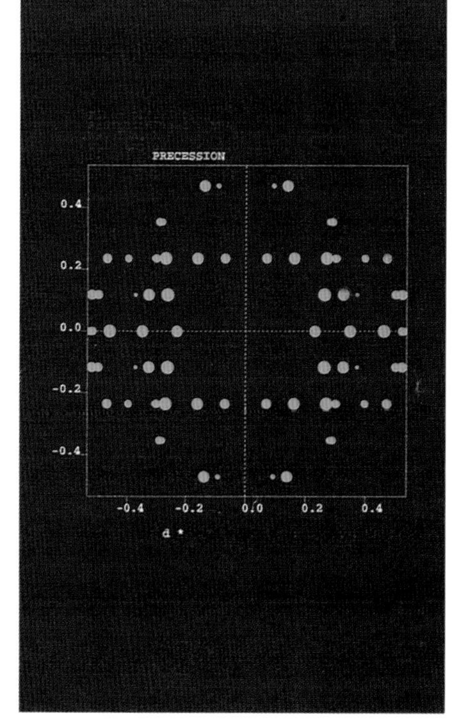

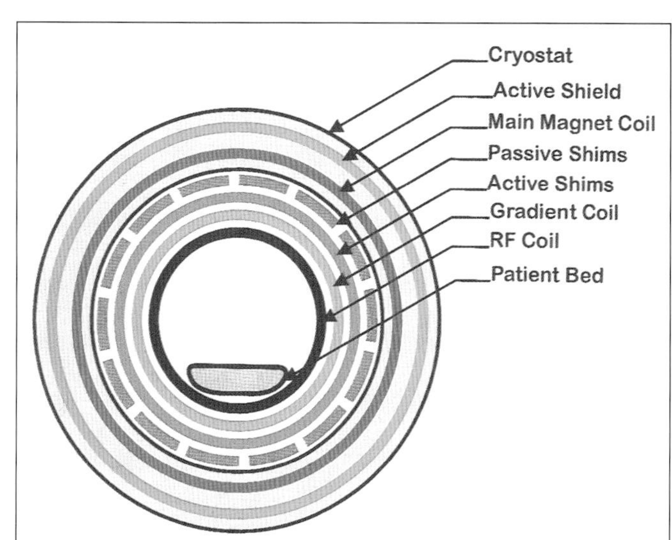

Plate 36 (above) The left-hand picture illustrates a powder diffraction pattern of S_2N_2 at a wavelength of 0.325 Å. The insert is the actual pattern which can be integrated and displayed as a function of *d* [d* = 2 sin q/1]. The right-hand picture shows the result if one single S_2N_2 crystal is rotated in the X-ray beam. The size of the spots illustrates the difference in diffracted intensities from the separated Bragg relections. *See Material Science Applications of X-ray Diffraction*. Courtesy of Svensson and Krick, 1998.

Plate 37 (left) Schematic cross-section through a typical superconducting clinical MR scanner. Within the cyrostat (light blue) are the superconducting coils of the primary magnet (red) and active shield (green). In the bore of the magnet there are passive shim rods (grey), active shim coils (orange), gradient set (blue), whole body RF coil (black) and patient bed. The tractable diameter is generally half the magnet bore diameter. *See MRI Instrumentation*. Courtesy of P. D. Hockings, J. F. Hare and D. G. Reid.

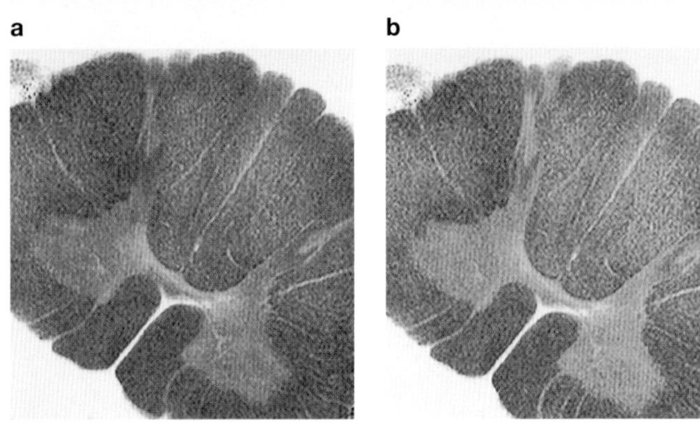

Plate 38 a and b Black and white & colourscale diffusion weighted spin echo magnetic resonance images of postmortem human cervical spinal cord using a micro-imaging probe at 600 MHz observation frequency. *See NMR Microscopy*. Reproduced with permission from Doty Scientific Inc.

Plate 39 (right) Small angle X-ray fibre diffraction pattern recorded form of the DNA double-helix. *See Nucleic Acids and Nucleotides Studied Using Mass Spectroscopy; Nucleic Acids Studied by NMR*. Courtesy of A. Mahendrasigam and W. Fuller.

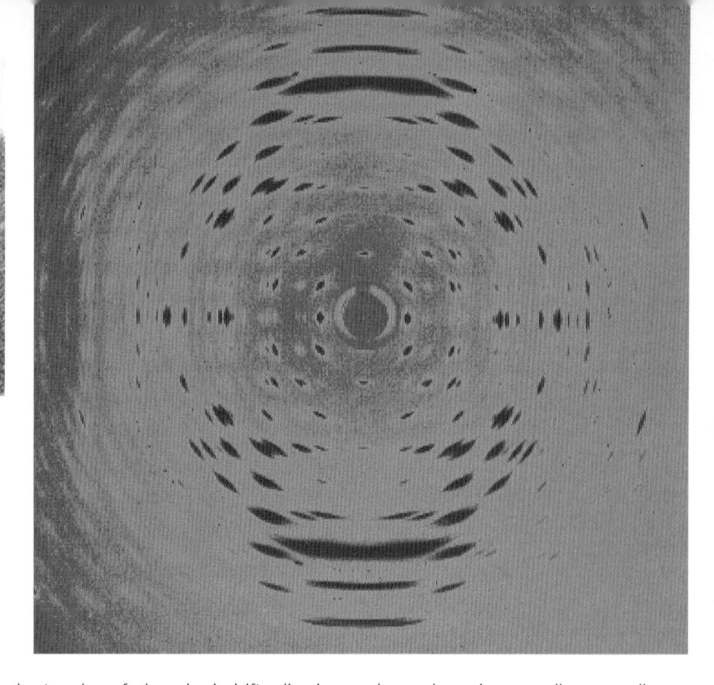

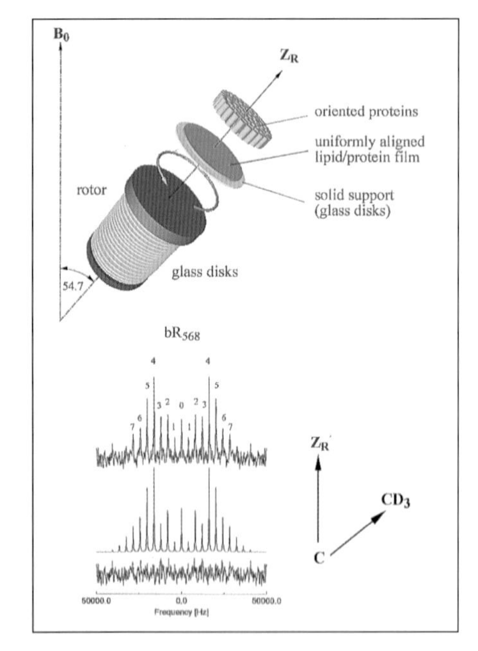

Plate 40 (left) The anisotrophy of chemical shift, dipolar and quadrupolar coupling usually broadens NMR spectra of very large, slowly tumbling complexes making a detailed structural analysis a difficult task. If however the sample is set spinning at an angle of 54.7° (MASS = magic angle sample spinning) with respect to the applied magnetic field, all anisotropies are scaled down and eventually collapse at sufficiently high speeds resulting in isotropic spectra. In a special version of MASS called MAOSS, oriented membranes are set spinning at moderate speeds (1–4 kHz) at the magic angle, and now the narrow spinning side bands, which arise from not-averaged anisotropic interactions, give information about the orientation of the group being observed (here a CD^3 group in deuterium NMR of a prosthetic group in a large protein embedded in membrane) at much higher sensitivity and resolution than could be obtained from a conventional spectrum of a statically positioned sample. Similar experimental condition for other nuclei (^{13}C, ^{15}N), permit both the magnitude and direction of the dipolar couplings to be extracted, to give high resolution structural details within a large membrane complex. (courtesy of Dr C. Glaubitz; adapted from Glaubitz and Watts, (1998) J. Mag. Res. 130; 305–316). *See NMR in Anisotropic Systems, Theory; Solid State NMR, Methods*.

a

b

Plate 41 (right) a A superconducting NMR magnet operating at 18.8T for ^1H NMR observation at 800 MHz demonstrating the size of these state-of-the-art magnets. Reproduced with permission from Bruker Instruments Inc., Billerica, MA, USA.
Plate 41 b A modern high-resolution NMR spectrometer. A superconducting magnet is shown at the rear, in this case providing a field of 18.8T corresponding to a ^1H observation frequency of 800 MHz. Behind the operator is the single console containing the RF and other electronics and the temperature-control unit. The whole instrument is computer controlled by the workstation shown at the right. Reproduced with permission from Bruker Instruments Inc., Billerica, MA, USA. *See NMR Spectrometers*.

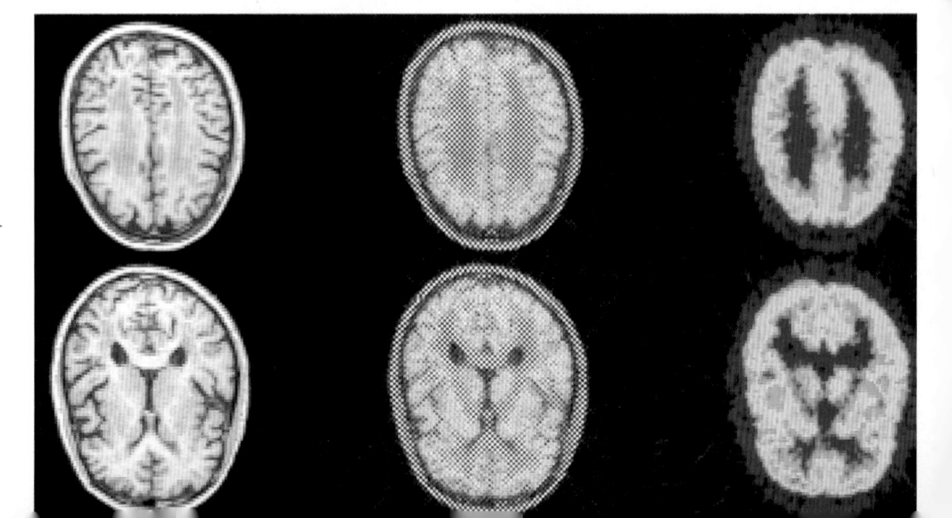

Plate 42 (right) PET functional images of glucose metabolic rate (MRG) (right), MRI images (magnetic resonance images, left) showing anatomical detail, and coregistered (overlaid) PET/MRI (centre). *See PET, Methods and Instrumentation.*

Liquid Crystals and Liquid Crystal Solutions Studied By NMR

Lucia Calucci and **Carlo Alberto Veracini**,
University of Pisa, Italy

MAGNETIC RESONANCE
Applications

Since the first publications on this subject in 1963, NMR in liquid crystalline systems has been a wide and active field of research in many branches of organic and physical chemistry. In fact, NMR spectroscopy has revealed a powerful means of probing molecular structure, anisotropic magnetic parameters and dynamic behaviour of solute molecules dissolved in liquid crystals. Moreover, this technique has been successfully employed to investigate properties of mesophases themselves, such as their orientational ordering, translational and rotational diffusion and their effects on nuclear relaxation, and molecular organization in different liquid crystalline phases.

NMR of different nuclei has been performed: ^{1}H, ^{2}H and ^{13}C NMR have been used extensively, but a number of studies are reported on ^{14}N, ^{19}F, ^{23}Na and ^{31}P also. Recently NMR of active isotopes of noble gases dissolved in liquid crystals has also been employed.

Liquid crystals

We define a liquid crystal as a state of aggregation that is intermediate between the crystalline solid and the amorphous liquid. In this state the molecules show some degree of orientational order with respect to a preferred direction, $\boldsymbol{n}$, called the director; long-range positional order does not exist or is limited to one or two dimensions. In several liquid crystal phases, termed uniaxial phases, only one director is required to describe the molecular order. In contrast, for biaxial phases the orientational order needs two directors for its description. The principal molecular axes are uniformly aligned only on average. The degree to which the molecules are ordered about the director is described by means of a number of parameters, called order parameters.

Substances that show a liquid crystalline phase, or mesophase, are called mesogens. Several thousands of compounds, both with low molecular mass and polymeric, are now known to form mesophases. They are mainly highly geometrically anisotropic in shape, rodlike or disclike (hence the terms calamitic and discotic liquid crystals), or they are anisotropic in solubility properties, like amphiphilic molecules and, depending on their detailed molecular structure, they can exhibit one or more mesophases between the crystalline solid and the isotropic liquid. Transitions to these intermediate states may be induced by purely thermal processes (thermotropic liquid crystals) or by the action of solvents (lyotropic liquid crystals). Each of these two categories can be further divided according to the structure of the mesophases and/or molecules; **Scheme 1** shows the classification of thermotropic mesophases.

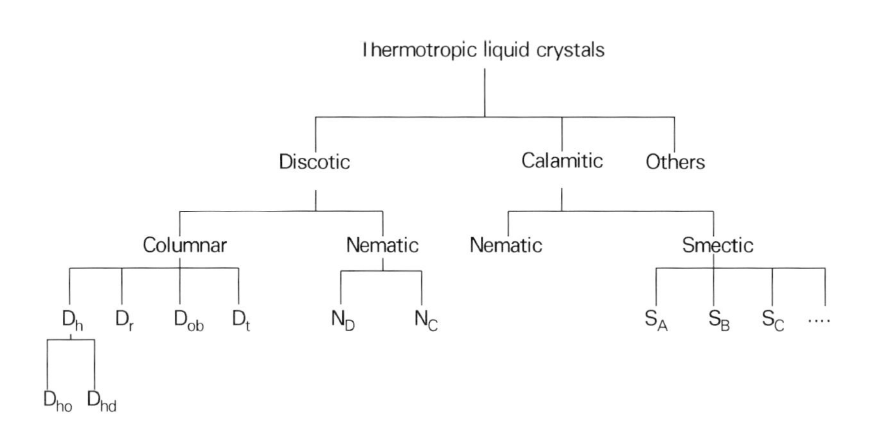

Scheme 1 Classification of thermotropic liquid crystals.

Macroscopic alignment

When subjected to an external magnetic field B_0 of sufficient magnitude (> 0.5 T) the molecules in a mesophase exhibit a macroscopic orientation so that the director $\boldsymbol{n}$ is uniformly aligned in the whole sample. The magnetic field exerts a torque on the molecules, which reorientate cooperatively; the induced reduction in free energy is $\Delta G = \Delta\chi\ B_0^2\ (3\cos^2\alpha - 1)/6$, where $\Delta\chi = \chi_{\parallel} - \chi_{\perp}$ is the difference between the component of the diamagnetic bulk susceptibility along ($\chi_{\parallel}$) and perpendicular to ($\chi_{\perp}$) the director and α is the angle between the director and the magnetic field. The direction of minimal G, i.e. the preferred orientation of $\boldsymbol{n}$, depends on the sign of $\Delta\chi$. Thus, for calamitic mesogens with aromatic cores, which have $\Delta\chi > 0$, $\boldsymbol{n}$ aligns parallel to B_0, while for those formed uniquely by aliphatic units $\Delta\chi < 0$ and $\boldsymbol{n}$ aligns perpendicularly to B_0. Most discotic liquid crystals have $\Delta\chi < 0$, and thus an orientation of $\boldsymbol{n}$ perpendicular to B_0 is observed.

The response time for the alignment depends on the relative values of the magnetic torque experienced by the molecules and of the opposing viscous torque. For most nematic liquid crystals it takes only a few seconds; for some lyotropic phases the alignment is relatively slow and takes minutes or hours. Smectic and columnar discotic liquid crystals are sometimes very viscous and a uniform alignment can only be obtained by reaching these mesophases on slow cooling from the nematic or isotropic phases. For cholesteric samples, the presence of magnetic fields can untwist the helical structure, giving rise to a nematic phase.

Molecules dissolved in liquid crystal solvents become partially oriented, the degree of orientation being dependent on the nature and concentration of the solute.

Observable interactions

The NMR spectra of molecules in a liquid crystalline environment, either mesogenic molecules or solutes, are dominated by second-rank tensorial properties (**T**) such as the shielding tensor (σ_i), the dipolar and indirect spin–spin interactions ($\mathbf{D}_{ij}$ and $\mathbf{J}_{ij}$) and, for nuclei with I>1/2, the quadrupolar interaction ($\mathbf{q}_i$). In most liquid crystals, molecular reorientations and internal motions reduce these interactions to a well-defined average, without the loss of all information. Moreover, molecular translational diffusion averages out to zero the intermolecular interactions, which thus do not complicate the partially averaged spectra. In the usual high-field approximation, in NMR experiments we measure T_{zz}, the component of the averaged interaction tensors resolved along the magnetic field direction (z in the laboratory frame).

The observed averaged quantities are related to the interactions in their principal axis systems (PAS), fixed on the molecules or on rigid molecular fragments, by several transformations of reference frames. Since the molecules are not fixed in a lattice, another frame, the liquid crystal (director) frame (LC), has to be defined. The transformation between the molecular and the LC frame is time dependent because of molecular reorientational motions and the use of the order parameters is needed in this step. Thus T_{zz} relates the NMR measured quantity to the molecular ordering in the mesophases by a number of order parameters depending on the symmetry of the molecule and phase. Measurements of partially averaged second-rank tensors give access only to second-rank order parameters, which are equal to 0 for completely disordered systems, and range from –0.5 to 1 for partially ordered systems, according to $\langle 3l_\alpha l_\beta - 1\rangle/2$, where l_α and l_β are the direction cosines between the PAS and the LC frames. The last step, from the LC frame to the laboratory frame, requires integration over all possible orientations in the case of a powder sample, or the use of a distribution around some preferred orientation in the case of an aligned sample.

Chemical shift anisotropy

NMR spectroscopy of oriented molecules provides a simple method for measuring chemical shift anisotropies, $\Delta\sigma_i$. The change in chemical shift with respect to a reference signal as a function of the molecular degree of order, which can be varied by changing phase and temperature, determines the elements of the shift tensor. In practice, however, the direct measurement of $\Delta\sigma_i$ poses problems related to the reference signal, which depends on changes in local and bulk effects produced by variable sample temperature, phase transitions and degree of alignment with temperature. The variable angle spinning (VAS) method can be used to avoid some of these problems.

Dipolar and indirect couplings

The direct dipolar coupling, D_{ijzz}, and J_{ij}^{aniso}, the anisotropic component of the indirect spin–spin interaction in the direction of the magnetic field, appear in the Hamiltonian in the same form. The splittings that arise in NMR spectra are the sum of these two quantities, called the experimental anisotropic coupling ($T_{ij} = 2D_{ijzz} + J_{ij}^{\text{aniso}}$). It is therefore important to know experimentally, or to evaluate theoretically, the values of the pseudo-dipolar coupling J_{ij}^{aniso} in order to accurately determine D_{ijzz}, which is related to the order parameter of the direction joining the two nuclei and to their distance, i.e. to molecular structure. On the other hand, the magnitude of J_{ij}^{aniso}, and in particular the components of J_{ij} expressed in a molecular axis system, are related to the molecular electronic structure. The measurement of a finite value of J_{ij}^{aniso} is not an unambiguous process, as it does not have a unique effect on the spectrum. Attempts to obtain values of J_{ij}^{aniso} for different pairs of nuclei have been made, and it has been concluded that some indirect couplings such as ^{1}H–^{1}H, ^{1}H–^{13}C, ^{15}N–^{1}H have small anisotropic contributions, while other couplings such as ^{13}C–^{13}C, ^{19}F–^{19}F, ^{1}H–^{19}F and ^{1}H–X (where X is Hg, Cd, Si and Sn) may have a significant contribution from the anisotropic term.

Quadrupolar interaction

Nuclei with spin $I > 1/2$ possess electric quadrupole moments which, in the presence of an electric field gradient, give an additional contribution to the spin Hamiltonian, termed the quadrupolar interaction. In anisotropic media this interaction produces line broadening and gives rise to additional lines in the spectrum. ^{2}H has a relatively small quadrupolar moment and is therefore particularly useful for studying anisotropic media. In an X–D bond the field gradients arise almost entirely from the charge on X, and it can be usually assumed that the X–D bond lies parallel to the principal axis for $\mathbf{q}_i$, the quadrupolar interaction tensor. Furthermore, for C–D bonds the quadrupolar coupling constants are found to have typical values depending on hybridization (see **Table 1**).

NMR of liquid crystalline solutions

Structural studies of rigid molecules

^{1}H NMR spectroscopy in oriented media has been widely employed for determining the structure of rigid molecules. Proton spectra in nematic solvents are dominated by homonuclear dipolar couplings, D_{ijzz}, whose magnitudes are large (up to 3–5 kHz) compared to the line width and the chemical shift differences and which are observable between all active nuclei in a molecule. The spectra can be very complex (see **Figure 1**) even for molecules with a small number of nuclei and various strategies have been devised for their analysis.

For simple spin systems, analytical solutions to the Hamiltonian can be found and transition frequencies and intensities can be written in closed forms. Spectra of complex systems may be simplified by reducing the number of nuclei contributing to them. This may be achieved by partial deuteration of molecules followed either by spin decoupling or spin-echo refocusing. Alternatively, simplified spectra may be obtained by VAS or multiple-quantum experiments. Complex spectra have been analysed by iterative and automatic methods using computer programs; both require good starting parameters, which can be obtained from simplified spectra, and an experienced operator.

Recently, methods for automatic analysis have been improved allowing the treatment of very complex systems; with these procedures it is possible to analyse spectra that depend on up to 27 spectral parameters.

The dipolar couplings obtained from spectral analysis are then used to determine molecular geometries, namely bond lengths and angles. Since these couplings depend on both geometrical factors and order parameters, they never define an absolute length but only relative distances. In contrast, angles can be measured absolutely.

Only for very simple molecular systems with favourable symmetry are there enough dipolar couplings for the determination of both the structural and the order parameters. In more complex cases, the solution must be found iteratively by computer

Table 1 Quadrupolar coupling constants for deuterium

C–D bond hybridization	*Quadrupolar coupling constant (kHz)*
sp	200 ± 5
sp^2	185 ± 5
sp^3	170 ± 5

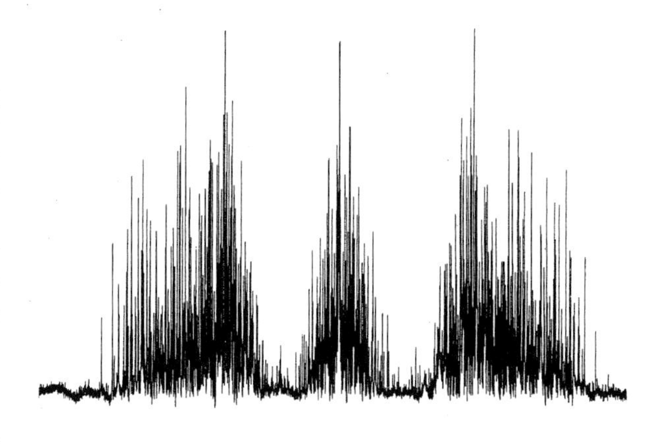

Figure 1 300 MHz proton spectrum of a sample of biphenyl dissolved in the nematic solvent ZLI1115.

calculations, and appropriate programs, such as SHAPE, have been written for this purpose. To get a unique and precise geometrical solution in the case of complex systems, it is highly recommended to record several spectra of the same molecule in different solvents. In addition, information on the carbon positions may be derived from the dipolar coupling constants D (^{13}C–H) observed as the ^{13}C satellites in the proton spectra, even though this method has two considerable disadvantages: the spectra are quite complex and the measured ^{13}C–H couplings are usually less precise than H–H couplings.

The presence of molecular vibrations makes internuclear distances and angles, and therefore the degree of ordering, time dependent. Thus the NMR structural data must be corrected for vibration. The harmonic correction can be computed relatively easily and a computer program, VIBR, is available; the structure corrected for these effects, called the r_α structure, has the advantage of being unaffected by shrinkage effects, and can be compared with the data obtained from different spectroscopies.

The determination of molecular structure by ^{1}H NMR spectroscopy in oriented media is apparently extremely precise: dipolar couplings of the order of 10^3 Hz can be measured with an error of 10^{-1} Hz, thus giving a precision of 10^{-5} in distance ratios. In practice, however, limitations to structure precision apply because of solvent effects, vibrational corrections and anisotropy of the indirect coupling constants, and the precision that can be easily reached is of the order of 10^{-2} Å in bond lengths and 10^{-1} degrees in angles. As far as solvent effects on the structure are concerned, correlations of molecular shape with angular orientation have been observed. For example, they are responsible for the observation of dipolar and quadrupolar splittings in NMR spectra of molecules that in their equilibrium conformation possess tetrahedral and higher symmetry. The use of 'magic' mixtures has been suggested in order to obtain minimal distortions. Theoretical evaluations of the deformations have been made and used to correct the NMR couplings in structural studies.

^{2}H NMR has been used to determine quadrupolar coupling constants and asymmetry parameters, once the molecular structure and orientational order parameters are known.

NMR in nematic solutions has also been applied in the study of weak molecular complexes. The interactions of systems such as pyridine, pyrimidine and quinoline with iodine have been investigated. ^{1}H and ^{13}C NMR spectra have been recorded and analysed in order to detect whether the complex formation measurably affects the molecular structure, i.e. the proton and carbon positions. Since these effects are small, vibrational as well as deformation corrections have been performed.

Structural studies of flexible molecules

The presence of large, low-frequency intramolecular motions (such as internal rotation and ring puckering) considerably complicates the interpretation of proton NMR spectra of flexible molecules in nematic solvents. In fact, in this case the effects of two simultaneous motional averagings on the magnetic interactions are present: the internal molecular motions and the overall motions of molecules in the mean potential due to the anisotropic environment. Moreover, owing to possible correlation between these motions, it is not possible to reach the same degree of precision on the orientational and geometrical parameters as can be obtained for rigid molecules.

Information about structure and orientational ordering can be obtained for individual rigid molecular fragments. Problems arise about the relationship between the data relating to different fragments and the interpretation of inter-fragment couplings. In early studies, several approximations were made, case by case, and their degree of soundness was tested by comparison of the results with those from other techniques. Later, systematic approaches were introduced for the analysis of dipolar data from flexible molecules; among which the additive potential (AP) method and the maximum entropy (ME) method have been found to be particularly useful.

Chiral recognition

A simple and convenient method for enantiomeric analysis by means of NMR has been proposed that consists of using a chiral lyotropic liquid crystal obtained by mixing poly(γ-benzyl-L-glutamate) (PBLG) and various organic solvents. (R) and (S) enantiomers interact with the chiral centres of PBLG and thus orient differently in the liquid crystal solvent, which implies that all the order-dependent NMR interactions are different for the two enantiomers. ^{2}H NMR has been applied successfully to a large number of partially deuteriated chiral molecules bearing various functional groups: enantiomeric excess (ee) up to 98% has been measured by signal integration. In many cases, chemical shift anisotropy of ^{13}C in natural abundance may be used to discriminate enantiomers and to measure ee through conventional integration with an accuracy of about 5%. In this case, isotopic enrichment is not needed and more sites in the same molecule may be observed in the same spectrum. On the other hand, chemical shift differences between enantiomers can be very small (a few hertz) and, in particular for hydrocarbons, the method may not work. At present the chiral recognition mechanism in PBLG is not fully understood so that it is not possible to assign absolute configurations.

Probe studies of liquid crystals

A well-developed approach to the study of structure and dynamics of liquid crystals is based on the use of probes, although the probe can disturb the mesophase. Generally, highly symmetric and rigid molecules are dissolved in mesophases, but, for special purposes, different probes can be chosen; for instance, for lyotropic systems the solvent itself can be used as a probe of the behaviour of aggregates.

The order and dynamic behaviour of probes, revealed by suitable NMR techniques, have been used to study phase transitions and pretransitional phenomena. They are also useful for revealing segregation effects: when highly ordered smectic or columnar discotic phases organize themselves, they may exclude added solutes or confine them in aliphatic or interlayer regions.

The anisotropic potential acting on a probe is a solute–solvent property and therefore the study of probes gives valuable insight into the mechanism governing solute–solvent interactions. It has been found that very often the probe and the solvent can interact more or less specifically and in some cases an exchange between two or more sites has been inferred from large effects on the apparent geometry of the dissolved molecule.

Noble gases have been used as probes and have proved to be very useful for obtaining information about physical properties of liquid crystals. In particular, the shielding of ^{129}Xe is highly responsive to the structure and temperature of the environment and ^{129}Xe NMR has been employed to monitor phase transitions of thermotropic and lyotropic liquid crystals and the formation of induced smectic phases. Quadrupolar noble gas nuclei are powerful probes for detecting electric field gradients (EFG); ^{131}Xe, ^{83}Kr and ^{21}Ne quadrupolar splittings and relaxation behaviour in various calamitic liquid crystals have given hints that there are two contributions to the EFG at the site of the nucleus: one is attributed to the distortion of the electron cloud of the atom from spherical symmetry and the other is the external EFG produced by the electric moments of the neighbouring molecules.

NMR of liquid crystals

Study of orientational order

Long-range orientational order is a fundamental character of liquid crystals. NMR is perhaps the most powerful technique for studying this property at the molecular level; its results are extremely useful for testing molecular theories that predict the variation of orientational order as a function of temperature, especially in uniaxial mesophases. The order parameters of different molecular segments in a liquid crystal can be obtained from ^{2}H quadrupolar couplings, ^{2}H–^{2}H, ^{2}H–^{1}H and ^{13}C–^{1}H dipolar couplings, ^{13}C chemical shifts and, for fluorinated mesogens, from ^{13}C–^{19}F dipolar couplings and ^{19}F chemical shifts.

The study of deuterium interactions is now a general method in liquid crystal research and, although it requires use of specially deuteriated compounds, it has been widely applied to both thermotropic and lyotropic mesogens. ^{2}H NMR spectra of aligned samples are well resolved and relatively simple (see **Figure 2**). An analysis of these spectra allows the magnitude of quadrupolar and, in some cases, dipolar splittings to be obtained, which are related to local order parameters for the ^{13}C–^{2}H, ^{2}H–^{2}H and ^{2}H–^{1}H directions.

^{13}C–^{1}H dipolar couplings of a large number of calamitic liquid crystals have been determined by SLF/VAS, a 2D ^{13}C NMR technique which combines separated local field spectroscopy (SLF) and variable angle spinning (VAS). In this method, mesogens with ^{13}C in natural abundance are used; however, special software and hardware as well as prolonged spectrometer time are required. It has been shown that the order parameters of the phenyl rings and the aliphatic C–H bonds in liquid crystals, determined by the SLF/VAS technique, are linearly related to their ^{13}C chemical shifts. The same linear correlation is applicable to different compounds in a homologous

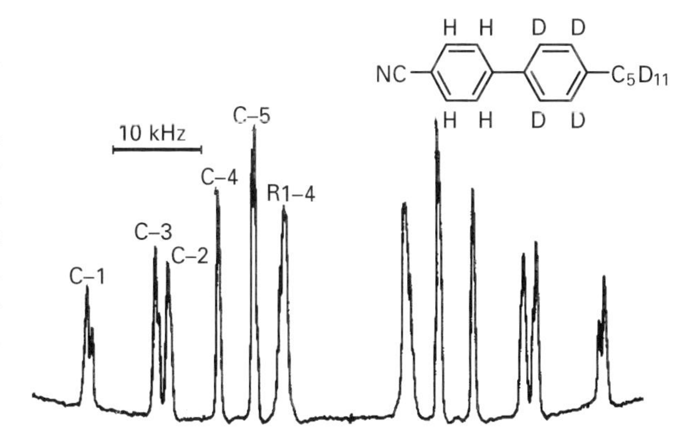

Figure 2 30.7 MHz ^{2}H NMR spectrum of 5CB-d_{15}. The peaks are labelled according to their assignment to sites in the molecule. Reproduced with kind permission from Kluwer Academic Publishers from Emsley JW (1985) In Emsley JW (ed) *NMR of Liquid Crystals*, p 397. Dordrecht: Reidel.

series: this provides a convenient means for the determination of the order parameters of liquid crystals from their ^{13}C chemical shift data, without the need to know the chemical shift tensors.

The local order parameters obtained by the various methods can be used to describe the orientational ordering of molecules, or fragments of molecules, through reference system transformations that require assumptions about geometry (i.e. bond length and angles) and conformational distributions. Usually, the order parameters for the aromatic core of mesogens can be determined quite accurately and can be assumed as a measure of the molecular ordering in the mesophases: values up to 0.6–0.7 are found for nematic phases, while values up to 0.9–1.0 are typical of highly ordered smectic and columnar discotic phases.

In fortunate cases the measured interactions also allow the calculation of structural parameters of molecular fragments, the structure of mesogenic molecules being too complex for a complete determination.

As far as side chains are concerned, the local order parameters relative to the various CH_2 and CH_3 groups give a measure of chain disorder, being dependent on chain flexibility and conformation–orientation correlations. Different theories of orientational ordering in flexible chains, such as the AP and the chord models, have been formulated and critically tested by comparison with experimental data (see **Figure 3**).

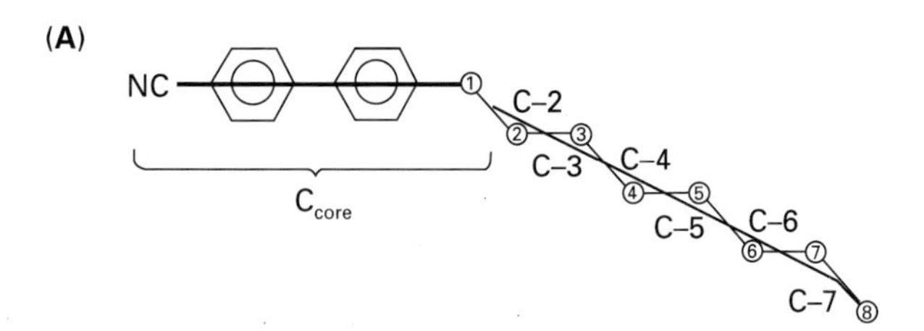

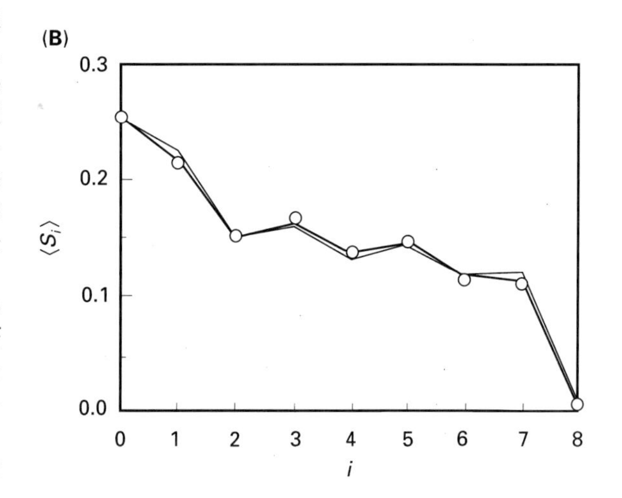

Figure 3 (A) Molecular geometry and interacting modules of the 8CB molecule in the chord model; (B) orientational order parameter profile of 8CB at $T_c - T$ = 4 K. The open circles represent experimental values; the bold line gives the calculated profile using the chord model, while the fine line gives the calculated profile using the AP model. The order parameter of the cyanobiphenyl para-axis (i = 0) is reduced by 50% for plotting convenience. Reproduced with permission of Springer-Verlag from Dong RY (1994) *NMR of Liquid Crystals*, p 107. New York: Springer-Verlag.

Phase structure determination

NMR spectroscopy can provide valuable information on the structure of mesophases. The most commonly used method for this is the analysis of line shapes in spectra of various nuclei, and, in particular, of ^2H. For a large number of mesophases, ^2H NMR spectra are motionally averaged and the spectral line shape reflects the phase symmetry. Line shape analysis of ^2H NMR spectra has been employed successfully to investigate the molecular organization and tilt angle of smectic C phases, as well as the structure of columnar discotic phases and it has been shown to be more sensitive than optical techniques to some aspect of biaxial ordering. Moreover, this technique has been shown to be extremely helpful in the discrimination of lyotropic phase symmetry. As examples, ^2H NMR spectra of some lyotropic phases are reported in **Figure 4**.

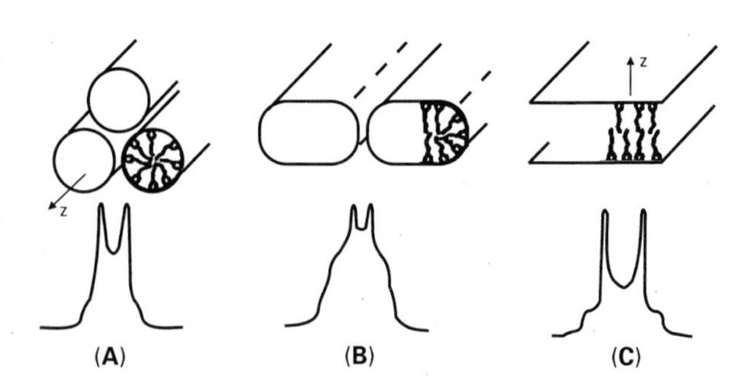

Figure 4 Schematic view of the cylindrical (A), ribbon (B) and lamellar (C) lyotropic aggregates along with their corresponding ^2H NMR spectral patterns recorded for a potassium palmitate-d_3 in a mixture with 7 wt% potassium laurate and 30 wt% water. Reproduced with kind permission from Kluwer Academic Publishers from Doane JW (1985) In: Emsley JW (ed) *NMR of Liquid Crystals*, p 414. Dordrecht: Reidel.

Other nuclei can be useful in determining structure of lyotropic liquid crystals. In fact, ^{14}N and ^{31}P NMR have been applied successfully to investigation of the structure of phospholipid aggregates in water.

Molecular dynamics

The complex dynamics of liquid crystals is characterized by a superposition of local and collective motions, comprising internal isomerization, overall rotational diffusion (rotation of the molecule about the long axis and reorientation of this axis) and translational diffusion, and collective order fluctuations. Different NMR techniques are designed to follow these motions and to differentiate the various motional modes on the basis of their timescale.

Overall molecular reorientations and internal motions take place in the 10^{-11}–10^{-6} s time window and information about them can be accessed using relaxation rate measurements. By far the best approach is to use ^{2}H NMR experiments where deuterium Zeeman and quadrupolar spin–lattice relaxation times are measured on selectively deuteriated mesogens or on deuteriated probes dissolved in the mesophase. Frequency, orientation and temperature dependence of the spectral densities obtained in these relaxation studies give indications about the various motional modes and are extremely useful for testing orientational diffusion and chain dynamics models. Different deuterium relaxation experiments can be employed to extend the observable dynamic range: the quadrupolar echo sequence gives spin–spin relaxation times sensitive to motions in the range 10^{-8}–10^{-4} s, while extremely slow motions (10^{-4}–10 s) are accessible by the Carr–Purcell–Meiboom–Gill pulse train.

Internal and overall motions that take place on a dynamic scale from 10^{-7} to 10^{-4} s can affect the line shape of ^{2}H NMR powder spectra. This is often the case for phenyl ring rotations or chain isomerizations in polymeric liquid crystals and for rotational diffusion in discotic mesophases. Line shape analysis allows different types of motions to be discriminated and the relative kinetic parameters to be obtained. Moreover multidimensional exchange NMR can be applied to characterize molecular details of dynamical processes when the correlation times range between 10^{-5} and 10^{2} s.

Fluctuations in the orientation of the director are better investigated by field cycling experiments that allow frequency-dependent T_1 measurements to be performed over a range from 100 Hz to 10 MHz. ^{1}H and ^{2}H spin relaxation studies have been carried out for numerous nematic and smectic liquid crystals to investigate the frequency dependence of director fluctuations, not accessible by experiments run on standard NMR spectrometers. It has been found that this motion is clearly observable by the relaxation rates only at low frequencies, i.e. below the MHz regime.

The translational diffusion motion is usually monitored on macroscopically aligned samples, by imposing a concentration or field gradient. The range of accessible diffusion rates is determined by the spectral or temporal resolution of the measuring technique. Moreover, translational diffusion constants may be determined indirectly by measuring spin–lattice relaxation times (T_1, T_{1D}, $T_{1\rho}$) and analysing them in terms of suitable theories.

^{1}H and ^{2}H NMR studies of dynamic processes such as isomerization, ring inversion, hindered rotation and intramolecular rearrangement have also been performed on solute molecules in nematic solvents. In the case of proton NMR the dominant anisotropic splittings in the spectra come from the dipole–dipole interaction, and dynamic processes in the range of 10–10^3 s^{-1} may be studied by line shape analysis. With ^{2}H NMR the dominant quadrupolar interaction allows motions in the range of 10^3–10^9 s^{-1} to be investigated. Besides the considerably larger dynamic intervals that may be achieved, NMR in liquid crystalline solvents is superior to that in isotropic solvents in revealing the details of reaction mechanisms. In this respect 2D exchange deuteron NMR techniques have also been found helpful.

List of symbols

B_0 = applied magnetic field; **D** = dipolar interaction tensor; ee = enantiomeric excess; G = free energy; I = nuclear spin; **J** = spin-spin interaction tensor; l = direction cosine; $\boldsymbol{n}$ = the director, preferred direction of orientation; **q** = quadrupolar interaction tensor; **T** = generic interaction tensor; T_1, T_{1D}, $T_{1\rho}$ = spin-lattice relaxation times; T_{zz} = component of averaged interaction tensor resolved along the magnetic field (z) direction; σ = shielding constant; χ = diamagnetic susceptibility.

See also: **^{13}C NMR, Methods; Chemical Exchange Effects in NMR; Chiroptical Spectroscopy, Oriented Molecules and Anisotropic Systems; Diffusion Studied Using NMR Spectroscopy; Enantiomeric Purity Studied Using NMR; ^{19}F NMR, Applications, Solution State; High Resolution Solid State NMR, ^{13}C; Nitrogen NMR; NMR Data Processing; NMR in Anisotropic Systems, Theory; NMR Pulse Sequences; NMR Relaxation Rates; ^{31}P NMR; Rigid Solids Studied Using MRI; Structural Chemistry Using NMR Spectroscopy, Organic Molecules; Xenon NMR Spectroscopy.**

Further reading

Chandrasekhar S (1992) *Liquid Crystals*, 2nd edn. Cambridge: Cambridge University Press.

Courtieu J, Bayle JP and Fung BM (1994) Variable angle sample spinning NMR in liquid crystals. *Progress in NMR Spectroscopy* **26**: 141–169

Diehl P and Khetrapal CL (1969) In: Diehl P, Fluck E and Kosfeld R (eds) NMR — *Basic Principles and Progress*, Vol. 1. Berlin: Springer-Verlag.

Doane JW (1979) NMR of liquid crystals. In: Owens FJ, Poole CP and Farach HA (eds) *Magnetic Resonance of Phase Transitions*. New York: Academic Press.

Dong RJ (1994) *Nuclear Magnetic Resonance of Liquid Crystals*. New York: Springer-Verlag.

Emsley JW (ed) (1985) *Nuclear Magnetic Resonance of Liquid Crystals*. Dordrecht: Reidel.

Emsley JW and Lindon JC (1975) *NMR Spectroscopy in Liquid Crystal Solvents*. Oxford: Pergamon Press.

Jokisaari J (1994) NMR of noble gases dissolved in isotropic and anisotropic liquids *Progress in NMR Spectroscopy* **26**: 1–26.

Luckhurst GR and Gray GW (eds) (1979) *The Molecular Physics of Liquid Crystals*. London: Academic Press.

Luckhurst GR and Veracini CA eds (1994) *The Molecular Dynamics of Liquid Crystals*. Dordrecht: Kluwer Academic.

Lithium NMR Spectroscopy

See **NMR Spectroscopy of Alkali Metal Nuclei in Solution.**

Luminescence Spectroscopy of Inorganic Condensed Matter

See **Inorganic Condensed Matter, Applications of Luminescence Spectroscopy.**

Luminescence Theory

Mohammad A Omary and **Howard H Patterson**,
University of Maine, Orono, ME, USA

ELECTRONIC SPECTROSCOPY
Theory

Introduction

Luminescence refers to the emission of light from an excited electronic state of a molecular species. In the case of photoluminescence, a molecule absorbs light of wavelength λ_1 decays to a lower energy excited electronic state, and then emits light of wavelength λ_2, as it radiatively decays to its ground electronic state. Generally, the wavelength of emission, λ_2, is longer than the excitation wavelength but in resonance emission $\lambda_1 = \lambda_2$. Luminescence bands can be either fluorescence or phosphorescence, depending on the average lifetime of the excited state which is much longer for phosphorescence than fluorescence. The relative broadness of the emission band is related to the relative difference in equilibrium distance in the excited emitting state versus the ground electronic state.

Photoluminescence of a molecular species is different from emission of an atomic species. In the case of atomic emission, both the excitation and the emission occur at the same resonance wavelength. In contrast, excitation of a molecular species usually results in an emission that has a longer wavelength than the excitation wavelength.

If a chemical reaction results in the production of a molecular species in an excited electronic state that emits light, this phenomenon is termed chemiluminescence. Chemiluminescence usually occurs in the gas or liquid phase. In contrast, photoluminescence can occur in the gas, liquid or solid phases.

The next two sections provide a discussion of the basic principles of luminescence spectroscopy, which include the electronic transitions and the important parameters determined from luminescence measurements. Then follow two sections that describe the general characteristics of luminescence measurements and one which provides two case studies for organic and inorganic luminophores. The remainder of the article covers more specific topics and phenomena in luminescence spectroscopy, namely quenching, energy transfer, exciplexes and chemiluminescence. The examples in this article were selected to cover multidisciplinary areas of science.

Electronic transitions and relaxation processes in molecular photoluminescence

The Jablonski energy level diagram

The radiative and nonradiative transitions that lead to the observation of molecular photoluminescence are typically illustrated by an energy level diagram called the Jablonski diagram. **Figure 1** shows a Jablonski diagram that explains the mechanism of light emission is most organic and inorganic luminophores. The spin multiplicity of a given electronic state can be either a singlet (paired electrons) or a triplet (unpaired electrons). The ground electronic state is normally a singlet state and is designated as S_0 in **Figure 1**. Excited electronic states are either singlet (S_1, S_2) or triplet (T_1) states.

When the molecule absorbs light an electron is promoted within 10^{-14}–10^{-15} s from the ground electronic state to an excited state that should possess the same spin multiplicity as the ground state. This excludes a triplet excited state as the final state of electronic absorption because the selection rules for

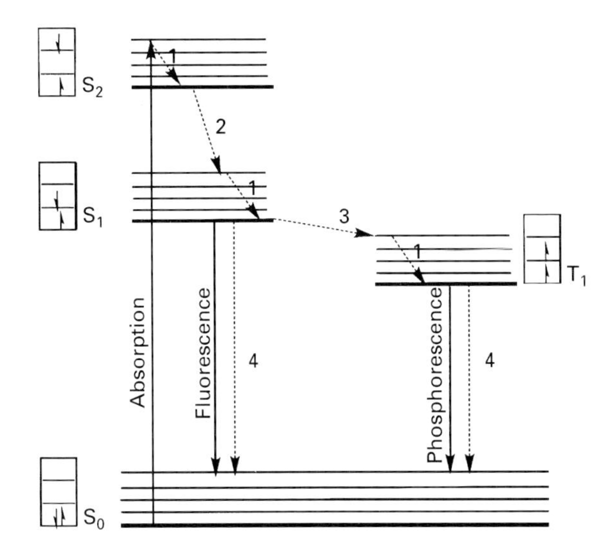

Figure 1 The Jablonski diagram. Radiative and non-radiative processes are depicted as solid and dashed lines, respectively.

electronic transitions dictate that the spin state should be maintained upon excitation. A plethora of non-radiative and radiative processes usually occur following the absorption of light en route to the observation of molecular luminescence. The following is a description of the different types of non-radiative and radiative processes.

Non-radiative relaxation processes

Vibrational relaxation Excitation usually occurs to a higher vibrational level of the target excited state (see below). Excited molecules normally relax rapidly to the lowest vibrational level of the excited electronic state. This non-radiative process is called 'vibrational relaxation'. Vibrational relaxation processes occur within 10^{-14}–10^{-12} s, a time much shorter than typical luminescence lifetimes. Therefore, such processes occur prior to luminescence.

Internal conversion If the molecule is excited to a higher-energy excited singlet state than S_1 (such as S_2 in **Figure 1**), a rapid non-radiative relaxation usually occurs to the lowest-energy singlet excited state (S_1). Relaxation processes between electronic states of like spin multiplicity such as S_1 and S_2 are called 'internal conversion'. These processes normally occur on a time scale of 10^{-12} s.

Intersystem crossing Non-radiative relaxation processes between different excited states are not limited to states with the same spin multiplicity. A process in which relaxation proceeds between excited states of different spin multiplicity is called 'intersystem crossing'. The relaxation from S_1 to T_1 in **Figure 1** is an example of an intersystem crossing.

Intersystem crossing is generally a less probable process than internal conversion because the spin multiplicity is not conserved. Because of the lower probability for intersystem crossing processes they occur more slowly ($\sim 10^{-8}$ s) than internal conversions. Intersystem crossing processes become more important in molecules containing heavy atoms such as iodine and bromine in organic luminophores and metal ions in inorganic luminophores (transition-metal complexes). Significant interaction between the spin angular momentum and the orbital angular momentum (spin–orbit coupling) becomes more important in the presence of heavy atoms and, consequently, a change in spin becomes more favourable. In solution, the presence of paramagnetic species such as molecular oxygen increases the probability of intersystem crossing. In transition metal complexes, intersystem crossing processes become increasingly important as one proceeds from complexes of the 3d-block to those of the 4d- and 5d-blocks.

Non-radiative de-excitation The excitation energy stored in the molecules following absorption must be dissipated owing to the law of conservation of energy. The aforementioned non-radiative processes occur very rapidly and only release very small amounts of energy. The rest of the stored energy will be dissipated either radiatively, by emission of photons (luminescence), or non-radiatively, by the release of thermal energy. The non-radiative decay of excitation energy which leads to the decay of the excited molecules to the ground electronic state is called 'non-radiative de-excitation'. These processes result in the release of infinitesimal amounts of heat that cannot normally be measured experimentally. The experimental evidence for non-radiative de-excitation processes is the quenching of luminescence. A major route for non-radiative de-excitation processes is energy transfer to the solvent or non-luminescent solutes in the solution. In the solid state, crystal vibrations (phonons) provide the mechanism for non-radiative de-excitation.

Radiative processes: fluorescence and phosphorescence

The spin selection rule for electronic transitions (both absorption and emission) states that 'spin-allowed' transitions are those in which the spin multiplicity is the same for the initial and final electronic states. Therefore, spin-allowed transitions are more likely to take place than spin-forbidden transitions. 'Fluorescence' refers to the emission of light associated with a radiative transition from an excited electronic state that has the same spin multiplicity as the ground electronic state. Fluorescence is depicted by the radiative transition $S_1 \rightarrow S_0$ in **Figure 1**. Since fluorescence transitions are spin-allowed, they occur very rapidly and the average lifetimes of the excited states responsible for fluorescence are typically $< 10^{-6}$ s.

Electronic transitions between states of different spin multiplicity are 'spin-forbidden' which means that they are less probable than spin-allowed transitions. However, spin-forbidden transitions become more probable when spin–orbit coupling increases. The factors that increase the probability of phosphorescence are the same factors discussed above that increase the probability of intersystem crossing. Therefore, if the triplet excited state is populated by intersystem crossing then luminescence might occur from the triplet state to the ground state. 'Phosphorescence' refers to the emission of light associated with a radiative transition from an excited electronic state that has a different spin multiplicity from that of the ground electronic state. Phosphorescence is depicted by the radiative transition $T_1 \rightarrow S_0$ in **Figure 1**. Since phosphorescence transitions are spin-forbidden, they occur slowly and the average lifetime of the excited states responsible for phosphorescence typically range from 10^{-6} s to several seconds. Therefore, some textbooks refer to phosphorescence as 'delayed fluorescence'. Photoluminescence refers to both fluorescence and phosphorescence.

Excited-state distortions and the Franck–Condon principle

Luminescence spectroscopy can be used to gain information about the geometry of a molecule in an excited electronic state. Such information provides an understanding of the difference in the bonding properties of an excited state relative to the ground state. These properties can be better understood when the electronic transitions are discussed in the context of the potential surfaces of the ground and excited states.

Electronic absorption of light occurs within 10^{-15} s. Since this time is extremely short the nuclei are assumed to be 'frozen' during the time scale of absorption. Therefore, the transitions between various electronic levels are depicted as 'vertical' transitions in energy level diagrams. This assumption of negligible nuclear displacement during electronic transitions is known as the 'Franck–Condon

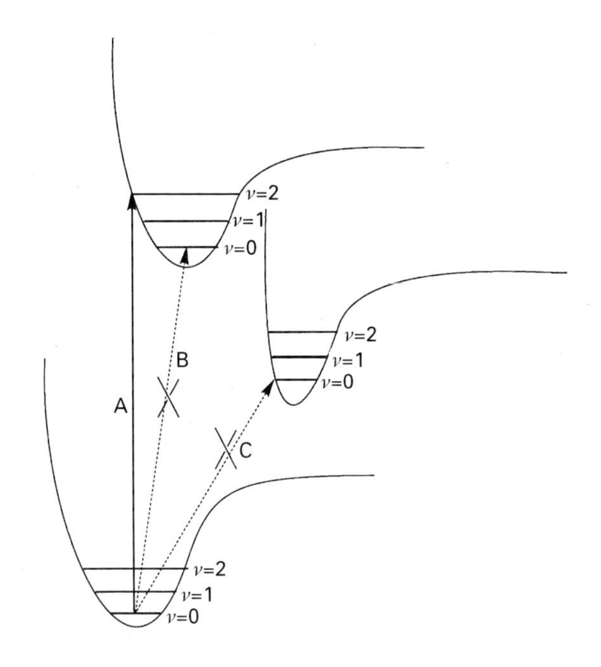

Figure 2 The Franck–Condon principle. Only vertical electronic transitions are allowed.

principle'. **Figure 2** illustrates the Franck–Condon principle. Electronic excited states usually have different geometries than the ground state and, consequently, different equilibrium distances. Since electronic transitions are vertical, only transition A in **Figure 2** occurs. Transition C involves an excited state that is largely displaced from the ground state and thus no vertical transition is possible to this state. Transition B, on the other hand, terminates in the lowest vibrational level of the excited state. **Figure 2** shows that this transition cannot occur vertically, either. The three transitions depicted in **Figure 2** explain why in the Jablonski diagram (**Figure 1**): (i) the absorption was depicted to a higher vibrational level of the S_2 excited state than the $\nu = 0$ level, and (ii) no direct excitation to the triplet excited state (T_1) was depicted.

Usually, the nuclei of excited molecules are displaced from their ground state positions. The displacement is caused by differences in the bonding properties between the molecular orbitals that represent the ground and excited states, respectively. The extent of the nuclear displacement varies from one case to another. **Figure 3** illustrates two cases (A) and (B) of excited state distortion. The transitions depicted in **Figure 3** are called 'vibronic transitions'. A vibronic transition refers to a transition that involves a change in both electronic and vibrational states.

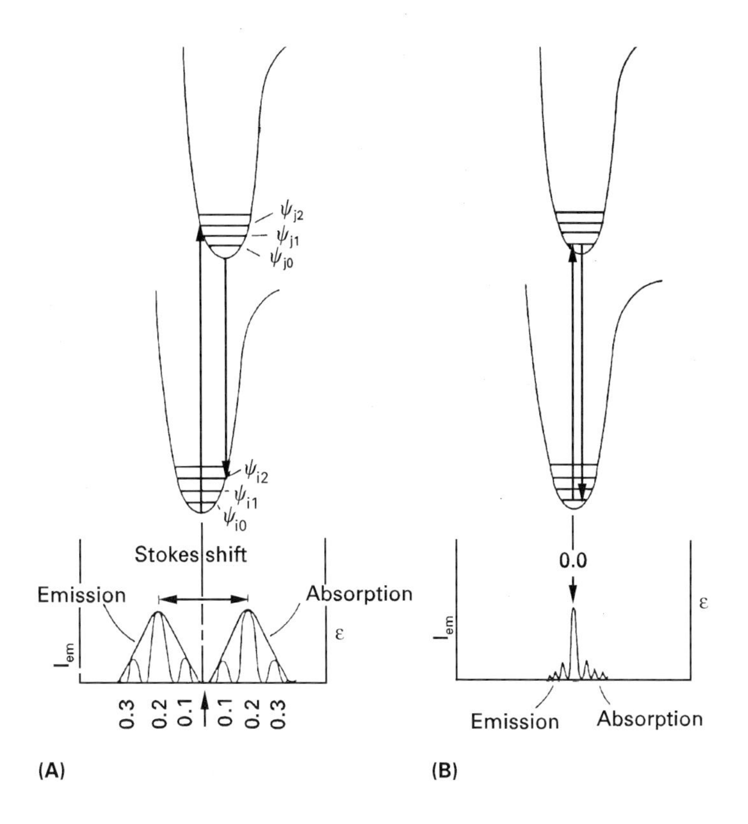

Figure 3 Potential energy diagrams for two cases of excited state distortions: (A) non-zero distortion and (B) zero distortion. The corresponding absorption and emission spectra are shown below. Reproduced with permission from Adamson AW and Fleischauer PD (1975) *Concepts of Inorganic Photochemistry*, p 4. New York: Wiley-Interscience.

The case shown in **Figure 3B** represents an extreme case in which the excited state distortion is virtually zero. As a result, the electronic transition for which $v = 0$ in both the ground and excited states (called the 0–0 vibronic transition) has the greatest probability and, thus, the strongest intensity in both absorption and emission spectra. When excited state distortion is significant, as in **Figure 3A**, the 0–0 transition becomes much less probable and so its intensity becomes much weaker than other 0–v vibronic transitions with higher v values.

The probability (P) of the occurrence of a vibronic transition is proportional to the square of the *Franck–Condon factor*, which is defined as the overlap integral of the vibrational wavefunctions, $\Psi(v)$. Therefore,

$$P \propto |\langle\Psi(v_1)|\Psi(v_2)\rangle|^2 \quad [1]$$

In many instances the absorption and emission maxima correspond to the same (v_1–v_2) vibrational pair. For example, **Figure 3A** shows that the 0–2 vibronic transition has the strongest intensity in both the absorption and the emission spectra. This is called the 'mirror image' rule and is followed by luminophores whose excited state distortion is zero or small. However, the mirror image rule may not apply for cases where large excited state distortion exists. Examples of each case are provided in the section dealing with organic luminophores.

Quantitative information about excited state distortion can be obtained from the luminescence spectra. For cases where vibronic structure is observed, the excited state distortion can be probed qualitatively simply by identifying the most intense vibronic transition: the larger the value of the vibrational level of the excited state, the larger the excited state distortion (as illustrated in **Figure 3**). Quantitatively, the displacement (Δq) of the excited state equilibrium distance from the corresponding ground state distance can be calculated. The procedure involves the calculation of the Franck–Condon factor as a function of Δq for each vibronic transition. This results in a calculated emission spectrum, which can be compared with the experimental emission spectrum as a function of Δq until a good fit is obtained. The actual Δq would be the value that gives the best fit.

The emission (and absorption) spectra in many practical cases do not show resolved vibronic peaks. This is especially the case in solutions of the luminophores at ambient temperatures. The interaction of excited molecules of the luminophore with solvent molecules (especially polar solvents) is responsible for this broadening. There are methods to quantify Δq in these cases with the aid of some computer programs but the description of these methods is beyond the scope of this article. In the section on inorganic exciplexes, nevertheless, we provide an example in which Δq is evaluated theoretically for a system that exhibits structureless emission bands.

Another quantitative measure of excited state distortion is the *Stokes shift*, defined as the energy difference between the emission and absorption peak maxima for the same electronic transition. The lower part of **Figure 3A** illustrates how the Stokes shift is evaluated in typical cases. The Stokes shift of the rare case shown in **Figure 3B** is zero because the absorption and emission spectra have the same peak positions. The band width can also be used to quantify excited state distortions. The band width is normally quantified in terms of the full-width-at-half-maximum (fwhm). The value of fwhm is calculated as the energy difference between the band positions that have intensities equal to one half the peak maximum, assuming a gaussian shape is obtained for the emission band. In the absence of excited state distortions the emission bands appear as sharp peaks with very small fwhm values (**Figure 3B**), whereas in more common cases the electronic bands are much broader because of excited state distortions (**Figure 3A**). In those cases where the emission spectra are devoid of vibronic structure, the fwhm values are calculated for the whole broad band. The Stokes shift and fwhm are parameters for excited state distortion because their values are larger for 0–v vibronic transitions with higher v values. The Stokes shift and fwhm are especially useful in cases where the emission spectra are structureless, because of the difficulty of carrying out the Δq calculation in such cases.

Emission and excitation spectra

In luminescence spectroscopy the emission of the luminophore is monitored. There are two different types of luminescence spectra that can be recorded with modern spectrofluorometers, emission and excitation spectra. Note that although the name implies that these instruments measure only fluorescence, spectrofluorometers can also measure phosphorescence, especially when special accessories are added. Spectrofluorometers contain both an excitation monochromator and an emission monochromator. In emission spectra, the excitation wavelength is fixed and the emission monochromator is scanned. The excitation is usually fixed at a wavelength at which the sample has significant absorbance. In excitation spectra, on the other hand, the emission wavelength is fixed and the excitation monochromator is scanned. The emission is normally fixed at a wavelength that corresponds to the emission peak of the sample.

Theoretically, the excitation spectra should mimic the absorption spectra of the luminophores. However, the lamp output is wavelength-dependent. For example, a common light source in modern spectrofluorometers is a xenon lamp. The output of this lamp is a continuum spectra in the range ~200–1200 nm. The radiation curve approximates blackbody radiation with a maximum near 550 nm and a sharp decline at short wavelengths. A correction is, therefore, needed for the lamp background in order for the excitation spectra to correlate with the absorption spectra. Corrected excitation spectra are usually obtained by using a *quantum counter*, a strong luminophore capable of absorbing virtually all incident light over a wide range of wavelength. Rhodamine B is a commonly used quantum counter. Concentrated solutions of rhodamine B absorb virtually all the incident light in the 220–600 nm range. The excitation spectrum of rhodamine B monitoring its emission maximum (633 nm) is determined only by the lamp output in the 220–600 nm range (**Figure 4A**).

Corrected excitation spectra of luminophores are obtained by dividing the uncorrected spectra by the excitation spectrum of the quantum counter. A common improvement in the corrected excitation spectra versus the uncorrected spectra is to eliminate the sharp 'spikes' of the xenon lamp between 450 and 500 nm (**Figure 4A**). This makes the excitation spectra of many luminophores very similar to their absorption spectra. However, more dramatic differences may exist, especially for luminophores which absorb in the UV region. An example of such a luminophore is the silver complex $[Ag(CN)_2^-]$. **Figure 4B** shows the corrected and uncorrected excitation spectra of $[Ag(CN)_2^-]$ doped in NaCl crystals. Note that the corrected excitation spectra are significantly different from the uncorrected spectra. This difference is attributed to the low output of the xenon lamp at wavelengths shorter than 280 nm (**Figure 4A**), at which $[Ag(CN)_2^-]$ species absorb strongly.

Absorption and excitation spectra are complementary. The necessity to obtain a correction for the excitation spectra represents a disadvantage. There are advantages, however, for excitation spectra over absorption spectra. The much higher sensitivity of luminescence techniques compared to absorption techniques is an obvious advantage for excitation spectra. The greater sensitivity of luminescence techniques stems from the fact that the luminescence intensity can be enhanced by increasing the intensity of the excitation source, which is not the case in absorption. The greater sensitivity of luminescence techniques is, however, accompanied by less precision

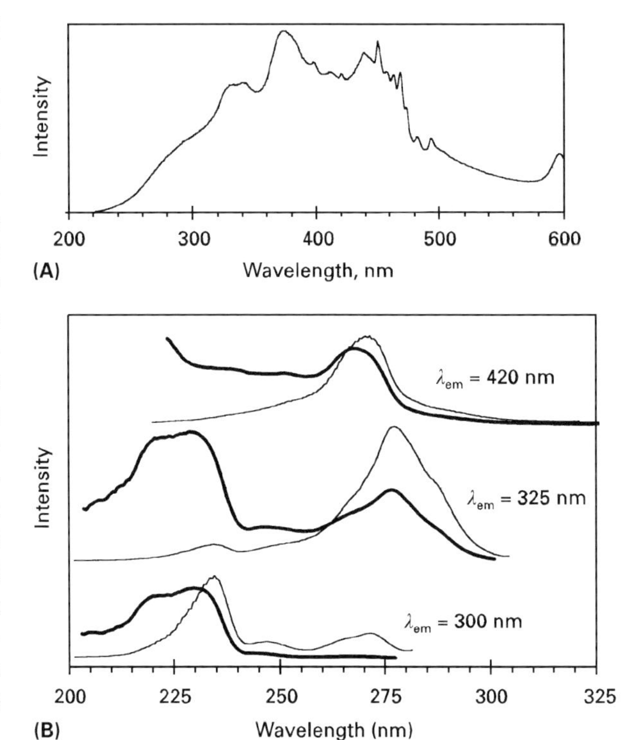

Figure 4 Correction of the excitation spectra by the quantum counter method: (A) excitation spectrum of rhodamine B; (B) excitation spectra of $[Ag(CN)_2^-]$/NaCl doped crystals monitoring the emission at different wavelengths. Corrected and uncorrected spectra are shown as thick and thin lines, respectively.

compared to absorption techniques. The relative ease of acquiring excitation spectra for some materials such as solids represents another advantage over absorption measurements. Finally, excitation spectra can provide valuable information about excited state processes such as energy transfer (see below) that cannot be obtained by absorption spectra.

Luminescence lifetimes

Theory

Consider the simplest case in which a molecule A absorbs light and is in an excited electronic state denoted by *A. If the molecule has a single pathway for decay, say fluorescence, we can write the following equation:

$$^*A \rightarrow A + h\nu \quad [2]$$

This is a first-order process whose rate can be expressed mathematically as:

$$-d[^*A]/dt = k_f[^*A] \quad [3]$$

where k_f is the rate constant of the fluorescence decay (Eqn [2]). The reciprocal of k_f is called the fluorescence lifetime τ_f ($\tau_f = 1/k_f$). Integration of Equation [3] gives:

$$[^*A] = [^*A]_o \exp(-t/\tau_f) \qquad [4]$$

Hence, a plot of ln [*A] versus time (t) should give a straight line with a slope of $1/\tau_f$. The value of [*A] is determined from the fluorescence intensity. Experimentally, lifetime measurements are obtained using a pulsed laser source. Pulsing leads to the population of the excited state of A, followed by emission of light by *A with a time profile according to Equation [4]. **Figure 5** shows a schematic description of a luminescence decay curve (A) and the plot used for the determination of the excited state lifetime (B).

Next, consider the case where *A can exhibit fluorescence (Eqn [2]) and also non-radiative decay:

$$^*A \rightarrow A + \text{vibrational energy} \qquad [5]$$

with a rate constant of k_{nr}. The decay of *A can now be described as:

$$-d[^*A]/dt = k_f[^*A] + k_{nr}[^*A] = (k_f + k_{nr})[^*A] \qquad [6]$$

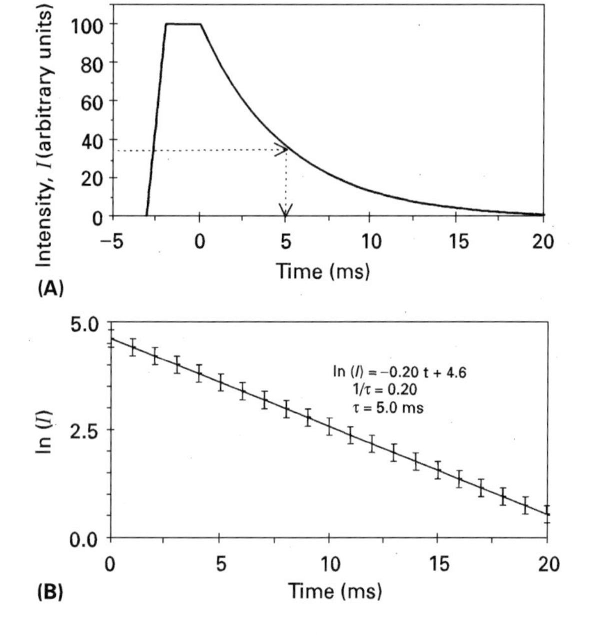

Figure 5 Experimental determination of excited state lifetimes: (A) a plot of the intensity (I) versus time after the laser pulse. The lifetime corresponds to the time at which the intensity decays to 1/e of its maximum value; (B) a plot of ln (I) versus time after laser pulse. The lifetime here can be calculated directly from the slope of the linear equation.

In this case the lifetime τ becomes:

$$\tau = 1/(k_f + k_{nr}) \qquad [7]$$

or the reciprocal of the sum of k_f and k_{nr}.

Generally, the non-radiative rate constant decreases at lower temperatures. As the temperature (T) approaches absolute zero, τ approaches $1/k_f$. A plot of $1/\tau$ versus T extrapolated to $T = 0$ K allows one to determine k_f as the value of the intercept (see the example in the case study on inorganic luminophores). Therefore, the value of the fluorescence lifetime (τ_f) is determined as $1/k_f$. Note that the treatment described in this section applies for both fluorescence and phosphorescence lifetimes and not just for fluorescence.

Now consider a final lifetime case in which two states are thermally populated at a temperature T. If excited state 1 has a lifetime τ_1 and excited state 2 has a lifetime τ_2, the observed lifetime τ_{obs} will be a weighted average of the two lifetimes:

$$\tau_{obs} = (n_1/N)\tau_1 + (n_2/N)\tau_2 \qquad [8]$$

with $N = n_1 + n_2$. Assuming that excited states 1 and 2 are non-degenerate and using the Boltzmann distribution function gives:

$$\tau_{obs} = \{[\tau_1 + \tau_2 \exp(-\Delta E/kT)]/[1 + \exp(-\Delta E/kT)]\} \qquad [9]$$

where ΔE is the energy difference between states 1 and 2 and k is the Boltzmann constant. Fitting τ_{obs} versus T to Equation [9] allows the determination of the value of ΔE as well as τ_1 and τ_2.

Fluorescence versus phosphorescence lifetimes

A major advantage of luminescence lifetime measurements is their use for spectral assignment. Specifically, the assignment of the luminescence bands as fluorescence or phosphorescence is primarily determined via luminescence lifetime measurements. As a rule of thumb, lifetimes on the order of microseconds and longer (milliseconds, seconds) are normally indicative of phosphorescence, while fluorescence lifetimes are normally on the sub-microsecond level (nanoseconds, picoseconds etc). It should be noted, however, that fluorescence and phosphorescence lifetime values vary from one case to another, depending on the system under study as well as other factors such as the extent of excited state distortion. This causes some subjectivity in the assignment, especially when the measured lifetimes are at borderline levels between the aforementioned levels of fluorescence and phosphorescence lifetimes. The clearest cases are those in which fluorescence and phosphorescence bands are both present in one system. In these cases, the phosphorescence bands exhibit lifetimes that are orders of magnitude longer than the lifetimes of the corresponding fluorescence bands. An example is given for such a case in the section on inorganic luminophores.

Luminescence quantum yields

The quantum yield is a luminescence property that is related to lifetimes. The luminescence quantum yield (Φ) is the ratio of the number of photons emitted to the number of photons absorbed. Therefore, the maximum value of Φ is 1. In practice, the quantum yield is less than unity for virtually all luminescent materials. The reason is the large number of non-radiative processes that lead to a decrease in the number of emitted photons. The quantum yield can be defined in terms of the rate constants of the radiative (k_{rad}) and non-radiative (k_{nr}) processes according to Equation [10]:

$$\Phi = \frac{\sum k_{\mathrm{rad}}}{\sum k_{\mathrm{rad}} + \sum k_{\mathrm{nr}}} \qquad [10]$$

The k_{rad} term includes the rate of fluorescence and phosphorescence while the k_{nr} term includes the rate constants of all the non-radiative processes described previously. Remembering that $\tau_f = 1/k_f$ and also using Equation [7], one can express the quantum yield of the fluorescence (Φ_f) in terms of the luminescence lifetimes as:

$$\Phi_f = \frac{k_f}{k_f + \sum k_{\mathrm{nr}}} = \frac{\tau}{\tau_f} \qquad [11]$$

A variety of factors are believed to influence the luminescence quantum yields of luminophores.

The transition type Because luminescence is a technique of electronic spectroscopy, the same selection rules apply for luminescence as those that apply for absorption. For organic compounds, the most common luminescence bands are due to π^*–π and π^*–n transitions. The σ^*–σ transitions, although strongly allowed, are not normally seen because of their high energies in most organic compounds. The spin selection rules also apply. Consequently, fluorescence in most luminescent organic compounds is much stronger than phosphorescence (see the example in the section on organic luminophores). For inorganic compounds, the luminescence transitions may involve the energy levels of the ligands (intra-ligand transitions), the metal ions (d-d transitions), or both the metal and the ligand (charge transfer transitions). Intra-ligand transitions are similar to the transitions discussed above for organic compounds. The d-d transitions are usually forbidden transitions, so the corresponding absorption and emission bands are generally weak. Charge transfer transitions are strongly allowed and can occur either from the ligand orbitals to the metal orbitals (ligand-to-metal charge transfer, LMCT) or vice versa (metal-to-ligand charge transfer, MLCT).

Structural rigidity Molecules that have *rigid* structures normally exhibit strong luminescence. Three examples are shown in **Scheme 1**. In each pair the fluorophore on the left has the more rigid structure, resulting in greater luminescence quantum yield. Note in the third example that the stronger luminescence intensity is due to the complexation with a metal ion (i.e. the complex is more rigid than the free ligand). The higher luminescence quantum yields for *rigid* luminophores are probably due to the

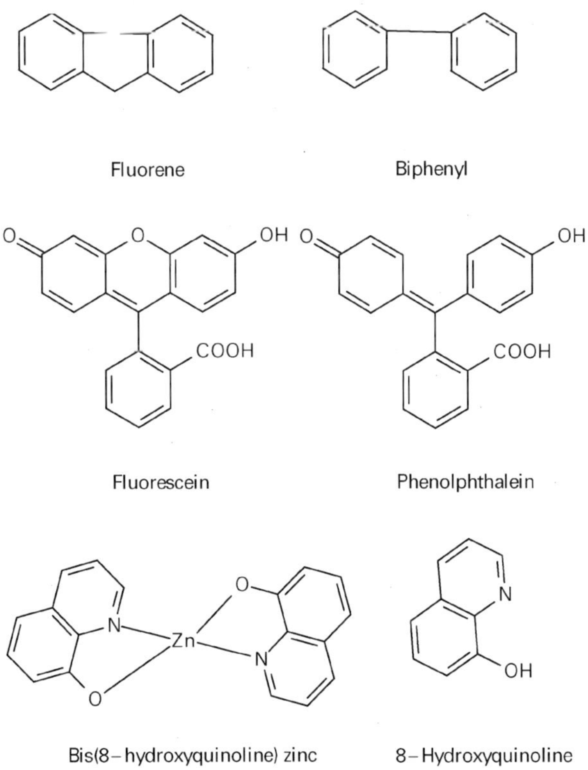

Fluorene

Biphenyl

Fluorescein

Phenolphthalein

Bis(8-hydroxyquinoline) zinc

8-Hydroxyquinoline

Scheme 1

inhibition of the internal conversion rates and vibrational motion in these compounds.

Substitution The nature of the substituents on, say, an aromatic ring may increase or decrease the quantum yield. For example, halogen and ketone substituents on anthracene generally decrease the fluorescence quantum yield. In contrast, the quantum yield of diphenyl anthracene is nearly unity (compared with $\Phi_f = 0.4$ for unsubstituted anthracene). Phosphorescence becomes an important factor if the substituents contain heavy atoms (for example halogens). In this case the reduction in fluorescence is accompanied by an increase in phosphorescence. This effect (called the *heavy atom effect*) is specially important in inorganic luminophores because of the involvement of the metal ions. The strong spin–orbit coupling in metal ions leads to the relaxation of the spin selection rules. Therefore, most inorganic luminophores exhibit strong phosphorescence (see below).

Other factors Many other factors affect luminescence quantum yields. Among these factors are temperature, solvent, phase and pH. A reduction in temperature suppresses non-radiative processes and thus increases Φ. This is why it is common to run luminescence experiments at cryogenic temperatures. The solvent is involved in many non-radiative processes. An increase in the viscosity of the solvent generally decreases the rate of non-radiative de-excitation. Also, the stretching frequency of the bonds of the solvent molecules is an important factor. For example, one of the common procedures to increase the quantum yield is to run the luminescence measurements in deuterated solvents (e.g. D_2O instead of H_2O). The quenching caused by the solvent may be removed by running the luminescence measurements in the solid state instead of solutions (examples are given later). Finally, a change in pH may strongly alter the Φ value because the structure of many luminophores may be different in acidic and basic media.

Case studies for photoluminescence

Organic luminophores

The luminescence properties of anthracene and its derivatives provide an excellent illustration of the relation between the excited state distortion and the profile of the luminescence spectra. **Figure 6** shows the luminescence and absorption spectra of anthracene. Note that the extinction coefficient for the singlet–singlet transition is 8 orders of magnitude higher than the value for the singlet–triplet transition. This observation suggests that the absorption and emission bands in the region with $\lambda < 500$ nm are due to a spin-allowed transition (singlet $\leftrightarrow$ singlet), hence the emission in this region is due to fluorescence. On the other hand, the bands in the region with $\lambda > 500$ nm are due to a spin-forbidden transition (singlet $\leftrightarrow$ triplet), hence the emission in this region is due to phosphorescence.

Figure 6 shows that the mirror image rule applies very well to the absorption and fluorescence bands of the $S_0 \leftrightarrow S_1$ transition. Note that among the vibronic bands, the 0–0 and 0–1 transitions have the strongest intensities. These observations suggest that the excited state distortion is very small in anthracene. **Figure 6** also shows that there is correlation even between the absorption and phosphorescence characteristic of the $S_0 \leftrightarrow T_1$ transition. However, the mirror image rule does not apply as strongly as it does for the $S_0 \leftrightarrow S_1$ transition. For example, the intensity is greater for the 0–1 transition than for the 0–2 transition in the phosphorescence band, but the opposite trend is seen in the corresponding absorption band. Moreover, while the absorption and emission peaks are nearly superimposed for the 0–0 vibronic peak of the $S_0 \leftrightarrow S_1$ transition, there is a greater separation between the corresponding peaks characteristic of the $S_0 \leftrightarrow T_1$ transition. These observations are consistent with the excited state distortion for the triplet excited state (T_1) being greater than the distortion of the singlet excited state (S_1).

Substitution of hydrogen atoms of anthracene may lead to a geometry change in the excited state. The extent of this change can be probed by luminescence spectroscopy. **Figures 7** and **8** provide an illustration. In **Figure 7**, the absorption and fluorescence spectra are shown for 9-anthramide. Similar geometries of the ground state and the fluorescent excited state of 9-anthramide are illustrated by: (i) the applicability of the mirror image rule, (ii) the peaks characteristic of the 0–0 and 0–1 transitions both being strong, (iii) the absorption and emission peaks being superimposed for the 0–0 transition. The situation is not the same when the substituent on anthracene is changed from an amide group to an ester group. **Figure 8** shows the absorption and emission spectra of cyclohexyl-9-anthroate. Note that the emission spectrum of cyclohexyl-9-anthroate is structureless and the mirror image rule is lost. Also note the large values of the Stokes shift (~6000 cm^{-1}) and fwhm (~5000 cm^{-1}). These observations suggest a largely displaced excited state for cyclohexyl-9-anthroate from the ground state geometry of the molecule. In conclusion, **Figures 7** and **8** show that the luminescent excited state is much more distorted for the ester derivative of anthracene (cyclohexyl-9-

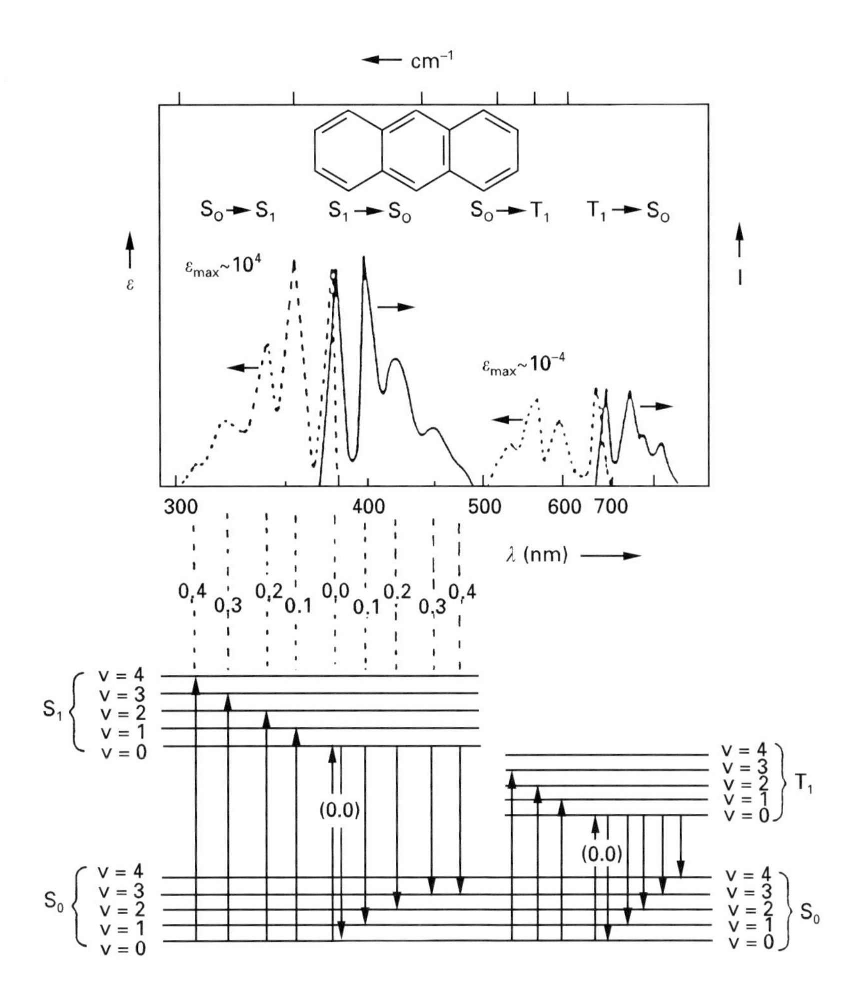

Figure 6 Emission (solid line) and absorption (dashed line) spectra of anthracene in solution. The assignments of the vibronic transitions are shown in the bottom portion of the figure. Reproduced with permission from Turro NJ (1978) *Modern Molecular Photochemistry*, p 94. Menlo Park: Benjamin/Cummings.

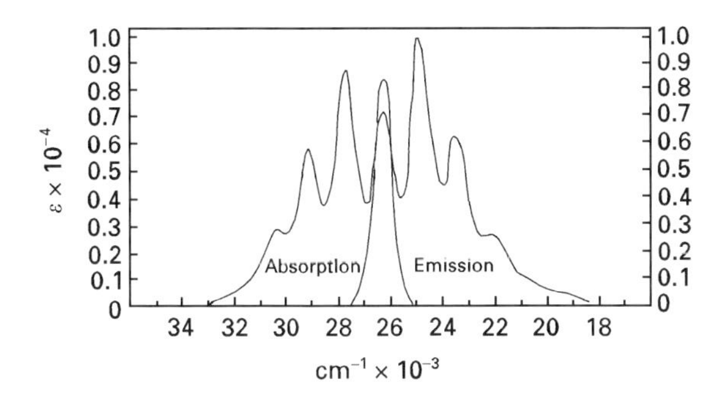

Figure 7 Emission (right) and absorption (left) spectra of 9-anthramide in tetrahydrofuran. Reproduced with permission from Shon RS-L, Cowan DO and Schmiegel WW (1975) Photodimerization of 9-anthroate esters and 9-anthramide. *The Journal of Physical Chemistry* **79**: 2087–2092.

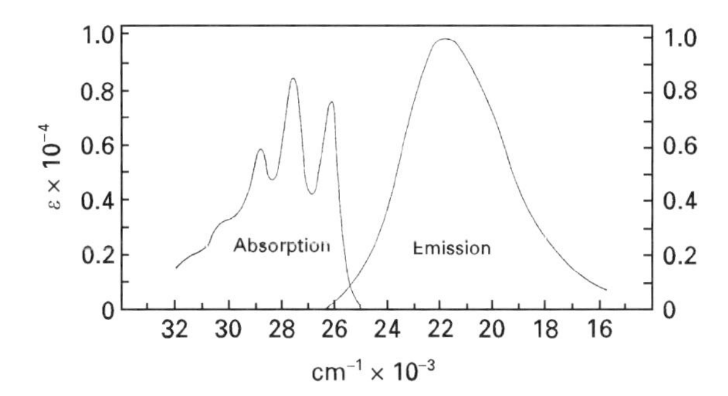

Figure 8 Emission (right) and absorption (left) spectra of cyclohexyl-9-anthroate in benzene. Reproduced with permission from Shon RS-L, Cowan DO and Schmiegel WW (1975) Photodimerization of 9-anthroate esters and 9-anthramide. *The Journal of Physical Chemistry* **79**: 2087–2092.

anthroate) than for the amide derivative of the same molecule (9-anthramide). One can therefore predict that the ester has a greater tendency to undergo photochemical reactions than the amide.

Inorganic luminophores

The compound $BaPd(CN)_4.4H_2O$ provides an excellent example for the differentiation between fluorescence and phosphorescence based on lifetimes and other luminescence properties. Inorganic compounds that exhibit chain structures show interesting luminescence properties in the solid state. The X-ray structure of $BaPd(CN)_4.4H_2O$ consists of $[Pd(CN)_4]^{2-}$ square-planar ions stacked in one-dimensional chains with an intra-chain Pd–Pd distance of 3.37 Å at room temperature. **Figure 9** shows the luminescence spectrum of $BaPd(CN)_4.4H_2O$ at 7 K using a nitrogen pulsed laser for excitation. Two emission bands are observed at ~ 19×10^3 and 26×10^3 cm^{-1} (~512 and 382 nm, respectively). These bands are designated the *lower energy* (LE) and the *higher energy* (HE) bands, respectively. In **Figure 10** the lifetime of the LE band is plotted versus temperature. A linear fit is obtained. From the resulting equation, the lifetime at 0 K is estimated to be ~ 2.5 ms and decreases by approximately a factor of 2 for each 30 K temperature rise. The shorter lifetimes at higher temperatures are due to the larger values of k_{nr} at higher temperatures (Eqn [7]). In contrast, the HE band has a lifetime shorter than the instrumental resolution of 10 ns. Thus, the lifetime data suggest that the HE band is fluorescence while the LE band is phosphorescence.

Polarized light can be used to study the luminescence bands for $BaPd(CN)_4.4H_2O$. The HE band has polarized emission with the phase of polarization of the fluorescence perpendicular to the $Pd(CN)_4^{2-}$ square planar plane (z-direction). In contrast, the LE band has polarized emission with the phase of the polarized light the same as the $Pd(CN)_4^{2-}$ molecular plane (xy-direction). The metal ions in one-dimensional chained compounds are oriented along one particular direction in crystals of layered compounds (along the xy-direction in $BaPd(CN)_4.4H_2O$). The absorption band of $BaPd(CN)_4.4H_2O$ solid is polarized along the xy-direction. The fact that the HE emission band has the same polarization as the absorption band provides further evidence that the HE band is fluorescence. This is because fluorescence takes place immediately after absorption ($\tau < 10$ ns) and from the same singlet excited electronic state as the one which absorption populates. In contrast, the LE band occurs from a triplet excited state long after absorption ($\tau = 2.5$ ms). Because the absorption and the LE emission bands have different excited states, these bands have different polarization. In accordance with the polarization of the HE band along the same direction as the Pd chains, the emission maximum of the HE band undergoes a progressive red shift (longer wavelength, lower energy) as the temperature is decreased. The red shift occurs because cooling leads to a thermal contraction of the intra-chain Pd–Pd distance, which results in a smaller HOMO–LUMO energy gap (HOMO = highest occupied molecular orbital, LUMO = lowest unoccupied molecular orbital). In contrast, the position of the LE band is independent of temperature because this band is not polarized along the Pd chains.

The preceding results can be compared with the selection rules for all the possible electronic transitions obtained from group theory, to give electronic assignments (term symbols) for each of the luminescence bands. The resulting electronic assignments are $^1A_{2u}$ and $^3A_{2u}$ for the HE and LE bands, respectively. A discussion of the concepts of group theory that lead

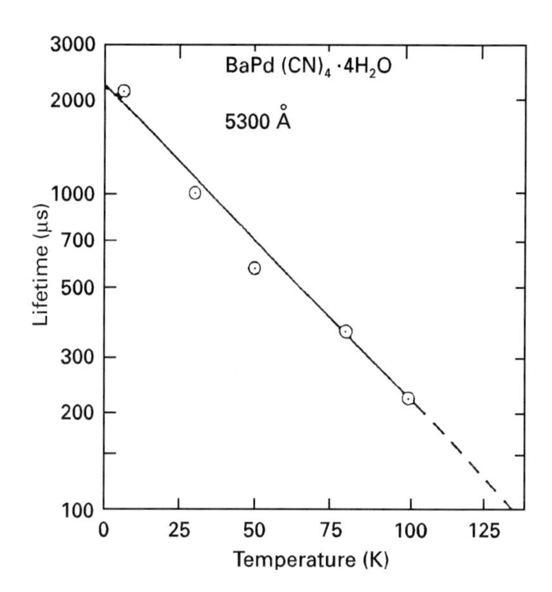

Figure 10 Lifetime of the 19×10^3 cm^{-1} luminescence band of $BaPd(CN)_4.4H_2O$ versus temperature. The dashed line is extrapolation to higher temperatures. Reproduced with permission from Ellenson WD, Viswanath AK and Patterson HH (1981) Laser-excited luminescence study of the chain compound $BaPd(CN)_4.4H_2O$. *Inorganic Chemistry* **20**: 780–783.

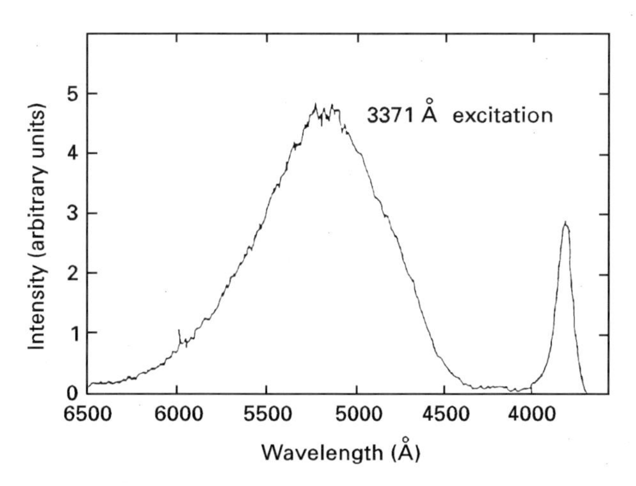

Figure 9 Luminescence spectrum of $BaPd(CN)_4.4H_2O$ at 7 K. A nitrogen laser was used for excitation. Reproduced with permission from Ellenson WD, Viswanath AK and Patterson HH (1981) Laser-excited luminescence study of the chain compound $BaPd(CN)_4.4H_2O$. *Inorganic Chemistry* **20**: 780–783.

to this assignment is beyond the scope of this article. However, it is enough to note that the term symbols of the HE and LE bands of $BaPd(CN)_4.4H_2O$ differ only in the spin multiplicity (superscripts indicate singlet and triplet, respectively). With this in mind, the difference in the excited state distortion between the HE and LE bands can be used to provide further evidence for the band assignment. The lowest energy absorption band of $BaPd(CN)_4.4H_2O$ solid has an energy of $\sim 31 \times 10^3$ cm^{-1}. This gives a Stokes shift of ~5000 and 11 500 cm^{-1} for the HE and LE bands, respectively. **Figure 9** shows that the LE band is much broader than the HE band. The fwhm values of the HE and LE bands are 900 and 3300 cm^{-1}, respectively. The higher values of the Stokes shift and fwhm for the LE band are consistent with the assignment that the LE band is phosphorescence while the HE band is fluorescence of the same electronic transition. This is the case because the excited state distortion of phosphorescence is greater than that of the fluorescence of the same electronic transition (with a different spin state). The Jablonski diagram shown in **Figure 1**) illustrates the higher Stokes shift for phosphorescence compared with fluorescence.

In summary, the HE and LE luminescence bands of $BaPd(CN)_4.4H_2O$ are assigned as fluorescence and phosphorescence, respectively. The basis of this assignment is the difference between the two bands in the lifetime values, polarization character, temperature dependence, and excited state distortion (Stokes shift and fwhm values).

Finally, an interesting aspect of the luminescence spectrum of $BaPd(CN)_4.4H_2O$ is the strong phosphorescence intensity (compared with the relative phosphorescence/fluorescence intensity of anthracene). This is a direct consequence of the strong spin–orbit coupling of palladium. This provides an illustration of the heavy atom effect which is generally more important in inorganic luminophores relative to their organic counterparts.

Quenching of emission

Theory

In luminescence spectroscopy, quenching refers to any process that leads to a reduction in the luminescence intensity of the luminophore. Static and dynamic quenching are the most common types of luminescence quenching and will be described in the following sections. A third type of quenching also exists, namely 'inner-filter quenching'. An inner-filter effect occurs when the total absorbance of the solution is high (greater than 0.1 au). This leads to a reduction in the intensity of the excitation radiation over the path length. Quenching of this type is not generally categorized among the major quenching process because it is a trivial type of quenching that is not really involved in the radiative and non-radiative transitions in luminescence spectroscopy.

Both static and dynamic quenching require contact of the luminophore with a quencher molecule. This requirement is the basis of the many applications of luminescence quenching. Because there are so many molecules that can act as luminescence quenchers, an appropriate quencher can be selected for any luminophore under study in order to investigate specific properties of the luminophore. An example of biochemical applications of luminescence quenching is given later in this section.

Static quenching

Static quenching occurs upon the complexation of the luminophore (L) with a quencher (Q):

$$\mathrm{L} + \mathrm{Q} \rightleftharpoons \mathrm{L} - \mathrm{Q} \qquad [12]$$

The total concentration of the luminophore, $[\mathrm{L}]_0$, is given by:

$$[\mathrm{L}]_0 = [\mathrm{L}] + [\mathrm{L} - \mathrm{Q}] \qquad [13]$$

The dependence of the luminescence intensity on the quencher concentration can be derived by considering the association constant for the formation of the L–Q complex, K_s:

$$K_s = \frac{[\mathrm{L} - \mathrm{Q}]}{[\mathrm{L}][\mathrm{Q}]} = \frac{[\mathrm{L}]_0 - [\mathrm{L}]}{[\mathrm{L}][\mathrm{Q}]} \qquad [14]$$

Rearrangement gives:

$$\frac{[\mathrm{L}]_0}{[\mathrm{L}]} = 1 + K_s[\mathrm{Q}] \qquad [15]$$

Static quenching occurs because the resulting ground-state complex is not luminescent and, therefore, the concentration of free L decreases upon its complexation. The ratio $[\mathrm{L}]_0/[\mathrm{L}]$ is a measure of the decrease of the luminescence intensity due to static quenching. In the absence of the quencher (Q), the luminescence intensity, I_0, is highest because $[\mathrm{L}]_0 = [\mathrm{L}]$. In the presence of Q, the luminescence intensity, I, decreases due to static quenching. Therefore, the ratio $[\mathrm{L}]_0/[\mathrm{L}]$ is the same as I_0/I. Substituting in Equation [15] gives the

Stern–Volmer equation:

$$I_0/I = 1 + K_s[Q] \quad [16]$$

A Stern–Volmer plot of I_0/I against [Q] should give a straight line with a slope equal to K_s, which is also known as the Stern–Volmer constant for static quenching.

Dynamic quenching (collisional quenching)

Dynamic or collisional quenching occurs when the lifetime of the fluorophore is reduced, thus reducing the luminescence quantum yield. The mechanism of this type of quenching involves a collision of the excited luminophore molecule (*L) with a quencher molecule (Q). As a result, *L returns to its ground state without emitting photons and the excitation energy is transferred to Q. The transfer of the excitation energy to Q leads to a reduction of the excited state lifetime of *L and, therefore, a reduction of the luminescence intensity. It can be shown that the reduction of the luminescence intensity due to dynamic quenching is the same as the corresponding reduction of the excited state lifetime:

$$I_0/I = \tau_0/\tau \quad [17]$$

where τ_0 and τ refer to the lifetimes in the absence and presence of the quencher, respectively. The Stern–Volmer equation that governs dynamic quenching can be derived on the basis of the kinetic model shown in **Figure 11**.

The radiative decay rate (k_{rad}) is the reciprocal of the excited state lifetime in the absence of the quencher ($1/\tau_0$). The rates of depopulation of *L in the absence and presence of Q are given in Equations [18] and [19], respectively:

$$d[^*L]/dt = w - k_{rad}[^*L]_0 = 0 \quad [18]$$

$$d[^*L]/dt = w - k_{rad}[^*L] - k_Q[^*L][Q] = 0 \quad [19]$$

Substituting ($1/\tau_0$) for K_{rad}, after rearrangement, gives Equation [20], which is the Stern–Volmer equation for dynamic quenching:

$$I_0/I = 1 + k_Q\,\tau_0[Q] \quad [20]$$

The term ($k_Q\,\tau_0$) is the Stern–Volmer quenching constant for dynamic quenching, K_D. Therefore, the Stern–Volmer equation for dynamic quenching can be re-written as:

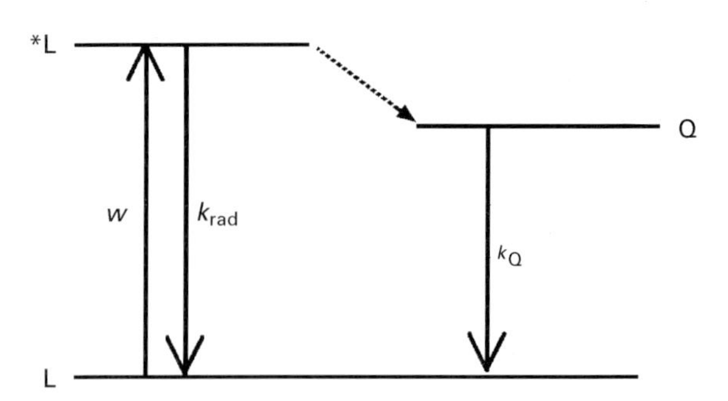

Figure 11 Kinetic model for dynamic quenching. In the notation used w. k_{rad} and k_Q refer to the absorption rate, radiative rate and quenching rate, respectively.

$$I_0/I = 1 + K_D[Q] \quad [21]$$

Note that Equations [16] and [21] are identical in form, and from both equations Stern–Volmer plots can be made in order to obtain the constants K_S, and K_D, respectively. This can be carried out simply by measurements of the luminescence intensity in the presence and absence of the quencher. However, such a study does not distinguish between the two mechanisms. The easiest and most definitive way to determine the quenching mechanism is to carry out a study of the lifetimes in the presence and absence of the quencher. The excited state lifetime decreases in the presence of quencher if dynamic quenching is the mechanism involved. Note that according to Equation [17] a Stern–Volmer plot can be obtained by plotting (τ_0/τ) on the y-axis instead of I_0/I:

$$\tau_0/\tau = 1 + K_D[Q] \quad [22]$$

In contrast, the excited state lifetime is invariant in the presence of the quencher if static quenching is the mechanism involved. Therefore, plotting (τ_0/τ) versus [Q] will simply yield a horizontal line parallel to the x-axis ($y = 1$).

Example

Tryptophan residues in a protein luminesce strongly whether they are on the surface of the protein or in its interior. When a quenching experiment is carried out with ionic quenchers, such as the iodide ion, the resulting quenching involves only surface-localized tryptophan residues. On the other hand, non-ionic quenchers, such as acrylamide, are able to penetrate into the interior of the protein and provide quenching constants that are really average constants and give information about both surface and interior tryptophan residues.

One way of separating the fluorescence quenching parameters associated with external and internal fluorophores is to use a double-quenching method in which two quenchers are applied simultaneously. The first quencher is used to selectively quench the fluorescence emission of exposed fluorophores. On the other hand, the second quencher is used to quench the fluorescence emission of both surface and interior fluorophores non-selectively.

Figure 12 shows Stern–Volmer plots of two samples of high density lipoproteins, HDL_1 and HDL_2 (referring to different densities of high density lipoproteins). These samples were obtained from rats raised on diets containing different amounts of the element manganese. A linear plot was obtained only for the manganese-deficient sample of HDL_2. From the equation of the straight line for this sample, the Stern–Volmer constant was determined as 0.88 M^{-1}. The plots were non-linear for the other samples, indicating the presence of more than one class of fluorophores, which are unequally accessible to the charged I^- quencher. In **Figure 13**, the fluorescence quenching of HDL_1 by iodide in the presence of various amounts of the non-ionc quencher acrylamide are shown using a modified Stern–Volmer method. The plots are all linear and provide fluorescence quenching parameters characteristic of interior and surface-exposed residues. **Table 1** shows a listing of the quenching constants for the different samples.

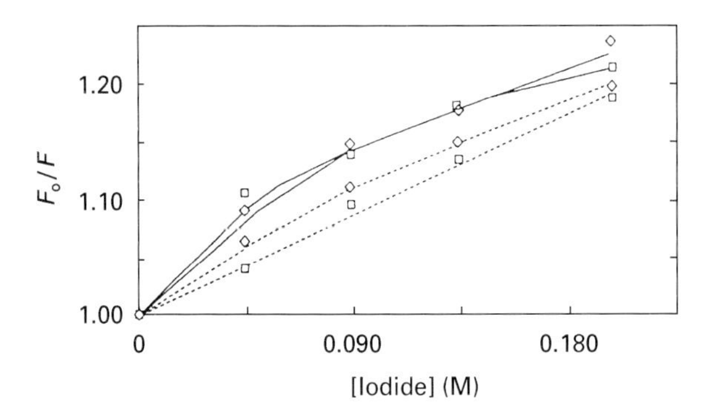

Figure 12 A Stern–Volmer plot for the fluorescence quenching of HDL_1 (solid lines) and HDL_2 (dashed lines). The different legends refer to samples obtained from rats fed with manganese-adequate (◊) and manganese-deficient (□) diet. The fluorescence was monitored at 338 nm (λ_{exc} = 295 nm) characteristic of tryptophan residues. (Reproduced with permission from Taylor PN, Patterson HH and Klimis-Tavatzis DJ (1997) A fluorescence double-quenching study of native lipoproteins in an animal model of manganese deficiency. *Biological Trace Element Research* **60**: 69–80.

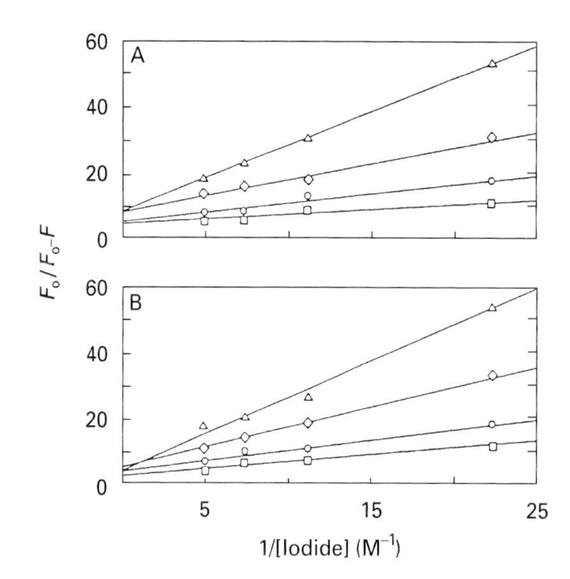

Figure 13 A modified Stern–Volmer plot for the fluorescence quenching of HDL_1 by iodide in the presence of varying amounts of acrylamide. The data are shown for HDL_1 samples obtained from rats fed with manganese-deficient (top) and manganese-adequate (bottom) diet. The fluorescence was monitored at 338 nm (l_{exc} = 295 nm) characteristic of tryptophan residues. Reproduced with permission from Taylor PN, Patterson HH and Klimis-Tavatzis DJ (1997) A fluorescence double-quenching study of native lipoproteins in an animal model of manganese deficiency. *Biological Trace Element Research* **60**: 69–80.

Table 1 Quenching constants for tryptophan residues in different high-density lipoprotein samples

Quenching constant / M^{-1}	*Lipoprotein*			
	HDL_1		HDL_2	
	MnA	MnD	MnA	MnD
K_I	9.14±0.44	17.18 ± 7.72	4.63±0.00	0.88±0.00
K_{A1}	5.21±0.88	6.15 ± 1.50	1.61±0.50	–
K_{A2}	4.27±0.11	5.96 ± 0.05	1.51±0.00	–

MnA, manganese-adequate sample; MnD, manganese-deficient sample; K_I, iodide quenching constant; K_{A1} and K_{A2}, acrylamide quenching constant for exposed and partly-exposed fluorophores, respectively.

A statistical analysis of the parameters in **Table 1** indicates that the acrylamide quenching constant for exposed fluorophores is significantly different in manganese-adequate HDL_1 compared with manganese-deficient HDL_1. In manganese-adequate HDL_2, there were two populations of fluorophores accessible to acrylamide, whereas in manganese-deficient HDL_2, all fluorophores were accessible to both quenchers. It was concluded based on these results that the local environments of the external fluorophores have different structures and charge distribution in manganese-adequate HDL_1 versus manganese-deficient HDL_1. For HDL_2 samples, it was concluded that one-third of the fluorophores were accessible to iodide and all external and internal fluorophores were accessible to acrylamide in manganese-adequate samples, whereas in manganese-deficient samples, all fluorophores were accessible to both quenchers.

Energy transfer

Theory

Energy transfer refers to a process in which an excited atom or molecule (*donor*) transfers its excitation energy to an *acceptor* atom or molecule during the lifetime of the donor excited state. **Figure 14** shows that as a result of energy transfer, the donor returns to its ground state while the acceptor is promoted to its excited state. If the acceptor is a luminescent species, it can emit by virtue of energy transfer, i.e. the acceptor luminesces as a result of the excitation of the donor. Such a luminescence is called 'sensitized luminescence', and some textbooks use the terms 'sensitizer' and 'activator' instead of 'donor' and 'acceptor'. Many applications in chemistry, physics, materials science and biochemistry are based on sensitized luminescence. A later section provides an example of the application of energy transfer processes in biological systems. Energy transfer processes occur via radiative or non-radiative mechanisms.

Radiative mechanisms This mechanism involves the absorption of light by a donor atom or molecule (D) followed by emission of a photon by the donor and the absorption of the emitted photon by another molecule called the acceptor (A). The process can be represented as follows:

$$D + h\nu \rightarrow D^* \quad [23]$$

$$D^* \rightarrow D + h\nu' \quad [24]$$

$$A + h\nu' \rightarrow A^* \quad [25]$$

The radiative mechanism is important if the acceptor A absorbs at the wavelength at which the donor emits. The efficiency of the process is determined simply by the quantum yield of the donor luminescence and the absorbance of the acceptor at the donor emission wavelength. No significant interaction between A and D is required in this mechanism and, therefore, radiative energy transfer can occur over extremely large separations of D and A. The radiative mechanism is not important if A and D are similar molecules because of the usually small overlap of the emission and absorption spectra in this case.

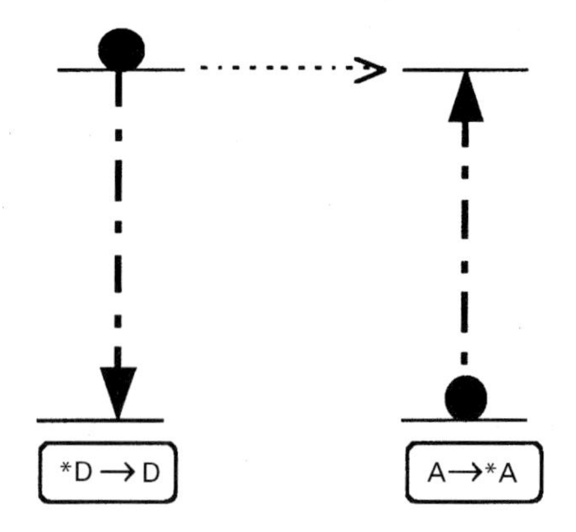

Figure 14 Energy transfer from a *donor* (D) to an *acceptor* (A).

Non-radiative mechanisms The radiative mechanism is a trivial case of energy transfer because it can be characterized by measuring the donor absorption and the acceptor emission separately from each other. In fact, most textbooks refer to the radiative pathway as a 'trivial mechanism' for energy transfer and do not really consider it as an energy transfer mechanism. Note in **Figure 14** that the donor does not emit light. Instead, the excitation energy is transferred non-radiatively to the acceptor. The non-radiative pathway shown in **Figure 14** is the mechanism that is typically used in most textbooks to represent energy transfer processes. Non-radiative energy transfer processes occur via two major mechanisms, the *Förster resonance mechanism* and the *Dexter exchange mechanism*.

The *Förster resonance mechanism* involves an electrostatic interaction between D and A. Such an interaction can occur over a long range, i.e. it does not require a very short contact between the donor and the acceptor. Energy transfer via the resonance mechanism may proceed over donor–acceptor distances as long as 50–100 Å. In order for energy transfer to be efficient via the resonance mechanism, the energies of the donor and acceptor transitions $D^* \rightarrow D$ and $A \rightarrow A^*$ must be nearly identical (hence the name '*resonance mechanism*'). Nevertheless, the presence of phonons (vibrational quanta) may provide assistance for energy transfer when small differences exist between the donor and acceptor excited states. Such processes are called *phonon-assisted* energy transfer processes (**Figure 15**). The energies of the phonons should be high enough to surmount the difference

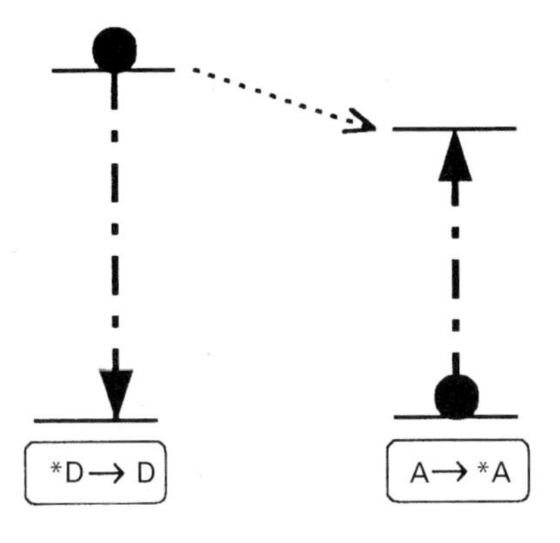

Figure 15 Phonon-assisted energy transfer. The energy difference between the excited states of D and A is provided by phonons.

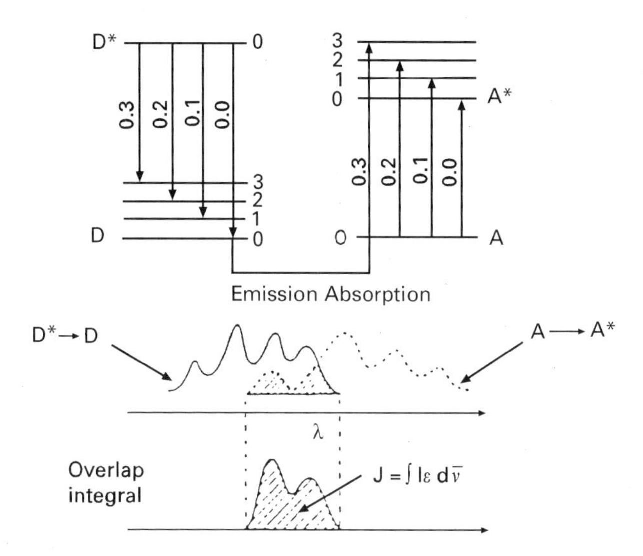

Figure 16 Experimental determination of the donor–acceptor spectral overlap from emission and absorption spectra. Reproduced with permission from Turro NJ (1978) *Modern Molecular Photochemistry*, p 299. Menlo Park: Benjamin/Cummings.

between the donor and acceptor excited states. However, too high phonon energies increase the likelihood of non-radiative de-excitation (see above), which would take place before energy transfer can take place.

The *Dexter Exchange Mechanism* involves a direct contact between the donor and the acceptor atoms or molecules. The exchange interaction between the donor D and acceptor A involves a transition state with a D–A distance that is close to the sum of the gas-kinetic collision radii of D and A, respectively. Therefore, information about the energy transfer mechanism can be gained by structural studies of the system under consideration. If the D–A distance is short (normally < 5 Å) then the exchange mechanism is the more likely, but if the D–A distance is long (>> 5 Å) then the resonance mechanism would be more likely.

The energy transfer efficiency is strongly dependent on the spectral overlap between the donor emission and the acceptor absorption. This is valid for both exchange and resonance mechanisms, as shown in the following expression for the energy transfer rate:

$$k_{\mathrm{ET}} = f(r_{\mathrm{D-A}}) \int I_{\mathrm{D}} \varepsilon_{\mathrm{A}} \, \mathrm{d}\bar{\nu} \qquad [26]$$

where the integral represents the spectral overlap between the donor emission I_D and the acceptor absorption ε_A. The factor $f(r_{D\text{-}A})$ is a function of the intermolecular distance between the donor and acceptor centres ($r_{D\text{-}A}$) and is governed by the relevant energy transfer mechanism (Förster or Dexter mechanism). The donor–acceptor spectral overlap can be determined experimentally by spectroscopic measurements, as illustrated in **Figure 16**.

Example

Trivalent lanthanide ions, Ln(III), can be substituted for certain metal ions in proteins to gain structural information about the sites of these metals. The ability of Ln(III) ions to substitute for metal ions such as Ca(II), Zn(II), and Mn(II) in proteins stems from the similarity between Ln(III) and these ions in ionic radii, coordination numbers (6–9) and preference to oxygen donor ligands. The higher charge density of Ln(III) makes these ions substitute with higher affinity than the metal ions in many metal-binding proteins. The quantum yield of luminescence is low for free Ln(III) but increases upon binding in close proximity to an aromatic amino acid within the protein: phenylalanine (Phe), tyrosine (Tyr) or tryptophan (Trp). When the protein is irradiated with wavelengths that correspond to the absorption maxima of these amino acids (~250–300 nm), the luminescence bands of the bound Ln(III) are enhanced by energy transfer. The strong sensitized luminescence of the Ln(III) makes these ions act as 'probes' of detailed and accurate structural information about the metal-binding sites in the proteins studied, as illustrated in the following example.

The protein α-lactalbumin is involved in the regulation of lactose synthesis. The binding of bovine α-lactalbumin (BLA) to Ca(II) is very strong (log K = 8–9). Eu(III) can bind apo-BLA (apo: metal-free) in the Ca(II) binding site. The appearance of sensitized luminescence for Eu(III) is illustrated in **Figure 17**. Note that owing to energy transfer, the

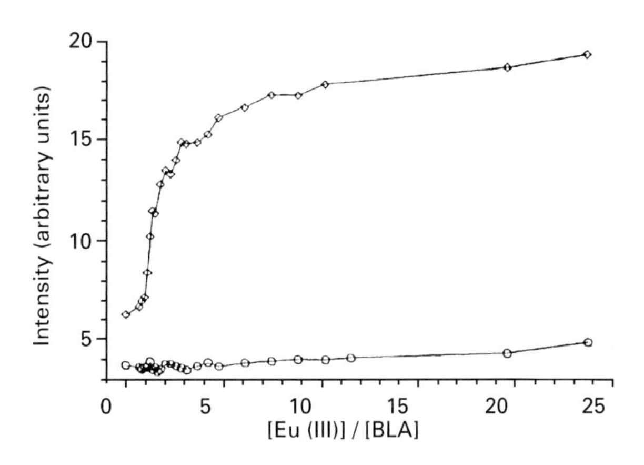

Figure 17 Luminescence titration of apo-α-lactalbumin (BLA) with europium chloride in D_2O. The luminescence intensity is monitored for the $^5D_0 \rightarrow ^7F_2$ emission line of Eu(III) with $\lambda_{exc} = 395$ nm. The upper curve shows the data for BLA-bound Eu(III) and the lower curve shows the data for $EuCl_3$ alone. Reproduced with permission from Bünzil JCG and Choppin GR (1989) *Lanthanide Probes in Life, Chemical and Earth Sciences*, p 279. Amsterdam: Elsevier.

luminescence intensity of BLA-bound Eu(III) (upper curve) is much stronger than the intensity of free Eu(III) (lower curve). **Figure 17** shows that the increase in the intensity of the $^5D_0 \rightarrow ^7F_2$ transition is most drastic when the ratio (R) of [Eu(III)/[BLA] is between 1 and 2, and the intensity continues to increase when $R > 2$. This is an indication of the binding of at least two Eu(III) ions to BLA. The Eu(III) luminescence is recorded in **Figure 17** for the $^5D_0 \rightarrow ^7F_2$ *hypersensitive* transition. A hypersensitive transition is a transition whose intensity is extremely sensitive to changes in the environment of the Ln(III) ion. Hypersensitivity is exhibited because of the $\Delta J = 2$ quadrupolar nature of the transition. The importance of this transition in the BLA-bound Eu(III) is illustrated in **Figure 18**. The excitation spectra of the Eu(III)-bound BLA are shown in **Figure 18** monitoring the $^5D_0 \rightarrow ^7F_0$ emission. The excitation spectra are resolved into several components, especially for R values > 1. The appearance of bands I and II indicate the presence of two different binding sites for Eu(III) in BLA. Band III is due to non-bonded (solvated) Eu(III) ions. Therefore, it is concluded that Eu(III) ions displace Ca(II) ions from BLA and bind into two sites.

Besides the determination of the number of metal binding sites, luminescence studies of Ln(III) ions in biological systems allow the determination of other structural properties of proteins such as the number of bonded water molecules, the sum of ligand formal charges and the site symmetry of the metal ion sites. For example, lifetime measurements in H_2O/D_2O mixtures allow the determination of the number of water molecules (n) that are coordinated to the metal ions in the different binding sites:

$$n = 1.05\,[1/\tau(H_2O) - 1/\tau(D_2O)] \qquad [27]$$

In the preceding example, the numbers of coordinated water molecules in sites I and II were determined as 2 and 4, respectively.

Excimers and exciplexes

Theory

Excimers and exciplexes are excited state complexes. An *exc*ited state di*mer* is called an *excimer*. Excimer formation can be represented by the following

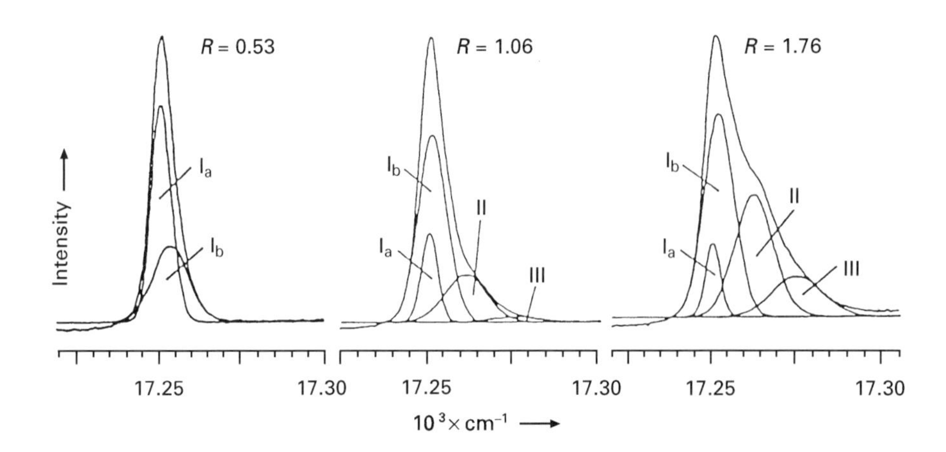

Figure 18 Curve-resolved Eu(III) excitation spectra of apo-BLA in D_2O. R represents the [Eu(III)]/[BLA] ratio. Reproduced with permission from Bünzil JCG and Choppin GR (1989) *Lanthanide Probes in Life, Chemical and Earth Sciences*, p 280. Amsterdam: Elsevier.

equation:

$$A + {}^*A \rightarrow {}^*[A - A] \quad [28]$$

where the asterisks refer to species in their excited states. According to Equation [28], an excimer is formed when an excited molecule interacts with another identical ground-state molecule. As a result of this interaction, an actual bond forms between the two monomer atoms in the excited state. Excited state interactions are not limited to the formation of homonuclear diatomic complexes. If interaction occurs between an excited molecule of one type and a ground state molecule of a different type, the resulting *exci*ted state com*plex* is called an *exciplex*:

$$A + {}^*B \rightarrow {}^*[A - B] \quad [29]$$

An exciplex can also form if more than two atoms are involved in the excited-state bond, whether these atoms are identical or different. Therefore, we can write a general equation to represent exciplex formation:

$$n\,A + m\,{}^*B \rightarrow {}^*[A_n - B_m] \quad [30]$$

From Equation [30], it is clear that an excimer is a special kind of exciplex where A = B and $n = m = 1$, so it is more appropriate to use the term *exciplex* when referring to excited state complexes in general, including both homoatomic and heteroatomic species.

It is important to recognize that exciplex formation is a physical phenomenon and not a chemical one. That is, bonding between the molecules only lasts as long as the excited state lifetime of the molecule, and the exciplex bond dissociates upon radiative or non-radiative de-excitation. Nevertheless, there are cases in which exciplex formation leads to photochemical reactions. For example, exciplexes are believed to be the reactive intermediates in the photochemical pathway of the Diels–Alder cycloaddition reactions of unsaturated organic compounds. Since photoluminescence is a photophysical process, the focus here will be on the photophysical aspect of exciplex formation instead of the photochemical reactions that result from exciplex formation.

Example of an organic exciplex

The characteristics of exciplex emission can be understood from potential energy diagrams of exciplex-forming species. **Figure 19** illustrates the spectroscopic features of exciplexes in relation to the potential surfaces of the ground and the excited electronic states. The spectra and potential surfaces are shown for pyrene, the classical example for excimer emission in organic compounds. According to **Figure 19**, the ground state is repulsive (no potential well) and the excited state is strongly bonding (deep potential well) along the pyrene–pyrene internuclear distance. This explains why the excimer bond forms only in the excited state.

Organic exciplexes are best identified by studying the variation of their emission spectra with concentration. This is illustrated in **Figure 19** for pyrene (py). At low pyrene concentrations ($\leq 10^{-5}$ M), the major luminescence band is due to the monomer. The characteristics of the monomer emission include: (i) vibronic structure is present, and (ii) the emission profile is independent of concentration at concentrations $\leq 10^{-5}$ M. As the pyrene concentration is increased above 10^{-5} M, the monomer emission is quenched and a new lower-energy emission band appears due to the formation of a *[py–py] excimer. The intensity of the excimer band increases with a concentration increase. The characteristics of the

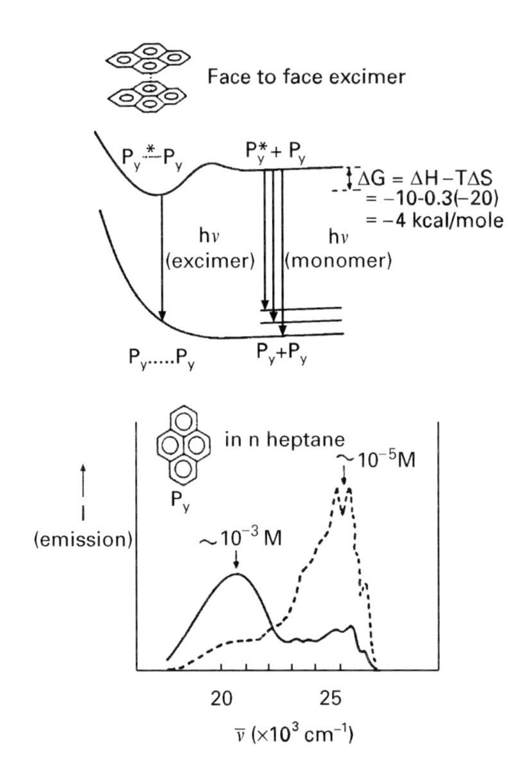

Figure 19 Excimer emission of pyrene. The upper curves show the potential surfaces of the ground and the luminescent excited electronic state (Reproduced with permission from Turro NJ (1978) *Modern Molecular Photochemistry*, p 141. Menlo Park: Benjamin/Cummings.

excimer emission band of pyrene in **Figure 19** are: (i) it is a structureless band with no vibronic structure, (ii) it has a lower energy than the monomer emission band and (iii) its intensity increases relative to the monomer emission band as the concentration is increased above a critical value (10^{-5} M for pyrene). These characteristics of the pyrene excimer bands are valid for the emission bands of organic exciplexes in general.

Examples of inorganic exciplexes

The formation of inorganic exciplexes has attracted attention only recently. Exciplexes formed in coordination compounds could be either *ligand-centred exciplexes*, or *metal-centred exciplexes*. Ligand-centred exciplexes are normally formed from coordinatively saturated complexes (where the metal ion has a high coordination number with no 'vacant' sites) and another species. In this situation, excitation of a ligand in the complex may result in the formation of an exciplex bond between the ligand and another molecule that exists in the solution (such as a solvent molecule). Metal-centred exciplexes, on the other hand, are normally formed from coordinatively unsaturated complexes and another species. In coordinatively unsaturated complexes, the potential coordination site(s) present in the metal ion can be filled by neutral electron-donor species (Lewis bases), anions or another metal ion. The latter type gives rise to the formation of *metal–metal bonded exciplexes*. This class of inorganic exciplexes is rather interesting because thus far all reported examples are luminescent.

The formation of the silver–silver bonded exciplexes $^*[Ag(CN)_2]_n$ ($n \geq 2$) will serve as an illustration. The formation of these exciplexes in the solid state has been reported by the authors of this article. The $[Ag(CN)_2^-]$ complex ion is a good candidate for the formation of metal–metal bonded exciplexes because its low coordination number (2) implies that several potential coordination sites are available.

Single crystals have been grown from a saturated solution of KCl that contains small amounts of $K[Ag(CN)_2]$. The crystals harvested from this solution are called $[Ag(CN)_2^-]$/KCl *doped crystals*, that is, $[Ag(CN)_2^-]$ guest ions are incorporated into (or *doped in*) the KCl host lattice. In these doped crystals, the Ag^+ and CN^- ions replace the K^+ and Cl^- ions, respectively, in some sites in the KCl lattice. Infrared measurements of single crystals of $[Ag(CN)_2^-]$/KCl have shown that multiple peaks exist in the ν_{C-N} region, indicating the presence of several local environments for the $[Ag(CN)_2]^-$ ions within the KCl lattice. This result has been explained in terms of the presence of monomers, dimers, trimers, etc of $[Ag(CN)_2^-]$ in the doped crystal.

The luminescence spectra of $[Ag(CN)_2^-]$/KCl doped crystals at 77 K are shown in **Figure 20.** It is interesting to note that at least four emission bands appear in the luminescence spectra of a single crystal of $[Ag(CN)_2^-]$/KCl. The most prominent bands are labelled as A, B, C and D in **Figure 20.** This observation is consistent with the conclusion based on the infrared spectra that the $[Ag(CN)_2^-]$ ions exist as monomers, dimers, trimers, etc in the KCl lattice. Therefore, $[Ag(CN)_2^-]_n$ oligomer ions in the KCl lattice with different values of n are responsible for the different luminescence bands. The strong dependence of the emission spectra on the excitation wavelength is unusual because emission spectra of luminescent materials are usually independent of the excitation wavelength. By controlling the excitation wavelength, a specific $[Ag(CN)_2^-]_n$ oligomer can be excited independently from the other oligomers so that the luminescence occurs primarily from this

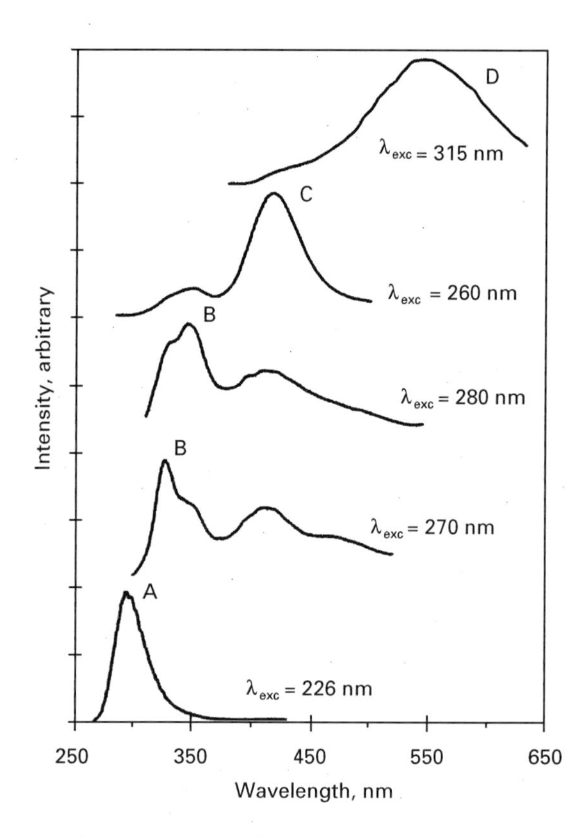

Figure 20 Photoluminescence spectra of $[Ag(CN)_2^-]$/KCl] doped crystals at 77 K. The letter assignment of the luminescence bands follows the notation used in **Table 2**. Reproduced with permission from Omary MA and Patterson HH (1998) Luminescent homoatomic exciplexes in dicyanoargentate (I) ions doped in alkali halide crystals: 1. Exciplex tuning by site-selective excitation. *Journal of the American Chemical Society* **120**: 7696–7705.

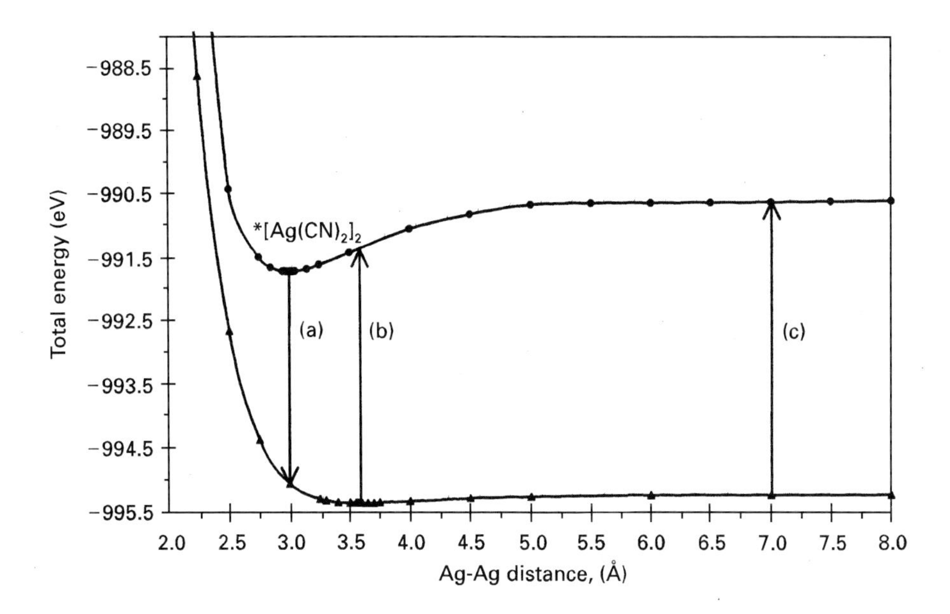

Figure 21 Potential energy diagram of the ground and the first excited electronic states of $[Ag(CN)_2^-]_2$ (eclipsed configuration) as plotted from extended Hückel calculations. The excimer $[Ag(CN)_2^-]_2$ corresponds to the potential minimum of the excited state. The optical transitions shown are (a) excimer emission, (b) solid state excitation and (c) dilute solution absorption. (Reproduced with permission from Omary MA and Patterson HH (1998) Luminescent homoatomic exciplexes in dicyanoargentate (I) ions doped in alkali halide crystals: 1. Exciplex tuning by site-selective excitation. *Journal of the American Chemical Society* **120**: 7696–7705.

excited oligomer. That is, different $[Ag(CN)_2^-]_n$ oligo-mers in the KCl lattice act as independent luminophores, each of which has a characteristic excitation wavelength. The emission is *tuned* to any of the bands A–D by changing the excitation wavelength. This is an interesting optical phenomenon that has been called *exciplex tuning*.

The luminescence bands of $[Ag(CN)_2^-]$/KCl are assigned, as shown in **Table 2**, to $^*[Ag(CN)_2]_n$ exciplexes that differ in the value of n and in their geometry. The exciplex assignment of these bands is based on both experimental and theoretical considerations. The following is an overview of the evidence used for this assignment:

Table 2 Assignment of the luminescence bands of $[Ag(CN)_2^-]$/KCl doped crystals

Band	λ_{max}^{em} *nm*	λ_{max}^{exc} *nm*	*fwhm/ (10^3 cm^{-1})*	*Assignment*
A	285–300	225–250	3.31	$^*[Ag(CN)_2^-]_2$ (excimers)
B	310–360	270–390	3.70	Angular $^*[Ag(CN)_2^-]_3$ (trimer exciplexes)
C	390–430	250–270	3.05	Linear $^*[Ag(CN)_2^-]_3$ (trimer exciplexes)
D	490–530	300–360	4.01	$^*[Ag(CN)_2^-]_n$ ($n \geq 5$, delocalized exciplexes)[a]

[a] It is assumed that the stabilization due to Ag–Ag interactions converges in $[Ag(CN)_2^-]_n$ oligomers with $n \geq 5$.

(1) The energies of the luminescence bands of $[Ag(CN)_2^-]$/KCl are extremely low relative to the absorption bands of dilute solutions of $[Ag(CN)_2^-]$ (which represent the $[Ag(CN)_2^-]$ monomer).

The absorption spectra of dilute solutions of $[Ag(CN)_2^-]$ show peak maxima with energies >50 000 cm^{-1}. The excitation and emission maxima are red-shifted by as much as 23 000 and 30 000 cm^{-1}, respectively, from the monomer transition. These large red shifts must be due to the oligomerization of $[Ag(CN)_2^-]$ units, as suggested by electronic structure calculations (see below).

The fact that the Ag(I) ion has a $4d^{10}$ closed shell electronic configuration forbids the formation of strong Ag–Ag bonds in the ground state. Excited states such as $4d^9$ $5s^1$, however, do not have a closed shell configuration. Hence, Ag–Ag bonding in the excited state is possible, which leads to the formation of $^*[Ag(CN)_2^-]_n$ exciplexes. **Figure 21** depicts the potential surfaces of the ground state and the first excited state of a $[Ag(CN)_2^-]_2$ dimer, as plotted from electronic structure calculations. The formation of a $^*[Ag(CN)_2^-]_2$ excimer is illustrated by the deep potential well in the excited state at a shorter Ag–Ag equilibrium distance than the corresponding ground state distance. The electronic transitions depicted in **Figure 21** explain the low emission and excitation energies (transitions (a) and (b)) relative to the monomer absorption (transition (c)).

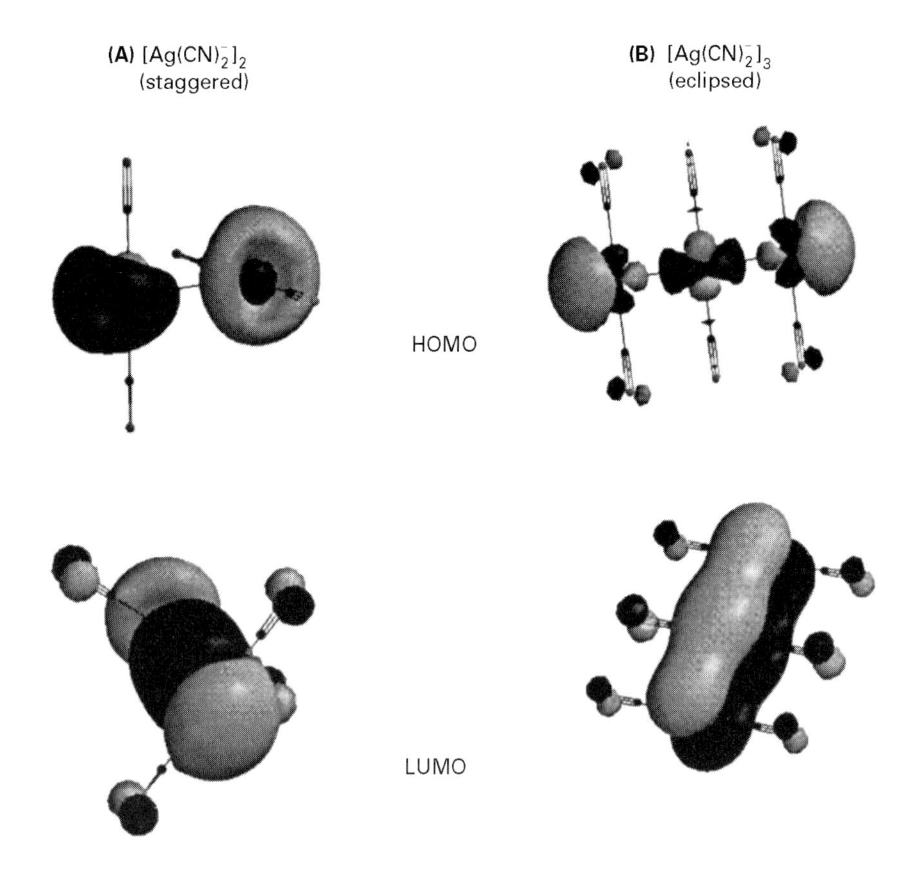

Figure 22 Surfaces of the highest occupied molecular orbital (HOMO) and the lowest unoccupied molecular orbital (LUMO) for $[Ag(CN)_2^-]_2$ (staggered isomer) and $[Ag(CN)_2^-]_3$ (eclipsed isomer), as plotted from *ab initio* calculations. Note the Ag–Ag antibonding character for the HOMO and the Ag–Ag bonding character for the LUMO in both cases.

(2) The excited states of $[Ag(CN)_2^-]_n$ oligomers are largely distorted from their ground states.

The Stokes shifts for the luminescence bands of $[Ag(CN)_2^-]$/KCl range from 7000 to 15 000 cm^{-1}. The fwhm values are in the 3000–4000 cm^{-1} range (**Table 2**), which are large values. As explained earlier, large excited state distortions are due to large differences in the bonding properties between the molecular orbitals that represent the ground and excited states, respectively. The ground state of an exciplex is antibonding while its first excited state is bonding. **Figure 22** illustrates this fact for the dimer $[Ag(CN)_2^-]_2$ and the trimer $[Ag(CN)_2^-]_3$. In both cases, the HOMO has a Ag–Ag antibonding character while the LUMO has a Ag–Ag bonding character. Similar results were obtained for other $[Ag(CN)_2^-]_n$ oligomers. Photoexcitations from antibonding HO-MOs to bonding LUMOs lead to the formation of $^*[Ag(CN)_2]_n$ exciplexes.

(3) The luminescence bands are generally lacking in structure.

Note that all the luminescence bands shown in **Figure 20** have no detailed structure. This is a similar feature to exciplex emission in organic compounds, as illustrated in **Figure 19** for the excimer band of pyrene. The apparent structure for band B is due to the existence of two different geometrical isomers of the $[Ag(CN)_2^-]_3$ trimer, not to vibronic structure. The absence of structured emission was obtained for $[Ag(CN)_2^-]$/KCl even though the luminescence measurements were carried out in the solid state for a doped crystal (1.1 mol% Ag) at cryogenic temperature (77 K). Doping and low temperatures normally reduce the chances of band broadening for luminophores that do not exhibit exciplex emission. Hence, the structureless emission of $[Ag(CN)_2^-]$/KCl must be due to exciplex formation.

Chemiluminescence

When exothermic chemical reactions occur, the product species are usually in their ground electronic states. However, some chemical reactions produce product species in electronically excited states which luminesce. This phenomenon is called *chemiluminescence*. Incidentally, chemiluminescence occurs in a number of biological systems (for example, the firefly), where it is called bioluminescence.

A common example of chemiluminescence is the reaction of luminol in basic solution with an oxidizing agent such as oxygen:

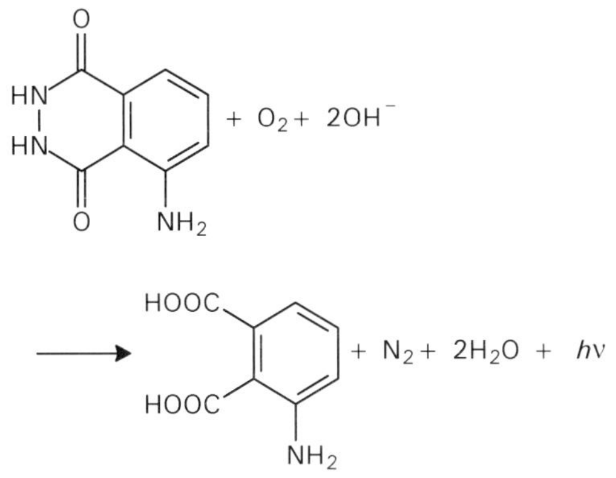

O, HN, HN, O, NH_2 $+ O_2 + 2OH^-$

$\longrightarrow$ HOOC, HOOC, NH_2 $+ N_2 + 2H_2O + h\nu$ [31]

The 3-aminophthalate product ion is in an excited electronic state and shows luminescence in the visible region. Reaction [31] is catalysed by the presence of certain metal ions such as Cr(III) and the intensity of the chemiluminescence has been found to be proportional to the concentration of the specific metal ion, usually in the concentration range of parts per billion (ppb). Thus, the luminol reaction can be used to determine the concentration of selected metal ion species at very low concentrations.

A second common example of chemiluminescence in the gas phase is the reaction of nitric oxide with ozone:

$$NO + O_3 \rightarrow O_2 + NO_2^*(\rightarrow NO_2 + h\nu) \qquad [32]$$

Here, the product species NO_2 is produced in an excited electronic state and emits light in the visible–near IR region. It has been found that the intensity of the chemiluminescence is proportional to the concentration of NO in the ppm–ppb range. Thus, the reaction shown in Equation [32] can be used as the basis for the development of a chemical sensor for NO. The detection of NO is important because nitric oxide is a chief environmental pollutant, and also because NO plays an important role in human biology.

See also: **Biochemical Applications of Fluorescence Spectroscopy; Laser Spectroscopy Theory**.

Further reading

Adamson AW and Fleischauer PD (1975) *Concepts of Inorganic Photochemistry*. New York: Wiley-Interscience.

Bünzil JCG and Choppin GR (1989) *Lanthanide Probes in Life, Chemical and Earth Sciences*. Amsterdam: Elsevier.

Cotton FA (1990) *Chemical Applications of Group Theory*, 2nd edn. New York: Wiley-Interscience.

Cowan DO and Drisko RL (1976) *Elements of Organic Photochemistry*. New York: Plenum.

Gilbert A and Baggot J (1991) *Essentials of Molecular Photochemistry*. Boca Raton: CRC Press.

Lakowicz JR (1983) *Principles of Fluorescence Spectroscopy*. New York: Plenum.

Omary MA and Patterson HH (1998) Luminescent homoatomic exciplexes in dicyanoargentate (I) ions doped in alkali halide-crystals: 1. Exciplex tuning by site-selective excitation. *Journal of the American Chemical Society* **120**: 7696–7705.

Skoog DA, Holler FJ and Nieman TA (1998) *Principles of Instrumental Analysis*, 5th edn. Philadelphia: Harcourt Brace.

Turro NJ (1978) *Modern Molecular Photochemistry*. Menlo Park: Benjamin/Cummings.

Wayne RP (1988) *Principles and Applications of Photochemistry*. Oxford: Oxford University Press.

Macromolecule–Ligand Interactions Studied By NMR

J Feeney, National Institute for Medical Research, London, UK

MAGNETIC RESONANCE
Applications

Introduction

NMR spectroscopy has proved to be a useful technique for studying interactions between proteins and other molecules in solution. Such interactions are important in biological molecular recognition processes and they have particular significance for studies of drug–receptor complexes where the results can assist in rational drug design. This article indicates how the appropriate NMR data can be extracted and analysed to provide information concerning interactions, conformations and dynamic processes within such protein–ligand complexes. For complexes of moderate size (up to 40 kDa), nuclear Overhauser effect spectroscopy (NOESY) measurements can often be used to determine the full three-dimensional structure of the complex, thus providing detailed structural information about the binding site and the conformation of the bound ligand. For larger complexes (typically up to 65 kDa), ligand-induced changes in protein chemical shifts, dynamic properties, amide NH exchange behaviour and protection from signal broadening by paramagnetic agents can all be used effectively to map out the ligand-binding sites on the protein by reporting on the nuclei influenced by ligand binding. In addition, NMR can sometimes be used to detect bound water molecules within the binding site and to monitor changes in water occupancy accompanying ligand binding.

NMR offers some advantages over X-ray crystallography in that it examines the complexes in solution, does not require crystals and provides a convenient method for defining specific interactions, monitoring changes in dynamic processes associated with these interactions, detecting multiple conformations and identifying ionization states of interacting groups within the protein–ligand complexes. However, unlike X-ray crystallography, NMR can provide full structural determinations only for moderately sized proteins (up to 40 kDa at the present time).

Equilibrium binding studies

The starting point for studies of protein–ligand interactions often involves determining the equilibrium binding constants for ligands binding reversibly to the protein. These measurements are sometimes made for a series of complexes where either the ligand or the protein is systematically modified in order to measure changes in the binding resulting from the introduction or removal of particular interactions in the complexes. Such investigations need to be accompanied by structural studies on the complexes to see whether the predicted effects have taken place and whether any major conformational perturbations have occurred in the rest of the system. These structural studies need large quantities of purified protein. For a typical sample size of 0.5 mL, the concentrations required vary from 10 µM for one dimensional spectra to 2 mM or greater for some multidimensional experiments. Large quantities of $^{13}C/^{15}N$-labelled proteins are usually prepared by cloning the appropriate gene into an overexpressing bacterial cell line and growing the cells using [^{13}C]glucose or [^{15}N]ammonium salts as the sole sources of carbon and nitrogen respectively.

Assignment of protein and ligand signals in the complex

Fast and slow exchange conditions

Before any detailed structural and dynamic information can be obtained from the NMR spectra of the complexes, the signals need to be assigned to specific nuclei in the ligand or protein. An important first step is to ascertain whether the bound and free species coexist under conditions of fast or slow exchange on the NMR timescale. For a nucleus with chemical shift frequencies ω_B and ω_F in the bound and free species respectively, separate signals are seen for the bound and free species for the case where the lifetime of the complex is long compared with $(\omega_B - \omega_F)^{-1}$: this is designated as the slow exchange condition. If the lifetime of the complex is short compared with $(\omega_B - \omega_F)^{-1}$, then conditions for fast exchange prevail and one observes a single averaged signal weighted according to the populations and chemical shifts in the bound and free forms. When the lifetime of the complex is of the same order as $(\omega_B - \omega_F)^{-1}$ then intermediate exchange conditions prevail, giving rise to spectra with broad, complex signals that are more difficult to analyse.

It is necessary to find out whether one is dealing with fast or slow exchange before further work can be attempted. The data can then be analysed to give the chemical shifts of the signals from the bound ligand/or protein. The line widths of the signals can sometimes provide information about the dissociation rate constants of the complex.

Assignment of protein signals

In making the assignments of the protein resonances, it is important to ensure that the protein is fully saturated with the bound ligand. Using multidimensional NMR methods in combination with ^{2}H-, ^{13}C- and ^{15}N-labelled proteins, it is now possible to obtain almost complete signal assignments for backbone and Cβ protons in proteins of molecular masses up to about 65 kDa. These resonances, once assigned, can be used to monitor ionization state changes, to characterize conformational mixtures and to provide conformational information from NOE measurements for the various complexes.

Assignment of ligand signals

Assigned signals for nuclei in the ligand are particularly important because these nuclei are obviously well placed to provide direct information about the binding site in the complex. It is easy to assign signals from bound ligands in fast exchange with free ligand if the assignments of the free ligand are known simply by following the progressive shift of the ligand signals during the ligand titration. It is more difficult to assign signals of nuclei in very tightly binding ligands ($K_a > 10^8$ M^{-1}) that are in very slow exchange with those in the free ligand. The usual method of assigning signals from tightly bound ligands is to examine complexes formed with isotopically labelled analogues (^{2}H, ^{3}H, ^{13}C and ^{15}N). Deuterated ligands can sometimes assist in making ^{1}H assignments by producing differences between ^{1}H spectra of complexes formed with deuterated and nondeuterated ligands, since signals from deuterated sites will disappear from the spectra. Complexes formed with ^{13}C- or ^{15}N-labelled ligands can also be examined directly by using ^{13}C or ^{15}N NMR: only the signals from nuclei at the enriched positions are detected, which simplifies their assignment. Protons directly attached to ^{13}C or ^{15}N can be detected using an appropriate editing or filtering pulse sequence. Heteronuclear multiple-quantum (or single-quantum) coherence (HMQC or HSQC) experiments allow the attached protons to be detected selectively and the X nuclei to be detected indirectly. A powerful extension of this approach is the 3D-NOESY-HSQC experiment, which allows selective detection of the NOEs from the ligand protons (attached to ^{15}N or ^{13}C nuclei) to neighbouring protons on the protein. The observed ^{1}H–^{1}H NOESY cross peaks are dispersed over the X-chemical shift frequency range. This considerably simplifies the NOESY spectrum at any particular X-frequency and is particularly useful for studying large complexes where there is extensive signal overlap in the normal NOESY spectra.

Complexes formed using less tightly bound ligands ($K_a < 10^6$ M^{-1}) can sometimes have spectra showing separate signals for bound and free species in slow exchange that are exchanging sufficiently rapidly to allow their signals to be connected using transfer of magnetization methods. Since the assignments for the free ligand are usually known, these methods give the assignments for the connected signals from the bound ligand.

Other nuclei can sometimes be used effectively for studying protein–ligand interactions. For example, the tritium (^{3}H) spectrum of a complex formed with a selectively tritiated ligand shows signals from the ligand only and the chemical shifts of these signals can be directly related to the corresponding protons in the nontritiated ligand. ^{19}F NMR measurements on complexes formed with fluorine-containing ligands or proteins can also provide useful information. Assignments of ^{19}F signals from the ligand are often straightforward, since usually only one or two sites are labelled. The simple spectra are ideal for monitoring multiple conformations and dynamic processes in the complexes. Making ^{19}F signal assignments for fluorine-containing proteins is more difficult, but they can be assigned by comparing ^{19}F spectra from different proteins where each fluorine-containing amino acid residue has been systematically replaced by a different amino acid using site-directed mutagenesis. Complexes formed with ligands containing phosphorus can be examined directly by ^{31}P NMR to provide detailed information about phosphate group ionization states and conformations in the bound state.

Determination of conformations of protein–ligand complexes

NMR is now able to provide full three-dimensional structures for protein–ligand complexes in solution. The general method involves first making the ^{1}H resonance assignments, then estimating the interproton distances from NOE measurements and dihedral angles from vicinal coupling constants and related data, and finally calculating families of structures that are compatible with both these distance and angle constraints and the covalent structure using some optimal fitting method usually, distance geometry-based and/or molecular dynamics simulated annealing-based calculations. Ideally, the structures of the unbound species as well as that of the complex should be determined.

Several workers have reviewed this area, particularly from the perspective of its value in drug design, and there have been many reported studies of ligand–receptor complexes where NMR has provided relevant structural information (see **Table 1**). This present overview will consider only a few examples chosen to illustrate particular aspects of protein–ligand interactions.

Table 1 Some examples of protein complexes studied by NMR

β-Lactamase with substrates
β-Lactoglobulin with β-ionone
Bcl-x(L) (survival protein) with Bak (cell death protein)
Calmodulin with peptides
Cyclosporin A with cyclophilin
Cytochrome P450 with substrate analogues
Dihydrofolate reductase with coenzyme and substrate analogues
Elastase with peptides
ETS domain of FLI-1 with DNA
FK506 binding protein with ascomycin
FKBP with immunosuppressants
GAT1 domain with DNA
Glutathione *S*-transferase with cofactor and substrate analogues
Homeodomain proteins with DNA
HPr phosphocarrier protein with phosphotransferase domain
Integration host factor (*E. coli*) with DNA
Lac repressor headpiece with DNA
Mu-Ner protein with DNA
P53 domain with DNA
Pepsin with inhibitors
Phospholipase with substrate analogues
Pleckstrin homology domain with phosphatidylinositol 4,5-bisphosphate
Protease with serpin
Protein G (streptococcal) domain with antibody fragment
PTB domain of insulin receptor substrate-1 (IRS-1) with phosphorylated peptide from IL-4 receptor
Rotamase enzyme FKBP with rapamycin
S100B with actin capping protein Cap 2
SHC SH_2 domain with tyrosine phosphorylated peptide
SRY with DNA
Staphylococcal nuclease with substrate analogues
Stromelysin domain with N-TIMP-2 inhibitor
Stromelysin with nonpeptide inhibitors
Thioredoxin with NFκβ peptide
Topoisomerase-I domain with DNA
Trp repressor with DNA
Trypsin with proteinase inhibitors
Urbs 1 with DNA

Many ligands that are flexible in solution adopt a single conformation when bound to a receptor protein. It is important to know the conformation of the bound ligand since this could provide the basis for designing a more rigid and effective inhibitor. Clearly, such information can be obtained directly once the full three-dimensional structure of the complex has been determined. However, in some cases the bound conformation of the ligand can be determined without determining the full structure of the complex if sufficient intramolecular distance and torsion angle constraints can be measured. Several methods based on measurements of intramolecular NOEs in the bound ligand have been proposed. One of these uses the transferred NOE (TrNOE) technique to provide conformational information about the bound ligand. In this method, cross-relaxation (NOE effect) between two protons in the bound ligand is transferred to the free molecule by chemical exchange between bound and free species. Under conditions of fast exchange, the negative NOEs from the bound state can thus be detected in the averaged signals for free and bound ligand. Transferred NOE effects can be detected in 2D-NOESY spectra and this approach has been used, for example, to obtain a set of intramolecular distance constraints between pairs of ligand protons in the tetrapeptide acetyl-Pro-Ala-Pro-Tyr-NH_2 bound to porcine pancreatic elastase and to determine the conformation of the bound peptide.

Other methods of determining the conformation of a bound ligand and details of its binding site involve using isotopically labelled proteins or ligands to simplify the NMR spectra. These approaches are particularly useful for studying tightly binding ligands where transferred NOE methods cannot provide any information. In such cases, it is necessary to measure directly the intramolecular NOEs within the bound ligand. The main problem is one of detecting the relevant NOEs in the presence of a large number of overlapping NOE cross-peaks from protons in the protein. There are several elegant techniques for measuring intra- and intermolecular NOEs in protein ligand complexes by isotopically labelling only one of the partners in the complex. One very direct strategy is to measure intramolecular ^1H–^1H NOEs in unlabelled ligands bound to perdeuterated proteins. Because only the ligand ^1H signals are detected, the 2D-COSY (correlation spectroscopy) and NOESY spectra are relatively simple. This approach has been used to examine cyclosporin A in its complex with perdeuterated cyclophilin. Another approach is to examine complexes of unlabelled protein with $^{13}C/^{15}N$-labelled ligands using NMR isotope-editing procedures that selectively detect only those NOEs involving ligand protons directly attached to ^{13}C or ^{15}N. In a ^{15}N-edited 2D-NOESY experiment on a pepsin/inhibitor (1:1) complex formed with ^{15}N-labelled inhibitors, NOE cross-peaks between the amide protons attached to ^{15}N in the ligand and their neighbouring protons in the protein could be detected. Isotope-editing methods have also been used to study ^{13}C- and ^{15}N-labelled cyclosporin A bound to cyclophilin. It is also possible to use NMR filter experiments to measure ligand–protein NOEs selectively for complexes containing nonlabelled ligand with ^{13}C-labelled proteins; this is a useful approach because it is usually easier to obtain labelled proteins than labelled ligands.

Specificity of interactions

Information about the groups on the protein and ligand that are involved in specific interactions can be obtained by determining the full three-dimensional structure of the complex in solution. More detailed information about specific interactions can often be deduced by monitoring the ionization states of groups on the ligand and protein and noting any changes accompanying formation of the complex. Further information about specific interactions comes from detecting characteristic low-field shifts for NH protons involved in hydrogen bonds.

Determination of ionization states

NMR is particularly effective for studying electrostatic interactions involving charged residues on the protein or ligand. A change in the charge state of an ionizable group is usually accompanied by characteristic changes in the electronic shielding of nuclei close to the ionizable group. Thus, NMR can monitor the ionization states of specific groups, measure their pK values and detect any changes that accompany protein–ligand complex formation.

The pK values of histidines in proteins are typically in the range 5.5 to 8.5 and they can easily be studied by carrying out pH titrations of the ^1H chemical shifts of the imidazole ε_1 protons over a suitable pH range and by fitting the data to the Henderson–Hasselbach equation. Ligand-induced changes in the pK behaviour of His residues have been used to monitor interactions in protein complexes formed with novel inhibitors. Protonation states of carboxylate groups in aspartic and glutamic acid residues in proteins have also been studied using ^{13}C NMR on suitably labelled proteins.

When the ionization state is a protonated species, it is sometimes possible to directly observe the proton involved in the protonation using NMR. If the protonation is at a nitrogen atom, then observation of the selectively labelled ^{15}NH group provides an unambiguous method of assigning the bonded proton. Such ^{15}NH proton signals have a doublet splitting (~90 Hz) characteristic of one bond ^{15}N–^1H spin coupling and they can be detected either directly in 1D experiments or by using 2D-HMQC (or HSQC) based experiments. In a ^1H NMR study examining ^{15}N-enriched trimethoprim in its complex with dihydrofolate reductase (DHFR), a 90 Hz doublet at 14.79 ppm in the spectrum could be assigned to the N-1 proton of bound trimethoprim (see structure in **Figure 1**). The ^{15}N chemical shift of the N-1 nitrogen is also characteristic of the protonated species (80 ppm different from the nonprotonated species). Earlier studies using [2-^{13}C]trimethoprim had already shown that the N-1 position is protonated in the bound state and that the pK value for this protonation is displaced by at least 2 units as a result of formation of the complex in which the protonated N-1 group interacts with the γ-carboxylate group of the conserved Asp-26 residue.

Ionization states of phosphate groups can be monitored using ^{31}P NMR and this approach has been used in studies of a coenzyme (nicotinamide–adenine

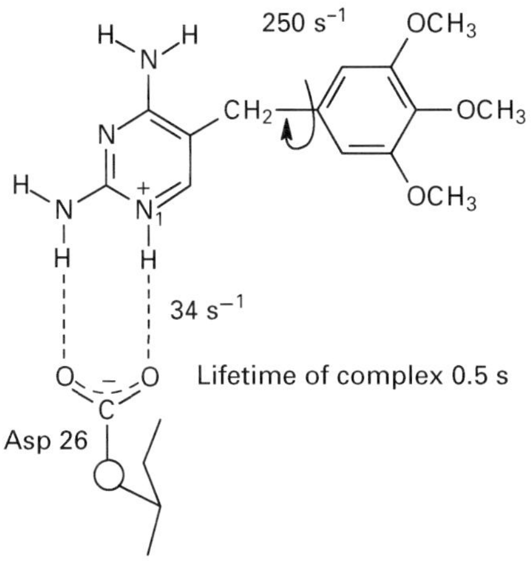

Figure 1 Dynamic processes in the complex of trimethoprim with *Lactobacillus casei* dihydrofolate reductase measured at 298 K. Reproduced with permission from Searle MS *et al.* (1988) *Proceedings of the National Academy of Sciences of the USA* **85**: 3787–3791.

dinucleotide, NADPH or NADP+) binding to dihydrofolate reductase. In each case the monophosphate group binds in the dianionic form with its p*K* value perturbed by at least 3 units compared to that of the free ligand.

Hydrogen-bonding interactions involving arginine residues

NMR has proved to be a very effective method for studying hydrogen bonding and electrostatic interactions involving side-chains of arginine residues in protein–ligand complexes. These studies are based on detection of ^{1}H and ^{15}N NMR signals from NH groups in ^{15}N-labelled proteins using gradient-enhanced two-dimensional $^{1}H/^{15}N$ HSQC NMR experiments where signals for the guanidino NH^{ε} and NH^{η} nuclei in arginine residues involved in protein ligand interactions can be detected. Such methods have been used on complexes of SH_2 domains formed with phosphopeptides to detect interactions between arginine NH^{η} hydrogens and phosphorylated tyrosines in the protein. Similar interactions have been studied in complexes of *Lactobacillus casei* dihydrofolate reductase formed with antifolate drugs such as methotrexate where four separate NH^{η} signals were observed for the Arg-57 residue, indicating hindered rotation in its guanidino group. Two of the NH^{η} signals had very low-field chemical shifts characteristic of NH hydrogen-bonded protons. From a consideration of the ^{1}H and ^{15}N chemical shifts it was possible to deduce that the central pair of NH^{η} protons in the guanidino group of Arg-57 interact with the α-carboxylate group of the glutamic acid moiety of methotrexate in an end-on symmetrical fashion (see **Figure 2**). The rates of rotation about the N^{ε}—C^{ζ} and C^{ζ}—N^{η} bonds were determined in the binary and ternary complexes of *L. casei* DHFR with methotrexate and NADPH, and their relative values compared with those in free arginine indicate correlated rotation about the N^{ε}—C^{ζ} bond of the Arg-57 guanidino group and the C'—C^{α} bond of the glutamate α-carboxylate group of methotrexate (**Figure 2**).

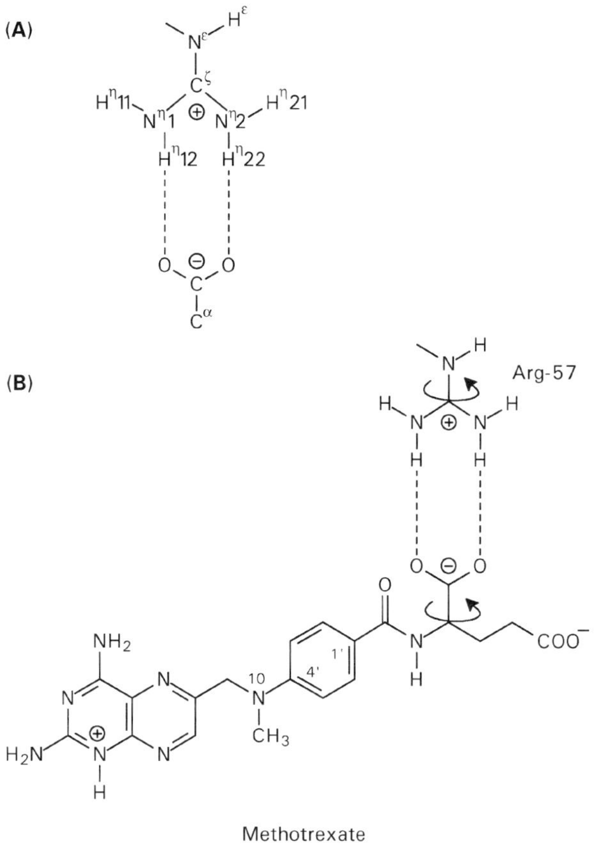

Figure 2 (A) Symmetrical end-on interaction of a carboxylate group with the guanidino group of an arginine residue. (B) Structure of methotrexate showing interactions of its α-carboxylate group of the glutamic acid moiety interacting in a symmetrical end-on manner with the guanidino group of Arg-57 of *Lactobacillus casei* dihydrofolate reductase and indicating the correlated rotation about the $N^{\varepsilon}C^{\zeta}$ bond of the Arg-57 guanidino group and the $C'C^{\alpha}$ bond of the glutamate α-carboxylate group of methotrexate, which allows the guanidino group to rotate without breaking its hydrogen bonds to the ligand. Reproduced with permission from Nieto PM, Birdsall B, Morgan WD, Frenkiel TA, Gargaro AR and Feeney J (1997) *FEBS Letters* **405:** 16–20. With kind permission of Elsevier Science-NL, Sara Burgerhartstraat 25, 1055 KV Amsterdam, The Netherlands.

Mapping binding sites by ligand-induced chemical shifts

A simple method of mapping the interaction sites in a protein–ligand complex involves measuring the ligand-induced chemical shifts accompanying complex formation using the ^{1}H and ^{15}N chemical shifts of backbone amide NH groups measured in ^{1}H/^{15}N HSQC spectra. This method indicates those residues that undergo a change in environment or conformation on complex formation and it works well, even for the case where the full assignments are available only for the uncomplexed protein. In such cases, lower limits for the shift changes can be estimated and these have proved to be adequate for mapping the binding sites. This method can be used for large protein–protein or protein–DNA complexes (up to 65 kDa).

Using these mapping procedures, a very elegant strategy for designing *de novo* ligands with high-affinity binding for selected target proteins has been developed (the so-called SAR (structure–activity relationships) by NMR approach). Large numbers of ligands were screened for their potential binding to target proteins by measuring ^{1}H/^{15}N HSQC spectra of the target ^{15}N-labelled protein in the presence of batches of ligands. These spectra could be collected relatively quickly and it was possible to screen up to 1000 compounds per day. This method identifies any ligands perturbing the ^{1}H/^{15}N chemical shifts (the binding, if any, usually results in conditions of fast exchange). Once a useful binding ligand has been identified, the protein is saturated with this ligand and the screening is continued to find another ligand that binds noncompetitively with the first one. When a suitable second candidate is found, detailed NMR structural work on the ternary complex is undertaken and, based on the structural information obtained, a strategy is developed for chemically linking the ligands to produce a high-affinity binding ligand. This approach has been used successfully to construct inhibitors with high binding affinity for metalloproteinases such as stromelysin.

Detection of multiple conformations

NMR spectroscopy has proved to be very useful for detecting the presence of different coexisting conformational states in protein–ligand complexes in solution. In some cases the different conformations are in slow exchange such that separate NMR spectra are observed for the different conformations. It is important to characterize the different conformations since each conformation offers a potentially new starting point for the design of improved inhibitors. Recognizing the presence of such conformational mixtures is also important when one is considering structure–activity relationships. NMR is the only method that can provide detailed quantitative information about such conformational equilibria in solution.

Several examples of multiple conformations have been uncovered in NMR studies of complexes of *L. casei* dihydrofolate reductase (DHFR). In many cases the different conformations correspond to a flexible ligand occupying essentially the same binding site but in different conformational states. For example, three conformational states have been detected in the NMR spectra of complexes of the substrate folate with DHFR. Two of the forms have the same pteridine ring orientation as bound methotrexate and their enolic forms can thus bind in a very similar way to the pteridine ring in methotrexate. The other form has its folate pteridine ring turned over by 180°.

Multiple conformations have been detected in several other complexes of *L. casei* DHFR (for example, with NADP^{+} and trimethoprim, and with substituted pyrimethamines) and also in complexes with *S. faecium* DHFR and *E. coli* DHFR: it seems likely that many other protein–ligand complexes will exist as mixtures of conformations. Of course, such conformations are more difficult to detect directly if they are in fast exchange.

Dynamic processes in protein–ligand complexes

NMR measurements can be used to characterize many of the dynamic processes occurring within a complex: this dynamic information complements the static structural information and provides a more complete description of the complex. Studies using NMR relaxation, line-shape analysis and transfer of magnetization have provided a wide range of dynamic information relating to protein–ligand complexes. The NMR-accessible motions range from fast ($>10^{9}$ s^{-1}) small-amplitude oscillations of fragments of the complex to slow motions ($1–10^{3}$ s^{-1}) involved in the rates of dissociation of the complexes, rates of breaking and reforming of protein–ligand interactions and rates of flipping of aromatic rings in the bound ligands; several illustrative examples, mainly from studies of dihydrofolate reductase complexes are considered below.

Rapid motions in protein–ligand complexes

Rapid segmental molecular motions ($>10^{9}$ s^{-1}) can be determined by measuring ^{13}C relaxation times and useful information about the binding can be obtained from the changes induced in the motions by the formation of the complex. Protein backbone dynamics are also frequently probed by making ^{15}N T_1, T_2, and $\{^1H\}^{15}N$ heteronuclear NOE measurements on ^{15}N-labelled proteins and analysing the data using the 'model-free' approach suggested by Lipari and Szabo.

Dissociation rate constants from transfer of saturation studies

If protons are present in two magnetically distinct environments, for example one corresponding to the ligand free in solution and the other to the ligand bound to the protein, then under conditions of slow exchange separate signals are seen for the protons in the two forms. When the resonance of the bound proton is selectively irradiated (saturated), its saturation will be transferred to the signal of the free proton via the exchange process and the intensity of the free proton signal will decrease. The rate of decrease of the magnetization in the free state as a function of the irradiation time of the bound proton can be analysed to provide the dissociation rate constant. This method has been used to measure the dissociation rate constant for the complexes of $NADP^+$·DHFR (20 s^{-1} at 284 K) and trimethoprim·DHFR (6 s^{-1} at 298 K). 2D-NOESY/EXCHANGE type experiments can also be used for such measurements.

Rates of ring flipping

Slow and fast rates of aromatic ring flipping have been characterized in ligands bound to proteins. Such studies are facilitated by using ^{13}C-labelled ligands. For example, ^{13}C line-shape analysis on the signals from the enriched carbons in [*m-methoxy*-^{13}C]trimethoprim and brodimoprim bound to DHFR has been used to measure the rates of flipping of the benzyl ring in the bound ligand. In all cases these rates are greater than the dissociation rates of the complexes and the flipping takes place many times during the lifetime of the intact complex. Thus the measured rate of flipping is indirectly monitoring transient fluctuations in the conformation of the enzyme structure that are required to allow the flipping to proceed.

Hydrogen exchange rates with solvent

Extensive NMR measurements of exchange rates between solvent and labile protons on protein or ligand have been reported. These are usually based on line shape analysis or transfer of magnetization methods. Such measurements have been made for the N-1 proton of bound trimethoprim in complexes of ^{15}N-labelled trimethoprim with DHFR. The line shape of the N-1 proton signal varies with temperature owing to changes in the exchange rate of this proton with the H_2O solvent. This line-width data can be analysed to estimate the exchange rate. This exchange can be considered as a two step process: in the first step the structure opens to allow access of the solvent, and in the second step the exchange process takes place. In this case, the N-1 proton forms and breaks a hydrogen bond with the carboxylate group of the conserved Asp-26 and the measured exchange rate (34 s^{-1} at 298 K) is thus the rate of breaking and reforming this hydrogen bonding interaction. This provides a further example of a very important interaction in the complex breaking and reforming at a rate much faster than the dissociation rate. Thus, individual protein interactions involving both the pyrimidine ring and the benzyl ring are involved in transient fluctuations during the lifetime of the complex (see **Figure 1**). If these structural fluctuations take place in close succession, they could form part of a sequence of events leading to complete dissociation of the complex.

Future perspectives

It is clear that advances in NMR methodology, particularly in multidimensional NMR experiments used in conjunction with isotopically labelled molecules, will provide even more detailed information about protein–ligand complexes in solution. Improved methods of structure determination will eventually allow the detection of smaller differences in structure between different complexes. The recently developed approaches for obtaining structural information from dipolar coupling contributions in the spectra resulting from orienting the molecules in solution (either by using high magnetic fields or by using liquid crystal solvents) could have an important impact on structural studies of large protein–ligand complexes. It seems likely that there will be increased input into structure–activity relationship (SAR) studies by use of the 'SAR by NMR' method for designing tightly binding ligands as inhibitors of important target proteins, particularly in industrial pharmaceutical laboratories where suitable libraries of compounds are readily available for screening. Future work should lead to an improved understanding of the implications of the dynamic processes taking place within ligand–protein complexes. Solid-state NMR studies on ligand complexes of membrane-bound proteins will be undertaken more frequently as the methodology and instrumentation become more widely available: although these studies require demanding isotopic labelling of the ligands, they can provide excellent information about distances and bond orientations that can be used to answer specific questions about the structures of protein–ligand complexes within lipid bilayers. The difficulty of obtaining such information by any other method provides a strong driving force for improving the solid-state NMR approach.

List of symbols

T_1 = spin–lattice relaxation time; T_2 = spin–spin relaxation time; ω_B (ω_F) = chemical shift frequency on the bound (free) species.

See also: **Drug Metabolism Studied Using NMR Spectroscopy; ^{19}F NMR Applications, Solution State; Hydrogen Bonding and other Physicochemical Interactions Studied By IR and Raman Spectroscopy; Nitrogen NMR; Nuclear Overhauser Effect; ^{31}P NMR; Proteins Studied Using NMR Spectroscopy.**

Further reading

Craik DJ (ed) (1996) *NMR in Drug Design.* Boca Raton, FL: CRC Press.

Emsley JW, Feeney J and Sutcliffe LH (eds) *Progress in NMR Spectroscopy*, Vols 18–33. Oxford: Elsevier. (See articles by C. Arrowsmith (32); M. Billeter (27); G.M. Clore (23); J.T. Gerig (26); A.M. Gronenborn (23); F. Ni (26); G. Otting (31); P. Rosch (18); B.J. Stockman (33); G. Wagner (22); G. Wider (32).)

Feeney J (1990) NMR studies of interactions of ligands with dihydrofolate reductase. *Biochemical Pharmacology* **40**: 141–152.

Feeney J and Birdsall B (1993) NMR studies of protein–ligand interactions. *NMR of Macromolecules* **7**: 183–215.

Fesik SW (1993) NMR structure-based drug design. *Journal of Biomolecular NMR* **3**: 261–269.

Fesik SW, Gampe RT Jr, Holzman TF, *et al* (1990) Isotope-edited NMR of cyclosporin A bound to cyclophilin: evidence for a *trans* 9,10 amide bond. *Science* **250**: 1406–1409.

Handschumacher RE and Armitage IM (eds) (1990) NMR methods for elucidating macromolecule–ligand interactions: an approach to drug design. *Biochemical Pharmacology* **40**: 1–174.

James TL and Oppenheimer NJ (eds) (1989, 1992) Nuclear magnetic resonance. In *Methods in Enzymology*, Vols 176, 177 (1989), Vol 239 (1992). London: Academic Press.

Jardetzky O and Roberts GCK (1981) *NMR in Molecular Biology*. London: Academic Press.

Markley JL (1975) Observation of histidine residues in proteins by means of nuclear magnetic resonance spectroscopy. *Accounts of Chemical Research* **8**: 70–80.

Roberts GCK (ed), (1993) *NMR of Macromolecules: A Practical Approach*. New York: Oxford University Press.

Shuker SB, Hajduk PJ, Meadows RP and Fesik SW (1996) Discovering high-affinity ligands for proteins – SAR by NMR. *Science* **274**: 1531–1534.

Watts A, Ulrich AS and Middleton DA (1995) Membrane protein structure: the contribution and potential of novel solid state NMR approaches. *Molecular Membrane Biology* **12**: 233–246.

Wüthrich K (1976) *NMR in Biological Research: Peptides and Proteins*. Amsterdam: North-Holland.

Macromolecules Studies By Solid State NMR

See **High Resolution Solid State NMR, ^{13}C.**

Magnetic Circular Dichroism, Theory

Laura A Andersson, Vassar College, Poughkeepsie, NY, USA

ELECTRONIC SPECTROSCOPY
Theory

Introduction and overview

Magnetic circular dichroism (MCD) spectroscopy is a type of electronic spectroscopy, also called the Faraday effect or the Zeeman effect, that can be a particularly useful and effective method for structural analysis. For example, MCD can be used to assign the transitions in the electronic absorption spectrum (UV-visible), with respect to details such as the molecular orbital origins of the transitions. Often, such transitions are not clearly observed in the UV-visible spectra, because they are spin-forbidden and weak, but upon application of the magnetic field, H_0, they can be detected. MCD spectroscopy can also be used to determine not only the spin state for a metal such iron, but also the coordination number at the metal.

There is an extensive body of detailed MCD structural data provided for a variety of different biological, organic, and inorganic systems. However, MCD has been surprisingly neglected, given its broad utility, ease of handling, and low sample-concentration requirements relative to many other spectroscopic methods. MCD spectroscopy has only recently begun to be utilized to its full potential.

Biological systems that have been studied by MCD include: (a) a haem (iron porphyrin-containing) proteins and enzymes such as oxygen transport proteins (haemoglobin and myoglobin), electron-transfer proteins (cytochromes), the diverse and ubiquitous P450 enzymes, and peroxide-metabolizing enzymes such as peroxidases and catalases; (b) other biological chromophores such vitamins B_{12} and chlorophylls; (c) tryptophan-containing proteins (this amino acid has a unique and distinguishing MCD signal); (d) non-haem iron proteins; (e) copper- and cobalt-containing proteins (natural or metal-substituted); and (f) a variety of other systems, too diverse to list here.

One particular advantage of MCD spectroscopy is the limited sample requirements, particularly relative to other experimental methods, even in these days of cloning and massive expression of samples. For example, as much as 500 µL of a 1–2 mM solution of haem protein must be used for NMR structural analysis. In contrast, to study the same sample by conventional (electromagnet) MCD, only 2.6 mL of an ~10–25 µM sample is required. Second, the ability to determine key structural information such as spin and coordination states at, or near, biological temperatures is also significant. Whereas the electronic absorption spectrum of a ferric haem protein can generally be used to distinguish high-spin from low-spin systems, more specific information concerning the coordination number was once routinely determined by EPR (also called ESR) spectroscopy. This method not only requires at least 250 µL of an ~250 µM sample, it also requires either liquid nitrogen or liquid helium cooling of the sample to gain the EPR *g* values and make the assignments. In contrast, MCD spectroscopy of a sample under biologically relevant conditions can provide highly detailed and specific data with respect to both the spin and the coordination states of the system (e.g. high-spin pentacoordinate haem vs high-spin hexacoordinate haem). This has recently been illustrated in the case of the haem catalases, which are among the most rapid of all enzymes, converting H_2O_2 to O_2 and H_2O with a turnover rate of ~100 000 per second per active centre (most catalases have four active centres). A novel set of X-ray crystallographic data for bacterial catalase were published, which were not only in conflict with previous X-ray data for a mammalian catalase, but also appeared inconsistent with the rapidity of normal enzymatic activity. Specifically, the ferric haem of the catalase was suggested to have a water molecule as its sixth ligand that was furthermore stabilized by participation in a hydrogen-bonding network. MCD spectral analysis of the identical bacterial catalase, as well as a mammalian, and a fungal catalase, clearly and unequivocally demonstrated that, under approximately biological conditions, all of the native catalases were always high-spin and pentacoordinate, with NO water ligated at the haem regardless of pH in the range 4–10. Again, this empty coordination site for the haem is of critical significance for the enzymatic reaction of the catalases, where the first step of the reaction requires H_2O_2 ligation at the haem.

In part, the increasing employment of MCD spectroscopy in structural analysis derives from a widening array of modifications that extend the diversity and accuracy of the method. A simple listing of such variations includes:

(1) method of field generation: electromagnet vs superconducting magnet;
(2) spectral region studied: near-infrared (near-IR; NIR) vs UV-visible (ultraviolet and visible);
(3) VT (variable temperature) and VTVH (variable temperature, variable field; $[H_0]$ MCD, using a superconducting magnet, with temperature variations to as low as ~1.5 K, and magnetic field variations from ~1 to 50 T (1T = 10 000 G);
(4) 'fast' MCD, which includes nanosecond and picosecond experiments, using an 'ellipsometric approach', also called TRMCD (time-resolved MCD);
(5) VMCD, vibrational MCD, particularly Raman;
(6) the newest modification, XMCD, X-ray detected magnetic circular dichroism.

Fundamentally, MCD spectroscopy can be defined as the differential absorption of left and right circularly polarized light, induced by an external magnetic field (H_0) that is parallel (or anti-parallel) to the direction of light propagation. This property is known as the 'Faraday effect', after Michael Faraday who observed (ca. 1845) that *any* substance, when placed in a magnetic field, will rotate the plane of polarized light. Indeed, it was the Faraday effect that was used to establish the electromagnetic nature of light. A general schematic of an MCD instrument is presented in **Figure 1**.

In the case of *degenerate* electronic transitions, for which the components are not resolved in the absorption spectrum, one has access to only limited structural information. However, in the presence of a magnetic field, these degeneracies are lifted (Zeeman effect) and now can be explored in more detail. Using ordinary (conventional) electronic absorption spectroscopy, no detectable spectral difference is observed for such a sample in the presence or absence of the applied external magnetic field. This is because the spectral line width (for most samples) is greater than the splitting of the energy levels. However, using circularly polarized light it is possible to measure and record the differences between these magnetically degenerate states (see *A* and *C* terms below). For a sample that has no nondegenerate energy levels, it is still possible to obtain an MCD spectrum if the nondegenerate energy levels undergo a magnetically induced mixing; this is the origin of MCD *B* terms. A more detailed analysis presented below.

MCD vs CD spectroscopy

Three aspects of MCD spectroscopy are clearly distinct from those of 'natural' circular dichroism (CD) spectroscopy:

(a) CD requires an optically active, chiral, molecule (essentially one of low molecular symmetry at the chiral centre lacking even a simple mirror plane), whereas MCD has no structural requirements, but rather is a property of *all* matter;
(b) chirality and optical activity (CD) are derived from the presence of *both* electric *and* magnetic dipole transition moments in the sample under study, which furthermore must be parallel (or anti-parallel) to one another, whereas for magnetic optical activity (MCD) only an electric dipole transition moment is required, with the external magnetic field supplying the magnetic component (see **Figure 2**);
(c) CD spectra are sensitive to molecular structure and perturbations of the chiral centre(s) by the physical environment, which is most clearly seen as asymmetry in the chromophore and/or its environment. MCD spectra are representative of the electronic structural properties of a given molecule, such as field-induced perturbations in energy levels. The latter, however, does *not* imply the absence of environmental sensitivity, but rather that molecular perturbations must directly affect the electronic properties. For example, this may include not only a concentration depend-

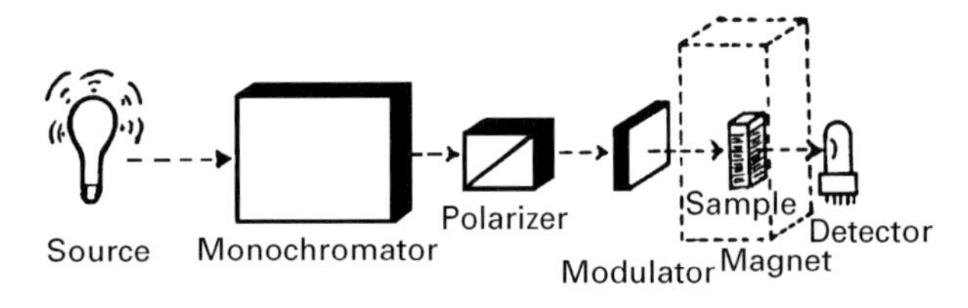

Figure 1 Optical components of a typical MCD/CD instrument. The modulator, now most commonly a piezoelectrically driven photoelastic device, converts linearly polarized light to a.c. modulated circularly polarized light.

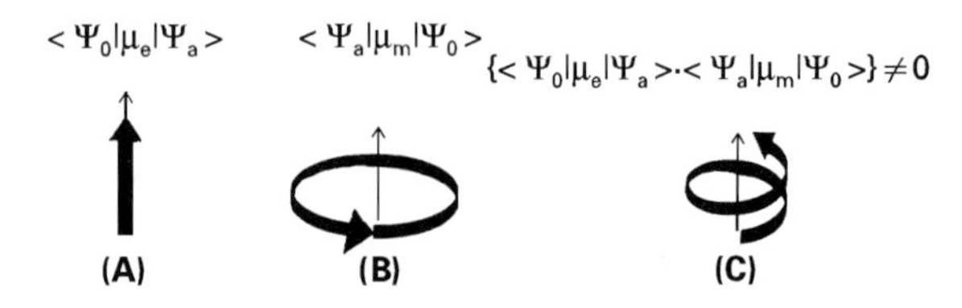

Figure 2 Cartoon illustrating the photon-induced transitions in a molecule. (A) Electronic absorption from ground to excited state is expressed as shown, where μ_e is the electric dipole moment operator; (B) magnetic absorption and the mathematical expression, where μ_m is the magnetic dipole moment operator; and (C) interaction of electronic and magnetic absorption, yielding optical activity.

ence, but also a sensitivity to structural variations, and to precise ligand geometry surrounding the chromophore (often a metal). MCD reflects electronic structural features such as spin and orbital degeneracies – information about spatial and coordination structure.

Note that in the simplest quantum mechanical expression, for a CD spectrum to be observed there must be both electric and magnetic dipole transition moments, for which the cosine between the two transition dipole moments must be non-zero (**Figure 2**). Essentially, this means that the transition dipole moments must have a parallel (or anti-parallel) relationship to one another. Without all three components (the electric dipole transition moment, the magnetic dipole transition moment, and their parallel relationship) there can be no optical activity.

Extensive theoretical discussions of CD spectroscopy focus on the specific origin of CD activity, such as the 'one electron model'. MCD differs specifically here from CD, in that MCD spectroscopy provides the external magnetic field, H_0, whereas chiral systems have their own magnetic transition dipole as a consequence of their very low symmetry at the chiral centre.

MCD experimental details

Fundamentally, then, both magnetic circular dichroism and circular dichroism are phenomena dependent upon the Beer–Lambert law (Eqn [1]) that is to say, upon the *concentration* of the sample, and upon the inherent 'responsiveness' of the sample under study to light, called the extinction coefficient:

$$A_\lambda = \varepsilon_\lambda cb \qquad [1]$$

where A = absorbance (unitless), λ = the specific wavelength, ε = the molar extinction coefficient (M^{-1} cm^{-1}); c = molar concentration (mol L^{-1}; M), and b = cuvette pathlength (cm). More specifically, this is written as shown in Equations [2] and [3] for circular dichroism and magnetic circular dichroism, respectively, where ε_l and ε_r are the specific extinction coefficients for left- and right-circularly polarized light (LCPL and RCPL, respectively):

$$A_l^0 - A_r^0 = \Delta A_{CD}^0 = \Delta A_\lambda^0 = (\varepsilon_l^0 - \varepsilon_r^0)c\,b \qquad [2a]$$

or

$$\Delta\varepsilon^0 = \varepsilon_l^0 - \varepsilon_r^0 \qquad [2b]$$

$$\Delta A_\lambda = A_l - A_r \qquad [3a]$$

$$= \Delta A_{\lambda(CD)}^0 + H\Delta A_{\lambda(MCD)} \qquad [3b]$$

and $\Delta\varepsilon_m = (\Delta\varepsilon - \Delta\varepsilon^0)/H_0 = (\Delta A_\lambda - \Delta A_\lambda^0)/cbH_0$ where A^0; $\Delta\varepsilon^0$ etc., with a superscript zero represent those values in the absence of a magnetic field. Thus, the actual MCD experiment requires collection of the MCD spectra for both sample and standard (buffer), and collection of the CD sample for both sample and standard.

The natural CD signal (sample minus buffer) is subtracted from the signal for the sample MCD minus buffer MCD, to yield the 'raw' MCD data. These data are then corrected for field strength (in tesla) and for the molar concentration of the sample under study. Note that the MCD intensity is actually dependent on the strength of the magnetic field, H_0, which is a key factor in the type of MCD experiment described as [3] above. This final correction means that it is actually the MCD 'extinction coefficient', $\Delta\varepsilon_M$, that is being reported, and thus one can directly compare the MCD data between different samples in a meaningful manner.

MCD *A*, *B* and *C* terms; MCD data analysis

In the case of 'natural' CD spectra, each CD spectral band is generally Gaussian in shape, and is associated with a single optically active transition. In contrast, a given electronic spectrum for a sample can result in several MCD spectral features, given the several different mechanisms by which the spectra feature may arise.

Under experimental conditions of temperature such that Zeeman energies are $\ll \kappa T$, and where the Zeeman splitting is small compared with absorption line width, there can be three separate contributions to net ellipticity, imaginatively called A, B and C terms. The magnetically degenerate ground or excited states are split by the application of the external field, H_0.

The MCD spectral magnitude is directly related to the difference between the LCP (left circularly polarized) and RCP (right-circularly polarized) light, where the (+) and (−) terms below refer to RCP and LCP, respectively:

$$\text{MCD} \propto (\Delta A_\pm \equiv A_- - A_+) \qquad [4]$$

Mathematically, MCD is defined as the difference between two transition moments. It is thus differential, and the observed MCD spectrum can be either positive or negative for a given absorption:

$$\Delta A_{\pm} = \gamma\beta H_0 c\, b[A_1(-\delta f/\delta E) + (B_0 + C_0/\kappa T)f(E) \quad [5]$$

where $\Delta A_{\pm}$ = change in absorption, γ = spectroscopic constant (defined as $N\pi^2\alpha^2\log e)/250hcn$), N is Avogadro's number, α is the proportionality constant between the electric field of the light and the electric field at the absorbing centre, h is Planck's constant, c is the speed of light, n is the refractive index, β is the Bohr magneton, H_0 is the magnetic field strength, κ is the Boltzmann constant, and T is the absolute temperature. The term $f(E)$ is a general Gaussian line shape function, and thus Equation [5] has contributions from both the line shape and derivative of the line shape. A_1, B_0 and C_0 are the MCD parameters (A, B and C terms) defining the amount of absorptive (B_0 and C_0) or derivative (A_1) signals in the MCD spectrum.

The A_1 term can *only* be non-zero if *either* the ground or the excited state is degenerate and is Zeeman split by longitudinal magnetic field; the derivative line shape function for A_1 in Equation [5] results in its characteristic shape. An A term is most easily described for degenerate excited state (**Figure 3**). Because there is a frequency shift between the two transitions, there cannot be cancellation of equal and opposite ellipticity, so the 'derivative shaped' A term is observed. (Note the important distinction that while this feature is shaped like a derivative, it is *not* one in actuality, being a difference spectrum.) Two diagnostic features of an A term are (1) that it is centred at ν_0, which means that the 'zero-point' where the MCD band goes from negative to positive or vice versa, corresponds exactly (in practice, very closely) with the electronic absorption spectral maximum of the sample, and (2) the intensity of the MCD A terms is invariant with temperature, and thus is independent of temperature.

The C_0 term is analogous to A_1, in that it is only non-zero if there is a degeneracy; however for C_0 this must be a ground-state degeneracy. The C_0 term is absorptive in nature and also has an associated temperature dependence, which means that the extent of its contribution to the MCD line shape varies with absolute temperature. Assuming thermal equilibrium for the electronic states, the populations are dominated by Boltzmann statistics. Thus as the temperature is lowered, the lower energy level of the degenerate ground state has an increasing population, and the C term increases in intensity. Indeed, the inverse correlation between MCD intensity and absolute temperature is a distinguishing feature of the C term. This is illustrated in **Figure 4**. Again, in comparison with the A term, the C term is not derivative shaped, nor does it have a zero-point that corresponds with the absorption spectral maximum of the correlated electronic transition. For both the C term and the B term (discussed below), the positions of the absorption maximum correlates with the MCD band position, as either a maximum (positive MCD band) or minimum (negative MCD band).

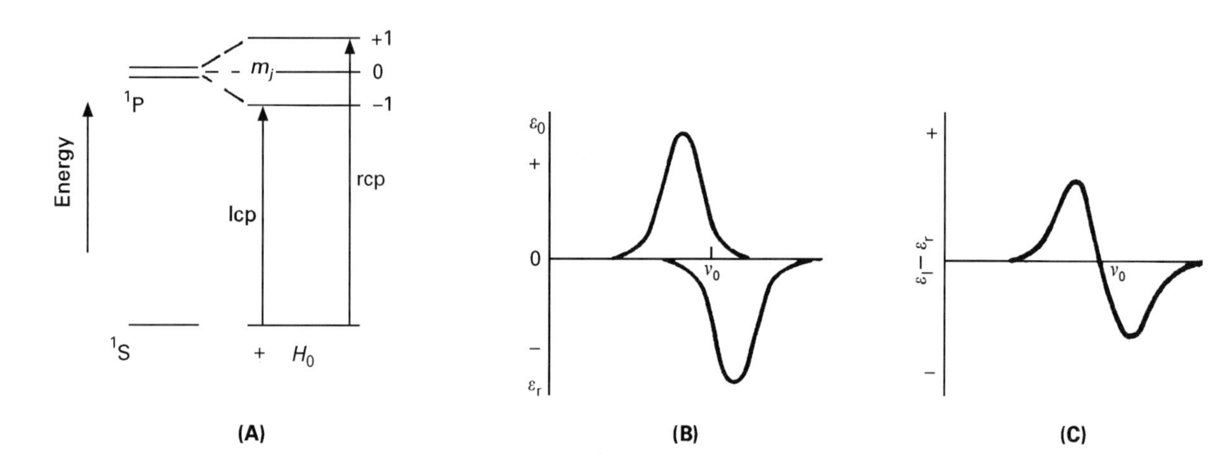

Figure 3 Origin and typical spectrum of an MCD A term. (A) Electronic transition from ground to degenerate excited state; on the right, in the presence of the magnetic field, H_0, the excited state degeneracy is split. This results in separate transitions, corresponding to the differential absorption of left or right circularly polarized light. (B) The separate left (ε_l) and right (ε_r) circularly polarized spectra; ν_0 is the position of the absorption maximum. By convention, the absorption of left circularly polarized light is positive and absorption of right circularly polarized light is negative. (C) The A term MCD spectrum, $\Delta\varepsilon = \varepsilon_l - \varepsilon_r$; note that the position of ν_0 is at the 'crossover' (zero-point) of the derivative-shaped A term. This change of sign for the MCD signal at the position of the absorption maximum is diagnostic for an A term.

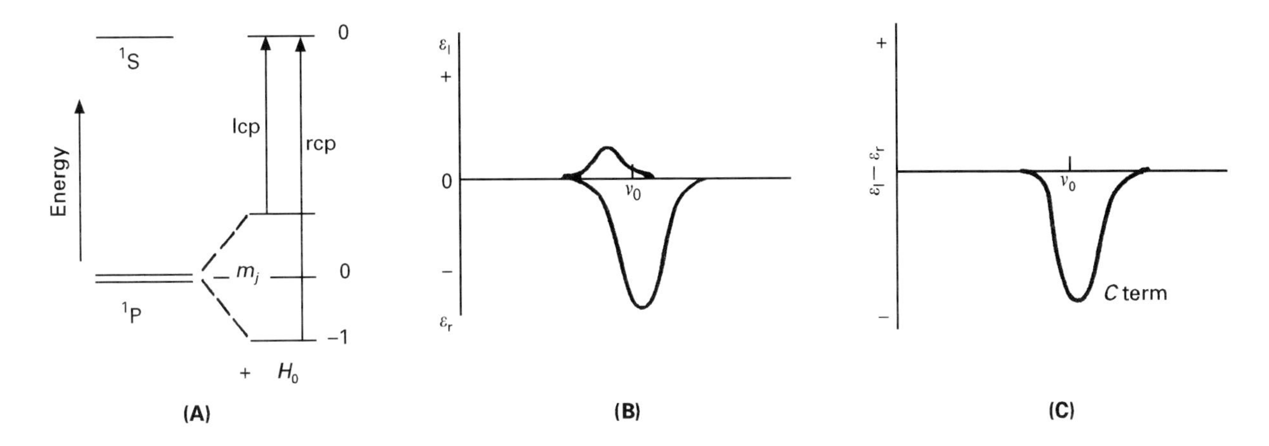

Figure 4 Origin and typical spectrum of an MCD '*C*' term. (A) Electronic transition from degenerate ground to excited state; on the right, the magnetic field, H_0, splits the excited state degeneracy. This results in separate transitions, corresponding to the differential absorption of left or right circularly polarized light. (B) The separate left (ε_l) and right (ε_r) circularly polarized spectra; ν_0 is the position of the absorption maximum. (C) The *C* term MCD spectrum, $\Delta\varepsilon = \varepsilon_l - \varepsilon_r$; note that the ν_0 absorption maximum is not in any particular relationship to the resulting *C* term spectrum. The *C* term intensity is inversely dependent upon temperature, due to Boltzmann population of energy levels; as the temperature decreases, more of the molecules populate the lower of the two degenerate orbitals, and thus the *C* term intensity increases as the temperature is lowered, until the saturation point.

Finally, there is the *B* term of the MCD spectra, which is a complex species deriving from the mixing of electronic states. The quantum-mechanical origin of the B_0 term is the magnetically induced mixing of nondegenerate states. The wide spacing of energy levels generally results in a small B_0 term. The B_0 term is, like the *C* term, Gaussian in shape. However, the *B* term is the temperature-independent portion of the normal absorptive line shape of Equation [5]. To some extent, the MCD spectrum can be considered to be an extension of electronic absorption spectroscopy, revealing a great deal more specific and precise information. This simplification, of course, neglects the information that MCD can provide with respect to magnetic parameters of a transition. As shown in **Figures** 3 and 4, the MCD spectrum not only has band position and intensity for each transition, but also *sign* hence there is more information 'content' provided.

One additional MCD term should be defined: D_0, the dipole strength, such that

$$A/E = \gamma D_0 f(E) \qquad [6]$$

Mathematically, the magnetic *g* values for the state involved in the electronic transitions are directly related to the ratios A_1/D_0 and C_0/D_0. Plotting $\Delta\varepsilon$ vs $\beta H/2\kappa T$ (this graph is called an MCD magnetization curve) permits the determination of the magnetic properties (*g* values) of the ground and excited states, C_0 and A_1. Although the *g* values determined in this manner are not as precise as those obtained by EPR spectroscopy, they have proved useful. Indeed, the use of group theoretical techniques to calculate MCD parameters is a major factor in the use of the MCD for assignment of energy states. A second key factor, discussed above, is the common ability to use sample concentrations that are an order of magnitude weaker than needed for other structural methods.

Experimental variations on MCD spectroscopy

The 'basic' MCD instrument is similar to that used for electronic absorption spectroscopy, with the exception that the light must be circularly polarized (see **Figure 1**). A photoelastic modulator (PEM) inputs linearly polarized light, and converts it into circularly polarized light, with RCP and LCP being generated alternatively. The detector observes a signal fluctuation at the ~ 50 Hz frequency and this a.c. signal is input into a sensitive amplifier, capable of detecting very small intensity differences. The sample itself is placed between the two poles of the magnet, and the magnet itself is oriented so that the direction of the field, H_0, is parallel (or anti-parallel) to the directions of propagation of the light.

Variations in method of field generation, electromagnet vs superconducting magnet

Some systems use a simple electromagnet, capable of producing magnetic fields up to 1.5 T (15 000 G). This method is simple and rapid, in that it only requires cooling water and temperature regulation for the sample. In contrast, other MCD systems use superconducting magnets, which have more demanding experimental constraints (such as liquid helium cooling and a longer 'setup' time prior to onset of experimental data collection), but are capable of generating magnetic fields as large as 50 T.

Variations in the spectral region studied

Near-infrared (near-IR, NIR) vs UV-visible (ultraviolet and visible): both electromagnets and superconducting magnets can be used successfully non only in the more common ~800–200 nm spectral regions (UV-visible), but also in the near-IR region. There has been considerable success in the assignment of the axial ligands to biological haem systems through study of MCD data for the near-IR region.

VT and VTVH MCD

These are variable-temperature MCD and/or variable temperature, variable-field MCD, respectively using a superconducting magnet, with temperature variations to as low as ~1.5 K and magnetic field variations from ~1 to 50 T.

VTVH MCD is used to study ground-state electronic structure such as ground-state splittings, especially for non-Kramers ions which do not always have EPR spectra. This approach is a generally useful probe of the ground-state electronic properties of paramagnetic metal ions. MCD is not a 'bulk property technique' like magnetic susceptibility, so is not prone to errors due to paramagnetic impurities. MCD spectroscopy, because it is a type of electronic absorption spectroscopy, can zero in on specific metal ions. VTVH can determine 'single-ion zero field splitting (ZFS)' and the exchange coupling constant, J. In comparison, EPR cannot be used in systems that are non-Kramers, even spin systems with large ZFS. Once the ground-state electronic properties are known, then one can determine changes in J or in ZFS. This approach, with examination of samples at very low temperatures, down to 1.5 K, is still an absorptive method, so the sample must be optically transparent. This is generally accomplished by dissolution of the sample in, e.g., 50:50 glycerol–buffer (aqueous media).

Low-temperature MCD spectroscopy is an extremely useful tool to distinguish between electronic transitions arising from a diamagnetic ($S = 0$) ground state and transitions arising from a paramagnetic ($S > 0$) ground state. This is because the $S = 0$ level is nondegenerate, and thus cannot provide a temperature-dependent C term in the presence of the magnetic field.

'Fast' MCD, also called TRMCD (time-resolved MCD)

For reviews of the methodology and applications of nano- and picosecond MCD experiments, see the Further Reading section. This approach has required extensive modification of the equipment used for sample analysis, in particular using elliptically polarized light. This work has led to exciting results, permitting the examination of transient molecular species. Applications of nanosecond MCD have focused primarily on ligand complexes of haem proteins and their photo-produced dissociation intermediates, particularly given the intense absorption maxima of haem systems (typical ε 100 000 M^{-1} cm^{-1}), and their strong MCD signals even at room temperature.

To date, the experimental focus has been on systems with unpaired spins (metal complexes), rotational symmetries (aromatic molecules such as the amino acid trytophan, porphyrins), and metalloporphyrins (haem proteins). An exciting application came from TRMCD of the photodissociated CO adducts of, e.g. haem proteins such as mammalian cytochrome *c* oxidase: the diamagnetic, low-spin, hexacoordinate Fe(II), of the ferrous–CO haem becomes a paramagnetic high-spin, pentacoordinate Fe(II), with a concomitant appearance of a new C term. In the case of picosecond TRMCD, picosecond lasers are used. One such application demonstrated that upon photodissociation of the CO ligand bound to the haem protein myoglobin, the change from a hexacoordinate to a pentacoordinate haem occurred, 'within the 20 ps rise time of the instrument'.

VMCD, vibrational MCD, particularly Raman

Magnetic Raman optical activity determines transitions of electrons among energy levels created by an applied external magnetic field; problems arise here owing to limitations in the field strength. MVCD (magnetic vibrational CD) splits degenerate levels of vibrational transitions and aids in the analysis of bonding.

X-ray detected magnetic circular dichroism (XMCD)

This technique has only recently evolved into an important method for magnetometry. This technique has unique strengths in that it can be used to determine quantitatively spin and orbital magnetic moments for specific elements, and can also be used to determine their anisotropies through analysis of the experimental spectra. For example, XMCD has been applied to the study of thin films of transient metal multilayers, such as Cu or Fe.

The XMCD method is one where the properties of $3d$ electrons are probed by exciting $2p$ core electrons to unfilled 3d states. The p → d transition dominates the L-edge X-ray absorption spectrum. L-edge X-ray spectroscopy of iron has proved to be useful because the transitions from the $2p$ ground state to $3d$ excited states are strong and dipole allowed, and the small natural line widths also indicate potentially strong MCD spectra. The intense L-edge XMCD spectra of the iron–sulfur protein rubredoxin and of the 2Fe–2S centre of *Clostridium pasteurianum* have also been studied. Both the XMCD sign, and its field dependence, can be used to characterize the type of coupling between magnetic metal ions and the strength of such coupling.

Conclusion

MCD spectra can profitably separate contributions from multiple metal centres to a protein electronic spectrum, be used to evaluate metallo-biological systems without complications from the protein 'milieu', determine zero-field splitting, assign electronic transitions, provide information about a chromophore's electronic structure, evaluate theoretical models, obtain magnetic properties (g values, spin states, magnetic coupling) and be used for structural comparison of 'model' and biological systems.

Modern MCD spectroscopy can only prove to be increasingly useful. Whereas the standard (electromagnetic) instruments available in the 1970s and 1980s could require up to ~45 min per single scan of the data (not counting the buffer, CD, and CD of buffer scans), modern multi-scanning capability permits a significant improvement in signal-to-noise ratio. This has a concomitant advantage in permitting careful and detailed studies to be performed.

Perhaps the greatest utility of MCD spectroscopy is in concert with other methods. No one spectroscopic or structural analysis method can have 'all the answers'. Only a consistent overall structural picture, provided by analysis of data from several methods, with awareness of the shortcomings of each, can lead us closer to the desired 'truth' with respect to the systems under study.

List of symbols

A_1, B_0, C_0 = MCD parameters (A, B and C terms) defining the amount of absorptive (B_0 and C_0) or derivative (A_1) signals in the MCD spectrum; A = absorbance (unitless); b = cuvette pathlength (cm); c = molar concentration (mol L^{-1}; M); c = speed of light; $f(E)$ = general Gaussian line shape function; g = EPR g values; h = Planck's constant; H_0 = external magnetic field; n = refractive index; N = Avogadro's number; T = absolute temperature; α = proportionality constant between the electric field of the light and the electric field at the absorbing centre; β = Bohr magneton; γ = spectroscopic constant; $\Delta A_{\pm}$, A_-, A_+ are the change in absorbance, the negative absorbance, and the positive absorbance, respectively; $\Delta\varepsilon_M$ = MCD 'extinction coefficient'; ε = molar extinction coefficient (M^{-1} cm^{-1}); ε_l and ε_r are the specific extinction coefficients for left- and right-circularly polarized light (LCPL and RCPL, respectively); κ = Boltzmann constant; λ = specific wavelength; ν_0 = the position, in nm, of the electronic absorption maximum for a given transition.

See also: **Near-IR Spectrometers; Vibrational CD, Applications; Vibrational CD Spectrometers; Vibrational CD, Theory.**

Further reading

Andersson LA, Johnson AK, Simms MD and Willingham TR (1995) Comparative analysis of catalases: spectral evidence against haem-bound water for the solution enzymes. *FEBS Letters* **370**: 97–100.

Ball DW (1990) An introduction to magnetic circular dichroism spectroscopy: general theory and applications. *Spectroscopy* **6**: 18–24.

Cheesman MR, Greenwood C and Thomson AJ (1991) Magnetic circular dichroism of hemoproteins. *Advances in Inorganic Chemistry* **36**: 201–255.

Dawson JH and Dooley DM (1989) Magnetic circular dichroism spectroscopy of iron porphyrins and heme proteins. In: Lever ABP and Gray, HB (eds) *Iron Porphyrins, Part III*, pp. 1–133. New York: V. V. H. Publishers.

Goldbeck RA, Kim-Shapiro DB and Kliger DS (1997) Fast natural and magnetic circular dichroism spectroscopy. *Annual Review of Physical Chemistry* **48**: 453–479.

Goldbeck RA and Kliger DS (1992) Natural and magnetic circular dichroism: spectroscopy on the nanosecond time scale. *Spectroscopy* **7**: 17–29.

Holmquist B (1978) The magnetic optical activity of hemoproteins. In: Dolphin D. (ed.) *The Porphyrins*, Vol. III, Chapter 5. New York: Academic Press.

Peng G, van Elp J, Janh H, Que L Jr, Armstrong WH and Cramer SP (1995) L-edge X-ray absorption and X-ray magnetic circular dichroism of oxygen-bridged dinuclear iron complexes. *Journal of the American Chemical Society* 117: 2515–2519.

Solomon EI, Machonkin TE and Sundaram UM (1997) Spectroscopy of multi-copper oxidases. In: Messerschmidt A (ed.) *Multi-Copper Oxidases*, pp. 103–127. Singapore: World Scientific.

Solomon EI, Pavel EG, Loeb KE and Campochiaro C (1995) Magnetic circular dichroism spectroscopy as a probe of the geometric and electronic structures of non-heme ferrous enzymes. *Coordination Chemistry Reviews* **144**: 369–460.
Stohr J and Nakajima R (1997) X-ray magnetic circular dichroism spectroscopy of transition metal multilayers. *Journal de Physique (Paris) IV* 7: C2–C47.
Sutherland JC (1995) *Methods in Enzymology* **246**: 110–131.

Magnetic Field Gradients in High-Resolution NMR

Ralph E Hurd, G.E. Medical Systems, Fremont, CA, USA

MAGNETIC RESONANCE
THEORY

Introduction

In the 1990s pulsed field gradients became a more common element in multiple-pulse high-resolution NMR methods. Gradients have been incorporated into these sequences to improve water suppression, to spoil radiation damping, to remove undesired signals, and to collect faster or higher-resolution multidimensional spectra. Although the potential for pulsed magnetic field gradients has been known since the early years of NMR, only recently has the performance of gradient systems been sufficient to take full advantage of this tool. There are essentially four ways in which gradients are used: coherence pathway selection, spatial encoding, diffusion weighting and spoiling are all used in modern high-resolution systems. These methods have common and differentiating elements. Coherence selection and diffusion weighting take advantage of the reversible behaviour of the pulse gradient effect. Spoiling is a subset of coherence pathway selection that requires no encoding gradient and hence no read or rephase gradient. Spatial encoding can be used to image and correct B_0 inhomogeneity, and can be used to restrict the detected sample volume. The basic elements of B_0 field gradients, as used in high-resolution NMR are described.

Basic properties of gradients

On a typical high-resolution NMR system, a B_o gradient probe can transiently generate a linear change in the otherwise homogeneous B_o field of ±1 mT or moreover the approximately 2 cm z sample length. Many gradient systems can also independently generate linear transverse (x and y) gradient fields of similar magnitudes. Linearity, switching speed and gradient recovery times are important gradient performance criteria. The switching time or recovery time was the most significant limitation of early gradient systems. In these early designs, the gradient field was not constrained inside the gradient cylinder, as shown in **Figure 1A**, and the process of generating a transient gradient interval induced undesirable currents in nearby conductors, especially the components of the magnet itself. These induced eddy currents in turn generate magnetic fields that perturb the NMR spectrum. These stray fields cause significant spectral distortions and last for hundreds of milliseconds. It was therefore impossible to maintain reasonable timing in multiple-pulse NMR experiments using this type of gradient. The invention of the actively shielded gradient coil in the late 1980s removed this limitation by constraining the gradient field inside the gradient cylinder, as illustrated in **Figure 1B**. This innovation and the development of dedicated high-resolution NMR gradient probes have made this technology readily available to NMR spectroscopists.

The gradient pulse effect

A gradient in the B_0 field across a sample will cause the spins in that sample to precess at spatially dependent rates. More specifically, a gradient pulse will add a reversible, spatially dependent, and coherence order-dependent, phase to the magnetization:

$$\varphi_r = \gamma prGt \qquad [1]$$

where γ is the magnetogyric ratio, r is the distance from gradient isocentre, G is the gradient amplitude, t is the gradient pulse duration and p is coherence

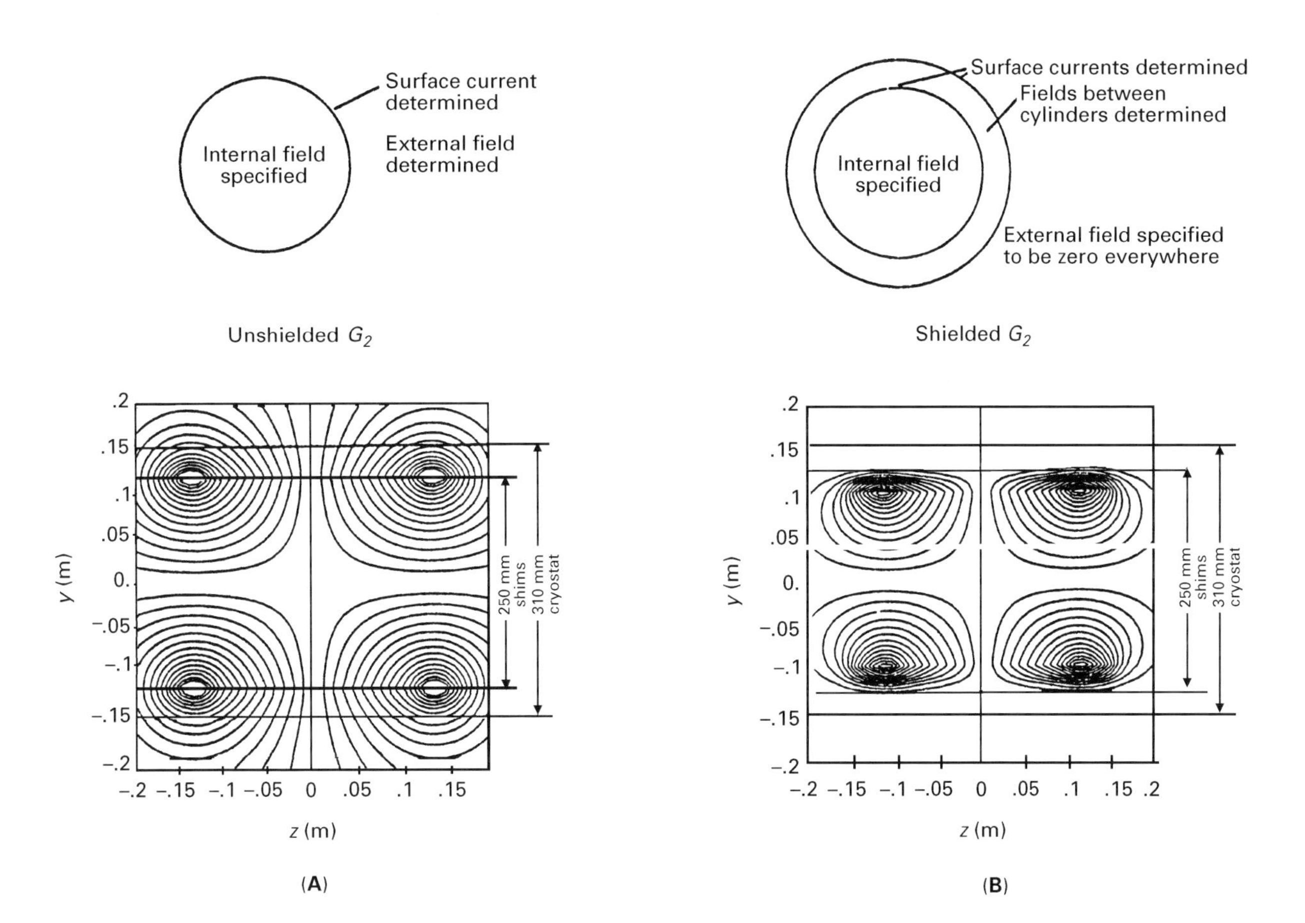

Figure 1 (A) Current diagram for conventional unshielded z gradient coil. (B) Current diagram for actively shielded z gradient coil.

order. If not resolved in space, or rephased, the impact of a spatially dependent phase across the detected volume is self-cancellation of signal. In the absence of B_1, radiofrequency magnetic field inhomogeneity and susceptibility shifts, in a detected region $\pm\frac{1}{2}\,r_{\max}$, the impact of a gradient pulse on pure x magnetization ($p = 1$) will be:

$$M_x(t) = 2\gamma G r_{\max} t \sin\left(\tfrac{1}{2}\,\gamma G r_{\max} t\right) \qquad [2]$$

Under these idealized conditions, perfect cancellation occurs at multiples of $\phi(r_{\max}) = 2\pi$ but practical dephasing requires many cycles of 2π, where residual signal can be approximated as, $2/(\gamma G t r_{\max})$. Thus, a typical 1 ms 0.25 T m^{-1} gradient pulse over a 1.5 cm B_1 sample volume would reduce the observable proton signal by a factor of about 1000. For pathway selection, it is the difference in gradient integrals that determine the level of suppression. Of course, practical matters such as gradient linearity (especially the fall-off at the ends of the sample volume), and B_1 homogeneity, will determine the actual suppression.

Coherence, coherence order and pathway selection

Coherence is a generalization of the idea of transverse magnetization. Coherence order is the quantum number difference associated with the z component of the rotation generated by the RF excitation, and can only be changed by another RF pulse. Thus, coherence order is conserved in the time periods separating RF pulses, during which the application of a gradient pulse will encode magnetization according to the coherence order of that interval. The route of the observed magnetization is referred to as the coherence transfer pathway. All pathways start at $p = 0$ (thermal equilibrium) and must end with single-quantum coherence to be detectable. Transverse magnetization is a specific type of coherence characterized by the single-quantum coherence levels $p = +1$ and -1. Both components are detected to distinguish positive and negative frequencies in a quadrature receiver. By convention, $p = -1$ represents the quadrature detected signal, $s^+(t) = s_x(t) + is_y(t)$.

Coherence transfer pathway diagrams are a good way to visualize the need for pathway selection in multiple-pulse NMR experiments. These pathways remind us that each RF pulse transfers magnetization to multiple coherence levels, only one or two of which must be retained to end up with the desired artefact-free spectrum. The traditional way to select a given pathway is to apply phase cycling. With phase cycling, the pulse sequence is repeated, using changes in the phase of the RF pulses, along with addition or subtraction of the corresponding complex signals to retain the desired pathway and cancel all the others. As a difference method, phase cycling can become a problem when the desired pathway is much smaller than the unwanted ones, as is the case in many multiple-quantum experiments. As a non-difference method, pulsed field gradient selection of the pathway is an advantage in these cases.

Coherence transfer pathways are also a convenient way to visualize the action of gradient pulses in a NMR sequence, since the spatial encoding of each interval in the pathway is directly proportional to the product of gradient integral, Gt, and coherence order, p. Any pathway in which the sum

$$\sum_i p_i (Gt)_i = 0 \quad [3]$$

will be passed. Pathways where this is not true will retain a spatially dependent phase and will self-cancel.

The pathway for homonuclear correlation spectroscopy (COSY) is shown in **Figure 2** and provides a simple example. The first pulse creates coherence with orders +1 and −1 and leaves some z magnetization as coherence order 0. Thus, there are three pathways by which the coherence can reach the receiver after the second RF pulse, namely $[0 \to 0 \to -1]$, $[0 \to +1 \to -1]$ and $[0 \to -1 \to -1]$. If the RF carrier is placed on one side of the F2 spectrum, all of the peaks in the 2D spectrum corresponding to the $[0 \to -1 \to -1]$ coherence pathway will lie on one side of $F_1 = 0$ and the peaks from $[0 \to +1 \to -1]$ will lie on the other side of $F_1 = 0$. The $[0 \to 0 \to -1]$ peaks will occur only at $F_1 = 0$. A single gradient pulse placed between the two RF pulses will spoil coherence that passes through both $p = +1$ and $p = -1$, and will select the $[0 \to 0 \to -1]$ pathway. The addition of a read gradient interval after the second RF pulse will allow one of the other two pathways to be selected. If the read gradient is equal in sign and integral to the first (encode) gradient, then the pathway that goes through the $[0 \to +1 \to -1]$ transfer will be selected, while the coherence that remains at −1 during evolution and acquisition $[0 \to -1 \to -1]$ will be selected by a gradient of equal integral but opposite sign.

Multiple-quantum coherence transfer selection

A common usage of pulsed field gradients is multiple-quantum coherence transfer selection, which takes advantage of the nondifference filtering of large unwanted signals from the small desired ones. The simplest homonuclear example is the three pulse sequence shown in **Figure 3**. Homonuclear scalar coupled spins will give rise to both double and zero quantum coherences in the mixing time (t_m)

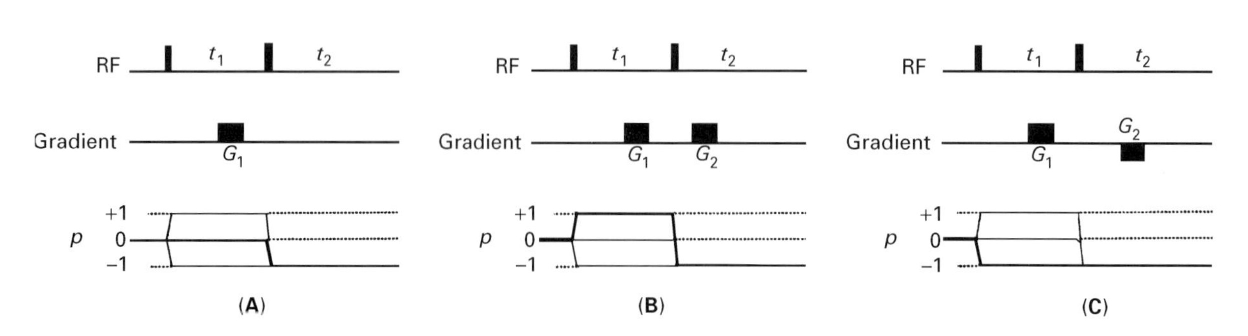

Figure 2 Coherence-transfer pathway diagrams for COSY, illustrating gradient selection of (A) the $F_1 = 0$ artefacts only, $[0 \to 0 \to -1]$; (B) N-type signals, $[0 \to +1 \to -1]$; and (C) P-type signals, $[0 \to -1 \to -1]$.

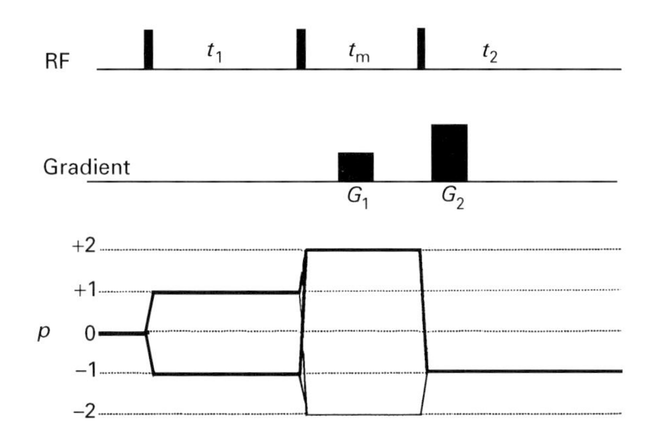

Figure 3 Coherence transfer pathway diagram for homonuclear double-quantum selection with gradients.

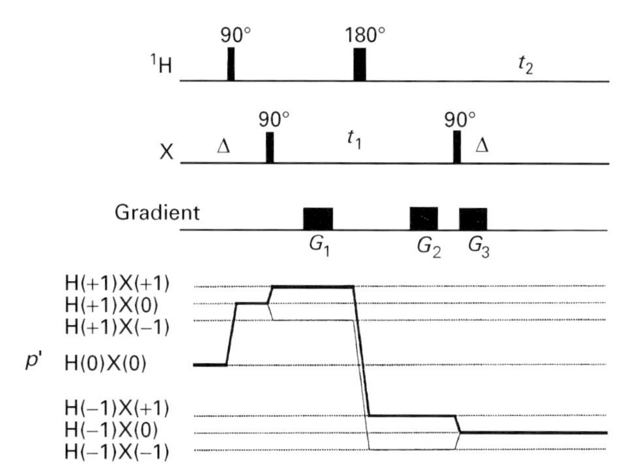

Figure 4 Coherence transfer pathway diagram for gradient-enhanced HMQC sequence. The pathway illustrated by the solid line selects the pathway through heteronuclear double-quantum [H(+1):X(+1)] and heteronuclear zero quantum [H(−1):X(+1)] levels. For X = ^{13}C, this pathway can be selected by any of the gradient ratios 4:0:5, 0:4:−3 or 4:4:2.

interval. Uncoupled spins, such as solvent water, will not and thus the use of a 1:2 ratio of gradients (G_1:G_2) will select only coupled spin coherence that goes through the [0 → +/−1 → +2 → −1] pathways. A spoiler gradient (G_1 only) during the mixing time will select for pathways that go though $p = 0$ in the mixing time, selecting the zero-quantum pathway plus any residual z magnetization. A single pathway [0 → +1 → +2 → −1] is selected using a 1:1:3 gradient sequence. This is achieved by adding a gradient of integral I during t_1 evolution time and increasing the read gradient at the start of the t_2 interval by the same area. These methods are all very good at suppressing the uncoupled water signal and reducing t_1-noise. However, both double-quantum and zero-quantum sequences may also pass water or other large solvent signals via the dipolar field effect unless the gradients are oriented at the magic angle 54.7° where triple axis gradients are used and ($G_x = G_y = G_z$). Heteronuclear multiple-quantum selections, often for protons attached to a lower magnetogyric ratio nucleus, are also very common applications of gradient selection. In this case it is often convenient to generate a combined coherence transfer pathway diagram for coupling partners and to use normalized heteronuclear coherence order p', scaled to the proton magnetogyric ratio. The resulting normalized coherence levels are then directly related to the sensitivity to pulsed field gradient integrals. The coherence pathway diagram for the gradient-enhanced heteronuclear multiple quantum correlation (HMQC) experiment is illustrated in **Figure 4**. For X = ^{13}C, the initial heteronuclear double-quantum level p' [H(+1): ^{13}C(+1)] = 1.25, and the initial heteronuclear zero-quantum level p' [H(−1): ^{13}C(+1)] = −0.75. In this example, gradient ratios of 4:0:5 or 0:4:−3 or 4:4:2 would all select for the same pathway through the double → zero-quantum transfer, and spoil the zero → double-quantum transfer pathway, as well as coherence pathways for proton spins not coupled to a ^{13}C nuclei.

As in the homonuclear case, this method provides excellent water suppression. The suppression of the t_1-noise artefacts is so good with these methods that data can be collected under conditions that are not possible with traditional phase cycled methods. This advantage has been exploited especially in long-range proton–carbon correlation studies of polymer branching, as illustrated in **Figure 5**, and for proton–proton correlation at the water chemical shift frequency.

Spin echoes and gradient pulses

Spin-echo selection with gradient pulses was the first and is probably now the most common use of gradients in magnetic resonance. This element is common to MR imaging, localized spectroscopy, diffusion measurements, water suppression and artefact reduction in multiple-pulse NMR. On high-resolution spectrometers, where all of the B_1 sample volume is normally detected, RF refocusing pulses produce a considerable fraction of non-π rotation. The placement of equal gradient pulses on either side of the π pulse, as illustrated in **Figure 6A**, filters out any coherence that does not refocus ($p \rightarrow -p$ transition). This is also an especially effective method for improving the performance of frequency-selective π pulses such as are used in the gradient-enhanced version of spin echo water suppression (SEWS). Gradients of equal integral, but opposite sign, placed on

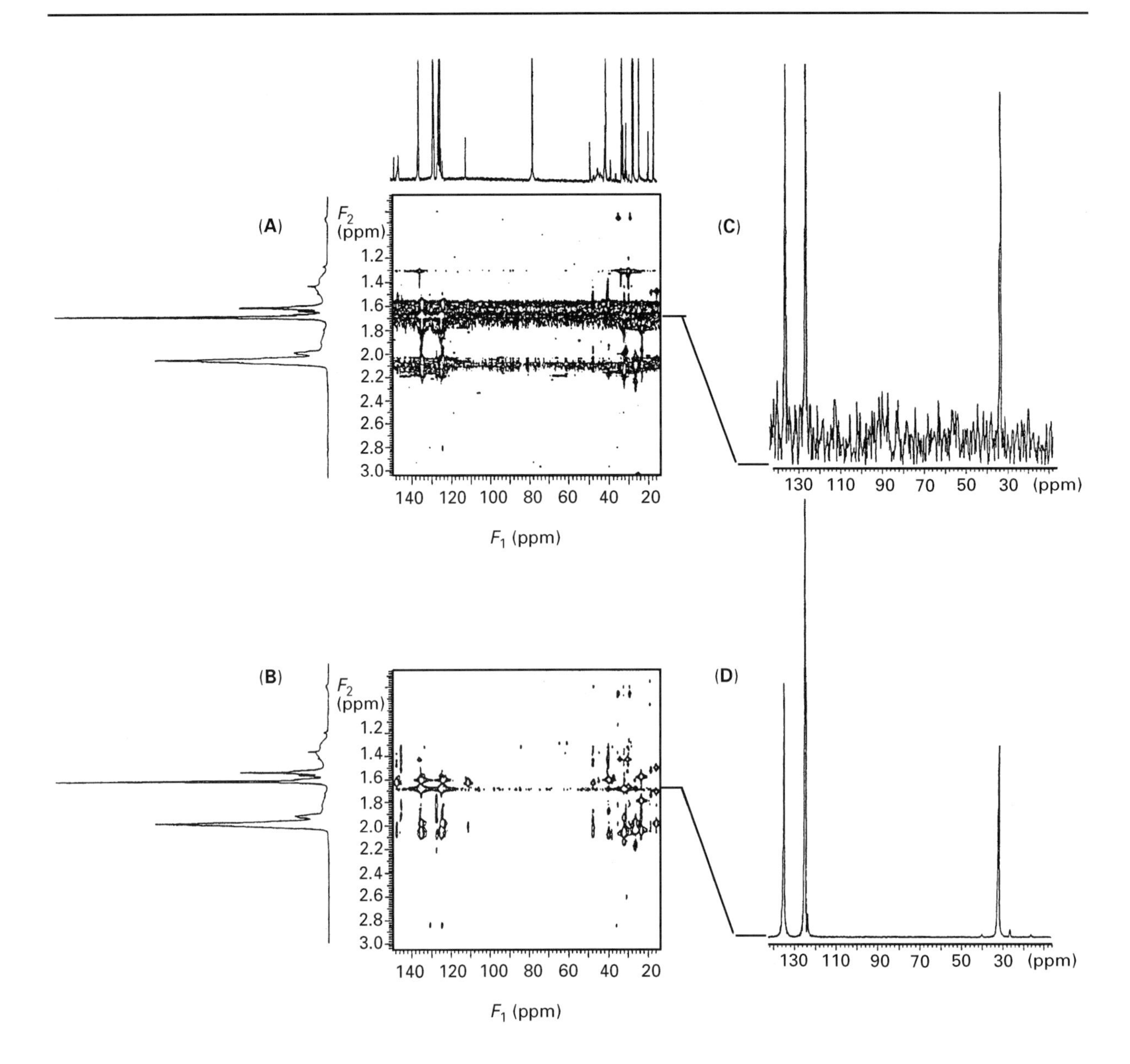

Figure 5 Gradient (B and D) versus phase-cycled (A and C) HMBC spectra of the polymer PI-b-PS. The comparative traces at $F_2 = 1.7$ ppm show the far superior signal-to-t_1-noise achieved by the gradient method (D) relative to the traditional phase-cycled approach (C). Reproduced with permission of The Society of Chemical Industry and John Wiley & Sons, from Rinaldi P, Ray DG, Litman V and Keifer P (1995) The utility of pulse-field gradient–HMBC indirect detection NMR experiments for polymer structure determination. *Polymer International* **36**: 177–185.

either side of a chemical shift selective refocusing pulse, such as the 1–1 binomial example shown in **Figure 6B**, are a powerful way to capture a selective, refocused ($p \rightarrow -p$ transition) bandwidth. This approach can be used to dramatically avoid residual out-of-band signal (e.g. water) relative to the phase-cycled method. Frequency-selective suppression using spin echoes and gradients has also proved very successful in methods such as WATERGATE (water suppression by gradient tailored excitation) and MEGA as illustrated in **Figures 6C and 6D**. In addition to $p \rightarrow -p$ transfer, π pulses invert z magnetization, $I_z \rightarrow -I_z$. In this case imperfect π pulses will generate transverse magnetization. To select for the $I_z \rightarrow -I_z$ transition and spoil both transverse magnetization and the $p \rightarrow -p$ refocused magnetization, nonequal gradients can be applied before and after a π inversion pulse (**Figure 6E**). Each gradient pulse will spoil any transverse magnetization during those intervals, and the nonequal integrals of the gradients will prevent the refocusing of the $p \rightarrow -p$ transition. The selection of $I_z \rightarrow -I_z$ transitions are also useful in multinuclear experiments, in which case the gradient dephasing of S coherences must be avoided. This can

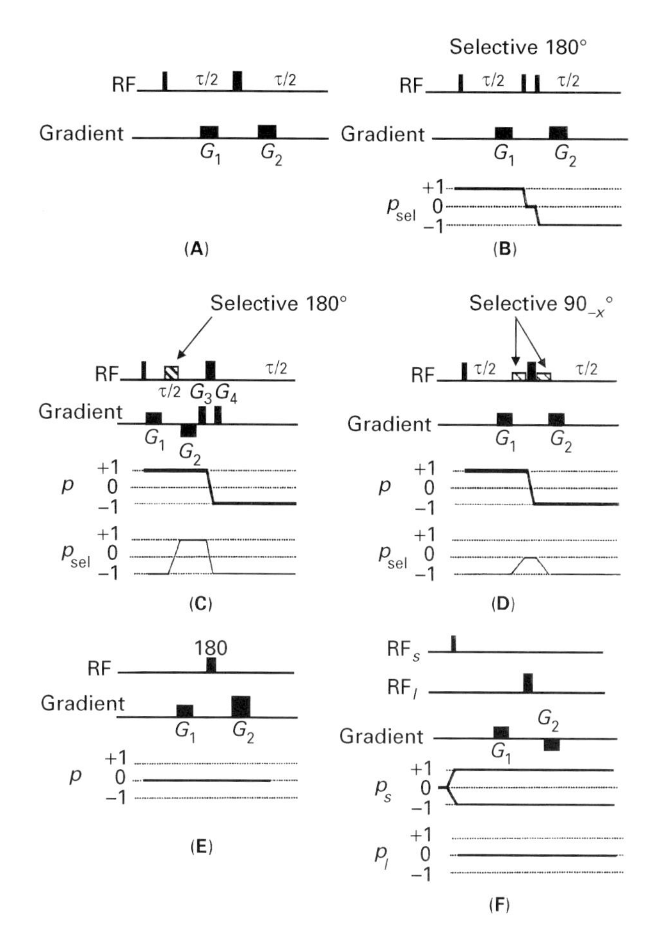

Figure 6 The use of gradients with RF π pulses. (A) Standard spin echo selection, $p \to -p$ transitions are selected. Any imperfection in the RF refocusing is cancelled. (B) Frequency-selective spin echo selection. Only the spins in the refocused bandwidth are selected. (C) Pathway for MEGA. Spins that are refocused by the selective RF refocusing pulse are dephased by the G_1:G_2 gradient pair. Outside the frequency-selective bandwidth, G_2 reverses the effect of G_1. (D) Pathway for WATERGATE. A net zero RF rotation leaves signals in the frequency-selective bandwidth dephased by the G_1:G_2 pair, while spins outside the selective bandwidth are rephased as in (A). (E) Selection of I_z to $-I_z$ ($p = 0 \to 0$) transitions uses nonequal gradients prior to the π pulse to eliminate any existing transverse magnetization, and after the π pulse to eliminate any transverse magnetization generated by RF pulse inhomogeneity. (F) selection of I_z to $-I_z$ ($p = 0 \to 0$) in a heteronuclear sequence while preserving any nonzero S coherence levels.

be accomplished by using gradients of equal integrals but opposite sign (**Figure 6F**). The second gradient will reverse any accumulated phase for the S spin caused by the first gradient, but will still spoil all I spin coherences except of the $I_z \to -I_z$ transitions.

Spoiling

A gradient spoiler pulse can be applied in intervals where the desired signal has coherence order $p = 0$. These applications include gradient-enhanced z and zz filters, stimulated echo selection, multiple-quantum suppression during NOESY (nuclear Overhauser effect spectroscopy) mixing times and the homonuclear zero-quantum methods as previously described. Two examples of this gradient element are illustrated in **Figure 7**.

The gradient-enhanced z filter is a pulse field gradient version of the multiple-acquisition nongradient method. In the original method, magnetization is stored as I_z, and multiple delay times are collected to allow non-I_z magnetization to evolve and self-cancel. The gradient method accomplishes this in a single step. As shown in **Figure 7A**, the two-pulse RF filter acts as a π pulse for the desired magnetization, which means the spoiler during the I_z interval can be combined with a spin-echo gradient pulse pair outside the I_z interval.

The gradient-enhanced zz filter selects for heteronuclear longitudinal spin order described by the density operator I_zS_z, and can be easily integrated into the heteronuclear single quantum correlation (HSQC) type sequences, or as a preparation period for HMQC methods. The gradient version of the zz filter as shown in **Figure 7B** also passes I_z magnetization and is not as efficient at rejecting unwanted pathways as coherence selection.

Spoiler gradients can also be used following frequency-selective excitation to eliminate a narrow band of chemical shift. This approach is often referred to as a chemical shift selective (CHESS) pulse. Optimum performance requires the tip angle of the selective excitation pulse be adjusted for water T_1 relaxation that occurs during the excitation–dephase intervals. Multiple excitation–dephase intervals can be concatenated to achieve a moderate level of B_1 and T_1 insensitivity. Alternatively, T_1- recovery time and water excitation flip angle can be adjusted to exploit differences in the solute and the water T_1 values and to allow significant recovery of the solute spins during the time it takes water to reach a null. Like many gradient methods, the T_1-delayed CHESS pulse inherently eliminates the radiation damping effect and makes it possible to take advantage of the true water T_1.

Diffusion-weighted water suppression

In any experiment where gradients are used to label spins with a spatially dependent phase, that are subsequently rephased with a second gradient pulse, there will be a loss of signal due to any movement of the spins during the time interval between labelling and rephasing. For a spin-echo (p +1 $\to -1$) transition, this loss of signal is related to translational

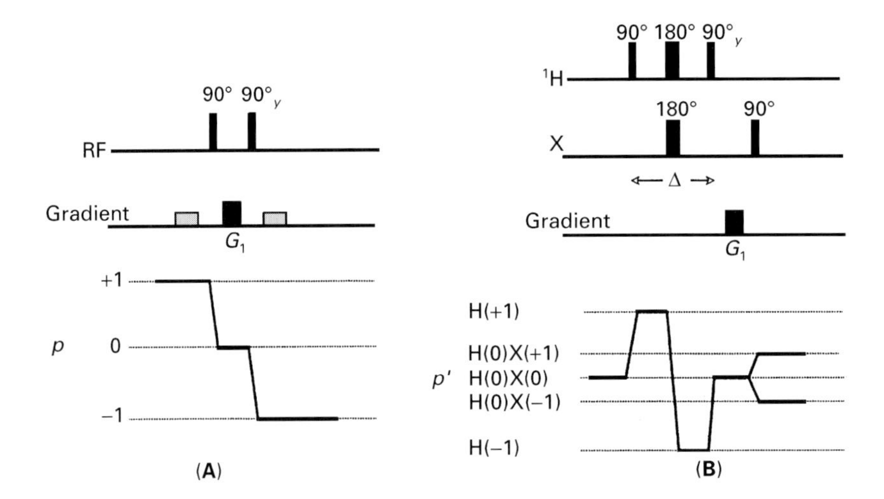

Figure 7 (A) Gradient-enhanced *z* filter and (B) *zz* filter. As in the inversion examples shown in **Figure 6**, these gradients dephase all but $p = 0$ coherence order.

diffusion by the Stejskal–Tanner equation:

$$\ln(S/S_0) = -\gamma^2 g^2 \tau^2 (\Delta - \left(\tfrac{1}{3}\tau\right) D \qquad [4]$$

where γ is the magnetogyric ratio, g is the strength, and τ the duration of the gradient pulse pair, Δ is the time between gradient pulses, and D is the diffusion coefficient. Normally, diffusion weighting is minimized by using modest gradient integrals (g and τ) and by keeping the separation (Δ) between the encode and rephase portion of a gradient pair small. However, by increasing both gradients integrals and separation (Δ), it is possible to take advantage of the significant differences in the translational diffusion of solvent water and large solute molecules such as proteins. This is the basis of the DRYCLEAN, diffusion reduced water signals in spectroscopy of molecules moving slower than water. With a modest 20-fold difference in diffusion constant, D, a gradient pair could be selected to preserve over 70% of the solute signal, while suppressing water by 1000-fold. It is important to note that, like multiple-quantum coherence pathway selection, this method is also independent of the width and shape of the water signal. The same basic gradient-selected spin-echo methodology is also used to study exchange processes in biomolecules.

Phase-sensitive methods

Modern multidimensional spectra are almost always recorded in pure absorption mode. The primary reasons are phase sensitivity, improved resolution, and a $\sqrt{2}$ factor increase in SNR compared with magnitude mode. Pure absorption phase is obtained from the amplitude-modulated signal in t_1, separating the frequencies of the two mirror image pathways, $p = +1$ and $p = -1$ in an evolution time analogue to quadrature detection. In methods without gradients, or in methods that use gradients only for spoiling, spin echo and/or I_z inversion selection, this is accomplished using a two-step phase cycle for each t_1 increment. Both steps contain $p = +1$ and $p = -1$ coherence, and the combination provides frequency discrimination at full signal intensity. Pure absorption line shape with gradient selection during evolution is also a two-step process, but each acquisition contains only $p = +1$ or $p = -1$, leading to a $\sqrt{2}$ factor loss in SNR relative to the phase-cycled selection of quadrature in F_1. The trade-off is that signal-to-t_1 noise is often better for the gradient methods, as illustrated in **Figure 5**, and in the instances where pure absorption line shape is not required the gradient selection methods are significantly faster. Unlike the single-step selection with gradients, phase-cycled methods require multiple steps to separate $p = +1$ or -1. The advantage of a reduction in required phase cycle steps is most evident in three- and four-dimensional NMR studies, where proper sampling of the evolution time alone generates more signal averaging than necessary. It is possible to collect separate $p = +1$ and $p = -1$ pathways in a single acquisition per t_1 time by using the switched acquisition time (SWAT) gradient method. In this method the two coherence pathways are alternately and individually acquired on alternate sampling points in the digitizer. Although a doubling in F_1 bandwidth still results in the $\sqrt{2}$ factor loss in SNR, this approach offers the ability to collect pure absorption multidimensional data in a minimum total acquisition time.

The method, however, is very demanding on gradient switching time.

Spatial selection

In the typical high-resolution NMR experiment, the entire B_1 volume contributes to the final result. The volume of spins in the transition band where the relatively linear B_1 field falls from maximum to zero can be significant. This inhomogeneity and line shape distortions from bulk susceptibility effects also found at the ends of the sample are among the reasons why gradient selection and phase cycling methods are so heavily used for artefact reduction in multiple-pulse NMR. A gradient-based method that can be used to reduce these end effects (transition band suppression or TBS), uses slice select–spoil intervals during the pulse sequence preparation period to avoid this difficult region of the sample. Combination of TBS with T_1-delayed CHESS pulses and gradient selection of double-quantum coherence makes it possible to study proton–proton correlations at the water chemical shift in both F_1 and F_2. The pulse sequence and an example are shown in **Figure 8**.

Another application of spatial localization in high-resolution NMR applies to the specialized field of high-performance liquid chromatography–NMR (HPLC-NMR). With NMR as the detector for a liquid chromatography system, it can be valuable to spatially resolve the NMR sample volume. This can be done by phase encoding, which allows the data to be retrospectively processed to eliminate end effects or to separate partially overlapping HPLC fractions.

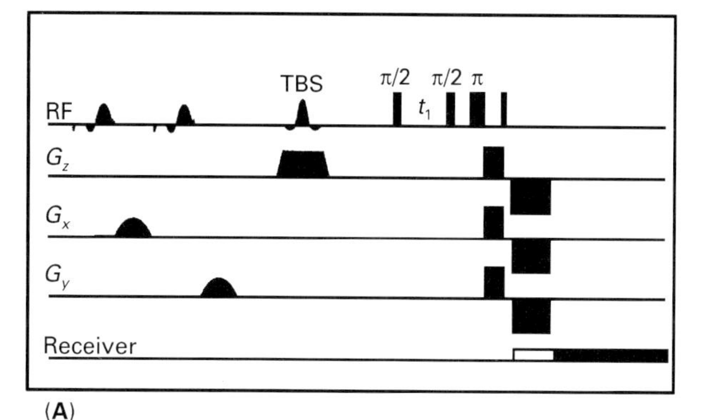

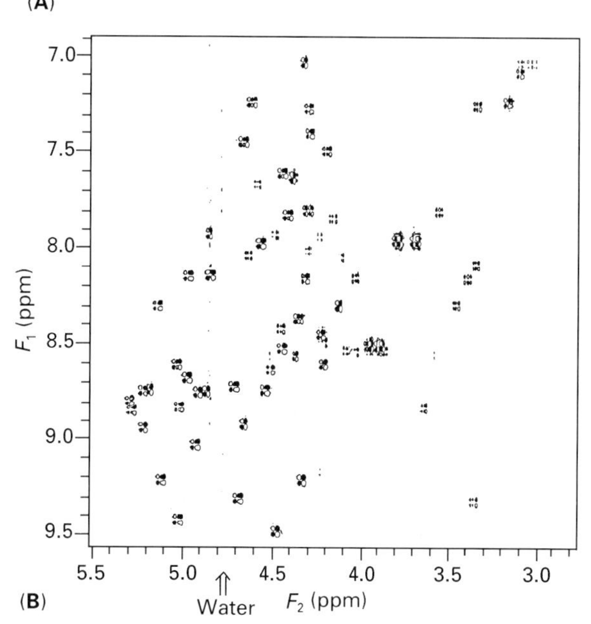

Figure 8 (A) Pulse sequence for phase-sensitive version of gradient-enhanced double-quantum correlation method incorporating T_1 delayed CHESS sequence and TBS. (B) Phase-sensitive contour plot of data for 1 mM ubiquitin in 90% H_2O–10% D_2O collected using this method. A water inversion null time of 200 ms was used to allow CαH protons at 4.8 ppm to recover fully as expansion near $F_2 = 4.8$ (water) illustrates. Reproduced from Hurd R, John B, Webb P and Plant P (1992) *Journal of Magnetic Resonance* **99**: 632–637 with permission of Academic Press.

Field maps and homogeneity adjustment

One relatively obvious application of three-axis gradients is to image and correct for any B_0 field inhomogeneity. Three-dimensional phase or frequency maps can be obtained and used to image the inhomogeneity of the sample, and with previously obtained maps for fixed offsets of the known shims a best-field solution can rapidly be made. Normally this approach works best with a strong solvent signal such as water, but in a limited way it can be accomplished using the deuterium-lock solvent signal.

Summary

Gradients are useful as an integral part of multiple-pulse NMR methods. High-resolution NMR systems and probes continue to incorporate these devices and accordingly the use of these tools continues to become more common. In many ways, gradients are a perfect partner for the limitations of the native high–resolution NMR B_1 fields, and also work complement only to crafted RF pulse methods. When used appropriately, gradients have the ability to enhance the quality of most multiple-pulse NMR results.

List of symbols

B_0 = applied magnetic field; B_1 = RF magnetic field strength; D = diffusion coefficient; F_1 = evolution frequency; g = strength of gradient pulse; G = gradient amplitude; I_z = z magnetization; p = coherence order; r = distance from gradient isocentre;

t = gradient pulse duration; γ = magnetogyric ratio; Δ = time between gradient pulses; τ = duration of gradient pulse; ϕ = magnetization phase.

See also: **Diffusion Studied Using NMR Spectroscopy; NMR Pulse Sequences; Product Operator Formalism in NMR; Solvent Suppression Methods in NMR Spectroscopy; Two-Dimensional NMR, Methods.**

Further reading

Freeman D and Hurd RE (1992) Metabolite specific methods using double quantum coherence transfer spectroscopy. In Diehl P, Fluck E, Günther H, Kosfeld R and Seelig J (eds) *NMR: Basic Principles and Progress*, Vol 27, pp 200–222. Berlin: Springer-Verlag.

Hurd RE (1995) Field gradients and their application. In Grant DM and Harris K (eds) *Encyclopedia of NMR*. Chichester: Wiley.

Hurd RE and Freeman D (1991) Proton editing and imaging of lactate. *NMR in Biomedicine* **4**: 73–80.

Keeler J, Clowes RT, Davis AL and Laue ED (1994) Pulsed-field gradients: theory and practice. *Methods in Enzymology* **239**: 145–207.

Zhu J-M and Smith ICP (1995) Selection of coherence transfer pathways by pulsed field gradients in NMR spectroscopy. *Concepts in NMR* **7**: 281–288.

Magnetic Resonance, Historical Perspective

J W Emsley, University of Southampton, UK

J Feeney, National Institute for Medical Research, London, UK

Introduction

NMR dates from 1938 when Rabi and co-workers first observed the phenomenon in molecular beams. This was followed in 1946 by the NMR work in the laboratories of Bloch and Purcell on condensed-phase samples. In the intervening 53 years there has been a wonderful revelation of how rich this spectroscopy can be, and only a flavour can be given here of the many significant developments. A very detailed account of the history is given in Volume 1 of the *Encyclopaedia of NMR*, which also includes biographies of many of those who created the subject as it is today. A shorter, but still very detailed, history can be found in five articles published in *Progress in NMR Spectroscopy*. Here we present a summary of the main developments under five headings: Establishing the principles; Solid-state and liquid crystal NMR; Liquid-state NMR; Biological applications of NMR; Magnetic resonance imaging. We also present three tables that give some of the important milestones in the development of NMR.

Establishing the principles

NMR arises because some nuclei may have an intrinsic spin angular momentum, which has the consequence that they also have a magnetic dipole moment. The existence of a magnetic dipole moment for hydrogen nuclei was established in 1933 by Gerlach and Stern, who observed the effect of an applied magnetic field gradient on a beam of hydrogen molecules. The trajectories of the molecules are changed if their nuclei have magnetic moments. Inducing transitions between nuclear spin states by the application of electromagnetic radiation at the appropriate resonance frequency was introduced by Rabi and co-workers, also using molecular beams. In this experiment the beam passed first through a field gradient, which deflected the atoms in a direction dependent on the value of m, the magnetic quantum number, then through a homogeneous field, where they were subjected to the electromagnetic radiation, and finally through another field gradient whose sign was opposite to that in the first region. If the nuclei in the atoms do not absorb the radiation, then the effect of the two field gradients cancels, and the beam is undeflected. Absorption or emission of radiation leads to a net deflection of the beam. This simple experiment therefore provided a foretaste of the use of gradients to create or destroy signals.

The first successful NMR experiments on condensed-phase samples were done in 1945, and published in 1946, separately by the group at Stanford led by Bloch, who observed the protons in water, and a group at Harvard led by Purcell, who also observed protons, but in solid paraffin. Unlike the beam experiments, in these the detection was of a net nuclear magnetization arising from the imbalance between states with different values of *m*, and it was crucial for their success that nuclear spin relaxation was occurring at a favourable rate. The first systematic experimental measurements of spin–lattice and spin–spin relaxation rates were published in 1948 by Bloembergen, Purcell and Pound, who also gave an interpretation of their magnitudes in terms of the dynamics of the molecules containing the nuclei.

The magnetic shielding of a nucleus from the applied field by the surrounding electrons was recognized to occur in atoms by Lamb, who published a method for calculating the effect in 1941. The aim of these calculations was to correct for the effect of the shielding on the resonance frequencies observed in molecular beam experiments and hence to obtain the true nuclear magnetic moment. Ramsey extended these calculations to nuclei in molecules in 1949–52, and in this same period the phenomenon was observed in the NMR of condensed-phase samples, first in the resonances of metals and metal salts by Knight in 1949, and in the following year by Proctor and Yu, who observed different resonances for ^{14}N in ammonium nitrate, and by Dickinson, who reported the same phenomenon for ^{19}F in various compounds (e.g BeF_2, HF, BF_3, KF, NaF, $C_2F_3CCl_3$). The physicists working on these problems thought these 'chemical shift' effects uninteresting and a nuisance, since they impeded the important task of measuring nuclear gyromagnetic ratios accurately! We can now pinpoint the years 1949–50 as the period when NMR ceased to be predominantly a technique of the 'physicists' and when the 'chemists' began to realize the potential usefulness of the 'chemical shift'.

It was also at this time that the effects produced by spin–spin coupling were first observed. Experimental results now preceded theory. Proctor and Yu observed a multiplet for the ^{121}Sb resonance in a solution containing the ion SbF_6^-. They observed only five lines of the seven-line multiplet, and so were sidetracked into attempting to explain the splitting as incomplete averaging of the internuclear dipolar coupling. Dipolar coupling had been observed in molecular beam experiments, and its origin was well understood. The problem facing Proctor and Yu was that this interaction, being entirely anisotropic, should vanish if the molecules are rotating rapidly and isotropically, as in an isotropic liquid sample. Gutowsky and McCall also observed spin–spin splittings, but this time in the ^{31}P and ^{19}F resonances in the compounds $POCl_2F$, $POClF_2$ and CH_3OPF_3. They were able to deduce that the number of lines is determined by the product of the *m* values of the coupled nuclei. Ramsey and Purcell published an explanation of the splitting as arising from a rotationally invariant interaction between nuclear spins that proceeds via the electrons in the molecule. Spin–spin splitting was also observed at the same time by Hahn and Maxwell as a modulation on a spin-echo signal, Hahn having discovered the spin-echo phenomenon in 1949.

By 1952 all the basic, important interactions that affect NMR spectra had been demonstrated, and their relationship with molecular structure had been explained (see **Table 1**). The challenge then, as now, was how to exploit the value of NMR for samples of varying degrees of complexity, and this proved to be an exciting and rewarding quest. There were still many new effects to be discovered, and these began to appear quickly as the early pioneers started to explore this new spectroscopy. In 1953 Overhauser predicted that it should be possible to transfer spin polarization from electrons to nuclei. He delivered this prediction to an initially sceptical audience at a meeting of the American Physical Society in Washington, DC. Overhauser was a postdoctoral worker in Illinois when he made this prediction, and he had interested Slichter in the possibility of enhancing NMR signals in this way. Slichter and Carver succeeded in demonstrating the enhancement in lithium metal and all doubts about the nuclear Overhauser effect (NOE) were put to rest.

The chemical shift and spin–spin coupling phenomena were clearly destined to be discovered as soon as magnets became sufficiently homogeneous. They might be classed as inevitable discoveries. The Overhauser effect is different, and it is conceivable that it would have lain undiscovered for many years without the perception of one individual. We might call this a noninevitable discovery. One of the remarkable features of NMR development has been the number of such noninevitable discoveries, some of which have been fully exploited only many years after their discovery. Another such example is the invention by Redfield in 1955 of spin locking, a technique that produces a retardation of spin–spin relaxation in the presence of a radiofrequency field. This not only led to a method of studying slow molecular motions, but also provided a method for transferring polarization between two nuclei that are simultaneously spin-locked, as ingeniously demonstrated by Hahn and Hartmann in 1962.

There were many developments going on in the period 1955–65, some of which we will discuss later. The successes of the early pioneers encouraged the development of commercial spectrometers, and this provided increased access to NMR for a wider

Table 1 Milestones in the development of NMR basic principles and solid–state and liquid-crystal NMR

Date	*Milestone*	*Literature citation*
1924–1939	Early work characterizing nuclear magnetic moments and using beam methods	Frisch and Stern, *Z. Phys.* **85**: 4; Esterman and Stern *Phys. Rev.* **45**: 761, Rabi *et al.*, *Phys. Rev.* **55**: 526
1938	First NMR experiment using molecular beam method	Rabi *et al.*, *Phys. Rev.* **53**: 318
1941	Theory of magnetic shielding of nuclei in atoms	Lamb, *Phys. Rev.* **60**: 817
1945	Detection of NMR signals in bulk materials	Bloch *et al.*, *Phys. Rev.* **69**:127; Purcell *et al.*, *Phys. Rev.* **69**: 37
1948	Bloembergen, Purcell and Pound (BPP) paper on relaxation	Bloembergen *et al.*, *Phys. Rev.* **73**: 679
1949	Hahn spin echoes	Hahn, *Bull. Am. Phys. Soc.* **24**: 13
1949	Knight shift in metals	Knight, *Phys. Rev.* **76**: 1259
1950	Discovery of the chemical shift	Proctor and Yu, *Phys. Rev.* **77**: 717; Dickinson *Phys. Rev.* **77**: 736
1951	Discovery of spin–spin coupling	Proctor and Yu, *Phys. Rev.* **81**: 20; Gutowsky and McCall, *Phys. Rev.* **82**: 748; Hahn and Maxwell, *Phys. Rev.* **84**:1246
1952	First commercial NMR spectrometer (30 MHz)	Varian
1953	Bloch equations for NMR relaxation	Bloch *et al.*, *Phys. Rev.* **69**: 127; Bloch, *Phys. Rev.* **94**: 496
1953	Overhauser effect	Overhauser, *Phys. Rev.* **91**: 476; Carver and Slichter, *Phys. Rev.* **102**: 975
1953	Theory for exchange effects in NMR spectra	Gutowsky *et al.*, *J. Chem. Phys.* **21**: 279
1953	Proton spectrum of a liquid crystal	Spence *et al.*, *J. Chem. Phys.* **21**: 380
1954	Carr–Purcell spin echoes	Carr and Purcell, *Phys. Rev.* **94**: 630
1955	Solomon equations for NMR relaxation	Solomon, *Phys. Rev.* **99**: 559
1955	Relaxation in the rotating frame	Redfield, *Phys. Rev.* **98**: 1787
1957	Redfield theory of relaxation	Redfield, *IBM J. Res. Dev.* **1**: 19
1958	Magic angle spinning for high-resolution studies of solids	Andrew *et al.*, *Nature* **182**: 1659; Lowe, *Phys. Rev. Lett.* **2**: 285
1963	Liquid crystal solvents used in NMR	Saupe and Engelert, *Phys. Rev. Lett.* **11**: 462
1964	Deuterium spectrum of a liquid crystal	Rowell *et al.*, *J. Chem. Phys.* **43**: 3442
1966	NMR spectrum shown to be Fourier transform (FT) of free induction decay (FID)	Ernst and Anderson, *Rev. Sci. Instrum.* **37**: 93
1971	Deuterium spectrum of a membrane	Oldfield *et al.*, *FEBS Lett.* **16**: 102
1976	Cross-polarization magic angle spinning for solids	Schaeffer and Stesjkal *J. Am. Chem. Soc.* **98:** 1030

scientific community. The first commercial spectrometer (30 MHz for ^{1}H) was marketed by Varian Associates in 1952, and many of the new early developments stemmed from Varian's research and development department. Sample spinning and field-frequency locking are just two examples that led to dramatic improvements in the quality of high-resolution spectra of liquids. However, the most significant development was the pulse Fourier transform (FT) method of acquiring spectra, which Anderson and Ernst realized at Varian, the first account of which appeared in 1966. At that time their spectrometer did not have an on-line, or even a close at hand computer on which to do the transform, and the exploitation of the method in a commercial spectrometer had to wait for the development of the on-line computer. In fact, the first commercial pulse FT spectrometer was marketed by Bruker in 1969.

Varian introduced superconducting magnets into NMR with a 200 MHz proton spectrometer, first produced in 1962, and whose field strength was soon increased so as to give proton resonance at 220 MHz.

By 1971 NMR was beginning to look like a mature spectroscopy with all the major developments in place. However, in that year Jeener suggested the idea of multidimensional spectroscopy, and in 1973 Lauterbur published his method for imaging of objects by applying magnetic field gradients. These two events stimulated Ernst and his collaborators to develop the first two-dimensional experiments, and a new age of rapid development in NMR began, leading to the marvellous portfolio of experiments in NMR spectroscopy and imaging that are available today.

Solid-state and liquid crystalline samples

The rapid and isotropic motion in normal liquids averages the anisotropic interactions to zero. The rapid motion also produces a long spin–spin relaxation time (T_2), and hence a very narrow NMR line. In most solids there is little or no motion and the NMR, lines may be split by very large anisotropic interactions, and will usually have a very short T_2, and hence broad lines. In fact, in the early days of NMR, studies of solids and liquids were seen to be quite different activities. Commercial spectrometers were produced mainly for liquid-state studies, since it was appreciated that the applications of NMR for mixture analysis and structure determination by chemists would be the major market. Spectrometers were usually designed either to obtain high-resolution spectra – that is, to resolve the small chemical shifts and spin–spin couplings exhibited by liquids – or for solid samples, where magnet homogeneity was not so important but special techniques were necessary in order to record the very broad line spectra. The NMR community was divided mainly into those working with liquids and those looking at solids.

We will restrict our description of the historical development of the NMR of solids and liquid crystals to showing how the gap between these two communities has narrowed, and indeed now overlaps. The first steps along this path were taken by Andrew, Bradbury and Eades in 1958, and by Lowe in 1959, who showed that rotation of a solid sample about an angle of 54.7° to the magnetic field can remove the second-rank, anisotropic contributions to NMR interactions for spin-$\frac{1}{2}$ nuclei. This means that, in principle, the dipolar interaction, which is entirely a second-rank, anisotropic interaction, and the anisotropic contribution to the chemical shift can be removed by using this 'magic angle' spinning (MAS) technique. The spectra obtained show spinning sidebands at the frequency of the rotation speed, and have intensities that depend on the relative magnitudes of the rotation speed and the magnitude of the interaction being averaged. The early experiments demonstrated that the spectral lines can be narrowed to reveal chemical shift differences, and even in some cases spin–spin couplings, but the samples that could be studied in this way were limited, and the method did not find wide application. The MAS experiment had to wait until 1976 before it was used to provide high-quality, high-resolution ^{13}C spectra from solid samples. Carbon-13 is a special case in being isotopically dilute at natural abundance and so the spectra can easily be simplified by proton decoupling. This produces spectra from a liquid sample that have a single line for each chemically equivalent group of carbons. The low isotopic abundance, however, also leads to a low signal-to-noise ratio, and in liquids it was not until the advent of the pulse FT method that ^{13}C spectra with a good signal-to-noise ratios could be obtained by time averaging. For a solid sample the time averaging is often inefficient because the ratio of the relaxation times T_1/T_2 is high. To overcome this problem, Schaefer and Stejskal used an idea proposed and demonstrated by Hahn and Hartmann in 1962 in which the ^{13}C and ^{1}H nuclei can be made to transfer polarization by subjecting each of them simultaneously to spin-locking radiofrequency fields.

In liquid crystalline samples the molecules move rapidly, so that the NMR interactions are averaged, but they do not move randomly, and this results in nonzero averaging of the dipolar couplings, the chemical shift anisotropies and the quadrupolar interactions. The first reported observation of a spectrum from a liquid crystalline sample was by Spence, Moses and Jain in 1953. It was the proton spectrum of a nematic sample, and consisted of a very broad triplet structure and had a low information content. Ten years later, Englert and Saupe recorded the ^{1}H spectrum of benzene dissolved in a nematic solvent and this consisted of a large number of sharp lines; its analysis gave three, partially averaged dipolar couplings whose values could be related to the relative positions of the protons and the orientational order of the sixfold symmetry axis of the benzene molecule. The study of liquid crystals themselves received a boost with the publication in 1965 by Rowell, Melby, Panar and Phillips of the spectrum given by the deuterons in a specifically deuterated nematogen. They obtained partially averaged quadrupolar splittings, which can be used to characterize the orientational order of the deuterated molecular fragments. The realization that NMR could give useful information about membranes and model membranes dates from the late 1960s and early 1970s, and the particularly valuable role of deuterium NMR in membrane studies stems from the publication in 1971 of a study by Oldfield, Chapman and Derbyshire.

Liquid-state NMR

Following the detection of the NMR phenomenon and the subsequent discovery of the chemical shift and spin–spin coupling, NMR emerged as one of the most powerful physical techniques for determining molecular structures in solution and for analysing complex mixtures of molecules. The potential of the method as a structural tool was almost immediately recognized. High-resolution NMR spectrometers were constructed in several laboratories (such as those of HS Gutowsky, RE Richards and JD Roberts, and JN Shoolery at Varian Associates) and the pioneering efforts of these scientists and others began to demonstrate the scope of applications of the technique in chemistry. The success of the method for chemists derives from the well-defined correlations between molecular structure and the measured chemical shifts and spin coupling constants. In retrospect, the achievements of the early workers were truly remarkable considering that they were working at such low magnetic fields (30/40 MHz for ^1H) so that spectral dispersion was poor and the sensitivity was three orders of magnitude less than in present-day instruments. The ingenious adaptations of their instruments to increase the stability and resolution (for example, field-frequency locking, homogeneity shim coils and sample spinning) were absolutely essential to allow them to make progress in their structural determinations. As time progressed, the sensitivity was boosted initially by increasing the field strengths and improving the radiofrequency (RF) circuitry and probe designs, and subsequently by using spectral accumulation and Fourier transform methods.

Most of the important milestones in the development of the NMR technique for studies of solution state NMR are given in **Table 2**. By 1957 NMR was emerging as a powerful nondestructive analytical technique capable of providing structural information about the environment of more than 100 known nuclear isotopes. Initially the technique was held back by its relatively low sensitivity and the complexity of the ^1H spectra of larger molecules. In the late 1950s, although many problems were identified for NMR study in areas such as polymer chemistry, organometallic chemistry and even biochemistry, the method was proving to be grossly inadequate for tackling them. For example, polymer scientists, acquired some of the early instruments hoping to determine stereotacticities and cross-linking in synthetic polymers; in fact it was not until several years later that improved instrumentation allowed such problems to be tackled successfully. Meanwhile, the method was enjoying considerable success in helping to solve molecular structures of moderately sized molecules (M_r < 400): it was particularly useful in natural product chemistry where it became possible to differentiate between several structures that satisfied the compositional data. It was also proving to be a very powerful method for defining stereochemical details of various structures, for example alkaloids and steroids. Not surprisingly, organic chemists were immediately attracted to this technique, which could reveal unresolved structural details about some of the molecules they had been studying for decades.

More challenging applications to larger molecules became possible only with the eventual improvements in sensitivity and spectral simplification. Although the manufacturers made steady progress in providing higher and higher field strengths, it was not until 1966 that a significant impact was made on the sensitivity problem with the arrival of Fourier transform methods and the use of dedicated computers for data acquisition. These methods also facilitated studies of less-sensitive nuclei and from 1966 to 1975 ^{13}C studies at natural abundance became routine not only for structural studies but also for investigating rapid molecular motions (obtaining correlation times from ^{13}C relaxation studies). During this period, structural determinations of fairly large molecules (M_r ~3500) became commonplace and measurements of nuclear Overhauser effects were frequently used to identify protons that were near to each other. Fortunately, while this rapid expansion in applications work was underway, a few research groups continued to concentrate on understanding the basic spin physics. Some of the novel multipulse techniques developed at this time (such as INEPT and HMQC for indirect detection of insensitive heteronuclei via proton signals) were to prove of far-reaching value in eventually simplifying complex NMR spectra from large macromolecules.

Biological applications of NMR

Biochemists became interested in the NMR technique long before it could provide them with the detailed information they were seeking. For example, the first ^1H spectrum of a protein was recorded in 1957 and proved to be almost featureless. From these unpromising beginnings, who would have predicted that 40 years later the technique would be used to fully assign the resonances of proteins as large as 30 kda and to determine their three-dimensional structures? Early workers such as M Cohn, O Jardetzky and RG Shulman had sufficient vision to recognize the eventual potential of the method when they began their pioneering studies on nucleotides, amino acids, peptides, proteins, paramagnetic ion effects and metabolic applications. In the early days, brave attempts were made to solve the problem of signal overlap by studying partially deuterated,

Table 2 Milestones in the development of solution-state NMR

Date	*Milestone*	*Literature citation*
1949–1950	Discovery of the chemical shift	Knight, *Phys. Rev.* **76**: 1259; Proctor and Yu, *Phys. Rev.* **77**: 777; Dickinson *Phys. Rev.* **77**: 736
	Discovery of spin–spin coupling	Proctor and Yu, *Phys. Rev.* **81**: 20; Gutowsky and McCall, *Phys. Rev.* **82**: 748; Ramsey and Purcell, *Phys. Rev.* **85**: 143, Hahn and Maxwell, *Phys. Rev.* **84**: 1246
1951	Discovery of ^{1}H chemical shifts	Arnold *et al.*, *J. Chem. Phys.* **19**: 507
1952	First commercial NMR spectrometer (30 MHz)	Varian
1953	Overhauser effect	Overhauser, *Phys. Rev.* **91**: 476
1953	Theory for exchange effects in NMR spectra	Gutowsky *et al.*, *J. Chem. Phys.* **21**: 279
1953–58	Sample spinning used for resolution improvement	Bloch, *Phys. Rev.* **94**: 496
	Field gradient shimming with electric currents	Golay, *Rev. Sci. Instrum.* **29**: 313
	Magnetic flux stabilization (Varian)	Shoolery, *Prog. NMR Spectrosc.* **28**: 37
	Variable temperature operation	Shoolery, *Prog. NMR Spectrosc.* **28**: 37
	Spin decoupling	Bloom and Shoolery, *Phys. Rev.* **97**: 1261
1957	Analysis of second-order spectra	Gutowsky *et al.*, *J. Am. Chem. Soc.* **79**: 4596; Bernstein *et al.*, *Can. J. Chem.* **35**: 65; Arnold, *Phys. Rev.* **102**: 136; Anderson, *Phys. Rev.* **102**: 151
1959	Blood flow measurements *in vivo*	Singer, *Science* **130**: 1652
1959	Vicinal coupling constant dependence on dihedral angle	Karplus, *J. Chem. Phys.* **30**: 11; **64**: 1793
1961	First commercial 60 MHz field/frequency locked spectrometer (Varian A 60)	Varian
1962	First superconducting magnet NMR spectrometer (Varian 220 MHz)	Varian
1962	Indirect detection of nuclei by heteronuclear double resonance (INDOR)	Baker, *J. Chem. Phys.* **37**: 911
1964	Spectrum accumulation for signal averaging	Ernst, *Rev. Sci. Instrum.* **36**: 1689
1965	Nuclear Overhauser enhancements (NOEs) used in conformational studies	Anet and Bourn, *J. Am. Chem. Soc.* **87**: 5250
1966	Fourier Transform (FT) techniques introduced	Ernst, *Rev. Sci. Instrum.* **36**: 1689; Ernst and Anderson, *Rev. Sci. Instrum.* **37**: 93
1969	First commercial FT NMR spectrometer (90 MHz)	Bruker
1969	Lanthanide shift reagents used in NMR	Sievers, *NMR Shift Reagents,* Academic Press
1970–75	^{13}C studies at natural abundance become routine	
1970	First commercial FT spectrometer with superconducting magnet (270 MHz)	Bruker
1971	Pulse sequences for solvent signal suppression	Platt and Sykes, *J. Chem. Phys.* **54**: 1148
1971	Two-dimensional (2D) NMR concept suggested	Jeener
1972	^{13}C studies of cellular metabolism	Matwiyoff and Needham, *Biochem. Biophys. Res. Commun.* **49**: 1158
1973	^{31}P detection of intracellular phosphates	Moon and Richards, *J. Biol. Chem.* **248**: 7276
1973	NMR analysis of body fluids and tissues	Moon and Richards, *J. Biol. Chem.* **248**: 7276; Hoult *et al.* *Nature* **252**: 285
1973	360 MHz superconducting NMR spectrometer	Bruker
1974	2D NMR techniques developed	Aue *et al.*, *J. Chem. Phys.* **64**: 229
1976	Early NMR studies on body fluids and tissues	Moon and Richards, *J. Biol. Chem.* **248**: 7276; Hoult *et al.*, *Nature* **252**: 285
1976–79	^{31}P studies of muscle metabolism	Burt *et al.*, *J. Biol. Chem.* **251**: 2584; Burt *et al.*, *Science* **195**: 145; Garlick *et al.* *Biochem. Biophys. Res. Commun.* **74**: 1256; Jacobus *et al.*, *Nature* **265**: 756; Hollis and Nunnally, *Biochem. Biophys. Res. Commun.* **75**: 1086; Yoshizaki, *J. Biochem.* **84**: 11; Cohen and Burt, *Proc. Natl. Acad. Sci.* **74**: 4271; Sehr and Radda, *Biochem. Biophys. Res. Commun.* **77**: 195; Burt *et al.*, *Annu. Rev. Biophys. Bioeng.* **8**: 1
1977	First 600 MHz spectrometer	Carnegie Mellon University
1979	Detection of insensitive nuclei enhanced by polarization transfer (INEPT)	Morris and Freeman, *J. Am. Chem. Soc.* **101**: 760

Table 2 *Continued*

Date	*Milestone*	*Literature citation*
1979	Detection of heteronuclear multiple quantum coherence (HMQC)	Mueller, *J. Am. Chem. Soc.* **101**, 4481; Burum and Ernst, *J. Magn. Reson.* **39**: 163
1979	500 MHz superconducting spectrometer	Bruker
1980	Surface coils used for *in vivo* NMR studies	Ackerman *et al., Nature* **283**: 167
1980	Pulsed-field gradients used for coherence selection	Bax *et al., Chem. Phys. Lett.* **69**: 567
1981	NMR used to diagnose a medical condition	Ross *et al., N. Engl. J. Med.* **304**: 1338
1981–83	Perfusion methods used for NMR studies of cell metabolism	Ugurbil *et al., Proc. Natl. Acad. Sci,* **78**: 4843; Foxall and Cohen, *J. Magn. Reson.* **52**: 346
1982	Full assignments obtained for small protein	Wagner and Wüthrich, *J. Mol. Biol.* **155**: 347
1983	First 3D-structures of proteins from NMR data	Williamson *et al., J. Mol. Biol.* **182**: 195; Braun *et al., J. Mol. Biol.* **169**: 921
1987	600 MHz superconducting spectrometer	Bruker; Varian; Oxford Instruments
1988	2D-NMR combined with isotopically labelling for full assignments of proteins	Torchia *et al., Biochemistry* **27**: 5135
1988	Whole-body imaging and spectroscopy at 4.0 T	Barfuss *et al., Radiology* **169**: 811
1989	3D-NMR on isotopically labelled proteins	Marion *et al., Biochemistry* **28**: 6150
1990	4D-NMR on isotopically labelled proteins	Kay *et al., Science* **249**: 411
1990	Pulsed-field gradients routinely incorporated into pulse sequences	Bax *et al., Chem. Phys. Lett.* **69**: 567; Hurd, *J. Magn. Reson.* **87**: 422
1992	750 MHz spectrometers	Bruker; Varian; Oxford Instruments
1995	800 MHz spectrometer	Bruker

large biological macromolecules at ever-increasing field strengths. However, a general solution to the signal overlap problem became available only with the arrival of multidimensional NMR methods.

The most important breakthrough came in 1975 with the development of the first two-dimensional (2D) NMR experiments, which had the capability of both simplifying complex spectra and also establishing correlations between nuclei connected either by scalar spin coupling through covalent bonds (COSY spectra) or by dipole–dipole relaxation pathways through space (NOESY spectra). These 2D experiments allowed the assignment of complex NMR spectra and provided distance information for use in structural calculations. The eventual demonstration of the full potential of these methods was made by Wüthrich and co-workers, which eventually led to the first determination of a complete structure for a globular protein in solution.

The extension of the multidimensional NMR approach to larger proteins was subsequently made possible by the development of 3D- and 4D-NMR techniques incorporating INEPT and HMQC pulse sequences that were applied to ^{13}C- and ^{15}N-labelled proteins. These latter developments were made at NIH by Bax and Clore and their co-workers. These multidimensional NMR methods provide the spectral simplification required to completely assign the spectra of proteins of up to 30 kDa and to determine their structures to a resolution similar to the 0.20 nm resolution X-ray structure (see the relevant milestone experiments in **Table 2**). Using the modern techniques, detailed structural and dynamic information can now be routinely obtained for complexes of proteins formed with nucleic acids and other ligands with overall molecular masses of ~30kDa.

In the early 1970s a completely new area of NMR was opened by reports (by Moon and Richard and by Hoult and co-workers) showing that it was possible to record high-resolution ^{31}P NMR spectra on cells and intact organs. This led to an exciting area of research into metabolic processes that allows the chemistry within living cells to be monitored directly. These methods have reached the stage where they can be used to diagnose disease, to monitor biochemical responses to exercise and stress, and even to follow the effects of drug therapy by using repeated non-invasive examinations. The possibility of combining this approach with spatial localization techniques in whole-body magnetic resonance imaging (MRI) presents enormous opportunities for future work.

Magnetic resonance imaging (MRI)

Many of us can recall the great intellectual excitement that accompanied the publication of the early NMR experiments in 1973 showing how spatial information can be encoded into NMR signals. In particular, the simple approach adopted by Paul Lauterbur of using field gradients to produce the spatial resolution required to give a two-dimensional image of water in glass tubes was a brilliant example of lateral thinking that provided a completely new way of viewing the NMR experiment. Even in the very early days, the pioneering workers in MRI (the word 'nuclear' having been dropped because it was thought that it would suggest to the patients that radioactivity was involved) realized that the technique would make its largest contribution in the area of noninvasive clinical imaging. By 1977 the first images of the human body were being reported, one of the earliest being that of a wrist showing features as small as 0.5 cm. At first the method was greeted with much scepticism because its sensitivity performance compared unfavourably with the well-established X-ray CT scanning methods: however, rapid instrumental advances soon allowed the MRI technique to show its full potential, particularly in the ability to provide high-contrast images for soft tissues and tissues in areas surrounded by dense bone structures. The development of the echo planar imaging (EPI) method by Mansfield and his co-workers allowed well-resolved images to be obtained from a single pulse and this opened up many new applications requiring short examination times, such as in heart, abdomen and chest imaging. Other important milestones in the development of the MRI technique are summarized in **Table 3**. There are now many applications where MRI is the favoured imaging method (such as brain scanning for detecting encephalitis or multiple sclerosis (MS) and for monitoring therapy treatment of MS). Most of the images examined are based on detecting ^{1}H nuclei. However, recent high-quality images of the airways in human lungs have been provided by helium or xenon images obtained after inhalation of the polarized inert gases by the patient. Another recent and exciting application, called functional MRI, attempts to study the working of the human brain; by stimulating the brain either through the

Table 3 Milestones in the development of magnetic resonance imaging (MRI)

Date	*Milestone*	*Literature citation*
1973	Spin-imaging methods proposed	Lauterbur, *Nature* **242**: 190; Mansfield and Grannell, *J. Phys. C* **6**: L422; Damadian, *NMR in Medicine*, Springer-Verlag
1973	NMR diffraction used for NMR imaging	Mansfield and Grannell, *J. Phys. C* **6**: L422
1973	Zeugmatography; first two-dimensional NMR image	Lauterbur, *Nature* **242**: 195
1974	Sensitive point imaging method	Hinshaw, *Phys. Lett.* **48**: 87
1974	2D NMR techniques developed	Aue *et al.*, *J. Chem. Phys.* **64**: 229
1975	Slice selection in imaging by selective excitation	Garroway *et al.*, *J. Phys.* C **7**: L457; Sutherland and Hutchinson, *J. Phys. E* **11**: 79; Hoult, *J. Magn. Reson.* **35**: 69
1975	Fourier zeugmatography	Kumar *et al.*, *J. Magn. Reson.* **18**: 69
1977–80	Spin-imaging of human limbs and organs	Wehrli, *Prog. NMR Spectrosc.* **28**: 87
1977	Echo-planar imaging	Mansfield and Pykett, *J. Magn. Reson.* **29**: 355
1977–78	Whole-body scanning	
1979	Chemical shift imaging	Cox and Styles, *J. Magn. Reson.* **40**: 209 Brown *et al.*, *Proc. Natl. Acad. Sci.* **79**: 3523 Maudsley *et al.*, *J. Magn. Reson.* **51**: 147 Mauldsley *et al.*, *Siemens Forsch. Entwickl-Ber.* **8**: 326
1980	Spin-warp imaging	Edelstein *et al.*, *Phys. Med. Biol.* **25**: 751
1980	3D-projection reconstruction	Lai and Lauterbur, *J. Phys. E* **13**: 747
1983	Whole-body imaging at 1.5 T	Hart *et al.*, *Am. J. Roentgenol.* **141**: 1195
1984–87	Gradient methods used for spatial localization	Bottomley, *US Patent* 480/228; Ordidge *et al.*, *J. Magn. Reson.* **60**: 283; Frahm *et al.*, *J. Magn. Reson.* **72**: 502
1984	Combined imaging and spectroscopy on human brain	Bottomley *et al.*, *Radiology* **150**: 441
1985	FLASH imaging	Haase *et al.*, *J. Magn. Reson.* **67**: 258
1985	Magnetic resonance (MR) angiographic images	Wedeen *et al.*, *Science* **230**: 946
1986	NMR microscopy imaging on live cell	Aguayo *et al.*, *Nature* **322**: 190
1987	Echo-planar imaging at 2.0 T	Pykett and Rzedian, *Magn. Res. Med.* **5**: 563
1988	Whole-body imaging and spectroscopy at 4.0 T	Barfuss *et al.*, *Radiology* **169**: 811
1991	Functional MR-detection of cognitive responses	Belliveau *et al.*, *Science* **254**: 716; Prichard *et al.*, *Proc. Natl. Acad. Sci.* **88**: 5829
1993	NMR microscopy using superconducting receiver coil	Black *et al.*, *Science* **259**: 793
1994	Use of polarized rare gases in spin-imaging	Albert *et al.*, *Nature* **370**: 199

senses or by thought processes, it is possible to detect changes in MRI images of the brain. These are related to changes in oxygen levels in the blood induced in specific locations of the brain. This type of experiment opens up exciting possibilities for studying the human brain in action.

MRI scanners are now increasingly being used not only in research hospitals but also in the general hospital environment. The high-profile use of MRI as a major health-care tool has certainly increased the public awareness of NMR and drawn proper attention to the versatility of this exceptional phenomenon.

List of symbols

m = magnetic quantum number; T_1 = spin–lattice relaxation time; T_2 = spin–spin relaxation time.

See also: **Cells Studied By NMR; *In Vivo* NMR, Applications, Other Nuclei; *In Vivo* NMR, Applications, ^{31}P; *In Vivo* NMR, Methods; Labelling Studies in Biochemistry Using NMR; Liquid Crystals and Liquid Crystal Solutions Studied By NMR; Macromolecule–Ligand Interactions Studied By NMR; Membranes Studied By NMR Spectroscopy; MRI Applications, Biological; MRI Applications, Clinical; MRI Instrumentation; MRI Theory; NMR in Anisotropic Systems, Theory; NMR of Solids; NMR Spectrometers; NMR Pulse Sequences; Nuclear Overhauser Effect; Nucleic Acids Studied Using NMR; Perfused Organs Studied Using NMR Spectroscopy; Proteins Studied using NMR Spectroscopy; Solid State NMR, Methods; Two-Dimensional NMR Methods.**

Further reading

Grant DM and Harris RK (eds) (1996) *Encyclopedia of NMR*. Chichester: Wiley.

Emsley JW and Feeney J (1995) *Progress in Nuclear Magnetic Resonance Spectroscopy* **28**: 1.

Manganese NMR, Applications

See **Heteronuclear NMR Applications (Sc–Zn).**

Mass Spectrometry in Food Science

See **Food Science, Applications of Mass Spectrometry.**

Mass Spectrometry, Historical Perspective

Allan Maccoll†, Claygate, Surrey, UK

MASS SPECTROMETRY
Historical Overview

Introduction

Mass spectrometry has made many notable contributions to chemistry from the chemical physics of small molecules to the structures of large biomolecules. It is an instrument in which ions in a beam are separated according to their mass/charge ratio (m/z). Its humble beginnings lay in the works of physicists at the turn of the century. Up to the Second World War mass spectrometry was the province of the physicists along with a small band of physical chemists. However, the demands for accurate evaluation of the composition of aircraft fuel during the Second World War led to its extensive application to hydrocarbon analysis. Heartened by the success in this area, operators were encouraged to put 'dirty' organic chemicals in their instruments and so organic mass spectrometry was born. These developments were largely owing to the manufacturers responding to the demands for instruments to meet the needs of the chemists. The introduction of high-resolution instruments led to the developments of ion chemistry. This took place in the decades 1950–1980. By this time the mass spectrometric study of large organic molecules had been achieved and the prevailing interest switched to biomolecules – a good source of financial support in view of their medical relevance.

One of the important aspects of the development of mass spectrometry was the camaraderie (occassionally blighted by periods of frustration) that existed between the users and the manufacturers. This was nurtured by the introduction of user's meetings by Associated Electrical Industries (an offshoot of Metrovick). The users would foregather with the engineers responsible for instrumental development to explain *their* problems and requirements for instrument development. The author remembers well the confidence he gained from learning that his problems were not unique – other users had them too! Instrumental development was stimulated by the demands of the users and if the suggested instrumentation could be satisfactorily produced it soon became available. This made it an exciting period to live through.

The beginnings

Thomson

The origins of mass spectrometry lie in the work done in the Cavendish Laboratory in Cambridge by JJ Thomson and his colleagues at the start of the twentieth century on electrical discharges in gases. The first relevant work was the discovery of the electron, using a cathode ray tube. The rays from the cathode pass through a slit in the anode (**Figure 1**) and after passing through another slit pass between two metal plates and on to the wall of the tube. This wall had been treated with a phosphorescent material which glows where the beam strikes it. The beam can be diverted by applying a potential difference between the plates and also by superimposing a magnetic field. By adjusting the two fields so that there is no displacement of the beam Thomson was able to show that the particles carried a negative charge of around 10^{11} C kg^{-1}. Goldstein in 1886, using a perforated cathode, was able to show that there was always a beam travelling in the opposite direction to that of the electrons – the so called kanalstrahlen. Later, Wien showed that these were positively charged particles and concluded that they were positive ions. Thomson decided to investigate these particles. His positive ray apparatus (1912) is shown in **Figure 1**. A is a discharge tube producing positive ions which then pass through the cathode B and after collimation in the narrow tube BN are subjected to superimposed electric and magnetic fields (M,M′ P,P′). The displaced beams then travel to the fluorescent screen G where their effect is observed.

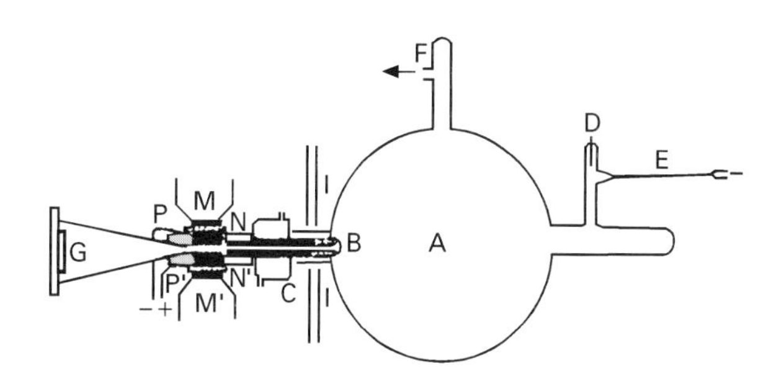

Figure 1 Thomson's cathode ray tube.

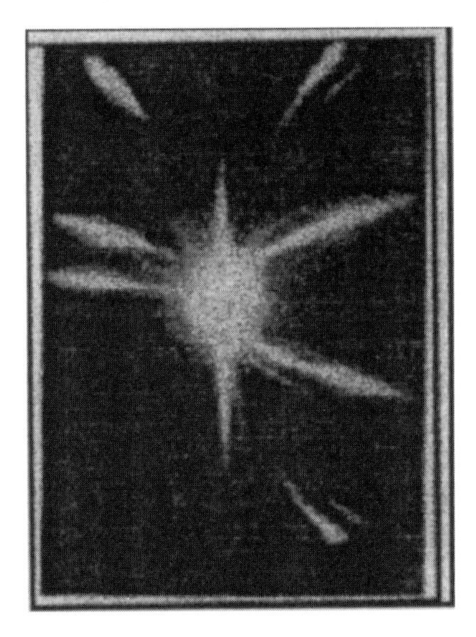

Figure 2 The parabolae.

In **Figure 2** the parabola formed by the top and bottom branches on the left-hand side are due to neon. Under better resolution they show the presence of isotopes at masses 20 and 22. Isotopes had previously been observed in studies of radioactivity. Thomson encouraged a research student in the Cavendish Laboratory, FW Aston, to build a mass spectrograph for further studies of stable isotopes. The research was interrupted by the war of 1914–1918 and so the work was not published until 1923.

Aston

His spectrograph is shown diagrammatically in **Figure 3**. A beam of ions passes through the collimating slits S_1, S_2 into an electric field P_1, P_2. It then enters a magnetic field centred upon M and the divergent beam is brought to focus on a photographic plate P. The geometry ensures that, irrespective of the velocity of the ions, they are brought to a sharp focus on the photographic plate. This is known as velocity focusing. By 1923 Aston had realized that deviations from integral values of the relative molecular masses (Prout's Rule) were of considerable importance for the study of nuclear structure. However, as will be seen later, they were of inestimable importance in the development of organic mass spectrometry. Current values for some atoms are shown in **Table 1** ($^{12}C = 12.000\ 000$).

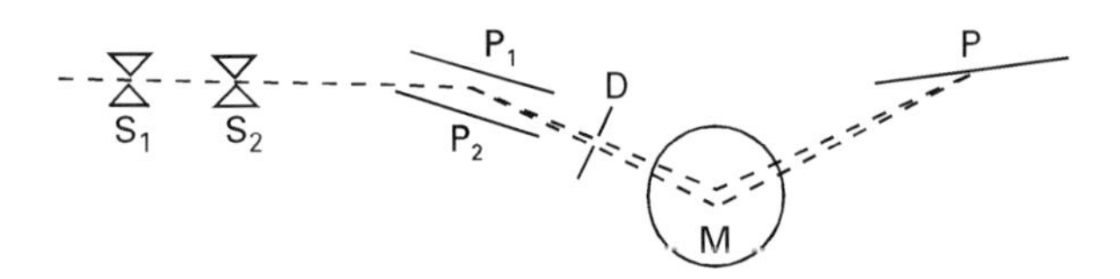

Figure 3 Aston's mass spectrograph.

Table 1 Accurate masses of some common atoms

Atom	*Relative atomic mass*
Hydrogen	1.007 825
Carbon	12.000 000
Nitrogen	14.003 074
Oxygen	15.994 915
Fluorine	18.888 405
Sulfur	31.972 074

Dempster

In 1918, a Canadian working in the University of Chicago (AJ Dempster) developed a different type of apparatus for investigating positive rays (**Figure 4**). It involved a 180° magnetic field. Ions produced by the filament in G are accelerated into the magnetic field through S_1 and pass through S_2 and hence to the collector E. Such a geometry gives rise to direction focusing – ions will arrive at the collector irrespective of the direction they enter the magnetic field. The experimental arrangement is described by the fundamental equation of sector mass spectrometry (Eqn [1]), namely

$$m/z = 2V/B^2R^2 \qquad [1]$$

where m is the mass of the ion, z its charge, B the magnetic field strength and R the radius of the magnetic field. A fundamental difference between

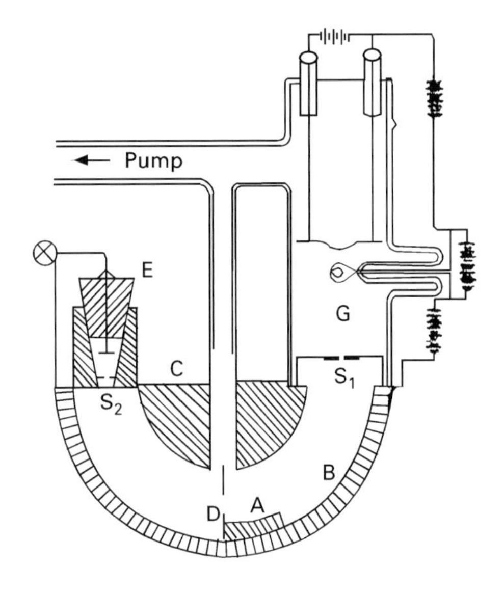

Figure 4 Dempster's mass spectrometer.

Aston's instrument and that of Dempster is that Aston's spectra are obtained instantaneously whereas Dempster's have to be scanned. This can be done simply in two ways, either by scanning the electric field at constant magnetic field or by scanning the magnetic field at constant electric field (more sophisticated methods of scanning have been developed, leading to a better understanding of mass spectrometric processes). Most sector mass spectrometers use the latter method.

Instrumental development

The basic mass spectrometer

In the mass spectrometer shown in **Figure 5** the sample is held in the reservoir and led into the ionization chamber via a leak. On ionization the ions are accelerated into the magnetic sector and eventually arrive at the collector. The current is amplified and recorded.

JJ Thomson was very percipient in predicting organic mass spectrometry in 1913. He wrote in his book *Rays of Positive Electricity and their Application to Chemical Analysis*. "I have described at some lengths the applications of positive rays to chemical analysis: one of the main reasons for writing this book was the hope that it might induce others and especially chemists, to try this method of analysis. I feel sure that there are many problems in chemistry which could be solved with much greater ease by this than by any other method. This method is surprisingly sensitive – more so even than that of spectrum analysis, requires an infinitesimal amount of material and does not require this to be especially purified; the technique is not difficult if appliances for producing high vacua are available . . .". It is a reflection upon the chemists of the period that it took thirty years for Thomson's predications to be verified.

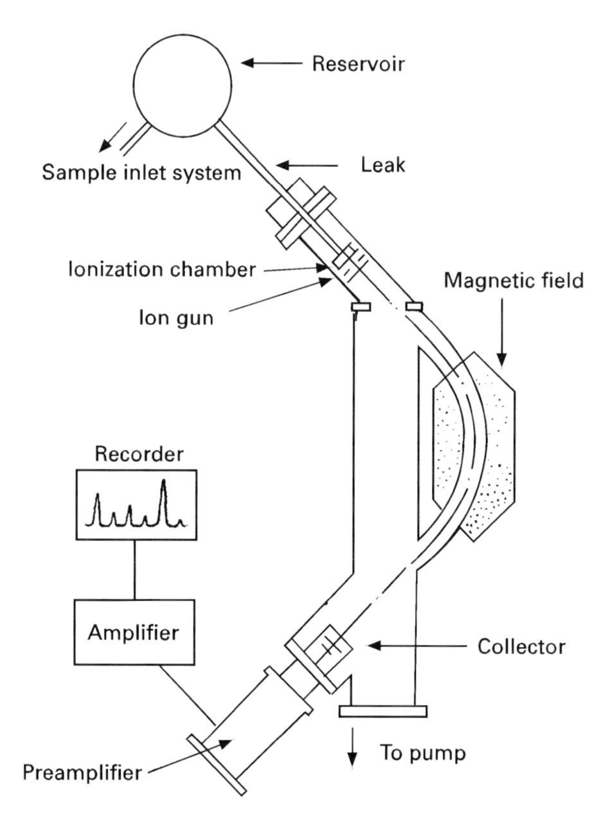

Figure 5 A single focusing mass spectrometer (MS2).

One difficulty was that the apparatus, simple to a physicist, appeared very complex to a chemist. The application of mass spectrometry of chemistry had to await the commercial production of instruments. The impetus came in the 1940s when the war effort demanded rapid and accurate hydrocarbon analysis in connection with aviation fuels. The next big step came in the 1950s when it was realized that in addition to quantitative analysis the technique could be used for the qualitative (structural) analysis of organic compounds. A certain resistance had to be overcome to induce mass spectrometrists to put 'dirty' compounds into their instruments rather than 'clean' hydrocarbons. This gave mass spectrometer manufacturers a further impetus to develop more and more advanced instruments and led to a new discipline – organic mass spectrometry.

Before the advent of the mass spectrometer the determination of the relative molecular mass (M_r) of an organic compound was performed by quantitative analysis (empirical formula) and a rough M_r was used to decide the number of empirical formula to make up the molecular formula. With the mass spectrometer the M_r could be determined directly: however, there was more to come. It was noted earlier that the relative atomic masses of atoms were slightly different from integer values. If the M_r of a compound could be accurately determined then there would be only one formula that would be consistent with it. So it was up to the manufacturers to produce instruments with sufficiently high resolution to be able to separate these values. A word about resolution or resolving power is appropriate here. Although there is no generally accepted definition, one that is widely used is the 10% valley definition. If two peaks of equal height are separated by Δm and the valley between them is 10% of the peak height then the resolving power is said to be $m/\Delta m$. If one considers the doublet at m/z 28 corresponding to C_2H_4 and N_2, Δm is (28.031 299 –28.006 158) = 0.025 141. Thus $m/\Delta m = 1114$ and a resolving power of about 1000 would be required to separate the two peaks. The search for higher and higher resolution led to the introduction of a double focusing mass spectrometer. It has been seen that while Aston's mass spectrograph gives velocity focusing, Dempster's mass spectrometer gave direction focusing. Nier and Roberts developed a geometry which ensured both velocity and direction focusing. This geometry formed the basis of the MS9 (Associated Electrical Instruments, AEI) (**Figure 6**) which for many years was the workhorse of the organic mass spectrometrists. Initially it had resolving power of 10000 but with modifications this value was raised tenfold. It became apparent that it would be advantageous if the ion beam could be selected before its subsequent analysis. This gave rise to the ZAB series of mass spectrometers (Vacuum Generators). These instruments also had the advantage of the ion beam being in the horizontal plane (the AEI instruments had the ion beam in the vertical plane) which made it much easier to add additional sectors when required.

Representation of mass spectra

The bar diagram

Mass spectra are usually represented by bar diagrams on which the relative intensity of peak or the relative abundance of an ion is plotted against the *m/z* value (**Figure 7**). The molecular peak $[M]^{\bullet+}$ is the one corresponding to the M_r of the compound and the base peak is the most intense one in the spectrum. A further alternative is the use of the fraction of the total ion current carried by the ion in question. In the early days of mass spectrometry the operator had to laboriously develop the trace recorded on photographic paper or equally laboriously plot the ion current against the *m/z* ratio. More recently the spectra are electronically recorded and can be plotted out according to the whim of the operator.

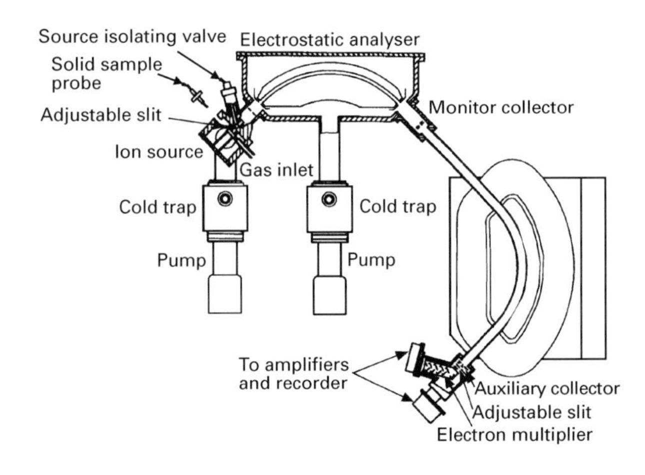

Figure 6 A double focusing mass spectrometer (MS9).

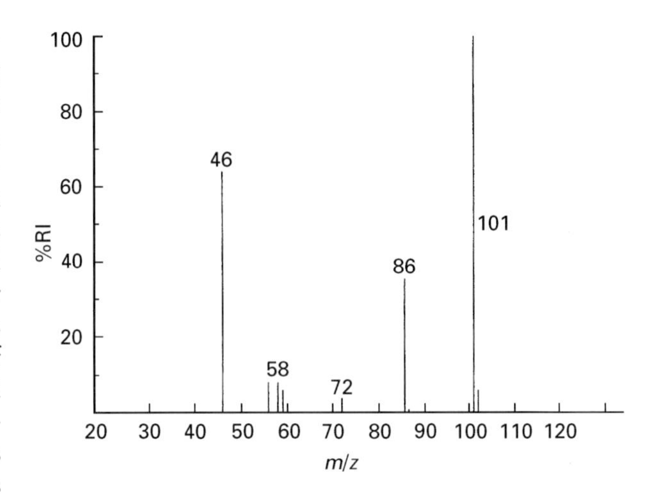

Figure 7 Mass spectrum of $[HCONHC(CH_3)_3]$.

The anatomy of a mass spectrometer

The components of a mass spectrometer

The mass spectrometer consists essentially of a source, which produces a beam of ions, an analyser which separates the beam according to the *m/z* ratio and a collector which determines the fraction of the total ion current carried by each of the ions.

Sector instruments

The source Probably the most widespread method of ion production is by electron impact. The other fundamental, though little used method, is that of photoionization. In recent years a number of other methods have been developed, such as fast atom bombardment (FAB) and electrospray (ES) both of which are known as 'soft' methods of ionization in that they transfer relatively little energy to the ion. A bonus with ESMS lies in the fact that multiply charged ions are produced, thus extending the mass range. Thus, for an ion m^{20+} the effective mass range will be 20 times that of a singly charged ion. These two techniques have had considerable application in biological and medical mass spectrometry. An alternative soft ionization method is to use low-energy electrons in impact ionization. If the measurements are also carried out using a cooled source the process produces what are known as LELT (low energy, low temperature) spectra.

In the electron impact source a beam of electrons (usually 70 eV) impacts the gaseous substrate under investigation and removes an electron from it, thus producing an ion. This is a drastic method since ionization energies are usually of the order of 10 eV and chemical energies of the order of a few volts. The processes occurring are:

$$e^- + M \rightarrow M^{\bullet +} + 2e^- \quad [2]$$

$$e^- + M \rightarrow F^+ + F^\bullet + 2e^- \quad [3]$$

$$e^- + M \rightarrow F^{\bullet +} + F + 2e^- \quad [4]$$

Process [2] represents ionization to form the molecular ion while [3] represents fragmentation to form an even electron ion and [4] represents fragmentation to form an odd electron ion. In writing equations for fragmentation it is essential that 'electron bookeeping' be maintained. What is shown here is primary fragmentation – the ions F^+ can further fragment to give secondary fragments and so on.

The analyser It has been seen that a transverse magnetic field can separate an unresolved beam of ions according to their *m*/*z* values (**Figure** 5). Such a system gives direction focusing. An electric field (see **Figure** 6) can give direction focusing and so lead to a double focusing mass spectrometer capable of high-resolution measurements.

The collector The usual collector is an electron multiplier which can give gains of 10^7 or more. The output is sent to a recorder or data system. An earlier form is the Faraday cup which collects the electrons – the current then being amplified and recorded.

The quadrupole

Originally the tool of physicists and physical chemists, now with improved electronics the quadrupole mass spectrometer has become an essential instrument for biological and biomedical research. Originally described as a mass filter, it operates by using a combination of a quadrupole static electric field and a radiofrequency field which combine to focus an ion beam on a collector.

The ion trap

This device is related to the quadrupole, being a three-dimensional quadrupole. The ion trap consists of a hyperbolic ring electrode (doughnut) and two hyperbolic end electrodes. To obtain a spectrum a variable amplitude radiofrequency is applied to the doughnut whilst the end plates are grounded. As the amplitude is increased ions of increasing *m*/*z* are collected.

The time-of-flight mass spectrometer

In this instrument ions produced in the source are accelerated to a given velocity. The unresolved beam is then injected into a field-free region and the ions drift towards the collector. The velocities will be inversely proportional to the square roots of the masses. This means that a pulse of ions will split up according to the ionic masses. The unresolved beam thus becomes resolved in time. Provided that the response time of the electronics is sufficiently fast a spectrum can be recorded. Obviously an average over many such pulses is necessary to provide a reliable signal. Once again the electronics lie at heart of this problem, which demands very fast amplifiers. Initially the time-of-flight mass spectrometer (TOF) was the province of physicists and later of chemists but, with the tremendous advance in electronics, instruments are now produced that are capable of routine operation by relatively untrained operators.

The ion cyclotron resonance mass spectrometer

An ion cyclotron resonance (ICR) spectrometer creates a pulse of ions in a magnetic field. These are brought into resonance by scanning the applied radiofrequency. From the cyclotron resonance frequency and the magnetic field strength the *m*/*z* ratio can be calculated. The use of a fast Fourier transform (FT-ICR) refines the method.

The energetics of ionization and fragmentation

The thermochemistry of ions

Just as the thermochemistry of neutral molecules has led to an understanding of the structure, stability and kinetics of chemical species, the thermochemistry of ions has led to a corresponding understanding of ionic species in the gas phase. Thus the enthalpy of formation ($\Delta_f H^o(M^{\bullet +})$ of the molecular ion is given by Equation [5].

$$\Delta_f H^\circ(M^{\bullet +}) = IP(M) - \Delta_f H^\circ(M) \quad [5]$$

where IE(M) is the ionization energy of the molecule and $\Delta_f H^o$ is the enthalpy of formation of the neutral molecules. Holmes and co-workers have published a very useful algorithm for estimating the enthalpies of

Table 2 The enthalpies of formation of *n*-alkane molecular ions

Molecule	$\Delta_f H°$ $(M^{\bullet +})$ (kJ mol^{-1})[a]	$\Delta_f H°$ $(M^{\bullet +})$ (kJ mol^{-1})[b]
CH_4	1142	1142
C_2H_6	1025	1021
C_3H_8	954	950
C_4H_{10}	891	895
C_5H_{12}	854	854
C_6H_{14}	816	816
C_7H_{16}	778	778

[a] Experimental value.
[b] Theoretical value.

formation of odd electron ions. Some typical values for hydrocarbons are shown in **Table 2**. The agreement between experimental and theoretical values is excellent. Often the enthalpies of formation of the substrate molecule are not known and so recourse has to be made to empirical methods such as that of Benson for estimation of the value.

In the case of the even electron ions one has, mainly, to have recourse to experimentally determined values. The enthalpies of formation of the even electron ions are given by Equation [6] where the appearance energy is represented by $AE(F^+)$, with $\Delta_f H°(F^{\bullet +})$, $\Delta_f H°(F^\bullet)$and $\Delta_f H°(M)$ being the enthalpies of formation of the ion, the radical and the molecule.

$$\Delta_f H°(F^{\bullet +}) \leq AE(F^+) + \Delta_f H°(F^\bullet) + \Delta_f H°(M) \quad [6]$$

In Equation [6] the inequality may be replaced by the equality in most instances. Some values for the primary carbonium ions are shown in **Table 3**. Values such as these can then be used in calculating ionization and appearance energies. These are, respectively, the lowest energy at which the molecular ion appears and the lowest energy at which a fragment ion appears. Thus the ionization energy is given by:

$$IE = \Delta_f H°(M^{\bullet +}) - \Delta_f H°(M) \quad [5a]$$

on rearranging Equation [5]. Similarly, the appearance energy is obtained by rearranging Equation [6].

$$AE(F^+) = \Delta_f H°(F^{\bullet +}) + \Delta_f H°(F^\bullet) - \Delta_f H°(M) \quad [6a]$$

Holmes and Lossing have developed an ingenious method of measuring the enthalpies of formation of neutrals by a further rearrangement of Equation [6]. This is extremely useful where the enthalpy of formation of the neutral has not been measured. The method depends on measuring the appearance energy of a fragment ion produced from different sources

$$\Delta_f H°(M) = \Delta_f H°(F^{\bullet +}) + \Delta_f H°(F^\bullet) - AE(F^+) \quad [6b]$$

and using the average value in Equation [6b].

Table 3 Some values of the enthalpies of formation of carbonium ions

Molecule	$\Delta_f H°$ $(F^{\bullet +})$ (kJ mol^{-1})
CH_3	1092
C_2H_5	916
C_3H_7	870
C_4H_9	841
C_5H_{11}	812[a]
C_6H_{13}	791[a]
C_7H_{15}	766[a]

[a] Estimated value.

Metastable ions

It will be seen in **Figure 6** that there are two important field-free regions (FFR) in the double focusing mass spectrometer, namely between the source and the electric analyser (FFRI) and between the electric and magnetic analysers (FFR2). It may so happen that in flight an ion decomposes in FFR2 in which case a diffuse peak appears in the mass spectrum at the position *m*/*z* given by Equation [7]

$$m^*/z = m_2{}^2/m_1 \quad [7]$$

for process [8]. A typical metastable peak is shown in **Figure 8** for the process

$$m_1{}^+ = m_2{}^+ + (m_1 - m_2) \quad [8]$$

The appearance of a metastable peak is confirmation of a fragmentation route, but absence of the peak does not indicate the absence of a fragmentation. The reason is that metastable ions are relatively long lived. If the fragmentation is rapid no metastable will be seen. A special scan, keeping *B*/*E* constant, will record all the daughter peaks resulting from a given parent ion. Equally, a scan keeping B^2/E constant will give all the progenitors of a given peak.

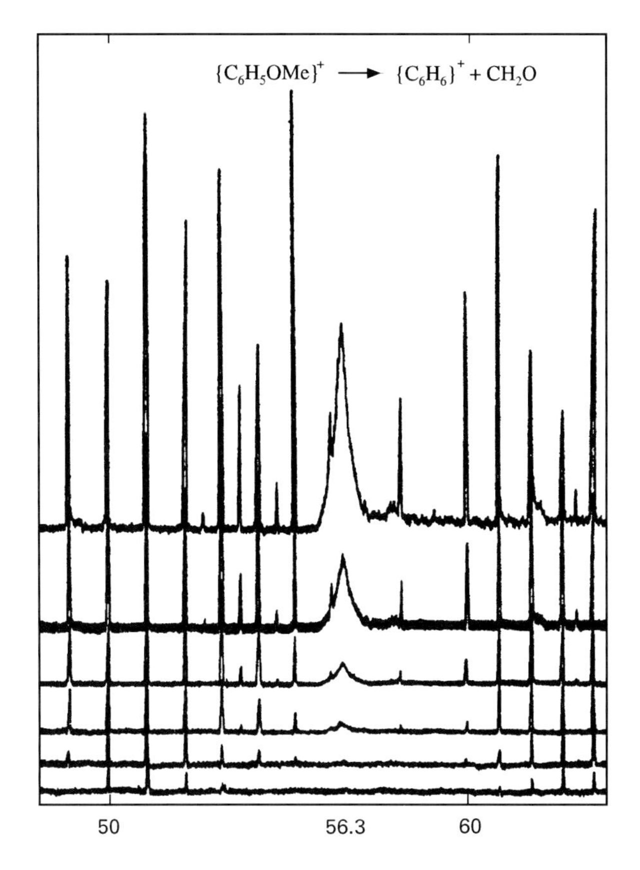

Figure 8 A metastable in the mass spectrum of anisole.

These scans are very useful in mapping out the fragmentation patterns of a given ion.

Collision induced dissociation

Another means of producing fragmentation involves collision processes – bimolecular as compared with the unimolecular processes previously discussed. In this method a beam of energetic ions is brought into collision with neutral molecules and fragmentation results – collision induced dissociation (CID). The spectra thus obtained were complex since they derived from an unresolved beam of ions. It was realized that it would be advantageous if the ions for collision were separated from the unresolved beam. This led to the development of a reversed geometry instrument – the ZAB, produced by Vacuum Generators. Finally there was the introduction of multisector instruments which gave rise to the technique of mass spectrometry–mass spectrometry (MS-MS). CID has proved very useful in assigning structures to fragment ions.

The future

Further developments in fundamental mass spectrometry will have to await for universities to return to their basic task – the pursuit of fundamental research. At the present time many workers in the field have to design their research to attract funds. This often leads to hack research – not always in the best interest of the subject or the scientists. It is to be hoped that the new millennium will see the universities of the world returning to their proper research areas, namely fundamental research. Only in this way will mass spectrometry develop in its fundamental aspects which in turn will lead to new and more powerful techniques.

The literature of mass spectrometry

1968 saw the first of the journals devoted to mass spectrometry. *Organic Mass Spectrometry* (OMS) and the *International Journal of Mass Spectrometry and Ion Physics* (IJMSIP). Later OMS spawned *Biomedical Mass Spectrometry* (BMS). IJMSIP has since changed its name to *The International Journal of Mass Spectrometry and Ion Processes* and latterly to the *International Journal of Mass Spectrometry*, while OMS and BMS have been incorporated in the *Journal of Mass Spectrometry*. The American Society for Mass Spectrometry has produced a Journal – *Journal of the American Society for Mass Spectrometry*. To facilitate rapid publication, *Rapid Communications in Mass Spectrometry* was born – the authors nominate their own referees.

List of symbols

B = magnetic field strength; m = mass of an ion; R = radius of the magnetic field; V = electric field strength; z = charge on an ion,; $\Delta_f H^o$ = enthalpy of formation.

See also: **Chemical Ionization in Mass Spectrometry; Fast Atom Bombardment Ionization in Mass Spectrometry; Fragmentation in Mass Spectrometry; Ion Structures in Mass Spectrometry; Ion Trap Mass Spectrometers; Ionization Theory; Ion Energetics in Mass Spectrometry; Ion Collision Theory; Metastable Ions; Quadrupoles, Use of in Mass Spectrometry; Sector Mass Spectrometers; Statistical Theory of Mass Spectra; Time of Flight Mass Spectrometers.**

Further reading

Aston FW (1924) *Isotopes*, 2nd edn. London: Edward Arnold.

Beynon JH and Morgan RP (1978) The development of mass spectrometry: an historical account. *International Journal of Mass Spectrometry and Ion Physics* 27: 1–30.
Thomson JJ (1898) *The Discharge of Electricity through Gases*. London: Archibald Constable.
Thomson JJ (1913) *Rays of Positive Electricity and their Application to Chemical Analyses*, p 56. London: Longmans and Green.

Mass Transport Studied Using NMR Spectroscopy

See **Diffusion Studied Using NMR Spectroscopy.**

Materials Science Applications of X-Ray Diffraction

Åke Kvick, European Synchrotron Radiation Facility, Grenoble, France

HIGH ENERGY SPECTROSCOPY
Applications

The X-ray diffraction technique is widely used in structural characterization of materials and serves as an important complement to electron microscopy, neutron diffraction, optical methods and Rutherford backscattering.

The early uses were mainly in establishing the crystal structures and the phase composition of materials but it has in recent years more and more been used to study stress and strain relationships, to characterize semiconductors, to study interfaces and multilayer devices, to mention a few major application areas.

One of the important advantages of X-ray diffraction is that it is a nondestructive method with penetration from the surfaces into the bulk of the materials. This article will outline some of the most important areas including some rapidly developing fields such as time-dependent phenomena and perturbation studies.

X-ray sources

X-rays are electromagnetic in nature and atoms have moderate absorption cross-sections for X-ray radiation resulting in moderate energy exchange with the materials studied, making diffraction a nondestructive method, in most cases.

Traditionally X-rays are produced by bombarding anode materials with electrons accelerated by a >30 kV potential. The collision of the accelerated electrons produces a line spectrum superimposed on a continuous spectrum called *bremsstrahlung*.

The line spectrum is characteristic of the bombarded anode material and has photon intensities much higher than the continuous spectrum. The characteristic lines are generated by the relaxation of excited electrons from the electron shells and are labelled K, L, M, etc. and signify the relaxation L to K, M to K, etc.

A table of available laboratory wavelengths is given in **Table 1**.

The increased importance of X-ray diffraction in materials science is coupled to the recent emergence of a new source of X-rays based on *synchrotron radiation* storage rings. The synchrotron radiation is produced by the bending of the path of relativistic charged particles, electrons or positrons, by magnets causing an emission of intense electromagnetic radiation in the forward direction of the particles. The

Table 1 Radiation from common anode materials

Radiation	*Wavelength (Å)*	*Energy (keV)*
Ag K_α	0.5608	22.103
Pd K_α	0.5869	21.125
Rh K_α	0.6147	20.169
Mo K_α	0.7107	17.444
Zn K_α	1.4364	8.631
Cu K_α	1.5418	8.041
Ni K_α	1.6591	7.742
Co K_α	1.7905	6.925
Fe K_α	1.9373	6.400
Mn K_α	2.1031	5.895
Cr K_α	2.2909	5.412
Ti K_α	2.7496	4.509
Synchrotron	~0.05–3	4.300

The value α is a mean of the $K_{\alpha1}$ and $K_{\alpha2}$ emissions. The synchrotron radiation is continuous and the range is the most commonly used. The range may be extended on both sides.

photons are generated over a wide energy range from very long wavelengths in the visible to hard X-rays up to several hundred keV.

The radiation is very intense and exceeds the available normal laboratory sources by up to 6 or 7 orders of magnitude. The synchrotron storage rings used for the radiation production, however, are large and expensive, with facilities characterized by storage rings with a circumference up to more than one thousand metres.

The main advantages of synchrotron radiation arc:

1. continuous radiation up to very high energies (>100 keV);
2. high intensity and brightness;
3. pulsed time structure down to picoseconds;
4. high degree of polarization.

Figure 1 illustrates a modern synchrotron facility with many experimental facilities in a variety of scientific areas from atomic physics to medicine.

Figure 1 Beam lines at the European Synchrotron Radiation Facility in Grenoble, France.

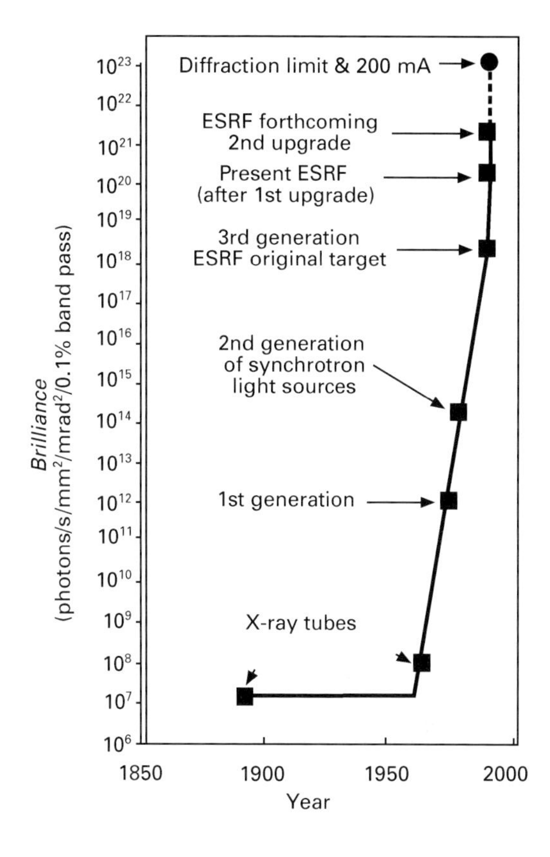

Figure 2 The brightness defined as photons/s/mm²/mrad²/0.1% energy band pass for conventional and synchrotron X-ray sources. ESRF denotes the European Synchrotron Radiation source in Grenoble, France.

Figure 2 compares the brightness of the available X-ray sources.

X-ray diffraction

The diffraction method utilizes the interference of the radiation scattered by atoms in an ordered structure and is therefore limited to studies of materials with long-range order.

The incoming X-ray beam can be characterized as a plane wave of radiation interacting with the electrons of the material under study. The interaction is both in the form of absorption and scattering. The scattering can be thought of as spheres of radiation emerging from the scattering atoms. If the atoms have long-range order the separate 'spheres' interfere constructively and destructively producing distinct spots, *Bragg reflections*, in certain directions.

The specific scattering angles, θ_{hkl}, carry information on the long-range ordering dimensions and the intensity gives information on the location of the electrons within that order.

The basis for all material science studies using X-ray diffraction is Bragg's law:

$$\lambda = 2d_{hkl}\sin(\theta_{hkl}) \quad [1]$$

where λ is the wavelength of the incoming radiation, d_{hkl} is the spacing of the (*hkl*) atomic plane and θ is the angle of the diffracting plane where constructive interference occurs. (see **Figure 3**).

Differentiation of Bragg's law gives the expression:

$$\Delta(\theta) = -\Delta(d)/d\tan(\theta) \quad [2]$$

which is an important formula relating the observed changes in scattering angles to structural changes in the material.

The penetration depth of the probing radiation is an important parameter in designing a diffraction experiment. The penetration depth is associated with the absorption of the radiation, which is a function of the absorption cross-section of the material under study. The absorption can be calculated by the formula:

$$I/I_o = \exp(-\mu t) \quad [3]$$

where I_0 is the intensity of the incident beam, I is the intensity of a beam having passed through t (cm) of material with an absorption coefficient of μ (cm^{-1}).

The absorption coefficient μ can be calculated as an additive sum over the different atomic species in the unit cell:

$$\mu = 1/V_c\,\Sigma(\sigma_n) \quad [4]$$

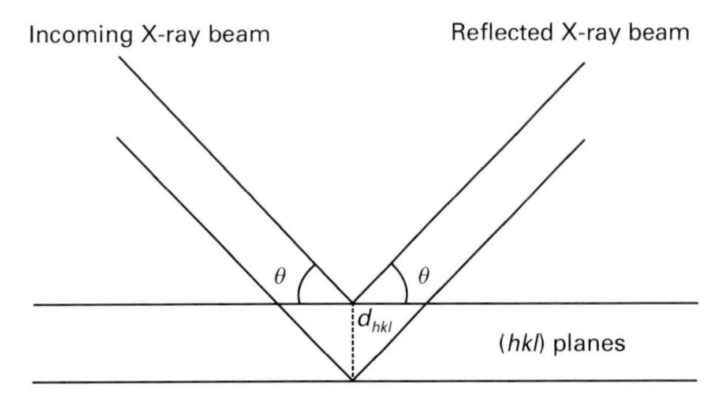

Figure 3 Reflection from the planes (*hkl*) with interplanar spacing d_{hkl}.

where V_c is the volume of the unit cell and σ_n is the absorption cross-section for component n. The absorption cross-sections vary as a function of the wavelength and can be calculated using the Victoreen expression:

$$\mu/\rho = C\lambda^3 - D\lambda^4 + \sigma_{K-N}NZ/A \quad [5]$$

where ρ is the density of the material with atomic number Z and the atomic weight A. The constants C, D and $\sigma_{K\text{-}N}N$ vary with the wavelength. Tabulations for various materials can be found in International Tables for Crystallography, Vol III, pp 161 ff.

It can be noted that the absorption drops off with decreasing wavelength and the penetration depth can thus be changed with a change in wavelength. A quantity called *penetration distance*, τ, is usually quoted for penetration depths and is defined as the distance where I/I_0 is reduced to 1/e. Penetration distances for a few elements are listed in **Table 2**, together with a comparison with other methods.

Structure determinations

Historically, and even today, the structure determination of crystalline materials is the most important application of X-ray diffraction in materials science. The relative intensities of Bragg reflections carry information on the location of the electrons in the solids and thus give precise information on the relative positions and thermal motion of the atoms. Even information on the bonding electrons may be obtained.

The scattered intensities from different planes (*hkl*) in a crystal are measured using precise diffractometers that orient the sample with respect to the incident X-ray beam for all the possible diffraction planes in the crystal. Intensities are measured using scintillation, semiconductor CCD or imaging plate detectors. The measured intensities are converted, after various geometric corrections, to the amplitude

Table 2 Penetration depth τ (1/e) in Al, Fe and Cu for various techniques in millimetres

	Al	*Fe*	*Cu*
Scanning electron microscope	<0.001	<0.001	<0.001
X-ray diffraction (Cu K_α)	0.14	0.007	0.005
Synchrotron X-rays (80 keV)	18	2.1	1.4
Synchrotron X-rays (300 keV)	36	11.5	10
Neutrons (cold)	97	8.3	10

structure factors $|F_{hkl}|$ which are proportional to the scattered intensities.

The factors can be expressed as:

$$F_{hkl} = \sum^{j} f_j \exp\{-2\pi i(hu_j + k\nu_j + lw_j)\} \quad [6]$$

where the f_j are the atomic scattering factors for atoms j and u_j, v_j and w_j are the fractional coordinates for atom j.

The electron density in the crystal unit cell can be determined from the structure factors $F_{hkl} = |F_{hkl}| \times e^{i\varphi hkl}$. The structure amplitudes ($|F_{hkl}|$) are obtained from the experiments and the phases φ can be obtained from a number of phasing procedures. The structure factors can then be converted to a mapping of the electron density distribution $\rho(xyz)$, which completely determines the crystal structure.

$$\rho(xyz) = 1/V_c \Sigma^{hkl} |F_{hkl}| e^{i\varphi hkl} \exp(-2\pi(hx + ky + lz)) \quad [7]$$

V_c is the volume of the unit cell and x,y,z are the positional coordinates in the unit cell.

The most precise structure determinations are performed using a single crystal sample where hundred to thousands of separate Bragg spots can be used to determine atomic positions to a precision of better than 0.001 Å. In materials science, however, quite frequently the material is in a polycrystalline form and powder diffraction methods have to be employed.

In the powder diffraction method the Bragg reflections from many thousands of microcrystals with different orientation overlap giving rise to 'powder' diffraction rings rather than distinct Bragg spots. Since the powder patterns are in the form of rings rather than distinct spots (see **Figure 4**) the observed intensities from planes with identical or closely similar scattering angles are analysed in terms of a deconvolution function:

$$Y_i = B_i + \Sigma^{hkl} I(hkl) Ng \quad [8]$$

where B_i is the background in point i, g is the powder peak shape function and N is a normalization factor. Once a structure model is obtained a refinement procedure proposed by Rietveld can be used to minimize the quantity

$$M = \Sigma^{i} W(Y_i(\mathrm{obs}) - Y_i(\mathrm{calc}))^2 \quad [9]$$

W is a weight factor based on counting statistics. Y(calc) may contain positional, thermal parameters, unit cells as well as peak shapes and these parameters are refined during the minimization.

The method is used for structure determinations and is well suited to handling multiphase systems. At present the method gives bond distances to within $\geq$0.005 Å precision, and *ab initio* structure determinations may handle systems with up to 200 parameters. **Figure 5** gives an example of a powder pattern collected at the European Synchrotron Radiation Facility of a two-phase powder sample.

The broadening of a diffraction line may give important information on changes occurring during the processing of the material. The line width is affected by instrumental resolution, source size, divergence of the radiation, particle size, microstrain components and stacking faults. After appropriate correction qualitative information on the micrograin size can be obtained by using the Scherrer formula. The grain size affects the line broadening as a function of wavelength and scattering angle as:

$$\Delta 2\theta = \lambda/(D \cos\theta) \quad [10]$$

where D is the particle size.

In order to separate particle broadening from strain effects two or more reflections may be used. The strain varies as function of tan θ and the different effects in line broadening may thus be separated by determining the broadening from different (hkl) diffraction lines.

Anomalous scattering

The absorption and scattering can change considerably around the absorption edges for the atoms. The absorption edge region is the atom specific energy region where the inner electrons are sufficiently excited by the X-rays to leave the atom or to be excited to an upper electronic shell. In this region the atomic scattering factor f_j is modified and is no longer a real property but has, in addition to an extra real term f'_j, an imaginary term f''_j.

The term in Equation [6] changes to:

$$f_j = f_j^0 + f'_j + i f''_j \quad [11]$$

These changes can be used to alter the absorption and consequently the penetration depth by changing the wavelength around the absorption edges. The changes may be considerable and can amount to the equivalence of tens of electrons.

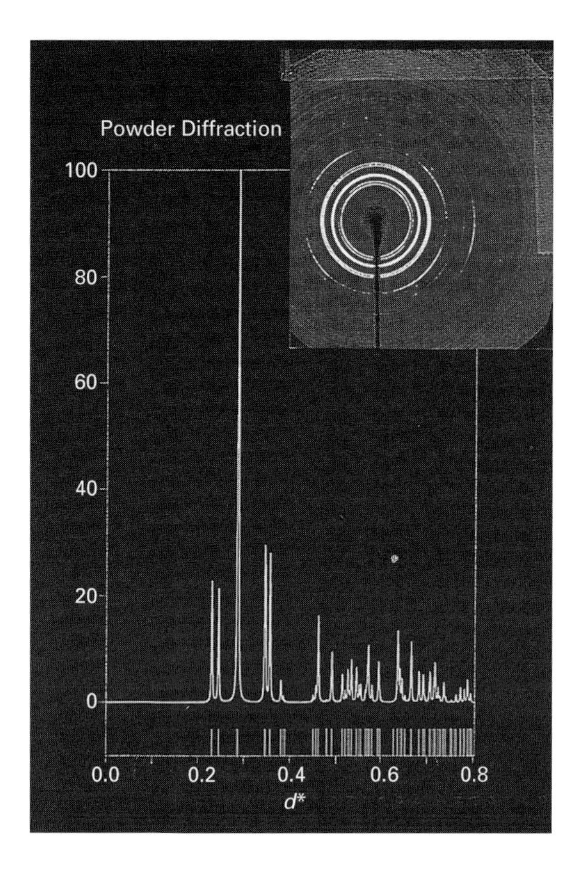

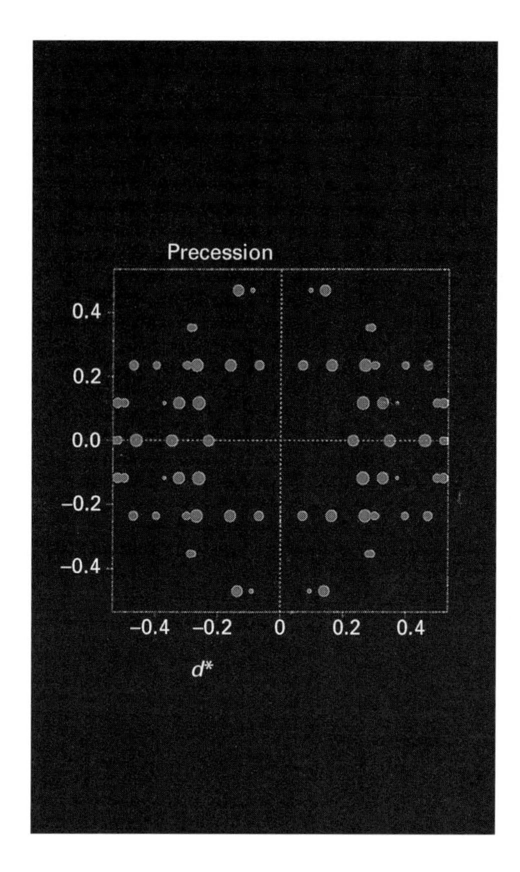

Figure 4 The left-hand picture illustrates a powder diffraction pattern of S_2N_2 at a wavelength of 0.325 Å. The insert is the actual pattern which can be integrated and displayed as a function of d^* [$d^* = (2 \sin \theta)/\lambda$]. The right-hand picture shows the result if one single S_2N_2 crystal is rotated in the X-ray beam. The size of the spots illustrates the difference in diffracted intensities from the separated Bragg reflections. Courtesy of Svensson and Kvick, 1998. (see Colour Plate 36).

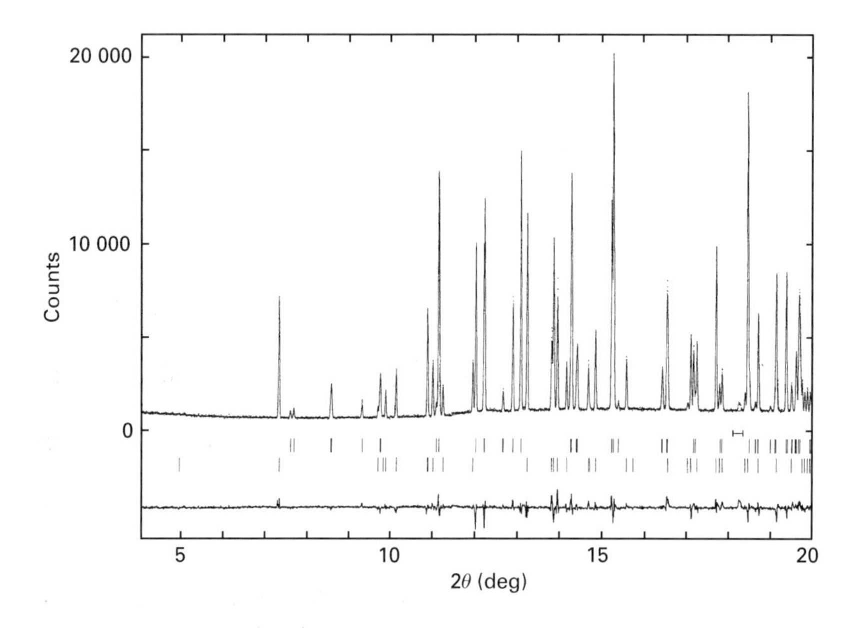

Figure 5 The powder pattern from a two-phase mixture of orthorhombic $(CH_3)_2SBr_{2.5}$ and monoclinic $(CH_3)_2SBr_4$ recorded at a wavelength of 0.94 718 Å at the BM16 beamline at the ESRF. The two lines of vertical bars show the location of diffraction lines for the two different phases. The bottom pattern shows the difference between the observed and refined intensities from the Rietveld refinement. Courtesy of Vaughan, Mora, Fitch, Gates and Muir, 1998.

In addition to changes in absorption there are fundamental differences in the scattering which may be used in materials science. Examples include the determination of absolute optical configuration in polar compounds.

Normally, the Friedel pairs of structure factors, i.e. $F(hkl) = F(-h-k-l)$, are identical. However, around the absorption edges this law breaks down since the imaginary term adds differently in the phase relation and the correct optical isomer can be determined.

The alternation of the scattering for a specific atom may also be used to resolve the partial contribution of that atom at a specific site by multiple wavelength differences studies.

The scattering contrast between almost isoelectronic elements such as Fe and Co may be enhanced sufficiently to allow precise determinations. Small chemical shifts in the energy of the edges depending on the valence state of the atom may also be used to differentiate between the valence states of the atoms in a compound.

The effect is particularly useful for structure determinations of macromolecules where multiple anomalous diffraction (MAD) experiments are combined to yield the phase factors necessary to obtain the electron density mapping.

The use of anomalous diffraction has gained in importance as synchrotron radiation sources have become available. The synchrotron sources provide easily tuneable radiation covering most of the atomic absorption edges.

Magnetic scattering

During the last few years a minute scattering component coming from the magnetic moments of atoms has been exploited. The scattering from the magnetic moments are smaller than the scattering from the electrons by orders of magnitude. However, successful studies of magnetic structures and even magnetic atomic overlayers are now possible by using the high brightness synchrotron sources.

The synchrotron radiation also makes it possible to tune the wavelength to the absorption edges of the magnetic atoms where the magnetic scattering is strongly enhanced. For a more detailed account of anomalous diffraction and magnetism the reader is referred to the book edited by Materlik and co-workers (1994).

Stress and strain relationships

Elastic X-ray stress analysis is based on recordings of the interplanar distances d_{hkl}. When stress is applied to a crystal the distances change from d_0 to $d_0 + \Delta d$ and the scattering in Bragg's law (Eqn [1]) will change accordingly. Using Equation [2] one obtains the relationship:

$$\Delta\theta = (\theta - \theta_0) = -\Delta d/d_0 \tan(\theta_0) = -\tan(\theta_0)\varepsilon \quad [12]$$

where θ_0 is the scattering angle from a strain free state, and ε is the elastic strain coefficient. θ is obtained from the diffraction experiment.

In strain analysis we can use the formalism given by Noyen and Cohen (1976) where the strain in the direction of a scattering plane (hkl) is measured as:

$$\varepsilon'(33)_{\phi\varphi} = (d_{\phi\varphi} - d_0)/d_0 \quad [13]$$

The plane spacing d_0 is measured from a strain free sample.

To obtain the strain in the sample system S one transforms the values from the laboratory coordinate system L.

The orthogonal coordinate systems S_1,S_2,S_3 and L_1,L_2,L_3, are defined as follows. S_3 is perpendicular to the sample surface and S_2 and S_3 are parallel to the sample surface. L_3 is the normal to the scattering plane hkl and makes the angle φ with S_3. The ϕ angle is the angle between the projection of L_3 onto the sample surface and vector S_1. The measured quantity $\varepsilon_{\phi\varphi}$ can then be converted to the sample coordinate system by the transformation:

$$\begin{aligned}\varepsilon'(33)_{\phi\varphi} &= \varepsilon_{11}\cos^2\phi\sin^2\varphi + \varepsilon_{12}\sin 2\phi\sin^2\varphi \\ &+ \varepsilon_{22}\sin^2\phi\sin^2\varphi\,\varepsilon_{33}\cos^2\varphi + \varepsilon_{13}\cos\phi\sin^2\varphi \\ &+ \varepsilon_{23}\sin\phi\sin^2\varphi \quad [14]\end{aligned}$$

This important equation is linear in ε_{11}, ε_{12}, ε_{22}, ε_{33}, ε_{13}, ε_{23} and these coefficients can thus be determined from six measurements at different angular values. When this is known the stress can be determined.

The stress (σ) and strains (ε) in the sample coordinate system are determined using Hooke's law:

$$\begin{aligned}\sigma_{ij} &= c_{ijkl}\,\varepsilon_{kl} \\ \varepsilon_{ij} &= s_{ijkl}\,\sigma_{kl} \quad [15]\end{aligned}$$

where the c_{ijkl} are the elastic stiffness constants and s_{ijkl} are the elastic compliances.

In the elastic case the observations may be expressed in terms of Young's modulus E and Poisson's ratio ν:

$$(d_{\phi\varphi} - d_0)/d_0 = (1 + \nu/E)\sigma_{ij} - (\delta_{ij}\nu/E)\sigma_{kk} \quad [16]$$

where δ_{ij} is the Kronecker delta.

This formalism is the basis for the commonly used $\sin^2\varphi$ method, which is usually limited to the surface region of the sample.

The availability of more elaborate diffractometers and the highly penetrating synchrotron radiation also gives possibilities of deeper penetration into the sample by rotation of the sample around the scattering vector direction, L_3, and by wavelength tuning or change in the scattering angle. Transmission studies permit the study of stresses in specific 'gauge' volumes in the bulk.

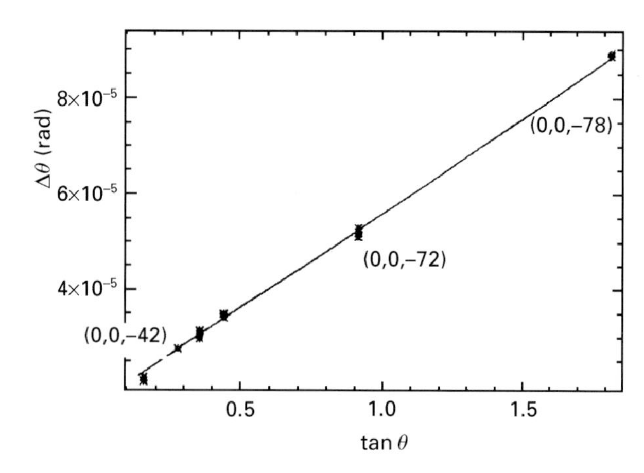

Figure 6 The shifts in $\Delta\theta$ for $(00l)$ reflections from a $LiNbO_3$ single crystal of 0.2 mm thickness subjected to an electric field of 50 kV cm^{-1} along the crystallographic c direction as measured using synchrotron radiation of a wavelength of 0.307 Å at the beamline 1D11 at the ESRF. The $d(33)$ element of the piezoelectric tensor can be evaluated to be $7.5(2) \times 10^{-12}$ CN^{-1}. Courtesy of Heunen, Graafsma, Kvick, 1997.

Externally perturbed systems

Deformations due to perturbations may also be followed by X-ray diffraction.

These perturbations may be caused by electric field or changes in temperature and thus piezoelectric effects and thermal expansion coefficients may readily be determined by using Equation [2].

In the case of changes occurring from the application of an external electric field, the converse piezoelectric effect, the induced strain coefficient ε_{ij} is related to the field by equation:

$$\varepsilon_{ij} = \sum_{k=1}^{3} d_{kij} E_k \quad [17]$$

where E_k is kth component of the electric field and d_{kij} is the kijth element of the third rank piezoelectric tensor.

Using Equation [2] Barsch has shown that the diffraction observable $\Delta\theta_r$ may be used to determine the piezoelectric tensor according to the formula (Coppens, 1992):

$$\Delta\theta_r = -E \tan\theta_r \sum_{k=1}^{3}\sum_{i=1}^{3}\sum_{j=1}^{3} e_k\, h_{r,i}\, h_{r,j}\, d_{kij} \quad [18]$$

where r refers to certain reflections (hkl) and e_k, $h_{r,i}$ and $h_{r,j}$ are, respectively, the components of unit vectors parallel to the electric field and the scattering vector for (hkl).

Figure 6 gives an illustration where the piezoelectric tensor element d_{33} has been determined from measurements of a series of $00l$ reflections with the electric field aligned along the $(00l)$ direction.

Ion-implantation effects

Ion implantation in layers of semiconductor material is an important process where Equation [2] may be very useful for assessing the effect on the implantation.

If dopants are introduced by ion implantation, for instance in silicon, the tetrahedral radius of the dopant (r_d) differs from that of the substrate and changes in the θ angle will be observed.

For cubic silicon material the corresponding change in the unit cell dimension can give information on the doping level according to Vegard's law:

$$\Delta a/a = (1 - r_d/r_s)KC_d N^{-1} \quad [19]$$

where r_d and r_s are the tetrahedral radii of the dopant and substrate respectively, C_d is the concentration of dopant and N is the number of substrate of atoms per unit volume.

The factor K takes account of the fact that only the lattice strain normal to the wafer is nonzero, whereas the in-plane strain is zero. For typical substrates with cubic structure the K values can be evaluated from the elastic stiffness constants.

Superlattice structures

Physical properties of materials may be changed by creating 'artificial' structural periodicity by depositing alternate thin layers of different materials. Structures of this type are commonly used in optics, electrooptical applications and coatings for corrosion protection or thermal barriers.

The production and performance of these structures can be characterized by X-ray diffraction. The diffraction patterns from the compound structures are characterized by main diffraction peaks interspersed by satellite peaks. The periodicity of superlattices, Λ, is given by the relationship:

$$\Lambda = (j - k)\ \lambda/2(\sin\theta_j - \sin\theta_k) \qquad [20]$$

where j and k represent the satellite order number and θ_j and θ_k are the observed scattering angles at wavelength λ. The periods may vary depending on the production and the variation can be monitored by the various Λ values obtained from different satellite pairs. These period variation may be interpreted as surface roughness.

Topography

Diffraction topography is an imaging technique based on Bragg's law (Eqn [1]) and ranges from a rather qualitative inspection method of the scattering power over a crystal to a much more complicated method employing dynamical scattering theory to elucidate microscopic strains in the crystal. Many phenomena of importance to a materials scientist, such as stacking faults, growth bands, strain around grain boundaries, twins, or even dynamical phenomena such as acoustic waves or magnetic domain formation, have been studied by this method.

The studies can be divided into two main areas:

(a) orientation contrast;
(b) extinction contrast.

The former studies (a) map the variation of the scattering power across the X-ray beam and detects misalignment of certain portions of the crystal where the misalignment is larger than the divergence of the monochromatic incident beam. The misalignment can be due to rotations or dilations of the lattice. When monochromatic radiation is used these effects are seen as distinct bands or patterns of *loss* of scattered radiation.

A richer pattern of intensity variations is seen if continuous wavelength radiation is used. In this case, due to divergence or convergence of the diffracted beams at the boundaries, *losses and gains* in intensity are observed.

The second phenomenon (b) is observed when the scattering power in the crystal is alternated by strain

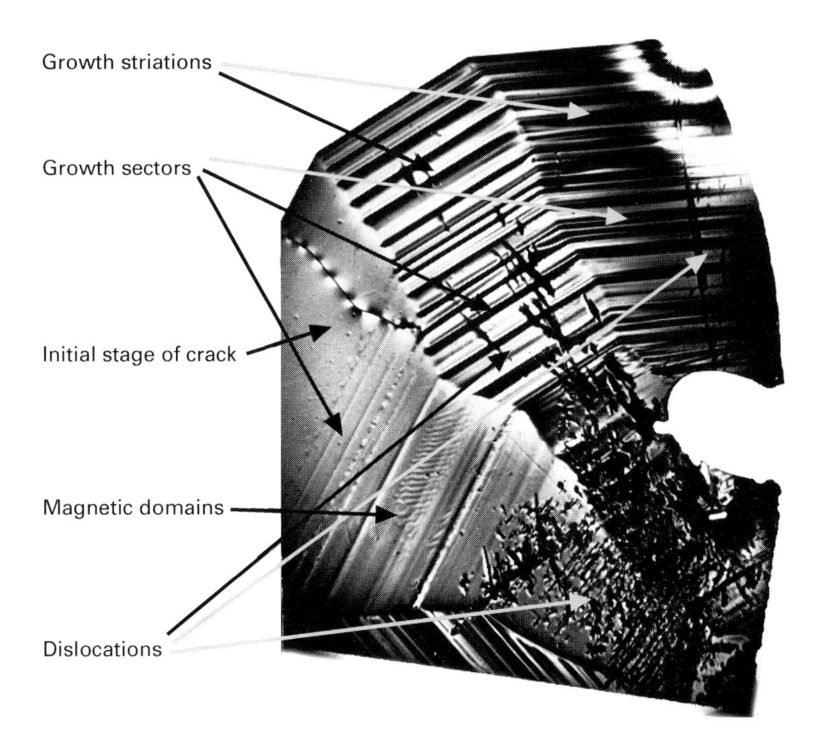

Figure 7 Transmission topograph of a flux grown Ga-YIG ($Y_3Fe_{5-x}Ga_xO_{12}$, $x \approx 1$) crystal plate, Mo $K_{\alpha 1}$-radiation (λ = 0.709 Å). Courtesy of J Baruchel, 1998.

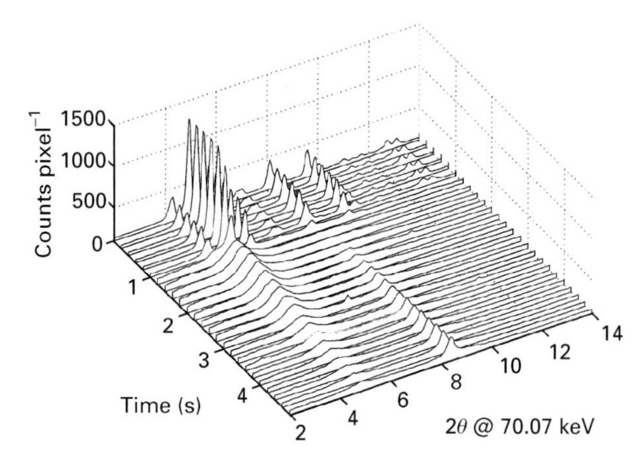

Figure 8 The time-evolution of the powder pattern of an exothermic reaction of a mixture of Al and Ni powders. The powder patterns were recorded at the 1D11 beam line at the ESRF every 250 ms during the self-propagating high-temperature synthesis. The wavelength 0.177 Å was chosen to allow penetration of the sample. It can be noticed that the original peak from Al and Ni disappear or change during the reaction and the formation of new phases and the nucleation can be followed. Courtesy of Kvick, Vaughan, Turillas, Rodriguez, Garcia, 1998.

fields around defects without major realignment in the crystal.

The observed contrasts are rather complicated to quantify and an understanding of the effects of the X-ray wavefields is necessary to interpret the observations in detail. For a detailed account of the theory the reader is referred to Tanner (1976).

Topography uses two different methods of observation; reflection topography, which maps the surface region of the samples, and transmission topography, which also samples the bulk of the material. In the latter case the radiation must be chosen so that the absorption is small enough to allow bulk penetration.

Different volumes in the crystal may be sampled if the collimation of the incident and diffracted beams is carried out carefully. This method is called section topography.

Figure 7 gives an example of the information one may obtain from a transmission topograph.

Time-resolved studies

Time-resolved X-ray diffraction has been used for a long time to study solid-state reactions. With the emergence of the new radiation sources it is now possible to follow solid-state reactions, phase transitions and physical changes caused by perturbations on much shorter time-scales.

Topography has already proved to be a suitable technique for studying magnetic domain formation or acoustic deformation down to time-scales of milliseconds.

The use of well collimated and high intensity synchrotron radiation beams is essential to reach the necessary time intervals without losing the statistical significance in the observed diffracted intensities. The white beam Laue technique has already been proven to facilitate studies down to the picosecond time regime for studies such as recombination of CO in myoglobin after flash photolysis. Nanosecond resolution has been obtained in a study of laser-annealing of defects in a silicon crystal.

The dynamics of reactions are either reversible or irreversible. Sufficient counting statistics can be obtained in the reversible processes through stroboscopic measurement where repeated measurements using short radiation time slices are performed. In the irreversible experiments the processes have to be followed by rapid consecutive exposures.

In many of the cases the reactions are often destructive to large single crystals and the powder method has to be employed. However, it has been proven that reactions such as polymerization and exothermic solid-state reactions may be followed down to the millisecond time-regime even when the reactions are irreversible.

Figure 8 exemplifies the time-evolution of an exothermic reaction between Al and Ni powders. In this time-resolved powder experiment one can follow the initial melting of Al and the formation of intermetallic phases as well as the crystallization process.

The rapid development of fast read-out CCD cameras in combination with the high brightness synchrotron sources promises to give X-ray diffraction a prominent role in the studies of dynamics of processes relevant to materials scientists.

List of symbols

A = atomic weight; B_i = background in point i; C_d = concentration of dopant; C, D, $\sigma_{N\text{-}K}$, N = constants; C_{ijkl} = elastic stiffness constants; d = atomic spacing; D = particle size; E = Young's modulus; E_k = kth component of electric field; f_j = atomic scattering factors; F_{hkl} = amplitude structure factors; g = powder peak shape function; I = intensity of beam; I_0 = incident intensity of beam; j,k = satellite order number; K = constant; N = normalization factor; N = number of atoms; R = tetrahedral radius; s_{ijkl} = elastic compliances; t = thickness of sample; u_j, v_j, w_j = fractional coordinates for atom j; ν = Poisson's ratio; V_c = volume of unit cell; W = weight factor; Y_i = deconvolution function; Z = atomic number; δ_{ij} = Kronecker delta; $\Delta\theta$ = change of angle of scattering; ε = elastic strain coefficient; θ = angle of scattering; λ = wavelength; Λ = periodicity of superlattices; μ = absorption coefficient; ρ = density; $\rho(xyz)$ = electron density distribution; σ_n = absorption cross-section for component n; σ = stress; τ = penetration distance; φ = angle; ϕ = phase.

See also: **Fibres and Films Studied Using X-Ray Diffraction; Powder X-Ray Diffraction, Applications; Scattering and Particle Sizing, Applications; Scattering Theory.**

Further reading

Authier A, Lagomarsino S and Tanner B (1996) *X-ray and Neutron Dynamical Diffraction*. New York and London: Plenum Press.

Coppens P (1992) *Synchrotron Radiation Crystallography*. London, San Diego: Academic Press.

International Tables for Crystallography, Vol I–III (1989) Dordrecht, Boston, London: Kluwer Academic Publishers.

International Tables for Crystallography, Vol A–C (1992) Dordrecht, Boston, London: Kluwer Academic Publishers.

Materlik G, Sparks CJ and Fischer K (1994) Amsterdam, London, New York, Tokyo: North-Holland.

Noyan IC and Cohen JB (1987) *Residual Stress*. New York, Berlin, Heidelberg, London, Paris, Tokyo: Springer-Verlag.

Tanner BK (1976). *X-ray Diffraction Topography*. Oxford: Pergamon Press.

Matrix Isolation Studies By IR and Raman Spectroscopies

Lester Andrews, University of Virginia, Charlottesville, VA, USA

VIBRATIONAL, ROTATIONAL & RAMAN SPECTROSCOPIES

Applications

Matrix isolation is a technique for maintaining molecules in an inert medium at very low temperature for spectroscopic study. This method is particularly well suited for preserving reactive species in a solid environment. Elusive molecular fragments, such as free radicals that may be important intermediates for chemical transformations used in industrial reactions, molecules that are in equilibrium with solids at very high temperatures, weak molecular complexes that may be stable at low temperatures, and molecular ions that are produced in plasma discharges or by high-energy radiation can all be observed and characterized using infrared absorption and laser-excitation spectroscopies. The matrix isolation technique enables spectroscopic data to be obtained for reactive molecular fragments, many of which cannot be studied in the gas phase.

Experimental apparatus

The experimental apparatus for matrix isolation experiments is designed with the methods of generating the molecular transient and performing the spectroscopy in mind. **Figure 1** is a schematic diagram of the laser-ablation matrix isolation apparatus for infrared absorption spectroscopy. Caesium iodide windows are typically employed. The rotatable cold window is cooled to 4–20 K by closed-cycle refrigeration or liquid helium. The matrix sample is introduced through the spray-on line at rates of 1–5 millimoles per hour; argon is the most widely used matrix gas, although neon, krypton, xenon and nitrogen are also used. In **Figure 1**, the Nd–YAG fundamental at 1064 nm in 5–50 mJ pulses of 10 ns duration is focused (spot size approximately 0.1 mm) onto a rotating metal target. Ablated metal atoms intersect the gas sample during co-deposition on the cold CsI window where collisions and reactions occur. The reactive species can be generated in a number of other ways: mercury arc photolysis of a trapped precursor molecule through the quartz window, evaporation from a Knudsen cell heated inside the chamber, chemical reaction of atoms evaporated from the Knudsen cell with molecules deposited through the spray-on line, and vacuum-ultraviolet photolysis of molecules deposited from the spray-on line by radiation from discharge-excited atoms. For laser excitation studies, the sample is deposited on a tilted copper wedge which is grazed by the laser beam, and light emitted or scattered at approximately 90° is examined by a spectrograph.

Free radicals, anions and cations

The first free radical stabilized in sufficient concentration for matrix infrared detection was formyl (HCO). Hydrogen iodide (HI) was deposited in a

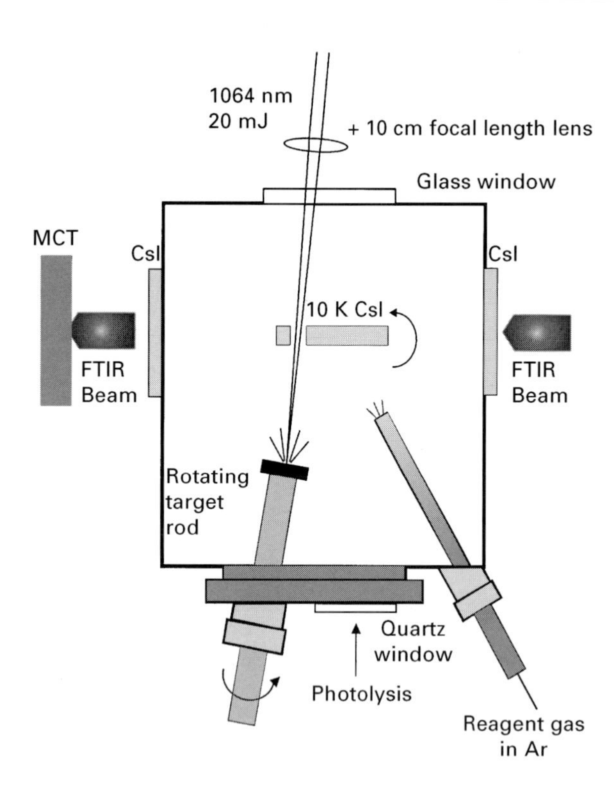

Figure 1 Vacuum chamber for infrared matrix isolation studies using laser ablation.

carbon monoxide (CO) matrix and photolysed with a mercury arc; hydrogen atoms produced by dissociation of HI reacted with CO in the cold solid to produce HCO. The infrared spectrum of HCO provided vibrational fundamentals and information about the chemical bonding in this reactive species. Free radicals have been produced in matrices using a variety of techniques such as ultraviolet photolysis (Eqn [1]); vacuum ultraviolet photolysis (Eqn [2]); lithium atom (Eqn [3]); hydrogen atom (Eqn [4]); and lithium atom (and electron) (reaction [5]).

$$OF_2 + \text{ultraviolet light} \rightarrow OF^{\bullet} + F^{\bullet} \quad [1]$$

$$CHF_3 + \text{vacuum ultraviolet light} \rightarrow CF_3{}^{\bullet} + H^{\bullet} \quad [2]$$

$$CCl_4 + Li^{\bullet} \rightarrow CCl_3{}^{\bullet} + LiCl \quad [3]$$

$$H^{\bullet} + O_2 \rightarrow HO_2{}^{\bullet} \quad [4]$$

$$Li^{\bullet} + O_2 \rightarrow (Li^+)(O_2{}^-) \quad [5]$$

The first molecular ionic species characterized in matrices, $Li^+O_2^-$, was formed by the co-condensation reaction of lithium (Li) atoms and oxygen (O_2) molecules at high dilution in argon. The infrared spectrum exhibited a weak $(O \leftrightarrow O)^-$ stretching vibration and two strong $Li^+ \leftrightarrow O_2^-$ stretching vibrations, as shown at the top of **Figure 2** for the ^7Li and $^{16,18}O_2$ isotopic reaction. The 1–2–1 relative-intensity oxygen isotopic triplets in this experiment showed that the oxygen atomic positions in the molecule are equivalent and indicated an isosceles triangular structure. The ionic model for the bonding in $Li^+O_2^-$ was confirmed by contrasting intensities between the laser-Raman and infrared absorption spectra shown in **Figure 2**. Infrared intensities depend upon a change in molecular dipole moment and Raman intensities depend on a change in molecular polarizability during the vibration. Hence, modes between ions ($Li^+ \leftrightarrow O_2^-$) will be strong in the infrared and modes within an ion ($O \leftrightarrow C^-$) will be strong in the Raman spectrum. **Table 1** lists the observed frequencies. The contrasting frequencies when Li^+ is replaced by Cs^+ verify the ionic model for the bonding in the $(M^+)(O_2^-)$ molecules.

High-temperature molecules like lithium fluoride (LiF) can be trapped in matrices by evaporating the molecule from the crystalline solid in a Knudsen cell at high temperature or by reacting lithium atoms with fluorine molecules during condensation. The latter method has been used to synthesize the calcium oxide (CaO) molecule from the calcium atom–ozone reaction, which gives the calcium ozonide ion pair $Ca^+O_3^-$ and CaO diatomic products.

Continuous exposure of a condensing sample to argon-resonance radiation during sample condensation has been used to produce molecular cations for spectroscopic study. In the case of fluoroform

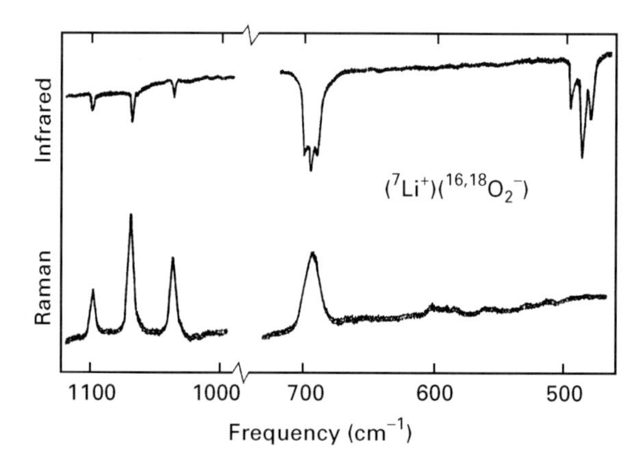

Figure 2 Infrared and Raman spectra of lithium superoxide, $Li^+O_2^-$, using lithium-7 and 30% $^{18}O_2$, 50% $^{16}O^{18}O$, and 20% $^{16}O_2$ at concentrations of $Ar/O_2 = 100$. The Raman spectrum was recorded using 200 mW of 488-nm excitation at the sample.

Table 1 Fundamental frequencies (cm^{-1}) assigned to the ν_1 intraionic and ν_2 and ν_3 interionic modes of the C_{2v} alkali metal superoxide molecules in solid argon

Molecule	ν_1	ν_2	ν_3
$^{6}LiO_2$	1097.4	743.8	507.3
$^{7}LiO_2$	1096.9	698.8	492.4
NaO_2	1094	390.7	332.8
KO_2	1108	307.5	–
RbO_2	1111.3	255.0	292.5
CsO_2	1115.6	236.5	268.6

(CHF_3), photolysis produced the $CF_3^{\bullet}$ radical, which may be photoionized by a second 11.6-eV photon to give CF_3^+. The infrared spectrum of CF_3^+ revealed a very high C–F vibrational fundamental, which indicates substantial pi bonding in the planar carbocation.

Complexes

Molecular complexes involving hydrogen fluoride (HF) serve as useful prototypes for the understanding of the important phenomenon of hydrogen bonding. An interesting chemical case is ammonia (NH_3) and HF, which on the macroscopic scale produce the salt ammonium fluoride but on the microscopic scale give the NH_3–HF complex. The infrared spectrum of this complex reveals vibrations for NH_3 and HF perturbed by their association in the complex. Fourier-transformed infrared spectroscopy is particularly advantageous in these studies because of the high vibrational frequencies for HF species and low sample transmission in this region. The ammonia symmetric bending motion is considerably blueshifted, and the hydrogen fluoride stretching fundamental is markedly redshifted. These shifts attest to a strong intermolecular interaction within the complex.

$$NH_3 + HF \rightarrow H_3N\text{- -}HF \quad [6]$$

Ozone (O_3) is a very important molecule in the upper atmosphere, as it absorbs harmful ultraviolet radiation. In the laboratory, this ultraviolet dissociation of ozone provides oxygen atoms for chemical reactions, but in complexes red light can induce an O-atom transfer. Photochemical reactions of elemental phosphorus (P_4), an extremely reactive molecule well suited for matrix isolation studies, and ozone produce a new low oxide of phosphorus, P_4O, which is involved as a reactive intermediate in the striking of a match. The infrared spectrum of P_4O shows a strong terminal PO bond stretching fundamental and characterizes the P_4O structure as tetrahedral P_4 with an O atom cap. Further photochemical reactions of phosphorus and oxygen produce and trap the reactive molecule $^{\bullet}PO_2$, isoelectronic with the pollutant molecule $^{\bullet}NO_2$, which can also be formed by laser evaporation of red phosphorus followed by reaction with a stream of oxygen gas.

$$P_4 + O_3 \rightarrow (P_4)(O_3)\,\text{complex} \xrightarrow{h\nu} P_4O + O_2 \quad [7]$$

High temperature molecules

Titanium dioxide is a white solid used as a pigment in paints, but at 2200–2400 K some TiO and OTiO molecules exist in equilibrium with the solid. The OTiO molecule is bent (113±3°) with two strong double bonds. The TiO and OTiO molecules have been prepared by reacting laser-ablated Ti atoms with O_2 and trapped in solid argon for infrared spectroscopic analysis. **Figure 3** shows the spectrum of a sample prepared by reacting laser-ablated Ti with O_2 (0.5%) in excess argon followed by condensation at 7 K. The weak 1039.5 cm^{-1} band is due to ozone which is made by combination of O_2 and O atoms produced in the ablation process. The weak 953.7 cm^{-1} band is due to O_4^- formed by combination of O_2 and O_2^-, the latter from capture of electrons by dioxygen. The strongest bands at 946.7 and 917.0 (TiO_2) and two weaker bands at 1012.8 (TiO^+) and 987.8 cm^{-1} (TiO) are important products. The reaction was repeated for $^{18}O_2$ and mixtures of $^{16}O_2$, $^{16}O^{18}O$, $^{18}O_2$. **Table 2** lists the isotopic for frequencies for these molecules. The 16/18 isotopic ratio is characteristic of the normal vibrational mode. Note that the 917.0 cm^{-1} band is strong enough to exhibit satellite adsorptions at 923.1, 919.9, 914.1 and 911.3 cm^{-1} that are due to the minor ^{46}Ti, ^{47}Ti, ^{49}Ti and ^{50}Ti isotopes in natural abundance around the major $^{48}TiO_2$ isotopic band.

The 946.7 and 917.0 cm^{-1} bands form triplet patterns in the statistical mixed isotopic experiment, which verifies the participation of two equivalent oxygen atoms, whereas the 1012.9 and 987.9 cm^{-1} bands form doublets, which shows that a single oxygen atom is involved.

The 946.7 and 917.0 cm^{-1} bands are due to the symmetric (ν_1) and antisymmetric (ν_3) Ti–O stretching fundamentals of OTiO. Titanium isotopic substitution provides basis for calculation of a lower limit (111±3°) to the OTiO valence angle whereas oxygen isotopic replacement gives a 115±3° upper limit to this angle owing to different anharmonicities. The true angle (113±3°) is the median of these limits.

The 16/18 oxygen isotopic ratios for the 987.8 and 1012.9 cm^{-1} bands are different from values for the

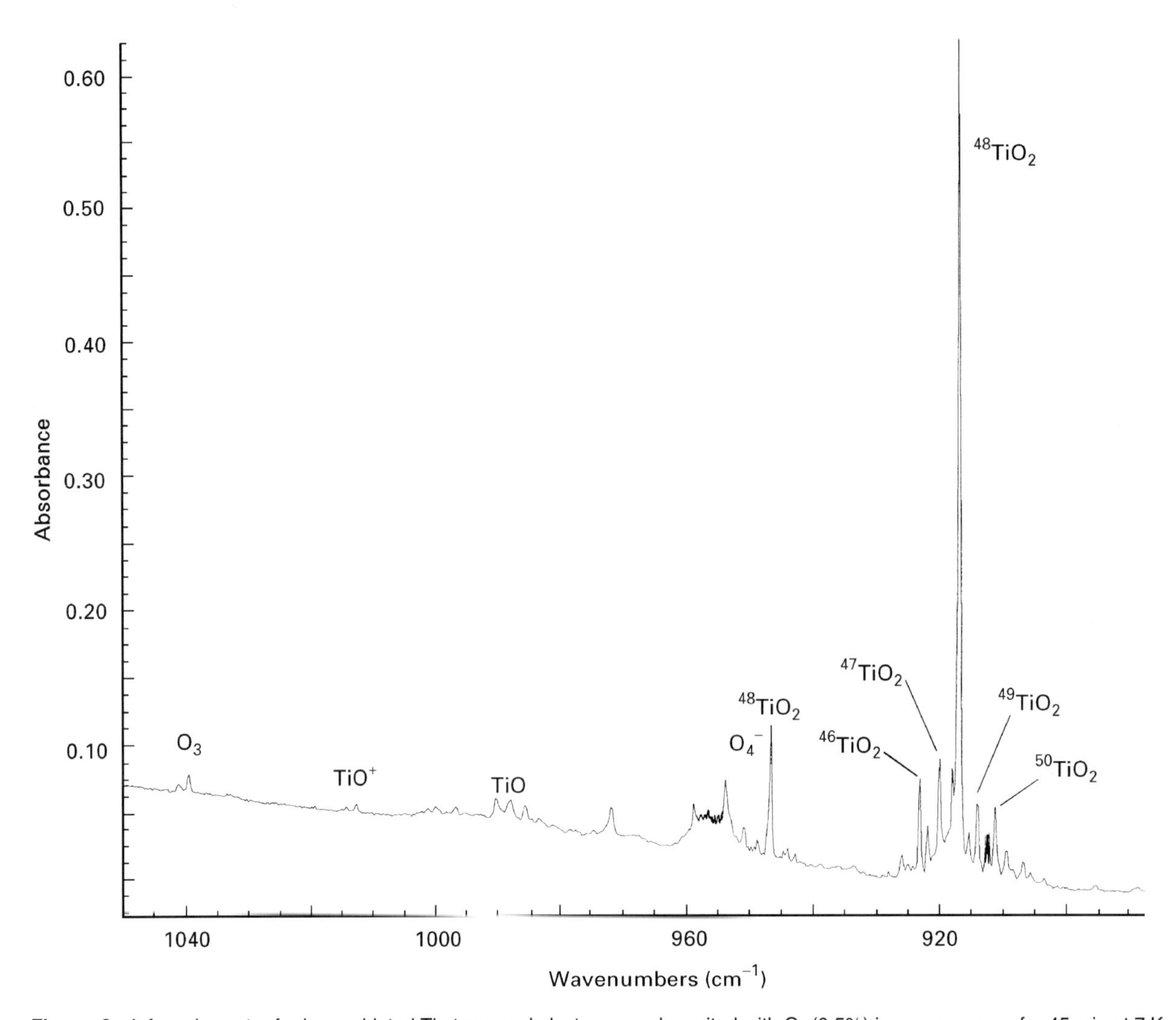

Figure 3 Infrared spectra for laser-ablated Ti atoms and electrons co-deposited with O_2 (0.5%) in excess argon for 45 min at 7 K.

above frequencies. These ratios (1.0442 and 1.0443) approach the harmonic value (1.0446) for diatomic TiO. The 987.8 cm^{-1} band is due to TiO in solid argon, which is redshifted from the 1000.0 cm^{-1} gas phase value, and the 1012.9 cm^{-1} band is due to TiO^+, which has not yet been observed by gas phase optical spectroscopy.

The use of pulsed-laser ablation to produce new chemical species for spectroscopic study is further

Table 2 Major infrared absorptions (cm^{-1}) observed for laser-ablated titanium and dioxygen reaction products isolated in solid argon

$^{16}O_2$	$^{18}O_2$	*Ratio* $^{16}O/^{18}O$	Identification
1039.6	982.3	1.0583	$O_3(\nu_3)$
1012.9	969.9	1.0443	TiO^+
987.8	946.0	1.0442	TiO
953.7	901.6	1.0578	O_4^-
946.9	904.5	1.0469	$^{48}TiO_2(\nu_1)$
917.1	881.7	1.0401	$^{48}TiO_2(\nu_3)$

illustrated for the vanadium atom reaction with N_2. Laser-ablated metal atoms are sufficiently energetic to dissociate molecular N_2 into N atoms for reaction to form metal nitrides. A sample of 2% $^{14}N_2$ + 2% $^{15}N_2$ in argon was reacted with laser-ablated V atoms, and spectra from this experiment are shown in **Figure 4**. Four very weak bands (A = absorbance = 0.002 to 0.001) were observed at 1026.2, 1010.3, 1014.4 and 1010.6 cm^{-1} from nitrogen-14 with an identical set at 999.4, 993.6, 987.9 and 984.2 cm^{-1} from nitrogen-15 on sample co-deposition at 10 K [**Figure 4A**]. Annealing to 25 and 30 K to allow diffusion and further reaction of trapped species slightly decreased the first band and increased the second, third and fourth bands in each set, and produced new weak bands at 997.8 and 971.4 cm^{-1} [**Figure 4B, C**]. Broadband photolysis decreased the lowest two bands and increased the second band in each set [**Figure 4D**]. Further annealing to 40 K decreased the first three bands, increased a

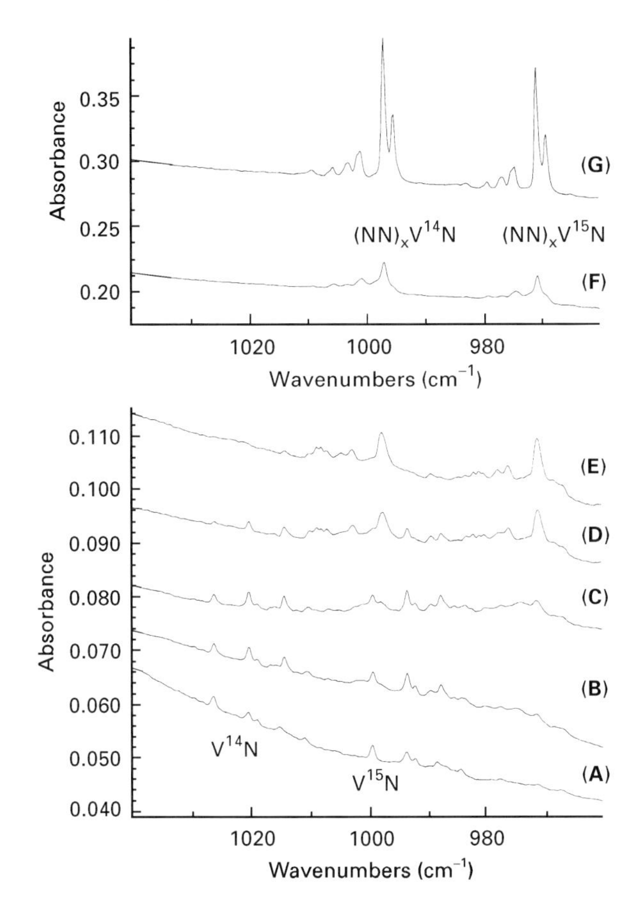

Figure 4 Infrared spectra in the 1040–960 cm^{-1} region for laser-ablated V atoms co-deposited with nitrogen. (A) 2% $^{14}N_2$ + 2% $^{15}N_2$ in argon co-deposited at 7 K for 1 h, (B) after annealing to 25 K, (C) after annealing to 30 K, (D) after annealing to 40 K, (E) after annealing to 43 K, (F) pure $^{14}N_2$ + $^{15}N_2$ co-deposited at 10–11 K for 1 h, and (G) after annealing to 30 K.

1002.8 cm^{-1} band (and 976.5 cm^{-1} counterpart) and increased the final 997.8 and 971.4 cm^{-1} bands [**Figure 4E**]. A final annealing to 43 K destroyed the first two bands and increased the last two bands in each set [**Figure 4F**]. One experiment was done with a pure dinitrogen $^{14}N_2$ + $^{15}N_2$ mixture and the spectra are shown at the top of **Figure 4** for deposition at 10 K and annealing to 30 K; note growth of the strong 997.0 and 970.6 cm^{-1} bands and satellites.

The VN example illustrates how the isolated VN molecule becomes successively ligated by dinitrogen to form $(NN)_xVN$ on annealing in the solid argon matrix or on deposition in pure dinitrogen matrix. The sharp weak new band at 1026.2 cm^{-1} in solid argon decreases on stepwise annealing while sharp bands increase at 1020.3, 1014.4 and ultimately 997.8 cm^{-1}. These bands exhibit sharp nitrogen-15 counterparts at 999.4, 993.6, 987.9 and 971.4 cm^{-1}, which define nitrogen 14/15 frequency ratios 1.0268, 1.0269, 1.0268 and 1.0272, respectively. These values are in excellent agreement with the harmonic VN diatomic ratio (1.0272) and attest to the pure V–N stretching character of these adsorptions. These bands showed no intermediate components with discharged (statistical) mixed isotopic dinitrogen, so a single nitrogen atom is involved in these vibrations. In solid dinitrogen, a sharp isotopic doublet at 997.0 and 970.6 cm^{-1} also increased on annealing and revealed the diatomic ratio (1.0272) and is only 0.8 cm^{-1} lower than the dominant nitrogen isotopic doublet surviving past 40 K annealing in solid argon (**Figure 4**).

The sharp 1026.2 cm^{-1} band is due to the VN diatomic molecule isolated in argon. The evolution of bands in **Figure 4** from 1026.2 cm^{-1} to 997.8 cm^{-1} on annealing in solid argon and the 997.0 cm^{-1} band in solid nitrogen provides a convincing picture for the attachment of dinitrogen ligands to the VN center. We note the appearance of five distinct bands for different ligated species. These are consistent with the eighteen electron rule in that these bands are due to $(NN)_xVN$ with $x = 1, 2, 3, 4, 5$ where the maximum ligated species has eighteen electrons in the valence shell about vanadium.

See also: **IR and Raman Spectroscopy of Inorganic, Coordination and Organometallic Compounds; IR Spectrometers; IR Spectroscopy, Theory; Nonlinear Optical Properties; Raman Spectrometers.**

Further reading

Andrews L (1984) Fourier transform infrared spectra of HF complexes in solid argon. *Journal of Physical Chemistry* **88**: 2940–2949.

Andrews L and Moskovits M (eds) (1989) *Chemistry and Physics of Matrix Isolated Species*. Amsterdam: Elsevier.

Andrews L and Smardzewski RR (1973) Argon matrix Raman spectrum of LiO_2. *Journal of Chemical Physics* **58**: 2258–2261.

Andrews L, Bare WD and Chertihin GV (1997) Reactions of laser ablated V, Cr and Mn atoms with nitrogen atoms and molecules. *Journal of Physical Chemistry A* **101**: 8417–8427.

Bondybey VE, Smith AM and Agreiter J (1996) New developments in matrix isolation spectroscopy. *Chemical Reviews* **96**: 2113–2134.

Chertihin GV and Andrews L (1995) Reactions of laser ablated Ti, Zr and Hf atoms with O_2 molecules in condensing argon. *Journal of Physical Chemistry* **99**: 6356–6366.

Zhou MF and Andrews L (1998) Matrix infrared spectra and density functional calculations of $Ni(CO)_x^-$, $x = 1, 2, 3$. *Journal of the American Chemical Society* **120**: 11499–11503.

Zhou MF and Andrews L (1999) Infrared spectra of the $C_2O_4^+$ cation and $C_2O_4^-$ anion in solid neon. *Journal of Chemical Physics* **110**: 6820–6826.

Medical Applications of Mass Spectrometry

Orval A Mamer, McGill University, Montréal, Québec, Canada

MASS SPECTROMETRY
Applications

Mass spectrometry has a breadth of application in medicine that is unequalled by any other spectroscopic technique. This is largely by virtue of having the uncommon ability to combine great sensitivity and specificity with near-perfect generality. Among the most common applications of mass spectrometry in medicine are the diagnosis and confirmation of known acquired and inherited metabolic disorders, characterization and investigation of those previously unknown, and the identification of intoxicants, whether inadvertently or deliberately administered. While this article will focus on the area of metabolic disease, the techniques described in brief here can be, and are, applied in a myriad other routine clinical situations. Some of these are measurement of serum homocysteine and methylmalonic acid as an indicator of increased risk for cardiovascular disease, and diagnosis of peptic ulcer, gastritis and gastric cancer caused by *Helicobacter pylori* by measurement of ^{13}C-labelled carbon dioxide released by *H. pylori* from an oral dose of ^{13}C-labelled urea. Mass spectrometry is the 'gold standard' in the clinical laboratory, and is at the heart of many reference methodologies. Serum cholesterol determinations are highly dependent on the natures of the wet chemistries used, and mass spectrometry with gas or liquid chromatographic or capillary electrophoresic inlets provide the only direct and unambiguous measurements available today. Metabolic disease investigation by mass spectrometry, then, is selected only as an example or model of how it may be applied routinely elsewhere in the clinical setting.

Metabolic diseases

The identification of accumulating metabolites characteristic of a metabolic disease employs mass spectrometry in a qualitative manner. Compounds are identified in a fingerprint sense by computer-comparison of the mass spectra of unknowns with a library of reference spectra either purchased with the instrument or accumulated in-house from reference compounds. Mass spectrometry may also be used in a quantitative sense for the purpose of monitoring or evaluating patients, usually as part of a treatment protocol.

The most significant metabolic disorders are life threatening in the neonatal period if left untreated and are usually the result of the inherited inability to catabolize normal substrates derived from diet or from normal tissue degradation and recycling. The typical catabolic pathway can be represented by Equation [1],

$$A \xrightarrow{E_A} B \xrightarrow{E_B} C \xrightarrow{E_C} \cdots\cdots \longrightarrow N \qquad [1]$$

where A, B, ..., N are precursors, intermediates and final products, and subscripted Es are the enzymes or enzyme complexes that effect the changes associated with each metabolic step. A metabolic disease is said to occur when one (or more) of these enzyme-mediated steps fails to produce the required transformation, and the blood and tissue concentrations of the substrate or precursor for that step (or earlier steps) increase to levels that are toxic, inhibit other enzyme systems or significantly alter the pH or other characteristics of blood or tissues and adversely affect the patient's well-being. Other disorders may be the result of the inability to synthesize a required substance or the inability to transport something essential across a membrane. Identification of accumulating or missing substrates is therefore crucial to understanding the nature of the disorder and the eventual identification of the faulty enzyme or enzyme system.

The list of inherited metabolic diseases continues to grow; at this writing, there are 517 distinct disorders that have been at least partially characterized, with the sequences of the responsible defective proteins known for many of them. Most of these disorders are very rare; the incidence of a given disorder may vary from 1 live birth in 500 000 to 1 in 750, and is frequently dependent upon the degree to which an isolated population is inbred. A paediatrician may spend a lifetime in practice and not encounter any but the most common of these rare diseases.

Diagnosis in the very early postnatal period is critically important if the newborn is to survive and not suffer toxic accumulations of metabolites that frequently lead to retardation of physical and intellectual development. Through correct early diagnosis of many diseases that result from errors of catabolism of dietary components, such as amino acids, diets restricted in these precursors may be instituted to prevent these accumulations.

Diagnosis *in utero* is often achieved by the mass spectral analysis of a small sample of the amniotic fluid for elevations of metabolites excreted by a possibly affected fetus. This invasive procedure is usually only applied when there is a reason to suspect that a fetus may be affected, such as a defect occurring in a prior birth. This provides a basis for an informed decision to be made either to terminate the pregnancy or to prepare for supportive intervention at birth.

The failure of an enzyme may be partial or complete, permanent or transitory. A genetic mutation that encodes for an enzyme that is partially or completely inactive will result in a permanent deficit. Some enzyme deficits that are due to mutations that lead to inefficient binding of a cofactor can be stimulated to a useful capacity by administration of large doses of the cofactor on a life-long basis.

Gas chromatography–mass spectrometry

The use of mass spectrometry in medicine was greatly facilitated by the development of coupled gas chromatography–mass spectrometry (GC-MS). The first commercially successful GC-MS, the LKB-9000, became available in the mid 1960s, and it is estimated that fully half of the inherited metabolic disorders initially described in the 1960s and 1970s were discovered and investigated with various versions of this instrument. Direct coupling of fused-silica capillary columns to a proliferation of small and inexpensive bench-top quadrupole and ion trap instruments has now made practical the screening for metabolic disease in every live birth in developed countries.

An example of an inherited disorder that is the result of a permanent enzyme deficit is methylmalonic acidaemia, in which methylmalonic acid (MMA), derived in large part from valine and isoleucine, accumulates. In one of the common forms of this disorder, methylmalonyl-coenzyme A mutase, the enzyme responsible for conversion of methylmalonyl-coenzyme A (CoA) to succinyl-CoA is partially or completely inactive, and the clinical result is a severe and life-threatening keto acidosis, accompanied by high concentrations of MMA in the blood and in tissues and the urinary excretion of up to several grams of MMA per day. In other forms, mutase activity can be increased by administration of large doses of vitamin B_{12} or one of its related cobalamins, which are necessary cofactors for the mutase. For diagnostic purposes, a sample of urine is collected from the patient and an extract is prepared containing the carboxylic acids present in the urine. The dried extract is treated with a chemical reagent that produces the trimethylsilyl (TMS) esters of the carboxylic acids, and this mixture is then analysed by capillary GC-MS. **Figure 1** is an organic acid profile obtained in this manner, and is an example of the use of GC-MS in the diagnosis of a mutase deficiency. Clearly evident is a very large GC peak due to

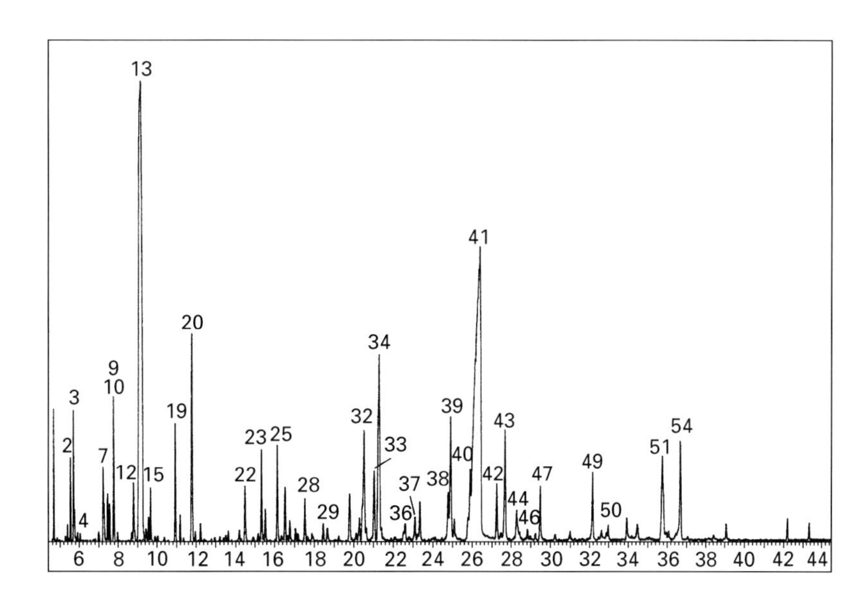

Figure 1 The electron ionization total ion current chromatogram for the trimethylsilyl derivatives of the organic acids extracted from the urine of a patient with an inherited error of methylmalonyl-CoA mutase. The major acidic component is methylmalonic acid. The annotated peaks are identified in **Table 1**.

bis(trimethylsilyl) methylmalonate. Other acids present and identified are normally found in human urine. **Table 1** lists the identities of the eluting peaks in this figure and in the other profiles presented below.

Integrated ion-current peak areas can be related to the quantities of metabolites present in the extract. While relative sensitivities for ionization must be known and taken into account for precise determinations, frequently one is forced to assume equality for all metabolites identified when these are unknown. The result is usually accurate to within a factor of 2 or 3 and is sufficient for screening of large numbers of samples.

When more precise measurements are required, stable-isotope dilution techniques are employed. In this approach to quantitation, a measured amount of a stable-isotope-labelled analogue of the compound of interest (internal standard) is added to the fluid to be analysed. The sample is then prepared for analysis and the mass spectrometer is used to report the ratio of the unlabelled to labelled analogues. With relatively few and minor corrections applied, the concentration of the unlabelled metabolite is easily determined. The technique is insensitive to losses in sample separations and variable derivatization yields as these will affect both analogues in the same proportion. As an example, **Figure 2** illustrates the use of this technique in the accurate determination of the concentration of MMA in human serum.

The internal standard is 20 μg of [*Me*-$^2H_3$]MMA, which was added to 1 mL of serum. The acidic extract was converted to the TMS derivatives and analysed in a GC-MS mode termed 'selected-ion monitoring', in which the quadrupole analyser is stepped discontinuously between selected ions and reports only their intensities. In this case, the ions selected has masses 218.2 and 221.2 Da, which are moderately intense, characteristic and distinguishing fragment ions in the spectra of the TMS derivatives of MMA and [*Me*-$^2H_3$]MMA, respectively. The data are obtained as plots of the intensities of these two ions as functions of GC retention time, and resemble independent gas chromatograms for these two fragment ions. The peak areas are integrated and a correction of the 218.2 area is made for isotopic impurity in the internal standard. The 221.2 area is also corrected for natural-abundance heavy isotope inclusion in the light ion that is present as an M+3 signal at 221.2. From these calculations one may then conclude that the serum sample was 20.06 μM in MMA, about 10-fold greater than the normal upper limit.

Another inherited error of valine, isoleucine and leucine metabolism is branched-chain α-keto aciduria, also known as maple syrup urine disease because the odour of urine of these patients resembles that of maple syrup. Branched-chain *α*-keto acid dehydrogenase, the enzyme complex that is used to oxidatively decarboxylate 2-ketoisovaleric, 2-keto-3-methylvaleric and 2-ketoisocaproic acids to the corresponding branched-chain CoA esters is defective, and very high concentrations of these keto acids accumulate in the blood and urine of these

Table 1 Identities of acid peaks annotated in **Figures 1, 3, 4 and 5**

1	Phenol	16	2-Hydroxyisocaproic	30	3-Phenyllactic	44	Vanillylmandelic
2	Lactic	17	(2*S*)-Hydroxy-3(*R*)-	31	Pimelic	45	Sebacic
3	2-Hydroxyisobutyric		methylvaleric	32	Hippuric (secondary	46	4-Hydroxyphenyllactic
4	Glycolic	18	(2*S*)-Hydroxy-(3*S*)-		derivative)	47	3-Indoleacetic
5	Oxalic		methylvaleric	33	4-Hydroxybenzoic	48	4-Hydroxyphenyl-
6	Glyoxylic oxime	19	Ethylmalonic	34	4-Hydroxyphenylacetic		pyruvic oxime
7	4-Cresol	20	Succinic	35	4-Hydroxybenzaldoxime	49	Palmitic
8	Pyruvic oxime	21	4-Hydroxybenzaldehyde	36	Phthalic	50	3-Hydroxysebacic
9	3-Hydroxyisobutyric	22	Glutaric	37	Suberic	51	4-Hydroxyhippuric
10	3-Hydroxybutyric	23	Methylmalonic (3-TMS)	38	Vanillic	52	*N*-Acetyltyrosine
11	2-Hydroxyisovaleric	24	2-Methoxybenzoic (IS)[b]	39	Homovanillic	53	5-Hydroxyindoleacetic
12	2-Methyl-3-hydroxybutyric	25	Capric (IS)[b]	40	Azelaic	54	Stearic
13	Methylmalonic (2-TMS)[a]	26	3-Hydroxyoctanoic	41	Hippuric	55	*N*-Acetyltryptophan
14	Benzoic	27	Mandelic	42	Citric		
15	2-Ethyl-3-hydroxy-	28	Adipic	43	3-(3-Hydroxyphenyl)-		
	propionic	29	3-Methyladipic		3-hydroxypropionic		

[a] TMS=trimethylsilyl.
[b] IS=Internal standard.

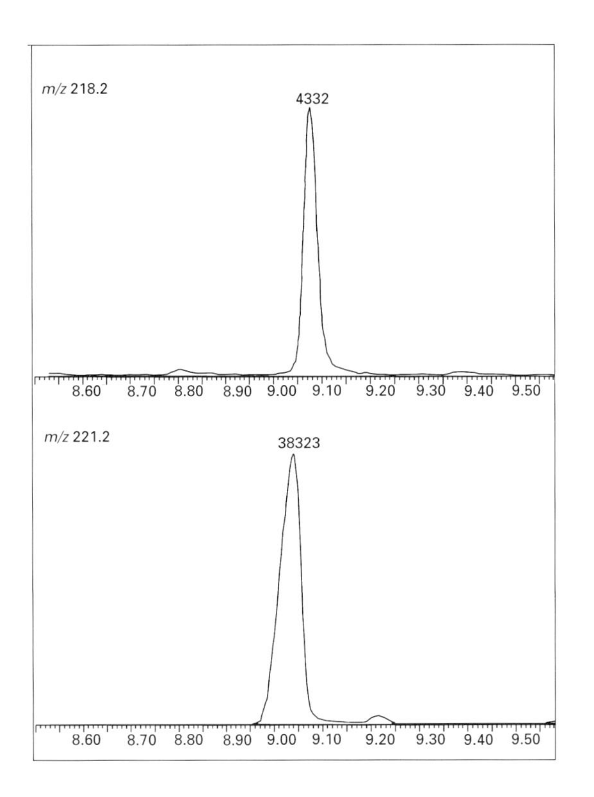

Figure 2 Selected ion monitoring analysis of serum for methylmalonic acid. The m/z 218.2 ion represents the $[M-CO_2]^{\cdot+}$ McLafferty fragment of the TMS derivative of endogenous unlabelled methylmalonic acid. The m/z 221.2 ion is the fragment derived by a similar process from the internal standard $[Me\text{-}^2H_3]$ methylmalonic acid. The areas integrated for the ions are related to the relative concentration of the derivatives (see text). The deuterium-labelled analogue has a slightly shorter retention time than the unlabelled analogue, a phenomenon perhaps unexpected but commonly observed in these circumstances.

patients. They also have in their fluids large elevations of the corresponding 2-hydroxy acids made by reduction of the keto acids, probably by lactate dehydrogenase. In sample preparation, addition of sodium borohydride to the urine will reduce the three 2-keto acids to the corresponding three 2-hydroxy acids and eliminate (E) and (Z) isomerism in the TMS derivatives of the enols of the 2-keto acids, thereby simplifying the gas chromatogram. If sodium borodeuteride is added instead, the 2-hydroxy acids that are produced will bear single deuterium substitutions on carbon-2 (Eqn [2]) and the labelled/unlabelled ratios may be determined easily for each of the 2-hydroxy acids.

$$\text{R—C(=O)—COOH} \xrightarrow{\text{NaB}^2\text{H}_4} \text{R—C(OH)(}^2\text{H)—COOH} \qquad [2]$$

These three ratios then represent the ratios of the 2-keto acids originally in the sample to their corresponding 2-hydroxy acids. **Figure 3** illustrates the urinary organic acid profile of one of these patients after reduction of the urinary 2-keto acids by sodium borodeuteride to the labelled 2-hydroxy acids. The keto/hydroxy ratios are clinically significant as they are related to the NAD^+/NADH status of these patients. (NAD^+/NADH = the oxidized/reduced forms of nicotinamide-adenine dinucleotide.)

A transitory enzyme deficit is one that arises commonly as the result of the ingestion of a toxic material that is or can be metabolized to a suicide substrate for that enzyme, which if depleted must be replaced by *de novo* synthesis. Another common temporary deficit is the result of an immature enzyme system in the newborn that spontaneously resolves itself within a few days.

As an example of the former, Jamaican vomiting sickness presents with severe hypoglycaemia and acidosis that resembles known types of inherited acyl-CoA dehydrogenase deficiencies. The disorder is precipitated by consuming the unripe fruit of the akee plant that grows in the Caribbean area. The protoxin is an amino acid, hypoglycine A (α-amino-β-(2-methylenecyclopropyl)propionic acid), which is metabolically converted to methylenecyclopropylacetic acid that irreversibly binds covalently to the flavin moiety of several acyl-CoA dehydrogenases and acts as a suicide substrate to permanently disable their active sites. Treatment is usually limited to supportive care during the period in which the inactivated enzyme system is replaced by new synthesis. The organic acid profile is dominated by very large concentrations of butyric, isovaleric and 2-methylbutyric acids, whose CoA esters require active dehydrogenases for further catabolism to crotonyl-CoA, 3-methylcrotonyl-CoA and 2-methylcrotonyl-CoA esters. Often these acids and other short-chain acids are also esterified to carnitine and excreted in the urine in easily detected amounts.

An example of a transient deficit that is the result of an immature enzyme system is a disorder in which the identification of increased concentrations of tyrosine and 4-hydroxyphenylpyruvic acid (4-HPPA) in urine of a premature newborn would suggest reduced activity of 4-HPPA oxidase. Administration of pharmacological doses of ascorbic acid, the cofactor required for activity of this enzyme, may overcome the temporary oxidase deficit and stimulate catabolism of the accumulating 4-HPPA. **Figure 4** is an illustration of this disorder, which is termed transient neonatal tyrosinaemia.

In addition to increased excretion of 4-HPPA, present here as the oxime, elevations of

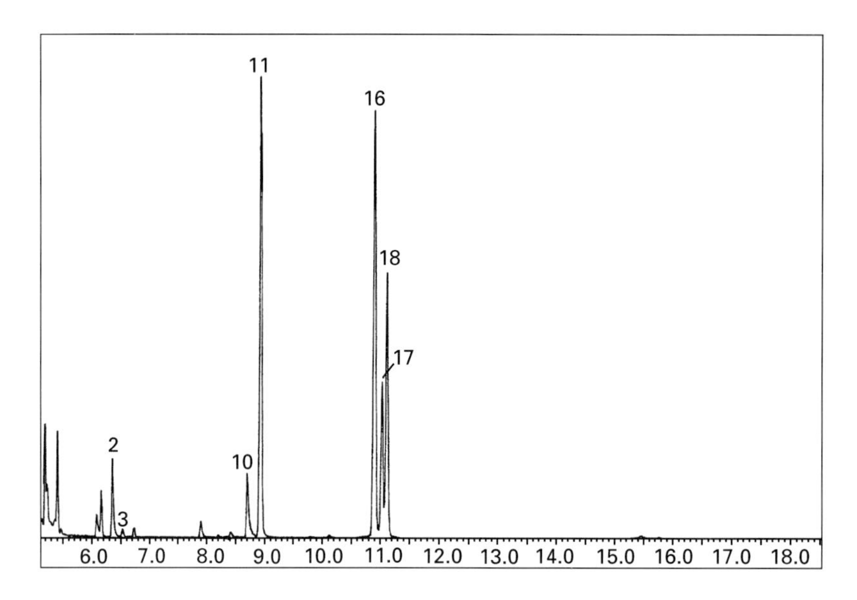

Figure 3 The total ion current chromatogram obtained for the organic acids isolated from the sodium borodeuteride-treated urine of a patient with maple syrup urine disease. The ratios of the ^{2}H-labelled to unlabelled analogues for 2-hydroxyisovaleric acid, 2-hydroxyisocaproic acid and the two 2-hydroxy-3-methylvaleric acid diastereomers are respectively 0.24, 34.5, 30.0 and 3.70.

4-hydroxyphenyllactic and 4-hydroxyphenylacetic acids are noted, and these are metabolites that are not important in normal catabolism and are derived from 4-HPPA by other enzymes. The oximes are made by addition of hydroxylamine hydrochloride to the urine during sample preparation for the purpose of detecting succinylacetone (see below)

Tyrosinaemia has an inherited form that is permanent and is due to inactive fumarylacetoacetate (FAA) hydrolase. As a result, FAA accumulates and is enzymatically converted into succinylacetoacetate (SAA), very low levels of which in turn severely inhibit 4-HPPA-oxidase. Many of the same accumulations seen in the transient neonatal form are measured as a result (**Figure** 5).

N-Acetyltyrosine is often noted in these organic acid profiles obtained for the hereditary form of tyrosinaemia, the result of *N*-acetylation occurring when blood levels of tyrosine rise to very high values. As a further complication, succinylacetone (SA), a

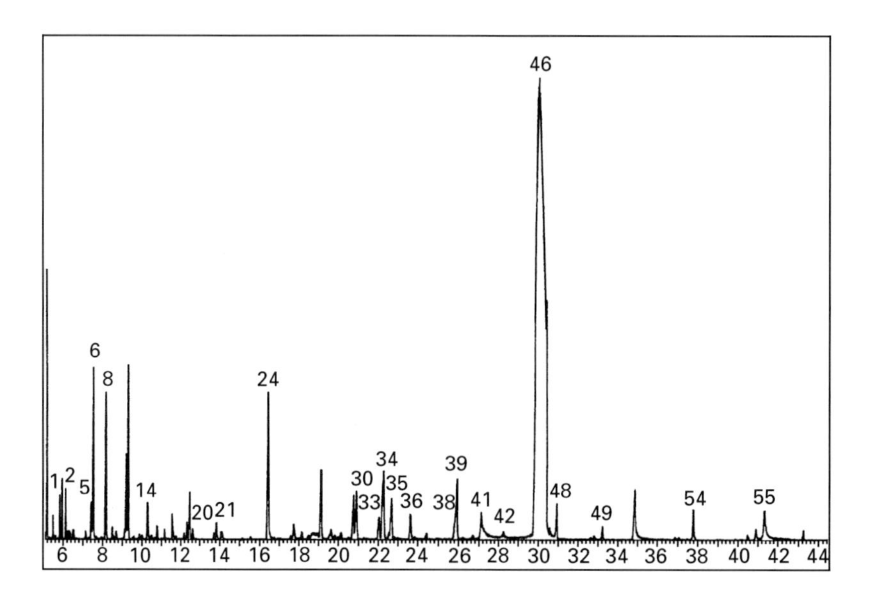

Figure 4 The urinary organic acids excreted by a premature infant with immature 4-hydroxyphenylpyruvic acid oxidase. This disorder, also known as transient neonatal tyrosinaemia, is characterized by excretion of elevated amounts of 4-hydroxyphenylacetic, 4-hydroxyphenyllactic and 4-hydroxyphenylpyruvic acids. The urine was pretreated with hydroxylamine hydrochloride, which converts the keto acids into the respective oximes.

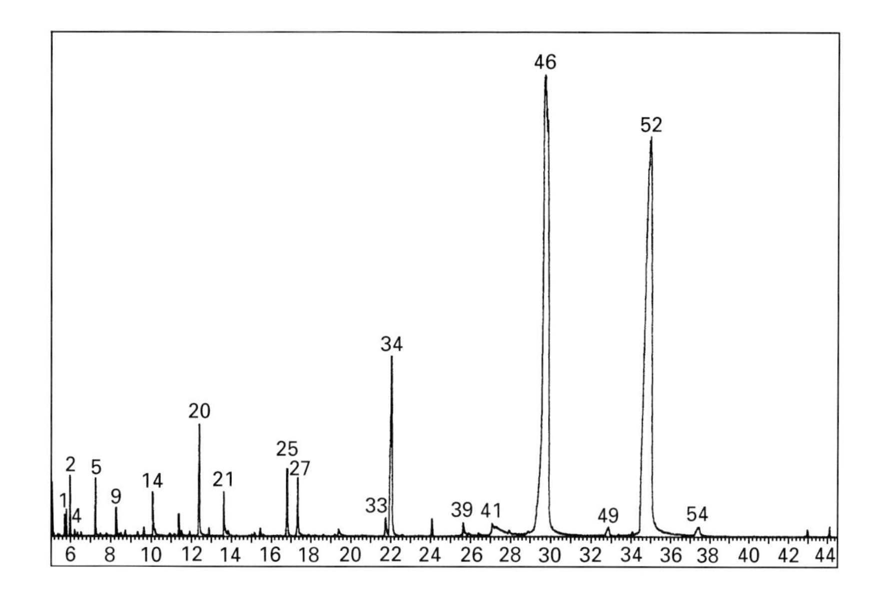

Figure 5 The organic acid profile obtained for a patient with inherited tyrosinaemia. Very large concentrations of 4-hydroxyphenyllactic acid and *N*-acetyltyrosine are in evidence.

spontaneous decomposition product of SAA, markedly inhibits porphobilinogen synthetase and other enzymes, which leads to the large number of clinical presentations seen in this disorder. To distinguish the transient and inherited forms, one needs only to measure the serum concentration of SA. SA is unstable to the usual sample preparation methods and, to avoid losses, the oxime is made by addition of hydroxylamine hydrochloride to the serum or urine sample. Since all of the hydrogen atoms in SA are easily exchanged with aqueous protons, no deuterium-labelled analogue of SA resistant to back-exchange can be made for use as an internal standard, and synthesis of a suitably ^{13}C-substituted analogue would be very difficult. This problem can be sidestepped by using 2-methoxybenzoic acid as the internal standard, as this does not occur in human metabolism and it elutes without interference by an endogenous metabolite. Ions selected for monitoring are m/z 209.1 and 212.1, the $[M-CH_3]^+$ fragments of the TMS derivatives of 2-methoxybenzoic acid and the positional isomers 3-[5-(3-methylisoxazolyl)] propionic acid and 3-[3-(5-methylisoxazolyl)] propionic acid. The latter two compounds are produced in the oximation of SA, and while they are separable on gas chromatography they are summed for quantitation. **Figure 6** illustrates an example of such an analysis of a urine sample.

Hence, this example represents the use of an unrelated compound as the internal standard, 2-methoxybenzoic acid in this case. It is necessary in these circumstances to carefully prepare calibration curves with samples of known composition to take into account the relatively different extraction and ionization efficiencies and appearance potentials for the ions selected.

Fast atom bombardment and electrospray ionization

Several diseases are known that result in elevations in tissues and fluids of various esters of carnitine and reduce the availability of free carnitine, which is normally synthesized by humans and is necessary for the transport of long-chain fatty acids into mitochondria for oxidation. In several disorders arising from acyl-CoA dehydrogenase deficiencies, the accumulation of the acyl-CoA substrate frequently sequesters coenzyme A and reduces its availability for other unrelated but important and otherwise competent pathways. Carnitine administration can displace and make available much of the coenzyme A that had been isolated, and stimulate the excretion of the accumulating acidic metabolites now esterified to carnitine. Detection of reduced levels of serum or urinary free carnitine and elevations of esterified carnitine is therefore useful for diagnosis of a variety of metabolic disorders, among them congenital inability to synthesize carnitine. In this disorder, carnitine must be supplied by a carnitine-enriched diet as it is, in effect, a vitamin.

$$(CH_3)_3\overset{+}{N}-CH_2-\underset{}{\overset{\overset{\text{O-X}}{|}}{CH}}-CH_2-COO^- \qquad [1]$$

Free carnitine (X=H)
Carnitine esters (X=acyl)

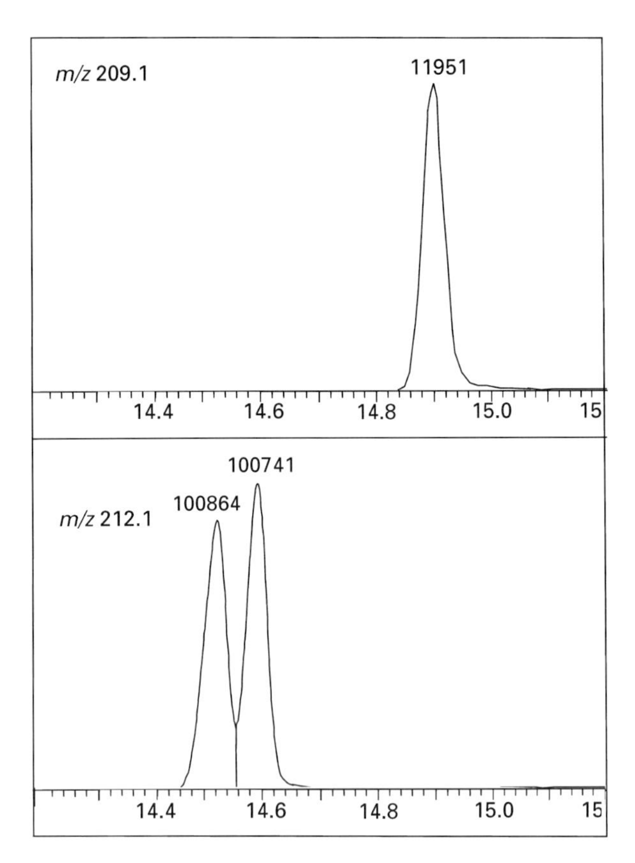

Figure 6 Selected-ion monitoring analysis of a urine sample from a patient with the inherited form of tyrosinaemia. Succinylacetone, which is distinctly elevated here, is produced from fumarylacetoacetic acid and is measured as an indication of fumarylacetoacetate hydrolase inactivity, the precipitating cause of this form of tyrosinaemia. The ions monitored are the $[M-CH_3]^+$ fragments of the TMS derivatives of the internal standard 2-methoxybenzoic acid (*m/z* 209.1) and the two positional isomers of the oxime derivative of succinylacetone (*m/z* 212.1).

Carnitine and its esters (see [1]) cannot be introduced to the mass spectrometer by gas chromatography, as they incorporate quaternary amine functions and will decompose in the attempt. Fast atom bombardment (FAB) and electrospray ionization (ESI) can use the formal charge on the quaternary amine function to advantage, as carnitine and its esters are very easily desorbed from glycerol on the FAB probe and from aerosol sprays in ESI. **Figure 7A** illustrates the use of FAB in the quantitation of carnitine and its esters excreted in the urine of a patient presenting with a severe dicarboxylic aciduria associated with medium-chain acyl-CoA dehydrogenase (MCAD) deficiency.

D,L-[Me_1,N-$^2H_3$]carnitine, D,L-[Ac-$^2H_3$]acetylcarnitine, and D,L-decanoyl[Me_1,N-$^2H_3$]carnitine are added as internal standards, and the sample is prepared as the trideuteromethyl ester for FAB analysis. Esterification of the carnitine carboxylate function with perdeuteromethanol removes the zwitterionic character of carnitine and its esters, and leaves them with a full positive charge. It also raises their ion mass by 17 Da to avoid confusion between them and their nonesterified homologues. High-resolution (10 000) spectra confirm the presence of the ions of interest free of interference from unrelated ions of the same nominal masses. Quantitations are then simply a matter of applying the usual calculations associated with a stable-isotope dilution assay. The identity of the annotated peaks is given in **Table 2**, along with the concentrations determined. Large concentrations of acetyl-, propionyl-, nonanoyl-, suberyl- and sebacylcarnitines are found that reflect elevations of the free acids in the urine. As a second illustration of the technique, urine from a patient with propionic acidaemia (**Figure 7B**) is shown to have a large concentration of propionylcarnitine (**Table 2**).

ESI has made possible many recent advances in the application of mass spectrometry to diagnostic medicine. Its great sensitivity and applicability to minuscule sample sizes together with its ability to analyse aqueous solutions form the basis of its utility. Because it is also a desorption technique, it is especially useful and sensitive for compounds that incorporate formal charges or chemical groups that are easily ionized.

Trimethylaminuria is an inherited disease related to the inability to convert trimethylamine (TMA) metabolized from dietary sources into trimethylamine *N*-oxide. The result is a disorder that is not clinically acute but has the unpleasant effect of producing a body odour resembling that of rotting fish in those affected. The social consequences of this are severely debilitating. While it is relatively easy to diagnose this disorder, objective means are needed to evaluate the efficacy of treatment protocols. TMA can be quaternized with [$^2H_3$]methyl iodide and, with ^{15}N-labelled TMA as the internal standard, an ESI method can be used to determine TMA concentrations in urine and blood. **Figure 8A** represents an example of the analysis of a normal urine for TMA. **Figure 8B** is obtained for a similar analysis of another normal urine spiked with 40 ng mL^{-1} of unlabelled TMA.

A known amount of [^{15}N]TMA hydrochloride is added to the urine, the pH of the solution is increased to 12 in an ice bath, an ether extract is made, and an excess of [$^2H_3$]methyl iodide is added. The resulting ether solution is added to water, which is warmed to drive off the ether and excess methyl iodide, and the remaining solution is infused by syringe through the ESI probe. Selected-ion monitoring with averaging or short scan modes

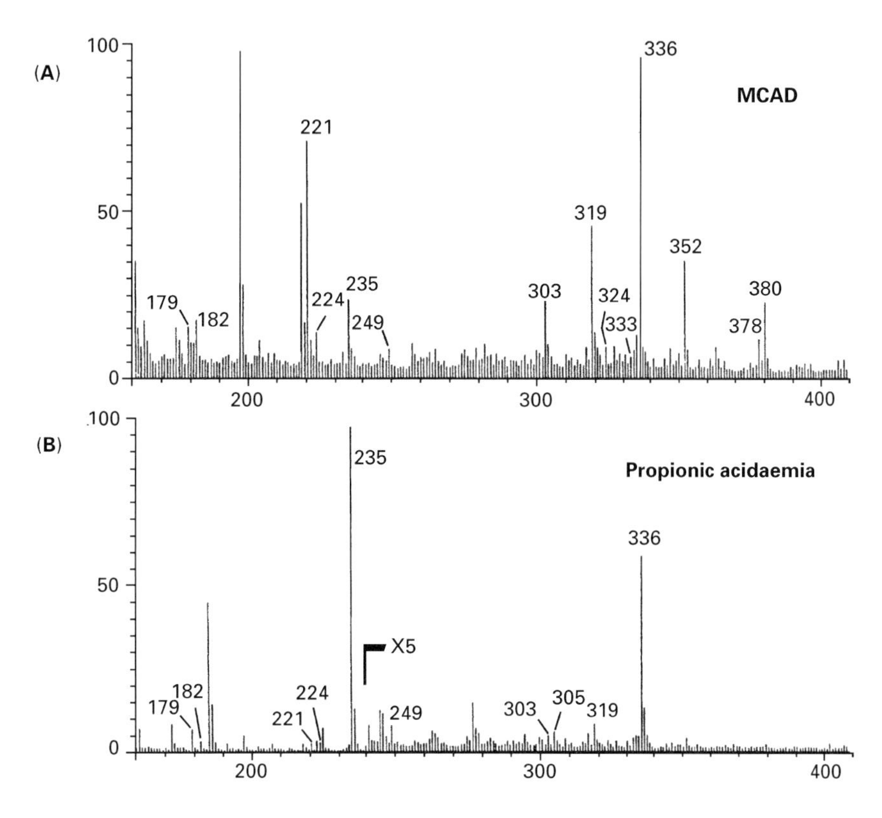

Figure 7 Fast atom bombardment mass spectra obtained in glycerol matrix for quantitation of free carnitine and carnitine esters in the urines of a patient with medium-chain acyl-CoA dehydrogenase deficiency (A) and a patient with propionic acidaemia (B). The identities and concentrations of the annotated ions are listed in **Table 2**. The composition of the ions has been confirmed by measurements at 10 000 resolution.

Table 2 Identities and concentrations of carnitine metabolites annotated in **Figure 7**

			Concentration (μmol/24 h)	
Carnitine	*Mass*	*Ion composition*	**Fig. 7A**	**Fig. 7B**
Free carnitine	179.1475	$C_8H_{15}{}^2H_3O_3N$	1688	2035
Carnitine IS[a]	182.1663	$C_8H_{12}{}^2H_6O_3N$		
Acetylcarnitine	221.1581	$C_{10}H_{17}{}^2H_3O_4N$	1625	210
Acetylcarnitine IS[a]	224.1769	$C_{10}H_{14}{}^2H_6O_4N$	25	4000
Propionylcarnitine	235.1737	$C_{11}H_{19}{}^2H_3O_4N$		
Butyrlcarnitine	249.1894	$C_{12}H_{21}{}^2H_3O_4N$	29	19
Octenoylcarnitine	303.2363	$C_{16}H_{27}{}^2H_3O_4N$	41	8
Octanolylcarnitine	305.2520	$C_{16}H_{29}{}^2H_3O_4N$	563	11
Nonanoylcarnitine	319.2676	$C_{17}H_{31}{}^2H_3O_4N$	67	16
Adipylcarnitine	324.2293	$C_{15}H_{26}{}^2H_6O_6N$	28	ND[b]
Decanoylcarnitine	333.2833	$C_{18}H_{33}{}^2H_3O_4N$	6	ND[b]
Decanolylcarnitine IS[a]	336.3021	$C_{18}H_{30}{}^2H_6O_4N$		
Suberylcarnitine	352.2606	$C_{17}H_{26}{}^2H_6O_6N$	50	ND[b]
Decenedioylcarnitine	378.2763	$C_{19}H_{28}{}^2H_6O_6N$	34	ND[b]
Sebacylcarnitine	380.2919	$C_{19}H_{20}{}^2H_6O_6N$	68	ND[b]

[a] Internal standard.
[b] Not detected.

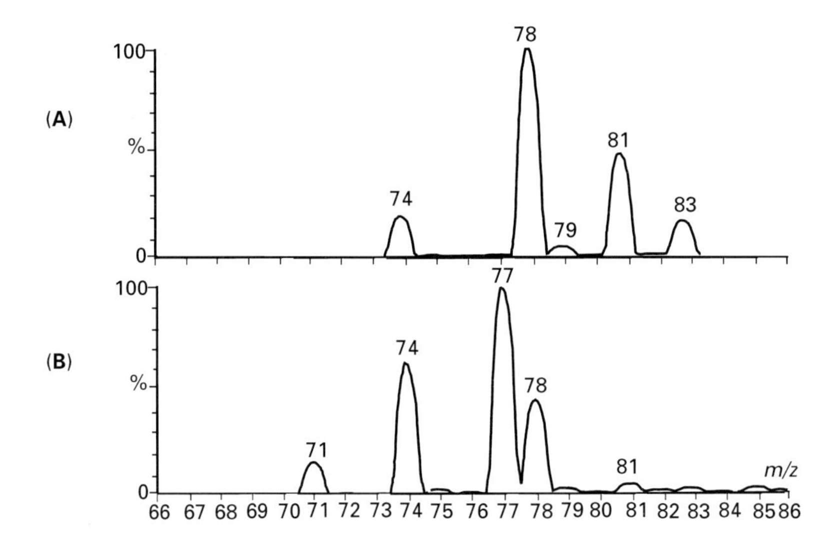

Figure 8 Positive ion electrospray spectrum of tetramethylammonium ions produced by quaternization of trimethylamine (TMA) excreted by a patient with trimethylaminuria. When the amine fraction is quaternized with [$^2H_3$]methyl iodide, endogenous TMA is measured at *m/z* 77, while the [^{15}N]TMA internal standard is measured at *m/z* 78. The ion signal at *m/z* 74 is unlabelled tetramethylammonium produced endogenously, which would interfere with the analysis if unlabelled methyl iodide were used for quaternization.

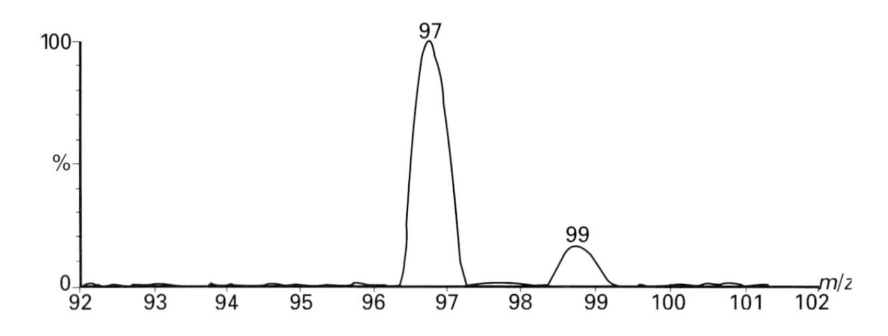

Figure 9 Negative ion electrospray collision-induced neutral loss of 17 Da (hydroxyl radical) analysis of plasma inorganic sulfate as $H^{32}SO_4^-$ (*m/z* 97) and $H^{34}SO_4^-$ (*m/z* 99). Phosphate, which occurs in plasma in much larger concentrations, also presents an intense ion with mass 97 Da ($H_2PO_4^-$), but can be distinguished from $H^{32}SO_4^-$ and excluded from the analysis as it eliminates 18 Da (water) under similar conditions. Heavy isotopes of hydrogen, oxygen and sulfur included in HSO_4^- in their natural abundances contribute an ion intensity at *m/z* 99 of about 5% of the intensity at *m/z* 97. Here the analysis shows that about 12% of the sulfate in the sample is derived from an oral load of ^{32}S-labelled sulfate.

in multichannel array detection may be used. Calculations usually associated with stable-isotope dilution analyses are then applied. [$^2H_3$]Methyl iodide is used as the quaternizing reagent to avoid interference by endogenous tetramethylammonium ions (mass 74 Da) usually present in urine and blood.

A second use of ESI is in the measurement of inorganic sulfate in blood and urine. Sulfate is extruded actively from cells and resides virtually exclusively in the extracellular fluid compartment. Sulfur has four stable isotopes, ^{32}S, ^{33}S, ^{34}S and ^{36}S (95.02%, 0.75%, 4.21% and 0.02%, respectively, although these values may vary somewhat with the nature of the source), and as sulfate it is an end-metabolite of sulfur-containing amino acids and other physiologically significant organosulfur compounds. Sulfate does not respond well in positive-ion FAB or ESI, but in negative-ion ESI the ion HSO_4^- is desorbed very efficiently from aqueous solutions. A stable-isotope dilution assay for sulfate based upon this fact uses highly enriched sodium [^{34}S]sulfate as the internal standard that is added to the biological fluid. The ratio of the $H^{32}SO_4^-$ analyte and $H^{34}SO_4^-$ reference ions could then be measured directly at *m/z* 97 and 99 if phosphate were not also present. $H_2PO_4^-$

also has mass 97 Da, and is present in concentrations much larger than that of sulfate. Collision-induced dissociation of HSO_4^- yields $SO_3^{\bullet -}$ (80 Da) while $H_2PO_4^-$ yields PO_3^-(79 Da), which distinguishes them in the tandem mass spectrometer. It is then a simple matter to measure the relative abundances of $H^{32}SO_4^-$ and $H^{34}SO_4^-$ by neutral loss of an OH radical (17 Da), producing ions at *m/z* 80 and 82, respectively, as $H_2PO_4^-$ dissociates exclusively with the loss of water (18 Da) and does not interfere with the sulfate measurement. An adaptation of this technique can be used to monitor sulfate produced from ^{34}S-containing amino acids in suspected errors of their catabolism. Radioactively-labelled sulfur-containing substrates have the drawback that ^{35}S (the longest-lived radioisotope) has a half-life of only 87 days, and they must be prepared shortly before each use. [^{34}S]sulfate may also be used in tracer studies to follow the excretion of label in the urine following an oral dose of the labelled sulfate. These experiments are usually conducted with radioactive [^{35}S]sulfate for the purpose of determining a patient's extracellular fluid volume. An example of this last use is presented in **Figure 9**.

While this article presents and concentrates on examples of the common and routine use of mass spectrometry in the diagnosis of metabolic disease, many other applications are important in the identification and quantitation of metabolites in other areas of clinical medicine.

See also: **Biochemical Applications of Mass Spectrometry; Biomedical Applications of Atomic Spectroscopy; Chromatography-MS, Methods; Fast Atom Bombardment Ionization in Mass Spectrometry; Isotopic Labelling in Mass Spectrometry.**

Further reading

Blasi F (ed) (1986) *Human Genes and Diseases*, Vol 8 in *Horizons in Biochemistry and Biophysics*. Chichester: Wiley.

Borum PR (ed) (1986) *Clinical Aspects of Human Carnitine Deficiency*. New York: Pergamon Press.

Burlingame AL and Carr SA (eds) (1996) *Mass Spectrometry in the Biological Sciences*. Totowa, NJ: Humana Press.

Chapman TE, Berger R, Reijngoud DJ and Okken A (eds) (1990) *Stable Isotopes in Paediatric Nutritional and Metabolic Research*. Andover, UK: Intercept.

Desiderio DM (ed) (1992) *Mass Spectrometry: Clinical and Biomedical Applications*, Vol 1. New York: Plenum Press.

Goodman SI and Markey SP (1981) *Diagnosis of Organic Acidemias by Gas Chromatography–Mass Spectrometry*. New York: Alan R. Liss.

Mamer OA (1994) Metabolic profiling: a dilemma for mass spectrometry. *Biological Mass Spectrometry*, 23: 535–539.

Matsumoto I (ed) (1992) *Advances in Chemical Diagnosis and Treatment of Metabolic Disorders*, Vol 1. Chichester: Wiley.

Matsumoto I, Kuhara T, Mamer OA, Sweetman L and Calderhead RG (eds) (1994) *Advances in Chemical Diagnosis and Treatment of Metabolic Disorders*, Vol 2. Kanazawa: Kanazawa Medical University Press.

Matsuo T (ed) (1992) *Biological Mass Spectrometry*. Kyoto: Sanei Publishing.

Weaver DD (1989) *Catalog of Prenatally Diagnosed Conditions*. Baltimore: Johns Hopkins University Press.

Wolfe RR (1984) *Tracers in Metabolic Research*. New York: Alan R. Liss.

Medical Science Applications of IR

Michael Jackson and **Henry H Mantsch**,
National Research Council Canada, Winnipeg, Manitoba, Canada

VIBRATIONAL, ROTATIONAL & RAMAN SPECTROSCOPIES
Applications

Traditionally a technique associated with astronomy and organic chemistry, IR spectroscopy has emerged as a powerful technique for the analysis and classification of human tissues and fluids. Such an application is possible because of the sensitivity of IR spectroscopy to changes in tissue biochemistry. In general, IR spectroscopy is sensitive to the structure and concentration of the macromolecules (nucleic acids, proteins, lipids) present in cells and tissues and relatively insensitive to low molecular mass metabolites (such as glucose, lactate and individual amino acids). Disease states that affect tissue ultrastructure are therefore potential targets for IR spectroscopic analysis. For example, cancer is often associated with the appearance of gross nuclear abnormalities, such as the presence of multiple nuclei in cells. As nucleic acids provide some of the most characteristic IR absorptions in biological materials, the changes in DNA associated with cancer result in changes to the spectral signature of malignant cells, and these changes are diagnostic. Changes in absorptions from other macromolecules have been reported to be diagnostic for a wide range of disorders affecting a variety of tissues.

A major advantage of IR spectroscopy is that a single instrument can in principle be used to characterize tissues affected by a wide range of disorders without the need for major reconfigurations of the instrument or the addition of probes (such as stains or other contrast-enhancing reagents) to the sample. In addition, valuable information concerning the molecular nature of the disease process is obtained. A general overview of available sampling methodologies and data analysis techniques will be presented, followed by illustrative examples.

Methodological aspects

General comments

Infrared absorption bands from biological materials are generally broad, largely owing to the presence of a heterogeneous population of chromophores. This heterogeneity has two sources. Firstly, functional groups are typically found in a variety of environments (e.g. polar/nonpolar, hydrogen-bonded/non-hydrogen-bonded), giving rise to slightly different but overlapping absorption profiles. Secondly, the complex biochemical nature of tissues, cells and biological fluids means that in many spectral regions overlapping absorptions from different functional groups are found (e.g. phosphate bands from nucleic acids overlap with collagen bands).

From a methodological viewpoint, this means that high-resolution measurements are not required. Typically, a spectral resolution of 2 or 4 cm^{-1} is sufficient. From an interpretational viewpoint, the presence of such broad absorption bands often means that application of band-narrowing techniques such as Fourier self-deconvolution or derivation is required before information can be reliably extracted from spectra. Application of second-derivative methods has the additional advantage that variations in baseline slope and offset are eliminated. While reducing the width of broad overlapping absorption bands, these techniques also disproportionately enhance the weak but narrow water vapour bands in the 1500–1800 cm^{-1} region, leading to many erroneous assignments in the literature. It is therefore essential that the spectrometer and any sampling accessories be well purged.

Sampling methods

A wide variety of techniques exists for obtaining spectra from tissues and fluids; the simplest of these are transmission methods. For biological fluids, small volumes (typically 5–10 μL) of sample are placed between a pair of CaF_2 or BaF_2 windows. In the mid-IR the pathlength must be limited to 10 μm or less owing to the presence of intense absorptions from water. Even under these circumstances, absorptions from water dominate the spectrum, and identification of dissolved/suspended materials is difficult. Subtraction of a water spectrum is possible, but it should be borne in mind that the structure of water is altered by dissolved solutes. This alters the spectrum of water, such that water in biological fluids gives rise to spectra that are subtly different from that of pure water. Problems associated with water can be removed by drying the sample to form a thin film. Drying of samples may result in the loss of volatile components from biological fluids, and also results in loss of information relating to hydration of materials. An alternative approach that minimizes sample preparation and prevents loss of information is to acquire spectra in the near-IR spectral region. The absorption bands due to combination and overtone vibrations seen in the near-IR spectral region are significantly reduced in intensity compared to those of the fundamental vibrations. In practice, this means that longer pathlengths may be utilized before water absorptions become a severe problem. However, it should be remembered that only overtones and combinations of X–H vibrations, (X = N, C, O) show significant absorption intensity, reducing the information content of the near-IR spectral region.

For tissue samples, transmission measurements are made in essentially the same way. Samples may either be small pieces of tissue (1 mm^3) dissected from a larger sample or microtomed thin sections. In either case, the sample is compressed between calcium fluoride windows. For some tissues (e.g. skin, muscle) such an approach is not possible owing to the high mechanical strength of the sample. For such tissues, specialized sampling techniques and accessories are recommended, such as the attenuated total reflection (ATR) measurements using a ‘split pea’ accessory. Using this approach, tissue is compressed with a reproducible force against a small hemispherical ATR element and spectra are acquired.

A major drawback of these approaches is that the physical integrity of the tissue is destroyed, preventing subsequent histological analysis. The investigator is therefore often uncertain of the exact nature of the sample being studied. This problem may be avoided with the use of an IR microscope. Using this technique, IR light is focused onto a small region of microtomed tissue and a spectrum is acquired. Spectra are sequentially acquired from the entire surface of the tissue by raster scanning with a computer-controlled stage. In this way a complete spectroscopic picture of the tissue section can be obtained with a high spatial resolution. The sample can then be stained and analysed by standard histological methods. This allows a direct correlation between spectra and sample histology, which aids in spectral interpretation.

Analysis of preserved samples

In all studies involving biological materials, questions arise of sample degradation over time. With cells and tissues, it is possible to prevent this degradation by fixation with preservatives such as 70% ethanol or formalin. However, these fixatives exhibit strong IR absorptions and are thus a source of potential artefacts. Furthermore, these fixatives preserve tissue by inactivation of degradative enzymes, which may also introduce artefacts into spectra if this inactivation is reflected in changes in protein conformation. Suspension of cultured tumour cells in 70% ethanol significantly alters the absorptions arising from cellular proteins compared to suspension in isotonic saline. Most noticeably, the observation of two protein absorption bands at 1625 and 1680 cm^{-1} is highly indicative of the formation of aggregates of protein molecules stabilized by intermolecular hydrogen bonding. This is generally associated with large-scale protein denaturation. The intensity of these absorptions is time dependent, increasing with prolonged suspension in the alcohol, owing to continued penetration of the alcohol into the cells. Fixation with ethanol also reduces the intensity of the ester C=O stretching vibration, suggesting a decreased lipid content in ethanol-fixed cells. This reduction in lipid content reflects solubilization of membrane lipids by ethanol.

Spectra of cells dried from formalin do not show the characteristic absorption associated with protein aggregates, indicating that formalin fixation does not induce protein denaturation. However, drying from formalin does result in the appearance of a series of sharp absorptions between 1000 and 1500 cm^{-1}. These narrow absorption bands arise from formaldehyde that is retained in salt crystals in the film. These distinctive absorptions are not present if the cell suspension is washed with isotonic saline before drying. Formalin should therefore be the fixative of choice for the vibrational spectroscopist, and cells and tissues should be rinsed with isotonic saline before spectral acquisition.

Spectral interpretation and data analysis methods

The group frequency approach

IR spectra have traditionally been interpreted by assigning absorptions that fall in particular frequency ranges to specific functional groups within molecules, in what is termed the functional group approach to spectral interpretation. Thus, absorption bands in the range 2800–3050 cm^{-1} are attributed carbon–hydrogen stretching vibrations of ethylene, methylene and methyl groups, while absorptions at 1700–1750 and 1600–1700 cm^{-1} are attributed to stretching vibrations of ester and amide C=O groups, respectively. A brief summary of the assignment of major absorptions in biological systems is presented in **Table 1**. Using such classical assignments, the biochemical and histological properties of tissues and cells may be inferred.

Functional group mapping

In an extension of the functional group approach to spectral analysis, data acquired by mapping tissues using an IR microscope may be analysed using an approach termed functional group mapping. The principle behind functional group mapping is illustrated in **Figure 1**. A grid first defines the area of the sample to be analysed. A spectrum is then sequentially acquired from each pixel in the grid, and the intensity or peak frequency of an absorption band arising from a functional group of interest is calculated for each spectrum. The intensity or frequency of the absorption band is then plotted as a function of position within the grid to produce a map of intensity or frequency distribution. In essence, therefore, this functional group mapping produces an image of the distribution of materials throughout the sample. For complex samples such as tissue sections, the distribution of important constituents such as collagen, proteins, acylglycerides and nucleic acids throughout the sample can be measured. These ‘chemical images’ can then be directly compared with the stained tissue to allow correlation of spectral and histological features.

Statistical and multivariate analysis

Spectroscopic differences between tissues or cell types are often subtle, leading to difficulties in

Table 1 Frequencies of biologically important IR absorptions. All absorptions cover a range of frequencies, and only representative frequencies are given here

Frequency (cm^{-1})	*Assignment*
3500	OH stretch; water, carbohydrates
3290	Amide A (N–H stretch): proteins
3050	Amide B (N–H bend first overtone): proteins
3010	Olefinic =CH stretch: lipids, cholesterol esters
2955	CH_3 asymmetric stretch: lipids, proteins, carbohydrates, nucleic acids
2925	CH_2 asymmetric stretch: lipids, proteins, carbohydrates, nucleic acids
2875	CH_3 symmetric stretch: lipids, proteins, carbohydrates, nucleic acids
2855	CH_2 symmetric stretch: lipids, proteins, carbohydrates, nucleic acids
2600	S–H stretch: proteins
2340	Solution phase and enclathrated CO_2
2058	SCN stretch: thiocyanate
1740	Ester C=O stretch: lipids, cholesterol esters
1655	Amide I (amide C=O stretch): proteins, α-helices
1635	Amide I (amide C=O stretch): proteins, β-sheet
1545	Amide II (amide N–H bend); proteins
1515	'Breathing' vibration of tyrosine ring (C–C/C=C stretching)
1467	CH_2 bend: lipids, proteins, cholesterol esters
1455	CH_3 asymmetric bend: lipids, proteins, cholesterol esters
1400	COO^- symmetric stretch: amino acid side chains, fatty acids
1380	CH_3 symmetric bend: lipids, proteins, cholesterol esters
1280	Amide III (C–N stretch) of collagen
1240	PO_2^- asymmetric stretch: phospholipids, nucleic acids; amide III (C–N stretch) of collagen
1204	Amide III (C–N stretch) of collagen
1170	CO–O–C asymmetric stretch: phospholipids, cholesterol esters
1150	C–O stretch, carbohydrates (glycogen)
1080	PO_2^- symmetric stretch: lipids, nucleic acids. C–O stretch, glycoproteins
1060	CO–O–C symmetric stretch: phospholipids, cholesterol esters
1035	C–O stretch, glycoproteins, carbohydrates

determining the significance of any observed differences. This is illustrated by a comparison of baseline-corrected mean spectra of cultured melanoma, lung adenocarcinoma and breast tumour cells (**Figure 2**). Spectra of the three cell types appear essentially identical, dominated by absorptions from proteins and nucleic acids. However, subtle differences are apparent throughout the spectra, including the region exhibiting absorptions from nucleic acid phosphate vibrations. Given the subtle nature of these spectral differences, the question arises as to the significance of the differences. Obviously, significance must be assessed using statistical techniques. The simplest way to assess whether significant differences exist between sets of spectra is to perform Student's *t*-test. The results of *t*-tests performed to compare the second-derivative spectra of the three groups of cells at each wavelength between 1000 and 1750 cm^{-1} are shown in **Figure 3**. For each comparison, the probability that differences between spectra at each wavenumber arose by chance is plotted. Thus, significant differences exist between spectra of breast and lung adenocarcinoma cells at a number of wavelengths. Similarly, significant differences exist between breast and melanoma cells and between melanoma cells and lung adenocarcinoma cells throughout the spectrum.

Such an analysis allows important conclusions to be drawn. For example, the greatest number of significant differences are found between spectra of breast tumour cells and melanoma cells, indicating that these two cell lines have the most different cellular biochemistry. In contrast, lung adenocarcinoma cells and melanoma cells appear to be the most similar as indicated by the high proportion of wavelengths at which no significant difference is found between spectra. Significant differences are found between all cell lines in the region of the nucleic acid phosphate vibrations, implying that the structure of the DNA in the three cell lines is different.

Analysis of differences between spectra by application of Student's *t*-test directly complements the traditional group frequency approach to spectral interpretation. Unfortunately, spectral regions in which significant differences exist between cell or tissue types are not necessarily the regions that allow optimal classification of spectra into groups based upon cell or tissue type. For example, significant differences exist between spectra of breast and melanoma cells between 1220 and 1240 cm^{-1}. However, lung adenocarcinoma cells also differ significantly from breast cells in this region. Thus, the only conclusion that can be drawn is that these cells are not breast cells. More definitive identification requires a comparison of a number of spectral regions, or even the entire spectral range.

Such definitive classification may be achieved with the aid of multivariate pattern recognition techniques such as hierarchical clustering, linear discriminant analysis (LDA) and artificial neural network analysis. Hierarchical clustering techniques compare sets of data (e.g. individually acquired spectra or spectra acquired by mapping of tissue) and group the data according to some measure of similarity. For mapping data, the application of cluster analysis

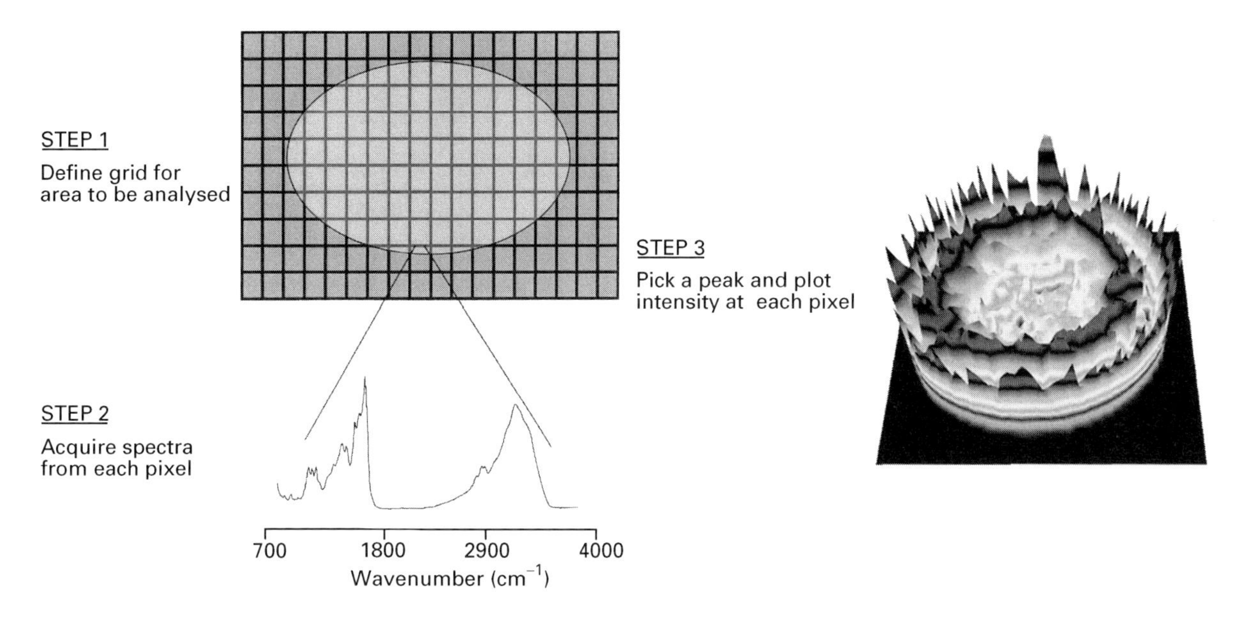

Figure 1 Pictorial description of the procedure for functional group mapping.

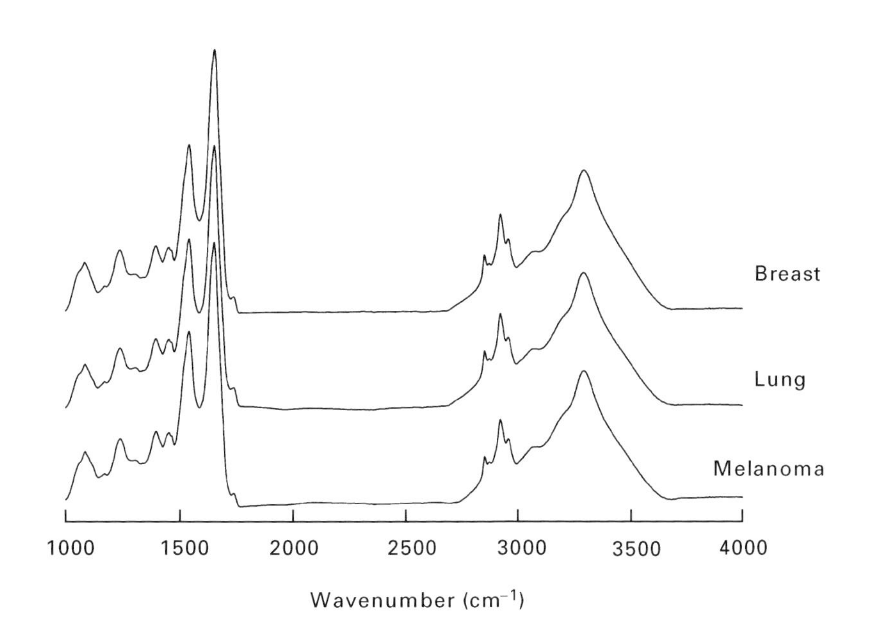

Figure 2 Infrared spectra of cultured human breast and lung tumour cells and murine melanoma tumour cells.

methods allows the identification of regions of the tissue that give rise to similar spectra and, by inference, have similar biochemistry. The advantage of such methods is that they perform a direct comparison of the spectral data (i.e. they are unsupervised, requiring no input from the investigator), and do not require assignment of spectral features.

More powerful multivariate methods such as LDA make use of the fact that the investigator is often in possession of class assignments (e.g. cell type). In such cases, so-called supervised methods such as LDA can be trained to detect patterns in the spectral data unique to each class. Unknown spectra can then be analysed using the trained algorithm to determine

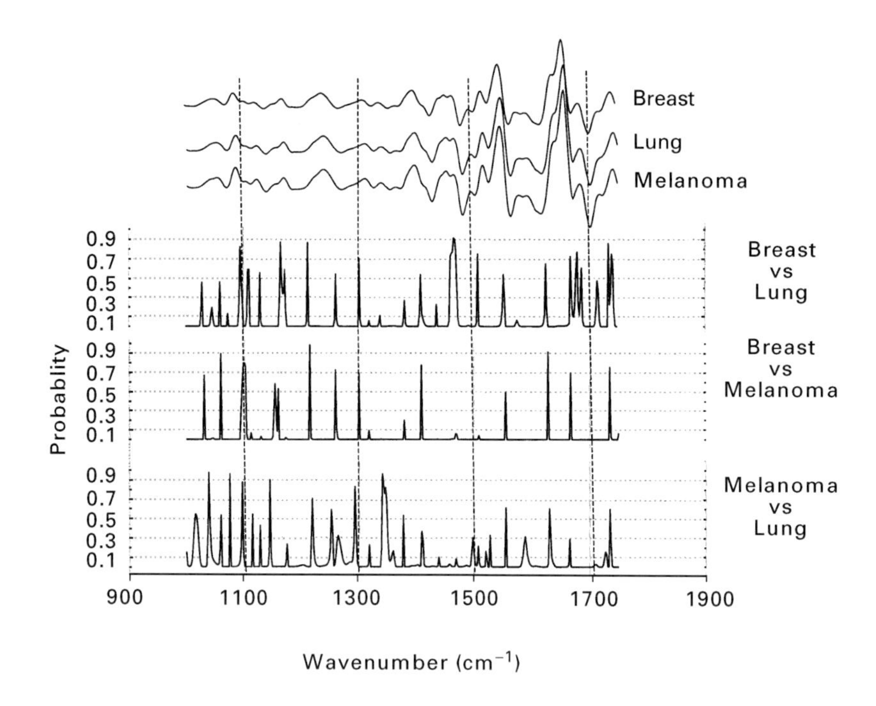

Figure 3 Results of Student's *t*-test performed to assess the significance of differences between spectra of cultured human breast and lung tumour cells and murine melanoma cells.

the appropriate class assignment based upon the spectral patterns present.

Applications of IR spectroscopy in medicine: Examples

With due regard to sampling, measurement and data analysis considerations, the application of IR spectroscopy to tissue, fluids and cells yields a remarkable amount of information. This information may relate to the concentration of particular analytes in biological fluids, the distribution of analytes within tissue, or the nature of biochemical changes associated with the disease process. In addition, information concerning metabolic processes in tissues and cells may be obtained. Examples will be presented to illustrate the general applicability of the technique.

Analyte determination

A major advantage of IR spectroscopy is that an IR spectrum is the summation of the spectra of all of the IR-active species present in the sample, weighted with respect to concentration. This means that the concentration of each of the major IR active analytes in a biological fluid can be predicted from a single IR spectrum. In practice this is achieved using such techniques as partial least squares (PLS) regression analysis, in which IR spectra are regressed against laboratory values for the concentration of analytes of interest to calibrate the method. This calibrated regression method can then be used to predict the concentration of the analytes from the spectrum of a fluid of unknown composition. Application of PLS methods to IR spectra can result in predictive accuracy as good as that of standard laboratory tests for many clinically relevant analytes such as glucose, urea, triglycerides and total protein (**Figure 4**).

Monitoring of metabolism

PLS techniques are required to decode the concentration information for most analytes present in biological fluids owing to the high degree of overlap between absorptions from most analytes. However, carbon dioxide exhibits a highly characteristic absorption in aqueous solution in a spectral region generally devoid of other absorptions (2340 cm^{-1}). This makes it relatively trivial to monitor CO_2 concentration. As CO_2 is a major by-product of cellular respiration, monitoring the intensity of this absorption in living systems should provide an indication of the rate of cellular metabolism. Obviously this cannot be done *in vivo* with current technology. However, monitoring the rate of respiration within suspensions of cells is possible. **Figure 5** shows the intensity of the absorption at 2340 cm^{-1} as a function of time for a suspension of yeast cells incubated with glucose. An increase in intensity with time as the glucose is metabolized is readily apparent. A

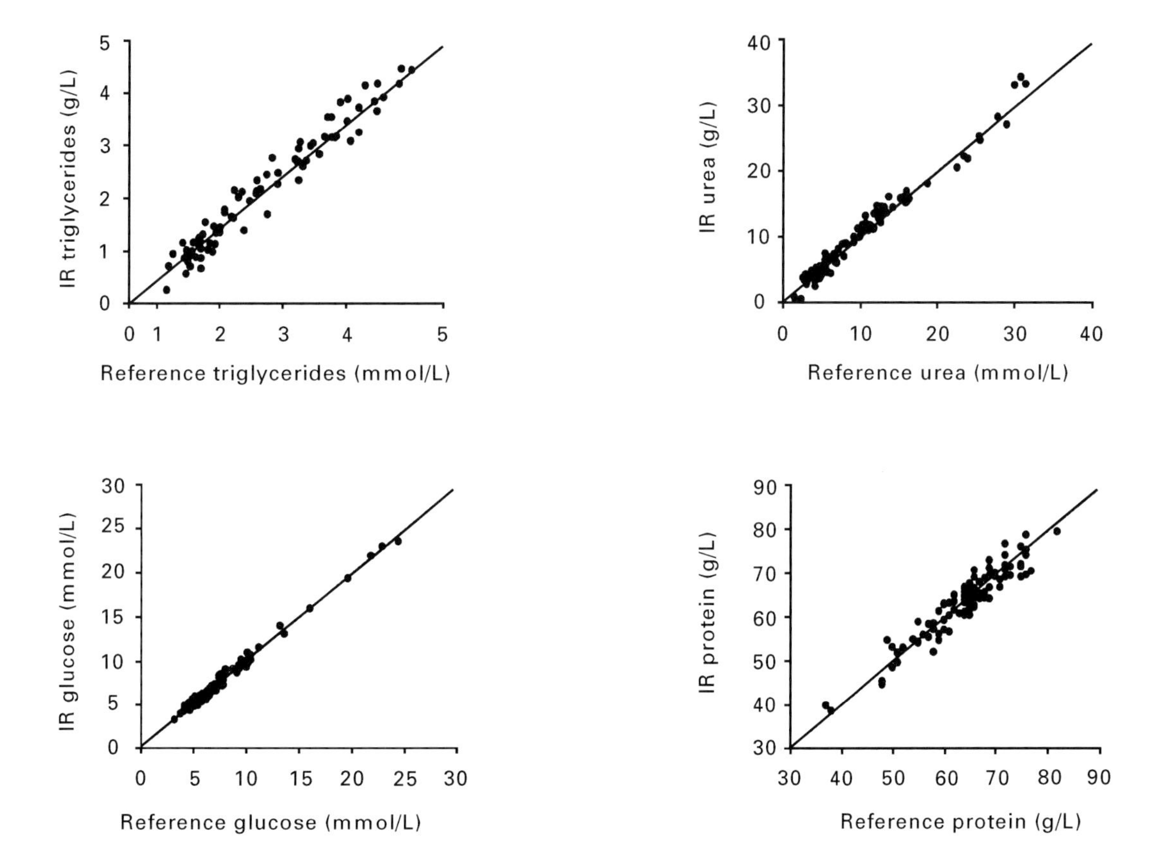

Figure 4 Concentration of triglycerides, urea, glucose and protein predicted by partial least-squares analysis of infrared spectra of serum plotted against reference values.

second, weaker absorption is apparent at 2275 cm^{-1}, attributed to the naturally occurring ^{13}C isotope of CO_2. The increase in intensity of this second absorption with time in the presence of ^{13}C-labelled glucose confirms that cellular respiration is indeed responsible for the increase in the intensity of the absorptions at 2275 and 2340 cm^{-1} with time. More importantly, the ability to detect metabolically produced $^{12}CO_2$ and $^{13}CO_2$ should allow IR spectroscopy to be used as a method for the elucidation of metabolic pathways in living cells using specifically labelled materials.

Grading of breast tumours

As discussed above, differences between spectra of cultured tumour cell lines may be subtle and difficult to discriminate from normal biological variability. Searching for diagnostic features in this relatively simple system may be compared to searching for the proverbial needle in a haystack, and multivariate pattern recognition methods are required. With this in mind, the task of identifying spectroscopic differences between sections of intact tissue becomes a daunting one. Not only must one find the subtle spectroscopic differences that exist between normal and diseased cells, but this information must now be extracted from the large signal arising from the matrix in which the cells sit. As an additional problem, the signal arising from the matrix may itself vary, either as a function of the disease or independently of the disease. The problem now becomes one of finding a needle in a field of haystacks.

The contribution of the extracellular matrix to spectra of breast tissue is shown in **Figure 6**. The spectra of low-grade breast tumours are vastly different from the spectra of cultured breast tumour cells. Breast tissue is composed predominantly of epithelial cells, connective tissue and adipose tissue. If we assume that cultured breast tumour cells and *in situ* breast tumour cells give rise to essentially similar spectra, then the differences seen between the spectra of tumours and cultured cells must reflect the presence of adipose and connective tissues. More specifically, a series of absorptions may be attributed to collagen and triglycerides. This example serves to illustrate not only the extent to which the extracellular matrix influences spectra, but also how spectroscopic studies of cultured cells can be used to identify such matrix effects.

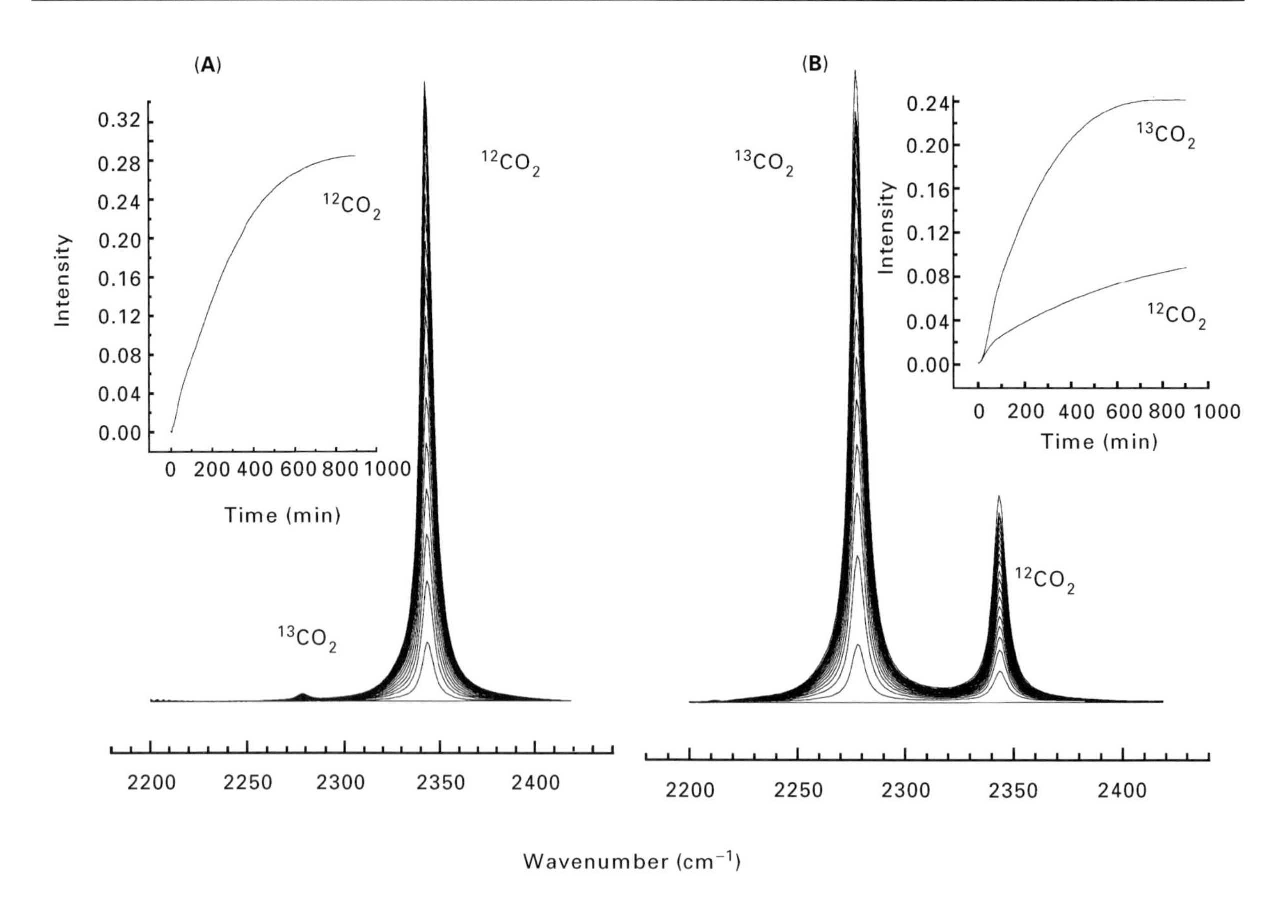

Figure 5 The intensity of the $^{12}CO_2$ and $^{13}CO_2$ absorption bands as a function of time in suspensions of yeast incubated with (A) ^{12}C and (B) ^{13}C labelled glucose.

To overcome the problem of matrix interferences and allow grading of tumours, spectra must be analysed by multivariate pattern recognition methods. Much of the spectral data is not related to changes associated with disease progression but results from normal spatial variations in breast histology. This information is therefore redundant, and only serves to increase computation time. Application of a genetic algorithm can be used to determine which regions of the spectrum carry the most useful diagnostic information. These regions can then be used as input for a pattern recognition strategy. Using this approach, tumours can be correctly classified as low, intermediate or high grade with an accuracy approaching 90%.

Microscopic studies of tumours

The problem of tissue homogeneity can also be addressed using IR microscopy and functional group imaging. Functional group maps showing collagen and adipose tissue distribution in a thin section of a subcutaneous murine tumour are shown in **Figure 7**. These maps are generated by plotting the intensity of characteristic absorptions from collagen and adipose tissue at each point within the map. The flat regions surrounding the tissue periphery correspond to the calcium fluoride substrate on which the tissue is supported. A low lipid content and a high collagen content characterize the periphery of the tissue, suggesting that this region of tissue is the epidermis and dermis. Underlying the epidermis/dermis, a large deposit of adipose tissue can be identified. A discrete band of collagen can then be seen forming a boundary layer around a central mass. This central mass is rich in protein and nucleic acids (as determined from functional group maps of the amide I and phosphate absorptions) but is low in collagen, identifying it as the main body of the tumour. The collagen found around the periphery of the tumour can therefore be recognized as part of a capsule surrounding the tumour. Functional group mapping thus allows identification of discrete histological structures within tissues.

In vivo spectroscopy and imaging

IR spectroscopy can be used to probe tissue biochemistry *in vivo*, using fibre optic accessories to bring the IR light to the patient. Such *in vivo* studies are performed almost exclusively in the near-IR region of the spectrum, owing to the enhanced penetration of

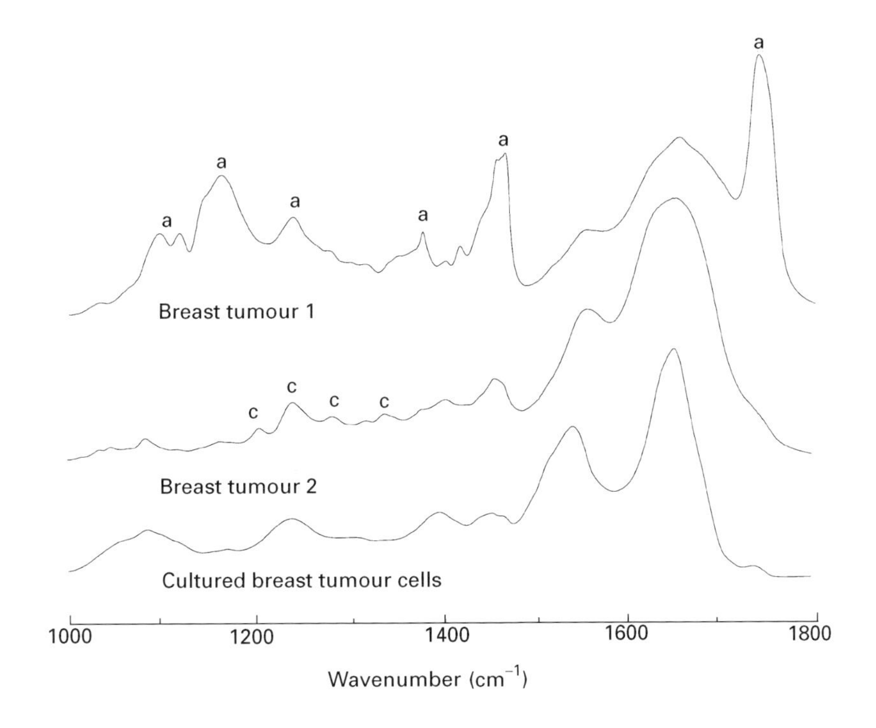

Figure 6 Infrared spectra of cultured human breast tumour cells and two low-grade human breast tumours. Absorptions marked (a) and (c) arise from adipose tissue and collagen, respectively.

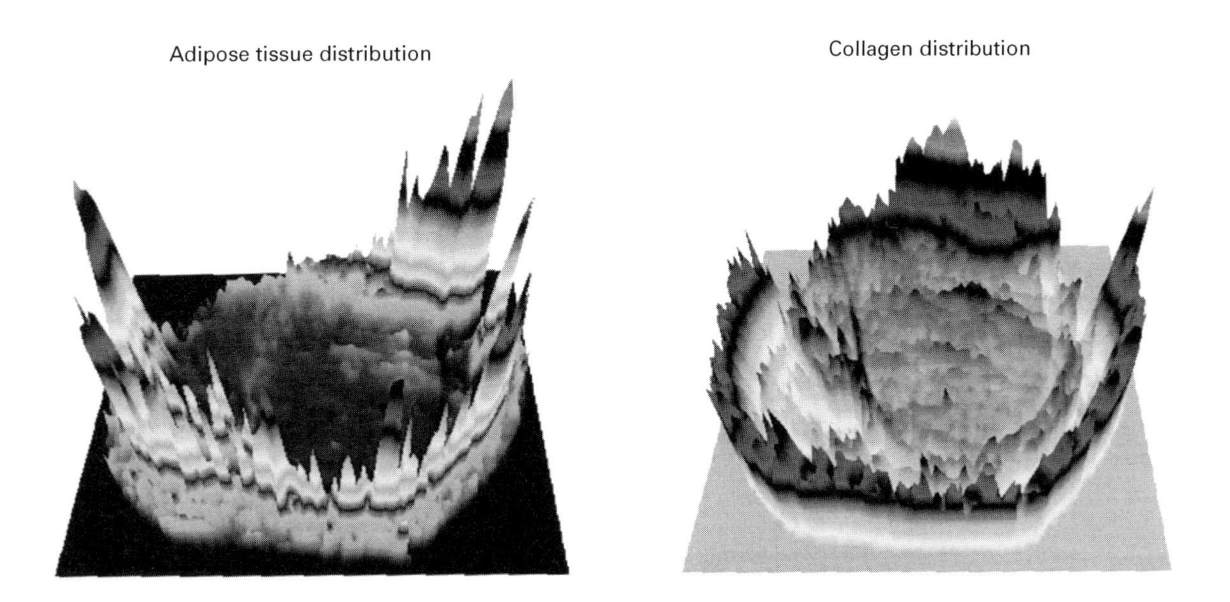

Figure 7 Functional group maps of a thin section of a subcutaneous tumour from a nude mouse showing adipose tissue and collagen distribution.

near-IR light into tissues. However, applications have been limited by the low information content of near-IR spectra. To date, most near-IR spectroscopic studies have focused on two main areas: blood glucose analysis and haemodynamic monitoring.

Attempts to develop methods for the noninvasive determination of blood glucose have met with little success, despite three decades of research. Major problems include: (i) the complex nature of skin; (ii) variations in the thickness and composition of skin within and between individuals and the consequent difficulties with variations in light scattering; (iii) the relatively low concentration of glucose in body fluids; (iv) the low intensity of the glucose signal in the near-IR; and (v) overlap of glucose absorption bands with absorption bands from water and other tissue components. Subtle effects such as changes in skin temperature and hydration further complicate the problem.

Haemodynamic monitoring has proved more successful. Haemodynamic monitoring using near-IR spectroscopy is possible owing to the occurrence of electronic transitions in the metal ions of haemoglobin and cytochrome species that are stimulated by absorption of near-IR light. Oxyhaemoglobin, deoxyhaemoglobin, oxidized cytochrome aa_3 and reduced cytochrome aa_3 exhibit absorptions in the near-IR region of the spectrum that are readily observed. Analysis of the relative intensities of these absorptions can therefore be used to assess tissue oxygenation and oxygen utilization within tissues. In addition, by combining an analysis of the haemoglobin and cytochrome absorptions with an analysis of tissue water absorptions, it is possible to extract information relating to haematocrit, tissue blood volume and blood flow.

The majority of such studies have been conducted using fibre optic-based systems. Spectra can be obtained from a small area of tissue by keeping to a minimum the distance between the fibre optic cables that bring the light to the tissue and those that return the light to the spectrometer. This approach provides information on a highly localized area, but does not provide gross haemodynamic information. Alternatively, the distance between the fibres may be increased and a large volume of tissue can then be probed. This approach provides information relating to the average haemodynamic parameters of the large volume of tissue. However, this approach does not allow localized haemodynamic information to be obtained. Recently, the availability of high-sensitivity charge-coupled-device (CCD) cameras that operate in the near-IR spectral region has been exploited to generate images that provide information concerning the haemodynamic parameters of large areas of tissue with a high spatial resolution. For example, an area of skin 10 cm × 10 cm can easily be imaged using a CDD camera with a 512 × 512 array detector. Spectra are obtained from each detector in the array, and each spectrum then corresponds to a region of the tissue approximately 200 μm × 200 μm. Haemodynamic parameters can be obtained from each spectrum, and these parameters can be used to produce an image. Thus, the global haemodynamic parameters of the 10 cm × 10 cm area of skin can be assessed, but information can also be obtained related to haemodynamic parameters with a spatial resolution limited only by diffraction.

The large amount of information generated in this manner (250 000 spectra) is ideally suited to multivariate analytical methods such as cluster analysis. Combining near-IR imaging and multivariate analysis allows one to determine nonsubjectively regions of tissue having similar haemodynamic properties (**Figure 8**). The applications of such techniques in areas such as wound healing are obvious.

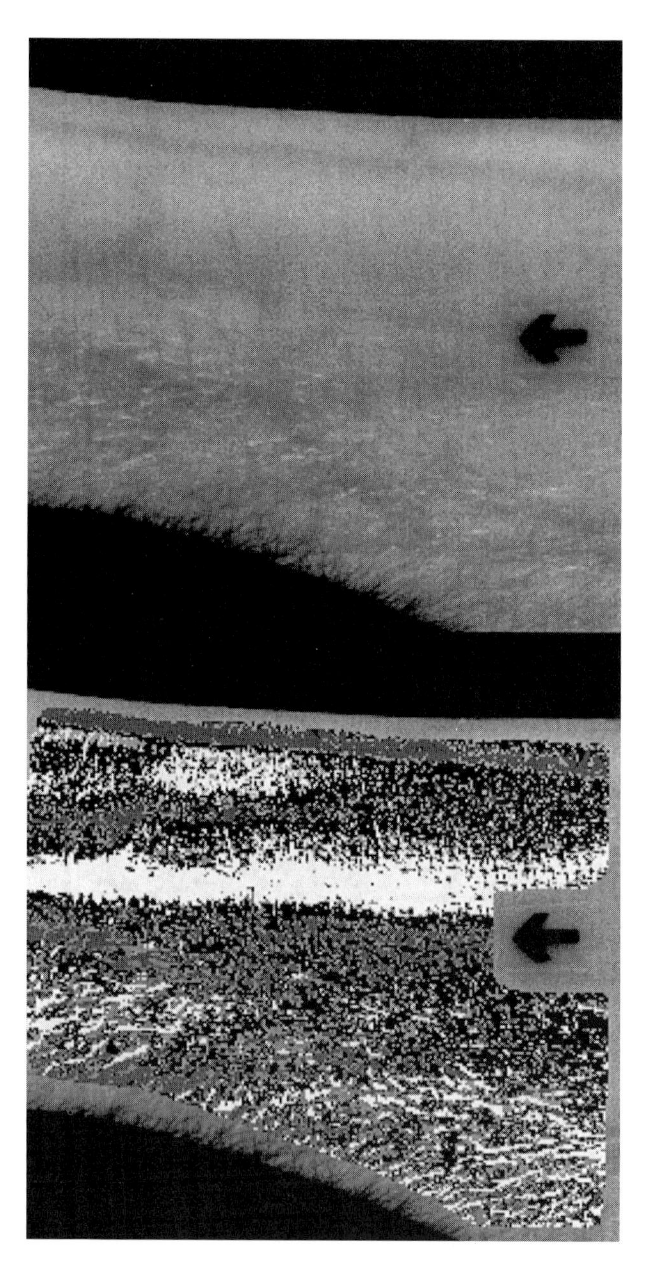

Figure 8 Noninvasive near-infrared imaging of the human forearm using a 512 × 512 array of CCD detectors. Visible image (upper panel) and results of cluster analysis on each pixel for the deoxyhaemoglobin absorption at 760 nm (lower panel). Areas with similar shading have similar deoxyhaemoglobin levels.

See also: **Biochemical Applications of Raman Spectroscopy; FT-Raman Spectroscopy, Applications; Multivariate Statistical Methods.**

Further reading

Fourier Transform Infrared (FT-IR) Microspectroscopy. A New Molecular Dimension for Tissue or Cellular Imaging and in situ Chemical Analysis. Cellular and Molecular Biology Special Issue (1998) Vol. 44, Issue no. 1.

Heise HM (1996) 'Near infrared spectroscopy for non-invasive monitoring of metabolites: state of the art'. *Hormone and Metabolic Research* **28**: 527–534.

Jackson M and Mantsch HH (1996) 'Biomedical IR spectroscopy'. In: Chapman D and Mantsch HH (eds) *FTIR Spectroscopy*, pp 311–340. New York: Wiley.

Jackson M and Mantsch HH (1996) 'FTIR spectroscopy in the clinical sciences'. In: Clark RJH and Hester RE (eds) *Biomedical Applications of Spectroscopy*, pp 185–215, Vol 25 in Advances in Spectroscopy. New York: Wiley.

Jackson M, Sowa M and Mantsch HH (1997) 'Infrared spectroscopy: a new frontier in medicine'. *Biophysical Chemistry* **68**: 109–125.

Manstch HH and Jackson M (eds) (1998) *'Infrared Spectroscopy: A New Tool in Medicine'*. Proc. SPIE 3257.

Membranes Studied By NMR Spectroscopy

A Watts, University of Oxford, UK
SJ Opella, University of Pennsylvania, Philadelphia, PA, USA

MAGNETIC RESONANCE
Applications

Introduction

Molecular motions in biomembranes are highly anisotropic. Since essentially all aspects of NMR spectroscopy are affected by molecular motions, the details of these motions strongly influence the spectroscopic approaches that can be applied to membrane samples. Although the motions of individual molecules or molecular groups may be very rapid (the molecular correlation time, τ_c, is in the ns range for lipid hydrocarbon chains and amino acid side chains of proteins), the overall tumbling of the lipid and polypeptide molecules in membrane bilayers is very slow and can be in the ms or s range. For this reason, conventional solution-state NMR, which relies on rapid overall molecular reorientation to narrow the resonance lines, has found little success in membrane research; even a lipid molecule with a relative molecular mass (M_r) of ~1000 or a small peptide (M_r ~ 3000) has, when in membrane bilayers, by solution-state NMR criteria a tumbling rate equivalent to a very large (M_r >> 30 kDa) macromolecule in free solution. Only by sonicating the sample to produce small (diameter 20–50 nm) bilayer vesicles can conventional solution-state NMR methods be used. However, not only are there very serious concerns about the status of the lipids and polypeptides embedded in highly distorted lipid bilayers, but also those portions of proteins embedded in the bilayers may still reorient too slowly to yield resolvable resonances. Solution-state NMR spectroscopy has been applied to peptides and proteins in organic solvents, detergent micelles and bicelles (**Figure 1**), and structures have been determined in these model membrane environments. While micelles certainly provide a more relevant environment for membrane proteins than do mixtures of organic solvents, fully hydrated lipid bilayers remain the definitive environment for structural and functional studies of membrane proteins.

As an alternative approach to solution-state NMR methods, which are ineffective with lipid bilayer samples, solid-state NMR methods have been refined sufficiently to permit structural details to be obtained for membrane-embedded peptides and proteins. This usually requires isotopic enrichment, either through chemical synthesis or biosynthetic incorporation in expressed peptides and proteins. In the absence of routine X-ray crystallographic structural studies for these molecules, solid-state NMR spectroscopy has the potential to be a powerful and unique approach to determining the structures and describing the dynamics and functions of membranes and membrane-bound proteins. In addition, solid-state NMR spectroscopy has been widely used to describe lipid structure, dynamics and phase properties. Thus, solid-state NMR experiments can be

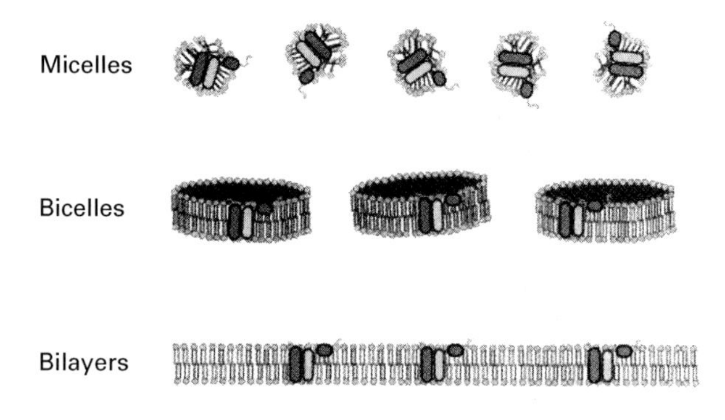

Figure 1 Representation of the types of sample preparations used for studying membrane lipids and proteins. Micelles and bicelles are usually small (diameters ~ 20–50 nm) structures and bilayers can be sonicated into small (20–50 nm diameter) vesicles, or produced as extended (diameters >> 100 nm) multi-bilayered or single bilayered closed or open structures, depending upon the method of preparation. Natural membranes are usually as large bilayer fragments or closed structures containing a complex and heterogeneous mixture of lipids and proteins and possibly carbohydrates.

applied to both the polypeptide and lipid components of membrane bilayers, without the need to disrupt the sample through sonication or the addition of organic solvents or detergents.

Nuclei used in membrane studies

With the exception of *J*-couplings, the major magnetic interactions (chemical shift, dipolar and quadrupolar couplings) for the nuclei exploited in biological NMR can be averaged with respect to the applied fields (B_0 ~ MHz and B_1 ~ kHz) by isotropic molecular motion of small molecules (**Table 1**). However, for biomembranes, any of these interactions may yield resonances with very broad lines and dominate the spectra, masking the resolution required for high resolution studies. Where these interactions can be exploited, their anisotropy (usually chemical shift, dipolar or quadrupolar) can give molecular orientational information from static samples, either oriented or as random dispersions (see below). Alternatively, magic angle spinning (MAS) of the sample can be used at spinning speeds (ω_r) which are either fast enough to average the interaction completely ($\omega_r >> \sigma$, D, Q) to give high resolution-like solid-state NMR spectra, or may be moderated either to recouple a dipolar interaction, such as in rotational resonance or REDOR, or provide orientionally dependent spinning spectral side-bands for nuclei which display chemical shift anisotropy (e.g. ^{31}P, ^{15}N).

Naturally occurring ^{13}C (natural abundance and with selective enrichment) and ^{31}P nuclei have been extensively exploited in membrane NMR studies

Table 1 Comparison of the strengths of the magnetic interactions in NMR and the ways in which they can be averaged to yield molecular information for lipids, peptides and proteins in membranes in liquid state and solid state approaches

Interaction	*Liquids* (Hz)	*Solids* (Hz)	*Methods*
σ	10–10^4	10–10^4	MAS
J	10^2	10^2	Decoupling
D	0	10^4	Decoupling, MAS
Q	0	10^5–10^6	MAS

Adapted with permission from Smith SO, Ascheim K and Groesbeck M (1996) Magic angle spinning NMR spectroscopy of membrane proteins. *Quarterly Review of Biophysics* **29**: 395–449. σ, Chemical shift anisotropy; J, J-coupling; D, dipolar coupling; Q, quadrupolar coupling.

(**Table 2**). However, replacement of ^{1}H by ^{2}H or ^{19}F, and ^{14}N by ^{15}N, has also found widespread application, although to date ^{17}O has not found application in these systems. Typical spectra for the more commonly exploited nuclei for lipids in bilayers are shown in **Figure 2**.

The need to average the strong dipolar coupling (~100 kHz) for ^{1}H to obtain high resolution spectra has, until now, excluded widespread observation of this nucleus in membranes. Extensive protein deuteration, to leave a minor ^{1}H density at a site of interest for observation in micellar suspensions, has been achieved. The realization that reorientation around the long molecular axis rotation of lipids and proteins in membrane bilayers in the liquid crystalline phase is sufficiently fast (at about 10^9 Hz for lipids and 10^6 Hz for proteins with radius ≤ 4 nm in fluid membranes) to average even homonuclear ^{1}H dipolar couplings, has opened a new avenue for membrane studies for most observable nuclei, including ^{1}H without the need for isotopic replacements. In addition, it is possible to perform magic angle oriented sample spinning (MAOSS) experiments to reap the benefits of both sample orientation and magic angle sample spinning in this situation.

Nature of the sample

Depending upon the kind of information desired, membrane bilayer samples can be prepared either oriented with respect to the applied field, or as random dispersions. For most studies, full hydration (>30 wt% of water) is desired, especially for protein studies where denaturation may occur and biological function be lost without sufficient amounts of water present.

Oriented membranes

Both natural and synthetic membranes can be effectively oriented and studied using NMR. In

Table 2 Properties, advantages and disadvantages of the commonly used nuclei in studies of membranes

Nucleus	*Relative sensitivity*	*Measured parameters*	*Advantages*	*Disadvantages*	*Common applications*
^{1}H	1 000	High resolution spectra Chemical shift, T_1, T_2	High sensitivity Natural abundance	Reasonable spectra with small vesicles, micelles, high speed MAS or MAS of oriented bilayers Several relaxation mechanisms Overlapping resonances	Dynamic properties Lipid diffusion
^{2}H	9	Powder spectra Quadrupole splitting T_1,T_2	Direct determination of order parameters and bond vectors Measurable in cells and dispersed lipids T_1 dominated by fast (ns) motions T_2 dominated by slow (μs– ms) motions Low natural abundance	Need for selective deutera tion Low sensitivity	Ordering properties of phospholipids Dynamic properties of phospholipids
^{13}C	16	High resolution spectra Chemical shift T_1 Dipolar couplings	Natural abundance T_1 dominated by one mechanism	Need MAS NMR to resolve spectra Without selective enrichment, overlapping resonances	Dynamic properties of phospholipids Lipid asymmetry Ligand–protein interactions Distance measurements
^{31}P	66	High resolution and powder spectra Chemical shift σ T_1 NOE	Natural abundance Chemical shift anisotropy is sensitive to headgroup environment and phase properties of the bulk lipids Measurable in cells and in dispersed lipids	Individual lipid classes can-not be resolved in mixed bilayer systems unless sonicated or MAS NMR is used	Quantitation of lipid composition Lipid asymmetry Phase properties
^{15}N	1.04	High resolution spectra Chemical shift σ T_1,T_2	Cost of labelling is low Can be incorporated in growth media Chemical shift sensitive to conformation	Low natural abundance, means of labelling required Overlapping resonance	Labelling of proteins and peptides Structural and dynamic studies
^{19}F	830	High resolution and powder spectra Chemical shift σ T_1	Chemical shift is sensitive to positional isomers Order parameters can be obtained High sensitivity Measurable in cells and in dispersed lipids	Need for selective fluorination Two factors contribute to the line shape, complicating the analysis High power proton decoupling is difficult May induce chemical perturbation compared to ^{1}H	Ordering properties of phospholipids

σ, Chemical shift anisotropy; T_1, spin–lattice relaxation time; T_2, spin–spin relaxation time; NOE, nuclear Overhauser effect; MAS, magic angle spinning.

general, reducing the hydration level of biomembranes supported and oriented on a substratum (glass or mica plates) improves their orientation, but if less than limiting levels of hydration are used (~<30 wt%), alterations may occur in the lipid phase from bilayer to isotropic or hexagonal phases. Fortunately, for all orientational studies of membranes, a good internal check for orientation can be made from the ^{31}P NMR spectrum of the bilayer phospholipids, since the line positions in the spectra recorded at 0° and 90° are separated by 40–50 ppm (**Figure 3**). The average mosaic spread can then be estimated from the spectral line-width.

Synthetic, model membranes, made from pure lipids, are best oriented on glass plates by drying down an organic solvent (usually $CHCl_3$/MeOH) solution of the lipid at 1–5 mg mL^{-1}. Overloading the lipids onto a substratum generally produces less good

Nucleus

^1H

6.0 4.0 2.0 0.0
Chemical shift (ppm)

^{13}C

180.0 140.0 100.0 60.0 20.0
Chemical shift (ppm)

^2H

$\Delta\nu_q$

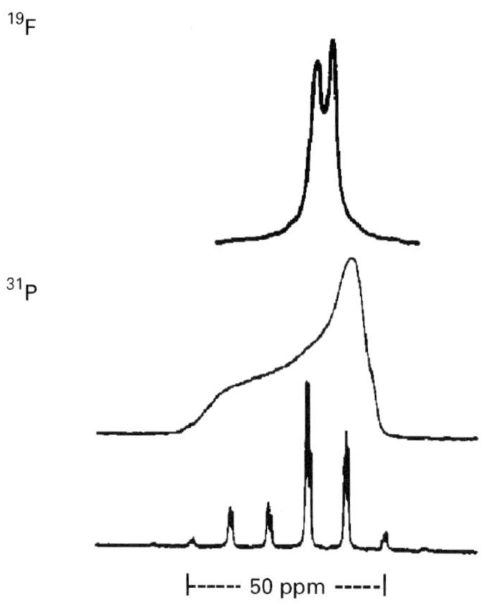

Figure 2 Typical spectra for various nuclei exploited in studies of lipids in bilayers. ^1H, a spectrum of phosphatidylcholine multi-bilayers, recorded under magic angle oriented sample spinning (MAOSS) conditions to give rise to narrow (~9 Hz) spectral lines; ^{13}C, proton-decoupled spectrum of sonicated, small (diameters <50 nm) phospholipid vesicles; ^2H, a typical deuterium NMR spherically averaged powder pattern for phospholipid bilayers deuterated in the choline head group, showing the way in which the quadrupole splitting ($\Delta\nu_q$) is determined; ^{19}F, spectrum of sonicated vesicles of lipids specifically ^{19}F-labelled in the 12′ position of both acyl chains; ^{31}P, a static, spherically averaged powder pattern from mixed, cardiolipin, phosphatidylcholine and phosphatidylethanolamine bilayers showing the lack of spectral resolution of the three lipid types (upper spectrum), and under MAS conditions to resolve the three individual phospholipids (lower spectrum).

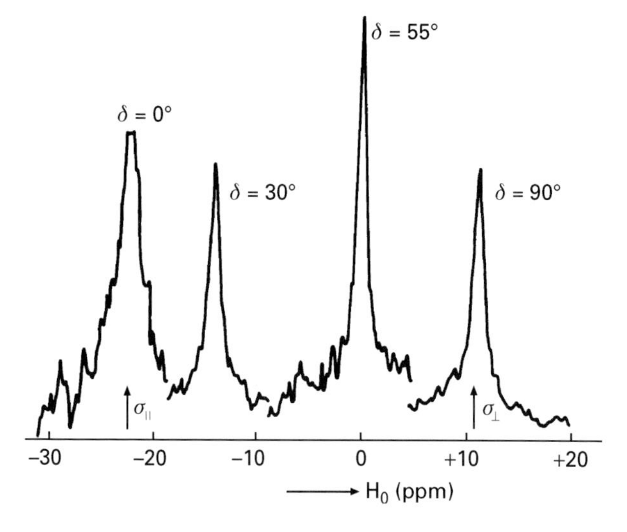

Figure 3 Orientational dependence of the ^{31}P NMR spectra (at 36.4 MHz) of planar multi-bilayers of phosphatidylcholine, where δ is the angle of the applied field with respect to the membrane normal. $T = 77°C$. Reproduced with permission from Seelig J and Gally HU (1976) Investigation of phosphatidylethanolamine bilayers by deuterium and phosphorus-31 nuclear magnetic resonance. *Biochemistry* **15**: 5199–5204.

orientation, and some lipids (notably zwitterionic ones such as phosphatidylcholines and phosphatidylethanolamines, both of which are major natural membrane components) orient better than others (anionic ones in particular), with cholesterol aiding orientation. Removal of solvent under high vacuum ($< 10^{-4}$ torr; 1 torr = 133.322 Pa) is followed by hydration, either by dropping buffer or water onto the film, or by incubating in a controlled atmosphere. It is also possible to orient hydrated random dispersions of lipids and proteins by applying pressure to the glass plates. In contrast, natural membranes are most readily oriented on glass plates using the isopotential spin dry ultracentrifugation (ISDU) method which involves centrifugation of a membrane dispersion followed by partial dehydration.

Membrane proteins prepared by solid-phase peptide synthesis or expression in bacteria can be reconstituted into lipids from detergents or organic solvents. The samples are placed on glass plates whose size and shape are determined by the radiofrequency coil with flat or square geometries for optimal filling factors in stationary experiments, or by the rotor for magic angle oriented sample spinning (MAOSS) NMR experiments. While most stationary experiments utilize samples arranged so that the bilayer normal is perpendicular to the direction of the applied magnetic field, it is possible to perform orientational dependent studies with the addition of a goniometer, which is usually rotated from the probe base without probe removal from the magnet. For MAOSS NMR experiments, the smallest diameter circular thin glass plates which can be handled are usually 4 mm, but for greater sensitivity, larger (up to 14 mm) rotor probes can be used to accommodate larger plates.

Random dispersions of membranes

Extensively sonicated dispersions of pure lipids usually form small single bilayer vesicles. These samples have been used extensively for ^{13}C NMR relaxation studies, which describe the motional properties of the various lipid groups, but have not been widely applied to proteins in membranes.

Natural membrane fragments or vesicles, large (diameter > 50 nm) liposomes, and hydrated unsonicated lipid dispersions are all indistinguishable (because of their slow tumbling) from the NMR perspective. They all give rise to anisotropically broadened spectral envelopes without the application of solid-state NMR techniques. For these systems, a spherically averaged powder pattern is observed which may be narrowed by molecular motion. Magic angle sample spinning (at a rate of ω_r) narrows many of the spectral features, but some potential orientational information content may be lost at high spinning rates (ω_r >> CSA, D or Q; **Table 1**) where there are few spinning side bands to analyse.

Experimentally, the amount of sample required is determined by the sensitivity of the nucleus to be observed. Typically, mM of the sample (5–10 mg of lipid; 0.5–1 mg of a small peptide; 1–4 mg of a large protein) in a volume of about 0.7 mL or on 10–50 glass plates, may be needed for less sensitive (^{2}H, ^{15}N) nuclei. For ^{1}H, ^{19}F, ^{31}P and ^{13}C in MAS NMR, somewhat less material is required. Cross-polarization from the abundant proton magnetization may improve sensitivity, and decoupling (5–10 kHz for ^{31}P, 80–100 kHz for ^{1}H) is routinely used to improve spectral shape and reduce spectral widths.

Micelles and bicelles

Micelles (SDS is often used) and bicelles (made of long and short chain phosphatidylcholines) tumble isotropically in solution and can accommodate peptides and proteins (**Figure 1**) to provide better mimetics for membrane proteins and peptides than organic solvents, which themselves can induce secondary structures in peptides and proteins. Indeed, good protein function and high resolution NMR spectra can be obtained from micellar-protein complexes, with sufficient resolution to permit structural analysis. Bicelles align in the applied field with the membrane normal perpendicular to the applied field, and by doping the system with lanthanides (Tm, Yb, Er or Eu) the orientation can be turned through 90°. Using paramagnetic chelates, direct protein–lanthanide interactions can be abolished leading to better spectral resolution although some hysteresis effects due to molecular reorganization may complicate their use.

Information content

Macroscopic structures

Phospholipids can, in aqueous dispersion, form a range of macroscopic structures including bilayers (predominantly for long chain derivatives, C_{12}–C_{24}), hexagonal, cubic and rhombic structures (**Figure 4**). ^{31}P NMR of static samples is a good diagnostic way of identifying such structures. Spherically averaged powder patterns (with a CSA of ~40–50 ppm) from bilayers are reversed and reduced in spectral widths (by 0.5) for hexagonal structures because of the added degree of freedom of molecular motion along the hexagonal cylinder. Isotropic spectra arise from small (diameter ≥ 50 nm) vesicles, cubic, rhombic and micellar molecular arrangements, as a result of molecular reorientation of the structure with respect to the applied field which is fast enough to average the frequency-dependent ^{31}P chemical shift anisotropy (τ_c < 3–10 kHz ~ = 40–50 ppm on 200–600 MHz instruments). Similar spectral changes are observed with ^{2}H NMR for deuterated lipids in similar structures.

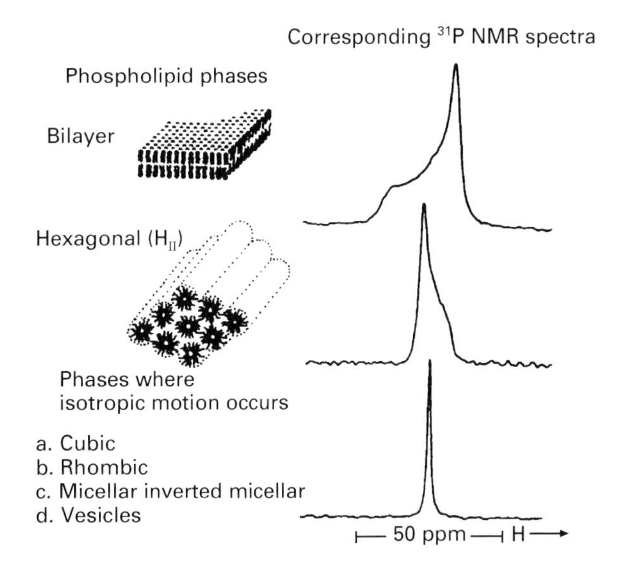

Figure 4 Three different types of phospholipid phases and their corresponding ^{31}P NMR spectra. From Cullis PR and Kruijff B (1979) Lipid polymorphism and the functional roles of lipids in biological membranes. *Biochimica et Biophysica Acta* **559**: 399–420.

A two-component bilayer and isotropic ^{31}P NMR spectrum from a membrane (usually a natural membrane) has been interpreted in terms of a major bilayer structure encompassing a much smaller (≤5%) population of inverted micelles. Although not the only explanation, the interpretation has many functionally attractive features (enhanced permeability, sites of membrane fusion, flip-flop regions, etc). Such isotropic spectral components are often produced by proteins interacting with the surface of lipid bilayers. The identity of the lipid type, in a mixed lipid membrane, which exists in this isotropic environment, can be determined using MAS ^{31}P NMR methods.

Molecular structure

Lipid bilayers are two-dimensional, axially symmetric structures with the normal to the plane of the bilayer being a reference axis. Lipids usually rotate quickly ($\tau_r \sim$ ns) around their long axis in fluid bilayers, but peptides and proteins may lack this motion which is determined by the association (controlled oligomerization or irreversible aggregation) behaviour of the various components, as well as the lipid dynamics.

Lipid orientational information By exploiting the anisotropic properties of the nuclear spin-interactions of ^{2}H, ^{13}C and ^{15}N (**Table 1**), placed at specific positions in peptides and ligand, rather precise bond vectors have been determined in oriented membranes.

Deuterium has low natural abundance and a low γ, but the quadrupole coupling can give rise to large splittings in the ^{2}H NMR spectrum ($\Delta\nu_q$(max) ~127 kHz), which is averaged by rotation of a methyl group around a C_3 axis (to 127/3 ~ 40 kHz). This high degree of orientational sensitivity has been exploited to determine the structure of specific residues (valines) involved in dimeric association for the gramicidin ion channel, and the structure of retinal within its binding site of bacteriorhodopsin and rhodopsin. Spectral simulations are necessary, as is some estimate of line broadening due to macroscopic mosaic spread (from ^{31}P spectral widths), but as a general method, the information gained is *ab initio* and, as such, model-independent. Most amino acids are available in a ^{2}H-labelled form for incorporation into peptides in solid phase synthesis, and a wide range of ^{2}H-precursors are available for specific labelling of ligands and prosthetic groups.

Peptide and protein orientation Many results have been obtained with specific, selective and uniform labelling of polypeptide sites with the spin $S = \frac{1}{2}$ nuclei ^{13}C and ^{15}N. These results have included the structure determination of gramicidin in bilayers, the geometrical arrangements of a variety of helical peptides in bilayers, the three-dimensional structures of peptides in bilayers, and the orientations of bound ligands and prosthetic groups. Uniform labelling of proteins with ^{15}N offers many advantages, and, indeed, was first developed for solid-state NMR spectroscopy before being applied to solution-state NMR where it has achieved nearly universal use. The nitrogen sites in the protein backbone are separated by two carbon atoms, leaving only minimal homonuclear ^{15}N dipolar effects, which is the essence of dilute-spin solid-state NMR spectroscopy. Each residue has one amide nitrogen in the backbone (with the exception of proline) and some have distinctive side-chain nitrogen sites. Further, uniform labelling allows the use of expressed proteins, and shifts the burden from sample preparation to spectroscopy, where complete spectral resolution is the essential starting point for structure determination.

Multidimensional solid-state NMR experiments have been shown to yield completely resolved spectra of uniformly ^{15}N labelled proteins in oriented lipid bilayers. In three-dimensional spectra, each amide resonance is characterized by three frequencies (^{1}H chemical shift, ^{15}N chemical shift and ^{1}H–^{15}N heteronuclear dipolar coupling), which provide the source of resolution among the various sites as well as the basic input for structure determination based on orientational constraints. The data shown in **Figure 5** are from a 50-residue protein in oriented lipid bilayers. More importantly, since the polypeptides are immobilized by the lipids on the relevant NMR time-scales, there can be no further degradation of line widths or other spectroscopic properties as the size of the polypeptide increases. Although larger proteins will have more complex spectra resulting from the increased number of resonances, there is no fundamental size limitation to solid-state NMR studies of membrane proteins.

In principle, a single three-dimensional correlation spectrum of an oriented sample of a uniformly ^{15}N labelled protein provides sufficient information in the form of orientationally dependent frequencies for each amide site to determine the complete structure of the polypeptide backbone. The two-dimensional ^{1}H–^{15}N heteronuclear dipolar–^{15}N chemical shift PISEMA spectrum in **Figure 5A** was obtained from a uniformly ^{15}N labelled sample of the 50-residue fd coat protein in oriented bilayers; it contains resonances from all of the amide backbone (and side-chain) nitrogen sites. The resonances in the box in the upper left are largely from residues in the

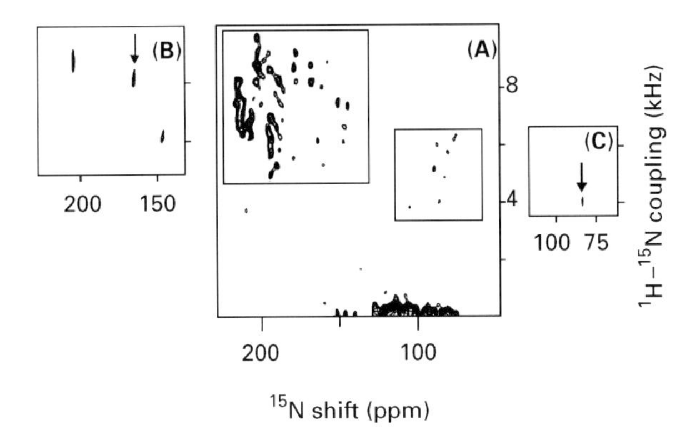

Figure 5 Multidimensional solid-state NMR spectra of uniformly ^{15}N labelled fd coat protein in oriented lipid bilayers. Panel (A) is the complete two-dimensional ^{1}H–^{15}N heteronuclear dipolar–^{15}N chemical shift spectrum. Panels (B) and (C) are spectral planes extracted from a three-dimensional correlation spectrum at specific ^{1}H chemical shift frequencies. The spectral regions in the planes correspond to the boxed regions of the complete two-dimensional spectrum with which they are aligned. Panel (B) contains resonances with ^{1}H chemical shift of 11.0 ppm. Panel (C) contains resonances with ^{1}H chemical shift of 11.6 ppm. The three orientationally dependent frequencies that can be measured for all of the resonances in the three-dimensional data set provide the input for structure determination. The arrow in panel (B) points to the resonance assigned to the amide of Leu 41 in the trans-membrane hydrophobic helix and that in panel (C) to the amide of Leu 14 in the in-plane amphipathic helix.

trans-membrane hydrophobic α-helix, and the resonances in the box on the right are from residues in the in-plane amphipathic α-helix. The two-dimensional PISEMA spectrum is generally well resolved, although there is some overlap among the resonances from residues in the trans-membrane α-helix because their peptide bonds have similar orientations, approximately parallel to the magnetic field. Two-dimensional planes extracted from a three-dimensional correlation spectrum of the same sample are shown in **Figures 5B** and **C**. The spectral regions in the planes correspond to the boxed regions of the complete two-dimensional spectrum with which they are aligned. There are very few resonances in the spectral regions in **Figures 5B** and **C** because only those resonances with a specific ^{1}H chemical shift, the third frequency in the three-dimensional spectrum, appear in each plane. These data illustrate how the three-dimensional correlation experiment contributes to these studies. First, it provides a substantial increase in resolution by separating resonances based on their ^{1}H chemical shift frequencies. Second, it enables the direct measurement of three orientationally dependent frequencies for each resolved resonances, the ^{1}H chemical shift, ^{1}H–^{15}N heteronuclear dipolar coupling and the ^{15}N chemical shift. Because the orientation of the protein-containing bilayers is fixed by the method of sample preparation, these frequencies can be used to determine the orientation of each peptide plane with respect to the applied magnetic field, since the magnitudes and orientations of the spin-interactions tensors for the amide sites in the molecular frame are known. Structures are then determined using the frequencies associated with individual resonances that have been assigned to specific residues as angular constraints. Using this approach the three-dimensional structure of the M2 trans-membrane segment from the acetylcholine receptor in oriented bilayers was determined. The ^{13}C chemical shift and ^{2}H quadrupolar coupling frequencies measured from selectively or specifically labelled samples in separate experiments also provide valuable structural information, and have been used together with the ^{15}N chemical shift and the ^{1}H–^{15}N heteronuclear dipolar coupling to determine the structure of the gramicidin channel at high resolution.

Amplitude of motion–order parameters Partial but fast ($\tau_c < \Delta\nu_{Q(\max)}$; Δ $\mathrm{CSA}_{(\max)}$) motional averaging of spectral anisotropy permits an order parameter, and hence an angle of motional amplitude with respect to a fixed axis (the membrane normal), to be determined. Direct measurements of order parameters from the powder patterns of random bilayers dispersions of deuterated lipids are conveniently made from the measured quadrupole splittings ($\Delta\nu_q$) (**Figure 2**), determined from the spectral maxima, corresponding to the 90°-orientational spectral components. Thus:

$$\Delta\nu_q = \left(\frac{e^2Qq}{h}\right)\left(3\overline{\cos^2\theta} - 1\right)$$

where $(e^2Qq/h) = 127$ kHz and thus

$$|S| = \frac{(\Delta\nu_q)_{\text{obs}}}{(\Delta\nu_q)_{\max}}$$

and so $1 > |S| > 0$.

The order parameter profile (S at various positions of measurement) usually observed for lipid acyl chains displays a plateau for lipid methylenes in the upper part of the bilayer (C_2–C_{10}; $S \sim 0.4$) as a result of water penetration and ordering of the upper part of the bilayer and then a decrease to the centre of the membrane (C_{10}–C_{16} gives S values of 0.4–0.1) (**Figure 6**). The discontinuity in S-profile at C_{10}–C_{12} corresponds to the main position of unsaturation in

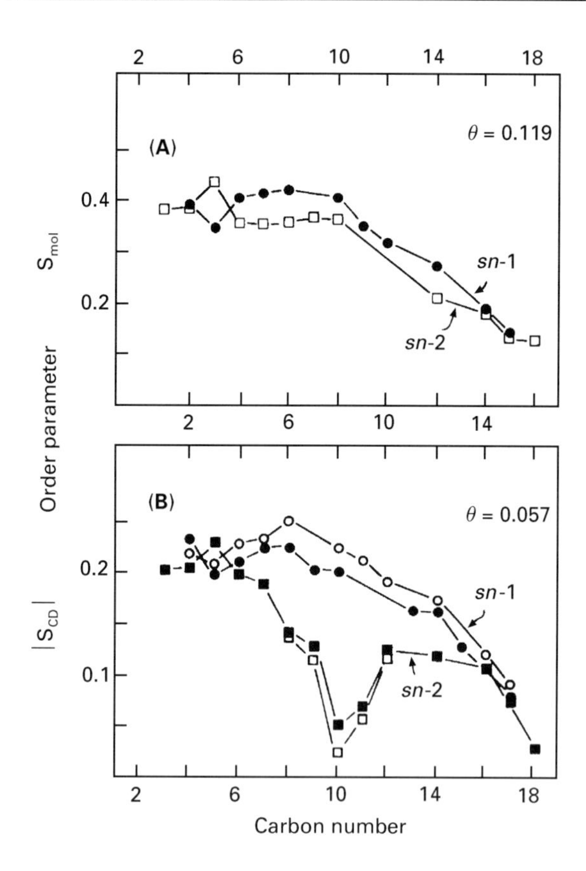

Figure 6 The measured deuterium order parameters, S_{CD} for the *sn*-1 (saturated) and *sn*-2 (unsaturated, C9=C10) bilayers of phospholipids specifically deuterated in their acyl chains, showing the consequence of the presence of double bonds causing the staggered conformation of the chains thus affecting the quadrupolar averaging (B), even though the molecular order parameter, S_{mol}, remains for each chain (A). Reproduced with permission from Seelig J and Seelig A (1977) Effect of single cis double bond on the structure of a phospholipid bilayer. *Biochemistry* **16**: 45–50.

natural membranes. At this position, the static kink in the chain contributes geometrically to the angle about which motional averaging occurs, thereby reducing the molecular order parameter (S_{mol} determined from $\Delta\nu_{q(obs)}$), since:

$$|S_{mol}| = |S(\text{orientation})| \times |S(\text{amplitude})|$$

Labelled ^{2}H lipid head groups give rise to small ($\Delta\nu_q$ < 10 kHz) quadrupole splitting, since their amplitude of motion with respect to the membrane normal is large, and their axis of averaging is not the membrane normal. In this case, the angle of molecular averaging cannot be uniquely determined, but the experimentally determined $\Delta\nu_Q$ values can be useful in studying ion binding (**Figure** 7), peptide and protein interactions, and electrostatic interactions, since the phospholipid; head group acts as a molecular voltmeter and sensor of interfacial pH at the bilayer surface.

Distance measurements within membranes In magic angle spinning (MAS) NMR, sample spinning averages out the dipolar couplings which are required for distance determinations. However, internuclear distance measurements can be made using rotational resonance (R^2) for homonuclear spins and REDOR for heteronuclear spins, in which the dipolar couplings are reintroduced into the spectrum under special spinning conditions. By spinning (at a speed of ω_r) the sample at multiples (n, where n = 1,2,3,4......) of the chemical shift difference (Δ in Hz) between a specific spin pair such that $\omega_r = n\Delta$, then transfer of magnetization occurs between the spin pair, and the dipolar interactions are recoupled. Now, the dependence of the NMR spectral intensity with mixing time shows a dependence on the distance between the spin pairs, and hence the internuclear distance (r) can be determined. This approach has been used to determine ^{13}C spin pair distances to sub-nm resolution in membranes between neighbouring lipids, between lipids and proteins (**Figure** 8), within a ligand at informative sites to give the structure of ligand at its site of action in a membrane bound target and of a prosthetic group (retinal) in its binding site of a receptor.

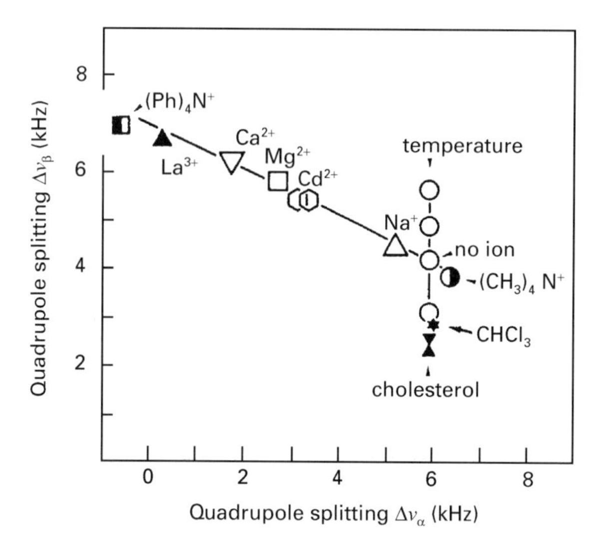

Figure 7 Deuterium NMR quadrupole splittings ($\Delta\nu_q$) of the β-CD_2 group plotted against the corresponding splitting for the α-CD_2 group of dipalmitoylphosphatidylcholine -$P(O_4)^-$ –αCD_2–βCD_2–$N^+(CH_3)_3$ in bilayers for a range of ions at constant ionic strength, I = 1.05 M and at 59°C, showing the sensitivity of lipid head group orientation to surface charge. Reproduced with permission from Akutsu H and Seelig J (1981) Interaction of metal ions with phosphotidylcholine bilayer membranes. *Biochemistry* **20**: 7366–7370.

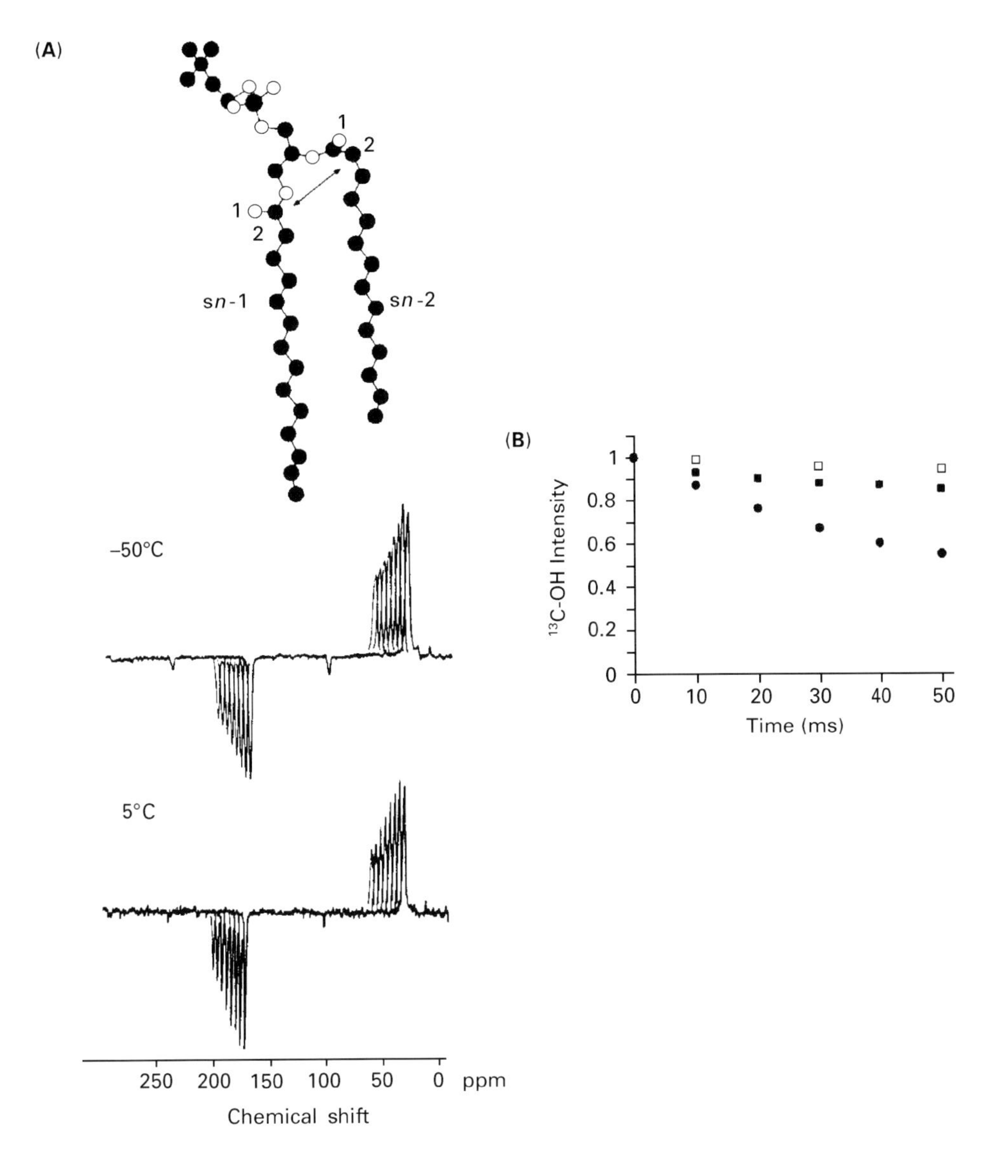

Figure 8 Rotational resonance ^{13}C MAS NMR has been used to determine both an intramolecular distance within a lipid and an intermolecular distance between a lipid and a protein in bilayers. A train of spectra are shown at different mixing times for the $n = 2$ resonance condition at a spinning speed of 6248 ± 5 Hz, at two different temperatures, giving a distance (from an analysis of the intensity changes with mixing time) from the C1 on the *sn*-1 chain to the C2 on the *sn*-2 chain of 4.0–5.0 Å at –50°C in dipalmitoylphosphatidylcholine bilayers (A). To determine the intermolecular peptide–lipid distance, the spectral intensity changes due to magnetization transfer were determined under conditions of no magnetization transfer, that is, at off-resonance, (□), with unlabelled lipid to show the contribution from natural abundance (1.1%) ^{13}C, (■), and with ^{13}C-1,2-[2-^{13}C] labelled lipid in bilayers containing glycophorin labelled at residue, ^{13}C–OH tyrosine-93 (●), to give a distance of 4.0–5.0 Å. Reproduced with permission from Smith SO, Hamilton J, Salmon A and Bormann BJ (1994) Rotational resonance NMR determination of intra- and intermolecular distance constraints in dipalmitoylphosphatidylcholine bilayers. *Biochemistry* **33**: 6327–6333).

Paramagnetic spectral broadening for a MAS NMR observed phosphate induced by a nitroxide spin-label, both at known positions in the primary sequence of membrane-bound bovine rhodopsin, a photoreceptor, has permitted some estimate of helix-loop distances to be made in the protein, supplementing other indirect structural details, even though the protein crystal structure is not available.

Membrane dynamics

Both spin–lattice (T_1) and spin–spin (T_2) relaxation times give motional information about specific nuclei in a membrane system. Faster ($\tau_c \sim 10^7$–10^9 s) motions associated with acyl chain *trans–gauche* rotational isomerisms and protein residue motions affect T_1 values whereas slower ($\tau_c \sim 10^3$–10^4 s) peptide backbone and membrane director fluctuations are inherently detected by T_2 values. Conventionally, high resolution NMR methods for measuring these relaxation times are used, the only difference being that broad spectral envelopes may show anisotropy in relaxation characteristics.

Membrane protein side chains (for example, Lys-$^{\varepsilon}CH_2$; Val-$^{\gamma}CH_2$) have been shown to possess fast (μs) motions from T_1 measurements of specifically deuterated residues in bacteriorhodopsin, even though the membrane environment was relatively rigid and crystalline. Similar approaches showed that the α-CH_3 group rotation is fast ($\tau_c \ll$ μs) for membrane-bound, specifically deuterated retinal in the same protein, even at −60°C. Somewhat slower motions (in the ms range) occur in peptide backbones of membrane proteins.

An enhancement of the T_1 for cardiolipin phosphates through direct contact of the haem group in cytochrome *c*, a peripherally bound protein, suggests that this protein undergoes considerable conformational distortions (molten globule) when at the bilayer surface, on the 10^{-6} s time-scale, a feature of peripherally bound proteins which may be general.

Slow (ms) director fluctuations (wobbling around the membrane normal) are induced in bilayer membranes by proteins, as shown by measurements of T_2. These motions are strongly coupled through the protein and membrane and may be significant in maintaining the protein in a dynamic equilibrium ready for function.

Ligand structure

Small molecules activate a range of cellular responses following binding to membrane-bound receptors. Solid-state NMR methods permit the structure and binding kinetics of such small ligands, using isotropic labelling to aid assignment and enhance sensitivity. The β-ionone ring orientation (6-S *trans* or 6-S *cis*) of ^{13}C-retinal in bacteriorhodopsin has been determined from direct measurements using rotational resonance MAS NMR, by comparison with the known crystal structure of retinal. Deuterated retinal in bacteriorhodopsin and mammalian rhodopsin has been observed and the structural changes induced upon light incidence determined to good precision using an *ab initio* approach. ^{13}C-labelled retinal has proved particularly successful for use in describing sugar binding to sugar transporters, drug binding to P-type ATPases (H^+/K^+-ATPase and Na^+/K^+-ATPase) and acetylcholine binding to the membrane-bound receptor. In the absence of crystals of this class of proteins at the present time, detailed structural information from solid-state NMR approaches and observation of a range of isotopes are helping to elucidate functional descriptions by this approach.

See also: **Cells Studied By NMR; ^{13}C NMR, Methods; ^{13}C NMR, Parameter Survey; Diffusion Studied Using NMR Spectroscopy; Liquid Crystals and Liquid Crystal Solutions Studied By NMR; Macromolecule–Ligand Interactions Studied By NMR; NMR in Anisotropic Systems, Theory; ^{31}P NMR; Proteins Studied Using NMR Spectroscopy; Solid State NMR, Methods; Solid State NMR, Rotational Resonance; Solid State NMR Using Quadrupolar Nuclei.**

Further reading

Cross TA and Opella SJ (1994) Solid-state NMR structural studies of peptides and proteins in membranes. *Current Opinion in Structural Biology* **4**: 574–581.

Glaubitz C and Watts A (1998) Magic angle-oriented sample spinning (MAOSS): a new approach toward biomembrane studies. *Journal of Magnetic Resonance* **130**: 305–316.

Gröbner G, Taylor A, Williamson, PTF *et al.* (1997) Macroscopic orientation of natural and model membranes for structural studies. *Analytical Biochemistry* **254**: 132–136.

Marassi FM, Ramamoorthy A and Opella SJ (1997) Complete resolution of the solid-sate NMR spectrum of a uniformly ^{15}N-labeled membrane protein in phospholipid bilayers. *Proceedings of the National Academy of Sciences* **94**: 8551–8556.

Opella SJ (1997) NMR and membrane proteins. *Nature Structural Biology* **4**(suppl.): 845–848.

Opella SJ, Stewart PL and Valentine KG (1987) Protein structure by solid-state NMR spectroscopy. *Quarterly Review of Biophysics* **19**: 7–49.

Opella SJ, Marassi FM, Gesell JJ *et al.* (1999) Three-dimensional structure of the membrane embedded M2 channel-lining segment from nicotinic acetylcholine receptors and NMDA receptors by NMR spectroscopy. *Nature Structural Biology* **6**: 374–379.

Pines A, Gibby MG and Waugh JS (1973) Proton-enhanced nmr of dilute spins in solids. *Journal of Chemical Physics*, **59**: 569–590.

Pinheiro TJT and Watts A (1994) Resolution of individual lipids in mixed phospholipid membranes and specific lipid-cytochrome *c* interactions by magic angle spinning solid-state phosphorus-31 NMR. *Biochemistry* **33**: 2459–2467.

Ramamoorthy A, Marassi FM and Opella SJ (1996) Applications of multidimensional solid-state NMR spectroscopy to membrane proteins. In: Jardetzky O and Lefevre J (eds) *Dynamics and the Problem of Recognition in Biological Macromolecules*, pp 237–255. New York: Plenum.

Reid DG (ed) (1997) Protein NMR techniques. In: *Methods in Molecular Biology*. Totawa, New Jersey: Humana Press.
Sanders CR II and Landis GC (1995) Reconstitution of membrane proteins into lipid-rich bilayered mixed micelles for NMR studies. *Biochemistry* **34**: 4030–4040.
Smith SO, Ascheim K and Groesbeck M (1996) Magic angle spinning NMR spectroscopy of membrane proteins. *Quarterly Review of Biophysics* **29**: 395–449.
Vold RR, Prosser RS and Deese AJ (1997) Isotropic solutions of phospholipid bicelles: A new membrane mimetic for high-resolution NMR studies of polypeptides. *Journal of Biomolecular NMR* **9**: 329–335.
Watts A, Ulrich AS and Middleton DA (1995) Membrane protein structure: the contribution and potential of novel solid state NMR approaches. *Molecular Membrane Biology* **12**: 233–246.
Watts A (1993) Magnetic resonance studies of phospholipid–protein interactions in bilayers. In: Cevc G (ed) *Phospholipids Handbook*, pp 687–740. New York: Marcel Dekker.
Watts A (1998) Solid state NMR approaches for studying the interaction of peptides and proteins with membranes. *Biochimica et Biophysica Acta* **1376**: 297–318.
Watts A (1999) NMR of drugs and ligands bound to membrane receptors. *Current Opinion in Biotechnology* **10**: 48–53.

Mercury NMR, Applications

See **Heteronuclear NMR Applications (La–Hg).**

Metastable Ions

John L Holmes, University of Ottawa, Ontario, Canada

MASS SPECTROMETRY
Theory

Metastable ions, their origins and observation

Metastable ions are those that dissociate en route from an ion source, through a mass analyser to an ion detection device. That ions can dissociate in flight was recognized by Hipple and Condon in 1946 when they provided the explanation for the presence of the diffuse peaks that they had observed in normal, electron ionization (EI) mass spectra obtained with a single-focusing magnetic sector mass spectrometer. It should be emphasized that the term metastable refers only to ions that are able to fragment in flight by virtue of internal energy that they acquired within the ion source, not after acceleration therefrom, nor by collisions with a target gas, nor by radiative excitation as they traverse the apparatus. They are thus unimolecular dissociations.

It was the striking and varied phenomenology produced by metastable ions that made them of particular interest in the 1960s and 1970s, when many laboratories had magnetic sector (B) instruments or double-focusing mass spectrometers consisting of an electric sector (E) and a magnetic sector arranged BE or EB (tandem mass spectrometers). Much of this article will describe observations made with tandem mass spectrometers.

First, however, consider a simple magnetic sector instrument. Singly charged molecular or fragment ions leaving the ion source will all have the same translational energy,

$$eV_{\mathrm{acc}} = \tfrac{1}{2} M_1 v_1{}^2 = \tfrac{1}{2} M_2 v_2{}^2 = \tfrac{1}{2} M_n v_n{}^2$$

independent of their mass M_n. The ion acceleration

potential is V_{acc} and v_n are the ion velocities. As the magnetic analyser field strength is changed, ions of different mass will be transmitted to the detector giving rise to the normal mass spectrum, a series of sharp, well-defined signals for each m/z value for the ions present. If, however, an ion $M_1^{\bullet+}$ is metastable and fragments in flight

$$M_1^{\bullet+} \rightarrow M_2^+ + M_3^\bullet$$

the fragment ion M_2^+ will have less translational energy than an ion M_2^+ produced in the ion source. The fragment ion's translational energy is $\frac{1}{2}M_2v_1^2$ and is equal to $\frac{1}{2}(M_2^2)v_2^2$ given that for source ions $\frac{1}{2}M_1v_1^2 = \frac{1}{2}M_2v_2^2$. The fragment ion M_2^+ from metastable $M_1^{\bullet+}$ will thus appear in the normal mass spectrum at an apparent mass (M_2^2/M_1).

For sector mass spectrometers, V_{acc} is typically 2–10 kV and the timescale for ions to pass from ion source to detector is typically in the range 5–50 μs for ions of $m/z \sim 100$.

The signals at (M_2^2/M_1) were observed in the majority of cases to be much wider than those of normal fragment ions and were typically of Gaussian profile.

Note that the presence of these diffuse peaks at nonintegral m/z values provides the observer with direct proof that a given fragmentation has indeed taken place, e.g.

$$\begin{array}{lll} C_5H_5^+ & \rightarrow C_3H_3^+ & + C_2H_2 \\ m/z\,65 & m/z\,39 & \end{array}$$

This metastable dissociation of $C_5H_5^+$ ions will produce a peak at apparent mass 23.4 Da. Clearly these phenomena can (and do) assist in the interpretation of mass spectra by unequivocally linking a fragment ion with a specific precursor.

The reason for the diffuse shape, i.e. Gaussian profile, of metastable peaks is that upon dissociation a fraction of the internal energy of $M_1^{\bullet+}$ is converted into translational degrees of freedom of the products M_2^+ and $M_3^\bullet$. This kinetic energy is released isotropically and results in the broadening of the metastable peak relative to the width of a beam of undissociating ions from the source. In a tandem mass spectrometer of BE geometry, under conditions of good energy resolution, the shape of a metastable peak may be examined in detail and physicochemical information derived therefrom. In such a tandem mass spectrometer (shown in **Figure 1**) the electromagnet (B) is used to select an ion $M_1^{\bullet+}$ of interest. Under normal operating conditions the electric sector potential (E_s) is set to transmit all the ions from the ion source. If however, $M_1^{\bullet+}$ is metastable during its flight through the (second) field-free region between the two analysers, the fragment ion M_n^+ (or ions) will only be transmitted through the electric sector if its potential is reduced in the ratio $M_n(E_s)/M_1$. Scanning this potential from zero to E_s will transmit all fragments from metastable $M_1^{\bullet+}$ ions at appropriate $M_n(E_s)/M_1$ values. Such a spectrum is known as a mass-selected-ion kinetic energy spectrum (MIKE).

Metastable peak shapes and their significance: kinetic energy release

As stated above, the majority of metastable peaks have a Gaussian-type profile (**Figure 2A**). Indeed many peaks can be described by a simple variant of the Gaussian formula, $h = \exp(-\alpha w^2)$ (where h is the fraction of the peak height at which the width, w, is

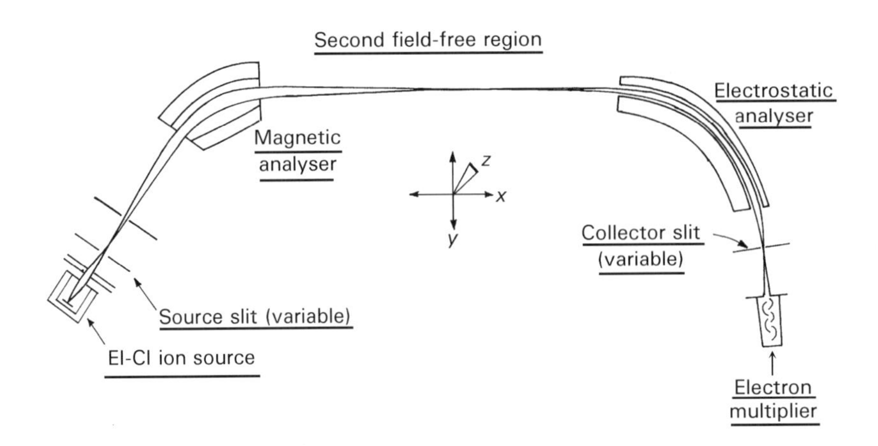

Figure 1 Schematic drawing of a tandem mass spectrometer of BE geometry.

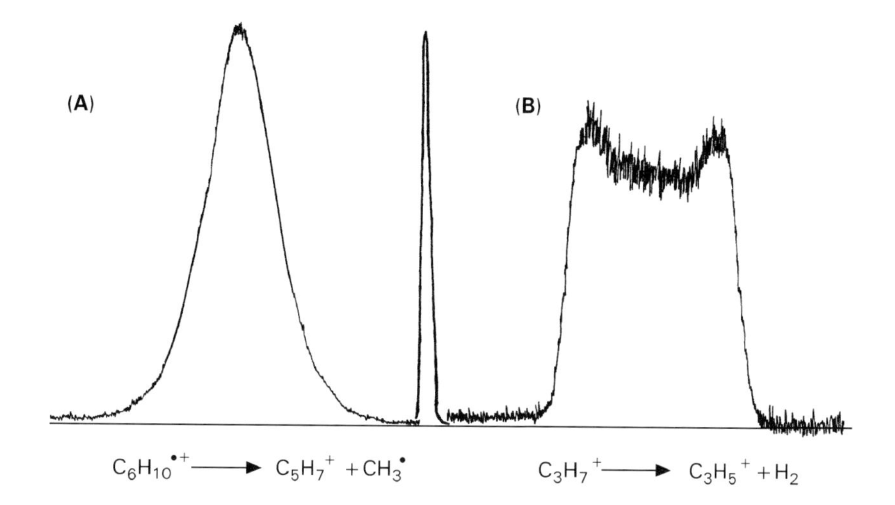

Figure 2 Simple Gaussian (A) and 'dished' (B) metastable peaks. For comparison a typical profile for an undissociating ion beam is shown between them.

measured and $\alpha = \ln 2$) where the index 2 is replaced by n, taking on a range of values from about 1.4 to 2.1. Note that the distribution of released energies derived from the exact Gaussian function is Maxwell–Boltzmann in form.

A second class of peak shapes are the so-called flat-topped or 'dished' metastable peaks (**Figure 2B**). The dish in such peaks is instrument-dependent and arises from the magnitude of the kinetic energy release (KER) and its distribution (see below). As the peak is recorded by sweeping the electric sector voltage over the appropriate range, all fragment ions having large components of translational energy released in the x- and y-axial directions (see **Figure 1**) will be recorded. Those having large z-axial components may, however, be unable to pass the collector slit because of its finite height. These 'lost' ions result in the observed dish. The effect is only important for dissociations in which the magnitude of the KER is large (see below).

The relationship between the metastable peak characteristics and the magnitude of the KER (T_h) is given by

$$T_\mathrm{h} = \frac{M_1{}^2}{16\, M_2 M_3 E_\mathrm{s}} (W_h{}^2 - W_\mathrm{m}{}^2)$$

where W_h is the width of the energy-resolved metastable peak at fractional height h, measured in electric sector volts; W_m is the corresponding width of the $M_1^{\bullet+}$ ion beam (also in sector volts). It has become customary to report KER values corresponding to half-height peak widths, $T_{0.5}$, they are usually given in meV. Under good energy-resolution conditions

$$\frac{W_{\mathrm{m}(0.5)}}{W_{0.5}} < 0.1$$

making the main beam correction small to negligible.

For Gaussian-type peaks, $T_{0.5}$ values lie in the range 0–50 meV, while for flat-topped peaks, $T_{0.5}$ is from ~200 meV to several volts.

It is believed that the great majority of Gaussian-type metastable peaks arise from ion fragmentations with no reverse energy barrier, E_{rev} (**Figure 3**),

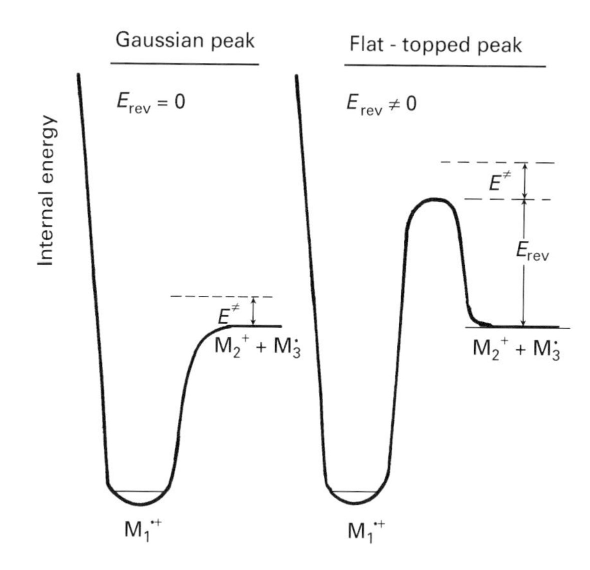

Figure 3 Potential energy profiles for ion fragmentations giving rise to Gaussian and flat-topped metastable peaks.

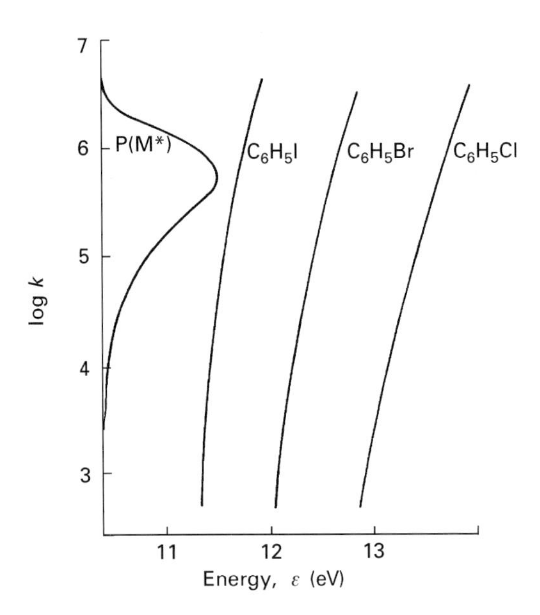

Figure 4 Log k vs. ε curves for the dissociation of the phenyl halides $C_6H_5X^{\bullet+} \rightarrow C_6H_5^+ + X^{\bullet}$. The metastable ion window is shown as P(M*). This curve represents the probability that an ion having the corresponding dissociation rate constant, k, will fragment in the metastable ion time-frame, here ~1–3 μs. At the maximum, about 11% of the beam is dissociating.

whereas flat-topped peaks are associated with a finite or large E_{rev}. This is also illustrated in **Figure 3**. The origin of the KER for the Gaussian peak is the excess energy, $E^{\ddagger}$, above the threshold for the fragmentation necessary to achieve the rate constant (k) for the process; typically k is ~10^5 s^{-1}. For the dissociation with the reverse energy barrier, E_{rev} (a fixed energy) is augmented by $E^{\ddagger}$ (sometimes referred to as the non-fixed energy).

The range of rate constants (k) accessed in the metastable time frame is easily calculated using the first-order rate equation. Direct experimental measurements of the way in which a fragmentation's rate varies with internal energy (E) are made by 'photoelectron photoion coincidence' methods. A typical plot of ln k vs. ε curves with the metastable ion window included is shown in **Figure 4**. In general, the fraction of $E^{\ddagger}$ that is released as translational energy is small, about 3–10%. For dished peaks, much larger fractions of E_{rev} may be released.

Composite metastable peaks are also known and these can arise from the following situations:

(a) A single ion structure $M_1^{\bullet+}$, dissociates metastably via two competing reactions to give isomeric fragment ions.
(b) A given ion $M_1^{\bullet+}$, has two stable isomeric forms (in an ion source) which metastably fragment via different transition states either to give fragment ions of different structure or to give fragment ions having only one structure.

Two examples are shown in **Figure 5**.

Collision induced dissociations

The early experiments on collision induced dissociations (CID) were performed using sector instruments such as that shown in **Figure 1**. A small constricted region or cell within the second field-free region close to the ion beam focus, with a diffusion pump placed thereunder, has a target gas admitted to it. The differential pumping of the region is such that

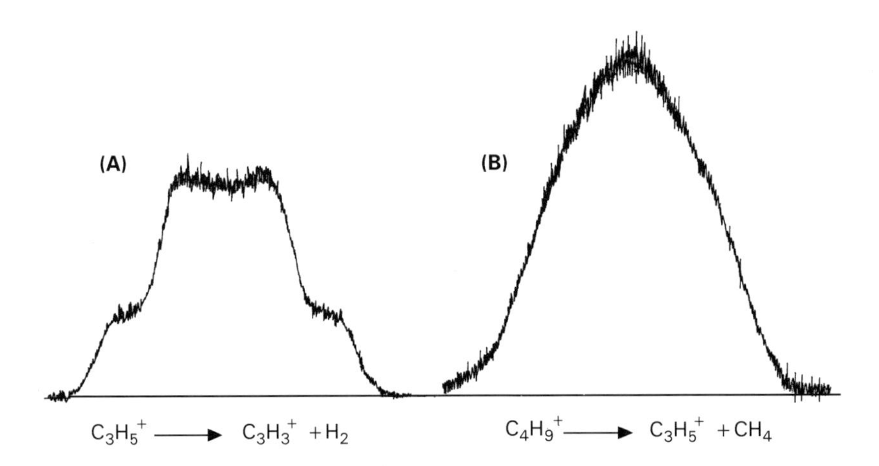

Figure 5 Composite metastable peaks. (A) Peak for $C_3H_5^+$ (m/z 41) ions dissociating to give isomeric fragment ions $C_3H_3^+$ (m/z 39). The broad signal corresponds to the production of the cyclopropenium cation, while the narrower component is for the generation of the propargyl isomer $[HCCCH_2]^+$. (B) Signal for isomeric $C_4H_9^+$ ions, fragmenting via different transition states, to produce by methane loss the 2-propenyl cation, $[CH_3{}^+CCH_2]$ (not the thermochemically more stable allyl isomer $[CH_2CHCH_2]^+$).

very little gas diffuses into the high-vacuum region outside the cell. The cell, 2–3 cm long, is only a small zone within the second field-free region, ~1 m long. When sufficient gas has been admitted to reduce the intensity of the mass-selected ion beam by 10% (a condition corresponding to predominantly single collisions) the ion kinetic energy spectrum is recorded. Unlike the metastable-ion (MI) mass spectrum, the CID mass spectrum contains many signals and is similar in appearance to the normal mass spectrum of $M_1^{\bullet+}$ (with M_1 being a stable neutral molecule).

The peaks in this CID mass spectrum are all broadened by kinetic energy release, making the spectrum rather poorly resolved. Nevertheless, as described elsewhere, the CID mass spectrum can be used as an invaluable tool in ion structure assignments. If a charge of say 500 V is placed on the collision cell (necessitating its electrical isolation from the apparatus), then MI and CID peaks may be separated. MI peaks remain undisplaced, whereas all CID peaks are shifted to higher energy (positive charge) or lower energy (negative charge).

Note that poor background pressures in the field-free regions can lead to unwanted CID contributions to MI mass spectra. In general, more than four peaks in a MI mass spectrum may be regarded as showing possible high vacuum failure. A consideration of the likely energy requirements for the observed fragmentations may also help to resolve this problem.

Metastable ions and ion structures

In the mid 1960s the first semiquantitative relationship between metastable-peak intensities and the structure of a fragmenting ion was proposed. This so-called 'metastable-peak abundance-ratio test' was based on the premise that when two (or more) competing fragmentations from the same ion give reasonably intense metastable peaks, then the ratio of their abundances may be used as a criterion for ion structure. Thus, if dissociating ions of a given molecular formula and *m*/*z* ratio, derived from a variety of precursor species, all display the same metastable peaks having similar relative abundances, then it can reasonably (but tentatively) be concluded that the fragmenting ions have the same structure. It is also possible that the fragmenting species consist of mixtures of common composition but having different structures.

The chief problem with this simple criterion was to decide what constitutes a 'significant' variation in metastable peak abundances. Indeed it was quickly realized that the ratio will be susceptible to the distribution of internal energies among the fragmenting ions. Notwithstanding these difficulties, it was observed that many even-electron hydrocarbon cations derived from a wide range of precursor molecules had closely similar metastable-peak abundance ratios, showing the ease with which isomeric hydrocarbon cations can rearrange prior to their metastable fragmentation. A good example are the C_6H_{10} isomers, 30 of which have been examined; they all, irrespective of their initial structure, produce the cyclopentenyl, $C_5H_7^+$, cation when their molecular ions lose a $CH_3^\bullet$ radical. The metastable ion behaviours of the $C_5H_7^+$ fragment ions are identical.

A more stringent test for ion structure comes from the examination of energy-resolved metastable peaks. **Figure 6** shows the signals for loss of a hydrogen atom from the isomeric $C_2H_4O^{\bullet+}$ ions, $CH_3CHO^{\bullet+}$, cy-$CH_2OCH_2^{\bullet+}$ and $CH_2CHOH^{\bullet+}$. The shapes are clearly structure-characteristic and so could, for example, serve as identifiers for $C_2H_4O^{\bullet+}$ *m*/*z* 44 fragment ions generated from different precursor molecules. Thus, for example, it has been shown that ionized 2-propanol, $CH_3CH(OH)CH_3^{\bullet+}$, loses CH_4 by two competing eliminations, one yielding $CH_3CHO^{\bullet+}$, and the other $CH_2CHOH^{\bullet+}$. The metastable peak for H loss from *m*/*z* 44 being composed of the two right-hand signals in **Figure 6**, the narrow component sitting atop the broad.

It is worth noting however, that the above observations are phenomena that only allow the above $C_2H_4O^{\bullet+}$ ions to be distinguished. The details of the physical chemistry of the H-loss processes are known and, for example, the structure from which $CH_2CHOH^{\bullet+}$ ions actually lose their H atom is the (uncommon) $C_2H_4O^{\bullet+}$ isomer, the carbene ion $CH_3^{\bullet+}COH$.

Isotopic labelling and metastable peaks

In studying ion fragmentation mechanisms, particularly molecular eliminations, it is important to discover which atoms are lost in the process. The use of isotopic labels to solve such problems is a long-established practice in organic chemistry. In general, the effect of label atoms on a normal EI mass spectrum does not provide definitive evidence for a reaction mechanism; this results from the possible multiple origins of fragment ions, especially those of lower *m*/*z* ratio. However, as described above, a metastable peak exactly defines an ion dissociation pathway and so here isotopic labelling has great utility. In particular, if MIKE mass spectra are examined, the incorporation of label in the sample molecule need not be complete; mass selection

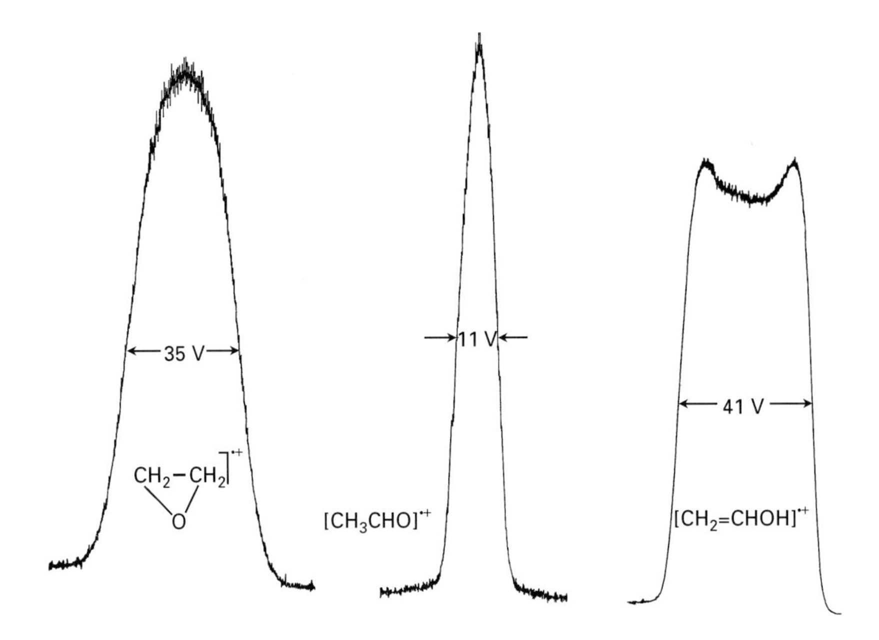

Figure 6 The energy-resolved metastable peaks for the loss of $H^{\bullet}$ from three isomeric $C_2H_4O^{\bullet+}$ ions. Note that the peak widths do not have a common scale. The half-height widths for the three signals are 35, 11 and 41 V from left to right.

allows the isolation of an ion containing the desired isotope(s). This is useful when for example, H,D exchange is performed in the mass spectrometer inlet system using D_2O to exchange hydroxyl, amino or even α-to-keto hydrogen atoms.

An important result of labelling experiments on metastable ions was the demonstration that in many species, especially hydrocarbon cations, complete loss of the positional identity of H and C atoms preceded fragmentation. This, of course, obscures all reaction pathways and so structures of fragmenting ions and their products have to be determined by other routes. Such statistical loss of labelled atoms is referred to as 'atom scrambling'. The butyl cations, $C_4H_9^+$, can serve as an example. $CD_3(CH_3)_2C^+$ loses methane as CD_3H, CD_2H_2, CDH_3 and CH_4 in the statistical random ratio of 6!/5! to 3!6!/2!4!2! to 3!6!/2!3!3! to 6!/4!2! or 2:15:20:5 and $(CH_3)_3{}^{13}C^+$ loses $^{13}CH_4$ to $^{12}CH_4$ in the ratio of 1:3. All $C_4H_9^+$ isomers behave similarly, showing that at energies below the dissociation limit for CH_4 loss, many isomeric structures are accessed. Isotopic labelling and metastable ion studies can also reveal unexpected mechanisms, hidden in the normal mass spectrum. Two examples should suffice. The mass spectrum of benzoic acid, C_6H_5COOH, is very simple with the only major fragments corresponding to simple bond cleavages; e.g. $C_6H_5COOH^{\bullet+}$, $C_6H_5{}^+CO$ and $C_6H_5^+$, are the dominant signals. Metastable $C_6H_5COOH^{\bullet+}$ ions lose $OH^{\bullet}$ as expected, but astonishingly, metastable $C_6H_5COOD^{\bullet+}$ ions lose $OD^{\bullet}$ *and* $OH^{\bullet}$, in the ratio of 1:2. That a hidden exchange between carboxyl and the two *ortho* H atoms has taken place was shown by metastable *ortho*-dideuterio benzoic acid ions losing $OD^{\bullet}$ and $OH^{\bullet}$ in exactly the inverse ratio.

The unexpected complexity of ion rearrangement processes is also well exemplified by the metastable loss of HCN from ionized benzonitrile $C_6H_5CN^{\bullet+}$. Here, only some 5–7% of the nitrile carbon is present in the HCN eliminated. The participation of linear isomers of the aromatic nitrile ion has been demonstrated.

Neutral species lost in metastable ion fragmentations

When metastable ions fragment, a neutral radical or molecule is generated. In most circumstances the identity of this neutral is not problematic and for example, is commonly a small molecule, H_2O, CH_4, CO, C_2H_4 or free radical $CH_3^{\bullet}$, $OH^{\bullet}$, $C_2H_5^{\bullet}$, $HC^{\bullet}O$, etc. The distinction between isobaric species (e.g. CO, C_2H_4, $C_2H_5^{\bullet}$, $HC^{\bullet}O$) may be made by the simple experiment described below.

If an ion-beam deflector electrode is placed in front of the collision cell in the second field-free region of a sector mass spectrometer (see **Figure 1**), then all mass-selected ions and their charged fragments may be deflected from the beam path, and only the neutral products from ion fragmentation will enter the collision cell. Neutral fragments having kV translational energies are readily ionized by collision with a target gas (e.g. He or O_2), producing the characteristic mass spectrum of the neutral species.

This technique has, for example, shown that HNC rather than HCN is lost from metastable aniline molecular ions and that ionized methyl acetate loses $^{\bullet}CH_2OH$ radicals rather than $^{\bullet}OCH_3$ as the lowest-energy fragmentation route.

List of symbols

e = electron charge; E = internal energy; E_s = electric sector potential; E_{rev} = reverse energy barrier; $E^{\ddagger}$ = excess energy; h = fraction of the peak height at which w is measured; $T_{0.5}$ = kinetic energy release calculated using the metastable peak width at half-height; v_n = ion velocities; V_{acc} = ion acceleration potential; w = peak width; W_h = width of energy-resolved metastable peak at fractional height h; W_m = width of the $M_1^{\bullet+}$ ion beam; $\alpha = \ln 2$.

See also: **Fragmentation in Mass Spectrometry; Ion Energetics in Mass Spectrometry; Ion Structures in Mass Spectrometry; Isotopic Labelling in Mass Spectrometry; Neutralization-Reionization in Mass Spectrometry; Photoelectron-Photoion Coincidence Methods in Mass Spectrometry (PEPICO); Sector Mass Spectrometers.**

Further reading

Burgers PC, Holmes JL, Mommers AA, Szulejko JE and Terlouw JK (1984) Collisionally induced dissociative ionization of the neutral products from unimolecular ion fragmentations. *Organic Mass Spectrometry* **19**: 442–447.

Cooks RG, Beynon JH, Caprioli RM and Lester GR (1973) *Metastable Ions*. Amsterdam: Elsevier, 1–296.

Derrick PJ and Donchi KF (1983) Mass spectrometry. In: Bamford CH and Tipper CFH (eds) *Comprehensive Chemical Kinetics* Vol 24, pp 53–247. Amsterdam: Elsevier.

Hipple JA, Fox RE and Condon EU (1946) Diffuse signals in mass spectra. *Physical Review* **69**: 347–351.

Holmes JL (1985) Assigning structure to ions in the gas phase. *Organic Mass Spectrometry* **20**: 169–183.

Holmes JL and Benoit FM (1972) Metastable ions in mass spectrometry. In: Maccoll A (ed) *MTP International Review of Science, Physical Chemistry, Series One*, Vol 5, pp 259–300. London: Butterworth.

Holmes JL and Terlouw JK (1980) The scope of metastable peak studies. *Organic Mass Spectrometry* **15**: 383–397.

Molenaar-Langeveld TA, Fokkens RH and Nibbering NMM (1986) An unusual pathway for the elimination of HCN from ionized benzonitrile. *Organic Mass Spectrometry* **21**: 15–22.

Microwave and Radiowave Spectroscopy, Applications

G Wlodarczak, Université des Sciences et Technologies de Lille, Villeneuve d'Ascq, France

VIBRATIONAL, ROTATIONAL & RAMAN SPECTROSCOPIES
Applications

Microwave spectroscopy covers, typically, the frequency range from 1–2 GHz to several THz. In this spectral region the rotational spectra of molecules which possess a permanent (or induced) dipole moment can be observed. The analysis of these spectra gives information on the geometrical structure of the molecule and the centrifugal distortion effects. Other information such as dipole moment, quadrupole coupling constants, spin rotation constants can also be determined. In the case of molecules that present large amplitude motions, like the internal rotation of a methyl group or ring deformation for cyclic compounds, the potential barrier is also accessible from the observation of line splittings. The large selectivity of microwave spectroscopy allows the diagnostic of interstellar medium, planetary atmospheres and minor components of the earth's atmosphere.

The number of molecules which have been studied by this method is increasing every year. Their complexity ranges from diatomic molecules to moderate size molecules (up to 30 atoms). Unstable species such as molecular ions, radicals, reactive molecules, molecular complexes, etc. can also be studied. The resulting molecular constants are regularly the subject of extensive compilations.

Spectrum measurements

Microwave spectroscopy uses tunable coherent sources of radiation such as microwave synthetizers, solid state oscillators (Gunn diodes) or electronic tubes (klystrons). These oscillators can be operated in their fundamental mode (up to 120 GHz) but harmonic generation is commonly realized with frequency multipliers up to 500 GHz, and has been used to reach 1 THz on occasions. Backward wave oscillators are available up to 1.2 THz in their fundamental mode. **Figures 1** and **2** show typical rotational spectra recorded with this type of sources. Different techniques can be used to work in the THz region:

- far-infrared laser tunable sideband generation,
- mixing of two infrared radiations (CO_2 lasers) and a microwave tunable radiation,
- photomixing of two diode lasers,
- far-infrared Fourier transform spectroscopy.

The detectors are generally Schottky diodes or InSb helium-cooled bolometers in the millimetre and submillimetre wave region.

Stark modulation has been widely used at microwave frequencies, while source modulation is the most common technique for millimetre and submillimetre spectroscopy. Recently a new technique, microwave Fourier transform spectroscopy, has been developed to obtain increased frequency resolution and sensitivity. A macroscopic polarization is created in the sample by a microwave pulse of appropriate strength and duration. The sample emits then a signal which decreases in time owing to relaxation

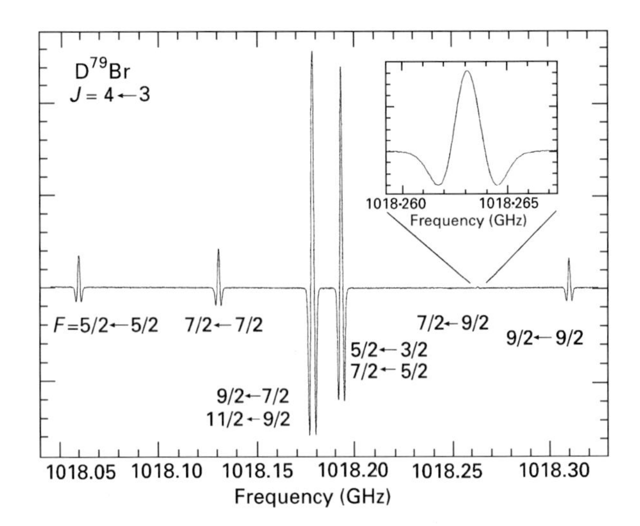

Figure 2 A 32-MHz scan showing the ^{79}Br hyperfine structure of the $J = 4 \leftarrow 3$ transition of $D^{79}Br$ at 1.018 THz. Reproduced with permission from Saleck AH, Klaus T, Belov S and Winnewisser G (1996) THz rotational spectra of HBr isotopomers in their v = 0,1 states. *Z. Naturforsch* **51A**: 898.

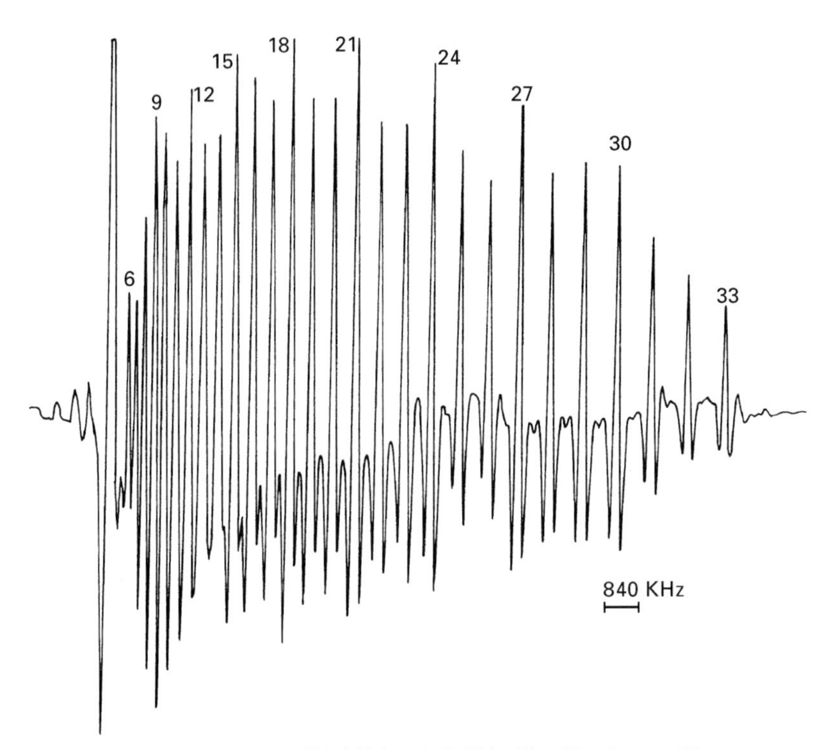

Figure 1 The $J = 34 \leftarrow 33$ transitions of trioxane [$(H_2CO)_3$] at 358 GHz. The $K = 3n$ transitions are easy to recognize and their quantum number K is indicated. Reproduced with permission from Gadhi J, Wlodarczak G, Boucher D and Demaison J (1989) The submillimeter-wave spectrum of trioxane. *Journal of Molecular Spectroscopy* **133**: 406.

effects. The Fourier spectrum of this decay signal contains the frequencies corresponding to rotational transitions of the sample. This technique is available between 1 and 40 GHz.

The sample is prepared in the gas phase, at low pressures (typically 10–30 mtorr) for stable or not too unstable species. Reactive species are prepared inside a free space cell, usually made of a pyrex glass tube, by different techniques: microwave discharge, a.c. or d.c. glow discharge, thermal decomposition or reaction, photochemical decomposition. This last process needs high power excimer or CO_2 lasers, but is much cleaner than electrical discharges. **Figure 3** shows spectral lines of linear C_3H, produced in a glow discharge of C_2H_2, CO and H_2 in a cell cooled to liquid nitrogen temperature to obtain stronger signals. Molecular ions can also be observed within electrical discharges: the sensitivity of the spectrometer is then increased by using the velocity modulation technique. This technique is based on the Doppler shift of ion spectra caused by the discharge electric field. An additive magnetic field usually produces an enhancement of the spectrum. For van der Waals or hydrogen-bonded complexes a supersonic expansion is generally needed. The monomers are diluted (several %) in rare gases (usually argon). The collisionless regime in molecular beams prevent the rapid destruction of the molecular complexes. **Figure 4** represents part of the spectrum of the phenol–water complex.

The isotopic species can be studied in natural abundance for ^{13}C, ^{15}N and most of the elements. In the case of deuterated isotopomers an enriched sample is generally needed. The line frequency measurements are made with a very high accuracy: 1 kHz for microwave Fourier transform measurements between 2 and 20 GHz to 50–300 kHz for far-infrared measurements up to 2–3 THz. The line shape is usually a Doppler or a Voigt profile if the collisional broadening becomes important. The Doppler half-maximum halfwidth is given by $\Delta\nu_D$ (MHz) $= 3.581\times10^{-7}$ $(T/M)^{1/2}$ ν_0, where T is the temperature in K, M the molecular weight in atomic mass units, ν_0 the frequency of the transition in MHz. For example the $J = 3–2$ transition of CO at 345.8 GHz has a Doppler width of 400 kHz.

The collisional broadening increases linearly with pressure: $\Delta\nu_L = \gamma_L P$, where $\Delta\nu_L$ is the collisional half-maximum halfwidth. Typical values for collisional

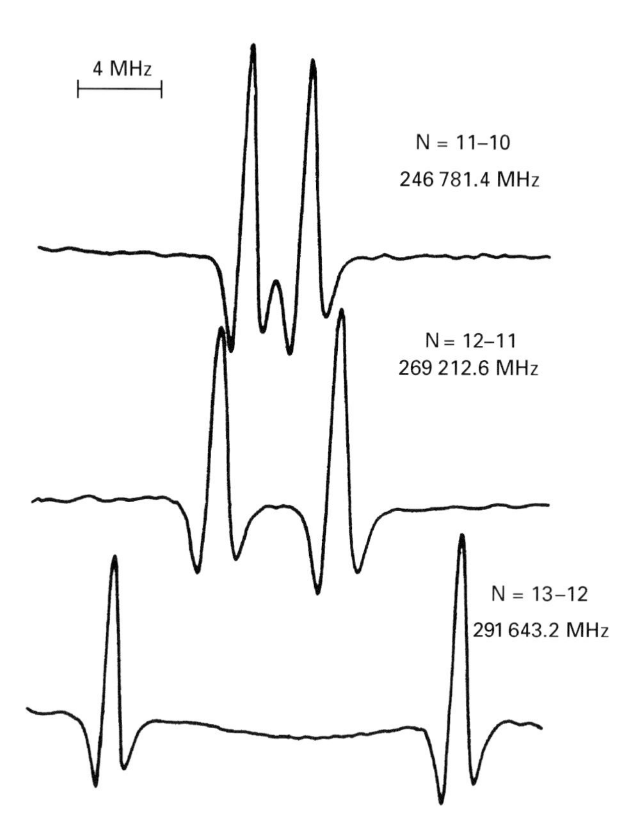

Figure 3 Typical traces of the spectral lines for the ν_4 ($^2\Sigma\mu$) state of C_3H (CCH bending mode). The centre frequencies of the doublet lines are given in the figure. The spin splitting becomes large as *N* increases. Reproduced with permission from Yamamuto S, Saito S, Suzuki H *et al* (1990) Laboratory microwave spectroscopy of the linear C_3H and C_3D radicals and related astronomical observation. *Astrophysical Journal* **348**: 363.

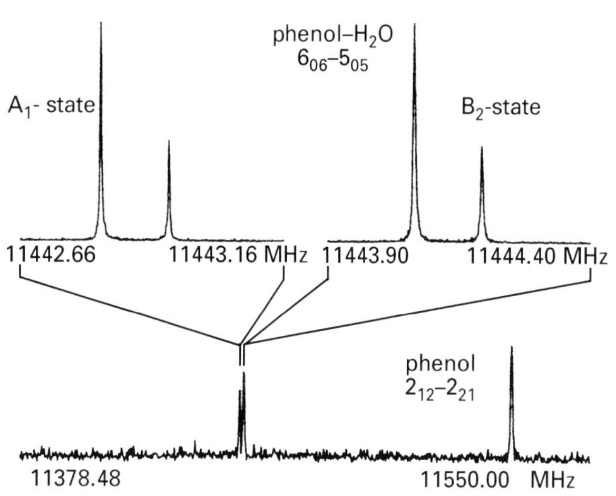

Figure 4 A 172 MHz broad-band scan (lower trace) and high-resolution spectrum (upper trace) of phenol and water in helium at a stagnation pressure of 100 kPa. Three lines can be recognized in the low-resolution spectrum: a doublet consisting of a strong and a weak component of the $6_{06} \leftarrow 5_{05}$ transition of the phenol–H_2O complex and the low frequency component of the $2_{12} \leftarrow 2_{21}$ internal rotation doublet of the phenol monomer. In the high-resolution spectrum, the lines appear as Doppler doublets. Reproduced with permission from Gerhards M, Schmitt M, Kleinemans K and Stahl W (1996). The structure of phenol–water obtained by microwave spectroscopy. *Journal of Chemical Physics* **104**: 967.

broadening parameters γ_L are 1–10 MHz torr^{-1}, as shown in **Figure 5**; the value of γ_L generally increases when the temperature decreases. At low pressures the line broadening is mostly due to the Doppler effect and is the general cause which limits the resolution of the spectrometer. The resolution can be increased by using sub-Doppler techniques such as saturated absorption spectroscopy (or Lamb dip spectroscopy) or by generating a microwave radiation perpendicular to the motions of the molecules when using a molecular beam.

Most often, molecules are studied in their ground electronic and vibrational state but rotational spectra in vibrationally excited states (up to 1000–1500 cm^{-1}) are commonly observed. **Figure 6** shows a stick diagram, reproducing the relative intensities of the lines, of the rotational spectrum of C_3S in the ground and various excited states: the vibrational energies for $v_5 = 1$ and $v_4 = 1$ are, respectively, 150 and 490 cm^{-1}. The observation of rotational spectra in excited electronic states is more difficult because of their short lifetime.

Rotational constants and geometrical structure

Most of the spectra recorded in the microwave region are rotational spectra. Among the molecular parameters used for the modelling of these spectra the most important are rotational constants. These constants are related to the principal moments of inertia by the relation $A = h/(8\pi^2 I_a)$. For light molecules such as hydrides, the spectrum lies in the millimetre or submillimetre wave region: for HCl the $J = 1–0$ transition, which is the lowest frequency transition, lies near 626 GHz. This is one of the reasons for the development of spectroscopy in the

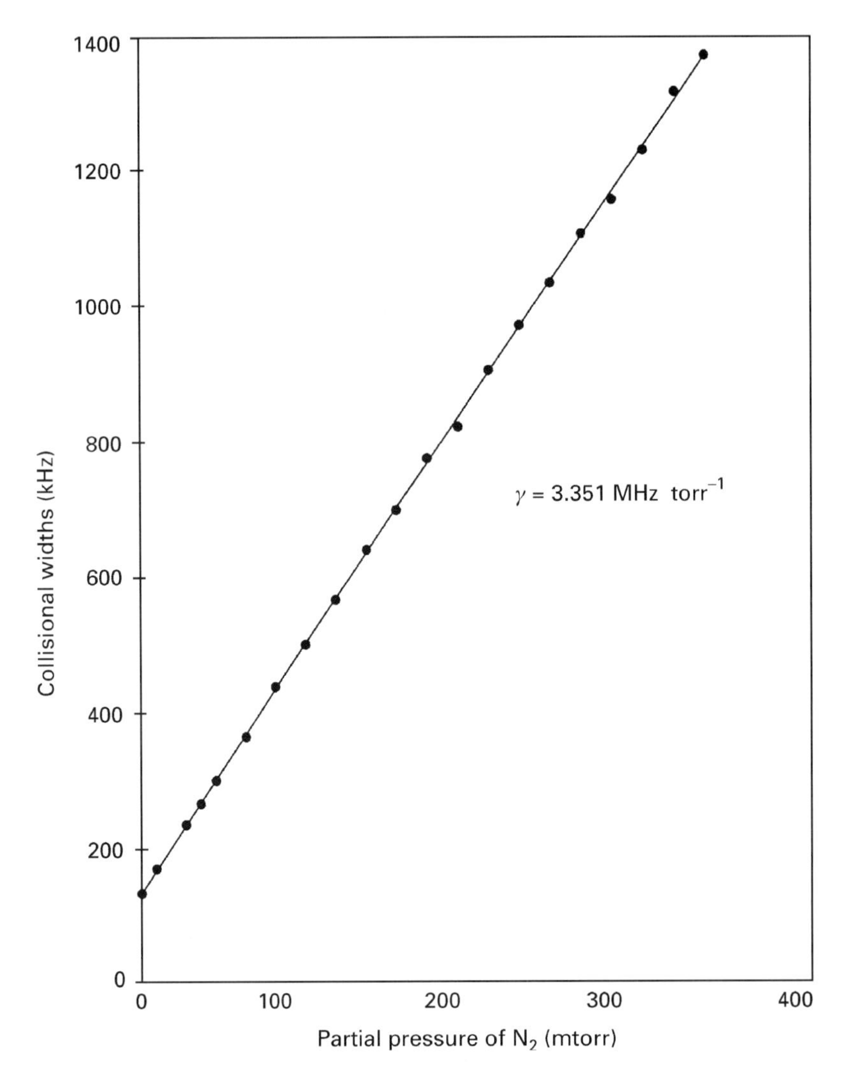

Figure 5 Nitrogen-broadened widths of the $J = 8 \leftarrow 7$ line of N_2O at $T = 295$ K; pressure of N_2O: 30 mtorr.

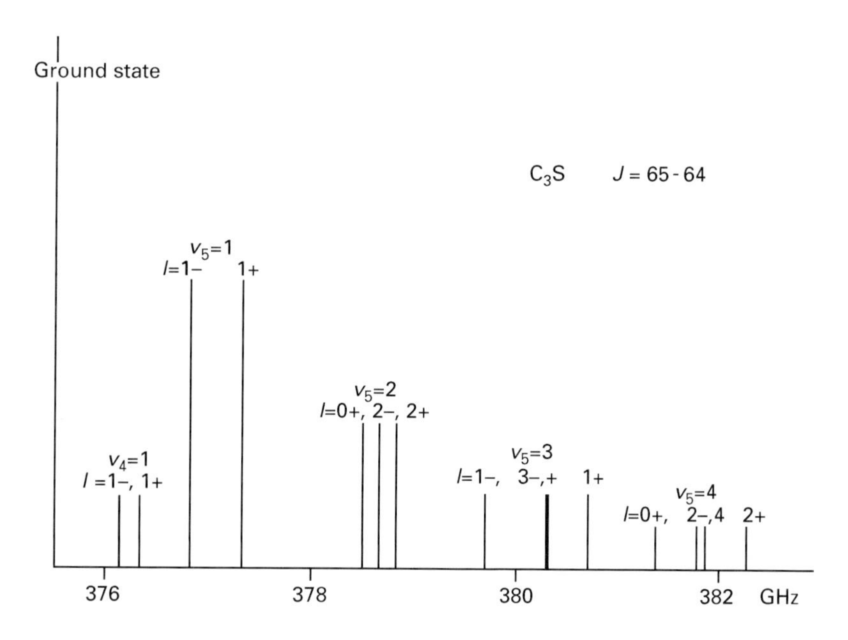

Figure 6 Spectral pattern and relative intensities of C_3S transition lines. Reproduced with permission from Tang J and Saito S (1995) Microwave spectrum of the C_3S molecule in the vibrationally excited states of bending modes v_4 and v_5. *Journal of Molecular Spectroscopy* **169**: 92.

terahertz domain. For heavy molecules the spectrum lies mostly in the microwave region. Nevertheless high J transitions are measured at millimetre and submillimetre wavelengths. This allows the analysis of the centrifugal distortion. The rotational Hamiltonian is generally expressed as a series expansion of the even powers of the angular momentum operator. The first term contains the rotational constants, the following terms contain, respectively, the quartic, sextic, octic, etc. centrifugal distorsion constants. The effects of these terms are important at high rotational excitation. This expansion is usually satisfactory except for floppy molecules, i.e. molecules which exhibit large amplitude motions (H_2O, CH_3OH, etc.). Quartic and sextic centrifugal distortion constants are usually determined during the analysis of the spectrum on a broad frequency range. They can be related to the force field and be estimated from *ab initio* calculations. Some empirical correlations were also found between these constants and the rotational constants: this allows also an estimation which can be useful at the beginning of the identification of a new spectrum. **Figure 7** shows an example of correlation found between a sextic centrifugal distortion constant and the rotational constant B for a class of symmetric top molecules.

The rotational constants are a source of information on the geometrical structure of the molecule. By combining the rotational constants of a parent molecule (usually the main species) and isotopically substituted daughter molecules it is now possible to determine a complete experimental structure which lies very close the equilibrium structure. These results mainly concern moderately sized molecules, containing between 3 and 8 atoms. A pure experimental equilibrium structure is difficult to obtain because the rotational constants in all fundamental

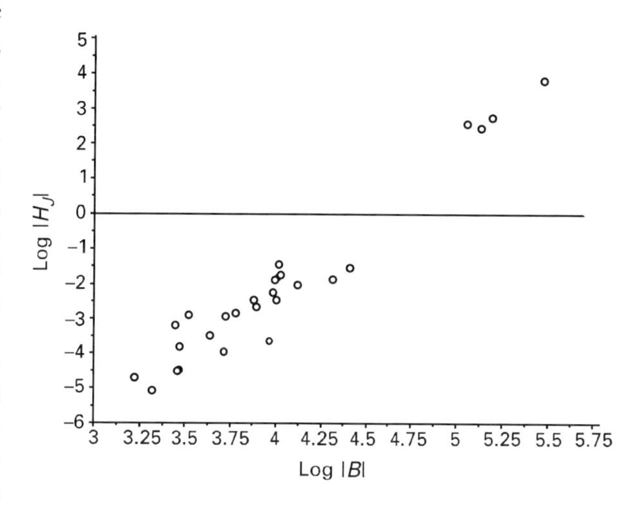

Figure 7 Plot of log $|H_J|$ against the function of log B for various C_{3v} symmetric tops: B is the rotational constant, H_J is one sextic centrifugal distortion constant. Reproduced with permission from Demaison J, Bocquet R, Chen WD, Papousek D, Boucher D and Bürger H (1994) The far-infrared spectrum of methylchloride: determination and order of magnitude of the sextic centrifugal distortion constants in symmetric tops. *Journal of Molecular Spectroscopy* **166**: 147.

vibrational states are needed to determine the equilibrium rotational constants. For the higher excited states the data are usually obtained from high resolution IR spectroscopy. In many cases a lot of interactions occur between these vibrational states (Fermi resonances, Coriolis interactions, etc.) which make it difficult to obtain unperturbed rotational constants.

Another type of problem which arises in structure determination is the ill-conditioning of the system of equations that link the moments of inertia and the structural parameters. To obtain reliable parameters we have to incorporate additional data obtained from other techniques: electron diffraction, *ab initio* calculations, empirical relations, etc. Bond distances with an accuracy of 0.001 Å and bond angles with an accuracy of 0.2° can then be determined.

For molecules that contain a greater number of atoms, different conformers are present and generally observed: their spectra are rather well separated and the comparison between their intensities gives an order of magnitude of their relative energy difference.

Determination of molecular parameters

Rotational constants and centrifugal distortion constants are not the only parameters accessible by microwave spectroscopy. In the case of linear and symmetric tops, doubly degenerate vibrationally excited states (bending modes for examples) are present and their spectra are slightly more difficult to analyse. More parameters are needed: l-type doubling constants, rotation–vibration interactions constants, etc. It should be noted that for symmetric tops the axial rotational constant is not accessible by pure rotational spectroscopy. Its determination remains a difficult problem, and needs the study of IR spectra (including hot bands and combination bands). Combined high-resolution microwave and IR studies have become the best way to determine a coherent set of molecular parameters for linear and symmetric tops. Owing to the growth in high quality experimental data, theoretical developments on the rovibrational Hamiltonian of symmetric tops have been made. The equivalence between different reductions of this Hamiltonian has been checked experimentally.

Large amplitude motions produces specific features in rotational spectra. Internal rotation of one methyl group induces a splitting of most of the lines into two components. This splitting depends on the barrier height and the geometry of the molecule. For high barriers the splitting is usually small but easily resolved by microwave Fourier transform spectroscopy. The splittings are larger in the excited torsional states but a complete theoretical description of the spectra has not yet been achieved. A prototype molecule in this field is acetaldehyde CH_3CHO.

Cyclic molecules containing an heteroatom in the cycle (O, S, etc.) also present large amplitude motions due to ring deformations, such as ring-puckering. The observation of rotational spectra in excited states is not a problem because these states are low in energy.

The behaviour of the rotational and centrifugal distortion constants with the vibrational quantum number can be related to the potential of the ring deformation (see **Figure 8**). Microwave measurements are then complementary to far-IR observations of the vibrational transitions (usually obtained at low resolution).

Hyperfine structure in rotational spectra is due to nuclei with a spin $I > \frac{1}{2}$ whose electric quadrupoles interact with the electric field gradient. The corresponding splittings depend on the nucleus involved in this interaction and its spin value. The spectra of numerous molecules containing ^{35}Cl, ^{37}Cl, ^{79}Br, ^{81}Br, ^{127}I, ^{14}N, ^{17}O, ^{33}S have been studied and the corresponding quadrupole coupling constants determined. The deuterium coupling constants have been studied more recently because the splittings are smaller (several tenths of kHz) and were observed only by very high resolving spectrometers (molecular beam maser, microwave Fourier transform spectrometers). These

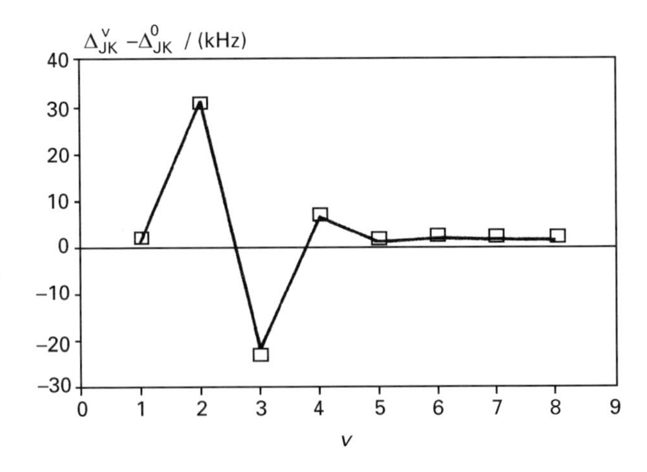

Figure 8 Comparison of observed (□) and calculated (—) variations of the quartic centrifugal distortion constant Δ_{JK} with the ring-puckering quantum number *v* for methylene cyclobutane. Reproduced with permission from Charro ME, Lopez JC, Alonso JL, Wlodarczak G and Demaison J (1993) The rotational spectrum of methylene cyclobutane. *Journal of Molecular Spectroscopy* **162**: 67.

spectrometers also allow the more or less complete resolution of the hyperfine structure due to two or more nuclei. Spin rotation and spin–spin coupling constants are also accessible by measuring the transitions involving the lowest values of the rotational quantum numbers.

Dipole moments are determined by applying an external electric field (Stark effect). The accuracy of the experimental dipole moments is about 0.001 D under good conditions. It is mainly limited by the homogeneity of the electric field. The calibration is generally done by using the OCS dipole moment as a reference. The vibrational dependence of the dipole moment can also be studied. In some cases (allene for example) the molecules possesses a vibrationally induced dipole moment and no permanent dipole moment in the ground state. In some spherical tops (CH_4, SiH_4, etc.) a very small dipole moment induced by centrifugal distortion has been measured (~10^{-5} D).

Atmospheric applications

The atmospheric transmission between 0 and 1 THz, at the ground level, is dominated by the absorption lines of water vapour and, to a less extent, by some absorption lines due to molecular oxygen (magnetic dipolar transitions), as shown in **Figure 9**. These strong, broad absorption lines are a limiting factor for the observations of other signals, i.e. absorptions due to minor components of terrestrial atmosphere or interstellar emissions. Nevertheless microwave sensors plays an important role in atmospheric measurements either in ground-based facilities or air- and spaceborne ones. The advantages of microwave sensor are the following:

- accurate measurements over the altitude range 0–100 km, mostly independent of clouds and aerosols,
- high frequency resolution and good sensitivity using superheterodyne receivers,
- accurate measurements of ozone profile and trace constituents of importance in catalytic ozone destruction cycles (ClO etc.)

In any event the data collected have to be analysed together with data obtained in the UV, visible and IR part of the electromagnetic spectrum for a reliable interpretation.

The frequency of the centre of the absorption line is not the only parameter which is necessary. The line shape is dominated by molecular collisions up to

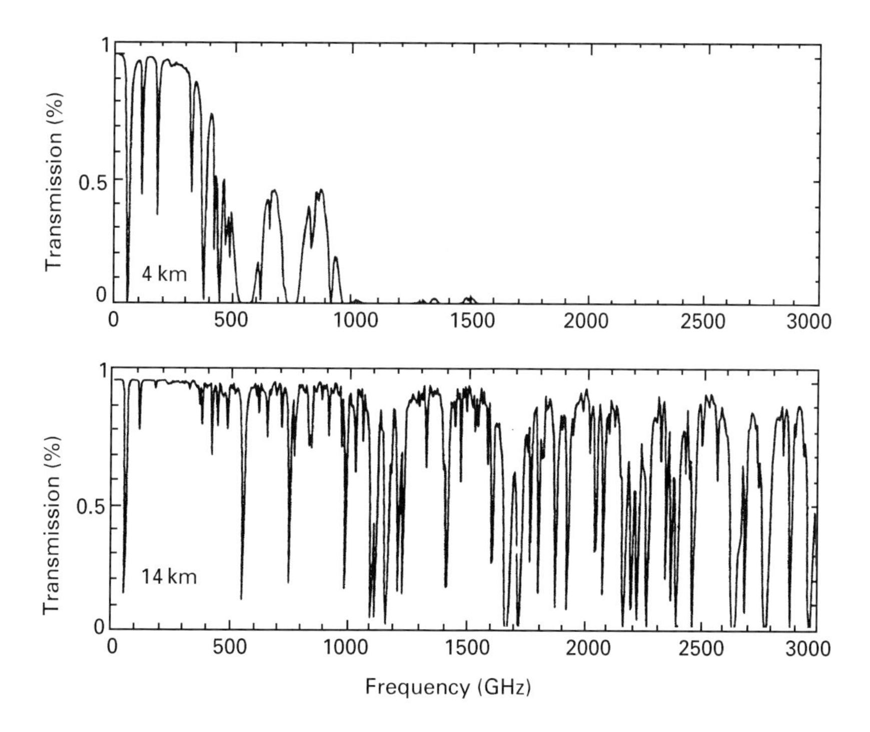

Figure 9 Atmospheric transmission in the submillimetre and far-IR from (top) a very good high-altitude ground-based site (Mauna Kea at 4.2 km) and from (bottom) an airborne observatory (e.g. KAO at 12 km). The blocked regions are mostly caused by molecular absorption. Reproduced with permission from Phillips TG and Keene J (1992) Submillimetre astronomy. *Proceedings of IEEE* **80**: 1662.

an altitude of 80 km. The collisional broadening parameters with N_2 and O_2, and their temperature dependence, are determined in the laboratory: they are of a crucial importance for data inversion. Experimental laboratory data with an accuracy of 2–3% are now obtained for the collisional broadening coefficients; the temperature dependance is usually determined with a greater uncertainty but this does not influence the data inversion too much. These laboratory data are also useful benchmarks for theoretical calculations and model testing.

Millimetre-wave sensors represent the only ground-based technique for the observation of stratospheric ClO, the abundance of which is fully correlated to ozone depletion. Moreover this technique allows a continuous observation of ClO, and the analysis of its diurnal cycle, as showing in **Figure 10**. The most frequently observed line is the $J = \frac{15}{2} - \frac{13}{2}$ transition at 278.632 GHz, which is the most intense one. This line is also one of the less blended lines, (interferences with ozone lines located in the neighbourhood are not too strong). This line is also broadened by the hyperfine components. The total line shape contains the contributions of the successive atmospheric layers, and its inversion leads to the vertical concentration profile of ClO.

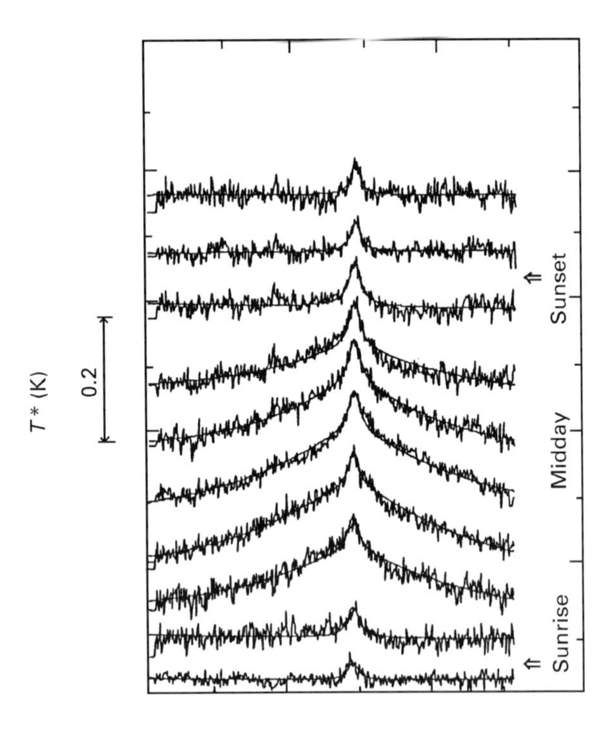

Figure 10 Diurnal variations of the stratospheric ClO lines shape over McMurdo Station, Antarctica, averaged over the period 20–24 September 1987. de Zafra RL, Jaramillo M, Barrett J, Emmons LK, Solomon P and Parrish A (1989) New observations of a large concentration of ClO in the springtime lower stratosphere over Antarctica and its implications for ozone-depleting chemistry. *Journal of Geophysical Research* **94**: 11423.

Another application of microwave spectroscopy is the analysis of pollutants. Recently, microwave Fourier transform spectrometers have been used to analyse polluted air samples, in the frequency range 10–26 GHz. The air sample is supersonically expanded in a Fabry–Perot resonator, the technique being the same as the one used for the study of molecular complexes. The difference is in the carrier gas which is now air instead of argon. Laboratory studies show that the sensitivity decreases by a factor of 30 when argon is replaced by air. Nevertheless, the sensitivity is still high enough to allow the detection of most of the polar constituents of the sample. Another advantage, already mentioned above, is the very high frequency resolution, which permits the unambiguous identification of a great number of pollutants.

Radioastronomy

One of the most fruitful application of laboratory microwave spectroscopy over the last twenty years is the analysis of the molecular content of interstellar clouds. These clouds contain gas (99% in mass) which has been mostly studied by radioastronomy, and dust, whose content has been analysed mostly by IR astronomy. The clouds rich in molecular content are dense or dark clouds (they present a large visual extinction), with a gas density of 10^3–10^6 molecules cm^{-3}, and temperatures of $T < 50K$. At these low temperatures only the low-lying quantum states of molecules can be thermally (or collisionally) excited, i.e. rotational levels. Spontaneous emission from these excited states occurs at microwave wavelengths. In some warm regions of dense clouds (star formation cores) the absorption of IR radiation produces rotational emission in excited vibrational states. Other rich chemical sources are the molecular clouds surrounding evolved old stars, such as IRC+10216, and called circumstellar clouds.

In the 1980s and 1990s a lot of radiotelescopes were built, with large antennas (diameter = 10–30 m) and sensitive receivers in the millimetre and submillimetre range. More than 100 different molecular species were found in the interstellar medium (see **Table 1**) and, for some of them, various isotopic species were also detected. The identification of interstellar species is not easy because of the high density of lines in the spectra of some interstellar clouds. A millimetre wave spectrum of the Orion nebula is shown in **Figure 11**. This is owing to the richness of the chemistry in these clouds and also to the improved sensitivity of the latest generation

Table 1 Interstellar molecules

Number of atoms										
2	*3*	*4*	*5*	*6*	*7*	*8*	*9*	*10*	*11*	*13*
H_2	H_2O	NH_3	SiH_4							
OH	H_2S	H_3O^+								
SO										
SO^+	N_2H^+									
NO	SO_2									
SiO	HNO									
SiS	SiH_2?									
SiN	H_2D^+									
NS	NH_2									
HCl										
HF										
NaCl										
KCl										
AlCl										
AlF										
PN										
NH										
CH^+	HCN	H_2CO	HC_3N	CH_3OH	HC_5N	$HCOOCH_3$	HC_7N	CH_3C_5N	HC_9N	$HC_{11}N$
CH	HNC	HNCO	C_4H	CH_3CN	CH_3CCH	CH_3C_3N	$(CH_3)_2O$	$(CH_3)_2CO$		
CC	C_2H	H_2CS	H_2CNH	CH_3NC	CH_3NH_2	CH_3COOH	CH_3CH_2OH			
CN	C_2S	HNCS	H_2C_2O	CH_3SH	CH_3CHO	C_6H_2	CH_3CH_2CN			
CO	SiC_2	C_3N	NH_2CN	NH_2CHO	CH_2CHCN	C_7H	CH_3C_4H			
CSi	HCO	l-C_3H	HCOOH	C_2H_4	C_6H		C_8H			
CS	HCO^+	*c*-C_3H	CH_4	C_5H	CH_2OCH_2					
CP	HOC^+	C_3O	*c*-C_3H_2	HC_2COH						
CO^+	OCS	C_3S	l-C_3H_2	l-H_2C_4						
	HCS^+	$HOCO^+$	CH_2CN	HC_3NH^+						
	CO_2	HCCH	C_4Si	C_4H_2						
	CCO	$HCNH^+$	HCCNC							
	MgNC	HCCN	HNCCC							
	MgCN	CH_2D^+	H_2COH^+							
	CaNC	H_2CN	C_5							
	C_3	SiC_3								
	NaCN									
	CH_2									

of radiotelescopes. The characterization of the molecules present in these dense cloud requires a knowledge of the laboratory spectra. In some cases (C_3H_2, HC_9N, etc.) the identification was first made in the interstellar medium, before laboratory evidence. Nevertheless in the case of $HC_{11}N$, the highest membrane of the cyanopolyine series, interstellar detection was claimed at the beginning of the 1980s. This molecule was recently produced in the laboratory and its rotational spectrum does not fit the interstellar line. A search for $HC_{11}N$ with the new experimental data was at first unsuccessful but, finally, a deeper search confirmed the presence of $HC_{11}N$ in the interstellar medium. A lot of laboratory studies have been devoted to this family of molecules: the rotational spectrum of $HC_{17}N$ has been observed, and numerous hydrocarbons of the type C_nH_m, with $n > m$, have been produced in discharges and their spectra analysed.

The detection of isotopomers in interstellar medium is a source of information on the elemental isotopic ratio. Molecules containing the following atoms have been detected: D, ^{13}C, ^{15}N, ^{17}O, ^{18}O, ^{33}S, ^{34}S and ^{36}S. The deuterated species are of particular interest because their abundances bring useful information on the chemical processes which take place in the peculiar conditions of the interstellar medium (isotopic fractionation).

Molecular hydrogen is the dominant molecule; the second most abundant molecule, CO, is four orders of magnitude less abundant. But H_2 has no strong transitions in the microwave regions, CO is mainly used to map interstellar clouds in our galaxy and others, and also in quasars. The observation of

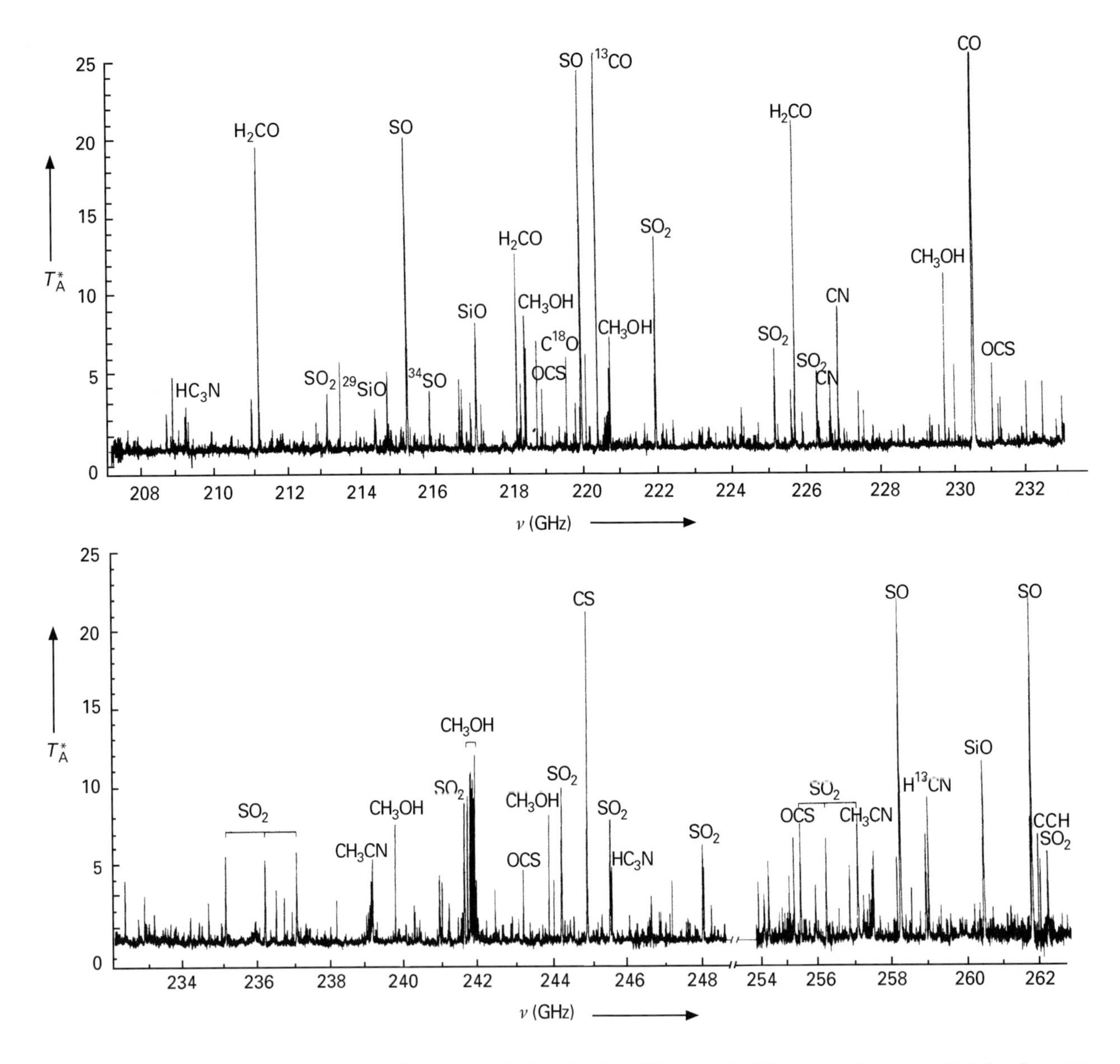

Figure 11 Millimetre wave spectrum of the Orion nebula in the direction of the so-called Kleinmann–Low area. Rotational spectra from many molecules are seen; ν = frequency and T_A^* = antenna temperature, a measure of emission intensity. Reproduced with permission from Blake GA, Sutton EC, Masson CR and Phillips TG (1987) Molecular abundances in OMC-1: the chemical composition of interstellar molecular clouds and the influence of massive star formation. *Astrophysical Journal* **315**: 621.

several lines of the same species gives information on the physical conditions in the interstellar cloud: temperature, molecular density. In the case of OH radical, the splitting of the observed microwave lines by the local magnetic field (Zeeman effect) is a way to evaluate its order of magnitude. Several molecular ions have been studied in the laboratory (H_2D^+, H_3O^+, CH_2D^+, etc.) because of their importance in interstellar chemistry, which consists mostly in gas phase ion–molecule reactions. But in many cases their reactivity prevents their interstellar detection. Radioastronomy has also been applied to the analysis of planetary atmospheres, together with infrared observations. Both CO and H_2O were detected in Mars and Venus, SO_2 in Io (a satellite of Jupiter), CO and HCN in Neptune. In Titan, a satellite of Saturn, HCN, HC_3N and CH_3CN were detected, indicating a complex photochemistry. More detailed mappings were undertaken more recently with interferometers working in the millimetre-wave region.

Millimetre astronomy has also been found to be a powerful tool for the physicochemistry of comets. This was fully demonstrated by the observations of two exceptional comets: Hyakutake (1996) and

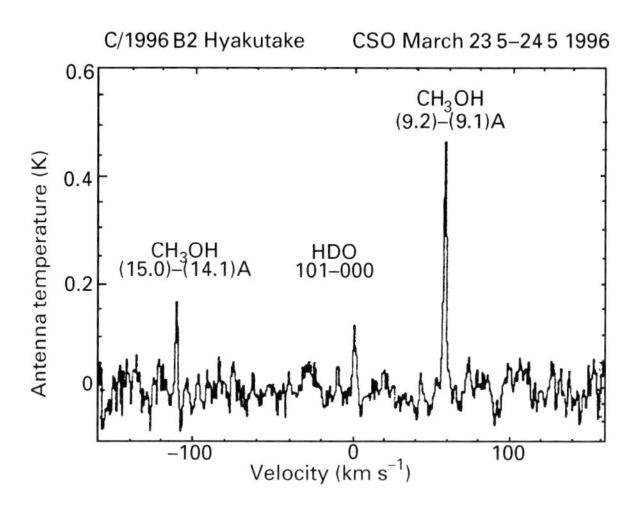

Figure 12 The $1_{10} \leftarrow 0_{00}$ HDO line at 465 GHz, observed at the Caltech Submillimetre Observatory, in comet Hyakutake. Two lines of methanol are present in the same spectrum. Reproduced by permission from Crovisier J and Bockelée-Morvan D (1997) Comets at the submillimetric wavelength in *ESA Symposium*, Grenoble, France.

Hale–Bopp (1996–1997). The newly detected molecules in these two comets are: CS, NH_3, HNC, HDO, CH_3CN, OCS, HNCO, HC_3N, SO, SO_2, HCCS, HCOOH, NH_2CHO, CN, CO^+, HCO^+ H_3O^+. This number is considerably bigger than the total number of molecules previously in comets. **Figure 12** shows the detection of HDO and methanol in the comet Hyakutake.

Increasing amounts of data are being obtained at higher frequencies, i.e. in the submillimetre region. A recent survey of Orion was made between 607 and 725 GHz, and another one between 780 and 900 GHz started. These spectral regions are well suited for the detection of light hydrides. They are limited by the atmospheric windows. A continuous coverage will be available with the future satellite observatories, which are planned for the beginning of the third millennium.

List of symbols

m = molecular weight; T = temperature (K); γ_L = collisional broadening parameter; $\Delta\nu_D$ = Doppler half-maximum halfwidth; ν_0 = transition frequency.

See also: **Atmospheric Pressure Ionization in Mass Spectrometry; Cosmochemical Applications Using Mass Spectrometry; Environmental Applications of Electronic Spectroscopy; Interstellar Molecules, Spectroscopy of; Microwave Spectrometers; Rotational Spectroscopy, Theory; Solid State NMR, Rotational Resonance; Vibrational, Rotational and Raman Spectroscopy, Historical Perspective.**

Further reading

Demaison J, Hüttner W, Tiemann E, Vogt J and Wlodarczak G (1992) *Molecular Constants mostly from Microwave, Molecular Beam, and Sub-Doppler Laser Spectroscopy*, Landolt–Börnstein, Numerical Data and Functional Relationships in Science and Technology (New Series) Group II, Vol 19. Berlin: Springer.

Encrenaz PJ, Laurent C, Gulkis S, Kollberg E and Winnewisser G (eds) (1991) *Coherent Detection at Millimetre Wavelengths and their Applications.* Les Houches Series. New York: Nova Science Publishers.

Gordy W and Cook CL (1984) *Microwave Molecular Spectra.* New York: Wiley.

Graner G, Hirota E, Iijima T, Kuchitsu K, Ramsay DA, Vogt J and Vogt N (1995) *Structure Data of Free Polyatomic Molecules*, Landolt–Börnstein, Numerical Data and Functional Relationships in Science and Technology (New Series) Group II, Vol 23. Berlin: Springer.

Kroto HW (1975) *Molecular Rotational Spectra.* London: Wiley.

Townes CH and Schawlow AL (1955) *Microwave Spectroscopy.* New York: McGraw-Hill.

Microwave Spectrometers

Marlin D Harmony, University of Kansas, Lawrence, KS, USA

Microwave radiation, defined roughly as electromagnetic radiation with a frequency in the range of 3000 to 300000 MHz (wavelengths from 10 to 0.1 cm), finds extensive use in chemistry and physics chiefly for two spectroscopic applications. The first of these involves the study of certain magnetic materials, especially paramagnetic substances, and is generally known as electron spin resonance spectroscopy. The second involves the spectroscopic study of the rotational energy states of freely rotating molecules in the gas phase. This latter field of investigation, properly known as rotational spectroscopy but universally and synonymously identified as microwave spectroscopy, is the subject matter of this article. Any instrument used to detect, measure and record the discrete and characteristic absorption of microwave radiation by gaseous molecular samples is thus commonly known as a microwave spectrometer.

General description

According to the well-known principles of quantum mechanics, the rotational energies of a rotating molecule, considered approximately as a rigid framework of atoms, are limited to certain discrete, quantized values E_i. Upon irradiation of a gaseous molecular sample by microwave radiation, an absorption of radiation is possible only if the frequency ν of the radiation satisfies the Bohr frequency relation

$$\frac{E_2 - E_1}{h} = \nu$$

where E_1 and E_2 are the initial and final rotational energies and h is Planck's constant (6.626×10^{-34} J s). When a molecule in the quantum state 1 absorbs radiation and is excited to the quantum state 2 we say a spectral transition has occurred. The spectral transitions permitted by the Bohr relation are further limited by other quantum mechanical rules known as selection rules. The net result is that a particular molecule will exhibit typically tens, hundreds or even thousands of relatively sharp, discrete, rotational absorption *lines* in the microwave spectral region. For gas samples at pressures of less than approximately 100 mtorr the frequency widths of the absorption lines are very narrow (typically 0.1–1 MHz) so the resolving power of a microwave spectrometer is very high.

Quantum mechanical and electromagnetic theory provide an additional extremely important restriction upon the occurrence of rotational transitions, namely, to a first and generally adequate approximation they can occur only for molecules having non-zero electric dipole moments. Thus, microwave spectra occur for the polar molecules of water, carbon monoxide and acetone but not for the non-polar moleculess of methane, carbon dioxide and benzene. It is worth stressing also that rotational spectra are produced only by gaseous molecules, not by liquids or solids. While this seems at first a serious limitation it should be noted that it is possible to vaporize even very refractory materials at elevated temperatures. Thus, the microwave spectrum of gaseous sodium chloride (NaCl) molecules is perfectly well known.

On the other hand, microwave spectroscopy is generally not useful for heavy molecules, i.e. those with molecular weights in excess of a few hundred atomic mass units. The reasons for this will be discussed later, but the result is that microwave spectroscopy tends to be far less generally applicable than other spectroscopic techniques such as IR or NMR spectroscopy.

Some detailed applications and theoretical aspects will be described later, but it is worthwhile noting in this general discussion that microwave spectroscopy clearly distinguishes molecular isotopic composition. Thus the microwave spectrum of carbonyl sulfide (OCS) exhibits distinct and easily identifiable spectral lines for various isotopomers such as $^{16}O^{12}C^{32}S$, $^{16}O^{13}C^{32}S$, $^{16}O^{12}C^{34}S$, $^{17}O^{12}C^{32}S$, and $^{18}O^{12}C^{32}S$ in natural abundance. This means that microwave spectra provide very specific information about the individual isotopomers rather than some molecule imagined to be composed of the elements with their average atomic masses or weights.

Experimental considerations

Microwave spectroscopic experimentation blossomed at the conclusion of World War II because of the military developments in microwave technology, especially the development of practical microwave generators such as the klystron (vacuum tube) oscillator, and of microwave detectors such as the silicon point contact mixer diode. Later developments led to the backward wave oscillator (BWO) and still more recent work in solid state electronics has led to the availability of a variety of entirely solid state microwave generators such as the Gunn diode. In accord with Maxwell's equations of electromagnetism, the wavelength of microwave radiation is perfectly adaptable for transmission in conducting metal tubing known as a waveguide or (depending upon the frequency) in specially designed coaxial cables. Microwave devices for attenuation, power splitting, impedance matching, frequency measurement and directional coupling are available. The Further reading section should be consulted for details of these and other rather specialized microwave components.

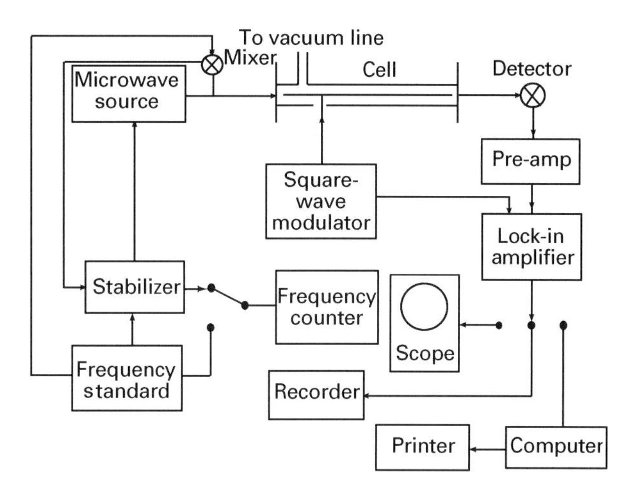

Figure 1 Block diagram of a conventional Stark-modulated microwave spectrometer.

Figure 1 presents a block diagram of a typical continuous wave (CW) microwave spectrometer. The gas sample is contained typically in a one to three metre length of standard rectangular waveguide fitted at each end with vacuum-tight windows that are transparent to the microwave radiation. The microwave generator is a klystron, BWO or solid state device, and has provision as shown for electronic apparatus for frequency stabilization and measurement. Modern microwave generators can have stabilities and accuracies as high as $1{:}10^7$ or $1{:}10^8$, which translates to 1 kHz or better. The microwave generator has provision through some electronic means for scanning the frequency over some appropriate range at selectable speeds. After passing through the sample cell, the microwave radiation is detected and further processed by a system of signal amplifiers. A critical aspect for obtaining high sensitivity is the use of square-wave electric field-modulation. This modulation, typically at a frequency of 5–100 kHz, is applied to a central electrode insulated from, and running the length of, the cell walls. The square-wave electric field, through the phenomenon of the Stark effect (see below), modulates the absorption signal at the square-wave frequency, thus permitting narrow band amplification and lock-in detection. Finally the resulting spectrum is commonly observed on either an oscilloscope synchronized with the sweep speed of the microwave generator or a strip chart recorder. Most modern instruments are computer interfaced, allowing powerful spectral manipulations and analyses. In this case the computer normally handles other tasks such as frequency range and sweep speed selection and control.

There are numerous variations to the basic design. In particular, the common rectangular waveguide gas-cell is often replaced with other structures for specialized experiments. For example, microwaves can be propagated through free space utilizing special microwave horns or antennae, so the metal surfaces can be largely eliminated for the study of reactive molecules. In this free-space design, the waveguide cell is thus replaced with a relatively large volume glass cylindrical enclosure fitted at its ends with transmitting and receiving horns. In still another design the microwaves are resonantly enclosed in a cavity whose physical size satisfies the boundary conditions for an electromagnetic standing wave according to Maxwell's equations. A particularly useful design for experiments requiring continuous high-speed pumping of unstable molecules is the microwave Fabry–Perot cavity. This design consists of two appropriately designed metal reflectors, typically circular discs with spherically machined reflecting surfaces. Microwave radiation is coupled into and out of the Fabry–Perot with appropriately designed coupling irises and the entire cavity is then enclosed in a large vacuum chamber attached to a high-speed pumping system. The unstable gas molecules of interest are produced by some means external to the cavity and are then rapidly injected into and pumped out of the cavity continuously.

The CW microwave spectrometer just described is a typical *frequency*-domain instrument. In the late 1970s it was demonstrated that *pulsed time*-domain microwave spectroscopy could be practically performed in analogy to the techniques already well known in other fields such as NMR spectroscopy. **Figure 2** depicts a block diagram of a modern version of a pulsed Fourier-transform microwave spectrometer. The particular instrument shown utilizes a Fabry–Perot cavity and a pulsed-gas nozzle, and is especially useful for detecting microwave spectra of molecular clusters in an expanding supersonic freejet.

Ignoring some of the details, which can be obtained from the Further reading section, the basic idea of the instrument is that a short pulse of monochromatic microwave radiation (approximately 1 μs in length) irradiates the gas sample in the cavity. If an appropriate transition exists within the bandwidth of the cavity (typically a few MHz), the radiation pulse produces a non-equilibrium ensemble of excited molecules which then immediately begin emitting radiation as they return to equilibrium after the pulse has dissipated. The resulting microwave emission is processed by a succession of coherent mixing processes which eventually yields a low-frequency signal for computer processing. Normally the experiment is repeated hundreds or thousands of times (at typically a 10 Hz repetition rate) to accumulate an observable signal. In accord with the theory of the coherent emission, Fourier transformation of the signal is found to produce the ordinary absorption spectral line. To scan a complete spectrum it is necessary to move the cavity resonance and microwave frequency along in small overlapping steps, repeating the entire signal accumulation process at each frequency.

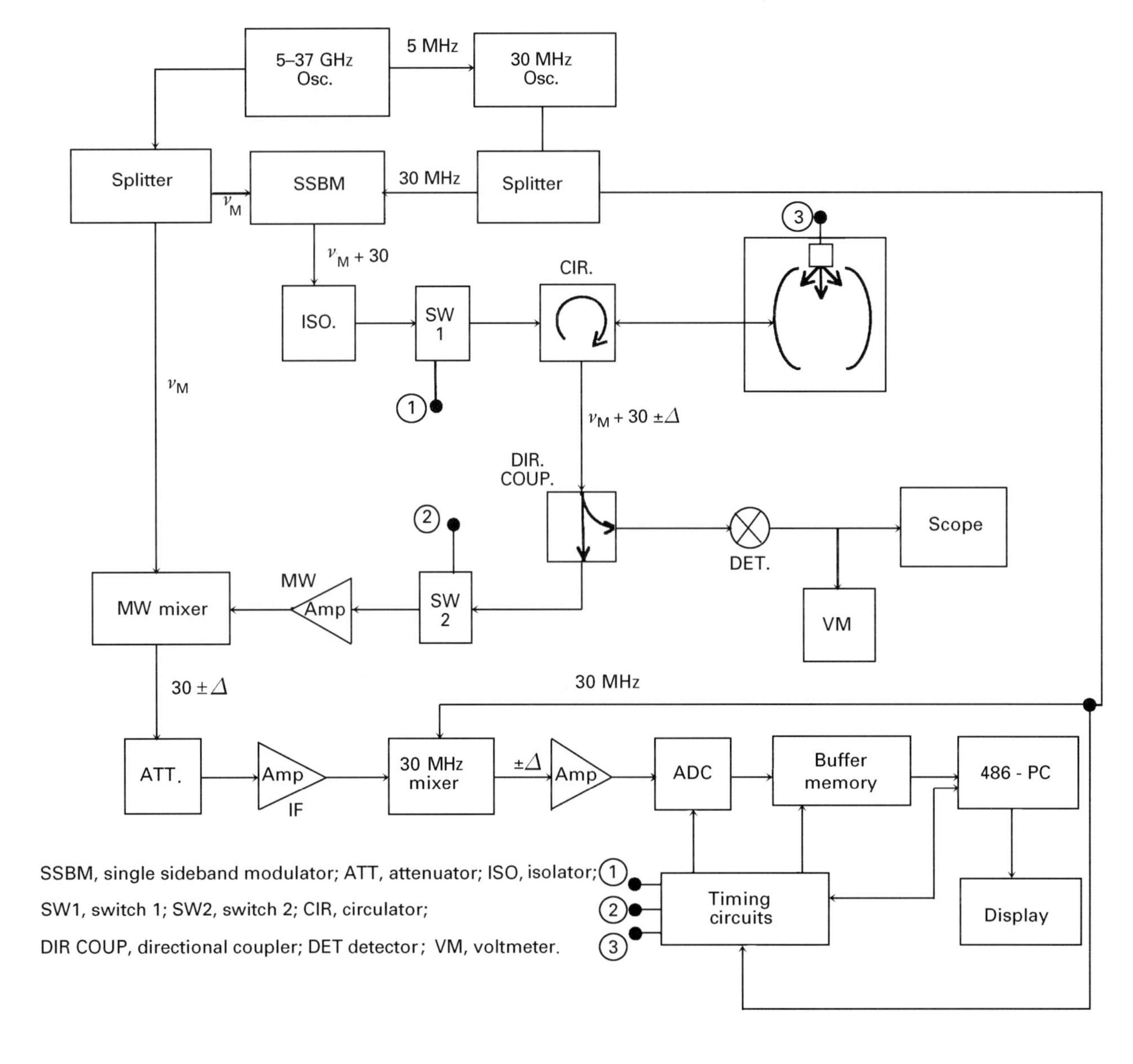

Figure 2 Block diagram of pulsed Fourier-transform microwave spectrometer. Reproduced with permission of the American Institute of Physics from Harmony MD, Beran KA, Angst DM and Ratzlaff KL (1995). A compact hot-nozzle Fourier transform microwave spectrometer. *Review of Scientific Instruments* **66**: 5196–5202. Copyright 1995, American Institute of Physics.

The result is that the FT-microwave spectrometer (FTMWS) produces the 'same' spectrum as the CW-spectrometer in a much more complex fashion. What are its advantages? As with all spectroscopic experiments carried out in the time domain, the data collection is inherently more efficient, so that the ultimate sensitivity of the FT-spectrometer is substantially higher (perhaps by a factor of 10–100 in practice). In addition, the FT instrument yields much narrower line widths than achievable in typical CW experiments, so the spectral resolution is even higher than for ordinary CW experiments.

Theoretical aspects of rotational spectra

The rotational quantum states of molecules are characterized by quantum numbers which specify the angular momentum of the rotating molecules. For all molecules, regardless of geometry, the quantum number J, with values, 0, 1, 2, . . ., specifies the total rotational angular momentum of the allowed energy states. (Note: we exclude from our discussion molecules having spin angular momentum, in which case a more careful specification of quantum numbers is necessary.) For all linear molecules this quantum number suffices to describe the rotation energy levels (aside from some special effects arising from vibrational motions) in the absence of additional applied fields. The permitted spectral transitions are limited by the selection rule $\Delta J = \pm 1$, i.e. transitions can occur only with a change of one unit of angular momentum. Thus, a typical observed microwave transition for $^{19}F^{12}C^{12}CH$ (in conventional notation) is the $J = 2 \leftarrow 1$, occurring at $\nu = 38824.64$ MHz. The notation means the molecule is excited from the lower $J = 1$ state to the higher $J = 2$ state. For the linear molecule, the rotational energy states in the simplest approximation are given by the expression

$$E = hBJ(J + 1)$$

where $B = h/8\pi^2 I$ and I is the classical moment of inertia of the molecule. The term B is known as the 'rotational constant'.

For non-linear molecules, additional quantum numbers (or labels) are necessary, and moments of inertia must be defined for three axes, conventionally labelled a, b, c. Molecules such as CH_3Cl or NH_3 can be shown to have $I_a < I_b = I_c$ and are known as prolate symmetric rotors. By convention the a-axis is chosen to lie along the molecular threefold (or higher) axis of symmetry. Then the theory for the rotating symmetric rotor shows that the energies depend now not only upon J but also upon the quantum number K which specifies the component of total angular momentum J lying along the a-axis. The value of K is limited to $-J, -J + 1, \ldots 0 \ldots J-1, J$. The energy levels are then (to the first approximation again) expressed by

$$E = hBJ(J + 1) + h(A - B)K^2$$

with definitions of the rotational constants as before, i.e. $B = h/8\pi^2 I_b$ and $A = h/8\pi^2 I_a$. The spectrum of the symmetric rotor is now determined by the selection rules $\Delta J = 0, \pm 1$ and $\Delta K = 0$. Note that the $\Delta K = 0$ rule leads to the result that the spectrum does not depend upon A at all. Moreover, $\Delta J = 0$, which is a formal rule according to theory, leads to no observable microwave transition. The net result is that the symmetric rotor microwave spectrum is essentially of the same structure as that of the linear molecule.

Non-linear or general polyatomic molecules (known as asymmetric rotors) with no threefold or higher axes of symmetry will generally have $I_a \neq I_b \neq I_c$. The rotational energy levels for this case have a complex pattern, depending upon the rotational constants A, B and C, the rotational angular momentum quantum number J, and two other pseudo-quantum numbers or labels related to K for the symmetric rotor. The spectrum is specified by the rules $\Delta J = 0, \pm 1$ again, and some additional symmetry rules involving the pseudo-quantum numbers and the dipole moment components μ_a, μ_b and μ_c. Some typical observed transitions for bicyclobutane (C_4H_6) are the $J = 1_{1,0} \leftarrow 0_{0,0}$ at $\nu = 26625.55$ MHz and the $J = 2_{2,1} \leftarrow 2_{1,1}$ at $\nu = 23995.38$ MHz. The transitions with $\Delta J = +1$ are known as R-branch lines while the $\Delta J = 0$ transitions are known as Q-branch lines.

The previous description has been based upon the so-called rigid-rotor approximation. In fact, molecules deform as they rotate, leading to the phenomenon known as centrifugal distortion. This produces small corrections to the previously described energy expressions, usually amounting to changes of less than 0.1%. Because of the very high precision of microwave measurements, such changes are, however, easily detectable and can be accounted for by appropriate theory. A number of other factors contribute to the finer details of microwave spectra. Some of these will be described in the next section and additional information can be obtained by consulting the Further reading section.

In addition to understanding the frequency axis (x-axis) of microwave spectra, it is important to have some knowledge about the intensity (or y-) axis. The theory describing the absorption of microwave radiation is complex, but it is worthwhile looking at some of the key factors. In a useful approximate theory for an asymmetric rotor, the intensity (a quantity proportional to the fraction of absorbed radiation) is given for a microwave transition by

$$\text{Intensity} \propto \sqrt{ABC}\, \mu_g^2 \nu^2$$

where the rotational constants have been defined previously, μ_g is the dipole moment along one of the axes $g = a$, b, c, and ν is the frequency of the transition. The expression leads to several key conclusions:

(1) Microwave intensities vanish (i.e. no radiation is absorbed) if $\mu_g = 0$, that is if the molecule is nonpolar as mentioned earlier. Conversely, the squared dependence of μ strongly favours very polar molecules. Thus, all other factors being equal, the spectral intensities of nitriles (such as C_2H_5CN) with μ values of typically 4 debye, will be approximately $(4/0.08)^2$, i.e. 2500, times greater than those of simple alkanes such as propane ($\mu \approx 0.085$ debye).
(2) Intensities are generally greater at high frequencies, according to the ν^2 dependence. Heavy molecules, with large moments of inertia and corresponding small rotational constants exhibit their transitions generally at low frequencies while the converse is true for light molecules. Thus heavy molecules tend to have ‘weak’ spectra while light molecules have ‘strong’ spectra.
(3) The factor $\sqrt{ABC}$ can be seen to emphasize the dependence upon molecular size and mass, or more precisely, upon moments of inertia. Small, light molecules are favoured because of their large rotational constants, while large, heavy molecules are discriminated against.

Applications of microwave spectroscopy

Structure determination

Microwave spectroscopy is the premier physical method for determining accurate and precise molecular structures, i.e. values of interatomic distances (bond distances) and angles (bond angles). This capability arises because the moments of inertia are directly related to the coordinates of the atoms as follows:

$$I_a = \sum_i m_i(b_i^2 + c_i^2)$$

with similar expressions for I_b and I_c. In this expression, m_i is the mass of the ith atom while b_i and c_i are the b- and c-axis coordinates of the atom. Assignment, measurement and analysis of microwave spectra yield precise values of rotational constants A, B and C and hence values of I_a, I_b and I_c. Thus the latter quantities provide equations which permit the evaluation of atomic coordinates, a_i, b_i and c_i. Once the coordinates are known, distances and angles are also known. Thus, the bond distance between atoms i and j is given by

$$R_{ij} = \sqrt{(a_i - a_j)^2 + (b_i - b_j)^2 + (c_i - c_j)^2}$$

A number of problems dealing with molecular non-rigidity must be considered if accurate and meaningful bond distances are to be obtained. Ideally, one would like to determine the coordinates (and hence structure) for the hypothetical vibrationless molecule. Methods for achieving this ideal (to various approximations) have been developed, so that numerous accurate structures have been determined from microwave spectral data. The Further reading section provides examples of such molecular structure determinations.

Molecular electric dipole moments

It has been mentioned that microwave intensities are determined by the size of the electric dipole moment, so one might suppose that accurate measurements of intensities might provide values of μ. This turns out not to be practical for various reasons. However, another very accurate procedure can be used. If an electric field is applied to a rotating molecule, a well-understood phenomenon known as the Stark effect splits the rotational transitions into a number of components. Precise measurements of these small splittings (typically several MHz) lead to very precise values of the electric dipole moment. Values of μ determined by this method refer to particular quantum states and are thus much more meaningful theoretically than those determined by classical bulk-gas relative permittivity (dielectric constant) measurements.

Hyperfine structure

Molecules containing nuclei whose nuclear spin values satisfy $I \geq 1$ exhibit splittings of the rotational transitions known as hyperfine structure. The predominant cause of these splittings (which for most common quadrupolar nuclei is typically several MHz or less) is the nuclear electric quadrupole interaction. Measurements of the splittings and application of appropriate theory lead to values of a quantity known as the quadrupole coupling constant, usually symbolized as eQq. In this expression Q is the nuclear quadrupole moment (a fundamental nuclear constant), e is the charge on the electron, and q is the electric field gradient at the nucleus produced by the surrounding electron and nuclear charges. Coupling constants have been extensively measured for nuclei such as ^{35}Cl ($I = \frac{3}{2}$), ^{14}N ($I = 1$) and D ($I = 1$) in a variety of molecules. The resulting values provide important information about the chemical bonding of the atom in question. Note that several very common nuclei, such as ^{1}H, ^{12}C and ^{16}O, have $I < 1$ and consequently produce no quadrupolar hyperfine splittings.

Internal rotation

Molecules such as propane, methanol or acetone have methyl groups which undergo large amplitude torsional oscillations or internal rotation. This internal rotation is hindered in general by a potential barrier, and the well-known quantum mechanical theory for the effect often leads to observable splittings (typically a doubling) of microwave spectral lines. In general, for high barriers (>1000 cm^{-1}) the splittings are small (typically several MHz or less) while for low barriers (~300 cm^{-1} or less) the splittings can be very large (100 MHz or greater). Because of these easily observed splittings microwave spectral measurements have led to a wealth of data on molecular internal rotation barriers. Several related phenomena, involving the puckering or inversion of four- or five-membered ring compounds, or the inversion about pyramidal nitrogen (as for NH_3), have also been extensively studied by microwave methods.

Interstellar microwave spectra

One of the most exciting applications since the 1970s has been the observation of microwave (rotational) spectra of interstellar molecules. Common species such as formaldehyde, ammonia and methylamine and more exotic species such as HCO and H–CC–C≡C–CN have been detected in various interstellar media. The experimental technique differs substantially from that outlined in **Figures 1** and **2**. In this case the interstellar molecular spectra are detected by collecting microwave emissions from interstellar space with large radio telescopes equipped with sensitive microwave receivers. An interesting feature of the interstellar spectra is that the spectral lines are generally Doppler-shifted from their laboratory ‘rest’ frequencies because the absorbing medium is moving rapidly relative to the background radiation source.

Multiple irradiation experiments

Microwave spectroscopy is often coupled with a second electromagnetic radiation field to perform specialized experiments. Thus microwave-optical double resonance (MODR) uses optical (say 400–800 nm) radiation simultaneously. The optical radiation transfers molecules to excited electronic states which are then probed by the microwave radiation before the excited molecules return to the normal ground state. Similar experiments utilizing infrared radiation (IRMDR) permit probing of excited vibrational states. Analogous experiments using two microwave fields (MMDR) and a microwave and radiofrequency field (RFMDR) are very commonly used to produce spectral simplification and to aid in spectral interpretation. The double resonance experiments have also been important for obtaining information about collisional energy transfer rates and mechanisms.

Studies of weakly-bound complexes

Since about 1980 there has been great interest in performing microwave studies of weakly bound species such as $(H_2O)_2$, ArHCl and (HC≡CH)HCl. These species are studied with the unique instrument shown earlier in **Figure 2**, known as a pulsed-nozzle Fourier-transform microwave spectrometer. The weakly-bound species are formed by pulsing a gas-mixture through a small nozzle such that it undergoes a supersonic free-jet expansion. Complexes are formed rather abundantly in such expansions and are stabilized by the low temperatures (<5 K) achieved in the expansion. Pulsed FTMWS (synchronized with the pulsed nozzle) is then used to sensitively observe and study the rotational spectrum. Such investigations will surely continue to be of great future interest because they provide information on van der Waals and hydrogen-bonding forces, both of which are of critical importance to understanding intermolecular potentials.

Analytical applications

The very high resolution and selectivity of microwave spectroscopy make it an excellent tool for qualitative analysis of gas-phase samples. Indeed, a substantial amount of effort has been placed by microwave spectroscopists in using the method to identify and characterize new chemical species, especially those which are unstable and hence difficult to study by more conventional techniques. Because microwave spectroscopy is readily adaptable to continuously flowing gas samples (with special cell designs as mentioned earlier) it is an ideal method for investigating the products of combustion, pyrolysis, photolysis or electric discharges. Examples of such studies include OH, CS, CH_2=NH, CF_2=C=C=O, HCO^+, HNN^+ and many others. Of course, the last section described the unique application of microwave spectroscopy to unstable molecular clusters and earlier the high selectivity for isotopic analyses was mentioned.

The chief disadvantage of microwave spectroscopy for gas-phase analytical applications is that its sensitivity is not as high as for some other methods (such as laser fluorescence or mass spectrometry). For low molecular weight polar species such as SO_2, NH_3 and NO_2, analytical detection sensitivities using FT-MWS instruments certainly extend into the parts per billion (ppb) range. However, as the molecular size and mass increase or the polarity decreases the sensitivities may fall more typically into the ppm range. Naturally, as with all spectroscopic methods, appropriate preconcentration or preselection schemes may lead to effectively improved detection limits.

From the above it is clear that quantitative measurements at high sensitivities are most useful for a variety of small polar molecules which are of concern from the atmospheric environmental pollution point of view. Thus a substantial amount of effort has been and continues to be placed upon the development of field operable, portable microwave spectrometers for trace gas monitoring using both CW and FT instrumentation. Although there are likely to be continued applications of microwave spectroscopy to pure analysis problems in the future, it seems likely that the microwave spectrometer will continue to find its most exciting applications in the chemistry and physics research laboratory.

List of symbols

A, B, C = rotational constants; b_i, c_i = b- and c-axis coordinates of the ith atom; e = charge on an electron; E = rotational energy; h = Planck's constant; I = moment of inertia, and the nuclear spin angular momentum quantum number; J = angular momentum quantum number; K = quantum number specifying component of J lying along the a-axis; m_i = mass of the ith atom; q = electric field gradient; Q = nuclear quadrupole moment; R_{ij} = bond distance between atoms i and j; μ = dipole moment; ν = frequency.

See also: **EPR Spectroscopy, Theory; Gas Phase Applications of NMR Spectroscopy; Microwave and Radiowave Spectroscopy, Applications; Rotational Spectroscopy, Theory; Solid State NMR, Rotational Resonance; Vibrational, Rotational and Raman Spectroscopy, Historical Perspective.**

Further reading

Balle TJ and Flygare WH (1981) Fabry–Perot cavity pulsed Fourier transform microwave spectrometer with a pulsed nozzle particle source. *Review of Scientific Instruments* **52**: 33–45.

Gordy W and Cook RL (1984) *Microwave Molecular Spectra*. New York: Wiley-Interscience.

Harmony MD (1981). In: Anderson, HL (ed) *AIP Physics Vade Mecum*, Chapter 15. New York: American Institute of Physics.

Harmony MD, Beran KA, Angst DM and Ratzlaff KL (1995) A compact hot-nozzle Fourier transform microwave spectrometer. *Reviews of Scientific Instruments*. **66**: 5196–5202.

Harmony MD, Laurie *et al* (1979) Molecular structures of the gas-phase polyatomic molecules determined by spectroscopic methods. *Journal of Physical Chemistry Reference Data* **8**: 619–721.

Harmony MD and Murray AM (1987). In: Rossiter BW and Hamilton JF (eds) *Physical Methods of Chemistry: Vol. IIIA – Determination of Chemical Composition and Molecular Structure*, Chapter 2. New York: Wiley.

Legon AC (1983) Pulsed-nozzle, Fourier-transform microwave spectroscopy of weakly bound dimers. *Annual Review of Physical Chemistry* **34**: 275–300.

Steinfeld JI and Houston PL (1978) In: Steinfeld JI (ed) *Laser and Coherence Spectroscopy*, Chapter 1. New York: Plenum.

Townes CH and Schawlow AL (1955) *Microwave Spectroscopy*. New York: McGraw-Hill.

Varma R and Hrubesh LW (1979) *Chemical Analysis by Microwave Rotational Spectroscopy*. New York: Wiley-Interscience.

Mineralogy Applications of Atomic Spectroscopy

See **Geology and Mineralogy, Applications of Atomic Spectroscopy.**

Molybdenum NMR, Applications

See **Heteronuclear NMR Applications (Y–Cd).**

Mössbauer Spectrometers

Guennadi N Belozerski, St.-Petersburg State University, Russia

HIGH ENERGY SPECTROSCOPY

Methods & Instrumentation

To obtain the Mössbauer spectrum the radiation from a Mössbauer source should be directed onto the sample under study. In Mössbauer experiments it is not the absolute energy of the γ-quanta which is determined but the energy shift of the nuclear levels. The energy scanning is carried out by the use of the Doppler effect and the energy parameters (Γ, δ) are expressed in velocity units, $v(E\ v/c)$. The Mössbauer spectrum is a measure of the dependence of the total intensity of radiation $I(v)$ registered by a detector in a definite energy region on the relative velocity v of the source.

A schematic diagram of a Mössbauer experiment and the spectrum is shown in **Figure 1**. If both the source and the absorber are characterized by single lines of natural width Γ_{nat}, δ being zero, the spectrum will show maximum absorption at $v = 0$. In this situation the intensity, $I(0)$, registered by the detector is minimized (**Figure 1C**). When the source moves at a certain velocity v, the emission line $J_M(E)$ is displaced relative to the absorption line $J^a(E)$. The overlap then decreases and the intensity increases. Finally, at a velocity that may be considered to be infinitely large ($v = \infty$), the spectrum overlap becomes so small that any further increase in velocity will not result in a significant increase in relative intensity. This value of intensity may be described as $I(\infty)$. The fact that the line shapes of the source and absorber are described by Lorentzians causes the experimentally observed line for a thin absorber to be Lorentzian, and its half-height width is the sum of the line widths of the source and the absorber.

A typical device for accumulating the Mössbauer spectrum is the multichannel analyser, where the count rate is a function of a definite value of the Doppler velocity. The count rate is normalized relative to the off-resonance count rate. Hence, for transmission-mode Mössbauer spectroscopy relative intensities are always less than unity (or 100%). In Mössbauer scattering experiments relative intensities always exceed 100% and can reach several hundred percent in the case of electron detection from samples with a high abundance of the resonant isotope. It is most often that the $-v_{max}$ value corresponds to the first channel and the $+v_{max}$ value to the last channel. The quality of a Mössbauer spectrometer is determined by how accurately the modulation of the γ-quanta energy follows the chosen mode of movement.

Typical Mössbauer spectrometers

The Mössbauer experiment may be in transmission mode, where γ-quanta are detected. The detector registers not only the γ-rays of the Mössbauer transition, but also the background noise. The main process competing with resonance interactions in the transmission mode experiments is the photoelectric effect. In transmission experiments there are three sources of background: (i) γ- and X-rays of higher energies which may be Compton-scattered; Bremsstrahlung produced outside the detector may contribute to this too; (ii) high-energy γ- and X-rays having lost only a part of their energy in the detector; (iii) X-rays that are not distinguished by the detector from the Mössbauer quanta.

In scattering Mössbauer spectroscopy the processes competing with Mössbauer scattering are the Compton effect, Rayleigh scattering and classical resonant scattering of γ-rays. The Compton effect is to be specially taken into account when the source emits high-energy γ-rays in addition to the Mössbauer radiation. The typical experimental arrangements are presented in **Figure 2**.

In Mössbauer spectroscopy the shape of the spectrum and its area are the 'signals' conveying quantitative information on a phase. When the shape is known to be Lorentzian, for example, the amplitudes and the line positions are often used as para-meters of the signal $-I(v_i)$ value at v_i $(i = 0, 1, \ldots)$.

Mössbauer scattering spectra obtained by detection of the γ-quanta or X-rays emitted out of the bulk of a material, convey information on the layer with a depth which is determined by the total linear absorption coefficient $\mu_a(E)$. The values of $\mu_a(E)$ for γ-rays and X-rays are generally different; therefore the Mössbauer spectra correspond to the layers, which are different in depth (from one to several μm). Backscattering Mössbauer spectroscopy is the most promising technique for applied research and industrial applications (see **Figure 2C**). The backscattering geometry is simple, efficient and suitable for any type of radiation. In such an experiment one can detect any radiation in different scattering channels. However, to detect γ-quanta, a special detector is needed. It has been shown by many experimentalists that the signal/noise ratio of the detection of γ-rays in the experimental geometry of **Figure 2B** is better than for detection of X-rays. At the same time the flat proportional counter has never been used to detect γ-quanta in the backscattering geometry of **Figure 2C**. Indeed, the direct Mössbauer radiation of an intensity which is 100 times as high

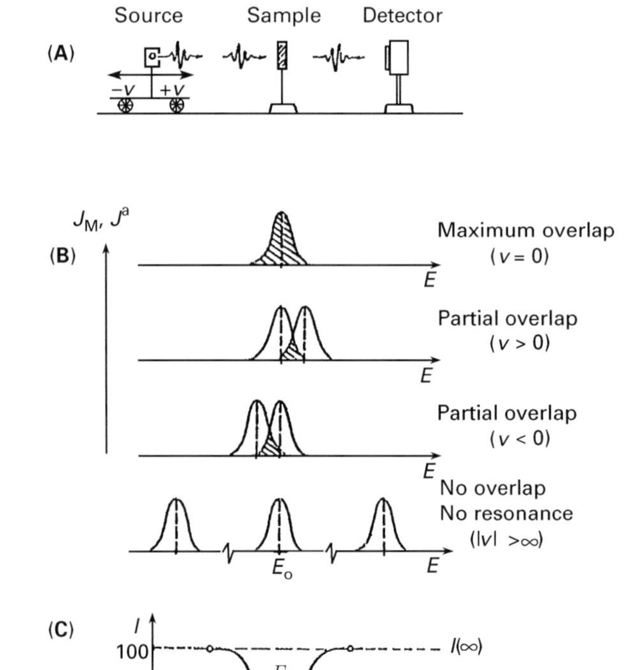

Figure 1 Schematic illustration of the experimental arrangement (A) used to obtain a Mössbauer spectrum (C) for a single Lorentzian line both in the source and in the sample (B).

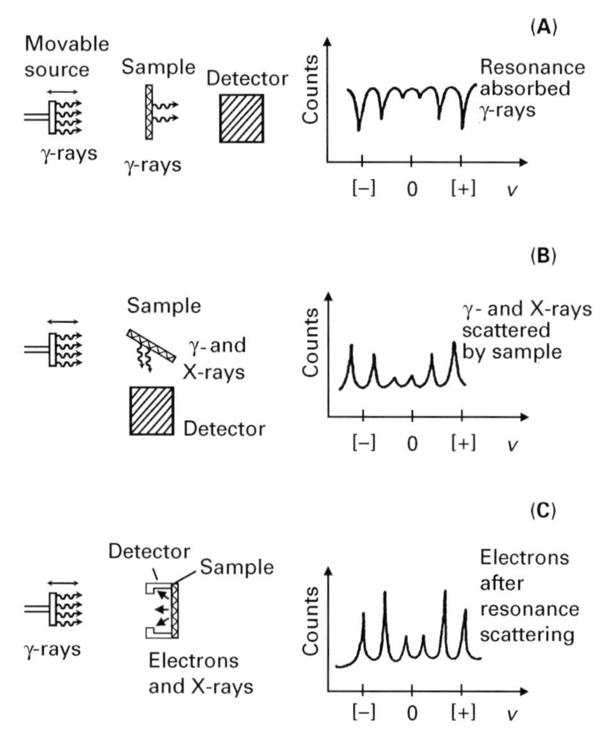

Figure 2 Experimental arrangements and Mössbauer spectra for a ^{57}Co (Cr) source and a sample of α-Fe: (A) transmission geometry, (B) scattering geometry with the detection of γ- or X-rays, (C) backscattering geometry with the detection of X-rays and electrons. The source moves at a velocity *v*.

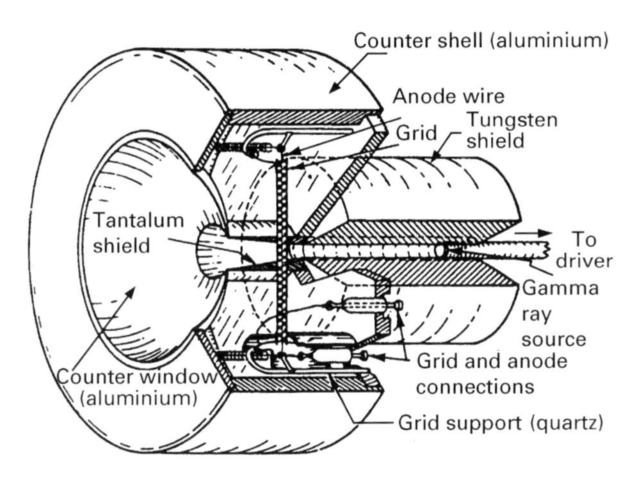

Figure 3 Spectrometer based on the toroidal detector.

as the scattered intensity also passes through the detector such that the effect would be very small. The need to detect the resonantly scattered γ-quanta in a solid angle close to 2π stimulated the search for a detector capable of sensing scattered photons with an energy of 10–20 keV and which would be insensitive to the direct primary beam of γ-quanta.

The requirements have been met by the use of toroidal detectors. The main problem has involved the necessity to create the inner electric field with circular equipotential lines around the anode. For this purpose, one uses cylindrical grid wires surrounding the anode. Electrons produced within the counter volume travel to the grid and through it to the anode wire. After filling with a krypton–methane mixture the resolution for the 14 keV line for such a counter is ~15%. A section of a Mössbauer spectrometer using the counter is shown in **Figure 3**. This toroidal proportional detector is easy to handle and can be usefully applied to surface studies with high efficiency.

Conversion electron Mössbauer spectroscopy

Mössbauer transitions are usually highly converted and are followed by the emission of characteristic X-rays and Auger electrons. (The total internal conversion coefficient is high. For most cases de-excitation of the nucleus is via the emission of conversion electrons followed by rearrangement of the excited atomic shell by X-ray emission and Auger processes. More than one electron is produced per resonant scattering event.) The detection of electrons has proved in many cases to be the most efficient means of observing the Mössbauer effect. The principal feature of Mössbauer spectroscopy based on the detection of electrons is that the average energy of an electron beam reaching thexx detector, and also the shape of the energy spectrum, depends on the depth x of a layer dx from which the beam has been generated. This provides interesting possibilities for layer-by-layer phase analysis. Various modifications of Mössbauer spectroscopy based on the detection of electrons have been developed including a technique which allows the Mössbauer signal from a very thin surface layer (~3 nm) of a homogeneous bulk sample to be distinguished. The techniques in this field of Mössbauer spectroscopy are classified as either integral or depth-selective.

The integral technique is called conversion electron Mössbauer spectroscopy (CEMS). In CEMS, of prime interest is the probability that electrons originating from a layer dx at a depth x with energy E_0 leave in a random direction from the surface with any energy and at any angle and will be registered by a detector. The electrons may be divided into several groups: conversion electrons, Auger electrons, low-energy electrons resulting from shake-off events and secondary electrons resulting from the re-emitted Mössbauer quanta and the characteristic X-rays. The energies and relative intensities of the first two groups of electrons for ^{57}Fe and ^{119}Sn are given in **Table 1**.

The development of CEMS as an independent analytical method came as a result of the development of gas-filled proportional counters for the detection of electrons. **Figure 4** illustrates the operating principle of such a CEM spectrometer. The proportional counter in CEMS detects all the electrons in the energy interval from about 1 keV up to the Mössbauer transition energy. In addition to the high efficiency, the proportional counters have an energy resolution allowing, if we need it, a certain depth selectivity to be obtained.

Phase analysis of multiphase mixtures, fine particles and disordered substances, as well as surface studies, require Mössbauer spectra to be recorded over a wide range of temperatures. The problem of

Table 1 Main radiation characteristics for the ^{57}Fe and ^{119}Sn de-excitation

	^{57}Fe		*^{119}Sn*	
Type	*Energy* (keV)	*Probability per de-excitation (C_i)*	*Energy* (keV)	*Probability per de-excitation (C_i)*
K-conversion	7.3	0.796		
L-conversion	13.6	0.09	19.6	0.83
M-conversion	14.3	0.01	23.0	0.13
KLL Auger	5.4	0.543		
LMM Auger			2.8	0.74

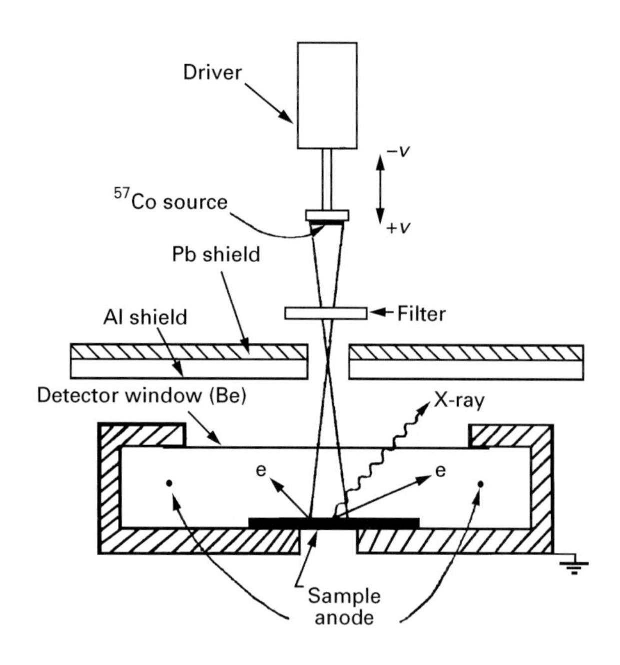

Figure 4 Schematic picture of a spectrometer for back-scattering studies.

the counter operation at temperatures other than ambient has received significant attention in Mössbauer spectroscopy. The counters can operate CEMS at low temperatures near 4.2 K and up to 1100 K.

Arrangements based on proportional counters which allow an independent and simultaneous recording of CEM spectra and X-ray Mössbauer spectra in backscattering geometry, and γ-ray absorption spectra in transmission, have been developed for industrial application purposes, see **Figure 5**. Due to the different escape or penetration ranges of the three radiations involved, the spectra give information on phases, depth and orientation. From a practical point of view the counters for γ-rays, X-rays and electrons must be separated and shielded to ensure independent detection.

In addition to the proportional counters, other types of gas-filled detectors are used in CEMS. First was the parallel-plate avalanche counter. In Mössbauer spectroscopy such detectors have been used as resonance detectors and at higher counting rates. These counters have found application in surface studies and are the effective tool for the registration of low-energy ($E < 1$ keV) electrons, which are practically impossible to detect with the proportional counter. Because of the high electron-detection efficiency this enables the measurement of reasonable spectra in a relatively short time for ^{57}Fe, ^{119}Sn, ^{151}Eu, ^{161}Dy and ^{169}Tm. Second, scintillation detectors may be considered. Thin organic (crystal or plastic) scintillators are used for detecting electrons. Gas scintillation proportional counters with a good energy resolution (e.g. $R \approx 8\%$ at 6 keV) may also be constructed for CEMS as well as semiconductor detectors.

Channeltrons, microchannel plates and window less electron multipliers constitute a special group of detectors for CEMS. These have no entrance

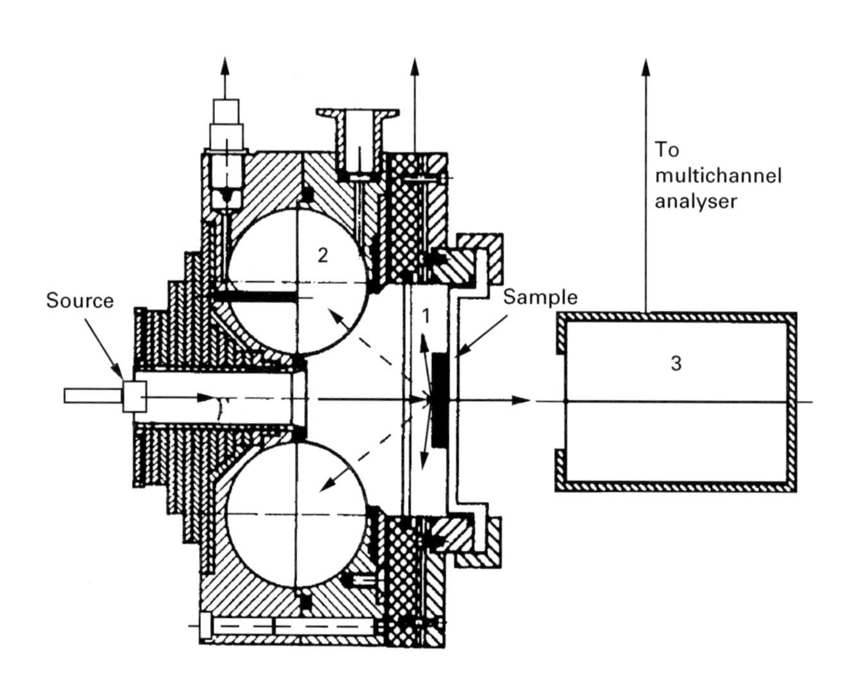

Figure 5 Set-up for simultaneous recording of CEM spectra (1), X-ray Mössbauer spectra (2) and transmission spectra (3).

windows and are designed for vacuum operation which can be used to advantage in CEM spectrometers operating both at high and low temperatures. A new method of surface study has recently appeared, CEMS based on the detection of very low-energy electrons. Detectors in this group have no energy resolution. The pulse–height distribution at the out-put of these detectors is similar to the noise distribution.

Advantages of the best CEM spectrometers with a channeltron include their easy sample access, high cooling rate, capability of simultaneous transmission measurements and adaptability to on-line experiments. To increase the count rate, detection efficiency or the effect value, a bias potential is sometimes applied to the sample or to the input of the channeltron. The statistical quality of spectra is, as a rule, nearly as good as for the gas-filled ionization detectors. The effective technique of collecting secondary electrons by applying a bias potential between the sample surface and a channeltron has been used to develop a spectrometer for low-temperature measurements (see **Figure 6**). The beam of γ-quanta is incident at 45° to the sample surface. The sample is the first electrode in a system of electrodes used to attract the secondary electrons to the entrance of the channeltron and to accelerate them to an energy corresponding to the maximum detection efficiency.

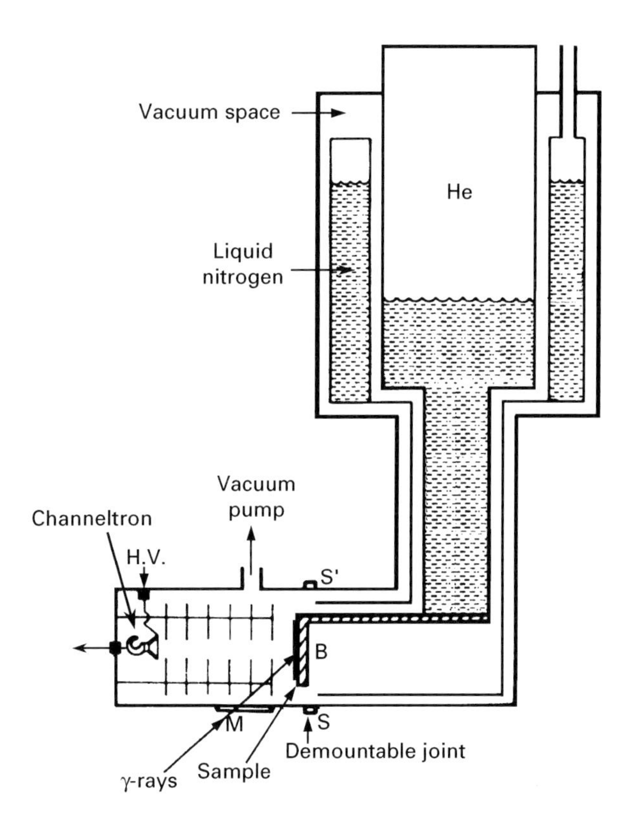

Figure 6 CEMS spectrometer used to operate at 4.2 K. M, mylar window, B, cold finger. The detection assembly is screwed on to the dewar at SS′.

Low energy electron Mössbauer spectroscopy (LEEMS)

Conversion electrons, KLL, KLM and KMM Auger electrons, photoelectrons and Compton-scattered electrons which are produced by γ-rays (with the energy above several hundred eV) in this context may be regarded as 'high energy' electrons emitted by the atom. Secondary electrons result from the interaction of the above electrons with matter. Also, there are electrons that are primarily produced with a very low energy. Two processes contribute to the intensity of the electrons. These are very low energy Auger electrons (LMM, MMM, MMN) and shake-off electrons.

Experimental data show a sharp peak in the number of electrons (related to Mössbauer events) at energies below 20 eV. These electrons supply information on a surface layer to a depth of ~5 nm. The detection of very low energy electrons offers the advantage of short data acquisition times (~77% of the electrons emitted from the Fe atom are low-energy Auger and shake-off electrons), and increases surface sensitivity compared to established procedures relying on the collection of electrons near 7.3 keV. CEMS detectors and techniques are summarized in **Table 2**.

Depth-selective conversion electron Mössbauer spectroscopy

The detection of electrons with energy E by a β-spectrometer with high-energy resolution gives a Mössbauer spectrum corresponding to the phase at a

Table 2 Detectors and electron detection techniques in CEMS

With energy resolution		*Without energy resolution*
Electron spectrometers	Magnetic	Parallel-plate avalanche counters
	Electrostatic	Channel electron multipliers
Ionization detectors	Proportional counters	Gas scintillation detectors
	Multiwire proportional counters	Microchannel plates
	Semiconductor detectors	Windowless multipliers
	Gas scintillation proportional counters	Organic scintillation detectors
	X-rays controlled proportional counters	Detection of light produced by microcharges
		Geiger-Müller counters

depth x_1 in the scatterer. If the known relationship between the energy of detected electrons and the depth of the layer through which they have passed is used, then depth-selective analysis of the surface layers can be performed. In depth-selective conversion electron Mössbauer spectroscopy (DCEMS) the group of electrons with a fixed energy leaving the surface at a certain angle within a small solid angle $d\omega$ are of interest.

There are a number of electrostatic and magnetic electron spectrometers that have been used, designed and developed for DCEMS. A schematic description of the DCEM spectrometer, based on the mirror analyser is depicted in **Figure 7**. Electrons, starting from inside the inner cylinder at angles close to 45° to the sample surface (1 cm^2), move out through slits in the inner grounded cylinder into a strong field region. The field bends the trajectories of the electrons back towards the inner cylinder. The group of electrons of interest passes through another slit, to be collected on the spectrometer axis. If a position-sensitive detector is placed on the axis, a series of Mössbauer spectra corresponding to different electron energies can be recorded simultaneously. Using three slits enables the simultaneous recording of K-, L- and M-conversion electron spectra.

An important methodology problem in DCEMS is the measuring time. For ^{57}Fe only the K-conversion electrons lead to a DCEMS spectrum. The thickness of the analysed layer in DCEMS is significantly less than in CEMS, being less than 80 nm for iron. It is dependent on parameters of the β-spectrometer. The resolution enhancement from 3% to 1% is significant for experiments involving the investigation of surface layers 0–5 nm thick. To investigate a thinner layer (0–2.5 nm thick), the β-spectrometer should detect separate groups of electrons in the 7.2–7.3 keV interval, and it is desirable to have R ≈ 0.5% and $\theta \approx 90°$. The maximum possible selectivity can probably be attained with electrostatic β-spectrometers whose accuracy of energy determination is about 1 eV and the half-width is –10 eV on the 7.3 keV line. To summarize, the experimentalist in DCEMS should try to use a detector with an efficiency close to 100%. There should be no window in the path of the electrons. The temperature of the sample may be varied in the range 4.2–1000 K.

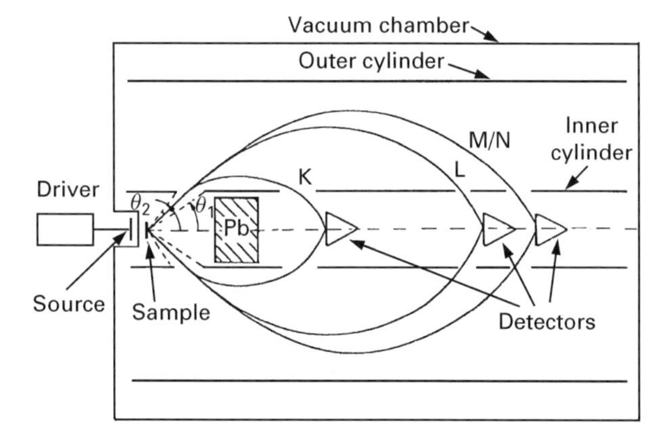

Figure 7 Schematic diagram of a DCEM spectrometer based on the electrostatic cylindrical mirror analyser. Forward scattering geometry is used. θ_1 and θ_2, minimal and maximal angles for the input slit edge positions; Pb, lead shielding.

Special Mössbauer spectrometers

There is a special situation where hyperfine interactions are present and a constant velocity v_i is chosen, so that the incident radiation is on resonance with the scatterer's line number 2 ($v_i = v_2$). There is, in this situation, no unique relation between the energies of the incident and scattered γ-quanta. The scattered quantum may have the energy of the incident quantum, as well as the energy belonging to the line number 4. The same is true if relaxation processes or very complicated hyperfine interactions occur in the sample. To study the phenomenon the incident γ-quanta energy should be fixed and the energy spectrum of scattered γ-quanta will show directly the energy change of γ-quanta on scattering.

To obtain the energy distribution, a γ-ray detector is needed with an energy resolution of approximately Γ_{nat}. For this purpose a resonant filter is placed in front of the conventional detector (see **Figure 8**). This filter is a 'single line' Mössbauer absorber. Driving the filter ('analyser') in the constant acceleration mode and detecting the outgoing radiation allows the $I(v,v_i)$ spectrum to be produced (see **Figure 9**). The observed effect is determined now by the two elastic resonant scattering processes (by four f factors). Two synchronized drive systems are necessary to observe the two scattering processes. This is known as selective-excitation double Mössbauer spectroscopy (SEDMS).

The method is demonstrated by considering the SEDM spectrum recorded for scattering at the energy corresponding to the $-\frac{1}{2} \rightarrow -\frac{1}{2}$ transition in a 9 μm thick ^{57}Fe foil (**Figure 9**). The Mössbauer spectrum consists of the second and fourth lines of the usual spectrum of α-Fe, i.e. the lines corresponding to the $-\frac{1}{2} \rightarrow -\frac{1}{2}$ transition as well as to the $-\frac{1}{2} \rightarrow +\frac{1}{2}$ transition.

The main advantage of SEDMS is that the method offers a direct means by which the relaxation processes between sublevels of the excited nucleus can be observed. Indeed, the experimental spectrum $I(v,v_i)$ gives direct information on time-dependent hyperfine interactions which determine the nuclear level

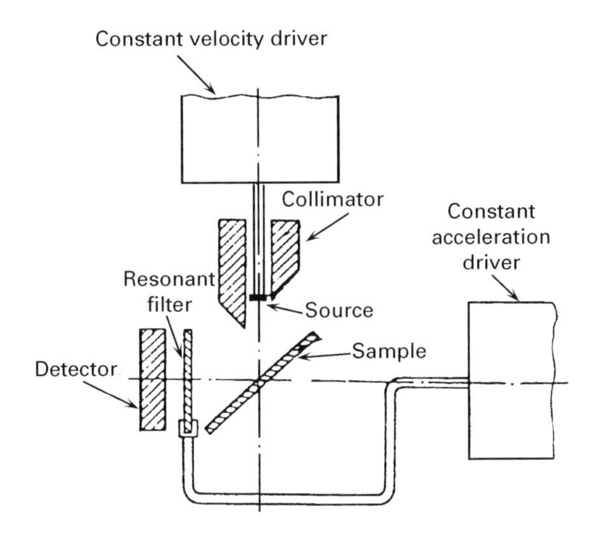

Figure 8 A schematic experimental arrangement used for selective-excitation double Mössbauer spectroscopy.

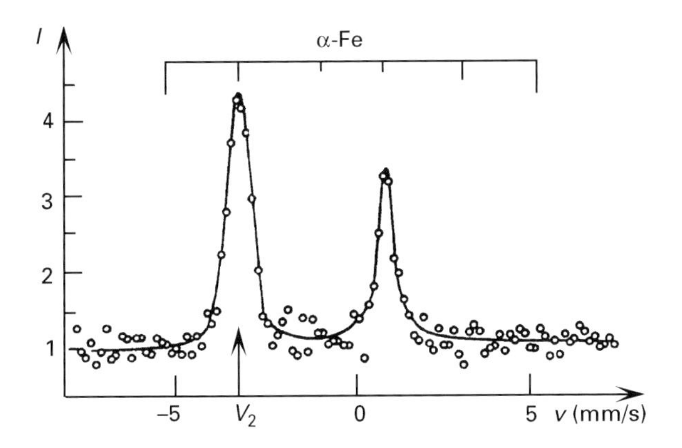

Figure 9 SEDM spectrum of α-Fe.

splitting. The relaxation times in the region of 10^{-7}–10^{-10}s are the most convenient to measure. Unfortunately, the necessity of having two successive resonant interaction processes results in a very low detected intensity. Indeed, the second part of a SEDMS experiment is a transmission experiment with the scatterer being the Mössbauer source.

Also, special Mössbauer spectrometers are used for total external reflection (TER) studies. On reflection at angles less than γ_{cr} the electromagnetic field intensity falls off rapidly (for the metal iron mirror, $\gamma_{cr} = 3.8 \times 10^{-3}$ sr). The penetration depth for the radiation (i.e. the thickness L of a layer under study) is taken to be equal to the depth at which the intensity is less by times e. If only the elastic scattering by electrons is considered, L is evaluated to be 1.3 nm for an iron mirror.

An experimental set-up is given for studies of TER of Mössbauer quanta in **Figure 10**. The design of the Mössbauer spectrometer for TER studies ensures: (1) simple and reliable setting and measurement of the grazing angle γ_{cr}; (2) convenience in the adjustment of the angular beam divergence; (3) sample replacement without affecting the experimental geometry; (4) reproducibility of all source–collimator–sample distances; (5) sample rotation in the range 0–90°. The spectrometer consists of the analytical unit and electronic system for control, acquisition and processing of spectrometric data.

The analytical unit of the spectrometer comprises a vibration damping platform suspended on shock-absorbers. Mounted on the platform are guides of the 'wedge slide' type, which carry the driver, shielding screens, collimator to form narrow directed plane-parallel radiation beams, proportional counter and scintillation detector. A narrow plane-parallel γ-ray beam from the source rigidly attached to the driver is formed by the slit collimator and, through the entrance window of the dual detector, falls on the sample. The γ-radiation is reflected from the sample surface and passed through the exit window of the dual detector and slotted mask (screen), and detected by the scintillation detector D1.

Although the analysed layer is very thin, the technique has not been widely used due to the very low luminosity. Of no less importance is the fact that

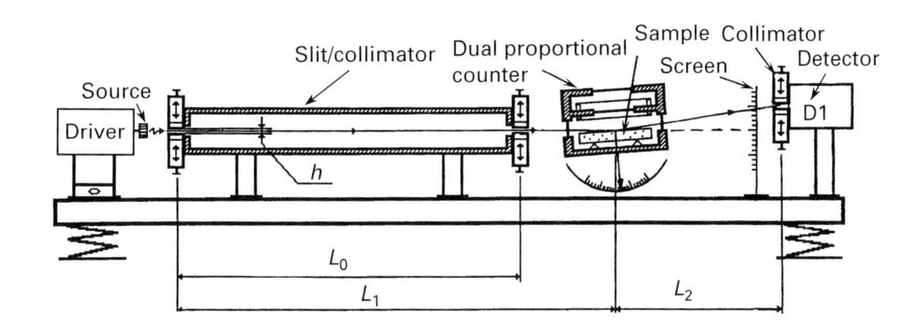

Figure 10 An experimental set-up for studies of total external reflection of Mössbauer quanta. D1, scintillation detector. $L_0 \sim 600$ mm, $L_1 \sim 700$ mm, $L_2 \sim 400$ mm, $h = (1 \pm 0.05)$ mm.

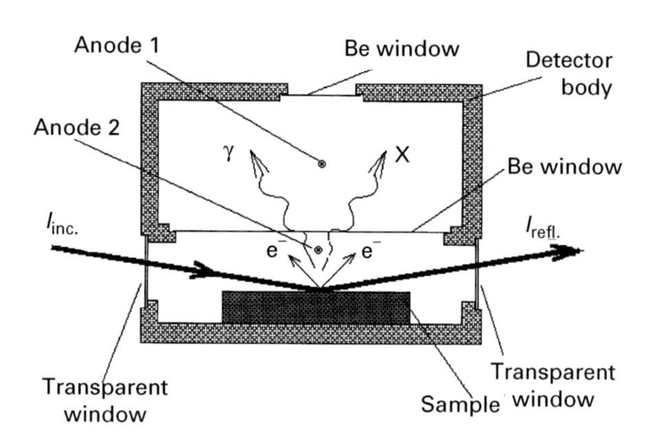

Figure 11 Part of an experimental set-up (see **Figure 10**): the dual proportional counter.

interference effects complicate the interpretation of the experimental data. Substantial progress is achieved by detecting not only the mirror-reflected γ-quanta, but all secondary radiation leaving the surface when Mössbauer radiation is incident at an angle that is less than critical. The key part of an experimental set-up is the dual proportional counter (see **Figure 10**). A schematic picture of the dual-chamber gas proportional counter is shown in **Figure 11**. The sample under investigation is inside the electron chamber of the detector. The gas mixture in the chamber is He + 8% CH_4. The gas mixture for detection of γ- and X-rays is Ar + 8% CH_4. Thus during a single run (preset γ value) one can obtain four Mössbauer spectra simultaneously: three from the combined detector and one from the scintillation detector (mirror-reflected γ-rays).

Spectrum quality and quantitative information from Mössbauer spectra

The amplitude of Mössbauer lines in scattering experiments can often be greater than in a transmission geometry. However, the intensity loss of the scattered radiation of about two orders of magnitude makes it necessary to compare both the sensitivity of the two methods and the quality of the two spectra obtained. For a thin sample characterized by a single Lorentzian and the effective thickness t_a, the quality of the spectrum in relation to the quantity of information on the t_a parameter (the information matrix element of interest), $[\hat{J}]_{t_a, t_a}$, is:

$$[\hat{J}]_{t_a, t_a} = \frac{\pi}{4}\varepsilon^2(0)I(\infty)\Gamma$$

where $\varepsilon(0)$ is the resonance effect magnitude. In order to increase $\varepsilon(0)$, the experimentalist needs to decrease the solid angle towards the detector and sample to prevent the source radiation from reaching the detector as a result of multiple nonresonant scattering in collimators and surrounding materials. This always gives a greater $\varepsilon(0)$ value, but the $I(\infty)$ value is decreased. The expression allows the evaluation of the limit when a further increase of $\varepsilon(0)$ values is no longer reasonable. After the optimal experimental conditions are chosen, the $\varepsilon^2(0)$ values are fixed for each sample under investigation. The quality of the spectrum is determined by the product $I(\infty)\Gamma$ and, as well as $I(\infty)$, it is also proportional to the measuring time.

In any spectroscopy, the intensity of the detected radiation may be written in the form:

$$I(\mathcal{E}) = C\int_{-\infty}^{\infty} L(E-\mathcal{E})\varphi(E)\mathrm{d}E + \xi(\mathcal{E})$$

where $\mathcal{E}$ is an energy parameter depending on the experimental setup, $L(E-\mathcal{E})$ is the instrumental line, $\varphi(E)$ is a function describing the response of the substance under investigation to monochromatic radiation, and $\xi(\mathcal{E})$ is the noise due to the stochastic processes.

In Mössbauer spectroscopy, any sample is characterized by $\mu_a(E)$. The simplest situation for recovering the $\mu_a(E)$ function from experimental data is in transmission spectroscopy, where $\varphi(E) = \exp[-\mu_a(E)d]$. There are two ways to find the $\mu_a(E)$ function. The first involves a hypothesis concerning the nature of this function. Analysis of an experimental spectrum amounts to the determination of the parameters characterizing $\mu_a(E)$ in accordance with the hypothesis.

The second way is connected with natural assumptions only on the nature of the $\mu_a(E)$ functions, for example, their smoothness. If there are no grounds for choosing a hypothesis, a certain initial assumption is made as to the nature of the required function. This often amounts to a search for an expression describing the response of the medium to monochromatic radiation, and sometimes 'an enhanced resolution of the method' is spoken of. The idea is that the best quality of the spectrum is attained using a source with a line shape described by the δ-function. Some methods of enhanced signal recovery have been developed for Mössbauer spectroscopy. As in sensitivity or resolution enhancement in other types of spectroscopy, a compromise has to be made between sensitivity and line width, as increasing the resolution always causes a decrease in sensitivity. Other types of data processing have been used to minimize distortion introduced by the measuring instruments.

List of symbols

E = energy; $\mathcal{E}$ = energy parameter depending on the experimental setup; E_0 = initial energy of electrons; $I(0)$ = intensity on resonance; $I(\infty)$ = intensity off resonance; $I(\mathcal{E})$ = intensity of the detected radiation; $I(v)$ = intensity at any velocity v; $I(v_i)$ = amplitude line at v_i position; $I(v,v_i)$ = experimental spectrum SEDMS; $[J]_{t_a,t_a}$ = information matrix element of interest; $J^a(E)$ = absorption line; $J_M(E)$ = emission line; $L(E-\mathcal{E})$ = instrumental line; R = energy resolution; t_a = effective thickness of the sample; v = relative velocity; γ_{cr} = angle of total reflection; Γ = full width at half maximum; Γ_{nat} = natural line width; δ = isomer (chemical) shift; $\varepsilon(0)$ = resonance effect magnitude; θ = direction electrons leaving the scatterer with an energy E; $\mu_a(E)$ = total linear absorption coefficient; $\xi(\mathcal{E})$ = noise due to the stochastic processes; $\varphi(E)$ = function describing the response of the substance under investigation to monochromatic radiation.

See also: **Calibration and Reference Systems (Regulatory Authorities); Mössbauer Spectroscopy, Applications; NMR Spectrometers; Quantitative Analysis.**

Further reading

Andreeva MA, Belozerski GN, Grishin OV, Irkaev SM, Nikolaev VI and Semenov VG (1993) Mössbauer total external reflection: A new method for surface layers analysis. I. Design and developing of the Mössbauer spectrometer. *Nuclear Instruments and Methods* **B74**: 545–553.

Atkinson R and Cranshaw TE (1983) A Mössbauer backscatter electron counter for use at low temperature. *Nuclear Instruments and Methods* **204**: 577–579.

Balko B (1986) Investigation of electronic relaxation in a classic paramagnet by selective excitation double-Mössbauer techniques: Theory and experiment. *Physical Review B* **33**: 7421–7437

Bäverstam U, Bohm C, Ekdahl T and Liljequist D (1975) Method for depth selective ME-spectroscopy. In: Gruverman IJ and Seidel CW (eds) *Mössbauer Effect Methodology*, Vol 9, pp 259–276. New York: Plenum Press.

Belozerski GN (1993) *Mössbauer Studies of Surface Layers*. Amsterdam: Elsevier Science.

Flin PA (1975) Mössbauer backscattering spectrometer with full data processing capability. In: Gruverman IJ and Seidel CW (eds) *Mössbauer Effect Methodology*, Vol 9, pp 245–250. New York: Plenum Press.

Lippmaa M, Tittonen I, Linden J and Katila TE (1995) Mössbauer NMR double resonance. *Physical Review B: Condensed Matter* **52(14)**: 10268–10277.

Meisel WP (1996) Surface and thin film analysis by Mössbauer spectroscopy and related techniques. In: Long GJ and Grandjean F (eds) *Mössbauer Spectroscopy Applied to Magnetism and Materials Science*, Vol 1, pp 1–30. New York: Plenum Press.

Nasu S (1996) High-pressure Mössbauer spectroscopy with nuclear forward scattering of synchrotron radiation. *High Pressure Research* **14(4–6)**: 405–412.

Pasternak MP and Taylor RD (1996) High pressure Mössbauer spectroscopy: The second generation. In: Long GJ and Grandjean F (eds) *Mössbauer Spectroscopy Applied to Magnetism and Materials Science*, Vol 2, pp 167–205. New York: Plenum Press.

Schaaf P, Kramer A, Blaes L, Wagner G, Aubertin F and Gonser U (1991) *Nuclear Instruments and Methods in Physics Research* **B53 (2)**: 184–188.

Weyer G (1976) Applications of parallels-plate avalanche counters in Mössbauer spectroscopy. In: Gruverman IJ and Seidel CW (eds) *Mössbauer Effect Methodology*, Vol 10, pp 301–320. New York: Plenum Press.

Mössbauer Spectroscopy, Applications

Guennadi N Belozerski, St.-Petersburg State University, Russia

HIGH ENERGY SPECTROSCOPY
Applications

The application of Mössbauer spectroscopy in diverse fields of qualitative and quantitative analysis is based on the ease with which hyperfine interactions can be observed. The information obtained from Mössbauer spectroscopy may be correlated with other methods by which HI can be examined such as NMR, EPR, ENDOR, PAC (perturbed angular correlations), nuclear orientation and neutron scattering. However, Mössbauer spectroscopy often proves to be experimentally simpler, more illustrative and an efficient method for studying applied problems. Mössbauer nuclei are ideal 'spies' supplying information on both the microscopic and macroscopic properties of solids.

Three factors may be identified as responsible for the widespread use of Mössbauer spectroscopy in both fundamental and applied research. First is the highest relative energy resolution $R \sim \Delta E/E$ and rather good absolute energy resolution $\Delta E \sim \Gamma_{nat}$ (the natural line width) (sometimes $\sim 10^{-9}$ eV). Secondly, the absolute selectivity of Mössbauer spectroscopy means that in each experiment a response is registered from only one isotope of the element. Thirdly, Mössbauer spectroscopy has a high sensitivity that is determined by the minimum number of resonant atoms needed to produce a detectable response. In transmission Mössbauer spectroscopy for ^{57}Fe, a response is given by a monolayer with an area of the order of 1 cm^2. Also important is the absence of any limitation on experimental conditions other than that the sample should be a solid.

Applications in physics

Mössbauer spectroscopy offers a resolution sufficient to measure the effect of differing gravitational potentials on frequency or time as predicted by Einstein. The sign of the effect can be reversed by inverting the sense of travel over a fixed vertical path. Pound and Rebka measured the gravitational red shift in a 22-metre tower and observed -5.1×10^{-15} shift in the γ-ray energy of ^{57}Fe. The source–detector setup was interchanged every few days to allow comparison of the results from a rising γ-ray beam to those from a falling one. When all of these measurements were combined, they yielded a result 0.9970 ± 0.0076 times the result predicted from the principle of equivalence.

There have also been some applications of Mössbauer spectroscopy in nuclear physics, to measure quadrupole moments of long-lived nuclear states by observing the orientation of a state at very low temperatures through the intensity ratios in a Mössbauer transition. The spectra for the case of the nuclear orientation in the $-\frac{11}{2}$ state of ^{119}Sn by a quadrupole interaction are shown in **Figure 1**. In this case no macroscopic orientation of the hyperfine fields is needed. At very low temperatures the hyperfine splitting of the $-\frac{11}{2}$ state leads to an alignment in the $+\frac{3}{2}$ state yielding different intensities of the two Mössbauer lines. The increase of the intensity of line A located at higher velocities is clearly seen in the 16 mK spectrum. This uniquely indicates that the quadrupole moments of the $-\frac{11}{2}$ state and of the $+\frac{3}{2}$ state

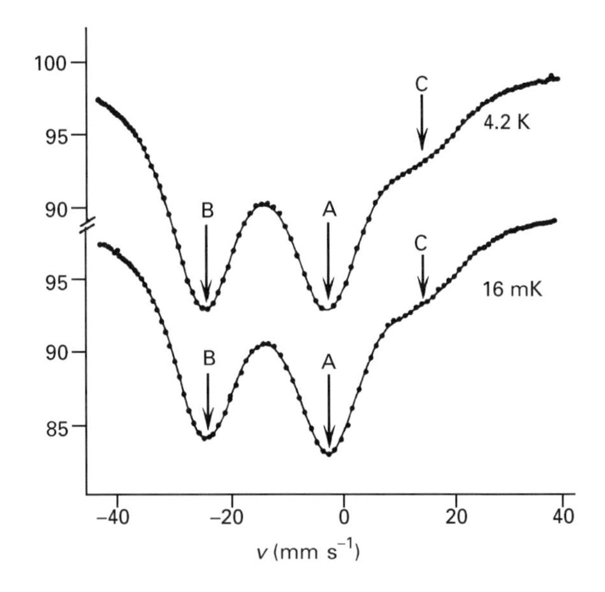

Figure 1 Mössbauer spectra of the 23.9 keV transition in ^{119}Sn with a source of $^{119}Sn^m(OH)_2$ (polycrystalline samples with one of the largest quadrupole splitting of ionic Sn^{2+} compounds) at 4.2 K and 16 mK and a ^{119}Sn:Pd (3 at% ^{119}Sn) absorber chamber at a temperature of about 1.3 K are shown. The, $-\frac{11}{2}$ state decays by an *M*4 transition to the $+\frac{3}{2}$ state of ^{119}Sn, which itself decays to the $-\frac{1}{2}$ ground state with the 23.9 keV *M*1 Mössbauer transition. The source was cooled inside the mixing chamber of the $^3He/^4He$ dilution refrigerator specially designed for Mössbauer experiments. The weak line C at 1.5 mm s^{-1} is attributed to Sn^{4+} impurities in the source.

have the same sign. From a theoretical fit to the data it was deduced that $Q_{11/2} = -0.13 \pm 0.04$b. The same principle is used for measurements of temperature below 100 mK, i.e. by a ^{151}Eu Mössbauer thermometer using an absorber of EuS.

More then 99.5% of all applications of Mössbauer spectroscopy are connected with hyperfine interaction parameters and structure factor determinations. An example of a sophisticated application is the study of the low temperature properties of magnetic impurities in metals that have an antiferromagnetic exchange interaction (Kondo effect). In order to study the very low temperature behaviour of a Kondo system, Mössbauer spectroscopy was used on two 'typical' Kondo systems, Fe:Cu and Fe:Au. The first system (Fe:Cu) showed expected Kondo-type properties – an extra polarization in the electron gas due to the correlations produced by the Kondo effect. The Fe:Au system, on the other hand, exhibited quite unexpected and striking results incompatible with those for Fe:Cu. In brief, in Fe:Au the temperature dependence of the susceptibility differed for $T > 10$ K from that for $T < 10$ K, yielding $\theta_1 = 10$ K and $\theta_2 = 0.5$ K for the Curie–Weiss θ of the two temperature regimes. Some data at low temperature are given in **Figure 2**. All data show that Fe:Au is clearly a system in which even at very low impurity concentrations the interaction effects are important at low temperatures and low external fields. Thus, it is difficult to extract parameters for the intrinsic Kondo behaviour of the isolated impurities from the experimental data on this system. The data show that the Kondo temperature for Fe:Au is most likely in the order of 10 K and not 0.5 K; the lower temperature seems to be connected with some kind of magnetic order.

Mössbauer spectroscopy is especially fruitful if the effective field, H_{eff} is one of the parameters determining the shape of the spectrum. H_{eff} is the resultant of the averaging of the hyperfine field H_n, but the way in which it is derived depends on the sample. The outcome is determined by comparing the frequency of the fluctuation of the H_n direction with the Larmor precession frequency of the nucleus in the hyperfine field. In a ferromagnetic material, the temperature dependence of H_{eff} is predicted to be the same as the temperature dependence of the expectation value of the atomic spin, and consequently of the magnetization. It thus provides a method for investigating magnetization without the application of external fields.

The study of the magnetic properties of materials and the dynamic aspects of magnetic interactions have been the most frequent applications of Mössbauer spectroscopy. The latter can be done

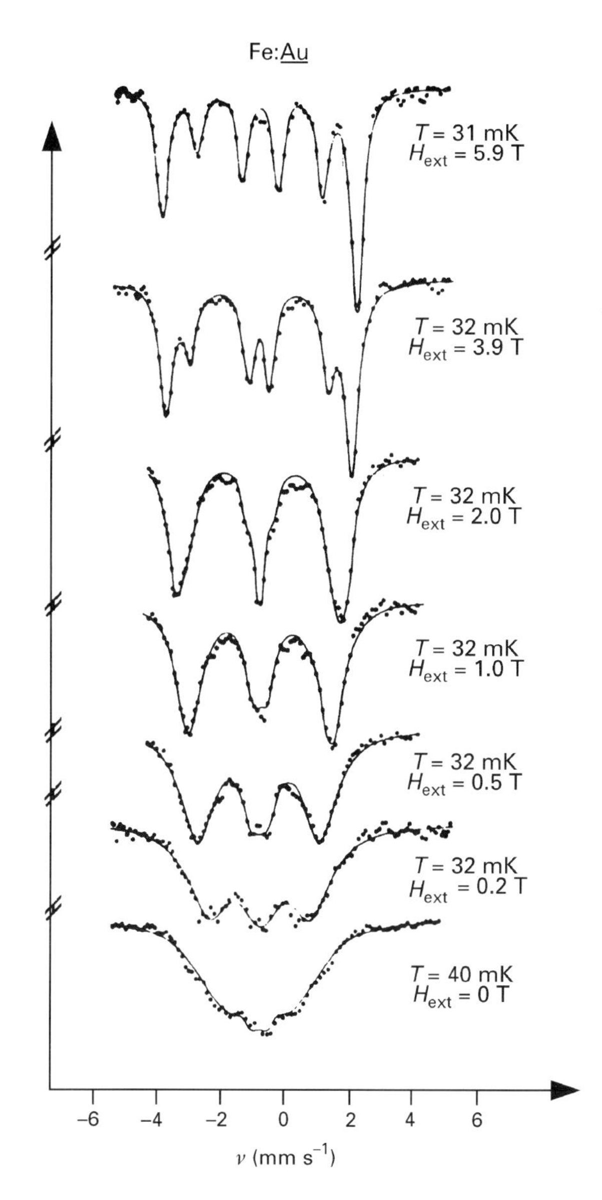

Figure 2 Mössbauer spectra of the source of Fe:Au, prepared by diffusing ^{57}Co into a foil of Au; the total impurity content (Co + Fe) was below 10 ppm. The source was attached to the mixing chamber of a ^{3}He/^{4}He cryostat. The absorber was iron potassium hexacyanide at 1.3 K in the same external field ($H_{ext} \leq 6.0$ T). The solid lines are fits to the spectra including line broadening due to relaxation effects or a distribution of hyperfine field.

particularly conveniently in the case of single-domain particles. A reduction in the dimensions of single-domain particles increases the probability of thermal fluctuations of the magnetization orientations and this leads to superparamagnetic phenomena. One can suppress superparamagnetism either by cooling the sample or by applying an external magnetic field H_{ext} (see **Figure 3**). Such studies permit the determination of the magnetic anisotropy, the alignment of magnetic domains, the particle size, the magnetic interaction among microcrystals, and surface effects.

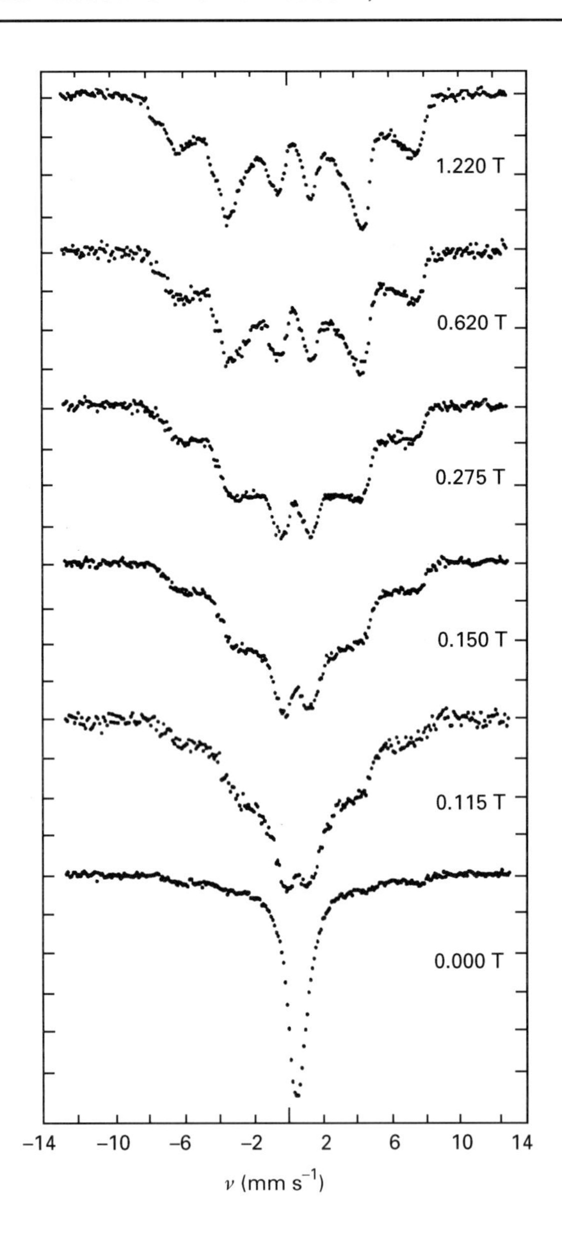

Figure 3 Spectra of microcrystals of Fe_3O_4 obtained at 260 K in various applied magnetic fields. The iron atoms at the octahedral and the tetrahedral sites in Fe_3O_4 have different hyperfine field with opposite directions, and the corresponding Mössbauer lines are not completely resolved even in external field resolved H_{ext} = 1.22 T.

A final illustration of the application of Mössbauer spectroscopy in physics is the study of spin texture in amorphous metals. Mössbauer spectroscopy provides information on both magnetic anisotropy and the texture of the principal electric field gradient (EFG) axes that reflects crystallographic features of the solid-state structure. It is known that spin texture in glassy metal ribbons, formed by high-speed quenching of fluid metals, is extremely sensitive to strains and stresses. This is partly consistent with the very low crystalline anisotropy in these materials. The results of a study of $Fe_{80}B_{20}$ are presented in **Figure 4**. The Mössbauer spectra for the contact with the quenching support and noncontact surfaces of the band are different. These differences in magnetic properties of the surfaces are due to the different cooling conditions during the band production. As the result of texture determination by Mössbauer spectroscopy, a distribution function $D(\theta_m, \Phi_m)$ (see **Figure 5**) was found for the contact and noncontact surfaces of the band, which describes the distribution of the relative volumes wherein H_{eff} or the EFG are oriented along the (θ_m, Φ_m) direction in a unit solid angle. Consideration of **Figure 4** shows that the magnetic anisotropy is along the *y*-axis. Hence, the magnetic dissipation fields on the alloy surface are significantly weaker. This is indicative of the high magnetic quality of the surface of the $Fe_{80}B_{20}$ alloy. This means that the local environment of Fe atoms is the same for the two sides of the band, i.e. there is no difference in the concentrations of Fe atoms and the metalloid atoms.

Applications in chemistry

The three components of the hyperfine interaction – the isomer shift, the nuclear quadrupole splitting and the nuclear magnetic dipole splitting – have immediate chemical application. The isomer shift δ measures the charge density of the atomic electrons at the nucleus and is therefore directly related to chemical bonding and covalency. The ligand field splitting energy Δ depends on the gradient of the electric field produced by the other ions in the lattice, i.e. it is related to the point symmetry of the lattice information. H_{eff} provides a sensitive tool for the detection of magnetically ordered states. Mössbauer spectroscopy is a major tool for studying the formation of new inorganic materials and probing structural and magnetic phase transformations in inorganic compounds. This can be demonstrated with some specific examples. As the first example, the participation of 3d and 4s electrons in the chemical bond of iron in a complex determines the total electron density at the Fe nucleus (see **Table 1**), and hence the isomer shift for the Fe complex. A change in the oxidation state influences the electron density at the nucleus in a major way, since such a change is usually associated with the removal (or the addition) of one or more valence electrons. Thus the two oxidation states differ in their outer valence electron number by an integer, and this results in a major difference in the isomer shift for the two oxidation states. This is illustrated in **Figure 6**, where the isomer shifts for monovalent, trivalent, and pentavalent gold

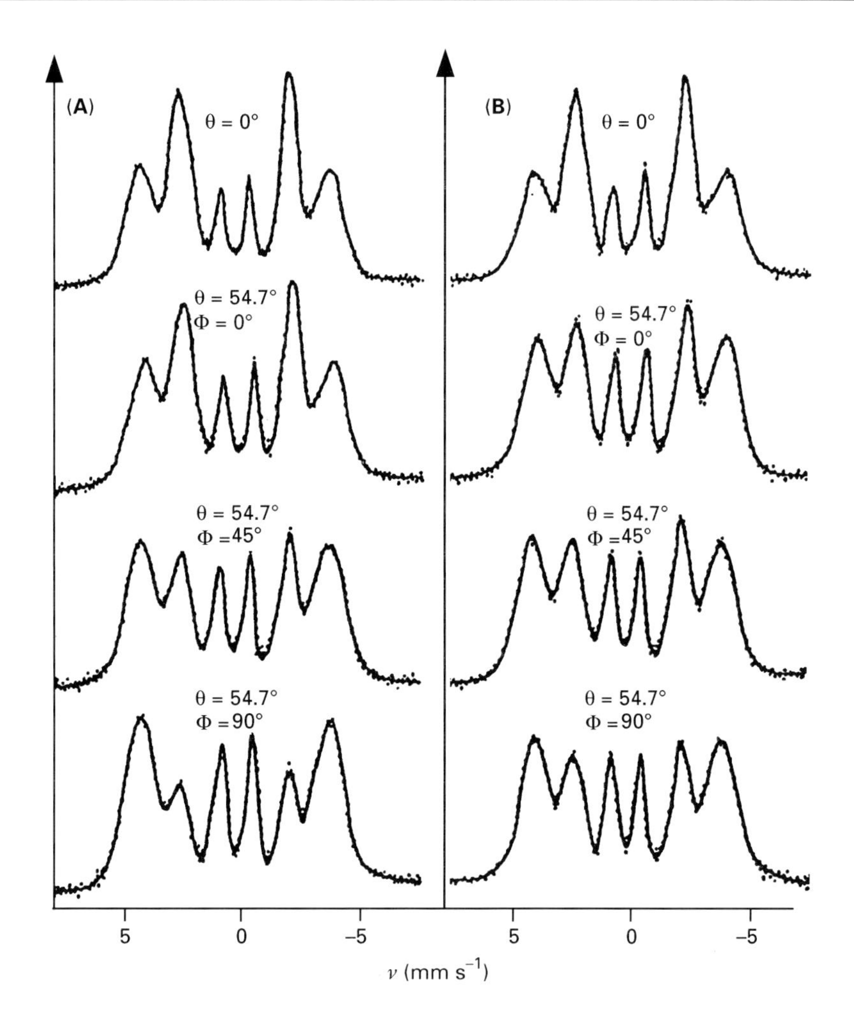

Figure 4 Mössbauer spectra from a band of an amorphous alloy: (A) contact surface, (B) noncontact surface. In accordance with the foil preparation, the *z* principal axis is perpendicular to the foil surface, the *x* axis is along the band and the *y* axis is perpendicular to the rolling direction. The direction of the incident Mössbauer quanta is specified by the angles θ and Φ.

complexes with various halide ligands are shown. The electron density $\rho(0)$ is the least in Au(I) compounds and the highest in Au(III) compounds. The spread in the isomer shift for the selected halide ligands is small within an oxidation state.

The ability to determine oxidation states is demonstrated in **Table 2** for the isomer shift data of fluorides of neptunium for various oxidation states represented by the $5f^n$ configuration.

Table 1 Isomer shifts, electron configurations, and value of the electron density difference $\Delta\rho(0)$ for matrix-isolated Fe ions

Ion	*Configuration outside Ar core*	*Isomer shift[a] (mm s⁻¹)*	*$\Delta\rho(0)$ (au)*
Fe^0	$3d^64s^2$	−0.75±0.03	16.0
Fe^+	$3d^7$	+1.77±0.08	0.0
Fe^+	$3d^64s$	+0.26±0.05	10.5

[a] Relative to Fe metal at 300 K.

The ice produced by the slow freezing of solutions that contain Mössbauer atoms, and the glass produced by their rapid freezing, are also suitable for investigation by Mössbauer spectroscopy. Systematic Mössbauer examinations of ices were started with investigations of the polymorphic transformation

Table 2 Isomer shifts for the various fluorides of Np and the values of $\rho(0)$ for electron configurations, $5f^n$, nominally representing the oxidation states in fluorides

$5f^n$	*Oxidation state*	*Electron density (au)*	*Isomer shift[a] (mm/s⁻¹)*
$5f^4$	Np(III)	6 965 162	+34(1)
$5f^3$	Np(IV)	6 965 343	−9(1)
$5f^2$	Np(V)	6 965 555	−37(1)
$5f^1$	Np(VI)	6 965 793	−63(2)
$5f^0$	Np(VII)	6 966 057	−77(2)[b]

[a] Relative to $NpAl_2$ at 4.2 K.
[b] For Li_5NpO_6.

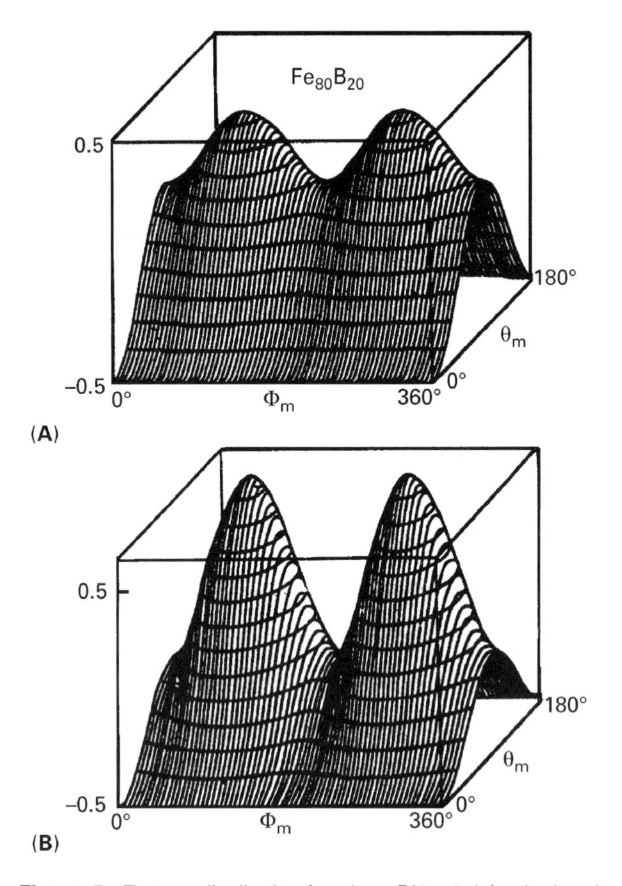

Figure 5 Texture distribution functions $D(\theta_m, \Phi_m)$ for the band of amorphous alloy $Fe_{80}B_{20}$: (A) contact surface, (B) noncontact surface. The direction of the quantization axis is determined by the angles θ_m and Φ_m.

that occurs in the ice as a result of changes in temperature. Detailed investigations showed how the structure of the ice formed depends on the rate of freezing, and how this affects the Mössbauer parameters measured on the frozen solution. If rapid freezing is applied to equilibrium solution systems in which the equilibrium shifts rapidly (within 1–2 s) and measurably as a result of the change in temperature, Mössbauer spectroscopy provides reliable information only if the temperature of thermostating before the freezing is close to the solidification point.

Research into nonaqueous solvents and solutions is also worth note. When metal salts are dissolved in various solvents, it is possible to study the coordination of the solvent molecules to the cation and the extent of association between the cation and its anion.

The phenomenon of spin transition (spin crossover), is another example of a fruitful application of Mössbauer spectroscopy in chemistry. The phenomenon is observed in transition metal complexes with d^4, d^5, d^6, d^7 and d^8 electron configurations. Depending on the ligand field splitting energy, Δ, relative to the mean spin pairing energy, E_p, the ground state of

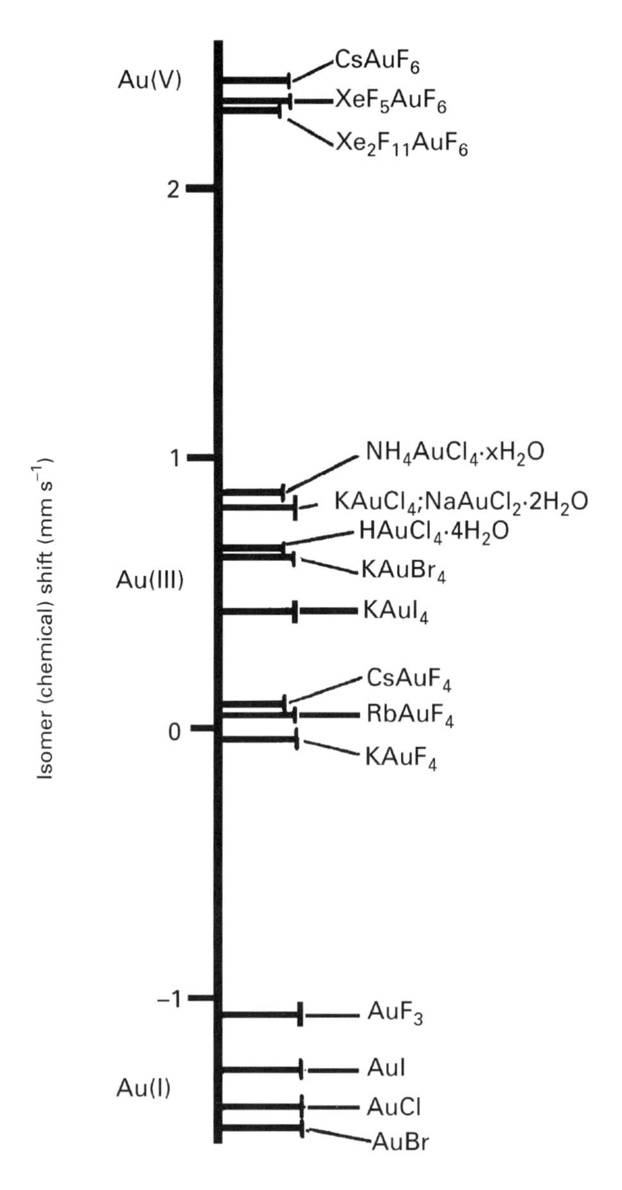

Figure 6 Isomer shift scale for Au(I), Au(III), and Au(V) halides.

a particular transition metal complex shows either high-spin (HS) behaviour or low-spin (LS) behaviour. If $|\Delta - E_p| \sim k_B T$, both HS and LS spin states may coexist in thermal equilibrium: HS $\rightleftarrows$ LS.

Mössbauer spectroscopy reveals distinct resonance lines for the individual spin states provided their individual mean lifetimes are comparable to or longer than the lifetime τ of the nuclear excited state. If one determines the area fractions of the spin states involved (which in most cases come close to the actual molar fractions of HS and LS molecules), one may follow quantitatively the spin conversion as a function of temperature. The simplest way to observe the transition is to change the temperature. An unusual way to affect the spin transition behaviour is to change the elastic properties of the crystal lattice, which are important for the spin transition behaviour. This was done by a partial isomorphous substitution of the central metal ions in mixed crystals of $[Fe_xMn_{1-x}(phen)_2(NCS)_2]$ by manganese ions, which differ in the volume and the electron configuration from the replaced metal ions (see **Figure** 7). The transition temperature shifts to lower temperatures upon dilution, and the residual paramagnetism at low temperatures increases systematically with decreasing iron concentrations x.

There are many other fields of application of Mössbauer spectroscopy in chemistry, including coordination chemistry, catalysis (heterogeneous), mixed-valence compounds, glasses, oxide and oxyhydroxide studies, and actinide chemistry.

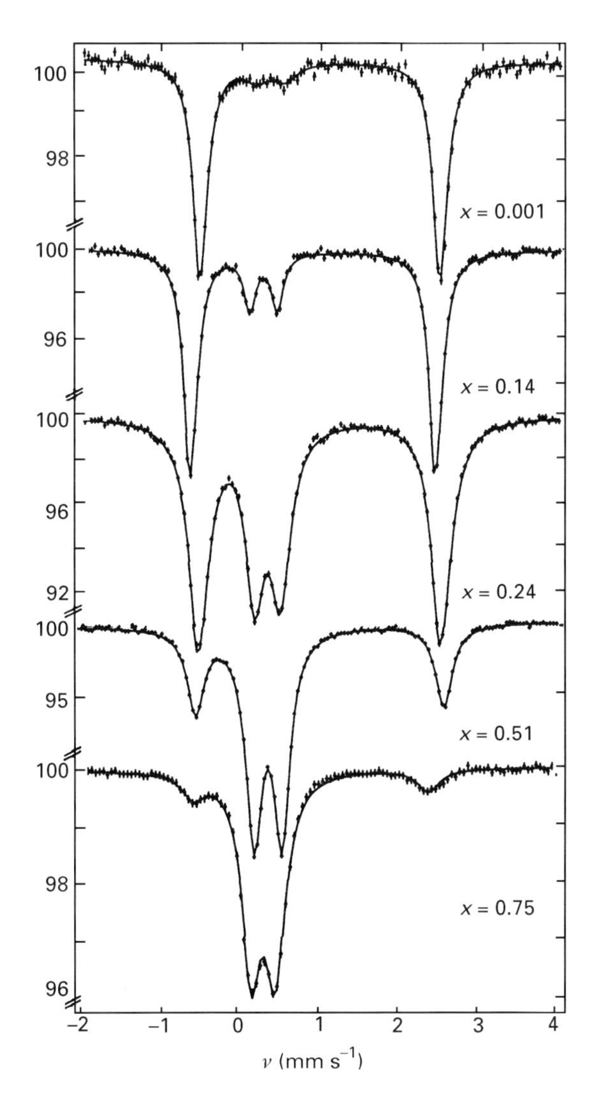

Figure 7 Mössbauer spectra of $[Fe_xMn_{1-x}(phen)_2(NCS)_2]$ at 5 K and various iron concentrations x. The spectra demonstrate that, with increasing dilution of iron by manganese, the intensity of the quadrupole doublet of the high-spin state of iron(II) (outer two lines) increases steadily at the expense of the low-spin quadrupole doublet of iron(II) (inner two lines).

Applications in biology and geology

Many biological molecules contain iron and Mössbauer spectroscopy is a useful tool for the study of proteins and enzymes. The measurement of the magnetic properties of transition metal elements in biological molecules by MS, NMR and EPR is an important way of characterizing the electronic state of the metal ion, and hence of providing a clue to the structure and function of the molecule. Mössbauer spectroscopy may be used to study their chemical state and bonding and to obtain qualitative data on the local structure and symmetry in their neighbourhood.

Haem proteins are the best-understood of these molecules, and the first systematic study of biological molecules using the Mössbauer effect was done on haemoglobin and its derivatives. Since then a great deal of work has been done on iron–sulphur proteins and on iron-storage proteins. Magnetic susceptibility data showed that the chemical state of the iron atom and its spin state are very sensitive to the nature of the sixth ligand. The Mössbauer spectra of these molecules have been valuable in confirming these earlier conclusions, and have yielded quantitative data on the way that the energy levels and wavefunctions of the iron atoms are affected by the ligand field and spin–orbit coupling in the protein. They have also been valuable in providing standard spectra for each of the four common states of iron, and a summary of the chemical shifts and quadrupole splittings measured at 195 K is shown in **Table** 3.

In disease, haemoglobin may become ferric (denoted Hi) and of low spin, e.g. in a haemoglobin cyanide HiCN, azide HiN_3 or hydroxide HiOH. The

Table 3 Isomer shifts δ and quadrupole splittings ΔE_Q for haemoglobin at 195 K. The symbol Hi is used for ferric haemoglobin and Hb for ferrous haemoglobin

Material	δ	ΔE_Q
HiF	(0.3)	(0.67)
HiH_2O	0.20	2.00
HiCN	0.17	1.39
HiOH	0.18	1.57
Hb	0.90	2.40
HiN_3	0.15	2.30
HiOH	0.18	1.57
Hb	0.90	2.40
HbNO[a]	–	–
HbO_2	0.20	1.89
HbCO	0.18	0.36

[a]Spectrum very broad.

most noticeable feature of the spectra is the appearance of magnetic hyperfine structure at 77 K. Spectra of HiCN at these temperatures are shown in **Figure 8**. At 4.2 K the magnetic hyperfine spectrum is well resolved, but is complex and asymmetrical owing to the anisotropy of the hyperfine interaction tensor.

Apart from haem proteins, there are studies of iron–sulfur proteins, iron transport and iron-storage compounds, iodine compounds and vitamin B_{12}. There are many publications connected with nitrogenase, oxygenase, hydrogenase and cytochrome P450–ferredoxin enzyme systems and medical and physiological applications.

Application of Mössbauer spectroscopy to crystallography, mineralogy and geology is very similar to chemical applications, with the main emphasis on phase analysis and structure determination. In view of the great importance of iron in the earth's crust and the widespread occurrence of this element in rock-forming minerals, earth scientists have naturally focused attention on applications of ^{57}Fe Mössbauer spectroscopy. One of the most important groups of rock-forming minerals are the silicates, in which particular lattice position is often occupied by more than two atomic species. In these cases, accurate site occupancy numbers for each species cannot be obtained by diffraction alone. Mössbauer spectroscopy has been a useful tool for investigating the local properties of iron sites in complex crystal structures, particularly when employed on minerals with carefully defined (well-known) positional parameters.

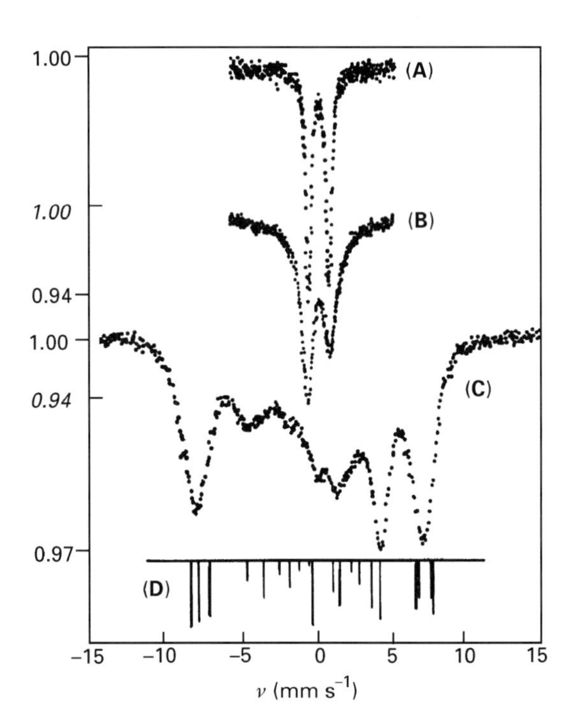

Figure 8 Mössbauer spectra of a low-spin ferric haemoglobin HiCN – at (A) 195 K, (B) 77 K and (C) 4.2 K. The broadening at 77 K relative to 195 K is due to the slower electron spin relaxation rate. At 4.2 K the relaxation is so slow that the hyperfine pattern is resolved, but complex.

Typical applications of Mössbauer spectroscopy to mineralogy and geology have been analysis of the oxidation state of iron at iron sites in minerals. The ferric to ferrous iron ratio reveals important information on the partial pressure of oxygen during crystallization, which is a parameter of geological significance. The study of area ratios of distinct hyperfine patterns has led to thermodynamic analyses of order–disorder phenomena in minerals. Application of Mössbauer spectroscopy to poorly crystallized materials of geological relevance may be fruitful. Studies have been made on hydrolysis and absorption in clay minerals and on coordination polyhedra of iron in glasses.

Mössbauer spectroscopy has made a modest contribution to the study of composite samples and separated mineral phases from the moon. The constituents of the lunar soil are of a highly complex nature. The primary aim has been to study the amount of iron in the soil, the distribution of iron over the soil constituents, and the particular valence states. The soils at the moon landing sites do not represent averages of the collected rocks. The spectrum of the soil (**Figure 9**) is the superposition of patterns of ^{57}Fe in silicate minerals, silicate glasses, ilmenite, metallic iron, and troilite.

The identification of the magnetic patterns of metallic iron and FeS and the quadrupole-split pattern of paramagnetic $FeTiO_3$ is generally no problem in

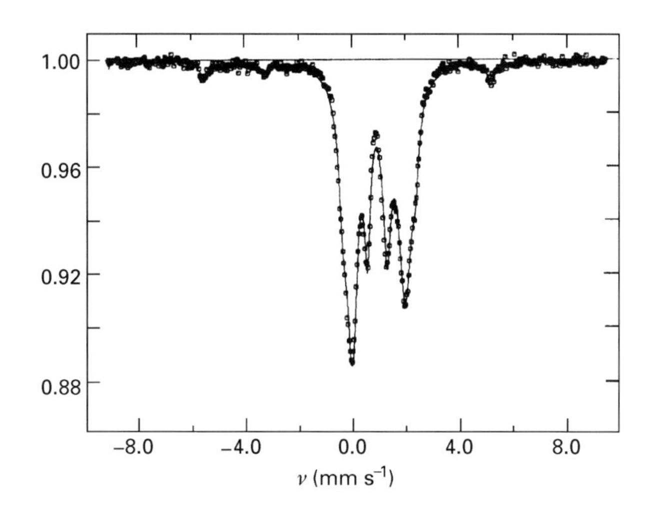

Figure 9 Resonant absorption spectrum of ^{57}Fe (295 K) in lunar soil from the Apollo 11 landing site at Mare Tranquillitatis. The absorption in the range between 0 and 0.3 mm s^{-1} results primarily from iron in silicate glass, pyroxene and olivine; the peaks at lower and higher velocities are due to metallic iron.

spectra from soil absorbers kept at room temperature (**Figure 9**). However, the paramagnetic silicates in the soil cannot be identified easily because of the substantial overlap of their quadrupole-split patterns.

Surface layer studies

Mössbauer spectroscopy has developed as one of the few methods available for investigation of solids differing in depth by several orders of magnitude. These can be layers with the depth of less than 1 nm (scattering at glancing angles) as well as layers that are about 10 μm deep (electron and X-ray detection). The technique is able to examine solids over a wide range of compositions, gaseous pressures, and temperatures and under conditions of practical interest. It should also be noted that Mössbauer spectroscopy is one of the best methods for *in situ* characterization of solid–solid and solid–solution interfaces. This lends itself to *in situ* studies of surfaces under various coatings and processes, surface magnetism and the effect of the gas phase on the properties of the surface layers and the structure and magnetic properties of epitaxially grown monolayers on the surface of oriented single crystals. The techniques of surface layer analysis find extensive applications in science and industry.

The thinnest layers to be studied in Mössbauer spectroscopy are those studied by GA DCEMS and (total external reflection) TER. For example, the TER method has been used to study the initial stage of oxidation of α-Fe foil. The substantial progress in surface studies using glancing angles was achieved by simultaneous detection of electrons, γ-rays or X-rays as well as mirror reflected γ-rays, which allows reconstruction of the process of surface layer formation. The sample under investigation was inside the electron chamber of the detector. The results of the calculation and the fitting of the experimental spectra are very sensitive to the model of surface layer structure and to the density of the top layer.

For phase modification of the surface layer, the initial sample (sample I) was oxidized at 150°C in air for 4 hours (to give sample II). From experimental spectra (see **Figure 10**) it can be concluded qualitatively that in the top layers iron is oxidized and is characterized by different spectra with and without magnetic splitting. The sample II spectra in the right column of **Figure 10** clearly illustrate the phase transformation on the top of the α-Fe film. The spectra differ mostly at the lower grazing angles, where the penetration depth is minimal. For example, the contribution of the doublet is much larger for the spectrum of sample II at θ = 2.2 mrad.

Thus even qualitative analysis clearly shows that the technique is more sensitive to the changes of chemical and magnetic states of ultra-thin surface layers than the usual CEMS techniques. Detailed information about distributions of different phases

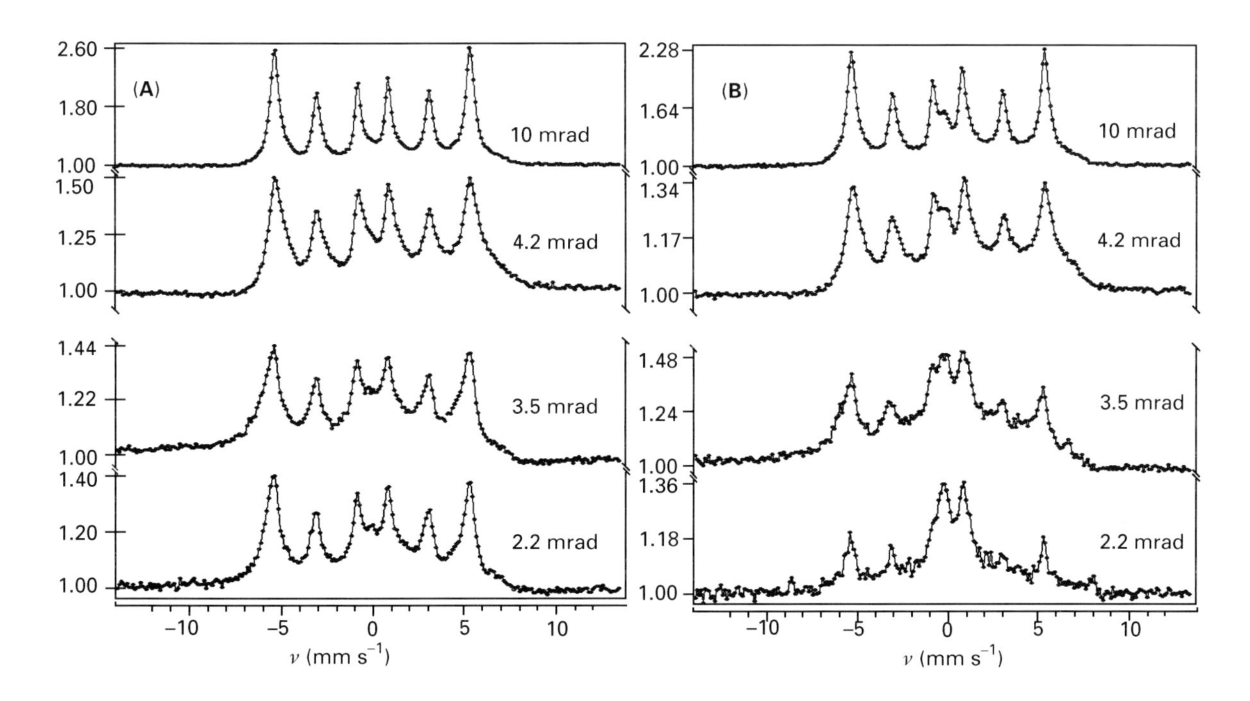

Figure 10 Sets of TER-CEM spectra (A) for the initial sample α-^{57}Fe (enriched up to 90% and ~ 50 nm thick), sputtered on a 10 mm thick beryllium disk (sample I) and (B) after oxidizing in air for 4 hours at 150°C (sample II). TER take place at grazing angle θ ~ 3 mrad.

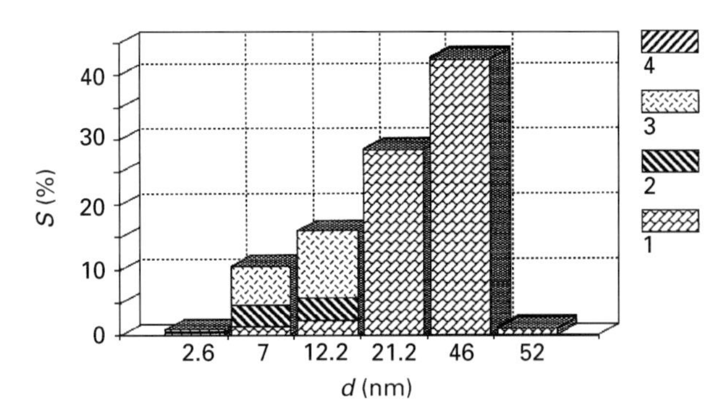

Figure 11 The histograms of iron content S in different phases of different sublayers (sample I) as function of the lower border of the sublayers. The sample is an iron film about ~ 50 nm thick, sputtered on the beryllium disk 10 mm thick. 1; sextet of α-Fe; 2, broadened sextet with smaller magnetic splitting than in α-Fe; 3, asymmetric strongly broadened sextet associated with hyperfine interaction parameters for α-FeOOH; 4, the doublet.

with depth may only be obtained after quantitative analysis of sets of spectra. The results of calculations for the initial sample are given in **Figure 11.** It is seen that the α-Fe phase is present in the top layer also. This points to the existence of an island phase structure of the top layer. The amount of FeOOH phase increases with depth but has a maximum value at ~10 nm. From analysis of histograms such as that shown in **Figure 11**, it can be seen explicitly how different oxide phases move inside the sample during oxidation processes.

Industrial applications

The scope of problems addressed is extremely wide and includes metals research of industrial significance, steel and steel alloys, amorphous alloys and glasses, corrosion, surface magnetism and superparamagnetism, catalysis and solid-state reactions on solid surfaces, minerals, and mineral processing, superconductivity including high-temperature superconductivity, nanocrocrystals, thin films, ion implantation and laser treatment of metals, the coal industry, amorphous alloys and glasses, and numerous other miscellaneous applications. Areas of application of Mössbauer spectroscopy to corrosion studies are listed in **Table 4**.

Physical metallurgy is a rather wide field of applications of Mössbauer spectroscopy and it is possible to enumerate only the main topics: phase analysis, order–disorder alloys, surfaces, alloying, interstitial alloys, steel, ferromagnetic alloys, precipitation, diffusion, oxidation, lattice defects; etc. Alloys are well represented by the iron–carbon system, the mechanism of martensite transformation, high-manganese and iron–aluminium alloys, iron–silicon and Fe–Ni–X alloys.

Table 4 Corrosion studies using Mössbauer spectroscopy

Processes studied	*Corrosive environments, objectives, applications*
Passivation	Solutions and electrode surfaces in solutions at various potentials
	Properties and structure of passive films in solutions, in gases or in vacuum
Corrosion in water and in aqueous solutions	Distilled water
	The effect of oxygen
	The effect of other admixtures in water
	The temperature effect in aqueous media
	Water vapour corrosion
	Corrosion in power plants
Corrosion in gases	Pure oxygen or dry air
	Atmospheric and water vapour corrosion
	Aggressive gases
	Combined action of both gas media and solutions
Corrosion in aggressive environments	Acids
	Alkalis
	Organic and natural media
	Corrosion in tubing and autoclaves
	Applications in agriculture
Specific corrosion processes	Stress corrosion
	Corrosion beneath lake and polymeric coatings
	Transformations of corrosion products to enhance the protective properties and to identify the corrosion products
Inhibition and passivation	The effect of special inhibiting or passivating admixtures on the on the composition and growth rate of protective films.
Corrosion of amorphous alloys	Materials science
	A check on the theory used to describe corrosion of amorphous alloys
Internal oxidation	Materials science

There have been a number of regional and international conferences devoted to industrial applications of Mössbauer spectroscopy or specific problems of materials science. The scope of problems addressed is extremely wide and ranges over all the above-mentioned problems. One can expect the contributions of Mössbauer spectroscopy in industry to divide into three areas: (1) as a research tool, (2) in quality control, and (3) for in-service evaluation. Unfortunately, there are still very few Mössbauer instruments in use for quality control, or for in-service evaluation of materials or surface. More than 99% of Mössbauer publications cover use of the technique as a research tool.

An example of the application of Mössbauer spectroscopy to metals research of industrial significance is the study of fractures and ruptures in steels (see **Figure 12**). The phase compositions of the metal microvolumes that are the sites for the propagation of fractures were determined. The spectra obtained from the fracture surfaces after various cyclic loadings are a superposition of a sextet with a single line in the centre (γ- and ε-phases). The central line is identical to that of the original sample, the sextet lines being broadened. Such line broadening may be attributed to different local environments of Fe atoms, or to significant local lattice distortions, or to both. It follows from **Table 5** that on increasing the loading rate the surface layer of the fracture is enriched in martensite. The appearance of the α-phase at the site of destruction and its predominance there explains the contradiction between the intercrystallite character of the steel destruction and the high energy that is required. Martensite in the fracture layer is formed in front of a crack to be nucleated and propagated. The predominance of martensite in the fracture layers testifies to the large number of cracks in this area and to a significant stress relaxation preceding the formation and development of the cracks. This is the reason for the high energy of destruction of steels with the initial ($\gamma + \varepsilon$) structure. A comparison of the CEMS and X-ray detection data shows that the processes responsible for the destruction of steels with the $\gamma + \varepsilon$ structure are localized in a zone with a width less than 5 μm. On static loading tests, the width increases up to 10 μm.

Concluding remarks

The technique has grown rapidly and is now an effective form of spectroscopy. Mössbauer spectroscopy using ^{57}Fe is so simple and straightforward that it is now an integral part of many undergraduate laboratory curricula.

An explosive growth in the application of Mössbauer spectroscopy in a broad range of

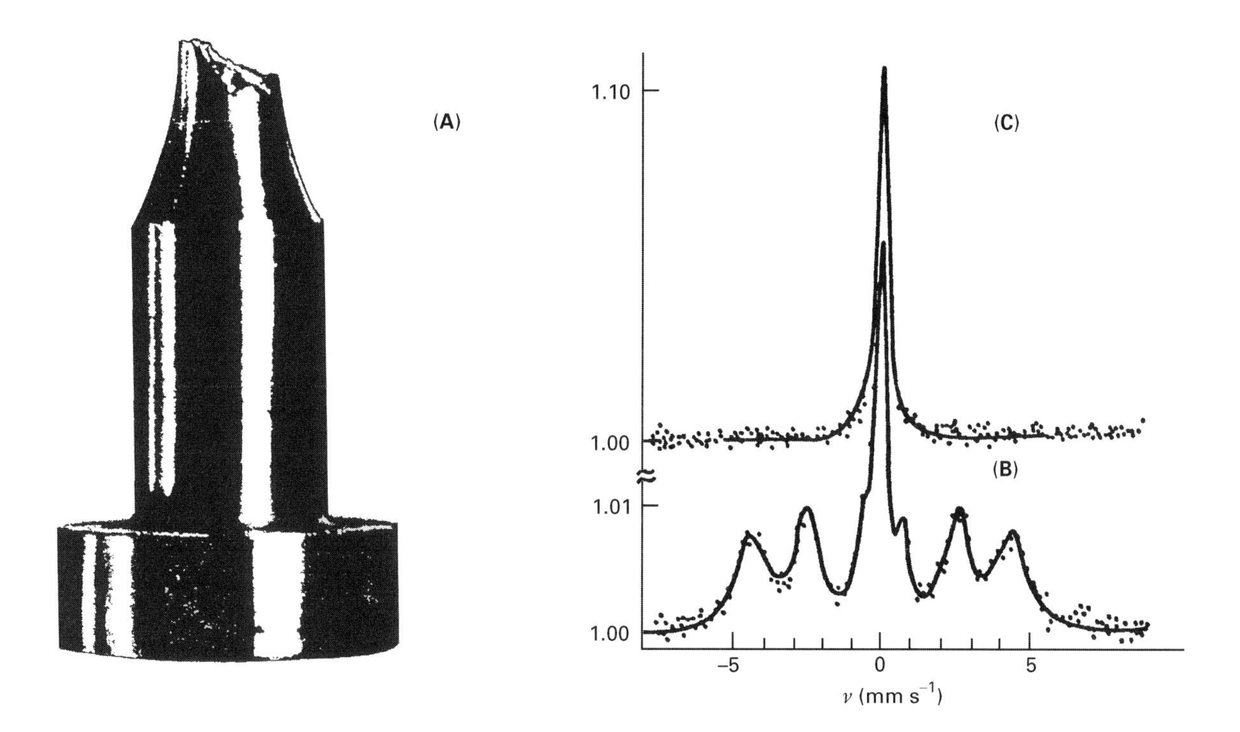

Figure 12 (A) An 8 mm fracture of steel with the initial $\gamma + \varepsilon$ structure (19.2% Mn, 1.67% Si, 0.9% Ti, 0.07% C). (B) CEM spectrum of the steel from the fracture surface shown in (A). (C) CEM spectrum after annealing at 1050°C for 6 hours, followed by cooling in air.

Table 5 Results of the phase analysis of iron-based alloy (19.2% Mn, 1.67% Si, 0.9% Ti, 0.07% C)

Loading	Detected radiation	*Parameters of the* $\gamma+\varepsilon$ *phases* $\delta_{\gamma+\varepsilon}$	$\eta_{\gamma+\varepsilon}$	*Parameters of the magnetic phase* $\langle H_{eff}\rangle$	δ_α	H_α	*Phase composition (saturation effect and rescattering are accounted for)* $C_{\gamma+\varepsilon}$	C_α
Low-cycle fatigue	Electrons	−0.12	28	27.3	−0.01	72	27.6	72.3
	X-rays	−0.14	29	27.5	−0.04	71	35	65
Fatigue tests at 600 cycles min^{-1}	Electrons	−0.12	10	27.5	−0.01	90	8.8	91.2
	X-rays	−0.14	26	27.5	−0.01	74	31.9	68.1
Impact tests	Electrons	−0.12	23	28.0	−0.01	77	21.4	78.6
	X-rays	−0.15	46	28.1	−0.05	54	53.2	46.8

$\langle H_{eff}\rangle$ is the average effective magnetic field at a ^{57}Fe nucleus in the α-phase; $\Delta\delta = \pm 0.01$ mm s^{-1}; η_i are the relative areas (%) under the spectrum ($i = \alpha, \gamma+\varepsilon$); C_i is the relative content of the *i*th phase (%), $\Delta\eta = \pm 5\%$, $C_\alpha = \eta_\alpha/(\eta_\alpha + 1.43\eta_\gamma)$. H_{eff} values are given in tesla, isomer shifts are in mm s^{-1}

scientific areas has resulted in publications in a diverse number of journals. From January 1978, the Mössbauer Data Center in Virginia USA has published the *Mössbauer Effect Reference and Data Journal* monthly. This journal has become an invaluable resource for information on Mössbauer spectroscopy.

List of symbols

C_i = relative content of *i*th phase (%); $D(\theta_m, \Phi_m)$ = distribution function; $\Delta\rho(0)$ = value of the electron density difference; E = energy; E_P = mean spin pairing energy; ΔE = absolute energy resolution; ΔE_Q = quadrupole splittings; $\langle H_{eff}\rangle$ = average effective magnetic field; H_{ext} = external field; H_n = hyperfine field; k_B = Boltzmann constant; R = relative energy resolution; T = temperature; Γ_{nat} = natural line width; Δ = ligand field splitting energy; δ = isomer shift; $\Delta\delta$ = inaccuracy of the δ determination; η_i = relative area (%) under the spectrum; $\Delta\eta$ = inaccuracy of the η determination; θ = grazing angle; τ = lifetime of nuclear excited state; $\rho(0)$ = electron density.

See also: **Chemical Applications of EPR; Chemical Shift and Relaxation Reagents in NMR; Cosmochemical Applications Using Mass Spectrometry; Geology and Mineralogy, Applications of Atomic Spectroscopy; Industrial Applications of IR and Raman Spectroscopy; Interstellar Molecules, Spectroscopy of; Materials Science Applications of X-Ray Diffraction; Mössbauer Spectrometers; Mössbauer Spectroscopy, Theory; Solid State NMR, Methods; Stars, Spectroscopy of; Surface Studies By IR Spectroscopy; Zeeman and Stark Methods in Spectroscopy, Applications.**

Further reading

Belozerski GN, Bohm C, Ekdahl T and Liljequist D (1982) A Mössbauer investigation of the surface of α-iron. I. Corrosion and passivation in H_2O_2 solution. *Corrosion Science* **22**: 831–844.

Cadogan JM (1996) Mössbauer spectroscopy and rare-earth permanent magnets. *Journal of Physics D: Applied Physics* **29**: 2246–2254.

Clark SJ, Donaldson JD and Grimes SM (1996) Mössbauer spectroscopy. In: *Spectroscopic Properties of Inorganic and Organometallic Compounds*, Vol 29, pp 330–417. Cambridge: The Royal Society of Chemistry.

Cranshaw TE (1986) Metal research of industrial significance by Mössbauer spectroscopy. In: Long GJ and Steven JG (eds) *Industrial Applications of the Mössbauer Effect*, pp 7–24. New York: Plenum Press.

Gütlich P (1984) Spin transition in iron complexes. In: Long GJ (ed) *Mössbauer Spectroscopy Applied to Inorganic Chemistry*, Vol 1, 287. New York: Plenum Press.

Greneshe JM (1997) Nanocrystalline iron-based alloys nsinvestigated by Mössbauer spectroscopy. *Hyperfine Interactions* **110**: 81–91.

Hafner SS (1975) Mössbauer spectroscopy in lunar geology and mineralogy. In: Gonser U (ed) *Mössbauer Spectroscopy*, pp 167–199. New York: Springer-Verlag.

Jonson CE (1975) Mössbauer spectroscopy in biology In: Gonser U (ed) *Mössbauer Spectroscopy*, pp 139–166. New York: Springer-Verlag.

Morup S (1986) Industrial applications of Mössbauer spectroscopy to microcrystals. In: Long GJ and Stevens JG (eds) *Industrial Applications of the Mössbauer Effect*, pp 63–82. New York: Plenum Press.

Shenoy GK (1984) Mössbauer-effect isomer shift. In: Long GJ (ed) *Mössbauer Spectroscopy Applied to Inorganic Chemistry*, Vol 1, pp 57–78. New York: Plenum Press.

Steiner P, Belozerski GN, Gumprecht D, Zdrojewski W and Hufner S (1974) Impurity–impurity interaction in a very dilute Fe:Au system. *Solid State Communications* 157–160.

Mössbauer Spectroscopy, Theory

Guennadi N Belozerski, St.-Petersburg State University, Russia

HIGH ENERGY SPECTROSCOPY
Theory

Mössbauer spectroscopy is concerned with the scattering and emission of γ-radiation by atomic nuclei in the condensed phase. The phenomenon was discovered in 1958 by the German physicist R. L. Mössbauer. It makes use of the probability that the state of a system will remain unchanged when γ-quanta are absorbed or emitted with an energy that is exactly equal to the nuclear transition energy E_0.

Hence the γ-spectrum $J(E)$ of a Mössbauer source may be represented by the sum of a line $J_R(E)$ that is displaced due to recoil effects and broadened by the Doppler effect, and a line $J_M(E)$ with its centre at the energy that is exactly equal to the transition energy, the half-width being close to the natural one, Γ_{nat}. The $J_M(E)$ part of the spectrum is of particular interest and reveals itself most strikingly when the source and the sample under study (absorber) are in the solid state, see **Figure 1**. The following normalization conditions may be assumed:

$$\int_{-\infty}^{\infty} J_R(E)\,dE = 1 - f; \qquad \int_{-\infty}^{\infty} J_M(E)\,dE = f \qquad [1]$$

A fraction $(1-f)$ of the disintegrations occurs with the energy transferred to the lattice, and the fraction f is recoilless.

The particular dependence $J_R(E)$ is of no interest but it should be noted that the centre of gravity of this distribution is shifted by the amount E_R relative to the transition energy E_s in the source. $E_R = E_0^2/2Mc^2$ is the recoil energy imparted to an isolated nucleus of mass M (where c = vacuum speed of light). The energy distribution of the Mössbauer γ-quanta $J_M(E)$ may be considered as Lorentzian $L(E)$ with the full half-width Γ such that $\Gamma = \Gamma_{nat}$. Owing to the normalization conditions, $J_M(E)$ may be written as

$$J_M(E) = \frac{f\Gamma_{nat}}{2}\frac{1}{(E-E_s)^2 + (\Gamma_{nat}/2)^2} = \frac{2f}{\pi\Gamma_{nat}}L(E) \qquad [2]$$

The theoretical spectrum of the 129 keV γ-ray of ^{191}Ir absorbed by an atom in iridium metal is shown in **Figure 1**. The Mössbauer spectrometer is sensitive only to the narrow, recoil-free line at zero energy

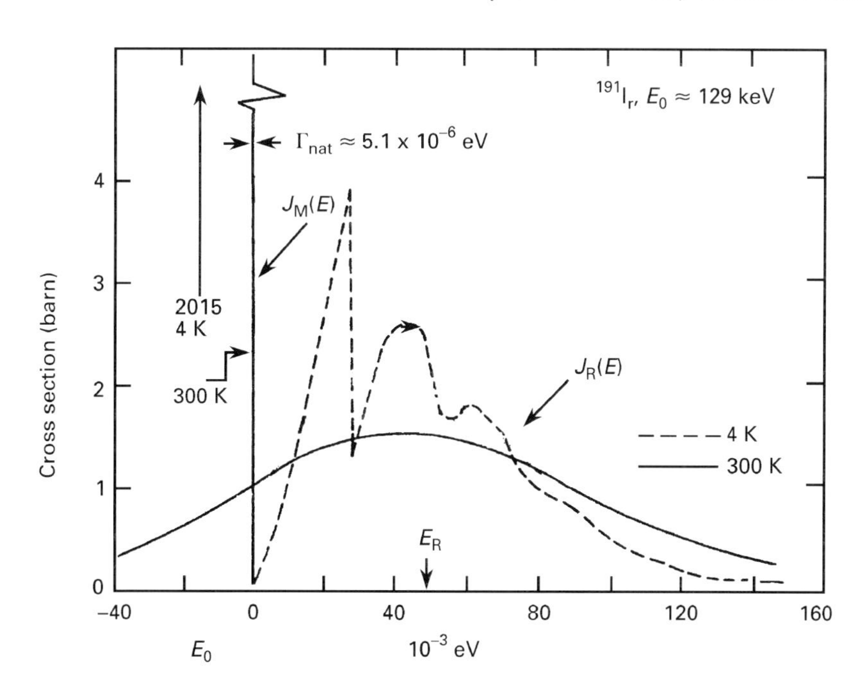

Figure 1 Absorption cross section for ^{129}Ir at 4 K and 300 K. The Debye model was used to calculate the lattice vibration.

shift, which contains 5.7% of the total area under the curve at 4 K for ^{191}Ir, $E_0 \approx 129$ keV.

There are two types of Mössbauer spectroscopic experiments based on scattering and transmission techniques. In transmission mode experiments, the resonant scattering leads to the sharp attenuation of the radiation intensity registered by a detector; it is therefore sometimes referred to as resonant absorption. The resonant absorption cross section is the total cross section of resonant scattering; the probability of detecting the scattered radiation in transmission spectroscopy may be neglected when the geometrical arrangement is appropriate. Scattering Mössbauer experiments involve the detection of either resonantly scattered γ-quanta or other radiation that is emitted in the process of resonant scattering or immediately after it.

The emission probability f and the absorption probability f' of recoilless γ-quanta and the temperature dependence of f and f' are determined by the γ-quantum energy, the mass of the nucleus, lattice vibrations and other properties of the sample. This is why the Mössbauer effect is not observed for all elements (see **Figure 2**).

The Mössbauer effect probability

To obtain information on chemical bonds of atoms in solids from experimental data, an explicit theoretical relation is needed to associate experimental f (or f') values with the phonon spectrum and the force constants of the crystal. Unfortunately, this seemingly rather simple approach produces a number of problems that primarily result from the limited information that is available on the phonon spectra of solids of practical interest. Only for a cubic crystal where interatomic forces may be assumed to be harmonic is there a simple relation for the probability of γ-quantum resonant emission with a wave vector $\boldsymbol{k}$ when a nucleus undergoes a transition from an excited state to the ground state:

$$f = \exp\left\{-k^2\,\overline{x^2}\right\}$$

where $\overline{x^2}$ is the mean square displacement of the Mössbauer atom from its equilibrium position at temperature T.

The Debye model is most widely used and is assumed irrespective of the lattice symmetry and the number of atoms in the unit cell. Hence, from an observed f value the effective characteristic temperature (Deby temperature) Θ may be evaluated and compared with the temperatures obtained for the substance by methods such as X-ray analysis or specific heat measurements. For very low and high temperatures (as compared with Θ),

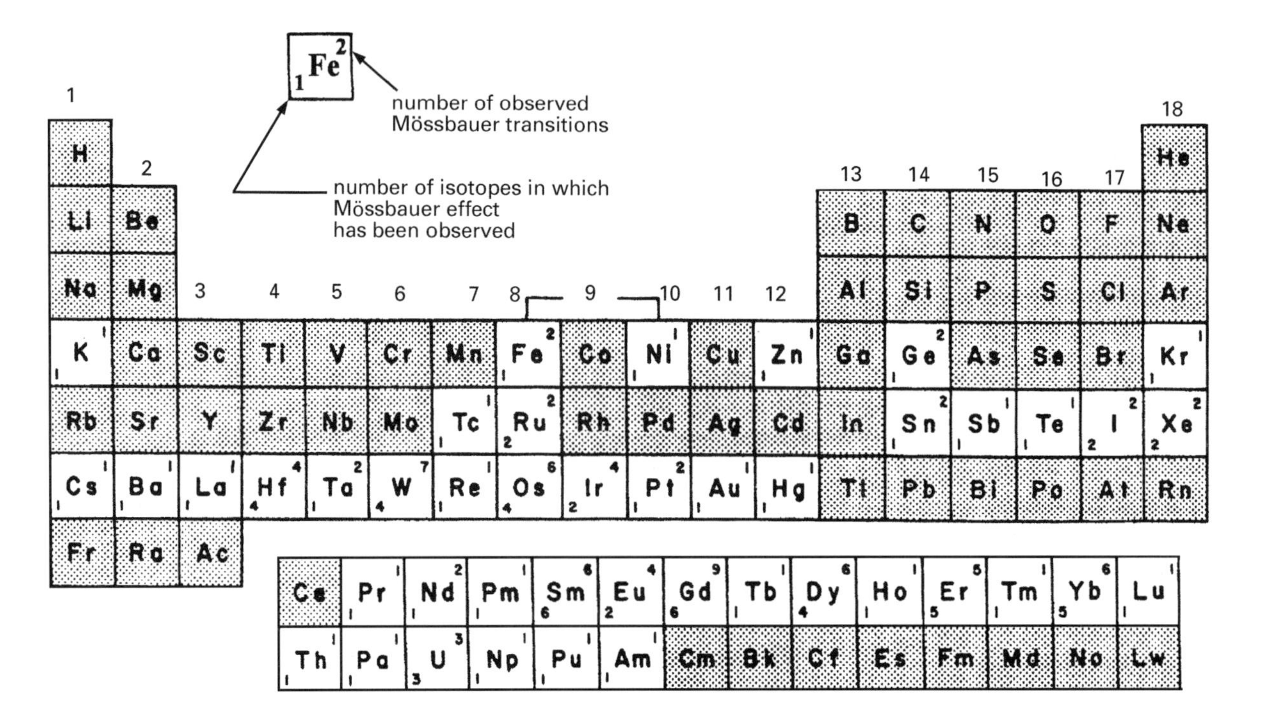

Figure 2 Mössbauer periodic table.

$$f = \exp\left(-\frac{3E_R}{2k_B\Theta}\right) \qquad \text{for } T \ll \Theta$$
$$f = \exp\left(-\frac{6E_R T}{k_B\Theta^2}\right) \qquad \text{for } T > \Theta \qquad [3]$$

where c is the velocity of light and k_B is the Boltzmann constant. Using resonance atoms as admixtures in various compounds extends the number of solids in which the Mössbauer effect is observable and this expands the usefulness of the method. Unfortunately, the theoretical treatment is much more complicated than that of the ordinary Mössbauer effect.

In general one needs to take into account the dependence of phonon excitation on the direction of the recoil momentum of the nucleus with respect to the crystallographic axes. This must be considered in particular when strong anisotropy is observed for laminar crystals. The anisotropy factor f' that is detected for some nontextured polycrystalline samples takes this into account in a phenomenon known as the Goldanskii–Karyagin effect.

Hyperfine interactions and line positions in Mössbauer spectra

The energy of a nucleus, as well as of any system of charges and currents, changes upon interaction with an external electromagnetic field by an amount E'. Using classic electrodynamics, the energy may be described by the multipole moments series as

$$E' = q\varphi_0 - \boldsymbol{p}\boldsymbol{E}_0 - \boldsymbol{\mu}\boldsymbol{H}_0 - \frac{1}{6}\sum_{i,k=1}^{3} \boldsymbol{Q}_{ik}(\partial^2\varphi/\partial x_i\,\partial x_k)_0 - \cdots \qquad [4]$$

where $\boldsymbol{E}$ and $\boldsymbol{H}$ are the electric and magnetic field strengths, respectively, φ is the electrostatic potential, $q = eZ$ is the nuclear charge, $\boldsymbol{p}$, $\boldsymbol{\mu}$ are vectors of electric and magnetic dipole moments, $\boldsymbol{Q}_{ik}$ is the tensor of the electric quadrupole moment. The subscript zero indicates that the quantity is that at the centre of the nucleus. Moments of higher orders may be neglected. Since nuclei do not have electric dipole moments, the second term of Equation [4] is zero and the energy of a nucleus in an external electromagnetic field is determined by the product of the nuclear (q, $\boldsymbol{\mu}$, $\boldsymbol{Q}_{ik}$) and electron (φ_0, $\boldsymbol{H}_0$, $(\partial^2\varphi/\partial x_i\partial x_k)$) factors. In solid-state physics and in fields of application the first factors are supposed to be known.

As seen from Equation [4], the Hamiltonian H, describing the interaction of a nucleus with effective fields, may be represented as a sum of the two Hamiltonians, one for interactions of the nucleus with electrons (H_Q) and the other for interactions with the magnetic field (H_M):

$$H = H_Q + H_M \qquad [5]$$

The Hamiltonian of the electrostatic interaction is

$$H_Q = e\sum_{p=1}^{Z} \varphi(\boldsymbol{r}_p) \qquad [6]$$

where $\boldsymbol{r}_p$ is the radius vector of the pth proton, $\varphi(\boldsymbol{r}_p)$ is the electric potential in the vicinity of the pth proton, and the summation is over all protons, $p = 1,\ldots, Z$. The coordinate system is chosen such that the origin is at the centre of the nucleus and the axes $x^i (i = 1, 2, 3;\ x^1 \equiv x;\ x^2 \equiv y;\ x^3 \equiv z)$ are directed along the principal axis of the tensor of the electric field gradient (EFG) acting on the nucleus. $\varphi_p(0)$ is the electric potential at the centre of the nucleus due to the pth proton and $r_p^2 = \Sigma_{i=1} (x_p^i)^2$.

Using the fact that the electrostatic potential φ satisfies the Poisson equation $\nabla^2\varphi = 4\pi\rho_e$, where $\rho_e = -e\ |\psi(0)|^2$ is the charge density at the nucleus ($r = 0$), and introducing the tensor of nuclear quadrupole moment Q_{ik} and the EFG tensor φ_{ik}, we can now rewrite the interaction of the nucleus with the electric fields as a sum of two interactions

$$H_Q = H_\delta + H'_Q = -\frac{2\pi}{3}e^2\sum_{p=1}^{Z} r_p^2|\psi(0)|^2 + \frac{e}{6}\sum_{i,k=1}^{3} \varphi_{ik}Q \qquad [7]$$

The nucleus is here considered to be a sphere with a mean-square radius $\overline{r^2} = \Sigma_{p=1}^{Z}\ \overline{r_p^2}/Z$ for the ground state (g) and excited state (e). As a rule, $\overline{r_g^2} \neq \overline{r_e^2}$ and the nuclear charge is uniformly distributed inside the sphere. The external electric field acting on such a spherical nucleus does not split the levels but shifts them by the quantity

$$\delta E_{g,e} = \frac{2\pi}{3}e^2 Z r_{g,e}^2|\psi(0)|^2 \qquad [8]$$

The shift due to Coulomb interactions is of the order of 10^{-12} of the transition energy. The value of the shift for every nuclear level depends on the chemical state of the atom. This is characterized by the $|\psi(0)|^2_{as}$ parameter which is the electron density at the nucleus in the absorber (a) or in the source (s). In a Mössbauer spectrum this part of the full electrostatic interaction manifests itself as the isomer (chemical) shift δ between the centre of gravity of the emission spectrum of the source and the centre of gravity of the absorption spectrum of the sample, which is called the absorber (**Figures 3A, B**). Thus the transition energy in the source E_s is different from the energy E_a in the absorber, both of them being different from the transition energy E^0 for $\overline{r^2} = 0$. It must be appreciated that in Mössbauer experiments it is not the absolute energy of the γ-quanta that is determined but the energy shift of the nuclear levels. The energy scanning is carried out by the use of the Doppler effect. Therefore, the energy parameters (Γ, δ) are expressed in velocity units, v. For a pair of source and absorbed nuclei we may write

$$\delta = \frac{2\pi}{3} Ze^2(\overline{r_e^2} - \overline{r_g^2})(|\psi(0)|_a^2 - |\psi(0)|_s^2 \qquad [9]$$

The charge density at the nucleus is mainly determined by s electrons and only partially by p-electrons. The main effect of the p electrons and d electrons and any other electron shells that do not contribute directly to the electron density $|\psi(0)|^2$, is to shield the s electrons. The determination of the scale factor $(\overline{r_e^2} - \overline{r_g^2})$ in Equation [9] is called the isomer shift calibration. The interpretation of isomer shifts in Mössbauer spectra involves the correlation of a given $\Delta\rho_e(0) = |\psi(0)|_a^2 - |\psi(0)|_s^2$ value (the electron density difference) with the known electronic structure of the Mössbauer atom or the change of the structure resulting from the examination of different samples. It should be noted that only one part in 10^{20} of the electrons in a solid directly participate in the isomer shift; the nuclear parameter $(\overline{r_e^2} - \overline{r_g^2})$ is of the order of 10^{-29} cm^2. The isomer shift is four orders of magnitude smaller than the Lamb shift caused by quantization of the electromagnetic field.

The second of the above-mentioned interactions is known as the electric quadrupole interaction, H'_Q. The value of Q_{zz} when the nucleus is in the state $m = I$ is conventionally called the nuclear quadrupole moment $eQ = \langle I, I\ |Q_{zz}|\ I, I\rangle$. The EFG tensor in the principal axes, taking into account LaPlace's equation, is determined by two independent parameters: first, φ_{zz}, commonly called 'the electric field gradient' or 'the principal component of the electric field gradient tensor' and sometimes written as $\varphi_{zz} = eq$; second $\eta = (\varphi_{xx} - \varphi_{yy})/\varphi_{zz}$, called the 'asymmetry parameter', the axes being chosen such that $|\varphi_{zz}| > |\varphi_{xx}| > |\varphi_{yy}|$ with $0 < \eta < 1$.

In Mössbauer spectroscopy it is necessary to evaluate the eigenvalues of the H'_Q Hamiltonian, that is the energies E_Q^m for the ground state and for the excited state, the transition from which is followed by the emission of a Mössbauer γ-quantum (see **Figure 4A**). The line positions in Mössbauer spectra

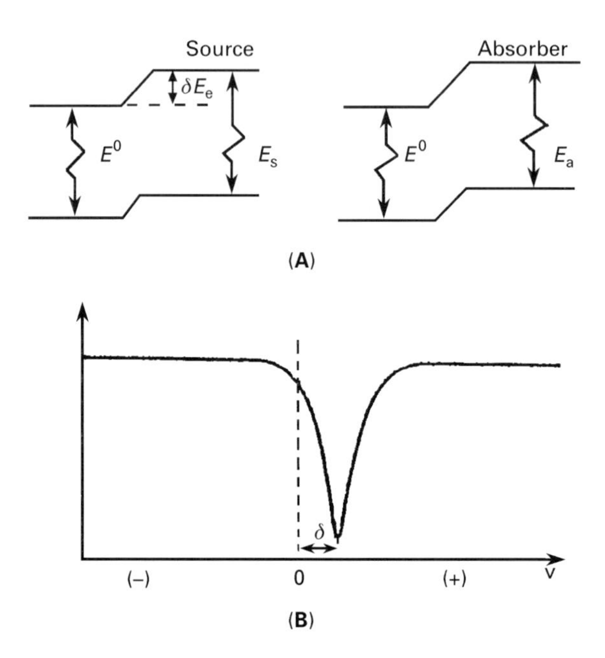

Figure 3 (A) Energy level shifts for a ^{57}Fe nucleus, resulting in the appearance of the isomer shift δ. (B) The corresponding Mössbauer spectrum.

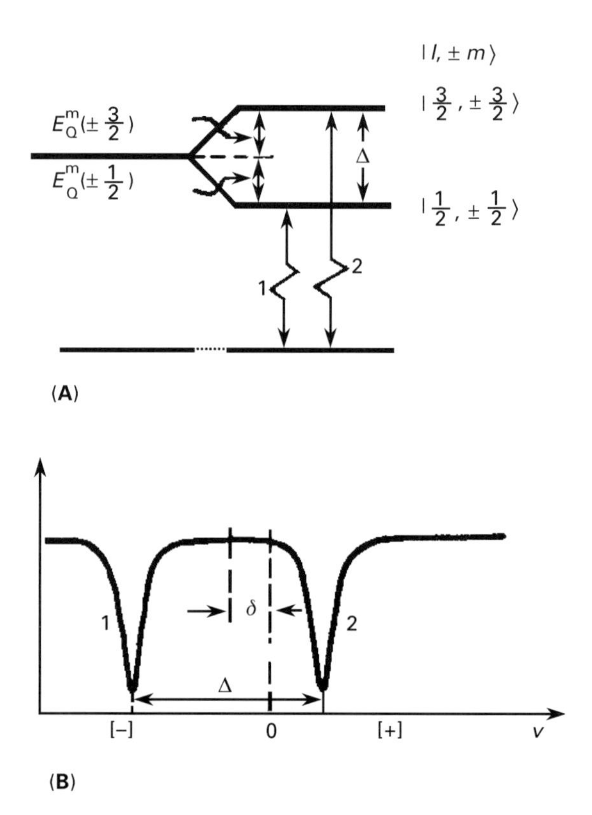

Figure 4 (A) The splitting of the excited level of a ^{57}Fe nucleus due to the electric quadrupole interaction Δ. (B) The corresponding Mössbauer spectrum.

are determined by the eigenvalues of the sum Hamiltonian H_Q for the nucleus in excited and ground states in the source and absorber, i.e. both δ value and E_Q^m. The intensities of the lines that provide valuable information on the structure of the surface layers are determined by the eigenvectors of the Hamiltonian H'_Q.

For the axially symmetric EFG tensor ($\eta = 0$) the degeneracy of the nuclear energy levels is not completely split, and the energy depends only on the absolute value of the spin projection. The energy level displacement is given by the expression

$$E_Q^m = \frac{e\varphi_{zz}Q}{4I(2I-1)}3m^2 - I(I+1)] \qquad [10]$$

where m is the value of the spin projection onto the quantization axis. If the nuclear spin is half integral, the quadrupole interaction will cause the levels to be at least twofold degenerate. If the spin values are integral, the level degeneracy for $\eta \neq 0$ may be completely lifted. For $\eta \neq 0$, the E_Q^m values may be found by solving a secular equation that has no general analytical solution for $I > 2$. Of special interest in Mössbauer spectroscopy are the transitions between states with spin quantum numbers $I = \frac{1}{2}$ and $I = \frac{3}{2}$. This is the case for ^{57}Fe, ^{119}Sn, ^{125}Te and many other nuclides. The quadrupole splitting, the distance between two lines, is equal to

$$\Delta = \frac{|e\varphi_{zz}Q|}{2}\left(\frac{1+\eta^2}{3}\right)^{1/2} \qquad [11]$$

According to Sternheimer, two primary sources of the EFG may be identified. First, charges on ions surrounding the nucleus (provided the symmetry of the surroundings is lower than cubic), and secondly, the unfilled valence shells (since filled shells possess a spherically symmetric charge distribution). The actual EFG at the nucleus is determined by the extent to which the electronic structure of the Mössbauer atom is distorted by electrostatic interactions with external charges. This leads to the so-called 'antishielding' effect, which is described by $1 - \gamma_\infty$.

The Hamiltonian for the interaction of the magnetic dipole moment of a nucleus with the effective magnetic field $\boldsymbol{H}_{\text{eff}}$ acting on it may be written

$$E_M = -g_I\mu_n \boldsymbol{I}\boldsymbol{H}_{\text{eff}} \qquad [12]$$

where μ_n is the nuclear magneton, g_I is the gyromagnetic ratio, $\boldsymbol{I}$ is the nuclear spin operator (the quantization axis coincides here with the direction of $\boldsymbol{H}_{\text{eff}}$). The degeneracy of the nuclear levels is completely split. **Figure 5** depicts the splitting of the nuclear energy levels and the corresponding Mössbauer spectrum. The shift of the levels is determined by the expression

$$E_M^m = -g_I\mu_n m H_{\text{eff}} \qquad [13]$$

(where m = spin projection onto the quantization axis). In ^{57}Fe, where the transition multipolarity of interest is $M1$, $m_e - m_g = 0, \pm 1$, and out of eight possible transitions in H_{eff} only six are present (**Figure 5A**).

Often all three interactions, i.e. the electric monopole, magnetic dipole and electric quadrupole interactions, occur simultaneously. If the quadrupole interaction is small compared with the magnetic interaction ($E_Q^m \ll E_M^m$), a correction to the interaction energy may be applied using first-order perturbation

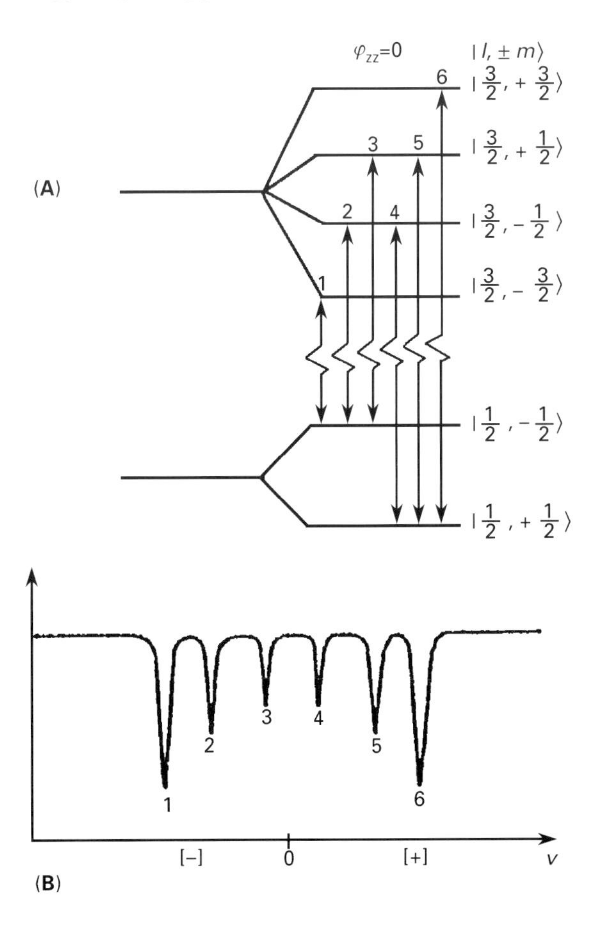

Figure 5 Effect of the magnetic dipole interaction on energy level splitting in ^{57}Fe. (A) Energy level diagram in the field $H_{\text{eff}} \neq 0$, $\varphi_{zz} = 0$. (B) The corresponding Mössbauer spectrum.

theory for a nondegenerate spectrum. For the case of an axially symmetric EFG tensor ($\eta = 0$) the level positions are given by

$$E_{\mathrm{M}}^{\mathrm{m}} = -g_I\mu_{\mathrm{n}}mH_{\mathrm{eff}} + \frac{eQ\varphi_{zz}}{4I(2I-1)} \left[3m^2 - I(I+1)\right]\frac{3\cos^2\theta - 1}{2} \qquad [14]$$

The splitting of the energy levels and the corresponding Mössbauer spectrum are shown in **Figure 6A** and **B**. If the z axis of the axially symmetric EFG is parallel to the magnetic field $E_{\mathrm{Q}}^{\mathrm{m}} \sim E_{\mathrm{M}}^{\mathrm{m}}$, the hyperfine structure is also described by Equation [14]. The sublevels are not equidistant. This results in an asymmetric magnetically split Mössbauer spectrum as depicted in **Figure 6B**. For the more general case, there is a dependence of the sublevels shift on the angle θ. If $E_{\mathrm{Q}}^{\mathrm{m}} \sim E_{\mathrm{M}}^{\mathrm{m}}$, $\eta = 0$ and $\theta \neq 0$, then the wavefunctions φ_m describing a nuclear state with a definite spin projection m onto the z axis are not the eigenfunctions of that Hamiltonian. The wavefunctions of the nuclear state with energies given by the roots of the secular equation $\mathrm{Det}(H_{mm'} - \varepsilon\delta_{mm'}) = 0$ will be a superposition of φ_m functions at different m ($H_{mm'}$ is the matrix element of the Hamiltonian H).

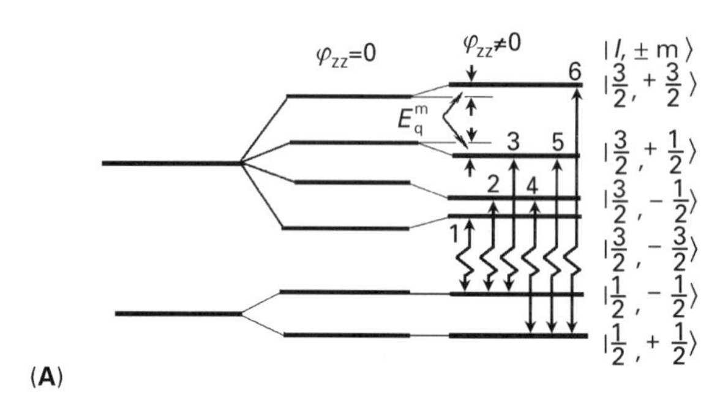

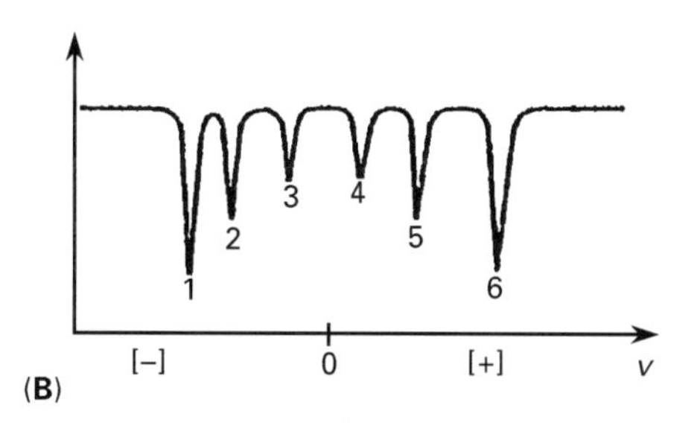

Figure 6 (A) Energy level splitting diagram with combined hyperfine interactions ($E_{\mathrm{Q}}^{\mathrm{m}} \ll E_{\mathrm{M}}^{\mathrm{m}}$) for ^{57}Fe. (B) The corresponding Mössbauer spectrum.

The superposition will cause the relative line intensities of the Mössbauer spectrum to be different from those characterizing a pure magnetic interaction. This effect may also give rise to the appearance of additional lines in the Mössbauer spectrum.

Relative intensities of spectral lines

In the absence of relaxation effects and saturation arising from finite sample thickness, the intensity of a spectral component is determined by the nuclear transition characteristics (see **Figures 3–6**). The most important of these are the spin and the parity of the excited and ground states of the Mössbauer nuclei, the multipolarity of the transition, and the direction of the wave vector $\boldsymbol{k}$ of the γ-quanta emitted with respect to a chosen direction which is specified, for example, by the magnetic field or by the electric field gradient that causes the nuclear level degeneracy to be lifted.

The probability P of the occurrence of a nuclear transition of multipolarity $M1$ from a state $|I_{\mathrm{e}}m_{\mathrm{e}}\rangle$ to a state $|I_{\mathrm{g}}m_{\mathrm{g}}\rangle$, equals

$$P(I_{\mathrm{g}}m_{\mathrm{g}}1M, I_{\mathrm{e}}m_{\mathrm{e}}; \theta, \varphi) = |G(m_{\mathrm{e}}, m_{\mathrm{g}})|^2 F_{\mathrm{L}}^{\mathrm{M}}(\theta, \varphi)|\langle I_{\mathrm{g}} \,||\, 1 \,||\, I_{\mathrm{e}}\rangle|^2 \qquad [15]$$

where θ, φ are the polar and azimuthal angles determining the direction of emitted γ-quanta in the coordinate system defined by the magnetic field direction, $M = m_{\mathrm{e}} - m_{\mathrm{g}}$; $G(m_{\mathrm{e}}, m_{\mathrm{g}}) = \langle I_{\mathrm{g}}m_{\mathrm{g}}LM \mid I_{\mathrm{e}}m_{\mathrm{e}}\rangle$, are the Clebsh–Gordan coefficients; $\langle I_{\mathrm{g}} \mid\mid 1 \mid\mid I_{\mathrm{e}}\rangle$ is the reduced matrix element which does not depend on the quantum numbers m_{g}, m_{e}. The angular function $F_L^{\mathrm{M}}(\theta, \varphi)$ is determined only by the transition multipolarity. The intensity of the Mössbauer line is proportional to the product of the Clebsh–Gordan coefficients and the $F_L^{\mathrm{M}}(\theta,\varphi)$ functions. Plots of angular dependence of the intensities of the spectral components are given in **Figure** 7. In the sample the purely magnetic hyperfine splitting of nuclear levels take place.

The effect of anisotropy of atomic vibrations in solids not only causes the Mössbauer effect probability f to be anisotropic in single crystals, but may also lead to anisotropy in f for nontextured polycrystalline samples consisting of randomly orientated crystallites. The relative line intensities of the Mössbauer spectrum (**Figures 3–6**) will be different for negative and positive velocities. Similar deviations may be caused by texture, that is by a preferred orientation of crystals in a polycrystalline sample.

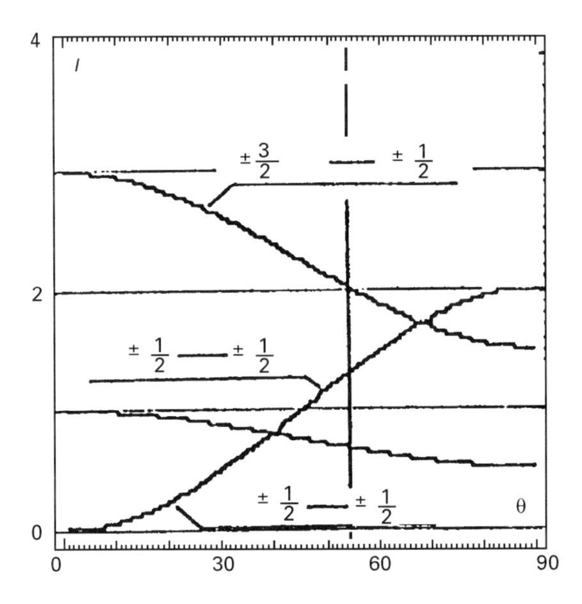

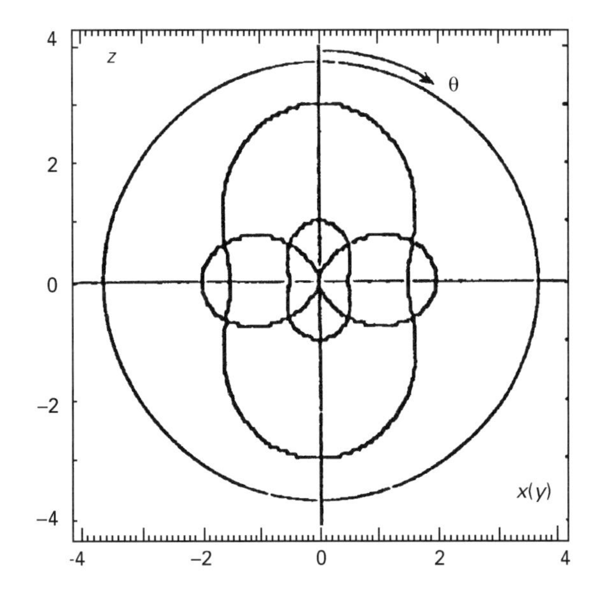

Figure 7 Angular dependences of relative intensities of the hyperfine structure components for the $I_e = \frac{3}{2}$, $I_g = \frac{1}{2}$ transition in ^{57}Fe, for magnetic dipole interaction. The polar angle θ, defining the wave vector $\boldsymbol{k}$ of the emitted γ-quantum, is the angle between the radiation direction and the quantization axis. The quantization axis z is parallel to H_{eff}.

Resonance fluorescence and interference effects

The resonantly scattered radiation may interfere with the radiation scattered by electrons of the atom. The characteristic time – the lifetime of nuclear excited state $\tau \propto \Gamma^{-1}$ – is longer by several orders of magnitude than the lattice vibration periods. There is no correlation here between the initial and final positions of the atom. Despite this, the scattered wave remains coherent with the incident one.

The energy distribution of the scattered γ-radiation may differ substantially from that of the incident radiation and is determined by convolution of the emission and scattering spectra. The Rayleigh scattering spectrum intensity effectively coincides with the emission spectrum. The intensity of the resonantly scattered radiation follows the usual Lorentzian curve, while the contribution of the interference term to the total intensity of the scattered radiation takes the form of a dispersion curve. The use of Bragg reflections in a single-crystal scatterer permits a substantial reduction in the contribution from incoherent scattering. The interference pattern in this case may be unambiguously connected with the crystallographic and electronic structure.

The directions of Mössbauer diffraction, when the hyperfine splitting is absent, generally coincide with the directions of Rayleigh coherent scattering. However, the angular dependences of diffraction line intensities from nuclear scattering and Rayleigh scattering are different. Since the Debye–Waller factor decreases with the scattering angle, it is necessary to use large scattering angles to increase the contribution of nuclear diffraction to the total spectrum.

When the hyperfine splitting is present, the diffraction pattern caused by resonant scattering is much more complicated. Magnetic fields at the different Mössbauer atoms may be not parallel. Only one of the spin subsystems will participate in the coherent scattering of the quantum and there will be no cancellation of the scattering amplitudes. This leads also to the observation of pure nuclear diffraction maxima. When the hyperfine interaction energies are sufficiently different, it should be possible to tune the incident radiation to select a particular chemical environment, and then measure the diffraction pattern from only these atoms. Only Mössbauer effect diffraction can provide independent autocorrelation functions for atoms in different chemical environments. Two physical problems must be given special mention. First, the diffraction of Mössbauer radiation is dynamic in nature. Secondly, the suppression of inelastic scattering channels requires attention. The resulting effect is the nuclear resonant analogue of the Borman effect and is realized when a thick perfect crystal containing Mössbauer nuclei is set up at a diffraction angle and the transmittance of the crystal increases when the source velocity is such that the system is brought into resonance.

Considerable interest in pure nuclear back-reflections arises also from application to γ-optical devices, such as the filtering of Mössbauer radiation from the white spectrum of synchrotron radiation.

Extremely narrow band with (10^{-6}–10^{-8} eV) and small angular width (0.4 arc second) have been obtained from the synchrotron radiation continuum. Progress in this technique has made it feasible to produce diffracted γ-quanta with intensities unattainable from conventional Mössbauer sources, thereby increasing interest in hyperfine spectroscopy. The standard experiment will be time-resolved observation of forward scattering from a polycrystalline target instead of the pure nuclear reflection from a single crystal that has been used to date. The use of synchrotron radiation may allow the Mössbauer effect to be observed in new isotopes. Such isotopes would need low-energy excited nuclear levels but need not have appropriate parent nuclei, and hence they are not given in **Figure 2**. The interference of the elastically scattered radiation gives rise to a mirror reflected wave. It is known that if electromagnetic radiation falls onto a mirror surface characterized by complex index of refraction $n = 1 - \sigma - i\beta$ at a glancing angle $\gamma \leq \gamma_{cr}$ the reflectivity R, i.e. the ratio of the reflected and incident intensities, becomes equal to unity. For real media there is always some absorption and the imaginary part of the index of refraction is not zero. However, if the R value rises sharply when γ becomes less than γ_{cr}, the situation is described as total external reflection (TER). The coherent amplification of the scattered wave under conditions of TER is analogous to diffraction on scattering from single crystals. The index of refraction depends only on the forward scattering amplitude and hence there is no phase shift between the waves scattered by various atoms and nuclei in the unit cell. In the presence of hyperfine splitting and of nonrandomly orientated quantization axes in the scatterer, polarization effects should also be taken into consideration. TER may be used for studies of very thin surface layers.

Relaxation phenomena in Mössbauer spectroscopy

The term 'relaxation' is used to indicate that time-dependent effects occur in the system under study. The γ-quantum scattering leads, as a rule, only to change of the nuclear state, while the electronic system remains unchanged. Sometimes, i.e. in paramagnets, the interaction of the electronic shell with the environment may be comparable to or much weaker than the hyperfine coupling. The atom follows a random, stochastic 'path' through its allowed states owing to time-dependent, extra-atomic interactions, and as a result of the hyperfine interaction the Mössbauer spectrum will be affected.

There are two main types of relaxation in Mössbauer studies: paramagnetic and superparamagnetic relaxation. As a rule, the observed spectra are quite complicated. The simple relaxation processes for electronic spin $S = \frac{1}{2}$ can be analysed in terms of a fluctuating hyperfine field $H_n(t)$ which takes on the values $+H_n$ and $-H_n$. If off-diagonal terms in the hyperfine Hamiltonian are absent, then

$$H_M = A_z I_z S_z(t) = g_n \mu_n H_n(t) I_z \qquad [16]$$

where A is the hyperfine coupling constant and $A_x = A_y = 0$. Under the influence of the electron–bath interaction, the electronic spin $S_z(t)$ fluctuates between the values $S_z = \pm\frac{1}{2}$ at some rate ν_R. Using either stochastic arguments or a rate equation approach, one can arrive at a closed-form expression for the line shape. The simplicity of the result obtained has made this model very popular. If off-diagonal terms are present in H_M, such simplifications are not possible.

Relaxation theory, as it applies to Mössbauer spectroscopy, has two main approaches: perturbation calculations and stochastic models. The stochastic approach is easily visualized and adapted to various physical situations. The approach generally proceeds by considering the system divided into two parts: the radiating system and the 'bath'. Depending on the particular model, the bath can induce fluctuations in the radiating system by providing unspecified 'hits' or by being represented by a fluctuating effective magnetic field. Blume formulated the effective-field, nonadiabatic model in a particularly useful way by introducing the superoperator (or Liouville operator) formalism. The superoperator formalism is extended to the case where the nucleus, and the atomic electrons are treated as a fully quantum-mechanically coupled system. It is possible to carry out good calculations of all experimental relaxation spectra.

List of symbols

A = hyperfine coupling constant; c = velocity of light; E = energy; E' = energy of interaction of a nucleus with an electromagnetic field; E_0 = energy of an excited state; $\mathbf{E}_0$ = electric field strength at the centre of the nucleus; E^0 = transition energy for $\overline{r^2} = 0$; E_a = transition energy in absorber; E_M^m = shift of levels due to magnetic interactions; E_Q^m = eigenvalues of the H'_Q Hamiltonian; E_R = recoil energy; E_s = transition energy in the source; $\delta E_{g,e}$ = value of the shift of a nuclear level; e = electron charge; e = excited state (index); $G(m_e, m_g)$ = Clebsh–

Gordan coefficient; g = ground state (index); g_l = gyromagnetic ratio; $f(f')$ = probability of recoilless emission (absorption), Lamb–Mössbauer factor; $F_L^M(\theta, \varphi)$ = angular functions; $\boldsymbol{H}$ = magnetic field strength; $\boldsymbol{H}_0$ = magnetic field strength at the centre of the nucleus; $\boldsymbol{H}_{eff}$ = effective magnetic field acting on the nucleus; H_n = hyperfine field; H = Hamiltonian; H_M = Hamiltonian for interaction with the magnetic field; H_Q = Hamiltonian for interaction with electrons; H_Q' = Hamiltonian; H_δ = Hamiltonian; $H_{mm'}$ = matrix element of the Hamiltonian H; I = nuclear spin; $\boldsymbol{I}$ = nuclear spin operator; $J_R(E)$, $J_M(E)$ = energy distributions of Mössbauer γ-rays; k_B = Boltzmann constant; $\boldsymbol{k}$ = wave vector; $L(E)$ = Lorentzian line; M = mass of nucleus; $M1$ = magnetic dipole transition; m = spin projection onto the quantization axes; $n = 1 - \sigma - i\beta$ = the complex index of refraction; $\boldsymbol{p}$ = vector of electric dipole moment; P = probability of a nuclear transition; Q_{ik} = tensor of the electric quadrupole; $q = eZ$ = nuclear charge; R = reflectivity; $\boldsymbol{r}_p$ = radius-vector of the pth proton; $\overline{r^2}$ = mean-square radius; S = electronic spin; T = temperature; v = velocity; $\overline{x^2}$ = mean square displacement of the Mössbauer atom; Z = number of protons in the nucleus; Γ = full width at half-maximum; Γ_{nat} = natural line width; γ = glancing angle; γ_{cr} = angle of total reflection (the critical angle); γ_∞ = antishielding factor; Δ = quadrupole splitting; δ = isomer (chemical) shift; ε = resonance effect magnitude; $\delta_{mm'}$ = Kronecker symbol; $\eta = (\varphi_{xx} - \varphi_{yy})/\varphi_{zz}$ = asymmetry parameter; Θ = Debye temperature; θ = polar angle, specifying $\boldsymbol{H}_{eff}$; μ_n = nuclear magneton; $\boldsymbol{\mu}$ = vector of magnetic dipole moment; ρ_e = charge density at the centre of the nucleus; τ = lifetime of nuclear excited state; φ = electrostatic potential; φ = azimuthal angle, specifying $\boldsymbol{H}_{eff}$; φ_m = wavefunction; $\varphi(\boldsymbol{r}_p)$ = electric potential in the vicinity of the pth proton; $\varphi_p(0)$ = electric potential at the centre of nucleus due to the pth proton; $\varphi_{xx,yy,zz}$ = x,y,z-component of the EFG tensor; $|\psi(0)|^2_{a,s}$ = the electron density at the nucleus in the absorber (a) or in the source (s).

See also: **Electromagnetic Radiation**; **Mössbauer Spectrometers**; **Mössbauer Spectroscopy, Applications**; **NMR Principles**; **Scattering Theory**; **X-Ray Spectroscopy, Theory.**

Further reading

Andreeva MA, Belozerski GN, Grishin OV, Irkaev SM and Semenov VG (1995) Mössbauer total external reflection. *Hyperfine interactions* **96**: 37–49.

Butz T, Ceolin M, Ganal P, Schmidt PC, Taylor MA and Troger W (1996) A new approach in nuclear quadrupole interaction data analysis: cross-correlation. *Physica Scripta* **54**: 234–239.

Deak L, Bottyan L, Nagy DL and Spiering H (1996) Coherent forward-scattering amplitude in transmission and grazing incidence Mössbauer spectroscopy. *Physical Reviews B: Condensed Matter* **53**: 6158–6164.

Gütlich P, Link R and Trautwein A (1978) *Mössbauer Spectroscopy and Transition Metal Chemistry*, p 280. Berlin: Springer-Verlag.

Hoy J (1997) Quantum mechanical model for nuclear resonant scattering of gamma-radiation. *Physics of Condensed Matter* **9**: 8749–8765.

Long GJ (ed) (1984–1989) *Mössbauer Spectroscopy Applied to Inorganic Chemistry*, Vols 1–3. New York: Plenum Press.

Long GJ and Grandjean F (eds) (1994) *Applications of the Mössbauer Effect*, International Conference on the Applications of Mössbauer Effect (ICAME-93), Vancouver, Vols I–IV. Amsterdam: Baltzer Science Publishers.

Mössbauer RL (1958) Kernresonanzfluoreszenz von Gammastrahlung in ^{191}Ir. *Zeitschrift für Physik* **151**: N1, 124–137.

Shenoy GK and Wagner FE (eds) (1978) *Mössbauer Isomer Shifts*, p 780. Amsterdam: North-Holland.

Smirnov GV (1996) Nuclear resonant scattering of synchrotron radiation. *Hyperfine Interactions* **97/98**: 551–588.

Thosar BV and Srivastava IK (eds) (1983) *Advances in Mössbauer Spectroscopy Application to Physics, Chemistry and Biology*, p 924. Amsterdam: Elsevier.

Wertheim GK (1964) *Mössbauer Effect: Principles and Applications*, p 145. New York: Academic Press.

MRI Applications in Food Science

See **Food Science, Applications of NMR Spectroscopy.**

MRI Applications, Biological

David G Reid, **Paul D Hockings** and **Paul GM Mullins**, SmithKline Beecham Pharmaceuticals, Welwyn, UK

Non-invasive MRI is at the forefront of clinical diagnostic imaging; its non-destructive nature also gives it great potential as a tool in biological research, involving animal models of disease. Because it is possible to scan the same animal as often, and over as long a period, as necessary, before and after experimental surgery and/or administration of test compounds, MRI is assuming increasing importance in longitudinal evaluation of novel pharmaceuticals and characterization of animal models of disease. Experiments can usually be designed so that each subject acts as its own control, increasing statistical power with smaller group sizes, and longitudinal studies are possible without killing groups of animals at each time point; two factors that, separately and in combination, offer dramatic sparing of laboratory animals. In general, measurement of anatomical features from MR images is much quicker than conventional invasive methodologies like tissue histology. It is often possible to acquire MRI data as three-dimensional images with isotropic resolution. These can be subsequently 'sliced' or rendered along arbitrary planes or surfaces to highlight irregular structures. The MR image is acquired *in situ*, so anatomy is undistorted by fixation, excision, sectioning and staining processes. Finally MRI methods developed to highlight features of animal disease models are often directly transferable to clinical trials and diagnoses.

MRI is so powerful because of the wide range of contrast mechanisms available to differentiate different organs, tissues and pathologies. The physicochemical basis of these contrast mechanisms, and the MR pulse sequences designed to exploit them, are treated in comprehensive standard works and other articles in this Encyclopedia. Important sources of MRI contrast are described below.

Differences in tissue water T_1 and T_2 relaxation times generally depend on differences in the extent to which water molecules interact with soluble macromolecules. Thus changes in the concentration of soluble proteins in oedema will usually cause changes in T_1 and T_2, so that MRI acquisition sequences weighted according to one (T1W or T2W) or both of these will distinguish oedematous from normal tissue. Water–macromolecule interactions are also the basis of magnetization transfer contrast (MTC), particularly effective at highlighting fibrous structures like cartilage. 'Pools' of water in which diffusion is more or less restricted by cell boundaries, or anisotropic environments, can be distinguished by diffusion weighted (DW) imaging. DWI is particularly effective at detecting cell swelling during ischaemic energy depletion, and in delineating the course of highly anisotropic microstructures like nerve cells. MR pulse sequences, which refocus magnetization using pulsed magnetic field gradients rather than spin echoes, produce images which are sensitive to differences in magnetic susceptibility between and within tissue, and are a function of the 'inhomogeneous T_2'', or T_2^*. Because paramagnetic deoxyhaemoglobin and diamagnetic oxyhaemoglobin affect the magnetic susceptibility of neighbouring tissues in very different ways, T_2^* weighted (T2*W) techniques can be used to define tissues where deoxy-haemoglobin has built up as a result of underperfusion, or where metabolic activation has increased oxygenated blood – Blood Oxygen Level Determination (BOLD). MR angiography (MRA) takes advantage of the different behaviours of moving and static nuclear spins, and can delineate vasculature and measure blood flow. Tissue perfusion can be measured using paramagnetic contrast reagents, usually stable chelates of gadolinium or manganese ions, or preparations of magnetic iron oxide particles, which reduce tissue relaxation times. T1W, T2W or T2*W images are obtained before and after administration (usually intravenously) of a contrast reagent; regions accessible to the reagent change in MRI intensity, and the time course of 'wash in' and 'wash out' gives a measure of perfusion status. Contrast reagents are widely used in animal models where the blood–brain barrier is compromised (such as demyelinating disorders and stroke), in studies of tumour perfusion, and as an alternative or adjunct to MRA.

Practicalities

Although useful work is possible in vertical magnets designed for high resolution spectroscopy, most animal MRI is done in horizontal superconducting magnets with field strengths ranging from 2 to 7 T (corresponding to ^{1}H resonances from 86 to 300 MHz) and clear magnet bore diameters ranging from about 20 to 40 cm. Concentric shim, gradient and RF coils reduce the useable diameter of 20 and 40 cm systems to about 8 to 20 cm respectively. Subjects must usually be anaesthetized with a suitable inhalation (e.g. isoflurane, halothane) or injectable (e.g. alphaxalone/alphadalone, fentanyl/fluanisone and midazolam) anaesthetic compatible with the animal model under study. It is often necessary to coordinate, or 'gate', the acquisition of NMR data with heart beat and breathing, which can be done by monitoring the animal's electrocardiogram (ECG) and respiration, and triggering data acquisition in synchrony with one or both. Tracheal intubation and mechanical ventilation allow respiratory gating on the ventilation cycle. Whether triggering is necessary or not, ECG and respiratory monitoring are essential for ensuring animal well-being and effective anaesthesia in the magnet. Other vital parameters like rectal temperature and blood pressure are often also monitored, and animal temperature can be controlled with thermostatted heating blankets or air conditioning. Radio frequency probe and animal holder design is the province of the on-site engineer in many institutions, but increasingly manufacturers are offering these items ready made. At the conclusion of an experiment it is still usual to compare *in vivo* MRI measurements with more conventional histological or organ weight measurements. **Figure 1** shows an unconscious rat supported in an animal holder and connected to ECG and respiratory monitoring systems (right), and about to be inserted into a typical horizontal laboratory MR scanner (left).

Applications

Central nervous system

MRI has been fruitfully applied to a number of animal models of CNS conditions, such as demyelination (as in for instance experimental allergic encephalomyelitis, EAE), excitotoxicity and neurotoxicity, identification of the spread of neuronal depolarization in the cortical spreading depression phenomenon, identification of neuroanatomical

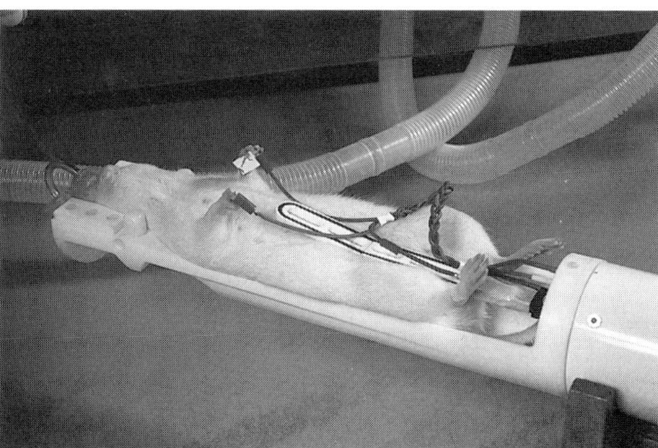

Figure 1 Right: Unconscious laboratory rat mounted in a non-magnetic holder for MR scanning. Note the face mask for delivery of inhalation anesthetic, conducting sticky pad electrodes on fore and hind paws for ECG signal detection, and the lever (containing a fibre optic cable) placed over the abdomen for respiratory monitoring. Incisor and ear bars are also built into the assembly for stereotaxic positioning if necessary. Left: The entire animal holder about to be inserted into a 7 T laboratory scanner. Although the notional diameter of the horizontal superconducting magnet is 18.3 cm the addition of concentric shim, pulsed field gradient and resonator coils reduces the useable diameter to about 7 cm – adequate for most small laboratory rodents. The gradient coils produce linear variations in the magnetic field of up to 150 mT m^{-1} (15 gauss/cm^{-1}) in each of three orthogonal directions; they are actively shielded to reduce induction of eddy currents in the magnet bore. ECG (electric) and respiratory (optical) signals are sent to monitors and triggering electronics outside the RF-impenetrable Faraday cage containing the magnet.

abnormalities in genetically modified animals, blood–brain barrier disruption using contrast reagents and localization of sites of action of psychoactive compounds using BOLD.

The versatility of MRI in this area is well illustrated by its application in models of stroke, where it has been widely used to study the evolution and properties of lesions produced by experimental cerebral ischaemia. Models investigated by MRI include permanent and transient versions of carotid artery, four vessel and middle cerebral artery occlusion (the latter commonly known as MCAO), in rats, mice, gerbils and larger animals like cats. The clarity of T2W images of ischaemic infarcts in some of these models makes this area extremely attractive for the development and implementation of 'high throughput' MRI screening strategies in testing neuroprotective treatments. MRI can exploit different sources of contrast to define the physiological events underlying ischemic injury. Thus **Figure 2** shows representative MR image slices through the brains of rats during an experiment to study the efficacy of an experimental neuroprotective treatment. The top row images are from a control subject, and the bottom row from a subject that received a neuroprotective treatment. Columns labelled (A) and (B) were acquired during a 100 minute period of MCAO using T2*W and DW respectively. In the ischaemic hemisphere (right hand side of each transverse brain image) buildup of paramagnetic deoxyhaemoglobin causes ischaemic regions with perfusion deficit to darken on the T2*W images due to T2* shortening. Additionally, cells swell and undergo cytoskeletal changes in response to energy depletion, which restricts the diffusion of tissue water. These regions show up bright relative to non-ischaemic tissue in DW imaging. Diffusibility changes are further emphasised if DW images are acquired using several different diffusion encoding gradient strengths, allowing a diffusion coefficient to be calculated for each pixel in the image, and the

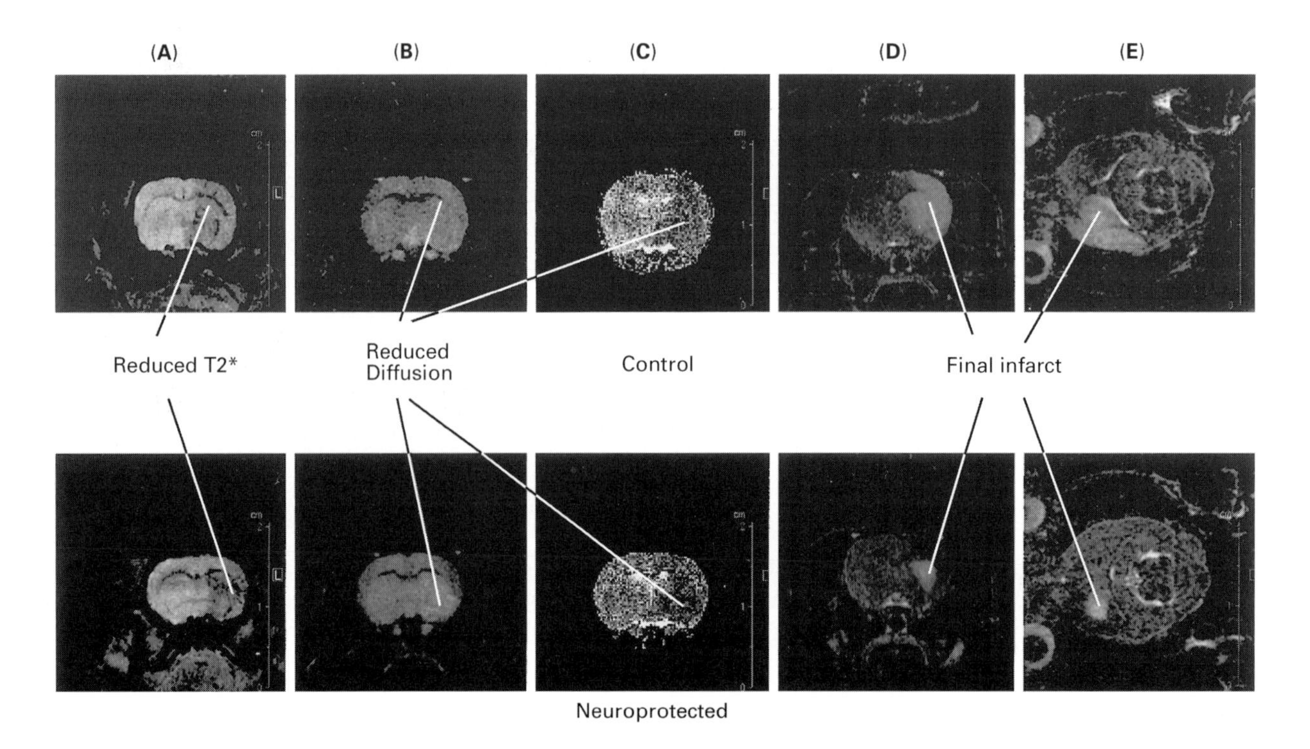

Figure 2 300 MHz MR images from the brains of rats subjected to temporary MCAO. The top row of images was acquired from a control animal, and the bottom row from an animal which received a prior neuroprotective treatment. Columns (A)–(D) show transverse images across the brain and column (E) shows a slice taken horizontally. The image columns show: (A) T2*W and (B) DW images acquired during the 100 min period of MCAO; areas of deoxyhaemoglobin buildup, and restricted diffusion, show up as dark and bright regions respectively in the affected (right) cerebral hemisphere; (C) Diffusion map plotting diffusion coefficients during the ischaemic period, calculated from images acquired with three different diffusion gradient strengths; areas of decreased diffusibility which show up bright in (B) manifest lower diffusion coefficients and hence appear dark in the map; Representative transverse (D) and horizontal (E) slices through 3D T2W images acquired 24 h after 100 min MCAO, in which oedema in infarcted regions appears bright. Pulse sequence conditions were: (A) Gradient echo technique, TE/TR = 13/1000 ms, flip angle $\alpha = 90°$; (B) TE/TR = 64/1200 ms, diffusion sensitization applied in vertical direction, *b* value = 11 370 s cm^{-2}; (C) Diffusion coefficient map calculated by exponential fitting the signal intensity decay to 3 *b* values of 0, 2350 and 11370 s cm^{-2}; (D) and (E) Interecho delay = 6.5 ms, which a repetition (RARE) factor of 16 converts to a $TE_{effective}$ of 54 ms, TR = 1500 ms.

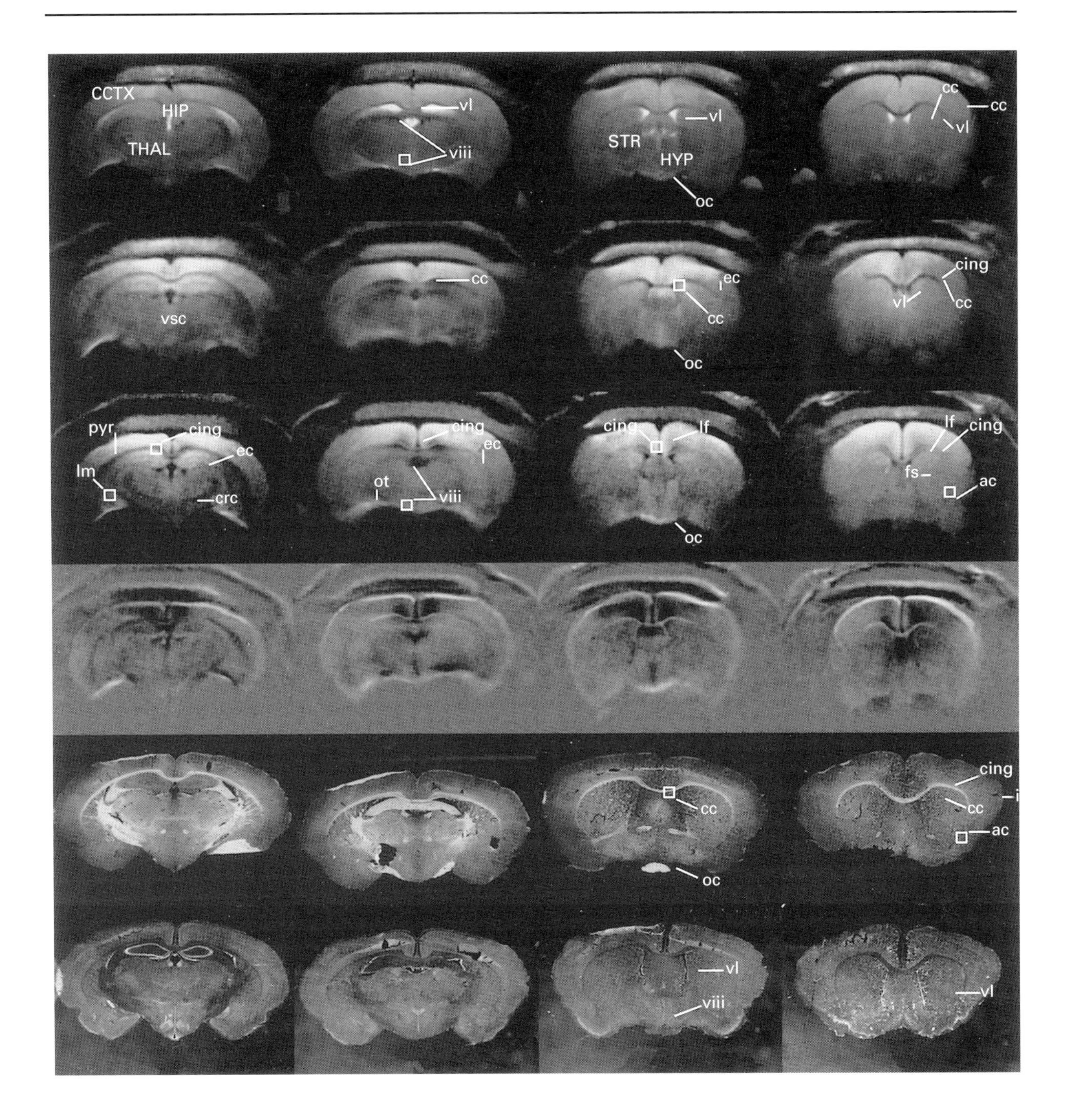

Figure 3 Transverse diffusion weighted MR images of a rodent brain acquired at four different levels (images progress from caudal ('back') to rostral ('front') from left to right). Images were acquired with TE/TR = 82/2000 ms, field of view (FOV) = 2 cm, and diffusion sensitization $b = 0$ (top row), and $b = 29600$ s cm^{-2} applied in a horizontal direction (2nd row) and orthogonal to the slice direction (3rd row). Further anatomical definition is apparent in difference images (4th row) calculated by digitally subtracting one diffusion sensitized image slice from another. Neuroanatomical structures delineated by the DWI method, and closely corresponding to structures identifiable using different histological stains (5th and 6th rows) are labelled as follows: CCTX – cerebral cortex; THAL – thalamus; HIP – hippocampus; cc – corpus callosum; STR – striatum; HYP – hypothalamus; ox – optic chiasm; ec – external capsule; 3v – third ventricle; LV – left ventricle; ot – optic tract; vsc – ventral spinocerebellar tract; ac – anterior commissure; cg – cingulum.

spatial dependence of the diffusion coefficient itself is displayed as a 'map' (column C). In contrast to DW and T2*W images, during and shortly after ischaemia no lesion is apparent using T2WI. However, 24 h after the transient MCAO oedema in the infarcted region manifests clearly as hyperintensity using T2W imaging; columns (D) and (E) show transverse and horizontal slices respectively through the same 3D T2W datasets at this time point. Apart from the distribution of oedema, which can be easily quantified, the high isotropic (~ 140 μm) resolution facilitates observation of a number of other neuroanatomical structures, such as the fluid-filled ventricles. Concerted application of different MRI acquisition modalities enables one to measure areas undergoing energy depletion, perfusion deficit and oedema, and make inferences regarding areas at risk at early time points which are destined to evolve into infarcts, and those which may be salvageable by neuroprotective intervention.

The strong directionality of nerve fibres makes DW imaging a very useful method for delineating neuroanatomy, and for studying models of neurodegenerative disorders like demyelinating diseases. **Figure 3** shows slices through a rodent brain acquired with diffusion sensitization in different directions. Nerve fibres which run parallel to the diffusion gradient direction show up dark (due to relatively unrestricted diffusion along the fibre) while those running orthogonal to the gradient direction manifest as bright because diffusion across the axon is relatively restricted, so signal loss during diffusion sensitization is minimal.

Cardiovascular system

Gating of the MRI acquisition to the cardiac and (preferably) respiratory cycles is essential in studies of cardiovascular anatomy and function. Images acquired at full systole, and diastole, enable one to measure the change in volume of the four chambers of the heart during a single contraction, and so calculate the ejection fraction. By preceding the MRI acquisition sequence with a selective presaturation method like DANTE, parts of the myocardium can be 'tagged' and their movement during the cardiac cycle mapped and correlated with cardiac dysfunctions.

Cardiac enlargement, or hypertrophy, is common in diseases like congestive heart failure, and can be a drug side effect. MRI is well suited to measuring changes in the cross-sectional area, or volume, of the chambers, and changes in wall thickness in response to hypertrophic stimuli. **Figure 4** shows transverse slices through the chests of three rats, orthogonal to the long axes of their hearts, acquired at diastole when the heart is fully distended. Image (A) is from a control animal, image (B) is from an animal which has received treatment which increases ventricle lumen size, and image (C) is from an animal after

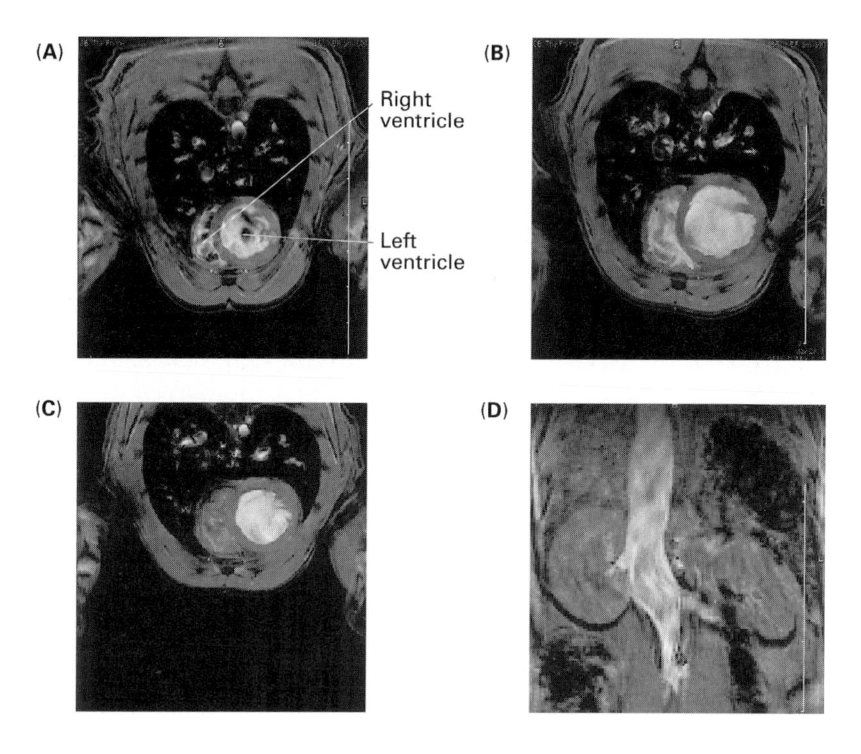

Figure 4 Transverse 300 MHz gradient echo (TE/TR = 4/1000 ms, flip angle = 90°) images through the chests of (A) a control rat, and (B) and (C), animals subjected to experimental treatments which increase the heart ventricular lumen size, and wall thickness, respectively; image acquisition was triggered on the QRS complex of the ECG signal obtain images with the hearts in diastole and hence fully dilated. Panel (D), a coronal 'bright blood' image obtained from the same animal shown in (B), depicts the enlargement of the great vessels in the abdomen provoked by an aorto-caval shunt operation.

administration of an agent to increase wall-thickness, respectively; both effects are quantifiable from the images. The coronal image (D) shows the aorto-venacaval shunt (AVS) which caused the lumen increase seen in (B); the success of the operation is obvious from the distension of the descending aorta and inferior vena cava in this bright blood image. Without MRI the success of the AVS could only be confirmed *post mortem*.

Detection of atherosclerotic plaque would be extremely useful in evaluating therapy to reduce deposition in blood vessels. Atherosclerosis models usually involve feeding appropriate animals diets rich in fat and cholesterol to induce plaque; this process may take many months so that conventional longitudinal studies use large groups of animals killed at a number of time points. Plaque detection by MRI obviates this need for large multi-group studies. **Figure 5** shows slices from 3D T2W datasets acquired around peak cardiac systole, in a transgenic atherosclerosis-prone mouse, from the region above the heart containing the aortic arch and branch points of the vessels supplying the upper body, shown in the coronal slice (A). Panel (B) is a transverse slice through the branching vessels before induction of atherosclerosis, while (C) and (D) were obtained after a few months on a high fat diet. Strong flow and turbulence around the aortic arch region make it a primary site for plaque deposition, but cardiac and respiratory motion here make good image acquisition challenging. Nevertheless the buildup of atherosclerotic plaque is clearly visible and quantifiable in, for instance, the innominate artery.

MR angiography (MRA) can be used to define vascular anatomy. **Figure 6** shows 3D images from the brain, and the upper abdomen, of a rat acquired with a fast gradient echo technique. Static water in the field of view undergoes saturation on account of the high pulse repetition rate, but blood flowing into the field of view during acquisition gives a strong signal. The delineation of the portal vasculature achieved by this technique is further enhanced by administration of a suitable contrast reagent; the cerebral vasculature is well delineated without any enhancing agent.

Liver

As the site of metabolism and toxicity of many xenobiotic compounds, non-invasive characterization of liver properties is of great interest. Many physiological and pharmacological interventions change liver size and morphology but its irregular shape can make quantification difficult; the use of pulse sequences giving adequate contrast between liver and

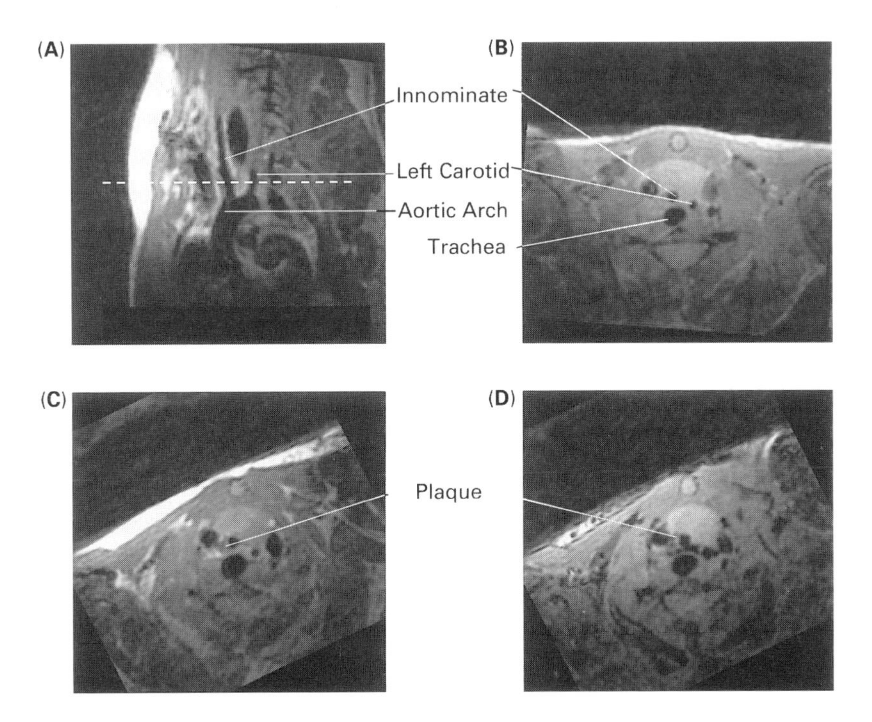

Figure 5 Coronal (A) and transverse (B) – (D) image slices through the aortic arch region of an atherosclerosis-prone mouse acquired before (A) and (B) and 17 weeks after (C) and (D) commencement of a high fat diet, selected from 300 MHz 3D T2W datasets. Plaque is arrowed in (C) and (D); perivascular fat is removed from the latter by a fat suppression procedure. In these spin echo (TE/TR = 13/1000 ms) images triggered in full systole 65 ms after the QRS wave, rapidly moving blood gives no NMR signal and so appears black.

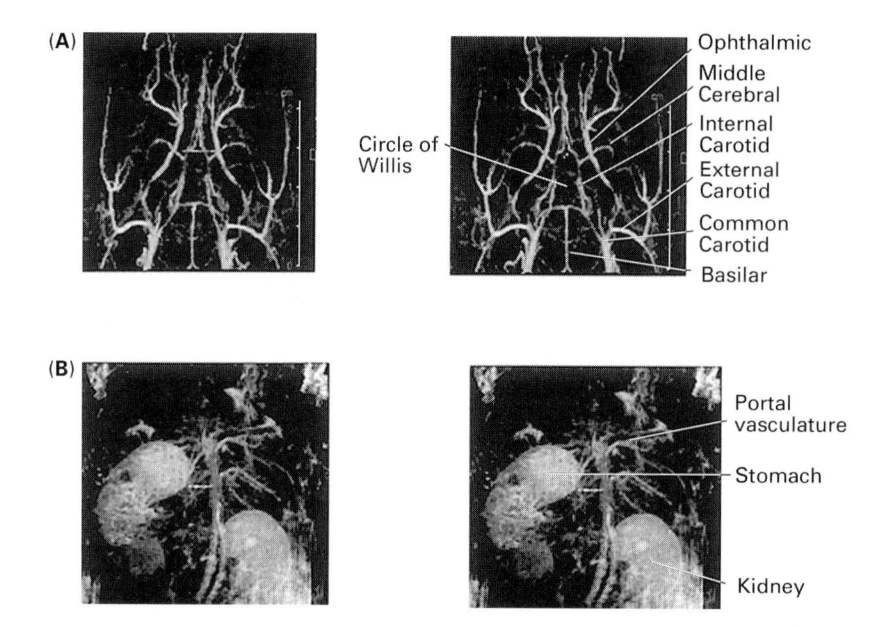

Figure 6 'Stereo pairs' of 'maximum intensity projection' bright blood MR angiograms acquired from rat brain (A) and abdomen (B). Contrast between flowing and static fluid was enhanced in (B) by administration of a colloidal magnetite contrast reagent which shortens T_2 and T_2^* of blood relative to static tissue. The 3D effect can be best appreciated by viewing the images through stereo viewing glasses.

surrounding tissues is essential. **Figure 7** shows a series of coronal slices from a 3D image from a rat abdomen acquired with a T1W method. Such data allows accurate liver volume quantification and measurement of changes induced by natural diurnal variations and feeding, and hypertrophic stimuli. Many stresses also cause changes in liver ultrastructure; although *in vivo* MRI cannot resolve microscopic necroses, these often manifest in changes in the gross MRI properties of the tissue reflected in changes in relaxation or diffusion contrast, or altered susceptibility to contrast reagents. Note the excellent delineation of other abdominal organs, particularly the stomach, kidneys and adrenals, and the abdominal aorta.

Kidney

Because it receives such a high proportion of the cardiac output, this organ is another important site of toxicity. T2W and proton density images delineate its anatomically and functionally distinct zones. **Figure 8** shows a slice along the median plane of a T2W 3D image of the kidney of a healthy rat. The divisions of the organ into outer and inner cortex, medulla and papilla, are obvious, as are neighbouring structures like the adrenal gland and fat pads. Treatment of the animal with regiospecific nephrotoxins produces characteristic changes in the MR images. Thus an inner cortical toxin brightens the corticomedullary boundary due to anomalous water buildup in this region. A papillary toxin evokes buildup of water in the inner zones of the organ leading to loss of medullary–papillary contrast, and swelling. Areas of anomalous MRI appearance correlate well with necrotic areas assessed by *post mortem* histology.

Musculoskeletal system

Articular cartilage is readily visible by MRI. **Figure 9A** shows spin echo T2W image slices through the long dimension of a tibio-tarsal (ankle) joint of a rat subjected to an arthrogenic procedure. Degradation, remodelling, and swelling of the joint as the disease progresses can be clearly seen. **Figure 9B** displays images acquired from joints excised *post mortem* from a control and an arthritic rat; they were acquired on an instrument custom modified to operate with an autosampler – an example of high throughput biological MRI data acquisition.

Oncology

MRI is a powerful technique for investigating the progression and properties of experimental tumours, as exemplified in **Figure 10**, which shows slices through a GH3 pituitary tumour implanted in a rat. Distinction of the tumour from surrounding tissue on the basis of relaxation time differences, and measurement of its volume, is straightforward. The left hand images were obtained with gradient (top) and spin echo (bottom) methodologies respectively while the rat breathed a normal air–anaesthetic mixture. The

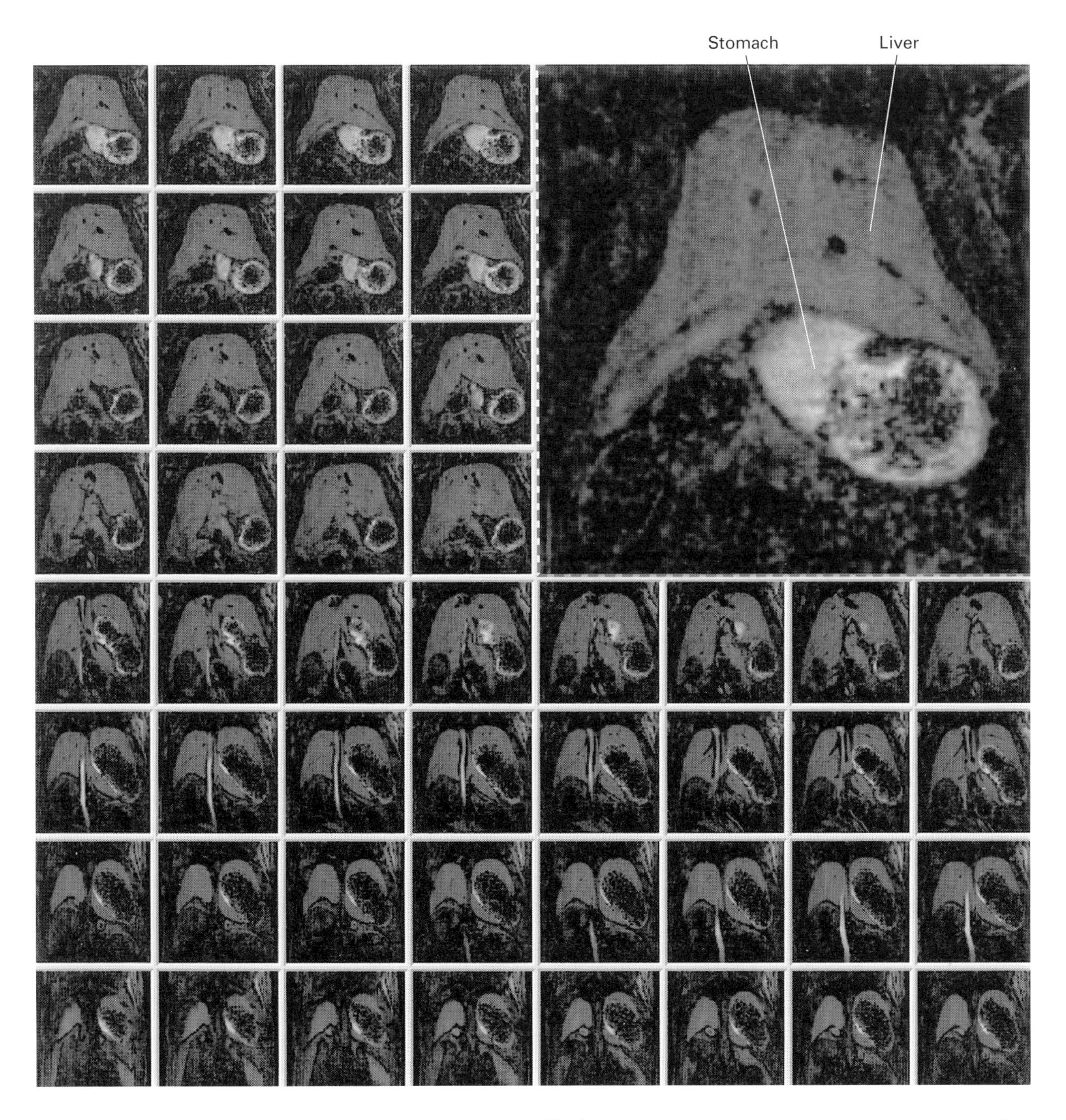

Figure 7 Contiguous coronal sagittal slices through a 3D dataset (300 MHz) acquired from the upper abdominal area of the rat. The acquisition method, combining inversion recovery (950 ms) and segmented (16, TE = 3.3 ms) low flip angle (~ 30°) fast gradient echo readout, was designed to optimize contrast between liver and surrounding structures, but note also the excellent definition of the kidneys, stomach, and moving (bright) blood in the descending aorta. Abdominal fat was suppressed by selective saturation 3.25 parts per million (975 Hz) upfigeld of the water signal before readout.

right hand images were obtained after increasing the CO_2 content of the breathing mixture to 5% – a powerful vasodilatory stimulus. Oxygenated haemoglobin increases in the tumour reducing T2* relaxation, producing more signal in the gradient echo T2*W image. This is a dramatic example of the use of BOLD to study functional activation.

Future developments

BOLD methodologies aided by fast techniques like echo planar imaging (EPI) in high field magnets promise the localization of the sites of action of neuroactive compounds. Cheaper actively shielded magnets will facilitate the use of MRI in biology,

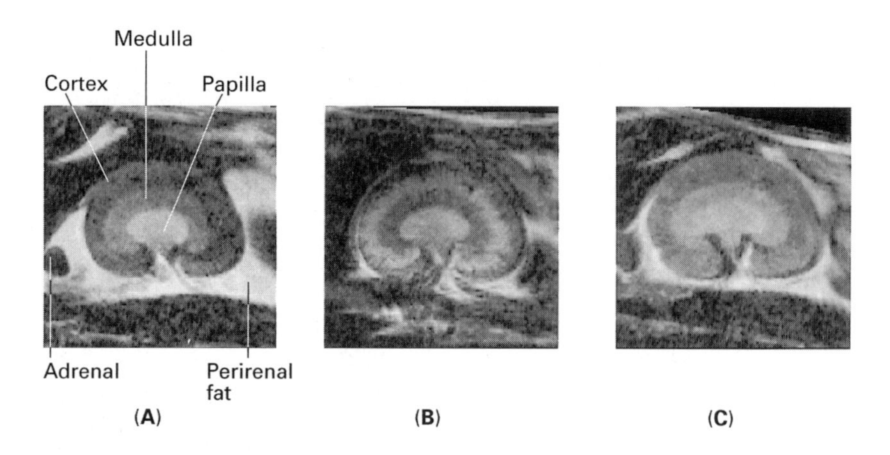

Figure 8 Slices through the median planes of 3D 300 MHz T2W images (TE = 6.5 ms, TR = 1.5 s, multiecho segmentation, or RARE, factor = 32, $TE_{effective}$ = 104 ms) of kidneys from a control rat (A), and from rats treated with an inner cortical (B) and papillary (C) toxin. Note the clear differentiation in the control kidney between cortex, medulla and papilla, and also the good definition of perirenal fat and adrenal glands. Note also the evolution of a hyperintense band in the cortical toxin-treated kidney reflecting derangement of renal tubular function and water buildup in this region. The papillary toxin evokes a loss of papillary–medullary contrast. Marked swelling of both treated kidneys is also obvious and easily quantifiable.

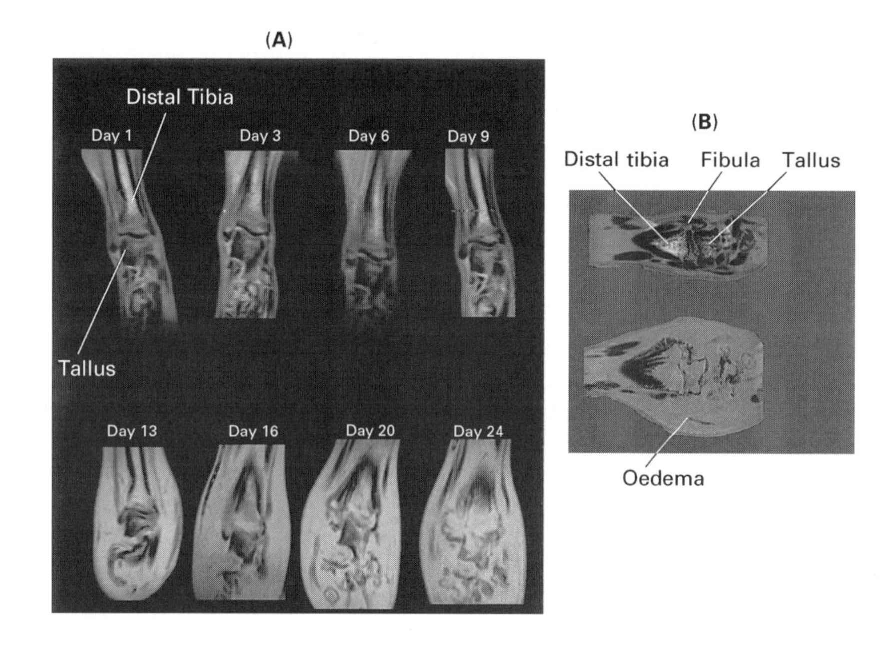

Figure 9 (A) MR images from longitudinal assessment of degeneration of the posterior tibio-tarsal joint of a rat, rendered arthritic by intra-venous injection of a *Mycobacterium butyricum* suspension at Day 1 (200 MHz, TE/TR = 9/2500 ms, 100 × 100 μm in plane resolution, 1 mm trans-plane resolution). (B) 400 MHz images of excised tibio-tarsal joints from control and adjuvant-arthritic rats, acquired using autosampler technology (TE/TR = 8/1000 ms, 70 × 70 × 250 μm resolution). Reproduced by permission of Dr. Rasesh Kapadia, SB Pharmaceuticals, Upper Merion, PA.

pharmacology and toxicology as the systems become less demanding of laboratory space. Robust acquisition and processing software will remove the routine conduct of biological MRI from the hands of the NMR expert and place it in those of the biologist. Complete automation of *in vivo* experiments is unlikely, but fully automated imaging of fixed tissue is already possible; 3D images of fixed tissue, which can be subjected to a battery of image analysis procedures, will become valuable complements to conventional fixed tissue histology. Image analysis is the rate-limiting step in many experiments.

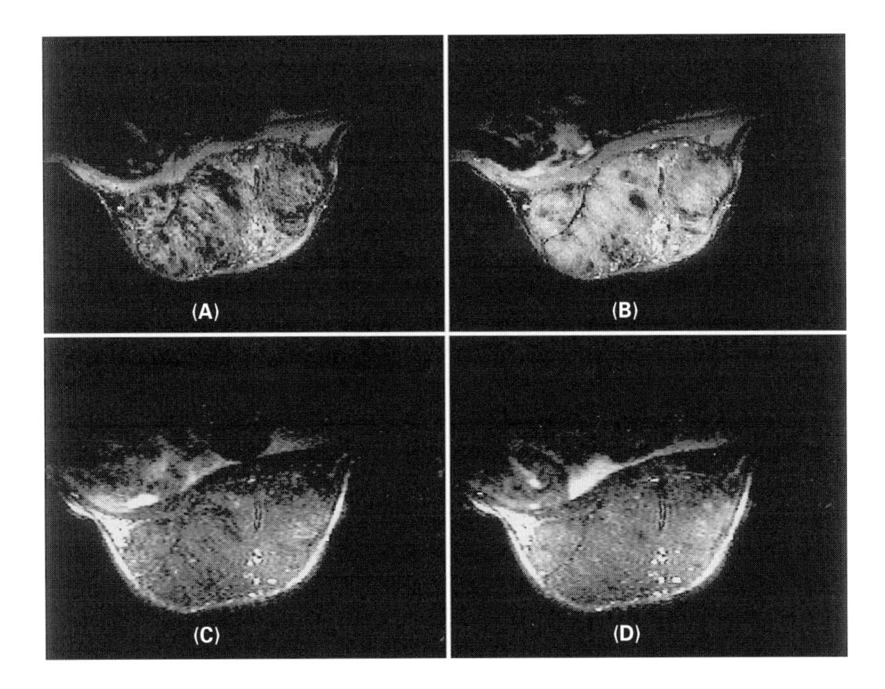

Figure 10 200 MHz MR images of a transplanted rat GH3 pituitary tumour. The top pair of images (A) and (B) were acquired with a T2*W gradient echo method (TE/TR = 20/80 ms, flip angle 45°) and the bottom pair (C) and (D) with T2W (TE/TR = 20/300 ms). The left hand images were acquired while the animal breathed normal air–anaesthetic gas mixture, while the right hand images were acquired shortly after switching the breathing mixture to carbogen (5% CO_2). Note the striking increase in intensity in the T2*W image as blood flow to the tumour increases due to vasodilation (B). This vasodilation is also reflected in the increase in T2W intensity in blood vessel cross-sections (D). Reproduced by permission of Dr Simon Robinson and Professor John Griffiths, St George's Hospital Medical School, London.

Perfection of automatic image coregistration and segmentation methods promise to break this logjam. MRI will be increasingly combined with *in vivo* spectroscopy, and other imaging methods like positron emission tomography (PET) to produce simultaneous anatomical, functional, metabolic and drug distributional information. Finally the interface between experimental and clinical MRI will strengthen as clinical trials are planned on the basis of laboratory protocols and *vice-versa*.

List of symbols

AVS = aorto-venacaval shunt; BOLD = blood oxygen level determination; CNS = central nervous system; DW = diffusion weighted; EAE = experimental allergic encephalomyelitis; ECG = electrocardiograph; EPI = echo planar imaging; FOV = field of view; MCAO = middle cerebral artery occlusion; MRA = magnetic resonance angiography; MTC = magnetization transfer contrast; RARE = rapid acquisition with repeated echo; RF = Radio frequency; T1W = T_1 weighted; T2W = T_2 weighted; T2*W = T_2* weighted; TE = echo time; TR = repetition time.

See also: **Chemical Shift and Relaxation Reagents in NMR; Diffusion Studied Using NMR Spectroscopy; *In Vivo* NMR, Methods; *In Vivo* NMR, Applications – ^{31}P; *In Vivo* NMR, Applications, Other Nuclei; MRI Applications, Clinical; MRI Applications, Clinical Flow Studies; MRI Instrumentation; MRI Theory; NMR Microscopy; NMR Relaxation Rates.**

Further reading

Anderson CM, Edelman RR and Turski PA (1993) *Clinical Magnetic Resonance Angiography*. New York: Raven Press.

Bachelard H (1997) *Magnetic Resonance Spectroscopy and Imaging in Neurochemistry*, New York: Plenum.

Bushong SC (1996) *Magnetic Resonance Imaging: Physical and Biological Principles*, St Louis: Mosby-Year Book.

Callaghan PT (1991) *Principles of Nuclear Magnetic Resonance Microscopy* , Oxford: Clarendon.

Chen C-N and Hoult DI (1989) *Biomedical Magnetic Resonance Technology*. Bristol: Institute of Physics.

Elster, AD (1994) *Questions and Answers in Magnetic Resonance Imaging*. St Louis: Mosby-Year Book.

Flecknell, P (1996) *Laboratory Animal Anaesthesia*. London: Academic.

Gadian, DG (1995) *NMR and its Applications to Living Systems*. Oxford: Oxford University Press.
Underwood R and Firmin D (eds.) (1991) *Magnetic Resonance of the Cardiovascular System*. Oxford: Blackwell Scientific.
Yuh W, Brasch R and Herfkens R (eds) (1997) *Journal of Magnetic Resonance Imaging (Special Edition – MR Contrast Reagents)* 7: 1–262.

MRI Applications, Clinical

Martin O Leach, The Institute of Cancer Research and The Royal Marsden Hospital, Sutton, Surrey, UK

MAGNETIC RESONANCE
Applications

In 1973 both Lauterbur, and Mansfield and Grannell, proposed that a shift in resonance frequency, induced by a spatially varying magnetic field, could be used to encode the spatial location of nuclear magnetic resonance signals. Developments in the late 1970s demonstrated the feasibility of magnetic resonance imaging and led to the construction of clinical instruments, with the first whole-body image published in 1977 by Damadian and colleagues. Initially a range of different imaging techniques were employed, with commercial developments at first using filtered back-projection. While this is the standard method of reconstruction in X-ray computed tomography (CT), the limited homogeneity of early magnets, together with the inherent variations in human magnetic susceptibility, gave rise to considerable image artefacts. These approaches were superseded by spin-warp imaging, introduced by Edelstein and colleagues in 1980, an extension of Kumar and colleagues' Fourier zeugmatography technique. Spin-warp imaging remains the method used for most clinical magnetic resonance imaging. Imaging technique are discussed in more detail by Morris and Leach as given in the Further reading section.

The ensuing 20 years saw an unprecedented development in the scope and quality of magnetic resonance imaging, compared with the growth of previous medical imaging techniques. Although the advent of CT revolutionized diagnosis by providing high-quality cross-sectional images, its use has generally been limited to the detection and measurement of anatomical abnormality and it provides limited functional information. As CT was well-established when MRI was introduced, MRI initially supplied supplementary information, particularly in neurological examinations where the increased soft-tissue contrast of MRI and lack of bony artefacts allowed better depiction of the brain and spinal cord, together with improved visualization of physiological processes. Hardware has progressively developed, with the introduction of superconducting magnets leading to more stable and homogeneous magnets and allowing the introduction of higher-field magnets of up to 1.5 T to many hospitals. Magnet field strengths now range from 0.2 T, often with open configurations (based on electromagnets or resistive coils), aiding orthopaedic, paediatric and interventional applications, through 0.5 T (superconductive or permanent, with open designs again possible) used for a wide range of applications, to 1.0 T and 1.5 T superconducting designs, with manufacturers now developing 'short-bore' magnets with flared apertures to increase patient acceptability. High-field magnets are used where signal-to-noise is a principal concern, for angiography, functional MRI (brain activation), cardiology and real-time imaging. At 1.5 T, magnetic resonance spectroscopy is also possible, with many instruments being capable of proton spectroscopy, and some also having facilities for broad-band spectroscopy. Research sites have installed higher-field magnets, with many 3.0 T installations, some at 4.0–4.7 T and recent installations at 7 T and 8 T. These systems are primarily used for spectroscopy and for brain activation studies.

There has been a range of further developments in hardware. These include shielded gradient coils, facilitating high-speed imaging by reducing eddy currents, and large increases in the strength and switching speed of gradients, allowing clinical implementation of snapshot imaging, echo planar imaging and similar real-time techniques. Circularly polarized and phased-array coils have significantly increased the sensitivity of measurements, with modern systems having a wide range of coils. Automatic shimming techniques have improved fat signal suppression, as well as aiding spectroscopy. Self-shielded magnets have eased the installation requirements for many clinical systems. These improvements have been accompanied by advances in pulse sequence design, versatility and reconstruction speed. Packages for specific clinical specializations are now available, providing pulse sequences and analysis techniques tailored to particular applications, e.g. functional neuroimaging and cardiac packages. In addition, a range of contrast agents with differing pharmaceutical properties have been developed that are leading to new clinical applications.

Anatomical imaging

Clinical applications of MRI primarily make use of the high soft-tissue contrast, which can be readily manipulated by appropriate choice of pulse sequences, to demonstrate cross-sectional anatomy at any arbitrary orientation. One of the initial motivations for developing clinical MR imaging instruments was the observation that tumours had long T_1 relaxation times. Although this was shown not to provide a unique discriminator for cancer, the different T_1 and T_2 relaxation times, together with other intrinsic properties affecting the MR signal, allow contrast to be changed between tissues by selecting appropriate pulse repetition times and flip angles (T_1 weighting) and echo times (T_2 weighting). This allows abnormal or distorted anatomy to be seen, and aberrant tissues can often be identified by different relaxation properties. T_1 and T_2 relaxation times reflect the environment and ease of movement of water and fat molecules. The greater the water content and the greater the freedom of movement, the longer are T_1 and T_2. When water is tightly bound, magnetization transfer imaging techniques can be used to interrogate this bound compartment by exploiting the short T_2 and broad line shape. An off-resonance (several kHz) irradiation suppresses the bound component, without directly affecting unbound water. However, the signal of the unbound water is subsequently reduced by exchange with the partially saturated bound component. A difference image reveals the degree of magnetization transfer.

A basic clinical examination will employ both T_1- and T_2-weighted multislice imaging sequences, chosen in a particular plane. Typically a set of scout images (very rapid T_1-weighted images in several orientations) will be acquired to aid the prescription of these images (orientation and number of slices, etc.). A fast spoiled gradient-echo image (e.g. fast low angle single-shot (FLASH)) might be chosen with a 300 ms repetition time (TR), a 12 ms echo time (TE) and a 70° flip angle (α), to provide T_1 weighting. A dual-echo spin-echo sequence with TR=2 s, TE=30 ms and 120 ms, and α=90° would provide, respectively, proton density and T_2-weighted images. The gradient-echo image is subject to signal loss in areas of magnetic field inhomogeneity, or variations in magnetic susceptibility, for example in the brain adjacent to air-filled sinuses or near sites of previous haemorrhage. The effect can be minimized by selecting a very short TE, or using a T_1-weighted spin-echo sequence (see **Figure 1**). With these conventional sequences, straightforward anatomical examinations can be performed in most parts of the body that are free from movement. A number of additional gradient-echo sequences are available that exploit the principle of steady-state free precession. FISP (fast imaging with steady state precession) maintains the steady-state signal, and does not suffer signal loss from flowing blood, providing high signal from long-T_2 fluids, with a signal that does not depend strongly on TR. This is valuable for generating MR myelograms or for angiography. PSIF (a time reversed FISP sequence also called CE-FAST) provides strong T_2 weighting that is a function of TR.

A further basic sequence that is widely used is the inversion recovery sequence, in which the magnetization is initially inverted, and then sampled with a 90° pulse at an inversion time (TI) after the 180° pulse. This can provide a greater range of T_1-weighted contrast, and has the particular property that it can be used to null signal from a particular tissue on the basis of its T_1 relaxation time, by selecting the TI to sample signal from that tissue as it recovers through zero longitudinal magnetization. A widely used variant of the inversion recovery sequence is the STIR (short τ inversion recovery) sequence, which is used to null the signal from fat, which usually has a bright signal on T_1-weighted images and can obscure important anatomical detail. Similar sequences can be used to null the cerebrospinal fluid (CSF) signal in spine imaging, allowing the spinal cord to be clearly seen. A further variant is the FLAIR (fluid attenuated IR) sequence, which nulls CSF in the brain, enhancing visualization of brain tissue. Alternative methods are available to obtain fat, or water, images using selective excitation with, for example, binomial pulses, or conventional frequency-selective pulses, or by employing a multi-acquisition method sensitive to the phase difference between fat and water (the Dixon method).

While providing excellent images in many parts of the body, acquisition times for these measurements are relatively long, reducing their value in moving tissues. In areas such as the abdomen, affected by respiratory and bowel movement, image quality can be improved by averaging, at the cost of longer measurement times. In the mediastinum, ECG triggering allows high-quality images at appropriate stages of the cardiac cycle to be obtained, despite the vigorous,

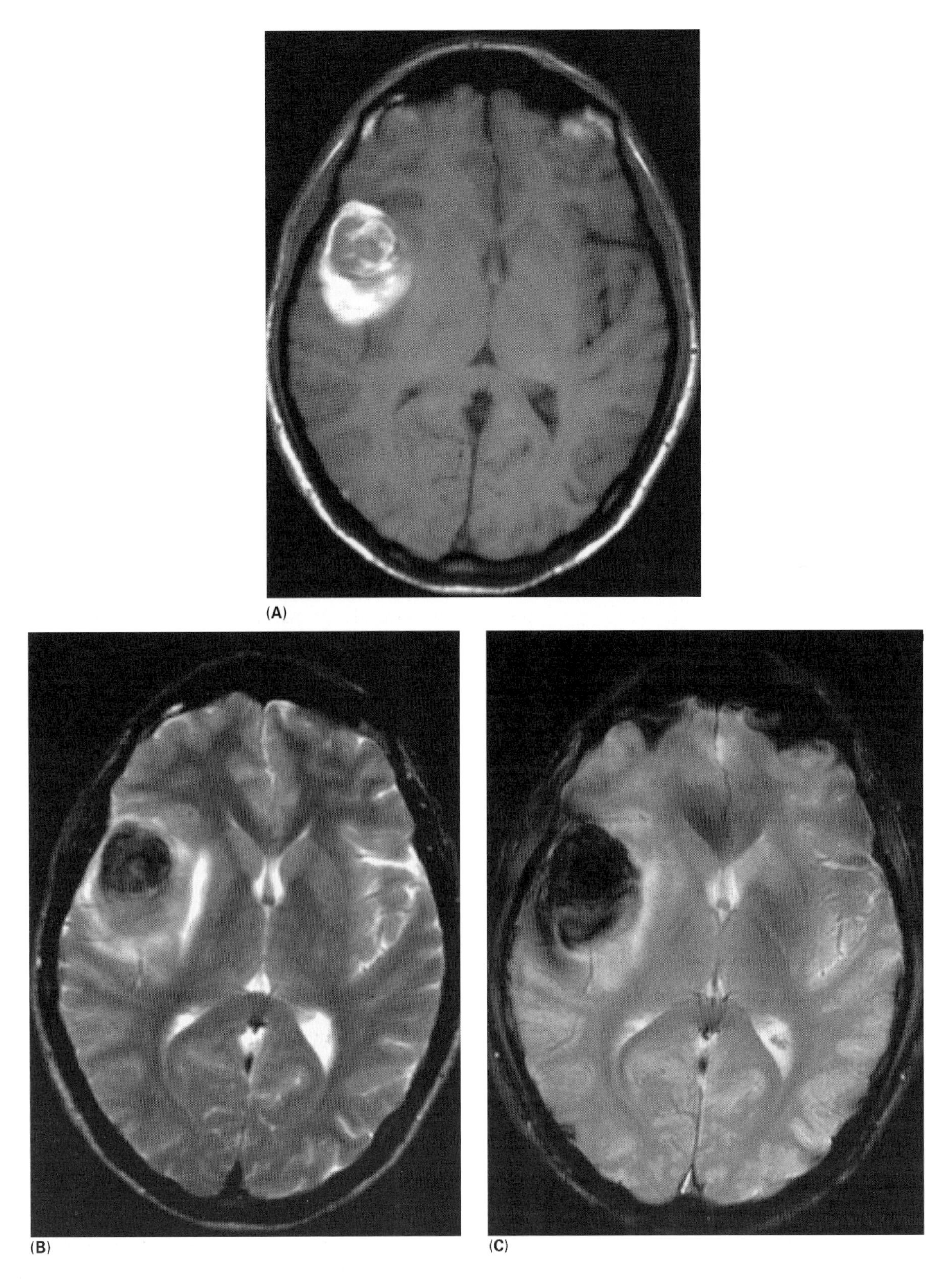

Figure 1 Transaxial images through the brain of a patient with a haemorrhagic melanoma metastasis. (A) T_1-weighted spin-echo image (TR = 665 ms, TE = 14 ms, $\alpha = 80°$) showing bright signal in the regions of recent haemorrhage. (B) T_2-weighted turbo spin-echo image (TR = 4500 ms, effective TE = 90 ms, $\alpha = 90°$) showing bright signal from cerebrospinal fluid and low signal arising from T_2 shortening due to melanin deposits in the tumour. (C) T_2^*-weighted FLASH image (TR=1604 ms, TE = 35 ms, $\alpha = 30°$) showing increased T_2^* signal loss within the tumour resulting from susceptibility changes due to melanin.

multidirectional motion. Image quality can be further improved by placing saturation slabs through moving high-signal regions, or by saturating in-flowing blood in adjacent planes. Respiratory gating, and methods of reordering phase encoding (ROPE) can also reduce motion effects. Although these techniques are still sometimes employed, major advances in imaging moving tissues, and in speeding up examinations, have been attained by a range of new rapid imaging techniques, made possible by recent advances in instrumentation. Where motion cannot be avoided, or where individual data sets building an image have to be acquired during movement, navigator echoes provide a way of accurately monitoring motion as well as providing the information necessary for correcting for the motion.

Turbo or magnetization prepared gradient-echo sequences have one or more preparation pulses, followed by a rapid succession of small flip angle pulses to interrogate the longitudinal magnetization, each encoding a different line in *k*-space, thus building up the image with only one preparation pulse. This sequence, and variants that further reduce the measurement time by reduced *k*-space sampling, provide rapid images that allow subsecond image acquisition, and a set of slices can be acquired within a breath-hold period. Preparation can include a large flip-angle pulse or an inversion pulse. Contrast and the relative weighting of spatial frequencies can be altered by changing the *k*-space sampling order. These techniques are often employed in 3D imaging sequences, to allow a 3D data set to be acquired in an acceptable time. A highly effective sequence providing rapid T_2-weighted measurements is the turbo spin-echo or RARE (rapid acquisition with relaxation enhancement) sequence. In this sequence, multiple echoes are acquired, each sampling a different line of *k*-space, thus speeding up acquisition of the image. A consequence of the many 180° pulses is a change in contrast in some tissues compared with spin-echo sequences, as well as increased power deposition. Contrast and resolution can also be varied by altering the *k*-space sampling scheme. Echo planar imaging uses a single-shot sequence to obtain a full image based on a single preparation or read-out pulse. This is one of the fastest imaging methods and places high demands on the gradient and acquisition system. It is now available on commercial systems, and is being applied to functional and physiological measurements, which are particularly sensitive to motion. A number of variants of the above techniques are in use, including GRASE (gradient and spin-echo), combining spin echo and gradient echo imaging and fast imaging with BURST RF excitation, which utilizes a sequence of RF pulses to generate images very rapidly.

Bone is not visible on MR images owing to the extremely short T_2 of hydrogen atoms in bone. The presence of bone can usually be inferred from the lack of signal, although estimation of bone volume is complicated by the relative shift in position of fat with respect to bone (the chemical shift artefact). In some areas of the body, signal voids from air spaces can also complicate interpretation. High resolution 3D imaging of joints can show excellent cross-sectional images of trabecular structure. The development of bone interferometry, based on the loss of signal in T_2*-weighted images from susceptibility effects, has provided a means of measuring changes in trabecular bone mineral mass in diseases such as osteoporosis. T_2* includes the contribution of local magnetic susceptibility.

MRI is widely used in musculoskeletal and orthopaedic examinations. The use of site-specific surface coils, combined with 3D or narrow slice imaging sequences, allows the detailed structure of joints to be visualized (see **Figure 2**). Tendons can be seen as regions of low signal, and there is good contrast between cartilage, synovial fluid and the meniscus. Open magnet designs associated with fast imaging techniques facilitate kinetic imaging of joints and tissues. Absence of radiation and the ability to freeze motion have also extended the application of MRI to resolving problems in pregnancy and examining the fetus.

Contrast agents are now widely used to enhance the appearance of pathology, separating it from normal tissues based on differential uptake of a labelled pharmaceutical. These agents principally affect T_1, as they are usually paramagnetic compounds with several unpaired electrons. These cause increased intensity on T_1-weighted images in areas of high uptake because of the reduced T_1. They can also be used to affect T_2 using superparamagnetic or very small ferromagnetic particles, causing a loss of signal on T_2-weighted images. The most commonly used agent is gadolinium, usually chelated to diethylene triamine pentaacetic acid (DTPA) or similar compounds. The agent is injected intravenously and diffuses rapidly into the extracellular space. Its first use was to demonstrate breakdown of the normal blood–brain barrier (see **Figure 3**), but it is now widely used to delineate pathology outside of the brain, exploiting differences in blood vessel density and vascular permeability. Diagnosis may be based on the standard enhanced images, but often contrast is improved by subtracting post-contrast from pre-contrast images, or by performing fat-suppressed imaging.

More complex approaches exploit the dynamic behaviour of contrast agents to obtain physiological information, discussed further below. In some tissues,

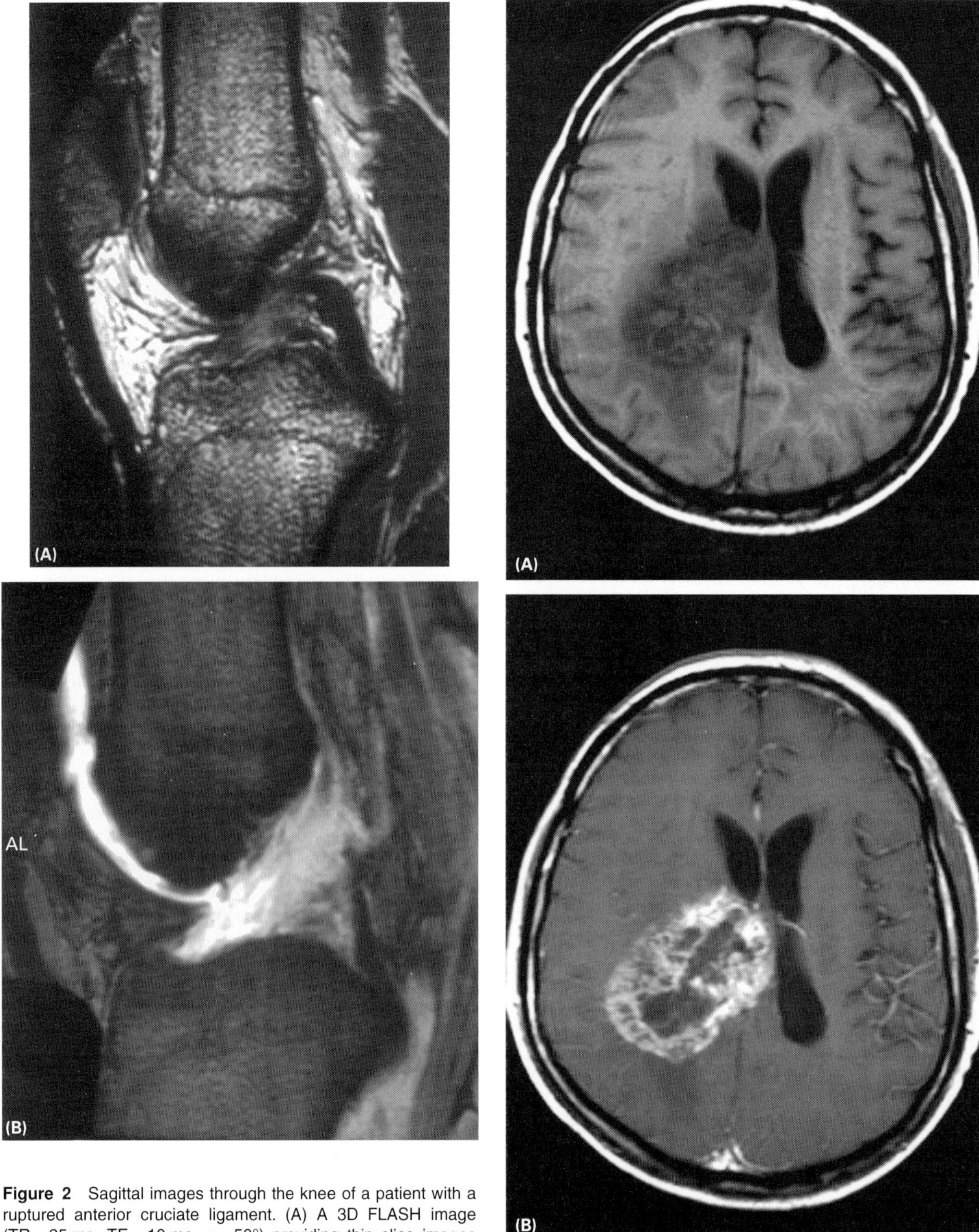

Figure 2 Sagittal images through the knee of a patient with a ruptured anterior cruciate ligament. (A) A 3D FLASH image (TR = 25 ms, TE = 10 ms, $\alpha = 50°$) providing thin slice images (1.5 mm) showing trabecular bone structure. (B) A turbo spin-echo image (TR = 4500 ms, effective TE = 96 ms, $\alpha = 90°$) showing synovial effusion and oedema (4mm slice thickness).

Figure 3 Transaxial images through the brain of a patient with a glioma. (A) T_1-weighted spin-echo sequence showing a large tumour in the deep cerebral white matter. (B) The same slice following injection with 0.1 mmol kg^{-1} of gadolinium contrast agent. In the latter slice, the sequence also had gradient moment rephasing to reduce artefacts from flowing blood, causing a slight change in white/grey matter contrast.

magnetization transfer techniques are employed to further improve contrast. While many agents are in development, the other major class of agent entering clinical practice is positive (T_1) liver agents such as gadoterate meglumine (Gd-DOTA) (taken up in tumour) and Mn-pyridoxal-5′-phosphate DPDP (taken up in normal liver cells) and negative (T_2*) liver agents such as superparamagnetic iron oxide particles (SPIO) which are taken up by the mononuclear phagacytosing system (Kupffer cells), reducing signal from normal liver (**Figure 4**).

Measuring physiology and function

The nature of magnetic resonance measurements confers sensitivity to a range of properties of water molecules that can be exploited to measure functional aspects of tissues and fluids. Probably the most widely used feature is the sensitivity of MR measurements to motion. When imaging static tissues, motion of fluids or of other tissues presents as a problem to be minimized so as to reduce artefacts. The most common effect is misregistered signals at the same frequency (same position in the read-out direction) but displaced in the phase-encoding direction. This effect can be reduced by the strategies discussed above or by the use of gradient motion rephasing sequences, where the phase gain resulting from movement in the gradient is cancelled by reversed-polarity gradients. Subtraction of pairs of images with and without these additional gradient-lobes results in images of the moving material. The flow of fluids can be measured by bolus-tracking techniques, where a slice is saturated and inflow is observed, or where a distant slice is tagged (by inversion, for example) and the appearance of the tagged blood in the slice of interest is observed. Alternative approaches make use of the phase gain occurring in moving fluids, allowing the speed and direction of flow to be calculated from phase maps (**Figure 5**). Specific sequences can directly measure flow profiles in any arbitrary direction. These techniques are used to make direct measurements of flow velocity, cardiac valve performance, vessel patency and the effects of obstruction, but can also be used to produce flow images.

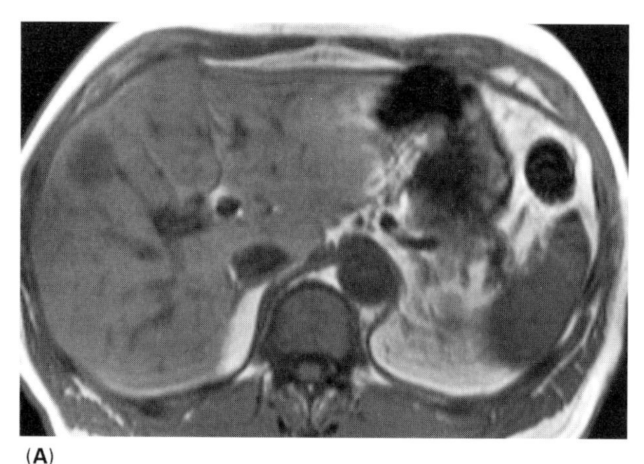

(A)

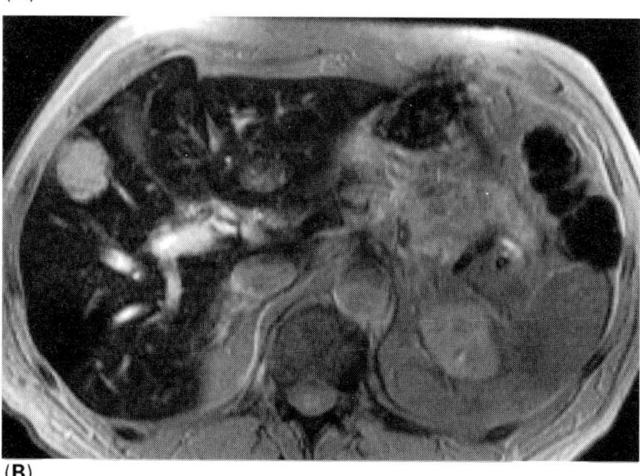

(B)

Figure 4 Transaxial images through the liver of a patient with hepatic metastases from colon cancer. (A) Breath-hold FLASH T_1 image (TR = 80 ms, TE = 4.1 ms, α = 80°) showing limited lesion contrast. (B) Proton density-weighted FLASH image (TR = 127 ms, TE = 10 ms, α = 40°) showing darkening of the liver from application of 15 μmol kg^{-1} of superparamagnetic iron oxide contrast agent, increasing the conspicuousness of the lesion.

Based on these flow-sensitive techniques, a major area of MRI development and application has been MR angiography. A range of time-of-flight and phase-contrast techniques are used to produce 3D data sets, or direct projection views, of vascular structure. 3D data sets are usually processed to produce a set of maximum intensity projections (MIPs), at different orientations, which can then be presented as a cine-loop display, giving apparent 3D visualization of vascular structures. The sensitivity of the measurement techniques has been improved with travelling saturation sequences and by the use of contrast agents and bolus-tracking approaches (**Figure 6**). A major advantage of MR angiography (MRA) is that registered high-resolution soft-tissue images can be obtained at the same time, aiding resolution of diagnostic problems. Initially the major area of interest was in carotid artery stenosis and in vascular abnormalities in the brain. Advances in technique now allow major vessels to be evaluated throughout the body, including the lung and peripheral vascular disease (**Figure 7**). It is now possible to use such approaches to replace expensive diagnostic angiography in application including screening for brain aneurysms and selection of donors for renal transplant.

While MRI is sensitive to bulk flow in vessels, it is also possible to assess the slower nutritive blood supply or perfusion of tissues, together with vascular permeability, and to measure the diffusion of water molecules within tissues. Diffusion is usually measured by determining the loss of signal resulting from the additional dephasing of magnetization experienced by spins moving in a magnetic field gradient. Initially this was achieved by strong pairs of gradients on either side of the 180° pulse in a spin-echo sequence, resulting in moving spins receiving a net dephasing, whereas the phase change cancelled for static spins. The loss in signal is proportional to the diffusion coefficient, but is also affected by the dimensions of the structures in which the spin can move in the time available, leading to the term apparent (or restricted) diffusion coefficient (ADC). Early measurements with this approach showed that by sensitizing the gradients in different directions, it was possible to demonstrate the orientation of white-matter tracts in the brain. Molecules travelling along the tracts could travel a considerable distance, leading to a large loss of signal, whereas those travelling across the tracts could not move far, resulting in little loss of signal. This provided a powerful tool for analysing brain structure *in vivo*, and for better understanding of anatomical distortion due to disease. In early machines, the techniques were very susceptible to eddy currents

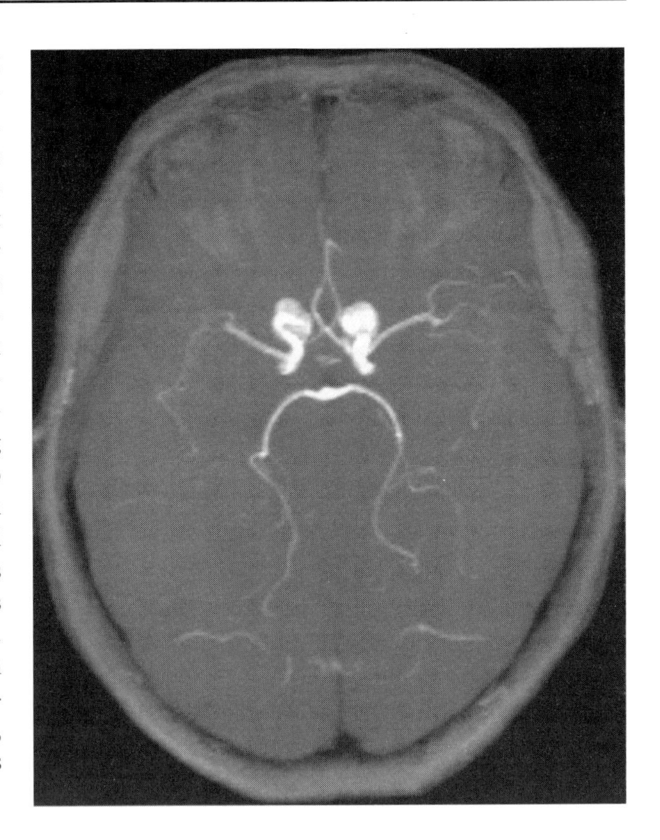

Figure 6 A maximum intensity projection of a set of MR time-of-flight angiography images, showing aneurysms on the circle of Willis (bright areas left and right of brain centreline).

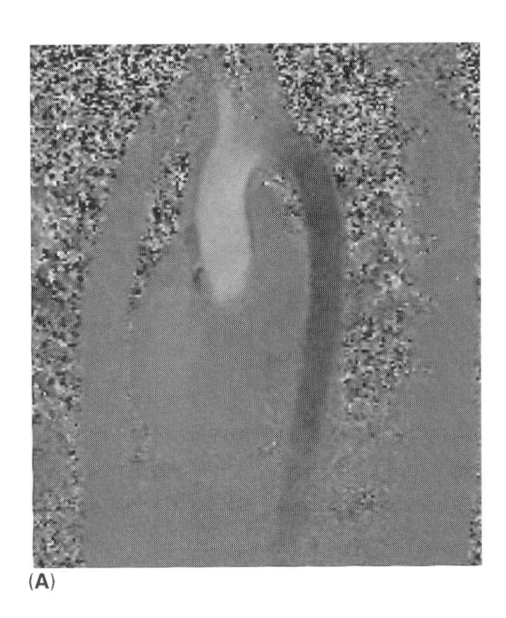

(A)

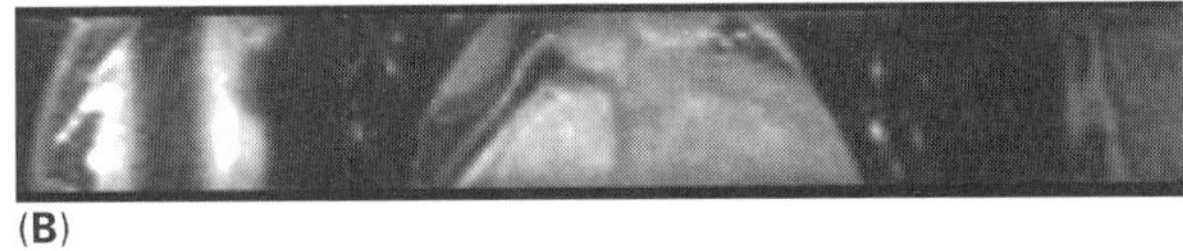

(B)

Figure 5 Flow-sensitive images of blood flow. (A) An oblique coronal phase-contrast image through the ascending and descending aorta, where white shows flow out of the heart and up the ascending aorta, dark shows flow downwards, through the descending aorta. (B) A 3D FLASH image using navigator echo techniques to remove motion effects, showing the right coronary artery just above the aortic arch (the thin white vessel seen against a dark background, centre left of image). Both images were acquired with ECG triggering.

induced in the magnet by the large gradient pulses, and by small bulk movements in tissues and fluids, which could give rise to much greater signal changes than the diffusion itself. These problems have been largely overcome by real-time imaging sequences and improved hardware. Diffusion measurements now commonly apply a set of six differently gradient-sensitized sequences to evaluate both the magnitude and spatial distribution of restricted diffusion, providing a diffusion tensor measurement. The method is now of considerable importance in the diagnosis of stroke and other ischaemic disease, where increased diffusion is an early and sensitive indicator of insult, providing the possibility of early and effective intervention before cell function is irreversibly lost.

Perfusion has also been measured using variants of the tagging or outflow techniques described above, where signal or apparent relaxation time changes occur as a result of the inflow or outflow of labelled spins. Contrast-enhanced studies provide a tracer, allowing the inflow or washout of the tracer, as seen on T_1 weighted images, to be used to derive perfusion. This approach is complicated for those positive contrast agents currently licensed for clinical use (gadolinium-labelled chelates) as they equilibrate rapidly with the extracellular space and also relax

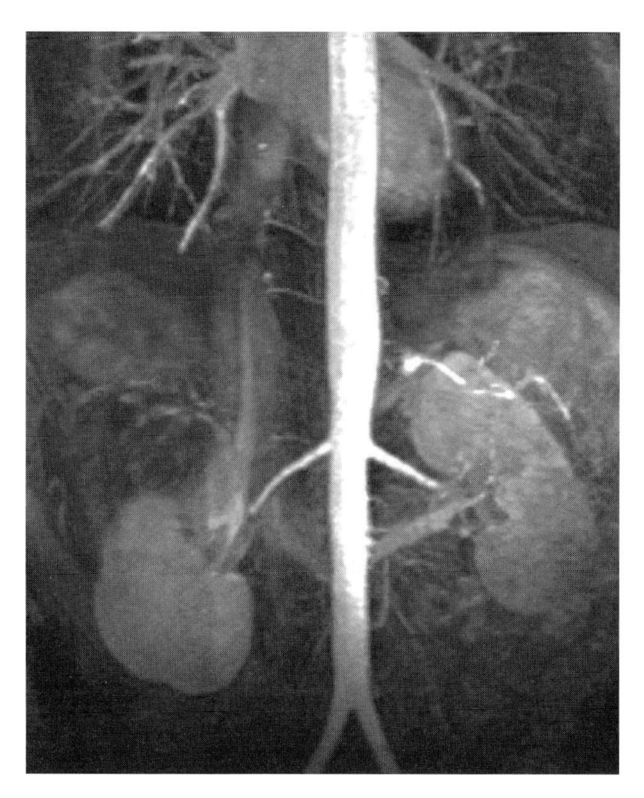

Figure 7 A maximum intensity projection of a set of MR contrast-enhanced angiography images obtained from a 3D FISP sequence, following administration of a gadolinium contrast agent. The image shows the renal arteries and descending aorta (bright centre right with downward-angled renal arteries) and more faintly the upward-angled renal veins and kidneys, draining into the inferior venacava (centre left). The right kidney (to the left of the image) is reduced in size owing to involvement of a renal carcinoma (dark outline visible). At the top of the image the pulmonary veins can be seen clearly.

water molecules that distribute between the extracellular space, intracellular space and vascular space. Future generations of blood pool agents will be more effective in measuring perfusion as their distribution will be limited to the vascular space. An alternative, and more effective, approach is to make use of the local change in magnetic susceptibility that occurs as a bolus of high-concentration contrast agent passes through the vascular bed. Prior to the contrast agent equilibrating between the intra- and extravascular space, there is a large susceptibility gradient around each capillary, which will result in signal loss due to dephasing on gradient-echo sequences. Using T_2*-weighted fast-imaging sequences, this transient phenomenon can be detected. It is proportional to the blood volume in the image, and the timing is related to blood flow (**Figure 8**).

By using T_1-weighted images with positive extracellular contrast agents, combined with modelling techniques, it is also possible to calculate the volume of the extravascular extracellular space, and to calculate the permeability–surface area product governing the rate of transfer of the contrast agent out of the vascular system. This measure is of particular interest, as developing tumour vasculature is characteristically leaky. Development of this new vasculature by tumour-initiated growth factors is believed to be a necessary condition for tumour growth above the limit at which nutritional requirements can be supplied by simple diffusion, and is a target for new generations of anti-angiogenic therapies. Permeability and vascular volume can be calculated on a pixel-by-pixel base as colour-mapped functional images and superimposed on anatomical images. Such measurements require quantitative imaging sequences. Much useful information can be obtained by characterizing the behaviour of contrast uptake and wash-out, and studies have shown that this can be of value in identifying and characterizing tumours, and in monitoring response.

MRI also provides a number of approaches by which tissue motion can be measured. In principle, phase maps or tagging can be employed, although, owing to slice thicknesses larger than or comparable to the motion, this is rarely done. A more widely used approach in cardiac wall motion studies is the application of a one- or two-dimensional criss-cross pattern of parallel signal-suppressed lines on the object. After a defined period, short compared with T_1 relaxation, an image is read out and the movement of tissue relative to the original grid can be deduced. Appropriate software can provide for sophisticated wall motion studies (**Figure 9**). Associated with techniques for monitoring ventricular function based on flow, tissue perfusion studies and assessment of cardiac artery patency (see **Figure 5**), these provide a powerful range of techniques for cardiology.

A recent area of development has been the generation and application of hyperpolarized gases. Both ^{3}He and ^{129}Xe can be prepared at high nuclear polarizations (10–50%) compared with ^{1}H (0.0006% at 1.5 T). This provides a very high signal, and initial measurements have shown the potential to image the lung air-spaces. This complements recent advances in fast very short echo-time sequences that have allowed the lung parenchyma to be imaged, as well as MRA approaches imaging the lung vasculature. Most measurements have been made with ^{3}He, which has low solubility in tissues. ^{129}Xe is of particular interest in measuring perfusion and other properties of tissue spaces, where it demonstrates a large tissue composition-dependent chemical shift. The potential for intravenous delivery using perfluorocarbon blood-substitutes and other suitable media is being evaluated. Studies using hyperpolarized gases require new imaging

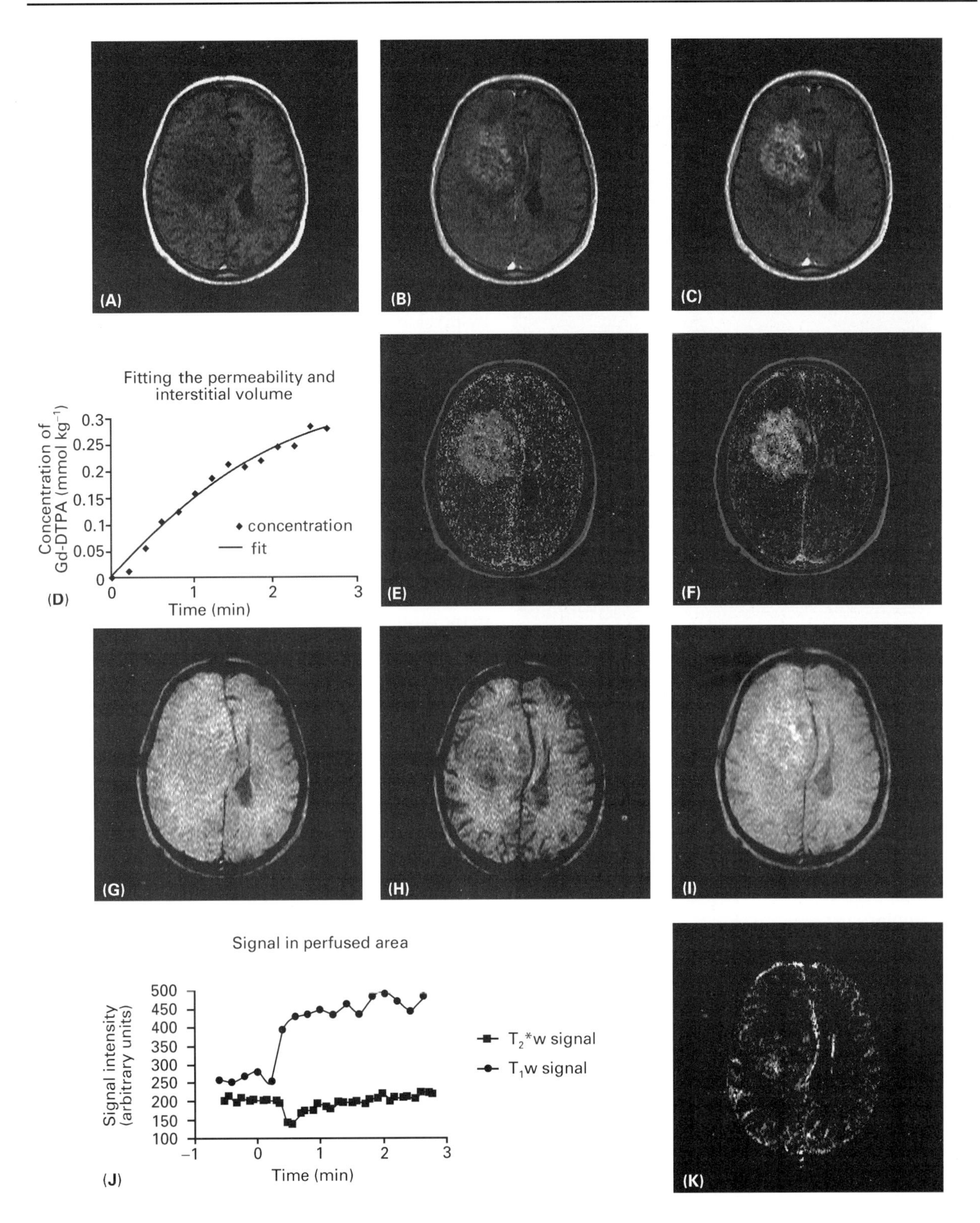

Figure 8 Figures showing quantitative measurements of permeability and blood volume in trans-axial images through the brain of a patient with a recurrent glioma being treated with chemotherapy. (A–C) rapid T_1-weighted images showing uptake of the contrast agent (Gd-DTPA) in the tumour (pre-contrast, 0.8 min and 2.6 min). (D) A graph of the calculated concentration of Gd-DTPA in a volume of interest (points) compared with a constrained fit to a multicompartment model used to derive physiological features. (E) Pixel-by-pixel map of vascular permeability. (F) Pixel-by-pixel map of interstitial volume. (G–I) T_2^* images obtained using the same sequences as for images (A–C) (pre-contrast, 0.28 min, 2.79 min), showing loss of signal due to the passage of contrast agent through the capillary bed; (J) Graph of signal intensity on T_1-weighted images, and on T_2^*-weighted sequences, where the integral of the signal drop on the latter curve is proportional to relative blood volume. (K) Pixel-by-pixel relative blood volume map. These images and calculated maps were obtained using sequences and methods developed by Ms I. Baustert and Dr G. Parker at the Royal Marsden Hospital/Institute of Cancer Research.

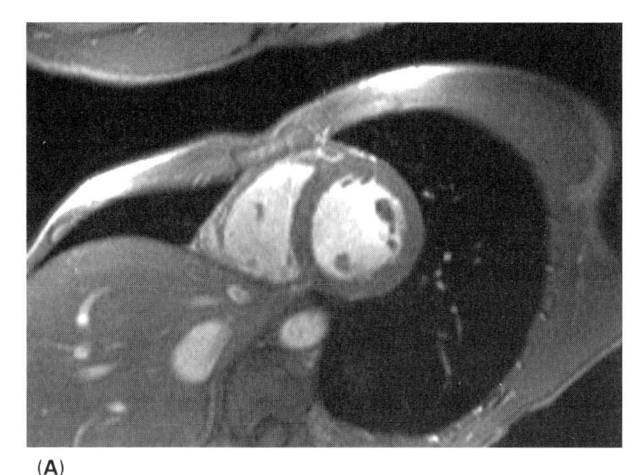

(A)

(B)

Figure 9 ECG-gated images through the heart showing bright blood and orientated to show left ventricle wall muscle. (A) Showing anatomy. (B) Tagged in one direction at early systole, to demonstrate myocardial wall motion.

approaches, as the polarization is exhausted by sampling and signal can only be restored by delivery of fresh hyperpolarized gas.

A major new area of functional MRI has been the discovery that brain activity associated with specific functional tasks causes a change in MR signal observable on T_2*-weighted imaging sequences. This is believed to result from brain activation causing increased local blood flow, which then provides an increased oxygen supply exceeding the increased demand. The blood thus contains proportionately less paramagnetic deoxyhaemoglobin, reducing the susceptibility between blood and surrounding tissues and thus reducing the susceptibility-induced signal loss. This approach provides higher-resolution images than the positron emission tomography techniques used previously, and allows functional activation measurements to be related to high-resolution images of local anatomy. Typically imaging is conducted with and without a stimulus, with subtraction or comparison of the two image sets to provide a difference image demonstrating the region of activation (**Figure 10**). Single-shot techniques are now being developed. The approach is being employed for basic neurological and psychiatric research, as well as in conditions affected by brain function. Signal-to-noise improves with field strength, and a number of centres are exploring the application of higher-field machines to improve the quality of these measurements. As with many of the more advanced techniques, motion and registration between measurements present problems, and sophisticated motion correction, image registration and mapping techniques are being developed.

MR spectroscopy (MRS) provides a complementary means of studying tissue function and metabolism. In the past, spectroscopic examinations have

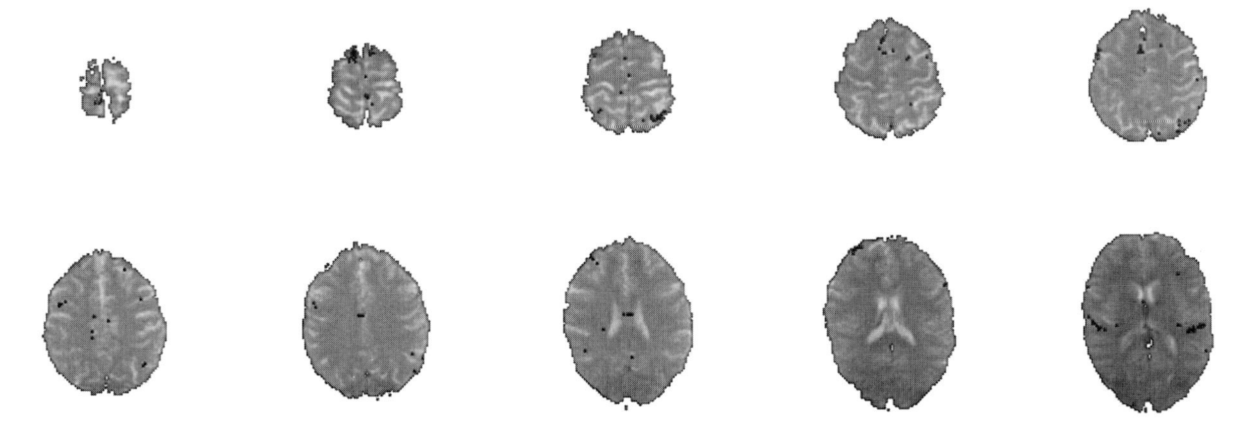

Figure 10 A set of processed image planes through the head of a volunteer showing (black) areas of significant neural activation following exposure to a pure audio tone. Activation data were obtained at The Royal Marsden Hospital by Mr D. Collins using a real-time echo planar imaging (EPI) sequence, and processed at the Institute of Psychiatry by Dr J. Suckling.

often been distinct from imaging studies, but the increase in imaging speed, increased automation and more robust instrumentation have allowed spectroscopy to be integrated with imaging examinations. This trend will continue, allowing specific metabolic pathways, tissue metabolism via ^{1}H or ^{31}P spectroscopy, and drug distribution studies to be integrated with measurements of perfusion, diffusion or activation.

Interventional techniques

The development of methods of guiding interventions or operations is a growing area of MRI. Following identification of suspicious lesions by MRI, it is often necessary to sample tissue to allow cytology or histopathology. Where MRI has provided better imaging, it is desirable to perform sampling using MRI, and eventually this might occur at the diagnostic visit. The design of most clinical MR systems using a cylindrical superconducting magnet design has limited access to the patient or biopsy site, presenting difficulties in performing biopsy or fine-needle aspirates in the magnet. A number of approaches are now being developed. MR-compatible biopsy tables designed for particular organs, often using specialist coils, are being designed for use with conventional systems. The breast is one such region, where MRI is demonstrating high sensitivity for the detection of breast cancer. Magnets have also been designed to provide open access, so that they can be used in the operating theatre or for more conventional image guided sampling. These systems employ either C-configuration magnets at about 0.2 T or a dual-doughnut superconducting design at 0.5 T, allowing access between the two superconducting rings. A particular objective of this latter design has been to enable interactive image guidance during neurosurgery. These approaches are requiring the development of a wide range of MR-compatible accessories, together with rapid imaging techniques and display technology.

Minimally invasive therapeutic approaches are also being piloted with MRI guidance and monitoring. These methods include high-intensity focused ultrasound, laser, electric current, RF hyperthermia and cryoablation. Areas of interest include breast, prostate and liver cancer. In principle, MR provides a valuable means of directly measuring temperature distributions in monitoring these treatments, although current techniques require a field strength of 1.5 T to provide adequate signal-to-noise ratio. An extension of these approaches is monitoring of intravascular or intra-gastrointenstinal tract using small surface coils providing high-resolution images local to the intervention.

Acknowledgements

I am grateful to Dr Anwar Padhani, Mrs Janet McDonald and colleagues in the Diagnostic Radiology Department for providing many of the illustrations shown. Images and data used to illustrate this article were obtained as part of the Cancer Research Campaign supported research in the Magnetic Resonance Unit of the Royal Marsden Hospital and Institute of Cancer Research.

List of symbols

T_1 = spin–lattice relaxation time; T_2 = spin–spin relaxation time; T_2^* = transverse relaxation including susceptibility effects; TE = echo time; TI = inversion time; TR = repetition time; α = flip angle.

See also: ***In Vivo* NMR, Methods; Magnetic Field Gradients in High Resolution NMR; MRI Applications, Biological; MRI Applications, Clinical Flow Studies; MRI Instrumentation; MRI of Oil/Water in Rocks; MRI Theory; MRI Using Stray Fields; NMR Microscopy; NMR Pulse Sequences; Xenon NMR Spectroscopy.**

Further reading

Edelman RR, Hesselink JR and Zlatkin MB (1996) *Clinical Magnetic Resonance Imaging*, 2nd edn. Philadelphia: WB Saunders.

Gadian DG (1995) *NMR and Its Application to Living Systems*. Oxford: Oxford University Press.

Glover GH and Herfkens RJ (1998) Future directions in MR imaging. *Radiology* 207: 289–295.

Grant DM and Harris RK (eds) (1996) *Encyclopaedia of Nuclear Magnetic Resonance*. Chichester: Wiley.

Higgins CB, Hricak H and Helms CA (1992) *Magnetic Resonance Imaging of the Body*, 2nd edn. New York: Raven Press.

Leach MO (1988) Spatially localised NMR. In: Webb S (ed) *The Physics of Medical Imaging*, pp 389–487. Bristol: IOP Publishing.

Morris PG (1986) *Nuclear Magnetic Resonance Imaging in Medicine and Biology*. Oxford: Clarendon Press.

MRI Applications, Clinical Flow Studies

Y Berthezène, Hôpital Cardiologique, Lyon, France

MAGNETIC RESONANCE
Applications

Magnetic resonance imaging (MRI) is a useful, versatile diagnostic tool that can achieve contrast among different tissues by taking advantage of differences in T_1 relaxation times, T_2 relaxation times and proton densities. In recent years there has been considerable interest in the development of MRI techniques as a noninvasive method of measuring blood flow and tissue perfusion in certain clinical conditions. MRI flow measurements have been applied particularly to the vascular system and compared with other techniques, such as ultrasound. MRI provides a noninvasive method for quickly measuring velocity and volume flow rates *in vivo* using readily available methods and equipment. Flow quantification by means of MRI does not require the use of ionizing radiation and/or contrast agents, as X-ray techniques do. Unlike ultrasound, MRI measurements are not hindered by the presence of overlying bone and air.

The two principal methods of velocity measurement use either 'time of flight' or 'phase contrast' techniques. Time-of-flight methods are well suited for determining the presence and direction of flow, and phase-based methods are well suited for quantifying blood velocity and volume flow rate.

Furthermore, MRI offers the opportunity to quantitatively assess properties of tissue, such as perfusion and blood volume. Use of such quantification potentially allows tissue to be characterized in terms of pathophysiology and to be monitored over time, during the course of therapeutic interventions.

Magnetic resonance flow measurements and MR angiography

Time of flight

On cine gradient echo images, blood flow is bright. The high signal intensity (bright signal) of vessels on cine gradient echo images is achieved by the entry of unsaturated protons into the image, a phenomenon called time of flight or flow-related enhancement. By displaying flowing blood as high signal intensity, cine gradient echo images generally provide a better signal-to-noise ratio within the blood pool than spin-echo images, which demonstrate the arterial lumen as a dark region of signal void. Because the signal in cine gradient-echo imaging is based on through-plane movement of protons, this technique is occasionally less sensitive to slow flow or flow within the imaging plane (in-plane flow). In cases where blood flow is slow, the vessel is tortuous or flow is primarily in-plane, there may be diminished signal or even complete signal saturation on cine gradient echo images.

However, signal loss on cine gradient echo images can be used as a diagnostic aid in special circumstances. In cases of haemodynamically significant stenosis (as in aortic coarctation or aortic stenosis), a dark, fan-shaped flow jet can be seen on cine gradient echo images. This area of intravoxel dephasing results from the turbulent flow typically seen distal to a significant vascular narrowing. Aortic insufficiency may also manifest as a flow jet on cine gradient echo images. Although the relative size of the jet has been shown to correlate with the clinical severity of the stenosis, the appearance of the jet is highly variable and can be greatly affected by a variety of factors (e.g. imaging plane, pulse sequence, echo time). The jet may be small or even absent despite the presence of a high-grade (haemodynamically significant) vascular narrowing.

Phase contrast (PC)

One good method for quantitatively measuring blood flow assesses the change in phase of the blood signal as the blood flows through a slice oriented perpendicular to the direction of flow. This method derives velocity from the phase of the MR signal, and calculates volume flow rate by multiplying the average velocity by the vessel's area. To determine flow velocity, cine PC imaging takes advantage of the phase shifts experienced by moving protons (within blood) as they move along a magnetic field gradient. Bipolar flow-encoded gradients are applied to measure these phase shifts. This technique requires the operator to prescribe a velocity encoding that determines the flow-encoding gradient strength and sensitivity to flow direction(s) (anterior-to-posterior, anterior-to-posterior and left-to-right), which dictate the plane(s) of the gradient application.

The vascular information from cine PC acquisitions may be displayed as simple angiographic images (similar to cine gradient-echo images) in which all flow is bright or as phase map images in which the flow directional information is coded as bright or dark and the flow velocity data are reflected in the signal intensity (relative brightness or darkness). With phase map cine PC imaging, blood flow can be quantified (millilitres per minute).

Ideally, if flow measurement is desired, one should choose an imaging plane perpendicular to the direction of the flow, select a velocity encoding at least as high as the fastest expected flow velocity, and prescribe the flow sensitivity to be in accordance with the direction of desired flow measurement. For measurement of normal flow within the ascending or descending aorta, for example, an axial cine PC prescription with a velocity encoding of 150 cm s^{-1} and superior-to-inferior flow direction is appropriate. If slow flow is expected or the goal is to visualize flow in a false lumen, a lower velocity encoding such as 50 cm s^{-1} may be more appropriate.

MR angiography (MRA)

Angiography is the imaging of flowing blood in the arteries and veins of the body. In the past, angiography was only performed by introducing an X-ray opaque dye into the human body and making an X-ray image of the dye.

Many techniques have been developed for MRA of the great vessels, including gradient echo time-of-flight and phase-contrast techniques. Both time-of-flight and phase-contrast MRA methods can be implemented as either a sequential 2D or a true 3D acquisition. The encompassed MRA volume is analysed by postprocessing with a maximum intensity projection (MIP) technique or with multiplanar reformatting (MPR). The MIP technique allows a rotational '3D' display of the vessel, viewed from different angles. MPR allows reconstruction of parallel thin slices in any orientation.

Three-dimensional gadolinium-enhanced MR angiography is a recently developed angiographic technique that can substantially improve the resolution, signal-to-noise ratio, speed and overall quality of vascular MRI. 3D gadolinium-enhanced MRA achieves its image contrast and hence its angiographic information from the T_1-shortening effect of gadolinium on blood. Because it is less dependent on inherent blood flow characteristics for the generation of vascular signal, 3D gadolinium-enhanced MRA is minimally degraded by flow-related artifacts. Three-dimensional gadolinium-enhanced MRA can be performed quickly (within a 20–40-s breath hold) on high-performance MR imagers. With a computer workstation, data from 3D gadolinium-enhanced MRA can be postprocessed to generate projection aortograms in any obliquity, that are similar to conventional angiograms (**Figure 1**).

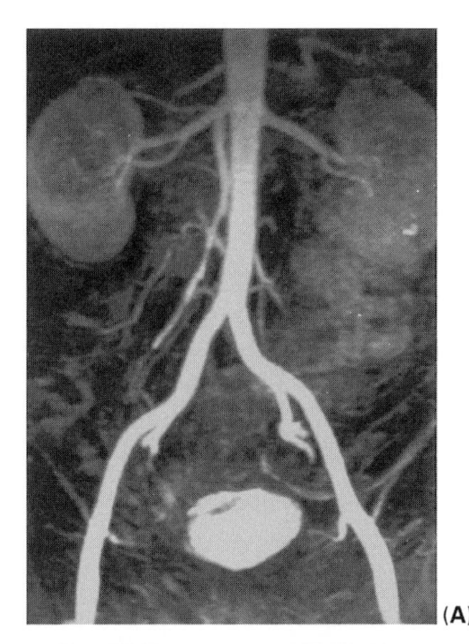
(A)

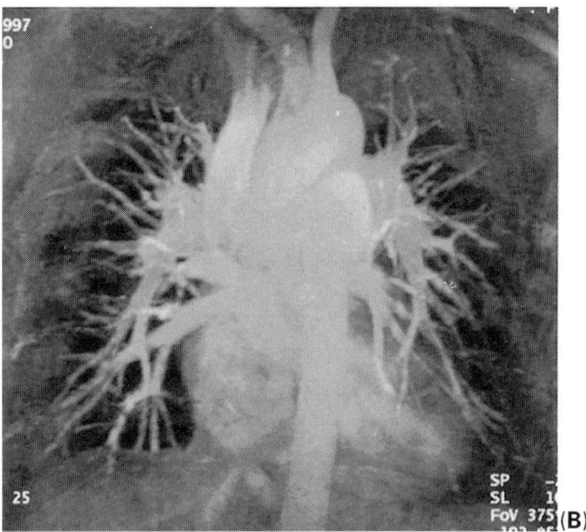

(B)

Figure 1 3D gadolinium-enhanced MR angiography of the abdominal aorta (A) and pulmonary vessels (B).

MRA is already used routinely in many centres for evaluation of the carotid arteries and intracerebral vasculature, aortography and assessment of the ileofemoral system. MRA of the coronary arteries is technically more difficult due to their relatively small size, their complex 3D anatomy and their constantly changing position within the thoracic cavity due to cardiac motion and respiration.

Tissue perfusion

In the broadest sense, perfusion refers to one or more of various aspects of tissue blood flow. Parenchymal blood flow is the ratio of blood volume to the transit time of blood through the tissue. The different techniques of MR perfusion typically deal with blood volume, transit times and blood flow as relative measures, although absolute quantification may also be possible. The two perfusion strategies are based either on induced changes in intravascular magnetic susceptibility (T_2* effect) or relaxivity (T_1 effect) and on tagging inflowing arterial spins.

Dynamic contrast-enhanced MR imaging

Dynamic contrast-enhanced MRI is a method of physiologic imaging, based on fast or ultrafast imaging, with the possibility of following the early enhancement kinetics of a water-soluble contrast agent after intravenous bolus injection.

Bolus tracking techniques have been used to measure tissue perfusion, notably in the kidney, heart and brain. These methods are based on fundamental MR contrast mechanisms that promote either T_1 or T_2/T_2* enhancement. Gadolinium chelates administered in low doses lead to predominantly T_1-weighted signal increases, mediated by water proton–contrast agent dipolar relaxivity interactions. Alternatively, the T_2/T_2* dephasing of spins, due to a locally heterogeneous high magnetic susceptibility environment, has been exploited by using higher doses of gadolinium. If the curve of concentration versus time can be plotted as a known quantity of a tracer passes through an organ, organ perfusion can be calculated from the area under the curve. Alternatively, if the tracer is wholly extracted by the organ, the principles described by Sapirstein enable perfusion to be measured from the amount of tracer trapped by the organ.

In the normal brain, tight junctions of nonfenestrated capillaries effectively prevent Gd chelates from leaking into the interstitial space. Thus during bolus-tracking experiments, Gd-DTPA behaves like a true intravascular contrast agent as long as there is no brain abnormality that causes blood–brain barrier disruption, and regional cerebral blood volume may be determined by integrating the time-versus-concentration curve. As opposed to perfusion imaging of the brain, in other tissues such as the breast the technique is hampered by the fact there is nothing like a blood–brain barrier. Accordingly, Gd-DTPA will not be a true intravascular contrast agent. Nevertheless, by treating it mainly as an extracted tracer, it is possible to measure perfusion from the peak tissue enhancement. The model assumes a linear relation between tracer concentration and signal enhancement. Now that echo planar and ultrafast gradient-echo imaging can provide at least one image for each cardiac cycle during the passage of the tracer, measurement of myocardial perfusion with high resolution is possible.

Arterial spin labelling

Blood flow imaging with MR by spin labelling, or spin tagging, of the water protons in the arterial source to a slice has the advantage that it is completely noninvasive, is a more direct assessment of blood flow, and may generate absolute blood flow quantification. Cerebral blood flow quantification has been accomplished by continuous adiabatic inversion of arterial spins and use of tracer kinetic models of cerebral blood flow determination. Qualitative cerebral blood flow mapping has also been described using echo planar sequences, a single inversion pulse to inflowing arterial spins, and subtraction of tagged and untagged echo planar images. In principle it is also quantifiable, to give absolute flow quantification.

Clinical applications

Brain

Hyperacute stroke Whereas conventional computed tomography and MRI are excellent modalities with which to detect and characterize central nervous system disease, they fail to depict acute ischaemia and infarction reliably at its earliest stages. Detection of cerebral infarction by dynamic MR contrast imaging is now possible. Some of the most promising work is being done with perfusion and diffusion imaging. Perfusion MRI characterizes how much brain tissue an occlusive blood clot has placed at risk (see **Figures 2** and **3**), whilst diffusion measurement shows how much tissue is already damaged or is possibly even dead.

Flow-restrictive lesions MR volume flow rate measurements have been used to evaluate the severity and haemodynamic significance of flow-restrictive lesions in the carotid, vertebral and intracranial vessels. A severe stenosis can result in a significant decrease in volume flow rate distal to the stenosis. Because brain perfusion relates directly to the volume of blood delivered, identifying an area of decreased volume flow rate distal to a stenosis may be of clinical importance.

Intracranial volume flow rate measurements are technically difficult using methods other than MRI, and for this reason the normal volume flow rates for intracranial vessels are not well established.

Vascular flow reserve Another evaluation process measures the change in volume flow rate in a given vessel before and after vascular challenge. In normal

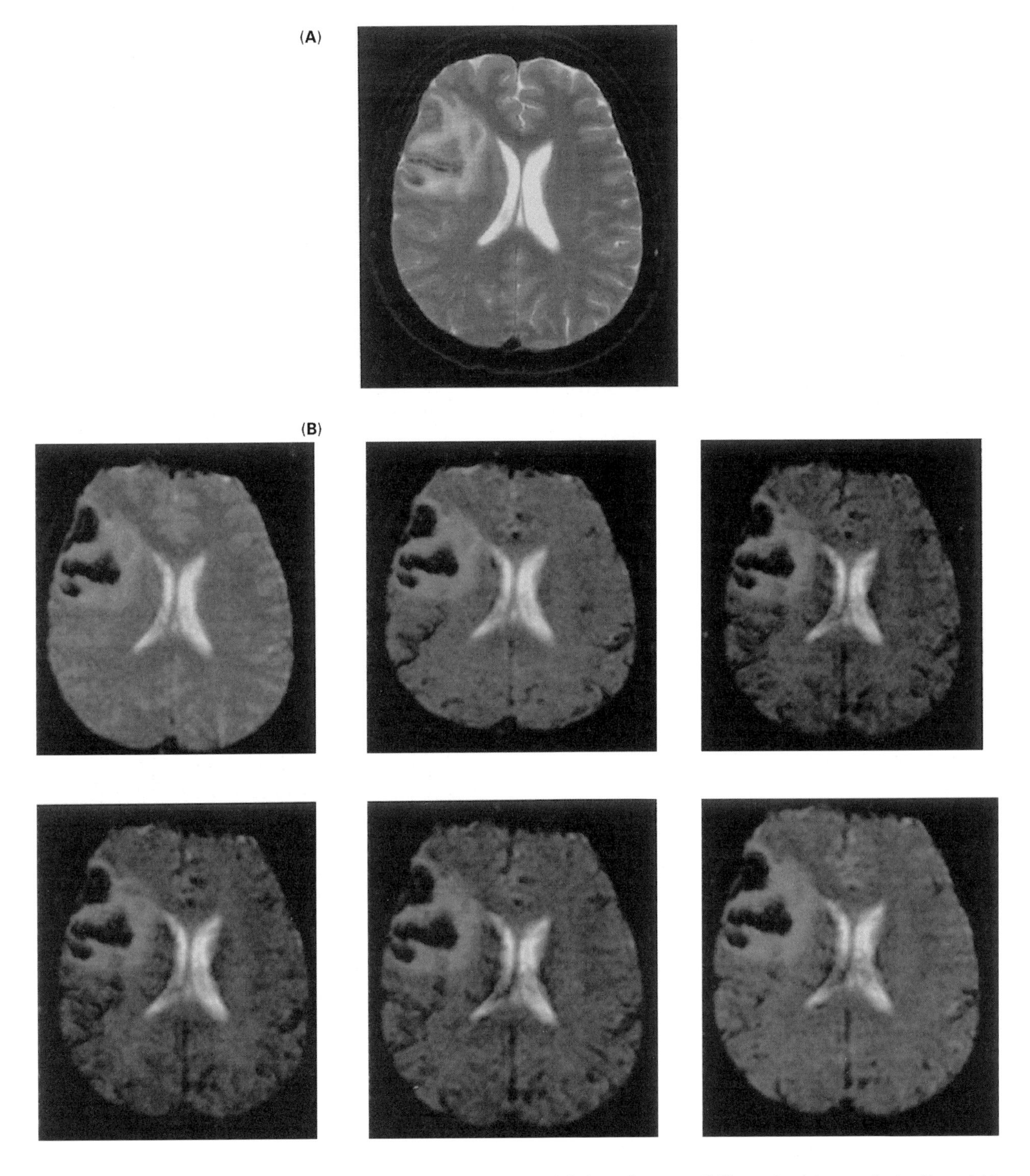

Figure 2 Brain axial spin-echo T_2-weighted image (A) and sequential dynamic susceptibility-contrast in a patient with a right infarct (B).

situations, inhalation of CO_2 or intravenous injection of a vasodilator (acetazolamide), causes intracranial arteries to dilate, leading to an increase in flow velocity and volume flow rates in these vessels. The difference between the flow rate under routine conditions and maximal flow rate after chemically induced vasodilatation is designated as the flow reserve. In human subjects and specifically in patients with cerebrovascular disease the acetazolamide test is performed to evaluate the decrease in cerebral perfusion pressure through the investigation of the vasomotor reactivity (VMR), which is thought

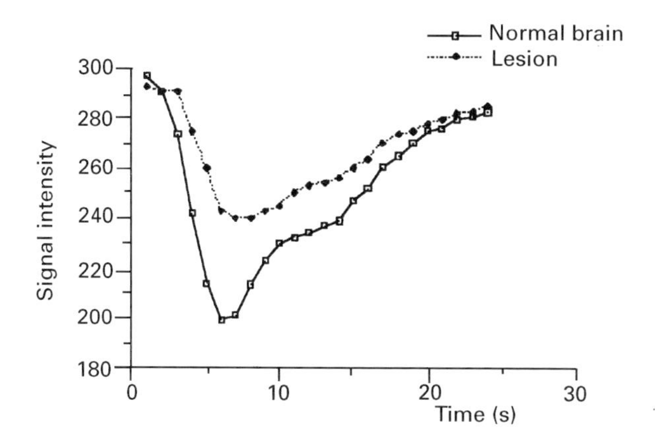

Figure 3 Change in signal intensity during a rapid bolus contrast injection (T_2* effect) comparing normal brain and ischaemic regions. The lesion shows a less dynamic decrease in signal intensity than the contralateral normal region.

to reflect compensatory vasodilatation. In patients with occlusion or stenosis of more than 90% of the internal carotid artery, diminished VMR was reported to be significantly associated with low flow infarctions and higher rate of future ipsilateral stroke compared with patients with a normal or only slightly disturbed VMR. The quantification of the response of the blood vessels to the stimulus can be obtained by measuring cerebral blood flow, cerebral blood volume or blood flow velocity.

Subclavian steal In this syndrome due to occlusion of the subclavian artery proximal to the origin of the vertebral artery, the blood flow is reversed in the vertebral artery and redirected from the basilar artery into the arm. Phase contrast MRI can be used to determine the direction of vertebral artery flow. This information is valuable for monitoring the progression of disease, for assessing the magnitude of the steal, and in the postoperative setting, for determining the efficacy of vascular reconstructive surgery.

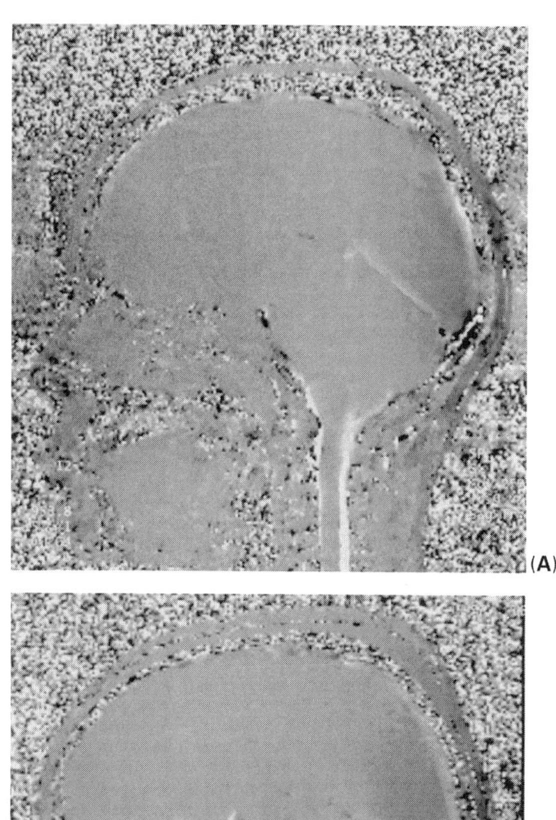

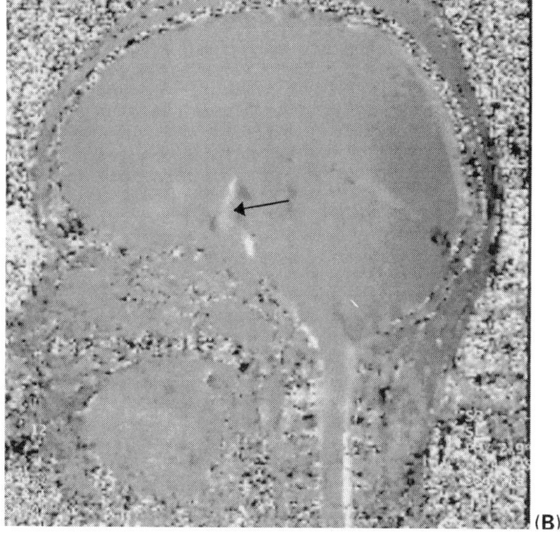

Figure 4 Phase contrast image in a patient with a brain tumour before (A) and after (B) surgery. Before surgery no flow is seen in the third ventricle because of tumour compression. After surgery flow can be seen in the floor of the third ventricle (arrow).

Cerebrospinal fluid flow Phase contrast methods have been used to measure velocity and volume flow rates of cerebrospinal fluid in healthy volunteers and in patients with various diseases. This method can be used to measure the flow rate of cerebrospinal fluid through ventriculo-peritoneal shunts in patients with hydrocephalus (**Figure 4**).

Thorax

Valvular heart disease The signal intensity of flowing blood during cine gradient echo imaging depends upon the nature of the flow. In general, flowing blood generates uniform high signal because of continuous replacement of magnetically saturated blood by fresh blood. Turbulence leads to loss of signal and so the turbulent jet of mitral regurgitation can be seen in the left atrium. The size of the signal void can be used as a semiquantitative measure of regurgitation but the signal void will vary with imaging parameters such as echo time. This is similar to colour flow Doppler where technical factors such as gain adjustment and filter setting are important. A more fundamental problem common to both is that

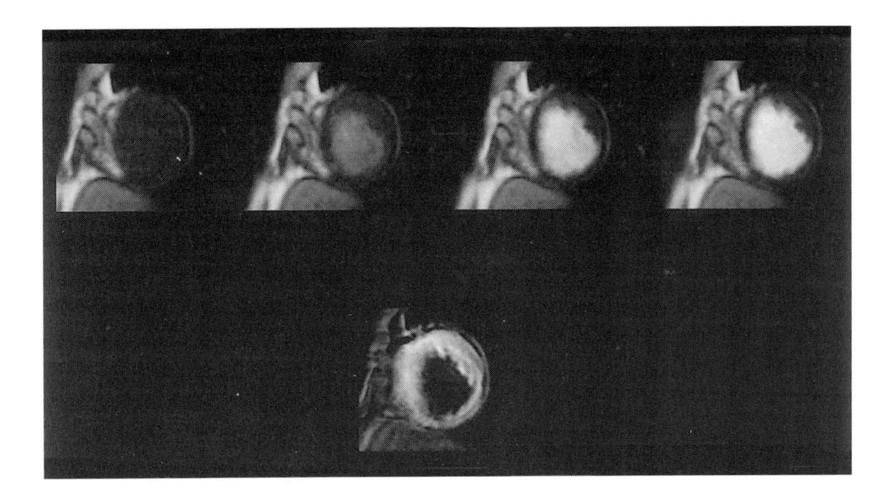

Figure 5 Selected T_1-weighted images, of a single short-axis section, illustrating myocardial transit of the contrast agent in the left ventricle (top images). Myocardial perfusion is difficult to assess visually. However, postprocessing the image (factor image) demonstrates myocardial enhancement (bottom image).

the size of the regurgitant jet is influenced by many factors in addition to the severity of regurgitation, such as the shape and size of the regurgitant orifice and the size of the receiving chamber.

Myocardial perfusion MRI can be employed to evaluate myocardial perfusion at rest and during pharmacological testing. Ultrafast MRI sequences with image acquisition at every heart beat provide the opportunity to acquire dynamic information related to the passage of a paramagnetic contrast agent through the myocardial microcirculation and thus provides an indirect measure of myocardial perfusion (**Figure 5**). A myocardial region supplied by a severely stenosed coronary artery can be detected by a delayed increase in signal intensity and a decreased peak signal intensity. Recently, several tomographic images could be acquired during a unique bolus of a small amount of paramagnetic chelate allowing the study of almost the entire myocardial volume compared to the previous situation where only one slice was available.

Great arteries

Aorta Next to congenital heart disease, the clinical utility of MRI has been most convincingly documented in patients with large vessel disease, and more specifically with acquired aortic disease. The wide field of view and the ability to freely adjust the orientation of imaging planes to the vessel direction do not only favour a clear depiction of the anatomy of the vessel lumen and vessel wall, but also facilitate the understanding of the relation to other anatomic structures within the chest and ensure highly accurate dimensional measurements.

Furthermore, it is relatively easy to combine the morphological information with functional aspects on blood flow, which can be assessed both qualitatively and quantitatively. The increased flow rate in arteries during systole, and in veins during both systole and diastole, enhances the contrast between intraluminal blood flow and vessel wall. Thus, a good image quality is usually obtained even without administration of intravenous MR contrast material.

Gradient echo techniques and phase velocity mapping are useful for demonstration and characterization of mural thrombus and for qualitative and quantitative assessment of aortic regurgitation associated with aneurysm of the ascending aorta.

There is substantial evidence demonstrating that from all the available modalities MRI has the highest sensitivity and specificity for detection of aortic dissection. MRI is not only well suited to identify an intimal flap, but can also detect aortic regurgitation and pericardial effusion with high accuracy.

The extent of aortic dissection is readily detected by NMR imaging and is displayed including involvement of other vessels. The entry and exit points are more difficult to localize, but there is no doubt that invasive investigation can be avoided with a combination of echocardiography and NMR imaging.

Pulmonary arteries The retrosternal position of central pulmonary arteries makes it difficult to assess pulmonary blood flow by Doppler echocardiography, especially in the presence of skeletal or lung abnormalities. NMR velocity imaging is not technically constrained and is capable of accurate blood flow measurement in any plane. Flow can be accurately quantified in the left and right main pulmonary artery with use of phase velocity mapping.

MR velocity mapping is an accurate technique to measure volumetric pulmonary flow after repair of congenital heart disease. The consequences of pulmonary regurgitation on right and left ventricular function can be comprehensively evaluated by the combined use of MR velocity mapping and gradient-echo MRI of both ventricles. This unique information may have prognostic and therapeutic implications for the management of patients with (repaired) congenital heart disease.

The flow pattern in the main pulmonary artery differs between normals and patients with pulmonary hypertension. The latter have lower peak systolic velocity and greater retrograde flow during end systole.

Early studies have already indicated the possible role of MRI in detecting central pulmonary emboli with the use of conventional MRI techniques. MRA using fast 2D time-of-flight gradient-echo techniques combined with maximum intensity projections showed good sensitivity but only moderate specificity. Better results may be obtained with the use of phased-array coils or 3D MRA. Gadolinium-enhanced MRA of the pulmonary arteries, as compared with conventional pulmonary angiography, had high sensitivity and specificity for the diagnosis of pulmonary embolism. This new technique shows promise as a noninvasive method of diagnosing pulmonary embolism without the need for ionizing radiation or iodinated contrast material.

Tumours

Dynamic contrast-enhanced MRI has been used as an additional imaging technique in various clinical applications, such as differentiation of benign from malignant lesions, tissue characterization by narrowing down the differential diagnosis, identification of areas of viable tumour before biopsy and detection of recurrent tumour tissue after therapy. This technique provides information on tissue vascularization, perfusion, capillary permeability and composition of the interstitial space.

Diagnosis in dynamic contrast-agent-enhanced breast MRI is primarily based on lesion contrast-agent-enhancement velocity, with breast cancers showing a faster and stronger signal intensity increase after contrast injection than benign lesions. The rapid enhancement seen in carcinomas is thought to be due to the angiogenic potential of malignant lesions.

While the dynamic technique proves very sensitive, specificity remains a problem: initial experiences with dynamic contrast-enhanced breast MRI suggest a clear-cut separation of benign and malignant lesions on the basis of their enhancement velocities. This concept has to be abandoned when more and more benign lesions have enhancement velocities comparable to or even higher than those of malignant tumours.

Kidneys

MRI has advantages over both computed tomography and nuclear scintigraphy for assessing renal function, because it combines high spatial resolution with information on perfusion and function. Quantification of flow rate by phase contrast in the renal arteries and veins has the potential to provide estimation of renal blood flow, which could prove useful in a number of clinical situations, especially for studying renal vascular disorders and the effects of treatment, and for assessing renal transplants. Evaluation of renal perfusion with MRI has become feasible with the development of rapid data acquisition techniques, which provide adequate temporal resolution to monitor the rapid signal changes during the first passage of the contrast agents in the kidneys. More recently, magnetically labelled water protons in blood flowing into kidneys has been used to noninvasively quantify regional measurement of cortical and medullary perfusion. Dynamic MRI demonstrates renal morphology and reflects the functional status of renal vasculature. The measurement of renal perfusion by MRI could provide a noninvasive diagnostic method for monitoring the status of renal transplants and renal ischaemic lesions.

Conclusion

With the above developments currently underway, the outlook for magnetic resonance flow measurements and contrast-enhanced MRI is bright. The opportunity to extract quantitative regional physiologic information in addition to anatomic information will definitely elevate MRI from 'anatomic imaging with soft tissue contrast' to 'a noninvasive technique for assessment of physiologic processes and tissue integrity with high spatial resolution', offering new power for diagnosis and treatment monitoring, and insights into the very mechanisms of disease physiopathology.

List of symbols

T_1 = spin–lattice relaxation time; T_2 = spin–spin relaxation time.

See also: **Contrast Mechanisms in MRI; MRI Applications, Clinical; MRI Instrumentation; MRI Theory.**

Further reading

Detre JA, Alsop DC, Vives LR, Maccotta L, Teener JW and Raps EC (1988). Noninvasive MRI evaluation of cerebral blood flow in cerebrovascular disease. *Neurology* **50**: 633–641.

Ho VB and Prince MR (1998) Thoracic MR aortography: imaging techniques and strategies. *Radiographics* **18**: 287–309.

Korosec FR and Turski PA (1997) Velocity and volume flow rate measurements using phase contrast magnetic resonance imaging. *International Journal of Neuroradiology* **3**: 293–318.

Mohiaddin RH and Longmore DB (1993) Functional aspects of cardiovascular nuclear magnetic resonance imaging. Techniques and application. *Circulation* **88**: 264–281.

Roberts TPL (1997) Physiologic measurements by contrast-enhanced MR imaging: expectations and limitations. *Journal of Magnetic Resonance Imaging* **7**: 82–90.

Sorensen AG, Tievsky AL, Ostergaard L, Weisskoff RM and Rosen BR (1997) Contrast agent in functional MR imaging. *Journal of Magnetic Resonance Imaging* **7**: 47–55.

MRI Contrast Mechanisms

See **Constrast Mechanisms in MRI.**

MRI Instrumentation

Paul D Hockings, **John F Hare** and **David G Reid**,
SmithKline Beecham Pharmaceuticals, Welwyn, UK

MAGNETIC RESONANCE
Methods & Instrumentation

Synopsis

Since 1973 when Paul Lauterbur published the first practical magnetic resonance imaging (MRI) method in Nature the one constant in this exciting area of science has been the rapid pace of change. Novel MRI methods forced the development of new technologies such as pulsed field gradients which have, again, opened the field to even more exciting pulse sequence developments. There has been a vast improvement in image quality over these years. Obviously many factors have contributed to this improvement and these will be discussed individually below. However, one factor stands pre-eminent and that is the improvement in pulsed magnetic field gradient technology. Improvements in gradient coil design have meant that gradients have become more linear and more sensitive, and the introduction of gradient shielding technology has reduced the problems of pre-emphasis and B0 correction to a thing of the past except for the most demanding methodologies. And, of course, there have been major innovations in gradient amplifier technology shortening rise times, increasing gradient strength and reducing gradient noise. Other major innovations of recent years that have significantly improved image quality have been the introduction of birdcage resonators and phased array coils. Oversampling of the receiver signal by the analogue-to-digital converter (ADC) has allowed the introduction of digital filtering techniques that prevent the folding of noise from outside the spectral width of interest back into the image. And, of course, there have been some technology improvements that have not contributed directly to improvements in image quality but have made the MRI technique easier to implement such as the introduction of self-shielded magnets and the enormous increase in computer power that has made 3D MRI techniques practical in terms of 3D Fourier transforms and image processing and display.

Introduction

NMR spectrometers can be converted into MR scanners by the addition of gradient handling capacity and gradient amplifiers. In the crudest configuration the output of the gradient amplifiers can be fed into the room temperature shim set to create the linear magnetic field gradients necessary for imaging. Thus, modern NMR spectrometers with triple axis gradient sets can be used for micro imaging.

However, for biological and clinical MRI applications there are a number of additional hardware items that need to be considered. The basic components of the typical clinical superconducting MR scanner can be seen in **Figures 1** and **2**. The individual components are described in more detail in the text, but briefly, the magnet cryostat is kept at liquid helium temperature and houses the windings of the primary magnet and, if active shielding is used, a second set of superconducting coils outside the primary coils to reduce the fringe field effect. Inside the magnet bore, clinical scanners will usually have a passive shim assembly, active shim coils, gradient set, RF whole body coil and patient bed. Typically, the RF coils will be tuned to the proton frequency; however, the addition of RF coils tuned to other nuclei and the appropriate RF amplifiers will allow such nuclei to be imaged if the signal-to-noise ratio is sufficient.

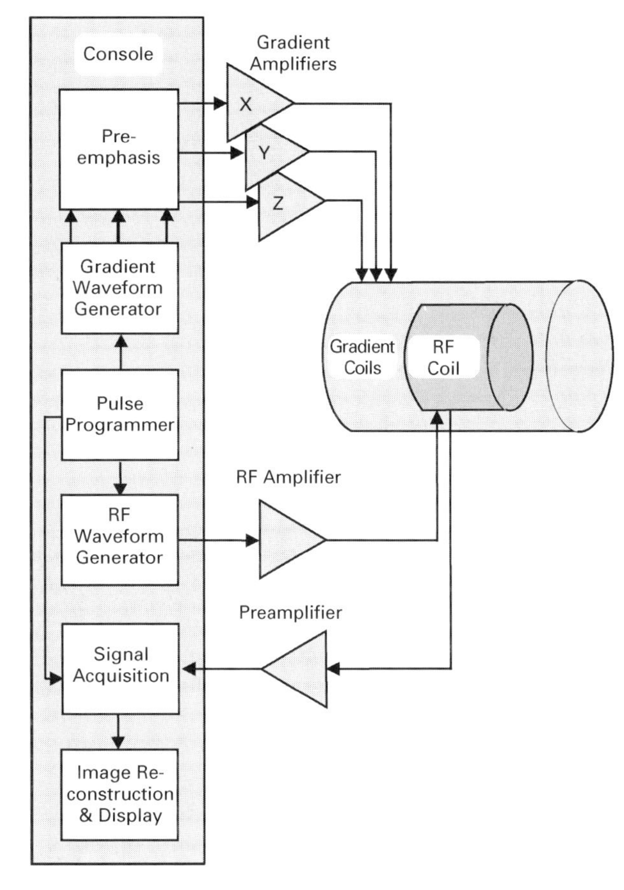

Figure 2 The essential components of a typical clinical MRI system.

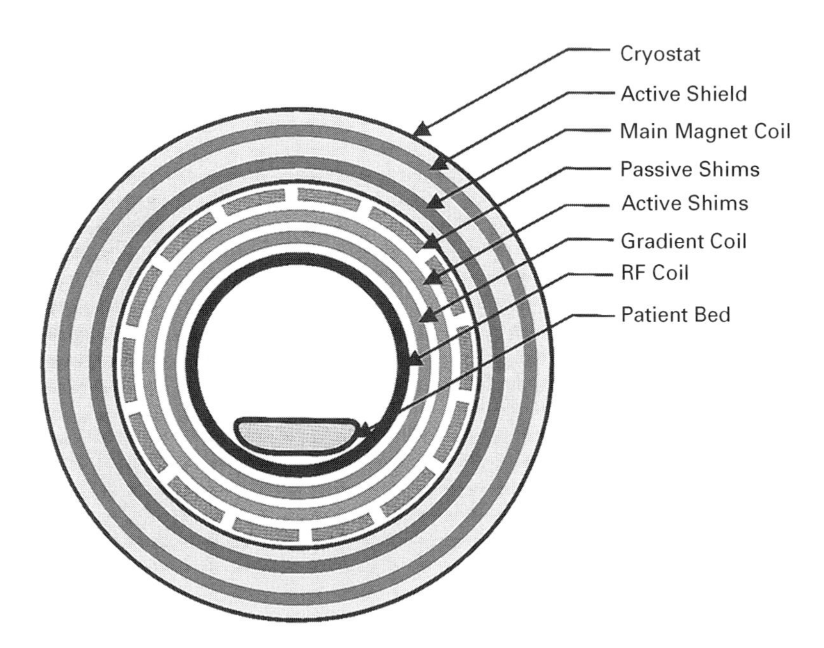

Figure 1 Schematic cross-section through a typical superconducting clinical MR scanner. Within the cryostat (light blue) are the superconducting coils of the primary magnet (red) and active shield (green). In the bore of the magnet there are passive shim rods (grey), active shim coils (orange), gradient set (blue), whole body RF coil (black) and patient bed. The tractable diameter is generally half the magnet bore diameter. (See Colour Plate 37).

Magnet

Bore

First, one needs to decide the largest patient size that needs to be scanned, as this will govern the magnet bore size. There is a roughly two to one relationship between bore size and tractable patient diameter. Most clinical whole body scanners have a 1 m bore, but specialist magnets exist for imaging larger patients. Head only scanners will typically have a diameter of 60 cm. For smaller animals such as rats and rabbits a range of smaller bore magnets exists.

Field

As with conventional high-resolution NMR, higher field strength produces more signal. However, for imaging applications this must be tempered by the consideration that differences in T_1 are generally greater at low field strengths and therefore images from lower field magnets intrinsically have more contrast. For micro-imaging applications where pixel size is approaching the distance water can diffuse during the application of the pulse sequence, practical experiments generally require field strengths in excess of 7 T. However, in the clinical realm superb images may be produced on systems with 0.5 T fields.

Type

There are three types of magnets used for MRI.

Superconducting Higher field magnets are superconducting (**Figure 3**). The magnet coil sits in a pool of liquid helium at 4.2 K. In most animal imaging systems this is surrounded by a secondary liquid nitrogen temperature dewar (77 K) to reduce heat transfer to the liquid helium dewar (liquid helium is much more expensive than liquid nitrogen). Typically, the liquid helium would need to be topped up at intervals of 3 to 12 months. Liquid nitrogen is usually filled weekly. Modern clinical systems dispense with the secondary cryostat in favour of a helium refrigerator. These systems still need periodic refilling with liquid helium, generally at yearly intervals. The superconducting magnet offers high field strength, stability and homogeneity; however, the initial cost of the magnet can be an order of magnitude more than the electromagnets and permanent magnets and the extensive fringe field can make finding an appropriate installation site difficult.

Electromagnets Resistive magnets have less extensive fringe fields than superconducting magnets but require up to 60 kW to produce fields of 0.3 T and consume large quantities of cooling water (**Figure 4**). The open access design can be ideal for interventional MRI applications. Their main drawback, aside from the limited field strength available, is field instability due to fluctuations in the power supply and temperature.

Permanent magnets Like the electromagnet, the field strength of the current generation of permanent magnets is restricted to 0.3 T. For many applications this will be sufficient and, given the insignificant fringe magnetic field and open access design, will prove an ideal solution for some installations, particularly where the power supply and/or supply of cryogens is unreliable. However, permanent magnets require very careful temperature regulation to prevent drifts in field and they can be extremely heavy.

Shielding

The stray fields emanating from superconducting magnets can pose a hazard to the surrounding environment. Unauthorized access within the 0.5 mT (5 gauss) field must be prevented to hinder entry of persons with cardiac pacemakers. In addition, fields as low as 0.1 mT can exert deleterious effects on colour computer monitors and analytical equipment such as scanning electron microscopes and mass spectrometers. When space is limited it may be necessary to shield the magnet to reduce its magnetic footprint. Passive shielding can be achieved by encasing either the magnet or the magnet room in ferromagnetic material. This iron shield can be both heavy and expensive. Alternatively, an active shield can be introduced by placing a second superconducting magnet outside the primary magnet and polarised in the opposite direction. The importance of this innovation to the whole body MRI market has been considerable, allowing magnets to be installed on sites throughout the world previously considered unsuitable or uneconomic and thereby contributing greatly to the overall market growth.

Shim set

As in high-resolution NMR spectroscopy, it is not sufficient just to have a magnetic field of a certain value in the centre of the magnet. The field also needs to be homogeneous over the volume being sampled. The requirements for imaging are not nearly as stringent as for high-resolution spectroscopy but as the volumes being sampled are generally much larger the demands on the magnet design are equally exacting. Shimming is the process of optimization of the

Figure 3 Actively shielded 2.0 T whole body superconducting magnet. Reproduced by permission of Oxford Magnet Technology, Oxford, UK.

magnetic field homogeneity and is a two-stage procedure. In the first stage, the homogeneity of the primary magnet field is optimized in the absence of a sample. Magnets will either have several cryoshim coils with windings of different designs inside the cryostat or a series of iron rods placed around the room temperature bore of the magnet to balance imperfections in the field. Generally, the cryoshim currents or passive iron shims need only be adjusted on installation and can thereafter be left unless the magnetic environment changes through, for example, building work.

However, MRI subjects also introduce their own inhomogeneities into the magnetic field as tissue has a different magnetic susceptibility than air. These sample-induced field disturbances can be partially removed by the active shims. Small bore and clinical research instruments will typically include an active shim set with perhaps a dozen shim windings. Adjustment of the current in each coil to optimize

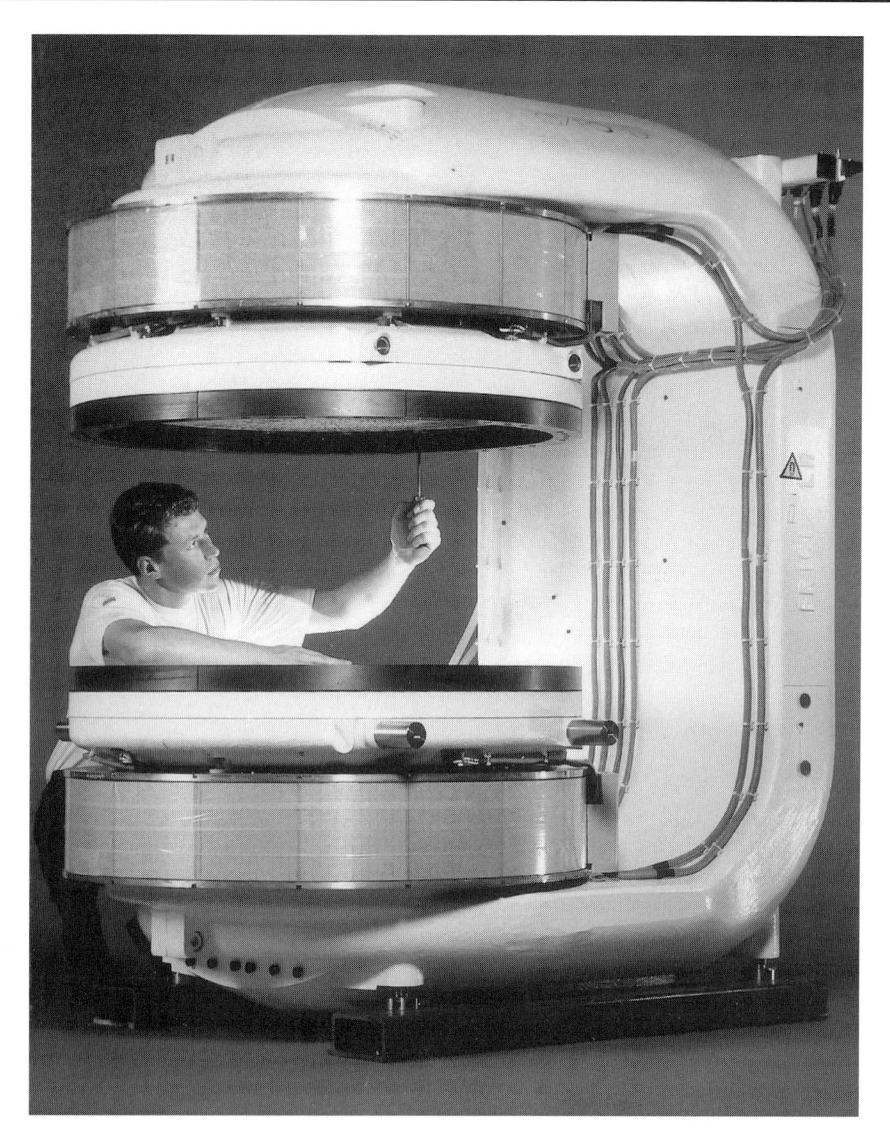

Figure 4 Open access 0.24 T resistive electromagnet without cladding. Reproduced by permission of Oxford Magnet Technology, Oxford, UK.

the magnetic field homogeneity of the sample may be done by hand or by using a simplex minimization routine. Alternatively, one may first map the field inhomogeneities using an imaging method and then calculate the currents necessary to counteract the inhomogeneity in the sample. Many clinical scanners do not have active shim sets but rely solely on DC currents through the gradient set to shim in the *X*, *Y* and *Z* directions.

Magnetic field gradients

Among the most critical components of an imaging system are the pulsed field gradients used to encode the images. Here, the characteristics that contribute to high quality images are the spatial linearity of the induced gradient pulses over the volume of interest and the decay characteristics of the gradient pulse. In the simplest system a linear gradient may be induced in the *Z*-axis by passing a direct current of opposite polarity through a Maxwell pair of coils wound on cylindrical formers. The greater the current the larger the linear field gradient imposed on top of the primary magnetic field.

Gradient set

As described above it is possible to make a *Z* axis gradient set by winding a pair of circular coils onto a cylindrical former and passing a DC current through the coils such that the polarity is opposed. *X* and *Y* gradients can be formed using saddle coils. Today, most gradient sets are no longer wire coils wound onto formers but are streamline patterns etched into copper sheet or cut into a copper plated cylinder. These have the advantage that the fabrication of complex current paths is easier and, generally, they are more compact. There are a number of conflicting parameters that must be considered when designing gradient coils. The sensitivity of the coil (in $T m^{-1} A^{-1}$), the region of acceptable linearity, the physical dimensions, the impedance and the shielding characteristics (more on this below) must all be weighed in the light of the proposed application. The coils will usually be embedded in epoxy resin to resist the torque generated when current is passed through the coils in the presence of the primary magnetic field. This torque would distort the shape of the coils and is the source of the drumming sound generated when the gradients are pulsed. Water cooling of the gradient set may be necessary for demanding applications with low field of view, thin slices and high duty cycle.

Amplifiers

The gradient pulse strength will be directly proportional to the current fed through the gradient coils. In modern clinical systems with echo planar imaging (EPI) capability the gradient amplifiers may need to produce 600 A. However, even for small animal systems in which the gradient amplifiers are more typically in the range of 50 A, current fed into the coil will take a finite time to reach the plateau value. That is to say that the gradient pulse will not be an ideal square function but will instead be trapezoidal. The duration of this rise time will depend on the inductance of the coil (hence low inductance coils are favoured for their short rise times) and on the voltage of the gradient amplifiers. Some systems are now provided with a 'booster' to raise the voltage and shorten the rise times. This 'booster' is basically a capacitor bank that discharges during the main amplifier switch on, increasing the voltage to drive the current through the coils. However, for fast imaging experiments such as EPI it is still important to minimize inductance in the design of the gradient coils. The other important criterion in selecting gradient amplifiers is low noise characteristics.

Preemphasis and active shielding

When the gradients are pulsed, residual fields called eddy currents are induced in the cryostat and other metallic structures. These fields decay with time constants typically in the order of tens of milliseconds, but for eddy currents in the cold cryostat vessel wall they may be hundreds of milliseconds long. Eddy currents can have a devastating effect upon image quality unless countered. Increasing the distance between the gradient coil and the magnet bore can reduce them, but as this will reduce the space available for the MRI subject it is often not an option. Eddy currents can be compensated for by overdriving the gradient waveform with a current that will itself counter the effect of the eddy currents. However, adjustment of this 'preemphasis' of the gradient pulse can be a tedious business as there are often eddy currents decaying with several different time constants.

Another approach to preventing eddy currents distorting the images is to shield the primary coil with a secondary coil placed outside the primary coil and connected to it in series. The secondary coil is designed to null the pulsed gradient field of the primary coil everywhere external to the coils but to have minimum effect in the centre of the coil. This approach has been almost universally adopted.

Radiofrequency

As in high-resolution NMR, the nuclei in the MR imaging experiment, be they the water protons of the typical anatomical imaging experiment or other nuclei such as ^{19}F, ^{31}P or ^{23}Na, must first be excited. The requirements for amplitude and phase control of RF pulses are similar to those in high-resolution NMR spectrometers, though the addition of phase coherent frequency switching can be an advantage for multislice fast spin–echo experiments.

RF amplifiers

Clinical MR scanners used for fast imaging experiments may have up to 15 kW RF amplifiers. These high powers are necessary to reduce pulse duration in fast spin–echo imaging sequences. However, care must be taken that the amplifiers are linear otherwise the shaped pulses necessary for slice selection will be distorted and the slice profile degraded. Of course, many manufacturers are aware of this problem and compensate their pulse shapes for the known distortions induced by the RF amplifier so that the final pulse shape delivered to the RF coil is optimal. If slice profiles are inadequate it is always worth checking for non-linearity in the RF amplifiers.

RF probes

The alternating current generated by the RF amplifier is fed into a probe to create an alternating magnetic field at the Larmor frequency in the sample. There are a number of basic probe types each with their own advantages and disadvantages.

Surface coils The simplest type of RF coil is the surface coil. These usually consist of a single loop of wire and give high signal-to-noise ratio for surface structures due to the close coupling of the nuclei in the region of interest and the surface coil. They are used where high signal-to-noise ratio is of primary importance such as in localized spectroscopy experiments, functional imaging and experiments with nuclei other than the proton. The main disadvantage of the surface coil is the loss of signal intensity with distance from the coil, which results in signal intensity variation across the image and a limited field of view.

Volume coils Both the Alderman and Grant probe and the birdcage resonator use distributed capacitance to produce a relatively homogeneous RF field in the centre of the probe and hence uniformity of signal intensity across the image (**Figure 5**). Also, these coils lend themselves to operation in quadrature mode which brings a $\sqrt{2}$ increase in transmission efficiency and a corresponding $\sqrt{2}$ improvement in the signal-to-noise ratio upon reception. Volume and surface coil use can be combined so that the volume coil is used for transmission to produce uniform excitation across the MRI subject and the surface coil is used for reception to increase the signal-to-noise ratio. However, care must be taken that the two coils do not couple to each other, either by ensuring that their fields are orthogonal (geometric decoupling) or by employing active decoupling using additional electronic circuits.

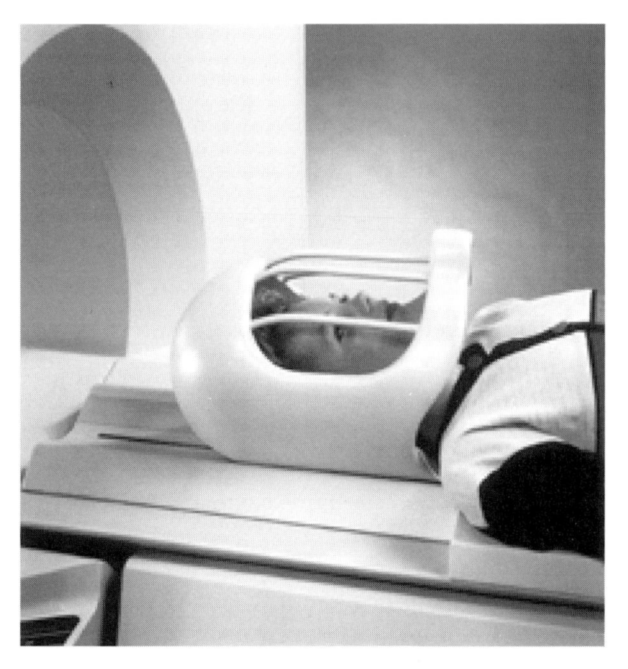

Figure 5 Clinical receive only volume coil. Reproduced by permission of Bruker Medical, Ettlingen, Germany.

Phased array coils In order to combine the signal-to-noise advantage of surface coils with the larger usable region obtained with volume coils, phased array coils can be used. These consist of an array of coils, each similar to a conventional surface coil, distributed over a surface. Each coil acts independently so that the required output signal can be obtained by combining the outputs from all or some of the elements. In order to reduce interaction between the adjacent coils, each one overlaps its immediate neighbours to minimize mutual inductance and is provided with its own preamplifier. The disadvantage of this approach is the relatively high price of the multiple amplifiers required.

Faraday cage

The antennas used to detect NMR signals will pick up extraneous signals from the environment unless they are shielded in some way. In a high-resolution NMR instrument the bore of the magnet acts as a waveguide, effectively shielding the RF coil from the outside world. However, in imaging systems the dimensions of the magnet bore are often of the same order as the wavelength of the RF frequency of interest and then it is necessary to introduce additional shielding measures. The most common solution is to enclose the entire magnet in a continuous sheet or mesh of copper or aluminium. All services to this Faraday cage must be electrically filtered to ensure they do not act as gateways for environmental RF.

Quality assurance

In addition to the physical hardware necessary to conduct an imaging experiment, every MR imaging lab will have a quality control process in place to identify spectrometer faults as they develop. In the clinical setting this will usually be included as part of the maintenance contract with the spectrometer manufacturer. Non-clinical labs will need to instigate their own procedures using standard phantoms. The parameters that need to be monitored are signal uniformity (RF coil homogeneity); signal-to-noise ratio; geometric linearity; spatial resolution; slice thickness; and relaxation time.

Patient monitoring

A description of ancillary equipment for the holding and positioning of animal and human patients is beyond the scope of this article. Similarly, monitoring equipment used for controlling animal or patient well-being such as pulse oximeters and blood pressure transducers will not be described. However, it is often necessary to monitor physiological parameters such as electrocardiogram (ECG) and/or respiration so that spectrometer acquisition can be synchronized with heart and/or abdominal motion. Many clinical systems have introduced optical transducers to convert the subject's ECG signal into optical signal for transfer via fibre optic lines to a monitoring device placed outside the magnet room. The advantage of the fibre optic line is that it cannot pick up extraneous RF and therefore does not need to be electrically filtered. Similarly, fibre optic and pneumatic devices are available for monitoring respiratory motion.

Computing

The same computers can be used for MRI applications as for high-resolution NMR. However, MRI systems can quickly generate large datasets requiring 2D and, these days, 3D Fourier transformations and, if there are animals or patients in the magnet, the operator will not want to wait for long periods during data reconstruction. Therefore, thought should be given to installing an adequate computer workstation to operate the spectrometer console. In addition, other workstations will be needed for off-line processing of images. The demands of multi-planar reformatting of 3D data, image segmentation, surgery planning and so on, can also be quite intensive and so these additional machines also need to be high-end machines.

Safety

Magnetic field

The static magnetic field of any NMR instrument poses a hazard to persons with surgical implants. The large bore and horizontal geometry of most MR superconducting scanners means that the stray field can emanate for several metres and provision must be made to prevent members of the public being exposed to a potentially lethal threat. Normally, this will consist of appropriate warning signs and restricted access to areas where the field is above 0.5 mT (5 gauss). However, in addition to the well-known dangers of static magnetic fields there is a potential hazard to patients and volunteers from peripheral nerve stimulation due to switched magnetic field gradients. This occurs when strong gradients are switched on very rapidly and results in muscular twitching and, possibly, pain. Clinical systems that can achieve such fast gradient switching should have a gradient supervision unit to ensure that they meet the requirements of the appropriate regulatory agencies, e.g. the Medical Devices Agency in the UK and the Food and Drug Administration in the US. The switching of magnetic field gradients also generates acoustic noise, which is a potential risk to both patients and staff.

RF

Pulse sequences that generate multiple 180° RF pulses can cause local tissue heating. Again the national regulatory agencies have laid down guidelines on RF power deposition in human subjects and a RF supervisor unit is necessary to ensure compliance.

Cryogens

Superconducting magnets may contain hundreds of litres of liquid helium. In the event of either a spontaneous or emergency quench of the main magnetic field, possibly due to someone being trapped against the magnet by an uncontrolled ferrous object, the energy stored in the superconducting coils of the magnet dumps into the cryogenic liquid. The expansion factor for liquid helium is 760:1 so a large amount of cryogen gas is released into the surrounding space in a very short time. Magnet manufacturers have designed their magnets to fail safe under these conditions. However, there is still the risk of asphyxiation as an opaque fog of helium and perhaps nitrogen gas replaces the air in the magnet room. All clinical systems and large bore animal scanners should be fitted with a quench vent to allow these gases to escape safely. In addition, clinical scanners will require an oxygen detector set to alarm should the oxygen level in the magnet room fall below safe levels.

Future trends

In the last few years MR functional imaging, in which activated regions of the brain can be visualized, and MR angiography, which visualizes flowing blood, have had a considerable impact on the specifications demanded of MR scanners. Both techniques benefit from high field strength and both rely on speed and hence gradient amplitude and switching speed. Combined with the inexorable drift to bigger and better magnets and magnetic gradient coils, there has been a move in the research MR field to follow clinical colleagues in demanding robust, easy to use scanners. In the pharmaceutical industry and in university laboratories it is often necessary to train relatively MR illiterate scientists and technicians in the routine operation of the scanner. Automated tuning, shimming and resonance frequency adjustment makes this task easier. The introduction of actively shielded magnets to the small bore end of the market will mean these systems can be installed on crowded sites and thus greatly expand the potential market. In short, the future looks bright for continued improvement and expansion in the MR scanner market.

See also: **Contrast Mechanisms in MRI; Magnetic Field Gradients in High Resolution NMR; MRI Applications, Biological; MRI Applications, Clinical; MRI Theory; NMR Spectrometers; NMR Microscopy; NMR Relaxation Rates; Radio Frequency Field Gradients, Theory.**

Further reading

Bushong SC (1996) *Magnetic Resonance Imaging: Physical and Biological Principles.* St Louis: Mosby.

Callaghan PT (1991) *Principles of Nuclear Magnetic Resonance Microscopy.* Oxford: Clarendon.

Chen C-N and Hoult DI (1989) *Biomedical Magnetic Resonance Technology.* Bristol and Philadelphia: Institute of Physics.

Fukushima E and Roeder SBW (1981) *Experimental Pulse NMR: a Nuts and Bolts Approach.* Reading, MA: Addison-Wesley.

Gadian DG (1995) *NMR and its Applications to Living Systems.* Oxford: Oxford University Press.

Lerski RA and de Certaines JD (1993) Performance assessment and quality control in MRI by Eurospin test objects and protocols. *Magnetic Resonance Imaging* **11**: 817-833.

Shellock FG and Kanal E (1996) *Magnetic Resonance Bioeffects, Safety, and Patient Management.* Philadelphia: Lippincott-Raven.

MRI of Oil/Water in Rocks

Geneviève Guillot, CNRS, Orsay, France

MAGNETIC RESONANCE
Applications

In recent years a large amount of basic and applied work on the application of NMR and MRI to the study of fluid distributions inside porous materials has appeared. With NMR one selectively observes one type of nucleus, by choosing the corresponding resonance frequency ω at a given static magnetic field intensity B_0 through the Larmor relationship

$$\omega = \gamma B_0 \qquad [1]$$

where γ is the gyromagnetic ratio of the examined nucleus. The proton, which is abundantly available in both water and oil, is the nucleus most frequently observed. This means that in contrast to other noninvasive visualization techniques NMR directly probes the fluid (liquid or gas) phases within opaque porous matrices. At the same time, the unique feature of NMR is that the signal is sensitive to the physicochemical environment of the fluid. Thus, characterization of the porous material itself is also possible.

Apart from the NMR signal intensity, which is proportional to the transverse magnetization, the main quantities of interest are the relaxation times, T_1 (longitudinal) and T_2 (transverse). It is through the modification of the relaxation properties of the fluid inside the solid phase that one can obtain physicochemical information on the porous matrix such as pore size, permeability or surface chemistry. Moreover, diffusion and flow, or more exactly fluid particle displacements, can be measured and visualized by NMR techniques using pulsed gradient techniques. The susceptibility contrast between the fluid and the solid phases, however, is usually very strong in rocks, and consequently a significant line-broadening is observed. Thus, spin-echo methods must be used, and in some cases more specialized solid-state methods are necessary.

From these principles, new instruments for the characterization of oil wells by NMR have been designed and are now routinely used to obtain rock porosity, water, oil and gas saturations, and other quantities of interest to the oil engineer. Applications have also appeared in other fields such as civil engineering (water in cement, bricks or clays), polymer engineering (solvent in solid polymers or polymer–polymer mixtures) or fluid mechanics. Specific methods or hardware are being developed for nonmedical applications of MRI and they may in turn find their way back into the medical field in the near future.

Relaxation properties of fluids in rocks

Surface effects

One usually observes faster relaxation rates for fluids inside a solid porous structure than for bulk fluids. This can be described as a surface relaxation effect, two or three fluid molecular layers having a specific relaxation rate much shorter than the bulk value. The origin of this shorter relaxation rate, for most mineral materials like rocks, is the presence of paramagnetic centres (usually iron). It is also considered that a reduction of molecular mobility or orientation could play a role. Whatever the origin of the surface relaxation, it can be shown that, under conditions of fast exchange, the relaxation rate measured for the fluid inside the pore space is proportional to the surface to volume ratio S/V, i.e. is inversely proportional to a characteristic pore size. The proportionality constant is the surface relaxation strength ρ characteristic of the solid–liquid pair under consideration, its order of magnitude for water in sandstone is 8×10^{-4} cm s^{-1}.

However, in many materials, pore sizes range over several orders of magnitude (from nanometres to hundreds of micrometres), and the experimental relaxation curves present a strong deviation from a monoexponential decay. A first approach is to use a stretched-exponential law to describe the relaxation curve, which has the advantage that a single relaxation parameter is obtained. Another approach is to calculate a relaxation time distribution from the relaxation curve by Laplace inversion; this is a mathematical task that presents some difficulties (the solution is not unique), but the inclusion of a regularization term, which is equivalent to favouring artificially smooth distributions, allows one to obtain reproducible results. With the assumption of fast exchange in each pore and slow exchange between different pores, the relaxation time distribution then gives the pore size distribution directly; this relationship is theoretically valid under the condition of uniformly distributed surface relaxation properties. The value of ρ must be obtained independently, usually by the use of mercury porosimetry.

Susceptibility contrast

The surface mechanism affects both T_1 and T_2. Another microscopic mechanism influences the apparent transverse relaxation and has important consequences in the methodology. This is the susceptibility difference between the fluid and the solid matrix. This difference creates an inhomogeneous magnetic field inside the fluid, and thus a broader line width. The main consequence is that it is almost always necessary to use spin-echo methods to observe fluids in a porous matrix.

Moreover, fluid molecular diffusion inside the field inhomogeneities causes decay of the transverse magnetization. The phenomenon cannot easily be described analytically, owing to the random character of molecular diffusion and the geometric complexity of porous media. Multi-echo sequences, such as CPMG (Carr Purcell Meiboon Gill), are employed to obtain the transverse relaxation curve, and it is considered that at interpulse spacing short enough (below 1 ms) and at low enough magnetic fields (below 0.2 T) the influence of field inhomogeneities is eliminated for many rock applications. One then recovers in liquids T_2 of the same order as T_1 within a factor 2 or so, that is to say equivalent physicochemical information. However, diffusion in gases being faster than in liquids, the apparent T_2 for gases in rocks can be shorter than for liquids. Because T_2 measurement times by CPMG are orders of magnitude faster than acquisition times for robust T_1 determinations, this method has become the standard protocol in the new logging instruments.

Laboratory applications

Methods

Standard imaging sequence The standard imaging sequence is the two-dimensional Fourier transform (2D FT) spin echo sequence, as described in **Figure 1**. It consists of a spin echo in coincidence with a gradient echo; the frequency encoding or read gradient pulses G_{read} and the phase encoding gradient pulses G_{cod} encode two orthogonal spatial directions; slice selection is obtained by the application of the gradient G_{sel} along the third orthogonal direction. The resolution within the image plane, or the voxel size δr, is fixed by the maximum applied gradient intensity G, and by the time duration of the gradient pulse T, through

$$\delta r = \frac{2\pi}{k} \qquad \text{with } k = \gamma GT \qquad [2]$$

The wave vector k represents the maximum length explored in the reciprocal space of the image. With gradient intensities G in the 20–100 mT m^{-1} range, k can be of order 10^4–10^5 m^{-1}, or equivalently δr can

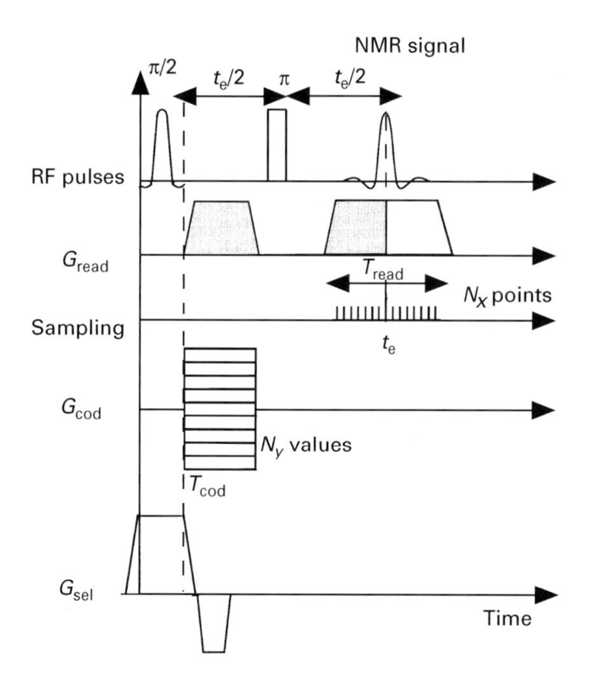

Figure 1 The two-dimensional Fourier-transform spin echo NMR imaging sequence. A $\pi/2$ radiofrequency pulse flips the magnetization into the transverse plane, where it is refocused into a spin echo at the echo time t_e by a π refocusing pulse applied $t_e/2$ after the first pulse. Spatial encoding is obtained (1) by using a shaped $\pi/2$ pulse, and simultaneously applying a selective gradient pulse G_{sel} to define a slice within the object; (2) by applying two read gradient pulses so as to form a gradient echo in coincidence with the spin echo, and by sampling N_{read} data points within the time T_{read} in the presence of the gradient G_{read}; (3) by repeating the acquisition for N_{cod} different gradient values applied during the time T_{cod} with a maximum amplitude G_{cod}. One then computes the two-dimensional Fourier transform of the resulting $N_{read} \times N_{cod}$ data points in order to obtain a two-dimensional image. The products $G_{read} \times T_{read}$ and $G_{cod} \times T_{cod}$ are chosen to achieve the desired resolution within the image plane (see text).

be a few hundred μm. Use of longer pulse gradients has a limited efficacy for resolution improvement in the case of heterogeneous porous media with susceptibility broadening, corresponding to a short lifetime of the NMR signal. Thus, δr is usually much larger than typical pore sizes in rocks. This also means that MRI will give images at a macroscopic scale of fluid distributions.

More complex imaging sequences The 2D FT sequence can be extended to three-dimensional imaging by using the phase encoding scheme instead of selection on the third axis. Chemical shift information can be extracted by adding a complementary chemical shift dimension; however, this procedure has rarely been used in practice for two main reasons: (1) four-dimensional data acquisition requires a prohibitive duration, and (2) the susceptibility effect spreads the spectra to the extent that the method is irrelevant in many practical situations. A more frequent approach is to extract 'water-only' and 'oil-only' images, with the simplification that the two chemical species are considered to give only single lines, and to use special protocols to eliminate local field inhomogeneities due to susceptibility as much as possible. This can be done only on rocks 'clean' enough and at magnetic field strengths above 1–2 T. Other relevant physicochemical information can be extracted by relaxation time imaging: the methods used are standard relaxation time measurement sequences combined with 2D FT or 3D FT imaging sequences.

Resolution: choice of magnetic field, different methods A usual rule in MRI is that a better resolution can be achieved at higher field intensity by an improvement of the signal-to-noise ratio. In the MRI of heterogeneous media, one must carefully examine the validity of that rule, since the spatial resolution is intrinsically limited by the line broadening due to the susceptibility contrast, which can be overcome only by increasing the gradient intensity. Since the susceptibility-induced field inhomogeneities are proportional to the magnetic field strength, the resolution achieved will be a compromise between the gradient intensity available from the instrument and the signal-to-noise ratio available. Orders of magnitude for the susceptibility internal gradients can be from 100 mT m^{-1} (pores of 100 μm in a 1 T field) up to a few T m^{-1} (pores of 1 μm in a 0.1 T field), comparable or much higher than the gradients available on large-scale imaging systems.

Other methods, which are usually considered as solid-state MRI methods because of their ability to obtain signals from samples with very short transverse relaxation times, are under development and offer very interesting possibilities for the exploration of small samples with very high controlled gradients. The first method uses fast oscillating gradients to obtain echoes at very short echo time; the second uses large static gradients and is called the STRAFI (Stray Field Imaging) method. The latter, which uses the very high gradients available in the stray fields of superconducting magnets (these can be as high as 10 to 100 T m^{-1}), probably offers the best possibility of going beyond the limit of large susceptibility gradients. However, these methods present the limitation that only objects of about 1 cm in size can be examined at the moment.

Laboratory measurements

Porosity, saturation The NMR signal amplitude gives a fairly straightforward measurement of porosity or fluid saturation when only one liquid (water or oil) is present in the porous sample, via simple calibration procedures such as reference measurements on bulk fluids in similar conditions. The relative accuracy achieved is usually of the order of 1% or better. Even so, extrapolation to zero time to eliminate relaxation weighting can be a difficult task in some iron-rich materials.

In diphasic cases (oil plus water), two simple techniques for the measurement of saturation have been suggested and used for laboratory applications. The first is to add to water a paramagnetic tracer, which effectively 'kills' the water signal by shortening its relaxation time below the observable limit; then only the oil signal is available. The second technique is to use NMR signals from other nuclei, such as deuterium (D_2O replacing H_2O) or fluorine ^{19}F (a fluorinated oil replacing the normal oil); the latter nucleus presents the advantage of resonating at a frequency only 0.94 times lower than the proton frequency. Chemical shift ^1H imaging should be the first choice technique, of course, and examples of chemical shift-resolved images have been obtained in various laboratories, at fields of about 2 T, of sandstone or dolomite samples saturated with water and dodecane (**Figure 2**). However, as discussed above, this technique can work only at static magnetic fields high enough to produce resolved water and oil resonance spectra, and with reasonably 'clean' samples in which line broadening does not cause their overlap. In addition, as detailed below, the analysis of relaxation spectra has also proved to yield fruitful information.

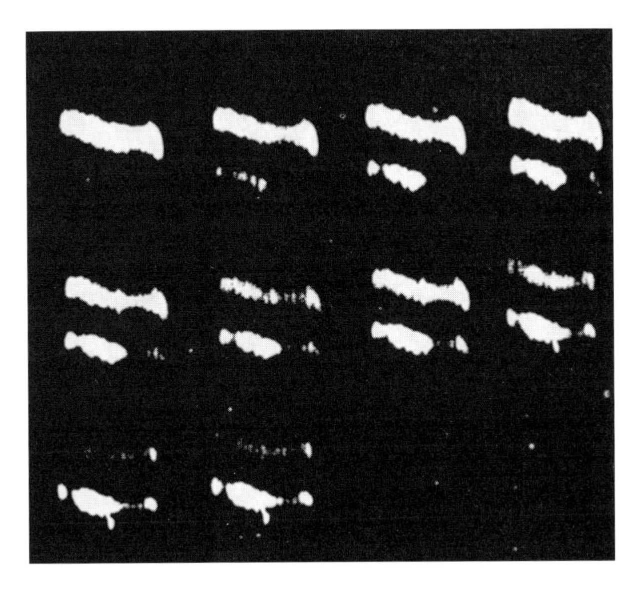

Figure 2 Chemical-shift imaging (CSI) in laboratory MRI of oil/water in rocks: time course CSI images of oil (upper line) displacement by water (lower line) in a Baker dolomite core sample, over 30 h, obtained at a 1.89 T magnetic field strength. The absolute intensities are not normalized from one image to the other, thus the change in the ratio of oil and water intensities with time (from left to right) is the meaningful parameter. The oil signal is initially (upper left) more intense, but a uniform decrease in the oil signal and increase in the water signal with time is observed. Reproduced with permission from Majors PD, Smith JL, Kovarik FS and Fukushima E (1990) *Journal of Magnetic Resonance* **89**: 470–478.

Information obtained from relaxation times The different physicochemical phenomena that influence relaxation times should be taken into account with some care when examining NMR images. At the same time, they can be exploited as specific contrast mechanisms.

The general theoretical picture described above relating pore size distribution and relaxation time spectra works satisfactorily for solid materials of reasonably uniform surface chemistry, such as many model porous systems (glass bead or particle packs), and most sandstones. A number of laboratory studies have used it to deduce pore size distributions from longitudinal or transverse decay curves in saturated porous systems. The fact that NMR and mercury porosimetry, which measure respectively the accessible surface and the throat dimensions, give comparable pore size distributions can be explained by the regular geometry of these systems. One should also mention that an empirical correlation between hydraulic permeability and some representative relaxation time value have been observed to be more or less satisfactory. In other rocks with more irregular geometry, the relationship between throats and pore dimensions does not hold systematically.

Changes in fluid arrangement with saturation have been followed, for example, in drying or centrifugation experiments. The displaced water tends to occupy smaller and smaller pores as its saturation decreases, and the corresponding relaxation time spectrum is generally observed to be displaced to lower values. Light oils present lower surface relaxation strengths than water in many rocks, presumably because of the natural water-wet character of the rocks. As in the drying experiments, saturation changes in immiscible situations (water plus oil) are most apparent on the water part of the relaxation spectrum, which tends to be displaced to lower values as water is displaced out, while the oil part is generally less affected.

Surface wettability also has an influence on the surface relaxation process: hydrophobic treatments of originally water-wet surfaces, by grafting of organic chains or by coating of surfactant layers, are known to increase the water-proton relaxation times. Images weighted in wettability have thus been obtained from T_1-weighted images in water-

Figure 3 Imaging of wettability contrast: images of water-saturated Fontainebleau sandstone samples of similar porosity (15%) and permeability, but with different surface treatments, obtained at a 0.1 T magnetic field strength. The right-hand sample is without treatment and naturally water-wet, and the left-hand sample was rendered oil-wet by chemical grafting of a silane chain. The image was acquired with a repetition time of 1 s, longer than the T_1 of the water-wet sample but shorter than the T_1 of the oil-wet sample; the resulting contrast due to different T_1-weighting by a factor 2 is much higher than the image signal-to-noise. Reproduced with permission from Guillot G, Chardaire-Rivière C, Bobroff S, Le Roux A, Roussel JC and Cuiec L (1994) *Magnetic Resonance Imaging* **12**: 365.

saturated rocks with different surface treatments (**Figure 3**). Moreover in mixed saturation (water plus oil) states the microscopic fluid arrangement depends on the surface wettability, and modifies the contribution to surface relaxation. Thus, NMR indices of wettability have been suggested from the shift of the water part of the relaxation spectrum at variable saturation. These indices are reasonably correlated to more traditional measurements of wettability properties. Fluid arrangement with respect to the solid surface has also been observed to influence the transverse relaxation for the wetting fluid via the susceptibility effect: indeed, the wetting fluid is in the vicinity of both a solid interface and the interface with the other fluid, while for the nonwetting fluid susceptibility effects play a role only on one fluid–fluid interface.

Another example of NMR relaxation weighting is the MRI study of mud filtration by rocks. Mud suspensions are used in oil-well drilling and their invasion into the surrounding rock is of importance for petroleum engineers. Water relaxation is faster in the presence of the mud particles, owing to their large surface area. Thus, the building of filtration cake has been followed quantitatively by MRI, as well as depth filtration of clay in natural rocks.

In many other potential application fields, the heterogeneous nature of the materials or their short transverse relaxation times cause similar difficulties in the collection of MRI images. Two strategies can be used. The first is to examine samples of realistic size (10–20 cm) at a moderate resolution, of the order of 1 mm, if T_2 values are long enough, typically longer than a few milliseconds: low-field equipment allows the collection of such images in many heterogeneous cases. When a finer resolution is necessary, other methods or specific equipment should be used.

For long enough transverse relaxation, images have been obtained by conventional liquid-state MRI sequences in different systems. From images of a solvent in a polymer matrix, quantitative measurements of solvent diffusion and possibly of matrix swelling have been performed in several systems, such as water–epoxy, water–nylon, methanol or chloroform–poly(methylmethacrylate). Elastomers are another example of samples with long enough T_2 and for which conventional MRI gives effective detailed information: the presence of voids in ill-cured elastomers is a spectacular source of contrast (corresponding to susceptibility defects), which can disappear with curing treatment. For building materials such as limestone and sandstone, the situation is comparable to that of oil-bearing rocks, and drying experiments have been monitored quantitatively. Similarly, the hardening of cement pastes is related quantitatively to the evolution of the water signal and of its longitudinal relaxation time (**Figure 4**).

For other samples, more solid-like or specific techniques should be used. Multipulse line-narrowing methods are well adapted to the case of solid polymers, such as adamantane, poly(methylmethacrylate) and polyacrylate. Fast gradient switching has been used to obtain one-dimensional images of water or solvent distribution in zeolite powders, with T_2 smaller than 1 ms. Moisture in building materials can cause spectacular damage and some groups have developed specific NMR instrumentation for moisture profile measurement at 1 mm resolution by point-to-point acquisition in bricks and mortars; these building materials are of very fine porosity and of an iron content (a few per cent) prohibitive for liquid NMR with conventional systems. But it is probably the STRAFI technique that will allow the finest resolution in solids to be achieved and that presents the highest efficiency for overcoming susceptibility broadening in heterogeneous materials.

Flow and diffusion

Methods The simplest and most straightforward method for flow imaging inside porous materials at a

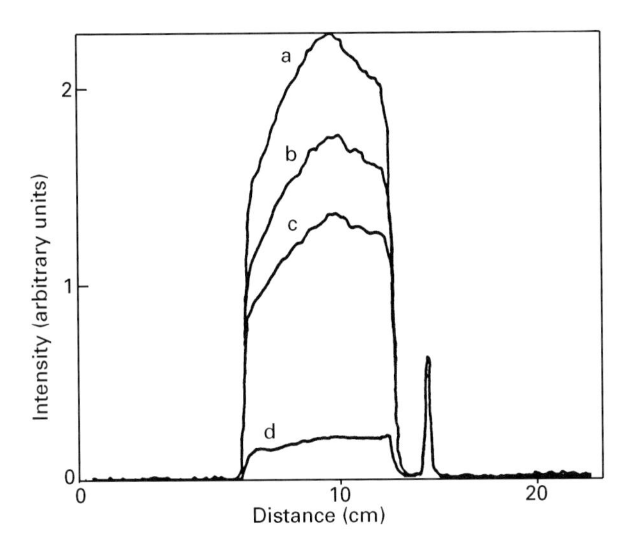

Figure 4 Thickening of a white cement paste monitored by MRI: time evolution of 1D FT images of water in white cement obtained at 0.1 T over 4 h (curve a, 1 h; b, 2 h; c, 2.5 h; d, 4 h). The acquisition duration of each profile is a few seconds; the sharp peak on the right corresponds to a water reference sample; during cement thickening, this peak maintains the same intensity, while the signal from the cement paste decreases, corresponding to the progressive immobilization of water as solid hydrates and to the shortening of the remaining liquid water T_2. Reproduced with permission from Guillot G and Dupas A (1994) In: Colombet P and Grimmer AR (eds) *Applications of NMR Spectroscopy to Cement Science*, p 313. Amsterdam: Gordon and Breach.

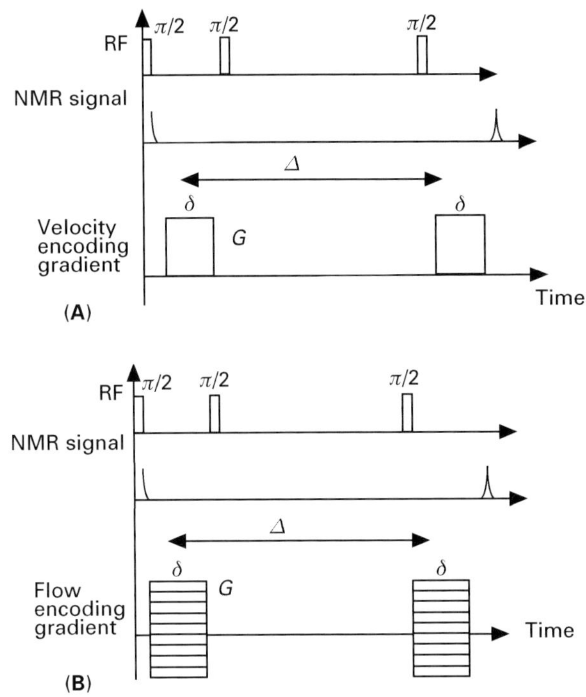

Figure 5 Sequences using stimulated echoes for flow velocity imaging (A) and for the measurement of the displacement distribution function without imaging (B). The use of stimulated echoes often proves very convenient since the NMR signal can be observed for longer delays Δ, limited only by the T_1 value. The pair of gradient pulses, when exactly matched, produce no phase shift for nonmoving spins, and for moving spins they induce a phase shift $\Delta\phi$ equal to the product of the wavevector $q = \gamma G\delta$, by the displacement $(r(\Delta) - r(0))$. For a uniform velocity field v, $(r(\Delta) - r(0)) = \Delta v$ everywhere in space and v can be found from the phase-shift measurement at a given value of q (A). In more complex flow situations, the full displacement distribution can be obtained from a Fourier transform analysis of data acquired with incremented G, and thus q, values (B).

macroscopic scale is to use paramagnetic solutions, which act as contrast agents just as in clinical applications of MRI. More refined techniques have been used and studies are currently in progress to study flow and diffusion. Their basis is generally the pulse field gradient-stimulated spin echo sequence. Susceptibility differences also create problems and can lead to an undervaluation of diffusion coefficients; multi-echo versions of this sequence derived from the CPMG echo train have been shown to compensate the susceptibility artefacts to a large extent.

Imaging of velocity is also possible at a macroscopic resolution. An appropriate gradient pulse pair causes a phase shift of the NMR signal. This phase shift is proportional to velocity (if all spins within each voxel move with the same velocity), via a controlled factor equal to the product of the wave vector q and the time delay Δ between the gradient pulses (see **Figure 5A**). One can combine this velocity encoding gradient with the imaging gradient pulses, so as to compute a velocity image from the phase-shift image.

Another interesting and powerful approach for obtaining detailed information on the flow field inside porous materials (without imaging) is to study the displacement distribution function, which can be obtained by Fourier transformation of the NMR signal acquired for incremented values of the wave vector q (**Figure 5B**). Of course it is also possible (but time-consuming) to make images of the displacement distribution. For these methods, the flow should be steady during the long data acquisition under the different gradient conditions, but this is a very realistic condition considering the low values of the Reynolds numbers normally encountered in the study of flow in porous media.

Results Some groups have mapped fluid velocity inside water-filled rocks, either sandstone or limestone samples, using the phase-shift method. The measured velocities have the expected order of magnitude and some reasonable correlation with rock porosity has been observed. However, one should be aware that in these studies the spatial

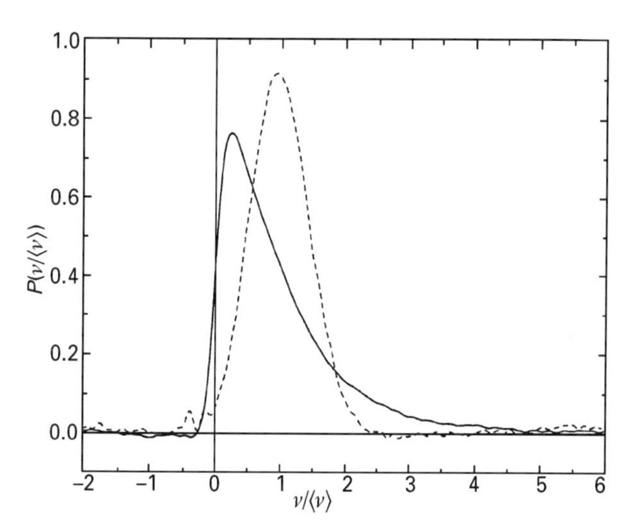

Figure 6 Velocity probability distribution $P(v/\langle v\rangle)$ as a function of $v/\langle v\rangle$, where $\langle v\rangle$ is the average velocity, for water flowing in a glass bead pack. For short Δ (solid line), corresponding to a displacement $\Delta\langle v\rangle = 0.08d$, where d is the bead diameter, the distribution can be described by an exponential decay, in agreement with the expected Stokes behaviour; here $d = 800$ µm, and $\Delta = 19$ ms. For long Δ (dashed line), corresponding to a displacement $\Delta\langle v\rangle = 7.3d$, the distribution can be described by a Gaussian law in agreement with the classical models of hydrodynamic dispersion in a porous media: here $d = 80$ µm, and $\Delta = 103$ ms. Courtesy of Lebon L, Leblond J and Hulin J-P PMMH, CNRS UMR 7636, ES-PCI, Paris, France.

resolution is usually larger than the pore sizes, i.e. larger than the scale of velocity variations, so the measured phase shift is related to some velocity value averaged in a complex way over the microscopic velocity distribution, in both space and time.

Other groups have measured, without imaging, the displacement distribution function for water in bead packs, and for water and oil in sandstone, and have examined its dependence on the time delay Δ. At short Δ, or at a mean displacement smaller than the pore size, the displacement distribution corresponds to the velocity distribution and has an exponential shape, in good agreement with numerical simulations of Stokes flow. At long Δ, or for a mean displacement larger than a few pore sizes, the distributions have more a Gaussian shape that reflects the hydrodynamic dispersion of the fluid particles in the velocity field (**Figure 6**).

Oil well logging

Logging tools

A new generation of logging tools for measurement in the severe conditions encountered in oil formations has appeared in recent years. This has been made possible by the use of permanent magnets, such as samarium–cobalt alloys with Curie temperatures above 200°C. The sample of interest is the rock formation surrounding the tool, in contrast to the usual laboratory NMR situation where the sample of interest is inside the magnet and the RF probe, and different designs for the static magnetic field and for the RF probe, adapted to this specific geometry, have been developed. Working static magnetic fields of 10–100 mT can be obtained, with sensitive volumes of toroidal shape that are typically of 20–1000 cm^3, at a distance of a few centimetres away from the borehole wall. From their specific designs, the static magnetic field for most logging tools is in fact a field gradient of about 100 mT m^{-1}. Another severe constraint is that the tool must move continuously in the bore, at speeds of several cm s^{-1}. Under these conditions, chemical shift is not obsservable, and the only NMR pulse sequence fit for use is the CPMG echo train. Typical logs consist of one-dimensional images (along the bore axis) of the NMR signal intensity and of the relaxation time distribution extracted from the CPMG measurement, at a resolution of about 0.2–1 m; from these data, rock porosity and various information on recoverable oil can be computed. **Figure 7** shows as an example a prototype logging tool that was designed to attain a finer spatial resolution of 2 cm.

Logging applications

The NMR signal intensity provides a measurement of the fluid-filled porosity. However, water in clay or shale has an apparent T_2 that is too short to be visible to the logging tools and the NMR signal comes mainly from water in larger pores and from oil. From the relaxation time distribution, an estimate of the movable fluid, called the free fluid index (FFI), is obtained by choosing a cutoff value, from *a priori* knowledge of the formation lithology. FFI is the proportion of fluid with relaxation times higher than this chosen value, or the proportion of fluid within pores larger than a given cutoff size. Laboratory measurements of centrifugeable water have shown a reasonable correlation with FFI measurements derived from NMR logs. An empirical estimate of permeability is also often calculated from FFI.

The differentiation of water from oil is based on the same general trends as presented above. The relaxation time spectrum can be separated into two parts: the water relaxation times are the shorter and change with the saturation state, whereas the light oils have the longer relaxation times, which are not strongly modified by confinement in the rock. It is also possible to detect the presence of gas, since at

Figure 7 Example of a NMR logging tool sensor prototype working at 4 MHz. The main magnetic field is produced by permanent magnets, plus V-shaped polar pieces to concentrate the magnet induction in the central plane, so as to define the measurement zone in the central area, with a spatial resolution along the tool axis of 2 cm. As the tool moves along the bore wall, the spins are prepolarized by the magnet induction before they arrive in the measurement zone; thus the standard logging speed can be as high as 15 cm s^{-1}. One can see the V-shaped main polar pieces between two cobalt–samarium permanent magnets, the RF antenna in the V space and the tuning capacitor. Courtesy of Locatelli M. LETI CEA-Technologies Avancées DSYS, Grenoble, France.

the high pressures in the reservoirs the corresponding density gives an NMR signal intensity only about 5 times lower than the signal from liquid oil or water. Gas can be distinguished from the other fluids through its specific relaxation behaviour: T_1 is a few seconds and T_2 is strongly influenced by diffusion effects in susceptibility-induced field inhomogeneities because of the higher diffusion coefficient of the gas. It has been shown that the amount of gas-filled porosity can be measured from T_2 acquisitions differently weighted in diffusion by changing the interpulse spacing in the CPMG sequence.

List of symbols

B_0 = applied static magnetic field strength; G = gradient pulse (G_{read}, frequency encoding; G_{cod}, phase encoding; G_{sel}, selective); k = reciprocal-space wave vector; q = wave vector; T = pulse duration; T_1 = longitudinal (spin–lattice) relaxation time; T_2 = transverse (spin–spin relaxation time); t_e = echo time; γ = gyromagnetic ratio; δ = pulse duration; δr = voxel size; Δ = delay between pulses; ρ = surface relaxation strength (relaxation for the fluid owing to relaxation centres on the solid surface); ω = resonance frequency.

See also: **Contrast Mechanisms in MRI; Diffusion Studied Using NMR Spectroscopy; Geology and Mineralogy, Applications of Atomic Spectroscopy; MRI Applications, Clinical Flow Studies; MRI Instrumentation; MRI Theory; MRI Using Stray Fields; NMR Microscopy; NMR of Solids; NMR Principles; NMR Pulse Sequences; NMR Relaxation Rates; Relaxometers; Solid State NMR, Methods.**

Further reading

Borgia GC (ed) (1991, 1994, 1996) *Proceedings of the International Meetings on Recent Advances in MR Applications to Porous Media:* Special Issues of *Magnetic Resonance Imaging* Vols **9**, **12**, **14**.

Brownstein KR and Tarr CE (1979) Importance of classical diffusion in NMR studies of water in biological cells. *Physical Review A* **19**: 2446–2453.

Callaghan PT (1993) *Principles of Nuclear Magnetic Resonance Microscopy*. Oxford: Oxford University Press.

Edelstein WA, Vinegar HJ and Tutunjian PN (1988) NMR imaging for core analysis. *Society of Petroleum Engineers* 18272.

Kleinberg RL (1996) Well logging. In: Grant DM and Harris RK (eds) *Encyclopedia of Nuclear Magnetic Resonance*, Vol **8**, pp 4960–4969. Chichester: Wiley.

Special Issue of the *Log Analyst* on NMR Logging. November–December 1996.

Watson AT and Chang CT (1997) Characterizing porous media with NMR methods. *Progress in NMR Spectroscopy* **31**: 343–386.

MRI of Rigid Solids

See **Rigid Solids Studied Using MRI.**

MRI Theory

Ian R Young, Hammersmith Hospital, London, UK

MAGNETIC RESONANCE
Theory

Introduction

In essence the theory of nuclear magnetic resonance peculiar to magnetic resonance imaging (MRI) alone is very simple and can be simply summarized as being a specialist application of multidimensional Fourier transformation NMR, in which the various frequency axes are related to spatial ones by assuming that the gradient magnetic fields applied to encode space produce equivalent frequency variations. In practice, the situation is very much more complicated, and involves reviewing a number of different aspects of the data recovery process.

In reality, the most complex differences between small-scale high-resolution studies and those involving human subjects lie not so much in the spatial encoding process but in the interactions between the RF coils and the subject and in the desirable targets for human studies. These are not, to anything like the same extent as in high-resolution studies, dictated by the need for quantitative accuracy of the measurement of a multiplicity of chemical components but are, rather, driven by the requirement to highlight certain structures of the body with respect to others. Imaging in general, and MRI in particular, is driven predominantly by issues of contrast. Whole-body magnetic resonance spectroscopy (MRS) is, similarly, driven by rather different factors from those affecting normal work in small-bore high-field spectrometers. In many ways MRS is closer to MRI in terms of its strategies and problems and both are, in effect, considered in this article (the former by implication only).

Relative to normal spectroscopy, both MRI and MRS rely to a very much greater extent on comparisons of results from regions of tissue considered to be normal and those felt to include more or less severely diseased structures. Biological diversity ensures that the reproducibility of results from one subject to another will not be as good as in most spectroscopic studies. On the other hand, much of what is attributed to this cause is due to ill-considered research strategies and artefactual results, but there is an undeniable level of difference between individuals in animal species of all kinds so that the spread of results will always be greater than that obtained from small, passive samples. Sensate beings move in complex and more or less uncontrollable ways, have highly nonreproducible sizes and shapes and are hugely complex, so that practically all data obtained from them are contaminated by significant partial volume effects (which means that data sampled from a region of tissue contains components of multiple structures and a variety of different tissue types).

This article discusses the basic processes of spatial localization and the formation of images; it considers how the form of the signal-to-noise relationship is affected by the large, conducting load that the body represents, and how its movement affects the quality of the data obtained from studies. The issues surrounding the creation of contrast, which is the prime clinical desire of MRI strategies, are addressed in a separate article on contrast mechanisms (q.v.). This also discusses the development of a number of important artefacts, the formation of which is a derivation from the strategies outlined in this article. The reader may find it convenient to treat both as being closely related topics in *in vivo* NMR.

In all of what follows, it is assumed that the reader is familiar with conventional NMR theory as described elsewhere in this encyclopedia (see Further reading), as this article examines only the extensions needed to the theory and practice that are the basis of whole-body MR.

Spatial localization

The basic concept of imaging is very simple. Gabillard first suggested the use of magnetic field gradients as a means of identifying positional data in NMR. However, it required the development of Fourier-transform NMR before these ideas could be usefully applied to imaging in its standard form. Mansfield and Grannell pioneered the application of gradients in FT-NMR, with the aim of achieving the analogue of optical diffraction, with resolution of lattice plane dimensions, while Lauterbur was the first to publish a two-dimensional image of an identifiable object.

In principle, assuming that B_0 is completely homogenous, application of a gradient G_r (the component of a field varying along the r axis parallel to the Z axis (parallel to B_0)), results in a divergence of the spin resonant frequencies represented by

$$\omega_i = \gamma(B_0 + r_i G_r) \qquad [1]$$

where $r_i G_r$ is the magnitude of the gradient field at position r_i along the gradient. If the object is multidimensional (as in the human body), the signal observed at frequency ω_i (S_i) is derived from all the spins in the plane orthogonal to the gradient axis through r_i.

If we have a series of planes normal to r with respective uniform proton densities P_i at distances r_i (each generating a signal S_i) then, after demodulation, the signal from the whole object (S_{obj}) is given by

$$S_{\text{obj}} = \sum_{i=1}^{n} S_i = \sum_{i=1}^{n} A_i P_i \exp(-\mathrm{i}\gamma r_i G_r t) \qquad [2]$$

where A_i measures the extent of the object in the plane at r_i. In the limit, this becomes

$$S_{\text{obj}} = \int_r P_r A_r \exp(-\mathrm{i}\gamma_r G_r)\mathrm{d}r \qquad [3]$$

ignoring all relaxation effects.

The above argument can be developed in two, or all three, dimensions to yield, for the latter, a relation of the form:

$$S = \int_x \int_y \int_z \rho_{ijk} \exp\left(-\frac{\mathrm{TE}}{T_{2ijk}} + \mathrm{i}\gamma\mathrm{TE}\right.$$
$$\left.(G_x x_i + G_y y_j + G_z z_k)\right)\mathrm{d}x\,\mathrm{d}y\,\mathrm{d}z \qquad [4]$$

where ρ_{ijk} is the proton density of the (ijk) voxel, and T_{2ijk} is its spin–spin relation time constant. This recognizes that there will be, at least, monoexponential transverse relaxation effects in the data, and that their magnitude will be those developed at a time TE (the echo time, the impact of which will be discussed later), which is the time at which it is assumed all spin dephasing due to the applied gradient is zero during the data acquisition.

In practice, the gradients require to be applied at different times to achieve the desired three-dimensional encoding, since the concurrent application of more than one gradient identifies a single axis, which is that determined by the vector sum of the applied fields.

Spatial resolution

It is easier, at the beginning, to discuss how resolution is controlled by the spatial encoding process by considering a single-axis experiment (as in Equation [3]). In order to achieve a resolution of n points along the direction of observation, the Nyquist criterion demands that there is a $n/2$ Hz spread of frequencies across the field of view being studied. Thus, the gradient G_r, the distance r which spans the object, and time t for which the signal is observed, are related by

$$n\pi = \gamma G_r r t \qquad (\gamma \text{ in Hz/T}) \qquad [5]$$

During the period t at least n data samples are required to identify the individual frequencies.

Conventionally, all imaging procedures acquire data in the presence of a gradient, depending on Equation [5]. In spatially localized spectroscopy, data are frequently measured in the absence of any gradient field and this approach could also, in theory, be useful in microscopy experiments, though in this case to minimize the impact of diffusion.

Essentially, there are only two fundamental imaging strategies, though with a growing number of variants of each. In one, the spins are excited and data are recovered along a direction determined by the vector sum of the applied gradients until enough has been collected, after which the magnetization is allowed to recover before it is excited again and data are acquired with a different gradient vector direction. The other strategy exploits the property that, after excitation spins retain the phase relationships into which they have been placed by the application of a short pulsed gradient field. In a good field, the relationships are held sufficiently well as the system relaxes for another gradient to be applied for long enough, in what is usually an orthogonal direction, while data are recovered. Subsequently, after time for the magnetization to recover and be excited again, another, different, gradient pulse is applied and more data are obtained. The process is then repeated sufficiently often for a complete set of the information needed for a two-dimensional image to be obtained.

The former technique was that known as filtered back-projection, which, as originally applied, was a direct MR analogue of the original translate–rotate CT–X-ray scanner developed by Hounsfield. The first version of the other strategy, involving Fourier transformation in two or more directions, was proposed by Ernst and his colleagues in a form that ultimately proved much less useful than the spin warp method that has become the basis of the vast majority of clinical MRI.

The aim of every image recovery procedure is to obtain enough good information to fill all the locations of '*k*-space' (i.e. spatial frequency space) so that when the data processing has been completed there are no artefacts in the image due to missing or corrupt data.

Figure 1 shows the sequence form (A,B) and coverage of *k*-space (C) developed during back-projection imaging. The various acquisitions form the spokes of a wheel and are produced in a plane (say the *X*/*Y* plane used as an example here) by the vector sums of the two gradients $x(=R\cos\theta)$ and $y(=R\sin\theta)$. Sufficient acquisitions are made to cover *k*-space at the density required. Suppose that the target of the image acquisition is an $n \times n$ matrix. The number of acquisitions needed to cover *k*-space is then $\pi n/2$. As will be discussed below, the efficiency of data acquisitions that fill *k*-space along orthogonal coordinates is better (as these require n acquisitions only) and this is one factor contributing to the unpopularity of back-projection acquisition.

It will be noted from **Figure 1** that if data were to be acquired as shown in **Figure 1A**, then regular sampling would result in oversampling at the start of the acquisition (as the spin frequency spectrum is spreading relatively slowly), though it may be just adequate later (after the gradients have flattened out and the full set of spin frequencies has developed). This form of acquisition was used in the early days of MRI by sampling the data nonlinearly (to account both for delays in amplifier response to command signals and for eddy currents arising from the changing gradients). Conjugate symmetry was also used to permit a significant reduction of the number of acquisitions as they only had to sample data though 180°.

Back-projection suffers from major problems in poor fields owing to those fields resulting, in effect, in the angular misplacement of points in *k*-space (which reconstruct to give streaks of intensity variation leaving the edges of structures). The centre of *k*-space is oversampled (and has an appropriately improved signal-to-noise ratio) as sampling can be continued as long as there is useful signal, though this results in angular undersampling at great distances from the centre. Nevertheless, the approach is not without merit. It is relatively impervious to the effects of motion (for reasons too complex to discuss in detail here) and it permits very high-resolution imaging of a local region in the body without the accompanying problem of aliasing that affects the spin warp method.

The spin warp technique and its development have, however, been the methods upon which practically all subsequent workers have based their work. The concept is shown in **Figure 2**. The block diagram of the data acquisition process is given in **Figure 2A** and the *k*-space strategy is indicated in **Figure 2B**. The data acquisition gradient is the same for each data recovery but the dotted lines indicate the varying nature of the phase-encoding pulses. The initial inverted gradient (G_t) (of the data acquisition) has the same value of

$$\int_0^{T_\mathrm{p}} G_t \,\mathrm{d}t$$

(where T_p is the width of the pulse) as the first half of the longer lower amplitude acquisition gradient (G_a) during the presence of which data are recovered.

Thus

$$\int_0^{T_\mathrm{ip}} G_\mathrm{it} \mathrm{d}t = \int_0^{T_\mathrm{ap}} G_{at} \mathrm{d}t \qquad [6]$$

where T_ip is the duration of the inverted gradient pulse and T_ap is half the duration of the acquisition gradient pulse.

During the warp gradient, spins are dephased, but they are then refocused at the centre of the data acquisition so that data sampling (which can be linear, as the gradient is constant throughout the data acquisition period) occurs through the echo peak. Thus, *k*-space is fully sampled for each line, and conjugate symmetry (as is needed for the technique in **Figure 1A**) is not necessary. The method is generally more robust than that in **Figure 1**, and in practice back-projection is also generally now

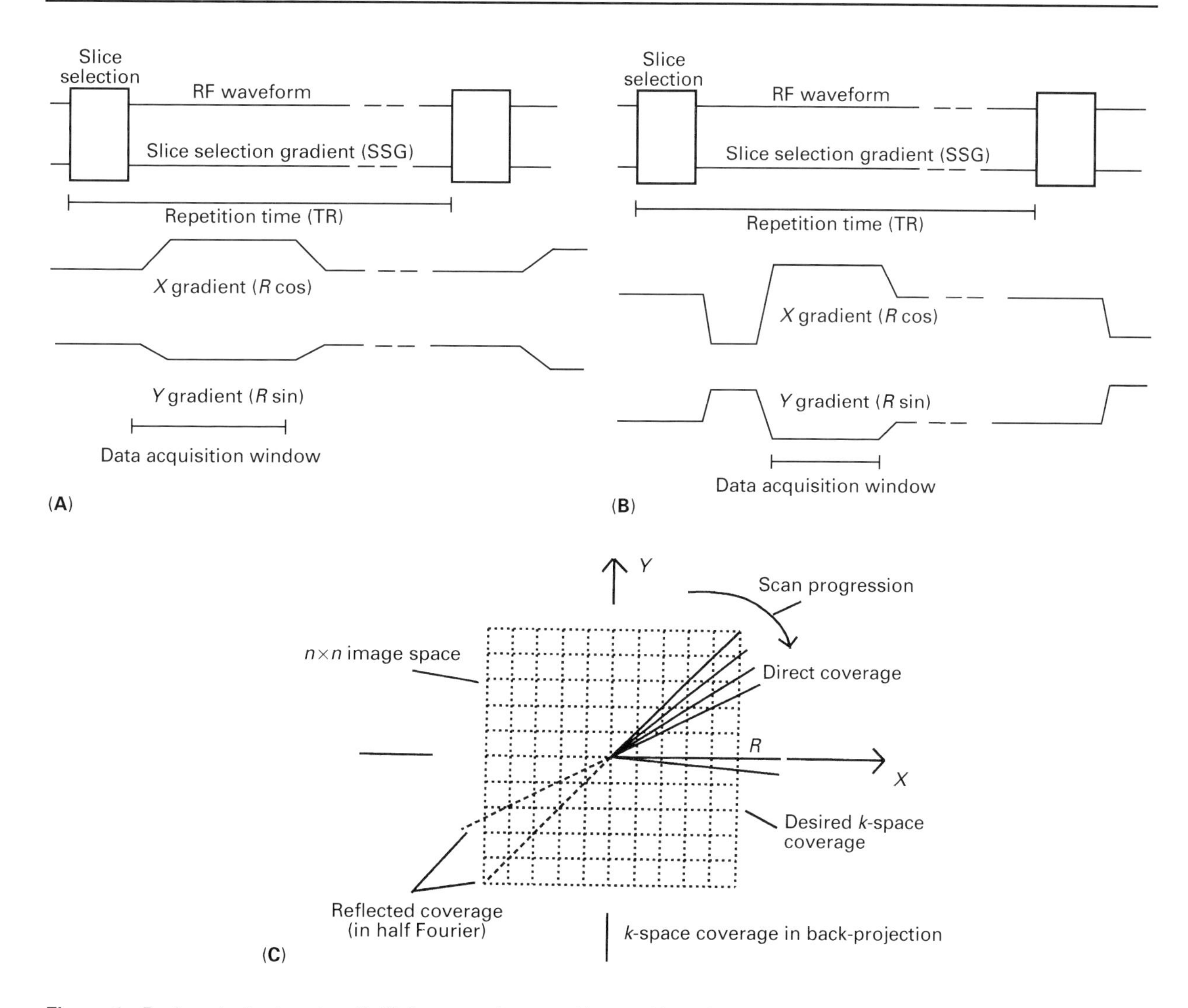

Figure 1 Back-projection imaging. (A, B) Sequence form used for acquiring a back-projection data set for an image in the *X*-*Y* plane. At this time the slice selection procedure is simply shown as a block. It will be described later in the text. The *X* and *Y* gradients are constrained to define a series of vectors (given by $\theta = \tan^{-1}(Y/X)$ of constant magnitude R $(= (x^2 + y^2)^{1/2})$: (A) shows the variant in the sequence used when conjugate symmetry is to be applied; (B) is the form of gradients used when an echo is to be formed during the flat regions of the gradients. (C) Coverage of *k*-space with back-projection. Workers reconstruct the data either using genuine back-projection algorithms (as in CT–X-ray) or by interpolating onto a two-dimensional grid, followed by a two-dimensional Fourier transformation.

implemented with echo formation (as in **Figure 2B**) except where it is desirable to acquire data as fast as is possible after excitation. The method in **Figure 2A** is a useful technique in the imaging of very short-T_2 proton moieties or nuclei such as sodium that also have short T_2 values.

In order to cover *k*-space completely, all its lines must be filled. Each line is obtained using a 'phase encoding' pulse (see **Figure 2A**), during which the spins precess at the frequencies dictated by the applied gradient. At the end of the gradient pulse, different groups of spins (isochromats) will have different phase relationships, which they retain in a perfect field along parallel lines even in the presence of another gradient applied orthogonal to the first. When enough phase-encoding pulses have been applied, each followed by a gradient in an orthogonal direction, the set of data generated will be given (ignoring relaxation and recovery effects) by

$$S_{x,y} = k\int_{0}^{m}\int_{-(n-1)}^{n} S_{p,t}\exp\left(\mathrm{i}\gamma(xG_x t + p\,\Delta\,G_y y\,T_\mathrm{p})\mathrm{d}p\,\mathrm{d}t\right) \quad [7]$$

where S_{xy} is the image dataset, assuming, in this case, that the readout direction is x, and the phase encode direction y. ΔG_y is the increment in the y gradient between samples; T_p is the phase encode pulse

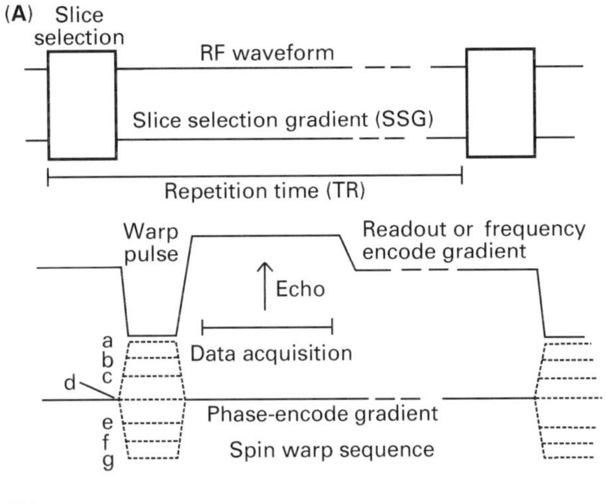

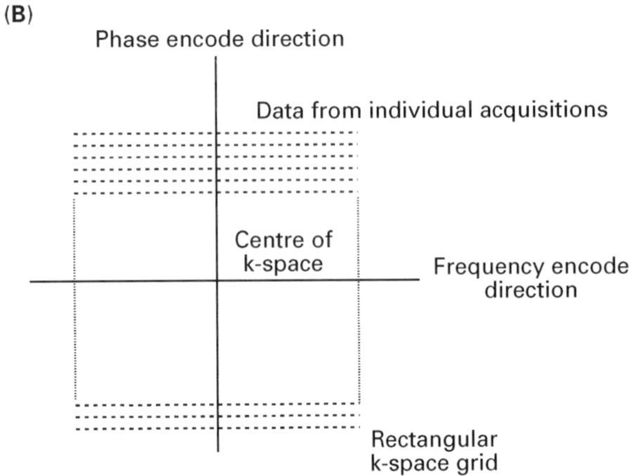

Figure 2 Spin warp imaging. (A) Sequence structure used in spin-warp imaging (again the slice selection component is shown as a block). The data acquisition gradient is fixed throughout the procedure; the phase encoding gradient is stepped uniformly from one extreme to the other, hence the difference from one excitation to the next. (B) The *k*-space average resulting from the spin-warp sequence.

duration. $S_{p,t}$ is the signal recovered at time t during recovery of the pth line of k-space.

A two-dimensional Fourier transform applied to the data results in a set of amplitudes associated with the set of positions x, y.

The process can be extended to three dimensions by adding another phase-encoding step in the third orthogonal direction. This is altered after the complete set of phase-encode steps in the second direction has been obtained. In order to obtain sufficient data to generate a volume data set with $n_i \times n_j \times n_k$ voxels, $n_j \times n_k$ acquisitions of the n_i points in the readout direction are needed. If $n_j = n_k = 128$ (a relatively modest resolution target), 16384 acquisitions are required. Even if the acquisitions are repeated at 20 ms intervals, the recovery of the data takes around 5.4 minutes. Extra time, to allow for greater recovery of signal after excitation, can quickly result in very extended, and practically unacceptable, durations. Signal-to-noise ratio in such imaging procedures can be very good, but the risk of patient movement, or even refusal to proceed, becomes much greater.

Slice selection

In most instances, single planes of data ('slices') are recovered during imaging. Slice selection is performed by selective excitation, in which an RF pulse is applied at the same time as a gradient. This selects a slice orthogonal to the direction of that gradient.

The process is illustrated in **Figure 3**. The RF waveform is modulated by a computer-generated pulse profile to give a burst of RF frequencies that are as uniform in amplitude as possible and that have minimal components outside the bandwidth wanted. The pulse profile ($B_1(t)$) results in a frequency spectrum $f_{B_1}(t)$ which, in turn, results in a range of magnetization flip angles α_t given by

$$\alpha_t = \gamma K_{B_1} f_{B_1}(t) t_p \qquad [8]$$

where K_{B_1} is a profile relating RF field intensity and the spectral content, and t_p is the pulse duration.

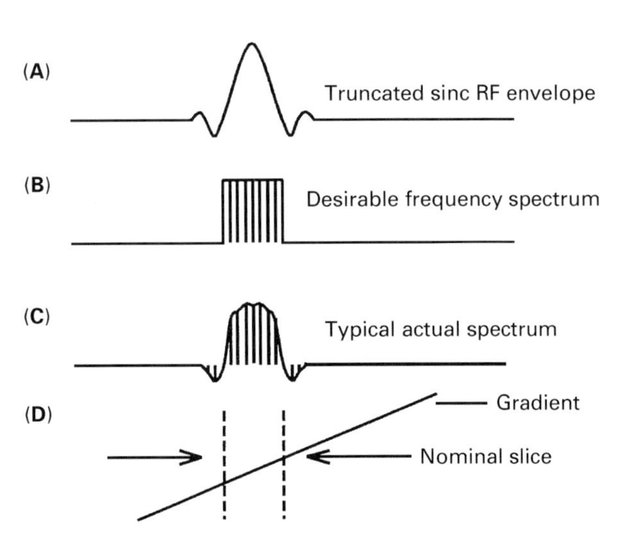

Figure 3 Slice selection process. (A) Envelope of the RF pulse (typical of the simpler pulses used). (B) Desirable burst of frequencies. (C) More typically achieved burst of pulses. The frequencies shown as negative are, of course, of opposite phase to those in the main block of frequencies. (D) Applied gradient with slice selected marked on it relative to the components of (A) to (C). The slice is shown as perfect, but actual performance is generally significantly poorer.

Even as the B_1 irradiation is present, and before the gradient is completely removed, the excited spins start dephasing relative to each other and, at the completion of the process, little or no signal may be obtained. Another, inverted, gradient is then used to refocus the spins and reform the signal.

The width of the slice is determined by the gradient amplitude (G, conventionally measured in mT m^{-1}) and is given by

$$W = \frac{f_{B_1}(t)}{\gamma G} \quad [9]$$

Initially RF pulses such as sinc pulses (with trapezoidal gradients) or high-order sinc pulses (with sinusoidal gradients) were used. Now much more sophisticated complex pulses (i.e. containing real and imaginary profile information) are used to obtain better and more exact slice selection.

Because tissue relaxation times are relatively long, scanning times are typically quite extended, with much apparently wasted time. Crooks and his colleagues followed an earlier suggestion and showed how multiple slices could be interleaved between each other, and acquired at the same time, by varying the operating frequency of the machine at successive acquisitions. Coincidentally, they also showed how to exploit the relatively long T_2 relaxation time of many tissues to recover more than one image from each slice excitation.

Signal-to-noise ratio (SNR)

Although the basic formulation for signal-to-noise ratio is the same as in classic NMR, there are very significant differences in what actually happens. These arise from the fact that the body loading on the coils (which can generally be ignored in typical high-field (small-bore) systems), can easily dominate the noise from the resistance of the coils, and the input stages of the preamplifier, in all except very low main fields. The following discussion concentrates on factors that are important (or controllable) in the whole-body experiment. Thus we express Hoult and Richards' form for the signal-to-noise ratio, which is

$$\mathrm{SNR} = \frac{N_\gamma^2 \hbar^2 I(I+1) B_0^2 B_1(w) V s^{1/2}}{2(k_B T l \xi)^{1/2} (T \mu_0 \mu \omega p)^{1/4}} \quad [10]$$

where N is the Avogadro number, $\hbar$ is the Planck constant/2π, I is the spin quantum number, V the volume of the sample, s the circumference and l the length of the windings of the receiver coil, k_B is the Boltzmann constant, T is the coil absolute temperature, ξ is the 'proximity factor' (dependent on things such as conductor spacing), ρ is the resistivity of the material of which the coil is made, μ is the relative permeability, $B_1(w)$ is the field at the sample due to a unit current flowing in the coil ($\omega_0 = \gamma B_0$), as assuming only that field magnitude and coil design parameters are variables.

$$\mathrm{SNR} \propto B_0^{7/4} B_1(w) \frac{s^{1/2}}{(l\xi)^{1/2}} \quad [11]$$

A unit volume is assumed as the target of the experiment. This derivation ignores any noise contributed by the object being studied, particularly one with large dimensions, and Hoult and Lauterbur later extended this formula to allow for the case where a large load is placed in the coil, to give the relationship

$$\mathrm{SNR} \propto \frac{B_0^2}{\left(k_d \frac{n^2}{\theta} (\xi \rho \mu \mu B_0)^{1/2} \left(k_b + k_c \frac{g}{a}\right) + k_d B_0^2 b^5 n^2 \frac{\sigma}{a^2}\right)^{1/2}} \quad [12]$$

where the k_c etc are numerical values. This describes the situation for a round 'headlike' object of radius b and conductivity σ in an n-turn saddle coil of included angle θ, radius a and length g.

The concept of the 'intrinsic SNR' was developed to demonstrate the sensitivity of system performance to changes in the main field in the situation where the coil is heavily loaded, giving it as

$$\mathrm{SNR} \propto B_0 \quad [13]$$

in whole-body systems in which the body noise is the dominant factor (which it is in all except the lowest fields, or when the coils in use are very small or poorly coupled to the target tissue).

However, it was pointed out that in the form of data recovery generally used, in which multiple experiments are needed, relaxation effects cannot be ignored. Tissue T_2 is effectively constant over the range of fields used in current whole-body studies but T_1 shows a dependence on B_0 for which the empirical relationship [14] was proposed,

$$T_{1A} = T_{1B}\left(\frac{B_{0A}}{B_{0B}}\right)^{\nu} \qquad [14]$$

where A and B are two fields levels, and υ is found, empirically, to be in the range of 0.3–0.4 for the majority of relevant tissues.

The signal in the most basic form of study is

$$S = S_0\left[1 - \exp\left(-\frac{\mathrm{TR}}{T_1}\right)\right]\exp\left(-\frac{\mathrm{TE}}{T_2}\right) \qquad [15]$$

where TR is the time between successive excitations, and TE is the time between the middle of the excitation RF pulse, and the time at which the central point of k-space is acquired. (The American College of Radiology published a glossary of terms for whole-body NMR quite early in its clinical application. This ignored the traditional NMR notation quite cavalierly, but has become so well established in the biomedical community that it has been used in this article. Only in a few instances (some diffusion weighting and magnetization transfer terms) does it align with standard NMR practice.) Where TR is short relative to T_1 (as in volume scanning), equation [15] reduces to

$$S = S_0^*\frac{\mathrm{TR}}{T_1} = kS_0^*B_0^{-\upsilon} \qquad [16]$$

where S_0^* includes the signal dependency on T_2, and so on.

In addition, data have to be acquired over a finite period of time, as the acquisition gradient disperses the spin frequencies. Further, it is easier to recover data with a reduced bandwidth in lower fields as magnet homogeneities are conventionally expressed as fractions of the main field (and are a function of the unit design and relatively independent of field level), while dephasing effects due to field inhomogeneity are a function of actual field variations. For any one form of data acquisition, the noise voltage ($\propto f^{1/2}$), where f is the bandwidth of the acquisition) is proportional to $B_0{}^{1/2}$.

Taking into consideration these latter two factors, the overall signal-to-noise of a practical imaging experiment is

$$\mathrm{SNR} \propto \left[1 - \exp\left(-\frac{\mathrm{TR}}{T_1}\right)\right](pB_0)^{1/2} \qquad [17]$$

where p is the number of acquisitions needed to recover the data. Volume scans (where p can be very large) can thus have excellent SNRs, even if the acquisition time is long. At the extreme where TR<<T_1, in a repetitive scanning situation,

$$\mathrm{SNR} \propto B_0^{0.2} \qquad [18]$$

(assuming $\nu = 0.3$) and the benefit of increasing the field is marginal. Note, however, that as TR increases, the performance at higher fields improves with respect to that at lower ones. In practice, in the majority of whole-body situations, where noise from the body dominates that from the coil and electronics, the limits of performance are set by artefacts.

Fast scanning, one of the most significant issues in imaging because extended scanning times are unpopular with patients, and uneconomic for hospitals, has followed two very different routes. In the first, a method much used for MR angiography and in volume acquisitions, short repetition times are used with a reduced flip-angle, selected using the relationship derived by Ernst and Anderson that relates flip angle (α), TR and T_1 for the optimum signal-to-noise ratio:

$$\alpha = \cos^{-1}\left(\exp\left(-\frac{\mathrm{TR}}{T_1}\right)\right) \qquad [19]$$

This is clearly only ideal for any one tissue in a complex system with multiple T_1 values, but, in spite of reservations about its contrast, it has been applied very successfully.

The alternative strategies depend on acquiring many lines of k-space consequent upon a single excitation. Mansfield developed echo planar imaging very early on, to acquire whole images in a single acquisition, while other popular forms of fast acquisition of k-space data imaging are more modest in their ambitions. All pay some price in signal-to-noise ratio, but all have their uses and particular capabilities.

Imaging of solids (or other systems with very short T_2 or T_2^*) involves special techniques to overcome the need to encode spatial data very quickly. The approaches are not dissimilar to those of medical MRI, but the demands on the machine are more extreme, and the topic has developed a specialized scientific community of its own.

Contrast-to-noise ratio (CNR)

Clinicians who use MRI (and, indeed, in reality, all users who view MRI images as the basis of their studies) are actually less concerned about the signal-to-noise ratio of the data they have than they are about the contrast-to-noise ratio. In X-ray terms contrast is usually defined as

$$C = \frac{S_A - S_B}{S_A} \quad [20]$$

where S_A and S_B are the signals from two components A and B. In NMR this definition has proved inappropriate as S_A can be zero or nearly so, as can easily occur, for example, with the inversion recovery sequence (see below), which actually delivers very high contrast. For MRI purposes, contrast-to-noise is better defined as

$$\mathrm{CNR} = \frac{S_A - S_B}{S_{0w} N} \quad [21]$$

where S_A and S_B are the signals as defined above, S_{0w} is the fully recovered signal from water (or other suitable reference) and N is the noise voltage. Much of medical MRI research has been devoted to studying how to maximize this ratio in a huge variety of diseases and circumstances.

Spectral spatial selectivity

It is not appropriate here to enter into a discussion of spectral spatial selection methods or their problems. However, it is worth making the point that these are much closer to the concepts (and difficulties) of MRI than they are to those of traditional high-resolution spectroscopy. In particular, problems of apparent low signal-to-noise ratio may have much more to do with artefacts than traditional noise. MRI has a surprising amount, largely ignored, to contribute to high-resolution spectroscopy.

Concluding remarks

The implementation of spatial localization in MRI (or MRS) is not hugely complex, nor is it difficult in practice. In many ways, it is deceptively easy in both instances, and it is necessary to be very aware of the range of artefacts that can rise from inadequacies of very many kinds in the design of hardware, in its implementation and in its use. It is hard to overemphasize the need for stability and accuracy in the equipment used, nor to underrate the complexity and subtlety of the effects that can arise from failure to achieve these. Unfortunately, providing even a simple discussion of these is beyond the scope of this article, though there is a brief reference to them in another article on contrast mechanisms in MRI (q.v.). Almost invariably, except in very low fields, results from whole-body experiments are more determined by artefact than they are by noise. Not infrequently, the presence of artefact is indicated only by an apparently excessive level of noise.

List of symbols

A_i = a measure of the extent of the object in the plane at r_i; B_0 = applied magnetic field strength; $B_1(t)$ = RF pulse profile; $B_1(w)$ = field at sample due to unit current in coil; C = contrast; CNR = contrast-to-noise ratio; f = bandwidth; $f(t)$ = frequency spectrum; G = gradient field amplitude; G_r = gradient field along r axis (parallel to z axis and to B_0); $\hbar$ = Planck constant/2π; I = spin quantum number k_B = Boltzmann constant; K = profile relating RF intensity and spectral content; l = receiver coil winding length; n = number of observation points; N = Avogradro number; N = noise voltage; P_i = proton density at position r_i; r_i = position along r axis; s = receiver coil winding circumference; S = signal strength; SNR = signal-to-noise ratio; t = observation time; t_p = RF pulse duration; T = temperature; TE = echo time; TR = time between successive excitations; T_1 = spin-lattice relaxation constant; T_{2ijk} = spin–spin relaxation constant for ρ_{ijk}; T_p = duration of phase-encoding pulse; T_{ip} = duration of inverted gradient pulse; T_{ap} = half duration of acquisition gradient pulse; V = ample volume; $\omega_0 = \gamma B_0$; W = width of slice; α = flip angle; γ = gyromagnetic ratio; μ = relative permeability; μ_0 = permeability of free space; ξ = proximity factor; ρ = resistivity of coil material; ρ_{ijk} = proton density of the ijk voxel; ω = angular frequency.

See also: **Contrast Mechanisms in MRI; Fourier Transformation and Sampling Theory; Magnetic Field Gradients in High Resolution NMR; MRI Applications, Biological; MRI Applications, Clinical; MRI Applications, Clinical Flow Studies; MRI Instrumentation; NMR Principles; Two-Dimensional NMR Methods.**

Further reading

Abragam A (1961) *Principles of Magnetic Resonance*. Oxford: Clarendon Press.

Budinger TF and Margulis AR (eds) (1986) *Medical Magnetic Resonance Imaging and Spectroscopy*. Berkely, CA: International Society for Magnetic Resonance in Medicine.
Edelstein WA, Hutchison JMS, Johnson G and Redpath T (1980) Spin warp NMR imaging and applications to human whole-body imaging. *Physics in Medicine and Biology* **25**: 751–756.
Foster MA and Hutchison JMS (eds) (1987) *Practical NMR Imaging*. Oxford: IRL Press.
Grant DM and Harris RK (eds) (1996) *Encyclopedia of NMR*. Chichester: Wiley.
Lauterbur PC (1973) Image formation by induced local interactions. Examples employing nuclear magnetic resonance. *Nature (London)* **242**: 191–192.
Mansfield P (1977) Multi-planar image formation using NMR spin echoes. *Journal of Physics C: Solid State Physics* **10**: L55–58.
Mansfield P and Morris PG (1982) *NMR Imaging in Biomedicine*. New York: Academic Press.
Slichter CP (1980) *Principles of Magnetic Resonance*. Berlin: Springer-Verlag.
Young IR (1984) Signal and contrast in NMR imaging. *British Medical Bulletin* **40**(2): 139–147.

MRI Using Stray Fields

Edward W Randall, Queen Mary and Westfield College, London and Instituto Superior Tecnico, Lisboa, Portugal

MAGNETIC RESONANCE
Applications

Introduction

MRI is accomplished using gradients in the magnetic field flux density (normally referred to as the magnetic field). These gradients are generally produced in conventional imaging with the aid of sets of gradient coils. The suggestion to use the gradients that are present in the stray field of a superconducting magnet for imaging purposes was made first by Samoilenko and colleagues in 1988. The method has the advantage that these field gradients are large, of the order of 50 T m^{-1}. This has proved to be very useful not only for the imaging of solids but also additionally for the imaging of liquids in solids, neither of which can be imaged very satisfactorily or even at all with conventional MRI techniques. The method was generalized from one spatial dimension to three by Samoilenko and Zick, working at Bruker Spectrospin, but the technique was exemplified only by proton and fluorine work in diamagnetic solids such as organic polymers. The extension to multinuclear work, even quadrupolar nuclides, was accomplished by Randall and colleagues in one-dimensional studies. They also showed that the method gave good images of diamagnetic crystalline solids, and even of paramagnetic crystalline solids. Thus now virtually any nuclide, even those with electric quadrupole moments, can be imaged in any solid, with the possible exception of ferromagnetics. Because metals conduct electricity, they are not penetrated by radiofrequency fields and therefore can be imaged only when in powder form. Distortions of the images produced by magnetic susceptibility are greatly reduced in proportion to the gradient strength.

The stray field and its large gradient have been of use for spectroscopic studies also, mainly for the investigation of relaxation times and of diffusion processes. Additionally, the basic spin physics in the stray field has been of considerable interest in its own right. In the case of quadrupolar nuclides, for example, solid-state NMR spectroscopy in the stray fields gives a new method of determining the electric quadrupolar coupling constant. This may now be scanned spatially for the first time.

Initially the acronym STRAFI was used to denote stray field imaging, but more recently it has been used by Randall and colleagues more generally to denote simply the stray field itself. Studies are not confined simply to imaging. The three types of work in the stray field may thus be referred to as STRAFI imaging (or STRAFI-MRI), STRAFI spectroscopy and STRAFI diffusion studies. All of these are covered in this article, since the technique is relatively new.

It is possible to combine the techniques and not only study *localized* free induction decays, relaxation times and self-diffusion coefficients but also to constructs maps of these properties.

The stray field

For superconducting magnets such as are used for NMR spectroscopy or for MRI studies, the central very homogenous field on the Z axis falls off rapidly as the distance from the centre of the magnet increases. Near the edge of the solenoid the gradient is particularly high. This is illustrated in **Figure 1**. The gradient can reach values of the order of 10–100 T m^{-1} typically, depending on the value of the central magnetic flux density, the size of the bore of the magnet, and the details of primary coils. **Table 1** shows some typical values for proton resonance frequencies that have been reported for STRAFI work on various instruments: the positions in the gradient have not always been optimized.

Even higher fields are available on resistive magnets such as the 24 T magnet at the National High Magnetic Field Laboratory at the University of Florida. The maximum STRAFI gradient is approximately 138 T m^{-1} at a flux density of about 17 T corresponding to a proton resonance frequency of 723 MHz. There is a 45 T system under construction.

High gradients give better spatial resolution in imaging experiments and allow the measurement of smaller diffusion constants. Additionally, they reduce the long-range effects of variations in magnetic susceptibility. Large fields are useful for the usual reasons, such as increased sensitivity, and for nuclides with low gyromagnetic ratios (γ) the higher resonance frequencies make probe ringing less of a problem. Thus the first solid-state STRAFI work on ^{14}N was accomplished at 32 MHz on the high-field, narrow-bore instrument shown in **Table 1**, in a stray field of 8.9 T.

Table 1 Typical values for the centre and stray fields and their proton resonance frequencies

Centre field (T)	*Centre frequency (MHz)*	*Bore (mm)*	*STRAFI (T)*	*Fringe frequency (MHz)*	*Gradient (T m^{-1})*
4.7	200	400	1.86	79	9.0
4.7	200	300	2.61	111	12.0
7.05	300	89	2.94	125	37.5
9.4	400	89	5.52	235	58.5
14.1	600	89	9.66	411	52.9
14.1	600	54	8.9	380	90.0
19.6	834	32	11.2	477	75.7

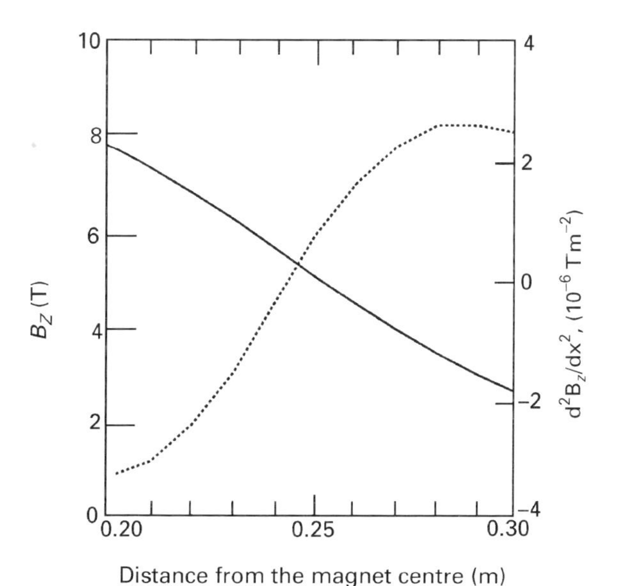

Figure 1 Plots of the field strength B_Z on the Z axis versus distance x from the magnet centre (solid line, left-hand abscissa) and of the second derivative of B_Z with respect to x (dotted line, right-hand abscissa). The plots are for a typical superconducting magnet (Magnex Scientific Ltd) of bore 89 mm and a centre field of 9.4 T. Reproduced with permission from: McDonald PJ (1997) Stray field magnetic resonance imaging. *Progress in NMR Spectroscopy* **30**: 69–99.

So far no magnet has been designed specifically for STRAFI work. This is a development that can be forecast confidently since it should be possible to obviate the expense associated with very high central field homogeneities. The exception is the portable system being developed in Aachen by Professor Blümich's group called the MOUSE, which uses an asymmetrical permanent magnet and employs the field outside the magnet. There are analogies with the systems developed for oil-well logging: although the fields are very low the *ratio* of the field strength to the gradient strength is similar to values in work on conventional superconducting magnets.

The spin physics and STRAFI spectroscopy

With gradients of the order of 50 T m^{-1} only a thin slice of the sample is excited for line widths of the order of a few kHz. The slice thickness for such narrow lines is governed (inversely) by the pulse duration, the value of γ, and the gradient strength. For 'isolated' protons in a 50 T m^{-1} gradient and a pulse duration of 10 μs, the slice thickness is 80 μm, which increases to 360 μm for ^{23}Na (even if the quadrupole coupling is zero).

For broader lines the line width of the sample in the absence of the gradient can become the governing factor. The line has, of course, two components, one due to homogenous broadening, governed by T_2, and a heterogeneous component, say for polycrystalline samples, that depends on the dominant interaction (apart from the gradient): dipolar interactions for dipolar solids and anisotropic liquids; the quadrupolar interaction for quadrupolar cases; or even the chemical shift anisotropy.

The STRAFI free induction signal for a single pulse has a decay that is governed by T_2^* (T_2^* is the apparent T_2 including inhomogeneity effects) and diffusion. Since the large gradient makes T_2^* very short, it is usual to use the spin-echo technique of Hahn to overcome dead-time problems in the detection. The width of the echo is determined by T_2^*. The gradient term is negligible at the top of the echo where there is perfect refocusing. If diffusion is negligible then T_2 may be determined by arraying the spin echo delay (τ) since the echo relaxation is then governed by T_2 only. The STRAFI echo is a true Hahn echo even for solids (because the strength of the gradient dominates other terms in the Hamiltonian), so that any two pulses will in general give an echo irrespective of their relative phase. In the absence of the large gradient, dipolar solids will give an echo only if the relative phase is different: this is the dipolar echo.

STRAFI Hahn echoes are very useful spectroscopically since they are obtained for both solids and liquids even in the same sample. Additionally, multiple primary Hahn echoes from just two pulses can be produced in the stray field even from solids.

It is usual to employ a train of pulses to produce a train of echoes. These are not to be confused with the multiple echoes mentioned above. Application of more than two pulses gives both primary and stimulated echoes. For a pulse angle of 90°, the sample response is as shown in **Figure 2** both for the EVEN sequence, $90_x - [\tau - 90_x - \tau - \text{echo} -]_n$, which has no phase change, and for the ODD sequence, which has $90_x - [\tau - 90_y - \tau - \text{echo} -]_n$. The two sequences produce the same intensity for the first echo, the relaxation properties of which are determined only by T_2, if diffusion is negligible. The echoes after the first one have more than one component, the number of which increases with the echo number. These contributions are all of the same sign for the ODD sequence, so that they augment each other – there is an analogy with constructive interference in diffraction – but not for the EVEN sequence, which produces echoes of opposite phase (destructive interference) and the up–down pattern shown in **Figure 2**. This has a (2 up, 2 down) pattern for 90° and an (n up, n down) pattern where n is an integer given by $180/\alpha$ when α, the pulse angle, is a submultiple of 180°. This is illustrated in **Figure 3**, which

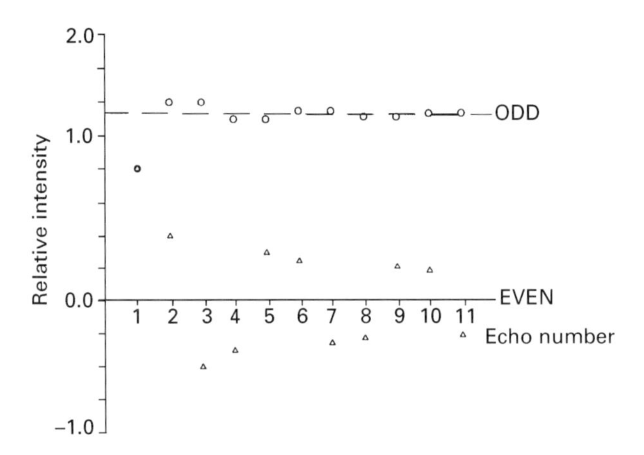

Figure 2 Plots of the calculated normalized echo intensity in the absence of relaxation and diffusion for the ODD sequence (top, circles) and the EVEN sequence (bottom, triangles) versus echo number following a train of 90° pulses. The first echo intensity is the same for each sequence and is governed only by T_2 in the absence of diffusion. Subsequent echoes have contributions from T_1. The EVEN repeating pattern has two points up and two points down for 90°. Reproduced with permission from: Bain AD and Randall EW (1996) Spin echoes in static gradients following a series of 90 degree pulses. *Journal of Magnetic Resonance* **A123**: 49–55.

shows the responses for pulse angles of 60° (3 up and 3 down) and 45° (4 up and 4 down). The EVEN sequence is therefore very useful for the setting of the pulse angle. In fact, α varies across the slice but, remarkably, the observed value corresponds to the value at the middle of the slice, as confirmed by detailed calculations based on the Bloch approach or on the density matrix.

The odd sequence has the advantage as in normal spectroscopy that it produces line narrowing if τ is short enough to produce some spin locking, in which case the T_2 contribution to the STRAFI-echo decay can be replaced by T_e (the effective decay time constant), which in the limit of spin locking may be approximated to $T_{1\rho}$. The line narrowing in the imaging experiments gives a greater spatial resolution. Even if there is no spin locking, extended echo trains may be produced since the last echoes have substantial contributions from T_1. Very long values of T_1 and $T_{1\rho}$, and hence T_e, can be obtained in many aprotic solids particularly if a low-γ nuclide is being observed. A good example is ^{31}P in bone. In this case a very large number of pulses can be applied in a long pulse train which is of great advantage for signal accumulation (long echo train summation or LETS). The same behaviour can be seen even for quadrupolar nuclides; indeed the first such observation was for ^{2}H in heavy ice, for which 9000

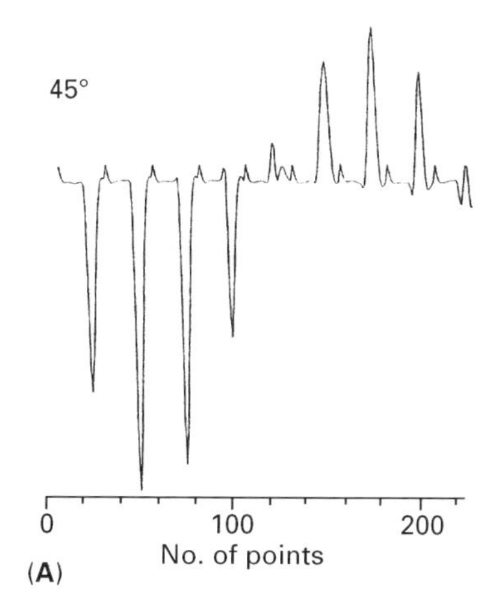

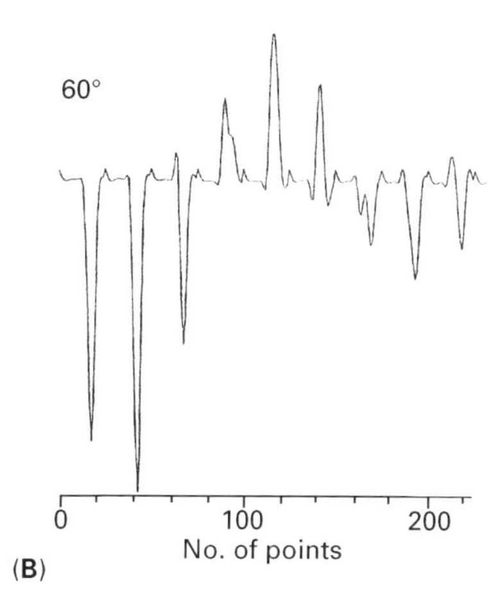

Figure 3 Experimental data of the EVEN sequence and a pulse angle of 45°, which gives a 4-up and 4-down repeating pattern, and 60°, which gives a 3-up and 3-down pattern. The signals were for ^{1}H at 237.5 MHz in the 58 T m^{-1} gradient of the Chemical Magnetics Infinity System at the University of Surrey for a silicone rubber sample. The τ value was 25 µs. Reproduced with permission from: Randall EW (1997) A convenient method for calibration of the pulse length in high field gradients using Hahn echoes. *Solid State NMR* **8**: 179–183.

echoes in each train have been produced. **Figure 4** shows the decay of the tops of these 9000 echoes. This decay is much longer than T_2 since echoes after the first have increasing degrees of T_1 governing their relaxation behaviour as the echo number increases. The decay is then said to be T_1-weighted. Of particular interest is the case of ^{14}N, which for unprotonated nitrogens gives very long echo trains even when the value of the quadrupolar interaction is large, such as in sodium nitrite for which the literature value is about 4.9 MHz.

Figure 4 The echo decay for every third echo of 9000 ^{2}H echoes from a sample of heavy water in a gradient of 74.2 T m^{-1} and resonance frequency of 27.5 MHz. The tail of the decay is governed mainly by self-diffusion. The pulse duration was 4 µs (tip angle 90°). A total of 569 echo trains were accumulated with a repetition time of 0.5 s. Courtesy of Drs Teresa Nunes, Geneviève Guillot and the author.

STRAFI spectroscopy resembles normal spectroscopy except that the chemical shift and dipolar interactions are too small to perturb the signal appreciably. Studies are therefore confined to the study of relaxation times and diffusion. For heterogeneous solid samples, or liquids in solids, discrimination between different motional components is possible, the contributions for components with fast motions being dominant for the last echoes in a long sequence.

For quadrupolar nuclides in solid samples that have a large electric quadrupole interaction, C_q, the echoes exhibit splittings and modulations that depend on the magnitude of C_q, which may thus be measured.

STRAFI-diffusion

Gradients have been used for the determination of diffusion coefficients since the earliest days of NMR. The advantages of pulsed gradients led to studies in fixed gradients being superseded. Now, however, the STRAFI technique, which is a fixed-gradient technique, has the advantage of employing large gradients that bring into measurable range smaller diffusion constants. Kimmich and colleagues were thus able to measure diffusion coefficients as small as 6×10^{-11} cm^2 s^{-1} in siloxane melts at 20.5°C in a gradient of 38.4 T m^{-1}. Even smaller diffusion coefficients should be accessible at gradients higher than this.

STRAFI-MRI

STRAFI-MRI is a slice-selective method since even very short pulses excite only a narrow portion of the sample: Samoilenko's 'sensitive slice'.

The free induction decay of the excited slice contains spatial information, which can be revealed by Fourier transformation as in most other conventional NMR techniques. If the sample is a thin film with a thickness less than the width of the excited slice, the STRAFI image obtained in one spatial dimension is the image across the whole film. Larger samples, however, must be scanned to get the whole 1D projection. One way is to sweep the frequency of the pulse, but the older STRAFI method is to move the sample through the field. The first possibility is analogous to the method of frequency sweep in continuous-wave NMR spectroscopy. Field sweep in STRAFI-MRI has also been tried.

If the scanning is accomplished by sample movement, Fourier transformation is not necessary: the magnetization and the signal are directly proportional to the nuclear spin density. The 1D profile is generally obtained by translation of the sample along the Z axis of the magnet. The motion can be continuous, if it is slow enough, or stepwise. The size of the step may become the major factor governing the spatial resolution if it is coarse. The slices may be interleaved so that there is no interference between the excitations of contiguous slices. Accumulation can be accomplished by repeating the motion from the starting position, since this allows relaxation to occur during the recycle time of the motion. An alternative for samples with a long T_1 is to use very long pulse trains as in the ^{31}P example (the LETS protocol); see **Figure 5**.

The profiling can be repeated for different orientations around the same axis by rotation of the sample to produce a 2D image. Then the third dimension can be scanned by a second rotation orthogonal to the first. Back-projection algorithms can be used to produce the images. 3D work is very time-consuming by this method, since there are not the usual gains from the Fourier technique, but it is possible with axially symmetric samples to design the sample (normally called the phantom in imaging studies) so that useful but short experiments can be conducted in one dimension only. Examples are liquids diffusing into solids, such as solvents into organic polymers (**Figure 6**), or water diffusing into cements or concrete (**Figure 7**). In each case, relaxation weighting can be used to advantage. The system shown in **Figure 6** depicts the diffusion of hexafluorobenzene into the polymer from right to left. The sensitive

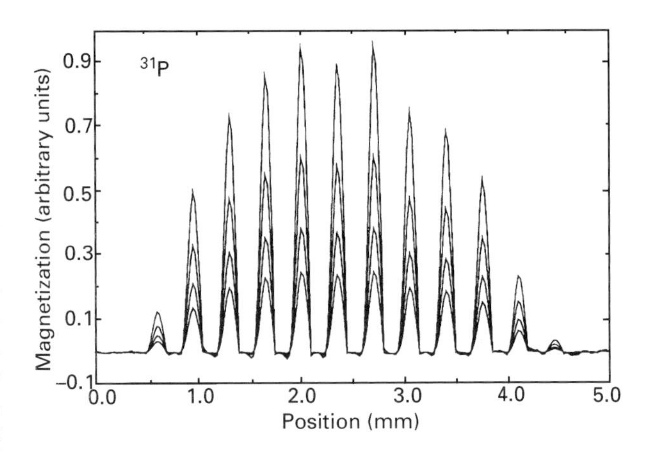

Figure 5 ^{31}P one-dimensional (semi-elliptical) profiles of a cylindrical sample containing ammonium hexafluorophosphate. At each position four summed echo trains are shown for long echo train summation (LETS). The highest intensity is for $n = 8192$ echoes. The gain in the signal-to-noise ratio is proportional to $n^{0.5}$ where n is the number of echoes in the train. The variation in n from the lowest to the highest intensity was 1024, 2048, 4096 and 8192. The odd sequence was used. Experimental details were resonance frequency 91 MHz; gradient strength 58 T m^{-1}; τ value 20 μs; pulse duration 3.6 μs; manual steps 0.35 mm; 64 averages of each echo train were used at 5.7 min per slice. Courtesy of Drs Duncan G. Gillies and Ben Newling.

planes for ^{1}H and ^{19}F are displaced one from the other because of the different values for the gyromagnetic ratios, and so are the images. The separation between the planes is about 4.9 mm in this example. The overall image therefore has a proton region (at the left) and a fluorine region at the right. In the sample these two regions overlap. There is a direct analogy with the chemical shift effect in conventional imaging.

The best position in the stray field is actually not where the gradient is maximum, because at this position the resonant slice is not planar since the magnetic field off-axis is not uniform. The optimum position is where the slice is planar. This occurs where the lines of equipotential lines are orthogonal to the Z axis. Nearer the centre of the magnet the equipotential lines are curved one way, but further from the centre they curve another way. Exactly at the optimum position a flat disc extends someway off-axis, typically about 1 cm for superconducting magnets with bores of 8.9 cm.

The distortion of this disc away from the optimum position is not large for small samples, which are normally cylinders about 1–2 cm long and less than 1 cm in diameter, so that positions away from the optimum are frequently used. This is part of the reason for the apparent irregularities in **Table 1**. Thus, the inserts, RF channels and tuning boxes used for the study of ^{31}P in the central field may be used for

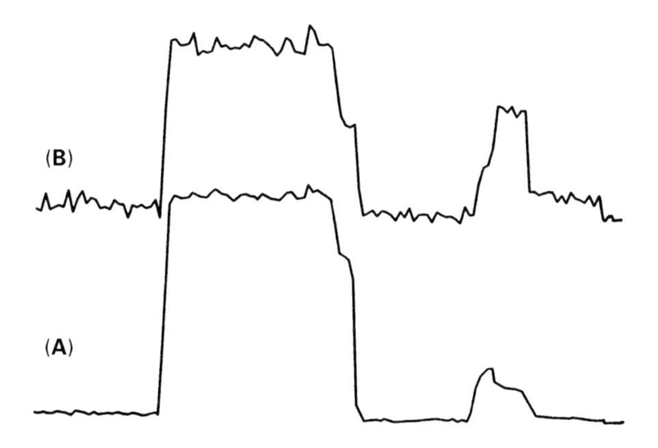

Figure 6 Profiles of a cylindrical phantom (5 mm thick, diameter 10 mm) of a polymer consisting of a 2:1 mixture of poly(methyl methacrylate) and poly(*n*-butyl methacrylate) following ingress of hexafluorobenzene (from the right): (A) projection produced from the first four echoes of 32 echoes; (B) projections from the last four echoes. The ^{19}F image is spatially separated at the right from the ^{1}H image at the left. The fluorine images exhibit spatially variable relaxation behaviour. The gradient strength was 37.5 T m^{-1} and the resonance frequency was 123.4 MHz. Reproduced with permission from: Randall EW, Samoilenko AA and Nunes T (1996) Simultaneous ^{1}H and ^{19}F stray field imaging in solids and liquids. *Journal of Magnetic Resonance* **A117**: 317–319.

protons in the stray field at a frequency that is rather lower than the optimum.

The gradient at the chosen position can be calibrated by measurement of an accurately known diffusion constant.

The spatial resolution has reached about 5 μm for ^{1}H in the case of line narrowing by spin locking for thin samples by use of the Fourier technique in a gradient of about 38 T m^{-1} in a 7 T magnet. For similar samples at the higher gradient strengths available on higher-field magnets, it should be possible to obtain a resolution of about 1 μm. Typically, however, for protons the resolution is about 50 μm or less.

For the ^{1}H and ^{19}F nuclides, the sensitive planes are spatially separated, with the fluorine-sensitive plane being at higher field close to the magnetic centre. The distance between these planes depends on both the field and the gradient but can be of the order of 5 mm. This means that both the ^{1}H and ^{19}F signal of a fluorohydrocarbon can be detected as the sample moves. For a small enough sample, for which the projection on the field axis is less than 5 mm, the two 1D profiles are completely separated, see **Figure 6**. This γ-displacement, similar to the chemical shift displacement in conventional imaging (which is too small generally to be seen with STRAFI work), is useful, otherwise the discrimination would have to be made with relaxation weighting alone. The γ-displacement itself can, however, be very small, so that overlap nearly always occurs for certain pairs of nuclides such as ^{23}Na and ^{27}Al, ^{63}Cu and ^{65}Cu and ^{75}As and ^{115}In (see **Table 2**).

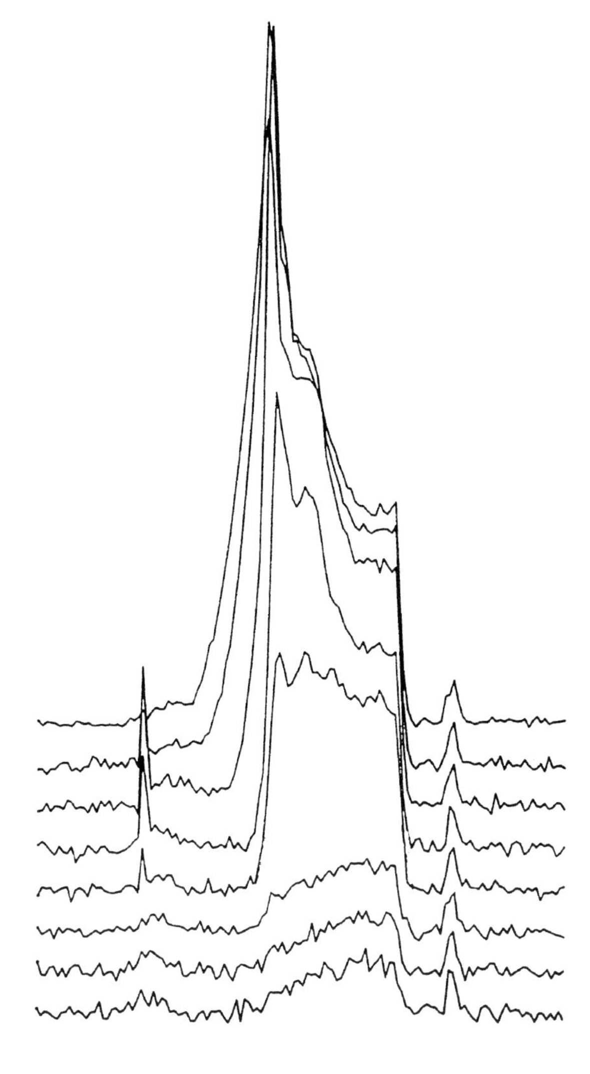

Figure 7 Hydration and curing of a Portland (type I) cement contained in a glass tube and covered with water without initial mixing of the components. The ^{1}H profiles were taken at various intervals thereafter: the top profile after 15 min; and the bottom one after 48 h. The peak at the right is from a plastic reference disc at the bottom of the vial, whereas the sharp peak at the left is from a parafilm top-cover and from water that initially condensed on it. The apparent loss of signal arises because of the growth of components with very short relaxation times as hardening occurs, which become invisible with the long echo times used (τ = 30 μs). The resonance frequency was 123.4 MHz and the gradient strength was 37.5 T m^{-1}. The ODD sequence was employed with trains of 64 echoes; 50 echo trains were accumulated for each point on each profile with a repetition time of 1 s. Courtesy of Dr Teresa Nunes.

For quadrupolar nuclides of half-integral spin there is a relatively narrow transition ($\frac{1}{2}$ to $-\frac{1}{2}$) and much wider transitions for the satellites. It can be that, as in spectroscopy, the latter are effectively not detected. In

Table 2 STRAFI-frequencies of some nuclei in the 2.9 T stray field of a Bruker MSL 300 and the proportion of the signal in the central transition

Nucleus	*I*	*Frequency* (MHz)	*Intensity of central transition (%)*[a]
^{1}H	$\frac{1}{2}$	123.4	100
^{19}F	$\frac{1}{2}$	116.1	100
^{7}Li	$\frac{3}{2}$	47.8	40
^{11}B	$\frac{3}{2}$	39.6	40
^{23}Na	$\frac{3}{2}$	32.2	40
^{65}Cu	$\frac{3}{2}$	32.2	40
^{27}Al	$\frac{3}{2}$	32.3	40
^{51}V	$\frac{7}{2}$	32.5	19
^{59}Co	$\frac{7}{2}$	29.2	19
^{115}In	$\frac{9}{2}$	27.0	15

[a] The intensity of the central transition expressed as a percentage of the whole signal.

this case the image is reduced in intensity by an amount that depends on the spin quantum number as shown in **Table 2**. The resolution then is determined by the line width of the central transition, which to first order is given by C_q^2/ν, where ν is the Larmor frequency. This can be of the order of kHz, so that the resolution is about the same as for dipolar-broadened ^{1}H in polymers, say. Curiously, if the signal-to-noise ratio increases, so that the satellite transitions are detected, the spatial resolution decreases.

It is possible to distinguish the loss of signal from this source from the case of a lower spin density by a comparison of the echo trains produced by the odd and even pulse sequences.

In the case of nuclides with integral values of the spin quantum number, there is no narrow transition and the slice thickness will be very large if the line width is big because of a large value for C_q. Thus the observed spatial resolution may be a factor of 10 worse than the correct image: a feature of a millimetre in the phantom may appear to be a centimetre or more in the image. This has been illustrated for ^{14}N in a series of samples in which the value of C_q was changed from 0 to about 5 MHz.

It can be noted here that an observed profile, *O*, is the convolution of the actual physical profile, *P*, and the line shape function *L*:

$$O = L \otimes P \qquad [1]$$

If *L* is known, say for a powdered sample consisting of one component containing a known dipolar or quadrupolar interaction, from which the line shape can be computed, this source of heterogeneous broadening may be removed from the observed image by deconvolution. The resolution of the image is then determined partly by the homogenous broadening only, which is governed by T_2 or $T_{1\rho}$ if there is spin locking and line narrowing.

Samples that are paramagnetic have proved to be easy to image, at room temperature at least, rather surprisingly perhaps for high-resolution spectroscopists. The reason, as shown as long ago as 1950 by Bloembergen, is that the electron relaxation times are so short that the NMR line widths are not greatly affected. For $CuSO_4{\cdot}5H_2O$ the proton line widths are determined mainly by the proton–proton dipolar interaction, and therefore the imaging problem is no different from that for a diamagnetic hydrate. The same hold true for other ligands and for more paramagnetic samples. An advantage is that although T_2 is not greatly affected, T_1 is shortened considerably, as illustrated in **Table 3**, which is an aid to accumulation when short pulse-trains are employed.

The compounds in **Table 3** have produced 1D profiles for both the ^{1}H and ^{19}F nuclides.

A most striking result is the observation of STRAFI echoes for the quadrupolar ^{51}V ($I = \frac{7}{2}$) nuclide in the paramagnetic vanadyl bis(acetonato) compound, which has two unpaired electrons on each vanadium atom. This result opens up the prospects of imaging for quadrupolar nuclides in paramagnetic situations in general.

Conclusions: Prospects for the future

It seems clear that STRAFI-MRI will become the method of choice for the imaging of solids, and samples consisting of solids and liquids. Any type of nucleus may be used, including quadrupolar nuclides, in any type of solid. All types of materials may be addressed. The method will be seen to best advantage when simple 1D profiles are sufficient for the particular study. Its worth has been shown already for water in cements (**Figure 7**) and concrete and other building materials, and for organic liquids

Table 3 Some values of proton relaxation times in paramagnetic compounds and values of the gram magnetic susceptibility χ_g

Formula[a]	$10^6\chi_g$	T_1/ms	T_2/μs
Eu(fod)^3	3.91	277	25.3
Dy(fod)^3	42.80	5.04	13.1
Ho(fod)^3	35.06	5.40	14.0
Yb(fod)^3	7.81	16.8	26.5

[a] fod is the 6,6-,7,7-,8,8,8-heptafluoro-2,2-dimethyl-3,5-octanedionato ligand.

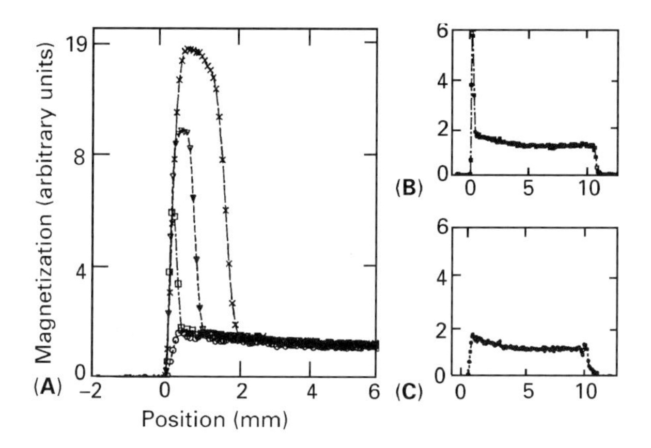

Figure 8 One-dimensional ^{1}H profiles for poly(vinyl chloride) polymer exposed at the left to acetone vapour of different activities for 48 h at 20°C. In (A) the circles are for an activity of 0.13, squares for 0.35, triangles for 0.60; and crosses for 0.82. The image of the largely undisturbed solid polymer is to the right. At the left the image has overlapping images of the ingressing acetone and the softening polymer. Part (B) shows the points for an activity of 0.35 in more detail, and those for an activity of 0.13 are shown in (C) The macroscopic diffusion in (C) is of case I type. The onset of case II diffusion at the higher activities is shown clearly in (B) and in (A) – the acetone in this case has a very sharp front that is absent in the case 1 example. STRAFI-MRI has the advantage that all components, solids as well as the liquid, can be imaged simultaneously. It is then possible to employ relaxation weighting to separate the components of the image. Reproduced with permission from: Perry KL, McDonald PJ, Randall EW and Zick K (1994) Stray-field magnetic resonance imaging of the diffusion of acetone into poly(vinylchloride). *Polymer* **35**: 2744–2748.

in polymers, for which it has the advantage of imaging the undisturbed polymer, the softening polymer and the liquid phase in the softened polymer as illustrated in **Figure 8**. The macroscopic diffusion may be followed easily.

STRAFI-diffusion studies of microscopic diffusion can bring into range very slow self-diffusion coefficients.

STRAFI-spectroscopy in the excited slice can yield details of the relaxation characteristics in general, and even the values of C_q for quadrupolar nuclides.

These various techniques may be combined for spatial mapping purposes, e.g. for relaxation times and quadrupole coupling constants.

There seems to be little doubt that STRAFI studies will be of great value in the general field of materials science.

As regards clinical work, *in vitro* studies have proved to be very valuable: for example, the study of materials and cements for dental work and hip-joint prostheses. One may predict that *in vivo* studies will emerge. Indeed, having the subject nearly outside the magnet has considerable advantages. For human patients, the question of the safety aspects of the large gradients will need to be addressed.

It is likely that special spectrometers will be designed exclusively for STRAFI studies in the near future.

List of symbols

C_q = quadrupolar interaction; T_1 = spin–lattice relaxation time; T_2 = spin–spin relaxation time; T_2^* = effective T_2 in the presence of inhomogeneities; T_e = effective decay constant for the STRAFI echo train; $T_{1\rho}$ = relaxation time in the rotating frame; α = tip angle; γ = gyromagnetic ratio; τ = spin echo delay; ν = Larmor frequency; χ_g = magnetic susceptibility.

See also: **Magnetic Field Gradients in High Resolution NMR; MRI of Oil/Water in Rocks; MRI Theory; NMR in Anisotropic Systems, Theory; NMR Microscopy; NMR of Solids; NMR Principles; NMR Pulse Sequences; Solid State NMR, Methods.**

Further reading

Bodart P, Nunes T and Randall EW (1997) Stray-field imaging of quadrupolar nuclei of half integer spin in solids. *Solid State NMR*, 8: 257–263.

Kimmich R and Fischer E (1994) One- and two-dimensional pulse sequences for diffusion experiments in the fringe field of superconducting magnets. *Journal of Magnetic Resonance* **A106**: 229–235.

McDonald PJ (1997) Stray field magnetic resonance imaging. *Progress in NMR Spectroscopy* **30**: 69–99.

McDonald PJ and Newling B (1998) Stray field magnetic resonance imaging. *Reports on Progress in Physics* **61**: 1441–1493.

Randall EW (1997) ^{1}H and ^{19}F MRI of solid paramagnetic compounds using large magnetic field-gradients and Hahn echoes. *Solid State NMR* **8**: 173–178.

Randall EW (1997) A convenient method for calibration of the pulse length in high field gradients using Hahn echoes. *Solid State NMR* **8**: 179–183.

MS–MS

See **Hyphenated Techniques, Applications of in Mass Spectrometry; MS–MS and MS^n.**

MS-MS and MS^n

WMA Niessen, hyphen MassSpec Consultancy,
Leiden, The Netherlands

MASS SPECTROMETRY
Methods & Instrumentation

Introduction

Tandem mass spectrometry (MS-MS) currently is an important technique, both in fundamental studies concerning the behaviour and structure of gas-phase ions, and in many analytical applications of MS. The history of MS-MS can be considered to go back to the observation and explanation of metastable ions by Hipple and Condon in 1945. Subsequently, the potential of metastable decompositions of ions was further investigated on magnetic sector instruments. MS-MS as a technique gained momentum in the mid-1970s. Reversed-geometry double focusing sector instruments were built in the laboratories of Cooks and of McLafferty and used for MS-MS studies, and the commercial ZAB range of sector mass spectrometers became widespread. The next major breakthrough was the introduction of triple-quadrupole instruments by the group of Yost in the early 1980s. Commercially available triple-quadrupole systems with an easy user interface stimulated the analytical use of MS-MS, which previously was primarily used in more fundamental studies. More recently, MS-MS applications have been implemented with other types of mass analysers, i.e. ion trap, time-of-flight (TOF) and Fourier-transform ion-cyclotron resonance (FT-ICR) mass spectrometers.

The principles, instrumentation and some (analytical) applications of MS-MS are reviewed in this article. More elaborate accounts on MS-MS may be found in the Further reading section.

Principles of MS-MS

A basic instrument for MS-MS consists of a combination of two mass analysers with a reaction region between them. While a variety of instrument setups can be used in MS-MS, there is a single basic concept involved: the measurement of the *m*/*z* of ions before and after a reaction in the mass spectrometer; the reaction involves a change in mass and can be represented as

$$m_p^+ \rightarrow m_d^+ + m_n$$

where m_p^+ is the precursor (or parent) ion, m_d^+ is the product (or daughter) ion, and m_n represents one (or more) neutral species. In terms of mass: $m_p = m_d + m_n$. The basic MS-MS experiment is the mass selection of the precursor ion in the first stage of analysis, the fragmentation of the precursor ion, e.g. metastable or by collision-induced dissociation (CID), and the mass analysis of the product ions in the second stage of analysis (product-ion scan).

The fragmentation of the precursor ion depends on the activation barrier of the reaction. The energy to overcome this barrier is due to the excess energy deposited in the precursor ion during ionization and transfer to the first mass analyser and, when applied, to the internal energy of the precursor ion gained by ion activation.

A metastable ion is an ion which during the ionization gained sufficient internal energy for fragmentation, but survived long enough to be extracted from the ion source. Such an ion may dissociate spontaneously during its flight from the ion source to the detector. Double focusing sector instruments can be used to detect the metastable ions.

However, in most cases, ion activation in a reaction region is applied to increase the internal energy of the ions transmitted from the ion source. Although ion activation by photodissociation and surface-induced collisions has been described, the most widely applied method is collisional ion activation. The CID process results from the conversion of translational energy of the precursor ion into internal energy by collisions with a neutral target gas, e.g. helium or argon, admitted to the collision cell.

In CID, two collision-energy regimes should be considered: low-energy and high-energy collisions, depending on the initial translational energy of the precursor ion upon collision. High energy collisions (kV energy) are applicable in magnetic sector instruments as well as in certain applications of post-source decay in TOF instruments (see below), while low energy collisions are applied in most other systems (triple-quadrupole, ion trap and FT-ICR).

While in principle any ion generated in the ion source by any ionization technique can be subjected to MS-MS in the product-ion mode, MS-MS has especially found analytical applications in combination with soft ionization techniques, where without MS-MS only information on the intact molecule is obtained and no fragmentation is observed. MS-MS is then required to achieve structure informative fragmentation. However, there are some limitations as well. First of all, the fragmentation in CID may not always lead to sufficient fragmentation to allow unambiguous structure assignment. Furthermore, MS-MS is limited in practice to ions up to m/z of ~2000 because with larger ions the number of degrees of freedom is so large that the internal energy gained in CID is readily dissipated over a large number of bonds and no single bond acquires sufficient energy for cleavage on the observational time-scale of the instrument.

In addition to the product-ion scan, most MS-MS instruments allow other scan modes, e.g. neutral-loss and precursor-ion scan modes. The modes are not only useful in the elucidation of the fragmentation pattern of a particular compound, but also in the screening for a series of structurally-related compounds in complex samples. In the product-ion scan mode, the first mass analyser selects a particular precursor ion, while the product ions obtained by CID of this precursor are analysed in the second mass analyser. In the precursor-ion scan mode, this process is virtually reversed: the first mass analyser transmits all ions in a preset m/z window to the collision cell, while the second analyser selects only the ions of one particular m/z, e.g. a particular structure informative fragment for a series of ions or compounds. An example of the use of the precursor-ion scan mode is the monitoring of phthalate plasticisers by means of the common fragment ion at m/z 149 due to protonated phthalic anhydride. In the neutral-loss scan modes, both mass analysers are operated in the scanning mode, but at a fixed mass (m/z) difference, corresponding to a characteristic neutral loss. An example of the use of the neutral-loss mode is the monitoring of the CO_2 loss from deprotonated carboxylic acids. In addition, selective reaction monitoring (SRM) can be applied to monitor a specific reaction in the mass spectrometer. Both mass analysers are operated in the selection mode, i.e. selecting a particular precursor ion in the first and a particular product ion in the second mass analyser. SRM is extremely useful in the quantitative analysis of compounds in complex matrices, as a significant gain in selectivity may be achieved, leading to improvement of detection limits.

Instrumentation for MS-MS

A wide variety of instruments for MS-MS are available.

MS-MS in sector instruments

Initially, the study of metastable ions was performed using sector instruments. In a double-focusing magnetic sector instrument, a variety of MS-MS related experiments may be performed. Two reaction regions may be used in the field-free regions, i.e. between the ion source and the first sector and in between the two sectors (RR1 and RR2 in **Figure 1A**). A fragmentation reaction in the first field-free region of an instrument with either geometry (EB or BE) can be monitored using linked scan experiments. In a linked scan, both the magnetic field strength $\boldsymbol{B}$ and the electrostatic sector voltage are scanned but maintained in a fixed relationship throughout the scan, e.g. B–E linked scans for product-ion scans and B^2–E linked scans for precursor-ion scans. A disadvantage of the linked scan procedures is the limitation in resolution (typically ~1000 for the precursor ion and ~5000 for the product ion). The product ions generated in the second field-free reaction region of a reversed-geometry instrument (BE) may be monitored via a mass-analysed ion kinetic energy (MIKE) scan, where the electrostatic sector voltage is scanned while maintaining a constant magnetic field and accelerating voltage.

More advanced possibilities for MS-MS in sector instruments can be achieved in three- and four-sector instruments. The four-sector instruments enable MS-MS with high-resolution selection/analysis for both precursor and product ions. CID is performed with high-energy collisions, i.e. in most cases by single high-energy collisions with helium. The four-sector instruments obviously are complex to operate. In addition to the three- and four-sector instruments, a

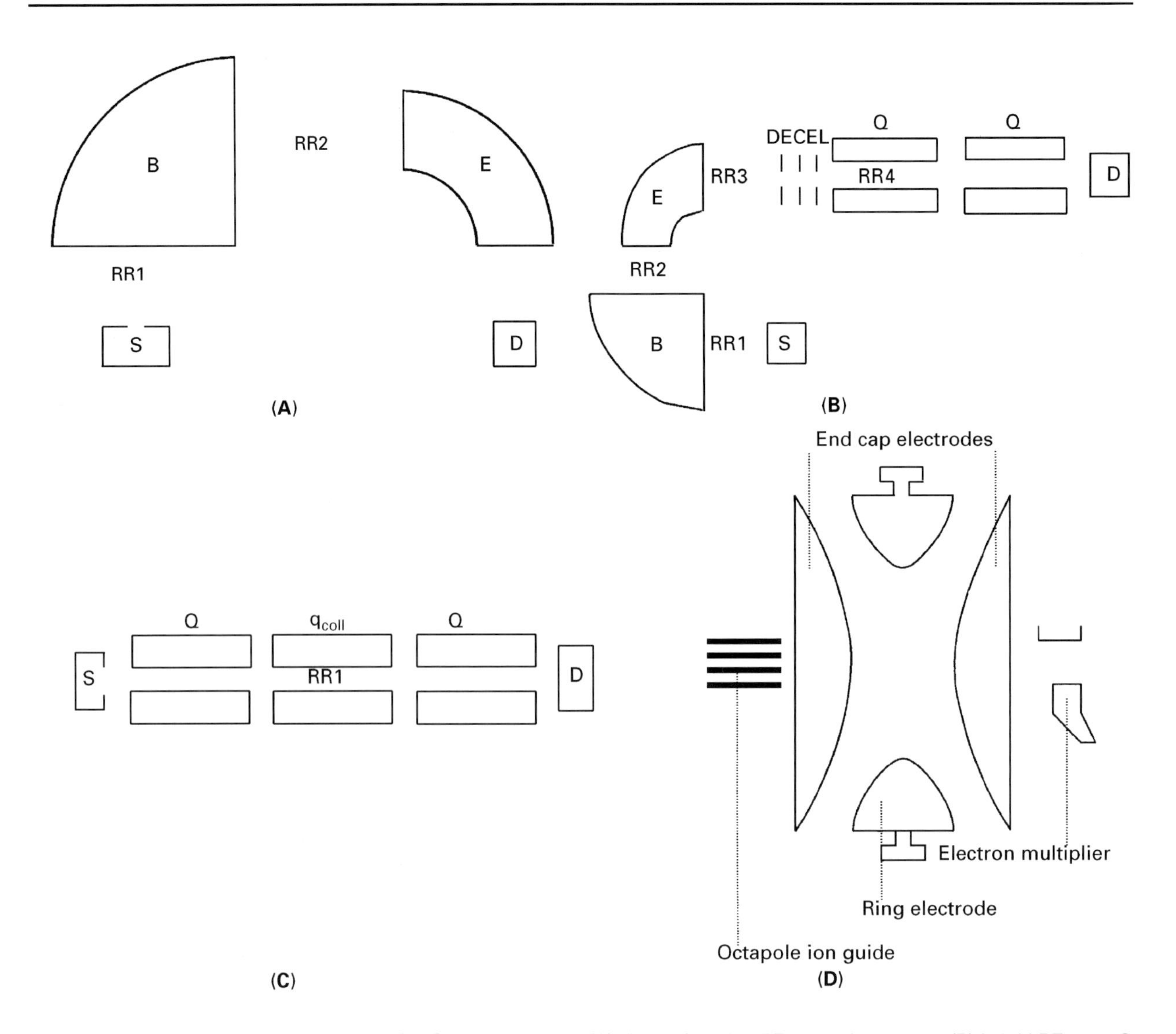

Figure 1 Schematic diagrams of some MS-MS instrumentation: (A) double-focusing BE sector instrument, (B) hybrid BE–q_{coll}–Q instrument, (C) triple quadrupole Q–q_{coll}–Q instrument and (D) ion-trap instrument.

number of hybrid sector-based MS-MS instruments have been developed, i.e. combination of a double-focusing sector instrument as the front-end and a quadrupole, an ion trap or a TOF analyser as the back-end. These hybrid instruments, such as the BE–q_{coll}–Q hybrid schematically drawn in **Figure 1B**, allow high-resolution precursor-ion selection, while collisions can be achieved in either the high-energy or the low-energy regime, i.e. in the third field-free region or in the quadrupole collision cell (RR4 in **Figure 1B**), respectively.

MS-MS in triple-quadrupole instruments

A triple-quadrupole instrument (Q–q_{coll}–Q) consists of two quadrupole mass analysers, connected by means of a collision cell, which is an RF-only quadrupole device (**Figure 1C**). The mass analysers can be used for rapid scanning or selection in various MS-MS scan modes. The collision cell transmits virtually all ions irrespective of their *m/z* without separation of different *m/z*, enabling efficient collection and refocusing of the product ions generated by low-energy, multiple collisions with argon. Good sensitivity is achieved with such an instrument. Two-fold improvements in sensitivity can be achieved by replacing the RF-only quadrupole with hexapole or octapole collision cells. In addition, some manufacturers position the three quadrupole elements in a banana shape to reduce the collisions of neutrals on the detector, thereby reducing the background noise. The triple-quadrupole instruments are currently the most widely used instruments for MS-MS, although most instruments are used for routine quantitative bioanalysis in SRM mode after liquid chromatography (LC).

Recently, an extremely powerful hybrid Q–TOF system has been introduced, based on a quadrupole analyser as the front-end and a TOF analyser as the back-end (Q–q_{coll}–TOF geometry). As a result of the orthogonal geometry of the quadrupole and TOF, the TOF analyser enables high-resolution operation and thus accurate mass determination (up to 5 ppm), which can be extremely useful in structure elucidation by MS-MS, e.g. in peptide sequencing and impurity profiling.

MS-MS and MS-MS^n in ion-trap instruments

MS-MS in an ion-trap instrument is fundamentally different from MS-MS in sector and triple-quadrupole instruments. While in the latter the various stages of the process, i.e. precursor ion selection, CID and product-ion mass analysis are performed in different spatial regions of the instrument, in ion-trap instruments these stages are performed consecutively within the ion trap itself. It is 'tandem-in-time' rather than 'tandem-in-space' mass spectrometry. A simplified diagram of an ion-trap system is shown in **Figure 1D**.

The measurement process consists of a series of steps: ions are either generated inside the ion-trap or injected into the trap from an external source. A suitable storage voltage applied to the ring electrode enables trapping of the ions generated or injected. Next, the precursor ion is selected by applying an ion isolation RF waveform voltage at the endcaps. At the same time the voltage on the ring electrode must be ramped to a new value to store both the precursor ion and its product ions. Subsequently, CID is achieved by applying a resonance excitation RF voltage at the endcaps. This induces faster and more extensive ion trajectories in the trap, resulting in ion activation and subsequent fragmentation. The helium bath gas, present in the ion trap for stabilization of the ion trajectories, serves as the collision gas. In a normal product-ion MS-MS experiment, at this stage the product ions are scanned out of the ion trap towards the detector. However, the system also allows multiple stages of MS-MS, i.e. one of the product ions may be selected by means of an ion isolation RF waveform voltage at the endcaps, and subsequently activated and dissociated. This process can be performed up to ten times in current commercially available ion-trap systems, or as long as a sufficient number of ions remains; in practice, this typically allows four or five stages of MS-MS for most applications.

Compared with triple-quadrupole and especially with sector instruments, the ion-trap instrument provides more efficient conversion of precursor ion into product ions. However, the CID process via resonance excitation, although quite efficient in terms of conversion yield, generally results in only one (major) product ion in the product-ion mass spectrum. While with a triple-quadrupole instrument a series of product ions is observed, only one major fragment ion is observed with the ion trap. Other fragment ions are detected in the ion-trap system after multiple stages of MS-MS. The product-ion mass spectrum of the protonated triazine herbicide propazine described by Hogenboom and co-workers may serve as an example. In the spectrum from a triple-quadrupole instrument, the product ions are found at m/z 188, 146, 110 and 79. With the ion-trap instrument, three stages of MS-MS are required to observe all these ions, i.e. m/z 188 and a minor 146 in the first stage, m/z 146 in the second stage, and m/z 110, 104, 86 and 79 in the third stage. In this experiment, the most abundant product ion in each stage is selected as the precursor ion for the next stage. The fragmentation in the ion trap appears to be softer, or more controllable, than that in triple-quadrupole instruments. This stepwise fragmentation can be extremely useful in structure elucidation, but is a serious limitation in cases where confirmation of identity should be based on the detection of a number of particular product ions.

With the current commercial availability of various ion-trap MS-MS^n systems, equipped with an external ion source and for use in both GC-MS and LC-MS, MS-MS in ion-trap systems can be expected to find more elaborate analytical applications.

MS-MS in time-of-flight instruments

MS-MS in TOF instruments has only recently been reported by Kaufmann and Spengler. A reflectron-TOF should be used for MS-MS. While the ion activation may take place during ionization or even after ion acceleration, the actual fragmentation takes place after the ions have left the ion source (post-source decay). As a result, the precursor ions p and product ions d will have the same velocity, but different kinetic energy. Therefore, the MS-MS experiments can be performed in two steps, outlined in **Figure 2**: first, the instrument is operated in the linear mode, with the flight times of precursor and product ion being the same. Second, the instrument is operated in the reflectron mode. In this case, the precursor ion, having the higher kinetic energy, takes a longer path than the product ion. By comparison of the mass spectra from the linear and the reflectron

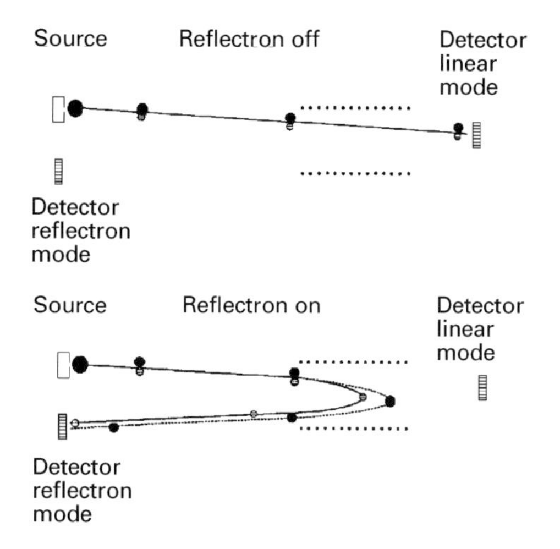

Figure 2 Schematic diagram of the procedure of MS-MS with post-source decay in a reflectron TOF instrument. Reproduced with permission of Masson éditeur and John Wiley & Sons from de Hoffmann E, Charette J and Stroobant V (1996) *Mass Spectrometry, Principles and Applications*. Paris: Masson éditeur. Chichester: John Wiley & Sons.

mode, the fragment ions can be identified. Recently, it was also demonstrated that, via electrostatic ion manipulation of the ion beam, precursor-ion selection before post-source decay can be used. These approaches are extremely useful for the advanced mass analysis of ions generated by matrix-assisted laser desorption/ionization (MALDI).

An interesting approach to MS-MS is the combination of an ion-trap storage device and a reflectron time-of-flight instrument. While initially the ion trap was used only to store ions from the continuous ion source before the pulsed acceleration to the TOF analyser, it has recently been demonstrated by the group of Lubman that MS-MS in the ion trap before TOF mass analysis results in additional possibilities.

MS-MS in FT-ICR instruments

In an FT-ICR instrument, which also is an ion-trapping device, MS-MS can be performed in a manner similar to MS-MS in ion-trap instruments. However, fragmentation by collisions is generally much less effective because of the significantly lower pressures in the FT-ICR cell; this is only partially compensated for by the longer ion residence times that are achievable.

In both the quadrupole ion traps and the FT-ICR instruments, MS-MS in the product-ion scan mode is feasible, but other scan modes such as neutral-loss and precursor-ion scans are not possible.

Applications of MS-MS

Numerous applications of MS-MS have been reported in the literature, in which the MS-MS instrumentation is either used as a stand-alone instrument for sample introduction by a probe, or column-bypass injection in a liquid stream, or used in on-line combination with GC or LC. Especially in the latter area, where soft ionization strategies are frequently applied, MS-MS plays an important role. In this section, a number of applications of different MS-MS instruments is briefly reviewed. Most attention is focused on the analytical applications of MS-MS.

Fundamental studies

The first studies with MS-MS concerned fundamental studies on the properties, structures and behaviour of gas-phase ions. This area continues to be important, as these fundamental studies lead to a better understanding of the processes involved in ion activation, fragmentation and CID. Fundamental studies can be subdivided into studies concerning the structures of ions in the gas phase, elucidation of reaction mechanisms and studies directed at obtaining thermochemical information from gas-phase species.

A currently important area of fundamental research in ion structures and reaction mechanisms, which also has significant analytical application, is the study of the fragmentation mechanisms of biomolecules in MS-MS. MS-MS is frequently applied to perform sequence analysis of biomolecules such as peptides, oligosaccharides and oligonucleotides. Fundamental studies are directed at elucidation of the formation of the various fragment ions observed, i.e. ions from backbone cleavages, side-chain specific ions and low mass ions such as immonium ions. In this respect, but also in other contexts, considerable research is directed at gaining a better understanding of the charge-remote fragmentations observed in the analysis of peptides, fatty acids and other compounds.

Structure elucidation

MS-MS in the product-ion scan mode is generally quite successful in structure elucidation of unknowns. There are ample examples available in the literature. However, the interpretation of the production mass spectrum is not always straightforward. When a protonated molecule is selected as precursor ion, the fragmentation rules significantly differ from the well-known fragmentation rules valid for molecular ions generated by electron ionization. Upon fragmentation, protonated molecules have a high tendency to rearrange and hydrogen shifts often take place. Knowledge of this type of fragmentation is less extensive and less systematically documented. Furthermore, the fragmentation in MS-MS may not result in a sufficient number of fragments to allow complete elucidation of the structure. In that case, additional strategies are required. In MS-MS studies with sample introduction via an electrospray interface for LC-MS, the combination of in-source CID and product-ion MS-MS of one of the fragment ions can be a useful tool to achieve further structure elucidation, as has been demonstrated, for example, by Bateman and co-workers for isomeric sulfonamides separated by capillary electrophoresis.

Because the experimental conditions for product-ion mass spectra are strongly instrument dependent, generally not very well standardized and difficult to exchange between instruments from different manufacturers, there are at present no generally applicable spectral libraries for MS-MS spectra that can assist in structure elucidation problems.

Two recent instrumental innovations can be applied to further assist in the interpretation of product-ion mass spectra. As discussed above, the $MS\text{-}MS^n$ capabilities of an ion-trap system allow the step-wise fragmentation of an analyte, facilitating the interpretation of the product-ion information and the fragmentation reactions involved. The use of a Q-TOF hybrid instrument allows accurate mass determination (at an accuracy of 5 ppm) of the product ions observed, which also facilitates the interpretation of the product-ion mass spectra.

Screening for structurally-related compounds

The combination of MS-MS and a soft ionization method is frequently applied in screening food or environmental samples for contaminants. The strategies involved are briefly illustrated with an example related to the detection of the sulfonamide antibiotic sulfamethazine (SDM) and its possible conversion products in meat samples.

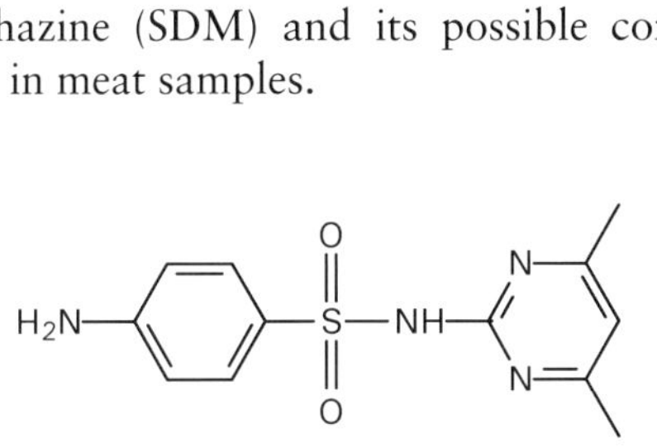

In the product-ion mass spectra of protonated SDM and one of its expected conversion products (deaminosulfamethazine, DAS), a common peak at *m*/*z* 124 is observed, corresponding to the protonated aminodimethylpyrimidine. This ion is complementary to the ion owing to the phenyl–SO_2^+ group, found at *m*/*z* 141 for DAS and at *m*/*z* 156 for SDM. To screen meat samples for the presence of SDM and its conversion products, a neutral-loss scan was applied with a common loss of 123 Da, corresponding to the loss of the aminodimethylpyrimidine. This provides a highly selective screening method for SDM-related compounds: only those compounds in the meat extract that show the neutral loss of 123 Da are detected. The result for a sausage extract, shown in **Figure 3**, indicates that both SDM and DAS are present in the meat extract, as well as a compound with *m*/*z* 280, most likely a hydroxy analogue, and a monochloro compound with *m*/*z* 298, where the chlorine attachment is due to the brine used in production of the sausage. In this particular case, the use of the precursor-ion scan mode with *m*/*z* 124 as the common product ion would have been successful as well. However, for many compounds the charge is preferentially retained at only one part of the molecule, and in CID the part of the molecule with the most-favourable ionization properties will be observed as an ion in the product-ion mass spectrum, while the complementary part is lost as a neutral and thus not detected. For example, this is the case with protonated nucleosides: after CID only the protonated nucleobase is observed, while no fragment ions due to the sugar ring are observed. The neutral-loss scan mode is now applicable to screen for modifications of the base, while possible modifications of the sugar ring remain undetected. The complementary precursor ion scan is not applicable in this case.

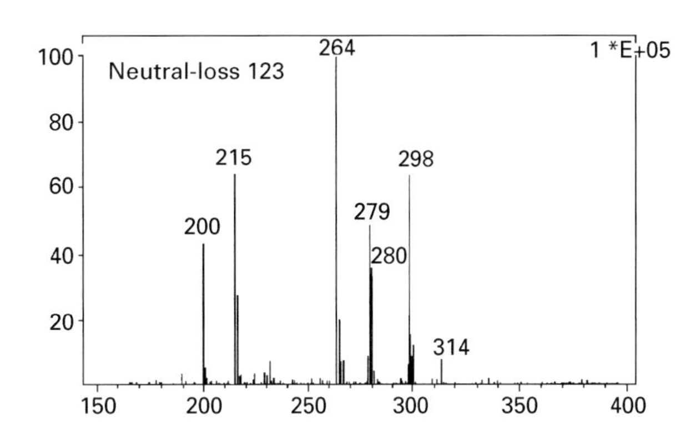

Figure 3 Neutral-loss 123 Da mass spectrum obtained from the introduction of a sausage extract into a thermospray-MS-MS system. The peaks detected at *m*/*z* 200 and 215 were also present in the blank.

Similar screening procedures may be applied to perform Phase II metabolic profiling, i.e. based on the neutral-loss scan mode with a loss of 176 or 80 Da for glucuronide and sulfate conjugates, respectively.

Quantitative bioanalysis

The use of MS-MS, especially in SRM mode, is an important application of MS-MS, especially in terms of instrument sales. Triple-quadrupole instruments are most widely used for this purpose, in most cases in an online LC-MS combination. In developing the method, the MS-MS parameters are optimized in such a way that only one or two intense fragment ions are preferentially observed. These fragment ions are subsequently used in the SRM procedure, selecting the precursor ion in the first mass analyser and the fragment ion in the second mass analyser. By means of SRM, a specific reaction in the gas phase is monitored. By careful selection of an appropriate reaction, excellent selectivity can be achieved, and the analyte of interest can be analysed at low levels in complex matrices. This approach is especially useful in high-throughput quantitative bioanalysis, required to acquire pharmacokinetic and pharmacodynamic data of new drugs and their metabolites during drug developmental studies. In such studies, an isotopically labelled internal standard is applied. This internal standard preferentially shows the same neutral loss as the analyte of interest, enabling the detection of the analyte and the internal standard at two different *m/z* values in the SRM procedure.

Peptide sequencing

A special application of structure elucidation by product-ion MS-MS is the sequence determination of amino acids in a peptide, which has been reviewed by Papayannopoulos. The low-energy CID of a protonated peptide in a triple-quadrupole instrument leads to a series of fragment ions that originate from backbone cleavages. The cleavage of the peptide bond primarily leads to either an acylium ion, when the charge is retained on the N-terminal side of the peptide, or a protonated peptide when the charge is retained at the C-terminal side of the peptide. According to the Roepstorff–Fohlmann nomenclature, these ions are indicated as b_n and y_n'' ions, respectively. The unknown amino acid sequence may be read from the two complementary series of ions in the product-ion mass spectrum. In high-energy CID, more advanced information can be obtained in peptide sequencing because, in addition to the backbone fragmentation, side-chain-specific fragment ions may be observed as well. This is especially important in the structure elucidation of isomeric amino acids, e.g. leucine and isoleucine. Given the extensive use of MALDI in peptide and protein mass analysis, the use of post-source decay processes to achieve peptide sequencing is of increasing importance.

See also: **Ion Trap Mass Spectrometers; Metastable Ions; Peptides and Proteins Studied Using Mass Spectrometry; Sector Mass Spectrometers; Surface Induced Dissociation in Mass Spectrometry; Time of Flight Mass Spectrometers.**

Further reading

Bateman KP, Locke SJ and Volmer DA (1997) Characterization of isomeric sulfonamides using capillary zone electrophoresis coupled with nano-electrospray quasi-MS/MS/MS. *Journal of Mass Spectrometry* **31**: 297.

Busch KL, Glish GL and McLuckey SA (1988) *Mass Spectrometry/Mass Spectrometry*. New York: VCH Publishers.

Cooks RG, Beynon JH, Caprioli RM and Lester GR (1973) *Metastable Ions*. Amsterdam: Elsevier.

Hogenboom AC, Niessen WMA and Brinkman UATh (1998) Rapid target analysis of microcontaminants in water by on-line single-short-column liquid chromatography combined with atmospheric-pressure chemical ionization ion-trap mass spectrometry. *Journal of Chromatography* A **794**: 201.

Hunt DF, Shabanowitz J, Harvey TM and Coates ML (1983) Analysis of organics in the environment by functional group using a triple quadrupole mass spectrometer. *Journal of Chromatography* **271**: 93.

Johnson JV and Yost RA (1985) Tandem mass spectrometry for trace analysis. *Analytical Chemistry* **57**: 758A.

McLafferty FW (ed) (1983) *Tandem Mass Spectrometry*. New York: John Wiley & Sons.

Papayannopoulos IA (1995) The interpretation of collision-induced dissociation tandem mass spectra of peptides. *Mass Spectrometry Reviews* **14**: 49.

Perchalski PJ, Yost RA and Wilder BJ (1982) Structural elucidation of drug metabolites by triple quadrupole mass spectrometry. *Analytical Chemistry* **54**: 1466.

Spengler B (1997) Post-source decay analysis in matrix-assisted laser desorption/ionization mass spectrometry of biomolecules. *Journal of Mass Spectrometry* **32**: 1019.

Wu J-T, He L, Li MX, Parus S and Lubman DM (1997) On-line capillary separations/tandem mass spectrometry for protein digest analysis by using an ion trap storage/reflectron time-of-flight mass detector. *Journal of the American Society of Mass Spectrometry* **8**: 1237.

Multiphoton Excitation in Mass Spectrometry

Ulrich Boesl, Technische Universität München, Germany

MASS SPECTROMETRY
Methods & Instrumentation

Multiphoton excitation in mass spectrometry is determined by the characteristics of the applied light source (wavelength and intensity). One dominant feature therefore is spectroscopy. This concerns spectroscopy as a means of obtaining a species-selective ion source as well as mass selection for species-selective spectroscopy in a mixture. The latter leads to a rich manifold of techniques in many spectroscopic fields such as UV spectroscopy, cation and anion spectroscopy and photoelectron spectroscopy. Another dominant feature is photodissociation. Here multiphoton excitation allows variation from exceptionally soft ionization to very hard dissociation. It enables high yields of metastable decay products, which are particularly valuable for kinetic and energetic studies of molecular ions, a field of basic research in mass spectrometry. A third aspect is good compatibility with other techniques, e.g. chromatography or laser desorption, creating connections to different types of molecular systems (e.g. biomolecules) as well as to other analytical techniques. The last feature of multiphoton excitation dealt with in this article is options in analytical chemistry. The combination of speed, selectivity and sensitivity in a multiphoton ion source has great potential for environmental trace analysis or process integrated analysis of industrial production procedures, to name only two fields of application.

Multiphoton excitation schemes

In **Figure 1** different types of multiphoton excitation are schematically presented.

Figure 1A. Resonance ionization spectroscopy is performed by monitoring molecular ions while tuning the laser wavelength. If the energy of one photon is in resonance with a neutral electronic excited state a second photon is able to be absorbed, giving rise to an ion current peak. Thus, the neutral UV-absorption spectrum is transferred to the ion current which can be recorded mass selectively (in opposition to the absorption). Thus UV spectroscopy and mass spectroscopy are combined as a two-dimensional technique.

Figure 1B. Instead of a fixed mass window, the laser wavelength may be kept constant and in resonance with a specific transition of one molecule in a mixture (e.g. molecule M_A) but out of resonance with other molecules (e.g. molecule M_B). Thus, a species selective ion source is achieved. This is unique in mass spectrometry and particularly useful for trace analysis. In addition, by selecting particular intermediate states, molecular cations may even be formed in a few vibrational or rotational levels of the ionic ground state. This is a very valuable feature for studying molecular structure, reaction kinetics or spectroscopy of molecular ions.

Figure 1C. Additional excitation of the molecular ions takes place at increased laser intensities, where photon absorption (in the ion and fragment ion manifold) and dissociation processes to fragment ions follow each other. Owing to this 'ladder switching' the degree of fragmentation is tuneable from soft ionization (mainly or solely molecular ions) to hard ionization (mainly small or even atomic fragment ions). At appropriate laser wavelengths and intensities metastable decay processes may dominate the mass spectrum. Multiphoton excitation and dissociation is also possible with delayed laser pulses and at different positions in space, giving rise to new forms of tandem mass spectrometry.

Figure 1D. If a tunable laser is used for secondary excitation of molecular ions, resonance dissociation spectroscopy of molecular cations may be performed. Here excited cationic levels are subject to laser spectroscopy. They serve as intermediate states for the process of resonance enhanced multiphoton dissociation. This is quite similar to resonance ionization spectroscopy of neutrals. The difference is that a dissociation instead of an ionization continuum is finally reached by multiphoton excitation. The advantage of this technique is that it is independent of high ion numbers (as necessary for absorption spectroscopy), fluorescence [necessary for laser-induced fluorescence (LIF)] or predissociation and therefore is fairly general. In addition, mass selectivity is intrinsic and one may benefit from state selective ion formation if resonance multiphoton ionization is used as an ion source.

Figure 1E. The degree of this state selective ion formation can be tested by analysing the kinetic energy of the emitted electrons while the laser

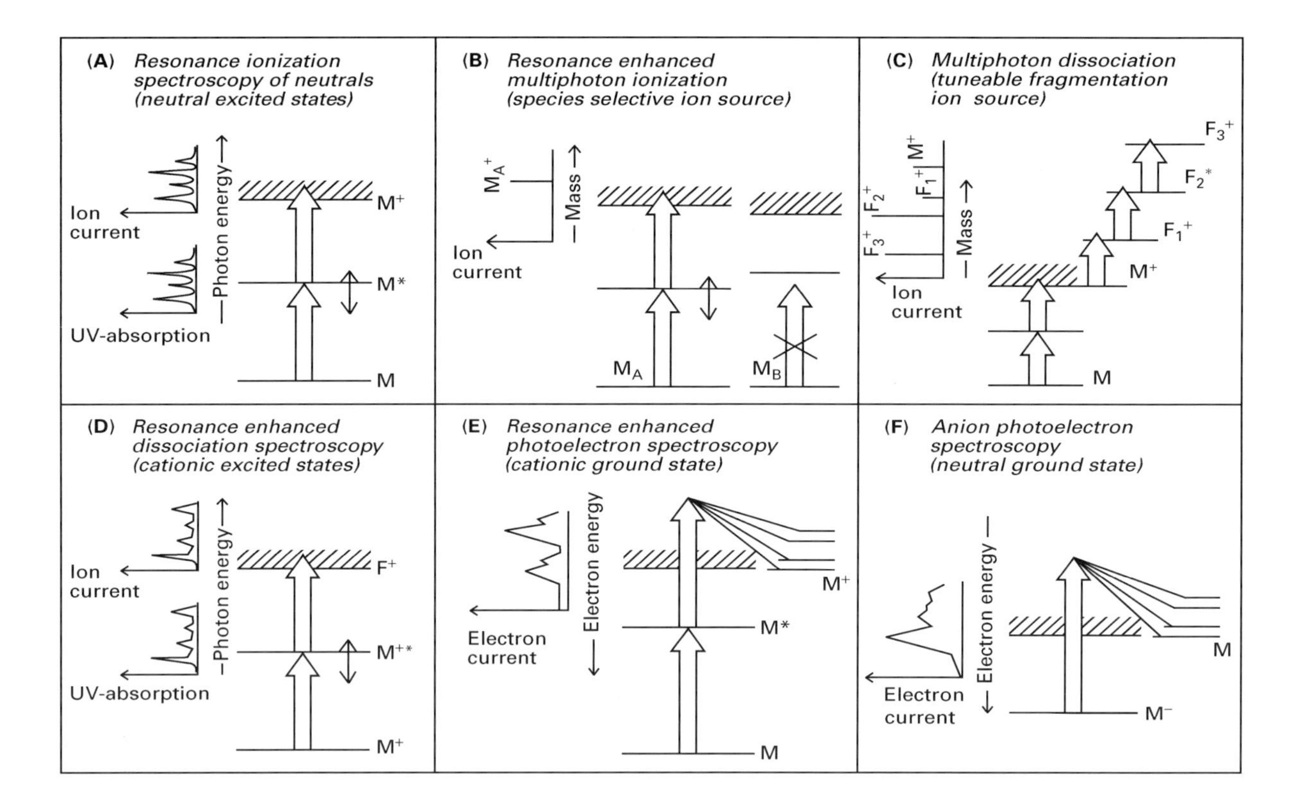

Figure 1 Different multiphoton excitation schemes applicable to molecular systems in the gas phase in mass spectrometry.

wavelength is kept in resonance with the selected intermediate levels. This resonance enhanced photoelectron spectroscopy gives information about cationic ground states. It also helps in studies of the nature of the neutral intermediate states due to selection or propensity rules valid at the ionization and photoelectron emission process. A special, high-resolution version of photoelectron spectroscopy is zero kinetic energy photoelectron spectroscopy (ZEKE). If an intermediate excited state is involved, the first photon has to be kept in resonance with it while the second one is tuned over so-called ZEKE states very near to ionization thresholds. They end up in cationic states after field ionization. Instead of electrons, the cations may be recorded, allowing the option of mass selectivity.

Figure 1F. Finally, anion photoelectron spectroscopy (or photodetachment photoelectron spectroscopy) is discussed (although mostly a one-photon process). Owing to electron affinities being much smaller than ionization energies one photon absorption and detachment with lasers is possible for most anions. Either conventional photoelectron spectroscopy (**Figure 1F**) with a fixed laser wavelength or anion-ZEKE spectroscopy (similar but not equal to ZEKE) with tuned laser wavelengths is possible. The intriguing feature of anion photoelectron spectroscopy is that it starts with ionic species (mass selection before spectroscopy) and ends up with neutral species (access to short-lived species, radicals, weakly bound molecular systems, traces in complex mixtures, etc.).

Experimental realization in a mass spectrometer

In **Figure 2** several experimental set-ups are summarized in a hypothetical instrument that consists of a central part (neutral source, ion sources, mass separator, ion detector) and several options for a variety of multiphoton excitation experiments (see **Figure 1**). The neutral inlet system or source may be narrow tubing, for an effusive molecular beam, or a pulsed nozzle for supersonic molecular beams. The advantage of the former is its very simple and inexpensive construction. By placing the end of the tubing between the electrodes of the ion extraction optics (ion source II) high gas densities may be achieved at the position of laser ionization without too high a load on the vacuum system. Supersonic beams, on the other hand, allow efficient cooling of the internal degrees of freedom of molecular motion. Well-structured spectra of even larger molecules are

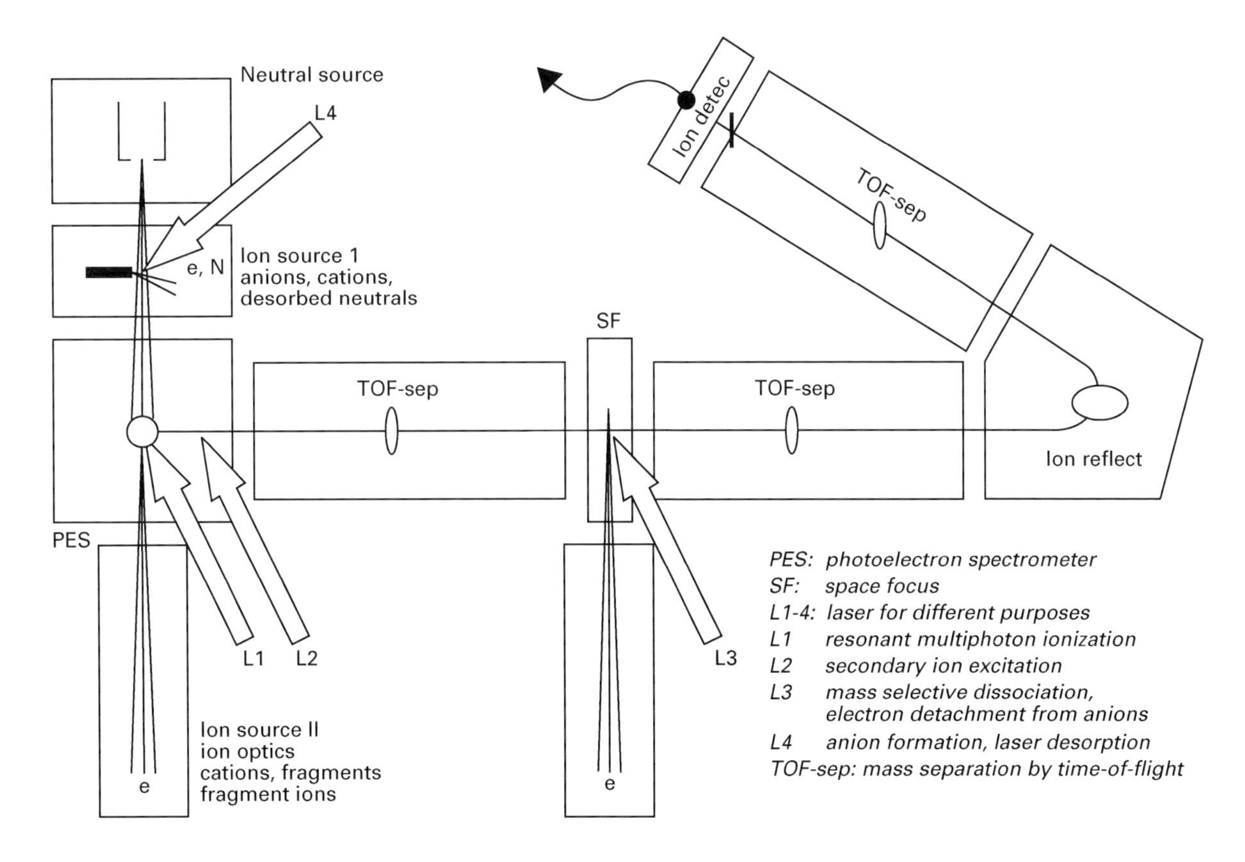

Figure 2 A hypothetical instrument that summarizes experimental arrangements where multiphoton excitation is involved.

thus achievable, containing rich spectroscopic information and enabling highly selective excitation and multiphoton ionization. Large involatile neutral molecules may be transferred into the gas phase by laser desorption. The insertion of a gas chromatographic capillary column in the neutral source even makes possible species selection before the gas inlet.

There exist a number of laser-induced techniques for ionizing these neutrals, such as resonance-enhanced multiphoton ionization (in ion source I or II, **Figures 1A** and **1B**), laser-induced vacuum UV ionization (e.g. by 118 nm from a 3×355 nm/Nd:YAG laser), electron attachment (in ion source I) and electron ionization (in ion source II). Electrons may be supplied by laser-induced photoelectron emission from metal surfaces, preferably from thin wires made out of material having a low work function (e.g. hafnium). Ions formed in source I have to drift into the ion optics of the mass spectrometer (together with the neutral molecular beam) and can be extracted by a pulsed electric field while ionization in source II may be performed within a static electric field with instantaneous ion extraction.

Extracted ions will enter the field-free drift region of a time-of-flight (TOF) mass separator. Time-of-flight mass analysers have turned out to be the ideal mass selection tools for pulsed laser-induced multiphoton excitation and ionization. A first mass selective detection may be performed in the so-called space focus of the ion source. Even for very short field-free drift regions (e.g. some 15 cm) a mass resolution of 200 to 300 is possible in routine operations. A considerable enhancement of mass resolution is achieved by adding a special ion reflector with further field-free drift regions. In such a reflectron TOF analyser the ion cloud in the space focus (SF) is imaged onto the ion detector. This preserves the flight time distribution Δt, but extends the total flight time t thus enhancing the mass resolution ($R_{50\%} = \frac{1}{2}t/\Delta t$).

Optional photoelectron spectrometers are included in **Figure 2**. These preferably are electric and magnetic field-free drift regions, allowing the analysis of electron kinetic energies by measuring electron flight times. They may be combined with ion source II for analysing photoelectrons emitted at resonance enhanced laser ionization (**Figure 1E**). A photoelectron TOF spectrometer can also be placed at the SF. Electrons due to photodetachment of anions may then be analysed (**Figure 1F**). Since in the space focus masses are already separated, this kind of anion photoelectron spectroscopy is intrinsically mass selective. For neutral ← cation photoelectron spectroscopy at ion source II, mass selection is only possible by additional photoion/photoelectron coincidence techniques.

In **Figure 2**, several laser beams are supposed to be used in combination or separately. By laser beam L1 resonance enhanced multiphoton ionization and spectroscopy may be performed (**Figures 1A** and **1B**). At higher intensities of laser L1, multiphoton dissociation additionally takes place (**Figure 1C**). Secondary multiphoton dissociation is possible by laser beam L2. Mass selective analysis of the manifold of secondary fragments results in tandem mass spectrometry. Recording the intensity of a selected fragment ion as function of laser L2 wavelength allows cation UV-visible spectroscopy (**Figure 1D**). Both experiments (tandem mass spectrometry or cation UV-spectroscopy) are also possible with laser beam L3 coupled into the SF. By the latter experimental arrangement considerably higher secondary mass selectivity is achieved, while the former arrangement may give better sensitivity. In addition, with laser L3 mass selective neutral ← anion photoelectron (**Figure 1F**) and anion-ZEKE spectroscopy can be performed as mentioned above. Finally, laser beam L4 is used for photoelectron emission from a wire for anion formation by electron attachment. In addition, desorption of involatile neutral molecules into a supersonic molecular gas beam is performed with laser L4.

Resonance multiphoton ionization: spectroscopy and selective ion source

Mass selective laser excitation was applied originally for the UV and VIS spectroscopy of neutral molecules. However, the benefits of both mass selected UV spectra and UV-resonance selective mass spectra were soon recognized.

In **Figure 3** (right-hand side) the resonance ionization spectra recorded mass selectively (at 106 and 107 amu) show the electronic origin of the $S_1 \leftarrow S_0$ transition of *p*-xylene (106 amu) and its $^{13}C_1{}^{12}C_7H_{10}$ natural isotopomer. An isotopic blue-shift of +2.7 cm^{-1} (corresponding to 0.02 nm) is found. On the left-hand side of **Figure 3** two UV-resonance selected mass spectra are shown, resulting from resonance enhanced multiphoton ionization at 272.17 nm and 272.14 nm. The latter is the band centre of the 107 amu selective laser spectrum and clearly gives rise to a strong relative enhancement of the heavier mass in the mass spectrum.

An extraordinary impact on molecular UV spectroscopy (in particular of large molecules) has been brought about by supersonic beam cooling of

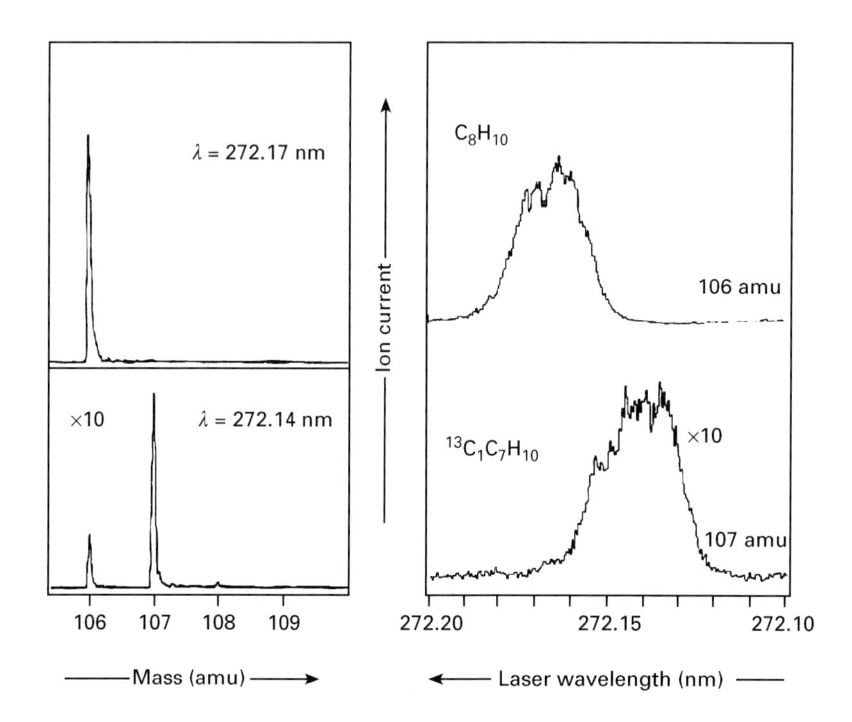

Figure 3 Two options for resonance enhanced multiphoton ionization concerning ion source or spectroscopy: UV-resonance selected mass spectra (on the left) and mass selected molecular UV spectra (on the right) for *p*-xylene and its $^{13}C_1$-isotopomers.

internal molecular motions. The narrow line width in **Figure 3** is due to this effect and allows isotopomer-selective ionization as illustrated in **Figure 3**. However, even for large involatile molecules well-resolved spectra with vibrational fine structure can be observed. To get these molecules into the gas phase, high temperatures or laser desorption were necessary. **Figure 4** shows the spectra of heated gas samples containing dibenzo-dioxins seeded in argon gas and cooled in a supersonic molecular beam. For dibenzodioxins only their UV spectra in solution were known before the application of mass selective resonance ionization spectroscopy. On the left-hand side of **Figure 4**, the cold spectrum of unsubstituted dibenzodioxin is represented. Single vibronic bands of the $S_1 \leftarrow S_0$ transition appear, revealing low frequency 'butterfly' and torsional modes excited in S_1. This and further arguments suggest a slight non-planarity of dibenzodioxin in the S_0 and planarity of its S_1 and ionic ground state.

The combination of supersonic beams with mass selective spectroscopy has a further advantage. While mass selection allows discrimination against impurities, fragments and differently substituted (e.g. halogen, isotope) species, the high spectroscopic resolution enables the discrimination of structural isomers having the same mass. **Figure 4** clearly shows the significant difference of spectra of dibenzodioxins with two chlorine atoms substituted at symmetrically equivalent positions (i.e. the 2, 3, 7, and 8 positions). They differ only by their position relative to each other [e.g. neighbouring (bottom right) or not]. Wavelengths can easily be found (e.g. between 32 790 and 32 805 cm^{-1}) where these isomers may be ionized selectively. Spectroscopic studies of many different substituted aromatics have been performed. Effects such as increasing ionization energies with increasing degree of halogenation, decreasing energy of the S_1 state (e.g. below half the ionization energy) with increasing molecular size and strongly differing lifetimes of the intermediate states have been experimentally observed or theoretically determined, and of course, these effects have to be considered for efficient ionization schemes.

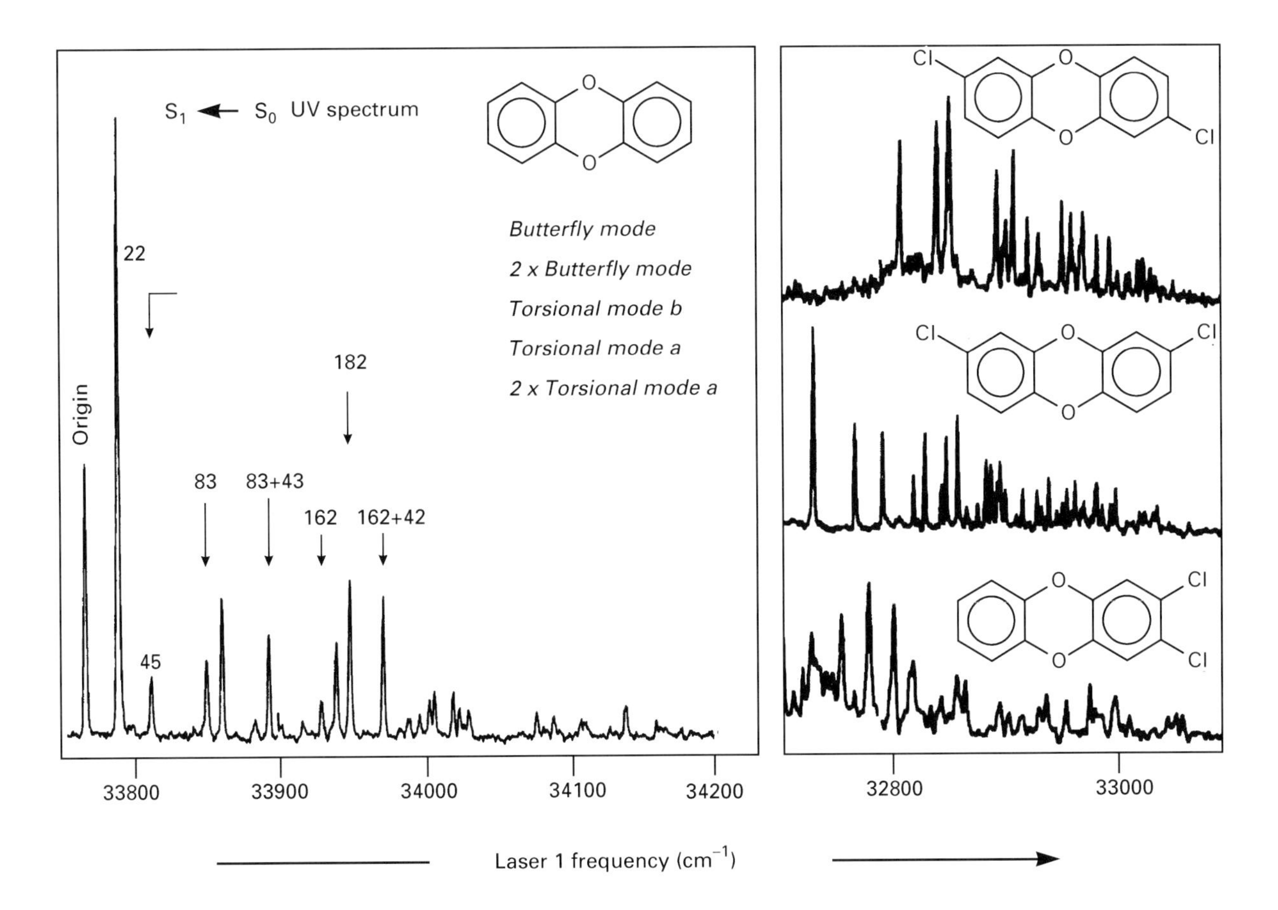

Figure 4 On the left: the cold (supersonic beam cooled) UV spectrum of dibenzo-dioxin measured via multiphoton ionization. On the right: The dichloro-congeners (with chlorine at the 3, 8, the 2, 8, and the 2, 3 positions) show similar well-structured spectra, giving wavelengths for congener selective ionization.

Multiphoton dissociation: tuneable fragmentation ion source

Species-selective ionization is one quality of multiphoton excitation in mass spectrometry. Another is the option to tune the degree of fragmentation from exceptionally low to very strong. The different possibilities of this feature are illustrated by the example of benzene in **Figure 5**. The variation of fragmentation is obtained by changing the laser intensity from 10^7 W cm^{-2} (corresponding to a UV-laser pulse with a 10 µJ pulse energy, a 10 ns pulse length and a 100 µm focus diameter) to 10^9 W cm^{-2}. While the first mass spectrum is nearly free of fragments, the last mass spectrum consists mainly of carbon ions at mass 12. Both extremes cannot be achieved by conventional electron ionization or other ion sources and are unique for multiphoton excitation. It should be mentioned that these mass spectra have been taken with a short TOF mass spectrometer consisting only of three electrodes and one ion detector.

The reason for the astonishing feature of tuneable fragmentation is the specific excitation mechanism of multiphoton absorption in comparison with electron ionization which is a one-step process (the total energy is deposited in the neutral molecule in one step, causing ionization and fragmentation). Multiphoton ionization is due to consecutive absorption steps which are interrupted if fast decay channels are reached and are continued within the decay products (so-called 'ladder-switching'). Increasing the intensity increases the absorption probability (less time per absorption step), allowing the whole process to reach higher steps of the 'ladder of fragment ions' during the laser pulse. This process is indicated schematically in **Figure 5**. However, if very intense and very short laser pulses are used (i.e. femtosecond lasers), absorption is faster than decay and even ionization process and the total absorbed multiphoton energy is deposited in the molecule as a single event.

Since fragmentation is a rich source of information in mass spectrometry, this feature of multiphoton

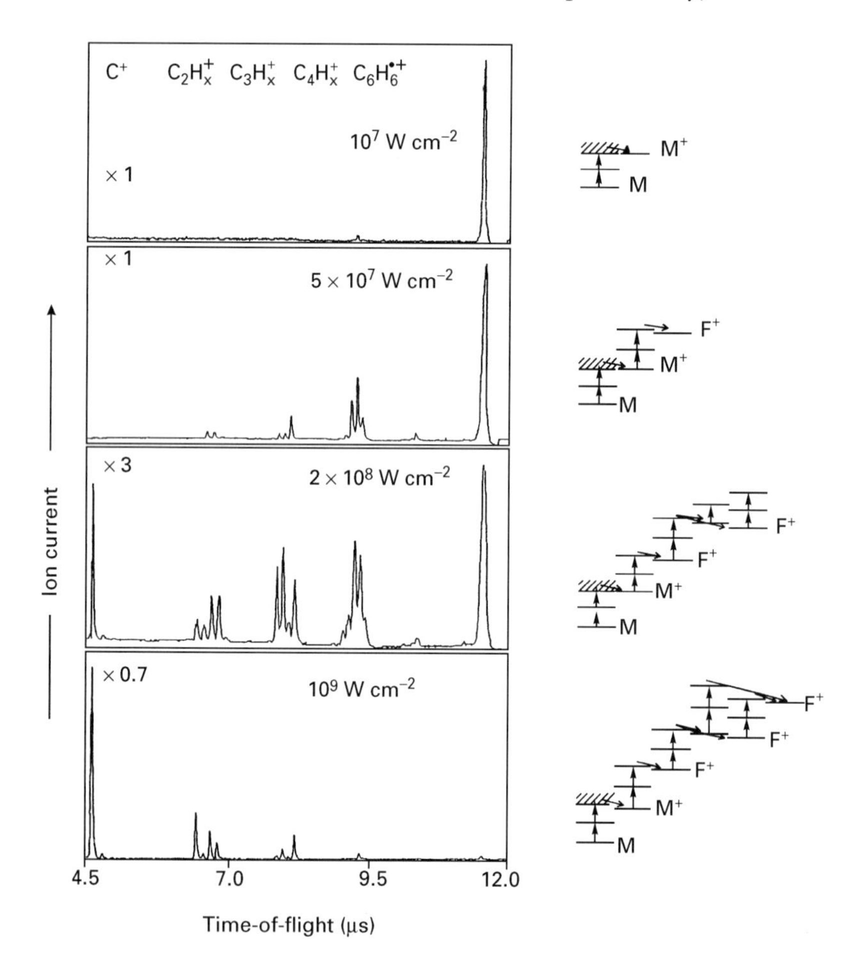

Figure 5 Multiphoton dissociation of benzene at different laser intensities, inducing different degrees of fragmentation (from fragmentation-free at the top to dominating atomic fragment ions at the bottom of the figure). On the right of the spectra the model of 'ladder switching' is schematically displayed and explains the large variation of fragmentation by multiphoton absorption processes.

excitation may give a new impetus to this field of research. For example, increasing the laser intensity may allow one to tune for specific decay channels (e.g. breakage of bonds to functional groups, typical substituents or bonds within the molecular framework) and thus supply a low level but inexpensive type of tandem mass spectrometry. However, multiphoton excitation also allows access to the sensitive tool of metastable decay, the most informative source of fragmentation in mass spectrometry. While in conventional mass spectrometry considerable effort is necessary to study tandem mass spectra (mass separation before secondary collision-induced excitation) multiphoton excitation may create controlled ion decay dissociation (e.g. by enhancing specific metastable decay channels) very efficiently already in the ion source. This is illustrated and explained in **Figure 6**.

Here, part of a multiphoton ionization mass spectrum in the region around the benzene radical cation is represented. While the 'linear TOF' spectrum only reveals the molecular ion and its $^{13}C_1$-isotopomer, the 'reflectron TOF' spectrum additionally shows the fragment ions $C_6H_5^+$ and $C_6H_4^{\bullet+}$ although the ion source conditions were identical. (The difference in mass resolution will not be discussed here, see the Further reading section and other articles in this Encyclopedia.)

To understand this effect one should remember that product ions formed in the field-free drift region of a linear TOF instrument cannot be distinguished from their parent ion (same ion velocity). A reflectron TOF mass analyser, however, corrects for ion kinetic energy deviations and shows all fragment ions of the same mass independent of their ion kinetic energy (at least for an energy mismatch of ± 10%) at the same flight time. Thus **Figure 6** demonstrates the existence and high yield of metastable decay induced by multiphoton excitation (in **Figure 5** the total fragmentation yield was small, main peak was $C_6H_6^{\bullet+}$). An explanation for this is given at the top of **Figure 6**. By resonance enhanced multiphoton ionization, molecular ions $M^{\bullet+}$ with small internal energy and a narrow energy distribution ΔE are produced (soft ionization, no absorption in the ion). The absorption of a further photon in the molecular ion $M^{\bullet+}$ will then result in excited ions which still have a narrow energy distribution ΔE^*. If the photon energy is in the range of the energy interval between the ion ground state and the threshold of a decay channel then the overlap of the metastable decay region (near F^+ decay threshold) and internal energy distribution ΔE^* will be large. In contrast to multiphoton excitation, electron ionization is usually performed with 70 eV electrons,

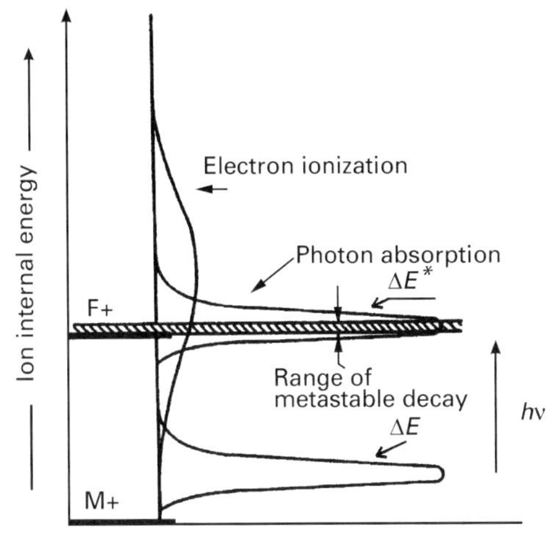

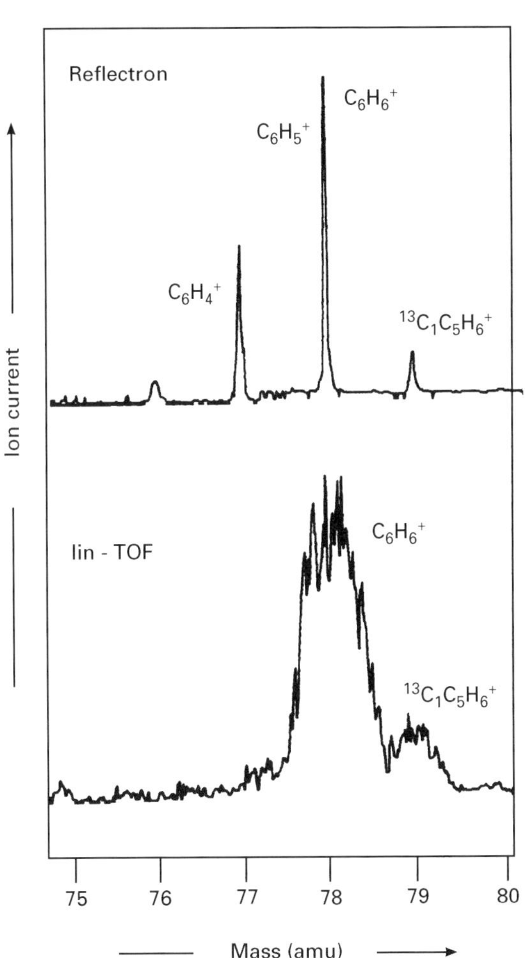

Figure 6 Experimental demonstration (bottom) and schematic explanation (top) of the high yield of metastable decay products available with multiphoton dissociation.

resulting in a broad internal energy distribution whose overlap with the metastable decay range is relatively small (few metastable decay products). The special features of multiphoton excitation concerning metastable decay have initiated a renaissance of ion kinetic decay studies.

Instead of a single colour excitation (laser L1, **Figure 2**) a second laser (L2 or L3) can be used. Secondary multiphoton excitation is then possible, which can be performed with different laser wavelengths, at different positions and different times from that of the ionizing laser L1. Thus, selective secondary excitation of a group of fragments (e.g. with L2 in ion source II) or even of single product ions with one well-defined mass (e.g. with L3 in the SF) is possible. This already is an approach to highly sophisticated tandem mass spectrometry with the benefit of well-defined internal ion energies and a high yield of secondary fragmentation. In **Figure 7**, a primary multiphoton mass spectrum of benzene and secondary mass spectra of four different fragment ions are shown. This experiment has been performed to elucidate multiphoton fragmentation and is shown here to illustrate its tandem mass spectrometry option. The primary mass spectrum is due to a laser wavelength of $\lambda_1 = 258$ nm (absorption maximum of benzene), while the secondary mass spectra of $C_4^{\bullet+}$, C_4H^+ and $C_4H_2^{\bullet+}$ have been performed with $\lambda_3 = 355$ nm. The C_4H_3+ fragment does not show fragmentation at $\lambda_3 = 355$ nm and has been excited

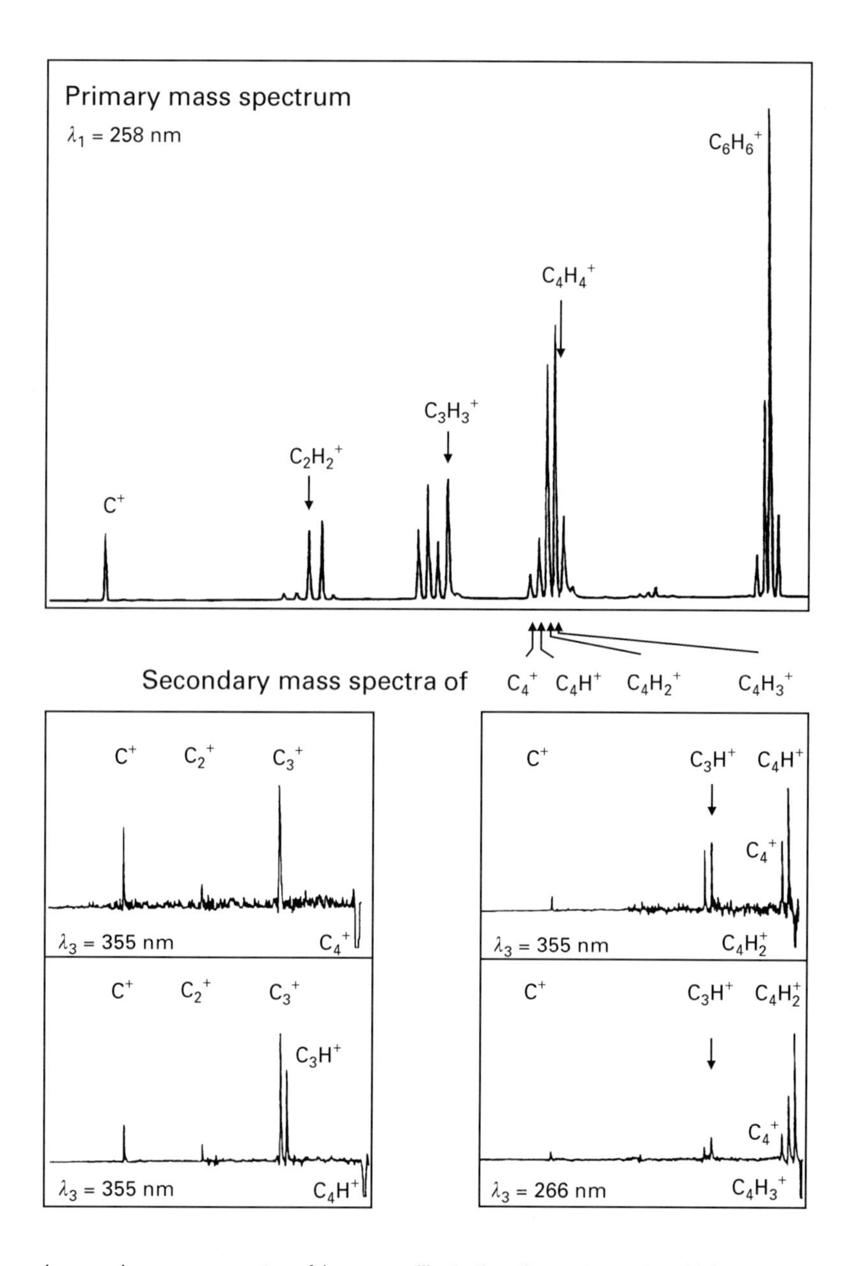

Figure 7 Primary and secondary mass spectra of benzene, illustrating the options of multiphoton excitation for tandem mass spectrometry.

with $\lambda_3 = 266$ nm. Furthermore, C_4H^+ does not exhibit a secondary $C_4^{\bullet+}$ product ion as $C_4H_2^{\bullet+}$ and $C_4H_3^+$ although it dissociates at 355 nm. These effects already indicate a complex multiphoton fragmentation tree with several decay channels. It is not the task of this article to go into more detail of that specific process, rather this example should illustrate the available options of multiphoton excitation for tandem mass spectrometry.

Multiphoton excitation for mass selective ion spectroscopy

There are several options for the use of multiphoton excitation for ion spectroscopy. One is resonance enhanced photoelectron spectroscopy. In contrast with conventional photoelectron spectroscopy, where single VUV photons are used, a neutral excited intermediate state is involved which allows the application of the low energy UV-photons available with lasers. Owing to small excess energies, the ionic ground state is most often studied by this technique. The involvement of different vibronic intermediate states results in a variation of the photoelectron spectrum and is a valuable means for obtaining additional information about the character of ionic and even of neutral intermediate states. Resonance enhanced photoelectron spectroscopy cannot be performed in a truly mass selective way. Either the neutral source consists only of one sort of molecule (which can be checked by synchronously observing the produced molecular ions) or photo-ion photoelectron coincidence spectroscopy is performed. But the high-resolution version of resonance enhanced photoelectron spectroscopy, ZEKE spectroscopy, can be performed in a true mass selective mode. Both methods are beyond the scope of this article (for further information see the Further reading section).

In addition to the study of the ionic ground state, resonance enhanced photoelectron spectroscopy allows one to characterize multiphoton ion sources with respect to the internal energy distribution of the produced ions. For favourable cases it even may supply information as to how to achieve state selective ion production. The example of fluorobenzene has been chosen in **Figure 8** to illustrate resonance enhanced photoelectron spectroscopy as well as further types of ion spectroscopy involving multiphoton excitation.

At the bottom of **Figure 8** the resonance ionization spectrum of fluorobenzene (cooled in a supersonic beam) is shown, exhibiting the electronic origin 0^0_0 as well as other vibronic transitions (e.g. excitation of one quantum of the 6b-vibration). In

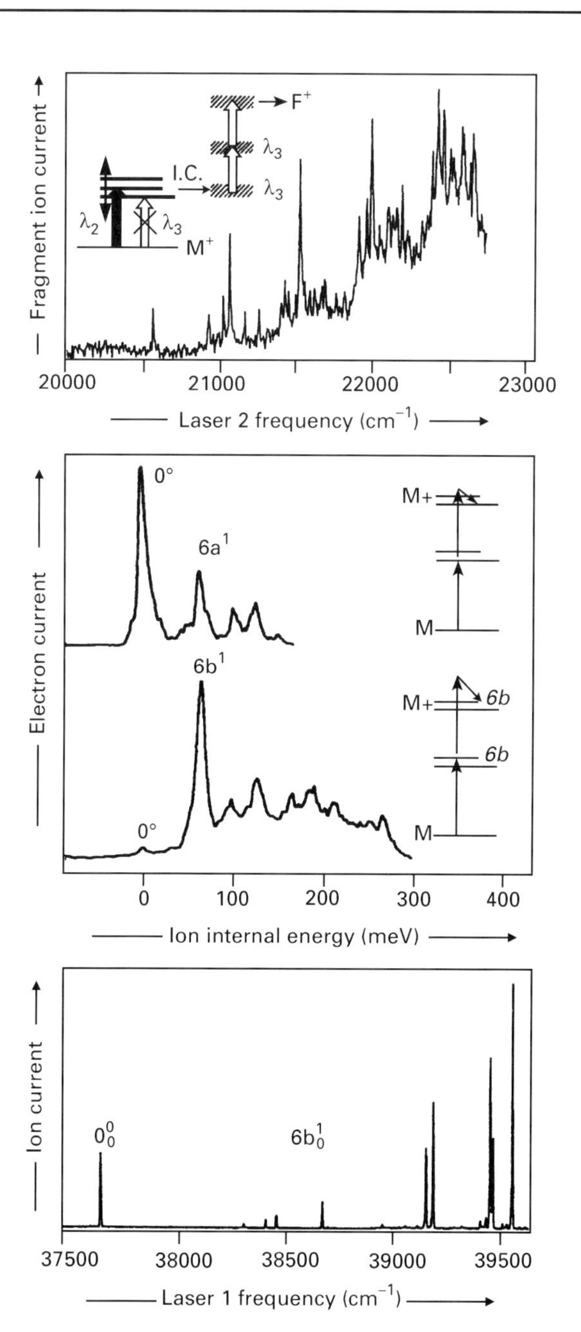

Figure 8 Multiphoton excitation and ion spectroscopy: Bottom spectrum: cold, mass selected UV spectrum ($S_1 \leftarrow S_0$ transition) of fluorobenzene, providing wavelengths for efficient and selective ionization. Middle spectra: photoelectron spectra induced by UV-resonance enhanced two-photon absorption. Choosing different intermediate states [$S_1(0,0)$ or $S_1(6b^1)$] results in different populations of the final fluorobenzene radical cations. Top spectrum: spectroscopy of the excited ionic state of the fluorobenzene radical cation measured by multiphoton dissociation spectroscopy. The ions have been prepared via the neutral 0^0_0 transition. A special excitation scheme has been used to optimize cation spectroscopy (for further details see text).

the middle of **Figure 8**, two photoelectron spectra are presented which involve different neutral intermediate states. A significant difference is observed, indicating that the excitation of the vibrationless S_1 or of the 6b-mode in the S_1 induces different populations in the molecular ion. This supports assignments in the photoelectron spectrum, and allows production of state-selected fluorobenzene ions for further experiments (ion kinetics, ion spectroscopy).

Such a secondary experiment is displayed at the top of **Figure 8**. After production of fluorobenzene radical cations via resonance enhanced multiphoton ionization, preferentially in the ionic ground state, multiphoton dissociation spectroscopy has been applied. However, in contrast to **Figure 1D**, an even more elaborate excitation scheme has been used here, exploiting the experimental options of TOF mass spectrometry. Laser L2 (ion source II) is used solely for spectroscopy; laser L3 (SF) is responsible for detection by multiphoton dissociation. The benefit of this scheme is that laser L2 can be optimized for spectroscopy while dissociation by laser L3 is highly mass selective, efficiently suppressing background signals from other molecular or fragment ions. Using modern double-pulse lasers, the requirements for the laser system and mass analyser are still reasonable. The result is a very well-resolved ion spectrum. One should keep in mind that in the case of fluorobenzene, as for many other halogen-substituted and the non-substituted benzene cations, fluorescence spectroscopy is not possible (to say nothing about absorption spectroscopy) and only low-resolution conventional photoelectron spectra exist.

When treating ion spectroscopy one should not forget anions. Similar spectroscopic techniques may be used as for cation spectroscopy. For instance dissociation spectroscopy is also possible for molecular anions. Since excited anionic electronic states mostly do not exist, one uses infrared multiphoton dissociation to study vibrational levels of the ground state. Another interesting technique is the photoelectron spectroscopy of anions (photodetachment photoelectron spectroscopy), which exhibit a very specific feature. This technique differs from cation ← neutral photoelectron spectroscopy in two respects: (i) the final state is a neutral one; thus anion photoelectron spectroscopy delivers information about neutrals rather than ionic systems. (ii) The initial state is anionic; thus mass selection before spectroscopy is possible. As a result, mass selective spectroscopic information of neutral molecular systems is supplied which otherwise is not accessible. This is of particular interest for neutral systems which are only available in complex mixtures or are short-lived intermediate reaction products or radicals.

One example is shown in **Figure 9**. Metal–carbon–hydrogen complexes are of great importance as intermediates in heterogeneous catalytic reactions of hydrocarbons on metal surfaces, but are not available as pure samples. At the bottom of **Figure 9** an anion mass spectrum of such complexes is presented. These complexes have been formed by a gas phase reaction in the high density region of a supersonic molecular beam (acetylene seeded in argon). Iron atoms as well as low-energy electrons are supplied by laser ablation from an iron wire which is positioned as near as possible to the nozzle of the supersonic beam valve. Clearly, $FeC_iH_n^-$ complexes appear in the spectrum together with carbon and carbon hydride clusters. Mass selected anion photoelectron spectra which supply information about the neutral complexes FeC_2, FeC_2H and FeC_2H_2 are presented at the top of **Figure 9**. Obviously, an increasing hydrogen content of the complex results (i) in a decreasing electron affinity (transition to state X), (ii) in a smaller change of molecular structure (Fe–C distance) between anion and neutral (Franck–Condon envelope of the Fe–C stretching frequency) and (iii) a strong decrease of energy gaps between higher electronic states (X, A, B, C) (a preliminary assignment in the case of FeC_2H_2). Although anion photoelectron spectroscopy is mostly owing to a one-photon process, it has been considered here to show the large variety of experimental possibilities when combining laser excitation and mass spectrometry.

Combination of multiphoton excitation with other neutral sources: laser desorption

Since multiphoton excitation in mass spectrometry takes place in the more or less tight laser focus, which can easily be shifted in space and time or be subject to other variations, it can be combined with different ion optical or mechanical arrangements (e.g. sources of neutral molecular systems) without the need for much additional hardware. Thus, by combination with chromatography (particularly gas chromatography), species selection has successfully been realized. Another very promising combination, which has frequently been applied in the recent past for the study of involatile molecules (e.g. polycyclic aromatics, biomolecules), is that of laser desorption of neutral molecules and resonance enhanced multiphoton ionization. All the benefits of multiphoton mass spectrometry, such as soft ionization, selective ionization, controllable fragmentation or secondary excitation for tandem mass spectrometry, may be used in this field.

In **Figure 10**, the example of a biologically relevant molecule is displayed. Porphyrins containing central metal atoms (e.g. Mg or Fe) represent the

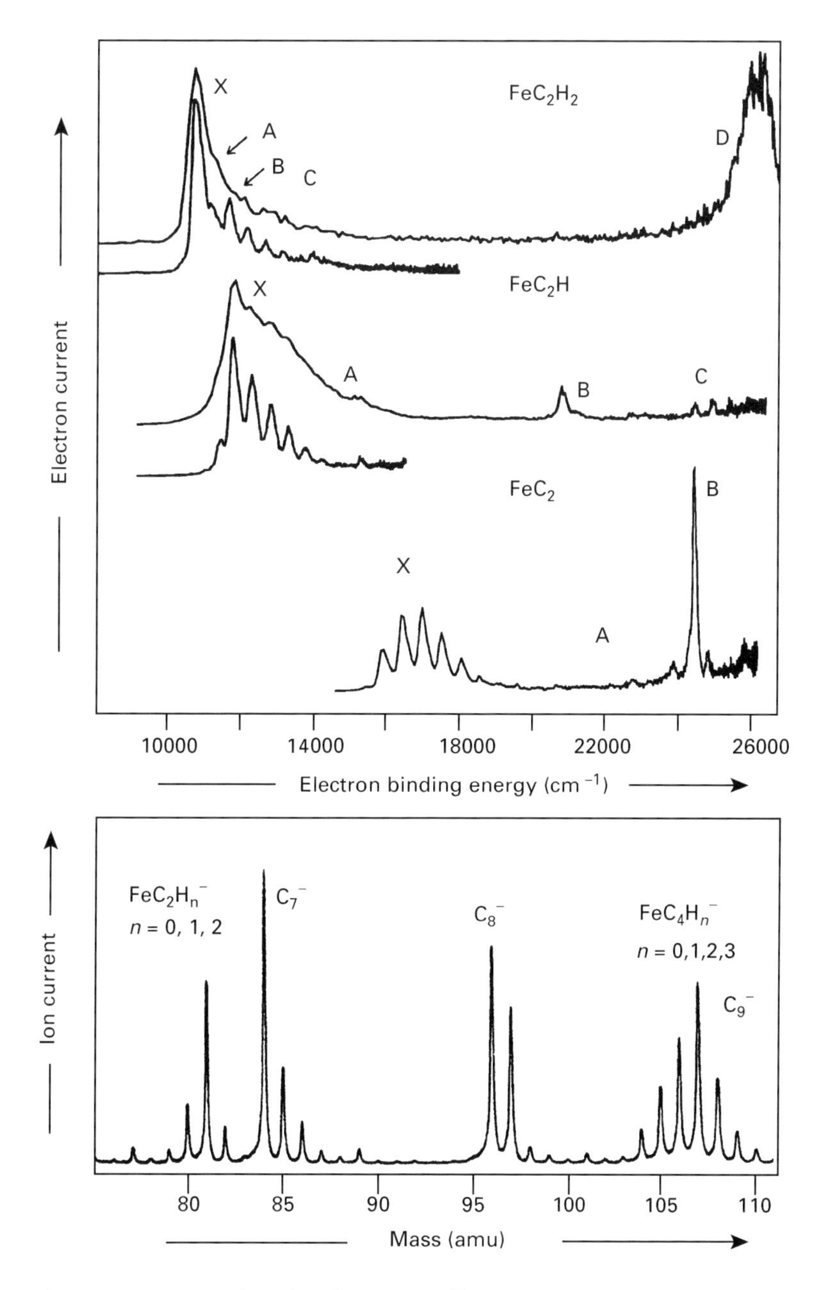

Figure 9 Anion photoelectron spectroscopy. Its unique features are (i) intrinsic mass selectivity and (ii) neutrals as final states. Here, as an example the results for compounds of iron, carbon and hydrogen are shown which exist in catalytic processes, high-temperature terrestrial or low-temperature astrophysical chemistry. Bottom spectrum: a primary anion mass spectrum containing anions of the complexes of interest. Top spectra: anion photoelectron spectra obtained by electron kinetic energy analysis after laser-induced photodetachment. They reveal the change of molecular structure and electronic energies for increasing numbers of hydrogen atoms in the complex.

molecular frame of important biomolecules such as chlorophyll or haemoglobin which mainly differ in their central metal atom and specific side-chains. In **Figure 10**, the TOF mass spectrum of zinc mesoporphyrin-IX dimethyl ester is shown, revealing a small degree of fragmentation. Inset A displays the molecular ion on an expanded mass scale and nicely reveals the typical isotopic pattern of Zn convoluted with that of ^{13}C-isotopomers (at the used wavelength of $\lambda_1 = 280$ nm and the internal molecular temperature of ≥ 300 K, ionization is not isotope selective). The fragment peaks near 155 μs consist of a narrow peak ($M^+ - 64$) and a group of peaks ($>M^+ - 73$). The latter results from the loss of a side-chain (Zn-

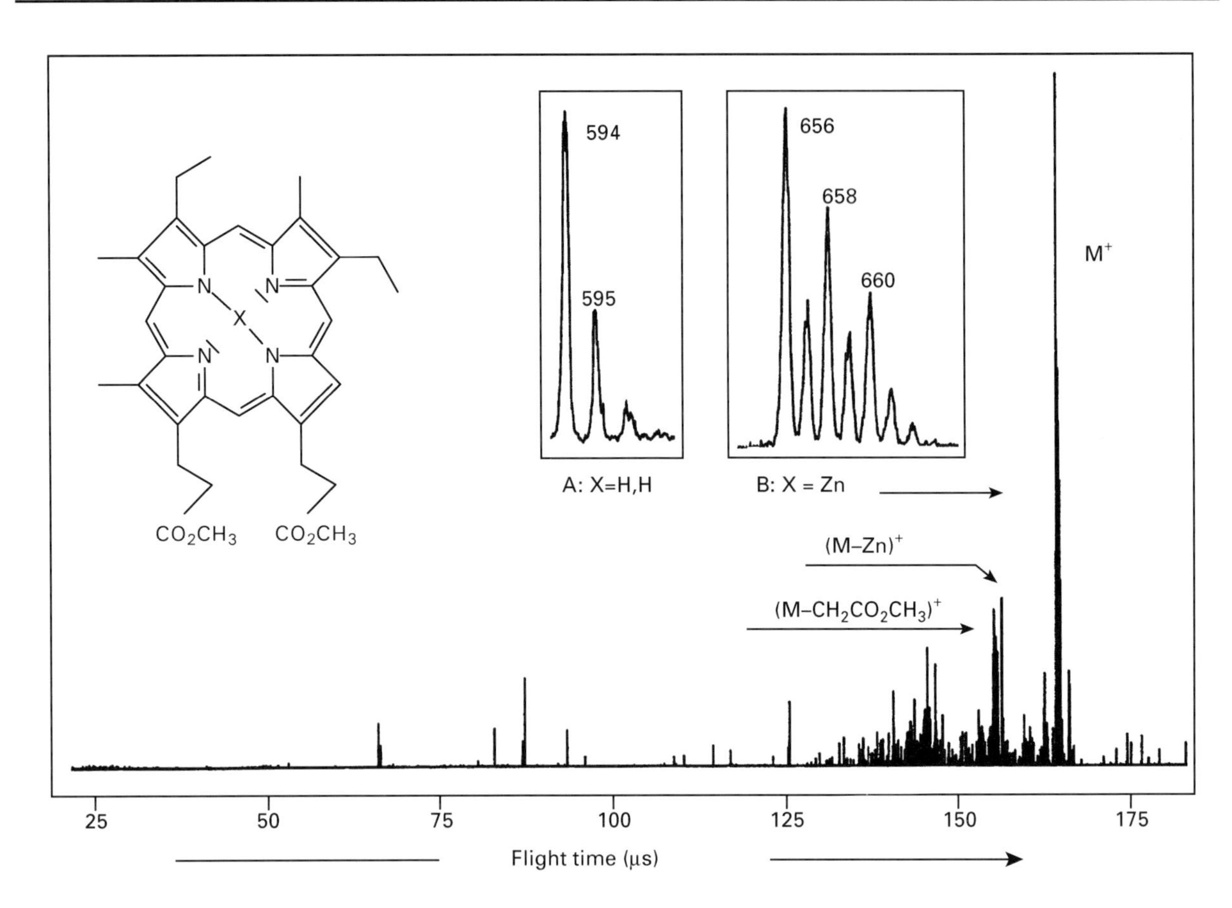

Figure 10 The combination of neutral laser desorption and resonance enhanced multiphoton ionization. A TOF mass spectrum of zinc mesoporphyrin-IX dimethyl ester is presented. Inset *A*: the isotropic pattern (zinc, carbon) of the molecular ion. Inset *B*: the molecular ion of mesoporphyrin-IX dimethyl ester (TOF mass spectrum not shown) which displays the same isotope pattern as peak (M–64) due to loss of a zinc atom.

isotope pattern preserved) while the former is owing to the loss of the central Zn atom (no Zn-isotope pattern remains). It shows a nearly identical isotope pattern (owing to ^{12}C–^{13}C distribution) to the mesoporphyrin-IX dimethyl ester without a central atom, whose molecular ion is shown in inset B of **Figure 10**.

Multiphoton mass spectrometry for chemical trace analysis

Both the high species selectivity and the soft ionization make resonance enhanced multiphoton excitation an excellent ion source for chemical trace analysis. In addition, the use of pulsed lasers combined with TOF mass spectrometry (also a pulsed technique) enables the recording of single mass spectra within a few milliseconds, depending on the repetition frequency of the laser pulses (typically 20 to 50 Hz). The unification of high speed, selectivity and sensitivity indicate that multiphoton mass spectrometry will become firmly established in the highly sophisticated analytical technology of the future. In **Figure 11**, the variety of applications is illustrated by examples from different fields, namely raw gas analysis of a waste incinerator, research for health care, production integrated analysis in the food industry and exhaust trace analysis of motor cars.

In **Figure 11A** (incinerator), traces of naphthalene (sub-ppb range) in the raw gas of a pilot waste incinerator have been recorded over 900 s. During this time a major failure of the combustion process (at 500 s) which gave rise to an increased concentration of polycyclic aromatics by many orders of magnitude has been studied. This event was preceded by small but significant fluctuations of the napthalene concentration, which were only detectable by a fast, but nevertheless selective and sensitive, method. Effects like these could be used in future for online control of incinerator plants. This could help to improve pollutant emission by reducing their formation rather than retaining them after the combustion process by expensive cleaning procedures. An example of

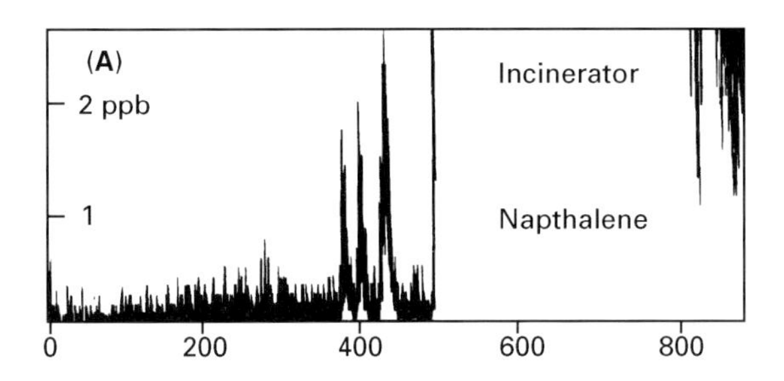

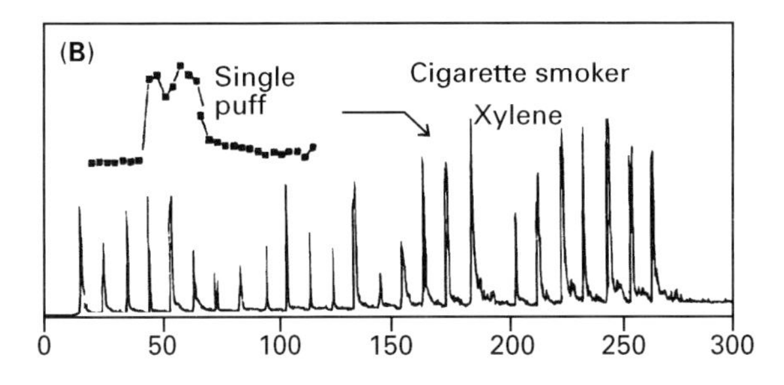

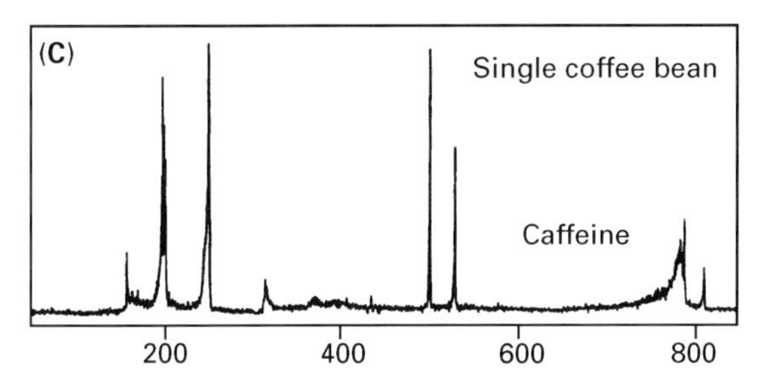

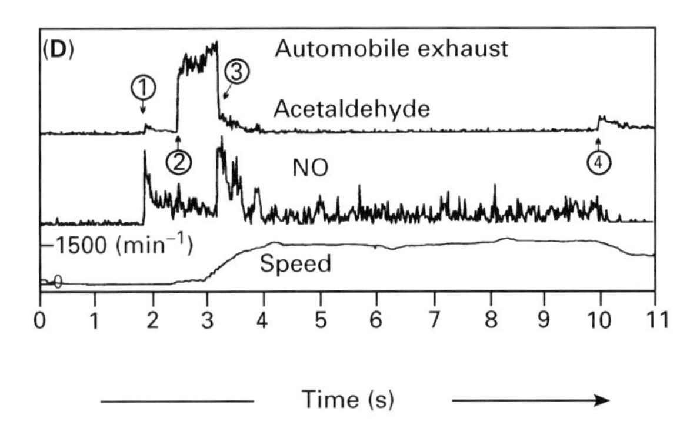

Figure 11 Several examples of applications of multiphoton ionization mass spectrometry for chemical analysis. (A) Incinerator: traces of naphthalene have been recorded in the raw gas of a pilot waste incinerator. Here a major failure of the combustion and preceding fluctuations of the naphthalene concentration were studied. (B) Cigarette-smoker: the xylene concentration in the mouth space of a cigarette-smoker. (C) Single coffee bean: a single coffee bean has been heated to untypically high temperatures. Single short ejection events of caffeine have been observed due to distinct cracks of the plant skin of the coffee bean. (D) Automobile exhaust: acetaldehyde due to incomplete combustion and NO (complete combustion) during the starting of a cold engine measured in the exhaust. Different behaviour is observed when the ignition is starting (1), the fuel mixture becomes rich (2) and lean (3) and the speed of idle running is reduced to its regular frequency (4).

even more direct health care research is shown in **Figure 11B** (cigarette smoker). Here the xylene content in air recorded in the mouth space of a cigarette smoker is recorded. The 50-fold expansion of a single puff demonstrates the high temporal resolution of less than 0.1 s.

The next spectrum shows fluctuations of caffeine concentration in the headspace above one single green coffee bean which has been suddenly heated up to high temperatures (far above typical roasting temperatures). The very short spikes of caffeine emission are induced by CO_2 (pyrolysis product) causing distinct cracks of the plant skin. At the single ejection events solvated caffeine is swept along with the CO_2. Of course, other molecular components (e.g. those which are used as indicators for the progress of coffee roasting) can be monitored thus enabling online control of the roasting process.

In **Figure 11** (automobile), the behaviour of acetaldehyde (a dehydrogenation product of ethanol which is typical for incomplete combustion) and nitric oxide in the exhaust of a motorcar is recorded during the first seconds of a cold engine start. Unleaded gasoline containing 10% of ethanol has been used as fuel. Both pollutants strongly support the formation of ozone in the troposphere (e.g. smog in large cities). Their reduction, e.g. by an electronically improved regulation of the motor, is therefore of major interest. Thus, the highly dynamic processes in combustion engines must be studied, but appropriate fast analytical techniques are not yet commercially available. In **Figure 11D** five different situations initiated by distinct fast events of the motor regulation can be recognized. From 0 to 2 s, no combustion is active and none of the two pollutants are emitted. At time (1) ignition starts, causing a high peak of the combustion product NO which drops off to a level of some 100 ppm. At time (2), a fuel-rich mixture is injected for a better start of the engine. A sudden increase of acetaldehyde is observed, indicating incomplete combustion due to a shortage of oxygen. At time (3) the speed of idle running is reached (faster in a newly started engine for stabilization of the combustion) and the fuel is changed to a lean mixture. Acetaldehyde, the product of incomplete combustion drops instantly, while NO, the product of complete combustion, increases strongly again. However, heavy fluctuations of NO still indicate some instability of the whole process. After stabilization (NO level now at 80 to 100 ppm) the idle running frequency is reduced to the regular speed at time (4). The NO level drops significantly, but a slight increase and then a gradual decrease of acetaldehyde is observed. This is consistent with a reduced combustion temperature at lower speed and desorption of fuel from the walls of the injection system for about half a second.

List of symbols

$R_{50\%}$ = mass resolution; t = time; ΔE = energy distribution; ΔE^* = internal energy distribution, λ = wavelength.

See also: **Environmental Applications of Electronic Spectroscopy; Fragmentation in Mass Spectrometry; Ion Dissociation Kinetics, Mass Spectrometry; Ion Energetics in Mass Spectrometry; Laser Applications in Electronic Spectroscopy; Metastable Ions; Multiphoton Spectroscopy, Applications; Negative Ion Mass Spectrometry, Methods; Photoelectron–Photoion Coincidence Methods in Mass Spectrometry (PEPICO); Photoionization and Photodissociation Methods in Mass Spectrometry; Spectroscopy of Ions; Time of Flight Mass Spectrometers.**

Further reading

Baer T (ed) (1996) Ion spectroscopy. *International Journal of Mass Spectrometry and Ion Processes (Special Edition)* **159**: 1–261.

Boesl U (1991) Multiphoton excitation and mass selective ion detection for neutral and ion spectroscopy. *Journal of Physical Chemistry* **95**: 2949–2962.

Boesl U, Weinkauf R and Schlag EW (1992) Reflectron time-of-flight mass spectrometry and laser excitation for the analysis of neutrals, ionized molecules and secondary fragments. *International Journal of Mass Spectrometry. Ion Processes (Special Edition)* **112**: 121–166.

Boesl U, Weinkauf R, Weickardt C and Schlag EW (1994) Laser ion sources for time-of-flight mass spectrometry. *International Journal of Mass Spectrometry. Ion Processes (Special Edition)* **131**: 87–124.

Boesl U, Heger HJ, Zimmermann R, Püffel PK and Nagel H (2000) Laser mass spectrometry in trace analysis. Submitted to *Encyclopedia of Chemical Analysis*. John Wiley & Sons. In press.

Gobeli DY, Yang JJ and El-Sayed MA (1985) Laser multiphoton ionization-dissociation mass spectrometry. *Chemical Reviews* **85**: 529–554.

Grotemeyer J, Boesl U, Walter K and Schlag EW (1986) A general soft ionization method for mass spectrometry: resonance-enhanced multiphoton ionization of biomolecules. *Organic Mass Spectrometry* **21**: 645–653.

Hayes JM (1987) Analytical spectroscopy in supersonic expansions. *Chemical Reviews* **87**: 745–760.

Kimura K (1987) Molecular dynamic photoelectron spectroscopy using resonant multiphoton ionization for photophysics and photochemistry. *International Reviews in Physical Chemistry* **6**: 195–226.

Letokhov VS (ed) (1985) *Laser Analytical Spectrochemistry*. Bristol: Adam Hilger.

Letokhov VS (1987) *Laser Photoionization Spectroscopy*. Orlando: Academic Press.

Lubman DM (ed) (1990) *Lasers and Mass Spectrometry*. New York: Oxford University Press.

Neusser HJ (1989) Lifetimes of energy and angular momentum selected ions. *Journal of Physical Chemistry* **93**: 3897–3907.

Vertes A, Gijbels R and Adams F (eds) (1993) *Laser Ionization Mass Analysis*. New York: John Wiley & Sons.

Zenobi R (1995) In situ analysis of surfaces and mixtures by laser desorption mass spectrometry. *International Journal of Mass Spectrometry Ion Processes* **145**: 51–77.

Zimmermann R, Lenoir D, Kettrup A, Nagel H and Boesl U (1996) On-line emission control of combustion processes by laser-induced resonance-enhanced multiphoton ionization mass spectrometry. *Twenty Sixth Symposium (International) on Combustion*, pp 2859–2868 Pittsburgh: The Combustion Institute.

Multiphoton Spectroscopy, Applications

Michael NR Ashfold and **Colin M Western**,
University of Bristol, UK

ELECTRONIC SPECTROSCOPY
Applications

Multiphoton spectroscopy involves the excitation of an atom or molecule from one electronic state A to another B by absorption of two or more photons in contrast to more conventional spectroscopies that involve just a single photon. We identify two distinct variants of multiphoton spectroscopy. **Figure 1A** depicts a sequential multiphoton absorption, i.e. an incoherent multiphoton excitation proceeding via successive allowed one-photon transitions:

$$\mathrm{A} \xrightarrow{h\nu_1} \mathrm{M} \xrightarrow{h\nu_2} \mathrm{B} \quad [1]$$

where the two photons may have the same (or, more generally, different) frequencies. Such schemes provide the basis for a wide range of double-resonance spectroscopies. By way of contrast, **Figure 1B** illustrates the simultaneous (i.e. coherent) absorption of two photons without the involvement of any resonant intermediate state M. The realization that an atom or molecule could undergo such coherent multiphoton excitation dates back to the early days of quantum mechanics, but the experimental demonstration of such multiphoton excitations had to await the advent of the laser to provide the necessary very high light intensities required to compensate for the inherently low multiphoton transition cross sections. With a conventional light source, a two-photon excitation might be 10^{20} times less probable than a one-photon absorption, but the I^n intensity dependence combined with the $> 10^{12}$ increase in power density available with lasers makes such transitions relatively straightforward to observe with modern pulsed lasers. The fraction of photons absorbed will always be small, so identifying that a multiphoton excitation has occurred almost always involves monitoring some consequence of the multiphoton excitation rather than observing the absorption itself. Three of the more common consequences are depicted in **Figure 1B**. If the excited state B fluoresces, the multiphoton excitation spectrum can be obtained simply by monitoring the (multiphoton) laser-induced fluorescence (LIF) as a function of excitation wavelength. Given the high light intensities required to drive the multiphoton absorption step, however, it will generally be the case that some of the molecules excited to state B will absorb one (or more) additional photons and ionize. This is termed resonance-enhanced multiphoton ionization (REMPI), and is the most widely used method for detecting multiphoton absorption by gas-phase species. The third possible process for the excited state, dissociation (or any other loss mechanism), will reduce the photons or ions detected, and is a potential limitation that is discussed further below. The other important condition is that the B ← A two-photon (or multi-photon) excitation has a nonzero transition probability; the selection rules depend on the number of photons and the differences between one- and two-photon transitions have many analogues with those distinguishing infrared and Raman vibrational spectroscopy. If the selection rules are satisfied, then the spectrum obtained by measuring the ion yield (or the yield of the accompanying photoelectrons) as a function of excitation wavelength will provide a signature of the B ← A two-photon transition of the neutral molecule; analysis can provide structural (and, in some cases, dynamical) information about the excited state B.

Multiphoton selection rules

As in one-photon spectroscopies, symmetry is crucial in determining multiphoton transition probabilities. A multiphoton transition between two states A and B is 'allowed' if the transition moment $\langle A | T^k_q(\hat{O}) | B \rangle$ is nonzero; i.e. if the product of the irreducible representations for the wavefunctions of state A and B and that of $T^k_q(\hat{O})$—the qth component of the spherical tensor of rank k representing the multiphoton transition operator $\hat{O}$—contains the totally symmetric representation. Symmetry considerations ensure that only spherical tensors of either odd or even rank will contribute to any one-colour multiphoton excitation. Thus, for example, whereas one-photon electric dipole transitions must be carried by components of rank 1, only components of rank $k = 0$ and/or $k = 2$ can contribute to two-photon transitions brought about using photons of identical frequency and polarization. The $k = 0$ component (a scalar) can only contribute to a two-photon transition connecting states of the same symmetry. Identification of $k = 0$ components in two-photon excitation spectra is generally rather straightforward since they are forbidden (and 'thus disappear') when the spectrum is recorded using circularly polarized light. Sensitivity to the polarization state of the exciting radiation is one important feature distinguishing one-photon and multiphoton transitions. As **Tables 1** and **2** show, for all but the least symmetric molecules, at least some of the $k \neq 1$ components will span representations different from (or additional to) those of the one-photon electric dipole moment operator.

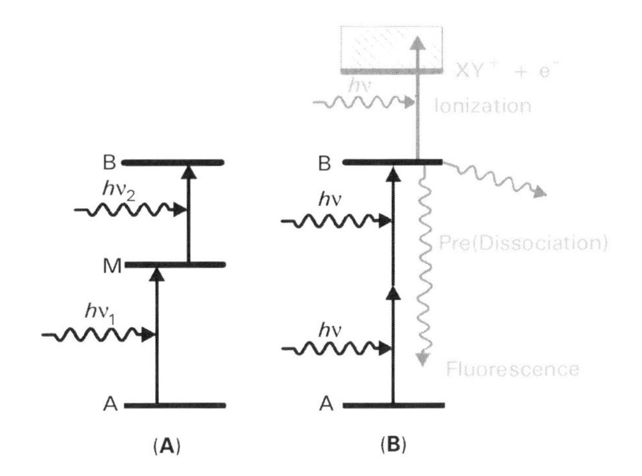

Figure 1 Illustration of (A) sequential and (B) simultaneous two-photon excitation from state A to state B. Also shown in (B) are three possible fates of the excited state B: fluorescence, dissociation and further photon absorption that ionizes the molecule. This latter process it termed 2+1 resonance-enhanced multiphoton ionization (REMPI).

Multiphoton excitations can thus provide a means of populating excited states via transitions that are 'forbidden' in traditional one-photon absorption spectroscopy. Two-photon spectroscopy has proved to be particularly valuable in this regard, especially in the case of centrosymmetric molecules, e.g. H_2, N_2, O_2, the halogens, ethyne and benzene. All of these molecules have *gerade* ground states. Thus, in each case, one-photon absorption provides a route to populating the *ungerade* excited states but the *gerade* excited states are inaccessible unless the excitation is carried out using an even number of photons. Further inspection of **Table 1** hints at the increased complexity of coherent multiphoton excitation spectra. While governed by the same spin-conservation requirements and vibrational (i.e. Franck–Condon) restrictions as one-photon spectra, an *n*-photon excitation can support changes in rotational quantum number $\Delta J \leq n$.

Experimental methods

The 'basic' multiphoton excitation experiment simply involves focusing tuneable laser radiation into a cell containing a low pressure (typically a few torr) of the atomic or molecular gas of interest and observing the resulting laser-induced fluorescence or, more commonly, the resulting ions or electrons (in the case of REMPI detection). In the latter experiment, the cell will typically be equipped with a pair of biased electrodes: an MPI spectrum is obtained simply by measuring the total ion, or the total photoelectron, yield as a function of excitation wavelength. The structure appearing in such a spectrum will reflect the resonance enhancements provided by the various rovibrational levels of the resonant intermediate electronic state(s) of the neutral, and may be analysed to provide spectroscopic (and thus structural) information about the excited neutral molecule. This 'basic' form of the REMPI experiment has limitations, and much of the recent experimental effort has been directed at improving both the selectivity and sensitivity of the technique.

One deficiency of this basic style of REMPI experiment is that all ions will be measured, irrespective of their masses, or that all photoelectrons will be counted, irrespective of their kinetic energies. A molecular MPI spectrum recorded in a static cell could therefore well include superimposed features associated with REMPI of the parent of interest, of neutral fragments arising from unintentional photodissociation of the parent molecule and of any other species present in the sample. Such potential ambiguities can usually be resolved by mass-resolving the resulting ions, and in most contemporary REMPI experiments this is achieved by time-of-flight (TOF) methods using either a linear TOF mass spectrometer, or a reflectron TOF mass spectrometer to provide enhanced resolution. A variety of fast charged particle detectors can be used, together with suitable time-gated signal-processing electronics, to monitor the REMPI spectrum associated with formation of any

Table 1 Allowed changes in some of the more important quantum numbers and symmetry descriptors for atoms and molecules undergoing one-colour multiphoton transitions involving one ($k = 1$), two ($k = 0$ and 2) and three ($k = 1$ and 3) photons

	Rank of transition tensor, k (number of photons)			
Quantum number / property of interest	*0 (2)*	*1 (1 or 3)*	*2 (2)*	*3 (3)*
(a) Atoms:				
Orbital angular momentum, *l* of electron being excited	$\Delta l = 0$	$\Delta l = \pm 1$	$\Delta l = 0, \pm 2$ (but s $\not\leftrightarrow$ s)	$\Delta l = \pm 1, \pm 3$ (but s $\not\leftrightarrow$ p)
(b) Linear molecules (case (a)/(b)):				
Axial projection of electronic orbital angular momentum, Λ	$\Delta\Lambda = 0$	$\Delta\Lambda = 0, \pm 1$	$\Delta\Lambda = 0, \pm 1, \pm 2$	$\Delta\Lambda = 0, \ldots, \pm 3$
linear molecules (case (c)):				
Axial projection of total electronic angular momentum, Ω	$\Delta\Omega = 0$	$\Delta\Omega = 0, \pm 1$	$\Delta\Omega = 0, \pm 1, \pm 2$	$\Delta\Omega = 0, \ldots, \pm 3$
(c) Centrosymmetric molecules:				
Inversion symmetry, u/g	u $\leftrightarrow$ u g $\leftrightarrow$ g	u $\leftrightarrow$ g	u $\leftrightarrow$ u g $\leftrightarrow$ g	u $\leftrightarrow$ g
(d) Atoms and molecules:				
Total angular momentum, *J*	$\Delta J = 0$	$\Delta J = 0, \pm 1$ (but $J = 0 \not\leftrightarrow J = 0$)	$\Delta J = 0, \pm 1, \pm 2$ (but $J = 0 \leftrightarrow J = 0,1$)	$\Delta J = 0, \ldots, \pm 3$ (but $J = 0 \not\leftrightarrow J = 0,1,2$; and $J = 1 \not\leftrightarrow J = 1$)
Total parity, +/–	$+ \leftrightarrow +$ $- \leftrightarrow -$	$+ \leftrightarrow -$	$+ \leftrightarrow +$ $- \leftrightarrow -$	$+ \leftrightarrow -$
Electron spin, *S*	$\Delta S = 0$	$\Delta S = 0$	$\Delta S = 0$	$\Delta S = 0$

Table 2 Representations of the spherical tensor components $T^{\kappa}_{q}(O)$ of the one-colour, ($n = 1$–3) transition operator

Number of photons, n	*k*	*q*	$D_{\infty h}$[a]	D_{6h}	D_{3h}[b]
1	1	0	Σ_u^+	A_{2u}	A_2''
	1	±1	Π_u	E_{1u}	E'
2	0	0	Σ_g^+	A_{1g}	A_1'
	2	0	Σ_g^+	A_{1g}	A_1'
	2	±1	Π_g	E_{1g}	E''
	2	±2	Δ_g	E_{2g}	E'
3	1	0	Σ_g^+	A_{2u}	A_2''
	1	±1	Π_u	E_{1u}	E'
	3	0	Σ_g^+	A_{2u}	A_2''
	3	±1	Π_u	E_{1u}	E'
	3	±2	Δ_u	E_{2u}	E''
	3	±3	Φ_u	$B_{1u} + B_{2u}$	$A_1'+A_2'$

[a] Assuming Hund's case (a) or (b) coupling. Ignore u/g labels for non-centrosymmetric linear molecules.
[b] A_1' and A_2'' reduced to A_1, A_2' becomes A_2, and E' and E'' both transform as E in C_{3v} molecules.

single, user-selected, ion mass. In this way it is usually possible to distinguish spectral features associated with the parent from those arising from REMPI of neutral photofragments, or to distinguish different isotopomers of the same parent. Mass-resolved REMPI spectroscopy necessarily requires use of collision-free conditions; the precursor of interest in such experiments is thus introduced into the mass spectrometer source region as a molecular beam.

It often proves useful to measure the kinetic energies (KEs) of the resulting photoelectrons also. Such measurements also require use of a molecular beam so that their KEs (which are usually measured by TOF methods in a spectrometer designed to minimize stray electric and magnetic fields) can be recorded under collision-free conditions; they provide the basis for a number of variants of photoelectron spectroscopy discussed below.

Applications

Less restrictive selection rules are just one of several benefits that can arise when using multiphoton excitation methods. Experimental convenience is another. A multiphoton excitation using visible or near-ultraviolet (UV) photons can often prove the easiest route to populating an excited state lying at energies that, in one-photon absorption, would fall in the technically much more demanding vacuum ultraviolet (VUV) spectral region. Other benefits derive from the fact that multiphoton excitations normally require the use of a focused pulsed laser because of the small excitation cross-section. The interaction is thus concentrated in a localized volume (the focal volume). The technique is therefore highly suitable for spatial concentration profiling, and well matched for use with supersonic molecular beams; many previously impenetrable molecular spectra have been interpreted successfully after application of multiphoton excitation methods to jet-cooled samples of the molecule of interest. This can be a huge benefit, especially in the case of REMPI where the resulting particles are charged and can be collected with far higher efficiency than could, for example, laser-induced fluorescence (LIF) from the excited state B. This benefit not only manifests itself in high sensitivity but, as we have seen, also offers additional species selectivity by allowing both mass analysis of the resulting ions and KE analysis of the accompanying photoelectrons.

Spectroscopy, structure and dynamics of excited state species

REMPI spectroscopy is typically used to probe high-lying electronic states, for which dissociation is always likely, but it will discriminate in favour of the more long-lived states because of the competition between ionization and dissociation (as in **Figure 1B**). There is one class of excited states that are often relatively long-lived — Rydberg states, which thus tend to dominate REMPI spectra. Molecular Rydberg states are conveniently pictured as a positive ion core, consisting of the nuclei and all but one of the valence electrons, with the remaining valence electron promoted to a state with a high principal quantum number, n. Such orbitals are large, spatially diffuse, and hence nonbonding, and are known as Rydberg orbitals. This is because the physical picture is very similar to that in the hydrogen atom, and the energy levels follow a modified Rydberg formula:

$$\tilde{\nu} = E_{\mathrm{i}} - \frac{R}{(n-\delta)^2} \qquad [2]$$

where R is the Rydberg constant. As a written, $\tilde{\nu}$, E_i and R must have the same units. E_i is the ionization limit of the ion core and δ is known as the quantum defect. It provides an indication of the extent to which the wavefunction of the Rydberg electron penetrates into the core region and its value is found empirically to be fairly constant for a given type of orbital. For molecules composed entirely of first-row atoms, typical values are $\delta = 1.0$–1.5 for s orbitals, $\delta = 0.4$–0.8 for p orbitals and $\delta \sim 0$ for all higher-l functions. Such qualitative ideas can be very useful for interpreting the patterns of excited states observed in many families of polyatomic molecules, though modifications due to configuration interaction (i.e. mixing between zero-order states sharing a common symmetry species but arising from different electronic configurations) can complicate such simple expectations. **Figure 2**, which shows a 2+1 REMPI spectrum of the NH radical serves to demonstrate several of these points. The spectrum is obtained by linearly polarized simultaneous two-photon absorption (at wavelengths ~271.2 nm) of NH radicals in their low-lying metastable excited $a^1\Delta$ state, followed by further one-photon excitation and detection of ions with m/z 15. Rotational analysis confirms that the spectrum is carried by a two-photon transition, linking states of $^1\Delta$ (lower state) and $^1\Pi$ symmetry, while the observation of neighbouring vibrational brands (including hot bands originating from the $v = 1$ level of the $^1\Delta$ state) verifies that this is an electronic origin band. Changes in rotational quantum number $\Delta J \leq 2$ are clearly evident, as anticipated in **Table 2**. Knowing the ionization limit of the NH radical ($108\,804 \pm 5\,\mathrm{cm}^{-1}$, measured relative to the $X^3\Sigma^-$ ground state), we can deduce a value for the quantum defect of this state ($\delta = 0.79$) which, taken together with its known symmetry, suggests that this transition should be associated with electron promotion from the highest occupied doubly degenerate 1π orbital (the little-perturbed $2p_x$ and $2p_y$ orbitals of atomic N) to a $3p\sigma$ Rydberg orbital. The spectrum appears red-degraded, indicating a 13% reduction in the effective B rotational constant upon electronic excitation. Multiphoton rotational line strengths (the multi-photon analogues of the Hönl–London line strength factors applicable to one-photon spectra) may be calculated, allowing derivation of the relative populations of the various initial quantum states contributing to the spectrum.

The simulated S branch contour shown in the top left part of **Figure 2** serves to illustrate another possible application of REMPI spectroscopy. Closer inspection of the experimental spectrum reveals that all transitions involving excited state levels with rotational quantum number $J' = 7$, 8 or 9 appear anomolously weakly. This is due to a very localized predissociation of the $v = 0$ level of the $f^1\Pi$ state of the NH radical. For these rotational levels in particular, the $f^1\Pi$ state predissociates at a rate that is comparable to, or greater than, the ionization rate; this competition leads to reduced ionization probability and a relative diminution of the eventual ion yield; multiphoton excitations proceeding via such predissociated levels thus appear with reduced relative intensity in the REMPI spectrum. In extreme cases the transitions involving such predissociated excited levels may show lifetime broadening as well. Clearly, in the case of more heavily predissociated excited states the REMPI signal is not only weaker (and thus harder to detect), but also less resolved, because of the increasing overlap of neighbouring lifetime-broadened spectral lines.

Figure 3 shows an example involving the SO radical where the predissociation is so severe that an alternative detection scheme must be used. The necessary two-colour sequential double-resonance excitation scheme is indicated in the figure; the first step is designed to populate a single rotational level in the $A^3\Pi$ state. The fluorescence from this state is monitored, and a drop is seen when the second laser is tuned to a frequency appropriate for further excitation of these state-selected molecules to the $D^3\Pi$ state. **Figure 3** shows the A state fluorescence intensity as the second laser is scanned, and reveals a very broad Lorentzian peak ($50\ \mathrm{cm}^{-1}$ full-width half-maximum). Use of the energy–time form of the uncertainty principle allows determination of the excited state lifetime (100 fs) from the width of the measured line shape. The D state is notionally a $4s\sigma$ Rydberg state, and its short lifetime is presumably indicative of significant mixing with a valence state, since the D state is the lowest-lying Rydberg state in SO.

This is just one of many instances where two-colour, double-resonance multiphoton spectroscopy can of great help in providing additional spectroscopic, structural and dynamical information about the

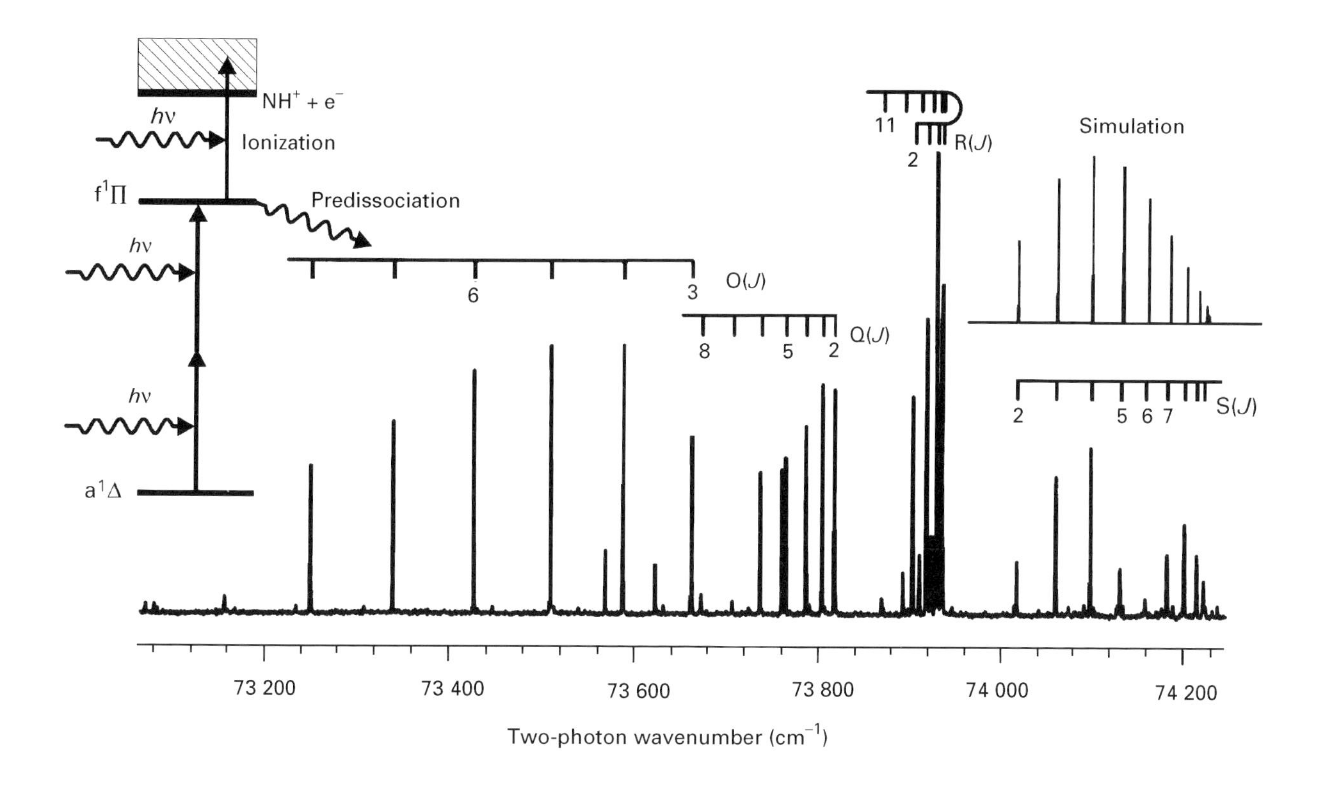

Figure 2 Two-photon resonant MPI spectrum of the origin band of the $f^1\Pi \leftarrow a^1\Delta$ ($3p\sigma \leftarrow 1\pi$) transition of the NH radical obtained using the excitation scheme shown at the top left and monitoring the *m/z* 15 ion mass channel as a function of the laser wavelength. Individual line assignments are indicated via the combs superimposed above the spectrum. The simulation of the S branch (top right) highlights lines that appear in the experimental REMPI spectrum with reduced intensity because of competing predissociation.

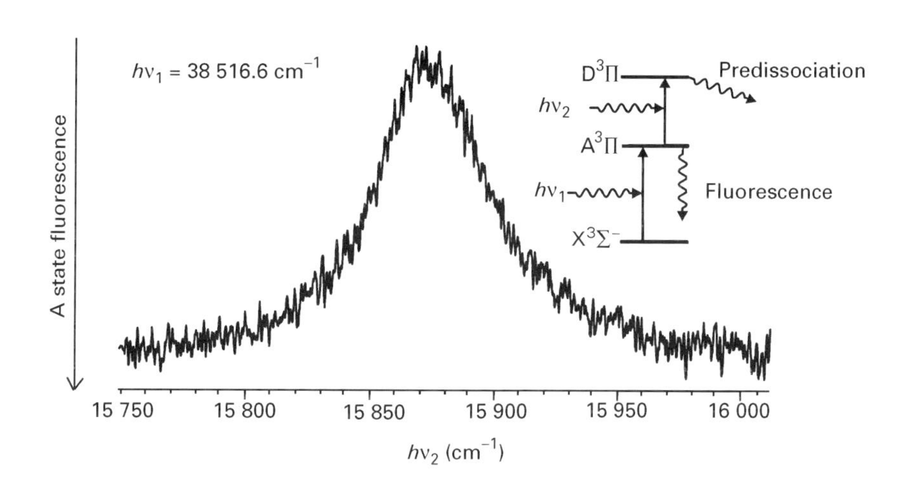

Figure 3 A two-colour fluorescence depletion spectrum of one rovibronic line associated with the $D^3\Pi \leftarrow A^3\Pi$ transition in SO. The two-colour excitation scheme used (upper right) is required because of the very short lifetime (100 fs) of the D^3P state. This results in the linewidth of 50 cm^{-1} shown in the spectrum.

excited states of small and medium-sized gas-phase molecular species. **Figure 4** shows the opposite extreme, where the final state is long lived (and ionization is used to detect that the multiphoton absorption has occurred), but double resonance is required to reach the states at all. The example involves states of the S_2 radical lying at energies around 75 000 cm^{-1} where, without the simplification of jet

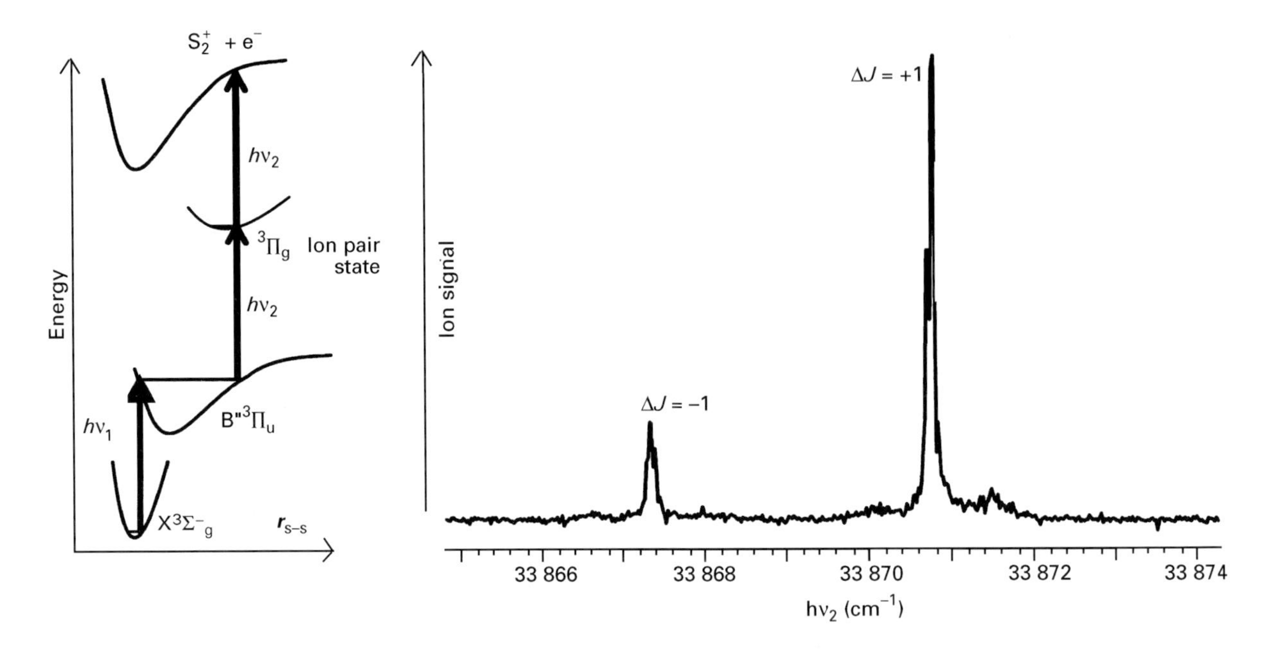

Figure 4 Two-colour ionization spectrum exciting levels of a $^3\Pi_g$ ion pair state of S_2, using the scheme shown in the inset. Note that a two step process is required, both to give a net g ← g excitation and to overcome the poor Franck–Condon factors for the transition.

cooling and the additional state selectivity afforded by double resonance methods, the S_2 spectrum would be impenetrably complex. Further, the excited state of interest and the ground state of S_2 both have *gerade* parity. Recalling the selection rules listed in **Table 1**, we see that the excited state can only be reached by a spectroscopy that involves use of an even number of photons. The final state in this case is an ion pair state of S_2, a state that is best described as a pair of oppositely charged ions, S^+S^-, rather than a covalently bound S=S. As for Rydberg states, most if not all molecular species will have such ion pair states but, to date, their observation remains quite rare. This is because the equilibrium bond length in an ion pair state is generally much larger than in the ground state, with the result that the two states show little Franck–Condon overlap. Double-resonance spectroscopy can provide a means of accessing such states if, as here, and as illustrated, the first step is to the inner turning point of the wavefunction associated with a high vibrational level of an intermediate valence excited state while the second step is arranged to excite from near the outer turning point of this same vibrational level to the ion pair state of interest. In the case of S_2, the ground and ion pair states have B rotational constants of 0.30 cm^{-1} and ~0.13 cm^{-1}, respectively, implying a 50% extension in equilibrium bond length when undergoing this double-resonant excitation. The versatility illustrated by these few examples serves to explain why multiphoton excitation methods in general, and REMPI in particular, continue to find widespread use as one of the most general and most sensitive species-selective methods of detecting atoms and small molecules (including radicals) in the gas phase.

REMPI–photoelectron spectroscopy (PES)

We now consider the photoelectrons formed in the MPI process, and the information they may carry. The measurement of their kinetic energies has become an established technique, thereby providing a means of performing photoelectron spectroscopy on excited electronic states. Such measurements can therefore give important clues as to the electronic and vibrational make-up of the excited state. They also allow determination of such details as the number of photons involved in the overall ionization process and the source of any fragment ions. For example, a given daughter ion, Y^+, seen in the TOF spectrum of the ions resulting from REMPI of a parent, XY, can arise from photodissociation of the neutral parent followed by one- (or more) photon ionization of the fragment, i.e.

$$XY + nh\nu \rightarrow X + Y \quad [3a]$$

$$Y + mh\nu' \rightarrow Y^+ \quad [3b]$$

or from MPI followed by photodissociation of the

resulting parent ion, i.e.

$$XY + nh\nu \rightarrow XY^+ \quad [4a]$$

$$XY^+ + mh\nu \rightarrow X + Y^+ \quad [4b]$$

or as a result of direct dissociative ionization of the parent, i.e.

$$XY + nh\nu \rightarrow (XY^+)^* \rightarrow X + Y^+ \quad [5]$$

Given the pulsed nature of the REMPI process, the electron KEs are almost always measured by TOF methods, either using a conventional (mu-metal-shielded) TOF spectrometer or a magnetic bottle photoelectron spectrometer. The latter offers the advantage of much higher collection efficiency, with comparatively little loss of ultimate KE resolution.

Recalling **Figure 1B** we note that when an MPI process is resonance enhanced by a bound excited state B, the vibrational structure in the resulting photoelectron spectrum will reflect the differences in the equilibrium geometries of state B and the parent ion, rather than between the ground state A and the ion as in traditional one-photon (e.g. He I) PES. Thus if the geometry and the vibrational level structure of the ion are already known, the vibronic structure evident in REMPI-PES can yield insight into the geometry of the resonance enhancing state B. If B is a pure Rydberg state, the electronic configuration of its core should be the same as that of the ionic state that lies at the convergence limit of the series to which it belongs. The Rydberg state and the ion will therefore be likely to have very similar geometries. Thus, by the Franck–Condon principle, we can anticipate that the final ionizing step in a REMPI process via such a Rydberg state will involve a $\Delta v = 0$ transition, leading to selective formation of ions with the same vibrational quantum number(s) as in state B. Since the photoionization is brought about using a (known) integer number of photons, the photoelectrons accompanying such state specific ion formation will have a narrow spread of KEs. In favourable cases the TOF spectrum of these photoelectrons can be resolved to the extent that individual rotational states of the ion are revealed, thus explaining the continuing appeal of REMPI-PES as a means of determining accurate ionization thresholds and of investigating photoionization dynamics in simple molecular systems.

Another form of PES has emerged that can provide a further order of magnitude improvement in energy resolution (i.e. cm^{-1} resolution). This technique is now generally referred to as zero kinetic energy (ZEKE)-PES. In conventional photoelectron spectroscopy, and in REMPI-PES, we learn about the energy levels of the ion by measuring the photoelectron KEs as accurately as possible. ZEKE-PES also reveals the energy levels of the cation but is based on a different philosophy. The principle of the method is illustrated in **Figure 5**. In the particular double-resonant variant shown, one laser is tuned so as to populate a (known) excited state M, and a second laser pulse is then used to excite this population to the energetic threshold for forming one of the allowed quantum states of the ion. Any energetic electrons (e.g. those formed via an autoionization process) will quickly recoil from the interaction region. The ZEKE (threshold) electrons can be detected by application of a suitably delayed pulsed extraction field. A ZEKE-PES spectrum is obtained by measuring the excitation spectrum for forming photoelectrons with zero kinetic energy; precise energy eigenvalues are obtained because the spectral resolution is determined, ultimately, by the bandwidth of the exciting laser. It is now recognized that this description, while appealing, actually oversimplifies the physics. The 'ZEKE' electrons detected in an experiment as described actually derive from pulsed-field ionization of very high Rydberg states belonging to series converging to the threshold of interest. As a result, the ionization thresholds determined via this type of experiment will all be subject to a small, systematic shift to low energy. However, the magnitude of this shift scales with the applied extraction voltage, so the true thresholds can be recovered by recording such spectra using a number of different pulsed extraction voltages and extrapolating the observed line frequencies to zero applied field.

Ion imaging

As REMPI is a very sensitive, selective and convenient means of detecting small localized concentrations of gas phase species, it is particularly suited to probing atomic or molecular products resulting from a gas-phase photodissociation or a crossed-beam reaction. Accurate knowledge of the energy disposal in such products, their recoil velocities and the angular distribution of these velocity vectors, the alignment of their rotational angular momenta, and the way all these quantities are correlated, can provide considerable insight into the detailed dissociation and/or reaction dynamics. Ion imaging is one way in which REMPI spectroscopy is being used to provide such information. The experiment is simple in concept. In the case of photodissociation, the precursor of interest, in a skimmed molecular beam, is photolysed to yield fragments, one of which is ionized selectively and in a

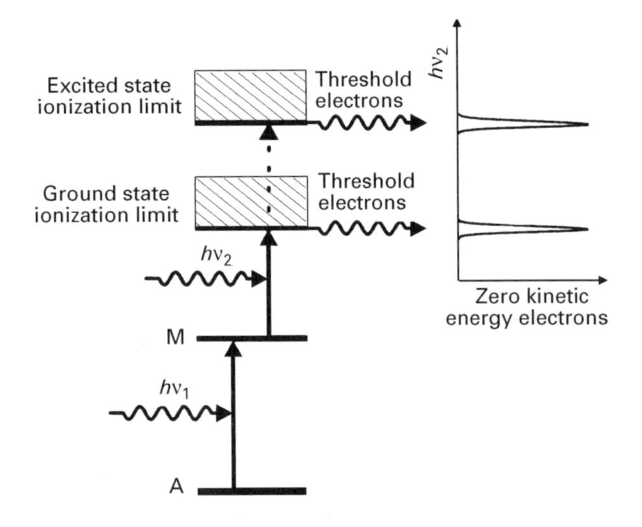

Figure 5 Illustration of two-colour two-photon ZEKE excitation scheme, in which the first photon is fixed so as to be resonant with a known M ← A transition and the frequency of the second photon is tuned. As shown in the inset, the peaks in a ZEKE spectrum correspond to the onsets of new ionization thresholds.

quantum state-specific manner, by REMPI, as soon as it is created. The resulting cloud of ionized fragments continues to expand with a velocity and angular distribution characteristic of the original photolysis event, but is simultaneously accelerated out of the interaction region and arranged to impact on a position-sensitive detector, e.g. a microchannel plate behind which is mounted a phosphor screen that is viewed using a gated image-intensified CCD camera. The result in a squashed two-dimensional projection of its initial 3D recoil velocity distribution. This can be reconstructed mathematically to yield the speed and angular distributions of the tagged fragment, in the particular quantum state defined by the REMPI excitation wavelength. The structure in the image provides information about the speed and angular distribution of the tagged fragments and also, by energy and momentum conservation arguments, the quantum state population distribution in the partner fragments; such knowledge can provide a uniquely detailed view of the parent photofragmentation dynamics. By way of illustration, **Figure 6** shows ion images obtained by ionizing ground-state ($^2P_{3/2}$) Br atoms resulting from photolysis of Br_2 molecules at three different wavelengths. It is clear from these that the velocity and angular distribution of these Br atoms (defined relative to the electric vector, ε, of the photolysis laser – vertical in **Figure 6** as indicated by the double-headed arrow) depends on the photolysis laser wavelength. The radii of the partial rings apparent in each image give the speed of the atoms and hence, by energy conservation, the energy of the other fragment. The middle image reveals that Br_2 photolysis at 460 nm yields ground-state Br($^2P_{3/2}$) atoms (the tagged species) in conjunction with both another ground-state atom (outer ring) and a spin–orbit excited-state ($^2P_{1/2}$) partner, the inner ring. These product sets show different recoil anisotropies. The relative intensities of the two partial rings provides a measure of the branching ratios into these two product channels. Clearly, dissociation via a perpendicular transition (ν_{recoil} perpendicular to ε) yielding two ground state Br

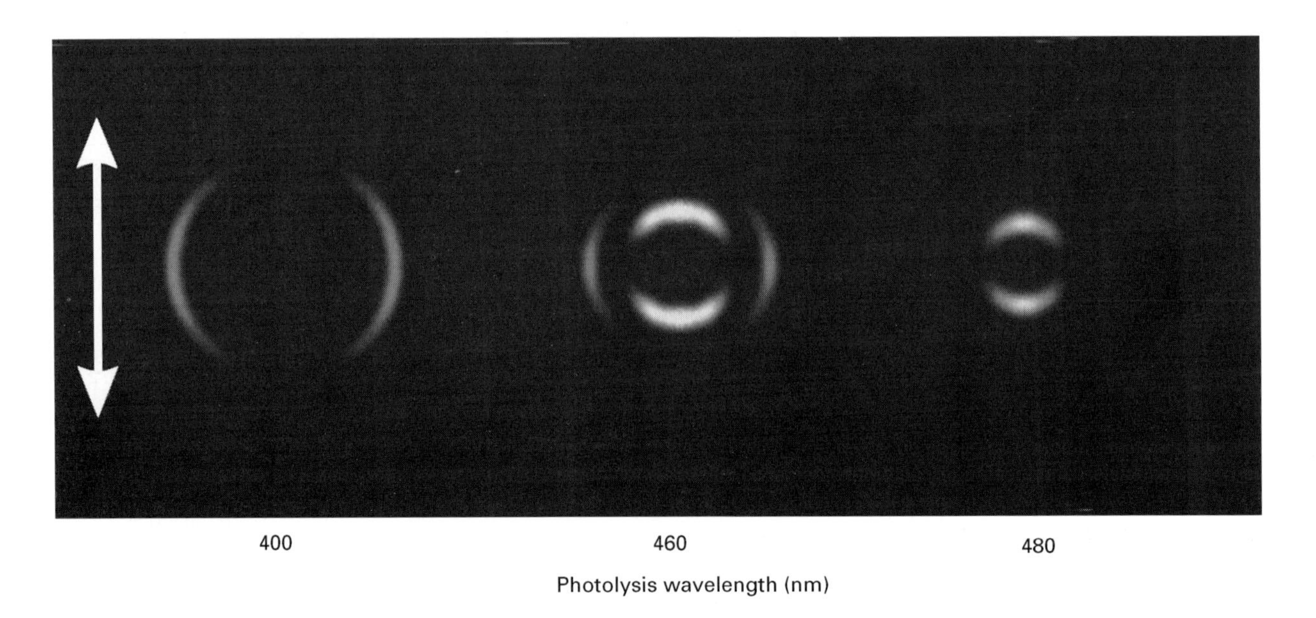

Figure 6 Ion images of ground-state Br ($^2P_{3/2}$) atoms resulting from Br_2 photolysis at the specified wavelengths. The double-headed arrow indicates the plane of polarization of the photolysis laser radiation.

atoms is the dominant decay mechanism at 400 nm, whereas at longer excitation wavelength (e.g. 480 nm) the dominant fragmentation is to one ground-state ($^2P_{3/2}$) and one spin–orbit excited-state ($^2P_{1/2}$) Br atom, following a parallel excitation process. Analysis of images like these, and their dependence on photolysis wavelength, can provide much insight into both the mechanism and the timescale of the dissociation process.

List of symbols

B = rotational constant; E_i = ionization limit; h = planck constant; J = rotational quantum number; n = principal quantum number; $\hat{O}$ = multiphoton transition operator; R = Rydberg constant; $T_q^k(\hat{O})$ = qth component of the tensor of rank k representing $\hat{O}$; δ = quantum defect; ε = electric vector; ν = frequency; v = vibrational quantum number; $\tilde{\nu}$ = wavenumber.

See also: **Ion Imaging Using Mass Spectrometry; Laser Spectroscopy Theory; Multiphoton Excitation in Mass Spectrometry; Multiphoton Spectroscopy, Applications; Photoelectron Spectrometers; Photoelectron Spectroscopy; Photoelectron-Photoion Coincidence Methods in Mass Spectrometry (PEPICO); Photoionization and Photodissociation Methods in Mass Spectrometry; Time of Flight Mass Spectrometers.**

Further reading

Ashfold MNR, Clement SG, Howe JD and Western CM (1993) Multiphoton ionization spectroscopy of free radical species. *Journal of the Chemical Society, Faraday Transactions* **89**: 1153–1172.

Ashfold MNR and Howe JD (1994) Multiphoton spectroscopy of molecular species. *Annual Review of Physical Chemistry* **45**: 57–82.

Heck AJR and Chandler DW (1995) Imaging techniques for the study of chemical reaction dynamics. *Annual Review of Physical Chemistry* **46**: 335–372.

Houston PL (1995) Snapshots of chemistry: product imaging of molecular reactions. *Accounts of Chemical Research* **28**: 453–460.

Kimura L and Achiba Y (1989) In Lin SH (ed) *Advances in Multiphoton Processes and Spectroscopy* Vol 5, pp 317–370. New Jersey: World Scientific.

Lin SH, Fujimara Y, Neusser HJ and Schlag EW (1984) *Multiphoton Spectroscopy of Molecules*. New York: Academic Press.

Müller-Dethlefs K and Schlag EW (1991) High resolution zero kinetic energy (ZEKE) photoelectron spectroscopy of molecular systems. *Annual Review of Physical Chemistry* **42**: 109–136.

Multivariate Statistical Methods

RL Somorjai, Institute for Biodiagnostics, National Research Council, Winnipeg, Canada

FUNDAMENTALS OF SPECTROSCOPY
Methods & Instrumentation

Introduction, basic ideas, terminology

Spectroscopic methods are increasingly becoming the methods of choice for analysing a variety of experimental data, in chemistry, biology, food industry, medicine, etc. This popularity is well deserved. The spectral methods are generally faster, more accurate and frequently much cheaper than conventional analytical techniques. Furthermore, in biomedical applications they provide the means for noninvasive or minimally invasive diagnosis. However, these obvious advantages are somewhat offset by the indirect relationship between spectral features and the measurables or observables of interest (such as analyte concentrations or disease class assignments). Consequently, we have to model a generally complex and frequently nonlinear relationship. This relationship can be represented by

$$\mathbf{Y} = \mathbf{F}(\mathbf{X}) \quad [1]$$

where $\mathbf{Y} = \{y, y_2, y_3, \ldots, y_K\}$ is the set of K measurables (responses, observables, targets) (e.g. the concentrations of K analytes or membership labels for K classes), $\mathbf{X} = \{x_1, x_2, x_3, \ldots, x_N\}$ is the collection of N

samples (objects, patterns), with x_k the kth sample. Every sample is represented as an L-vector, with each of its L elements corresponding to one of the L spectral features (attributes, predictor variables, measurements) (such as wavelengths or frequencies), i.e. $x_k = \{x_k(1), x_k(2), x_k(L)\}$, and $\mathbf{F}$ is the model (function) that couples $\mathbf{Y}$ with $\mathbf{X}$ and whose parameters we are to optimize, implicitly or explicitly. Any practical scheme of optimization must necessarily be data-driven. Hence the larger N, the number of samples, the more reliable any prediction based on Equation [1]. In spectroscopy (whether infrared, IR, magnetic resonance, MR, or other), K (the number of analytes or the number of classes) typically ranges from one to about a dozen, L (the number of spectral wavenumbers or frequencies) is in the hundreds or thousands, and N ought to be at least in the hundreds. Since in spectroscopic applications L, the dimension of the feature space, is invariably greater than one, and typically large, multivariate methods of analysis are necessary. Furthermore, L is frequently larger than N, the sample space dimension. As this causes numerical problems, special approaches are needed and precautions have to be taken. These we can handle via some preprocessing procedure. Its most important goal is data reduction (compression). In practice, data compression is divided, somewhat artificially, into feature extraction or feature subset selection.

Exploratory data analysis and representation is usually the first step in probing the functional relationship between $\mathbf{Y}$ and $\mathbf{X}$. This falls into the purview of unsupervised pattern recognition, with clustering methods the most common representatives. Hierarchical clustering is the older, having its roots and initial applications in taxonomy, psychology and the social sciences. Its limitations are due, paradoxically, to its flexibility: by changing the merging (or splitting) criterion and/or the dissimilarity measure, we can create almost any grouping of the samples. The final number of clusters depends on a user-selected threshold; hence it is subjective. The fact that the samples are partitioned into mutually exclusive (non-overlapping) groups must be regarded as unrealistic.

The other major variant of clustering minimizes some objective function of the intersample distances. The results are also dependent on user-selected parameters (e.g. the number of clusters and the type of the distance measure used). However, these results become more realistic if we allow overlap between clusters, i.e. if we accept that the samples can be fuzzy, having memberships in all clusters. Bezdek's fuzzy c-means clustering algorithm is the most popular of such clustering methods. Neither clustering variant can be expected to outperform methods based on supervised pattern recognition.

Although calibrating or classifying spectra are merely two different applications of the mathematical/statistical technique of regression, historically they have evolved independently, with different goals and requirements. Whenever possible attempts will be made to connect the two by drawing attention to common concepts and methods that only differ in their terminology and emphasis. Occasionally, for the sake of simplicity in presentation, discussion will cover certain aspects and peculiarities of the two disciplines separately.

Calibration *is* regression, but with a specific connotation: it involves the development of a quantitative, and generally linear relationship between, say, the concentrations of the analytes present and their spectral manifestations. Significantly, this relationship is continuous. In contrast, classification assigns class labels to the samples and thus establishes a categorical correspondence between the samples and their spectral features, a generally discrete relationship.

Both calibration and classification belong to the group of supervised pattern recognition methods. Supervision means that the model parameters in Equation [1] must be optimized using information provided by and extracted from the available data. To do this so that prediction is reliable and robust, leads naturally to the concept of partitioning the samples into training (design, learning) and validation (test, prediction) sets.

This article will focus on concepts and ideas, without delving into detailed description of individual methods. The interested reader is directed to Further reading for more in-depth accounts and discussions where there are also listed some of the most likely journal sources for the latest advances.

Preprocessing

The characteristics of spectra demand that we use preprocessing to guarantee that the predictions from the optimized Equation [1] are reliable and robust. In fact, extensive experience suggests that if the preprocessing part of the analysis is done properly, then even the simplest calibration or classification methods will succeed.

At its most straightforward, Equation [1] describes some simple functional transformation of the original data. This might be as elementary as mean-centring and scaling ('whitening'), i.e. subtracting the overall sample mean from each individual sample and dividing by the sample variance. Other useful examples include smoothing (filtering), normalization (e.g. by the overall area under the spectra), monotonic nonlinear operations (e.g. using the logarithms or powers of the spectral intensities), and analysing the (numerical) derivatives of the original spectra. For instance, the assignment of mid-IR spectra of biofluids into different disease classes is most successful when the classifier uses the first derivatives of the spectra.

Extracting some intrinsic structure from the data, while in the original L-dimensional feature space, is rarely successful because the majority of the L spectral features are either redundant (typically owing to correlation) or represent irrelevant information ('noise'). Any of these will usually mask the discriminating features. We then interpret preprocessing as either a procedure that removes irrelevant features (a type of filtering) or one that finds the optimal subspace in which the data can be best analysed (a form of projection).

The number of spectral features L is generally large and frequently larger than N, the number of samples. Furthermore, (adjacent) spectral features are often strongly correlated, i.e. they are not independent. If $L > N$, linear dependency or near-dependency (called multicollinearity by the calibration community) could lead to numerical instabilities, with consequent unreliable and non-reproducible results. For instance, when classifying spectra via linear or quadratic discriminant analysis (LDA or QDA), the occurrence of singular or near-singular covariance matrices causes matrix inversion problems. The overall conclusion, supported by extensive experience, is that we must somehow reduce the number of features. Although no general theoretical proof is available, in practice the ratio $N{:}L$ should be at least 10, but preferably 20, for reliable classification or calibration results. Instead, for spectra this ratio is more likely to be 0.1 or even less.

Data reduction can be achieved by feature extraction, which first transforms the original L variables $\mathbf{X}$ into L new variables $\mathbf{Z} = \mathbf{G}(\mathbf{X})$. $\mathbf{G}$ is nonlinear in general. However, principal component analysis (PCA), the most commonly used transformation, is linear: $\mathbf{Z} = \mathbf{AX}$, where $\mathbf{A}$ is a matrix. PCA is an unsupervised method, applied to the entire available data. It is carried out by diagonalizing the sample covariance matrix $\mathbf{S}$. If we sort the resulting eigenvalues in decreasing order, then the corresponding eigenvectors (principal components, PCs) point, in the original L-dimensional feature space, in directions of successively decreasing variance. The PCs are orthogonal and uncorrelated. Geometrically, PCA corresponds to a rotation of the original L coordinate axes to L new orthogonal axes formed by the L PCs. We achieve the data reduction by retaining only the first $M << L$ PCs ($\mathbf{Z}$s) for further analysis. (Recall that the PCs are different linear combinations of all L original features.) However, the first M PCs, even if they account for a large fraction of the total variance in the data, are not necessarily the most discriminating directions along which classes may be separated. In fact, it is more useful to select those M PCs that give rise to the lowest classification error. These are generally not the first M in the ordered list. Similarly, in calibration, it makes more sense to select those M PCs that show maximum correlation with the $\mathbf{Y}$. Such arguments lead naturally to the other data reduction method, feature subset selection. As a first application, we could select from the L PCs the M that best satisfy the above requirements. This is always beneficial. However, the general approach has several attractive characteristics which don't depend on a prior PCA, especially for features originating in spectra. The foremost of these is that the features of the optimal subset retain their spectral identity. This is, of course, important for interpretability, especially in MR spectroscopy, where the spectral peaks are relatively narrow and have distinct chemical identities. Another significant practical advantage is that once we identify a smaller subset of the complete set of features we never need to determine or measure the remaining features again.

Assume that we wish to select the $M << L$ best features, with 'best' defined appropriately for the given application. The naive approach of combining the individually best M features rarely produces the best M-feature subset. Examining all possible M-feature subsets to find the best is usually prohibitive computationally. A possible solution is to restrict the search to a subspace of the complete space of all feature sets. The standard stepwise methods proceed along these lines. The forward selection (FS) version is computationally the least demanding. FS starts by selecting the M individually best features. This is repeated, at each stage adding that feature which, when combined with those already chosen, gives the best result. The backward elimination (BE) algorithm starts with the complete set of features and sequentially removes the one that least degrades performance. The BE method is computationally prohibitive when the starting set of features is large, as is the case for spectra. Both have been generalized (e.g. more than one feature can be added or removed at a time). FS and BE can also be combined. Neither method is guaranteed to find the globally optimal M-feature subset. Branch-and-bound methods, which would guarantee the best possible subset, are rarely applicable to spectra. Most recently, optimal feature selection techniques based on genetic algorithms (OFS-GA) have been used with great success. The approach is natural for preprocessing spectra or time series, for which the feature set originates from sequential discrete sampling from or measurement of an underlying continuum. Consequently, the dimensionality L of the original feature space is large. The inherent order (along time or wavelength) implies that adjacent features will be correlated, a characteristic that causes redundancy. Unlike BE, OFS-GA can start with the original feature space. It maps this onto a bit string (zeroes and ones). GA operations, such as mutation and crossover are used on a population of bit strings. The performance criterion is the minimization of an objective (fitness) function F. If F is the mean square error between the training set classification results and the a priori class indicator, then the best F simultaneously maximizes the overall accuracy for the training set and the crispness of the class assignments. The input to the algorithm is M, the ultimate number of features (distinct subregions) required, and the type of feature space-reducing operation or transformation to be carried out in these subregions. Typical operations include (but are not confined to) replacing the individual features in a subregion by their average or variance. Selecting an $M << L$ and one of these operations lead jointly to substantial feature space reduction.

Training versus test sets and crossvalidation

Suppose we knew exactly the multivariate distributions from which we drew our samples. Then, at least in principle, we could calculate the true error (called the Bayes error, e_B, which is the smallest achievable error given the exact distributions). In practice, we have a finite, limited number of N samples. When the samples are random and N is sufficiently large to be representative of the underlying distributions, the entire data set can be used to optimize the calibration or classifier parameters. Then predicting the concentrations of the K analytes present in a new sample (calibration), or assigning this sample to one of the K classes (classification), is analogous to interpolation. In contrast, for the typical case of small or moderate N, prediction, or class assignment corresponds to the much less reliable process of extrapolation. Thus, in most practical situations two contradictory requirements confront us. One the one hand, for reliability we would like to use all N samples. When this is done, it is called the resubstitution (plug-in) method in the classification literature. However, the resubstitution or apparent error estimate e_R is generally overoptimistic. The danger of overfitting threatens, without any guarantee that new samples, not used for training, would be correctly predicted or assigned. The consequence of overfitting is that the classifier or regressor will have poor generalizing power. On the other hand, let us partition the N available samples into independent training and validation sets, with N_T and N_V ($N = N_T + N_V$) samples respectively. Then we may satisfy validation requirements but reliability suffers because N_T is not sufficiently large.

Crossvalidation techniques may be used to compensate for the smallness of the sample set (unfortunately far too frequent in biomedical applications). Lachenbruch proposed the leave-one-out or delete-1 ($D_{[1]}$) method (a close relative of Tukey's jackknife). For classification problems this proceeds by successively deleting each of the N samples from the data set, training a classifier on the remaining $N - 1$, and assigning the deleted sample to one of the classes. ($N_T = N - 1$ and $N_V = 1$). Thus we create and test N $(N - 1)$-sample classifiers, each with performance only very slightly worse than that of the complete N-sample classifier. (The entire process is identical for calibration, with 'classification' and 'classifier' replaced by 'regression/calibration' and 'regressor'.) The result is a much-reduced bias for the classification (calibration) error, $e_{D[1]}$. Unfortunately, for small or even moderate N the variance can be large. The delete-d ($D_{[d]}$) crossvalidation strategy can reduce the variance. The complete delete-d ($CD_{[d]}$) version produces $N!/d!(N - d)!$ classifiers, a huge number even for moderate N. (If $N = 50$ and $d = 10$, this number is $\sim 1 \times 10^{10}$). Thus, in practice we select uniformly a random subset of size B out of these. As long as $N/B \to 0$, e.g. with $B = N^{\delta}$, $\delta > 1$, the $D_{[d]}$ strategy retains the efficiency of the $CD_{[d]}$. The expected classification error is the average of the B individual errors, as measured on the B sets of d deleted, i.e. test samples.

The above are examples of nonparametric resampling (RS) methods, without replacement. The deservedly popular bootstrap method is also a resampling method, but with replacement. Because of replacement, some objects may appear several times in any of the B bootstrap samples. This could cause difficulties for those classifiers that we optimize by inverting covariance matrices. A simple remedy is to perform the sample replacement only after we have selected for the current training set a particular subset of $N - d$ distinct samples. Again, we secure the end result by averaging the B individual classifier outcomes.

Resampling, based on a random (Monte Carlo) subset selection is a very powerful crossvalidation approach. It is not constrained by the computational limitations of the $CD_{[d]}$, even for $d \approx N/2$. Furthermore, RS methods are equally effective in providing nonparametric confidence intervals/regions for any computed estimate.

(Supervised) classification methods

These fall into two categories: parametric and nonparametric. The former assumes that the samples for the K classes derive from some known distribution (usually multivariate normal), whereas the latter is distribution-free. There is no best method: the choice of the classifier depends both on the problem and on the property that we want to optimize. Most frequently, we minimize the classification error, in particular e_V for the validation set, but this need not be the only choice. Our current concern is with the classification of spectra. In the following, we shall assume that the original, L-dimensional feature space was already reduced to an M-dimensional space, $M << L$ (M-space).

After 60 years, Fisher's LDA method is still the workhorse of the parametric methods. Of all the statistical classifiers, we understand LDA's theoretical properties best. We can derive it from ordinary (linear) least-squares (OLS) regression. LDA creates those linear combinations of the M features that best separate the classes. For each class k there is one of these discriminant functions d_k. The decision surfaces $g_{jk} = d_j - d_k = 0$ between any pair (j,k) of classes are also linear, i.e. they are hyperplanes in M-space. The properties of LDA are optimal in the Bayes sense if all K distributions are multivariate normal and have a common covariance matrix **S**. In practice, these conditions are rarely satisfied. Nevertheless, LDA is surprisingly robust against even moderate deviations from normality and/or from the equality of the covariance matrices. LDA requires the optimization of $M + 1$ parameters for a M-dimensional feature space, the minimum number for a parametric statistical classifier.

If the covariance matrices $\mathbf{S}_k$ are different, then the classifier created is called quadratic discriminant analysis (QDA). The hypersurfaces separating the classes become quadratic (curved). Unfortunately, this additional flexibility comes at a cost: we have to optimize $O(M^2)$ parameters instead of the $O(M)$ in LDA. Usually $O(M^2) \gg N$, leading to overfitting, numerical instabilities, etc. Furthermore, QDA is less reliable, because it uses the K $\mathbf{S}_k$s, each an estimate of the (generally unknown) population covariance matrix, whereas LDA uses the more reliable single, pooled sample covariance matrix.

Friedman suggested regularized discriminant analysis (RDA) to alleviate some of these problems. RDA introduces a two-parameter estimate of the sample covariance matrix. If $\mathbf{S}_k$ is the sample covariance matrix for class k, and **S** their pooled version, then we construct a two-parameter covariance matrix $\mathbf{S}_k(\lambda,\gamma)$, $0 \le \lambda \le 1$, $0 \le \gamma \le 1$. The four corners defining the extremes of the λ,γ parameters represent well-known classifiers; ($\lambda = 0$, $\gamma = 0$) corresponds to QDA, ($\lambda = 1$, $\gamma = 0$) to LDA, whereas ($\lambda = 1$, $\gamma = 1$) gives the nearest-means classifier (assigning a sample to the class with the nearest training set mean, using Euclidean distance). When ($\lambda = 0$, $\gamma = 1$) a variance-weighted version of the nearest-means classifier is generated. The optimum (λ, γ) pair is usually determined via crossvalidation. Excellent results have been reported even when the 'number of samples' to 'number of features' ratio N:M was unfavourably low.

RDA modifies (shrinks) the eigenvalues of the original covariance matrices. Other methods exist with similar goals. Amongst these is DASCO (discriminant analysis with shrunken covariances). Although popular with the chemometric community, DASCO is quite complex, yet does not outperform RDA or the simpler ridge method. The latter also modifies the sample covariance matrix to be used with LDA from **S** to $\mathbf{S} + \alpha\,\mathbf{I}$, where α is a parameter to be optimized, **I** the unit matrix.

Among the nonlinear methods, neural net (NN) classifiers are relatively recent. Conceptually attractive, they are frequently used uncritically. A common danger is overfitting: because of the essentially unlimited flexibility of multilayer NNs, the training set can be classified perfectly (by creating an arbitrarily convoluted and complicated decision surface). Unfortunately, this high classification accuracy does not generally carry over to the validation set.

Recursive partitioning (decision tree) methods had found early use in numerical taxonomy, the social sciences and medicine. More recently, researchers in both artificial intelligence and statistics have substantially advanced these methods. There are two major types. Binary decision trees would be relevant in medical diagnosis, where the features are symptoms, with only yes/no answers for the decisions. If the features are continuous (spectral features qualify) then the decision trees are more complex, requiring the partitioning of each feature. The partitioning criteria and the sequence in which the features are probed are aspects of the tree design. The major advantage of the decision tree methods is ease of interpretability. To apply these methods to spectra, prior feature space reduction seems mandatory.

The nearest neighbour (nn) methods are nonlinear, nonparametric classification methods. They estimate probabilities at x using the classes of adjacent training set points. The estimates will have smaller bias if only close neighbours are used, hence nearest neighbours. In applying these methods a distance measure, e.g. Euclidean, is needed. Among the advantages are simplicity and an upper bound on their asymptotic misclassification error: $e_{nn} \leq 2e_B$. A major disadvantage in a high-dimensional feature space is the unacceptably large number of training samples required for accuracy.

Calibration techniques

A major desideratum in most calibration work is linearity. If the relation between **Y** and **X** (Eqn [1]) appears nonlinear, then either the **X** or the **Y** (or both) are transformed until linearity is achieved. This is important for accurate, quantitative predictions of continuous responses such as analyte concentrations. This preoccupation with linearity guided the development of the most important calibration methods.

Because of their common origin (i.e. regression), calibration methods share many characteristics with classification methods. Nevertheless, we can introduce most calibration techniques more simply in terms of their original formulation. Furthermore, several of these methods treat data reduction and calibration (fitting) simultaneously. Wold's partial least squares (PLS) regression reduces the number of features on which the OLS regression is carried out. This is done by feature extraction, i.e. from the raw features the algorithm produces 'derived features' (DFs, the partial least-squares). The first DF is the linear combination of the L original features that has maximum covariance (rather than correlation) with **Y**. Subsequent DFs are found in the same way, subject to being uncorrelated with previous ones. PLS can be described in terms of covariance regularization (maximization).

Principal component regression (PCR) is intimately related to PCA. If the spectral decomposition (diagonalization) of the cross-product matrix **X′X** is,

$$\mathbf{S} = \sum_{k=1}^{L} E_k \mathbf{u}_k \mathbf{u}_k'$$

then in PCR the matrix

$$\mathbf{S}^{-1} = \sum_{k=1}^{M} E_k^{-1} \mathbf{u}_k \mathbf{u}_k'$$

is used, where $M << L$. $\mathbf{u}_k$ is the kth PC and the M PCs are usually chosen to have the highest correlation with **Y**. There are other methods, e.g. shortest least-squares, but the above two are used most frequently.

Discussion

Is there a best method for classification or regression? The answer must be: 'It depends on the data and the criterion to be optimized'. Nevertheless, some general recommendations can be made. The first is that preprocessing the data is probably the most important step to be taken for reliable, reproducible results. The second is that crossvalidation is essential if we want reliable predictions for new samples. Finally, as many samples should be collected as possible. If at least the first two are carried out properly, then even the simplest classifiers or regressors will work well. However, the most important message is: know your data!

Some important topics could not be dealt with because of space limitation. Outlier detection is amongst these. Every attempt should be made to find outliers before classification or calibration is carried out, otherwise the results will be distorted.

List of symbols

$CD_{[d]}$ = complete delete-d; d = number of samples deleted; e_B = Bayes error; e_R = resubstitution error; $e_{D(1)}$ = classification error; e_{nn} = classification error; L = number of variables; N = number of samples; N_T = number of samples in training sets; N_V = number of validation sets; **S** = covariance matrix; **X** = collection of N samples; **Y** = set of K measurables; LDA = linear discriminant analysis; QDA = quadratic discriminant analysis; IR = infrared; MR = magnetic resonance; PC = principal component; PCA = principal component analysis; RDA = regularized discriminant analysis; OFS-GA = optical feature selection-genetic algorithm; FS = forward selection; BE = backward elimination; DASCO = discriminant analysis with shrunken covariances; NN = neutral net; nn = nearest neighbour; PLS = partial least squares; PCR = partial least squares.

See also: **Biofluids Studied By NMR; Calibration and Reference Systems (Regulatory Authorities); Computational Methods and Chemometrics in Near-IR Spectroscopy; Fourier Transformation and Sampling Theory; Laboratory Information Management Systems (LIMS).**

Further reading

Bezdek JC (1981) *Pattern Recognition with Fuzzy Objective Function Algorithms*. New York: Plenum Press.

Bishop CM (1995) *Neural Networks for Pattern Recognition*. Oxford: Clarendon.

Breiman L, Friedman JH, Olshen RA and Stone CJ (1984) *Classification and Regression Trees* Belmont, CA: Wadsworth International Group.

Cook RD and Weisberg S (1982) *Residuals and Influence in Regression*. London: Chapman & Hall.
Deming SN and Morgan SL (1987) *Experimental Design: A Chemometric Approach*. New York: Elsevier.
Efron B and Tibshirani RJ (1993) *An Introduction to the Bootstrap*. New York: Chapman & Hall.
Fukunaga K (1990) *Introduction to Statistical Pattern Recognition*, 2nd edn. Boston: Academic Press.
Hand DJ (1997) *Construction and Assessment of Classification Rules*. Chichester: Wiley.
Jackson JE (1991) *A User's Guide to Principal Components*. New York: Wiley.
Kowalski BR (ed) (1984) *Chemometrics: Mathematics and Statistics in Chemistry* Dordrecht: Reidel.
Martens H and Naes T (1991) *Multivariate Calibration*. Chichester: Wiley.
McLachlan GJ (1992) *Discriminant Analysis and Statistical Pattern Recognition*. New York: Wiley.
Rencher AC (1995) *Methods of Multivariate Analysis*. New York: Wiley.
Ryan ThP (1997) *Modern Regression Methods*. New York: Wiley.
Sharaf MA, Illman DL and Kowalski BR (1986) *Chemometrics*. New York: Wiley.

Relevant journals *Analytical Chemistry, Analytical Clinica Acta, Applied Statistics, Biometrics, Biospectroscopy, Chemometrics, Chemometrics and Intelligent Laboratory Systems, IEEE Transactions on Pattern Analysis and Machine Intelligence, International Journal of Pattern Recognition and Artificial Intelligence, Journal of Chemometrics, Journal of Classification, Journal of Multivariate Analysis, Journal of Royal Statistical Society, Neural Computation, Neural Networks, Neurocomputing, Pattern Recognition, Pattern Recognition Letters, Statistical Methods in Medical Research, Technometrics.*

Muon Spin Resonance Spectroscopy, Applications

Ivan D Reid, Paul Scherrer Institute, Villigen PSI, Switzerland
Emil Roduner, Universität Stuttgart, Germany

MAGNETIC RESONANCE
Applications

Spin-polarized positive muons, when injected into matter, can serve as magnetic probes for the investigation of various properties. The evolution of muon spin polarization rests on the same basis as conventional magnetic resonance techniques, but the methods involved in exploiting these probes are considerably different. The basic elements of muon spin resonance spectroscopy are described, with an emphasis on spectroscopic and kinetic applications in chemistry.

The positive muon, a magnetic probe

Muon spin resonance spectroscopy is an offshoot of the experiment that proved parity nonconservation in pion and muon decay. The observation that the initial amplitudes and the relaxation rates of the muon precession signals depended on the nature of the stopping medium led to a new analytical method that is closely related to Fourier transform magnetic resonance.

The positive muon μ^+ is an elementary particle that in the present context is best regarded as a light proton with a mass one-ninth that of the proton. Like the proton it is a spin-$\frac{1}{2}$ particle, but its magnetic moment is 3.18 times μ_p. It is available for experiments as spin-polarized beams that can be implanted into most environments and then used to probe the local magnetic field, either as static spectators or by sampling a certain region as a mobile species. Because of its low mass and low linear energy transfer during the thermalization process, the muon causes relatively little damage near its stopping site. Nevertheless, it can probe radiation-chemical effects near the end point of its thermalization track.

In some environments the positive muon can capture an electron to form a hydrogen-like atom with the μ^+ as a nucleus. Called muonium (Mu), this atom behaves chemically as a light hydrogen isotope. It is a paramagnetic species that can also serve as a static or diffusing magnetic probe. Some of the properties of muons and muonium are summarized in **Table 1**. Chemical reaction of Mu with unsaturated molecules leaves the muon as a polarized spin label in organic free radicals, for example

$$C_6H_6 + Mu \rightarrow C_6H_6Mu \quad [1]$$

Such species are routinely observed and can be characterized spectroscopically.

In comparisons of muons with protons and of muonium with hydrogen atoms, pronounced quantum effects occur whenever dynamics are involved. In this way, muons have been utilized to probe a large variety of properties and materials: insulators, semiconductors, metals, superconductors, insulators, gases, liquids, crystalline and amorphous solids, static and dynamic magnetic properties of all kinds, electron mobility, quantum diffusion, chemical reactivity and molecular structure and dynamics. The term adopted for the broad field of muon spin spectroscopy techniques, μSR, emphasizes the analogy with other types of magnetic resonance; for example EPR. 'μS' represents 'muon spin', and 'R' in a more general sense stands simultaneously for 'rotation', 'relaxation' and 'resonance'.

Muon production

Muons are produced artificially from pions formed by the collision of energetic protons with low-Z targets. The pions decay with a lifetime of 26 ns:

$$\pi^+ \rightarrow \mu^+ + \nu(\mu) \quad [2]$$

The neutrino, ν, has negative helicity (angular momentum antiparallel to linear momentum), so from conservation laws the muon must also have negative helicity. By momentum selection, muon beams are obtained with a polarization close to 100%.

Two types of positive muon beams are possible: (1) surface muons, arising from pions decaying at rest close to the surface of the production target, with a momentum of 28 MeV/c and a stopping range of 140 mg cm^{-2} (corresponding to a water layer 1.4 mm thick); and (2) decay muons, from pions decaying in flight, having a selectable momentum normally in the range 70–130 MeV/c, and requiring a stopping range up to several centimetres of water. Surface muons are often preferred since they allow the use of thinner samples. They are also amenable to in-flight spin rotation where the crossed electric and magnetic fields of a Wien filter can be used to orientate their polarization with respect to their momentum.

Muon beams can be further classified on the basis of their time structure. Quasi-continuous muon sources allow high-precision measurement of the time interval between the arrival of the muon and its decay. However, the requirement in most experiments that each muon decay before the next is admitted to the sample limits the usable flux to about 10^5 muons per second. Pulsed beams avoid this pile-up limit since a large number of muons are stopped simultaneously, followed by a long delay during which the decay of most muons can be accumulated, leading in principle to background-free histograms and a longer time window. Pulsed beams are also ideally suited for experiments that require coincidence with other pulses, e.g. laser or radiofrequency excitation.

Table 1 Properties of μ^+ and Mu

	Property	*Symbol*	*Value*[a,b]
μ^+	Mass	m_μ	$1.883\,53 \times 10^{-28}$ kg = $0.1126\ m_p = 206.7864\ m_e$
	Life time	τ_μ	2.197 134 μs
	Spin	I	$\frac{1}{2}$
	Magnetic moment	μ_μ	3.183 344 μ_p
	Gyromagnetic ratio	γ_μ	135.54 MHz T^{-1}
Mu	Mass	m_{Mu}	0.0113 15m_H
	Bohr radius	a_0 (Mu)	0.053 17 nm = 1.0043a_0(H)
	Ionization potential	I (Mu)	13.539 eV = 0.9957I(H)

[a] μ_p = magnetic moment of the proton.
[b] H = hydrogen.

Detection of muon polarization

Conventional magnetic resonance uses radiofrequency techniques to detect the time evolution of spin polarization. This is impossible in muon spin resonance because of the extremely dilute concentration of the probes within the sample (often no more than one muon at a time). However, the decay of the muon

$$\mu^+ \rightarrow e^+ + \nu(e) + \bar{\nu}(\mu) \quad (\tau_\mu = 2.197\ \mu s) \quad [3]$$

does not conserve parity, leading to an anisotropic decay in which the positron (e^+) is emitted preferentially along the instantaneous direction of the muon spin at the moment of decay. Detection of the decay positron (and also of the muon's arrival in the case of time-differential experiments at a continuous-beam facility) involves standard particle-physics techniques, typically scintillating detectors coupled to photomultiplier tubes. Collection of an ensemble of correlated muon–positron events as a function of time and/or magnetic field in a given direction from the sample allows the evolution of muon polarization to be monitored.

Thus the basic methods of muon spin resonance differ considerably from those of conventional magnetic resonance spectroscopy. First, the probe is delivered to the sample with its spin polarization already prepared, and second the evolution of the polarization is monitored statistically by collecting information on the decays of individual probes.

Experimental techniques of muon-spin resonance

Transverse-field muon spin rotation

The transverse-field μSR (TF-μSR) technique is shown schematically in **Figure 1A**. Muons enter the sample S either perpendicular to the magnetic field (as shown in **Figure 1A**) or (for a spin-rotated surface beam) along the field axis such that their polarization is perpendicular to the field. Each arrival is detected by a muon counter that starts a clock to measure the individual muon lifetime (at pulsed-beam facilities the start signal is synchronized to the muon pulse). The clock is recorded when the corresponding decay positron is detected in a positron counter, e.g. f or b, and the event is stored in a histogram $F(t)$ of decay positrons as a function of time spent by the muon in the sample. In the absence of any time evolution of the muon polarization, $F(t)$ is simply the radioactive decay curve (dashed line in **Figure 1A**) that corresponds to the muon decay. In a nonzero local magnetic field the muon will undergo a free induction decay (FID), similar to the response after a $\pi/2$ pulse in conventional magnetic resonance. Owing to the decay anisotropy (Eqn [3]) the FID appears in the histogram superimposed on the muon decay curve. $F(t)$ takes the shape indicated schematically in **Figure 1A**.

In many cases, different muons find themselves in different magnetic environments in the same sample, resulting in an FID with multiple frequencies. In this case it is convenient to work with the Fourier transform (FT) of the time-domain data, just as for conventional magnetic resonance. This is sometimes called FT-μSR.

Zeeman precession of the diamagnetic muon occurs at $\nu_\mu = \gamma_\mu B_i$ where the local magnetic field B_i can deviate from the applied external field as a result of Knight shifts. Muonium, on the other hand, is observed mostly in low fields where two characteristic intra-triplet precession frequencies of the combined muon–electron systems occur. Below ~2 mT they become degenerate at $\nu_{Mu} = 13.94$ MHz mT$^{-1} \times B$. The two transitions are indicated in the Breit–Rabi diagram for Mu in **Figure 2** (ν_{Mu}, solid arrows). Two further allowed transitions are of very high frequency

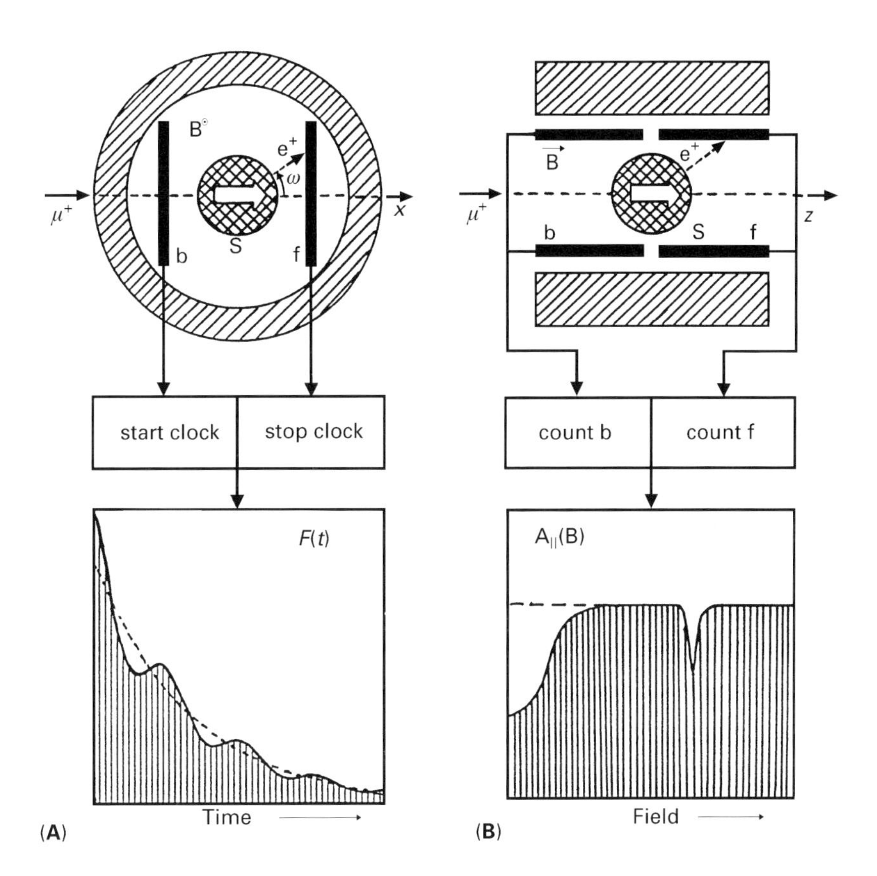

Figure 1 Schematic drawing of the experimental apparatus used for muon spin resonance spectroscopy: (A) time-resolved transverse-, longitudinal- and zero-field studies; (B) time-integrated longitudinal fields (ALC).

and not normally observed (indicated by broken arrows). Muonated organic free radicals, however, are studied more easily in high fields where the muon transitions degenerate into two, independent of any other magnetic nuclei that may be present, corresponding to the first-order ENDOR frequencies,

$$\nu_{\mathrm{R}\pm} = \left|\tfrac{1}{2}A_\mu \pm \nu_\mu\right| \quad [4]$$

These are indicated for Mu in **Figure 2**. The muon hyperfine coupling constant A_μ is obtained directly as the sum of the two observed frequencies (or the difference, in fields where $[\frac{1}{2}A_\mu - \nu_\mu]$ becomes negative). In general A_μ is in fact the component of a tensor $\tilde{\mathbf{A}}_\mu$ so its value depends on the orientation of the radical to the magnetic field B; however, most experiments are carried out with fluid samples where motion of the radical averages out the anisotropy, so just the isotropic value is observed.

Zero-field and longitudinal-field muon spin relaxation (ZF-μSR and LF-μSR)

These experiments are performed in time-differential mode using the same setup as TF-μSR, except that the external field is absent or is applied parallel to the muon spin polarization in the beam (either by rotating the magnet or by removing spin rotation).

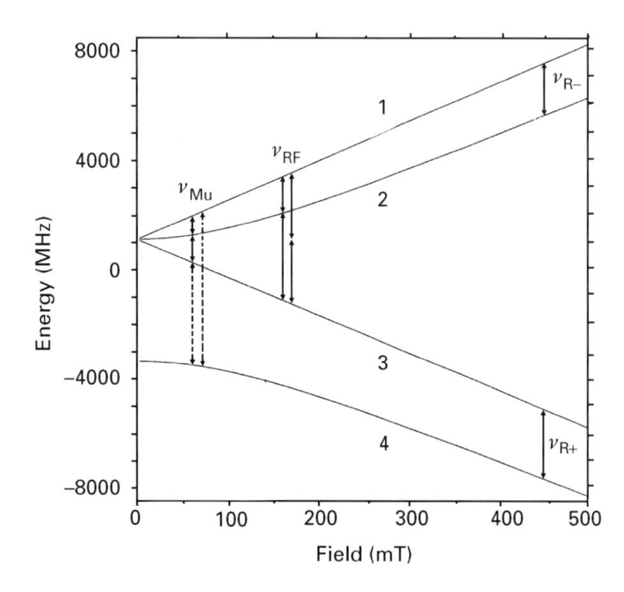

Figure 2 Breit–Rabi diagram for muonium with transverse-field transitions indicated by arrows. The zero-field splitting corresponds to the muon hyperfine coupling constant (A_μ = 4463 MHz for Mu). In high fields, the four eigenstates become pure Zeeman states ($|1\rangle = |\alpha^\mu \alpha^e\rangle$, $|2\rangle = |\beta^\mu \alpha^e\rangle$, $|3\rangle = |\beta^\mu \beta^e\rangle$, $|4\rangle = |\alpha^\mu \beta^e\rangle$).

Avoided-level-crossing resonance (ALC-μSR)

A basically different type of μSR takes advantage of the effect of avoided crossings of magnetic energy levels and is commonly performed in a time-integral mode as a function of longitudinal magnetic field (**Figure 1B**). Decay positrons are counted in the 'forward' (N_f) and in the 'backward' (N_b) direction with respect to the incoming muon beam. Of interest is the decay asymmetry

$$A_{\|}(B) = \frac{N_\mathrm{f} - N_\mathrm{b}}{N_\mathrm{f} + N_\mathrm{b}} \quad [5]$$

In sufficiently high fields, the eigenstates of any spin system are pure Zeeman states, and muons that are aligned with the external field are in an eigenstate where there is no time evolution of spin polarization, and $A_{\|}(B)$ is constant. Coupling of the muon to an unpaired electron, as in Mu or muonated radicals, leads to eigenstates that are mixtures of Zeeman states in low fields. The prepared non-eigenstate oscillates between eigenstates, resulting in partial depolarization of the muon and a strong field dependence of the detected decay asymmetry at low fields. The same effect operates at those high fields where there is an avoided crossing of two levels and eigenstates are mixtures of two Zeeman states. As **Figure 1B** indicates, sharp resonances are observed at these fields.

The resonances are distinguished by the selection rules $|\Delta M| = 0$, 1, 2, where M is the quantum number for the z component of the total spin. They appear to first order at resonance fields

$$B_\mathrm{r} = \left|\frac{A_\mu + (\Delta M - 1)A_k}{2[\gamma_\mu + (\Delta M - 1)\gamma_k]}\right| \quad [6]$$

Thus, if A_μ is known, the nuclear hyperfine couplings A_k are readily obtained, as are their relative signs, usually a difficult result in conventional magnetic resonance. The technique has been applied to semiconductors and to insulators, where dipolar and quadrupolar terms are determined in addition to the isotropic hyperfine couplings of Equation [6].

Radiofrequency resonance techniques (RF-μSR)

RF-μSR resembles conventional CW (continuous wave) magnetic resonance, as it relies on resonance between an exciting RF field and a transition between two magnetic energy levels. The technique is of advantage, for example, for studying radicals in dilute solutions where the phase coherence of muon precession is lost in transverse fields during the formation process. The experiment is carried out in a longitudinal external field with a transverse RF field B_1. At high RF power a narrow double-quantum transition is observed in addition to the two power-broadened lines. The experiment can be conducted at constant field and variable RF frequency, or perhaps more conveniently at constant frequency and variable applied magnetic field as demonstrated nicely in a report on endohedral Mu and exohedral Mu adducts of fullerenes.

Applications of muon spin resonance spectroscopy

To illustrate some of the experimental techniques outlined above, a selection of experiments using muon spin resonance spectroscopy are described.

Spectroscopy of muonium-labelled organic free radicals

The prototypical sample used in μSR spectroscopy is benzene (see Eqn [1]). A raw histogram of the TF-μSR data obtained in a benzene sample at an applied field of 0.2 T is shown in **Figure 3A**. The precession

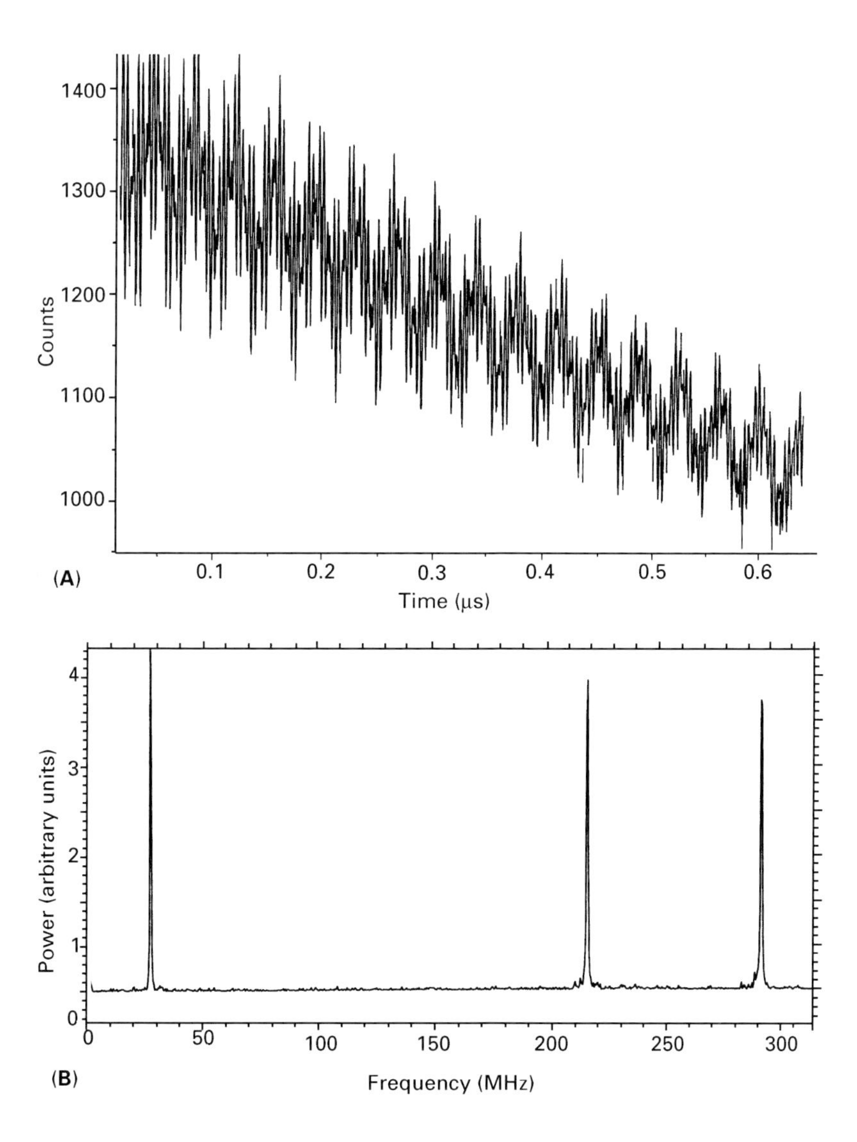

Figure 3 TF-μSR spectra for positive muons implanted into benzene: (A) the raw time-differential histogram recorded for the experiment; (B) the Fourier transform of (A) showing the two transitions due to the C_6H_6Mu radical (218.18 and 295.85 MHz, giving a hyperfine coupling constant of A_μ of 514.03 MHz) and the signal from muons in diamagnetic environments (27.1 MHz).

signal at 27 MHz arising from muons that have thermalized in a diamagnetic environment is easily seen, as is the overall decay curve due to the finite muon lifetime. **Figure 3B** shows the Fourier transform of the histogram, revealing not only the muon precession but also the ENDOR-like precession signals from muons which have reacted to form the free radical C_6H_6Mu.

Benzene is a simple example, since all addition sites are equivalent. However, consider 6,6-dimethylfulvene (**Figure 4**). Here there are four possible addition sites and thus potentially four radicals can be formed. **Figure 5A** shows the FT-μSR signal obtained in this compound at 0.3 T in ether solution. Two pairs of radical signals are seen, with hyperfine coupling constants of 105.2 and 203.5 MHz, respectively. To elucidate which radicals were observed, ALC spectra were also obtained for this compound (**Figure 5B**, lower trace). The resonances indicate where the muon's energy level has become degenerate with that of one of the H atoms in the radical (i.e. $\Delta M = 0$). Since the A_μ values are known from the FT-μSR, it is possible to devise a simulation to model the ALC spectrum, obtaining the various A_p values within the

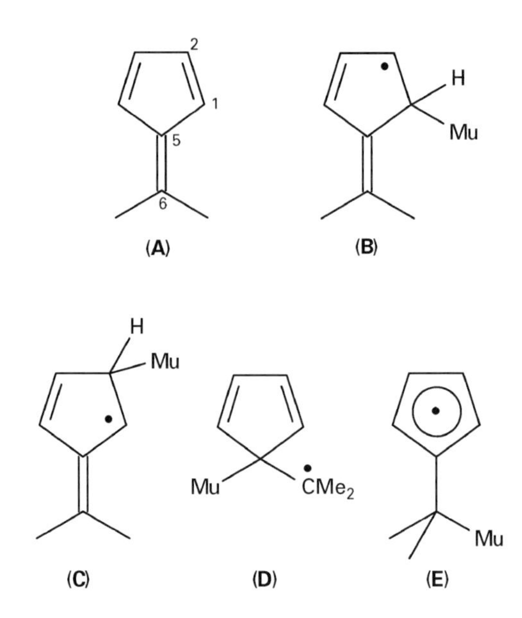

Figure 4 The structure of 6,6-dimethylfulvene (A) and the four radicals (B–E) that may be derived from it by Mu addition. Reproduced with permission of the Royal Society of Chemistry from Rhodes CJ, Roduner E, Reid ID and Azuma T (1991) *Journal of the Chemical Society, Chemical Communications* 208–209.

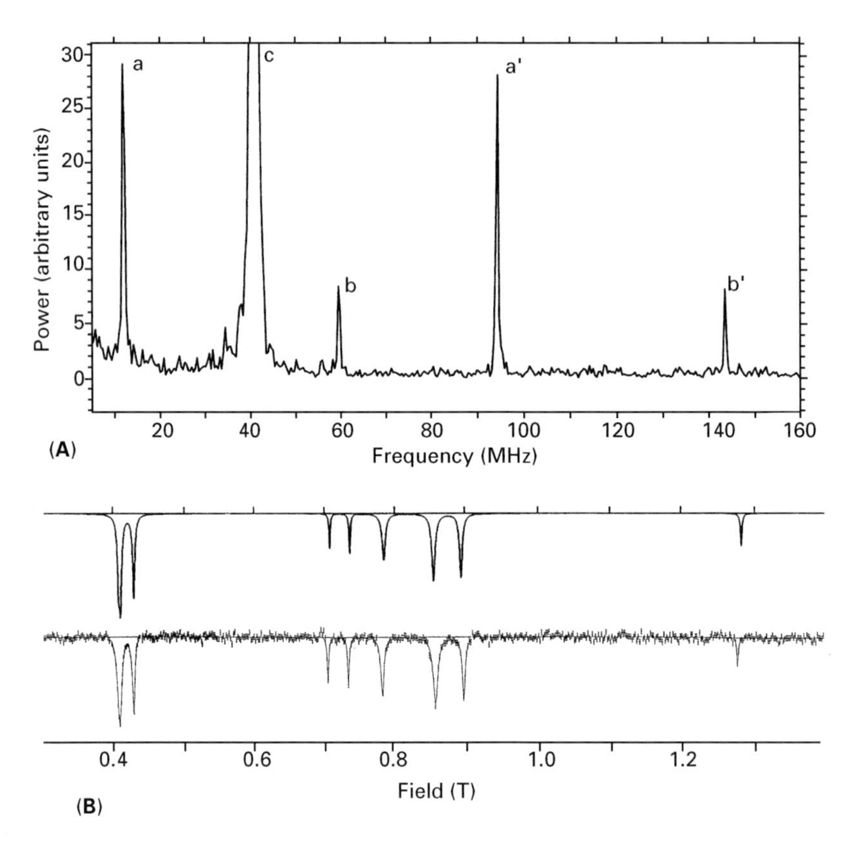

Figure 5 Muon spin spectra from positive muons stopped in 6,6-dimethylfulvene: (A) FT-μSR spectrum showing the pairs of signals for each radical (a,a′; b,b′) and that from muons stopped in diamagnetic environments (c); (B) ALC spectra: lower trace, experimental data; upper trace, numerical simulation. Reproduced with permission of the Royal Society of Chemistry from Rhodes CJ, Roduner E, Reid ID and Azuma T (1991) *Journal of the Chemical Society, Chemical Communications* 208–209.

radicals, and to assign these values to specific sites. The results indicate that the two radicals are those formed by addition at the 1 and 2 positions in the ratio 1.4:1. The derived coupling constants and spin populations for these radicals are given in **Figure 6**.

Note that the ratio of the reduced muon coupling A'_μ ($A'_\mu = A_\mu$ (μ_p/μ_μ)) to the proton coupling at the same site is somewhat higher than unity. This is typical for muonated organic free radicals, the isotope effect normally lying between +5% and +15%. This can be understood as a simple consequence of the larger zero-point vibrational amplitude of the lighter isotope. Since the potential well for vibration is anharmonic, the average bond length is also larger. In other words, the muon and electron are moved towards the configuration of a free Mu atom, where A_μ is much higher than in the radical, and so their hyperfine coupling constant is correspondingly increased.

As noted earlier, the hyperfine coupling constant A_μ is in fact the component of a tensor $\tilde{\mathbf{A}}_\mu$ with a value dependent on the orientation of the radical to the magnetic field B. Whereas in solution the anisotropic part of the tensor is averaged out, in a crystalline solid the full anisotropic nature can be seen. **Figure 7** shows FT-μSR spectra from positive muons in naphthalene. The spectrum in (A) was obtained in acetone solution; the four lines are readily attributed to α and β radicals as indicated in the figure. The spectrum in (B) results from a large (~25 mm) single crystal – it is quite evident that the simple solution spectrum has been replaced by a much more complicated system. Indeed, the four lines have been split into 32 because there are two molecules per unit cell, because the crystal environment halves the symmetry of the parent molecules, and because additions from above and below the molecule are also not symmetric. None the less, the angular variation of the signals as the crystal was rotated about three orthogonal axes was mapped out and the hyperfine tensor was calculated for each of the 16 radicals. From symmetry and physical arguments, each was assigned to a specific addition site within the crystal. **Table 2** gives the results obtained for the tensor components.

The orientation of the radicals within the crystal was deduced to be very similar to that of the parent radicals, but effects from the crystal field on the hyperfine coupling constants were quite evident, particularly for the β radicals; note particularly the differences in the isotropic components for addition from both sides of the molecule. Comparison with an ENDOR study of α-hydronaphthyl radicals showed that the earlier work had not detected all the possible radicals and that the two identified arose from addition at only one of the α sites; the ENDOR study did note see any β radicals. The reduced hyperfine coupling constants were ~10–20% larger than the proton values, although the comparison is not rigorous since the two studies were performed at greatly different temperatures.

As well as studies in condensed phases, μSR can also be carried out in the gas phase using either surface muons at moderate pressures or decay muons at higher pressures. Experiments with ethylene gas at room temperature and pressures of 25, 35 and 50 atm revealed FT-μSR signals that can readily be ascribed to the $CH_2Mu\dot{C}H_2$ radical. The resultant spectra are shown in **Figure 8**. The hyperfine coupling is 329.7 MHz, quite close to that found in liquid ethylene, and does not vary with pressure. The two radical signals are not of equal strength owing to the finite lifetime of the Mu precursor, which leads to dephasing of the muons and a corresponding loss of polarization. Calculations from the ratio of the polarizations of the two signals indicate a reaction rate coefficient of 16×10^{-12} cm^3 molecule^{-1} s^{-1}, compared to a value of 6×10^{-12} cm^3 molecule^{-1} s^{-1} from relaxation measurements on the Mu signal in ethylene in a moderator gas at 295 K. The higher reaction

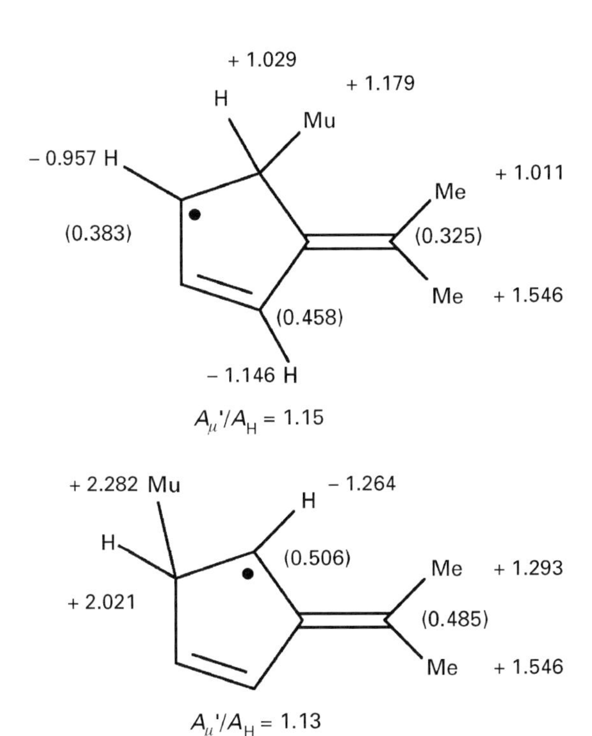

Figure 6 Coupling constants in mT and (spin populations) for the two Mu adducts to 6,6-dimethylfulvene. Spin populations were obtained using Q = –2.6 mT for α-protons and Q = +2.925 mT for Me protons. Reproduced with permission of the Royal Society of Chemistry from Rhodes CJ, Roduner E, Reid ID and Azuma T (1991) *Journal of the Chemical Society, Chemical Communications* 208–209.

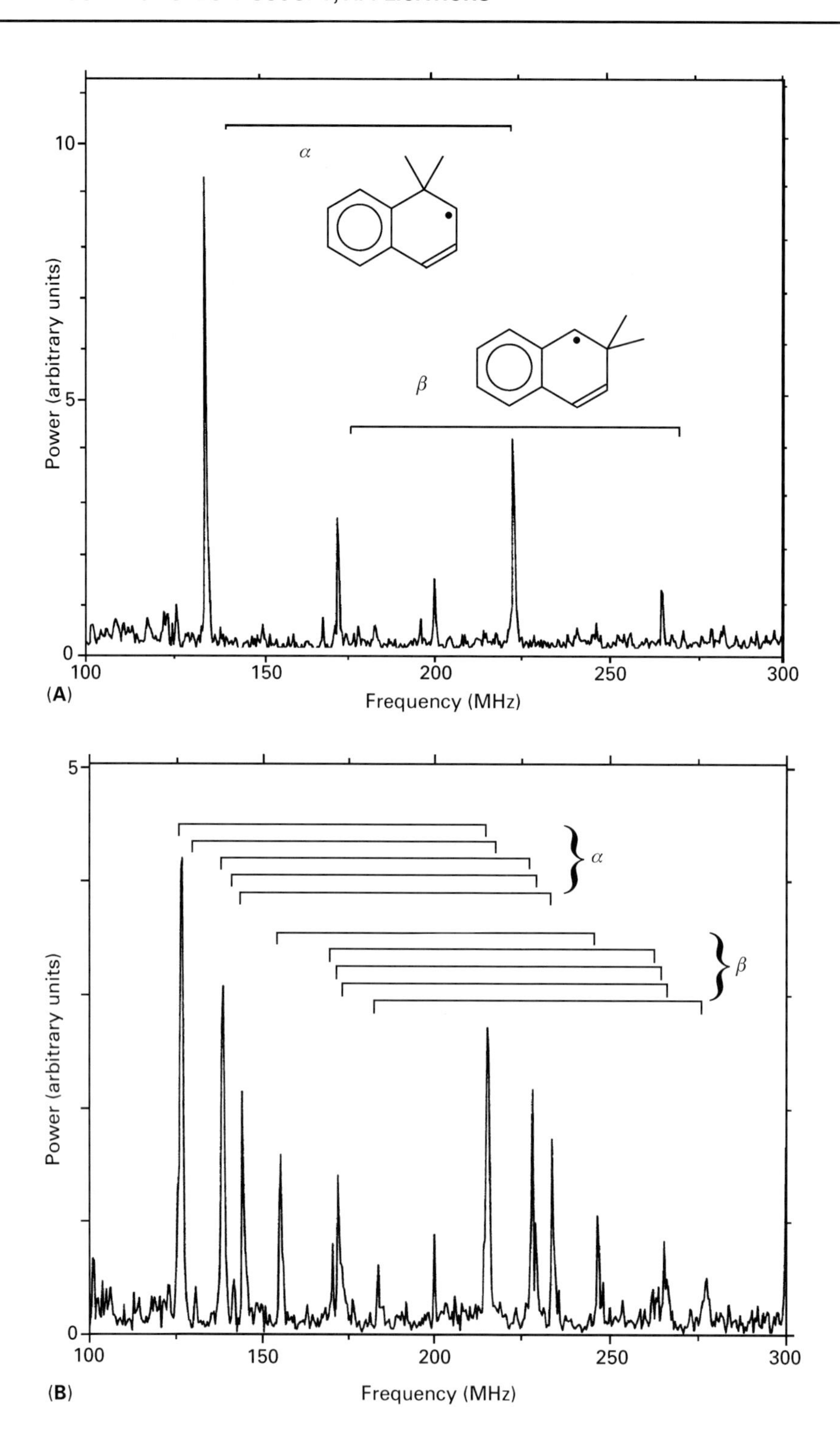

Figure 7 FT-μSR spectra obtained for positive muons implanted in naphthalene samples: (A) in saturated acetone solution; (B) in a large single crystal. Reproduced with permission from Reid ID and Roduner E (1991) *Structural Chemistry* **2**: 419–431.

rate coefficient implies a Mu temperature of around 500 K, showing that it has not completely thermalized before the reaction with the ethylene. The broad nature of the radical signals has been explained in terms of the spin–rotational interaction within the radical as it undergoes collisions within the gas.

Muonium and radical kinetics and dynamics

The mass ratio between Mu and H is unparalleled and pronounced isotope effects are to be expected. In reaction kinetics experiments, for example, considerable kinetic isotope effects (KIE) may be found.

Consider isotopic variations of the simple abstraction reaction

$$H + X_2 \rightarrow HX + X \quad [7]$$

First, comparing the reaction with Mu instead of H, one should expect a trivial KIE of about 3 owing to the lower mass and hence higher thermal velocity of Mu. However, the lower mass can also lead to greater zero-point vibrational energy effects in the activated complex and on the product side, causing a higher barrier for the Mu reaction so that reaction of the heavier isotope may be faster. Finally, quantum-mechanical tunnelling is often considered important in H-atom reactions, so that the lighter mass of Mu should favour it in crossing reaction barriers.

Table 2 Principal values of reduced hyperfine coupling tensors for Mu-substituted radicals in single-crystal naphthalene ($A'_{ii} = A'_{iso} + B'_{ii}$). Also given are the isotropic solution couplings and the tensors obtained for the CH_2 protons of analogous hydrogen radicals by ENDOR spectroscopy.

Nucleus[a]	A'_{iso} (MHz)	B'_{11} (MHz)	B'_{22} (MHz)	B'_{33} (MHz)
CHMu muon α_a	115.68	−2.61	−1.09	3.70
CHMu muon α'_a	108.75	−2.64	−1.06	3.70
CHMu muon α_b	117.98	−2.94	−0.95	3.89
CHMu muon α'_b	108.83	−2.85	−0.74	3.59
CHMu muon α_{soln}	112.17			
CHMu muon β_c	128.71	−2.82	−1.69	4.51
CHMu muon β'_c	143.54	−3.11	−1.27	4.38
CHMu muon β_d	138.90	−2.61	−1.75	4.36
CHMu muon β'_d	136.56	−2.86	−1.73	4.59
CHMu muon β_{soln}	137.14			
CH_2 proton 1	101.78	−2.80	−1.28	4.10
CH_2 proton 1′	90.43	−3.03	−1.25	4.27

Reproduced with permission from Reid ID and Roduner E (1991) *Structural Chemistry* **2**: 419–431.
[a]The subscripts indicate the various inequivalent addition positions; a prime indicates addition from the opposite side of the molecular plane.

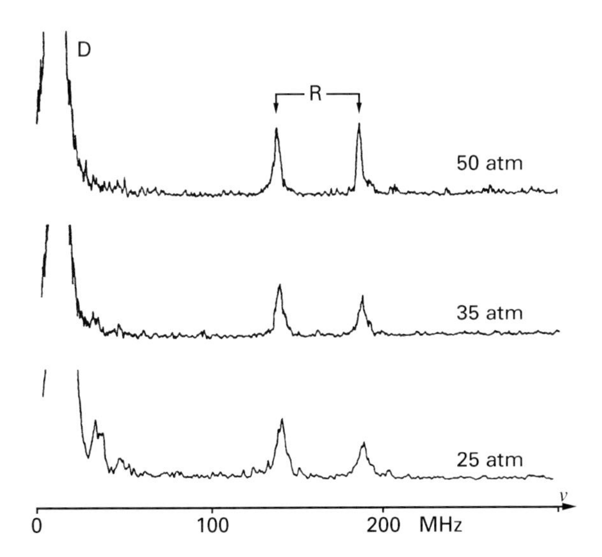

Figure 8 FR-μSR spectra for ethylene gas at various pressures. D-muons in diamagnetic environments; R-muonated ethyl radical. Reproduced with permission of Baltzer Science Publishers from Roduner E and Garner DM (1986) *Hyperfine Interactions* **32**: 733–739.

Compared to H-atom reaction studies, Mu reactivity experiments are reasonably simple. Positive muons are stopped, in a low (<0.8 mT) transverse field, in a moderator that produces a high yield of Mu, doped with varying concentrations $[X_2]$ of the reactant. Exponential relaxation of the coherent Mu precession signal is attributed to randomly occurring chemical reactions that transfer the muon to a different magnetic state where its precession frequency is significantly different. There is thus a linear relationship between the relaxation rate λ and $[X_2]$, leading directly to the thermally averaged bimolecular reaction rate constant k:

$$\lambda([X_2]) = \lambda(0) + k[X_2] \quad [8]$$

where $\lambda(0)$ is the relaxation rate for $[X_2] = 0$.

Many studies have been undertaken of such reactions, for example reaction of Mu with the halogen gases (F_2, Cl_2 and Br_2). For both F_2 and Cl_2 the data showed a significantly shallower slope in the Arrhenius plots of k vs T at temperatures below 200 K, clear evidence for tunnelling dominating the reaction in this region. This feature is absent from comparable H-atom reaction data. For Br_2, on the other hand, a negative slope was observed, indicating a negative activation energy E_a. In these highly exothermic reactions, the barrier is early on the reaction path and so the transition state approximates a slightly perturbed reactant molecule whose vibrations are only weakly dependent on the isotopic substitution. There may, however, still be a significant effect in the transmission coefficients for tunnelling, $\kappa_{H,Mu}$. **Table 3** summarizes some of the results obtained in this study and shows that there are indeed significant tunnelling effects even at the higher temperatures.

On the other hand, there are significant zero-point energy effects in the reaction of Mu with H_2 and D_2, since these reactions are highly endothermic ($\Delta H_R^0 = 32$ and 38 kJ mol^{-1}, respectively) so the vibrationally adiabatic barrier is late. The potential energy surface of the H_3 system is known to a high accuracy, so comparison of the results for these reactions with various theoretical treatments should help elucidate the extent to which the above effects are important and how well they can be predicted theoretically. Such measurements have been carried out over the temperature ranges 473–843 K and 598–843 K, respectively. Comparison with theoretical results showed that 3D quantum coupled states (CS) calculations of the reactions fit the data much better than improved canonical variation theory treatments, verifying the validity of both the potential energy surface and the CS treatment.

Studies of hydrogen KIE often involve reactions where a reagent attacks a H—R/D—R bond, transferring the isotope atom, whereas in Mu reactivity studies muonium is normally the attacking species. One example where Mu is thought to be transferred is the reaction of cyclohexadienyl radicals with 2,3-dimethyl-1,3-butadiene (DMBD).

In both pure benzene and pure DMBD, just one radical is observed via TF-μSR, with narrow lines (λ_0 <0.5 μs^{-1}). However, in binary mixtures of the compounds or in three-component mixtures with cyclohexane, the cyclohexadienyl lines broaden with increasing DMBD concentration while the line width of the allyl-type radical remains narrow. This obviously indicates a reaction of the cyclohexadienyl radical with DMBD. From the variation of line width with concentration (Eqn [8]) a reaction rate constant k_{Mu} of 0.95×10^6 M^{-1} s^{-1} was obtained at 293 K. The protiated cyclohexadienyl radical has been observed in time-resolved EPR experiments, but no significant reaction with DMBD was seen, with an upper limit to k_H of ~12 M^{-1} s^{-1}. This yields an enormous KIE: $k_{Mu}/k_H \gtrsim 7.5 \times 10^4$ at room temperature. For such an affect Mu must be directly involved in the reaction and it is believed the scheme is:

Reaction of C_6H_6Mu is favoured over C_6H_7 primarily because of its higher zero point vibrational energy, but tunnelling is also involved.

Other μSR methods can also be used to probe reactivity. For example, muon relaxation in longitudinal fields has recently been used to study the gas-phase reaction of the muonated ethyl radical with oxygen. Analysis of the results enabled a separation of the effects of chemical reaction and spin-exchange, leading to values of $k_{ch} = 8.4(3) \times 10^{-12}$ cm^3 $molecule^{-1}$ s^{-1} and $k_{ex} = 2.8(2) \times 10^{-10}$ cm^3 $molecule^{-1}$ s^{-1}, respectively, at room temperature. It is striking that the chemical reaction rate is so much smaller than the spin-exchange rate, but this is explained by the anisotropy of the potential energy surface for chemical reaction leading to successful bond formation.

Dynamic effects in radicals also can be studied by μSR. Particularly useful is the $\Delta M = 1$ ALC-μSR resonance, which is only observable when there is anisotropy in the hyperfine coupling. Phenomenologically it can be thought of as occurring when the lower frequency ν_{R-} in Equation [4] passes through zero, i.e. a field where $\nu_\mu \sim \frac{1}{2}A\mu$ (Eqn [6]). Here a component of the hyperfine field is nullified by the external field and any anisotropy is seen by the muon as a transverse field, which causes it to precess and thus depolarize, leading to an observable resonance. Since the hyperfine tensor is anisotropic, this resonance should be seen unless the radicals are reorientating so fast as to average out the anisotropy. This was found to be the case for cyclohexadienyl radicals in a benzene monolayer on silica, for example, where the lack of the resonance down to 139 K was taken as evidence that the radicals remained mobile on the surface.

Table 3 Reaction rate coefficients, kinetic isotope effects, and relative transmission coefficients for the reactions of H and Mu with the halogen gases.

	F_2	Cl_2	Br_2
k_{298}(Mu) (10^{-11} cm^3 $molecule^{-1}$ s^{-1})	2.62±0.06	8.50±0.14	56.0±0.9
k_{298}(H) (10^{-11} cm^3 $molecule^{-1}$ s^{-1})	0.16±0.01	2.10±0.1	8.2±3.8
KIE_{298}(Mu/H)	16.4	4.0	6.8
κ_{Mu}/κ_H (298 K)	5.7	1.4	2.3
κ_{Mu}/κ_H (250 K)	8.0	2.1	
E_a (Mu) (kJ mol^{-1})	3.1± 0.3	2.7±0.2	−0.4±0.8
E_a (H) (kJ mol^{-1})	9.2± 0.3	5.0±0.4	

Reproduced with permission from Gonzalez AC, Reid ID, Garner DM, Senba M, Fleming DG, Arseneau DJ and Kempton JR (1989) *Journal of Chemical Physics* **91**: 6164–6176.

A further study used this resonance to ascertain the anisotropic components of the hyperfine coupling in the radicals formed by irradiating bicyclo[2.2.1]hept-2-ene (norbornene) in its plastic phase. The tensor had already been derived from a single-crystal FT-μSR investigation but the ALC experiment was able to determine the (axial) anisotropy in a powder sample from the asymmetry of the shape of the resonance. The results agreed well with the FT-μSR data and confirmed effective partial averaging above T_c (129 K) and a rapid approach to complete isotropy at T_m (320 K).

Potential and limitations of the muon as a probe

Spin polarization and usable field range

Clearly, for a magnetic resonance type technique, the availability of a probe with a polarization close to 100% under all conditions is an immense advantage. Most techniques must work with Boltzmann populations, corresponding to polarizations of typically 10^{-5} for protons and 10^{-3} for electrons in commonly available magnetic fields and at room temperature. Muons have their full polarization independently of sample temperature and magnetic field and are therefore ideal probes, especially at low or zero field and at high temperatures.

Longitudinal fields up to about 7T have been used for ALC-μSR experiments. TF-μSR has been limited by the routinely available time resolution to ~3–4 T, corresponding to a Zeeman frequency of about 500 MHz. An important advantage over conventional magnetic resonance, however, is the availability of a fully polarized probe in *zero* field where additional information becomes accessible.

Time resolution

Magnetic resonance techniques in time-resolved mode normally create transverse magnetization by application of a preparation pulse. This limits time resolution to typically 50 ns in EPR, and to several hundred nanoseconds in NMR, although a combination with optical techniques can break this limitation.

Since muons can be injected with their spins transverse to the external field, a preparation pulse is not necessary and time resolution is limited only to the accuracy with which t_0, the entrance time of the muon into the sample, can be measured. This is typically half a nanosecond when working with individual muons, and if needed it can be improved by a factor of about 5 at the cost of counting statistics.

Frequency resolution

The muon lifetime of 2.2 μs sets a natural limit of 0.45 MHz on the widths of μSR lines. The nominal frequency resolution is given by the inverse length of the histogram. Using pulsed machines it is not uncommon to observe the FID over a time window of 20 μs, which leads to a nominal resolution of 0.05 MHz.

Time window

The accessible time window for processes that can be observed directly extends between typically 10^{-9}s and 10^{-5}s. The limits are determined by the same parameters as time and frequency resolution. The windows accessible to NMR and EPR are thus extended considerably towards shorter times.

The time window can be extended to even shorter times if there is a muonium precursor state. Evolution of spin polarization in Mu occurs partly at frequencies near the Mu hyperfine frequency (4463 MHz in vacuum, broken arrows in **Figure 2**), which sets the timescale for loss of phase coherence during formation of the observed muonated species. The formation process can be studied indirectly in transverse fields by interpretation of shifts in the initial phase and concomitant loss of amplitude. Thus, processes occurring on a timescale down to 10 ps can be analysed, but the results rely on the validity of the underlying model.

The muon as a local probe

In the same sense as nuclei in conventional magnetic resonance techniques, the muon is a local probe. While, for example, magnetization measurements give an integral response, the local magnetic probe may precess at different frequencies when it sits at different sites. The advantage of the muon is that it can be implanted in any material, including those that contain no other suitable NMR-active nuclei. For organic muonated radicals, the stopping site can be derived with confidence by analogy with the known chemistry of hydrogen atoms. In organic solids it is in general an interstitial site – this may be a nonbonding electron pair if oxygen or nitrogen are present. In semiconductors, some muons come to rest at bond centre sites. There are, however, many examples of materials where a firm determination of the stopping site is difficult.

It is important to consider to what extent the muon is also a perturbing probe. In comparison with the analogous proton or hydrogen defects, the differences are certainly small. The structure and properties of muonated radicals are very similar to those of their hydrogen analogues and therefore well understood. On the other hand, a radical could be viewed as a strongly perturbed diamagnetic precursor molecule, which demonstrates how much the probe can change the electronic structure. In crystalline solids the properties measured are often not the same as those in the absence of the probe. It is clear, for example, that in a metal the positive charge of the muon causes charge polarization of the surroundings, leading to some structural relaxation.

Sensitivity

Both the high spin polarization of the muon and the single-event detection technique lead to a high sensitivity of the muon spin resonance methods. The total number of muons needed for a routine μSR experiment in transverse fields is of the order of 10^7, while X-band ESR at room temperature needs about 10^{11} unpaired electrons, and single-scan proton NMR needs on the order of 10^{17} spins to allow the observation of a narrow, unsplit signal. This sensitivity is particularly crucial for the observation of surface-adsorbed organic free radicals – it is difficult to maintain the minimum concentration for EPR detection under conditions where the radicals are mobile since they disappear via bimolecular termination reactions, whereas in TF-μSR one has a single muonated radical at a time in the entire sample. This allows adsorbed radicals to be observed up to temperatures where they desorb or undergo catalytic reactions. Termination reactions also represent a severe limitation to kinetic work in liquid solutions and in the gas phase, so the fact that the μSR technique guarantees ideal pseudo-first-order kinetics is clearly advantageous.

List of symbols

A_k = nuclear hyperfine coupling constant; A_μ = muon hyperfine coupling constant; A'_μ = reduced muon hyperfine coupling constant; $\tilde{\mathbf{A}}_\mu$ = muon hyperfine coupling tensor; $A_\parallel$ = decay asymmetry; B = magnetic field strength; B_{ii} = anisotropic components of the hyperfine coupling; e^+ = positron; E_a = activation energy for reaction; k = reaction rate constant; k_{ch} = chemical reaction rate constant; k_{ex} = spin-exchange reaction rate constant; M = spin quantum number; N_b = 'backward' decay positron count; N_f = 'forward' decay positron count; T_c = critical temperature for phase change; T_m = melting point; ΔM = change in spin quantum number; ΔH_R^0 = reaction enthalpy; γ_μ = muon gyromagnetic ratio; κ = tunnelling transmission coefficient; λ = relaxation rate; μ^+ = positive muon; μ_p = proton magnetic moment; μ_μ = muon magnetic moment; $\nu(e)$ = electron neutrino; $\nu(\mu)$ = muon neutrino; $\bar{\nu}(\mu)$ = muon antineutrino; ν_μ = muon precession frequency; ν_{Mu} = muonium precession frequency; ν_R = radical precession frequency; τ_μ = muon lifetime.

See also: **Chemical Applications of EPR; EPR, Methods; EPR Spectroscopy, Theory; Spin Trapping and Spin Labelling Studied Using EPR Spectroscopy.**

Further reading

Cox SFJ (1987) *Journal of Physics C: Solid State Physics* 20: 3187–3319.

Davis EA and Cox SFJ (1996) (eds) *Protons and Muons in Materials Science*. London: Taylor & Francis.

Roduner E (1986) *Progress in Reaction Kinetics* 14: 1–42.

Roduner E (1988) *The Positive Muon as a Probe in Free Radical Chemistry. Potential and Limitations of the μSR Techniques*, Vol. 49 in Lecture Notes in Chemistry. Heidelberg: Springer.

Roduner E (1993) *Chemical Society Reviews* 22: 337–346.

Roduner E (ed) (1997) *Applied Magnetic Resonance* 13.

Schenck A (1985) *Muon Spin Rotation Spectroscopy*. Bristol: Hilger.

Walker DC (1983) *Muon and Muonium Chemistry*. Cambridge: Cambridge University Press.

N

Near-IR Spectrometers

R Anthony Shaw and **Henry H Mantsch**, Institute for Biodiagnostics, National Research Council of Canada, Winnipeg, Manitoba, Canada

VIBRATIONAL, ROTATIONAL & RAMAN SPECTROSCOPIES
Methods & Instrumentation

Introduction

While near-infrared spectroscopy was for many years a sleeping giant, the diversity of instrumentation available today attests to its new-found and widespread acceptance as an analytical method. The aim of this article is to outline the measurement principles for each of the various types of instruments that are commercially available as of this writing, and to outline the strengths of each. Because analysis is by far the most common application, we also include accounts of specialized measurement techniques that have emerged for on-line, in-situ, and remote spectral measurements, as well as briefly outlining dedicated analysers founded upon near-infrared technology.

The near-infrared (near-IR) falls between the visible and mid-infrared regions, with corresponding vibrational frequencies in the terahertz range (**Figure 1**). It has become customary to report near-IR absorption positions in units of wavelength, either micrometres or nanometres, although it is becoming more common to see positions measured in wavenumbers (cm^{-1}). Simply the inverse of the wavelength (in cm), the wavenumber scale has become standard for mid-infrared spectroscopy and is becoming more commonly used for near-IR work. The advantages most often cited are that the scale is linear in energy and that it provides values of a convenient order of magnitude, particularly in the mid-infrared.

It seems odd at first glance that the near-IR region remained largely an afterthought for many years, given the relatively long and productive histories of both UV-visible and mid-infrared spectroscopy. However, the weak bands in the intervening near-IR region were long viewed as being superfluous; with broad, weak absorptions corresponding almost exclusively to overtones and combinations of vibrational modes, it was (and still is) generally considered that the structural information latent in the spectra is far more readily available from the mid-infrared spectra.

Times change. The 1980s and 1990s have witnessed revolutionary growth in the use of near-IR spectroscopy as an *analytical* technique. Near-IR spectrometers may now be found in applications ranging from agriculture to food processing and pharmaceuticals. Estimated at $4 500 000 in 1977, near-IR instrumentation revenues exceeded the $100 000 000 mark in 1997 and continue to grow at a compound rate of 20% annually. This demand has spurred the development of a variety of near-IR spectrometers ranging from dedicated analysers to research-grade systems of enormous versatility. Remote measurements may be carried out in conjunction with fibre optics, and, from a vantage point aboard the Hubble telescope, even more distant measurements are being carried out as the near-IR camera and multi-object spectrometer 'NICMOS' yields a new vision of the universe. Application of the same spectroscopic-imaging technology is yielding new insights here on Earth, and the next 20 years will be eventful ones as this technique begins to live up to its potential.

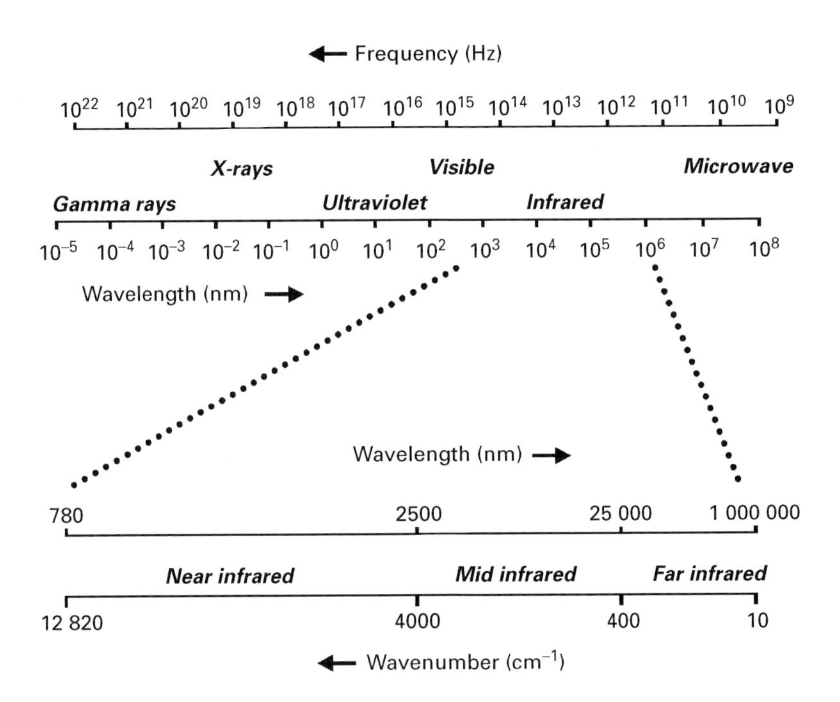

Figure 1 The near-infrared lies between the visible and mid-infrared regions, ranging from 780 to 2500 nm.

Instrumentation: an overview

Historical development

In 1800, William Herschel assembled a spectrometer in which the Sun served as the source of radiation, a prism as the dispersing element, and a thermometer as the detector. Having first measured the distribution of radiant heat across the visible region, he then took the unprecedented step of moving the thermometer to a position beyond the red end of the spectrum. To his surprise, the thermometer not only registered heat, but a higher temperature than that found in the visible region. This signalled the discovery of a 'non-visible spectrum', and coincidentally of the first near-IR spectrometer.

The idea of a spectrometer dedicated to near-IR spectroscopy is relatively recent, and the early commercial near-IR instruments were simply UV-visible (or mid-infrared) spectrometers fitted with an additional detector and occasionally a second grating blazed for the near-IR. This equipment provided the basis for the pioneering work of Kermit Whetzel and Wilbur Kaye, who were largely responsible for laying the foundation of analytical near-IR spectroscopy. The development of modern near-IR instrumentation was spurred by research at the US Department of Agriculture; Karl Norris discovered that no commercial spectrometer of the time could provide diffuse reflectance measurements of the quality he required, and developed his own computerized near-IR spectrometer for meat analysis. The original aim of this work was to develop a convenient means to monitor the water content of agricultural products. The research was complicated by the observation that components other than water contributed to the absorption profiles; however, this observation was quickly turned to advantage when it was found that protein could be quantified accurately from the near-IR spectrum of wheat. Under the direction of Phil Williams, the near-IR method was adopted by the Canadian Grain Commission to replace routine Kjeldahl testing. It worked, and spurred the widespread adoption of near-IR spectroscopy as an analytical method, and with it the birth of dedicated near-IR instrumentation.

What is a near-IR spectrometer?

It is easy to find a pair of near-IR spectrometers that would not be recognizable as belonging to the same species even placed side by side. This diversity has arisen largely as a result of the tremendous variety of applications, with each unique application satisfied by a unique set of design and performance characteristics.

Some of the spectral criteria that characterize and differentiate near-IR spectrometers are wavelength accuracy and precision; photometric accuracy, precision, and linearity; signal-to-noise; and resolution (or spectral bandwidth). Among these, the attributes

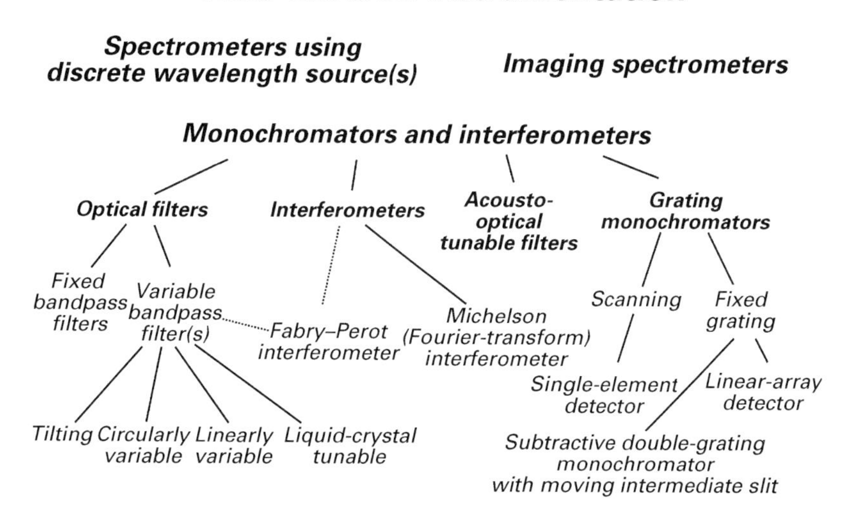

Figure 2 Near-infrared instrumentation.

that are most important for analytical work are wavelength precision, photometric precision, and signal-to-noise. While the last may occasionally be sacrificed, it is generally the case that the signal-to-noise is the limiting factor governing the analytical accuracy; spectra are commonly measured with noise at single-digit micro-absorbance levels.

Practical considerations almost always play a role in choosing a near-IR spectrometer, reflecting specific demands that may be placed by a specific application. Among these considerations are size and/or portability; speed; susceptibility to extreme (e.g. industrial) environments; cost; measurement type; and flexibility (e.g. can the spectrometer be used only for transmittance measurements, or are reflectance and/or fibre optic measurements feasible?). Manufacturers consider all of these factors. For example, it is common to find commercial instrumentation housed in enclosures that are ruggedized to meet specific industry standards (National Electrical Manufacturers Association – NEMA).

For convenience, the spectrometers discussed in this article have been divided into the six design categories and variations illustrated in **Figure 2**. The instrumentation will be discussed after first introducing two essential components, the source and the detector.

Near-IR sources

The most common broadband light source used for near-IR spectroscopy is the tungsten-halogen lamp, typically powered by a stabilized DC power supply. This source provides high energy throughout the near-IR, and has a long life with the halogen acting as a scavenger to deposit tungsten vapour back onto the filament.

Detectors

No single detector covers the entire 780–2500 nm near-IR range. The detectors most commonly used for near-IR spectroscopy are lead sulfide (or lead selenide) photoconductors and silicon photodiodes, with PbS(Se) covering the region 1100 to 2500 nm and Si the visible to 1100 nm range. A huge advantage of these detectors is that both provide good signal-to-noise operation at room temperature. **Figure 3** illustrates the specific detectivities (D^*) as a function of wavelength for PbS, PbSe, Si, and other detectors. As illustrated, both PbS and PbSe become more sensitive at lower temperatures, and some spectrometer manufacturers do capitalize on this. The alternative detector most commonly used at wavelengths longer than 1100 nm is indium gallium arsenide (InGaAs), which extends to 1800 nm with a substantially better D^* than either PbS or PbSe. At the expense of some sensitivity, the range of the InGaAs detector may be extended still further to 2200 nm, or even 2600 nm. Thermoelectric cooling is commonly used to increase D^*, but this improvement comes at the expense of long-wavelength response (**Figure 3**).

Fixed-grating spectrometers require a linear detector array for the simultaneous detection of the spectral elements dispersed along a line, and imaging applications require a two-dimensional detector array for spatial resolution. While both types of array are available for all of the detector materials included in **Figure 3**, those most commonly used are silicon, InSb and InGaAs.

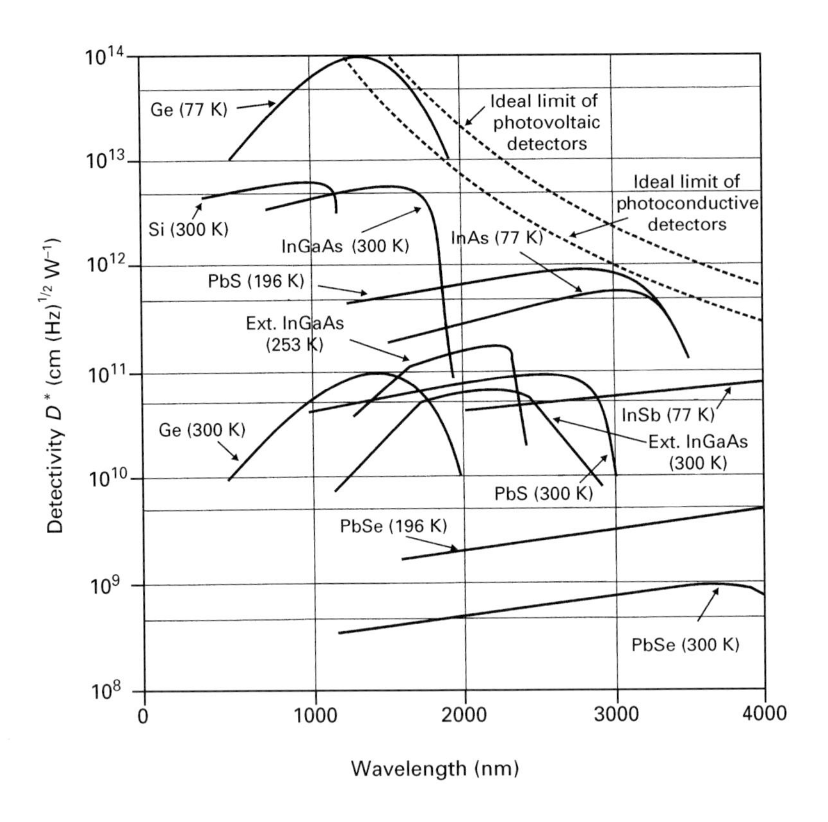

Figure 3 Specific detectivities and wavelength ranges for near-infrared detectors.

Near-IR instrumentation

The following sections are intended to illustrate the measurement principles for commercial near-IR spectrometers, and to provide an understanding of the strengths inherent in each design. Because near-IR spectroscopy and chemical analysis are inextricably linked, particular emphasis is placed on features relevant to this application.

Grating instruments

Scanning-grating spectrometers In this category, it is important to distinguish spectrometers that are designed explicitly for quantitative analytical applications from those that are not. Several manufacturers offer conventional scanning single-monochromator spectrometers that extend from the ultraviolet region through the visible and near-IR. The main attraction of these instruments is the wide-wavelength coverage. A second monochromator may be included, which reduces the stray light and increases spectral resolution; these instruments have extraordinary photometric range, with linear response to 6 absorbance units and beyond. For analytical work however, these UV-visible near-IR spectrometers do not match the performance – in particular the combination of speed and signal-to-noise – that dedicated near-IR instrumentation provides.

On the other hand, commercial rapid-scanning monochromators that incorporate both PbS and Si detectors are capable of scanning the range 400–2500 nm in about 1 s. Typically 30 to 60 scans are co-added to provide a spectrum with very low noise levels (typically less than 30 microabsorbance units) within a minute or so. There are a wide variety of applications that require this combination of speed and accuracy, and the relatively high cost of these instruments is fully justified for these applications. These spectrometers are also found in research laboratories. The role of the ‘high-end’ research spectrometer is to test the feasibility of new analytical applications and, if successful, to provide a basis to judge how best to implement the method in the field. Is the full spectrum required, or will discrete wavelengths suffice? What spectral regions (and hence detectors) are essential? What is the minimum signal-to-noise that will still yield acceptable analytical results? It is often the case that the only way to answer these questions is to carry out a careful study using a full-range research-grade spectrometer.

Fixed-grating spectrometers The near-IR spectrum may be obtained by using a grating spectrometer with no moving parts at all. Since the grating disperses the spectrum of the source along an axis, the spectrum may be measured by an array of detector elements spaced along that axis (**Figure 4**). The most common detectors are silicon photodiode (PDA) and

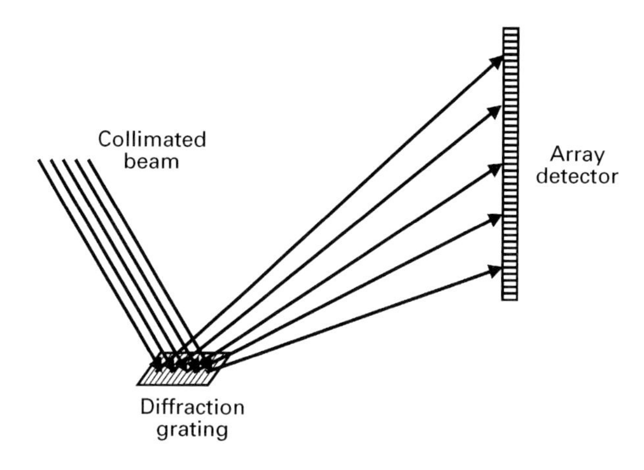

Figure 4 The fixed-grating array detector. The grating spreads the spectrum across the array of detector elements, so that each element senses a different wavelength.

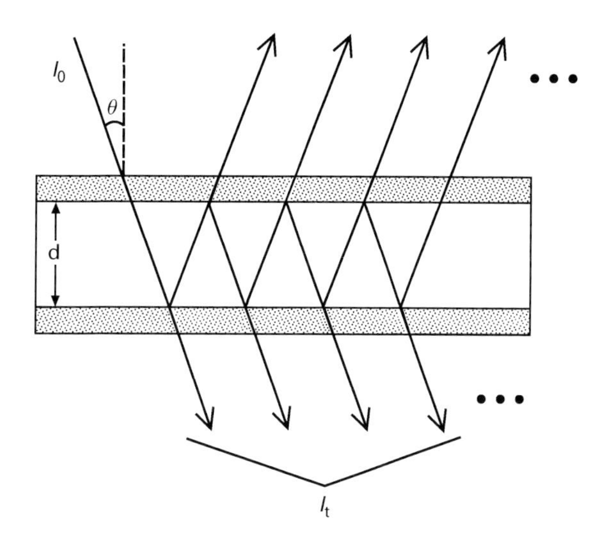

Figure 5 The interference filter. The profile of the transmitted intensity I_t is an Airy function whose bandwidth is governed by the reflectivity of the dielectric-coated reflective layers.

charge coupled device (CCD) arrays, with thermoelectrically cooled InGaAs arrays available for operation above 1100 nm. One advantage of this design is that it may be miniaturized to the point where the spectrometer is truly portable; these 'mini spectrometers' typically accept light via a single strand fibre optic cable of 10–1000 μm diameter, with resolution of 0.5 to 10 nm depending on the choice of grating and slit. They are inexpensive enough that several configurations, optimized for various resolutions and spectral bandwidths, may be linked in parallel by running several fibres from the same source.

It is possible to combine the mechanical stability of fixed gratings with a single-element detector. One manufacturer provides a subtractive double-monochromator system with the wavelength scanned by displacement of movable intermediate slits. Using slits cut in a rotating disk, this arrangement is capable of acquiring up to 1000 spectra per second. While the design was developed primarily for kinetic studies in the visible region, extension to the near IR is readily achieved.

Interference filters and the Fabry–Perot interferometer

A number of commercial instruments make use of a set of interference filters to send discrete bands of radiation successively through the sample. In their simplest form, these filters appear as illustrated in **Figure 5**. Flat plates coated with dielectric reflecting layers are placed in parallel. Incident light undergoes multiple reflections in the region between the two plates, and the transmitted components interfere constructively or destructively as a function of (i) incident angle θ, (ii) gap thickness 'd', (iii) refractive index of the material between the reflective plates, and (iv) wavelength 'λ'. If the light is incident normal to the surface of the filter the transmission profile has a maximum centred at a wavelength λ_{max} equal to twice the gap thickness 'd', with higher-order transmission maxima at $\lambda_{max}/2$, $\lambda_{max}/3$, $\lambda_{max}/4$, etc. The desired order is selected using appropriate high-pass and low-pass filters.

Quite elaborate filters may be built upon this principle, with bandwidths ranging from 0.25 up to 10% of the centre wavelength. A typical instrument includes several filters mounted on a rotating filter wheel (**Figure 6**). The wheel may include a set of filters with bandpasses tailored to fit the requirements of a specific application or, if the instrument is to be used for a variety of applications, a set of filters to survey the near-IR region of interest.

Because of the angular dependence of the transmission profile, the peak transmitted wavelength may be fine tuned by tilting the filter relative to the incident beam. In fact, this property has been capitalized on to scan small spectral regions using a single filter, and one commercial instrument used seven tilting filters of appropriate centre wavelengths to cover the 1400–2400 nm region. However with the advent of inexpensive holographic gratings, the tilting-filter instrument is no longer manufactured commercially.

The same interference principle may be applied to make a circularly-variable filter. The cavity thickness, and hence the centre wavelength, varies systematically around the circumference of the filter. Rotating the disk in a position between the source

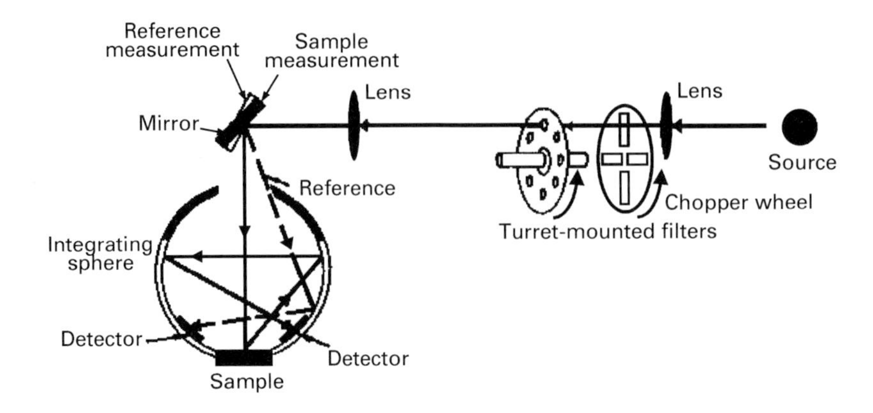

Figure 6 A filter-based instrument operating in reflectance mode. The sample is probed with the mirror in the 'sample' position, and the beam is directed to the internal surface of the integrating sphere to measure the background. Courtesy of Bran+Luebbe Analyzing Technologies Ltd.

and the detector scans the wavelengths. Similarly, the cavity thickness may vary systematically along a line to produce a linearly variable filter suitable for use with an array detector. While neither of these has found wide application for analytical work, both find niches in applications where a broad bandpass is not an issue.

The Fabry–Perot spectrometer is an interference filter in which the centre transmitted wavelength is scanned either by varying the gap between the parallel plates, or by varying the refractive index of the medium separating them. Again, this design is not widely used for analytical work, primarily due to practical difficulties in manufacturing devices that work in first order across the near-IR region. Devices operating at higher order may have extraordinarily high resolution; however the free spectral range is correspondingly narrow.

The optical path of the liquid-crystal tunable filter comprises a series of polarizers and liquid crystal elements whose birefringent properties are electronically controlled. These devices can select wavelengths ranging from the visible to 1100 nm, with a bandwidth as low as 5 nm. The wide, circular aperture makes it particularly attractive as a means to select wavelength regions for transmission to a near-IR camera.

The Michelson interferometer

The Michelson interferometer does not measure the infrared spectrum directly. Rather, an 'interferogram' is measured, and converted to a single-beam spectrum via Fourier transformation. Because of the critical role of this transformation, the method is generally referred to as Fourier-transform infrared spectroscopy, or 'FT-IR'. Instruments using this design have recently proven to be suitable for a number of near-IR applications.

The optical layout is illustrated in **Figure 7**. As the moving mirror is displaced, there is an increasing difference in the optical pathlength travelled by the two components incident on the detector. This results in interference between the two merging beams, and for a monochromatic source of wavelength λ_1 the resulting AC signal (a mirror is moving!) is a cosine wave. The Fourier transform of this interferogram is an infinitely narrow 'band' (delta function) at a position $1/\lambda_1$. (FT-IR spectra are typically reported in

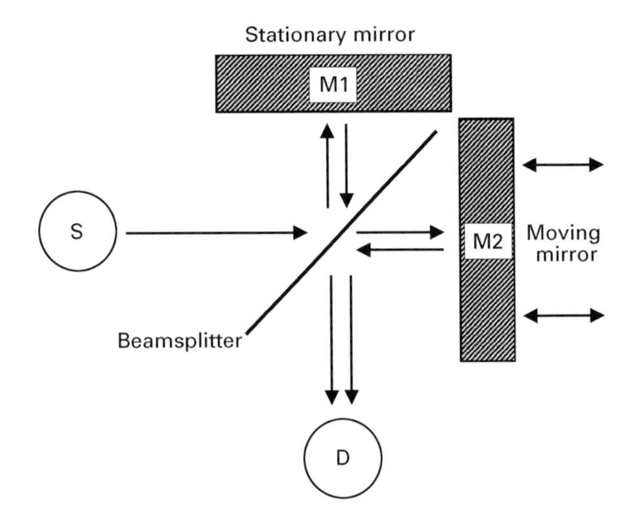

Figure 7 The Michelson interferometer. The beamsplitter splits light from the broadband source S into two components that travel separate paths to mirrors M1 and M2. The recombined beams incident on the detector D move progressively further and further out of phase as the moving mirror is scanned. The interferogram is a plot of the AC component of the detector signal *vs* the path difference.

wavenumbers (cm^{-1}), the inverse of the wavelength (and, not coincidentally, the inverse of the unit of length measuring the optical-path-difference that defines the horizontal axis of the interferogram). The scale is linear in energy, and the resolution is a constant in these units.) For a polychromatic source the ideal interferogram is a superposition of cosine waves, one for each 'colour' in the spectrum of the source; the Fourier transform converts this to the emission spectrum of the source. The absorption spectrum is acquired by placing a sample before the detector in the optical path, repeating the measurement and ratioing the two single-beam spectra.

Although they have long been the design of choice for mid-IR spectroscopy, FT-IR instruments have only recently become popular in analytical near-IR applications. This emergence is due in large part to improvements in the interferometer design and materials, resulting in modern instruments that are very stable both in the short and the long term. Because the position of the moving mirror and the digitization interval are monitored very accurately by a parallel helium–neon laser interferometer, the FT-IR spectrometer is also characterized by superb wavelength (wavenumber) accuracy and precision.

The Michelson interferometer combines very good signal-to-noise with high spectral resolution. The resolution is governed by the maximum displacement of the moving mirror, and a resolution of 4 cm^{-1} or better is readily achieved. For comparison, a bandpass of 10 nm is typical for grating near-IR spectrometers; the relatively wide slit is preferred in order to transmit enough energy to quickly record spectra of good signal-to-noise. **Figure 8** graphically illustrates the resolution advantage of a Michelson interferometer operating at 4 cm^{-1} or even at 16 cm^{-1} nominal resolution. While the practical advantages of the higher resolution remain largely unexplored, there is little doubt that the enhanced ability to distinguish closely spaced absorptions will prove beneficial in some analytical applications.

Two final points specific to FT-IR spectrometers should be mentioned. First, the signal-to-noise can often be improved by using an optical bandpass filter to isolate the spectral region of interest. Second, this approach is particularly well suited for measurements where the light flux on the detector is low, for example for measurements using fibre optics where the light-gathering efficiency may be quite poor.

Acousto-optical tunable filters

If an acoustic wave is produced in a suitable crystal, the result is that the refractive index varies periodically across the crystal. Intuition would then suggest that the crystal might act as a grating with diffraction properties governed by the wavelength of the sound wave.

In fact the crystal does act as a monochromator, but the mechanism underlying this phenomenon is more complex than the simple diffraction grating analogy would indicate. In fact the process is

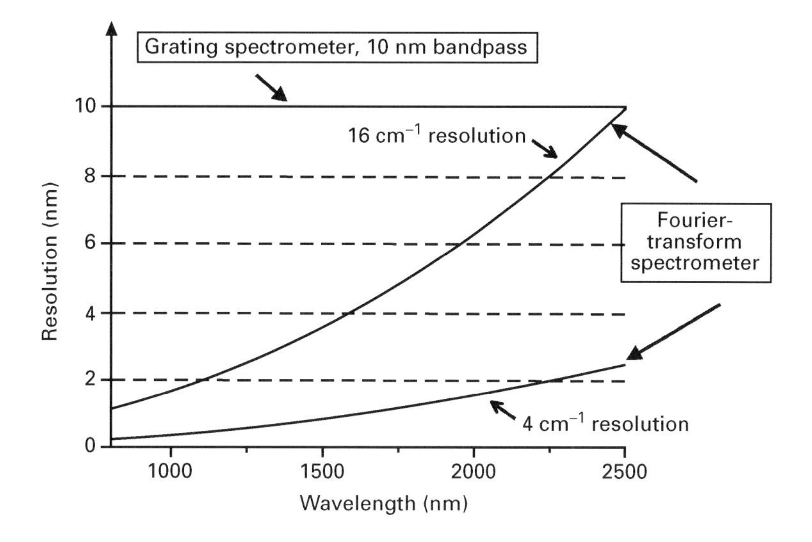

Figure 8 Comparison of the resolution of a FT-IR (Michelson) spectrometer to that of a grating spectrometer. The solid lines represent the resolution of a grating spectrometer with a bandpass of 10 nm and the resolution (in nm) for a FT-IR spectrometer operating at a nominal resolution of either 4 cm^{-1} or 16 cm^{-1} (because noise levels rapidly increase for narrower slitwidths, a 10 nm bandpass is typical of analytical rapid-scanning grating spectrometers). For example, a FT-IR spectrometer operating at 4 cm^{-1} resolution has an effective bandpass of 2.5 nm at 2500 nm, falling to less than 0.5 nm at 800 nm.

governed by a phonon–photon scattering mechanism the details of which lie beyond the scope of this article. The result, however, is that a tellurium oxide crystal with one or more piezoelectric transducers (PZTs) bonded to it acts as a monochromator (**Figure 9**). Wavelength selection is achieved by varying the driving frequency of the PZTs, and the intensity may also be varied by varying the driving power. The resolution is governed by the physical size of the crystal (or, more precisely, by the length of the path over which the light/crystal interaction takes place).

Two distinctive features of this design are the fact that it has no moving parts and that there is 'random access' to any pre-selected wavelength or set of wavelengths. Wavelength stability, precision, and accuracy are also very good, since the wavelength selection is controlled by the RF drive. But the most striking feature of the AOTF design is the speed of measurement.

Spectrometers using discrete-wavelength sources

Light-emitting diodes (LEDs) are available with centre wavelengths throughout the NIR range, and are used as the basis of near-IR instruments with no moving parts. Because LEDs typically have bandwidths of 50 nm or larger, a small interference filter is required to select the desired centre wavelength and bandwidth. An appropriate set of filtered LEDs may then be assembled to span the wavelength range of interest, and the spectrum acquired by cycling through the set. Two advantages of this design are that the instrument can easily be miniaturized, and that the LED sources are stable for decades. Commercial instruments are in essence portable analysers that happen to be based upon near-IR spectroscopy, with built-in calibrations to give direct read-out of analyte levels.

Imaging spectrometers

Every picture tells a story. This is as true in the near-IR as it is of the visible, and the combination of imagery and spectroscopy is a marriage with a short track record but enormous potential.

In its simplest form the measurement requires a near-IR camera, typically equipped with a two-dimensional silicon array detector, with some means to select wavelengths. Some of the advantages of spectroscopic imaging may be realized by simply using fixed bandpass filters to capture images at two or more discrete wavelengths; two judiciously chosen wavelengths often complement one another to provide information that is not available from either of the individual images.

The full potential of near-IR imaging spectroscopy is realized by replacing the fixed bandpass filters with a monochromator having a continuously variable bandpass. Several possibilities exist. The Fabry–Perot interferometer is suitable for scanning over small wavelength ranges at high resolution, and is used primarily for astronomical observations. Two elements most commonly used for imaging across wide spectral regions are the acousto-optic tunable filter and the liquid-crystal tunable filter. **Figure 10** illustrates the kind of data that can be measured

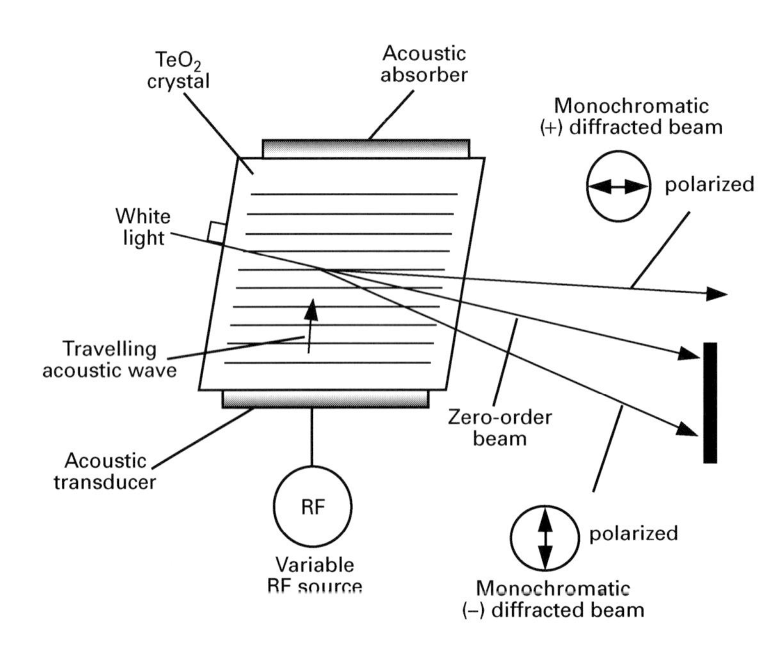

Figure 9 Acousto-optical tunable filter. Courtesy of Brimrose Corporation.

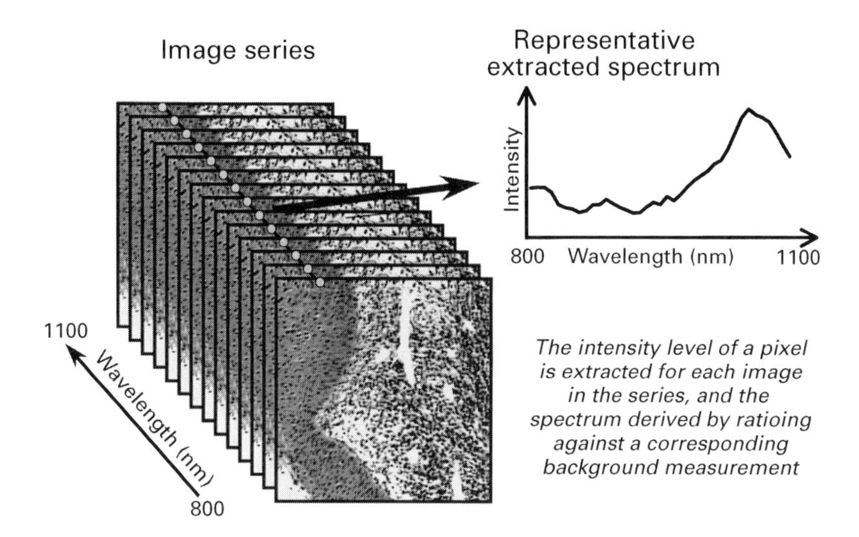

Figure 10 Near-infrared spectroscopic imaging. A series of *N* images is acquired using a CCD-array camera, using optical filtering to step successively through the near infrared (800–1100 nm in this illustration). An *N*-point near-infrared spectrum may then be reconstructed for each pixel in the image. Courtesy of Jim Mansfield.

using this type of arrangement; a series of images is acquired at discrete wavelengths, and a spectrum then constructed for each pixel by extracting the corresponding intensity value for each image in the series. The analogous experiment has been carried out using a Fourier-transform spectrometer, using an infrared microscope in combination with an InSb focal-plane array detector. In that instance, an interferogram is collected at each pixel and transformed to yield the near-IR spectra.

Sampling methods: from the laboratory to on-line

Diffuse reflectance

Analytical applications of near-IR spectroscopy can only succeed if the sample can be presented to the spectrometer in a reproducible fashion. This is simple enough for liquids, but less straightforward for solid, physically inhomogeneous samples. The most common solution is to measure radiation that is diffusely reflected from the sample. **Figure 11** illustrates a typical configuration for this measurement, with two detector elements at 45° to the surface. Another way to collect the reflected radiation is to use an integrating sphere, as illustrated in **Figure 6**. In either event, the reflectance spectrum of a coarse, inhomogeneous sample is influenced to some degree by the surface terrain. This influence is typically minimized either by spinning the sample or translating the surface of the sample stepwise across the beam. It should be mentioned that the Michelson interferometer is not well suited for reflectance measurements of coarse moving samples. The moving sample modulates the intensity of the reflected light, giving rise to a noise component that superimposes on the interferogram. The result upon Fourier transformation is a noisy spectrum.

Fibre optics, on-line and in-situ measurements

It is often desirable to measure near-IR spectra for samples that for reasons of convenience or necessity cannot be transported to the spectrometer for 'conventional' transmission or reflectance measurements. Examples range from processes such as fermentation and refining, where on-line monitoring is desirable,

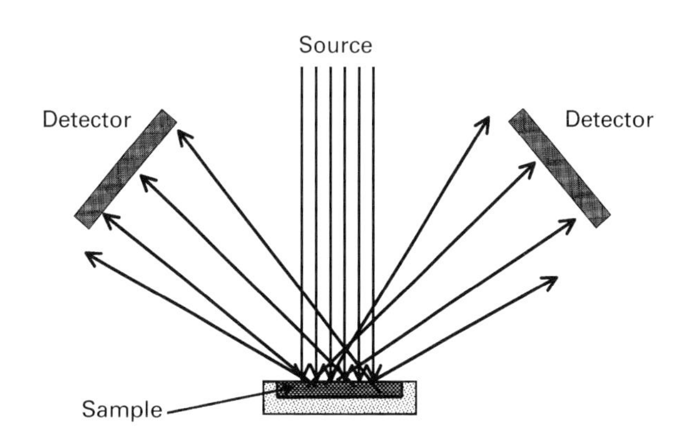

Figure 11 Typical geometry used for diffuse-reflectance spectroscopy. The detectors may be the same materials, or two different materials (e.g. Si and PbS) for wider wavelength coverage.

to the analysis of pharmaceutical blends and incoming raw materials. A common solution is to use fibre optics or fibre-optic bundles to carry the spectral information from the sample to the spectrometer. Silica fibres are ideal for this purpose, as they are transparent from the visible through the near-IR.

One way to use fibre optics is for transmission measurements, with one bundle carrying near-IR radiation to the interface of a transmission cell, and a second bundle gathering the transmitted radiation and carrying it back to the spectrometer. Specialized fibre-optic probes exist for in-situ transmission measurements with the 'sending' and 'receiving' fibres in the same bundle. The immersion tip includes built-in optics to create a reproducible transmission path; light that has travelled through the sample is reflected back into the receiving fibres by a mirror built into the probe tip itself.

While transmission measurements are suitable for liquids, powders and other highly scattering samples require a different approach. One measurement simply uses a fibre-optic bundle with a measurement tip that exposes both sending and receiving fibres to the specimen. While many of the photons entering the sample are lost, a fraction of them are scattered in a path that leads them back into one of the receiving fibres. The background spectrum is typically reflectance from a ceramic standard. The 'interactance' spectrum is obtained by ratioing the single-beam powder spectrum against the single-beam ceramic reference spectrum, and reflects both the absorption properties of the sample and the wavelength dependence of the scattering efficiency. Probes based upon this general principle are suitable for checking the identity/purity of batch powder samples. The operator simply inserts the probe tip into the sample and waits for the analytical result to appear before moving onto the next measurement.

Remote reflectance

For many production-line applications, the most convenient way to measure the near-IR spectrum is by 'remote' diffuse reflection. The source and specialized collection optics are combined in a single device and simply aimed at the target, for example chemicals or food products moving along conveyor lines or webs and sheets in the production process. Just as is the case for laboratory measurements, these instruments vary in sophistication from single-wavelength devices to full grating spectrometers – the choice of the appropriate system is dictated by the difficulty of the analysis. Because water has extraordinarily strong absorptions in the near-IR, one of the more common on-line applications is monitoring moisture content; more than one manufacturer offers near-IR filter-based analysers exclusively for this purpose.

Summary

From its original niche as a convenient method for grain analysis, near-IR spectroscopy has evolved to the point where new applications appear almost daily. The parallel evolution in hardware has led to modern instruments that could not have been contemplated by the founding fathers of near-IR analysis. Many devices faithfully produce analytical results running unattended around the clock, while others are used by operators who need not even be aware that near-IR spectroscopy is involved. In short, the technology has reached a certain maturity.

What does the future hold? In addition to ever-faster and more accurate instruments, one exciting prospect is the full realization of the marriage of spectroscopy with imaging. Practical roadblocks to the widespread adoption of this technique are rapidly disappearing as both imaging arrays and high-throughput tunable filters become more widely accessible, and the realm of possibilities seems limitless. This and the other extraordinary instruments that appear ever-more frequently are vital to our gathering and understanding the information that nature has to offer, and the next twenty years will surely be as eventful as the past twenty.

List of symbols

D^* = detectivity; I = intensity; λ = wavelength.

See also: **Fibre Optic Probes in Optical Spectroscopy, Clinical Applications; IR Spectrometers; IR Spectroscopy Sample Preparation Methods; IR Spectroscopy, Theory; Medical Science Applications of IR.**

Further reading

Burns DA and Ciurczak EW (eds) (1992) *Handbook of Near-Infrared Analysis*. New York: Marcel Dekker.

Davies AMC and Williams P (eds) (1996) *Near Infrared Spectroscopy: The Future Waves*. West Sussex, UK: NIR Publications.

Davies AMC (ed) *Journal of Near-Infrared Spectroscopy*. (This journal regularly includes research articles reporting refinements and advances in near-IR instrumentation.)

Griffiths PR and de Haseth JA (1986) *Fourier Transform Infrared Spectroscopy*. Toronto: Wiley-Interscience.

McLure WF (1994) Near-Infrared spectroscopy: the giant is running strong. *Analytical Chemistry* **66** 43A–53A.

Murray I and Cowe IA (eds) (1992) *Making Light Work: Advances in Near Infrared Spectroscopy*. Weinheim: VCH.
Osborne BG, Fearn T and Hindle PH (1993) *Practical NIR Spectroscopy with Applications in Food and Beverage Analysis*. Harlow, England: Longman Scientific & Technical.
Wetzel DL (1998) Contemporary near infrared instrumentation. In: Williams P and Norris K (eds) *Near-Infrared Technology in the Agriculture and Food Industries*, 2nd edn. St. Paul, MN: The American Association of Cereal Chemists.

Negative Ion Mass Spectrometry, Methods

Suresh Dua and **John H Bowie**, The University of Adelaide, Australia

MASS SPECTROMETRY
Methods & Instrumentation

Introduction

Investigations of both negative-ion and positive-ion mass spectrometry as aids to structure determination commenced in earnest in the middle of the twentieth century. The ease of formation of molecular cations, together with their characteristic fragmentations, led to the development of positive-ion mass spectrometry as one of the most important of all analytical techniques. In contrast, initial difficulties with the formation of suitable anions together with the early nonavailability of commercial mass spectrometers that could readily detect negative ions delayed the analytical development of negative-ion mass spectrometry for several decades. Modern instrumentation allows the measurement of the mass spectra of molecular anions ($M^{\bullet-}$), and of $(M-H)^-$ ions and multiply charged anions. The analytical applicability of the fragmentations of $(M-H)^-$ species will be outlined. Mechanistic information that can be obtained from gas phase studies with respect to rearrangement reactions of anions, and the collision-induced conversion of negative ions into neutrals and positive ions and the application of these techniques will also be described. Selected examples will be used to illustrate each section. References to reviews that provide further information concerning aspects of anion chemistry are listed at the end of the article.

Analytical applications of negative ions

The fragmentations of $(M-H)^-$ species

$(M-H)^-$ ions may be formed by deprotonation of organic molecules in the chemical ionization (NICI) source of a mass spectrometer using a strong base (e.g. HO^- from H_2O or NH_2^- from NH_3), or using fast-atom bombardment (FAB), electrospray

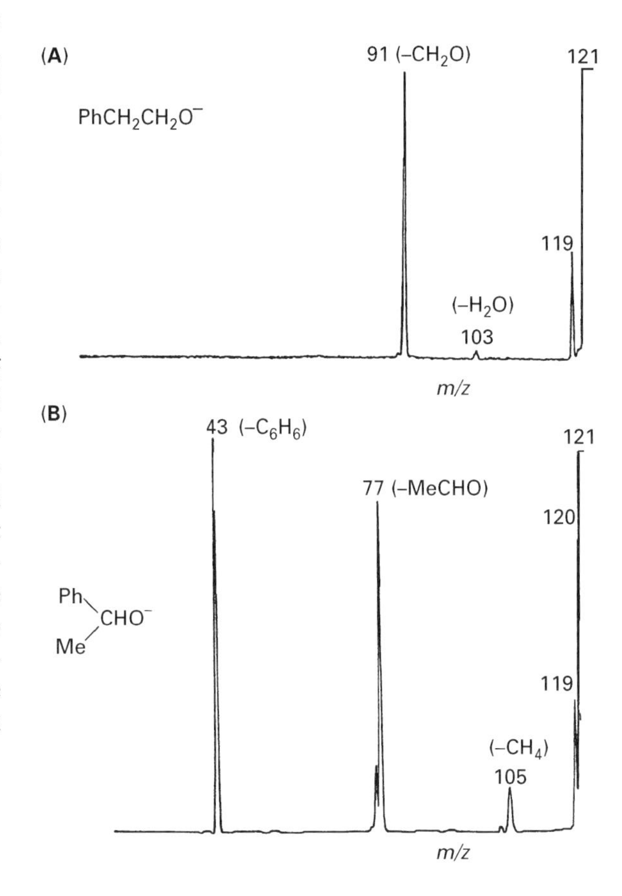

Figure 1 Collision-induced HO^- negative-ion chemical ionization tandem mass spectra (MS/MS) of (A) $PhCH_2CH_2O^-$ and (B) $PhCH(Me)O^-$. ZAB 2HF instrument. Argon collision gas: pressure of gas in first collision cell adjusted so that the reduction in the main beam is 10%.

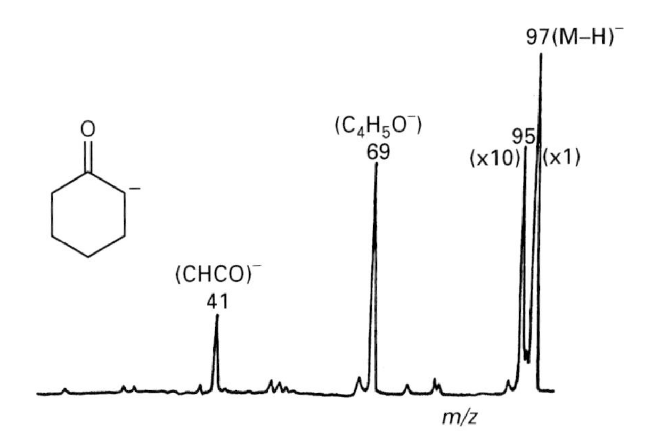

Figure 2 Collision-induced HO^- negative-ion chemical ionization tandem mass spectra (MS/MS) of deprotonated cyclohexanone. ZAB 2HF instrument. Details as for **Figure 1**.

→ C_2H_4 [5]

$[(\text{cyclohexenone})H^-]$ → + H_2 [6]

→ + H_2 [7]

Scheme 2

ionization (ESI), atmospheric pressure ionization (API), matrix-assisted laser desorption ionization (MALDI) mass spectrometry, and a number of associated ionization techniques. The $(M–H)^-$ ions are generally formed with little excess energy and consequently undergo little decomposition. Molecular mass information can be obtained in this way for the majority of organic compounds, including those that do not form detectable molecular cations or where peaks from such cations are of small abundance in the positive-ion spectra (examples of the latter category include many long-chain compounds such as alcohols, ketones, acids and esters, and also some peptides and polysaccharides). Fragmentation data of $(M–H)^-$ ions may be obtained using collisional activation or some other method of energy activation of the parent anion. Ionization and energy activation techniques mentioned above are described in other articles.

Table 1 Characteristic negative-ion fragmentations of side chains of amino acid residues from $(M–H)^-$ ions of peptides

Residue	*Loss (or formation)*	*Mass*
Phe	$PhCH_2^-$	91
Tyr	*p*-$HOC_6H_4CH_2^-$	107
	$O{=}C_6H_4{=}CH_2$	106
Trp	C_9H_7N	129
Ser	CH_2O	30
Thr	MeCHO	44
Cys	H_2S	34
Met	MeSH	48
	MeSMe	62
	$^{\bullet}CH_2CH_2SMe$	75
Asp	H_2O	18
Glu	H_2O	18
Asn	NH_3	17
Gln	NH_3	17

$$Ph{-}CH_2{-}CH_2{-}O^- \longrightarrow Ph{-}CH_2^- + CH_2O \quad [1]$$

$$Ph{-}C(Me)(H){-}O^- \longrightarrow [(MeCHO)Ph^-] \longrightarrow Ph^- + MeCHO \quad [2]$$

$$[(MeCHO)Ph^-] \longrightarrow [CH_2CHO]^- + PhH \quad [3]$$

$$\longrightarrow [(PhCHO)Me^-] \longrightarrow (C_6H_4)^-{-}CHO + CH_4 \quad [4]$$

Scheme 1

The negative-ion fragmentations of $(M–H)^-$ ions derived from organic molecules are often simple and provide useful structural information. Fragmentations involving particular functional groups are often characteristic of those groups. Consider first the spectra of the two isomeric alkoxide anions shown in **Figure 1**. These spectra are collision-induced HO^- NICI tandem mass (MS/MS) spectra that have been measured in a reverse sector mass spectrometer. The negative-ion fragmentations are rationalized in **Scheme 1**. They include simple cleavage (Eqns [1] and [2]), together with reactions of an anion within an anion neutral complex (Eqns [3] and [4]). A second example is shown for the $(M–H)^-$ parent ion of cyclohexanone in **Figure 2** cleavage mechanisms are outlined in **Scheme 2**. Fragmentations include a retro cleavage of the ring (Eqn [5]) together with two competitive losses of dihydrogen (Eqns [6] and [7]).

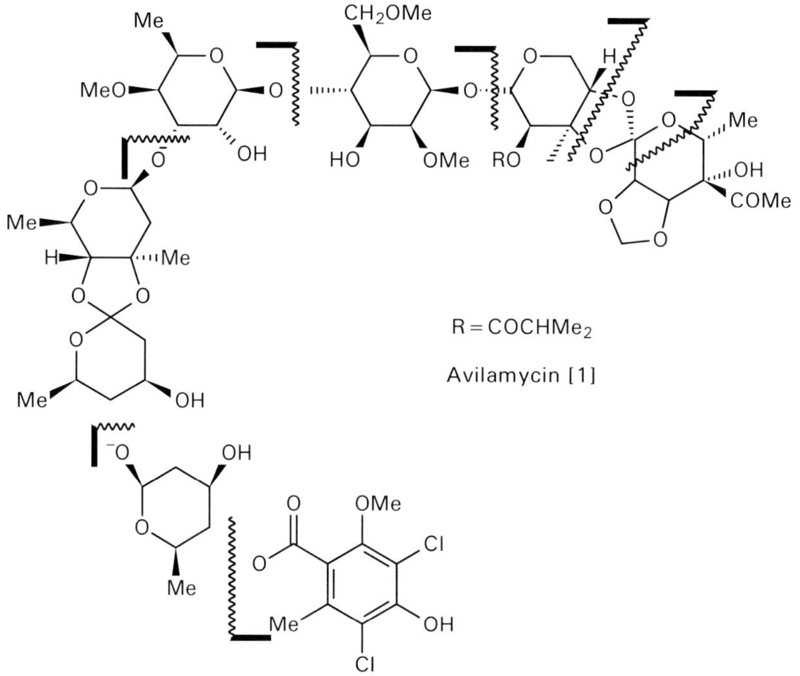

R = $COCHMe_2$

Avilamycin [1]

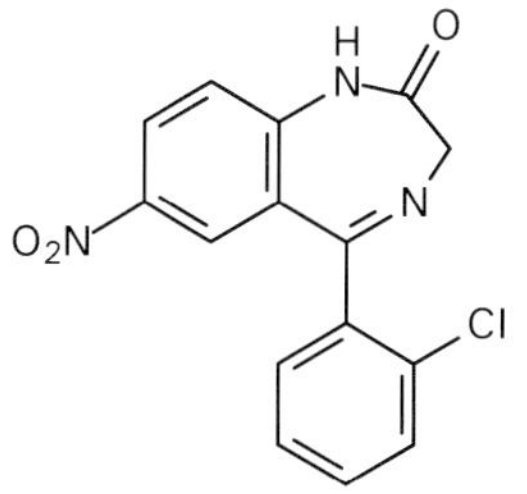

[2] Clonazepam

Many biologically important molecules form $(M-H)^-$ parent ions, and these often undergo characteristic fragmentations. For example, $(M-H)^-$ ions from molecules containing C–O and P–O bonds (Such as nucleosides, nucleotides and molecules containing saccharide linkages) undergo collision-induced cleavage reactions to yield stable $-O^-$ fragment anions. Negative-ion mass spectrometry is often the analytical method of choice for such molecules. A particular example is shown above for avilamycin [**1**], where C–O bond cleavages (the bold lines indicate the direction of charge retention) provide sequence information.

Positive-ion mass spectrometry has traditionally been the MS method of choice for sequencing peptides and proteins. However, the fragmentations of both $(M-H)^-$ and $(M-nH)^{n-}$ ions of peptides and proteins can also provide useful structural information. The negative-ion spectra of $(M-H)^-$ ions show two types of cleavages: (i) characteristic fragmentation of the side chains of some amino acid residues (these data are summarized in **Table 1**), and (ii) cleavages of the peptide backbone. As an illustration, MS/MS data from the $(M-H)^-$ parent ion of a 12-residue peptide are shown in **Figure 3**. The sequence of this peptide can be determined by the α and β negative-ion backbone cleavages shown in **Figure 3** (see **Scheme 3** for the nomenclature and mechanisms of α and β backbone cleavages). In addition, the ion formed by side-chain cleavage of CH_3CHO from the Thr side chain (cf. **Table 1**) also produces backbone cleavage ions: these are indicated by arrows in **Figure 3**.

Further negative-ion backbone cleavages are initiated by the enolate anion of an Asp or Asn side chain. This is illustrated in the spectrum of the non-apeptide shown in **Figure 4**, with the characteristic Asp backbone cleavages rationalized in **Scheme 4**. The usual α and β cleavage ions are indicated on the formula shown in **Figure 4**.

Radical anions

Parent radical anions may be formed following low-energy electron capture by a neutral if the electron affinity of the neutral is suitably positive. Particular examples are conjugated systems (such as α-diketones, α, β-unsaturated carbonyl systems, quinones and flavones, etc.) together with molecules containing specific functional groups that readily capture electrons (e.g. nitro, sulphonyl, phosphate esters, etc). For example, the limit of detection of the clonazepam [**2**] molecular anion is about 25 times lower than that of the corresponding molecular cation.

In some cases derivatization of a neutral with a functional group that readily accepts a low-energy electron can provide molecular mass information: for example, a long-chain alcohol can be converted into a perfluorobenzoate ester that readily yields a molecular anion.

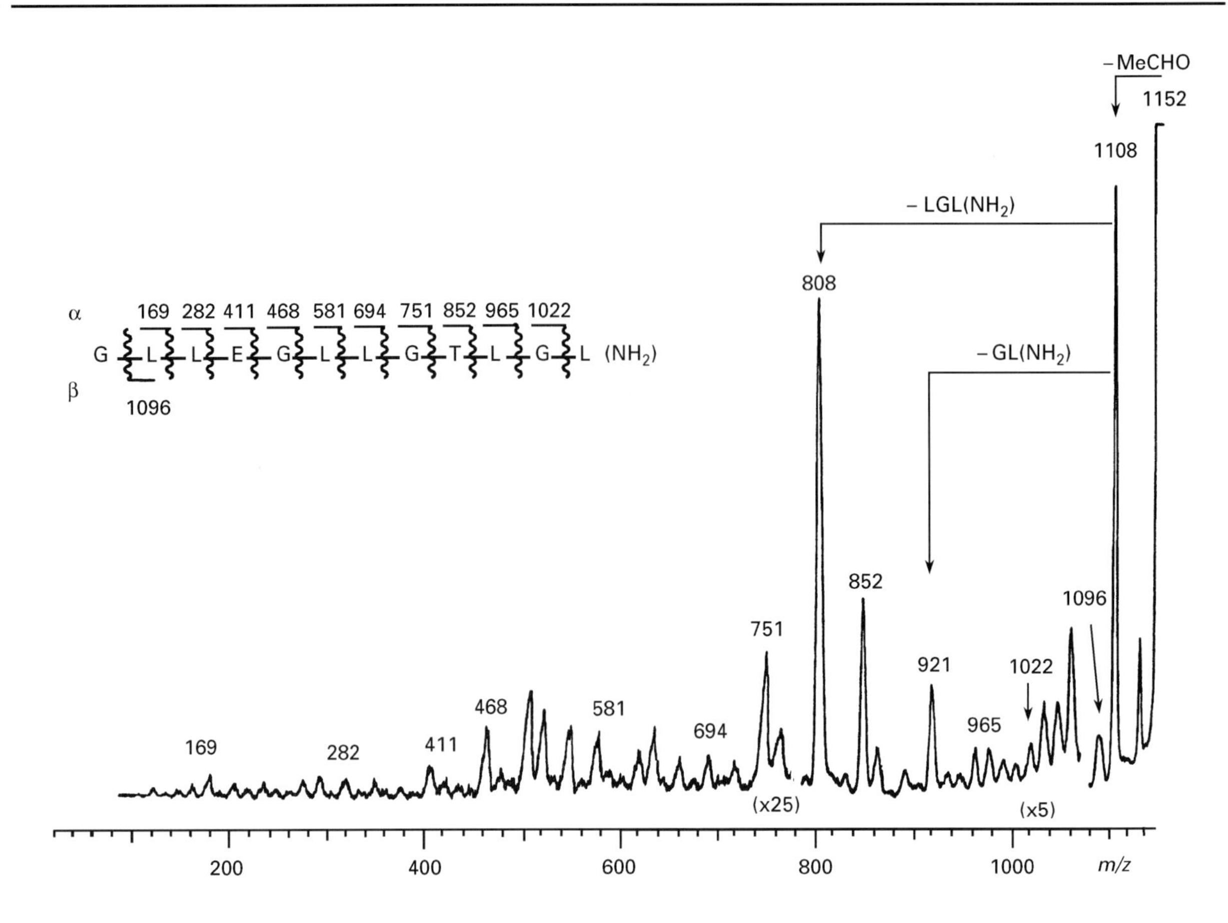

Figure 3 Collision-induced negative-ion fast-atom bombardment tandem mass spectrum (MS/MS) of GLLEGLLGTLGL(NH_2). ZAB 2HF instrument. Glycerol was used as matrix. Other details as for **Figure 1**. For the mechanisms of the backbone cleavages, see **Scheme 3**.

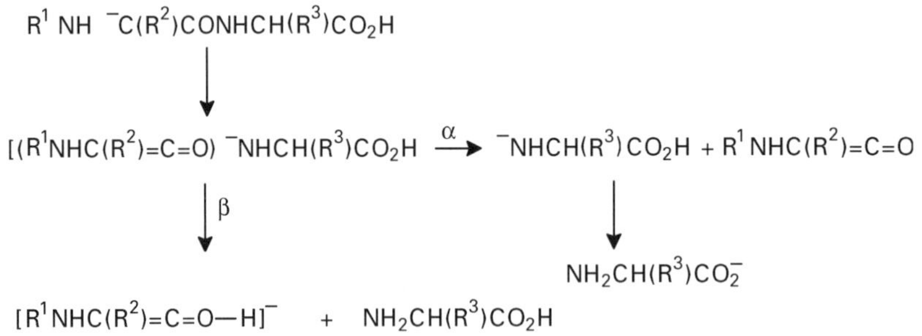

$$R^1NH\ {}^-C(R^2)CONHCH(R^3)CO_2H$$

$$\downarrow$$

$$[(R^1NHC(R^2){=}C{=}O)\ {}^-NHCH(R^3)CO_2H \xrightarrow{\alpha} {}^-NHCH(R^3)CO_2H + R^1NHC(R^2){=}C{=}O$$

$$\downarrow \beta \qquad\qquad \downarrow$$

$$NH_2CH(R^3)CO_2^-$$

$$[R^1NHC(R^2){=}C{=}O{-}H]^- + NH_2CH(R^3)CO_2H$$

Scheme 3

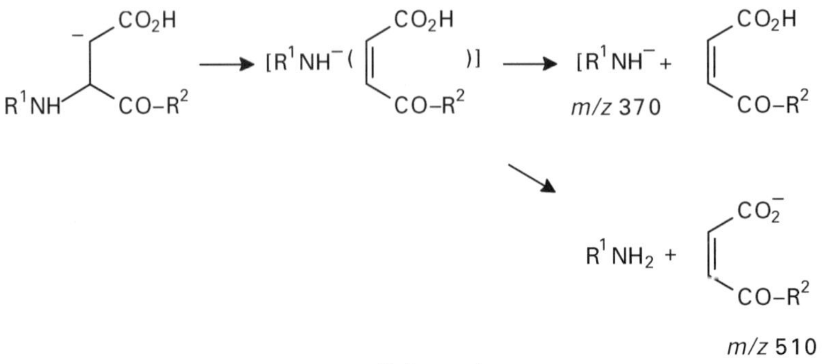

$$R^1NH{-}{}^-CH(CO_2H){-}CH_2{-}CO{-}R^2 \longrightarrow [R^1NH^-(CO_2H{-}CH{=}CH{-}CO{-}R^2)] \longrightarrow [R^1NH^- + CO_2H{-}CH{=}CH{-}CO{-}R^2$$

m/z 370

$$\searrow\ R^1NH_2 + CO_2^-{-}CH{=}CH{-}CO{-}R^2$$

m/z 510

Scheme 4

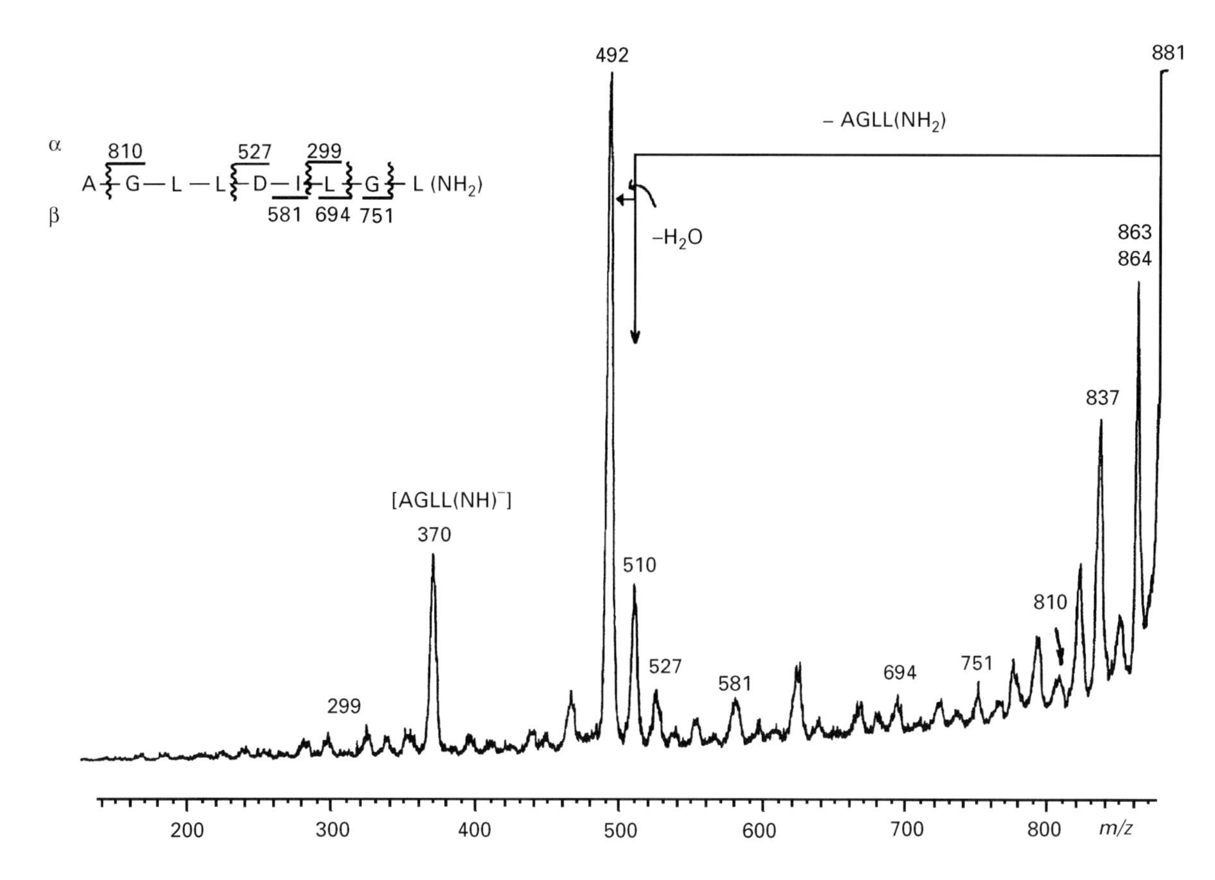

Figure 4 Collision-induced negative-ion fast-atom bombardment tandem mass spectrum (MS/MS) of AGLLDILGL(NH_2). ZAB 2HF mass spectrometer. Glycerol was used as matrix. Other details as for **Figure 1**. For the mechanisms of the backbone cleavages, see **Schemes 3** and **4**.

Mechanistic studies of negative ions

The study of intramolecular rearrangements of negative ions

There are many rearrangement reactions of anions in the condensed phase in which the products and rates of reaction are dependent on the solvent and counterion used. The gas phase is thus the best medium for the investigation of the fundamental reactivity of the rearranging anion and the mechanism of the rearrangement. Many such rearrangements have been studied including well-known 1,2 anionic rearrangements such as the Wittig, Wolff, acyloin, negative-ion pinacol and Favorskii reactions.

The pinacol–pinacolone rearrangement (Eqn [8]) is arguably the most famous of all acid-catalysed rearrangements and involves a simple Whitmore 1,2 methyl shift. Base-catalysed analogues of the pinacol rearrangement are not common, but the rearrangement does occur for deprotonated β-chlorohydrins. For example, base-catalysed rearrangement of *cis*-2-chlorocyclohexanol yields formylcyclopentane. The negative-ion pinacol rearrangement has been studied in the gas phase using MeO^- as the leaving group. This is illustrated for the reactions shown in Equations [9] and [10]. The *trans* isomer (Eqn [9]) does not undergo the pinacol rearrangement, instead, an S_Ni cyclization gives an epoxide, which ring-opens as shown to yield deprotonated cyclohex-2-en-1-ol and methanol. In contrast, the *cis* isomer (see Eqn [10]) undergoes the pinacol rearrangement as shown. The product anions of Equations [9] and [10] are identified from a comparison of their negative-ion mass spectra with those of independently synthesized anions.

The base-catalysed rearrangements of α-halo ketones are classical examples of the reactions of ambident enolate anions in solution. The extent of each of the two reactions shown in Equations [11] and [12] is principally a function of the type of solvent used. A protic solvent solvates more strongly at the oxygen centre of the ambident anion and thus reaction proceeds through the carbanion centre to yield the Favorskii species as the major product (Eqn [11]). In marked contrast, the Favorskii rearrangement does not occur in the gas phase. Here,

$$Me_2C(OH)C(^+OH_2)Me_2 \longrightarrow Me_2C(OH)^+CMe_2 + H_2O$$
$$\longrightarrow Me^+C(OH)CMe_3$$
$$\longrightarrow MeCOCMe_3 + H^+ \quad [8]$$

O⁻ OMe → (H, ⁻OMe) O → O⁻ + MeOH [9]

O⁻ OMe → [(CHO) MeO⁻] → O ⊖ + MeOH [10]

nucleophilic attack occurs exclusively via the more electron-rich alkoxide centre to form an allene oxide adduct that fragments as shown in Equation [13].

Charge reversal and neutralization of negative ions

When a negative ion with high translational energy undergoes a soft collision with an inert gas atom (such as He or Ar) in a collision cell, some translational energy may be converted into internal energy. The energized ion rids itself of this internal energy in any one of a number of different ways; for example it may (i) radiate, (ii) cleave to give fragment negative ions (see, e.g. **Figures 1–4**), (iii) eject one electron to yield a neutral, or (iv) eject two electrons (either sequentially or synchronously) to yield a positive ion. These processes may occur for either parent or daughter anions, irrespective of whether they are radical anions or even-electron anions.

Charge reversal is that process by which an anion loses two electrons to form the corresponding cation. The process is best explained by reference to the schematic diagram shown in **Figure 5**. The particular polyatomic anion under study is selected using the analyser system of a sector mass spectrometer; it then proceeds into the first collision cell, which contains sufficient gas to effect single-collision conditions. The parent cations so produced have a range of internal energies and generally produce a mass spectrum containing peaks corresponding to the parent (cation) (the recovery signal: sometimes this is absent) together with a variety of fragment cations. The first application of charge reversal is that it is possible to make positive ions that cannot be formed by conventional ionization procedures. For example, *m/z* 31 from methanol in the positive mode is $CH_2{=}^+OH$: CH_3O^+ is not normally formed. However, CH_3O^+ can be formed by charge reversal of CH_3O^-. Similarly, RCO_2^+ is not a species formed readily by decomposition of any positive ion, but it can be formed by charge reversal of RCO_2^-.

The second application is analytical. If an $(M–H)^-$ ion undergoes little or no fragmentation following excitation, another option is to charge-reverse the $(M–H)^-$ to give a spectrum of positive ions that

H_2C O ⊖ Me X Me — Favorskii → O ; $-X^-$(X=Cl,Br,I) ; NuH → Me_3CCONu [11]

→ O ; NuH → $MeCOC(Nu)Me_2$ + $NuCH_2COCHMe_2$ [12]

(X=OMe) → [(O) MeO⁻] → O ⊖ H_2C + MeOH [13]

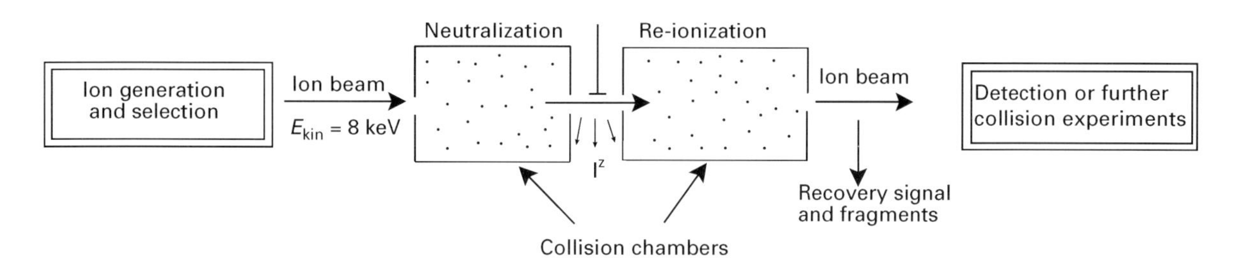

Figure 5 Schematic representation of the two collision cells in a reversed sector mass spectrometer. Reproduced with permission of The American Chemical Society from Goldberg N and Schwarz H (1994) *Accounts of Chemical Research* **27**: 347–352.

should provide structural information. Alternatively, the structure of some fragment anion may need to be determined. Comparison of the charge-reversal spectrum of this anion with the charge-reversal spectra of anions of known structure can often identify the unknown. Suppose the major peak in a negative-ion spectrum corresponds to $C_2H_3O_2$ (*m/z* 59). The unknown is either $MeCO_2^-$ or $^-OCH_2CHO$. The charge-reversal spectra of these anions are reproduced in **Figure 6** each is characteristic and readily identifies the anion precursor.

Co-occurring with the charge-reversal process is that which produces an (even-electron) neutral by ejection of an electron from a radical anion, or a radical by loss of an electron from an even-electron anion. If a potential is applied after the first collision cell (see **Figure 5**) to deflect all ions, the only species proceeding into the second collision cell are neutrals formed in the first collision cell. If the second cell contains a gas (normally oxygen) that can ionize the neutrals, a composite positive-ion spectrum of all neutrals formed in the first collision cell can be obtained. This is called neutralization–reionization mass spectrometry (NRMS). The technique has a number of applications. First, the neutral formed in a negative-ion fragmentation may be identified. More importantly, interesting neutrals can be synthesized in the mass spectrometer, and their structures probed both experimentally (from their mass spectra) and theoretically (using computational methods). Fred McLafferty, a pioneer in the development and application of NRMS has stated that 'NRMS has made its largest contribution to research areas outside mass spectrometry in the characterization of unstable and radical species'. Several examples of the application of this technique are outlined below.

The radical $HSO^•$ is believed to be implicated in ozone depletion in the upper atmosphere by the reactions shown in Equations [14] and [15]. This species and its isomer $HOS^•$ have been made in the mass spectrometer by electron loss from the precursor anions HSO^- and HOS^-. The HOS^- precursor anion may be prepared by the collision-induced process summarized in Equation [16], while HSO^- is

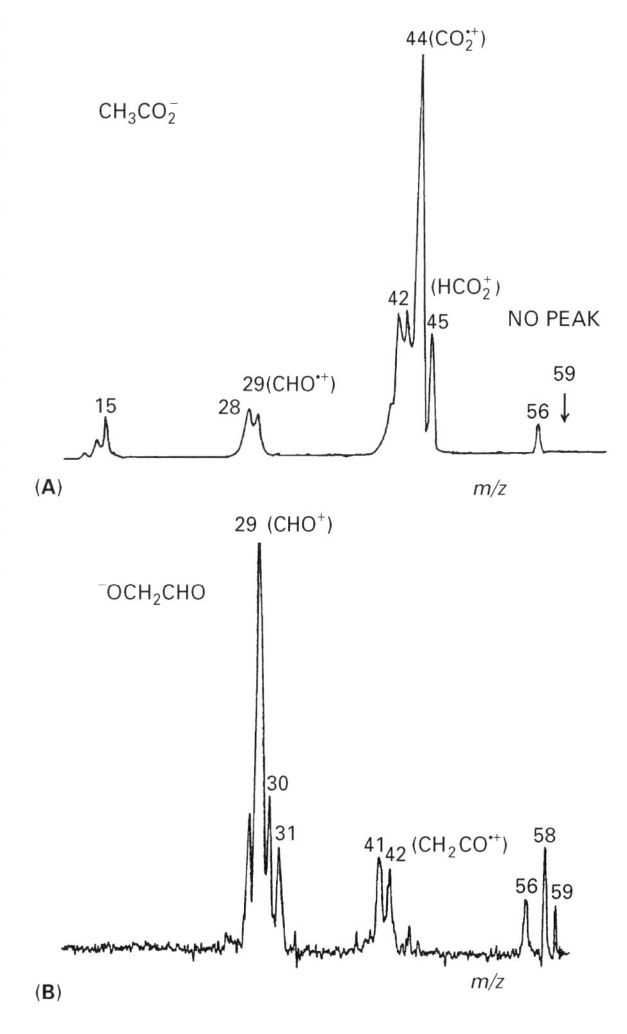

Figure 6 Charge-reversal (positive-ion) tandem mass spectra of (A) $CH_3CO_2^-$ and (B) $^-OCH_2CHO$. ZAB 2HF mass spectrometer. Argon collision gas: pressure adjusted in the first collision cell so that the reduction in the main beam is 10%.

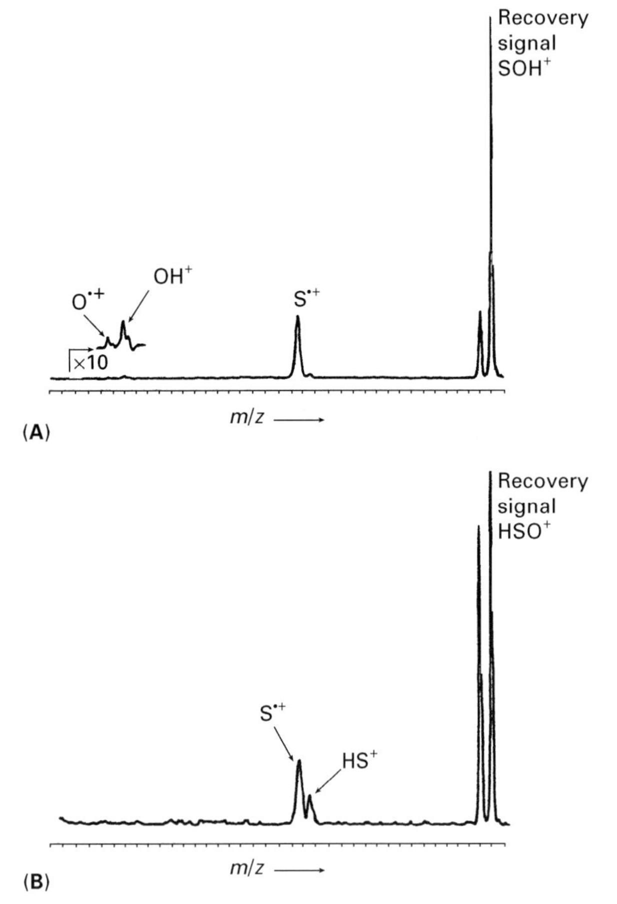

Figure 7 Neutralization–reionization mass spectra of (A) HOS⁻ and (B) HSO⁻. Modified ZAB four-sector mass spectrometer of BEBE configuration (B = magnet sector; E = electric sector). Oxygen used as collision gas used in each collision cell (see **Figure 5**). The pressure of the gas in each cell was adjusted so the reduction in the main beam is 20% for each collision event. Reproduced with permission of ACS from Goldberg N and Schwarz H (1994) *Accounts of Chemical Research* **27**: 347–352.

prepared by admixture of H_2S and N_2O under conditions of negative-ion formation, perhaps by the reaction shown in Equation [17]. The reionization positive-ion spectra of the two non-interconverting radicals are shown in **Figure** 7.

$$HS^{\bullet} + O_3 \longrightarrow HSO^{\bullet} + O_2 \quad [14]$$

$$HSO^{\bullet} + O_3 \longrightarrow HS^{\bullet} + 2O_2 \quad [15]$$

$$[CH_3SOCH_2]^- \longrightarrow HOS^- + C_2H_4 \quad [16]$$

$$HS^- + N_2O \longrightarrow HSO^- + N_2 \quad [17]$$

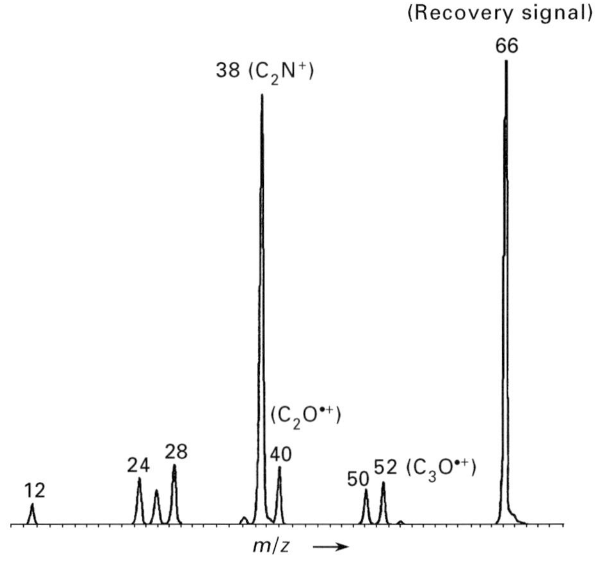

Figure 8 Neutralization–reionization mass spectrum of N≡C–C≡C–O⁻. Details as in the legend to **Figure** 7. Reproduced with permission of Elsevier from Muedas CA, Sülzle D and Schwarz H (1992) *International Journal of Mass Spectrometry and Ion Processes* **113**: R17–R22.

There is much current interest in the formation of small cumulene and heterocumulene anions and their neutrals. Two examples are mentioned. The anion N≡CC≡C–O⁻ can be prepared by collisional activation of the precursor anion shown in Equation [18]. The corresponding radical is made by charge stripping. The positive-ion reionization spectrum of NC_3O^- is shown in **Figure** 8.

$$NC{-}CH^-{-}CO_2Me \longrightarrow [(NC{-}CH{=}C{=}O)\ ^-OMe] \longrightarrow N{\equiv}C{-}C{\equiv}C{-}O^- + MeOH \quad [18]$$

Neutral CH_2C_2: is formed following electron loss from radical anion $CH_2C_2^{\bullet-}$. The anion is made by the reaction between allene and the monooxygen radical anion as shown in Equation [19].

$$O^{\bullet-} + CH_2{=}C{=}CH_2 \longrightarrow CH_2{=}C{=}\dot{C}{:}^- + H_2O \quad [19]$$

Conclusions

Negative-ion mass spectrometry is a useful analytical technique that is complementary to more ubiquitious positive-ion method. There are particular classes of compounds where the spectra of (M–H)⁻ ions provide more structural information than the corresponding positive-ion spectra.

Studies of intramolecular reactions of negative ions can provide useful mechanistic information.

Electron stripping of negative ions can be used to produce both neutrals and positive ions not available via other synthetic pathways.

See also: **Chemical Ionization in Mass Spectrometry; Fragmentation in Mass Spectrometry.**

Further reading

Born M, Ingemann S and Nibbering NMM (1997) Formation and chemistry of radical anions in the gas phase. *Mass Spectrometry Reviews* **16**: 181–201.

Bowie JH (1984) The formation and fragmentation of negative ions derived from organic molecules (mainly radical anions). *Mass Spectrometry Reviews* **3**: 161–207.

Bowie JH (1990) The fragmentations of even-electron organic negative ions. *Mass Spectrometry Reviews* **9**: 349–379.

Bowie JH (1994) The fragmentations of $(M-H)^-$ ions derived from organic compounds. An aid to structure determination. In: Russell DH (ed) *Experimental Mass Spectrometry*, pp 1–39. New York: Plenum Press.

Eichinger PCH, Dua S and Bowie JH (1994) Review. A comparison of skeletal rearrangement reactions of even-electron anions in solution and in the gas phase. *International Journal of Mass Spectrometry and Ion Processes* **133**: 1–12.

Goldberg N and Schwarz H (1994) Neutralisation–reionisation mass spectrometry: a powerful "laboratory" to generate and probe elusive molecules. *Accounts of Chemical Research* **27**: 347–352.

Neutralization–Reionization in Mass Spectrometry

Chrys Wesdemiotis, University of Akron, OH, USA

MASS SPECTROMETRY
Methods & Instrumentation

Neutralization–reionization mass spectrometry (NRMS) involves the synthesis and characterization of neutral species inside a tandem mass spectrometer *via* gas-phase redox reactions. The neutrals are produced by neutralization of the corresponding mass-selected cations or anions (precursor ions) and are subsequently characterized by the mass spectra arising after reionization to positive or negative ions. The successive neutralization and reionization events are effected by collisions with gaseous targets at high kinetic energy. Equations [1] and [2] illustrate a neutralization–reionization (NR) sequence that starts with a positive precursor ion and reionizes the intermediate neutral to cations. These choices generally proceed with higher yields than alternative charge permutations and, therefore, have been employed in most NRMS studies reported so far; nonetheless, any charge combination of precursor ion and final product ions is possible and has been documented.

NRMS is uniquely suitable for the study of highly reactive neutrals that cannot be isolated and characterized in condensed phases, where interactions with neighbouring molecules would cause their immediate decomposition. By preparing the neutrals in the gas phase of the mass spectrometer, such destructive intermolecular collisions are avoided. Furthermore, the ionized forms of reactive neutrals are often stable, well known ions that can easily be generated in the mass spectrometer to serve as the starting material for the neutrals' synthesis. For example, carbene ion $H{-}C{-}NH_2^{\bullet+}$, which is available *via* electron ionization (EI) of cyclopropyl amine, can be used to synthesize the highly reactive prototype carbene $H{-}C{-}NH_2$ (a putative interstellar species); similarly, the distonic ion $^{\bullet}CH_2CH_2OCH_2^+$, which is formed by EI of 1,4-dioxane, can serve as the precursor for the hypovalent 1,4-diradical $^{\bullet}CH_2CH_2OCH_2^{\bullet}$ i.e. ring-opened oxetane (pyrolysis intermediate). NRMS has so far enabled the first experimental characterization of a large variety of neutral intermediates, in particular species with weak bonds, extra or missing valences, unpaired electrons, or a high tendency toward intermolecular isomerization (see below).

Knowledge of the intrinsic properties of such unusual neutrals is desirable for many reasons. They are postulated key intermediates in organic and biological reactions; they appear in the atmosphere, in interstellar clouds, and in other cosmic environments; they are involved in plasma and combustion chemistry and in photochemical processes and they play an important role in industrial and biological catalysis.

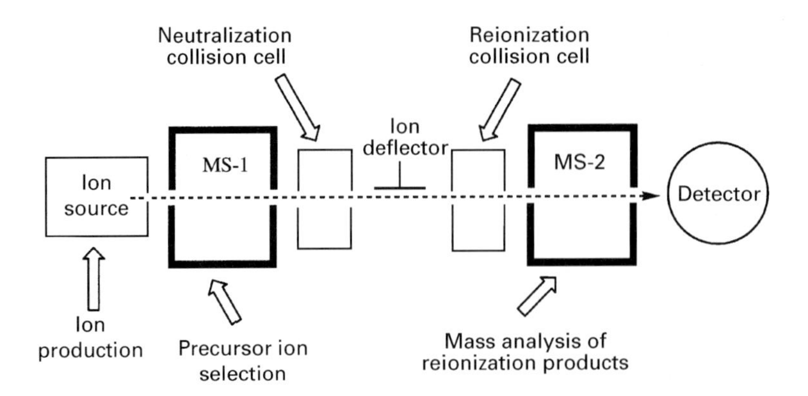

Figure 1 Basic components of a mass spectrometer for NRMS.

NRMS instrumentation

NRMS experiments are conducted with beam tandem mass spectrometers consisting of two mass-analyzing devices (MS-1 and MS-2) and equipped with at least two collision cells and an intermediate ion deflector in the field-free region between MS-1 and MS-2 (**Figure 1**). The desired precursor ion, *i.e.* the cationic or anionic form of the reactive neutral to be synthesized, is generated by the appropriate ionization method in the ion source, accelerated to a keV kinetic energy (3–10 keV), and separated from all other ions simultaneously produced in the ion source by MS-1. The mass-selected precursor ion beam is then partly neutralized by charge exchanging collisions with the neutralization target (Eqn [1]) in the collision cell following MS-1 (neutralization cell). The mixture of neutrals and ions exiting the cell passes the ion deflector, where the ions are removed from the beam path by electrostatic deflection. The remaining neutral beam subsequently enters the next collision cell (reionization cell) where it is ionized by collisions with the reionization target (Eqn [2]). The newly formed product ions are then dispersed according to their mass-to-charge ratio (m/z) by MS-2 to yield the NR spectrum of the precursor ion under study.

MS-1 and MS-2 may be simple, one-stage mass analyzers or tandem mass spectrometers. In the latter case, it is possible to form precursor ions by MS/MS of *specific* source-generated ions, and/or to analyze by MS/MS a *specific* product ion arising in the reionization event. Generally, the lifetimes of the ions subjected to NRMS experiments have been in the order of tens of microseconds. Depending on the distance between neutralization and reionization cells, the lifetimes of the neutrals generated in NRMS experiments range between a few tenths of a microsecond to a few microseconds. The field-free region between MS-1 and MS-2 may contain a third collision cell and a second intermediate ion deflector. Such a configuration allows one (a) to vary the lifetime of the neutral intermediates (by changing the neutralization and/or reionization cell) and (b) to probe the collisionally activated dissociation (CAD) of the neutrals *before* they are reionized.

The neutralization and reionization events

Positive or negative precursor ions can be subjected to neutralization and the resulting neutral species may be reionized to cations or anions. Customarily, the charges of precursor and final ions are indicated by superscripts to the NR acronym. The most frequently performed experiment is $^+NR^+$; it involves neutralization of a cation followed by reionization of the neutral intermediate(s) to cations, *i.e.* a gas phase reduction–oxidation succession.

Cations with kinetic energies in the keV range can be neutralized with metal vapours (e.g. Na or Hg), atomic or molecular gases (e.g. Xe or $(CH_3)_3N$), and organic compound vapours (e.g. CH_3SCH_3). Electron exchange takes place during the collision encounter of the fast moving precursor ion with the essentially stationary target atom/molecule. This encounter lasts only femtoseconds, a time much shorter than typical rovibrational periods. As a result, such neutralization is *vertical*, generating the incipient neutral in the geometry of its ionic precursor (Franck–Condon process, see **Figure 2**).

The neutralization yield and internal energy of the neutral species generated in the neutralization step (Eqn [1]) depend on several variables, including the internal energy of the precursor ion, the equilibrium geometries of precursor ion and neutral, and the target used for neutralization. One important factor determining the internal energy of the incipient neutral species emerging upon vertical neutralization is the thermochemistry of the charge exchange process.

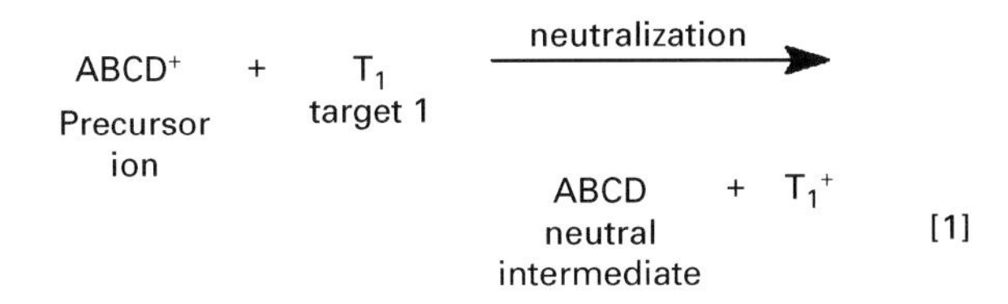

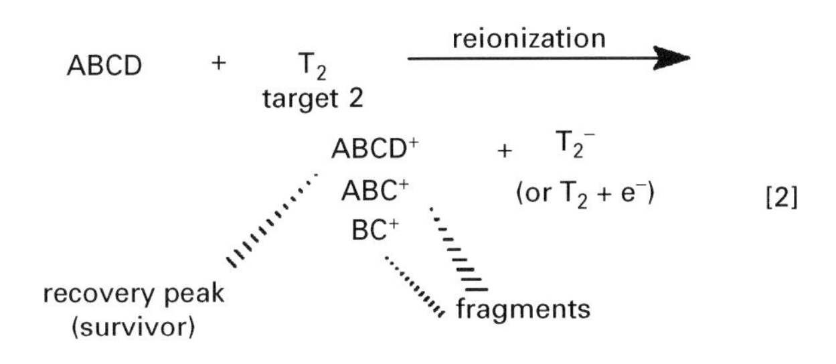

The enthalpy change of this reaction, ΔH_N, is given in Equation [3] where $IE_v(T_1)$ is the vertical ionization energy of the gaseous target oxidized and $RE_v(ABCD^+)$ the vertical recombination energy of the precursor ion reduced.

$$\Delta H_N = IE_v(T_1) - RE_v(ABCD^+) \qquad [3]$$

Exothermic neutralization ($\Delta H_N < 0$) usually leads to an excited species which may fragment if the energy deposited (ΔH_N) exceeds the dissociation threshold. On the other hand, thermoneutral ($\Delta H_N = 0$) or endothermic ($\Delta H_N > 0$) neutralizations normally yield ground-state neutrals. With endothermic encounters, the energy deficit is supplied by the kinetic energy of the precursor ion. In general, thermoneutral or nearly thermoneutral reactions exhibit the highest yields.

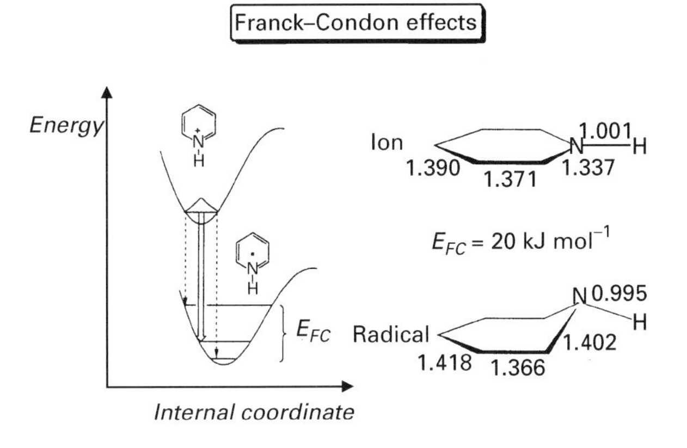

Figure 2 Franck–Condon energies in vertical neutralization of pyridinium ion. Reprinted with permission of John Wiley and Sons from Turecek F (1998) Modelling nucleobase radicals in the mass spectrometer. *Journal of Mass Spectrometry* **33**: 779–795.

It is believed that neutralization involves glancing collisions with a relatively large projectile–target distance ('impact parameter'). An endothermic neutralizing collision may provide some extra internal energy to the newly formed neutral; the degree of activation in the nascent neutral increases with ΔH_N, presumably because of the decreased impact parameter needed to convert some kinetic into internal energy, so that charge exchange becomes thermochemically possible. Additional excitation may be imparted by Franck–Condon effects; thus, if the equilibrium geometries of $ABCD^+$ and ABCD differ substantially, ABCD can emerge with rovibrational excitation, even if it is formed by a thermoneutral or endothermic process. For example, endothermic neutralization of pyridinium cations produces vibrationally excited pyridinium radicals due to the different ground-state structures of cation and radical (**Figure 2**). The average internal energy imparted to the radical by Franck–Condon effects (E_{FC}) is ~20 kJ mol^{-1} in this case. This modest amount is not sufficient to cause N–H bond cleavage in the radical (D = 108 kJ mol^{-1}), but increases the dissociation extent after reionization. In some systems, Franck–Condon effects can be very large; this is true for the neutralization of ground-state $(CH_3)_2Si^+OH$, which produces the radical $(CH_3)_2Si^{+\bullet}OH$ with enough internal energy for O–H or C–Si bond cleavage; therefore, such radicals do not survive intact until reionization, although $(CH_3)_2Si^\bullet OH$ is a bound species in its ground state. When molecular targets are used for neutralization, Franck–Condon effects play a more significant role than the thermochemistry of the neutralization reaction in determining the energy state of ABCD; further, in exothermic neutralizations, the excess ΔH_N can end up as excitation of the target ion (T_1^+ in Eqn [1]) and not in the incipient neutral ABCD. The energy state of ABCD also depends on the internal energy of the precursor ion $ABCD^+$; an activated precursor ion can give rise to an activated or vibrationally cool neutral, depending on the Franck–Condon overlap of Equation [1]. The former case is true for NH_4^+ neutralization, while the latter behaviour has been observed for neutralized $(CH_3)_2OH^+$, $C_2H_5OH_2^+$, and $(CH_3)_2Si^+OH$. Finally, it must be kept in mind that the neutralizing collision (Eqn [1]) may also cause some CAD of the precursor ion. The extent of concomitant CAD is minimized by avoiding strongly endothermic neutralizations whose small impact parameters may excite and dissociate the precursor ion (see above).

In the second $^+NR^+$ step (Eqn [2]), the neutral intermediate (ABCD) is collisionally reionized. This process, which resembles ionization by electron impact, produces a molecular ion as well as fragment ions (see Eqn [2]). Customarily, the molecular ion ($ABCD^+$) has been called the 'survivor ion', as it originates from ABCD that survived intact. The survivor ion gives rise to the 'recovery peak' in the NR spectrum, *i.e.* the peak appearing at the same *m*/*z* value as the precursor ion. Regarding the reionization target (T_2 in Eqn [2]), O_2 and He are the most widely employed choices. O_2 yields more abundant recovery peaks and He more fragments, justifying their classification as 'softer' and 'harder' reionization gas, respectively. Other, less frequently used soft reionization gases are NO and NO_2. Reionization efficiencies depend significantly on the structure of the neutral and rise with the neutral's kinetic energy but decline with its internal energy. Furthermore, kinetic and internal energies of the neutral intermediate also affect how much internal energy is deposited on collisional ionization; here, the average internal energy gained increases with both.

With anionic precursors, the neutralization step entails electron removal, *i.e.* an oxidation reaction, as does reionization of a neutral to cations. Therefore, O_2 which is the best reionization target (see above) also is the target of choice for the neutralization of anions. Conversely, cation neutralization targets, which effect a gas-phase reduction, are most suitable for reionization to anions. Widely used targets for this purpose have been xenon and trimethylamine. Reionization to anions suffers from substantially poorer yields than reionization to cations, often by ca. 10 times or more; therefore, it has been employed to a much lesser extent. However, in certain cases it can provide superior structural information; for example, the unequivocal identification of oxygen-centered radicals, such as the carbonate radical $CO_3^{\bullet-}$ and the diradical $^\bullet CH_2CH_2O^\bullet$ relied on their reionization to anions.

The study of reactive neutrals *via* NRMS

The stability and unimolecular reactivity of the neutral generated in the neutralization step are characterized by the mass spectrum arising after reionization. The presence of a recovery peak in this spectrum provides evidence that the neutral intermediate has survived intact (*i.e.* undissociated) for microsecond(s). Whether it has retained the connectivity of the precursor ion is judged by comparison of the NR spectrum to the CAD spectrum of the precursor ion or the NR spectra of other, usually stable and known isomers. Both these strategies are presented below with representative examples.

The hypermetallic dilithium fluoride (Li_2F), a compound violating the octet rule, can be formed in the gas phase by neutralizing the known cation $[Li–F–Li]^+$, which is produced abundantly upon fast atom bombardment ionization of lithium trifluoroacetate. Neutralization with trimethylamine (TMA) and subsequent reionization with O_2 lead to the NR spectrum of **Figure 3A**. The CAD spectrum of $[Li–F–Li]^+$ (using O_2) is displayed in **Figure 3B**. The significant recovery peak in the NR spectrum convincingly shows that Li_2F has survived intact. It is, thus, a stable species and bound in respect to Li + LiF. The experimentally documented stability is supported by theory which predicts that the reaction $Li_2F \rightarrow Li + LiF$ must overcome a barrier of 130–140 kJ mol^{-1}. The fragmentation patterns in NR and CAD spectra of **Figure 3** are similar. In both spectra, the most intense fragments are $LiF^{\bullet+}$(*m*/*z* 26) and F^+ (*m*/*z* 19), and the peak widths at half-height ($w_{0.5}$) follow the order $w_{0.5}$ ($LiF^{\bullet+}$, Li^+) < $w_{0.5}$ ($Li_2^{\bullet+}$) < $w_{0.5}$(F^+). These common trends strongly suggest that the majority of NR fragments arise from decomposing survivor ions, not from dissociation of the inter-

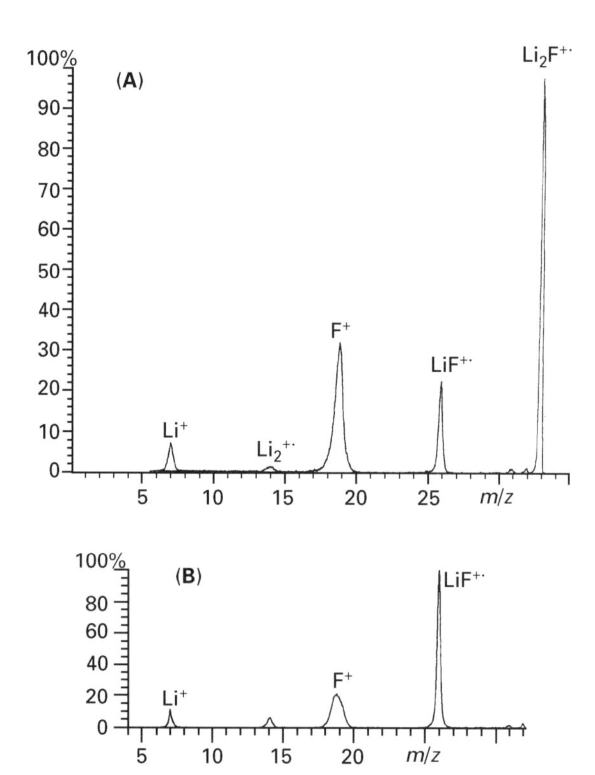

Figure 3 (A) Neutralization–reionization ($^+NR^+$) and (B) collisionally activated dissociation (CAD) spectra of Li_2F^+. Reprinted with permission of Elsevier from Polce MJ and Wesdemiotis C (1999) Hypermetallic dilithium fluoride, Li_2F, and its cation and anion: a combined dissociation and charge permutation study. *International Journal of Mass Spectrometry* **182/183**: 45–52.

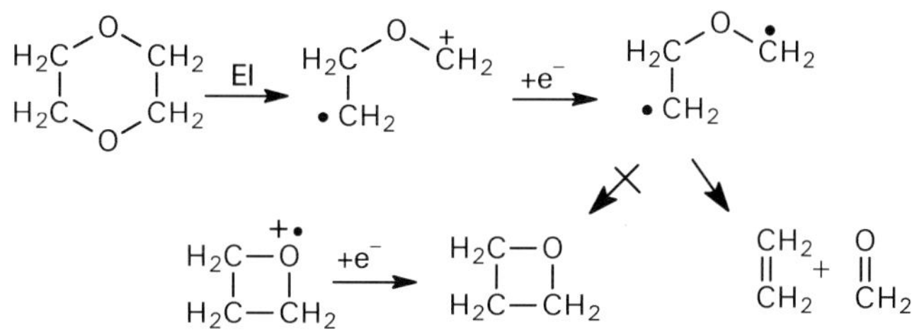

Scheme 1

mediate Li_2F, in keeping with the considerable thermodynamic stability of this hypermetallic compound.

A reactive neutral that dissociates extensively in the time span between neutralization and reionization is diradical $^{\bullet}CH_2CH_2OCH_2^{\bullet}$ (C–C ring-opened oxetane). This species is available through neutralization of the distonic ion $^{\bullet}CH_2CH_2OCH_2^+$ which is generated by electron ionization of 1,4-dioxane (**Scheme 1**). When Xe and O_2 are used for neutralization and reionization, respectively, the NR spectrum of **Figure 4A** is obtained; it is dominated by peaks at $m/z \leq 30$, verifying that the incipient diradical emerging in the neutralization step decomposes to ethylene (28 Da) and formaldehyde (30 Da). A small fraction remains, however, undissociated, as indicated by the presence in **Figure 4A** of a survivor ion (m/z 58) and fragment ions of $m/z > 30$. The surviving diradicals could cyclize to the thermodynamically more stable oxetane molecule (cf. **Scheme 1**). Whether this happened is best determined by comparison of the NR spectra of $^{\bullet}CH_2CH_2OCH_2^+$ (**Figure 4A**) and ionized oxetane (**Figure 4B**). Consistent with the high stability of neutral oxetane, the corresponding NR spectrum (**Figure 4B**) contains a substantial recovery peak (m/z 58); further, the $C_3H_{0-3}^+$ fragments in this spectrum (m/z 36–39) are diagnostic for oxetane's cyclic structure which has three adjacent C atoms. These latter fragments are minuscule in the NR spectrum of $^{\bullet}CH_2CH_2OCH_2^+$ (**Figure 4A**), which shows that the diradical $^{\bullet}CH_2CH_2OCH_2^{\bullet}$ does not cyclize to oxetane within the time available between neutralization and reionization. Hence, the diradical is kinetically stable in respect to both isomerization and dissociation and exists as a bound, high-energy intermediate.

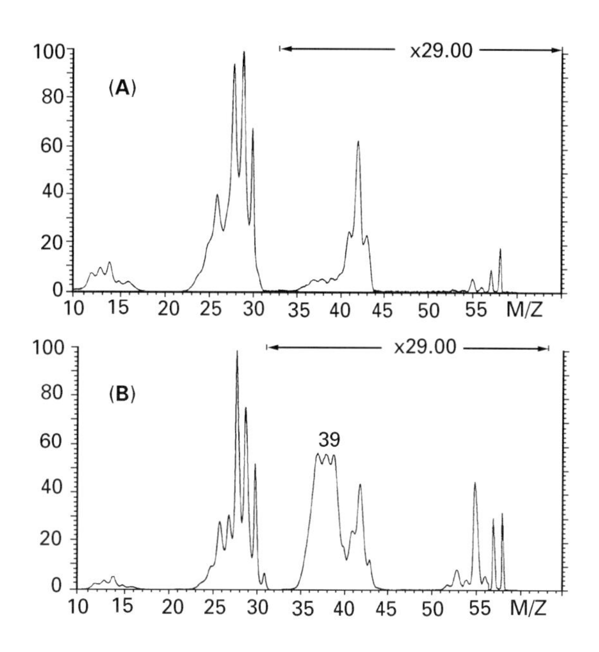

Figure 4 Neutralization–reionization (+NR+) spectra of (A) $^{\bullet}CH_2CH_2OCH_2^+$ and (B) oxetane cation. Reprinted with permission of the American Chemical Society from Polce MJ and Wesdemiotis C (1993) The unimolecular chemistry of the 1,4-diradical $^{\bullet}CH_2CH_2OCH_2^{\bullet}$ in the gas phase. Comparison to the distonic radical ions $^{\bullet}CH_2CH_2OCH_2^+$ and $^{\bullet}CH_2CH_2OCH_2^-$ *Journal of the American Chemical Society* **115**: 10849–10856.

Auxiliary NRMS methods

When the precursor ion is not isomerically pure or the intermediate neutrals rearrange partly or dissociate to products that are isobaric with those arising after reionization of the surviving neutrals, a straightforward characterization from the NR spectrum alone may be impossible. This problem has led to the development of an array of auxiliary NRMS methods that can be used to explore the stability and unimolecular reactivity of transient neutrals with complex and/or ambiguous NR spectra. A brief overview of these experimental approaches follows, along with examples demonstrating their usefulness.

NRMS-CAD studies

The isomerization proclivity of a reactive intermediate can be interrogated by measuring the tandem mass spectrum of the survivor ion. This capability requires that the instrumentation available be equipped with an additional mass analyzing device after the one used for the mass separation of the NR products.

The NRMS-CAD method is illustrated with the CH_3NO tautomers depicted in **Scheme 2**.

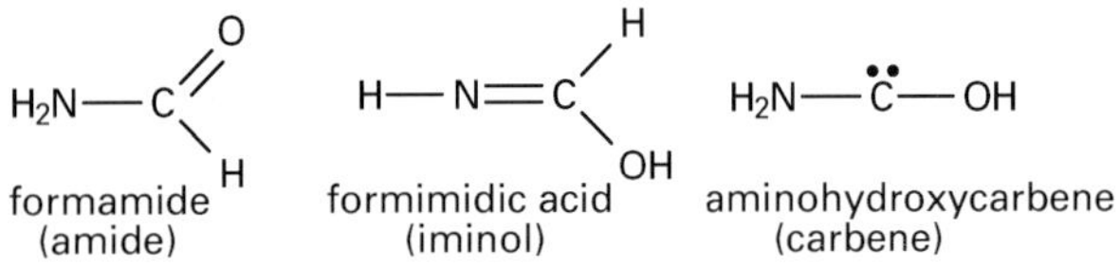

$H_2N—C(=O)H$ formamide (amide)   H—N=C(H)OH formimidic acid (iminol) $H_2N—\ddot{C}—OH$ aminohydroxycarbene (carbene)

Scheme 2

Formimidic acid and aminohydroxycarbene, both tautomers of formamide, are reactive molecules that do not exist in condensed media. Using NRMS, they can be synthesized in the gas phase from the corresponding radical cations, which are formed *via* dissociative electron ionization of *N*-formylhydrazone (yields the iminol ion) and oxamide (yields the carbene ion). After reionization, all three tautomers give rise to abundant survivor ions, but the NR spectra are not distinctive enough to unequivocally determine whether the less stable iminol and carbene neutrals underwent isomerization to the most stable formamide molecule. This question can be answered by subjecting the three survivor ions to CAD and comparing the resulting spectra to the CAD spectra of authentic (i.e. source-generated) amide, iminol, and carbene ions (**Figure 5**). The ions are known not to interconvert.

The NR-CAD and CAD spectra of formamide and the carbene are essentially indistinguishable, providing strong evidence that the amide and carbene tautomers retain their structures upon NR and do not undergo any rearrangements. In sharp contrast, NR-CAD and CAD spectra of the iminol ion differ markedly from each other; the NR-CAD spectrum can be interpreted as a combination of the CAD spectra of genuine amide and iminol ions, in agreement with a partial tautomerization of the neutral iminol to the amide.

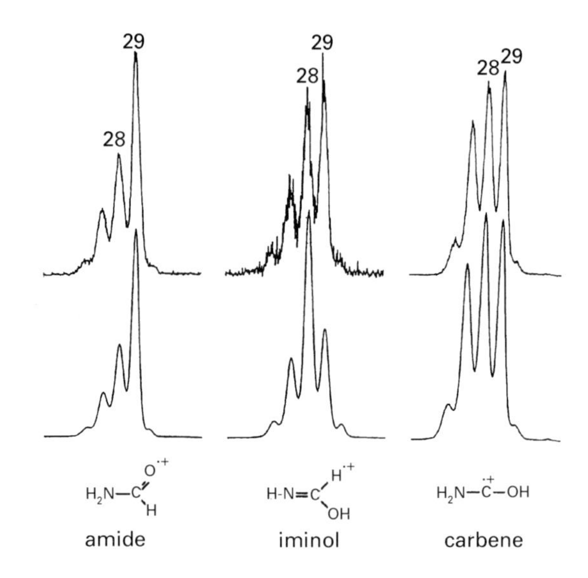

Figure 5 Partial CAD spectra (using He) of $^+NR^+$ survivor ions (top) and CAD spectra of source-generated ions (bottom) of H_2N-$CH(=O^{\bullet+}$ (left), $HN=C(H)OH^{\bullet+}$ (centre), and $H_2N–C–OH^{\bullet+}$ (right). Reprinted with permission of Elsevier from McGibbon GA, Burgers PC and Terlouw JK (1994) The imidic acids H–N=C(H)–OH and CH_3–N=C(H)–OH and their tautomeric carbenes H_2N–C–OH and CH_3–N(H)–C–OH: stable species in the gas phase formed by one-electron reduction of their cations. *International Journal of Mass Spectrometry and Ion Processes* **136**: 191–208.

CAD of the neutral intermediate (NCR studies)

Instruments with three successive collision cells and two intermediate ion deflectors allow for CAD of the neutral intermediate before reionization. With such an arrangement, the neutral is produced in the first cell, undergoes CAD in the second cell, and is reionized (along with its dissociation and/or isomerization products) in the third collision cell. The first deflector removes unneutralized precursor ions and the second one any ions formed during the neutral CAD step. This procedure gives rise to an NCR spectrum (neutralization – CAD – reionization) which, when compared to the corresponding NR spectrum, reveals insight about the favoured fragmentation and/or isomerization of the neutral species under study.

As an example, **Figure 6** shows the NR and NCR spectra of the distonic ion $^+CH_2CH_2O^\bullet$ whose neutralization creates the elusive diradical $^\bullet CH_2CH_2O^\bullet$ (ring-opened oxirane, cf. **Scheme 3**). The increased relative abundances of *m/z* 43, 29, 28, and 12–15 in the NCR spectrum support the occurrence of the top channel in **Scheme 3** (dissociation through acetaldehyde). Corroborative evidence that collisionally activated $^\bullet CH_2CH_2O^\bullet$ partially interconverts to $CH_3CH=O$, but does not ring-close to oxirane, is obtained by the NR-CAD spectrum of $^+CH_2CH_2O^\bullet$ (see previous section), which is consistent with a mixture of $^+CH_2CH_2O^\bullet$ and $CH_3CH=O^{\bullet+}$ and significantly different from the reference CAD spectrum of oxirane ion.

Photodissociation/photoionization of the intermediate neutrals

Laser light can be used, in place of the reionization gas, to probe the reactivity and electronic state(s) of the neutrals produced in the neutralization step. Due to the low neutral currents generated in NRMS experiments (~ 10^6 molecules per second), a coaxial arrangement of the neutral and laser beams and continuous laser light are necessary to ensure detectable photodissociation or photoionization yields. An

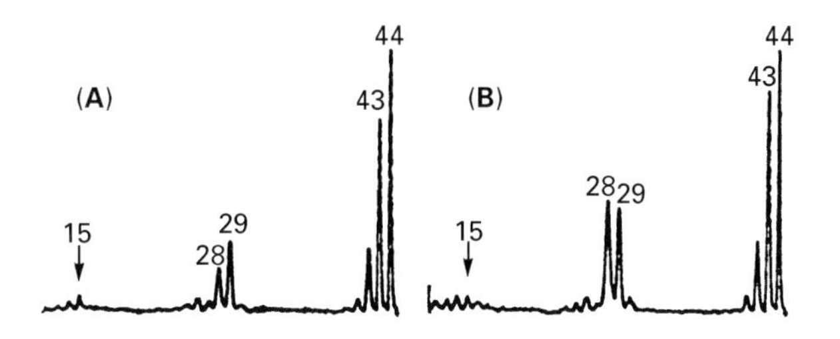

Figure 6 (A) Neutralization-reionization ($^+$NR$^+$) and (B) neutralization–CAD–reionization ($^+$NCR$^+$) spectra of $=^+CH_2CH_2O^{\bullet}$. Hg was used for neutralization, O_2 for reionization, and He for CAD. Reprinted with permission of the American Chemical Society from Wesdemiotis C, Leyh B, Fura A and McLafferty FW (1990) The isomerization of oxirane. Stable $^{\bullet}CH_2OCH_2^{+\bullet}$, $^{\bullet}CH_2CH_2O^{\bullet}$, and $:CHOCH_3$ and their counterpart ions. *Journal of the American Chemical Society* **112**: 8655–8660.

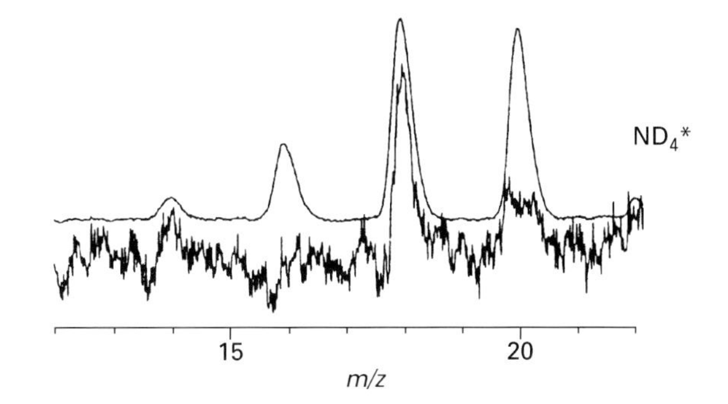

Figure 7 Neutralization–photodissociation/photoreionization spectrum of $ND_4^{\bullet}$. Top: total reionization, bottom: after background subtraction. $ND_4^{\bullet}$ was formed by ND_4^+ neutralization with dimethyl disulfide. Reprinted with permission of Elsevier from Sadilek M and Turecek F (1996) Laser photolysis of $ND_4^{\bullet}$ and trimethylamine formed by collisional neutralization of their cations in the gas phase. *Chemical Physics Letters* **263**: 203–308.

argon laser, which emits intense lines at 488 nm (2.54 eV) and 514.5 nm (2.41 eV), is suitable for such experiments. The procedure is illustrated for the hypervalent radical $ND_4^{\bullet}$.

Collisional neutralization of ND_4^+ followed by laser irradiation of $ND_4^{\bullet}$ gives rise to the NR spectrum of **Figure** 7, top. A second NR spectrum is acquired with the laser off, so that the contribution of the background gases can be assessed. Subtraction of these sequential spectra gives the net change due to interaction with photons (**Figure** 7, bottom). The background-corrected spectrum includes a very weak positive contribution to the ND_4^+ survivor ion at *m/z* 22, indicating that the $ND_4^{\bullet}$ beam contains a minor fraction of long-lived excited electronic states with ionization energies ≤ 2.54 eV; for comparison, the IE of ground-state $ND_4^{\bullet}$ is 4.6 eV. The corrected NR spectrum also shows increased intensities of $ND_3^{\bullet+}$ (*m/z* 20) and ND_2^+(*m/z* 18), which are attributed to photodissociation of ground-state $ND_4^{\bullet}$ and subsequent collisional reionization with the background gases. The alternative scenario, namely collisional reionization to ND_4^+ followed by photodissociation of ND_4^+ to $ND_3^{\bullet+}$ and ND_2^+ is energetically impossible with the wavelengths used.

Variable-time NRMS

A given NR product may arise from dissociation of the survivor ion or from dissociation of the neutral intermediate (see above). As explained above, NCR helps identify the predominant neutral fragmentations. An alternative approach for distinguishing overlapping neutral and ion dissociations is to systematically alter the time scales of these processes, using variable-time NRMS; here, the position of the reionization cell is varied, in order to change the *lifetimes* ('observation times') of the neutrals (ABCD) and reionized ions (ABCD$^+$ plus fragments). The cell need not be moved physically; instead, a segmented reionization cell can be used, in which the segments are individually floatable so that ions formed inside them are either prevented or permitted passage to MS-2, depending on the potentials applied. This

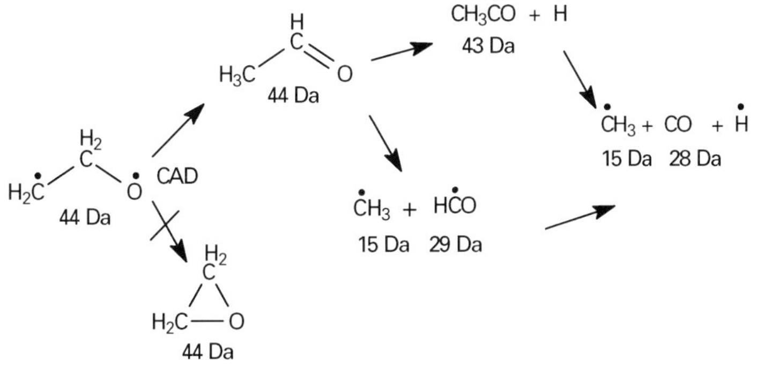

$\dot{C}H_2CH_2\dot{O}$ (44 Da) —CAD→ CH_3CHO (44 Da) → $CH_3\dot{C}O + \dot{H}$ (43 Da) → $\dot{C}H_3 + CO + \dot{H}$ (15 Da, 28 Da)

CH_3CHO (44 Da) → $\dot{C}H_3 + H\dot{C}O$ (15 Da, 29 Da) → $\dot{C}H_3 + CO + \dot{H}$

$\dot{C}H_2CH_2\dot{O}$ (44 Da) ✕→ oxirane H_2C–O–CH_2 (44 Da)

Scheme 3

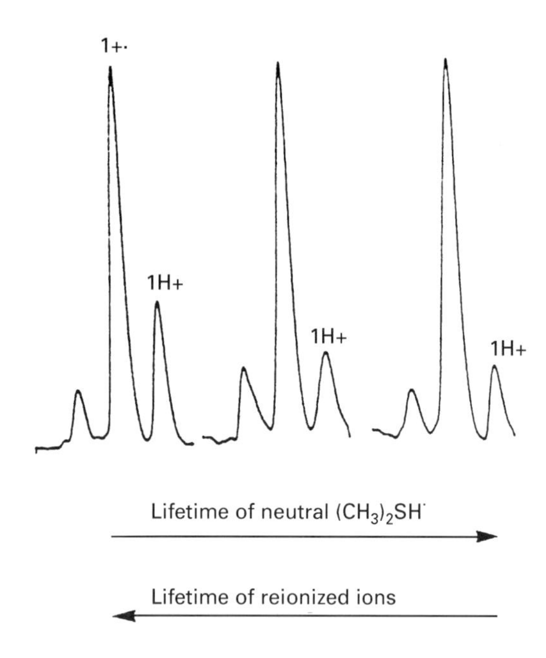

Figure 8 Variable-time $^{+}NR^{+}$ spectra of $(CH_3)_2SH^+$ formed by chemical ionization of $(CH_3)_2S$ with $(CH_3)_2CH{=}OH^+$. Dimethyl disulfide and oxygen served as the neutralization and reionization targets, respectively. The observation times of the neutrals are 0.35 µs (left), 1.05 µs (centre), and 1.76 µs (right). The corresponding observation times of the ions are 3.44 µs (left), 2.73 µs (centre), and 2.02 µs (right). Reprinted with permission of Elsevier from Sadilek M and Turecek F (1999) Metastable states of dimethylsulfonium radical, $(CH_3)_2SH^\bullet$: a neutralization–reionization mass spectrometric and ab initio computational study. *International Journal of Mass Spectrometry* **185/186/187**: 639–649.

essentially corresponds to moving the entrance of the reionization cell; if the segments are floated such that the entrance moves towards (away from) MS-2, the observation time of the neutrals increases (decreases). Since the field-free region housing of the NR cells has a fixed length, increasing the observation time of the neutrals decreases that of the reionized ions (time elapsing between their formation and MS-2 entry) and *vice versa*.

Figure 8 shows partial variable-time NR spectra of the dimethylsulfonium cation, $(CH_3)_2SH^+$, in which the lifetime of neutral $(CH_3)_2SH^\bullet$ (a hypervalent radical) gradually increases from 0.35 to 1.76 µs while that of the reionized ions decreases from 3.44 to 2.02 µs. The relative abundance of the survivor ion (designated $1H^+$) *decreases* upon *increasing* the observation time of the *neutrals* (equivalent with time available for *neutral dissociation*). This clearly points out that a fraction of radical $(CH_3)_2SH^\bullet$ dissociates in the microsecond time scale and that the peak labelled $1^{+\bullet}$ primarily originates through the neutral dissociation $(CH_3)_2SH^\bullet \rightarrow (CH_3)_2S + H^\bullet$ and subsequent reionization of $(CH_3)_2S$.

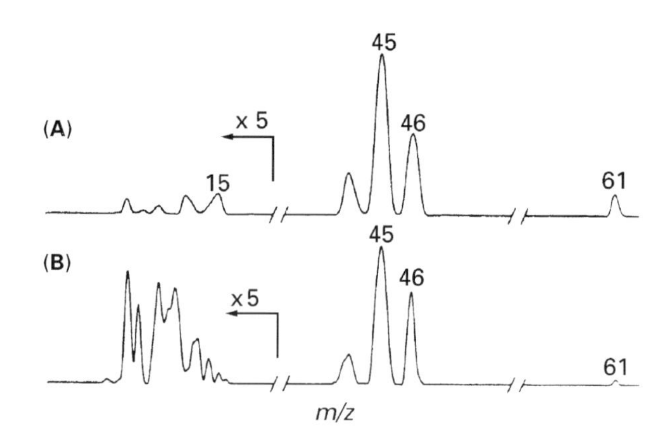

Figure 9 Variable time $^{+}NR^{+}$ spectra of $CH_3SCH_2^+$ using $(CH_3)_3N$ for neutralization and O_2 for reionization. The observation times for $CH_3SCH_2^\bullet$ and reionized $CH_3SCH_2^+$ are (A) 2.44 and 1.70 µs and (B) 0.36 and 3.78 µs, respectively. Reprinted with permission of the American Chemical Society from Kuhns DW, Tran TB, Shaffer SA and Turecek F (1994) Methylthiomethyl radical. A variable-time neutralization-reionization and ab initio study. *Journal of Physical Chemistry* **98**: 4845–4853.

Another example is illustrated in **Figure 9**, which depicts variable-time NR spectra of the methylthiomethyl cation, $CH_3SCH_2^+$. In this case, the observation times for $CH_3SCH_2^\bullet$ and its reionization products are 2.44 and 1.70 µs, respectively, in the top and 0.36 and 3.78 µs, respectively, in the bottom spectrum. Now, the relative abundance of the survivor ion exhibits a 6-fold *decrease* upon *decreasing* the minimum neutral lifetime from 2.44 to 0.36 µs while increasing that of reionized $CH_3SCH_2^+$ from 1.70 to 3.78 µs. Thus, the depletion of the survivor ion in the lower spectrum must be due to an increased fraction of reionized $CH_3SCH_2^+$ that dissociates. This, in turn, strongly suggests that (a) the fraction of $CH_3SCH_2^\bullet$ dissociating within microseconds is negligible and (b) collisional reionization of $CH_3SCH_2^\bullet$ is accompanied by substantial energy deposition, thereby extending the unimolecular decomposition of the reionized ions into the microsecond time window.

Neutral and ion decomposition difference (NIDD)

In this method, a neutral produced from an *anionic* precursor is reionized to *cations*, and one produced from a *cationic* precursor is reionized to *anions*, to thus obtain a $^{-}NR^{+}$ or $^{+}NR^{-}$ spectrum, respectively. This is then compared to the corresponding charge reversal spectrum ($^{-}CR^{+}$ or $^{+}CR^{-}$) of the precursor ion, where charge inversion is afforded by a two-electron transfer in a *single* collision (either one of the two collision cells in **Figure 1** can be used for this purpose). Under *single* collision conditions, the CR signals originate solely through ionic fragmentations of the charge-inverted ions. In contrast, the NR signals, which result from two spatially separated collisions involving neutral intermediates, may arise *either* from the neutral *and/or* the reionized ions. Normalization of the spectra to the sum of all fragments followed by subtraction of the CR from the NR intensities (Eqn [4]) results in the NIDD spectrum, in which processes due to the fragmentation of the neutrals have positive intensities, while ionic contributions (of the reionized ions) show up as negative peaks. Hence, the neutral and ionic processes of NR are readily deconvoluted.

$$I_i(\mathrm{NIDD}) = [I_i(\mathrm{NR})/\Sigma_i I_i(\mathrm{NR})] - [I_i(\mathrm{CR})/\Sigma_i I_i(\mathrm{CR})] \qquad [4]$$

where I_i is the intensity of the ith ion.

NIDD is most suitable for the study of anionic precursors, which easily undergo two-electron loss in one step. The method has been applied extensively to alkoxide anions to decipher the chemistry of the corresponding radicals and cations, both of which are more reactive and less readily accessible than the anions. As an example, **Figure 10** shows the $^-$NR$^+$, $^-$CR$^+$, and $^-$NIDD$^+$ spectra of *n*-butoxide anion, $CH_3CH_2CH_2CH_2O^-$. Three prominent features are worth mentioning. (a) The recovery peaks are weak, consistent with the high reactivity expected for the oxenium cation $CH_3CH_2CH_2CH_2O^+$. (b) The favoured dissociations of this elusive cation are identified by the negative signals in **Figure 10C** and primarily lead to $C_3H_5^+$, $C_3H_3^+$, and $C_2H_5^+$ and $C_2H_3^+$; most likely, these products arise from elimination of CH_2O or C_2H_4O (inductive cleavages), followed by one or more dehydrogenations. (c) The fragments generated from the radical $CH_3CH_2CH_2CH_2O^{\bullet}$ are revealed by the positive signals in **Figure 10C.**

Table 1 Examples of elusive neutral intermediates that have been synthesized and characterized by NRMS.

Intermediate type	*Examples*
Ylides	$H_2N^+{=}N^-$ (*iso*-diazine), $^-HC{=}N^+{=}NH$ (nitrileimine), $^-HC{=}N^+{=}CH_2$ (*N*-methylide of HCN), C_3H_3NS (thiazole-2-ylide), C_5H_5N (pyridine-2-ylide)
Carbenes	$H{-}C{-}NH_2$ (aminocarbene), $H{-}C{-}NO_2$ (nitrocarbene), $H_2N{-}C{-}NH_2$ (diaminocarbene), HO–C–OH (dihydroxycarbene), :C=C=O (carbonyl carbene), $CH_3O{-}C{-}OCH_3$, $H{-}C{-}OC_2H_5$ (ethoxycarbene), :C=C(H)CN (cyanovinylidene), $C_3H_4N_2$ (imidazol-2-ylidene and imidazol-4-ylidene), C_3H_5NO (2-oxazolidinylidene), H_2CCCC:, $CH_3O{-}C{-}OCH_2CF_3$
Cumulenes	O=C=C=O (ethylenedione), $NNCCN^{\bullet}$, $NCNCN^{\bullet}$, $NCCO^{\bullet}$, $CNCO^{\bullet}$, $CCNO^{\bullet}$, $OCCN^{\bullet}O$, $OCN^{\bullet}CO$, $NCCO_2^{\bullet}$, NCNCS, O=C=C=C=O (carbon suboxide), $NCCCO^{\bullet}$, $O{=}C{=}C{=}CH{-}CO_2H$
Radicals	$HOSO_2^{\bullet}$ (bisulfite radical), $HOCO_2^{\bullet}$ (bicarbonate radical), $FC^{\bullet}(OH)_2$, $^{\bullet}CH_2NHNH_2$, $^{\bullet}CH_2CH_2Br$, $^{\bullet}CH_2N(H){-}COOH$, $H_2NCH_2C^{\bullet}(OH)_2$, $(CH)_3)_2NCH_2^{\bullet}$, $CH_3CH_2C^{\bullet}(OH)_2$, $C_4H_6N^{\bullet}$ (neutral form of C–2 and C–3 protonated pyrrole), $C_4H_6N^{\bullet}$ (neutral form of N-3 protonated imidazole), $C_4H_5N_2^{\bullet}$ (neutral form of protonated pyrimidine), $C_4H_6N_3^{\bullet}$ (neutral form of protonated 2- or 4-pyrimidinamine)
Hypervalent radicals	$CH_2{=}CH(OH)_2^{\bullet}$; $C_2H_5OH_2^{\bullet}$; $(C_2H_5)_3O^{\bullet}$; *n*-$C_7H_{13}O^{\bullet}H(CH_3)$, $CH_3O(CH_2)_nOH^{\bullet}(CH_3)$ ($n = 2,4$), $(CH_3)_2N^{\bullet}H_2$ (several isotopomers), $H_3N^{\bullet}CH_2CO_2H$, $(CH_3)_3N^{\bullet}H$, $(CH_3)_4N^{\bullet}$, *n*–$C_6H_{11}N^{\bullet}H_3$, $C_6H_5CH_2N^{\bullet}H_3$, $(C_2H_5)_4N^{\bullet}$, *n*–$C_6H_{11}N^{\bullet}H(CH_3)_2$, $C_6H_5CH_2N^{\bullet}H_2(CH_3)$, *n*–$C_7H_{15}N^{\bullet}H(CH_3)_2$, $C_6H_5CH_2N^{\bullet}H(CH_3)_2$, $C_6H_5N^{\bullet}(CH_3)_3$, $C_6H_5CH_2N^{\bullet}(CH_3)_3$, $C_6H_5CH_2N^{\bullet}(C_2H_5)_3$, $HN(CH_2CH_2)_2N^{\bullet}H(CH_3)$, $N(CH_2CH_2)_3N^{\bullet}H$, $CH_3N(CH_2CH_2)_2N^{\bullet}H(CH_3)$, $(CH_3)_2N(CH_2)_nN^{\bullet}H(CH_3)_2$ ($n = 2,3,4,6$), $(NO_2)_2C_6H_4CH_2N^{\bullet}H(CH_3)_2$, $C_6F_5CH_2N^{\bullet}H(CH_3)_2$, $(C_2H_5)_2X^{\bullet}$ (X = Cl,Br,I), $H_3S^{\bullet}$ (various isotopomers)
Diradicals	$^{\bullet}CH_2OSi^{\bullet}$, $^{\bullet}CH_2NHCH_2^{\bullet}$, $^{\bullet}CH_2COO^{\bullet}$ (acetoxy), $^{\bullet}CH_2CH_2OCH_2^{\bullet}$, $^{\bullet}CH(CH_3)OCH_2^{\bullet}$, $^{\bullet}CH_2CH_2CH(OH)^{\bullet}$, $^{\bullet}CH_2CH_2SCH_2^{\bullet}$, $^{\bullet}CH_2CH_2CH_2CH(OH)^{\bullet}$
Intermediates in atmospheric chemistry	NH_2NO (nitrosamine), $HSO^{\bullet}$, $SOH^{\bullet}$, HSOH (hydrogen thioperoxide), SOH_2 (thiooxonium ylide), $O{=}S^{\bullet}OH$ (hydroxysulfinyl radical), $O_2S^{\bullet}H$ (hydrogensulfonyl radical), HOSOH (sulfinic acid), $CH_3SCH^{\bullet}$, $(CH_3)_2S^{\bullet}OH$
Clusters	NSS, SNS (nitrogen disulfide), SSNS, O_2SOSO (sulfur dioxide dimer), He@C_{60}
Reactive closed-shell species	HOSSOH (dihydroxy disulfide), FCO_2H (fluoroformic acid), RCONO (R = H, CH_3, CF_3), HC≡NS (thiofulminic acid), $CH_3O{-}P{=}O$, $CH_3S{-}P{=}O$, CH_3PS_2, $CH_3S{-}P{=}S$, $CH_3C{\equiv}NS$, CH_3H_7NO (3-methyl-2-aziridinone), $CH_3CH{=}CH{-}SH$, $C_6H_5C{\equiv}NS$ (benzonitrile–sulfide), $C_6H_5PS_2$, $C_6H_5S{-}P{=}S$
Organometallic species	$Fe(O_2)$ (peroxide), OFeO (dioxide), Fe(CO), $Fe(C_2H_4)$,$C_5H_5{-}Fe{-}R$ (R = halogen, O, OH, OCH_3, C_6H_5, H), $C_5H_5FeC_5H_4{=}X$ (X=O, CH_2, CO), $SiNH_2^{\bullet}$, Si_2O_2, Si_3N, $D_2Si{=}CH_2$ (silaethene), $SiC_nH^{\bullet}$ (n=4,6) (silicon carbon hydrides), $(CH_3)_3SiC^{\bullet}{=}CH_2$, $(CH_3)_3SiCH{=}CH^{\bullet}$, PrF_n(n = 1-2), $(C_5H_5)_2Zr$ (zirconocene), $F_3CCSeNSeCCF_3^{\bullet}$

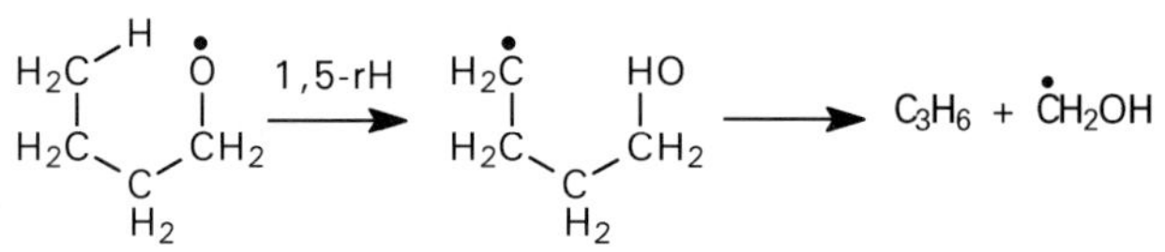

$\xrightarrow{1,5\text{-rH}}$ $\longrightarrow C_3H_6 + {}^{\bullet}CH_2OH$

Scheme 4

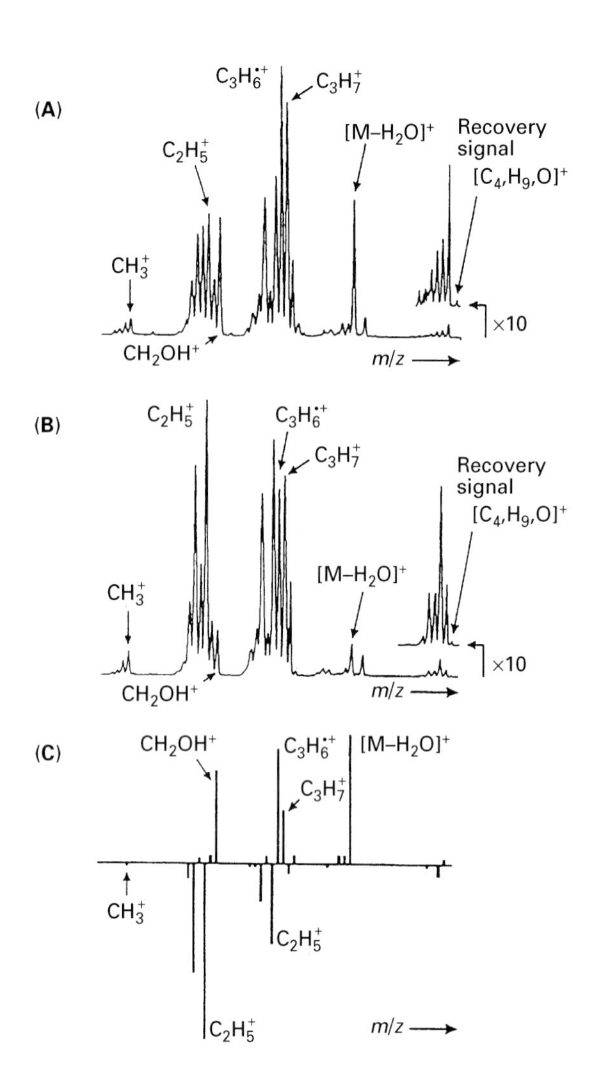

Figure 10 (A) Neutralization–reionization ($^-NR^+$) spectrum of *n*-butoxide anions using O_2 for neutralization and reionization. (B) Charge reversal ($^-CR^+$) spectrum of *n*-butoxide anions using O_2. (C) $^-NIDD^+$ spectrum of *n*-butoxide anions. Reprinted with permission of Wiley-VCH from Hornung G, Schalley CA, Dieterle M, Schröder D and Schwarz H (1997) A study of the gas-phase reactivity of neutral alkoxy radicals by mass spectrometry: α-cleavages and Barton-type hydrogen migrations. *Chemistry: a European Journal* **3**: 1866–1883.

Among them are the α-cleavage products $C_3H_7^{\bullet}$+CH_2O (weak); the intramolecular hydrogen transfer products C_3H_6+$^{\bullet}CH_2OH$; as well as the product of H_2O loss which indicates the occurrence of a double hydrogen transfer to oxygen. Repeating these experiments with $CD_3CH_2CH_2CH_2O^{\bullet}$ reveals that the intramolecular H-rearrangement involves a 1,5-hydrogen migration (**Scheme 4**), similar to the Barton reaction observed in solution. On the other hand, the water elimination from $CD_3CH_2CH_2CH_2O^{\bullet}$ is found to proceed with partial H/D scrambling, leading to the losses of H_2O, HDO, and D_2O in the ratio 1:10:5.

Reactive neutrals investigated and outlook

The basic NRMS method and the auxiliary techniques outlined have enabled the discovery and characterization of a large variety of reactive neutrals that had eluded experimental studies due to their unusual structures and reactivities. The species studied so far are summarized in **Table 1** and include radicals, diradicals, carbenes, nitrenes, cumulenes, ylides, hypervalent species, and weak intermolecular complexes.

NRMS has primarily been a qualitative method. It determines whether a reactive neutral species is bound and to what it rearranges or dissociates, but it does not easily yield *quantitative* insight about bond dissociation energies and isomerization barriers or about the thermochemistry of the neutral under study; these data have generally been obtained by parallel theoretical calculations. NRMS also does not allow the assessment of the *bimolecular* reactivity of neutral intermediates, which is of paramount interest in organic and biological chemistry and in catalysis. However, in light of the new knowledge NRMS has yielded in less than two decades thus far (**Table 1**), the shortcomings mentioned provide the motivation for future efforts to overcome them.

See also: **Ion Dissociation Kinetics, Mass Spectrometry; Ion Energetics in Mass Spectrometry; Ion Imaging Using Mass Spectrometry; Ion Molecule Reactions in Mass Spectrometry; Ion Structures in Mass Spectrometry; Ion Trap Mass Spectrometers.**

Further reading

Busch KL, Glish GL and McLuckey SA (1988) *Mass Spectrometry/Mass Spectrometry*. New York: VCH.

Goldberg N and Schwarz H (1994) Neutralization-reionization mass spectrometry: a powerful "laboratory" to generate and probe elusive neutral molecules. *Accounts of Chemical Research* **27**: 347–352.

Holmes JL (1989) The neutralization of organic cations. *Mass Spectrometry Reviews* **8**: 513–539.

McLafferty FW (1990) Studies of unusual simple molecules by neutralization-reionization mass spectrometry. *Science* **247**: 925–929.

McLafferty FW (1992) Neutralization-reionization mass spectrometry. *International Journal of Mass Spectrometry and Ion Processes* **118/119**, 221–235.

Polce MJ, Beranova S, Nold MJ and Wesdemiotis C (1996) Characterization of neutral fragments in tandem mass spectrometry: a unique route to mechanistic and structural information. *Journal of Mass Spectrometry* **31**: 1073–1085.

Schalley CA, Hornung G, Schröder D and Schwarz H (1998) Mass spectrometric approaches to the reactivity of transient neutrals. *Chemical Society Reviews* **27**: 91–104.

Terlouw JK and Schwarz H (1987) The generation and characterization of molecules by neutralization-reionization mass spectrometry (NRMS). *Angewundte Chemie, International Edition; English* **26**: 805–815.

Turecek F (1992) The modern mass spectrometer: a chemical laboratory for unstable neutral species. *Organic Mass Spectrometry* **27**: 1087–1097.

Wesdemiotis C and McLafferty FW (1987) Neutralization-reionization mass spectrometry (NRMS). *Chemical Reviews* **87**: 485–500.

Zagorevskii DV and Holmes JL (1994) Neutralization-reionization mass spectrometry applied to organometallic and coordination chemistry. *Mass Spectrometry Reviews* **13**: 133–154.

Neutron Diffraction, Instrumentation

AC Hannon, Rutherford Appleton Laboratory, Didcot, UK

HIGH ENERGY SPECTROSCOPY
Methods & Instrumentation

Neutron diffraction is a very powerful technique for investigating the structure of condensed matter. Crystal structures can be studied, either by diffraction from a polycrystalline powder or from a single crystal sample. Neutron diffraction is also a very important tool for the study of the structure of noncrystalline forms of matter, such as liquids and glasses. This article considers in detail the instrumentation for studying isotropic samples, either polycrystalline powders or non-crystalline materials. Neutrons may be obtained either from a nuclear reactor or from an accelerator-based source. They may be detected using either gas detectors or scintillator detectors. Typical neutron diffractometers at both reactor and accelerator-based sources are described, and a consideration of the resolution is given in each case.

The neutron

The neutron was discovered by Chadwick in 1932. It is a neutral subatomic particle which has a finite mass, $m_n = 1.0087$ amu, similar to that of the proton. It has a spin of $\frac{1}{2}$ and a magnetic moment of −1.9132 nuclear magnetons. A neutron with speed v has a de Broglie wavelength given by

$$\lambda = \frac{h}{m_n v} \quad [1]$$

and thus the neutron exhibits wavelike behaviour including diffraction.

Normally neutrons only exist within the atomic nucleus and hence a nuclear reaction of some sort must be used in order to produce a beam of neutrons for neutron diffraction. Chadwick first produced neutrons by the interactions in beryllium of alpha particles from the decay of natural polonium. The first experimental demonstrations of the phenomenon of neutron diffraction were performed in 1936 and these also used radioactive sources. However, the neutron flux available from the decay of a radioactive source is very weak, and since neutron scattering is an intensity-limited technique, all neutron diffraction experiments soon came to be performed using neutrons produced by either a nuclear reactor or an accelerator-based neutron source, both of which produce a much greater flux. The properties of the neutron source have important consequences both for the way in which a neutron diffraction experiment is performed, and for the results obtained, and hence it is worthwhile to discuss below the two important types of neutron source.

Neutron sources

Reactor sources of neutrons

Conventionally, from the 1940s onwards, neutron diffraction experiments were performed using a beam of neutrons derived from a nuclear reactor in which the neutrons are produced by the fission of ^{235}U nuclei. The cross-section for neutron-induced fission of ^{235}U is only high for slow neutrons with energies in the meV range, whereas the fast neutrons produced by fission have high energies in the MeV range. Hence, in order to sustain the fission process, a reactor includes a component, known as a moderator, which slows down the neutrons. The neutrons undergo inelastic collisions with the nuclei in the moderator so that they are in thermal equilibrium at the temperature of the moderator. The moderator normally contains large numbers of low mass nuclei (usually H or D) because the energy transferred in the inelastic collisions is maximized when the mass of the colliding nucleus is as close as possible to the neutron mass. The peak flux within the moderator is at a neutron speed ν_p given by

$$\frac{1}{2}m_n\nu_p^2 = k_B T \qquad [2]$$

where T is the temperature of the moderator. For example, a temperature of 290 K corresponds to a neutron energy E of 25 meV, a neutron wavelength λ of 1.8 Å (0.18 nm), or a neutron speed v of 2200 m s^{-1}. In order to perform a neutron diffraction experiment to study the atomic structure of condensed matter it is necessary to use neutrons whose wavelength is of a similar order of magnitude to the interatomic separations in materials. It is thus fortuitous that the process of moderation produces neutrons which, as well as being slowed down for maintaining the fission reaction, also have a wavelength suitable for performing neutron diffraction experiments.

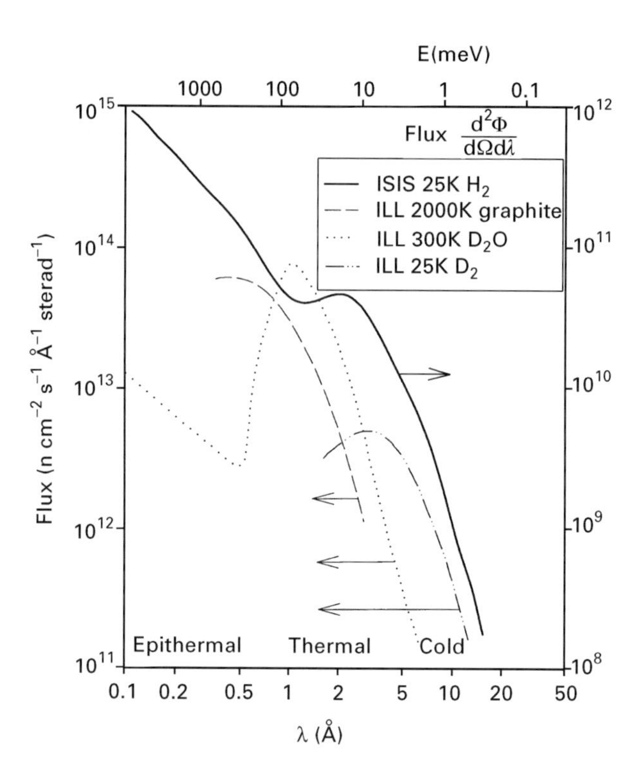

Figure 1 The neutron flux distribution for three different moderators at the ILL reactor and for the liquid hydrogen moderator at the ISIS accelerator. The accelerator flux distribution is adjusted by a factor 10^3 to represent the increased efficiency for time-of-flight experiments due to the pulse structure. Modified with permission from Price DL and Sköld K (1986) Introduction to neutron scattering in: *Neutron Scattering, Part A*, Sköld K and Price DL (eds). Orlando: Academic Press.

A neutron diffractometer uses a beam of neutrons which is obtained by viewing a moderator through a beam-tube or neutron guide which passes through the shielding around the neutron source. Note that, in practice, the moderator used as a source of neutrons for neutron diffraction experiments at a reactor may be separate from the moderator used to slow the neutrons in order to maintain the fission reaction. **Figure 1** shows the neutron flux for three different moderators at the world's preeminent reactor source of neutrons, the Institut Laue–Langevin (ILL) in Grenoble, France (see **Figure 2**). Reactor neutron sources produce a high flux of thermal neutrons (E~25 meV, T~290 K) and cold neutrons (E~1 meV, T~12 K), but they have little flux at higher epithermal energies (E~1 eV, T~12 000 K). This is a consequence of the fact that a reactor can only produce neutrons which are in thermal equilibrium with a moderator and there are practical limitations on the maximum temperature of the moderator.

The neutron flux produced by a normal nuclear reactor is unchanging with time and covers a wide

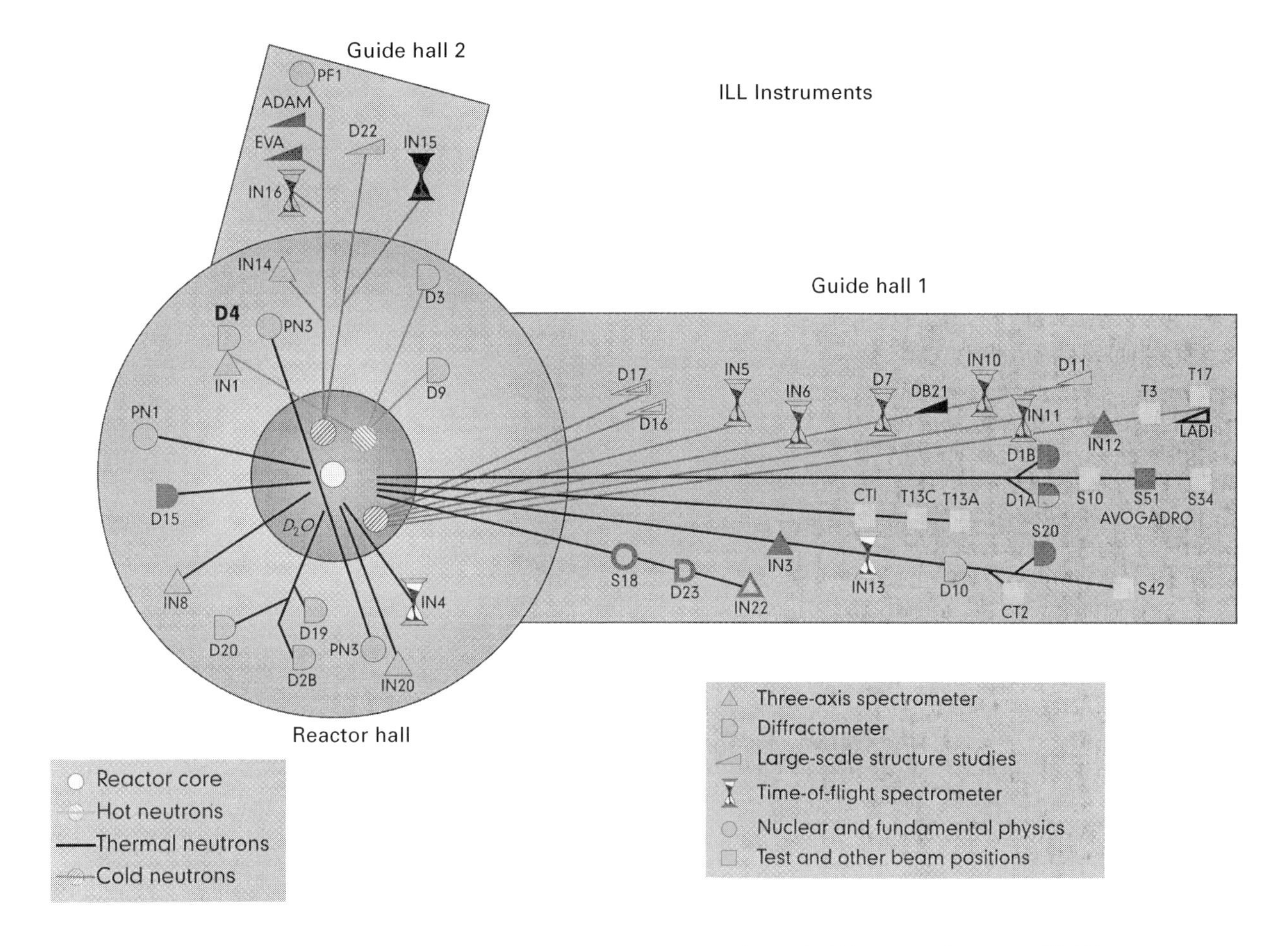

Figure 2 The layout of the reactor and the neutron scattering instrumentation at the ILL reactor. Courtesy of H. Büttner.

range of neutron wavelengths. In order to perform a neutron diffraction experiment it is thus necessary to monochromate the neutron beam from a reactor so that it covers a narrow range of neutron wavelengths and the vast majority of the flux from the source is lost at this stage.

Accelerator-based sources of neutrons

In more recent years neutron diffraction experiments have increasingly come to be performed using sources of neutrons which are based on a particle accelerator. A beam of charged particles is accelerated to a high energy and then fired at a target. Interactions between the particle beam and the nuclei in the target produce high energy neutrons which are then slowed down by a moderator.

The earlier accelerator-based neutron sources from the 1960s and 1970s use an electron linac to accelerate an electron beam to relativistic energies (~50 MeV). The electron beam is fired at a dense target made of a heavy element, usually uranium, and neutrons are produced by a two-stage process. First, the electrons are slowed down extremely rapidly due to the strong interaction with the electromagnetic field of the target nuclei, producing a cascade of bremsstrahlung photons. Secondly, some of these photons go on to produce neutrons by photoneutron reactions where the photon excites a target nucleus which subsequently decays with the emission of a neutron. About twenty electrons must be accelerated for each neutron produced.

More recent accelerator-based neutron sources use a linear accelerator, usually together with a synchrotron, to accelerate a beam of protons to a high energy (~800 MeV). The proton beam is fired at a heavy metal target (made, for example, of tantalum, uranium or tungsten) and neutrons are produced by the spallation process. Spallation is a violent interaction between the proton and the target nucleus which results primarily in the emission of neutrons, but also a variety of light nuclear fragments. In effect, the protons chip pieces off the target nuclei and each proton produces of the order of fifteen neutrons for a nonfissile target (or about twenty-five neutrons for a fissile target).

Accelerator-based sources are usually pulsed (typically with a pulse repetition rate of the order of

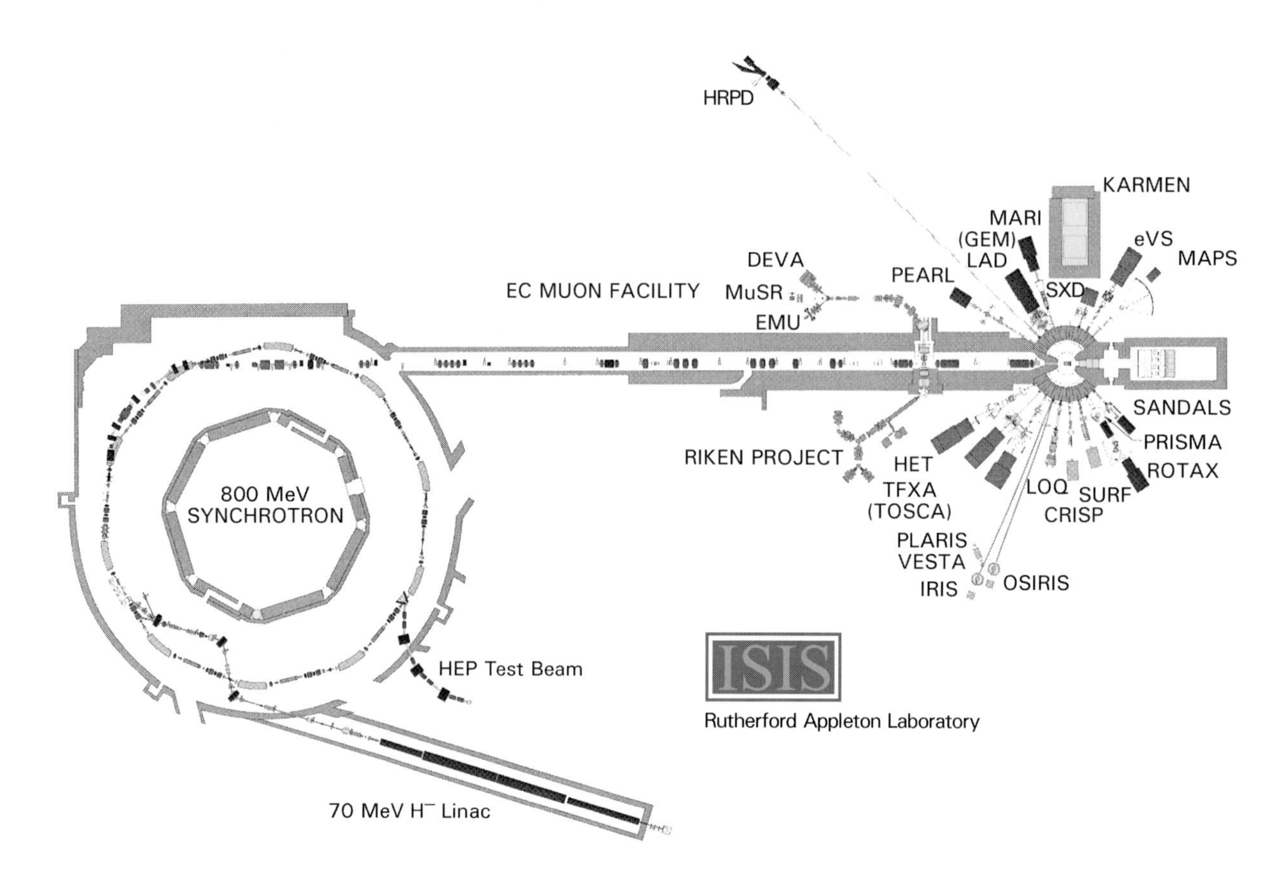

Figure 3 The layout of the neutron source and the neutron scattering instrumentation at the ISIS spallation neutron facility.

50 Hz) and so they produce a pulsed neutron flux which is ideally suited to the time-of-flight neutron diffraction technique. (Accelerator-based neutron sources which are quasi-steady-state or intensity-modulated also exist. Furthermore, pulsed neutrons have also been produced in Russia by using a pulsed reactor.) This technique involves measuring the time-of-flight, t, taken for a neutron to travel the total flight path, L, from the moderator to the detector, via the sample. On the assumption of elastic scattering (i.e. initial and final neutron energies are the same, $E_i = E_f$) then

$$t = \frac{m_n}{h} L\lambda \qquad [3]$$

(or $t = 252.82L\lambda$ in convenient units of μs, metres and Ångstroms, respectively) and it is thus straightforward to determine the neutron wavelength. The use of the time-of-flight technique removes the need to monochromate the neutron beam and thus, even though the raw flux produced initially by an accelerator-based source is much less than that produced by a reactor source, the final flux available for neutron diffraction is of a comparable order of magnitude (see **Figure 1**).

The moderator at an accelerator-based neutron source is used to slow the neutrons down so that they have suitable wavelengths for neutron diffraction, in the same way as for a reactor neutron source. However, in order that the moderation process does not broaden the pulsed time structure of the neutron flux too much, the moderator must be relatively small. (Also note that, unlike a reactor, the process of moderation plays no role in the production of neutrons at an accelerator-based source.) This has the consequence that the neutrons produced by an accelerator-based source are under-moderated and there are many more epithermal neutrons than for a reactor source. **Figure 1** also shows the neutron flux for a moderator at the world's most intense pulsed neutron source, the ISIS spallation neutron source at the Rutherford Appleton Laboratory, UK (see **Figure 3**).

Neutron detectors

The single most important component of a neutron diffractometer is the detector array. The lack of a charge on the neutron means that it cannot directly create an electrical signal in a detector. Instead the neutron must be made to undergo a nuclear reaction which subsequently leads to an electrical signal.

The three most important reactions for neutron detection involve the ^{3}He, ^{10}B and ^{6}Li nuclei:

$$\begin{aligned} {}^3_2\text{He} + {}^1_0\text{n} &\rightarrow {}^3_1\text{H} + {}^1_1\text{p} + 0.770\ \text{MeV} \\ {}^{10}_5\text{B} + {}^1_0\text{n} &\rightarrow {}^7_3\text{Li} + {}^4_2\text{He} + 2.3\ \text{MeV} \\ {}^6_3\text{Li} + {}^1_0\text{n} &\rightarrow {}^4_2\text{He} + {}^3_2\text{He} + 4.79\ \text{MeV} \end{aligned} \qquad [4]$$

Generally, the efficiency of detection of neutrons of wavelength, λ, by a detector of thickness x, containing a number density, n_v, of absorbing atoms with neutron absorption cross-section, $\sigma_{abs}(\lambda)$, may be expressed as

$$\text{Efficiency} = \varepsilon(1 - \exp(-n_v x \sigma_{abs}(\lambda))) \qquad [5]$$

where ε is the fraction of absorption events which result in an output pulse from the detector. Usually the absorption cross-section used in detection is proportional to the neutron wavelength (a so-called $1/v$ cross-section):

$$\sigma_{abs}(\lambda) = \lambda \sigma^0_{abs} \qquad [6]$$

where σ^0_{abs} is a constant. For good performance the ideal neutron detector should have a high neutron detection efficiency, a low intrinsic detector background, a low sensitivity to non-neutron events (particularly gamma rays), a good stability and a short dead-time.

Gas detectors

A simple gas detector consists of a sealed metal tube, which contains either ^{3}He or BF_3 gas under pressure (typically 10 bar; 10^6 Pa), with a high voltage thin wire anode along its axis (see **Figure 4A**). When a neutron is absorbed (Eqn [4]) the energetic recoil particles produce ionization in the gas. The anode then collects electrons from the ionization and an electrical signal is produced. Although BF_3 is cheaper than ^{3}He, it is little used now because it is less efficient and the gas is toxic. An advantage of using ^{3}He gas is that it is self-regenerating because the triton produced when a neutron is absorbed undergoes beta decay with a half-life of 12.33 years to produce another ^{3}He nucleus:

$${}^3_1\text{H} \rightarrow {}^3_2\text{He} + \text{e}^- + \bar{\nu}_e + 18.6\ \text{keV} \qquad [7]$$

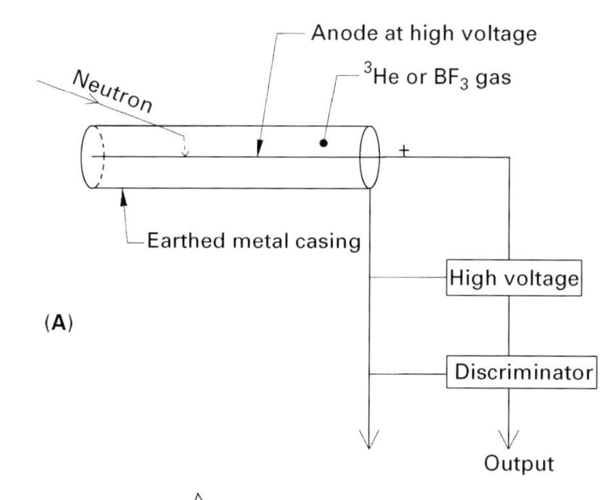

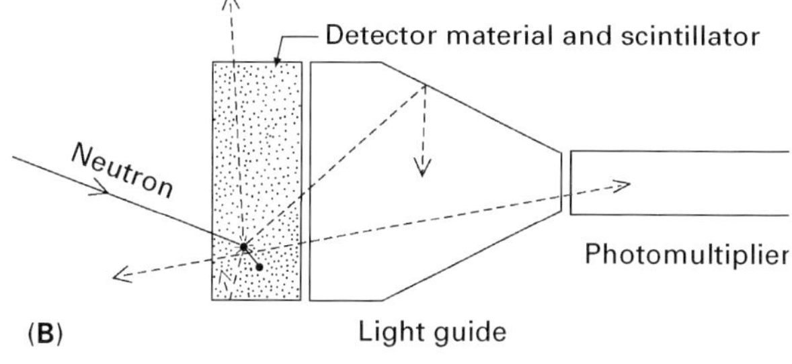

Figure 4 Schematic of (A) a gas detector and (B) a scintillator detector (modified with permission from Windsor CG (1981) *Pulsed Neutron Scattering*. London: Taylor and Francis).

Gas detectors have a pulse height distribution in which the signal due to gamma rays is well separated from the neutron signal. Hence, the detector electronics can be made to discriminate very well against gamma rays, leading to a very low gamma ray sensitivity (~10^{-8} at 1 MeV). Furthermore, gas detectors have a very low instrinsic detector background (at best ~3–5 counts per hour) and a good stability. In practice, modern neutron diffractometers often use multiwire detectors, where a large number of anodes are contained within a single envelope, in order to cover a larger solid angle and hence obtain a higher count rate. Also resistive-wire position-sensitive detectors are sometimes used for applications where it is useful to be able to determine the position along the length of the detector where a neutron is detected.

Gas detectors are relatively expensive and it is difficult to make the active element take up more complex geometries and shapes. Furthermore, it is difficult to make the active elements small, which is advantageous for good diffractometer resolution.

The most recent development in gas detector technology to be applied is the microstrip detector. In a microstrip detector the metal anode is replaced by a semiconducting glass plate on which very thin

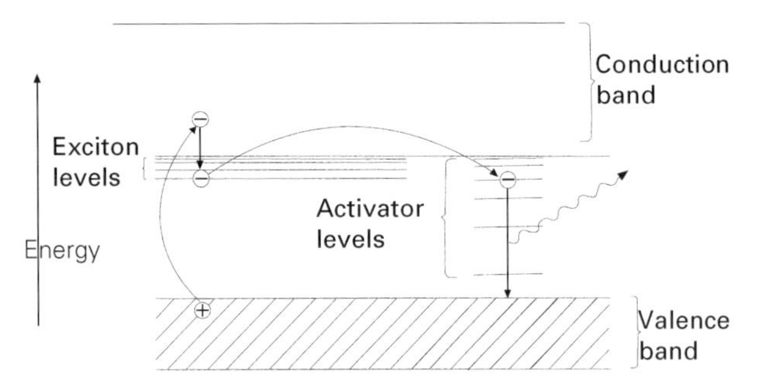

Figure 5 The physical principles behind the operation of a scintillator material. Reproduced with permission from Bennington SM, Hannon AC and Forsyth JB (1993) *Rutherford Appleton Laboratory Report* RAL-93-083.

metallic strips are engraved by photolithographic means. In this way, the limitations on the size and shape of gas detector elements may be overcome in the future.

Scintillator detectors

The aim with a scintillator detector is to combine the neutron absorbing element intimately with a scintillating phosphor so that the reaction products from the capture of a neutron strike the phosphor and produce a light flash, which is then detected by a photomultiplier tube (see **Figure 4B**).

There are two types of scintillator material in common use for neutron detection: ^{6}Li/Ce glass scintillator and granular ZnS(Ag)/^{6}Li scintillator. **Figure 5** illustrates the physical principles behind the operation of an ideal scintillator material. The absorption of a neutron by a ^{6}Li nucleus causes a large amount of energy to be deposited in the scintillator material, resulting in the excitation of electrons from the valence band into the conduction band. An excited electron can bind together with a hole to form an exciton which propagates through the lattice. When the exciton reaches an impurity activator atom (Ce for the glass scintillator, or Ag for the ZnS scintillator) it becomes trapped and then it recombines to emit a flash of light. Since the activator atoms have levels within the band gap of the scintillator material, the photon emitted has an energy less than the band gap and can pass through the material without absorption.

The earlier scintillator detectors used at ISIS were based on the lithium glass material. Such detectors have a high neutron detection efficiency (comparable to or even exceeding that of a gas detector), but also a very high gamma sensitivity (0.02 at 1 MeV), a large instrinsic detector background (~150 counts per minute) and a poor stability. Thus, neutron diffraction results obtained using lithium glass scintillator detectors can suffer from background and absolute normalization problems. More recent scintillator detectors are almost always based on ZnS. This gives a high neutron detection efficiency (comparable to, or even exceeding, that of a gas detector), a low gamma sensitivity (at best ~ 10^{-8} at 1 MeV), a low intrinsic detector background (at best ~ 12 counts per hour) and a much improved stability.

One of the advantages of scintillator detectors is that a high efficiency can be obtained in a small thickness of material because a higher density of absorbing atoms can be achieved than is possible with a gas detector (cf. Eqn [5]). This can be of particular importance for achieving a large efficiency for the detection of higher energy neutrons, for which the absorption cross-section is relatively small (cf. Eqn [6]). Another advantage of scintillator detectors is that a high degree of flexibility can be achieved in terms of the size and geometry of the active component. For example, this allows the detectors to be designed to be narrow and to follow the Debye–Scherrer cone so that the resolution width for the diffractometer is minimized.

Neutron diffraction instrumentation

Instrumentation – general principles

The purpose of a total neutron diffractometer is to measure the differential cross-section

$$\left(\frac{d\sigma}{d\Omega}\right)_{tot} = \frac{R_{tot}}{N\Phi d\Omega} \qquad [8]$$

where R_{tot} is the rate at which neutrons of wavelength λ are scattered into the solid angle dΩ in the direction $(2\theta, \phi)$ (see **Figure 6**), regardless of whether or not they are scattered elastically (i.e. total diffraction). N is the number of atoms in the sample and Φ is the flux of neutrons of wavelength λ which is incident on the sample.

In general, the differential cross-section depends upon the scattering vector

$$\boldsymbol{Q} = \boldsymbol{k}_i - \boldsymbol{k}_f \qquad [9]$$

where $\boldsymbol{k}_i$ and $\boldsymbol{k}_f$ are the neutron wave vector before and after scattering. The term momentum transfer (i.e. the momentum transferred to the sample) is commonly used for Q, although strictly speaking this term should be used for $\hbar Q$. Diffraction data

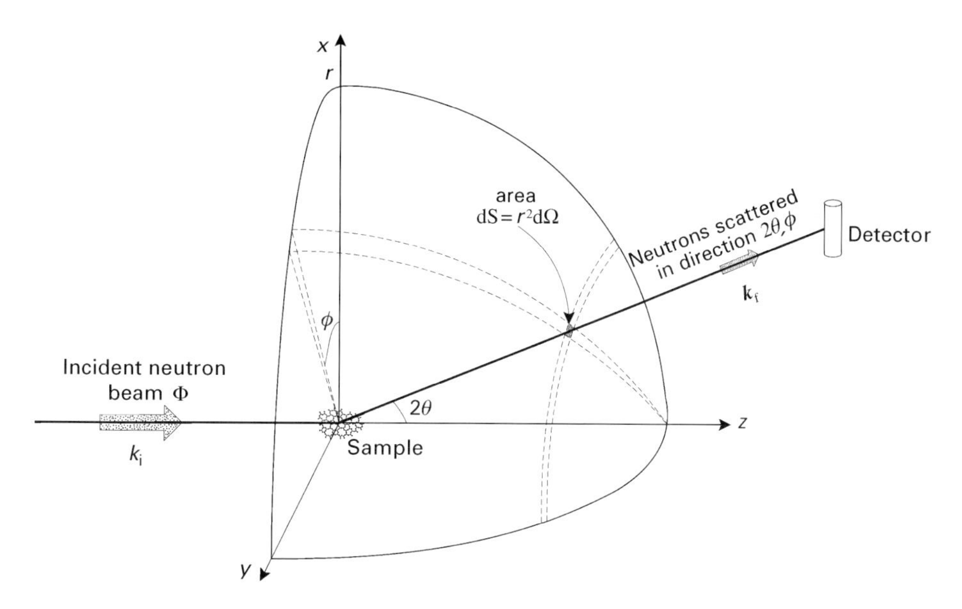

Figure 6 The geometry for a neutron diffraction experiment.

may be treated by considering the scattering to be totally elastic so that the magnitudes of the initial and final neutron wave vectors are the same,

$$|\boldsymbol{k}_{\mathrm{i}}| = |\boldsymbol{k}_{\mathrm{f}}| \qquad [10]$$

The most frequently used type of neutron diffraction involves the study of the atomic structure of isotropic samples (polycrystalline powders, glasses, liquids, etc.) and hence this article gives greatest emphasis to the instrumentation for this type of experiment. For an isotropic sample, it is only the magnitude of the momentum transfer which is important (the direction is not important), and in this case the differential cross-section is a function of a single variable

$$Q = |\boldsymbol{Q}| = \frac{4\pi \sin\theta}{\lambda} \qquad [11]$$

For neutron diffraction experiments on polycrystalline powders it is usually more convenient to treat the differential cross-section as a function of d-spacing, $d(=2\pi/Q)$, defined according to Bragg's law by

$$2d\sin\theta = \lambda \qquad [12]$$

Bragg peaks are then observed in the differential cross-section whenever the d-spacing satisfies

$$d = d_{hkl} \qquad [13]$$

where d_{hkl} is a d-spacing between atomic planes in the crystal for which the structure factor is non-zero.

In order to produce high quality data, a neutron diffractometer must satisfy several requirements: The data must have a good statistical accuracy, which is obtained by having a high count rate. This is achieved by such factors as a bright source and a large total detector solid angle. The corrections which must be made to the data must be small. In particular, the background must be small, featureless and unchanging. The range in Q must be as wide as possible, in order to provide high resolution in real-space. The reciprocal-space resolution must be as narrow as possible (see below).

Reactor source instrumentation

Figure 7 shows a schematic of the layout of a typical neutron diffractometer at a reactor source. The neutron beam coming from the moderator at a reactor covers a wide range of wavelengths and is unchanging with time. It is thus necessary to use a single crystal monochromator so as to produce a monochromatic beam. The general principle of operation of a neutron diffractometer at a steady state source is the same as for a conventional X-ray diffractometer. The differential cross-section is measured as a function of Q by moving the detector to different scattering angles, 2θ, and measuring the scattered count

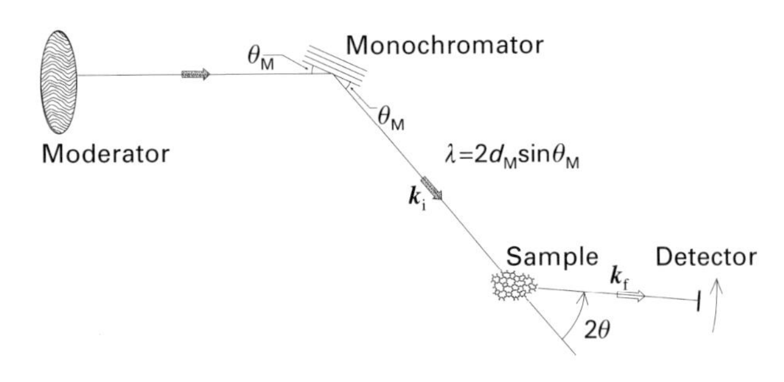

Figure 7 Schematic of a neutron diffractometer for a steady-state (reactor) source.

rate. That is to say, Q (or d) is scanned by varying 2θ whilst keeping the neutron wavelength λ constant (cf. Eqn [11]).

Figure 8 shows the D4 diffractometer at the ILL reactor which, for many years, has been the most successful reactor-based diffractometer for studying the structure of liquids and amorphous materials. The diffractometer uses neutrons from a hot graphite moderator at a temperature of 2400 K because this produces neutrons with short wavelengths and hence a high maximum Q can be achieved, leading to good real-space resolution. A copper monochromator is used to produce neutrons with a wavelength of 0.7 Å, 0.5 Å or 0.35 Å, depending upon which copper reflection is selected. Most of the neutron flight path is evacuated in order to minimize background due to the scattering of neutrons by air. Two 64-wire ^3He gas multidetectors are used in order to provide a large detector solid angle and hence a high count rate. The multidetectors are moved on compressed air across a marble floor in order to cover the full range of scattering angles. With a wavelength of 0.5 Å the Q-range of D4 extends from 0.3 to 24 Å^{-1}. (A higher maximum Q may be attained at $\lambda = 0.35$ Å, but the flux available at this wavelength is too low for regular use.) A development program is currently underway to replace the multidetectors with microstrip detectors, increasing the count rate by an order of magnitude.

Pulsed source instrumentation

Figure 9 shows a schematic of the layout of a typical neutron diffractometer at a pulsed neutron source. The time-of-flight technique (see Eqn [3]) is used to determine the wavelength of the detected neutrons and hence a monochromator is not needed. The differential cross-section is measured as a function of Q with the detector at a fixed scattering angle, 2θ, and Q (or d) is scanned by varying the neutron wavelength λ (cf. Eqn [11]). The time-of-flight technique is thus a dispersive technique and a white beam covering a wide range of wavelengths is incident on the sample. For powder diffraction, pulsed source data have the simplifying property that time-of-flight

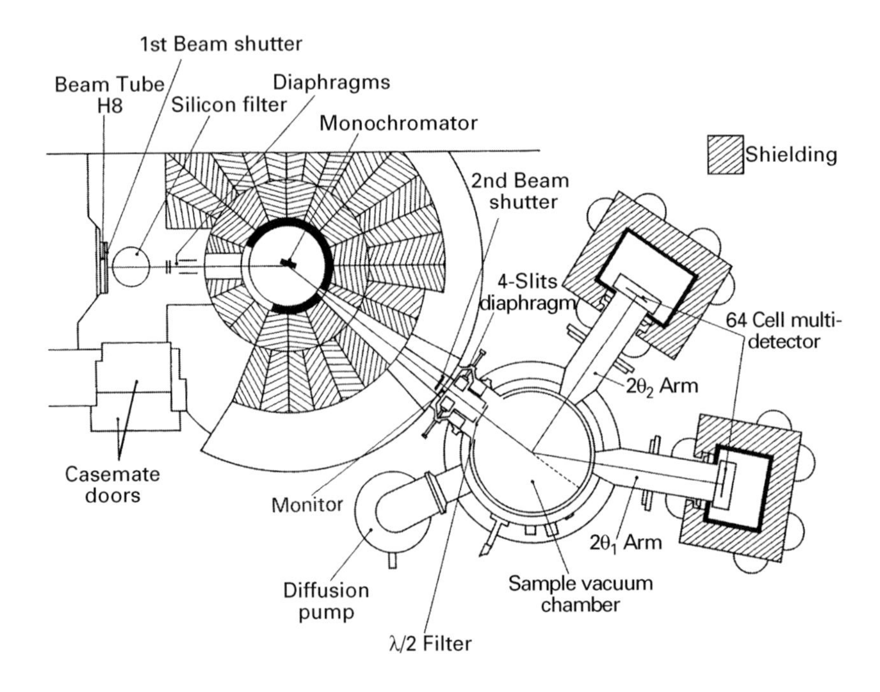

Figure 8 The D4 neutron diffractometer at the ILL reactor. Reproduced with permission from American Institute of Physics, Clare AG, Etherington G, Wright AC, et al. (1989) A neutron-diffraction and molecular-dynamics investigation of the environment of Dy^{3+} ions in a fluoroberyllate glass. *Journal of Chemical Physics* **91**: 6380–6392.

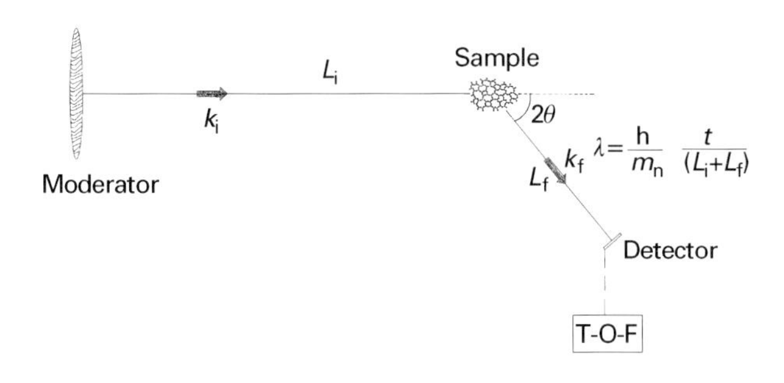

Figure 9 Schematic of a neutron diffractometer for a pulsed (accelerator-based) source.

is proportional to d-spacing;

$$t = \frac{2m_n}{h} Ld \sin\theta \qquad [14]$$

(or $t = 505.64Ld \sin\theta$ in convenient units of μs, metres and Ångstroms, respectively).

A noteworthy advantage of the ability to measure a full diffraction pattern at a single fixed scattering angle is that complex sample environment equipment can be used (e.g. for high pressures) with two well-collimated flight paths for the incident and scattered beams so that background from the equipment is minimized.

In the analysis of data from a pulsed source diffractometer, it is necessary to normalize out the flux distribution $\Phi(\lambda)$ arising from the moderator. This is done by measuring the time-of-flight spectrum for a vanadium standard, as well as for the sample. Vanadium is used for this purpose because the scattering from vanadium is almost completely incoherent and the Bragg peaks are extremely small (nevertheless, it is necessary to remove the Bragg peaks from the data by some kind of smoothing process). In practice, the differential cross-section for the sample is determined by performing the following operation with the measured time-of-flight spectra:

$$\text{Normalized sample spectrum} = \left(\frac{\text{sample} - \text{empty container background}}{\text{vanadium} - \text{background}}\right) \qquad [15]$$

(A fully corrected dataset will also have been corrected for such effects as detector dead-time, attenuation and multiple scattering.) **Figure 10** illustrates the effect of flux normalization for pulsed diffraction data, using data from the LAD diffractometer described below. The vanadium time-of-flight spectrum is closely related to the flux distribution $\Phi(\lambda)$ arising from the moderator. The upturn in $\Phi(\lambda)$ at short times is due to high energy epithermal neutrons whilst the broad peak at

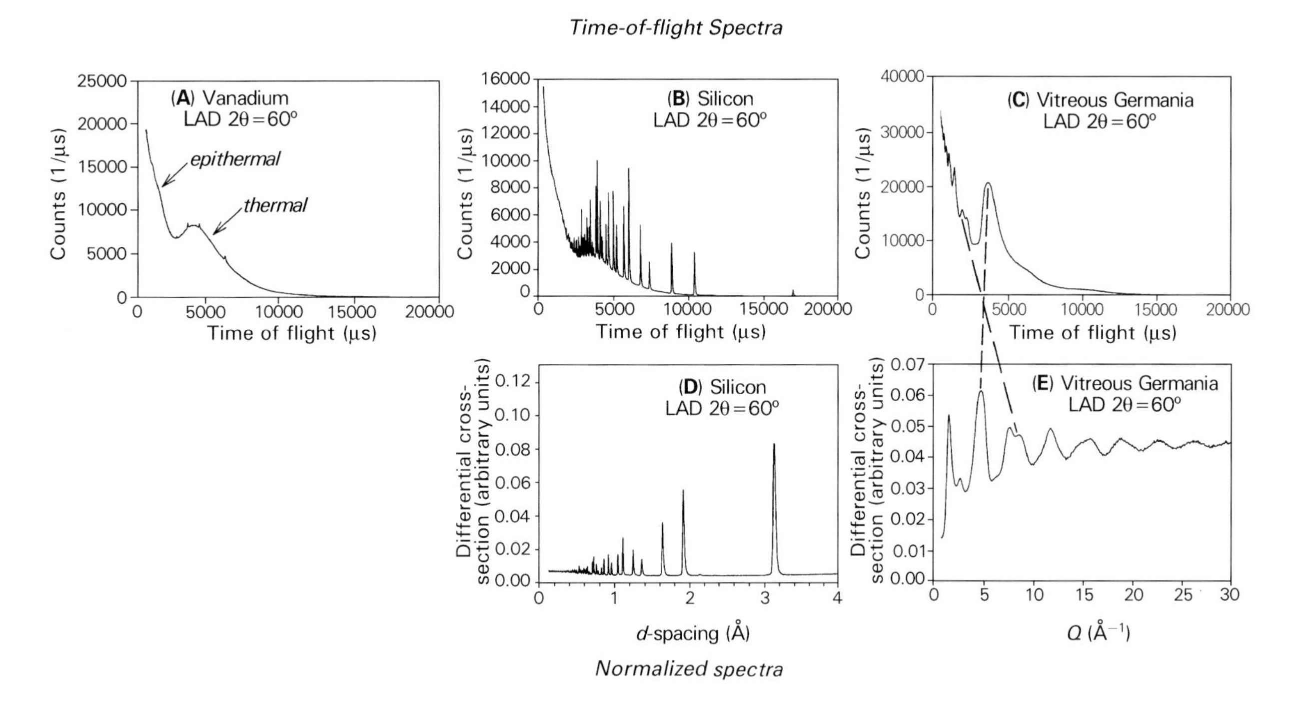

Figure 10 Time-of-flight spectra for (A) vanadium, (B) polycrystalline silicon and (C) vitreous germania. Also shown are the normalized spectra for (D) polycrystalline silicon and (E) vitreous germania.

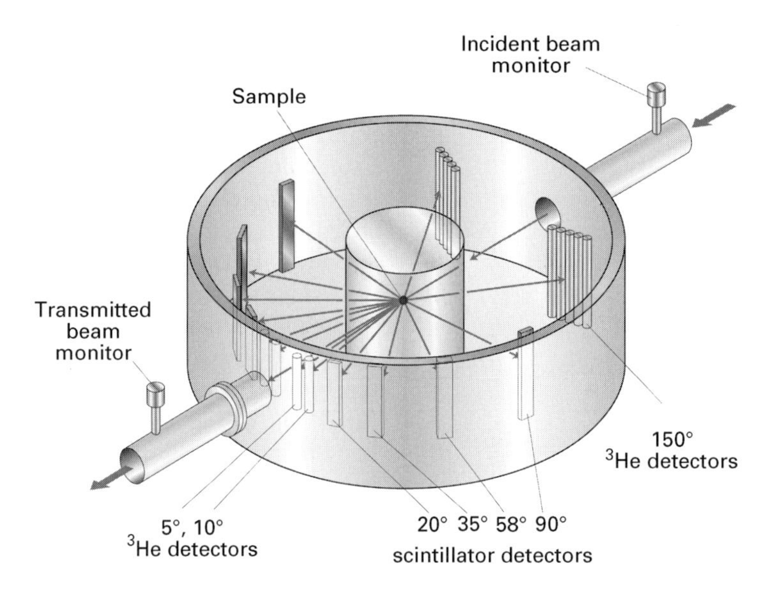

Figure 11 The LAD diffractometer at the ISIS spallation neutron source.

intermediate times is the peak of a Maxwellian distribution whose position depends on the moderator temperature (cf. Eqn [2]). The spectra measured on a reactor diffractometer do not need to be normalized to a flux distribution, but a vanadium standard may still be used in an experiment in order to achieve an absolute normalization of the differential cross-section.

Figure 11 shows the LAD (liquids and amorphous diffractometer) instrument at the ISIS spallation neutron source. LAD has been the main ISIS diffractometer for the study of disordered materials in the decade since the start of operations at ISIS and its design is typical for an early pulsed source diffractometer. The neutron beam comes from a liquid methane moderator at a temperature of 110 K. A cooled moderator is used in order to reduce the correction for inelasticity effects. The incident flight path, L_i, is 10.0 m long, leading to a moderately high reciprocal-space resolution. In practice, a pulsed source diffractometer has several different detector banks at different scattering angles so as to extend the Q-range of the data. LAD has detector banks at seven different scattering angles, some of which use ^{3}He gas detectors, whilst the others use lithium glass scintillator detectors. The detectors are all in the horizontal plane. In order to minimize background there are large amounts of shielding between the detector banks. Thus the solid angle subtended by the relatively low detector area is small and hence the count rate of the instrument is not very high. The best resolution is obtained in the backward angle detectors ($2\theta = 150°$) with $\Delta Q/Q$ ~0.5–0.6%.

Figure 12 shows the GEM (general materials) diffractometer at ISIS which is being constructed to replace the LAD diffractometer. This will be a state-of-the-art pulsed source diffractometer to be used for both non-crystalline diffraction and for powder crystallography. The incident flight path, L_i, will be relatively long at 17.0 m, giving a very small flight path contribution to the resolution. Thus the resolution in the backward angle detectors will be very good with $\Delta Q/Q$ ~0.2–0.3%. All of the detectors will be ZnS scintillators using narrow 5 mm active elements in order to minimize the angular contribution to the resolution. The detectors will cover a very large solid angle (area ~10 m^2, maximum azimuthal angle ~45°) so as to achieve a high effective count rate. A nimonic t_0 chopper at a distance 9.3 m from the moderator will be used to close off the beam at $t = 0$ and thus prevent very fast neutrons and prompt gamma rays from reaching the sample. This will prevent high energy neutrons from thermalizing in large pieces of sample environment equipment (e.g. high pressure) and then giving rise to a substantial background. In addition, two disc choppers will be used at flight paths of 6.5 m and 9.5 m to define a restricted wavelength range for the beam reaching the sample. This will be done so as to avoid frame overlap, which can be a significant problem for a diffractometer with a longer flight path. Frame overlap occurs when slower neutrons from a pulse of the source are overtaken by faster neutrons from the subsequent pulse. If the flux of the slower neutrons is significant, then a diffraction peak, which in reality is detected at the long time-of-flight, t, appears to be detected at the earlier time, $t - \tau_0$ (where τ_0 is the period of the source), and this leads to spurious peaks in the data.

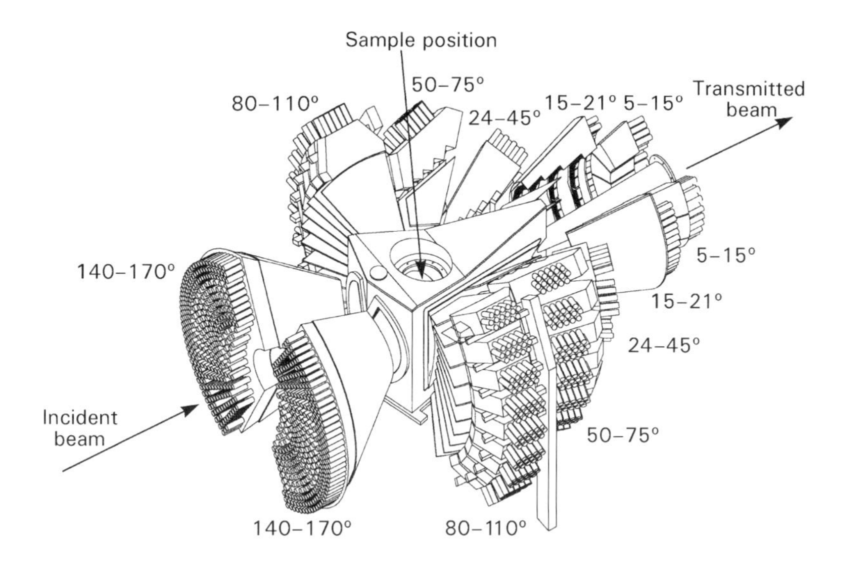

Figure 12 The new GEM diffractometer at the ISIS spallation neutron source.

Reciprocal-space resolution

A certain point in a diffraction pattern picks out the scattering at some momentum transfer Q. However, this Q is never completely defined due to uncertainties in the various parameters α_i of the diffractometer used to determine Q, leading to an overall uncertainty ΔQ. In practice the parameters α_i may be highly dependent, but for a simple discussion of Q-resolution it may be assumed that the parameters all act independently so that

$$\Delta Q = \left[\sum_i \left(\frac{\partial Q}{\partial \alpha_i} \Delta \alpha_i \right)^2 \right]^{1/2} \qquad [16]$$

The parameters which are important in determining the resolution will depend on the details of the particular diffractometer being used and thus only an outline of some of the general principles and dependencies for diffraction from an isotropic sample can be given here.

Resolution of a reactor diffractometer

For a reactor diffractometer the resolution width, ΔQ, arises because the acceptance angle of the collimation around the monochromator, the size of the sample and the acceptance angle of the detector system are all greater than zero and also because the monochromator crystal has a mosaic spread. For a treatment of the Q-resolution in terms of independent parameters, these factors may be considered to lead to a geometrical uncertainty in angle, $\Delta\theta$, and an uncertainty in wavelength, $\Delta\lambda$. Differentiating the definition of Q (Eqn [11]) thus gives

$$\frac{\Delta Q}{Q} = \frac{\Delta d}{d} = \left[(\cot\theta \Delta\theta)^2 + \left(\frac{\Delta\lambda}{\lambda} \right)^2 \right]^{1/2} \qquad [17]$$

where the semi-angle, θ, is in radians, not degrees.

Figure 13 shows the experimental Q-resolution of the D4 reactor diffractometer (measured at $\lambda = 0.7$ Å for powder samples of YIG, Si, Ni and Fe with diameter 7 mm and height 45 mm, and with the detector at 1.455 m). For the range of angle shown, $Q\cot\theta$ changes little and hence the Q-resolution may be expressed approximately as

$$\Delta Q = (C_\theta + C_\lambda Q^2)^{1/2} \qquad [18]$$

where C_θ and C_λ are constants. This simple independent-parameter expression predicts the general trend of **Figure 13**. However, the detailed behaviour, and in particular the upturn in ΔQ at low Q, is not predicted, and a full analysis of the Q-resolution must treat the parameters as being dependent.

The resolution of a reactor diffractometer may be improved by reducing the uncertainties in angle and wavelength. However, this can only be achieved at the expense of a corresponding decrease in the count rate of the diffractometer. For non-crystalline samples the Q-resolution is of less importance than for polycrystalline samples because the diffraction peaks are relatively broad (cf. **Figure 10**). The D4 diffractometer is used mainly to study non-crystalline samples and hence it has been

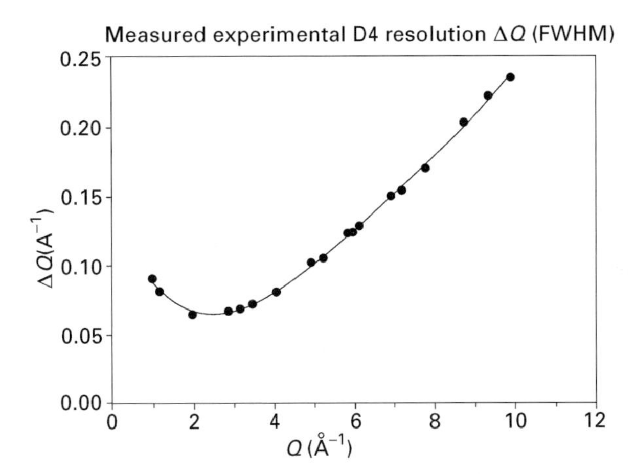

Figure 13 The measured experimental resolution (full width at half maximum) of the D4 reactor diffractometer. Courtesy of HE Fischer.

optimized primarily to have a high count rate. Thus the Q-resolution of D4 is not the best that can be achieved at a reactor neutron source.

Resolution of a pulsed source diffractometer

For a time-of-flight diffractometer the three major contributions to the Q-resolution are due to the geometrical uncertainty in angle, $\Delta\theta$, the uncertainty in flight path, ΔL, and the uncertainty in time-of-flight, Δt, leading to the following expression for the total combined resolution width:

$$\frac{\Delta Q}{Q} = \frac{\Delta d}{d} = \left[(\cot\theta\Delta\theta)^2 + \left(\frac{\Delta L}{L}\right)^2 + \left(\frac{\Delta t}{t}\right)^2\right]^{1/2} \quad [19]$$

The angular uncertainty for a time-of-flight diffractometer, $\Delta\theta$, is similar to the reactor case in that it arises from the acceptance angle of the collimation in front of the moderator, the size of the sample and the acceptance angle of the detector. However, there is a major difference in its effect because of the way in which a single detector measures a complete diffraction pattern without moving to different scattering angles, 2θ, so that the angular contribution to $\Delta Q/Q$ is independent of Q. This contribution to the resolution is ideally suited to powder diffraction because it has a Δd which is small for short d-spacing where Bragg peaks are very close together, and only becomes large at high d-spacing where Bragg peaks are well separated.

The flight path uncertainty, ΔL, arises from the finite sizes of the moderator, sample and detector. Its contribution to the resolution is symmetric and may be approximated by a Gaussian, as is also the case for the angular contribution. The contribution to $\Delta d/d$ due to the flight path uncertainty may be minimized simply by using a very long flight path, L, although this may only be achieved at the expense of a corresponding decrease in the count rate.

At low and medium scattering angles, $\cot\theta$ is relatively large and the angular contribution to the resolution tends to dominate. However, at backward angles ($2\theta \rightarrow 180°$) this contribution becomes very small so that a diffraction pattern can be measured with a very narrow angular resolution across its entire range. The time-of-flight uncertainty, Δt, is then of greater importance. Its main component is the time uncertainty which arises from the moderation process. This gives an exponential contribution to the resolution with a decay time, τ, which is the mean time spent in the moderator for neutrons of a particular wavelength. **Figure 14** shows the resolution of the LAD time-of-flight diffractometer at backward

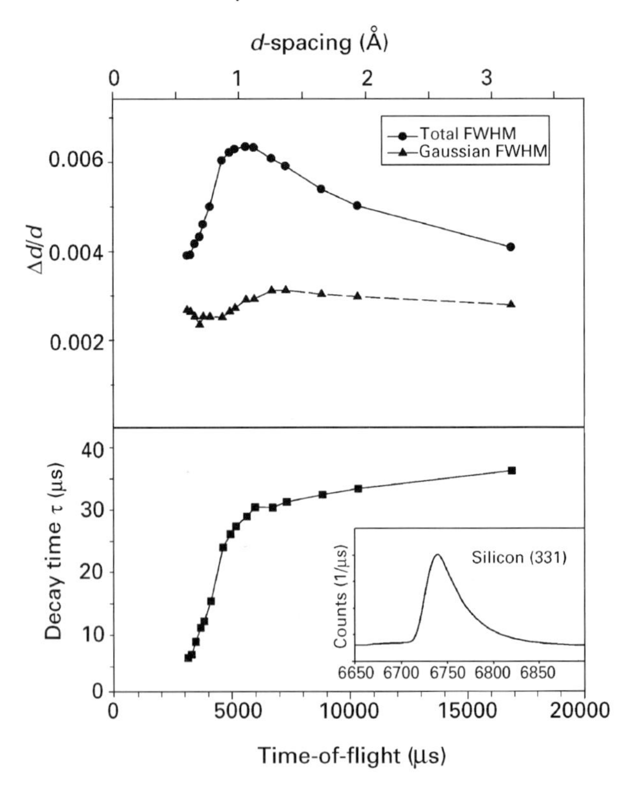

Figure 14 The measured experimental resolution of the LAD pulsed neutron diffractometer at backward angle, $2\theta = 146°$. The total full width at half maximum (FWHM) and the FWHM for the Gaussian contribution are shown in the form of $\Delta d/d$. Also shown is the decay constant τ for the exponential contribution to the resolution. The inset shows the peak shape measured for a typical Bragg peak (the (331) reflection for polycrystalline silicon).

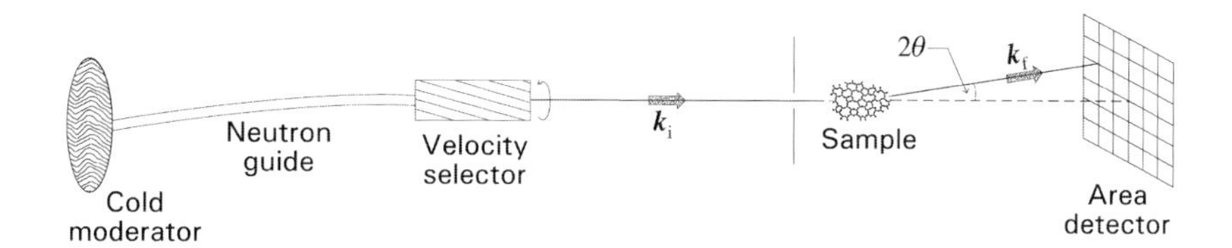

Figure 15 Schematic of a small angle neutron scattering (SANS) diffractometer for a steady-state (reactor) source.

angle, obtained by fitting the convolution of a Gaussian and an exponential to Bragg peaks in experimental data. The Gaussian contribution to $\Delta d/d$ is almost independent of d-spacing, as predicted. The detailed behaviour of the decay time, τ, depends upon the design of the moderator, but is generally small at short times (i.e. for high energy neutrons) and tends to a large value at long times (i.e. for low energy neutrons). The overall peak shape, Gaussian-exponential convolution (see inset to **Figure 14**), has a very sharp leading edge which can be advantageous in resolving overlapping peaks.

Some other types of neutron diffraction instrumentation

In this article the greatest emphasis is given to wide angle neutron diffraction from isotropic samples. However, a range of other neutron diffraction techniques is also available and it is only possible to describe some of them briefly here. (One of the major advantages of neutrons is that polarized beams can be used and these are especially useful for the study of magnetic structures. However, a discussion of polarized neutron diffraction instrumentation is beyond the scope of this article.)

Small angle neutron scattering at a reactor

A small angle neutron scattering (SANS) diffractometer is used to study structures whose dimensions range from 10 to 1000 Å (e.g. defects, voids, micelles, macromolecules, etc.) and as such their Q-ranges are in the region of 0.005 to 0.5 Å^{-1}. In order to access these low values of Q the detector should be at low scattering angle, 2θ, and a cold moderator should be used so as to provide long wavelength neutrons (cf. Eqn [11]).

Figure 15 shows a schematic of the layout of a typical small angle neutron scattering (SANS) diffractometer at a reactor source. A large two-dimensional multiwire area detector is placed at a low angle so as to cover as large a solid angle as possible and thus maximize the count rate. The detector is either placed symmetrically around the transmitted beam (as shown in **Figure 15**) to maximize the count rate for the lowest Q, or is placed off-axis to extend the Q-range upwards. The neutron beam is monochromated by a velocity selector which consists of a rotating cylinder with a set of helical grooves cut into its outer surface. The angular contribution to the resolution is unavoidably large at low angle because cot θ becomes large (cf. Eqn [17]). Hence the monochromation produced by the velocity selector is relatively coarse ($\Delta\lambda/\lambda$~5–20%) so as to match the angular contribution and provide as large an incident flux as possible. The D11 diffractometer at the ILL a prime example of a reactor SANS instrument.

Single crystal diffraction at a plused source

In an experiment on a single crystal sample the differential cross-section must be measured as a function of the magnitude and the direction of the momentum transfer vector, $\mathbf{Q}$. Bragg peaks are then observed in the differential cross-section whenever $\mathbf{Q}$ satisfies

$$\mathbf{Q} = \boldsymbol{\tau} \qquad [20]$$

where $\boldsymbol{\tau}$ is a reciprocal lattice vector of the crystal.

Figure 16 shows a schematic of the layout of a typical pulsed source single crystal diffractometer. A goniometer is used to provide full control of the orientation of the sample in space so that any region of reciprocal-space of interest can be studied. An area

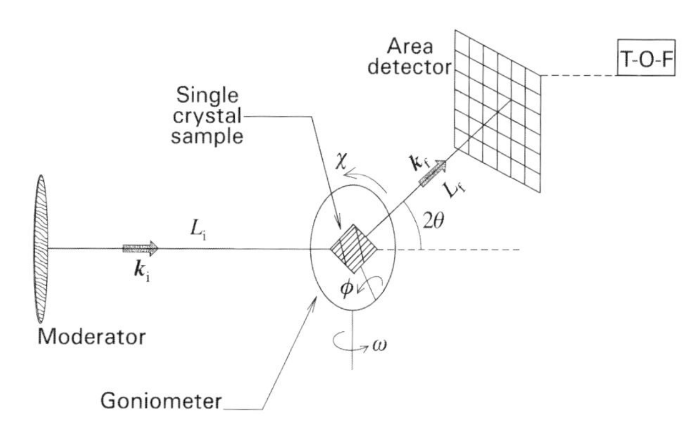

Figure 16 Schematic of a single crystal diffractometer on a pulsed (accelerator-based) source.

detector combined with the use of multiwavelength data provides a three-dimensional sampling of reciprocal-space that may contain hundreds of Bragg reflections. This wide sampling for a pulsed source diffractometer is advantageous because it may reveal unexpected Bragg reflections or satellites and can show diffuse scattering between the Bragg peaks. The SXD diffractometer at ISIS is a prime example of a pulsed source single crystal diffractometer.

List of symbols

C_x = a constant; d = d-spacing; d_{hkl} = interlayer spacing between crystal planes; $(d\sigma/d\Omega)_{tot}$ = differential cross-section for total neutron diffraction; E = neutron energy; E_f = neutron energy after scattering (final); E_i = neutron energy before scattering (initial); h = Planck's constant; $\hbar$ = Planck's constant divided by 2π; k_B = Boltzmann's constant; $\mathbf{k}_f$ = neutron wavevector after scattering (final); $\mathbf{k}_i$ = neutron wavevector before scattering (initial); L = total flight path of neutron; L_f = flight path of neutron after scattering (final); L_i = flight path of neutron before scattering (initial); m_n = neutron mass; N = number of atoms in sample; n_v = atomic number density; Q = magnitude of momentum transfer; $\mathbf{Q}$ = momentum transfer vector; R_{tot} = rate at which neutrons are scattered into the specified solid angle; t = time-of-flight of neutron; T = temperature; v = neutron speed; x = detector thickness; α_i = diffractometer parameter; Δd = resolution uncertainty for d-spacing, d; ΔQ = resolution uncertainty for momentum transfer, Q; $\Delta\theta$ = resolution uncertainty for semi-scattering angle, θ; $\Delta\lambda$ = resolution uncertainty for neutron wavelength, λ; ε = output pulse fraction for detector; 2θ = scattering angle; λ = neutron wavelength; $\sigma_{abs}(\lambda)$ = neutron absorption cross-section at neutron wavelength, λ; σ^0_{abs} = neutron absorption cross-section at neutron wavelength of 1 Ångstrom; $\boldsymbol{\tau}$ = reciprocal lattice vector of crystal; τ = exponential decay time; τ_0 = period of the pulsed source; ϕ = polar coordinate; Φ = neutron flux; Ω = solid angle.

See also: **Inelastic Neutron Scattering, Applications; Inelastic Neutron Scattering, Instrumentation; Inorganic Compounds and Minerals Studied Using X-Ray Diffraction; Materials Science Applications of X-Ray Diffraction; Neutron Diffraction, Theory; Powder X-Ray Diffraction, Applications; Structure Refinement (Solid State Diffraction).**

Further reading

Bacon GE (1969) *Neutron Physics*. London: Wykeham.

Bacon GE (1975) *Neutron Diffraction*, 3rd edn. Oxford: Oxford University Press.

Brown PJ and Forsyth JB (1973) *The Crystal Structure of Solids*. London: Edward Arnold.

Carpenter JM and Yelon WB (1986) Neutron Sources. In: (eds) Sköld K and Price DL. *Neutron Scattering, Part A*, pp 99–196. Orlando: Academic Press.

Egelstaff PA (ed.) (1965) *Thermal Neutron Scattering*. London: Academic Press.

Newport RJ, Rainford BD and Cywinski R (eds) (1988) *Neutron Scattering at a Pulsed Source*. Bristol: Adam Hilger.

Stirling GC (1973) Experimental techniques. In: Willis BTM (ed.) *Chemical Applications of Thermal Neutron Scattering*, pp 31–48. London: Oxford University Press.

Williams WG (1988) *Polarized Neutrons*. Oxford: Clarendon Press.

Willis BTM (1970) *Thermal Neutron Diffraction*. Oxford: Clarendon Press.

Windsor CG (1981) *Pulsed Neutron Scattering*. London: Taylor and Francis.

Windsor CG (1986) Experimental techniques. In: Sköld K and Price DL (eds) *Neutron Scattering, Part A*, pp 197–257. Orlando: Academic Press.

Neutron Diffraction, Theory

Alex C Hannon, Rutherford Appleton Laboratory, Didcot, UK

HIGH ENERGY SPECTROSCOPY
Theory

The cross-section measured in a neutron scattering experiment on a non-magnetic material depends in general upon a correlation function involving the positions in time and space of the atomic nuclei in the sample. For a total neutron diffraction experiment the diffraction pattern depends upon a correlation function involving the instantaneous interatomic vectors between the nuclei. However, for an elastic diffraction experiment the diffraction pattern depends upon a correlation function involving the time-averaged interatomic vectors between the positions of the nuclei. The correlation function is obtained by performing a suitable Fourier transform of the diffraction pattern and its features indicate which interatomic distances occur more or less frequently in the sample. The neutron correlation function is thus a powerful tool for the study of the atomic structure of glasses, liquids and disordered crystalline materials. For a crystalline sample the symmetry of the atomic structure gives rise to sharp Bragg peaks in the diffraction pattern involving only elastically scattered neutrons. The positions of the Bragg peaks and their systematic absences depend upon the symmetry of the crystal lattice. Meanwhile the intensities of the Bragg peaks depend upon the positions of the atoms in the unit cell. A single crystal neutron diffraction experiment yields both the symmetry of the lattice and the contents of the unit cell, whilst a powder neutron diffraction experiment is usually only used to reveal the unit cell contents for a crystal whose symmetry is already known.

Fundamental theory

The interaction between a neutron beam and a sample

If a sample is placed in a beam of neutrons then some of the neutrons will be removed from the transmitted beam due to scattering. There are two interactions which give rise to this scattering. The first of these is the nuclear force between a neutron and the nuclei of the sample. The second interaction is that between the magnetic moment of the neutron and the unpaired electrons of magnetic ions in the sample. The treatment of magnetic scattering (and polarized neutron scattering) is beyond the scope of this article. (There are two key differences between magnetic and nuclear scattering: magnetic scattering involves a form factor, and it also has a directional dependence (on magnetic moment) which does not arise for nuclear scattering). The basic properties of the neutron and their consequences for diffraction are summarized in Table 1.

Table 1 The properties of the neutron–sample interaction and their consequences for diffraction

Property	*Consequences*
No form factor for scattering	Data can be measured to high momentum transfer, leading to high real-space resolution
	Absolute normalization of results is relatively straightforward
Scattering length is different for different isotopes	Isotopic substitution can be used to extract element-specific information
Scattering length varies haphazardly across the periodic table	The positions of light atoms (especially hydrogen) can be determined in the presence of heavy atoms
	The positions of atoms of elements which are close in the periodic table can be distinguished
Interaction is relatively weak	Neutrons are highly penetrating and hence results are characteristic of the bulk of a sample, not the surface
	Bulky equipment can be used to subject the sample to a wide range of environments
Magnetic interaction between neutron and magnetic ions	Magnetic structures can be determined

Neutron cross sections

Consider a monochromatic beam of neutrons of energy E_i, wave vector $\boldsymbol{k}_i$ and flux Φ which is incident on a sample of N atoms. The scattering of the neutrons may then be defined experimentally by the double differential cross section:

$$\frac{d^2\sigma}{d\Omega\, dE} = \frac{R_{inel}}{N\Phi\, d\Omega\, dE} \qquad [1]$$

where R_{inel} is the number of neutrons scattered per unit time into the solid angle $d\Omega$ in the direction $(2\theta, \phi)$ (see **Figure 1**) with final energy E_f in the range E_i–E to E_i–$(E + dE)$. The energy E $(= E_i$–$E_f)$ is that transferred by the neutron to the sample in a given scattering event. The double differential cross-section is the fundamental quantity for the scattering of neutrons and the cross-section for a given experimental arrangement (e.g. diffraction or transmission) is obtained from it by performing suitable integration(s).

In a total diffraction experiment all scattered neutrons are detected regardless of their final energy. In this case the relevant cross-section is the differential cross-section defined by

$$\left(\frac{d\sigma}{d\Omega}\right)_{tot} = \frac{R_{tot}}{N\Phi\, d\Omega} = \int_{-\infty}^{\infty} \left(\frac{d^2\sigma}{d\Omega\, dE}\right) dE \qquad [2]$$

where R_{tot} is the number of neutrons scattered per unit time into the solid angle $d\Omega$ in the direction $(2\theta, \phi)$. This is to be contrasted with the case of elastic diffraction where only neutrons whose energy is unchanged (i.e. they are scattered elastically and have $E = 0$) are detected. In this case the relevant cross-section is

$$\left(\frac{d\sigma}{d\Omega}\right)_{el} = \frac{R_{el}}{N\Phi\, d\Omega} = \left.\frac{d^2\sigma}{d\Omega\, dE}\right|_{E=0} \qquad [3]$$

where R_{el} is the number of neutrons scattered elastically per unit time into the solid angle $d\Omega$ in the direction $(2\theta, \phi)$. The elastic differential cross-

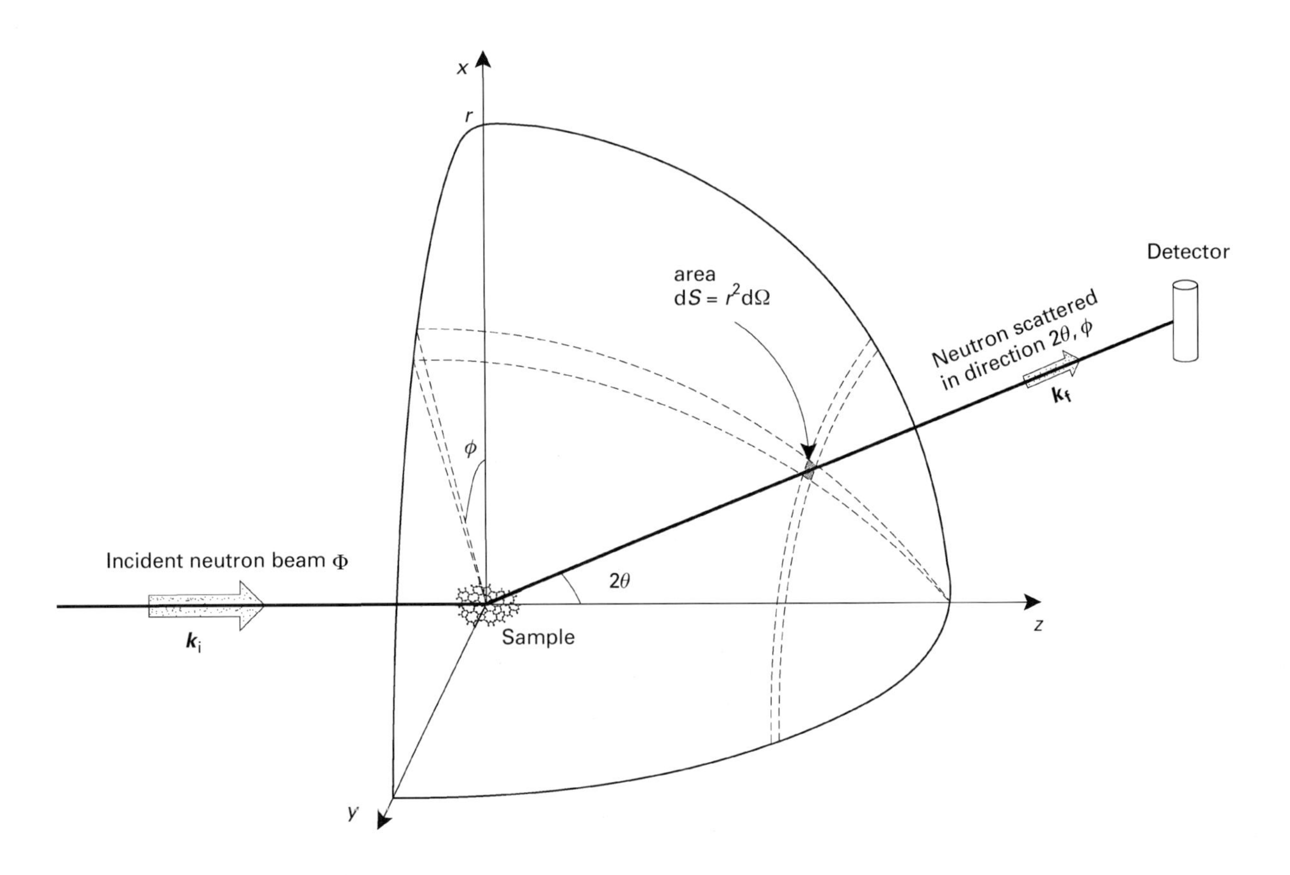

Figure 1 Geometry for a neutron scattering experiment.

section is given by evaluating the double differential cross-section at $E = 0$.

Neutron scattering length

To perform a neutron diffraction experiment, the neutron wavelength needs to be of similar magnitude to interatomic distances, say of the order of 1 Ångström (1 Å ≡ 10^{-10} m – Ångström units are almost universally used in the field of diffraction). This corresponds to a neutron energy of 25 meV, and thus the energies of neutrons used for diffraction are very much lower than typical X-ray energies used in diffraction. Nuclear forces are very short range (~10^{-15} m), operating over much shorter distances than neutron wavelengths. Hence nuclei can be treated as point-like scattering centres which give rise to an isotropic scattered neutron wave (i.e. only s-wave scattering need be considered). The wavefunction of a neutron scattered by a single nucleus at the origin is thus expressed as

$$\Psi_{\mathrm{f}} = -\frac{b}{r}\exp(\mathrm{i}kr) \qquad [4]$$

where k is the magnitude of the wave vector for the scattered wave and r is a polar coordinate. b is a constant, known as the (bound atom) scattering length, which has units of length. A positive scattering length corresponds to a phase change of π between the scattered and incident waves. Most scattering lengths are positive, but there is a minority of cases where the scattering length is negative. The neutron scattering length corresponds to the form factor in X-ray diffraction, but with the important difference that it is a simple constant without any wave vector dependence.

Fermi pseudo potential and the general expression for cross-section

The interaction between a neutron and a sample may be represented by the Fermi pseudo potential:

$$V(\boldsymbol{r}) = \frac{2\pi\hbar^2}{m_{\mathrm{n}}}\sum_{j=1}^{N} b_j\delta(\boldsymbol{r} - \boldsymbol{R}_j) \qquad [5]$$

where m_{n} is the neutron mass, and the summation is taken over the N nuclei whose position vectors and scattering lengths are $\boldsymbol{R}_j$ and b_j respectively. Use of the Fermi pseudo potential together with Fermi's golden rule (sometimes termed the Born approximation) eventually results in the following general expression from which all results for the nuclear scattering of neutrons may be derived:

$$\frac{\mathrm{d}^2\sigma}{\mathrm{d}\Omega\,\mathrm{d}E} = \frac{1}{N}\frac{k_{\mathrm{f}}}{k_{\mathrm{i}}}\frac{1}{2\pi\hbar}\sum_{j,j'} b_j b_{j'}\int_{-\infty}^{\infty}\left\langle\exp\{-\mathrm{i}\boldsymbol{Q}.[\boldsymbol{R}_j(0) - \boldsymbol{R}_{j'}(t)]\}\right\rangle\exp(-\mathrm{i}Et/\hbar)\mathrm{d}t \qquad [6]$$

The summations j and j' are both taken over the N nuclei of the sample and the angular brackets denote a thermal average at the temperature T of the sample. $\boldsymbol{R}_j(t)$ is the position of the jth nucleus at time t and $\boldsymbol{Q}$ is the scattering vector, defined by

$$\boldsymbol{Q} = \boldsymbol{k}_{\mathrm{i}} - \boldsymbol{k}_{\mathrm{f}} \qquad [7]$$

where $\boldsymbol{k}_{\mathrm{i}}$ and $\boldsymbol{k}_{\mathrm{f}}$ are the incident and scattered neutron wave vectors respectively. The term momentum transfer (i.e. the momentum transferred to the sample) is commonly used for $\boldsymbol{Q}$, although strictly speaking this term should be used for $\hbar\boldsymbol{Q}$.

Total diffraction

The static approximation

In the static approximation it is assumed that the incident neutron energy E_{i} is large compared with the excitation energies of the sample. This means that $k_{\mathrm{f}} \approx k_{\mathrm{i}}$ so that the scattering vector may be taken to have its elastic value:

$$Q = |\boldsymbol{Q}| = 2k\sin\theta = \frac{4\pi\sin\theta}{\lambda} \qquad [8]$$

where λ is the wavelength of the neutrons. In the static approximation Equation [6] may be integrated according to Equation [2] to obtain Equation [9]. (In practice the assumptions inherent in the static approximation do not hold exactly and as a consequence an inelasticity correction must be made to experimental total diffraction data).

$$\left(\frac{\mathrm{d}\sigma}{\mathrm{d}\Omega}\right)_{\mathrm{tot}} = I(\boldsymbol{Q}) = \frac{1}{N}\sum_{j,j'} b_j b_{j'} \times\left\langle\exp\{-\mathrm{i}\boldsymbol{Q}.[\boldsymbol{R}_j(0) - \boldsymbol{R}_{j'}(0)]\}\right\rangle \qquad [9]$$

Thus the total diffraction pattern $I(Q)$ depends upon the vectors $R_j - R_{j'}$ between the atoms, with a weighting according to scattering length. This is the basis for the use of total diffraction to measure the atomic structure of a sample. Total diffraction depends upon the positions of the atoms in the sample at an arbitrary time zero. Thus total diffraction yields an instantaneous 'snapshot' of the interatomic vectors.

Coherent and incoherent scattering

For a consideration of coherent and incoherent scattering it is convenient to write Equation [9] in the form

$$I(\boldsymbol{Q}) = \sum_{j,j'} b_j b_{j'} \langle j, j' \rangle \qquad [10]$$

The value of b_j is not the same for all the nuclei of a single element, owing to spin and isotopic incoherence, as will be explained below. Hence, to obtain a useful result, Equation [10] is averaged over all possible distributions of scattering length, making the assumption that there is no correlation between the values of b_j for any two nuclei. Now the average value of $b_j b_{j'}$ differs, depending on whether $j = j'$ is satisfied;

$$\begin{aligned} \overline{b_j b_{j'}} &= \left(\overline{b}\right)^2, \; j \neq j' \\ \overline{b_j b_{j'}} &= \overline{b^2}, \quad j = j' \end{aligned} \qquad [11]$$

where $\overline{b}$ is the average scattering length (usually known as the coherent scattering length) for all nuclei of a particular element, whilst $\overline{b^2}$ is the average of the squared scattering length for the relevant element. The double summation of Equation [10] may thus be separated into $j \neq j'$ (distinct) terms and $j = j'$ (self) terms:

$$I(\boldsymbol{Q}) = i(\boldsymbol{Q}) + \sum_l c_l \overline{b_l^2} \qquad [12]$$

where $c_l = N_l$, N is the atomic fraction for element l and the distinct differential cross-section is

$$i(\boldsymbol{Q}) = \sum_{l,l'} \overline{b_l}\,\overline{b_{l'}} \sum_{\substack{j=1 \\ j \neq j'}}^{N_l} \sum_{j'=1}^{N_{l'}} \frac{1}{N} \times \left\langle \exp\{-\mathrm{i}\boldsymbol{Q}.\left[\boldsymbol{R}_j(0) - \boldsymbol{R}_{j'}(0)\right]\right\rangle \qquad [13]$$

The l and l' summations are over the elements in the sample (e.g. for $Li_2Si_2O_5$, l = Li, Si, O). The j (or j') summations are then over all the N_l (or N_l') atoms of element l (or l'), excluding terms where j and j' refer to the same atom. By the algebraic trick of adding and subtracting a $\Sigma_j \overline{b}_j^2 \langle j, j \rangle$ term the differential cross-section for total diffraction may be separated into its coherent and incoherent parts:

$$I(\boldsymbol{Q}) = \left\langle \overline{b}^2 \right\rangle_{\mathrm{av}} S(\boldsymbol{Q}) + \frac{\sigma_{\mathrm{inc}}}{4\pi} \qquad [14]$$

where the structure factor is given by

$$S(\boldsymbol{Q}) = \frac{1}{\left\langle \overline{b}^2 \right\rangle_{\mathrm{av}}} \sum_{l,l'} \overline{b_l}\,\overline{b_{l'}} \sum_{j,j'} \frac{1}{N} \langle \exp\{-\mathrm{i}\boldsymbol{Q}.[\boldsymbol{R}_j(0) - \boldsymbol{R}_{j'}(0)]\} \rangle = 1 + i(\boldsymbol{Q}) \Big/ \left\langle \overline{b}^2 \right\rangle_{\mathrm{av}} \qquad [15]$$

in which self terms ($j = j'$) are now included in the summation. The incoherent cross-section of the sample is

$$\sigma_{\mathrm{inc}} = 4\pi \left(\left\langle \overline{b^2} \right\rangle_{\mathrm{av}} - \left\langle \overline{b}^2 \right\rangle_{\mathrm{av}} \right) \qquad [16]$$

in which average values for the sample are defined as

$$\begin{aligned} \left\langle \overline{b}^2 \right\rangle_{\mathrm{av}} &= \sum_l c_l \overline{b}_l^2 \\ \left\langle \overline{b^2} \right\rangle_{\mathrm{av}} &= \sum_l c_l \overline{b_l^2} \end{aligned} \qquad [17]$$

($4\pi \langle \overline{b^2} \rangle_{\mathrm{av}}$ is the coherent cross-section for the sample). The interpretation of the coherent differential cross-section $\langle \overline{b^2} \rangle_{\mathrm{av}} S(Q)$ is that this is what would be measured from a sample for which all nuclei of element l had a scattering length of $\overline{b}_l$. **Figure 2** shows the total diffraction pattern $I(Q)$ for a glass (for an isotropic sample, such as a glass, the direction of the Q vector is not important). It is the coherent contribution to the differential cross-section that contains interference information relating to the positions of atoms in the sample. Also shown in **Figure 2** are the self and incoherent contributions – these are featureless and contain no interference information. For most elements the incoherent cross-section is

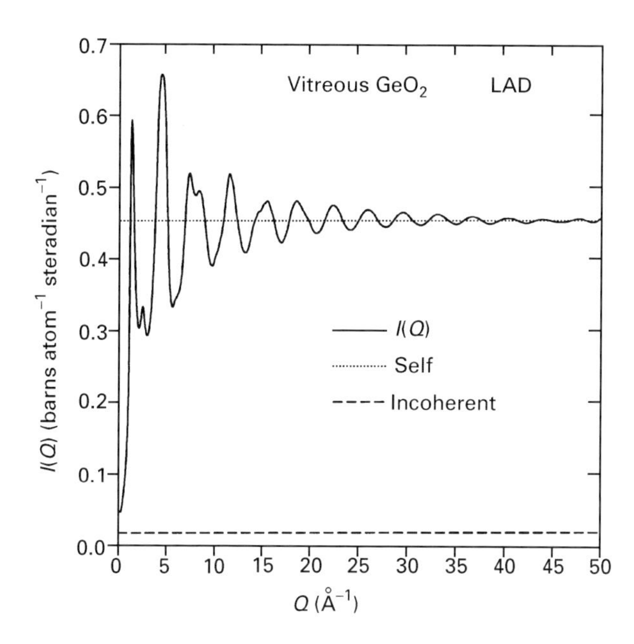

Figure 2 The total diffraction pattern, $I(Q)$, for GeO_2 glass, shown together with the self and incoherent contributions.

relatively small, but a notable exception to this is hydrogen for which the incoherent cross-section is very large, so that the measured experimental diffraction pattern is dominated by the incoherent contribution.

The sources of incoherence

The scattering length b_j for an individual nucleus (i.e. the amplitude of the neutron wave which it scatters) is not the same for all nuclei of a particular element due to two factors. These are spin incoherence and isotopic incoherence.

Spin incoherence is due to the fact that a neutron and a nucleus of spin I can form two different compound nuclei of spin $I \pm \frac{1}{2}$; the amplitude of the neutron wave scattered by the nucleus, and thus the scattering length, is generally different for the two different compound nuclei. Isotopic incoherence arises as a result of the presence of more than one isotope of a particular element.

Total diffraction from a disordered sample

The total diffraction pattern measured in reciprocal-space may be interpreted in terms of a correlation function in real-space, as is discussed below. Although this approach to neutron diffraction was developed primarily for non-crystalline systems (liquids and glasses), it is increasingly being used for powder diffraction from disordered crystalline systems as well. For a disordered crystal, the correlation function method has the advantage that it reveals the local structure, rather than the structure averaged over all unit cells that is measured by Bragg diffraction (see below).

For an isotropic sample, the diffraction pattern depends only on the magnitude Q of the momentum transfer $\mathbf{Q}$, and not its direction. In this case the distinct scattering, $i(Q)$ (Eqn [12]), is related to a neutron correlation function, $T(r)$, by a Fourier transform:

$$T(r) = T^0(r) + \frac{2}{\pi}\int_0^\infty Qi(Q)\sin(rQ)\,\mathrm{d}Q \qquad [18]$$

$T^0(r)$ is the average density contribution to the correlation function, given by

$$T^0(r) = 4\pi r g^0\left(\sum_l c_l\bar{b}_l\right)^2 \qquad [19]$$

where g^0 is the average atomic number density in the sample, c_l is the atomic fraction for element l, and the l summation is over elements in the sample (e.g. for $Li_2Mo_2O_7$, l = Li, Mo, O). For example, **Figure 3** shows the neutron correlation function, $T(r)$, for GeO_2 glass, obtained by Fourier transformation of the diffraction data shown in **Figure 2**. Also shown is the average density contribution $T^0(r)$.

The total correlation function is a weighted sum of partial correlation functions:

$$T(r) = \sum_l \sum_{l'} c_l\bar{b}_l\bar{b}_{l'}t_{ll'}(r) \qquad [20]$$

Each partial correlation $t_{ll'}(r)$ is related to a generalized van Hove distinct correlation function by

$$t_{ll'}(r) = 4\pi r G^{\mathrm{D}}_{ll'}(\boldsymbol{r}, 0) \qquad [21]$$

where

$$G^{\mathrm{D}}_{ll'}(\boldsymbol{r}, t) = \frac{1}{N_l}\sum_{\substack{j=1\\ j\neq j'}}^{N_l}\sum_{j'=1}^{N_{l'}}\int\langle\delta(\boldsymbol{r}' - \boldsymbol{R}_j(0)) \times \delta(\boldsymbol{r}' + \boldsymbol{r} - \boldsymbol{R}_{j'}(t))\rangle\mathrm{d}\boldsymbol{r}' \qquad [22]$$

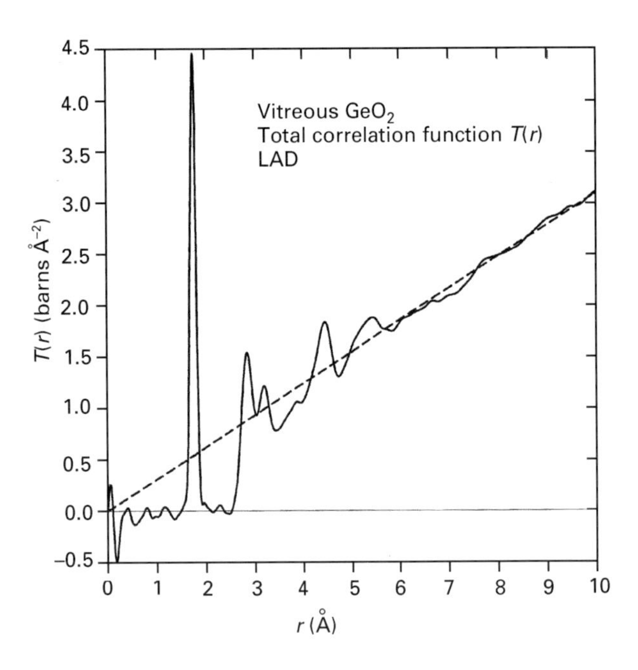

Figure 3 The neutron correlation function, $T(r)$, for GeO_2 glass, obtained from the data in **Figure 2**. The dashed line indicates the average density contribution, $T^0(r)$.

Although the formal definition of the partial correlation function, $t_{ll'}(r)$, may appear complicated, its interpretation is simple: $rt_{ll'}(r)\mathrm{d}r$ is the average number of atoms of element l' which are located in a spherical shell of radius r to $r+\mathrm{d}r$, centred on an atom of element l. For example, **Figure 4** shows how the first two peaks in the correlation function, $T(r)$, arise for an A_2X_3 material with a network structure. For the experimental data for GeO_2 glass shown in **Figure 3**, the first peak in the correlation function arises from the nearest neighbour Ge–O bonds, whilst the second peak arises from the non-bonded O–O distance in the GeO_4 tetrahedra which connect together to form the random network structure.

In the harmonic approximation the contribution to $t_{ll'}(r)$ due to a single interatomic distance r_{jk} with a root mean square thermal variation in distance of $\langle u_{jk}^2\rangle^{1/2}$ is

$$t_{jk}(r) = \frac{n_{jk}}{r_{jk}\left(2\pi\langle u_{jk}^2\rangle\right)^{1/2}} \exp\left(-\frac{(r-r_{jk})^2}{2\langle u_{jk}^2\rangle}\right) \quad [23]$$

where n_{jk} is the coordination number of k atoms around the atom j. In reciprocal-space this corresponds to

$$i_{jk}(Q) = n_{jk}\bar{b}_j\bar{b}_k \frac{\sin(Qr_{jk})}{Qr_{jk}} \exp\left(\frac{-\langle u_{jk}^2\rangle Q^2}{2}\right) \quad [24]$$

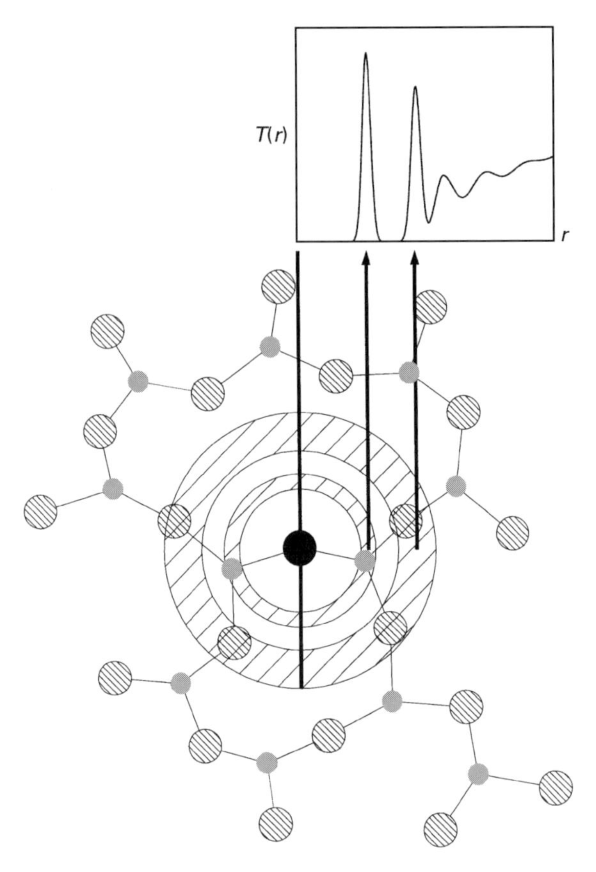

Figure 4 A simulated neutron correlation function, $T(r)$, together with a fragment of an A_2X_3 network, showing how the peaks in the correlation function arise from the interatomic distances.

so that the total distinct scattering is given by

$$i(Q) = \sum_j \sum_k i_{jk}(Q) \quad [25]$$

Small angle neutron scattering

Small-angle neutron scattering (SANS) involves the study of diffraction in the regime where the magnitude of the momentum transfer, Q, is small compared with the position of the first peak in the structure factor or the highest d-spacing Bragg peak. Thus SANS is used to study structures with dimensions of tens or hundreds of Ångströms which are large compared with interatomic distances in condensed matter. The individual scattering centres in a sample are not resolved and hence a continuous scattering length density may be introduced:

$$\rho_b(\boldsymbol{r}) = \sum_l \bar{b}_l \sum_{j=1}^{N_l} \delta[\boldsymbol{r} - \boldsymbol{R}_j(0)] \quad [26]$$

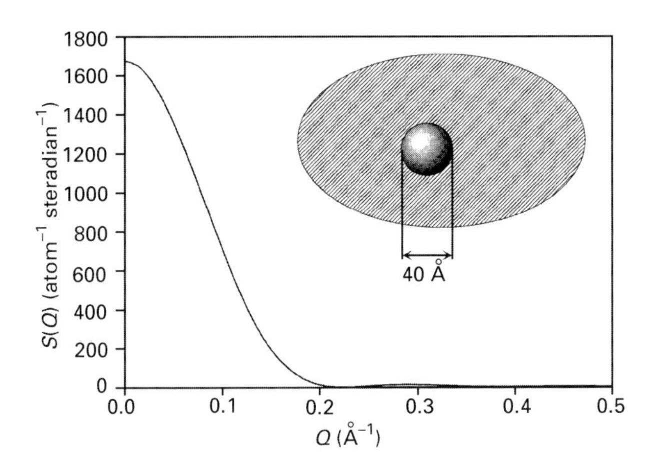

Figure 5 The structure factor, $S(Q)$, calculated for a homogeneous sphere of radius 20 Å and uniform density 0.05 atoms Å^{-3}, surrounded by empty space.

The coherent differential cross-section observed in a SANS experiment is then

$$\left(\frac{d\sigma}{d\Omega}\right)_{coh} = \frac{1}{N}\left|\int\left[\rho_b(\boldsymbol{r}) - \rho_b^0\right]\exp(i\boldsymbol{Q}.\boldsymbol{r})\right|^2 \quad [27]$$

where ρ_b^0 is the average scattering length density for the sample. A SANS diffraction experiment is thus sensitive to deviations of the scattering length density from the average value. To analyse data from a SANS experiment it is usually necessary to employ a model for the structures under investigation. For example, **Figure 5** shows the structure factor for a sphere with a homogenous scattering length density, surrounded by free space. As a consequence of the square modulus in Equation [27], SANS is only sensitive to the contrast in scattering length density. Thus the same structure factor as that shown in **Figure 5** would be obtained for an empty void of the same size embedded in a homogenous medium.

Elastic diffraction from a crystalline sample

Single crystal diffraction

An ideal crystalline material has an atomic structure which can be described in terms of infinite repetitions in three-dimensional space of a unit cell. If the sides of the unit cell are described by the three lattice vectors $\boldsymbol{a}$, $\boldsymbol{b}$ and $\boldsymbol{c}$, then repeating the unit cell ad infinitum by translations made up of all the possible (whole number) combinations of $\boldsymbol{a}$, $\boldsymbol{b}$ and $\boldsymbol{c}$ will generate the ideal crystal structure.

For a single crystal the coherent elastic contribution to the differential cross-section is

$$\left(\frac{d\sigma}{d\Omega}\right)_{coh\ el} = \frac{(2\pi)^3}{\nu_0}\sum_{\tau}\delta(\boldsymbol{Q} - \boldsymbol{\tau}_{hkl})|F(\boldsymbol{Q})|^2 \quad [28]$$

where ν_0 is the volume of the unit cell, given by

$$\nu_0 = \boldsymbol{a}\,.\,(\boldsymbol{b}\times\boldsymbol{c}) \quad [29]$$

τ_{hkl} is a reciprocal lattice vector, given by

$$\tau_{hkl} = h\boldsymbol{a}^* + k\boldsymbol{b}^* + l\boldsymbol{c}^* \quad [30]$$

where h, k and l are integers, and $\boldsymbol{a}^*$, $\boldsymbol{b}^*$ and $\boldsymbol{c}^*$ are the unit cell vectors of the reciprocal lattice, defined by

$$\boldsymbol{a}^* = \frac{2\pi}{\nu_0}\boldsymbol{b}\times\boldsymbol{c};\quad \boldsymbol{b}^* = \frac{2\pi}{\nu_0}\boldsymbol{c}\times\boldsymbol{a};\quad \boldsymbol{c}^* = \frac{2\pi}{\nu_0}\boldsymbol{a}\times\boldsymbol{b} \quad [31]$$

$F(Q)$ is known as the structure factor:

$$F(\boldsymbol{Q}) = \sum_d \bar{b}_d \exp\left(i\boldsymbol{Q}\,.\,\overline{\boldsymbol{R}}_d\right)\exp(-W_d) \quad [32]$$

where the d summation is taken over the atoms in the unit cell and $\overline{\boldsymbol{R}}_d$ is the equilibrium position of the dth atom in the unit cell. The Debye–Waller factor, W_d, is a 'damping' term which depends upon the thermal atomic displacements. A complete treatment of the Debye–Waller factor is beyond the scope of this article, but its general behaviour is well illustrated by the simple case of a cubic Bravais lattice for which

$$2W = \frac{1}{3}Q^2\langle u^2\rangle \quad [33]$$

where $\langle u^2\rangle$ is the mean square thermal displacement of an atom from its equilibrium position.

The delta function of Equation [28], $\delta(Q-\tau_{hkl})$, shows that coherent elastic scattering only occurs when the following condition is satisfied

$$\boldsymbol{Q} = \boldsymbol{\tau}_{hkl} \quad [34]$$

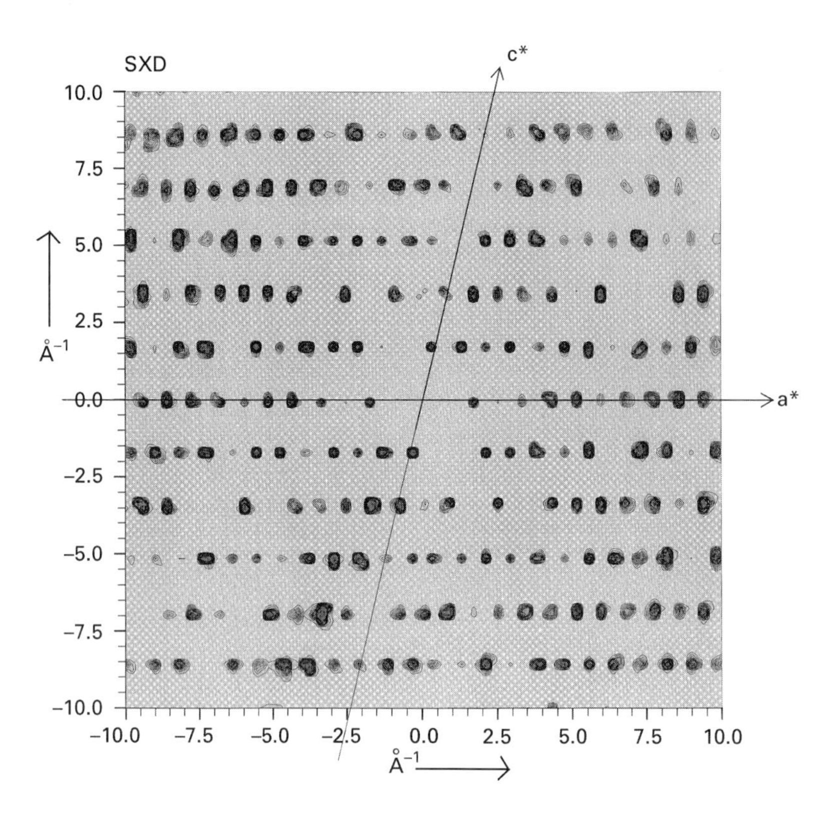

Figure 6 A single crystal diffraction pattern for $KDCO_3$ ($a = 15.2$ Å, $b = 5.6$ Å, $c = 3.7$ Å, $\beta = 104.8°$). Reproduced courtesy of Fillaux F, Cousson A and Keen DA.

That is to say, peaks are observed in the diffraction data when the momentum transfer vector is equal to a vector of the reciprocal lattice. Such peaks are known as Bragg peaks and Equation [34] is Bragg's law, for diffraction from a single crystal. **Figure 6** shows a typical single crystal diffraction pattern; for every peak in the observed diffraction pattern the momentum transfer vector, Q, satisfies Bragg's law. Each and every Bragg peak may be identified by a unique combination of indices (h, k, l).

The diffraction pattern of a single crystal contains a wealth of information in three-dimensional reciprocal-space and, if a sufficient number of Bragg peaks is observed, a complete description of the crystal structure may be determined. The symmetry of the Bragg peaks observed in the diffraction pattern depends upon the symmetry of the crystal structure. Thus the space group of the crystal can be determined from the symmetry observed in the diffraction pattern. The positions of the Bragg peaks in reciprocal-space depend upon the lattice parameters a, b, c, α, β and γ (i.e. the magnitudes of the lattice vectors $\boldsymbol{a}$, $\boldsymbol{b}$ and $\boldsymbol{c}$, and the angles between them). Consequently the dimensions of the unit cell can be determined from the positions of the observed Bragg peaks. The intensity of each Bragg peak depends upon its structure factor, which itself depends on the positions of the atoms in the unit cell. Thus the contents of the unit cell can be determined from the integrated intensities of the observed Bragg peaks. In this way a complete description of the crystal structure may be developed from the single crystal diffraction pattern.

It should be noted that the structure factor, $F(Q)$, used in crystallography (Eqn [32]) has a different meaning and definition to the structure factor, $S(Q)$, used in studying diffraction from disordered samples (Eqn [15]). In addition, there is a fundamental difference between the Bragg scattering observed for crystals and the total diffraction cross-section in that the Bragg scattering is purely elastic, whereas total diffraction involves both elastically and inelastically scattered neutrons. This has the consequence that the correlation function which is addressed by the study of Bragg diffraction is the time-averaged correlation function $G^D_{ll'}(\boldsymbol{r}, \infty)$ (see Eqn [22]). Hence, for a crystal, the Bragg scattering provides information which is in principle complementary to total diffraction, which yields the instantaneous correlation function $G^D_{ll'}(\boldsymbol{r}, 0)$.

Table 2 The planar d-spacings for each crystal system

Crystal system	*d-spacing*
Cubic $a=b=c$ $\alpha=\beta=\gamma=90^\circ$	$\frac{1}{d_{hkl}^2}=\frac{h^2+k^2+l^2}{a^2}$
Tetragonal $a=b$ $\alpha=\beta=\gamma=90^\circ$	$\frac{1}{d_{hkl}^2}=\frac{h^2+k^2}{a^2}+\frac{l^2}{c^2}$
Orthorhombic $\alpha=\beta=\gamma=90^\circ$	$\frac{1}{d_{hkl}^2}=\frac{h^2}{a^2}+\frac{k^2}{b^2}+\frac{l^2}{c^2}$
Monoclinic $\alpha=\gamma=90^\circ$	$\frac{1}{d_{hkl}^2}=\frac{1}{\sin^2\beta}\left(\frac{h^2}{a^2}+\frac{k^2\sin^2\beta}{b^2}+\frac{l^2}{c^2}-\frac{2hl\cos\beta}{ac}\right)$
Hexagonal $a=b$ $\alpha=\beta=90^\circ$, $\gamma=120^\circ$	$\frac{1}{d_{hkl}^2}=\frac{4}{3}\left(\frac{h^2+hk+k^2}{a^2}\right)+\frac{l^2}{c^2}$
Rhombohedral $a=b=c$ $\alpha=\beta=\gamma$	$\frac{1}{d_{hkl}^2}=\frac{(h^2+k^2+l^2)\sin^2\alpha+2(hk+kl+lh)(\cos^2\alpha-\cos\alpha)}{a^2(1+2\cos^3\alpha-3\cos^2\alpha)}$
Triclinic $a\neq b\neq c$ $\alpha\neq\beta\neq\gamma$	$\frac{1}{d_{hkl}^2}=\frac{\frac{h^2\sin^2\alpha}{a^2}+\frac{k^2\sin^2\beta}{b^2}+\frac{l^2\sin^2\gamma}{c^2}+\frac{2hk}{ab}(\cos\alpha\cos\beta-\cos\gamma)+\frac{2kl}{bc}(\cos\beta\cos\gamma-\cos\alpha)+\frac{2lh}{ac}(\cos\gamma\cos\alpha-\cos\beta)}{1+2\cos\alpha\cos\beta\cos\gamma-\cos^2\alpha-\cos^2\beta-\cos^2\gamma}$

Powder diffraction

Although single crystal diffraction is the most powerful method for studying the structure of crystals, the powder diffraction method is also frequently used to reveal structural information for crystalline materials. In this case the sample is not a single crystal, but is instead a powder which contains numerous microcrystals, such that all orientations are equally likely. The observed diffraction pattern is then an average of Equation [28] over all possible relative orientations between the momentum transfer vector Q and the crystal lattice. The measured diffraction pattern does not depend on the orientation between the sample and the neutron beam, and it is usual to identify each Bragg peak in the diffraction pattern by its d-spacing, d_{hkl}, defined by

$$|\tau_{hkl}|=\frac{2\pi}{d_{hkl}} \qquad [35]$$

Combining this definition with Equations [8] and [34] leads to Bragg's law for powder diffraction:

$$2d_{hkl}\sin\theta=\lambda \qquad [36]$$

Bragg's law indicates the combinations of θ and λ for which Bragg peaks are observed in the powder diffraction pattern of a crystalline material. **Table 2** gives the formulae for the possible d-spacings of each of the crystal systems. Note that some of the d-spacings predicted by these formulae may not occur as peaks in the diffraction pattern, depending on the symmetry of the crystal structure; some Bragg peaks may have structure factors equal to zero, owing to the crystal symmetry, with the result that they are not allowed. The d-spacing of a Bragg peak may be interpreted as being the spacing for a set of planes of atoms in the crystal such that reflection from these planes gives rise to the Bragg peak (see **Figure 7**).

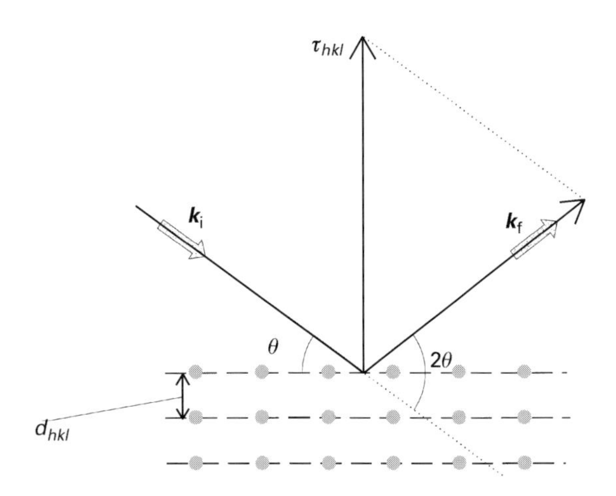

Figure 7 Bragg planes with spacing d_{hkl} which give rise to the *hkl* Bragg peak.

Figure 8 shows the relatively simple powder diffraction pattern of polycrystalline silicon, measured by time-of-flight neutron diffraction. Each allowed Bragg peak for the crystal structure is observed as a sharp peak in the diffraction pattern. A fundamental difference between single crystal and powder diffraction is that the single crystal diffraction pattern has the full directional information described by Equation [28], whereas for the powder diffraction pattern this information has been collapsed down into a one-dimensional function. This difference has two important consequences: firstly, the loss of directional information makes it very much more difficult to deduce the symmetry of the crystal lattice from powder diffraction data. For this reason, powder diffraction is more commonly used to study the contents of the unit cell for crystals whose symmetry is already known; secondly, the Bragg peaks are much more likely to suffer from significant overlap in a powder diffraction pattern than in single crystal data. This means that it may not be possible to determine the structure factors of the reflections by integrating the intensities of individual peaks and instead the profile refinement method must be used to analyse the data. The profile refinement method involves fitting the whole of the diffraction pattern simultaneously in order to extract structural information. This method requires an explicit knowledge of the resolution function of the neutron diffractometer.

List of symbols

a = lattice parameter; $\boldsymbol{a}$ = lattice vector; $\boldsymbol{a}^*$ = reciprocal lattice vector; b = lattice parameter; $\boldsymbol{b}$ = lattice vector; $\boldsymbol{b}^*$ = reciprocal lattice vector; b_j = (bound atom) neutron scattering length for nucleus j; $\overline{b}_l$ = coherent scattering length for element l; $\overline{b_l^2}$ = mean square scattering length for element l; $\langle b^2 \rangle_{av}$ = average value of $\overline{b^2}$ for a sample; $\langle \overline{b}^2 \rangle_{av}$ = average value of $\overline{b}^2$ for a sample; c = lattice parameter; $\boldsymbol{c}$ = lattice vector; $\boldsymbol{c}^*$ = reciprocal lattice vector; c_l = atomic fraction for element l; d_{hkl} = d-spacing; dS = elemental area; $\mathrm{d}^2\sigma/\mathrm{d}E$ = double differential cross-section; $\mathrm{d}\sigma/\mathrm{d}\Omega$ = differential cross-

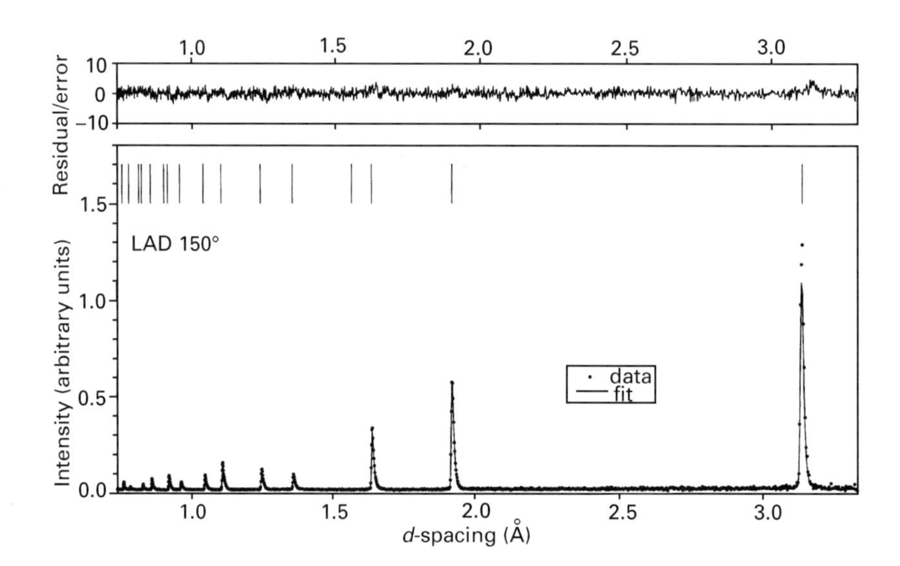

Figure 8 The neutron diffraction pattern for polycrystalline silicon together with a fit obtained by profile refinement. Also shown is the residual divided by experimental error for the fit and vertical bars which indicate the positions of the allowed reflections under the space group $Fd\overline{3}m$.

section; E_i = energy of neutrons in incident monochromatic beam; E_f = energy of neutron after scattering; E = energy transferred from neutron to sample; $F(Q)$ = structure factor in crystallography; g^0 = average atomic number density; $G_{ll'}^D(\boldsymbol{r},t)$ = generalized van Hove distinct correlation function; h = index of Bragg reflection; $i(Q)$ = distinct scattering; I = nuclear spin; $I(Q)$ = differential cross-section for total diffraction; k = index of Bragg reflection; $\boldsymbol{k}$ = neutron wave vector; $\boldsymbol{k}_i$ = wave vector of incident neutron; $\boldsymbol{k}_f$ = wave vector of scattered neutron; l = index of Bragg reflection; m_n = neutron mass; n_{jk} = coordination number; N = number of atoms in sample; N_l = number of atoms of element l in sample; Q = momentum transfer; r = polar coordinate; R_x = the rate at which neutrons are scattered into a given final state; $\mathbf{R}_j(t)$ = position of jth nucleus at time t; $\overline{R}_d$ = equilibrium position of dth atom in unit cell; $S(Q)$ = structure factor for disordered materials; t = time; T = temperature of sample; $t_{ll'}(r)$ = partial neutron correlation function; $T(r)$ = neutron corelation function; $T^0(r)$ = average density contribution to neutron correlation function; $\langle u^2 \rangle$ = mean square thermal displacement of atom from equilibrium; $\langle u_{jk}^2 \rangle$ = mean square thermal variation in interatomic distance; $V(r)$ = Fermi pseudo-potential; W_d = Debye–Waller factor; α = lattice parameter; β = lattice parameter; γ = lattice parameter; $\delta(x)$ = Dirac delta function; 2θ = scattering angle; λ = neutron wavelength; v_0 = volume of unit cell; $\rho_b(r)$ = scattering length density; ρ_b^0 = average scattering length density; σ_{inc} = incoherent cross-section; τ_{hkl} = reciprocal lattice vector; ϕ = polar coordinate; Φ = flux in incident neutron beam; Ψ_f = wave-function of scattered neutron; Ω = solid angle.

See also: **Fourier Transformation and Sampling Theory; Inelastic Neutron Scattering, Applications; Inelastic Neutron Scattering, Instrumentation; Inorganic Compounds and Minerals Studied Using X-Ray Diffraction; Material Science Applications of X-Ray Diffraction; Neutron Diffraction, Instrumentation; Powder X-Ray Diffraction, Applications; Structure Refinement (Solid State Diffraction).**

Further reading

Bacon GE (1969) *Neutron Physics*. London: Wykeham.

Bacon GE (1975) *Neutron Diffraction*, 3rd edn. Oxford: Oxford University Press.

Brown PJ and Forsyth JB (1973) *The Crystal Structure of Solids* London: Edward Arnold.

Hannon AC, Howells WS and Soper AK (1990) ATLAS: A suite of programs for the analysis of time-of-flight neutron diffraction data from liquid and amorphous samples. *IOP Conference Series* 107: 193–211.

Lovesey SW (1984) *Theory of Neutron Scattering from Condensed Matter. Volume 1: Nuclear Scattering*. Oxford: Oxford University Press.

Marshall W and Lovesey SW (1971) *Theory of Thermal Neutron Scattering*. London: Oxford University Press.

Price DL and Sköld K (1986) Introduction to neutron scattering. In: Sköld K and Price DL, (eds) *Neutron Scattering, Part A*, pp 1–97. Orlando: Academic Press.

Squires GL (1996) *Introduction to the Theory of Thermal Neutron Scattering*. Mineola: Dover Publications.

Wright AC (1974) The structure of amorphous solids by X-ray and neutron diffraction. *Advances in Structure Research by Diffraction Methods* 5: 1–120.

Wright AC and Leadbetter AJ (1976) Diffraction studies of glass. *Physics and Chemistry of Glasses* 17: 122–145.

Nickel NMR, Applications

See **Heteronuclear NMR Applications (Sc–Zn).**

Niobium NMR, Applications

See **Heteronuclear NMR Applications (Y–Cd).**

Nitrogen NMR

GA Webb, University of Surrey, Guildford, UK

MAGNETIC RESONANCE
Applications

Introduction

Nitrogen has two stable isotopes, ^{14}N and ^{15}N, both of which are NMR active. In addition nitrogen is one of the most important atoms in organic, inorganic and biochemistry due to its occurrence in a variety of valence states with various types of bonding and stereochemistry. Thus the nuclear shieldings, spin–spin couplings and relaxation data obtained from nitrogen NMR studies are of widespread interest to molecular scientists. The relatively large range, in excess of 1000 ppm, found for the nitrogen chemical shifts of organic compounds indicates that the shifts are sensitive to relatively small electronic changes in the environment of the nitrogen atom. Consequently a change of solvent or substituent can have a very significant effect on nitrogen chemical shifts.

Solvent effects can be produced by specific interactions, such as protonation or hydrogen bonding and nonspecific interactions which may arise from solvent polarity effects. Both of these types of interaction are found in studies of solvent effects on nitrogen nuclear shieldings. The extent of substituent effects depends upon the position of substitution and the electronic nature of the substituent.

Most spin–spin couplings involving nitrogen nuclei are relatively small, and both positive and negative couplings are found. Both the sign and the magnitude of the couplings are influenced by the orientation of the nitrogen lone pair electrons with respect to the direction of the bond through which the coupling occurs.

^{14}N nuclear relaxation is usually dominated by the quadrupolar mechanism which results in broad lines both in the nitrogen NMR spectrum and in the spectra of nuclei spin–spin coupled to nitrogen. The relaxation of ^{15}N is controlled by one, or more, of the less efficient processes.

Nuclear properties

In natural abundance nitrogen exists in two isotopic forms, the most common is ^{14}N which is 99.635% abundant and has a spin of 1, the other is ^{15}N which has an abundance of 0.365% and a spin of $\frac{1}{2}$. Thus both isotopes are NMR active. The first report of a ^{14}N NMR spectrum appeared in 1950 when NMR was still very much in its infancy. However, both nitrogen isotopes have low NMR sensitivities relative to that of a proton in the same applied magnetic field; these are 0.00101 and 0.00104 for ^{14}N and ^{15}N respectively. Coupled with the relatively high cost of ^{15}N isotopic enrichment, the low sensitivities ensured that nitrogen NMR studies were not too widespread until the mid-1960s. Since then improvements in both NMR instrumentation and experimental techniques have generated a wide and growing interest in nitrogen NMR spectroscopy. Today the NMR spectra of both isotopes are normally studied at natural abundance due to the increased sensitivity which is currently available in high-field NMR spectrometers.

The fact that the ^{14}N nucleus has a spin of 1 implies that it has an electric quadrupole moment, which is able to interact with local electric field gradients in the molecule and produce rapid nuclear relaxation. A consequence of this is that ^{14}N NMR signals can be very broad, as much as a few kilohertz, and overlapping. Since the quadrupolar relaxation mechanism is independent of the magnitude of the magnetic field used in the NMR experiment, with a high-field spectrometer the ^{14}N signals appear to be relatively sharper and less overlapping than they do in a lower field instrument. Hence the widespread availability of high-field spectrometers has led to a renaissance of interest in ^{14}N NMR studies both from the point of view of increased sensitivity and from the diminishing in importance of the effects of quadrupolar relaxation in high-resolution NMR spectroscopy.

Chemical shifts

The most widely used nitrogen chemical shift standard is neat nitromethane. This is used as an external reference by typically employing a set of coaxial tubes with the reference material in the inner one. Ideally an internal chemical shift standard is desirable since this removes the necessity of correcting for any magnetic bulk susceptibility difference between the sample and the standard. However, in nitrogen NMR this is usually not an option since the nitrogen shielding of a chemical shift standard may vary by up to 40 ppm due to molecular interactions in liquids and solutions. A possible candidate for the role of internal nitrogen chemical shift standard is molecular nitrogen, N_2. This is present in nearly all solutions and gives a clear sharp signal in ^{14}N NMR spectra. However, its shielding is not entirely immune from solvent effects; a range of about 2 ppm is found for solvent-induced variations. As shown in **Figure 1** the overall range of nitrogen chemical shifts for organic compounds is in excess of 1000 ppm.

Only approximate ranges are given as a general guide, and in practice most of them may be further divided into subsections to reflect the finer points of variation in molecular structure.The extent of the chemical shift range shown in **Figure 1** shows clearly that spectral assignments are usually very straightforward in nitrogen NMR. In general those nitrogen environments with σ-bonding only, and no nitrogen lone-pair electrons, are the most highly shielded ones, e.g. alkylamines. In contrast, when the nitrogen atom is involved in π-bonding and lone-pair electrons are available, then the shielding is decreased by a contribution from the paramagnetic term in the shielding expression, e.g. nitrites and nitroso compounds. In a molecular system where nitrogen is involved in multiple bonding an increase in shielding or decrease in chemical shift is observed if the nitrogen lone-pair is replaced by a covalent bond, e.g. in passing from pyridine to the pyridinium ion.

Solvent effects on nitrogen shielding

Solvent effects on nitrogen shieldings can be remarkable and need to be carefully considered in the interpretation of nitrogen chemical shifts in terms of molecular structure. So far the largest variation reported in nitrogen chemical shift for a neutral molecular species, as a function of solvent, is about 50 ppm for pyridazine (1,2-diazine). This is shown in **Table 1** together with some other examples of typical solvent effects on nitrogen shieldings.

Table 1 Examples of the range of solvent effects on nitrogen shieldings

Molecule	*Range of solvent effects on nitrogen shielding (ppm)*
Pyridine	38
Pyridazine (1,2-diazine)	49
Pyrimidine (1,3-diazine)	18
Pyrazine (1,4-diazine)	16
1,3,5-Triazine	11
Pyridine *N*-oxide	30
Indolizine and azaindolizine systems	
Bridgehead nitrogen	1–3
Pyridine-type nitrogen	26–32
Alkyl cyanides (nitriles)	22–26
Nitromethane	11
Methyl nitrate	5
t-Butyl nitrite	26
Methyl isothiocyanate	10
Azo bridge	2
Dinitrogen	1.3

These data are quoted from systematic studies on dilute solutions where bulk susceptibility effects have been taken into account and the solvents used were chosen to encompass a broad range of properties.

In ionic species the largest solvent-induced variation of nitrogen shielding has been observed for the nitroso moiety of $[C(NO)(CN)_2]^-K^+$ in water and a number of alcohols. The range is about 200 ppm and is related to the pK_a values of the solvents. The data reported in **Table 1** have been analysed in terms of the Kamlet–Taft system of solvent properties. This can be represented by the following expression:

$$\sigma(i,j) = \sigma_0 + a_i\alpha_j + b_i\beta_j + s_i(\pi_j^* + d_i\delta_j)$$

where i denotes a particular nitrogen atom in the molecule examined, j denotes the solvent, σ is the relevant nitrogen shielding, σ_0 is the shielding observed in a solution using cyclohexane as a standard inert solvent, α_j is the hydrogen-bond donor strength of the solvent j, β_j is the corresponding hydrogen-bond acceptor strength of the solvent, π_j^* represents the solvent polarity/polarizability and δ_j is a correction for the 'superpolarizability' of aromatic and highly chlorinated solvents. The terms a, b, s and d are the relevant nitrogen shielding responses to the individual bulk solvent properties. If a solute nitrogen atom is protonated by the solvent then a large increase in shielding, often in excess of 100 ppm, occurs. Hydrogen-bonding from solvent to a solute nitrogen atom usually also produces a smaller increase in nitrogen shielding. In terms of ppm per scale unit of α the increase is about 21 for the nitrogen atoms of pyridine and pyridazine, about 17 for pyridine-type nitrogen atoms in the five-membered ring moieties of azaindolizines and about 10 for covalent cyanides. The magnitudes of these solvent-induced shifts reflect the relative strengths of the hydrogen-bonds concerned.

Since the d term in the above equation is usually negligibly small in practice, nitrogen shielding responses to changes in solvent polarity are normally represented by the s term. In terms of ppm per unit scale of π^* the values found for s may be of either sign. A positive sign indicates an increase in nitrogen shielding as the polarity of the medium increases. For pyridazine the value is about 13, whereas for other pyridine-type nitrogen atoms in azines and

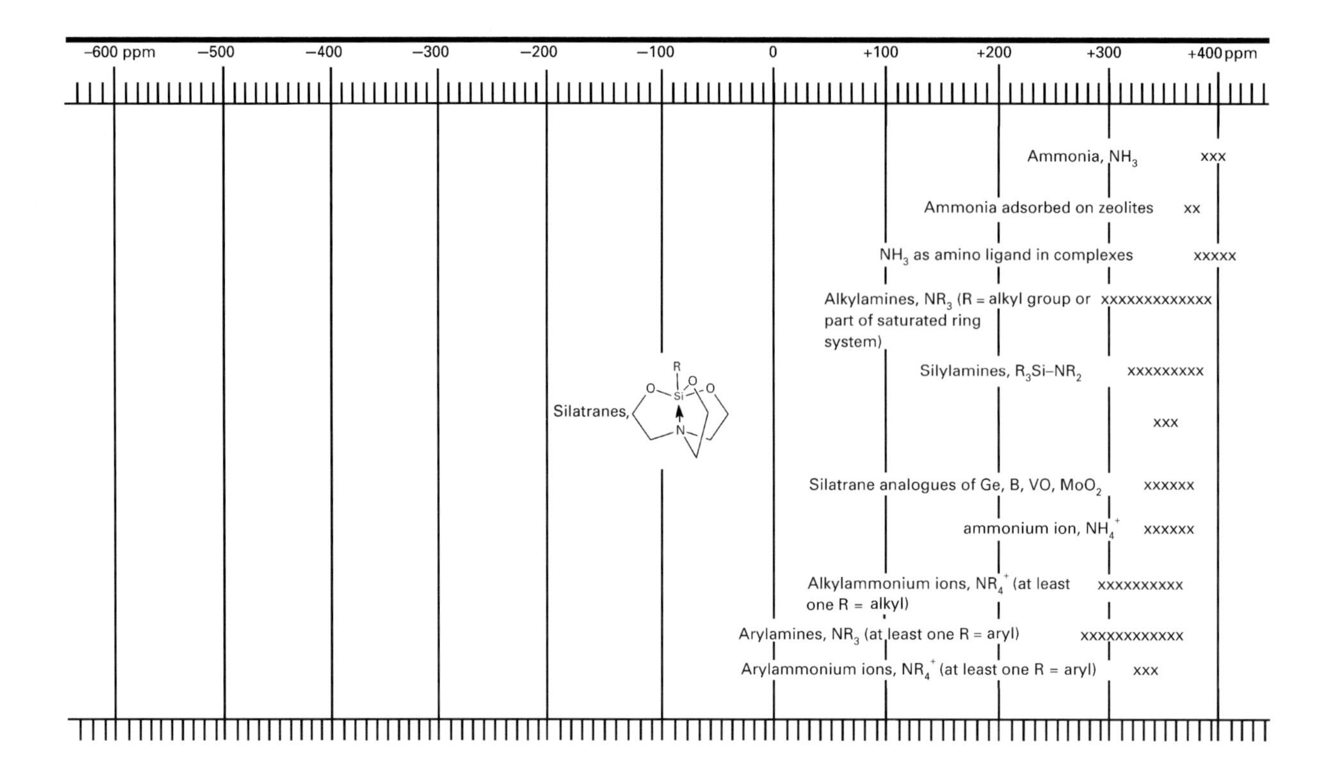

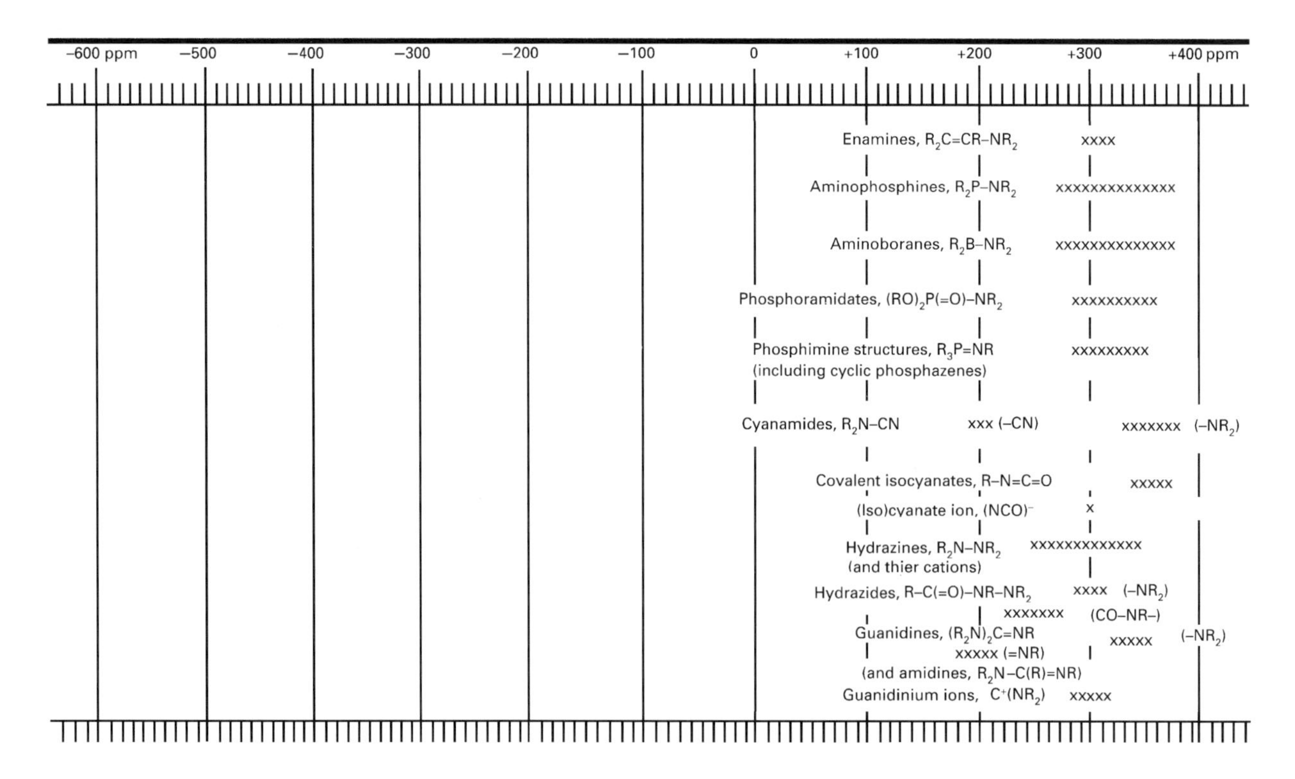

Figure 1 *Continued*

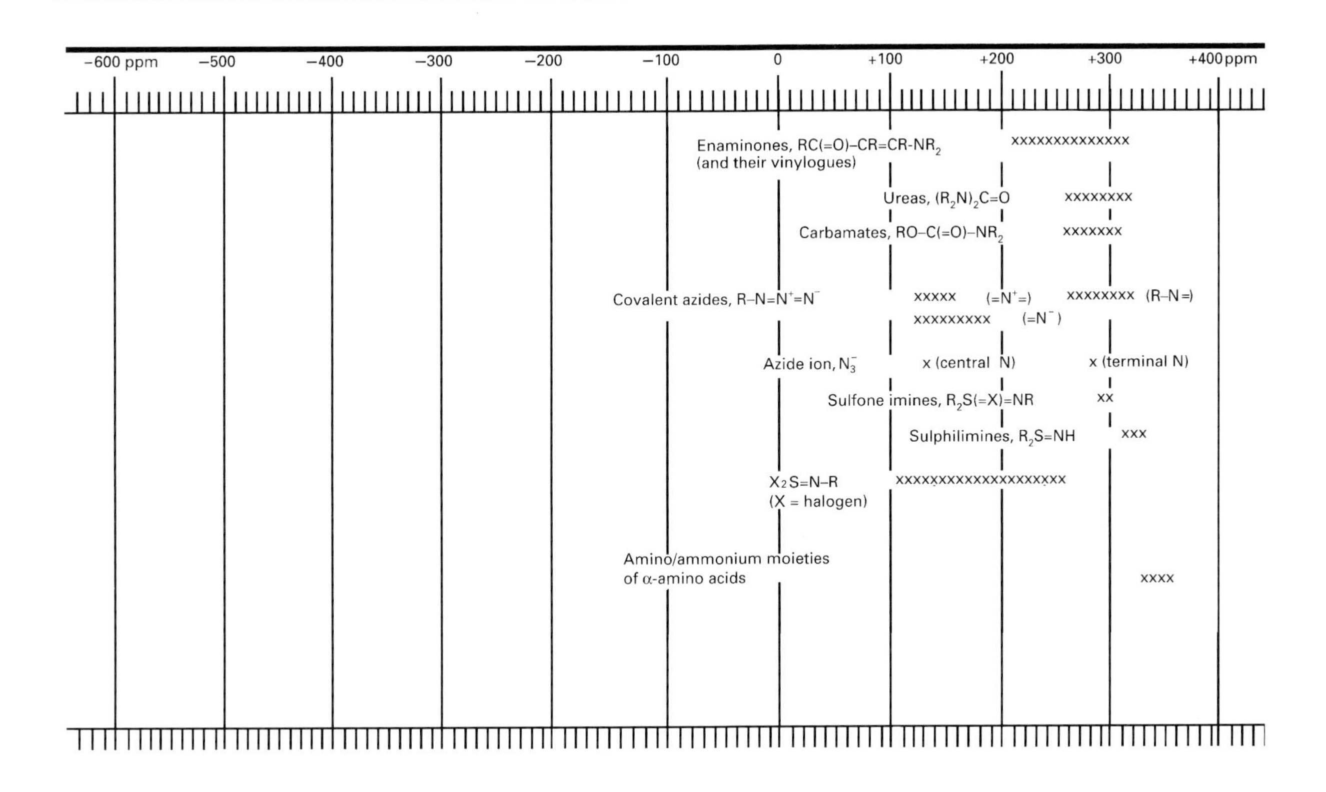

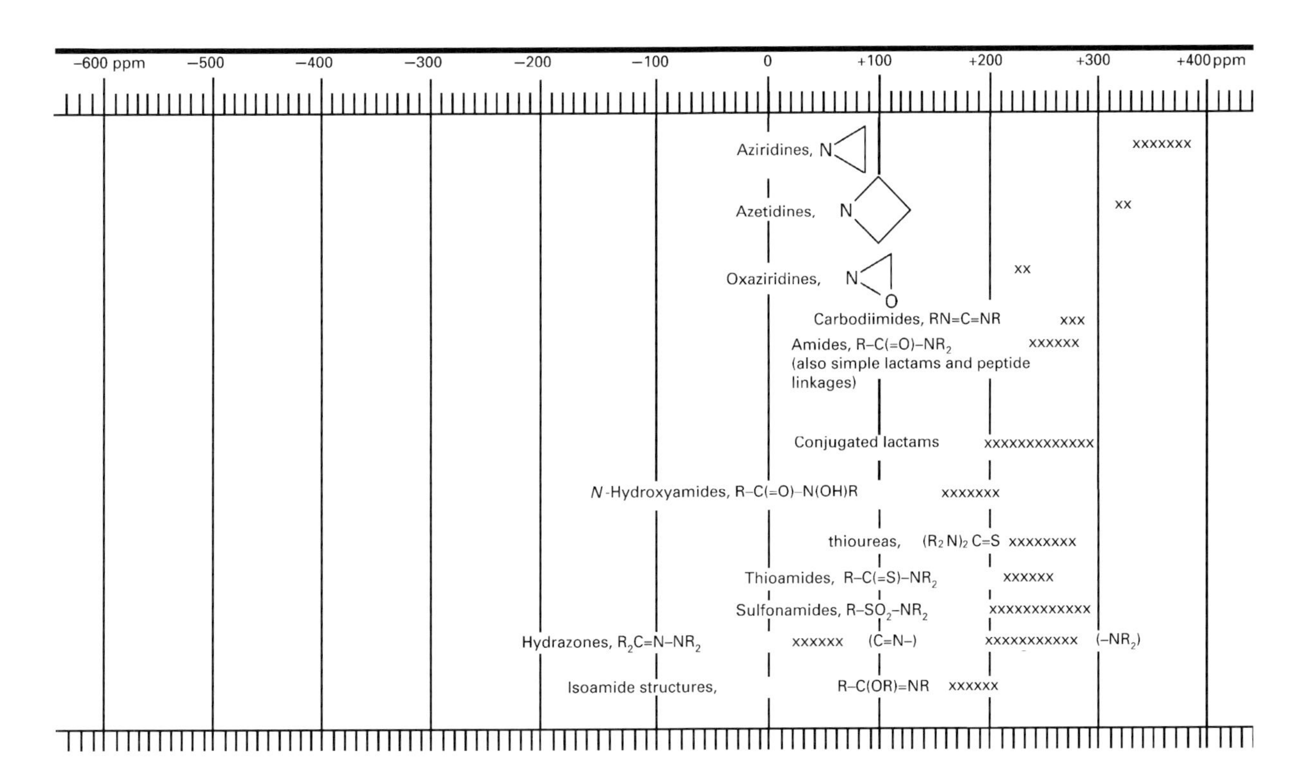

Figure 1 *Continued*

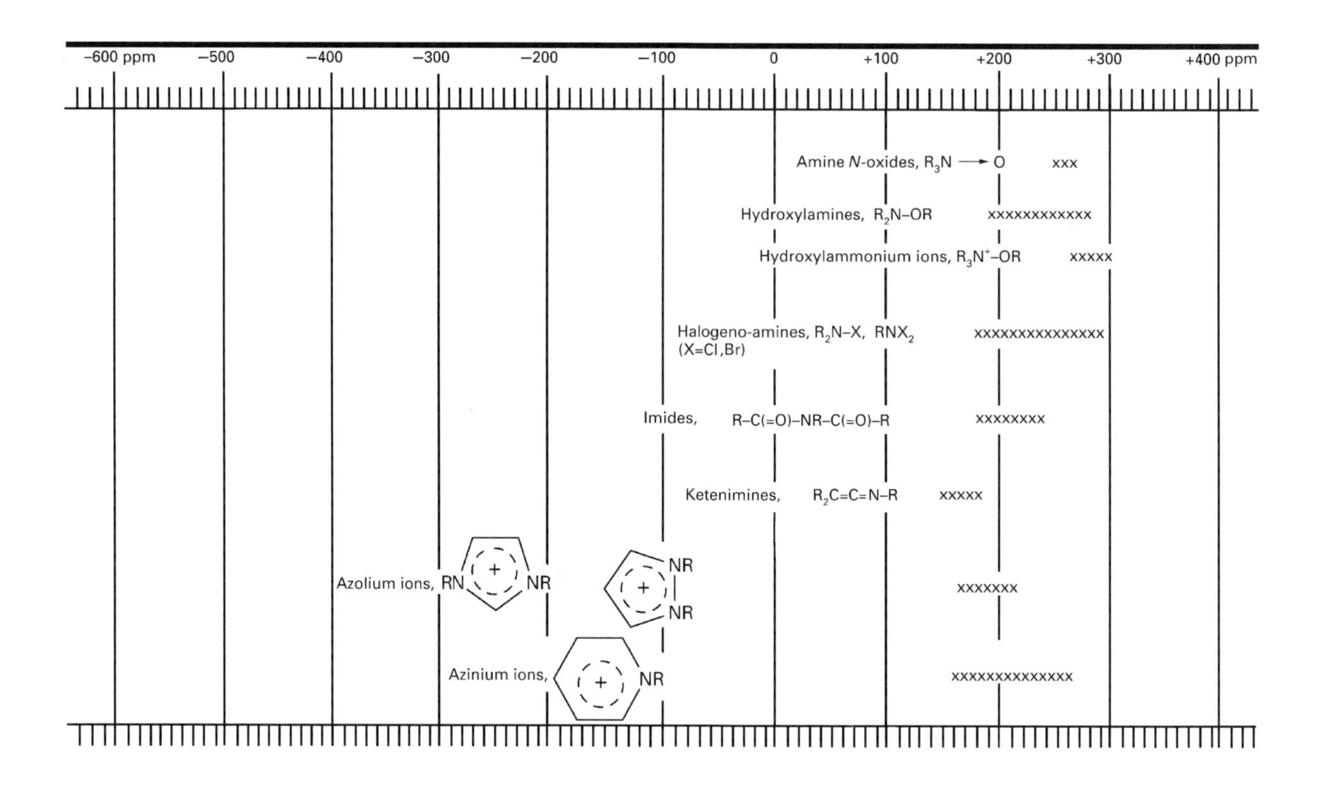

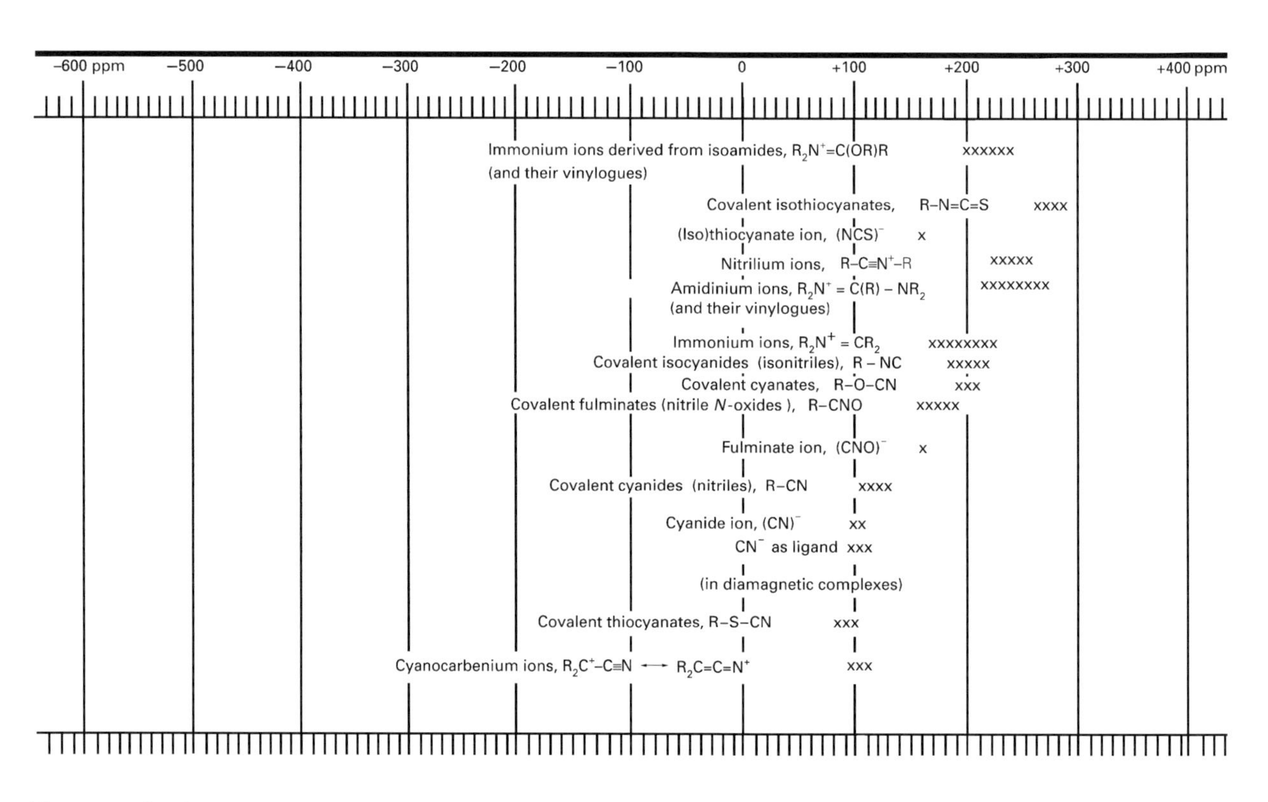

Figure 1 *Continued*

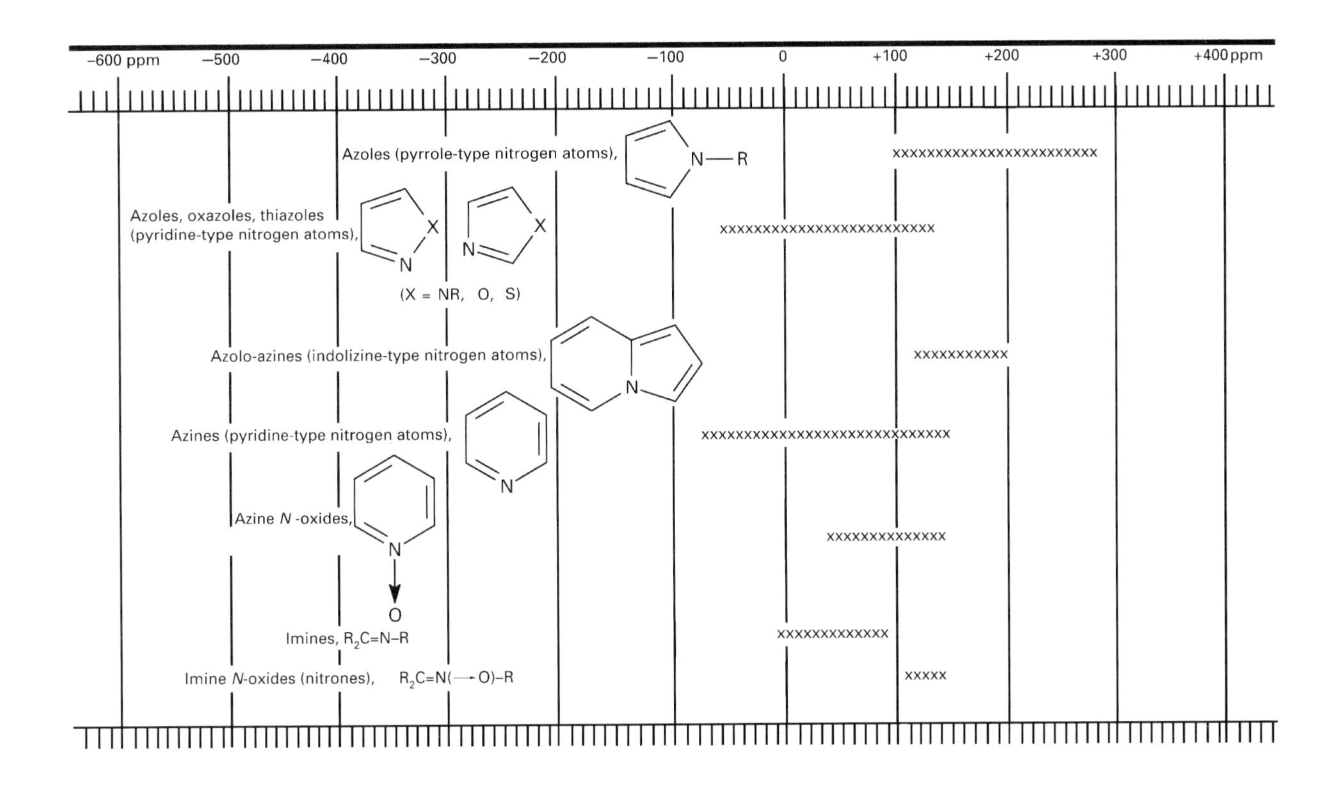

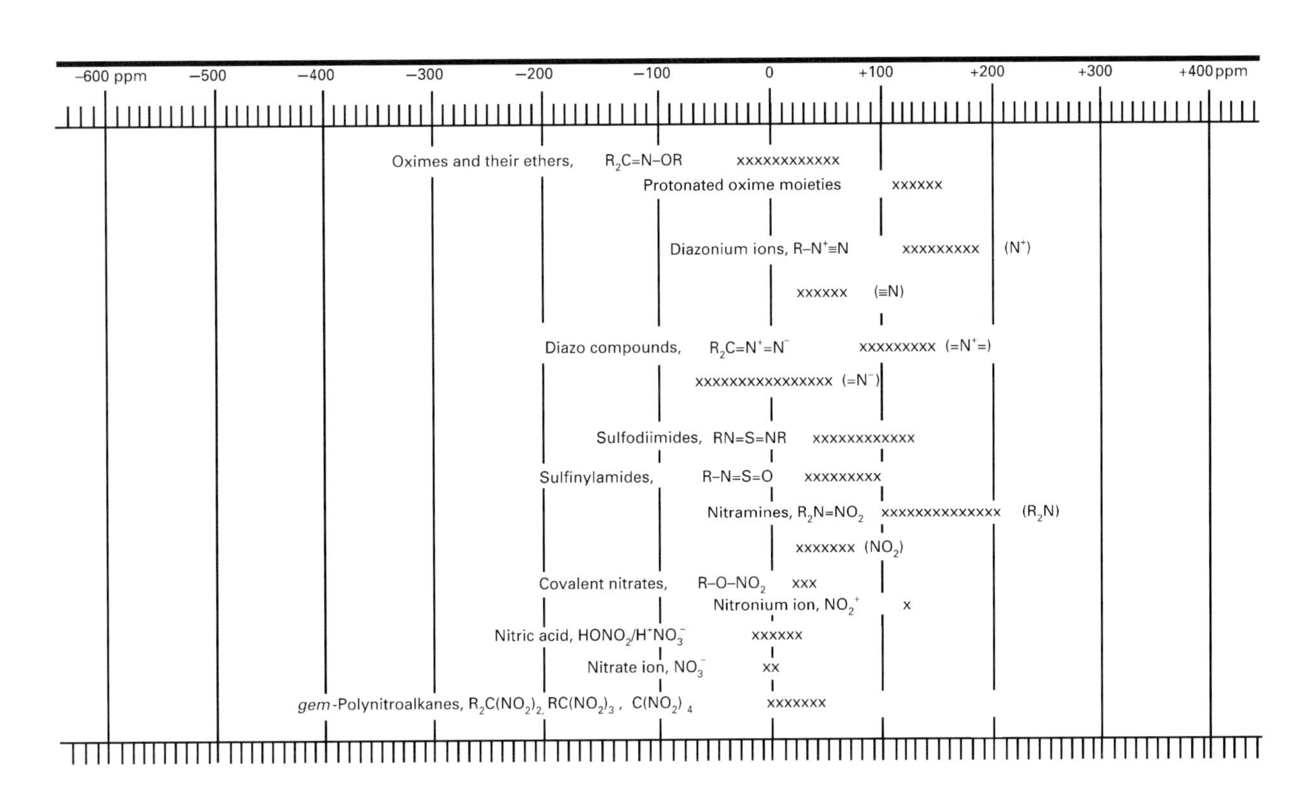

Figure 1 *Continued*

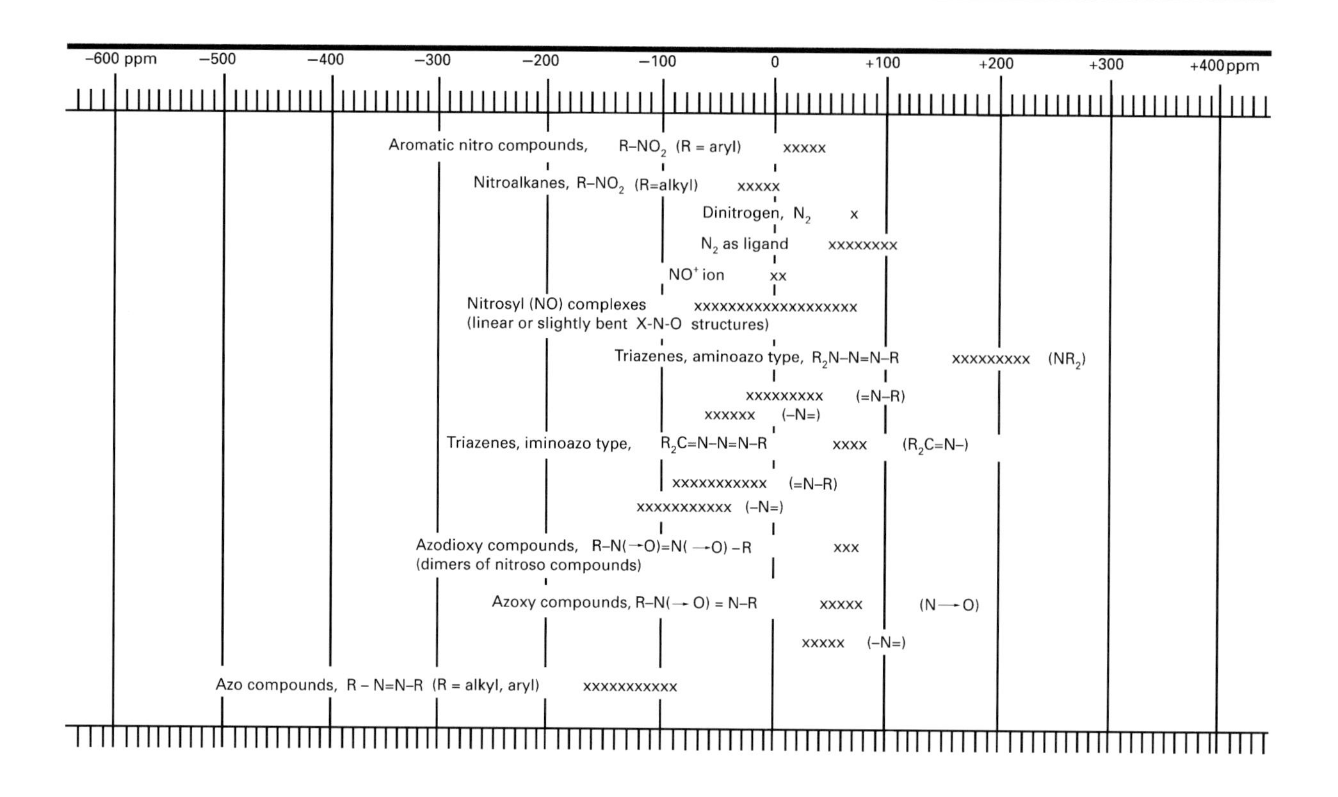

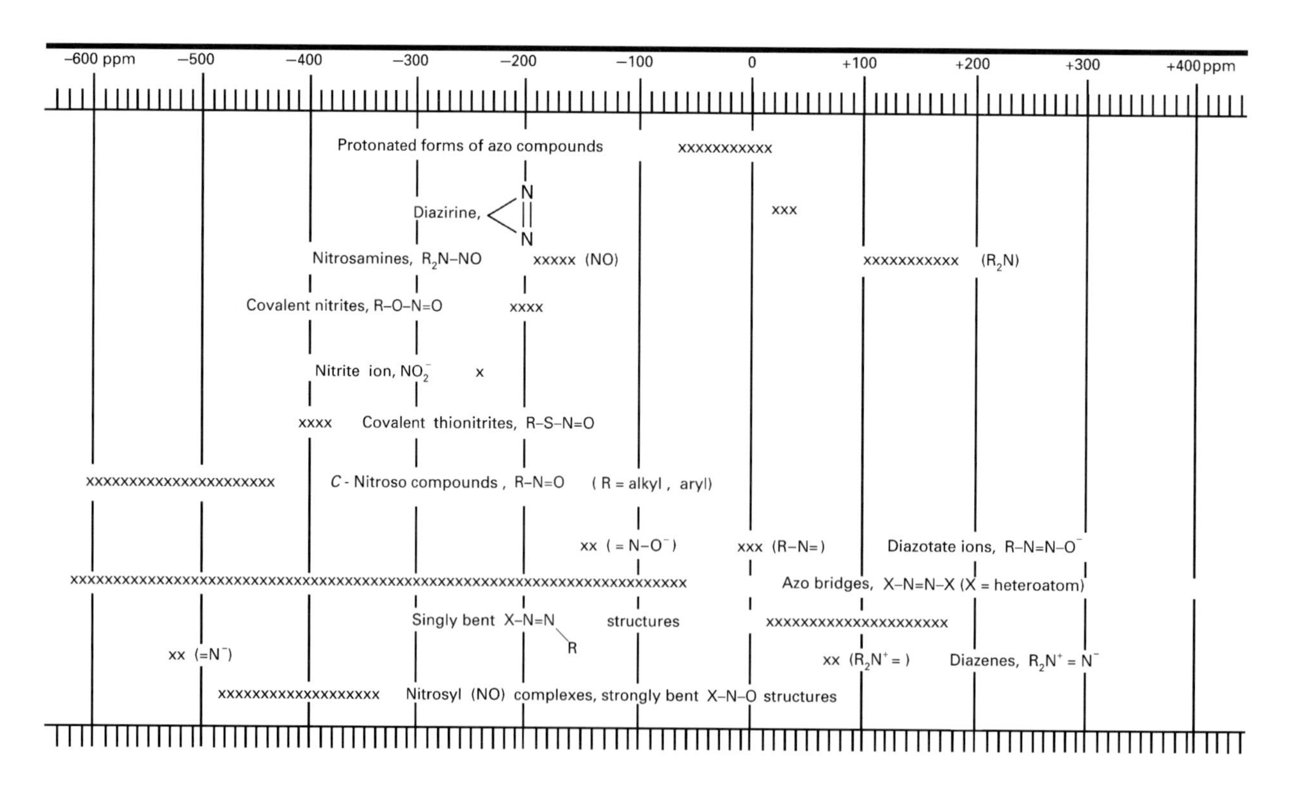

Figure 1 Characteristic nitrogen shielding ranges for various classes of molecules and ions (referred to external neat nitromethane)

azaindolizines it is in the region of 5. The value of s is also found to be positive for the nitrogen atoms of covalent cyanides, whereas it is negative for covalent nitrites, isothiocyanates, C-nitro and O-nitro groups. The sign of s is related to the orientation of bond dipoles in the vicinity of the nitrogen atoms concerned. This point of view is supported by molecular orbital calculations incorporating the Solvaton model.

In general the *trans*-azo bridge nitrogen atoms reveal only a very small shielding variation as the solvent is changed. An exception occurs when trifluoroethanol is used as solvent. This is probably due to a hydrogen-bonding interaction with the π-electrons of the N=N bond. The nitrogen shieldings of nitroso groups show a diversity of solvent-induced effects. The observed range for Bu^t-NO is about 10 ppm. The major contribution to this range of solvent-induced shielding changes arises from variations in solvent polarity. This is probably due to the fact that the bulky *tert*-butyl group prevents any significant amount of solvent-to-solute hydrogen bonding in this case. In contrast, the O-nitroso groups in covalent nitrites show much larger solvent-induced nitrogen shielding ranges which include significant contributions from solvent-to-solute hydrogen bonding effects.

Substituent effects on nitrogen shielding

A general trend in the values of the nitrogen shielding, which is observed for a large variety of structures, is for the shielding to decrease along the sequence, N–CH_3, N–CH_2R, N–CHR_2, N–CR_3, where R is an alkyl group. This is referred to as the β effect and is useful for locating nitrogen-containing functional groups within a hydrocarbon chain. Other comparable effects may arise when alkyl groups replace hydrogen atoms at other positions in an aliphatic chain, as shown in **Figure 2**.

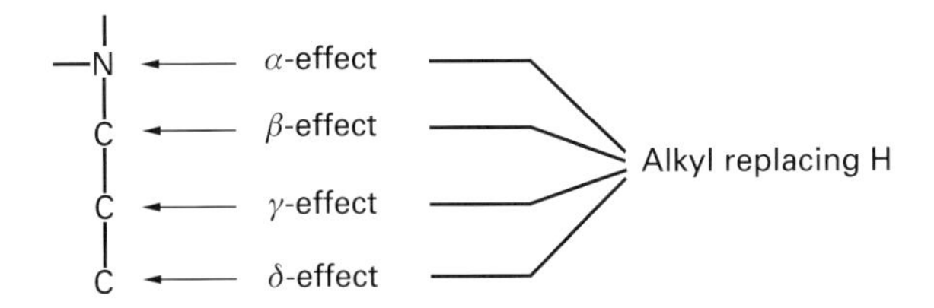

Figure 2 Effect on nitrogen chemical shifts caused by replacing hydrogen atoms by alkyl groups in an aliphatic chain.

For aliphatic amines the nitrogen shielding decreases by 7.8 ppm when an alkyl group is introduced into the α position. It decreases by 18.6 ppm when the introduction occurs at the β position. If the alkyl group appears at the γ position the opposite effect is observed and the shielding increases by 2.4 ppm, whereas the introduction of the alkyl group at the δ position produces a further decrease of nitrogen shielding by 2.5 ppm. Local molecular interactions can also be monitored by means of changes in nitrogen chemical shifts. Interactions of protein fragments provide an example of this, as shown in **Figure 3**.

Escherichia coli thioredoxin has characteristic sites which are affected by the transformation between its reduced and oxidized forms. Variations in the nitrogen chemical shifts of the amino acid fragments have been used in a study of the binding of Mg^{2+}, Ca^{2+} and Ba^{2+} to *E. coli* ribonuclease HI.

Spin–spin couplings

Due to the normally rapid quadrupolar relaxation of ^{14}N nuclei, spin–spin coupling interactions between ^{14}N and other nuclei are only seldom observed. Thus most couplings involving nitrogen nuclei are usually measured from the NMR spectra of ^{15}N-coupled nuclei. Spin–spin couplings between a nucleus X and ^{14}N and between X and ^{15}N are interconvertible by means of the following equation:

$$J(^{15}N{-}X) = -1.4027\ J\ (^{14}N{-}X)$$

Spin–spin couplings involving ^{15}N play an important role in the application of nitrogen NMR to the structure determination of nitrogen-containing molecules, since their values are often characteristic of the type and number of intervening bonds between the coupled nuclei.

Values of $^{1}J(^{15}N{-}^{1}H)$ are negative in sign and considerably larger in magnitude than spin–spin couplings between this pair of nuclei when separated by more than one bond. If proton exchange occurs within a given molecule then the observed value of the $^{15}N{-}^{1}H$ coupling represents a weighted average

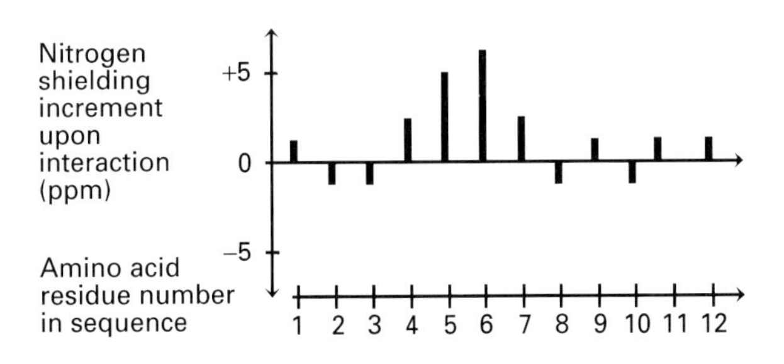

Figure 3 Effect on nitrogen shielding due to interactions of protein fragments.

which includes coupling with the proton at all its sites of residence. Such a situation arises with some porphyrins, for example octaethylporphyrin in $CDCl_3$. At −53°C the ^{15}N NMR spectrum consists of a singlet and a doublet split by 98 Hz; this latter value is typical of $^1J(^{15}N–^1H)$ in pyrrole-type systems. At +28° C only a quintet is observed split by 24 Hz. This indicates that the NH protons are exchanging amongst the four nitrogen atoms at higher temperatures and that the long-range N–H couplings are close to zero.

If major structural differences between molecules are considered then values of $^1J(^{15}N–^1H)$ often show a reasonable correlation with the s-characters of the N–H bonds involved. However, notable exceptions are known and thus it may be unsound to try to estimate the s-character of a given N–H bond from the observed value of the $^1J(^{15}N–^1H)$ interaction. Any such correlation would have to rely on the dominance of the contact term to the spin–spin coupling interaction. In cases where this dominance occurs then a simple distinction may be made between a variety of structures by means of $^1J(^{15}N–^1H)$ values.

Since absolute values of $^1J(^{15}N–^1H)$ are normally of the order of 100 Hz, the collapse of multiplet patterns in the ^{15}N NMR spectra of NH moieties can be used to monitor proton exchange processes which occur at rates of around 100 s^{-1}. Some examples where this technique has been used are ureas, lactams, arginine and histidine. Values of $^2J(^{15}N–^1H))$ across a saturated carbon atom are normally small in magnitude and positive in sign. Such couplings reveal a dependence of the orientation of the lone-pair electrons on the nitrogen atom with respect to the C–H bond. The largest values are observed when the bond is *cis* to the lone pair. If the intervening carbon atom is tricoordinate, then the value of $^2J(^{15}N–^1H))$ is much larger, about 15 Hz. When the coupling occurs in an imino-type moiety the value of $^2J(^{15}N–^1H)$ depends critically upon the bond structure at the nitrogen atom in question. If the nitrogen atom has a lone pair of electrons, the coupling is usually large in magnitude and negative in sign. If protonation of the lone pair electrons occurs, then the absolute value of the coupling decreases significantly. An even more dramatic change occurs when an N-oxide is formed; the resulting coupling may then have a small positive value. Similar observations have been made for cyclic lactams of the uracil type; if the nitrogen has a lone pair of electrons the observed value of $^2J(^{15}N–^1H)$ is large, e.g. the anion of 3-methyluracil. The coupling is considerably reduced when a hydrogen atom is attached to the nitrogen.

Values of $^3J(^{15}N–^1H)$ in saturated systems are often larger than the corresponding two-bond couplings. However, the three-bond couplings depend upon the dihedral angle between the N–C and C–H bonds and attain maximum values for the *cis* and *trans* arrangements and a minimum value when the dihedral angle is around 90°. Thus for the *gauche* configuration, where the dihedral angle takes a value of about 60°, rather small absolute values of $^3J(^{15}N–^1H)$ are expected. For unsaturated systems, e.g. aza-aromatic structures, the values of $^3J(^{15}N–^1H)$ do not differ appreciably in absolute magnitude from those found in saturated systems. This is in contrast to the situation found for values of $^2J(^{15}N–^1H)$.

In general, few values have been found for $^{15}N–^1H$ couplings across more than three bonds. Those reported for pyridine, its cation and its N-oxide are small and positive in sign, whereas those for nitrobenzene are small and negative in sign.

Usually the values of $^1J(^{15}N–^{13}C)$ are negative in sign and less than ~35 Hz. However, the range of values reported is from +4.9 Hz for an oxaziridine derivative to −77.5 Hz for 2,4,6-trimethylbenzonitrile *N*-oxide. Some attempts have been made to relate $^1J(^{15}N–^{13}C)$ data to the amount of s-character of the N–C bonds. Such a correlation would depend upon the dominance of the contact interaction to the spin–spin coupling and this is not always the case in practice. In general, nitrogen–carbon couplings across one bond are larger than those across more bonds. However, there are exceptions. These usually arise when the nitrogen atom in question has a lone pair of electrons in an orbital with high s-character. Examples are provided by pyridine-type nitrogen atoms in azines and azoles and imino-type nitrogen atoms in imines and oximes. For pyridine the value of $^1J(^{15}N–^{13}C)$ is +0.62 Hz, that for $^2J(^{15}N–^{13}C)$ is +2.53 Hz and for $^3J(^{15}N–^{13}C)$ it is −3.85 Hz. In general it pays to be cautious in N–C spin–spin coupling data to determine the presence, or absence, of specific N–C bonding arrangements.

Values of $^2J(^{15}N–^{13}C)$ are usually smaller then those across one bond in saturated systems. If the coupling is across a carbonyl carbon atom then the absolute value of $^2J(^{15}N–^{13}C)$ increases significantly to about 4–12 Hz. Thus a means is provided for distinguishing between saturated and unsaturated molecular fragments. If the nitrogen atom involved in the coupling has a lone pair of electrons in an orbital with significant s-character then the sign and magnitude of $^2J(^{15}N–^{13}C)$ depend critically upon whether the coupled carbon nucleus is *cis* or *trans* to the lone-pair electrons.

Three-bond $^{15}N–^{13}C$ couplings are usually negative in sign and less than 5 Hz. For saturated systems attempts have been made to relate values of $^3J(^{15}N–^{13}C)$ to the dihedral angle between the C–C and N–C bonds. Usually the rather small values observed for the couplings give rise to considerable errors in any such angle determination.

Couplings between ^{15}N and ^{13}C across more than three bonds are usually less than 1 Hz in absolute magnitude. A lone exception is a derivative of 1,2,4-triazine where a value of 3.9 Hz has been reported for the four-bond coupling between 1-N and a methyl group attached to a vinyl substituent.

The largest $^{1}J(^{15}N{-}^{15}N)$ values are found for *N*-nitrosoamines and the smallest for hydrazino moieties, nitramines and molecular N_2. Generally, values occur within the range of 10–20 Hz for N=N moieties. Exceptions are found when the moieties are charged, as they are in diazo compounds and azides, where the coupling is less than 10 Hz. The values of $^{2}J(^{15}N{-}^{15}N)$ across a nitrogen atom are close to zero in azides but may be as large as 11 Hz in some isomers of imino-type triazenes. Two-bond $^{15}N{-}^{15}N$ couplings across a carbon atom can also be significant, particularly when the carbon atom in question is tricoordinate.

Values of $^{1}J(^{31}P{-}^{15}N)$ cover a broad range and may be either positive or negative in sign. Hence, if only the absolute value is known their interpretation is not always straightforward. It seems that the value observed depends critically upon the dihedral angle between the nitrogen and phosphorus lone pairs of electrons. A significant algebraic decrease in the value of $^{1}J(^{31}P{-}^{15}N)$ occurs upon passing from tricoordinate to tetracoordinate phosphorus atoms.

A number of values of $^{2}J(^{31}P{-}^{15}N)$ have been reported for couplings across metal atoms in various complexes. For square-planar complexes a clear distinction is found in the values of the couplings depending upon whether the ligands in question are *cis* or *trans* to each other.

Values of $^{1}J(^{19}F{-}^{15}N)$ are positive in sign and usually between 150 and 460 Hz. Nitrogen–fluorine couplings across two or more bonds are also significant in magnitude.

$^{1}J(^{195}Pt{-}^{15}N)$ couplings have large absolute values, usually ranging from 100 to 580 Hz. They are found to be sensitive to the nature of the ligands, as well as to their arrangement in square-planar complexes. Normally the largest influence on the value of the coupling is exerted by the ligand that is *trans* to the nitrogen atom involved.

Nitrogen nuclear relaxation

Since the ^{14}N nucleus has an electric quadrupole moment, this may interact with local electric field gradients in the molecule, to produce rapid nuclear relaxation and broad NMR transitions. The electric field gradient may vary due to molecular motion and this variation may be studied by means of ^{14}N NMR. A combination of the ^{14}N quadrupolar relaxation rate and $^{13}C{-}^{1}H$ dipolar relaxation data can be used to determine the rotational correlation times for motion about each principal molecular axis. This approach has been applied to pyrimidine, pyridazine and pyrazine. ^{14}N relaxation data for formamide are found to be very sensitive to the presence of both cations and anions. These results provide direct evidence for specific ion–amide interactions and a tentative model for the interaction of electrolytes in liquid formamide.

^{14}N NMR line widths may be used as an aid to the assignment of nitrogen chemical shifts. In particular, a nitrogen atom bearing a partial positive charge tends to have a relatively small local electric field gradient associated with it, consequently its ^{14}N signal will be rather sharp. This approach has been extensively used in studies on meso-ionic compounds.

^{15}N nuclear relaxation usually occurs through varying contributions from the dipole–dipole, spin-rotation, chemical shielding anisotropy and scalar coupling mechanisms. If the ^{15}N nucleus is directly bonded to a proton then the dipole–dipole interaction is normally dominant. In the case of *trans*-azobenzene, the ^{15}N relaxation occurs by a combination of the spin-rotation, dipole–dipole and chemical shielding anisotropy interactions. The relative proportions of these mechanisms contributing to the relaxation are found to vary considerably over the temperature range from 5°C to 80°C. The ^{15}N relaxation times of a number of aldoximes and ketoximes are between 25 s and 50 s. In general the dipole–dipole interaction is the major contributor but other mechanisms can account for more than half of the measured ^{15}N relaxation in these molecules.

See also: **Chemical Shift and Relaxation Reagents in NMR; NMR Relaxation Rates; Relaxometers; Solvent Suppression Methods in NMR Spectroscopy; Spin Trapping and Spin Labelling Studied Using EPR Spectroscopy; Structural Chemistry using NMR Spectroscopy, Inorganic Molecules; Structural Chemistry using NMR Spectroscopy, Organic Molecules.**

Further reading

Witanowski M and Webb GA (1972) *Annual Reports on NMR Spectroscopy* **5A**: 395–464.

Witanowski M and Webb GA (eds) (1973) *Nitrogen NMR*. London: Plenum Press.

Witanowski M, Stefaniak L and Webb GA (1977) *Annual Reports on NMR Spectroscopy* **7**: 117–244.

Witanowski M, Stefaniak L and Webb GA (1981) *Annual Reports on NMR Spectroscopy* **11B**: 1–502.

Witanowski M, Stefaniak L and Webb GA (1986) *Annual Reports on NMR Spectroscopy* **18**: 1–761.

Witanowski M, Stefaniak L and Webb GA (1993) *Annual Reports on NMR Spectroscopy* **25**: 1–480.

NMR Data Processing

Gareth A Morris, University of Manchester, UK

MAGNETIC RESONANCE
Methods & Instrumentation

Computers play a central part in modern NMR spectroscopy. Their use for the real-time control of pulsed NMR experiments has enabled the development of multiple pulse techniques such as two-dimensional NMR; this article deals with the part played by computers in the acquisition, processing and presentation of experimental NMR data.

All NMR experiments rely on the excitation of a nuclear spin response by a radiofrequency magnetic field, usually in the form of a short pulse or pulse sequence. This generates a rotating nuclear magnetic moment, which induces a small oscillating voltage in the probe coil: a 'free induction decay' (FID). The spectrometer receiver amplifies this voltage, shifts it down into the audio frequency range, and filters out high-frequency components before it is digitized by an analogue-to-digital converter (ADC). In modern instruments two receiver channels with phases 90° apart are used ('quadrature detection'), allowing the relative signs of frequencies to be distinguished. From this point onwards the signal is handled digitally until it is presented to the experimenter as a printed spectrum or an interactive spectral display. The three main stages of processing are first, the acquisition of a filtered, averaged time-domain recording of the nuclear spin response; second, the generation of a frequency-domain spectrum, usually but not always by Fourier transformation; and third, postprocessing of the spectrum to aid its interpretation. The three stages are summarized briefly below, followed by more detailed discussions of the techniques used and some practical illustrations.

Data acquisition

NMR experiments generally require the coaddition of a number of recordings of the nuclear spin response ('transients'), often using different permutations of radiofrequency pulse and receiver phases ('phase cycling'). The data are recorded as a set of complex numbers which sample the in-phase and quadrature NMR signals as a function of time. Early Fourier transform spectrometers had limited word length and required some scaling of the data before coaddition, but this is no longer necessary and successive transients are simply added to memory. Many recent spectrometers interpose a stage of digital signal processing before data summation: a wide receiver band width is used, and data are digitized very rapidly. These oversampled data are then filtered digitally before downsampling and addition to memory, giving better filtration of unwanted noise and signals, better effective ADC resolution, and less baseline distortion.

In two-dimensional (2D) NMR a series of FIDs is acquired using a pulse sequence containing a variable evolution period, which is incremented regularly to map out the behaviour of the nuclear signal as a function both of real time t_2 during the FID, and of the evolution time t_1. A typical 2D NMR experiment might acquire 512 FIDs of 1024 complex points, which would then be doubly Fourier transformed as a function of the two time variables. 3D and 4D NMR extend the principle to two and three evolution periods respectively.

Spectrum generation

To make the raw experimental data interpretable they must be converted into a frequency-domain spectrum. The classical, and commonest, method is discrete Fourier transformation. This is a linear operation: the information content of the data is unchanged. Alternative methods such as maximum entropy reconstruction and linear prediction are nonlinear, changing the information content. In Fourier processing a weighting function is usually applied to the time-domain data before transformation; a DC correction based on the last portion of the data may also be performed. After weighting and any zero-filling (see below), the fast Fourier transform (FFT) algorithm is used to produce the frequency spectrum. The resultant spectrum consists of a set of complex numbers sampling a defined frequency range at equal intervals. In general the real and imaginary parts of the spectrum will both be mixtures of absorption mode and dispersion mode signals; a spectrum suitable for display and analysis is obtained by taking linear combinations of the real and imaginary data ('phasing'), or, if that is not possible, by taking the modulus ('absolute value mode') or square modulus ('power mode').

In 2D NMR, weighting and zero filling are carried out on the FIDs as normal, but after Fourier transformation with respect to t_2 to give a series of spectra $S(t_1, F_2)$ the data matrix is transposed to give a set of 'interferograms' $S(F_2, t_1)$. A second Fourier transformation, with respect to t_1, yields the 2D spectrum $S(F_1, F_2)$. Because most coherence transfer processes cause amplitude rather than phase modulation as a function of t_1, information on the signs of F_1 frequencies is usually missing; it can be recovered by making two measurements using different phase cycles. If the two sets of measurements are combined before Fourier transformation, converting amplitude modulation into phase modulation, absolute value mode presentation is generally required because individual signals will show a 'phase-twist' line shape; this is also the case with simple experiments which use pulsed field gradients for coherence transfer pathway selection. If the two data sets are recombined appropriately after the first Fourier transformation, both phase cycled and pulsed field gradient 2D experiments can be made to yield pure double absorption mode line shapes. Two common recombination schemes are the 'hypercomplex' method of States, Haberkorn and Ruben, and the time-proportional phase incrementation (TPPI) method of Marion and Wüthrich.

Postprocessing

Many chemical questions can be answered with a simple spectrum, but to extract all the useful information from an NMR experiment it can require a wide range of data processing methods. Integration (usually after baseline correction) and peak picking allow the relative numbers of spins responsible for different multiplets to be found, and chemical shifts and scalar couplings to be measured. Better measures of signal intensity and line shape can be found by least-squares fitting, if necessary in conjunction with line shape correction by reference deconvolution. Computers can also aid in the analysis of strongly coupled spectra and the spectra of dynamic systems.

Zero-filling

Data acquisition produces a set of N complex points, sampled at equal time intervals Δt, which describe the nuclear FID as $2N$ independent pieces of information. Discrete Fourier transformation of these data will produce a spectrum of N complex points, at frequency intervals $1/(N\Delta t)$. A final absorption mode spectrum will contain just N independent pieces of information, only half the amount originally acquired; the remaining N points form the dispersion mode spectrum. Full use of the experimental data can be achieved if N complex zeroes are appended to it before Fourier transformation ('zero-filling'). Transforming N data points plus N zeroes generates a spectrum of $2N$ complex points, at frequency intervals $1/(2N\Delta t)$ Hz. The $2N$ real points are independent, and contain the same information as the $2N$ imaginary points: the real and imaginary data are correlated, so the total information content of the spectrum is unchanged. Zero-filling thus improves the digital resolution of the frequency spectrum twofold, as can be seen from **Figure 1**. Appending more than N zeroes before transformation cannot increase the information content of the spectrum: the extra data points obtained simply interpolate between those produced by a single zero-filling.

Time-domain weighting

All experimental NMR signals decay, sooner or later: most well-designed experiments sample the signal until it has decayed close to zero. The later stages of the recorded data contain less signal, but the noise remains more or less constant. Recording for too short a time will lose valuable data; recording for too long will emphasize the noise relative to the signal. The widths of spectral lines depend on the rate of decay of the NMR signal: resolution can be improved if the natural decay of the signal is counteracted. Both issues can be addressed by weighting the time-domain signal before Fourier transformation. The choice of weighting function determines the compromise between resolution and signal-to-noise ratio in the resultant spectrum.

The signal-to-noise ratio of a spectrum can be optimized by multiplying the experimental signal by a weighting function which matches the experimental decay envelope: 'matched filtration'. The Fourier transform of an exponential decay with time constant T is a Lorentzian line shape with a full width at half height of $1/(\pi T)$ Hz. Thus to obtain the best signal-to-noise ratio for a Lorentzian line of width W Hz the experimental data should be multiplied by a decaying exponential of time constant $1/(\pi W)$ s. This

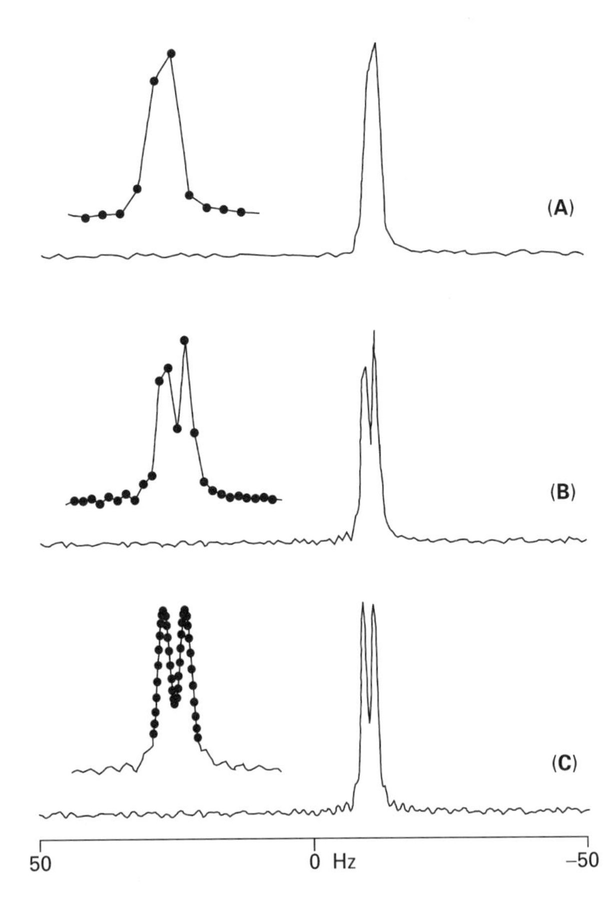

Figure 1 Spectra of a doublet with splitting 2 Hz, centred at –10 Hz, calculated for a 64 complex point FID with (A) no zero filling; (B) one zero filling, to 128 complex points; and (C) four zero fillings, to 1024 complex points. The splitting becomes visible after one zero filling; further zero filling is equivalent to interpolation between the data points with a sin(x)/x function.

gives a spectrum with optimum signal-to-noise ratio, at the expense of a doubling of the line width to $2W$ Hz. Where a spectrum with poor signal-to-noise ratio contains lines with a range of widths it can be helpful to try exponential weighting with several different time constants.

Time-domain weighting is equivalent to frequency-domain convolution. The convolution theorem states that the Fourier transform of the product of two functions $a(t)$ and $b(t)$ is the convolution of the two individual transforms $A(\nu)$ and $B(\nu)$:

$$\begin{aligned} \mathrm{FT}^{-}[a(t) \times b(t)] &= \int_{-\infty}^{\infty} a(t) \times b(t) \exp(-2\pi i \nu t)\, \mathrm{d}t \\ &= \int_{-\infty}^{\infty} A(\nu') \times B(\nu - \nu')\, \mathrm{d}\nu' \\ &= \mathrm{FT}^{-}[a(t)] \otimes \mathrm{FT}^{-}[b(t)] \\ &= A(\nu) \otimes B(\nu) \end{aligned} \qquad [1]$$

where $\mathrm{FT}^{-}[\,]$ indicates Fourier transformation and $\otimes$ denotes convolution. Thus time-domain exponential weighting is equivalent to convolution, or smoothing, with a Lorentzian lineshape in the frequency domain. Matched filtration corresponds to smoothing the raw experimental spectrum with a function which matches the experimental line shape, as **Figure 2** illustrates.

Time-domain weighting is also extensively used for resolution enhancement. Since this emphasizes the later part of the experimental signal, the noise energy is increased with respect to the signal energy, and resolution enhancement reduces the signal-to-noise ratio of the resultant spectrum. The aim of resolution enhancement is to reduce line widths without degrading the signal-to-noise ratio unacceptably. The natural decay of individual NMR signals is normally exponential, but countering this decay by multiplication with a rising exponential would lead to steeply rising noise. To stop the exponential rise in noise a further weighting using a function with a steeper decline is required. A weighting function $W(t)$ composed of a rising exponential with time constant t_e and a falling Gaussian with time constant t_g

$$W(t) = \exp(+t/t_e) \exp[-(t/t_g)^2] \qquad [2]$$

is generally the method of choice for resolution

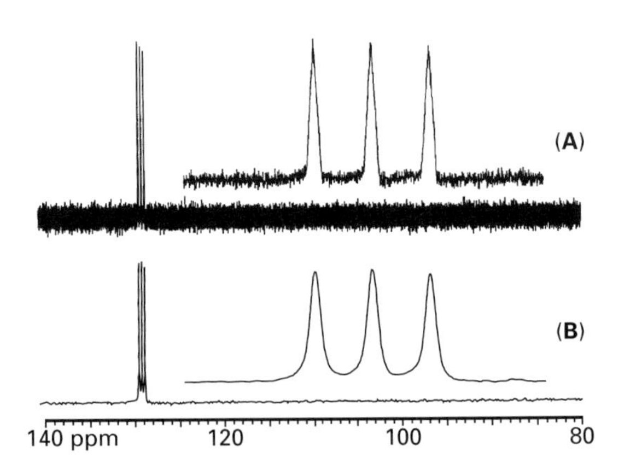

Figure 2 75.4 MHz spectra of the ^{13}C triplet of deuteriobenzene in the ASTM (American Society for Testing and Materials) sensitivity test sample (60% deuteriobenzene/40% dioxane), with and without matched filtration. The unweighted spectrum (A) shows a signal-to-noise ratio of 18:1; the same data given an exponential multiplication with a time constant $1/(3\pi)$ s before Fourier transformation, corresponding to a 3 Hz Lorentzian line broadening, show (B) a signal-to-noise ratio of 92:1. An acquisition time of 5.462 s was used, with a spectral width of 12000 Hz and one zero filling. The insets show expansions of the triplet signal, illustrating the broadening of the lines and the smoothing of the noise caused by the exponential multiplication.

enhancement. $W(t)$ can also be written as a time-shifted Gaussian

$$W(t) = \exp(+t_s^2/t_g^2)\exp[-(t-t_s)^2/t_g^2] \quad [3]$$

where $t_s = t_g^2/(2t_e)$. If t_e is equal to the decay constant of the experimental NMR signal, then multiplication by $W(t)$ before Fourier transformation converts a Lorentzian lineshape of width $1/(\pi t_e)$ Hz to a Gaussian of width $(2/\pi t_g)\sqrt{\log_e 2}$ Hz. Since spectra normally contain a range of line widths, it is usually necessary to experiment with t_e and t_g to find the best values for a given region of a spectrum. Because instrumental effects such as field inhomogeneity make experimental line shapes non-Lorentzian, resolution enhancement is best combined with reference deconvolution. **Figure 3** shows the application of Lorentz–Gauss resolution enhancement to a proton multiplet.

Even where neither sensitivity nor resolution enhancement is sought, time-domain weighting is desirable where some NMR signal survives at the end of the sampled data. Such a truncated dataset is equivalent to the full, untruncated signal multiplied by a window function; the convolution theorem shows that the resultant spectrum will contain the true line shapes convoluted by a 'sinc' [sin(x)/x] function, giving rise to 'wiggles' on either sides of lines. Applying a weighting function $W(t)$, which brings the time-domain data smoothly to zero ('apodization'), can reduce or suppress such undesirable artefacts.

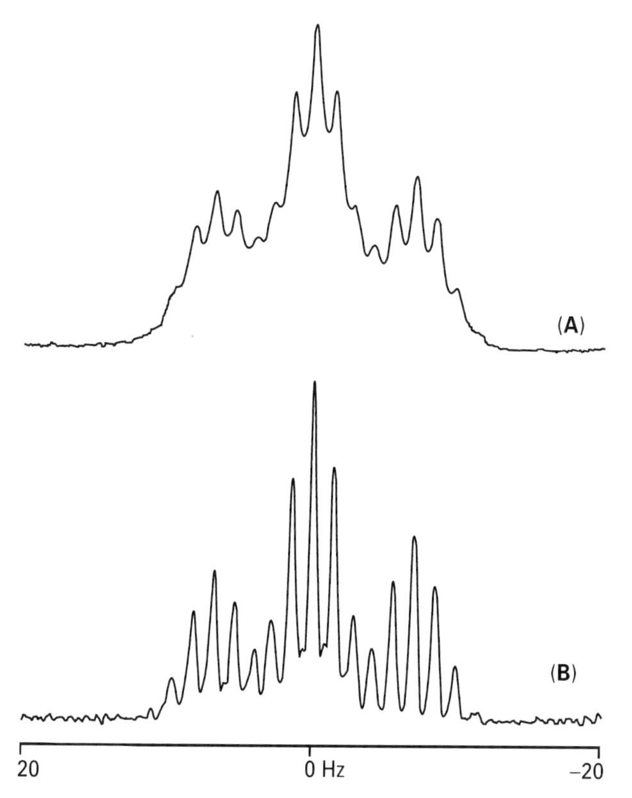

Figure 3 Expansions of the multiplet at 5.1 ppm in the 400 MHz proton spectrum of geraniol in deuteriomethanol: (A) raw spectrum; and (B) spectrum after Lorentz–Gauss conversion using rising exponential weighting with a time constant of $1/\pi$ s and Gaussian weighting with a time constant of 1 s.

Most NMR spectra are presented in phase-sensitive mode, but this is not appropriate where the phases of signals vary rapidly or unpredictably with position, as in some magnetic resonance imaging and multidimensional NMR experiments. The modulus of a complex Lorentzian line shape shows a very broad base because of the contribution from the (imaginary) dispersion mode component. This can be suppressed if time-domain weighting is used to force the experimental signal into a form that is time-symmetric, for example using the function $W(t)$ above with a small t_e ('pseudo-echo' weighting) or using a half sine-wave ('sine-bell' weighting). Although it is common to arrange for such weighting to leave the maximum of the weighted signal at the midpoint of the experimental data, this is neither necessary nor always desirable. Absolute value mode presentation is the norm where phase cycling or pulsed field gradients are used to produce signal phase modulated as a function of t_1 in 2D NMR; it is to be avoided where possible because overlapping peaks are distorted by interference between their dispersion mode parts.

Fourier transformation

The classical frequency domain spectrum $S(\nu)$ is the Fourier transform of the FID $s(t)$:

$$S(\nu) = \int_0^{t_a} s(t)\exp(-2\pi i\nu t)\,dt \quad [4]$$

where the integration limits reflect the fact that the FID starts at time zero and is recorded for a time t_a. Practical spectrometers use digital technology, so the FID $s(t)$ is digitized at regular intervals Δt to give a time series of M points $s_k = s[(k-1)\Delta t]$, where $(M-1)\Delta t = t_a$, and a discrete Fourier transform (DFT) is carried out using the Cooley–Tukey FFT algorithm. The DFT of a time series of N complex points with spacing Δt generates a frequency spectrum which is a series of N complex points with spacing $1/(N\Delta t)$ Hz:

$$S_n = \frac{1}{\sqrt{N}}\sum_0^{N-1} s_k \exp(-2\pi ikn/N) \quad [5]$$

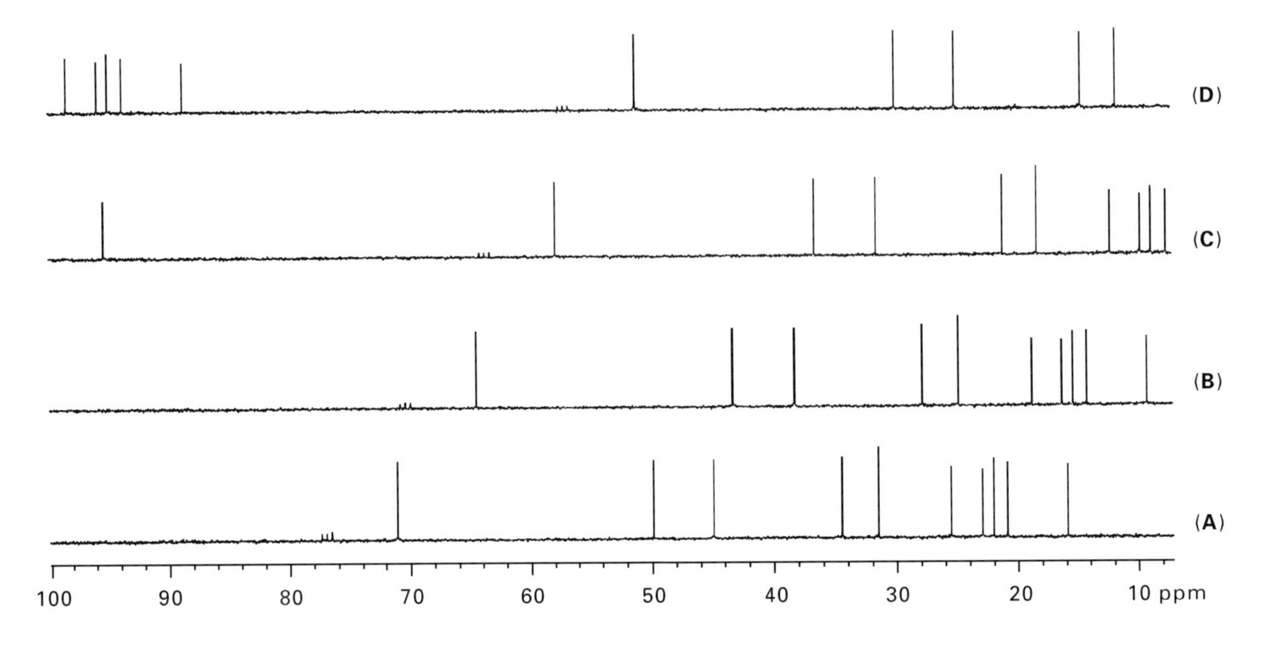

Figure 4 75.4 MHz proton-decoupled ^{13}C spectra of 30% menthol in deuteriochloroform, (A) recorded with all signals within the spectral window, and (B)–(D) with the transmitter displaced to high field in 500 Hz steps. Spectra (C) and (D) show the aliasing of the high-field signals to reappear at the low-field end of the spectrum.

where the frequency of the nth point is $(n-1)/(N\Delta t)$ Hz. Both the continuous and the discrete Fourier transform can use several different sign and normalization conventions; those given here are widely used, but others are equally valid. Spectrometers are almost invariably restricted by the FFT to Fourier transforming numbers of points N which are powers of 2.

Discrete sampling in the time domain introduces an ambiguity into the frequency domain: signals at frequencies separated by multiples of $1/(\Delta t)$ Hz are indistinguishable, since their relative phases only change by multiples of 2π between sampling points. NMR signals which lie outside the range 0 to $(N-1)/(N\Delta t)$ will be 'aliased' by adding or subtracting the spectral width $1/\Delta t$ until they lie within this window, as seen in **Figure 4**. Since the signals that emerge from the spectrometer receiver may have positive or negative frequencies, the output of the DFT needs to be rotated by $N/2$ points $[1/(2\Delta t)$ Hz$]$ so that the digitized spectrum covers the range $-1/(2\Delta t)$ to $(N/2-1)/(N\Delta t)$ Hz. By convention, the resultant spectrum is plotted with frequency increasing from right to left, so that the most shielded nuclei (those with lowest chemical shift) lie at the right. Some older spectrometers sample the real and imaginary receiver channels alternately rather than simultaneously, using a real rather than a complex Fourier transform; signals outside the spectral width then 'fold' back into the spectrum by reflection about the frequency limits $\pm 1/(2\Delta t)$.

Phasing

The measurement of NMR data must wait until the radiofrequency pulse and its after-effects have died away, which can take several tens of microseconds; analogue or digital filtration also delays the arrival of the NMR signal at the receiver output. The effect on an NMR signal of a delay time δ is to add a phase shift $\exp(2\pi i\nu\delta)$ to a signal of frequency ν, causing the signal phase to vary linearly across the spectrum. In addition, the relative phases of the receiver reference signals and the transmitter pulse are arbitrary, so both a zero- and a first-order phase correction are needed to bring all signals into absorption mode. The phased spectrum $S_p(\nu)$ can be written

$$S_{\mathrm{p}}(\nu) = S(\nu)\exp[-\mathrm{i}(\phi_0 + \phi_1\nu\Delta t)] \qquad [6]$$

where the zeroth-order and first-order phase shifts ϕ_0 and ϕ_1 are normally determined either automatically, or by the spectrometer operator using an interactive display. **Figure 5** shows a typical spectrum before and after phasing.

First-order phase correction has one insidious effect, baseline distortion. A frequency-dependent phase shift cannot make up for the data that were lost during δ; the baseline error is just the DFT of the missing data. However, provided δ is small compared to Δt, this baseline curvature can easily be corrected during postprocessing of the spectrum. In 2D

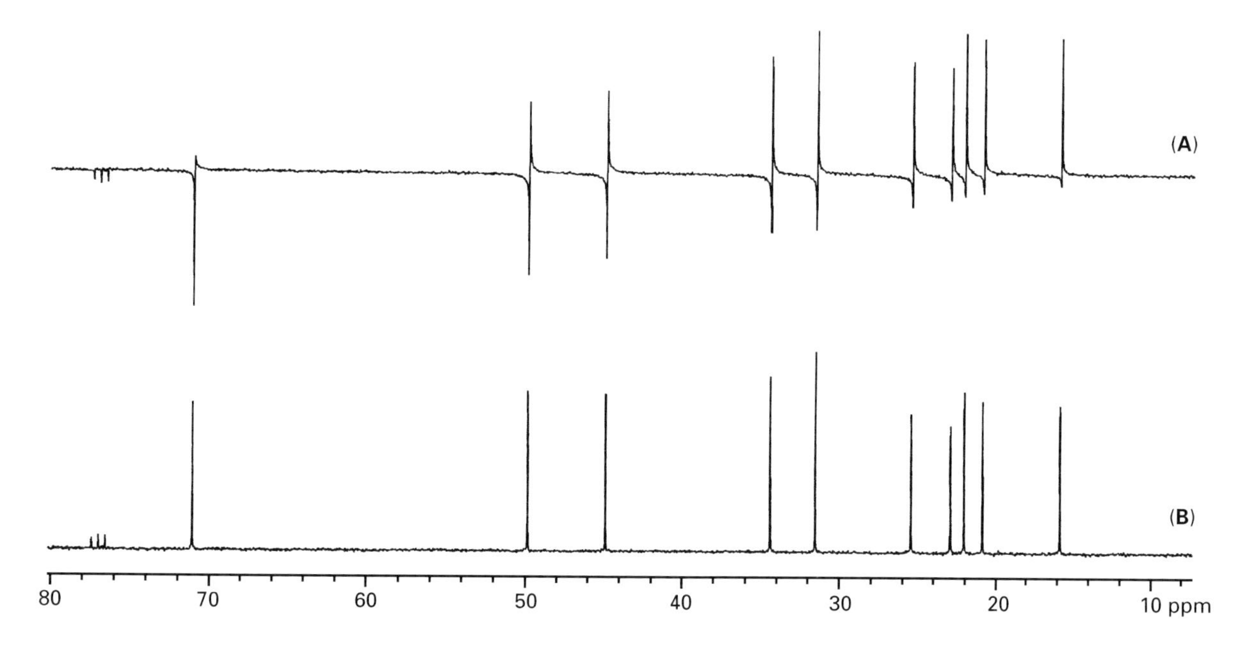

Figure 5 75.4 Mhz proton -decoupled 13c spectra of 30% menthol in deuteriochloroform, (A)b before and (B) after zero- and first-order phase correction.

NMR, there will be different time delays δ_1 and δ_2 in the two time dimensions; phasing again is normally carried out using an interactive display.

Linear prediction

A FID and its Fourier transform contain exactly the same information, but sometimes this is insufficient to give a readily interpretable spectrum. Where there is adequate internal evidence within the FID, it may be possible to extrapolate the NMR signals forwards and/or backwards in time to synthesize missing data and hence create a time-domain signal that transforms to a clearer spectrum. The two commonest uses of such an extrapolation are backwards in time to replace data lost during the time δ, and forwards in time to improve resolution. A typical digitized experimental FID contains a series of n exponentially damped, complex signals, plus a background of random noise. The NMR signal can be written:

$$s_k = \sum_{j=1}^{n} \alpha_j \beta_j^k \qquad [7]$$

where the complex number $\alpha_j = A_j \exp(i\phi_j)$ defines the phase ϕ_j and amplitude A_j of the jth signal, and $\beta_j = \exp(2\pi i \Delta t \nu_j) \exp(-\Delta t / T_j)$ is determined by the frequency ν_j and decay constant T_j. The contribution made by component j to point s_k is just the contribution to s_{k-1} multiplied by β_j. Linear prediction (LP) algorithms take a FID of M points and fit this time series with a set of m complex coefficients a_j

$$s_k = \sum_{j=1}^{m} a_j s_{k-j} \qquad [8]$$

so that point k is expressed as a linear combination of the previous m points (forward prediction), or of the subsequent m points (backward prediction)

$$s_k = \sum_{j=1}^{m} a_j s_{k+j} \qquad [9]$$

A variety of algorithms exist for finding the coefficients a_j and multipliers β_j, with which the experimental data can be extrapolated forwards or backwards; all share some common problems. Linear prediction has difficulty distinguishing between positive and negative decay constants T_j, and so is best suited to time series in which all the decay constants are either positive or negative, allowing spurious β values to be rejected. The number m of coefficients a_j to be used has to be decided by the experimenter: too few, and peaks will be missed; too many and noise will be treated as signal. Non-Lorentzian line shapes make exponential damping a poor approximation, increasing the number of coefficients needed.

Linear prediction is (despite its name) a nonlinear method and can produce very misleading results, but with care it can greatly ease interpretation of poorly digitized spectra. It is particularly useful in 2D NMR, where signals are routinely truncated in the t_1 dimension.

Maximum entropy reconstruction

Although linear prediction can be used to extract spectral data directly from a FID ('parametric LP'), it is commonly used to extrapolate the experimental time-domain data, which are then weighted and transformed as normal. Maximum entropy reconstruction, in contrast, seeks to fit the experimental FID with a model function that contains the minimum amount of information consistent with fitting experiment to within the estimated noise level. The criterion of minimum information corresponds to the maximum Shannon informational entropy $S(p)$, which for a probability distribution p is defined as

$$S(p) = -\sum_{n} p_n \log_e p_n \qquad [10]$$

Maximum entropy methods have generated considerable controversy; they have been described (a little unkindly) as generating more heat than light. Their results can show spectacular improvements in signal-to-noise ratio, but this should not be confused with sensitivity of detection. Maximum entropy methods successfully pick out those signals that are above a defined threshold, but miss those below it; the signal amplitude estimates produced are comparable to those obtainable by simply fitting a model line shape to the Fourier transform spectrum. Thus for well-sampled experimental data the advantages of maximum entropy methods are largely cosmetic, and come at a high computational cost. Where such methods can be very valuable is with data sets that are damaged, incomplete, not sampled at uniform time intervals, or require deconvolution.

Postprocessing techniques

The simplest and most widely used form of postprocessing is integration of the signal intensity. The integral of a resonance is proportional to the number of spins contributing to it; thus the relative numbers of nuclei in different chemical groupings can be found by comparing the integrals of their signals under appropriate experimental conditions. Accurate integration almost always requires operator intervention to carry out baseline correction, varying according to need from simple offset and slope correction through to the subtraction of a baseline calculated by spline or polynomial fit to operator-defined regions of empty baseline.

A second common example of postprocessing is the listing of signal heights and positions ('peak picking'), for the measurement of chemical shifts and coupling constants, or as the first step in the extraction of parameters such as relaxation times, rate constants or diffusion coefficients. In principle, the integral of a signal should give a better estimate of signal amplitude than peak height, being independent of line shape; in practice, baseline errors and signal overlap mean that peak height measurements are usually preferred for intensity comparisons between corresponding signals in different spectra.

Where signals overlap, neither integration nor peak picking gives accurate signal intensities. Here iterative fitting can be used to decompose the experimental spectrum into contributions from individual lines, typically assumed to have Lorentzian or Gaussian shapes; the positions, amplitudes and widths of the theoretical line shapes are varied to minimize the sum of the squares of the differences between the experimental and calculated spectra. Such least-squares fitting is easily perturbed by instrumental distortion of the line shapes, for example as a result of static field inhomogeneity, so the best results require either great care with shimming or some form of compensation for instrumental line shape contributions.

One effective way to compensate for many instrumental sources of error is reference deconvolution. Since most instrumental errors (e.g. static field inhomogeneity and magnetic field instability) affect all signals equally, multiplying the experimental FID by the complex ratio of the theoretical and experimental signals for a reference resonance leads to a spectrum in which such errors have been corrected. This technique can be used to ensure that all lines are basically Lorentzian, and also to enforce strict comparability between different spectra in a series, for example to correct t_1 noise in multidimensional NMR.

Many other data processing techniques are used to extract useful information from experimental NMR spectra. Signal intensities compiled by peak picking may be fitted to an exponential or Gaussian function, as in the determination of relaxation times and in pulsed field gradient spin-echo measurements of diffusion coefficients. The latter experiment can be extended to the construction of a pseudo-2D spectrum in which signals are dispersed according to chemical shift in one dimension and diffusion coefficient in the other ('diffusion-ordered spectroscopy'). In many spectra the extraction of chemical shift and coupling constant values is hindered by second-order effects ('strong coupling'); the analysis of strongly coupled spectra is most effectively carried out using quantum-mechanical simulation. This can be partially automated in favourable cases, the experimental spectrum being used as a target for least-squares fitting in which the variable parameters are the chemical shifts and coupling constants rather than simply the signal positions, amplitudes and widths. The extraction of kinetic parameters from exchange-broadened band shapes ('line shape analysis') can be similarly automated.

List of symbols

a, b = time-domain functions; A, B = frequency-domain structures; F, ν = frequency; $\mathrm{i} = \sqrt{-1}$; m = number of complex coefficients; M, N = number of complex points; p = probability distribution; s_k = kth point in free induction decay; S_n = nth point in spectrum; $S(p)$ = Shannon informational entropy; $S_p(\nu)$ = phased spectrum; $s(t)$ = free induction decay; t_1 = evolution time; t_2 = real time; t_e = exponential time constant; t_g = Gaussian time constant; t_s = time shift of Gaussian; T = decay constants; W = line width; $W(t)$ = weighting function; α_j = jth complex amplitude; β_j = jth complex exponential; δ = delay time; ϕ = phase shift.

See also: **Fourier Transformation and Sampling Theory; Laboratory Information Management Systems (LIMS); NMR Principles; NMR Pulse Sequences; NMR Spectrometers; Two-Dimensional NMR, Methods.**

Further reading

Freeman R (1997) *Spin Choreography*. Oxford, UK: Spectrum.

Freeman R (1987) *A Handbook of Nuclear Magnetic Resonance (2/e)*. Harlow, UK: Longman.

Hoch JC and Stern AS (1996) *NMR Data Processing*. New York: Wiley-Liss.

Lindon JC and Ferrige AG (1980) Digitisation and data processing in Fourier transform NMR. *Progress in NMR Spectroscopy* **14**: 27–66.

Morris GA, Barjat H and Horne TJ (1997) Reference deconvolution methods. *Progress in NMR Spectroscopy* **31**: 197–257.

Rutledge DN (ed) (1996) Signal treatment and signal analysis in NMR. *In*: Vol 18 of *Data Handling in Science and Technology*. Amsterdam: Elsevier.

NMR in Anisotropic Systems, Theory

JW Emsley, University of Southampton, UK

MAGNETIC RESONANCE
Theory

The simplest anisotropic system is a single crystal and in this case the spin interactions depend upon the orientation of the molecules with respect to the direction of the applied magnetic field, B_0, of the spectrometer. The anisotropies of the various interactions that affect the spectrum can be measured by changing the orientation of the crystal in the field. The spin interactions in polycrystalline or noncrystalline solids are still anisotropic, but now the spectra are invariant to sample orientation. Molecular motion can have a dramatic effect on the NMR spectra observed for anisotropic systems, and characterizing such motion is one of the principal uses of this spectroscopy.

The most complex anisotropic systems are the various liquid crystalline phases; now the molecular motion is similar to that in a normal, isotropic liquid in that there is rapid rotation and diffusion of the molecules. These phases differ from isotropic phases in that the motion is not random, and this has a profound affect on their NMR spectra.

The spectra produced by anisotropic systems are more complex than those from isotropic phases, but they contain more information. However, the spectral complexity is often too great for all the information to be obtained, and for solid samples it is usually preferable to dissolve the material in order to simplify the spectrum. Various other methods have been developed that enable some, if not all, of the information to be extracted from the NMR spectrum of an anisotropic sample, and these are described elsewhere. The aim of this article is to show what can, in principle, be obtained by a successful analysis of the NMR spectra obtained from solid and liquid crystalline samples.

General considerations

We will consider only experiments that use a strong magnetic field, that is those using mainstream spectrometers in which B_0 is greater than about one tesla. In this case, for nuclei with spin $I = \frac{1}{2}$, the largest magnetic interaction is that between μ, the magnetic dipole moment of the nucleus, and the field, the Zeeman interaction. This leads to the field direction being, to a good approximation, a unique axis of quantization for the nuclear spins; and the experiments yield only the component, $T_{B,q}$, of the qth interaction along $\boldsymbol{B}_0$, and so these are the parameters that may be obtained from the spectra. For nuclei with $I > \frac{1}{2}$ the interaction of the nuclear electric quadrupole moment, eQ_N, of a nucleus of isotopic species denoted by N with the electric field gradient, $\boldsymbol{V}_i$, at the site i may be larger than the Zeeman interaction, and the spins will no longer be quantized along $\boldsymbol{B}_0$. We will not deal with this situation, and the interested reader is referred to the article on the theory of nuclear quadrupole resonance. The present discussion is therefore confined to spin $\frac{1}{2}$ nuclei, or to those with small quadrupole moments, or at sites of small field gradients, such that the Zeeman term still dominates the nuclear spin Hamiltonian.

For each of the spin interactions that affect the spectrum of an anisotropic sample, the value of $T_{B,q}$ may have both a part, $T_{B,q}$(isotropic), that is independent of the orientation with respect to B_0 of the molecule containing the spin, and a part, $T_{B,q}$(anisotropic), that is orientation dependent. Thus,

$$T_{B,q} = T_{B,q}(\text{isotropic}) + T_{B,q}(\text{anisotropic}) \qquad [1]$$

Four interactions may affect the spectrum of an anisotropic sample. The shielding (or chemical shift), σ_i, of the nucleus at a site i always has a finite value of $T_{B,q}$(isotropic), σ_B^0, in an anisotropic sample. There will be a nonvanishing contribution $T_{B,q}(\text{anisotropic}) = \sigma_B^a$ provided that the symmetry of the site i is lower than tetrahedral. The electron-mediated spin–spin coupling, J_{ij}, between a pair of nuclei in an anisotropic sample will also in general have two nonvanishing contributions, J_{ij}^0 and J_{ij}^a. Both terms are negligible when the two nuclei are in different molecules. For nuclei within one molecule, the anisotropic term ranges from being negligible to being very large depending on the nuclear isotopes involved. The dipolar spin–spin coupling, D_{ij}, has only an anisotropic term, as does the quadrupolar coupling, $eQ_N V_i$. We will return to the individual anisotropic interactions after discussing the features they have in common.

Relationship between spectral parameters $T_{B,q}$ and molecular properties

To understand what information is provided on properties of the molecules, we need to know how to relate $T_{B,q}$ to components $T_{\alpha\beta,q}$ of the qth interaction in a reference frame fixed within a molecule. This relationship is a general one since all four interactions are second-rank tensors, and so:

$$T_{B,q} = \sum\nolimits_{\alpha,\beta} T_{\alpha\beta,q} \cos\theta_\alpha \cos\theta_\beta \qquad [2]$$

$T_{\alpha\beta,q}$ is a component in the (xyz) molecular axes, and θ_α is the angle $\boldsymbol{B}_0$ makes with the axis α. There are nine components of $T_{\alpha\beta,q}$:

$$\begin{matrix} T_{xx,q} & T_{xy,q} & T_{xz,q} \\ T_{yx,q} & T_{yy,q} & T_{yz,q} \\ T_{zx,q} & T_{zy,q} & T_{zz,q} \end{matrix}$$

The sum $T_{xx,q} + T_{yy,q} + T_{zz,q}$ is independent of the choice of the axes. For some systems $T_{\alpha\beta,q} \neq T_{\beta\alpha,q}$ (for a discussion see the paper of Smith, Palke and Gerig listed in Further reading), but the effect of this asymmetry can usually be neglected, and in this case there is a special set of axes (abc) such that only the diagonal elements are nonzero. These are known as the principal axes of the interaction tensor and the $T_{aa,q}$, $T_{bb,q}$ and $T_{cc,q}$ are principal components of $\boldsymbol{T}_q$. It is important to note that the different interactions will not usually share a common set of principal axes. In principal axes, Equation [2] is simplified to

$$T_{B,q} = T_{aa,q} \cos^2\theta_a + T_{bb,q} \cos^2\theta_b + T_{cc,q} \cos^2\theta_c \qquad [3]$$

There are obvious advantages in using principal axes, but their location in a molecule is not always known. This does not prevent us from working in principal axes when doing algebraic manipulations on Equation [2]. Thus, Equation [3] might seem inconsistent with Equation [1], but that this is not so can easily be appreciated by replacing $\cos^2 \theta_\alpha$ by $(1 - \sin^2 \theta_\alpha)$ to give

$$T_{B,q} = T_{aa,q} + T_{bb,q} + T_{cc,q} - T_{aa,q} \sin^2 \theta_a - T_{bb,q} \sin^2 \theta_b - T_{cc,q} \sin^2 \theta_c \quad [4]$$

There is, however, a more useful way of expressing $T_{B,q}$ as a scalar plus an anisotropic part, and this is derived by noting that when the molecules move rapidly and randomly, as in an isotropic liquid or gas, $T_{B,q}$ is averaged (denoted by the brackets $\langle\,\rangle$) to produce $\langle T_{B,q}\rangle_{\text{iso}}$:

$$\langle T_{B,q}\rangle_{\text{iso}} = \sum\nolimits_{\alpha,\beta} T_{\alpha\beta,q} \langle \cos\theta_\alpha \cos\theta_\beta \rangle \quad [5]$$

Using principal axes, and noting that $\langle \cos^2 \theta_\alpha \rangle = \frac{1}{3}$

$$\langle T_{B,q}\rangle_{\text{iso}} = \tfrac{1}{3}(T_{aa,q} + T_{bb,q} + T_{cc,q}) = T_q^0 \quad [6]$$

Remembering that the sum of the diagonal elements of $\boldsymbol{T}_q$ is independent of the axes, then we can see that Equation [6] is true for all choices of molecular axes even though we derived it in principal axes. Rearranging Equation [3] as

$$T_{B,q} = T_q^0 + \sum\nolimits_\alpha T_{\alpha\alpha,q}(\cos^2\theta_\alpha - \tfrac{1}{3}) \quad [7]$$

which leads to

$$T_{B,q} = T_q^0 + \tfrac{2}{3}\sum\nolimits_\alpha T_{\alpha\alpha,q}(3\cos^2\theta_\alpha - 1)/2 \quad [8]$$

Rearranging gives

$$T_{B,q} = T_q^0 + \tfrac{2}{3}[T_{aa,q} - \tfrac{1}{2}(T_{bb,q} + T_{cc,q})](3\cos^2\theta_a - 1)/2 + \tfrac{1}{2}(T_{bb,q} - T_{cc,q})(\cos^2\theta_b - \cos^2\theta_c) \quad [9]$$

Equation [9] has the advantage of showing clearly what happens if the tensor $\boldsymbol{T}_q$ is axially symmetric in the molecular principal axes, that is, when $T_{bb,q} = T_{cc,q}$.

Effect of molecular motion on the interactions in anisotropic systems

We have seen already that rapid, random motion averages $T_{B,q}$(anisotropic) to zero, but what is the effect of motion that is rapid but not random? The definition of rapid is that the rate of the motion is much greater than the magnitude of $T_{B,q}$(anisotropic). Slower motion will contribute to the relaxation rates and is not discussed here. The effect of motion is to produce $\langle T_{B,q}\rangle$, an averaged value of $T_{B,q}$, which can be thought of as a sum of contributions, $T_{B,q}(n)$ from n configurations of the system, each with a normalized probability $P(n)$:

$$\langle T_{B,q}\rangle = \sum_n P(n)[T_q^0(n)(\text{isotropic}) + T_{B,q}(n)(\text{anisotropic})] \quad [10]$$

For most systems the isotropic contribution will not have a strong dependence on n, and so we will concentrate on the averaging of the anisotropic term.

For solids it is better to consider the different kinds of motion that occur, and to see how they affect the averages.

Rotation of whole, rigid molecules on lattice sites in crystals Rotation about an n-fold axis, or hopping between n equivalent sites, produces an averaged tensor, one of whose principal axes, r, lies along the rotation axis. The location of the other two principal axes, s and t, will be in the plane orthogonal to r, and they are fixed with respect to the lattice and not the molecules. Thus,

$$\langle T_{B,q}\rangle = \langle T_q^0\rangle + \tfrac{2}{3}[T_{rr,q} - \tfrac{1}{2}(T_{ss,q} + T_{tt,q})](3\cos^2\theta_r - 1)/2 + \tfrac{1}{2}(T_{ss,q} - T_{tt,q})(\cos^2\theta_s - \cos^2\theta_t) \quad [11]$$

The relationship between $T_{rr,q}$ and the components of $\boldsymbol{T}_q$ in the (abc) principal frame is given by

$$T_{rr,q} = T_{aa,q}\cos^2\theta_{ar} + T_{bb,q}\cos^2\theta_{br} + T_{cc,q}\cos^2\theta_{cr} \quad [12]$$

The angles θ_{ar}, θ_{br} and θ_{cr} are between the axes a, b, c and r. For two-site hopping, the location of s and t will depend upon the structure of the molecule.

When $n \geq 3$ the averaged tensor is axially symmetric about r, so that Equation [11] simplifies to

$$\langle T_{B,q}\rangle = \langle T_q^0\rangle + \tfrac{2}{3}[T_{rr,q} - \tfrac{1}{2}(T_{ss,q} + T_{tt,q})](3\cos^2\theta_r - 1)/2 \quad [13]$$

and $T_{rr,q}$ and $\frac{1}{2}(T_{ss,q}+T_{tt,q})$ can be replaced by $T_{\parallel r,q}$ and $T_{\perp r,q}$.

These relationships apply to all the spin interactions when the molecule rotates as a whole because in this case the internuclear distances within a molecule do not change.

Rotation of a group of nuclei within a molecule The values of $\langle T_{B,q} \rangle$ for interactions involving nuclei only in the group that is moving relative to the crystal lattice (and hence relative to $\boldsymbol{B}_0$) are affected in the same way as described for whole-molecule rotation except that now the axes s and t are fixed in the nonrotating part of the molecule. The spin–spin interactions between nuclei in the rigid and moving parts of the molecules will be averaged in a way that depends on the structure of the molecules.

Hopping (diffusion) of molecules between lattice sites This kind of motion will usually occur in combination either with whole-molecule rotation on a lattice site or with internal motion in the molecules. The diffusion process will have a large effect on dipolar interactions between nuclei in different molecules. The exact magnitude of the effect will depend on the crystal structure, but it will always lead to a reduction in the intermolecular contribution to dipolar coupling.

Molecular motion in liquid crystalline samples In liquid crystalline samples the molecules rotate and diffuse rapidly but not randomly. The diffusive motion means that the distance between nuclei in different molecules is changing, with the result that the intermolecular contribution to the dipolar coupling is zero. The averaging produced by the rotational motion of whole, rigid molecules depends to some extent on the nature of the phase and the effect on it of the magnetic field of the spectrometer. We will consider here the simplest case only, which is when there is axial symmetry about $\boldsymbol{B}_0$. With uniaxial symmetry the result for a rigid molecule is to average the angular factors in Equation [9], which is usually expressed in slightly different way; thus

$$T_{B,q} = T_q^0 + \tfrac{2}{3}[T_{aa,q} - \tfrac{1}{2}(T_{bb,q} + T_{cc,q})]S_{aa}^B + \tfrac{1}{3}(T_{bb,q} - T_{cc,q})(S_{bb}^B - S_{cc}^B) \quad [14]$$

where

$$S_{\alpha\alpha}^B = \langle 3\cos^2\theta_\alpha - 1)/2 \rangle \quad [15]$$

θ_α is the angle between $\boldsymbol{B}_0$ and axis α. Note that now both the $T_{\alpha\alpha,q}$ and the $S_{\alpha\alpha}^B$ are unknown parameters that may be obtained from an analysis of the NMR spectra, and that their absolute values cannot be obtained separately. In practice, for chemical shifts and quadrupolar coupling it is usually possible to use values of the $T_{\alpha\alpha,q}$ obtained from the spectrum of a single-crystal sample to determine the $S_{\alpha\alpha}^B$. For dipolar coupling it is necessary to assume just one internuclear distance in order to obtain the separate values of the $S_{\alpha\alpha}^B$ and the D_{ij}.

The relationship between the values of the $S_{\alpha\alpha}^B$ and the orientational order of the molecules in the liquid crystalline phase can be derived by first introducing the concept of a director, $\boldsymbol{n}_i$. This is a unit vector at point i in the sample which defines the preferred orientation of the anisotropically shaped molecules. This can be quantified by a function, $P_{LC}(\beta, \gamma)$ which describes the probability that the director makes angles between β and $\beta + d\beta$ and γ and $\gamma + d\gamma$ in a frame (abc) fixed in the molecules, as shown in **Figure 1**. This probability density function has a maximum when β is zero, whose magnitude grows towards a value of unity as the molecules become ordered by reducing temperature. The values of $\langle T_{B,q} \rangle$ can be thought of as arising from rapid motion of the molecules about the director giving an average, $\langle T_{n,q} \rangle$, of the $\boldsymbol{T}_q$ along $\boldsymbol{n}$, and then using the relationship

$$\langle T_{B,q} \rangle = \langle T_{n,q} \rangle (3\cos^2\theta_{Bn} - 1)/2 \quad [16]$$

where θ_{Bn} is the angle between $\boldsymbol{n}_i$ and $\boldsymbol{B}$. If the directors are distributed relative to $\boldsymbol{B}_0$ then the spectrum will be a sum of spectra and will be like that from a polycrystalline powder. The averages $\langle T_{n,q} \rangle$ are given by

$$\langle T_{n,q} \rangle = T_q^0 + \tfrac{2}{3}[T_{aa,q} - \tfrac{1}{2}(T_{bb,q} + T_{cc,q})]S_{aa} + \tfrac{1}{3}(T_{bb,q} - T_{cc,q})(S_{bb} - S_{cc}) \quad [17]$$

where the $S_{\alpha\alpha}$ are known as Saupe order parameters. They are defined as

$$S_{aa} = \int \tfrac{1}{2}(3\cos\beta - 1)P_{LC}(\beta, \gamma)\sin\beta\, d\beta \quad [18]$$

$$S_{bb} - S_{cc} = \tfrac{3}{2}\int \sin^2\beta \cos 2\gamma P_{LC}(\beta, \gamma)\sin\beta\, d\beta\, d\gamma \quad [19]$$

The important point to note is that they can act as tests for models of $P_{LC}(\beta, \gamma)$. This function can be

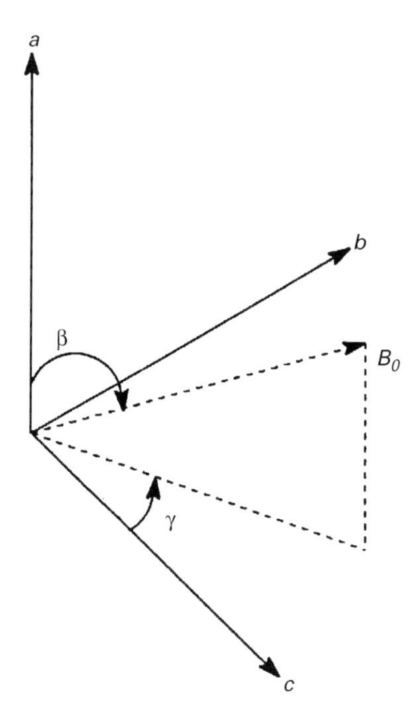

Figure 1 Angles defining the orientation of the magnetic field, $\boldsymbol{B}_0$, in axes (*abc*) fixed in a molecule.

used to define an effective orienting potential, or potential of mean torque, $U_{ext}(\beta, \gamma)$, thus

$$P_{LC}(\beta, \gamma) = Z^{-1} \exp[-U_{ext}(\beta, \gamma)/k_B T] \quad [20]$$

with

$$Z = \int \exp[-U_{ext}(\beta, \gamma)/k_B T] \sin\beta \, d\beta \, d\gamma \quad [21]$$

where T = temperature, k_B = Boltzmann constant.

In the absence of a magnetic field, the directors vary in orientation relative to some space-fixed axes in a completely random way. The directors may, however, be aligned uniformly by a magnetic field. If the molecules comprising the liquid crystal phase have a positive anisotropy, $\Delta\chi$, in their magnetic susceptibility, then the directors align parallel to the field, so that θ_{Bn} is 0, and $\langle T_{B,q}\rangle = \langle T_{n,q}\rangle$. If $\Delta\chi$ is negative the directors align in a plane perpendicular to the field, making $\theta_{Bn} = 90°$ and $\langle T_{B,q}\rangle = -\frac{1}{2}\langle T_{n,q}\rangle$. In both cases the spectra are like those for a single crystal except the lines are much narrower because of the absence of broadening from intermolecular effects.

The effect of internal motion on $\langle T_{n,q}\rangle$ for liquid crystalline samples Fast internal motion produces an additional averaging of the $T_{n,q}$, in a similar way to the case of solid samples, but with the very important difference that there is a cooperative effect of changes in molecular shape and orientational order. For example, consider rotation about a bond through an angle ϕ. The probability distribution becomes dependent on ϕ as well as β and γ, and so too does the mean potential energy of the molecules in the phase. Thus, we revise Equations [20] and [21] to

$$P_{LC}(\beta, \gamma, \phi) = Z^{-1} \exp[-U(\beta, \gamma, \phi)/k_B T] \quad [22]$$

$$Z = \int \exp[-U(\beta, \gamma, \phi)/k_B T] \sin\beta \, d\beta \, d\gamma \, d\phi \quad [23]$$

The mean potential may be divided into a purely anisotropic part, $U_{ext}(\beta, \gamma, \phi)$, which vanishes in the isotropic phase, and a part, $U_{int}(\phi)$, which does not:

$$U(\beta, \gamma, \phi) = U_{ext}(\beta, \gamma, \phi) + U_{int}(\phi) \quad [24]$$

Because the potential of mean torque has become dependent on the internal motion, then so too do the order parameters; that is, Equations [18] and [19] become

$$S_{aa}(\phi) = \int \tfrac{1}{2}(3\cos\beta - 1) P_{LC}(\beta, \gamma, \phi) \sin\beta \, d\beta \quad [25]$$

$$S_{bb}(\phi) - S_{cc}(\phi) = \tfrac{3}{2} \int \sin^2\beta \cos 2\gamma \, P_{LC}(\beta, \gamma, \phi) \sin\beta \, d\beta \, d\gamma \quad [26]$$

Various strategies have been suggested for obtaining $P_{LC}(\beta, \gamma, \phi)$ from observed values of $\langle T_{B,q}\rangle$.

Having obtained $P_{LC}(\beta, \gamma, \phi)$, it is then possible to determine $P_{LC}(\phi)$, the probability that the molecule in the liquid crystalline phase is in a conformation defined by the angle ϕ, as

$$P_{LC}(\phi) = \int P_{LC}(\beta, \gamma, \phi) \sin\beta \, d\beta \, d\gamma = \frac{\int \exp[-U(\beta, \gamma, \phi)/k_B T] \sin\beta \, d\beta \, d\gamma}{\int \exp[-U(\beta, \gamma, \phi)/k_B T] \sin\beta \, d\beta \, d\gamma \, d\phi} \quad [27]$$

Note that in principle this differs from $P_{iso}(\phi)$, the conformational probability distribution for the molecule in the isotropic phase, which is

$$P_{\mathrm{iso}}(\phi) = \frac{\exp[-U_{\mathrm{iso}}(\phi)/k_{\mathrm{B}}T]}{\int \exp[-U_{\mathrm{iso}}(\phi)/k_{\mathrm{B}}T]\mathrm{d}\phi} \qquad [28]$$

The anisotropic nuclear spin interactions

As noted earlier, we will confine our discussion to the case when the interaction of the spins with the magnetic field, the Zeeman interaction, is dominant so that the direction of the applied, static field determines the axis of quantization and the experiments yield the component $T_{B,q}$ of the various interactions in this direction. It is now convenient to define this as the Z direction. A very detailed discussion of the spin interactions, and their contribution to the nuclear spin Hamiltonian has been given by Smith, Palke and Gerig.

The Zeeman interaction

The contribution, $\mathcal{H}_{\mathrm{Zeeman}}$, to the nuclear spin Hamiltonian, in units of hertz, has the form:

$$\mathcal{H}_{\mathrm{Zeeman}} = -\frac{1}{2\pi}\sum_i \gamma_i(1 - \sigma_{ZZi})B_0 I_{Zi} \qquad [29]$$

where γ_i is the magnetogyric ratio of the ith nucleus, and I_{Zi} is the component along Z of the nuclear spin angular momentum operator, I_i. The important feature is that this Hamiltonian has an identical form in both isotropic and anisotropic systems, and hence the spectral consequences are exactly the same. That is, the resonance is shifted in energy depending on the electronic environment at the ith site. The difference between these two kinds of system lies in the contributions to σ_{ZZ} in the two situations. Thus for a rigid molecule Equation [9] becomes

$$\sigma_{ZZ} = \sigma^0 + \tfrac{2}{3}[\sigma_{aai} - \tfrac{1}{2}(\sigma_{bbi} + \sigma_{cci})](3\cos^2\theta_{ia} - 1)/2 + \tfrac{1}{2}(\sigma_{bbi} - \sigma_{cci})(\cos^2\theta_{bi} - \cos^2\theta_{ci}) \qquad [30]$$

The isotropic term, σ^0, is much greater than the anisotropic contribution and so that there is only a small (~kHz) difference between the resonance frequency of a nucleus in an isotropic and anisotropic environment.

Note that the location of the principal axes (abc) depends on the site occupied in a molecule by the nucleus. The site symmetry in a rigid molecule determines whether $\sigma_{aai} = \sigma_{bbi} = \sigma_{cci}$ (isotropic site symmetry), $\sigma_{aai} \neq \sigma_{bbi} = \sigma_{cci}$ (axial site symmetry), or $\sigma_{aai} \neq \sigma_{bbi} \neq \sigma_{cci}$ (site asymmetry). The nature of the site symmetry is in fact revealed most clearly and easily by recording the spectra of polycrystalline samples, if possible. This is particularly true when the Zeeman term is the only one determining the observed spectrum. This situation can easily be realized, for example for ^{13}C, ^{15}N or ^{31}P nuclei.

Figure 2 shows the shape of spectra obtained for nuclei at sites of the three different symmetries.

Electron-mediated spin–spin coupling

The contribution to the Hamiltonian, in hertz, is

$$\mathcal{H}_J = \sum_{i<j}\{J_{ij}^0\,[I_{Zi}I_{Zj} + \tfrac{1}{2}(I_{+i}I_{-j} + I_{-i}I_{+j})] + J_{ij}^a\,[I_{Zi}I_{Zj} - \tfrac{1}{4}(I_{+i}I_{-j} + I_{-i}I_{+j})]\} \qquad [31]$$

J_{ij}^0 ($\equiv T_{B,q}$(isotropic) in Equation [1] is the scalar contribution to coupling, which is the only contribution in isotropic environments. J_{ij}^a is the anisotropic contribution, and its components in the principal, molecular frame can be obtained by substitution in Equation [9] for a rigid crystal, and Equation [14] for a rigid molecule in a liquid crystalline phase.

The operators I_{+i} and I_{-i} are defined as:

$$I_{+i} = I_{Xi} + \mathrm{i}I_{Yi} \quad \text{and} \quad I_{-i} = I_{Xi} - \mathrm{i}I_{Yi} \qquad [32]$$

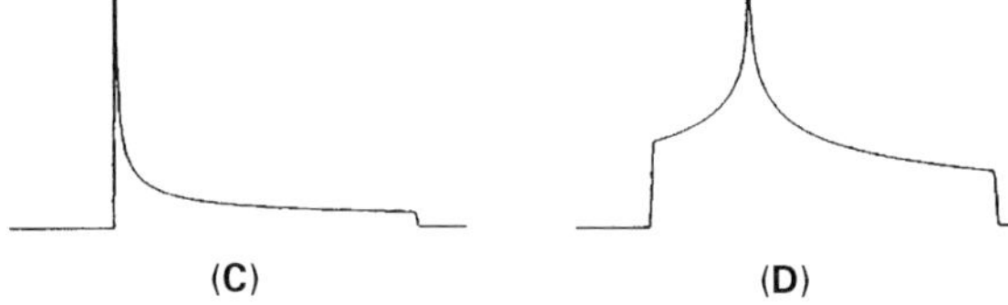

(A)

(B)

(C)

(D)

Figure 2 Spectra for a polycrystalline sample for nuclei with different site symmetries and subject only to the Zeeman interaction: (A) $\sigma_{aa} = \sigma_{bb} = \sigma_{cc}$; (B) $\sigma_{aa} = \sigma_{bb} > \sigma_{cc}$; (C) $\sigma_{aa} = \sigma_{bb} < \sigma_{cc}$; (D) $\sigma_{aa} \neq \sigma_{bb} \neq \sigma_{cc}$.

The magnitude of J^a_{ij} is predicted to be much smaller than either J^0_{ij} or D_{ij}, the dipolar coupling between the same nuclei, when one of the nuclei is a proton. For other pairs of nuclei, J^a_{ij} may not be negligible. Values have been obtained experimentally from the spectra of molecules dissolved in liquid crystalline phases.

Dipolar coupling

This is the through-space coupling between a pair of nuclear magnetic dipoles. It is a purely anisotropic interaction and is also symmetric about $\mathbf{r}_{ij}$, the internuclear vector. It contributes a term to the Hamiltonian, in hertz, of

$$\mathcal{H}_D = \sum_{i<j} 2D_{ij}[I_{zi}I_{zj} - \tfrac{1}{4}(I_{+i}I_{-j} + I_{-i}I_{+j})] \quad [33]$$

Note that this has identical spin operators to those involving J^a_{ij} in Equation [31]. This means that the spectra of anisotropic systems depend on $(2D_{ij} + J^a_{ij})$, and that these two interactions cannot be determined separately from the spectra. For this reason, J^a_{ij} is often referred to a pseudo-dipolar coupling.

Dipolar coupling is axially-symmetric about $\mathbf{r}_{ij}$, which is a principal axis, so that from Equation [9] there is just one contribution to D_{ij}:

$$D_{ij} = D_{aaij}(3\cos^2\theta_{ij} - 1)/2 \quad [34]$$

θ_{ij} is the angle between $\mathbf{r}_{ij}$ and $\mathbf{B}_0$. D_{aaij} for a fixed internuclear distance is given, in hertz, by

$$D_{aaij} = -[\mu_0 h\gamma_i\gamma_j/(16\pi^3)]/r_{ij}{}^3 \quad [35]$$

where $\mu_0 = 4\pi \times 10^{-7}$ and is the permeability of free space. This makes it possible to determine molecular structure from dipolar couplings.

The quadrupolar interaction

Remembering that we are restricting our discussion to the cases when the Zeeman term determines the axis of quantization of the nuclear spins, then the quadrupolar interaction contributes a term, in hertz, to the spin Hamiltonian of

$$\mathcal{H}_Q = \sum_i [eQ_N V_{ZZi}/(4hI_{Ni}(2I_{Ni} - 1)] \times [3I_{Zi}{}^2 - I_{Ni}(I_{Ni} + 1)] \quad [36]$$

I_{Ni} is the nuclear spin quantum number of spin i of type N, which has quadrupolar moment eQ_N, and is at a site with an electric field gradient $\mathbf{V}_i$. It is a purely anisotropic interaction, and so for a rigid molecule in a solid crystalline sample,

$$V_{ZZi} = V_{aai}(3\cos^2\theta_a - 1)/2 + \tfrac{1}{2}(V_{bbi} - V_{cci})(\cos^2\theta_b - \cos^2\theta_c) \quad [37]$$

List of symbols

a, b, c = molecule-fixed principal axes; $\mathbf{B}_0$ = applied magnetic field, $B_0 = |\mathbf{B}_0|$; D_{ij} = dipolar spin–spin coupling; h = Planck constant; $\mathcal{H}$ = Hamiltonian; I = nuclear spin quantum number; $\mathbf{I}_i$ = spin angular momentum operator; I_{Zi} = component of $\mathbf{I}_i$ along Z; J_{ij} = electron-mediated spin–spin coupling; k_B = Boltzmann constant; $\mathbf{n}_i$ = director (unit vector at point i); $P_{LC}(\beta,\gamma)$ = probability that director angle is in β+dβ, γ+dγ; $P(n)$ = probability of configuration n; Q_N = nuclear quadrupole moment; r, s, t = lattice-fixed principal axes; $\mathbf{r}_{ij}$ = internuclear vector; $S_{\alpha\alpha}$ = Saupe order parameters; T = temperature; $T_{B,q}$ = component of qth interaction along $\mathbf{B}_0$; $\mathbf{T}_q$ = interaction tensor; $T_{\alpha\beta,q}$ = component of qth interaction with respect to molecule fixed axes α and β; U_{ext} = potential of mean torque; $\mathbf{V}_i$ = electric field gradient at site i; x, y, z = molecular axes; γ_i = magnetogyric ratio of ith nucleus; θ_i = angle of $\mathbf{B}_0$ with axis i; μ = nuclear magnetic dipole moment; μ_0 = permeability of free space; σ = shielding constant; ϕ = angle of rotation about bond; χ = magnetic susceptibility.

See also: **Chiroptical Spectroscopy, Oriented Molecules and Anisotropic Systems; Diffusion Studied Using NMR Spectroscopy; Liquid Crystals and Liquid Crystal Solutions Studied By NMR; Solid State NMR, Methods; Solid State NMR Using Quadrupolar Nuclei; Solid State NMR, Rotational Resonance.**

Further reading

Emsley JW (ed) (1985) *NMR of Liquid Crystals*, Dordrecht: Reidel.

Schmidt-Rohr K and Spiess HW (1994) *Multidimensional Solid State NMR and Polymers*. New York: Academic Press.

Smith SA, Palke WE and Gerig JT (1992) *Concepts in Magnetic Resonance* 4: 107; (1992) 4: 181; (1993) 5: 151; (1994) 6: 137.

NMR Microscopy

Paul T Callaghan, Massey University, Palmerston North, New Zealand

MAGNETIC RESONANCE
Methods & Instrumentation

Introduction

A microscope is generally regarded as an assembly of lenses used to give an image of small objects at high spatial resolution. By high spatial resolution is meant a resolution finer than can be resolved by the naked human eye, in other words better than 0.1 mm. The well-known 'optical microscope' uses either reflected or transmitted light to present an image and the resolution of this instrument is determined by the wavelength of the electromagnetic radiation, around 0.5 μm. There are many other forms of microscopy that use different radiations, such as the electron microscope, whose resolution is much finer because of the shorter wavelength of the electron beam, or the acoustic microscope which is very effective at probing near-surface properties using high-frequency sound waves. X-ray microscopes have been developed in recent years, although these are rather tricky to use because of the need to generate relatively monochromatic soft X-rays that can then be focused by Fresnel lenses. This article describes a very different microscope, based on the use of radio waves, the so-called 'nuclear magnetic resonance microscope'.

If one were to present the concept of a radio wave microscope in the context of our usual perspectives on such devices, three factors would emerge. First, one would expect that, like the X-ray microscope, the device would have excellent powers of penetration, since radio waves pass easily through most matter, excluding good conductors such as metals. Second, one would expect that it would be almost ideally noninvasive, because of the very low energy of the radiofrequency photon. In this regard it could be contrasted with electron microscopy and X-ray microscopy, which are both capable of breaking covalent bonds between atoms. Finally, it might be guessed that such a device would have hopelessly poor spatial resolution because of the very long wavelengths (several metres!) of the radiation.

This last point would be certain to render radio wave microscopy useless were it not for the phenomenon of nuclear precession exhibited by atomic nuclei with nonzero spin when placed in a magnetic field. This property means that those nuclei (the 'spins') have associated with them a special radio frequency that depends precisely on the local magnetic field strength, and if that magnetic field strength is varied from place to place in the sample, then the measurement of the frequency of each nucleus will indicate where its parent atom or molecule is positioned. By this means one could build up an image of the atomic distribution, avoiding altogether the normal wavelength (or 'diffraction') limit to resolution. The technique used to measure the spatially dependent spin precession frequency is nuclear magnetic resonance (NMR).

The prospects for such a method seem almost too good to be true, but there is an Achilles heel. NMR relies on the measurement of the nuclear frequency by means of the exchange of radiofrequency photons whose energies are so weak that they must compete with the thermal noise that exists in the detection circuitry. It is this latter effect that determines the ultimate resolution of such an instrument, and, using the best possible stable nucleus (the proton) in thermal equilibrium at room temperature, and with a receiving antenna made from room-temperature metals, the limit is volume elements around $(10\ \mu m)^3$ for the image, an exceptionally poor resolution by comparison with all other microscopies. Indeed the NMR microscope is only just a microscope in the normal sense and one would hardly bother with it at all were it not for a few remarkably useful properties. It is exceptionally noninvasive as intimated earlier, and this makes it of special interest in *in vivo* biological applications. It is highly penetrating and can work perfectly well with optically opaque materials with minimal problems of 'transparency', and it is especially well suited to the study of liquid phases, a rather unusual property in the context of other microscopies. Most important of all, it enables the imaging of a number of molecular and atomic properties to which NMR is particularly suited. These include the local chemical composition, the local molecular order and rotational dynamics, and the local molecular translational motions. It is for these reasons that NMR microscopy, despite its rather poor spatial resolution, has found a number of important applications in science, technology and medicine.

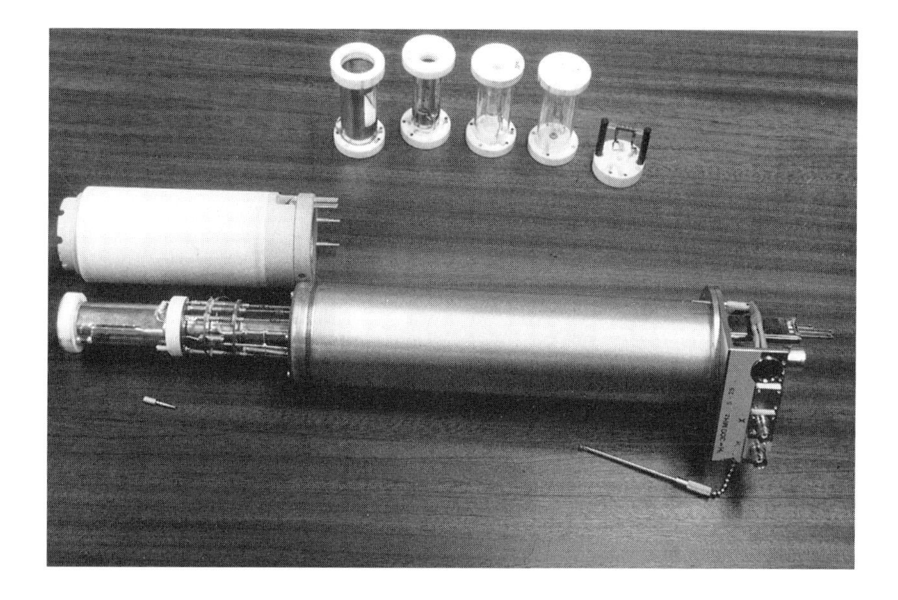

Figure 1 Radiofrequency probe, gradient coils set and RF coil inserts used in NMR microscopy. The probe assembly is placed in the 89 mm diameter vertical bore of a superconducting magnet. The set of RF coils and resonators enable samples of different sizes to be inserted, and have diameters ranging from 25 mm diameters down to 2 mm. Photograph courtesy of Bruker Analytische Messtechnik, Karlsruhe, Germany.

Nuclear magnetic resonance imaging

Proton NMR

Nuclear magnetic resonance was first detected in 1945, although its extension to imaging applications was to wait until 1973. NMR is an enormous field of research and the subject of very many textbooks in its own right. This article will attempt to give only a cursory introduction to the phenomenon, and the reader should look to some of the articles elsewhere in this Encyclopedia for further information. An isolated proton, when placed in a magnetic field, B_0, occupies one of two quantum states with respect to the field direction and the quantum phases of those states rotate at the frequency $\omega = \gamma B_0$ (for example $\omega/2\pi = 300$ MHz for a proton immersed in a 7 T field), where γ is the gyromagnetic ratio of the proton, the factor that determines the ratio of its magnetic properties to its spin properties. This phase rotation arises from the combined magnetic and spin properties of the nucleus and is known as precession. The proton has a very high value of γ (and hence radiofrequency photon energy) relative to other nuclei and it is highly abundant in most materials, as the nucleus of atomic hydrogen. It is thus the prime candidate for NMR microscopy.

In practice we deal with large ensembles of nuclei whose phases, in thermal equilibrium, are randomized so that the net magnetization presented by the sample is directed along B_0, and results from the slight preponderance of spin-up over spin-down (typically around 1 in 10^5). Detection of the underlying precession frequency therefore requires the intervention of a small, transverse, oscillatory magnetic field, B_1 (the so-called resonant radiofrequency field, generated by a coil surrounding the sample). This field is applied as a pulse for a finite duration t_p at the same frequency as the underlying precessional circulation, thus keeping its effect on the spins in step with their motion. As a result the spins gradually re-align themselves with respect to the combined effects of B_0 and B_1, the result being a reorientation of the net magnetization vector to an angle $\gamma B_1 t_p$ with respect to B_0. A '90°' RF pulse would be one that left the magnetization precessing in the transverse plane. The same sample coil is used to excite the spin system from equilibrium (relaxation time T_1). Thus the NMR spectrometer consists of a magnetic field, a coil as antenna, and a radio transceiver. In the case of a microimaging system, an additional requirement is a set of magnetic field gradient coils. Typical NMR microscope RF coils and gradient coils are shown in **Figure 1**.

Position encoding using field gradients

The imaging principle is as follows. Suppose we now apply, in addition, a magnetic field gradient $\mathbf{G} = \nabla B_0$ (this can be achieved with specialized coil designs capable of generating quite uniform gradients along three orthogonal axes), then the precession frequency will depend on spin location and can be written

$$\omega = \gamma B_0 + \gamma \mathbf{G} \cdot \mathbf{r} \quad [1]$$

Because the receiver uses heterodyne phase-sensitive detection with reference frequency γB_0, the additional term, $\gamma \boldsymbol{G}\cdot\boldsymbol{r}$, shows up as a difference oscillation, generally in the audiofrequency part of the spectrum. As a consequence, the heterodyne signal detected via the NMR coil at time t after the spins have precessed in the gradient has the mathematical form

$$S(\boldsymbol{k}) = \int \rho(\boldsymbol{r}) \exp(\mathrm{i}2\pi \boldsymbol{k}\cdot\boldsymbol{r})\, \mathrm{d}\boldsymbol{r} \qquad [2]$$

where $\exp(\mathrm{i}2\pi\boldsymbol{k}\cdot\boldsymbol{r})$ is the local spin phase factor for the spins at position $\boldsymbol{r}$, $\boldsymbol{k} = (2\pi)^{-1}\gamma \boldsymbol{G}t$, and $\rho(\boldsymbol{r})$ is the density of spins at position $\boldsymbol{r}$ and is the quantity we seek to image. The remarkable fact about Equation [2] is that it is a simple Fourier transform in which $\boldsymbol{k}$ is the reciprocal-space wave vector conjugate to the spatial dimension $\boldsymbol{r}$. Thus the acquisition of a signal in a domain of '$\boldsymbol{k}$-space' leads to direct computation of $\rho(\boldsymbol{r})$ by means of Fourier inversion, a fact that is made possible by the ability of the NMR experiment to provide full phase information (i.e. the complex number $S(\boldsymbol{k})$ via the acquisition of both in-phase (real) and quadrature-phase (imaginary) signals.

Generally one acquires an image of the spin distribution in a 2-dimensional planar 'slice', with a frequency-selective RF pulse being used to excite only spins within a preselected plane, and the $\boldsymbol{k}$-space encoding being applied independently for two orthogonal directions within the plane, one direction via a fixed time duration variable-magnitude gradient pulse (phase-encoding) and the other direction via a fixed gradient pulse with the signal being sampled at successive time points during the evolution (read-encoding). A typical pulse sequence that performs these tasks along with the corresponding $\boldsymbol{k}$-space map, is shown in **Figure 2**. Note that this sequence employs the spin-echo method in which a separate 180° RF pulse is used to invert spin phases. This allows us to traverse both negative and positive regions of $\boldsymbol{k}$-space. The spin echo has important applications in the measurement of transverse relaxation, and of molecular flow and diffusion.

Note also that the finite time needed to encode the NMR magnetization in $\boldsymbol{k}$-space requires that the nuclear spin relaxation time, T_2, be sufficiently long. T_2 is determined predominantly by internuclear interactions, which in turn are motionally averaged by molecular tumbling. This means in effect that molecules in the liquid state have long proton T_2 values (10–1000 ms) while those in the solid state may have relaxation times as short as 10 μs. This results in a sharp discrimination of the signal in which the solid state component is entirely filtered out, unless special rapid-encoding methods are used. Most NMR microscopy images therefore arise from the liquid phase alone. Readers interested in the rarer solid-state options should consult references given under Further reading.

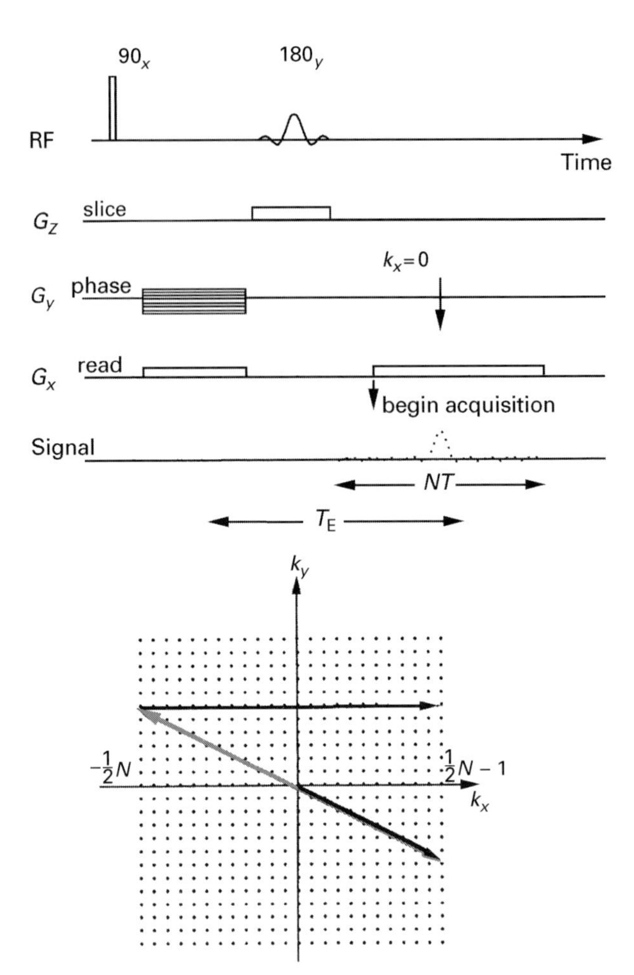

Figure 2 RF and magnetic field gradient pulse sequence used in NMR microscopy along with the associated trajectory through $\boldsymbol{k}$-space. The frequency-selective 180° pulse is used to select a layer of spins (the slice plane) to participate in the spin echo and hence to contribute to the image. NT=acquisition time; T_E=*i* echo time.

Limits to resolution

The fundamental spatial resolution for NMR microscopy is limited by the three factors of intrinsic signal-to-noise, molecular self-diffusion and diamagnetic susceptibility effects. For protons at room temperature and with an RF coil at room temperature, the signal-to-noise limit is easily calculated under optimal experimental conditions and depends directly on the ratio of T_1 to T_2, the available experimental time for signal averaging, and the magnetic field strength. Typical resolution limits for superconducting magnet systems for imaging times of the order of 30 min are (10 μm)^3 and (20 μm)^3 for T_1/T_2 values of unity and 100, respectively. These estimates assume 256 × 256 pixels with the sample exactly filling a solenoidal RF coil, a consequence of which is that the RF coil dimensions must progressively decrease with increasing resolution, thus limiting the highest resolution to very small samples. Note that resolution in one dimension can be traded against another so that it is possible to observe finer details (in-plane resolution) if a thicker slice can be tolerated. One of the highest-resolution examples yet reported, at (4.5 μm)^2 in-plane and 70 μm slice thickness, was obtained using a sample of onion cells contained with an RF coil of 1 mm diameter (see **Figure 3**).

In principle, the resolution could be further improved if the receiver coil temperature and/or its electrical resistance could be lowered. Some progress has been reported on the use of liquid nitrogen-cooled superconducting ceramics in which an effective signal-to-noise increase was obtained. The need for inductive coupling and the problem of poor filling factor means in effect that no major advantage has resulted to date.

The fact that the NMR imaging signal arises from nuclei whose parent molecules exist in a liquid state implies that imaging resolution will be strictly limited by self-diffusion. A rule of thumb is that diffusion limits will be important when the rms Brownian displacements over the ***k***-space encoding time are comparable with the pixel dimension. The usual consequence is severe signal attenuation, an effect that is apparent in **Figure 4**, which shows an image from water contained in a rectangular capillary of 100 μm wall spacing. Here only the layer of molecules near the wall, whose motions are impeded by the boundary, fails to suffer attenuation and therefore presents a bright edge effect. Such phenomena can provide potentially useful contrast. Another effect that also has its origin in boundaries associated with structural heterogeneity is illustrated in **Figure 5** which shows high-resolution NMR-microscope images of a section of geranium stem at two different read gradient strengths for which the acquisition times, NT, and echo time T_E (see **Figure 2**) are different by a factor of 5. The image obtained at longer acquisition and echo time (**Figure 5A**) suffers from attenuation and distortion effects due to the differing diamagnetic susceptibility across cell wall boundaries. Note that the pixel dimension is 10 μm while the slice thickness is 500 μm. The subtle interplay between diffusion and susceptibility effects can also result in characteristic bright image features seen in **Figure 5A**. It is apparent that the susceptibility and diffusion effects can provide image features that are useful in discriminating boundaries between fluid regions. These complex phenomena and their elucidation are discussed in more detail in references listed at the end of the article.

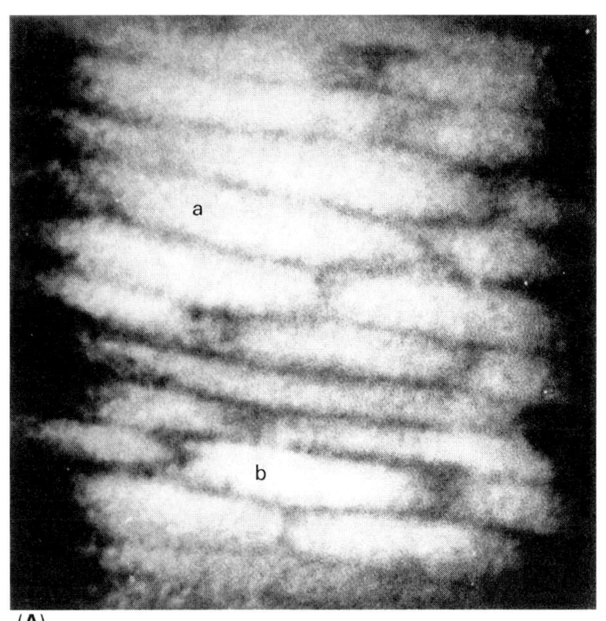

(A)

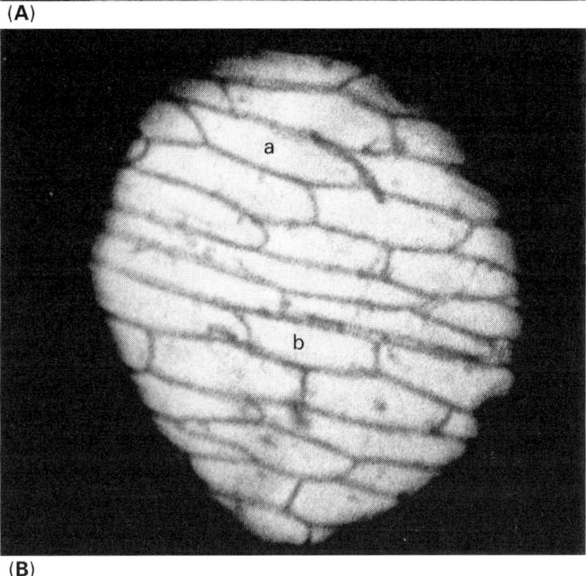

(B)

Figure 3 NMR image (A) and corresponding optical micrograph (B) of the epidermal cells of *Alium cepa*. The pixel size is 5 μm with a slice thickness of 70 μm. Reprinted with permission of the International Society of Magnetic Resonance in Medicine. From Glover PM, Bowtell RW, Brown GD and Mansfield P (1994) *Magnetic Resonance in Medicine* **31**: 423–428.

Image contrast

The particular utility of NMR microscopy lies in the contrasts that are available. These include, in

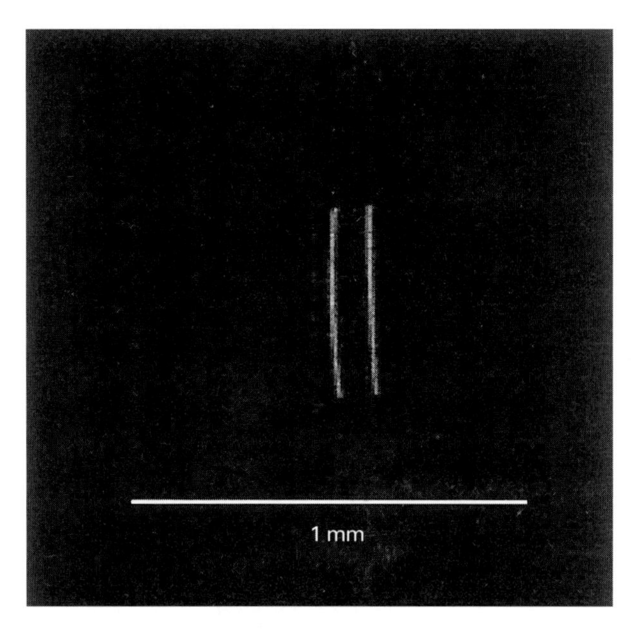

Figure 4 NMR micrograph obtained from water in a rectangular glass capillary whose walls are spaced by 100 μm. The pixel dimension is 8 μm, comparable with the distance diffused by the water molecules over the echo time T_E, and hence the image intensity is severely attenuated, except at the walls where the molecular Brownian motion is restricted. Courtesy of S.L. Codd and the author.

addition to the usual spin density (or, by implication, molecular density) the susceptibility effects discussed above, the chemical shift (the small changes in frequency due to different electronic environments of the nuclei), the nuclear spin relaxation times (sensitive to the rotational dynamics of parent molecules and hence ideally suited to discriminating solids, liquids and semisolid phases), the molecular translational self-diffusion coefficient and the molecular translational flow rate. Each contrast, if it is to be appropriately quantified or enhanced, requires a particular modification to the basic imaging pulse sequence shown in **Figure 2**, details of which can be found in the book by Callaghan. For example, in the case of chemical shift imaging, the initial 90° RF pulse can be made frequency-selective so as to excite only those spins residing in a particular chemical site. For relaxation contrast the echo time can be altered by adjusting delay times or pulse repetition times. However, because the imaging pulse sequence is based on the use of magnetic field gradients in conjunction with spin echoes, this can make the image intensity especially sensitive to the effects of molecular self-diffusion when one is close to the diffusion limit. This is the reason for the bright edge effects seen in **Figure 4**.

A commonly used approach is to leave the parameters of the imaging spatial encoding (the part to the right of the grey line in **Figure 2**) unchanged and to add the pulse sequence needed for contrast encoding as a precursor. For example, in the case of

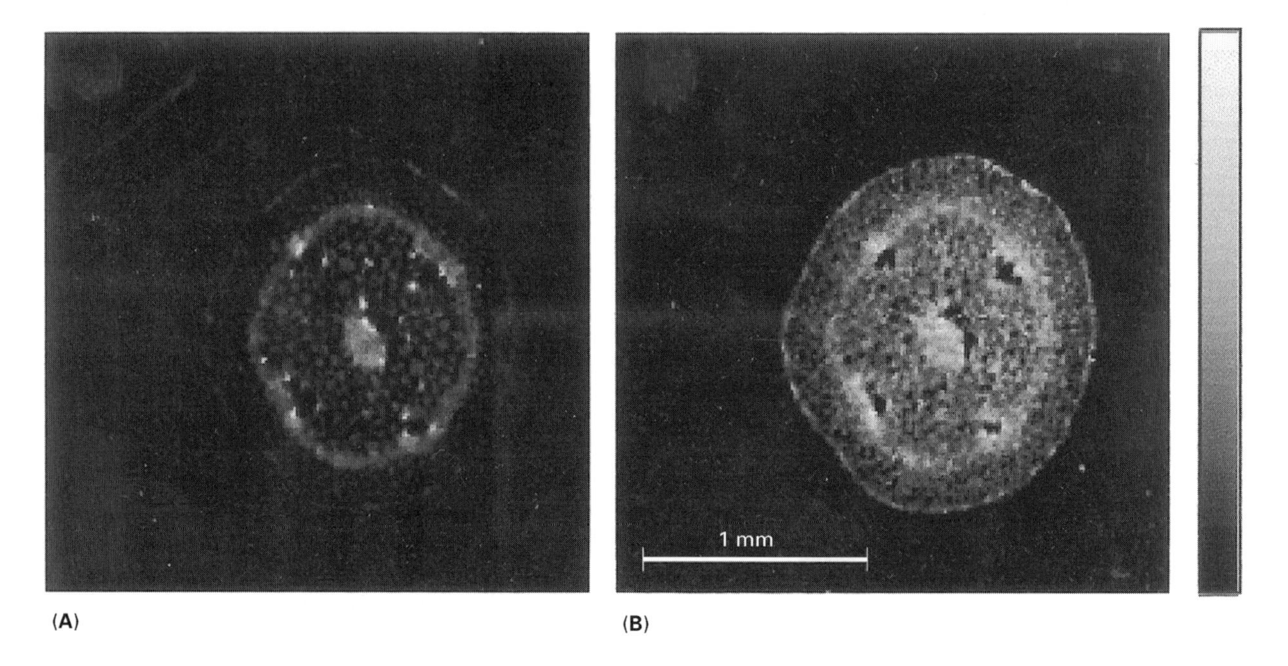

Figure 5 Two images (256^2 pixels of $(10\ \mu m)^2$, slice thickness 500 μm) obtained with the spin-echo pulse sequence of **Figure 2** from a geranium stem in which different read gradients and echo times are employed: (A) 20 kHz with $T_E = 13.5$ ms; (B) 100 kHz with $T_E = 3.2$ ms. Reprinted with permission from Rofe CJ, Van Noort J, Back PJ and Callaghan PT (1995) *Journal of Magnetic Resonance B* **108**: 125–136.

T_2 contrast this would take the form of a prior 90_x–τ–180_y–τ spin-echo segment where the remaining echo amplitude used for spatial encoding depends on the relaxation that occurs over the delay time τ. By acquiring separate images at different delay times τ, one generates a three-dimensional data set (two for the image within the slice plane and one additional for τ). Analysis in the third dimension enables one to display an image of relaxation rate. Another higher-dimensional encoding is that needed for molecular self-diffusion or flow analysis. This method is so important to NMR microscopy that it requires a separate discussion.

Pulsed gradient spin-echo NMR

Encoding for translational motion

The spin echo allows one to perform a useful trick in the encoding of spin positions. Consider the pulse sequence shown in **Figure 6**. At the echo maximum (the point where $\boldsymbol{k} = 0$), the phase excursions of the spins return to zero, unless, that is, any of the parent molecules happen to have moved along the direction of the gradient. In this case the subtraction performed by the echo is imperfect and the residual phase shifts provide a signature for motion.

The pulsed gradient spin echo (PGSE) experiment of **Figure 6**, uses two narrow gradient pulses of amplitude g, duration δ and separation Δ. These pulses effectively define the starting and finishing point of spin translational motion over the well-defined timescale, Δ. A spin that moves by a distance $\boldsymbol{R}$ over time Δ will acquire a phase shift, $\gamma\delta\boldsymbol{g}\cdot\boldsymbol{R}$. Indeed the resulting phase encoding can be expressed in the same reciprocal space language that we have already seen; the difference now is that the wave vector, q ($q = (2\pi)^{-1}\gamma g\delta$), is conjugate to molecular displacements over the time Δ between the gradient pulses, rather than in the case of imaging where it is conjugate to the actual spin positions. The analogue to Equation [2] is given by the echo amplitude,

$$E(\boldsymbol{q},\Delta) = \int \overline{P_s}(\boldsymbol{R},\Delta)\exp[\mathrm{i}2\pi\boldsymbol{q}\cdot\boldsymbol{R}]\,\mathrm{d}\boldsymbol{R} \qquad [3]$$

where the 'average propagator', $\overline{P_s}(\boldsymbol{R},\Delta)$, gives the probability that a spin in the ensemble examined displaces by $\boldsymbol{R}$ over the encoding time Δ. Just as inverse Fourier transformation of $S(\boldsymbol{k})$ in Equation [2] returns an image of the spin density $\rho(\boldsymbol{r})$, so inverse Fourier transformation of $E(q, \Delta)$ with respect to q returns an image of the average propagator, $\overline{P_s}$.

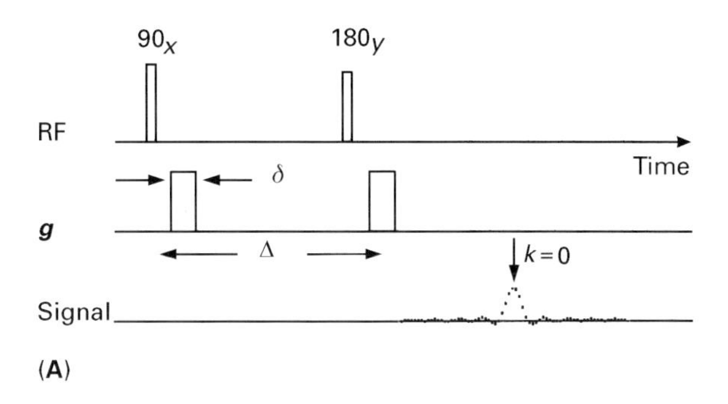

(A)

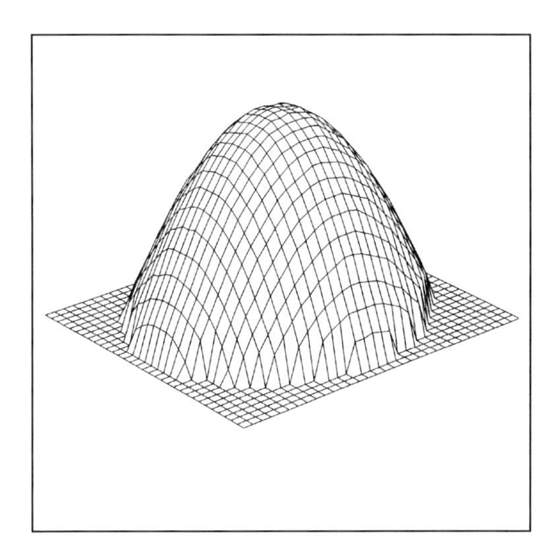

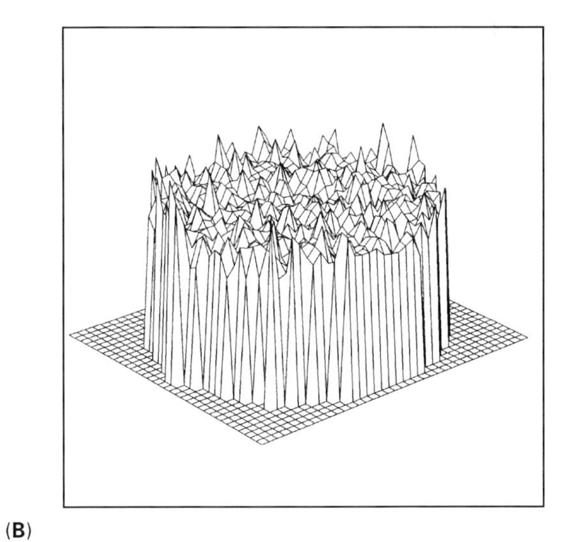

(B)

Figure 6 (A) Pulsed gradient spin echo sequence used to encode spin magnetization phase for molecular translational motion. (B) Velocity and diffusion maps for a water molecule flowing through a 2 mm diameter capillary. The images are shown as stackplots. The velocity profile is Poiseuille while the diffusion map is uniform. Courtesy of RW Mair, MM Britton and the author.

Diffusion and flow

In the case of two simple examples of motion of importance in NMR microscopy, namely self-diffusion and flow, the propagator takes the form

$$\overline{P_s}(\boldsymbol{R},\Delta) = (4\pi D_s\Delta)^{-3/2}\exp\left(-\frac{(\boldsymbol{R}-\boldsymbol{v}\Delta)^2}{4D_s\Delta}\right) \qquad [4]$$

where $\boldsymbol{v}$ is the constant velocity and D_s is the self-diffusion coefficient. When the PGSE encoding is used as a higher contrast dimension in conjunction with NMR imaging, the signal acquired is effectively modulated both in $\boldsymbol{k}$-space and q-space. Generally the experiment is performed with two dimensions of $\boldsymbol{k}$ (the slice planes) and one dimension of q (where $q = |\boldsymbol{q}|$), so that the signal is

$$\int(\boldsymbol{k},q) = \int \rho(\boldsymbol{r})E(q,\Delta)\exp(\mathrm{i}2\pi\boldsymbol{k}\cdot\boldsymbol{r})\,\mathrm{d}\boldsymbol{r} \qquad [5]$$

Inverse Fourier transformation of $S(\boldsymbol{k},q)$ with respect to $\boldsymbol{k}$ returns a set of 2-dimensional images modulated by $E(q,\Delta)$ while transformation with respect to q returns, for every image pixel, $\overline{P_s}(Z,\Delta)$, the average propagator for displacements Z along the gradient axis at position $\boldsymbol{r}$ within the image. By appropriate processing of these average propagators, details of the local motion can be calculated. For example, the width of $\overline{P_s}(Z,\Delta)$ is determined by the rms Brownian motion $(2D_s\Delta)^{1/2}$ whilst the displacement of $\overline{P_s}(Z,\Delta)$ along the Z axis is determined by the flow displacement $v\Delta$ where v is the local molecular velocity. In this manner maps of $D_s(\boldsymbol{r})$and $v(\boldsymbol{r})$ may be constructed, examples of which are shown in **Figure 6B**.

Applications of NMR microscopy

Numerous factors militate against the widespread use of NMR microscopy: the resolution is poor by optical standards, the apparatus is expensive, the technique requires a high level of scientific expertise and the arrangements for sample loading are inconvenient and restrictive. Set against these are the uniquely non-invasive character of the method, its sensitivity to fluid phases, its unique ability to measure specific molecular properties and the especially powerful insights it can provide regarding fluid dynamics. Studies in which NMR microscopy has been able to provide unrivalled information include those concerned with membrane filtration, flow and dispersion in porous media, non-Newtonian flow in viscoelastic fluids, nonequilibrium phase transitions, electrophoresis and electroosmosis, interdiffusion of fluids, and the monitoring of chemical wave propagation. In biology the method has provided new insights into plant physiology *in vivo*, multicellular tumour development and angiogenesis in tumours, while the method enables detailed MRI investigations concerning rat and mouse physiology in the study of diseases and their treatment. It should be noted that while the majority of applications of the method have concerned ^{1}H NMR, there are a number of important examples of its applications using other nuclei, including ^{19}F, ^{31}P, ^{13}C, ^{2}H and ^{17}O. These rarer nuclei allow site- or molecular-specific labelling, and provide the opportunity to investigate new contrast schemes, for example the mapping of pH in the case of ^{31}P.

A few examples selected from this range of applications are mentioned here. First, the use of NMR microscopy as a chemical mapping tool in plant physiology is illustrated in **Figure 7**, where two images of a castor bean stem cross section are illustrated alongside the total NMR spectrum from the stem. The spectral regions in the initial excitation pulse are indicated by the vertical line in each case and the images correspond to (A) water and (B) fructose. The sugar image is intense precisely at the phloem regions within the vascular bundles. There are many interesting applications of relaxation time weighting in order to obtain useful image contrast, some of which can be seen in chapter 5 of the book by Callaghan. One example from that chapter concerns profiles taken across several growth rings of a wood segment in which the water components associated with early wood, late wood and cell walls are separated by different spin relaxation times. This illustrates the potential of NMR microscopy to resolve differing aqueous components in porous media, in biological tissue and in food products. Another example concerns the sensitivity of T_1 and T_2 to the oxidation state of ions, an effect that is put to use in the imaging of chemical waves associated with the Belusov–Zhabotinsky reaction.

Figures 8 and **9** show applications of NMR microscopy in the rheological investigation of complex viscoelastic fluids. In **Figure 8** comparative velocity and diffusion profiles are shown across the diameter of a 700 μm diameter capillary though which is pumped a solution of high-molecular-mass polymer undergoing laminar flow. The velocity profile is distinctly non-Poiseuille, consistent with shear thinning, while the polymer self-diffusion coefficients exhibit a dramatic enhancement once the shear rate (the velocity gradient) exceeds a characteristic value. This value corresponds to the slowest relaxation rate of the molecule τ_d^{-1}, where τ_d is the so-called tube disengagement time. τ_d indicates the time

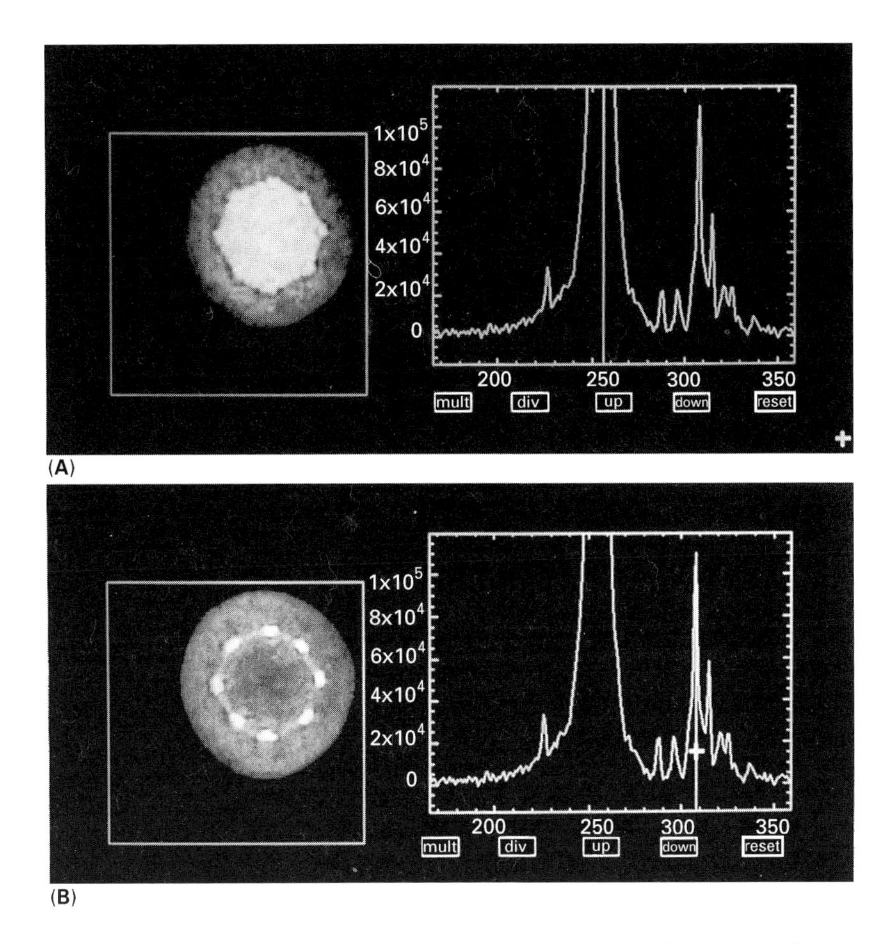

Figure 7 Chemical shift-selective proton NMR images of castor bean stems along with the corresponding proton NMR spectra from the whole stem. The region of the spectrum used to generate the corresponding image is shown by the vertical line. The upper image corresponds to water distribution while the lower corresponds to fructose distribution. Note the confinement of the sugars to the site of the phloem. Courtesy of W. Koeckenberger.

taken for the polymer to completely reconfigure its conformation. This diffusion enhancement phenomenon is associated with the breakdown of chain entanglements under shear and illustrates the changes that occur when an externally imposed rate of strain competes with the natural Brownian dynamics of a molecular system. In the example shown here, the crossover in the competition can be observed because the diameter of the capillary is sufficiently small that high shear rates can be observed, thus emphasizing the importance of the microscopic length scale.

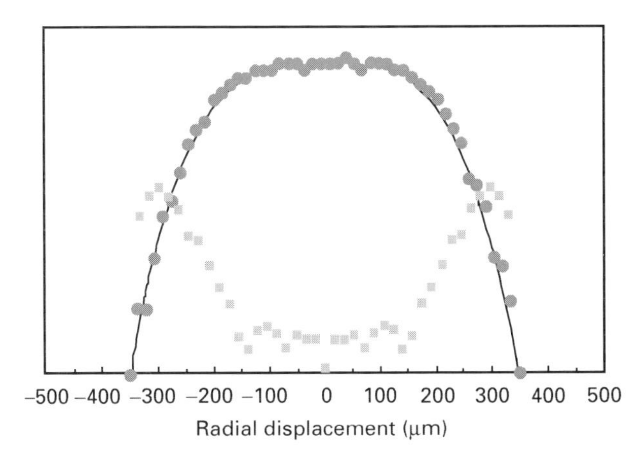

Figure 8 Velocity (●) and diffusion (■) profiles taken across a diametral slice for a solution of 5% 1.6 MDa poly(ethylene oxide) in laminar flow through a 700 mm diameter capillary. Courtesy of Y Xia and the author.

The second example, shown in **Figure 9**, also illustrates some of the strange phenomena apparent when competitive crossover occurs in systems that exhibit flow instability or are close to a phase transition. The image shows the velocity profile obtained for a wormlike micelle solution sheared in the gap of a cone-and-plate rheometer, along with the resultant shear rate of the fluid. Flow instability phenomena have led to distinctive shear banding, in effect a first-order phase transition of the fluid maintained under non-equilibrium conditions. Microscopy has been able to provide unique insight in this emerging area of condensed-matter physics.

(A)

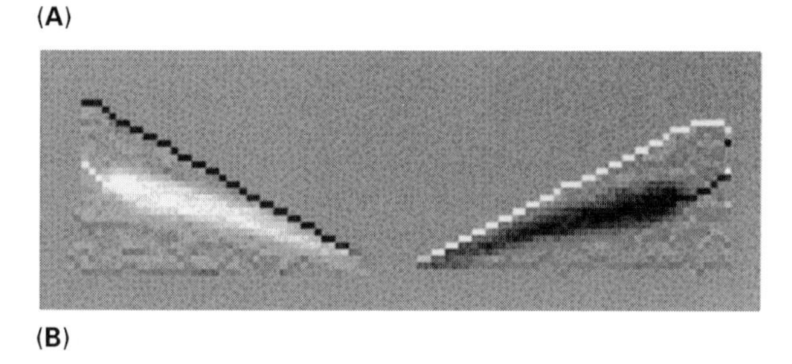

(B)

Figure 9 Stacked profile map (A) of the fluid velocity across the gap of a 7° cone and plate system containing a semidilute wormlike micelle solution, and shear rate image (B) obtained by taking the derivation of the velocity across the gap. Dramatic shear banding effects are apparent. Note the use of an expanded field of view across the gap achieved by using different magnitudes of spatial encoding gradients. Courtesy of MM Britton and the author.

Diffraction phenomena and higher spatial resolution

The diffractive physics at the heart of NMR imaging means in effect that any degree of structural homogeneity can be used to advantage in improving resolution. For example, in a periodic structure, the raw signal $S(\boldsymbol{k})$ will contain coherence peaks whose spacing bears an inverse relationship to the separations of 'scattering centres' in real space. Shorter-range spatial correlations can be investigated using Patterson function ($|S(\boldsymbol{k})|^2$) analysis. An entirely different but related diffraction phenomenon arises in the PGSE NMR experiment from the diffusion or dispersion of fluid within pores, where the boundary restrictions to molecular translation result in distinctive coherences in $E(\boldsymbol{q})$. All these approaches have the potential to greatly extend the resolution by which NMR field gradient methods can be used to probe liquid-phase morphology in materials science and biology.

Conclusion

NMR microscopy is an expensive and sophisticated technique that requires specialist insight in order to gain maximum advantage. Every new application requires a significant degree of spectroscopic optimization (e.g. RF and gradient pulse sequence design) and hardware optimization (e.g. RF antenna and sample holding design) if the best results are to be achieved. The method has now been realistically assessed and is likely to find extensive use in chemical physics and in chemical engineering in the study of fluid dynamics, multiphase flow and dispersion, restricted diffusion and flow in porous media, the rheology of soft condensed matter, phase separation and chemical instability, pH and temperature mapping, permeation and interdiffusion, and in research concerning electrophoresis and current imaging. In biological applications the method has unrivalled capability in studying the distribution and transport of specific molecular species in plant and insect physiology *in vivo*. In medical research it is proving an important tool in rat and mouse studies *in vivo* and it holds considerable promise as a complementary method in histology.

While the intrinsic insensitivity of NMR confines both the spatial and temporal resolution possible in NMR microscopy, a number of important new technical developments may yet extend those limits. These include the use of superconducting RF coils, and the use of indirect optical pumping techniques to greatly enhance nuclear polarization. Furthermore, in systems exhibiting a degree of structural homogeneity, the limits to resolution may be significantly enhanced by taking advantage of diffraction methodology. These latter approaches, while scientifically challenging, provide a nice link between the field of NMR spectroscopy and the wealth of scattering techniques that are so widely used in the study of condensed matter.

List of symbols

B_0 = applied magnetic field strength [flux density]; B_1 = applied transverse magnetic field strength [flux density]; D_s = self-diffusion coefficient; E = echo amplitude; $\boldsymbol{g}$ = gradient pulse amplitude; $\boldsymbol{G}$ = magnetic field gradient; $\boldsymbol{k}$ = reciprocal-space wave vector; P_s = average propagator; $\boldsymbol{q}$ = wave vector conjugate to molecular displacement; $\boldsymbol{r}$ = position vector; $\boldsymbol{R}$ = distance moved by spin over time Δ; S = signal amplitude; t_p = pulse duration; T_1 = spin–lattice relaxation time; T_2 = spin–spin relaxation time; v = local molecular velocity; γ = gyromagnetic ratio; δ = gradient pulse duration; Δ = gradient pulse separation; $\rho(\boldsymbol{r})$ = density of spins at $\boldsymbol{r}$; τ = delay time; ω = angular frequency of applied radiation/nuclear precession.

See also: **Contrast Mechanisms in MRI; Diffusion Studied Using NMR Spectroscopy; EPR Imaging; Fourier Transformation and Sampling Theory; Magnetic Field Gradients in High Resolution NMR; MRI Applications, Biological; MRI Instrumentation; MRI of Oil/Water in Rocks; MRI Theory; NMR Principles; NMR Pulse Sequences; NMR Relaxation Rates.**

Further reading

Blümich B and Kuhn W (eds) (1992) *Magnetic Resonance Microscopy: Methods and Application in Materials Science, Agriculture, and Biomedicine*. Weinheim: VCH.

Callaghan PT (1991) *Principles of NMR Microscopy*. Oxford: Oxford University Press.

Callaghan PT (1996) NMR imaging, NMR diffraction and applications of pulsed gradient spin echoes in porous media. *Magnetic Resonance Imaging* **14**: 701–709.

Callaghan PT and Stepisnik J (1996) Generalised analysis of motion using magnetic field gradients. *Advances in Magnetic and Optical Resonance* **19**: 325–388.

Kimmich R (1997) *NMR Tomography, Relaxometry and Diffusiometry*. Berlin: Springer.

Mansfield P and Morris PG (1982) *NMR Imaging in Biomedicine*. New York: Academic Press.

NMR of Solids

Jacek Klinowski, University of Cambridge, UK

MAGNETIC RESONANCE
Applications

Introduction

NMR spectra cannot normally be measured in solids in the same way in which they are routinely obtained from liquids. For example, the width of the ^{1}H NMR line in the spectrum of water is ~0.1 Hz, while the line from a static sample of ice is ~100 kHz wide. The reason for this is the existence of net anisotropic interactions which in the liquid are exactly averaged by the rapid thermal tumbling of molecules. A typical high-resolution spectrum of an organic compound in solution contains a wealth of information. The frequency of the radiation absorbed by the various non-equivalent nuclei in the molecule depends subtly on their chemical environments, giving rise to very sharp spectral lines. The parameters derived from such a spectrum (positions, widths, intensities and multiplicities of lines, relaxation mechanisms and rates) provide detailed information on the structure, conformation and molecular motion. This is not the case in a solid, where the nuclei are static and a conventional NMR spectrum is a broad hump which conceals most structural information. Although certain solids have sufficient molecular motion for NMR spectra to be obtainable without resorting to special techniques, we are concerned here with the general case, where there is no motion of nuclei and where conventional NMR, instead of sharp spectral lines, yields a broad hump which conceals information of interest to a chemist. Although the study of moments of such spectra and of various temperature-dependent parameters can still yield information on the degree of crystallinity, interatomic distances and molecular motion ('wide-line NMR'), we shall be primarily interested in ways of achieving high-resolution spectra, i.e. spectra which enable magnetically non-equivalent nuclei of the same spin species (e.g. ^{13}C) to be resolved as individual lines.

The interactions to be considered in the solid state and their Hamiltonians are as follows:

(1) Zeeman interaction with the magnetic field, $\mathcal{H}_Z$;
(2) chemical shielding, $\mathcal{H}_{CS}$;
(3) dipolar interaction, $\mathcal{H}_D$;
(4) *J*-coupling, $\mathcal{H}_J$;
(5) quadrupolar interaction, $\mathcal{H}_Q$.

The total Hamiltonian is a sum of all these contributions:

$$\mathcal{H} = \mathcal{H}_Z + \mathcal{H}_{CS} + \mathcal{H}_D + \mathcal{H}_J + \mathcal{H}_Q \qquad [1]$$

with the quadrupolar term $\mathcal{H}_Q$ non-zero only for nuclei with $I > \frac{1}{2}$. In general, $\mathcal{H}_Z$, $\mathcal{H}_{CS}$, $\mathcal{H}_D$ and $\mathcal{H}_Q$ are much larger than $\mathcal{H}_J$. *J*-Coupling is rarely observed in solids, so that $\mathcal{H}_J$ will henceforward be neglected.

The interaction Hamiltonians have the general form

$$\mathcal{H} = \boldsymbol{I} \cdot \mathbf{A} \cdot \boldsymbol{S} = [I_x\ I_y\ I_z] \begin{bmatrix} A_{xx} & A_{xy} & A_{xz} \\ A_{yz} & A_{yy} & A_{yz} \\ A_{zx} & A_{zy} & A_{zz} \end{bmatrix} \begin{bmatrix} S_x \\ S_y \\ S_z \end{bmatrix} \quad [2]$$

where *I* and *S* are vectors and **A** is a second-rank Cartesian tensor. We shall consider the various interactions in turn.

The Zeeman interaction

The Zeeman Hamiltonian, which determines the resonance frequency of an NMR-active nucleus in the magnetic field $\boldsymbol{B}_o$ is

$$\mathcal{H}_Z = \boldsymbol{I} \cdot \mathbf{Z} \cdot \boldsymbol{B}_o \quad [3]$$

where $\mathbf{Z} = -\gamma\hbar\mathbf{1}$, $\boldsymbol{I} = [I_x, I_y, I_z]$, $\boldsymbol{B}_o = [B_x, B_y, B_z]$ and **1** is a unit matrix. When the magnetic field is aligned with the *z*-axis of the laboratory frame of reference, $\boldsymbol{B}_o = [0, 0, B_o]$. The Zeeman interaction, which is directly proportional to the strength of the magnetic field, is thus entirely under the operator's control.

Magnetic shielding

The effect known as the chemical shift, central to the application of NMR in chemistry, is caused by simultaneous interactions of a nucleus with surrounding electrons and of the electrons with the static magnetic field $\boldsymbol{B}_o$. The field induces a secondary local magnetic field which opposes $\boldsymbol{B}_o$ thereby 'shielding' the nucleus from its full effect. The shielding Hamiltonian is

$$\mathcal{H}_{CS} = -\gamma\hbar\, \boldsymbol{I} \cdot \boldsymbol{\sigma} \cdot \boldsymbol{B}_o \quad [4]$$

The shielding is anisotropic, which is quantified in terms of a second-rank tensor σ ('the chemical shielding tensor'):

$$\boldsymbol{\sigma} = \begin{bmatrix} \sigma_{xx} & \sigma_{xy} & \sigma_{xz} \\ \sigma_{yx} & \sigma_{yy} & \sigma_{yz} \\ \sigma_{zx} & \sigma_{zy} & \sigma_{zz} \end{bmatrix} \quad [5]$$

In strong magnetic fields σ is axially symmetric. When transformed into its principal reference system (PAS) by using rotation matrices, the tensor is described by three principal components σ_{ii} (i = 1, 2, 3):

$$\begin{bmatrix} \sigma_{xx} & \sigma_{xy} & \sigma_{xz} \\ \sigma_{yx} & \sigma_{yy} & \sigma_{yz} \\ \sigma_{zx} & \sigma_{zy} & \sigma_{zz} \end{bmatrix} \xrightarrow{\text{Rotation}} \begin{bmatrix} \sigma_{11} & 0 & 0 \\ 0 & \sigma_{22} & 0 \\ 0 & 0 & \sigma_{33} \end{bmatrix}$$

and three direction cosines, $\cos\theta_i$, between the axes of PAS and the laboratory frame.

The observed shielding constant, σ_{zz}, is a linear combination of the principal components:

$$\sigma_{zz} = \sum_{i=1}^{3} \sigma_{ii}\cos^2\theta_i = \frac{1}{3}\mathrm{Tr}\,\boldsymbol{\sigma} + \frac{1}{3}\sum_{i=1}^{3}(3\cos^2\theta_i - 1)\sigma_{ii} \quad [6]$$

where Tr σ stands for the trace of the tensor. Since the average value of each $\cos^2\theta_i$ is $\frac{1}{3}$, the average value of σ_{zz} in the NMR spectra of liquids (where there is random molecular tumbling) is the isotropic value:

$$\overline{\sigma_{zz}} = \frac{1}{3}\mathrm{Tr}\,\sigma = \sigma_{\text{iso}} \quad [7]$$

In solids the angle-dependent second term on the right of Equation [6] survives, giving rise to a spread of resonance frequencies, i.e. line broadening.

Dipolar interactions

The Hamiltonian for the dipolar interaction between a pair of nuclei *i* and *j* separated by the internuclear vector *r* is given by

$$\mathcal{H}_D = R\boldsymbol{I}_i \cdot \mathbf{D} \cdot \boldsymbol{I}_j = R\,[I_{ix}\ I_{iy}\ I_{iz}] \times \begin{bmatrix} r^2 - 3x^2 & -3xy & -3xy \\ -3xy & r^2 - 3y^2 & -3yz \\ -3xz & -3yz & r^2 - 3z^2 \end{bmatrix} \begin{bmatrix} I_{jx} \\ I_{jy} \\ I_{jz} \end{bmatrix} \quad [8]$$

where $R = \gamma_i\gamma_j\hbar\mu_o/4\pi r^3$ is the dipolar coupling constant, γ the nuclear gyromagnetic ratio and **D** the dipolar interaction tensor. In the PAS of the tensor, with the internuclear vector aligned along one of the coordinate axes, we have $xy = yz = zx = 0$, $r^2 = x^2+y^2+z^2$ and the tensor becomes

$$\mathbf{D} = \begin{bmatrix} 1 & 0 & 0 \\ 0 & -2 & 0 \\ 0 & 0 & 1 \end{bmatrix} \qquad [9]$$

It is clearly traceless (Tr $\mathbf{D}$ = 1−2+1 = 0).

The truncated dipolar interaction Hamiltonian may be written in the form

$$\mathcal{H}_\mathrm{D} = \frac{\gamma_i \gamma_j \hbar^2}{2r^3}[\boldsymbol{I}_i \cdot \boldsymbol{I}_j - 3I_{iz}\, I_{jz}](3\cos^2\theta - 1) \qquad [10]$$

where θ is the angle between $\boldsymbol{r}$ and the external magnetic field $\boldsymbol{B}_\mathrm{o}$. Since the average value $\overline{\cos^2\theta_{ij}} = \frac{1}{3}$, the isotropic average of the Hamiltonian is $\overline{\mathcal{H}_\mathrm{D}} = 0$, so that the dipolar interaction does not affect the NMR spectrum in solution. In the solid the interaction remains, greatly increasing the spectral line width.

Quadrupolar interactions

Some 74% of all NMR-active nuclei have $I > \frac{1}{2}$, so that, in addition to magnetic moment, they possess an electric quadrupole moment brought about by non-spherical distribution of the nuclear charge. The quadrupole interaction broadens and shifts the NMR lines, and also affects their relative intensities.

When the quadrupolar Hamiltonian is considered as a perturbation on the Zeeman Hamiltonian, there is no general analytical solution for the eigenvalues of $\mathcal{H}_\mathrm{Z}$ in the (very rare) case when $\mathcal{H}_\mathrm{Z}$ and $\mathcal{H}_\mathrm{Q}$ are of comparable magnitude. When $\mathcal{H}_\mathrm{Q} >> \mathcal{H}_\mathrm{Z}$, the splitting of the nuclear states is very large and 'pure quadrupole resonance' (NQR) is observed even in the absence of a magnetic field. In the usual 'high field' case, $\mathcal{H}_\mathrm{Z} >> \mathcal{H}_\mathrm{Q}$, the quadrupole Hamiltonian in the PAS of the electric field gradient tensor is

$$\mathcal{H}_\mathrm{Q} = \frac{e^2qQ}{4I(2I-1)}[3I_z^2 - I^2 + \eta(I_x^2 - I_y^2)] \qquad [11]$$

where η is the asymmetry parameter which describes the symmetry of the electric field gradient. The definitions of η and of the 'quadrupole frequency', ν_Q, which describes the magnitude of the interaction, are

$$\eta = \frac{V_{xx} - V_{yy}}{V_{zz}} \qquad \nu_\mathrm{Q} = \frac{3e^2qQ}{2I(2I-1)h} \qquad [12]$$

Perturbation theory allows us to calculate the energy levels $E_m^{(0)}$, $E_m^{(1)}$ and $E_m^{(2)}$ (superscripts denote the order). Because of the first- and second-order shifts in energy levels, instead of a single (Larmor) resonance frequency $\nu_\mathrm{L} = [E_{m-1}^{(0)} - E_m^{(0)}]$, as with spin-$\frac{1}{2}$ nuclei, there are now several resonance frequencies:

$$\nu_m = \frac{E_{m-1} - E_m}{h} = \nu_\mathrm{L} + \nu_m^{(1)} + \nu_m^{(2)} \qquad [13]$$

Detailed calculations reveal that:

(1) The first-order frequency shift is zero for $m = \frac{1}{2}$, so that the central transition for non-integer spins (such as ^{27}Al with $I = \frac{5}{2}$) is not affected by quadrupolar interactions to first order. It is thus advantageous to work with such nuclei, especially since the central transition is normally the only one which is observed: other transitions are so broadened and shifted as to be unobservable.
(2) The first-order shift is scaled by $\frac{1}{2}(3\cos^2\theta - 1)$.
(3) The second-order shift increases with ν_Q^2 and is inversely proportional to the magnetic field strength. Since the dispersion of the chemical shift, which is what we normally wish to measure, is proportional to $\boldsymbol{B}_\mathrm{o}$, it is advantageous to work at high fields, where the chemical shift effects make the maximum contribution to the spectrum. As the second-order frequency shift is always present for all transitions, the feasibility of obtaining useful spectra depends on the magnitude of ν_Q.

The very small quadrupole interactions of ^{2}H and their sensitivity to molecular motion at a wide range of frequencies make this integer spin nucleus very useful for chemical studies. ^{2}H NMR experiments normally use static samples, and dynamic information is extracted by comparing spectra measured at different temperatures with model computer simulations.

Magic-angle spinning

Magic-angle spinning (MAS) is by far the most powerful tool in solid-state NMR. The technique averages anisotropic interactions by acting on the factor $(3\cos^2\theta - 1)$ in the Hamiltonians, which in solids is not averaged to zero by rapid molecular motion. MAS was first introduced to deal with the dipolar interaction. It can be shown that when the sample is rapidly spun around an axis inclined at the angle β to the direction of the magnetic field, the time-averaged value of the angle θ, which an arbitrary internuclear vector makes with $\boldsymbol{B}_o$, is

$$\overline{3\cos^2\theta - 1} = \frac{1}{2}(3\cos^2\beta - 1)(3\cos^2\chi - 1) \qquad [14]$$

where χ, is the angle between the internuclear vector and axis of rotation, is constant for each vector, because the solid is rigid. The result is that the term $\frac{1}{2}(3\cos^2\beta - 1)$ scales the spectral width, and that for $\beta = \cos^{-1}/\sqrt{3} = 54.74°$ (the 'magic angle'), $\overline{3\cos^2\theta - 1} = 0$. The dipolar Hamiltonian in Equation [10] is averaged to zero.

For MAS to be effective, the sample must be spun at a rate greater than the static spectral width expressed in Hz. As the homonuclear ^{1}H–^{1}H interactions may lead to spectra which are as much as 50 kHz wide, it is not possible to spin the sample fast enough. Thus high-resolution solid-state ^{1}H spectra of most organic compounds, where protons are generally close together, cannot be obtained with the use of MAS alone, but require the additional use of multiple-pulse techniques (see below). However, MAS is successful in removing homonuclear interactions for ^{13}C, ^{31}P and nuclei of small gyromagnetic ratios.

The chemical shift anisotropy is also reduced by MAS, because the tensor interactions controlling all anisotropic interactions in solids all have a common structure and may be expressed in terms of Wigner rotation matrices which are scaled by MAS.

High-power decoupling

When dilute spins, such as ^{13}C, interact via the dipolar interaction with ^{1}H or other abundant nuclei, the large heteronuclear broadening of an already low-intensity spectrum is a considerable problem. High-power decoupling, used to remove heteronuclear coupling effects, applies a continuous, very-high-power pulse at the ^{1}H resonance frequency in a direction perpendicular to $\boldsymbol{B}_o$. The ^{13}C pulse is then applied, and the ^{13}C free induction decay measured while continuing the ^{1}H irradiation. The powerful decoupling pulse stimulates rapid ^{1}H spin transitions, so rapid that the ^{13}C spins experience only the time-average of the ^{1}H magnetic moment, i.e. zero. Since the technique relies on selective excitation of the abundant and dilute nuclei, it can only remove heteronuclear interactions.

Cross-polarization

Dilute nuclei, such as ^{13}C and ^{15}N, are more difficult to observe than abundant nuclei, such as ^{1}H or ^{31}P, particularly when they also have a low gyromagnetic ratio. However, the dilute and abundant nuclei are often in close proximity, and coupled via the dipolar interaction. Cross-polarization (CP) exploits this interaction to observe dilute nuclei, at the same time overcoming two serious problems often encountered in solid-state NMR: (i) because of a very small population difference in the polarized sample, NMR actually observes very few dilute spins and consequently the sensitivity of the experiment is low; (ii) spin–lattice relaxation times of spin-$\frac{1}{2}$ nuclei in solids are often very long, so that long delays are required between experiments and the spectral signal-to-noise ratio is poor.

The sequence of events during the ^{13}C–^{1}H CP experiment is as follows. After the end of the 'preparation period', during which the sample polarizes in the magnetic field, a $\pi/2$ pulse is selectively applied to ^{1}H along the x-axis of the rotating frame, aligning the ^{1}H magnetization with the y-axis. A long pulse of amplitude B_{1H} is then applied along the y-axis. Since the ^{1}H magnetization is now aligned with the effective field in the rotating frame, it becomes 'spin locked' along this direction. At the same time, a long pulse of amplitude B_{1C} is selectively applied to ^{13}C along the x-axis. The amplitudes B_{1H} and B_{1C} are adjusted so as to satisfy the Hartmann–Hahn condition:

$$\gamma_H B_{1H} = \gamma_C B_{1C} \qquad [15]$$

The energies of ^{1}H and ^{13}C in the rotating frame are thus equal, and the two spin reservoirs can transfer magnetization in an energy-conserving manner during the 'contact time'. Finally, the ^{13}C radiofrequency field is turned off and a free induction decay observed in the usual way. During the observation time the ^{1}H field is still on, but serves as the high-power decoupling field to reduce the ^{1}H–^{13}C dipolar broadening.

Detailed arguments show that the magnetization of ^{13}C nuclei is theoretically increased by the factor of $\gamma_H/\gamma_C \approx 4$. After the ^{13}C free induction decay signal has been measured, the magnetization of carbons is again almost zero, but the loss of proton magnetization is small. The CP experiment can be repeated without waiting for the carbons to relax. The only limitations are the gradual loss of polarization by the ^1H spin reservoir, and the decay of the ^1H magnetization during spin locking. The latter process proceeds on a time-scale ('spin–lattice relaxation in the rotating frame') which is much shorter than the ^{13}C spin–lattice relaxation time.

Multiple-pulse line narrowing

Although homonuclear dipolar couplings are in principle removable by MAS, with abundant nuclei they are often very strong. For example, the removal of the ^1H–^1H interaction in most organic compounds requires spinning rates far in excess of what is practically feasible. The alternative to MAS is to manipulate the nuclear spins themselves using 'multiple-pulse line narrowing' so as to average the dipolar interaction. The method uses specially designed sequences of pulses with carefully adjusted phase, duration and spacing. The result is that, when the signal is sampled at a certain moment during the sequence, the dipolar interaction is averaged to zero.

WAHUHA, the simplest multiple pulse sequence, is composed of four 90° pulses:

$$(\mathrm{P}_x - 2\tau - \mathrm{P}_{-x} - \tau - \mathrm{P}_y - 2\tau - \mathrm{P}_{-y} - \tau)\ n \text{ times} \qquad [16]$$

where P_i represents rotation about the particular i-axis of the rotating frame and τ is the time interval between pulses. Over the sequence, the magnetic moments spend equal amounts of time along each of the three principal axes. The NMR signal is sampled in one of the 2τ windows. Sequences have been developed involving from 4 to as many as 52 pulses. The entire sequence must be short relative to the relaxation time T_2, and the pulses themselves must also be very short.

Multiple pulse sequences average the dipolar Hamiltonian, but also affect other Hamiltonians to an extent which depends on the particular sequence. For example, the WAHUHA sequence scales chemical shift anisotropies by a factor of $1/\sqrt{3}$.

Moments of an NMR line

Even when the dipolar ^1H–^1H interaction is not removed from the spectrum, the method of moments can provide important structural information. The nth moment of the line shape $f(\omega)$ about ω_0 is defined as

$$M_n = \frac{\int_0^\infty (\omega - \omega_0)^n f(\omega)\, \mathrm{d}\omega}{\int_0^\infty f(\omega)\, \mathrm{d}\omega} \qquad [17]$$

where

$$M_0 = \int_0^\infty f(\omega)\, \mathrm{d}\omega$$

is the area under the line (the zeroth moment). For a normalized function $M_0 = 1$. The second moment is physically analogous to the moment of inertia of an object with the same shape as the line. If $f(\omega)$ is an even function of ω, $M_n = 0$ for all odd values of n. It is convenient to calculate moments about the centre of gravity of the line shape, i.e. the value of ω_0 for which the first moment is zero.

The second moment can be calculated from the interatomic distances in the solid containing pairs i, j of dipolar-coupled nuclei. Van Vleck has shown that, for a polycrystalline powder composed of randomly oriented crystals in which we observe identical spin-$\frac{1}{2}$ nuclei, the second moment is

$$M_2^{\mathrm{homo}} = \frac{9}{16}\gamma^4 \hbar^2 \left(\frac{\mu_0}{4\pi}\right)^2 \sum_j \frac{1}{r_{ij}^6} \qquad [18]$$

while for pairs of unlike nuclei the second moment is different:

$$M_2^{\mathrm{hetero}} = \frac{1}{4}\gamma_i^2 \gamma_j^2 \hbar^2 \left(\frac{\mu_0}{4\pi}\right)^2 \sum_j \frac{1}{r_{ij}^6} \qquad [19]$$

Thus, even when the interacting nuclei have very similar gyromagnetic ratios, the homonuclear second moment is larger by a factor of $\frac{9}{4}$ than the heteronuclear moment. This is because dipolar coupling between unlike spins cannot lead to an energy conserving mutual spin flip. The second moment is thus very sensitive to the kind of neighbour.

The method of moments has further advantages. First, since the second moment is inversely proportional to the sixth power of the internuclear distance, it is a very sensitive means of determining interatomic distances. Second, it can provide insights into the structure. For example, it was used to demonstrate the presence of groups of three equivalent protons in solid hydrates of strong acids, thus proving the presence of oxonium ions, H_3O^+. Third, it is useful for the study of motion, because the moments are dramatically reduced when the dipolar interaction is partly or completely averaged out by an onset of a specific motion.

DOR, DAS and MQ-MAS

We have seen that the second-order quadrupolar interaction, which affects all quadrupolar nuclei, is reduced, but not removed, by MAS. Its complete removal is the most important current problem in solid-state NMR. Three different techniques have been proposed to achieve this aim.

When the second-order quadrupole interaction is expanded as a function of Wigner rotation matrices, and we consider the case of a sample rapidly rotated about an angle β with respect to $\boldsymbol{B}_o$, the average second-order quadrupolar shift of the central transition becomes

$$\nu_{\frac{1}{2}}^{(2)} = \frac{\nu_Q^{(2)}}{\nu_L}\left[I(I+1) - \frac{3}{4}\right][A_0 + B_2 P_2(\cos\beta) + B_4 P_4(\cos\beta)] \quad [20]$$

where ν_Q is the quadrupole frequency, ν_L is the Larmor frequency, A_0 and B_0 are constants and the $P_n(\cos\beta)$ terms are the Legendre polynomials

$$P_2(\cos\beta) = \frac{1}{2}(3\cos^2\beta - 1)$$
$$P_4(\cos\beta) = \frac{1}{8}(35\cos^4\beta - 30\cos^2\beta + 3) \quad [21]$$

There is no value of β for which both the $P_2(\cos\beta)$ and the $P_4(\cos\beta)$ terms can be zero, so that the angle-dependent terms cannot be averaged by spinning about a single axis. Instead, in the ingenious 'double-rotation' (DOR) experiment the sample is spun *simultaneously* about two different axes β_1 and β_2, so that

$$P_2(\cos\beta_1) = 0 \qquad P_4(\cos\beta_2) = 0 \quad [22]$$

with solutions $\beta_1 = 54.74°$ (the conventional magic angle) and $\beta_2 = 30.56$ or $70.12°$. As a result, only the A_0 term remains in Equation [20]. This is accomplished by a rotor-within-a-rotor probehead in which the centres of gravity of the two rotors, each spinning at a different angle with respect to $\boldsymbol{B}_o$, exactly coincide. Although the daunting engineering problems posed by the design of a DOR probehead have been overcome, it is very difficult to spin the two rotors simultaneously at sufficiently high spinning speeds, and the spinning rates are at present limited to ~6 and 1 kHz for the inner and outer rotors, respectively, compared with ~30 kHz achievable with MAS. This is an unfortunate limitation, since multiple spinning sidebands appear in the spectra if the rate of the rotation is lower than the strength of the quadrupolar interaction.

The technique known as 'dynamic-angle spinning' (DAS) adopts an alternative approach to DOR: the sample is rotated *sequentially* about two different axes, β_1' and β_2', which are chosen so that

$$P_2(\cos\beta_1') = -P_2(\cos\beta_2')$$
$$P_4(\cos\beta_1') = -P_4(\cos\beta_2') \quad [23]$$

with the solutions $\beta_1' = 37.38°$ and $\beta_2' = 79.19°$. The rotation axis is switched very rapidly, which poses technical problems, given that the minimum time required for changing the spinning angle must be shorter than the relaxation time of the nucleus being observed. As a result, DAS often cannot be applied to many nuclei, including ^{27}Al, and is limited to the study of nuclei with long relaxation times (for example in amorphous samples, such as glasses).

Yet another solution to the problem, known as 'multiple-quantum magic-angle spinning' (MQ-MAS) relies on the fact that B_2 and B_4 are functions of I, p, η, α and β, where p is the order of the multiquantum coherence and α and β are the Euler angles corresponding to the orientation of each crystallite in the powder with respect to the rotor axis. Under fast MAS, the chemical shift anisotropy, heteronuclear dipolar interactions and the term proportional to P_2 in Equation [20] are removed, so that

$$\nu_{\frac{1}{2}}^{(2)} = \frac{\nu_Q^{(2)}}{\nu_L}\left[I(I+1) - \frac{3}{4}\right][A_0 + B_4 P_4(\cos 54.74°)] \quad [24]$$

Although the second term, proportional to P_4, still causes substantial line broadening, it can be eliminated by using p-quantum transitions. A p-quantum transition (with $p = 3$ or 5 for ^{27}Al) is excited and the signal allowed to evolve during time t_1. As multiple quantum transitions are not directly observable by

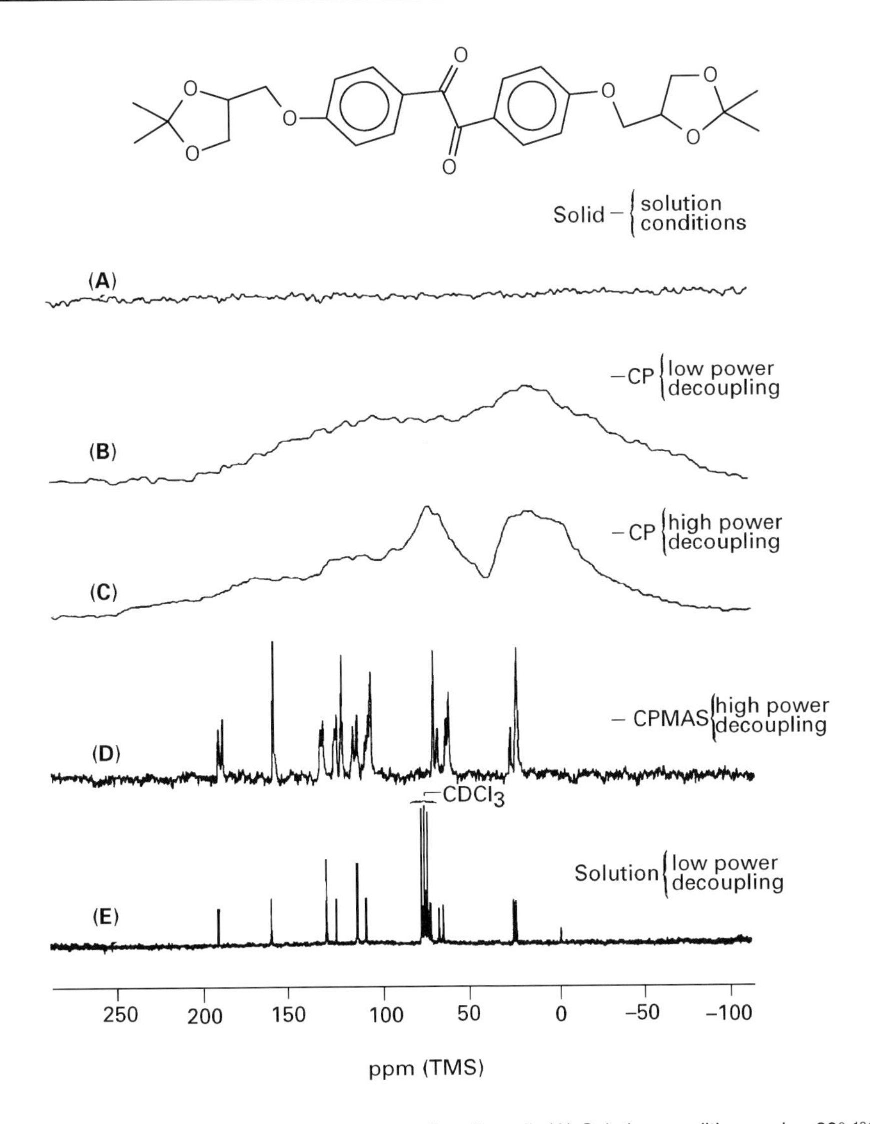

Figure 1 ^{13}C NMR spectra of solid 4,4′-bis[(2,3-dihydroxypropyl)oxy]benzil. (A) Solution conditions using 60° ^{13}C pulses and 10 s recycle delays; (B) as in (A) but with ^{1}H–^{13}C cross-polarization, low-power proton decoupling and 1 s recycle delays; (C) as in (B) but with high-power proton decoupling; (D) as in (C) but with the addition of magic-angle spinning; (E) high-resolution spectrum of a solution in $CDCl_3$ with the same NMR parameters. Reproduced with permission of the American Chemical Society from Yannoni CS (1982) *Accounts of Chemical Research* **15**: 201–208. Copyright 1982 American Chemical Society.

NMR, a second pulse converts the signal into a single-quantum transition, which is observable. The technique enables a two-dimensional representation of the spectra, with a regular increment of t_1 providing a 'p-quantum dimension', free of quadrupolar interactions. Although the optimal conditions for MQ-MAS are difficult to establish, the technique is being increasingly used for the study of quadrupolar nuclei of half-integer spin, such as ^{27}Al, ^{85}Rb, ^{23}Na, ^{11}B and ^{93}Nb.

Note that DOR, DAS and MQ-MAS do not remove the A_0 term in Equations [20] and [24]. Thus the position of the line in the spectrum, however narrow, does not correspond to the pure chemical shift, but includes the effect of the quadrupole interaction.

Modern solid-state NMR

Magic-angle spinning has greatly enhanced our knowledge of a wide range of materials used in chemical, physical, biological and earth sciences and in the technology of glass and ceramics. It took nearly twenty years, since its discovery in 1958, for MAS to become a routine tool of structural investigation. The reasons were the difficulty of spinning the sample at the very high speeds required and the insufficiently high magnetic fields. However, the introduction of Fourier-transform NMR, cross-polarization and superconducting magnets during the 1960s and 1970s greatly improved the sensitivity of the spectra and enabled virtually all NMR-active nuclei to be observed in solids. ^{1}H MAS NMR was used to examine polymers as early as 1972, and Schaefer and Stejskal were the first to combine CP and MAS in ^{13}C NMR studies of organics. Much important work, at first mostly with ^{13}C but later with other nuclei, has been done since. Since the early 1980s great progress has been made in the study of ^{29}Si and ^{27}Al in natural and synthetic molecular sieve catalysts and minerals, which is particularly significant since nearly a half of all known minerals are silicates or aluminosilicates.

High-resolution spectra of solids are now routinely obtained using a combination of CP and MAS (see **Figure 1**), and it is fair to say that CP-MAS has revolutionized materials science. The otherwise weak signals from dilute nuclei (such as ^{13}C or ^{29}Si) are enhanced by cross-polarization, heteronuclear dipolar interactions are removed by high-power decoupling, chemical shift anisotropy and the weak dipolar interactions between dilute nuclei are averaged by fast MAS, and the signal-to-noise ratio is increased further thanks to the more frequent repetition of the experiment and the availability of high magnetic fields. Although the line widths is in such high-resolution spectra are still greater than these measured in liquids, the various non-equivalent nuclei can in most cases be separately resolved.

List of symbols

$\boldsymbol{B}_0$ = magnetic flux density; $\mathbf{D}$ = dipolar interaction tensor; $\mathcal{H}$ = interaction Hamiltonian; p = order of multiquantum coherence; P_i = rotation about the i-axis; R = dipolar coupling constant; T_1, T_2 = relaxation times; γ = nuclear gyromagnetic ratio; η = asymmetry parameter; ν_Q = quadrupole frequency; ν_L = Larmor resonance frequency; σ_{zz} = shielding constant; τ = time interval between pulses.

See also: **^{13}C NMR, Parameter Survey; ^{13}C NMR, Methods; High Resolution Solid State NMR, ^{13}C; High Resolution Solid State NMR, ^{1}H, ^{19}F; Magnetic Field Gradients in High Resolution NMR; NMR Principles; NMR Pulse Sequences; Solid State NMR, Methods; Solid State NMR, Rotational Resonance.**

Further reading

Abragam A (1983) *The Principles of Nuclear Magnetism*. Oxford: Clarendon Press.

Andrew ER (1981) Magic angle spinning. *International Reviews of Physical Chemistry* **1**: 195–224.

Engelhardt G and Michel D (1987) *High-Resolution Solid-State NMR of Silicates and Zeolites*. Chichester: John Wiley.

Fukushima E and Roeder SBW (1981) *Experimental Pulse NMR – A Nuts and Bolts Approach*. Reading, MA: Addison-Wesley.

Fyfe CA (1983) *Solid State NMR for Chemists*. Ontario: CFC Press.

Mehring M (1983) *High-Resolution NMR Spectroscopy in Solids*, 2nd edn. New York: Springer-Verlag.

Slichter CP (1989) *Principles of Magnetic Resonance*, 3rd edn. New York: Springer-Verlag.

Stejskal EO and Memory JD (1994) *High Resolution NMR in the Solid State. Fundamentals of CP/MAS*. Oxford: Oxford University Press.

NMR Principles

PJ Hore, Oxford University, UK

MAGNETIC RESONANCE
Theory

Nuclear magnetic resonance spectroscopy is an extraordinarily powerful source of information on the structure and dynamics of molecules. Almost every molecule one can think of has at least one magnetic nucleus already in place, exceedingly sensitive to its surroundings but interacting very weakly with them. As such, nuclear spins are ideal probes of molecular properties at the atomic level.

NMR spectra of molecules in liquids contain essentially five sources of information: the intensities of individual resonances (which depend on the number of nuclei responsible), chemical shifts (the interaction of nuclear spins with an applied magnetic field), spin–spin coupling (their interactions with one another), spin relaxation (the restoration of thermal equilibrium), and chemical exchange (the effects of conformational and chemical equilibria).

Spin angular momentum and nuclear magnetism

Most atomic nuclei have an intrinsic angular momentum known as *spin*. Like the angular momentum of a gyroscope, nuclear spin is a vector quantity – it has both magnitude and direction. Unlike classical angular momentum, however, nuclear spin is quantized. Its magnitude is

$$\sqrt{I(I+1)}\hbar \qquad [1]$$

where I is the *spin quantum number* of the nuclide in question and $\hbar$ is Planck's constant h divided by 2π. I may be zero, or a positive integer or half-integer:

$$I = 0, \tfrac{1}{2}, 1, \tfrac{3}{2}, 2, \cdots \qquad [2]$$

Table 1 gives the spin quantum numbers of some popular NMR nuclei.

The projection of the angular momentum vector $\boldsymbol{I}$ onto an arbitrary axis (labelled z) is also quantized:

$$I_z = m\hbar \qquad [3]$$

where the *magnetic quantum number*, m, can have values between $+I$ and $-I$ in integral steps:

$$m = +I, +I-1, \cdots, -I+1, -I \qquad [4]$$

Table 1 Nuclear spin quantum numbers of some popular NMR nuclides

I	*Nuclide*					
0	^{12}C	^{16}O				
$\frac{1}{2}$	^{1}H	^{13}C	^{15}N	^{19}F	^{29}Si	^{31}P
1	^{2}H	^{14}N				
$\frac{3}{2}$	^{11}B	^{23}Na	^{35}Cl	^{37}Cl		
$\frac{5}{2}$	^{17}O	^{27}Al				
3	^{10}B					

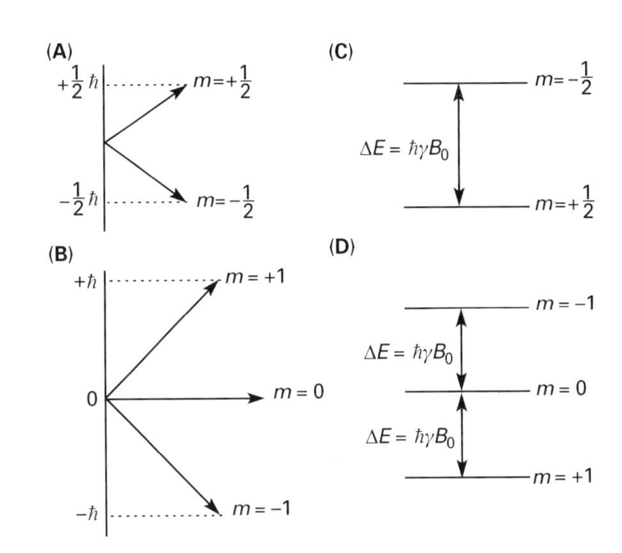

Figure 1 Space quantization and energy levels of spin-½ and spin-1 nuclei. (A) and (C) spin-½; (B) and (D) spin-1. The energy level splittings produced by an applied magnetic field depend on the value of the gyromagnetic ratio, γ (here taken as positive).

The spin of a nucleus with $I = \frac{1}{2}$ (e.g. ^{1}H) has magnitude $(\sqrt{3}/2)\,\hbar$ and z component $I_z = \pm\frac{1}{2}\hbar$; for $I = 1$ (e.g. ^{2}H), the spin angular momentum is $\sqrt{2}\hbar$, and $I_z = 0$ or $\pm\hbar$ (**Figures 1A** and **1B**). According to the uncertainty principle, the other two (x and y) components of the angular momentum cannot be known once the magnitude and the z component of $\boldsymbol{I}$ have been specified.

Closely associated with nuclear spin is a *magnetic moment* μ

$$\boldsymbol{\mu} = \gamma \boldsymbol{I} \quad [5]$$

which is parallel or sometimes antiparallel to $\boldsymbol{I}$, with a proportionality constant γ called the *gyromagnetic ratio*. As a consequence, both the magnitude and orientation of μ are quantized. In the absence of a magnetic field, all $2I + 1$ states of a spin-I nucleus are degenerate, and the direction of the quantization axis is arbitrary.

In an applied magnetic field $\boldsymbol{B}_0$ with strength B_0, the spins are quantized along the field direction (the z-axis) and have an energy

$$E = -\boldsymbol{\mu} \cdot \boldsymbol{B}_0 = -\mu_z B_0 \quad [6]$$

where $\boldsymbol{\mu} \cdot \boldsymbol{B}_0$ is the scalar product of the two vectors, and μ_z is the projection of μ onto $\boldsymbol{B}_0$. Since $\mu_z = \gamma I_z$ and $I_z = m\hbar$, it follows that

$$E = -m\hbar\gamma B_0 \quad [7]$$

That is, the $2I + 1$ states are split apart in energy, with a uniform gap $\Delta E = \hbar\gamma B_0$ between adjacent levels (**Figures 1C and 1D**).

The NMR experiment involves applying electromagnetic radiation of the correct frequency ν to 'flip' spins from one energy level to another, according to the selection rule $\Delta m = \pm 1$, i.e.

$$h\nu = \Delta E = \hbar\gamma B_0 \quad [8]$$

which may be rearranged to give the *resonance condition*

$$\nu = \frac{\gamma B_0}{2\pi} \quad [9]$$

The NMR frequency of a nucleus is proportional to its γ and to the strength of the field; the $2I$ allowed transitions of a spin-I nucleus have identical frequencies (e.g. **Figure 1D**). Typical magnetic fields used in modern NMR spectroscopy are in the range 4.7–20.0 T, giving proton (^{1}H) resonance frequencies of 200–850 MHz, falling in the *radiofrequency* region of the electromagnetic spectrum. **Table 2** gives the gyromagnetic ratios, resonance frequencies in a 9.4 T field, and natural isotopic abundances of some commonly studied NMR nuclei.

Table 2 Gyromagnetic ratios, NMR frequencies (in a 9.4 T field), and natural isotopic abundances of selected nuclides

	γ (10^7 T^{-1} s^{-1})	ν (MHz)	Natural abundance (%)
^{1}H	26.75	400.0	99.985
^{2}H	4.11	61.4	0.015
^{13}C	6.73	100.6	1.108
^{14}N	1.93	28.9	99.63
^{15}N	−2.71	40.5	0.37
^{17}O	−3.63	54.3	0.037
^{19}F	25.18	376.5	100.0
^{29}Si	−5.32	79.6	4.70
^{31}P	10.84	162.1	100

The intensity of the observed NMR signal depends on the difference between the numbers of nuclei in the states involved in the transition. At thermal equilibrium the fractional difference in populations, of a spin-$\frac{1}{2}$ nucleus with positive γ, is given by the Boltzmann distribution:

$$\frac{n_\alpha - n_\beta}{n_\alpha + n_\beta} = \frac{e^{\Delta} - e^{-\Delta}}{e^{\Delta} + e^{-\Delta}} \cong \Delta = \frac{\hbar\gamma B_0}{2kT} = \frac{h\nu}{2kT} \quad [10]$$

where α and β denote the $m = +\frac{1}{2}$ and $m = -\frac{1}{2}$ levels, k is the Boltzmann constant, and T is the temperature in kelvin. The approximation made in Equation [10] is that the NMR energy gap $\hbar\gamma B_0$ is tiny by comparison with kT, which is the situation in essentially all NMR experiments. For protons (^{1}H) in a 9.4 T field, $\nu = 400$ MHz, so that $\Delta = 3.2 \times 10^{-5}$, giving a population difference of about one part in 31 000.

Chemical shifts

Although the resonance frequency of a nucleus in a magnetic field is determined principally by γ, it also depends, slightly, on the immediate surroundings of the nucleus. This effect, the *chemical shift*, is of crucial importance for chemical applications of NMR because it allows one to distinguish nuclei in different environments. For example, the ^{1}H spectrum of liquid ethanol (**Figure 2**) shows clearly that there are three types of protons (methyl, methylene and hydroxyl).

The chemical shift exists because the applied magnetic field $\boldsymbol{B}_0$ causes electrons in atoms and molecules to circulate around the nuclei. Somewhat like an electric current in a loop of wire, the swirling electrons generate a small local magnetic field that

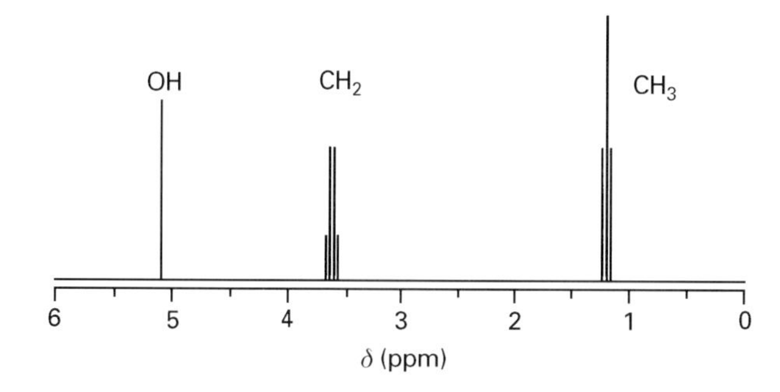

Figure 2 Schematic ^{1}H NMR spectrum of liquid ethanol, C_2H_5OH. The three multiplets, at chemical shifts of 1.2, 3.6 and 5.1 ppm arise from the CH_3, CH_2, and OH protons. The multiplet structure (quartet for the CH_2, triplet for the CH_3) arises from the spin–spin coupling of the two sets of protons. Splittings are not normally seen from the coupling of the OH and CH_2 protons, because the hydroxyl proton undergoes rapid intermolecular exchange, catalysed by traces of acid or base.

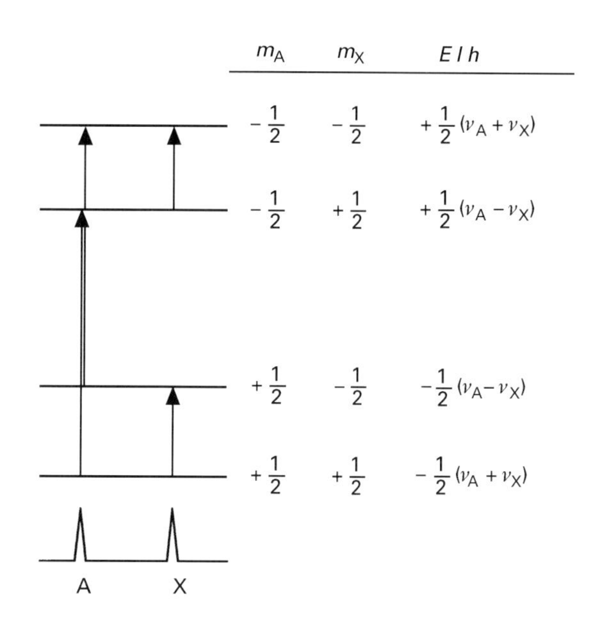

Figure 3 Energy levels and NMR spectrum of a pair of spin-$\frac{1}{2}$ nuclei, A and X. m_A and m_X are the magnetic quantum numbers, ν_A and ν_X are the two resonance frequencies, and E is the energy. The spin–spin coupling J_{AX} is zero.

augments or opposes $\boldsymbol{B}_0$. This induced field $\boldsymbol{B}_{\text{ind}}$ is proportional in strength to $\boldsymbol{B}_0$ and, in atoms, is antiparallel to it. The net field $\boldsymbol{B}$ experienced by the nucleus is thus slightly different from $\boldsymbol{B}_0$:

$$B = B_0 - B_{\text{ind}} = B_0 - \sigma B_0 = B_0(1 - \sigma) \qquad [11]$$

where the proportionality constant σ is known as the *shielding* or *screening constant*. The resonance condition, Equation [9], thus becomes

$$\nu = \frac{\gamma B}{2\pi} = \frac{\gamma B_0}{2\pi}(1 - \sigma) \qquad [12]$$

The shielding constant is determined by the electronic structure of the molecule in the vicinity of the nucleus: ν is thus characteristic of the chemical environment.

The relation between the energy levels of a pair of spin-$\frac{1}{2}$ nuclei A and X,

$$\frac{E}{h} = -\frac{m_A \gamma B_0(1 - \sigma_A)}{2\pi} - \frac{m_X \gamma B_0(1 - \sigma_X)}{2\pi} = -m_A\nu_A - m_X\nu_X \qquad [13]$$

and the NMR spectrum is shown in **Figure 3**.

The chemical shift is customarily quantified by means of a parameter δ, defined in terms of the resonance frequencies of the nucleus of interest and of a reference compound:

$$\delta = 10^6 \times \left(\frac{\nu - \nu_{\text{ref}}}{\nu_{\text{ref}}}\right) \qquad [14]$$

δ is dimensionless and independent of B_0; values are usually quoted in parts per million (ppm). The most commonly used reference compound for ^{1}H and ^{13}C NMR is tetramethylsilane, $(CH_3)_4$ Si.

NMR spectra are displayed with δ increasing from right to left, with the reference compound at $\delta = 0$. As a consequence, nuclei with higher resonance frequencies (i.e. those that are less shielded) appear towards the left-hand side of the spectrum. Although spectra are now normally recorded at a fixed field strength, the old terms 'upfield' and 'downfield', meaning 'more shielded' and 'less shielded', dating from the days of field-swept NMR, are still in common use.

Chemical shifts are easily converted into frequency differences using Equation [14]. For example, the chemical shifts of the methyl and methylene signals of ethanol (**Figure 2**) are 1.2 and 3.6 ppm, respectively, giving a difference in resonance frequencies in a 9.4 T field of $(3.6 - 1.2) \times 10^{-6} \times 400$ MHz $= 960$ Hz.

The relative intensities of the signals in an NMR spectrum are proportional to the population differences (Eqn [10]), and therefore to the numbers of nuclei responsible for each signal. The CH_3, CH_2, and OH resonances of ethanol (**Figure 2**), for example, thus have integrated areas in the ratio 3:2:1.

Spin–spin coupling

Magnetic nuclei interact not only with applied and induced magnetic fields, but also with one another. The result, for molecules in liquids, is a fine structure known as *spin–spin coupling*, *scalar coupling* or *J-coupling*, illustrated by the ^{1}H spectrum of ethanol in **Figure 2**.

The effect of spin–spin coupling on a pair of nuclear spins A and X is to shift their energy levels by amounts determined by the two magnetic quantum numbers and by the parameter that quantifies the strength of the interaction, the *spin–spin coupling constant*, J_{AX}. Thus, Equation [13] becomes

$$\frac{E}{h} = -m_A \nu_A - m_X \nu_X + J_{AX} m_A m_X \quad [15]$$

For spin-$\frac{1}{2}$ nuclei, the energies are raised or lowered by $\frac{1}{4} J_{AX}$ according to whether the spins are parallel ($m_A m_X = +\frac{1}{4}$) or antiparallel ($m_A m_X = -\frac{1}{4}$). Equation [15] leads to the modified resonance condition for spin A:

$$\nu = \nu_A - J_{AX} m_X \quad [16]$$

i.e. the resonance frequency of A is shifted from its chemical shift position by an amount that depends on the orientation of the X spin to which it is coupled. Since X has in general $2I + 1$ states, the A resonance is split into $2I + 1$ uniformly spaced lines, with equal intensities (because the different orientations of X are almost exactly equally likely). The effect that spin–spin coupling has on the energy levels of two spin-$\frac{1}{2}$ nuclei is shown in **Figure 4**. Each nucleus now has two NMR lines (a doublet).

The origin of spin–spin coupling is not the direct, through-space *dipolar interaction* of two magnetic moments: being purely anisotropic, this interaction is averaged to zero by the rapid end-over-end tumbling of molecules in liquids. Rather, the nuclei interact via the electrons in the chemical bonds that connect them. The interaction usually falls off rapidly as the number of intervening bonds increases beyond 3, so that the existence of a scalar coupling between two nuclei normally indicates that they are close neighbours in a molecular framework.

Equation [16] can easily be extended to describe more than two nuclei:

$$\nu = \nu_A - \sum_{i \neq A} J_{Ai} m_i \quad [17]$$

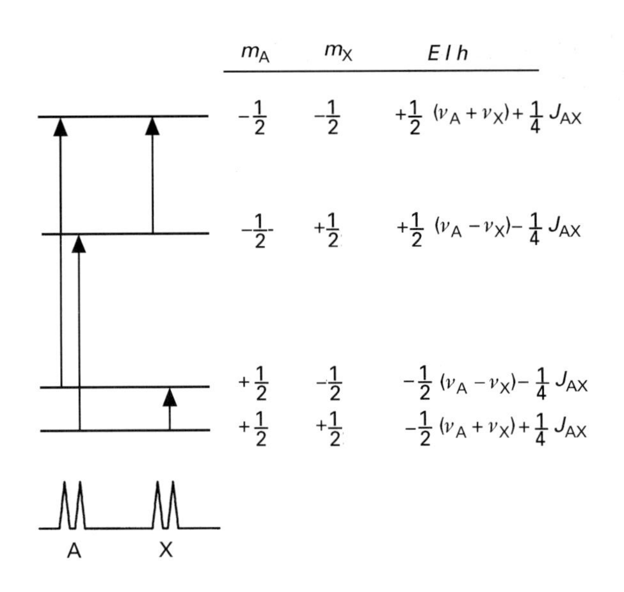

Figure 4 Energy levels and NMR spectrum of a pair of spin-$\frac{1}{2}$ nuclei, A and X. m_A and m_X are the magnetic quantum numbers, ν_A and ν_X are the two resonance frequencies, J_{AX} is the spin–spin coupling constant, and E is the energy.

where the sum runs over all spins to which A has an appreciable coupling. If A is coupled to N identical spin-$\frac{1}{2}$ nuclei (e.g. the three protons in a methyl group), it can be seen from Equation [17] that its resonance is split into $N + 1$ equally spaced lines with relative intensities given by the binomial coefficients

$$\binom{N}{i} = \frac{N!}{i!(N-i)!}, \quad i = 0, 1, 2, \cdots N \quad [18]$$

Thus, the CH_2 and CH_3 resonances in ethanol (**Figure 2**) are respectively a 1:3:3:1 quartet and a 1:2:1 triplet.

This discussion of the *multiplet* (i.e. doublet, triplet, quartet, ...) structure arising from spin–spin coupling is valid in the *weak coupling* limit, i.e. when the difference in resonance frequencies of the coupled nuclei $|\nu_A - \nu_X|$ is much larger than their interaction $|J_{AX}|$. When this is not the case (*strong coupling*), the positions and intensities of the lines are modified, as illustrated in **Figure 5**. The origin of these effects lies in the NMR transition probabilities. As the coupling becomes stronger, the outer line of each doublet in **Figure 5** becomes weaker relative to the inner line. In the limit that the chemical shift difference is zero, the transitions leading to the two outer lines become completely forbidden, and the two inner lines coincide, so that only a single line is

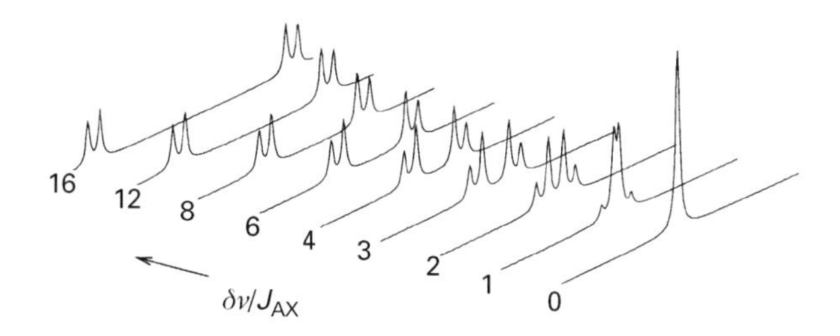

Figure 5 Calculated NMR spectra of a pair of spin-$\frac{1}{2}$ nuclei for a range of $\delta\nu = \nu_A - \nu_X$ values between $16J_{AX}$ and zero.

observed. This is a general result: *spin–spin interactions between protons in identical environments do not lead to observable splittings.*

Vector model of NMR

Considerable insight into the operation of simple NMR experiments may be derived from a straightforward *vector model.* It relies on the fact that while the individual nuclear magnetic moments behave quantum mechanically, the net magnetization of a large collection of nuclear spins obeys classical mechanics.

The motion of a classical magnetic moment $\boldsymbol{M}$, possessing angular momentum, in a magnetic field $\boldsymbol{B}$ is described by the differential equation

$$\frac{\mathrm{d}\boldsymbol{B}}{\mathrm{d}t} = -\gamma \boldsymbol{B} \times \boldsymbol{M} \qquad [19]$$

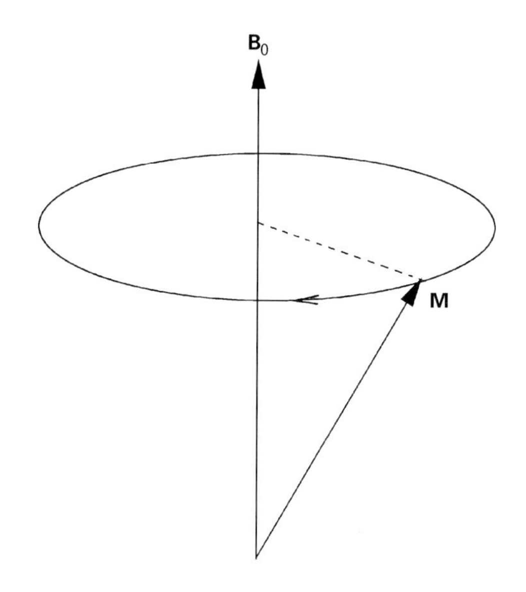

Figure 6 The motion of a magnetization vector $\boldsymbol{M}$ in a magnetic field $\boldsymbol{B}_0$. $\boldsymbol{M}$ precesses around the field direction rather like the axis of a spinning gyroscope.

or equivalently,

$$\begin{aligned}
\frac{\mathrm{d}M_{x'}}{\mathrm{d}t} &= -\gamma B_{y'} M_{z'} + \gamma B_{z'} M_{y'} \\
\frac{\mathrm{d}M_{y'}}{\mathrm{d}t} &= +\gamma B_{x'} M_{z'} - \gamma B_{z'} M_{x'} \qquad [20] \\
\frac{\mathrm{d}M_{z'}}{\mathrm{d}t} &= -\gamma B_{x'} M_{y'} + \gamma B_{y'} M_{x'}
\end{aligned}$$

where $\boldsymbol{B} \times \boldsymbol{M}$ is the vector product of $\boldsymbol{B} = (B_{x'}, B_{y'}, B_{z'})$ and $\boldsymbol{M} = (M_{x'}, M_{y'}, M_{z'})$, and the (x', y', z') coordinate system is called the *laboratory frame.*

These expressions describe the precession of $\boldsymbol{M}$ around $\boldsymbol{B}$ at *angular* frequency $\omega = \gamma B$, as may be seen by taking $\boldsymbol{B} = \boldsymbol{B}_0$, along the z'-axis:

$$\begin{aligned}
\frac{\mathrm{d}M_{x'}}{\mathrm{d}t} &= +\gamma B_0 M_{y'} \\
\frac{\mathrm{d}M_{y'}}{\mathrm{d}t} &= -\gamma B_0 M_{x'} \qquad [21] \\
\frac{\mathrm{d}M_{z'}}{\mathrm{d}t} &= 0
\end{aligned}$$

(see **Figure 6**). This motion is known as *Larmor precession*, and it occurs at the NMR frequency of the nuclear spins in the field $\boldsymbol{B}_0$:

$$\omega_0 = \gamma B_0 = 2\pi\nu \qquad [22]$$

An NMR experiment involves the application of a brief, intense burst of radiofrequency radiation, known as a 'pulse', along, say, the x' axis in the laboratory frame. The frequency of this field, ω_{RF} is very close to the Larmor frequency ω_0. Regarding this linearly oscillating field as the sum of two counter-rotating fields, we may ignore the component that rotates in the opposite sense to the Larmor precession because, being $2\omega_{\mathrm{RF}}$ off-resonance, it has a negligible effect on the spins. The other component is

$$\boldsymbol{B}_1 = (B_1 \cos \omega_{\mathrm{RF}} t, -B_1 \sin \omega_{\mathrm{RF}} t, 0) \qquad [23]$$

The nuclear spins thus experience the sum of two magnetic fields: a strong static field $\boldsymbol{B}_0$ along the z' axis, and a much weaker, time-dependent field $\boldsymbol{B}_1$

rotating in the $x'y'$ plane. $\boldsymbol{M}$ therefore precesses around the time-dependent vector sum of $\boldsymbol{B}_0$ and $\boldsymbol{B}_1$ (**Figure 7A**). To make this complicated motion easier to visualize, Equation [20] is transformed into the *rotating frame* (x, y, z), a coordinate system rotating around the z' axis at frequency ω_{RF}, in which the radiofrequency field appears stationary. In this frame, the components of the bulk magnetization are

$$\begin{aligned} M_x &= M_{x'} \cos \omega_{\mathrm{RF}} t - M_{y'} \sin \omega_{\mathrm{RF}} t \\ M_y &= M_{x'} \sin \omega_{\mathrm{RF}} t + M_{y'} \cos \omega_{\mathrm{RF}} t \\ M_z &= M_{z'} \end{aligned} \qquad [24]$$

Differentiating Equations [24], and using Equations [20] with $\boldsymbol{B} = (B_1 \cos\omega_{\mathrm{RF}}\, t, -B_1 \sin\omega_{\mathrm{RF}}\, t, B_0)$ gives

$$\begin{aligned} \frac{\mathrm{d}M_x}{\mathrm{d}t} &= +\gamma \Delta B M_y \\ \frac{\mathrm{d}M_y}{\mathrm{d}t} &= +\gamma B_1 M_z - \gamma \Delta B M_x \\ \frac{\mathrm{d}M_z}{\mathrm{d}t} &= -\gamma B_1 M_y \end{aligned} \qquad [25]$$

or, more compactly,

$$\frac{\mathrm{d}\boldsymbol{M}}{\mathrm{d}t} = -\gamma \boldsymbol{B}_{\mathrm{eff}} \times \boldsymbol{M} \qquad [26]$$

where $\boldsymbol{B}_{\mathrm{eff}} = (B_1, 0, \Delta B)$, and $\Delta B = B_0 - \omega_{\mathrm{RF}}/\gamma$. Equation [26] describes the precession of $\boldsymbol{M}$ about a static field $\boldsymbol{B}_{\mathrm{eff}}$ at frequency $\gamma \boldsymbol{B}_{\mathrm{eff}} = \gamma\sqrt{B_1^2 + \Delta B^2}$ in the rotating frame (**Figure 7B**). $\gamma\Delta B = \omega_0 - \omega_{\mathrm{RF}} = \Omega$ is the offset of the radiofrequency field from resonance. To include chemical shifts, γB_0 should be replaced by $\gamma B_0(1 - \sigma)$.

Radiofrequency pulses

At equilibrium, in the absence of a radiofrequency field, the bulk magnetization of the sample $\boldsymbol{M}_0$ is parallel to the $\boldsymbol{B}_0$ direction (z axis) with a magnitude proportional to the population difference ($n_\alpha - n_\beta$, for a spin-$\frac{1}{2}$ nucleus). If the radiofrequency field strength is much larger than the resonance offset ($B_1 >> \Delta B$) then $\boldsymbol{B}_{\mathrm{eff}} \cong \boldsymbol{B}_1$ and the effective field lies along the x axis in the rotating frame. The pulse therefore causes $\boldsymbol{M}_0$ to rotate in the yz plane at frequency γB_1 (**Figure 8A** and **8B**). In this way a short, intense monochromatic burst of radiofrequency

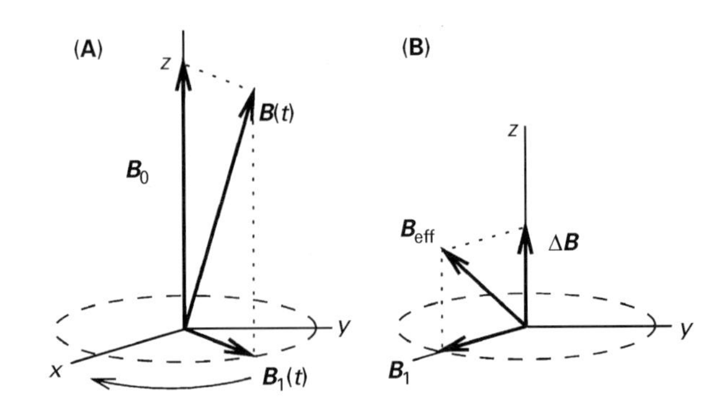

Figure 7 The magnetic fields present in an NMR experiment in (A) the laboratory frame and (B) the rotating frame. $\boldsymbol{B}_0$ is the strong static field, $\boldsymbol{B}_1$ is the much weaker oscillating radiofrequency field, $\Delta\boldsymbol{B}$ and $\boldsymbol{B}_{\mathrm{eff}}$ are respectively, the offset and effective fields in the rotating frame, and $\boldsymbol{B}(k)$ is the resultant of $\boldsymbol{B}_0$ and $\boldsymbol{B}_1$ in the laboratory frame.

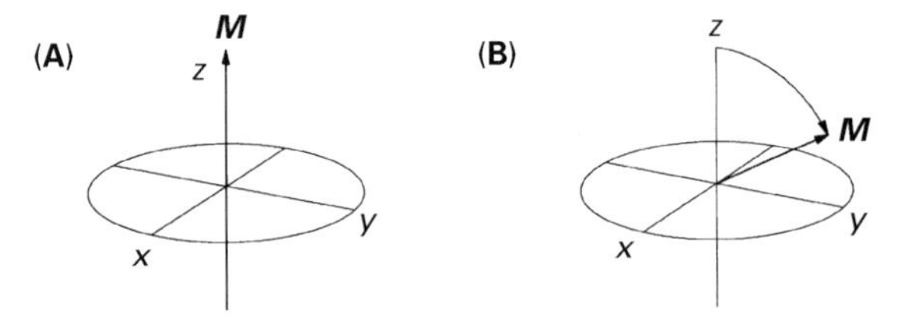

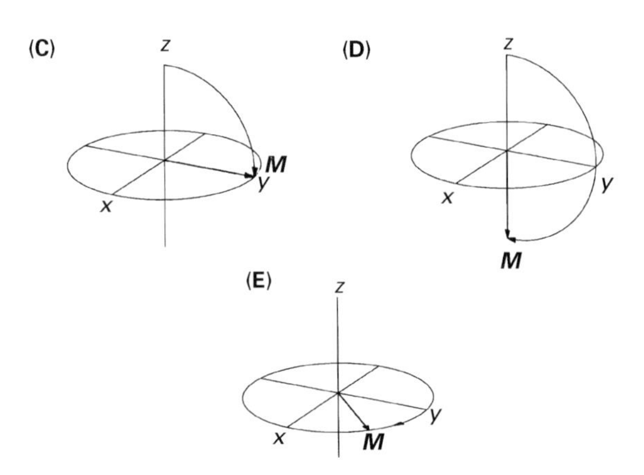

Figure 8 The effect of radiofrequency pulses (in the rotating frame). (A) At thermal equilibrium, the net magnetization of the sample is parallel to the $\boldsymbol{B}_0$ direction. (B) A pulse along the x axis, whose strength $\boldsymbol{B}_1$ is much greater than the offset field $\Delta\boldsymbol{B}$, causes $\boldsymbol{M}$ to rotate in the yz plane at angular frequency γB_1. (C) A 90° pulse, of duration t_p, $\gamma B_1 t_p = \pi/2$), rotates the magnetization from the 'north pole' (z axis) to the 'equator' (y axis) (D) A 180° pulse, of duration t_p ($\gamma B_1 t_p = \pi$), rotates the magnetization from the 'north pole' to the 'south pole' ($-z$ axis). (E) Following a 90° pulse, the magnetization precesses around the 'equator' in the rotating frame at frequency $\Omega = \gamma\Delta B$. Relaxation is ignored throughout.

radiation can excite spins uniformly over a range of resonance frequencies, provided their offset frequencies Ω are much smaller than γB_1. If the field is switched off after a time t_p, given by $\gamma B_1 t_p = \pi/2$, **M** is turned through 90° and is left along the y axis (**Figure 8C**). A radiofrequency pulse with this property is known as a 90° pulse. If t_p is twice this duration, the magnetization is *inverted* (a 180° pulse, **Figure 8D**); this is equivalent to exchanging the n_α and n_β populations of a spin-$\frac{1}{2}$ nucleus.

Free precession

Equation [26] may also be used to predict what happens *after* a 90° pulse. Setting $B_1 = 0$, the effective field is $\mathbf{B}_{\text{eff}} = (0, 0, \Delta B)$ and **M** precesses in the xy plane at angular frequency $\gamma\,\Delta B = \Omega$, i.e. at the offset frequency determined by ω_{RF} and the chemical shift (**Figure 8E**):

$$\begin{aligned} M_x &= M_0 \sin \Omega t \\ M_y &= M_0 \cos \Omega t \\ M_z &= 0 \end{aligned} \qquad [27]$$

where t is now the time after the end of the pulse. When several nuclei with different chemical shifts have been excited by the pulse, the xy magnetization of the sample is the sum of several oscillating terms of the form of Equation [27].

Free induction decay

Up to this point it has been assumed that the nonequilibrium state produced by the radiofrequency pulse does not relax back towards equilibrium. This is a reasonable approximation during the very short pulse. However, to describe the behaviour of the spins during the period of free precession that follows the pulse, relaxation must be included. This is traditionally done by allowing M_x and M_y to decay exponentially back to zero with a time constant T_2, while M_z grows back to M_0 with a time constant T_1:

$$\begin{aligned} \frac{\mathrm{d}M_x}{\mathrm{d}t} &= +\gamma \Delta B\, M_y - \frac{M_x}{T_2} \\ \frac{\mathrm{d}M_y}{\mathrm{d}t} &= +\gamma B_1 M_z - \gamma \Delta B M_x - \frac{M_y}{T_2} \\ \frac{\mathrm{d}M_z}{\mathrm{d}t} &= -\gamma B_1 M_y - \frac{(M_z - M_0)}{T_1} \end{aligned} \qquad [28]$$

T_1 and T_2 are the *spin–lattice* and the *spin–spin relaxation times*. These expressions are known as the *Bloch equations*. With relaxation included, Equation [27] becomes

$$\begin{aligned} M_x(t) &= M_0 \sin \Omega t \exp(-t/T_2) \\ M_y(t) &= M_0 \cos \Omega t \exp(-t/T_2) \\ M_z(t) &= M_0[1 - \exp(-t/T_1)] \end{aligned} \qquad [29]$$

The two components M_x and M_y represent the detectable signal in an NMR experiment – the *free induction decay* (**Figure 9**). Fourier transformation of the free induction decay gives the NMR spectrum.

Spin relaxation

Relaxation processes allow nuclear spins to return to equilibrium following a disturbance, e.g. a

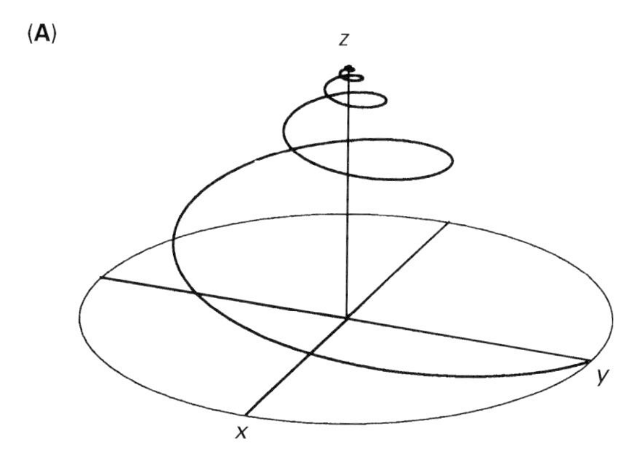

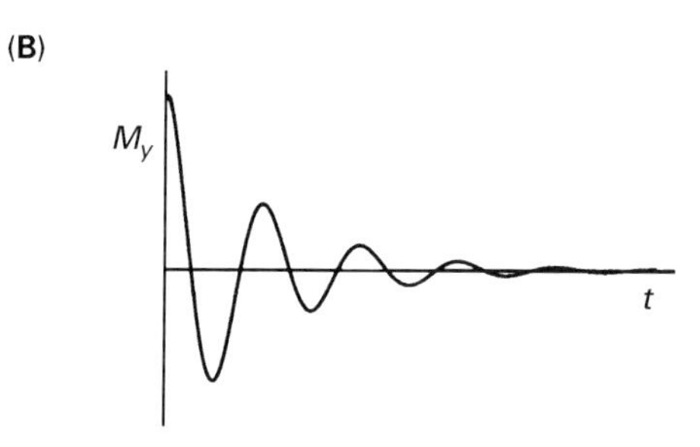

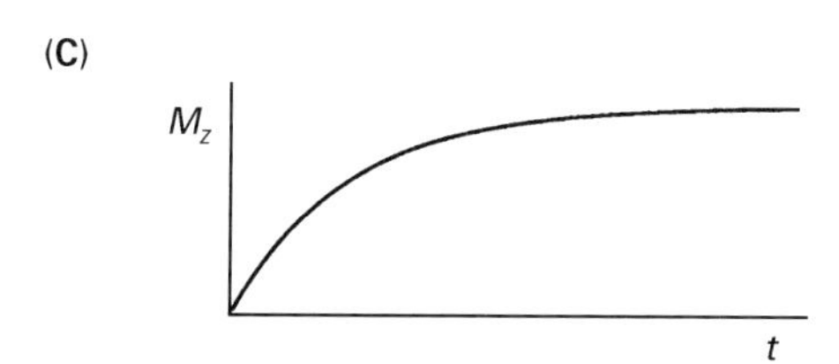

Figure 9 Following a 90° pulse, the magnetization precesses around the z axis and at the same time returns to its equilibrium position at the 'north pole' (A). The transverse components of ***M*** decay to zero with time constant T_2, the spin–spin relaxation time (B). The z component of ***M*** grows back to $\boldsymbol{M}_0$ with time constant T_1, the spin–lattice relaxation time (C).

radiofrequency pulse. The relaxation times T_1 and T_2 characterize the relaxation of, respectively, the *longitudinal* and *transverse* components of the magnetization $\boldsymbol{M}$, respectively parallel and perpendicular to $\boldsymbol{B}_0$. Equivalently, T_1 is the time constant for the return to equilibrium of the populations of the spin states, while T_2 is the time constant for the dephasing of the *coherence* between spin states. In the absence of any significant spatial inhomogeneity of $\boldsymbol{B}_0$, or other sources of line broadening such as chemical exchange, the width of the NMR line (in hertz) is $1/\pi T_2$.

Spin–lattice relaxation is caused by randomly fluctuating local magnetic fields. A common source of such fields is the dipolar interaction between pairs of nuclei, modulated by molecular tumbling in a liquid. The component of these fields that oscillates at the resonance frequency can induce transitions between the spin states, so transferring energy between the spin system and the 'lattice' (i.e. everything else) and bringing the spins into equilibrium with their surroundings. In the simplest case, T_1 depends on the mean square strength of the local fields $\langle B_{\mathrm{loc}}^2 \rangle$, and the intensity of the fluctuations at the resonance frequency ω_0

$$\frac{1}{T_1} = \gamma^2 \langle B_{\mathrm{loc}}^2 \rangle J(\omega_0) \qquad [30]$$

where

$$J(\omega) = \frac{2\tau_c}{1 + \omega^2 \tau_c^2} \qquad [31]$$

is the *spectral density* function, and τ_c is the *rotational correlation time* (roughly the average time the molecule takes to rotate through 90°).

Spin–spin relaxation has two contributions:

$$\frac{1}{T_2} = \tfrac{1}{2}\gamma^2 \langle B_{\mathrm{loc}}^2 \rangle J(\omega_0) + \tfrac{1}{2}\gamma^2 \langle B_{\mathrm{loc}}^2 \rangle J(0) \qquad [32]$$

The first is closely related to spin–lattice relaxation, and arises from the finite lifetime of the spin states, through the uncertainty principle. The second term is due to the loss of coherence caused by local fields of very low frequency (hence the $J(0)$ factor), which augment or oppose $\boldsymbol{B}_0$ and so give rise to a spread of resonance frequencies, and hence the dephasing of transverse magnetization. **Figure 10** shows the dependence of T_1 and T_2 on τ_c.

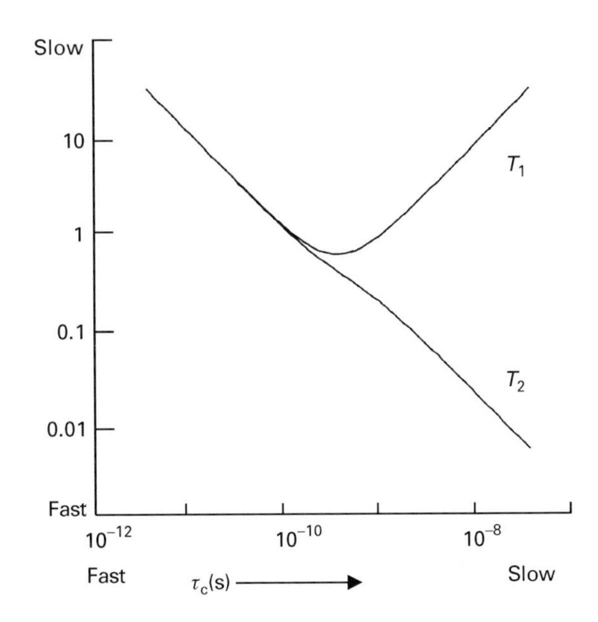

Figure 10 The dependence of T_1 and T_2 on the rotational correlation time τ_c, using $\gamma^2\langle B_{\mathrm{loc}}^2\rangle = 4.5 \times 10^9\ \mathrm{s}^{-2}$ and $\omega_0/2\pi = 400$ MHz. The units for the vertical axis are seconds.

Relaxation times contain information on both $J(\omega)$ (i.e. on molecular motion) and $\langle B_{\mathrm{loc}}^2 \rangle$, (i.e. on molecular structure via, for example, the r^{-3} distance dependence of the dipolar interaction). A further relaxation phenomenon that provides important information on internuclear distances is the *nuclear Overhauser effect*.

Chemical exchange

In addition to chemical shifts, spin–spin coupling and spin relaxation, NMR spectra are affected by, and may be used to study, chemical and conformational equilibria. Consider an equilibrium

$$\mathrm{A} \rightleftharpoons \mathrm{B} \qquad [33]$$

which exchanges the chemical shifts of two nuclei, with equal forward and backward rate constants, k. At low temperature, the NMR spectrum comprises two sharp resonances at frequencies ν_A and ν_B. As the temperature is raised, the following sequence of events occurs: the two lines broaden and move towards one another until they coalesce into a broad flat-topped line which then narrows into a sharp single resonance at the average chemical shift $\frac{1}{2}(\nu_A + \nu_B)$ (**Figure 11**).

The mid-point of this process, when the two lines just merge into one, occurs when

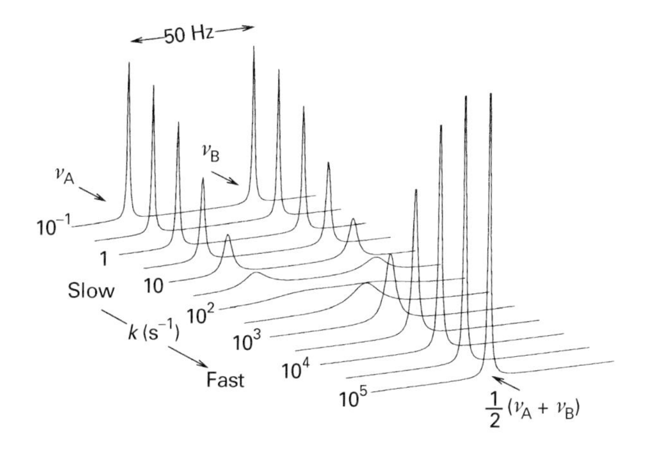

Figure 11 Calculated NMR spectra for a pair of nuclei exchanging between two sites with equal populations. Spectra are shown for a range of values of the exchange rate k. The difference in resonance frequencies of the two sites, $\delta\nu$, is 50 Hz.

$$k = \frac{\pi(\nu_A - \nu_B)}{\sqrt{2}} = \frac{\pi\delta\nu}{\sqrt{2}} \quad [34]$$

For *slow exchange*, the exchange broadening of the two separate resonances is

$$\Delta\nu = \frac{k}{\pi} \quad [35]$$

while for *fast exchange*, the single line has an extra width

$$\Delta\nu = \frac{\frac{1}{2}\pi(\delta\nu)^2}{k} \quad [36]$$

Related but more complex expressions are found if the forward and backward rate constants differ, or if there are more than two exchanging species.

List of symbols

$\boldsymbol{B}$ = magnetic field vector; B = magnitude of $\boldsymbol{B}$; $\hbar$ = Planck constant $(h)/2\pi$; $\boldsymbol{I}$ = nuclear spin angular momentum vector; I = nuclear spin quantum number; J = spin–spin coupling constant; m = nuclear magnetic quantum number; $\boldsymbol{M}$ = classical (macroscopic) magnetization vector; T_1 = spin–lattice relaxation time; T_2 = spin–spin relaxation time; γ = gyromagnetic ratio; μ = nuclear magnetic moment; $\mu_z = z$ component; δ = chemical shift; σ = shielding (screening) constant; τ_c = rotational correlation time; ω = angular frequency; ω_0 = Larmor frequency.

See also: **Chemical Exchange Effects in NMR; Fourier Transformation and Sampling Theory; NMR Relaxation Rates; NMR Spectrometers; Nuclear Overhauser Effect; Parameters in NMR Spectroscopy, Theory of.**

Further reading

Carrington A and McLachlan AD (1967) *Introduction to Magnetic Resonance*. New York: Harper and Row.

Ernst RR, Bodenhausen G and Wokaun A (1987) *Principles of Nuclear Magnetic Resonance in One and Two Dimensions*. Oxford: Clarendon Press.

Freeman R (1997) *A Handbook of Nuclear Magnetic Resonance*, 2nd ed. Harlow: Longman.

Günther H (1995) *NMR Spectroscopy*, 2nd edn. Chichester: Wiley.

Harris RK (1983) *Nuclear Magnetic Resonance Spectroscopy*. London: Pitman.

Hore PJ (1995) *Nuclear Magnetic Resonance*. Oxford: Oxford University Press.

McLauchlan, KA (1972) *Magnetic Resonance*, Oxford: Clarendon Press.

Sanders JKM and Hunter BK (1993), *Modern NMR Spectroscopy*, 2nd edn. Oxford: Oxford University Press.

NMR Pulse Sequences

William F Reynolds, University of Toronto, Ontario, Canada

MAGNETIC RESONANCE
Theory

Introduction

The single most important development in nuclear magnetic resonance (NMR) spectroscopy since the initial observation of the NMR phenomenon in bulk phases in 1945 was undoubtedly the introduction of pulse Fourier transform NMR by Anderson and Ernst. This technique provided greatly increased sensitivity per unit time, making it feasible to obtain spectra for low sensitivity/low abundance nuclei such as ^{13}C. More importantly, it allowed the development of a wide variety of sophisticated and powerful multipulse experiments which have revolutionized the use of NMR spectroscopy in studies of molecular structure and dynamics. This article provides an overview of pulse sequence experiments. Many individual experiments are discussed in other articles.

The classical vector model of NMR and the basic one-pulse Fourier transform experiment

Many NMR pulse sequences can be described either by a classical model describing the motions of magnetic vectors or by quantum mechanical models of different levels of sophistication. The attractive feature of the classical vector model is that it provides simple physical pictures of many of the basic pulse sequences. However, it does not work for many multipulse experiments that involve multiple quantum coherence. These experiments can only be described by quantum mechanical methods. Because of the insights which the vector model provides into many of the basic sequences, I will use this model wherever possible.

The fundamental magnetic properties of nuclei are well described elsewhere. I will begin with the bulk magnetization vector $\boldsymbol{M}$ for a series of nuclei of the same type. This is parallel to the external magnetic field $\boldsymbol{B}_0$ and is the resultant of individual magnetic moment vectors μ, precessing about $\boldsymbol{B}_0$ with the Larmor angular velocity

$$\omega = -\gamma B_0 \tag{1}$$

where γ is the magnetogyric ratio of the nucleus (**Figure 1**). For a nucleus with spin quantum number $I = \frac{1}{2}$ $\boldsymbol{M}$ actually results from the slight excess of nuclei in the α spin state ($m_I = +\frac{1}{2}$) over those in the β spin state ($m_I = -\frac{1}{2}$).

Now consider the effect of a 'pulse' of electromagnetic radiation of frequency corresponding to the Larmor frequency $\omega/2\pi$. This is applied so that the oscillating magnetic component of the pulse is in a plane at right angles to $\boldsymbol{B}_0$. This oscillating component can be resolved into two rotating components of angular velocity $\pm 2\pi\nu$. Only the component rotating in the same direction as the magnetic moments need be considered since the opposite component has no net effect on the nuclear magnetization. The former

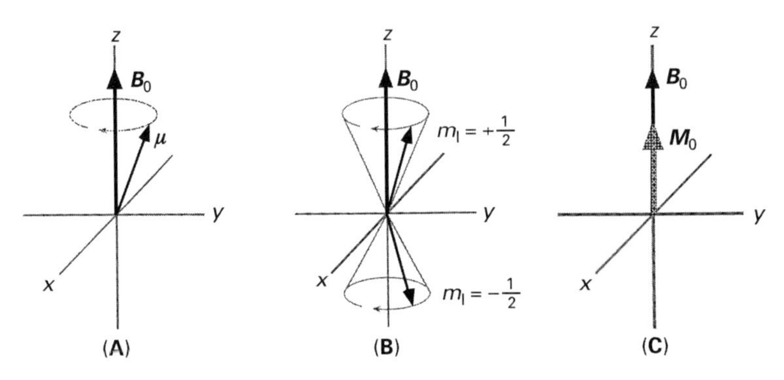

Figure 1 (A) The precession of an individual magnetic moment μ about the external magnetic field $\boldsymbol{B}_0$. (B) The precession of magnetic moments in the α ($m_I = +\frac{1}{2}$) and β ($m_I = -\frac{1}{2}$) spin states. (C) The resultant magnetic moment $\boldsymbol{M}$ of a large number of nuclei of the same kind, reflecting the small excess population of nuclei in the α spin state.

component is represented by a magnetization vector, B_1, rotating in the x–y plane at frequency ν. However, to simplify the visualization, the Cartesian coordinate system is also assumed to be rotating at frequency ν, called the rotating frame model (**Figure 2**). This allows us to concentrate on frequency differences rather than absolute frequencies in considering NMR experiments.

During the pulse, the individual magnetic moments and consequently the bulk magnetization vector precess about B_1 with angular velocity γB_1. If B_1 is taken as defining the x-axis in the rotating frame, M rotates towards the y-axis through an angle:

$$\alpha\,(\text{radians}) = \gamma B_1 \tau \qquad [2]$$

where τ is the pulse duration in seconds. Thus if the duration of the pulse is just sufficient to rotate M through $\pi/2$ radians, it is called a 90° pulse. The resultant magnetization generated in the x–y plane can then be detected by a receiver.

Now consider the one-pulse Fourier transform experiment. The pulse sequence is illustrated in **Figure 3**. The pulse does not excite a single frequency but rather a range of frequencies whose width (in Hz) is inversely proportional to the pulse duration (in s). The frequency excitation profile is provided by taking the Fourier transform of the time profile of the pulse (**Figure 3**). Quadrature detection distinguishes positive and negative frequencies, allowing one to position the transmitter frequency at the midpoint of the spectral window. Modern high resolution spectrometers typically have 90° pulses of duration 10 µs or less. While this allows excitation over a 200 kHz spectral window, it is important to have near uniform excitation over the entire spectral window. A 10 µs pulse provides near uniform excitation over ~25 kHz, which is adequate for most high resolution applications. However, solid state spectra have much wider spectral windows, requiring much shorter pulses.

After the pulse generates x–y magnetization, the return of this magnetization to equilibrium is sampled as a function of time. This response is called the free induction decay (FID) signal. In the vector model, it can be regarded as the resultant of a series of individual magnetization vectors, each precessing in the x–y plane at some frequency $\Delta\nu$ relative to the transmitter frequency and decaying exponentially with a time constant T_2, characteristic of the return to equilibrium of x–y magnetization. Each vector corresponds to a specific signal in the frequency spectrum, and thus the Fourier transform of the FID yields the frequency spectrum:

$$F(\nu) = \int_{-\infty}^{+\infty} f(t)\exp(2\pi i\nu t)\mathrm{d}t = \int_{-\infty}^{+\infty} f(t)[\cos(2\pi\nu t) + i\,\sin(2\pi\nu t)]\mathrm{d}t \qquad [3]$$

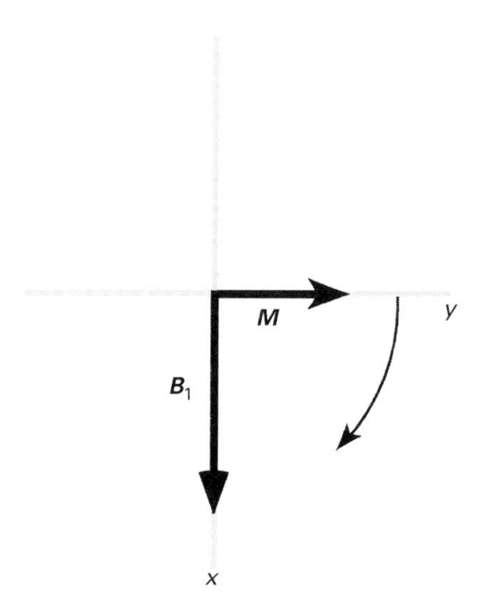

Figure 2 The rotating frame coordinate system. The coordinate system is assumed to be rotating at the same frequency as B_1 and consequently B_1 appears to be stationary along the x-axis. The x,y magnetization M, generated after a pulse, will be stationary along the y-axis if the Larmor precession frequency is equal to the pulse frequency or rotating at a frequency $\Delta\nu$, corresponding to the frequency difference.

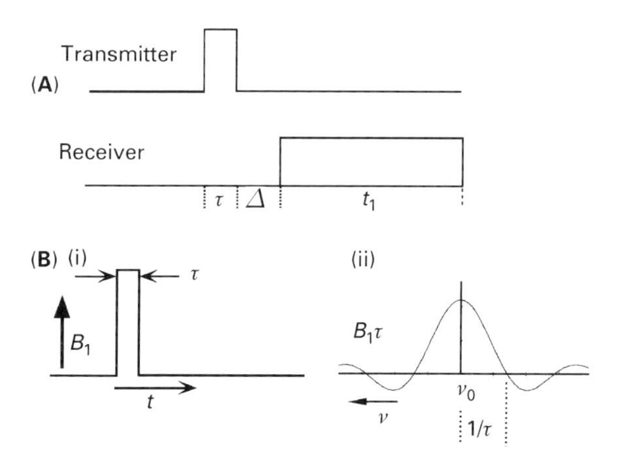

Figure 3 (A) The basic pulse Fourier transform sequence; τ represents the pulse duration and Δ is a small delay, comparable to τ, to ensure that the pulse is not detected by the receiver. Note that in this and subsequent pulse sequences, the duration of the pulse is exaggerated. The actual pulse duration is ~10 µs compared with an acquisition time, t_1, of ~1 s. (B) (i) The time profile of the pulse and (ii) the frequency excitation profile due to the pulse. The frequency profile is the Fourier transform of the time profile.

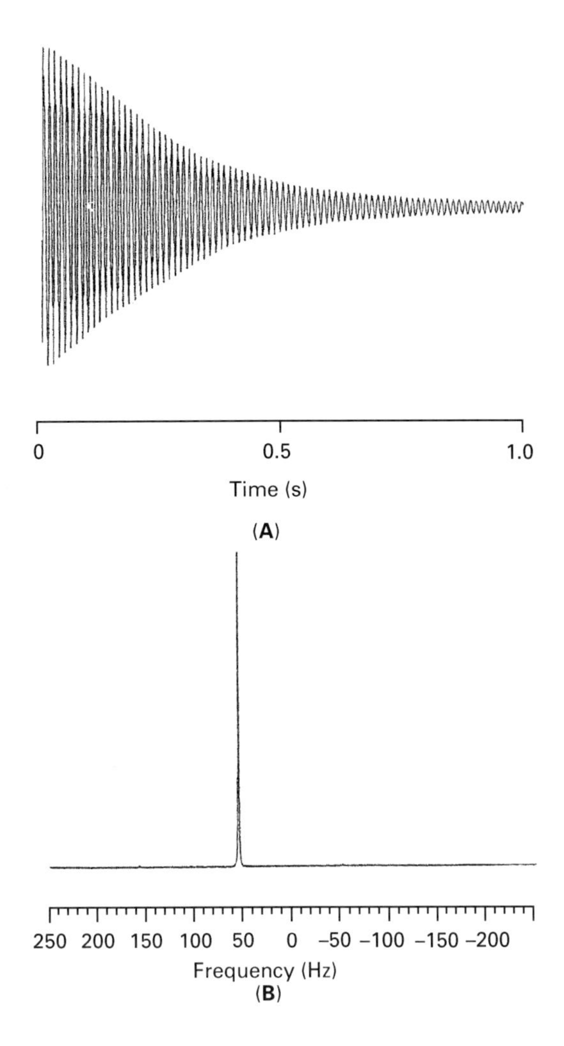

Figure 4 The FID time signal (A) and resultant ^{1}H frequency spectrum (B) for a single off-resonance peak.

This is illustrated in **Figure 4** for a spectrum of a single off-resonance peak. **Figure 5** shows the FID and frequency spectrum for the aliphatic region of kauradienoic acid [1].

[1]

The two basic types of signals that can be detected in an NMR experiment (**Figure 6**) are an absorption signal (owing to detection of a signal at right angles to $\boldsymbol{B}_1$) and a dispersion signal (owing to a signal parallel to $\boldsymbol{B}_1$), corresponding to the cosine and sine terms in Equation [3]. The phase for an on-resonance (i.e. $\Delta\nu = 0$) peak can easily be adjusted to give an absorption signal via a zero frequency phase adjustment.

However, off-resonance peaks undergo additional phase shifts due to vector evolution during the finite pulse and the delay between the pulse and gating on the receiver (see **Figure 3**). For example, for a total time before acquisition of 20 μs, a peak at $\Delta\nu = 10\ 000$ Hz will rotate through an angle:

$$\begin{aligned}\alpha &= 2\pi(2.0 \times 10^{-5})(1.0 \times 10^{4}) \text{ radians}\\ &= 0.40\pi \text{ radians} = 72^\circ\end{aligned} \qquad [4]$$

introducing a significant dispersive component. Fortunately, this phase shift varies linearly with $\Delta\nu$ and thus can be corrected by applying a phase correction which varies linearly with frequency.

The acquisition time, t_1, is determined by the time required for x–y magnetization to decay to near zero as well as by the desired data point resolution. Typical values for one-dimensional spectra range from 0.5 to 5 s. The ability to excite and acquire all signals for a given nucleus simultaneously provides a major sensitivity advantage over the older continuous wave (CW) method which involved slowly sweeping through the spectral window, exciting one signal at a time. Typically, one can acquire at least 100 FID signals in the time taken to acquire a CW spectrum. Since the signal-to-noise increases as the square root of the number of scans, this provides at least a 10-fold increase in sensitivity.

However, the acquisition of multiscan spectra introduces a new problem. Ideally, $\boldsymbol{M}$ should have returned to its equilibrium position along the $+z$-axis before the next pulse. Otherwise, the residual magnetization will fractionally decrease with each scan, a phenomenon known as saturation. Compounding the problem is the fact that the time constant for return to equilibrium along the z-axis, T_1, can be longer than T_2 (see below). One solution is to introduce an additional relaxation delay between the end of each acquisition and the next pulse. The second is to use a shorter pulse duration so that the rotation angle of $\boldsymbol{M}$, α, is $< 90°$. Richard Ernst conclusively demonstrated that the second approach gives a superior signal-to-noise ratio. The ideal pulse flip angle, α_E, called the Ernst angle, is given by:

$$\cos\alpha_E = \exp(-t_1/T_1) \qquad [5]$$

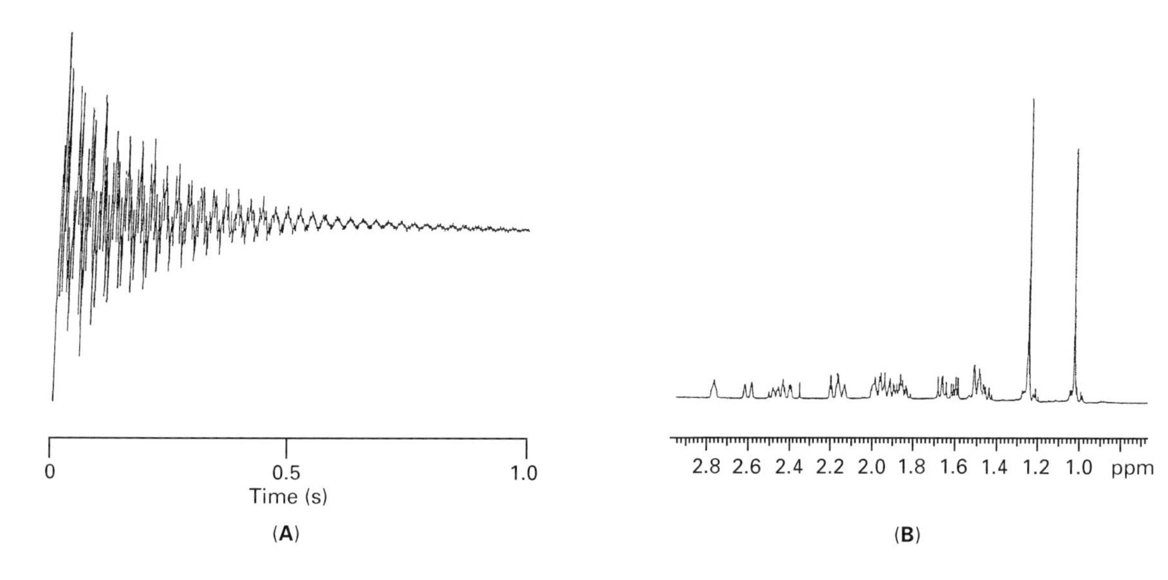

Figure 5 The FID time signal (A) and ^{1}H frequency spectrum (B) for the aliphatic region of kauradienoic acid [1]. The scale along the bottom of the frequency spectrum is the δ scale [chemical shift in parts per million relative to $(CH_3)_4Si$].

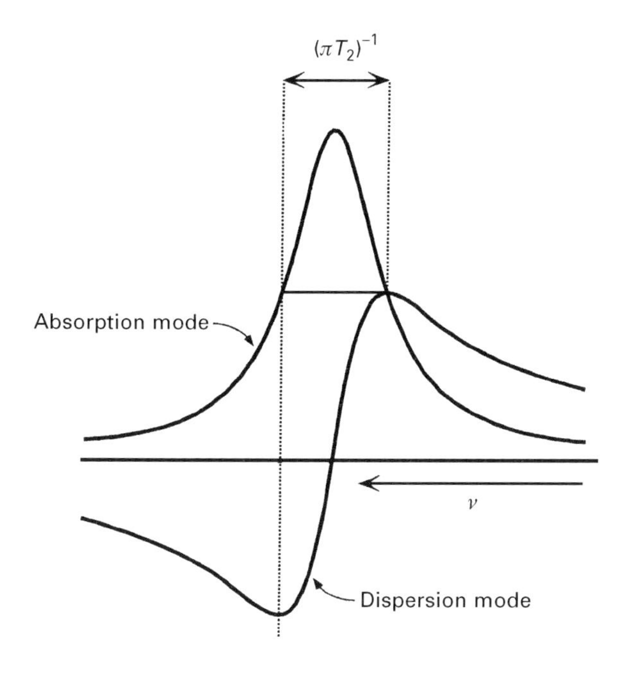

Figure 6 Comparison of absorption (ν mode) and dispersion (u mode) signal shapes. Spectra are usually phase corrected to give pure absorption mode peaks.

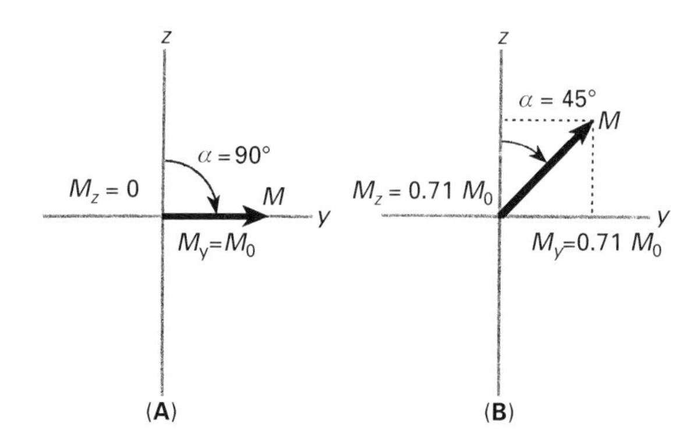

Figure 7 Comparison of z (M_z) and y (M_y) magnetization immediately following (A) a $90°_x$ pulse and (B) a $45°_x$ pulse. The latter generates 71% of the amount of M_y magnetization (and therefore 71% of the signal) while retaining 71% of equilibrium z magnetization, compared with 0° z magnetization after a 90° pulse. This allows most, if not all, of the equilibrium z magnetization to be restored during the acquisition time, t_1, after a 45° pulse while a 90° pulse typically will require a lengthy delay after t_1 to restore equilibrium z magnetization. If T_1 is very long compared with t_1, an even smaller pulse angle must be used (see Eqn [5]).

This is illustrated in **Figure 7**, using a simple trigonometric argument. However, T_1 may be significantly different for different peaks in a spectrum (e.g. a ^{13}C spectrum of a molecule containing protonated and non-protonated carbons). This requires a compromise choice of α and relative peak areas may no longer be quantitative.

Finally, the analogue voltage signal detected by the receiver must be digitized for computer storage and processing. This puts some constraints on data acquisition, as discussed in any of the texts listed in the Further reading section.

Measurement of T_1 and T_2 relaxation times

The classical equations for the return of magnetization to equilibrium along the z- and y-axes are respectively:

$$\frac{dM_z}{dt} = -(M_z - M_0)/T_1 \text{ and } \frac{dM_y}{dt} = -M_y/T_2 \quad [6]$$

where M_0 is the magnitude of equilibrium z magnetization. The spin–lattice or longitudinal relaxation time, T_1, reflects the effect of the component of the random fluctuating magnetic field (arising from the thermal motion of dipoles in the sample) at the Larmor frequency. However, the spin–spin or transverse relaxation time, T_2, is also affected by static magnetic field components, including any inhomogeneity of the magnetic field over the region of the sample. Consequently, transverse relaxation is in principle faster than longitudinal relaxation, i.e. T_2 can be smaller than T_1.

The inversion–recovery sequence can be used to measure T_1 (**Figure 8**). The 180° pulse inverts the equilibrium magnetization **M**. During the delay t_1, the magnetization begins to return to equilibrium. A $90°_x$ pulse then samples the magnetization remaining after t_1. Since the final pulse is a 90° pulse, it is necessary to include a relaxation delay, Δ, to allow for return to equilibrium between scans. Ideally $\Delta \geq 5T_1$. However, it has been found that accurate values of T_1 can still be obtained using shorter values of Δ, known as the fast inversion–recovery method. After a sufficient number of scans, n, have been collected to achieve adequate signal-to-noise, the experiment is repeated, systematically varying t_1 from small values out to $\sim 2T_1$. The intensity of each peak exponentially returns to equilibrium as t_1 increases (**Figure 8**). T_1 can be determined from a least-squares fit of the equation

$$\ln[S_\infty - S(t_1)] = \ln 2 + \ln S_\infty - t_1/T_1 \qquad [7]$$

This gives a linear plot of slope $-1/T_1$. However, this approach is particularly sensitive to errors in S_∞ (obtained with $t_1 \geq 5T_1$). The alternative, more reliable, approach is to carry out an exponential fit to each relaxation curve. This does not require an accurate value of S_∞ and is well suited to the fast inversion–recovery method.

The value of T_2 can be determined from the line width of a signal at half of its maximum height:

$$\Delta\nu_{1/2} = (\pi T_2)^{-1} \qquad [8]$$

However, this includes any contribution to the line width from magnetic field inhomogeneity. A 'true' T_2, independent of contributions from field inhomogeneity, can be obtained with the aid of a spin-echo or refocusing pulse sequence (**Figure 9**). This pulse sequence also forms a key component of many other multipulse experiments. Consider a single magnetization vector which is off resonance by $\Delta\nu$ Hz, either owing to chemical shift effects or field

(A)

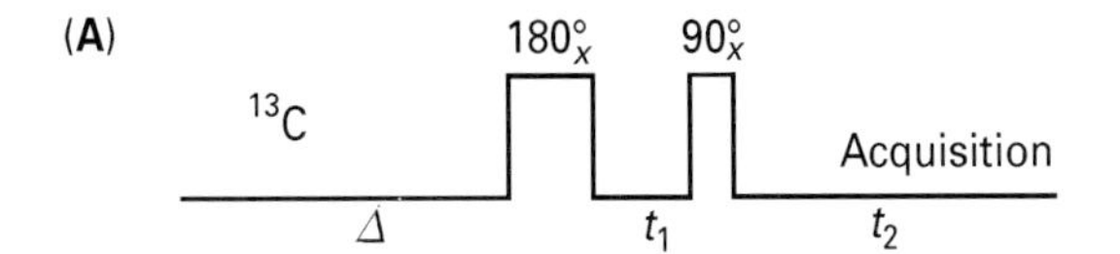

(B)

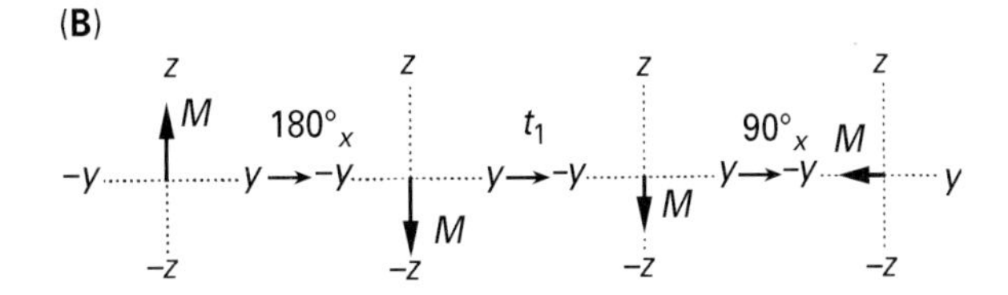

(C)

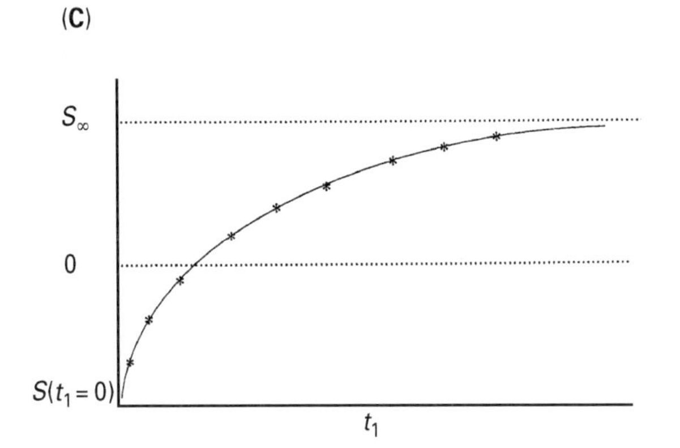

Figure 8 (A) The inversion–recovery sequence used to measure T_1. The experiment is repeated with a number of different values of t_1. (B) The behaviour of a magnetization vector during the inversion–recovery pulse sequence. (C) A plot of signal intensity (s) versus t_1, illustrating the exponential return to the equilibrium value, S_∞, as t_1 increases.

inhomogeneity. After the initial $90°_x$ pulse rotates it to the y-axis, it precesses during $t_1/2$ at angular velocity $2\pi\Delta\nu$ rad s^{-1}, rotating through an angle α. The $180°_y$ pulse 'flips' it from the positive to the negative x region (or vice versa), so that it is now at an angle $-\alpha$ with respect to the y-axis. During the second $t_1/2$ period it again rotates through α, returning the vector to the y-axis, i.e. it is refocused (see **Figure 9**). The FID is then collected. The spin-echo sequence can be repeated a number of times, systematically varying t_1. Alternatively, one can generate an echo train in a single experiment, applying a $180°_y$ pulse at $t_1/2$, $3t_1/2$, $5t_1/2$, etc. and sampling the FID at the peaks of the echoes produced at t_1, $2t_1$, $4t_1$, etc. In either case, an exponential fitting process can be used to determine T_2 from the variation in signal intensity as a function of t_1.

Spectral editing pulse sequences

These pulse sequences, which are important in ^{13}C NMR spectroscopy, allow assignment of individual

(A)

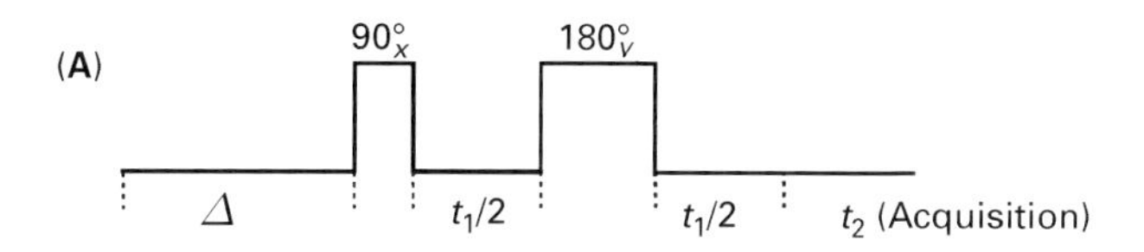

(B)

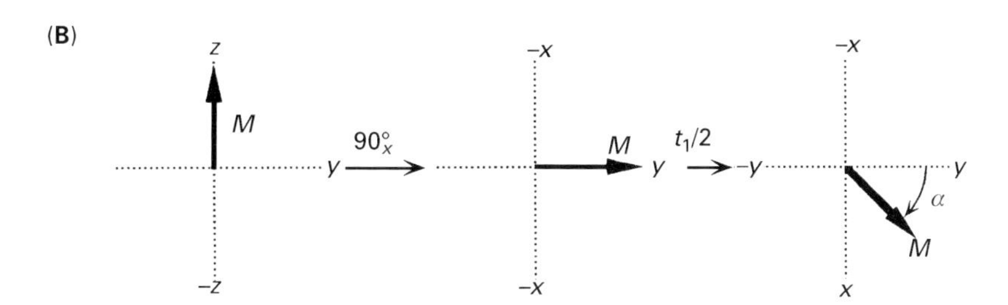

(C)

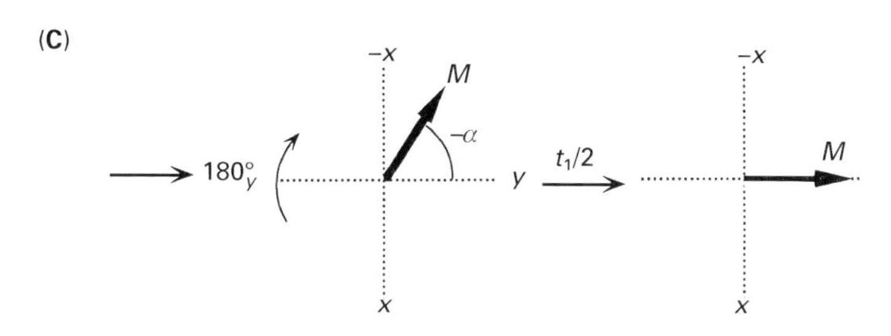

Figure 9 (A) The spin-echo or refocusing sequence for measuring T_2. (B) Behaviour of a magnetization vector, corresponding to an off resonance ($\Delta\nu \neq 0$) signal, during the spin-echo sequence. The vector returns to the initial position after t_1, producing an 'echo'.

peaks in a heteronuclear spectrum in terms of the number of bonded hydrogens. There are three basic sequences which fall into two distinct classes. The first is the APT sequence which involves initial ^{13}C excitation and ^{13}C detection. **Figure 10** illustrates the multiplet patterns expected for ^{13}C peaks coupled to 0, 1, 2 and 3 hydrogens. Now consider the effect of the APT pulse sequence (**Figure 11**) upon a ^{13}C–^{1}H spin system. In the vector model, the ^{13}C magnetization is excess α spin. This can be divided into two nearly equal magnetic vectors, corresponding to ^{13}C α spin bonded to either ^{1}H α or β spins. Following the initial 90°_x pulse, these two vectors begin to precess in the x–y plane with angular velocities $2\pi(\Delta\nu \pm J/2)$ where $\Delta\nu$ is the chemical shift (in Hz), relative to the transmitter, and $J = {}^1J_{CH}$, the one-bond ^{13}C–^{1}H coupling constant. After $t_1/2$, the 180°_y ^{13}C pulse flips the vectors about the y-axis. The simultaneous 180° ^{1}H pulse inverts the equilibrium (z) ^{1}H magnetization, interchanging ^{1}H α and β spin states. The two vectors continue to precess during the second $t_1/2$ period. At the end of this period, they

Table 1 Vector evolution for various CH_n (n = 0–3) multiplets and resultant vectors at the end of t_1 for the APT sequence with various values of t_1

	t_1									
	0		1/4J[a]		1/2J		3/4J		1/J	
n[b]	$\alpha(^0)$[c]	$\langle J\rangle$[d]	$\alpha(^0)$	$\langle J\rangle$	$\alpha(^0)$	$\langle J\rangle$	$\alpha(^0)$	$\langle J\rangle$	$\alpha(^0)$	$\langle J\rangle$
0	0	1.00	0	1.00	0	1.00	0	1.00	0	1.00
1	0, 0	1.00	$\pm\pi/4$	0.71	$\pm\pi/2$	0.00	$\pm 3\pi/4$	−0.71	$\pm\pi$	−1.00
2	0, 0, 0	1.00	0, $\pm\pi/2$	0.50	0, $\pm\pi$	0.00	0, $\pm 3\pi/2$	0.50	0,$\pm 2\pi$	1.00
3	0, 0, 0, 0	1.00	$\pm\pi/4$, $\pm 3\pi/4$	0.35	$\pm\pi/2$, $\pm 3\pi/2$	0.00	$\pm 3\pi/4$, $\pm 9\pi/4$	−0.35	$\pm\pi$, $\pm 3\pi$	−1.00

[a] $J \equiv {}^1J_{CH}$, the one-bond ^{13}C–^{1}H coupling constant for CH_n.
[b] n = number of hydrogens directly bonded to carbon.
[c] Angles of rotation (relative to y-axis) of coupling vectors for the different peaks in each multiplet. These are calculated from the frequencies in **Figure 10** plus Equation [12].
[d] Vector average, relative to y-axis, of coupling vectors relative to an initial value of 1.00. Effects of T_2 relaxation during t_1 are not included.

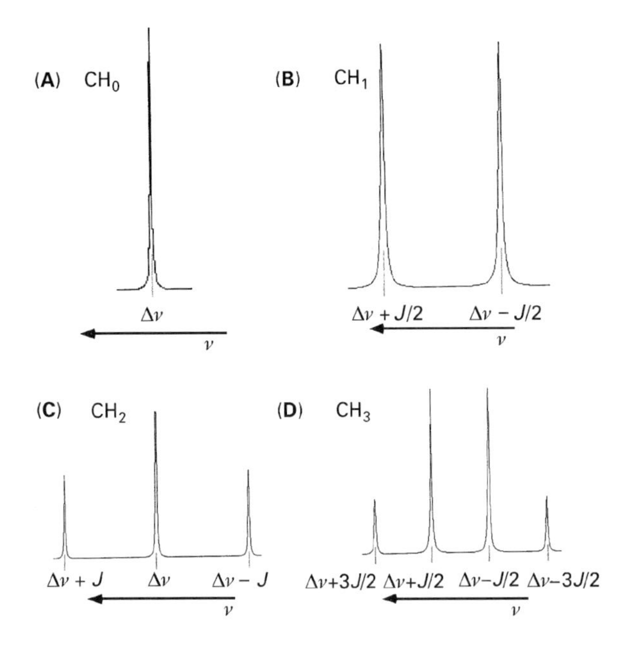

Figure 10 Multiplet patterns arising from to $^1J_{CH}$ (one-bond ^{13}C–^1H coupling constant) in the ^{13}C spectra of CH_n groups ($n = 0$–3).

are at angles with respect to the y-axis given by:

$$\alpha = \pm(2\pi J/2)t_1 = \pm\pi J t_1 \qquad [9]$$

Thus, the ^{13}C chemical shift is refocused by the ^{13}C 180° pulse while the pair of 180° pulses allow $^1J_{CH}$ to evolve through t_1. Applying ^1H decoupling during acquisition rapidly scrambles ^1H spin states, producing a single, averaged vector which initially is along the y-axis but precesses at a frequency $\Delta\nu$ during FID acquisition. This results in a peak at $\Delta\nu$ in the frequency spectrum with an intensity determined by the vector average at the end of t_1. **Table 1** summarizes the results for CH_0, CH_1, CH_2 and CH_3 peaks for different values of t_1. Only a CH_0 peak is observed at $t_1 = 1/2J$ since the coupling vectors for the other multiplets average to zero. For $t_1 = 1/J$, CH_0 and CH_2 peaks are positive (upright) while CH_1 and CH_3 peaks are negative (inverted). This allows partial assignment of carbons in terms of numbers of attached protons. The main weakness of this approach is that it is sensitive to variations in $^1J_{CH}$ and thus may give unreliable results for compounds which have a wide range of $^1J_{CH}$ for carbons.

The other two spectral editing sequences are insensitive nuclei enhanced by polarization transfer (INEPT) and distortionless enhancements by polarization transfer (DEPT) (see **Figure 12**). Both of these sequences involve ^1H excitation, followed by magnetization transfer to the heteronucleus by a mechanism called polarization transfer. This provides heteronuclear signal enhancement by a factor of γ_H/γ_X, e.g. ~4 for ^{13}C. In the case of INEPT, this can be adequately described by the vector model in terms of selective inversion of the populations of ^1H energy levels corresponding to carbons in the α spin state. However, although the sequences appear similar, DEPT can only be explained by a quantum mechanical model.

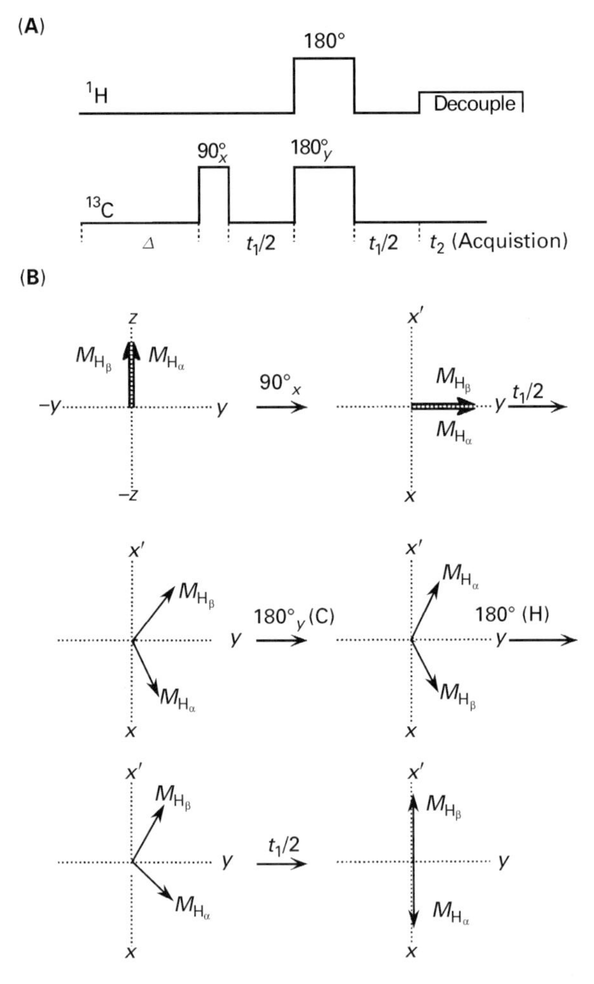

Figure 11 (A) The APT (attached proton test) pulse sequence. (B) Behaviour of the ^{13}C magnetization due to a ^{13}C–^1H spin pair during the APT sequence. The two components, corresponding to ^{13}C coupled to ^1H in α or β spin states, precess at frequencies $\Delta\nu \pm J/2$ (where $J = {}^1J_{CH}$). The spin-echo sequence refocuses the chemical shift ($\Delta\nu$) but not J (see Eqn [9]). This figure illustrates the result when $t_1 = J/2$ with vectors rotating through $\alpha = \pm\pi/2$.

The two sequences use different forms of spectral editing. With INEPT, this is done by the choice of the final delay. If $\Delta_3 = J/2$, only CH carbons appear while $\Delta_3 = 3J/4$ produces a spectrum with CH and CH_3 up and CH_2 down. With DEPT, editing is carried out by varying the angle of the final ^1H pulse;

$\Theta = 90°$ yields only CH carbons while $\Theta = 135°$ yields CH and CH_3 up and CH_2 down. Because it relies on a pulse angle rather than a delay, DEPT is less sensitive than INEPT to variations in $^1J_{CH}$ and is thus the sequence of choice. Note that residual ^{13}C magnetization is suppressed by phase cycling in each case (see below) and thus non-protonated carbons are not observed with either sequence.

Phase cycling for artifact suppression

Before the development of pulsed field gradient sequences (see below), most NMR pulse sequences included a phase cycle in which the phases of at least one pulse and the receiver were varied systematically. This was needed for one or more of several reasons, e.g. suppression of unwanted signals, suppression of artifacts due to hardware imperfections and/or incomplete return to equilibrium between scans and coherence pathway selection in multidimensional NMR. One example of each the first two kinds of phase cycle is briefly discussed below.

In the INEPT sequence (**Figure 12**), the final 90°_y ^1H pulse sets up selective inversion of the populations of a pair of levels within the coupled AX (^1H–^{13}C) spin system. The 90°_y ^{13}C pulse then generates two antiphase magnetization vectors of relative intensity +4 and −4 (relative to equilibrium ^{13}C magnetization) along the $\pm x$-axes owing to magnetization (polarization) transfer from ^1H arising from the selective population inversion. However, the ^{13}C 90°_y pulse also generates a magnetization vector from the initial ^{13}C magnetization. It is desirable to eliminate the latter component to avoid complications with spectral editing. This is done by alternating the phase of the final ^1H pulse 90°_y, 90°_{-y} while alternately adding and subtracting FID signals. With the 90°_{-y} pulse, the antiphase ^{13}C vectors become −4, +4 but this is converted back into +4, −4 by subtracting this FID. However, the ^{13}C 90°_{-y} pulse always generates a signal of the same phase owing to ^{13}C magnetization and thus is cancelled by the alternate addition and subtraction of FID signals.

Quadrature detection involves the use of two receivers. If the two receivers have different gains, 'quadrature image' peaks are generated at $-\Delta\nu$ for every true peak at $+\Delta\nu$. This can be eliminated by using a four-step CYCLOPS phase cycle in which the phase of a transmitter is cycled through relative phases x, y, $-x$, $-y$ along with the receiver. Derome (see Further reading) gives a very clear account of quadrature images and how they are suppressed by CYCLOPS phase cycling.

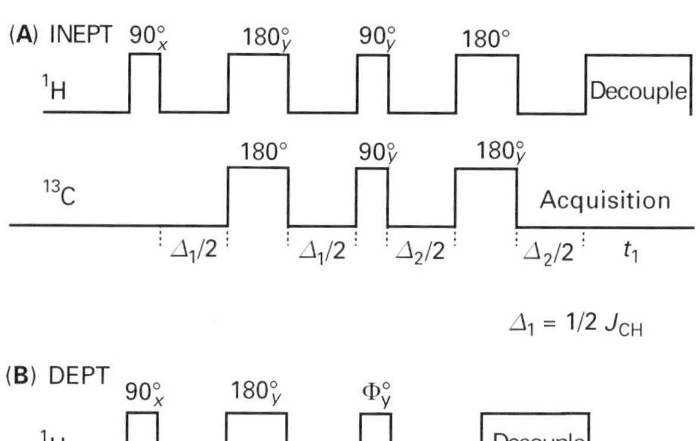

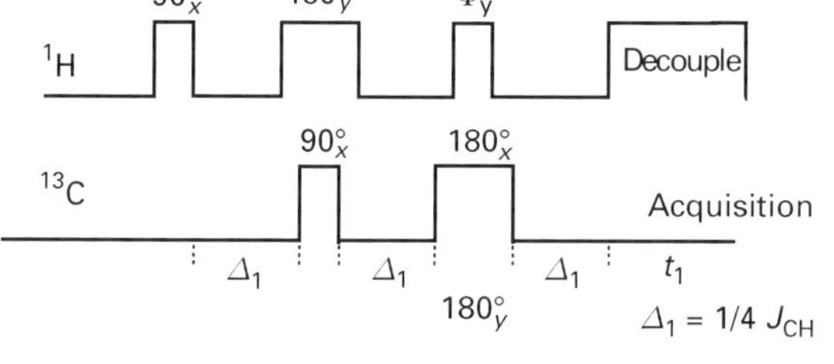

Figure 12 (A) INEPT pulse sequence. (B) DEPT pulse sequence. The article on product operator formalism describes the behaviour of the DEPT sequence while the texts by Harris and Günther (see Further reading section) describe the behaviour of INEPT in terms of vector diagrams and energy levels.

Quantum mechanical methods for understanding pulse sequences

The ultimate approach for interpreting multipulse sequences and their resultant spectra is a full density matrix treatment. While this approach is ideal for simulating the spectrum generated by a multipulse experiment, the calculations are complex and do not provide obvious physical insights. A very useful and widely used simplified quantum mechanical approach involves product operator formalism. This focuses on the components of the density matrix which are directly relevant to the experiment. Product operator descriptions of several of the pulse sequences discussed here are given in another article.

Mastery of this approach is essential for anyone desiring to design new pulse sequence experiments and valuable for anyone wishing to understand modern NMR experiments.

Multidimensional NMR experiments

Multidimensional NMR experiments have revolutionized the use of NMR spectroscopy for the structure determination of everything from small molecules to complex proteins. Since most of the 3D and 4D experiments are essentially combinations of two-dimensional (2D) experiments, this section will focus on 2D NMR. Only a basic overview will be given since many specific multidimensional experiments are discussed elsewhere.

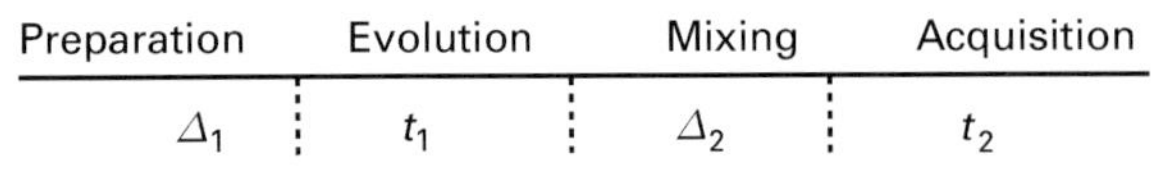

Preparation	Evolution	Mixing	Acquisition
Δ_1	t_1	Δ_2	t_2

Figure 13 A general two-dimensional NMR pulse sequence. Data are acquired during t_2 for a series of spectra in which t_1 is regularly incremented from 0 to some maximum value. Fourier transformation with respect to t_2 and then t_1 generates a spectrum with two difference frequency axes.

Two-dimensional NMR spectroscopy

A generalized two-dimensional experiment is illustrated in **Figure 13**. The preparation time is usually a relaxation delay followed by one or more pulses to start the experiment. The evolution period establishes the second frequency dimension. A series of FID signals are collected with t_1 regularly incremented from 0 up to the desired maximum value, t_1 (max). The number of increments and t_1 (max) depend on the desired spectral width and data point resolution along the time-incremented axis. Depending on the experiment, the evolution period may contain one or more pulses, most commonly a spin-echo sequence. The mixing period, which is not required in some sequences, can be a 90° mixing pulse, a fixed delay, a more complex pulse such as a spin lock or isotropic mixing pulse or some combination of these. Finally, data is acquired during t_2, as in a 1D experiment. Double Fourier transformation, with respect to t_2 then t_1, yields a spectrum with two orthogonal frequency scales.

2D NMR experiments are designed to generate different kinds of frequency information along the two axes. The principle behind this frequency separation is most easily seen by considering modification of the APT sequence (**Figure 11**) to produce a 2D sequence, the heteronuclear *J*-resolved sequence. This involves replacing the constant t_1 period by an incremented t_1 period. Each ^{13}C signal in f_2 is then modulated by the evolution of $^1J_{CH}$ coupling vectors as t_1 is incremented (e.g. see **Table 1**). Fourier transformation with respect to t_1 at each f_2 frequency then produces a 2D spectrum where a cross section through each ^{13}C peak in f_2 will give a $^1J_{CH}$ multiplet pattern similar to one of those shown in **Figure 10**. Similarly, the INEPT sequence can be converted into a 2D sequence (the heteronuclear shift correlation sequence or HETCOR) by inserting a spin-echo sequence, $t_1/2$–180°(C)–$t_1/2$ immediately after the initial ^{1}H 90° pulse (see **Figure 12**). The extent of polarization transfer from ^{1}H to ^{13}C is then modulated by ^{1}H chemical shift evolution as t_1 is incremented, with the resultant 2D experiment having ^{13}C chemical shifts along f_2 and ^{1}H chemical shifts along f_1.

Table 2 Characteristics of several commonly used 2D NMR pulse sequences

Sequence	*Display mode*[a]	*f_1, f_2*[b]	*Transmission*[c]
COSY	D, OD	δ_H, δ_H	J_{HH}
DQCOSY[d]	D, OD	δ_H, δ_H	J_{HH}
TOCSY[e]	D, OD	δ_H, δ_H	$J_{HH} \rightarrow J_{HH}$
NOESY	D, OD	δ_H, δ_H	H–H dipolar
ROESY[f]	D, OD	δ_H, δ_H	relaxation
EXSY[g]	D, OD	δ_H, δ_H	$H_A \rightleftharpoons H_B$
HETCOR[h]	f_1, f_2	δ_H, δ_X	$^1J_{CH}$
COLOC[h]	f_1, f_2	δ_H, δ_X	$^nJ_{CH}$
FLOCK[h]	f_1, f_2	δ_H, δ_X	$^nJ_{CH}$
HMQC[i]	f_1, f_2	δ_X, δ_H	$^1J_{CH}$
HSQC[i]	f_1, f_2	δ_X, δ_H	$^1J_{CH}$
HMBC[i]	f_1, f_2	δ_X, δ_H	$^nJ_{CH}$
INADE-QUATE	DQ, SQ	$\delta_{C(1)}+\delta_{C(2)}$, $\delta_{C(1)}$	$^1J_{CC}$

[a] D, OD: spectrum along diagonal with off-diagonal peaks between correlated protons, e.g. see COSY spectrum (**Figure 14**), f_1, f_2: different chemical shift scales along f_1, f_2, e.g. see HSQC spectrum (**Figure 15**), DQ, SQ: double quantum frequencies along f_1 (sum of frequencies of coupled ^{13}C peaks, relative to transmitter), regular (single quantum) ^{13}C spectrum along f_2.

[b] Chemical shift information appearing along each axis. {δ scale ≡ chemical shift in parts per million relative to internal reference [$\delta_1(CH_3)_4$ for ^{1}H, ^{13}C and ^{29}Si]}. Note that the ^{1}H-axis normally also shows multiplet structure owing to J_{HH}.

[c] Parameter by which information is transmitted to establish correlations between the spectra on the two frequency axes: J_{HH} = ^{1}H–^{1}H coupling constant, $H_A \rightleftharpoons H_B$ is chemical exchange between different sites, J_{CH} = one-bond ^{13}C–^{1}H coupling constant, $^nJ_{CH}$ = *n*-bond (n = 2 or 3) ^{13}C–^{1}H coupling constant, $^1J_{CC}$ = one-bond ^{13}C–^{13}C coupling constant.

[d] DQCOSY ≡ double quantum filtered COSY. This suppresses strong singlets (e.g. solvent peaks) and gives well-resolved off-diagonal peaks with up–down intensity patterns for coupled protons, i.e. those giving rise to the off-diagonal peak.

[e] TOCSY (also called HOHAHA) relays information among sequences of coupled protons. A cross section through the f_2 frequency of a specific proton shows f_1 peaks for all of the protons within the coupled sequence.

[f] ROESY ≡ NOESY in the rotating frame.

[g] The EXSY sequence is identical to the NOESY sequence but detects cross peaks between chemically exchanging hydrogens. Both EXSY and NOESY peaks may appear in the same spectrum.

[h] X nucleus (usually ^{13}C) detected heteronuclear shift correlation sequences.

[i] ^{1}H detected heteronuclear shift correlation sequences. These are more sensitive than the earlier X-nucleus detected sequences [by $(\gamma_H/\gamma_X)^{3/2}$] but have more limited resolution along the X(f_1) axis.

High-resolution 2D NMR pulse sequences can be based on information transfer via homonuclear or heteronuclear scalar coupling, dipolar relaxation or chemical exchange while solid-state 2D NMR experiments normally use dipolar coupling in place of scalar coupling. There are three basic modes of spectral

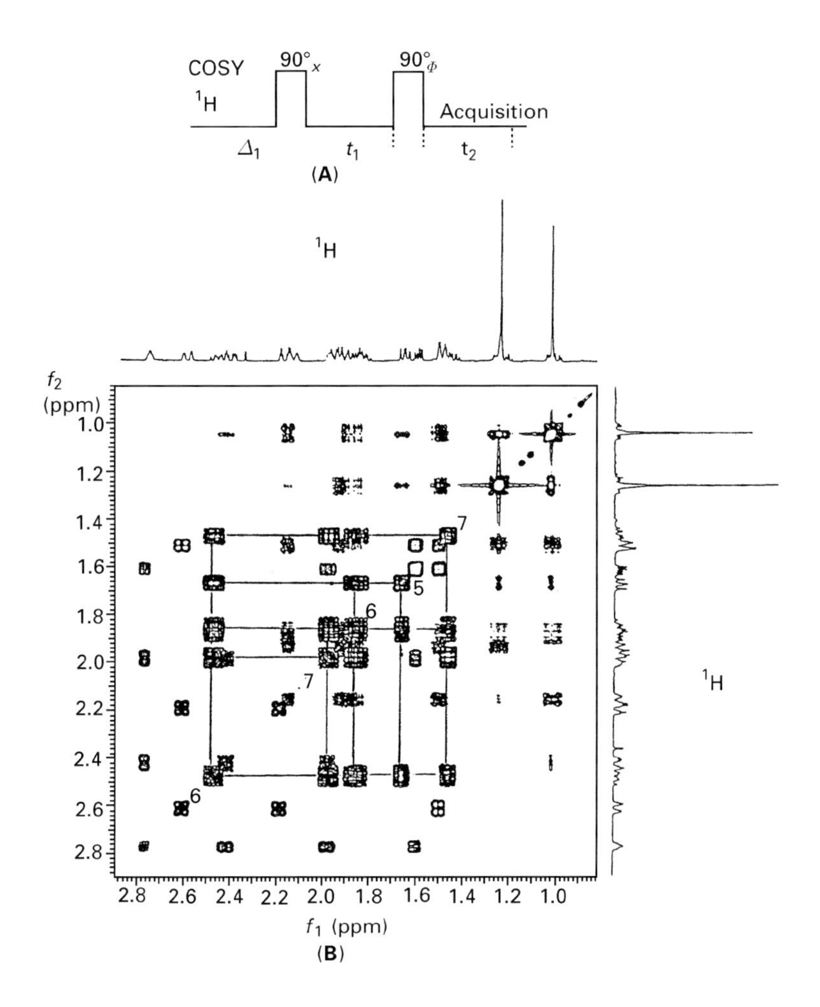

Figure 14 (A) The COSY pulse sequence. (B) The COSY spectrum for the aliphatic region of [1], showing the ^{1}H spectrum along the diagonal and symmetric off-diagonal peaks between coupled protons. The connections for one molecular fragment [C(5)H–C(6)H_2–C(7)H_2)] are traced out.

display: with the normal spectrum along the diagonal and off-diagonal peaks for correlated signals (see COSY spectrum, **Figure 14**), different chemical shift information along the two axes (e.g. ^{1}H and ^{13}C or ^{15}N, see **Figure 15**) and spectra with single quantum frequencies along f_2 and multiple quantum frequencies along f_1. The characteristics of many of the common high resolution 2D sequences are summarized in **Table 2**.

However, the real strength of multidimensional NMR is not in the information provided by a single experiment but rather in the synergy provided by carrying out several different experiments on the same molecule. This is illustrated below for kauradienoic acid [1], one of the very first molecules where combined 2D methods were used for spectral assignment. The ^{1}H–^{1}H COSY spectrum (**Figure 14**) and the ^{1}H–^{13}C shift correlation spectrum (**Figure 15**) for [1] allow assignment of molecular fragments involving sequences of protonated carbons. Further experiments allow completion of the structural and spectral assignments (see caption to **Figure 15**).

Absolute value versus phase sensitive 2D spectra

Many of the original 2D sequences gave spectra which could not be phased since they involved different mixtures of absorption and dispersion modes for different peaks. To simplify displays, these spectra were plotted in absolute value mode, $(u^2 + v^2)^{1/2}$, where u refers to dispersion mode and v to absorption mode. While the individual absolute value mode peaks appear to be properly phased, they are distorted from Lorentzian shape with broad tails. Better resolution and sensitivity can be obtained if spectra are obtained in a manner which provides pure absorption mode peaks. There are now phase sensitive versions of most 2D sequences. There are two requirements for obtaining phase sensitive spectra. First, any fixed delay must include a spin-echo sequence to prevent chemical shift evolution. Second, one must acquire two separate data sets with

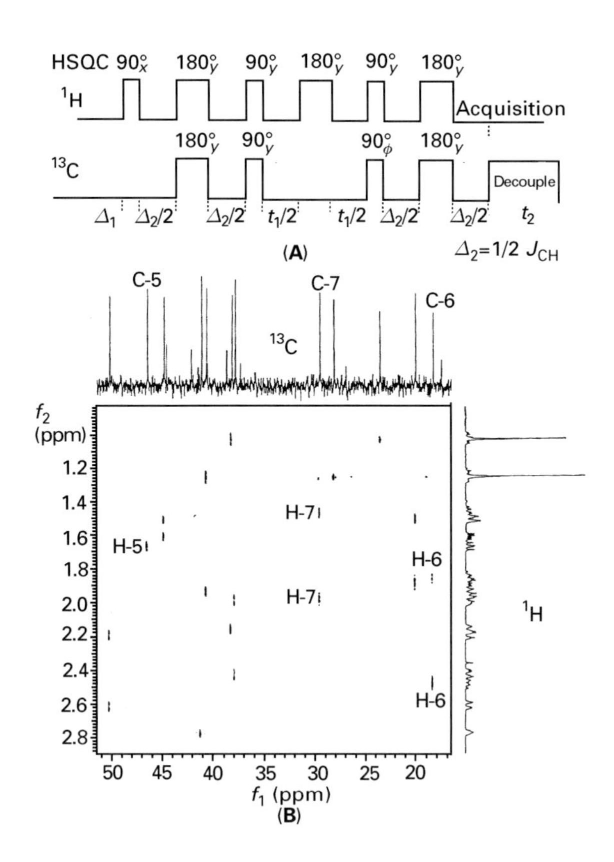

Figure 15 (A) The basic HSQC (heteronuclear single quantum coherence) pulse sequence. (B) The HSQC spectrum of the aliphatic region of [1] with the ^{13}C along f_1 and the ^{1}H spectrum along f_2. The ^{13}C–^{1}H connectivities are marked for the same molecular fragment as in **Figure 14**. Other sequences of protonated carbons can be determined from the same spectrum while an n-bond ($n = 2,3$) ^{13}C–^{1}H shift correlation spectrum such as HMBC, COLOC or FLOCK (see **Table 2**) can identify non-protonated carbons and tie together the molecular fragments into a complete structure.

one of the pulses having a 90° phase difference for the two spectra or one must increment the phase of one of the pulses by 90° with each time increment while doubling the number of time increments collected.

Phase cycling for coherence pathway selection in 2D NMR

This important topic can only be understood in quantum mechanical terms. Owing to space limitations, only a very brief introduction can be given here.

The COSY sequence (**Figure 14**) will be used to illustrate the concept. A coherence level diagram for this sequence is given in **Figure 16**. Equilibrium z magnetization is defined as having a coherence level of 0. The 90° pulse acts as a raising or lowering operator, i.e. it can change the spin quantum number of an individual nucleus by ±1, resulting in the generation of observable x,y magnetization. This evolves during t_1 at frequencies determined by the chemical shift and homonuclear coupling. The second 90° pulse then can generate further coherence level changes, including level changes of 0 to ±2 associated with a pair of coupled nuclei. The receiver can only detect single quantum coherence and is chosen to be at coherence level +1. Thus, a pair of ideal pulses can generate a COSY signal by two paths with coherence level changes +1, 0 or −1, +2, respectively called P (or antiecho) and N (echo) signals. If both paths are detected, signals will occur at both $+\Delta\nu_1$ and $-\Delta\nu_1$ along the f_1 axis. However, phase cycling allows one to choose one path, rejecting the other. The change in coherence phase associated with a pulse is given by:

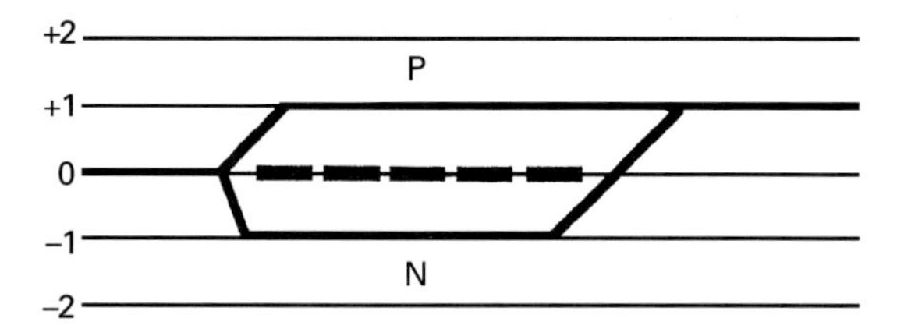

Figure 16 Coherence level diagram for the COSY sequence. The symbols N and P designate the N and P pathways.

$$\Delta\theta = \Delta p \cdot \Delta\phi \qquad [10]$$

where $\Delta\theta$ is the change in coherence phase (in increments of $\pi/2$), Δp is the change in coherence level and $\Delta\phi$ is the change in pulse phase (in units of $\pi/2$). **Table 3** shows how this can be used to design two-step phase cycles for a coherence pathway section with the COSY sequence.

The dotted line indicates a third possible coherence pathway. If the initial 90° pulse is imperfect, there will be some residual z magnetization which will be raised to coherence level +1 by the second 90° pulse. Suppression of this pathway requires two extra steps, yielding a four-step phase cycle. Finally, if one also wishes to incorporate a CYCLOPS cycle for f_2 quadrature image suppression, the total phase cycle is $4 \times 4 = 16$ steps. The number of scans in a 2D experiment should be some whole number multiple of the number of steps in the phase cycle. However, with COSY, the sensitivity is high enough that 16 scans are usually more than necessary to acquire good spectra and thus the phase cycle determines the minimum time for the experiment. Fortunately, pulsed field gradient sequences have overcome this problem.

Table 3 Alternative two-step phase cycles for N-type pathway selection and P-pathway suppression for a COSY spectrum plus a four-step phase cycle which selects the N-pathway while suppressing both P and Z paths

	N		*P*			*N*		*P*	
Scan	1	2	1	2	Scan	1	2	1	2
ϕ(P1)	0[a]	0	0	0	ϕ(P1)	0	3	0	3
$\Delta\theta$(P1)	0	0	0	0	$\Delta\theta$(P1)	0	1[b]	0	3
ϕ(P2)	0	1	0	0	ϕ(P2)	0	0	0	0
$\Delta\theta$(P2)	0	2	0	0	$\Delta\theta$(P2)	0	0	0	0
$\Delta\theta$(P1 + P2)	0	2	0	0	$\Delta\theta$(P1 + P2)	0	1	0	3
ϕ(R)[c]	0	2	0	2	ϕ(R)	0	1	0	1

	N				*P*				*Z*			
Scan	1	2	3	4	1	2	3	4	1	2	3	4
ϕ(P1)[d, e]	0	3	2	1	0	3	2	1	0	3	2	1
$\Delta\theta$(P1)	0	1	2	3	0	3	2	1	0	0	0	0
ϕ(P2)	0	0	0	0	0	0	0	0	0	0	0	0
$\Delta\theta$(P2)	0	0	0	0	0	0	0	0	0	0	0	0
$\Delta\theta$(P1 + P2)	0	1	2	3	0	3	2	1[e]	0	0	0	0[e]
ϕ(R)[c]	0	1	2	3	0	1	2	3	0	1	2	3

The symbols N and P indicate the coherence level change from the first pulse and are respectively negative (–1) and positive (+1) for the two paths. The third path, from an imperfect initial 90° pulse, has zero coherence level change in the initial pulse and is thus given the symbol Z.

[a] Pulse and receiver phases *x*, *y*, $-x$ and $-y$ are, respectively, given as 0, 1, 2 and 3, corresponding to the number of $\pi/2$ phase increments relative to a $90°_x$ pulse.

[b] $\Delta p = -1$ for the coherence level change in the N pathway while $\Delta\theta(\text{P1}) = 3$. From Equation [10], $\Delta p \Delta\theta = (-1)\,(3) = -3$. However, since a –270° phase shift corresponds to a −90° phase shift, $-3 \equiv 1$.

[c] $\phi(\text{R}) \equiv$ receiver phase. When the sums of coherence phase changes in different scans match the receiver phase cycle, successive scans add, while when the relative phases change 0, 2 successive scans cancel. Thus in each case, the signals from the N path add while signals from the P path cancel.

[d] An alternative four-step phase cycle for N-path selection involves a $\phi(\text{P1}) = 0, 1, 2, 3$ and $\phi(\text{R}) = 0, 3, 2, 1$. For P-type selection ϕ(P1) and ϕ(R) should either both be 0, 1, 2, 3 or both be 0, 3, 2, 1. Another alternative for N-pathway selection is to expand the first two-step phase cycle to a four-step cycle with $\phi(\text{P2}) = 0, 1, 2, 3$ and $\phi(\text{R}) = 0, 2, 0, 2$.

[e] In this four-step phase cycle, the P-pathway is cancelled in steps 1 + 2 and in steps 3 + 4 while the Z pathway is cancelled in steps 1 + 3 and steps 2 + 4.

Gradient pulse sequences

Many of the pulse sequences discussed above now have versions which incorporate magnetic field gradient pulses that can be used to replace phase cycling. They allow one to acquire a spectrum in greatly reduced time and/or with greatly reduced artifacts. For example, applying two identical field gradient pulses before and after the final 90° pulse in COSY selects the N (echo) path while suppressing the other paths.

Pulse sequences which replace single pulses

A number of pulse 'sandwiches' have been developed to replace single pulses in specific cases. These include the following.

Composite 180° pulses

The quality of the spectra obtained with many pulse sequences is strongly dependent on the precision of 180° pulses, particularly in the case of 180° inversion pulses for heteronuclei with broad spectral windows. Problems can arise due to mis-set pulses, inhomogeneity in pulses over the sample or incomplete excitation at large frequencies relative to the transmitter. These problems can be minimized by the use of composite 180° pulses, e.g. a 90°_x, 180°_y, 90°_x composite pulse in place of a 180°_x pulse (**Figure 17**).

BIRD (bilinear rotating decoupling) pulses

BIRD pulses act as selective 180° ^{1}H pulses either for protons directly bonded to ^{13}C or not bonded to ^{13}C, while simultaneously providing a ^{13}C 180° pulse (**Figure 18**). Earlier uses of these pulses included partial ^{1}H–^{1}H decoupling in HETCOR and optimization

Figure 17 (A) Vector diagram illustrating the effect when a nominal 180° pulse is mis-set, resulting in only 170° rotation. (B) Illustration of how a composite $90°_x$, $180°_y$, $90°_x$ compensates for the effect of a mis-set pulse. Compensation is less complete for off-resonance signals.

of performance of long-range ^{13}C-detected ^{13}C–^{1}H correlation sequences such as COLOC and FLOCK. The most common current use is for suppression of ^{1}H–^{12}C magnetization in ^{1}H-detected one-bond ^{13}C–^{1}H correlation sequence, i.e. HMQC and HSQC. Although BIRD pulses can be explained by vector diagrams (**Figure 18**), a full understanding of these pulses requires a quantum mechanical treatment.

Frequency selective pulses

The ability to selectively excite a narrow spectral region is important both for solvent suppression and because it often allows one to replace a full 2D experiment by a limited number of 1D experiments. A 'soft' (i.e. low power, long duration) pulse can be used for selective excitation but this does not generate

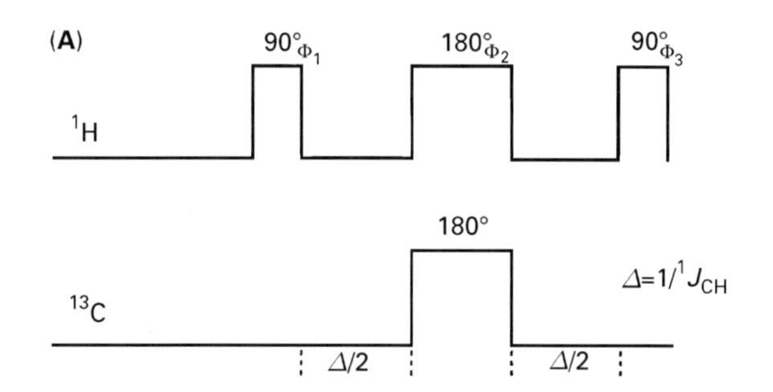

Φ_1	Φ_2	Φ_3	^{1}H signal inverted
x	*y*	*x*	^{1}H - ^{12}C
x	*y*	−*x*	^{1}H - ^{13}C
x	*x*	−*x*	^{1}H - ^{12}C
x	*x*	*x*	^{1}H - ^{13}C

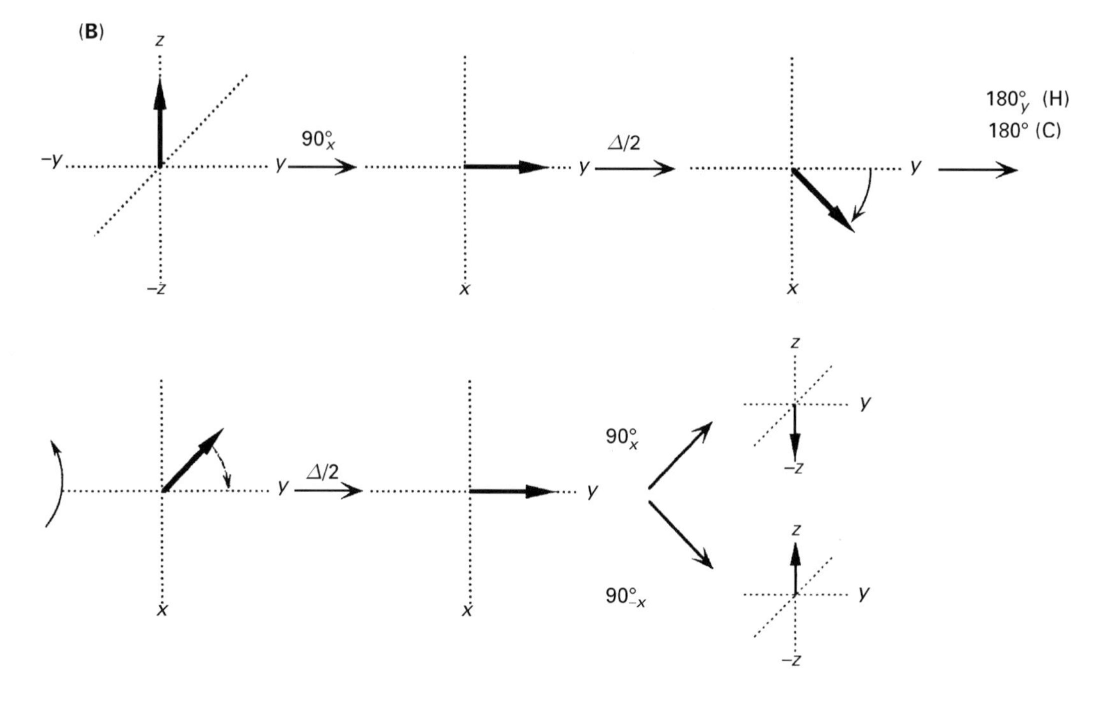

Figure 18 (A) The BIRD pulse sequence and effects of different combinations of phases within the BIRD pulse. (B) Vector diagram for a ^{12}C–^{1}H spin system, illustrating how $90°_x$, $180°_y$, $90°_x$ and $90°_x$, $180°_y$, $90°_{-x}$ BIRD pulses, respectively, act as 180° and 0° ^{1}H pulses. Since the BIRD pulse corresponds to the APT sequence (with ^{1}H and ^{13}C pulses interchanged) up to the point of the final 90° pulse, the effect of these two BIRD pulses on a ^{1}H–^{13}C pair can be deduced from the data in **Table 1** for $n = 1$. With $\Delta = 1/J_{CH}$, the vectors associated with the ^{1}H–^{13}C pair are refocused along the −*y*-axis and a $90°_x$ pulse will rotate them back to the *z*-axis (0° pulse) while a $90°_{-x}$ pulse rotates them to the −*z*-axis (180° pulse).

uniform excitation (see **Figure 3**). Better results are obtained from multiple pulse sequences. Modern spectrometers allow the generation of 'shaped' pulses whose time profiles are designed to produce the desired excitation profiles. A series of pulses of controlled amplitude, duration and phase (without intervening delays) are used which provide the desired profile. For example, generating a pulse with a time profile similar to the frequency profile in **Figure 3** will give a narrow square wave excitation profile. Both 90° and 180° pulses can be generated as well as pulses which simultaneously irradiate at two or more chosen frequencies. The bandwidth of each pulse can be adjusted to selectively irradiate a chosen signal or to cover a specific spectral region (e.g. irradiation of the amide $^{13}C{=}O$ region in a 3D or 4D protein spectrum).

Broad-band decoupling pulse sequences

Multiple pulse sequences can also be used to provide effective broad-band decoupling. The original pulse sequence of this kind was the WAHUHA sequence of Waugh and co-workers which was designed to minimize broadening arising from homonuclear dipolar coupling in solid-state spectra.

For high-resolution NMR, the main interest has been in heteronuclear broad-band decoupling. Initially, the interest was in broad-band ^1H decoupling while acquiring heteronuclear (e.g. ^{13}C) spectra. More recently, with the increasing use of ^1H-detected 2D, 3D and 4D sequences involving ^1H–X chemical shift correlation, the emphasis has been on decoupling of heteronuclei (e.g. ^{13}C, ^{15}N). This is much more demanding owing to the much wider heteronuclear chemical shift window. Increasingly effective decoupler pulse sequences have been developed with acronyms such as MLEV, WALTZ, GARP, DIPSI and WURST. Most are based on a composite 180° decoupler pulse which is subjected to a series of phase cycles. For example, WALTZ is based on a 90°_x, 180°_{-x}, 270°_x composite pulse (which in shorthand form is designated 1 $\bar{2}$ 3, justifying the name WALTZ).

Summary

This article has given an overview of the many different multiple pulse experiments which have developed from the original pulse Fourier transform experiment. These experiments, along with major improvements in spectrometer instrumentation, have dramatically increased the range of structural and dynamic problems that can be studied by NMR spectroscopy.

List of symbols

B_0 = external magnetic field vector; B_1 = rotating magnetic field vector arising from RF electromagnetic radiation; u = dispersion mode; M = resultant of individual magnetic moment vectors; v = absorption mode; α,β = spin states corresponding to allowed values of m_I; α = angle of rotation of M with respect to initial axis; γ = magnetogyric ratio for nucleus, i.e. the ratio of magnetic moment/spin angular momentum; Δ = fixed delay; θ = phase of coherence; ν = frequency (s^{-1}); $\Delta\nu$ = frequency difference; ϕ = phase of pulse or receiver; ω = angular velocity (rad s^{-1}).

See also: **^{13}C NMR, Methods; ^{13}C NMR, Parameter Survey; Fourier Transformation and Sampling Theory; High Resolution Solid State NMR, ^1H, ^{19}F; Magnetic Field Gradients in High Resolution NMR; NMR Principles; NMR Spectrometers; Product Operator Formalism in NMR; Proteins Studied Using NMR Spectroscopy; Solvent Suppression Methods in NMR Spectroscopy; Structural Chemistry using NMR Spectroscopy, Inorganic Molecules; Structural Chemistry using NMR Spectroscopy, Organic Molecules; Structural Chemistry Using NMR Spectroscopy, Peptides; Structural Chemistry Using NMR Spectroscopy, Pharmaceuticals; Two-Dimensional NMR, Methods.**

Further reading

Derome AE (1987) *Modern NMR Techniques for Chemistry Research*. Oxford: Pergamon Press.

Ernst RR (1992) Nuclear magnetic resonance Fourier transform spectroscopy (Nobel lecture). *Angewante Chemie* **31**: 805–823.

Freeman R (1997) *A Handbook of Nuclear Magnetic Resonance*, 2nd edn. Harlow: Addison Wesley Longman.

Freeman R (1998) Shaped radio frequency pulses in high resolution NMR. *Progress in NMR Spectroscopy* **32**: 59–106.

Günther H (1995) *NMR Spectroscopy*, 2nd edn. Chichester: Wiley.

Harris RK (1986) *Nuclear Magnetic Resonance Spectroscopy: A Physiochemical View*. Harlow: Longman.

Keeler J (1990) Phase cycling procedures in multiple pulse NMR spectroscopy of liquids. In: Granger P and Harris RK (eds) *Multinuclear Magnetic Resonance in Liquids and Solids*. Dordrecht: Kluwer.

Levitt M (1986) Composite pulses. *Progress in NMR Spectroscopy* **18**: 61–122.

Parella T (1998) Pulsed field gradients: a new tool for routine NMR. *Magnetic Resonance in Chemistry* **36**: 467–495.

Shaka AJ and Keeler J (1987) Broadband spin decoupling in isotropic liquids. *Progress in NMR Spectroscopy* **19**: 47–129.

NMR Relaxation Rates

Ronald Y Dong, Brandon University, Manitoba, Canada

MAGNETIC RESONANCE
Theory

How a nuclear spin system achieves thermal equilibrium by exchanging energy with its surrounding medium or the 'lattice' is governed by the NMR relaxation rates. The lattice consists of all degrees of freedom, except those of the nuclear spins, associated with the physical system of interest. Pulsed NMR provides a highly versatile and flexible tool to determine spin relaxation rates, which can probe the entire spectrum of molecular motions. These include molecular rotation, translational self-diffusion, 'coherent' rotational motion, and the internal motion in nonrigid molecules. Physical systems investigated by NMR range from condensed matter phases to dilute molecular gases. The theory of nuclear spin relaxation is now well understood, and the details are given in the classical treatise by Abragam. An elementary treatment of the same material can be found in the text by Farrar. In relating the measured spin relaxation rates to molecular behaviours, there are severe limitations and difficulties that a newcomer can often fail to appreciate. Several nuclear interactions may simultaneously all contribute to the relaxation of a spin system. These may include the magnetic dipole–dipole interaction, the quadrupole interaction, the spin–rotation interaction, the scalar coupling of the first and second kind, and the chemical-shift anisotropy interaction. Due to the need of estimating certain nuclear couplings and/or correlation times associated with molecular motions, considerable uncertainty may exist in identifying and separating these contributions. The semiclassical relaxation theory of Redfield is outlined in this short review to give expressions of spin relaxation rates in terms of spectral densities of motion. The treatment is semiclassical simply because it uses time correlation functions which are classical.

The most difficult problem in any relaxation theory is the calculation of correlation functions or spectral densities of motion. It is often possible to determine the mean square spin interaction $\langle H_q^2(t)\rangle$, where $H_q(t)$ is a component of the spin Hamiltonian which fluctuates randomly in time owing to molecular motions. The time dependence of the correlation function $\langle H_q(t)H_{q'}(t-\tau)\rangle$ can often be approximated by an exponential decay function of τ, i.e.

$$\langle H_q(t)H_{q'}(t-\tau)\rangle = \langle H_q(t)H_{q'}(t)\rangle \exp(-\tau/\tau_c) \qquad [1]$$

where the angle brackets denote an ensemble average, and the correlation time τ_c for the motion can be determined with the help of experiments. There are many examples of exponentially decaying correlation functions. For instance, thermal motion of molecules in liquids was first treated in the classical BPP paper by Bloembergen. Spectral density calculations for liquids normally use a classical picture for the lattice. Quantum calculations of spectral densities are feasible for spin relaxation due to lattice vibrations or conduction electrons. Such calculations are, in general, impossible since the eigenstates of the lattice are often unknown. Molecular motions that are too fast ($\omega_0\tau_c << 1$) or too slow ($\omega_0\tau_c >> 1$) with respect to the inverse of the Larmor frequency ω_0 are not amenable to nuclear spin–lattice relaxation (T_1) studies. Fortunately, measurements of spin–spin relaxation time (T_2) or spin–lattice relaxation time ($T_{1\rho}$) in the rotating frame can be used. Both T_1 and T_2 appear in the phenomenological Bloch equations, which describe the precession of nuclear magnetization in an external magnetic field. In NMR, the coupling between the lattice and the Zeeman reservoir of the nuclear spin system is magnetic in all cases except one. The exception is the quadrupole coupling between the nuclear quadrupole moment (for spin angular momentum $I > ½$) and the lattice via an electric field gradient, which is electrical in nature. When this coupling exists, it is generally more efficient than any magnetic coupling. Relaxation of a quadrupolar nucleus of spin $I = 1$ (i.e. ^2H) will be explicitly addressed. The deuteron has a small quadrupole moment with a coupling constant e^2qQ/h typically 150–250 kHz, large enough so that relaxation is dominated by the quadrupole interaction and small enough so that perturbation theory is applicable. In liquids, the couplings between nuclear spins are greatly reduced by rapid thermal motions of molecules. Since these couplings are weak and comparable to the coupling of the spins with the lattice, one can consider relaxation of individual spins or, at most, groups of spins inside a molecule. For deuterated molecules in liquids, the dipole–dipole coupling between deuterons is much weaker than the quadrupole interaction. As a consequence, one can normally consider a collection of isolated deuteron spins in liquid samples.

Theory

Suppose that an assembly of N identical spin systems is considered. This allows a quantum statistical description of a spin system, for example the kth spin system in the ensemble. If the spin system is in a state with wavefunction or ket $|\psi_k\rangle$, the expectation value of a physical observable given by its operator Q is

$$\langle Q\rangle_k = \langle \psi_k \mid Q \mid \psi_k\rangle \quad [2]$$

NMR spectroscopy deals with the observation of macroscopic observables rather than states of individual spin systems. Thus, one needs to perform an average over the members of the ensemble:

$$\langle Q\rangle = \sum_{k=1}^{N} \langle Q\rangle_k / N \quad [3]$$

In general, the ket $|\psi_k\rangle$ is time dependent and may be expanded using a complete orthonormal basis set of m stationary kets $|\phi_\beta\rangle \equiv |\beta\rangle$:

$$|\psi_k\rangle = \sum_{\beta=1}^{m} C_\beta^k(t)|\beta\rangle \quad [4]$$

where the expansion coefficients C_β^k are time dependent. This leads to

$$\langle Q\rangle = \sum_{\alpha,\beta} \langle \beta \mid Q \mid \alpha\rangle \sigma_{\alpha\beta}(t) \quad [5]$$

where the matrix elements of a density operator σ are defined by

$$\sigma_{\alpha\beta} = \langle \alpha \mid \sigma \mid \beta\rangle = \overline{C_\alpha(t) C_\beta^*(t)} \quad [6]$$

and the bar denotes an ensemble average. Now the density operator is Hermitian and has real eigenvalues. In particular, its diagonal elements $\sigma_{\alpha\alpha}$ represent the probabilities of finding ket $|\alpha\rangle$ (or populations of $|\alpha\rangle$) in $|\psi\rangle$. The equation of motion for σ can easily be obtained from the Schrödinger equation for $|\psi\rangle$,

$$\frac{\mathrm{d}}{\mathrm{d}t}|\psi\rangle = -\mathrm{i}H\,|\psi\rangle \quad [7]$$

where H is an appropriate spin Hamiltonian (in angular frequency units) for the spin system. The result is the Liouville–von Neumann equation for the time dependence of the density operator σ:

$$\frac{\mathrm{d}\sigma}{\mathrm{d}t} = -\mathrm{i}[H, \sigma(t)] \quad [8]$$

A spin system with the Hamiltonian given by

$$H = H_0 + H'(t) \quad [9]$$

is now taken, where H_0 is the static Hamiltonian and $H'(t)$ represents time-dependent spin–lattice coupling. H' is a random function of time with a vanishing time average [i.e. $\overline{H'(t)} = 0$], and H_0 includes the Zeeman interactions, static averages of dipolar and quadrupole couplings, and time-dependent radiofrequency (RF) interactions. Writing σ and H' as $\tilde{\sigma}$ and $\tilde{H}'$ in the *interaction* representation and using the second-order perturbation theory, the time evolution of the density operator can be shown to obey

$$\frac{\mathrm{d}\tilde{\sigma}}{\mathrm{d}t} = -\int_0^t \mathrm{d}t' \overline{[\tilde{H}'(t), [\tilde{H}'(t-t'), (\tilde{\sigma}(t) - \sigma_{\mathrm{eq}})]]} \quad [10]$$

where the bar is now used to indicate an average over all identical molecules in the sample. Using the eigenket basis of the static Hamiltonian H_0 (i.e. $H_0|\alpha\rangle = \alpha|\alpha\rangle$), Redfield has obtained a set of linear differential equations:

$$\frac{\mathrm{d}}{\mathrm{d}t}\tilde{\sigma}_{\alpha\alpha'} = \sum_{\beta\beta'} \exp[-\mathrm{i}(\omega_{\alpha'\alpha} - \omega_{\beta'\beta})t] \times R_{\alpha\alpha'\beta\beta'}[\tilde{\sigma}_{\beta\beta'}(t) - \tilde{\sigma}_{\beta\beta'}(\infty)] \quad [11]$$

where $\omega_{\alpha\beta} = \alpha - \beta$ ($\equiv E_\alpha - E_\beta$), $\tilde{\sigma}_{\beta\beta'}(\infty)$ corresponds to the matrix elements $\tilde{\sigma}_{\beta\beta'}$ at thermal equilibrium, and **R**, the Redfield relaxation supermatrix, is given by

$$R_{\alpha\alpha'\beta\beta'} = U_{\alpha\alpha'\beta\beta'} + U_{\beta'\beta\alpha'\alpha} - \delta_{\alpha'\beta'} \sum_\gamma U_{\gamma\gamma\beta\alpha} - \delta_{\alpha\beta} \sum_\gamma U_{\beta'\alpha'\gamma\gamma}$$

This treatment is closely related to the relaxation theory of Wangsness and Bloch. The U functions are further simplified by examining, for example, $U_{\alpha\alpha'\beta\beta'}$,

$$U_{\alpha\alpha'\beta\beta'} = \int_0^{\Delta t} \mathrm{d}\tau \exp(-\mathrm{i}\omega_{\alpha'\beta'}\tau) G_{\alpha\beta\alpha'\beta'}(\tau) \times \left[\frac{\exp\{\mathrm{i}(\alpha-\beta-\alpha'+\beta')(\Delta t-\tau)\}-1}{\mathrm{i}(\alpha-\beta-\alpha'+\beta')\Delta t}\right] \quad [12]$$

where $G_{\alpha\alpha'\beta\beta'}(\tau)$ denote time correlation functions of a stationary random function $H'(t)$, which is by definition independent of the origin of time, and

$$G_{\alpha\beta\alpha'\beta'}(\tau) = \overline{H'^{*}_{\alpha'\beta'}(t) H'_{\alpha\beta}(t+\tau)} \quad [13]$$

where $H'_{\alpha\beta}(t) = \langle \alpha | H'(t) | \beta \rangle$. Note that the integrand is large only if $\tau << \tau_c$, the correlation time for $G_{\alpha\beta\alpha'\beta'}(\tau)$. Thus, the upper limit (Δt) of the integral can be set to infinity. The bracket in the integrand is a constant ($\Delta t >> \tau$) and equals 1. In this limit, $U_{\alpha\alpha'\beta\beta'}$ become the spectral densities $J_{\alpha\beta\alpha'\beta'}(\omega_{\beta'\alpha'})$ given by

$$J_{\alpha\beta\alpha'\beta'}(\omega_{\alpha'\beta'}) = \int_0^{\infty} \exp(-\mathrm{i}\omega_{\alpha'\beta'}\tau) G_{\alpha\beta\alpha'\beta'}(\tau) \mathrm{d}\tau \quad [14]$$

and the relaxation matrix elements are now given by

$$R_{\alpha\alpha'\beta\beta'} = J_{\alpha\beta\alpha'\beta'}(\omega_{\alpha'\beta'}) + J_{\alpha\beta\alpha'\beta'}(\omega_{\alpha\beta}) - \delta_{\alpha'\beta'} \sum_{\gamma} J_{\gamma\beta\gamma\alpha}(\omega_{\gamma\alpha}) - \delta_{\alpha\beta} \sum_{\gamma} J_{\gamma\alpha'\gamma\beta'}(\omega_{\gamma\alpha'}) \quad [15]$$

Because of the large heat capacity of the lattice relative to that of the nuclear spins, the lattice may be considered at all times to be in thermal equilibrium, while the time-varying spin states, in the absence of a RF field, evolve to thermal equilibrium because of the spin–lattice interactions. When the exponential argument $[(\omega_{\alpha'\alpha} - \omega_{\beta'\beta})$ in Equation [11]] is significantly larger than the spin relaxation rates, the exponential term oscillates rapidly in comparison with the slow variation in the density matrix due to relaxation. As a consequence, the impact of these terms becomes zero. The so-called secular approximation ($\omega_{\alpha'\alpha} = \omega_{\beta'\beta}$) effectively simplifies the equation of motion to

$$\frac{\mathrm{d}}{\mathrm{d}t}\tilde{\sigma}_{\alpha\alpha'} = \sum_{\beta\beta'}{}' R_{\alpha\alpha'\beta\beta'}[\tilde{\sigma}_{\beta\beta'}(t) - \tilde{\sigma}_{\beta\beta'}(\infty)] \quad [16]$$

where the prime on the summation indicates that only terms that satisfy $\omega_{\alpha'\alpha} = \omega_{\beta'\beta}$ are kept. Now the exponentials in front of those $R_{\alpha\alpha\beta\beta}$ terms in Equation [11] are clearly secular. These $R_{\alpha\alpha\beta\beta}$ parameters control the spin–lattice relaxation and are associated with the diagonal elements $\tilde{\sigma}_{\alpha\alpha}$, which specify the probabilities (P_α) that spin states $|\alpha\rangle$ are occupied. The exponentials in front of $R_{\alpha\beta\alpha\beta}$ are also secular. These $R_{\alpha\beta\alpha\beta}$ parameters control the spin–spin relaxation. When only spin–lattice relaxation is considered, the important Redfield terms in the eigenbase representation are limited to the following two types:

$$R_{\alpha\alpha\beta\beta} = 2J_{\alpha\beta\alpha\beta}(\omega_{\alpha\beta}) - 2\delta_{\alpha\beta} \sum_{\gamma} J_{\gamma\beta\gamma\alpha}(\omega_{\gamma\beta})$$
$$R_{\alpha\alpha\alpha\alpha} = -\sum_{\gamma\neq\alpha} 2J_{\gamma\alpha\gamma\alpha}(\omega_{\gamma\alpha}) = -\sum_{\gamma\neq\alpha} R_{\gamma\gamma\alpha\alpha} \quad [17]$$

Now $H'(t)$ in Equation [9] determines what is called the spin relaxation mechanism. As an example, the dipole–dipole Hamiltonian or quadrupolar Hamiltonian with an axially symmetric ($\eta = 0$) electric field gradient tensor is given by

$$H'_{\lambda}(t) = \sum_{m_L} A_{2,m_L} \left\{ D^2_{m_L,0}[\Omega(t)] - \overline{D^2_{m_L,0}} \right\} \quad [18]$$

where the time dependence arises via Euler angles Ω in the Wigner rotation matrices $D^2_{m,n}(\Omega)$, and A_{2,m_L} is defined by

$$A_{2,m_L} = C_{\lambda}\rho_{2,0} T_{2,m_L}$$

with $C_{\lambda}\rho_{2,0} = \sqrt{3/8}\,(e^2qQ/\hbar)$ for the quadrupole ($I = 1$) terms, and for the dipolar Hamiltonian this is $-\sqrt{3/8}\,(\mu_0\gamma_i\gamma_j\hbar/\pi r^3_{ij})$, where γ is the gyromagnetic ratio of a nuclear spin, r_{ij} is the internuclear distance between the spin pair, and μ_0 is the magnetic vacuum permeability. T_{2,m_L}, the spin operators in the laboratory frame, are given for a deuteron by

$$T_{2,0} = \frac{1}{\sqrt{6}}(3I_z^2 - I^2)$$
$$T_{2,\pm1} = \mp\frac{1}{2}(I^{\pm}I_z + I_zI^{\pm})$$
$$T_{2,\pm2} = \frac{1}{2}(I^{\pm})^2$$

A similar set of equations can be written for the case of a pair of $I = \frac{1}{2}$ spins. When the cross-products between spin Hamiltonian matrix elements of different

m_L values can be ignored (e.g. in liquids) where m_L is the projection index of a rank L (=2) interaction Hamiltonian, the spectral densities of Equation [14] become

$$J_{\alpha\beta\alpha'\beta'}(\omega_{\alpha\beta}) = \sum_{m_L} \langle \alpha \mid A_{2,m_L} \mid \beta \rangle \times \langle \alpha' \mid A_{2,m_L} \mid \beta' \rangle^* J_{m_L}(\omega_{\alpha\beta}) \quad [19]$$

where

$$J_{m_L}(\omega) = \int_0^\infty G_{m_L}(\tau) \exp(-\mathrm{i}\omega\tau)\mathrm{d}\tau \quad [20]$$

with

$$G_{m_L}(\tau) = \left\langle \left\{ D^2_{m_L,0}[\Omega(t)] - \overline{D^2_{m_L,0}} \right\} \times \left\{ D^{2*}_{m_L,0}[\Omega(t-\tau)] - \overline{D^{2*}_{m_L,0}} \right\} \right\rangle \quad [21]$$

It should be noted that the $J_{m_L}(\omega)$ are quantities that are obtained from experiments without reference to any molecular dynamics model. Now, Equation [11] can be transformed back to the Schrödinger representation:

$$\frac{\mathrm{d}}{\mathrm{d}t}\sigma_{\alpha\alpha'}(t) = \mathrm{i}[\sigma, H_0]_{\alpha\alpha'} + \sum_{\beta\beta'} R_{\alpha\alpha'\beta\beta'}[\sigma_{\beta\beta'}(t) - \sigma_{\beta\beta'}(\infty)] \quad [22]$$

The first term on the right-hand side describes spin precessions and is only important for spin–spin relaxation. According to Redfield, the above equation is valid provided that the relaxation elements are small in comparison to the inverse correlation time τ_c^{-1} of the thermal motion, i.e.

$$\frac{1}{R_{\alpha\alpha'\beta\beta'}} \gg \Delta t \gg \tau_c$$

where Δt represents the time interval over which the density matrix of the spin system has not appreciably changed. Different nuclear spin relaxation mechanisms (i.e. quadrupole, dipole–dipole, spin–rotation, chemical-shift anisotropy, and scalar spin–spin relaxation) are surveyed below. The list of relaxation mechanisms is by no means inclusive, but contains the most commonly discussed mechanisms.

Quadrupole relaxation

Let us apply the Redfield theory to a deuteron with its quadrupole moment experiencing a fluctuating electric field gradient arising from anisotropic molecular motions in liquids. When the static average of quadrupole interaction is nonzero, i.e. $\overline{H_Q} \neq 0$, it can be included in the static Hamiltonian H_0. The density operator matrix for a deuteron spin is of the dimension 3 × 3 and the corresponding Redfield relaxation supermatrix has the dimension $3^2 \times 3^2$. When only nuclear spin–lattice relaxation is considered, the spin precession term in Equation [22] is set to zero and the diagonal elements $\sigma_{\alpha\alpha}$ (α = 1, 2, 3) satisfy

$$\frac{\mathrm{d}}{\mathrm{d}t}P_\alpha(t) = \sum_\beta R_{\alpha\beta}[P_\beta(t) - P_\beta(\infty)] \quad [23]$$

where $P_1 \equiv P_1$, $P_2 \equiv P_0$ and $P_3 \equiv P_{-1}$ are the populations in spin states $|1\rangle$, $|0\rangle$ and $|-1\rangle$, respectively (see **Figure 1**), and $R_{\alpha\beta} \equiv R_{\alpha\alpha\beta\beta}$ given in Equation [17]. $R_{\alpha\beta}$ represents the transition probability per second from the spin state β to the spin state α and $R_{\alpha\beta} = R_{\beta\alpha}$. Thus, nuclear spin–lattice relaxation involves transitions induced between nuclear states of different energies by the time-dependent part of quadrupolar interactions $H_Q(t) - \overline{H}_Q$. Solving Equation [23] in terms of linear combinations of the eigenstate populations P_α gives

$$\begin{pmatrix} 1 & 0 & 0 \\ 0 & \exp(-t/T_{1Z}) & 0 \\ 0 & 0 & \exp(-t/T_{1Q}) \end{pmatrix} \times \begin{pmatrix} P_1(0) + P_0(0) + P_{-1}(0) \\ P_1(0) - P_{-1}(0) \\ -P_1(0) + 2P_0(0) - P_{-1}(0) \end{pmatrix} \quad [24]$$

where the deuteron spin–lattice relaxation times T_{1Z} and T_{1Q} for relaxation of the Zeeman and quadrupolar orders, respectively, are

$$T_{1Z}^{-1} = K_Q[J_1(\omega_0) + 4J_2(2\omega_0)] \quad [25]$$

$$T_{1Q}^{-1} = 3K_QJ_1(\omega_0) \quad [26]$$

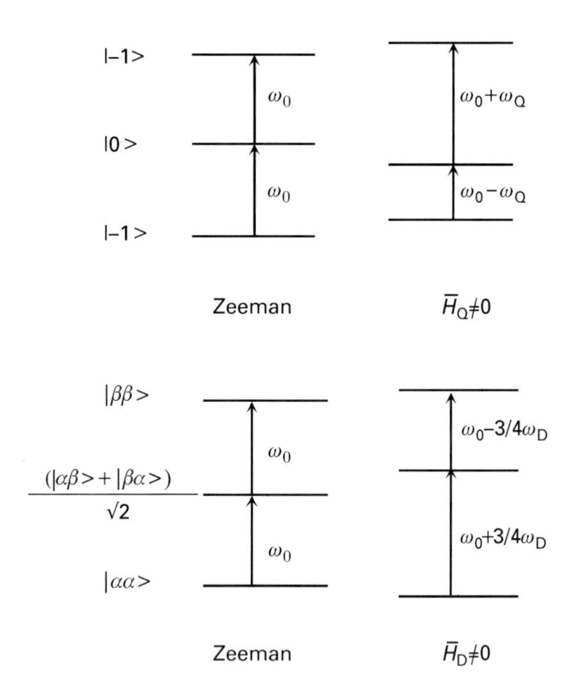

Figure 1 Energy level diagram for a deuteron spin ($\eta = 0$) and for a pair of protons ($I = 1$ triplet; $I = 0$ is not shown) in an external magnetic field. $\omega_0/2\pi$ is the Larmor frequency.

where $K_Q = (3\pi^2/2)(e^2qQ/h)^2$. The asymmetry parameter η of the quadrupolar coupling is assumed to be zero here. T_{1Z} can be measured using an inversion–recovery pulse sequence, while T_{1Q} can be obtained using the Jeener–Broekaert pulse sequence $90^\circ_x - \tau - 45^\circ_y - t - 45^\circ_y$.

When considering spin–spin relaxation, it is necessary to examine the off-diagonal elements $\sigma_{\alpha\beta}$ of the density operator matrix and there are three independent spin–spin relaxation times (T_{2a}, T_{2b}, and T_{2D}):

$$T_{2a}^{-1} = K_Q\left[\frac{3}{2}J_0(0) + \frac{5}{2}J_1(\omega_0) + J_2(2\omega_0)\right]$$
$$T_{2b}^{-1} = K_Q\left[\frac{3}{2}J_0(0) + \frac{1}{2}J_1(\omega_0) + J_2(2\omega_0)\right]$$
$$T_{2D}^{-1} = K_Q[J_1(\omega_0) + 2J_2(2\omega_0)] \qquad [27]$$

The quadrupolar (solid) echo pulse sequence ($90^\circ_x - \tau - 90^\circ_y$) allows measurement of the spin–spin relaxation time T_{2a}. The double-quantum spin–spin relaxation rate T_{2D}^{-1} can be determined using a double quantum spin–echo pulse sequence $90° - \tau - 90° - t_1/2 - 180° - t_1/2 - 90°$. The first two 90° pulses create the double-quantum coherence, which is refocused by a 180° pulse, and the spin–echo is detected by the last monitoring 90° pulse.

Dipole–dipole relaxation

Treatment of spin–lattice relaxation of an isolated spin-$\frac{1}{2}$ pair by an intramolecular dipole-dipole interaction is identical to that for a spin-1 system (see **Figure 1**). Two like spin-$\frac{1}{2}$ nuclei separated by an internuclear distance r are considered. The longitudinal or Zeeman spin–lattice relaxation time T_{1Z} is given by Equation [25], but with K_Q replaced by a different multiplicative constant $K_D = \frac{3}{2}(\mu_0\gamma^2\hbar/4\pi r^3)^2$ which determines the dipolar coupling strength. In solids with isolated spin-$\frac{1}{2}$ pairs, the Jeener–Broekaert sequence can be used to determine the dipolar spin–lattice relaxation time T_{1D} which is the counterpart of T_{1Q} described above. Now T_{1Z} depends on the spectral density of the dipolar interaction fluctuations at the Larmor frequency and twice the Larmor frequency, whereas T_{1D} depends in addition on the spectral density of dipolar fluctuations in the low-frequency region around the line width of dipolar couplings. Thus T_{1D} can be quite sensitive to slow motions, which can significantly contribute to the spectral density at low frequencies. Similarly, the transverse or spin–spin relaxation rate for a spin-$\frac{1}{2}$ pair is, according to T_{2a}^{-1} in Equation [27], given by

$$T_2^{-1} = K_D\left[\frac{3}{2}J_0(0) + \frac{5}{2}J_1(\omega_0) + J_2(2\omega_0)\right] \qquad [28]$$

The spin–lattice relaxation rate ($T_{1\rho}^{-1}$) in the rotating frame is given by

$$T_{1\rho}^{-1} = K_D\left[\frac{3}{2}J_0(2\omega_1) + \frac{5}{2}J_1(\omega_0) + J_2(2\omega_0)\right] \qquad [29]$$

where ω_1/γ is the spin-locking field B_1.

Now suppose the motional process (e.g. rotational Brownian motion in normal liquids) can be described by a single exponential correlation function of the form given in Equation [1]. The corresponding spectral density, which appears in the BPP theory, is a Lorentzian function:

$$J(\omega_0) = \frac{1}{5}\left(\frac{\tau_c}{1 + \omega_0^2\tau_c^2}\right) \qquad [30]$$

where τ_c is a correlation time for the rotational motion. Hence,

$$T_1^{-1} = \frac{1}{5}K_D\left(\frac{\tau_c}{1 + \omega_0^2\tau_c^2} + \frac{4\tau_c}{1 + 4\omega_0^2\tau_c^2}\right) \qquad [31]$$

$$T_2^{-1} = \frac{1}{5}K_D\left(\frac{3}{2}\tau_c + \left(\frac{5}{2}\right)\frac{\tau_c}{1+\omega_0^2\tau_c^2} + \frac{\tau_c}{1+4\omega_0^2\tau_c^2}\right) \quad [32]$$

In **Figure 2**, a sketch of these two equations as a function of τ_c is shown. As seen in this Figure, $T_1 = T_2$ in the extreme narrowing limit ($\omega_0\tau_c << 1$), which is observed in isotropic liquids. Note also that T_1^{-1} goes through a maximum at $\omega_0\tau_c \sim 1$.

The above equations are developed for the rotational motion which modulates *intra*molecular dipole–dipole interactions in isotropic liquids. Translational motion can also cause spin relaxation by modulating *inter*molecular dipole–dipole interactions. Again for a pair of like spin-$\frac{1}{2}$ nuclei on two different molecules, the relaxation rate T_{1T}^{-1} due to translation involves different correlation times and somewhat different spectral densities:

$$T_{1T}^{-1} = \frac{9}{8}(\mu_0\gamma^2\hbar/4\pi)^2[J_1(\omega_0) + 4J_2(2\omega_0)] \quad [33]$$

This relaxation mechanism is important when ^{1}H NMR is used to study protonated samples and samples that contain paramagnetic impurities like oxygen. As the gyromagnetic ratio of an electron is about 1000 times greater than that of a proton, a small amount of paramagnetic impurity can drastically reduce the T_1 values in a sample.

For the case of an unlike spin-$\frac{1}{2}$ pair in a molecule (e.g. ^{13}C–^{1}H), the Zeeman spin–lattice relaxation time for the resonant (I) spin depends on the spectral density at its Larmor frequency (ω_I), at the sum ($\omega_I + \omega_S$) and difference ($\omega_I - \omega_S$) of the two Larmor frequencies for the spins I and S:

$$T_1^{-1} = \frac{1}{3}K'_D[J_0(\omega_I - \omega_S) + 3J_1(\omega_I) + 6J_2(\omega_I + \omega_S)] \quad [34]$$

where $K'_D = \frac{3}{2}(\mu_0\gamma_I\gamma_S\hbar/4\pi r^3)^2$.

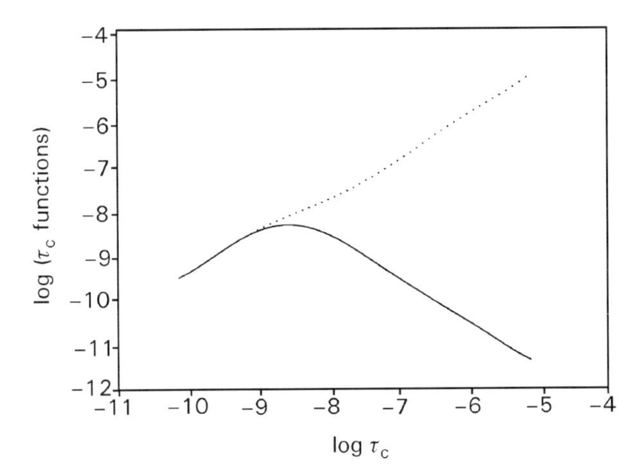

Figure 2 Plot of the τ_c functions in Equation [31] (solid curve) and Equation [32] (dotted curve) as a function of the correlation time τ_c for $\omega_0/2\pi = 100$ MHz.

Spin–rotation relaxation

A rotating molecule or a rotating group of atoms is a rotating charge system, thereby it can generate a magnetic field at the resonant nucleus. The fluctuating magnetic field depends on the magnitude and/or orientation of the angular momentum vector of the rotating molecule. Spin–rotation interactions can be an effective relaxation mechanism if the timescale of the fluctuating magnetic field at the nucleus is comparable to the inverse of the Larmor frequency. This is often encountered as the collisions among small molecules become infrequent. This mechanism is, therefore, important for gases, small molecules, and freely rotating groups ($-CH_3$, $-CF_3$). For molecules undergoing isotropic rotation in liquids, the longitudinal relaxation rate due to the spin–rotation interaction is

$$T_1^{-1} = \frac{2\pi kT}{\hbar^2} I_m C_{eff}^2 \tau_J \quad [35]$$

where k is the Boltzmann constant, T the absolute temperature, I_m the moment of inertia of the molecule, τ_J the angular momentum correlation time, and $C_{eff}^2 = \frac{1}{3}(C_{\parallel}^2 + 2C_{\perp}^2)$ with $C_{\parallel}$ and $C_{\perp}$ being the principal elements of the spin–rotation interaction tensor **C** parallel and perpendicular to the symmetry axis of the molecule, respectively. While the correlation time τ_c for molecular rotation decreases with increasing temperature, τ_J becomes longer. They are related by the simple relation:

$$\tau_c\tau_J = \frac{I_m}{6kT} \quad [36]$$

Hence it is a characteristic feature for the spin–rotation relaxation to show an opposite temperature dependence to that observed for the other relaxation mechanisms.

Chemical-shift anisotropy relaxation

The magnetic field at a resonant nucleus depends on the applied B_0 field, as well as the screening of the applied field at the nucleus by the surrounding electrons. As the shielding (or chemical shift) tensor is anisotropic, its magnitude fluctuates with time due to modulations by molecular rotations. The fluctuating magnetic field at a nucleus from the chemical-shift anisotropy can cause spin relaxation. This relaxation mechanism depends on the square of the applied B_0 field and is therefore important at high fields. It also tends to dominate for nuclei exhibiting large chemical-shift ranges. For molecules with axial symmetry, one can obtain:

$$T_1^{-1} = \frac{2}{15}\gamma^2 B_0^2(\sigma_{\|} - \sigma_{\perp})^2 \frac{\tau_c}{1 + \omega^2\tau_c^2} \quad [37]$$

$$T_2^{-1} = \frac{2}{15}\gamma^2 B_0^2(\sigma_{\|} - \sigma_{\perp})^2 \left[\frac{2\tau_c}{3} + \left(\frac{1}{2}\right)\frac{\tau_c}{1 + \omega^2\tau_c^2}\right] \quad [38]$$

where $\sigma_{\|}$ and $\sigma_{\perp}$ are the principal components of the chemical-shift tensor parallel and perpendicular to the symmetry axis of the molecule, respectively. In contrast with the dipole–dipole relaxation, the longitudinal and transverse rates do not tend to the same value in the limit of short τ_c, as seen in the above equations.

Scalar relaxation

For spin–spin (scalar) coupling between spins I and S, where $I = \frac{1}{2}$ and $S \geq \frac{1}{2}$, the interaction Hamiltonian involves the scalar coupling tensor **J**. The scalar relaxation of the nucleus I can arise from either a time-dependent S or a time-dependent **J** ('first kind'). If the relaxation time ($T_1(S)$) of the nucleus S is short compared with $1/J$, where J is the scalar coupling constant, the nucleus I sees the average of the spin–spin interaction and does not show the expected multiplet, but a single line. The scalar relaxation correlation time τ_s is then equal to $T_1(S)$, and the scalar spin–spin contributions to the longitudinal and transverse relaxation times of the I nucleus are given by

$$T_1^{-1} = \frac{8\pi^2 J^2}{3} S(S+1)\frac{\tau_s}{1 + (\omega_I - \omega_S)^2\tau_s^2} \quad [39]$$

$$T_2^{-1} = \frac{4\pi^2 J^2}{3} S(S+1)\left[\tau_s + \frac{\tau_s}{1 + (\omega_I - \omega_S)^2\tau_s^2}\right] \quad [40]$$

Scalar relaxation of the first kind involves collapse of spin multiplets. When $T_1(S)$ is primarily due to the quadrupole relaxation for $S > \frac{1}{2}$, this is often referred to as scalar coupling of the second kind. In the above equations, the denominators involving $\omega_I - \omega_S$ become very large when the Larmor frequencies ω_I and ω_S are very different. In this case, the scalar relaxation becomes unimportant for T_1, but still exists for T_2 due to the frequency-independent term in Equation [40].

Spectral density of motion

As mentioned above, the evaluation of correlation functions or spectral densities is a daunting task for any relaxation theory. To further complicate the matter, different motional (or relaxation) processes can simultaneously occur in the material being studied by nuclear spin relaxation. However, the observed relaxation rate can often be given by

$$T_1^{-1} = T_{1a}^{-1} + T_{1b}^{-1} + T_{1c}^{-1} + \cdots \quad [41]$$

provided that different relaxation mechanisms labelled by the subscripts a, b, c occur at very different timescales. Otherwise, possible couplings between these processes may also exist and their contributions to relaxation must be properly treated.

In the above discussion of the quadrupole and dipole–dipole relaxation, the relaxation rates are written in terms of spectral densities for general applications. As an example, the reorientation correlation functions ($g_{mn}(t)$) for molecules rotating in an anisotropic medium are calculated using a rotational diffusion model. The rotational diffusion equation, which involves a rotational diffusion operator (Γ) and also contains the pseudopotential for reorienting molecules, must first be solved to get the conditional probability that a molecule has a certain orientation at time t given it has a different orientation at time $t = 0$. This, together with the equilibrium probability for finding the molecule with a certain orientation, is required to work out $g_{mn}(\tau)$. In general, the orientational correlation functions can be written as a sum of decaying exponentials:

$$g_{mn}(t) = \sum_K (\beta_{mn}^2)_K \exp[(\alpha_{mn}^2)_K t] \quad [42]$$

where m and n represent the projection indices of a rank 2 tensor in the laboratory and molecular frames, respectively; $(\alpha_{mn}^2)_K/\rho$, the decay constants, are the eigenvalues of the rotational diffusion Γ matrix and $(\beta_{mn}^2)_K$, the relative weights of the exponentials, are the corresponding eigenvectors. In this model, the decay constants contain the model parameters $D_{\parallel}$ and $D_{\perp}$ specifying rotational diffusions of the molecule about its long axis and perpendicular to the long axis. The spectral densities for a deuteron residing on the rigid part of a uniaxial molecule are the Fourier transform of the orientational correlation functions ($m = 0$, 1, or 2) to give

$$J_m(m\omega) = \sum_n [d^2_{n,0}(\beta_{M,Q})]^2 \sum_K \frac{(\beta^2_{mn})^2_K(\alpha^2_{mn})^2_K}{(\alpha^2_{mn})^2_K + m^2\omega^2} \quad [43]$$

These can now be substituted into Equations [25–27] to obtain deuteron relaxation rates.

By fitting the experimental spectral densities with the predictions from a certain motional model, its model parameters can then be derived. However, the derived motional parameters are model dependent. It is a price one normally has to pay when using NMR relaxation rates. Justification of NMR model parameters may be obtained by comparing them with those observed by other spectroscopic techniques.

List of symbols

B_1 = RF field along an axis of the rotating frame; C = spin–rotation interaction tensor; $d^l_{mn}(\Omega)$ = reduced Wigner rotation matrix elements; $D^l_{mn}(\Omega)$ = Wigner rotation matrix elements; eQ = nuclear electric quadrupole moment; eq = electric field gradient at nucleus; $g_{mn}(t)$ = reduced time correlation function; $G_m(t)$ = time correlation function of spin coupling tensor; H_0 = static spin Hamiltonian; $H'(t)$ = zero-average, time-dependent spin Hamiltonian; $H_q(t)$ = qth component of time-dependent spin Hamiltonian; I = nuclear spin angular momentum; I_m = moment of inertia; J = scalar coupling constant; $J_n(n\omega_0)$ = spectral density of the nth component of a fluctuating coupling tensor at frequency $n\omega_0$; P_α = population of the spin state $|\alpha\rangle$; r_{ij} = internuclear distance; $\mathbf{R}$ = Redfield relaxation supermatrix; $R_{\alpha\alpha'\beta\beta'}$ = Redfield relaxation supermatrix elements; T_{1Z} = longitudinal or Zeeman spin–lattice relaxation time; T_{1D} = dipolar spin–lattice relaxation time; T_{1Q} = quadrupole spin–lattice relaxation time; T_{1T} = longitudinal relaxation time due to translation; $T_{1\rho}$ = rotating frame spin–lattice relaxation time; $T_{2,m}$ = spin operator tensor; T_2 = transverse or spin–spin relaxation time; γ = nuclear gyromagnetic ratio; η = asymmetry parameter of electric field gradient tensor; σ = chemical-shift tensor; σ = density operator; $\tilde{\sigma}$ = density operator in *interaction* representation; τ_c = correlation time; τ_J = angular momentum correlation time; τ_s = scalar relaxation correlation time; ψ = wavefunction; ω_D = average dipolar coupling expressed as a frequency; ω_0 = Larmor precession frequency; ω_Q = average quadrupole coupling expressed as a frequency; Ω ($\frac{1}{2}\alpha, \beta, \gamma$) = Euler angles; 90°_x, 45°_y = RF pulses producing rotations of 90°, 45° about the x, y axes of the rotating frame.

See also: **Chemical Shift and Relaxation Reagents in NMR; Liquid Crystals and Liquid Crystal Solutions Studied By NMR; NMR in Anisotropic Systems, Theory; NMR Principles; Nuclear Overhauser Effect.**

Further reading

Abragam A (1961) *The Principles of Nuclear Magnetism.* Oxford: Clarendon.

Bloembergen N, Purcell EM and Pound RV (1948) *Physical Review* 73: 679.

Cowan B (1997) *Nuclear Magnetic Resonance and Relaxation.* Cambridge: Cambridge University Press.

Dong RY (1997) *Nuclear Magnetic Resonance of Liquid Crystals,* 2nd edn. New York: Springer.

Farrar TC (1989) *Introduction to Pulse NMR Spectroscopy.* Madison: Farragut Press.

Goldman M (1988) *Quantum Description of High-Resolution NMR in Liquids.* Oxford: Clarendon.

Jeener J and Broekaert P (1967) *Physical Review* **157**: 232–240.

Redfield AG (1965) *Advances in Magnetic Resonance* **1**: 1–32.

Slichter CP (1990) *Principles of Magnetic Resonance,* 3rd edn. New York: Springer.

NMR Spectrometers

John C Lindon, Imperial College of Science, Technology and Medicine, London, UK

MAGNETIC RESONANCE
Methods & Instrumentation

Introduction

After the first observation of nuclear magnetic resonance in bulk phases in 1946 and the realization that it would be useful for chemical characterization, which first came with the discovery of the chemical shift in 1951, it was only a few years before commercial spectrometers were produced. By the end of the 1950s a considerable number of publications on the application of NMR to chemical structuring and analysis problems had appeared, and then during the 1960s and later it became clear that useful information could be obtained in biological systems. Since then, the applications and the consequential instrument developments have diversified and now NMR spectroscopy is one of the most widely used techniques in chemical and biological analysis. The very high specificity, the exploratory nature of the technique without the need to preselect analytes and its nondestructive nature have made it very useful despite its lower sensitivity compared to some spectroscopic methods.

A general description is given of the way in which a modern NMR spectrometer operates, of the various components that go into making a complete system and of the particular role that they play. A block diagram of the components of a high-resolution NMR spectrometer is given in **Figure 1**.

Components and principles of operation of NMR spectrometers

Continuous wave (CW) and Fourier transform (FT) operation

For many years, all commercial NMR spectrometers operated in continuous wave mode. This type of operation required a sweep of the NMR frequency or the magnetic field over a fixed range to bring each nucleus into resonance one at a time. These scans for ^{1}H NMR spectroscopy would take typically 500 s to avoid signal distortion. Since most NMR spectra consist of a few sharp peaks interspersed with long regions of noise, this was a very inefficient process. A fundamental paper by Ernst and Anderson in 1966 pointed out the favourable gain in efficiency that could be obtained by simultaneously detecting all signals. This is achieved by the application of a short intense pulse of RF radiation to excite the nuclei, followed by the detection of the induced magnetization in the detector coil as the nuclei relax. The decaying, time-dependent signal, known as a free induction decay (FID) is then converted to the usual frequency domain spectrum by the process known as Fourier transformation (FT). For speed of implementation, in NMR computers this requires the data to have a number of values that is a power of 2, typically perhaps 16K points for modest spectral widths, up to 128K or even 256K points for wide spectral widths on high-field spectrometers (1K is 1024 or 2^{10} points). Acquisition of a ^{1}H FID requires typically a few seconds and opens up the possibility of adding together multiple FID scans to improve the spectrum signal-to-noise ratio (S/N), since for perfectly registered spectra the signals will co-add but the noise will only increase in proportion to the square root of the number of scans. The S/N gain, therefore, is proportional to the square root of the number of scans. This, for the first time, made routine the efficient and feasible acquisition of NMR spectra of less sensitive or less abundant nuclei such as ^{13}C.

The magnet

The most fundamental component of an NMR spectrometer is the magnet. Originally, this would have been a permanent or electromagnet and these provided the usual configurations for field strengths up to 1.41 T (the unit of magnetic flux density is the tesla (T) equivalent to 10 000 gauss), corresponding to a ^{1}H observation frequency of 60 MHz. Because the sensitivity of the NMR experiment is proportional to about the 3/2 power of the field strength, denoted B_0, there has been a drive to higher and higher magnetic fields. This led the commercial NMR manufacturers to develop stronger electromagnets for NMR spectroscopy that took the highest field strengths to 2.35 T, i.e. 100 MHz for ^{1}H NMR observation. Materials suitable for electromagnets have a maximum saturation field strength at about this value and at this field the current used and the consequential water cooling required was a considerable running expense.

Because of the continued need for even higher strengths, NMR manufacturers have collaborated closely with magnet developers to produce high-resolution magnets based upon superconducting solenoids. The magnetic field is generated by a current circulating in a coil of superconducting wire immersed in a liquid helium dewar at 4.2 K. This bath is shielded from ambient temperature by layers of vacuum and a jacket of liquid nitrogen at 77 K, which is usually topped up at a weekly interval. A liquid helium refill is carried out at approximately 2-month intervals depending on the age and field strength of the magnet. The initial development of superconducting magnets was at 5.17 T, corresponding to 220 MHz for ^{1}H and operated in continuous wave (CW) mode (q.v.)

Until about 1972, this represented the highest field strength, but then at regular intervals the available field strength gradually increased along with the emergence of wider-bore magnets, enabling the incorporation of larger samples. Thus, a 270 MHz spectrometer was produced along with a wide-bore 180 MHz machine, and subsequently the field was increased to allow ^{1}H observation at 360 MHz, 400 MHz, 500 MHz, 600 MHz, 750 MHz; the observation frequency limit of any machine yet delivered to a customer is 800 MHz (mid 1999). The development of such magnets has required new technology in which part of the liquid helium bath is kept at about 2K by an adiabatic cooling unit, thereby allowing higher current to be used in the coils. This approach should lead to higher field strengths being available in

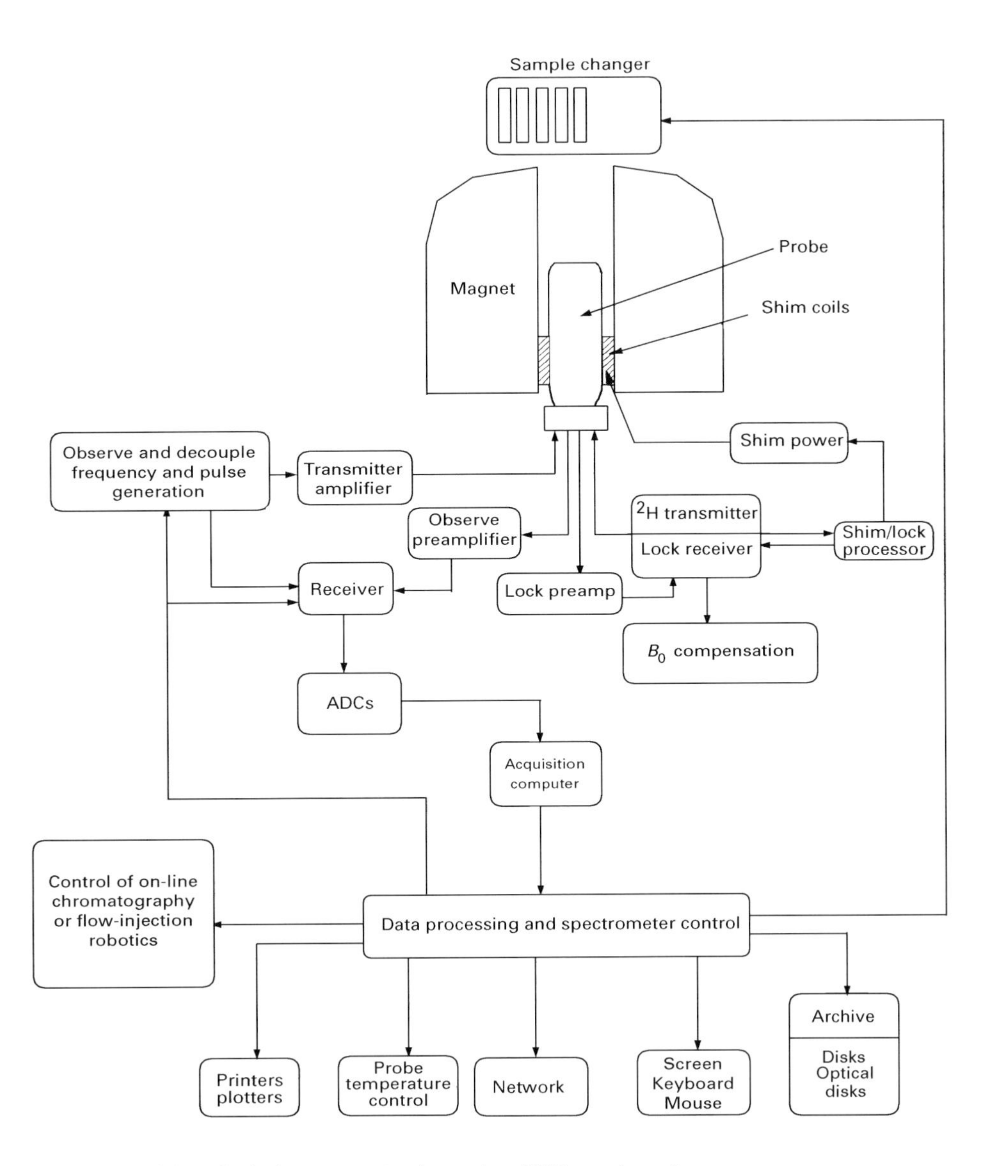

Figure 1 A block diagram of the principal components of a modern NMR spectrometer.

the near future. A photograph of a superconducting magnet designed to give a field of 18.8 T and ^{1}H NMR spectra at 800 MHz is shown in **Figure 2**, indicating the size that such magnets have reached. A modern, recently installed high-field NMR spectrometer using this type of superconducting magnet is shown in **Figure 3**. Nowadays, apart from very basic routine low-field spectrometers used, for example, for monitoring chemical reactions, all NMR spectrometers are based on superconducting magnets.

NMR magnetic field optimization, signal detection and sample handling

Inserted into the magnet is the NMR detector system or probe. High-resolution NMR spectra are usually measured in the solution state in glass tubes of standard external diameters; 5 mm is the most common, but larger ones (10 mm) are used where improved sensitivity is required and sample is not limited. Also a range of narrow and specially designed tubes is available for limited sample studies, including 4 mm, 3 mm, 1.5 mm diameter and even smaller specially shaped cavities such as capillaries or spherical bulbs, plus tubes containing limited-volume cavities where the glass has a magnetic susceptibility tailored to be the same as that of a specific NMR solvent such as D_2O. The probe contains tunable RF coils for excitation of the nuclear spins and detection of the resultant signals as the induced magnetization decays away. A capability exists for measuring NMR spectra over a range of temperatures, typically 125–475 K.

Although modern high-resolution magnets have very high field stability and homogeneity, this is not sufficient for chemical analysis, in that it is necessary to resolve lines to about a width of 0.2 Hz; at 800 MHz this represents a stability of one part in 4×10^9. This performance is achieved in three ways: first by locking the magnetic field to the RF to ensure that successive scans are co-registered; second by improving the homogeneity of the magnetic field; and finally by sometimes spinning the sample tube.

Deuterated solvents are usually used for NMR spectroscopy to avoid the appearance of solvent peaks in the ^{1}H spectrum. Deuterium is an NMR-active nucleus and the spectrometer will contain a ^{2}H channel for exciting and detecting the solvent resonance. Circuitry is provided in the spectrometer for maintaining this ^{2}H signal exactly on resonance at all times by detecting any drift from resonance caused by inherent magnet drift or room temperature fluctuations and for providing an error signal to bring the magnet field back on resonance by applying small voltages through subsidiary coils in the magnet bore. This is known as a 'field-frequency lock' and it means that successive scans in a signal accumulation run are registered exactly.

To improve the homogeneity of the magnet, an assembly of coils is inserted into the magnet bore (shim coils). These consists of about 20–40 coils specially designed so that adjustable current can be fed through them to provide corrections to the magnetic field in any combination of axes to remove the effects of field inhomogeneities. The criterion of the best

Figure 2 A superconducting NMR magnet operating at 18.8 T for ^{1}H NMR observation at 800 MHz demonstrating the size of these state-of-the-art magnets. Photograph courtesy of Bruker Instruments Inc., Billerica, MA, USA. (See Colour Plate 41a).

homogeneity is based upon the fact that when the ^{2}H lock signal is sharpest (i.e. at the most homogeneous field) the signal will be at its highest. The currents in the shim coils are therefore usually adjusted to give the highest lock signal. Alternatively, it is possible, although less common, to 'shim' on the ^{1}H NMR signal. It is possible to 'map' the field inhomogeneities using MRI methods involving magnetic field gradients prior to automatic compensation. This whole process is now largely computer-controlled in modern spectrometers.

NMR spectra are sometimes measured with the sample tube spinning at about 20 Hz to further improve the NMR resolution. This can introduce signal sidebands at the spinning speed and its harmonics, and on modern high-field machines with improved resolution, this is becoming less necessary and is undesirable in some cases.

In analytical laboratories where large numbers of samples have to be processed, automatic sample changers can play a large part in improving efficient use of the magnet time. These devices allow the measurement of up to about 120 samples in an unattended fashion with insertion and ejection of samples from the magnet under computer control. Automatic lock detection and optimization of sample spinning, NMR receiver gain and shimming are also standard. The data are acquired automatically and can be plotted and stored on backing devices. As an additional aid in routine work, it is possible to purchase an automated work bench that will produce the samples dissolved in the appropriate solvent in an NMR tube starting from a solid specimen in a screw-capped bottle and which will also dispose of samples safely and wash the NMR tube. It is possible to foresee the demise of the glass NMR tube in laboratories requiring high sample throughout. This can now be achieved using a flow probe type of NMR detector and automatic sample handling robots taking samples from 96-well plates. This is an extension of the technology used for direct coupling of chromatography, such as HPLC, to NMR spectroscopy.

Figure 3 A modern high-resolution NMR spectrometer. A superconducting magnet is shown at the rear, in this case providing a field of 18.8 T corresponding to a ^{1}H observation frequency of 800 MHz. Behind the operator is the single console containing the RF and other electronics and the temperature-control unit. The whole instrument is computer controlled by the workstation shown at the right. Photograph courtesy of Bruker Instruments Inc., Billerica, MA, USA. (See Colour Plate 41b).

Excitation, detection and computer processing of NMR signals.

The RF signal is derived ultimately from a digital frequency synthesizer that is gated and amplified to provide a short intense pulse. Pulses have to be of short duration because of the need to tip the macroscopic nuclear magnetization by 90° or 180° and at the same time to provide uniform excitation over the whole of the spectral range appropriate for the nucleus under study. Thus for ^{13}C NMR, for example, where chemical shifts can cover more than 200 ppm, this requires 25 kHz spectral width on a spectrometer operating at 500 MHz for ^{1}H, which corresponds to 125 MHz for ^{13}C. To cover this range uniformly requires a 90° pulse to be < 10 μs in duration.

The RF pulse is fed to the NMR probe, which contains one or more coils that can be tuned and matched to the required frequency, this tuning changing from sample to sample because of the different properties of the samples such as the solution dielectric constant. The receiver is blanked off during the pulse and for a short period afterwards to allow the pulse amplifier to recover. The receiver is then turned on to accept the NMR signal that is induced in the coil as the nuclei precess about the field and decay through their relaxation processes. The detection coil is wound on a former as close as possible to the sample to avoid signal losses and is oriented with its axis perpendicular to the magnetic field. In a superconducting magnet the sample tube is aligned along the field, and this coil axis is therefore at right angles to the field and a simple solenoid, which would provide the best S/N, is not possible. Consequently most detector coils are of the saddle type.

The weak NMR signal is amplified using a preamplifier situated as close to the probe as possible, and then also in the main receiver unit where it is mixed with a reference frequency and demodulated in several stages to leave the FID as an oscillating voltage in the kHz range. This signal is then fed to an analogue-to digital converter (ADC) and at this point the analogue voltage from the probe is converted into a digital signal for data processing. ADCs are described in terms of their resolution, usually in terms of the number of bits of resolution: a typical high-field NMR FID is digitized to a resolution of 16 bits or one part in 2^{16} or 65536. This digital signal can then be manipulated to improve the S/N ratio or the resolution by multiplying the FID by an appropriate weighting function before the calculation of the digital Fourier transform.

If only one ADC is used to collect the NMR FID, it is not possible to distinguish frequencies that are positive from those that are negative with respect to the pulse frequency. For this reason, the carrier frequency used to be set to one edge of the spectral region of interest to make sure that all of the NMR frequencies detected were of the same sign. This had the disadvantage of allowing all of the noise on the unwanted side of the carrier to be aliased onto the noise in the desired spectral region, hence reducing the final S/N by $\sqrt{2}$. To overcome this problem it is general practice now to collect two FIDs, separated in phase by 90°, either using two ADCs or multiplexing one ADC to two channels. This approach allows the distinction of positive and negative frequencies and means that the carrier can be set in the middle of the spectrum and the hardware filters can be correspondingly reduced in width by a factor of 2, giving an increase in S/N by $\sqrt{2}$. This process is termed quadrature detection.

In modern NMR spectrometers, the electronics are largely digital in nature, thus providing greater opportunities for computer control and manipulation of the signals. This includes the use of oversampling and digital filtering to improve the dynamic range of the signal acquisition.

Modern NMR spectrometers usually have two separate computer systems. One is dedicated to the acquisition of the NMR FID and operates in the background so that all necessary accurate timing requirements can be met. The FID is transferred, either at the end of the acquisition or periodically throughout it to enable inspection of the data, to the host computer for manipulation by the operator. These computers are based on modern operating systems such as UNIX. The computer software can be very complex, using multiple graphics windows on remote processors, and can, like any modern package, take advantage of networks, printers and plotters. Typical operations include manipulations of the signal-averaged FID by baseline correction to remove DC offset; multiplication by continuous functions to enhance S/N or resolution; Fourier transformation; phase correction; baseline correction of the frequency spectrum; calculation and output of peak lists; calculation and output of peak areas (integrals); and plotting or printing of spectra. It is common to have a separate computer workstation solely for data inspection and manipulation, networked to the host computer. This may be the same model as the host computer but is often an industry-standard model from a third-party supplier. NMR data processing software can also be purchased from a number of companies other than the instrument manufacturers, and these often have links to document production software or provide output of NMR parameters for input into other packages such as those for molecular modelling. A number of approaches alternative to the use of FID weighting functions for improving the quality of the NMR data have been developed and are available from software suppliers. These include such methods as maximum entropy and linear prediction, and indeed it is now possible to purchase these as supplementary items from some NMR manufacturers.

Multiple-pulse experiments and multidimensional NMR

Everything described so far applies to the basic one-dimensional NMR experiment in which the nuclear spin system is subjected to a 90° (or less) pulse and the FID is collected. A wide variety of experiments are reported in the literature and are routinely applied to measure NMR properties such as relaxation times T_1, T_2 and $T_{1\rho}$, which can be related in some cases to molecular dynamics. These experiments involve the use of several pulses separated by timed variable delays and are controlled by pulse programs written in a high-level language for ease of understanding and modification. The computer system will have software to interpret the data and calculate the relaxation times using least-squares fitting routines.

Such pulse programs are also used to enable other special one-dimensional experiments such as saturation or nonexcitation of a large solvent resonance (these are different in that the former method will also saturate NH or OH protons in the molecules under study through the mechanism of chemical exchange), or the measurement of nuclear Overhauser enhancement (NOE) effects which are often used to provide distinction between isomeric structures or to provide estimates of internuclear distances. Pulse programs are also used for measuring NMR spectra of nuclei other than ^{1}H and sometimes in order to probe connectivity between protons and the heteronucleus. In this case, pulses or irradiation can be applied on both the heteronucleus and ^{1}H channels in the same experiment. The commonest use is in ^{13}C NMR where all spin–spin couplings between the ^{13}C nuclei and ^{1}H nuclei are removed by ‘decoupling’. This involves irradiation of all of the ^{1}H frequencies while observing the ^{13}C spectrum. In order to cover all of the ^{1}H frequencies, the irradiation is provided as a band of frequencies covering the ^{1}H spectral width; this is consequently termed noise decoupling or broad-band decoupling. Alternatively, it is possible to obtain the effect of broad-band decoupling more efficiently by applying a train of pulses to the ^{1}H system, this being known as composite pulse decoupling.

Recently, a whole family of experiments have been developed that detect low-sensitivity nuclei such as ^{13}C or ^{15}N indirectly by their spin coupling connectivity to protons in the molecule. This involves a series of pulses on both ^{1}H and the heteronucleus but allows detection at the much superior sensitivity of ^{1}H NMR. Special probes have been developed for such ‘indirect detection’ experiments in which the ^{1}H coil is placed close to the sample, and the heteronucleus coil is placed outside it, the opposite or ‘inverse geometry’ to a standard heteronuclear detection probe.

The one-dimensional NMR experiment is derived from measuring the FID as a function of time. If the pulse program also contains a second time period which is incremented, then a second frequency axis can be derived from a second Fourier transform. This is the basis for ‘two-dimensional’ NMR and its extension to three or even four dimensions. For example, a simple sequence such as

$$90^\circ(^1\text{H})-t_1-90^\circ(^1\text{H})-\text{collect FID for time } t_2$$

where t_1 is an incremented delay, results after double Fourier transformation with respect to t_1 and t_2 in a spectrum with two axes each corresponding to the ^{1}H chemical shifts. This is usually viewed as a contour plot with the normal 1D spectrum appearing along the diagonal and any two protons that are spin coupled to each other giving rise to an off-diagonal contour peak at their chemical shift coordinates. This simple experiment is one of a large family of such correlation experiments involving either protons alone or heteronuclei. The extension to higher dimensions has already been exploited to decrease the amount of overlap by allowing spectral editing and the spreading of the peaks into more than one dimension. Hardware and software in modern NMR spectrometers allows this wide variety of experiments.

The increasingly complex pulse sequences used today rely on the ability of the equipment to produce exactly 90° or 180° pulses or pulses of any other angle. One way to do this is to provide trains of pulses that have the desired net effect of, for example, a 180° tip but which are compensated for any mis-setting. An example of such a ‘composite pulse’ is $90^\circ_x - 180^\circ_y - 90^\circ_x$, which provides a better inversion pulse than a single 180° pulse. Many complex schemes have been invented both for observation and for decoupling (especially for low-power approaches that avoid heating the sample). A universal approach to removing artefacts caused by electronic imperfections, and one which is also used to simplify spectra by editing out undesired components of magnetization, is the use of 'phase cycling'. This allows the operator to choose the phase of any RF pulse and of the receiver, and cycling these in a regular fashion gives control over the exact appearance of the final spectrum.

So far only pulses that excite the whole spectrum (hard pulses) have been described. For spectral editing purposes or to prove some NMR spin connectivity, it can be very convenient to perturb only part of spectrum, possibly only that corresponding to a given chemical shift or even one transition in a multiplet. This approach is achieved by using lower-power pulses applied for a longer period of time (e.g. a 10 ms 90° pulse will only cover 25 Hz). Such selective pulses are often not rectangular as are hard pulses but can be synthesized in a variety of shapes such as sine or Gaussian because of their desirable excitation frequency profiles. Modern research spectrometers can include such selective, shaped pulses in pulse programs.

Instruments for special applications

NMR of solids

Although ^{1}H high-resolution NMR spectroscopy is possible in the solid, most applications have focused on heteronuclei such as ^{13}C. High-resolution studies rely on very short pulses, so high-power amplifiers are necessary. Similarly, because of the need to decouple ^{1}H from ^{13}C and thereby to remove dipolar interactions not seen in the liquid state, high-power decoupling is required. However, the major difference between solution and solid-state high-resolution NMR studies lies in the use of 'magic-angle spinning' (MAS) in the latter case. This involves spinning the solid sample packed into a special rotor at an angle of 54°44′ to the magnetic field. This removes broadening due to any chemical shift anisotropies that are manifested in the solid-state spectrum and any residual ^{1}H–^{13}C dipolar coupling not removed by high-power decoupling. Typical spinning speeds are 2–6 kHz or 120000–720000 rpm although higher speeds up to 25 kHz, at which the rotor rim is moving at supersonic velocity, are possible and necessary in some cases.

For nuclei with spin $> \frac{1}{2}$, MAS is insufficient to narrow the resonances and more complicated double angle spinning (DAS) or double orientation rotors (DOR) are necessary.

NMR imaging

A whole new specialized subdivision of NMR has arisen in the allied disciplines of NMR imaging (magnetic resonance imaging or MRI) and NMR spectroscopy from localized regions of a larger object. MRI applications range from the analysis of water and oil in rock obtained from oil exploration drilling to medical and clinical studies, and spectroscopic applications include the possibility of measuring the ^{1}H or ^{31}P NMR spectrum from a particular volume element in the brain of a living human being and relation the levels of metabolites seen to a disease condition. Some experiments on smaller samples can be carried out in the usual vertical-bore superconducting magnets, but studies are more often performed in specially designed horizontal-bore magnets with a large, clear bore capable of taking samples up to the size of adult human beings. Because of their large bore, they operate at lower field strengths compared to analytical chemical applications, and typical configurations are 2.35 T with a 40 cm bore or 7.0 T with a 21 cm bore. Clinical imagers generally utilize magnetic fields up to 2 T with a 1 m bore. Imaging relies upon the application of magnetic field gradients to extra coils located inside the magnet bore in all three orthogonal axes including that of B_0 and excitation using selective RF pulses. Virtually all clinical applications of MRI use detection of the ^{1}H NMR signal of water in the subject, with the image contrast coming from variation of the amount of water or its NMR relaxation or diffusion properties in the different organs or compartment being imaged. Very fast imaging techniques have been developed that allow movies to be constructed of the beating heart or studies of changes in brain activity as a result of visual or aural stimulation to be conducted.

Benchtop analysis

Specialist tabletop machines can be purchased and these are used for routine analysis in the food and chemical industries. They operate automatically, typically at 20 MHz for ^{1}H NMR, using internally programmed pulse sequences, and are designed to give automatic printouts of analytical results such as the proportion of fat to water in margarine or the oil content of seeds.

Future trends

NMR spectroscopy has shown a ceaseless trend in improvements in S/N, field strength, new types of pulse experiments and computational aspects. This trend is not slowing down and, with the rapid advances in computers, it is probably accelerating; it is therefore difficult to predict NMR developments in the long term. However, some recent research developments mentioned below will certainly break through into commercial instruments.

Higher magnetic field strengths 800 MHz detection for ^{1}H NMR is the current (mid 1999) commercial limit and the first machines at this field have now been delivered and 900 MHz systems are being developed. Higher fields must be on the way and clearly an emotive figure would be the 1 GHz ^{1}H NMR spectrometer. This development will require the design of transmitter and detection technology working at or beyond the limit of RF methods and investigation of new superconducting materials for the magnets. Although the higher field strengths provide greater spectral dispersion and yield better sensitivity, it may be that some applications involving heavier nuclei are less suited to such fields because of the field dependence of certain mechanisms of nuclear spin relaxation, which could cause an increased line broadening and hence lower peak heights and delectability. It has been demonstrated that cooling the NMR detector to liquid helium temperature has the effect of improving the S/N by up to about 500%. This will have an even more dramatic effect on sensitivity than higher magnetic fields.

New NMR pulse experiments Four-dimensional experiments are reported in the literature and developments, through such approaches as selective excitation, allow the reduction of the enormous data matrices that result. This also means that new methods of detecting only the desired information in complex spectra are becoming possible through such approaches as the detection of ^{1}H NMR resonances only from molecules containing certain isotopes of other nuclei. Transfer of the use of pulsed magnetic field gradients has occurred from the MRI field to the high-resolution NMR area and this provides new ways of editing complex spectra with improved data quality and acquisition speed. This technology will find widespread application in the near future, for example in the measurement of diffusion coefficients and other forms of molecular mobility.

Novel data processing The advent of more widespread application of the maximum-entropy technique, where any prior knowledge about the system can be used to advantage, is imminent as the method becomes more widely available. It will probably gain more credence when careful benchmarking and comparisons have been completed. Undoubtedly, it will find application in all areas of NMR spectroscopy.

Coupled techniques The recent coupling of HPLC to NMR has been shown to be of great use in separating and structuring components of complex mixtures such as drug metabolites in body fluids. This technique has been extended to other chromatographic techniques such as supercritical fluid chromatography (SFC) and to the use of nuclei other than ^{1}H or ^{19}F which form the basis of most studies so far because of their high NMR sensitivity. The direct coupling of capillary electrophoresis (CE) and capillary electrochromatography (CEC) to NMR has also been developed and commercial systems based on these approaches will become available. The hyphenation of HPLC with both NMR spectroscopy and mass spectrometry has been achieved and the first commercial systems are now being produced. It is expected that a wealth of applications based on these technologies, such as the identification of drug metabolites, will be forthcoming. Finally, the technology that has led to the direct coupling of separation to NMR spectroscopy is leading to the demise of the glass NMR tube for high-throughput applications and its replacement by flow-injection robots.

List of symbols

B_0 = magnetic field strength [flux density]; T_1 = spin–lattice relaxation times; T_2 = spin–spin relaxation time; $T_{1\rho}$ = spin–lattice relaxation time in the rotating frame.

See also: **Diffusion Studied Using NMR Spectroscopy; Fourier Transformation and Sampling Theory; Magnetic Field Gradients in High Resolution NMR; MRI Theory; NMR Data Processing; NMR Principles; NMR Pulse Sequences; NMR Relaxation Rates; Solid State NMR, Methods; Solvent Suppression Methods in NMR Spectroscopy; Two-Dimensional NMR, Methods.**

Further reading

Ernst RR and Anderson WA (1996) Applications of Fourier transform spectroscopy to magnetic resonance. *Review of Scientific Instruments* 37: 93–102.

Lindon JC and Ferrige AG (1980) Digitisation and data processing in Fourier transform NMR. *Progress in NMR Spectroscopy* **14**: 27–66

Sanders JKM and Hunter BK (1993) *Modern NMR Spectroscopy. A Guide for Chemists*, 2nd edn. Oxford: Oxford University Press.

NMR Spectroscopy in Food Science

See **Food Science, Applications of NMR Spectroscopy.**

NMR Spectroscopy of Alkali Metal Nuclei in Solution

Frank G Riddell, The University of St Andrews, UK

MAGNETIC RESONANCE
Applications

The alkali metals, lithium, sodium, potassium, rubidium and caesium all possess NMR active nuclei, all of which are quadrupolar.

- Lithium has two NMR active isotopes ^{6}Li (7.4%) and ^{7}Li (92.6%), of which ^{7}Li is the isotope of choice due to its higher magnetogyric ratio and natural abundance. Both isotopes are available in isotopically enriched form making NMR tracer studies relatively easy.
- Sodium has only one NMR active nucleus, ^{23}Na (100%).
- Potassium has two NMR active isotopes ^{39}K (93.1%) and ^{41}K (6.9%), of which ^{39}K is the isotope of choice due to its much greater natural abundance and ^{41}K is observable only with the greatest difficulty.
- Rubidium has two NMR active isotopes ^{85}Rb (72.15%) and ^{87}Rb (27.85%), of which ^{87}Rb is the isotope of choice due to its much higher magnetogyric ratio despite its lower natural abundance.
- Caesium has only one NMR active nucleus, ^{133}Cs (100%).

Lithium is important as the treatment of choice for manic depressive psychosis and this has provoked a wide variety of NMR studies in an endeavour to probe its mode of action. Organolithium compounds are used extensively in synthetic organic chemistry and as industrial catalysts, especially in polymerization reactions.

Both sodium and potassium are essential for life. Potassium is the major intracellular cation in most living cells, with sodium having the second highest concentration. These concentrations are generally reversed in the extracellular fluids. The concentration differences across the cellular membrane are maintained by ion pumps, the most important of which is Na/K/ATPase. This enzyme pumps three sodium ions out of the cell and two potassium ions in for the consumption of one molecule of ATP. This enzyme consumes about one-third of the ATP produced in the human body, emphasizing the importance for life of maintaining the concentration gradients of these ions. In addition, large numbers of enzymes require the presence of sodium or potassium for them to function by mechanisms such as symport or antiport. The human need for sodium chloride as a part of the diet is recognized in many proverbs and sayings in common use, and in the word ‘salary’ which is a reminder that salt has in the past been used as a form of payment.

Although the chemistry of rubidium is close to that of potassium it cannot be used as a substitute for potassium in biological systems *in vivo*, although it has been used in studies of perfused organs and cellular systems. The same applies for similar reasons to caesium. These metals can be taken into biological systems where they generally replace potassium, but the ingestion of large amounts of the salts of either metal has severe physiological consequences leading in extreme cases to death.

Many reasons exist, therefore, for the development and implementation of NMR methods for the study of the alkali metals.

Nuclear properties

The nuclear properties of the NMR active isotopes of the alkali metals are presented in **Table 1**.

Table 1 Nuclear properties of the alkali metals

Isotope	*Spin, I*	*Natural abundance (%)*	*Magnetogyric ratio, $\gamma/10^7$(rad T^{-1} s^{-1})*	*Quadrupole moment $Q/10^{-28}(m^2)$*	*NMR frequency, Ξ (MHz)*	*Relative receptivity, D^c*
^{6}Li	1	7.42	3.937	-8×10^{-4}	14.716	3.58
^{7}Li	3/2	92.58	10.396	-4.5×10^{-2}	38.864	1.54×10^3
^{23}Na	3/2	100	7.076	0.12	26.451	5.25×10^2
^{39}K	3/2	93.1	1.248	5.5×10^{-2}	4.666	2.69
^{41}K	3/2	6.88	0.685	6.7×10^{-2}	2.561	3.28×10^{-2}
^{85}Rb	5/2	72.15	2.583	0.247	9.655	43.0
^{87}Rb	3/2	27.85	8.753	0.12	32.721	2.77×10^2
^{133}Cs	7/2	100	3.509	-3×10^{-3}	13.117	2.69×10^2

Ξ is the observing frequency in a magnetic field in which ^{1}H is at 100 MHz.
D^c is the receptivity relative to ^{13}C.
Quadrupole moments Q are the least well determined parameters in this Table.
Data taken from: *NMR and the Periodic Table* (1978) Harris RK and Mann BE (eds) London: Academic Press.

Quadrupolar relaxation and visibility

The NMR spectra of the alkali metals are dominated by the fact that all the isotopes are quadrupolar. Effective use of alkali metal NMR requires an understanding of the resulting quadrupolar interactions and the best ways to make use of them and to avoid their pitfalls. Many of the problems that arise and solutions adopted are similar to those involved with the halogens. Quadrupolar nuclei have an asymmetric distribution of charge which gives rise to an electric quadrupole moment. Apart from when the nucleus is in an environment with cubic or higher symmetry, the quadrupole moment interacts with the electric field gradient (EFG) experienced by the nucleus, giving rise among other things to quadrupolar relaxation. The strength of the quadrupolar interaction between the quadrupole moment (eQ) and the electric field gradient (eq) is given by the quadrupolar coupling e^2qQ/h. This can take from very small values to hundreds of MHz, depending on the magnitudes of Q and q.

In solution, modulation of the EFG at the quadrupolar nucleus by isotropic and sufficiently rapid molecular motions (where $\omega\tau << 1$) leads to relaxation according to the expression:

$$\frac{1}{T_1} = \left(\frac{3}{40}\right)\left[(2I+3)/I^2(2I-1)\right] \times \left(1+\eta^2/3\right)\left(e^2qQ/\hbar\right)^2\tau_c$$

where η is the asymmetry parameter associated with the EFG.

The alkali metal ions in solution are subject to relatively low quadrupolar interactions. This is particularly true for ^{6}Li and ^{7}Li and for ^{133}Cs, which have inherently low quadrupole moments. Indeed ^{6}Li and ^{133}Cs have the two lowest known quadrupole moments and ^{6}Li is often referred to as an 'honorary' spin $\frac{1}{2}$ nucleus. In aqueous solution the ions are solvated by charge dipole interactions with the water molecules. At any one instant the pattern of water molecules around the cation does not have spherical symmetry but is always close to it. Thus the quadrupolar couplings are low but are never zero. Typically, in aqueous solution and in the absence of extraneous influences, both isotopes of lithium show Li^+ line widths of < 1 Hz, Na^+ and K^+ show a line width of *ca.* 12 Hz, both rubidium isotopes show line widths of *ca.* 140–150 Hz and ^{133}Cs shows a line width of *ca.* 1 Hz.

Because of the low values of the quadrupole moments for both isotopes of lithium, dipolar relaxation becomes important. In aqueous solution at ambient temperature dipolar relaxation accounts for over 75% of the relaxation of ^{7}Li ($T_1 \sim 20$ s) and almost 100% of ^{6}Li ($T_1 \sim 170$ s). In D_2O solution with no ^{1}H available for dipolar relaxation and virtually no quadrupolar mechanism available, the relaxation time of ^{6}Li becomes very long ($T_1 \sim 830$ s). In contrast, it appears that despite the low quadrupole moment of ^{133}Cs, dipolar relaxation does not contribute significantly.

In cases where molecular motion is restricted ($\omega\tau$ is not << 1) the situation is more complex. Such cases arise when the alkali metal is bound to the surface of a large molecule such as a protein or membrane surface and thereby has its motion restricted. The quadrupolar interaction with the nucleus shifts the energies of the Zeeman levels according to the square of the quantum number to a first approximation. Thus, the energy level splittings for a nucleus with $I = \frac{3}{2}$ (e.g. ^{7}Li, ^{23}Na, ^{39}K and ^{87}Rb) become as illustrated in **Figure 1**. With rapid isotropic motion, as described above, the multiple line pattern will

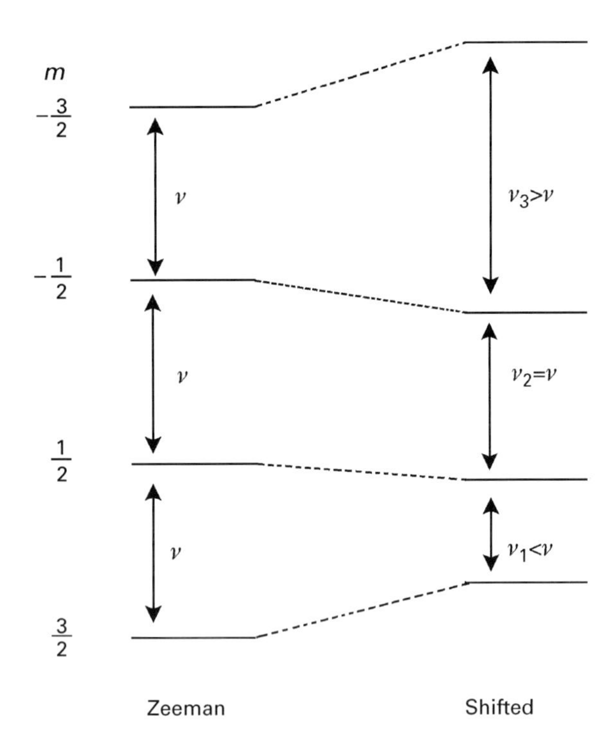

Figure 1 Changes in the energy levels for a spin $\frac{3}{2}$ nucleus subject to a quadrupolar interaction. Note: The shifts in energy levels shown are exaggerated for clarity

collapse into a single line. In the absence of rapid isotropic motion the relaxation rate of the outer transitions, which combine to have 60% of the total intensity, is different from that of the inner transitions, which has 40% of the total intensity. The frequencies of the outer transitions also shift from the inner transition (dynamic frequency shift). There are three principal consequences of these changes for cases where the motion of the cation is restricted. First, line shapes become a double and not a single Lorentzian; secondly, two relaxation times are apparent; and finally, where the more rapid relaxation time becomes very short, partial or total invisibility of the signal from the outer transitions may occur. An excellent review of quadrupolar relaxation effects is given in the review by Springer given in the Further reading section.

Biological applications

Contrast reagents

One of the main problems in using NMR to study the alkali metals in biological systems is that the chemical shifts of the aqueous ions are essentially independent of the ion's surroundings in all cases except for $^{133}Cs^+$, making differentiation of intra- and extracellular cations difficult. This problem can be met by using a contrast reagent (either a shift or a relaxation agent) in one of the compartments, normally extracellular. A large number of aqueous shift reagents have been employed. They all work on the same principle, that a paramagnetic lanthanide, typically dysprosium, is enclosed in a complex by a ligand or ligands, and the resulting complex has several negative charges. With the overall charge on the complex being negative, the alkali metal cations are attracted to the negatively charged species and thereby brought into a region where the paramagnetic interaction induces a chemical shift change. Typical shift reagents for the alkali metals are given in **Table 2**. The resonances of the cations are also broadened by the process, but this broadening has been shown to be largely due the quadrupolar interaction of the cation with the reagent and not due to paramagnetic relaxation. For rubidium, which has a substantial line width that is comparable to the shifts capable of being induced by the best shift reagents, it is preferable to employ a relaxation agent to relax the signal from the extracellular Rb^+ into the baseline noise.

The first important application of alkali metal shift reagents was the use of dysprosium bistripolyphosphate $(DyP_3O_{10})_2^{7-}$ ($DyPPP_2$) to differentiate between intra- and extracellular ^{23}Na in human erythrocytes. This was soon followed by a similar experiment revealing the intracellular signal from ^{39}K. In both cases it seems as if the intracellular metal ions in human erytrocytes are essentially 100% visible. The spectra obtained for the ^{39}K experiment are shown in **Figure 2**.

The maximum shift generated by the shift reagents varies with the alkali metal according to the number of shells of electrons shielding the nucleus from the paramagnetic centre. For example the maximum shifts available from $DyPPP_2$ are approximately: $^7Li^+$, 40 ppm: $^{23}Na^+$, 20 ppm; $^{39}K^+$, 10 ppm; $^{87}Rb^+$, 4 ppm; $^{133}Cs^+$, 2 ppm.

The shift reagent $DyPPP_2$ is commonly used for *in vitro* systems such as vesicles or with isolated erythrocytes. However, it displays considerable toxicity for *in vivo* systems, in which cases the shift reagent $TmDOTP^{5-}$ (see **Table 2**) is preferred. For example during *in vivo* studies of rat kidneys using $TmDOTP^{5-}$, three $^{23}Na^+$ signals were resolved, corresponding to intracellular Na^+, vascular Na^+ and intraluminal Na^+.

Multiple quantum filtration

It has been shown that ^{23}Na double-quantum-filtered NMR spectroscopy can be used to detect anisotropic motion of bound sodium ions in biological systems. The technique is based on the formation of

Table 2 Shift reagents for alkali metal cations

Shift reagent	*Acid anion*
$[Dy(PPP)_2]^{7-}$	Tripolyphosphate $(P_3O_{10})^{5-}$
$[Dy(DPA)_3]^{3-}$	Dipicolinate
$[Dy(NTA)_3]^{3-}$	Nitrilotriacetate $[N(CH_2COO)_3]^{3-}$
$[Dy(CA)_3]^{6-}$	Chelidamate
$[Dy(THHA)]^{3-}$	Triethylenetetraminehexaacetate
$TmDOTP^{5-}$	Thulium 1,4,7,10-tetraazacyclodecane-1,4,7,10-tetrakismethylenephosphonate

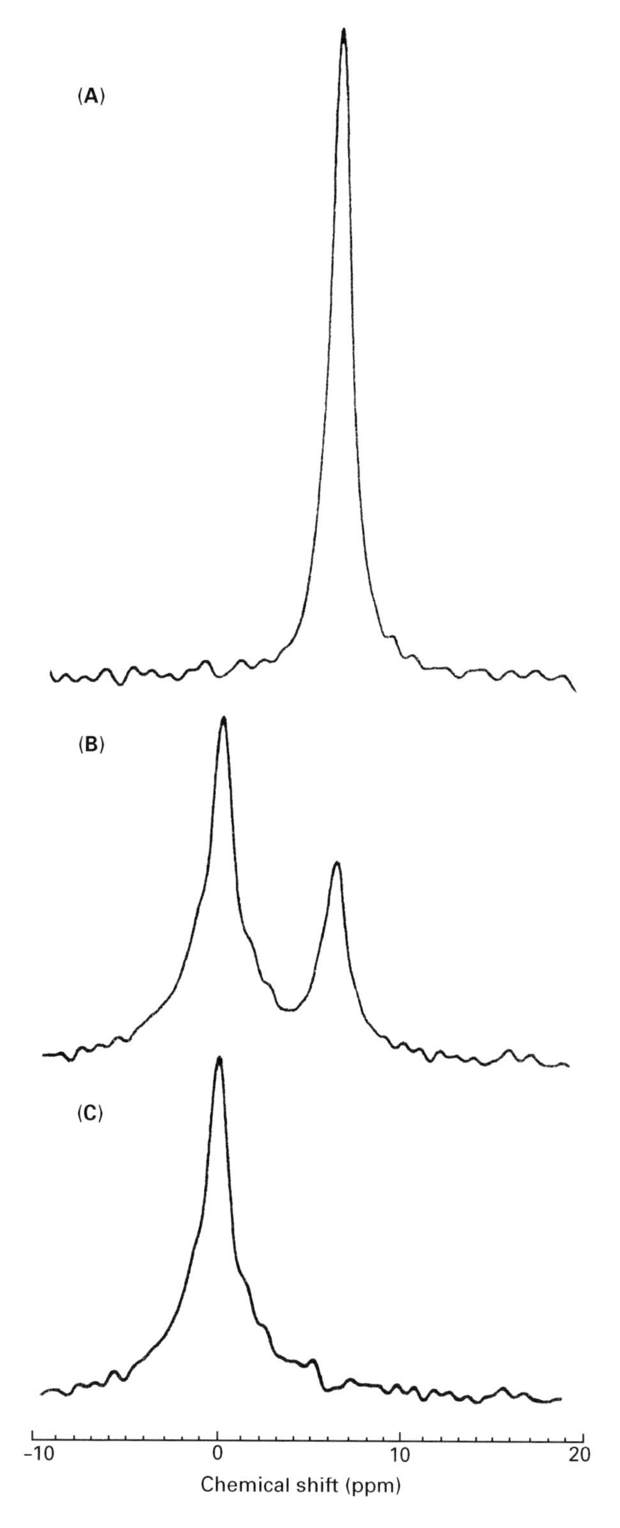

Figure 2 ^{39}K NMR spectra recorded at 16.8 MHz of (A) resuspension medium containing 60 mM K^+, 6 mM Dy^{3+} and 15 mM tripolyphosphate: (B) human erythrocytes in the same medium; and (C) difference spectrum after the subtration of 0.3 of the intensity of spectrum (A) from spectrum (B). For each spectrum 20 000 free induction decays were collected in approx. 20 min. Reproduced with permission of the Biochemical Society from Brophy PJ, Hayer MK and Riddell FG (1983) *Biochemical Journal* **210**: 961.

the second-rank tensor when the quadrupolar interaction is not averaged to zero. Isotropically tumbling ^{23}Na, free in aqueous solution, is not seen by these methods. Such techniques allow, for example, the detection of $^{23}Na^+$ bound to macromolecules such as proteins or membranes or the detection of intracellular $^{23}Na^+$ if its motion is partially restricted. Triple-quantum-filtered spectra can also be used for similar purposes.

Multiple-quantum-filtered NMR offers the possibility of monitoring the intracellular Na^+ content in the absence of shift reagents provided that three criteria are met: (1) the contribution from intracellular $^{23}Na^+$ to the multiple-quantum-filtered spectrum is substantial, (2) that it responds to a change in intracellular $^{23}Na^+$ content and (3) that the amplitude of the extracellular multiple-quantum-filtered component remains constant during a change in intracellular $^{23}Na^+$ content.

Lithium NMR

The use of Li^+ salts as the preferred treatment for manic depressive psychosis has spurred on the use of ^7Li NMR in biological systems, particularly work on cellular systems. The use of the shift reagent $DyPPP_2$ for ^{23}Na and ^{39}K to separate intra- and extracellular signals in human erythrocytes was rapidly followed by similar experiments with $^7Li^+$. The object of these experiments was to determine lithium transport rates across the erythrocyte membrane as a model for the blood–brain barrier. Comparisons were made of the transport rates of $^7Li^+$ into and out of the erythrocytes of manic depressive patients being treated with Li^+, with those of normal controls. These experiments have demonstrated that at extracellular concentrations ranging from 50 to 2 mM the efflux rate from the erythrocytes of the patients was significantly slower than for those of the controls. Moreover, the experiments have shown that the abnormal transport rate is a consequence of Li^+ treatment and is not a marker for the illness. Similar experiments have been carried out with other cellular systems including astrocytomas, neuroblastomas, rat hepatocytes and cultured Swiss Mouse 3T3 fibroblasts. The work on astrocytomas, an immortalized cell line from a human brain cancer, allowed visualization of Li^+ inside the cells which were supported on microcarrier beads (**Figure 3**) and showed that there is an active Li^+ extrusion pump present in these cells and, therefore, that there must also be a Li^+ pump present in astrocytomas in the brain.

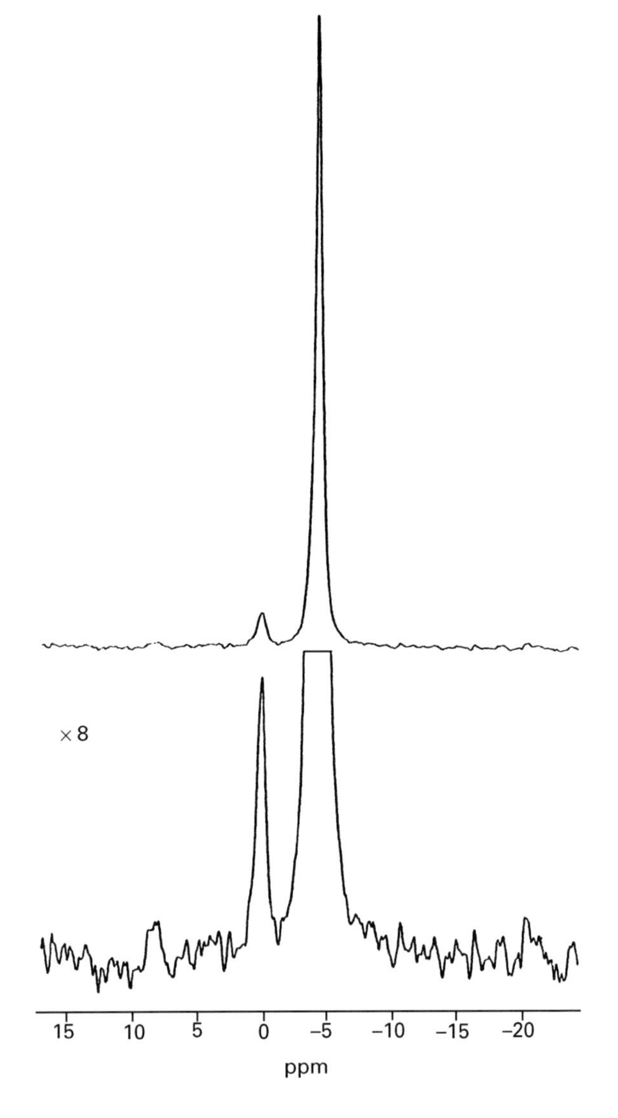

Figure 3 ^7Li NMR spectra of astrocytomas on microcarrier beads in a buffer containing dysprosium tripolyphosphate shift reagent. Each spectrum is the sum of 48 acquisitions recorded at 194 MHz, $[Li^+]_{out}$ = 10 mM. Reprinted from Bramham J, Carter AN and Riddell FG *Journal of Inorganic Biochemistry* **61**: 273–284, copyright 1996, with permission from Elsevier Science.

It is widely believed that the enzyme interacting with Li^+ when it acts to control manic depressive psychosis is inositol monophosphatase. $^7Li^+$ NMR signals from Li^+ bound to the inositol monophosphatase enzyme have been observed. This suggests that $^7Li^+$ NMR could make an important contribution to the study of this and other lithium-sensitive enzymes.

NMR imaging techniques have been applied to study the concentrations and the pharmacokinetics of Li^+ in various parts of the human body. Such experiments have shown that the concentration of Li^+ in the brain and muscle is lower than that in the blood serum. Interestingly, they have also shown that after ingestion of Li^+ there is lag in the uptake of Li^+ into the brain. Maximum concentrations of Li^+ in the brain occur several hours after the concentration maximum in the serum has been reached. Such experiments point towards better treatment regimes for patients.

Sodium, potassium, rubidium and caesium NMR

A very substantial body of literature exists on the use of ^{23}Na and ^{39}K NMR to study cardiac function. Such experiments are typically performed on isolated perfused rat hearts, although hearts from other species including guinea pigs and dogs have also been used. Often the experiments involve shift reagents, although more recently multiple quantum filtration has been extensively used to differentiate between pools of ions. During ischaemia (low or zero blood flow which mimics a heart attack) there is an accumulation of intracellular sodium, cellular swelling and energy deficiency that participate in the transition to irreversible ischaemic injury. Such changes can be followed readily by a combination of ^{23}Na and ^{31}P NMR techniques. These experiments have provided valuable information on the behaviour of hearts under conditions of ischaemia and their recovery afterwards during reperfusion. They provide valuable information on methods for the resuscitation of ischaemic hearts and their protection against hypoxic injury. Although not present in normal biological systems in other than trace amounts, Rb^+ and Cs^+ have been used on several occasions as K^+ analogues in the above experiments, thus extending the range of nuclei available for study. Other similar experiments have been performed on hearts from genetically hypertensive rats. Similar experiments have been performed on other organs such as kidney and liver from small animals.

Studies of $^{23}Na^+$ in cellular systems have been performed on cells such as superfused isolated rat cardiomyocytes, *Methanobacterium thermoautotrophicum*, porcine vascular endothelial cells, the halotolerant bacterium *Brevibacterium* sp., *Escherichia coli*, murine TM3 Leydig and TM4 Sertoli cell lines, and mouse 3T3 fibroblasts. These experiments have been employed to determine the NMR visibility of $^{23}Na^+$, its intracellular concentration, membrane transport properties and dynamics and its ionic mobility inside the cells. ^{23}Na NMR studies have contributed to the study of Na/K/ATPase.

Since the principal cytoplasmic cation is K^+, NMR experiments on $^{39}K^+$ in cellular systems should give valuable information on the intracellular environment. That they have been used much less frequently than experiments on $^{23}Na^+$ is because of the lower receptivity of ^{39}K. The utility of $^{39}K^+$ studies is shown by work on $^{39}K^+$ from *E. coli* after plasmolysis. The $^{39}K^+$ signals are 100% visible and show biexponential relaxation, with both components relaxing very rapidly. The result was attributed to a substantial interaction between the $^{39}K^+$ and the polynegatively charged surface of the ribosomes.

The uptake of Rb^+ into human erythrocytes has been studied by ^{87}Rb NMR using the relaxation agent $LaPPP_2$ to contrast the two pools of Rb^+. Uptake was linear over a 24 h period.

With $^{113}Cs^+$ NMR there is no need for a contrast reagent to separate the intra- and extracellular signals since the chemical shifts of the intra- and extracellular signals are well separated. Variations of the phosphate concentration in the suspension buffer are sufficient for this purpose. Uptake of $^{113}Cs^+$ into human erythrocytes was observed to be linear with a rate of 0.33 mM h^{-1} at an extracellular Cs^+ concentration of 10 mM. When the cells were removed to a Cs^+-free buffer they retained the Cs^+, indicating that there is no transport mechanism available for the removal of Cs^+ from the cells. Cs^+ was shown to replace K^+ inside the cell. The favourable properties of $^{113}Cs^+$ as indicated above, primarily its chemical shift range without the use of shift reagents and its low quadrupolar interactions, have led to its use as an analogue of K^+ in several studies of its tissue compartmentation.

Mediated membrane transport

A variety of NMR methods exists for the study of the mediated transport of alkali metal ions through model biological membranes. Substrates that mediate the transport include the ionophoric antibiotics such as monensin [1], channel forming peptides such as gramicidin and the peptaibols, other channel forming substrates such as amphotericin and the brevitoxins, or synthetic carriers, for example, those based around crown ether-like skeletons such as [2]. For such experiments large unilamellar vesicles (LUV) formed from phospholipid are prepared and a chemical shift difference between the intra- and extravesicular compartments is established by means of a shift reagent. For rapid exchange of ions across the membrane ($k > 10\ s^{-1}$) dynamic line broadening provides information on the transport kinetics (**Figure 4**). For cases where the transport rate is comparable to the relaxation rate a magnetization transfer technique can be employed. In this experiment one of the two signals, normally the extracellular signal, is inverted by a simple pulse sequence. Chemical exchange then causes a time-dependent reduction in the signal of the other resonance. Analysis of signal intensities against time gives the transport kinetics. For relatively slow exchange ($k < 10^{-3}\ s^{-1}$) isotope exchange is used, e.g. $^6Li/^7Li$ or $^7Li/^{23}Na$. In such experiments the concentration gradients of the cations form the driving force for the transport. Such experiments have provided extremely valuable

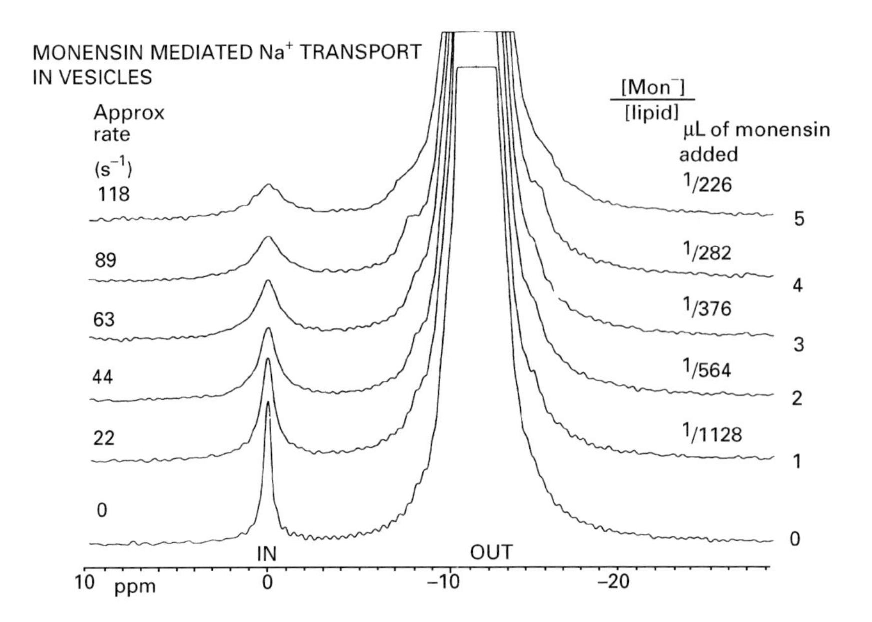

Figure 4 Changes in the ^{23}Na NMR spectra recorded at 21.16 MHz of LUV containing 120 mM NaCl on addition of increasing microlitre amounts of a dilute solution of monensin in methanol at 303 K. The surrounding solution contains 10 mM $Na_5P_3O_{10}$, 70 mM NaCl and 4.0 mM $DyCl_3$. Reprinted from Riddell FG and Hayer MK *Biochimica Biophysica Acta* **817**: 313–317, Copyright © 1985, with kind permission of Elsevier Science-NL, Sara Burgerhartstraat 25, 1055 KV Amsterdam, The Netherlands.

insights into the kinetics and mechanism of mediated transport. For the ionophoric antibiotics such as monensin they have shown that one ionophore transports one alkali metal and that the rate limiting step is almost invariably release of the alkali metal ion at the membrane surface. However, several synthetic ionophores, e.g. ([2], $n = 3$, $R_1 = R_2 = C_{10}H_{21}$), which transports Na^+ at a rate comparable to that of monensin, exhibit diffusion as the rate limiting step. For gramicidin for example, these experiments have confirmed that two molecules are required to come together to form a pore. For the peptaibols the transport has been shown to occur by 'barrel stave' assembly of peptide molecules across the membrane forming a pore inside the 'barrel' that is of sufficiently long duration to allow complete exchange of the intra- and extravesicular media. For the brevetoxins, selectivity for various ions was probed by changing the ions involved in the concentrations gradients. The dependence on cholesterol incorporation in the membrane was studied.

So-called 'bouquet molecules', based on a central crown ether or cyclodextrin unit equipped with pendant arms that are also capable of complexing cations and are long enough to traverse a lipid bilayer, have been studied in vesicles with a Na^+ / Li^+ gradient across the membrane using both ^7Li and ^{23}Na NMR. Such systems show a one-for-one exchange of Na^+ for Li^+ (antiport). These molecules were found to transport Na^+ at similar rates in fluid- and gel-state membranes; this suggests that ion passage occurs preferentially by a channel mechanism and not by the carrier mechanism. Monensin, known to operate as a carrier, was shown to transport at a slower rate in a gel-state membrane.

Another interesting aspect of these experiments is their ability to probe the effect of changes in the membrane composition on the transport kinetics. Thus, placing positive and negative charges on the membrane surface causes changes in ionophore mediated transport rates, and the introduction of pharmaceuticals such as chlorpromazine and imipramine cause changes in the nigericin mediated Na^+ transport rates.

Chemical applications

Covalently bound lithium

Lithium covalently bound to carbon may be observed by ^7Li NMR in lithium alkyls. In such molecules ^7Li has a small chemical shift range (~12 ppm). Tables of chemical shifts and coupling constants are to be found in the review by Günther.

Metal ion complexation studies

Although biological applications have been the major use of alkali metal ion NMR, it has also proved to be valuable for the study of complexation of the alkali metals in host–guest systems by suitably designed ligands, e.g. the crown ethers and cryptands. Two parameters are important in detecting complexation: chemical shift changes and decreases

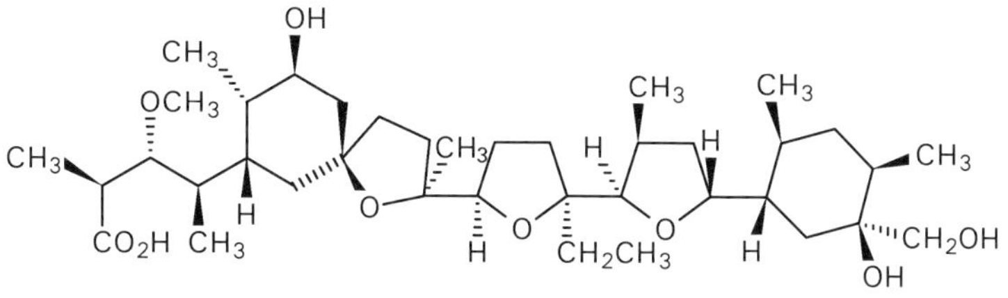

[1]

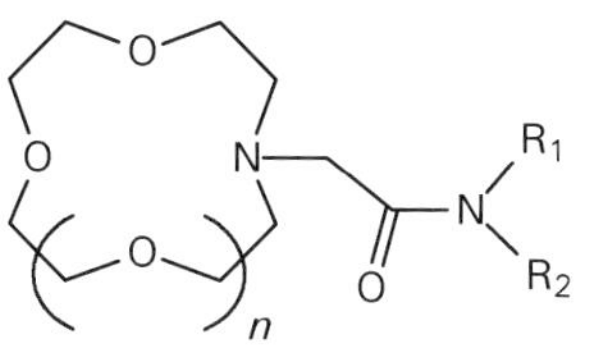

[2]

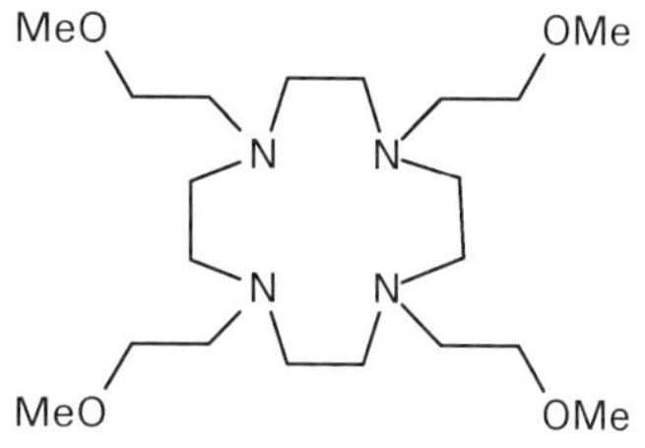

[3]

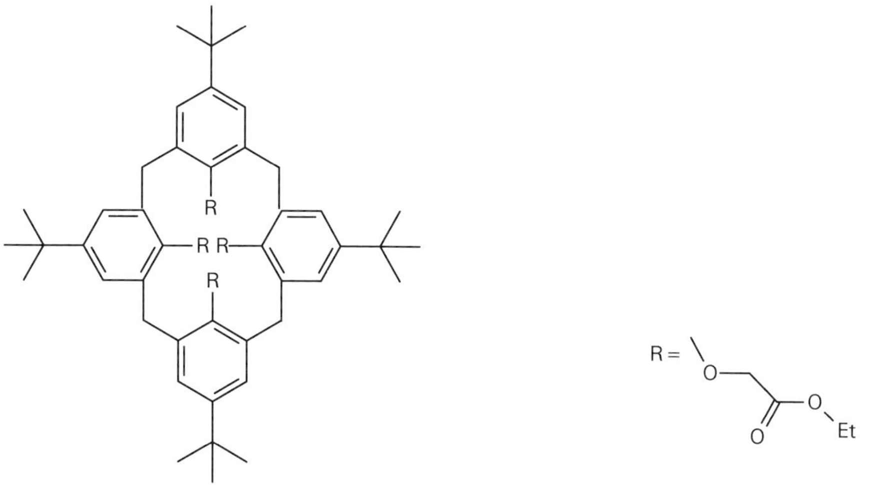

[4]

in relaxation times as a result of enhanced quadrupolar interactions.

Often dynamically broadened alkali metal NMR spectra can be seen as a result of exchange between the free and complexed cation. A good example is provided by the dynamic ^{7}Li and ^{23}Na spectra for the interaction of Li^{+} and Na^{+} with the pendant arm macrocycle 1,4,7,10-tetrakis(2-methoxyethyl)-1,4,7,10-tetraazacyclododecane [3]. The dynamic ^{7}Li spectra are shown in **Figure 5**. Evidence of a slowly exchanging 1:1 complex and of a 2:1 complex in rapid equilibrium with the 1:1 complex between calixarene [4] and Na^{+} is provided by studies of this system by ^{23}Na and ^{1}H NMR.

Frequently, when the alkali metal ion is exchanging between the complex and the solution, the temperature variation of T_1 and/or T_2 for the metal can give information about the exchange kinetics. The complexed ion has a much shorter T_1 (and T_2) value due to strong quadrupolar interactions with the ligand. In the slow exchange limit the observed T_1 value approaches that of the ion free in solution, whilst in the rapid exchange limit the T_1 value is an average of the values for the complexed and free

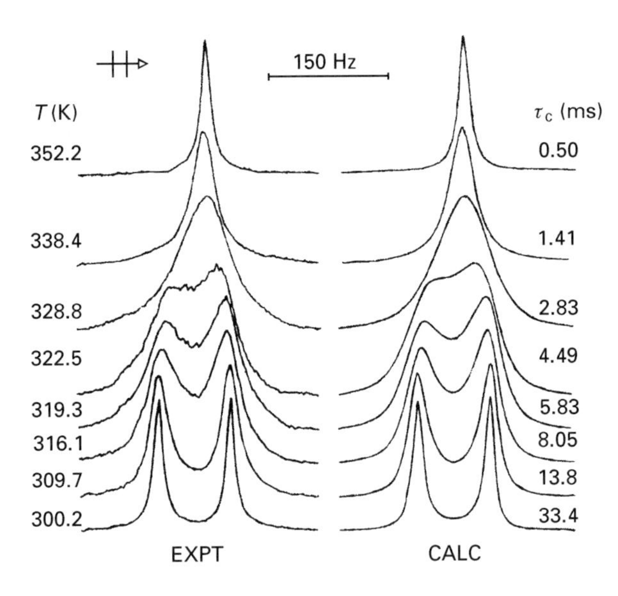

Figure 5 Typical exchange-modified ^7Li NMR spectra recorded at 116.59 MHz of a dimethylsulfoxide solution of solvated Li^+ and [3]. Experimental temperatures and spectra appear on the left of the figure and the best fit calculated line shape and lifetime values on the right. Complexed $^7Li^+$ appears as the left-hand side, high frequency, peak. Reprinted with permission from Stephens AKW, Dhillon RS, Madbak SE, Whitbread SL and Lincoln SF *Inorganic Chemistry* **35**: 2019–2024, copyright 1996, American Chemical Society.

ions. In between these extremes T_1 follows a sigmoid curve when plotted against $1/T$.

Alkali metal anions in solution: alkalide ions

Under conditions of the most rigorous purity and using high vacuum techniques in dipolar aprotic and similar solvents, the alkali metals yield metal anions (M^-) known as the alkalides. Thus sodium in hexamethylphosphoric triamide solutions gives rise to sodide (Na^-) ions. Sodium and rubidium in 1,4,7,10-tetraoxacyclododecane (12-crown-4) give sodide and rubidide ions (**Figure 6**). These anions are most readily identified by their NMR spectra which occur substantially to low frequency of the chemical shift standards of the alkali metal salts in D_2O solution. The alkalide ions are formed by the addition of one electron to the partially filled outermost S orbital. This is expected to lead to a substantial shielding increase, the observation of which is a good indication of the ion formation. Chemical shifts of the alkalide anions vary slightly with solvent and temperature but are near the following values: Na^-, −62 ppm, K^-, −103 ppm, Rb^-, −191 ppm, Cs^-, −280 to −300 ppm.

The spectra of the sodide ion and the potasside ion at low temperatures show relatively sharp line widths, indicative of low ion–solvent interactions suggesting that these ions in solution resemble those

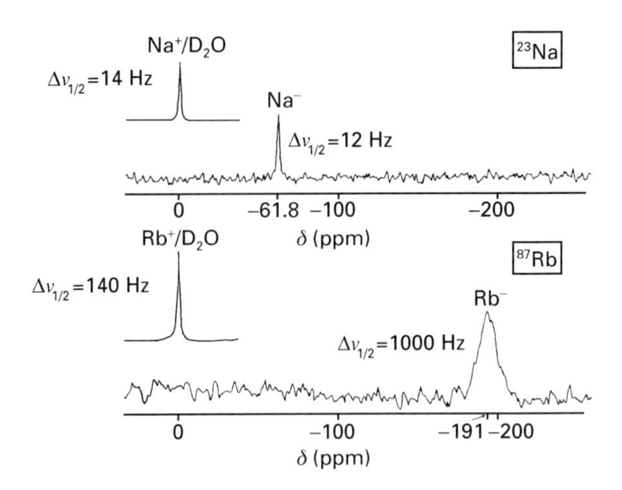

Figure 6 ^{23}Na and ^{87}Rb NMR spectra of solutions of sodium and rubidium in 1,4,7,10-tetraoxacyclododecane (12-crown-4). Negative chemical shift values correspond to a decrease in resonance frequency and an increase in nuclear shielding. Reproduced with permission of The Royal Society of Chemistry from Holton DM, Edwards PP, Johnson DC, Page CJ, McFarlane W and Wood B (1984) *Journal of the Chemical Society, Chemical Communications*, 740–741.

in the gas state. On the other hand, at room temperature the line width of the rubidide ion (~1000 Hz vs. 140 Hz for Rb^+ in H_2O) indicates that there is quadrupolar broadening and the observed shift falls short of that calculated for a gaseous-like ion, unlike the shifts of the sodide and potasside ions. This strongly suggests that there are interactions between the solvent and the rubidide ion. In the case of caesium dissolved in crown ethers the species Cs^+e^- has also been observed.

List of symbols

eq = electric field gradient strength; e^2qQ/h = quadrupolar coupling; eQ = quadrupole moment strength; I = spin quantum number; T_1 = longitudinal relaxation time; T_2 = transverse relaxation time; τ = correlation time for molecular motion (s); ω = Larmor frequency (rad s^{-1}).

See also: **Biofluids Studied By NMR; Cells Studied By NMR; Halogen NMR Spectroscopy (excluding ^{19}F); *In Vivo* NMR, Applications, ^{31}P; *In Vivo* NMR, Applications, Other Nuclei; Membranes Studied By NMR Spectroscopy; NMR Relaxation Rates; Perfused Organs Studied Using NMR Spectroscopy; Relaxometers.**

Further reading

Bramham J and Riddell FG (1994) Cesium uptake studies on human erythrocytes, *Journal of Inorganic Biochemistry* 53: 169–176.

Brophy PJ, Hayer MK and Riddell FG (1983) Measurement of intracellular potassium ion concentration by NMR, *Biochemical Journal* **210**: 961–963.

Edwards PP, Ellaboudy AS and Holton DM (1985) NMR spectrum of the potassium anion K^-, *Nature (London)* **317**: 242–244.

Edwards PP, Ellaboudy AS, Holton DM and Pyper NC (1988) NMR studies of alkali anions in non-aqueous solvents. *Annual Reports on NMR Spectroscopy* **20**: 315–366.

Günther H (1996) Lithium NMR, In: Grant DM and Harris RK (eds) *Encyclopedia of Nuclear Magnetic Resonance*, p. 2807, Chichester: Wiley.

Laszlo P (1996) Sodium-23 NMR, *In*: Grant DM and Harris RK (eds) *Encyclopedia of Nuclear Magnetic Resonance*, p. 4551. Chichester: Wiley.

Lindman B and Forsén S (1978) The Alkali Metals, *In*: Harris RK and Mann BE (eds) *NMR and the Periodic Table*, London: Academic Press.

Mota de Freitas D (1993) Alkali metal nuclear magnetic resonance, *Methods in Enzymology* **227**: 78–106.

Riddell FG (1998) Studying biological lithium using nuclear magnetic resonance techniques, *Journal of Trace and Microprobe Techniques* **16**: 99–110.

Sherry AD and Geraldes CFGC (1989) Shift reagents in NMR spectroscopy in lanthanide probes, *In*: Bünzli J-CG and Chopin GR (eds) *Life, Chemical and Earth Sciences, Theory and Practice*, Amsterdam: Elsevier.

Springer CS (1996) Biological systems, spin-3/2 nuclei. *In*: Grant DM and Harris RK (eds) *Encyclopedia of Nuclear Magnetic Resonance*, p. 940. Chichester: Wiley.

NMR Spectroscopy, Applications

See **Diffusion Studied Using NMR Spectroscopy; Drug Metabolism Studied Using NMR Spectroscopy; Structural Chemistry Using NMR Spectroscopy, Pharmaceuticals; Biofluids Studied By NMR; Carbo-hydrates Studied By NMR; Cells Studied By NMR; Structural Chemistry Using NMR Spectroscopy, Peptides; Proteins Studied Using NMR Spectroscopy; Nucleic Acids Studied Using NMR; Structural Chemistry Using NMR Spectroscopy, Inorganic Molecules; Structural Chemistry Using NMR Spectroscopy, Organic Molecules.**

NOE

See **Nuclear Overhauser Effect.**

Nonlinear Optical Properties

Georges H Wagnière, University of Zurich, Switzerland
Stanisław Woźniak, A. Mickiewicz University, Poland

ELECTRONIC SPECTROSCOPY
Theory

Nonlinear optical phenomena manifest themselves as special forms of light scattering and refraction. Ordinary, linear, light scattering may be viewed as a quasi-simultaneous absorption and re-emission of a photon of same frequency ω. The event occurs on a very short timescale; for radiation in the UV-visible region, of the order of 10^{-15} s, or 1 fs. Light scattering is not to be confounded with resonant absorption and emission. Atoms and molecules absorb and emit light at particular, selected frequencies; they scatter light within the whole spectrum of electromagnetic radiation. Resonant absorption and emission are connected with a change of energy state of the atom or molecule; light scattering is an elastic process, leaving the atom or molecule in the same state after as before the event. Under conditions that will be specified in more detail below, forms of elastic scattering may also occur in which more than one incident photon are involved. For instance, it may happen that two photons of frequency ω incident on a molecule merge to form a single re-emitted photon of frequency 2ω. This process is called second-harmonic generation and is a nonlinear optical phenomenon.

In the following, the most important nonlinear optical effects are reviewed. The emphasis is laid on the material systems in which they have been observed. Nonlinear optical properties of both organic and inorganic materials are presented in tabular form and commented on.

Basic quantities

The semiclassical description of light scattering attributes the effect to a light-induced (electric) dipole moment $\boldsymbol{p}^{(1)}$ in the molecule that oscillates with the frequency of the incident radiation ω and becomes the source of quasi-immediate re-emission at the same frequency. Mathematically, this is expressed as

$$\boldsymbol{p}^{(1)}(\omega) = \alpha^{(1)}(-\omega;\omega)\cdot\boldsymbol{E}(\omega) \qquad [1a]$$

$E(\omega)$ is the electric field strength of the incident light at the frequency ω, and $\alpha^{(1)}(-\omega;\,\omega)$ the molecular *frequency-dependent polarizability* of first order. The descriptive parenthesis $(-\omega;\,\omega)$ indicates incidence at frequency ω and quasi-simultaneous re-emission at frequency ω. For a macroscopic sample we have the corresponding equation:

$$\boldsymbol{P}^{(1)}(\omega) = \varepsilon_0\,\chi^{(1)}(-\omega;\omega)\cdot\boldsymbol{E}(\omega) \qquad [1b]$$

where $\boldsymbol{P}^{(1)}(\omega)$ is the volume polarization and $\chi^{(1)}(-\omega;\,\omega)$ the macroscopic susceptibility. χ is the appropriate sum of the molecular contributions α.

If two separate radiation frequencies, ω_1 and ω_2, act on a molecule at the same time, we will mainly have separate and distinct scattering at these two basic frequencies. However, if the coherence and intensity of the incident radiation are sufficiently high, we may observe the generation of 'overtones' of frequency $\omega_1+\omega_2$:

$$\boldsymbol{p}^{(2)}(\omega_1+\omega_2) = \alpha^{(2)}(-\omega_1-\omega_2;\omega_1,\omega_2):{}^1\boldsymbol{E}(\omega_1)\,{}^2\boldsymbol{E}(\omega_2) \qquad [2]$$

We notice that the induced polarization $\boldsymbol{p}^{(2)}$ for this second-order effect is proportional to the product of the field strengths ${}^1\boldsymbol{E}(\omega_1)$ and ${}^2\boldsymbol{E}(\omega_2)$. The parenthesis $(-\omega_1-\omega_2;\,\omega_1,\omega_2)$ indicates incidence at frequencies ω_1 and ω_2 and scattering at frequency $\omega_1+\omega_2$.

For $\omega_1=\omega_2$, sum frequency generation becomes tantamount to second-harmonic generation (SHG), i.e. incidence at frequency ω and quasi-simultaneous re-emission at the doubled frequency 2ω:

$$\boldsymbol{p}^{(2)}(2\omega) = \alpha^{(2)}(-2\omega;\omega,\omega):\boldsymbol{E}^2(\omega) \qquad [3]$$

We encounter a situation for which the induced polarization is no longer linearly but now quadratically dependent on the field strength of the incident radiation. Correspondingly, the intensity of the radiation generated at frequency 2ω is also quadratically dependent on the intensity $I(\omega)$. In contrast, the intensity of 'ordinary' light scattering at frequency ω is linearly dependent on $I(\omega)$.

As well as sum frequency generation, the nonlinear optical effect of difference frequency generation is also possible, $\omega_1 - \omega_2$ being generated from ω_1 and ω_2. For $\omega_1 = \omega_2$, we then have optical rectification, namely, the induction of a static electric polarization of frequency 0, by the radiation field:

$$p^{(2)}(0) = \alpha^{(2)}(0; \omega, -\omega):\ {}^{1}E(\omega)\ {}^{2}E(-\omega) \qquad [4]$$

(Mathematically, $E(-\omega)$ is expressed as the complex conjugate of $E(\omega)$.)

Tables 1 and **2** and **Figure 1** summarize a number of distinct nonlinear optical effects, in particular including those of third order arising from the combined influence of three frequencies, ω_1, ω_2, ω_3. These effects are collectively called four-wave mixing, as three incident waves combine coherently to give a fourth resulting one of frequency $\omega_1 \pm \omega_2 \pm \omega_3$. The radiation-induced polarization depends to third order on the electric field strength of the incident radiation, namely, on the triple product ${}^{1}E(\omega_1)\ {}^{2}E(\pm\omega_2)\ {}^{3}E(\pm\omega_3)$. In the particular case where $\omega_1 = \omega_2 = \omega_3$, we may have third-harmonic generation, proportional to $E^3(\omega)$. Correspondingly, the intensity of the third harmonic radiation $I(3\omega)$ depends on $I^3(\omega)$.

The variety of nonlinear optical phenomena is indeed vast, but with increasing order their observation becomes more and more difficult. Conventional, thermal light sources do not produce radiation of sufficient coherence and intensity to induce observable nonlinear effects. The field of nonlinear optics has been made accessible by the laser. Laser light is generated by induced emission and can be pictured as consisting of wave trains oscillating in phase. This allows the generation of tightly bundled, sharply focusable beams of high intensity.

The key quantity to understanding the nonlinear optical response of a given molecule is the corresponding generalized polarizability $\alpha^{(n)}(\ldots)$, sometimes also called molecular susceptibility or, for $n > 1$, hyperpolarizability, where n denotes the order of the effect. The first-order polarizability $\alpha^{(1)}$ is a second-rank tensor. The tensor is symmetric and, therefore, in general has six independent elements, assumed to be defined in a symmetry-adapted molecular reference frame x, y, z. These six different tensor elements manifest themselves in point groups belonging to the triclinic symmetry system: $xx(\equiv \alpha^{(1)}_{xx})$, yy, zz, $xy = yx$, $yz = zy$, $zx = xz$. In the monoclinic system, there occur four independent elements. In higher systems, by symmetry the tensor becomes diagonal. In the cubic system, all three diagonal elements are the same. For an isotropic medium, we obtain a single scalar average.

The second-order polarizabilities $\alpha^{(2)}(\ldots)$ are third-rank tensors. Such tensors in general vanish in centrosymmetric media, as they are parity-odd. In the triclinic symmetry system, there are $3^3 = 27$ independent tensor elements. In a molecule of higher symmetry, some elements become zero, others may

Table 1 Overall classification of nonlinear optical effects

Frequency of incident radiation	Frequency of scattered radiation	Order of effect (n)	Rank of susceptibility tensor	Name/description
ω	ω	1	2	Rayleigh scattering, ordinary refraction
ω_1, ω_2	$\omega_1 + \omega_2$ $\omega_1 - \omega_2$	2	3	Sum-frequency generation Difference-frequency generation
ω_1, ω_2, ω_3	$\omega_1 + \omega_2 + \omega_3$ $\omega_1 + \omega_2 - \omega_3$ $\omega_1 - \omega_2 + \omega_3$ $\omega_1 - \omega_2 - \omega_3$	3	4	Four-wave mixing
ω_1, ω_2, ω_3, ω_4	$\omega_1 \pm \omega_2 \pm \omega_3 \pm \omega_4$	4	5	Five-wave mixing

Table 2 Some important nonlinear optical effects

Frequencies of interacting electric fields	Effect	First experiment
$\omega + \omega \rightarrow 2\omega$	Second-harmonic generation (SHG)	a
$\omega - \omega \rightarrow 0$	Optical rectification (OR)	b
$2\omega - \omega \rightarrow \omega$	Parametric amplification (PA)	c
$\omega + \omega + 0 \rightarrow 2\omega$	Electric field-induced second- harmonic generation (EFISH)	d
$\omega + \omega + \omega \rightarrow 3\omega$	Third-harmonic generation (THG)	d
$\omega + \omega + \omega - \omega \rightarrow 2\omega$	Second-harmonic generation by five-wave mixing	e

[a] Franken PA, Hill AE, Peters CW and Weinreich G (1961) *Physical Review Letters* **7**: 118.
[b] Bass M, Franken PA, Ward JF and Weinreich G (1962) *Physical Review Letters* **9**: 446.
[c] Giordmaine JA and Miller RC, (1965) *Physical Review Letters* **14**: 973.
[d] Terhune RW, Maker PD and Savage CM (1962) *Physical Review Letters* **8**: 404.
[e] Shkurinov AP, Dubrovskii AV and Koroteev NI (1993) *Physical Review Letters* **70**: 1085.

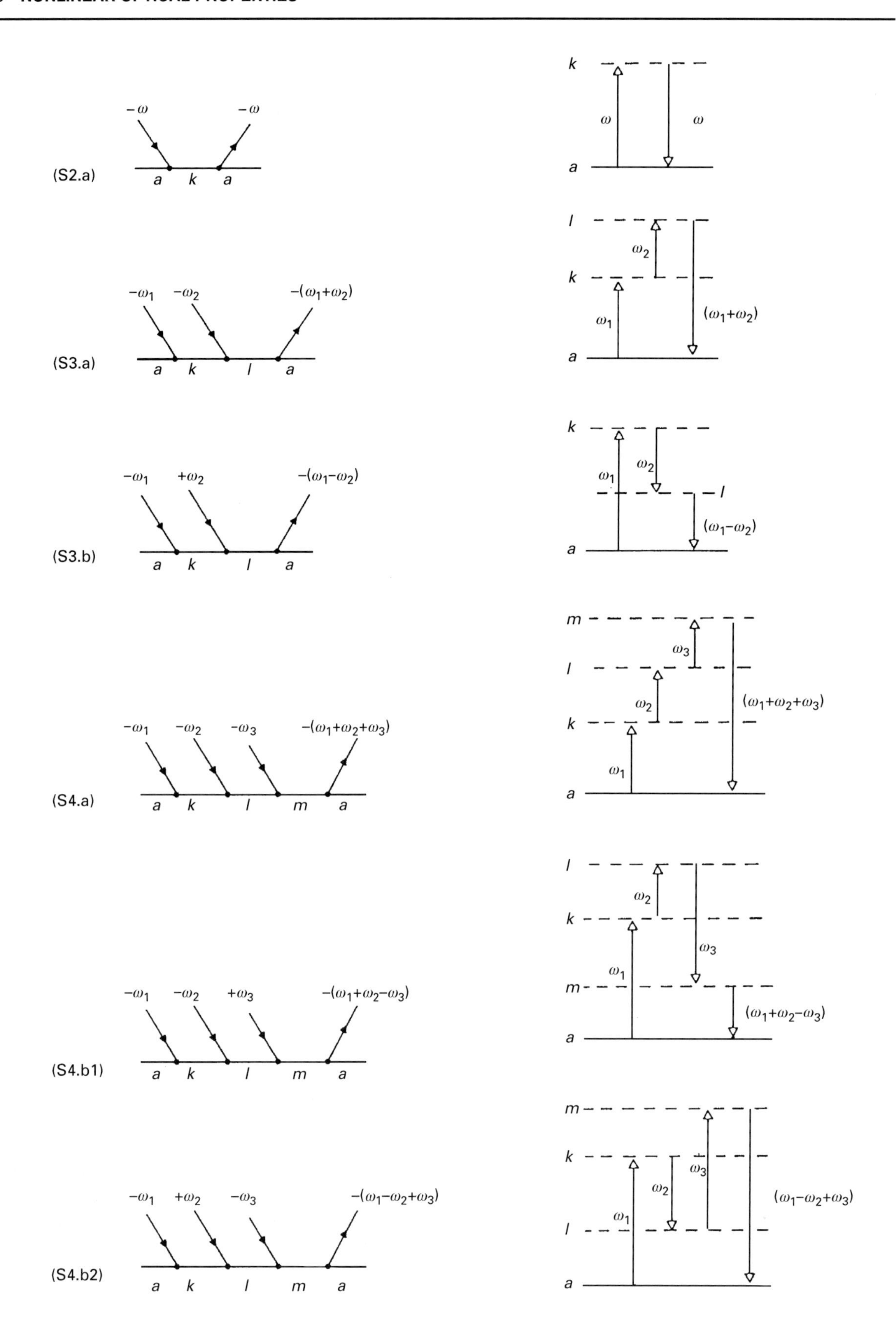

Figure 1 Ward graphs (at left) and ladder graphs (at right) for linear (S2.a), second-order nonlinear (S3.a, b), and third-order nonlinear (S4.a, b1, b2) elastic scattering processes. The broken horizontal lines in the ladder graphs represent virtual, nonstationary states of the molecular system. Reproduced with permission from Wagnière GH (1993) *Linear and Nonlinear Optical Properties of Molecules.* Basel: Verlag Helvetica Chimica Acta.

Table 3 Independent nonvanishing elements of $\chi^{(2)}(-\omega_1-\omega_2;\, \omega_1, \omega_2)$ for crystals of given symmetry classes

Crystal system	*Crystal class*	*Nonvanishing tensor elements*
Triclinic	1	All elements are independent and nonzero
	$\bar{1}$	Each element vanishes
Monoclinic	2	*xyz, xzy, xxy, xyx, yxx, yyy, yzz, yzx, yxz, zyz, zzy, zxy, zyx* (twofold axis parallel to $\hat{y}$)
	m	*xxx, xyy, xzz, xzx, xxz, yyz, yzy, yxy, yyx, zxx, zyy, zzz, zzx, zxz* (mirror plane perpendicular to $\hat{y}$)
	2/*m*	Each element vanishes
Orthorhombic	222	*xyz, xzy, yzx, yxz, zxy, zyx*
	*mm*2	*xzx, xxz, yyz, yzy, zxx, zyy, zzz*
	mmm	Each element vanishes
Tetragonal	4	*xyz* = −*yxz*, *xzy* = −*yzx*, *xzx* = *yzy*, *xxz* = *yyz*, *zxx* = *zyy*, *zzz*, *zxy* = −*zyx*
	$\bar{4}$	*xyz* = *yxz*, *xzy* = *yzx*, *xzx* = −*yzy*, *xxz* = −*yyz*, *zxx* = −*zyy*, *zxy* = *zyx*
	422	*xyz* = −*yxz*, *xzy* = −*yzx*, *zxy* = −*zyx*
	4*mm*	*xzx* = *yzy*, *xxz* = *yyz*, *zxx* = *zyy*, *zzz*
	$\bar{4}2m$	*xyz* = *yxz*, *xzy* = *yzx*, *zxy* = *zyx*
	4/*m*, 4/*mmm*	Each element vanishes
Cubic	432	*xyz* = −*xzy* = *yzx* = −*yxz* = *zxy* = −*zyx*
	$\bar{4}3m$	*xyz* = *xzy* = *yzx* = *yxz* = *zxy* = *zyx*
	23	*xyz* = *yzx* = *zxy*, *xzy* = *yxz* = *zyx*
	*m*3, *m*3*m*	Each element vanishes
Trigonal	3	*xxx* = −*xyy* = −*yyz* = −*yxy*, *xyz* = −*yxz*, *xzy* = −*yzx*, *xzx* = *yzy*, *xxz* = *yyz*, *yyy* = −*yxx* = −*xxy* = −*xyx*, *zxx* = *zyy*, *zzz*, *zxy* = −*zyx*
	32	*xxx* = −*xyy* = −*yyx* = −*yxy*, *xyz* = −*yxz*, *xzy* = −*yzx*, *zxy* = −*zyx*
	3*m*	*xzx* = *yzy*, *xxz* = *yyz*, *zxx* = *zyy*, *zzz*, *yyy* = −*yxx* = −*xxy* = −*xyx* (mirror plane perpendicular to $\hat{x}$)
	$\bar{3}$,$\bar{3}m$	Each element vanishes
Hexagonal	6	*xyz* = −*yxz*, *xzy* = −*yzx*, *xzx* = *yxy*, *xxz* = *yyz*, *zxx* = *zyy*, *zzz*, *zxy* = −*zyx*
	$\bar{6}$	*xxx* = −*xyy* = −*yxy* = −*yyx*, *yyy* = −*yxx* = −*xyx* = −*xxy*
	622	*xyz* = −*yxz*, *xzy* = −*yxz*, *zxy* = −*zyx*
	6*mm*	*xzx* = *yzy*, *xxz* = *yyz*, *zxx* = *zyy*, *zzz*
	$\bar{6}m2$	*yyy* = −*yxx* = −*xxy* = −*xyx*
	6/*m*, 6/*mmm*	Each element vanishes

Reproduced with permission from Boyd RW (1992) *Nonlinear Optics.* Boston: Academic Press.

be equal to each other (see **Table 3**). For instance, in the chiral cubic point group O, the only nonvanishing elements are: $xyz = yzx = zxy = -xzy = -yxz = -zyx$; in the achiral point group T_d: $xyz = yzx = zxy = xzy = yxz = zyx$. For these cases, a second-order nonlinear response can only be detected if ${}^1E$ and ${}^2E$ are nonparallel to one and the same symmetry-adapted coordinate axis. As will be seen in more detail in the next section, in an isotropic medium, the averaged value of $\alpha^{(2)}$ only fails to vanish if the individual molecules of which the medium is composed are chiral, and if the frequencies ω_1 and ω_2 of the incident radiation are different.

The third-order polarizabilities $\alpha^{(3)}$ (…) are parity-even fourth-rank tensors. In the triclinic system, there occur $3^4 = 81$ independent and nonzero tensor elements. The presence of symmetry leads to corresponding simplifications.

The same symmetry considerations, which here are stated for individual molecules, may of course also be applied to macroscopic systems, in particular to crystals. The determining aspect is here the overall crystal symmetry, and instead of the molecular polarizabilities $\alpha^{(n)}$, we consider the bulk susceptibilities of the crystal $\chi^{(n)}$.

Organic media

Organic liquids and solutions

The nonlinear optical properties of solutions of organic molecules have been investigated extensively, although the selection rules for second-order nonlinear optical effects in isotropic liquids are quite restrictive. In order to be noncentrosymmetric, a fluid must consist of, or contain, chiral molecules. Such a chiral medium is 'optically active' and not superposable on its mirror image. Although sum and difference frequency generation are then possible, the important special cases of second-harmonic generation and optical rectification are still forbidden. The respective molecular polarizabilities $\alpha^{(2)}(-2\omega;\, \omega, \omega)$ and $\alpha^{(2)}(0;\, \omega, -\omega)$ vanish upon isotropic averaging.

Second-harmonic generation may be induced in any liquid medium (or gas) if an external static electric field is applied to it, whereby the medium loses its centrosymmetry and the conditions for second-harmonic generation are fulfilled. The generalized polarizability leading to this effect may be expressed as $\alpha^{(3)}(-2\omega; \omega, \omega, 0)$ and is described by a fourth-rank tensor. With electric field strengths applicable in the laboratory, the effect is in general quite small. However, if the liquid is composed of polar molecules (not necessarily chiral), the applied electric field will also partially align them. This then leads to an additional, temperature-dependent, contribution to second-harmonic generation that can be stronger. It is proportional to the molecular dipole moment μ and to an average value of the tensor $\alpha^{(2)}(-2\omega; \omega, \omega)$, often denoted in the literature by β. Electric field-induced second-harmonic generation, in the literature sometimes abbreviated as EFISH, has been widely applied to study solutions of polar organic molecules in nonpolar solvents. To allow extraction of significant molecular data, the interaction between solute molecules should be negligible and the influence of the nonpolar solvent must be taken into account as an averaged correction.

Although second-harmonic generation attained by the EFISH effect is in general weak, the method has been applied widely and successfully. The molecular data so obtained serve as a point of departure for the interpretation of the nonlinear optical properties of molecular crystals and arrays and for the design of novel systems. One observes (see **Tables 4** and **5**) that particularly large quantities for β are found in molecules containing one or more electron-donor substituent(s), such as $-NH_2$ (amino), one or more electron-acceptor substituent(s), such as $-NO_2$ (nitro), bound to a polarizable π electron system (containing conjugated C=C double bonds).

The tensor elements of the molecular quantity $\alpha^{(2)}(-2\omega; \omega, \omega)$ and, therefrom, the averaged quantity β may in principle be calculated quantum mechanically. $\alpha^{(2)}$ may be expressed in terms of the energy levels of the molecule and the electric dipole transition moments between the corresponding quantum states. Exact (*ab initio*) calculations are very cumbersome, but a number of simplified procedures (semiempirical calculations) have been applied to this problem and their results allow a reasonably successful interpretation of the measured results, in particular where strong charge-transfer effects come into play.

Third-order nonlinear optical effects, such as third-harmonic generation or other kinds of four-wave mixing phenomena, occur in all media, irrespective of their symmetry. This follows from the parity-even property of the corresponding tensors. Consequently, third-harmonic generation can be observed both in liquids and gases. Some results on organic molecules are given in **Tables 4** and **5**. In general γ, the dominant component of $\alpha^{(3)}(-3\omega; \omega, \omega, \omega)$, is a small quantity leading to correspondingly small effects.

Organic layers and crystals

Any surface or interface breaks the inversion symmetry and is therefore a possible source of second-order effects. Owing to their surface sensitivity, second-harmonic generation measurements have developed into a very useful tool for probing the orientation of organic molecules in well-structured monolayers, such as those obtainable by the Langmuir–Blodgett technique (see **Table 6**). The surface susceptibility may in general be written as

$$\chi_{\mathrm{S}}^{(2)} = \chi_{\mathrm{SM}}^{(2)} + \chi_{\mathrm{SB}}^{(2)} \qquad [5a]$$

where $\chi_{\mathrm{SM}}^{(2)}$ stands for the part arising from the adsorbed molecules and $\chi_{\mathrm{SB}}^{(2)}$ for the background contribution of the adjoining media. In order to obtain strong signals, the molecules in the layer must themselves be noncentrosymmetric. Often one chooses the adjoining bulk media to be centrosymmetric (air, water, glass; see **Figure 2**). Then

$$\chi_{\mathrm{S}}^{(2)} \approx \chi_{\mathrm{SM}}^{(2)} \qquad \chi_{\mathrm{SM}}^{(2)} \gg \chi_{\mathrm{SB}}^{(2)} \qquad [5b]$$

Among organic crystals, one of the most frequently used for second-harmonic generation is urea, composed of noncentrosymmetric molecules arranged in a noncentrosymmetric fashion, according to the tetragonal space group $P\bar{4}2_1m = D_{2d}^3$ (see **Table 7**). Much attention has been devoted to the design and fabrication of even more efficient media, based on large β values obtained from EFISH experiments. In some cases, such as that of *p*-nitroaniline, β is large, but the molecules crystallize in a centrosymmetric space group, rendering the crystal useless. One strategy to overcome such difficulties consists in making the molecules chiral, thereby forcing them into a noncentrosymmetric crystal structure. From a theoretical point of view, one is interested in relating the bulk susceptibility of the crystal $\chi_{ijk}^{(2)}$ to the susceptibilities of the individual molecules $\alpha_{xyz}^{(2)}$ in their respective positions and orientations in the unit cell. Neglecting intermolecular interaction, this may be written as a sum

$$\chi_{ijk}^{(2)} = \frac{1}{V} L_{ijk} \sum_{s} \sum_{x_s y_s z_s} C_{ix_s} C_{jy_s} C_{kz_s} \alpha_{x_s y_s z_s}^{(2)} \qquad [6]$$

Table 4 Properties of *para*-disubstituted benzenes: Y–C$_6$H$_4$–X

X	Y	Solvent	λ_{max} (nm)[a]	μ (10^{-30} cm)[b]	$\alpha^{(1)}$ (10^{-40}J m^{2}V^{-2})	$\alpha^{(2)} \equiv \beta$ (10^{-50}J m^{3}V^{-3})	$\alpha^{(3)} \equiv \gamma$ (10^{-60}J m^{4}V^{-4})
NO	NMe_2	*p*–Dioxane	407	20.7	23.3	4.44	
NO_2	Me	*p*–Dioxane	272	14.0	17.8	0.78	0.99
NO_2	Br	*p*–Dioxane	274	10.0	20.0	1.22	
NO_2	OH	*p*–Dioxane	304	16.7	16.7	1.11	0.99
NO_2	OPh	*p*–Dioxane	294	14.0	28.9	1.48	1.11
NO_2	OMe	*p*–Dioxane	302	15.3	16.7	1.89	1.23
NO_2	SMe	*p*–Dioxane	322	14.7	21.1	2.26	2.10
NO_2	N_2H_3	*p*–Dioxane	366	21.0	20.0	2.81	1.11
NO_2	NH_2	Acetone	365	20.7	18.9	3.41	1.85
NO_2	NMe_2	Acetone	376	21.3	24.4	4.44	3.46
NO_2	CN	*p*–Dioxane		3.0	18.9	0.22	0.86
NO_2	CHO	*p*–Dioxane	376	8.3	18.9	0.07	0.86
$CHC(CN)_2$	OMe	*p*–Dioxane	345	18.3	26.7	3.63	3.70
$CHC(CN)_2$	NMe_2	$CHCl_3$	420	26.0	31.1	11.85	

[a] λ_{max} denotes the wavelength of the lowest electronic transition; [b] μ denotes the ground-state dipole moment; the other quantities are explained in the text.

Data from Cheng L-T, Tam W, Stevenson SH, Meredith GR, Rikken G and Marder SR (1991) *Journal of Physical Chemistry* **95**: 10631; converted therefrom into SI units (see **Table 9**).

Table 5 Properties of 4,4′-disubstituted stilbenes: Y–C$_6$H$_4$–CH=CH–C$_6$H$_4$–X

X	Y	Solvent	λ_{max} (nm)[a]	μ (10^{-30} cm)[b]	$\alpha^{(1)}$ (10^{-40}J m^{2}V^{-2})	$\alpha^{(2)} \equiv \beta$ (10^{-50}J m^{3}V^{-3})	$\alpha^{(3)} \equiv \gamma$ (10^{-60}J m^{4}V^{-4})
CN	OH	*p*–Dioxane	344	15.0	35.6	4.81	6.42
CN	OMe	$CHCl_3$	(340)	12.7	37.8	7.04	6.67
CN	$N(Me)_2$	$CHCl_3$	382	19.0	43.3	13.33	15.43
NO_2	H	*p*–Dioxane	345	14.0	32.2	4.07	7.53
NO_2	Me	*p*–Dioxane	351	15.7	38.9	5.56	9.51
NO_2	Br	*p*–Dioxane	344	10.7	42.2	5.19	12.10
		$CHCl_3$	(356)	11.3	36.7	6.67	5.56
NO_2	OH	*p*–Dioxane	370	18.3	36.7	6.30	12.84
NO_2	OPh	*p*–Dioxane	350	15.3	46.7	6.67	9.88
NO_2	OMe	*p*–Dioxane	364	15.0	37.8	10.37	9.75
		$CHCl_3$	(370)	15.0	37.8	12.59	11.48
NO_2	SMe	*p*–Dioxane	374	14.3	43.3	9.63	13.95
		$CHCl_3$	(380)	14.3	42.2	12.59	12.35
NO_2	NH_2	$CHCl_3$	402	17.0	35.6	14.81	18.15
NO_2	$N(Me)_2$	$CHCl_3$	427	22.0	37.8	27.04	27.78

For footnotes, see **Table 4**.

where i, j, k denote the coordinate system of the crystal, x_s, y_s, z_s that of the molecule s in the unit cell. L_{ijk} is a local-field correction, V the volume of the unit cell. The trigonometric factors C_{ix_s} relate the molecular coordinate systems to the crystal. This purely additive orientated gas model presents a useful first approximation for the interpretation of data on organic molecules. To refine it, intermolecular interaction in the crystal must be included in the calculation. For crystals of strongly polar molecules, methods based on the dipole–dipole approximation have been successful.

From harmonic generation to parametric amplification

Conservation of photon energy

The photons involved in a nonlinear optical process must fulfil the requirement of energy conservation. For a three-wave mixing effect in which the incident photons are of frequency ω_1, ω_2, leading to an outgoing photon of frequency ω_3, this implies

$$\omega_1 + \omega_2 = \omega_3 \qquad [7]$$

For sum-frequency generation, where $\omega_3 = (\omega_1 + \omega_2)$, this is automatically fulfilled. For difference-frequency generation, where $\omega_3 = (\omega_1 - \omega_2)$, the above equation as such evidently cannot be satisfied; we must write

$$\omega_1 + \omega_2 = \omega_3 + 2\omega_2 \qquad [8]$$

This means that for each incident photon of frequency ω_2 there are two outgoing photons of the same frequency. Simultaneously with the generation of a new wave of frequency $\omega_1 - \omega_2$, the incident wave of frequency ω_2 is parametrically amplified. If the nonlinear medium is placed between two mirrors reflecting at the frequencies ω_2 and (or) ω_3, this parametric effect may be increased. One calls such a device a parametric oscillator (see **Figure** 3). From this point of view, $\omega_1 \equiv \omega_P$ corresponds to the so-called pump wave, $\omega_2 \equiv \omega_S$ to the (amplified) signal wave, and $\omega_3 \equiv (\omega_1 - \omega_2) \equiv \omega_I$ to the idler wave. Equation [8] may be simplified to

$$\omega_1 = \omega_2 + \omega_3, \qquad \text{or} \qquad \omega_P = \omega_S + \omega_I \qquad [9]$$

The fundamental process then appears to be the conversion of a photon of higher frequency ω_P into two photons of lower frequency ω_S and ω_I. Interestingly, this process may go on in a parametric oscillator merely as a result of sending in a pump wave. The signal photons are first generated inside the cavity by spontaneous emission and then coherently amplified. Carried out in this manner, the intensity of

Table 6 Surface susceptibility $\chi^{(2)}(-2\omega;\omega,\omega)$ and molecular second-order nonlinear polarizability $\alpha^{(2)}(-2\omega;\omega,\omega)$ for organic monomolecular layers on water

Molecule	$\chi^{(2)}_{jkj}$ (10^{-20} m V^{-1})	$\alpha^{(2)}_{zzz}$ (10^{-50} J m^3 V^{-3})
$C_8H_{17}(C_6H_4)_2CN$	46[a]	9.2
$C_9H_{19}(C_6H_4)_2CN$	46	9.2
$C_{10}H_{21}(C_6H_4)_2CN$	46	9.2
$C_{12}H_{25}(C_6H_4)_2CN$	46	9.2
$C_{14}H_{29}COOH$	0.21	0.030
$C_{17}H_{35}COOH$	0.17	0.026
$C_{22}H_{45}COOH$	0.17	0.026
$C_{17}H_{35}CH_2OH$	0.25	0.041
$C_{12}H_{25}(C_{10}H_6)SO_3Na$	0.75	0.28
$C_8H_{17}(C_6H_4)_2COOH$	12[b]	2.2
$C_7H_{15}(C_4N_2H_2)C_6H_4CN$	8	3.0
$C_5H_{11}(C_6H_4)_3CN$	15	2.8

Data from Rasing Th, Berkovic G, Shen YR, Grubb SG and Kim MW (1986) *Chemical Physics Letters* **130**: 1 and Berkovic G, Rasing Th and Shen YR (1987) *Journal of the Optical Society of America B* **4**: 945.
Fundamental wavelength $\lambda = 532$ nm.
[a] For surface density 3.0×10^{18} molecules m^{-2}.
[b] For surface density 2.5×10^{18} molecules m^{-2}.

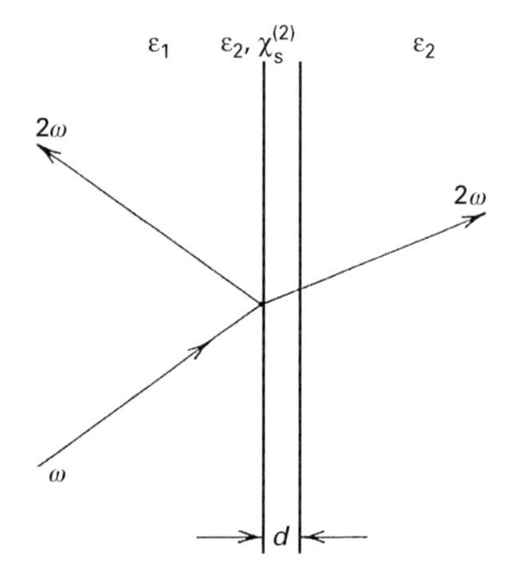

Figure 2 Sketch of second-harmonic generation from an interface between two isotropic media. The interfacial layer of thickness d is specified by a linear dielectric constant ε_2 and a second-order surface nonlinear susceptibility $\chi^{(2)}_s$. Reproduced with permission of John Wiley and Sons from Shen YR (1984). *The Principles of Nonlinear Optics.* New York: © 1984 John Wiley and Sons.

Table 7 Experimental second-order nonlinear optical susceptibilities d_{il} of organic crystals

Crystal	Symmetry	d_{il} (10^{-12} m V^{-1})		Reference
MBBCH (2,6-bis(*p*-methylbenzylidene)-4-*t*-butylcyclohexanone)	Orthorhombic	d_{31}	15	a
	$mm2 = C_{2v}$	d_{32}	12	
		d_{33}	4	
		d_{eff}	12 (I)	
BBCP (2,5-bis(benzylidene)cyclopentanone)	$222 = D_2$	d_{14}	7	a
m-NA (*m*-nitroaniline)	$mm2 = C_{2v}$	d_{31}	13.05	b
		d_{32}	1.09	
		d_{33}	13.72	
		d_{eff}	10.35 (I)	
5NU ($P2_12_12_1$; 5-nitrouracil)	$222 = D_2$	d_{14}	8.7	c
POM (3-methyl-4-nitropyridine-1-oxide)	$222 = D_2$	d_{14}	9.6	d
MNA (2-methyl-4-nitroaniline)	Monoclinic	d_{11}	167.6	e, j
	$m = C_s$	d_{12}	25.1	
		d_{33}, d_{13}, d_{31}	$\sim 10^{-3} d_{11}$	
		d_{eff}	20.8 (I)	
L-PCA (*L*-pyrrolidone-2-carboxylic acid)	Orthorhombic	d_{14}	0.22	f
	$222 = D_2$	d_{eff}	0.20 (I)	
MAP (methyl-(2,4-dinitrophenyl)amino-2-propanoate)	Monoclinic	d_{21}	16.8	g, j
	$2 = C_2$	d_{22}	18.4	
		d_{23}	3.7	
		d_{25}	–0.54	
		d_{eff}	16.3 (I)	
		d_{eff}	8.8 (II)	
Urea ($CO(NH_2)_2$)	Tetragonal	d_{14}	1.4	h, i
	$\bar{4}2m = D_{2d}$			

Fundamental wavelength $\lambda = 1.064$ µm. *Data for different frequencies available.
(I) For type I phase-matched SHG; (II) for type II phase-matched SHG.
[a] Kawamata J, Inoue K and Inabe T (1995) *Applied Physics Letters* **66**: 3102.
[b] Huang G-F, Lin JT, Su G, Jiang R and Xie S (1992) *Optical Communications* **89**: 205.
[c] Puccetti G, Perigaud A, Badan J, Ledoux I and Zyss J (1993) *Journal of the Optical Society of America B* **10**: 733.
[d] Zyss J, Chemla DS and Nicoud JF (1981) *Journal of Chemical Physics* **74**: 4800.
[e] Levine BF, Bethea CG, Thurmond CD, Lynch RT and Bernstein JL (1979) *Journal of Applied Physics* **50**: 2523.
[f] Kitazawa M, Higuchi R, Takahashi M, Wada T and Sasabe H (1995) *Journal of Applied Physics* **78**: 709.
[g] Oudar JL and Hierle R (1977) *Journal of Applied Physics* **48**: 2699.
[h] Catella GC, Bohn JH and Luken JR (1988) *IEEE Journal of Quantum Electronics* **24**: 1201.
[i] Halbout J-M, Blit S, Donaldson W and Tang CL (1979) *IEEE Journal of Quantum Electronics* **QE-15**: 1176.
[j] Nicoud JF and Twieg RJ (1987) In: Chemla DS and Zyss J (eds) *Nonlinear Optical Properties of Organic Molecules and Crystals*, Vol.1, pp 227–296. London: Academic Press.

the signal wave becomes linearly dependent on the intensity of the incident pump wave.

Evidently, a photon of frequency ω_P may break up into two photons of lower frequency in an infinity of ways, depending on the relative frequencies ω_S and ω_I. In order to select which frequency ω_S should be amplified, the parametric oscillator must be correspondingly tuned. The most important and practical way to achieve this tuning is by phase matching in a crystal.

Conservation of photon momentum: phase matching

To optimize the intensity of a coherent nonlinear optical effect, there must be conservation of photon momentum. For sum frequency generation this requirement is expressed as

$$\boldsymbol{k}_1 + \boldsymbol{k}_2 = \boldsymbol{k}_3 \qquad [10a]$$

Table 8 Experimental second-order nonlinear optical susceptibilities d_{il} of inorganic crystals

Materials	*Symmetry*		d_{il} *(10^{-12}m V^{-1})*	d_{il} λ(μm)	*Reference*
Quartz (α-SiO_2)	$32 = D_3$	d_{11}	0.46	1.06	a
		d_{14}	0.009	1.0582	b
$LiIO_3$	$6 = C_6$	d_{31}	6.43	2.12	a
			7.11	1.06	
			8.14	0.6943	
		d_{33}	6.41	2.12	
			6.75	1.318	
			7.02	1.06	
$LiNbO_3$	$3m = C_{3v}$	d_{31}	5.77	1.15	a
			5.95	1.06	
		d_{33}	29.1	2.12	
			31.8	1.318	
			34.4	1.06	
		d_{22}	3.07	1.0582	b
$KNbO_3$	$mm2 = C_{2v}$	d_{31}	– 15.8	1.064	g
		d_{32}	– 18.3		
		d_{33}	– 27.4		
		d_{24}	– 17.1		
		d_{15}	– 16.5		
$Ba_2NaNb_5O_{15}$	$mm2 = C_{2v}$	d_{31}	– 14.55	1.0642	b
		d_{32}	– 14.55		
		d_{33}	– 20		
$BaTiO_3$	$4mm = C_{4v}$	d_{15}	–17.2	1.0582	b
		d_{31}	– 18		
		d_{33}	– 6.6		
$NH_4H_2PO_4$(ADP)	$\bar{4}2m = D_{2d}$	d_{14}	0.48	0.6943	b
		d_{36}	0.485		
KH_2PO_4(KDP)	$\bar{4}2m = D_{2d}$	d_{14}	0.49	1.0582	b
		d_{36}	0.599	1.318	a
			0.630	1.06	
			0.712	0.6328	
KD_2PO_4(KD*P)	$\bar{4}2m = D_{2d}$	d_{14}	0.528		c
		d_{36}	0.528		
GaP	$\bar{4}3m = T_d$	d_{14}	35	3.39	b
		d_{36}	58.1	10.6	a
			77.5	2.12	
			99.7	1.06	
GaAs	$\bar{4}3\ m = T_d$	d_{14}	188.5	10.6	b
		d_{36}	151	10.6	a
			173	2.12	
$AgGaSe_2$	$\bar{4}2m = D_{2d}$	d_{36}	57.7	10.6	a
			67.7	2.12	
$AgSbS_3$	$3m = C_{3v}$	d_{31}	12.6		c
		d_{22}	13.4		
Ag_3AsS_3	$3m = C_{3v}$	d_{31}	15.1		c
		d_{22}	28.5		
CdS	$6mm = C_{6v}$	d_{33}	36.0		c
		d_{31}	37.7		
		d_{36}	41.9		
CdSe	$6mm = C_{6v}$	d_{15}	31	10.6	h
		d_{31}	28.5		
		d_{33}	55.3	10.6	a

Table 8 *Continued*

Materials	*Symmetry*		d_{il} (10^{-12} m V^{-1})	d_{il} λ (μm)	*Reference*
			65.4	2.12	
Te	$32 = D_3$	d_{11}	5×10^3	10.6	b
β-BaB_2O_4(BBO)	$3m = C_{3v}$	d_{11}	1.6	1.064	d
		d_{22}, d_{31}	< 0.08		
$LaBGeO_5Nd^{3+}$		d_{eff}	0.296	1.064	e
$KTiOPO_4$	$mm2 = C_{2v}$	d_{15}	1.91	1.064	h
(KTP)		d_{24}	3.64		
		d_{31}	2.54		
		d_{32}	4.35		
		d_{33}	16.9		
$RbTiOPO_4$	$mm2 = C_{2v}$	d_{15}	6.1	1.064	f
		d_{24}	7.6		
		d_{31}	6.5		
		d_{32}	5.0		
		d_{33}	13.7		
ZnO	$6mm = C_{6v}$	d_{31}	2.1	1.0582	b
		d_{15}	4.3		
		d_{33}	–7.0		

[a] Absolute values: Choy MM and Byer RL (1976) *Physical Review* **B14**: 1693
[b] Shen YR (1984) *The Principles of Nonlinear Optics*. New York: Wiley
[c] Boyd RW (1992) *Nonlinear Optics*. Boston: Academic Press
[d] Eimerl D, Davis L, Velsko S, Graham EK and Zalkin A (1987) *Journal of Applied Physics* **62**: 1968
[e] For type I phase-matched SHG; Capmany J and Garcia Sole J (1997) *Applied Physics Letters* **70**: 2517
[f] Zumsteg FC, Bierlein JD and Gier TE (1976) *Journal of Applied Physics* **47**: 4980
[g] Biaggio I, Kerkoc P, Wu L-S, Günter P and Zysset P (1992) *Journal of the Optical Society of America B* **9**: 507
[h] Vanherzeele H and Bierlein JD (1992) *Optics Letters* **17**: 982.

k_i denotes the wave vector of the corresponding beam. To avoid reduction of effective beam interaction length due to finite cross-sections, collinear phase matching is aimed at. One then may write equation [10a] in scalar form

$$k_1 + k_2 = k_3 \quad [10b]$$

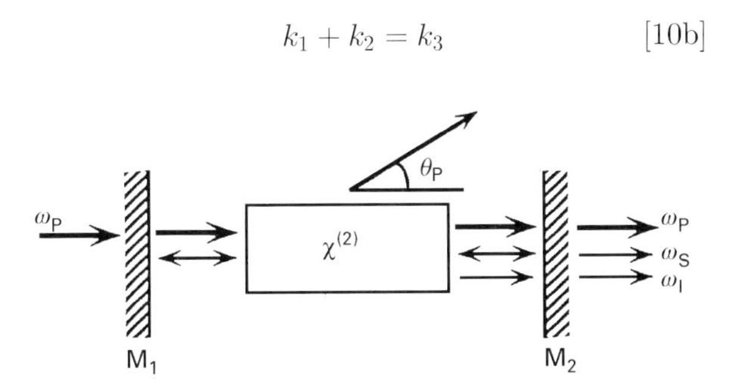

Figure 3 Schematic representation of a singly-resonant optical parametric oscillator. Pump wave of frequency ω_P, (reflected) signal wave of frequency ω_S, idler wave of frequency ω_I. The signal wave ω_S becomes amplified. θ_P denotes the angle of orientation of the direction of propagation with respect to the crystal optic axis. Adapted with permission from Tang CL and Cheng LK (1995) *Fundamentals of Optical Parametric Processes and Oscillators*. Amsterdam: Harwood Academic Publishers.

With $k_i = n_i 2\pi/\lambda_i$, where $n_i \equiv n(\omega_i)$ stands for the refractive index of the medium at frequency ω_i and λ_i for the vacuum wavelength, this may be expressed as

$$\omega_1 n_1 + \omega_2 n_2 = \omega_3 n_3 \quad [10c]$$

In a lossless medium, $n(\omega)$ in general increases monotonically with ω owing to normal dispersion. In an isotropic medium such as a liquid, $n(\omega)$ is independent of beam polarization. It can then easily be shown that for $\omega_1 \le \omega_2 < \omega_3$, Equation [10c] cannot be satisfied.

In a uniaxial birefringent crystal, excluding propagation along the optic axis, an incident beam may, depending on its polarization, be made ordinary or extraordinary. The ordinary (o) and extraordinary (e) rays, perpendicularly polarized with respect to each other, will each experience a different index of refraction, $n^{(o)}(\omega) \neq n^{(e)}(\omega)$. According to the crystalline medium, the frequency ω and the angle of incidence with respect to the optic axis, situations may be found where the phase-matching condition is fulfilled.

For sum-frequency generation in a positive uniaxial crystal, in which $n^{(e)} > n^{(o)}$, the phase-matching condition may be satisfied in two different ways:

$$n_1^{(e)}\omega_1 + n_2^{(e)}\omega_2 = n_3^{(o)}\omega_3 \qquad \text{Type I} \qquad [11a]$$

$$n_1^{(o)}\omega_1 + n_2^{(e)}\omega_2 = n_3^{(o)}\omega_3 \qquad \text{Type II}$$

or

$$n_1^{(e)}\omega_1 + n_2^{(o)}\omega_2 = n_3^{(o)}\omega_3 \qquad [11b]$$

Similarly for the parametric effect in a negative uniaxial crystal, in which $n^{(e)} < n^{(o)}$ (see **Figure 4**):

$$n_P^{(e)}\omega_P = n_S^{(o)}\omega_S + n_I^{(o)}\omega_I \qquad \text{Type I} \qquad [12a]$$

$$n_P^{(e)}\omega_P = n_S^{(e)}\omega_S + n_I^{(o)}\omega_I \qquad \text{Type II}$$

or

$$n_P^{(e)}\omega_P = n_S^{(o)}\omega_S + n_I^{(e)}\omega_I \qquad [12b]$$

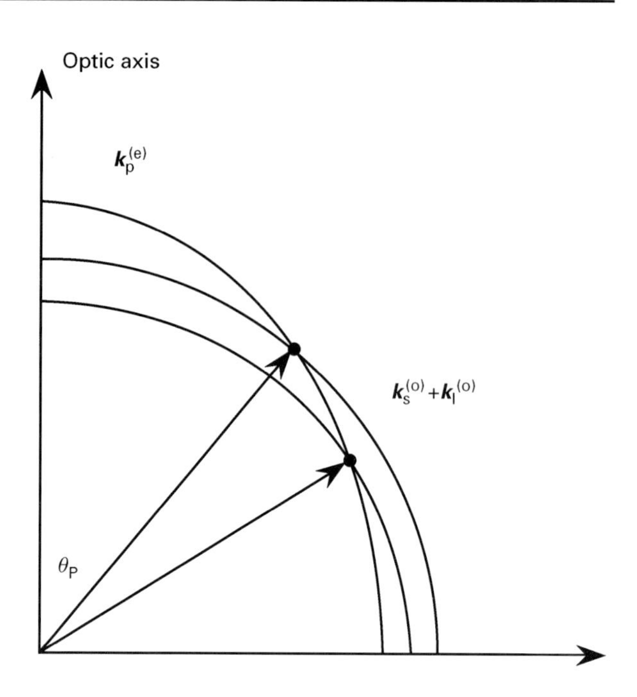

Figure 4 Phase-matching in an optical parametric process to achieve photon momentum conservation is based on the use of birefringence to compensate for normal material dispersion. In an uniaxial crystal, the ordinary wave (o) is polarized perpendicularly to the plane defined by the direction of propagation and the optic axis. The corresponding value of $k^{(o)}$ (or $n^{(o)}$) is independent of θ_P, the angle of orientation of the direction of propagation with respect to the optic axis. $k_S^{(o)}$ and $k_I^{(o)}$ therefore lie on a circle. The extraordinary wave (e) is polarized in the plane defined by the direction of propagation and the optic axis. The value of $k_P^{(e)}$ (or $n_P^{(e)}$) in its dependence on θ_P is described by an ellipse. In a negative uniaxial crystal, and for given values of $\omega_P = \omega_S + \omega_I$, the ellipse for $k_P^{(e)}$ may intersect the circle for $k_S^{(o)} + k_I^{(o)}$. At the corresponding angle θ_P there is phase matching. Rotation of the crystal relative to the direction of propagation of the waves correspondingly leads to tuning of the frequencies of the signal and idler waves. Adapted with permission from Tang CL and Cheng LK (1995). *Fundamentals of Optical Parametric Processes and Oscillators.* Amsterdam: Harwood Academic Publishers.

Crystals belonging to the cubic crystal system are isotropic, and therefore unsuited for phase-matching. Tetragonal and trigonal crystals are uniaxial; those of the orthorhombic, monoclinic and triclinic symmetry are biaxial. The description of phase-matching in biaxial crystals is somewhat more complicated than in uniaxial crystals, but it essentially rests on the same principles.

The search for birefringent crystals with good phase-matching properties is of great technical importance in nonlinear optics. Although phase-matching has been achieved in organic crystals (see **Table** 7), inorganic materials appear so far to offer a greater variety of possibilities.

Inorganic media

Noncentrosymmetric crystals

Inorganic crystals are widely applied for second-harmonic generation and for optical parametric processes. Some frequently used materials: for second-harmonic generation from the near-IR into the visible and beyond KH_2PO_4 (KDP), KD_2PO_4 (KD*P) and more recently β-BaB_2O_4(BBO). For optical parametric amplification into the mid-IR: $AgGaSe_2$, GaSe; the visible and near-IR: $LiNbO_3$, $KTiOPO_4$, $KNbO_3$; and into the visible and UV: β-BaB_2O_4 and LiB_3O_5.

Table 8 shows experimental second-order nonlinear optical susceptibilities for different tensor components d_{il} and various fundamental wavelengths. The quantities d_{il} are defined as follows:

$$\chi_{ijk}^{(2)}(2\omega) = 2d_{ijk}(2\omega) \qquad [13]$$

The second and third indices of d_{ijk} are then replaced by a single symbol l according to the piezoelectric contraction:

$jk:$	11	22	33	23, 32	31, 13	12, 21
$\ell:$	1	2	3	4	5	6

The nonlinear susceptibility tensor can then be represented as a 3×6 matrix containing 18 elements. In the transparent region, i.e. outside of absorption bands, one may assume the validity of the Kleinman symmetry condition, which states that the indices *i*, *j*, *k* may be freely permuted:

$$d_{ijk} = d_{kij} = d_{jki}, \quad \text{etc.} \qquad [14a]$$

One then finds, for instance,

$$d_{123} = d_{312}, \quad d_{14} = d_{36} \qquad [14b]$$

In this case there are only 10 independent elements for d_{il}.

Table 8 shows that the values for d_{il} may vary over several orders of magnitude, and that it is not necessarily the crystals with the highest values that are most commonly used. The technical applicability is partly also determined by other qualities, such as phase-matching properties, ease of crystal growth, mechanical strength, chemical inertness, temperature stability and light-damage threshold.

A quantity often used to characterize the optical properties of nonlinear optical materials is the Miller index:

$$\delta_{ijk} = \frac{d_{ijk}(2\omega)}{\chi^{(1)}_{ii}(2\omega)\,\chi^{(1)}_{jj}(\omega)\,\chi^{(1)}_{kk}(\omega)\cdot\varepsilon_0} \qquad [15]$$

$\chi^{(1)}(2\omega)$ represents the linear susceptibility for the doubled frequency 2ω, $\chi^{(1)}(\omega)$ that for the fundamental frequency ω. One finds that for most materials δ is not far from a mean value of about 2×10^{-2} m^2 C^{-1}, suggesting that in a given substance nonlinear and linear susceptibilities are closely related.

Noncentrosymmetric crystals show other properties in addition to frequency conversion, for instance the linear electro-optic or Pockels effect: the linear change of the refractive index induced by an applied DC electric field. Furthermore, the point groups C_n and C_{nv} allow for the existence of a permanent electric dipole moment. Indeed, crystals such as $LiNbO_3$ (C_{3v}) and $BaTiO_3$ (C_{4v}) are well-known for their ferroelectric properties. Crystals transforming according to point groups containing only rotations, such as C_n, D_n, T and O are chiral and therefore optically active. In **Table 8** we find quartz α-SiO_2(D_3), $LiIO_3$(C_6) and Te(D_3).

Harmonic generation in metal vapours

Third-harmonic generation can in principle occur in all matter, as it is not tied to the condition of noncentrosymmetry. While the effect has been investigated in liquids and solids, the use of gases, in particular alkali metal vapours, has proved particularly interesting. In spite of the relatively low density of atoms, the third-harmonic generation efficiency can become quite high, up to 10%. The limiting laser intensity in gases is orders of magnitude higher than in condensed matter. Furthermore, the sharper transitions in gases allow strong enhancement of $\chi^{(3)}$ near *resonances*, especially three-photon resonances, which are electric dipole-allowed with respect to the atomic ground state. In sodium vapour this corresponds to transitions $3s \rightarrow 3p$, $3s \rightarrow 4p$, etc. Enhancement may in principle also occur via intermediate one-photon resonances, of same symmetry as three-photon resonances; or by two-photon resonances at transitions of symmetry $3s \rightarrow 4s$, $3s \rightarrow 5s$, or $3s \rightarrow 3d$, etc. The resonance enhancement of $\chi^{(3)}(3\omega)$ will evidently be diminished by concurrent multiphoton (or single-photon) absorption. In tuning ω, a compromise must be sought, whereby the anomalous dispersion of $\chi^{(3)}$ is maximized in comparison to energy dissipation through absorption.

The anomalous dispersion of $\chi^{(3)}(3\omega)$ near resonances may also be used to achieve phase matching,

$$n(\omega) = n(3\omega) \qquad [16a]$$

which in a normally dispersive isotropic medium would be impossible. Considering an alkali atom A, and assuming ω to be below, and 3ω to be above a strong $s \rightarrow p$ transition, we find

$$n_A(\omega) > n_A(3\omega) \qquad [16b]$$

Phase matching may be achieved by admixture of a buffer gas B. Such an inert gas must be transparent at frequency 3ω and above; then

$$n_B(\omega) < n_B(3\omega) \qquad [16c]$$

The relative concentration of the inert gas is adjusted, so as to have for the mixture M,

$$n_M(\omega) = n_M(3\omega) \qquad [16d]$$

High conversion efficiencies have, for instance, been achieved with the mixtures Rb:Xe (10%) and Na:Mg (3.8%).

Four-wave mixing

Beside third-harmonic generation, there exists a large variety of four-wave mixing effects. Depending on the combination of frequencies, on the occurrence of intermediate resonances and on the polarization of the light beams involved, the manifestation of these phenomena may be very different. We limit our considerations to a few selected examples.

Coherent Raman spectroscopy

In coherent anti-Stokes Raman spectroscopy (CARS) two beams of frequency ω_1 and ω_2 are mixed in the sample to generate a new frequency $\omega_s = 2\omega_1 - \omega_2$. If there is a Raman resonance at $\omega_1 - \omega_2 = \Omega$, an amplified signal is detected at the anti-Stokes frequency $\omega_1 + \Omega$ (see **Figure 5**). The corresponding susceptibility $\chi^{(3)}(-\omega_4; \omega_1, \omega_2, \omega_3)$ may be written $\chi^{(3)}(-\omega_1 - \Omega; \omega_1, -\omega_1 + \Omega, \omega_1)$. The major experimental advantage of CARS and of other coherent Raman techniques is the large, highly directional signal produced, of the order of 10^4 times more intense than would be obtained for conventional spontaneous Raman scattering. Usually, CARS experiments are performed with pulsed lasers delivering a peak power of the order of 10 – 100 kW. High frequency-resolution measurements with CW lasers are also possible. CARS experiments have been performed in gases, liquids and solids and on a variety of substances, ranging from diamond to aqueous solutions of biological macromolecules. Of particular interest is the use of CARS for combustion diagnostics. The coherent Raman signals can easily be separated from the luminescent background in flames.

Other, related coherent Raman effects are also represented in **Figure 5**, such as the case (C) where the signal beam is detected at the Stokes frequency. The Raman-induced Kerr effect (B) may be interpreted as the quadratic influence of an electric field of frequency ω_2 on the elastic scattering of radiation at a frequency ω_1, or vice versa. In this case the phase-matching (or wave-vector-matching) condition is fulfilled for any angle between beams 1 and 2, while in cases (A) and (C) it may only be met for certain angles of the beams with respect to each other.

Degenerate four-wave mixing

The process governed by the third-order susceptibility $\chi^{(3)}(-\omega; \omega, -\omega, \omega)$ is called degenerate four-wave mixing. It may lead to a variety of highly interesting effects, one of them being that the index of refraction $n(\omega)$ becomes dependent on the incident light intensity $I(\omega)$:

$$n(\omega) = n_0(\omega) + n_2(\omega)I(\omega) \qquad [17]$$

For a single-mode laser beam with a Gaussian transverse intensity distribution, the index of refraction at the centre of the beam will then be larger than at its periphery, provided $n_2(\omega)$ is positive. Thereby the medium will act as a positive lens, tending to bring the incident beam to a focus at the centre on the beam. However, only if the intensity of the laser beam is sufficiently large will this self-focusing effect be able to counteract the beam spread due to ordinary diffraction.

An effect that may also occur with other nonlinear optical phenomena, but that has been extensively

Table 9 Conversion from CGS-esu to SI units for *n*th order optical quantities

Conversion factor for n ≥ 1	*[SI] ← [CGS-esu]*	*Dimension in SI units*
$\chi^{(n)}[\mathrm{SI}] = \dfrac{4\pi}{(3\times 10^4)^{n-1}}\,\chi^{(n)}[\mathrm{esu}]$		$\left(\dfrac{\mathrm{m}}{\mathrm{V}}\right)^{n-1}$
$\alpha^{(n)}[\mathrm{SI}] = \dfrac{10^{-7}}{(3\times 10^4)^{n+1}}\,\alpha^{(n)}[\mathrm{esu}]^*$		$\mathrm{J}\left(\dfrac{\mathrm{m}}{\mathrm{V}}\right)^{n+1*}$
* The case $n = 0$ corresponds to the conversion factor for a permanent electric dipole moment:		
$\mu[\mathrm{SI}] = \dfrac{1}{3}\times 10^{-11}\mu\,[\mathrm{esu}]$		$\mathrm{J}\dfrac{\mathrm{m}}{\mathrm{V}} = \mathrm{cm}$

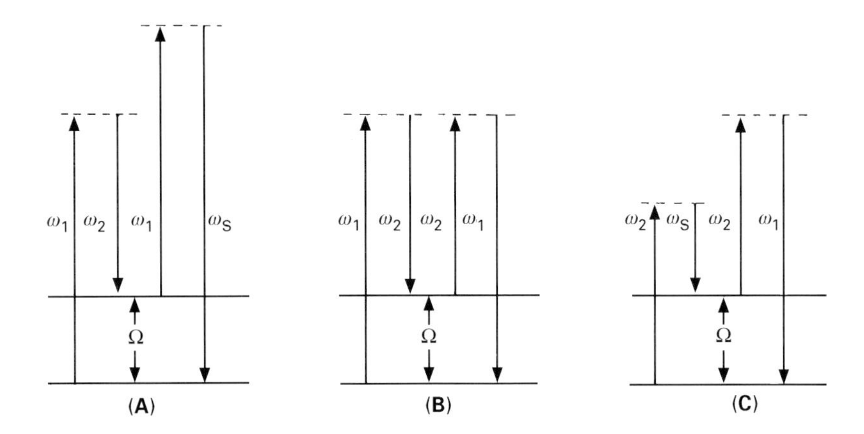

Figure 5 Ladder graphs for four-wave mixing effects containing Raman processes. In all cases there is assumed an intermediate Raman-type resonance at the frequency Ω. (A) The coherent anti-Stokes Raman (CARS) process. (B) The process responsible for stimulated Raman spectroscopy (SRS) as well as the Raman-induced Kerr effect (TRIKE). (C) The coherent Stokes Raman spectroscopy (CSRS). Adapted with permission from Levenson MD (1982), *Introduction to Nonlinear Laser Spectroscopy*. New York: Academic Press.

studied in the frame of degenerate four-wave mixing, is phase conjugation. Here we consider not a single beam of frequency ω, but four different beams: the collinear counterpropagating pump beams 1 and 2 interfere in the $\chi^{(3)}$-active medium to form an induced static grating. From this grating a signal wave 3, incident at a given angle with respect to 1 and 2, is scattered and reflected. The coherent reflected wave 4 is phase conjugate with respect to 3. For instance, if 3 is a forward-travelling plane wave

$$E_3(z,t) = A_3 \cos(\omega t - kz - \phi_0) \qquad [18a]$$

the corresponding phase-conjugate wave 4 will be

$$\begin{aligned} E_4(z,t) &= A_4 \cos(-\omega t - kz - \phi_0) \\ &= A_4 \cos(+\omega t + kz + \phi_0) \end{aligned} \qquad [18b]$$

It will travel backwards and behave as if the time t had been replaced by $-t$. A nonlinear medium susceptible to degenerate four-wave mixing can thus be used as a phase-conjugate mirror. A left circularly polarized incident beam will be reflected as a left circularly polarized beam, and not as a right circularly polarized one as would be the case upon ordinary reflection. The phase conjugation process can be thought of as the generation of a time-reversed wavefront. If the input signal wave in passing through a medium before entering the phase-conjugate mirror suffers a wavefront distortion, the phase conjugate wave reflected back through the medium will remove this distortion. The phenomenon of phase conjugation can, for instance, be used to correct for aberrations induced by amplifying media.

Particular aspects of nonlinear optics

Higher order electromagnetic effects

The interaction energy of a molecular system with the radiation field may formally be expanded into a multipole series. The first term in this expansion contains the electric dipole–electric field interaction; in the second term appear the magnetic dipole–magnetic field interaction and the electric quadrupole interaction with the electric field gradient of the radiation, and so on. If the wavelength of light is large compared to the molecular dimensions, the higher multipole effects tend to be small and are often negligible from an experimental standpoint. The discussion until now has therefore considered only dominant electric dipole contributions to the molecular polarizability or bulk susceptibility. However, depending on molecular symmetry, there are situations where magnetic dipole and electric quadrupole interactions may become measurable. For instance, owing to these, weak second-harmonic generation may also be observed in some centrosymmetric crystals. Furthermore, the interplay of electric dipole, magnetic dipole and electric quadrupole interactions in chiral media leads to natural optical activity and to related higher-order nonlinear circular differential effects.

Particular nonlinear optical phenomena arise also when static electric or magnetic fields are applied. The molecular states and selection rules are thereby modified, leading, for instance, to higher-order, nonlinear-optical variants of the linear (Pockels) and quadratic (Kerr) electro-optical effect, or of the linear (Faraday) and quadratic (Cotton–Mouton) magneto-optical effect.

Incoherent higher-harmonic scattering

We have seen that coherent second-harmonic generation is forbidden in liquids, even in chiral ones. This is due to the fact that the relevant molecular quantity $\alpha^{(2)}(-2\omega; \omega, \omega)$ vanishes when averaged over all possible molecular orientations:

$$\langle \alpha^{(2)}(-2\omega; \omega, \omega) \rangle = 0 \qquad [19a]$$

However, the inhomogeneity of the liquid at the molecular level and the fact that every molecule is an individual scatterer of radiation are not fully taken into account. The superposition of this molecular scattered radiation is partly incoherent. It consists mainly of 'ordinary' Rayleigh scattering at the basic frequency ω; but if the molecules are noncentrosymmetric, some incoherent radiation of frequency 2ω is also generated. This hyper-Rayleigh scattering, though weak, is clearly detectable with pulsed lasers of megawatt peak power. Its intensity is proportional to the square of $\alpha^{(2)}(-2\omega; \omega, \omega)$, which upon averaging over all spatial orientations in the liquid does not vanish:

$$\langle |\alpha^{(2)}(-2\omega; \omega, \omega)|^2 \rangle \neq 0 \qquad [19b]$$

From the directional dependence and the depolarization ratios of the scattered radiation, information may be gained on particular tensor elements of $\alpha^{(2)}(-2\omega)$. The method has the advantage over EFISH measurements that it is also applicable to noncentrosymmetric molecules that do not posses a permanent dipole moment, in particular to, 'octopolar' molecules of symmetry D_{3h} (such as tricyanomethanide $[C(CN)_3]^-$) or of symmetry T_d (such as CCl_4). It is to be expected that progress in laser technology and light detection systems will further improve the applicability of the method.

List of symbols

C_{ix_s} etc. = trigonometric factors; (e) refers to the extraordinary ray; E = electric field strength of incident radiation; $I(\omega)$ = intensity of incident/scattered radiation; i,j,k = coordinate system of crystal; k_i = wave vector of beam i; L_{ijk} = local-field correction; n = order of nonlinear effect; n_i = refractive index of medium at ω_i; (o) – refers to the ordinary ray; $p^{(n)}$ = molecular induced electric dipole moment (nth-order effect); $P^{(1)}$ = volume polarization; V = volume of unit cell; x_s, y_s, z_s = coordinate system of molecules; $\alpha^{(n)}$ = molecular polarizability of nth order; δ_{ijk} = Miller index (see equation [15]); ε_0 = permittivity of free space; λ = wavelength; μ = static molecular dipole moment; ϕ = phase angle; $\chi^{(1)}$ = macroscopic susceptibility; $\chi_S^{(2)}$ = surface susceptibility; ω = photon frequency.

See also: **Electromagnetic Radiation; Laser Applications in Electronic Spectroscopy; Laser Spectroscopy Theory; Linear Dichroism, Theory; Multiphoton Spectroscopy, Applications; Optical Frequency Conversion; Raman Optical Activity, Applications; Raman Optical Activity, Spectrometers; Raman Optical Activity, Theory; Raman Spectrometers; Rayleigh Scattering and Raman Spectroscopy, Theory; Symmetry in Spectroscopy, Effects of.**

Further reading

Andrews DL (1993) Molecular theory of harmonic generation. Modern nonlinear optics, Part 2. *Advances in Chemical Physics* 85: 545–606.

Bloembergen N (1965) *Nonlinear Optics*. NewYork: WA Benjamin.

Boyd RW (1992) *Nonlinear Optics*. Boston: Academic Press.

Chemla DS and Zyss J (1987) *Nonlinear Optical Properties of Organic Molecules and Crystals*, Vols 1 and 2. London: Academic Press.

Clays K, Persoons A and De Maeyer L (1993) Hyper-Rayleigh scattering in solution. Modern nonlinear optics, part 3. *Advances in Chemical Physics* 85: 455–498.

Flytzanis C (1975) Theory of nonlinear susceptibilities. In: Rabin H and Tang CL (eds) *Quantum Electronics*, Vol. I, Nonlinear Optics, part A. New York: Academic Press.

Lalanne JR, Ducasse A and Kielich S (1996) *Laser–Molecule Interaction*. New York: Wiley.

Levenson MD (1982) *Introduction to Nonlinear Laser Spectroscopy*. New York: Academic Press.

Shen YR (1984) *The Principles of Nonlinear Optics*. New York: Wiley.

Tang CL and Cheng LK (1995) *Fundamentals of Optical Parametric Processes and Oscillators*. Amsterdam: Harwood Academic.

Wagnière GH (1993) *Linear and Nonlinear Optical Properties of Molecules*. Basel: Verlag HCA, VCH.

Yariv A (1975) *Quantum Electronics*. New York: Wiley.

Zel'dovich BY, Pilipetsky NF and Shkunov VV (1985) *Principles of Phase Conjugation*. Berlin: Springer-Verlag.

Nonlinear Raman Spectroscopy, Applications

W Kiefer, Universität Wurzburg, Germany

VIBRATIONAL, ROTATIONAL & RAMAN SPECTROSCOPIES
Applications

Linear, spontaneous Raman spectroscopy is a powerful tool for structural analysis of materials in the gaseous, liquid or solid state. Its scattering cross-section can be increased considerably by resonance excitation, i.e. irradiation in spectral regions where there is strong absorption or by applying surface enhanced methods like SERS (surface enhanced Raman scattering). Also, the scattering volume, determined by the dimensions of a focused laser beam, can be as small as a few μm^2 if a microscope is incorporated in a Raman spectrometer. There are, however, cases where ordinary Raman spectroscopy has limitations in allowing the derivation of the desired information. For example, particular vibrational modes of specific symmetry are neither allowed in linear Raman scattering nor in infrared absorption, but their vibrational bands show up in what is called a hyper-Raman spectrum, because there is a nonvanishing contribution from the nonlinear part of the induced dipole moment. Also, fluorescence simultaneously excited with visible laser light, may obscure the Raman scattered light. This can often be overcome by near-infrared laser excitation. Another way is to apply nonlinear coherent Raman techniques like CARS (coherent anti-Stokes Raman spectroscopy). In general, nonlinear optical properties of materials can only be obtained using nonlinear optical methods. One of the major advantages of nonlinear coherent Raman spectroscopy is its possible high resolution of up to three orders of magnitude better than its linear counterpart. In addition, these methods allow spectral information to be obtained from scattering systems which produce a high light background like flames, combustion areas, etc. In recent years there has been a dramatic development in time-resolved linear and nonlinear Raman spectroscopy due to the availability of commercial pico- and femtosecond lasers which allows direct insight into the dynamics of molecules in their ground or excited electronic state. After a short description of the various nonlinear Raman techniques, typical applications will be given for these methods.

A short description of nonlinear Raman techniques

Spontaneous nonlinear as well as coherent nonlinear Raman methods are considered here. These are based on the contributions of the nonlinear part of the induced dipole moment (spontaneous effects) or the induced polarization (coherent effects) to the intensity of the frequency shifted light. In the first case, the Raman signal is generated in a spontaneous, incoherent but nonlinear optical process, whereas in the second case the Raman information is contained in a coherent laser beam whereby the nonlinear polarization acts as a coherent light source.

Hyper-Raman effect

Generally, the induced dipole moment $\boldsymbol{p}$ in a molecular system is written as

$$\boldsymbol{p} = \alpha \boldsymbol{E} + \frac{1}{2}\beta \boldsymbol{EE} + \frac{1}{6}\gamma \boldsymbol{EEE} + \ldots \qquad [1]$$

where α is the polarizability, β the hyperpolarizability and γ the second hyperpolarizability. $\boldsymbol{E}$ is the incident electric field. The nonlinear terms in Equation [1] are usually small compared to the linear term which gives rise to normal, linear Raman scattering. However, when the electric field is sufficiently large, as is the case when a high-powered laser is focused on the sample, contributions from the second term in Equation [1] are sufficiently intense to be detected. This scattering is at an angular frequency $2\omega_L \pm \omega_R$, where ω_L is the angular frequency of the exciting laser beam and $-\omega_R$ and $+\omega_R$ are the Stokes and anti-Stokes hyper-Raman displacements, respectively. Scattering at $2\omega_L \pm \omega_R$ is called hyper-Raman scattering.

The hyper-Raman effect is a three-photon process involving two virtual states of the scattering system. The level scheme for Stokes hyper-Raman scattering is presented in **Figure 1**.

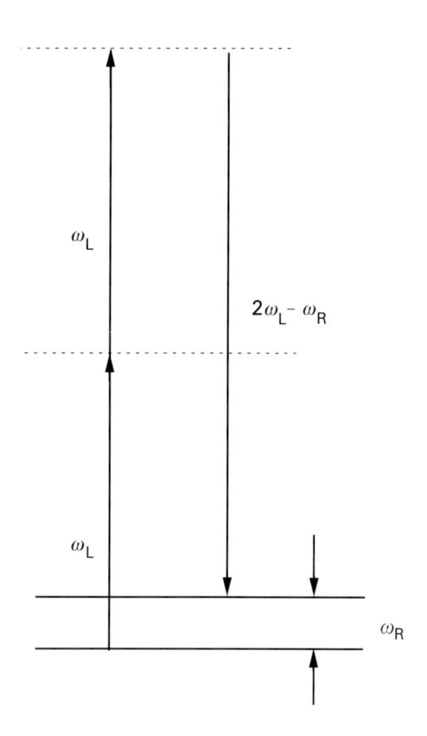

Figure 1 Schematic level diagram for Stokes and hyper Raman scattering.

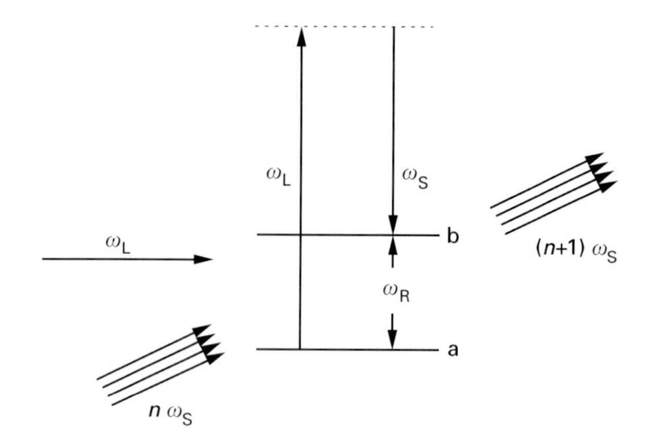

Figure 2 Schematic diagram for stimulated Raman scattering as a quantum process.

The importance of the hyper-Raman effect as a spectroscopic tool results from its symmetry selection rules. It turns out that all infrared active modes of the scattering system are also hyper-Raman active. In addition, the hyper-Raman effect allows the observation of 'silent' modes, which are accessible neither by infrared nor by linear Raman spectroscopy.

Stimulated Raman effect

The stimulated Raman process is schematically represented in **Figure 2**. A light wave at angular frequency ω_S is incident on the material system simultaneously with a light wave at angular frequency ω_L. While the incident light beam loses a quantum ($\hbar\omega_L$) and the material system is excited by a quantum $\hbar\omega_R = \hbar(\omega_L - \omega_S)$, a quantum $\hbar\omega_S$ is added to the wave at angular frequency ω_S, which consequently becomes amplified. It can be shown theoretically that a polarization at Stokes angular frequency ω_S is generated via the third-order nonlinear susceptibility $\chi^{(3)}$. Including a degeneracy factor, the polarization oscillating at angular frequency ω_S is given by Berger and co-workers (1992):

$$\boldsymbol{P}^{(3)}(\omega_S) = \frac{6}{4}\varepsilon_0\chi^{(3)}(-\omega_S;\, \omega_L, -\omega_L, \omega_S) \times |\boldsymbol{E}(\omega_L)|^2\, \boldsymbol{E}(\omega_S) \qquad [2]$$

where ε_0 is the permittivity constant of vacuum.

Optimum gain for this effect is found at the centre of the Raman line where $\omega_R = \omega_L - \omega_S$. There, the gain constant for stimulated Raman scattering at Stokes frequency is given by

$$g_S \propto \left(\frac{d\sigma}{d\Omega}\right) \cdot \frac{1}{\Gamma} \qquad [3]$$

where $(d\sigma/d\Omega)$ is the differential Raman cross-section and Γ represents the line width of the molecular transition (ω_R). From Equation [3] we immediately recognize that in stimulated Raman scattering processes where only *one* input laser field with frequency ω_L is employed a coherent Stokes wave is generated for those Raman modes which have the highest ratio between differential Raman cross-section and line width Γ. The distinctive feature of stimulated Raman scattering is that an assemblage of coherently driven molecular vibrations provides the means of coupling the two light waves at angular frequencies ω_L and ω_S by modulating the nonlinear susceptibility.

Nonlinear Raman spectroscopies based on third-order susceptibilities

From the discussion on stimulated Raman scattering it is clear that during this nonlinear process coherently driven molecular vibrations are generated. In what is usually called the stimulated Raman effect only one input field (ω_L) is used for this type of excitation. We have seen that only particular Raman modes, i.e. those with highest gain factors, give rise to stimulated Stokes emission. Thus, for molecular spectroscopy in which we are interested in determining *all* Raman active modes, excitation with one strong laser field would not serve the purpose, although it would

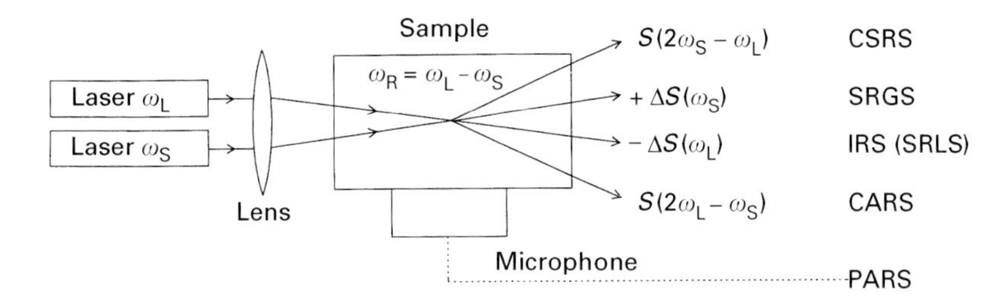

Figure 3 Schematic diagram for a few techniques in nonlinear (coherent) Raman spectroscopy (CSRS: Coherent Stokes Raman Spectroscopy; SRGS: Stimulated Raman Gain Spectroscopy; IRS: Inverse Raman Spectroscopy (= SRLS: Stimulated Raman Loss Spectroscopy); CARS: Coherent anti-Stokes Raman Spectroscopy; PARS: Photoacoustic Raman Spectroscopy).

provide very high signals in the form of a coherent beam, but unfortunately, only at one particular vibrational frequency. However, the advantages of stimulated Raman scattering, being high signal strength and coherent radiation, can be fully exploited by a very simple modification of the type of excitation. The trick is simply to provide the molecular system with an intense external Stokes field by using a second laser beam at Stokes angular frequency ω_S instead of having initially the Stokes field produced in the molecular system by conversion of energy from the pump field. Thus, by keeping one of the two lasers, e.g. the laser beam at Stokes angular frequency ω_S tunable, one is now able to excite selectively coherent molecular vibrations at any desired angular frequency ω_R assuming the transitions are Raman allowed. A variety of nonlinear Raman techniques based on this idea have been developed, which combine the wide spectroscopic potentials of spontaneous Raman spectroscopy and the high efficiency of scattering, strong excitation and phasing of molecular vibrations in a macroscopic volume of substance, that are the features inherent to stimulated Raman scattering.

The following acronyms of some of these nonlinear coherent Raman techniques have been widely used: CARS, CSRS (coherent Stokes Raman spectroscopy), PARS (photoacoustic Raman spectroscopy), RIKE (Raman induced Kerr effect), SRGS (stimulated Raman gain spectroscopy), IRS (inverse Raman scattering) also called SRLS (stimulated Raman loss spectroscopy). A schematic diagram of these methods is illustrated in **Figure 3**. The common physical aspect is the excitation of Raman active molecular vibrations and/or rotations in the field of two laser beams with angular frequencies ω_L and ω_S in such a way that their difference corresponds to the angular frequency of the molecular vibration ω_R $(= \omega_L - \omega_S)$. The strong coupling between the generated coherent molecular vibrations with the input laser fields via the third-order nonlinear susceptibility $\chi^{(3)}$ opens the possibility for various techniques.

The most powerful of these methods is CARS since a new coherent, laser-like signal is generated. Its direction is determined by the phase-matching condition

$$\boldsymbol{k}_{AS} = 2\boldsymbol{k}_L - \boldsymbol{k}_S \qquad [4]$$

where $\boldsymbol{k}_{AS}$, $\boldsymbol{k}_L$ and $\boldsymbol{k}_S$ are the wave vectors of the anti-Stokes signal, pump and Stokes laser, respectively. The laser-like anti-Stokes signal is therefore scattered in one direction, which lies in the plane given by the two laser directions $\boldsymbol{k}_L$ and $\boldsymbol{k}_S$ and which is determined by the momentum vector diagram shown in **Figure 4**. Therefore, CARS is simply performed by measuring the signal $S(2\omega_L - \omega_S) = S(\omega_L + \omega_R)$, which is a coherent beam emitted in a certain direction. These coherent signals with anti-Stokes frequencies are generated each time the frequency difference of the input laser fields matches the molecular frequency of a Raman active transition.

The mixing of the two laser fields can also produce radiation on the Stokes side of the ω_S-laser. The direction of this coherent Stokes Raman scattering (CSRS) signal is again determined by a corresponding momentum conservation diagram, which leads to a different direction (see **Figure 3**), labelled by $S(2\omega_S - \omega_L)$. Since the CSRS signal is in principle weaker than the CARS signal, and because the former may be overlapped by fluorescence, the CARS technique is more frequently used.

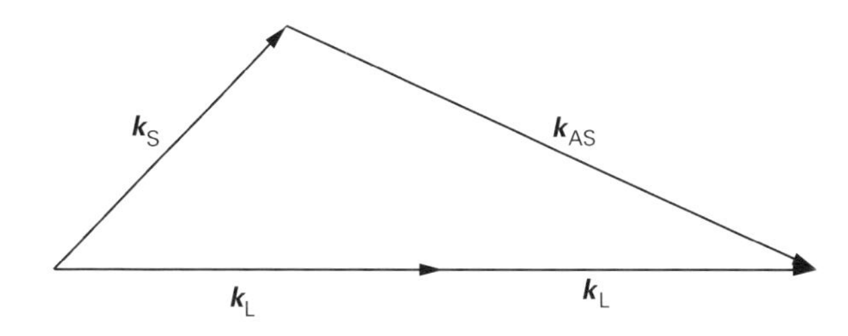

Figure 4 Momentum conservation for CARS (representation of Equation 4).

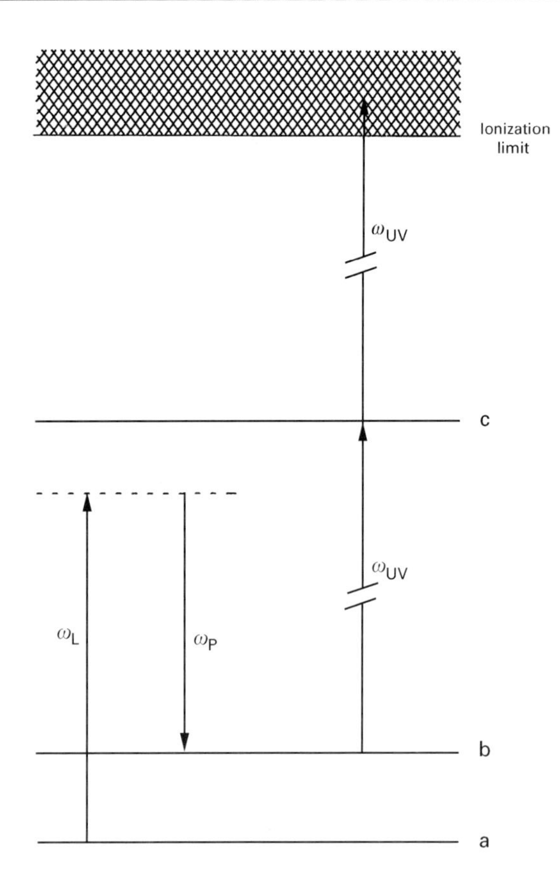

Figure 5 Energy-level diagram illustrating the two excitation steps of Ionization Detected Stimulated Raman Spectroscopy (IDSRS).

The interaction of the electric fields of the two ω_L and ω_S lasers with the coherent molecular vibrations yields also a gain or a loss in the power of the lasers. The method where the gain at the Stokes frequency (labelled in **Figure 3** by $+\Delta S(\omega_S)$) is measured is generally referred to as 'stimulated Raman gain spectroscopy' whereas the 'inverse Raman scattering' (IRS) is the terminology commonly used to designate the induced loss at the pump laser frequency (**Figure 3**, $-\Delta S(\omega_L)$). IRS is also often called stimulated Raman loss spectroscopy (SRLS). In order to get full Raman information of the medium, it is necessary to tune the frequency difference $\omega_L - \omega_S$; then, successively all Raman-active vibrations (or rotations, or rotation–vibrations) will be excited and a complete nonlinear Raman spectrum is then obtained either by measuring newly generated signals (CARS, CSRS) or the gain (SRGS) or loss (SRLS) of the pump or the Stokes laser, respectively. In what is called broadband CARS, the Stokes (ω_S) is spectrally broad, while the pump laser (ω_L) is kept spectrally narrow, resulting in the simultaneous generation of a broad CARS spectrum. For the detection of the latter a spectrometer together with a CCD camera is needed.

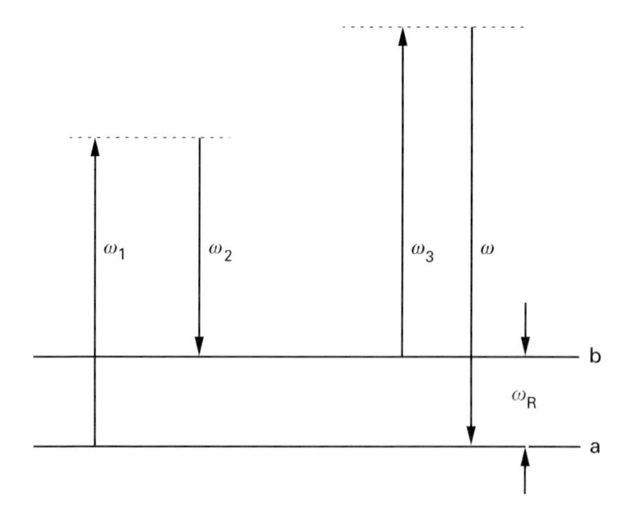

Figure 6 Schematic diagram representing the four-wave mixing process: a polarization is generated at the frequency $\omega = \omega_1 - \omega_2 + \omega_3$.

In photoacoustic Raman spectroscopy (PARS), due to the interaction of the two input laser fields (ω_L, ω_S) a population of a particular energy level (ω_R) of the sample is achieved. As the vibrationally (or rotationally) excited molecules relax by means of collisions, a pressure wave is generated in the sample and this acoustic signal is detected by a sensitive microphone.

A technique which combines the high sensitivity of resonant laser ionization methods with the advantages of nonlinear coherent Raman spectroscopy is called IDSRS (ionization detected stimulated Raman spectroscopy). The excitation process, illustrated in **Figure 5**, can be briefly described as a two-step photoexcitation process followed by ion/electron detection. In the first step two intense narrow-band lasers (ω_L, ω_S) are used to vibrationally excite the molecule via the stimulated Raman process. The excited molecules are then selectively ionized in a second step via a two- or multiphoton process. If there are intermediate resonant states involved (as state c in **Figure 5**), the method is called REMPI (resonance enhanced multi-photon ionization)-detected stimulated Raman spectroscopy. The technique allows an increase in sensitivity of over three orders of magnitude because ions can be detected with much higher sensitivity than photons.

The nonlinear Raman techniques discussed above are special cases of a general four-wave mixing process, which is schematically illustrated in **Figure 6**. Here, three independent fields with angular frequencies ω_1, ω_2 and ω_3 may be incident upon the matter. A fourth field, which is phase coherent relative to the input fields, is then generated at angular frequency $\omega = \omega_1 - \omega_2 + \omega_3$. When the angular frequency

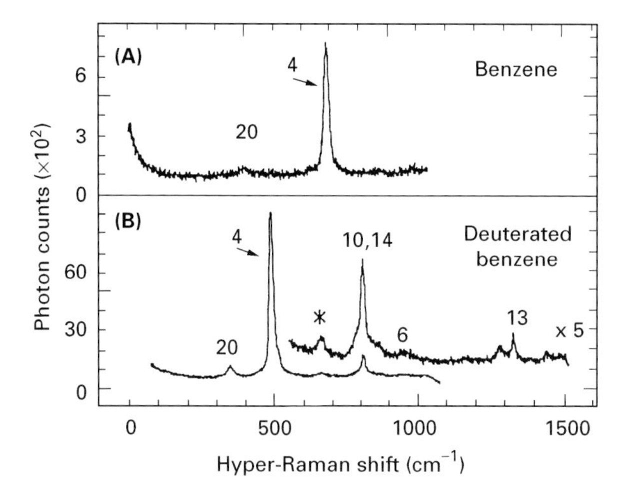

Figure 7 Hyper-Raman spectra of C_6H_6 excited with a Nd:YAG laser (λ_0 = 1.064 nm) Q-switched at 1 kHz (A) and of C_6D_6 in the lower spectrum with the laser Q-switched at 6 kHz (B). Reproduced by permission of Elsevier Science from Acker WP, Leach DH and Chang RK (1989) Stokes and anti-Stokes hyper Raman scattering from benzene, deuterated benzene, and carbon tetrachloride. *Chemical Physics Letters* **155**: 491–495.

difference $\omega_1 - \omega_2$ equals the Raman excitation angular frequency ω_R, the signal wave at ω is enhanced, indicating a Raman resonance. For example, a CARS signal is Raman resonantly generated when $\omega_1 = \omega_3 = \omega_L$, $\omega_2 = \omega_S$ and $\omega_L - \omega_S = \omega_R$.

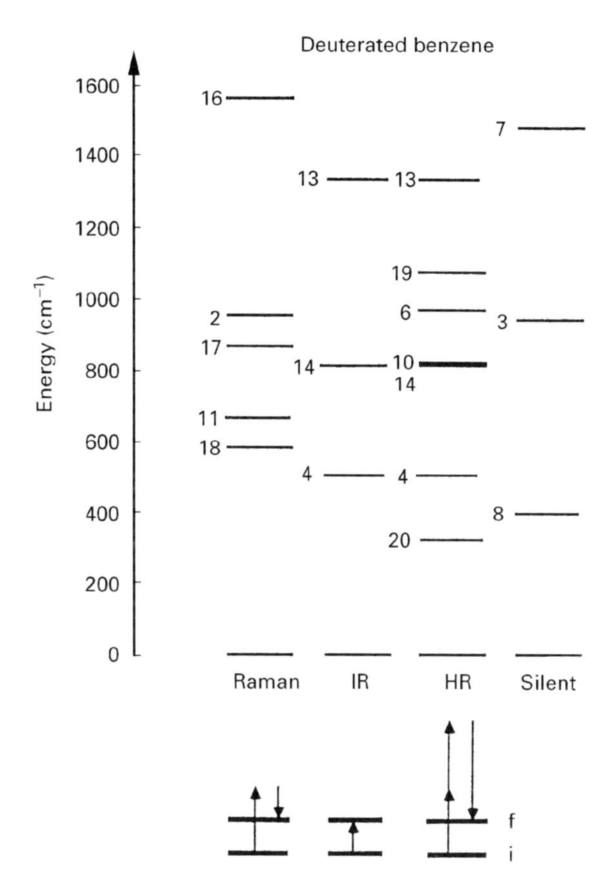

Figure 8 Vibrational energy levels of C_6D_6 (energy < 1600 cm^{-1}) grouped by their activity from the ground state, i.e. Raman, IR, or hyper-Raman (HR). Modes which are not active in Raman, IR, or hyper-Raman are grouped. Reproduced by permission of Elsevier Science from Acker WP, Leach DH and Chang RK (1989) Stokes and anti-Stokes hyper Raman scattering from benzene, deuterated benzene, and carbon tetrachloride. *Chemical Physics Letters* **155**: 491–495.

Applications

Applications of spontaneous nonlinear Raman spectroscopy (Hyper-Raman scattering)

Since its discovery in 1965, hyper-Raman spectra have been observed in all three states of aggregate. However, reasonable signal-to-noise ratios could only be obtained for a convenient measurement time after the development of fast pulsed, high power lasers and highly sensitive detectors (multichannel diode arrays or charge-coupled devices (CCDs)). Before that time only a few gases had been studied which included ethane, ethene and methane. Only vibrational spectra of modest resolution have been obtained in these studies. A number of group IV tetrahalides have been studied in the liquid phase. Other liquids whose Raman spectra have been reported include water and tetra-chloroethene. Probably most hyper-Raman work was performed in crystals: NH_4Cl, NH_4Br, calcite, $NaNO_2$, $NaNO_3$, $LiNbO_3$, $SrTiO_3$, caesium and rubidium halides, rutile, PbI_2, CuBr, diamond and quartz. Stimulated hyper-Raman scattering has been observed from sodium vapour, resonance hyper-Raman scattering from CdS and surface enhanced hyper-Raman scattering from SO_3^{2-} ions adsorbed on silver powder.

Technological advances, i.e. CW pumped acoustooptically Q-switched Nd:YAG lasers with repetition rates of up to 5 kHz combined with multichannel detection systems have increased the ease of obtaining hyper-Raman signals. By making use of this advanced technology, hyper-Raman spectra of benzene and pyridine could be obtained. Spectra from benzene, deuterated benzene and carbon tetrachloride have been measured with high signal-to-noise ratios. As examples, we show in **Figure 7** the hyper-Raman spectra of benzene and deuterated benzene. The observed hyper-Raman bands are labelled by numbers (4, 6, 10, 13, 14, 20) and correspond to the ν_4 (A_{2u}), ν_6 (B_{1u}), ν_{10} (B_{2u}), ν_{13} (E_{1u}), ν_{14} (E_{1u}) and ν_{20} (E_{2u}) vibrations of C_6D_6, respectively. **Figure 8** shows the low-lying vibrational energy levels for C_6D_6 grouped by their activity involving transition from the ground state, i.e. Raman, IR, hyper-Raman (HR) and none of the above which are grouped as silent. Note that in the third column four modes with energy below 1500 cm^{-1} are only hyper-Raman active and three modes of symmetry A_{1u} and E_{1u} are both IR and hyper-Raman active. Except for the ν_{19} (E_{2u}) mode all hyper-Raman active modes can be found in the spectrum displayed in **Figure 7**. The modes of class B_{2g} are active in the second hyper-Raman effect which is controlled by the fourth rank second hyperpolarizability tensor γ.

Hyper-Raman scattering under resonance conditions for molecules in the gas phase was observed in 1993. High quality rotational resonance hyper-Raman spectra of NH_3 were obtained using blue incident radiation at half the $\tilde{X} \rightarrow \tilde{A}$ transition energy. Also hyper-Raman scattering of methyl iodide for excitation with a laser line which has been tuned through the two-photon resonance with the absorption band of a predissociative Rydberg transition in the VUV (175–183 nm) was reported. Similarly to linear resonance Raman scattering, overtones or combination bands can also be observed for resonantly excited hyper-Raman sc attering. An example is given in **Figure 9** where several higher order modes of methyl iodide can be observed.

The use of CW pumped acoustooptically Q-switched Nd:YAG lasers (repetition rates of 5 kHz), synchronously gated photomultiplier tubes, and synchronously gated two-dimensional single-photon counting detectors has improved the signal-to-noise ratio of hyper-Raman spectra. Considerable further improvements have been obtained with mode-locked pulses (at 82 MHz) from a Nd:YAG laser to observe the surface-enhanced hyper-Raman signal from pyridine adsorbed on silver. In these studies, hyper-Raman signals were observed with intensities close to spontaneous Raman scattering. It was shown that surface enhanced hyper-Raman scattering (SEHRS) has become a useful spectroscopic technique. In view of the recent advances in laser and detector technology, significant improvement in SEHRS sensitivity will come rapidly from the use of an intensified CCD camera for hyper-Raman signal detection and the use of a continuously tunable mode-locked Ti:sapphire laser as the excitation source.

Applications of coherent anti-Stokes Raman spectroscopy (CARS)

The advantages of CARS, i.e. high signal strength, very high spectral or temporal resolution, discrimination against fluorescence, etc., have opened new ways to study molecular structure. In the following

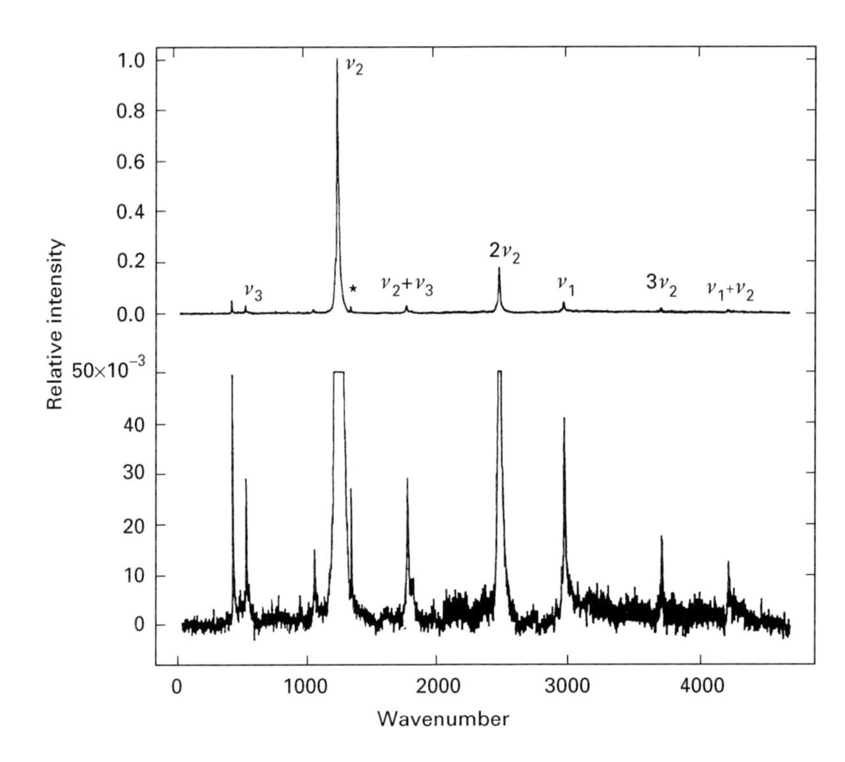

Figure 9 Resonance hyper Raman spectrum of CH_3I vapour excited at 365.95 nm. Reproduced by permission of Elsevier Science from Campbell DJ and Ziegler LD (1993) Resonance hyper-Raman scattering in the VUV. Femtosecond dynamics of the predissociated C state of methyl iodide. *Chemical Physics Letters* **201**: 159–165.

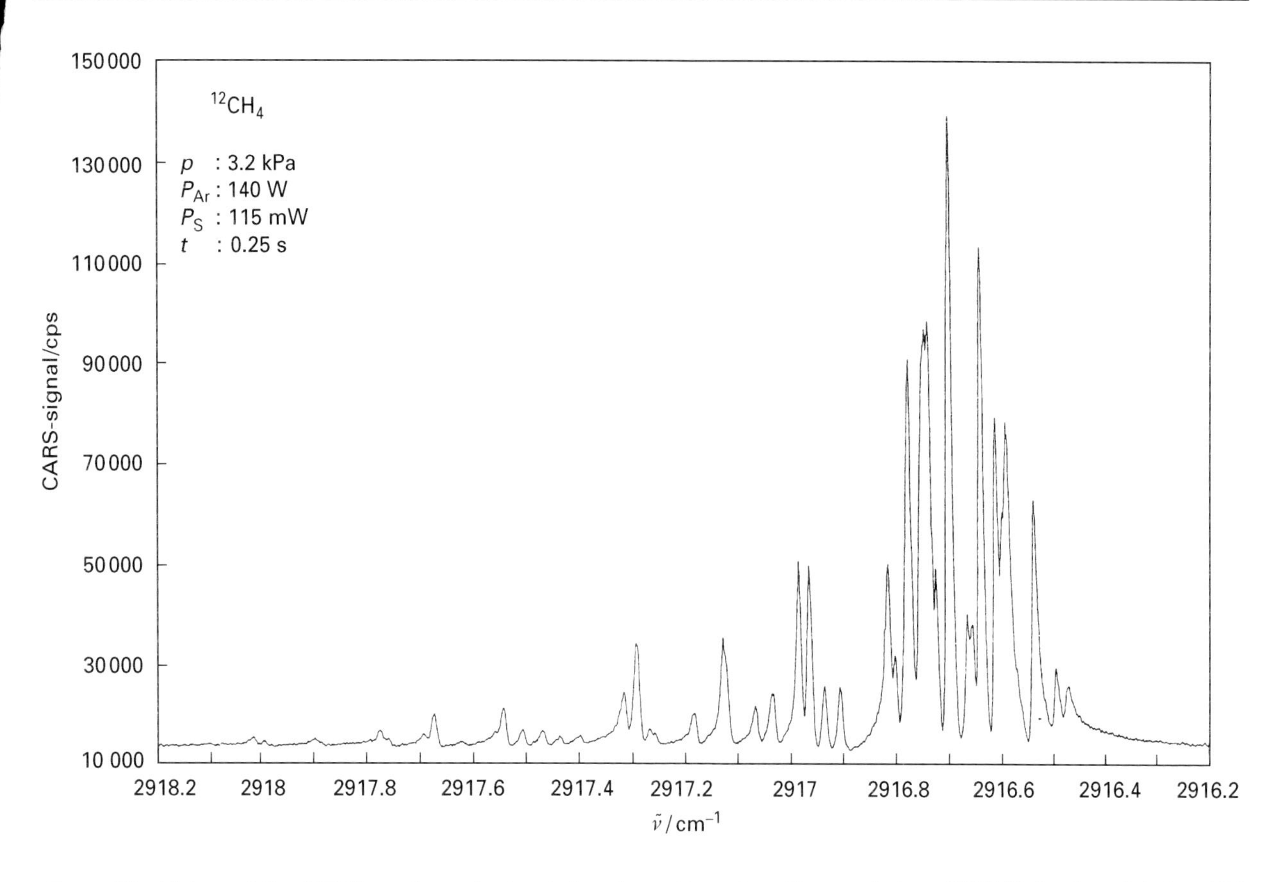

Figure 10 High-resolution CARS spectrum of ν_1 band of methane. Reproduced by permission of VCH Verlag from Schrötter HW (1995) Raman spectra of gases. In: Schrader B (ed.) *Infrared and Raman Spectroscopy*, pp 277–297. Weinheim: VCH Verlag.

some selected examples will be given to demonstrate the capability of this nonlinear coherent technique.

The 1980s and 1990s have seen a remarkable growth in the number of CARS applications to molecular and physical properties, particularly in the field of gas-phase systems. The latter are challenging because of low sample densities and the narrow transition line widths make them attractive for high resolution studies. Gas-phase CARS spectra have been obtained so far at pressures down to a few pascal, at temperatures ranging from a few K to 3600 K, and at a resolution better than about 10^{-3} cm^{-1}.

Mainly, the Q-branches of simple molecules, like di-, tri-, and four-atomic as well as spherical XY_4 top molecules have been studied. As an example **Figure 10** shows the Q-branch of methane. The complicated rotational structure seen there has been resolved by applying this powerful nonlinear Raman technique. This very high resolution of the order of 10^{-3} cm^{-1} allows us to study in detail collisional effects, which is of particular importance as a basis for the determination of temperatures and pressures.

One very active area of the gas-phase CARS technique has been the remote sensing of temperature and species in hostile environments such as gas discharges, plasmas, flames, internal combustion engines, and the exhaust from jet engines. The high signal intensity and the excellent temporal and spectral resolution of CARS make it a favourite method for such studies. For example, CARS has been used to measure state populations and changes in discharges of H_2, N_2 and O_2 at pressure ranging from a few kPa down to 0.6 Pa. Also, gas-phase CARS can be employed to monitor SiH_2 intermediates in their investigations of silane plasmas commonly used in amorphous silicon deposition processes. Many laboratories are engaged in combustion research. Combustion studies in engines include thermometry in a diesel engine, in a production petrol engine, and thermometry and species measurements in a fully afterburning jet engine. Investigations on turbulent and sooting flames were performed.

Temperature information from CARS spectra derives from spectral shapes either of the Q-branches or of the pure rotational CARS spectra of the molecular constituents. In combustion research it is most common to perform thermometry from nitrogen since it is the dominant constituent and present everywhere in large concentration despite the extent

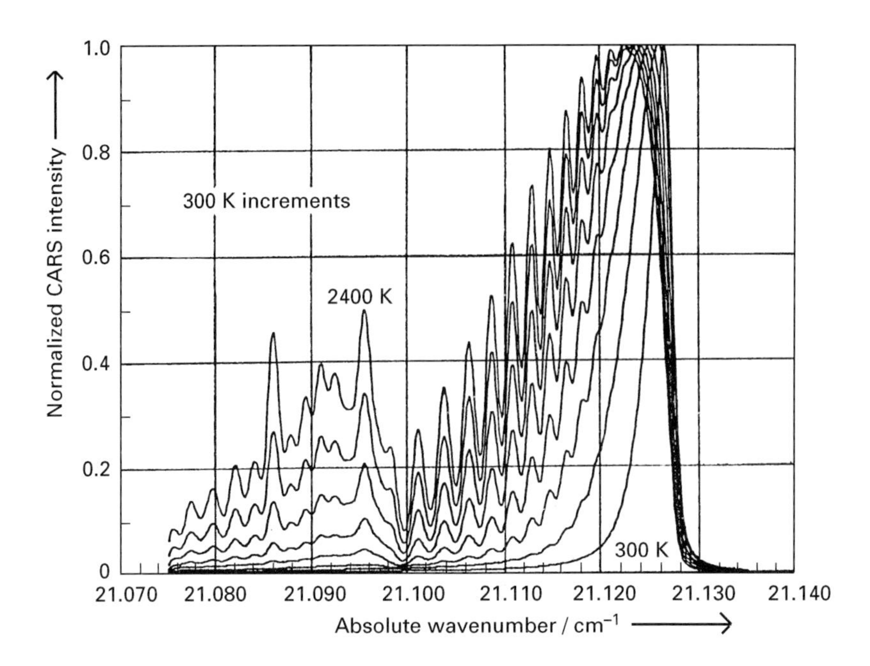

Figure 11 Temperature dependence of N_2 CARS spectrum from 300 to 2400 K in 300 K increments (Hall and Eckbreth, 1984).

of chemical reaction. The Q-branch of nitrogen changes its shape due to the increased contribution of higher rotational levels which become more populated when the temperature increases. **Figure 11** displays a calculated temperature dependence of the N_2 CARS spectrum for experimental parameters typically used in CARS thermometry. Note that the wavenumber scale corresponds to the absolute wavenumber value for the ~2320 cm^{-1} Q-branch of N_2 when excited with the freqency doubled Nd:YAG laser at 532 nm (≅ 18 796 cm^{-1}), i.e. $\tilde{\nu}_{AS}$ = 18 796 + 2320 = 21 126 cm^{-1}. The bands lower than about 21 100 cm^{-1} are due to the rotational structure of the first vibrational hot band.

For the case that there are not too many constituents in the gas under investigation, the use of the pure rotational CARS technique may be superior to vibrational CARS thermometry since the spectra are easily resolvable (for N_2 the adjacent rotational peaks have a spacing of approximately 8 cm^{-1}) compared with the congestion of the rotational lines in the vibrational bands of the Q-branch spectra (see **Figure 11**). An experimental comparison of rotational and vibrational CARS techniques, under similar conditions has been made that demonstrates that rotational CARS may be viable for flame-temperature measurements up to 2000 K. Of course, the pure rotational approach cannot be applied for spherical molecules which have no pure rotational CARS spectrum. An elegant method, using Fourier analysis based on the periodicity of pure rotational CARS spectra has been introduced recently.

In addition to temperature measurements the gas-phase CARS technique also provides information on the fluctuating properties occurring for instance in turbulent combustion systems. However, concentration measurements are more difficult to perform than temperature ones because the absolute intensity is required, while temperature measurements are only based on the shape of the spectrum. Simultaneous information on the relative concentrations between several species are easier to obtain.

Quantitative gas-phase CARS spectroscopy has also been applied to probing species in a laboratory chemical reactor and to temperature measurements inside incandescent lamps. Another interesting area is that of CARS applied to free expansion jets. The key benefits of this technique are the spectral simplification of cold molecules and the increased concentrations of small van der Waals complexes obtained under the non-equilibrium jet conditions.

CARS is also used for the study of samples in the condensed phase. The major experimental advantage of CARS (and most nonlinear coherent Raman techniques) is the large signal produced. In a typical CARS experiment in a liquid or a solid, the applied laser power of the pump and Stokes laser (10^4–10^5 W) generates an output power of up to 1 W, while conventional Raman scattering would give a collected signal power of ~10^{-4} W with the same lasers. Since the CARS output is directional, the collection angle can be five orders of magnitude smaller than that needed in spontaneous scattering. Taken together, these two factors imply that CARS is nine orders of

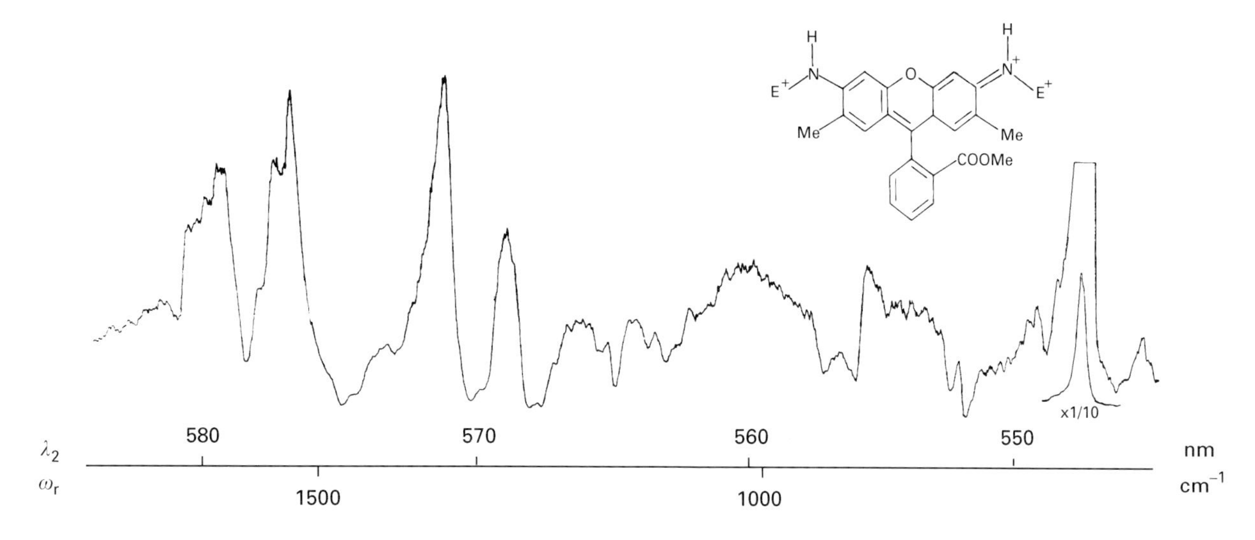

Figure 12 CARS spectrum of rhodamine 6G in solution (Carreira and Horovitz, 1982).

magnitude less sensitive to sample fluorescence than spontaneous scattering. The advantage is actually even greater since the CARS signal is at higher frequency than any of the input laser frequencies.

While it is nearly impossible to obtain Raman spectra of highly luminescent materials, e.g. dye solution, it was the CARS technique which first overcame this problem because of the reasons mentioned above. As an example the CARS spectrum of a rhodamine 6 G (R6G) water solution is displayed in **Figure 12**. The vibrational modes of the strongly luminescent R6G molecule can be seen. It should be mentioned at this point that by long wavelength excitation, i.e. for example excitation with the 1.064 μm line of a CW Nd:YAG laser or by making use of the SERS effect luminescence-free linear Raman spectra can be obtained. Since the latter methods are in any case much easier to perform than CARS or other nonlinear Raman techniques, they are to be preferred. However, if one is interested in obtaining structural as well as electronic properties of absorbing materials through resonance excitation, there are many cases where linear resonance Raman spectroscopy is limited because of the mentioned strong luminescence.

On the other hand, many, particularly organic, substances show considerable third-order nonlinear susceptibilities $\chi^{(3)}$, as for example polyacetylenes, polydiacetylenes or chlorophyll. For such systems, resonance CARS spectroscopy is a suitable tool to obtain resonance Raman information via the anti-Stokes, coherent spectroscopic method. However, in performing resonance CARS spectroscopy in solids one must realize that this technique results in a fairly complicated arrangement between the sample and the coherent beams. First, the phase-matching conditions (Eqn [4], **Figure 4**) have to be obeyed, where the momentum vectors depend also on the refractive index of the solid media. Therefore a continuous adjustment of the crossing angle between the incident laser beams ($\boldsymbol{k}_L$, $\boldsymbol{k}_S$) as well as of the angle between the pump laser beam and the CARS beam ($\boldsymbol{k}_L$, $\boldsymbol{k}_{AS}$) is required during the scan of the CARS spectrum. Secondly, in order to excite particular phonons in the crystals, the difference between the pump and the Stokes beam wave vectors must coincide with the wave vector of the coherently excited phonon in the crystal ($\boldsymbol{k}_L - \boldsymbol{k}_S = \boldsymbol{k}_{phonon}$). Depending on the strength of absorption and sample thickness, CARS in solids is either performed in transmission or in reflection (backscattering CARS).

As an example of resonance CARS studies in solids, for which a linear resonance Raman study has been impossible to perform because of simultaneous strong luminescence, we considered here investigations on colour zones in substituted diacetylene crystals originating from partial polymerization.

For a long time it has been known that diacetylene monomer single crystals undergo, upon thermal annealing or exposure to high-energy radiation, topochemical solid-state polymerization. From this reaction, polymer chains are formed which have a substantial π-electron delocalization, forming a pseudo-one-dimensional electronic system. Colour zones occur in such crystals due to different chain lengths and CARS studies were performed on these zones in crystals with low polymer content, where the polymer chains were embedded in the monomer matrix. As mentioned, resonance Raman excitation within the strong absorption of the polymer chains, i.e. within the absorption of the colour zone, produced high luminescence levels which obscured the bands in linear Raman spectroscopy. In contrast,

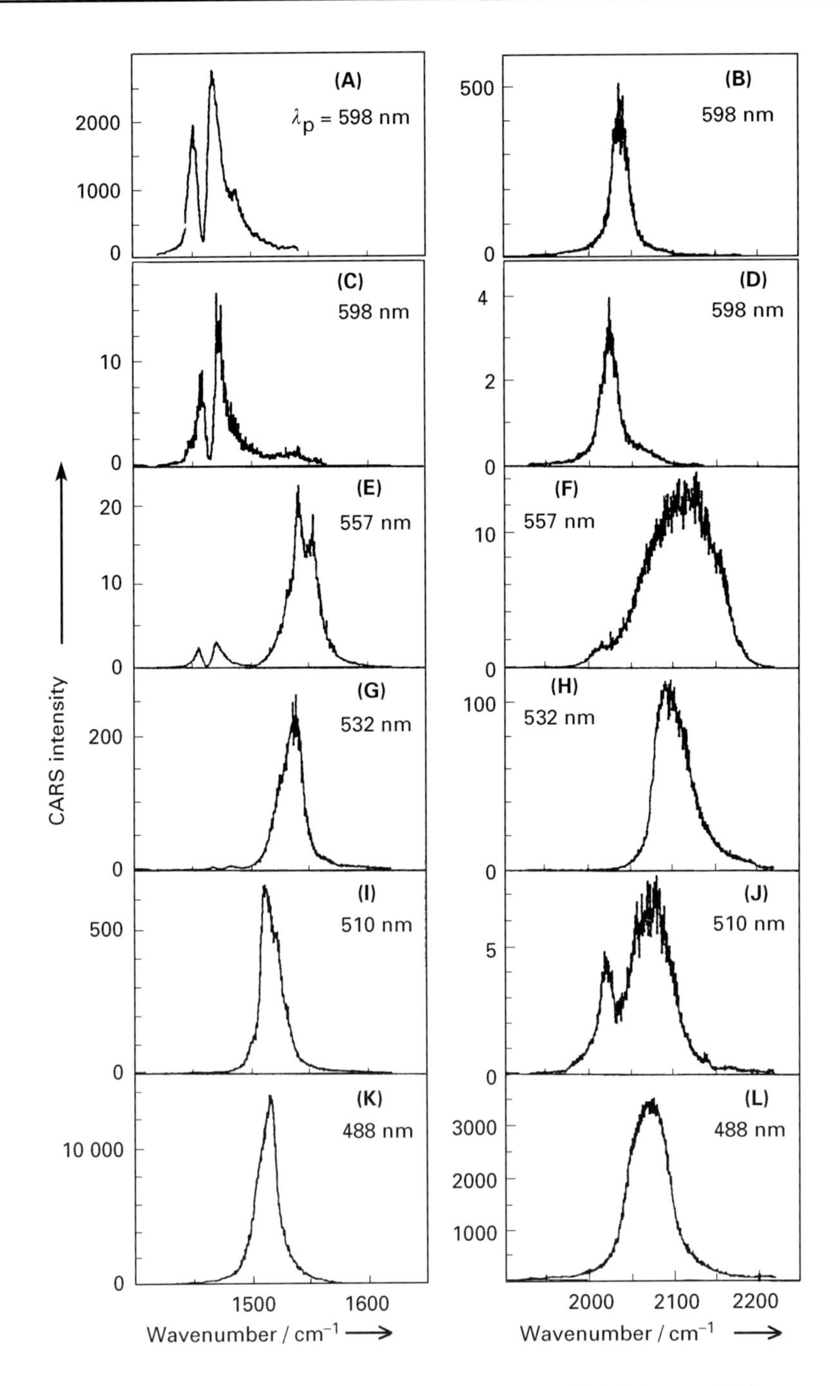

Figure 13 Resonance CARS spectra of a substituted diacetylene single crystal (FBS-DA) at 10 K. The pump wavelength λ_p used is labelled for each spectrum. (A) and (B) show CARS spectra of the P-colour zone, and (C)–(L) those for the Y-colour zone. Spectra on the left side correspond to the C=C stretching region, and those on the right side to the C≡C stretching region. For further details, see text. Reproduced by permission of John Wiley & Sons from Materny A and Kiefer W (1992) Resonance CARS spectroscopy on diacetylene single crystals. *Journal of Raman Spectroscopy* **23**: 99–106.

luminescence-free resonance CARS spectra can be obtained, as shown in **Figure 13** for the case of an FBS DA crystal at 10 K (FBS = 2,4-hexadiynylene-di-p-fluorobenzene sulfonate, DA = diacetylene). On the left and right panels of **Figure 13** CARS spectra are displayed for the region of the C=C and C≡C stretching region around 1500 cm^{-1} and 2100 cm^{-1}, respectively. Spectra (A) and (B) are those of the P colour zone (P = principal) and (C) – (L) those of the Y-colour zone (Y = yellow). Note the very different CARS intensities as well as band shapes for the various excitation wavelengths of the pump laser (λ_p, which corresponds to ω_L of the CARS process as outlined above) which are due to different resonant

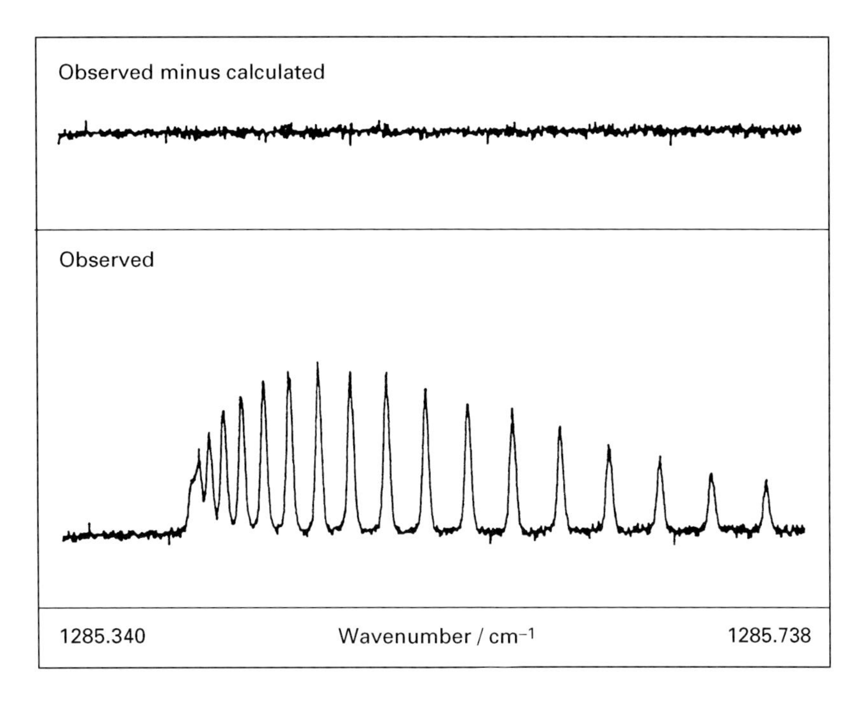

Figure 14 High resolution multi-pass stimulated Raman gain spectrum (SRGS) of the Q-branch of the lower component of the Fermi resonance diad of $^{12}C^{16}O_2$ at a pressure of 200 Pa (1.5 torr). Reproduced by permission of John Wiley & Sons from Saint-Loup R, Lavorel B, Millot G, Wenger C and Berger H (1990) Enhancement of sensitivity in high-resolution stimulated Raman spectroscopy of gases. *Journal of Raman Spectroscopy* **21**: 77–83.

enhancements. Comparing spectrum (K) with (C), for example, shows in addition the very high dynamic range (at least four orders of magnitude) inherent in this type of spectroscopy. Analysing the CARS spectra together with the absorption spectra of several substituted DA crystals, one is able to derive important structural as well as electronic properties of this type of crystal.

It should be mentioned that there are some disadvantages of CARS: (i) an unavoidable electronic background nonlinearity that alters the line shape and can limit the detection sensitivity; (ii) a signal that scales as the square of the spontaneous scattering signal (and as the cube of the laser power), making the signals from weakly scattering samples difficult to detect; and (iii) the need to fulfil the phase matching requirements. While other techniques avoid these difficulties, CARS still remains the most popular coherent nonlinear technique.

Applications of stimulated Raman gain and inverse Raman spectroscopy (SRGS, IRS)

The advantages of SRGS and IRS are that (in contrast to CARS) the signal is linearly proportional to the spontaneous Raman scattering cross-section (and to the product of the two laser intensities), and that the phase-matching condition is automatically fulfilled. The fact that the resolution of the nonlinear Raman techniques is limited only by the laser line widths gives the stimulated Raman techniques particular appeal under conditions where interference from background luminescence is problematic or in situations where very high resolution is required. The main disadvantage of these techniques, however, is that they are quite sensitive to laser noise. The latter requires high stability in laser power.

Due to complexity, only a few stimulated Raman gain and loss spectrometers with a main application in high resolution molecular spectroscopy have been built since the fundamental developments around 1978.

Here, we present an instructive example for each of the two techniques (SRGS, IRS) emphasizing the high resolution capability of these methods.

The Q-branches of numerous molecules, particularly of linear and spherical top molecules have been analysed by means of SRGS and IRS. As an example of a recent high resolution SRGS spectrum we show in **Figure 14** the spectrum of the Q-branch of the lower component of the Fermi resonance diad of $^{12}C^{16}O_2$ at 1285 cm^{-1}. The spectrum has been recorded at a pressure of 200 Pa (1.5 torr). The excellent agreement with a calculation assuming Voigt line profiles is demonstrated by the residual spectrum in the upper trace.

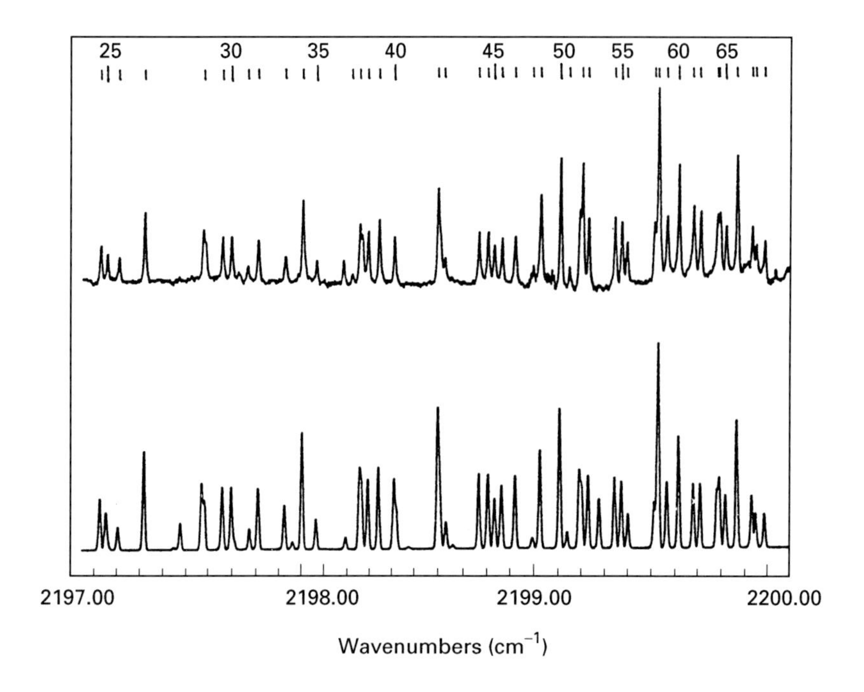

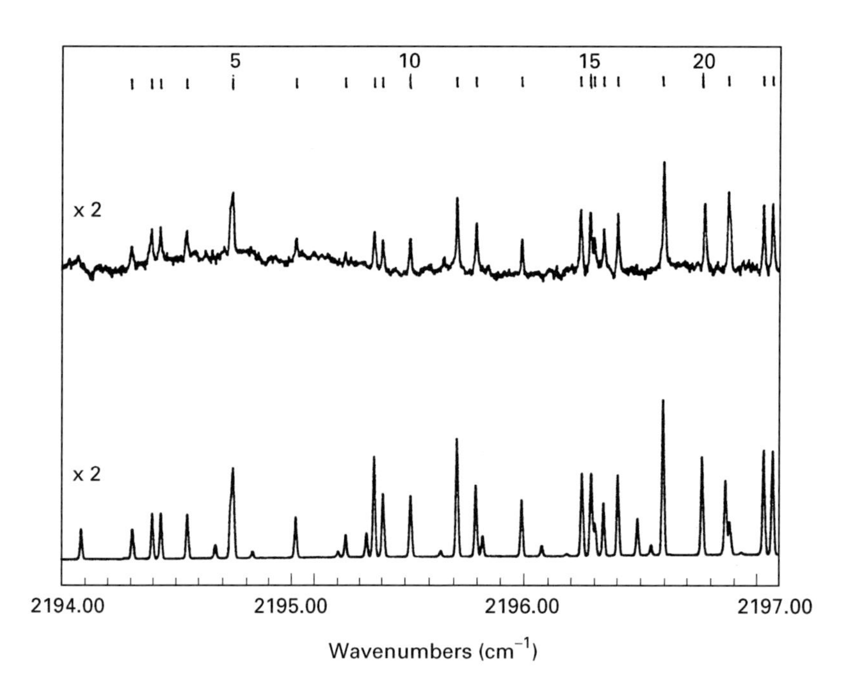

Figure 15 High resolution inverse Raman Spectrum of the ν_2 Q-branch of CH_3D between 2194 and 2200 cm^{-1}. Upper traces : Observed, lower traces: calculated spectra. Reproduced by permission of John Wiley & Sons from Bermejo D, Santos J, Cancio P *et al* (1990) High-resolution quasicontinuous wave inverse Raman spectrometer. Spectrum of CH_3D in the C-D stretching region. *Journal of Raman Spectroscopy* **21**: 197–201.

An example for high-resolution IRS is given in **Figure 15**, where the ν_2 Q-branch of CH_3D is displayed. This spectrum represents a Doppler-limited spectrum of the C–D stretching band. The authors were able to assign the observed transitions by performing a theoretical fit to the observed data which allowed them to refine some of the rotational–vibrational constants.

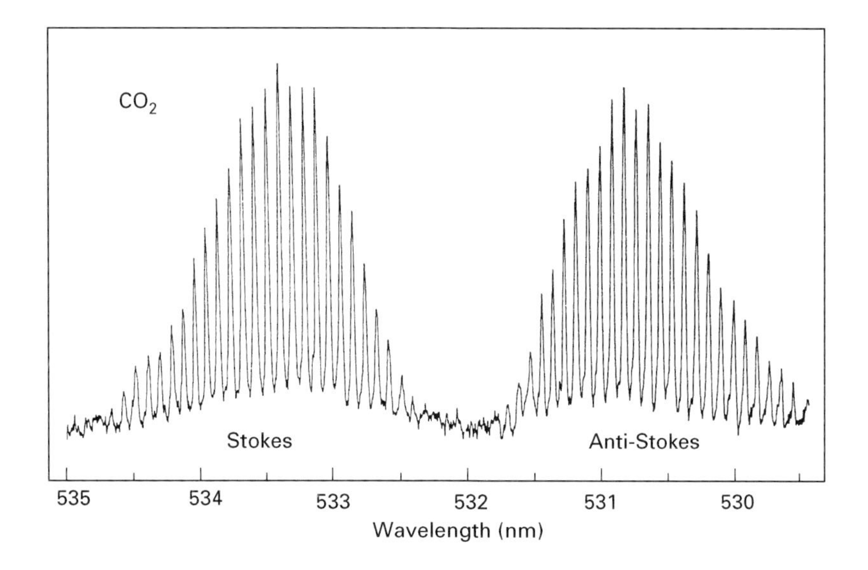

Figure 16 The pure rotational photoacoustic Raman (PARS) spectrum of CO_2 gas at a pressure of 80 kPa (600 torr); pump laser wave length at 532 nm. Note the complete absence of any acoustical signal due to Rayleigh scattering (at 532 nm). Reproduced by permission of Academic Press from Barrett JJ (1981) Photoacoustic Raman Spectroscopy. In: Harvey AB (ed) *Chemical Applications of Nonlinear Raman Spectroscopy*, pp 89–169. New York: Academic Press.

Applications of photoacoustic Raman spectroscopy (PARS)

As discussed above in photoacoustic Raman spectroscopy (PARS) the energy deposited in the sample by excitation of, for example, a vibration by the stimulated Raman process leads to pressure increases through relaxation to translational energy and can therefore be detected by a sensitive microphone.

When the pump (ω_L) and Stokes (ω_S) beams have only small frequency differences, as can be achieved, for example, by using a frequency-doubled Nd:YAG laser for ω_L and a dye laser with amplifier pumped by the third harmonic of the same Nd:YAG laser for ω_S, the recording of pure rotational PARS spectra becomes possible. Such a spectrum at medium resolution is shown in **Figure 16**. The striking feature of this spectrum is the absence of a strong Rayleigh component at the pump wavelength (532 nm) because at that wavelength no energy is deposited in the sample.

The PARS technique has been extended to study vibrational–rotational transitions with high resolution (~0.005 cm^{-1}). For example, a high resolution PARS spectrum of the lower component of the Fermi resonance diad of CO_2 at a pressure of 1.6 kPa (= 11 torr) could be obtained with high signal-to-noise ratio.

In another PARS study it was shown that photoacoustic Raman spectroscopy is a sensitive technique for obtaining Raman spectra of hydrogen-bonded complexes in the gas phase. PARS spectra of the CN stretching ν_1 region of HCN as a function of pressure revealed bands which could be assigned to HCN dimers and trimers.

Applications of ionization detected stimulated Raman spectroscopy (IDSRS)

Above we have discussed how the sensitivity in determining Raman transitions can be enormously increased by employing nonlinear Raman schemes in which the shifts in vibrational state populations due to stimulated Raman transitions are probed by resonance-enhanced multiphoton ionization. As ions can be detected with much higher sensitivity than photons, the signal-to-noise ratio in the nonlinear Raman spectrum of, for example, NO could be improved by a factor of 10^3 by this method. In fact, one can obtain sufficient sensitivity to characterize the Raman transitions of species even in molecular beams. The high sensitivity of IDSRS made it, for instance, possible to investigate the degenerate Fermi doublet of benzene in such a molecular beam experiment. The two Fermi subbands could be recorded separately by selectively tuning the UV laser into resonance with electronic transitions from one of the two states. When the Stokes laser is tuned, then the rovibrational structure of only one Raman transition is recorded. **Figure 17** shows in the upper part the lines belonging to ν_{16} and in the lower part those assigned to $\nu_2 + \nu_{18}$ in the same spectral region.

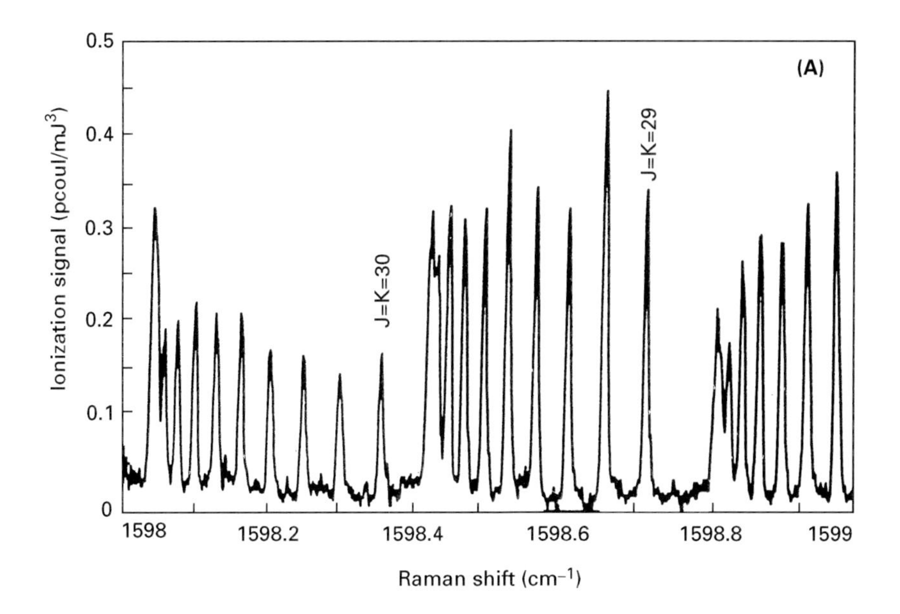

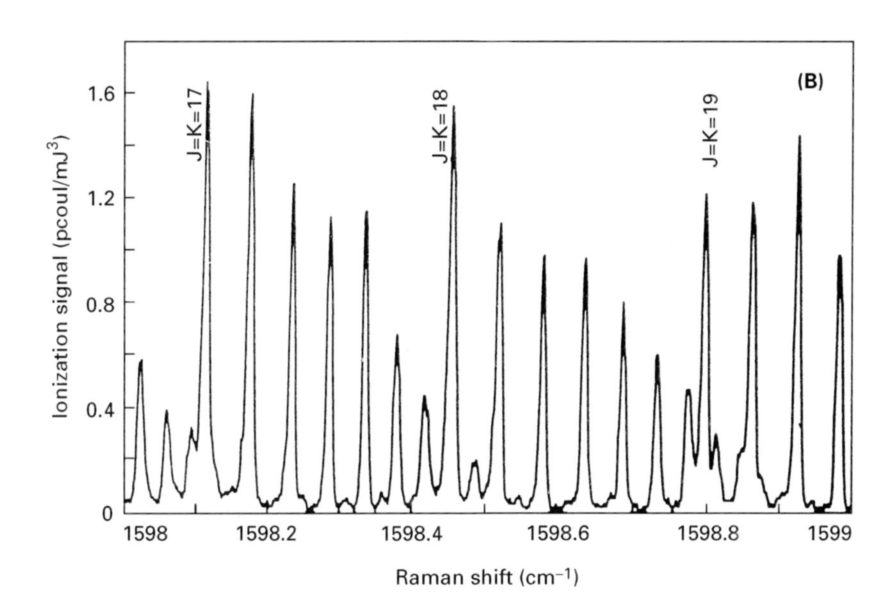

Figure 17 Ionization detected stimulated (IDSRS) spectra of benzene in the region of overlap between O-branch transitions of ν_{16} and the S-branch transitions of $\nu_2 + \nu_{18}$. (A) UV laser tuned to 36 467 cm^{-1}; (B) UV laser tuned to 36 496 cm^{-1}. Reproduced with the permission of the American Institute of Physics from Esherick P, Owyoung A and Pliva J (1985) Ionization-detected Raman studies of the 1600 cm^{-1} Fermi diad of benzene. *Journal of Chemical Physics* **83**: 3311–3317.

List of symbols

$d\sigma/d\Omega$ = differential Raman cross-section; E = electric field; g_S = gain constant; $\mathbf{k}$ = wave vector; $\mathbf{p}$ = dipole moment; α = polarizability; β = hyperpolarizability; γ = 2nd hyperpolarizability; Γ = line width; ε_0 = permittivity of vacuum; λ = wavelength; $\chi^{(3)}$ = 3rd order nonlinear susceptibility; ω_L = angular frequency of exciting beam; $+\omega_P$ = anti-Stokes hyper-Raman displacement; $-\omega_R$ = Stokes hyper-Raman displacement.

See also: **Matrix Isolation Studies By IR and Raman Spectroscopies; Nonlinear Optical Properties; Nonlinear Raman Spectroscopy, Instruments; Nonlinear Raman Spectroscopy, Theory; Photoacoustic Spectroscopy, Theory; Raman Optical Activity, Applications; Raman Optical Activity, Theory; Rayleigh Scattering and Raman Spectroscopy, Theory; Surface-Enhanced Raman Scattering (SERS), Applications.**

Further reading

Acker WP, Leach DH and Chang RK (1989) Stokes and anti-Stokes hyper Raman scattering from benzene, deuterated benzene, and carbon tetrachloride. *Chemical Physics Letters* **155**: 491–495.

Barrett JJ (1981) Photoacoustic Raman spectroscopy. In: Harvey AB (ed) *Chemical Applications of Nonlinear Raman Spectroscopy*, pp 89–169. New York: Academic Press.

Berger H, Lavorel B and Millot G (1992) In: Andrews DL (ed.), *Applied Laser Spectroscopy*, pp 267–318. Weinheim: VCH Veilagsgesellschaft.

Bermejo D, Santos J, Cancio P *et al* (1990) High-resolution quasicontinuous wave inverse Raman spectrometer. Spectrum of CH_3D in the C-D stretching region. *Journal of Raman Spectroscopy* **21**: 197–201.

Campbell DJ and Ziegler LD (1993) Resonance hyper-Raman scattering in the VUV. Femtosecond dynamics of the predissociated C state of methyl iodide. *Chemical Physics Letters* **201**: 159–165.

Carreira LA and Horovitz ML (1982) Resonance coherent anti-Stokes Raman spectroscopy in condensed phases. In: Kiefer W and Long DA (eds) *Nonlinear Raman Spectroscopy and its Chemical Applications*, pp 429–443. Dordrecht: D. Reidel Publishing Company.

Esherick P, Owyoung A and Pliva J (1985) Ionization-detected Raman studies of the 1600 cm^{-1} Fermi diad of benzene. *Journal of Chemical Physics* **83**: 3311–3317.

Hall RJ and Eckbreth A (1984) Coherent anti-Stokes Raman spectroscopy (CARS): Application to combustion diagnostics. In: Ready F and Erf RK (eds) *Laser Applications* **5**: 213–309. New York: Academic Press.

Harvey AB (1981) *Chemical Applications of Nonlinear Raman Spectroscopy*. New York: Academic Press.

Kiefer W and Long DA (1982) *Nonlinear Raman Spectroscopy and its Chemical Applications*. Dordrecht: D. Reidel.

Kiefer W (1995) Nonlinear Raman Spectroscopy. In: Schrader B (ed.) *Infrared and Raman Spectroscopy*, pp 162–188. Weinheim: VCH Verlag.

Kiefer W (1995) Applications of non-classical Raman spectroscopy: resonance Raman, surface enhanced Raman, and nonlinear coherent Raman spectroscopy. In: Schrader B (ed.) *Infrared and Raman Spectroscopy*, pp 465–517. Weinheim: VCH Verlag.

Materny A and Kiefer W (1992) Resonance CARS spectroscopy on diacetylene single crystals. *Journal of Raman Spectroscopy* **23**: 99–106.

Saint-Loup R, Lavorel B, Millot G, Wenger C and Berger H (1990) Enhancement of sensitivity in high-resolution stimulated Raman spectroscopy of gases. *Journal of Raman Spectroscopy* **21**: 77–83.

Schrötter HW (1995) Raman spectra of gases. In: Schrader B (ed.) *Infrared and Raman Spectroscopy*, pp 227–297. Weinheim: VCH Verlag.

Nonlinear Raman Spectroscopy, Instruments

Peter C Chen, Spelman College, Atlanta, GA, USA

VIBRATIONAL, ROTATIONAL & RAMAN SPECTROSCOPIES
Methods & Instrumentation

Introduction

When exposed to large electric fields generated by intense sources of light (e.g. a laser), the charges in a material exhibit a nonlinear response. The resulting induced polarization of charge P is described by the series expansion

$$P = \chi^{(1)}E + \chi^{(2)}EE + \chi^{(3)}EEE + \ldots \quad [1]$$

where the second-, third-, and higher-order terms account for the nonlinear contribution. The coefficients $\chi^{(n)}$ are separate complex susceptibility tensor elements that describe the magnitude of the nonlinear contribution. The Es are the applied electric fields from the lasers with the form $E = A\exp[-i(\boldsymbol{k}x - \omega t)]$, where the $\boldsymbol{k}$ is the propagation or wave vector, ω is the angular frequency, and x and t indicate space and time, respectively. Since $\chi^{(1)} >> \chi^{(2)} >> \chi^{(3)}$, lasers with sufficiently large Es are required in order for the second and third terms to be significant. Most nonlinear Raman techniques rely on the third-order term to drive the induced polarization that generates an intense output beam.

Nonlinear spectra are produced by monitoring the intensity of the output beam while varying some parameters, such as the frequency ω of one or more of the laser beams. When the difference in frequency between two laser beams matches the frequency of a Raman-active mode, the resulting resonance enhances the nonlinear optical effect, causing a change in the intensity of the output beam. The result is a peak in the nonlinear Raman spectrum. Some nonlinear Raman techniques that use this approach are given in **Table 1**.

Table 1 Comparison of some nonlinear Raman techniques

Technique	*Comment*	*Variable*	*Output*	*Advantages*	*Disadvantages*
CARS	Most popular form of nonlinear Raman	ω_1 or ω_2	Intensity of newly generated light at $\omega_4 = \omega_1 - \omega_2 + \omega_3$	Fluorescence-free, intense signal at new wavelength	Phase matching required owing to dispersion, nonresonant background, complex line shape
CSRS	Nonparametric version of CARS	ω_1 or ω_2	Intensity of newly generated light at $\omega_4 = \omega_1 - \omega_2 + \omega_3$	Intense signal at new wavelength, can be used to observe dephasing effects	Susceptible to fluorescence, phase matching required due to dispersion, nonresonant background, complex line shape
SRG	Induced amplification of ω_2	Modulation of ω_1	Increase in intensity of ω_2 when $\omega_1 - \omega_2 = \omega_{Raman}$	No phase matching, no nonresonant background, linear with concentration	Sensitivity limited by stability of probe laser, difficult to multiplex
SRL	Induced reduction in intensity of ω_1	Modulation of ω_2	Decrease in intensity of ω_1 when $\omega_1 - \omega_2 = \omega_{Raman}$	No phase matching, no nonresonant background, linear with concentration	Sensitivity limited by stability of probe laser, difficult to multiplex
RIKES	Raman-induced birefringence	Modulation of ω_2, ω_1 is CW	Induced change in ω_1 polarization for $\omega_1 - \omega_2 = \omega_{Raman}$	Nonresonant background can be suppressed, no phase matching	Limited sensitivity, susceptible to turbulence and birefringence from windows, optics, sample
DFWM	Laser-induced grating		Beam of light at $\omega_4 = \omega_1 = \omega_2 = \omega_3$	Very sensitive, no phase matching	Multiple mechanisms (local and nonlocal), not a Raman technique
IRSFG	$\chi^{(2)}$ process, IR and Raman active		Intensity of newly generated light at $\omega_3 = \omega_1 + \omega_2$	Surface-specific	Requires tunable coherent IR source, relatively low signal intensity

Coherent anti-Stokes Raman spectroscopy (CARS)

Perhaps the best-known and most widely used form of nonlinear Raman spectroscopy is CARS. One of its attractive properties is that it generates an intense beam of light at a new frequency that is anti-Stokes (blue shifted) and spectrally separable from the input beams. Therefore, CARS is not susceptible to fluorescence (red shifted) and mechanisms (e.g. non-local effects) that can affect elastic scattering of the input beams.

CARS relies on the third-order term from Equation [1] which can be expanded as

$$\chi^{(3)} A_1 A_2 A_3 \exp[-i((\pm \boldsymbol{k}_1 \pm \boldsymbol{k}_2 \pm \boldsymbol{k}_3)\, x - (\pm\, \omega_1 \pm \omega_2 \pm \omega_3)t)] \qquad [2]$$

where the subscripts are labels for three input laser fields. This term can cause new light to be generated at frequency combinations corresponding to $\pm\,\omega_1 \pm \omega_2 \pm \omega_3$. CARS involves the generation of light at the specific output frequency $\omega_4 = \omega_1 - \omega_2 + \omega_3$. Raman-like peaks in the spectra are obtained when ω_1–ω_2 is tuned to the frequency of Raman-active vibrations or rotations. Judicious selection of the three input fields so that ω_1 or ω_4 matches the frequencies of coupled higher lying electronic levels can lead to the same type of enhancement observed in resonance Raman spectroscopy.

The intensity of the generated CARS beam can be written as

$$I_4 = \frac{144\pi^4 \omega_4 |\chi^{(3)}|^2 I_1 I_2 I_3 L^2}{c^4\, n_1\, n_2\, n_3\, n_4} \left\{ \frac{\sin(L\Delta k/2)}{L\Delta k/2} \right\}^2 \qquad [3]$$

where the *I*s correspond to the intensities of the beams, *n* is the refractive index, and *c* is the speed of light. This equation indicates that the output beam intensity varies as the product of the input laser intensities and the square of their overlap length *L* in the sample. The squared sinc function on the right is equal to 1 when the phase of the input and output beams are matched (i.e. phase matched).

Peaks in the nonlinear Raman spectrum are produced when $\chi^{(3)}$ changes while varying the frequencies of the input beams. The intensity of the CARS process varies as the squared modulus of the nonlinear susceptibility:

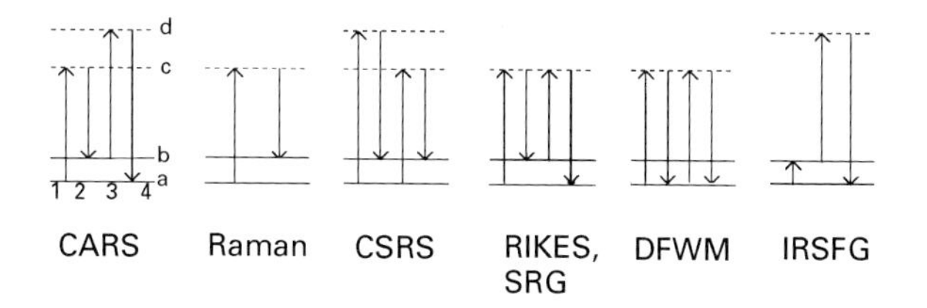

Figure 1 Energy level diagram for CARS, Raman, CSRS, DFWM, RIKES/SRG, and IRSFG. The dotted horizontal lines represent virtual levels and the solid horizontal lines represent ground, rotational, or vibrational levels. The output frequency corresponds to the downward arrow furthest to the right in each diagram. Electronic enhancement may be achieved if the virtual levels are replaced by real levels.

which is a fourth-ranked tensor that is summed over all possible states. *N* is the concentration, and the μs are transition dipole moments. The three products in the denominator are resonant terms that approach a minimum value of $i\Gamma$ (the dephasing linewidth) when a laser combination frequency matches the frequency of a level. The labels for the transition moments and the angular frequencies correspond to those shown in the CARS energy level diagram in **Figure 1**.

Equation [4] can be used to compare spectra from incoherent Raman and CARS. First, while conventional Raman varies linearly with the sample concentration, the CARS signal varies as $|\chi^{(3)}|^2$ and is therefore proportional to the square of the *N*. Furthermore, $\chi^{(3)}$ is a summation of terms, including both resonant and nonresonant contributions ($\chi^{(3)} = \chi_{\mathrm{res}}{}^{(3)} + \chi_{\mathrm{nr}}{}^{(3)}$). Contributing terms near resonance are primarily imaginary ($i\Gamma$ dominates), while non-resonant terms are primarily real ($i\Gamma$ is negligible). The nonresonant contributions result in a nonzero background, which determines the detection limits of the technique to around 0.1% in the condensed phase and 10 ppm in the gas phase. Therefore, although the nonlinear signal is more intense, the sensitivity of CARS for trace analysis is not necessarily higher than that of more conventional techniques. Finally, since the observed signal goes as $|\chi^{(3)}|^2$, the cross-product between the $\chi_{\mathrm{res}}{}^{(3)}$ and $\chi_{\mathrm{nr}}{}^{(3)}$ can contribute dispersion-like character. Therefore, CARS peaks often have asymmetric line shapes, especially when the nonresonant background is large relative to the resonant peak.

$$\chi^{(3)} \propto N \sum \frac{\mu_{ac}\mu_{cb}\mu_{bd}\mu_{da}}{(\omega_c - \omega_1 - i\Gamma_{ac})(\omega_b - (\omega_1 - \omega_2) - i\Gamma_{ab})(\omega_d - (\omega_1 - \omega_2 + \omega_3) - i\Gamma_{ad})} \qquad [4]$$

Other nonlinear Raman techniques

In addition to CARS, other closely related but less commonly used nonlinear Raman techniques have been developed. The energy level diagram for coherent Stokes Raman spectroscopy (CSRS) is shown in **Figure 1**. Unlike CARS, CSRS is nonparametric; the final state is not the same as the initial state. Therefore, CSRS spectra may exhibit extra peaks due to coherence dephasing. Furthermore, the CSRS output beam is generated to the Stokes (lower frequency) side of ω_3. CSRS is therefore more susceptible to spectral interference from fluorescence and Rayleigh scattering of the input beams.

Other Raman-based forms of nonlinear spectroscopy include stimulated Raman gain (SRG) or stimulated Raman scattering, stimulated Raman loss (SRL) or inverse Raman spectroscopy, and Raman induced Kerr effect spectroscopy (RIKES). Some information on these techniques are provided in **Table 1**. Many of these other forms do not produce light at wavelengths that are different from the input lasers, do not involve phase matching, and may be susceptible to multiple effects that may interfere with the measurement. Consequently, these techniques have not been as widely used as CARS.

Other nonlinear techniques

Several other forms of nonlinear spectroscopy have been developed that are not strictly based on Raman-active vibrations or rotations. Degenerate four-wave mixing (DFWM) is a $\chi^{(3)}$ technique where all input and output frequencies are identical. Because it does not involve the generation of light at new frequencies, it can rely on non-local mechanisms other than the local electronic polarizability (e.g. electrostriction). The selection rules for DFWM are closely related to those of one-photon techniques (e.g. absorption). DFWM using infrared beams is therefore used to probe infrared absorbing transitions instead of Raman-active transitions.

Finally, other nonlinear techniques can be used to obtain spectra that are both infrared and Raman active. Infrared sum frequency generation (IRSFG) is a surface-specific nonlinear technique that relies on $\chi^{(2)}$. The coherently generated output beam has a frequency of $\omega_3 = \omega_1 + \omega_2$, where ω_1 is in the infrared region. The selection rules for IRSFG require that the medium be anisotropic and that the transition be both IR and Raman active.

Although the remainder of this article will focus primarily on CARS, the described advances in instrumentation and methods typically also benefit other forms of nonlinear spectroscopy.

Experimental setup

Conventional Raman spectroscopy involves the collection and spectral analysis of light that is incoherently scattered in many directions. Nonlinear Raman spectroscopy requires careful alignment and overlap of multiple laser beams in order to produce a coherent output beam. Phase matching is also required for CARS and some other closely related nonlinear techniques (e.g. CSRS).

Overlap

Nonlinear Raman spectroscopy requires spatial and temporal overlap of the input beams. All beams should be spatially overlapped, which can be achieved by ensuring that all beams are parallel or collinear as they enter the lens that focuses them into the sample. Spatial overlap at the sample position can then be verified by temporarily placing a knife edge or a small pinhole into the focal point overlap region. If the spatial properties of the beams (e.g. divergence and diameter) are poor or not well matched, spatial filters and additional lenses may be used to improve the quality of the overlap. Temporal overlap of the incoming beams at the sample is also essential, since the response times of some mechanisms (i.e. the local electronic polarizability) are on the order of femtoseconds. Therefore, most CARS systems use a single fixed-wavelength laser to pump all tunable lasers. Temporal overlap is then optimized using optical paths that delay any beams that would otherwise arrive at the sample prematurely. Temporal overlap may be confirmed by scattering light at the overlap region into a fast photodiode when working with nanosecond pulses. For shorter pulses, temporal overlap may involve the use of an autocorrelator. Finally, the frequency and polarization of input beams and the detection system should be adjusted as needed. Polarization optics may be inserted both in the pump beams and in the detection system.

Phase matching

Phase matching is required for CARS experiments in normally dispersive media (i.e. condensed phase samples). The exponential terms from Equation [2] can be written as $\omega_1 + \omega_3 = \omega_2 + \omega_4$ (conservation of energy) and $\mathbf{k1} + \mathbf{k3} = \mathbf{k2} + \mathbf{k4}$ (conservation of momentum). The magnitude of each k vector is $|\mathbf{k}| = n\omega/c$, and the direction corresponds to the direction of the beam as it propagates through the sample. In a dispersionless material (the refractive index is constant for all wavelengths of light) both conditions may be satisfied using collinear alignment of all beams (see **Figure 2A**). For most materials,

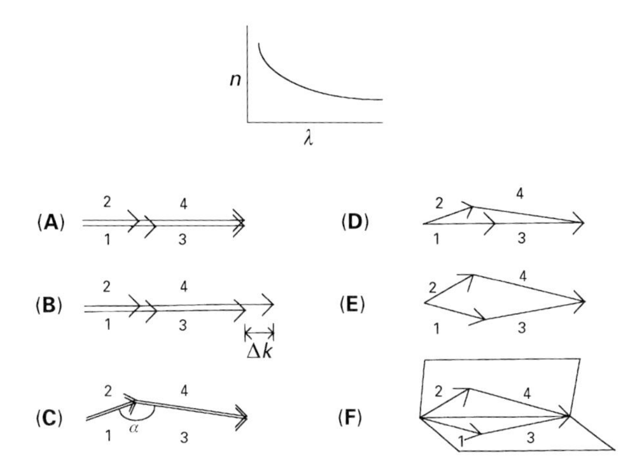

Figure 2 The refractive index in a sample with normal dispersion increases with decreasing wavelength. The phase matching diagrams are as follows: (A) collinear phase matching in the gas phase where dispersion is negligible (e.g. gas phase); (B) phase mismatch Δk encountered when using collinear geometry in a sample with normal dispersion; (C) possible arrangement in RIKES, SRG, and SRL where the angle α between beams is not critical and phase matching calculations are not needed; (D) conventional phase matching in condensed phase ($\Delta k = 0$); (E) BOXCARS phase matching; and (F) folded BOXCARS phase matching.

however, the refractive index increases with the frequency of light. Since ***k4*** has the highest frequency and therefore the largest refractive index, it is disproportionately long, causing $|\mathbf{k2}| + |\mathbf{k4}|$ to be greater than $|\mathbf{k1}| + |\mathbf{k3}|$ (see **Figure 2B**). The discrepancy in length indicates the presence of a phase mismatch Δk between the beams. The result is a loss in the efficiency of the output beam, described by the squared sinc function in Equation [3] and shown in **Figure 3**.

Figure 2D shows how this problem can be fixed by introducing an angle between ***k2*** and ***k4*** to match the phases of the beams. The fact that ***k4*** is emitted along its own unique trajectory provides the ability to separate spatially the CARS output beam from the pump beams, other nonlinear processes, or other sources of spectral interference. Additional spatial discrimination may be achieved using BOXCARS phase matching, where an angle is introduced between ***k1*** and ***k3*** to increase further the angle between ***k4*** and ***k2*** (see **Figure 2E**).

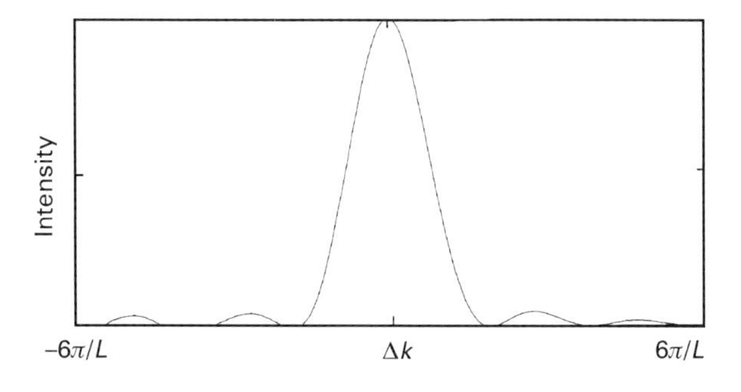

Figure 3 Effect of the phase matching Δk on the intensity (I) of the CARS output beam.

In the gas phase, dispersion may be negligible, making collinear phase matching possible (see **Figure 2A**). However, the BOXCARS approach is often preferred because it allows spatial discrimination between the input and output beams. Additional spatial discrimination may be achieved using a three-dimensional form called folded BOXCARS (see **Figure 2F**).

Unfortunately, the angles required for phase matching often vary when the laser frequencies change. The magnitude of each k vector depends upon both its frequency ω and the frequency-dependent refractive index n. Changing the frequency of any one of the four beams forces one other beam frequency to change. Therefore, the scanning of beam frequencies while producing spectra usually requires adjustment of the phase matching angles in order to avoid a phase mismatch. Without correction, the growing phase mismatch can be approximated by

$$\Delta k \approx \frac{n\Delta\omega}{c}\sqrt{2 - 2\cos\theta} \qquad [5]$$

where n is the approximate refractive index, $\Delta\omega$ is the change in frequency, and θ is the angle between the two beams with changing frequencies. Therefore, this phase mismatch problem can also be minimized by reducing the angle θ between changing k vectors.

Instrumentation

In conventional Raman spectroscopy, the required instrumentation includes (1) a fixed-wavelength narrowband laser, (2) a filter, monochromator, or some other means for rejecting Rayleigh scattering, and (3) a detection system for spectrally analysing and measuring the intensity of the scattered light. Factors such as spectral resolution and scan range depend primarily upon the detection system. For CARS and other forms of nonlinear Raman spectroscopy, however, the scanning of wavelengths is often performed by the laser instead of the detection system. Therefore, the quality of the spectra depends primarily upon the lasers.

Lasers

Most nonlinear Raman spectrometers include a fixed-wavelength laser that pumps one or more continuously tunable lasers. Some common pump lasers include the Nd:YAG laser, excimer laser, nitrogen laser, and argon ion laser. The traditional source for broad tunability has been the dye laser, which is tunable over several tens of nanometres. For example, a common configuration involves the second harmonic of an Nd:YAG laser ($\lambda = 532$ nm) split into two beams, one for pumping a dye laser (ω_2) and the other for both ω_1 and ω_3 (see **Figure 4**). Tuning of the dye laser frequency causes the frequency difference $\omega_1 - \omega_2$ to pass through Raman-active rotations or vibrations.

In recent years, however, dye lasers have been replaced or enhanced by sources that are more broadly tunable or that allow extension of wavelength into regions that are inaccessible by dyes. Difference frequency generation, sum frequency generation, and stimulated Raman scattering are nonlinear optical processes that can generate tunable light in the infrared and UV regions. Ti:sapphire lasers, tunable over a range of roughly 700–900 nm, are widely commercially available in both CW and pulsed (mode-locked) versions. Optical parametric devices such as the optical parametric oscillator (OPO) and the optical parametric amplifier (OPA) are nonlinear devices that are continuously tunable over wide regions of the spectrum. For example, optical parametric oscillators (OPOs) pumped by the third harmonic ($\lambda = 355$ nm) of an Nd:YAG laser can produce tunable signal and idler beams that cover a range of roughly 450–1800 nm.

The temporal behaviour of the laser source is also an important factor to consider. CW dye lasers can have low noise and extremely narrow bandwidths (10^{-4} cm^{-1}) for high-resolution work. However, their peak powers are low (~watts), making their use with CARS possible for only the strongest Raman

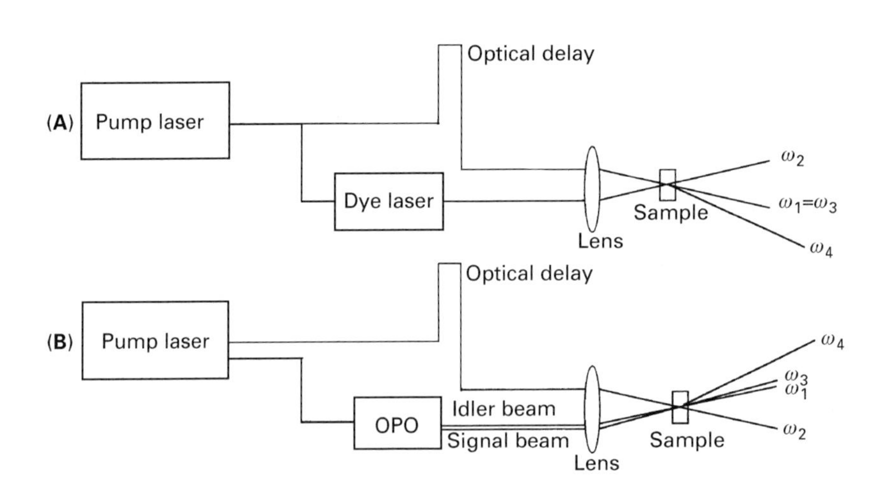

Figure 4 Experimental setups, illustrating two possible CARS spectrometers using (A) a single dye laser where $\omega_1 = \omega_3$, and (B) an optical parametric oscillator for single-wavelength detection.

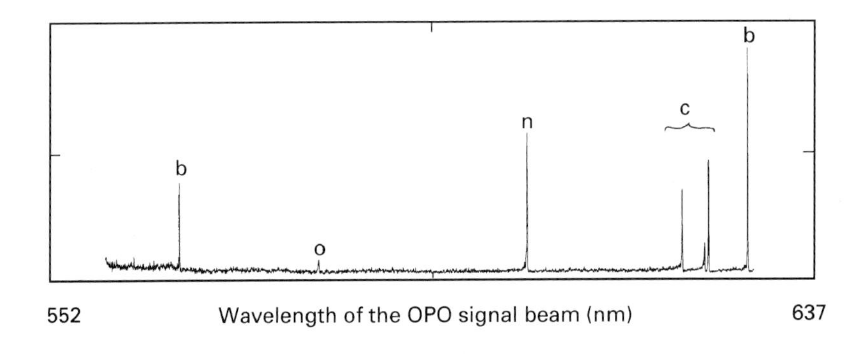

Figure 5 A CARS vibrational spectrum produced by monitoring the output beam intensity (at ω_4) while wavelength scanning an OPO (see Figure 4(B)). This spectrum shows Raman-active peaks from benzene (b), oxygen (o), nitrogen (n), and cyclohexane (c) covering a range from 681 cm^{-1} ($\lambda_{OPO} = 552$ nm) to 3098 cm^{-1} ($\lambda_{OPO} = 637$ nm). Zero frequency shift corresponds to $\lambda_{OPO} = 532$ nm.

transitions. Unless extremely high resolution work is needed, Q-switched lasers are preferable because they can generate high peak power (MW to GW) nanosecond pulses with sufficiently narrow bandwidth (0.01–0.2 cm^{-1}) for most Raman applications. The noise due to these lasers may be problematic, given the relatively low repetition rate (<100 Hz) and shot-to-shot fluctuations of several per cent. For example, the mode-beating effects in Q-switched Nd:YAG lasers can cause huge shot-to-shot variations in the nonlinear optical effect. Fortunately, this mode-beating problem can be addressed by using an injection seeded Nd:YAG. Mode-locked lasers provide incredibly high peak powers (many gigawatts) in an extremely short pulse (femtoseconds to picoseconds). These lasers typically also have high repetition rates (kHz or MHz) which is convenient for signal averaging. Owing to the inverse relationship between bandwidth and pulse duration, however, mode-locked lasers generate spectrally broad pulses that may not permit adequate resolution for the system under study. Mode-locked lasers are often used for time-domain CARS.

Although narrow bandwidth is often preferred for high-resolution studies, some applications benefit from the use of a laser having a broad bandwidth in conjunction with multiwavelength detection. Spectra generated by wavelength scanning a narrowband laser can contain unwanted effects when working with samples or environments that change rapidly (e.g. turbulent combustion systems). On the other hand, broadband dye lasers with broad bandwidths (>100 cm^{-1}) can be used with multiwavelength detection to perform single-shot CARS spectroscopy.

The following are some key properties of a laser system:

- tuning range
- linewidth
- pulse length
- pulse energy (peak power)
- coherence
- polarization
- stability and reproducibility
- practical issues (e.g. cost, maintenance, convenience).

Detection system

The primary functions of the detection system are to reject unwanted light, to measure the intensity of the output signal, and to analyse spectrally or temporally the output signal if needed. Collection optics (e.g. lenses, optical fibres), wavelength separation devices, detectors, and associated electronics are common components of the detection system.

Possible sources of unwanted light include fluorescence, Rayleigh scattering and ambient room light. A well-designed system will provide three means for rejecting this unwanted light and for minimizing potential damage to optics, slits, and detectors from intense beams of light. Spectral rejection may be accomplished using a combination of wavelength separation devices such as filters, prisms, gratings, or monochromators. Temporal rejection may be achieved using electronic gating, optical gating, or lock-in amplification if the signal is driven by pulsed or modulated lasers. Spatial filters that interrupt the input beams can be incorporated into the detection system if phase matching can cause the output beam to leave the sample at a different angle than that of the input beams.

Measurement of the intensity may be performed using a broad range of photoemissive or semiconductor detectors. Common issues to consider include wavelength sensitivity, damage or saturation threshold, linearity, and noise. Photoemissive detectors such as photomultiplier tubes are fast and sensitive in the UV and visible regions. They are, however, insensitive in the infrared region and easily damaged by high levels of light. Semiconductor photodiodes are less sensitive but more rugged, and may be used for more intense signals (>10^7 photons per pulse). Photoemissive and semiconductor detectors may also be used in multichannel form for multiwavelength detection. Examples include charge-coupled devices (CCDs) and photodiode arrays with or without microchannel plate intensifiers.

If needed, spectral analysis of light from the output beam can be achieved using a simple monochromator with a multiwavelength detector (CCD or diode array). The spectral resolution is determined by the size of the monochromator, the width of the entrance slit, the density and order of the grating, and the distance between individual elements in the detector array. Fast temporal analysis may be achieved using a fast detector such as a streak camera with picosecond or subpicosecond temporal resolution.

Most detection systems operate in one of four possible modes: single-wavelength detection, scanning detection, multiwavelength detection, and time-resolved detection. The simplest of these, single-wavelength detection, involves the detection of light at one fixed wavelength with rejection of light at all other wavelengths. The detection system may be as simple as a narrowband dielectric filter in front of a photodetector, although the use of a monochromator allows more flexibility for control of bandwidth and selection of wavelength. Scanning detection is needed for measurement of an output beam that is changing in wavelength. It typically involves wavelength scanning of a monochromator with a photodetector, although broadband filters may be used if the change in wavelength of the output beam is small. Multiwavelength detection is required when the output beam contains multiple frequency components that form a complete spectrum. The equipment for this mode is briefly discussed in the preceding paragraph. Time-resolved detection provides temporal information when the temporal behaviour of the output beam provides information such as the response of a system to an externally controlled stimulus.

The following are some figures of merit for detectors:

- time response and resolution
- wavelength response, discrimination, and resolution
- spatial discrimination
- sensitivity
- stability and noise
- practical issues – cost, convenience
- multichannel vs single channel
- saturation and damage threshold
- linear response and dynamic range.

Techniques

Acquisition of spectra

Spectral information may be acquired in three ways. The first method is the conventional approach where one or more of the laser fields frequencies are tuned to match Raman-active resonances. The second approach is to use a broadband source that allows the spectral information to be obtained in a single shot. The third is to use a time-resolved approach, where time between the pulses is varied and the sample response is measured as a function of the delay.

Scanned CARS

In conventional frequency-domain CARS, either ω_1 or ω_2 is scanned so that $\omega_1 - \omega_2$ passes through Raman-active resonances. As the difference in frequency between these two beams is tuned to each resonance, a resonance enhancement of the nonlinear optical effect occurs, leading to a peak in the intensity of the output beam. Spectra are produced by plotting the intensity of the output beam as a function of $\omega_1 - \omega_2$.

The output beam is typically monitored using a scanning detection system because $\omega_4 = \omega_1 - \omega_2 + \omega_3$ varies as the input frequencies are varied. However, single-wavelength detection may be accomplished if ω_4 is held constant by simultaneously tuning ω_3 to compensate for the changes in $\omega_1 - \omega_2$. One way to accomplish this compensation is to let ω_1 and ω_3 be generated by an OPO idler and signal beam (see **Figure 4**). As the OPO beams are tuned, $\omega_1 - \omega_2$ changes, but ω_2 and ω_4 remain constant. This approach also reduces the phase mismatch during a scan because the angles between the scanned beams (θ in Equation [5]) may be reduced to zero.

Shot-to-shot noise in the laser system can degrade the quality of the spectra for scanned CARS. Since the signal depends on the product of three input intensities, relatively small noise in the pump laser can result in a much greater noise in the output beam intensity. This problem is especially problematic in Q-switched Nd:YAG lasers that are not injection seeded. Furthermore, shot-to-shot temporal jitter between pulses in a system that does not have a single pump laser can result in noisy spectra. Such noise problems may be corrected by simultaneously monitoring and dividing the signal by the individual pump beam intensities. Alternatively, parts of the input beams may be focused into a separate reference cell to simultaneously generate a non-resonant signal to correct for fluctuations.

Single-shot CARS

Single-shot CARS may be accomplished by using one or more broadband lasers in addition to one or more narrowband lasers for the input beams. Each frequency element of the broadband laser(s) can independently mix with the narrowband frequency, contributing a separate frequency element to the output beam. This approach, called multi-colour CARS, multiplex CARS, or single-shot CARS, typically uses a broadband dye laser and multiwavelength detection in order to capture simultaneously a region of a few hundred wavenumbers of a rotational and/or vibrational spectrum. For example, dual broadband CARS involves the use of a single broadband dye laser for ω_1 and ω_2, and a fixed narrowband frequency beam for ω_3 (e.g. the pump beam for the dye laser). The resulting technique provides a relatively simple way to obtain single-shot rotational spectra in the range 0–150 cm^{-1}. Unlike scanned CARS, the spectral resolution for this technique is often determined by the detection system.

This approach is especially useful in the analysis of gas-phase combustion and other systems where turbulence may be a problem. In the condensed phase, the range of coverage may be limited by phasematching.

Time-resolved nonlinear Raman

Time-domain CARS involves the use of short picosecond or femtosecond pulses to generate the nonlinear Raman signal. Up to three separately timed excitation pulses may be combined in the sample, resulting in the generation of a pulse of light called a photon echo. Measurement of size of the photon echo as a function of the delay time between pulses can be used to determine values of both the energy relaxation times T_1 and the phase relaxation times T_2. Time resolution of several femtoseconds is possible.

Another option for performing time-resolved nonlinear Raman spectroscopy is to use a fast detector such as a streak camera. By combining short picosecond or femtosecond pulses with longer nanosecond pulses, a generated signal can be produced that evolves over time. This approach can be used to obtain simultaneously both frequency and time domain information.

List of symbols

c = speed of light; E = applied electric field; I = beam intensity; $\boldsymbol{k}$ = propagation or wave vector; L = overlap length; N = concentration; n = refractive index; P = polarization of charge; t = time; x = space; θ = angle between beams; μ = transition dipole moment; ω = angular frequency; $\chi^{(n)}$ = complex susceptibility tensor element.

See also: **Laser Applications in Electronic Spectroscopy; Light Sources and Optics; Multiphoton Spectroscopy, Applications; Nonlinear Optical Properties; Nonlinear Raman Spectroscopy, Applications; Nonlinear Raman Spectroscopy, Theory; Optical Frequency Conversion; Raman Spectrometers.**

Further reading

Bloembergen N (1992) *Nonlinear Optics*. Redwood City, CA: Addison-Wesley.

Boyd RW (1992) *Nonlinear Optics*. San Diego, CA: Academic Press.

Eckbreth AC (1996) *Laser Diagnostics for Combustion Temperature and Species*, 2nd edn. Amsterdam: Gordon and Breach.

Levenson MD and Kano SS (1988) *Introduction to Nonlinear Laser Spectroscopy*, revised edition. San Diego, CA: Academic Press.

Mukamel S (1995) *Principles of Nonlinear Optical Spectroscopy*. New York: Oxford University Press.

Shen YR (1984) *The Principles of Nonlinear Optics*. New York: Wiley.

Wright JC (1996) Nonlinear laser spectroscopy. *Analytical Chemistry* **68**: 600A–607A.

Wright JC (1982) Applications of lasers in analytical chemistry. In: Evans TR (ed) *Techniques of Chemistry*, Vol 17, pp 35–179. New York: Wiley.

Yariv A (1989) *Quantum Electronics*, 3rd edn. New York: Wiley.

Zinth W and Kaiser W (1993) Ultrafast coherent spectroscopy. In: Kaiser W (ed) *Topics in Applied Physics*, 2nd edn, Vol 60, pp 235–277. Berlin: Springer-Verlag.

Nonlinear Raman Spectroscopy, Theory

J Santos Gómez, Instituto de Estructura de la Materia, CSIC, Madrid, Spain

VIBRATIONAL, ROTATIONAL & RAMAN SPECTROSCOPIES
Theory

Introduction

In a typical spontaneous Raman experiment, an incident, nonresonant photon of energy $\hbar\omega_P$ interacts with the molecule and is scattered into a photon of energy $\hbar(\omega_P \pm \omega_R)$ where ω_R is the frequency of a vibrational mode. The molecule undergoes a transition that balances the gain or loss of field energy. The spectroscopic information is extracted by measuring the energy change of the scattered photon.

From a classical point of view, the molecule is polarized by its interaction with the input field at ω_P and an oscillating dipole at frequency ω_P is induced. As the molecular polarizability itself is modulated by internuclear motion at frequency ω_R, lateral bands appear at combination frequencies. The molecular dipole oscillating at $\omega_P \pm \omega_R$ radiates a field at these frequencies. Even with a coherent input field, as provided by lasers, the output field is incoherent because the phases of individual scatterers are not correlated.

In a typical nonlinear Raman experiment the molecule interacts with two strong coherent fields at frequencies ω_P (pump) and ω_S (Stokes). As we will see below, for strong field the response of the system is nonlinear and the molecular polarizability at a given frequency is periodically modulated at the driving frequencies and at new frequencies that are linear combinations of these $2\omega_P$, $2\omega_S$, $\omega_P \pm \omega_S, \ldots$. If we focus on the oscillating polarizability component at $\omega_P - \omega_S$, the interaction with either of the input fields or with a third input field at ω_0 gives rise as before to a component of the induced dipole at a combination frequency, such as $\omega_P + (\omega_P - \omega_S) = \omega_{AS}$, $\omega_P - (\omega_P - \omega_S) = \omega_S$, $\omega_0 + (\omega_P - \omega_S)$, whose amplitude depends on intrinsic molecular properties and on the product of three field amplitudes. The input fields being temporal and spatially coherent, there is a definite phase relationship between driven dipole oscillations of different molecules in the interaction volume, giving rise to a coherent macroscopic polarization in the medium.

The created polarization acts as a source and a new coherent field at frequency ω_σ grows to some extent or, if $\omega_\sigma = \omega_P$, ω_S or ω_0, the corresponding amplitude increases or decreases depending upon the relative field-polarization phase.

The above behaviour is quite general and will be observed for any molecule and frequency combination, provided that some macroscopic symmetry requirements are met. The interaction just described is a four-wave process: a macroscopic polarization at a signal frequency ω_σ builds up through a nonlinear frequency mixing of three frequency components of the electromagnetic field. In a general (n+1)-wave process, n frequency components are mixed to produce an oscillating polarization at $\omega_\sigma = \pm(\omega_1 \pm \omega_2 \pm \omega_3 \pm \cdots \pm \omega_n)$. The magnitude of the nonlinear polarization depends on the product of n field amplitudes and an nth-order nonlinear susceptibility $\chi^{(n)}(\omega_\sigma; \omega_1, \omega_2, \ldots \omega_n)$, which is a material property.

The connection of this nonlinear optical effect with spectroscopy lies in the fact that $\chi^{(n)}$, and hence the signal strength, will be enhanced whenever a linear combination of a subset of $m \leq n$ frequency components approaches the energy difference of two molecular states of proper symmetry. This m-photon resonant enhancement — which can in principle be observed even in the absence of real transitions, as will be the case for states with no thermal population — has been exploited to develop a number of nonlinear spectroscopic techniques that differ in order n, order of the used resonance m, number of colours (i.e. different actual laser fields that provide the frequency components), spectral and temporal resolution and the actual method used to detect the resonances by monitoring either the power generated at ω_σ or the change in amplitude, polarization or phase at some of the input frequencies.

The simplest m-photon resonance is obtained for $m = 2$ when $\Omega_{fg} = (E_f - E_g)/\hbar \simeq \omega_1 \pm \omega_2$. If we consider ω_1 and ω_2 to be typical optical frequencies, the + sign correspond to two vibronic states g and f, from different electronic states, in resonance with $2\omega_1$, $2\omega_2$ or $\omega_1 + \omega_2$: two-photon absorption-like resonance (TPA).

Most nonlinear Raman techniques correspond to the − sign above: $\Omega_{fg} \simeq \omega_P - \omega_S$ (Raman-like resonance), where frequency subscripts have been converted to the usual Raman convention. These resonances can contribute to enhance nth-order processes for $n \geq 2$. For isotropic media, all even-order susceptibilities vanish and Raman techniques involve 3rd, 5th, 7th,... order nonlinear susceptibilities. The ordering comes from a perturbative development of the field–molecule interaction and, as far as the perturbative approach can be applied, successive orders correspond to much smaller terms. Hence we can study the main spectroscopic features with only the lowest-order nonvanishing term, and most nonlinear Raman techniques are 3rd-order four-wave processes, which will constitute the main topic in this article. These techniques are indeed readily implemented with present laser technology in a variety of media, including gases at low pressure.

We can perform Raman spectroscopy by monitoring different properties of the macroscopic field as we scan ω_P or ω_S in such a way that $\omega_P - \omega_S \simeq \Omega_{fg} \equiv \omega_R$, leading to several techniques that have usually been identified by acronyms. The energy level diagrams for the main nonlinear Raman techniques are depicted in **Figure 1**.

With only two colours we can monitor the intensity of the generated beam at the anti-Stokes frequency $\omega_{AS} = \omega_P + \omega_R$ (coherent anti-Stokes Raman spectroscopy, CARS) or the Stokes frequency respect to ω_S, $\omega_{CSRS} = \omega_S - \omega_R$ (coherent Stokes Raman spectroscopy, CSRS). Alternatively, we can monitor the intensity increase at ω_S (stimulated Raman gain spectroscopy, SRG) or the intensity decrease at ω_P (stimulated Raman loss, SRL, also known as inverse Raman spectroscopy, IRS). If we focus on the polarization of the created field or the change in polarization state of input fields, we arrive at different polarization variants of the above techniques, such as polarization CARS, CARS ellipsometry or Raman-induced Kerr effect (RIKES). The input frequencies can be tuned to match additional one-photon resonances that can enhance the signal by several orders of magnitude, leading to techniques such as resonance CARS. A third colour ω_0 can be used for different purposes such as fine tuning of additional one-photon resonances or shifting the signal frequency $\omega_\sigma = \omega_0 + \omega_R$ to a more convenient region; this expands the possibilities, leading to Raman resonant four-wave mixing (FWM). The different choices of input and signal beams are collected

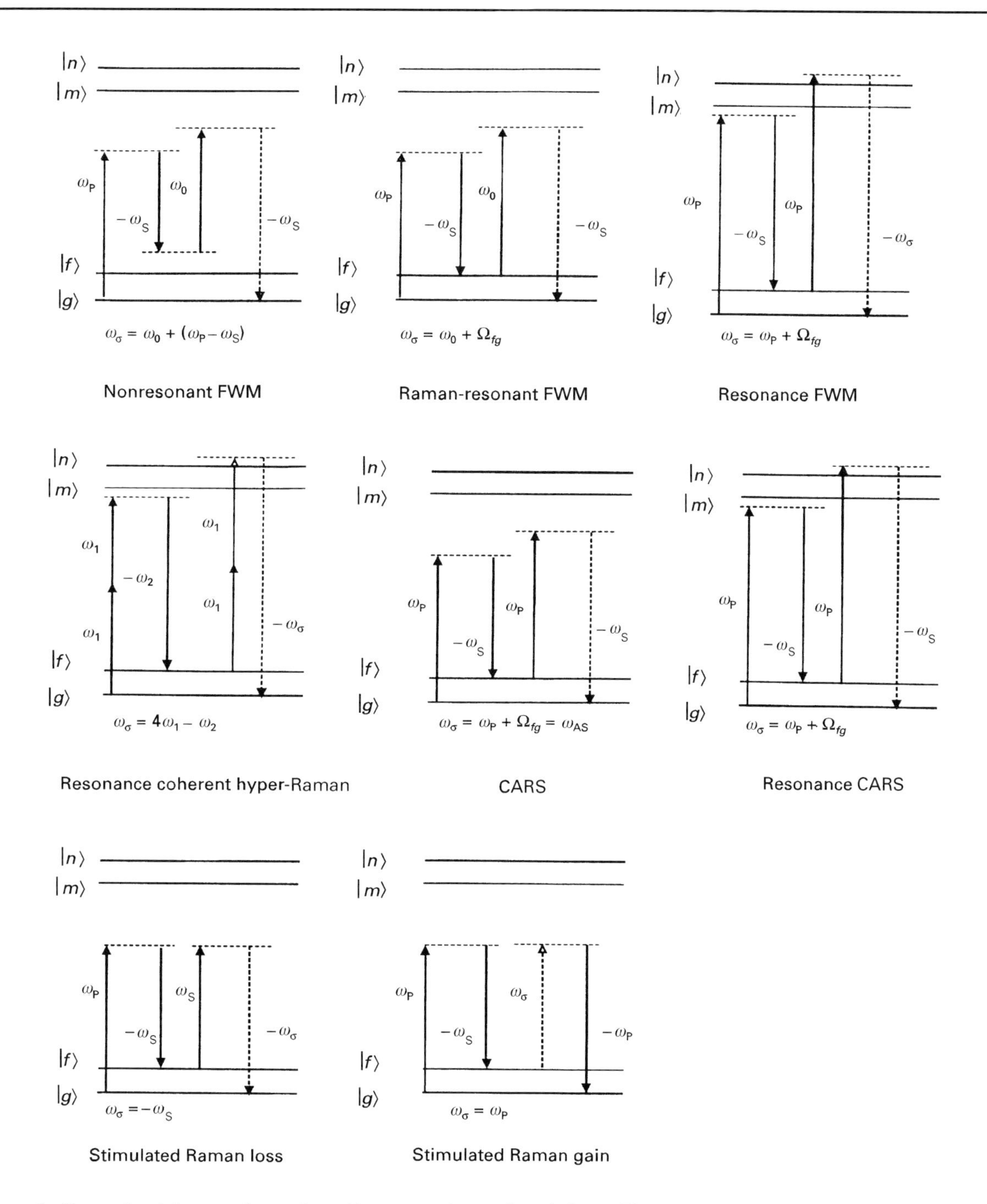

Figure 1 Energy level diagrams for nonlinear Raman spectroscopic techniques. The input fields are shown as arrows pointing upwards for positive frequency (downwards for negative) in the arguments of the nonlinear susceptibility, and the output field as a dashed arrow, which must close the diagram. Schematic molecular energy levels are represented to show the main Raman resonance and additional one-photon resonances. Horizontal position does not imply time ordering. The energy balance involves the field and the material excitation in different ways for different techniques. In pure nonresonant processes, the energy of created photons balances that of those destroyed. In SRL and SRG, each scattering event leads to an excited molecule. In Raman-resonant FWM and CARS, the energy of destroyed photons is shared among material excitation and created photons depending on the relative magnitude of the imaginary and real parts of the nonlinear susceptibility.

in **Table 1**, along with the names of the associated nonlinear Raman spectroscopic techniques.

The spectroscopic use of three-photon resonances has been demonstrated, although these usually require additional enhancement through one-photon electronic resonance, or the use of condensed media. Closely related to Raman spectroscopy is coherent hyper-Raman scattering (see **Figure 1**), the stimulated analogue of the spontaneous hyper-Raman effect, for which a three-photon resonant

Table 1 Nonlinear Raman spectroscopic techniques

	FWM	*CARS*	*RIKES*
Name	Four-wave mixing	Coherent anti-Stokes Raman spectroscopy	Raman-induced Kerr effect
Monitored effect	Intensity at $\omega_\sigma = \omega_0 + \omega_R$	Intensity at $\omega_{\text{anti-Stokes}} = \omega_P + \omega_R$	Intensity at ω_S polarized $\perp$
Input/output field	$\omega_0, \omega_P, \omega_S / \omega_0 + \omega_R$	$\omega_P, \omega_S / \omega_A$	$\omega_P, \omega_S / \omega_S$
Input/output polarization	All possibilities with paired Cartesian index	$E_{\omega_P x}, E_{\omega_S x} / E_{\omega_A x}$ $E_{\omega_P x}, E_{\omega_S y} / E_{\omega_A y}$	E_{ω_P} circular, $E_{\omega_S x} / E_{\omega_S y}$
Effective susceptibility	$\chi^{(3)}_{\mu\alpha\beta\gamma}(-\omega_\sigma; \omega_0\omega_P - \omega_S) = \chi^{(3)NR}_{\mu\alpha\beta\gamma} + \frac{1}{6\epsilon_0\hbar} \times \sum_{g,f}(N_g - N_f)\alpha_{\alpha\mu}\alpha_{\gamma\beta}\left(\frac{\Delta_{fg}}{\Delta_{fg}^2 + \Gamma_{fg}^2} + i\frac{\Gamma_{fg}}{\Delta_{fg}^2 + \Gamma_{fg}^2}\right)$	$\chi^{(3)}_{\mu\alpha\beta\gamma}(-\omega_\sigma; \omega_P\omega_P - \omega_S) = \chi^{(3)NR}_{\mu\alpha\beta\gamma} + \frac{1}{3\epsilon_0\hbar} \times \sum_{g,f}(N_g - N_f)\alpha^2_{x\alpha}\left(\frac{\Delta_{fg}}{\Delta_{fg}^2 + \Gamma_{fg}^2} + i\frac{\Gamma_{fg}}{\Delta_{fg}^2 + \Gamma_{fg}^2}\right)$	$\chi^{(3)}_{\text{eff}}(-\omega_S; \omega_P - \omega_P\omega_S) = \frac{i}{2}\left(\chi^{(3)}_{yyxx} - \chi^{(3)}_{yxyx}\right) \approx \frac{i}{2}\left(\chi^{(3)R}_{yyxx} - \chi^{(3)R}_{yxyx}\right)$
Signal[a]	$I_{\omega_\sigma} = \frac{9}{4\varepsilon_0^2 c^4 n_\sigma n_0 n_P n_S}\omega_\sigma^2 \left\lvert\chi^{(3)}_{\mu\alpha\beta\gamma}\right\rvert^2 I_{\omega_0} I_{\omega_P} I_{\omega_S} L^2 \operatorname{sinc}^2(\Delta kL/2)$	$I_{\omega_A}(L) = \frac{9}{16}\frac{\omega_A^2}{\varepsilon_0^2 c^4 n_A n_P^2 n_S}\left\lvert\chi^{(3)}_{\alpha x x\alpha}\right\rvert^2 I^2_{\omega_P}(0) I_{\omega_S}(0) L^2 \times \operatorname{sinc}^2(\Delta kL/2)$	$\propto \left\lvert\chi^{(3)\text{RIKES}}_{\text{eff}}(-\omega_S; \omega_P - \omega_P\omega_S)\right\rvert^2 I_P^2 I_S$

	OHD-RIKES	*SRL*	*SRG*
Name	*Optically heterodyne detected RIKES*	Stimulated Raman loss or inverse Raman spectroscopy	Stimulated Raman gain
Monitored effect	Intensity at ω_S polarized $\perp$	Intensity decrease at ω_P	Intensity at ω_S
Input/output fields	$\omega_P, \omega_S / \omega_S$	$\omega_P, \omega_S / \omega_P$	$\omega_P, \omega_S / \omega_S$
Input/output polarization	E_{ω_P} circular, $E_{\omega_S x} / E_{\omega_S y}$	$E_{\omega_P x}, E_{\omega_S x} / E_{\omega_P x}$	$E_{\omega_P x}, E_{\omega_S x} / E_{\omega_S x}$
Effective susceptibility	$\chi^{(3)}_{\text{eff}}(-\omega_S; \omega_P - \omega_P\omega_S) = \frac{i}{2}\left(\chi^{(3)}_{yyxx} - \chi^{(3)}_{yxyx}\right) \approx \frac{i}{2}\left(\chi^{(3)R}_{yyxx} - \chi^{(3)R}_{yxyx}\right)$	$\chi^{(3)}_{x\alpha x\alpha}(-\omega_P; \omega_S\omega_P - \omega_S) = \chi^{(3)NR}_{x\alpha x\alpha} + \frac{1}{6\epsilon_0\hbar} \times \sum_{g,f}(N_g - N_f)\alpha^2_{\alpha x}\left(\frac{\Delta_{fg}}{\Delta_{fg}^2 + \Gamma_{fg}^2} + i\frac{\Gamma_{fg}}{\Delta_{fg}^2 + \Gamma_{fg}^2}\right)$	$\chi^{(3)}_{x\alpha x\alpha}(-\omega_S; \omega_P - \omega_P\omega_S) = \chi^{(3)NR}_{x\alpha\alpha x} + \frac{1}{6\epsilon_0\hbar} \times \sum_{g,f}(N_g - N_f)\alpha^2_{x\alpha}\left(\frac{\Delta_{fg}}{\Delta_{fg}^2 + \Gamma_{fg}^2} - i\frac{\Gamma_{fg}}{\Delta_{fg}^2 + \Gamma_{fg}^2}\right)$
Signal[a]	$\propto \operatorname{Im}\left(\chi^{(3)\text{RIKES}}_{\text{eff}}\right) I_P I_S$ or $\propto \operatorname{Re}\left(\chi^{(3)\text{RIKES}}_{\text{eff}}\right) I_P I_S$	$I_{\omega_P}(L) = I_{\omega_P}(0)\left[1 - \frac{3\omega_P}{\varepsilon_0 c^2 n_S n_P}\operatorname{Im}\left(\chi^{(3)\text{SRL}}_{x\alpha x\alpha}\right) I_{\omega_S}(0) L\right]$	$I_{\omega_S}(L) = I_{\omega_S}(0)\left[1 - \frac{3\omega_S}{\varepsilon_0 c^2 n_S n_P}\operatorname{Im}\left(\chi^{(3)\text{SRG}}_{x\alpha\alpha x}\right) I_{\omega_P}(0) L\right]$

[a] Signal expression for plane monochromatic input waves. L is the interaction length.

enhancement $\Omega_{fg} \simeq 2\omega_1 - \omega_2$ can be detected in $\chi^{(5)}$ in a six-wave process, for example by monitoring the intensity of the generated field at $\omega_\sigma = 4\omega_1 - \omega_2 \simeq 2\omega_1 + \omega_R$. We will not explicitly consider these higher-order resonances in the following, but the theory involved closely follows that for the simplest two-photon resonant four-wave case.

From a fully quantum-mechanical view, the system matter+field undergoes a transition from an initial state $| n_P = n, n_S = 0, g \rangle$, for which the input field has n photons, the output field is at the vacuum state and the molecule in an internal state g, to a final state $| n_P = n - 1, n_S = 1, f \rangle$. Owing to the low Raman scattering cross section, the probability of finding one scattered photon within the interaction region at the time of the next scattering event is vanishingly small at low input intensity. As a consequence, the initial state of the output mode is always the vacuum state and the number of scattered photons is linearly dependent on the number of photons in the input field alone. At high input intensity or in the presence of a second external field at ω_S, $n_S \neq 0$ and the matrix element depends on the product of $n_P \times n_S$. Therefore, spontaneous and nonlinear Raman are closely connected. In fact, the spontaneous Raman effect cannot be considered among linear optical phenomena such as absorption, diffraction or stimulated emission because the scattered field has a different frequency.

The dependence of the spatial, spectral and temporal signal properties on those of the input beams, along with the broad range of lasers available, opens many possibilities ranging from high-resolution studies of gases at Doppler-limited resolution using narrow-bandwidth lasers to high-temporal-resolution studies of intramolecular and intermolecular dynamics and relaxation processes at the femtosecond scale in condensed media. Although the underlying theory is essentially the same, high-resolution studies are better described in the frequency domain, looking at the Raman resonances in the nonlinear susceptibility of the medium, allowing for interpretation of line position, intensity and (with the inclusion of relaxation terms) line width. Broadband ultrashort pulses lead to an impulse response, where the molecule evolves freely after the sudden excitation involving many states and is probed at later times. This situation is better described in the time domain, looking at the time evolution of the nonlinear response function, which shows relaxation and dynamic processes, directly and still contains information about the spectrum in the form of oscillations or beating among different excited modes. Midway between these limits, lasers with small bandwidth and short duration can be used to excite a selected set of molecular levels and study their evolution.

Theory

The theory relating first principles to practical signal expressions for an arbitrary material system can be considered to be well understood. We present here a sketch of its main aspects, indicating the approximations needed and its scope. Our goal is to give a practical signal expression in terms of Raman scattering tensor components, field intensities and polarizations. Our model system is an isotropic medium composed of polarizable units, small compared with the wavelength, interacting only through their coupling to a common thermal bath, such as a gas or dilute solution, in the presence of a superposition of quasi-monochromatic ($\delta\omega_i << \omega_i$) laser pulses.

Nonlinear interaction can be studied on a full quantum basis by expressing the Hamiltonian in terms of quantum operators for the material and the radiation degrees of freedom, but the most fruitful treatment has been the semiclassical approach in which the electric field operator is replaced by its expectation value and considered as a function, though quantum operators are used for the material. Only electric dipole polarization will be addressed.

Owing to the coherence, we need to consider the macroscopic evolution of the field in a medium that shows a macroscopic polarization induced by the field–matter interaction. This will be done in three steps. First, the polarization induced by an arbitrary field will be calculated and expanded in power series in the field, the coefficients of the expansion being the material susceptibilities $\chi^{(n)}$ (frequency domain) or response function $\mathbf{R}^{(n)}$ (time domain) of nth-order. Nonlinear Raman effects appear at third order in this expansion. Second, the perturbation theory derivation of the third-order nonlinear susceptibility $\chi^{(3)}$ in terms of molecular eigenstates and transition moments will be outlined, leading to a connection with the spontaneous Raman scattering tensor components. Last, the interaction of the initial field distribution with the created polarization will be evaluated and the signal expression obtained for the relevant techniques of **Table 1**.

Macroscopic nonlinear polarization

We consider a unit volume that contains many molecules but is small compared to the wavelength. The electric field can be considered uniform inside this volume. By expanding the polarization at a given time in power series in the field, we obtain for a given point:

$$\boldsymbol{P}(t) = \boldsymbol{P}^{(0)}(t) + \boldsymbol{P}^{(1)}(t) + \boldsymbol{P}^{(2)}(t) + \boldsymbol{P}^{(3)}(t) + \cdots \quad [1]$$

where $\boldsymbol{P}^{(0)}$ stands for the static polarization, $\boldsymbol{P}^{(1)}$ is linear in the field, $\boldsymbol{P}^{(2)}$ quadratic and so on (bold typefaces represent a vector, tensor or matrix magnitude).

The next approximation is to consider only the *local response*: the polarization at the point $\boldsymbol{r}$ depends only on the field at this point. In condensed media, the local field sensed by the molecule includes the depolarizing effect of the surrounding molecules, and it does depend on the field values at different points. In an anisotropic medium the material response shows *spatial dispersion* and is sensitive to both frequency and wavevector of the driving fields.

Nonlinear Raman processes are governed by the third-order polarization, which can be expressed in the time domain, for a generic electric field, in component form:

$$P_{\mu}^{(3)}(t) = \epsilon_0 \int_{-\infty}^{\infty} \mathrm{d}\tau_1 \int_{-\infty}^{\infty} \mathrm{d}\tau_2 \int_{-\infty}^{\infty} \mathrm{d}\tau_3 \; R_{\mu\alpha\beta\gamma}(\tau_1, \tau_2, \tau_3)\, E_{\alpha}(t-\tau_1)\, E_{\beta}(t-\tau_2)\, E_{\gamma}(t-\tau_3) \quad [2]$$

where summation over repeated indices is assumed.

$\mathbf{R}^{(3)}(\tau_1, \tau_2, \tau_3)$ is a fourth-order tensor ($\mathbf{R}^{(n)}$ has rank n+1) of components $R^{(3)}_{\mu\alpha\beta\gamma}(\tau_1, \tau_2, \tau_3)$ where $\mu\alpha\beta\gamma$ represent Cartesian coordinates describing the field polarization. It is a real function of n time variables (because $\boldsymbol{E}$ and $\boldsymbol{P}$ are physical observables) and vanish whenever any of its time arguments are negative because the polarization at time t cannot depend on the field values at later times. Owing to the implicit summation, the above equation is invariant to any permutation of the pairs (α, τ_1), (β, τ_2) etc. (intrinsic permutation symmetry).

To study narrow resonances, the frequency domain allows for a better description of the material response in terms of susceptibilities. To obtain the corresponding expressions we take the Fourier transform of $\boldsymbol{E}(t)$ and $\boldsymbol{P}(t)$.

$$\boldsymbol{E}(t) = \int_{-\infty}^{\infty} \mathrm{d}\omega\, \boldsymbol{E}(\omega) \exp(-\mathrm{i}\omega t)$$

$$\boldsymbol{E}(\omega) = \frac{1}{2\pi} \int_{-\infty}^{\infty} \mathrm{d}\tau\, \boldsymbol{E}(\tau) \exp(\mathrm{i}\omega\tau)$$

$$\boldsymbol{P}(t) = \int_{-\infty}^{\infty} \mathrm{d}\omega\, \boldsymbol{P}(\omega) \exp(-\mathrm{i}\omega t)$$

$$\boldsymbol{P}(\omega) = \frac{1}{2\pi} \int_{-\infty}^{\infty} \mathrm{d}\tau\, \boldsymbol{P}(\tau) \exp(\mathrm{i}\omega\tau) \quad [3]$$

By substituting Equation [3] into Equation [2] and rearranging, we arrive at

$$P_{\mu}^{(3)}(\omega_\sigma) = \epsilon_0 \int_{-\infty}^{\infty} \mathrm{d}\omega_1 \int_{-\infty}^{\infty} \mathrm{d}\omega_2 \int_{-\infty}^{\infty} \mathrm{d}\omega_3 \; \chi^{(3)}_{\mu\alpha\beta\gamma}(-\omega_\sigma; \omega_1\omega_2\omega_3) E_{\alpha}(\omega_1) E_{\beta}(\omega_2) E_{\gamma}(\omega_3) \delta(\omega - \omega_\sigma) \quad [4]$$

where $\omega_\sigma = \omega_1 + \omega_2 + \omega_3$, and the third-order nonlinear susceptibility $\chi^{(3)}_{\mu\alpha\beta\gamma}$ is defined by

$$\chi^{(3)}_{\mu\alpha\beta\gamma}(-\omega_\sigma; \omega_1\omega_2\omega_3) = \int_{-\infty}^{\infty} \mathrm{d}\tau_1 \int_{-\infty}^{\infty} \mathrm{d}\tau_2 \int_{-\infty}^{\infty} \mathrm{d}\tau_3 \; R^{(3)}_{\mu\alpha\beta\gamma}(\tau_1\tau_2\tau_3) \exp\left(\mathrm{i} \sum \omega_i \tau_i\right) \quad [5]$$

$\chi^{(3)}$ is a field- and time-independent fourth-rank tensor, and a function of three frequency arguments. Its general properties can be derived from that of the nonlinear response: $\boldsymbol{E}(\omega)$ and $\boldsymbol{P}^{(3)}(\omega)$ include both positive and negative frequencies, related by the conditions $\boldsymbol{E}(-\omega) = \boldsymbol{E}^*(\omega)$, $\boldsymbol{P}(-\omega) = \boldsymbol{P}^*(\omega)$. The reality condition of $\mathbf{R}^{(n)}$ implies that $[\chi^{(3)}(-\omega_\sigma; \omega_1\omega_2\omega_3)]^* = \chi^{(3)}(\omega_\sigma; -\omega_1 - \omega_2 - \omega_3)$, $\chi^{(3)}_{\mu\alpha\beta\gamma}(-\omega_\sigma; \omega_1\omega_2\omega_3)$ is invariant under the permutation of the pairs (α, ω_1), (β, ω_2).

The integration in Equation [4] considers the existence of several frequency combinations matching the condition $\omega_\sigma = \Sigma_j\, \omega_j$, within the bandwidth of the applied field. A single laser whose bandwidth is large enough to include frequency components that match $\omega_a - \omega_b = \omega_R$ can drive the nonlinear response. However, more frequently the laser bandwidth is smaller than ω_R and different beams with distinct centre frequencies at needed to match the Raman resonance. Let us now consider a field composed of a superposition of quasi-monochromatic beams, i.e. the amplitude at a carrier frequency is modulated by a complex envelope that defines its spectral shape. The central frequencies are chosen to match the Raman resonant four-wave mixing scheme (see **Table 1**), which can be later adapted to other colour choices. Three laser beams are present of frequencies ω_0, ω_P, ω_S, with $\omega_P - \omega_S = \omega_R$ as the only resonant frequency combination.

The field can be expressed as

$$\boldsymbol{E}(t) = \sum_{\omega_j = \omega_0, \omega_P, \omega_S} \frac{1}{2} [\boldsymbol{E}_{\omega_j}(t) \exp(-\mathrm{i}\omega_j t) + \boldsymbol{E}_{-\omega_j}(t) \exp(\mathrm{i}\omega_j t)] \quad [6]$$

The envelope $E_{\omega_j}(t)$ is now time-dependent, and contains both amplitude and phase information, thus defining the spectrum (centred at ω_j) of each field. It is implicit in the above considerations that $E_{\omega_j}(t)$ changes smoothly in the timescale of $(\omega_j)^{-1}$. By substituting Equation [6] into Equation [4], we found that $P^{(3)}(t)$ can be recast as a sum of quasi-monochromatic components at central frequency ω_σ. The envelope $P^{(3)}_{\omega_\sigma}(t)$ will vanish due to the fast oscillating exponential factors except when ω_σ is close to any linear combination of three frequencies chosen among the input frequencies, leading to

$$\boldsymbol{P}^{(3)}(t) = \sum_{\omega_\sigma} \frac{1}{2}\left[\boldsymbol{P}^{(3)}_{\omega_\sigma}(t)\exp(-\mathrm{i}\omega_\sigma t) + \boldsymbol{P}^{(3)}_{-\omega_\sigma}(t)\exp(\mathrm{i}\omega_\sigma t)\right] \quad [7]$$

Only combinations including $\pm(\omega_P - \omega_S)$ will show resonant enhancement. Among these, we are interested in the term that oscillates at a centre frequency $\omega_\sigma = \omega_0 + \omega_P - \omega_S$, which corresponds to a field ω_0 being scattered into an output field $\omega_0 + \omega_R$ due to the driving effect of the resonant pair ω_P, ω_S. This term is

$$\begin{aligned}\boldsymbol{P}^{(3)}_{\omega_\sigma}(t) = \epsilon_0 \frac{D}{4} \int_{-\infty}^{\infty} \mathrm{d}\tau_1 \int_{-\infty}^{\infty} \mathrm{d}\tau_2 \int_{-\infty}^{\infty} \mathrm{d}\tau_3 \\ \times\ \boldsymbol{\Phi}^{(3)}_{-\omega_\sigma;\,\omega_0\,\omega_P-\omega_S}(t-\tau_1, t-\tau_2, t-\tau_3) \\ \times\ \boldsymbol{E}_{\omega_0}(\tau_1)\ \boldsymbol{E}_{\omega_P}(\tau_2)\boldsymbol{E}^*_{\omega_S}(\tau_3)\end{aligned} \quad [8]$$

where $\boldsymbol{\Phi}^{(3)}(t-\tau_1, t-\tau_2, t-\tau_3) = \mathbf{R}^{(3)}(t-\tau_1, t-\tau_2, t-\tau_3)\exp(i\,\Sigma_j\,\omega_j\tau_j)$ is an *envelope response function* that relates the envelope of the third-order polarization at frequency ω_σ with the envelopes of the fields when their central frequency is tuned to the values given by the subscripts of $\boldsymbol{\Phi}^{(3)}$. The relationship between $\chi^{(3)}$ and $\boldsymbol{\Phi}^{(3)}$ is:

$$\chi^{(3)}(-\omega_\sigma; \omega_0\omega_P - \omega_S) \int_{-\infty}^{\infty} \mathrm{d}\tau_1 \int_{-\infty}^{\infty} \mathrm{d}\tau_2 \int_{-\infty}^{\infty} \mathrm{d}\tau_3\ \boldsymbol{\Phi}^{(3)}_{-\omega_\sigma;\omega_0\omega_P-\omega_S}(\tau_1, \tau_2, \tau_3) \quad [9]$$

The D factor above comes from the identification of terms in the summation that are identical owing to the intrinsic permutation symmetry. We must now compare the characteristic time scale of the material T_1 (population relaxation) and T_2 (phase relaxation) with that of the laser pulse τ_p, taken as the period of the fastest amplitude change in the field (similar to the pulse duration for a smooth, narrow-bandwidth pulse, but close to the inverse of the full bandwidth for a multimode laser).

For $\tau_p << T_2$ each pulse appears to the system as a δ impulsion at fixed time t_i, A $E_{\omega_a}(t) = A_{\omega_a}\delta(t - t_i)$, where A_{ω_a} is the complex pulse area. The polarization is

$$\boldsymbol{P}^{(3)}_{\omega_\sigma}(t) = \epsilon_0 \frac{D}{4} \boldsymbol{\Phi}^{(3)}_{-\omega_\sigma;\omega_0\omega_P-\omega_S}(t-t_0, t-t_P, t-t_S)\ \boldsymbol{A}_{\omega_0}\boldsymbol{A}_{\omega_P}\boldsymbol{A}^*_{\omega_S} \quad [10]$$

$\boldsymbol{\Phi}^{(3)}$ can now be directly measured by changing the times t_i at which the pulses are applied. A typical implementation of this scheme is to excite a Raman mode by applying ω_P and ω_S simultaneously at $t = 0$ and provide some variable delay for the response to evolve before applying the field at ω_0 that is scattered into the signal field ω_σ.

On the other hand, for $\tau_p >> T_2$ the macroscopic polarization relaxes very fast, the system loses memory of previous field values, and $P^{(3)}_{\omega_\sigma}(t)$ follows adiabatically the field envelope, leading to

$$\boldsymbol{P}^{(3)}_{\omega_\sigma}(t) = \epsilon_0 \frac{D}{4} \boldsymbol{\chi}^{(3)}(-\omega_\sigma; \omega_0\omega_P - \omega_S)\ \boldsymbol{E}_{\omega_0}(t)\ \boldsymbol{E}_{\omega_P}(t)\ \boldsymbol{E}^*_{\omega_S}(t) \quad [11]$$

This expression is the starting point for the vast majority of nonlinear Raman experiments. One special case that enters into this category is CW experiments with monochromatic waves ($\tau_p \to \infty$ and the amplitudes of incoming fields and nonlinear polarization become constant). The corresponding expression is

$$P^{(3)}_{\omega_\sigma\mu} = 2^{1-n}\epsilon_0 D \chi^{(3)}_{\mu\alpha\beta\gamma}(-\omega_\sigma; \omega_0\omega_P - \omega_S)\ E_{\omega_0\alpha} E_{\omega_P\beta} E^*_{\omega_S\gamma} \quad [12]$$

where individual vector and tensor components are used. D is the number of distinguishable permutations; thus $D = 6$ for processes involving three different input frequencies but $D = 3$ for two-colour processes (i.e. $\omega_0 = \omega_P$ or $\omega_0 = \omega_S$), which necessarily include two identical frequency arguments, whose permutation does not produce a distinguishable term. In this context ω_j and $-\omega_j$ must be considered different. The 2^{1-n} factor is linked to the $\frac{1}{2}$ factors appearing in Equations [6] and [7].

Although the order of frequencies in $\chi^{(3)}_{u\alpha\beta\gamma}(-\omega_\sigma; \omega_0\omega_P - \omega_S)$ has no special meaning itself, owing to the intrinsic permutation symmetry, we will keep it fixed (Raman convention), using ω_P and $-\omega_S$ as second and third arguments. Besides the intrinsic permutation symmetry, the medium macroscopic symmetry imposes further restrictions on the tensor index of nonvanishing, independent components $\chi^{(3)}_{u\alpha\beta\gamma}$. A very important result is that the *n*th-order susceptibility vanishes for even *n* in media showing inversion symmetry and $\chi^{(3)}$ contributes the lowest-order nonlinearity. For isotropic media, it can be shown that $\chi^{(3)}_{u\alpha\beta\gamma}$ vanishes if some Cartesian index appears an odd number of times in the subscript.

The symmetry properties of $\chi^{(3)}_{u\alpha\beta\gamma}$ in any medium can be established considering that it transforms as a fourth-rank polar tensor under the macroscopic symmetry operations. The symmetry of the microscopic polarizable units can be used to simplify the microscopic expression of $\chi^{(3)}_{\mu\alpha\beta\gamma}$.

Microscopic origin of the nonlinear Raman susceptibility

We must now relate $\chi^{(3)}$ to molecular eigenstates and transition moments. This can be accomplished through standard quantum-mechanical perturbation methods. Only a sketch of the steps involved will be given here.

The microscopic origin of the nonlinear response is the distortion induced in the molecular charge distribution due to the electrical field. The presence of a microscopic dipole produces a macroscopic polarization in the unit volume $\boldsymbol{P} = N\langle \varepsilon r\rangle$, where N is the number density of polarizable units and $\langle \varepsilon r\rangle$ the expectation value of the dipole moment induced in each unit. In order to evaluate $\langle \varepsilon r\rangle$ we will use the density matrix formalism, because it is the easiest way to relate microscopic properties to macroscopic ones and to cope with macroscopic coherence effects. In the absence of fields, the medium is supposed to be described by an unperturbed Hamiltonian $\mathbf{H}_0$ and to be at equilibrium. When the fields are applied, the field–matter interaction contributes a time-dependent term $\boldsymbol{V}(t) = -\boldsymbol{E}(t)\boldsymbol{P}(t)$ to the global energy. The evolution of the system under this perturbation can be described through the equation of motion of the density operator:

$$\mathrm{i}\hbar\frac{\partial \boldsymbol{\rho}(t)}{\partial t} = [\mathbf{H}(t), \boldsymbol{\rho}(t)] - \mathrm{i}\hbar\boldsymbol{\Gamma} \qquad [13]$$

where $\mathbf{H}(t) = \mathbf{H}_0 + \boldsymbol{V}(t)$ and the term $\mathrm{i}\hbar\Gamma$ is introduced to include relaxation. The evolution of the matrix density elements is given by:

Diagonal elements:

$$\mathrm{i}\hbar\frac{\partial \rho_{ii}(t)}{\partial t} = [\boldsymbol{V}(t), \boldsymbol{\rho}(t)]_{ii} - \mathrm{i}\hbar\Gamma_{ii}\rho_{ii}(t)$$

Nondiagonal elements

$$\mathrm{i}\hbar\frac{\partial \rho_{ij}(t)}{\partial t} = [\boldsymbol{V}(t), \boldsymbol{\rho}(t)] + \hbar\rho_{ij}(t)(\omega_{ij} - \mathrm{i}\Gamma_{ij}) \qquad [14]$$

where $\omega_{ij} = (E_i - E_j)/\hbar$ and $G_{ij} = \frac{1}{2}(G_{ii}+G_{jj}) + G_{ij}^{\mathrm{proper}}$, G_{ij} being the homogeneous width of level *i*. G_{ij}^{proper} includes the effect of dephasing processes that destroy the coherence between states *i* and *j* without altering their populations. Once we solve Equation [14] for $\rho_{ij}(t)$, the dipole moment can be calculated as $\langle \varepsilon r\rangle = e\mathrm{Tr}(\rho(t)\boldsymbol{r})$.

Diagonal elements $\rho_{ii}(t)$ represent the fractional population of state *i*. Equation [14] shows that the population factors change from their initial values owing to the perturbation, while the energy relaxation mechanisms tend to restore them to the equilibrium distribution with a rate constant $\Gamma_{ii} \sim 1/T_1$.

Nondiagonal terms in Equation [14] produce a macroscopic polarization; $\rho_{ij}(t)$ departs from zero at equilibrium and includes an oscillatory term at the frequency ω_{ij}. The existence of a nondiagonal term $\rho_{ij}(t) \neq 0$ is linked to the creation of a coherent superposition of states *i* and *j* due to the perturbation. This coherence is broken through the interaction with the molecules of the bath, with a rate constant of the order of $\Gamma_{ij} = 1/T_2$. Usually this dephasing is much faster than the energy relaxation and $T_2 << T_1$.

Two distinct strategies exist for solving Equation [13]. The best-suited to our previous treatment starts by expanding the density matrix in a power series in the perturbation. By identifying terms we arrive at

$$\boldsymbol{P}^{(3)}(t) = Ne\ \mathrm{Tr}\ (\boldsymbol{\rho}^{(3)}(t)\ \boldsymbol{r}) \qquad [15]$$

from which we can evaluate $\boldsymbol{P}^{(3)}(t)$ once we have obtained $\rho(t)$ iteratively, after expanding ρ in a basis of eigenstates of $\mathbf{H}_0$. In dense media local field factors must be included to relate the microscopic field at the molecule site to the macroscopic field of Equation [2].

This approach correctly describes the nonresonant nonlinear response and, with the inclusion of decay terms, allows treatment of Raman resonances as far as the density matrix remains close to its equilibrium value.

In the case of strong resonance, i.e. high-intensity fields closely tuned to narrow transitions, the perturbative series does not converge and this approach will fail. An alternative solution uses a simplified two-level or three-level atom model, along with effective multiphoton operators, and solves Equation [14] without using perturbation techniques. The result is a field-dependent *effective* susceptibility, that can describe population transfer and saturation effects. It can be related to $\chi^{(n)}$ by including correction factors that are a function of the amplitude, the transition moments and the detuning from exact resonance.

The final result for the perturbative approach is given without proof. Using a time-ordered, double diagrammatic technique, it can be shown that $\chi^{(3)}$ includes a sum of 48 terms, each corresponding to a diagram that specifies a time-ordered sequence of individual interactions. Quantum interference among these paths leads to a partial cancellation whenever proper dephasing is not relevant.

Among the remaining 24 terms, we select those with Raman-resonant denominators. The other terms can be considered a weakly dispersive nonresonant background for the Raman process. Typical resonant and nonresonant terms are shown in **Table 2**.

Let us now suppose that only two states g and f are connected through a Raman-type resonance ($\Omega_{fg} - (\omega_P - \omega_S) \simeq 0$), the rest of the denominators being well removed from zero. By inspection, it is possible to collect terms that start at either of these two levels and contribute to the spectral features of $\chi^{(3)}$. After referring all the frequencies to the lower g level, we arrive at

$$\begin{aligned}
\chi^{(3)}_{\mu\alpha\beta\gamma}(-\omega_\sigma;\omega_0\omega_P - \omega_S) &= \chi^{(3)NR}_{\mu\alpha\beta\gamma} \\
&+ \frac{N}{6\epsilon_0\hbar^3}\sum_{g,f}\frac{1}{\Omega_{fg} - (\omega_P - \omega_S)} \\
&\times \sum_j\left(\frac{R^\mu_{gj}R^\alpha_{jf}}{\Omega_{jg} - \omega_\sigma} + \frac{R^\alpha_{gj}R^\mu_{jf}}{\Omega^*_{jg} + \omega_0}\right) \\
&\times \sum_k\left[\rho^{(0)}_{gg}\left(\frac{R^\beta_{gk}R^\gamma_{kf}}{\Omega_{kg} - \omega_P} + \frac{R^\gamma_{gk}R^\beta_{kf}}{\Omega_{kg} + \omega_S}\right)\right. \\
&\left. -\rho^{(0)}_{ff}\left(\frac{R^\beta_{gk}R^\gamma_{kf}}{\Omega^*_{kg} - \omega_P} + \frac{R^\gamma_{gk}R^\beta_{kf}}{\Omega^*_{kg} + \omega_S}\right)\right]
\end{aligned} \qquad [16]$$

where N is the total number density of the chromophore. The summation in j and k runs over all molecular levels, whereas that in g, f runs over all pairs close to Raman resonance, which give rise to close-lying lines. When there is more than one chromophore, an additional summation over species should be included.

If an optical frequency ω_a is close to one-photon electronic resonance with an intermediate state j, the terms including the denominator $\Omega_{jg} - \omega_a \simeq 0$ will be dominant and we obtain a good approximation by retaining only these (resonant nonlinear Raman spectroscopy). In the absence of such additional electronic resonances, we can neglect the damping coefficient for all denominators whose absolute value is far from zero and write them in terms of real Raman scattering tensor components for the transition from

Table 2 Some terms in the microscopic expression of the third-order non-linear susceptibility for Raman resonance

$$\begin{aligned}
\chi^{(3)}_{\mu\alpha\beta\gamma}(-\omega_\sigma;\omega_0\omega_P - \omega_S) &= \frac{N}{6\varepsilon_0\hbar^3}\sum_{gkfj}\rho^{(0)}_{gg} \\
&\times\left\{\frac{1}{[\Omega_{fg} - (\omega_P - \omega_S)](\Omega_{kg} + \omega_S)}\right. \\
&\times\left(\frac{(R_{gk})_\gamma(R_{kf})_\beta(R_{fj})_\alpha(R_{jg})_\mu}{\Omega_{jg} - \omega_\sigma} + \frac{(R_{gk})_\gamma(R_{kf})_\beta(R_{fj})_\alpha(R_{jg})_\mu}{\Omega^*_{jg} + \omega_0}\right) \\
&+\frac{1}{[\Omega_{fg} - (\omega_P - \omega_S)](\Omega_{kg} - \omega_P)} \\
&\times\left(\frac{(R_{gk})_\beta(R_{kf})_\gamma(R_{fj})_\alpha(R_{jg})_\mu}{\Omega_{jg} - \omega_\sigma} + \frac{(R_{gk})_\beta(R_{kf})_\gamma(R_{fj})_\mu(R_{jg})_\alpha}{\Omega^*_{jg} + \omega_0}\right) \\
&+ \text{other resonant terms}^a \\
&+\frac{1}{[\Omega^*_{fg} + (\omega_0 + \omega_P)](\Omega^*_{jg} + \omega_P)} \\
&\times\left(\frac{(R_{gk})_\mu(R_{kf})_\gamma(R_{fj})_\alpha(R_{jg})_\beta}{\Omega^*_{kg} + \omega_\sigma} + \frac{(R_{gk})_\gamma(R_{kf})_\mu(R_{fj})_\alpha(R_{jg})_\beta}{\Omega_{kg} + \omega_S}\right) \\
&+\frac{1}{[\Omega^*_{fg} + (\omega_0 + \omega_P)](\Omega^*_{jg} + \omega_0)} \\
&\times\left(\frac{(R_{gk})_\mu(R_{kf})_\gamma(R_{fj})_\beta(R_{jg})_\alpha}{\Omega^*_{kg} + \omega_\sigma} + \frac{(R_{gk})_\gamma(R_{kf})_\mu(R_{fj})_\beta(R_{jg})_\alpha}{\Omega_{kg} + \omega_S}\right) \\
&\left. + \text{other nonresonant terms}^b\right\}
\end{aligned}$$

$\Omega_{ab} = (E_a - E_b)/\hbar - i\Gamma_{ab}$

$(R_{ab})_\mu$ = component along μ of the transition dipole matrix element $\langle b|er|a\rangle$

[a] For $\omega_P - \omega_S \approx \omega_R$ as the only frequency combination resonant with a molecular transition, Raman resonant terms dominate the nonlinear response and its change across the resonance defines the spectral shape. Additional one-photon resonance is possible through the denominators such as $(\Omega_{kg} - \omega_P)$ whenever input frequencies approach electronic transitions. The spectral behaviour in this case is dominated by a few terms in the summation.

[b] The sum of all nonresonant terms constitutes a weakly dispersive background. Some terms are nonresonant for a Raman transition but can give rise to two- or three-photon absorption for a different set of input frequencies.

g to f excited by the laser at frequency

$$\omega - (\alpha_{\delta\epsilon}(\omega))_{gf} = \frac{1}{\hbar}\sum_j \left(\frac{R^{\epsilon}_{gj}R^{\delta}_{jf}}{\omega_{jg}-\omega} + \frac{R^{\delta}_{gj}R^{\epsilon}_{jf}}{\omega_{jf}+\omega}\right)$$

We can further simplify by ignoring the weak dispersion of $(\alpha_{\delta\varepsilon}(\omega))_{gf}$ and dropping the gf subscript. The final expression is shown on **Table 1**. We see that, in the presence of resonances, the third-order susceptibility is generally complex:

$$\chi^{(3)} = \chi^{(3)\text{NR}} + \chi'^{(3)\text{R}} + \text{i}\chi''^{(3)\text{R}} \qquad [17]$$

The non-resonant part $\chi^{(3)\text{NR}}$ is real and includes the contribution of every species present. It will usually show a weak dispersion, being approximately constant for extended frequency ranges. In this case, the permutation of the frequency arguments alone does not alter much the numerical value of the susceptibility. This is the *Kleinman's symmetry conjecture* which, combined with the strict intrinsic permutation symmetry allows to freely permute the Cartesian subscripts in $\chi^{(3)\text{NR}}$. This property, although not exact, is of great relevance for the polarization nonlinear Raman techniques.

$\chi^{(3)\text{R}}$ has both real and imaginary parts, and is additive if different transitions from the same or other chromophore overlap. Disregarding multiplicative factors, the dispersion of $\chi^{(3)\text{R}}$ has the form

$$\chi'^{(3)\text{R}} \sim \sum_{g,f} \alpha^2(\rho^{(0)}_{gg} - \rho^{(0)}_{ff})\frac{\Delta_{fg}}{\Delta^2_{fg}+\Gamma^2_{fg}}$$
$$\chi''^{(3)\text{R}} \sim \pm\sum_{g,f} \alpha^2(\rho^{(0)}_{gg} - \rho^{(0)}_{ff})\frac{\Delta_{fg}}{\Delta^2_{fg}+\Gamma^2_{fg}} \qquad [18]$$

The imaginary part has a Lorentzian profile of halfwidth Γ_{fg}, centred at exact resonance. This Lorentzian line shape comes from the assumed form for the damping terms, and more realistic models should be used in Equation [14] when studying line shapes. The spectral shape is the same as for spontaneous Raman lines. The real part of $\chi^{(3)\text{R}}$ is multiplied by the detuning, and shows a dispersive line shape. $\chi^{(3)\text{R}}$ depends on the population difference, instead of just the population of the initial state as does spontaneous Raman. This dependence can lead to saturation whenever appreciable population is transferred to the excited level.

Generation, evolution and detection of the signal field

The last step is to analyse how the induced nonlinear polarization creates a new wave or interferes with the existing ones to generate the nonlinear Raman signal. The nonlinear wave propagation equation (taking $\boldsymbol{P}^{(2)} = 0$) is

$$\nabla \times \nabla \times \boldsymbol{E}(\omega) = \frac{\omega^2}{c^2}\boldsymbol{\varepsilon}(\omega)\boldsymbol{E}(\omega) + \omega^2\mu_0\boldsymbol{P}^{(3)}(\omega) \qquad [19]$$

where $\boldsymbol{\varepsilon}(\omega) = 1 + \boldsymbol{\chi}^{(1)}(-\omega;\omega)$ is the dielectric tensor that accounts for the linear propagation. We look for solutions in the form of running plane waves propagating collinearly along the z axis, orthogonal to the electric field vector, and suppose that the amplitude of the signal field will change slowly along z. We will restrict ourselves to the small-gain regime, in which neither appreciable depletion of the input beams nor population transfer occurs, leading to an amplitude rate change given by

$$\frac{\partial E_{\omega_\sigma\mu}(z)}{\partial z} = \frac{\text{i}\omega_\sigma}{2cn_\sigma}\left(\frac{D}{4}\right)\chi^{(3)}_{\mu\alpha\beta\gamma}(-\omega_\sigma;\omega_0\omega_\text{P} - \omega_\text{S})$$
$$E_{\omega_0\alpha}(z)E_{\omega_\text{P}\beta}(z)E^*_{\omega_\text{S}\gamma}(z)\exp(\text{i}\Delta kz) \qquad [20]$$

where $\Delta k = (k_0 + k_\text{P} - k_\text{S}) - k_\sigma$ is the phase mismatch, i.e. the difference between the wave vector of the 'polarization wave' at frequency ω_σ and that of the signal field at the same frequency. The former value is a combination of the refractive index at the input frequencies, whereas the latter depends on the refractive index at ω_σ. Owing to linear dispersion, these two values are in principle different. If $\Delta k \neq 0$ the change rate of the signal amplitude shows an oscillatory behaviour, whereas for $\Delta k = 0$ (perfect phase match) a monotonic change is obtained (increase or decrease depending upon the argument of the complex susceptibility) leading to a higher signal. In the small-signal regime we can integrate Equation [20] and evaluate the intensity at the detector, given by $I\omega_j = \frac{1}{2}\varepsilon_0 cn_j |E_j|^2$.

We are at last able to give a plane wave signal expression for the different nonlinear Raman techniques, collected in **Table 1**.

In several techniques we monitor the intensity at a new frequency (normal CARS, Raman resonant four-wave mixing), a new polarization (RIKES) or both (polarization CARS). The generated wave does not interfere with the input waves and the output intensity depends on the product of three input intensities, the square of the interaction length, within the coherence length of lasers, and the square of the effective susceptibility. If $\omega_\sigma = \omega_P$ or ω_S, the phase matching is automatic; in other cases it must be set by controlling the refraction index or crossing the beams.

The quadratic dependence with $\chi^{(3)}$ leads to interference among neighbouring lines, from the same or different species, and with the nonresonant background in dense media, which can lead to strongly distorted line-shapes. The different tensorial properties of $\chi^{(3)R}$ and $\chi^{(3)NR}$ can be exploited to reduce background interference, by using input waves with circular polarization or linear along a direction chosen to cancel the nonresonant part.

In other techniques, we monitor the change in intensity for the pump or Stokes beam. We can solve Equation [20] considering the input field as variable, leading to an exponential amplitude change that can be approximated by a linear change in the small signal regime, or keeping it constant as before. For the latter method we must include the interference of signal and input fields at the detector, which shows a close connection with heterodyne detection techniques. Both methods lead to the same result in the small-gain regime, showing a decrease in the pump intensity (SRL) and an increase in the Stokes intensity (SRG), linear with the imaginary resonant part of the susceptibility $\chi^{(3)\prime R}$, i.e. proportional to the spontaneous Raman spectrum and free from interference effects, automatically phase-matched and dependent on the product of pump and Stokes intensities.

The signal expressions in **Table 1** allow for an analysis of the spectral properties of the relevant techniques. In actual experiments, beams are focused and the detailed geometry must be considered.

List of symbols

A_ω = complex pulse area; D = number of distinguishable permutations of the pairs (ω, α); $\langle er \rangle$ = expectation value of the molecular dipole operator; E_i = energy of level i; $\boldsymbol{E}(t)$ = electric field vector; $E_\mu(t)$ = component along μ of the electric field vector; $\boldsymbol{E}(\omega)$ = amplitude of the Fourier component of $\boldsymbol{E}$ at frequency ω; $\boldsymbol{E}_\omega$ = amplitude of the monochromatic wave at frequency ω; $E_{\omega\mu}$ = component along μ of $\boldsymbol{E}_\omega$; $\boldsymbol{E}_\omega(t)$ = envelope of the quasi-monochromatic field at central frequency ω; $\mathrm{H}(t)$ = time-dependent total Hamiltonian; H_0 = unperturbed Hamiltonian in the absence of electromagnetic fields; I_ω = intensity of the monochromatic wave at frequency ω; N = number density of molecules; n_σ = refractive index at frequency ω_σ; $\boldsymbol{P}(t)$ = macroscopic polarization vector; $\boldsymbol{P}^{(n)}$ = n-th order term in the series expansion of $\boldsymbol{P}$ in powers of the applied field; $P^{(3)}_\mu$ = component along μ of $\boldsymbol{P}^{(3)}$; $\boldsymbol{P}(\omega)$ = amplitude of the Fourier component of $\boldsymbol{P}$ at frequency ω; $P^{(3)}_\mu(\omega)$ = amplitude of the monochromatic polarization wave at frequency ω; $\boldsymbol{P}^{(3)}_\omega(t)$ = envelope of the quasi-monochromatic third-order polarization component at central frequency ω; $P^{(3)}_{\omega\mu}$ = component along μ of $P^{(3)}_\omega$; $\mathbf{R}^{(n)}(\tau_1, \tau_2, \tau_3)$ = third-order nonlinear response function tensor; $R^{(3)}_{\mu\alpha\beta\gamma}(\tau_1, \tau_2, \tau_3)$ = component of $\mathbf{R}^{(3)}$ with Cartesian index $\mu\alpha\beta\gamma$; R^μ_{ab} = component along μ of the transition dipole matrix element for transitions from a to b; T_1 = population relaxation characteristic time; T_2 = phase relaxation characteristic time; $V(t)$ = interaction operator; $(\alpha(\omega))_{gf}$ = Raman scattering tensor for transitions from state g to state f; $(\alpha_{\delta\varepsilon}(\omega))_{gf}$ = component $\delta\varepsilon$ of the Raman scattering tensor; Γ = relaxation matrix; Γ_{ij} = matrix element of Γ; $\Gamma^{\text{proper}}_{ij}$ = pure dephasing term associated with element ρ_{ij} of the density matrix; $\Delta\boldsymbol{k}$ = phase mismatch vector; $\rho(t)$ = density matrix of the material system; $\rho_{ij}(t)$ = matrix element of ρ_k; $\rho^{(3)}(t)$ = third order term in the development of the matrix density in power series of the interaction; τ_p = characteristic time of the radiation (it represents the period of the fastest amplitude change in a laser pulse); $\Phi^{(3)}_{-\omega_\sigma;\,\omega_0\omega_\mathrm{p}-\omega_\mathrm{s}}(\tau_1, \tau_2, \tau_3)$ = third-order nonlinear envelope response function; $\chi^{(n)}(-\omega_\sigma; \omega_1\omega_2\omega_3)$ = nth-order nonlinear susceptibility tensor; $\chi^{(3)}_{\mu\alpha\beta\gamma}(-\omega_\sigma; \omega_1\omega_2\omega_3)$ = component of $\chi^{(3)}$ with Cartesian index $\mu\alpha\beta\gamma$; $\chi^{(3)NR}$ = nonresonant part of the third-order nonlinear susceptibility; $\chi'^{(3)R}$ = real resonant part of the third-order nonlinear susceptibility; $\chi''^{(3)R}$ = imaginary resonant part of the third-order nonlinear susceptibility; ω = frequency of the radiation, $E = \hbar\omega$; ω_R = frequency corresponding to a Raman transition; $\omega_{ij} = (E_i - E_j)/\hbar$ = frequency corresponding to the energy gap for levels i and j; $\Omega_{ab} = \omega_{ab} - \mathrm{i}\Gamma_{ab}$ = compact complex notation for detuning and relaxation; $\hbar$ = Planck constant/2π.

See also: **Nonlinear Optical Properties; Nonlinear Raman Spectroscopy, Applications; Nonlinear Raman Spectroscopy, Instruments; Raman Optical Activity, Applications; Raman Optical Activity, Theory; Raman Spectrometers.**

Further reading

(This work is mainly based on Santos (1996) and Butcher (1990.)

Bloembergen N (1965) *Nonlinear Optics*. Reading, MA: WA Benjamin.

Butcher PN and Cotter D (1990) *The Elements of Nonlinear Optics*. Cambridge: Cambridge University Press.

Clark RJH and Hester RE (eds) (1988) *Advances in Nonlinear Spectroscopy*. Chichester: Wiley.
Eesley GL (1981) *Coherent Raman Spectroscopy*. Oxford: Pergamon Press.
Lee D and Albrecht AC (1993) In: Prigogine I and Rice SA (eds) *Advances in Chemical Physics*, Volume LXXXIII, pp 43–87. New York: Wiley.
Levenson D and Kano SS (1988) *Introduction to Nonlinear Laser Spectroscopy*. San Diego: Academic Press.
Mukamel S (1995) *Principles of Nonlinear Optical Spectroscopy*. New York: Oxford University Press.
Prior Y (1984) A complete expression for the third-order susceptibility. Perturbative and diagrammatic approaches. *IEEE Journal of Quantum Electronics* QE-20: 37–42.
Santos Gómez J (1996) Coherent Raman spectroscopy. In: Laserna JJ (ed) *Modern Techniques in Raman Spectroscopy*, pp 305–340. Chichester: Wiley.

NQR, Applications

See **Nuclear Quadrupole Resonance, Applications.**

NQR, Spectrometers

See **Nuclear Quadrupole Resonance, Instrumentation.**

NQR, Theory

See **Nuclear Quadrupole Resonance, Theory.**

Nuclear Overhauser Effect

Anil Kumar and **R Christy Rani Grace**,
Indian Institute of Science, Bangalore, India

MAGNETIC RESONANCE
Theory

Introduction

The Nuclear Overhauser Effect (NOE) has become one of the key effects for obtaining the structures of molecules, especially biomolecules, in solution by NMR spectroscopy. The effect was first enunciated by its discoverer, Overhauser, as a large polarization of nuclear spins in metals on saturation of electrons to which they are coupled. Neither its discoverer, nor the famous people in the audience at the conference where he first presented his calculation in 1953, nor Charles Slichter, who took upon himself the task of experimentally verifying Overhauser's assertions, could have imagined that the effect would become the backbone of modern NMR. The effect remained as a curiosity and was sparingly used until its transformation into the nuclear–nuclear Overhauser effect, in which one of the nuclear spins is saturated and the nearby nuclear spin shows changes in the intensity of its NMR signal. For small molecules, in the fast motion limit, the Overhauser effect is positive for nuclear spins with the same sign of γ (the magnetogyric ratio) and negative for those with opposite sign.

The nuclear–nuclear Overhauser effect, in its early days, was used for enhancing the polarization of less-sensitive nuclei, such as natural abundant ^{13}C or ^{15}N and in double-resonance experiments for obtaining additional information on relaxation of molecules. It was utilized, in conjunction with the spin–lattice relaxation of various carbons, for the characterization of anisotropy of molecular reorientations. However, its largest utility continued to be for obtaining information on molecular conformations, by selectively saturating the magnetization of a specific proton in the molecule and monitoring the changes in the intensity of resonances of nearby protons.

In continuous wave (CW) NMR, one of the problems is the separation of the Overhauser effect from the coherent effects of radiofrequency (RF) irradiation. For example, even at very low RF field (~ 0.02 Hz) CW double resonance experiments, the observed intensities can only be explained by including coherent effects, as splitting or broadening of directly connected transitions. In higher power experiments, the lines are split and several coherent effects are simultaneously present. The development of one-dimensional (1D) Fourier transform NMR immediately freed the nuclear Overhauser effect from the shackles of such coherent effects. The irradiation can be time-shared, such that the saturating RF field is applied before the observation pulse. Since, during observation, the radiofrequency field and hence the coherent effects are absent, the spectrum contains information only on relaxation. The development of 1D Fourier transform NMR, with its added sensitivity advantage, opened the field of NMR, to the study of proton NMR of biomolecules. As soon as such spectra were recorded, the assignment of the large number of resonances became a major problem. Selective decoupling and NOEs were employed, respectively, to assign and to obtain information on the secondary structures of peptides and proteins. In small peptides, it became a matter of routine to pick-up a few characteristic NOEs and assign the secondary structure. For larger peptides, containing 20–50 amino acid residues, efforts were directed towards obtaining a large number of (50–100) selective decoupling and NOE experiments, to assign the spectra and to obtain the three-dimensional structure of the molecules. However, it was hard work to manually plot, measure and analyse these large number of 1D spectra. The development of two-dimensional (2D) NMR removed this obstacle. COSY NMR spectroscopy with its many improvements, is used for identifying coherently coupled spins, while the 2D NOESY is used for obtaining large-scale information on distances between nearby spins. The success of this algorithm prompted workers to look at even larger molecules (of relative molecular mass 10–15 kDa), crowding even the 2D spectra, and necessitating addition of a third or even fourth dimension to remove overlap. This was achieved by labelling the biomolecules with either ^{13}C and/or ^{15}N and spreading the 2D information into a third or fourth dimension, as a function of attached carbon or nitrogen chemical shifts. Selective polarization transfers (from a selected heteronucleus to a selected proton) followed by NOE are being developed to enhance the resolution and to reduce the burden of 2D or 3D NMR experiments.

Theory

In the NOE, the non-equilibrium magnetization of a spin is shared by nearby spins through mutual dipole–dipole relaxation. Any other relaxation mechanism, if present, attenuates the effect. For example, dissolved paramagnetic oxygen has a deleterious effect on the NOE. It is therefore necessary to have clean solvents and often it becomes necessary to remove the dissolved oxygen for a quantitative estimate of NOEs.

The source spin is brought to a non-equilibrium state, for example by selectively saturating its magnetization by a low power, long pulse. During this irradiation, the transfer of magnetization to the other nearby spins also takes place and this can be monitored as a function of the irradiation time. The irradiation can be performed for a sufficiently long time, so that a steady-state is reached between the irradiation and the relaxation and this gives rise to what is known as 'steady-state NOE'. However, such experiments, particularly in case of slowly reorienting molecules, give NOEs between fairly distant spins in the molecule, through migration of magnetization over several intervening spins (the phenomenon is known as 'spin-diffusion'). This has the disadvantage that the NOE loses selectivity of information. A simple alternative is to reduce the irradiation time or to monitor the growth of the NOE, as a function of irradiation time. This has been called 'truncated-driven-NOE'. In both these experiments, the NOE transfer takes place in the presence of the RF field. In another experiment, the source spin is selectively inverted (by a selective π pulse) and the NOE is monitored in the absence of the RF field. This is known as 'transient-NOE', and was first suggested by Solomon.

The 2D NOE (NOESY) experiment is formally equivalent to a large number of transient NOE experiments. Each cross-section of a 2D NOESY spectrum is equivalent to a 1D selective transient NOE experiment, in which the peak corresponding to the diagonal is selectively inverted. The NOE transfer in both NOESY and 1D transient NOE takes place in the absence of the RF field and, theoretically, the two experiments are identical (except for a factor of 2 in the intensities of 1D experiments which will be explained later). The theory of the NOE, given in the following subsections is divided into three parts, (i) steady-state NOE, (ii) transient NOE and (iii) 2D NOE (NOESY). Later sections describe the NOE in the rotating frame, heteronuclear NOE and transferred NOE.

In this article the NOE between various spins will be discussed, ignoring the spin–spin coupling between various spins and the associated splitting of lines into multiplet components. A spin will be assumed to be irradiated (saturated or inverted) as a whole and detected as a whole. This description leaves out many interesting effects, such as selective saturation of various transitions, direct pumping effects of irradiation or inversion, multiplet effects between various transitions of a spin arising from cross-correlations and strong coupling effects.

Steady-state NOE

Consider a two spin-$\frac{1}{2}$ system, with spins I and S relaxation coupled to each other. Relaxation of these two spins in the presence of mutual cross-relaxation is described by coupled rate equations known as Solomon's equations:

$$\frac{\mathrm{d}I_z(t)}{\mathrm{d}t} = -\rho_I(I_z(t) - I_z^{\mathrm{eq}}) - \sigma_{IS}(S_z(t) - S_z^{\mathrm{eq}}) \qquad [1]$$

$$\frac{\mathrm{d}S_z(t)}{\mathrm{d}t} = -\sigma_{SI}(I_z(t) - I_z^{\mathrm{eq}}) - \rho_S(S_z(t) - S_z^{\mathrm{eq}}) \qquad [2]$$

where $I_z(t)$ and $S_z(t)$ describe the instantaneous values of the longitudinal magnetization of the spins I and S and I_z^{eq} and S_z^{eq} represent their equilibrium values respectively. The terms ρ_I and ρ_S are the self-relaxation rates of the two spins, respectively, and $\sigma_{IS} = \sigma_{SI}$ is the mutual cross-relaxation (NOE) rate between the spins. All these rates depend on the mechanics of relaxation operative in the spin system and the state of mobility of the molecule in which they are embedded.

On selective saturation by RF irradiation at the resonance frequency of one of the spins (say S), it is generally assumed that the magnetization of the irradiated spin is fully saturated at all times, that is $S_z(t) = 0$, for all values of t. The steady state solution of the Solomon's equations is then obtained by making the time derivatives on the left-hand-side of Equations [1] and [2] as equal to zero, yielding the steady state value of I_z from Equation [1] as

$$I_z^{SS} - I_z^{\mathrm{eq}} = \frac{\sigma_{IS}}{\rho_I} S_z^{\mathrm{eq}} \qquad [3]$$

Since $I_z^{\mathrm{eq}}/S_z^{\mathrm{eq}} = \gamma_I/\gamma_S$, one obtains the well-known steady-state-NOE as

$$\eta = \frac{I_z^{SS} - I_z^{\mathrm{eq}}}{I_z^{\mathrm{eq}}} = \frac{\gamma_S}{\gamma_I}\frac{\sigma_{IS}}{\rho_I} \qquad [4]$$

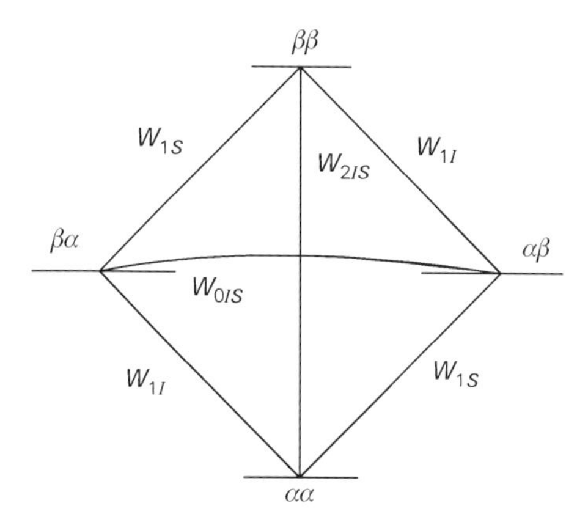

Figure 1 Energy-level diagram of a two spin-$\frac{1}{2}$ system, showing the transition probabilities and the spin states. The designation αα means both the spins are in the $+\frac{1}{2}$ state, while $\alpha\beta$ means the first spin (I) is in the $+\frac{1}{2}$ state and the second spin (S) is in $-\frac{1}{2}$ state. W_{1I} gives the single quantum transition probability for the I spin when it changes its spin state. W_0 and W_2 are the zero and double-quantum transition probabilities, respectively, by which both the I and S spins simultaneously change their spin states.

For two relaxation coupled spins, each spin-$\frac{1}{2}$, the energy level diagram consists of four levels (**Figure 1**).

From this diagram it is seen that W_{1I} and W_{1S} are the single quantum transition probabilities for the spin I and S respectively. The term W_{0IS} is the zero quantum transition probability for the flip-flop process, in which the two spins are in opposite spin states and exchange their spin states; W_{2IS} is the double quantum transition probability, by which the two spins in the similar states simultaneously change their spin states. In homonuclear systems, these probabilities are proportional to the spectral densities at 0, ω and 2ω frequencies, respectively, for W_0, W_1 and W_2, where ω is the Larmor frequency. Using population rate equations of the four levels of the two-spin system and using these transition probabilities connecting the four levels, it is straightforward to show that

$$\begin{aligned}\sigma_{IS} &= W_{2IS} - W_{0IS}\\ \rho_I &= W_{0IS} + 2W_{1I} + W_{2IS}\\ \rho_S &= W_{0IS} + 2W_{1S} + W_{2IS}\end{aligned} \qquad [5]$$

Using Equation [5] in Equation [4] the NOE on spin I, on saturating spin S, is obtained as

$$\eta_I\{S\} = \left(\frac{\gamma_S}{\gamma_I}\right)\frac{W_{2IS} - W_{0IS}}{W_{0IS} + 2W_{1I} + W_{2IS}} \qquad [6]$$

The origin of the NOE can now be explained in the following way. After the selected spins absorb a certain amount of energy from the RF field, they pass this energy to the other spins and to other degrees of freedom, known as the lattice. In other words, they start to relax. Mutual dipolar interaction between the nearby spins provides an effective mechanism for exchange of energy among the spins and to the lattice. However, if these spins are embedded in a rigid solid, the dipolar interaction is time-independent and does not couple the spins to the lattice, except in the case of like spins (same γ), the dipolar interaction in rigid solids does provide a mechanism of rapidly passing the energy from one spin to the next (via its energy conserving flip-flop, W_0, terms). This process, known as spin-diffusion in solids, is responsible for migration of magnetization across the whole sample and to relaxation sinks such as rapidly rotating methyl groups or rapidly relaxing paramagnetic centres. On the other hand, if the interacting spins are on molecules in solution, which are rapidly reorienting, the dipolar interaction becomes time dependent and, if there are molecular motions at the Larmor or twice the Larmor frequency, it couples the spins to the rotational degrees of freedom of the molecule and transfers the spin energy to the lattice. The lattice is assumed to have infinite heat capacity and remains at thermal equilibrium at room or sample temperature. The coupling of the spin to the lattice at the Larmor frequency is provided additionally by several other mechanisms, such as modulation of chemical shift anisotropy, quadrupolar interaction and paramagnetic relaxation centres. All these contribute to W_1 but not to W_0 and W_2, and attenuate the NOE, as seen from Equation [6]. It is only the dipolar interaction that has two spin terms, such as I_+S_-, I_-S_+ and I_+S_+, I_-S_-, which contribute respectively, to W_0 and W_2 and in turn to the NOE. The first two are flip-flop terms or zero-quantum terms, in which the two spins exchange energy among each other without requiring exchange of energy with the lattice (for $\gamma_I = \gamma_S$). Such processes dominate for slowly reorienting molecules such as biomolecules; and therefore the NOE is very significant in such cases, as hardly any energy is lost to the lattice. Here W_0 dominates over all the other terms, and as seen from Equation [6], the NOE is large and negative (−1). The third and the fourth terms, I_+S_+ and I_-S_- are the two quantum terms and the exchange energy of the spins with the lattice at twice the Larmor frequency. In this case both the spins flip simultaneously either up or down. If molecular motions are fast enough to have spectral densities at 2ω, then these processes dominate. It can be shown that in such circumstances the NOE is positive and small, with a maximum of +0.5.

The spectral densities for the various motional regimes are schematically sketched in **Figure 2**.

In the slow motion or long correlation time limit ($\tau_c >> 1/\omega_0$), the spectral densities are maximum near the zero frequency and minimum at ω_0 and $2\omega_0$, while for fast reorienting molecules $\tau_c << 1/\omega_0$, the spectral density is almost equal for all the frequencies till $\omega_c = 1/\tau_c$, where τ_c is the correlation time for molecular reorientations. For the dipolar interaction between spins *I* and *S*, the transition probabilities for isotropic molecular reorientations are obtained as

$$\begin{aligned} W_{0IS} &= \frac{1}{10}k^2\frac{\tau_c}{1+(\omega_I-\omega_S)^2\tau_c^2} \\ W_{1I} &= \frac{3}{20}k^2\frac{\tau_c}{1+\omega_I^2\tau_c^2} \\ W_{1S} &= \frac{3}{20}k^2\frac{\tau_c}{1+\omega_S^2\tau_c^2} \\ W_{2IS} &= \frac{6}{10}k^2\frac{\tau_c}{1+(\omega_I+\omega_S)^2\tau_c^2} \end{aligned} \qquad [7]$$

where $k = (\mu_0/4\pi)\gamma_A\gamma_X\hbar r_{IS}^{-3}$.

Homonuclear case When the two spins are of the same species, i.e. $\gamma_I = \gamma_S$, then the steady state homonuclear NOE for selective saturation of one of the spins with the other spin having a well-resolved chemical shift is obtained by substituting in Equation [6] $\omega_I = \omega_S = \omega$,

$$\eta_{\max} = \frac{5+\omega^2\tau_c^2-4\omega^4\tau_c^4}{10+23\omega^2\tau_c^2+4\omega^4\tau_c^4} \qquad [8]$$

This curve is plotted in **Figure 3**, and has the value +0.5 for fast reorientation with respect to ω (short correlation limit $\omega\tau_c << 1$) and –1 for slow reorientations (long correlation limit, $\omega\tau_c >> 1$).

These can also be seen for homonuclear spins, by direct substitution of *W*s in the two limits in Equation [6]. For example, for the short correlation time limit, $W_0 : W_1 : W_2 :: 2 : 3 : 12$, yielding $\eta = \frac{1}{2}$. For the long correlation time limit, W_1 and W_2 are negligible and $\eta = -1$. From Equation [8] and from **Figure 3** it is seen that $\eta = 0$, for $\omega\tau_c = 1.118$. This is often called the critical correlation time limit. In this limit, $W_0 = W_2$, $\sigma = 0$ and the laboratory frame NOE is zero. Experiments have been designed in which both the spins are spin-locked along an axis

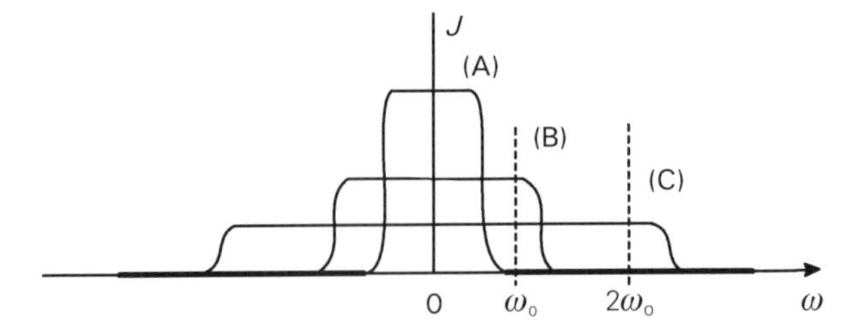

Figure 2 Spectral density $J(\omega)$ for three values of correlation time, plotted as a function of frequency ω. The spectral density has a cutoff frequency $\omega_c = 1/\tau_c$, where τ_c is the correlation time of molecular reorientations. As molecular reorientations become faster, τ_c decreases and the spectral density dispersion becomes flatter. The terms T_1, T_2 and NOE depend on the value of the spectral densities at 0, ω_0 and $2\omega_0$, where ω_0 is the Larmor frequency. (A) Spectral density for slowly reorienting molecules which have long correlation times ($\tau_c >> 1/\omega_0$). In such cases the spectral density has a negligible value at ω_0 and $2\omega_0$, but large values at low frequencies. (B) Spectral density for intermediate values of correlation times, for which $\tau_c \approx 1/\omega_0$. (C) Spectral density for small molecules undergoing fast reorientation, which have short correlation times ($\tau_c << 1/\omega_0$) and the spectral density has nearly equal values from 0 to $2\omega_0$. Since the area under the curves is constant, the spectral density has different magnitudes at each frequency in the above three cases.

making an angle θ from the **z**-direction, yielding a non-zero NOE. This will be described in the section on 'rotating-frame NOE'. The short correlation time limit, $\omega\tau_c << 1$ is generally applicable to small molecules (relative molecular mass ≤ 1 kDa) at low spectrometer frequencies, as was often the case in the early days of NMR. For larger size molecules, which reorient slowly ($\omega\tau_c >> 1$) the NOE is negative, but very useful in obtaining information on the proximity of the spins.

The observation of negative NOE among the spins with the same sign of γ and in the short correlation time limit, however, gave rise to some excitement during the late 1960s. It was soon found that the negative NOE was owing to what was called a 'three-spin-effect'. The explanation is as follows. When the first spin is saturated, the second spin is enhanced in intensity. By logical extension this means that the third spin is reduced in intensity. This of course requires that the three spins are in almost a linear configuration, such that the direct positive NOE from the first spin to the third is less than the transmitted NOE via the second spin. The observation of the negative three-spin-effect for homonuclear spins in the short correlation time limit is thus a signature of the linearity or near-linearity of the three spins. The other observation of a negative NOE between two protons, without the intervention of a third spin in a polypeptide by Balaram and coworkers, was the first evidence of molecules

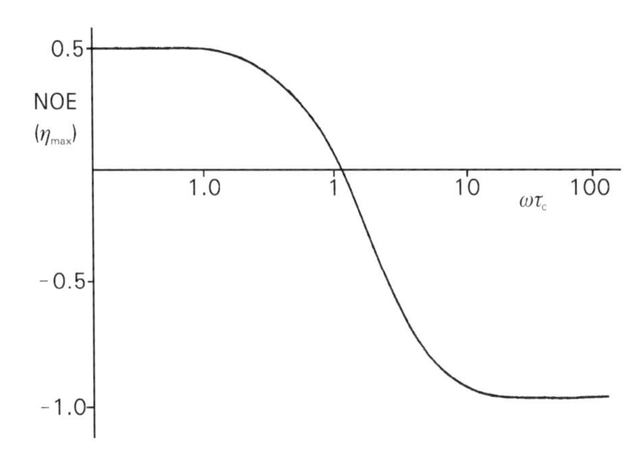

Figure 3 Variation of homonuclear NOE enhancement, Equation [8], plotted as a function of $\omega\tau_c$. Note the logarithmic scale of $\omega\tau_c$. For small molecules with short τ_c, the limiting value for η_{max} is +0.5. In practice, since relaxation mechanisms other than dipolar are also efficient in this extreme narrowing limit, positive enhancements as large as this are rarely observed. For large molecules with long τ_c, the limiting value of η_{max} is −1. Biomolecules and small molecules in viscous solvents come into this category and generally give significant NOEs. In the central region where η_{max} varies rapidly with $\omega\tau_c$, the NOE enhancements depend on the spectrometer frequency and the molecular tumbling rate.The value of η_{max} passes through a null for $\omega\tau_c \approx 1$.

tumbling at rates slower than the Larmor frequency. Ever since then, larger molecules have been studied by NMR spectrometers operating at higher frequencies, and negative NOEs between protons have become the backbone of NMR research. There is an additional advantage of negative NOEs, which becomes apparent in the transient NOE experiment, described in the next section.

Transient NOE

In the transient NOE experiment, the perturbed spin is selectively inverted rather than saturated. Since this can be done in times short compared with T_1 and T_2 of the spins, the NOE during the pulse is neglected and the migration of the magnetization is observed after the pulse, in the absence of the RF field. The time evolution of magnetization is obtained using Solomon's Equations [1] and [2]. Substituting the initial condition, $S_z(0) = -S_z^{eq}$, one obtains a biexponential time evolution for I_z and S_z magnetization, assuming $\rho_I = \rho_S = \rho_{IS}$, as

$$\frac{S_z(t) - S_z^{eq}}{S_z^{eq}} = -e^{-(\rho_{IS}+\sigma_{IS})t} - e^{-(\rho_{IS}-\sigma_{IS})t} \qquad [9]$$

$$\frac{I_z(t) - I_z^{eq}}{I_z^{eq}} = -e^{-(\rho_{IS}+\sigma_{IS})t} + e^{-(\rho_{IS}-\sigma_{IS})t} \qquad [10]$$

While the recovery of the inverted spin S_z to its equilibrium value is biexponential, that of I_z magnetization shows an initial growth and then a decay. Equation [10] can be rewritten as

$$\frac{I_z(t) - I_z^{eq}}{I_z^{eq}} = [1 - e^{-2\sigma_{IS}t}]e^{-(\rho_{IS}-\sigma_{IS})t} \qquad [11]$$

This gives the NOE on spin I, which is positive for positive s and negative for negative s. The NOE on spin I grows, reaches a maximum and then decays to zero. The initial rate of growth is obtained by differentiating Equation [11] with respect to time and taking the limit $t \to 0$, the so-called initial rate approximation, yielding

$$\frac{d}{dt}\left(\frac{I_z(t) - I_z^{eq}}{I_z^{eq}}\right)_{t\to 0} = 2\sigma_{IS} \qquad [12]$$

The initial rate of growth of NOE thus gives a direct measure of the cross-relaxation rate σ_{IS} and by inference the distance r_{IS} (σ_{IS} is proportional to r_{IS}^{-6})

The advantage of the transient NOE experiment is that the transport of magnetization takes place in the absence of RF irradiation and also the dynamics of Solomon's equations are identical to the 2D NOE experiment, described in the next section. The driven experiment has no 2D analogue and the solution given earlier in Equation [4] has the limitation that the details of saturation are not included. In fact if one uses Equation [2] instead of Equation [1], for the steady-state solution, by substituting $dS_z(t)/dt = 0$, $S_z(t) = 0$, one obtains a wrong result

$$\frac{I_z(t) - I_z^{eq}}{S_z^{eq}} = \frac{\rho_S}{\sigma_{IS}} \qquad [13]$$

Boulat and Bodenhausen earlier and recently Karthik have shown how this anomaly can be removed by describing the details of the saturation process of spin S.

2D NOE (NOESY)

Selective saturation or inversion of each transition out of a large number of closely spaced transitions of various protons of a protein is both tedious and difficult. The development of the two-dimensional nuclear Overhauser effect (2D NOE or NOESY) experiment was therefore a turning point in the application of NMR for the study of biomolecules. The 2D NOE experiment, **Figure 4A**, uses three 90°

pulses. The first pulse flips the magnetization of all the spins in the molecule to the transverse plane, which are then allowed to evolve during a frequency labelling period t_1. The second pulse flips the magnetization to the longitudinal direction. This non-equilibrium magnetization is allowed to relax and gives NOEs according to Equations [1] and [2] during the mixing period τ_m. The state of the spin system is read by the third 90° pulse with the signal being recorded as a function of the time variable t_2. A complete set of data $s(t_1, t_2)$, after Fourier transformation with respect to both t_1 and t_2 yields a 2D spectrum (**Figure 4B**).

The magnetization components, which have the same frequencies in time domains t_1 and t_2, lie along the diagonal of the NOESY spectrum, while those magnetization components which have crossed over from one spin to another spin during the mixing time τ_m, owing to the NOE, lie on both sides of the diagonal and are called the cross-peaks. Indeed, it has been shown that a cross-section parallel to ω_2 is identical, except for a factor of 2, to a 1D transient difference NOE experiment in which the peak on the diagonal is selectively inverted. Suppression of transverse magnetization during τ_m and the growth of longitudinal magnetization during τ_m (giving rise to an axial peak in the 2D spectrum) can be achieved either by phase cycling or by the use of gradients. The factor of 2 difference arises between the 1D and 2D experiment owing to the fact that the axial magnetization does not contribute in the 2D, but it does in the 1D experiment. However, the rate of transfer and hence the information content is identical in the two experiments. The first 2D NOE spectrum of a small protein, basic pancreatic trypsin inhibitor (BPTI) is shown in **Figure 5**.

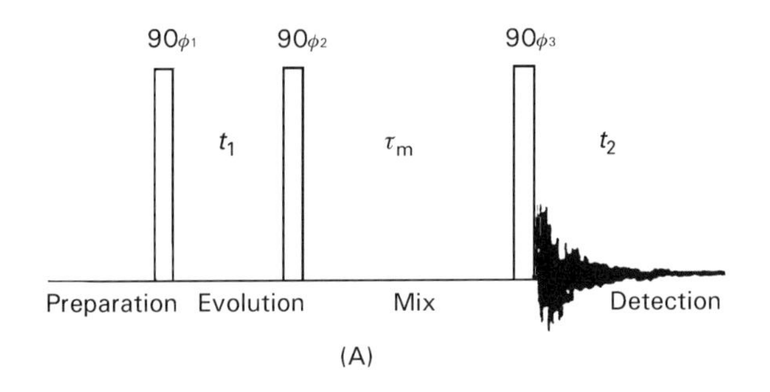

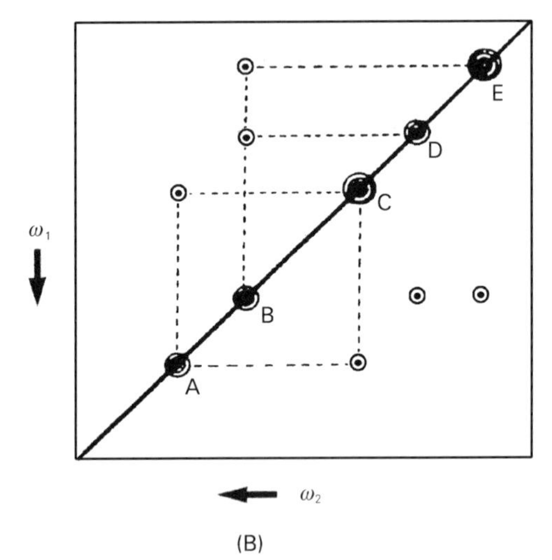

Figure 4 (A) The 2D NOESY pulse sequence, which uses three 90° pulses. The times t_1 and t_2 are the evolution and detection periods, respectively; τ_m is the mixing time during which only longitudinal magnetization is retained, either by gradients or by cycling the phases (ϕ_1, ϕ_2, ϕ_3) of the pulses. (B) Schematic NOESY spectrum, showing that in such spectra the NOEs are manifested as cross-peaks between the various spins, the resonances of which lie on the diagonal.

A large number of cross-peaks are observed, each indicating a NOE or exchange between the protons of the corresponding diagonals. Exchange also gives rise to cross-peaks identical to the negative NOE (same sign as the diagonal) and can only be distinguished from NOE by the use of the rotating frame NOE method, to be discussed later. The data in the NOESY experiment are analysed by measuring the peak volume of a cross-peak in a series of 2D experiments with various mixing times τ_m. The initial rate of growth of the NOE is directly proportional to $1/r^6$. To obtain the proportionality constant, the rate is compared with some known distance. However, this procedure is strictly valid for only two relaxation coupled spins. Since there are in general, several spins simultaneously coupled by relaxation to each other, the two-spin problem is generalized in the following manner.

Generalized Solomon's equations If there are more than two spins, relaxation coupled to each other, then the two-spin Solomon's Equations [1] and [2] can be generalized, in the following manner

$$\frac{\mathrm{d}(I_{zi}(t) - I_{zi}^{\mathrm{eq}})}{\mathrm{d}t} = -\rho_i(I_{zi} - I_{zi}^{\mathrm{eq}}) - \sum_{j \neq i} \sigma_{ij}(I_{zj} - I_{zj}^{\mathrm{eq}}) \qquad [14]$$

which states that there are n ($i = 1 \ldots n$) coupled equations describing the self-relaxation of each spin via the term ρ_i and the cross-relaxation with the other spins via σ_{ij}. This is a straightforward extension of the pairwise interaction and it neglects any cross-terms (cross-correlations) that may be present between the relaxation of various spins. It has been shown that the effect of cross-correlations on the total NOE (the average NOE, neglecting differences in the intensities of various transitions of a spin) are

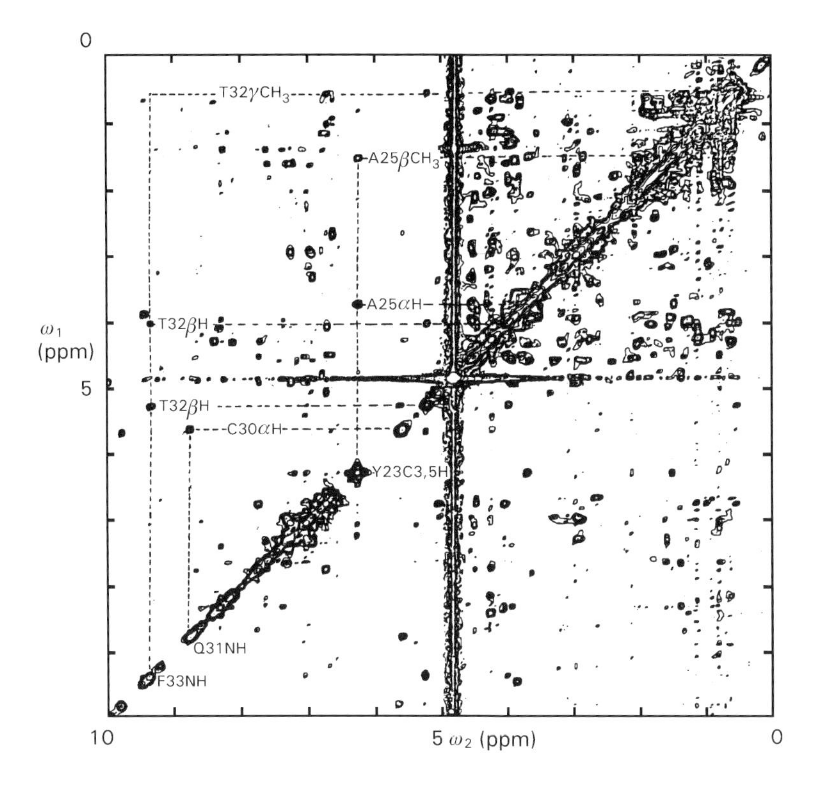

Figure 5 Contour plot of the ¹H NOESY spectrum at 360 MHz of the basic pancreatic trypsin inhibitor. The protein concentration was 0.02 M, solvent D_2O, pD = 3.8, $T = 18°C$. The spectral width was 4000 Hz; 512 data points were used in each dimension; 56 transients were accumulated for each value of t_1. The mixing time τ_m was 100 ms. The absolute value spectrum, obtained after digital filtering in both dimensions with a shifted sine bell, is shown. NOE connectivities for selected amino acid residues are indicated by the broken lines. Reproduced with permission of Academic Press from Kumar A, Ernst RR and Wüthrich K (1980) *Biochemistry and Biophysics Research Communications* **95**: 1.

generally small. In this review, the effect of cross-correlation on NOE will not be dealt with and the reader is referred to several articles on this field, including a recent review by the authors. The general solution of Equation [14] is a multiexponential time evolution of magnetizations which are coupled to each other. Once the geometry of the spins is known, it is possible to calculate the various rates of Equation [14] and compute the expected auto- and cross-peak intensities of the NOESY experiment. These computed intensities are then iteratively fitted to the observed intensities, to converge on possible structure(s) consistent with the observed intensities. Often there are differences between the computed intensities and the observed intensities that arise from internal motions, which in turn when built into the calculations give information on the internal motions. Anisotropy of reorientation of the molecules also plays a role and can also be built into the NOE calculations.

Three-dimensional structures of a large number of biomolecules (proteins, peptides, oligonucleotides and oligosaccharides) have been obtained using information derived from the NOESY experiments. The reader is referred to the 1986 book by Wüthrich and the *Encyclopedia of Nuclear Magnetic Resonance* for an exhaustive review up to 1996.

ROESY

For intermediate size molecules for which $\omega\tau_c \approx 1$, the zero-quantum (W_{0IS}) and double-quantum (W_{2IS}) transition probabilities are nearly equal and the cross-relaxation rate σ_{IS} approaches zero. In such cases there is no NOE. Bothner-By came up with the fascinating idea of doing cross-relaxation in the transverse plane by spin-locking the magnetization, using RF fields. He named the technique as CAMEL-SPIN (cross-relaxation appropriate for minimolecules emulated by locked spins), but it is now known as ROESY (rotating frame NOESY). Both 1D and 2D versions are known, and are shown schematically in **Figure 6**. The method will be explained using the 1D experiment; the 2D logic is identical. The first 90° pulse (**Figure 6A**) flips the magnetization to the

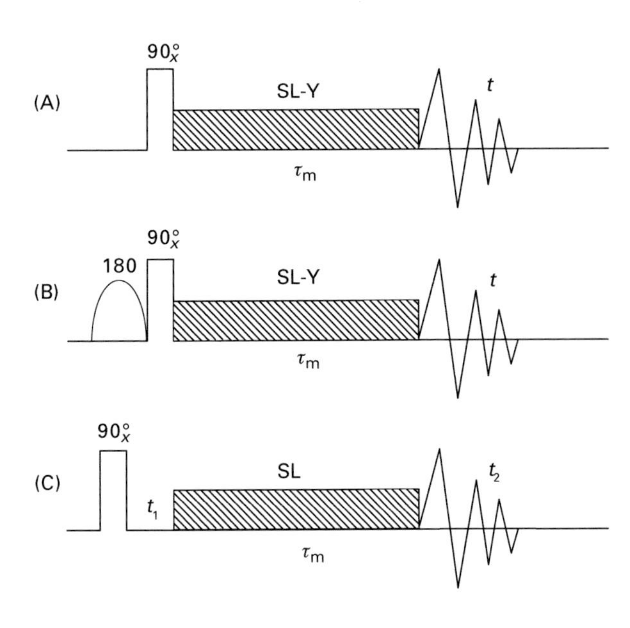

Figure 6 Rotating frame NOE pulse sequences. The 1D experiment requires two sequences, represented by (A) and (B). (A) is the reference experiment in which a 90°_x non-selective pulse is applied on all the spins, followed by a spin-lock along the *y*-direction for a time τ_m and the state of the spin system is detected. (B) The control experiment in which a selective 180° pulse, inverts the magnetization of the spin from which the NOE is to be observed before the 90°_x pulse and the experiment is continued as (A). The 1D NOE spectrum is the difference between the spectra obtained with the sequence (A) and (B). (C) The 2D ROESY sequence. The times t_1 and t_2 are the evolution and detection periods and τ_m is the mixing time. SL refers to the low power spin-locking RF field.

transverse plane, followed by a spin-lock using a 90° phase-shift.

The spin-locked magnetization of the two spins, which differ in chemical shifts, decay and cross-relax according to the rate equations,

$$\frac{dm_I(t)}{dt} = \rho_I m_I(t) - \sigma_{IS} m_S(t)$$
$$\frac{dm_S(t)}{dt} = \sigma_{SI} m_S(t) - \rho_S m_S(t) \qquad [15]$$

where m_I and m_S are the transverse magnetization of spins *I* and *S* spin-locked along the RF field. For the 1D case, two experiments (reference and control) are performed. For the reference experiment (**Figure 6A**), the initial condition is $m_I(0) = m_S(0) = 1$, while for the control experiment (**Figure 6B**), in which the magnetization of spin *S* is selectively inverted just before non-selective spin-lock, the initial condition is $m_I(0) = 1$, $m_S(0) = -1$. The solution of Equation [15] for these two cases can be written, respectively, (for $\rho_I = \rho_S = \rho$ and $\sigma_{IS} = \sigma_{SI} = \sigma$) as

$$m_I^r(t) = e^{-(\rho+\sigma)t}$$
$$m_I^c(t) = e^{-(\rho-\sigma)t} \qquad [16]$$

The NOE is the difference of the two experiments and is therefore given by

$$\eta = e^{-(\rho-\sigma)t} - e^{-(\rho+\sigma)t} \qquad [17]$$

η_{max} is obtained as

$$\eta_{\max} = \left(\frac{\rho+\sigma}{\rho-\sigma}\right)^{-\frac{(\rho-\sigma)}{2\sigma}} - \left(\frac{\rho+\sigma}{\rho-\sigma}\right)^{-\frac{(\rho+\sigma)}{2\sigma}}$$
$$t_{\max} = \frac{1}{2\sigma}\ln\left(\frac{\rho+\sigma}{\rho-\sigma}\right) \qquad [18]$$

ρ and σ in the above equations for homonuclear spins are obtained as

$$\rho = \frac{3}{4}\gamma^4\hbar^2\left[\frac{3}{8}J(0) + \frac{15}{4}J(\omega) + \frac{3}{8}J(2\omega)\right]$$
$$\sigma = \frac{3}{4}\gamma^4\hbar^2\left[\frac{1}{6}J(0) + \frac{3}{2}J(\omega)\right] \qquad [19]$$

For isotropic Brownian motion, the spectral densities are obtained as

$$J(0) = \frac{1}{r^6}\frac{24}{15}\tau_c,$$
$$J(\omega) = \frac{1}{r^6}\frac{4}{15}\left(\frac{\tau_c}{1+\omega^2\tau_c^2}\right)$$
$$J(2\omega) = \frac{1}{r^6}\frac{16}{15}\left(\frac{\tau_c}{1+4\omega^2\tau_c^2}\right) \qquad [20]$$

Using these values for the spectral densities, the expressions for ρ and σ in Equation [19] reduce to

$$\rho = \frac{1}{20}\left(\frac{\gamma^4\hbar^2}{r^6}\right)\left\{5\tau_c + \frac{9\tau_c}{1+\omega^2\tau_c^2} + \frac{6\tau_c}{1+4\omega^2\tau_c^2}\right\}$$
$$\sigma = \frac{1}{20}\left(\frac{\gamma^4\hbar^2}{r^6}\right)\left\{4\tau_c + \frac{6\tau_c}{1+\omega^2\tau_c^2}\right\} \qquad [21]$$

The maximum NOE for these conditions is plotted in **Figure 7**, which has a maximum value of 38.5% for $\omega\tau_c << 1$ and 67.5% for $\omega\tau_c >> 1$.

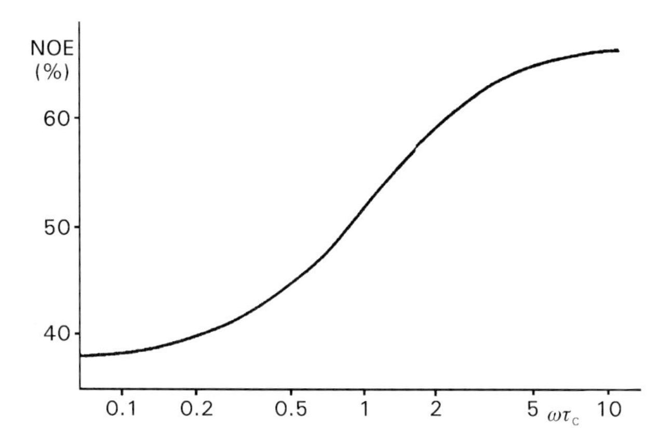

Figure 7 Plot of maximum rotating frame NOE (from Equation (18)) for a homonuclear two spin-$\frac{1}{2}$ system as a function of $\omega\tau_c$. The rotating frame NOE is positive for all values of correlation time.

The NOE is positive and does not have a null for any correlation time. This result holds for the 2D experiment as well, with the diagonal and the cross-peaks having opposite signs (positive NOE). A typical 2D ROESY spectrum is shown in **Figure 8**.

ROESY has several advantages and disadvantages. The advantages are (i) the NOE is finite (non-zero) for all sizes of the molecules, (ii) a positive NOE means that there is leakage to the lattice present and hence the magnetization does not migrate over long distances. The limited spin-diffusion helps in selecting the nearest neighbours. (iii) The three-spin effect yields a negative NOE; this can be looked at as an advantage or a disadvantage. However, the disadvantages are (i) The ROESY intensities are sensitive to the magnitude of the spin-locking RF field and to their resonance offsets. (ii) There is also a coherence transfer due to *J*-coupling, known as TOCSY (total correlation spectroscopy). In fact an identical pulse scheme can also be used for obtaining a TOCSY spectrum, with which one identifies all the resonances which are *J*-coupled through bonds. For example, all the resonances of an amino acid residue could be

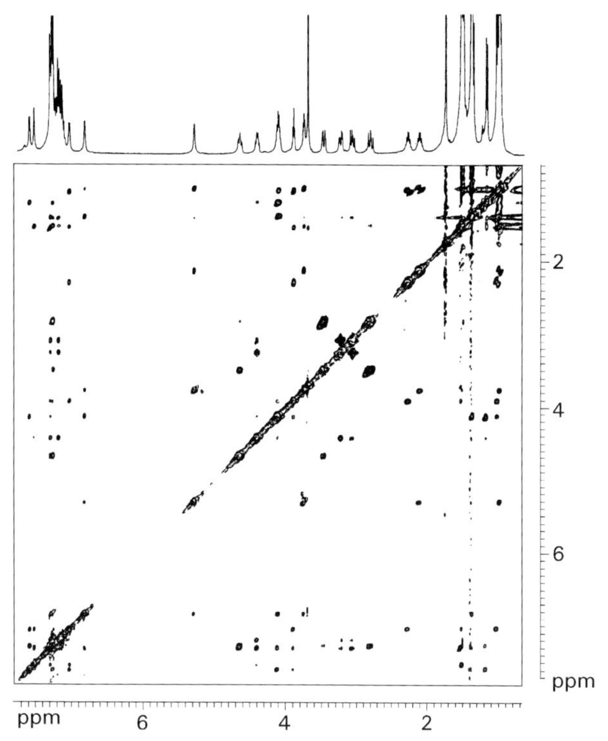

Figure 8 Contour plot of the 2D ^{1}H ROESY spectrum of a 0.5 M solution of Boc-Val-Ala-Phe-Aib-Val-Ala-Phe-Aib-OMe in $CDCl_3$, recorded at 400 MHz. A 2.25 kHz spin-lock field has been used during the 300 ms mixing period. 64 scans were performed for every t_1 value and 512 × 1k data were acquired. Zero filling was used to give a 1k × 1k size of the displayed absorptive part of the spectrum. The diagonal drawn is negative and the cross-peaks are positive. Unpublished results by Das C, Grace RCR and Balaram P.

identified by taking a cross-section at either the NH or αH position. The salient features of ROESY, TOCSY and NOESY are listed in **Table 1**.

Some of these differences are therefore utilized for differentiating the TOCSY and ROESY peaks, in particular the strength of the spin-locking field and the mixing time, as well as the sign of the cross-peak. Experiments have also been designed for obtaining clean TOCSY as well as clean ROESY spectra.

Table 1 Comparison of the salient features of ROESY, TOCSY and NOESY

Feature	*ROESY*	*TOCSY*	*NOESY*
Net transfer	Yes	Yes	Yes
Pure absorptive	Yes	Yes (almost)	Yes
Sign with respect to the diagonal	Opposite (+ve NOE)	Same	Opposite for $\omega\tau_c << 1$ Same for $\omega\tau_c >> 1$
Mixing time	Large (> 100 ms)	Small (<100 ms)	<500 ms
RF field amplitude needed	Small	Large	Not applicable

Heteronuclear NOE

So far we have discussed the NOE between nuclei having the same γ. The maximum use of these homonuclear NOEs has been for proton–proton NOEs. However, the heteronuclear NOE also has a long history. In the early days of NMR, the heteronuclear NOE was basically utilized for the enhancement of the intensities of low γ nuclei. For example, the steady state NOE, on spin I on irradiation of spin S when spins I and S are mutually dipolar-relaxation coupled is given by Equations [4] and [6]. For $\omega\tau_c << 1$, $W_0:W_1:W_2::2:3:12$, $\sigma = \frac{1}{2}\rho$, yielding

$$\eta_I\{S\} = \frac{1}{2}\frac{\gamma_S}{\gamma_I} \qquad [22]$$

Thus for ^{13}C the NOE is 2, while for ^{15}N it is –5. The ^{13}C magnetization could thus be enhanced by 200% by adding the NOE to carbon equilibrium magnetization. For ^{15}N, the enhancement was –4, which is like a 400% increase in the signal. These enhancements have been and are routinely utilized in heteronuclear NMR experiments.

As $\omega\tau_c$ increases, the heteronuclear NOE decreases in magnitude (approaching zero for $\omega\tau_c >> 1$) since in the heteronuclear case the flip-flop process becomes non-energy conserving and requires spectral densities at a frequency equal to the difference in the Larmor frequencies of the two spins. Even the rotating frame NOE for heteronuclei is very small as σ approaches zero. Therefore heteronuclear NOE has been used only for small molecules in the $\omega\tau_c << 1$ limit, and there is even a 2D version known as HOESY which uses identical logic as NOESY, except that the first two pulses are applied on the source spin and the third pulse on the target spin, with the detection on the target spin.

Transferred NOE (TRNOE)

NOE measurements are also made on spin systems undergoing chemical exchange in which a small ligand molecule (peptide, drug molecule) exchanges between free and bound states with a large host molecule (protein). Under this condition, the rotational correlation time of the small molecule changes from a regime of $\omega\tau_c << 1$ in the free state to $\omega\tau_c >> 1$ in the bound state. While sharp resonances of the ligand are observed only from its free state, the bulk of the NOE takes place when the ligand is in the bound state, giving rise to negative NOE between the various protons of the ligand. This is known as the transferred NOE. The observed NOE is related to the various exchange rates, concentration ratios of the two molecules and the binding constants. The transfer NOE experiment is thus often exploited for obtaining information on the exchange kinetics of ligand–protein interactions.

Conclusions

The nuclear Overhauser effect arises from mutual dipolar relaxation of nearby spins and is utilized for identifying spins which are close in proximity in reorienting molecules. The NOE thus provides information on the three-dimensional structures of the molecules in solution and has led to an alternate methodology of obtaining structure of biomolecules in solution as compared with those obtained by single crystal X-ray crystallography. Since the NOE is extremely sensitive to the modulation of the distance between the spins owing to internal motions, as well as to the anisotropy of molecular reorientations, this methodology has led to detailed information on internal motions, conformational flexibility, local order parameters, exchange between chemically inequivalent sites and anisotropic reorientations of the molecules, some of which are correlated to their biological functions. The field is growing rapidly with the help of modelling software, molecular mechanics calculations, development of newer experimental protocols such as isotopic labelling of spins (^{13}C and ^{15}N), measurement of J-coupling between heteronuclei, leading to conformational information, and the utilization of information contained in cross-correlations.

Acknowledgement

RCR Grace acknowledges CSIR for financial support.

List of symbols

$I_z(S_z)$ = longitudinal magnetization of spin I (and S); J = coupling constant, m_I = transverse magnetization of spin I; T_1, T_2 = relaxation times; $W = T_1/T_2$ transition probability; γ = magnetogyric ratio; η = NOE; ρ = self-relaxation rate; σ_{IS} = mutual cross-relaxation rate; τ_c = correlation time; τ_m = mixing time; ω = Larmor frequency.

See also: **Chemical Exchange Effects in NMR; Macromolecule–Ligand Interactions Studied By NMR; Magnetic Resonance, Historical Perspective; NMR Pulse Sequences; NMR Relaxation Rates; Nucleic**

Acids Studied Using NMR; Proteins Studied Using NMR Spectroscopy; Structural Chemistry Using NMR Spectroscopy, Organic Molecules; Structural Chemistry Using NMR Spectroscopy, Peptides; Structural Chemistry Using NMR Spectroscopy, Pharmaceuticals; Two-Dimensional NMR Methods.

Further reading

Abragam A (1989) *Time Reversal, an Autobiography*, p 164. Oxford: Oxford University Press.

Balaram P, Bothner-By AA and Dadok JJ (1972). *Journal of the American Chemical Society* **94**: 4015.

Boulat B and Bodenhausen G (1992). *Journal of Chemical Physics* **97**: 6040.

Chiarpin E, Pelupessy P, Cutting B, Eykyn TR and Bodenhausen G (1998) *Ang Chemie* (in the press).

Dalvit C and Bodenhausen G (1990) *Advances in Magnetic Resonance* **14**: 1.

Ernst RR, Bodenhausen G and Wokaun A (1987) *Principles of Nuclear Magnetic Resonance in One and Two-dimensions*. Oxford: Clarendon Press.

Karthik G (1999) *Journal of Chemical Physics* **110**: 4992.

Kumar Anil, Grace RCR and Madhu PK (1998) *Progress in NMR Spectroscopy* (in the press).

Levy GC, Lichter R and Nelson GL (1980) *Carbon-13 Nuclear Magnetic Resonance for Organic Chemists*. NewYork: Wiley Interscience.

Nageswara Rao BDN (1970) Nuclear spin relaxation by double resonance. *Advances in Magnetic Resonance* **4**: 271.

Neuhaus D and Williamson M (1989) *The Nuclear Overhauser Effect in Structural and Conformational Analysis*. NewYork: VCH.

Noggle JH and Schirmer RE (1971) *The Nuclear Overhauser Effect – Chemical Application*. London: Academic Press.

Overhauser AW (1996) In: Grant DM and Harris RK (eds) *Encyclopedia of Nuclear Magnetic Resonance*, Vol 1, p 513. Chichester: Wiley.

Sanders JKM and Hunter BK (1987) *Modern NMR Spectroscopy*. Oxford: Oxford University Press.

Slichter CP (1978) *Principles of Magnetic Resonance*. NewYork: Springer-Verlag.

Wüthrich K (1976) *NMR in Biological Research: Peptides and Proteins*. Amsterdam: North Holland.

Wüthrich K (1986) *NMR of Proteins and Nucleic Acids*. NewYork: Wiley.

Nuclear Quadrupole Resonance, Applications

Oleg Kh Poleshchuk, Tomsk Pedagogical University, Russia
Jolanta N Latosińska, Adam Mickiewicz University, Poznan, Poland

MAGNETIC RESONANCE
Applications

The NQR frequencies for the various nuclei vary from several kHz up to 1000 MHz. Their values depend on quadrupole moments of the nucleus, the valence electrons state and the type of chemical bond in which the studied atom participates. Using the NQR frequencies, the quadrupole coupling constant (QCC) and asymmetry parameter (η) can be calculated with either exact or approximate equations, according to the spin of the nuclei. For a polyvalent atom, the NQR frequencies depend on the coordination number and hybridization. The NQR frequency shifts for a single-valent atom in different environments can be classified as follows:

1. The greatest shifts of the NQR frequencies are determined by the valence electron state of the neighbouring atoms and can reach 1200–1500%. The changes of the NQR frequencies depend on ionic character, and for example for chlorine the lowest frequency corresponds to a chloride ion, whereas the highest corresponds to the chlorine atom in ClF_3.
2. Changes within the limits of one type of chemical bond (with the same kind of atom) can reach 40–50%. Thus the frequency changes of C–Cl bonds vary from 29 to 44 MHz.
3. The range of possible shifts of the NQR frequencies is reduced to 10–20% when only one class of compound is studied. In this case the shifts are determined by the surroundings and donor–acceptor properties of the substituents. For example, for halogen substituted benzenes, the changes of the frequencies for the C–Cl bonds are about 9% for C–Br bonds ~ 12% and for C–I bonds ~ 18%.
4. The changes of NQR frequencies caused by the occurrence of intramolecular and intermolecular interactions lie within the limits of 3–40%.

5. The shifts of NQR frequencies caused by crystal field effects in molecular crystals reach a maximum of 1.5–2%. However within a series of similar compounds this effect, as a rule, does not exceed 0.3%.

Such classification of NQR frequency shifts allows the determination of structural non-equivalencies from NQR spectra. Understanding how the frequency shifts depend on chemical non-equivalence and how they are determined by distinctions in distribution of electronic density in a free molecule is very important. Moreover, crystallographical non-equivalences are observed, and are a result of distinct additional contributions of the crystal field to the electric field gradients at a nucleus. It is obvious that the division of structural non-equivalencies into molecular and crystal types is not relevant in coordination and ionic crystals, in which there are no individual molecules. NQR, being highly sensitive to subtle changes in electron density distribution, provides diverse information on the structural and chemical properties of compounds.

Molecular structure studies using NQR

When applied to structural investigations, NQR spectra may prove an effective tool for the preliminary study of crystal structure in the absence of detailed X-ray data. Such parameters as spectroscopic shifts, multiplicity, spectroscopic splitting, resonance line width, the temperature dependence of resonance frequencies and relaxation rates all afford useful structural information and provide insight into the factors determining the formation of certain structural types.

The violation of chemical equivalence of resonance atoms due to a change in chemical bonding, such as, for example, dimerization in group IIIA halogenides, leads to a significant splitting of the spectroscopic multiplet caused by a difference in the electronic structure of bridging and terminal atoms.

Crystallographic structures

For crystallographically non-equivalent atoms the corresponding components of the electric field gradient (EFG) at the respective sites differ from each other in magnitude and direction due to the crystal field effect. This generally includes a contribution to the EFG of electrostatic forces between molecules, dispersion forces, intermolecular bonding and short-range repulsion forces. Physically non-equivalent sites differ from each other only in the direction of the EFG components, their magnitudes being identical. To distinguish between such sites, Zeeman analysis of the NQR spectrum is required.

Table 1 The angles (°) between the directions of the C–Cl bonds and the crystal axes in 1,3,5-trichlorobenzene according to NQR and X-ray results

Chlorine position	*Crystal axes*					
	a		*b*		*c*	
	NQR	*X-ray*	*NQR*	*X-ray*	*NQR*	*X-ray*
Cl-1	119.1	118.8	149.5	150	81.7	80.9
Cl-3	64.1	65.0	31.3	30.2	73.7	74.3
Cl-5	24.8	24.0	90.4	90.5	114.8	114.0

The intensities of spectroscopic lines are also important. They reflect the relative concentration of resonance nuclei at certain sites although one also has to take into account the transition probabilities and lifetimes of the energy states of the system investigated. The correspondence between the number and intensities of frequencies and the number of non-equivalent sites occupied by a resonant atom in a crystal lattice is very helpful in a preliminary structure study made with the use of NQR.

NQR single-crystal Zeeman analysis can provide information about special point positions occupied by the quadrupole atoms. This Zeeman analysis determines the orientation of the EFG components with respect to the crystal axes, which essentially facilitates the most difficult and time-consuming stage of X-ray analysis. **Table 1** gives a comparison of the angle values between the crystal axes and the EFG z-axis at the chlorine atoms in 1,3,5-trichlorobenzene (determined by the two methods).

As one can see from **Table 1** the angle values obtained by the two methods show rather good numerical agreement. To illustrate more completely the type of structural information that can be obtained by NQR spectroscopy we consider in more detail an NQR study of $BiCl_3$, whose structure is known. Two chlorine atoms are involved in bridging to two other Bi atoms while another chlorine atom is involved in bridging to only one other Bi atom. The ^{35}Cl and ^{209}Bi NQR parameters of $BiCl_3$ measured by means of the single-crystal Zeeman method are listed in **Table 2**.

The assignment of ^{35}Cl resonances is unambiguous owing to direct correspondence of the lower frequency line of double intensity to the two Cl atoms which are crystallographically nearly equivalent. The asymmetry parameter for the chlorine atoms has been determined, using the zero-splitting cone. According to the results the $BiCl_3$ crystal belongs to the Laue symmetry of D_{2h} and therefore to the orthorhombic crystal class. The Cl(1) and Bi atoms

Table 2 ^{35}Cl and ^{209}Bi NQR spectra of $BiCl_3$

Isotope	T (K)	Transition frequencies (MHz) $\frac{1}{2}-\frac{3}{2}$	$\frac{3}{2}-\frac{5}{2}$	$\frac{5}{2}-\frac{7}{2}$	$\frac{7}{2}-\frac{9}{2}$	e^2Qqh^{-1} (MHz)	η (%)	Assignment
^{35}Cl	291	15.952(2)	–	–	–	30.960	43.1	Cl(1)
		19.173(1)				38.145	17.8	Cl(2)
^{209}B	294	31.865	25.132	37.362	51.776	318.900	55.5	

occupy special point positions in the lattice since only two non-equivalent directions of the corresponding EFG components have been detected for them, while four orientations have been found for the EFG components at the site of Cl(2,3) atoms. The number of observed ^{35}Cl resonances suggests that the molecule possesses a mirror plane and therefore the centrosymmetric group P_{nma}. The lower frequency resonance of double intensity is then assigned to the chlorines out of the mirror plane while the higher frequency line corresponds to one lying in the mirror plane. The shorter (i.e. more covalent) bond length of the latter corresponds to the higher resonance frequency.

Another example is provided by the ^{35}Cl results for phosphorus pentachloride. In the gas phase this is known to be a trigonal bipyramid but the usual solid PCl_5 is likewise known to be an ionic crystal, $(PCl_4)^+(PCl_6)^-$. In accordance with this, and the detailed crystal structure, there are four resonances at high frequency that correspond to the $(PCl_4)^+$ group and six at a lower frequency that correspond to the $(PCl_6)^-$ group (**Table 3**).

It has recently been observed that quenching of the vapour of PCl_5 gives rise to a new metastable crystalline phase which can be preserved essentially indefinitely at low temperatures. The ^{35}Cl spectrum of this has two low frequencies, corresponding to the axial chlorine atoms, and three identical higher frequencies, corresponding to the equatorial substituents, which is strong evidence that this new phase is the corresponding molecular solid. Another example is taken from the chemistry of antimony pentachloride. It is known that the molecule of antimony pentachloride at 210 K has a trigonal bipyramid structure, and at 77 K it is a dimer. In **Table 3** the experimental ^{35}Cl NQR frequencies and their assignment to equatorial and axial chlorine atoms are given. In the dimer molecule, bridging chlorine atoms have much lower NQR frequencies, as in dimer molecules of group IIIA compounds. From **Table 3** it is also clear that in the monomer the NQR frequencies of equatorial chlorine atoms are greater than the axial ones.

In the dimer, the ratio of ^{35}Cl NQR frequencies is in agreement with the results of *ab initio* calculations. In dimers of transition elements, such as $NbCl_5$ and $TaCl_5$, the ^{35}Cl NQR frequencies of equatorial chlorine atoms are higher than those of axial, whilst those of bridging chlorine atoms are higher, on average, than those of terminal chlorines. A similar inversion of NQR frequencies in dimers of transition and non-transition elements is explained by a significant multiplicity of the metal–halogen bonds and hence by electron transfer from p-valent orbitals of the halogen atoms to the vacant d-orbitals of the central atom.

The fact that the difference between chemically non-equivalent atomic positions is readily revealed by NQR spectroscopic splitting may be utilized to identify geometric isomers. Octahedral complexes of tin tetrachloride, $[SnCl_4{\cdot}L_2]$ exist as either *cis* or *trans* isomers. In the *cis* isomers the axial and equatorial non-equivalence produces considerable splitting in the NQR spectra. In the *trans* complexes, all four chlorine atoms are chemically equivalent with identical electron density distribution. Splitting in the NQR spectra of these isomers arises therefore from the crystallographic non-equivalence of the chlorine positions. Indeed, the observed NQR splitting of two complexes (**Table 4**) provides evidence

Table 3 ^{35}Cl quadrupole resonance frequencies for phosphorus and antimony pentachlorides

Compound	$\nu\,^{35}Cl$ (MHz)	Assignment
$(PCl_4)^+(PCl_6)^-$	28.395, 28.711, 29.027,	$(PCl_6)^-$
	30.060, 30.457, 30.572	$(PCl_6)^-$
	32.279, 32.384, 32.420	$(PCl_4)^+$
	32.602	$(PCl_4)^+$
PCl_5	29.242, 29.274	Axial
	33.751	Equatorial
$SbCl_5$	30.18	Equatorial
	27.85	Axial
Sb_2Cl_{10}	27.76	Equatorial
	30.18	Axial
	18.76	Bridging

Table 4 ^{35}Cl Quadrupole resonance frequencies and asymmetry parameters of some [$SnCl_4{\cdot}2L$] and [$SbCl_5{\cdot}L$] complexes

Complex	*ν ^{35}Cl (MHz)*	*η (%)*	*Assignment*
[$SnCl_4{\cdot}(NC_5H_5)_2$]	17.644	15.4	–
	17.760	15.4	–
[$SnCl_4{\cdot}(OPCl_3)_2$]	19.030	11.7	Equatorial
	19.794	11.1	Equatorial
	21.132	2.3	Axial
[$SbCl_5{\cdot}OPCl_3$]	24.399	11.3	*Cis*
	25.821	2.5	*Trans*
	26.119	4.7	*Cis*
	27.314	4.9	*Cis*
[$SbCl_5{\cdot}NCCCl_3$]	26.008	6.2	*Cis*
	26.313	10.3	*Cis*
	26.409	2.2	*Trans*
	27.297	11.2	*Cis*

for the *cis* configuration of [$SnCl_4{\cdot}(OPCl_3)_2$] and the *trans* configuration of [$SnCl_4{\cdot}(NC_5H_5)_2$], which is confirmed by X-ray data.

However, it is not always possible to assign equatorial and axial chlorine atoms solely on the basis of the splitting of the ^{35}Cl NQR frequencies. For *cis* isomers, the ratio of the NQR frequencies of equatorial and axial chlorine atoms is fixed by several factors that determine the optimum crystal structure, among which is the influence of donor molecules, L is not a major contribution. In practically all structural investigations of complexes that exhibit this type of equatorial Sn–Cl bond it is usually noted that axial chlorines have lower relative NQR frequencies compared with equatorial atoms (**Table 4**). However, interpretation in structure terms is difficult because of the large values of the asymmetry parameters of chlorine atoms, which (**Table 4**) differ considerably for axial and equatorial atoms. On the other hand, donor ligands, L, influence the change in electron density of equatorial and axial Cl atoms and cause a relative lowering of frequencies of axial atoms in comparison with those of equatorial chlorines. Such a relation of frequencies in *cis* isomers is explained from the point of view of the 'mutual ligand influence' concept in non-transition element complexes. In *cis* complexes $SnCl_4L_2$ the interaction of –Sn–L bonds will be stronger with Sn–Cl bonds which are in a *trans* position, than with *cis*-Sn–Cl bonds. For these complexes, the mutual ligand influence establishes the greater *trans* effect, and leads to a redistribution of the electronic density on the chlorine atoms. In $SbCl_5{\cdot}L$ complexes (**Table 4**) to assign axial and equatorial chlorine atoms signals, even allowing for a knowledge of frequencies, splittings and asymmetry parameters, it is necessary in a number of cases to use the temperature dependences of NQR frequencies. In this case the NQR frequency is described by a square-law dependence: $\nu(T) = A + BT + CT^2$. A positive sign of the C coefficient indicates that a NQR frequency arises from an axial chlorine atom. Thus it appears that in some complexes the frequency of an axial chlorine atom lies above those of equatorial atoms. Apparently, in these complexes the spatial influence of the donor molecules on the NQR frequencies of the equatorial chlorine atoms is dominant.

The width of the NQR signal also provides structural information. In molecular crystals of high order and purity the line width is not much different from the value determined from the sum of the spin–lattice relaxation and spin–spin relaxation times. In the majority of inorganic compounds, the lines are, however, inhomogeneously broadened by lattice imperfections such as defects, vacancies, admixtures and dislocations, so that their widths are mainly determined by the crystal inhomogeneity. A systematic study of spectroscopic shifts and broadening produced by a continuous change of the relevant sources of that broadening is an effective approach to the investigation of problems concerned with the distribution of mixtures over a matrix, the nature of their interaction with the matrix, the mechanisms of disorder and the local order in vitreous compounds.

Studies of bonding using NQR spectra

The Townes–Dailey approximation

The most widely used approach to provide a meaningful account of bonding trends within a series of related compounds is that formulated by Townes and Dailey for the interpretation of nuclear quadrupole interactions (NQI). The electric field gradient at a quadrupole nucleus (q_{zz}) arises mainly from electrons of the same atom. To a first approximation, it is possible to consider that the internal electrons will form a closed environment with spherical symmetry and, consequently, do not contribute to the EFG. Actually, polarization of the internal electrons is taken into account through the Sternheimer antishielding factor (γ_∞). However, if the comparison of NQI is only for the purpose of chemical interpretation and is not accompanied by discussion of their absolute meanings, the polarization of the internal electrons can be neglected. Among valent electrons, those that are on s orbitals with spherical symmetry do not contribute to the EFG and the main contribution is caused by p electrons; the contribution of d and f electrons is much less significant because of their greater distance from the nucleus and their smaller participation in hybridization.

The quantitative consideration of the contributions to the EFG results in expressions of the following type, applied here to nuclei of chlorine in chloroorganic compounds:

$$q_{zz}{}^{\mathrm{Cl}} = [(1 - s^2 + d^2 - I_{\mathrm{B}} - \pi) + I_{\mathrm{B}}(s^2 + d^2)]q_{\mathrm{at}}{}^{\mathrm{Cl}} \quad [1]$$

where s^2 and d^2 are contributions from s and d orbitals to the hybridization of the chlorine atom bonding orbital, I_B is the ionic character of the bond (the chlorine atom carries a partial negative charge) and $q_{at}{}^{Cl}$ is the gradient of p electron density on the chlorine atom. The valent orbitals can be represented by somewhat modified expressions in that some treatments include the three nuclear p orbital populations, and the axes x, y and z are usually defined so that the z-axis coincides with the bond direction of the considered halogen or with an axis of symmetry in the molecule. These approaches can be more convenient for discussing bonding and the contribution of lone pair electron orbitals.

On the basis of the above interpretation NQI can be unequivocally represented in terms of the population of orbitals. Actually, NQR spectroscopy allows the determination, at best, of two parameters (e^2Qqh^{-1} and η) and in many cases ($I = \frac{3}{2}$) only e^2Qqh^{-1} can be obtained. However, the above-mentioned Equation [1] contains four parameters (s, d, I_B, π) which cannot be determined from one or two experimental parameters. It is therefore necessary to include approximations, which neglect the d orbital participation in hybridization, and also, in some cases, p bonding and to consider that the s orbital hybridization is a small contribution and remains constant in a series of compounds. Thus changes of e^2Qqh^{-1} are directly related to bond ionic character or p electron charge transfer in the case of a hydrogen bond.

In the case of nitrogen, whose nucleus has a spin $I = 1$, the situation is more favourable, as the experiment allows the determination of both the nuclear quadrupole coupling constant and the asymmetry parameter, and it is possible to make conclusions about sp hybridization, if the molecular geometry is known. However, the meaning of nuclear NQI for ^{14}N p electrons is not known with as much accuracy as for chlorine, bromine or iodine, but the estimation of a 9–10 MHz contribution can be considered reliable, as it is based on the analysis of a large number of experimental data. An even more difficult situation is the case of the antimony atom, for which very little reliable NQI data exist.

Donor–acceptor interactions

In addition to interpreting experimental NQI values in terms of orbital populations, a role is also played by structural, dynamic and simple contributions, which change the NQI such that their experimental meanings differ from those expected for a hypothetical molecule or complex in an isolated condition at rest. In the case of molecular complexes there is an additional contribution which results in NQI changes caused by a change in hybridization owing to a change in the donor and the acceptor molecule geometry. A typical example is given by MX_3 complexes, where M is a group IIIA or VB element and X is a halogen or methyl group. With complex formation, the X–M–X angle and appropriate hybridization change, and these result in changes in NQI interpretation even in the absence of any charge transfer. This complicates unequivocal interpretation of experimental NQR data.

It is necessary to answer the question: are the shifts of the NQR frequencies caused by a transition from a pure, non-complexed mixture of initial substances to a complex, or because of a charge transfer or for other reasons? This rather important question depends on several factors. With a small charge transfer the NQR frequency shifts are defined mainly by crystal or solid-state effects, which are caused by distinct effects in the crystal environment of molecules as the result of the transition from individual components to the complex (crystal electrical field, intermolecular interactions, thermal movement); the complexation shifts can reach several hundred kHz (in the specific case of a resonance on a chlorine nucleus a shift of the order 200 kHz is typical). When the observable shifts have larger values (one to several MHz) they cannot be considered as caused by crystal effects and it is then possible with confidence to attribute them to electronic effects arising from a charge transfer. However, it is necessary to take into account other contributions, such as hybridization changes. The hybridization change accompanying deformation of a flat $AlMe_3$ molecule to a pyramidal form formally results in an NQI change on the aluminium atom, even if there is no electron population change or change in ionic character of the bonding orbitals; thus, the shift of the NQR frequency in this case can be determined as having both a charge and a hybridization contribution.

A more difficult situation exists when, in the free compounds, there are strong intermolecular interactions: the perturbation of these interactions by complex formation can result in an increased NQI in a complex which contradicts the usual, simple prediction of an NQI reduction upon transition to a complex. Such situations are met in complexes of mercury halogenides such as $HgBr_2$·dioxan and HgI_2·dioxan. More complete interpretation of the experimental results can be achieved if in a complex there is present a number of quadrupole nuclei; this allows a comparison of shifts for each of them. In addition, there is often useful structural data available from X-ray diffraction. Finally for an estimation of a relative role of the various contributions to observable NQR frequency shifts, one can resort to theoretical calculations.

Intermolecular interactions

Another example arises in the situation where a quadrupole halogen atom makes a symmetrical bridge between two metal atoms, which is often the case in polymeric metal halides of composition MX_n (n = 3–5). Dimers of metal halides, e.g. $AlCl_3$, $GaBr_3$, $TaCl_5$, $SbCl_5$ or $[Al_2Br_7]^-$, have thus been attractive for NQR investigators, because of the differences between the resonance frequencies of bridging (X_b) and terminal (X_t) halogen atoms. The dimers have structures with either two bridging M–X_b–M bonds of about equal length, as in $GaBr_3$, or two bonds of significantly different length, as in complexes of oxygen donor ligands with mercuric halides, or one bridging bond as, e.g. in the aluminium heptabromide anion $[Al_2Br_7]^-$. **Figure 1** presents the structure of one type of symmetric bridging dimeric halide.

The spatial structure of the MX_3-type dimer is shown in **Figure 1A** whereas **Figure 1B** presents the same structure projected onto the plane of the bridging bonds. The halogen atoms involved in the bridges are, together with metal atoms (M), in one plane which is perpendicular to the plane of the other terminal halogen atoms (X_t) and the metal atoms.

Analysis of the Zeeman splitting of the NQR spectra of halides of non-transition metals from group IIIA of the periodic table permits a determination of the directions of the main axes of the EFG tensor on the bridging halogen atoms. These directions are marked in **Figure 1B**. The axis of the greatest gradient of the electric field on the bridging halogen atom is perpendicular to the plane that contains the metal atoms and the bridging halogen atoms. The same axis, but on the terminal halogen atoms, lies along the metal–halogen bond. The orientation of the main

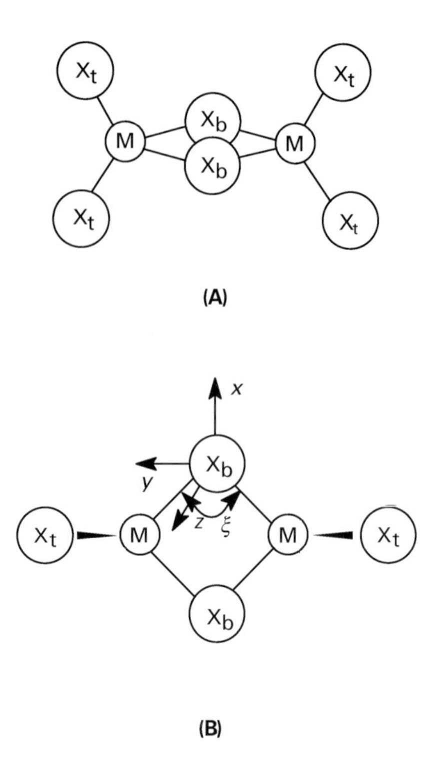

Figure 1 (A) Spatial structure of MeX_3 type halides with two bridging bonds. (B) The same structure projected onto the plane of bridging bonds. The directions of the main axes of the EFG tensor on the bridging halogen atoms are marked, in the case when $\xi < 109°28'$.

axes of the EFG tensor in the case of other types of bridging dimeric halides is similar.

We have indicated that the NQR frequencies of the bridging halogen atoms in non-transition metal compounds are lower than those of the terminal atoms and that this relationship is reversed in the case of transition metal compounds. **Table 5** includes the NQR frequencies of the bridging and terminal halogen atoms, the values of the EFG asymmetry parameter for the bridging and terminal halogen atoms, as well as the angles at the bridging atoms (see **Figure 1B**) for some non-transition and transition metal dimers.

X-ray structural and electron diffraction data indicate that, independent of the nature of the central metal atom, the length of the bridge is always greater than the distance between the terminal atoms. In non-transition metal compounds the effective negative charge on the bridging halogen atoms is smaller than that on the terminal atoms. The opposite situation is found in transition metal halides. In these compounds the effective negative charge on the bridging atoms is greater than that on the terminal ones. The explanation for this lies in the nature of the metal valence shells (p–d transfer). In the

Table 5 NQR spectral parameters of some non-transition and transition metal dimer halogenides

Compound	*$\nu(X_b)$ (MHz)*	*η_b (%)*	*$\nu(X_t)$ (MHz)*	*η_t (%)*	*ξ(°)*
$GaCl_3$	14.667	47.3	19.084	8.9	86.0
			20.225	3.4	
GaI_3	133.687	23.7	173.650	0.9	94.5
			174.589	2.8	
$AlBr_3$	97.945	–	113.790	–	82.0
			115.450		
AlI_3	111.017	18.1	129.327	0	93.5
			129.763		
InI_3	122.728	29.7	173.177	1.1	94.5
			173.633		
$NbCl_5$	13.290	58.8	7.330	–	101.3
$NbBr_5$	105.850	58.8	59.500	–	101.3
WBr_5	114.580	44.9	81.660	–	98.6

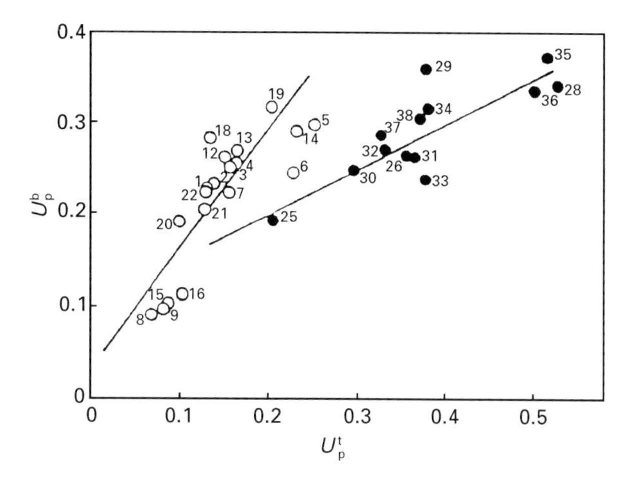

Figure 2 U_p^b versus U_p^t for halogen atoms in halides of transition (o) and non-transition (•) metal elements.

framework of the Townes and Dailey theory this means an effective lowering of the resonance frequency owing to the decrease in the p_x and p_y orbital population of the terminal halogens.

Figure 2 presents a correlation found by the linear regression method between the values of unpaired p-electron densities for the bridging (U_p^b) and terminal (U_p^t) halogen atoms for a large number of compounds. The existence of these correlations is connected with the influence of the p-electron distribution on NQR frequencies, and the bridging and terminal halogen atoms are connected through the central atom, which acts as a buffer. One can see from **Figure 2** that the experimental points can be divided into two groups, corresponding to non-transition and transition metal compounds. This is in good agreement with the difference in NQR frequencies for the bridging and terminal halogen atoms for all the compounds studied. As a consequence, the unpaired electron density differs for the compounds of both groups, although U_p^b increases with increasing U_p^t for all compounds studied. The observed correlations are approximate, which points to the influence of other factors, such as crystallographic uncertainty and the position of the molecules in the unit cell of the crystal.

Charge distribution in biological molecules

NQR spectroscopy can be also used to provide detailed information on the structure and conformation of biologically active systems. It offers a unique possibility of determining the quadrupole coupling constants and, as a consequence, effective charges, and in this way allows a determination of the electronic structure of the molecule. NQR spectroscopy appears to offer a powerful tool for the investigation of various chemical effects in the solid phase of many nitrogen-containing compounds. Analysis of the quadrupole coupling constants of nitrogen atoms allows an estimation of the electron density distribution on the nitrogen nuclei and enables the analysis of charge distribution in chemical bonds involving nitrogen.

For example, the effect of substitution at the 1-H position of the 2-nitro-5-methylimidazole [2] ring can be analysed (**Table 6**), Hitherto, different imidazole derivatives have been studied by ^{14}N NQR, by the continuous wave method, by ^{13}C NMR and ^{35}Cl NQR (pulse methods). However, the problem of the effect of a substituent at the 1-H position on the

Table 6 Chemical names and substituents of the compounds studied

No.	*Compound*	*R^1*	*R^2*	*R^3*	*R^4*
[1]	Imidazole	H	H	H	H
[2]	2-Nitro–5-methylimidazole	H	NO_2	H	CH_3
[3]	1-(2-Hydroxyethyl)-2-nitro-5methylimidazole (metronidazole)	CH_2CH_2OH	NO_2	H	CH_3
[4]	1-(2-Carboxymethyl)ethyl-2-nitro-5-[2-(P-ethoxy-phenyl)ethenyl] imidazole	$CH_2CH_2OCOCH_3$	NO_2	H	$CH{=}CHPhCH_3O$

Table 7 Quadrupole coupling constants and asymmetry parameters for imidazole derivatives

		Nitrogen Nucleus					
		−N=		−NR		$-NO_2$	
No.	T (K)	$e^2Qq_{zz}h^{-1}$ (MHz)	η	$e^2Qq_{zz}h^{-1}$ (MHz)	η	$e^2Qq_{zz}h^{-1}$ (MHz)	η
[1]	291	3.222	0.119	1.391	0.930	–	–
	77	3.253	0.135	1.418	0.997	–	–
[2]	296	3.243	0.250	1.569	0.821	1.225	0.356
	193	3.249	0.249	1.546	0.878	1.244	0.360
[3]	296	3.299	0.150	2.467	0.317	0.936	0.381
	193	3.320	0.156	2.479	0.320	0.950	0.381
[4]	296	3.755	0.038	2.566	0.238	0.921	0.239
	193	3.779	0.039	2.565	0.238	0.931	0.230

electron distribution in 2-nitro-5-methylimidazole [2] derivatives has not been considered. The results of NMR-NQR double resonance studies on a series of imidazoles (**Table 6**) are collected in **Table 7**.

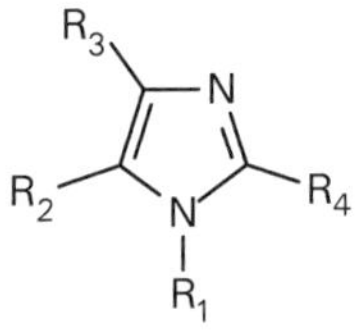

[1-4]

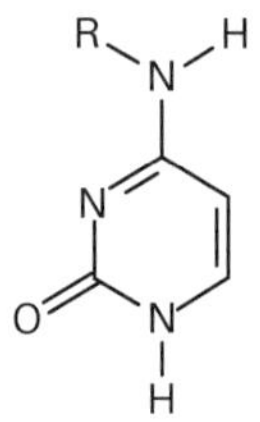

[5]

As follows from the data in **Table 7**, the introduction of NO_2 and CH_3 at positions 2 and 5 of the imidazole ring, respectively, leads to an increase in nuclear quadrupole coupling constants on both nitrogens of the ring and a decrease in the value of this parameter on the nitrogen from the NO_2 group. Results of a bond population analysis carried out according to the Townes–Dailey method for imidazole and its derivatives are displayed in **Table 8**.

According to the assumed notation n_{NC} and n_{NR} stand for the population of N–C and N–R bonds, n_a stand for the lone pair population of an N atom, while π is the π-electron density. A comparison of the data collected in **Table 8** shows that in the case of a 3-substituted nitrogen atom, the lone pair electron density and NR bond population increase with increasing length of a substituent at 1-H position in the imidazole ring, both at 193 and 269K. The substitution of an NO_2 group at the 2-position and a methyl group at the 5-position of the imidazole ring leads to an insignificant redistribution of electron density (of the order of 2%) relative to that in pure imidazole. Only the introduction of a substituent at the 1-H position of the ring causes considerable changes. Similarly for a 2-substituted nitrogen, the bond population changes in a characteristic way – the π-electron density and population of the N–C bond decrease with elongation of the substituent at 1-H position of the imidazole ring (**Table 8**). Interestingly, a qualitatively similar but much more effective phenomenon is the electron redistribution on the nitrogen atoms of the NO_2 groups. Therefore it can be concluded that, in the case of 2-nitro-5-methyl derivatives of imidazole, with increasing length of the chain of the aliphatic substituent at the 1-H position of the imidazole ring, the electron density of the π-orbital and σ-bond of the 2-substituted atom, as well as the π-orbital and σ-bond of the NO_2 group towards the 3-substituted nitrogen, undergoes redistribution. Of essential importance is the character of substituents, i.e. CH_2CH_2OH and $CH_2CH_2OCOCH_3$ groups are electron density acceptors, while for 1H-imidazole and 1-acetylimidazole, the opposite tendency was observed, i.e. the π-electron density and the population of the σ-bond increased. The results also indicated that the effect of temperature on the bond populations in the imidazole derivatives studied is negligible.

The 4-N-derivatives of cytosine [5], R-substituent R = H, CH_2Ph, CH_2CH_2Sh, CH_2CH_2Ph and naphtha-

Table 8 Population of the –NH–, –N=, $-NO_2$ bonds for imidazole and its derivatives

	−NH−			−N=		$-NO_2$		
No.	n_a	n_{NC}	n_{NR}	p	n_{NC}	p	n_{NC}	T(K)
[1]	1.330	1.120	1.330	1.640	1.260	–	–	77
	1.340	1.140	1.330	1.640	1.270	–	–	293
[2]	1.350	1.129	1.330	1.657	1.556	1.072	1.182	193
	1.361	1.139	1.330	1.652	1.556	1.070	1.178	296
[3]	1.672	1.250	1.368	1.449	1.390	1.042	1.128	193
	1.669	1.250	1.366	1.452	1.394	1.041	1.126	296
[4]	1.648	1.250	1.340	1.430	1.384	1.045	1.117	193
	1.648	1.250	1.340	1.430	1.384	1.044	1.116	296

lene, have aroused significant interest mainly because of their biological significance. 4-*N*-methyl and 4-*N*-acetylcytosine have been found in nuclei acids as rare bases. The results of the analysis of bond populations for cytosine and its derivatives performed according to the Townes–Dailey theory are described below. On the NH nitrogen, the changes in electron density distribution induced by substitution are insignificant. The symmetry of charge distribution at the NH nitrogen in 4-*N*-thioethylcytosine is the lowest while in 4-*N*-naphthalenecytosine it is the highest. On the other hand, the greatest value of the z-component of the EFG tensor was detected in the vicinity of the NH in pure cytosine while the smallest was in naphthalenecytosine, which implies that the electrons are drawn away from NH by the system of bonds depending on the substituent. Thus, it can be concluded that the introduction of a substituent at the amine group of cytosine at 4-N does not cause any changes when the substituent contains a chain such as CH_2CH_2 which separates the aromatic system from the cytosine ring. However, when the aromatic system is separated only by one CH_2 group, or is not separated at all, there occurs a strong inductive effect which is responsible for drawing π-electron density from the NH nitrogen. For a 2-substituted nitrogen the symmetry of charge density at –N= is 50% lower. This is if one compares that of 4-*N*-phenylmethylcytosine with that of 4-*N*-thioethylcytosine, and about 20% lower than that of 4-*N*-naphthalenecytosine. On the other hand, the quadrupole coupling constant and thus the z-component of the EFG tensor at the –N= nitrogen, is much higher in the derivatives whose substituents are not separated by the CH_2CH_2 chain. The difference between the population of the NC bond for pure and substituted cytosine is small –0.0001 for 4-*N*-thioethylcytosine but as great as 0.132 for 4-*N*-phenylmethylcytosine, which is still much less than the difference between the populations of σ and π bonds. Such a situation indicates that the changes in σ-bond population are dominant. In 4-*N*-naphthalenecytosine the change in π-electron density is dominant. This implies that –N= in phenylmethylcytosine and naphthalenecytosine plays the role of a buffer. In 4-*N*-phenylmethylcytosine and 4-*N*-naphthalenecytosine the electron density of a free electron pair at the nitrogen NHR is significantly delocalized. The electron density on the NH bond in 4-*N*-naphthalenecytosine, 4-*N*-phenylmethylcytosine and 4-*N*-thioethylocytosine relative to that in pure cytosine changes by 0.053, 0.038 and 0.005, respectively, while for 4-*N*-phenylethylcytosine there is no change. Thus, the amine group which acts as π-electron acceptor in the majority of molecular systems, becomes the electron donor in phenylcytosine and naphthalenecytosine. The aromatic rings which usually compensate changes in electron density in cytosine act as electron acceptors. When the aromatic substituents are separated by the CH_2CH_2 chain, the density redistribution is reduced, which is in agreement with the tendency observed for chlorobenzenes. Investigations of cytosine and its derivatives have shown that the cytosine derivatives with an aliphatic substituent do not show an anticancer activity, but those with an aromatic substituent do. However, the mechanism of this activity has not been explained. The results suggest that in the search for anticancer drugs from the group of 4-*N*-cytosine derivatives, the choice should be those in which the aromatic substituents are not separated by a CH_2CH_2 chain, whose presence induces a significant redistribution of π-electron density. The ^{14}N NQR frequencies recorded for cytosine derivatives were practically the same at ambient and liquid nitrogen temperatures, which means that no essential redistribution of electron density occurs in this temperature range.

The ^{14}N nucleus, because of its widespread occurrence in all types of systems (especially biologically active systems), is of particular interest in studying electron density distribution, molecular reorientations and intermolecular time-dependent interactions. It seems that such studies will acquire more and more importance in the future and will occur more frequently, especially with the availability of double resonance spectrometers and new data processing techniques such as the maximum entropy method. The examples discussed do not, of course, exhaust the potential of NQR as a tool for structure and chemical bonding. These are only simple illustrations of the applied aspects of NQR spectroscopy.

Application of NQR to the detection of explosives, contraband etc.

The increasing use of plastic explosive and drugs has found experts and research facilities scrambling for new detection methods. One of these 'new' methods is NQR. In the past five years a number of researchers around the world have independently begun to reconsider NQR as a possible solution for the detection of plastic explosives and started developing it specifically for bomb and narcotics detection. The noninvasive nature of NQR (closely connected with the absence of magnets) gives it some advantages over other methods.

Pulsed-RF NQR produces single, or nearly single, peak signals at specific frequencies that depend on the specific bond environment surrounding an element in a given compound, usually a crystalline solid. Because the resonance frequency is almost unique to each compound, NQR exhibits great specificity for analytes such as explosives and narcotics. The most useful elements to monitor by NQR are ^{14}N, ^{35}Cl and ^{37}Cl. Most high explosives contain 30–40% N and a large number of drugs such as cocaine and heroin are prepared as chloride salts. Most pure explosive such as RDX, HMX, TNT, C-4 are crystalline or semicrystalline compounds embedded in a polymer matrix, rather than pure polymeric compounds, so that they are immobilized and relatively ordered and as such give good NQR. Of course liquids and polymers are too disordered to give an NQR signal, although some monomers have shown detectable resonances.

NQR can be also used to differentiate between explosives, narcotics and benign nitrogen-containing compounds such as polyurethane foam or nylon. False alarms from these compounds are not the problem for NQR whereas, they might be for neutron activation or other generic nitrogen-detecting methods. It is well known that commercial and military explosives are physical mixtures of pure explosive compounds with some additive plasticizer or binder. Because NQR is so compound-specific, physical additives do not interfere with the signal for a target compound so NQR can be used to identify explosives that are not in a pure form. Moreover, it does not matter what form the explosive or drug is in – whether it be tin sheets or small pellets. NQR may eventually be used to detect bombs or narcotics with spatial resolution, in the same way as X-ray metal detectors.

NQR is inherently less flexible than NMR but when it works it is extremely attractive because of its specificity. NQR can work with slurries, aggregates and possibly even emulsions, as long as the molecular dynamics are slower than the NQR method time scale (the MHz range). The ongoing use of NQR as analytical tool for measuring solid-state phase transitions and order–disorder in materials may also be of interest.

Summary

NQR is not as extensively useful at present as NMR spectroscopy. The best results on light nuclei, such for example as those on ^{27}Al nuclear coupling constants in mineral samples, have been made using NMR. Here, the changes in the NMR spectra were considered as a function of the orientation of a single crystal in external magnetic field. NQR could, however, directly measure the same nuclear coupling constant data using neither single crystals nor an external magnetic field. NQR measurements possess high spectral resolution, precision, specificity and speed of measurements. The reason for the relatively limited practical application of NQR seems to lie in the lack of sufficiently sophisticated equipment.

NQR applications can be divided into four groups:

1. Studies of the electron density distribution in a molecule – changes in orbital populations under substitution, and complexation.
2. Studies of molecular motions – reorientations, rotations, hindered rotations.
3. Studies of phase transitions.
4. Studies of impurities and mixed crystals.

List of symbols

$e^2Qq_{zz}h^{-1}$ = quadrupole coupling constant; $e^2Qq_ph^{-1}$ = quadrupole coupling constant; I = nuclear spin quantum number; I_B = ionic character of a bond; q_p = the electric field gradient produced by one unbalanced p electron; q_{zz} = electric field gradient (EFG) at a quadrupole nucleus; U_p^b, U_p^t = unpaired p-electron density of a bridging and terminal, halogen, respectively; η = asymmetry parameter; γ = Sternheimer antishielding factor.

See also: **Mossbauer Spectrometers; Mossbauer Spectroscopy, Applications; Mossbauer Spectroscopy, Theory; NQR, Theory; Nuclear Quadrupole Resonance, Instrumentation.**

Further reading

Buslaev JA, Kolditz L and Kravcenko EA (1987) *Nuclear Quadrupole Resonance in Inorganic Chemistry*. Berlin: VEB Deutsche Verlag der Wissenschaften, 237 pp.

Das TP and Hahn EI (1958) *Nuclear Quadrupole Resonance Spectroscopy. Solid State Physics*, suppl. I. New York: Academic Press, 223 pp.

Gretschischkin WS (1973) *Yadernye Kwadrupolnye Vzaimodejstviya v Tverdych Telach*. Nauka: Moskva, 264 pp.

Lucken EAC (1969) *Nuclear Qudrupole Coupling Constants*. London: Academic Press, 360 pp.

Safin IA and Osokin D (1977) *Yadernyj Kvadrupolnyj Rezonans v Soedinieniach Azota*. Moskva: Nauka, 255 pp.

Semin GK, Babushkina TA and Yakobson GG (1975) NQR Group of INEOS AN SSSR, *Nuclear Quadrupole Resonance in Chemistry*, (English edition). London: Wiley, 334 pp.

Smith JA (1974–1983) *Advances in Nuclear Quadrupole Resonance*, Vol. 1–5. London: Heiden & Sons.

Townes CH and Dailey BP (1949) Determination of electronic structure of molecules from nuclear quadrupole effects. *Journal of Chemical Physics* **17**: 782–796.

Townes CH and Dailey BP (1955) The ionic character of diatomic molecules. *Journal of Chemical Physics* **23**: 118–123.

Nuclear Quadrupole Resonance, Instrumentation

Taras N Rudakov, S.E.E. Corporation Ltd., Bentley, WA, Australia

MAGNETIC RESONANCE
Methods & Instrumentation

Nuclear quadrupole resonance (NQR) is a modern research method for the analytical detection of chemical substances in the solid state. NQR is a type of radiofrequency (RF) spectroscopy, and is defined as the phenomenon of resonance RF absorption or emission of electromagnetic energy. It is due to the dependence of a portion of the energy of the electron–nuclear interactions on the mutual orientation of asymmetrically distributed charges of the atomic nucleus and the atomic shell electrons, as well as those charges that are outside the atomic radius. Thus, all changes in the quadrupole coupling constants and NQR frequencies are due to their electronic origin. The nuclear electric quadrupole moment eQ interacts with the electric field gradient eq, defined by the asymmetry parameter η. Therefore the nuclear quadrupole coupling constant e^2Qq and the asymmetry parameter η, which contain structural information about a molecule, may be calculated from the experimental data. The main spectral parameters in NQR experiments are the transition frequencies of the nucleus and the line width Δf. In addition, measurement of the spin–lattice relaxation time T_1, spin–spin relaxation time T_2 and line-shape parameter T_2^* (inversely proportional to Δf) is also of great value. These parameters must also be taken into consideration when choosing the experimental technique and equipment.

NQR methods

In contrast to nuclear magnetic resonance (NMR) methods, NQR can operate without a strong external DC magnetic field. This technique is known as 'pure NQR', or direct NQR detection, and has many advantages for some applications, such as identification of specific compounds and remote NQR. In turn, direct NQR detection techniques can be subdivided into two main areas: oscillator-detector methods [continuous wave (CW) and superregenerative techniques] and pulsed methods. At present the pulse method is the most widely used direct NQR method. The use of various multipulse sequences permits a considerable increase in the sensitivity, and a decrease in the duration of the experiment.

Besides direct methods of detection, indirect NQR detection methods have also been developed. As a rule, they are used at low frequencies or in cases when the concentration of quadrupolar nuclei is not high, i.e. when the sensitivity of the direct methods is not sufficient. For most experiments, when using these methods, a constant magnetic field is needed. To apply these methods, it is necessary for the sample to have two spin systems connected by dipole–dipole interactions. Indirect methods can be conventionally divided into double-resonance techniques and the cross-relaxation spectroscopy method.

A block diagram summarizing the composition of the main technologies for NQR methods is given in **Figure 1**.

Direct NQR detection (pure NQR) techniques and equipment

Direct methods are the basis of experimental methods in NQR. They are used in a wide frequency range from hundreds, sometimes even tens of kHz to hundreds of MHz. Historically these methods started to be applied earlier than indirect methods. The first observations of NQR were undertaken using CW and superregenerative techniques, which are simple and inexpensive. These methods and techniques were developed during the 1960s to 1980s. At present it is the pulsed techniques that are more widely used, as they permit a better sensitivity and are more convenient for measuring NQR parameters of the sample over a wide frequency range. The modern pulsed techniques use the latest signal processing methods, including fast Fourier transformation (FFT), digital filtering, and others.

Oscillator-detector methods

Continuous wave techniques CW NQR spectrometers are based on the use of oscillator-detectors, which are built around the circuits of a marginal oscillator or a limited oscillator (Robinson oscillator). Such an oscillator-detector includes a tank circuit with a coil, into which the studied sample is inserted. When the frequency of the oscillator-detector coincides with the NQR frequency in the sample, the

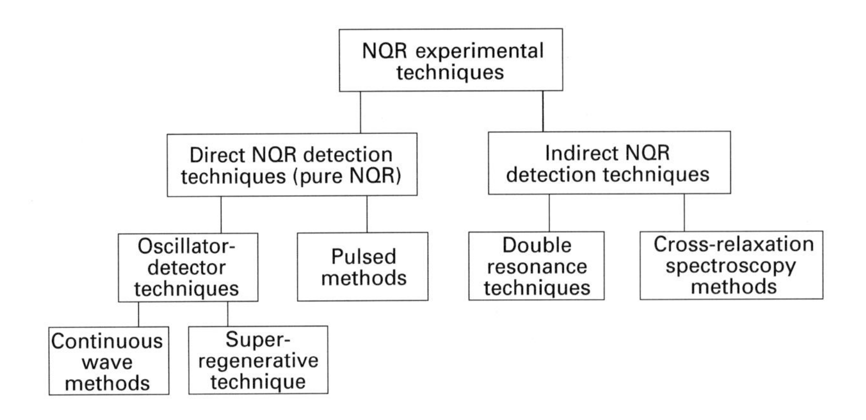

Figure 1 Main categories of NQR experimental techniques.

sample starts to absorb energy and the amplitude of oscillations in the circuit decreases. This change in the amplitude is detected on the output of the oscillator-detector. The sensitivity of such spectrometers can be high and depends on the Q-factor of the coil, on the filling factor and on the noise temperature of the oscillator. Normally these oscillator-detectors cannot work at high levels of RF voltages and require accurate tuning of the electric circuit parameters. To avoid saturation of the sample during the detection process, they use either frequency modulation or Zeeman modulation, usually bisymmetric. Frequency modulation is easier to perform from the technical viewpoint, but it has the drawbacks of giving base-line drift and spurious signals.

In the literature one can find many descriptions of CW NQR spectrometers and oscillator-detectors for various purposes and frequency ranges. Particular interest is evoked by limited oscillators, which have a low noise level similar to that of marginal oscillators but do not need continual critical adjustment of circuit parameters. **Figure 2** shows a simplified block diagram of a CW spectrometer, containing an oscillator-detector with a Zeeman modulation system and a computer for signal processing and frequency control.

Superregenerative technique Superregenerative oscillators (SRO) have been widely used as oscillator-detectors in the study of NQR because of their high sensitivity at high RF levels, reliability of operation and simplicity of construction. The phenomenon of NQR was discovered by Demelt and Kruger using a SRO spectrometer. SRO is actually an oscillator-detector, with oscillation periodically quenched either with the help of an external quench oscillator (quencher mode), or with the help of internal circuits (self-quenched mode). A high sensitivity of the SRO is provided during its coherent work, when the oscillations are not quenched completely, i.e. not to the level of noise. Besides, SRO performs very well in logarithmic mode, when RF pulses are built up to a limiting amplitude. Detection of the NQR signal in SRO occurs as follows. Using RF pulses, the SRO excites an NQR signal in a sample. This signal, which occurs in the intervals between these pulses, is added up with the resilient voltage. In this case oscillations in the next pulse will start on a higher level of voltage and will reach the limiting amplitude sooner, i.e. the length of the pulse will increase. This change in the length of the RF pulses of the SRO is detected as the signal.

Usually the SRO has a number of significant disadvantages: (i) poor frequency stability; (ii) the RF spectrum contains components (sidebands) spaced at integral multiples of the quench frequency on either side of the fundamental frequency, each of them being capable of exciting the NQR signal; (iii) poor fidelity of line shape reproduction; and (iv) dependence of the centre frequency on the quench frequency. The sideband suppression is achieved by using the quench frequency modulation. The other disadvantages can be eliminated by using the quench frequency modulation. The other disadvantages can be eliminated by using the phenomenon of frequency (or injection) and phase-locking. In this case the frequency stability of the SRO will be determined by the stability of the external locking oscillator. To stabilize the shape of lines of the detected NQR signal, phase-locking is used. A simplified block diagram of such an NQR spectrometer is given in **Figure 3**. The frequency locking of the SRO is done by introducing a small voltage from the external adjustable oscillator, which switches SRO from the regime of noncoherent work to coherent. An injection signal can be introduced with the help of a small capacitor or by

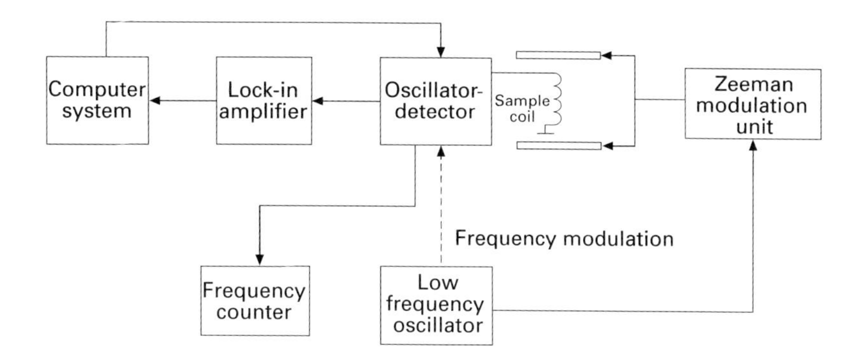

Figure 2 Simplified block diagram of a CW NQR spectrometer.

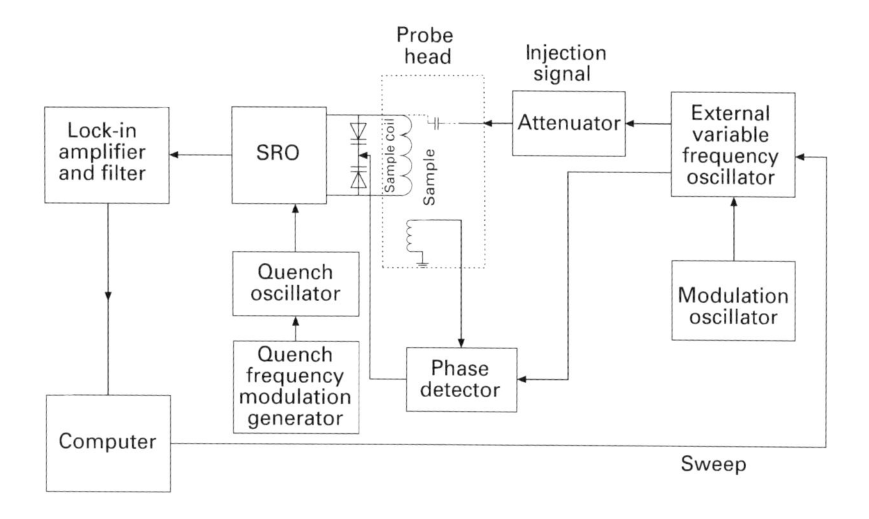

Figure 3 Simplified block diagram of the frequency and phase-locked NQR spectrometer based on a SRO-type oscillator-detector.

simply using the inductance of the connection. As the SRO frequency is determined by the frequency of the external oscillator, the frequency modulation is applied to that oscillator. In the mode of frequency locking the change of the capacity of the SRO circuit leads to a change in the phase of the generator frequency, which causes a change in the line shape of the detected signal. Therefore to stabilize the shape of the NQR signal line on the output of the SRO, so-called phase locking is carried out, using an additional phase detector. Such a phase detector permits the control of a required phase shift between the SRO frequency and the external oscillator frequency by automatic tuning of the capacity of the SRO circuit.

Pulsed methods

The essence of the pulse method approach consists of irradiating the spin-system by RF pulses with frequencies equal or close to the NQR transition frequency and followed by detection of signals induced by this spin-system. Direct pulsed NQR methods are of great interest for industrial, medical, chemical and biochemical investigations. They also seem to be an effective technique for detecting the presence of prohibited goods (narcotics), landmines and plastic bombs. This technique is very convenient and commonly used for the measurement of line widths and relaxation times. CW and SRO spectroscopy are quite inefficient with regard to measuring time and in particular for samples with long T_1 and short T_2. Nowadays, CW and SRO spectrometers have almost completely been replaced by pulsed FFT NQR instruments.

Multipulse techniques, widely used in magnetic resonance, are also very common in NQR spectroscopy. They are effectively used for increasing sensitivity, reducing the duration of the experiment, and for measuring relaxation times in the sample. In NQR such well-known sequences as spin-echo, Carr–Purcell (CP), Meiboom–Gill-modified CP, spin-locking sequence and phase-alternated pulse sequence are widely used. There is a large practical

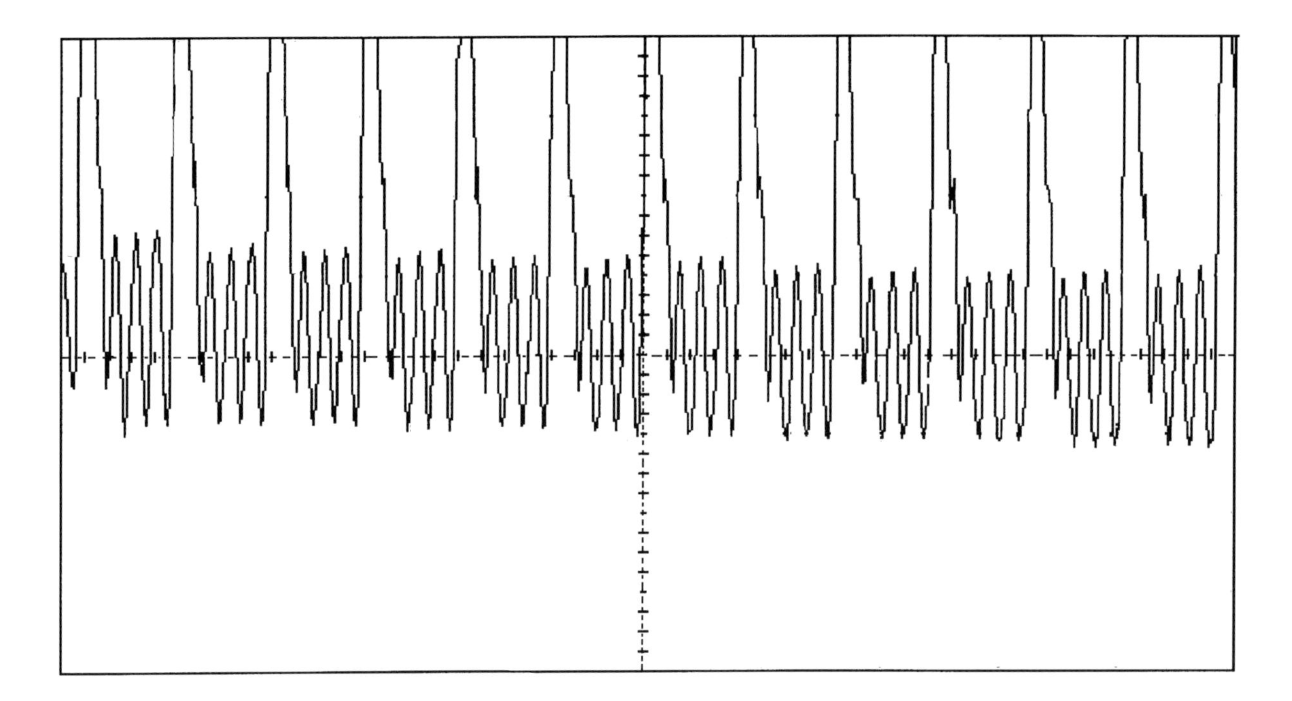

Figure 4 The response from nitrogen-14 in a sample of HMX to the SSFP pulse sequence with offset –5 kHz. The resonance frequency is 5.301 MHz at 295 K.

interest at present in pulse sequences of the steady-state free precession type (SSFP). The simplest version of this sequence is well known in NQR as 'strong off-resonance comb'. When certain requirements are complied with, these sequences give a stationary signal that does not die down while the sequence is in action. This method is very convenient for achieving fast coherent accumulation (averaging) of the signal.

At present, various modifications of the pulsed SSFP-type sequences have been developed or are being developed to increase sensitivity and eliminate some undesirable effects, such as intensity and phase anomalies. Typical results for ^{14}N responses to SSFP-type sequences in HMX ($C_4H_8O_8N_8$) are shown in **Figure 4**. To carry out multipulse experiments, the pulsed NQR spectrometer must have a programmable pulse generator with a variety of possibilities for changing pulse length and spacing, and a phase shifter, which permits a large variation of the phase of the RF pulse.

Now, pulse Fourier spectroscopy is the preferred experimental technique in NQR. It was made possible by the introduction of inexpensive computers and by the development of FFT algorithms. Note that the principles of Fourier spectroscopy are well developed and widely used in other fields of spectroscopy, including magnetic resonance, electron paramagnetic resonance etc. Several monographs are available on this subject. The reader is referred to the Further reading section for details of Fourier spectroscopy, including its application in NQR. To illustrate the FFT method in pulsed spectroscopy, **Figure 5** shows a typical ^{14}N NQR spectrum of powder RDX ($C_3H_6O_6N_6$), obtained from a free induction decay (FID).

One of the most important characteristics of a wide-band and pulsed NQR spectrometer is its

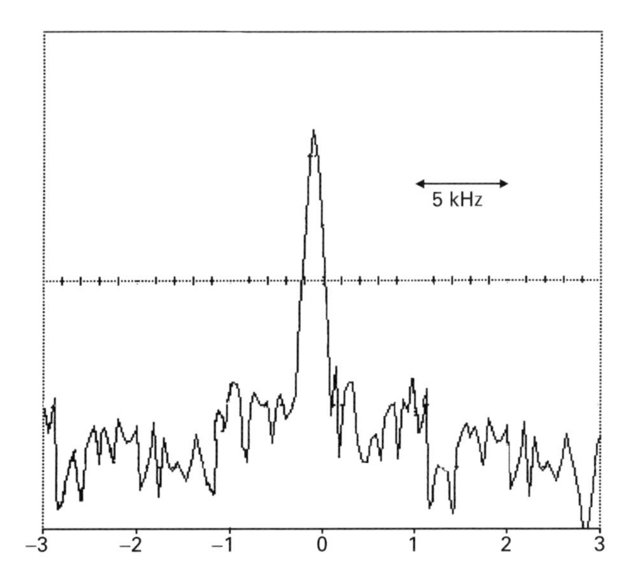

Figure 5 A typical nitrogen-14 NQR spectrum of powder RDX at 297 K, obtained from a FID. The resonance frequency is ~5.193 MHz.

sensitivity. This is because, in most cases, the intensity of NQR signals is very low. It is especially true for the low-frequency range, less than 10 MHz, but is also significant for higher frequencies. Therefore it is necessary for the receiving system of the spectrometer to have a low noise factor and to ensure matching of its input impedance with the circuit. Without this matching the noise factor can increase considerably. The signal-to-noise (S/N) ratio can also be increased by increasing the Q-factor of the circuit. The S/N ratio is proportional to $\sqrt{Q}$, because with the increase in quality, the amplitude of the signal voltage increases Q times, while the effective noise voltage increases by $\sqrt{Q}$ only. However, with the increase of the quality of the circuit, transient processes known as 'ringing', after the irradiating signal, become longer. At low frequency the length of transient processes can reach hundreds of microseconds, which creates considerable difficulties for detecting the inductance signal and even spin-echo signals in samples with short T_2 and T_2^* times. The problem of ringing can be solved by several methods. Most often Q-switching is used, i.e. during transmission the Q factor is low, about 5–10, but during receiving it is high, typically > 100. In a number of practical applications of NQR, e.g. in remote NQR, or detection using large volumes, it is desirable to have a high Q-factor in the transmit regime as well. In this case methods of active damping of transient processes in the circuit are used, i.e. a short-term decrease of Q directly after the irradiation pulse. There are many descriptions of circuits for this purpose in the literature. Good results are achieved when short antiphase RF pulses are used immediately after pulses of RF irradiation. This effect is shown in **Figure 6**.

Besides ringing, transient processes in the receiver module can also impede the observation of NQR signals in both high- and low-frequency parts of the receiver. To eliminate this, it is possible to use wide-band amplifiers in the high-frequency part, and to close the input of the receiver and the input of the synchronous detectors and switch off the reference frequency transmission during the excitation pulse. High sensitivity of an NQR spectrometer is achieved when using optimum filtering, which corresponds best to the parameters of the detected signal. For this purpose Bessel and Butterworth filters are usually used. As different substances have resonance lines of different width, it is preferable to have a receiver with a variable bandwidth. Most often receivers of NQR spectrometers are built on the basis of superheterodyne techniques. These techniques are very efficient for wide frequency band spectrometers operated at 10 MHz. At low frequencies though (0.5–10 MHz) straight receiver systems are successfully used too. They consist of an RF amplifier, synchronous detectors, low-pass filter and an output video amplifier. As a rule, the principle of quadrature detection is used in all these receivers. Digital receivers, where analogue-to-digital conversion is carried out either at the intermediate frequency, or directly on the resonance frequency, have recently become more widely used. Evidently in the future this will be the most promising direction in the development of receiver systems of NQR spectrometers.

The system of exciting the NQR signal in a sample is also an important part of a spectrometer. Besides a power amplifier, it contains a means of generating RF pulses (gate), a programmable pulse generator (sequencer) and a stable variable oscillator (synthesizer). To form various multipulse sequences it is also necessary to use phase shifters. A standard block diagram for a pulsed NQR instrument is given in **Figure 7**.

In the literature more complicated NQR spectrometers with frequency conversion in both the receiver and the transmitter (irradiation) modules have been described. It should be noted that in wide-band NQR spectrometers several receivers and power amplifiers are used, each of them intended for work in a specific frequency range. Many laboratories involved in NQR research use spectrometers that are either home-made or made to order by specialized companies. However, companies specializing in the technology for magnetic resonance also often produce small batches of high-quality NQR equipment. For example, the equipment produced by Tecmag, Inc. is extremely good for carrying out the most complicated NQR experiments. It contains devices for forming all kinds of common and rarely used pulse sequences. It also contains such vital modules as a digital signal processor, a pulse programmer, a signal averager with complex memory, a special frequency synthesizer, a digital receiver and the latest model of a PC with specialized software.

The development of electronics and new experimental techniques has allowed the considerable broadening of the areas of practical application for pulsed NQR. Important NQR applications include the detection of drugs, landmines and plastic explosives based on signals from the nuclei of nitrogen and chlorine, in which the NQR signal at low frequencies of 0.5–6 MHz is detected. The weakness of NQR signals hinders further development of this method, especially in this frequency range. Two different NQR techniques are being applied to detect these substances, detection using large ordinary coils and remote detection using special coils (antennae). The first technique is used to scan packages. It is a development of ordinary pulsed NQR techniques for

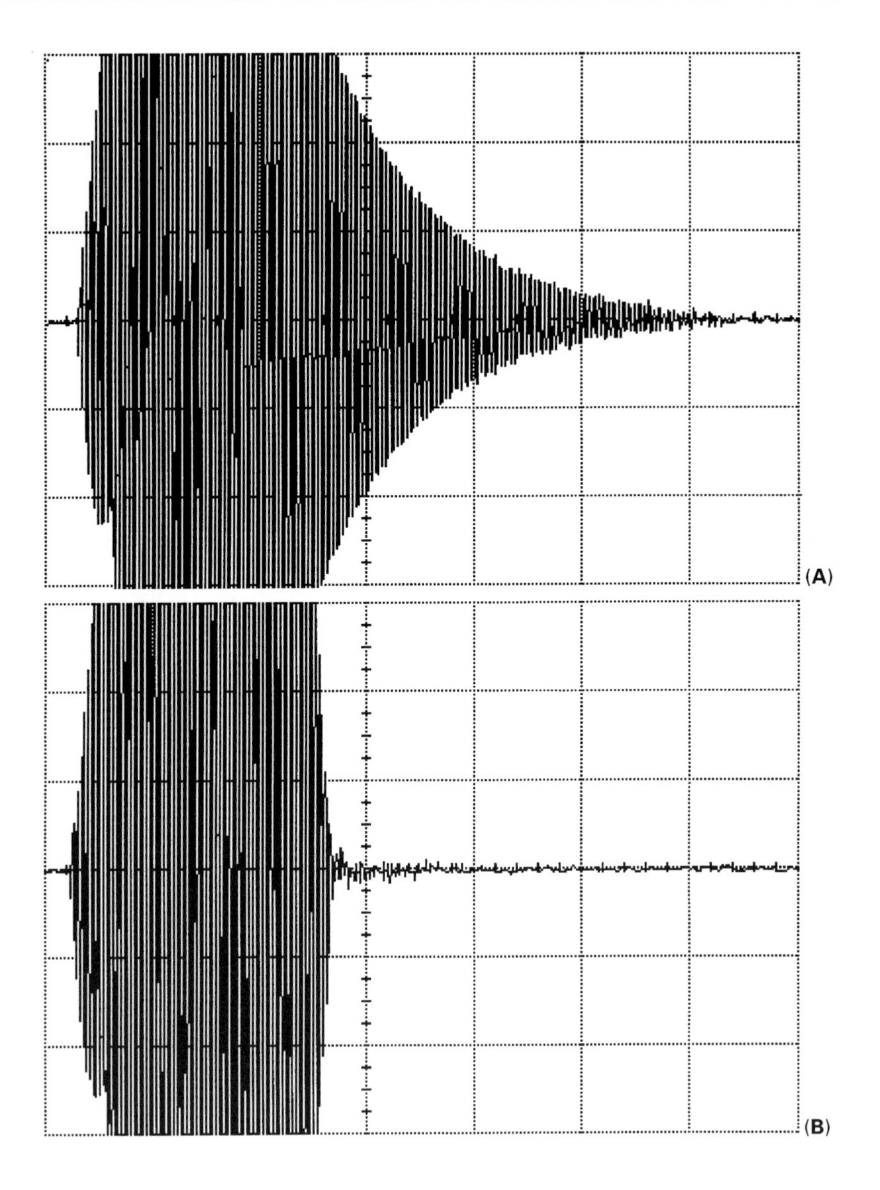

Figure 6 An RF pulse in the NQR spectrometer coil: (A) without ring damping, (B) after ring damping.

use with large volumes and cannot be utilized to detect samples beyond a certain distance from the spectrometer coil (e.g. landmines). At the same time, the remote NQR technique is more universal. However, besides extremely complicated highly sensitive receiving equipment, it requires the use of specific signal processing methods for eliminating external interference. It is also necessary to establish the required RF field at a definite distance from the coil. In remote NQR, special flat-surface coils of various designs are used, with irradiating and receiving coils (antennae) sometimes separated.

Indirect NQR detection technique

The indirect NQR detection technique was developed as a result of wide interest in important chemical compounds containing light nuclei. As a rule, these are organic compounds that are interesting from the point of view of biology and medicine. As NQR frequencies of light nuclei are located in the low-frequency range (less than 1 MHz), the intensity of the signals from these nuclei when using direct techniques is very low. The indirect NQR technique permits high sensitivity for detecting many light elements, which is very convenient for locating unknown resonances. This technique cannot replace completely the direct method for NQR detection, but is important as a complementary technique.

Double resonance

The principle of double resonance (DR) consists of the detection of a weak signal from one type of

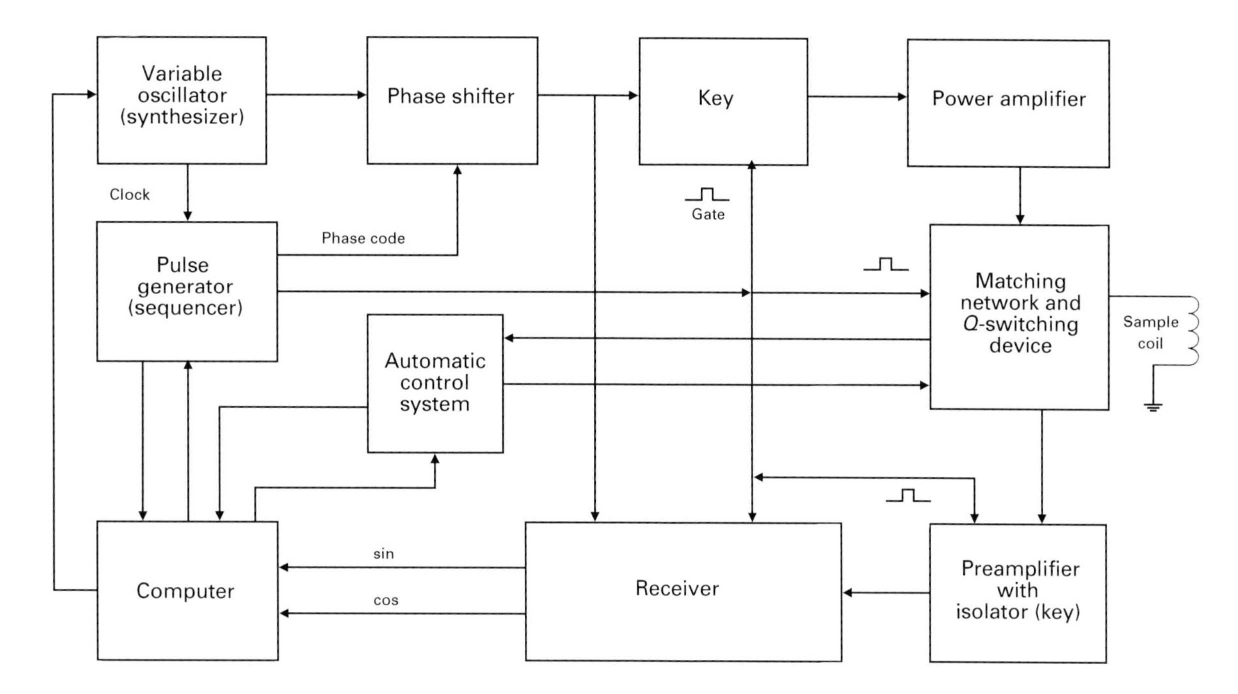

Figure 7 Block diagram of a pulsed NQR instrument.

nucleus when directly detecting changes in the strong signal from another type of nucleus. Thus, the studied sample must contain two types of nuclei, N_Q and N_P, with one of them (e.g. N_P) having a strong nuclear resonance signal. Spin systems N_Q and N_P must be connected by dipole–dipole coupling, so that the exposure of one system is reflected in the state of the other. Nuclei with a strong NQR signal could be used as the N_P nuclei, but most often nuclei with a strong NMR signal, e.g. protons are used for this purpose.

The DR techniques allow the successful detection of the NQR signal from such nuclei as ^{2}H, ^{10}B, ^{11}B, ^{14}N, ^{17}O, ^{23}Na, ^{25}Mg and ^{27}Al, which can be present in biologically important molecules. The DR methods have developed and improved quickly, and now a whole range of methods has appeared. Usually DR methods for NQR are divided into the following main groups: DR in the rotating frame, DR in the laboratory frame, DR by level crossing, DR by continuous coupling and DR by solid-state effect. The areas and details of the use of these methods can be found in the reviews given in the Further reading section. It should be noted that the choice of a specific method depends on the type of the studied nuclei and the parameters of the sample, with relaxation times of both spin systems being very important.

In spite of a wide variety of different DR methods, the main stages of the experiment can very briefly be described as follows, using the widely used concept of spin temperature in magnetic resonance. Initially the N_P spins are prepared in a polarized state. Then the polarization of the N_Q spins occurs through thermal contact between the two spin systems. Another method widely used is resonant heating of the N_Q spins by irradiating them with a RF field. As a result of the new thermal contact, the spin temperature of the N_P spins changes, and this change is then registered as the NQR signal.

The experimental equipment for NQR DR consists first of all of a spectrometer for detecting nuclear resonance signals from the nuclei of N_P. As a rule, this is an ordinary NMR spectrometer and a DC magnet, which permits the detection of FID signals with 90° pulses. In addition, a DR spectrometer contains a sample-transfer system, an irradiating system (for the N_Q nuclei), a cryostat, a controlling and automatic signal processing unit, including a PC computer, and also some additional modules depending on the chosen methods of detection, e.g. a source of additional magnetic fields.

Figure 8 shows a typical block diagram of an NMR/NQR DR spectrometer. Such spectrometers require a special sample transit system. Systems of various design can be used for this purpose, including those with linear induction motors, step motors or compressed air systems. As a source of DC magnetic field, permanent magnets or electromagnets are used. Initially the sample is located in the field of this magnet, and the NMR signal is detected (in the P-coil), then the sample is transferred into the Q-coil,

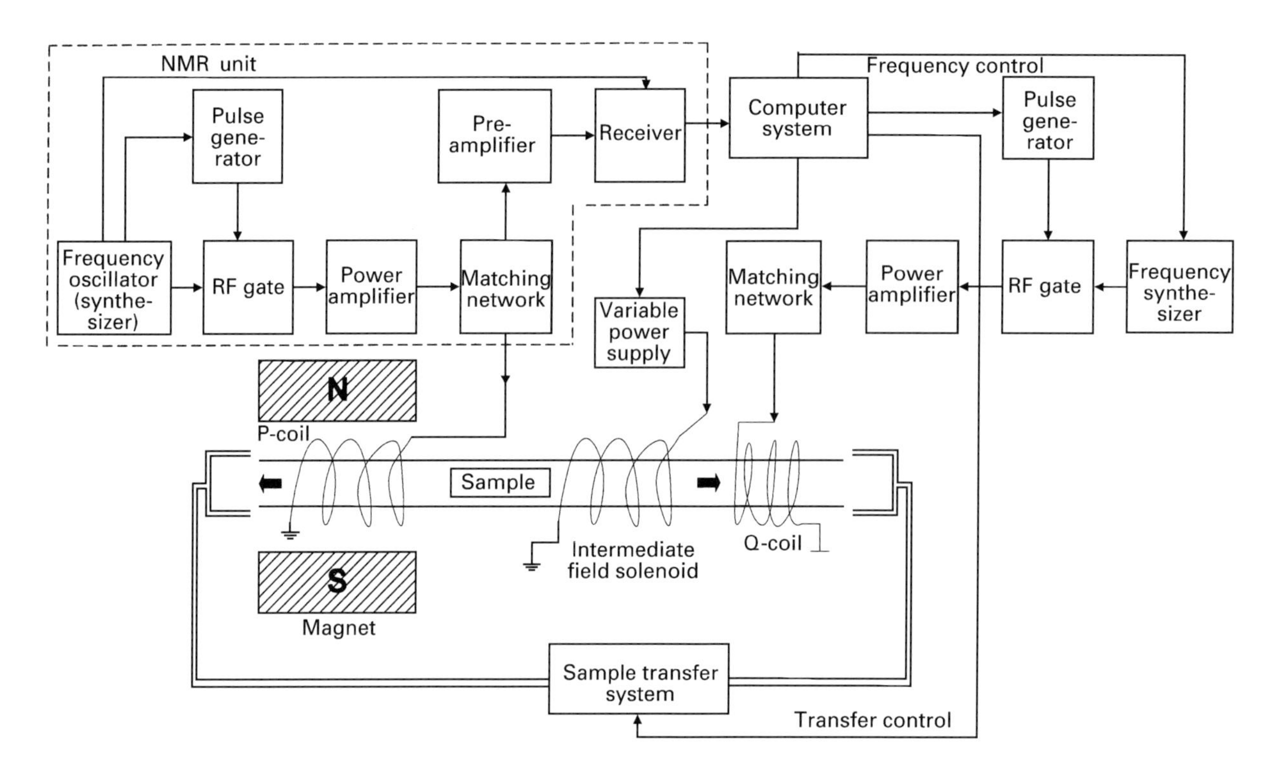

Figure 8 Simplified block diagram of the double resonance (NQR/NMR) spectrometer.

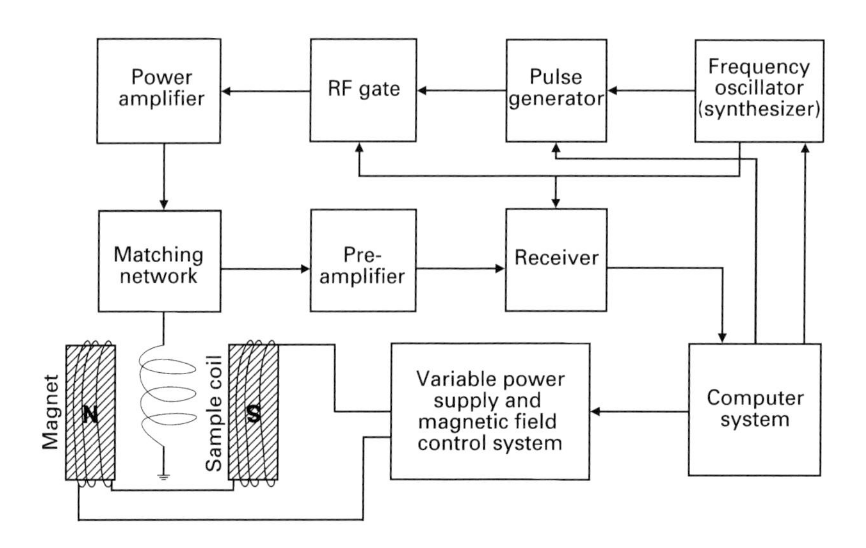

Figure 9 Simplified block diagram of a cross-relaxation NQR spectrometer.

where it is irradiated with RF pulses at the frequency of quadrupole nuclei. After this, the sample is returned into the magnetic field, where the NMR signal is detected again. The change in the amplitude of the NMR signal detects the NQR. For some experiments an intermediate magnetic field can be used. It is also possible instead of the mechanical transfer of the sample from the magnetic field, to switch the field. For experiments in quadrupole (NQR/NQR) DR the external magnetic field need not necessarily be used. This refers to spin-echo DR, or DR in the rotating frame, for example.

Cross-relaxation spectroscopy method

The cross-relaxation (CR) method is well known in magnetic resonance spectroscopy and can be successfully used for the detection of weak NQR signals in the low-frequency area in particular. A characteristic of this method is the absence of irradiation of the spin system at NQR frequencies. In principle this method can also be regarded as a version of DR. A block diagram of a spectrometer for implementing the method of CR spectroscopy, is given in **Figure 9**. It includes an NMR spectrometer, a controlling unit for regulating the magnetic field, and a computer system. The experiment is usually carried out in the following way. At first, as when using an ordinary DR method, the sample is acted on by a strong DC magnetic field, and NMR signals from the N_P nuclei are detected. Then the magnetic field is changed to satisfy the conditions of CR between the levels of the N_P and N_Q nuclei. If the conditions of CR are satisfied, the amplitude of the NMR signal detected at the end of the measurement cycle will change, and these changes are then registered as the NQR signal.

The use of amplifiers based on a DC SQUID

One of the new methods that permits effective detection of weak NQR signals at low frequencies, is the use of amplifiers based on a DC superconducting quantum interference device (SQUID). The operation of this is based on the use of two superconducting phenomena: flux quantization and Josephson tunnelling. In fact a SQUID represents a flux-to-voltage transducer with a very low noise temperature as compared with regular amplifiers, which is particularly important for low frequencies. The SQUID can be used in various methods for NQR detection, including CW methods as well as pulsed methods. Also, a SQUID amplifier does not have an input resonance circuit, and therefore permits the detection of signals over a broad bandwidth, from close to zero. Readers interested in the principles of operation, design and specificity of application for SQUIDs, can find more information in the literature, included in the Further reading section. NQR spectrometers using SQUIDs, are of great interest for many practical applications, particularly in areas where the use of indirect methods is complicated from the technical point of view. This technique is developing rapidly at present.

List of symbols

eq = electric field gradient; eQ = nuclear electric quadrupole moment; e^2Qq = nuclear quadrupole coupling constant; Q = quality factor; T_1 = spin–lattice relaxation time; T_2 = spin–spin relaxation time; T_2^* = line-shape parameter; Δf = line width; η = asymmetry parameter.

See also: **Fourier Transformation and Sampling Theory; NMR in Anisotropic Systems, Theory; NMR of Solids; NMR Pulse Sequences; NMR Spectrometers; Nuclear Quadrupole Resonance, Applications; Solid State NMR, Methods.**

Further reading

Blinc R (1975) Double resonance detection of nuclear quadrupole resonance spectra. In: Smith JA (ed) *Advances in Nuclear Quadrupole Resonance*, Vol 2, pp 71–90. London: Heyden.

Clarke J (1993) SQUIDs: Theory and practice. In: Weinstock H and Ralston RW (eds) *The New Superconducting Electronics*. Dordrecht: Kluwer Academic.

Edmonds DT (1977) Nuclear quadrupole double resonance. *Physics Reports (Section C of Physics Letters)* **29**: 233–290.

Ernst RR, Bodenhausen G and Wokaun A (1987) In: Rowlinson JS, Green MLH, Halpern J, Mukaiyama T, Schowen RL, Thomas JM and Zewail AH (eds) *Principles of Nuclear Magnetic Resonance in One and Two Dimensions*. Oxford: Clarendon Press.

Greenberg Ya S (1998) Application of superconducting quantum interference device to nuclear magnetic resonance. *Reviews of Modern Physics* **70**: 175–222.

Klainer SM, Hirschfeld TB and Marino R (1982) Fourier transform nuclear quadrupole resonance spectroscopy. In: Marshall AG (ed) *Fourier, Hadamard and Hilbert Transforms in Chemistry*, pp 147–182. New York: Plenum Press.

Read M (1974) A frequency and phase-locked super-regenerative oscillator spectrometer for nuclear quadrupole resonance at 200 MHz. In: Smith JA (ed) *Advances in Nuclear Quadrupole Resonance*, Vol 1, pp 203–226. London: Heyden.

Semin GK, Babushkina TA and Iakobson GG (1975) Pick AJ (ed) *Nuclear Quadrupole Resonance in Chemistry*. New York: Wiley.

Smith JAS (1986) Nuclear quadrupole interactions in solids. *Chemical Society Reviews* **15**: 225–260.

Nuclear Quadrupole Resonance, Theory

Janez Seliger, University of Ljubljana and
'Jožef Stefan' Institute, Ljubljana, Slovenia

MAGNETIC RESONANCE
Theory

Introduction

Nuclear quadrupole resonance (NQR) offers a unique means for the study of less structure, dynamics and chemical bonding in solids. A number of atomic nuclei possess nonzero electric quadrupole moments in the ground state. The interaction of a nuclear quadrupole moment with the local inhomogeneous electric field removes the degeneracy of the nuclear ground state. The transition frequencies between the nuclear quadrupole energy levels, which are typically in the MHz frequency region, depend on the nuclear quadrupole moment and on the electric field gradient (EFG) tensor at the nucleus. The electric quadrupole moment is a well-defined property of a nucleus in its ground state, whereas the EFG tensor depends on the electric charge distribution around the nucleus. The NQR frequencies thus indirectly provide valuable information on the local structure around the observed atom and its chemical bonding. Thermal motions in solids partially average out the EFG tensor, whereas thermal motions in isotropic liquids average the EFG tensor to zero. The temperature dependence of the NQR frequencies thus provides valuable information on the thermal behaviour of solids.

Nuclear constants and the natural abundance of naturally occurring quadrupolar nuclei are listed in **Table 1.**

The Hamiltonian

We start by considering a nuclear electric charge distribution $\rho(\boldsymbol{r})$ placed in an external inhomogeneous electric field with the potential $V(\boldsymbol{r})$. The origin, $\boldsymbol{r} = 0$, is at the centre of gravity of the nucleus. The interaction energy E is given by

$$E = \int \rho(\boldsymbol{r}) V(\boldsymbol{r})\, \mathrm{d}^3\boldsymbol{r} \qquad [1]$$

where the integral is taken over the volume of $\rho(\boldsymbol{r})$. When the electrostatic potential varies weakly over the nuclear electric charge distribution, it can be expanded in a Taylor series about the origin to give the familiar multipole expansion of the energy:

$$E = V(0) \int \rho(\boldsymbol{r})\, \mathrm{d}^3\boldsymbol{r} + \sum_k \left(\frac{\partial V}{\partial x_k}\right)_0 \int x_k \rho(\boldsymbol{r})\, \mathrm{d}^3\boldsymbol{r} + \frac{1}{2}\sum_{k,l} \left(\frac{\partial^2 V}{\partial x_k \partial x_l}\right)_0 \int x_k x_l \rho(\boldsymbol{r})\, \mathrm{d}^3\boldsymbol{r} + \cdots \qquad [2]$$

The first term is the energy of the nuclear 'monopole' and does not depend on its orientation. It is thus of no interest to us. The second term is the dipole contribution, which vanishes because a nucleus in its ground state has definite parity and therefore zero electric dipole moment. In fact all odd nuclear electric multipole moments vanish by the same argument. The strongest term representing the orientation-dependent energy of a nucleus is thus the third, quadrupole, term in Equation [2]. The symmetric traceless second-rank tensor $(\partial^2 V/\partial x_k \partial x_l)_0 = V_{kl}$ is called the electric-field-gradient (EFG) tensor. Introducing a symmetric traceless second rank tensor Q_{kl},

$$Q_{kl} = \int (3x_k x_l - r^2 \delta_{kl}) \rho(\boldsymbol{r})\, \mathrm{d}^3\boldsymbol{r} \qquad [3]$$

we may rewrite the quadrupole term E_Q in Equation [2] as

$$E_Q = \frac{1}{6}\sum_{kl} Q_{kl} V_{kl} \qquad [4]$$

Here Q_{kl} are the elements of the nuclear electric quadrupole moment tensor. Equation [4] is of course a classical expression. A quantum-mechanical expression for the quadrupole Hamiltonian H_Q is obtained from Equation [4] by replacing the tensor elements Q_{kl} by the operators $Q_{kl}^{(\mathrm{op})}$:

$$H_Q = \frac{1}{6}\sum_{kl} Q_{kl}^{(\mathrm{op})} V_{kl} \qquad [5]$$

The degeneracy of the nuclear ground state with a spin I is $2I + 1$. H_Q may be treated as a small

Table 1 Nuclear constants and natural abundance of naturally occurring quadrupolar nuclei

Isotope	Natural abundance (%)	Spin	Magnetic dipole moment (μn)	NMR Frequency in $B = 1$ T (MHz)	Electrical quadrupole moment Q ($10^{-28}m^2$)
^{2}H	0.0156	1	+0.8574	6.54	0.00286
^{6}Li	7.43	1	+0.8220	6.27	−0.00083
^{7}Li	92.57	3/2	+3.2564	16.55	−0.0406
^{9}Be	100	3/2	−1.1778	5.98	+0.053
^{10}B	18.83	3	+1.8006	4.58	+0.08472
^{11}B	81.17	3/2	+2.6886	13.66	+0.04065
^{14}N	99.64	1	+0.4038	3.08	+0.0193
^{17}O	0.037	5/2	−1.8938	5.77	−0.02578
^{21}Ne	0.257	3/2	−0.6618	3.36	+0.103
^{23}Na	100	3/2	+2.2175	11.26	+0.1006
^{25}Mg	10.05	5/2	−0.8555	2.61	+0.201
^{27}Al	100	5/2	+3.6415	11.09	+0.150
^{33}S	0.74	3/2	+0.6438	3.27	−0.076
^{35}Cl	75.4	3/2	+0.8219	4.17	−0.08249
^{37}Cl	24.6	3/2	+0.6841	3.47	−0.06493
^{39}K	93.08	3/2	+0.3915	1.99	+0.049
^{41}K	6.91	3/2	+0.2149	1.09	+0.060
^{43}Ca	0.13	7/2	−1.3176	2.87	−0.049
^{45}Sc	100	7/2	+4.7565	10.33	−0.22
^{47}Ti	7.75	5/2	−0.7885	2.40	+0.29
^{49}Ti	5.51	7/2	−1.1042	2.40	+0.24
^{50}V	0.24	6	+3.3457	4.25	+0.209
^{51}V	99.76	7/2	+5.1487	11.19	−0.052
^{53}Cr	9.54	3/2	−0.4754	2.41	−0.15
^{55}Mn	100	5/2	+3.4532	10.55	+0.33
^{59}Co	100	7/2	+4.627	10.10	+0.404
^{61}Ni	1.25	3/2	−0.7500	3.81	+0.162
^{63}Cu	69.09	3/2	+2.2233	11.29	−0.211
^{65}Cu	30.91	3/2	+2.3817	12.09	−0.195
^{67}Zn	4.12	5/2	+0.8755	2.66	+0.150
^{69}Ga	60.2	3/2	+2.0166	10.22	+0.168
^{71}Ga	39.8	3/2	+2.5623	12.98	+0.106
^{73}Ge	7.61	9/2	−0.8795	1.49	−0.173
^{75}As	100	3/2	+1.4395	7.29	+0.314
^{79}Br	50.57	3/2	+2.1064	10.67	+0.331
^{81}Br	49.43	3/2	+2.2706	11.50	+0.276
^{83}Kr	11.55	9/2	−0.9707	1.64	+0.253
^{85}Rb	72.8	5/2	+1.3534	4.11	+0.23
^{87}Rb	27.2	3/2	+2.7518	13.39	+0.127
^{87}Sr	7.02	9/2	−1.0936	1.85	+0.335
^{91}Zr	11.23	5/2	−1.3036	3.96	−0.206
^{93}Nb	100	9/2	+6.1705	10.41	−0.32
^{95}Mo	15.78	5/2	−0.9142	2.77	−0.022

Table 1 *Continued*

Isotope	Natural abundance (%)	Spin	Magnetic dipole moment (μn)	NMR Frequency in $B = 1$ T (MHz)	Electrical quadrupole moment Q ($10^{-28}m^2$)
^{97}Mo	9.60	5/2	−0.9335	2.83	+0.255
^{99}Ru	12.81	5/2	−0.641	1.95	+0.079
^{101}Ru	16.98	5/2	−0.7188	2.19	+0.457
^{105}Pd	22.23	5/2	−0.642	1.96	+0.660
^{113}In	4.16	9/2	+5.5289	9.31	+0.799
^{115}In	95.84	9/2	+5.5408	9.33	+0.86
^{121}Sb	57.25	5/2	+3.3634	10.25	−0.36
^{123}Sb	42.75	7/2	+2.5498	5.52	−0.49
^{127}I	100	5/2	+2.8133	8.52	−0.79
^{131}Xe	21.24	3/2	+0.6919	3.49	−0.120
^{133}Cs	100	7/2	+2.5820	5.59	−0.00371
^{135}Ba	6.59	3/2	+0.8379	4.23	+0.160
^{137}Ba	11.32	3/2	+0.9374	4.73	+0.245
^{138}La	0.089	5	+3.7136	5.62	+0.45
^{139}La	99.911	7/2	+2.7830	6.01	+0.20
^{141}Pr	100	5/2	+4.2754	13.04	−0.0589
^{143}Nd	12.20	7/2	−1.065	2.32	−0.63
^{145}Nd	8.30	7/2	−0.656	1.43	−0.33
^{147}Sm	15.07	7/2	−0.8148	1.77	−0.26
^{149}Sm	13.84	7/2	−0.6717	1.46	+0.075
^{151}Eu	47.77	5/2	+3.4717	10.59	+0.903
^{153}Eu	52.23	5/2	+1.5330	4.64	+2.412
^{155}Gd	14.68	3/2	−0.2591	1.32	+1.30
^{157}Gd	15.64	3/2	−0.3398	1.73	+1.36
^{159}Tb	100	3/2	−2.014	10.23	+1.432
^{161}Dy	18.73	5/2	−0.4803	1.46	+2.507
^{163}Dy	24.97	5/2	+0.6726	2.05	+2.648
^{165}Ho	100	7/2	+4.132	9.00	+3.58
^{167}Er	22.82	7/2	−0.5639	1.23	+3.565
^{173}Yb	16.08	5/2	−0.6799	2.07	+2.80
^{175}Lu	97.40	7/2	+2.2327	4.86	+3.49
^{176}Lu	2.60	7	+3.1692	3.45	+4.92
^{177}Hf	18.39	7/2	+0.7935	1.73	+3.365
^{179}Hf	13.78	9/2	−0.6409	1.09	+3.79
^{181}Ta	99.99	7/2	+2.3705	5.16	+3.28
^{185}Re	37.07	5/2	+3.1871	9.59	+2.18
^{187}Re	62.93	5/2	+3.2197	9.68	+2.07
^{189}Os	16.1	3/2	+0.6599	3.31	+0.856
^{191}Ir	38.5	3/2	+0.1507	0.81	−0.816
^{193}Ir	61.5	3/2	+0.1637	0.83	+0.751
^{197}Au	100	3/2	+0.1457	0.73	+0.547
^{201}Hg	13.24	3/2	−0.5602	2.85	+0.385
^{209}Bi	100	9/2	+4.1106	6.84	−0.37
^{235}U	0.71	7/2	−0.38	0.83	+4.55

perturbation that at least partially removes the degeneracy of the nuclear ground state. To calculate the energies of the nuclear quadrupole energy levels and the corresponding NQR frequencies, it is necessary to evaluate the matrix elements $\langle I,m \mid H_Q \mid I,m' \rangle$. Here I is the nuclear spin and m and m' are the magnetic quantum numbers. To evaluate the above matrix elements of H_Q, it is necessary to evaluate the matrix elements

$$\left\langle I, m | Q_{kl}^{(\mathrm{op})} | I, m' \right\rangle \quad [6]$$

of the quadrupole moment tensor. Using the Wigner–Eckart theorem, we may replace the matrix elements given by Equation [6] by the matrix elements

$$\langle I, m | \tilde{Q}_{kl} | I, m' \rangle = \left\langle I, m \middle| C \left(\frac{3}{2}(I_k I_l + I_l I_k) - \delta_{kl} \boldsymbol{I}^2 \right) \middle| I, m' \right\rangle \quad [7]$$

of a symmetric, traceless, second rank tensor $\tilde{Q}$ composed of the components of the nuclear angular momentum vector $\boldsymbol{I}$. We define a scalar constant eQ called the nuclear quadrupole moment as $eQ = \langle I,I \mid \tilde{Q}_{zz} \mid I,I \rangle$ and express the proportionality C as $C = eQ/I(2I-1)$. The quadrupole Hamiltonian may thus be expressed as

$$H_Q = \frac{eQ}{6I(2I-1)} \sum_{kl} V_{kl} \left[\frac{3}{2}(I_k I_l + I_l I_k) - \delta_{kl} \boldsymbol{I}^2 \right] \quad [8]$$

The elements V_{kl} depend on the choice of the coordinate system. The simplest situation arises in a coordinate system where the coordinate axes point along the principal directions of the EFG tensor. In this coordinate system the EFG tensor is diagonal. Let us denote the three principal values of the EFG tensor as V_{XX}, V_{YY} and V_{ZZ} ($\mid V_{XX} \mid \leq \mid V_{YY} \mid \leq \mid V_{ZZ} \mid$)| and the corresponding three principal directions as X, Y and Z. The three principal values of the traceless EFG tensor are not independent. They can be described by two parameters:

$$eq = V_{ZZ} \quad \text{and} \quad \eta = (V_{XX} - V_{YY})/V_{ZZ} \quad [9]$$

The quantity eq is thus the largest principal value of the EFG tensor measuring its magnitude, whereas the asymmetry parameter η measures the departure of the EFG tensor from axial symmetry. The asymmetry parameter η ranges between 0 and 1. Let us further choose the principal axis Z as the quantization axis and rewrite the quadrupole Hamiltonian as

$$H_Q = \frac{e^2 qQ}{4I(2I-1)} \left[3I_Z^2 - \boldsymbol{I}^2 + \frac{\eta}{2}(I_+^2 + I_-^2) \right] \quad [10]$$

The quantity e^2qQ divided by the Planck constant h is called the quadrupole coupling constant. It measures the magnitude of the nuclear quadrupole interaction. The NQR frequencies of a given nucleus depend on its spin I and on two parameters, e^2qQ/h and η, related to the EFG tensor.

In solid-state NMR of quadrupolar nuclei, H_Q often represents a perturbation of the Zeeman interaction. It is then natural to choose the quantization axis z along the direction of the external magnetic field and rewrite Equation [8] in a convenient form:

$$H_Q = \frac{eQ}{4I(2I-1)} \left[V_0(3I_z^2 - \boldsymbol{I}^2) + V_{+1}(I_- I_z + I_z I_-) + V_{-1}(I_+ I_z + I_z I_+) + V_{+2} I_-^2 + V_{-2} I_+^2 \right] \quad [11]$$

Here

$$V_0 = V_{zz}, \; V_{\pm 1} = V_{xz} \pm \mathrm{i} V_{yz}$$

and

$$V_{\pm 2} = (V_{xx} - V_{yy})/2 \pm \mathrm{i} V_{xy}$$

Nuclear quadrupole energy levels and resonance frequencies

It is obvious from Equation [10] that H_Q mixes the states with $\Delta m = 2$ when $\eta \neq 0$ and the quantization axis points along the Z principal axis of the EFG tensor. We shall from now on assume that the quantization axis points along the Z principal axis of the EFG tensor and expand the eigenstates $\mid \psi \rangle$ of H_Q in the representation of the eigenstates $\mid I, m \rangle$ of I_Z. The set of states $\mid I, m \rangle$ can first be divided into two subsets: (a) $\mid I, I \rangle$, $\mid I, I-2 \rangle$, etc. and (b) $\mid I, I-1 \rangle$, $\mid I, I-3 \rangle$, etc.

Each eigenstate of H_Q is represented either by the states of the subset (a) or by the states of the subset (b). H_Q is also invariant to time inversion. Let $\mid \psi, t \rangle$ be an eigenstate of H_Q. Owing to the time-inversion symmetry of H_Q the state $\mid \psi, -t \rangle$ obtained after the inversion of time is also an eigenstate of H_Q with the same energy as $\mid \psi, t \rangle$. A state $\mid I, m \rangle$ transforms under the time inversion into the state $(-1)^{I-m} \| I, -m\rangle$. In the case of an integer spin nucleus the states $|I\ m\rangle$ and $|I, -m\rangle$ belong to the same subset of states and $|\psi,-t\rangle$ is generally equal to $|\psi, t\rangle$ multiplied by a phase factor. The energy levels are thus nondegenerate, except for $\eta = 0$. In the case of half-integer spin nuclei the states $|I, m\rangle$ and $|I, -m\rangle$ belong to two different subsets of states and $|\psi, -t\rangle$ is different from $|\psi, t\rangle$. The energy levels are generally doubly degenerate. This is called Kramers degeneracy. The inhomogeneous electric field thus completely removes the degeneracy of the nuclear ground state only for integer spin nuclei when $\eta \neq 0$. In case of a half-integer spin nucleus with a spin I, we observe only $I + 1/2$ doubly degenerate nuclear quadrupole energy levels.

We shall first consider nuclei with an integer spin 1, 2 and 3. In all these cases the energies of the nuclear quadrupole energy levels can be calculated analytically.

Spin 1

The energies E of the three nuclear quadrupole energy levels and the expansion coefficients c^m of the corresponding eigenstates $|\psi\rangle = \Sigma_m C^m |1, m\rangle$, in the representation of the eigenstates of I_z are given in **Table 2**.

The upper energy level in the case of $\eta = 0$ is doubly degenerate and only one NQR frequency, $\nu_Q = 3e^2qQ/4h$, is observed. For a nonzero η, the three nuclear quadrupole less resonance frequencies ν_+, ν_- and ν_0 $(\nu_+ \geq \nu_- > \nu_0)$ are

$$\begin{aligned}\nu_+ &= \frac{e^2qQ}{4h}(3+\eta)\\ \nu_- &= \frac{e^2qQ}{4h}(3-\eta)\\ \nu_0 &= \nu_+ - \nu_- = \frac{e^2qQ}{2h}\eta\end{aligned} \quad [12]$$

The energies of the nuclear quadrupole energy levels and the NQR frequencies as functions of η are shown in **Figure 1**.

Table 2 Energies E in units of $e^2qQ/4$ and the expansion coefficients c^m of the eigenstates of $\boldsymbol{H}_Q$ for a nucleus with $I = 1$ in the representation of the eigenstates of $\boldsymbol{I}_z$

E	c^1	c^0	c^{-1}
$1+\eta$	$1/\sqrt{2}$	0	$1/\sqrt{2}$
$1-\eta$	$1/\sqrt{2}$	0	$1/\sqrt{2}$
−2	0	1	0

None of the NQR transitions is generally forbidden except for special orientations of the radiofrequency magnetic field. The NQR line at $\nu = \nu_+$ vanishes when the resonant radiofrequency magnetic field points perpendicularly to the principal direction X. Similarly, the NQR line at $\nu = \nu_-$ vanishes when the resonant radiofrequency magnetic field points perpendicularly to the principal direction Y, and the NQR line at $\nu = \nu_0$ vanishes when the resonant radiofrequency magnetic field points perpendicularly to the principal direction Z. In a powder sample all three NQR lines can be observed, whereas in a single crystal the orientation dependence of the intensity of the NQR lines may be used to determine the orientation of the principal axes of the EFG tensor in a crystal-fixed coordinate system. The quadrupole coupling constant is usually calculated from the NQR frequencies as $e^2qQ/h = 2(\nu_+ + \nu_-)/3$, and the asymmetry parameter η is equal to $\eta = 3(\nu_+ - \nu_-)/(\nu_+ + \nu_-)$.

Spin 2 and 3

The energies of the nuclear quadrupole energy levels and expansion coefficients c^m of the corresponding eigenstates of H_Q in the representation of the eigenstates of I_Z, $|I, m\rangle$ for spins 2 and 3 are given in **Table 3** and **Table 4**, respectively.

Nuclei with an integer spin larger than 1 are seldom observed in practice. **Tables 3** and **4** are therefore included only for completeness.

Spin $\frac{3}{2}$

As seen from **Table 1**, a half-integer nuclear spin is in practice much more common than an integer nuclear spin. Nuclei with a half-integer spin are often observed in practice. As already mentioned, the nuclear quadrupole energy levels of the half-integer spin nuclei are generally doubly degenerate. The two eigenstates of H_Q, $|\psi_+\rangle$ and $|\psi_-\rangle$, corresponding to the same doubly degenerate energy level are generally expressed as

$$\begin{aligned}|\Psi_+\rangle &= \sum_{k=0}^{I-1/2} c^k |I, I-2k\rangle\\ |\Psi_-\rangle &= \sum_{k=0}^{I-1/2} c^k |I, -I+2k\rangle\end{aligned} \quad [13]$$

The energies of the nuclear quadrupole energy levels and the corresponding eigenstates of H_Q can in the general case $(\eta \neq 0)$ be expressed analytically

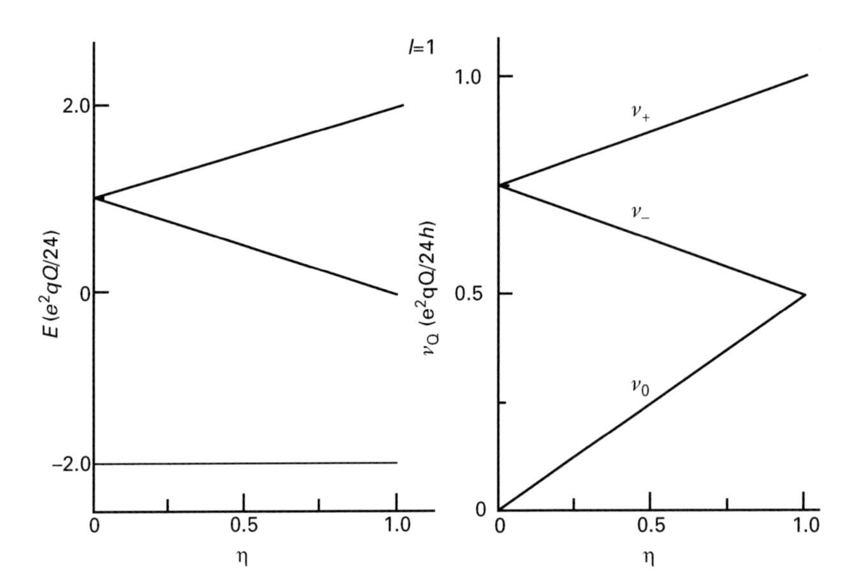

Figure 1 Energy levels and NQR frequencies for $I = 1$.

only for $I = \frac{3}{2}$, where

$$E_{\pm 3/2} = \pm \frac{e^2qQ}{4}\sqrt{1 + \frac{\eta^2}{3}} \quad [14]$$

and the eigenstates of H_Q are

$$\left|\Psi_{\pm 3/2}\right\rangle = \sqrt{\frac{z+3}{2z}}\left|\frac{3}{2}, \pm\frac{3}{2}\right\rangle + \sqrt{\frac{z-3}{2z}}\left|\frac{3}{2}, \mp\frac{1}{2}\right\rangle$$

$$\left|\Psi_{\pm 1/2}\right\rangle = -\sqrt{\frac{z-3}{2z}}\left|\frac{3}{2}, \pm\frac{3}{2}\right\rangle + \sqrt{\frac{z+3}{2z}}\left|\frac{3}{2}, \mp\frac{1}{2}\right\rangle$$

$$z = \sqrt{9 + 3\eta^2} \quad [15]$$

Only one NQR frequency ν_Q,

$$\nu_Q = \frac{e^2qQ}{2h}\sqrt{1 + \frac{\eta^2}{3}} \quad [16]$$

is observed in this case. The quadrupole coupling constant e^2qQ/h and the asymmetry parameter η cannot be determined separately from the NQR frequency. The problem is usually solved by the application of a weak magnetic field or by the application of two-dimensional NQR techniques.

Spin $\frac{5}{2}$

The energies E of the three nuclear quadrupole energy levels are obtained from the secular equation

$$x^3 - 7(3 + \eta^2)x - 20(1 - \eta^2) = 0 \quad [17]$$

Here energy E is given as $E = (e^2qQ)/20x$, where x is a solution of the secular equation. The energies are usually labelled as E_m, where m is the magnetic quantum number which can be assigned to a given energy level when $\eta = 0$. The three NQR frequencies are labelled as $\nu_{5/2-1/2}, \nu_{5/2-3/2}$ and $\nu_{3/2-1/2}$ ($\nu_{5/2-1/2} > \nu_{5/2-3/2} \geq \nu_{3/2-1/2}$). The energies E_m, and the NQR frequencies are shown in **Figure 2**.

The NQR line at the frequency $\nu_{5/2-1/2}$, $\nu_{5/2-1/2} = \nu_{5/2-3/2} + \nu_{3/2-1/2}$ is generally weaker than the other two NQR lines and cannot be observed when $\eta = 0$. The asymmetry parameter η is in practice calculated from the ratio $R = \nu_{3/2-1/2}/\nu_{5/2-3/2}$ which ranges from $R = 0.5$ for $\eta = 0$ to $R = 1$ for $\eta = 1$. When η is known, the quadrupole coupling constant can be calculated from any NQR frequency, most precisely from the highest NQR frequency $\nu_{5/2-1/2}$.

Table 3 Energies of the nuclear quadrupole energy levels in units of $e^2qQ/8$ and the expansion coefficients c^m for $I = 2$

E	c^2	c^1	c^0	c^{-1}	c^{-2}
$2z$	$\sqrt{\frac{z+1}{4z}}$	0	$\sqrt{\frac{z-1}{2z}}$	0	$\sqrt{\frac{z+1}{4z}}$
2	$\frac{1}{\sqrt{2}}$	0	0	0	$-\frac{1}{\sqrt{2}}$
$-(1-\eta)$	0	$\frac{1}{\sqrt{2}}$	0	$\frac{1}{\sqrt{2}}$	0
$-(1+\eta)$	0	$\frac{1}{\sqrt{2}}$	0	$-\frac{1}{\sqrt{2}}$	0
$-2z$	$-\sqrt{\frac{z-1}{4z}}$	0	$\sqrt{\frac{z+1}{2z}}$	0	$-\sqrt{\frac{z-1}{4z}}$

$z = \sqrt{1 + \eta^2/3}$

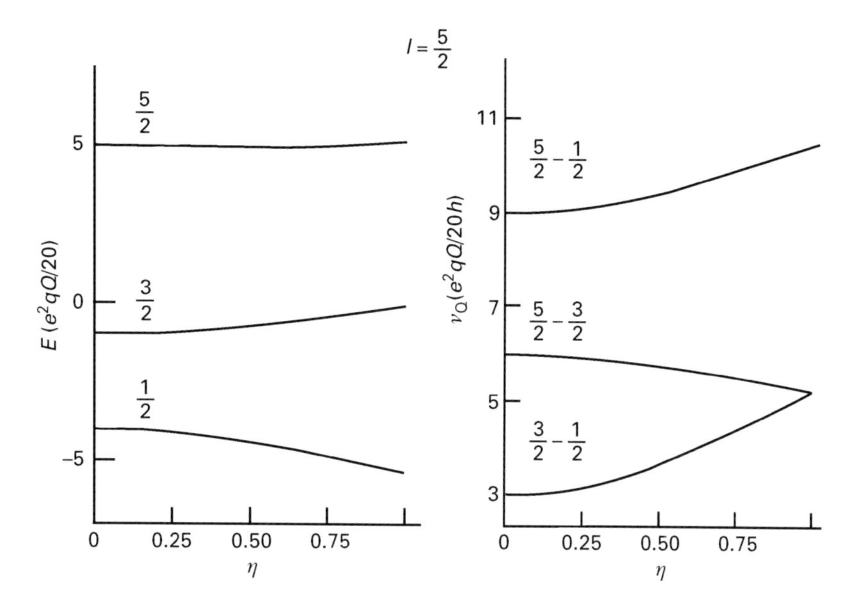

Figure 2 Energy levels and NQR frequencies for $I = \frac{5}{2}$.

Spin $\frac{7}{2}$

The four energies E of the nuclear quadrupole energy levels are calculated from the secular equation

$$x^4 - 42\left(1+\frac{\eta^2}{3}\right)x^2 - 64(1-\eta^2)x + 105\left(1+\frac{\eta^2}{3}\right)^2 = 0 \qquad [18]$$

where $E = e^2qQx/28$. They are again labelled as E_m, $m = \frac{7}{2}, \frac{5}{2}, \frac{3}{2}, \frac{1}{2}$. The dependence of the energies E_m and of the NQR frequencies $\nu_{m-(m-1)} = (E_m - E_{m-1})/h$ on the symmetry parameter η is shown in **Figure 3**.

The three NQR frequencies corresponding to $\Delta m = 1$ give the strongest NQR signals. The NQR signals at the frequencies corresponding to $\Delta m = 2$ and $\Delta m = 3$ are also observed for large values of η, but their intensities are lower than the intensities of the NQR lines corresponding to $\Delta m = 1$. As seen from **Figure 3**, the NQR frequency $\nu_{3/2-1/2}$ depends strongly on η, whereas the η-dependence of the

Table 4 Energies of the eigenstates of H_Q in units of $e^2qQ/20$ and the expansion coefficients c^m for $I = 3$

E	c^3	c^2	c^1	c^0	c^{-1}	c^{-2}	c^{-3}
$1+\eta+4x$	$\sqrt{\frac{4x+4-\eta}{16x}}$	0	$\sqrt{\frac{4x-4+\eta}{16x}}$	0	$\sqrt{\frac{4x-4+\eta}{16x}}$	0	$\sqrt{\frac{4x+4-\eta}{16x}}$
$1-\eta+4y$	$\sqrt{\frac{4y+4+\eta}{16y}}$	0	$\sqrt{\frac{4y-4-\eta}{16y}}$	0	$-\sqrt{\frac{4y-4-\eta}{16y}}$	0	$-\sqrt{\frac{4y+4+\eta}{16y}}$
$2z-2$	0	$\sqrt{\frac{z+1}{4z}}$	0	$\sqrt{\frac{z-1}{2z}}$	0	$\sqrt{\frac{z+1}{4z}}$	0
0	0	$\frac{1}{\sqrt{2}}$	0	0	0	$-\frac{1}{\sqrt{2}}$	0
$1+\eta-4x$	$-\sqrt{\frac{4x-4+\eta}{16x}}$	0	$\sqrt{\frac{4x+4-\eta}{16x}}$	0	$\sqrt{\frac{4x+4-\eta}{16x}}$	0	$-\sqrt{\frac{4x-4+\eta}{16x}}$
$1-\eta-4y$	$-\sqrt{\frac{4y-4-\eta}{16y}}$	0	$\sqrt{\frac{4y+4+\eta}{16y}}$	0	$-\sqrt{\frac{4y+4+\eta}{16y}}$	0	$\sqrt{\frac{4y-4-\eta}{16y}}$
$-2-2z$	0	$-\sqrt{\frac{z-1}{4z}}$	0	$\sqrt{\frac{z+1}{2z}}$	0	$-\sqrt{\frac{z-1}{4z}}$	0

$x = \sqrt{1-\frac{\eta}{2}+\frac{\eta^2}{6}}$, $y = \sqrt{1+\frac{\eta}{2}+\frac{\eta^2}{6}}$, $z = \sqrt{1+\frac{5\eta^2}{3}}$

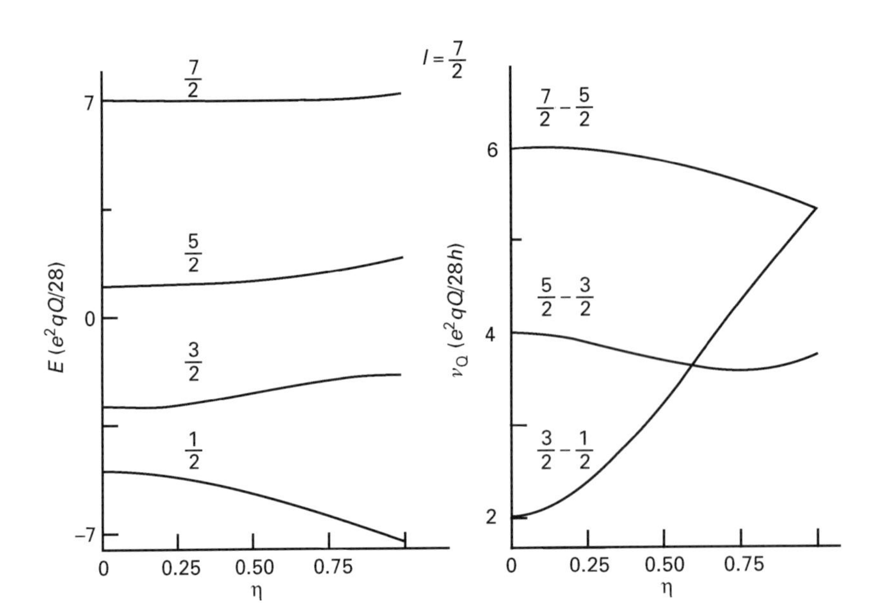

Figure 3 Energy levels and NQR frequencies $\nu_{m-(m-1)}$ for $I = \frac{7}{2}$.

other two NQR frequencies is weaker. The asymmetry parameter η is in practice determined either from the ratio $\nu_{3/2-1/2} / \nu_{5/2-3/2}$ or from the ratio $\nu_{3/2-1/2}/\nu_{7/2-5/2}$. When η is known, the quadrupole coupling constant is calculated from any NQR frequency, most precisely from the highest NQR frequency observed.

Spin $\frac{9}{2}$

The highest half-integer nuclear spin of a stable nucleus is $I = 9/2$. The energy E of a nuclear quadrupole energy level is given as $E = e^2qQx/24$, where x is a solution of the secular equation

$$x^5 - 11(3+\eta^2)x^3 - 44(1-\eta^2)x^2 + \frac{44}{3}(3+\eta^2)^2x + 48(3+\eta^2)(1-\eta^2) = 0 \quad [19]$$

The energies of the nuclear quadrupole energy levels are again labelled as E_m, with m being the magnetic quantum number assigned to a quadrupole energy level when $\eta = 0$. The dependence of the energies E_m and of the NQR frequencies $\nu_{m-(m-1)}$ on the asymmetry parameter η is shown in **Figure 4**.

The lowest NQR frequency $\nu_{3/2-1/2}$ also in this case exhibits the strongest dependence on η. The asymmetry parameter η is in practice determined from a ratio of the NQR frequencies, say $\nu_{3/2-1/2} / \nu_{5/2-3/2}$. When η is known, the quadrupole coupling constant e^2qQ/h is calculated from any NQR frequency.

Application of a weak magnetic field: Zeeman perturbed NQR

A weak static magnetic field is often used in NQR. In a powder sample it may cause broadening of a NQR line and consequently the disappearance of a NQR signal. In a single crystal a weak external magnetic field removes the degeneracy of the doubly degenerate quadrupolar energy levels. In the case of a half-integer quadrupolar nucleus, each NQR line splits into a quartet. The splitting depends on the orientation of the external magnetic field in the principal coordinate system of the EFG tensor. The orientation dependence of the splitting of the NQR lines gives the orientation of the principal axes of the EFG tensor in a crystal-fixed coordinate system and, for the case $I = 3/2$, also the value of the asymmetry parameter η. When I is integer, the external magnetic field slightly shifts the resonance frequencies. The orientation dependence of the frequency shift makes it possible to determine the orientation of the principal axes of the EFG tensor in a crystal-fixed coordinate system. In both cases the multiplicity of the resonance lines in nonzero magnetic field gives the number of magnetically nonequivalent nuclei in the crystal unit cell. Here we treat in detail only the situation for two nuclear spin systems $I = 1$ and $I = \frac{3}{2}$.

Spin 1

The Hamiltonian is

$$H = H_Q - h\nu_L(\boldsymbol{In}) \quad [20]$$

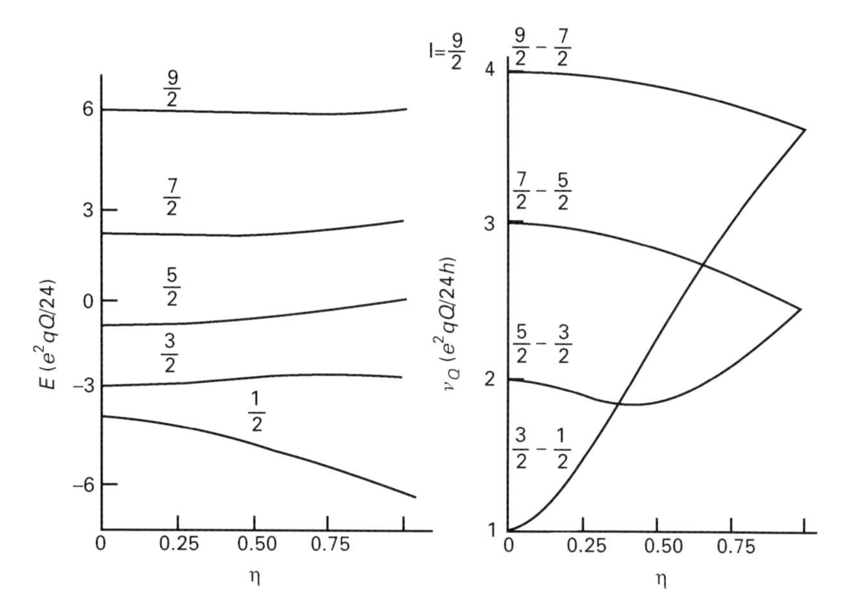

Figure 4 Energy levels and NQR frequencies $\nu_{m-(m-1)}$ for $I = \frac{9}{2}$.

Here H_Q is given by Equation [10], $\nu_L = \gamma B/2\pi$ is the Larmor frequency of a nucleus in the external magnetic field $\boldsymbol{B}$ and $\boldsymbol{n}$ is a unit vector in the direction of $\boldsymbol{B}$. We assume that the second term in equation [20] may be treated as a perturbation and that $\nu_L << \nu_0$. The eigenstates of H_Q and their energies are given in **Table 2** for the case of $I = 1$. The expectation value $\langle\psi|\boldsymbol{I}|\psi\rangle$ for any nondegenerated eigenstate of H_Q is identically equal to zero when I is integer. The first-order perturbation corrections of the energies are thus equal to zero. The lowest-order nonzero terms are thus the second-order terms. Let us denote the resonance frequencies of H as ν_+^*, ν_-^* and ν_0^*. Using the second-order perturbation theory, we obtain

$$\begin{aligned}\nu_+^* &= \nu_+ + \nu_L^2\left\{\frac{1}{\nu_0} - \sin^2(\theta)\left[\left(\frac{1}{\nu_0} - \frac{2}{\nu_+}\right)\right.\right.\\ &\quad \left.\left.+\left(\frac{2}{\nu_+} - \frac{1}{\nu_-}\right)\sin^2(\theta)\right]\right\}\\ \nu_-^* &= \nu_- + \nu_L^2\left\{-\frac{1}{\nu_0} + \sin^2(\theta)\left[\left(\frac{1}{\nu_0} + \frac{1}{\nu_+}\right)\right.\right.\\ &\quad \left.\left.+\left(\frac{2}{\nu_-} - \frac{1}{\nu_+}\right)\sin^2(\theta)\right]\right\}\\ \nu_0^* &= \nu_0 + \nu_L^2\left\{\frac{2}{\nu_0} - \sin^2(\theta)\left[\left(\frac{2}{\nu_0} - \frac{1}{\nu_+}\right)\right.\right.\\ &\quad \left.\left.+\left(\frac{1}{\nu_+} + \frac{1}{\nu_-}\right)\sin^2(\theta)\right]\right\}\end{aligned} \quad [21]$$

Here θ is the angle between the unit vector $\boldsymbol{n}$ and the principal axis Z of the EFG tensor, whereas ϕ is the angle between the projection of $\boldsymbol{n}$ on the X–Y plane and the principal axis X. It is clear from Equation [21] that the magnetic shifts of the NQR frequencies depend on the orientation of the external magnetic field with respect to the principal axes of the EFG tensor. In general in a magnetic field a NQR line splits into more lines. The multiplicity of the resonance line gives the number of crystallographically equivalent but magnetically nonequivalent positions of the studied atom in the unit cell. The crystallographically equivalent nuclei occupy positions where the principal values of the EFG tensor are equal. The magnetically nonequivalent sites differ in the orientation of the principal axes of the EFG tensor. An orientation dependence of the frequency shifts makes it possible to determine for each nuclear position the orientation of the principal axes X, Y and Z in a crystal-fixed coordinate system.

In the case of a low value of η with $\nu_0 << \nu_+, \nu_-$ when the lowest NQR frequency ν_0 may be comparable to the Larmor frequency ν_L, the first-order perturbation calculation gives

$$\begin{aligned}\nu_\pm^* &= 3K \pm \sqrt{\eta^2K^2 + \nu_L^2\ \cos^2\theta}\\ \nu_0^* &= 2\sqrt{\eta^2K^2 + \nu_L^2\ \cos^2\theta}\end{aligned} \quad [22]$$

Here $K = e^2qQ/4h$. A splitting of the upper energy level, and consequently three resonance frequencies, are observed even when $\eta = 0$ if $B \neq 0$.

Spin $\frac{3}{2}$

The Hamiltonian is given by Equation [20] where the second term is treated as a perturbation. The energies and eigenstates of the unperturbed Hamiltonian H_Q

are given by Equations [14] and [15]. By first-order perturbation theory, the upper energy level splits into two energy levels with the energies $E_{3/2} \pm h\delta_{3/2}$ where

$$\delta_{3/2} = \frac{\nu_L}{2z}\sqrt{\begin{array}{l}(z+6)^2 + 3(z^2 - 6z - 18)\sin^2\theta) \\ +6\eta(z-3)\sin^2(\theta)\cos(2\phi)\end{array}} \quad [23]$$

Here $z = \sqrt{(9 + 3\eta^2)}$. The lower energy level splits into two energy levels with the energies $E_{1/2} \pm h\delta_{1/2}$ where

$$\delta_{1/2} = \frac{\nu_L}{2z}\sqrt{\begin{array}{l}(z-6)^2 + 3(z^2 + 6z - 18)\sin^2(\theta) \\ -6\eta(z+3)\sin^2(\theta)\cos(2\phi)\end{array}} \quad [24]$$

The NQR line splits a magnetic field into four lines. In a crystal a set of four lines is observed for each magnetically nonequivalent atomic site. From the orientation dependence of the splitting it is again possible to determine the orientation of the principal axes of the EFG tensor in a crystal-fixed coordinate system.

In practice, NQR spectroscopists often locate zero splitting where the frequency of two inner resonance lines is equal to ν_θ. This happens when $\delta_{3/2} = \delta_{1/2}$ or, consequently,

$$3\cos^2(\theta_0) - 1 + \eta\sin^2(\theta_0)\cos(2\phi) = 0 \quad [25]$$

For zero η, zero splitting is observed when $\boldsymbol{B}$ lies on the surface of a cone around the principal axis Z with the angle $\theta_0 = -\boldsymbol{B}, Z = 54.44°$.

For nonzero η, the directions of the external magnetic field which leave an unsplit component at ν_θ form an elliptical cone around the Z axis. The angle θ_0 is maximum when $\phi = 0°(\boldsymbol{B}\perp Y)$ and minimum when $\phi = 90°(\boldsymbol{B}\perp X)$. From an experimentally determined elliptical cone it is then easy to find the directions of the principal axes of the EFG tensor. The knowledge of $\theta_0(\phi = 0°)$ and $\theta_0(\phi = 90°)$ makes it possible to calculate the asymmetry parameter η:

$$\eta = \frac{3[\sin^2\theta_0(0°) - \sin^2\theta_0(90°)]}{\sin^2\theta_0(0°) + \sin^2\theta_0(90°)} \quad [26]$$

Many of the features, of the $I = \frac{3}{2}$ case also appear in cases of $I = \frac{5}{2}, \frac{7}{2}$ and $\frac{9}{2}$. However, in these cases the asymmetry parameter η is determined simply from the ratio of the NQR frequencies, whereas the external magnetic field is used to determine the number of magnetically nonequivalent nuclei in the unit cell and the orientation of the principal axes of the EFG tensor in a crystal-fixed coordinate system.

List of symbols

c^m = expansion coefficients of eigenstates of the quadrupolar Hamiltonian; E_m = nuclear quadrupolar energy levels; I = nuclear spin quantum number; m = nuclear magnetic quantum number; Q_{kl} = nuclear quadrupole moment tensor; η = asymmetry parameter; $V(\boldsymbol{r})$ = electric potential; V_{kl} = electric field gradient tensor; ν = NQR frequency; ν_L = Larmor frequency; $\rho(\boldsymbol{r})$ = nuclear electric charge distribution.

See also: **NMR Principles; NQR Applications; Nuclear Quadrupole Resonance, Instrumentation.**

Further reading

Das TP and Hahn EL (1958) Nuclear quadrupole resonance spectroscopy. In F. Seitz and D. Turnbull (eds) *Solid State Physics* vol. 5. New York: Academic Press.

Grechishkin VS (1973) *Jadernie Kvadrupolnie Vzaimodejstvija v Tverdih Telah*. Moscow: Nauka.

Lucken EAC (1969) *Nuclear Quadrupole Coupling Constants*. New York: Academic Press.

Raghavan P (1989) Table of nuclear moments. *Atomic Data and Nuclear Data Tables* **42**: 189–291.

Schempp E and Bray PJ (1970) Nuclear quadrupole resonance spectroscopy. In Henderson D (ed) *Physical Chemistry*, pp 521–632. New York: Academic Press.

Semin GK Babushkina TA and Jakobson GG (1972) *Primenenie Jadernogo Kvadrupolnogo Rezonansa v Himii*. Leningrad: Himija. [English translation (1975): *Nuclear Quadrupole Resonance in Chemistry*. New York: Wiley.

Slichter CP (1963) *Principles of Magnetic Resonance*. New York: Harper & Row.

Nucleic Acids and Nucleotides Studied Using Mass Spectrometry

Tracey A Simmons, **Kari B Green-Church** and **Patrick A Limbach**, Louisiana State University, Baton Rouge LA, USA

Mass spectrometry is a powerful tool for the characterization of biomolecules including nucleotides, oligonucleotides and nucleic acids. The advantages of mass spectrometry are high sensitivity, high mass accuracy and, more importantly, structural information. Historically, oligonucleotides and nucleic acids have proved difficult to characterize using mass spectrometry. Problems often arise with impure samples, low ion abundance for analysis owing to inefficient ionization processes and low mass accuracy for higher-molecular-mass compounds. Advances in the development of electrospray ionization and matrix-assisted laser desorption/ionization now permit the analysis of oligonucleotides and nucleic acids with high sensitivity and good mass accuracy. These improvements allow the use of mass spectrometry for the identification of unknown oligonucleotide structures and are suitable for applications focused on acquiring sequence information on the samples of interest.

Nucleic acids and nucleotides

Nucleic acids are high-molecular-mass biopolymers composed of repeating units of nucleotide (nt) residues (**Figure 1**). The three major substituents of a nucleotide residue are the heterocyclic base, a sugar and a phosphate group. The five most common heterocyclic bases are adenine (Ade), cytosine (Cyt), guanine (Gua), thymine (Thy) and uracil (Ura). Cytosine, thymine and uracil are classified as pyrimidine bases, and adenine and guanine are classified as purine bases. Adenine, cytosine, guanine and thymine are the major bases found in deoxyribonucleic acids (DNA), and adenine, cytosine, guanine and uracil are the major bases found in ribonucleic acids (RNA). Nucleic acids are identified as either RNA or DNA, depending on the identity of the sugar. The sugar is a 2′-deoxy-D-ribose in DNA and a D-ribose in RNA. The phosphate group is usually attached through the 5′ or 3′ hydroxyl groups of the sugar.

Mononucleotides are typically represented by a shorthand notation that identifies the sugar, the base and the number of phosphate groups. For example, dNMP represents any 2′-deoxynucleoside monophosphate, dCDP represents 2′-deoxycytidine diphosphate and GTP represents guanosine triphosphate. Unfortunately, this convention is rarely followed in the mass spectrometry literature, where nucleosides and mononucleotides are denoted by their single letter base abbreviation and a preceding or following 'p' to represent the location of the phosphate group on the mononucleotides (for example, dC, dCp, pT and T for 2′-deoxycytidine, 2′-deoxycytidine 3′-monophosphate, thymidine 5′-monophosphate and thymidine respectively). This designation may be a source of confusion for those not familiar with this nomenclature and care should be used whenever such designations are used, especially when one wishes to distinguish nucleobases, nucleosides and nucleotides.

Oligonucleotides are most commonly joined together through the 3′ and 5′ sites of each nucleoside by a phosphodiester linkage. The primary sequence of an oligonucleotide is by definition determined from the 5′ to the 3′ end. Usually the sequence is written in shorthand notation, using the single letter abbreviations for the nucleobases; e.g., 5′-d(pTCAG)-3' as in **Figure 2**. In the case where the base composition is known, but the primary sequence is not, the unknown sequence regions are enclosed in parenthesis; e.g. 5′-dACT(GGCT)AAT-3′. Oligonucleotides of a specific length are typically referred to as '*n*-mers', where *n* is the number of oligonucleotide residues.

Modified nucleosides are designated in the overall sequence by their chemical symbol (e.g., Am for 2′-O-methyladenosine). Unknown modified nucleosides often are designated by an 'X' in the overall sequence. Modified linkages are generally identified by describing the type of modified oligonucleotide followed by the usual sequence notation: a methylphosphonate 10-mer d(ACACGTTGAC) or a phosphorothioate 12-mer d(GCGCATATGCGC). Occasionally, phosphorothiates are identified by an 's' internucleotide linkage, such as d(AsCsTsAsG).

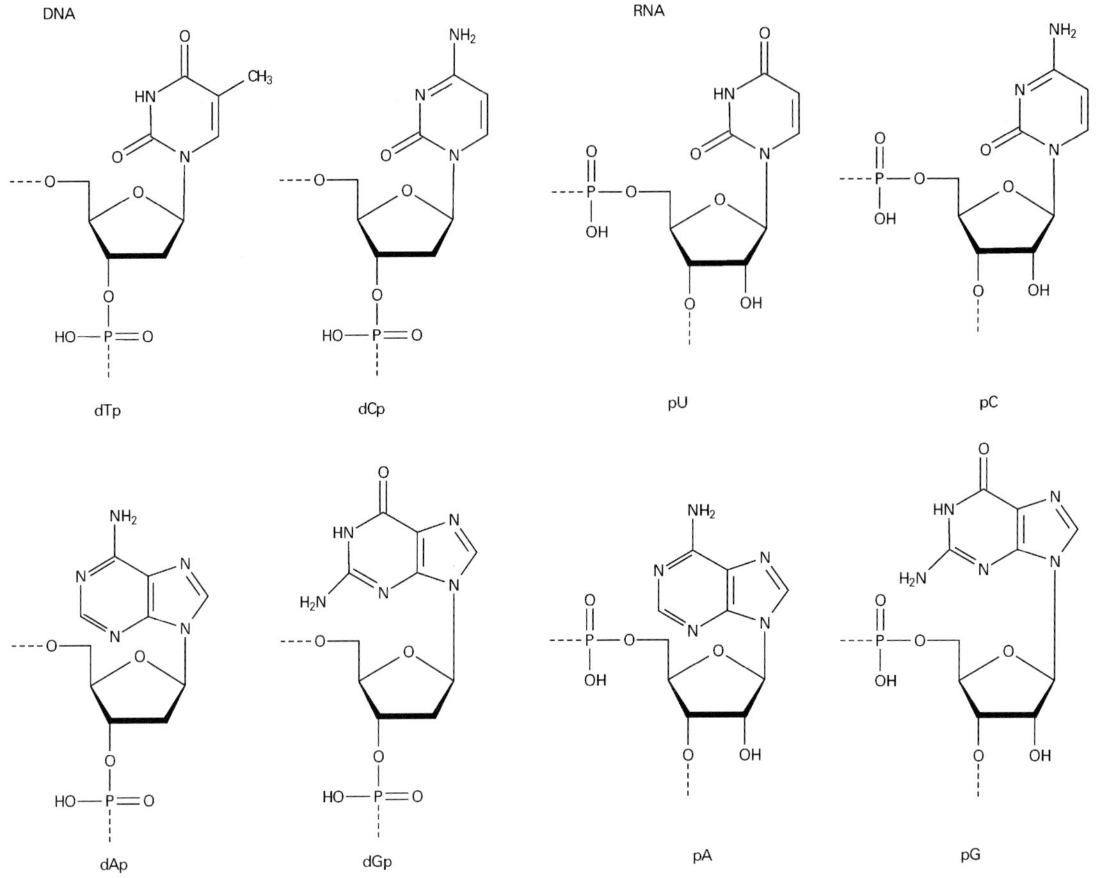

Figure 1 Subunit structures of the major nucleotides from deoxyribonucleic acid (DNA) and ribonucleic acid (RNA). T, C and U nucleobases are pyrimidine derivatives, and A and G nucleobases are purine derivatives.

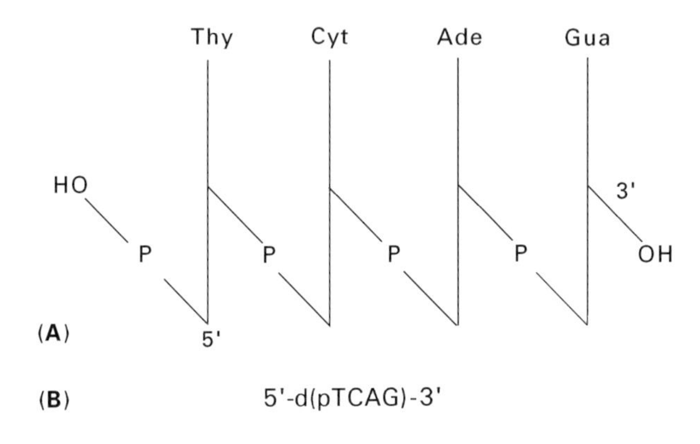

Figure 2 Shorthand notations of oligonucleotides: (A) line notation of oligonucleotides, and (B) one-letter sequence notation.

Ionization and mass analysis

Mass spectrometric measurements of oligonucleotides and nucleic acids involve the determination of molecular mass, or of the masses of dissociation products of gas-phase ions, which can then be related to various structural properties, such as sequence. In either case, the crucial step lies in the conversion of liquid- or solid-phase solutions of the analytes into gaseous ions. Common ionization sources used in mass spectrometry for nucleic acid and nucleotide experiments are fast-atom bombardment (FAB), electrospray ionization (ESI) and matrix-assisted laser desorption/ionization (MALDI). FAB was historically the ionization method of choice, but has largely been replaced by ESI or MALDI. ESI- and MALDI-based methods have particularly aided the analysis of nucleic acids because these techniques accomplish the otherwise experimentally difficult task of producing gas-phase ions from solution species that are both very highly solvated and highly ionic. Mass analysis of oligonucleotides and nucleic acids up to the 100-mer level has been demonstrated using ESI-MS. Molecular ions from significantly larger oligonucleotides (up to the 400-mer level) can be generated using MALDI-MS, but realistic mass analysis is limited to oligonucleotides and nucleic acids at the 125-mer level.

Fast-atom bombardment

Oligonucleotide samples used in FAB must be soluble in a liquid matrix and are spotted onto a probe tip. Matrices selected are liquids with low vapour pressures and include glycerol, or an equal volume mixture of dithioerythritol and dithiothreitol (the 'magic bullet' matrix). Oligonucleotide samples are typically prepared as saturated solutions in water, and are spotted on the probe tip with neat matrices, prior to instrument insertion. FAB-MS of oligonucleotides is typically performed on a sector or quadrupole mass spectrometer and is limited to oligonucleotides that are no greater than 5000 Da in mass (**Figure** 3).

Electrospray ionization

ESI requires sample introduction in liquid form and thus is convenient for the analysis of oligonucleotides. Samples are introduced into the instrument through a narrow capillary into a strong electrostatic field. Mixtures of an organic solvent and water are used in ESI-MS of oligonucleotides. Typical organic solvents are isopropanol, methanol or acetonitrile. Oligonucleotide sample concentrations for ESI experiments are prepared at the micromolar and nanomolar levels depending on the type of electrospray apparatus utilized. A mass spectrum of an oligonucleotide generated using ESI is characterized by a series of multiply charged negative ions that differ from one another by the removal of a single proton. The multiple charging effect from ESI-generated ions is advantageous for mass spectral analysis because mass analysers with limited upper mass ranges can be used (**Figure** 4).

Matrix-assisted laser desorption/ionization

For MALDI-MS, oligonucleotide samples are mixed with a UV-absorbing matrix, spotted on a sample plate, and allowed to air dry. The common oligonucleotide MALDI matrices are 3-hydroxypicolinic acid, 2,4,6-trihydroxyacetophenone and 6-aza-2-thiothymine. Sample solutions for MALDI are at the micromolar or nanomolar level, with only a small fraction of the sample utilized in the MALDI experiment. A MALDI-generated mass spectrum of an oligonucleotide typically contains a singly negatively charged peak at the mass of the deprotonated ion. The analysis of MALDI-generated oligonucleotides is limited to either time-of-flight (TOF) mass spectrometers, which have an extended upper mass range but limited mass accuracy, or Fourier transform ion cyclotron resonance (FT-ICR) mass spectrometers, which have a limited upper mass range but high mass accuracy (**Figure** 5).

Measurement of molecular mass

In ESI, both DNA and RNA are typically converted to gas-phase negative ions. Although the ionization efficiency is different for different homopolymers ($dT_{10} > dA_{10} \approx dC_{10} \approx dG_{10}$), both these and mixed bases are readily detectable. In MALDI, both DNA and RNA can be converted into gas-phase positive or negative ions. The quality of mass spectra from

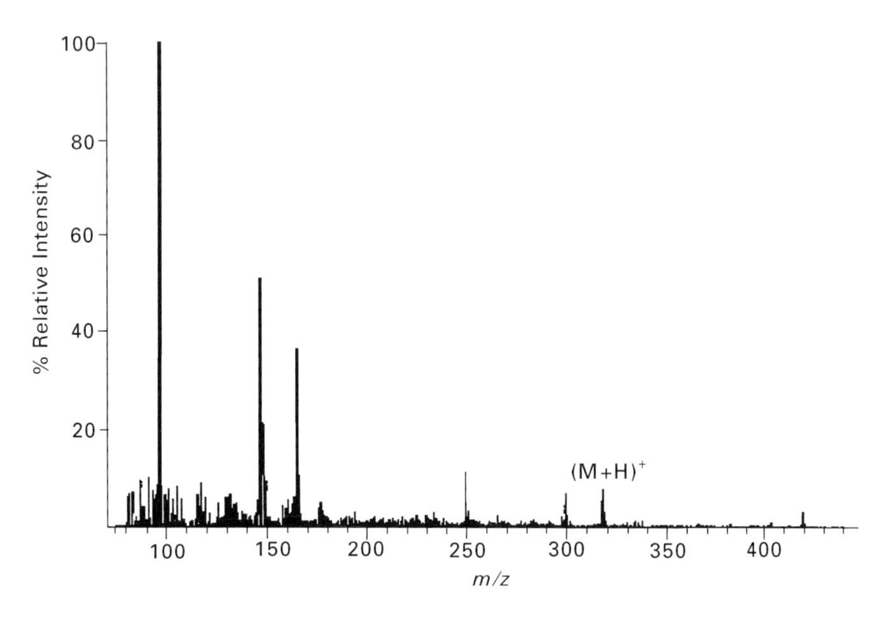

Figure 3 Positive ion mode FAB mass spectrum using a double focusing sector mass analyser. The sample is dpC with piperidine in nitrobenzyl alcohol matrix. The parent ion mass, $(M + H)^+$, is 307.3 Da.

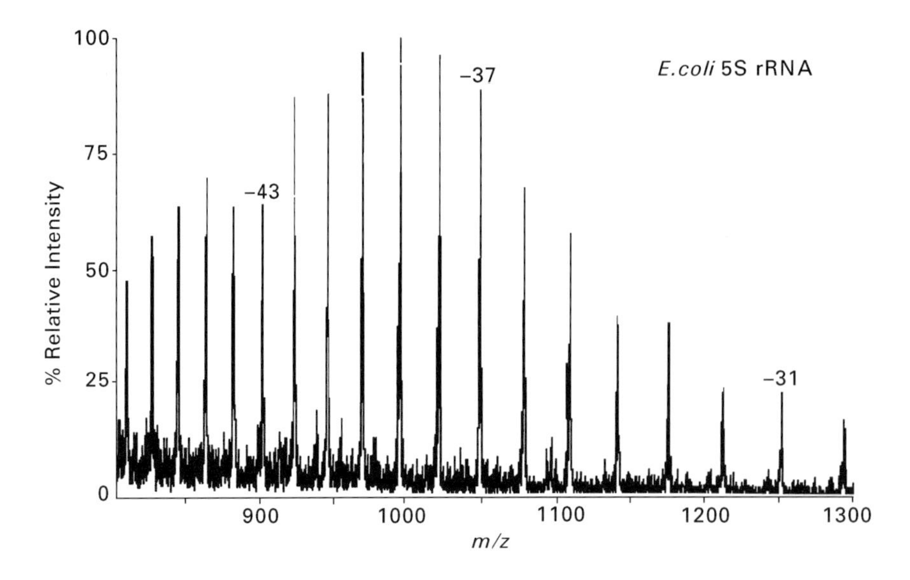

Figure 4 Negative ion mode ESI spectrum of *Escherichia coli* 5S rRNA treated with *trans*-1,2-diaminocylohexane-*N,N,N′,N′*-tetraacetic acid (CDTA) and triethylamine (TEA) additives.

MALDI analyses is profoundly influenced by base composition. Thus, polythymidylic acids are readily analysed, but polydeoxyguanylic acids are a distinct challenge and mixed-base oligomers represent an intermediate situation. As a consequence of the increased stability of the glycosidic bond in RNA, it is more readily analysed than DNA.

Relative molecular mass is an intrinsic molecular property, which, when measured with high accuracy, becomes a unique and unusually effective parameter for characterization of synthetic or natural oligonucleotides. Mass spectrometry-based methods can be broadly applied not only to normal (phosphodiester) nucleotides but also to phosphorothioates, methylphosphonates and other derivatives. The level of mass accuracy will depend on the capabilities of the mass analyser used. Quadrupole and TOF instruments yield lower mass accuracies than sector or FT-ICR instruments. High mass accuracy is necessary not only for the qualitative analysis of nucleotides present in a sample but to provide unambiguous peak identification in a mass spectrum. For example, uridine monophosphate and cytidine monophosphate are separated by 1 Da (306.26 u, and 305.25 u, respectively) and oligonucleotides differing by the number of uridine and cytidine

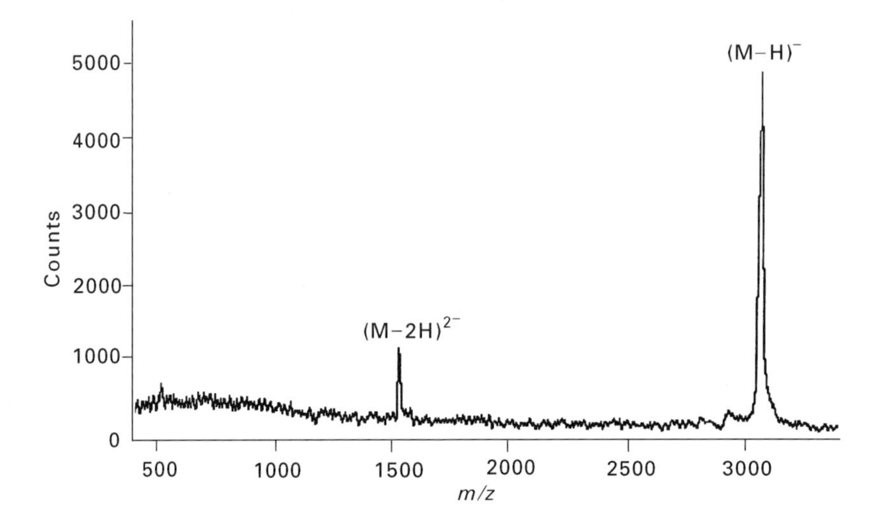

Figure 5 MALDI-TOF mass spectrum, in negative ion mode, of dA_{10} in 2,4,6-trihydroxyacetophenone matrix. The parent mass is seen as both singly and doubly charged ions.

nucleotides may be difficult to distinguish in a mixture. In such cases, the confidence of assigning a putative oligonucleotide base composition from the molecular mass measurement is dependent on the type of mass analyser used.

In ESI, mass measurement errors of 0.03% and lower are typical and, with appropriate sample clean-up, values lower than 0.01% can be obtained routinely. In contrast, for MALDI, typical measurement errors are between 0.03% and 0.05% and, with use of internal standards, mass measurement errors on the order of 0.01% can be obtained for small oligonucleotides.

The primary challenge to accurate molecular mass measurements of oligonucleotides and nucleic acids is reduction or complete removal of salt adducts. In solution, the phosphodiester backbone is completely ionized at pH > 1 and the solvent acts as a dielectric shield to reduce the repulsive Coulombic charging. During the ionization process, this Coulombic protection is lost, which eventually results in the adduction of any cations that may be present in solution. These cations (usually Na^+) reduce the polarity of this highly charged backbone, improving the production of gas-phase ions. It is these adducts that shift the centroid to higher values and limit the ability to determine molecular masses accurately.

Ammonium (or tetraethylammonium) are preferred salt forms for both ESI and MALDI. High-performance liquid chromotography (HPLC) purification is a suitable approach for sample purification, especially when ammonium buffers are used. Solvent additives, such as ethylenediaminetetraacetic acid and triethylamine, can also be added to the sample solution in ESI or MALDI to help alleviate the interference of alkali salt ions from the spectra (**Figures** 4 and 5). In MALDI, cation-exchange resin beads are effective at generating the ammonium salts of the oligonucleotides prior to mass analysis.

Determination of nucleotide sequence by gas-phase sequencing

General fragmentation processes

Dissociation of oligonucleotides can occur as a result of the excess energy that is imparted to the analyte during the desorption/ionization process. This dissociation is relatively fast (i.e. timescales shorter than the total mass spectral analysis time), resulting in ions that are generally difficult to identify accurately. Electrospray ionization produces stable, intact molecular ions except when the source conditions are adjusted to impart excess translational energy into the analyte, leading to its dissociation in the electrospray interface region (so-called nozzle–skimmer dissociation, discussed below). Thus, most dissociations that are desorption/ionization-induced are seen only in MALDI or FAB spectra. There are essentially four differing timescales for desorption/ionization-induced dissociations: prompt, fast, fast metastable and metastable.

- Prompt dissociations occur on a timescale equal to or less than the desorption event.
- Fast dissociations occur after the desorption event, but before or at the beginning of the acceleration event.
- Fast metastable decays occur on the timescale of the acceleration event.
- Metastable decays occur after the acceleration event, during the field-free flight time of the ion.

In theory, dissociations occurring during any one of these timescales will generate fragment ions that could be used to determine the sequence of the oligonucleotide. In practice, unless certain instrumental parameters are manipulated, most of these fragments result in a broadening of the molecular ion peak with a concomitant loss of resolution and sensitivity.

Owing to these drawbacks and because such a method for sequencing provides little control over the extent of fragmentation, there are few reports of using desorption/ionization-induced fragmentation to determine oligonucleotide sequences. Prompt fragment ions will be detected at their appropriate *m/z* values, using linear and reflectron time-of-flight mass spectrometers, with the reflectron geometry yielding higher resolution than the linear geometry, presumably owing to the interference of fast fragment ions in the latter case. Also, it is evident that formation of prompt fragment ions is matrix- and sequence-dependent, but is independent of the laser wavelength.

Figure 6 illustrates the possible fragments one might see during the dissociation of an oligonucleotide. B_1, B_2 and B_3 represent nucleobases numbered consecutively from the 5′-terminus of the molecule. Fragment ions containing the 5′-terminus of the original molecule are labelled a, b, c and d, and differ by the site of bond cleavage along the oligonucleotide backbone. Fragment ions containing the 3′-terminus of the original molecule are labelled w, x, y and z, and, as the 5′-fragments, they differ by the site of bond cleavage along the oligonucleotide backbone. The sequence location of bond cleavage for the a–d and w–z fragment ions is denoted by a subscript reflecting the number of nucleobases remaining in the fragment ion (**Figure** 7).

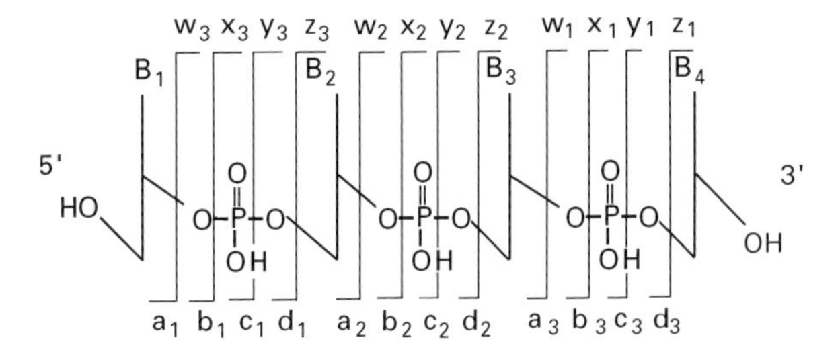

Figure 6 Representative fragmentation pattern of oligonucleotides.

Intentional dissociation methods

Intentional fragmentation of oligonucleotides may be caused by several methods including nozzle–skimmer dissociation (NS), photodissociation (PD), and collision-induced dissociation (CID). Intentional fragmentation can be made to occur within or near to the ionization source, or within a specified region of the mass analyser. NS dissociation occurs in ESI-MS when the translationally excited molecular ion collides with the background neutrals that are present in the high-pressure ESI source interface. NS dissociation generates fragment ions similar to those found in traditional CID spectra for small ($n \leq 15$) oligonucleotides. NS dissociation can also be used to obtain sequence information on larger ($n \geq 50$) oligonucleotides. However, complete sequence information on oligonucleotides of this size is not possible using this technique. Infrared multiphoton dissociation (IRMPD) has been used to characterize the sequence of small and large oligonucleotides. As with NS dissociation, complete sequence information can be obtained for small oligonucleotides, but this technique does not yield complete sequence information from larger oligonucleotides.

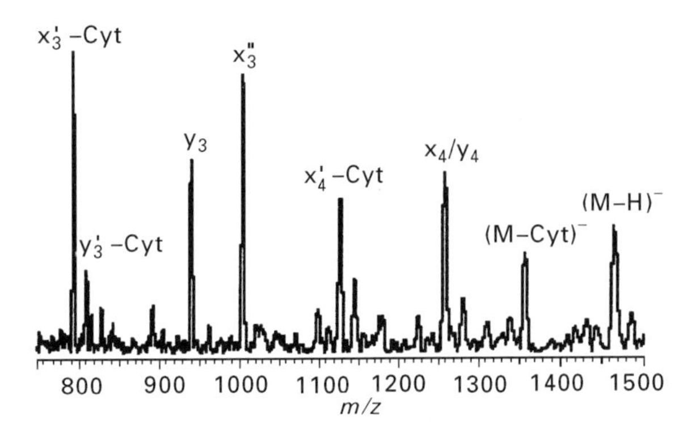

Figure 7 MALDI mass spectrum of d(CATCG) with 2,4,6-trihydroxyacetophenone as the matrix. The laser-induced fragment ions are labelled corresponding to the fragmentation nomenclature shown in **Figure 6.**

Tandem mass spectrometry (MS/MS)

An important goal in the mass spectrometric analysis of oligonucleotides has been to develop methods to determine structural information (such as sequence) through the dissociation of a molecule in the gas phase and interpretation of the resulting fragmentation pattern. The development of FAB-MS increased greatly the ability to perform such sequence analysis. However, it was not until the later development and application of primarily ESI-MS (and to a small extent, MALDI-MS) that meaningful tandem mass spectrometry studies were performed on oligonucleotides. In general, the fragmentation patterns of oligonucleotides are similar in spite of the different ionization techniques or mass analysers used.

Mass accuracy is essential for sequencing of unknowns using MS/MS experiments. If the oligonucleotide length is known, possible combinations of the four nucleotides may be calculated corresponding to the parent ion mass, and reasonable fragment ions can be assigned. If the length of the oligonucleotide is not known, working with high mass accuracy can help to reduce the possible sequence combinations calculated to match the parent ion mass. This technique is powerful because the specific fragment ions for a specific molecule can be identified and interference from other ions is eliminated. MS/MS also helps to reduce the problem of increased combination sequences as a result of increased sample size. When using mass analysers such as FT-ICR-MS or quadrupole ion traps, subsequent collisions and mass analysis (MS^n) may be conducted repeatedly, increasing the structural information obtained with each step.

Determination of nucleotide sequence by solution-phase sequencing

Mass ladders

The utility of any mass spectrometric sequencing method that relies on consecutive backbone cleavages depends on the formation of a mass ladder. The sequence information is obtained by determining the mass difference between successive peaks in the mass spectrum. In the case of oligodeoxynucleotides, the expected mass differences between successive peaks will correspond to the loss of dC = 289.25 u, dT = 304.26 u, dA = 313.27 u and dG = 329.27 u (all values are atomic mass-based). With oligoribonucleotides, the mass differences will be rC = 305.25 u, rU = 306.26 u, rA = 329.27 u and rG = 345.27 u (all values are atomic mass-based).

As is the case for all sequence determination methods that rely on the mass measurements of successive *n*-mers, oligodeoxynucleotides are easier to characterize owing to the relatively large differences in mass among the four oligodeoxynucleotide residues. The ribonucleotide residues, owing to the small mass difference between U- and C-containing nucleotides (a mass difference of only 1 Da), require a higher mass accuracy measurement to correctly distinguish U from C. Further, all mass ladder methods have a distinct advantage for sequence determination because it is the difference in two mass measurements that results in the desired information – the identity of the nucleotide residue. Because the accuracy of each individual oligonucleotide mass measurement is not critical, provided that the same bias exists for all ions detected, this procedure can be performed routinely on instruments with inherently lower mass accuracies; in particular, on TOF mass spectrometers.

Failure sequence analysis

One of the simplest methods for characterizing the chain length and sequence of synthetic oligonucleotides is the detection of failure sequences from the original synthesis step. This procedure takes advantage of the fact that automated solid-phase synthesis of oligonucleotides, especially those that contain modified internucleotide linkages such as methylphosphonates or phosphorothioates, is not 100% efficient. For example, the solid-phase synthesis of phosphodiester-linked oligonucleotides generally produces a yield of 99–100% for each synthesis cycle. Synthesis of a 10-mer at a 99.5% yield per cycle would result in an overall yield of 95.6%, with the remaining 4.4% being failure products that terminated before completion of the synthesis. Synthesis of larger oligonucleotides, naturally, will result in lower overall yields. For example, the synthesis of a 50-mer oligonucleotide at the 99.5% yield per cycle level would result in an overall yield of only 78.2%; nearly one-quarter of the sample would be failure sequences.

One potential drawback to this method is the reduction in yield of the failure sequences of larger oligonucleotides. For example, assuming that the synthesis yield per cycle is 95%, the 5-mer failure product will be present in the final solution at a concentration four times greater than a 30-mer failure product. As the coupling yields increase, this difference becomes smaller. If the coupling yield increases to 99%, the 5-mer will be only 1.28 times as concentrated as the 30-mer. However, the total amount of failure sequence products will also decrease in this case. These factors place some upper limit on the size of the oligonucleotide that is amenable to sequencing and to the applicability of this approach for very efficient synthesis protocols.

Sequence determination of oligonucleotides by an analysis of the failure sequences is an extremely simple and straightforward method. The mass spectrum will contain a series of peaks that corresponds to the final product and to each one of the failure sequences, each of which differs in mass by the appropriate nucleotide residue value. The sequence of the oligonucleotide is determined in the 5′ to 3′ direction from the mass ladder of the synthesis failure products.

Exonuclease digestion

The use of exonucleases to generate mass ladders of oligonucleotides that are suitable for analysis by mass spectrometry is now a standard method for sequencing small to moderate-length oligonucleotides. Unlike the failure sequence analysis method described above, this approach is suitable for the sequence analysis of naturally occurring oligonucleotides. Enzymes that hydrolyse the oligonucleotide consecutively from either the 5′- or 3′-terminus may be combined with unknown nucleotide samples and, over the reaction time, oligonucleotide ion signals are monitored by mass spectrometry. As the reaction progresses with the loss of one nucleotide at a time, the parent ion decreases in abundance and a ‘mass ladder’ is generated with each peak of lower mass corresponding to the sequential loss of a terminal nucleotide. Oligonucleotides up to the 40-mer level can be characterized readily by this approach.

Chemical digestion

An alternative to the enzymatic approach is the use of chemical agents to cleave the original oligonucleotide into smaller components for mass analysis. Although the number of applications of chemical digestion methods for sequence determination of oligonucleotides is far smaller than the number of applications of the failure sequence method or enzymatic approach, there are several important cases where neither of the latter two approaches is suitable for oligonucleotide analysis. These cases generally occur when the oligonucleotide backbone has been modified, such as in the case of antisense oligonucleotides, where modification is chosen specifically to reduce the susceptibility of the oligonucleotide to enzymatic digestion. The principles of analysis for the majority of the chemical digestion protocols are identical to those in failure sequence analysis and enzymatic digestion in that the goal is to generate a mass ladder of peaks that differ from one another by a nucleotide residue. As such, all of the general concerns for sequence determination by formation of mass ladders will apply to these methods.

We can classify chemical cleavages by their specificity for oligonucleotide base composition and/or linkage identity. Nonspecific chemical cleavages are reactions such as acid and/or base hydrolysis and alkylation reactions. In each of these reactions, the cleavage of the phosphodiester backbone does not depend on the base composition of the oligonucleotide. Acid hydrolysis is more specific for DNA, base hydrolysis is more specific for RNA and methylphosphonate-linked oligonucleotides, and alkylation is more specific for phosphorothioate-linked oligonucleotides. However, there is little to no base specificity for these chemical approaches.

The generation of a mass ladder of oligonucleotides for sequence determination using a chemical digestion approach can be complicated by the nonspecific nature of the chemical cleavage reaction. Any linkage site can potentially be cleaved by the chemical agent. If a single cleavage site is generated randomly for each oligonucleotide, then two sequence-specific fragments are produced, one from the 5′- and another from the 3′-terminus. If each of these fragments is detected in the resulting mass spectrum, then there is twice the necessary amount of information needed for the sequence to be determined. Although this additional information may facilitate sequence determination, especially if complete sequence coverage is unobtainable in any one direction, these two ion series can be a source of confusion.

See also: **Biochemical Applications of Mass Spectrometry; Fast Atom Bombardment Ionization in Mass Spectrometry; Fragmentation in Mass Spectrometry; Metastable Ions; Quadrupoles, Use of in Mass Spectrometry; Sector Mass Spectrometers; Time of Flight Mass Spectrometers.**

Further reading

Chapman JR (1993) *Practical Organic Mass Spectrometry*, 2nd edn. Chichester: Wiley.

Dienes T, Pastor SJ, Schurch S, *et al* (1996) *Mass Spectrometry Reviews* **15**: 163–211.

Limbach PA (1996) *Mass Spectrometry Reviews* **15**: 297–336.

Limbach PA, Crain PF and McCloskey JA (1995) *Current Opinion in Biotechnology* **6**: 96–102.

McCloskey JA (ed) (1990) *Methods in Enzymology*, Vol 193. San Diego: Academic Press.

McNeal CJ (ed) (1986) *Mass Spectrometry in the Analysis of Large Molecules*. Chichester: Wiley.

Nordhoff E, Kirpekar F and Roepstorff P (1996) *Mass Spectrometry Reviews* **15**: 67–138.

Saengar W (1984) *Principles of Nucleic Acid Structure*. New York: Springer-Verlag.

Nucleic Acids Studied Using NMR

John C Lindon, Imperial College of Science, Technology and Medicine, London, UK

MAGNETIC RESONANCE
Applications

Introduction

The methods used to analyse and assign the NMR spectra of nucleotides and nucleic acids (DNA and RNA) are very similar to those for peptides and proteins. These are the subject of other articles in this Encyclopedia. Individual residues in DNA comprise deoxyribose sugar units connected at C-1′ to either cytosine (C) or thymine (T) pyrimidine bases, or to adenine (A) or guanine (G) purine bases. For RNA, the corresponding sugar is ribose and uracil (U) is found instead of thymine. The residues are connected *via* phosphate linkages between the 3′- and 5′-positions of the sugar rings.

Unlike proteins where use is generally made of ^{1}H, ^{13}C and ^{15}N NMR data, the NMR spectroscopy of nucleotides and nucleic acids uses ^{1}H, ^{13}C and ^{31}P NMR spectra. In protein NMR studies, it is usual nowadays to prepare the protein extensively labelled with both ^{13}C and ^{15}N and also sometimes with ^{2}H. These techniques have not been applied so extensively in nucleic acid NMR spectroscopy.

Similar sample preparation methods are used for nucleic acids as for proteins. Nucleic acids are typically dissolved in 90% H_2O – 10% D_2O with the corresponding use of appropriate solvent resonance suppression methods for the measurement of ^{1}H NMR spectra.

Assignment of NMR spectra

The usual approach used for proteins is also applied to NMR spectra of nucleic acids. 1D ^{1}H NMR spectra can be supplemented with COSY or TOCSY spectra to aid spin system connectivities. Separate signals for the exchangeable amino and imino protons can often be observed and the properties of these signals can be very indicative of nucleotide structure. For example, imino protons involved in Watson–Crick base pairing such as A with T generally resonate in a distinctive window between δ13.0 and δ14.5. The ^{1}H chemical shifts in duplex structures can be different to those for bases in single strands and thus the unwinding of a duplex can be monitored using these shift changes.

Because base protons and sugar protons are separated by a minimum of four bonds, spin couplings are not usually observed between these units and recourse is made to the use of NOE measurements often as 2D NOESY studies. Thus, for example, NOEs observed between the sugar anomeric proton and H-6 and H-8 of a base serve to identify the base and sugar units of a single residue. NOE measurement can also be used to gain information on the sequence of residues in a nucleic acid.

In addition, heteronuclear coupling between ^{1}H and ^{31}P can be used to make sequential connectivities between residues. This is possible because there is a continuous relay of a series of homonuclear and heteronuclear couplings along the nucleotide backbone.

Variable-temperature studies can be very informative as they give information on the melting and denaturation of duplex structures.

Structural aspects of nucleic acids from NMR spectroscopy

It is possible to extract information on a wide variety of structural features using NMR spectroscopy of nucleic acids. These include characterization of base pairing, the conformation of the ribose and deoxyribose sugar rings, the torsion angles between sugar rings and the phosphate groups, the orientation of the sugar rings relative to the nucleotide bases, discrimination of right- and left-handed helices and the overall helix structure. In addition, information can be obtained on the location, dynamics and stoichiometry of binding of drugs to nucleic acids. Also, for duplex structures, it is possible to investigate mismatched pairings.

The overall aim is usually to determine the complete 3D structure of a nucleic acid and this is achieved in the same fashion as for protein structures. Thus, the NMR spectra are interpreted in terms of qualitative or semi-quantitative distance information which is then used as a set of constraints for a theoretical calculation of the structure generally based on distance geometry or restrained molecular dynamics calculations. One of the main problems with this approach, unlike for proteins, is the lack of long-range distance constraints for double-helix structures.

Despite these limitations, a combination of NMR spectroscopy and modelling has been used to derive structures for oligonucleotides, both DNA and RNA fragments, RNA–DNA hybrids and drug–DNA complexes. The review by Rizo and Bruch in the Further reading section provides references to a wide range of structural and dynamic studies of nucleotides and nucleic acids using NMR spectroscopy.

See also: **Carbohydrates Studied By NMR; Macromolecule–Ligand Interactions Studied By NMR; Nucleic Acids and Nucleotides Studied Using Mass Spectrometry; Peptides and Proteins Studied Using Mass Spectrometry; Proteins Studied Using NMR Spectroscopy; Solvent Suppression Methods in NMR Spectroscopy; Two-Dimensional NMR, Methods.**

Further reading

Davies DB and Veselkov AN (1996) Structural and thermodynamic analysis of molecular complexation by ^{1}H NMR spectrocopy – intercalation of ethidium bromide with the isomeric deoxytetra-nucleoside triphosphates 5′-d (GpCpGpC) and 5′-d (CpGpCpG) in aqueous solution. *Journal of the Chemical Society, Faraday Transactions* 2, **92**: 3545–3557.

Pardi A, Walker R, Rapoport H, Wider G and Wuthrich K (1983) Sequential assignments for the ^{1}H and ^{31}P atoms in the backbone of oligonucleotides by two-dimensional nuclear magnetic resonance. *Journal of the American Chemical Society* **105**: 1652–1653.

Patel D, Pardi A and Itakura K (1982) DNA conformation, dynamics and interactions in solution. *Science* **216**: 581–590.

Rizo J and Bruch MD (1996) Structure of biological macromolecules. In: Bruch MD (ed.) *NMR Spectroscopy Techniques*, Chapter 6, p. 285. New York: Marcel Dekker.

Searle MS, Hall JG, Denny WA and Wakelin LPG (1988) NMR studies of the interaction of the antibiotic nogalamycin with the hexadeoxyribonucleotide duplex D(5′-GCATGC)$_2$. *Biochemistry* **27**: 4340–4349.

Index

NOTE

Bold page number locators refer to complete articles on the various topics covered by this encyclopedia. Illustrations, including spectra, are indicated by *italic* page numbers.

Text and tables are located by page numbers in normal print.

Cross references, prefixed by *see* and *see also*, are also listed at the end of each article.

A

B

C

D

E

F

G

H

I

J

K

M

N

O

P

Q

R

S

T

U

V

W

X

Y

Z